HANDBOOK OF ELECTRONIC PACKAGING

OTHER McGRAW-HILL HANDBOOKS OF INTEREST

AMERICAN INSTITUTE OF PHYSICS · American Institute of Physics Handbook
BAUMEISTER AND MARKS · Standard Handbook for Mechanical Engineers
BEEMAN · Industrial Power Systems Handbook
BLATZ · Radiation Hygiene Handbook
BRADY · Materials Handbook
BURINGTON AND MAY · Handbook of Probability and Statistics with Tables
COCKRELL · Industrial Electronics Handbook
CONDON AND ODISHAW · Handbook of Physics
COOMBS · Printed Circuits Handbook
CROFT AND CARR · American Electricians' Handbook
ETHERINGTON · Nuclear Engineering Handbook
GRUENBERG · Handbook of Telemetry and Remote Control
HAMSHER · Communication System Engineering Handbook
HENNEY · Radio Engineering Handbook
HENNEY AND WALSH · Electronic Components Handbook
HUNTER · Handbook of Semiconductor Electronics
HUSKEY AND KORN · Computer Handbook
IRESON · Reliability Handbook
JASIK · Antenna Engineering Handbook
JURAN · Quality Control Handbook
KLERER AND KORN · Digital Computer User's Handbook
KOELLE · Handbook of Astronautical Engineering
KORN AND KORN · Mathematical Handbook for Scientists and Engineers
KURTZ · The Lineman's and Cableman's Handbook
LANDEE, DAVIS, AND ALBRECHT · Electronic Designers' Handbook
MACHOL · System Engineering Handbook
MARKUS · Electronics and Nucleonics Dictionary
MARKUS · Handbook of Electronic Control Circuits
MARKUS AND ZELUFF · Handbook of Industrial Electronic Circuits
PERRY · Engineering Manual
SHEA · Amplifier Handbook
TERMAN · Radio Engineers' Handbook
TRUXAL · Control Engineers' Handbook

HANDBOOK OF ELECTRONIC PACKAGING

CHARLES A. HARPER, EDITOR
Westinghouse Electric Corporation
Baltimore, Maryland

McGRAW-HILL BOOK COMPANY
New York St. Louis San Francisco London Sydney
Toronto Mexico Panama

HANDBOOK OF ELECTRONIC PACKAGING

 Library of Congress Catalog Card Number 68-11235

26671

1234567890 MAMB 754321069

CONTRIBUTORS

BAUM, JAMES R. *Mechanical Engineering Laboratory, Motorola, Incorporated, Government Electronics Division:* 11. THERMAL-DESIGN CONSIDERATIONS FOR PACKAGING ELECTRONIC EQUIPMENT (CONTRIBUTING AUTHORS: R. J. JIMINEZ *Universidad de Costa Rica,* J. R. WELLING, P. DICKERSON, and V. P. DUFFY *Motorola)*

BROACHE, EUGENE W. *Process Engineering, Westinghouse Electric Corporation, Aerospace Division:* 3. SOLDERING AND MECHANICAL INTERCONNECTIONS

CARTER, H. G. *Mechanical Design & Development Engineering, Westinghouse Electric Corporation, Aerospace Division:* 8. PACKAGING WITH CONVENTIONAL COMPONENTS

COUCH, W. R. *International Business Machines Corporation, Electronics Systems Center:* 15. COMPUTER-AIDED DESIGN

FULLER, W. D. *Lockheed Missile & Space Company:* 14. PACKAGING FOR SPACE ELECTRONICS

HARPER, CHARLES A. *Materials Engineering, Westinghouse Electric Corporation, Aerospace Division:* 7. MATERIALS FOR ELECTRONIC PACKAGING

HEIDLER, G. R. *General Electric Company, Missile and Space Division:* 5. DEPOSITIONS ON MICROELECTRONICS

KALB, ROBERT M. *Sperry Rand Corporation, Univac Division:* 12. PACKAGING FOR COMPUTER APPLICATIONS (CONTRIBUTING AUTHORS: EDWARD MASSELL *Electronic Associates, Incorporated,* LEE R. CARLSON *Rosemount Engineering Company,* and LARRY D. SLY *Fabri-Tek, Incorporated)*

McCAULEY, D. O. *Development Engineering, Westinghouse Electric Corporation, X-ray Division:* 10. ELECTRICAL FACTORS OF ELECTRONIC PACKAGE DESIGN

SAUNDERS, RALPH *Burroughs Corporation, Defense, Space & Special Systems Group:* 6. CONNECTORS AND INTERCONNECTING DEVICES

SAWYER, H. F. *Walter V. Sterling, Incorporated, Consultants, Claremont, California:* 4. WELDING AND METAL BONDING TECHNIQUES

SCHNORR, DONALD P. *Central Engineering, Defense Electronic Products, Radio Corporation of America:* 1. DESIGN AND APPLICATION OF RIGID AND FLEXIBLE PRINTED WIRING

SCHUH, ARTHUR G. *Materials Engineering Laboratories, Martin Marietta Corporation, Orlando Division:* 2. WIRES AND CABLES

STALEY, W. W. ***Mechanical Design & Development Engineering, Westinghouse Electric Corporation, Aerospace Division:*** **9. PACKAGING OF MICROELECTRONIC AND HYBRID SYSTEMS**

STALLER, JACK J. ***Microsystems Technology Corporation, Burlington, Massachusetts:*** **13. PACKAGING FOR MILITARY APPLICATIONS (CONTRIBUTING AUTHORS: E. KOVAL, J. THOMPSON, H. LAKE, ARNOLD BUCKMAN** ***Sylvania,*** **and L. BRICKER** ***Grumman Aircraft Engineering Company)***

PREFACE

Within the span of a few short years, electronic packaging has matured from merely a phrase to a broadly accepted discipline in the electronic and electrical industries. Today, new though the discipline is, most designers and manufacturers of electronic equipment have incorporated electronic packaging into, and as a vital part of, their functional organizational structures. It is this maturing which dictated the need for this *Handbook of Electronic Packaging*. And so, this Handbook was born.

In spite of the rapid technological advances and the growth of a discipline, electronic packaging is not easily defined in terms of some already existing technology. Rather, it is an overlapping of disciplines, which requires a breadth of knowledge rarely encountered before in emerging disciplines. While there may be no perfect definition of electronic packaging, it may, for practical purposes, be considered as the conversion of electronic or electrical functions into optimized, producible, electromechanical assemblies or "packages." True, electromechanical assembly is not new, as electrical equipment has always required some mechanical form factor; however, the binding of sound interdisciplinary principles into a single discipline of electronic packaging is new. The *Handbook of Electronic Packaging* is devoted to supplying useful information, data, and guidelines for all of those involved in any aspect of this new discipline of electronic packaging. This includes engineers and manufacturers who design and produce any type of electronic packages, as well as suppliers of parts, materials, and equipment.

Preparation of this Handbook presented many challenges. With electronic packaging having an interdisciplinary base, challenging questions arose on both content and organization. It is hoped that, based on the extensive considerations which led to the ultimate decisions in content and organization, an optimized balance of material is presented. The reader will find this book full of practical and useful data and guidelines on materials, components, processes, connection and interconnection techniques and devices, mechanical layouts, electrical factors, thermal design, and all other important aspects of fundamental electronic packaging. Both microelectronic and conventional electronic packaging are thoroughly covered. Further, much valuable data and many guidelines are presented for major application areas, such as military, space, and commercial industries, including computers.

The chapter content of this Handbook illustrates the breadth of material coverage which has been developed. The first six chapters cover connection, interconnection, and termination systems such as printed and conventional wiring, deposited films, soldering, welding, and connectors. Then comes a chapter on materials, followed by two chapters on packaging techniques and two chapters on design considerations, electrical and thermal. The next two chapters cover the important application areas of military and space packaging. Following this is a chapter specifically covering computer packaging, and a final chapter on computer-aided design.

In addition to the broad presentation of data and guidelines, several other features of this Handbook should be noted. First, a set of definitions and terms is presented at the beginning of most of the chapters. Such definitions can frequently be invaluable. Next, the reader will find tradeoff tables spread generously throughout the book. These will be very useful in making decisions among various possible approaches to given problem areas. Last, every attempt has been made to provide a very complete and thoroughly cross-referenced index. It is suggested that each reader acquaint himself with this index.

Length and coverage of a Handbook of this magnitude are necessarily measured compromises. Inevitably, varying degrees of shortages and excesses will exist, depending on the needs of each individual reader. Then too, the time required to complete such a major work as this necessarily demands that some most recent data may not be fully covered. Further, in spite of the tremendous efforts involved, some errors or omissions may exist. While every effort has been made to minimize such shortcomings, it is my greatest desire to improve each successive edition. Toward this end, any and all reader comments will be welcomed and appreciated.

Charles A. Harper

CONTENTS

Chapter 1

DESIGN AND APPLICATION OF RIGID AND FLEXIBLE PRINTED WIRING

By

DONALD P. SCHNORR

Radio Corporation of America
Camden, New Jersey

INTRODUCTION

Electronic packaging, by its very nature, has evolved into a combination of all the physical sciences; for those involved in this technology this fact has proved to be unavoidable. Elements of chemistry, physics, metallurgy, mechanical and electrical engineering, photography, and many other sciences and even skilled arts combine to assist the packaging engineer in his task. Probably nowhere is this case more true than in the critical subdivision of packaging interconnections. The problems involved with interconnecting functional elements and assemblies in modern electronic packaging well deserve this specialized treatment; interconnections remain one of the major problem areas in modern electronic packaging. Indeed, next to perhaps the functional devices (active and passive), the vastly complex task of "hooking them up" in a reliable fashion (in a manner not to degrade critical circuit parameters) remains a major preoccupation of the modern packaging engineer.

The development of printed wiring was an inevitable event in the history of electronic packaging. Without the density capability and reliability of printed wiring, along with its bulk-processing potential, many modern electronic developments and much of the production would have proved to be impossible, if not impractical. Furthermore, basic improvements on original printed-wiring innovations, i.e., multilayer wiring, flexible wiring, etc., have resulted in developments which may have been impossible with conventional interconnection techniques.

Since this subject is influenced by so many and such diverse disciplines, the author has divided the chapter into a number of functional sections. The first three sections are concerned with definitions and an explanation of the types of rigid and flexible printed wiring currently available; the next two sections are concerned with materials and processes used for producing printed wiring; and these are followed by discussions of design consideration, artwork, applications, and printed-wiring design parameters.

Although primarily intended as a reference source and design guide for those using printed wiring, this chapter may also provide some direction to those involved in other areas of packaging; often the problems shared are common, and an understanding of the fundamentals of printed wiring by those in other phases of packaging is important if an optimum design is sought.

GLOSSARY OF DEFINITIONS

A glossary of printed-wiring terms has been compiled in Table 1 to assist in a better understanding of the topic matter covered in the various sections in this chapter. Taken from multiple sources and personal experience, most of these terms occupy an important position in defining printed-wiring art today. More definitions will be included in individual sections of this chapter, pertinent to the subject matter there.

TABLE 1. Terms and Definitions

Accelerator. A chemical additive which hastens a chemical reaction under certain conditions; it is often used with a hardener and catalyst.

Acid copper. Copper electrodeposited using a copper sulfate–sulfuric acid bath.

Adhesion. The degree of retention by surface attachment of materials with each other.

Alumina. Aluminum oxide, Al_2O_3. An insulating ceramic with very good mechanical characteristics at elevated temperatures.

Annular ring. A ring, ideally concentric, described by a pad or a printed-wiring board and the hole drilled through it.

Artwork. A working tool for reproduction of precision patterns which is a glass plate or film with an enlarged or full-scale pattern of the circuit board desired.

B-stage. See Prepreg.

Bond strength. See Peel strength.

Checking. A surface condition evidenced by fine hairline cracks in the resin surface.

Circuit board. An etched blank of final dimensions with the desired copper circuit pattern on one or both sides.

Coordinatograph. A precision drafting instrument used for preparation of highly accurate artwork.

Current-carrying capacity. The maximum current which can be continuously carried without causing degradation of electrical or mechanical properties of the printed-wiring board or attached components.

Delamination. A separation between any of the layers of the base laminate and/or between the metal cladding and the laminate to or from hole or board edges.

Diallyl phthalate (DAP). A type of unsaturated polyester resin, usually cured with a peroxide catalyst; it is used with mineral or glass-fiber or mat fillers for printed-wiring substrate laminates.

Dielectric constant. That property of a dielectric which determines the electrostatic energy stored per unit volume for unit potential gradient. The rates of the capacitance of the material of given dimensions are comparable with that of air for the same thickness and area.

Dielectric strength. The root-mean-square voltage gradient between two electrodes at which electrical breakdown or rupture occurs. It is expressed in volts per mil. It is often expressed as breakdown voltage, short time or step by step.

Dissipation factor. Tangent of the loss angle ($\tan \delta$); the ratio of the loss current to the charging current.

Dwell time. A period of time in a preheated laminating press at contact pressure that allows B-stage prepreg bonding layers to soften, devolatilize, and further polymerize reducing resin loss, prior to application of cure pressure.

Electroless plating. Electroless plating consists of the controlled autocatalytic reduction of the desired metal ion to the corresponding metal on certain catalytic surfaces.

Emulsion side. As applied to artwork, the side of the film or glass on which the photographic image is present.

Epoxy. A straight-chain thermoplastic and thermosetting resin based on ethylene oxide, its derivative or homolog. A normally thermosetting organic polymer usually containing ether linkages and aromatic groups, it has exceptional adhesive properties and is popularly used as the reinforcement bonder in laminates.

Etch factor. In etching-foil patterns, the ratio of the depth of etch to lateral etch.

Flame retardant (FR). The ability of a material to extinguish itself once ignited after the source of ignition is removed. Nonburning is not implied.

Flexural strength. The strength of a material in bending.

Gel time. The time required at a specific temperature, usually the prepreg curing temperature, for prepreg resin to gel. Application of cure pressure prior to resin gelation is essential to obtain a sound laminate. Gel time is a time-temperature function and will vary with heat input.

Grid. A grid is a two-dimensional network consisting of a set of equally located parallel lines superimposed upon another set of equally located parallel lines which are mutually perpendicular, thereby forming square areas. These intersections provide the basis for an incremental location system.

Ground plane. A single conductive current-carrying layer of essentially unetched copper used as a primary ground bus to complete a circuit loop and provide transmission-line capabilities to a signal line.

In-process corrective procedures. Correction of printed-wiring manufacturing anomalies prior to completion of the board-manufacturing process.

Interlayer connection. A metal conductor which provides a communication to or between layers.

Jumper. A direct electrical connection which is not a portion of the conductor pattern between two points on a printed-wiring plane.

Laminate. The process of bonding together by simultaneous application of heat and pressure individually etched printed-circuit layers of a multilayer printed-circuit board. Also refers to a complete laminated board.

Lay-up. The process of registering and stacking individual layers of the multilayer in preparation for the laminating cycle. A lay-up also refers to the completed, registered stack of layers, usually in a lay-up, or laminated fixture.

Measling. A condition existing in the base laminate in the form of discrete white spots or crosses under the laminate surface, indicating a separation of glass-cloth fibers from resin at weave intersections.

Melamine. Melamine formaldehyde resin, commonly known as melamine, is a thermosetting plastic used as a molding compound for electrical insulation or with reinforcement for electrical insulation. It is not normally clad with metal foil for printed-circuit applications.

Microstrip line. A printed-wiring transmission-line configuration which consists of a printed conductor on a dielectric substrate with a ground plane on one side and air dielectric on the outside.

Migration. The tendency for some metals, especially silver, in the presence of a dc field and absorbed moisture to migrate across the intervening dielectric to lower the dielectric strength.

Multilayer board. A product consisting of layers of electrical conductors separated from each other with an insulating support and fabricated into a solid mass. Interlayer connections are used to establish continuity between various conductor patterns. This connection may be a plated-hole, solid-pillar, or plug, or a hollow-pin insert.

Overhang. The inverted shell of plating formed when conductor material is selectively removed from under the plating.

Peel strength. The force, measured in pounds per inch of width of cladding, necessary to maintain steady separation of the metal cladding from the laminate to which it is bonded when pulled at a prescribed rate at a right angle to the board. This process is normally performed at room temperature ("cold peel") but may be required at elevated temperature for specific applications ("hot peel").

Phenolic. A synthetic thermosetting resin produced by the condensation of phenol with formaldehyde. It is the least expensive printed-circuit-board type but has the disadvantage of poor arc and tracking resistance. It is usually used in punching grades with paper reinforcement.

Polyamide. A polymer in which the structural units are lined by amide or thioamide groupings. Nylon is an example of the normally thermoplastic type of resin system.

Polyester. A resin formed by the reaction between a dibasic acid and a dihydroxy alcohol. These resins are generally inexpensive but have found little use in printed-circuit-board laminates because of inability to B-stage them conveniently as with the epoxies. The chemical characteristics of this class of polymers is a carboxylate group.

Polyethylene. A thermoplastic material composed of polymers of ethylene.

Prepreg. This is generally glass fabric which has been impregnated with an epoxy resin system and partially cured to a nontacky condition (B stage). The resin is both fusible and soluble in organic solvents in this state and has a definite, limited shelf life.

Pyrophosphate copper. Copper electrodeposited using a copper pyrophosphate electrolyte bath.

Resist. A protective material (ink, paint, metallic plating, etc.) used to shield the desired portions of the printed conductive pattern from the action of the etchant, solder, or plating.

Repair. Correction of a printed-wiring defect after completion of board fabrication to render it functionally as good as a perfect board.

Solderability. The ability of the conductive pattern in a printed-circuit board to be wet by solder.

Steatite. A ceramic insulation, the multicomponent crystalline phase of the magnesia alumina silica group. The formula is $MgOSiO_2$.

Strip transmission line. A printed-wiring transmission-line configuration which consists of a printed conductor surrounded by dielectric material and shielded by copper ground planes on both sides.

Substrate. A material on whose surface an adhesive substance is spread for bonding or coating. Any material which provides a supporting surface for other materials used to support printed-circuit patterns.

Test coupon. A sample, or test, pattern, usually made as an integral part of a board, and on which selected tests may be performed to evaluate design or process control without destroying the parent board.

Thermosetting. A classification of a resin system that can be repeatedly softened or melted by heating and rehardened by cooling.

Thieving. The use of auxiliary material to "steal away" high current density from critical portions of the circuit pattern and to promote even plating over the entire workpiece.

Throwing power. The ability to deposit plating in recesses and low-current-density areas (without regard to the thickness of the printed-wiring board).

Undercut. The amount of reduction of the cross section of a metal-foil conductor caused by the removal of metal from under the edge of the resist by the etchant.

Warp and twist. Warp is the deviation from a plane surface measured across the length or width. Twist is defined as the deviation measured from one corner to the diagonally opposite corner.

Weave exposure. A surface condition in which the unbroken, woven-glass-cloth filaments are not uniformly encapsulated by resin.

TYPES OF PRINTED-WIRING PLANES

Although printed wiring since its inception during World War II has been predominantly of the rigid type, a later development of flexible wiring planes forced the classification of printed wiring into two general categories: *rigid printed-circuit boards* and *flexible printed wiring.* Another important subclassification is involved with the number of planes, or layers of wiring, which constitute the total wiring assembly, or structure.

The printed-wiring art has been particularly susceptible to inventions and processes of almost infinite variety and novelty; in spite of this apparent diversity a relatively small number are still employed to an extent. In this section, we shall discuss the most common processes. These are classed under the general heading of chemical and mechanical methods but will be broken down into single-sided, double-sided, and multilayer, the latter at present being produced predominantly by the chemical processes.

Rigid Printed-circuit Boards. Generically the first, and currently the most widely used, form of printed wiring, the rigid printed-circuit board was early recognized not only to provide a conductive wiring path but also to allow support and protection of components it connected, as well as to supply a heat sink to aid in thermal management of the total package. The chemical methods used in producing rigid printed-circuit boards are the subtractive, or etched-foil, type, and the additive, or plated-up, approach. The mechanical methods include stamped wiring, metal-sprayed wiring, embossed wiring, and molded wiring.

Single-sided Printed-circuit Boards. Referring to the fact that there is wiring on one side of the insulating substrate only, this type of board represents by far the largest volume of printed-circuit boards currently produced. Single-sided boards are used for relatively unsophisticated and simple circuitry, where circuit types and speeds do not place unusual demands on wiring electrical characteristics.

SUBTRACTIVE, OR ETCHED-FOIL, TYPE. Illustrated by Fig. 1, the etched-foil process begins with a base laminate composed of a variety of insulators clad with copper or some other metallic foil. Following suitable cleaning and other preparation, a pattern of the desired circuit configuration is printed using a suitable negative-resist pattern for photoresist or ink resist, a positive-resist pattern if plating is to be used as an etchant resist. In the case of the latter, gold or solder plating is applied in the nonphotoresist areas. In the next step, etching, all copper not protected by the resist material is removed by the etchant. Following etching, the printed-wiring board is *stripped,* or subjected to removal of resist, and otherwise cleaned to ensure that no etchant remains. The board is then ready to fabricate by drilling, trimming, etc.

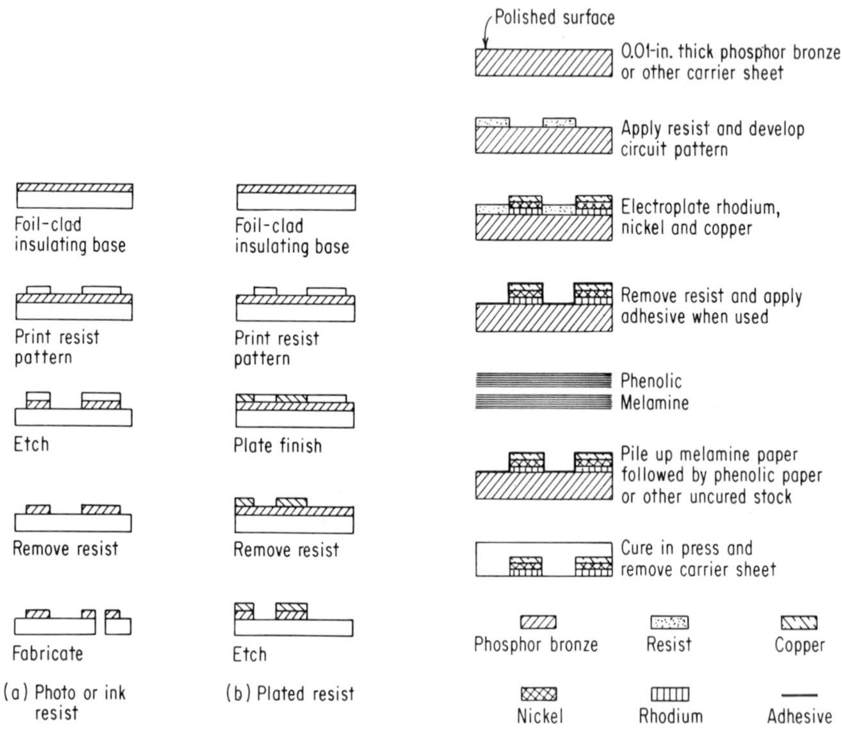

FIG. 1. Etched-foil processes.

FIG. 2. Electroplating-transfer process.[1]

ADDITIVE, OR PLATED, TYPE. Figure 2 illustrates a method for producing a printed circuit by using the additive, or plated, technique. Also called the *electroplating-transfer process,* this method was developed, and remains primarily used, for the fabrication of flush circuit boards. A carrier sheet is utilized; the sheet is copper, phosphor bronze, or aluminum if it is desired to remove it chemically after the process and stainless steel if it is desired to remove it mechanically. Resist is applied directly on the carrier sheet, and a pattern is exposed; thus when the sheet is developed, the circuit areas are exposed, free to accept subsequent electroplating. Although the figure shows a combination of rhodium, nickel, and copper characteristic of flush applications, other combinations or plain copper may be used. Following plating, the resist is removed and adhesive is applied if needed, depending upon the insulating material selected. The insulator "lay-up" is then placed over the electroplated circuitry and cured, or molded. The carrier sheet is then removed following this operation. When stainless steel is used and the carrier plate is mechanically stripped, adhesive is not used and glass-based "prepreg" material is used. Distortion caused by stripping may introduce some tolerance problems in this case. The other variation, using phosphor bronze, aluminum, or copper, uses an appropriate acid to dissolve the carrier plate; however, these solutions are only feasible when gold or rhodium are used for the first plating step.

STAMPED OR DIE-CUT WIRING. In this process a die is required which carries an image of the wiring pattern desired in the final product. As illustrated in Fig. 3, the die may be either photoengraved using conventional photoresist techniques or machine-engraved. The die in either case is heated, and when forcing the adhesive-coated foil into the base material, effects a bond as well as cuts out the circuit pattern and embeds it into the base at the conductor edges. Blanking and/or piercing of the board may be done at the same time. Popular in Europe (especially in France, where it is known as *appliqué wiring*), its main advantage, high production, is rapidly losing ground to more conventional techniques.

EMBOSSED WIRING. This technique uses a photoengraved die bearing the circuit pattern to push an adhesive-coated foil blank into a partially cured or thermoplastic base. Obviously a high-production technique, this step cures or sets the base material, imprints the circuit pattern, and bonds the foil to the insulator. A subsequent machining operation removes the unwanted embossed, or raised, areas, leaving the actual circuitry slightly recessed.

MOLDED WIRING. A thorough description of these techniques is given by Schlabach and Rider.[1] Three common methods are illustrated in Fig. 4; these are termed the *standard method*, the *dinked-conductor method*, and the *sintered-conductor method*. Molded-wiring processes differ in detail but hold out the common promise of a

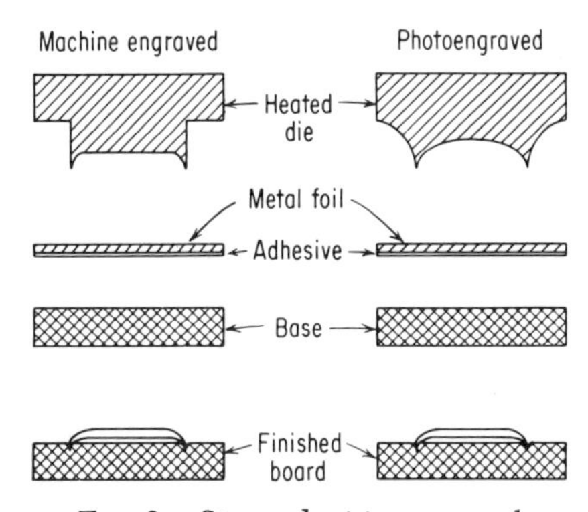

FIG. 3. Stamped-wiring process.[1]

three-dimensional configuration of wiring plus substrate. In one approach, the base insulator is premolded with a recessed-wiring pattern and truncated holes for desired through connections or component lead wires. Adhesive is sprayed on where needed, and the base, including holes, is made conductive by electroless-plated copper or nickel. The wiring paths are built up by the electrodeposition of copper, followed by solder. Finally, the adhesive is set by heat, and the surfaces ground to remove the unwanted conductor material.

A second method die-stamps the wiring pattern onto a resilient semicured preform using an adhesive-coated foil blank and a heated die. Following this, the wire preform is molded with the desired holes, bosses, etc., in matched metal dies. The method described by Bell,[2] illustrated in Fig. 4*a*, is based on the use of metal foil containing a thin layer of 325-mesh bronze balls sintered onto one surface. This foil blank is placed, ball side up, over a metal die containing the raised-wiring pattern and then molded to the desired substrate. The molding pressure causes the foil to conform to the die surfaces. After molding, the unwanted raised portions are removed by grinding to leave a flush or slightly recessed wiring pattern. Alternatively, the ball-coated foil can be sheared into a die with a recessed-wiring pattern using a rubber pressure pad and then molded to give a substrate with a raised wiring pattern as shown in Fig. 4*b*. This leads to a raised wiring pattern and eliminates the grinding step. A substantial degree of mechanical interlocking is obtained, and such bonds possess excellent resistance to thermal, chemical, and mechanical action.

A related process based on the use of sintered copper conductors is shown in Fig. 4*c*. In the first step, copper powder (approximately 200 mesh) is compacted into a hardened steel die containing recessed conductors. The compacting is done at about 50,000 psi with a relatively hard rubber pad. Following this, the compacted powder conductors in the die are sintered at 500°C for about 15 min in a reducing atmosphere. Finally the substrate is molded to the sintered-conductor system, and the board is ready for further processing.

OTHER PROCESSES. As stated before, the technology of printed circuits from its

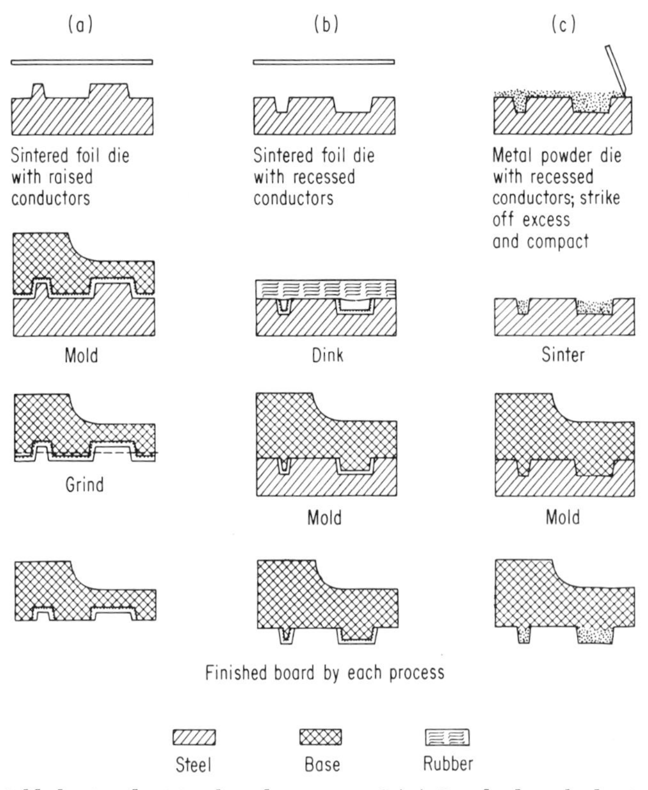

FIG. 4. Molded printed-wiring-board processes:[1] (*a*) Standard method using sintered ball-coated foils. (*b*) Modification based on a dinked conductor formed from a sintered ball-coated foil. (*c*) Sintered-conductor method based on the use of copper powder.

inception has evolved through a large number of processes, some of them practical, many of them not. Eisler,[3] terming this the "confused start of a new technology," lists 26 known methods for producing printed wiring of the single-sided, or most elementary, form. It is felt that the more practical and currently popular methods are covered above. There are several other methods, however, which are worthwhile noting. The metal-spray technique involves the deposition of molten metal by spraying through a stencil onto a roughened ceramic base. The Shoop process is one way to produce this atomized spray; this involves the melting of a metal or alloy in wire form in a gas flame, atomization, and propulsion by compressed air. The base layer, which may also be applied by a spray, is commonly produced by the fluidized bed-coating process. In this way the strength and desirable thermal characteristics of a metal-core board may be utilized. This technique is illustrated in Fig. 5. Ceramic wiring boards utilize printing in the true sense of the word, with conductive inks or paint coatings applied to an insulating substrate by hand painting, stencil spraying, decalcomania, and screening. Following application, the coating is baked or dried and fired to a high temperature, producing a conductor pattern with good adhesion and conductivity depending upon the material used. The conductor material is present as a pigment, e.g., silver, dispersed in a binder and solvent. In a sense, ceramic printed-wiring boards by this process are only an extension of the art of metallizing ceramics, an art which can be traced back to ancient craftsmen.

Although the electroplating-transfer process, previously covered, produced a flush printed-circuit board, it is also possible to produce flush circuitry by a variation of the subtractive, or etched-foil, method. Fig. 6*a* illustrates the pressure-flushing method which utilizes a flushing-grade laminate wherein the surface layer of insulation (next to cladding) is a semicured B-stage material. Following the etching of

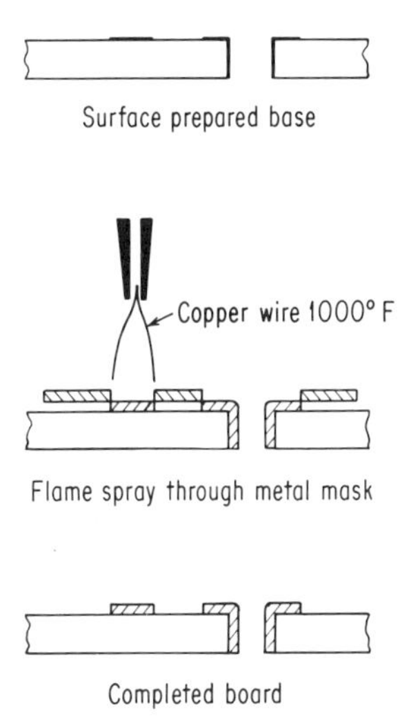

FIG. 5. Flame-sprayed process.

the clad copper as in the conventional etched-foil process, the etched circuit pattern is "pushed," or flushed, into the base laminate by means of a heated platen. Fig. 6*b* shows the etch-and-fill method which differs in that the flushing operation results in the application of liquid resin by squeegee or some other pressure method and then a cure cycle. The latter may require some machining to remove excess resin; this may damage surface plating unless protective measures are taken. Pressure flushing results in a board which may be sensitive to temperature since, upon exposure to temperatures in the laminating range, the structure "relaxes" and the etched foil returns to some degree of its nonflush condition.

Double-sided Printed-circuit Boards. When more than one layer of wiring is needed, circuit patterns are placed on both sides of a printed-wiring board, and a necessity arises for a means for interconnection between the two wiring layers. This is usually accomplished by a connection which goes through the board rather than around the edge, resulting in the name *through connection.* In many cases the "hole" serves a dual purpose: to accommodate a component lead and to provide a location for some method of interconnection between sides. The many processes for effecting interconnection between circuit layers may be generalized as including some plating process or method requiring a mechanical interconnection technique.

PLATED-THROUGH PROCESSES. There are two variations of the plated-through process, both utilizing a plated conductor through the hole to make the connection. For purposes of comparison we shall call them the *conventional plated-through-hole process* and the *plating-only-through-hole process*. Both are illustrated in Fig. 7.

The conventional plated-through-hole process starts with a double-clad laminate with a series of holes placed in it corresponding to the locations where a through connection is needed. Holes are drilled and deburred, and an electroless coating of copper applied over the entire board surface, including the holes. Copper is now electrodeposited to the exposed copper foil and sensitized walls of the hole, usually to a thickness of 0.001 in. A negative, or plating-resist, pattern is then applied, registered to both sides of the material. Resist covers all areas of foil where base copper conductor is not required; this will subsequently be etched off. The next plating step is to electrodeposition a thin layer of a suitable etch-resist plating, usually solder or gold. The original plating resist (photoresist) is removed, and the

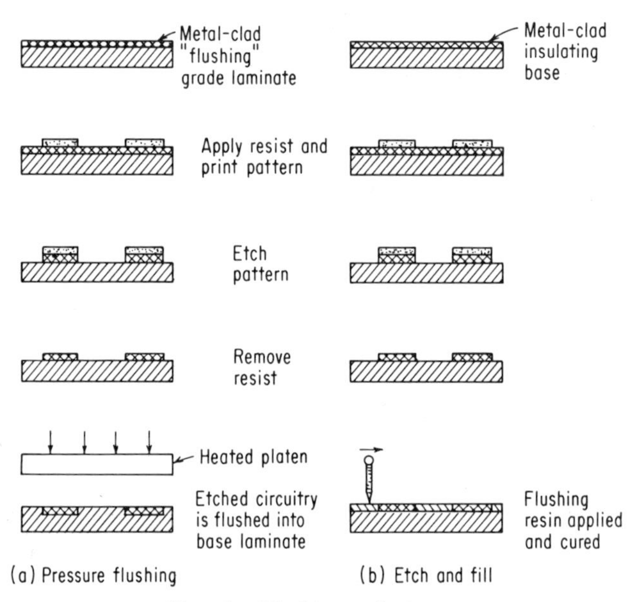

FIG. 6. Flushing techniques.

circuit pattern is defined by etching away the exposed copper in a suitable etchant, depending upon the type of plating resist used.

The plating-only process differs from the conventional plating-through-hole process in that no etching is required and the circuit pattern is defined at the same time the through connection is made. Holes are again drilled at desired connection points, and a thin layer of a suitable adhesive is applied as illustrated in Fig. 7*b*. Electroless deposition of copper sensitizes the entire surface and holes and is followed by a "flash" electroplate.

After the registered printing of a plating-resist pattern to both sides of the board, copper is electrodeposited to the desired thickness on exposed areas. Resist is removed, unwanted electroless copper is "flash-" etched off, and excess adhesive taken off with appropriate solvents. The last step is the curing of the adhesive by subjecting it to a heat-and-pressure cycle.

Although any discussion of through-hole processes should consider the built-up, or plated, stud technique discussed under multilayer methods, this technique is not practical and is rarely used for double-sided boards. The height to which it is feasible to plate a stud produces a board too thin to be useful unless it has another board for support; this removes it from the category of a double-sided board.

MECHANICAL PROCESSES. The nonplating techniques, illustrated in Fig. 8, preceded through-hole-plating processes for double-sided boards until the reliability of the latter was proved and the economic import of such mass-interconnection techniques realized. While not to be construed as an integral part of the board-fabrication procedure (as represented by the plated-through hole) these mechanical interconnections, actually board-assembly techniques, serve the purpose of connecting the two sides of a double-sided board.

A simple and easily made interconnection is represented by the clinched jumper wire, illustrated in Fig. 8*a*. A formed, uninsulated, solid-lead wire is placed through the hole and clinched and soldered to the conductor pad on each side of the board. Part-lead wires are not normally considered as interfacial connections. Three types of eyelets are also commonly used for double-sided-board interconnection; these

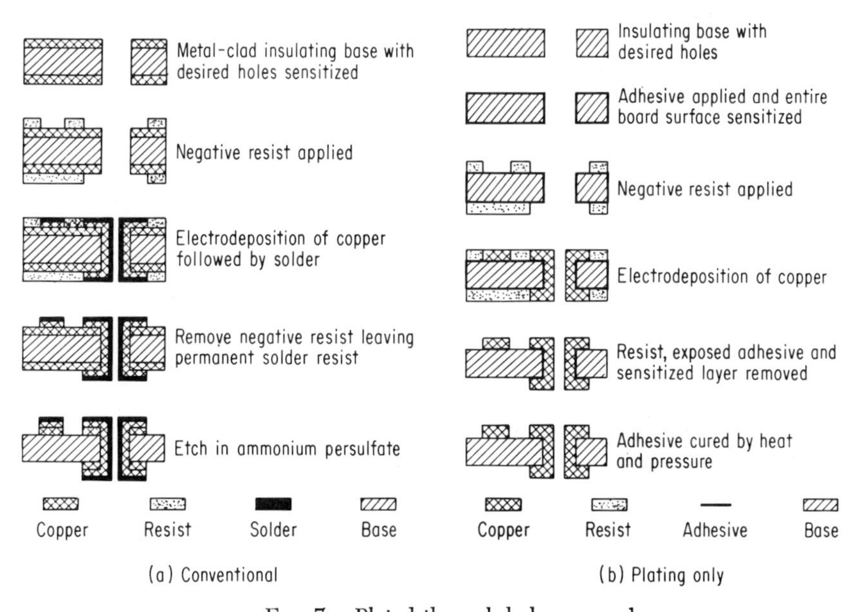

FIG. 7. Plated-through-hole process.[1]

are shown in Fig. 8*b* through *d*. Funnel-flanged eyelets are soldered to the terminal areas on the component side of the board prior to insertion of component leads. The other connection is made at assembly when the boards are dip-soldered. The *funnel flange* by definition shall have an included angle between 55 and 120°. Split-funnel-flanged eyelets differ only in the split and the fact that they need not be soldered to the terminal area on the part side of the board before insertion of part leads.

Many eyelets are machine-inserted, which increases uniformity of the connection and reduces the cost of assembly. Fig. 8*e* shows a recently developed interfacial through connection for double-sided printed-wiring boards called the *compliant-redundant through connection.* This technique utilizes a "sock" of multiple strands of wire braided around a rubber core. The unrestrained diameter of the rubber with braid is larger than the hole in the board; when the assembly is placed in the hole, the rubber exerts enough force on hole walls to keep solder from wetting strands within the hole. This phenomenon ensures that the strands are maintained compliant after soldered assembly. Advantages and disadvantages of these processes will be covered later in this chapter.

Multilayer Printed-wiring Boards. The problem of providing for increased wiring density required in many present electronic packaging applications could only be met by the use of more than two planes of wiring, resulting in the multilayer printed-wiring board. A multilayer printed-wiring board is a series of individual circuit layers bonded to produce a thin, monolithic assembly with internal and external connections to each level of the circuitry determined by the system wiring diagram. As an extension of the section on terminology, this discussion of printed-wiring processes will provide definitions only, with design and application information to follow later in the chapter.

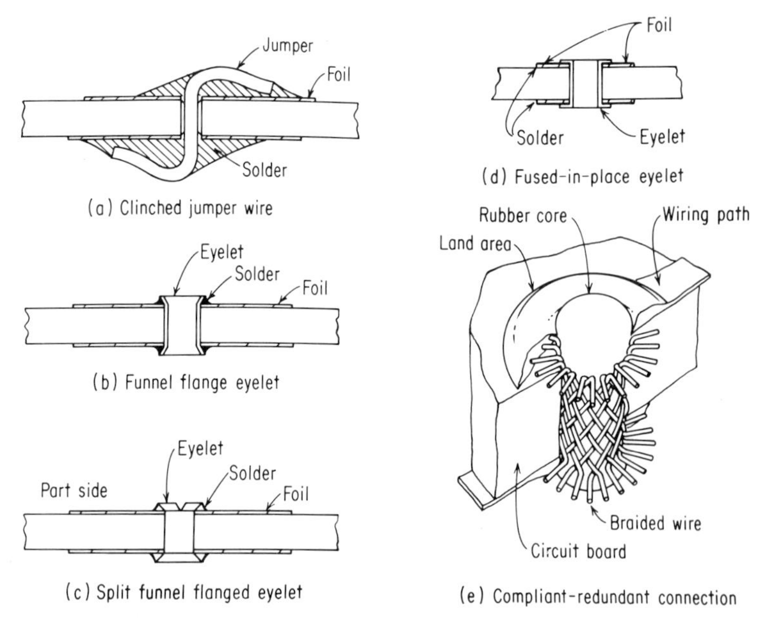

FIG. 8. Mechanical interconnections. (*Part e, Western Electric Co.*)

The processes for fabricating multilayer boards are basically extensions of methods used for single- and double-sided boards; therefore they will be described in less detail. As identified by board-fabrication process, multilayer printed-wiring boards can be broken down into two basic types: *laminated multilayer printed-wiring boards* and *built-up multilayer printed-wiring boards.* Other than the difference between the laminated approach and the screened or deposited insulation used in the built-up approach, these various methods differ only in the method in which interlayer connections are made. As in single-side printed-wiring technology, there are countless variations of these basic multilayer processes. Only the methods identified with significant process departures will be discussed here. Multilayer concepts classified as laminated printed-wiring boards are the clearance-hole process, the plated-hole process, the optical-chemical blanking technique, and the brazed-tubelet process. Techniques reasonably classified as built-up multilayer boards are the transfer process, the Beck process, and the metal-spray process. The merits of all of these methods when compared with each other are discussed later under design parameters, or trade-offs.

THE CLEARANCE-HOLE PROCESS. The clearance-hole multilayer board, illustrated in Fig. 9, consists of a number of layers of thin printed-circuit boards with a number of holes in graduated diameters arranged on the registered layers so that connections may be made directly to the desired pad through the clearance hole, since there are no interlayer connections. The clearance holes must be cut or formed into the layers of circuitry prior to lamination, and actual connection is made to the pad by the component lead after lamination by flow or dip soldering.

THE PLATED-HOLE PROCESS. Basically an extension of the plated-hole process previously described for double-sided boards, this technique lends itself well to multilayer construction. As shown in Fig. 10, the plated-through-hole multilayer printed-wiring board consists of a number of layers of thin circuit boards which are stacked one above the other and laminated into one monolithic assembly. At the points of proposed interconnection, holes are drilled which pass through pads on the conductors of the inner layers which are larger than the drilled hole. The drilling exposes a rim of copper around the entire circumference of the hole to which the copper on the individual layers in the through hole is connected by

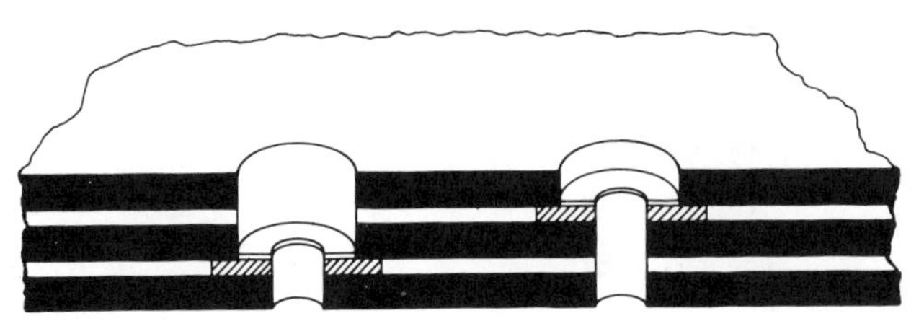

FIG. 9. The clearance-hole multilayer board.

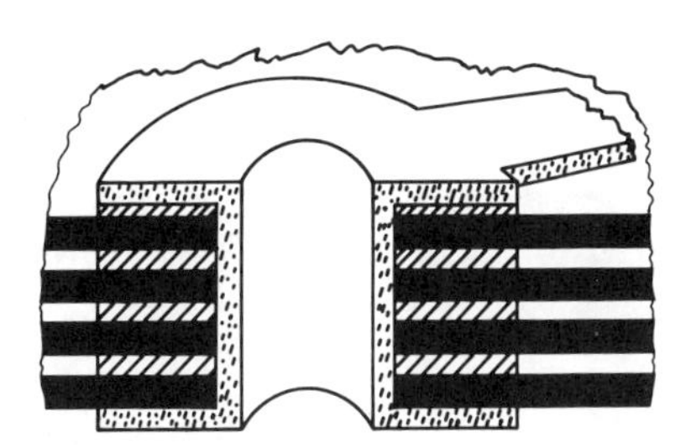

FIG. 10. Plated-through-hole multilayer board.

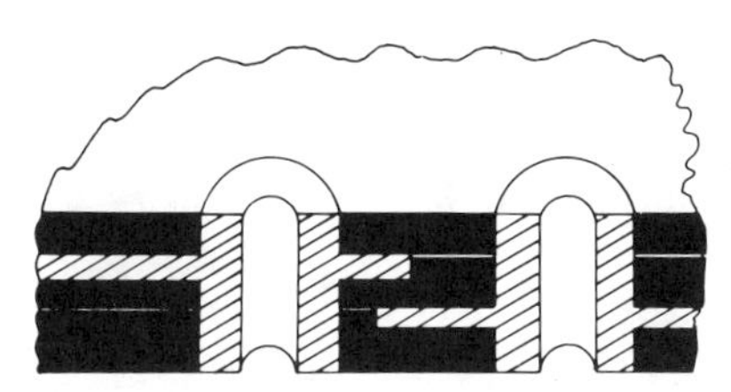

FIG. 11. The optical-chemical blanking process.

plating. The interconnection between the different layers is accomplished by plating through the holes connecting circuitry on individual layers with each other and to the surfaces of the board. Fabrication is by printing and etching of individual internal planes; registration and lamination; drilling; plating of circuit lines and holes with a copper and then a plated-etch resist, e.g., gold, solder; and then etching of outside circuitry.

THE OPTICAL-CHEMICAL BLANKING PROCESS. The optical-chemical blanking process, shown in Fig. 11, consists in laminating a board from alternate layers of insulating and conducting material which has been chemically etched so that the conducting material forms the circuit pattern and the insulating material locates and supports the circuitry. This process is workable without interlayer connection, with interlayer connections which are etched into the circuitry from one piece of conducting material, or with welded interlayer connections.

THE BRAZED-TUBELET PROCESS. This technique, illustrated in Fig. 12, involves the stacking and laminating into one board of a number of layers of thin printed-wiring planes printed and etched in the conventional manner. Locations where interconnections are to be made are drilled through such that pads on inner layers larger than the diameter of the hole are hit and a rim of copper around the entire circumference of the hole is exposed. Copper pads are interconnected by means of a silver-plated tubelet which is inserted into the hole and brazed in place, fusing to all the exposed copper rings within the hole. A variation of this technique uses a solder-coated eyelet or pin and establishes connection with internal layers by solder reflow in the presence of adequate heat.

THE TRANSFER PROCESS. One of the more popular of the built-up techniques which also results in flush circuitry, the transfer process is in reality alternate layers of insulation and plated conductive material which are applied to a metal base, or *transfer plate*. The first layer of circuitry is selectively plated to the metal baseplate by printing the plating-resist pattern directly on the metal baseplate. This plating has a weak bond to the baseplate. As illustrated in Fig. 13, insulation is then screened over the circuit, leaving only the through-connection pads exposed. Through connections are then plated to the pads above them and the second layer of circuitry is selectively plated to the insulation and first-layer pads. The procedure is repeated for the desired number of layers, and the entire structure is then transferred from the metal baseplate to an insulating backing material, which is needed for mechanical strength and rigidity.

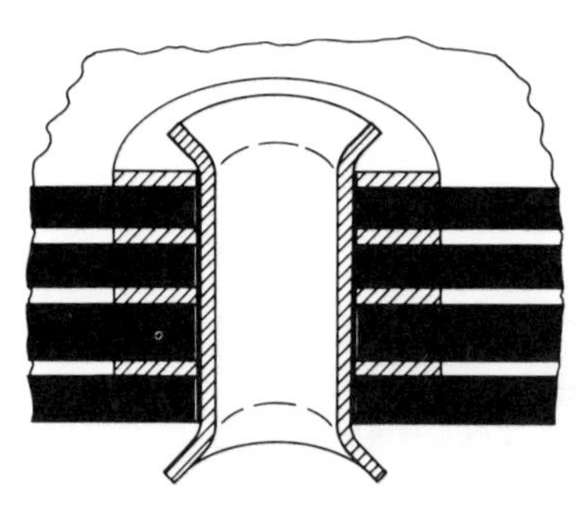

FIG. 12. The brazed-tubelet process.

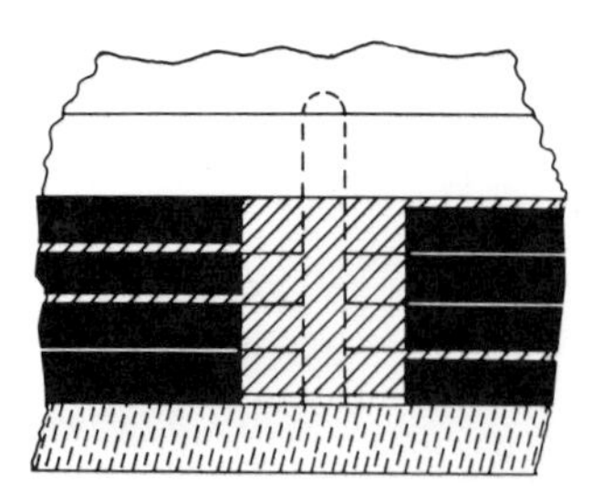

FIG. 13. The transfer process.

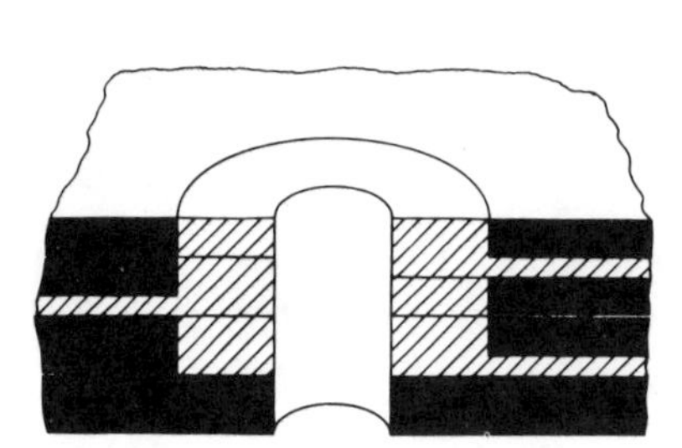

FIG. 14. The Beck process.

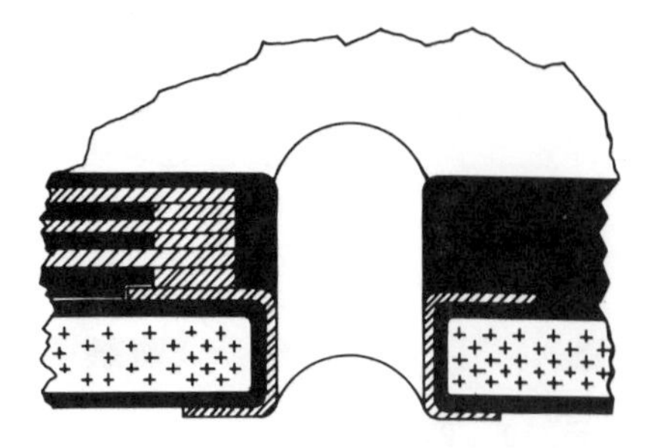

FIG. 15. The metal-spray process.

THE BECK PROCESS. This technique, illustrated in Fig. 14, differs from the transfer process in that the metal baseplate is replaced by a single-sided board and the circuitry is obtained by selective etching rather than selective plating. Etching resist is applied to the pad area on a single-sided board, and a portion of the copper is etched away, leaving the interlayer connections. Etching resist is added to the conductor area, and the balance of the copper is etched away, leaving the circuitry of one thickness and the interlayer connection pads of a greater thickness. Following the screening of an insulating material to the tops of the pads, copper is plated to the entire surface, and this procedure is repeated for the successive layers of circuitry.

THE METAL-SPRAY PROCESS. Illustrated in Fig. 15, this technique is initiated by drilling a metal or composition board with the holes which are required for interlayer connections. The entire board-and-hole surface is coated with an insulating material by the fluidized-bed or spray technique. Molten copper (or other metal) is then sprayed onto the dielectric-covered substrate through stainless-steel masks which define the circuit pattern. Areas for interlayer connections are then masked

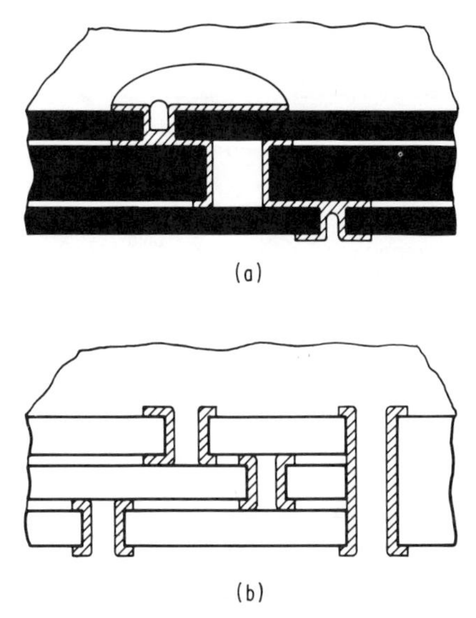

FIG. 16. Sequential-lamination techniques.

off, and insulting material is again applied for the next layer. The multilayer assembly is therefore built up by alternately spraying molten copper and applying insulating materials on the original-core substrate.

SEQUENTIAL-LAMINATION TECHNIQUES. Demands placed on multilayer printed wiring for greatly increased density have reached the point where conventional multilayer techniques are inadequate. Two attempts of designers to meet this need are the sequential-lamination methods illustrated in Fig. 16. One variation of the method, shown in Fig. 16*a*, is fabricated by sequentially laminating layers of a single-sided printed-wiring board onto each side of a two-sided, conventional printed-wiring board. Connections between layers are made by plating from the boards recently laminated to a land on the next internal layer during the process of board buildup. These interlayer connections do not need through-hole drilling in order to make internal connections. Fig. 16*b* shows another variation of the sequential-lamination process wherein a series of double-sided boards with plated-through interconnections are laminated together, with connections between these three boards established by plating a hole drilled through the entire laminated-board structure.

Flexible-flat-conductor Wiring. According to the Institute of Printed Circuits, flexible-flat-conductor wiring consists of a number of conductors laminated between two or more layers of plastic insulation and are of two basic types.

1. *Flexible flat cable.* Flexible, multiconductor, flat wire cable is used most often with connectors at each end in the same way as conventional, round-cable wiring.

2. *Flexible printed wiring.* Multiconductor, etched, flexible flat-encapsulated wiring is generally custom-designed for a special application with regard to an electronic-systems design and connected by conventional techniques.

Either of these generic types may be piled up in the same fashion as multilayer circuits in order to obtain the required circuit density. This kind of wiring is characterized by its flexible, conforming nature and by its flat, thin shape.

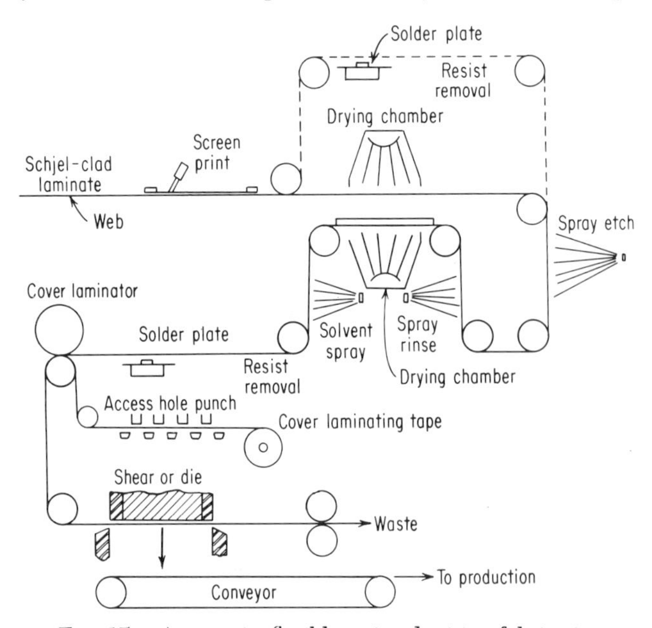

FIG. 17. Automatic, flexible, printed-wiring fabrication.

Single-layer Flexible Wiring. Whether classified as cable or wiring as in types 1 and 2 above, the constructional features are of little difference. In general, however, the cable is usually supplied with an encapsulating layer of plastic insulation on the outside surfaces since the usual mode of connection is through the ends. The process outline is very similar to that for rigid single-sided boards of the etched-foil variety, Fig. 1. Except for rare instances, flexible printed wiring of the built-up or electroplating-transfer process is not fabricated or used. The manufacturing process, as illustrated in Fig. 17, is predominantly a continuous or automatic one and consists in printing an acid-resist (or plating-resist) image in the appropriate form using either screening or photoresist methods. If the etch resist is used, the conductors are then etched. If plating resist is utilized, the conductors are plated with an acid-resistant metal and then etched. As explained above, an encapsulating layer may be applied and termination points made.

Double-sided and Multilayer Flexible Wiring. The etched-foil techniques are used almost exclusively for these varieties of flexible wiring, with the only options available in the area of interconnection between layers. For double-sided or very thin multilayer configurations, plated-through holes are rarely used. Interfacial connections in double-sided planes and interlayer connections in multilayer planes are made predominantly by the clearance-hole technique as illustrated in Fig. 9, one of the mechanical interconnection methods shown in Fig. 8, or the brazed-tubelet technique as illustrated in Fig. 12.

Shielded, Flexible Printed Wiring. A number of different methods of shielding may be used which will not interfere with flexibility. Some degree of shielding may be obtained in a single-sided plane merely by grounding adjacent conductors. The most efficient shielding, however, is accomplished by a three-layer construction where the outer layers are either of a solid-shield material or a lattice structure running across the conductors at a 90° angle. Arranging the lattice structure at an angle of 45° to the conductors permits sharper bending of the wiring. The constructional features outlined above for multilayer wiring apply. Effectiveness of the various types of shielded cables is outlined later in this chapter in the application section, page 1-100.

MATERIALS FOR PRINTED-CIRCUIT WIRING

Aside from the metallic foil or plating employed as the electrical conductor of a printed-wiring board, the only other major material category to be considered is that of insulating these base materials. These materials may be classed as organic and inorganic, e.g., ceramic, with organic materials accounting for more than 95 percent of all applications at the present time. Organic materials are typically rigid, reinforced thermosetting plastics or flexible, unsupported thermoplastics. Rigid-printed-wiring materials will be discussed first, followed by the more commonly used of the flexible materials. While the thinner epoxies (thermosetting plastics) are currently available in a flexible form, they have neither the mechanical strength, when unsupported, nor the flex-life characteristics with glass-cloth reinforcement to be classed with the flexible materials as commonly used.

Although, technically, printed-circuit base materials are composed of an organic or ceramic molding compound and some type of filler, or reinforcing material, for the sake of our discussion we shall consider only the end product of the material supplier's effort—the flat printed-circuit base, with or without the bonded metallic-foil facing or cladding. In the case of the reinforced thermosetting plastic materials, this is invariably a paper or glass-cloth reinforced laminate since molded plastics with or without fillers are rarely used in the industry. Non-woven-glass and glass-paper-mat reinforcements are, however, finding increasingly popular application as low-cost alternates to woven-fabric reinforcements.

Dielectric, or Base, Materials. Except for a small percentage of ceramic substrates, the materials most widely used to make insulating substrate materials are thermoplastic or thermosetting polymers. Referring to our definitions, it is obvious that thermoplastic materials may be rendered unserviceable if subjected to the temperature, e.g., in laminating composite structures and soldering, often experienced in rigid-printed-wiring-board fabrication. For this reason, primary emphasis for generally used rigid materials will be placed on reinforced thermosets and high-temperature thermosetting plastics, although some mention is necessarily made of unsupported thermoplastics such as polyolefins, polystyrenes, polyethylene terephthalates, and acrylics because of their specialty applications. Most of the films used for flexible-printed-wiring applications are thermoplastic. These will also be covered in this section. Properties of rigid and flexible substrate or base materials, most of which are used in printed wiring, are given in Tables 2 and 3, respectively.

Reinforcing Materials. Rigid-printed-wiring bases, or substrate support, in common use today are invariably reinforced laminates in which a number of sheets of a partially cured, resin-impregnated mat or fabric have been bonded under heat and pressure into a fully cured board, although resins such as melamines and polyesters which are difficult to B-stage (see Table 1) may be combined with a reinforcement by a wet lay-up method, i.e., pouring the liquid resin on the reinforcement and laminating. This "binder," or plastic, is largely responsible for properties such as dielectric and electrical characteristics, dimensional stability, and heat resistance. Organic reinforcements include kraft or rag paper, cotton and linen cloth, nylon fabric, and synthetic-fiber papers. Inorganic reinforcements are woven-glass-cloth–glass-fiber papers and nonwoven-glass-mat, flake-glass, and asbestos paper. By far the most important reinforcing materials currently in use in the printed-

TABLE 2. Properties of Laminated and

NEMA minimum or maximum average values, sheets*

NEMA Grade	Min. flexural strength, 1/16 in., 10^3 psi		Min. Izod impact strength, ft-lb/in., notch, edge		Min. bond strength, lb	Max. water absorption, 1/16 in., %	Min. dielectric strength,§ kv	Max. dielectric const., 1 MHz, 1/32 in. or more	Max. dissipation factor, 1 MHz, 1/32 in. or more	Min. arc resistance, sec
	LW‡	CW‡	LW	CW						
X	25	22	0.55	0.50	700	6.00	...	...	...	...
XP	13	11	...	...	...	3.60	40	...	...	...
XPC	10	8	...	...	...	5.50	...	...	...	...
XX	15	14	0.40	0.35	80	2.00	40	5.5	0.045	...
XXP	14	12	...	...	0	1.80	60	5.0	0.040	...
XXX	13.5	11.8	0.40	0.35	950	1.40	50	5.3	0.038	...
XXXP	12	10.5	...	...	...	1.00	60	4.6	0.035	...
XXXPC	12	10.5	...	...	...	0.75	60	4.6	0.035	...
ES-1	13.5	13.5	0.25	0.22	...	2.50	...	...	...	...
ES-2	...	...	0.25	0.22	...	...	...	...	...	...
ES-3	13.5	13.5	0.25	0.22	...	2.50	...	...	...	...
C	17	16	2.10	1.90	1,800	4.40	15	...	...	...
CE	17	14	1.60	1.40	1,800	2.20	35	...	...	...
L	15	14	1.35	1.10	1,600	2.50	15	...	...	...
LE	15	13.5	1.25	1.00	1,600	1.95	40	5.8	0.055	...
A	13	11	0.60	0.60	700	1.50	5	...	...	...
AA	16	14	3.60	3.00	1,800	3.00	...	...	...	...
G-2	20	16	4.5	3.3	1,000	1.50	30	5.5	0.025	...
G-3	20	18	6.5	5.5	850	2.70	...	...	...	...
G-5	50	40	7.5 (to 1/2 in.)	5.5	1,570	2.70	23	7.8	0.020	180
G-7	20	18	6.5	5.5	650	0.55	32	4.2	0.003	180
G-9	60	40	13.0	8.0	1,700	0.80	60	7.5	0.018	180
G-10	55	45	7.0	5.5	2,000	0.35	35	5.2	0.025	...
G-11	55	45	7.0	5.5	1,600	0.35	35	5.2	0.025	...
N-1	10	9.5	3.0	2.0	1,000	0.60	60	3.9	0.038	...
FR-2	12	10.5	...	...	...	0.75	60	4.6	0.035	...
FR-3	20	16	...	...	...	0.65	60	4.6	0.035	...
FR-4	...	...	7.0	5.5	2,000	...	35	5.2	0.025	...
FR-5	...	...	7.0	5.5	1,600	...	35	5.2	0.025	...
GPO-1	18	18	8.0	8.0	850	1.00	40	4.3†	0.03†	100
GPO-2	18	18	8.0	8.0	850	0.9	40	...	...	100

* See NEMA Publication No. LI 1-1965, "Standards Publication for Industrial Laminated Thermosetting Products," for test methods and conditions, etc.
‡ LW and CW indicate lengthwise and crosswise, respectively.
§ Parallel to lamination, step by step, 1/16 in. thick.

Reinforced Plastics [5]

Typical values, sheets†												
Tensile strength, 10^3 psi		Compressive strength, 10^3 psi		Rockwell hardness, M scale	Specific gravity, psi	Dielectric strength perpendicular to lamination, volts/mil		Thickness range, in.		Base material	Resin	NEMA Grade
LW	CW	Flat	Edge			Short time	Step by step	Min.	Max.			
20	16	36	19	110	1.36	700	500	0.010	2	Paper	Phenolic	X
12	9	25	...	95	1.33	650	450	0.010	1/4	Paper	Phenolic	XP
10.5	8.5	22	...	75	...	600	425	1/32	1/4	Paper	Phenolic	XPC
16	13	34	23	105	1.34	700	500	0.010	2	Paper	Phenolic	XX
11	8.5	25	...	100	1.32	700	500	0.015	1/4	Paper	Phenolic	XXP
15	12	32	25.5	110	1.32	650	450	0.015	2	Paper	Phenolic	XXX
12.4	9.5	25	...	105	1.30	650	450	0.015	1/4	Paper	Phenolic	XXXP
...	...	25	...	105	1.31	650	450	1/32	3/16	Paper	Phenolic	XXXPC
...	...	...	...	...	1.58	...	...	3/64	1/4	...	Melamine¶	ES-1
...	...	...	...	...	1.46	...	...	0.085	1/4	...	Phenolic¶	ES-2
...	...	...	...	...	1.48	...	...	3/64	1/4	...	Melamine¶	ES-3
10	8	37	23.5	103	1.36	150	...	1/32	10	Cotton	Phenolic	C
9	7	39	24.5	105	1.33	500	300	1/32	2	Cotton	Phenolic	CE
13	9	35	23.5	105	1.35	150	...	0.010	2	Cotton	Phenolic	L
12	8.5	37	25	105	1.33	500	300	0.015	2	Cotton	Phenolic	LE
10	8	40	17	111	1.72	225	135	0.025	2	Asb. paper	Phenolic	A
12	10	38	21	103	1.70	...	...	1/16	2	Asb. fabric	Phenolic	AA
13.7	10	38	15	105	1.50	500	360	1/32	2	Staple glass	Phenolic	G-2
23	20	50	17.5	100	1.65	700	500	0.010	2	Cont. glass	Phenolic	G-3
37	30	70	25	120	1.90	350	220	0.010	3½	Cont. glass	Melamine	G-5
23	18.5	45	14	100	1.68	400	350	0.010	2	Cont. glass	Silicone	G-7
40	25	65	...	...	1.90	400	350	...	...	Cont. glass	Melamine	G-9
35	30	70	30	110	1.75	700	500	0.010	1	Cont. glass	Epoxy	G-10
35	30	70	30	110	1.75	700	500	0.010	1	Cont. glass	Epoxy	G-11
8.5	8	28	...	105	1.15	600	450	0.010	1	Nylon	Phenolic	N-1
12.4	9.5	25	...	105	1.30	650	450	0.030	1/4	Paper	Phenolic	FR-2
12	9	28	...	...	1.45	600	500	1/32	1/4	Paper	Epoxy	FR-3
35	30	70	30	110	1.75	700	500	0.010	1	Cont. glass	Epoxy	FR-4
35	30	70	30	110	1.75	700	500	0.010	1	Cont. glass	Epoxy	FR-5
12	10	30	20	100	1.5–1.9	400	...	1/16	2	Glass mat	Polyester	GPO-1
10	9	30	20	100	1.5–1.9	...	...	1/16	2	Glass mat	Polyester	GPO-2

† Typical values from a number of sources—do not use for specifications or standards; consult manufacturers.

¶ Usually made with these resins.

TABLE 3. Typical Flat-cable Insulation [6]

	TFE Teflon	TFE Teflon Glass Cloth	FEP Teflon	FEP Teflon Glass Cloth	Kapton*	KEL-F†	PVF Tedlar‡	Polypropylene	Polyester	Polyvinyl chloride	Polyethylene	Mylar
Specific gravity	2.15	2.2	2.16	2.2	1.38	2.10	1.37	0.905	1.38	1.25	0.93	
In.^{2} of 1 mil. film per pound	12,800	13,000	12,800	13,000	20,000	12,000	20,000	31,000	20,000	22,000	30,100	19,850
Service temp. °C: Min.	−70	−70	−70	−70	Not known	−70	−55	−55	−55	−40	−20	−60
Max.	250	250	200	250	>400	150	125	125	100	85	60	150
Flammability	Nil	Nil	Nil	Nil	Nil	Nil	Yes	Yes	Yes	Slight	Yes	Slight
Appearance	Translucent	Tan	Clear bluish	Tan	Amber	Clear	Clear	Clear	Clear	Translucent	Clear	Clear
Thermal expansion, 10^6 in./in./°F	70	Low§	50	Low§	Not known	45	24	61	...	...	...	11
Bondability with adhesives	Good¶	Good¶	Good¶	Good¶	Good	Good¶	Good¶	Poor	Good	Good	Poor	Good
Bondability to itself	Good	Poor	Good	Good	Poor	Good	Good	Good	Poor	Good	Good	Good
Tensile strength, psi, 77°F	3,000	20,000	3,000	20,000	20,000	4,500	8,000	5,700	20,000	3,000	2,000	23,000
Modulus of elasticity, psi	80,000	§	50,000	§	400,000	200,000	220,000	170,000	500,000	...	50,000	
Electrical resistivity, ohms	2×10^{16}	$>10^{16}$	28×10^{16}	$>10^{16}$	10^{17}	1×10^{15}	3×10^{13}	10^{16}	4×10^{15}	1×10^{10}	1×10^{16}	1×10^{12}
Dielectric constant, 10^2–10^8 Hz	2.2	2.5–5§	2.2	2.5–5§	3.5	2.5	6.8	2.0	2–3	3–4	2.2	3.0–3.5
Dissipation factor, 10^2–10^8 Hz	0.0002	0.0007–0.001§	0.0002	0.0001–0.001§	0.1	0.015	0.013	0.0002–0.0003	0.01	0.14	0.0006	0.002–0.016
Dielectric strength (5 mils thickness), volts/mil	800	650–1,600	1,000	650–1,600	6,000	2,000	2,000	(0.125 in. thick) 750	5,000	800	1,500	7,000
Chemical resistance	Excellent	Excellent	Excellent	Excellent	Excellent	Excellent	Good	Excellent	Excellent	Good	Excellent	Excellent
Water absorption, %	0	0.10–0.68	0	0.18–0.30	0	0	0.10	<0.01	0.5	0.10	0	0.8
Sunlight resistance	Excellent	Excellent	Excellent	Excellent	Excellent	Excellent	Excellent	Low	Fair	Fair	Low	Low

* Trademark, E. I. du Pont de Nemours & Co., Inc.
† Trademark, Minnesota Mining and Manufacturing Co.
‡ Trademark, E. I. du Pont de Nemours & Co., Inc.
§ Depends on percent of glass cloth.
¶ Must be treated.

wiring industry at this time are wood-pulp or rag paper and glass cloth. Nylon and cotton fabrics and flake or "flock" glass find some use, while other reinforcing materials are called on for special applications. Few flexible-printed-wiring films are produced with reinforcement.

Thermosetting Polymers. *Phenoplasts or Phenolics.* Historically one of the oldest plastic-insulation materials used by the electrical industry, phenolic resins remain the largest-volume thermoset, having also a large number of applications outside the printed-wiring industry. The original single-stage phenolic resin, the product of a reaction between phenol and formaldehyde, represents one of the lowest production costs of any of the thermosets. Two-stage resin phenolics, however, are the most commonly used in printed-wiring manufacture because they are more stable, have better electrical characteristics, and lower volatile content (which may cause blistering during soldering) than the single-stage phenolics. Because of their chemical nature, two-stage phenolic resins continue to evolve minute quantities of ammonia even after curing, one of the disadvantages of using this material. Although the resins of the phenolic family possess good dimensional stability and reasonably good electrical properties, they are deficient with regard to arc resistance. The National Electrical Manufacturers Association (NEMA) classification for phenolic laminates is X, XX, or XXX; which means that they contain at least 35, 45, or 58 percent by weight, respectively, of phenolic resin. A P nomenclature indicates they can be punched hot, and PC indentifies a cold-punch grade. In the following listing, "MIL Grade" refers to unclad material only:

Grade XXP* (MIL Grade PCB). This paper-based grade may be used for ordinary electrical applications and is supplied plasticized for hot punching from 200 to 250°F. It is reasonably good in mechanical properties and fair in electrical properties, except for poor arc resistance.

Grade XXX (MIL Grade PBE). Another paper-based phenolic, this laminate has slightly better mechanical properties than XX but is not recommended for punching. It is suitable for radio-frequency work and high-humidity applications.

Grade XXXP (MIL Grade PBE-P). Suitable for the majority of electrical applications of XXX, XXXP has the advantage of being plasticized for hot punching. Electrical properties permit its use up to 10 MHz. It can be obtained with a modification that produces self-extinguishing characteristics when ignited.

Grade XXXPC (MIL Grade PBE-P). Similar in electrical properties to XXXP, this grade is suitable for punching at lower temperature. This grade is recommended for applications requiring high insulation resistance and low dielectric losses under severe humidity conditions.

Grade FR-2 (MIL Grade PBE-P). This grade is the same as XXXPC, except that FR-2 is self-extinguishing. The latter is suitable for punching at lower temperatures than XXXP. With good punching practice, sheets up to and including 0.062 in. may be punched at 75°F and up to 0.125 in. when heated to 140°F. It has generally superior electrically insulating characteristics.

Grade G-2. The only phenolic covered here with a glass-base (staple-fiber type) glass cloth, this grade is noted for its electrical properties and heat resistance. This material has good electrical properties under high-humidity conditions, and although mechanically the weakest of all the glass-cloth grades, it is also the lowest in dielectric losses of all glass-based thermosets except the silicones. In addition, it has good dimensional stability.

Epoxies. The superior insulating properties, low dielectric loss and shrinkage, and excellent mechanical strength, all maintained under wide temperature and moisture conditions, have made epoxies the most popular and fastest growing in the printed-wiring industry today. This polymer is chemically the reaction product of epichlorohydrin and bisphenol A. A multifunctional amine is generally employed in the curing process. Epoxies are highly chemical-resistant, being most affected by strong oxidizing acids and ketone-type solvents. They are characterized by excellent adhesion, cure capability at low molding or laminating pressures, excellent dimensional stability, and electrical properties. The superior mechanical properties and adhesion of the epoxy group are often attributed to the hydrogen bonding to

* The designations used are in accordance with ASTM and NEMA.

substrates produced by hydroxyl groups in the resin, and the low shrinkage is accounted for by factors involved in the crosslinked molecular structure.

Grade FR-3 (MIL Grade PEE). A paper-base material with epoxy-resin binder, this laminate has low dielectric-loss properties with good stability of electrical properties under conditions of high humidity. It has good punching characteristics; sheets up to and including 0.062 in. may be cold-punched (at 80°F, and sheets from 0.062 to 0.125 in. may be cold-punched when warmed to 150°F). It has high flexural strength (better than XXXPC), and it is suitable for electrical applications through the medium-frequency range. It is the only common paper-based epoxy used for printed circuits.

Grades G-10 and FR-4 (MIL Grade GEE). G-10 is a continuous-filament-type glass cloth with epoxy-resin binder. It has extremely high mechanical strength (flexural, impact, and bonding) at room temperature. Good dielectric-loss and dielectric-strength properties under both dry and humid conditions foster its use for most electrical applications up through the high-frequency range. Bond strengths of the copper-clad laminate are as high as any other common grade, and it is probably the most widely used of any base material for printed circuits, especially multilayer. It is available as a self-extinguishing type, FR-4.

Grades G-11 and FR-5 (MIL Grade GEB). Materials and properties of G-11 are similar to those of G-10 at room temperature, but with the ability to retain 50 percent of its initial flexural strength when measured at 150°C after 1 hr at that temperature. The material is somewhat more difficult to machine than G-10 or FR-4 and is more expensive. Again, the self-extinguishing type is FR-5.

Silicones. Chemically these partially inorganic resins are characterized by alternating atoms of silicone and oxygen, with organic groups attached to the silicon atoms. The extent and type of the organic substitution and molecular weight determine whether the resin is liquid or solid. Electrical properties include high arc resistance and dielectric strength, low dielectric constant and dissipation factor, and resistance to corona and electrical breakdown. These properties hold at elevated temperatures also, ranging from 350 to 750°F, and are resistant to degeneration due to weathering and oxygen. At the present time application of silicone resin laminates is somewhat limited because of the poor adhesion obtainable between foil cladding and base laminate and the poor interlaminar bond strength. New developments are expected to minimize these disadvantages, however.

Grade G-7 (MIL Grade GSG). A continuous-filament-type glass cloth with silicone-resin binder, this laminate has extremely good dielectric loss and insulation resistance under dry conditions and good electrical properties under humid conditions. It is suitable for printed-wiring applications in the very-high-frequency range up to 500 MHz. The material is difficult to obtain in the copper-clad form because of the bonding problems previously mentioned.

Amino Plastics. Although two plastics in this class, the urea formaldehyde and melamine formaldehyde resins, are both thermosetting, only the latter is suitable for printed-circuit use since it is relatively unaffected by processing chemicals. The material performs satisfactorily over a temperature range of −70°F to a continuous operating temperature of 210°F. While the glass-reinforced material has excellent arc characteristics, it is not generally popular as a printed-circuit material because of relatively poor loss characteristics.

Grade G-5 (MIL Grade GMG). A continuous-filament-type glass-cloth-with-melamine resin binder, this laminate has a high mechanical strength and is one of the hardest laminated grades. It has good flame resistance and is second only to silicone in heat and arc resistance.

Grade G-9 (MIL Grade GME). This laminate is similar to the preceding G-5, except that it has additional heat resistance.

Other Rigid-base Materials. Diallyl phthalate (DAP) molding compounds may also be obtained with various mineral or glass fillers or polyester fiber reinforcements. They are unsaturated polyester resins without volatile monomers or reactive diluents that often characterize polyesters, and are usually cured with a peroxide catalyst after formulation with a suitable filler or reinforcement. They are good where maximum insulation resistance at high humidity and good dimensional sta-

of accessory minerals by some sort of beneficiation, and are graded and supplied on the basis of the amount and type of impurities present. In the case of ceramics made from single oxides such as alumina, magnesia, or beryllia, these raw materials' complexities are nearly absent since the starting materials are essentially pure oxides obtained by chemical means. Actually, with the advent of precision ceramics for electronic applications, there is a growing need and tendency to start with chemically pure raw materials in all cases.

The selection of raw materials for a given ceramic composition is made in light of the forming method to be used and the solid-state reactions and phase transformations which can occur during sintering, or firing, at high temperatures. Most of the common ceramic compositions have a predominant crystalline phase or matrix which contains the other phase features, and numerous authors have discussed the systems in some detail. Various ceramic insulators and their chemical makeup which can be considered popular for circuit base materials are listed in Table 4, together with related minerals and compounds which are important in any discussion of ceramics.

TABLE 4. Selected Ceramic Raw Materials and Constituents of Fired Bodies [1]

Material	*Composition*
Kaolinite (kaolin, clay)	$Al_2O_3 \cdot 2SiO_2 \cdot 2H_2O$
Quartz (flint, silica)	SiO_2
Feldspar	$Na_2O/K_2O \cdot Al_2O_3 \cdot 6SiO_2$
Talc (steatite)	$3MgO \cdot 4SiO_2 \cdot H_2O$
Magnesite	$MgCO_3$
Magnesia	MgO
Corundum (alumina, sapphire)	Al_2O_3
Steatite (protoenstatite)	$MgO \cdot SiO_2$
Forsterite	$2MgO \cdot SiO_2$
Cordierite	$2MgO \cdot 2Al_2O_3 \cdot 5SiO_2$
Mullite	$3Al_2O_3 \cdot 2SiO_2$
Porcelain	$3Al_2O_3 \cdot 2SiO_2 + SiO_2$
Spinel	$MgO \cdot Al_2O_3$
Ferrite spinel	$Ni_xZn_{1-x}Fe_2O_4$
Zircon	$ZrO \cdot SiO_2$
Wollastonite	$CaO \cdot SiO_2$
Spodumene	$Li_2O \cdot Al_2O_3 \cdot 4SiO_2$
Barium titanate	$BaTiO_3$
Rutile (titania)	TiO_2
Beryllia	BeO
Boron nitride	BN

Although a number of ceramics are listed, only a relative few in number have proved capable of providing properties useful in printed wiring, and these only in specialized applications. Among ceramic materials currently used are steatite, alumina, beryllia, forsterite, glass-bonded mica, and photosensitive glass ceramics. The chemical, dimensional, thermal, and electrical stability of most ceramics usually cannot be equaled by conventional organic material. Ceramics, therefore, find their widest acceptance as printed-wiring material where precise control of properties is required and mechanical properties are of lesser importance (partly because of this limitation, ceramic-based circuits are generally found in relatively small physical dimensions). Significant properties of ceramic materials used for insulation are outlined in Table 5.

Conductive Materials. Printed-wiring materials in use today invariably come with a conductive metal foil bonded to one or both sides of the dielectric sheet,

TABLE 5. Properties of

Material	Dielectric properties at 1 kHz, room temp.			Dielectric properties at 1 MHz, room temp.			Dielectric properties at 10 MHz, room temp.		
	tan δ	Dielectric constant	Loss factor	tan δ	Dielectric constant	Loss factor	tan δ	Dielectric constant	Loss factor
Porcelain	......	......	......	0.008–0.020	5.0–6.5	0.10–0.05	0.02	7.0	0.14
Zircon	......	......	......	0.001–0.0014	8.0–10.5	0.013–0.009	0.0027	8.4	0.23
Alumina	0.0014–0.0042	8.1–9.5	0.012–0.04	0.0003–0.002	8.2–11.2	0.002–0.03	0.00014–0.0027	8.1–9.3	0.0013–0.022
Steatite	0.0007–0.009	5.1–5.6	0.0035–0.05	0.0008–0.0035	5.9–6.1	0.004–0.03	0.0009–0.0019	5.5–5.8	0.005–0.01
Forsterite	0.001	6.2	0.006	0.0004–0.001	5.8–6.7	0.002–0.008	0.001	5.8	0.0058
Cordierite	......	......	......	0.003–0.007	4.1–5.4	0.008–0.010	0.003	5.0	0.015
Spinel	......	......	......	0.0004	7.5	0.003	......	......	......
Mullite	......	......	......	0.004–0.005	6.2–6.8	0.025–0.034	......	......	......
Magnesia	......	......	......	0.001	8.2	0.002–0.01	......	......	......
Quartz (Fl-3)	......	......	......	0.0004	4.4	0.002	......	......	......
Quartz (single crystal)	......	......	......	0.0002	4.44	0.0011			
$BaO \cdot MgO \cdot Al_2O_3 \cdot SiO_2$(K-6)	......	......	......	0.0018	5.5	0.008–0.016	......	......	......
Wollastonite (E-16)	......	......	......	0.00038	6.6	0.0026	0.0013	6.3	0.008
BeO	0.0084	4.5	0.004	0.001	5.8	0.006	0.0005	4.2	0.002
Lead alumina silicate (P-2)	......	......	......	0.001	5.8	0.006	......	......	......
Spodumene (Sp-1)	......	......	......	0.004	6.4	0.03	......	......	......
$Li_2O \cdot Al_2O_3 \cdot SiO_2$	......	......	......	0.005	5.3	0.03–0.10	......	......	......
Boron nitride	0.001	4.2	0.004	0.001	4.2	0.004	......	......	......
Zirconia	......	......	......	0.01	12.0	0.12	......	......	......
Thoria (99.9% ThO_2)	......	......	......	0.0003	13.5	0.004	......	......	......
Pyroceram	......	......	......	0.0017–0.013	5.5–6.3	0.01–0.07	......	......	......
Glass-bonded mica	......	......	......	0.0015–0.003	6.4–9.2	0.023–0.011	......	......	......
Mica	......	......	......	......	5.4–8.7	......	......	......	......
Glass (gen'l)	......	......	......	0.0005–0.01	3.8–15.0	0.0019–0.14	......	......	......
Glass (quartz)	......	......	......	0.0003	3.5–4.0	0.0015	......	......	......
Porous ceramics	......	......	......	......	1.5	0.0015–0.10	......	......	......
Air or vacuum	......	......	......	......	1.0				
Copper	......	......	......	......	......	......	......	......	......
Steel	......	......	......	......	......	......	......	......	......
Kovar†	......	......	......	......	......	......	......	......	......
Hafnia	......	......	......	0.01	12	0.12	......	......	......
Ceria (99 + % CeO_2)	......	......	......	0.0007	15	0.011	......	......	......

* Te value is temperature at which 1 cm^3 has a resistance of 1 megohm.
† Trademark, Westinghouse Electric Corp.

Ceramic Insulation [15]

Dielectric strength, volts/mil	Te,* °C	Sp. gr.	Resistivity at 25°C, ohm-cm	Thermal conductivity, Btu/(hr)(ft²)(°F)(ft)	Thermal coef. of expansion 10^{-6} in./(in.)(°C)	Tensile strength, psi, t_0	Transverse strength, psi, t_0	Compressive strength psi, t_0	Thermal shock resistance
55–300	350	2.4	10^{14}	1.5	6.0	6,000	11,000	50,000	Fair
60–290	870	3.7–4.3	$>10^{14}$	3.0–5.0	3.5–5.5	13,000	24,600	75,000	Good
250–400	1070	3.6	10^{16}	10.0	8.0	25,000	60,000	500,000	Good
200–350	840	2.8	10^{17}	2.0	7.8–10.4	10,000	20,000	90,000	Moderate
200–300	1040	2.8	10^{17}	0.9–2.4	10.6	10,000	20,000	85,000	Poor
140–230	780	2.0–2.9	10^{16}	1.8	2.3	5,000	16,000	35,000	Excellent
300	1170	2.8	10^{12}	4.4	6.6	......	14,000	247,000	Fair
....	600	2.5	....	1.5	4.3–5.0	4,000	18,000	56,000	Fair
....	1300	2.5–3.3	$>10^{14}$	23.0	12.8	12,000	19,000	150,000	Fair
....	....	....	....	....	10.0				
....	....	....	....	....	2.4	......	......	....	Good
....	....	2.9	....	1.5	7.0	7,000	18,000	80,000	Moderate
240–350	....	2.8–2.9	$>10^{16}$	30.0–125.0	8.4	14,000	35,000	225,000	Good
....	....	....	....	....	5.1	......	6,000	....	Fair
....	....	2.4	....	....	2.0	......	7,000	....	Good
....	....	......	10^{12}	1.0	−0.04–+0.5	4,000	8,000	56,000	Excellent
900–1,400	....	2.1	10^{14}	16.6	4.3	3,500	7,500	40,000	Good
....	....	5.6	10^{8}	14.3	3.0–8.3	18,000	26,000	300,000	Poor
....	....	9.7	10^{10}	8	5.3–9.0	18,000	18,000	220,000	Poor
250–300	....	2.4–2.6	....	1.1–2.1	0.2–4	......	35,000	....	Good
270–600	....	2.6–3.8	10^{14}	....	10.5	7,000	16,000	25,000	Fair
1,000–2,000	....	2.6–3.8	....	0.2–0.4	18–27				
....	....	2.0–8.0	10^{12}	0.4–0.8	0.8–1.3	$<4{,}000$	......	100,000	Fair
400–410	....	2.2	10^{14}–10^{18}	0.8	0.3	7,000	......	$>160{,}000$	Excellent
....	....	1.9	....	0.05–0.5	6.3				
....	....	8.9	....	196–226	17.0	35,000			
....	....	7.8	....	21–38	15.0	50,000–300,000			
....	....	8.1	....	7.8–10.3	5.1	80,000			
....	....	9.0	10^{8}	1.0	6.5	......	15,000	200,000	Poor
....	....	7.0	10^{9}	7.0	10.0	......	15,000	200,000	Poor

which is masked and chemically etched into the circuit pattern. A small number of processes or applications involve plating or metal deposition of the conductor in some fashion on an insulating substrate. The metallic material must have suitable electrical and chemical properties to qualify it for its intended use and total material system. Conductive materials for printed-wiring boards are most commonly a bonded metallic foil of the rolled or electrodeposited variety (although plated-on metals and conductive inks and adhesives are also used); the properties of some of these metals are listed for consideration in Table 6.

TABLE 6. Metallic-foil Properties

Metal	Symbol	Electromotive potential, volts	Resistivity, microhms/cm²	Relative conductivity, Cu = 100	Oz/ft²/ 0.001 in.	Thermal coefficient of expansion, 10^{-5} in./(in.)(°F)	Mod. elast. in tension, 10^6 psi	Brinell hardness No.
Aluminum	Al	+1.67	2.665 at 20°C	65	0.22	1.33	10	15
Nickel	Ni	+0.25	6.84 at 20°C	25	0.74	0.76	30	110
Tin	Sn	+0.14	11.50 at 20°C	15	0.60	1.30	6	5.2
Lead	Pb	+0.13	20.65 at 20°C	7.7	0.94	1.60	2.6	3.9
Copper	Cu	−0.34	1.726 at 23°C	100	0.74	0.91	16	42
Silver	Ag	−0.80	1.59 at 20°C	104	0.79*	1.05	11	95
Rhodium	Rh	−0.82	9.83 at 0°C	33	0.95*	0.43	21	100
Gold	Au	−1.68	2.19 at 0°C	70	1.47*	0.80	12	28

* Troy oz.

Metallic Foils. The predominant metal used as foil cladding by the printed-wiring industry is 99.5+ percent pure copper. Other foils such as aluminum, silver, phosphor bronze, beryllium copper, nickel, Kovar, and magnetic alloys are also used for special applications, but to a much smaller extent. Therefore this section will treat only copper, although other materials and their properties are tabulated in Table 6 for comparison. The primary usefulness of Kovar and nickel is their weldability, which allows component mounting on base materials of much lower heat resistance than those subject to traditional soldering operations. Historically, rolled foil was the first form of copper available for cladding which was combined with laminated insulators in the early post World War II period. Its main disadvantage, availability only in relatively narrow widths, led to the development of the currently more popular electrodeposited form. Copper foil used in printed wiring, whatever the method of producing it, should have specific properties as defined in MIL-P-13949C.[7] This specification states: "Copper foil shall have a minimum purity of 99.5 percent and a maximum resistivity of 0.15940 ohm-gram per meter squared at 20°C before or after laminating to the base material." Copper-cladding thickness tolerances are given in Table 7.

TABLE 7. Copper-cladding Thickness Tolerances

Nominal thickness		Nominal weight		Tolerances		
in.	mm	oz/ft²	g/mm²	By weight, %	By gauge, in.	By gauge, mm
0.0007	0.0178	½	1.5	±10	±0.0002	±0.0051
0.0014	0.0355	1	3.06	±10	+0.0004 −0.0002	+0.0102 −0.0051
0.0028	0.0715	2	6.12	±10	+0.0007 −0.0003	+0.0178 −0.0076
0.0042	0.1065	3	9.18	±10	±0.0006	±0.0152
0.0056	0.1432	4	12.24	±10	±0.0006	±0.0152
0.0070	0.1780	5	15.30	±10	±0.0007	±0.0178
0.0084	0.2130	6	18.36	±10	±0.0008	±0.0204
0.0098	0.2460	7	21.42	±10	±0.001	±0.0254
0.014	0.3530	10	30.6	±10	±0.0014	±0.0355
0.0196	0.4920	14	43.2	±10	±0.002	±0.0508

ROLLED COPPER FOIL. Conventional metal-rolling techniques using electrolytically refined ingots of copper are used to produce rolled foil. Finish of this variety is characterized by relatively smooth surfaces on both sides. Unannealed, it has approximately twice the tensile strength of the electrodeposited form and is harder than the latter due to the work hardening induced in the rolling operation. Rolled copper has less inclusions and pinholes than the electrodeposited type and can be produced in a copper-alloy form to give it special characteristics. The superior fatigue resistance, i.e., flex-failure resistance, of this variety leads to its utilization in some flexible cables where vibration is encountered. Its main drawbacks, however, are its unavailability in roll widths of over 24 in. (for 0.001 in. thick) and its smooth surface requiring special treatment by the fabricator for bonding.

ELECTRODEPOSITED COPPER FOIL. Produced in widths up to 5 ft, electrodeposited foil is made by plating copper from a copper sulfate or copper cyanide bath onto a slowly turning, circular, burnished stainless-steel or lead drum. It is stripped from the drum on a continuous basis and cleaned and dried before cladding. The granular structure of the foil may be varied by controlling the bath composition and/or the plating parameters. As produced, the side toward the drum is relatively smooth and shiny and is used for resist printing or electrical contacts; the other surface is grainy or matte, which is welcome for bonding purposes. This surface is conventionally conditioned with a chemically oxidized surface, or A* condition, in varying degrees dependent upon the application.

PRINTED-WIRING PLATING MATERIALS. For a number of reasons a variety of metals other than copper are used in the fabrication of printed wiring, including an etchant resist for processing, e.g., gold, protection against corrosion, and enhancement of conductor trace hardness and resistance to abrasion. Copper is frequently plated to existing foil in board processing. This will be covered in a later section. Silver, once used in the range of 0.0001 to 0.002 in. thick for static-pressure contacts and switching surfaces, has more or less dropped from the printed-wiring picture owing to its propensity for tarnishing and migration across dielectric materials under certain conditions of humidity. In paragraph 5.9.1.1 of MIL-STD-275B,[8] entitled "Printed Wiring for Electronic Equipment," its position is clarified by the statement: "Silver plating, whether underplating or final plating, shall not be used."

Nickel of the low-stress variety is utilized to a minimum thickness of 0.0005 in., usually as a plating between copper and rhodium where the latter is used as a contact surface. Because it has a wide range of hardness, depending upon the plating bath and conditions, nickel has also been used as a reinforcing plating in plated-through-hole fabrication. Gold is widely used because of its excellent resistance to corrosion and processing chemicals and acceptable conductivity. Although deposits as low as 5 millionths inch thick serve as a tarnish-resistant coating, current industry practice is to use gold ranging in thickness from 50 to 100 millionths (0.00005 to 0.00010) in. as suggested by MIL-STD-275. Recently, it has been determined that gold in thicknesses over 80 millionths in. is detrimental to good soldering practice. Although tin and tin-lead alloys may be used for contact materials, a more important current use in the printed-wiring process is to provide a solderable finish on the completed board. In the case of plated-through-hole processes and for eyelet use, the thickness control of electrodeposited solder has distinct advantages over that found with the hot-tinning process. MIL-STD-275 states that tin-lead plating shall be a minimum of 0.0003 in. thick and shall contain a minimum of 50 percent, and a maximum of 70 percent, tin. A more typical industry requirement is 65 percent tin, 35 percent lead. A requirement for a hard-plated surface for printed-circuit boards with sliding contacts is usually met by rhodium, a wear-resistant finish which has a Brinell hardness number of 100.

Other Conductive Materials. At its inception, the printed-circuit industry, as pointed out by Eisler,[3] used a great many processes calling for deposition, in some form or another, of the conductive areas. This is still done at the present time, but mainly for special applications, for ceramic substrates, or for repair of conventional

* Trademark, Circuit Foil Corp., Bordentown, N.J.

printed wiring. The conductive material most commonly used for plastic-based substrates is almost always silver, either in a liquid or paste form, and this is usually applied by screening, using a suitable artwork with an image of the circuit pattern desired. There are many proprietary compounds currently on the market for this purpose, but the majority of these use silver for the conductive component and epoxy for the binder. These materials have fairly low baking or curing temperatures (165 to 345°F), which most plastic, laminated materials are able to withstand. A copper oxide paste discussed by Kohl[9] has also been developed which can be screened on a surface and which, by firing in a reducing atmosphere, will produce a reasonably good conductor. For ceramic substrates, where firing temperatures can approach the melting point of many metals, the designer has a wider choice of conductive materials. A commonly used material is a composition of molybdenum and manganese, which is applied by screening or printing in a slurry form and fired to produce a conductive pattern. Other processes for metallizing ceramic substrates are by vapor deposition or cathode sputtering and the use of active metal and metal hydride films. A wide variety of metals and alloys may be utilized when employing the latter two techniques.

Protective Coatings. Although the subject of protective coatings for printed wiring will receive more extensive treatment in Chap. 7, this important subject bears repetition and will be covered here as a basic printed-wiring material. In addition, the effect of protective coatings on printed-wiring electrical characteristics will be covered in a later section of this chapter.

The obvious and probably most important reason for using protective coatings is to protect the printed wiring from its environment and all that this implies. Among these conditions, a protective coating less than 5 mils thick is necessary to shield the electronic components and base dielectric from salt spray, humidity, and contamination from any cause including physical abuse and handling. The protective coatings covered here are not, then, of the temporary type, such as an agent to prevent copper oxidation during storage, but of the permanent type which may be removed only during repair. Another type of protective coating is the relatively thick conformal coating, 5 to 10 mils and over thick, whose increased bulk acts as an embedment. This protective coating is used to add mechanical strength to the component board joint and alleviate the effects of vibration. A third general reason for using protective coatings is similar to the first: to protect against solder bridging in the case of dip soldering. This makes a solder-resist coating a special case; however, its properties include many of the others found in coatings not used for this specific purpose. These include ease of application, good insulation resistance, abrasion resistance, low-temperature curing (under 170°F, which is typical of state-of-the-art integrated-circuits' thermal capability), thermal shock resistance, noncorrosiveness, adhesion, moisture block, transparency, and ability to be repaired. In addition, the coating should be of a material having properties similar to the rest of the printed-circuit system, should not crack in thermal shock, and should provide the design flexibility for easy application in various thicknesses. Flexible printed wiring is usually a special case in that a bondable film, usually of a material similar to the base substrate, is often applied over the cable or etched wiring.

There are almost an infinite number of coatings applicable to printed-circuit assemblies. Predominant are the epoxies, polyurethanes, silicones, acrylics, polystyrenes, and varnishes of the MIL-V-173 type. These coatings are closely allied to embedding resins of the same chemical type. Similarly, the number of combinations and variations of these coatings is quite variable. The most widely used are covered by a military specification on conformal coatings, MIL-I-46058.[10] This specification lists the following classification of protective coatings:

Class A: General-purpose coating
  Type ER: epoxy
  Type PUR: polyurethane
Class B: Heat-resistant coating
  Type SR: silicone
Class C: Low-loss dielectric coating
  Type PO: polystyrene

Schlabach and Rider[1] have some general information regarding coatings and state that "properly formulated epoxy coatings appear to offer real promise on most counts." Acrylics, with excellent electrical and moisture-resistance properties, are generally of the air-dry variety. Silicone resins, fluids, and greases, as well as polystyrenes, are quite permeable to moisture vapor. Polyurethanes can be difficult to apply uniformly because of their moisture sensitivity, which may cause pinholing and bubbling. Vinyls, alkyds, and polyesters frequently prove corrosive and the latter have poor adhesion and thermal shock characteristics. There are relatively few current references which can provide sufficient detail to provide a general guide in this important area; therefore the design engineer, in arriving at a sound decision, should first become conversant with the pertinent aspects of the fabrication and repair processes, materials, and the nature of the environment against which the coating is designed to protect.

Process Variables of Printed-Wiring Materials. Printed-wiring raw-material variables, e.g., physical strength, thickness, dielectric and electrical constant, dissipation factor, etc., are known to have a significant effect on the end-product quality. Other variables, such as process variables and artwork, will be covered later. It is important to recognize that any tolerances in any of these significant characteristics introduced at the raw-material stage will be carried through to the end product; i.e., the finished piece can be no better than the quality of workmanship used in its production. Since the rigid raw materials are composed both of, in sheet form, layers of resin-impregnated reinforcement clad with metal and of flexible material, i.e., metal-clad plastic film, it is important to understand something of their method of manufacture, how these variations arise, and what these tolerances are.

The base-laminate insulating material is most often produced by impregnating the dry reinforcement material with a suitable resin in a *treater*. The treater, or impregnating or coating machine, is essentially a series of rollers and velocity-control mechanisms that partially cures the resin-saturated reinforcement web to prepreg, or B-stage (this is a nontacky condition and greatly facilitates handling). Paper, fabric, or glass cloth in widths ranging up to 48 in. is impregnated with a liquid-resin system or varnish, by dip or roller coating, to a controlled thickness. Controlled drying drives the desired degree of volatiles out of the impregnated web, leaving the material still fusible under heat and pressure. Cut into sheets, stacks of the dry, partially cured prepreg are piled up to the desired number of plies, which determines thickness of the resultant laminate, and pressed between stainless-steel plates in heated hydraulic press. The curing cycle decides the nature, and to a degree, the final thickness, of the laminate. The final cured product, or C-stage laminate, is infusible and insoluble. This process, then, contains variables not only of the curing cycle (time, pressure, at what point pressure should be applied, and temperature) but also of composition and quality. Multilayer laminates, in particular, are subject to entrapment of air and volatile products of the resin. These variables are customarily optimized by the materials fabrication. For this reason the end-product laminate characteristics will vary, and this should be expected.

Metal foil, usually copper as previously discussed, is laminated to the base material at the same time the prepreg is pressed. Copper thickness is specified in ounces (ounces of copper per square foot) and 0.0014 in. (1 oz), 0.0028 in. (2 oz), and other thicknesses up to 0.0098 in. (7 oz) are common. Tolerances on copper-foil thickness and weight are set forth in Institute of Printed Circuits Specification IPC-CF-150.[11] The thicknesses given in this specification, shown in Table 7, also agree with those found in MIL-P-13949.[7] In the cladding of copper to the base-laminate surface, considerable care is taken to assure a suitable bond between the foil and the insulating base. This is done either by some type of chemical or electrolytic modification of the bonding surface or by oxidation which usually results in a copper oxide or other copper complex which fosters good adhesion to the prepreg resin. Because of their excellent adhesive properties, epoxy laminates do not normally require additional adhesive. The cladding is usually coated with an adhesive for most other grades, however. Adhesives usually used for bonding copper to rigid-board materials are vinyl-modified phenolics and modified epoxies although nitrile phenolics are also used. Many of the flexible materials, especially thermoplastics such as the polyolefins, polystyrenes, and the polyhalocarbons, e.g., polyvinyl chloride, can be heat-bonded without the use of adhesives. Modifying copolymers are often used to bond copper to films in the Teflon family.

At the present time there are about two dozen standard grades of resin-reinforced laminates as listed in NEMA and ASTM specifications,[12, 13] and these materials usually come in nominal thicknesses from 0.010 to 1.25 in., with flexible materials generally identified as under 0.020 in. and without glass reinforcement. Most manufacturers produce copper-clad laminates in sheet sizes of 36 by 42, 36 by 48, and 36 by 72 in., with standard copper weights of 1, 2, or 3 oz on one or both sides. Table 8 lists some properties of several materials currently available.

PROCESSES FOR PRODUCING PRINTED-WIRING BOARDS

In order to design and apply printed wiring most effectively, it is important to understand the manufacturing steps taken to produce it. With the exception of artwork, the most important cause for variation in printed-wiring characteristics may be reasonably charged to the various processes used to manufacture it. Another important cause, covered in the last section, is variation in the raw materials themselves.

As noted before, a large number of processes, additive and subtractive in nature, have been proposed and used for the manufacture of printed wiring. Many have some appeal for a specific application, such as flush boards. The majority of printed circuits processed at the present time, however, use the subtractive method, by etching copper-clad dielectric substrate. Patterns are generated by means of (1) the photoresist technique with suitable artwork, (2) photoresist in combination with a plated-etch resist such as gold or solder, or (3) screen or offset printing with acid-resist inks. Patterns for single-sided boards can be generated using any of the techniques above; patterns for double-sided boards can be generated by using a plating resist to produce plated-through holes; and patterns for multilayer boards can be generated by etching individual single-sided boards, laminating, drilling through, and plating in the holes, followed by a plated-etch resist in the holes and outside pattern to define that circuitry. A flow chart of the processes used for each of the above three cases may be found in Figs. 18 through 20. Additive techniques are not widely used. These methods require the application of conductive inks alone or in combination with electroplating, the use of negative masks over chemically prepared surfaces with electroplating, or the direct chemical plating of positive

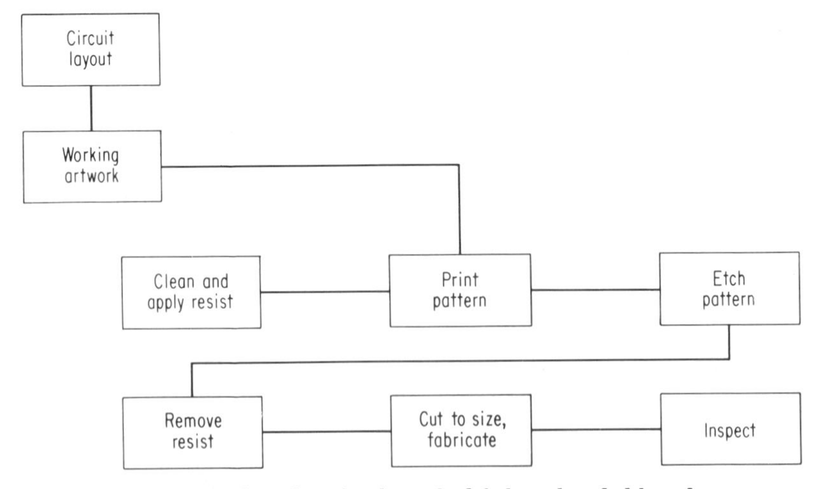

FIG. 18. Flow chart for the etched-foil single-sided board.

sensitized ink patterns. In addition to the explanation found in a previous section under types of printed-wiring planes, a more detailed explanation has been given by Schlabach and Rider.[1]

It is the intended purpose of this section to familiarize the reader with the processes commonly utilized in the fabrication of subtractive, etched-foil, printed-wiring boards of the single, double, and multilayer variety. Although processing tolerances will be tabulated later, this section will cover the nature of the processes as they affect printed-wiring design and application. Process details will not be discussed since they are adequately covered in the references.

Rigid Printed Wiring. The processes for single, double, and multilayer boards of the etched-foil variety are very similar in the elemental techniques used for the various steps in their fabrication. Single-sided boards and internal layers of multilayer boards require the application of a pattern and then the etching of the unwanted copper to define the circuit. Double-sided boards and external layers of multilayer boards must have protective plating applied over the circuit pattern

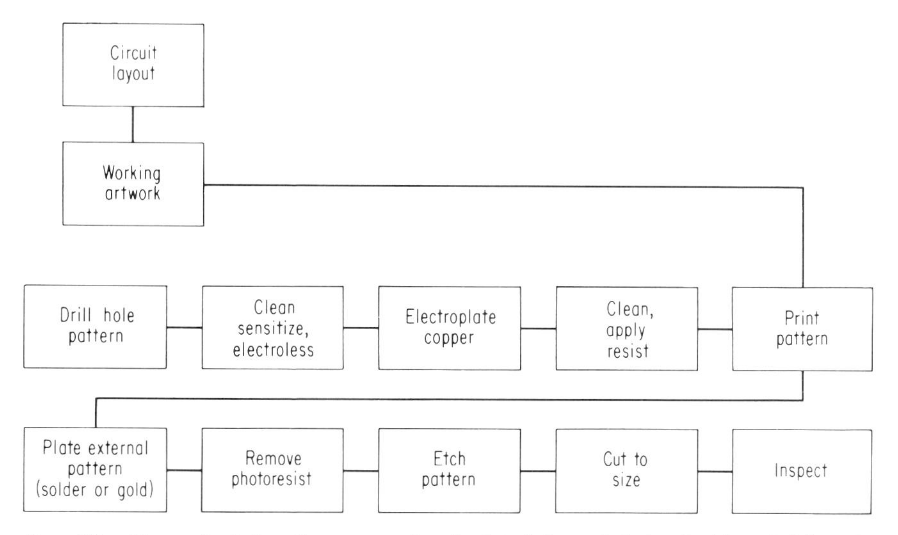

FIG. 19. Flow chart for the conventional plated-through-hole double-sided-board process.

and in the holes to protect this conductive material during the final etching operation. The elemental techniques described below are different only for different materials used and for the degree of complicity or sophistication required by the designer for his end product. For example, narrow, close tolerance lines require different etch-resist pattern techniques than do wider lines. Although a brief explanation must necessarily be given for these processes, the author will strive to explain reasons for process variations; the reader is left to uncover process details in the references cited.

Pattern Printing. This operation is perhaps the most important and key step in the production of printed wiring because this is the way in which the working "tool" of printed wiring, or artwork, is applied to the raw material. The finished product can be no better than the quality of the image or pattern applied to the raw material to produce it; the variations which may be introduced in pattern printing may have a significant effect in determining end-product quality. There are three principal methods currently used for pattern printing: screening, photosensitive-resist techniques, and offset printing. All these methods apply resist material to the foil in the designated pattern.

TABLE 8. Properties of Copper-clad Laminates [14]

Characteristics		XXXP-790 paper-base phenolic	XXXP-770 paper-base phenolic FR-2	EXXXP-845 paper-base epoxy FR-3	G-10-733 glass epoxy	LL-161 paper-base FR-4	G-10-900 glass-epoxy FR-4
		High electrical values for use in exacting electronic applications. Low moisture absorption (unnecessary to treat for fungus resistance). Cold-punching grade furnished as copper clad only. Base material approved to MIL-P-3115PBE-P. Natural color only.	Flame-resistant. Excellent electrical properties. Excellent cold punching. Excellent dimensional stability. Base material approved to MIL-P-3115 PBE-P. Dark amber color only.	Flame-resistant. Excellent electrical properties. Excellent cold punching. Approved to MIL-P-22324 and to MIL-P-13949, type PX. Cream color.	Excellent electrical properties. Low moisture absorption. Fungus-resistant. Excellent mechanical and machining properties. Base material approved to MIL-P-18177, type GEE. Copper clad approved to MIL-P-13949, type GE. Translucent, light-green color.	A paper-base laminate incorporating a newly developed polymer system. Designed especially to meet the requirements of electronic equipment demanding thermally stable dielectric properties. Hot-punch grade.	Flame-resistant. Glass epoxy with improved electrical properties over Grade G-10-773. Developed especially for cold punching and shearing. Approved to MIL-P-18177, type GEE. NEMA Grade FR-4. Approved to MIL-P-13949, type GF. Translucent, blue-green color.
Flexural strength, psi, condition A	LW	16,000	18,000	25,000	72,000	14,500	82,000
	CW	14,000	14,000	23,000	60,000	11,140	60,000
Tensile strength, psi, condition A	LW	10,500	12,000	20,000	49,000	7,600	57,000
	CW	9,500	8,500	16,000	34,000	6,400	44,000
Izod impact, ft-lb/in. of notch, condition E-48/50	LW	0.40	0.43	0.56	11.6	0.495	12.4
	CW	0.35	0.35	0.54	7.0	0.420	9.1
Rockwell hardness, condition A		M-90	M-80	M-97	M-114	M-105	M-111
Water absorption, %, condition E-1/105 D-24/23		0.54	0.051	0.32	0.08	0.50	0.11
Dissipation factor, 1 MHz	A	0.029	0.029	0.029	0.014	0.027	0.018
	D-24/23	0.031	0.034	0.030	0.015	0.027	0.019
Dielectric constant, 1 MHz	A	4.4	4.6	4.3	5.2	3.81	4.7
	D-24/23	4.5	4.8	4.4	5.3	3.93	4.7

Surface resistance, megohms, condition 96/35/90	141,500	70,700	177,000	353,000	2×10^{12}	3.5×10^{6}
Volume resistivity, megohms/cm, condition 96/35/90	6.3×10^{6}	12×10^{5}	24×10^{6}	14×10^{7}	2×10^{14}	3.3×10^{8}
Copper bond strength #/in. 2 oz, condition A	8	8	9	10	9.5	11.5
Solder blister, sec/@ 500°F, condition A	7	8	10	20	5	30
Flame resistance per NEMA LP 1-7.07E, sec, condition A	......	9	6	......	......	6
Minimum thickness, in.	1/32	1/32	1/32	0.0075	1/32	0.0075

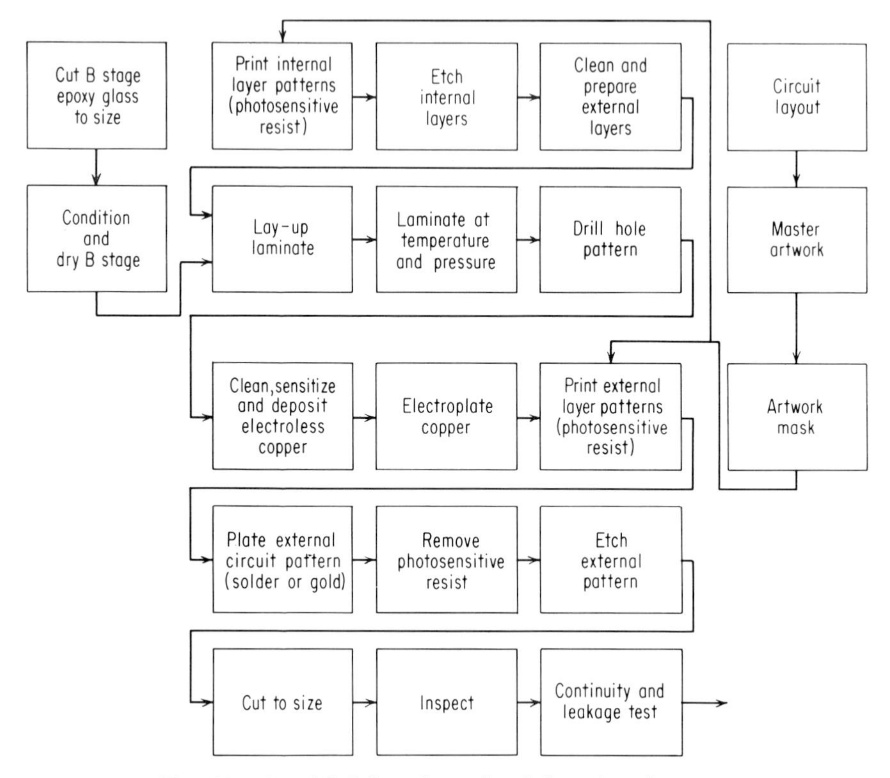

FIG. 20. Simplified flow chart of multilayer-board process.

SCREENING. Using a technique borrowed from the graphic arts industry, a screen provides a means for selectively transferring resist material to the foil surface through the open or "clear" areas of the screen. The pattern may be produced in photostencil paper using a variety of artwork methods and then transferred to the screen. Screens are also provided with the photographic emulsion on the surface or completely impregnating the screen; these screens are photographically processed. The screen material, either silk or metal, is stretched tightly over a metal frame. It is registered and placed above the workpiece and contacts the surface when the ink or resist material is squeegeed through the screen. Screening inks are available in oil, cellulose, asphalt, vinyl, or resin base. The image produced is a heavy ink layer which frequently bears a screen impression owing to the weave of the screen. Screen printing lends itself readily to production methods (usually 25 or more pieces); however, dimensional tolerances cannot be easily controlled, resulting in registration and pattern-configuration problems.

PHOTOSENSITIVE RESISTS. A *photosensitive resist* is a light-sensitive polymer which, when hardened, has very great chemical resistance. After the board or foil surface has been adequately cleaned (usually by pumice scrubbing and then chemical cleaning it) a thin layer of photoresist is applied by dipping, spraying, spinning, or roller coating. Available from a number of suppliers, these proprietary compounds may be classed as negative photoresist and positive photoresist. The former, which is most commonly used, is of the light-hardening variety; i.e., exposure to light produces a chemical reaction in the polymer, causing it to harden. Positive photoresist works in the opposite manner; the polymer is caused to soften upon impingement of light through a suitable mask or artwork. Usual light sources are utilized for exposure; however, a popular practice with current photoresist materials is to use light rich in the ultraviolet region. Following exposure, the unexposed photoresist is removed by a variety of methods, the most common being a trichloroethylene spray-vapor degreaser.

Baking of the photoresist following development is done if an unusually strong etchant is used; baking is almost always done if the photoresist is used as a plating resist. During subsequent etching, the hardened photoresist protects the conductive foil where desired, and all other superfluous metallic material is etched away. In cases where a plated coating, e.g., gold or solder, is used as an etch resist, photoresist is again used to determine where the plated coating is applied. Metal which is to be etched away is protected by resist; the desired conductor pattern is then left uncovered when the photoresist is printed and developed. Plating of the etchant-resistant metal takes place, the unexposed resist is removed with solvent, and following etching, the desired pattern remains with the plating covering the conductors. An example of the photoresist-printing technique is shown in Fig. 21.

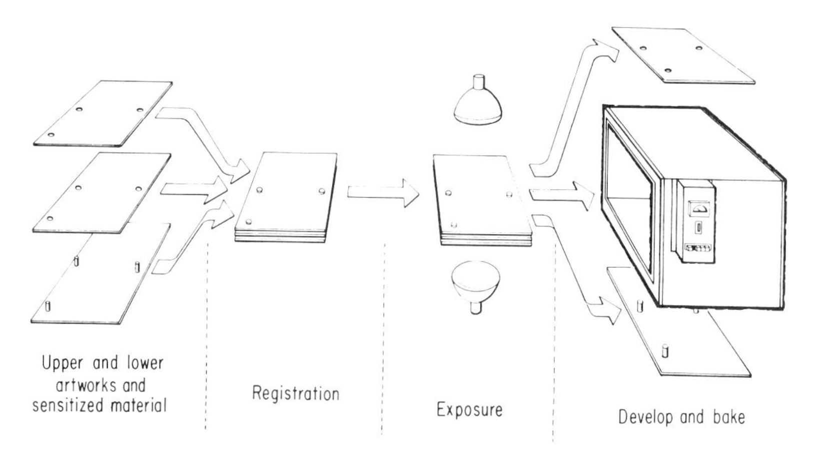

FIG. 21. Photosensitive resist-pattern printing.

OFFSET PRINTING. Using conventional printing methods to place the circuit-pattern image on the foil surface, this technique shows some promise as a high-volume-production technique. A master pattern in the form of a plate etched in relief (about 0.006 in. deep) is used as the printing plate. Ink is transferred from an ink roller via the raised-image areas on the pattern to a rubber "blanket" cylinder. The ink image is then transferred to the copper-plate printing surface with unusually good reproduction capability because the pressure of pickup and transfer can be low enough to prevent squeeze-out or smudging. Although the ink thickness from one pass is only a few ten-thousandths of an inch thick, the unusually good registration found in this method permits several impressions to be made, resulting in the application of a dense ink film of the required thickness suitable for either an etchant resist or a plating resist. Although tolerances are dependent upon the artwork and film masters which are used, with proper equipment it is possible to achieve a tolerance of ± 0.001 in. over the blanket, i.e., pattern plate to board. An offset printer of the type used for printed wiring is illustrated in Fig. 22.

Etching. The object in printing a pattern of the desired circuitry with a protective coating is to etch away foil material which is not covered. This is done in a variety of ways with a number of chemicals. Methods of etching are concerned

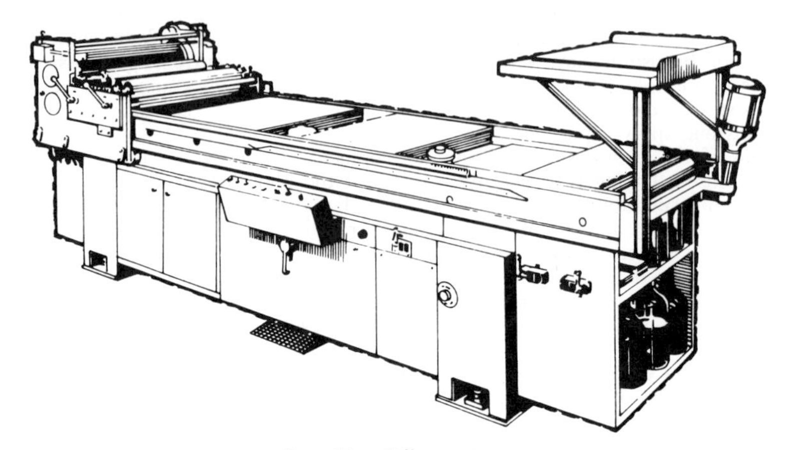

FIG. 22. Offset printer.

FIG. 23. Horizontal-spray etcher. (*Radio Corp. of America.*)

with the way in which the etchant is applied to the workpiece. In addition to dipping, feasible techniques are splash and spray etching, both of the latter oriented in either the vertical or horizontal position. It is important that a continuous supply of fresh etchant be brought in contact with the work area; hence the popularity of splash and spray etching. Only four main types of etchants are currently used in the printed-wiring industry: ferric chloride, chromic acid, ammonium persulfate, and cupric chloride. Details on these operations may be found in Schlabach and Rider.[1] The parameters conventionally considered for etching are type and thickness of resist used, type and thickness of foil, degree of undercutting, and type of etchant used. These will be tabulated later in this chapter under Design Considerations, page 1-46. A horizontal-spray etching machine, of the type commonly used in printed-wiring fabrication, is illustrated in Fig. 23.

Laminating. Referring again to Fig. 20, we see that the next step in the manufacture of multilayer boards is lamination, i.e., bonding all the layers of wiring planes to form a complete laminated assembly. A laminating press of the type illus-

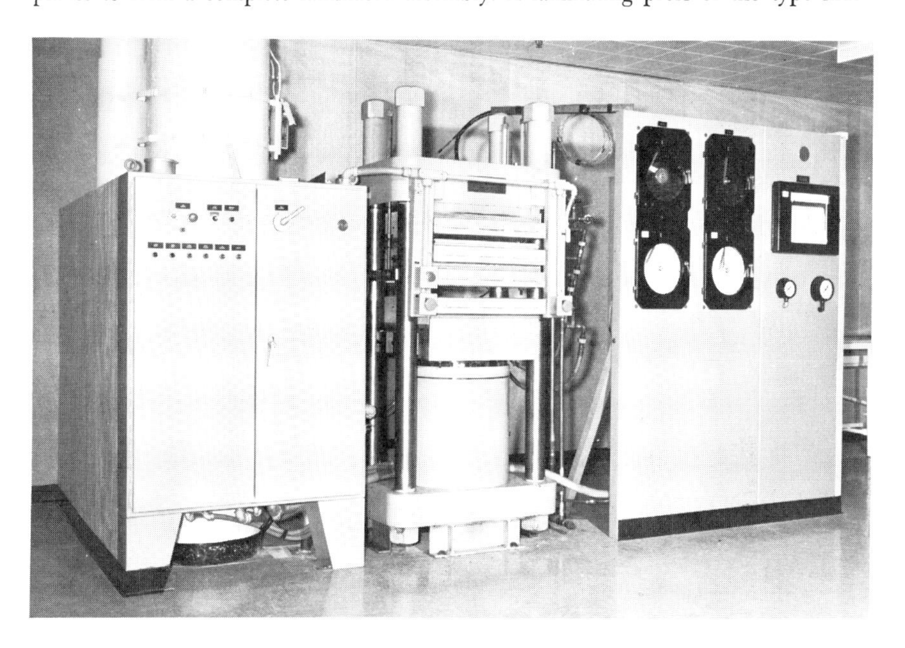

FIG. 24. Laminating press. (*Radio Corp. of America.*)

trated in Fig. 24 is used for this operation. With steam or electrically heated platens flat and parallel to within ±0.001 in. and capable of controlled heat in the range of 300 to 500°F, the press should be capable of exerting pressures in excess of 500 psi over the entire circuit area. In "laying up" or assembling the "stack" of sheets to be laminated, layers of fully cured circuit layers are placed over registration pins, alternating with adhesive; this laminating layer is known as B stage, or prepreg. This material is partially cured and is similar to the circuit layer material, except that it has higher resin content and flow characteristics to ensure good circuit encapsulation. Typical bonding materials reach the optimum cure in 40 min at 180 or 335°F. Bonding time is usually kept to a minimum to shorten expensive press time, but care must be taken to circumvent incomplete curing of the resin since the resin would soften during drilling. Good bonding practice is important for many reasons; e.g., the variety of subsequent manufacturing operations including drilling and subjection to harsh cleaners, chemicals, and hot solder would surely result in contamination or delamination without proper laminating procedure. Also, proper registration of circuit layers, established during lamination, is important to assure circuit interconnection following drilling and plating of holes.

Hole Fabrication. Except in the case of techniques where through-hole plating is used, hole fabrication is not an in-process step; i.e., it is usually done when the board is completed. For single-sided boards, holes are either drilled or punched, depending upon the volume of production and the number of holes required. Punching is more often done with paper-, fabric-, and asbestos-base grades, which are suitable for hot or cold punching as described previously in the materials discussion. Dies for punching are designed similarly to those used for metal, except smaller clearances should be allowed between punch and die. The minimum clearance between individual punchings and between punchings and the edge of plate should be two to three times the thickness of the plate. Since the diameter of the punched hole shrinks when the punch is removed, the size of the punch should be 0.001 in. larger than the desired diameter of the hole for every 0.020-in. thickness of material punched cold and 0.001 in. larger for every 0.015-in. thickness punched hot.

Laminates for hot punching should be preheated in an adequate oven to give uniform heat throughout. Temperatures of 100 to 120°C are recommended but only until the material reaches oven temperature; a delay may cause brittleness. Punching practices for glass-based materials are similar to those for the other laminates just described, but the operation with this material is much harder on the tools. Although the use of tungsten carbides and other special die steels increases die life, it does not circumvent a frequent trouble—haloing around holes and occasional delamination when punching near the recommended maximum thickness of 3/32 in. At the present time, holes for double-sided and multilayer printed-wiring boards using the plated-through-hole technique are drilled since this is the only process producing the quality of holes needed for plating. By this we mean holes clean, free of burrs and epoxy smears. For drilling non-glass-based laminates, a standard high-speed drill with recessed lips to provide ample clearance is satisfactory. Carbide drills, however, are used for production since they are used almost exclusively for glass-based laminates. The feed should be light and uniform, and the speed of the drill should be considerably in excess of that used for soft steel. With tungsten-carbide tips, speeds in excess of 15,000 rpm are frequently used. Where possible, the material being drilled should be backed up with scrap or other soft material to prevent chipping out.

Because of the nature of plastic material, the diameters of holes drilled in laminates are usually 0.002 in. under the drill size. Therefore, the drill selected should be at least 0.002 in. larger than the designed diameter of the hole. These materials are usually drilled dry with a good exhaust system to remove dust. Among the various methods for drilling printed-circuit boards are the following: (1) drilling by eye, (2) drilling with a jig-bored drill plate, (3) drilling with optical aids, and (4) automated drilling with a tape-programmed machine. The automated technique is oriented to high production; an example of this method is illustrated in Fig. 25.

Plating.[1] Following the drilling of holes in the board structure and the mechanical and/or chemical cleaning of the holes and board surface, the board is plated for a variety of reasons. Plating is used to establish interconnection between board sides and internal layers of a multilayer board. Plating may also be used as a final finish to protect against corrosion and furnish a hard surface or as a resist against etching during board processing. Before a board is plated, however, it must be sensitized, or made conductive. Sensitizer is sorbed at polar sites on the piece to be plated, then reduced in place to a thin metallic film. Tin and palladium formulations are most widely used in printed wiring.

Electroless plating is ordinarily used in printed wiring in preparation for electroplating. It is based upon the controlled autocatalytic reduction of the metal ion to the corresponding metal on certain catalytic surfaces. It differs from chemical reduction methods since with electroplating, thick coatings, dense and continuous, can be built up. Methods are currently available for the deposition of nickel, copper, cobalt, gold, and palladium, with nickel and copper most widely used, the latter preferred for printed-wiring boards having plated-through holes.

The conducting film formed by chemical reduction or electroless plating previously described is fragile and easily destroyed by electroplating unless care is taken. Intermediate flash plating is used to prevent dissolution of the film in the plating bath or adhesion failure from plating stresses. Following flash plating, the board may be transferred to a conventional plating bath for plating to the desired thickness.

Electrodeposition, or *electroplating*, as it is commonly known, consists in the

FIG. 25. Multispindle, tape-controlled drill. (*Radio Corp. of America.*)

deposition of metal on a conductive surface when the surface is made the cathode in a suitable, usually liquid, electrolyte. In this case, the plating is produced by the reduction of the metal ion in solution through electrons supplied to the cathode from an external dc source. The thickness of the electroplate obtained is proportional to the current density, usually expressed in amperes per square foot, and the time of plating. For a given thickness, plating time may be reduced by increasing current density, solution temperature, concentration, or degree of agitation. These parameters, as well as the electrolytic bath used, affect physical characteristics of the plating such as grain structure and ductility. Shorter plating times are desirable in alkaline baths which can damage both resist material causing breakdown and bonding adhesives. A large number of metals and some alloys may be plated in aqueous solutions but the deposits most widely used in printed-circuit applications are copper, silver, gold, tin, nickel, rhodium, and tin-lead alloys. An industrial printed-wiring plating facility is illustrated in Fig. 26. More detailed information on processes and conditions for plating may be obtained by consulting general texts on this subject.[15, 16]

FIG. 26. Printed-wiring plating facility. (*Radio Corp. of America.*)

Machining Processes. In addition to punching and drilling, already discussed, laminated materials are usually machined to cut raw stock and to cut or blank to the desired size or outline. Non-glass-based materials are machined on standard machine tools such as those used for wood or metal work. For most machining operations, ordinary high-speed steel tools are adequate. However, where production quality, production quantity, production speed, or finish is needed, carbide-tipped tools are usually employed. Cutting tools must be kept extremely sharp for best results. Thermosetting plastics are ordinarily machined dry. Cooling by air is preferable to the use of liquid coolants, which are difficult to remove from finished parts and may cause swelling or delamination. During machining, temperature of the workpiece may be controlled by air to keep material below 150°C, at which temperature distortion may be caused. For glass-based materials, diamond or tungsten-carbide tools give best results. Laminates of all grades can be cut by circular or band saws, usually using carbide or diamond impregnated grades of cutting wheels or bands for the glass-based varieties.

For shearing, standard shears suitable for sheet metal may be used, except that for glass-based laminates, thicknesses up to 3/32 in. only can be cut. The knife blade should be kept sharp and the material held rigid with suitable fixtures. Most paper laminates up to 1/16 in. thickness and canvas laminates up to ⅛ in. thickness may be sheared at room temperature (70°F min). Thicker stock may be sheared by heating the laminate at temperatures up to 250°F. Tolerances which are usually associated with various board-cutout methods are listed in Table 9.

TABLE 9. Machining Tolerances for Laminates

Machining method	Tolerance, in.	Edge finish
Shear	0.018	Very rough
Band saw	0.018	Rough
Diamond wheel	0.012	Smooth
Routing or milling	0.005	Smooth
Punching or blanking* . .	0.004	Rough

* With heat-resistant G-11, some delamination may result.

Material Process Variables. The very nature of the materals used in the fabrication of printed wiring deems that changes may be expected in the printed-wiring board during its manufacture. Whether the base material is organic or inorganic, changes due to thermal expansion, shrinkage behavior, and dimensional-stability characteristics must be known and anticipated in the initial design. In this way, printed-circuit materials are a good deal more affected by the manufacturing process than are metals because the former are subjected to a variety of chemical (cleaning and etching), heat, and pressure (lamination), as well as regular machining, processes in their fabrication. Considering the physical behavior of the material, it is evident that these processes must introduce some change in the material which the engineer must consider in his initial design.

THERMAL EXPANSION. Because of the nature of the materials comprising a laminated, reinforced plastic material and the method of its manufacture causing it to be anisotropic, it behaves strangely when subjected to heat. This is demonstrated by the fact that it has a different coefficient of linear thermal expansion in lengthwise, crosswise, and thickness directions. Typical values are illustrated in Table 10. The larger coefficients for organic materials are apparent, as well as the effect of filling with a different material. The differences in expansion due to temperature in these three directions may be directly attributable to the manufacture

TABLE 10. Mean Coefficient of Linear Thermal Expansion of Selected Materials

Material	Coeff., 10^{-5} in./(in.) (°C)	Material	Coeff., 10^{-5} in./(in.) (°C)
Thermoplastics (−30 to +30°C):		Laminates (0 to 60°C) (*cont.*):	
Polyethylene, 0.95 density	15	FR-3, epoxy paper:	
Polypropylene	10	Lengthwise	3
Fluorinated ethylene-propylene copolymer	9	Crosswise	4
Polystyrene	7	Thickness	10
Polycarbonate	7	G-10, epoxy glass:	
Polytrifluorochloroethylene	7	Lengthwise	1.1
Polytetrafluoroethylene	5.5	Crosswise	1.5
Molded thermosets (−30 to +30°C):		Thickness	6
Phenolic:		Ceramics (25 to 200°C):	
Wood-flour-filled	3.7	Steatite	0.78
Asbestos-filled	2.0	Forsterite	1.0
Glass-filled	1.1	Cordierite	0.23
Diallyl phthalate:		Mullite	0.50
Acrylic-fiber-filled	5.5	Alumina, 99%	0.67
Asbestos-filled	4.0	Photosensitive glass ceramic*	1.0
Glass-filled	3.2	Glasses (25 to 200°C):	
Alkyd:		Borosilicate (Pyrex†)	0.5
Asbestos-filled	3.6	Fused quartz	0.06
Glass-filled	3.6	Metals (25 to 100°C):	
Laminates (0 to 60°C):		Aluminum	2.4
XXXP, phenolic-paper:		Copper	1.7
Lengthwise	2	Silver	2.0
Crosswise	3	Gold	1.4
Thickness	10	Solder (60/40)	2.4
XXXPC, phenolic-paper:		Tin	2.4
Lengthwise	2	Nickel	1.3
Crosswise	4	Rhodium	0.8
Thickness	20		

* Fotoceram 8603, trademark, Corning Glass.
† Trademark, Corning Glass.

of the laminate. Reinforcement is stronger in the lengthwise direction than crosswise, and the material is compressed at pressures up to 1,000 psi while it is laminated, resulting in a larger movement of the material in the thickness direction when heated. In higher temperature ranges of 260 to 300°F, the coefficient of linear thermal expansion of most laminates is two to six times greater in the thickness direction than in the range from 32 to 140°F. When compared with metals in Table 10, the effect of this phenomenon on plated-through holes is obvious.

ETCH SHRINKAGE. Usually, in the cladding operation, when metal foil is clad to the dielectric material, a difference in the coefficient of thermal expansion of the metal foil and the base laminate results in a "locked-up stress" when the material cools from the bonding temperature. When the metal foil is subsequently processed and some of it is etched away, the stress is relieved and a dimensional change (usually expansion) occurs. Because of the inherent nonuniformity (directionality) of woven fabrics, dimensional changes are generally unpredictable and follow no set pattern. It appears that the net shift of a given point is a vector sum of random shifts of all layers (plies of glass cloth) subject to varying degrees of restraint and residual stress imposed during fabric weaving, impregnation, and laminating. Dimensional changes will be larger with large, thin substrates of low modulus, and warping may also occur on double-sided clad substrates where more foil is removed from one side than from another. In order to circumvent the above problems where critical tolerances are essential, every attempt should be made to process material close to room temperature. If this is not feasible, the designer should specify registration of critical dimensions in his pattern in the direction of low-shrinkage change of the substrate indicated by the manufacturer. Dimensional changes ranging from 0.002 in./in. shrinkage to 0.010 in./in. expansion are not unusual for present commercially available copper-clad laminate.

OTHER PROCESS VARIABLES. Laminate materials are susceptible to moisture absorption which will result in a dimensional change, or swelling, of the material. Paper-based phenolics are particularly hygroscopic, with water absorption ranging from 0.35 to 5.0 percent over a 24-hr period at 95 percent relative humidity. This can result in changes for a 1/16-in. XXXP material having dimensions of 0.003 in./in. (lengthwise), 0.004 in./in. (transverse), and 0.024 in./in. (thickness). Glass-based materials, especially epoxies, are not as liable to change under these conditions; figures for glass-epoxy materials range from 0.13 to 0.40 percent for the conditions cited above. Since these changes are directly dependent on moisture absorption, the designer must either select unaffected materials and conformally coat them to help shield them from moisture or specify the environment the equipment is to operate in.

Because of the way laminates are fabricated, they are capable of being deformed or compressed, especially when heated. Thermosets are much more stable than thermoplastics (to be discussed later), but they may nevertheless deform in the range of from 0.2 to 2.3 percent when subjected to a load of 3,000 psi at 125°F.

Flexible Printed Wiring. The processes for producing flexible printed wiring and cables do not differ greatly from those used for their rigid counterpart. Again, the predominant technique used in the fabrication of flexible printed wiring is the etched-foil, or subtractive, technique, using similar resisting methods to those found for rigid materials. The predominant difference in the processing method between the two material classes is that flexible cabling, at least, is processed on a continuous basis, in a manner illustrated in Fig. 17. Other flexible wiring may be processed in individual sheets, in batches, or on a continuous basis, as above, depending upon the volume of circuitry required.

The methods for pattern printing for flexible materials are dependent upon the accuracy of the pattern desired. Usually, screen-printed asphalt-base inks are used, but for precision or high-tolerance etching, conventional photoresist techniques are employed. Care must be taken in using photoresists on flexible materials; a crack in the photoresist caused by rough handling or flexing will result in a break in the copper at that point.

Etching is accomplished in a similar manner to that for conventional, rigid materials, and etchants are mainly determined by whether the clad copper is standard electrodeposited copper or copper with an oxidized surface to enhance adhesion. Since etching, resist, resist development, and cleaning chemicals can all affect the base material, manufacturers' recommendations should invariably be followed in processing the many types of materials which are included in the flexible-wiring category. A typical data sheet of this type, illustrated in Table 11, gives processing information supplied for a copper-clad polyester material by the manufacturer.

Lamination of flexible printed wiring applies only to heat sealing or film encapsulation of finished circuitry. Although fabrication with plated-through holes is possible with some materials, this technique is rare, and flexible, plated-through multilayer fabrication using thermoplastic materials is not widely available. The

TABLE 11. Flexible-material Processing Information [4]

Processing information	Standard electrodeposited copper[a]	Treated electrodeposited copper oxide or precleaned rolled copper[a]
Cleaning agents, surface	Pumice and water Sulfuric acid (20%) Pumice-alumina-water Ammonium persulfate Household cleanser: Comet[b]	HCl (dilute) and pumice Sulfuric acid (20%) Pumice and water Ammonium persulfate Household cleanser: Comet
Silk-screen resists	Warnow Process Paint Company (Los Angeles) Nazdar Company (Chicago) Advance Screen Supply (Chicago) Engravers Ink Company (Montebello, Calif.)	
Photosensitive resist	Kodak KPR[c] Lith-Kem-Ko[d] Kodak metal-etch resist	
Etching solutions	Ferric chloride Ammonium persulfate Hunt RCE[e]	Ferric chloride Ammonium persulfate Hunt RCE Cupric chloride Chromic acid
Resist removal (screen type)	Depends on screen resist used. Follow manufacturer's recommendations. Avoid using pure halogenated hydrocarbons with high solvating power, i.e., 1,1,2-trichloroethane, methylene chloride.	
Resist removal (KPR type)	Toluene methylene chloride 75/25 by volume Trichloroethylene (will soften adhesive at 100°F but strength will recover) Methyl isobutyl ketone	
Plating solutions	Plating not recommended, unless reverse printing system used. Use large border (1 in.) where using plating as resist.	Gold, acid-modified (Sel Rex, Nutley, N.J.) (Wildberg, Los Angeles) (Technic, Providence) Solder (Baker-Adamson Div. of Allied Chemical)

[a] Use all solutions per manufacturer's recommendations.
[b] Trademark, Procter and Gamble.
[c] Trademark, Eastman Kodak.
[d] Trademark, Lithographic Chemical Supply, Chicago.
[e] Trademark, Phillip A. Hunt Chemical Corp., Palisades Park, N.J.

properties of flexible printed wiring contradicts the requisites for environmental protection of plated-through-hole connections. Heat sealing is an ideal environmental protection for printed wiring supported by thermoplastic materials and is commonly employed. It is conventionally done with heated rollers or platens and cooled under pressure to minimize wrinkling, distortion, and local delamination or blistering. In most cases, a minimum heat-sealing temperature and pressure are preferred since these minimize conductor displacement and shrinkage accountable to the readjustment of internal stresses.

Hole fabrication for flexible materials is usually by means of punching; without reinforcement most thermoplastic materials punch easily. In many cases, flexible plastics are punched prior to the cladding step in order that the foil may be accessible from both sides. Blanking and other cutting operations are usually done by a shear or die.

Material process variables for flexible materials are more dependent upon the material or plastic itself than the method of its manufacture, as with laminated materials. The coefficient of linear thermal expansion is somewhat greater for thermoplastic materials than for the rigid types, as illustrated in Table 10. Some flexible materials, notably the polyesters, because of a unique production process, are *prestressed,* i.e., have locked-up stresses which give them exceptional strength. When the film is heated, these stresses are relieved, causing shrinkage, called *residual shrinkage.* Residual shrinkage is an irreversible phenomenon, but the material is still subject to the same rules as for ordinary thermal expansion. When flexible materials are clad with metal foil, usually copper, a machine stress, or "stretch," is put on the material in the direction of sheet travel. During the etching of the material, the material reverts to its former size or dimension. This *etch shrinkage,* or release of machine-induced stresses, has a wide range of values for thermoplastics currently popular; e.g., the values for polyester measured at 10°F are 0.3 percent in the machine direction and 0.15 percent in the transverse direction. Again, it is wise for the designer to register his pattern so that critical tolerance dimensions fall in the transverse direction of the material.

Moisture-absorption properties of some flexible materials are listed in Table 3. As a rule, thermosetting materials are exceptionally stable to changes induced by moisture or are amenable to coating or cladding with a material that is stable to this kind of change. Thermoplastic materials, on the other hand, are particularly prone to cold flow, or deformation under load, and the designer must be aware of this property to avoid a possible short in a critical area. This is one property which minimizes the use of plated holes in flexible substrates. Some typical deformations of thermoplastics at 3,000 psi and 125°F are the following:

Thermoplastic	*Deformation, %*
Nylon 6–6	1.3
Polypropylene	4.8–5.6
Polytrifluorochloroethylene	33
Polytetrafluoroethylene	25
Polycarbonate	0.1

DESIGN CONSIDERATIONS FOR PRINTED WIRING

Following a discussion of the various techniques for producing a printed-wiring board and the materials of its manufacture, the designer may find he still has a large number of factors to take into consideration in his quest for an optimum design. The objective of this section is to explain how the design of the printed-wiring board is affected by a variety of factors, from the basic materials and manufacturing processes to environment, end use, and other packaging trade-offs. The reader may, for a number of reasons, at this point favor certain processes and procedures for making and assembling boards in the system. It is important to remember, however, that there are a variety of factors which will influence his approach to board fabrication, i.e., dimensional tolerances, cost, effect of environment, electrical properties, and repair feasibility.

Probably the most important considerations in the design of electronic equipment are reliability, satisfactory performance, and maintainability. These factors are not inherent in printed wiring per se; however, by adequate design and proper selection of materials and manufacturing techniques, they can be introduced into the system. Initially, printed wiring was developed to adapt this interconnection method to mass-production and assembly schemes and economize on weight and space in military equipments. It was soon apparent, however, that other advantages would accrue from its use. The close control over electrical parameters and their reproducibility by using a certain printed-wiring design was soon recognized as an important and valuable attribute. Wiring bulk was greatly reduced. Wiring errors, except those in the initial design, were all but eliminated. Printed wiring was soon found to have the design flexibility to provide characteristics (electrical and mechanical) compatible with all types of electronic equipment. The potential of printed wiring to automation both in fabrication and use results in cost savings, but more importantly, facilitates control which can only result in enhanced reliability of the system. It goes without saying that the designer can realize the maximum benefit of printed wiring if the rules of design or guidelines, some of which are listed in Table 12, are utilized to their maximum possible extent.

TABLE 12. Required Artwork Dimensions for Etched-copper Foil *

Resist system	Lines and pads, in.	Spaces, in.
Photoresist:	*Negative artwork:*	*Negative artwork:*
½ oz (0.0007 in.)	X	X
1 oz (0.0014 in.)	$X + 0.0005$	$X - 0.0005$
2 oz (0.0028 in.)	$X + 0.0015$	$X - 0.0015$
Solder plate (pattern plate):	*Positive artwork:*	*Positive artwork:*
½ oz (0.0007 in.)	$X + 0.00025$	$X - 0.0025$
1 oz (0.0014 in.)	$X + 0.001$	$X - 0.001$
2 oz (0.0028 in.)	$X + 0.0025$	$X - 0.0025$
Gold plate (pattern plate):		
½ oz (0.0007 in.)	$X + 0.002$	$X - 0.002$
1 oz (0.0014 in.)	$X + 0.0035$	$X - 0.0035$
2 oz (0.0028 in.)	$X + 0.0065$	$X - 0.0065$

* Desired etched dimension is denoted by X.

Environment. Printed wiring may normally be expected to exhibit the same properties as the base materials, i.e., laminate and cladding, from which it is constructed. Environmental requirements, therefore, relate back to these elements, except in the case of monolithic assemblies of printed-wiring boards, which are by nature of the lamination and interconnection procedure additionally affected by the environment. This section will include, under the general topic of environment, expected performance and properties of printed-wiring boards. Although integrally associated with the base-material properties, electrical characteristics will by definition not be considered as environmental in nature and will necessarily be covered in a later section.

In addition to serving as an electrical connection media, printed-wiring planes also provide mechanical support for the active or passive components they are interconnecting. In this way, they become an integral part of the package, or assembly, and must therefore be able to withstand the environment stresses associated with the entire structure.

In order to categorize capabilities of printed wiring, i.e., what may be expected of printed-wiring planes with regard to environment, the author will treat subclasses of printed-wiring properties, mechanical, thermal, atmospheric, and processing. As a generic class, printed wiring of both the rigid and flexible types will exhibit various capabilities with regard to these attributes. The following is a brief summary of what these values may be expected to be.

Mechanical Factors. Although perhaps secondary in importance to the function of electrically interconnecting various circuit components, the rigid-printed-wiring board is also a mechanical member of the packaging system; it must provide mechanical support and protection to the components and, in many cases, assist in thermal management. For flexible printed wiring, this is not always the case, and other support may be needed.

Tables 2, 3, and 8 give the reader some knowledge of the mechanical properties. Tensile strength of paper-based phenolics ranges from 10,000 to 20,000 psi, and is usually greater in the lengthwise direction, a property associated with the fabrication of the reinforcing material. Flexible materials, which are usually thermoplastic, range from 2,000 to more than 30,000 psi for polyesters. Probably the strongest material used for printed wiring is glass epoxy; with a tensile strength ranging from 30,000 to 82,000 psi, it is often used as a structural material, and in many cases it approaches certain metals in strength. Glass-fiber-based epoxy materials are used almost exclusively in the fabrication of multilayer printed wiring since its superior tensile and flexural strength and modulus of elasticity enable it to withstand laminating and drilling operations without losing its dimensional stability. Flexural strength, modulus of elasticity, and impact strength, tabulated in the tables cited, are several times greater for glass epoxy than for the other grades. Of the thermoplastics, the polyesters seem to have the edge on strength.

Since the laminate is a composite structure, it may be expected to warp (between parallel sides) and twist (on the diagonal). Acceptable values for this are either 1 percent or 3 percent, according to Institute of Printed Circuits Specifications IPC-ML-950.[17] Printed-wiring materials are well able to withstand most any vibration force the package will experience. Vibration can deteriorate the component connection to the printed-wiring board, however, and it is estimated that 20 percent of component failures are caused by shock and vibration. Although shipboard installations may see vibrations only in the range of from 5 to 35 Hz, it is not uncommon to find a vibration spectrum of from 40 to 25,000 Hz in missile applications. *Peel strength,* a measure of the adhesion of the cladding to the base material, is considerably greater for rigid materials than for flexible films. As indicated in Table 8, this ranges from 8 lb for a 1-in.-wide strip on XXXP to 11.5 lb for FR-4. Untreated copper may be expected to yield one-half these values.

Thermal Factors. Apart from the ceramic materials, materials of printed wiring are at times severely restricted in service because of their relatively low resistance to heat. Maximum continuous-duty service temperature of some representative rigid materials are as follows:

NEMA grade	MIL Spec. type	Max. service temperature, °F
XX	PBG	250
XXP	. . . .	250
XXXPC	PBE–P	250
N–1	NPG	165
G–5	GMG	300
G–7	GSG	350
G–10	GEE	250
G–11	GEB	300+

Temperature characteristics of thermoplastic, flexible materials may be found in Table 3. Exposure to elevated temperatures tends to cause a permanent continuing change in properties of plastic laminates with increased effect at higher temperatures, partly the result of continued curing of the resin. Young[18] has conducted an exhaustive study on this subject and found that flexural strength, tensile strength, and elastic modulus all decrease with increasing temperature. Selecting G-10 material as an example, one finds that these values of decline in pounds per square inch per degree Fahrenheit are 283, 95, and 2,500, respectively. Flame-resistance properties are listed in Tables 3 and 8, and it is obvious that they must be given careful consideration in applications where a combustion hazard may exist.

Atmospheric Factors. By *atmospheric factors,* we mean the behavior of the printed-wiring board as affected by its immediate environment, or surroundings. Corrosive atmospheric factors might include the effect of salt spray which may be experienced in shipboard applications, high-humidity conditions, or a high-altitude application which affects the dielectric breakdown resistance. Where wiring is encapsulated in internal layers, the problems are somewhat alleviated, except in the case of moisture absorption which can permeate over a period of time. Individual suppliers must be contacted regarding resistance to salt spray. Data for moisture absorption are listed in the tables cited, and it may readily be seen that the values for the paper phenolic grades are several times higher than those listed for glass epoxies. Most flexible films are relatively insensitive to moisture except in the case of glass reinforcement, where moisture absorption can run between 0.10 and 0.68 percent.

Processing Factors. Although many of these factors will be covered in the next section, Manufacturing Processes, it is true that the processes which the board may see in its use and assembly must rightly be termed *processing factors.* For instance, delamination may occur or foil bond be destroyed by soldering and resoldering cycles commonly seen during assembly of components to the printed-wiring board. Foil-clad laminates experience a degradation of peel strength of printed conductors which is directly proportional to the elapsed time to which it is subjected to a given soldering temperature. Soldering temperatures may cause a blistering of the laminate surface as indicated in Table 8, which shows that XXXP can take a 500°F solder bath for 7 sec, G-10 for 20 sec. Many of the thermoplastic materials such as the polyesters, polyvinyl chloride, and polyethylene cannot withstand immersion in a solder bath, and other connection techniques must therefore be used. Although terminal pull is technically a function of the plated-through-hole integrity, it is somewhat dependent upon the strength characteristics of the material itself. In this case the glass-epoxy grades are best and exemplify the reasons for their almost exclusive use in multilayer fabrication.

Manufacturing Processes. Although board-fabrication techniques and its individual processes were previously described, little was said about how variations in manufacturing tolerances affected the original design. Printed wiring must be designed to consider all the factors affected by material, method of manufacture, and ultimate use and environment.

Pattern Printing and Etching. Pattern printing, whether it is used to define an etch-resist or plating-resist pattern, is probably one of the most critical steps in the production of printed wiring since it is here that the precision of the artwork is transferred to the workpiece. As will be developed later, the conventional artwork methods (large-scale ink drawings or taped patterns on drafting film reduced photographically) are not generally satisfactory for fine-line patterns, especially over large areas. In these cases, an automatic-pattern-generation coordinatograph (see pages 1-77 to 1-78) has proved to be most practical.

Changes in etched-line dimensions from the artwork dimensions vary according to the type and thickness of the positive-resist material used and the thickness of the copper (laminate and plating) and etching medium. Negative photoresist materials produce the same amount of change in the printed lines, pads, and spaces. Line, pad, and space dimensions with plated resists vary according to the type of metal plating and the thickness of the plating used.

Resist used for etching is either plating (which is usually applied using a negative-photoresist pattern) or acid-resisting inks or photoresist. Proper surface preparation is of essential importance since resist failure, causing defects in the pattern, most frequently occurs because of inadequate adherence of the resist or from dirt or inclusions in the resist. Variation in line width due to etching is mostly attributed to the fact that the foil material etches at approximately the same rate in the lateral as it does in the vertical direction. Among the common etchants for copper are ferric chloride, cupric chloride, ammonium persulfate, chromic sulfuric acid, and ferric chloride with additive agents. The quality of etching, etching rate, and slope of the sides of the lines are affected by the type of etchant used. It appears that the most effective etching control occurs when the etchant is applied to the copper surface with force, as in splash or spray etching. Table 12 gives artwork dimensions required for a given width of copper line.

Plating. In printed-wiring fabrication, plating is often used as a method for establishing connection between conductive patterns on the two sides or internal layers of the board. In addition, plating is used in the additive method, where lines are built up with electroless or electrolytic plating of copper or unclad, but sensitized, insulating boards. Lastly, plating is often used as an etch resist and always used as a last resist step in plated-through-hole methods, where protective gold or solder are plated in the holes and over the outside circuit patterns to protect the circuit during the final etching operation. Good design practice calls for a copper-plating thickness in the holes of at least 0.001 in. *Panel plating* means that the same thickness of copper which is plated in the holes is also plated on the entire board. Gold or solder is then selectively plated as an etch resist in the holes and on circuit lines only. In *pattern plating,* the copper plated in the holes is also plated on the circuit lines or pattern only, followed by gold or solder and the final etch. The same vertical-lateral relationship holds in plating as in etching; i.e., metal builds up as well as out. For example, artwork must be smaller than the finished line by two times the thickness of copper desired; for 0.003-in. artwork lines on unclad boards a copper thickness of 0.001 in. is maximum for a final line width of 0.005 in. Plated etch resists cause more severe undercutting than ordinary photoresist patterns. Solder-plated patterns are not as troublesome as gold-plated patterns in this aspect. It is believed that greater side etching of gold-plated boards results from a local galvanic action between gold and the etched foil, with the etchant acting as an electrolyte.

Laminating. There are several aspects of the laminating operation in which the design engineer should have an interest. Among the laminating factors which may be of critical importance in the final design are the following:

1. Registration of layers
2. Thickness between layers and total thickness
3. Warp and twist characteristics
4. Interlaminar bond

Registry of circuit planes is frequently of paramount concern, especially where critical electrical characteristics are desired. Multilayer circuit layers are stacked or "layed up" in laminating platens having individual locating pins to control registry. Because of the clearance between pins and hole, and possibly the shift of copper on the substrate during the bonding cycle, misregistration between layers for 12-in.^{2} boards of from ±0.004 to ±0.008 in. is not uncommon. For precise applications and for large boards with hole spacings less than 0.100-in. layer-to-layer registration of ±0.003-in. may be obtained with guide pins at 4-in. centers. In extreme cases, precision of registry between layers is so critical that the board is built up by layers in several separate laminating operations.

Thickness between layers is controlled by the amount of prepreg used. Adequate prepreg, or B-stage, adhesive must be used to prevent any voids or "resin starvation" in the multilayer laminate. As a rule of thumb, the minimum thickness of prepreg to be used is twice the total thickness of copper between adjacent circuits to be bonded. Consulting NEMA or MIL standards for thickness tolerances (which range on the average of 8 percent of sheet thickness), it is easy to see that precise planning is required to control laminate thickness. Where precise thickness tolerance is required, it is best to press a trial lay-up and by measuring, adjust the buildup by changing the thickness of either the circuit laminations or bonding layers. Overall thickness is very important if an edge-board connector is to be used, as will be discussed later.

To minimize warp and twist in the multilayer operation, it is important that all circuits and layers of bonding sheet be oriented in the same yarn-warp direction. This is evident if the material comes rolled; otherwise it can be supplied by the material vendor. Flexural strength is assured only by proper bonding conditions and may be expected to range from 55,000 psi and upward for typical glass-epoxy material.

Hole Fabrication. Tolerance capability for hole formation is dependent upon the way in which the holes are made. Although holes may be punched in some rigid materials (especially paper-phenolic punching grades) and most flexible materials with a positional accuracy of ±0.003 in., this technique is limited to single-sided wiring only. For double-sided and multilayer boards, holes are drilled in a variety of ways previously described. Tolerances on various methods of drilling holes are as follows:

Method	*Tolerance, in.*
Drilling by eye	±0.010
Jig-bored drill plate	±0.005
Drilling with optical aid	±0.005
Tape-controlled drill, 8-spindle.....	±0.003
Tape-controlled drill, 1-spindle.....	±0.001

Hole size for single-sided planes should be large enough to accommodate the stitch, wire, or component lead which is to be inserted through it. For plated holes, a good soldering relationship results when the hole is 0.006 in. larger than the lead it is to accept. For plated via holes, where no lead is used, there is no such restriction, but another limitation exists. At the present time practical manufacturing techniques can produce reliable plated-through holes with diameters down to one-fourth of the overall board thickness. This relationship between thickness and hole diameter (less than 4:1) limits either the hole size or the overall thickness of the board if conventional multilayer techniques are used. For holes which are to accept plating, drilled-hole diameter after expected shrinkage should be minus two times the plating thickness planned for the hole walls.

Projected Use in the System. The designer must, above all, be aware of the capabilities and limitations of the printed wiring he is specifying. It is the purpose of this section to outline application properties of rigid and flexible printed wiring and the way in which they can and do affect the designer's task.

Electrical Properties. The functional characteristics of the circuit and the environment to which it is subjected will largely decide which insulating substrate the designer will use. Since the electrical properties of the printed-wiring board

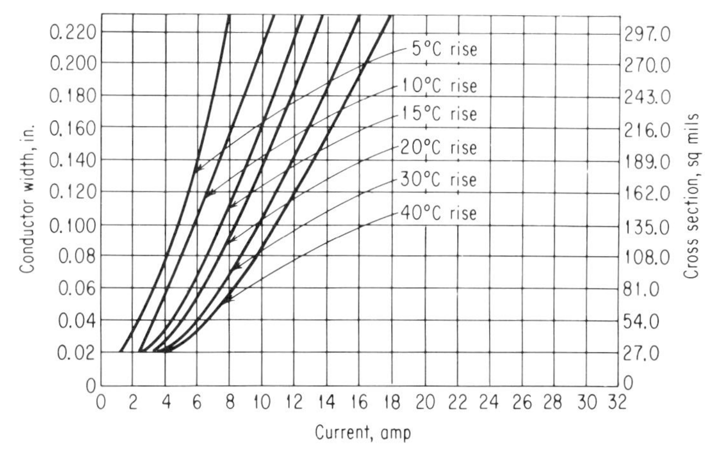

FIG. 27. Temperature rise versus current for 1-oz copper.

are integrally associated with the base-material properties, a clear understanding of these electrical properties may have a decisive effect on the design. The most important electrical properties of the printed-wiring board for dc as well as the rather low-frequency ac systems are current-carrying capacity, arc and tracking resistance, insulation resistance, and dielectric withstanding voltage. For high-frequency applications the most important characteristics are characteristic impedance, dielectric constant, capacitance (and crosstalk) dissipation factor, and propagation delay.

CURRENT-CARRYING CAPACITY. The current-carrying capacity of etched-copper conductors for rigid boards is given in Figs. 27 through 29. For 1- and 2-oz con-

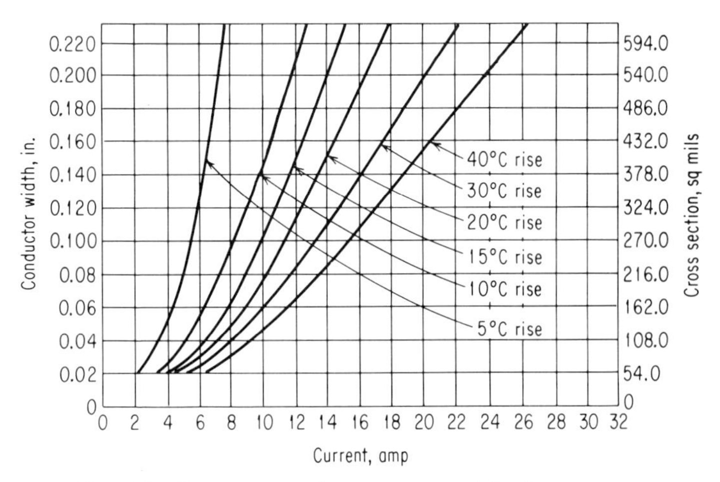

FIG. 28. Temperature rise versus current for 2-oz copper.

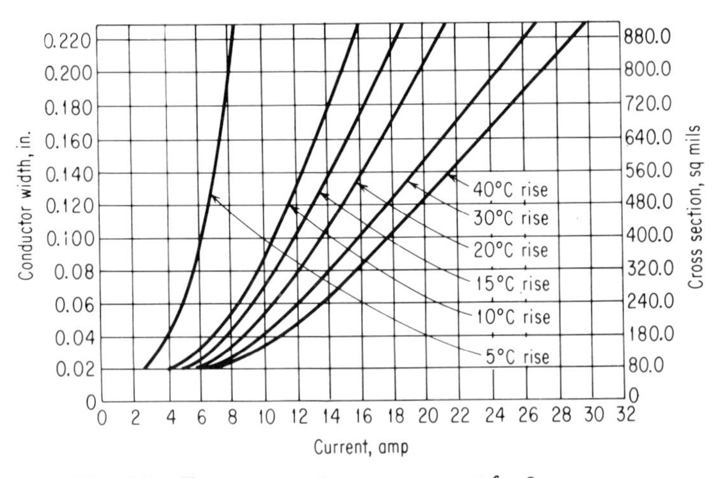

FIG. 29. Temperature rise versus current for 3-oz copper.

ductors, allow a nominal 10 percent derating (on a current basis) to provide for normal variations in etching methods, copper thickness, and thermal differences. Other common derating factors are 15 percent for conformally coated boards (for base material under 0.032 in. and copper over 3 oz) and 30 percent for dip-soldered

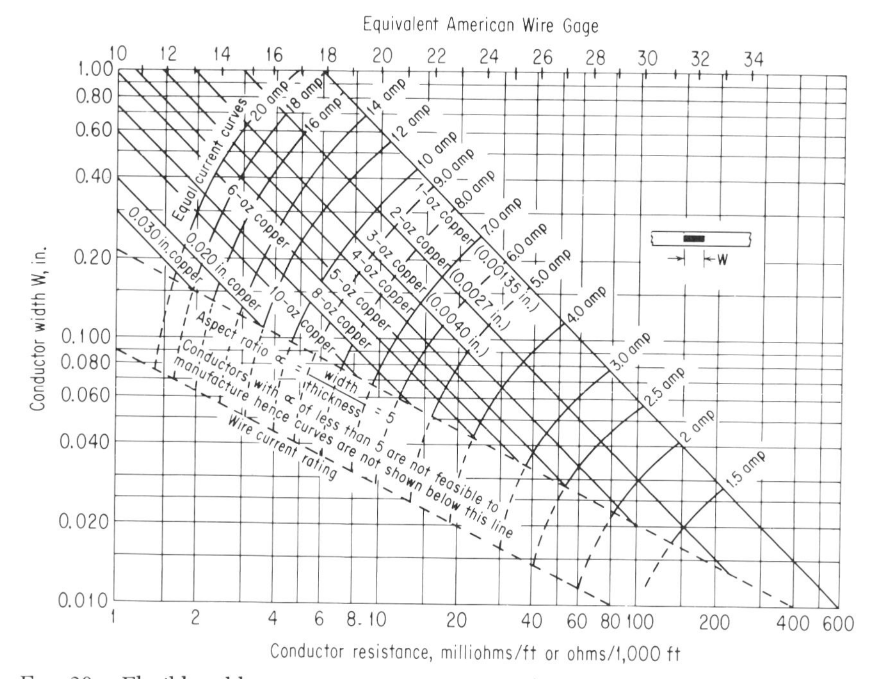

FIG. 30. Flexible-cable current-carrying capacity design chart for single, flat conductors with 0.010-in. insulation, assuming 10-in. allowable temperature rise at 20°C ambient temperature.[27]

boards. A convenient guide to the current-carrying capacity of flexible cable is given in Fig. 30.

CONDUCTOR RESISTANCE. Closely related to current-carrying capacity, the resistance of any metal varies, of course, with its absolute temperature. In the case of copper, if the resistance R_1 at any temperature T_1 is known (refer to Fig. 32 for resistance at 25°C), the resistance R_2 at any other temperature T_2 may be determined by the formula

$$R_2 = R_1 \qquad 1 + A\ (T_2 - T_1)$$

where R_1 = resistance of copper at 25°C (Fig. 32)
R_2 = unknown resistance at T_2
T_1 = 25°C
A = temperature coefficient of copper (Fig. 31)

A convenient empirical equation for standard annealed copper is

$$R_2 = R_1 \frac{234.5 + T_2}{234.5 + T_1}$$

where R_1 = resistance of copper at T_1
R_2 = resistance of copper at T_2

ARC AND TRACKING RESISTANCE. *Arc resistance* may be defined as the ability of an insulating substrate to resist the formation of a conducting path when subjected to electrical arcs over the material surface. *Track resistance* is the property which enables a material to resist the formation of creepage paths under the continuous application of electrical stress. These properties are only desirable when high

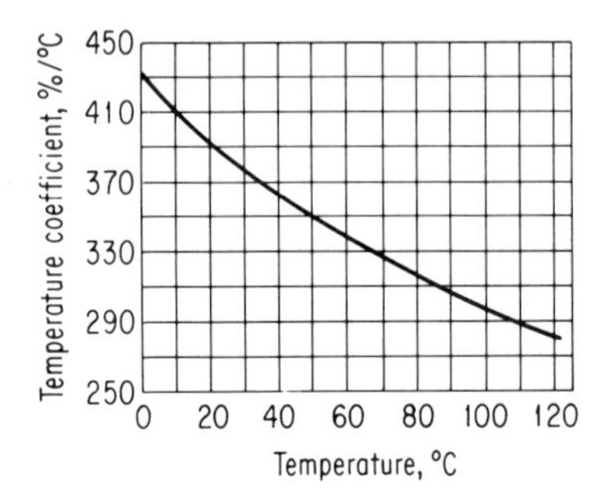

FIG. 31. Temperature coefficient of copper.

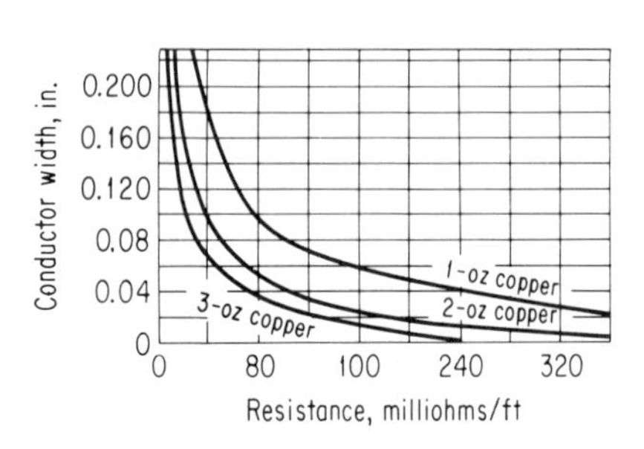

FIG. 32. Conductor resistance versus conductor width of 1-, 2-, and 3-oz copper conductors.

voltages or high-voltage pulses are expected in the packaging system. Although individual suppliers should be consulted for these values when the application warrants it, some comparative figures for arc resistance are as follows: XXXP, 5 to 10 sec; G-10, 60 to 80 sec; PTFE, 180 sec; and G-7 (silicone), 200 to 250 sec.

INSULATION RESISTANCE. *Insulation resistance* is probably the most important single parameter of the processed laminate from the standpoints of both design and material engineers. Insulation-resistance tests are important ways of determining raw-material properties as well as of controlling product quality since they are excellent indicators of the propensity of the material for moisture, ions, and other contaminants commonly found in the manufacturing and handling of the printed-wiring board. The test is usually performed by making insulation-resistance tests between conductors on a layer (or between layers in the case of double-sided and multilayer boards) after a specified time in a humidity and temperature cycle. A "polarization" voltage, usually 100 volts ac, is applied between conductors during the test and resistance measurements made immediately following the test. Typical resistance values are 1,000 megohms per 0.001-in. separation between conductors. Among the design features which may minimize low-insulation-resistance problems are the following:

1. Keep the conductors at least 0.10 in. away from the edge of the board.
2. Machine (mill or saw) the border rather than shearing it.
3. Avoid contamination during handling and processing.
4. Employ a conformal coating.

DIELECTRIC WITHSTANDING VOLTAGE. It appears that this property, also called *flashover strength,* is dependent only on the air density and conductor spacing and seems independent of moisture content or dielectric constant of the material or conductor-surface finish. An important factor for the design engineer to consider is the potential at which flashover occurs in the air above the conductor traces. Table 13, taken from MIL-STD-275, may be used for a design guide in this case. As indicated, increasing altitude lowers the insulating properties of air (causes it to ionize more readily), and, accordingly, greater spaces are called for. Other than this, high-altitude design may call for pressurization of equipment or conformal coating or encapsulation of wiring as found in multilayer wiring. Table 13 is not valid for alternating current above 400 Hz.

DIELECTRIC CONSTANT AND DISSIPATION FACTOR. The *dielectric constant* of an insulating material is defined as the ratio of the capacitance of a capacitor containing that particular material to the capacitance of the same electrode system, with air replacing the insulation as the dielectric medium. It may also be defined as that property of an insulation which determines the electrostatic energy, stored within the solid material, i.e., the ability of the insulating material to store energy.

The *dissipation factor* is defined as the ratio of the equivalent series resistance to the equivalent series capacitive reactance of a dielectric circuit. These values are especially important for high-frequency applications since the impedance of a printed-wiring transmission line is proportional to the frequency, and the dielectric constant for most known board materials decreases with frequency. Tables 2 and 3 give values for these factors for rigid and flexible materials, respectively. For

TABLE 13. Spacing of Conductors for Various Voltages [8]

Uncoated boards sea level to 10,000 ft		Uncoated boards over 10,000 ft and coated boards		
Voltage between conductors, dc or ac peak, volts	Minimum space, in.	Voltage between conductors, dc or ac peak	Minimum space, in.	
			Above 10,000 ft uncoated	Conformally coated all altitudes
0–150	0.025	0–50	0.025	0.010
151–300	0.050	51–100	0.060	0.015
301–500	0.100	101–170	0.125	0.020
Above 500	0.0002 (in./volt)	171–250	0.250	0.030
		251–500	0.500	0.060
		Above 500	0.001 (in./volt)	0.0012 (in./volt)

the lower values ordinarily found in printed-wiring materials, dissipation factor is the practical equivalent of the power factor. Q is the quality factor, or *Q factor,* which is equivalent to the reciprocal of the dissipation factor and is sometimes used to rate a dielectric. If the power losses in a dielectric are to be minimized, the loss factor must be low since the mean power developed in the dielectric is proportional to the loss factor. The interrelation of these values is given by the equation

$$\epsilon'' = \epsilon_r \tan \delta$$

where ϵ_r = dielectric constant
ϵ'' = loss factor
$\tan \delta$ = loss tangent or dissipation factor

For printed-wiring substrates, a very low dissipation factor and a low dielectric constant are usually the characteristics most in demand.

CAPACITANCE. The phenomena associated with capacitance, crosstalk, characteristic impedance, and propagation or time delay are normally important only for higher-speed circuit applications requiring the use of double-sided, or, more frequently, multilayer printed wiring. Multilayer boards are also needed to provide the density of wiring usually needed for the higher-speed circuits in order to reduce signal delay. Capacitance for double-sided boards is determined by the formula[20]

$$C = \frac{0.2249\epsilon_r A}{d}$$

where C = capacitance, pf/in.
ϵ_r = dielectric constant (relative to air = 1)
A = area of smallest electrode, in.^{2}
d = thickness of dielectric, in.

Under normal conditions, lines parallel and separated by 0.062 in. on 0.062-in.-thick G-10 will exhibit a capacitance of 0.5 pf/in., and 0.3 pf/in. for 0.062-in.-thick Teflon.

In multilayer boards, capacitance is calculated by a different formula although the conductors are still capacitively coupled when they physically approach each other in the horizontal or vertical direction. An approximate value for the amount of capacitive coupling may be calculated by the equation[20]

$$C = \frac{W \epsilon_r}{h\ 4.45}$$

where ϵ_r = effective dielectric constant
h = height, mils
W = width, mils
(See Fig. 33.)

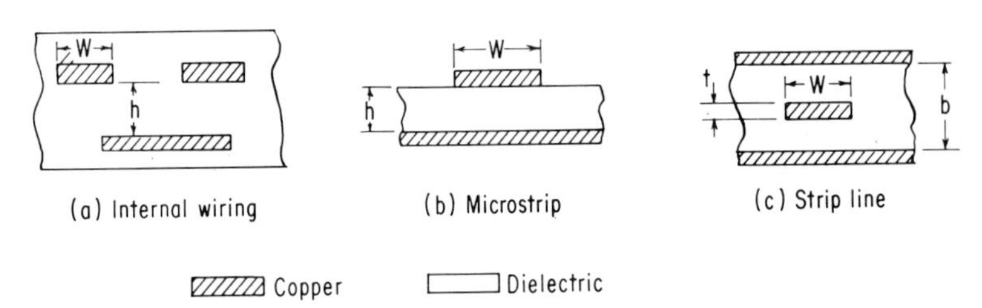

FIG. 33. Printed-wiring configurations.

The two methods of achieving transmission-line capability in wiring are illustrated in Fig. 33*b* and *c*. Microstrip may be used as a single wiring layer or the outside layer of a multilayer board. Strip transmission line may be used either as one or a multitude of wiring layers. Capacitance of strip transmission line, referring to Fig. 33*c*, can be approximated by the equation[20]

$$C = 0.9 \frac{W/b}{1-(t/b)} \epsilon_r$$

CROSSTALK.[19] Crosstalk is the undesirable coupling of energy between signal paths. This unwanted transfer of energy between lines results from capacitive and inductive coupling between lines and is a function of length of the lines and space between them, and the dielectric constant. This phenomenon is often a limiting factor in high-speed digital systems, where coupling between lengths of parallel lines may be the only alternative to the high wiring density which is required. Crosstalk between an active line, i.e., the line carrying the voltage pulse, and a passive line in the vicinity places a spurious pulse on the passive line. If the latter pulse is of sufficient amplitude, it could falsely switch circuits on the passive line. Traditionally, before crosstalk could be calculated it was necessary to calculate or measure mutual inductance, mutual capacitance, propagation velocity, and impedance of the lines. Analytical methods are unwieldy because of the effect of air on the dielectric constant and the uncertainty of fringe field effect. As indicated by Kaupp,[21] a new method is now available which depends only on the geometric cross section of the lines and is independent of time, length, and voltage.

For the case of microstrip, then, induced voltage values due to crosstalk may be calculated by use of the formulas

$$V_B(t) = K_B[V_i(t) - V_i\ (t - 2l\sqrt{LC})]$$

$$V_F(t) = K_F l \frac{dV_i}{dt} (t - l\sqrt{LC})$$

where l = length of coupled region
V_B = induced backward voltage } as seen at the ends of the
V_F = induced forward voltage } coupled region
V_i = input voltage
K_B = backward crosstalk constant
K_F = forward crosstalk constant

K_B is always positive with respect to the driving signal. K_F is negative for microstrip and will be zero for homogeneous dielectric media. To determine crosstalk constants for any system, use line lengths where $2T_d$ is significantly greater than the rise time and use the following formulas:

$$K_B = \frac{V_B}{V_i} \quad \text{and} \quad K_F = \frac{V_{F\,\max}}{dV_i/dt_{\max}}\left(\frac{1}{l}\right)$$

where l = length of coupled region
V_B = induced backward voltage
V_F = induced forward voltage
V_i = input voltage

Empirical values for K_B and K_F are given in Figs. 34 and 35. A summary of crosstalk values for flexible printed wiring is given in Fig. 36.

TIME DELAY OR PROPAGATION DELAY. For the microstrip configuration, Fig. 33*b*, signal propagation-velocity delay is calculated by means of the following formula:

$$T_d = 1.015\sqrt{\epsilon_r'}\ \text{nsec/ft}$$

where T_d is the time delay in nanoseconds

ϵ_r' = effective dielectric constant

CHARACTERISTIC IMPEDANCE. An important parameter for high-speed, digital-interconnection design, control of impedance is important to avoid signal reflections which result from the passage of fast pulses through an impedance discontinuity, or a mismatch at the load end of the line. It is dependent upon the line width, dielectric thickness, and dielectric constant of the insulating medium.

The general equation for characteristic impedance of any high-frequency transmission line is

$$Z_0 = \sqrt{\frac{R + j\omega L}{G + j\omega C}}$$

where the parameters are

Z_0 = impedance, ohms
R = resistance per unit length of line
L = inductance per unit length of line
G = conductance per unit length of line
C = capacitance per unit length of line

If the line is lossless, the equation is simplified:

$$Z_0 = \sqrt{\frac{L}{C}}$$

A specific formula for microstrip (Fig. 33*b*) is the following:

$$Z_0 = \frac{h}{W}\,\frac{377}{\epsilon_r}$$

where h and W are given in the diagram and ϵ_r is the effective dielectric constant of the material (considering the effect of air). This analytical method disregards fringing effects and leakage flux and gives validity to the notion that more reliable values may be obtained by measurement. To account for the fringing effects of microstrip transmission line, and if the analytical method must be used, the following formula is recommended:[20]

$$Z_0 = \frac{h}{W}\,\frac{377}{\sqrt{\epsilon_r}\,\{1 + (2h/\pi W)\,[1 + \ln(\pi W/h)]\,\}}$$

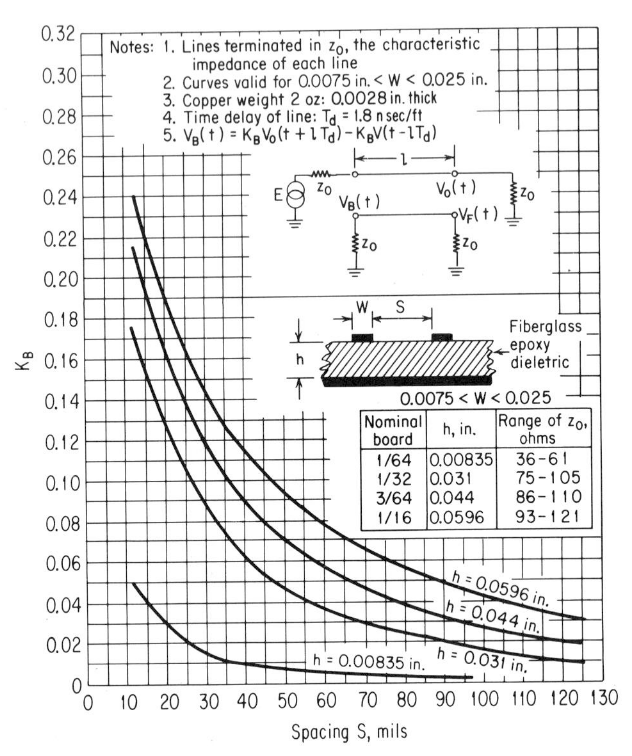

FIG. 34. Back crosstalk constant as a function of spacing.[1]

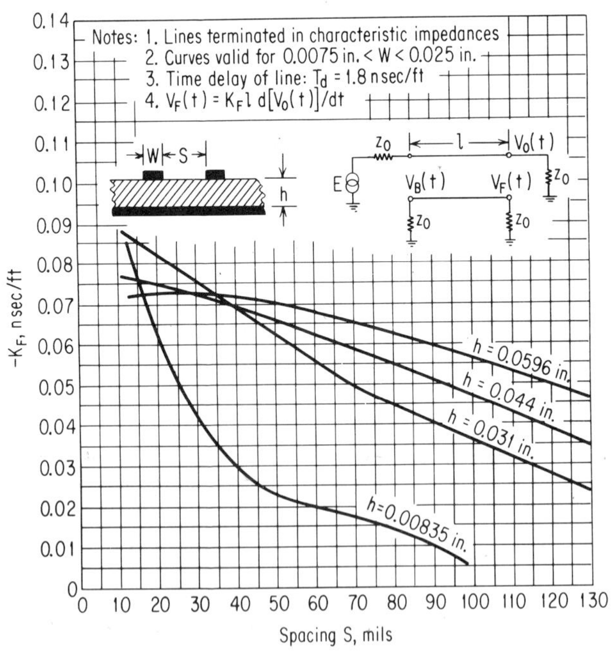

FIG. 35. Forward crosstalk as a function of spacing.[21]

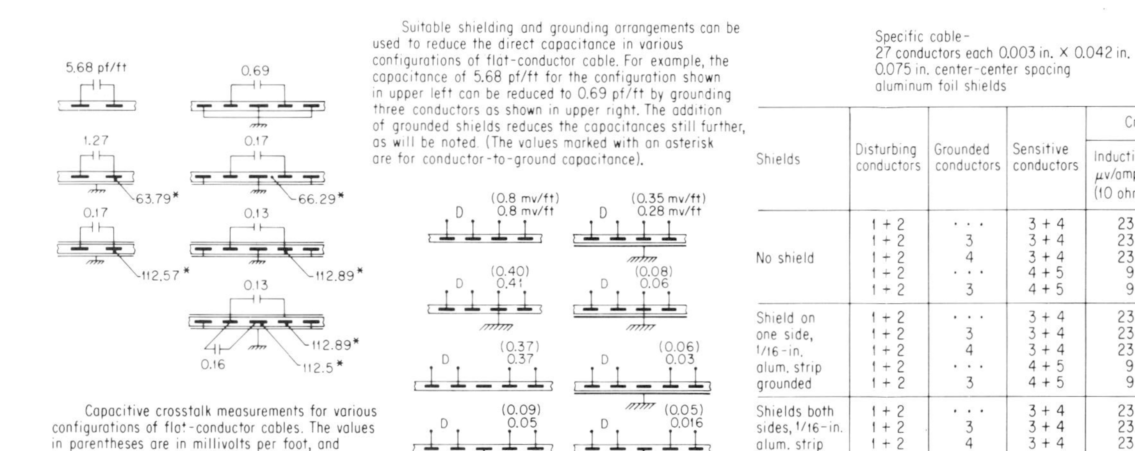

Capacitive crosstalk measurements for various configurations of flat-conductor cables. The values in parentheses are in millivolts per foot, and represent the pickup voltage caused by a current of 4 amp at 400 cps in the disturbing circuits. (D = 120 volts, 460 cps)

Suitable shielding and grounding arrangements can be used to reduce the direct capacitance in various configurations of flat-conductor cable. For example, the capacitance of 5.68 pf/ft for the configuration shown in upper left can be reduced to 0.69 pf/ft by grounding three conductors as shown in upper right. The addition of grounded shields reduces the capacitances still further, as will be noted. (The values marked with an asterisk are for conductor-to-ground capacitance).

Specific cable–
27 conductors each 0.003 in. × 0.042 in.
0.075 in. center-center spacing
aluminum foil shields

Shields	Disturbing conductors	Grounded conductors	Sensitive conductors	Crosstalk, μv	
				Inductive, μv/amp/ft (10 ohms)	Capacitive, μv/v/ft (5,000 ohms)
No shield	1 + 2	. . .	3 + 4	23	6.9
	1 + 2	3	3 + 4	23	3.4
	1 + 2	4	3 + 4	23	16.8
	1 + 2	. . .	4 + 5	9	3.0
	1 + 2	3	4 + 5	9	0.4
Shield on one side, 1/16-in. alum. strip grounded	1 + 2	. . .	3 + 4	23	2.3
	1 + 2	3	3 + 4	23	0.5
	1 + 2	4	3 + 4	23	5.2
	1 + 2	. . .	4 + 5	9	0.2
	1 + 2	3	4 + 5	9	0.1
Shields both sides, 1/16-in. alum. strip alum. foil grounded	1 + 2	. . .	3 + 4	23	0.4
	1 + 2	3	3 + 4	23	0.0
	1 + 2	4	3 + 4	23	0.8

FIG. 36. Summary of crosstalk measurements.[6]

Typical values of characteristic impedance values of microstrip on epoxy-glass material where the conductor is exposed to air are given in Fig. 37.

The characteristic impedance of strip transmission line, illustrated in Fig. 33*c*, when $W/b >> 0.35$ is

$$Z_o = \frac{60}{\sqrt{\epsilon_r}} \ln \left(4b/d_o\right)$$

where d_o, the effective wire diameter for square configuration, is $0.567W + 0.67t$. Representative values of Z_o for strip transmission line are shown in Fig. 38.

A complementary discussion of these printed-wiring electrical-design factors is given in Chap. 10.

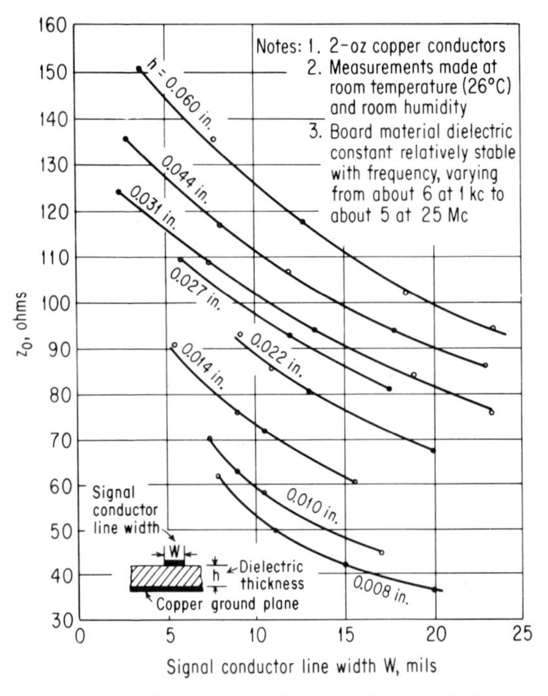

FIG. 37. Characteristic impedance versus line widths and dielectric thickness: G-10 epoxy-glass dielectric-microstrip transmission line.[20]

Layout and Mounting Considerations. Although many of the printed-wiring layout fundamentals will necessarily be covered in the Artwork section of this chapter, page 1-75, there are some items which concern initial planning of the printed-wiring system and mounting considerations which have a decided effect on the projected use of the printed-wiring board in the total system. More information on packaging and layout design may be found in Chaps. 8, 9, and 12 through 14 of this handbook.

Basically, the printed-wiring board electrically interconnects the active and/or passive devices mounted on it, lends them mechanical support, and assists in removing heat generated by them out of the system. In addition to these factors then, the designer must consider the probable number of layers required as well as the method in which it is anticipated to mount the components to the printed-wiring board.

GENERAL LAYOUT FACTORS. The printed-wiring board is the basic building element of a variety of packaging techniques. It is commonly used singly or in combination with a number of other boards. Combinations of single printed-wiring boards may also be made to form more complex units, such as planar and stacked modules, which are covered later in this chapter. Printed-wiring boards may also be utilized as logic-interconnect boards without components. For single-sided boards the printed wiring is on the obverse side from the components, and crossovers are accomplished by component placement. Components are placed on one side only, and leads are extended through holes in the base and clinched to the circuit pad or land. Electrical connection is then made to the copper pattern on the etched side, usually by soldering. Double-sided boards may have components on both sides of the board, and interconnection between sides is made by a plated hole, eyelet, com-

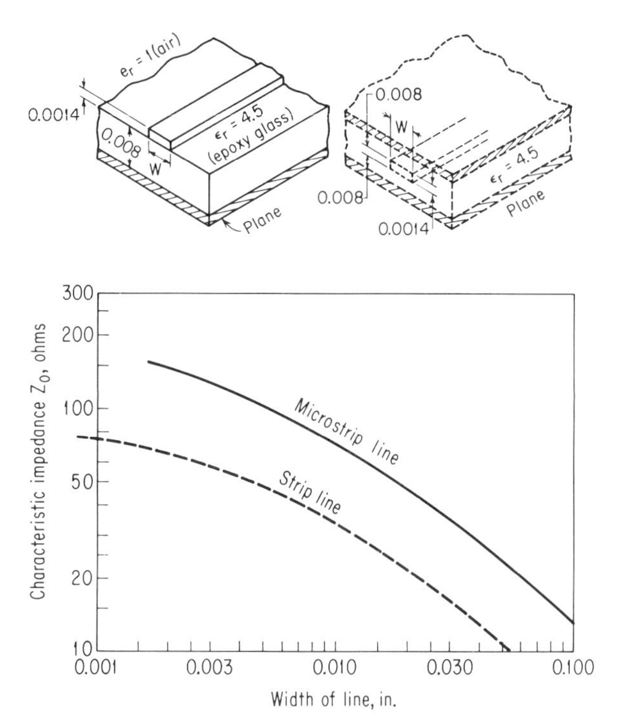

FIG. 38. Microstrip and stripline impedance comparison. (*Reprinted from Electronics*, July 11, 1966, copyright McGraw-Hill, Inc., 1966.)[22]

ponent lead, or "stitch." Multilayer boards provide wiring density not available by conventional techniques, and layout of these is complicated to the point of requiring computer-assisted routing and placement of modules.

Many printed-wiring designs are based on *modular design*, which means that the electronic package is subdivided into functional units or modules. Printed wiring is particularly adaptable to this philosophy (which yields a standard configuration of the same shape, dimensions, and construction), whose primary advantage is standardization. The many implications of optimum module size, minimum throwaway cost, and maintainability philosophy are usually the responsibility of the system designer.

The printed-wiring designer must be in a position to provide the flexibility of design required. In general, expendable or throwaway packages will be more of a "universal" logic, composed of high-quantity, low-cost circuitry of a fixed design. Repairable modules are more complex, low-quantity circuits, usually of the "logic-on-board" variety. Many of these considerations will be treated in later chapters on packaging, especially in Chap. 13.

ELECTRICAL DESIGN FACTORS. Electrical factors which affect the layout of the printed-wiring board are essentially the electrical requirements of the system of which it becomes a part. As illustrated previously, the conductor width and spacing are primarily a function of voltages, currents, and environment of the entire electronic package. Conductor spacing, location, and routing may also be affected by the types of signals which are carried when these signals may interfere with each other or be affected by stray capacitance. This factor may call for transmission-line capabilities, the characteristics of which have already been discussed. Because multilayer wiring is a number of individually designed and etched circuit boards, the same basic layout factors apply to the individual layers that apply to conventional one- and two-sided circuitry. The width of any conductor is a function of the current carried and the maximum-allowable heat rise due to resistance. The minimum conductor width, however, shall not be less than that required by the detail drawings. The conductor width should be as generous as possible, and 0.040 in. is a practical figure for lines and spaces. This configuration can tolerate the normal amount of undercutting caused by etching, as well as nicks and scratches in the artwork from careless handling, or caused by the manufacturing process. As explained previously, the amount of undercutting, or line reduction, in mils, is equal to twice the copper thickness in ounces, except for 3 and 5 oz copper, where the undercutting may average 0.007 and 0.012 in., respectively, depending upon the resist system used. Original papers should be consulted for specific details on finer-line etching. Line widths of 0.010 in. for 1 oz copper and 0.005 in. for ½ oz are not uncommon at the present time. The required spacing between conductors is influenced by several factors. These are generally

1. Peak voltage
2. Atmospheric pressure
3. Use of a coating
4. Capacitive coupling parameters

The distance between layers in a multilayer board is influenced by

1. Available material
2. Achievable board thickness, i.e., hole-diameter ratio
3. Insulation-resistance and dielectric-withstanding voltage requirements
4. Strip-transmission-line geometry, if used

The logician, or circuit designer, may place certain restrictions on board layout which must be heeded. These are usually called *wiring rules*. For instance, critical impedance or high-frequency components are usually placed very close together to reduce critical stage delay. Transformers and inductive elements should be isolated to prevent coupling, and inductive signal paths should cross at nominally right angles. Components which may produce any electrical noise from movement of magnetic fields should be isolated or rigidly mounted to prevent excessive vibration.

The reader may be aware of many published guidelines regarding rules for terminal-area spacing and conductor width. A conservative approach to this problem for single- and double-sided printed-wiring boards was outlined by Keonjian[23] and is reproduced in Tables 14 and 15. Although these data are based on standard industry and U.S. Army practice, a "tighter" design may result if the designer chooses to use different materials or sophisticated manufacturing procedures. Table 16 outlines design practice for multilayer printed wiring as suggested by the Institute of Printed Circuits. A composite picture of the interconnection density capabilities for all printed-wiring systems is given by Table 17. Although maximum, theoretical termination densities are quite high, they are not practical because this case would allow zero conductor-interconnection freedom. Practical termination densities allow for a reasonable number of current-carrying connections between various terminations.

TABLE 14. Terminal-area Spacings for Single- and Double-sided Printed Wiring (No Through Conductors)[23]

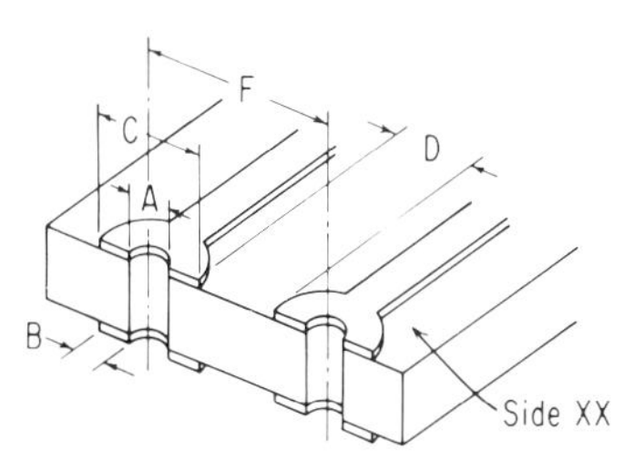

Dimension	Single-sided printed wiring, side XX, in.		Double-sided printed wiring, in.	
	300 volts	100 volts	100 volts	50 volts
A: Mounting-hole dia., max.	0.060	0.045	0.035	0.020
B: Annular-ring width, min.	0.040	0.025	0.015*	0.015*
C: Terminal-area dia., min. (A + 2B) ..	0.140	0.095	0.065	0.050
D: Electrical spacing (coated), min.	0.030	0.020	0.020	0.015
E: Tolerances	0.030	0.035	0.015	0.010
F: Recommended mounting-hole spacing (C + D + E)	0.200	0.150	0.100	0.075

* Reduced annular ring made possible by use of plated-through hole for increased solder-joint strength.

MECHANICAL DESIGN AND LAYOUT FACTORS. Although the printed-wiring board lends mechanical support to the components, it should not be used as a structural member of the overall equipment. Support should be provided at peripheral intervals of at least 5 in. at the board edge. An analysis of the board for any equipment shock and vibration requirements should reveal whether or not additional support is required. Board warpage should be considered in design of the mechanical mounting arrangements. Where the edge-board type of connector is to be used for the printed-wiring board, a careful design is required to assure that the overall board thickness remains within the tolerance capability of the connector. The design of board mounting means slides, handles, and other mechanical design features shall be such that close tolerances are not required on the board dimensions. Additional information on mechanical design features of printed wiring will be found in Chaps. 8, 9, and 13.

It goes without saying that the printed conductor should travel by the shortest path between components within the limits of the wiring rules imposed. The latter may restrict parallel runs of conductors owing to coupling between them. Good design dictates that the minimum number of layers of wiring be used, and also the maximum line-width and pad diameter (the area of connection between the component lead and the printed-circuit conductor is called the *pad*, or *land*) that is commensurate with the density of packaging required. Since rounded corners and smooth fillets avoid possible electrical as well as mechanical problems, sharp angles and bends in conductors should be avoided.

The pad should, ideally, symmetrically surround the mounting hole which envelops the component lead, and the mounting holes should be located at the grid intersection of a modular grid system, usually 0.025, 0.050, or 0.100 in. The annular ring around the hole should be at least 0.020 in. for nonminiaturized designs,

TABLE 15. Layout Design for Single- and Double-sided Printed Wiring [23]

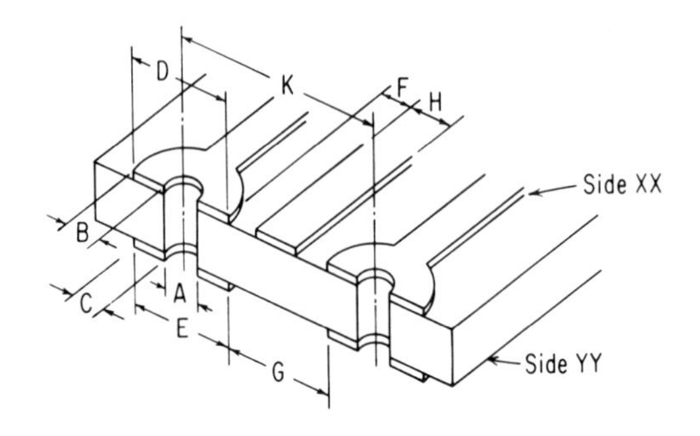

Dimension	Single-sided printed wiring, side XX, in.		Double-sided printed wiring, in.	
	300 volts	100 volts	100 volts	50 volts
A: Mounting-hole diameter, max.	0.045	0.035	0.035	0.020
B: Annular-ring width, min., side XX . .	0.025	0.020	0.005*	0.005*
C: Annular-ring width, min., side YY . . .	NA†	NA	0.015*	0.015*
D: Terminal-area diameter, min., side XX (A + 2B)	0.095	0.075	0.045	0.030
E: Terminal-area diameter, min., side YY (A + 2C)	NA	NA	0.065	0.050
F: Electrical spacing, min., side XX ‡ . .	0.030	0.020	0.015	0.010§
G: Electrical spacing, min., side YY ‡ . .	NA	NA	0.025	0.015§
H: Conductor width, min., side XX	0.030	0.020	0.015	0.015
I: Tolerances, side XX	0.015	0.015	0.010	0.010
J: Tolerances, side YY	NA	NA	0.010	0.010
K: Recommended terminal-area spacing with one conductor (D + 2F + H + I)	0.200	0.150	0.100	0.075

* Reduced annular ring made possible by use of plated-through hole for increased solder-joint strength.
‡ Protective coating over circuitry.
§ Minimum spacing available for 0.075 terminal-area spacing.
† NA means not applicable.

and pad diameters in these cases may run to 0.100 to 0.125 in. For a good solder fillet, the clearance hole for the electrical connection should not exceed the component-lead diameter by more than 0.012 in., unless the lead is to be clinched and soldered to the foil. For miniaturized applications, such as those used for integrated-circuit packaging, clearance-hole–lead-diameter relationships are given in Table 18. These dimensions apply to plated-through holes only.

MOUNTING CONSIDERATIONS. Printed-wiring boards are primarily used to interconnect various components, and the method of mounting these components to the board is of critical importance. Most conventional components such as diodes, capacitors, and resistors are mounted parallel to the printed-circuit board, with a minimum spacing of 0.015 in. The clearance between the lead on one component and the body of the next should be 0.06 in. for manual insertion and 0.03 in. for automatic assembly. Usually, the only fastening holding most components to the board are the leads themselves, as illustrated in Fig. 39. All components should be mounted as close to the board as practical unless some additional support is re-

TABLE 16. Layout Design for Multilayer Printed Wiring with Crossover Conductors [23]

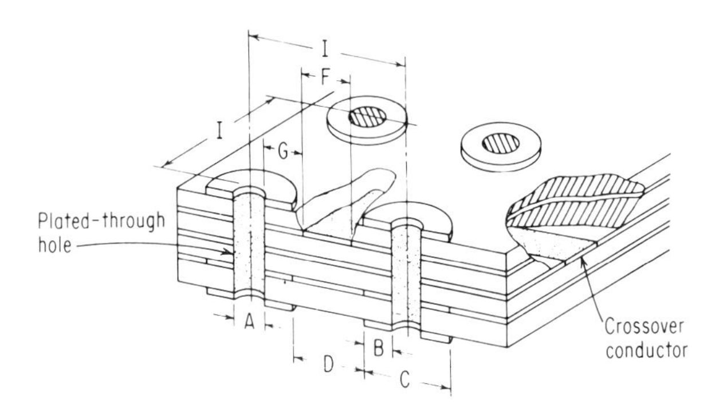

Dimension	Inches			
	300 volts	200 volts	100 volts	50 volts
A: Mounting-hole dia., max.	0.036	0.036	0.026	0.021
B: Annular-ring width	0.020	0.020	0.017	0.0025*
C: Terminal-area dia., max. (A + 2B) ..	0.076	0.076	0.060	0.026
D: Electrical spacing between terminal areas (coated), min.	0.030	0.020	0.011	0.020
E: Terminal-area tolerances (ICD)	0.044	0.004	0.004	0.004
F: Conductor width, min.	0.030	0.020	0.020	0.010
G: Electrical spacing between conductor and mounting hole (coated), min. ..	0.030	0.015	0.011	0.005
H: Conductor-to-mounting-hole tolerances	0.024	0.014	0.007	0.009
I: Mounting-hole spacing (A + F + 2G + H)	0.150†	0.100†	0.075†	0.050†

* Reduced annular ring made possible by use of plated-through hole for increased solder-joint strength.

† The dimensions given for 0.150-, 0.100-, and 0.075-in. mounting-hole spacings apply for either the plated-through or clearance-hole multilayer fabrication techniques. Dimensions for the 0.075- and 0.050-in. mounting-hole spacings apply to the "plated-through" and "clearance-hole" techniques, respectively.

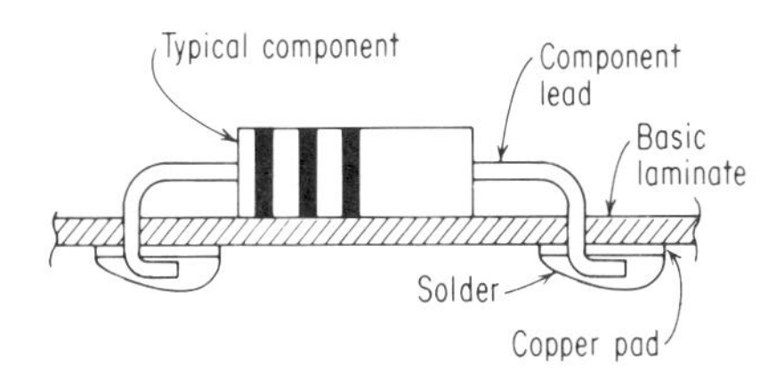

FIG. 39. Conventional component mounting.

TABLE 17. Comparison of Printed-wiring Terminations and Integrated Capabilities [23]

Type of printed wiring	Adjacent terminal spacing, in.	Minimum line width and electrical spacing, in.	Termination density terminations/in.²		Theoretical conductor density conductors/in.		Interconnection capability[a]
			Theoretical	Practical	Parallel	Crossover	
Single-sided ...	0.200[b] (300 volts)	0.030	25	5	4	0	100
	0.150[b] (100 volts)	0.20	45	10	7	0	315
Double-sided ..	0.100[b] (50 volts)	0.20	100	18	10	10	1,000
	0.075[b] (50 volts)	0.20	177	40	13	0 / 6.5[c]	2,300 / 1,700
Multilayer (six layers) ..	0.150[d] (300 volts)	0.20	45	25	14	14	2,260
	0.100[d] (300 volts)	0.20	100	50	20	20	4,000
	0.075[e] (100 volts)	0.20	177	88	26	26	9,200
	0.050[f] (50 volts)	0.005 for spacing / 0.010 for line	400	200	40	40	32,000

[a] Interconnection capability (a figure of merit) = terminations/in.² × conductors (parallel + crossover)/in.
[b] Production capability.
[c] Crossover capability for 88 terminations/in.²
[d] Crossover capability for 50 terminations/in.²
[e] Pilot-production capability.
[f] Development-model capability.

TABLE 18. Clearance Holes and Pad Diameters for Various Lead Sizes

Lead diam. (or max. dimensions), in.	Nominal hole diam., in.*	Design pad diam., in.		
		Preferred	Acceptable	Min.
0.014	0.020	0.060	0.050	0.040
0.019	0.025	0.065	0.055	0.045
0.024	0.030	0.070	0.060	0.050
0.029	0.035	0.080	0.070	0.060
0.034	0.040	0.100	0.080	0.070

* After plating.

quired, as in the case of heavier components or where unusual vibration is expected. MIL-STD-275B[8] states: "All parts weighing ¼ ounce per lead or more, shall be mounted by clamps or other means of support, such as embedment, which ensures that the soldered joints are not relied upon for mechanical support." Parts shall be mounted so as not to obscure the terminations of another part. In general, heavy components, even though clamped, should be over or near supported areas. Whenever possible, the mounting scheme should provide the flexibility for automatic assembly except where breadboarding or short production runs are expected.

Figure 40 illustrates the three predominant methods for attachment of components to a printed-wiring board: the surface-pad, post, and obverse-side, or through-hole, method. Recommended dimensions from the component body to the terminal area for various types of components are indicated. Special lead-bending recommendations for integrated circuits are given in Fig. 41.

An important aspect of printed wiring as it affects component mounting is the configuration used for attaching components to the board. Illustrated in Figs. 42 and 43 for soldering and welding, respectively, these techniques can only be utilized if the interconnection medium is compatible. The usual recommended soldered configuration, the clinched method, is illustrated in Fig. 42*b*. This technique provides good mechanical support and is adaptable to bulk-processing techniques such as automatic assembly and flow or dip soldering. In the welding category, Fig. 43, the surface-pad and post techniques are most used, although the latter must have a

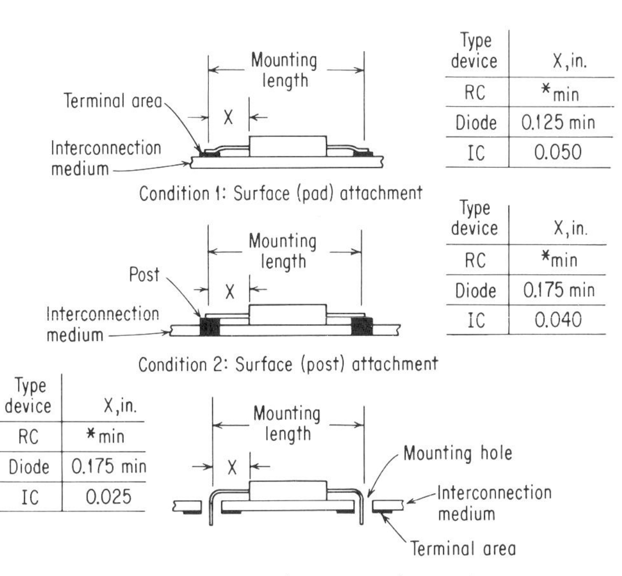

Condition 1: Surface (pad) attachment

Type device	X, in.
RC	*min
Diode	0.125 min
IC	0.050

Condition 2: Surface (post) attachment

Type device	X, in.
RC	*min
Diode	0.175 min
IC	0.040

Condition 3: Obverse side (mounting hole) attachment

Type device	X, in.
RC	*min
Diode	0.175 min
IC	0.025

FIG. 40. Printed-wiring component mounting.

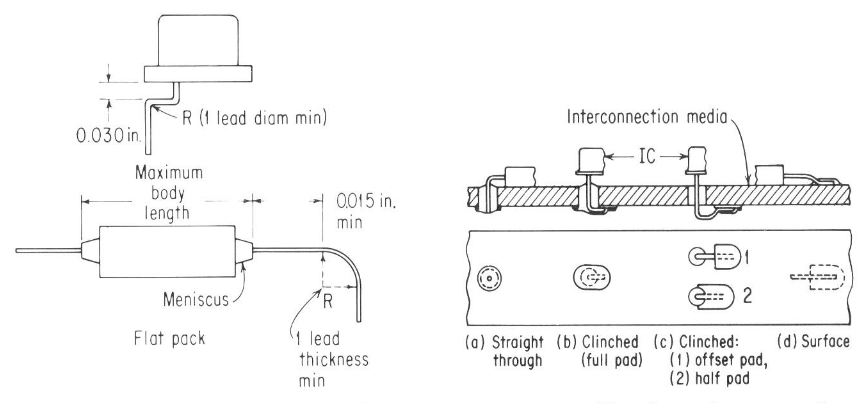

FIG. 41. Integrated-circuit lead forming requirements.

FIG. 42. Soldered-attachment configurations.[24]

board with a plated-up "pillar," such as is produced by the built-up method. Details on soldering and welding as applied to printed wiring may be found in Chaps. 3 and 4, respectively.

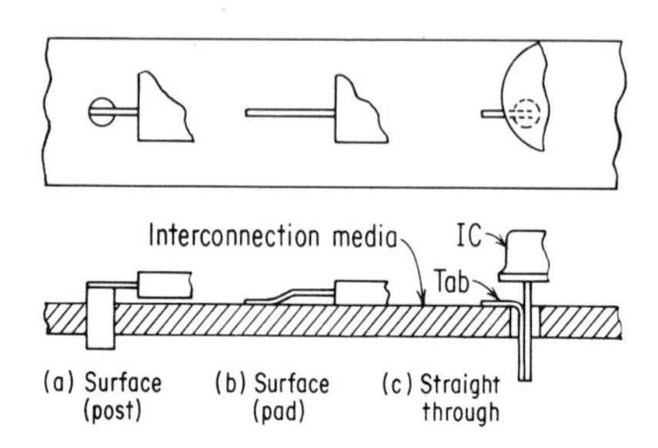

FIG. 43. Welded-attachment configurations.[24]

TABLE 19. Average Thermal Conductivity *K* of Various Materials from 0–100°C (32–212°F) [1]

Material	Specific gravity	Btu/(hr)(ft²)(°F)(ft)
Metals and alloys:		
Copper	8.50	218–224
Aluminum	2.71	117–119
Brass (70% Cu, 30% Zn)	8.53	56–60
Magnesium alloy (9% Al, 1% Zn)	1.81	26–28
Steel (mild)	7.85	26–28
Ceramics:		
Beryllium oxide	2.97	104–130
Magnesium oxide	3.2	18–21
Aluminum oxide	3.7	10–15
Photosensitive glass ceramic	2.46	1.3
Thermoplastics:		
Polyethylene	0.92–0.96	0.1–0.4
Polytetrafluoroethylene	2.15–2.25	0.1–0.2
Molded thermosets:		
Phenolic, wood-flour-filled	1.32–1.45	0.10–0.19
Phenolic, mineral-filled	1.65–1.92	0.24–0.34
Diallyl phthalate, acrylic fiber	1.31–1.45	0.18–0.19
Laminates, perpendicular-to-face:		
XXXP, paper-phenolic	1.3–1.4	0.04–0.12
G-7, silicone-glass	1.6–1.8	0.07–0.17
G-10, epoxy-glass	1.7–1.8	0.10–0.17
PTFE glass-cloth	2.1–2.2	0.02–0.05
Casting resins and foams:		
Epoxy, unfilled	1.16	0.13–0.20
Epoxy, 73% alumina by weight	......	0.82
Epoxy, 50–55% silica by weight	1.6–1.7	0.29–0.53
Epoxy, hollow phenolic spheres	0.86	0.16
Epoxy, hollow glass spheres	0.95	0.38
Polyester, unfilled	1.23	0.10–0.15
Polyester, 50% silica by weight	1.6	0.19
Polyurethane foam (10 lb/ft³)	0.16	0.02–0.03

* To obtain: Btu/(hr)(ft²)(°F)(in.), multiply by 12.
g-cal/(sec)(cm²)(°C)(cm), multiply by 0.00413.
watts/(cm²)(°C)(cm), multiply by 0.0173.

Thermal Problems. The printed-wiring board is an important constituent in the thermal management of the total electronic system. It provides a ready avenue of escape for heat generated by components to the board, through the board to the supporting member, and from there outside the package. Although all three principal modes of heat transfer—conduction, convection, and radiation—find use in cooling electronic packages, conduction is more directly involved in printed wiring since it is the point-to-point transfer of heat along bodies in direct contact with each other. As far as printed-wiring boards are concerned, this conductive heat transfer can be made more effective by using high-conductivity materals (see Table 19), using a conductive surface path as in Fig. 44, and using good thermal bonding between parts in the thermal path. Conductive cooling demands the use of large cross-sectional paths for heat flow and a small number of thermally conductive joints. Some general hints on methods for enhancing heat dissipation in printed-wiring assemblies are given in Fig. 45. A more comprehensive guide to cooling and thermal considerations for electronic package design is given in Chap. 11, which includes special consideration for printed-wiring applications.

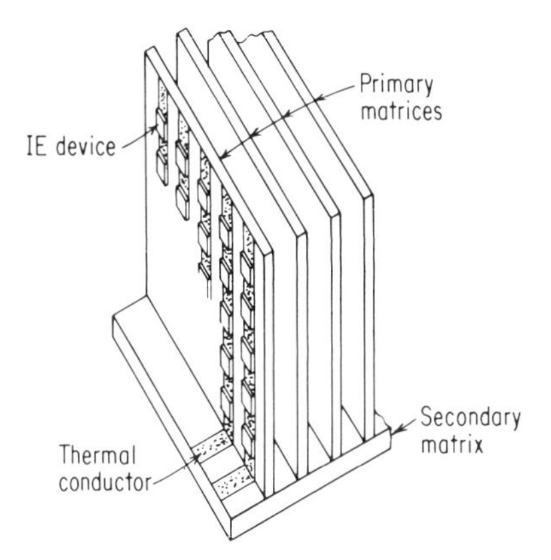

FIG. 44. Conductive-heat removal.

Termination Considerations. A critical link in any electronic packaging system is in the connector, i.e., the method for providing a continuous circuit path between electronic subassemblies. The implications of reliability, cost, maintainability and performance trade-offs for connectors are reasonably well known. Since a more complete discussion of connector types, application and technical description of connectors and interconnecting devices will be given in Chap. 6, the author will orient this section on terminations to the ways in which design and use of connectors is affected by the printed-wiring medium itself.

RIGID-PRINTED-WIRING CONNECTORS. Although many different types of connectors and connector schemes have been used for terminating printed wiring, by far the most important widely used are the card-edge type and the pin-socket type. Schematic illustrations of these are found in Fig. 46, and some practical examples are shown in Fig. 47. Card-edge-type connectors utilize a portion of the conductive pattern as a male contact and may have a common or separate connection to each side of the board as well as a mate on one or both sides. MIL-STD-275 states that edge-type connector plugs shall have copper contacts which are the thickness of the remaining conductive pattern and they shall be plated with rhodium (20 to 50 millionths in.) over low-stress nickel (0.0005 in.) or tin nickel with an average thickness of 0.0003 or 0.0001 in. gold or 0.00005 in. gold over low-stress nickel. Although the roughness of the surface finish should never exceed 32 μin., longer contact life may be achieved if a finish of 2 to 4 μin. is approached. As noted in the illustrations, the board edge must be chamfered to facilitate insertion into the

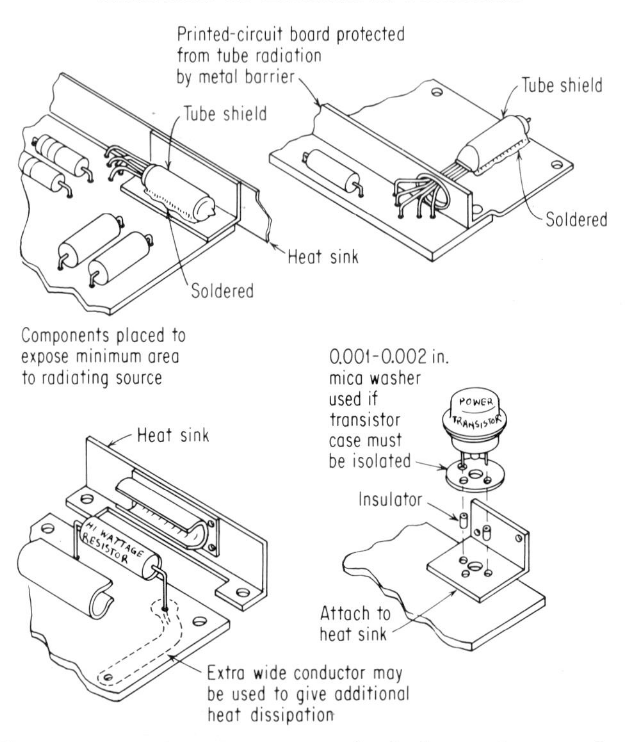

FIG. 45. Component-mounting arrangements for facilitating heat transfer in printed-wiring-board assemblies.[1]

connector. The small size of the edge-card connector and the convenience of direct board insertion into the connector is offset by disadvantages which include critical board-thickness tolerance required, special conductor finish required (above), contact wear, board-warpage problems, and problems in connector standardization with different board thicknesses. Edge-card connectors are available with contact center-to-center spaces down to 0.050, which calls for critical circuit tolerance to prevent interference between contacts. Board-thickness control is also an important factor for good connector performance. On a nominal 0.062-in. laminate a thickness tolerance of ±0.0075 in. may be expected; for multilayer boards of this thickness the tolerance may approach ±0.012 in.

For pin-and-socket-type connectors, the printed-wiring designer is somewhat removed from the design of the connector per se, and the connector may be treated as another component which must be fastened to the board. Costly board-plating problems are alleviated when this connector type is used, and the connector choice is not dependent upon board thickness. Although this connector type adds strength to the board and may reduce warpage, it is bulky, adds an extra part to the connector system, and requires two soldered or welded joints instead of the one required for the card-edge type. The pin-and-socket-type connector is somewhat better in vibration resistance than the card-type, however.

FLEXIBLE PRINTED WIRING. Until quite recently, the lack of good, flexible-cable-connector techniques and standard-connector types somewhat restricted the growth of flexible printed wiring to its full potential. Presently, except for heat limitations of some flexible base materials which preclude mass soldering techniques, similar connection techniques are used for flexible printed wiring as for their rigid counterpart, and a wide selection of standard connectors are available.

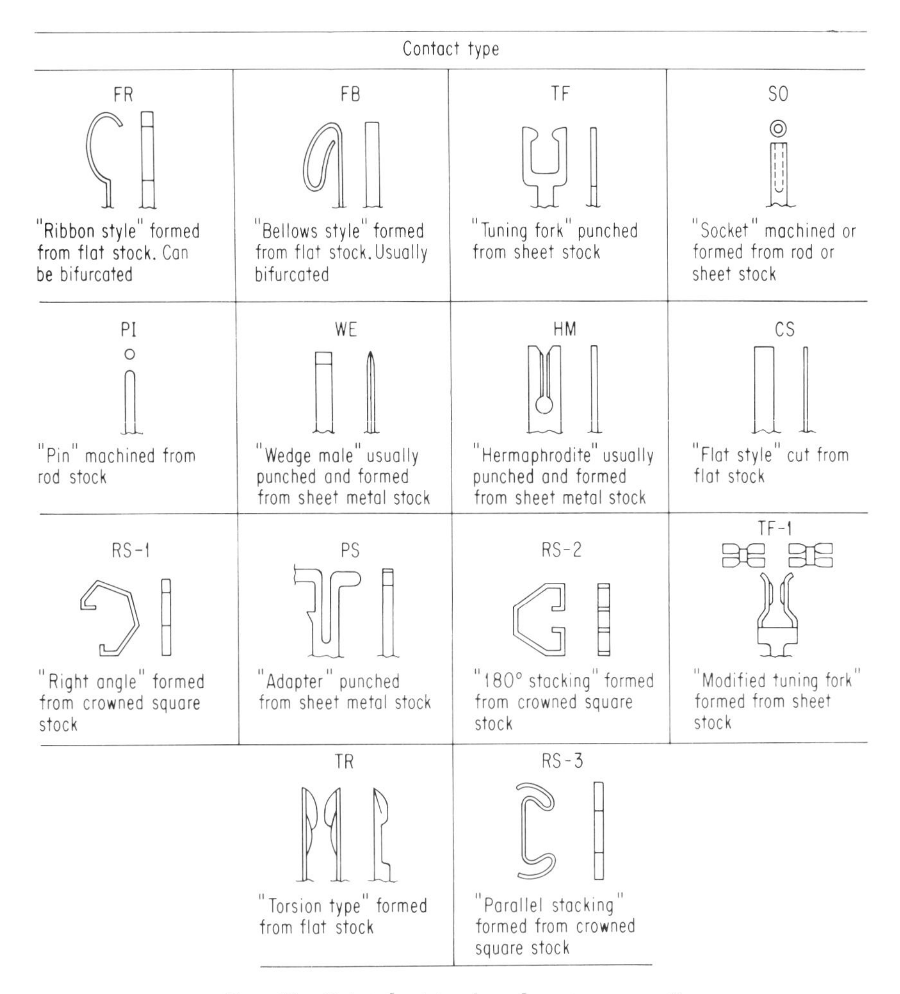

FIG. 46. Printed-wiring-board contact types.[26]

FLEXIBLE-PRINTED-WIRING CONNECTORS.[27] Any standard connector with square- or round-terminal pins can be used with a flexible cable. Connectors with broad-blade pins can be altered to become usable. One or more layers of flexible circuitry can be soldered to the pins of these connectors in any of the three basic ways illustrated in Fig. 48.

The standard connectors used with flexible circuits can be grouped into five categories: (1) rectangular connectors, (2) round connectors, (3) card-edge connectors, (4) bulkhead connectors, headers, and tube sockets, and (5) eyelets, etc. Rectangular connectors can have either regular-row or staggered-row pin geometry as shown in Fig. 49.

Right-angle address to the pins of a multipin connector such as those shown requires either two layers of flexible circuitry or a single layer with conductors passing between connector pins. In this latter case, it is good practice to use a connector having at least 0.150-in.-on-center pin spacing. Naturally, the heavier the current and necessary conductor size, the greater the pin spacing required. The most com-

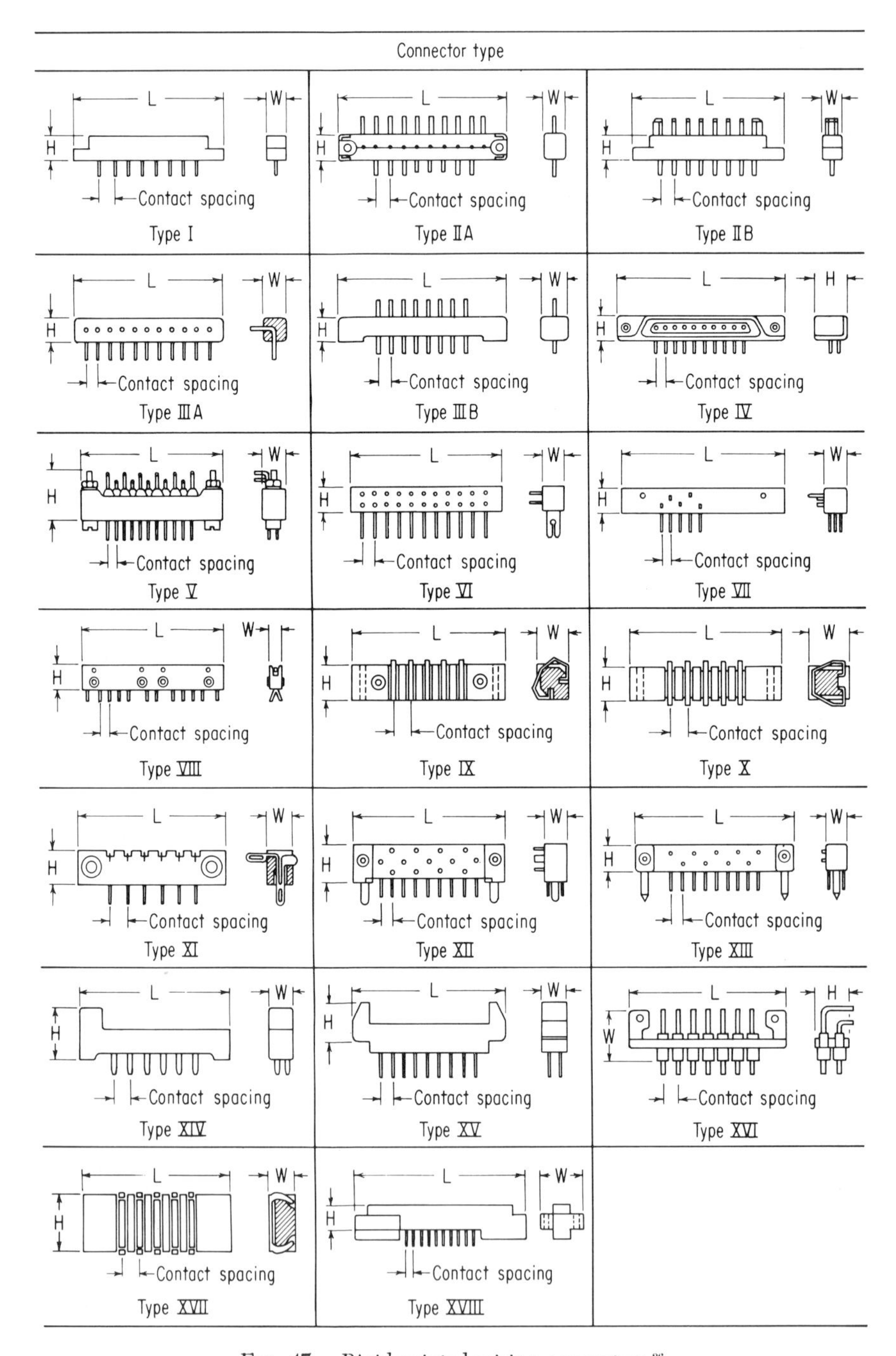

FIG. 47. Rigid-printed-wiring connectors.[26]

mon pin spacings on connectors used with flexible cables are 0.156, 0.150, 0.125, and 0.100 in. When the clearance between pins is less than 0.100 in., it is better not to pass conductors between pins. However, if it is mandatory to pass conductors between such closely spaced pins, precision photoetching methods may be used with a corresponding increase in the cost of the flexible circuits.

Round connectors have either straight-row or radial-row pin geometry as shown in Fig. 50. Round connectors with straight rows can be connected in several ways as shown in Fig. 51.

Card-edge and integral-cable connectors can be made by exposing a cable's flat conductors and fusing them onto an epoxy board for rigid support. Fig. 52 shows a male plug made in this way. This particular cable was made for a low-signal-level application. Since the ends of the cable are their own connectors, no solder joints are required, and the cable transmits low-level signals with the highest-possible

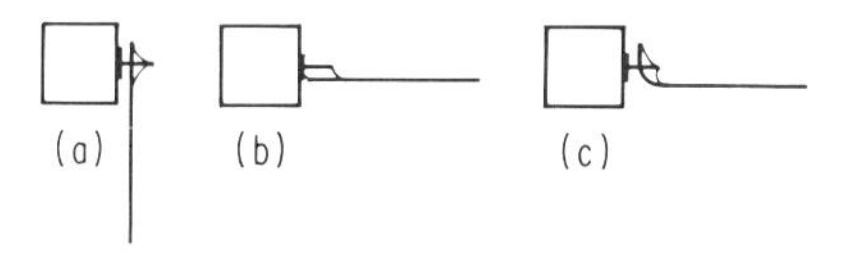

FIG. 48. Three standard methods of approaching pins on connectors.[27] (*a*) Right-angle address refers to the 90° angle formed by the conductor and pin. (*b*) In-line address refers to the parallel geometry of pin and conductor. (*c*) Everted address refers to the fact that even though the solder pad is connected in a fashion similar to right-angle address, the flexible circuit is bent to leave the pin in a nearly in-line manner.

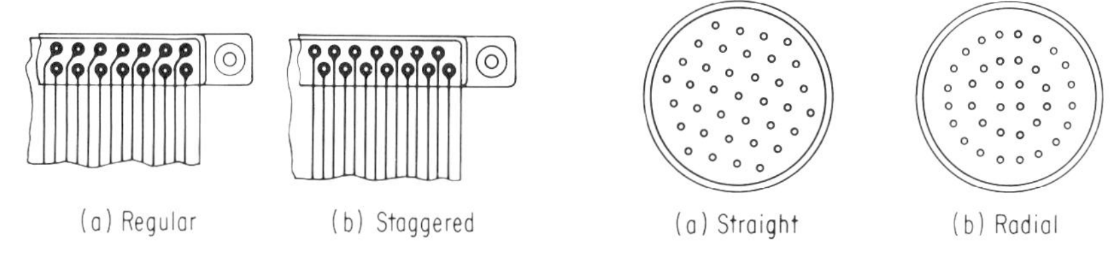

FIG. 49. Rectangular connectors with regular and staggered-pin geometry.[27]

FIG. 50. Round-connector geometries.[27]

signal-to-noise ratio. The exposed conductors are gold-plated to assure a good electrical contact even in somewhat corrosive atmospheres. The stack of circuits shown in Fig. 52 contains 112 conductors (seven layers of 16 conductors each) and is small enough to pass through a slot only ⅛ by ⅞ in., an example of the space savings possible with flexible circuits.

There are a number of other techniques, some proprietary and using a welding method, for terminating flexible wiring with an edge-card connector. Female receptacles can also be made using the exposed conductors of a flexible circuit. Headers, tube sockets, switches, etc., require various solder-pad terminations to fit their various pin designs. Flexible circuits can be made with the optimum solder pads for virtually any shape pin or terminal. Thus connection to headers, sockets, switches, etc., is generally no problem.

Figure 53, for example, shows flexible circuitry connected to a two-deck wafer switch. Note the slots in each solder pad which just fit the switch terminals. Also note that the cable is split to allow easy attachment to an assembled switch. Eyelets of all types can be used with flexible circuits. Figure 54 illustrates their use.

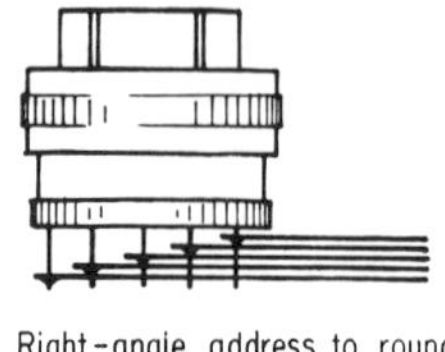

Right-angle address to round connectors with straight rows

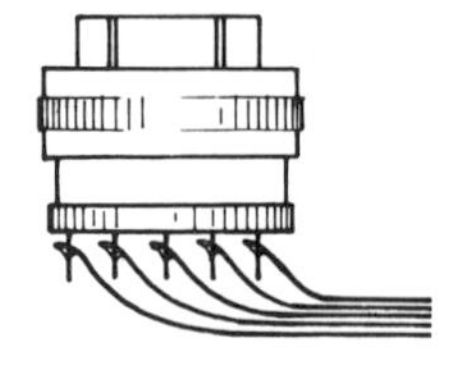

Everted address to round connectors with straight rows using cables grouped horizontally

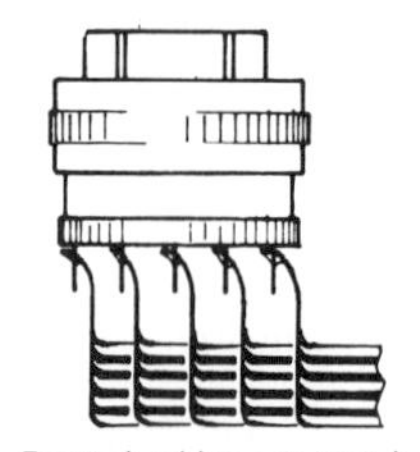

Everted address to round connectors with straight rows using cables having a 90° bend

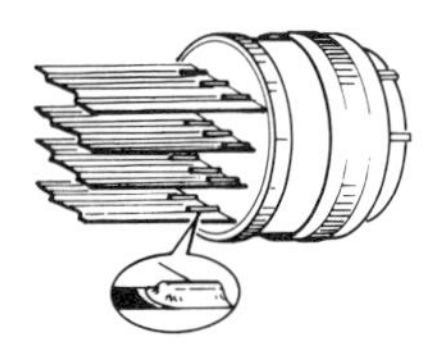

In-line address to round connectors with straight rows

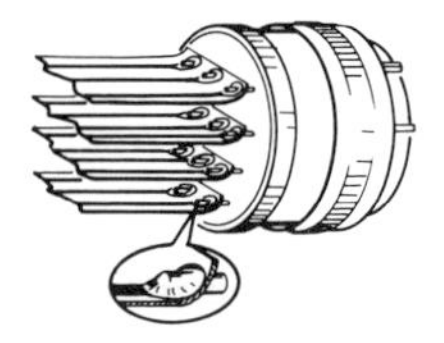

Everted address to round connectors with straight rows

FIG. 51. Round-connector configurations.[27]

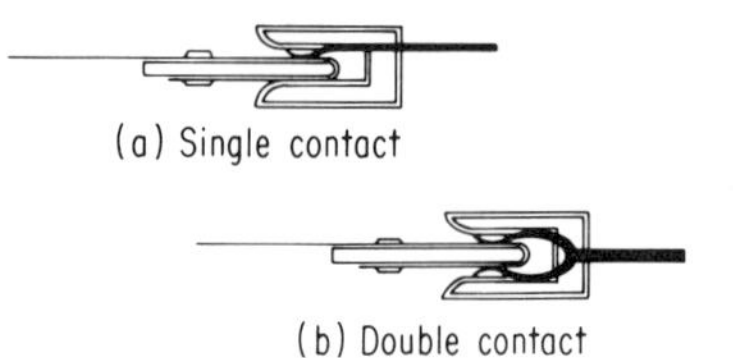

(a) Single contact

(b) Double contact

Typical types of female receptacles used with Flexprint male plugs

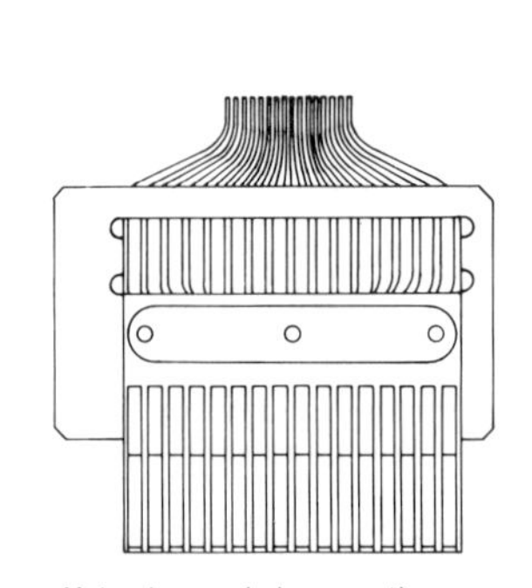

Male plug made by mounting exposed flexible circuit on a rigid board

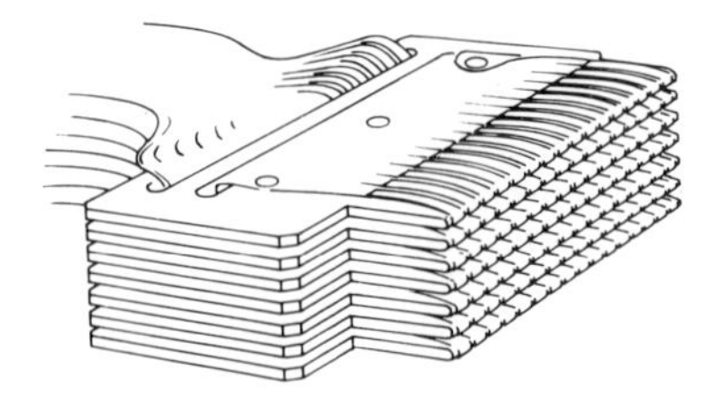

Stack of Flexprint male plugs showing high achievable wiring density

FIG. 52. Card-edge-connection techniques.[27] (U.S. Patent No. 2,951,112.)

ARTWORK

The most critical and sensitive operation in the entire printed-wiring manufacturing procedure is in the preparation of artwork. Except where a circuit pattern is generated directly on a photosensitized, copper-clad laminate, artwork is necessary to transfer the required pattern from the circuit designer's diagram to the end-product raw material. Since this usually involves all or a combination of steps including manual drafting or taping, photoreduction, composition, registration, and handling, it is imperative that inaccuracies introduced by these steps be held to an absolute minimum. The conductor pattern desired on the finished board cannot be any more accurate than the master artwork which was used to produce it; it is invariably less accurate due to manufacturing tolerances. Indeed, any tolerance deviations or errors generated at the artwork stage may be multiplied during the ensuing production process.

Preliminary Design Information. Before the artwork is actually made, a number of other operations and/or decisions must necessarily take place. Basic decisions involving numbers of layers of wiring, end-product environment, conductive and insulating materials, line width and spacing, manufacturing method, component-attachment techniques, and volume of production desired should have already been made. All of these aspects vitally affect the printed-wiring artwork. For instance, an intimate knowledge of photoresist and etching tolerances and variations must be in evidence if this method is used, and these factors will serve to dictate artwork requirements. Thermal-management problems, type of components used, and system-packaging hardware are other factors which the design engineer must realize and consider in the artwork stage.

Input to the drafting or artwork generation area should, then, reflect the factors above, as well as those mentioned elsewhere in this chapter and throughout the entire book. Based on the electrical connections he wishes to make (which in turn are dependent upon the raw logic information or circuit diagram), the engineer will provide information for artwork which may be a pencil sketch or a to-from wiring list. In the case of computer-generated wiring- and module-placement programs, this input will be in the form of a paper or magnetic tape or punched cards with which to operate an automatic artwork generator or to produce a hard-copy to-from list.

Artwork-production Methods. In the course of development of printed wiring, a large number of artwork-production techniques have been used, some almost as

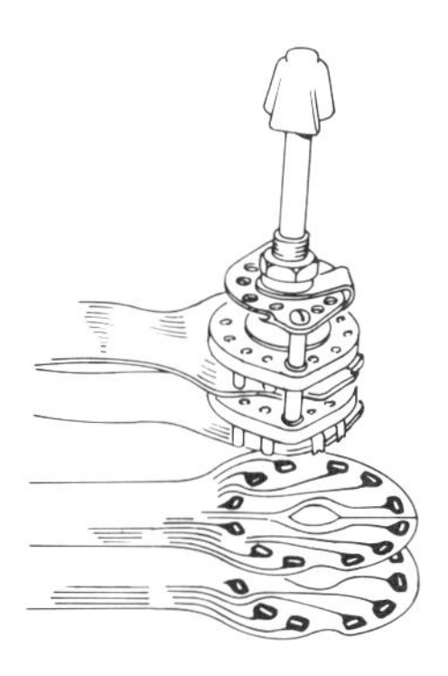

FIG. 53. Typical cable for connection to wafer switch.[27] (U.S. Patent No. 2,946,877.)

FIG. 54. Eyelet terminations.[27] (U.S. Patent No. 2,977,521.)

varied as the processes used to produce the boards. Only a few still remain. The main idea is to produce an accurate, durable pattern defining the circuit pattern which can be used directly or photo reduced with a minimum loss of accuracy.

Ink Method. Historically the first method used to produce artwork, the ink method depends upon use of a good, dense ink, which is compatible with the art-work-base film. The process is slow and requires complete remaking for minor changes. It has largely given way to other methods, but ink is still used for titles, dimensioning, marking for other reasons, and touch-up.

Tape Method. Most printed-circuit artwork is currently produced using black adhesive-paper tape or using red transparent tape on a white or transparent Mylar material, usually 0.004 or 0.007 in. in thickness. Mylar with blue lines delineating the standard grid system is most often used although many times a photographically reproduced "pad master" is the base sheet for tape application. Grids with increments of 0.100 and 0.050 in. are commonly used. The grid provides a

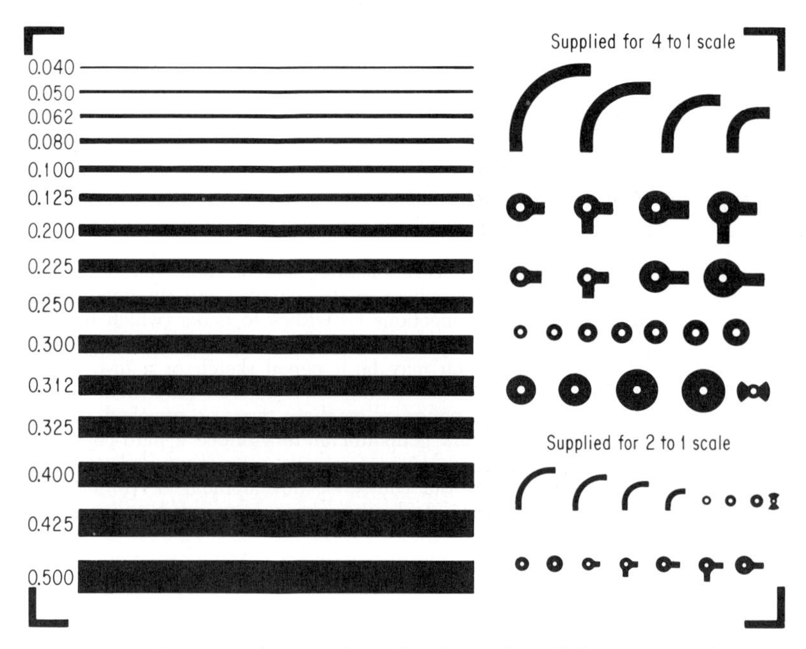

FIG. 55. Typical precut lines, bends, pads, and datum points.[1]

standard base for location of component mounting holes and minimizes misregistration pattern to hole, front to back, and layer to layer, with multilayer boards. In either case, the draftsman has accurate guides in laying out and taping the circuit pattern. Tape for lines as well as standard pad patterns and other geometric shapes are available from standard drafting-supply houses for this purpose. Some sample lines and pads are illustrated in Fig. 55.

When applying tape, the proper procedure is to lay it, not stretch it, on the base. Curving of the tape around bends is done by crimping the tape on the inside radius. Curves are also produced by cutting wide tape to the desired configuration. If the tape is stretched, subsequent shrinking causes the pattern to deform and leave an adhesive residue which will reproduce as a ragged edge on the conductor. Tape-tape and tape-to-ink joints should overlap to preclude separation from occurring, but should not exceed two layers. Red transparent tape produces finer edge-line definition than black adhesive-paper tape. Tape should be selected, which, when reduced to proper size, will produce the desired conductor width, as dictated by design and/or manufacturing tolerance requirements. Using the conventional taping or stripping method, the draftsman can locate pads and lines accurately to within 0.012 in. with the unaided eye. This tolerance can be reduced by using an accurate grid.

In order to simplify the task of laying out artwork accurately and achieving sharply defined conductor patterns, artwork should be produced from two to eight times actual size and photographically reduced to achieve the working master artwork.

Cut-and-strip Method. The strip-coat, or cut-and-strip, method utilizes a polyester-base material coated with a strippable lacquer film. A set of special tools must be used, tools capable of accurately cutting the peelable surface with single lines or mutually parallel double lines. One of the advantages of this method is that it normally produces a negative, and if made on a one-to-one scale, may be used directly for making the board or silk screen. However, the edges are not uniformly smooth when they are peeled off, and errors or changes are difficult or impossible

Fig. 56. Manual coordinatograph. (*Haag-Streit AG, Bern, Switzerland.*)

to handle. A variation on this material is a glass material, either with an adhesive-backed red plastic tape for stripping or with colloidal graphite coating.

Coordinatograph. This highly accurate drafting instrument, an example of which is illustrated in Fig. 56, provides mechanized assistance to the draftsman in manually producing artwork for printed wiring. Used almost exclusively for making masks for thin-film deposition and in the manufacture of integrated circuits, this instrument has a typical plotting precision of ±0.0012 in. Available for the rectilinear, and circular or radial modes, it operates on the principle of locating a point in a plane, referenced to a zero point by its rectangular and/or polar coordinates. A wide variety of tools for these coordinatographs permit their use with either strippable Mylar film or cut-and-strip or coated-glass working media.

Automatic Artwork Generation. Production of printed-wiring artwork by manual methods is usually adequate where moderate tolerances are required and a noncritical work load exists. Current electronic-packaging requirements place an added burden on the interconnection media, brought on by the extensive use of integrated circuits and multilayer printed-wiring boards with controlled electrical characteristics. Pin densities ranging from 70 to 110 per square inch are common. Line- and pad-width tolerances of ±0.0005 in. and pad-line location of ±0.001 in. over areas up to 20 by 20 in. are not uncommon requirements. The composite tolerance problems connected with these requirements are beyond the range of conventional artwork-preparation techniques.

Born of necessity, the automatic artwork generator is essentially the same as the coordinatograph previously described except that it is computer-controlled, and therefore much faster, more versatile, and accurate. A machine of this type is capable of automated production of computer-designed artworks where critical circuit paths and wiring rules are predetermined by means of a computerized data-processing routine.

One example of a precision plotter is illustrated in Fig. 57. This particular machine is capable of producing rectilinear or curvilinear line patterns with equal accuracy over a maximum work area of 48 by 60 in. with a true position accuracy of ±0.001 in. (0.0005 in. repeatability) at 60 in./min traverse speed. The controlling data for the artwork generator can originate from either of three different sources: a

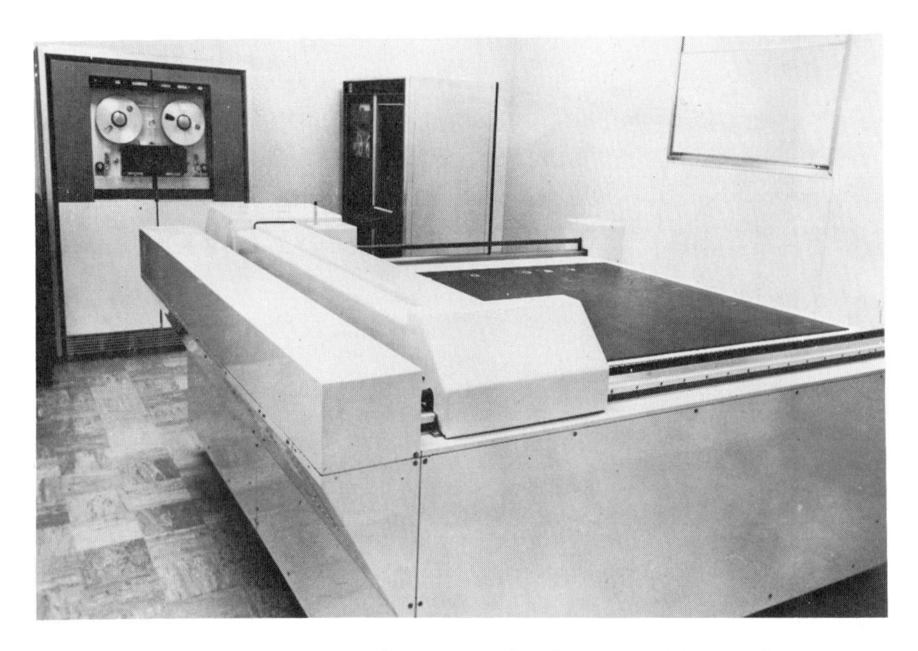

Fig. 57. Precision artwork generator. (*Radio Corp. of America.*)

manual keyboard, magnetic tape, or punched paper tape. Four different drawing modes are available, as illustrated in Fig. 58. These are inking, scribing, routing, or photoexposing. Photoexposure is most commonly used and has the advantage of generating various pad configurations in one exposure by means of special apertures. Twenty-four of these apertures are positioned for automatic sequencing in the output head. A machine of this type may also be used for generating a pattern directly on the photosensitized surface of clad laminate, eliminating artwork as it is conventionally used. A test sample pattern illustrating the capability of the machine illustrated in Fig. 57 is shown in Fig. 59.

Conventional (Master-dot-master) Method. In double-sided and multilayer printed-wiring boards, the most critical pattern to locate is that of the terminal pad since this is one item which determines the width of the annular ring upon drilling. Dot-master patterns are prepared either by very accurate conventional-type pad placement, or by step-and-repeat machine methods. Made on either

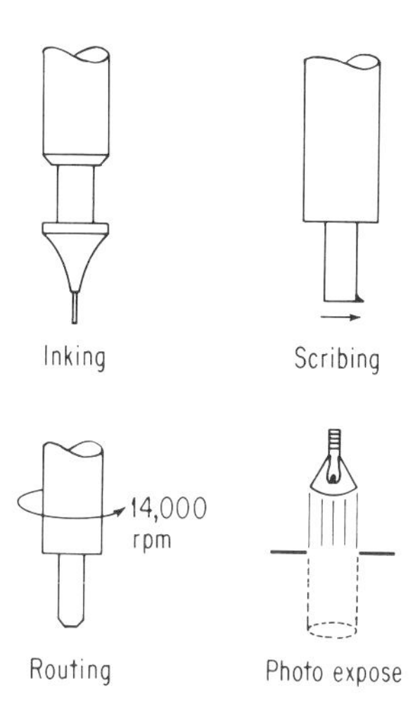

FIG. 58. Artwork-generation modes.

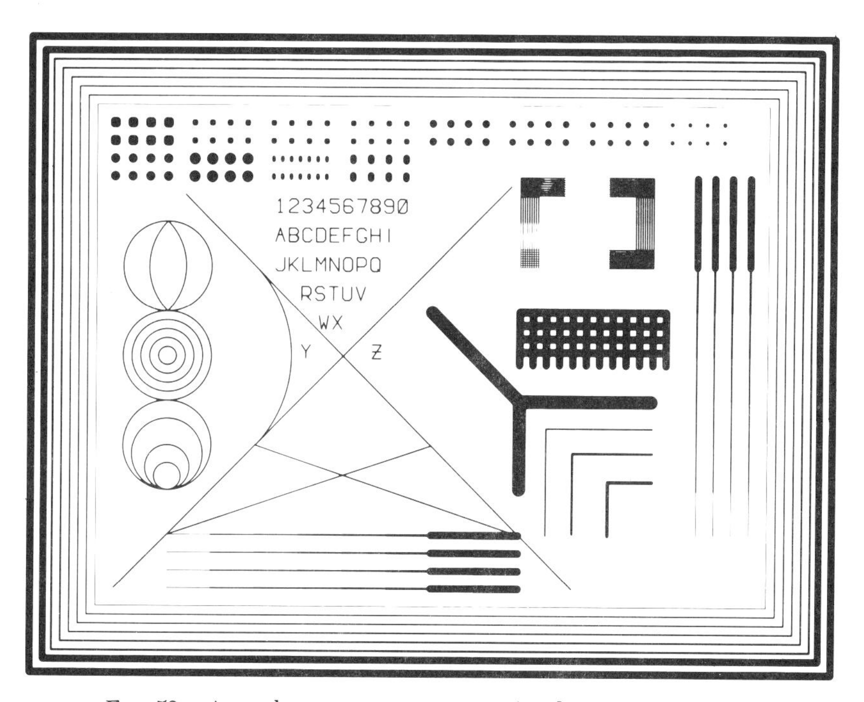

FIG. 59. Artwork-generator test pattern. (*Radio Corp. of America.*)

stabilized film or glass, this artwork characteristically has a scale easy to work with but which will provide required accuracy when reduced 1:1. A number of working dot-master layout sheets are then produced from the same master dot master; these sheets equal in number the number of layers in the board. Tape is applied to the working dot masters to define the circuit pattern, and this is then reduced if needed

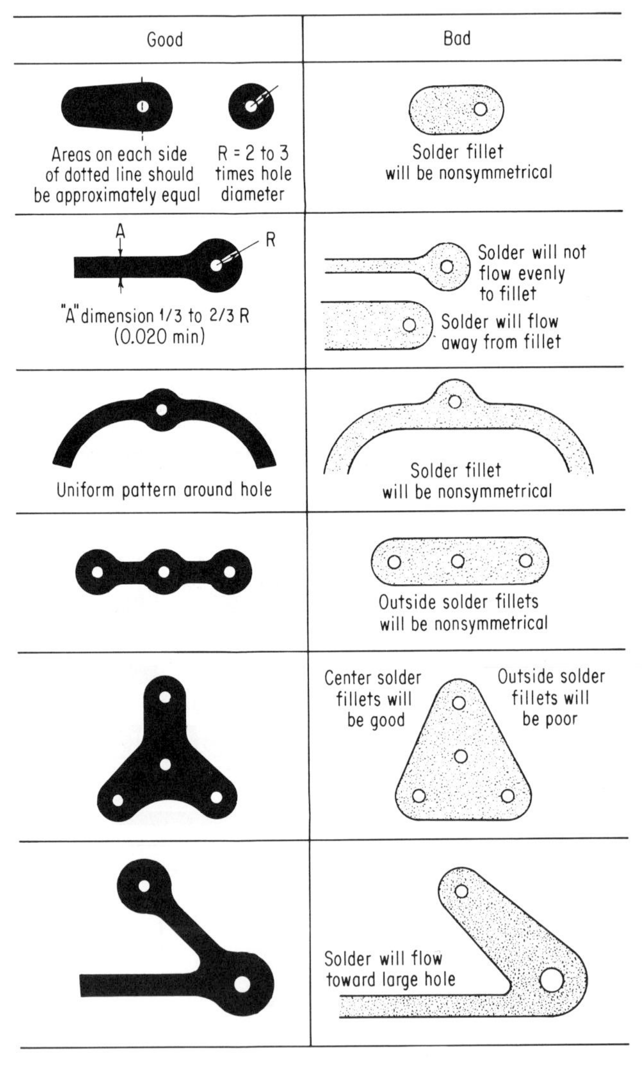

Fig. 60. Recommended artwork forms.[1]

to provide the working master artwork. Since all working dot overlays and working dot artwork layers are produced from the original dot master, it is obvious that each successive pad pattern on each layer will inherit the accuracy of original master-dot-master pattern placement.

Artwork Preparation Details. A major portion of printed-wiring artwork is still prepared by the taping method, and the technician or draftsman must have some guidelines to follow in his task. Most organizations, recognizing the need for a standardized procedure in this respect, have published guidelines in this area. The general features covered by instructions or standards of this type are as follows.

GENERAL DRAFTING REQUIREMENTS. The detail drawing provides the engineering requirements as well as manufacturing information on the required size and shape of the printed-wiring board. Material, board identification, and requirement notes are specified as well as the location of holes, conductor paths, and lands. Usually the engineering drawing reads as viewed from the wiring, or soldering, side, but in the case of double-sided and multilayer boards, it is usually right-reading, emulsion-side-down orientation. Master-pattern artwork for double-sided boards are usually two sheets of the same drawing termed *Front* and *Back* and shall be so marked in the title block. Grid locators, or "bull's-eyes," as are used for registration marks must be placed in the same relative physical location, all patterns comprising the same board to assure good registration. One example of a bull's-eye, or grid locator, is illustrated in Fig. 62*a*. In addition, position of these marks must be such that a

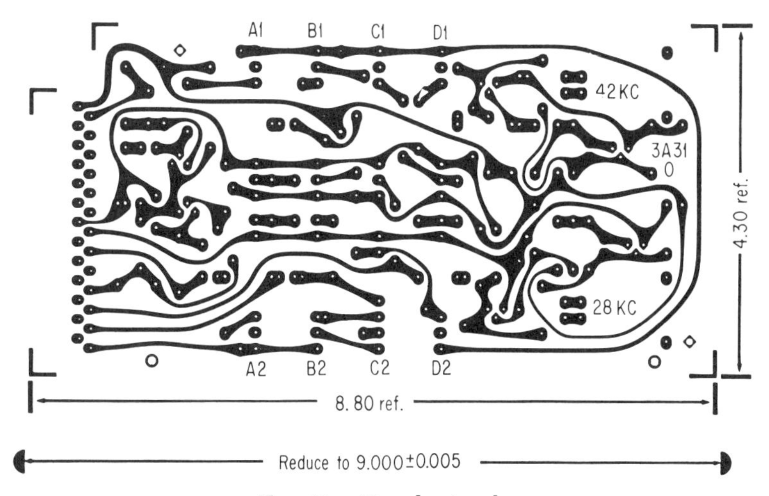

FIG. 61. Taped artwork.

match will be achieved only when the various patterns are placed in their proper relationship.

A drawing or an artwork is also needed for the purpose of marking the printed-wiring board; this is usually required in preparation of a silk screen to do this job. Marking may include reference designators, the polarity of polarized components, terminal numbers, board part number, and, in some cases, corner marks.

DETAIL DRAFTING REQUIREMENTS. The methods for preparation of artwork have already been covered; in the case of taping, precut lands, bends, and terminals are usually available as illustrated in Fig. 55. Recommended practices in pattern configuration are shown in Fig. 60. A typical artwork with proper tooling and registration marks is shown in Fig. 61.

Corner marks are required to indicate the corners of the board. It is necessary that those corner marks which correspond with the board edge be accurately located on the master-pattern drawing. These corner marks shall be applied outside the board outline and are illustrated in Fig. 62*b*. To assist the phototechnician in the reduction of the master-pattern drawing, reduction patches similar to those illustrated in Fig. 62*c* must be used. They should be applied so that the inner edges are straight, parallel to each other, and accurately spaced. The reduction dimension should be a whole number, at least as large as the longest edge of the pattern, and should have the tolerance indicated. The reduction patches should be spaced within 0.005 in. (before reduction) above the corner marks as illustrated in Fig. 62*c*.

Production Artwork. Since the artwork is an essential tool, or pattern in production of printed wiring, it is essential that it be compatible to the type and volume of production which is required. In addition, there must be some system for registering the artwork to the workpiece so that the proper pattern orientation results.

Registration Procedure. One of the most critical steps in the production of double-sided and multilayer printed-wiring boards is the registration of the pattern on the artwork to the workpiece. It is this step which assures pad-pad and pad-line relationships on the various layers and which, if badly miscalculated, could result in short circuits or the loss of a plated connection at a hole location. In offset printing and silk screening, the registration is accomplished by mechanical means during setup of the equipment, usually by means of the grid locators on the artwork. Film and glass master artworks are conventionally registered to an accurate glass grid plate (with an accurately marked grid pattern), which contains precision-

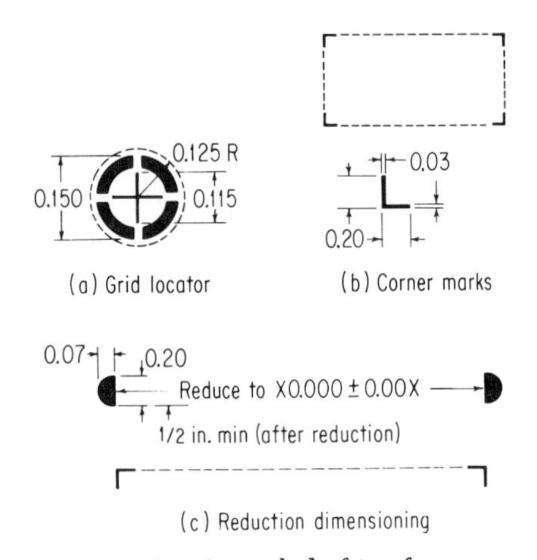

FIG. 62. Artwork drafting forms.

bushed registration or tooling holes. When the film or glass artwork is accurately positioned over predetermined positions on the grid plate, the registration or tooling-hole locations in the grid plate are transferred to the working artwork. A minimum of three holes should be used. For film, this means punching the film through the grid-plate registration hole. For glass artwork, an accurately bored drill bushing is affixed in a predrilled hole in the artwork glass and potted in place either with a low-melting-point metal or with a proprietary epoxy compound.

Types of Production Artwork. When a production artwork medium is chosen, the designer must consider a number of factors:

1. The precision requirements of the job, including pattern-configuration location and tolerance, and registration tolerance
2. Volume of production expected
3. Dimensional stability of artwork with respect to changes in temperature and relative humidity
4. Its ability to withstand handling, its durability, and its resistance to tearing and scratching
5. Its compatibility to photographic processes, an essential factor

Screen printing is ordinarily used for etch-resist patterns where unusual tolerance requirements are not present or where production volume averages over 25 pieces.

Offset printing, also suitable for etch resists, is used for slightly larger volume since it utilizes an etched master plate made from a photosensitive polymer. Precisions approaching those with photoresist are currently found in the development stage. For photoresist processes, either film or glass artwork are used. Characteristics of these materials are covered below. For high production, the more dimensionally stable and durable glass medium is used, either as a photographic emulsion protected with a clear epoxy or acrylic coating or as a metallized glass master with a proprietary material. This material is much thinner and more abrasion-resistant than photographic emulsion and is also covered with a protective coating for production purposes.

Artwork Tolerances.[28] Artwork for double-sided and multilayer printed-wiring boards could be produced by conventional, single-sided printed-wiring artwork procedures except that since there are a number of layers of artwork and they must all register, the problems of dimensional tolerances are multiplied by each additional layer required. In addition, as printed wiring becomes more and more miniature, dimensional tolerances become increasingly difficult to meet and/or maintain because of the decrease in the available board space; pad sizes are reduced almost to the size of the hole in the pad. Because of these dimensional problems, special precautions and techniques must be used in order to cope with these cumulative-type tolerances.

Artwork tolerances can be categorized as follows:

1. Mechanical layout tolerances (production of circuit configuration)
2. Material tolerances (temperature and humidity dimensional stability)
3. Photographic printing and reduction tolerances (optical, chemical, and mechanical processes)

Since mechanical layout tolerances have already been discussed and indicate a wide variance dependent upon methods or individuals used in layout, only material and photographic tolerances will be discussed here.

Material Dimensional Problems. Generally speaking, master-pattern (MIL-STD-429)[28] materials for multilayer printed wiring are photographic in nature and photographic films and plates are made to produce several layers of extremely thin, carefully coated materials. These materials may, of course, also be used for single- and double-sided applications. A transparent-film or glass base serves as a support for the thin, light-sensitive emulsion. The materials of most photographic films have extreme clarity and relatively good dimensional stability. One film, for example, has an "ortho" line emulsion on a polyester-base material. Glass plates are available that have an orthoemulsion on specially selected glass ranging in thickness from 0.060 to 0.0190 in., depending on the size of the plates required. The dimensional stability of photographic film depends on many factors, such as coating on film, treatment during manufacture, and storage conditions. Dimensional variation may be either temporary (reversible) or permanent (irreversible).

Temporary expansion or retraction is due to either (1) the loss or gain of moisture (relative humidity of air in contact with film) or (2) change in temperature.

The expansion of film owing to moisture in the air is generally greater than that owing to thermal expansion since the relative humidity of most photographic laboratories is more variable than the temperature.

Permanent shrinkage in photographic film is most often caused by loss of residual solvents and plasticizer (softening agent) from the base, plastic flow (shrinking caused by the contraction of the emulsion), and release of mechanical strain. Permanent shrinkage during storage prior to exposure can generally be discounted since there is no image on the film. Film swells during processing and contracts again during drying. If the film is brought to equilibrium with the same relative humidity after processing as exists before, a small net shrinkage called the *processing shrinkage* is usually found. In most cases, this processing shrinkage can be discounted.

Most film materials are in equilibrium with air at approximately 40 percent relative humidity at the time of shipment. In order to control dimensional stability of drafting and photographic films, laboratories and workrooms should be provided with temperature controls and humidity controls that provide temperature ranges between 68 and 75°F and 40 to 60 percent relative humidity. Too low a humidity increases static electricity and attraction of dust to unexposed or exposed films, causing defects in printed patterns.

A chart of the thermal and relative humidity coefficients of expansion of various materials is shown in Table 20. This chart shows what change can take place in a 30-in. dimension on a piece of material with a 10°F shift in temperature and a 20 percent shift in relative humidity.

Photographic Tolerances. The final accuracy of the 1:1 negative (master pattern) used to print conductor patterns on a printed-wiring conductor layer is a function of material dimensional tolerances, pattern-placement tolerances, and photoreduction tolerances. The final negative accuracy is also determined by the scale at which the original artwork was produced. A misregistration of 0.010 in. on an artwork produced at a magnification of 4:1 produces a misregistration on the 1:1 negative of 0.0025 in. At 2:1, this misregistration would be 0.005 in. Since large-scale artwork can be

TABLE 20. Representative Dimensional Changes Due to Thermal Conditions [20]

Material	Thermal and humidity coefficients of expansion		Possible representative change of 30 in. with 10°F rise in temperature and 20% relative-humidity rise
	Thermal coefficient, in./(in.)(°F)	Humidity coefficient, in./(in.) (% relative humidity)	
0.004 in. Cronaflex* drafting film	1.5×10^{-5}	1.15×10^{-5}	0.0114
0.004 in. Cronaflex contact film	1.5×10^{-5}	1.4×10^{-5}	0.0129
0.007 in. Cronaflex contact film	1.5×10^{-5}	1.2×10^{-5}	0.0117
0.004 in. Cronar Ortho A* projection film	1.5×10^{-5}	1.7×10^{-5}	0.0147
0.007 in. Cronar Ortho A projection film	1.5×10^{-5}	1.3×10^{-5}	0.0123
0.004 in. Cronar Ortho S* projection film	1.5×10^{-5}	1.4×10^{-5}	0.0129
0.007 in. Cronar Ortho S projection film	1.5×10^{-5}	1.04×10^{-5}	0.0108
Glass plates	0.45×10^{-5}	0	0.0013
Kodak Estar-base† films	1.5×10^{-5}	1.3×10^{-5}	0.0123
Aluminum	1.2×10^{-5}	0	0.0036
Sheet steel	0.6×10^{-5}	0	0.0018

* E. I. du Pont de Nemours & Co., Inc.
† Eastman Kodak Company.

used to produce the final negative for the smaller boards, the smaller the final product-board requirements, the more accurate is the final negative that can be produced. Generally, photographic-reduction cameras are capable of taking copy to the 30 by 40 or 36 by 42 size. If the printed-wiring-board size is larger than 8 by 8 in., it is no longer possible to produce 4:1 artwork that can be reduced by standard photographic equipment; i.e., $8 \times 4 = 32$ in. This is one of the principal reasons *large* multilayer printed-wiring cards require larger artwork tolerances than do *small* multilayer printed-wiring cards. This principle is the same for conventional printed-wiring artworks also. In addition, dimensional changes on the artwork base material become larger as the material size becomes larger.

In general, good, accurate camera reductions can produce a master pattern within ±0.001 in. of the copy provided.

Institute of Printed Circuits Standard Tolerances. A compilation of dimensional and mechanical tolerances as issued by the Institute of Printed Circuits (IPC-D-300)[29] is considered to be so pertinent to this handbook that the author has decided to publish it in its entirety. This has been done in Tables 21 to 32. Representing

TABLE 21. Dimensional Tolerances for Terminal Holes

Class 1 generally indicates low-cost tooling and/or material.

Classes 2, 3, and 4 indicate progressive upgrading of tooling and/or material, resulting in higher costs.

All dimensions and/or tolerances in parentheses are in millimeter equivalents.

Hole (either unplated or prior to plating): Diameter tolerances are for material thicknesses up to and including 0.0625 in. (1.59). For thicknesses greater than 0.0625 (1.59) up to and including 0.125 in. (3.18), add ±0.001 (0.025).

Classes	Hole-to-hole centers	Nominal conductor widths and spacings
1 and 2	Over 0.100 in. (2.54)	Over 0.015 in. (0.381)
3 and 4	0.100 in. or less (2.54 or less)	0.015 in. or less (0.381)

The following hole tolerances indicate total spread of tolerance and may be split up from nominal to satisfy individual requirements:

EXAMPLE:	±0.004 (±0.10)	or +0.005 −0.003 (+0.13) (−0.08)	or +0.006 −0.002 (+0.15) (−0.05)

Inches	Class 1	Class 2	Class 3	Class 4
0–0.032 diameter	±0.003	±0.002	±0.001	±0.001
0.032–0.063 diameter	±0.004	±0.003	±0.002	±0.001
0.063–0.188 diameter	±0.005	±0.004	±0.003	±0.002
Slots and notches up to 2.000	±0.005	±0.003		

Millimeters	Class 1	Class 2	Class 3	Class 4
0–0.81 diameter	±0.08	±0.05	±0.025	±0.025
0.81–1.60 diameter	±0.10	±0.08	±0.05	±0.025
1.60–4.78 diameter	±0.13	±0.10	±0.08	±0.05
Slots and notches up to 50.80	±0.13	±0.08		

Holes, slots, notches, etc., may have larger tolerances where space and component configuration permit. If radii are not provided in slots and notches and plating is required, extra cost will be incurred. Shown below is a table of hole-diameter to board-thickness ratios (unplated).

Class	Ratio	Min. hole diameter, %
1	2:3	66
2	3:6	50
3	3:10	30
4	3:12	25

METHOD OF MEASUREMENT:

Hole diameters and slot and notch widths up to 0.749 in. (19.02).

A: Taper gauge. The gauge must be used perpendicular to the plane of the board within 5°.

B: Round-plug gauge.

The length of slots and notches is determined by a vernier caliper or optical comparator.

The depth of notches is determined by a vernier caliper, optical comparator, or depth micrometer.

TABLE 22. Dimensional Tolerances for Plated-through Holes—Finished Diameter

Plating thicknesses may be specified as minimums only; otherwise a tolerance of −0, +100 percent is generally accepted.

Finished plated-hole tolerance is calculated as the unplated hole tolerance + (2 × the minimum plating thickness).

All dimensions and/or tolerances in parentheses are in millimeter equivalents.

EXAMPLE:

	Plated-hole tolerance		Plated-hole size	
Hole size: 0.052-in. (1.32) diameter (Table 21)	±0.004	(0.10)	0.052	(1.32)
Min. copper plating: 0.001 in. (0.025)	−0.002	(−0.25)		
Min. tin-lead plating: 0.0005 in. (0.013)	−0.001	(−0.025)	0.003	(0.08)
			Total min. plating	
Total tolerance	+0.004	(+0.10)	0.049	(1.24)
	−0.007	(−0.18)	+0.004	(+0.10)
	Finished hole		−0.007	(−0.18)

On boards with plated holes, plating thickness may build up on printed contact (plug-in fingers) over and above the specified minimum on each side as follows:

Classes	Thickness over min./side, %
1 and 2	200
3 and 4	100

If terminal areas (lands) are not provided on the back side of single-sided boards with plated-through holes, extra cost will be incurred.

METHOD OF MEASUREMENT:

The method used to measure plated holes is the same as that used to measure unplated holes.

The method of measurement for plating thickness is as follows:

A: Electronic-plating-inspection device (where applicable).

B: Approved microsection or micrograph techniques similar to ASTM A219, "Standard Method of Test for Local Thickness of Electrodeposited Coatings."

TABLE 23. Dimensional Tolerances for Centerline to Datum (True Position Tolerance) (See Table 30)

NOTE: Stated tolerance is total diameter of permissible movement around the true position (applied at maximum material condition). See Table 31 for explanation.

All dimensions and/or tolerances in parentheses are in millimeter equivalents.

Dimension from datum points	Class 1	Class 2	Class 3	Class 4
Greatest dimension less than 6 in. (152.40)	0.020 (0.50)	0.014* (0.36)	0.010* (0.25)	0.006 (0.15)
Greatest dimension over 6 in. (152.40)	0.028 (0.71)	0.020 (0.05)	0.014* (0.36)	0.010* (0.25)

* This tolerance conforms to MIL-STD-275.

METHOD OF MEASUREMENT:

A: Pocket comparator for distances up to and including 0.500 in. (12.70).

A: Optical comparator for distances greater than 0.500 in. (12.70).

B: Coordinatograph.

TABLE 24. Dimensional Tolerances for Terminal Area (Land) Centerline to Datum (True Position Tolerance) (See Table 30)

NOTE: The stated tolerance is the total diameter of permissible movement around the true position (applied regardless of feature size).

All dimensions and/or tolerances in parentheses are in millimeter equivalents.

With registration datum, either side				
Dimension from datum points	Class 1	Class 2	Class 3	Class 4
Greatest dimension less than 6 in. (152.40)	0.020 (0.5)	0.014* (0.36)	0.010* (0.25)	0.006 (0.15)
Greatest dimension over 6 in. (152.40)	0.028 (0.71)	0.020 (0.5)	0.014* (0.36)	0.010* (0.25)

Without registration datum, either side			
Dimension from datum points	Class 1	Class 2	Classes 3 and 4
Dimension less than 6 in. (152.40)	0.030 (0.8)	0.020 (0.5)	Use registration datum
Dimension over 6 in. (152.40)	0.040 (1.02)	0.028 (0.71)	

* This tolerance conforms to MIL-STD-275.

The use of above tolerances can result in twice the stated tolerance when measured from front pattern to back pattern if the datum is not a hole (indexing hole).

Front-to-back registration may be controlled by the minimum annular ring (edge-of-hole to edge-of-terminal area) specification if no other feature requires closer tolerancing.

When printed-contact fingers are used, selection of tolerance must be compatible with tolerances of mating connector, taking into consideration accumulation of tolerances due to processes, material, and front-to-back registration.

METHOD OF MEASUREMENT:

A: Pocket comparator.
A: Optical comparator.
B: Coordinatograph.

TABLE 25. Dimensional Tolerances for Terminal Area (Land) Size

Classes 1, 2, and 3. MIL-STD-275 and MIL-P-55110 set accepted limits for minimum terminal area around holes (annular ring) for various conditions.

Class 4. Finished minimum edge-of-hole to edge-of-terminal area (annular ring) 0.002 in. (0.05). The designer must consider total accumulation of tolerances when selecting terminal size in reference to hole size.

EXAMPLE:

Assumption: Plated-through holes, class 4.

Minimum terminal (land)

$$\text{diameter} = 2\left(\frac{A}{2} + \frac{B}{2} + \frac{C}{2} + \frac{D}{2} + E\right)$$

$$= 2\left(\frac{0.030}{2} + \frac{0.011}{2} + \frac{0.006}{2} + \frac{0.004}{2} + 0.002\right) = 0.055$$

$$= 2\left(\frac{0.76}{2} + \frac{0.28}{2} + \frac{0.15}{2} + \frac{0.10}{2} + 0.05\right) = 1.39\text{ mm}$$

where $A =$ maximum plated-hole diameter (assumed to be 0.030 in.); see Table 22.

$B =$ hole-position tolerance (applied at maximum condition); see Table 23 (see Table 31 for explanation).

$C =$ terminal area-position tolerance (applied regardless of feature size); see Table 24.

$D =$ total possible reduction on the diameter due to process class 4 (terminal areas will be reduced in diameter in proportion to conductors); see Table 24.

$E =$ minimum edge-of-hole to edge-of-terminal area (annular ring), class 4; see Table 25.

METHOD OF MEASUREMENT

A: Pocket comparator.
A: Optical comparator.
B: Coordinatograph.

TABLE 26. Dimensional Tolerances for Conductor Width and Spacing

Final product drawings and/or specifications should call out only minimums for conductors and spacings. Artwork should be done on a magnified scale suitable to produce required tolerances (normally 4:1). The following table is an example, by classes, using worst-case process reduction.

Conductor width	Class 1	Class 2	Class 3	Class 4
Design minimum	0.031 (0.79)	0.015 (0.38)	0.010 (0.25)	0.005 (0.13)
Finished minimum*	0.021 (0.53)	0.010 (0.25)	0.006 (0.15)	0.003 (0.08)

* These minimums do not make allowance for nicks, pinholes, and scratches. These imperfections are normally acceptable providing the line is not reduced by more than 20 percent.

The following table represents process tolerances which might be expected and are given as a guide only. Specific process tolerances should be ascertained by the supplier.

Conductor width	Class 1	Class 2	Class 3	Class 4
No plating	+0.006 (+0.15) −0.010 (−0.25)	+0.004 (+0.10) −0.005 (−0.13)	+0.002 (+0.05) −0.004 (−0.10)	+0.001 (+0.025) −0.002 (−0.05)
With plating	+0.015 (+0.38) −0.010 (−0.25)	+0.008 (+0.20) −0.005 (−0.13)	+0.004 (+0.10) −0.004 (−0.10)	+0.002 (+0.05) −0.002 (−0.05)

Conductor spacing	Class 1	Class 2	Class 3	Class 4
No plating	+0.010 (+0.25) −0.006 (−0.15)	+0.005 (+0.13) −0.004 (−0.10)	+0.004 (+0.10) −0.002 (−0.05)	+0.002 (+0.05) −0.001 (−0.025)
With plating	+0.010 (+0.25) −0.015 (−0.38)	+0.005 (+0.13) −0.008 (−0.20)	+0.004 (+0.10) −0.004 (−0.10)	+0.002 (+0.05) −0.002 (−0.05)

The above line and space tolerances can be applied to plated-through-hole boards where the laminate is not over 0.0625 in. (1.59) thick. Where thicker laminates are used, and hole-diameter to laminate-thickness ratio is less than 2:3, plating may further expand conductor widths.

Stated tolerances are based on 1 oz copper. For each 0.001 in. (0.025) additional copper thickness, an additional 0.001 in. (0.025) reduction per conductor side can be expected.

METHOD OF MEASUREMENT:

A: Pocket comparator, gap gauge.

B: 40X microscope or optical-projector comparator.

CAUTION: Measuring a tin-lead-plated conductor width does not give a true reading of the remaining copper width owing to undercutting in the process.

TABLE 27. Dimensional Tolerances for Edge of Board to Datum Point

Class 1	Class 2	Class 3	Class 4
±0.015 (±0.38)	±0.010 (±0.25)	±0.008 (±0.20)	±0.005 (±0.13)

All dimensions and/or tolerances in parentheses are in millimeter equivalents.

METHOD OF MEASUREMENT:

A: Vernier caliper.

B: Height gauge with dial indicator or optical comparator.

TABLE 28. Dimensional Tolerances for Overall Outside Dimensions (Wherever practical these shall coincide with lines of the 0.100-in. grid)

Class 1	Class 2	Class 3	Class 4
±0.030 (±0.76)	±0.020 (±0.50)	±0.016 (±0.41)	±0.010 (±0.25)

All dimensions and/or tolerances in parentheses are in millimeter equivalents.

METHOD OF MEASUREMENT:

A: Vernier caliper.

B: Height gauge with dial indicator.

TABLE 29. Dimensional Tolerances for Warp and Twist—Finished Part

Thickness		Paper base		Glass base	
in.	mm	in./in.	mm/mm	in./in.	mm/mm
Pattern one side					
0.0625	(1.59)	0.025	(0.025)	0.015	(0.015)
0.0937	(2.38)	0.020	(0.02)	0.010	(0.01)
0.125	(3.18)	0.012	(0.012)	0.008	(0.008)
0.250	(6.35) and up	0.008	(0.008)	0.006	(0.006)
Pattern two sides					
0.0312	(0.79)	0.020	(0.02)	0.015	(0.015)
0.0625	(1.59)	0.015	(0.015)	0.010	(0.01)
0.0937	(2.38)	0.010	(0.01)	0.007	(0.007)
0.125	(3.18)	0.007	(0.007)	0.005	(0.005)

NOTE: A gross differential of conductor area and distribution may increase the above tolerance.

Closer warp tolerances may limit selection of raw materials or make necessary unusual manufacturing operations or shipping procedures.

Closer tolerances may be necessary for miniaturized applications where space is at a premium.

All dimensions and/or tolerances in parentheses are in millimeter equivalents.

METHOD OF MEASUREMENT:

If a flat surface plate is used, three corners should touch the plate. An indicator reading may be taken at the high point of the board showing the true deviation of the overall diagonal length.

Similarly, as shown in the diagram below, if three-point suspension is used, the points should be selected at three corners and the deviation measured at the high point of the diagonal thus formed. Move one support to check the deviation along the other diagonal.

NOTE: Sag may occur with thin material and large panels.

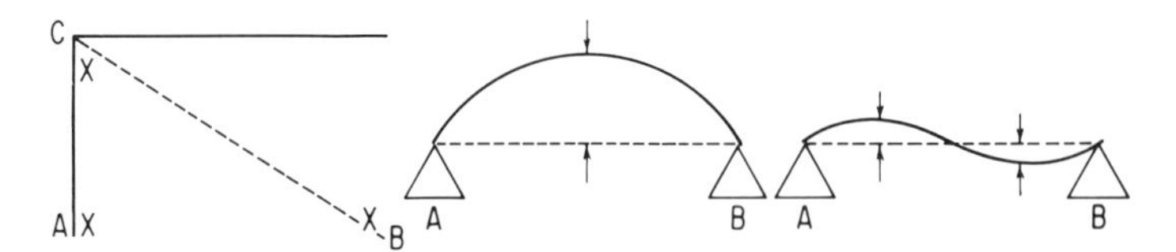

TABLE 30. True Position Tolerance*

It is recommended that *X* and *Y* datum lines be established. (These can be established through (3) holes with terminal area for printing registration if desired.) Point locations should be tabulated by *X* and *Y* coordinates and the tolerance specified by the total diameter within which the actual center may move from its true position.

For holes and terminal area, the true-position tolerance is the diameter of the tolerance zone within which the axis of the feature must lie, the center of the tolerance zone being at the true position.

Positional tolerances should be dimensioned using a modular grid system. Basic modular units of length should be 0.100 (2.54), 0.050 (1.27), or 0.025 in. (0.64), in that order of preference.

All dimensions and/or tolerances in parentheses are in millimeter equivalents.

Class	X basic	Y basic	Position tolerance diam. Class 3
1	0.100 (2.54)	0.200 (5.08)	0.010 (0.25)
2	0.200 (5.08)	0.400 (10.16)	0.010 (0.25)
3	0.300 (7.62)	0.100 (2.54)	0.010 (0.25)
4		Etc.	

Class	X	Y	Position tolerance diameter
1	0.100	0.200	0.020
2	0.200	0.400	0.020
3	0.300	0.100	0.020
4		Etc.	

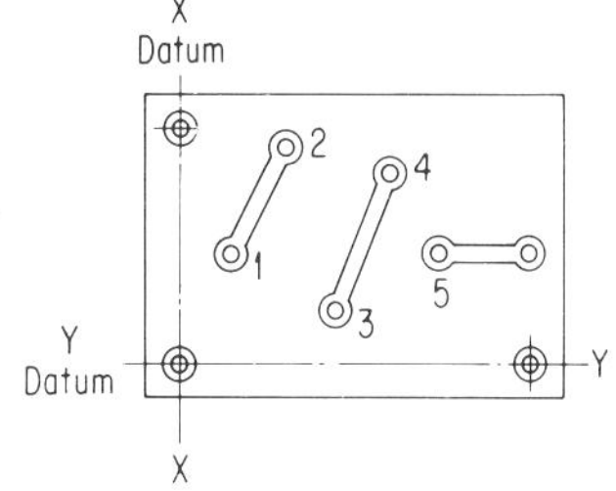

Holes and pads optional

NOTE: Frequently components must be mounted rigidly on printed-circuit boards. Usually, the component must be located to register with a pattern feature, but closer tolerances may be required within the mounting-hole pattern. In such cases, true-position tolerances can be reduced for those affected holes and still be dimensioned in relationship to the X and Y datum. The location of the component on the board may be shown by locating one mounting hole from the X and Y datum lines, and the pattern relationship will be dimensioned and toleranced directly on the drawing. (The latter method is mandatory to preclude tolerance accumulations when direct coordinate tolerancing is used.)

Also, the true-position tolerancing is only one recommended system. Direct coordinate tolerancing using master grid patterns is also used.

* MIL-STD-8.

TABLE 31. Explanation of B in Example of Table 25

All dimensions and/or tolerances in parentheses are in millimeter equivalents.

Hole-position tolerance can be found in Table 23. Dimension less than 6 in. (152.40), class 4, is 0.006 in. (0.15), except that this tolerance applies at maximum material condition only. Since the minimum terminal area is calculated using the largest hole, the 0.006 in. (0.15) must be increased the same amount the hole size departs from its smallest size after plating (when dimensions apply after plating). Therefore, this increase is equal to the class 4 hole tolerance from Table 21, 0.0002 in. (0.05), and the minimum plating buildup from Table 22, 0.003 in. (0.08), which results in a hole-location tolerance of 0.006 (0.15) + (0.05) + 0.003 (0.08) = 0.11 in. (0.28). Thus

$$B = 0.011 \text{ in. or } (0.28)$$

A graphical explanation of this equation is given in Table 25.

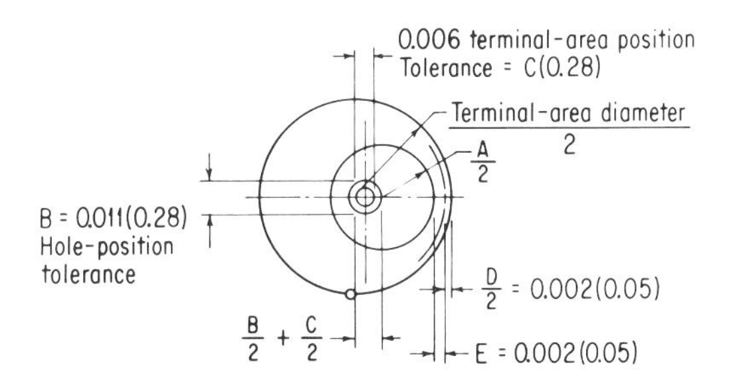

TABLE 32. Comparison Chart of True-position Tolerances to Rectangular Tolerances *

EXAMPLES:

Given, in.—	Find Equivalent	Chart Interpretation
± 0.010 Rectangular tolerance	True-position dia.	See X_1 and Y_1 on chart. Answer: 0.028-in. true-position dia.
+ 0.005 − 0.003 Rectangular tolerance	True-position dia.	See X and Y on chart. Answer: 0.012-in. true-position dia.
0.014 True-position dia.	Rectangular tolerance	See 45° line on chart. This line intersects 0.014-in. true position dia. at 5. Answer: ±0.005-in. rectangular tolerance.

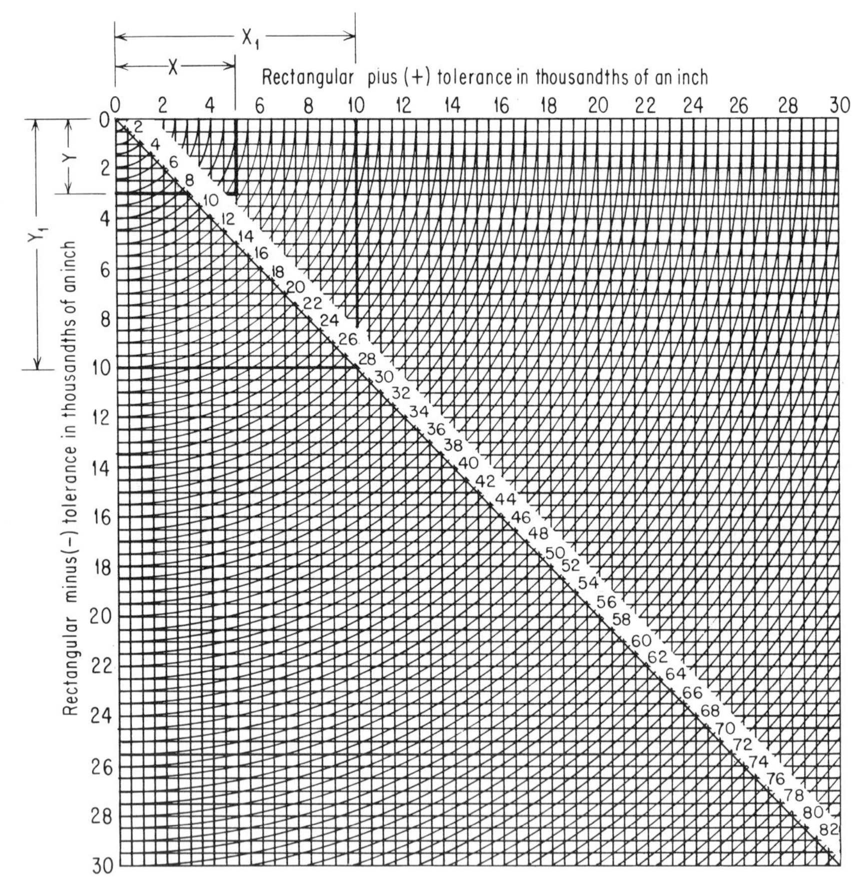

* This chart may be used for larger tolerances by assuming all figures doubled.

Checklist on Design Techniques for Reliability

1. Conductor thickness and width are determined on the basis of the current-carrying capacity required and allowable temperature rise.

CONSIDER:

a. Safe operating temperature of the laminate.
b. Maximum ambient temperature in operating location.
c. Spacing of parallel conductors versus adjacent free-panel area.

NOTE: For reference charts for cross-sectional area, temperature rise, and spacing recommendations, MIL-STD-275.

2. Conductors shall contain no exterior corners having less than a 90° included angle.
3. Conductor lengths shall be held to a minimum between various terminal areas.
4. A distance of not less than the board thickness is preferred between the edge of conductors and the edge of the board.
5. Use single-sided circuit boards where practical.
6. When conductors are required on both sides of the boards, parts are placed on one side only. This is done to avoid moisture traps between components and bridged conductors. Electrical interference should be avoided.
7. Conductive patterns larger than 0.5 in. shall be relieved to prevent blistering or warpage in soldering.
8. A hole within a terminal area shall be provided for each component part lead to be mounted on the printed-wiring board.
9. A distance of not less than the board thickness is preferred between the edges of component or lead holes as well as between the edges of component or lead holes and the edge of the board.
10. Use heat sinks where necessary to avoid hot spots.
11. Check compatibility of electrical, mechanical, and chemical characteristics of materials in the specific application as well as UL and MIL STDs., where applicable.
12. Consider space required for support and mounting of the board as well as weight, size, and location of components to prevent fracture or loosening resulting from flexing, vibration, and shock.
13. The manufacturing process must be considered in the master drawing (artwork) in that most processes will affect line widths, spacing, and terminal area.

the consensus of opinion among "user" companies in the industry, these tolerances are intended as a guide to engineers and designers to ensure that their dimensional requirements can be met by industry and to prevent a choice of tolerances which may make the cost of the resulting printed circuit excessive. These dimensional requirements are categorized by classes 1 through 4, indicating progressive upgrading of tooling and/or material resulting in higher costs.

APPLICATIONS

It took a pressing need brought on by wartime emergencies to push printed wiring from a laboratory curiosity to a practical technique for interconnecting electronic circuits, and its first large-volume application was for circuitry controlling an artillery proximity fuse. Although miniaturization, ruggedness, and ease of manufacture were the predominant influences in this case, it soon became evident that there were many other benefits to be gained from its use.

Printed wiring has always represented a cost saving over most other methods of wiring when a number of identical assemblies are needed. It is readily adaptable to automated production techniques as well as to mechanized assembly, i.e., dip or flow soldering. A beneficial by-product of the production technique used is repeatability; removal of the human element in wire layout and attachment of wires to terminals was eliminated, resulting in a more predictable and reliable product. Electrical as well as mechanical characteristics could be predicted, controlled, and repeated for each and every assembly. Although printed-wiring designers take it for granted, the printed-wiring plane does much more than merely electrically interconnect the circuit elements. It provides physical support and environmental protection to the various components used as well as to act as a conductive path leading to the removal of much of the heat generated.

Whether because of the impossible bulk of hand wiring, unpredictable electrical characteristics of hand wiring or impracticality of a nonautomatable fabrication scheme, many electronic packaging systems would be nonexistent were it not for printed wiring. It is the intended purpose of this section to cover some of the applications of printed wiring. Since advantages and disadvantages will be covered later, this discussion will be confined to describing typical packaging concepts which have resulted from the introduction of printed circuits.

Rigid Printed Wiring. In general, printed wiring replaces the point-to-point wiring historically used for interconnecting electronic components where conductive paths were reproduced with photochemical techniques on the surface of an insulating substrate. Rigid printed wiring was generically the first used, and remains the predominant version used to date. The combination of conductor paths and end-terminal areas or pads on modular grids leads to a major advantage of rigid printed wiring—its adaptability to automated component insertion. In general, the breakdown of rigid-printed-wiring applications closely parallels that of its hand-wired counterpart, except that printed wiring offers many options (mainly attributed to its miniaturization capability) which are not attainable with point-to-point hookup.

In order to describe rigid-printed-wiring applications in terms of packaging terminology so that it would be understandable to those concerned with all types of interconnections, it was decided to identify the application with the packaging level to which it logically belongs. Any electronic equipment using printed wiring may be theoretically partitioned into a number of packaging levels. These levels are essentially hierarchical arrangements of basic components. In our case, each of these levels will use some type of printed wiring, and each packaging level characteristically represents components, subassemblies, and/or assemblies interconnected by some type of rigid printed wiring. These levels are equally applicable regardless of the end-product environment, i.e., commercial, military, space, etc.

Level 1: The Module Element. Level 1 is a logical-function module of intraconnected circuits, otherwise identified with the component level. This is the smallest-field-maintainable, throwaway item. Fig. 63 illustrates some examples of this category and the way in which printed wiring is utilized. Actually, integrated circuits contain in themselves a form of printed wiring on the surface of the monolithic silicon chip, where aluminum interconnection traces are commonly used to connect logic elements on the chip. Usually single-sided circuitry is used for this application, with a wide range of organic and inorganic substrate media prevalent. A popular member of this category is the *sandwich,* or *"cordwood," approach,* a technique in which the components are sandwiched between printed-wiring boards in a cordwoodlike array. This rugged package is ideally suited to pigtail

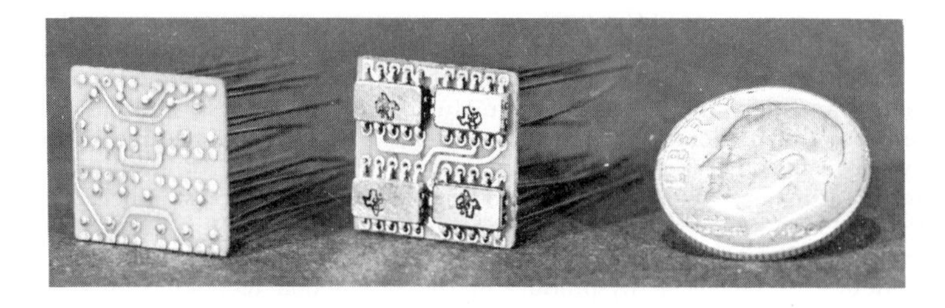

FIG. 63. Level 1 (component). (*Radio Corp. of America.*)

components or integrated circuits and can be encapsulated for environmental protection and enhancement of heat transfer. A variation of this is the on-end cordwood package, where only one printed-wiring board is used.

Level 2: Plug-in, or "Daughter," Board. This generally refers to the assembly level and is the first grouping of basic components or a subsystem of interconnected level-1 modules which are usually easily replaced. Commonly known as the *card level* when printed wiring is used, some examples of this level are illustrated in Fig. 64. Almost any type of printed wiring previously described may be used for this level, and connection is conventionally made with edge- or pin- and socket-type connectors previously explained. Among the parameters which may affect the application at this level are the type of component (level 1) used and how it will be mounted and attached as well as the type of level-3 assembly which will be used (nest, drawer, rack, etc.).

Level 3: Platter, or "Mother," Board. A more recent entry on to the list of printed-wiring applications is that of the large interconnecting plane, or platter, which is rapidly becoming a popular method for interconnecting level-2 assemblies. An example of this category is illustrated in Fig. 65; boards of this type are most often multilayer and range from as small as 6 by 6 to as large as 18 by 18 in.

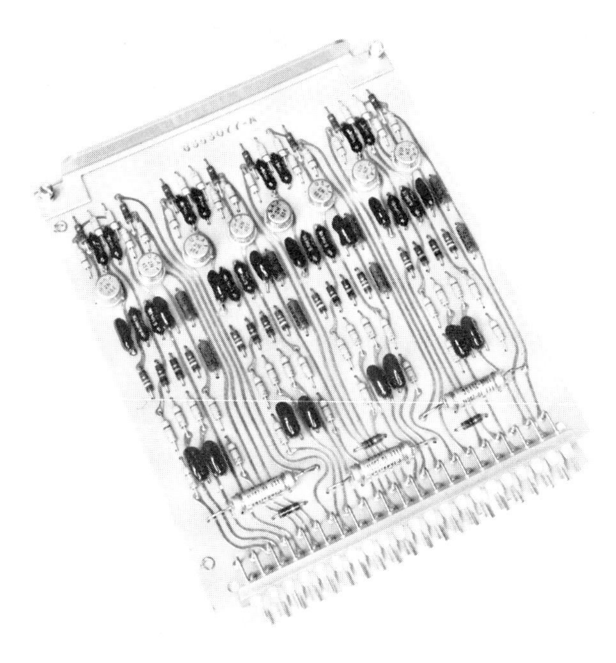

FIG. 64. Level 2 (card). (*Radio Corp. of America.*)

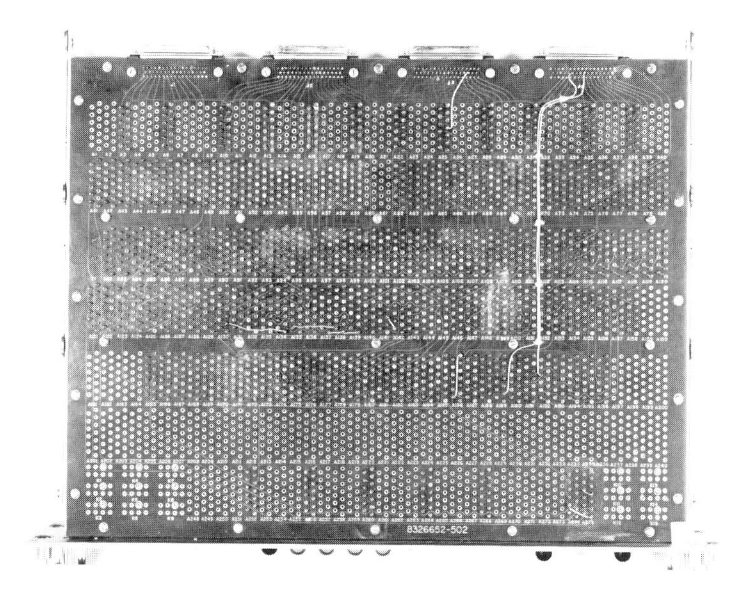

FIG. 65. Level 3 (platter). (*Radio Corp. of America.*)

It is often impossible to accommodate more than 90 to 95 percent of all wiring needed on the platter, and the remaining interconnections are made by some type of discrete wire hookup, e.g., point-to-point wiring by soldering, welding, or wire wrap. This type of wiring plane often includes a transmission line of some type (microstrip or strip transmission line) and usually has one or more copper planes devoted to ground- and/or voltage-supply purposes.

Other Levels. The recent popularity of computer-aided design in electronic packaging (covered more thoroughly in Chap. 15) has resulted in the use of printed wiring in levels above 3 as just defined. Many designers are discovering that as contrasted with other methods, there is a definite cost and performance advantage in some cases in favor of printed wiring for use above level 3. It is obvious that printed wiring at this level is not practical for prototype or small-quantity production or where a great many changes may be required for developmental logic changes or special customer requirements. Automated wiring (computer-assisted-design or design-automation) procedures are very helpful in implementing printed wiring for application, as well as for computer simulation, of machine characteristics. Automated artwork-preparation techniques become essential as pin density or the number of layers increase; both factors make dimensional tolerances increasingly difficult to maintain. In general, printed wiring looks attractive at any level when the terminal density approaches 50 terminals per square inch since conventional techniques such as point-to-point wiring become extremely unwieldy in these cases.

Other Rigid-printed-wiring Applications. It is safe to say that the variety of applications of printed wiring are limited only by the imagination of the designer; printed wiring has the natural capability of fulfilling a wide variety of interconnection, as well as component, needs. It is impossible to cover all the *possible* applications of printed wiring here; however, the author will attempt to point out major categories within which, understandably, there remain many variations.

FLUSH PRINTED WIRING. A printed circuit whose upper conductor surface is in the same plane as the surface of the insulating material is known as a *flush printed circuit.* The methods of producing flush boards, explained in some detail previously, include (1) use of the built-up approach, (2) etching circuitry on flush-grade laminate and "pushing" it down to flush, and (3) etching the circuitry and filling with a resin compatible with the base resin. Although originally designed to provide a switching surface for high-velocity wiping contacts where contact bounce and accompanying noise cannot be tolerated, many other applications have developed. Code wheels, which are analog-to-digital conversion devices, use them almost exclusively. A printed-circuit motor has been developed using a printed-wiring armature. This device, with low armature inertia, is capable of high pulse torque and controlled "stepping" to preferred armature positions. This motor is used in a wide variety of applications such as in tape recorders, computer peripheral equipment, and positioning and detenting mechanisms.

PRINTED-WIRING COMPONENTS. Printed-wiring conductors in themselves may be said to share a dual classification. As normally used, they interconnect circuits, or logic elements. They are, however, resistors of very low value. When materials other than the conventional copper are used, the resistance may be raised to some discrete value. In actual practice this is done by sputtering or vacuum evaporation of metals such as titanium, chromium, or tantalum; by etching of cladding of this type; or by screening of some conductive material such as silver acrylic or silver epoxy. Capacitors are also produced by photochemical processes; more information on this type of component will be found in Chap. 5, which discusses deposited-film techniques. A type of memory has been produced using a double-sided board with registered capacitor plates on top and bottom; this is another illustration of the versatility of printed wiring. Although somewhat wasteful of printed-board area, inductive elements are occasionally produced on printed-wiring boards by spiral configurations; commonly used in early radio and television applications of printed wiring, they have largely given way to discrete elements.

MEMORIES. Many memory systems depend on conducting read, write, and sense currents around or through a predetermined array of magnetically retentive materials, such as ferrite or various metallic oxides. Printed wiring has demonstrated its capacity for doing this, and a number of memory systems are currently operating which employ some form of printed wiring in their operation. Another variation is to use photochemical or printed-wiring techniques for deposition or etching of the actual memory elements; this is currently the case in the design of some superconductive memory systems.

UNIVERSAL-GRID WIRING. Most often used for breadboard designs, wiring systems of the universal-grid type are available in either the etched or unetched form. The unetched form is supplied as a prepunched insulating sheet with copper cladding on both sides. A circuit pattern is applied and etched, complete with conductors and terminals. Holes are presented when the cladding over them is etched away; they are then used for wire jumpers or eyelets to interconnect between sides. The other approach provides a standardized, prearranged wiring pattern on a dielectric base with conductors arranged mutually perpendicular to each other on opposing sides. The desired pattern is obtained by the addition of connections or interconnections via welding or eyelets, and unwanted conductive areas can be cut away or punched out. Universal-grid wiring is perhaps suitable for breadboard or prototype work, but by nature of its universality it is frequently unattractive for production since it fails to optimize space and cost factors.

Flexible-flat-conductor Wiring. In all flat-conductor wiring, two or more flat conductors are encapsulated between layers of film-type insulating material. The cables are relatively flexible because the flat conductors lie in a neutral-axis plane of the cable. For this reason tensile loads are distributed across the entire width of the cable, and the engineer is able to select conductors on the basis of their electrical characteristics rather than on strength alone.

As defined previously, the difference between flexible-flat-cable and flexible printed wiring is determined by the type of conductors laminated within the cable, the manufacturing process, and the fact that flexible printed wiring is generally custom-designed for a special application with regard to an electronic-systems design.

Flexible flat cable uses parallel, flat conducting ribbons as conductors. Different conductive materials may be used, but copper is standard. The conductors may be etched to shape, or stamped or slit. Uniform conductor sizes are usually employed, and widths and spacings recommended are as specified in the industry standard, NAS 729,[30] duplicated in Table 33. Flexible flat cable, like the conventional cable it replaces, is most commonly used to interconnect separate or remote component packages. Flexible printed wiring almost always has etched conductors of copper and occasionally of other metals such as Kovar, nickel, or bronze. In some mass-production applications such as automotive or telephone, the wiring may be more economically produced by die molding or die stamping. Conductors are not necessarily parallel and may run in any direction.

In many ways even more versatile in application than its rigid counterpart, flexible printed wiring has found an almost unlimited variety of uses since its appearance on the market. In many cases it gives way to rigid printed wiring only because it cannot lend structural support and many of the flexible materials are thermoplastic and require specialized soldering and joining techniques. In other cases, however, the flexible nature of this wiring media constitutes one of its principal advantages.

Flexible Flat Cable. As defined previously, this is flexible, multiconductor, flat-wire cable, used most often with connectors at each end in the same way as conventional, round-cable wiring. The largest application of this variety has always been in cables and harnesses, where its advantages are readily apparent. Used for power and signal distribution, it is generally applied to interconnecting separate or remote component packages. It is especially attractive where a conformal configuration is required or where the components have to move in relation to each other. In terms of electrical performance, flat-conductor cables are more efficient than round-wire cables because they expose proportionately more surface area than round conductors, transfer heat more efficiently, and hence carry a greater current load than round conductors with the same cross-sectional area.

TABLE 33. Standard Cables per NAS 729 [30]

Cable sizes	Conductor spacing 0.050 in. Conductor sizes 0.002 × 0.025 (No. 32 AWG) 0.003 × 0.026 (No. 30 AWG) Resistance at 20°C (ohms/1,000 ft) 0.002 × 0.025–188.0 ohms 0.003 × 0.026–119.0 ohms	Conductor spacing 0.075 in. Conductor sizes 0.002 × 0.025 (No. 32 AWG) 0.003 × 0.026 (No. 30 AWG) 0.003 × 0.042 (No. 28 AWG) Resistance at 20°C (ohms/1,000 ft) 0.002 × 0.025–188.0 ohms 0.003 × 0.026–119.0 ohms 0.003 × 0.042–78.7 ohms	Conductor spacing 0.100 in. Conductor sizes 0.003 × 0.062 (No. 26 AWG) 0.005 × 0.063 (No. 24 AWG) Resistance at 20°C (ohms/1,000 ft) 0.003 × 0.062–48.4 ohms 0.005 × 0.063–28.9 ohms	Conductor spacing 0.150 in. Conductor sizes 0.005 × 0.100 (No. 22 AWG) Resistance at 20°C (ohms/1,000 ft) 0.005 × 0.100–18.8 ohms
1 in. Conductors →	17	12	9	6
2 in. Conductors →	37	25	19	12
3 in. Conductors →	57	38	29	19

SOURCE: Aerospace Industries Association of America, Inc., ©. Reprinted by permission.

The most frequent use of flexible flat cable is in replacing the conventional round cable normally used. For many applications, substantial increases in density and reliability have resulted, with no decrease in performance. Flat cable is often used as an interconnection means, or harness, between rigid boards, computer- or communications-equipment back planes, and as flexible harnesses in many aerospace applications such as the Polaris and Titan missiles. Flat cable has also found some use in low-voltage domestic wiring systems, where the 28-volt restriction enforced by the building codes limit their application to intercom systems, fire and burglar alarms, bells, clocks, and low-voltage house-current switching systems. They have also been utilized for low-temperature heater applications, but these are usually in custom-designed configurations. Typical applications can include miniature ovens and drying chambers, ambient-temperature controls, and heaters for refrigerator defrosting systems.

There are a number of physical attributes of flexible flat cable which are particularly useful. One is its *memory*, or ability to return to a predetermined shape. This makes it very useful for retractable cables for drawers or doors, when the window-shade or accordion construction may be used.

The number of distinctive applications is limited only by the ingenuity of the engineer using it since flexible-flat-conductor cable is adaptable to a wide range of operating conditions and uses. Adhesive-backed flat cables, for instance, may be stuck to the inside of aircraft compartments, space chambers, and even to baseboards, walls, and other building compartments.

Custom, Flexible Printed Wiring. Although some flat cables are made on a custom basis, for our purpose we are defining them as essentially off-the-shelf items, which can be used as is, except for terminations which may be required. Most flexible printed wiring, on the other hand, requires a custom-designed artwork to pattern it to a particular application. As discussed before, this artwork is essentially the same as artwork which is used for rigid printed wiring. Flexible printed wiring is most applicable to electrical interconnection systems within a particular component package.

Figure 66 illustrates a typical example of a flexible-printed-wiring application within a drawer-type assembly. In this case, the wiring, several layers deep, is used to interconnect pins on connectors which accommodate plug-in cards.

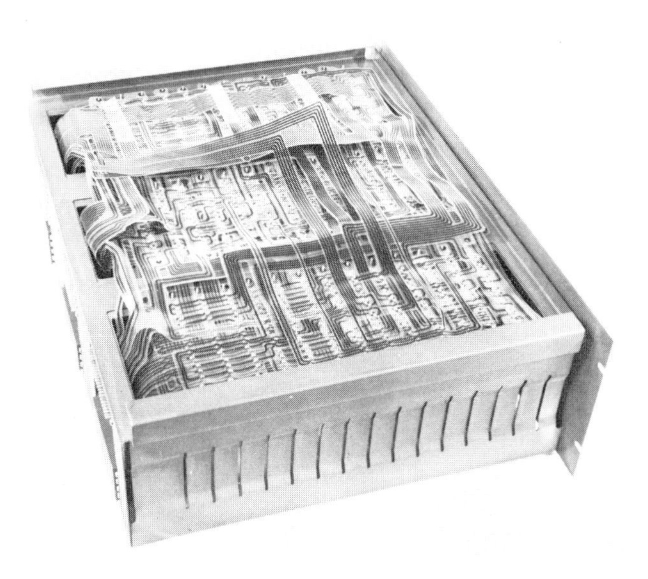

Fig. 66. Flexible cable wiring. (*Radio Corp. of America.*)

Among the many specialized applications of flexible printed wiring is its use in the fabrication of computer memories where it is utilized to conduct logic and sense signals around various types of memory elements. Flexible wiring is also used for security grids; in security grids a fine grid-type pattern is utilized to provide a secure network which, if any portion is broken, will actuate a signal or alarm. This technique is widely used in industrial, military, and space applications. For partial shielding applications, flexible printed wiring is produced either in a sandwich configuration with two latticelike windings simulating a twisted pair or, for complete shielding, in a fashion similar to strip transmission line.

PRINTED-WIRING DESIGN AND APPLICATION TRADE-OFFS

To have ventured this far into the chapter, it must be assumed that the reader is currently using printed wiring or is considering doing so. In arriving at his decision concerning which printed-wiring technique he will choose, the designer may indicate a number of reasons for his selection. Many times, customer preference or military requirements can be charged with the choice. More often than not, however, a choice is made primarily because a number of design goals exist for a task, and this interconnection technique must completely fulfill these design goals. In order to determine this, however, the engineer must consider the alternatives, or *trade-offs,* and take into account important design parameters such as cost, miniaturization, maintainability, and reliability, to mention a few.

Design and application trade-offs are much more comprehensive than the above list would indicate and do not mean merely choosing printed wiring in lieu of other interconnection techniques. Once an interconnection technique is chosen (in our case, assuming it is printed wiring) a number of other options affect its use. There are many insulating and conductive materials, for instance, which may be chosen for the application. The processes used to make the wiring plane have an important bearing on its design. The constructional technique employed can radically affect the overall packaging system. A wiring plane is normally used to interconnect functional elements to carry out system goals. The method by which these elements are connected to the interconnection plane (soldering, welding, etc.) can also affect the design and use of the packaging system. The number of layers of wiring which are required is another aspect requiring careful consideration in full light of all the facts.

Before discussing the details affecting the trade-offs in printed-wiring design, it is necessary to examine the factors affecting the choice of printed wiring in the first place.

Trade-offs between Printed-wiring and Other Wiring Techniques. Modern electronic systems have placed a number of problems before the packaging engineer, the most serious and provocative of which is that of an adequate interconnection method. There can be no doubt that the most important factors in the overall packaging of electronic equipment are reliability, satisfactory performance under all environments, and ease of serviceability. As previously mentioned, other design parameters also play an important role. These are producibility, or ease of manufacture, weight, miniaturization potential, and cost. Keeping these factors in mind then, the author will arbitrarily set up a list of requirements, a so-called "yardstick," against which to compare these various interconnection techniques.

1. Wiring accuracy
2. Reliability under adverse environments
3. Miniaturization (size and weight)
4. Ease of manufacture; bulk-processing potential for manufacture and assembly
5. Design flexibility
6. Maintainability and repairability
7. Mechanical and thermal support capability
8. Reproducible performance, especially electrical

Rigid Printed Wiring. When measured against all eight of the above requirements, printed wiring, as a general class, can offer satisfactory performance in varying degree depending upon the type of printed-wiring plane used. Wiring accuracy may be substantially enhanced by using computer-generated wiring lists and automated artwork production. Reliability is probably the best of any wiring system employed today. In addition, the rigid-printed wiring system has the potential of extreme miniaturization and is naturally amenable to mass production not only in the printed-wiring-plane manufacturing phase but also, if a flow soldering technique is used, in assembly of components. If a flow-soldering technique is used, design flexibility is in assembly of components. Design flexibility is relatively good before a design is committed; thereafter, of course, it requires a change in artwork, and changes are not easy to make. In the same manner, printed-wiring boards are not easy to change after their fabrication is completed. Advantages and disadvantages of rigid printed wiring are covered in Table 34.

TABLE 34. Advantages and Disadvantages of Rigid-printed-wiring (All Classes) Interconnections

Advantages	*Disadvantages*
1. Lowest cost for large-quantity production	1. High cost for small-quantity production
2. Good weight and space minimization	2. High tooling and artwork costs
3. Provides structural support for components, versatility in package design	3. Difficult to repair—especially multilayer
4. Rapid installation, can be mass-soldered	4. Design must be fixed to be feasible on a cost basis
5. Multilayer forms can provide voltage and ground buses	5. Reliability of interlayer connection varies with fabrication technique
6. Can provide transmission-line capability	

Flexible Printed Wiring. Essentially the same characteristics exist for the flexible classes of printed wiring as for the rigid ones. Obviously, the capability for mechanical support does not exist; some other means must usually be provided to support components which are to be interconnected. Flexible cable has the distinct and unique advantage of immediate availability; it is usually stocked in many sizes and may be merely cut to size and connected into the system. However, flexible wiring has always been difficult to terminate, and this continues to present many unique problems in its application. When compared with conventional point-to-point cabling systems, it must be stated that flexible wiring suffers from its inability to provide branching and crossovers as readily as conventional wiring. Crossovers may be achieved by double-sided flexible planes, but mostly by stitches, jumpers, eyelets, or other mechanical means; plated-through holes have not been established as practical in flexible wiring. The outstanding advantage of flexible wiring over rigid wiring is, of course, in its construction, which allows the designer to specify visually any three-dimensional shape or structure for any application. Flexible wiring can reduce space requirements by a factor of 7:1 over conventional cabling. A summary of the advantages and disadvantages of flexible printed wiring are covered in Table 35.

TABLE 35. Advantages and Disadvantages of Flexible Printed-wiring Interconnections

Advantages	*Disadvantages*
1. Possesses capability to be bent or flexed	1. High cost for small-quantity production
2. Possible to produce consistently uniform wiring assemblies	2. High tooling costs
3. Low cost for large-quantity production	3. Not easily repairable
4. Rapid installation	4. Reliability in military environment not proved; also, not UL approved
5. Weight and space reduction where voltage drop is not critical	5. Usually needs structural support for components
	6. Limited multilayer interlayer connection methods
	7. Soldering temperature critical (usually below 400 to 450°F)

Conventional Wiring (Point-to-point Hookup). Since components at the first and second levels of assembly must be physically supported anyway, there is hardly any reason for not using rigid printed wiring at this stage; thus it is preponderantly advantageous to do so. In radio and television work, where vacuum tubes are used, components are wired between socket pins, making conventional wiring somewhat feasible for moderate-run production. The replacement of many vacuum-tube circuits with solid-state devices, however, established yet another advantage for printed wiring at this level. Other than in back-plane wiring of large equipments such as computers, the one place where conventional point-to-point hookup is still in popular use is in cabling between points in an assembly or from points between assemblies. In terms of wiring accuracy, errors are liable to occur as with all nonautomated systems, although methods have been devised to provide mechanized assistance to check on accuracy of wiring. Although providing few restrictions in design flexibility, this method does not offer the size and weight savings which printed wiring does. Maintainability and repairability, on the other hand, are simplified when this method is used, and design changes are more readily incorporated than when printed wiring is employed. This method of wiring is most often employed in small-quantity or prototype production, and may use soldering, welding, or proprietary-connector techniques for the actual joint. Advantages and disadvantages of conventional point-to-point hookup are given in Table 36.

TABLE 36. Advantages and Disadvantages of Conventional Wiring (Hand-soldering) Interconnections

Advantages	*Disadvantages*
1. Low cost for small-quantity or prototype production	1. High costs in large-quantity production
2. Relatively easy to make repairs and/or changes	2. Limited to one point-to-point connection at a time
3. Minimum or no tooling required	3. Needs rack or chassis to support components
4. Well-known technique, inspectable joints	4. Bulky (heavy and space consuming)
5. Can use coaxial (transmission-line) techniques	5. Limited wiring density.
6. Flexible design, good for breadboarding	

Wire-wrap and Other Automated Techniques. Although these "joint-making" techniques may be accomplished on an individual basis and follow the same general rules as defined under conventional wiring, this section will describe the automated-wiring techniques currently available for wire-wrap and Termi-point* methods. In wire wrapping, multiple turns of wire are wound tightly on a pin-type terminal possessing a minimum of one sharp edge. Uniform and dependable gas-tight joints result when an automated tool is used, and an entire panel may be wrapped or connected if a programmable table is used. Termi-point is a similar connection method, where a clinched spade-type terminal is used, but likewise may also be connected into a programmed system for automated production. Wiring accuracy and design flexibility when using these techniques is probably equal to those of printed wiring, and design changes or repairs are relatively simple, especially with Termi-point, where connections may be "pushed down" on a post and need not be removed. When compared with printed wiring, which is the ultimate in bulk processing, it must be stated that these techniques, although automated, do not possess the manufacturing economy potential of printed wiring. These techniques, however, are commonly used for adding interconnections to printed-wiring planes and fill an important need in this fashion. Information on wire-wrap termination and interconnection capability is given in Table 37. Advantages and disadvantages of automated wire wrap are given in Table 38.

* Registered trademark, AMP, Inc.

TABLE 37. **Wire-wrap Termination and Interconnection Capability** [23]

Adjacent terminal spacing, in.	Terminal dimensions		Wire diameter over insulation		Turns per pin	Termination density, terminations/in.^{2}		Conductor density, conductors/in.^{2}		Interconnection capability*
	Cross section, mils	Pin length, in.	B&S wire size No.	Wire diameter, in.		Theoretical	Probable	Parallel	Crossover	
0.200†	20 × 60	0.75	24	0.060	7	25	21	25	25	1,250
0.150†	45 × 45	0.75	26	0.050	7	45	38	35	35	3,150
0.100‡	22 × 25	0.75	28	0.048	7	100	85	50	50	10,000
0.075§	25 × 25	0.5	30	0.043	7	177	150	65	65	23,000

* Interconnection capability (a figure of merit) = terminations/in.^{2} × conductors (parallel + crossover)/in.
† Automatic production capability in 1962.
‡ Hand-tool production capability in 1962.
§ Prototype of hand-tool production capability (1963 or 1964).

Table 38. Advantages and Disadvantages of Wire-wrap Interconnections

Advantages

1. Can be automated, increasing wiring accuracy
2. Features a reliable, gas-tight connection without plating or other materials
3. Can use component lead as connection point
4. Compatible to rack-panel and drawer-packaging schemes
5. Limited human-factor inputs needed, especially if automatic wire wrap is used

Disadvantages

1. Solid wire only; not stranded
2. Transmission-line techniques unwieldy, system-frequency limited
3. Close process control needed for reliable product
4. Only effective testing technique is destructive
5. Changes difficult to make
6. Expensive equipment, tooling, and programming for automatic wire wrap

Welded-wire Planes. *Welded-wire planes* represent an attempt to capitalize on the potential reliability of welded joints in the fabrication of interconnecting planes. A plane may be composed of a number of *matrices,* which by definition are two parallel planes of mutually perpendicular sets of wires. In use comparison with printed wiring, welded-wire planes are virtually insignificant. They do, however, represent an important attempt by the designer to achieve his design goals. From a standpoint of card for card, one welded-wire matrix medium is from 1½ to 3 times heavier and from 2 to 4 times thicker than the plated-through-hole, multilayer wiring-board medium. These results are graphically indicated in Fig. 67, which illustrates the attendant sacrifices in size and weight when designing with multilayer welded-wire matrices as opposed to multilayer printed-wiring boards. Because of its construction, the welded-wire plane increases in multiples of two layers, resulting in the step curve shown. Experience has shown that approximately 40 percent more welded connections must be made for a welded-wire matrix when compared with an equivalent printed board. The printed multilayer board lends itself to bulk-processing techniques; the welded-wire matrix tends to require an individual-connection technique, making the latter inherently more expensive. It can be concluded, therefore, that the use of the welded-wire matrix-interconnection concept can prove detrimental toward achieving equipment design goals particularly when miniaturization (size and weight) is a primary packaging factor. It is equally difficult to repair as is the printed-wiring board, and although welding is an inherently reliable technique, the integrity of each joint may vary and be subject to possible operator inconsistencies.

Trade-offs between Various Materials. An important option open to the packaging designer is in the choice of material he uses. Although this choice is often process-dependent, the printed-wiring designer has a comprehensive list of materials with diverse mechanical and electrical properties from which to choose, both in the dielectric and conductor category.

Insulating Materials. Although materials have been discussed to some extent previously in this chapter, it may be helpful to point out some of the advantages and disadvantages of these various dielectrics. This may already be apparent from a close look at Tables 2 and 8, but some factors are not altogether evident. In general, most rigid materials are reinforced thermosetting compounds, while flexible materials are usually nonreinforced thermoplastics with some thermosetting types. The advantage of thermosets is obvious; any flow soldering operation requires a substrate of this type. Tables 2 and 3 give the pertinent mechanical and electrical properties of the commonly used dielectric materials. Although paper- and glass-reinforced phenolics are still widely used in commercial and entertainment applications, the superior electrical characteristics, strength, and resistance to moisture absorption of the glass-reinforced epoxy grades have clearly indicated that these are the best materials to use where high performance is required; thus these materials are almost exclusively used in multilayer applications. Other resin systems indicated in Table

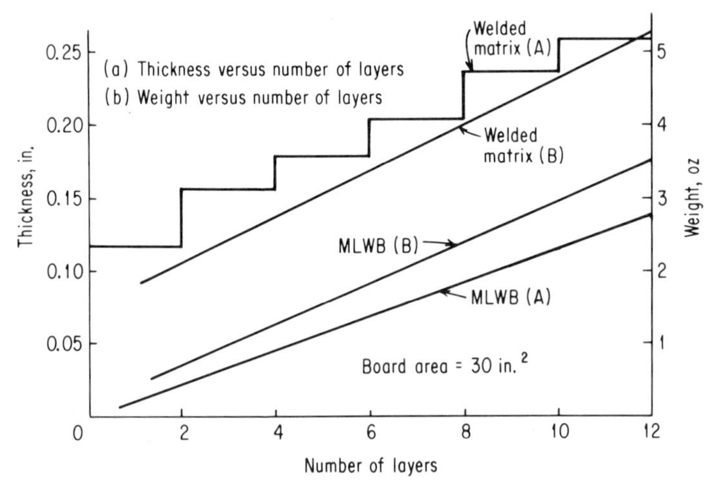

FIG. 67. Multilayer-board–welded-plane comparisons.

2 are also used as well as other reinforcing materials such as asbestos, polyester, cotton, mica and ceramic; however, these are special cases and always applicable. More important are newer dielectric materials such as polyphenylene oxide, irradiated polyolefin, and crosslinked styrene copolymers, which, by reason of their low relative dielectric-constant and loss factor, are desirable to use for high-frequency interconnection applications, especially in computers and communications equipment.

Conductive Materials. In view of the fact that more than 95 percent of all printed-wiring boards are made using copper (either electrodeposited or rolled) as a clad material, it appears that there are a few alternatives to consider for the actual circuit conductor. The recent advent of welded, surface-mounted components has popularized nickel and Kovar (an alloy of nickel, iron, and cobalt) for these applications. Although copper is a good electrical conductor, it also conducts heat away from the weld area too fast to permit good welding practice; consequently nickel, Kovar, and other proprietary alloys and/or clad compositions are currently available as conductors for printed wiring. A compromise must be accepted in this case regarding conductivity, however; referring to Table 6, the conductivity of nickel is shown to be one-quarter that of copper. Nickel, however, has strength advantages; it has a modulus of elasticity in tension almost twice that of copper.

Some manufacturers of multilayer wiring, for this reason, favor the use of some nickel plating in the hole to enhance the strength characteristics of the plated-hole system.

Although electrodeposited foil is most widely used as cladding, freedom from pinholes is one reason for the employment of rolled foil. Rolled foil is work-hardened in its manufacture; therefore it has a tensile strength twice that of electrodeposited foil and exhibits superior fatigue-resistance characteristics when compared with electrodeposited foil. For this reason the rolled type is often used for flexible cables and printed wiring.

These types will be discussed in some detail in Chap. 5. Conductive epoxies and conductive acrylics using silver as the conductor material have conductivities ranging from approximately 0.03 to 0.06 ohms/sq. When compared with copper, the relative conductivity of these materials would range from 1 to 3.

Adhesive Materials. It is particularly advantageous, in cladding of copper to an insulating substrate or in laminating boards, to use an adhesive system having a resin or active component which is compatible with the base-laminate resin system. In addition, every attempt is made not to use an adhesive which would degrade the electrical characteristics (impedance, loss factor, etc.) of the base laminate. Adhesion is enhanced when using any adhesive system, if oxide-treated copper is employed in the cladding operation; adhesion during etching is also improved when the oxide-treated foil is used.

Rubber-modified adhesive, particularly nitrite phenolics, have proved superior in peel strength but somewhat degrade the electrical properties of the material. Phenolics modified with vinyl appear to be best for phenolic materials, modified epoxy adhesives for epoxy substrates. Although many thermoplastic flexible materials are their own adhesive upon heating, some special copolymers such as auroethylene copolymers are used as long as they are closely physically and electrically compatible with the base material.

Processing Trade-offs. Although processes do not enter directly into the design cycle, it is important that the designer acquaint himself with the trade-offs in this area since these parameters may affect the initial design or selection of methods or materials. A good design of any product should normally be affected by the method intended for its manufacture; this is the best way to ensure "getting the most for your money." Although in some cases, e.g., etching, a number of alternative methods are presented, in others, e.g., laminating, the process variety is much more limited, resulting in a requirement that the engineer design with some consideration for the trade-offs and restrictions presented by the manufacturing process.

Etching. The etching process has probably more variables than any other process used to produce printed wiring. There are several etch-resist systems currently employed by the industry; in addition, it is common practice to use one of a variety of etchants applied in one of three modes.

RESIST SYSTEM. Ironically, etching is not so hard on the resist used on the clad-copper board as is the plating process. Plating solutions are many times quite hard on board material and photoresist as well, owing in part to their chemical composition as well as to the elevated temperature required for best plating results. A number of commercially available positive and negative photoresists are satisfactory for this operation. Among the plated resists, solder is usually preferred. Apart from other considerations solder is often required on the board for assembly of components, and it has the advantage of providing protection for the edges of the conductors when it is heated and reflowed. Gold plating is also used as an etch resist, particularly in the manufacture of boards with plated-through holes. When compared with other resist systems, however, severe undercutting occurs. As mentioned previously, in the case of some etchants, such as chromic sulfuric compounds, it is thought that this undercutting is caused by a galvanic action between the gold and copper in the etching bath, which acts as an electrolyte.

ETCHANTS. Etchants commonly used, together with some pertinent trade-offs which may affect design, include the following:

Ferric chloride base. This base is used where the acid-resist image is either a photographic-resist, screen or offset-printed resist, or a gold-plated resist. A major disadvantage is the violent agitation which must be used. The main problem in using this etchant is in removing the acid from the work following the etch cycle. Subsequent humidity- and insulation-resistance tests may reject the boards if acid residue is allowed to contaminate the board.

Cupric chloride. In combination with hydrochloric acid, this etchant may be used with any resist. It is easily washed from boards, and provides the advantage of continual regeneration by selected additions of chemicals.

Chromic-acid base. Often applied with sulfuric-acid or nitric-acid additives, this etchant is usually used for solder-plate-resist etching. This etchant is the most expensive and potentially the most dangerous of any covered here. Some staining of board substrate is caused by this etchant. Along with ferric chloride with certain addition agents, this etchant provides the most uniform etching rate, least undercutting of copper lines, and least slope to line sides.

Ammonium persulfate. This etchant can be used for all photographic-, screen-, and offset-printed resists as well as for both solder- and gold-plated circuits. It is clean and easy to handle, except for temperature-control problems caused by its exothermic nature. The etching solution is somewhat unstable, a distinct disadvantage in production.

ETCHANT APPLICATION. Perhaps as important as the etchants used is the way in which they are applied to the workpiece. Tank etching is very simple since it involves merely dipping the workpiece in a still or agitated supply of etchant. It is extremely difficult to produce consistent- or even minimum-quality work in this manner. Splash etching does not produce uniform quality but can achieve acceptable quality where conditions are closely controlled. Spray etching (either vertical or horizontal) produces consistently uniform and high-quality work with a minimum of operator skill. It is ideally adaptable to automatic production.

Plating. A number of variables having an important bearing on the quality of the plated-through hole are presented in the selection of a copper-plating system. Plain acid-copper sulfate baths have poor throwing power, produce a coarse-grained structure, and are not the most popular plating baths. Cyanide copper baths produce the best copper in terms of tensile strength (average 100,000 psi) but the strong alkaline solution used for the electrolyte may deteriorate both the insulating material as well as the bond. Fluoborate or pyrophosphate copper baths are most satisfactory in terms of throwing power and produce fairly fine-grained, smooth deposits. Pyrophosphate-plated copper has a tensile strength of from 35,000 to 70,000 psi and is the most ductile of all the platings discussed above.

Laminating. In laminating, close control of the heat in the platens is of major importance, and in some cases they should be heated rapidly. In general, steam heating is preferred in that it greatly decreases time necessary to heat platens and provides the most uniform platen-temperature distribution, $\pm 1°F$ for steam as opposed to $\pm 5°F$ for electrical methods. A temperature of from 100 to 400°F can be obtained in under 5 min with an adequate boiler; this same change would require 17 min with an electrical setup. Obviously, steam would provide a more rapid turnover in production. Steam has the disadvantage, however, of requiring expensive boilers, controls, and piping. As far as registration is concerned, high-temperature-bond-strength materials should be used to circumvent clad-copper shift, and registration pins should be placed on at least 4-in. centers to maintain good registration during lamination.

Hole Formation. Of the two hole-fabrication methods commonly used, punching is more often used where plating in the hole is not required. Drilling is the only technique which will provide adequate hole quality for the plated-hole processes. If hole quality is not adequate in the drilled condition, it appears that no amount of cleaning of any type will make the holes acceptable for plating. Although all drilled holes must be deburred or cleaned in some way, there is an open question regarding the cleaning technique (mechanical or chemical) to use. Mechanical methods include dry and wet abrasive deburring, vapor blasting, and sanding. Chemical methods are two in number: (1) the use of sulfuric acid to remove epoxy "smear"—called *chemical cleaning*, and (2) the *etch back* method, which employs combinations of sulfuric and hydrofluoric acids to remove epoxy material and glass fiber in the hole.

Trade-offs in the Number of Layers. In conventional point-to-point wiring, when the interconnection requirement increases, the wireman merely adds more wires, or interconnectons. When printed wiring is used, the designer attempts to add more printed lines on the layer. If the space on that layer is not adequate, a decision must be made to use other means (discrete wiring) or add another layer. In many cases, the decision is a simple one: density restrictions or electrical considerations demand that additional wiring layers be used. However, owing to increasing cost, and reduced design and maintenance flexibility as the number of layers is increased, the designer is usually committed to using the smallest number of layers possible.

Single-sided Boards. The simplest and most common of all printed-wiring media, this configuration also has the lowest pin density. Components may be surface-mounted but are more often mounted through holes (usually punched) and mass-soldered to the wiring. It is the most reliable and also the most easily maintained of all the board types. See Table 39 for comparative advantages and disadvantages.

TABLE 39. Advantages and Disadvantages of Rigid Printed Wiring

Single-sided

Advantages	*Disadvantages*
1. Lowest cost	1. Lowest density capability
2. Easily maintained	2. Crossovers must be hand-wired
3. Most reliable	

Double-sided

Advantages	*Disadvantages*
1. Increased density capability	1. Increased cost
2. Ground plane availability	2. Maintainability more difficult
3. Attachment to both sides available	3. Decreased reliability
4. Conductive heat removal possible	

Double-sided Boards. The double-sided board must have some interside connection method, resulting in a loss of reliability to some extent. It provides more density capability than the single-sided board; however, it is two to six times more costly. Although this board permits the attachment of components to both sides of the board, attachment is made to one side only if flow soldering is to be used. Refer to Table 39 for advantages and limitations.

Multilayer Boards. The ultimate in printed-wiring density, this technique is required where circuitry requirements dictate a high interconnection density. Depending upon the particular application involved, space savings over single-sided and double-sided printed circuitry of 15 percent to 70 percent can be realized. However, contrasted to a comparable double-sided board, the cost can be expected to increase at 25 percent plus 10 to 12 percent per additional layer. The many advantages and restrictions of this type of wiring plane are given later.

Board-fabricating Techniques. The trade-offs involved in the various printed-wiring-board fabricating techniques are not always the designer's for the choosing since the trade-offs may include volume considerations, severe environments requiring resistant materials, or even customers' demand. Although the many methods and variations of techniques employed when printed wiring was first developed may still be used for unusual cases in unique applications, the great bulk of printed wiring is still being manufactured using one of the techniques described in the beginning of this chapter.

Single-sided Boards. At the present time, single-sided boards represent the largest volume of production of any board type, and there are relatively few important techniques currently used for producing them. These methods are concerned mostly with expected volume of production, material used, and ultimate end-environment application.

The etched-foil process is at present the most widely used method for the production of printed wiring. Although the etched-foil process is the predominant method used in the production of single-sided boards (both rigid and flexible), it may also be used for double-sided boards and for internal layers of multilayer boards. This method is adaptable to small- or large-scale production, depending upon artwork type and resist application, and is adaptable to a wide variety of packaging applications. It is available with a protective overplate when a plated resist (gold, solder, etc.) is provided as an etch resist.

The additive or plated type is somewhat more expensive to produce than the above, and has the predominant advantage of supplying flush circuitry. A problem exists in producing holes since they must usually be produced by drilling rather than by the more economical punching procedure, and the board itself must usually be reinforced in some way to give it greater rigidity and strength than that provided by the screened epoxy insulation conventionally used.

The mechanical processes for producing printed wiring (stamped, embossed, sprayed, and molded) are historically among the first processes used and are very popular in Great Britain and Europe. They are almost all, however, applicable mostly to high-volume production and in general depend upon the use of tool-made dies bearing an image of the wiring pattern to obtain the desired product.

Though all these mechanical methods have individual capabilities and restrictions, there are certain general characteristics which they all share. These are high-production methods; however, for large volumes of repetitive circuitry important cost savings may be realized. A good example is in automotive-dashboard flexible circuitry, where stamped or die-cut circuitry is used almost exclusively. The mechanical processes do not use chemicals and as such are not subject to possible contamination or subsequent deterioration. These are generally custom-designed processes, which work optimally with certain conductor and base combinations and are not suitable for breadboard or prototype work. Certain other advantages associated mainly with the individual process details exist for some of the mechanical methods. For instance, the sintered-conductor method provides a conductor-insulator bond possessing excellent resistance to thermal, chemical, and mechanical action. In the case of flame-sprayed wiring, any base material may be used, including the actual metal-equipment chassis or drawer (with a suitable sprayed insulation), and this fact also enhances the heat-transfer characteristics of the printed-wiring media.

Double-sided Boards. As far as fabrication trade-offs are concerned, the double-sided board is in a hybrid class between single-sided boards and multilayer boards. The basic board-fabrication process, i.e., definition of the circuitry pattern, is the same as that found with single-sided boards. The main difference is that double-sided boards, like multilayer boards, require some technique or method of interconnecting circuitry between sides. This is simpler for double-sided than for multilayer boards; for this reason the following trade-off discussion of double-sided boards will be concerned with methods of establishing the interfacial connection only.

Although double-sided boards were originally used with mechanical interfacial connections similar to the type illustrated in Fig. 8, it soon became evident that the plated-through hole had much more to offer in terms of bulk-processing potential and, as processes improved, reliability. Much of the present reliability of plated-through holes for double-sided boards may be attributed to new materials, more rigid process control, and the mechanized nature of the process. Although the hole may actually be used as it is, it is usually filled with solder, providing a redundant electrical connection not always found in the mechanical interfacial connections. Should the plated-through hole become defective, it is simply repaired by inserting, clinching, and soldering a jumper wire.

Many industries still favor the mechanical interconnection methods for a number of reasons, not all associated with the relative reliability of the plated-through hole. The most simple of these connections is the clinched jumper wire (Fig. 8).

Although a wire jumper is used for a via hole, it is easy to see that a component lead, serving a dual purpose, could be used just as well. As in most of the mechanical processes, the integrity of the joint depends almost entirely upon the quality of the solder joint. Although fused-in-place eyelets (Fig. 8*d*) have been used successfully in epoxy-glass wiring boards they were found to be subject to cracking in paper-phenolic boards. The most recent entry into mechanical-type interfacial connections is the compliant-redundant connection, Fig. 8*e*. This technique is the most reliable of all of the mechanical types and is recommended where plated holes are not to be employed. However, it suffers from the same disadvantage characterized by all the mechanical interfacial connections; it is not a bulk-processing technique, and careful consideration must be given to any advantages it may have over plated-through holes if the economics are to be justified.

Multilayer Printed-wiring Boards. It is the intention of the author to cover here not only the trade-offs which must be considered in multilayer printed wiring in general but also the advantages and disadvantages of the various types of printed wiring currently employed in the industry. Multilayer printed wiring was a natural extension of double-sided printed wiring; the density and crossover capability were not available and additional layers of wiring were laminated to provide a functional but monolithic wiring plane. In general, multilayer printed-wiring boards may reduce space and weight, provide for greater density of plug-in components, and assure enhanced reliability and cost savings resulting from closely controlled mass processing.

Specifically, the pertinent trade-offs applying to printed wiring as an interconnection technique are as follows:

Advantages:

1. It provides a wiring-density capability not matched by any other method.
2. Once a master artwork has been checked, any number of boards may be produced which are exactly identical, not only in wire routing but also in electrical characteristics.
3. As with other printed-wiring boards, structural support of components and thermal-management assistance are provided.
4. Multilayer circuits may be used to run wiring to and from conventional connectors where there is no space for conventional harnessing.
5. The incorporation of the ground plane in conjunction with closely controlled dielectric thickness between layers and exact conductor width provides transmission-line capability, yielding controlled-impedance lines and crosstalk protection.
6. Separate planes of a multilayer board may be used for signal lines, ground planes, or power bus, thereby providing design flexibility.

Disadvantages:

1. At the present time there are no satisfactory methods for repairing the internal layers of a multilayer printed-wiring board.
2. Considerable artwork and tooling costs are necessary on prototype or short-run production.
3. Inner-layer connections are susceptible to damage when components are soldered or unsoldered.
4. A simple design change may cause an entire board to require costly artwork change and manufacture or be scrapped altogether.

Certainly the many multilayer-board-fabrication processes described early in this chapter are not all equally suitable for a specific application. Their individual constructional features make them particularly adaptable for some jobs and unsuitable for others. The following are some advantages of the techniques previously discussed.

THE CLEARANCE-HOLE MULTILAYER BOARD (FIG. 9)

Advantages:

1. For short-run production costs are lower than costs for plated-hole techniques.
2. Modules can be unsoldered for repair or replaced with less danger of damage to the board than with plated-hole techniques.
3. No interlayer connections depend on critical processes as is so with plated-hole techniques, thereby resulting in greater reliability.

4. All component connections to internal wiring can be visually inspected from the surface of the board.

Disadvantages:

1. A restriction on the density capability exists in the minimum distance between the holes on the board; currently this is 0.075 in.
2. It is difficult to use conductors in conjunction with ground planes for constant impedance for high-frequency applications.
3. The maximum number of layers is six to eight, which is insufficient for many applications.
4. There is no known method of reliably making interlayer connections at any one mounting hole, thereby limiting the number of possible layouts.
5. The large clearance holes required reduce the available wiring area on each layer above the layer to which the termination is required.
6. For long-run production, costs are higher than costs for plated-hole techniques.
7. Modifying designs for future production is more difficult than with plated-hole techniques owing to lack of design flexibility.

THE PLATED-THROUGH-HOLE MULTILAYER BOARD (FIG. 10)

Advantages:

1. This technique has a desirable density capability permitting hole centers of 0.050 in., which is satisfactory for miniaturized applications.
2. This technique incorporates the ability to use ground and shielding planes in printed-circuit form.
3. Transmission-line properties can be incorporated using strip-transmission-line techniques.
4. The method can be modified to provide for soldering, welding, or wire wrapping of component leads.
5. The maximum number of layers is 30, which is more than enough for most applications.
6. Production costs are low for long-run production owing to mass production of plated holes.
7. Several interlayer interconnections are provided at one hole, thereby allowing greater circuit density on fewer layers than clearance-hole techniques.

Disadvantages:

1. Interlayer connections are susceptible to damage when modules are unsoldered for repair or replacement.
2. The reliability of interlayer connection may be questioned because it depends upon a butt connection between the copper conductor at each layer in the hole and the plated copper, a critical process which is difficult to control.
3. Large tolerance of thickness of boards with a large number of layers results because the tolerances of layer are additive.
4. Connections between any two layers requires extension of the connection (plated hole) to both external surfaces, cutting down on the possibility of connections between other layers at that point and reducing further possible miniaturization.
5. Minimum hole size is approximately one-quarter of board thickness, or 0.018 in., whichever is larger owing to state-of-the-art plating capability.
6. Repair of faults in inner layer or connections between layers is not practical and often impossible.

THE OPTICAL-CHEMICAL BLANKING PROCESS (FIG. 11)

Advantages:

1. Very high reliabilities are possible in boards with no interconnections. High interconnection reliabilities are possible using welding between the layers.
2. Heat sinks with thick metal layers can be incorporated into the board.
3. Many materials may be used for the conductors; among these are the weldable alloys of nickel.
4. The process is very flexible, and many modifications of the technique are available for special applications, such as those for welding and wire wrap.
5. There are no interconnections at component holes to be damaged when modules are unsoldered for repair or replacement.

6. Connections between two layers can be made at a point other than the component hole and can be accomplished by welding or soldering. Interlayer connections need not extend to the surface on both sides.
7. Repair of interlayer connections is possible if the connections are on the surface of the board.
8. There is no minimum hole size.
9. Locating slots for modules can be incorporated into the board.

Disadvantages:
1. Production costs are very high owing to both the great number of steps in the process and fact that some steps are hard-to-control.
2. No interconnections are provided for at the hole, severely reducing the circuitry density possible.
3. There is an extremely limited source of supply for boards of this type.

THE BRAZED TUBELET (FIG. 12)

Advantage:
1. This process is suitable for interfacing between rigid- and flexible-printed wiring media.

Disadvantages:
1. The potential reliability is very poor owing to the small area of solder connection between the pad ring and eyelet at each layer.
2. Interlayer connection is more subject to damage when modules are removed because the interlayer solder connection is liquefied every time the component lead and pad solder connection is liquefied; thus the possibility of a break or cold solder joint is increased.
3. If a different fusing metal than solder is used, glass-epoxy-board material is not practical.

THE TRANSFER PROCESS (FIG. 13)

Advantages:
1. There is no minimum hole size.
2. This process provides for blind subsurface layer-to-layer connections, increasing the miniaturization capability.
3. Blind subsurface interconnections increase the design flexibility, increasing the ability to make modifications in existing designs for future production.
4. Interlayer connection is less subject to damage when modules are unsoldered for repair or replacement than with plated-hole boards.
5. Tolerance on total thickness of board is not as difficult to control as in laminated-board processes because tolerances of individual layers can be compensated for.

Disadvantages:
1. Reliability of the board is questionable. The screened insulation may have pinholes, with possible short circuiting between layers.
2. Costs are high owing to long fabrication process in production quantities.
3. The number of materials which can be used is limited, restricting design applications.
4. The exposed layer of circuitry has poor peel strength.
5. The maximum number of layers is 12, which is not satisfactory for some applications.
6. A hookup or reinforcing board must usually be added to give the assembly structural strength.

THE BECK PROCESS (FIG. 14)

Advantages:
1. There is no minimum hole size.
2. The process provides for blind, subsurface layer-to-layer connections, increasing the miniaturization capability.
3. Many insulation materials can be used, such as ceramic, for high-temperature operation.
4. Blind subsurface interconnections increase the design flexibility, increasing the ability to make modifications in existing designs for future production.
5. Tolerance on the total thickness of the board is not as difficult to control because tolerance of individual layers can be compensated for.

Disadvantages:

1. Reliability of board is questionable. The screened insulation may have pinholes with possible short circuiting between layers.

2. Costs are high owing to the long fabrication process in production quantities.

3. The mechanical strength is less because glass cloth or other reinforcing materials cannot be used in the individual layers.

4. Transmission-line capability is restricted since maximum insulation thickness is less than 0.008 in. owing to plating characteristics.

THE METAL-SPRAY PROCESS (FIG. 15)

Advantages:

1. The process provides for blind subsurface interconnections which increase design flexibility, increasing the miniaturization capability.

2. Interlayer connections are not subject to damage when modules are unsoldered for repair or replacement because they are not located at the component holes.

3. Reliability of interlayer connection should be higher; this, owing to insufficient data, however, has not been proved.

4. Blind subsurface connections increase the design flexibility, increasing the ability to make modifications in existing designs for future production.

5. Large number of materials can be used for conductor paths, resulting in more flexibility of application.

6. The process may be applied directly to equipment chassis.

Disadvantages:

1. Transmission-line properties cannot be easily incorporated owing to roughness of surfaces of substrate, affecting distance between the signal line and ground plane.

2. Reliability of the board is questionable. The insulation may have pinholes, with possible short circuiting between layers.

3. Line-width control is sloppy.

4. Not suitable for edge-board connection schemes.

A trade-off matrix on multilayer printed-wiring processes may be found in Table 40 (see also Fig. 68). This table, like the previous discussion on advantages and disadvantages, is useful in helping the designer to examine his application critically in order to select the most optimum technique.

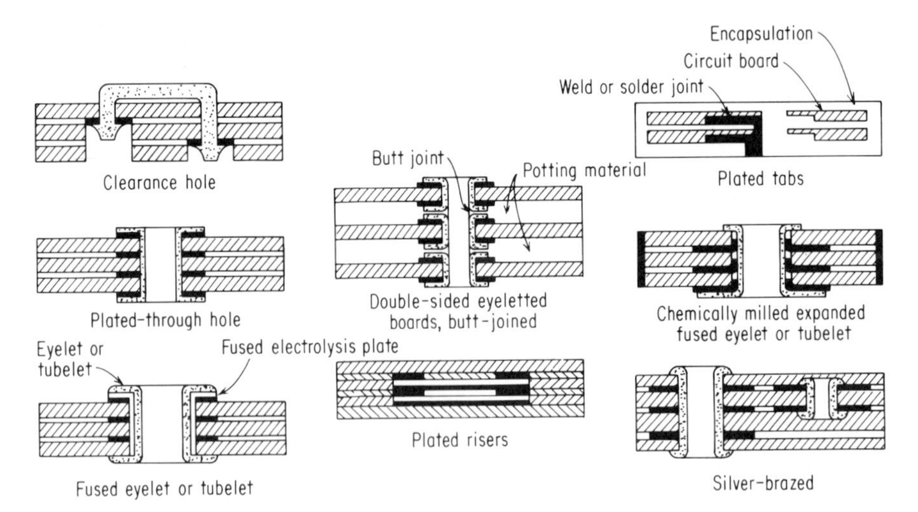

FIG. 68. Multilayer technique comparison.[31]

Component-attachment Techniques. At the present time, the two most commonly used methods for attaching components to printed-wiring boards are soldering and welding. Details on these two processes will be presented in Chaps. 3 and 4. It is important here, however, to emphasize the particular details of component attachment which are affected by the type of interconnection plane utilized as well as by the attachment configuration selected.

Before considering some of these connection techniques in detail, it may be well to examine briefly the pros and cons of soldering and welding as they apply to printed wiring. From the viewpoint of miniaturization, welding can be said to offer greater potential since weld joints can be formed closer to the device body without danger of thermal damage. Examining the cost aspect, it is seen that welding at this time is not amenable to bulk-processing techniques; therefore, the advantage must go to soldering, where the cost savings of dip-soldering techniques are readily apparent. Much reliability information is available supporting the integrity of soldered joints; this has been documented in the course of typical large-volume-production contracts such as Minuteman. Reliable soldering depends primarily on the solderability of the surfaces to be joined; this singular parameter is contrasted by the multiplicity of parameters requiring close control for reliable welding. The nature of soldering permits visual inspection of the joint. The welding process, however, basically holds promise of much greater reliability; the reliability of any joint tends to increase as the joint becomes more permanent. This very fact, however, leads us to maintainability considerations; the welded joint is impractical to repair at any level; solder-joint repair techniques are well developed. Welding, of course, requires a weldable surface; this rules out copper foil for the most part, necessitating the use of nickel, which is a relatively poor conductor, or Kovar, with its attendant plating problems.

In comparing printed-wiring techniques as they accommodate various attachment configurations, the overall connection system must be examined, i.e., the metal-joining method and configuration and the printed-wiring media and its relationship toward satisfying system-packaging goals. The trade-offs which influence the component printed-wiring interface, that is, the attachment technique, may well be the most important in the entire packaging system since this technique has been determined to be the most critical junction in the entire interconnection network. In the selection of an attachment technique, i.e., soldering, welding, etc., the reliability-maintainability trade-off must be considered. Experience has indicated that the reliability of a connection improves in direct relationship to its permanency. Therefore, the maintainability at the device level is negatively affected by any attempts to make the device attachment more permanent.

Soldered-attachment Configurations. The following is a brief explanation and trade-off analysis of the three attachment configurations which account for the majority of printed-wiring soldering at this time.

STRAIGHT-THROUGH ATTACHMENT (FIG. 42*a*). With this type of attachment, the actual connection is accomplished by some form of dip soldering, which is basically the application of molten solder to the workpiece. The straight-through approach is obviously an effective means for effecting a mass-interconnection program, where many components can be attached to a board at one time. For effective implementation, however, some method must be devised for forming leads and positioning and holding the device during the soldering operation. Attention is called to the mounting criteria previously discussed and to the lead-hole dimensional differences if this attachment technique is to be considered. A plated-through hole is recommended for this procedure; otherwise clinching should be used.

CLINCHED (LEAD) ATTACHMENT (FIG. 42*b*). A plain hole or a plated-through hole may be used. Similar to the straight-through approach, the clinched configurations feature a bend in the wire to ensure mechanical attachment prior to the soldering operation. The clinching operation is usually accomplished by "wiping" all the protruding leads in the same direction, and the printed-wiring-pad configuration will correspond to provide coverage in the proper areas. The offset

TABLE 40. Printed-wiring Trade-offs [31]

Characteristic	Clearance hole	Plated-through hole	Fused eyelet or tubelet	Double-sided eyeletted boards, butt-joined	Plated risers	Plated tabs	Chemically milled, expanded fused eyelet	Silver-brazed
Reliability of joint	Excellent	Fair	Good	Poor	Excellent	Excellent	Good	Good
Single- or double-sided circuits	Single	Both	Both	Both	NA	NA	Both	Both
Soldered circuits	Easy	Easy	Easy	Easy	NA	Easy	Easy	NA
Welded circuits	Easy	Hard	Hard	Hard	NA	Easy	Difficult	NA
Number of layers (max.)	8	20	10	No limit	8	10	10	No limit
Registration	Not critical	Not critical	Not critical	Not critical	Critical	Not critical	Not critical	Critical
Hole drilling	Not critical	Not critical	Not critical	Not critical	NA	NA	NA	Critical
Mechanical strength	Good	Good	Good	Poor	Fair	Fair	Good	Good
Environmental resistance	Poor	Fair; good if conformal-coated	Fair; good if conformal-coated	Poor	Excellent	Good	Fair; good if conformal coated	Fair to excellent
Number of process steps	Few	Many	Few	Few	Many	Few	Many	Few
Packaging density	Poor	Good	Good	Poor	Excellent	Good	Good	Fair to excellent
Weight	Heavy	Light	Medium	Heavy	Light	Light	Medium	Light to medium
Thickness	Thick	Thin	Medium	Thick	Thin	Thin	Medium	Thin to medium
Sockets or added terminals	Easy	Easy	Hard	Hard	Hard	Not possible	Hard	Easy
Flexible circuits	Easy	Easy	Easy	Hard	Hard	Hard	Hard	Easy
Design changes	Easy	Hard	Hard	Easy	Hard	Hard	Hard	Hard
Cost	Low	Low	Low	Low	High	High	High	Low
Production time	Low	Low	Low	Low	High	High	High	Low
Automation possible	Yes	Yes	Yes	Yes	Yes	No	Yes	Yes
Equipment readily available	Yes	No	Yes	Yes	No	Yes	No	Yes
Visual inspection	Easy	Not possible	Hard	Easy	Easy	Easy	Hard	Hard
Repair	Easy	Hard	Moderately hard	Easy	Hard	Easy	Moderately hard	Hard
Proprietary process	No	Yes	No	No	Yes	Yes	Yes	Yes

variation evolved as a means of making the clinched soldered joint more maintainable and is recommended for multiple lead components such as integrated circuits where the greater lead length is available and in systems where the extra lead length will not adversely affect performance. The printed-wiring substrate and foil bond should, as in the above case, withstand the temperature of flow soldering. The basic advantages of the clinched lead (Fig. 42*b*) are that mechanical support is provided and the electrical connection is more subject to straightforward visual inspection. Although the clinching procedure does eliminate the need for any jigs or holding fixtures for the components during the soldering operation, it also involves an extra mechanical operation, which may add to the cost. The clinched types also require slightly more board area for pads and serve to negate absolute-density capability. The maintenance action for the in-line clinched configuration is somewhat complicated by the fact that the lead wire must be completely straight to be withdrawn.

SURFACE ATTACHMENT (FIG. 42*d*). At the present time commonly used in the attachment of all types of components to be printed-wiring boards, this configuration is accomplished by reflow or conventional soldering. This is the only soldered attachment which will accommodate the placement of components on both sides of the wiring plane and which also features accessibility of the device and joint area on the same side of the printed-wiring board. This configuration is more adaptable to use with the flat-pack integrated circuit than the Joint Electron Device Engineering Council outline; the latter may require an elaborate lead-forming device to facilitate application to the board surface. This is the most easily maintainable of all the soldered-attachment configurations. Here the connection is made on the component-mounting side of the interconnection plane. The lead is attached by soldering to form a "lap" connection. The solder is applied by conventional methods or reflowed by application of heat. Device removal is accomplished by heating the junction and individually removing the leads.

Already very popular as a method for attaching miniaturized components such as integrated circuits, the surface-soldered connection is well known, has much published reliability information, and provides the ready maintainability so often required for integrated-circuit application. It is often used for a breadboard requirement where the device must be used again. It lends itself to maximum-density packaging applications since devices can be attached to both sides of the board and can be used with virtually any type of interconnection media. For maintenance action where the original connection was made by another technique, this is often useful for attaching the replacement package.

Welded-attachment Techniques. The basic problem in welding to printed-circuit boards is that a weldable material must be provided; copper, since it is such a good conductor of heat, is not usually used because of the difficulty in determining a suitable, versatile, and workable weld schedule. Although the post-attachment method is effective, it is restrictive in view of the few interconnection planes offering this configuration.

SURFACE (POST) ATTACHMENT (FIG. 43*a*). The basic welding process here is axial-electrode or crossed-wire welding as defined above. With this method of attachment, the component lead is welded to the butt, or top portion, of a post which is connected to the interconnection media. In implementing this attachment type, it is important that the materials be compatible with regard to composition, surface-finish, and bulk properties. The actual operation can be performed by a number of commercially available equipments. It is also possible to utilize parallel-gap welding in this configuration. Current reliability information indicates that the crossed-wire weld is the more reliable of the two. Surface (post) attachment is most adaptable to attaching components to interconnection media which present a post as an integral part of the media (such as the welded-wire matrix). The post attachment is also advantageous if maintenance action is anticipated since the lead can be taken off the post and the connection area resurfaced.

In application then, this attachment technique is best-suited to interconnection media which by nature present a weldable pin for connection. It appears to be a good method, for instance, for attaching components to pins of welded-wire planes. For dense packaging applications, this presents the best approach since a minimum lead length is required to make the connection; however, this approach is offset by the fact that components can be attached to one side of the plane only. Although soldering could be used for initial attachment and/or repair, caution should be exercised that the post material is either compatible with a soldering operation or can be treated to prepare for it.

SURFACE (PAD) ATTACHMENT (FIG. 43*b*). For practical reasons, the parallel-gap technique is most commonly used for this physical configuration. Since the other applicable methods such as ultrasonic, laser, and electron beam are all in various stages of development, they will not be considered here. In using this technique, the component lead is formed parallel to the interconnection terminal pad, both electrodes are applied to the lead, and the joint is made at the lead-pad interface halfway between the electrodes. Presently, the surface- (pad-) attachment configuration is used primarily for attachment of flat-pack integrated circuits to the interconnection media, utilizing the planar configuration. It is not likely that bulk-processing techniques can be used to reduce manufacturing costs for this method.

STRAIGHT-THROUGH (FIG. 43*c*). The straight-through welded connection features an inherent compatibility with crossed-wire welding which, supposedly, can be accomplished with greater consistency and thereby with improved reliability than can be accomplished with parallel-gap welding. Conventional interconnection media cannot be utilized with this concept unless they are modified to provide a tab extension to which the component lead can be attached. At least two products, currently commercially available, provide welded "tabs" which are inseparably associated with the interconnection medium. This method is used mostly for small "cordwood-" type modules and is characterized by its easy maintainability, which is accomplished by clipping the leads. It is not widely used.

MILITARY AND INDUSTRY GUIDES

Specifications are often written by an organization in order to qualify or set forth their exact needs in a product or service. These specifications, in turn, help the contracting agency, or buyer, to monitor or judge the acceptability of the material and/or services supplied, and put both the buyer and supplier on "common ground." For products which are ordered on a continual basis, or grouped in a general class, such as printed wiring, specifications and standards have been written which continue in force and may be used either above or supplemented by other documentation or written orders. The most common specifications written on printed wiring, as in many other fields, have been issued by governmental agencies on behalf of the military services. Industry associations also publish specifications and guides; this is particularly true in case of printed wiring. Industry-guide inputs come from manufacturers as well as users of printed wiring. Manufacturers are primarily concerned with regulating unfair competition and standardizing to foster industry acceptance and use of their product; users, on the other hand, are mostly interested in standardizing quality to assist in product assurance and acting as guides in application and design of in-house products.

Specifications and guides pertaining to printed wiring may be divided into three general categories: those specifying design of various types of printed wiring, those concerned mostly with material and acceptability criteria for same, and those setting forth the performance requirements of printed wiring and associated items.

Although it is not possible to include all of the applicable guides written on printed wiring, the most common are covered briefly below to acquaint the reader with this type of design and application assistance.

MIL-STD-275B: *Printed Wiring for Electronic Equipment.* This document is basically a design standard on single- and double-sided printed wiring with specific information on interfacial connections. The scope of this standard states: "This standard establishes design principles governing the fabrication of printed wiring, and the mounting of parts and assemblies thereon, used in electronic equipment. It does not apply to printed wiring boards used within microelectronic assemblies, multilayer and flexible wiring used in electronic equipment. The requirements do not apply to the fabrication of parts, such as resistors, inductors, capacitors or transmission lines fabricated using these techniques."

This standard specifies general requirements for the design of printed wiring, including dimensioning procedure and the master-pattern format as well as detail requirements delineating the conductive pattern configuration, recommended conductor widths and spaces, and design rules and limitations for seven types of interfacial connections. Assembly of components on the printed-wiring board is also covered, including mounting provisions, part spacing, and connectors. Finally, this standard spells out the recommended materials to use as well as platings and coatings.

IPC-ML-910: *Rigid, Multilayered Printed-wiring Boards.* This document is issued by the Institute of Printed Circuits, which is an industry association comprised of companies that are manufacturers as well as users of printed wiring. This is primarily a design specification for rigid multilayer printed wiring.

Specifying requirements for the design and construction of these boards, guidelines include information on materials (base material, bonding agent, and plating), and dimensional requirements (board thickness, conductor dimensions, spacing, terminal areas, hole diameter, and location and plating thickness). Brief requirements on acceptability are given, as well as packaging needs. Useful tables concerned with suggested dimensions and practical tolerances are included in this specification.

MIL-P-13949: *Plastic Sheet, Laminated, Copper Clad (for Printed Wiring).*[7] In this specification the scope covers "the requirements for copper-clad, laminated, plastic sheets (paper base and glass base), to be used primarily for the fabrication of printed wiring for electrical and electronic circuits." It includes laminates of nine different types, employing either paper or glass reinforcement and either epoxy or polytetrafluoroethylene resins.

Detailed requirements for the appearance and physical characteristics of the foil are given (dimensions, tolerances, warp and twist, uniformity, machinability, and workmanship). Very complete tables on mechanical and electrical properties are included. A major portion of this specification, given over to quality assurance provisions, is complete with recommended sampling plans and test methods and specimens. Although it is primarily a material-acceptance document, designers may find useful test methods detailed within enabling them to follow the authorized procedure to determine physical properties of their intended material.

IPC-CF-150: *Copper Foil for Printed Circuit Applications.*[11] Also issued by the Institute of Printed Circuits, this standard includes both electrodeposited and rolled foil suitable for circuit applications. This is basically an acceptance document, giving information on mechanical and electrical qualities, finish, copper defects, peel strength, inclusions, and solderability. Standard procedures for packing and marking are also outlined to assure uniformity within the industry on these important procedures.

MIL-I-46058: *Insulating Compound, Electrical (for Coating Printed Circuit Assemblies).* This specification covers three classes of uniformal coatings which are suitable for application to printed-circuit assemblies by dipping, brushing, and spraying procedures designed to ensure satisfactory performance over long storage periods and also under conditions of high humidity. Including four types of materials (epoxies, polyurethanes, silicones, and polystyrenes), this standard sets forth requirements, mechanical and electrical, and enters into some detail with regard to the quality assurance provisions governing acceptance of the material. A test pattern outlined to incorporate the qualification procedure and all tests included in the qualification inspection are spelled out in detail. Four different inspection and acceptance procedures, depending on the end-product requirements, are outlined as group A, B, C, or D inspection.

MIL-P-55110: *Printed-wiring Boards.* One of the most comprehensive and significant performance specifications currently invoked in the field of printed wiring, this specification covers "printed wiring boards consisting of a conductor pattern on the surface of one or two sides on an insulating base." Stating that the design features of the printed-wiring board shall conform to the requirements of MIL-STD-275, this standard delineates the detail requirements considered necessary to produce a printed-wiring board of acceptable quality. In addition to requirements relating to visual examination (holes, edges, surfaces, conductor patterns, and edges), standard acceptance tests for printed wiring are designated. These are continuity, thermal shock, dielectric withstanding voltage, bond strength, plating, moisture resistance, insulation resistance, peel strength, solderability, and workmanship. Three inspection groups are given: A, B, and C. The first two are acceptance tests to which all the boards are usually subjected, the last is essentially a qualification test used to qualify a production procedure initially and at periodic intervals during the production of a given design of boards. It is important to remember that MIL-P-55110 does not cover multilayer printed wiring.

IPC-FC-240: *Flexible Printed Wiring.* The scope of this standard states: "This specification establishes the requirements for the manufacture, performance, and inspection of flexible printed wiring, consisting of a conductor pattern on a flexible insulating base. A flexible, insulating cover sheet may be applied over the conductors, and conductors may be laminated to both surfaces of the base." This document covers seven types of currently used base materials and specifies the type and standards of copper foil to be used for cladding. Under quality assurance provisions, this specification delineates a preproduction inspection which is identical with group B inspection; this is intended to ensure continual qualification of the process. These group B tests include insulation resistance, peel strength, shrinkage, flexure, thermal shock, solder dip, vibration, solderability, and dielectric strength. Test specimens and fixtures are described and illustrated.

IPC-ML-950: *Multilayer Printed Circuit Boards.*[17] This Institute of Printed Circuits document is the result of a concerted effort on the part of representatives from many organizations, industrial, academic, and governmental, to produce a workable performance specification for multilayer boards. It establishes "qualifications and acceptance requirements" for rigid, multilayer printed-circuit boards, and designates three categories of decreasing severity, A, B, and C. The specification illustrates the recommended six-layer board pattern designed to provide the most suitable samples for consistent testing with the procedures given.

The detailed test requirements (many of which set forth class A, B, or C levels) are warp and twist, hot-oil test, plating adhesion, shock and vibration, flexural strength, machinability, water absorption, flammability, terminal pull, interconnection resistance, insulation resistance within and between layers, dielectric withstanding voltage, current-carrying capacity, internal short circuits, and cross sectioning. The recommended test-board design has designated sections for individual tests, and a testing summary and sequence are given. Because of the functional board design and test-sequence arrangement, it is possible to run several of the tests concurrently.

REFERENCES

1. Schlabach, T. D., and D. K. Rider: "Printed and Integrated Circuitry," McGraw-Hill Book Company, New York, 1963.
2. Bell, R. M.: *Electronics,* vol. 30, pp. 266, 268, 270, 272, and 274, 1957.
3. Eisler, P.: "The Technology of Printed Circuits," Academic Press, Inc., New York, 1959.
4. Schjeldahl, G. T.: Technical bulletin entitled "Schjeldahl Flexible Electrical Laminates Handbook."
5. *Insulation Directory/Encyclopedia Issue,* vol. 12, no. 6, June/July, 1967.
6. Hughes Aircraft Co.: Technical bulletin entitled "Design Guide, Flat Conductor Cable Flexible Cable," 1963.
7. MIL-P-13949 "Plastic Sheet, Laminated Copper Clad (for Printed Wiring)."
8. MIL-STD-275 "Printed Wiring."

9. Kohl, W. H.: "Materials and Techniques for Electron Tubes," Reinhold Publishing Corporation, New York, 1960.
10. MIL-I-46058 "Insulating Compound, Electrical (for Coating Printed Circuit Assemblies)."
11. IPC-CF-150 "Copper Foil for Printed Circuit Applications."
12. LP 1-1966 "NEMA Standards for Industrial Laminated Thermosetting Products."
13. ASTM D 1867 "Tentative Specifications for Copper-clad Thermoset Laminates for Printed Wiring."
14. Spaulding Fiber Co.: "Product Bulletin," 1966.
15. Gray, A. G.: "Modern Electroplating," John Wiley & Sons, Inc., New York, 1953.
16. Graham, A. K.: "Electroplating Engineering Handbook," Reinhold Publishing Corporation, New York, 1955.
17. IPC-ML-950 "Performance Specification, Multilayer Printed Circuit Boards," Institute of Printed Circuits, Chicago, Ill., 1966.
18. Young, M. G.: How Temperature Affects Plastic Laminates, *Prod. Eng.*, July, 1964, pp. 67–72.
19. Sanders Associates: Technical bulletin FT-169 entitled "Flexprint Circuit Design Handbook."
20. Geshner, R. A., and Messner, G., Institute of Printed Circuits: "IPC Multilayer Printed Circuit Boards Technical Manual."
21. Kaupp, H. R.: Pulse Crosstalk between Microstrip Transmission Lines, *Intern. Electronic Circuit Packaging Symp.*, Los Angeles, Calif., 1966.
22. Garth, E. C., and I. Catt: Ultrahigh-speed IC's Require Shorter, Faster Interconnections, *Electronics,* July 1966, pp. 103–110.
23. Keonjian, E.: "Microelectronics," McGraw-Hill Book Company, New York, 1963.
24. Szukalski, E. A., and D. P. Schnorr: Designing for Maintainability in Integrated Circuit Packaging, *Electronic Packaging and Production,* December, 1964, pp. 40-48.
25. Schnorr, D. P.: Factors Influencing the Selection and Use of Integrated Circuit Packages, *Elec. Design News (Packaging Reference Issue),* October, 1964.
26. IPC Connector Information Committee: "Printed Circuit Connector Availability Chart," Institute of Printed Circuits, Chicago, Ill., 1964.
27. Sanders Associates: "Flexprint Wiring Handbook of Design Information."
28. MIL-STD-429 "Printed Circuit Terms and Definitions."
29. IPC-D-300 "Dimensional Tolerances for Printed Circuit Boards," Institute of Printed Circuits, Chicago, Ill., 1966.
30. NAS 729 "Cable, Electrical, Flat Conductor, Flexible, 300 Volts, Copper," Aerospace Industries Association of America, Inc., Washington, D.C., 1963.
31. Rigling, W. S.: Designing and Making Multilayer Printed Circuits, *Electro-Technology,* May, 1966, pp. 54–57.

Chapter 2

WIRES AND CABLES

By

ARTHUR G. SCHUH

Materials Engineering Laboratories
Martin Marietta Corporation
Orlando, Florida

INTRODUCTION

The selection of wires or cables is often neglected within the total design concept. Hookup and interconnection wiring are frequently the last areas of consideration. More and more, engineers realize the importance of selecting conductors, insulation, shielding, jacketing, and cabling that offer the best combination of size, weight, environmental protection, and handling resistance, coupled with lowest cost and ease of availability and maintenance. If the intended environment is abnormal, proper wire and cable selection should be verified by adequate environmental testing for complete success in application. In addition, the designer must adequately evaluate the environment and mechanical stress and consider the compatability of his materials with all possible encapsulants, fuels, chemicals, explosives, and the new and varied types of associated hardware and fabrication techniques.

The purpose of this section is to provide the designer with sufficient background information and guideposts for proper wire and cable selection. Definitions of various terms which will be used appear in Table 1.

TABLE 1. Terms and Definitions

Abrasion resistance. Ability to resist surface wear.

Aging. The change in properties of a material with time under given conditions.

Ambient temperature. The temperature of the surrounding atmosphere.

Attenuation. The power or signal loss in a circuit, expressed in decibels (db).

AWG (American Wire Gauge). A standard for copper wire sizes. The diameters of successive sizes vary in geometric progression.

Braid. A woven outer covering. It may be composed of any filamentary material such as fiber, plastic, or metal.

Braid angle. The angle between the axis of cable and the axis of any one member or strand of the braid.

Breakdown. A disruptive discharge through insulation.

Bunch lay. The twisting of strand members in the same direction without regard to geometrical arrangement.

Capacity. That property of a system of conductors and dielectrics which permits the storage of electricity when potential differences exist between conductors. Its value is expressed as the ratio of a quantity of electricity to a potential difference.

Coat. To cover with a finishing, protecting, or enclosing layer of any compound.

Coax. Abbreviation for coaxial cable.

Concentric lay cable. A concentric lay cable is composed of a central core surrounded by one or more layers of helically wound insulated conductors.

Concentric lay stranding. Composed of a central core surrounded by one or more layers of helically laid wires.

Conductor. A slender rod or filament of metal or group of such rods or filaments not insulated from one another, suitable for transmitting an electric current.

Corona. A luminous discharge caused by the ionization of the gas surrounding a conductor around which exists a voltage gradient exceeding a certain critical value.

Corona resistance. The time for which insulation will withstand a specified level field of intensified ionization that does not result in the immediate complete breakdown of the insulation.

Crosslinking. The setting up of chemical bonds between molecular chains.

Crosstalk. Undesirable electromagnetic coupling between adjacent signal-carrying conductor pairs.

Cut-through resistance. Resistance of a solid material to penetration by an object under conditions of pressure, temperature, etc.

Decibel (db). Unit employed to express the ratio between two amounts of power, voltage, or current between two points.

Density. Weight per unit volume of a substance.

Dielectric. A nonconducting material or a medium having the property that energy required to establish an electric field is recoverable, in whole or in part, as electric energy.

Dielectric constant. That property of a dielectric which determines the electrostatic energy stored per unit volume for unit potential gradient.

Dielectric loss. The time rate at which electric energy is transformed into heat in a dielectric when it is subjected to a charging electric field.

Dielectric-loss factor. The product of its dielectric constant and the tangent of its dielectric-loss angle.

Dielectric strength. The voltage which an insulating material can withstand before breakdown occurs, usually expressed as a voltage gradient (such as volts per mil).

Direction of lay. The direction in which individual members of a multiconductor cable or stranded conductor spiral over the top of the cable in a direction going away from an observer who is standing behind the twisting device.

Drain wire. An insulated, stranded or solid conductor which is located directely under and in intimate contact with a shield.

Elastomer. A material which at room temperature stretches under low stress to at least twice its length and snaps back to its original length upon release of stress.

Elongation. The fractional increase in length of a material stressed in tension.

Extrusion. Compacting a natural or synthetic material and forcing it through an orifice in a continuous fashion.

Farad. Unit of capacitance. The capacitance of a capacitor which, when charged with a coulomb, gives a difference of potential of one volt.

Filler (cable). Fillers are used in multiconductor cable to occupy space or interstices formed by the assembled conductors. Fillers are employed to obtain circularity.

Flame resistance. Ability of a material to extinguish flame once the source of heat is removed.

Flammability. Measure of a material's ability to support combustion.

Flex life. The life of a material when subjected to continuous bending.

Flexural strength. The strength of a material in bending.

Foamed insulation. Resins in flexible or rigid sponge form with cells closed or interconnected.

Hard-drawn. Refers to the temper of conductors that are drawn without annealing or that may harden in the drawing process.

Hydrolysis. Chemical decomposition of a substance involving the addition of water.

Hygroscopic. Tending to absorb moisture.

Impact resistance. Relative susceptibility of material to fracture by shock.

Impedance. The apparent resistance in a circuit to the flow of an alternating current, analogous to the actual resistance to a direct current.

Insulation resistance. The insulation resistance of an insulated conductor is the electrical resistance offered by its insulation to an impressed current tending to produce a leakage current. Normally measured in megohms per 1,000 ft for insulated wire.

Insulator. A material of such low electrical conductivity that the flow of current through it can usually be neglected.

Interstice. A minute space between one thing and another, especially between things closely set or between the parts of a body.

Jacket. A protective sheath or outer covering applied over an insulated conductor or cable.

Layer. Consecutive turns of a coil lying in a single plane.

Lay length. The lay length of any helically wound strand or insulated conductor is the axial length of one turn of the helix.

Magnet wire. Insulated wire intended for use in windings on motor and transformer coils.

Migration of plasticizer. Loss of plasticizer from an elastomeric plastic compound, with subsequent absorption by an adjacent medium of lower plasticizer concentration.

Moisture resistance. The ability of a material to resist absorbing moisture from the air or when immersed in water.

Nylon. The generic name for synthetic fiber-forming polyamides.

Ohm. Unit of electrical resistance. Resistance of a circuit in which a potential of one volt produces a current of one ampere.

Ozone. A faintly blue, gaseous, allotropic form of oxygen, obtained by the silent discharge of electricity in ordinary oxygen or in air.

Plastic. High-polymeric substances, including both natural and synthetic products, but excluding the rubbers, that are capable of flowing under heat and pressure at one time or another.

Plasticizer. A chemical agent added to plastics to make them soft and more flexible.

Polyamide. A polymer in which the structural units are linked by amide or thiamide groupings.

Polychloroprene. The chemical name for neoprene, a polymer of chloroprene, a combination of vinyl acetylene and hydrogen chloride.

Polyethylene (PE). A thermoplastic material composed of the polymers of ethylene.

Polypropylene. A plastic made by the polymerization of high-purity propylene gas in the presence of an organometallic catalyst at relatively low pressures and temperatures.

Polyurethane. A copolymer of urethane similar in properties to neoprene.

Polyvinyl chloride (PVC). A family of thermoplastic insulating compounds composed of polymers of polyvinyl chloride or its copolymer, vinyl acetate, in combination with certain plasticizers, stabilizers, fillers, and pigments.

Polyvinylidene fluoride. A thermoplastic-resin crystalline high-molecular-weight polymer of vinylidene fluoride.

Primary insulation. A nonconductive material placed directly over a current-carrying conductor whose prime function is to act as an electrical barrier for the applied potential.

Quad. A four-conductor cable.

Relative humidity. Ratio of the quantity of water vapor present in the air to the quantity which would saturate it at any given temperature.

Resistance. Property of a conductor that determines the current produced by a given potential difference.

Resistivity. The ability of a material to resist passage of electrical current either through its cross section or on the surface. The unit of volume resistivity is the ohm-centimeter; of surface resistivity, the ohm.

rf. Abbreviation for the term "radio frequency." Usually considered the frequency spectrum above 10,000 hertz.

rms. Abbreviation for "root mean square." When the term is applied to voltages and currents, it means the effective value, i.e., that it produces the same heating effect as a direct current or voltage of the same magnitude.

Rope lay. In a rope-lay conductor or cable, stranded members are twisted together with a concentric lay; the stranded members themselves may have a bunched, concentric, or rope lay.

Serve. The spiral application of a material, as opposed to braid.

Shield. A metallic sheath placed around an insulated conductor or group of conductors to protect against extraneous currents and fields.

Shielded conductor. An insulated conductor which has been shielded to reject extraneous electrical fields.

Silicone. Polymeric materials in which the recurring chemical group contains silicon and oxygen atoms as links in the main chain.

Sintering. Forming articles from fusible powders at a temperature below melting point.

Solvent. A liquid substance which dissolves other substances.

Specific gravity. The density (mass per unit volume) of any material divided by that of water at a standard temperature.

Stabilizer. An ingredient used in some plastics to maintain physical and chemical properties throughout processing and service life.

TABLE 2. Bare and Coated Conductor Properties

Conductor material and coating	Tensile strength, psi	Elong, %	Flex. life	Conductivity min., %	Cont. oper. temp. max., °C	Oxidation resist.	Galv. corrosion resist.	Solderability	Availability	Cost	Specifications
Bare copper (ann.)	34,000	10	Fair	100	150	Poor	Good	Fair	Good	Low(1)	QQ-W-343 ASTM B3
TC copper (ann.)	34,000	10	Fair	100	150	Good	Good	Good	Good	Low(2)	QQ-W-343 ASTM B33
SC copper	34,000	10	Fair	102	200	Good	Poor	Good	Good	Medium	ASTM B298
NC copper	34,000	10	Fair	96	260	Good	Good	Poor	Fair	Medium	ASTM B355
Aluminum	20,000–32,000	15	Poor	61	150	Poor	Good	Poor	Good	Low	MIL-W-7071 (insulated)
SC alum.	20,000–32,000	15	Poor	63	150	Good	Poor	Good	Fair	Medium	
Silver	25,000	15	Fair	104	260	Fair	Good	Good	Poor	High	
SC copper-clad steel (HD)	110,000–125,000	1	Poor	30–40	200	Good	Medium	Good	Fair	Medium	QQ-W-345
SC copper-clad steel (ann.)	55,000–65,000	8	Good	30–40	200	Good	Medium	Good	Fair	Medium	
SC cadmium copper (HD)	90,000	1	Poor	80	200	Good	Poor	Good	Fair	Medium	
SC cadmium copper (ann.)	55,000	5	Good	80	200	Good	Poor	Good	Fair	Medium	
SC zirconium	55,000	5	Good	85	200	Good	Poor	Good	Fair	Medium	
SC chrome (ann.)	55,000	6	Good	85	200	Good	Poor	Good	Fair	Medium	ASTM B268
SC cadmium-chrome	60,000	5	Good	84	200	Good	Poor	Good	Fair	Medium	

TC = tin-coated. SC = silver-coated. NC = nickel-coated. HD = hard-drawn. ann. = annealed.

Strand. A single metallic conductor.

Tape. A relatively narrow woven or cut strip of fabric, paper, or film material.

Tear strength. Force required to initiate or continue a tear in a material under specified conditions.

Tensile strength. The pulling stress required to break a given specimen.

Thermoplastic. A classification of resin that can be readily softened and resoftened by repeated heating.

Volt. Unit of electromotive force. It is the difference of potential required to make a current of one ampere flow through a resistance of one ohm.

Volume resistivity. The electrical resistance between opposite faces of a 1-cm cube of insulating material, commonly expressed in ohm-centimeters.

VSWR. The abbreviation for voltage standing-wave ratio, the ratio of voltage maximum to voltage minimum in a transmission line.

Water absorption. Ratio of the weight of water absorbed by a material to the weight of a dry material.

Wire. A conductor of round, square, or rectangular section, either bare or insulated.

Working voltage. The recommended maximum voltage of operation for an insulated conductor.

CONDUCTORS

Materials. Following is a brief discussion of conductor materials most widely used in the aerospace and electronics industry; Table 2 presents a summary of conductor materials and coatings.

Copper. ELECTROLYTIC TOUGH PITCH (ETP). ETP copper makes up the majority of conductor strands used in industry. ETP copper is controlled in composition, conductivity, and purity by ASTM B3. Properties tinned of individual strands, such as tensile strength, elongation, and dimensions, are covered by ASTM B33 or QQ-W-343.

OXYGEN-FREE HIGH CONDUCTIVITY (OFHC). OFHC differs from ETP copper in its fabrication. Electrolytic slabs are melted into bars for extrusion or rolling in an inert atmosphere excluding oxygen. OFHC copper, covered by ASTM B170, has improved properties at temperatures above 1000°F and is resistant to hydrogen embrittlement.

Copper Alloys. Pure copper has poor mechanical characteristics; breakage will occur from tensile pull, flexing, and vibration. As a result of progressive miniaturization, various high-strength copper alloys have been developed. Figure 1 presents a comparison of flex endurance between high-strength copper alloys and copper. Since drawing and annealing techniques used by different suppliers play a significant part in the final physical and electrical properties of high-strength alloys, the recommended practice is to specify performance requirements of the conductor rather than require a specific alloy. One exception to this practice would be the use of zirconium alloy for high-temperature applications.

Because of high annealing temperatures during processing, high-strength alloys are available only with high-temperature coatings such as silver or nickel. Tin-coated high-strength-alloy conductors are still under development. A description of the more popular high-strength alloys follows.

CADMIUM-COPPER ALLOY (CADMIUM BRONZE). This is used in fire wire applications for increased strength; ASTM B105 covers hard-drawn temper. No specification is available for temper application to high-strength-alloy conductors of higher elongation. This alloy is not recommended for temperature applications exceeding 400°F.

ZIRCONIUM-COPPER ALLOY. Required physical and electrical characteristics of the high-strength-alloy conductors as defined are difficult to meet with zirconium alloy; however, zirconium alloy shows best retention of physical and electrical properties at temperatures up to 400°C.

CHROMIUM-COPPER ALLOY. This provides the best conductivity of the high-strength-alloy conductors; however, it is the most difficult alloy to process consistently. Chrome-copper alloy is a precipitation-hardened alloy; if improperly heat-treated, the chromium is not retained in solution, causing inconsistency in physical properties.

CADMIUM-CHROMIUM-COPPER ALLOY. This most recent addition to high-strength alloys appears to present the most consistent and best physical properties, although minimum conductivity is slightly less than in chrome alloy.

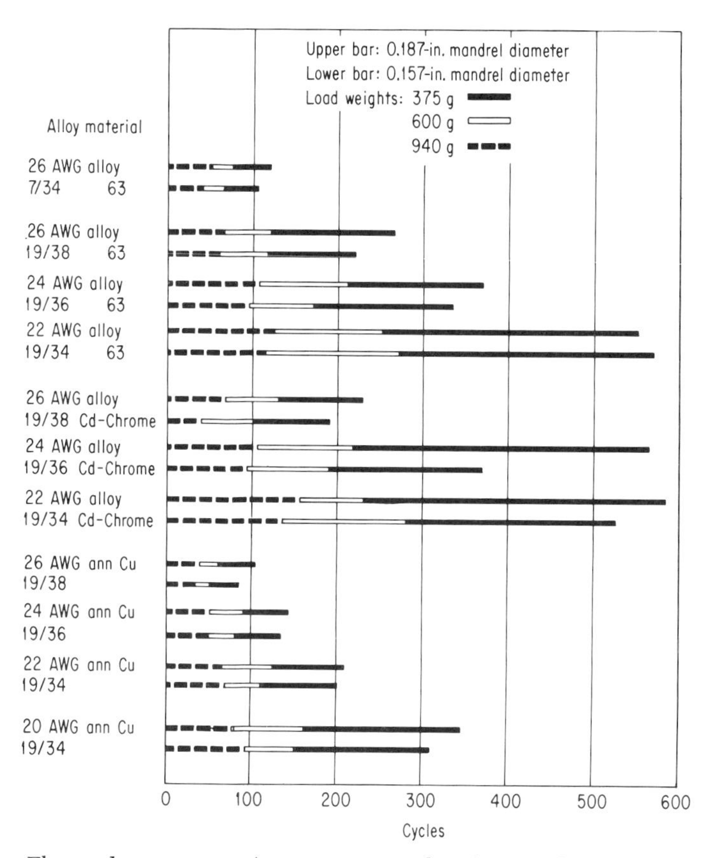

FIG. 1. Flex endurance tests (concentric strandings), annealed copper vs. high-strength alloys.[1]

Copper-covered Steel. This conductor is available in two consistencies: hard-drawn for high strength (125,000 to 150,000 psi) or annealed (50,000 to 80,000 psi) for greater ductility and two conductivities, 40 and 30 percent. At high frequencies, either 30 or 40 percent copper-covered steel has a conductance almost equivalent to that of copper; therefore, it is commonly used in coaxial cables. At low frequencies or with direct current, the conductance is 30 or 40 percent of an equivalent size of copper conductor. Table 3 presents the physical properties of the small sizes of copper-covered steel wire. Specifications for copper-covered steel wire are contained in QQ-W-345.

Aluminum. The major use of aluminum conductors has been in large-gauge power conductors. Aluminum offers a weight advantage of approximately 3:1, possessing approximately 60 percent conductance of an equivalent copper conductor. Compared with copper, weight savings of approximately 2:1 can be attained for

TABLE 3. Physical Properties of Copper-clad Steel Wire, Small Sizes [2]

Size, AWG	Nominal dia., in.	Weight		Average dc resistance, ohms/1,000 ft at 68°F		Nominal strength, lb				Size, AWG
						Annealed wire (soft), avg.		Hard-drawn wire, min.		
		lb/1,000 ft	ft/lb	40%	30%	40%	30%	40%	30%	
13	0.072	14.4	70	5.1	6.8	200	280	390	530	13
14	0.064	11.4	88	6.4	8.6	160	230	310	440	14
15	0.057	9.04	110	8.1	10.8	130	180	240	330	15
16	0.051	7.17	140	10.2	13.7	100	140	190	270	16
17	0.045	5.68	180	12.9	17.2	80	110	150	200	17
18	0.040	4.51	220	16.3	21.7	64	89	120	170	18
19	0.036	3.58	280	20.5	27.4	51	71	96	130	19
20	0.032	2.84	350	25.9	34.5	40	56	76	110	20
21	0.028	2.25	440	32.6	43.5	32	45	60	81	21
22	0.025	1.78	560	41.2	54.9	25	35	48	64	22
23	0.023	1.41	710	51.9	69.2	20	28	38	51	23
24	0.020	1.12	890	65.5	87.3	16	22	30	40	24
25	0.018	0.89	1,100	82.5	110	13	18	24	32	25
26	0.016	0.71	1,400	104	139	10	14	19	25	26
27	0.014	0.56	1,800	131	175	7.9	11	15	20	27
28	0.013	0.44	2,300	166	221	6.3	8.8	12	16	28
29	0.011	0.35	2,800	209	278	5.0	7.0	9.5	13	29
30	0.010	0.28	3,600	263	351	3.9	5.5	7.5	10	30
31	0.0089	0.22	4,500	332	442	3.1	4,4	5.9	7.9	31
32	0.0080	0.18	5.700	418	558	2.5	3.5	4.7	6.3	32
33	0.0071	0.14	7,200	528	703	2.0	2.8	3.7	5.0	33
34	0.0063	0.11	9,100	665	887	1.6	2.2	3.0	4.0	34
35	0.0056	0.87	11,000	839	1,120	1.2	1.7	2.4	3.1	35
36	0.0050	0.069	14,000	1,060	1,410	1.0	1.4	1.9	2.5	36
37	0.0045	0.055	18,000	1,330	1,780	0.8	1.1	1.5	2.0	37
38	0.0040	0.044	23,000	1,680	2,240	0.6	0.9	1.2	1.6	38
39	0.0035	0.035	29,000	2,120	2,830	0.5	0.7	0.9	1.2	39
40	0.0031	0.027	36,000	2,670	3,570	0.4	0.5	0.7	1.0	40

equivalent current-carrying capacity. The major disadvantages of aluminum conductors versus copper are shorter fatigue life, poor corrosion resistance, and decreased solderability.

The military specification for insulated aluminum wire is MIL-W-5088.

Stainless Steel. Although very seldom utilized by itself as conductor because of low conductivity, stainless steel (in strands) has been employed to reinforce stranded copper conductors. This reinforcement is normally used in small hookup wire (No. 24 AWG and smaller) and in ground support cable (No. 18 AWG and smaller).

Silver. Recent observations on the corrosion problems of silver-plated copper conductors have resulted in investigations into the use of pure silver and silver-alloy conductors. The physical properties of silver materials available to date are covered in Table 2. Investigations are being carried out at Battelle Memorial Institute under contract from NASA.

Coatings. Bare copper conductor is rarely used in the aerospace and military electronics industry, since copper will oxidize from exposure to the atmosphere. Corrosion will impair the conductors and reduce solderability. Some insulating materials also tend to corrode bare copper. Following is a list of the most widely used conductor coatings:

Tin. For temperatures below 150°C, the least expensive protective coating is tin. In addition to protecting copper from oxidation, it aids soldering. Tin should not be used as a protective coating above 150°C, as it will rapidly oxidize.

TIN DIP. Protective tin coating can be applied to a copper conductor by two methods: tin dip or electroplating. Tin dip is the process of passing strands through a molten tin bath. The major disadvantage of tin dip is poor control of coating thickness; thickness can vary from 20 to 200 μin. Specifications for tin-coated wire are covered in ASTM B33 and QQ-W-343.

Several variations of the standard tin-coated copper conductor have been utilized as a fabrication aid to eliminate strand twisting and pretinning. Following are some of the variations available:

Heavy tin is the application of a minimum tin thickness of 100 μin. to No. 31 AWG and smaller, and of 150 μin. to No. 30 AWG and larger. Utilized with the aid of high-frequency induction heaters, tin is melted and flowed, tacking the strands together in the area where the conductor is to be stripped.

Prefused conductor utilizes heavy tin-coated strands that are melted and fused together for the entire length of the insulated conductor. Although fabrication costs may be reduced, conductor flexibility is impaired. This process defeats the purpose of stranded conductors for increased flex life.

Overcoated conductor consists of individually tinned copper strands followed by an overcoating of tin over the conductor, bonding all strands together. Conductor flexibility is reduced significantly.

Top-coated conductor has bare, untinned copper strands followed by an overcoating of tin. The finished conductor is bonded along its entire length; it is less expensive than the overcoated conductor but has less corrosion resistance.

ELECTROPLATED TIN. The electroplating process applies to the copper strand a tin coating of controlled thickness and is becoming increasingly important for automated stripping and termination equipment. There is no individual specification covering the application of tin coating by electroplating.

Silver. This coating is utilized for continuous service of conductors in excess of 150°C to a maximum of 200°C and as required for the application of certain high-temperature insulation materials. Silver-coated conductors are widely used with fluorocarbon, polyimide, and silicone rubber insulation. Silver is very readily soldered; this at times is considered a disadvantage in that solder flows under the insulation and potentially reduces flex life at the conductor termination.

Silver-coated copper is susceptible to corrosion caused by the galvanic interaction between copper and silver. Microscopic breaks in the silver coating, caused by stranding or coating porosity, allow galvanic reaction to take place between bare copper and silver in the presence of moisture or other electrolytes. Considerable work is under way to solve this problem. Protective coatings applied over the silver, increased silver thickness, and intermediate barrier platings are the subject of current research. These approaches have been successful in reducing but not eliminating this potential problem.

ASTM B298 contains the specifications for silver-coated copper conductors.

Nickel. This is a good high-temperature protective coating, suitable for continuous use up to 300°C, and not susceptible to the corrosion potential of silver. The main disadvantage of nickel is its poor solderability. Even with activated fluxes and high soldering temperatures good solder terminations are extremely difficult to achieve. Nickel coating can be applied to copper conductors by two methods: cladding and electroplating.

Cladding is the application of a relatively thick nickel outer coating under heat and pressure over the copper billet. The process is similar to the fabrication of copper-covered steel wire. Cladding offers superior protection at extremely high-temperature applications (1000°F maximum, continuous). Cladding is not advantageous for lower temperatures because of the higher resistance of nickel.

Electroplating is applied by standard electroplating techniques, normally with 50 μin. minimum thickness.

Construction. Conductor construction plays a significant role in the proper functioning and reliability of any selected conductor. Solid wire, while low in cost and weight, will not withstand even mild flexure without breaking because of work

TABLE 4. Properties of Solid Conductors[4]

Gauge, AWG or B. & S.	Nominal diameter, in.	Area, cir mils	Weight, lb/1,000 ft	Length, ft/lb	Resistance at 68°F		
					ohms/1,000 ft	ft/ohms	ohms/lb
0000	0.4600	211,600	640.5	1.561	0.04901	20,400	0.00007652
000	0.4096	167,800	507.9	1.968	0.06180	16,180	0.0001217
00	0.3648	133,100	402.8	2.482	0.07793	12,830	0.0001935
0	0.3249	105,500	319.5	3.130	0.09827	10,180	0.0003076
1	0.2893	83,690	253.3	3.947	0.1239	8,070	0.0004891
2	0.2576	66,370	200.9	4.977	0.1563	6,400	0.0007778
	0.250	62,500	189.1	5.286	0.1659	6,025	0.000877
3	0.2294	52,640	159.3	6.276	0.1970	5,075	0.001237
4	0.2043	41,740	126.4	7.914	0.2485	4,025	0.001966
	0.188	35,344	106.98	9.425	0.2934	3,407	0.00276
5	0.1819	33,100	100.2	9.980	0.3133	3,192	0.003127
6	0.1620	26,250	79.46	12.58	0.3851	2,531	0.004972
7	0.1443	20,820	63.02	15.87	0.4982	2,007	0.007905
8	0.1285	16,510	49.98	20.01	0.6282	1,592	0.01257
9	0.1144	13,090	39.63	25.23	0.7921	1,262	0.01999
10	0.1019	10,380	31.43	31.82	0.9989	1,001	0.03178
11	0.09074	8,234	24.92	40.12	1.260	794	0.05053
12	0.08081	6,530	19.77	50.59	1.588	629.6	0.08035
13	0.07196	5,178	15.68	63.80	2.003	499.3	0.1278
14	0.06408	4,107	12.43	80.44	2.525	396.0	0.2032
15	0.05707	3,257	9.858	101.4	3.184	314.0	0.3230
16	0.05082	2,583	7.818	127.9	4.016	249.0	0.5136
17	0.04526	2,048	6.200	161.3	5.064	197.5	0.8167
18	0.04030	1,624	4.917	203.4	6.385	156.5	1.299
19	0.03589	1,288	3.899	256.5	8.051	124.2	2.065
20	0.03196	1,022	3.092	323.4	10.15	98.5	3.283

21	0.02846	810.1	2.452	407.8	12.80	78.11	5.221
22	0.02535	642.4	1.945	514.2	16.14	61.95	8.301
23	0.02257	509.5	1.542	648.4	20.36	49.13	13.20
24	0.02010	404.0	1.223	817.7	25.67	38.96	20.99
25	0.01790	320.4	0.9699	1,031	32.37	30.90	33.37
26	0.01594	254.1	0.7692	1,300	40.81	24.50	53.06
27	0.01420	201.5	0.6100	1,639	51.47	19.43	84.37
28	0.01264	159.8	0.4837	2,067	64.90	15.41	134.2
29	0.01126	126.7	0.3836	2,607	81.83	12.22	213.3
30	0.01003	100.5	0.3042	3,287	103.2	9.691	339.2
31	0.008928	79.7	0.2413	4,145	130.1	7.685	539.3
32	0.007950	63.21	0.1913	5,327	164.1	6.095	857.6
33	0.007080	50.13	0.1517	6,591	206.9	4.833	1,364
34	0.006305	39.75	0.1203	8,310	260.9	3.833	2,168
35	0.005615	31.52	0.09542	10,480	329.0	3.040	3,448
36	0.005000	25.00	0.07568	13,210	414.8	2.411	5,482
37	0.004453	19.83	0.06001	16,660	523.1	1.912	8,717
38	0.003965	15.72	0.04759	21,010	659.6	1.516	13,860
39	0.003531	12.47	0.03774	26,500	831.8	1.202	22,040
40	0.003145	9.888	0.02993	33,410	1,049	0.9534	35,040
41	0.00280	7.8400	0.02373	42,140	1,323	0.7559	55,750
42	0.00249	6.2001	0.01877	53,270	1,673	0.5977	89,120
43	0.00222	4.9284	0.01492	67,020	2,104	0.4753	141,000
44	0.00197	3.8809	0.01175	85,100	2,672	0.3743	227,380
45	0.00176	3.0976	0.00938	106,600	3,348	0.2987	356,890
46	0.00157	2.4649	0.00746	134,040	4,207	0.2377	563,900

hardening and fatigue. Bunch stranding offers extreme flexibility. While it offers low cost and fatigue resistance, it does not provide a consistent circular cross section; therefore, it is not recommended for thin-wall extruded insulated wire construction. Major conductor constructions are discussed below.

Solid. Table 4 contains the characteristics of solid conductors. Solid-conductor usage is limited for reasons given above. The major application of solid wire, magnet wire excepted, is for short jumper and bus wires not subjected to vibration or flexing.

Stranded. Besides increased flexibility, construction characteristics of stranded conductors may limit the application of insulating materials. Table 5 contains detailed characteristics of stranded conductors that meet military specifications. Following is a brief description of the various stranding constructions.

TRUE CONCENTRIC. This construction presents the most consistent circular cross section of all available strandings. Alternate layers, applied in opposite directions with increasing lay, hold strands in place and prevent "strand popping" and high strands. Insulated wire manufacturers prefer this stranding for conductors with extruded thin-wall insulation. Construction of concentric stranded conductors is shown in Fig. 2.

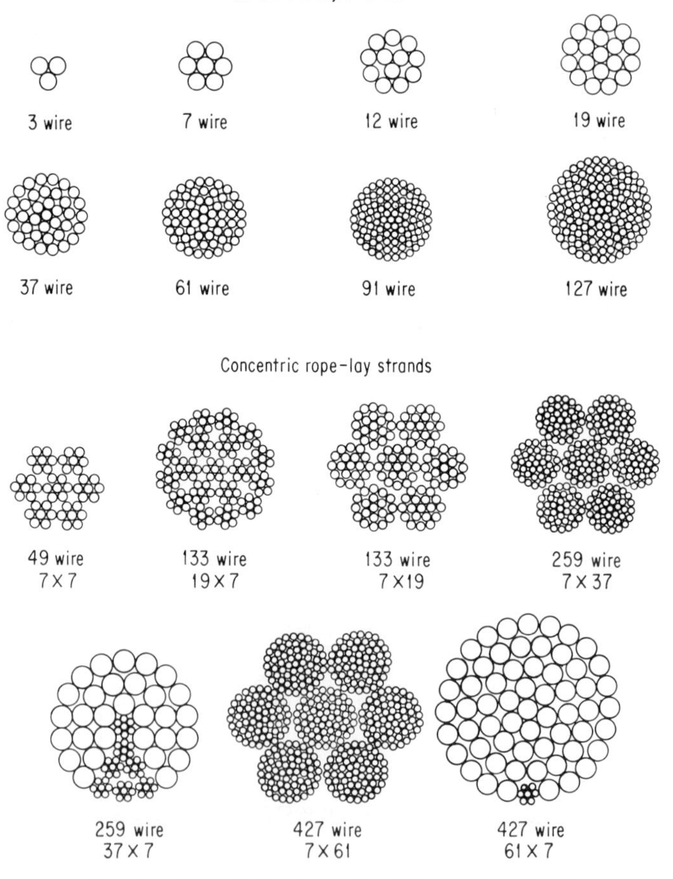

FIG. 2. Stranded wire and cable construction.[3]

TABLE 5. Details of Stranded Conductors[5]

Size designation, AWG	Nominal conductor area, cir mils	Number of strands	Allowable no. of missing strands	Nominal diameter of individual strands, in.	Max diameter of stranded conductor, in.	Max resistance of finished wire, ohms/1,000 ft at 20°C			
						Tin-coated copper	Silver-plated copper	Nickel-plated copper	Silver-plated high-strength copper alloy
30	112	7	0	0.0040	0.013	107.0	101.0	109.0	116.0
28	175	7	0	0.0050	0.016	67.6	62.9	68.3	72.2
26	304	19	0	0.0040	0.021	39.3	36.2	40.1	41.5
24	475	19	0	0.0050	0.026	24.9	23.2	25.1	26.6
22	754	19	0	0.0063	0.033	15.5	14.6	15.5	16.8
20	1,216	19	0	0.0080	0.041	9.70	9.05	9.79	10.4
18	1,900	19	0	0.0100	0.052	6.08	5.80	6.08	6.65
16	2,426	19	0	0.0113	0.060	4.76	4.54	4.76	5.23
14	3,831	19	0	0.0142	0.074	2.99	2.87	3.00	3.30
12	7,474	37	0	0.0142	0.102	1.58	1.48	1.59	1.70
10	9,361	37	0	0.0159	0.118	1.27	1.20	1.27	1.38
8	16,983	133	0	0.0113	0.176	0.700	0.661	0.680	0.760
6	26,818	133	0	0.0142	0.218	0.436	0.419	0.428	0.483
4	42,615	133	0	0.0179	0.272	0.274	0.263	0.269	0.302
2	66,500	665	2	0.0100	0.345	0.179	0.169	0.174	0.194
0	104,500	1,045	3	0.0100	0.432	0.114	0.105	0.109	0.123

UNIDIRECTIONAL CONCENTRIC. This construction differs from true concentric in that the lay of successive layers is applied over a core in the same direction; it does not alternate directions. Unidirectional lay is more flexible and has greater flex endurance than true concentric. A comparison of flex endurance among true concentric, unidirectional concentric, bunch, and unilay conductor constructions is shown in Fig. 3.

BUNCH. Since bunch stranding consists in twisting a group of strands with the same lay length in the same direction, without regard to geometric arrangement, the strands are susceptible to movement and circular cross section is not ensured. If conductor circularity is important, as demanded by extrusion of thin-wall insulating materials, bunch conductor is not recommended.

UNILAY. The advantages of unilay are a smaller diameter, flexibility approaching bunch construction, and superior flex endurance (Fig. 3).

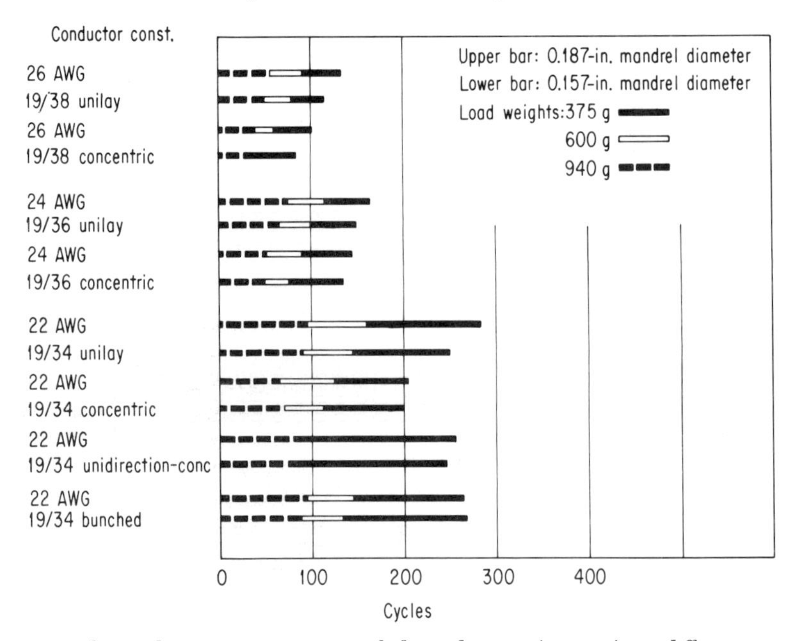

FIG. 3. Flex endurance tests on annealed conductors (copper) in different stranding configurations.[1]

EQUILAY. This is a variation of concentric construction that has reversed lay of layers but equal length of lay and a stiffer construction than unidirectional concentric lay.

ROPE LAY. Rope lay is basically a large-gauge (No. 10 AWG and larger) conductor construction that consists of a central-core stranded member with one or more layers of stranded members surrounding the core. The stranded members may be bunch-stranded (ASTM B172) or concentric (ASTM B173). Rope construction shown in Fig. 2 provides a uniform, circular cross section with good flexibility.

WIRE AND CABLE INSULATION

Materials. The properties of insulating and jacketing materials used with wire and cable can be divided into three main categories: mechanical, electrical, and chemical.

The mechanical and physical properties to consider are: tensile strength, elongation, specific gravity, abrasion resistance, cut-through resistance, and mechanical temperature resistance (cold bend and deformation under heat).

The electrical properties are: dielectric strength, dielectric constant, loss factor, and insulation resistance.

The chemical properties are: fluid resistance, flammability, temperature resistance, and radiation resistance.

Tables 6 through 8 give these major properties for the most commonly employed wire insulating and jacketing materials. Following is a brief description of the insulating materials.

Polyvinyl Chloride (PVC). This is a low-cost primary insulating and jacketing material available in many formulations tailored to meet applications. Two varieties of PVC, most widely used in aerospace and electronics, are (1) plasticized PVC, 105°C operating temperature applicable under most military specifications that require PVC insulation; and (2) semirigid PVC, a less flexible but more cut-through

TABLE 6. Mechanical and Physical Properties of Insulating Materials

Insulation	Common designation	Tensile strength, psi	Elongation, %	Specific gravity	Abrasion resist.	Cut-through resist.	Temperature resistance (mechanical)
Polyvinyl chloride	PVC	2,400	260	1.2–1.5	Poor	Poor	Fair
Polyethylene ..	PE	1,400	300	0.92	Poor	Poor	Good
Polypropylene	....	6,000	25	1.4	Good	Good	Poor
Crosslinked polyethylene	IMP	3,000	120	1.2	Fair	Fair	Good
Polytetrafluoroethylene	TFE	3,000	150	2.15	Fair	Fair	Excellent
Fluorinated ethylene propylene	FEP	3,000	150	2.15	Poor	Poor	Excellent
Monochlorotrifluoroethylene	Kel-F[*]	5,000	120	2.13	Good	Good	Good
Polyvinylidine fluoride	Kynar[†]	7,100	300	1.76	Good	Good	Fair
Silicone rubber	Silicone	800–1,800	100–800	1.15–1.38	Fair	Poor	Good
Polychloroprene rubber	Neoprene	150–4,000	60–700	1.23	Good	Good	Fair
Butyl rubber ..	Butyl	700–1,500	500–700	0.92	Fair	Fair	Fair-good
Fluorocarbon rubber	Viton[‡]	2,400	350	1.4–1.95	Fair	Fair	Fair-good
Polyurethane ..	Urethane	5,000–8,000	100–600	1.24–1.26	Good	Good	Fair-good
Polyamide	Nylon	4,000–7,000	300–600	1.10	Good	Good	Poor
Polyimide film	Kapton[‡]	18,000	707	1.42	Excellent	Excellent	Good
Polyester film	Mylar[‡]	13,000	185	1.39	Excellent	Excellent	Good
Polyalkene	....	2,000/7,000	200/3,001.2	1.76	Good	Good	Fair-good
Polysulfone	....	10,000	50–100	1.24	Good	Good	Good
Polyimide-coated TFE	TFE/ML[‡]	3,000	150	2.2	Good	Good	Good
Polyimide-coated FEP	FEP/ML[‡]	3,000	150	2.2	Good	Good	Good

* Trademark, 3M Company.
† Trademark, Pennsalt Chemicals Corporation.
‡ Trademark, E. I. du Pont de Nemours & Co.

TABLE 7. Electrical Properties of Insulation Materials

Insulation	Common designation	Property			
		Dielectric strength, volts/mil	Dielectric constant, 10^3 Hz	Loss factor, 10^3 Hz	Volume resistivity, ohm-cm
Polyvinyl chloride	PVC	400	5–7	0.02	2×10^{14}
Polyethylene	PE	480	2.3	0.005	10^{16}
Polypropylene	....	750	2.54	0.006	10^{16}
Crosslinked polyethylene	IMP	700	2.3	0.005	10^{16}
Polytetrafluoroethylene	TFE	480	2.1	0.0003	10^{18}
Fluorinated ethylene propylene	FEP	500	2.1	0.0003	10^{18}
Monochlorotrifluoroethylene	Kel-F	431	2.45	0.025	2.5×10^{16}
Polyvinylidine fluoride	Kynar	1,280 (8 mils)	7.7	0.02	2×10^{14}
Silicone rubber	Silicone	575–700	3–3.6	0.003	2×10^{15}
Polychloroprene rubber	Neoprene	113	9.0	0.030	10^{11}
Butyl rubber	Butyl	600	2.3	0.003	10^{17}
Fluorocarbon rubber	Viton	500	4.2	0.14	2×10^{13}
Polyurethane	Urethane	450–500	6.7–7.5	0.055	2×10^{11}
Polyamide	Nylon	385	4–10	0.02	4.5×10^{13}
Polyimide film	Kapton	5,400 (2 mils)	3.5	0.003	10^{18}
Polyester film	Mylar	2,600	3.1	0.15	6×10^{16}
Polyalkene	....	1,870	3.5	0.028	6×10^{13}
Polysulfone	....	425	3.13	0.0011	5×10^{16}
Polyimide-coated TFE	TFE/ML	480	2.2	0.0003	10^{18}
Polyimide-coated FEP	FEP/ML	480	2.2	0.0003	10^{18}

resistant version used primarily with automated termination devices (wire wrap, Termi-point,* etc.).

Polyethylene (PE). This possesses excellent properties for wire and cable insulation. Electrically, because of its dielectric constant, loss factor, and insulation resistance, PE can be matched only by some fluorocarbons. Solvent resistance, moisture resistance, and low-temperature performance are superior. PE has three distinct disadvantages: it is flammable, its maximum operating temperature is only 80°C, and it is stiff. Flame retardants can be added to render PE self-extinguishing. Crosslinking can be achieved either by chemical methods or by irradiation methods increasing the continuous operating temperature to the 135 to 150°C range and reducing its susceptibility to crack under thermal and environmental stresses.

Polypropylene. A member of the polyolefin family, as is polyethylene, polypropylene exhibits good electrical and chemical properties. Mechanically, polypropylene has superior properties to polyethylene: its abrasion and cut-through resistance is superior and comparable with that of nylon. Until recent improvements, polypropylene could cause copper poisoning (embrittlement of conductor). Polypropylene has poor low-temperature (cold-bend) characteristics; it is not recommended for low-temperature applications. Its main usage to date has been in telephone and communication wire and cable insulation.

Polytetrafluoroethylene (TFE). This is undoubtedly the best known and most widely used material in the fluorocarbon family. It has excellent electrical, chemical, and thermal properties. TFE is not a particularly tough material; its abrasion and cut-through resistance are rated only fair. However, it maintains mechanical resistance at temperatures exceeding 200°C. TFE can withstand short-time contact

* Trademark, AMP, Inc.

TABLE 8. **Chemical Properties of Insulation Materials**

Insulation	Common designation	Property: Fluid resistance	Flammability	Radiation resistance, rads gamma exposure	Temperature resistance, °C	Comments
Polyvinyl chloride	PVC	Good	Slow to self-extinguishing	10^6–10^7	−55–105	
Polyethylene	PE	Excellent	Flammable	10^8	−65–80	
Polypropylene	....	Good	Self-extinguishing	10^8	−20–125	
Crosslinked polyethylene	IMP	Good	Self-extinguishing	10^8	−65–150	
Polytetrafluoroethylene	TFE	Excellent	Nonflammable	10^6	−80–260	
Fluorinated ethylene propylene	FEP	Excellent	Nonflammable	10^6	−80–200	
Monochlorotrifluoroethylene	Kel-F	Good	Nonflammable	10^6	−80–200	Fluids tend to permeate at high temperature
Polyvinylidine fluoride	Kynar	Good	Self-extinguishing	10^8	−65–130	
Silicone rubber	Silicone	Poor	Flammable	10^8	−65–200	
Polychoroprene rubber	Neoprene	Good oil resistance	Self-extinguishing	10^7	−55–80	
Butyl rubber	Butyl	Good	Flammable	10^6	−55–85	Poor resistance to hydrocarbons
Fluorocarbon rubber	Viton	Excellent	Self-extinguishing	10^6–10^7	−40–200	Poor resistance to oxygenated alcohols
Polyurethane	Urethane	Good	Flammable	10^7–10^8	−55–85	
Polyamide	Nylon	Good	Self-extinguishing	10^7	−55–105	Soluble in alcohol
Polyimide film	Kapton	Excellent	Nonflammable	10^9	−80–260	
Polyester film	Mylar	Good	Flammable	10^8	−65–120	
Polyalkene	....	Good	Self-extinguishing	10^8	−65–135	
Polysulfone	Polysulfone	Fair	Self-extinguishing	....	−65–150	Soluble in chlorinated hydrocarbon
Polyimide-coated TFE	TFE/ML	Excellent	Nonflammable	10^6	−80–260	
Polyimide-coated FEP	FEP/ML	Excellent	Nonflammable	10^6	−80–260	

with a hot soldering iron without damage. TFE-insulated wire is in wide use because of its solder resistance characteristics: it reduces the problem of insulation overheating at solder terminations and allows soldering of high-density wiring where soldering without contact of nearby insulation is unavoidable. TFE insulating and jacketing applications are high-temperature interconnection and hookup wires and cables (also multiconductor) and coaxial cable. TFE is suitable neither for high-voltage applications, because of poor corona resistance, nor for radiation environments, because of mechanical degradation at levels in excess of 10^6 rads. TFE's radiation resistance improves in oxygen-free atmosphere.

Fluorinated Ethylene Propylene (FEP). A melt-extrudible counterpart of TFE, FEP has almost identical electrical and chemical properties; but it melts to a fluid at 290°C, and thus its solder iron resistance and temperature resistance are inferior. FEP is widely used as jacketing material, since it is melt-extrudible by thermoplastic techniques (TFE cannot be melt-extruded). Used as primary insulation, FEP offers the following advantages: lower cost (in quantities), longer continuous lengths, and easier identification (hot stamp).

Monochlorotrifluoroethylene (Kel-F). As part of the fluorocarbon family, this is less temperature-resistant but substantially tougher than TFE or FEP. It has good electrical, chemical, and mechanical properties but is susceptible to crystallization manifested in poor shelf aging. Kel-F has seen limited use as wire insulation in the military, aerospace, and electronics industries.

Vinylidene Fluoride (Kynar). Kynar is the most recent addition to the fluorocarbon family of wire and cable insulation materials. It is extremely tough and has excellent abrasion and cut-through resistance. Its electrical, thermal, and chemical properties are inferior to those of TFE and FEP; however, Kynar has superior radiation resistance. Kynar has been utilized by itself as insulation, in combination with polyethylene (Polyalkene), and as jacketing. Some problems have been encountered on insulation shrinkage and flaring while soldering, although the problem can be corrected by controlled extrusion techniques.

Silicone Rubber (Silicone). Silicone is a good, flexible, high-temperature, elastomeric insulator; however, it is not noted for its mechanical toughness or fluid resistance. Its electrical properties are good and its radiation resistance superior for an elastomer but generally inferior to most plastic insulations. Excellent corona resistance makes it useful as a high-voltage insulation. Applications include shipboard wire and cable, nuclear cable, high-voltage cable, and aircraft wire and cable. Silicone rubber is flammable but has the unique feature of leaving a white nonconductive ash which, if contained, forms good insulation.

Neoprene Rubber (Neoprene). Neoprene finds major usage in the aerospace and electronic industry as jacketing material. Its electrical insulating properties are inferior: it is suitable only for low-voltage low-frequency applications. Its resistance to abrasion and to mechanical bending and impact in the −55 to 70°C range is excellent. Neoprene is extruded as a jacket for ground-support cable applications. Neoprene tubing may be used as harness jacketing where extremely heavy use is anticipated.

Butyl Rubber (Butyl). Butyl rubber has better moisture resistance and electrical properties than neoprene. It is used as insulation for larger ground-support cable sizes. More flexible than polyethylene, butyl is used in thick-wall insulation. As a jacket material butyl does not possess the mechanical toughness of neoprene but has the advantage of compatibility with some of the more exotic missile fuels.

Fluorocarbon Rubber (Viton). This has some outstanding characteristics: high-temperature, oil, and fuel resistance. Its mechanical properties are inferior to those of neoprene, and its low-temperature cold-bend resistance is poor. Electrically it is not recommended as a primary insulation except for low-voltage low-frequency application.

Polyurethane Elastomers (Urethane). Two basic urethane elastomers are utilized in the wire and cable industry: polyether-based and polyester-based. The polyether-based material possesses superior low-temperature characteristics and humidity resistance. The polyester-based material exhibits superior high-temperature resistance but is susceptible to hydrolysis under high-humidity high-temperature conditions. Major application of urethanes has been as jacketing material, presenting an extremely tough abrasion- and tear-resistant covering.

Polyamide (Nylon). This is used almost exclusively as jacketing owing to its high moisture absorption. Nylon is a tough, abrasion-resistant material. Because of its stiffness and susceptibility to cracking it is not recommended for extruded jacketing over cores greater than 0.210-in. diameter. Nylon provides a good protective jacket over thin-wall thermoplastic insulations compatible with its temperature characteristics. However, nylon cracking has also been observed when the material is used in a low-humidity environment or if stored for a length of time in a formed condition.

Polyimide. Two forms of polyimide resin are at present in use in the wire and cable industry: (1) a coating material, applied as an insulation on magnet wire and as a mechanical barrier on TFE and FEP, enhancing their mechanical strength; (2) a film which, when coated with FEP, becomes heat-sealable. The film can be applied over a conductor by tape wrapping, where the final sintering provides a fused homogeneous insulation. Polyimide possesses exceptionally high heat resistance, excellent mechanical properties, high radiation resistance, and very good chemical resistance.

As a supplementary coating applied to TFE or FEP in 0.0005- to 0.001-in. thickness, polyimide more than doubles abrasion and cut-through resistance. Film construction provides a thin wall of approximately 0.007 in., which saves space and weight without sacrifice of mechanical strength or abrasion and cut-through resistance, compared with many larger and heavier insulations.

Polyester (Mylar). Mylar is available to date only in yarn or film, although several extrudible polyester-based resins have been applied to conductors experimentally. Polyester film is widely used in cable fabrication as a separator or binder wrap. A heat-sealable construction is available with either PE or PVC coating. The tape is applied by wrapping, as with the FEP-coated polyimide film previously mentioned. When applied and heat-sealed, the coated polyester film exhibits similar mechanical properties to the coated polyimide film. Heat resistance of polyester-coated film is inferior to that of polyimide. Major application of the coated polyester film insulation has been to date in the computer industry for tough, thin-walled, insulated wire used in conjunction with wire wrap and other automated wiring methods.

Polyalkene. Another recent development in insulation materials, this exhibits good properties for thinner-walled, lighter-weight wire constructions. Polyalkene wire is a dual extrusion of polyolefin and polyvinylidene fluoride. Both these materials are crosslinked for increased heat resistance and greater mechanical strength. Combined use of the two insulating compounds mutually offsets their individual disadvantages. Polyvinylidene fluoride provides mechanical toughness not inherent in polyolefin, while its main disadvantage, high dielectric constant, is tempered by the excellent electrical properties of polyolefin.

Polysulfone. This is manufactured by Union Carbide and is an extrudible resin used mostly for thin-walled wires and cables in computer applications. Polysulfone is tough, suitable for high-density computer wiring; its electrical properties are adequate; and it is self-extinguishing.

Construction and Application Methods. Reference has been made to extrusion, tape wrapping, and coating as methods of applying specific materials to a conductor. Certain materials are available only in a tape or solution form and cannot be obtained in extruded form. Following is a brief description of methods of application:

Extrusion. Many materials covered in the preceding section are thermoplastic; i.e., they are heated to softness, formed into shape, then cooled to become solid again. Thermoplastics are shaped around a conductor by the extrusion process. This consists in forcing the plastic material under pressure and heat through an orifice. The suitability for extrusion presents certain advantages to both the manufacturer and the user: longer continuous lengths of insulated conductor, smooth outer surfaces for easier connector sealing, homogeneous insulation, fast processing, quick delivery, and, usually, lower cost due to simpler processing.

Tape Wrap. Tape wrapping is the application of insulation in the form of a thin film or tape. The tape is normally applied to the conductor with minimum overlap to ensure wire flexibility without baring the conductor. Layers of tape can be built up to achieve the desired insulation wall thickness. Successive layers are wrapped in the opposite direction, ensuring coverage of tape overlaps. Finally, after the tape is applied, the wire is sintered, cured, or fused with sufficient heat to seal the layers of tape into a homogeneous mass. Tape wrap provides good control of wall thickness; there is no concentricity problem with tape construction as opposed to extruded insulation. Tape wrap has the disadvantage of slower fabrication. Problems of poor adhesion between tape layers have also been encountered. Details of representataive tape-wrapped Kapton* film insulated construction are shown in Figs. 4 and 5 .

Coating. Insulation by dip coating is limited almost exclusively to magnet wire. However, for small-diameter Teflon,* TFE, and FEP applications, a polyimide overcoating of less than 0.001 in. significantly enhances the mechanical strength, abrasion protection, and solder resistance of the insulation. The coating of fluorocarbon base is a difficult process; it requires complete surface preparation to promote good adhesion of the polyimide coating material.

* Trademark, E. I. du Pont de Nemours & Co.

SHIELDING

Materials and Construction. Although there is no real substitute for braided round copper as a general-purpose shield, certain applications may allow the use of a lighter, less bulky shield construction. The purpose of a shield is (1) to prevent external fields from adversely affecting signals transmitted over the center conductor, (2) to prevent undesirable radiation of a signal into nearby or adjacent conductors, or (3) to act as a second conductor in matched or tuned lines. Following is a brief description of various available shield types. Table 9 presents the shielding effectiveness at tested frequency ranges of the systems described.

Braided Round Shield. Braided round copper shielding is the most commonly utilized by industry today as flexible coaxial or shielded cable. The braided-shield

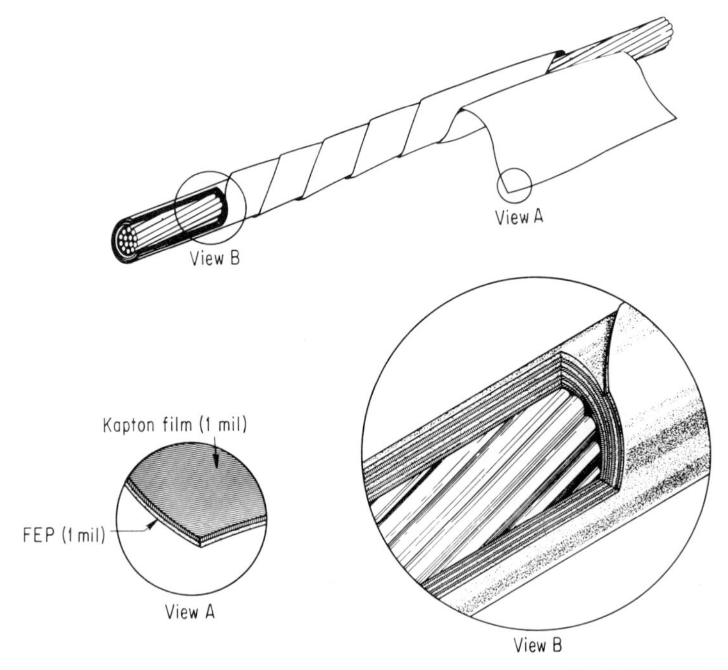

Fig. 4. Tape-wrap detail, single-wrap construction: ⅔ overlap.[6] [Note: Tape—single-faced 1:1 HF film (1 mil Kapton film, 1 mil FEP one side only).]

technique is highly effective in the frequency range below 0.5 Hz. Above this frequency range the gaps in the interstices of the shield braid cause loss in effectiveness. Braiding consists in interweaving groups (carriers) of strands (ends) of metal over an insulated conductor. The braid angle at which the carriers are applied with reference to the axis of the core, the number of ends, the number of carriers, and the number of carrier crossovers per inch (picks) determine the percent coverage, which is a measure for the shielding gap. As reference, a solid metal tube is equivalent to 100 percent coverage.

Braided-shield strands normally consist of coated copper. The coating must be compatible with the insulation material employed: tin coating is used for low-temperature insulations such as PVC, PE, and polyalkene; silver or nickel coating is used with high-temperature insulations.

Silver-plated aluminum strands are employed for lighter weights, although a significant galvanic corrosion potential exists. Stainless steel or other metals can be employed for added strength if conductivity is not the prime requisite.

Flat Braided Shield. The application and considerations are the same as with round shielding, except for shape: the individual strands are not round but flat. Flat braid combines the advantages of reduced shield buildup and weight with good coverage. Termination is a problem solved satisfactorily with a shrinkable solder sleeve.

Metal Tape. Metal tapes of either copper or aluminum with a minimum overlap of 10 percent can be employed as an effective shield. Because of their stiffness, metal tapes are normally employed only in large single-conductor or multiconductor cable construction.

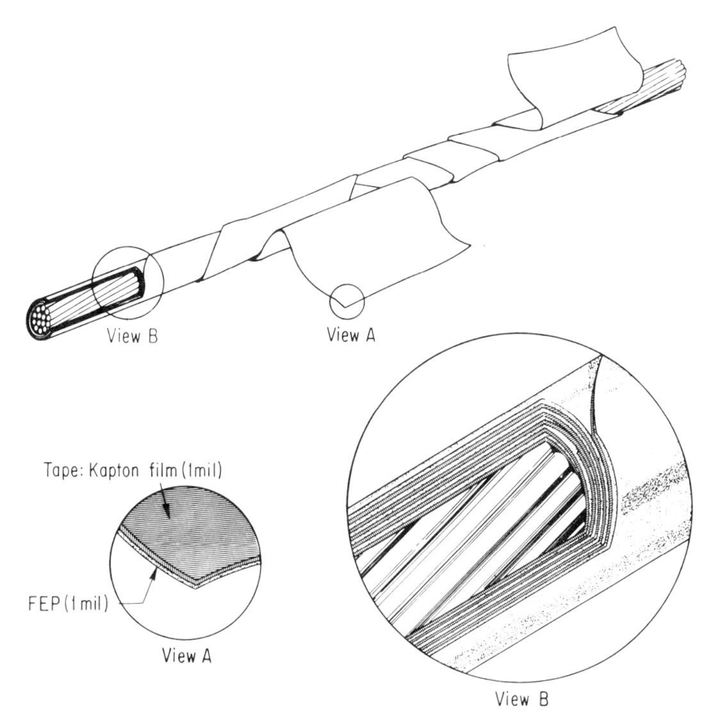

FIG. 5. Tape-wrap detail, double-wrap construction: ½ overlap.[6] [NOTE: Tapes—single-faced 1:1 HF film (1 mil Kapton film, 1 mil FEP one side only).]

Solid Shield. Solid or tubular shields may be applied by tube swaging or by forming interlocked, seam-welded or soldered tape around the dielectric. These methods offer 100 percent shielding and double as protective armor. Corrugated tapes facilitate cable bending, but their flexibility is poor. Solid shield is usually applied to buried cables or where 100 percent shielding is mandatory.

Served Shield. A served, or spiral, shield consists of a number of metal strands wrapped flat as a ribbon over a dielectric in one direction. Weight and size of shielding are approximately one-half of braided shields, as there is no strand crossover or overlap. A served shield is known to be effective in the audio-frequency range. Above audio frequencies the possibility of inductive effects caused by strands spiraling in one direction should be examined.

TABLE 9. Shield Effectivity in Volts Peak to Peak [7]

Sample description	Frequency									
	30 Hz	100 Hz	300 Hz	1 kHz	3 kHz	10 kHz	30 kHz	100 kHz	300 kHz	1 MHz
Control unshielded	<0.600	<1.6	4.0	10.0	12.0	12.5	13.5	12.5	12.5	14.0
Braid, No. 36 AWG, tinned copper, 90% coverage	<0.001	<0.001	0.0025	0.005	0.006	0.00625	0.0075	0.0075	0.007	0.008
Aluminum tape, No. 22 AWG drain wire	<0.001	<0.001	0.001	0.002	0.002	0.002	0.005	0.0085	0.012	0.014
Semiconductive PVC, No. 26 AWG drain wire	<0.001	<0.001	<0.00325	0.020	0.060	0.120	0.240	0.450	0.540	1.85
Serve—No. 36 AWG tinned copper, 90% coverage, 4 ends reversed	<0.001	<0.001	0.001	0.002	0.0025	0.0025	0.004	0.006	0.0065	0.013
Braid—8 carriers, No. 36 AWG tinned copper, 8 carriers, conductive glass yarn	<0.001	<0.001	<0.001	0.001	0.00125	0.00125	0.002	0.003	0.005	0.012
Braid—No. 36 AWG flat ribbon silver-plated copper	0.00125	0.003	0.008	0.018	0.024	0.024	0.028	0.028	0.027	0.029
Serve—No. 36 AWG tinned copper	<0.001	0.003	0.008	0.018	0.026	0.026	0.029	0.029	0.030	0.031
Solid—cadmium-plated copper	<0.001	<0.001	<0.001	<0.001	<0.001	<0.001	<0.001	<0.001	<0.001	<0.001

Foil Shield. Foil tape materials such as copper or aluminum-coated Mylar have been employed in such techniques as normal overlap taping and longitudinal wrapping with various interlocking approaches. Shielding effectiveness at low frequencies is poor, and termination is difficult. The shield is terminated with a single strand of solid wire (drain wire) in contact with the metal coating under the tape.

Conductive Plastics. PVC and PE compounds have been formulated with conductive additives for the purpose of shielding. Effectiveness has been poor, normally limited to the low audio-frequency range. As with foil, termination is a problem; a drain wire can be employed for terminating.

Conductive Yarns. Conductive yarns, such as impregnated glass, provide weight reduction, but lack of conductivity and difficulty in termination make them suited only for special applications.

Metal Shield and Conductive Yarn. An effective marriage between metal strands and conductive yarns has been developed and tested which provides effective shielding when braided and which reduces shield weight. Terminations can be accomplished by normal techniques.

SHIELD JACKETS

Materials and Construction. It is the basic function of a shield jacket to insulate a shield from ground (structure), other shields, or conductors. Secondarily, a shield jacket may serve: as a lubricant in multiconductor cables, enabling free movement during bending; as an insulator, to allow varying potentials between adjacent shields; and as a moisture and abrasion barrier. A few of the materials described in the section on Wire and Cable Insulation are appropriately used as jacketing materials, either because of superior mechanical strength or because of compatibility with the chosen primary insulation system.

Extrusion, braiding, and taping techniques are employed in the application of shield jackets. Fused tapes and extruded materials are the only reliable method for assuring a moistureproof jacket. Consistent wall thickness is best maintained with tapes; however, reliable fusion across interstices of multiple shielded conductors is frequently a problem. Lacquered fiber braids allow a large conductor or multiple conductors more flexibility than extruded or taped jackets but offer limited moisture and humidity resistance.

Table 10 lists the characteristics of extruded and taped jackets suited for low- and medium-temperature applications; Fig. 6 provides shield-jacket abrasion-resistance data. In comparing abrasion-resistance values, differences in wall thickness should be considered. Following is a discussion of available shield-jacket materials.

Polyamide (Nylon). Nylon has excellent mechanical properties within its temperature capabilities (120°C max.). Electrical properties, relatively unimportant in jacket materials, are poor owing to rather high moisture absorption. Nylon jacket is likely to crack when heated and bent at less than a 10 D mandrel. Jacketing is applied by extrusion or in the form of braided nylon yarn. Extrusion is not recommended over cores greater than 0.210 in. because of stiffness and cracking.

Polyvinyl Chloride (PVC). This provides good electrical insulation but requires substantial wall thickness to achieve adequate mechanical protection. PVC has seen much use as a coaxial cable jacket formulated with a noncontaminating compound containing a nonbleeding plasticizer. It is applied by extrusion with practically no limitation to core diameter.

Polyethylene (PE). Polyethylene is used rarely as a shield jacket because of poor high-temperature characteristics, stiffness, and inadequate mechanical resistance in thin walls. Crosslinked PE with improved temperature and mechanical properties is used with compatible primary insulation materials. PE jackets are applied by extrusion with minimum limitation on core diameter.

Teflon (TFE). This is supplied as a shield jacket by extrusion or taping. Tape jackets are the only solution for larger shields owing to practical extrusion limitations. Close control must be exercised by the manufacturer in the curing of TFE to prevent the flow of inner primary conductor insulation.

Teflon (FEP). As outlined in the section on Wire and Cable Insulation, FEP is much easier to fabricate as a jacket material than TFE. However, the lower tem-

TABLE 10. Shield-jacket Properties

No.	Sample description	OD, in.	Wall, in.	Thermal[a] shock	Aging stability[b]		Heat resistance[c]		Cold[d] bend	Cut through[e] 120°C	Cracking test[f]	
					Visual	Dielectric, kv	Visual	Dielectric, kv			Visual	Dielectric, kv
1	Polyester-coated Mylar tape	0.104	0.0045	Not applicable	Not applicable	....	Cracked	No test	Passed	120 hr	Passed 3X	3–4
2	Polyethylene-coated Mylar tape	0.092	0.005	Not applicable	Not applicable	....	Passed	4–5	Passed	120 hr	Passed 3X	4-5
3	Polyethylene-coated Mylar tape	0.095	0.006	Not applicable	Not applicable	....	Passed	6.5–7.5	Passed	120 hr	Passed 3X	3.5–4
4	Polyethylene-coated Mylar tape	0.112	0.005	Not applicable	Not applicable	....	Cracked	No test	Passed	120 hr	Passed 3X	Failed up to 9X
5	Kynar jacket	0.092	0.005	Not applicable	Not applicable	....	Passed	9.5	Passed	120 hr	Passed 5X	4
6	NAS-702-22SC9 nylon jacket	0.090	0.0055	Not applicable	Not applicable	....	Cracked	No test	Passed	120 hr	Failed up to 9X	No test
7	NAS-702-20SC9 nylon jacket	0.107	0.005	Not applicable	Not applicable	....	Cracked	No test	Passed	120 hr	Failed up to 9X	No test
8	Caprolactam nylon jacket	0.125	0.0075	Not applicable	Not applicable	....	Passed	4.5–5	Failed	6 min	Failed up to 9X	No test
9	Plasticized nylon jacket	0.118	0.009	Not applicable	Not applicable	....	Cracked	No test	Passed	120 hr	Passed 9X	5.5–6
10	Mylar wrap 0.002 nylon jacket	0.116	0.0075	Not applicable	Not applicable	....	Cracked	No test	Passed	120 hr	Failed up to 9X	No test
11	NAS-702 18SC7 nylon jacket	0.112	0.008	Not applicable	Not applicable	....	Cracked	No test	Passed	120 hr	Failed up to 9X	No test
12	2 cond. cable nylon jacket	0.129	0.010	Not applicable	Not applicable	....	Cracked	No test	Passed	120 hr	Failed up to 9X	No test

13	Tough PVC jacket . . .	0.077	0.010	Not applicable	Passed	11.5–14	Not applicable	. . . .	Passed	5 min	Passed	8–8.5
14	Polyurethane (Estane) jacket	0.145	0.010	Jkt. softened and flowed	Passed	5.5–10	Not applicable	. . . .	Passed	5 min	Jkt. softened flowed	No test
15	PVC 0.008 in. . . . . . . Nylon 0.003 in.	0.124	0.011	Not applicable	Not applicable	. . . .	Passed	11–12.5	Passed	120 hr	Passed 3X	10–11.5
16	22SCI PVC jacket . . .	0.108	0.012	Failed	Passed	2–15	Not applicable	. . . .	Passed	5 min	Passed 3X	5–6.5
17	20SC1 PVC jacket . . .	0.114	0.012	Passed	Passed	16–17.5	Not applicable	. . . .	Passed	5 min	Passed 3X	5–6
18	Tough PVC(B)	0.146	0.015	Passed	Passed	14	Not applicable	. . . .	Passed	5 min	Passed 3X	6–7.5
19	Tough PVC (A′) . . .	0.153	0.016	Passed	Passed	12.5–13.5	Not applicable	. . . .	Passed	5 min	Passed 3X	11.5–12
20	Kynar jacket 3 cond. .	0.143	0.021	Not applicable	. . . .	. . . .	Passed	. . . .	Passed	120 hr	Passed 3X	15–20

[a] Conducted in accordance with MIL-C-27500.

[b] Conducted in accordance with MIL-C-27500, with the addition of 1,000 volts (rms) for 1 min, then raised to breakdown.

[c] Conducted in accordance with MIL-C-27500—dielectric breakdown noted.

[d] Conducted in accordance with MIL-C-27500.

[e] Specimen subjected to 120°C oven temperature, 800-g load, over a 0.052-in. mandrel, 1,000 volts, rms applied continuously—time of breakdown noted. Test discontinued at 120 hr.

[f] Specimen conditioned 24 hr at 150°C wrapped around mandrel 3, 5, 7, and 9 times cable diameter, cooled in a desiccator, cable straightened and visually inspected, then subjected to 1,000 volts rms for 1 min, raised to breakdown after 1 min.

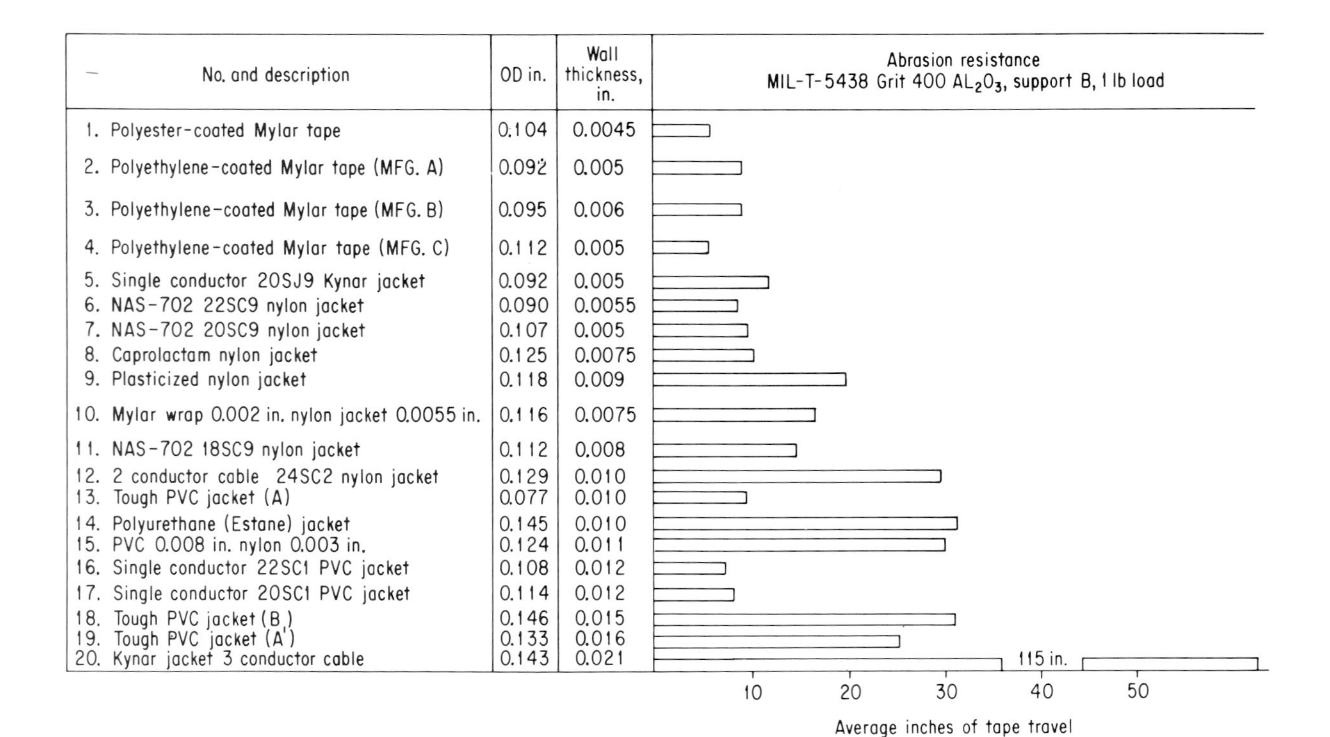

FIG. 6. Shield jacket—abrasion resistance.

perature rating of FEP must be kept in mind; the material should not be used above 200°C if subjected to any appreciable stress.

Kynar. This is a good, tough jacket material with mechanical resistance and superior crack resistance comparable with those of nylon.

Polyethylene-coated Mylar. Excellent for shield jacketing and applied by taping, this causes minimum size buildup. Some problems are encountered in obtaining a good seal over irregular multiconductor cores.

Dacron Braid. This has good temperature resistance (150°C) and good abrasion resistance. Braid construction is flammable and prone to moisture absorption.

Glass Braids. Glass braids are utilized for high-temperature nonflammable applications. Glass braids have poor abrasion resistance and are prone to fraying because satisfactory lacquering is difficult to achieve. Among glass-braid constructions, Teflon-coated glass braids appear most satisfactory where individual fiber strands are coated with a Teflon dispersion, then braided and cured to form a more homogeneous nonfraying construction.

DESIGN CONSIDERATIONS

Wire-gauge Selection. The following factors must be considered for proper wire-gauge selection:

1. Voltage drop
2. Current-carrying capacity
3. Circuit-protector characteristics

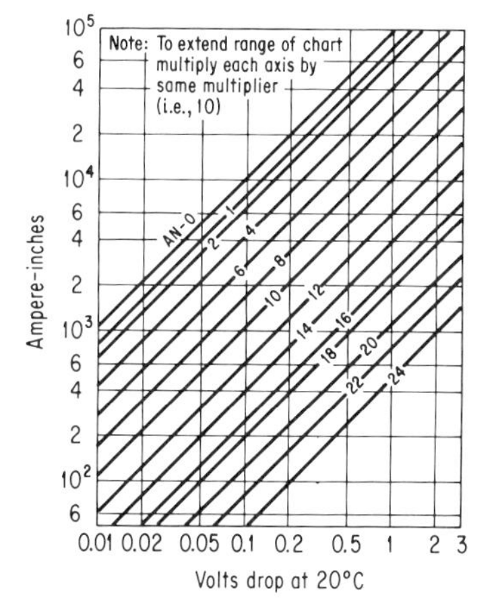

FIG. 7. Direct-current voltage drop of copper wire.[8]

Voltage drop is a major factor in low-voltage systems, except where leads are very short. At high ambient temperatures or high-voltage installations current-carrying capacity controls selection. If several loads are supplied by a single protector, the circuit protector becomes the significant factor. If voltage drop is the major consideration, a single wire should be used to save weight; but if current-carrying capacity is significant, two or more parallel wires will generally weigh less than a single wire of the same total current-carrying capacity. Parallel wires should be of the same gauge and length for even current split.

Voltage Drop. Voltage-drop calculations should be based on anticipated load current at nominal system voltage. Voltage drop through an aluminum structural ground return can be considered zero for all practical purposes. Normal voltage-drop limits do not apply to starting currents of equipment such as motors. The voltage at load during startup should be considered to ensure proper operation of equipment. Figure 7 may be used to select wire gauge. *Ampere-inches* are the product of wire length between terminations in inches and wire current in amperes.

Current-carrying Capacity. The following factors must be considered in determining the current-carrying capacity of a wire:

1. Continuous-duty rating
2. Short-time rating
3. Effect of ambient temperature
4. Effect of wire grouping
5. Effect of altitude

CONTINUOUS-DUTY RATING. This applies if a wire is to carry current for 1,000 sec. or more. The continuous current-carrying capacity of copper and aluminum wire in amperes for aerospace applications is shown in Table 11. The following criteria apply to the ratings in the table:

1. Ambient temperature is
57.2°C (135°F) for 105°C insulated wire
92°C (197.6°F) for 135°C insulated wire
107°C (225°F) for 150°C insulated wire
157°C (315°F) for 200°C insulated wire
2. "Wire bundled" indicates 15 or more wires in a group.
3. The sum of all currents in a bundle is not more than 20 percent of the theoretical capacity of the bundle, which is calculated by adding up the bundle ratings of the individual wires.

SHORT-TIME CURRENT RATINGS. These apply when a wire is to carry a current for less than 1,000 sec. The short time rating is generally applicable to starter lead applications. Figure 8 presents curves for various wire gauges in harness (bundle).

TEMPERATURE-CURRENT RELATIONSHIP. The following equation provides a means of rerating the current-carrying capacity of wire and cable at any anticipated ambient temperature,

$$I = I_r \sqrt{\frac{t_c - t}{t_c - t_r}}$$

where I = current rating in ambient temperature t
I_r = current rating in rated ambient temperature t_r (Table 11)
t = required ambient temperature
t_r = rated ambient temperature (Continuous-duty Rating, above)
t_c = temperature rating of insulated wire or cable

Figures 9 and 10 present curves showing the effect of temperature on the current-carrying capacity of copper wire with 10- and 15-mil insulation, respectively.

EFFECTS OF WIRE GROUPING. When wires are grouped (bundled, harnessed) together, their current ratings must be reduced owing to restricted heat loss. Table 11 and Fig. 6 take into account reduced ratings based on grouping 15 or more wires. If a harness or wire bundle contains fewer than 15 wires, the allowable capacity may be increased toward the rating of a single wire in free air. In grouping wires, it is good practice to use wires of the same gauge in one bundle. Grouping wires of widely differing gauges sacrifices the current capacity of smaller gauges.

EFFECTS OF ALTITUDE. Air density decreases with increasing altitude. Since lower density reduces the dielectric properties of air, trapped air between conductors and insulators represents a problem. In addition, air is retained in voids that occur primarily in stranded conductors, although voids cannot be eliminated from solid insulated wire either. A direct result of increased altitude on insulated wire is lower corona threshold, resulting in lower peak operating voltage. Wire insulated with organic materials should always be operated below the corona extinction voltage, as corona has a degrading effect on all organic materials. Operating below corona threshold but above corona extinction is risky, for a surge may start corona, which will continue until the voltage is lowered to the extinction level.

Figure 11 presents curves on insulation breakdown voltage of air as a function of altitude. In addition to proper derating of operating voltage on insulated wires and cables, derating of termination spacing in air must be considered (see Table 21).

Circuit-protector Characteristics. Selection of the circuit protector must be con-

TABLE 11. Maximum Current Capacity, Amperes

Size, AWG	MIL-W-5088				National Electrical Code	Underwriters Laboratory		American Insurance Association	500 cir mils/amp
	Copper		Aluminum						
	Single wire	Wire bundled	Single wire	Wire bundled		+60°C	+80°C		
30	...	...	...	...	...	0.2	0.4	...	0.20
28	...	...	...	...	...	0.4	0.6	...	0.32
26	...	...	...	...	...	0.6	1.0	...	0.51
24	...	...	...	...	...	1.0	1.6	...	0.81
22	9	5	...	...	...	1.6	2.5	...	1.28
20	11	7.5	...	...	...	2.5	4.0	3	2.04
18	16	10	...	...	6	4.0	6.0	5	3.24
16	22	13	...	...	10	6.0	10.0	7	5.16
14	32	17	...	...	20	10.0	16.0	15	8.22
12	41	23	...	...	30	16.0	26.0	20	13.05
10	55	33	...	...	35	...	...	25	20.8
8	73	46	58	36	50	...	...	35	33.0
6	101	60	86	51	70	...	...	50	52.6
4	135	80	108	64	90	...	...	70	83.4
2	181	100	149	82	125	...	...	90	132.8
1	211	125	177	105	150	...	...	100	167.5
0	245	150	204	125	200	...	...	125	212.0
00	283	175	237	146	225	...	...	150	266.0
000	328	200	...	...	275	...	...	175	336.0
0000	380	225	...	...	325	...	...	225	424.0

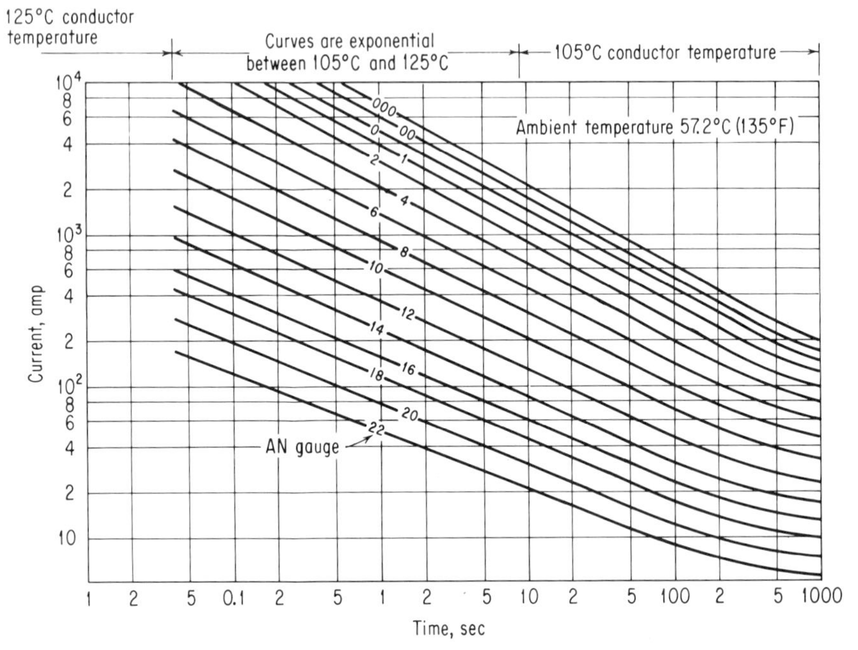

FIG. 8. Short-time working curves for 105°C insulated copper wire in bundles.[8]

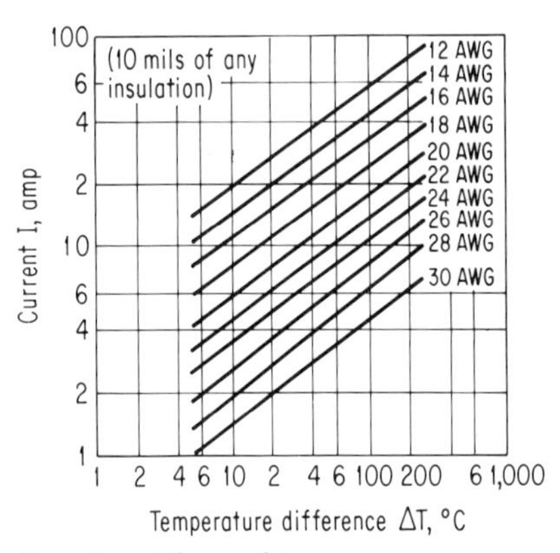

FIG. 9. Effects of temperature on current-carrying capacity of copper conductors (10-mil insulations).[9]

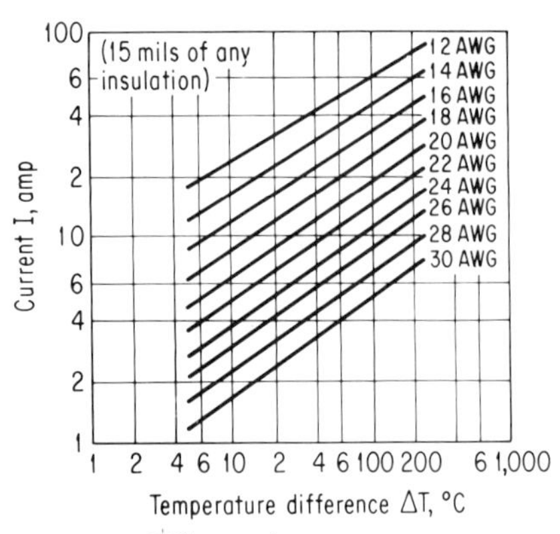

FIG. 10. Effects of temperature on current-carrying capacity of copper conductors (15-mil insulation).[9]

sidered carefully. The circuit protector rating must be low enough to protect the smallest gauge wire connected to it against damage from overheating, smoke, or fire from short circuits.

Insulation-material Selection. Factors governing selection of an optimum insulation material cover a wide range and are less theoretical in nature than wire gauge selection. In addition to meeting specific system electrical and environmental requirements, related fabrication techniques must be considered. The insulation must withstand mechanical abuse and the heat from soldering or application of associated shrinkable devices. Further, the insulation must be compatible with encapsulants, potting compounds, conformal coating, and adhesives.

Following is a designer's checklist for insulation selection:

Insulation Selection Checklist

Requirement	*Considerations*
Environment:	
Temperature extremes	
Continuous-operating	Refer to Tables 6 and 8
Short-term-operating	May require test that simulates specific application
Fabrication temperatures	Check for soldering iron resistance in high-density packaging; cure temperatures of encapsulant; compatibility with shrinkable devices, if employed
Storage	Check for embrittlement, long-term storage, low humidity conditions
Altitude	
Outgassing	Weight loss, smoke, condensation
Corona	Maintain voltage below corona level, especially with insulations susceptive of erosion
Radiation	Refer to Table 8
Weather	Moisture resistance, aging, ultraviolet radiation
Flame	Refer to Table 8
Fluids	Refer to Table 8
Electrical:	
Capacitance	$C = \frac{7.36K}{\log(10D/d)}$ where C = capacitance, pf/ft, K = dielectric constant (Table 7), D = insulated wire diameter, in., d = conductor diameter, in.
Dielectric strength	Refer to Table 7
Volume resistivity	Refer to Table 7
Loss factor	Refer to Table 7
Mechanical:	
Installation and handling	Check for minimum bend radius, special tooling, clamping stresses, chaffing. Refer to Table 6 for abrasion, cut-through, and mechanical resistance
Operating	Refer to Table 6
Size	Refer to applicable specifications for outside dimensions
Weight	Refer to applicable specification for maximum weight. If not listed, use the following equation for insulation weight: $W = \frac{D \times d^2}{2} \times K \times G$ (lb/1,000 ft), where D = diameter over insulation, in., d = diameter over conductor, K = 680, G = specific gravity of insulation (Table 6)

INTERCONNECTION AND HOOKUP WIRE

Definitions. For purposes of this discussion, the difference between interconnection and hookup wire is determined by the amount of mechanical stress applied to the wire. Interconnection wire is used to connect electrical circuits between pieces of equipment; it must withstand rough handling, the abrasion of pulling through conduit, and potential accidental damage during installation that results from the slip impact of hand tools. Hookup wire is used to connect electrical circuits within a unit of equipment or *black box*. This type of equipment may range from miniaturized airborne equipment to a massive ground installation. With today's emphasis on miniaturization, many applications require the use of hookup wire where interconnection wire was used in the past.

Interconnection Wire. As stated, interconnection wire is by nature bulky; it must withstand severe mechanical abuse; and since it has major applications in the aircraft industry, it must have minimum weight and size to allow greater payload. Table 12 presents a listing of interconnection wires most widely used in industry. The tabulation is made on the basis of No. 22 AWG insulated conductor so that weight and size can be analyzed and compared for the different types.

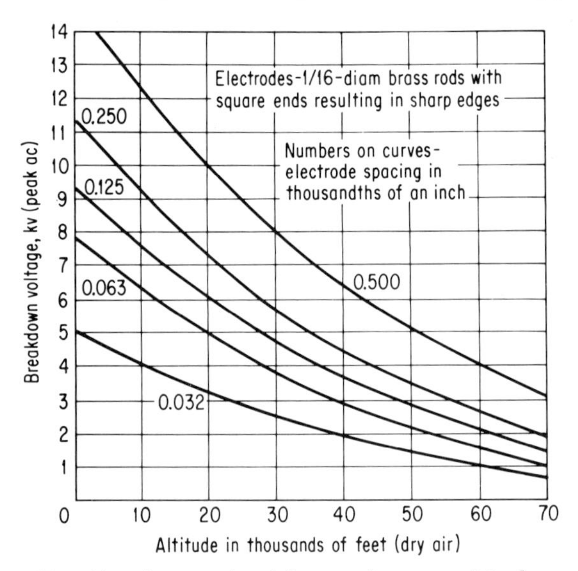

FIG. 11. Average breakdown voltage vs. altitude.

Figure 12 presents thermal life curves for five military specification wires and one nonmilitarized wire construction with 0.018 in. wall thickness of extruded polyvinylidene fluoride (Kynar).

Hookup Wire. Although the name hookup wire is likely to become a catchall for all wire constructions, it refers to the area where by far the highest footage of military and electronic wire is consumed. Under the demands of miniaturization, many aerospace and missile designs call for hookup wire rather than the heavier, bulkier, interconnection wire.

Table 13 presents a summary of hookup wire types, including recent thin-walled insulations.

Table 14 is a tabulation of the properties and characteristics of a specialized hookup wire used with automated termination techniques, such as wire wrap and Termi-point. These termination techniques require a special set of criteria. The major conductor size formerly was No. 24 AWG conductor with a nominal 10-mil wall of insulation. Most recently, however, a No. 30 AWG conductor with a nominal 5-mil wall of insulation is becoming more widespread. Important considerations for automated termination wire are:

1. Stiffness. It is undesirable to have wire "take a set" and "pop up" on wiring panels.

2. Cut-through. To achieve a satisfactory wire wrap, termination pins have very sharp edges.

3. Long lengths. Long, uninterrupted wire lengths are desirable to increase efficiency of operation.
4. Strippability. Machine automatically strips insulated wire. Wall thickness and concentricity must be controlled to close tolerances.

COAXIAL CABLES

Design Considerations. Coaxial cable consists of a center conductor, insulation, shield, and, usually, an outer jacket. It is essentially a shielded and jacketed wire. The term *coaxial* not only implies construction but also connotes usage at radio frequencies.

Background. The purpose of a coaxial cable is to transmit radio-frequency energy from one point to another with minimum loss (attenuation). Loss of radio-frequency energy in a coaxial cable can occur (1) in the conductor, which is a power loss due to heating caused by currents passing through a finite resistance, and (2) in the

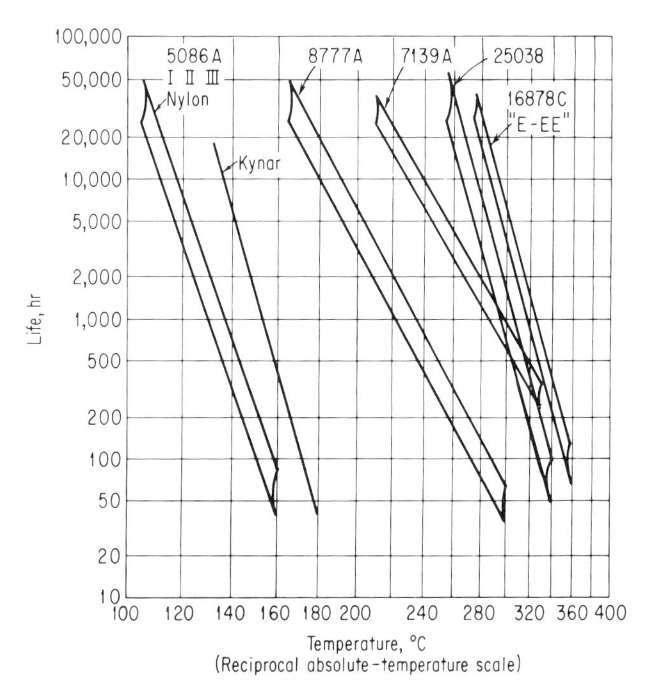

FIG. 12. Summary graph: ranges of life-temperature curves of mil-specification wires.[10]

dielectric, caused by the use of materials with high power factor. Unlike a shielded and jacketed wire used at low frequencies where the major loss is incurred in resistance of the conductor, high-frequency transmission invokes a phenomenon called *skin effect*, where currents travel on the outer surface (skin) of a conductor and partly through the adjacent insulating material; hence, loss in the insulation itself becomes more significant.

Electrical. In addition to loss, other important electrical characteristics of coaxial cables are velocity of propagation, impedance, capacitance, and corona extinction point. These are discussed in detail below.

VELOCITY OF PROPAGATION. Velocity is a function of the insulation dielectric constant, $V = 1/K$, and is expressed in percentage of the speed of light.

TABLE 12. Interconnection Wire Data (Copper Conductor)

Basic specification	Class type or MS No.	Size range, AWG	Conductor coating	Primary insulation	Jacket	Voltage rating, rms	Temperature rating, °C	Diameter rating, max.	Weight rating, max.
MIL-W-5086	MS-25190 Ty 1	22–12	Tin	PVC	Nylon	600	—55 to 105	1.0	1.0
	MS-25190 Ty 2	22–4/0	Tin	PVC	Glass nylon	600	—55 to 105	1.1	1.07
	MS-25190 Ty 3	22–4/0	Tin	PVC	Glass PVC nylon	600	—55 to 105	1.25	1.3
MIL-W-8777	MS-25190 Ty 4	22–16	Tin	PVC	Nylon	3,000	—55 to 105	1.46	1.64
	MS-25471	22–2/0	Silver	Silicone rubber	Dacron braid	600	—55 to 200	1.25	1.32
MIL-W-22759	MS-27110	22–4	Silver	Silicone rubber	FEP	600	—55 to 200	1.25	1.61
	MS-17331	22–8	Silver	TFE	Asbestos	600	—55 to 200	1.1	1.25
	MS-17332	22–8	Nickel	TFE	Asbestos	600	—55 to 260	1.1	1.25
	MS-17411	24–4	Silver	Reinforced TFE	....	600	—55 to 200	1.25	1.82
	MS-17412	24–4	Nickel	Reinforced TFE	....	600	—55 to 260	1.25	1.82
	MS-18000	24–4	Silver	Reinforced TFE	....	600	—55 to 200	1.04	1.36
	MS-18001	24–4	Nickel	Reinforced TFE	....	600	—55 to 260	1.04	1.36
	MS-90294	22–2/0	Silver	TFE and glass tape	Glass braid FEP	600	—55 to 200	1.07	1.3
MIL-W-81044	/1	24–4	Silver	Crosslinked polyalkene	Crosslinked Kynar	600	—55 to 135	1.0	0.98
	/2	24–4	Tin	Crosslinked polyalkene	Kynar	600	—55 to 135	1.0	0.98
MIL-W-25038	MS-27125	22–4/0	Nickel clad	Asbestos glass TFE tape	Glass braid	600	—55 to 288	1.62	2.26
MIL-W-7139	Class 1	22–4/0	Silver	TFE and glass	Glass braid	600	—55 to 200	1.25	1.82
	Class 2	22–4/0	Nickel	TFE and glass	Glass braid	600	—55 to 260	1.25	1.82
MIL-W-81381	/3	26–2	Silver	Polyimide FEP film	FEP dispersion	600	—55 to 200	0.84	0.89
MIL-W-81381	/4	26–2	Nickel	Polyimide FEP film	FEP dispersion	600	—55 to 260	0.84	0.89

TABLE 12. Interconnection Wire Data (Copper Conductor) *(Continued)*

Basic specification	Class type or MS No.	Duty rating	Cost rating	Availability	Mechanical properties	Electrical properties	Chemical resistance	Processing characteristics			
								Solder	Bondability	Strippability	Marking
MIL-W-5086	MS-25190 Ty 1	M	1.0	RA	Fair	Fair	Good	Poor	Good	Good	Good
	MS-25190 Ty 2	M	1.5	RA	Fair	Fair	Good	Poor	Good	Fair	Good
	MS-25190 Ty 3	H	1.7	RA	Good	Fair	Good	Poor	Good	Fair	Good
MIL-W-8777	MS-25190 Ty 4	M	1.3	RA	Fair	Fair	Good	Poor	Good	Good	Good
	MS-25471	H	9.6	LS	Good	Good	Fair	Fair	Good	Fair	Poor
MIL-W-22759	MS-27110	H	9.1	LS	Good	Good	Good	Fair	Poor	Good	Good
	MS-17331	M	10.5	LS	Good	Good	Good	Good	Fair	Fair	Poor
	MS-17332	M	12.0	LS	Good	Good	Good	Good	Fair	Fair	Poor
	MS-17411	H	10.0	RA	Excellent	Good	Excellent	Good	Poor	Good	Poor
	MS-17412	H	11.0	RA	Excellent	Good	Excellent	Good	Poor	Good	Poor
	MS-18000	M	7.8	RA	Good	Good	Excellent	Good	Poor	Good	Poor
	MS-18001	M	8.6	RA	Good	Good	Excellent	Good	Poor	Good	Poor
	MS-90294	M	9.3	LS	Good	Good	Good	Good	Poor	Fair	Good
MIL-W-81044	/1	M	5.6	LS	Good	Good	Good	Fair	Good	Good	Good
	/2	M	4.8	LS	Good	Good	Good	Fair	Good	Good	Good
MIL-W-25038	MS-27125	H	15.0	LS	Good	Fair	Good	Good	Poor	Fair	Poor
MIL-W-7139	Class 1	H	12.2	RA	Excellent	Good	Good	Good	Fair	Fair	Fair
	Class 2	H	12.3	RA	Excellent	Good	Good	Good	Fair	Fair	Fair
MIL-W-81381	/3	M	6.0	RA	Good	Good	Excellent	Good	Fair	Fair	Fair
MIL-W-81381	/4	M	6.7	RA	Good	Good	Excellent	Good	Fair	Fair	Fair

L = light. M = medium. H = heavy. RA = readily available. LS = limited sources (fewer than 4).

TABLE 13. Hookup Wire Data (Copper Conductor)

Basic specification	Type class or MS No.	Size range, AWG	Conductor coating	Primary insulation	Jacket material	Voltage rating, rms	Temperature rating, °C	Diameter rating	Weight rating	Duty rating
MIL-W-16878/1	Type B	32–14	Tin	PVC	....	600	−55 to 105	1.0	1.0	M
MIL-W-16878/1	Type B/N	32–14	Tin	PVC	Nylon	600	−55 to 105	1.15	1.15	H
MIL-W-16878/2	Type C	26–12	Tin	PVC	....	1,000	−55 to 105	1.28	1.22	M
MIL-W-16878/3	Type D	24–1/0	Tin	PVC	....	3,000	−55 to 105	1.81	1.88	M
MIL-W-16878/4A	Type E	32–10	Silver or nickel	TFE	....	600	−55 to 200 or 260	1.02	1.22	M
MIL-W-16878/5A	Type EE	32–8	Silver or nickel	TFE	....	1,000	−55 to 200 or 260	1.21	1.48	M
MIL-W-16878/6A	Type ET	32–20	Silver or nickel	TFE	....	250	−55 to 200 or 260	0.86	0.97	L
MIL-W-16878/7	Type F	24–4/0	Tin, silver, or nickel	Silicone rubber	....	600	−55 to 200	1.21	1.1	M
MIL-W-16878/8	Type FF	24–4/0	Tin, silver, or nickel	Silicone rubber	....	1,000	−55 to 200	1.83	1.75	M
MIL-W-16878/10A	Type J	24–4/0	Tin	Polyethylene	....	600	−55 to 75	1.13	0.98	M
MIL-W-16878/11	Type K	32–10	Silver	FEP	....	600	−55 to 200	1.02	1.22	M
MIL-W-16878/12	Type KK	32–8	Silver	FEP	....	1,000	−55 to 200	1.21	1.48	M
MIL-W-16878/13	Type KT	32–20	Silver	FEP	....	250	−55 to 200	0.86	0.97	L
MIL-W-22759	MS-18104	28–12	Silver	TFE	Polyimide dip-coated	600	−55 to 200	0.98	1.15	H
MIL-W-22759	MS-18105	28–12	Nickel	TFE	Polyimide dip-coated	600	−55 to 260	0.98	1.15	H
MIL-W-22759	MS-21985	28–12	Silver	TFE	....	600	−55 to 200	0.98	1.24	M
MIL-W-22759	MS-21986	28–12	Nickel	TFE	....	600	−55 to 260	0.98	1.24	M
MIL-W-22759	MS-18113	28–8	Silver	TFE	....	1,000	−55 to 200	1.17	1.48	M
MIL-W-22759	MS-18114	28–8	Nickel	TFE	....	1,000	−55 to 260	1.17	1.48	M
MIL-W-81044	/3	30–12	Silver	Crosslinked polyalkene	Crosslinked Kynar	600	−55 to 135	0.92	0.99	H
MIL-W-81044	/4	30–12	Tin	Crosslinked polyalkene	Crosslinked Kynar	600	−55 to 135	0.92	0.99	H
MIL-W-81381	/1	26–10	Silver	Polyimide/ FEP film	FEP dispersion	600	−55 to 200	0.95	1.0	H
MIL-W-81381	/2	26–10	Nickel	Polyimide/ FEP film	TFE dispersion	600	−55 to 260	0.95	1.0	H

TABLE 13. Hookup Wire Data (Copper Conductor) *(Continued)*

Basic specification	Type class or MS No.	Cost rating	Avail- ability	Mechanical properties	Electrical properties	Chemical properties	Iron resistance	Processing characteristics		
								Soldering-bond-ability	Strip-pability	Marking
MIL-W-16878/1	Type B	1.0	RA	Poor	Fair	Fair	Poor	Good	Good	Good
MIL-W-16878/1	Type B/N	1.2	RA	Good	Fair	Good	Poor	Good	Good	Good
MIL-W-16878/2	Type C	1.1	RA	Fair	Fair	Fair	Poor	Good	Good	Good
MIL-W-16878/3	Type D	1.5	RA	Fair	Fair	Fair	Poor	Good	Good	Good
MIL-W-16878/4A	Type E	6.4	RA	Fair	Excellent	Excellent	Excellent	Poor	Fair	Poor
MIL-W-16878/5A	Type EE	8.8	RA	Fair	Excellent	Excellent	Excellent	Poor	Fair	Poor
MIL-W-16878/6A	Type ET	6.4	RA	Poor	Excellent	Excellent	Excellent	Poor	Fair	Poor
MIL-W-16878/7	Type F	6.4	LS	Fair	Good	Poor	Fair	Fair	Good	Fair
MIL-W-16878/8	Type FF	9.0	LS	Fair	Good	Poor	Fair	Fair	Good	Fair
MIL-W-16878/10A	Type J	1.1	RA	Poor	Excellent	Good	Poor	Poor	Good	Good
MIL-W-16878/11	Type K	5.5	RA	Poor	Excellent	Excellent	Poor	Poor	Good	Fair
MIL-W-16878/12	Type KK	7.8	RA	Poor	Excellent	Excellent	Poor	Poor	Good	Fair
MIL-W-16878/13	Type KT	5.3	RA	Poor	Excellent	Excellent	Poor	Poor	Good	Fair
MIL-W-22759	MS-18104	16.1	LS	Good	Excellent	Excellent	Excellent	Fair	Fair	Poor
MIL-W-22759	MS-18105	18.1	LS	Good	Excellent	Excellent	Excellent	Fair	Fair	Poor
MIL-W-22759	MS-21985	6.2	RA	Fair	Excellent	Excellent	Excellent	Poor	Fair	Poor
MIL-W-22759	MS-21986	6.2	RA	Fair	Excellent	Excellent	Excellent	Poor	Fair	Poor
MIL-W-22759	MS-18113	8.9	RA	Fair	Excellent	Excellent	Excellent	Poor	Fair	Poor
MIL-W-22759	MS-18114	9.2	RA	Fair	Excellent	Excellent	Excellent	Poor	Fair	Poor
MIL-W-81044	/3	5.1	LS	Good	Good	Good	Fair	Good	Good	Good
MIL-W-81044	/4	4.2	LS	Good	Good	Good	Fair	Good	Good	Good
MIL-W-81381	/1	8.3	RA	Good +	Good	Excellent	Good	Fair	Fair	Fair
MIL-W-81381	/2	9.0	RA	Good +	Good	Excellent	Good	Fair	Fair	Fair

L = light. M = medium. H = heavy. RA = readily available. LS = limited sources (fewer than 4).

TABLE 14. Automated-termination Wire Data
(No. 30 AWG conductor—nominal 0.005 in. wall thickness)

Insulation material	TFE Teflon	TFE/ML	FEP/ML	Vinylidene fluoride	Polyethylene-coated Mylar	Polysulfone	Polyalkene + Kynar	Kapton
Conductor coating	Silver or nickel	Silver or nickel	Silver	Tin	Tin	Tin	Tin or silver	Silver or nickel
Temperature rating, °C	200 or 260	200 or 260	200	135	125	125	135	200 or 260
Cut-through resistance	Poor	Fair	Fair	Good	Good	Excellent	Good	Excellent
Abrasion resistance	Poor	Fair	Fair	Good	Good	Good	Good	Excellent
Dielectric constant	2.1	2.1	2.1	7.7	2.8	3.2	3.4	3.2
Dielectric strength	Good	Good	Good	Good	Good	Good	Good	Good
Flexibility (stiffness)	Good	Good	Good	Fair	Fair	Fair	Fair	Fair
Chemical resistance	Excellent	Excellent	Excellent	Fair	Good	Poor	Fair	Excellent
Cost	Medium	Medium to high	Medium	Low	Low	Low	Low to medium	High
Availability	RA	LS	LS	RA	LS	LS	LS	DR
Long lengths	Poor	Poor	Fair	Good	Fair	Good	Good	Fair

RA = readily available. LS = limited sources (fewer than 3 manufacturers). DR = development required.

IMPEDANCE. The three common impedance values for coaxial cables are 50, 75, and 95 ohms. Impedance can be determined by the following formula,

$$Z_0 = \frac{138}{K} \log \frac{D}{d}$$

where Z_0 = characteristic impedance, ohms
D = diameter over the insulation, in.
d = diameter over the conductor, in.
K = dielectric constant of the insulation

CAPACITANCE. It is usually desirable to have minimum capacitance for minimum coupling and crosstalk. Capacitance, as is impedance, is a logarithmic function of dimensions and is also dependent on the dielectric constant. The equation for calculating capacitance is

$$C = \frac{7.36K}{\log (D/d)}$$

where C = capacitance, pf/ft
D = diameter over the insulation
d = diameter over the conductor
K = dielectric constant of the insulation

CORONA EXTINCTION POINT. This determines the maximum voltage at which a coaxial cable may be operated. The corona extinction point of a cable is determined experimentally by gradually raising the voltage on a sample of cable until corona is detected, then lowering the voltage until no further ionization is present. If the cable is operated consistently below this level, corona will not occur within the cable. Corona can cause noise at higher frequencies and eventual degradation of certain insulating materials.

Mechanical. The dependence of significant electrical properties, such as attenuation, capacitance, and impedance, on the relative size of conductor and insulation has ramifications in respect to the mechanical strength of coaxial cables. Attenuation can be reduced by increasing conductor size, which in turn forces an increase in insulation wall thickness if capacitance and impedance are to be maintained. An additional means of maintaining low capacitance with increased conductor size is through foamed or air dielectrics. Introduction of air bubbles in an insulation material or the use of air itself reduces the dielectric constant. The introduction of air into a solid insulation material such as polyethylene or FEP Teflon can reduce the dielectric constant to as low as 1.4.

Other coaxial constructions make use of a spiral thread or a web of insulation to center the conductor, which is essentially surrounded by air as the dielectric. From a mechanical standpoint, skin effect aids the coaxial-cable user in that he may use a stronger conductor core coated with a thin layer of copper without significant attenuation but with significant increase in mechanical strength. Copper-coated steel conductors are widely used in coaxial-cable applications. For applications where the cable is to be flexed, an annealed conductor is recommended for greater flex life.

Coaxial cables must be treated with care in handling and installation, since important electrical properties are dependent on conductor-to-insulation dimensions. Any flow or movement of the conductor can seriously affect electrical properties. For installation, the minimum allowable bend radius should be at least 10 times the cable diameter in order to minimize stresses and preclude any cable deformation.

Environmental. Coaxial cables are normally fabricated with low-loss low-dielectric-constant insulation materials covered in the section on Wire and Cable Insulation. Teflon (TFE and FEP), PE, and irradiated PE, including foamed versions, are the most common dielectric materials. For coaxial-cable application, one additional requirement is imposed on shield-jacket materials: they must be noncontaminating. Shield jackets must withstand the required environment without allowing any contamination of the dielectric which might affect its loss characteristics. Con-

TABLE 15. Power Ratings of Coaxial Cable[4]

RG/U cable	Maximum input power rating, watts at frequencies, MHz									
	1.0	10	50	100	200	400	1,000	3,000	5,000	10,000
5, 5A, 5B, 6, 6A, 212	4,000	1,500	800	550	360	250	150	65	50	25
7	4,100	1,550	810	540	370	250	140	70	50	30
8, 8A, 10, 10A, 213, 215	11,000	3,500	1,500	975	685	450	230	115	70	
9, 9A, 9B, 214	9,000	2,700	1,120	780	550	360	200	100	65	40
11, 11A, 12, 12A, 13, 13A, 216	8,000	2,500	1,000	690	490	340	200	100	60	
14, 14A, 74, 74A, 217, 224	20,000	6,000	2,400	1,600	1,000	680	380	170	110	40
17, 17A, 18, 18A, 177, 218, 219	50,000	14,000	5,400	3,600	2,300	1,400	780	360	230	
19, 19A, 20, 20A, 220, 221	110,000	28,000	10,500	6,800	4,200	2,600	1,300	620	410	
21, 21A, 222	1,000	340	160	115	83	60	35	15		
22, 22B, 111, 111A	7,000	1,700	650	430	290	190	110	50		
29	3,500	1,150	510	340	230	150	95	50	35	
34, 34A, 34B	19,000	7,200	2,700	1,650	1,100	700	390	140	80	
35, 35A, 35B, 164	40,000	13,500	5,500	3,800	2,500	1,650	925	370	210	
54, 54A	4,400	1,580	675	450	310	210	120	60	40	
55, 55A, 55B, 223	5,600	1,700	2,700	480	320	215	120	60	40	
57, 57A, 130, 131	10,000	3,000	1,250	830	570	370	205	95		
58, 58B	3,500	1,000	450	300	200	135	80	40	20	
58A, 58C	3,200	1,000	425	290	190	105	60	25	20	
59, 59A, 59B	3,900	1,200	540	380	270	185	110	50	30	
62, 62A, 71, 71A, 71B	4,500	1,400	630	440	320	230	140	65	40	15
62B	3,800	1,350	600	410	285	195	110	50	31	15
63, 63B, 79, 79B	8,200	3,000	1,300	1,000	685	455	270	130	75	35
87A, 116, 165, 166, 226, 227	42,000	15,000	6,250	4,300	3,000	2,050	1,200	620	480	250
94	62,000	15,500	5,900	4,300	2,900	1,900	1,400	650	480	200
94A, 226	64,000	18,000	9,600	6,800	4,600	3,300	1,750	775	540	250

108, 108A	1,300	360	145	100	70	45	30	15	5	
114, 114A	5,300	1,350	475	345	230	150	85	40	25	15
115, 115A, 235	33,000	9,900	4,200	2,900	2,000	1,380	830	600	450	170
117, 118, 211, 228	200,000	66,000	25,000	19,000	12,800	8,500	4,800	2,200	1,400	490
119, 120	100,000	31,000	13,000	9,000	6,100	4,100	2,400	1,100	770	250
122	1,000	240	100	65	45	30	15	10	5	
125	8,500	2,300	910	620	435	285	165	75	45	
140, 141, 141A	19,000	6,300	2,700	1,700	1,200	830	450	220	140	65
142, 142A, 142B	19,000	5,700	2,600	1,800	1,300	900	530	265	175	100
143, 143A	26,000	8,700	3,750	2,600	1,800	1,250	750	390	275	160
144	51,000	17,000	7,500	5,400	3,700	2,500	1,400	700	440	20
149, 150	7,100	1,900	740	485	315	200	105	45	25	
161, 174	1,000	350	160	110	80	60	35	15	10	
178, 178A, 196	1,300	640	330	240	180	120	75	40		
179, 179A, 187	3,000	1,400	750	480	420	320	190	100	73	
180, 180A, 195	4,500	2,000	1,100	800	570	400	240	130	90	50
188, 188A	1,500	770	480	400	325	275	150	80	55	
209	180,000	55,000	22,000	15,000	8,500	6,000	3,400	1,600	1,000	310
281	150,000	47,000	19,000	13,500	8,800	6,000	3,300	1,650	1,150	625

Power–rating conditions: ambient temperature 104°F.
Center–conductor temperature 175°F with polyethylene dielectric.
Center–conductor temperature 400°F with Teflon dielectric.
Altitude–sea level.

TABLE 16. Attenuation Ratings of Coaxial Cables[4]

RG/U cable	Nominal attenuation, db/100 ft at frequencies, MHz									
	1.0	10	50	100	200	400	1,000	2,000	5.000	10,000
5, 5A, 5B, 6, 6A, 212	0.26	0.83	1.9	2.7	4.1	5.9	9.6	23.0	32.0	56.0
7	0.18	0.64	1.6	2.4	3.5	5.2	9.0	18.0	25.0	43.0
8, 8A, 10, 10A, 213, 215	0.15	0.55	1.3	1.9	2.7	4.1	8.0	16.0	27.0	>100.0
9, 9A, 9B, 214	0.21	0.66	1.5	2.3	3.3	5.0	8.8	18.0	27.0	45.0
11, 11A, 12, 12A, 13, 13A, 216	0.19	0.66	1.6	2.3	3.3	4.8	7.8	16.5	26.5	>100.0
14, 14A, 74, 74A, 217, 224	0.12	0.41	1.0	1.4	2.0	3.1	5.5	12.4	19.0	50.0
17, 17A, 18, 18A, 177, 218, 219	0.06	0.24	0.62	0.95	1.5	2.4	4.4	9.5	15.3	>100.0
19, 19A, 20, 20A, 220, 221	0.04	0.17	0.45	0.69	1.12	1.85	3.6	7.7	11.5	>100.0
21, 21A, 222	1.5	4.4	9.3	13.0	18.0	26.0	43.0	85.0	>100.0	>100.0
22, 22B, 111, 111A	0.24	0.80	2.0	3.0	4.5	6.8	12.0	25.0	>100.0	>100.0
29	0.32	1.20	2.95	4.4	6.5	9.6	16.2	30.0	44.0	>100.0
34, 34A, 34B	0.08	0.32	0.85	1.4	2.1	3.3	5.8	16.0	28.0	>100.0
35, 35A, 35B, 164	0.06	0.24	0.58	0.85	1.27	1.95	3.50	8.6	15.5	>100.0
54, 54A	0.33	0.92	2.15	3.2	4.7	6.8	13.0	25.0	37.0	>100.0
55, 55A, 55B, 223	0.30	1.2	3.2	4.8	7.0	10.0	16.5	30.5	46.0	>100.0
57, 57A, 130, 131	0.18	0.65	1.6	2.4	3.5	5.4	9.8	21.0	>100.0	>100.0
58, 58B	0.33	1.25	3.15	4.6	6.9	10.5	17.5	37.5	60.0	>100.0
58A, 58C	0.44	1.4	3.3	4.9	7.4	12.0	24.0	54.0	83.0	>100.0
59, 59A, 59B	0.33	1.1	2.4	3.4	4.9	7.0	12.0	26.5	42.0	>100.0
62, 62A, 71, 71A, 71B	0.25	0.85	1.9	2.7	3.8	5.3	8.7	18.5	30.0	83.0
62B	0.31	0.90	2.0	2.9	4.2	6.2	11.0	24.0	38.0	92.0
63, 63B, 79, 79B	0.19	0.52	1.1	1.5	2.3	3.4	5.8	12.0	20.5	44.0
87A, 116, 165, 166, 225, 227	0.18	0.60	1.4	2.1	3.0	4.5	7.6	15.0	21.5	36.5
94	0.15	0.60	1.6	2.2	3.3	5.0	7.0	16.0	25.0	60.0
94A, 226	0.15	0.55	1.2	1.7	2.5	3.5	6.6	15.0	23.0	50.0

108, 108A	0.70	2.3	5.2	7.5	11.0	16.0	26.0	54.0	86.0	>100.0
114, 114A	0.95	1.3	2.1	2.9	4.4	6.7	11.6	26.0	40.0	65.0
115, 115A, 235	0.17	0.60	1.4	2.0	2.9	4.2	7.0	13.0	20.0	33.0
117, 118, 211, 228	0.09	0.24	0.60	0.90	1.35	2.0	3.5	7.5	12.0	37.0
119, 120	0.12	0.43	1.0	1.5	2.2	3.3	5.5	12.0	17.5	54.0
122	0.40	1.7	4.5	7.0	11.0	16.5	29.0	57.0	87.0	>100.0
125	0.17	0.50	1.1	1.6	2.3	3.5	6.0	13.5	23.0	>100.0
140, 141, 141A	0.30	0.90	2.1	3.3	4.7	6.9	13.0	26.0	40.0	90.0
142, 142A, 142B	0.34	1.1	2.7	3.9	5.6	8.0	13.5	27.0	39.0	70.0
143, 143A	0.25	0.85	1.9	2.8	4.0	5.8	9.5	18.0	25.5	52.0
144	0.19	0.60	1.3	1.8	2.6	3.9	7.0	14.0	22.0	50.0
149, 150	0.24	0.88	2.3	3.5	5.4	8.5	16.0	38.0	65.0	>100.0
161, 174	2.3	3.9	6.6	8.9	12.0	17.5	30.0	64.0	99.0	>100.0
178, 178A, 196	2.6	5.6	10.5	14.0	19.0	28.0	46.0	85.0	>100.0	>100.0
179, 179A, 187	3.0	5.3	8.5	10.0	12.5	16.0	24.0	44.0	64.0	>100.0
180, 180A, 195	2.4	3.3	4.6	5.7	7.6	10.8	17.0	35.0	50.0	88.0
188, 188A	3.1	6.0	9.6	11.4	14.2	16.7	31.0	60.0	82.0	>100.0
209	0.06	0.27	0.68	1.0	1.6	2.5	4.4	9.5	15.0	48.0
281	0.09	0.32	0.78	1.1	1.7	2.5	4.5	9.0	13.0	24.0

tamination is usually associated with PVC compounds, which contain plasticizers that could migrate into the dielectric material. In coaxial-cable applications, moisture resistance of the cable jacket is important. If water, which has a relatively high dielectric constant, penetrates the core, cable performance can be seriously affected.

Cable Selection. A guide to military coaxial cable selection is MIL-HDBK-216. Specific cable types are documented in MIL-C-17, which covers requirements for approximately 150 different cable configurations. Power ratings of MIL-C-17 cable types are covered in Table 15. Table 16 presents nominal attenuation figures for MIL-C-17 cables at specific frequencies ranging from 1 to 10,000 MHz. MIL-C-23806 and MIL-C-22931 are recent specifications covering semiflexible cables with foamed dielectric and air-spaced dielectric.

MULTICONDUCTOR CABLES

Multiconductor cables fall into four major categories: airborne, ground electronics, ground support, and miscellaneous.

Airborne Cables. The primary design considerations for airborne multiconductor cables are size and weight. Airborne cables are often fabricated by a user who selects the appropriate interconnection or hookup wires, lays the insulated wires in a bundle or harness, then laces, spot-ties, and applies insulating tubing over the wiring assembly. Several variations in this type of harness construction have been utilized in an effort to reduce size and weight and increase mechanical protection. In one variation, small-diameter hookup wire is encapsulated in a hard, physically tough epoxy compound. This approach offers greater wire density, reduced overall weight, and effective mechanical protection. Its disadvantages are limited flexibility, reparability, and wire interchangeability. In a second variation, a braided fibrous covering or jacket impregnated with resin for greater abrasion and fray resistance is applied over the cabled wires.

Ribbon cable consisting of conventional wires arranged in single or multiple parallel layers offers particular geometric advantages in many applications. Tape cable, consisting of copper-foil conductors laminated between a dielectric, offers even greater advantages of space savings in one plane. However, the designer must remember that by the use of tape cabling, versatility (as offered by shielding, twisted pair, shielded pair, termination, etc.) is sacrificed in favor of space savings. Table 17 contains a comparative evaluation of the mechanical, electrical, and application features of conventional harness-, ribbon-, and tape-cabling concepts in airborne or ground-electronic applications.

Specifications MIL-C-7078 and MIL-C-27500 cover multiconductor cables utilizing interconnection and hookup wires in a round configuration. These specifications include single shielded and up to seven multiconductor cables with or without an overall shield and with or without an overall jacket. All conductors must be of the same gauge; no individually shielded conductors are permissible. Various shield and jacket options are available offering compatibility with the chosen primary wire. In the interest of minimum weight and size, fillers are not used. Table 18 presents recommended MIL-C-27500 options, including construction details and a mechanical-usage rating.

Ground-electronics Cables. The term *ground-electronics cabling* embraces rack and panel interconnection, equipment cabling installed in conduit (as used with fixed computer and data-processing installations), or cabling placed beneath flooring that is not subjected to extreme mechanical abuse or environment.

Military-specification coverage for multiconductor cables in this area of usage is extremely poor; MIL-C-7078 and MIL-C-27500 are frequently used for lack of adequate specification coverage. MIL-C-27072 is broad enough in scope to cover heavy-duty usage in this area but has had little industrial application to date. The most commonly used heavy-duty construction for this category of cabling contains PVC-insulated nylon-jacketed primary conductors. These are cabled and jacketed with a PVC sheath. Variations to this general construction are unjacketed PVC-insulated conductors (not recommended for individually shielded members because of potential shield-end puncture); shielded or unshielded single conductors, pairs, triplets, quads, etc.; shield-jacketed with PVC or nylon material; equipped with overall shield, if applicable, PVC-sheathed.

Ground-support Cables. Ground-support cables for tactical systems should receive early attention from the designer. These cables are tailored to system needs, require considerable lead time for delivery, and can amount to a considerable system cost if no effort at standardization is made. A major manufacturing cost in the production of this type of multiconductor cable is cabling-machine setup. Many cable manufacturers require a minimum order of 500 to 1,000 ft of cable for a given configuration. After setup additional cable footage can be produced more economically. A prime example of effective cable standardization is the Hawk Missile System employing a minimum number of ground-support cables. Following is a brief discussion of ground-support multiconductor cable applications:

Permanent Installation. Cables that are buried or placed in conduits, open ducts, troughs, or tunnels are considered permanent cables. These cables are not handled, flexed, reeled, or dereeled except at the time of installation.

For permanent installation either neoprene or polyethylene cable sheaths are preferred. Polyethylene offers greater moisture and water protection and has a lower coefficient of friction for easier installation in conduit where the cable must be pulled for substantial distances. Neoprene offers good mechanical protection and greater flexibility, affording easier termination. Cable-pulling compounds can be applied to neoprene-sheathed cable for easier installation in conduits.

Portable Installation. Portable heavy-duty multiconductor ground-support cables are designed to withstand installation and use under the following conditions:

1. Rocky, uneven, or sandy terrain
2. Mechanical abuse such as frequent reeling and dereeling; heavy vehicle traffic; twisting, kinking, jerking, and impact by heavy objects
3. Operating temperature —65 to +165°F, sand, dust, water immersion, high humidity, coastal (salt-water) atmosphere, and ultraviolet radiation

MIL-C-13777 is used as the basis for design of heavy-duty portable cables. The basic construction of this cable is as follows:

1. Conductor: tin-coated, annealed copper, stainless-steel-reinforced, No. 18 AWG and smaller
2. Insulation: polyethylene
3. Jacket: extruded nylon or nylon braid for No. 10 AWG and larger
4. Binder: Mylar wrap
5. Separator: braided cotton
6. Sheath: double-layer reinforced neoprene
7. Component shielding: No. 36 AWG tin-coated copper braid
8. Shield jacket: extruded or braided nylon
9. Overall shield: No. 34 AWG tin-coated copper braid or fused PE-coated Mylar

A working example of utilization of recent techniques in thin-wall insulation, high-strength conductors, and high-strength sheath materials is the lightweight small-diameter portable cable developed for the Pershing Weapons System. Conventional heavy-duty cable was not compatible with the Pershing concept of a quick-reaction weapon system, not only from the standpoint of ease of handling, but also from that of the size and weight of associated hardware and components such as cable reels and storage facilities. Cable size and weight reduction have been accomplished by incorporation of higher-strength conductors, tougher insulation, shield innovations, and tougher sheath materials discussed in the previous sections. For example, a conventional 60-conductor No. 20 AWG MIL-C-13777 cable would have a nominal diameter of 1.121 in. and a cable weight of 890 lb/1,000 ft, as compared with the lightweight cable having a diameter of 0.518 in. and a weight of 370 lb/1,000 ft.

Miscellaneous Cables. *Shipboard Cables.* Specifications MIL-C-915 and MIL-C-21984 cover electrical cables for installation in fixed wireways on combat ships. The available configurations are covered by referenced specifications. In general, shipboard cable design has the following features:

TABLE 17. Cabling-concept Comparison[11]

Conventional cable. Separate wires laced or tied into bundles or jacketed	Ribbon cable. Conventional wires bonded together in ribbon form	Tape cable. Copper-foil conductors sandwiched between thin sheets of dielectric
	Mechanical Features	
Greatest variety of conductor configurations: Single wires–all gauges Shielded wires—all gauges Twisted pair, triples—all gauges Twisted and shielded—all except very large gauges Coaxial cables—all sizes Jacketed overall for heavy duty, if needed	Wide variety of conductor configurations: Single wires—limited to smaller gauges Shielded wires—limited to smaller gauges Twisted pair, triples—limited to smaller gauges Twisted and shielded—limited to smaller gauges Coaxial cables–limited to smaller gauges Not suited for jacketed cables	Limited variety of conductor configurations—some are compromises: Single conductor—smaller gauges only Shielded conductor—smaller gauges only Twisted pair—simulated by zigzag crossovers in sandwich Twisted and shielded—simulated in sandwich construction Coaxial cable not practical—twin-line construction must be used Not suited for jacketed cables
Round-bundle configuration is relatively self-supporting	Flat-bundle configuration requires moderate support	Tape configuration requires continuous support
Stranded wire usually required for increased mechanical life	Stranded wire is desirable. On larger quantities of conductors per cable, solid wire can be used owing to the mutual mechanical effects of adjacent wires	Foil is ordinarily used. Some modifications have used stranded wire configured to a rectangular cross section
Termination preparation—simple mechanical or hot-wire stripper	Termination preparation—simple mechanical or hot-wire stripper after separating wires from each other	Termination preparation—minimum practical production process exists for removal of insulation to expose conductors. Certain types of insulation respond to new welding techniques which do not require insulation removal

Rugged insulation achieved by selective combination of dielectric materials in layers	Rugged insulation achieved by selective choice of primary dielectric plus an overall protective ribbon skin. Cables without this skin permit undesired separation of conductors even with careful handling	Edges of tape insulation or complete laminate should be reinforced because, once nicked, the cable can be torn across conductors with relatively small force. A deep scratch or cut in the outer layer of the cable insulation will allow conductors to fracture with sharp bend at the scratch
	Electrical Features	
Standard voltage-drop characteristics; use standard current ratings. Bundled circuits require derating because of heat rise if circuits are operated simultaneously	Standard voltage-drop characteristics; use standard current ratings. Derating not normally necessary	Voltage-drop characteristics will vary more widely than standard. Broad, tinsellike conductors permit much higher current limits compared with the equivalent cross section of circular conductors. In general this permits use of much less copper for a comparative current requirement if the increased voltage drop can be tolerated
Requires no heat sink in normal application	Requires no heat sink in normal application	Required *continuous* heat sink in normal application to utilize the high current-carrying capability
Interconductor capacitance is high; spacing generally random; capacitance values are unpredictable; capacitance to structure relatively low	Interconductor capacitance is high, spacing fixed, capacitance predictable. Circuits can be selectively spaced to minimize effects. Capacitance from cable to metal support or shielding is high	Interconductor capacitance is low, spacing fixed, capacitance predictable. Circuits can be selectively spaced to minimize effects. Shielding needed in stacked cables. Capacitance from cable to metal support or shielding foil is high
Crosstalk is uncontrolled because of random spacing; individual shielding usually required	Crosstalk can be controlled by conductor placement; shielded conductor not usually required. CAUTION: In stacking cables, overall interlayer shielding may be needed	Crosstalk can be controlled by circuit placement. Overall interlayer shielding may be necessary when cables are stacked

TABLE 17. Cabling-concept Comparison *(Continued)*

Conventional cable. Separate wires laced or tied into bundles or jacketed	Ribbon cable. Conventional wires bonded together in ribbon form	Tape cable. Copper-foil conductors sandwiched between thin sheets of dielectric
	Application Features	
Permanently installed interconnections between units and subsystems; also for field cables when jacketed:	Permanently installed interconnections:	Permanently installed interconnections:
Circular or oval cross-section bundles	Flat-ribbon cross section—ribbons are stackable	Flat tape cross section—tapes are stackable
Follows three-dimensional contour without folding or special fabrication, except for very sharp bends	Lies flat against two-dimensional contour, even with sharp bends. Must be folded over or looped for third dimension	Lies flat against two-dimensional contour, even with sharp bends. Foldover preferred over loop for third dimension
Requires narrow-width path; can be installed with minimum structural design preparation	Requires moderate width, preferably flat path. Needs detailed structural preparation in initial design	Requires excessive width of continuously flat path. Needs elaborate structural preparation in detail design. During initial design period structure must be designed in detail around cable requirements
Requires only air heat sink without special provisions	Requires only air heat sink—higher current ratings achieved with heat sink	Requires continuous heat sink to take advantage of the higher current ratings attainable
Fixed installation requires ties, lacing, or vinyl jacket	Requires no supplementary bundle ties	Requires no supplementary bundle ties
Portable application requires heavy-duty jacket	Not recommended for portable application	Not recommended for portable applications
Installation clamps and brackets required; quantity and spacing to suit application	Can be cemented in place for permanent installation or fastened with flat clamps	Should be cemented in place; clamps not recommended unless cable is derated

Each conductor requires color coding or stamped identification number. Controlled conductor-to-conductor orientation is not practical	Conductor orientation in cable relative to conductor orientation in an end item *should* be planned carefully and coordinated during all design phases to achieve proper electrical interface. This is particularly significant when rectangular connectors or grid-spaced terminals are used. Individual conductors can be split apart from the cable as an emergency procedure. Such practice is not recommended	Conductor orientation in cable relative to conductor orientation in an end item *must* be carefully planned and coordinated during all design phases to achieve proper electrical interface. Individual conductors cannot be reliably split apart from the cable
Can be used with largest variety of connectors; i.e., circular, rectangular or special. Termination by crimp, solder, weld, pressure, or other techniques	Can be used with large variety of connectors, but the inherent configuration is most compatible with rectangular connectors with terminals spaced on a grid. Grid spacing need not be exactly matched, since the cable conductors may be locally separated one from the other	Variety of connectors is very limited. Grid spacing of connector terminals must match grid spacing of cable conductors. The tolerance control requirements are very exacting. Industry standard is still nebulous
Cable replacement is relatively easy: requires loosening or removing clamps and installing new cable without special tools or processes; field repair of individual conductors by simple splice techniques	Cable installed with clamps can be replaced as readily as conventional cables. Use of cable cemented in place is not desirable if field replacement or repair is required. Procedures for removal, cleaning, and re-installation, and the need for special materials unduly complicate field repairs	Cable installed with clamps can be readily replaced; cable cemented in place cannot be serviced readily. Individual repair is impractical

TABLE 18. Multiconductor Cable Options—Recommended MIL-C-27500

Basic primary wire specifications	Specification symbol	Size range, AWG	Shielded		Jacketed		Shield-jacketed		Voltage rating, rms	Temperature rating, °C
			Shield style	Jacket style	Shield style	Jacket style	Shield style	Jacket style		
MIL-W-5086:										
MS-25190, TY I	A	22–12	T	O	U	1 or 3	T	1, 2, or 3	600	–55 to 105
MS-25190, TY II	B	22–4/0	T	O	U	3	T	1 or 3	600	–55 to 105
MS-25190, TY III	C	22–4/0	T	O	U	3	T	1 or 3	600	–55 to 105
MS-25190, TY IV	P	22–16	T	O	U	3	T	1 or 3	3,000	–55 to 105
MIL-W-7139, CL 1	D	22–4/0	S	O	U	7	S	6 or 7	600	–55 to 200
MIL-W-7139, CL 2	E	22–4/0	N	O	U	7	N	6 or 7	600	–55 to 200
MIL-W-8777:										
MS-25471	H	22–2/0	S	O	U	4	S	4	600	–55 to 150
MS-27110	F	22–4	S	O	U	5	S	5	600	–55 to 200
MIL-W-22759:										
MS-17411	V	24–4	S	O	U	6 or 7	S	6 or 7	600	–55 to 200
MS-17412	W	24–4	N	O	U	6 or 7	N	6 or 7	600	–55 to 260
MS-18000	S	24–4	S	O	U	6 or 7	S	6 or 7	600	–55 to 200
MS-18001	T	24–4	N	O	U	6 or 7	N	6 or 7	600	–55 to 260
MS-18113	LA	28–8	S	O	U	6	S	6	1,000	–55 to 200
MS-18114	LB	28–8	N	O	U	6	N	6	1,000	–55 to 200
MS-21985	R	28–12	S	O	U	6	S	6	600	–55 to 200
MS-21986	L	28–12	N	O	U	6	N	6	600	–55 to 260
MS-90294	N	22–2/0	S	O	U	6 or 7	S	6 or 7	600	–55 to 200
MIL-W-25038:										
MS-27125	J	22–4/0	F	O	U	7	F	7	600	–55 to 750
MIL-W-81044/1	M	24–4	S	O	U	4* or 5	S	4* or 5	600	–55 to 135
MIL-W-81044/2	MA	24–4	T	O	U	4 or 5	T	4 or 5	600	–55 to 135
MIL-W-81044/3	MB	30–12	S	O	U	4 or 5	S	4 or 5	600	–55 to 135
MIL-W-81044/4	MC	30–12	T	O	U	4 or 5	T	4 or 5	600	–55 to 135

* A jacket compatible with primary insulation system not available to date.

Kynar, polyethylene-coated Mylar, and crosslinked Kynar are proposed additions to specification.

TABLE 18. Multiconductor Cable Options—Recommended MIL-C-27500 *(Continued)*

Basic primary wire specifications	Specification symbol	Mechanical duty rating	Conductor	Construction details: Primary insulation	Shield	Jacket
MIL-W-5086:						
MS-25190, TY I	A	Medium	Tinned copper	PVC/nylon	Tinned copper	(1) PVC
MS-25190, TY II	B	Medium; fire-resistant	Tinned copper	PVC/glass nylon	Tinned copper	(2) Extruded nylon
MS-25190, TY III	C	Heavy	Tinned copper	PVC/glass PVC/nylon	Tinned copper	(3) Nylon braid
MS-25190, TY IV	P	Medium	Tinned copper	PVC/nylon	Tinned copper	
MIL-W-7139, CL 1	D	Heavy	Silver-coated copper	Teflon TFE	Silver-coated copper	(6) Taped TFE Teflon
MIL-W-7139, CL 2	E	Heavy	Nickel-coated copper	Tapes and glass braid	Nickel-coated copper	(7) Glass braid
MIL-W-8777:						
MS-25471	H	Heavy	Silver-coated copper	Silicone rubber	Silver-coated copper	(4) Dacron braid
MS-27110	F	Medium	....	....	....	(5) Extruded FED Teflon
MIL-W-22759:						
MS-17411	V	Heavy	Silver-copper	Mineral-filled Teflon (TFE)	Silver-copper	(6) Taped TFE Teflon
MS-17412	W	Heavy	Nickel-copper		Nickel-copper	(7) Glass braid
MS-18000	S	Medium	Silver-copper	....	Silver-copper	
MS-18001	T	Medium	Nickel-copper	....	Nickel-copper	
MS-18113	LA	Light	Silver-copper	Extruded TFE Teflon	Silver-copper	(6) Taped TFE Teflon
MS-18114	LB	Light	Nickel-copper	....	Nickel-copper	
MS-21985	R	Light	Silver-copper	....	Silver-copper	
MS-21986	L	Light	Nickel copper	....	Nickel-copper	
MS-90294	N	Medium	Silver-copper	TFE-glass-FED	Silver-copper	(6) Taped TFE Teflon (7) Glass braid
MIL-W-25038:						
MS-27125	J	Heavy; fire-resistant	Nickel-clad copper	TFE tapes and glass braid	Stainless steel	(7) Glass braid
MIL-W-81044/1	M	Medium	Silver-copper	Polyalkene and Kynar (cross-linked)	Silver-copper	(4) Dacron braid*
MIL-W-81044/2	MA	Medium	Tinned copper	....	Tinned copper	(5) Extruded FED Teflon
MIL-W-81044/3	MB	Light	Silver-copper	....	Silver-copper	
MIL-W-81044/4	MC	Light	Tinned copper	....	Tinned copper	

1. Conductors are uncoated copper made watertight by filling the strands with a flexible material that is compatible with the insulation.
2. Insulation is either extruded silicone rubber or silicone-rubber-treated glass tape.
3. Fillers are used to obtain a well-rounded cable core.
4. A qualified binder may be used at the manufacturer's option.
5. A separator is employed only with specific approval.
6. An impervious sheath, normally neoprene rubber, is applied over the cable core.
7. A braided metal armor is applied over the sheath.
8. Finally, the metal armor braid is painted for corrosion protection.

TABLE 19. Building-wire Insulations*

Trade name	Type letter	Insulation	Outer covering
Code	R	Code rubber	Moisture-resistant, flame-retardant, nonmetallic covering†
Heat-resistant	RH RHH	Heat-resistant rubber	Moisture-resistant, flame-retardant, nonmetallic covering†
Moisture-resistant	RW	Moisture-resistant rubber	Moisture-resistant, flame-retardant, nonmetallic covering†
Moisture and heat-resistant	RH-RW	Moisture and heat-resistant rubber	Moisture-resistant, flame-retardant, nonmetallic covering†
Moisture and heat-resistant	RHW	Moisture and heat-resistant rubber	Moisture-resistant, flame-retardant, nonmetallic covering†
Heat-resistant latex rubber	RUH	90% unmilled, grainless rubber	Moisture-resistant, flame-retardant, nonmetallic covering
Moisture-resistant latex rubber	RUW	90% unmilled, grainless rubber	Moisture-resistant, flame-retardant, nonmetallic covering
Thermoplastic	T	Flame-retardant, thermoplastic compound	None
Moisture-resistant thermoplastic	TW	Flame-retardant moisture-resistant thermoplastic	None
Heat-resistant thermoplastic	THHN	Flame-retardant, heat-resistant thermoplastic	Nylon jacket
Moisture and heat-resistant thermoplastic	THW	Flame-retardant, moisture- and heat-resistant thermoplastic	None

TABLE 19. Building-wire Insulations* (*Continued*)

Trade name	Type letter	Insulation	Outer covering
Moisture- and heat-resistant thermoplastic	THWN	Flame-retardant, moisture- and heat-resistant thermoplastic	Nylon jacket
Thermoplastic and asbestos	TA	Thermoplastic and asbestos	Flame-retardant, nonmetallic covering
Thermoplastic and fibrous braid	TBS	Thermoplastic	Flame-retardant, nonmetallic covering
Synthetic heat-resistant	SIS	Heat-resistant rubber	None
Mineral-insulated metal-sheathed	MI	Magnesium oxide	Copper
Silicone-asbestos..	SA	Silicone rubber	Asbestos or glass
Fluorinated ethylene propylene	FEP	Fluorinated ethylene propylene	None
	FEPB	Fluorinated ethylene propylene	Glass braid Asbestos braid
Varnished cambric	V	Varnished cambric	Nonmetallic covering or lead sheath
Asbestos and varnished cambric	AVA and AVL	Impregnated asbestos and varnished cambric	AVA-asbestos braid or glass AVL-lead sheath
Asbestos and varnished cambric	AVB	Impregnated asbestos and varnished cambric	Flame-retardant cotton braid (switchboard wiring)
			Flame-retardant cotton braid
Asbestos	A	Asbestos	Without asbestos braid
Asbestos	AA	Asbestos	With asbestos braid or glass
Asbestos	AI	Impregnated asbestos	Without asbestos braid
Asbestos	AIA	Impregnated asbestos	With asbestos braid or glass
Paper	.....	Paper	Lead sheath

* Excerpted from table 310-2(b), National Electrical Code 1965, USAS C1-1965.
† Outer covering is not required over rubber insulations which have been specifically approved for the purpose.

Shipboard cables are designed for watertightness, to prevent water transmission between sealed bulkheads, and for heat and flame resistance, to prevent flame travel if exposed to fire or current overload.

Rubber-insulated Cables. IPCEA Publication S-19-81 was combined with NEMA Publication WC3 for the IPCEA-NEMA Standards Publication entitled "Rubber-insulated Wire and Cable for the Transmission and Distribution of Electrical Energy." These standards apply to materials, construction, and testing of rubber-insulated wires and cables for installation and service for indoor, aerial, underground, portable, or submarine applications.

Construction and materials covered in the standards vary with application. General features are as follows:

1. Conductors are annealed, coated or uncoated copper or aluminum; sizes AWG 8 and smaller are normally solid; stranded for specific applications.
2. Insulations are vulcanized natural or synthetic rubber compounds.
3. Insulation shielding is metallic, nonmagnetic, and made of tape, braid, serve, or tubular.
4. Coverings—single-conductor wires or cables, AWG 8 and smaller, rated at 600 volts or less, have at least one covering applied over the insulation.

All other constructions have at least two coverings applied; i.e., two braids, a braid and a serve, or a tape and two servings. Coverings shall be compound-filled tape, cotton yarn, rubber, or thermoplastic jackets.

Building Wires and Cables. Building wire and cable applicable to the wiring of buildings or apparatus installed in buildings are described in the National Electrical Code as published by Underwriters' Laboratories, Inc.

There are 26 different types of building wire and cable as listed in the 1965 edition of the National Electrical Code. Table 19 presents the various constructions available as listed in Table 310-2(b) of the National Electrical Code. Federal Specifications J-C-580 and J-C-129 partly cover these types.

Design Considerations. Design of the cabling installation is an integral part of the mechanical design of any equipment or system requiring electrical interconnection. Design planning for electrical installation should be concurrent with the layout of the mechanical design. Quality installation design requires that each of the following major considerations be thoroughly evaluated and a positive design approach determined.

Environment. Vibration, acceleration, and shock are dynamic environments which are controlling factors from a design viewpoint. Acceleration places a load on cables, supports, brackets, connectors, and mounting points for black boxes. Wherever practicable, these items must be so designed as to be in compression against a structural member. Connectors should be oriented so that the possibility of inadvertent disconnection is minimized. Consideration shall also be given to deceleration forces.

Vibration sets up varying stresses in cables and supports proportional to the mass being supported. Shock, including acceleration and deceleration, also contributes to an overstressed condition.

Other environmental conditions include high dynamic pressure, elevated temperatures, and lowered atmospheric pressure. High temperature has the most severe effect on network installations. Insulation against high temperature and selection of materials capable of withstanding high temperature are the basic approaches for controlling problems caused by temperature extremes. Low atmospheric pressure is insignificant except in outer-space applications. Here the effects can be significant: outgassing and deterioration of plastics with time.

Several different phenomena are generated by a nuclear burst: radiation, heat pulse, shock wave, and electromagnetic pulses (EMP). The initial radiation may be of such type and intensity and over a sufficient time interval so that it may seriously degrade the quality of materials, including metals. The heat pulse and shock wave generate conditions similar to those already discussed above. The EMP can produce voltages and currents through the structure. Thus, installation hardware can become electrically energized with electrical stresses far higher than those normally encountered.

Ground environmental factors include high and low temperatures, humidity, and dynamic parameters due to handling and transportation. The levels of these ground environmental factors are usually far less severe than flight environment. However, despite the lower level, the duration of these stresses is far in excess of normal flight time. The accumulated stress or degradation under these conditions may be quite appreciable.

Routing and Grouping. Interconnecting cables and networks should be designed and installed to minimize the adverse effects of electromagnetic interference and to control crosstalk between circuits. To eliminate the adverse effects, special grouping, separation, and shielding practices should be followed, for which the following general guidelines are recommended:

1. Direct-current supply lines. Use twisted pair. Separate from ac power and control lines.
2. Alternating-current power lines. Use twisted lines. Separate from susceptible lines. Shield ac circuits in which switching transients occur, and ground the shield at both ends.
3. Low-level signals. Use shielded twisted pair, and ground the shield at one end.
4. High-level signals. Use shielded twisted pair, and ground the shield at both ends.
5. Provide adequate filtering to prevent conducted-noise problems.
6. Follow a single-point ground concept where possible. Analyze flow of parasitic chassis currents, and design ground conductor for worst case.
7. Plan the separation of signal and power circuits with maximum distance between runs.
8. Hold wire and cable length to minimum.
9. Locate high-heat-generating wires on the outside.
10. Plan the cable routing in coordination with the structural-design effort. Plan cable runs and tie-down points in the early design phase for incorporation into the structure. Plan for minimum length. Attempt to optimize cable installation; compromise only in the solution of installation and maintenance problems.
11. Hold bend radius of coaxial and multiconductor cables to at least 10 times the cable outside diameter.

MAGNET WIRE

The field of film-insulated wire is so vast and, in general, unrelated to insulation materials and to practices connected with hookup and interconnection wires that some descriptions of conductors and insulation must be restated in this section in a form applicable to magnet wire.

Conductors. *Materials and Construction.* COPPER. The most common conductor is bare, round, solid annealed copper wire in accordance with USA Standards Specification C7.1. Square and rectangular copper wire is available as described in C7.9. Copper strip can be obtained in accordance with QQ-C-576. Rounded edges should be specified to preclude any roughness or sharp projections.

Bare copper will oxidize rapidly at temperatures approaching 200°C; at these temperatures copper should be protected with silver or nickel coating. Nickel plating can safely be used for oxidation protection of copper up to 260°C. Above 260°C and up to 400°C nickel cladding can be used on a continuous basis with permissible time exposures up to 700°C. Hollow copper conductors through which a coolant is forced are also available. Stainless-steel-clad copper has been used up to 650°C with little oxidation effect; above 650°C alloys such as Inconel are recommended.

ALUMINUM. Shortage of copper has resulted in the increasing application of aluminum conductors. Unfortunately, in most applications, direct substitution of copper is impossible without significant change in design because aluminum conductors have lower conductivity and increased brittleness. A significant area for the application of aluminum is in insulated strip conductors for distribution and power transformers. Anodization (surface oxidation) of aluminum conductors presents a unique approach to both mechanical protection and electrical insulation. Figure 13 is a chart for the dielectric strength of various anodized-aluminum surface thicknesses. Aluminum oxide is inorganic and therefore possesses many desirable electrical insulation properties, such as resistance to radiation, to aging at high temperatures (melting point 3600°F), and to chemical attack. The film is somewhat porous, but sealing treatments for protection against moisture are available.

CONDUCTORS FOR HIGH-TEMPERATURE APPLICATIONS. The usable temperature range of copper conductors (bare and with protective coatings) is evaluated on the basis of oxidation, melting point, grain growth, and solid-state diffusion. Temperature ranges have been established in the form of a spectrum as follows:

1. Very low temperature (VLT) 70 to 90°C (160 to 195°F)
2. Low temperature (LT) 90 to 120°C (195 to 250°F)
3. Medium temperature (MT) 120 to 170°C (250 to 340°F)
4. High temperature (HT) 170 to 250°C (340 to 480°F)
5. Very high temperature (VHT) 250 to 400°C (480 to 750°F)
6. Ultrahigh temperature (UHT) 400 to 650°C (750 to 1200°F)
7. Superhigh temperature (SHT) 650 to 1000°C (1200 to 1830°F)
8. Extremely high temperature (EHT) 1000 to 1500°C (1830 to 2730°F)

Table 20 presents bare conductors, coated conductors, and magnetic materials suitable for use at each of the eight temperature ranges.

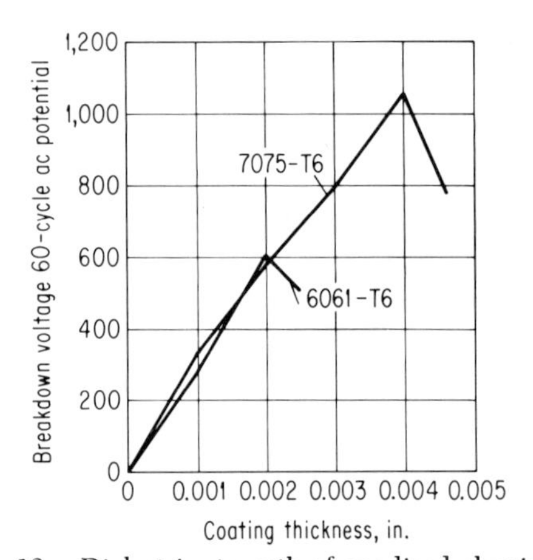

FIG. 13. Dielectric strength of anodized aluminum.

Insulations. *Film Insulation.* ACRYLIC. Rated for 105°C temperature operation, acrylic coating is available with modifiers to produce an enamel with solderable characteristics. The basic unmodified resin is resistant to refrigerants and solvents. Major usage of the insulation has been in hermetically sealed motors. The non-solderable version is covered by NEMA Standard MW4.

CERAMIC. Ceramic insulation may be used for temperatures as high as 650°C depending on the conductor utilized. Ceramic insulation exhibits good resistance to radiation but is difficult to handle and has poor moisture resistance. At present there is no specification covering ceramic insulation.

CERAMIC WITH OVERCOAT. Ceramic insulation with overcoats of polyimide, silicone, or polytetrafluoroethylene for moisture barrier and crack protection is available. Since it decomposes at temperatures below that of the ceramic, the overcoat material is the temperature-limiting factor. None of the overcoats is specifically covered in MIL-W-583. NEMA Standards MW8 and MW7 provide coverage for the Teflon and silicone overcoated constructions, respectively.

EPOXY. Epoxy enamels may be rated as high as 130°C. They exhibit good moisture, chemical, and corona resistance. A cement-coated epoxy magnet wire is available offering a self-bonding coating that can be used at elevated temperatures. This type of construction is suitable for self-supported coils. The cement-coated epoxy can be bonded by oven heating, resistance heating, or solvent activation. Epoxy-insulated magnet wire meets the requirements of MIL-W-583, Class 130, Types

TABLE 20. Thermal Spectrum for Electrical Insulation, Conductors, and Magnetic-circuit Elements from 70 to 150°C[12]

Class of materials	VLT (very low temperature) 70°C ← → 90°C 160°F 195°F	LT (low temperature) 90°C ← → 120°C 195°F 250°F	MT (medium temperature) 120°C ← → 170°C 250°F 340°F	HT (high temperature) 170°C ← → 250°C 340°F 480°F	VHT (very high temperature) 250°C ← → 400°C 480°F 750°F	UHT (ultrahigh temperature) 400°C ← → 650°C 750°F 1200°F	SHT (superhigh temperature) 650°C ← → 1000°C 1200°F 1830°F	EHT (extremely high temperature) 1000°C ← → 1500°C 1830°F 2730°F
Electrical insulation	Untreated cotton, paper, silk, etc.	Oil-filled or varnished cotton, paper, silk, enamels such as Formvar, nylon, etc.	Varnished glass and mica, Mylar and other polyester films, enamels such as polyurethane, polyester, epoxy, and combinations	Polyimide, silicone, TFE-fluorocarbon, silicone-varnished glass fibers, mica, etc., resins plus ceramic, polyimide plus glass fiber	Ceramic-coated wires, glass-bonded fiber glass, glass enamel, glass-bonded mica and asbestos, etc.	Glass-bonded fibers, glass plus ceramic, glass-bonded synthetic mica, glass enamel, etc.	Glass plus refractories, crystallized glass, quartz, ceramic fibers plus glass	Pure refractory oxides, sapphire, beryllia, magnesia
Conductors	Copper, aluminum	Copper, aluminum	Copper, aluminum	Nickel-plated copper, aluminum copper at 180°C	Nickel-plated copper, nickel-clad copper, aluminum	Nickel-clad copper, stainless-steel-clad copper, Cufenic (nickel-iron-clad copper), nickel-clad silver	Inconel plus barrier over Inconel-clad silver	Platinum
Magnetic materials	Iron	Iron	Iron	Iron	Iron	Iron (to 500°C), cobalt alloys	Cobalt alloys, cobalt	None available

B, B_2, B_3, and B_4. NEMA Standards MW14 and MW9 cover rectangular (square wires) and round wires, respectively. To date, there is no specification coverage of the cement-coated epoxy.

POLYAMIDE (NYLON). Nylon is used mainly for overcoating of magnet wire. It provides a tough, smooth surface for better windability, solvent resistance, abrasion resistance, and varnish compatibility. It can be soldered through with resin-alcohol flux and tin-lead solder. Its moisture resistance is poor, which may preclude its use where high insulation resistance is required.

Nylon is covered by MIL-W-583, Class 105, Types T_1, T_2, T_3, and T_4, and NEMA Standard MW6.

OLEORESINOUS. This film, consisting of a cured varnish made with a natural resin and a drying oil, is the oldest of the enameled magnet wires. During recent years natural resins have been replaced with synthetic materials. Oleoresinous coating is preferred for paper-filled coils and where low initial cost is important.

Oleoresinous enamels are covered by MIL-W-583, Class 105, Types E and E2. The NEMA Standard Classification is MW1.

POLYAMIDE-POLYIMIDE (AMIDE-IMIDE). One of the most recent advances in magnet-wire insulations, Amide-Imide presents a host of desirable properties such as (1) toughness, smoothness, and the abrasion resistance of nylon, (2) good dielectric strength in a humid environment, (3) resistance to deformation at high temperatures, (4) operating temperature range up to 220°C, (5) resistance to solvents, (6) compatibility with insulating varnishes and encapsulants, and (7) good radiation resistance, withstanding gamma exposure up to 3×10^9 rads. Amide-Imide insulation will meet the requirements of MIL-W-583, Class 220, Types M_1, M_2, M_3, and M_4. No NEMA standard has been prepared to date.

POLYESTER. Polyester is one of the oldest insulating-film materials. Significant improvements in the heat resistance of the coating have been made: temperature ratings range from 130 to 200°C. Polyester insulation is generally provided with an overcoat for improved physical performance. The insulation is susceptible to hydrolysis; encapsulation is recommended. Polyester-polyimide, a basic polyester material that has been modified with polyurethane and polyimide, was recently introduced; it is tougher and more heat-resistant than its predecessors. Polyester materials offer good solvent resistance and compatibility with insulating varnishes and encapsulating materials. Specifications MIL-W-583, Classes 155, 180, and 200, depending on formulation and overcoating, are applicable for Types L1, L2, L3, and L4; H1, H2, H3, and H4; and K, K1, K2, K3, and K4. NEMA Standards are MW5 for Class 155, Round Wire, MW13, Class 155, Rectangular and Square Wire, and MW25, Class 180, Round Wire.

POLYTETRAFLUOROETHYLENE (TEFLON). Teflon is a high-temperature insulation that may be used up to 260°C maximum. Silver- or nickel-coated conductors are recommended for use above 200°C. The insulation is applied as a dispersion coating, then cured to remove the carrier. Teflon exhibits good electrical properties, flexibility, fair abrasion resistance and plasticity, but poor adhesion properties unless treated. Teflon magnet wire meets the requirements of MIL-W-583, Class 200, Types K1, K2, K3, and K4. NEMA Standard MW10 is applicable to coated round copper wire.

POLYURETHANE. Polyurethane polymers, a family of relatively recent magnet-wire insulation materials, have captured a significant portion of the market. Polyurethane-coated wire can be soldered without prior removal of insulation. The film presents a tough coating that has good chemical, moisture, and corona resistance.

Polyurethane-coated, round copper magnet wire is covered by NEMA Standard MW2, thermal classification 105°C. (MIL-W-583 does not specifically cover polyurethane-coated wire.)

Polyurethane-insulated magnet wire is available with various overcoats: a friction-surface overcoat is applied containing inorganic materials, which allows winding of basket as universal weave coils without the use of adhesives; polyvinyl butyral resin overcoat is applied to achieve a bondable wire with coating that may be activated by heat or solvent; a combination overcoat of nylon and polyvinyl butyral is available for increased cut-through resistance in a bondable 130°C version; nylon overcoat is available for improved windability and varnish compatibility. The construction is solderable without prior removal of the insulation. NEMA Standard MW28 is applicable to polyurethane-nylon-coated round copper in thermal class 130°C. The construction is not specifically covered by MIL-W-853 but will meet the requirements of Class 130, Types B, B2, B3, and B4.

POLYVINYL FORMAL (FORMVAR). This resin is the "old reliable" of film-insulated wire. Though challenged by many newly developed resins, Formvar is still the most widely used. Formvar insulation is suitable for use in Class A electrical equipment. It has high dielectric strength, good abrasion resistance, and good windability characteristics and is compatible with most electrical insulations and varnishes. The thermal classification for polyvinyl formal is 105°C. NEMA Standards MW15 and MW18 cover the insulation for round and rectangular (square) wires, respectively. MIL-W-583, Class 105, Types T, T2, T3, and T4, is applicable.

Polyvinyl formal has been modified with isocyanates for resistance to fluorinated refrigerants in hermetic use. NEMA Standard MW27 (proposed) covers this specific construction for round wire. Polyvinyl formal is also available with a nylon or a polyvinyl butyral overcoat for improved windability of self-bonding. NEMA Standards are MW17 for nylon-coated polyvinyl formal and MW19 for self-bonding overcoat, round wire. Nylon-overcoated construction meets the requirements of MIL-W-583, Class 105, Types T, T2, T3, and T4. Self-bonding overcoat construction is not covered by military specifications.

POLYIMIDE (PYRE-ML). The development of Pyre-ML coating is an outstanding advance for magnet-wire insulation. Pyre-ML is the only enameled wire which has a 220 thermal classification. It is chemically inert and therefore is compatible with all varnishes and encapsulating compounds. Polyimide coating provides an extremely tough, abrasion-resistant film which exhibits high nuclear-radiation resistance, excellent thermal resistance, and good windability. NEMA Standard MW16 covers polyimide-coated round wire. MIL-W-583, Class 220, Types M, M2, M3, and M4, is applicable to polyimide-coated round, square, and rectangular wire.

Textile and Composite Insulation. ASBESTOS FIBER. Asbestos-fiber-covered magnet wire is available with phenolic or asphaltic impregnation for 130°C usage and with silicone for 180°C applications. There are no NEMA standards for these constructions. MIL-W-583, Class 130, Type AV, covers asbestos varnish construction.

CELLULOSE ACETATE FIBER. Insulation is applied by one or more servings of fiber. There is no NEMA standard. Construction meets the requirements of MIL-W-583, Class 90,* Types F and F2.

COTTON. Cotton-insulated wire is available with a single or double serve; impregnated or unimpregnated, it is the oldest such construction. Primary insulation is achieved through space separation. NEMA standards are MW11, Class 90 or 105 for (impregnated) round copper wire, and MW12 for rectangular wire (MW12 has been omitted from the latest proposed standards). MIL-W-583, Class 90,* Types C and C2, covers cotton insulation.

GLASS FIBER. Glass-fiber insulation is available with a wide variety of impregnating and bonding agents and in combination with polyester fibers. These constructions are employed when long service at high temperatures and high cut-through strength are required. Glass insulation provides insulation by spacing where extreme temperatures may drive out binders. Glass, used in combination with polyester fibers which are subsequently fused, provides a smooth surface and prevents fraying. Glass-fiber combinations are available in the following forms:

1. Glass-fiber-covered and impregnated round copper wire, Class 155, NEMA Standard MW41. Square or rectangular wire, MW42. Construction is covered by MIL-W-583, Class 130, Types GV and G2V.

2. Glass-fiber-covered, silicone-treated, round copper wire, Class 180, NEMA Standard MW44. Rectangular and square wires are covered by MW43. Construction is in accordance with MIL-W-583, Class 200, Types GH and G2H.

* Obsoleted by Revision C. Not to be used for new design.

3. Polyester-glass-fiber-covered round copper wire, Class 155, NEMA Standard MW45. Rectangular and square wires are covered by MW46. MIL-W-583, Class 130, Types DG and DG2, cover the construction.

4. Polyamide- (nylon) fiber insulation is used in small rotating machines, in instruments, and where more spacing of conductors is needed than can be realized with enameled insulation. Nylon-fiber insulation is covered by NEMA Standard MW22 and MIL-W-583, Class 90,* Types F and F2.

5. Paper insulation is used primarily in oil-filled transformers. NEMA standards for round and rectangular conductors are MW31 and MW33, respectively. MIL-W-583 Class 90,* Types P and P2, is applicable to paper-insulated round wire.

6. Silk insulation is applied with one or two serves of yarn. NEMA Standard MW21 and MIL-W-583, Class 90, Types S and S2, are applicable.

Design Considerations. Table 21 contains the characteristics of most commonly used magnet wires. Following is a brief discussion of environmental, electrical, and mechanical considerations:

Environmental Requirements. In contrast to interconnection and hookup wire, the environmental requirements for magnet wire are largely self-imposed. High-temperature extremes are imposed by demands for increased efficiency, higher operating temperatures, and overload protection. Insulation compatibility is required to cope with a multitude of varnishes, encapsulants, and system fluids.

In recent years, radiation resistance has become a significant environmental consideration in industrial use and in military applications.

In addition to radiation, moisture is a significant external factor in the environment. High humidity can cause insulation degradation with the use of materials that are susceptible to hydrolysis or high moisture absorption.

Electrical Requirements. The electrical requirements of magnet wire are not normally as stringent as those of exposed wire systems. A continuous insulation film is certainly desirable; however, the chance of breakdown caused by two discontinuities lined up in adjacent windings and facing each other is extremely small. The likelihood of this mode of failure is further decreased by coil impregnation, encapsulation, or potting.

Mechanical Requirements. Mechanical property considerations for magnet wire insulation and conductors are as follows:

Conductor properties:

1. Malleability
2. Dimensional uniformity
3. Solderability

Insulation properties:

1. Surface condition of insulation—windability
2. Abrasion resistance
3. Flexibility
4. Resistance to flow
5. Solderability

Conductor malleability is an important quality in magnet wire. A conductor that has been hardened by excessive tension will not form an evenly wound package and will tend to spring out in unreeling and coil winding. In addition, if a conductor has been hardened, reduced cross section and higher resistance are indicated.

Dimensional uniformity results in proper winding buildup and minimum "hot spots." The surface condition of the conductor, such as the lack of oxidation prior to the application of a conductive or insulating coating, has a significant effect on solderability.

WIRE AND CABLE TERMINATIONS

Terminating Hardware. Wire and cable terminations are of major importance to design reliability. Careful selection of proper terminating hardware and the reduction

* Obsoleted by Revision C. Not to be used for new design.

TABLE 21. Film-insulated Wire Characteristics [13]

Insulation type	Class	Snap	Maximum flexibility	Abrasion		Dielectric strength, volts/mil	Solubility					Completeness of cure	Cut-through, °C	Solderability, °F	Remarks
				Repeated, strokes	Single, g		Naphtha	Toluol	Alcohol	Mild acid	Mild alkali				
Black enamel (oleo resins)	A	OK	15% + 1 time	1–2	900	1,200–1,500	OK	OK	OK	OK	OK	Fails	160	No	Minimum abrasion, no solder
Formvar	A	OK	Snap + 1 time	60	2,700	2,000–2,500	OK	OK	OK	OK	OK	OK	225	1000	Excellent windability, excellent moisture resistance
Nyform	A	OK	Snap + 1 time	40	2,400	2,000–2,500	OK	OK	OK	OK	OK	OK	220	1000	Same as Formvar, slight moisture absorption
Thermoplastic-overcoated Formvar	A	OK	Snap + 1 time	30	No test	2,000–2,500	OK	OK	Softens for self-bond	OK	OK	No test	200	1000	Self-bonding when alcohol is applied
Solderable Acrilac	A	OK	Snap + 1 time	50	2,000	2,000–2,500	OK	OK	OK	OK	OK	OK	200	850	Good solderability, slight moisture absorption
Polyurethane	A	OK	Snap + 3 times	40	2,300	2,000–2,500	OK	OK	OK	OK	OK	OK	220	680	Excellent solderability
Thermoplastic-overcoated polyurethane	A	OK	Snap + 1 time	30	No test	2,000–2,500	OK	OK	Softens for self-bond	OK	OK	No test	220	680	Excellent solderability, self-bonding
Epoxy	B	OK	Snap + 3 times	25	1,400	2,000–2,500	OK	OK	OK	OK	OK	Fails	175	900	Chemically inert
Nylon over polyurethane	B	OK	Snap + 1 time	30	2,300	2,000–2,500	OK	OK	OK	OK	OK	OK	240	680	Excellent solderability and windability
Thermoplastic-overcoated epoxy	B	OK	Snap + 3 times	25	1,500	2,000–2,500	OK	OK	Softens for self-bond	OK	OK	Fails	175	900	Self-bonding
Polyester	F	OK	Snap + 1 time	50	2,000	2,000–2,500	OK	OK	OK	OK	OK	OK	300	No	No solder
Linear polyester over polyester	F	OK	Snap + 1 time	60	1,700	2,000–2,800	OK	OK	OK	OK	OK	OK	300	No	No solder, excellent abrasion resistance
Polythermalax	H	OK	Snap + 2 times	60	1,700	2,000–2,800	OK	OK	OK	OK	OK	OK	300	No	No solder, high abrasion resistance and thermal properties
Polyamide	250C	OK	Snap + 3 times	25	1,500	2,000–3,000	OK	OK	OK	OK	OK	OK	Over 500	No	No solder, excellent thermal properties

of the number of terminations to a minimum should be primary design goals. Refer to Chaps. 3, 4, and 6 for further details on terminations.

The following conditions must be evaluated in the selection and use of terminations: (1) termination life; (2) connection density; (3) compatibility; (4) environment; (5) preparation; (6) mass production; (7) process control; (8) inspectability; (9) current and voltage and resistance limits; (10) maintenance tools; (11) repairability, including time and skill required; (12) contractual constraints.

A number of wire-attachment methods are used in wire terminations such as (1) crimping, (2) soldering, (3) clamping, (4) welding, (5) wrapping, and (6) friction.

Terminating devices normally used in electrical installations are (1) studs, (2) lugs (crimp), (3) terminal posts (solder or wrap), (4) connectors (with solder or crimp-type contact terminal), (5) splices (crimp or solder), (6) compression screw lugs, (7) screw terminals (usually limited to use on barrier strips), (8) ferrules, (9) taper pins, and (10) pads and eyelits.

Terminals. TERMINAL LUGS. These are designed to establish electrical connection between a wire and a connection point such as a stud.

TERMINAL POSTS. Terminal posts are used on terminal boards, in the assembly type of wiring, and on many components, such as electrical connectors (solder-type) relays, transformers, lampholders, switches, etc.

DESIGN GUIDES FOR TERMINALS. Following are the "dos" and "don'ts" of terminals and wire terminations:

Do:

Use special, prebussed connector terminals where required.

Apply supplementary insulation sleeving over axial terminations where continuous insulation is not provided between adjacent terminations.

Ensure that electrical spacings between terminals conform to Table 22.

TABLE 22. Allowable Voltage between Terminals, Volts

Min. air space, in.	Creepage distance, in.	At sea level			At 50,000 ft			At 70,000 ft		
		Flash-over, rms	Working		Flash-over, rms	Working		Flash-over, rms	Working	
			dc	ac		dc	ac		dc	ac
*	3/64	800	280	200	300	100	75	200	70	50
1/32*	1/16	1,400	490	350	500	190	125	375	125	90
3/64	5/64	2,000	700	500	700	210	175	500	175	125
1/16	7/64	2,500	840	600	900	315	225	600	210	150
5/64	1/8	3,000	1,050	750	1,050	360	260	675	230	165
3/32	5/32	3,600	1,260	900	1,200	420	300	750	260	185
1/8	3/16	4,500	1,550	1,100	1,400	490	350	900	310	225
3/16	1/4	6,100	2,000	1,500	1,800	630	450	1,100	375	275
1/4	5/16	7,300	2,500	1,800	2,000	700	500	1,300	455	325
5/16	3/8	8,500	2,900	2,100	2,300	810	575	1,420	500	355

NOTE: The allowable voltage is determined by the actual creepage distance or the minimum air space, whichever provides a lower rating. At 70,000 ft visible corona has been recorded by voltages as low as 350 volts rms. Consequently, at these elevations corona may be the limiting factor rather than flashover.

* Continuous insulation should be provided between electrical connections of 1/32 in. or less.

Don't:

Use solder cap adaptors to accommodate additional connectors or larger gauges in connector terminals.

Connect more than three leads to one terminal.

Twist multiple wires or leads to effect terminations.

Terminate more than one wire in a connector terminal.

Terminal Boards. These are used for junctions or terminations of wire or cable assemblies as an aid to installation and maintenance.

STUD TERMINAL BOARD. The *stud terminal board* is generally a threaded post with the axial portion of its body firmly anchored into a mounting panel; it requires the use of tools for attachment of wire lugs. The features of the stud terminal board are listed below:

1. Provisions for four wires per stud
2. Greater mechanical strength than the barrier type
3. Adaptability to bus connections
4. Suitability for larger wire gauges
5. Vibration resistance by proper choice of locking nut or vibration-proof washer

BARRIER TERMINAL BOARD. The *barrier terminal board* is molded of insulating material; it has integral raised barriers between pairs of screw terminals. Its features are listed below:

1. Longer leakage path than the study type between adjacent terminals
2. More connections in a given length of board
3. Limited current-carrying capacity
4. Poor adaptability to applications with high levels of dynamic stress

Maximum wire sizes which can be connected to these terminal boards are shown in Table 23.

TABLE 23. Maximum Wire Size for Connection to Terminal Boards

	Stud-type terminal board			Barrier-type terminal board	
	Stud size			Screw size	
	6	10	¼–⅜	5/40	6/32
Copper wire (AWG) . .	12	6	2/0	18	18
Aluminum wire (AWG)	None	6	1/0	None	None

TAPER-PIN TERMINAL BLOCK. This is composed of molded insulating material containing metal inserts designed to hold taper pins. Its features are listed below:

1. High-density construction
2. Provisions for mutual connection of four wires (in the dual-insert type)
3. Limited range of wire gauges
4. Special tooling requirement for pin insertion and pin-to-wire crimping

SPECIFICATIONS. Terminal boards should be installed in accordance with MIL-E-7080 for aircraft and MIL-E-25366 for missiles unless other specific requirements are established.

Splices. Permanent splices, available for both shielded and unshielded cables, should be used only when absolutely required. Conductor splices in interconnecting wiring should be grouped and located in designated areas selected for ready access. Where leads from electrical equipment are spliced into a cable assembly, the splice area should be located as near to the equipment as practical. Nonpermanent splices should be avoided; however, certain very special applications may require their use.

Shield Wiring Terminations. There are two basic shield terminations: terminated to a shield common, and floating.

Shield termination merits careful consideration from the design phase through production. A judicious grouping of wires and careful examination of the need for shielding will alleviate termination problems. Following are generalized design recommendations:

1. Minimize the use of shielded wiring.
2. Avoid shielding of leads less than 4 in. long.
3. Provide lead segregation instead of shielding where this is practical.
4. Become familiar with all facets of shield termination techniques (see subsequent paragraphs, pages 2-64 to 2-65) to ensure that the techniques fully satisfy the design environment.

Some of the above is, of course, not applicable to ac, pulsed, or rf leads and cables with significant radiation potential. The following shield terminations are in general practice:

1. Direct shield termination (pigtail)
2. Ferrule termination (crimp attachment)
3. Solder sleeve termination (solder attachment)

DIRECT SHIELD TERMINATION. An established practice is to form the shield braid into a pigtail as in Fig. 14. No special tooling is required, but the pigtail must be

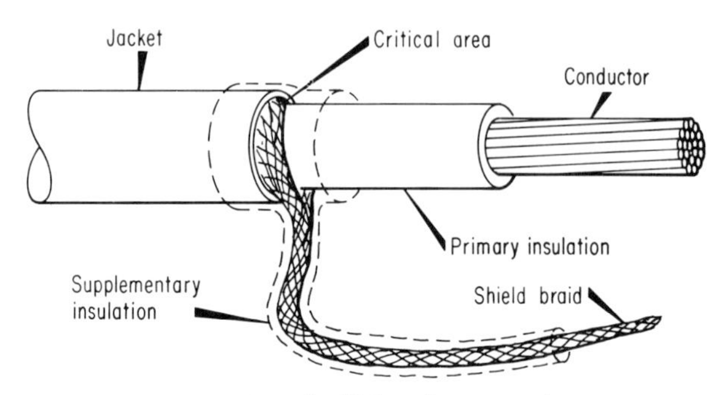

FIG. 14. Shield braid in pigtail.

prepared with special care to preclude damage to the primary wire insulation beneath the braid at the breakout point or the shield braid strands. No external pressure should be applied to the breakout point by clamp, tie, or flexure. Supplementary insulation (sleeving) over the breakout of the braid and the braid itself is required.

SHIELD FERRULE TERMINATION. Shield termination ferrules are available in two basic types: two-piece preinsulated and single-piece uninsulated. Both types require careful design control to assure proper application. The preinsulated type satisfies applications where the shield braid must be electrically isolated from other shields. The uninsulated ferrule would be unsuitable for this application. While ferrules are lightweight, they add to the bulk of the harness or cable trunk. The bulk problem can be remedied by the staggering of ferrule positions back along the harness or cable trunk. However, this practice leads to degradation of overall shielding effectiveness. A typical application of two-piece insulated ferrule is shown in Fig. 15.

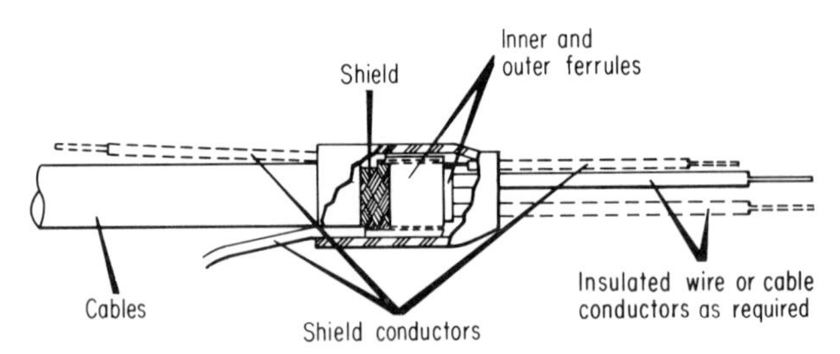

FIG. 15. Insulated ferrule shield termination.

SOLDER SLEEVE TERMINATION. Solder sleeve is another shield termination device. Prior to its selection, compatibility between solder melting temperature, insulating sleeve shrink temperature, and the temperature resistance of the primary wire insulation, as well as that of the jacket over the shield, must be determined. Grouping of shield conductors for solder sleeve applications is shown in Fig. 16. The use of solder sleeves permits the use of center-strip shield terminations to minimize the bulk of shield terminations at connector back shells and to allow continuation of shields closer to the point of termination for the shielded conductor.

Further design considerations involve shield grouping and collection. With direct shield termination the shield conductor (pigtail) may be formed by the twisted braid terminated in a crimped lug for connection to stud or screw. In certain applications the pigtail may be directly soldered to a suitable terminal post.

Multiple shield braids or multiple shield conductors extending from shield ferrules or solder sleeves may be collected into a single shield conductor as shown in Fig. 16, which also gives typical shield-conductor grouping schemes.

In floating terminations, a preferred practice, a short length of the shield is folded back over the outer sheath, and a short length of close-fitting insulating sleeving is applied over the fold-back. The heat-shrinkable tubing will provide snugly fitting insulation that is safely retained in place. This method should be used for floating shield terminations with all three of the shield terminations previously discussed.

RF CABLE (COAXIAL) TERMINATION. Terminations for rf cables may be selected from MIL-HDBK-216. Straight rf connectors of the TNC (threaded coupling) type are desirable. The right-angle type of rf connectors and adaptors should be avoided

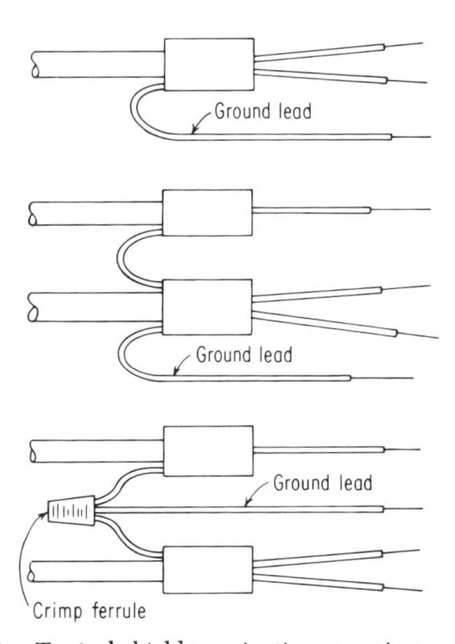

FIG. 16. Typical shield-termination grouping practice.

because of the inherent mechanical weakness of many designs which use brazed metal housings.

Identification. Identification of wiring and cabling includes the marking or coding of individual wire leads, harnesses, cables, and termination devices. Wire identification facilitates design control and traceability (wiring diagram to hardware, etc.), manufacturing efficiency, and maintenance (trouble shooting). Identifying markings on harness and cable assemblies usually provide usage information, interconnection instructions, and part number. The marking may also include serial number, source, assembly date, lot number, etc. Marking is also useful for inventory control and supply stock records.

Wire Marking. MIL-STD-681 is applicable to various wire-marking methods. Color stripes, bands, and numbers are acceptable. Numbering can be a relatively simple, sequential matter, beginning with number 1 and progressing consecutively to the highest number required for an assembly.

Certain military requirements specify a coded marking for individual wires, which includes (1) unit number, (2) equipment identity or circuit function, (3) wire number, (4) wire segment letter, (5) wire gauge, and (6) ground, phase, or thermocouple letter. Hot impression stamping or color banding of the required wire identification is a practical production marking method. Some wire-insulation materials are difficult to mark by this method. In these cases, a short length of close-fitting insulating sleeving or shrinkable tubing with the marking is slipped over the wire, adjacent to its point of termination.

Harness Marking. Wire-harness assembly marking should be as simple as possible to convey the information required. A simple method is to use a short length of close-fitting insulating sleeving over the harness trunk, adjacent to each termination. Identification can be applied by hot impression stamping of the thermoplastic sleeve.

Cable Marking. Sheathed cables are identified in a manner similar to wiring-harness assemblies. The most significant difference between the two involves the materials used in the actual marking. A preferred cable-marking method for production cables uses a reflective label with pressure-sensitive, adhesive backing. Other types of cable marking used in military electrical applications have specific disadvantages:

1. Embossed metal bands or straps are difficult to read under poor lighting conditions. Retention is difficult. The band may constitute a hazard to cable sheath and personnel.

2. Hot-impression-stamped thermoplastic sleeves, the type used on wiring harnesses, are difficult to retain in position. From experience, this type of marking on sheathed cables may discolor under field conditions and the marking information may thus be obliterated.

Associated Hardware. *Insulation Sleeving and Tubing.* GENERAL. Insulation sleeving and tubing serve a multiple purpose in electrical assembly and harness fabrication. They are used for insulating; protection from chafing or abrasion; jacketing; strain relief; thermal or chemical protection; and identification. For best selection of sleeving a thorough knowledge of the physical and chemical properties of the materials employed and of the available construction configurations is required. Following is a review of sleeving and tubing characteristics.

MATERIALS. Extruded tubings are made from all the plastic and rubber materials discussed in the section on Wire and Cable Insulation. Material characteristics described there apply to insulating tubing as well and will not be reviewed here.

A major category not previously covered is braided insulating sleeving. Braided sleeving is made from basic uncoated yarns, lightly treated yarns, or yarns heavily coated with various insulating varnishes or resins. Although practically any yarn can be used for braiding, the materials most frequently used are fiber glass, cotton, rayon, and asbestos. Untreated braided sleevings have excellent flexibility, but the only space advantage they offer is with respect to electrical insulation. Sleevings may be coated with various resin compounds such as PVC, oleoresinous varnish, acrylic varnish, polyester, polyurethane, epoxy, polyimide, polytetrafluoroethylene, silicone varnish, and rubber. Such coating reduces flexibility, however; elasticity of the finished sleeving will thus largely depend on the choice of coating material. Elastomeric materials, such as silicone rubber, yield the highest flexibility.

MILITARY SPECIFICATIONS. MIL-I-631, MIL-I-22076, and MIL-I-7444 are the military specifications for extruded polyvinyl chloride tubing. Refer to Table 24 to determine the characteristics for proper selection.

Extruded polytetrafluoroethane (TFE) tubing is covered by MIL-I-22129 for thin-wall electrical applications and by MIL-P-22296 for heavier-wall, larger-size-range tubing designed mainly for mechanical applications.

Neoprene tubing for harnessing applications that require severe abrasion resistance is covered by MIL-R-6855.

Federal Specification ZZ-R-765 covers silicone rubber tubing for extraflexible heat-resistant harness applications.

Specifications for braided sleeving with optional coatings are available in MIL-I-3190, MIL-I-18057, and MIL-I-21557. Refer to Table 25 to determine the proper selection for a specific application.

TABLE 24. Extruded Tubing

Tubing type	Size range	Wall thickness nominal, in.	Temperature range, °C	Colors	Flammability	Applicable specifications	Characteristics
Polyvinyl chloride (PVC)	AWG 20 to 2 in.	0.016–0.060	−20 to +105	Clear and colors	Self-extinguishing	MIL-I-631 Gr. C; Cl. I, Cat. I	Fungus- and flame-resistant
High-temperature	AWG 24 to 2 in.	0.012–0.060	−10 to +105	Clear and colors	Self-extinguishing	ASTM D922, Gr. C., UL–105	Flame-resistant
Polyvinyl chloride, general-purpose	AWG 20 to 2 in.	0.016–0.060	−30 to +80	Clear and colors	Self-extinguishing	MIL-I-631, Gr. a, Cl. I, Cat. I	Fungus- and flame-resistant
	AWG 24 to 2 in.	0.012–0.060	−30 to +60	Colors only	Self-extinguishing	ASTM D922 Grade A	Good printability
Polyvinyl chloride (PVC) low-temperature	AWG 20 to 2 in.	0.016–0.060	−46 to +80	Clear and colors	Self-extinguishing	MIL-I-631, Gr. 6, Cl. I, Cat. I	Fungus- and flame-resistant, good dielectric
	AWG 24 to 2 in.	0.012–0.060	−55 to +80	Clear and colors	Self-extinguishing	MIL-I-22076	Good low and high temp., noncorrosive, flame- and fungus-resistant
	AWG 24 to 2 1/2 in.	0.012–0.070	Cl I: −68 to +80 Cl. II: −55 to +80	Clear and colors	Self-extinguishing	MIL-I-7444	Corrosion-resistant, good low temperature
Polytetrafluoroethylene (TFE)	AWG 30 to 0	0.009–0.020	−70 to 250	Natural and colors	Non-burning	MIL-I-22129	Excellent dielectric, chemically inert
	1/16 to 3 in.	As specified	−70 to 250	Natural	Non-burning	MIL-P-22296	Mechanical applications, heavy wall, abrasion-resistant
Polychloroprene (neoprene)	1/8 to 1/8 to	Light, 3/64–1/16 Heavy, 3/32–1/8 Extra heavy, 3/16–1/4	−40 to 75	Black	Self-extinguishing	MIL-R-6855, Class II	Mechanical applications, oil-resistant
Silicone rubber . .	AWG 24 to 2 in.	0.012–0.060	−75 to 200	White	Self-extinguishing	ZZ-R-765, Class III, Grade 60	Tear-resistant, good flexibility

TABLE 25. Braided Sleeving

Tubing type	Size range	Temperature range, °C	Dielectric strength, volts	Wall thickness, in.	Color	Specifications	Characteristics
Varnished cotton, rayon, or nylon	AWG 24 to 1 in.	−10 to 105	To 7,000	......	Natural, yellow, and black std. for classes A and B	MIL-I-3190, Cl. A NEMA vs 1, Ty 1 ASTM D372	Low moisture absorption; oil- and acid-resistant; high tensile strength; good flexibility
Varnished glass	AWG 24 to 1 in.	−10 to 130	To 7,000	......	Natural, yellow, and black std.	MIL-I-3190, Cl. B, NEMA vs 1, Ty 2 ASTM D372	Chemical-resistant flexible; tear and moisture-resistant
Silicone-varnished glass	AWG 24 to 1 in.	−60 to 200	To 7,000	......	Natural	MIL-I-3190, Cl. H, NEMA vs 1, Ty 4	Heat-resistant, compatible with magnet wire coatings
Vinyl-coated glass	AWG 24 to 1 in.	−10 to 130	To 8,000	......	....	MIL-I-21557, NEMA vs 1, Ty 3	Oil- and solvent-resistant, hot-spot temperature 130°C
Acrylic-coated glass	AWG 24 to 1 in.	−10 to 155	To 7,000	......	Natural, yellow, and black std.	NEMA vs 1, Ty 6 ASTM D372	Tough; abrasion-resistant; chemical-resistant
Silicone-rubber-coated glass	AWG 24 to 1 in.	−70 to 200	To 8,000	0.030–0.075	Natural	MIL-I-19057, NEMA vs 1, Ty 5	Extra-flexible; resists bending and flexing; radiation resistant
Polytetrafluoroethylene-coated glass	AWG 24 to 1 in.	−80 to 250	....	......	Natural	Not covered to date—tentative NEMA Ty 4	Excellent resistance to fluids; high heat resistance; nonflammable
Polyimide-coated glass	AWG 24 to 1 in.	−70 to 250	....	......	Natural	Not covered to date—tentative NEMA Ty 4	Tough; high heat resistance; nonflammable

INDUSTRIAL SPECIFICATIONS. ASTM D922, Grades A, B, and C, covers general-purpose, low-temperature, and high-temperature extruded PVC tubing. NEMA VS-1 and ASTM D372 cover coated braided sleeving.

Shrinkable Tubing. Shrinkable tubing is based on the theory of elastic memory. Under specific thermal and mechanical conditions, molecules of certain materials may be overexpanded, then frozen in place in a strained condition; finally, when heated, the material tends to return to its original shape and size as strains are released.

MATERIALS. Heat-shrinkable tubing is available in many of the plastics and elastomers listed in the section on Wire and Cable Insulation. The properties of these heat-shrinkable materials are comparable with those of conventional materials, Heat-shrinkable tubing is at present available in:

1. Polyvinyl chloride (PVC)
2. Polyolefin (irradiated)
3. Polyvinylidene fluoride (Kynar)
4. Polychloroprene (neoprene)
5. Silicone rubber
6. Butyl rubber
7. Polytetrafluoroethylene (TFE)
8. Fluorinated ethylene propylene (FEP)

Table 26 presents typical properties of heat-shrinkable tubings.

SPECIFICATIONS. Military Specification MIL-I-23053 covers heat-shrinkable polyvinyl chloride, polyolefin, and polytetrafluoroethylene materials. The specification has six classes, which are given in Table 27.

Industrial SAE Specifications AMS-3636, -3637, -3638, and -3639, respectively, cover the following heat-shrinkable polyolefin tubing: pigmented, flexible; clear, flexible; pigmented, semirigid; and clear, semirigid.

Shrinkable Devices. MELTABLE LINER. A variation of the standard shrinkable tubing is of dual wall construction. The inner wall is composed of noncrosslinked material, and the outer of crosslinked material. With the proper amount of heat the inner wall melts, and the outer shrinks, forcing the melted material into voids. The meltable-liner construction is applicable to tubing and end-cap devices, offering the additional advantage of moisture sealing.

MOLDED SHAPES. A great variety of shrinkable molded shapes (boots, etc.) that offer bend and strain relief are available in polyolefin, neoprene, silicone, and butyl materials. In addition to connector boots, shrinkable molded breakouts are available for harnessing applications. Adhesives are used to attach the molded shapes to connector shell and cable jacket. The interior of the boot or breakout may be potted to ensure complete encapsulation and moisture resistance.

SOLDER SLEEVE. This shield termination device consists of a crosslinked, shrinkable plastic tube containing a flux-cored, preformed solder ring at the center, with thermoplastic rings at either end. The device is placed over a cable shield and ground wire; then heat is applied to melt the solder and the thermoplastic rings and to shrink the outer sleeve, all in one operation. Softening of the primary thermoplastic insulation and flow of the conductor must be prevented by careful heating.

Tubular Zipper Tubing. Tubular jacketing with zipper closure is available in polyvinyl chloride, polyethylene (plain or irradiated), polyvinyl chloride-impregnated nylon cloth, or fiberglass and polytetrafluoroethylene-impregnated glass cloth. Combinations of the above materials are available in conjunction with shielding materials of aluminum, conetic, or netic foil, for shielding from magnetic or low-frequency interference.

Zipper tubing offers the unique advantage of applying a harness jacket without regard to size, shape, and quantity of conductor or size of conductors. Reworking or repair of cable harness can be accomplished with minimum complexity. The zipper track may be permanently sealed with an adhesive for optimum moisture and water protection. Special configurations such as breakouts, multiple channels, or boots are available for specific applications.

Wire and Cable Mounting and Spacing Hardware. Good installation-design prac-

TABLE 26. Typical Heat-shrinkable Tubing Properties

Properties	Irradiated polyolefin					Irradiated PVF_2	Flexible PVC	Flexible irradiated PVC	Semirigid irradiated PVC	Neoprene rubber	Silicone rubber	Butyl rubber	PTFE
	Flexible opaque	Flexible clear	Semirigid opaque	Semirigid clear	Dual wall								
Tensile strength, psi	2,500	2,500	3,000	3,000	2,000	7,000	3,000	3,000	5,000	1,900	900	1,600	4,500
Ultimate elongation, %	400	400	400	400	400	300	300	300	250	220	300	350	250
Brittleness temperature, °C	−60	−85	−60	−90	....	−73	−20	−20	−20	−40	−75	....	−90
Hardness	98A	90A	....	....	....	....	85A	....	....	85A	70A	80A	....
Specific gravity	1.3	0.93	1.3	0.95	0.94	1.76	1.4	1.35	1.4	1.4	1.2	1.2	2.2
Water absorption, %	0.05	0.01	0.05	0.01	0.1	0.1	....	0.6	0.6	0.5	0.5	0.1	0.01
Dielectric strength, volts/min	1,300	1,300	1,300	1,300	1,100	1,500	750	750	900	300	300	130	1,200
Volume resistance, ohm-cm	10^{15}	10^{17}	10^{15}	10^{17}	10^{16}	....	10^{12}	10^{12}	$>10^{13}$	10^{11}	10^{15}	10^{12}	10^{18}
Dielectric constant	2.7	2.3	2.7	2.4	2.4	....	5.4	....	....	....	3.3	....	2.1
Power factor	0.003	0.0003	0.003	0.0003	0.0005	....	0.12	....	....	....	....	....	0.0002
Fungus resistance	Inert	Inert	Inert	Inert	Inert	Inert	Inert	Inert	Inert	Inert	Inert	Inert	Inert
Fuel and oil resistance	Excellent	Excellent	Excellent	Excellent	Excellent	Excellent	....	R exc.	Excellent	Good	Fair	Fair	Excellent
Hydraulic fluid resistance	Excellent	Excellent	Excellent	Excellent	Excellent	Excellent	....	Excellent	Excellent	Fair	Poor	Good	Excellent
Solvent resistance	Good	Good	Good	Good	Good	Excellent	....	Excellent	Excellent	Fair	Fair	Fair	Excellent
Acid and alkali resistance	Excellent	Excellent	Excellent	Excellent	Excellent	Excellent	....	Excellent	Excellent	Good	Good	Good	Excellent
Flammability	Self-extinguishing	Burns slowly	Self-extinguishing	Burns slowly	....	Non-burning	Self-extinguishing	Self-extinguishing	Self-extinguishing	Self-extinguishing	Self-extinguishing	Burns slowly	Non-burning

tice requires adequate space, not only for wiring and cabling, but also for the supporting hardware (clamps, sleeving, grommets, guides, etc.). Space is also needed for manipulating tools during initial installation as well as during maintenance and replacement.

HARDWARE. Cable mounting with the MS type of cable clamp is a proved method. Many variations of this clamp are available from specialty suppliers. The principal advantages of MS clamps are low cost, light weight, high strength, ready adaptability, and ease of installation and servicing. Clamps can be installed on any structure or skin of adequate strength that can be drilled. If the structure or skin cannot be drilled, bonding is recommended. One technique is to bond a cable-supporting device or pad to the supporting area, then strap, tie, or clamp the cable to it. An alternative technique is to bond the entire cable to the supporting area for a very secure installation. Since this technique inhibits design flexibility, it is best used with firm design or under more flexible conditions where spare wires can

TABLE 27. Classification of Heat-shrinkable Tubing Materials in Accordance with MIL-I-23053

Classification	Material	Consistency	Flame resistance	Pigmentation	Temp. rating, °C	Colors
Class 1 . . .	Polyolefin (irradiated)	Flexible	Flame-retarded	Yes	135	Black, red, yellow, blue, white
Class 2 . . .	Polyolefin (irradiated)	Flexible	Non-flame-retarded	No	135	Clear
Class 3 . . .	Polyolefin (irradiated)	Semirigid	Flame-retarded	Yes	135	Black, red, yellow, blue, green, slate, brown, white
Class 4 . . .	Polyolefin (irradiated)	Semirigid	Non-flame-retarded	No	. . . .	Clear
Class 5 . . .	Polyvinyl chloride (irradiated)	Flexible	. . . .	Yes	105	Black
Class 6 . . .	Polytetrafluoroethylene	. . . .	. . . .	No	250	Clear

be installed. In areas where large spaces must be spanned without available support, either an integral cable support or an auxiliary structure must be installed. Integral cable supports can be tubes or geometric extruded shapes in the center of the cable. A variation of this method is to mold the entire cable into a rigid mass, which, however, greatly reduces design flexibility. If bonding is the only possible means of attachment, MS nylon, reinforced nylon, or Kynar harness straps, mounting plates, and a compatible bonding material can be used. Specification MIL-S-23190 covers adjustable plastic cable straps in military use.

SUPPORT SPACING. The spacing of clamps and other cable-support tie-down devices can be determined from experience and development mock-ups. Applicable electrical system specifications generally establish bundle tie-down spacing by stating a maximum distance between supports (MIL-W-8160 maximum spacing 24 in.). For adequate design of electric cable installations in missiles and space vehicles, the spacing must be resolved analytically and tested for verification. The spacing will be determined by the dynamic environment in which the cabling must reliably perform. Therefore, the installation designer must coordinate his cable tie-down spacing with the dynamics specialist for mechanical stability under extreme dynamic conditions. Complete design coordination must exist between the structures, dynamics, and electrical-installations engineers in order to meet system requirements. Dynamic tests on development hardware are recommended early in a program for verifying the installation.

Drastic changes in cable stiffness or section size caused by the ending or branching of wires may lead to points of dynamic weakness. Firm support is recommended on both sides immediately adjacent to these points, regardless of the spacing of other tie-downs or clamps.

The spacing of supporting devices on a high-acceleration missile system can be determined by the formula

$$F = \Sigma LANG$$

where F = design load of attachment device, lb
L = unsupported length, in.
A = unit weight per length, in., for each wire size
N = number of each wire size in bundle
G = maximum dynamic environmental load in g's

A sample calculation to illustrate a design example is given below.

Given harness parameters are:

10 unshielded wires size 20 AWG
10 shielded and jacketed cables size 20 AWG
10 twisted, shielded, and jacketed pairs size 26 AWG
G load = 150 g
F = 50 lb
$A_{20u} = 4.02 \times 10^{-4}$ lb/in.
$A_{20s} = 6.36 \times 10^{-4}$ lb/in.
$A_{26tws} = 4.98 \times 10^{-4}$ lb/in.

F is indicated as 50 lb. However, a safety factor of 2:1 changes this value to

$$F = \frac{50}{2} = 25 \text{ lb}$$

From this,

$$L = \frac{25}{150} \frac{1}{10(4.02 \times 10^{-4}) + 10(6.36 \times 10^{-4}) + 10(4.98 \times 10^{-4})}$$

$$L = \frac{1}{6} \frac{1}{1.536 \times 10^{-2}}$$

$$L = 10.8 \text{ in.}$$

The sample harness must be clamped or attached every 10.8 in. to satisfy the given conditions.

The above details indicate a technique evolved under a specific set of requirements and are presented as a guide only.

SUPPORT AND CLAMPING OF CABLES TO CONNECTORS. The cable clamp associated with a multipin connector is used primarily to support the wire(s) or the cable terminating at the connector and also to relieve strain from the terminations. Soft telescoping bushings (in accordance with Specification AN3420) are available for cables smaller than the cable clamp opening. The bushings permit the cable to be centered and anchored securely without excessive padding. Selection of the proper size of bushing permits easy assembly of the clamp. There should be adequate clamping pressure without bottoming the two halves of the clamp. Clamp screw-thread engagement should be equal to ⅔ to 1½ times the major nominal screw diameter.

Major differences in size between cable and connector can be corrected with step-up or step-down telescoping extension sleeves instead of bushings.

REFERENCES

1. Schuh, A. G., and J. Penkacik: High Strength Alloy Conductors, *Symp. Commun. Wire and Cables,* December, 1964.
2. Copperweld Steel Company: "Copperweld Wire for Electronic Applications."
3. Rome Cable Division: "The Rome Cable Manual of Technical Information," Copyright 1957, Rome Cable Corp.
4. Amphenol Corporation: "Cable Products Catalog ACD-5."
5. MIL-W-81044.
6. Bigelow, N. R.: Development and Evaluation of a Lightweight Airframe and Hookup Wire for Aerospace Applications, *Bur. Naval Weapons Symp.,* October 13 and 14, 1964.
7. Martin Marietta Corp.: Extra Flexible Tactical Cable—Report No. 3 Dec. 1964 —adapted from:
8. The Martin Company: "Electrical Design," Copyright 1958.
9. Reed, J. C.: Save Space by Hookup Wire Insulated with Teflon, *J. Teflon,* 1964.
10. Campbell, F. J., C. L. Baggett, R. J. Flaherty, and J. A. Kimball: Wire Insulation Thermal Life Studies, *Bur. Naval Weapons Symp.,* Oct. 12 and 13, 1965.
11. French, E. M.: Electrical Interconnections and Cabling, TOS 521 *Study Rep.,* December, 1964.
12. Pendleton, W. W.: Advanced Magnet Wire Systems, Electro-Technol., October, 1963.
13. Martin, G. C.: Insulated Magnet Wire Characteristics, *EDN,* August, 1963.
14. "Insulation Directory Encyclopedia Issue," May/June, 1966, Lake Publishing Corporation.

Chapter 3

SOLDERING AND MECHANICAL INTERCONNECTIONS

By

EUGENE W. BROACHE

Process Engineering
Westinghouse Electric Corporation
Aerospace Division
Baltimore, Maryland

INTRODUCTION

The American Welding Society defines *soldering* as a joining process in which coalescence is produced by heating (below 800°F) and by using a nonferrous filler metal that has a melting point below that of the base metal(s). The 800°F temperature is the dividing line between soldering and brazing (above 800°F). This definition will be used in this chapter to set the upper limit of temperature for soldered joints. However, practically all solders used for electrical connections melt below 600°F.

Soldering has been a major means of making electrical connections since the discovery of electricity, and today soldered connections are used in all phases of electronic packaging, including interconnecting wiring, terminal boards, modules, printed circuits, and thin- and thick-film circuitry. All types of components from conventional axial lead resistors to bare integrated circuit chips are soldered.

The inherent simplicity of the soldering process, which can be accomplished with simple low-temperature equipment, has allowed this process to compete favorably with other more sophisticated methods on a reliability and economy basis for most applications. Soldering has been applied to many mass-production and automated processes, as well as to individual operations.

The increasing complexity of electronic systems, some of which contain millions of connections, has in turn caused an increase in the requirements for reliable joints which can be produced at high production rates. Modern circuitry techniques and high-density packaging have brought the solder joints in close proximity to heat-sensitive materials and/or components. Thinner leads, special lead materials, and thin conductors are being used. All these factors place new demands on solders, fluxes, and techniques of heating in order to make joints reliably, economically, and with minimum damage to adjacent materials or components.

Soldering has successfully met this challenge with improved solders, fluxes, and equipment as well as better operator training, process control, and inspection techniques. Increasing attention has been paid to design of soldered joints and the increasing knowledge of solder and fluxes has been applied to the selection of these materials to meet changing requirements. There is every reason to believe that soldering will remain a major means of making electrical connections in future electronic packages.

Definitions of terms to be used in this chapter are given in Table 1.

TABLE 1. Definitions

Alloy. A substance, with metallic properties, that is composed of two or more chemical elements of which at least one is an elemental metal.

Amalgam. An alloy of mercury with one or more other metals.

Atomic percentage. The number of atoms of an element in a total of 100 representative atoms of a substance; often written a/o.

Base metal. (1) The metal present in the largest proportion in an alloy; brass, for example, is a copper-base alloy. (2) The metal to be brazed, cut, or welded. (3) After welding, that part of the metal which was not melted.

Binary alloy. An alloy containing two component elements.

Butt joint. A joint between two abutting members lying approximately in the same plane.

Capillary attraction. The combination force, adhesion and cohesion, which causes liquids, including molten metals, to flow between very closely spaced solid surfaces even against gravity.

Cast structure. The internal physical structure of a casting evidenced by shape and orientation of crystals and segregation of impurities.

Clearance. (1) The gap or space between two mating parts. (2) Space provided between the relief of a cutting tool and the surface cut.

Creep. Time-dependent strain occurring under stress. The creep strain occurring at a diminishing rate is called *primary creep;* that occurring at a minimum and almost constant rate, *secondary creep;* that occurring at an accelerating rate, *tertiary creep.*

Creep strength. (1) The constant nominal stress that will cause a specified quantity of creep in a given time at constant temperature. (2) The constant nominal stress that will cause a specified creep rate at constant temperature.

Crystal. A solid composed of atoms, ions, or molecules arranged in a pattern which is repetitive in three dimensions.

Crystallization. The separation, usually from a liquid phase of cooling, of a solid crystalline phase.

Ductility. The ability of a material to deform plastically without fracturing, being measured by elongation or reduction of area in a tensile test, by height of cupping in an Erichsen test, or by other means.

Elastic deformation. Change of dimensions accompanying stress in the elastic range, original dimensions being restored upon release of stress.

Elastic limit. The maximum stress to which a material may be subjected without any permanent strain remaining upon complete release of stress.

Elastic modulus. See Modulus of elasticity.

Elasticity. That property of a material by virtue of which it tends to recover its original size and shape after deformation.

Electrode potential. The potential of a half cell as measured against a standard reference half cell.

Electromotive series. A list of elements arranged according to their standard electrode potentials. In corrosion studies the analogous but more practical galvanic series of metals is generally used. The relative position of a given metal is not necessarily the same in the two series.

Equilibrium. A dynamic condition of balance between atomic movements where the resultant is zero and the condition appears to be one of rest rather than change.

Equilibrium diagram. A graphical representation of the temperature, pressure, and composition limits of phase fields in an alloy system as they exist under conditions of complete equilibrium. In metal systems, pressure is usually considered constant.

Eutectic. (1) An isothermal reversible reaction in which a liquid solution is converted into two or more intimately mixed solids on cooling, the number of solids formed being the same as the number of components in the system. (2) An alloy having the composition indicated by the eutectic point on an equilibrium diagram. (3) An alloy structure of intermixed solid constituents formed by a eutectic reaction.

Fatigue. The phenomenon leading to fracture under repeated or fluctuating stresses having a maximum value less than the tensile strength of the material. Fatigue fractures are progressive, beginning as minute cracks that grow under the action of the fluctuating stress.

Fatigue life. The number of cycles of stress that can be sustained prior to failure for a stated test condition.

Fatigue limit. The maximum stress below which a material can presumably endure an infinite number of stress cycles. If the stress is not completely reversed, the value of the mean stress, the minimum stress, or the stress ratio should be stated.

Fatigue strength. The maximum stress that can be sustained for a specified number of cycles without failure, the stress being completely reversed within each cycle unless otherwise stated.

Fatigue-strength reduction factor (K_f). The ratio of the fatigue strength of a member or specimen with no stress concentration to the fatigue strength with stress concentration. K_f has no meaning unless the geometry, size, and material of the member or specimen, and its stress range, are stated.

Filler metal. Metal added in making a brazed, soldered, or welded joint.

Fillet. (1) A radius (curvature) imparted to inside meeting surfaces. (2) A concave cornerpiece used on foundry patterns.

Fluidity. The ability of liquid metal to run into and fill a mold cavity.

Flux. In brazing, cutting, soldering, or welding, material used to prevent the formation of, or to dissolve and facilitate removal of, oxides and other undesirable substances.

Galvanic corrosion. Corrosion associated with the current of a galvanic cell consisting of two dissimilar conductors in an electrolyte or two similar conductors in dissimilar electrolytes. Where the two dissimilar metals are in contact, the resulting reaction is referred to as *couple action.*

Galvanic series. A series of metals and alloys arranged according to their relative electrode potentials in a specified environment. Compare with Electromotive series.

Grain. An individual crystal in a polycrystalline metal or alloy.

Hardness. (1) Resistance of metal to plastic deformation, usually by indentation. However, the term may also refer to stiffness or temper, or to resistance to scratching, abrasion, or cutting. Indentation hardness may be measured by various hardness tests, such as Brinell, Rockwell, and Vickers. (2) For grinding wheels, same as *grade.*

Humidity test. A test involving exposure of specimens at controlled levels of humidity and temperature.

Hypereutectic alloy. Any binary alloy which has a composition that lies to the right of the eutectic on an equilibrium diagram, and which contains some eutectic structure.

Impact energy (*impact value*). The amount of energy required to fracture a material, usually measured by means of an Izod or Charpy test. The type of specimen and the testing conditions affect the values and therefore should be specified.

Intermetallic compound. An intermediate phase in an alloy system, having a narrow range of homogeneity and relatively simple stoichiometric proportions, in which the nature of the atomic binding can vary from metallic to ionic.

Lap joint. A joint made with two overlapping members.

Liquids. In a constitution or equilibrium diagram, the locus of points representing the temperatures at which the various compositions in the system begin to freeze on cooling or to finish melting on heating.

Magnetostriction. The characteristic of a material that is manifest by strain when it is subjected to a magnetic field; or the inverse. Some iron-nickel alloys expand; pure nickel contracts.

Mechanical properties. The properties of a material that reveal its elastic and inelastic behavior where force is applied, thereby indicating its suitability for mechanical applications; for example, modulus of elasticity, tensile strength, elongation, hardness, and fatigue limit.

Melting point. The temperature at which a pure metal, a compound, or a eutectic changes from solid to liquid; the temperature at which the liquid and the solid are in equilibrium.

Metal. (1) An opaque, lustrous elemental chemical substance that is a good conductor of heat and electricity and, when polished, a good reflector of light. Most elemental metals are malleable and ductile and are, in general, heavier than the other elemental substances. (2) As to structure, metals may be distinguished from nonmetals by their atomic binding and electron availability. Metallic atoms tend to lose electrons from the outer shells, the positive ions thus formed being held together by the electron gas produced by the separation. The ability of these free electrons to carry an electric current, and the fact that the conducting power decreases as temperature increases, establish one of the prime distinctions of a metallic solid. (3) From the chemical viewpoint, an elemental substance whose hydroxide is alkaline. (4) An alloy.

Micrograph. A graphic reproduction of the surface of a prepared specimen, usually etched, at a magnification greater than 10 diameters. If produced by photographic means it is called a *photomicrograph* (not a *microphotograph*).

Microstructure. The structure of polished and etched metals as revealed by a microscope at a magnification greater than 10 diameters.

Modulus of elasticity. A measure of the rigidity of metal. Ratio of stress, within proportional limit, to corresponding strain. Specifically, the modulus obtained in tension or compression is Young's modulus, stretch modulus, or modulus of extensibility; the modulus obtained in torsion or shear is modulus of rigidity, shear modulus, or modulus of torsion; the modulus covering the ratio of the mean normal stress to the change in volume per unit volume is the bulk modulus. The tangent modulus and secant modulus are not restricted within the proportional limit; the former is the slope of the stress-strain curve at a specified point; the latter is the slope of a line from the origin to a specified point on the stress-strain curve. Also called *elastic modulus* and *coefficient of elasticity*.

Monotectic. An isothermal reversible reaction in a binary system, in which a liquid, on cooling, decomposes into a second liquid of a different composition and a solid. It differs from a eutectic in that only one of the two products of the reaction is below its freezing range.

Notch sensitivity. A measure of the reduction in strength of a metal caused by the presence of stress concentration. Values can be obtained from static, impact, or fatigue tests.

Notch strength (notch tensile strength). The ratio of maximum load to the original minimum cross-sectional area in notch tensile testing.

Peritectic. An isothermal reversible reaction in which a liquid phase reacts with a solid phase to produce another solid phase on cooling.

Phase. A physically homogeneous and distinct portion of a material system.

Physical properties. The properties, other than mechanical, that pertain to the physics of a material; for example, density, electrical conductivity, heat conductivity, thermal expansion.

Plastic deformation. Deformation that does or will remain permanent after removal of the load which caused it.

Proportional limit. The maximum stress at which strain remains directly proportional to stress.

Reflowing. The melting of an electrodeposit followed by solidification. The surface has the appearance and physical characteristics of being hot-dipped (especially tin or tin alloy plates).

Residual elements. Elements present in an alloy in small quantities, but not added intentionally.

Segregation. Nonuniform distribution of alloying elements, impurities, or microphases.

Shear strength. The stress required to produce fracture in the plane of cross section, the conditions of loading being such that the directions of force and of resistance are parallel and opposite although their paths are offset a specified minimum amount.

Soldering. Similar to brazing, with the filler metal having a melting temperature range below an arbitrary value, generally 800°F. Soft solders are usually lead-tin alloys.

Solidification shrinkage. The decrease in volume of a metal during solidification.

Solidus. In a constitution or equilibrium diagram, the locus of points representing the temperatures at which various compositions finish freezing on cooling or begin to melt on heating.

Stress raisers. Changes in contour or discontinuities in structure that cause local increases in stress.

Stress-rupture test. A tension test performed at constant load and constant temperature, the load being held at such a level as to cause rupture. Also known as *creep-rupture test*.

Substrate. A layer of metal underlying a coating, regardless of whether the layer is the base metal.

Tarnish. Surface discoloration of a metal caused by formation of a thin film of corrosion product.

Tensile strength. In tensile testing, the ratio of maximum load to original cross-sectional area. Also called *ultimate strength*.

Ternary alloy. An alloy that contains three principal elements.

Thermal shock. The development of a steep temperature gradient and accompanying high stresses within a structure.

Thermal stresses. Stresses in metal, resulting from nonuniform temperature distribution.

Tin pest. A polymorphic modification of tin which results in crumbling of the tin into a powder known as *gray tin.* The reaction can occur below 32°F but does not proceed rapidly unless the metal is much colder. Maximum rate is −54°F.

Tinning. Coating metal with a very thin layer of molten filler metal.

Torsion. A twisting action resulting in shear stresses and strains.

Toughness. Ability of a metal to absorb energy and deform plastically before fracturing. It is usually measured by the energy absorbed in a notch impact test, but the area under the stress-strain curve in tensile testing is also a measure of toughness.

Ultimate strength. The maximum conventional stress—tensile, compressive, or shear—that a material can withstand.

Wetting. A phenomenon involving a solid and a liquid in such intimate contact that the adhesive force between the two phases is greater than the cohesive force within the liquid. Thus, a solid that is wet, on being removed from the liquid bath, will have a thin continuous layer of liquid adhering to it. Foreign substances such as grease may prevent wetting. Other agents, such as detergents, may induce wetting by lowering the surface tension of the liquid.

Yield point. The first stress in a material, usually less than the maximum attainable stress, at which an increase in strain occurs without an increase in stress. Only certain metals exhibit a yield point. If there is a decrease in stress after yielding, a distinction may be made between upper and lower yield points.

Yield strength. The stress at which a material exhibits a specified deviation from proportionality of stress and strain. An offset of 0.2 percent is used for many metals.

MECHANISM OF SOLDERING

The most important requirement for a soldered joint is that there be metallic continuity between the solder and the surface which is soldered. The atoms of the solder actually form a metallic bond with the atoms of the metal being soldered. The layer of alloy may be only a few atoms thick in some cases while in other instances an appreciable thickness of alloy layer will be present (see Fig. 1). Excess thickness

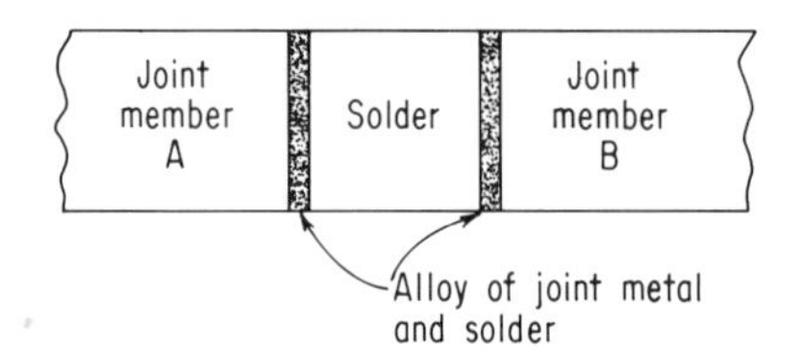

FIG. 1. Schematic of alloy layer.

of this layer, which is cause for problems in some alloy systems, is discussed later.

In order to attain this unity between the solder and the metal being joined, the joint and solder must be heated to a temperature sufficient to melt the solder and raise it to soldering temperature. In addition, all the surfaces to be soldered must be free of contaminants so that the solder may alloy with the atoms of the surface. The packaging engineer must select a solderable metal, the solder alloy, and an appropriate flux, and then design the joint so that it can be adequately heated in order to assure the wetting or alloying which will assure a reliable joint.

In designing for soldering, consideration of manufacturing processes is as important in attaining a satisfactorily alloyed joint as are the physical and mechanical properties of the materials used.

CONSIDERATIONS IN THE DESIGN OF SOLDERED JOINTS

Many factors must be considered simultaneously when designing soldered electrical assemblies. The choices of flux, solder, solderable surfaces, and heating methods interact so that a decision on any one of these factors may profoundly affect the decision on another. A listing of the major factors to be considered follows.

Solder Composition. The operating temperature of the equipment will be a major consideration in selection of the solder since the composition determines the melting point. The melting point determines the soldering temperature and thus must be considered in view of the temperature tolerance of the surrounding materials. The solder alloy also determines strength and physical properties.

Flux. The selection of flux will influence the selection of materials to be soldered. Most soldering of electrical assemblies must be performed using a mild flux which is electrically nonconductive, thus requiring very solderable surfaces. Strong fluxes, which by nature promote corrosion and have low electrical resistivity, are sometimes necessary. In instances where strong fluxes are used, the assembly must be capable of being washed free of all flux residue.

Solderability of Surfaces. Solderability, as mentioned above, depends to a large extent on the flux used. However, easily solderable surfaces may be ruined by coatings applied during manufacturing operations. Solderability may deteriorate due to oxidation. Paints, potting compounds, silicone mold releases, and various organic compounds are typical examples of substances that destroy solderability. Clean, solderable surfaces must exist at the time of soldering.

Heating Methods. The packaging engineer should know the method of heating for soldering. For instance, furnace soldering heats the whole assembly; therefore, all components must be capable of withstanding the temperature cycle for soldering. Other methods heat locally.

Thermal Properties of Materials and Components in the Assembly. Since electrical assemblies often contain insulation, and characteristically insulation has drastically different thermal expansion rates, thermal conductivity, etc., from metals, consideration must be given to the effect of the soldering temperature on such materials. Also, some components are damaged by elevated temperature, thus requiring control of soldering temperature, localized application of heat, use of heat sinks, and other measures to prevent damage.

Joint Design. Joint design can have major effects on strength of joints, manufacturability, and inspectability as well as on thermal damage. Extremely large clearances between leads and holes in printed-circuit boards can result in incomplete joints. Inaccessibility for soldering and inspection can raise costs and jeopardize reliability.

It should be the aim of the designer to present the manufacturing operation with a selection of materials, conditions, and design which can be fabricated with least problems.

SELECTION CRITERIA FOR SOLDERS

In selecting a solder a number of considerations must be taken into account, including:

1. *Liquidus temperature.* At this temperature the solder is completely molten. For most soldering the soldering temperature should exceed the liquidus temperature by 100°F. Components and materials must be able to survive the temperatures resulting from this soldering temperature.
2. *Solidus temperature.* Above this temperature part or all of the alloy is liquid and possesses practically no strength. In addition, solders decrease in strength rapidly as the solidus is approached; therefore, service temperature should be well below this temperature.
3. *Strength.* While electrical joints are not usually expected to carry loads, many joints will experience loads as a result of differential thermal expansion and vibration; therefore, this is an important factor.
4. *Corrosion resistance.* In corrosive atmospheres, corrosion may become a problem, especially where the solder is galvanically dissimilar to adjacent metal.
5. *Electrical conductivity.* While joint area is usually many times larger than the conductor, extremely low conductivity (high resistance) may present a problem.
6. *Thermal conductivity.* This may be a factor in dissipating heat from small components.
7. *Thermal expansion.* Where ceramics or glasses are present, this may become a major consideration as a source of stress.

Solder Alloys. While tin-lead solders in the range of 40 to 70% tin account for a large percentage of the solder used in electronics, a number of other solder alloy systems are available. Alloys of lead-silver-tin, lead-indium, tin-silver, tin-indium, and others are used to meet requirements of high or low temperature, process needs, corrosion resistance, and increased strength.

TABLE 2. List of Typical Solder Alloys

Solidus, °F	Liquidus, °F	Composition, percent by weight					
		Tin	Lead	Indium	Silver	Antimony	Other
158	165	12.5	25	...	...	...	Bismuth 50, cadmium 12.5
243	243	48	...	52			
321	321	....	...	100			
361	361	63	37				
361	370	60	40				
390	390	91	...	...	...	...	Zinc 9.0
430	430	96.5	...	...	3.5		
450	450	100					
450	465	95	...	...	...	5	
579	579	....	97.5	...	2.5		
588	588	1.0	97.5	...	1.5		
572	597	....	95	5			
640	740	....	...	...	5	...	Cadmium 95
720	720	....	...	...	...	...	Zinc 95, aluminum 5

Table 2 lists selected alloys with their compositions, their melting points, and some physical properties. To supplement this information, the phase diagrams of some of the common two-component binary alloy systems are shown in Figs. 2 to 5. These phase diagrams can be used to determine liquidus and solidus points for any composition in the binary system. Further alteration of the properties of binary systems can be made by adding other elements. While such three-component systems can be represented in ternary diagrams, such representation is beyond the scope of this chapter. Sufficient data on specific alloys are available from solder manufacturers.

Using the tin-lead diagram in Fig. 2 we can gather considerable information

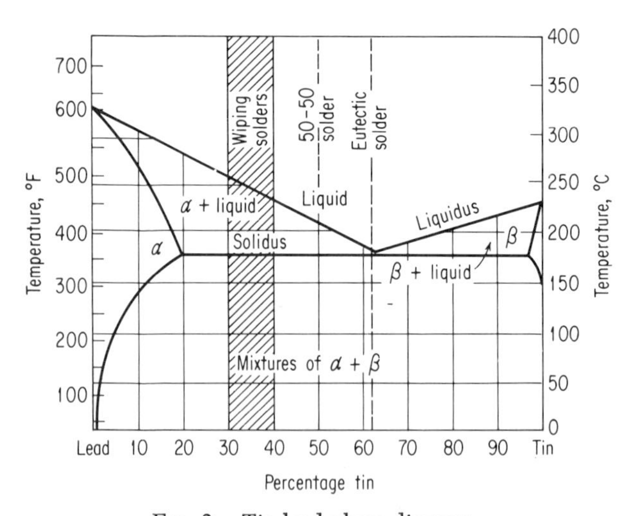

Fig. 2. Tin-lead phase diagram.

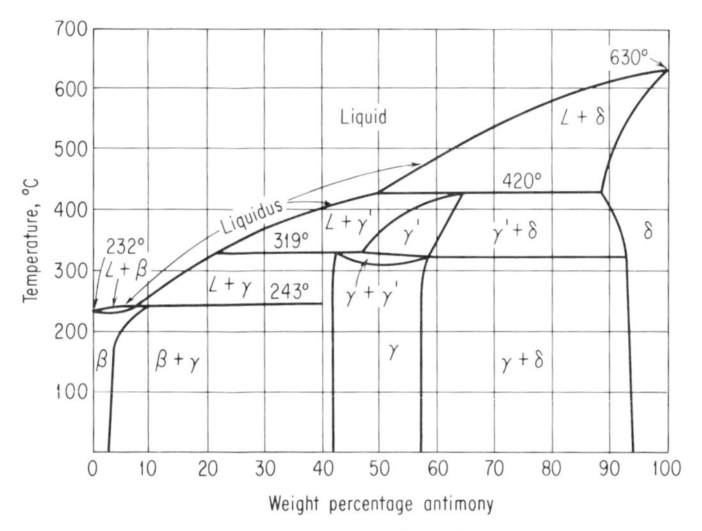

FIG. 3. Tin-antimony phase diagram.

about the melting and freezing characteristics of the alloy system. First, it is evident that adding tin to pure lead decreases the melting point until we reach 63% tin. Further increase in tin raises the melting point until we reach 100% tin. It can also be seen that pure tin, pure lead, and the 63% tin–37% lead compositions have a sharp melting point; that is, their respective melting temperatures and freezing temperatures

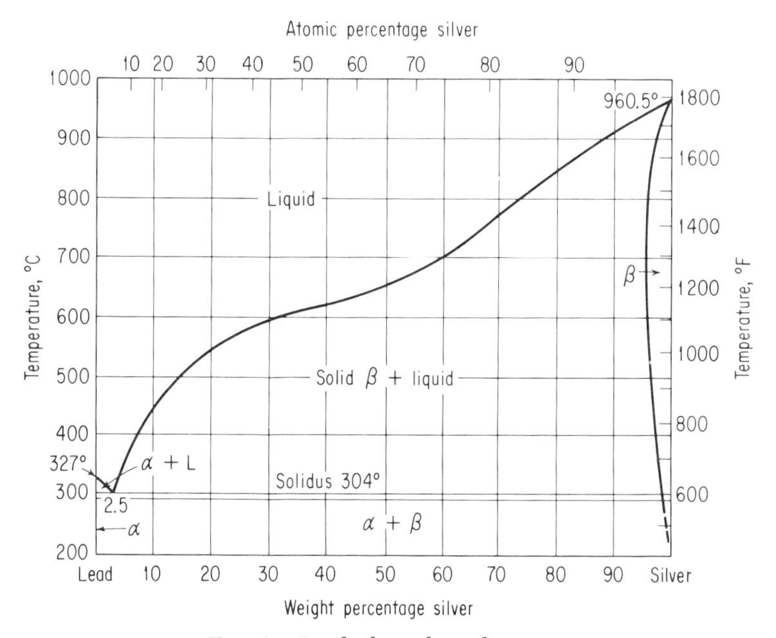

FIG. 4. Lead-silver phase diagram.

are the same. Sharp melting (or freezing) points are a characteristic of pure metals and of eutectic compositions. The 63% tin–37% lead composition is a eutectic.

The compositions other than those mentioned (the pure metals and the eutectic) have a temperature range in which liquid and solid coexist. This is the so called plastic, or pasty, or semiliquid, range. The important thing about this range is that actual liquid exists and therefore the solder has essentially no strength.

While the tin-indium, tin-antimony diagrams are of a different shape, the same type of information on solidus, liquidus, and solid-liquid coexistence may be found in them.

Tin-Lead Alloys. The tin-lead alloys in the range of 50 to 70% tin are used for making the large majority of the soldered electrical connections, and the 60% tin–40% lead composition is the most frequently used. A survey of the properties of the 63% tin–37% lead (eutectic) alloy will reveal why this is so.

The melting point (and freezing point) is 361°F. This allows soldering at temperatures low enough to prevent damage to components or materials in most cases. Also, at such low temperatures simple, low-cost, nonhazardous tools may be used. Fluxes

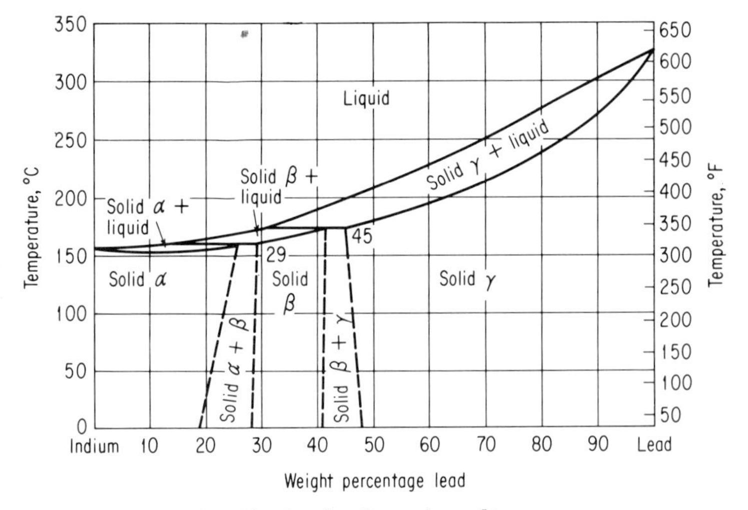

FIG. 5. Lead-indium phase diagram.

are active at the melting point but do not tend to deteriorate quickly. Even such a low melting point is adequate for the service temperature experienced by most electronic systems.

Eutectic tin-lead solder flows well and wets most surfaces well, leaving a bright, thin film of solder when applied correctly. The brightness and the small fillets are two good criteria for visual inspection.

The cost of tin-lead is low compared to most other low-melting-point metals or alloys.

The advantages mentioned for eutectic tin-lead alloys apply to a slightly lesser degree to any tin-lead alloy containing between 50 and 70% tin. Raising or lowering the tin content will raise the liquidus temperature and reduce solderability, while the service temperature at which it can be used will not be raised. Lowering the tin will decrease solder cost but will probably raise process cost since the temperature of soldering is raised.

Thus we can see that from an ease of manufacturing and inspecting standpoint the tin-lead solders of around 60% tin are the most satisfactory. It is generally accepted that a process that lends itself to ease and simplicity of manufacture also leads to highest reliability. It is little wonder that 60% tin–40% lead solder has become the standard for the majority of the electronics industry.

Short-time Mechanical Properties. The engineering properties of tin-lead solders are affected by composition and temperature. Figure 6 shows a plot of strength versus composition on which it can be seen that the strength at room temperature peaks at about 65% tin, or close to the eutectic composition, which is also the lowest-melting alloy. Fortunately, the tin-lead alloy possessing best manufacturability also has the highest strength. Shear strength also varies with composition but not to so great an extent. In normal engineering materials, shear strength is usually about 60 to 70% of the ultimate strength. The available data, including that in Fig. 6, shows that there is not a constant shear strength to ultimate strength ratio.

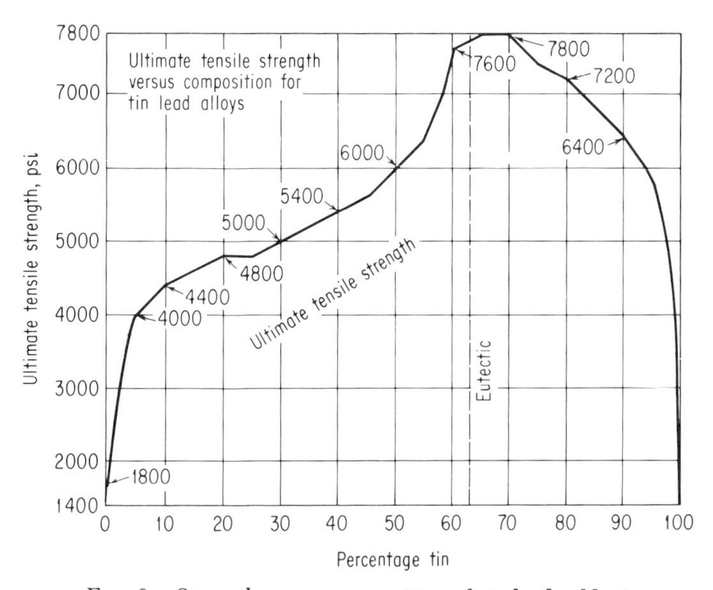

FIG. 6. Strength versus composition of tin-lead solder.[7]

Long-time Mechanical Properties (Creep). Solder under constant stress well below its tensile strength will permanently creep and eventually fail if stress and time are great enough. Creep is commonly thought of as a high-temperature phenomenon. If we use the criteria of high temperature being above one-half the melting point on the *absolute scale,* then room temperature *is* high temperature for solder. For most compositions of tin-lead the stress required to produce an 0.0001 in./in. strain per day is on the order of 100 to 300 psi at room temperature. Increase of temperature to 175°F decreases the stress required to the order of 25 to 50 psi. It can be seen that constant loads of very little magnitude can cause appreciable plastic deformation in relatively short times. Some selected data on creep appear in Table 3.

TABLE 3. Creep Data

Solder	Creep resistance,* psi		
	Steel	Copper	Brass
Commercial 44 (Sn 44%, Cu 0.09%)	310	310	470
Antimonial 40 (Sn 40%, Sb 2%, Cu 0.09%) ..	325	390	470
Pure 44 (Sn 44%, Cu 0.009%)	210	310	470

* Maximum stress capable of being sustained for 500 days.

Effect of Tin-Lead Ratio on Physical Properties. The variation of electrical conductivity and elastic modulus with tin-lead composition is shown in Fig. 7.

Thermal conductivity decreases directly as lead content decreases, being 0.157 for pure tin and 0.083 for pure lead (cgs units) (see Table 4).

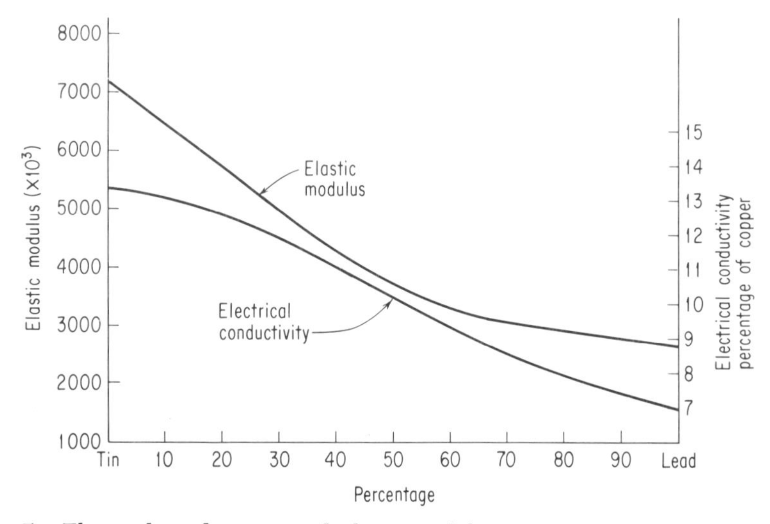

FIG. 7. Electrical conductivity and elastic modulus versus composition of tin-lead solders.

TABLE 4. Thermal Conductivity of Tin-Lead Solders

Sn, %	Pb, %	Thermal conductivity	
		Temperature, °C	(g)(cal)(cm)/cm²
100	0	Room	0.083
		Room*	0.118
		40*	0.1178
		70*	0.1177
62	38	102*	0.1125
		125*	0.1150
		139*	0.1115
		236 liquid*	0.0553
		310*	0.0610
		420*	0.071
0	100	Room	0.155

* W. B. Brown, *Phys. Rev.*, vol. 22, p. 171, 1923.

Effect of Alloying Elements on Tin-Lead Solders. Other elements are often present in tin-lead solders. Some elements are intentionally added for specific purposes, others exist as impurities from the refining process, while others are present due to alloying with the metal being soldered during making of the joint. Both beneficial and deleterious effects are brought about by alloying additions. The most common elements found in tin-lead solders are discussed below.

ANTIMONY. Antimony increases the strength of tin-lead alloys. The elevated-temperature tensile strength and creep strength are more drastically improved than are the room-temperature properties (see Fig. 8).

An additional advantage of antimony is that it suppresses the transformation of tin to gray tin or tin pest at low temperatures (see the discussion in the tin solder section). Federal specification QQ-S-571 specifies a minimum antimony content for tin-lead solders for this purpose.

Antimony decreases the spreading qualities of solders and thus may make soldering somewhat more difficult. It will also decrease the luster of joints and thus may cause questioning of joints by inspection personnel accustomed to pure tin-lead joints.

SILVER. The most common reason for adding silver to tin-lead solders is to slow down the alloying of the solder with thin coatings of silver on nonsolderable base materials such as (1) vacuum-deposited silver on glass or (2) fired-on silver paints on ceramics.

Thin electroplated silver on metals that are difficult to solder such as aluminum or stainless steels may present similar problems. A 3% silver alloy is available from most solder manufacturers and has been widely used for solving the above difficulties.

Silver lowers the melting point of tin-lead alloys to an insignificant degree. An improvement in creep strength is brought about by the addition of small amounts of silver.

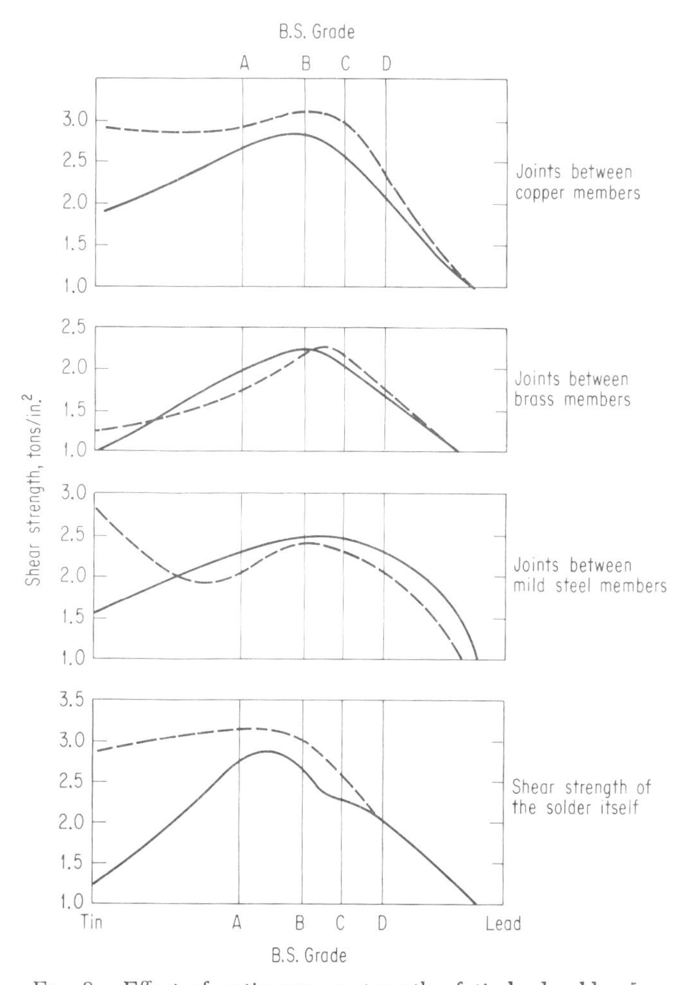

FIG. 8. Effect of antimony on strength of tin-lead solders.[5]

Joints made with silver containing solder will show less luster than tin-lead solders such as antimony-containing solders. Since tin-lead will alloy readily with silver, connections made with tin-lead solder on silver-plated surfaces may show some decrease in luster, especially at the edges of the joint where the solder fillets into the silver surface. The amount of alloying will increase with increasing time and increasing temperature.

COPPER. Copper is seldom added intentionally, although solder alloys with copper surfaces much as silver does. One manufacturer marketed a tin-lead solder with 2 to 3% copper, in which the purpose of the copper was to decrease erosion of copper soldering tips. This solder has been successfully used in soldering to vacuum-deposited copper on glass.

At present, most solder specifications consider copper to be an impurity and limit its presence in solder to a very low level. However, small amounts of copper will raise creep strength of solder.

Tin-lead solders containing appreciable amounts of copper will soldify with a rough-looking dull surface which is difficult to inspect since it gives the appearance of being a cold joint.

GOLD. Gold is seldom, if ever, added to tin-lead alloys intentionally. However, due to its rapid alloying with both tin and lead, considerable gold may exist in finished solder joints made on gold or gold-plated surfaces. Both gold-tin alloys and gold-lead alloys form intermetallic compounds at certain compositions. Intermetallic compounds are brittle by nature, and when the joint structure consists of appreciable amounts of these compounds, brittleness will result. The tin-gold intermetallic compound seems to be the most detrimental of the two alloys formed by tin and lead with gold. About 5% gold in a tin-lead joint will cause the whole joint to be brittle. At 3% gold the composite material retains appreciable toughness and ductility. Several investigations have been run on this subject.[1,2]

Due to the successful use of gold-plated terminals over many years and billions of soldered joints, it seems premature to condemn all joints made on gold surfaces. Some suggested precautions for minimizing gold alloy problems are:

1. Keep gold coatings thin (about 0.00005 in.).
2. Keep soldering times short.
3. Keep soldering temperatures low.
4. Pretin surfaces in a large solder pot prior to soldering, in order to remove most of the gold before making the final joint.
5. Pretin surfaces and then vacuum off solder prior to making the final joint.

Some specifications require removal of gold by erasing it from the soldering area as a means of eliminating the pickup of gold in the solder joint.

ALUMINUM. Small percentages of aluminum in solder (as little as 0.001%) are reported to cause poor soldering. Therefore, efforts should be made to keep aluminum out of solder pots used for making electrical connections. No advantages of aluminum as an alloying element are known.

ARSENIC. Small, closely controlled amounts of arsenic are present in solder from the raw material.

BISMUTH. Bismuth is present in solder in small amounts and exerts some beneficial effects on the spread of solder. No pickup of bismuth occurs during soldering.

CADMIUM. Although clean cadmium can be soldered readily and although cadmium is used as a major constituent in some low-temperature and high-temperature alloys, the presence of cadmium in tin-lead alloys is considered contamination. It imparts a sluggish property to the solder. Since no beneficial effects are imparted by cadmium, it should be kept to a low level.

IRON. As little as 0.1% iron can cause grittiness in solder. Solder does not alloy with iron readily at soldering temperatures unless fluxed with strong fluxes. Therefore, cast-iron pots can be used for solder pots. If temperatures exceed 800°F, alloying with iron may occur.

When soldering to iron alloys with strong flux, a layer of alloying with the base metal at the solder interface will occur. The extent of this alloying will increase with increasing time and temperature. Metallographic study of soldered joints on ferrous alloys shows the existance of this alloy layer.

No useful beneficial effects of iron in solder are reported.

NICKEL. Since nickel neither alloys greatly with solder nor dissolves in solder, little problem is found with nickel in solder. No detrimental effects are reported.

SULFUR. Sulfur is reported to be detrimental above 0.03% by weight.

ZINC. Minute amounts of zinc are reported to be detrimental to soldering, as it causes grittiness, lack of adhesion, etc. Brasses and galvanized parts are sources of zinc.

OXIDES. Proprietary solder is now available with low entrained oxide. Such solder solidifies in thin coatings with a bright smooth surface. Advantages claimed are better wetting and better inspectability. While oxides have not been reported as detrimental, the stated advantages have been accepted by a large portion of industry and this special grade of solder is in widespread use.

SUMMARY. It can be seen that some elements are purposely added to tin-lead to improve specific properties. These elements are specified within given limits in applicable specifications. Also, known impurity elements are limited to maximum contents.

When it comes to unintentional elements which are picked up by the solder when it is molten, during soldering operations, the limits are less clear. It has been the writer's experience that the first evidence of unacceptability of contaminated solder shows up as finished joints with less luster than shown by joints made with non-contaminated tin-lead alloy. When using tin-lead as a solder a major inspection criterion is that the joint members be covered with a thin film of smooth lustrous solder whose edges fair into the surfaces being joined. Contaminated solder generally results in joints exhibiting variation from the desired appearance such as heavy solder coatings, rough surfaces, frosty- or grainy-looking surfaces, and abruptly ending fillets. Joints of similar appearance could be caused by low soldering temperatures, lack of flux, poor wetting, or other deficient soldering practices. Therefore, one of the most important reasons for purity of solder is inspectability. Other reasons, as mentioned above, are brittleness of joints which are highly alloyed and poor wetting characteristics imparted by some contaminants.

Elevated-temperature Solders. Applications arise where temperatures experienced by equipment exceed the temperature tolerance of tin-lead solders, thus requiring a solder with a higher solidus temperature. Several alloy systems satisfy this requirement while maintaining adequate properties to serve as good solders. Some of the solder alloy systems adequate for elevated-temperature joints are given in Table 5.

TABLE 5. Elevated-temperature Solders

ASTM grade	Sn, %		Sb, %		Pb, %	Ag, %		Completely solid		Completely liquid	
	Min.	Max.	Min.	Max.	Max.	Min.	Max.	°C	°F	°C	°F
95TA	94.5	95.5	4.75	5.25	0.07	...	...	236	457	243	469
5A	4.75	5.25	...	0.10	Remainder	...	...	270	518	312	594
1.5S	1.0	1.5	...	0.10	Remainder	1.4	1.6	309	588	310	590

100% TIN. With a sharp melting point (typical of pure metals) of 450°F, tin would appear to be a good "next-step-up" solder for elevated-temperature connections and is used as such by some manufacturers. It has good wettability for most surfaces.

There is concern over two phenomena of tin which have caused doubt about the advisability of using it. The allotropic transformation of tin to "gray tin" at low temperatures, referred to as *tin disease,* is a definite possibility although extremely rare. Also, the occurrence of "tin whiskers" on some tin surfaces has been the source of electrical shorts between closely spaced terminals. While these phenomena occur rarely and under special conditions, they are definite possibilities which cannot be ignored. Usually slight alloying will reduce their occurrence. Consultation with manufacturers of solders is recommended.

TIN-ANTIMONY. The 95% tin–5% antimony alloy with a melting range of 457 to 469°F is a suitable alloy with no whisker or phase transformation problem. It should be avoided on brass due to the possibility of embrittlement of that alloy. Some reduction in solderability will be experienced over tin, but should not be objectionable.

TIN-SILVER. Tin-silver, usually 3.5 to 5% silver, has a solidus of 430°F and offers some improvement in elevated-temperature service over tin-lead. This alloy has a color which matches closely that of stainless steel.

The addition of silver to tin reduces the dissolving of silver, and thus this alloy may be beneficial when soldering to thin silver coatings.

LEAD-SILVER. Lead alloyed with silver in the amount of 1.5 to 5% silver gives an alloy with a solidus point of 589°F. Lead-silver alloys have poor wetting characteristics and poor corrosion resistance. Addition of tin in percentages from 1.0 to 10% has been used to enhance wetting and improve corrosion resistance with slight effect on melting points. Most lead-silver alloys in use have some tin added.

CADMIUM-SILVER, CADMIUM-ZINC, ZINC-ALUMINUM. These three alloys are special ones not normally used for electrical connections. Their melting points are as follows:

Alloy	*Solidus, °F*	*Liquidus, °F*
95% Cd–5% Ag	640	740
40% Cd–60% Zn	509	635
95% Zn–5% Al	720	720

The 95% Zn–5% Al is probably the most successful alloy for soldering aluminum, while the cadmium-silver alloy has good structural characteristics at temperatures up to 425°F. The cadmium-zinc is useful for aluminum soldering.

LEAD-INDIUM. Lead alloyed with 5% indium has found use as an elevated-temperature solder. Its melting temperature is 600°F. These alloys have exceptional wetting characteristics for high-melting-point solder.

TIN-INDIUM PLUS. The tin-indium solders offer low melting points running as low as 117°F for the 8.3% tin, 19.1% indium, 44.7% bismuth, 22.6% lead, 5.3% cadmium alloy. Other alloys have a solidus as high as 243°F.

Pure indium is also used as a solder. Its melting point is 313°F. An alloy of indium with 3% silver has been successfully used for fluxless soldering of thin silver surfaces. See Table 6 for properties of indium solders.

TABLE 6. Properties of Indium Solders

Composition						Melting point, °C	Tensile strength, psi
In, %	Sn, %	Pb, %	Cd, %	Ag, %	Cu, %		
50	50	....	...	...	...	117	1,720
25	37.5	37.5	...	...	...	135–181	
75	....	....	25	...	...	123	
80	....	15	...	5	...	157	2,550
90	....	....	...	10	...	230	1,650
100	....	....	...	...	...	156.7	515
25	37.5	37.5	...	...	...	138	5,260
5	....	92.5	...	25	...	280–285	4,560
50	....	50	...	...	...	215	4,670
12	70	18	...	...	...	150–174	5,320
25	....	75	...	...	...	230	5,450
5	....	95	...	...	...	315	4,330
5	....	90	...	5	...	292	5,730
15	....	....	...	61	24	630–685	

LEAD-BISMUTH-TIN PLUS (FUSIBLE ALLOYS). This group includes alloys with solidus as low as 158°F. Generally these solders require strong fluxes and are brittle. However, some use has been made of these alloys as solders.

Physical Forms of Solder. Solder is available in a wide variety of forms and shapes.

Wire solder is one of the most common forms in use. Both solid wire and flux-cored wire are available in a variety of sizes. Rosin, organic (nonrosin), and chloride fluxes are available in cored solder. The amount of flux in the core may also be varied. The percentage of flux (percentage of flux by weight compared to the flux plus metal weight) in the core may be obtained as low as 0.5% and as high as 3.5%. Most flux cores fall between 2 and 3%.

Rod solders in specific lengths are also available.

Ingot solder for use in solder pots is furnished in various ingot weights.

Foil, sheets, strips, and ribbons of solder are available in almost any combination of thickness, width, and length desired.

Preforms to fit a large variety of joint shapes are furnished by suppliers either in standard shapes or to customer requirements. Such forms may be made from wire (solid or cored), or may be punched or formed from flat forms. Economy of manufacture can be realized in many instances by use of preforms.

Solder in the form of balls or pellets or specific sizes is available for use where carefully measured amounts of solder are required.

Powdered or granulated solders have been in use for some time. Much of this solder is sold as a mixture of solder and flux described as solder "pastes" or solder creams. Some of these products may be locally applied by silk-screen techniques. Others are applied with spatulas, brushes, etc.

Some solders may be applied by electroplating. The most common ones are tin, alloys of tin-lead, and indium.

Solder and flux rings inside a heat-shrinkable tubing is on the market. The tubing is put over the joint area and heated with hot air, infrared heat, or resistance heat to shrink the tubing and make the solder joint in one operation.

SELECTION OF FLUXES

Fluxes are used in most soldering operations primarily as a means of providing a clean metallic surface to which the solder can alloy. In some few cases the cleaning for soldering is accomplished by mechanical removal of surface films utilizing ultrasonic energy or abrasive means, such as wire brushing or rubbing while the solder is molten. The overwhelming majority of electrical joints are made using a fluxing agent.

Flux is used to clean the surface to be soldered, to protect it from reoxidation while hot, and to aid in the spreading of solder. The flux required will be influenced by the metal being soldered and the nature of the surface film to be removed.

In electrical joints the flux used is also determined by its electrical and corrosive properties. For this reason rosin or modified rosins are almost always specified for electrical joints since they are the only fluxes which have residues sufficiently inert to maintain the necessary insulation resistance and which do not promote corrosion.

In special instances fluxes stronger than rosin are used but only after it has been determined that all flux residues can be removed from the assembly after soldering.

Electronic equipment has many flux traps which are difficult to clean. The insulation on wire, component clips, and component bodies on printed circuits, and the holes in printed circuits, are examples. Thus, fluxes whose residues are harmless are generally the only fluxes allowed on electrical assemblies.

Strong fluxes are often used for applying hot-dipped coatings to leads of components. Flux residues are then removed by complete washing. Under these circumstances there are no flux traps, and the washing liquid, usually water, can reach all flux residue.

Classes of Fluxes. Three general classes of fluxes are available. They are (1) rosin base, (2) organic (nonrosin base), and (3) inorganic.

Rosin Base. Fluxes based on rosin are the only ones universally approved for use on electrical and electronic equipment. Within this group there are varying grades of activity with a resulting varying degree of risk from the flux residue. For purposes of this discussion we will further divide the rosin fluxes into three categories as follows:

1. Water-white rosin
2. Moderately activated rosin
3. Highly activated rosin

These categories are based on Federal Specification QQ-S-571 and MIL-F-14256. See Table 7 for requirements.

TABLE 7. Rosin Flux Requirements

Flux class	Test requirements		
	Resistivity of water extract (min)	Copper mirror corrosion test	Polarized wire corrosion test
Water-white rosin (*a*)	100,000 ohm-cm	Meet standards of MIL-F-14256	Not required
Mildly activated rosin (*b*) ..	100,000 ohm-cm	Same as for water-white rosin	Not required
Highly activated rosin (*c*) ..	45,000 ohm-cm	Not required	Meet standards of QQ-S-571

Synopsis of tests and requirements for rosin fluxes as specified in government specifications (see QQ-S-571 and MIL-F-14256 for test procedures): (*a*) QQ-S-571 Type R and MIL-F-14256 Type W; (*b*) QQ-S-571 Type RMA and MIL-F-14256 Type A; (*c*) QQ-S-571 Type RA.

WATER-WHITE ROSIN. Water-white rosin is a natural product of pine sap, being the nonsteam volatile fraction.[3] It is the safest flux from an electrical resistivity standpoint, and its residue does not promote corrosion. Therefore, it may be safely left on electrical assemblies. However, it must be remembered that in some instances the flux may remain tacky, may pick up dust, fibers, and other contamination which in the presence of moisture will possibly lower insulation resistance. The safest practice is to remove all flux residue that tends to bridge insulation between conductors.

MODERATELY ACTIVATED ROSIN. Moderately activated rosin is a slightly activated revision of rosin. The activator increases the cleaning power of the flux and thus increases efficiency of soldering. It can also be argued that the activator increases the probability of good soldered joints since it will allow soldering to surfaces that would be of marginal solderability with rosin.

The activators used in the rosin are proprietary but it is relatively certain they are selected from among the organic (nonrosin) fluxes. The claim is made that the activator in the flux is driven off with soldering heat and only the rosin residue remains. However, if the residue is "driven off" by volatilization, then it may redeposit on the equipment when it condenses on cooling. Thus with such fluxes it is always possible that activator residues are present on the assembly.

MIL-F-14256 Type A covers moderately activated liquid flux and QQ-S-571 Type RMA covers moderately activated flux in the core of wire solder. These specifications do not specify the chemical composition of the flux or activator. Instead, they rely on performance tests for corrosion and electrical resistivity (see Table 7).

HIGHLY ACTIVATED ROSIN. Highly activated rosin is the highest activation rosin covered by a specification. It is covered by QQ-S-571 Type RA for cored solder and again is controlled by performance tests for electrical resistivity and corrosion (see Table 7).

It will be noted that only the cored solder specification covers this highly activated flux. The reasoning behind this is reported to be as follows. With cored flux the flux must experience the soldering heat in order to be released, and thus the activator will be driven off. With liquid flux it is impossible to assure heating of all flux and it is virtually impossible to remove all flux residue. Therefore, the use of highly activated liquid flux is not approved for most military electronic soldering.

Choosing a Rosin Flux. There is no clear-cut method for choosing among the rosin-base fluxes. Well-known, reputable companies have chosen fluxes at either end of the scale. Some general rules are:

1. If water-white rosin is used, all surfaces to be soldered must be clean and of solderable materials.

2. If liquid, activated fluxes are used, a thorough cleaning cycle with a solvent or solvents that remove both organic and inorganic compounds should be used.

3. Assemblies with small insulation clearances which flux may bridge require special attention. The same applies to high-impedance circuitry and assemblies with many flux traps.

Organic (Nonrosin Base). Organic fluxes cover a wide variety of compounds including organic acids, organic halogens, and amines and amides.

These fluxes have better activity than the rosin-base fluxes, thus allowing the soldering of nickel and other materials which are not normally solderable even with activated rosin on a routine basis.

A major interest has been shown in fluxes from this group due to their water solubility. Some electronics manufacturers have used them on printed-circuit boards. After soldering, the complete assembly was thoroughly washed by soaking in water, rinsing in hot water several times, and then finally rinsing in deionized water. So far as the author can determine, these practices have declined in the past several years, in favor of activated rosin fluxes.

The low residue from these fluxes after soldering is another major advantage.

Any use of nonrosin fluxes on assemblies should be considered only when a washing cycle which completely removes flux residue can be accomplished.

At present government specifications do not allow the use of these fluxes on electronic equipment.

Inorganic Fluxes. The major components of the fluxes in this class are ammonium and zinc chloride. They are highly conductive and corrosive materials and should not be used on electronic assemblies.

The major use of chloride fluxes is in pretinning leads or conductors for subsequent assembly. All residues must be removed by thorough washing. In some cases the vehicle in which the flux is contained is an organic material such as petrolatum. In order to remove such flux, both an organic solvent and an inorganic solvent, usually water, must be used.

Flux Vehicles. The fluxing agents are generally furnished dissolved in or mixed with a vehicle, especially when they are not present in the core of wire solder.

The rosin-base fluxes are primarily furnished dissolved in alcohol. Various solids contents are available; 35 percent solids is one popular composition. Proprietary flux-solvent mixtures utilizing solvents with higher boiling points than alcohol are available for use where high soldering temperatures are used.

The organic fluxes are generally dissolved in water, as are the zinc-chloride-ammonium fluxes. Use of polyethylene glycol as a flux vehicle for these fluxes has been reported; its advantage is that it does not boil and spatter at soldering temperature. Polyethylene glycol is water-soluble and thus can be removed by water rinsing.

Chloride fluxes are normally dissolved in water. They are also available in vehicles such as petrolatum or other heavy greaselike vehicles. While these substances retard corrosion in the joint area, it is important that all residue be removed from electronic and electrical equipment. Since the organic residue is present, use of an organic solvent followed by water rinse is mandatory on such fluxes.

Utilization of Fluxes. From the above discussion some general guidelines may be arrived at on the use of fluxes in electronic equipment.

The chloride fluxes must be confined to uses where all residues can be positively removed. They should be confined to (1) pretinning leads on components which have no flux traps, (2) solder-coating terminals, (3) pretinning areas on chassis prior to assembly, (4) other operations of this type. All such operations should take place outside the electronic assembly area due to the possibility of transfer of flux on the hands of personnel, on fixtures, or from benches on which flux has spilled. Again, all flux residue must be removed by proper washing with water where the flux vehicle is water, or with organic solvent followed by water where organic vehicles are used. Caution must be exercised that the final water rinse is *clean water.*

The organic fluxes offer somewhat the same problems as the chloride fluxes. But since their residue is somewhat easier to remove, they have been used on printed-circuit assemblies where the whole assembly can be soaked in water without component damage. The most successful use of this flux has been where the washing was done in a commercial dishwasher, which involves use of detergent and several hot water rinses. A final rinse in deionized or distilled water is highly desirable. Use of these fluxes in complex assemblies containing insulated wires or in other flux traps is by no means desirable. Even if the assembly could be water-washed, it is difficult if not impossible to assure removal of flux residue which has wicked under insulation in stranded wires.

Rosin-base fluxes are the only fluxes whose residue can be safely left on electrical and electronic assemblies. Water-white rosin, of itself, is a good insulator and will not decrease resistivity of most insulating materials. With the activated rosins, the matter becomes one of degree of activation. For a point of reference the residue from the Type RMA flux core of QQ-S-571 solder is considered nonharmful, while the type RA requires removal of residue. While there is no clear delineation of the degree of activation that will cause insulation and/or corrosion problems, QQ-S-571 has established a guideline.

Whether a flux is harmful is a function of several things. For example, if the environmental conditions are dry the effect of flux will be lessened. The insulation distance between circuits, the quality of insulation, and the operating conditions of the circuit will figure in the determination of the effect. High-impedance circuitry will tax the capability of most insulations without the further effects from flux, and therefore such circuitry should have all traces of flux removed from it.

The harmful effects of flux will depend also on how it is applied. If a cored flux is used and a proper soldering operation is done there should be only a thin film of dry flux on the joint area. Under such conditions little harm can be done by even Type RA flux. On the other hand, if liquid flux is applied sloppily so as to bridge conductors, it may leave a sticky flux residue which is prone to picking up and retaining contaminants that will lower insulation resistance. Thus, even water-white rosin could be harmful. Such conditions are particularly bad for activated fluxes since not all the activating agents are driven off by the heat of soldering.

Removal of rosin fluxes then becomes a question of a number of compromises. If flux residue is to be removed, there are several factors to be considered. The most common method of removing flux residues from hand-soldered joints is brushing the flux with a solvent. This is a questionable practice which essentially dissolves and spreads the flux. It may be argued with validity that it is more reasonable to leave the flux on the joint where it can do the least harm from an electrical resistivity standpoint. To remove water-white rosin flux residues any number of conventional and proprietary solvents are available. The activated fluxes require an organic solvent plus an inorganic solvent in order to assure solution and removal of both the rosin and the activator. Many practices are used. The most complete cycle appears to be a chlorinated solvent followed by an alcohol rinse followed by deionized or distilled water rinse.[4] Alcohol soaking plus a fresh alcohol rinse has also been used with some success. It is stated that normal isopropyl alcohol has sufficient water to dissolve the activators. One point seems worthwhile emphasizing: that any removal of flux residue (or any soil residue) requires a rinsing with a clean solvent in order to get complete residue removal.

Flux Residue Tests. No standard spot test for effectiveness of flux residue removal exists. However, silver nitrate and silver chromate have been suggested as means for detecting chloride residues from fluxes. The silver nitrate test has been used for many years to check for brazing flux residue and has a high sensitivity to the presence of chlorides. It may be used by placing a drop of 5% silver nitrate in water on the surface in question and checking for a white precipitate which indicates chloride. Another method is to soak the assembly in distilled or deionized water and then to check the water by dropping in the silver nitrate solution. Sometimes the dissolving water is boiled down to reduce volume and raise chloride concentration prior to testing.

Paper impregnated with silver chromate is being considered as a chloride detection method. The paper is wet with water and touched to a surface. If chlorides are present a positive indication is present on the paper.

Other more elaborate tests can detect harmful flux residues. Electrical resistivity of water in which assemblies have been soaked has been suggested. Also, assemblies such as printed-circuit boards could be checked for insulation resistance. Such techniques are not suitable for routine production testing. They seem more useful in setting up standards of acceptability for chemical spot checks such as the silver nitrate or silver chromate tests.

SELECTION OF HEAT SOURCES

The engineer does not necessarily select the heating method for soldering. However, his design may determine the method to be used. Some knowledge of the heating methods is necessary so that a reliable, economical method will not be excluded by a relatively unimportant design feature. Again cooperation between engineering and manufacturing is mandatory.

Practically any source of heat that can be used to raise the temperature of the part to the melting point of the solder being used can be used for soldering. These methods include: conduction soldering iron, electrical resistance, molten solder pots (including baths, waves, jets, cascades), hot gas (including air, nitrogen, hydrogen), ovens, hot plates, flame, light (infrared, laser, microscopic), hot oil and induction.

In selection of a heating method for soldering electrical or electronic joints, all the ordinary considerations of a production process must be considered, such as economy, design of parts, accessibility of joints to heat source, availability of equipment, and production rate. Additional requirements are introduced for electrical joints due to the possibility of thermal damage to the surrounding material and components. A rapid, localized heating method is often required to minimize heat effects on heat-sensitive components and materials in close proximity to the joint. In other cases it is necessary to maintain molten solder conditions for some extended length of time (such as in repair of printed circuits, removal of modules from printed circuits, joining multiple leads of integrated circuit chips to circuitry on a metallized substrate), which necessitates a uniform, controlled, safe temperature which can be accurately maintained over a period of time.

Actually selection of a heating method is dependent upon the combination of time and temperature. With few exceptions a rapid heating time is desired for the following reasons:

1. Flux tends to degenerate on heating and may be "used up" before the solder melts if the heating rate is slow.
2. The surface may oxidize and become difficult to solder during prolonged heating.
3. Solder may dissolve thin metal coatings or fine wires if the time at a high temperature is extended.
4. Components may be degraded if allowed to come to soldering temperature.

Slower, more uniform heating is sometimes desired when materials sensitive to thermal shock are employed in the assembly. Some ceramics and glasses may crack when rapidly heated, especially when they are bonded to metals.

A compromise solution, which has the advantages of slow and of rapid heating, is preheating. In this procedure the part is preheated to some lower temperature than that required to melt the solder. Then a faster heating method is used to rapidly raise the temperature to soldering temperature. Thus the time in the most damaging temperature range is minimized, and there is less thermal gradient, which minimizes thermal shock.

Soldering Irons. The soldering iron is a relatively simple device consisting of nothing more than a tip for carrying heat to the joint and a handle cool enough for the operator to hold comfortably. The tip of course must be heated by some means. Present-day tips are almost always heated by a built-in electrical resistance unit. However, in the past (and to a very limited extent today), the tips were heated by flame or by dipping into a molten-solder pot. We will primarily deal with electrically heated irons here.

The primary commercial classification of electrical soldering irons is by wattage rating, or the heat output of the heating resistance coil to the tip. The wattage rating alone is not a sufficient basis for selecting a soldering iron. The important factors in soldering are the heat input to the joint, the rate at which it is delivered, and the temperature to which the joint is heated. With a given wattage these factors will be determined by the equilibrium temperature of the tip, the mass of the tip and the associated heat reservoir, the distance of the tip from the heating element, and the resistance to thermal flow from the heater to the joint (including the thermal resistance of the materials and the area of the path).

Three important functional characteristics of soldering irons are as follows:

1. *Maximum tip temperature.* The maximum tip temperature must be capable of bringing the joint to soldering temperature in the desired time. This temperature is the equilibrium temperature of the soldering iron system when the heat being generated by the coil is being dissipated so that no temperature change is occurring.
2. *Recovery rate.* The recovery rate is the time necessary for a material to return to equilibrium temperature after being cooled, as occurs when making a joint.
3. *Heat content of system.* The heat content of a system depends on the mass and heat contents of the tip and associated materials.

The type of work being soldered and the conditions of soldering will determine the characteristics of the soldering irons used. Some of the applications are:

1. *High production rates on repetitive work with skilled operators with good accessibility to joints.* This type of work should be done with high-temperature irons with tip temperatures of 800 to 900°F. Heat reservoirs should be large with a short, thinner tip. Recovery rate must be rapid. It may be necessary to decrease tip temperature to 700 to 800°F when soldering printed circuits or when new operators are being broken in.
2. *Repair work, intermittent soldering, removal of components from printed circuits.* Lower-temperature irons are required here. If work can be reached easily for soldering a 700 to 800°F tip, temperature with low heat content is desirable. If long dwell time is required, as in removing a crimped lead from a printed-circuit board, decreasing the tip temperature to about 500°F may be desirable to prevent damage to insulating materials.
3. *Soldering fine wires, making joints to thin films.* A small hot iron which will give up heat quickly and recover quickly is desirable. The idea is to make the joint quickly and avoid long soldering times, which allow alloying of solder with the wire. The mass of the system should be small.
4. *Soldering connections close to heat-sensitive components of materials.* Two approaches can be taken. One approach is to heat quickly with a high-temperature iron with low mass and high recovery rate. Thus there is low total heat input to the joint and heat is confined to the joint. On the other hand, if the tolerable temperature is only slightly above the soldering temperature, the tip temperature should be controlled to a temperature just below the tolerable temperature.

Tip Materials. Copper and its alloys are the almost universal material for soldering iron tips. Due to the alloying of solder with copper, bare copper tips are quickly eroded away in soldering operations. To avoid this, many tips are plated with iron, nickel, or various alloy coatings which can be wet with solder but which do not alloy as quickly as copper does. Such tips should be kept well tinned at all times. They should never be filed or abraded since the coating will be removed.

Electrical Resistance Heating Devices. Resistance heating units differ from electrically heated soldering irons in that the current-carrying elements of resistance units actually contact the joint area. Heating may be accomplished by two means with these units. The most common method is passing electrical current through the joint area and creating heat due to the resistance of the joint to the flow of the current. The resistance is a combination of the resistance of the joint material, the resistance of the interface(s) between the electrodes and the joint, and the resistance of the electrodes themselves. The other method of resistance heating relies upon the heat created by a shorting bar or shunt between the two electrodes. This shunt is of a high-resistance metal which heats quickly when current flows. The heated shunt is in contact with the joint, which is then heated by conduction.

Primary advantages of resistance heaters are: (1) The heating cycle can be controlled. (2) The unit is hot only when the operator desires. (3) Higher temperatures than those supplied by soldering irons can be used. (4) The joint itself is a part of the heating circuit and must therefore be heated in order to melt solder. (This does not apply to the shunt method.)

Resistance Units Using Joint Resistance Heating. Two electrode configurations are used in this type.

Single electrode. In the single-electrode type the joint is made ground. Touching the single electrode to the joint causes current to flow and results in heating.

Double electrode. The double-electrode unit is the most common type. It has two electrodes which are put into contact to the joint. Current flows between the electrodes through the joint material and develops heat. A more sophisticated approach to this technique is the split electrode or the parallel-gap electrode. This method uses a thin layer of insulation between the two electrodes to give an electrode pin of small size.

Control of Electrical Resistance Heating. Equipment is supplied with varying degrees of control, depending upon the application.

For hookup wire to terminals, axial lead components to terminals, hookup wire to solder cups, and such work, the heating units may be relatively uncontrolled. A variable-power input which allows the operator to adjust the power level for various sizes of work is often the only control means. Touching the electrodes to the work completes the circuit. The operator watches for solder flow and then removes the electrode.

The next level of control is the adding of a switch which the operator actuates to cause heating. With the switch he can locate the electrodes as desired before applying heat and cut off the heat without having to remove his electrodes.

Still further control is available by use of a timer which cuts the power after a given elapsed time. On repetitive joints this is a decided advantage since a schedule of power level and time can be developed and evaluated. Use of this schedule should aid in joint consistency. Operator skill in applying the electrodes to the joint consistently is still required.

Planar layout of flat packs on printed-circuit boards where the leads of the flat pack form a lap joint on the printed-circuit conductor provided a further standardization of conditions for soldering. In this packaging scheme all joints are of the same material and the same size and are in the same horizontal plane. This configuration made possible a further refinement in resistance soldering in that the electrodes could be in the same plane each time they fired. Resistance soldering equipment is now available where the joint on the printed-circuit board is brought under the electrodes and the electrodes are lowered by a lever or foot switch to contact the joint. When a given electrode force is reached, the current flows for a set time at a given power level. Solder is preplaced on the leads and on circuitry by electroplating or hot dipping, or may be provided as shims, pastes, or creams. This method allows for control of electrode pressure and placement, power level, time, and amount of solder. Providing that the solderability of surfaces is consistent, this method can be considered a highly controlled process.

The next level of control which may be used is programmed heat cycles with resistance units. Resistance heating machines may be controlled to have preheat cycles of low power, followed by higher power levels to raise the joint to soldering temperature, followed by postheating cycles to cause slower cooling. Such controls are available on resistance welders today. The need for gradual heating and/or cooling may be useful for soldering to thin films on glass and on other joints where thermal shock may cause damage.

Shorting-bar Resistance Units. The simplest of these units is a hand-held soldering iron which is activated by a switch. The shorting bar is heated and is touched to the joint to raise it to soldering temperature.

Recently an adaptation of this method has been used for soldering flat packs to printed circuits in a planar layout. A shorting bar is put between the electrodes of a standard welding machine head. The electrode gap is wide enough to accommodate all the leads of a flat pack. The shorting bar is lowered to contact the flat-pack leads which are in contact with the printed-circuit conductors. Power is supplied to the shorting bar, causing it to heat and flow the solder. All the controls of time, power level, preheat, and postheat, mentioned above, may be used in this equipment. Solder is preplaced by electroplating or dipping or by pastes, shims, etc.

SUMMARY. Resistance soldering has many possibilities of control. It can provide high temperature quickly, thus allowing localized heat. The heating tips can be applied to the joint cold, thus separating placement time from heating time. Control of solderability is necessary to assure good results with this process as well as any other process.

Molten Solder Pots. The widespread use of printed circuits has resulted in a widespread use of a relatively old process in soldering. Molten solder pots have proved to be an extremely high production process for soldering printed circuits and multilayer boards. Because all joints are located on a planar surface, contact may be made with molten solder with the result that solder alloys with the solderable surfaces, and sufficient solder remains on the joint to join the members.

In its simplest form, dip soldering consists of floating a printed-circuit board on the surface of a molten solder pot for sufficient time to allow soldering to be accomplished (several seconds) and then removing the assembly. The technique of entry into and exit from the pot will have a major effect on soldering results. For best results the leading edge of the board should contact the solder and then the balance of the surface should be lowered until flat contact is effected. After the prescribed soldering time, the leading edge is lifted and the board is brought up to about a 15° angle before the trailing edge is lifted from the solder surface. This procedure will allow the gas generated during the flux decomposition to escape without creating gas pockets under the board. The gradual lifting of the leading edge during exit will allow excess solder to be pulled from the metal surfaces on the board, resulting in a thin layer of solder free from peaks and bridges. Mechanized devices to accomplish soldering by the pot method are available.

Wave Soldering. Systems in which molten solder is pumped to give a wave or ripple have become very popular primarily due to two factors:

1. Mechanization of the process can be accomplished with a straight conveyor since the solder can be pumped to an elevation above the walls of the container or solder pot.
2. Since fresh solder is constantly being pumped, no dross removal is needed prior to soldering.

In most wave soldering the solder is pumped through a rectangular orifice to form a standing wave over which the boards may be passed. In other systems the solder is allowed to cascade over a plate which contains ripples that cause waves to be formed. The surface to be soldered is passed through the peaks of these waves to be heated and soldered.

While there is no dross removal to be accomplished with wave soldering, the production of dross is very high since fresh solder is constantly being exposed to air. To combat this problem various covers have been tried. Usually these are high-temperature oils or graphite or mixtures of the two. In at least one device, oil is pumped along with the solder, thus minimizing oxidation of the wave surface. A smoothing effect is also claimed for this system.

Hot Gases. Systems utilizing hot gases have been used for soldering. They are especially suitable where it is desired to avoid pressure on fragile joints or where the joints are not readily accessible for contact with a solid conduction tip. The gas is usually heated by passing it over an electrically heated coil. The heated gas is blown on the joint through a small orifice such as a hypodermic needle. The gases generally used are air, nitrogen, and hydrogen. Air of course is the least expensive, but it has the disadvantage of being strongly oxidizing and having a relatively low heat capacity. Nitrogen is inert, thus overcoming the oxidizing problem if sufficient flow is provided, but it also has a low heat capacity. Sufficient flow of either air or nitrogen to provide adequate heat to melt solder rapidly requires high flow rates which may tend to blow solder away from the joint. Hydrogen has two major advantages: (1) It has a high heat capacity. (2) It is a reducing atmosphere and thus prevents oxidation and in some cases actually acts as a flux. In use of hydrogen it must be kept in mind that mixtures of hydrogen and air in certain concentrations are explosive. Adequate exhausting and other safety measures must be taken into account.

Flame. Flame heating is not often used in electronics; however, with the development of miniature flame devices interest in such devices has revived. The concentrated heat of oxyhydrogen flames from hypodermic needle torches offers a cheap means of providing heat to joints which cannot be reached by conduction heating units.

Light. Light rays of various kinds are used to heat joints to soldering temperature. Heating by infrared, laser, and ordinary light focused through lenses has been utilized on varying scales. While laser heating is the most glamorous of the methods, infrared heating is in wide use as a source of elevating temperatures. It has been found to be an excellent means of soldering multiple-lead components since with proper focusing of the beams by reflectors, a shaped beam capable of heating multiple joints can be attained. Soldering all leads of a flat pack, or even lines of flat packs, at one pulse of infrared energy is possible. Focused light has the advantage of not requiring contact with the joint in order to develop soldering temperature. No products of combusion are involved so that contamination is minimized. Unless sharp focus is provided, to focus sharply on the joint, the energy will be absorbed by the surrounding materials, and temperature-sensitive materials such as printed-circuit laminate may be damaged. Shields may be necessary to prevent this damage.

Mechanization of infrared soldering is relatively simple because no contact is required with the joint, and therefore a horizontal movement beneath the light source is the only motion required. Temperature can be controlled by power level to the lamp and by "time on."

Hot Oil (or Nonmetal Liquid). Dip soldering with a hot oil bath at soldering temperature as the source of heat is not widely used in electronics work. However, reflowing of solder with hot oil is fairly routine. Various oils including palm oil, vegetable shortenings, and mineral oils are used. Control of the acid number of oil is desirable in order to assure a fluxing action by the oil.

Induction Heating. Induction heating utilizes the induced energy from high-frequency alternating currents to heat by eddy currents and magnetic hysteresis losses. It is a fast means of heating, and since the heating is within the metal, it is reasonably certain that the metal surface will reach the temperature of soldering before the solder melts, a desirable condition for reliable soldering.

Since consistent heating is attained by a proper size and shape of induction coil, induction soldering can be quite expensive unless a large quantity of identical joints are to be soldered. Due to the bulk of the equipment required, the work must be brought to the machine. Also, coils are comparatively large, and joints in closely packed areas cannot be readily soldered.

Solderability. The selection of metals or finishes for joint members must be made with the purpose in mind of supplying the manufacturing operation with surfaces which will readily alloy with molten solder when applied with the specified flux. Not only must a basically solderable surface be specified but procedure must be controlled to assure that the surfaces are not contaminated by materials which cannot be removed by the fluxes. It must be emphasized that an acceptable joint requires alloying between the solder and the joint members, and this can occur only when clean molten solder contacts the clean metal of the joint members.

A solderable surface must possess the following qualities:

1. Thermodynamic ability to alloy with the solder, forming a thin alloy layer without harmful phases.
2. Freedom from chemical compounds on the surfaces which cannot be cleaned off by the flux. Examples of such compounds are oxides and sulfides which form on certain metals exposed to the atmosphere.
3. Freedom from externally applied contaminating films such as plastic coatings, paints, silicone mold releases, and various other contaminants which are not cleaned off by flux.

Foreign material contamination usually takes the form of organic compounds such as paints, resins, and silicone mold releases which are used in the manufacture of electronic components. Such contaminants are not removed by the fluxes used in soldering, and as a result the solder cannot come into contact with clean metal to form the metallic bond required of a true soldered joint.

Chemical reaction between oxygen in the air and the surface of such metals as chromium and aluminum accounts for the extremely poor solderability of these metals and the alloys containing them. Such stable oxides resist fluxing action, and thus it is not possible to bring about a metal-to-molten-solder interface so that alloying may occur.

The thermodynamic properties of some alloy systems make formation of alloys difficult. Lead, for instance, does not wet either iron or copper under ordinary conditions of soldering. Rhodium, though free from oxide, is still somewhat difficult to solder with rosin fluxes. Fortunately, most commonly used metals are capable of being wet by the commonly used solders if correct flux is used.

The wide variety of metals and alloys required for electronic parts and assemblies and the processes used in their fabrication make soldering of metals with the above difficulties inevitable. Since final soldering must be nearly always performed with rosin fluxes, the problem becomes even more difficult. However, procedures are available for handling the difficulties.

The cleaning of organic residues should be accomplished prior to soldering by cleaning with a solvent which will dissolve the contaminant where practical. Some contaminants such as epoxies and silicones are impractical to remove with solvents and must be mechanically removed by such methods as pencil erasers, abrasive blasting, emery cloth, or fiber-glass brushes.

Most metals which form tenacious oxides or other surface chemical compounds that inhibit soldering may be soldered with strong fluxes to remove these compounds. These fluxes are generally corrosive and electrically conductive and must not be used in assemblies. If the parts can be thoroughly washed of the residue, strong fluxes can safely be used for pretinning prior to being assembled into equipment. Again, caution should be exercised in assuring either that traps for flux residues are not present or that the washing cycle is such as to penetrate, dissolve, and rinse away all residue.

A more common and much safer approach is to coat the difficult-to-solder metal with a solderable coating either by hot dipping or by electroplating during manufacture of the part.

Solderability is a complex subject complicated by the presence of foreign contaminants, natural contaminants (oxides, etc.), and inherent metal characteristics. The flux used also drastically affects solderability. Comments are made below based on solderability with the least active flux which will promote wetting.

Metals Solderable with Water-white Rosin Flux and Moderately Activated Rosin Flux. CADMIUM. Can be soldered with rosin but is not generally recommended due to formation of brittle intermetallics with solder. Usually found as an electroplated finish.

COPPER. Needs to be freshly cleaned for good solderability with rosin flux. When tarnished, activated rosin will be required. Copper is used as a base metal or as electroplate.

GOLD. Excellent solderability when purity is high. Alloy gold plates may give poorer performance. Does not oxidize; however, when thin electroplates are used, underlying metals may diffuse through porous coatings and react with environment to form nonsolderable surfaces. Gold is nearly always present as an electroplate. Thickness of 0.00005 in. minimum to 0.0001 in. is usually specified. Gold has a strong tendency to dissolve in tin and form intermetallics which are brittle. (Further discussion is given later on.)

SILVER. Excellent solderability when clean. Tends to react with sulfur, forming poorly solderable surfaces. Often overcoated with gold to minimize sulfur reaction. Silver is most often present as an electroplate, although coin silver (92% silver–8% copper) is sometimes used for parts.

SOLDER. Excellent solderability when used as a coating on other metals. Applied by either electroplating or hot dipping. Hot-dipped coatings are generally superior to electroplates due to lack of porosity and superior adhesion to base metal. Tolerance of solder coatings for foreign contaminants is good since coating is melted during soldering allowing surface films to float away.

TIN. Very good solderability for same reasons as solder. Thin coatings may oxidize causing difficulty in soldering. This is especially true of electroplates. A minimum of 0.0003-in.-thick electroplating is required. Hot-dipped coatings are preferred since porosity will be eliminated.

TIN-ZINC. Very good solderability in the composition range of 80% tin–20% zinc. Resistance to solderability deterioration with age is good. Applies as an electroplated coating.

Metals Solderable with Highly Activated Rosin Flux or with Organic Acid Fluxes. COPPER ALLOY (BRASS, BRONZE, BERYLLIUM COPPER, GERMAN SILVER, etc.). Freshly cleaned, these alloys may be soldered, in most cases, with highly activated fluxes. The organic acid fluxes will be required under any but ideal soldering conditions. Where items made from these alloys are to be used in electronic assemblies, use of a finish selected from Group I above is recommended to render parts solderable with Group I fluxes.

NICKEL. Nickel plate and A nickel may be soldered with activated rosin or organic acid fluxes. In some cases freshly cleaned surfaces may be soldered with rosin, but the procedure is not reliable.

Metals Solderable with Modified Chloride Fluxes. ALNICO, CHROMIUM, INCONEL, other than stainless require the zinc-ammonium chloride fluxes. Normally these alloys are furnished with solderable finish.

IRON-NICKEL ALLOY. (Kovar,* Rodar,† and other glass-sealing alloys) Alloys of iron-nickel with other alloying elements are widely used for component leads when glass sealing is required. Bare surfaces require use of strong flux. However, components are normally furnished with gold-plated leads or hot-solder-tipped leads.

Metals Solderable with Modified Chloride Fluxes. ALNICO, CHROMIUM, INCONEL, MONEL, STAINLESS. These materials are made up of metals which form tenacious oxides of chromium, titanium, and/or aluminum. They require chloride fluxes modified with other fluxing agents in order to remove these oxides and allow alloying.

Metals with Special Solderability Problems. ALUMINUM. Strong fluoride-type fluxes are available for soldering aluminum and its alloys. Fluxless soldering with ultrasonics is done also, as is "rub tinning," in which a high-zinc alloy is rubbed on the heated aluminum surface. Corrosion resistance of soldered aluminum joints is poor and must be carefully considered. Plating with a solderable metal is often used to achieve solderability.

MAGNESIUM. Soldering magnesium is possible under special conditions but is rarely done. Plating has been done with electroless nickel plus tin to achieve solderability.

Many of the metals and alloys used in electronic equipment are not inherently solderable and must be finished with a plated or hot-dipped finish to allow soldering with the low-activity rosin fluxes. This is apparent from studying the above summary of solderability.

Coatings for Solderability. There are several factors to be considered when selecting a coating system for solderable electronic parts, including:

1. Solderability of coating
2. Method of coating application
3. Thickness of coating
4. Solubility of coating in solder, and solderability of base-metal components
5. Reaction of coating with storage and service environments

The coating, of course, must be solderable, with solder and flux specified. The method of application of the coating is important in that it determines the type of bond that the coating has with the base metal. A hot-dipped coating, applied with a proper flux, is alloyed with the base metal and thus should remain as a solderable coating even if the coating melts during soldering. However, low-melting electro-

* Trademark, Westinghouse Electric Corp.

† Trademark, Wilbur B. Driver Co.

plates such as cadmium or tin on a difficult-to-wet metal such as aluminum can melt during soldering. Often this will cause the coating to ball up, leaving a bare, nonsolderable base-metal surface. Use of a solderable, high-melting-point electroplate such as copper or silver under the low-melting-point coating will generally give satisfactory solderability.

The coating thickness must be such as to prevent the environment from attacking the base metal. It must prevent diffusion of elements from the base metal through the coating to the surface where it may form unsolderable surface films. This may occur with brass, where zinc from the base alloy diffuses through thin, porous platings and reacts with oxygen to form a nonsolderable film. Also, silver protected by a thin film of gold may be sulfided by sulfur that diffuses through the thin gold coating. A barrier plate on 0.0003 in. of copper prior to application of the solderable coating will usually solve the zinc-diffusion problem with brass. Dense plating of gold 0.00005 in. thick is normally sufficient to protect silver from sulfiding.

Another area where thickness is critical occurs in hot-dipped coatings. Coatings are applied by hot-dipping the alloy with the base metal and, depending on the alloy system, on time, and on temperature of dipping, may form appreciable alloy layers. These layers are often intermetallic compounds which are difficult to wet. If severe wiping is carried out after dipping while the coating is molten, the unsolderable layer may be exposed. Inspection for a smooth, lustrous coating will generally assure good solderability of hot-dipped coating.

Most easily soldered coatings are soluble to some extent in solder. Thus when specifying coatings for solderability it is necessary to consider the solubility of the coating in the solder used. Thin films of gold, silver, or copper on ceramic, glass, or nonsolderable metal substrates may be completely dissolved in the solder when making a connection. The higher the temperature and the longer the time of soldering, the more dissolving occurs. Gold dissolves readily in tin-containing solders, as does silver. Copper also will dissolve in solders but at a lesser rate than gold and silver. Standard tin-lead solders with approximately 3% silver, which partially saturates the solder with silver, are available for use in soldering to thin silver plates. Copper-containing solders have been used for the same purpose on thin copper surfaces. Because of the large solubility of gold in tin, the increase in melting point of solder, and the formation of brittle intermetallic compounds, the use of gold additions to decrease solubility of solder has not been successful. Using higher-lead-content solders will reduce dissolving but at the expense of higher soldering temperature. Low-temperature, fast soldering probably offers the best means of lowering solution of plating.

Another factor in considering the solubility of the metal to be soldered is the resulting alloy phases formed between the solder and the surface being soldered. As mentioned above, the readily soldered metals generally dissolve to some extent in solder. Where joints are made rapidly with temperature approximately 100°F above the solder melting point, the alloying is confined to a layer on the order of 0.001 in. thick or less at the solder–base-metal interface. For copper and silver, this is the extent normally found, and the only visual evidence externally is a slight dull line around the periphery of the joint where the solder fillet joins the base metal.

Gold dissolves more readily in solder and quickly diffuses through the molten metal. Tin-lead solder joints made on gold surfaces will generally be less lustrous than those made on other surfaces, especially if appreciable time is used to make the joint. Considerable study has been made of the gold-solder microstructure as well as the mechanical properties of the alloys formed. From the phase diagram it can be predicated that alloying gold with solder will result in the formation of brittle intermetallic compounds. At about 5% gold, the amount of intermetallic becomes such a large percentage of the joint that it drastically reduces the impact strength of the solder. Joints made by normal practices on surfaces of 0.00005-in.-thick gold plate seldom reach this concentration of gold. Heavier gold plus long, high-temperature soldering practices may result in embrittlement of joints. Field experience does not indicate that joints soldered to gold offer a great reliability problem, since gold plating has been standard practice on many connector solder cups for many years. However, some specifications now require erasing of gold from the surfaces to be soldered.

Another technique for reducing the gold content of solder joints is pretinning leads in a solder pot, which tends to dissolve away some of the gold. Still another technique used on printed-circuit boards is pretinning the surface and then vacuuming off the solder prior to final soldering.

It would appear from the evidence that use of thin gold platings in the order of 0.0005 to 0.0001 in. thick combined with short soldering times results in reliable joints to gold without prior treatment.

Solderability of Fine Wires. The soldering of fine wires results in the same problems in solution of base metals as does the soldering of thin platings. With wires in the range of 0.001 in. diameter, complete dissolving of gold and copper wires may occur during soldering. Rapid soldering with high-lead-content solder seems to be one practical approach and has been suggested for copper wires. Resistance soldering where heating time can be held to a minimum will help in minimizing alloying of gold with solder.

DESIGN OF SOLDERED ELECTRICAL CONNECTIONS

Soldered electrical connections are seldom purposely designed to carry service structural loads. In fact, most specifications for electrical connections require that they not carry loads. However, the solder used to make electrical connections is subjected to loading in many applications due to the operating environment and due to manufacturing processes used in assembly, testing, and installation. Sources of mechanical loading in various types of joints are discussed below.

Hookup Wire to Terminals. The major source of load on such joints comes from bending stresses. The source of the bending may come from vibration of equipment during its operation or from vibration of a vehicle in which it is installed. In either case, the wire or wires may move from the inertia of its weight. Since the terminal is usually tied down securely to a chassis or other relatively solid object, the wire moves relative to the terminal. Another type of bending comes from handling during assembly or inspection, or from plugging or unplugging a connector. In the case of vibration the movement is generally small and the bending stresses are low, giving a condition of low-stress fatigue which may cause failure only after millions of cycles. In the case of the handling type of bending, the stress is much higher. It is not uncommon to encounter 90° sharp radius bends. In such severe bending, failure of solid conductors can occur in several bends due to work hardening or to high-stress, low-cycle fatigue.

Generally neither of the above conditions causes failure in the solder joint. Instead the failure occurs in the wire adjacent to the solder fillet. This is a logical place for failure since (1) the wire is stripped of insulation in this area, thus creating a notch, (2) the wire can be bent around the sharpest radius in this area since no insulation is present to force a large radius, and (3) the wire terminates at a much stiffer member (the terminal), thus increasing the notch effect.

The fact that a large percentage of hookup wires fail in the stripped area adjacent to the solder joint has led to much attention being paid to the wicking of solder along stranded wire. Wicking of solder is a condition in which solder has traveled along the strands, causing them to form essentially a single strand. Wicking is more prevalent on silver-plated wires than on other wires since solder readily wets and spreads on clean silver. Specification requirements on wicking vary among various government agencies and civilian companies. Some specifications prohibit wicking beyond the solder joint while others require that the wicking be present to a given distance beyond the joint. Other specifications do not mention wicking as an inspection criterion. Reports have been issued covering tests which show wicking to be harmful, while other tests show it to be helpful. With such conflicting data, it is difficult to come to a concrete conclusion. The following seem to be reasonable conclusions based on the evidence reported:

1. Wicked wire will fail more readily than wire with free strands when a sharp bend is forced to take place in the uninsulated area.

2. Wicked wire will outperform unwicked wire when the movement of the wire is small, as in vibration. NOTE: This generalization is limited to a condition in which

the length of the wicked portion is small compared to the unsupported length of wire.

The above conclusions may be explained by the fact that a wicked wire acts as a single strand, and when it is bent around a sharp radius, the yield strength of the wire is exceeded, causing work hardening. Several bends will harden the wire to a point at which it is not ductile enough to withstand further bending, and it then fractures. However, if the applied stress is low, as in vibration, the bending will take place in the unwicked portion of the wire which is encased by the insulation. The insulation keeps the wire from taking a sharp radius and, therefore, prevents high bending stresses from being applied to the wire. The author's observation of unwicked wire during low-load-type flex testing has shown that the strands tend to spread out under flexing and the load is taken by individual strands which then break one by one. Wicked wire under the same type of loading does not flex in the stiffened portion.

The major consideration in design of hookup wire to terminal joints is prevention of movement of the wire in relation to the terminal. Tiedowns close to the terminal with adequate wire supports which keep loads off the wire can result in high reliability. Training of inspectors, assembly line personnel, and equipment installers to exercise care in handling wires, especially in joint areas, will result in reduction of wire breakage.

Joints on Printed-circuit Boards. A first look at printed-circuit boards will lead to a conclusion that the loads on printed-circuit joints are minimal. Calculation of the load put on a joint by the weight of a component under heavy g loading still shows an entirely adequate safety factor. However, there are several sources of mechanical loads in printed-circuit boards that may cause joint failure.

Flexing. Flexing of boards under vibration can be quite severe if the boards are large and the supporting points are widely separated. A bending load can occur which may result in a fatigue failure. The presence of the component lead hole is a stress concentration point which can raise the stress in this area.

Thermal Expansion. The thermal expansion of the insulating boards used for printed circuitry is considerably higher than that of metals. The value of thermal expansion of epoxy glass has been found to be as high as 10 times that of copper when measured in the thickness direction of the board. On double-sided boards with plated-through holes, this difference in expansion has caused cracking of the plating in the hole during soldering.

Two types of failures attributed to differential coefficient of expansion have been encountered in eyeletted printed circuits. In thermal cycling of the type usually specified for military equipment (—55 to +125°C), failures have occurred in the solder joint around the periphery of the eyelet. Usually a number of cycles is required to cause this type of failure indicating a low-cycle, high-stress fatigue failure mode. The other type of failure occurs during the soldering process. In this case the board and eyelet expand due to the heat of solder. The solder solidifies, completing the joint. As cooling continues, the contraction of the board exceeds that of the eyelet and sets up a stress sufficient, in some cases, to fracture the solder joint.

Wraps for Terminals. The usual requirement for wrapping hookup wire and component lead wires onto terminals is for "mechanically secure" wrap. The main purposes of this requirement are to prevent loads from being imposed on solder, and to prevent movement of the joint during the solidification of solder. One complete wrap has been the rule in most cases. Recently deviations from the mechanical wrap have been proposed and sometimes specified. These have been proposed as reducing costs and improving reliability. The use of slotted terminals in which component leads could be laid and soldered have been demonstrated to be capable of withstanding most military environments. From the reliability standpoint it has been argued that the biggest reliability problem is joints which were never soldered. With mechanically secure joints, joints that are missed may have sufficient continuity to pass in-plant tests but fail under dynamic conditions encountered in service. This argument has been used in favor of using nonmechanically secure wraps on terminals.

Figure 9 shows the wrap requirements of United States of America Standards

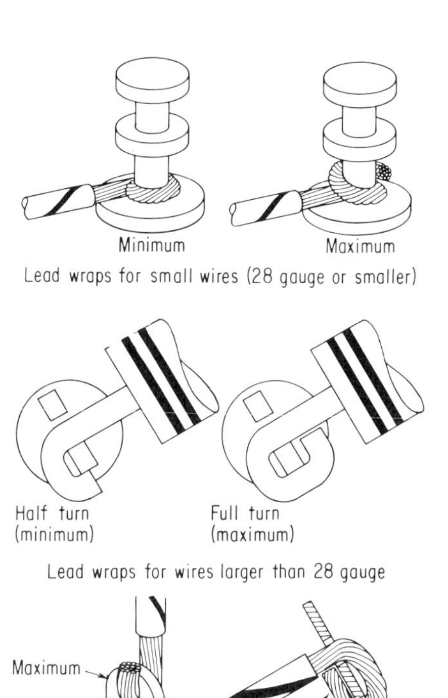

FIG. 9. Wrap requirements for hookup wire attachment.

Association Standard C-99. These wraps are not universally accepted but represent one group's opinion of good practice. Specific requirements varying in some details from these are called out in government specifications.

Joint Clearances. The clearance between the members of a soldered joint will affect the strength of the joint. Usually the thinner the layer of solder, the stronger the joint. But the joint clearance must be wide enough to allow solder to enter the space during soldering. In addition, the clearance should not be so great as to prevent filling of the joint by capillary action. Figure 10 shows the effect of clearance on strength of soldered joints made with three metals. Note that clearances of 0.003 to 0.005 in. show maximum strength. Such clearances are satisfactory for manufacturing

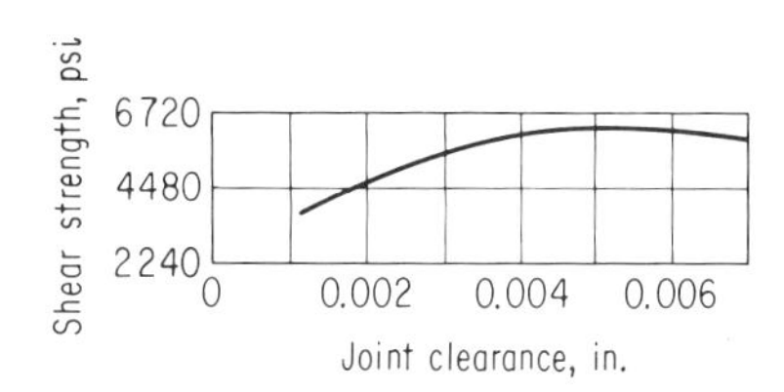

FIG. 10. Effect of clearance on strength of joints on copper.[5]

purposes since they are large enough to allow solder to enter the joint by capillary action and to be held there after entering.

It can be noted from Fig. 10 that joints made on copper had the highest strength. Two factors may account for this circumstance. Since copper alloys readily with solder and gives a strengthening affect, copper may give a sounder joint with more alloy (and resultant strengthening) than the other metals.

With conventional printed circuits and other joints where a lead is inserted into a hole it is important to consider the relative size of the lead and the hole. This clearance will affect the quality of the joints from the standpoint of joint soundness and completeness as well as final strength. The use of clearances of 0.003 to 0.008 in. is recommended for complete fillets from dip soldering.

QUALITY ASSURANCE OF SOLDERED ELECTRICAL CONNECTIONS

The final inspection of soldered connections is primarily done by visual means, usually with relatively low magnification in the range of 3 to 10 diameters. The amount of inspection is a matter of weighing a number of factors, including: (1) the consequences of failure, (2) confidence in the process control, (3) the difficulty of making the joints, (4) the environment in which the equipment will operate.

In a missile application, where one electrical connection failure can cause failure of the mission, 100 percent inspection is usually required. On the other hand, a $3.98 radio can hardly bear the cost of 10 to 15 min required for 100 percent inspection. On the other hand, neither the missile nor the radio can tolerate defective work. The missile manufactured with a multitude of defective or marginal joints is unreliable since no amount of inspection can detect every possible defect or weak point. The low-price radio cannot economically compete where production yield is low or extensive repair is required.

It is important that trained personnel, proper materials, processes, equipment, and design be utilized to assure that the joint is made right the first time. Old adages that apply especially to soldering are: "An operator is the best inspector of his work." "Quality must be built into, not inspected into equipment." The operators must be trained in the skills of their jobs. Design must allow proper access for the heating process with controlled joint clearances and mechanical security during solidification of the joint. Surfaces must be readily solderable with the flux specified. Fluxes must be safe for the application. If fluxes are to be removed, the proper cleaning solvents and rinsing procedures must be used—and the assembly must be cleanable, i.e., no inaccessible flux traps can be present. Heating equipment must be capable of supplying heat at the temperature level and at the rate required. Solder composition must be maintained within specification requirements.

Quality assurance then means much more than inspection. With proper control of materials and processes and proper training of operators, the inspection of soldered electrical connections can become less burdensome and more positive.

Visual Requirements for Soldered Joints. The purpose of visual inspection is to assure that wetting has occurred on all surfaces required to be joined. That is, the surface members of the joint should be alloyed with the solder. Several well-known visual characteristics give assurance that wetting or alloying has occurred. These are:

1. Spreading out of the solder into a thin film on the solderable surface with the fillets feathering into the surface. Conversely, a nonwetting condition will result in the solder's forming a ball or bead on the surface, much as water beads form on a well-waxed surface.

2. A smooth surface which indicates that the joint was heated sufficiently high to allow the solder to have good fluidity. The smooth surface also indicated that no appreciable metallic or nonmetallic impurities were present. Such a condition also indicates that adequate flux was present to cover the molten surface preventing oxidation. When tin-lead solder is used, the surface should generally be bright and shiny. However, in some cases with tin-lead, such as when soldering to gold or cadmium, some lack of brightness may be encountered due to alloying with these

surfaces. This will not indicate poor joints if the requirements of No. 1 above are met. Tin-antimony, tin-silver, tin-lead-silver, and other alloys may deviate from the bright shiny luster of tin-lead joints. When using these alloys special inspection standards should be set up to distinguish such joints from tin-lead joints.

3. The contours of the joint members are visible. While extra solder obscuring the joint does not necessarily indicate a nonwet joint, it does prevent visual inspection for condition No. 1 and thus prevents the use of the most valuable visual inspection criteria.

Some of the rejection criteria are discussed below:

1. *Solder points or peaks.* Electrically such points act as concentration points for electrical arcs. They also indicate that temperature may have been on the low side or that insufficient flux was used.

2. *Pits, holes, or porosity.* Such defects may occur due to gases generated in soldering or from solidification shrinkage. Such visible surface defects may be indications of subsurface imperfection. Gases may occur from organic contaminants, moisture in insulation, or breakdown of insulation. Gas holes may be due to insufficient heating time to allow the flux to escape from the solder.

3. *Excessive solder.* As mentioned above, solder that obscures the outline of the joint members destroys inspectability. The configuration of the joint may influence the shape of the solder fillet. On small printed-circuit pads, for instance, there is not room for the solder to spread out and feather into the pad. Thus the solder may assume a semiball shape. In such cases it is important that some projection of the component lead be visible to assure that visual inspection for lead wetting can be accomplished.

4. *Cold solder.* Joints made with insufficient heat will not flow out. They generally do not flow out and feather into the surfaces. Usually these joints will show wrinkled or rough surfaces and a tendency to form a ball. The cause is insufficient time and/or temperature to accomplish alloying.

5. *Rosin joints.* This condition is evidenced by a film of rosin flux between the solder and the surface to be joined. The probable cause is too little time or too low temperature.

6. *Dewetting.* This defect is usually evidenced by a dull surface with islands of thicker shiny solder. The surface has been wet by the solder, but then pulls away as if the wet surface were unsolderable. It is the author's opinion that the new alloy formed by the solder and the surface during heating is thermodynamically incompatible with solder, thus resulting in rejection of the solder or dewetting. Another defect which gives a similar appearance occurs when the solderable coating on an unsolderable base metal is dissolved by the liquid solder, leaving the base metal bare and no longer solderable. This condition sometimes occurs with gold-plated Kovar or gold-plated nickel. This is an especially difficult inspection problem since in many cases the joints feather into the lead surface much as a perfect joint does. Careful inspection at relatively high magnification is necessary to detect this condition. A corrective action for this phenomenon may be use of hot solder dip applied directly to the lead, using chloride flux, of course, followed by washing to remove the flux before final assembly.

7. *Disturbed solder.* This condition is evidenced by a rough surface. It is sometimes difficult to distinguish between disturbed solder, cold solder, and contaminated solder. Movement of a member of the joint during solidification of the solder causes a disturbed solder. Cracking and partial separation may result. Proper mechanical support, accessibility of tools, and operator skill are factors in the occurrence of this condition.

The visual characteristics specified in most soldering specifications and outlined above are primarily based on tin-lead alloys with a tin content near the eutectic (63% tin) composition. This is especially true of the requirement for a smooth, shiny surface. Tin-lead alloys with appreciably lower or appreciably higher tin content will not exhibit the same degree of luster as do the near-eutectic compositions. Also, since all tin-lead alloys except the eutectic pass through a temperature range in which liquid and solid coexist during solidification, the surface of the solidified joint will be less smooth. For the same reason fillets will be heavier. This does not imply that the joints are inferior, but they may be suspected of being "cold" or contaminated when compared to joints of near-eutectic composition.

Many of the other alloys, such as tin-antimony, lead-indium, and lead-silver, will exhibit appearances similar to those described above. Therefore, when using such alloys the visual characteristics for inspection that are relied on for near-eutectic tin-lead joints are lost. It becomes necessary to set up standards of acceptance for the particular alloy being used in soldering.

A suggested method is to make joints simulating production joints. These joints can be tested for electrical and mechanical properties. In addition, metallographic examination for internal structure should be performed to assure that alloying has occurred. When satisfactory mechanical, physical, and metallurgical properties are attained, visual standards can be set up to serve as standards for inspection.

Nondestructive Tests. Use of nondestructive testing, such as radiography, penetrant inspection, ultrasonics, and thermography, has been proposed for inspection of soldered electrical connections. However, to the author's knowledge no production system exists for checking final assemblies. Some use has been made of nondestructive techniques on an investigative basis.

Radiography has been used to check for proper filling of solder cups on connectors. By this means the completeness of filling of the cup can be determined. Penetration of the wire in the cup can also be determined by x-ray. This method has also been used in some cases to check the soundness of joints on printed-circuit boards.

It has been suggested that use of thermographic means for checking the temperature rise of a soldered joint under electrical power could detect marginal joints. It is reasoned that an incomplete joint would have a high resistance, thus causing a rise in temperature due to I^2R heating. Infrared temperature measurement and heat-sensitive materials have been suggested as sensors. By the same reasoning, checking of the resistance of solder joints could furnish the same information. However, no workable scheme for applying these techniques on a production basis has been developed.

The primary means of inspection of soldered joints remains that of visual observation.

MECHANICAL CONNECTIONS FOR ELECTRICAL PURPOSES

In a great number of electrical joints, the electrical continuity is attained through intimate metal-to-metal contact between the surfaces of the joint members. One of the most frequent uses of this method is in plug-in connectors. The subject of connectors is covered in another chapter and will not be dealt with here. The use of mechanical connections for individual joints to terminals is discussed, however.

Many of the techniques of mechanical connections have caused considerable controversy over reliability performance. While no valid comparison with soldered and welded joints is known to the writer, mechanical connections have performed satisfactorily in commercial and military equipment when designed within the limitations of the process.

The list of techniques for making electrical joints by mechanical means is quite long. Three general techniques are considered in this chapter: (1) wire-wrap type, (2) wire-to-clamp type, (3) crimp type.

Wire Wrap. The wirewrap process consists of wrapping a *solid* wire conductor around a post with sharp corners. The wire is wrapped while under tension, causing deformation of the wire at the sharp corners of the post. The post is also deformed under the wire. The resulting joint gives a gas-tight interface between the wire and the post.

Two types of wraps are used. One type has all hard wire wrapped on the post while the other type has one insulated turn (the last turn) wrapped on the post. The insulated turn prevents deformation of the wire at the last turn. Under dynamic conditions, the deformed wire acts as a stress concentration notch with resulting low fatigue life.

The solderless wire-wrap technique is limited to round solid wire since the properties of solid wire are needed to assure the intimate and uniform contact which assures continuity. Use of wire-wrap tools to install stranded wire onto terminals for subsequent soldering has been reported.

The wire-wrap process is an extremely economical process which combines the wrapping of wire on the terminal with the making of the joint. No intermediate hardware is needed.

Materials. Control of material properties of wire, terminals, and coatings on these materials is necessary to assure reliable wire-wrap joints. It is also necessary to control the shape of both wires and terminals. Some general requirements are discussed below:

WIRE. The wire generally used is round soft copper similar to QQW-343 Type S, Soft. When copper alloy is used it should be of a similar hardness. The elongation of the wire should be controlled so as to assure sufficient deformation in wrapping. Elongation (in 10 in.) normally specified is 15 percent minimum for AWG sizes 24 and 26 and 20 percent minimum for AWG sizes 18, 20, and 22.

WIRE FINISH. Tin, tin-lead alloy, silver, and gold with a minimum thickness of 0.000050 in. are suitable finishes for wire to be used in the wire-wrap process.

TERMINALS. Square or rectangular terminals or wrap posts are used as members on which the wire is wrapped. The edge radius of the post should not exceed 0.003 in. Typical sizes are 0.045 in. square, 0.025 in. square, and 0.06 × 0.03 in. rectangular.

Copper alloys are used for terminals. Gold plating is often used for a finish to assure low contact resistance.

Tools. Special tools for wire wrapping are available on the market. Hand-held tools for making of joints by individual operators are widely used. Most of these tools are also available for lab operation. Automated systems programmed to wire complicated circuits, such as back-panel wiring, are available for wire wrapping.

Joint Characteristics and Controls. WIRE TURNS. The number of turns of wire wrapped on the terminal is typically four turns for 18 and 20 AWG wire size, five turns for 22 and 24 wire, and six turns for 26 wire. In addition the class A wraps have an additional one-half wrap minimum of insulated wire in contact with the terminal.

STRIPPING FORCE. The completed wire-wrap joint must withstand a stripping force which tends to slide the wrap off the terminal. Forces specified by MIL-STD-1130 are shown in Table 8.

GAS TIGHTNESS. Another test to assure a good electrical joint is the testing to assure gas tightness between the edge terminal and the wire. Joints are first exposed to aqua regia fumes, then to ammonium sulfide gas. Upon unwrapping, 75 percent of the contact areas (except the first and last turn) must be free of evidence of the sulfide presence.

VOLTAGE DROP. Voltage drop across the wrapped connections must not exceed 4 mv.

Control of Wire-wrapped Joints. Quality assurance in wire wrapping is highly dependent on the control of tools. Tools are qualified regarding their ability to make joints that meet the stripping force, gas tightness, and voltage drop requirements. Periodic checks are made for one or more of these characteristics, as in process control.

Visual checks of joint characteristics are also made. The number of turns is an im-

TABLE 8. Strip Force Limits for Wire Wrap

Size designation, AWG	Diameter, in.	Minimum strip force, lb.
26	0.0159	6
24	0.0201	7
22	0.0253	8
20	0.0320	8
18	0.0403	15

portant quality requirement. Crossing of the wires from adjacent turns is not allowed. Neither is the overlapping of the wires from one wrap to another when two wraps are made on the same post.

General requirements for wire wrap are given in MIL-STD-1130.

TERMI-POINT* Connections. TERMI-POINT is a mechanical system for making electrical connections. The system is applicable to both solid and stranded wire. These connections can be made by hand tools or by programmed equipment which strips the wire and makes the connection.

The components of the connection are a post, a clip, and the wire. The posts are mounted in suitable bases which may be connectors, bus bars, or large back panel boards. The stripped portion of the wire is laid flat against the post face and the spring clip is applied over the post and the wire. The spring action of the clip holds the wire and post in intimate contact. The resulting contact is said to be gas tight.

Removal of clips and subsequent application of new connections without disturbing other connections is possible using simple tools.

Major advantages of TERMI-POINT are its adaptability to automated programmed wiring, ease of replacement, and, compared to wire wrap, its usefulness with stranded wire.

Crimp Connections. A wide variety of crimp-type terminals are available and are in wide use in both military and industrial applications. The connection to the wire is made by deforming the tubular shape of the terminal, which has been placed over the wire. The connection to the terminal is then made by any number of mechanical means such as by screw pressure, taper pins, or taper tabs.

The application of crimp connections to wires can be mechanized or can be by hand tools.

REFERENCES

1. Foster, F. G.: Gold from Plated Surfaces May Embrittle Solder, report from Bell Telephone Laboratories Inc., Murray Hill, N.J.
2. Braun, J. D.: An Improved Soft Solder for Use with Gold Wire, *Trans. ASM*, vol. 57, 1964.
3. Manko, H.: "Solders and Soldering," McGraw-Hill Book Company, New York, 1964.
4. ASTM Publication No. 189, "Symposium on Solder," June, 1956.
5. Lewis, W. R.: "Notes on Soldering," Tin Research Institute, Middlesex, England, 1959.
6. Barber, Clifford L.: "Solder, Its Fundamentals and Usage," Kester Solder Co., Chicago.
7. Alpha Metals, Inc., Jersey City, N.J., Technical Data Bulletin A-103B.

*Trademark of AMP, Inc.

Chapter 4

WELDING AND METAL BONDING TECHNIQUES

By

H. F. SAWYER

Walter V. Sterling, Inc.
Consulting Engineers
Claremont, California

INTRODUCTION

Joining of lead wires that carry electrical information for electronic circuits is the subject of this chapter. The significance of the multiple of techniques, of analyses, and of cost may all be measured against whether the electronic system performs under the required environments. Specifically, two conditions must be met: (1) electrical conductivity and (2) adequate physical strength. Both these conditions must be acceptable at a very high reliability because of the large number of joints typical in the complex systems of concern. Assuming weld-joint resistance to be negligible compared to the lead resistance, the only requirement for welded joints is that they do not break under specified environments. As a result, a minimum strength is the most realistic requirement for any proposed material combination or joining method. The object of any joint testing program on particular process settings is to predict the percentage of the parent population that will be below this minimum strength. For high reliability requirements needed in high-density electronics, strength predictions are involved far out on the lower tail of the probability density function. Out in the tails is precisely the area in which a "fitted" distribution (such as the normal) can be seriously in error, even though it passes a "goodness-of-fit" test at a high significance level. Large errors, which may take on factors rather than percentages, have been experienced in statements about minimums using normal statistics. All joining methods are subject to estimates of strength based upon test samples since it is not possible (yet) to perform nondestructive measurements on each joint made, nor would economics permit the expense of such extensive measurement. Reliance on process controls is mandatory and when carefully done will provide the necessary minimum strength guarantees. Each application must be considered as a separate case to determine what minimum strength will *always* survive the manufacturing and use environment. The minimum is elusive if one tries to compute each joint in each application; therefore, it is recommended that a *worst-case* minimum be established by good engineering analysis, and then a margin applied to extrapolate the measured probability to the desired reliability. A good margin empirically determined is a factor of 5. An example is resistance-welded component leads, where $P = 0.003$ at 90 percent CC, where P is the probability of having a strength lower than the lowest in the sample. A sample of 778 is used; a minimum of 5 lb is established by the observation that in mixed designs joints with lower than 1-lb strength gave frequent failures and the use of a margin factor of 5 to guarantee a *much* higher reliability *necessary* to prevent joint failure from becoming a predominant cause of module failure.

All other joining methods should be treated similarly, and controls should be effected to maintain *above*-minimum strength in production. The information which follows in this chapter is intended to provide an insight into each type of joint so that analytical means can be applied to establish the process, set up the critical process controls, and monitor the result.

Types of Welding. Several methods of welding electronic connections are being used with good success. The degree of success depends primarily upon the suitability of the welding method to the application. Each method has inherent advantages and disadvantages for a particular application, which must be considered when making a choice of method. It is the intent of this section to provide the designer with a working knowledge of how each type of welding is achieved. Various packaging applications and their relationship to the methods of welding are discussed in a later section of this chapter.

TYPES OF WELDING

Welding of metals is not a new art but one which has a wealth of technical information in the literature particularly relating to the knowledge of metallurgy. The current interest in welding of electronic packaging involves two new developments: (1) specialized machines and (2) electronic components with leads of known and controlled materials. The advantages of welding may be obtained, then, by an understanding of the new methods and machines in an old subject. A number of welding methods are described in the following pages. New techniques and machines will be developed which are not described and which the reader will need to research for himself. There is a wide variety of techniques available and described, and great versatility therefore exists, permitting welded joints for any application where welds are needed.

Resistance Welding. Resistance welding is a weld system wherein the welding heat is generated by an electric current flowing through the pieces or one of the pieces being joined. The heat is produced by the resistance characteristics of the materials themselves, where $W = I^2R = E^2/R$. It is important to recall that for welding considerations R is not a constant because of the temperature coefficient. This temperature coefficient increases with temperature for most metals, which decreases the current (for a constant-voltage welder) or increases the power (for a constant-current welder). There are many developed resistance welding methods to account for different applications; the important ones are described.

Power Source. The source of electric power has an effect on the weld due to the effects of polarity and waveform, and therefore the weld schedule must take into account the parameters of the source and must provide setting information and control.

DIRECT CURRENT. The most popular and commonly used dc power is the capacitor discharge welder. This machine provides energy which is stored in a capacitor bank. The stored energy is adjusted for a weld schedule by setting the voltage of the bank. The energy in the bank, W, is $W = \frac{1}{2} CE^2$, where C is the capacitance and E is the voltage. The energy is seen to be dependent upon voltage and capacity, but it is commonplace to use the voltage as a measure of energy by meter scale conversion. It is incorrect, obviously, to equate the energy of two banks of the same voltage but different capacitance. The energy of a capacitor bank is transferred to the work through a step-down transformer to increase the current and decrease the time. A short time is desirable to prevent excessive heat loss by conduction during the weld cycle. The capacitor discharge results in a single current direction and a single current discharge cycle.

OTHER POWER SUPPLY FORMS. Many other discharge forms are possible and available in equipment such as polarity reversal dc, battery source switched dc, ac-controlled pulse welders with many electric pulse forms, and feedback-controlled welders which sense and control workpiece resistance. These electives complicate the choice of what to use for a particular application but enhance the opportunity of effecting a high-reliability joint.

Electrode Configurations. Several electrode configurations have been used to fit particular applications. These are described briefly with their important characteristics.

SERIES. In series welding, both electrodes are on one of the two pieces being joined. Current passes from one electrode through the top workpiece and out through the other electrode, as shown in Fig. 1. The procedure results in two welds, one under each electrode. The welder electrodes are usually separated by more than 0.025 in., and each electrode is separately loaded. The process is particularly suitable where the bottom workpiece is inaccessible and where two welds are desirable. This kind of weld is subject to polarity effects since unidirectional weld current flows through each weld in opposite directions. This frequently makes one stronger than the other. A dual polarity dc weld or an ac weld will overcome this disadvantage.

Wires up to 0.040 in. and ribbons up to 0.015 in. thick × 0.050 in. wide can be welded in this series welding process with the conventional equipment. It is essential that the base metal have the capacity for weld current between the electrodes or the

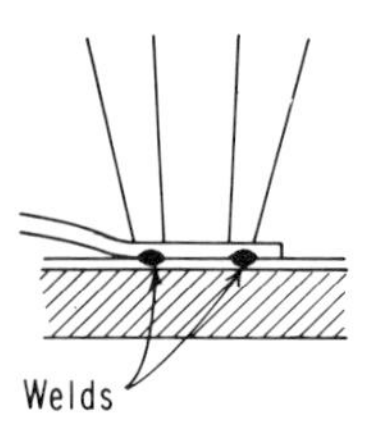

FIG. 1. Resistance microwelding—series configuration.[1]

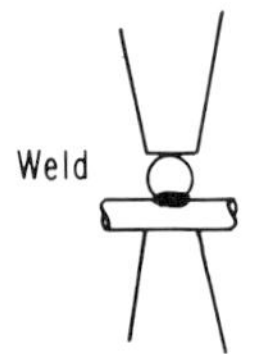

FIG. 2. Resistance microwelding—opposed-electrode configuration.[1]

top workpiece will overheat and "blow out." Wide variations in thickness of the two wires to be welded (over 2:1) cannot be tolerated in series welding.

OPPOSED ELECTRODE. The principal feature of opposed-electrode resistance welding is the fact that the electrodes approach the parts to be welded from directly opposite sides (Fig. 2). In this type of arrangement the surfaces of the electrodes which come in contact with the work are directly opposed to each other. No electrode springback may be expected, so that "follow-up" dynamics must be provided by low-inertia electrode holders and head moving parts. Some instances of burning due to lack of follow-up have been experienced. A necessary requirement in opposed-electrode welding is access for the electrodes to approach the workpiece from opposite sides.

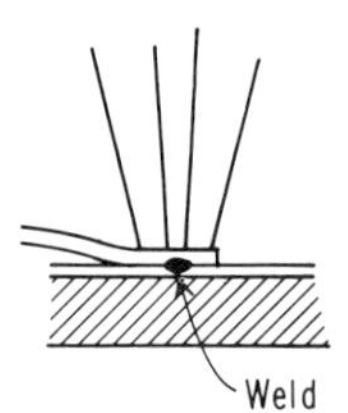

FIG. 3. Resistance microwelding—parallel-gap configuration.[1]

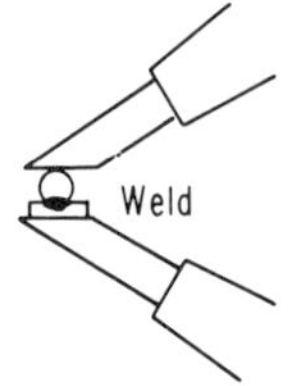

FIG. 4. Resistance microwelding—pincer or angular configuration.[1]

Relatively high forces are easily obtained in this configuration, which enhances welding of comparatively large-diameter wires. The process is one of the simplest to set up and perform.

PARALLEL GAP. Parallel-gap welding is a method whereby a single weld joint can be made where only one side of the work is accessible. The electrodes are positioned very close together (Fig. 3) and separately loaded, and they may or may not be separated by insulation. Weld current flows from one electrode through the work and back out through the other electrode. The weld size is a function of the electrode separation and energy. The weld surface must be well supported and *exactly* normal to the electrodes. Better welds result from using high-resistance-type materials on top with lower-resistance materials beneath. The electrodes should be wider than the material so that positioning of the electrodes by the operator is not critical. Split-tip welding is a variation of the parallel-gap system in which the tip is solidly constructed with insulation laminated between the conducting halves. The split-tip system is particularly suited for very small wires and metal films. Both split-tip and parallel-gap welding are insensitive to polarity.

ANGULAR ELECTRODE. The angular electrode configuration (Fig. 4) is the most important for construction of cordwood welded modules because of the accessibility it provides for interconnecting on a plane surface where access is only from one side and lead wire projections are to be joined. The most popular included angle is 70°, a compromise between spring force limits of the electrodes and accessibility to the closely spaced lead wires. Any angle may be used between a minimum to clear the holders (20°) and opposed (180°). Frequently the tips of the electrodes are reduced in diameter in order to provide additional clearance. If this is done, straight rather than tapered reduction has the advantage, since the electrode face area remains constant as the electrode wears down due to resurfacing. The clearance needed is 0.050 in. between adjacent projections to weld with 0.012- × 0.030-in. nickel ribbon. The angular configuration creates a spring force applied through the moving and stationary weld head and holders. This condition helps in weld follow-up and requires that electrode holders be pinned or securely clamped to prevent rotation. The electrode welding surfaces should be flat and parallel to each other. The polarity of the weld current using a dc single-cycle capacitor-discharge power supply is a factor in the weld schedule. As we have seen, dissimilar materials exhibit different resistance depending upon the direction of current flow.

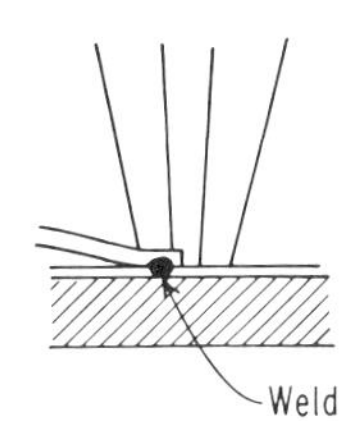

FIG. 5. Resistance microwelding—step-series configuration.[1]

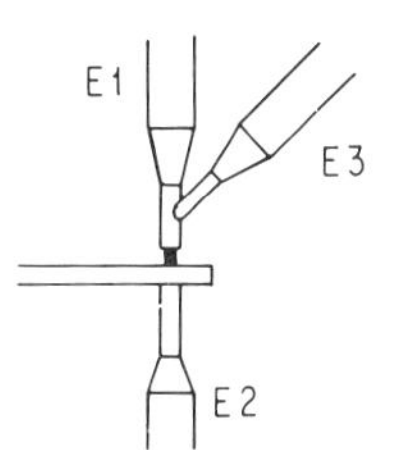

FIG. 6. Resistance microwelding—through insulation with hot third electrode.[1]

STEP SERIES. In step-series resistance welding, one electrode is in contact with the base metal while the other is in contact with the ribbon or wire being welded, as shown in Fig. 5. Step series is used primarily on materials that cannot be welded by parallel-gap methods. The process is especially suitable for welding high-resistance leads to base metals. A big advantage in this process is that high-resistance leads can be welded from one side of the workpiece. Base metals must be thick enough to carry the electrical current needed for welding heavy leads. Step-series welds on thin films are not very practical because of the low current-carrying capacity of the base metal. The electrode spacing is not critical; the weld is sensitive to polarity when using a dc power supply.

HOT ELECTRODE (THREE). (As shown in Fig. 6.) Insulated wire size 0.0005 to 0.015 in. This process permits welding insulated wire without prestripping the insulation. The insulation is vaporized by the generation of heat at the point of weld. The third electrode (E3) passes some of the weld current, heating one of the other welding electrodes (E1). The two electrodes are made of high-resistance material for this purpose. The type of material recommended for welding is copper: the insulation material is most critical.

HOT ELECTRODE (TWO). (As shown in Fig. 7.) The use of a special, heated top electrode is acceptable and permits the use of conventional equipment except for a special electrode holder or electrode heater and heater power supply. Most common varnish-insulated wires and plastic-insulated wires may be welded without prior insulation removal. Typically, magnet wires of various sizes may be welded using this technique; the insulation types which are successful are polyurethane, nylon, Nyleze, and Formvar. The parameter of concern for effecting insulation breakdown is the electrode temperature (50°C or less). Wire sizes possible range from 0.005

to 0.025 in. (at this time), with limits fixed only by the welder and heater power supplies.

The hot-electrode welding system presents a point of danger for the weld operator and the work in that the hot electrode can accidentally burn a hole in the Mylar* (or positioner) or injure an operator's hand. The training and protection guards,

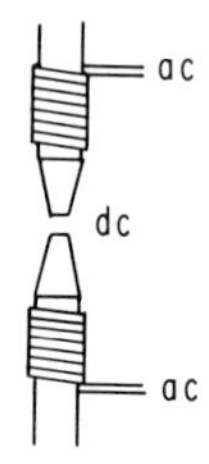

FIG. 7. Resistance microwelding—through insulation with continuously heated electrode.[1]

however, are a small price to pay for the added capability provided. Welds made this way do not suffer any degradation of strength and are sometimes made with the same schedule as is used with the uninsulated combination. Typical insulated wire welding combinations are listed in Table 1.

* Trademark, E. I. du Pont de Nemours & Co., Inc.

TABLE 1. Insulated Wire Welding

Wire, diameter in inches	Insulation	Interconnection	Electrode	Electrode orientation, and strength
0.014 Ni	Formex [a]	0.018 Kovar [b]	Tungsten	Opposed electrode 6 lb
		0.020 Dumet	Tungsten	Opposed electrode 6 lb
		0.025 OF Cu	RWMA-2 Tungsten	Opposed electrode 7 lb
		0.020 Ni	Tungsten	Opposed electrode 7 lb
0.020 Cu	Nyleze [c]	0.010 × 0.031 Ni	RWMA-2 Tungsten	Opposed electrode 6.5 lb
0.010 Cu	Nyleze	0.025 OF Cu	Tungsten	Opposed electrode 7 oz
0.006 Cu	Nyleze	0.031 × 0.031 OF Cu	Brass, tin-plated	Opposed electrode 13.5 oz
0.004 Cu	ML [d]	0.028 × 0.040 OF Cu	Brass, tin-plated	Opposed electrode 13.0 oz
0.005 Cu	Poly-thermaleze [e]	0.020 dia. OF Cu	Dumet	Opposed electrode 4.5 oz
0.0075 Cu	Varnish	0.014 × 0.050 OF Cu	Copper, tin-plated	Opposed electrode 5.0 oz
0.011 Cu	Heavy Formvar [f]	0.045 × 0.065 OF Cu	Cold-rolled steel	Opposed electrode 9 oz

[a] Trademark, General Electric Co.
[b] Trademark, Westinghouse Electric Corp.
[c] Trademark, Phelps Dodge Magnet Wire Corp.
[d] Trademark, E. I. du Pont de Nemours & Co., Inc.
[e] Trademark, Phelps Dodge Magnet Wire Corp.
[f] Trademark, Monsanto Co.

Percussive Arc. Percussive-arc welding is the joining of two workpieces where a flash of arc melts both surfaces locally. Percussive welding can effectively join wires of widely differing physical properties, including fine wire (0.002 to 0.015 in. in dia.). There are three variations of importance described below. Many advantages are realized by using this process to terminate connectors and terminals: true fused welds exist in place of diffused welds; stranded wire as well as solid can be used; a terminal needs only a simple butt end—no pot or cavity—providing greater miniaturization. The contact may be equal in diameter to the wire it is terminating. The joint may be cut and rewelded many times since only the very ends are involved in the weld. Percussive-arc welding offers advantages in joining delicate electronic components, as follows: (1) high heat concentration, which permits a wire to be butt-welded to a large surface such as a ground wire to a chassis; (2) low pressure, which permits butt welding of stranded wires or any material of low columnar strength; (3) low sensitivity to conductivity and low melting point of the workpieces, which minimizes the dependence upon precise welded settings; (4) joining of dissimilar metals including various surface finishes; (5) the short weld pulse does not permit significant contamination of the reactive metals or permit the formation of intermetallic low-strength areas in the joints.

Percussive Welding. Percussive welding employs a charged capacitor bank connected across the two pieces to be joined, as shown in Fig. 8. The two workpieces are brought together rapidly, and just before they meet, a flash of arc melts the colliding surfaces. The molten surfaces are then squeezed together by the collision,

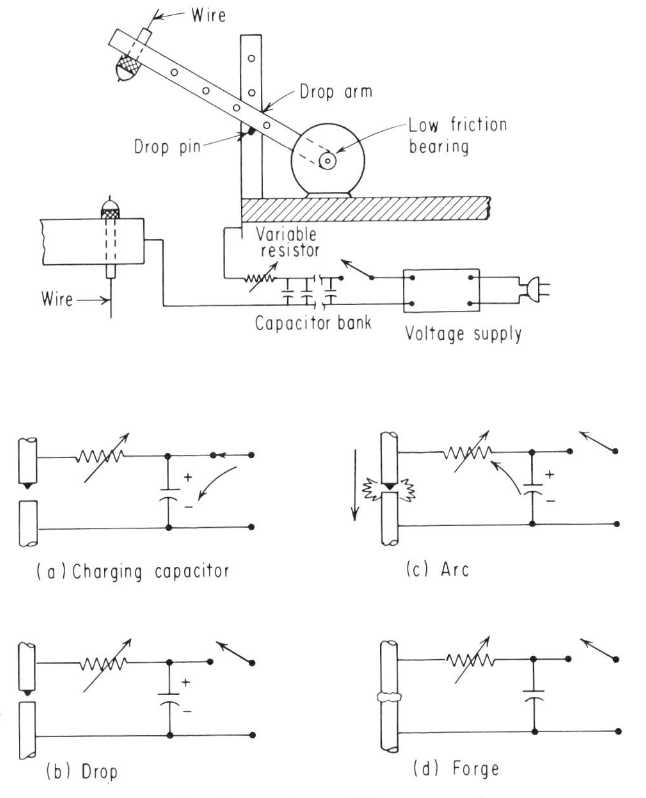

FIG. 8. Percussive welding system.[2]

and some of the metal is forced out to the sides of the joints. Precision controls are required for the charging voltage, the total capacitance, the forging force and speed of the approaching parts, and the value of the series resistance.

The voltage and capacity control the energy stored in the system and therefore also control the heating effect of the arc. The approach speed determines how much time the arc has to heat the parts, and the series resistance regulates the discharge time. The mechanical parts of the welder must be able to provide forging forces and positioning. Some representative examples of materials which have been joined by this system are given in Table 2. When tested for strength, they are stronger than

TABLE 2. Typical Percussive Weld Combinations [2]

Material	Material size, in.	Tensile strength, 1,000 psi	Location of failure
Chromel to alumel	0.015 diam (both)	77.3	Alumel
Copper to Nichrome	0.015 diam (both)	39.1	Copper
Copper wire to type 304 plate	0.015 diam (both)	40.0	Copper
Nichrome wire to type 304 plate	0.015 diam (both)	145.5	Joint
Chromel-alumel wire to type 304 plate	0.015 diam (both)	65.0	Joint
Thorium to thorium	0.040 sq (both)	Not measured	Wire
Thorium to Zircaloy-2	0.040 sq (both)	Not measured	Zirc wire

the weaker base metal, and in all cases, the failure in the base metal is far from the weld.

Percussive-arc Welding. The percussive-arc weld utilizes the two workpieces to be joined as the anode and cathode. The workpieces are separated by a gap which, at the initiation of the welding cycle, is ionized by a burst of rf energy. A capacitor bank then discharges across the gap, generating an arc. This arc heats the two surfaces so that on each a thin layer of metal is brought to welding temperature. Simultaneously with striking the arc, an electromagnetic actuator is energized that accelerates one workpiece toward the other. When the workpieces are at the proper heat, the propelling force of the actuator percussively joins them together with proper forging action to make the weld. The driving force hot-forges the weld and expels any surface oxide films from the interface, which results in a strong, clean weld. The driving force and the inertia of the actuator can be adjusted so that conductors of low columnar strength such as stranded copper wire can be welded without distortion.

The basic component of the percussive-arc welding system is the capacitor discharge power supply. The capacitor banks are charged to the desired voltage through an adjustable output ac-dc rectifier. Usually the bank is kept connected to the power supply until a weld pulse is triggered at which time the charging circuit is opened and the weld circuit is simultaneously energized. The capacitor discharge current passes through an auto transformer which steps the voltage down to a safe value. From the auto transformer, the weld pulse passes through a toroid, generating a high-frequency signal which precedes the dc weld pulse and ionizes the gas between the electrodes. This provides the low-resistance path for the relatively low-voltage weld current. The members to be joined are the electrodes which discharge the capacitor banks through the arc. One member is held by a fixture, while the other is secured to the welding head. Horizontal and vertical alignment adjustment is required, and an adjustment for the gap is required. For welding conductors to large members where accessibility is more of a problem than alignment accuracy, a hand tool can be used.

Pulse Arc Welding.* In Pulse Arc welding, members to be joined are held together, and an arc is struck by cycling a tungsten rod to the work and back a preset distance while the workpieces complete the circuit to the capacitor bank as shown in Fig. 9. When an rf voltage precedes the welding current, the arc is struck without contact between the tungsten electrode and the work. The capacitor bank is the same as described above for the percussive-arc welder.

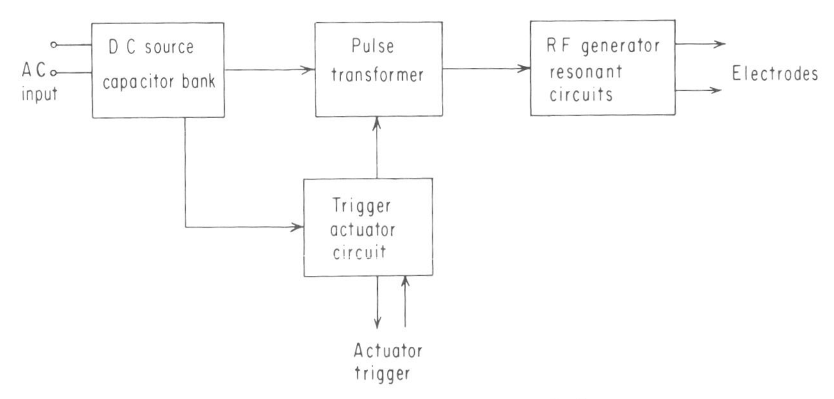

FIG. 9. Percussive arc/Pulse Arc block diagram.[2]

Process Variables. The important process variables are described as applicable.

RF TIMING. The rf timing controls the point at which rf pulse initiates the welding energy. Control settings which provide less delay to the rf timing circuit initiate the arc at an earlier point in the cycle. This provides more heating time on the weldments. The optimum setting for rf timing is determined by developing a weld schedule.

ACTUATOR VELOCITY. The actuator velocity control is required to provide adjustment of acceleration to the workpiece, thereby creating more or less force at impact. Generally speaking, the lower velocities are used for small materials and the higher velocities for larger wire sizes.

WELD ENERGY. Polarity may or may not affect the weld. In cases where the polarity has an obvious effect on the operation, incorrect polarity will produce either an excessively explosive weld or no weld at all. In some cases the differences may be less obvious; therefore, it is recommended that welds be made at each polarity and tested for strength. As an example of this, tests have shown that when copper is being welded to another metal, the copper generally should be at + polarity.

VOLTAGE. The weld-energy voltage control determines the amount of voltage applied to the weld capacitors. The optimum setting for the voltage control is determined by developing a weld schedule.

SLIPPAGE. Slippage applies to percussive-arc welding only. When the arc is struck, the ends of the workpieces become molten. One of the pieces is accelerated toward the other, fusing the two molten ends to form the weld. If one or the other of the workpieces is allowed to slip on impact, a sufficient amount of forging action will take place to expel oxides, inclusions, etc., from the interface. If there is no slippage, the actuator velocity setting will be more critical or some of the molten metal may be splattered from the interface, eliminating the weld bead and causing a weaker weld joint. The exact amount of slippage is not so important as is ensuring that the same amount of slippage occurs with each weld.

GAP SETTING. The gap or air space separating the workpieces varies with the type and size of materials. Therefore, it is quite difficult to establish a fixed rule. Generally speaking, the most useful gap lies between 0.020 and 0.040 in., although occasionally

* Trade name, Pulse Arc Welder Co.

the gap may be outside this range. The gap setting, along with the actuator velocity, determines the amount of force on impact. For a critical weld schedule, the gap setting, the rf timing, and the weld voltage require simultaneous optimization.

ESTABLISHING A WELD SCHEDULE. To establish a weld schedule, weld strength is plotted as a function of the two most important variables, rf timing and weld-energy voltage. As was seen in the discussion of process variables, slippage, actuator velocity, polarity, gap, and weld-energy capacity can be set after a few exploratory welds and then are no longer considered to be variables.

Laser. *General.* A laser beam may be used as a source of energy for welding metals together. The most significant features of this welding method are listed in Table 3, showing the advantages and disadvantages.

TABLE 3. **Laser Welding Advantages and Disadvantages**

Advantages

Welds small leads to large leads.
Welds dissimilar materials easily.
Welds through insulation.
Weld parameters are not critical.
Joining surfaces do not require contact.
Joint is inspectable.
Welds easily in inaccessible places.
Does not require special atmosphere.
Joints usually stronger than parent material.
Adjacent heat damage nonexistent.
Welds in very small spots.
Low cost of fixturing.

Disadvantages

Cost of equipment.
Shallow heat penetration.

Operation. The laser welder shown in Fig. 10 consists of a power supply, an energy-storage and pulse-forming network, a pumping system, and an optical system for focusing the beam and viewing the work. The equipment's features are as follows:

1. Power supply: ac to dc adjustable power rectifier system
2. Energy storage and pulse forming: Capacitor-discharge stored energy system
3. Pumping system: Usually a helical xenon flash lamp
4. Laser rod: Typically a synthetic ruby or doped crystalline material
5. Optics: Focuses monochromatic parallel light to a spot as small as 0.0001 in. but usually between 0.005 and 0.040 in. in diameter. Viewing optics are incorporated by a sliding prism or mirror to permit precise observation of the work before and after welding.

The parameters and typical setting ranges are:

Parameters	*Ranges*
Energy output	to 5 joules
Pulse length	to 4 msec
Repetition rate	to 20 pulses/min
Spot size	0.0001 to 0.020 in. in diameter

The two surfaces to be welded are brought together so that they just touch, using simple tooling. The energy is set by adjusting the power supply and the spot size. Too much energy will vaporize the material and create a crater, while insufficient energy will not result in fusion. Between these two extremes there is a large number of possible energy-per-unit-area combinations. The extent of melting depends upon the exposure time, the intensity, and the thermal properties of the materials to be welded. The laser intensity is dependent upon the laser output energy, pulse duration, and spot size in the following relationship:

$$I = E/tA$$

where E = energy in joules or watt-seconds
t = pulse duration in milliseconds
A = area of focused spot in square centimeters
I = intensity in watts per square centimeters

Weld Schedule Factors. Thin materials require small spot diameters, while thick materials require proportionately larger spot diameters, as shown in Fig. 11. The optimum welding conditions are determined by setting the parameters for the cleanest looking fusion puddle with no craters and by a satisfactory metallurgical combination as examined by sectioning and staining. Pull tests should be used to determine the average strength and the range. This type of weld should routinely develop average strengths above 80 percent of the wire; breaks usually occur next to the weld because of annealing around the fusion zone. Higher strengths occur for shorter pulse lengths since there is less time for annealing. Electronic component leads ranging in thickness from 0.001 to 0.020 in. require pulse durations from 1 to 6 msec. Many different

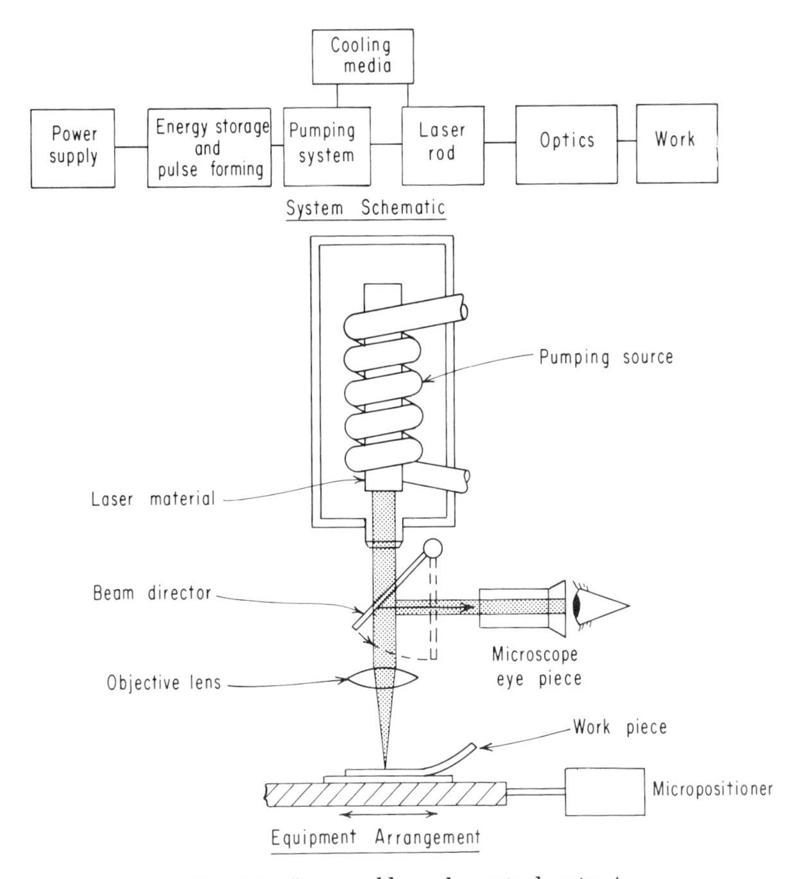

FIG. 10. Laser welder, schematic drawing.[4]

wire-to-wire materials may be welded with the same schedule and relate generally as shown in Fig. 12.

Material Properties. The number of materials that can be welded is almost limitless, although some weld better than others. A partial listing of materials with quality classifications is given below:

Material	*Quality*
Titanium	Excellent
Zirconium	Excellent
Columbium	Excellent
Tantalum	Excellent
Stainless 300 and 400	Excellent
Steel alloy 4130 and 4340	Excellent
Aluminum alloys	Good
Copper	Good
Bronze	Fair
Tungsten	Fair
Molybdenum	Fair
Steel 303	Fair
Sulfur-bearing metals	Poor
Selenium-bearing metals	Poor

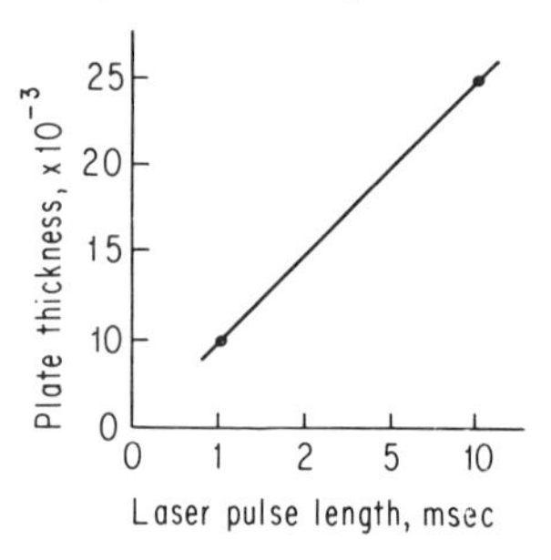

FIG. 11. Laser pulse length vs. plate thickness for copper.[4]

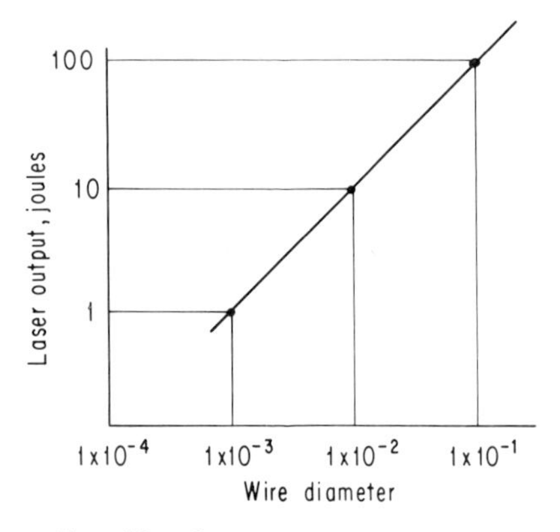

FIG. 12. Laser output requirement for various wire diameters.[4]

Joint Design. The most frequently used joint designs are shown in Fig. 13.

Thermocompression. Thermocompression welding is defined as a process of joining in which the application of appropriate pressure and temperature causes diffusion of the parent materials without exceeding their melting points. Various references to the process add to the confusion since they all describe this mechanism. Thermocompression bonding, diffusion bonding, thermal bonding, and thermocompression joining are a few names applied to this form of welding. Many equipments are available to effect the welding process, and in fact it is frequently difficult to describe processes without detailed reference to specific equipment.

WELD PHYSICS. The following factors are involved in the process:

1. Nature of alloying; i.e., solid solution, intermetallic compound
2. Surface characteristics when exposed to weld conditions; i.e., temperature, atmosphere
3. Diffusion rate of materials at the welding temperature
4. Magnitude of strain in the joint area
5. Material concentration gradient across interface

The choice of materials is a significant part of the joint design and will define most of the properties. Soft materials commonly in use are aluminum, gold, platinum, copper, and silver.

The diffusion rate for a particular material combination will determine the extent to which bands will be seen that correlate with their phase equilibrium diagram. A desirable metallurgical property is a low recrystallization temperature because there is then a tendency for grains to grow across and erase interface lines. Table 4 gives the recrystallization temperatures for some materials.

A good contact between surfaces to be welded is necessary before diffusion can occur. Although pressure does affect diffusion rate, it is important to obtain an adequate area of surfaces in contact. Surface reactions with the environment may limit the diffusion possible and degrade the weld quality. Aluminum, nickel, and silver are particularly sensitive to surface reactions. Some surface oxide layers have a lesser effect on weldability (e.g., copper). It is necessary to maintain a clean film of minimum oxide and wire of similar surface condition to obtain consistent weld results. Figures 14 to 16 show phase diagrams of three typical material combinations, and Figs. 17 to 19 show the resulting joint cross sections. The gold-silver combination demonstrates solid solubility in all proportions with minimum oxide formation for both materials. An excellent diffusion joint will result from this combination. The copper-silver combination shown in Figs. 15 and 18 shows a eutectic mixture formation. This low-temperature eutectic compound will form at temperatures above the eutectic melting point and below the parent material melting points. Some other typical eutectic formations are with gold and germanium, lead and tin, silver and germanium, silver and silicon, and gold and silicon. The gold and aluminum combination is shown in Figs. 16 and 19, where a homogeneous intermetallic band is formed. Joints of this type may be mechanically satisfactory, but they have the additional requirement that the intermetallic compounds have the desired properties or the combination should not be used.

(a) Butt joint – nearly vertical fusion

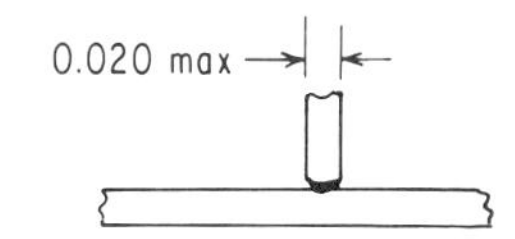

(b) T joint – weld from one side only

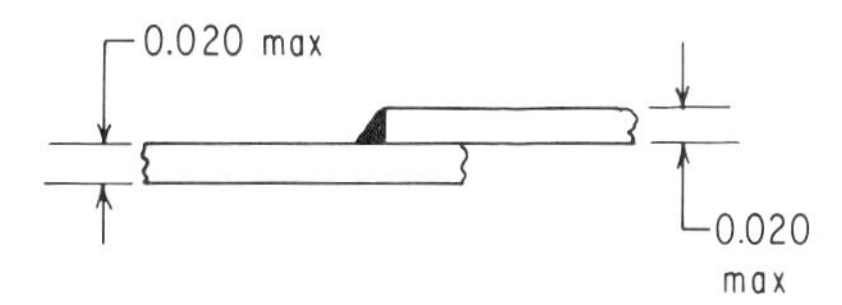

(c) Lap joint – melt top member into bottom

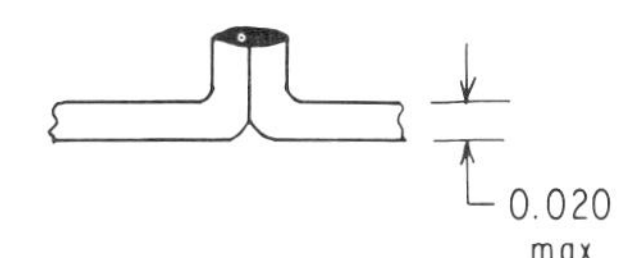

(d) Flange joint – puddle standing ends

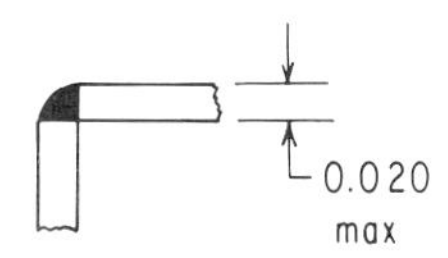

(e) Corner joint – puddle corner, starting at top

FIG. 13. Recommended joint designs for laser welding.[5]

Metallurgical Compatibility. An important consideration in the lead attachment process is the metallurgical compatibility of the various constituents in the bonding system. This is especially important in silicon semiconductor systems where high-temperature operation and storage are important, for some bimetallic systems (i.e., gold-aluminum) are quite unstable at high temperatures, especially in the presence of certain other materials which may act as catalysts to compound formation. The gold-aluminum system, in particular, has stimulated considerable interest in the industry because of the widespread use of aluminum as a contact material for

TABLE 4. Recrystallization Temperatures of Some Metals

Material	Recrystallization temperature (approx.), °F *
Gold	350
Copper	550
Silver	300
Platinum	1,000
Nickel	900
Aluminum	600

* Dependent on degree of cold work. Temperature given is for 30% deformation short-time heating (less than 10 min).
SOURCE: "Metals Handbook," 8th ed., vol. 1, American Society for Metals.

silicon and the great desirability of gold as a bonding wire on account of its ductility, electrical conductivity, and corrosion resistance. When gold and aluminum are placed in intimate contact, as in a weld or thermocompression bond, and the combination is heated to modest temperatures (greater than 200°C), chemical changes take place. These changes can be further complicated by the presence of silicon and oxygen. If a pure gold-aluminum system is heated to 300°C, a purplish material begins to appear at the interface between the two metals. This material has been analyzed and been found to be the chemical compound $AuAl_2$. This compound, which is a good electrical conductor and is strong mechanically, has been widely referred to as the *purple plague.*

Initially, it was felt that the $AuAl_2$ compound was responsible for the failure of bonds on silicon transistors metallized with aluminum, but it is now believed that this is not the case. To support this contention, an interesting experiment (Fig. 20) was recently described. In the experiment a silicon wafer was heavily oxidized to form a thick layer of SiO_2. The SiO_2 layer was then removed from half of the wafer. Small aluminum hemispheres were placed over the entire wafer, and the system was heated

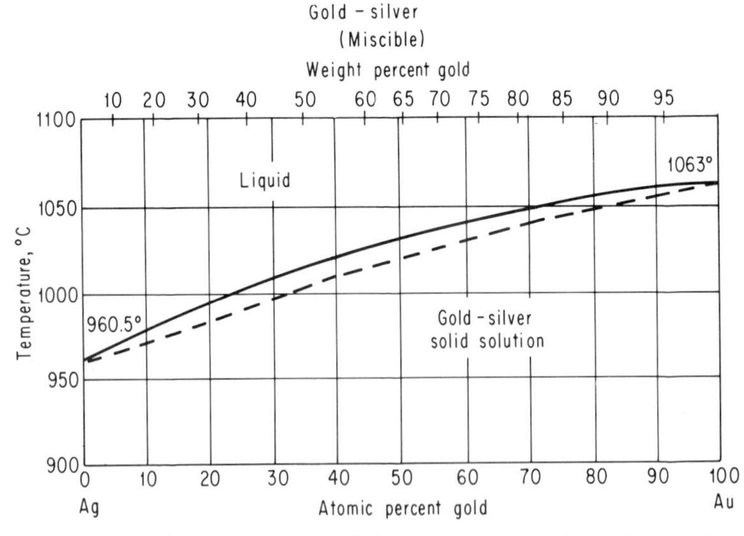

FIG. 14. Thermocompression joining—gold-silver phase diagram.[6]

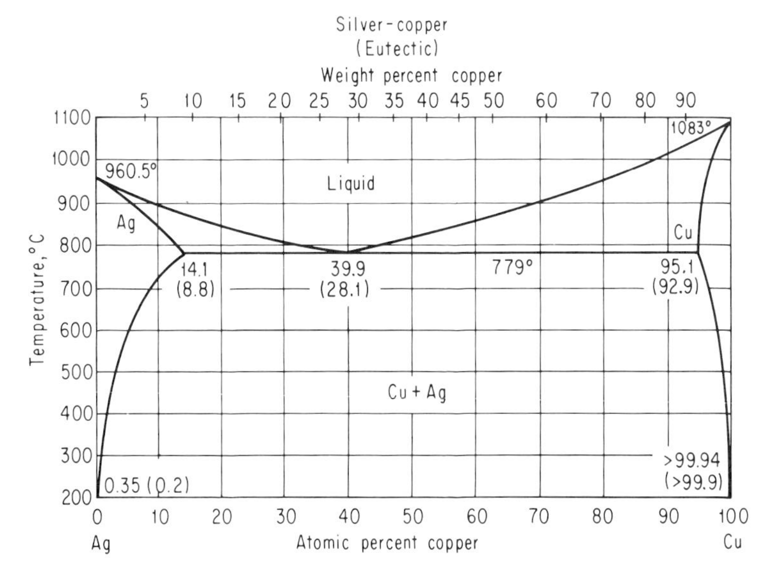

Fig. 15. Thermocompression joining—silver-copper phase diagram.[6]

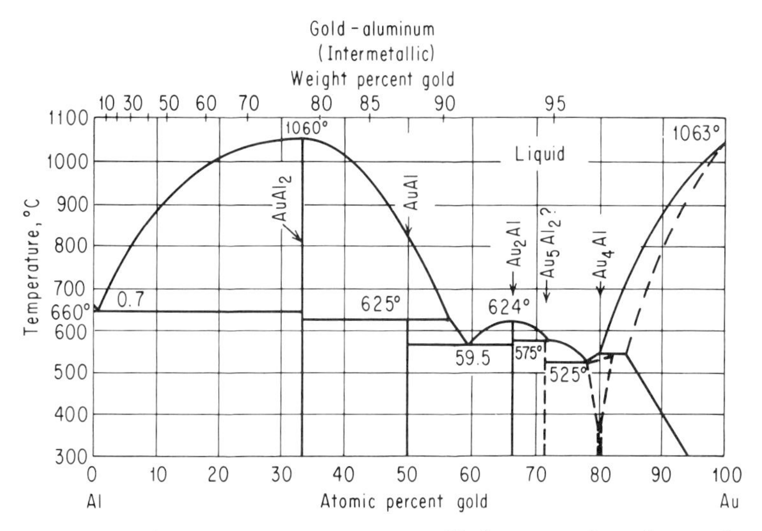

Fig. 16. Thermocompression joining—gold-aluminum phase diagram.[6]

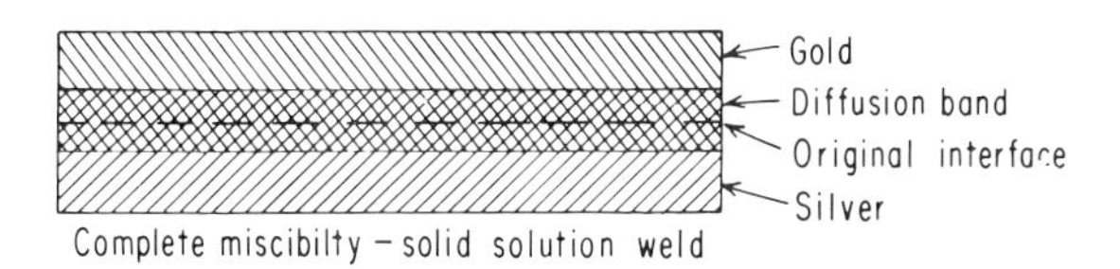

Fig. 17. Thermocompression joining—gold-silver diffusion joint.[6]

beyond the melting point of aluminum (660°C). On the half of the wafer on which the oxide remained, the aluminum merely melted and resolidified. On the other half of the wafer, the aluminum alloyed into the silicon, changing the composition of the hemisphere to the aluminum-silicon eutectic composition, 89 percent aluminum and 11 percent silicon. After cooling, a thin film of pure gold was evaporated over the entire wafer. After this, the wafer was heated to 300°C for approximately 10 hr and then cooled. A surface grinding operation was performed to cut through the gold film and halfway through the aluminum hemispheres. Upon examination of this section, it was found that on the side of the wafer retaining the oxide, the familiar purple phase

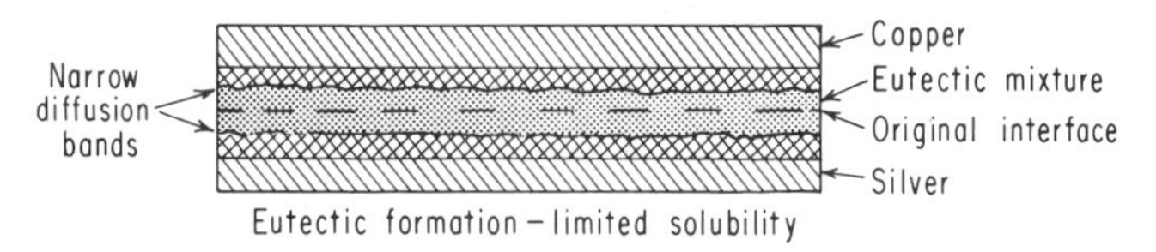

Fig. 18. Thermocompression joining—copper-silver diffusion joint.[6]

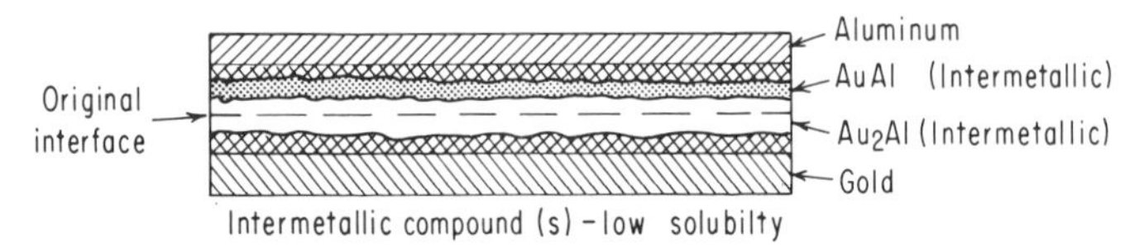

Fig. 19. Thermocompression joining—gold-aluminum diffusion joint.[6]

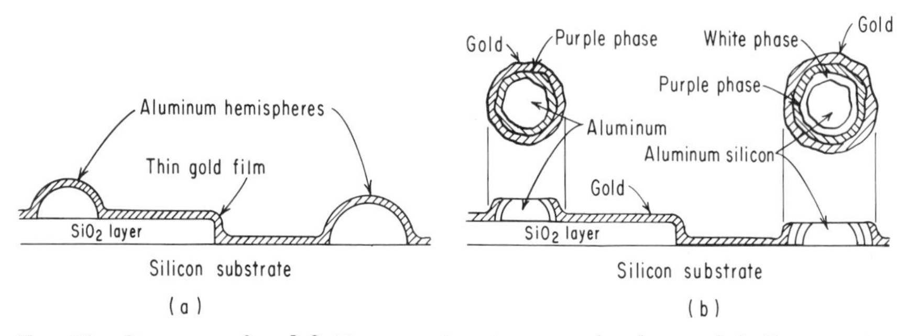

Fig. 20. Summary of a definitive experiment on purple plague. (*a*) Cross section of specimen as originally prepared. (*b*) Cross section of specimen after heating to 660°C. Both regions show the relatively strong and conductive purple phase. The brittle and resistive white phase has occurred where silicon was available.[22]

existed. On the other side of the wafer a new white phase existed in addition. This white phase was analyzed and was found to be Au_2Al. Ostensibly, this new compound had formed in the presence of silicon as a catalyst. This compound was found to be a poor electrical conductor and was extremely fragile. Thus it is this *white plague* which is believed to be responsible for the failure of bonds.

In order to circumvent these problems, there has been a general trend toward changing from the commonly used gold-wire-to-aluminum-contact-area bonding system to the more compatible aluminum-wire-to-aluminum-contact-area system. Today, the use of aluminum-alloy bonding wires is widespread in the industry. However, the suppliers of the fine bonding wires have found it necessary to introduce a trace of silicon into the aluminum melt in order to draw the wire to the fine diameters required. This means that when such aluminum wires are used for bonding to gold or gold-plated regions as on the package lead wires, there could be a plague problem. However, it has been found experimentally that the quantity of silicon in the wire is small enough that the white plague formation is usually negligible.

Joint Types. Three joint types are discussed briefly with their important control parameters.

WEDGE. Generally, heating is applied to the wedge or tip as shown in Fig. 21. The base is often heated. The base may be heated without the tip being heated. The variables controlled here are: (1) element's heated temperature(s) (tip, base, or both), (2) time, (3) force, and (4) wire and film properties.

BALL OR SCISSOR. Ball bonding employs a glass capillary (Fig. 22), through which the wire passes. The ball is formed by a hydrogen gas flame cutoff operation which fuses a ball on both ends of the severed wire. The ball on the glass capillary is larger than the diameter by at least 2 times, and the capillary transmits the force of joining against the ball. The substrate or base is heated. A variation in this form of bonding substitutes mechanical cutting for flame cutting. The welding parameters are: (1) base temperature, (2) force, (3) time, and (4) wire and film properties.

RESISTANCE. Resistance welding is generally accomplished with a split-tip or parallel-gap welding apparatus. The wire only is heated by the passage of electric current through the wire between two closely spaced electrodes, as shown in Fig. 23. It is possible to arrange pre- and postheating cycles by current control in the welding power supply. These features permit control of stresses by applying heat changes gradually and avoiding thermal shocks. The welding parameters are: (1) the current settings (energy), (2) time, (3) pressure, (4) electrode configuration, (5) electrode spacing, and (6) electrode cleaning.

Weld Schedule Development. A weld schedule results in experimenting with these parameters of welding on the materials chosen to obtain their best combination, using weld strength as the controlling parameter. There are frequently other restricting factors such as temperature-sensitive parts which dictate limits rather than weld results. General procedures for optimizing parameters against weld strength should

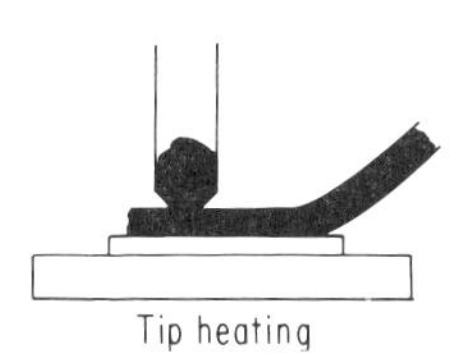

FIG. 21. Thermocompression weld methods—weld-tip heating.[6]

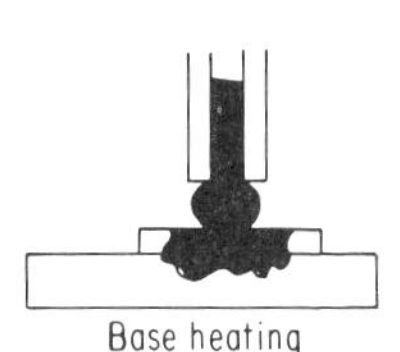

FIG. 22. Thermocompression weld methods — weld-base heating.[6]

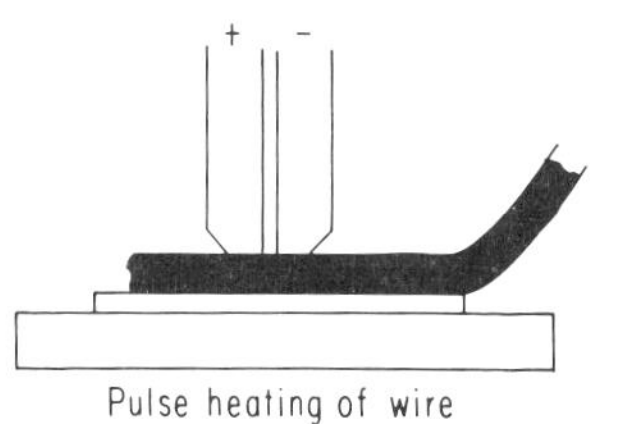

FIG. 23. Thermocompression weld methods—parallel-gap current heating.[6]

be followed as discussed later in this chapter in procedures for weld schedule development. The weld joint strength is measured by tensile testing. The most discerning, the peel test shown in Fig. 24, is recommended. Other tests such as shear are not so effective in determining the best weld schedule because of lack of repeatability and sensitivity to critical joint properties. Minimum strength limits are often arbitrary and empirical. Joint property requirements should be established for the particular application. Some typical peel and shear strengths are shown in Table 5.

Ultrasonic Welding. In *ultrasonic welding*, components to be joined are clamped together between a welding tip and a supporting member or anvil, with only sufficient static pressure to hold them in intimate contact. High-frequency vibratory energy is transmitted to the joint through the tip for a brief interval. Thus high-strength metallurgical bonds can be made in many similar and dissimilar metal combinations

TABLE 5. Thermocompression Welding—Joining Strength Data [6]

Wire or ribbon material, dimensions in inches	Film substrate	Peel strength (percentage of ultimate)	Shear strength (percentage of ultimate)
Heated tip			
0.003 diam gold	Copper-aluminum * on glass	84	92
0.003 diam gold	Copper-Nichrome † on glass	84	86
0.003 diam gold	Copper-Nichrome on alumina	88	93
0.003 diam gold	Silver on glass ‡	10 §	89
0.005 diam gold	Copper-Nichrome on glass	50–86 §	Over 70
0.001 × 0.014 gold	Copper-aluminum on glass	42	74
0.001 × 0.014 gold	Copper-Nichrome on glass	62	74
0.001 × 0.014 gold	Copper-Nichrome on alumina	41	77
0.001 × 0.014 gold	Silver on glass	15	73
0.001 gold plated on 0.001 diam nickel	Copper-Nichrome on alumina	65	82
0.003 diam platinum	Silver on glass	No peel §	58
0.003 diam platinum	Copper-Nichrome on alumina	38	56
0.0026 diam copper	Copper-Nichrome on glass	19	66
0.0026 diam copper	Silver on glass	10	68
0.005 diam aluminum	Copper-Nichrome on glass	13	20
Split tip			
0.003 diam gold	Copper-Nichrome on glass	74	89
0.003 diam gold	Copper-Nichrome on alumina	80	95
0.003 diam gold	Silver on glass	12 §	93
0.001 × 0.014 gold	Copper-Nichrome on glass	36	63
0.001 × 0.014 gold	Silver on glass	13	66
0.001 × 0.014 gold	Copper-Nichrome on alumina	32	78
0.003 diam platinum	Copper-Nichrome on glass	19	35
0.003 diam platinum	Silver on glass	4.2 §	52
0.0026 diam copper	Silver on glass	17 §	46

* 50 Å aluminum, 5,000 Å copper.
† 50 Å Nichrome, 5,000 Å copper.
‡ 2.5×10^{-4}-in. silver.
§ Dependent on film adhesion.

without applying external heat, without melting weld metal, without using fluxes or filler metal, and without passing electrical current through the joint. Many metal combinations are weldable by ultrasonic means which are not by other techniques. Materials which have been successfully joined are shown in Fig. 25. There are two

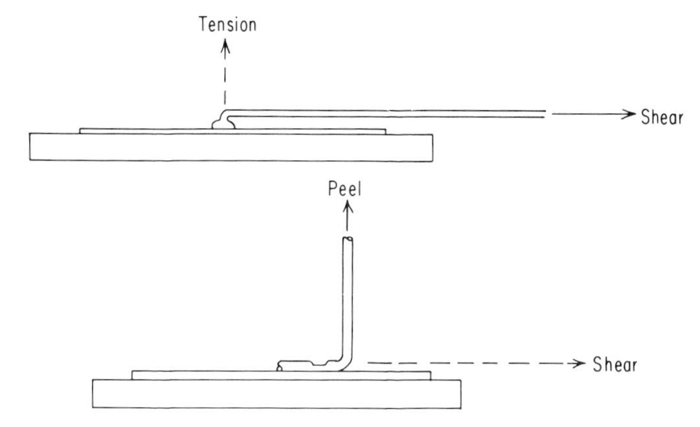

FIG. 24. Thermocompression weld test methods.[6]

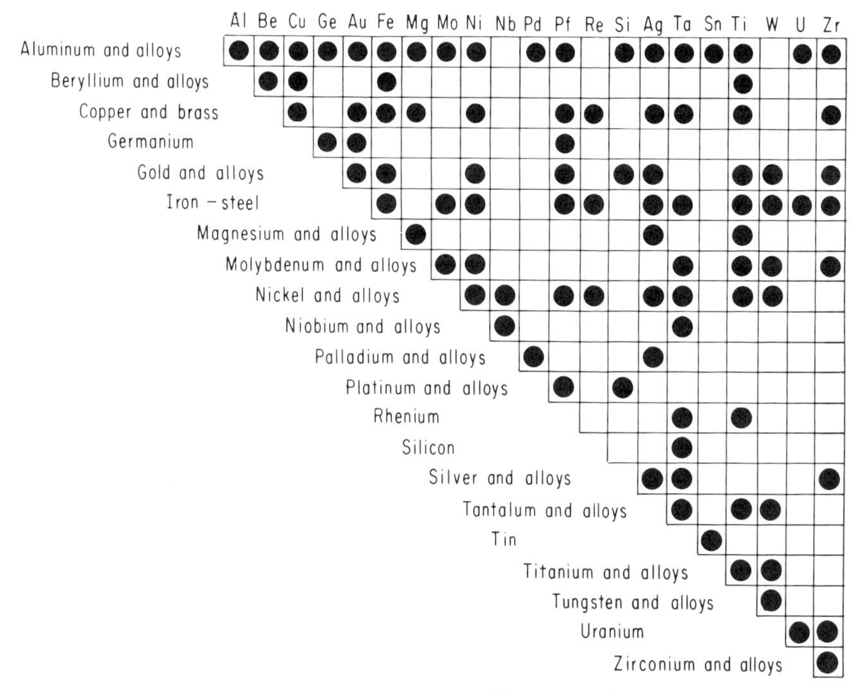

	Al	Be	Cu	Ge	Au	Fe	Mg	Mo	Ni	Nb	Pd	Pt	Re	Si	Ag	Ta	Sn	Ti	W	U	Zr
Aluminum and alloys	●	●	●	●	●	●	●	●	●		●	●		●	●	●	●	●		●	●
Beryllium and alloys		●	●			●												●			
Copper and brass			●		●	●	●		●			●	●		●	●		●			●
Germanium				●	●							●									
Gold and alloys					●	●			●			●		●	●			●	●		●
Iron – steel						●		●	●			●	●		●	●		●	●	●	●
Magnesium and alloys							●								●			●			
Molybdenum and alloys								●	●							●		●	●		●
Nickel and alloys									●	●		●	●		●	●		●	●		
Niobium and alloys										●						●					
Palladium and alloys											●				●						
Platinum and alloys												●		●							
Rhenium																●		●			
Silicon																●					
Silver and alloys															●	●					●
Tantalum and alloys																●		●	●		
Tin																	●				
Titanium and alloys																		●	●		
Tungsten and alloys																			●		
Uranium																				●	●
Zirconium and alloys																					●

FIG. 25. Ultrasonically weldable metals.[7]

distinct methods of using ultrasonic energy in metal joining. One method is to apply ultrasonic energy to metal being soldered, brazed, or welded. This aids the joining process, although heat is still needed to bring the filler material or base metal to its melting temperature. This might be called *ultrasonically assisted fusion welding.*

The second method is the conventional one, as follows: Since the atoms at the surface of a metal are not completely surrounded by neighboring atoms, they are capable of bonding to those in another surface, providing the two surfaces are perfectly clean and smooth and are brought close enough together for the formation of atomic bonds. In reality, such surfaces do not exist and surface atoms of most metals combine with oxygen to form a metal oxide layer, which may be 20 to 200 molecules thick. The oxide layer, in turn, generally has a strong affinity for water, so that a layer of water covers the layer of oxide, as shown in Fig. 26. In ultrasonic welding, high-frequency vibrations applied in a plane parallel to the weld surfaces shatter the contaminating layers and reduce the yield strength of the metal so that only a small clamping force is necessary to bring the metals into intimate contact.

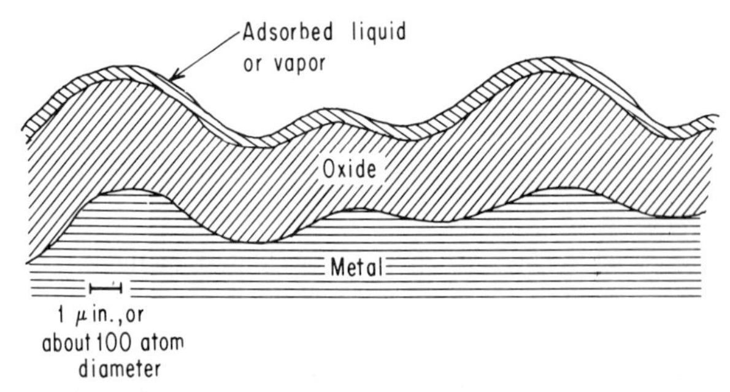

FIG. 26. Surface schematic of a smooth clean metal.[7]

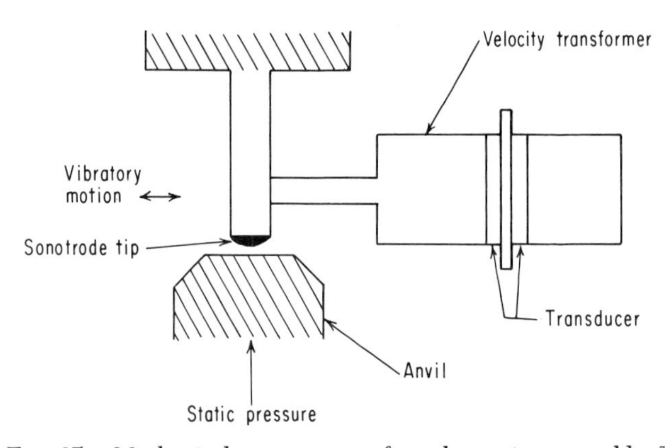

FIG. 27. Mechanical arrangement of an ultrasonic spot welder.[7]

Equipment. There are two major components: (1) a power source or frequency converter which converts ordinary 60-cycle line voltage into high frequency, (2) the transducer-coupling system which converts the high-frequency electrical power into vibratory energy and delivers it to the weld zone. The welding head provides the static clamping force to the workpieces. Maintenance of the equipment is not complex. Welding tips may deteriorate with use and will require sanding and burnishing periodically. The cleaning frequency depends upon the materials being joined, but on soft materials cleaning may not be required before several million welds. The essential parts of the welding apparatus are diagramed in Fig. 27.

Primary Variables. The variables which must be specified and controlled are (1) power, (2) clamping force, and (3) pulse time. These variables can be present so that the weld cycle may be completed automatically. A foot switch or other triggering mechanism lowers the welding head, applies clamping force, introduces the ultrasonic energy for the prescribed length of time, and retracts the welding tip upon completion of the energy application. The welding tip is a critical variable which must be properly designed for the materials being joined. For example, joining flat sheets of foil to metal deposited on a substrate requires a flat-faced anvil with the opposing tip contoured to a spherical or crown radius. No exact design rules can be stated for other geometry and material, leaving the user to "design" each optimum tip and anvil configuration.

In setting up the weld schedule, the clamping force is not critical and can be made effective over a rather broad range. At any clamping force there is a minimum power level for making good welds. In addition, there is an *optimum* clamping force range within which good welds can be produced with least power input. This is determined by making weld samples at various settings for integers or families of clamping force and plotting the results against the weld strength. The weld time during which vibratory energy is transmitted to the workpieces depends in part on the power and is usually in the range of 0.005 to 1.0 sec. Longer weld times usually indicate insufficient power. High-power, short-time welds generally produce better results than low-power, long-time welds. The effects of weld times should be explored (within limits) initially and then considered fixed at the best value unless large variations in weld strength result from small changes. Surface preparation of the material to be joined is not critical and no postcleaning is required. Large-scale and excessive lubrication should be removed prior to welding, however. It is possible to weld through a number of surface deposits or coatings, although higher energy levels are required. Welding has been effected through anodized aluminum and through plastic films such as polyvinyl chloride or polyethylene.

Joint Strength Testing. Visual examination provides some indication of weld quality but it is not possible to judge each weld absolutely by visual methods. Good welds are generally accompanied by a slight roughening of the surface. The thickness deformation is usually less than 5 percent even in the soft metals. Each weld should be examined metallurgically when establishing the schedule, to assure proper design, but metallurgical examination is not required subsequently. Destructive testing is used extensively on representative welds to establish weld schedules and to monitor production. The strength tests on thin-gauge material are peel type and usually provide good correlation. The weld failure should occur in the base metal outside the weld area, leaving the nugget intact. For thick materials, torsion shear tests are recommended.

Electron Beam. It is convenient to group joining into three classifications: (1) fusionless joining, typical in thermocompression bonding, (2) fusion welding, in which metals in the interface are melted and recrystallized as in laser welding, and (3) soldering, in which an interface material is melted below the temperature of either of the pair being joined. Electron beam welding is a fusion joining method characterized by its control, efficiency, reproducibility, and high reliability. The ease of programming the beam and the weld energy provide an additional attraction of this welding method. Electron beam welding produces an extremely rapid melting and quenching. This condition results in a much finer grain structure than is found in conventional resistance welding. The smaller grain size results in greater weld strength and a limited amount of grain growth adjacent to the weld nugget. The extent of penetration is dependent upon the time during which the material surrounding the nugget is at welding temperature. Due to the short time required for beam welding (typically 10 msec), the alloy zones are small. Generally a welding cycle is accomplished by setting the machine parameters so that the weld is formed by deflecting the beam over the pieces at their overlap. The inherent accuracy of the programmed beam spot permits operation down to extremely fine wires with the required reproducibility.

EQUIPMENT. The basic equipment consists of an electron beam gun, a magnetic lens assembly, a deflection coil, a vacuum chamber, an optical viewing system, and an x and y table for positioning the work. The vacuum system must be capable of holding a lower pressure than 10^{-3} torr; the power supply for the beam typically ranges between 50 and 150 k. There are equipments available which do not require the work to be carried out in a vacuum lower than 1 to 3 torr. To prevent air molecules from scattering the electron beam, the workpieces are welded using a helium shield. Alternately, the electrons are emitted from an electrically heated tungsten rod accelerated and focused by electric and magnetic fields. With the nonvacuum environment the weld head can be brought to the work.

Process Variables. The following process variables must be considered in establishing weld schedules:

1. *Beam accelerating voltage (the cathode-to-anode accelerating potential).* The work is usually at ground potential. Lower beam accelerating voltages require correspondingly higher beam currents. Penetration of the top material, and therefore its thickness, controls the minimum voltage. If the voltage is too low, the welds will not be consistent; if too high, the material will vaporize.

2. *Beam current (weld pulse height).* The beam current can be controlled independently by applying a gun grid bias. The voltage requirement depends upon the gun and is typically 1 to 2 k. The beam current will be reduced as the accelerating voltage is increased for the same welding energy.

3. *Beam on-time (weld pulse width).* The beam on-time is defined as the time during which electrons from the source are directed to the weld area. The on-time must be systematically varied while maintaining other parameters fixed to "determine" the desired welding heat.

BEAM REPETITION RATE. The beam can be energized manually or adjusted to operate automatically at a preselected rate.

BEAM DEFLECTION. Weld temperature distribution will be enhanced if the beam is programmed over a prescribed pattern unless the nugget is of the same size as the spot. A sinusoidal scan has been employed, is easy to program, and provides adequate joint heat distribution. The excursions are independently adjustable.

SPOT SIZE. The spot size is required to be small with respect to the weld nugget size in order to prevent too large a melt zone which will jeopardize the area adjacent to the weld by reducing the cross section. The spot size is typically 0.0005 to 0.10 in. in diameter.

OTHER ADJUSTMENTS. In addition to normal schedule adjustment for welding, the machine should have an x and y control for the work table which is compatible with the size and precision desired on the work. In addition, this x-and-y-controlled table is required to work in the required vacuum. The versatility and precision of the electron beam equipment make it suitable for other duties, including serving as an energy source for change of state of matter such as solid-state diffusion, recrystallization, alloying, selective etching, and milling (sublimation and vaporization).

MATERIALS

To maintain high-reliability welded connections, it is necessary to use materials which offer consistently good welding capabilities. The feasibility of minimizing the number of different lead sizes and materials is highly desired and this can only be done effectively when a planned program exists. The most logical and practical approach is to utilize what is already in use with part manufacturers so as to not introduce a new unknown factor into the part reliability. There are sufficient available parts with desired materials to permit good selections for all electronic parts. Interconnection materials must be selected with the knowledge of their ability to be joined with all common lead materials with consistently high reliability. Each of the joining methods described above has a different set of parameters, including materials; therefore, what makes an ideal weld combination for one does not necessarily make a universally ideal combination.

Interconnect Material. Interconnect materials hopefully will have two basic properties idealized: (1) lowest electrical resistance and (2) weldability with high consistency with all lead materials. There are two prevalent lead-joining techniques which are singled out because their use is larger than all others combined.

Wire Bonding. This form of joining is being used to interconnect semiconductor chips to header pins or to interconnect pads within a microassembly. The welding methods are thermocompression bonding and ultrasonic bonding. The most prevalent materials here are round gold wire from 0.0005 to 0.002 in. and round aluminum wire from 0.001 to 0.005 in. These interconnect materials are used because of their welding properties with semiconductor device terminating pads and with external leads. The thermal heat conductance of these connections is low because of the small cross section, and therefore heat from a chip is removed through the die bonding interface. The nonmagnetic properties of both gold and aluminum provide an advantage in applications where a system must use nonmagnetic materials.

Ribbon Welding. The much larger electronic assemblies which involve interconnecting leads between 0.015 and to 0.060 in. in diameter usually employ a rectangular (ribbon) cross-section wire which is ordinarily resistance-welded to part leads. There are suitable combinations which use round wire, although the rectangular cross section provides the better and more consistent weld. The bus material in highest use is pure soft nickel wire ranging from 0.007 × 0.015 in. to 0.012 × 0.030 in. in dimension; square cross sections have also been used successfully. Nickel has excellent weldability with nearly all component lead materials; however, nickel is significantly more resistive than copper and has poorer thermal conductivity. Nickel is also magnetic and sometimes may be excluded from a system for this reason alone. Note that nickel is a good material to use in resistance welding because of the following:

1. It has enough resistance for heat to generate due to welding current.
2. It has a lower thermal conductivity than copper.
3. It is easily controlled for its physical and chemical properties.
4. It does not require corrosion protection.
5. Quality rectangular and round wire are available.

Where nonmagnetic bus material is needed, the nickel interconnect is unacceptable, and alloy 90 or alloy 180 is frequently used instead. These materials are cuprous-nickel alloys and have good welding properties. Both will harden and sometimes crack due to the heat application and quenching when welding, but alloy 90 is less susceptible to cracking and is preferred. The two alloys may be obtained in the same size ranges as the nickel.

Resistance Weldable. The welding parameters selected by the operator for optimum welding will differ, of course, with the type of welding to be performed. The type of welding may be selected because of the particular welding problem. Assuming that the joint will be made via a resistance welding method, then it is desired that the heat-affected zone be at the interface between the parts being joined. Ordinarily this ideal is not achieved and the hottest spot will be within one or the other of the materials being joined. The heat-affected zone can be "adjusted" by making one of the weld electrodes more thermally resistant than before; this is done by use of a molybdenum electrode in place of a copper electrode. Other ways of adjusting the heat-affected zone are: (1) vary the bus or wire thickness; (2) vary electrode pressure; (3) vary lead and/or ribbon dimensions; (4) vary the rate of application of heat (pulse length and height). It is important to note that laser, beam, and ultrasonic welding methods can apply heat exactly at the weld interface. The choice of a material with an appreciable thermal resistivity is desired because the energy applied to the joint must result in the interface surfaces being at least in the plastic condition; if the materials are thermally low in resistance, then heat may flow away faster than it is applied and very large amounts of energy are needed to make a weld. Copper needs very short cycles (1 msec) to resistance weld for this reason. Copper is particularly difficult where parallel-gap welding is to be applied to printed-circuit boards; here the copper becomes so hot that the printed-circuit board traces are lifted due to overheating of the plastic cement. In this case, some other printed-circuit-board metal is desirable; gold-plated Kovar and nickel, plated over copper, have been used successfully for this purpose.

The high thermal conductivity of copper is responsible for the high temperature achieved in the copper-to-board interface during the welding cycle. Kovar and nickel are better materials because of their higher thermal resistance. Copper can be welded easily with electron beam and laser welding methods.

Welding with Copper. Pure copper material has been shown by metallurgists to be more weldable than copper with traces of oxygen. Copper, as used in electronic part leads, has been usually electrolytic tough pitch (ETP) because of its stiffness. Three types of copper will be found in use as component leads and sometimes as printed-circuit P.C. board laminates:

1. Oxygen-free, high-conductivity copper (OFHC)
2. Deoxidized copper, low phosphorous residue (OF)
3. "Silver-bearing" coppers (three general grades)

The copper content of all three is roughly 99.95 percent. However, the different trace elements present as a consequence of the different manufacturing processes result in widely divergent weldability.

The maximum allowable oxygen for type OF is 0.005 percent as described by the ASM Metals Handbook, which also specifies ETP copper to have oxygen in the range 0.020 to 0.070 percent. Excellent welding results of OF copper versus ETP copper are obtained with oxygen content as high as 0.015 percent. Measurement of oxygen content is particularly difficult but it is readily determined by the vacuum fusion method. Other methods which have been tried and are not effective are:

1. Micrographs
2. Emission spectrograph
3. Tensile properties
4. Elongation

MIL-STD-1276 has included the necessary description and control for copper as a lead material. There are two very significant advantages to the use of OFHC coppers for a material where welding is to be the joining method.

1. The resulting strength distributions will be considerably better than with most other materials, as shown in Fig. 28.

2. The weld schedule (for resistance welding particularly) is grossly broader than for ETP, resulting in higher strengths with a wider energy range (see Figs. 29 to 31).

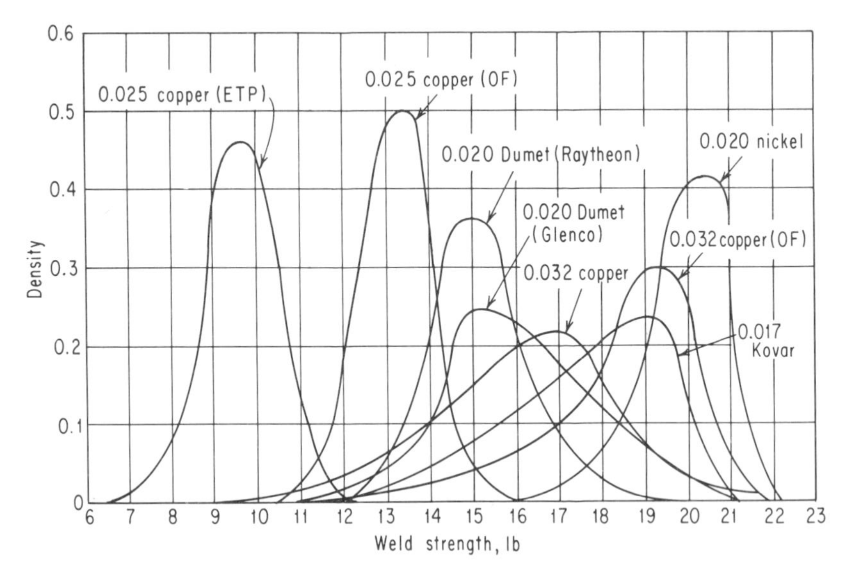

FIG. 28. Normalized crosswire weld-strength distribution shapes for various metal combinations.[8]

This quality makes copper the best material for use in electronics, where the following properties are desired:

a. High electrical conductivity
b. High thermal conductivity
c. High joint reliability
d. Nonmagnetic properties
e. Workability
f. Low cost
g. Availability
h. Universality

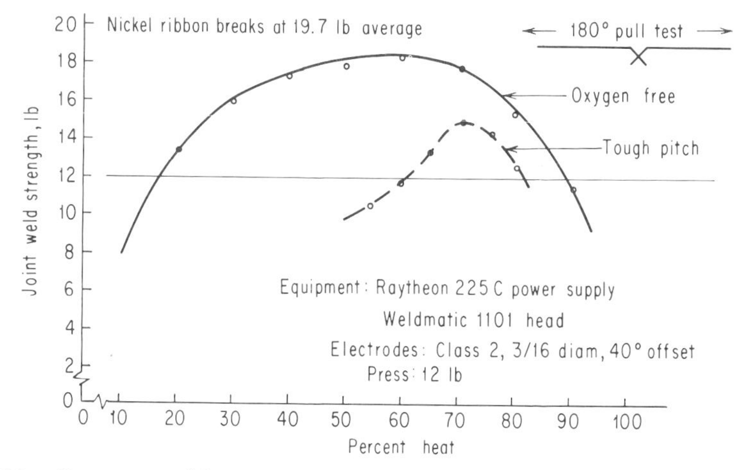

FIG. 29. Crosswire weld-strength profile, 0.033 diam OFHC copper to 0.010 × 0.032 nickel ribbon.[9]

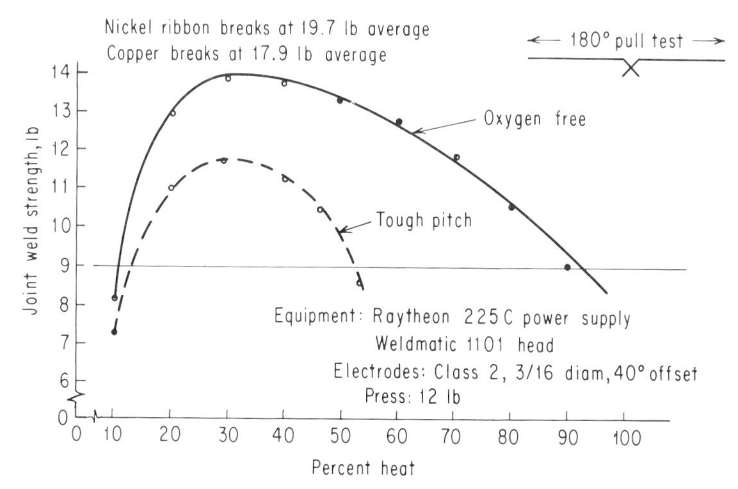

FIG. 30. Crosswire weld-strength profile, 0.025 diam OFHC copper to 0.010 × 0.032 nickel ribbon.[9]

Component Leads. Due to the large number of component manufacturers currently supplying industry, a wide range of component lead diameters exists. It has become desirable to limit the number of different lead diameters available to effect reliable welding at reasonable costs. Unless such a limiting control is placed on component lead diameters, many costly and unnecessary control data must be generated, such as weld schedules and metallurgical, chemical, and statistical analyses.

Satisfactory results can be experienced for most applications by limiting the lead diameters to a few sizes in a specified range.

The preferred nominal diameters are 0.016, 0.020, 0.025, and 0.032. The above lead sizes will accommodate nearly all components currently being manufactured for modular application.

Welding of microelectronic parts and assemblies is a different situation. Here "parts" are things like semiconductor chips and leads are things like 0.001- to 0.005-in.-diameter gold. The joining technique is necessarily selected for the type of work to be done. Microelectronic component parts are available with leads, but the trend and recommendation is to supply this class of component leadless.

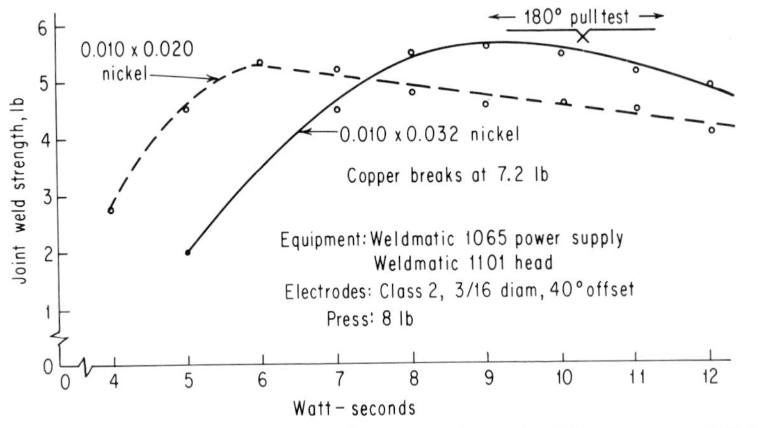

FIG. 31. Crosswire weld-strength profile, 0.015 diam OFHC copper to 0.010 × 0.032 nickel ribbon.[9]

For the modular cordwood assemblies, component lead materials are recommended to be as follows, for the reasons given:

Material	*Reason*
Nickel 200	High-temperature coefficient of resistivity, excellent welding properties, good stability
Copper—OFHC	High thermal conductivity, excellent welding, low electrical resistivity, nonmagnetic
Kovar*	Compatible with glass for sealing
Dumet	High electrical conductivity, high thermal conductivity, compatible with glass for sealing

A listing of these four materials, together with their minimum strengths, is given in Table 6. The strongest joints are the most reliable and where possible these should be used first. The minimum strengths given are shown together with the sample size. Comparisons cannot be accurately made with minimums at other sample sizes. Large

* Trademark, Westinghouse Electric Corp.

sample sizes are needed to obtain even modest precision. The use of normal or parametric statistics is sometimes grossly in error.

TABLE 6. **Resistance Welding—Joint Strength Data**

Material	Diam, in.	y_l, lb	x_{min}, lb	$\bar{x}/y_l$, %	$x_{min}/\bar{x}$, %
Copper ETP	0.025	18	6.0	62	58
Copper ETP	0.025	18	6.5	56	64
Copper OF	0.025	20	9.0	67	67
Copper	0.032	31.5	5.5	48*	50
Copper	0.032	28	9.0	68*	57
Copperweld	0.020	19	11.0	89	65
Copperweld	0.025	28	11.5	74*	68
Copperweld	0.032	45	9.5	70*	59
Dumet	0.020	23.7	10.0	73*	60
Dumet	0.020	23	14.0	83	74
Dumet	0.025	40	10.5	81*	57
Dumet	0.032	60	12.0	68*	77
Kovar	0.017	20	9.5	78	61
Kovar	0.025	34	11.0	83*	58
Kovar	0.040	110	14.5	78*	81
Nickel	0.016	15.5	5	70	51
Nickel	0.020	21.5	8	70	53
Nickel	0.025	34	13.5	80*	73
Nickel	0.032	52	11.0	86*	56

NOTE: y_l = lead wire average breaking load, in pounds.
$\bar{x}$ = average weld breaking load of sample, in pounds.
x_{min} = lowest-strength coupon where $n = 770$ (no exclusions).
Welds were made with Weldmatic 1065 machine—long pulse setting.
* Weakest of pair is 0.012 × 0.030 Ni ribbon, 23 lb average.

WELD SCHEDULE DEVELOPMENT

General Considerations. There are several steps to be taken in determining the quality of a welded connection. These steps include mechanical testing, visual inspection, and metallurgical analysis. Each of these steps must be included in the development of an optimum weld schedule.

It is incorrect to generalize the requirements of such factors as deformation and type of interface structure. Optimization of each material combination should include the optimum physical characteristics which are *discovered* by metallurgical analyses by a professional metallurgist. Metallography should be employed for each material and geometric combination for the purpose of photographically recording the micrographic appearance of optimum welds. The procedure involves determination of optimum metallurgical appearance for each material combination, then confirmation by comparison with the appearance of other catalogued similar material combination.

The basic considerations found applicable to all resistance weld processes involve (1) heat balance (influence of electrode materials and electrode force) and (2) the extent of the heat-affected zone. Limitations here are applied where (1) an insufficiently large heat zone causes low weld strength while the structure may be "excellent" and (2) an excessively large heat-affected zone may degrade the properties of the parent metal cross section out of the weld zone. Excess heat can cause undesirable electrode contamination.

In welding process control, duplication of a condition previously determined as acceptable is sought, rather than "optimum" structure. For example, the basic objects of a Dumet-to-nickel weld are (1) to expel the copper sheath, exposing a nickel-to-Dumet core interface, and (2) to limit overheating of the copper sheath (of Dumet) to prevent electrode contamination due to skin melting. Metallographic analysis coupled with mechanical analysis will corroborate object 1 above; similarly, weld strength consistency should corroborate object 2.

Weld Coupon Testing. The most discerning of the mechanical test procedures available is the torsion-shear test in which the coupon is stressed as shown in Fig. 32. This places the weld coupon under combined tension, torsion, and shear stresses. The manner in which the coupon fails is significant, and this information, as well as the numerical value obtained from the pull tester, is recorded. A tension failure indicates that the weld and adjacent areas are stronger than the parent material; a torsion failure shows insufficient fusion, and a shear-torsion failure indicates a weakness in the parent material adjacent to the weld. The most desirable condition is usually obtained when coupons from the same group fail from a combination of causes, such as tension and shear-tension failures.

The pull test rates must be consistent for all tests where data comparisons are made. The change in resultant strength readings resulting from different rates on a Hunter Spring Co. tensile tester over the range from 1 to 60 in./min is linear, 0.01 lb increase for each inch per minute increase in rate. Microjoint testing will be different depending upon the weld method, and tubelet testing will also differ. For example, peel tests are generally used for bonds to thin films.

Weld Data Charting. A weld profile method of establishing an optimum weld schedule is recommended as a means of providing an efficient and simplified approach. The weld profile is a plot of weld breaking strength versus energy at constant electrode pressures, as shown in Fig. 33.

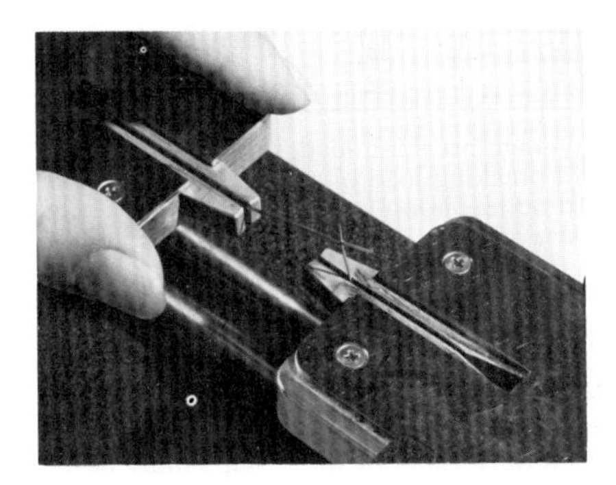

FIG. 32. Crosswire weld—torsion-shear testing coupon.[10]

The parameters to be controlled were discussed earlier in this chapter where each welding technique was described. All weld methods have more than two variables which require simultaneous optimization. The easiest method of chart optimization is to use weld strength as the dependent variable, the most critical parameter as the independent variable, and all others as fixed families.

Through the use of a weld profile it is possible to completely analyze the average weld strength and the range of strengths at each energy setting. A line drawn through the average strengths will usually establish a plateau of high average strength values. The plateau is then examined and a point is selected which shows a combination of high average strength and low spread from the high to the low value within the sample. A sample of 10 welds is usually made at each energy setting on the chart, and the profile is plotted as the joints are made and pulled so that additional points may be called for as indicated by the plotted curves.

Another method of charting, the isostrength diagram, will be found in common use particularly in resistance welding for cordwood modules. This method is a plot of electrode pressure versus electrical energy plus the associated strength for each point. The strength of each point on the isostrength diagram is found by averaging the strengths of 60 welds made at the corresponding energy and pressure settings. This method is not recommended because the optimum weld parameters are not readily observed.

Other Considerations.

CLEANLINESS. Weld locations along component leads must be free of coatings other than lead plating. The maximum dimensions of a component body must include the meniscus along the leads. Welding will not be consistent in the region which includes the transition between the controlled lead and the component package. Provision should be made in the product design for this need.

Resistance welding is relatively insensitive to many forms of contamination and, in fact, even benefits from some. Ordinary oils and finger acids do not interfere with optimum welding conditions. Insulating materials such as molding epoxies will cause blowouts or no welds, however. Other welding techniques require exceptionally clean conditions for any degree of consistency. Thermocompression bonding, for example, requires very clean surfaces. Reference to the discussions on the welding techniques for parameters to be controlled should provide guidance for contamination control.

COATINGS. Component lead coating is used (1) to provide an aid for soldering, (2) to protect the base metal from corrosion, or (3) to provide an interface material for bonding. Except where interface bonding is intended, plating is unnecessary for

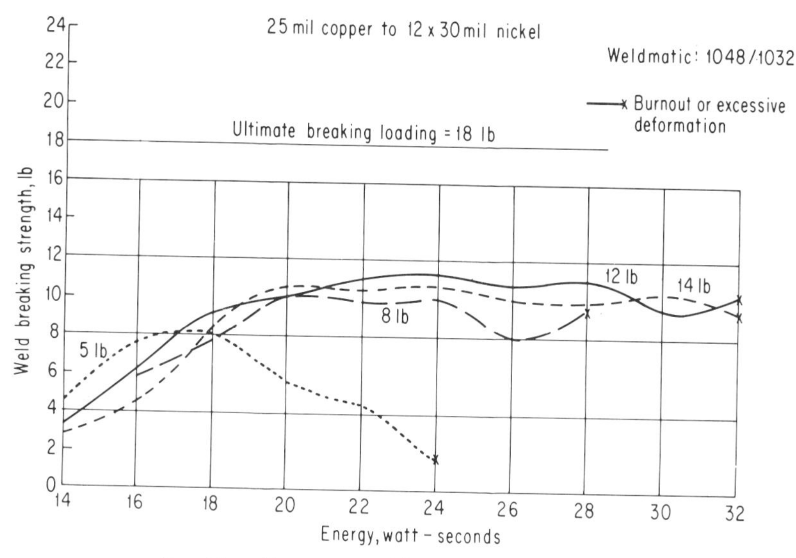

FIG. 33. Crosswire weld-strength profile chart.[11]

effecting a reliable weld. If plating is deliberately included in leads or interconnect buses, the thickness must be controlled in order to assure that welds will always be made through the plating to the core materials and not in the plating. Occasionally, plating materials will permit the formation of undesirable intermetallics under the expected welding conditions. A metallurgically acceptable weld schedule must include an examination of the effects of plating materials. Tin and lead platings are undesirable for resistance welding because of the tip pickup which degrades the electrode surfaces. Except for the rather minor problem that results with such surface pickup, solder-coated component leads can be routinely welded with high reliability. Gold coating is treated as any other metal coating material, and it possesses the added advantage of permitting consistent bonding results without requiring core involvement. Small-diameter gold (0.001 in.) is commonly bonded to the gold plating on (0.017-in.-diameter) gold-plated Kovar wire.

ELECTRODE SHORTING. Plating materials are troublesome in parallel-gap welding due to tip pickup. In this welding technique, where the two electrodes are separated by approximately 0.010 in. of insulation, a continuous degradation occurs via leakage between electrodes, which shunts some welding current. Frequent tip cleaning is required, is sometimes necessary between each weld, and is dependent upon the plating material and thickness.

EFFECTS OF DESIGN

Cordwood Modules. The cordwood assembly of electronic parts has been employed to increase the density of packaging over the two-dimensional printed-circuit board assembly approach for the same parts. Welding of the component leads is done to effect a higher-strength joint under certain temperature environments. Where necessary, an extensive parameter measurement system during each weld cycle can provide excellent quantitative knowledge of each joint. The cordwood module con-

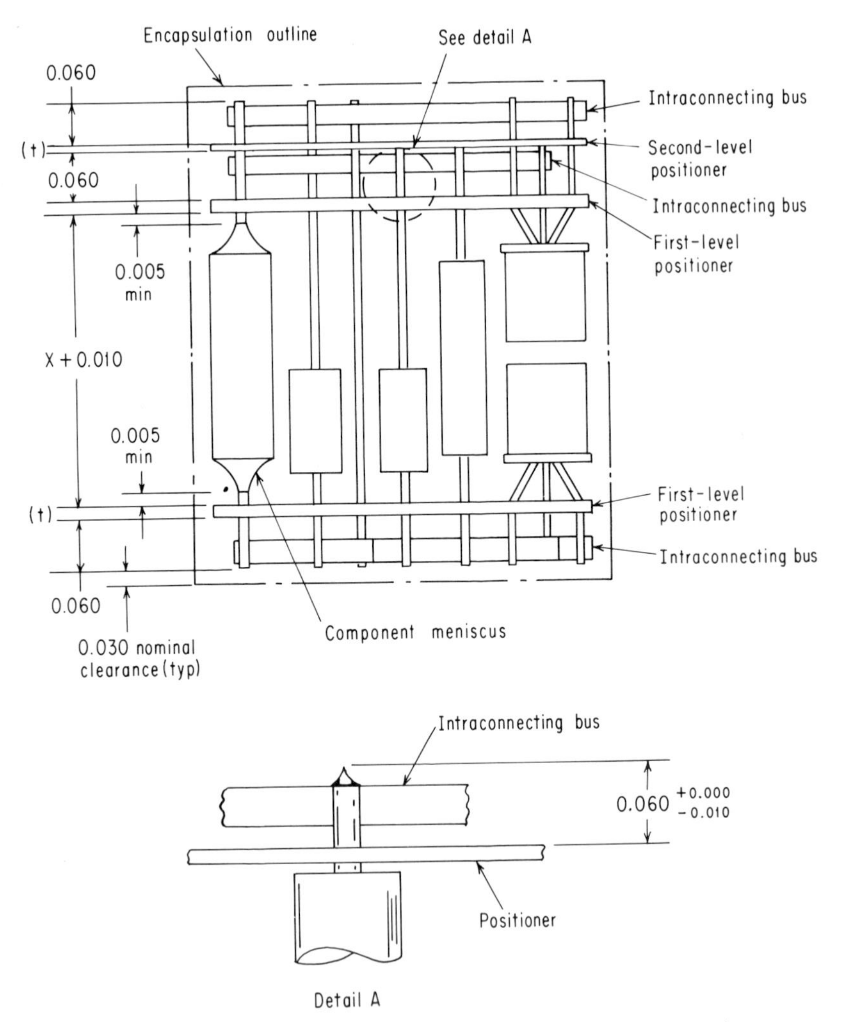

FIG. 34. Typical cordwood welded module dimensional tolerances.[12] (*Note:* X = the maximum length, including meniscus, of the longest component in the module; t = thickness of positioner.)

sists of two end plates (called positioners) in addition to the stacked parts. Certain design instructions are given to obtain the best welding conditions.

Component Positioners or Positioning Films. Component positioners are used within the module for several reasons, the most important being the retention of components in the desired position throughout the assembly and welding processes. Figure 34 shows a sectional view of a module with the top and bottom positioners in edge view. The component leads extend through the positioners which provide a structure for the components after welding. Perpendicular location of components held between the positioners may be optional as long as sufficient clearance is maintained between the positioners and the maximum-size components. In case of multiple-lead components, the bodies are located away from the positioner so that the leads may be spread to allow for welding accessibility. Such spreading should occur on a generous radius and not require bending the leads more than approximately 30°. No coating or holding material should be used to position parts because of the possibility of contaminating the material to be welded. All component bodies should be located within the positioner boundaries. A plan view of associated clear-

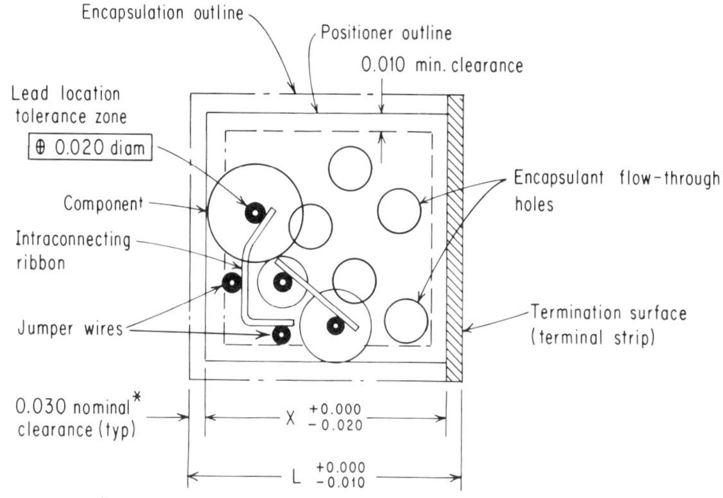

FIG. 35. Typical cordwood welded module positioner boundary.[12]

ances and tolerances related to positioner boundaries, encapsulation outline, and lead locations are shown in Fig. 35. The location of component lead holes should allow for a 0.02-in.-diameter true position tolerance. The outer limits of this tolerance zone should not be closer than 0.01 in. to positioner edge. Encapsulation flow-through holes should be provided in the positioner, where space permits, to improve the encapsulation integrity of the finished module. These holes should be at least 0.094 in. in diameter and located at a density of four to eight holes per square inch, depending on the availability of unused positioner area.

The component positioners may be oriented parallel or perpendicular to the module termination surface. When they are oriented perpendicular to the termination surface, they should be parallel to the rows of terminations so that the interconnecting bus and the terminations are in the same plane. This provides good welding access when attaching a termination strip (if used) to the welded assembly. When orienting the positioners parallel to the termination surface, additional jumper wires may be required between the positioners. Figure 36 illustrates the methods of positioner orientation and welded assembly attachment to the module terminations.

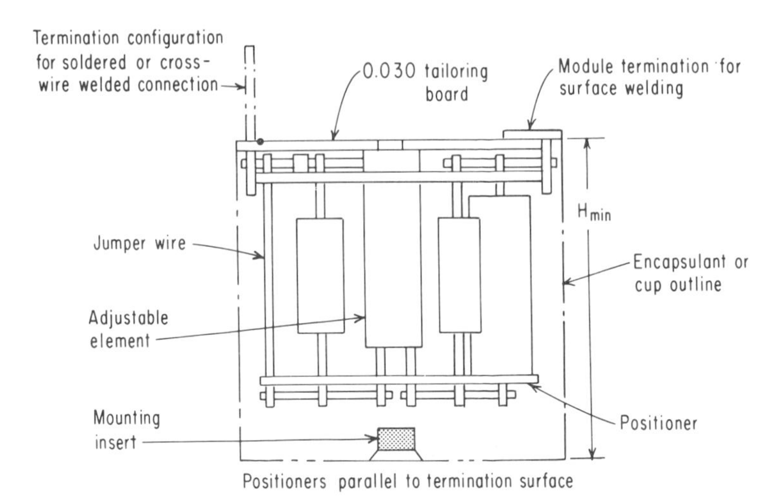

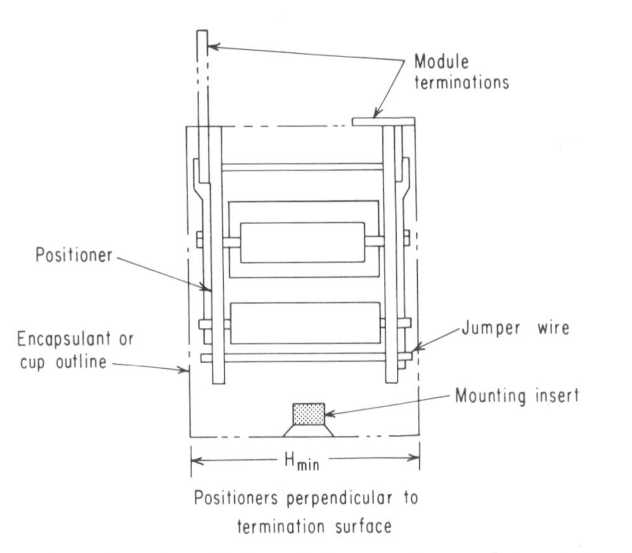

FIG. 36. Typical cordwood welded module mounting and termination.[12] (*Note:* $H_{min} = 0.070 + L + N(P + R) + M + B$, where 0.070 = component-to-positioner clearance [0.010] + weldment-to-surface clearance [2(0.030)], L = large component maximum length, N = number of positioner levels, P = positioner thickness, R = lead height beyond positioner [0.050], M = mounting hardware [if any], and B = termination or tailoring board [if any].)

Some of the restrictions which may be imposed on positioner orientation are as follows:

1. An adjustable component generally requires that the positioner be oriented parallel to the module top surfaces.
2. A component size or configuration may require that the positioner be oriented parallel to the top surface.
3. Heat-flow rods, when required, may dictate the direction of positioner orientation, depending on the heat-sink arrangement and the basic module configuration.
4. Module repairability becomes more difficult when header boards are an integral part of the module and the positioners are placed parallel to the termination surface.

The configuration of the positioner depends upon the orientation of the components and the module envelope. The positioners are defined as being located on the top and bottom with respect to the component axes. The top and bottom positioners may be in more than one layer depending upon the complexity of interconnections. The second layer need only be as large as necessary to show additional second-level bus routing and provide insulation between layers. If a partial positioner is used, it must be the outside layer.

Output Pins. The terminations to a cordwood module should be of the same size and material and should be rugged and weldable. For this purpose, 0.020- to 0.025-in. nickel 200 is a good choice. Although there are advantages to using component leads as terminations, there are many disadvantages, as follows:

1. Variation in size and material
2. Restricted location of pins
3. Interference during module welding
4. Variations in corrosion resistance
5. Frequently inadequate sealing conditions
6. Exclusion of premade header

Ideally the location of the module terminations should be determined by the system wiring design in order to reduce cabling complexity. Unfortunately the normal case is to locate module pins arbitrarily without system considerations and to require the cabling to do all the organizing. This approach adds substantial cost to a system. The type of output pin depends upon the interconnection system to be employed so that there is optimum compatibility for the joining technique. If the modules are to be soldered to a mother board, then the pins must be both weldable (for intramodule connection) and solderable (for intermodule connection). The best material for solderable and weldable pins is nickel, which has been properly hot-tin-dipped. Other considerations, such as module mounting to a connector, will dictate materials which are best for the connector and which can be welded for internal connections. Gold plating is recommended for this type of termination.

Jumpers. Components must be placed within the module in a manner which makes the interconnections easy, as short as possible, and with as few jumpers between positioners as possible.

A simple method of achieving this is to label, on the schematic, each termination of each component either "top" or "bottom" (Fig. 37). As each component lead is labeled, care must be taken that leads that must be connected together will enter the same positioner. If a situation exists such that after completing a circuit loop a contradiction occurs, then a jumper can resolve the incompatibility of two adjacent loops.

An alternate method for jumper optimization uses a sketch (Fig. 38) of the component connections to provide the designer with the placement sequence. The sketch is used only while making the module design and does not become a part of the final documentation. The diagram is constructed by taking each component in signal path order sketched to agree with the lead positions, adjustments, and any other component case features. Starting at the left edge of both the individual module schematic and this diagram, locate component positions in order of their interconnections as shown. The bottom positioner is indicated at the upper and lower part of the diagram. The top positioner is indicated in two places in the center of the diagram. The ground bus is drawn in at the upper and lower part of the diagram, which signifies that ground appears only at the bottom of the module. The components are evenly divided, insofar as practical, between the upper and lower part of the diagram, and are placed in sequence until all are accounted for. Component rearrangements should be tried until the simplest connection diagram results. When the designer is satisfied with this step, he can visualize, roughly, the module part placement by folding the sketch at the center and simultaneously compressing the length. This sketch may be retained by the designer as a checking aid when making the final component arrangement. Additional levels of interconnects may be necessary when crossovers occur. The number of crossovers must be mini-

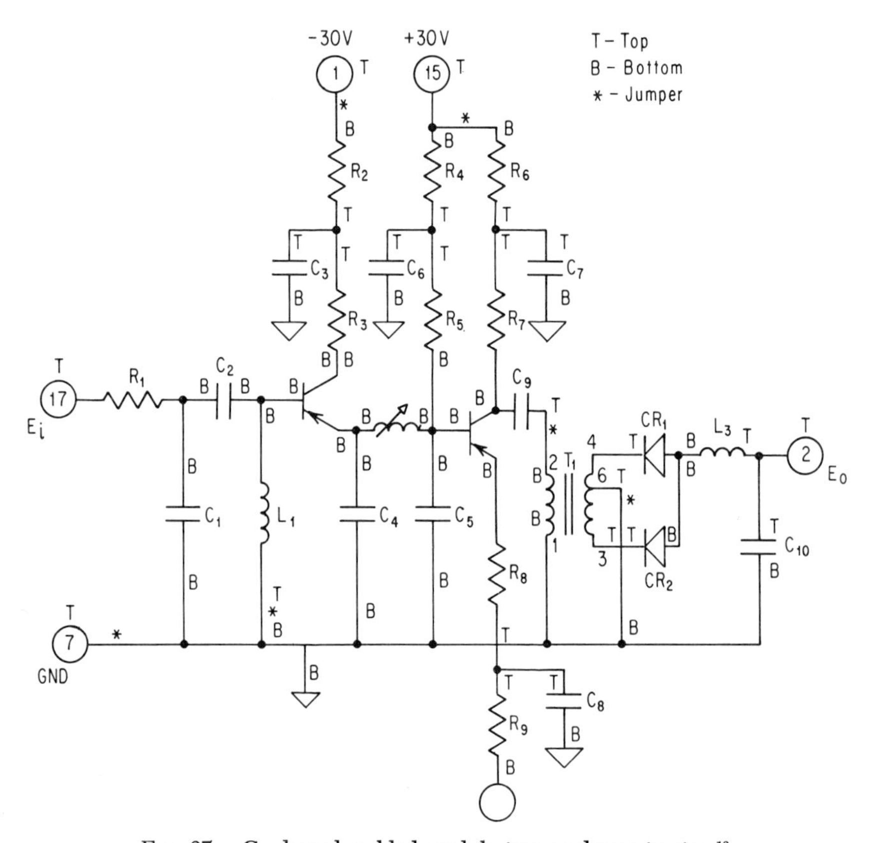

Fig. 37. Cordwood welded module jumper determination.[12]

mized when determining the optimum arrangement of components. Jumper material should be of round nickel wire and should be located such that there is no possibility of accidental internal shorting. Jumper wires do not require assembly between positioners as do other cordwood parts, since they can be inserted easily through one positioner into the other after the assembly has been completed.

Back-to-back Components. Components which have all leads terminating on one side should be mounted and welded in place prior to cordwood assembly. Efficient use of volume is obtained when parts of this type are distributed on both positioners in a back-to-back arrangement, as shown in Fig. 34.

Insulating Components. Insulation is required wherever there is a possibility that shorts can occur due to close packaging or tolerance drifts. Insulating parts is an added task that increases cost and usually can be avoided. The most frequent source of trouble by shorting is where semiconductor "cans" come in accidental contact with adjacent lead wires. A shrinkable plastic sleeve placed over the transistor prior to assembly completely resolves this problem. Large-diameter metal-bodied parts are also a potential source of trouble, and where there is danger of contact due to part density, these part bodies should be insulated.

Insulating Leads and Ribbon. Leads and bus ordinarily do not require insulation; however, whenever there is an unsupported long bus run, insulation will prevent accidental electrical contact. Sometimes a single crossover is needed to complete the connections, which would require a second bus level. In this case, the most efficient solution is to insulate a section of bus wire and perform this crossover on the first level. Part leads should not be insulated since most parts are larger in diameter than their lead wires, and therefore the space allotted to the component body provides adequate clearance to other leads. Occasionally a jumper

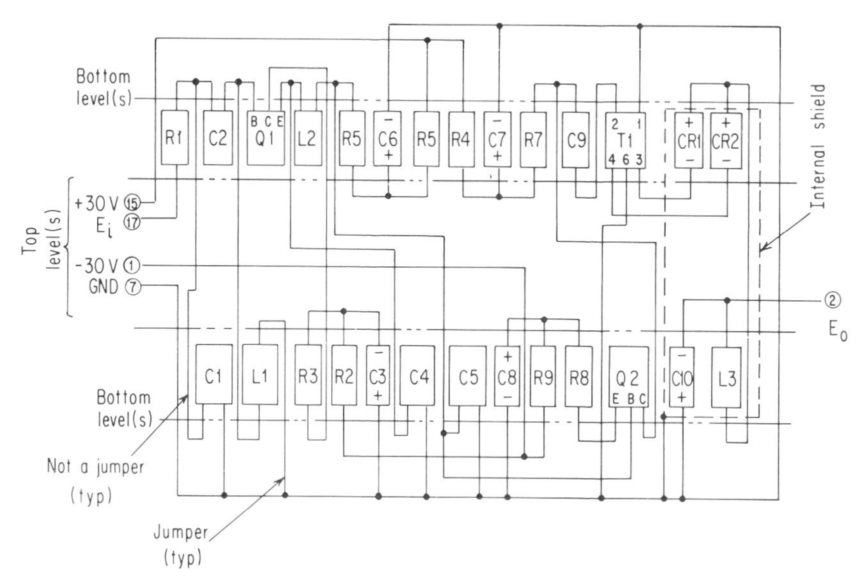

FIG. 38. Alternate method—cordwood welded module jumper determination.[12]

is so close to components that using an uninsulated wire is unsafe. The ease of inserting the jumper after assembly may still be achieved by using jumper wire holes with clearance over the insulation. This provision also assures that the insulation is long enough to provide the desired protection and, in addition, that the assembly is inspectable.

REPAIRABILITY

The cordwood module is not considered repairable, although some repair is possible during fabrication. The module reliability is seriously degraded due to repair operations, even when done carefully. Welds are repairable, however, and when the assembly configuration permits, part replacements may be made with ease.

Modules.

Part Replacement. Part replacement in a cordwood module is accomplished by cutting a hole in one positioner, clipping the leads and removing the part. The new part is inserted and the bus work patched back if possible. Repairs of this kind are not feasible with matrix positioners or with any positioner where there is more than one level. Other types of positioners can permit replacement by the use of ribbon for bus patching. Splicing buses by lap welding is the accepted practice. Splicing onto a plated trace requires the trace to be cleared and bent up to provide a surface for welding. Welding to a lead where a weld is already existent cannot be done reliably because weld schedules cannot account for the variability due to the third element and the material already welded.

Bus Repairs. Various defects can be corrected on the welding planes with good confidence. A bus blowout can be corrected by cutting away the blown area and patching another bus over the old one by lap welding. A missed weld is simply rewelded. A broken weld may be rewelded under some circumstances. The circumstances are too extensive to provide herein, but an operator instruction and schedule is derived by performing repairs and testing these joints essentially in the same manner as that in which the original weld schedules were developed.

Two-dimensional Assemblies.

Part Replacement. In most cases the parts, being readily accessible, may be removed by clipping their leads. If the leads are bonded through an interface material such as gold, the leads should be peeled and the new parts welded over the old. In many cases there is enough area provided on the printed-circuit-board trace to find a new area for the new part termination. Where a core-to-core weld has been made, such as may be made to a Kovar taper pin, the old part lead should be left and the new lead welded on top. A proofed weld schedule should be developed for this type of weld.

Rewelds. When it is desirable to repair a bad weld, it is not acceptable to reweld in the same location. Preferably, a new weld should be made—closer to the component—and possibly the circuit trace should be patched if it has been damaged by the bad weld. If a bond is not acceptable, it may be rebonded in the same location, providing the lead reduction due to bonding force is not reduced below a predetermined dimension.

Encapsulation. Encapsulation of electronic assemblies is needed for environmental protection. Many circuits require a uniform dielectric material to maintain consistent performance. Since cordwood modules present a collection point for dust and dirt, they benefit from encapsulation by enclosing the interpart volume. Cordwood assemblies are structurally weak until encapsulated, to the extent that handling damage will result prior to, or without, encapsulation. The usual argument against encapsulation is that repair is impossible after encapsulation. In fact, it is questionable whether a cordwood module should be repaired, encapsulated or not, because of degraded reliability. In any case, it is difficult to diagnose the failure of an encapsulated module, and materials are available that can be removed if desired. Planar and cordwood assemblies can be conformally coated for the desired environmental protection.

BASIC INTRACONNECTION TECHNIQUES

Various intraconnection welding techniques have been developed to improve the reliability and cost effectivity of electronic assemblies. The high part reliability now achievable places a new emphasis on the assembly yield and reliability. A few intraconnection techniques are described in the following paragraphs.

Cordwood Modules.

Point-to-point. The cordwood modules intraconnection design having the least necessary joints is where point-to-point bus routing is employed on each end, as shown in Fig. 39. The minimum joint count is not automatically achieved since part arrangements can radically change the routing complexity. As described earlier under the heading *Jumpers,* the circuit being packaged determines the routing limitations. Point-to-point routing requires either hand routing to a pattern printed on the positioner, or hand placing of precut and prebent ribbon. This technique is subject to assembly errors and cannot be considered an efficient mass production method.

Matrix. Another method of intraconnecting cordwood parts, shown in Fig. 40, is called matrix positioner. In this technique a film is punched with a predetermined hole pattern. Wiring on one side of the film is at 90° from wires on the other and on established grid dimensions. Crossed wires are welded together where there are holes in the film, and additional holes off the grid but alongside the wires are provided for later welding to the component leads. This system can be easily premade either manually or automatically, and in addition, the intraconnections can be pretested prior to use in module assembly. The intraconnecting wires are usually nickel of 0.020-in. diameter on both sides of the matrix. The matrix wires are welded using opposed electrodes and a dc capacitor discharge power supply. The undesired connections are cut out after fabrication. The component part leads are welded with pincer electrodes to the round matrix wire leads.

Intercon. The Intercon* system (Fig. 41) has tabs which are integral parts of the intraconnecting wiring pattern and which project into the third dimension beside each component. There is no nickel wire ribbon or wire to route between com-

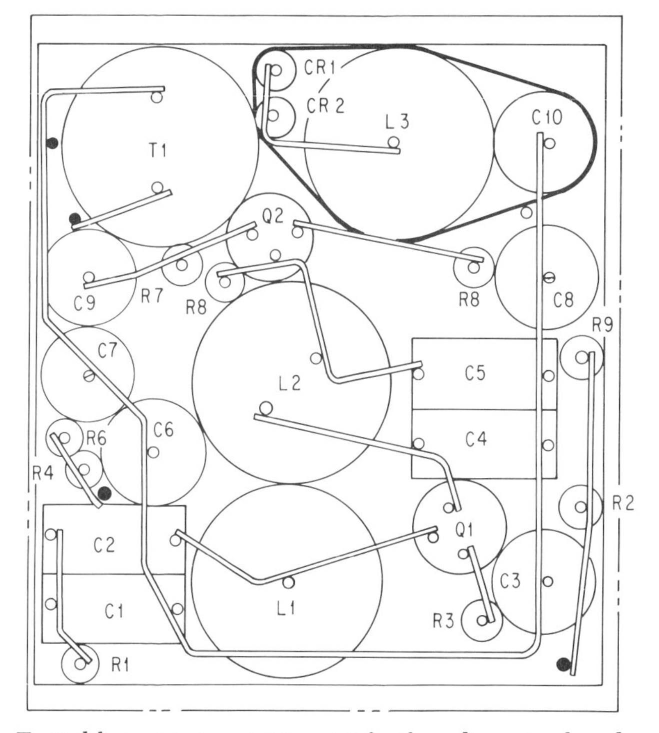

FIG. 39. Typical bus wiring—point-to-point for three-dimensional cordwood welded modules.[12]

ponents. The premanufactured positioner is of plated nickel where the tabs are subsequently bent up and where the circuitry is embedded in the insulation. The process is photochemical, so that once an intraconnection design is perfected all production articles will be identical. The system employs one layer of intraconnection with point-to-point wiring. The widest dimension of the ribbon is parallel to the positioner, with thickness and width controlled by the plating parameters. Welds are made by either a cross-wire configuration where the tab is L shaped or a lap-type weld where the lead is alongside the bent-up tab.

Tubes. Plated nickel tubes (Fig. 42) will weld with good strength and strength consistency when the tube dimensions and metal properties are precisely controlled.

* Trademark, Amphenol-Borg Electronics Corp.

A positioner with tubes and their intraconnecting circuitry was first developed by Tume. In this concept, all holes for locating parts by their leads are predrilled and then the tubes and intraconnecting circuitry are simultaneously plated. The tubes permit welding at any position around the tube and eliminate the need for weld polarity control. There are two welds made for each lead, one on each side of the

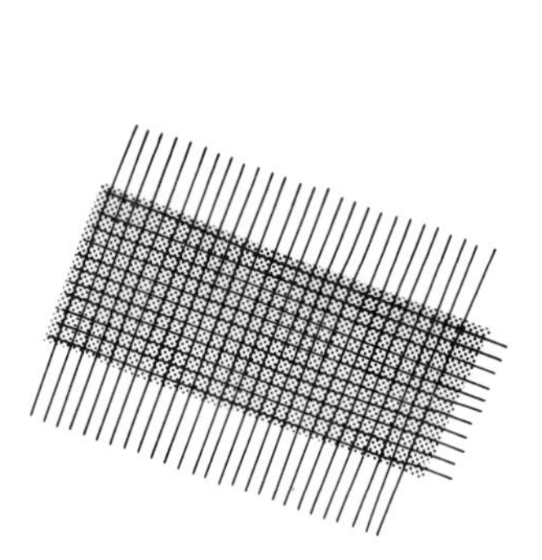

FIG. 40. Typical bus wiring—matrix method for three-dimensional cordwood welded modules.[13]

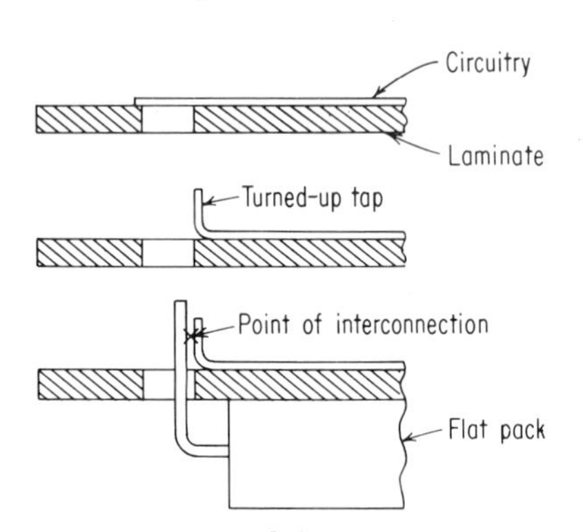

FIG. 41. Typical bus wiring—Intercon method for three-dimensional cordwood welded modules.[14] (*Amphenol-Borg Electronics Corp.*)

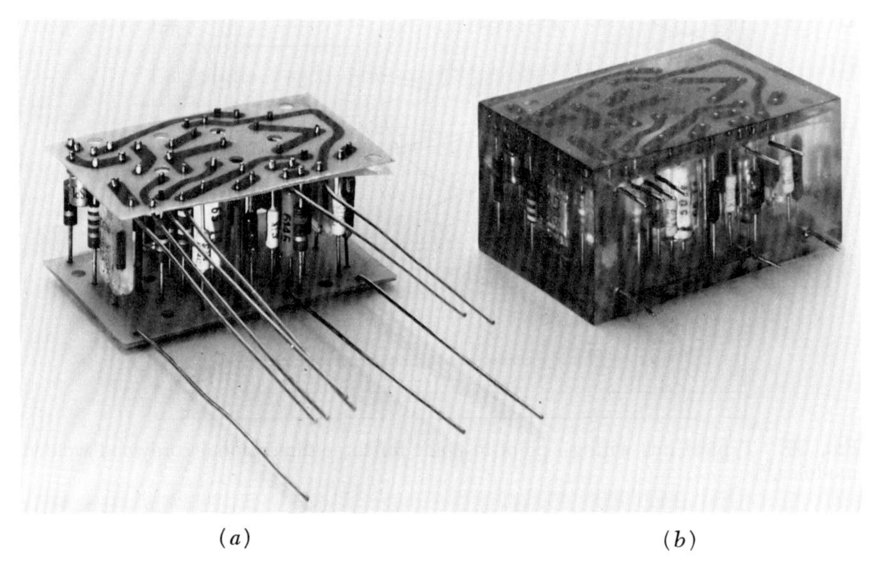

(*a*) (*b*)

FIG. 42. Typical bus wiring of cordwood welded modules using tubelet circuitry: (*a*) assembled module, (*b*) embedded module.[15] (*Litton Industries.*)

lead as the tube is forged onto the lead under electrode pressure. The precision required for the tube dimensions and the metal properties results in a somewhat costly process; however, since the processes are many but not complex and control measures are understood, the process should be attractive where large-quantity production is involved. The nominal wall thickness is 0.004 in., and the tube height is 0.030 in. Insulator sheets are required to prevent shorting to the circuit paths.

The method permits mounting and welding 10-pin integrated-circuit enclosures without lead spreading. The lead diameters, in the recommended ranges of 0.016 to 0.041 in., require four different-sized tubes. Weld testing is performed by welding leads with a portion of the lead extending beyond the tube. The part is cut off, the board is supported, and the tension is applied in the lead-tube axis. Average strengths are higher than cross-wire welds tested by the torsion-shear method.

Weld Preforms. Preformed circuit elements are manufactured out of sheet stock in a single operation. Included in the completed element, as an integral part, are all required conductors, terminal pins, terminal extensions, heat sinks, shielding, and alignment fixtures.

The conducting elements are created within a framework in the sheet stock. This framework, through supporting wires, maintains proper spatial relationships for the individual traces until the components are welded into place. The supporting wires are then removed along with the frame. Tooling holes in the frame correspond to similar holes in the assembly fixture to ensure accurate alignment during assembly.

The combination of preformed circuitry with etched epoxy laminates permits fabrication of a relatively high-density circuit. By placing the preformed circuitry into corresponding grooves etched in an epoxy-glass laminate, the exact positioning of a weldable circuit on a dielectric is achieved. Because the traces are locked in position there can be no movement in the potting operation allowing shorts.

Applying this to module manufacture substantially improves system reliability. For example, in point-to-point wiring, using wire or ribbon, the assembler wires the circuitry according to guidelines printed on an insulating support structure. Tremendous amounts of time are involved in designing and assembling, with ample opportunity for human errors. Using preformed circuitry, all the trace lines are prepositioned from a master layout. The result is no time for routing traces and the elimination of the human error of a misdirected trace. The terminals for welding the component leads are accurately positioned and the positioning is reproducible for any given module design. Inspection time is greatly reduced to only a surveillance of the condition of the weld joint.

An inherent feature is that there is no change in the fundamental properties of the metal trace. Thus, when weld schedules are established for a particular trace metal size and configuration, there will be no changes due to variations in the process. Where required for quality assurance purposes, it is possible to include in the design a test tab on which the weld schedule may be reaffirmed before the circuit itself is welded in place.

Another advantage is that the only connections are those made attaching the component leads. There are no internal electrical or mechanical connections. Inspection of the final part is simplified without sacrificing reliability.

The success of this approach to the manufacture of modules depends upon the consistent production of weldable traces from the appropriate sheet stock. Weld schedules have been determined for etched traces from nickel and phosphor-bronze sheets in thicknesses from 0.007 to 0.020 in. There is, of course, a difference in the configuration of the etched trace as compared with the ribbon. A compromise between the etched-trace cross section and the ribbon cross section is obtained by electropolishing the chemically blanked trace.[17]

Welded Insulated Wire. There is a tremendous amount of good usage history on magnet wire such that highly reliable, intricate, and yet flexible intraconnections can be constructed using this material. Welding through the insulation is accomplished using conventional capacitor discharge welder power supplies. Terminals spaced on 0.050-in. centers can be connected in series randomly to as many points as needed with one continuous wire. The reliability is equal to conventional cross-wire welds with minimum strengths at 50 percent of the wire. AWG 34 Formvar* has been used in this application, as shown in Fig. 43.

Printed-circuit Board—Two-dimensional Assemblies.

Tapered Pins. Plated tapered pins have been used in printed-circuit boards

* Trademark, Monsanto Co.

to obtain a raised solid post through the board for weld terminations of flat leads like those on integrated-circuit flat packs. A plated-through hole in a conventional printed-circuit board receives a gold-plated Kovar pin wedged in place to effect a diffusion bond with the hole plating. Subsequent mounting of the flat pack permits opposed electrode welding of the leads to the post. Accuracy of hole location and diameter are the only critical conditions that must be controlled for automatic assembly and welding.

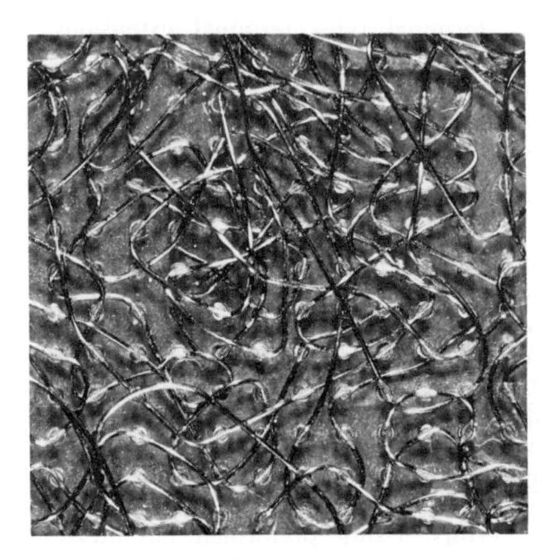

FIG. 43. Welded through insulation using magnet wire for complex interconnection designs.[16]

Kovar Cladding. Kovar-clad printed-circuit boards provide a much more weldable material if the leads are to be welded directly to the circuit-board traces. Kovar is somewhat active and requires a plating for protection. Gold plating is recommended unless the board is also to be machine-soldered. In some cases a gold-to-gold bond using only the plating on the Kovar and the lead may be desired. Kovar, being a magnetic material, may present some problems.

Copper-Nickel-Gold. Another good weldable combination where high conductivity is required is the conventional copper laminate with an added nickel plate. The nickel provides the material to be used for welding, and the copper provides electrical conductivity. The small amount of nickel reduces the magnetic effects, and gold plating can be applied over the nickel for gold-to-gold bonds if desired.

Substrate Connection—Face Bonding. The connections from semiconductor devices to other circuit connection points have traditionally involved the use of fine wires to bridge the chip pads to the circuit terminations. The fine wire terminations have been made by ultrasonic or thermocompression methods. This operation is by far the highest source of failure and consumes a high share of labor in the construction of microcircuits. The industry's concentration on improvement has resulted in the development of a technique called *face bonding* which avoids the use of fine wire altogether. Face bonding may be accomplished in a number of ways, and precision equipment is available to perform the operation efficiently.

Devices with Projections. The most efficient and reliable configuration for the face-bonding intraconnection is to form the projections on the semiconductor by vacuum deposition during its manufacturing while it is still in the wafer form. The manufacturing processes of metallization are already in existence and are used in the process cycle for other connection requirements. The metal systems are either aluminum or gold with projection dimensions limited to 0.003 in. maximum in height and with position accuracy equal to normal pad location requirements. The availability of such semiconductor-chip devices is not a restriction as at least one semiconductor manufacturer produces a major part of his integrated-circuit devices in this form. An example of a chip with projection terminators is shown in Fig. 44. There are other methods of obtaining projections on devices which are also satisfactory. Projections can be obtained by electroplating or electroless plating, or by a solder system using microscopic metal balls, or by screening conductive pastes.

Bonding Techniques. The ultrasonic bonding of projections on chips to mating pads on a substrate provides an excellent and reliable joining method. With well-controlled projection dimensions there is no difficulty in obtaining simultaneous low-resistance joints for 14 or more projections per device. Substrate packages are being marketed which provide required dimensional accuracy at a cost of $0.15 to $0.20 each. The bonding operation requires a special optical arrangement to assure accurate positioning prior to energy application. When a suitable optical and mechanical system is provided the die position can be adjusted with precision to mate with the pads. A modified Hughes Model 2902 ultrasonic flip-chip die mounter

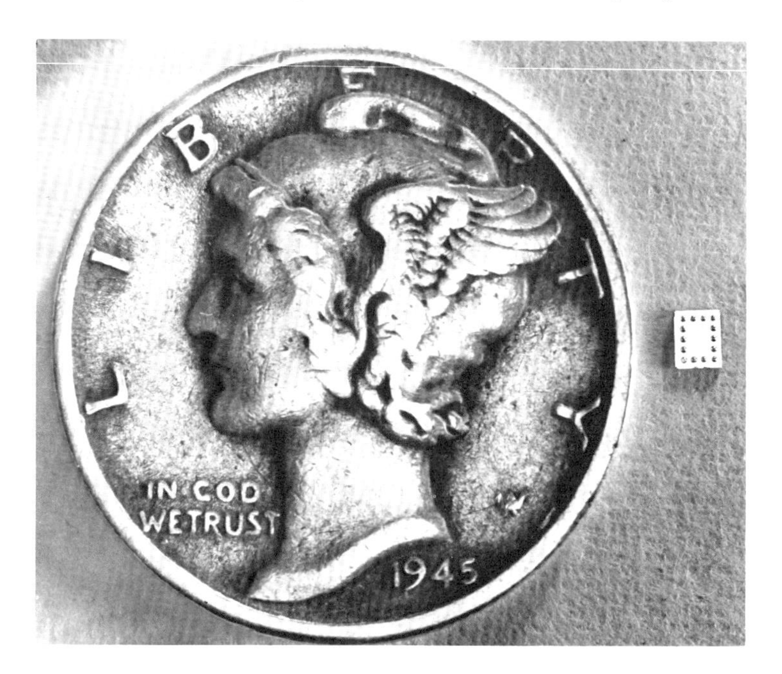

FIG. 44. Example of chip with projection terminators, or bumps.

shown in Fig. 45 illustrates the arrangement and positioning features. A vacuum chuck is employed to hold the chip in position while presenting it to the substrate. The unmodified equipment was designed for three-projection devices. A Unitek Model 8-142-01 flip-chip bonder is shown in Fig. 46. This machine employs a half-silvered mirror and a positioning mechanism with three degrees of freedom. In this way the part can be viewed while precise location adjustment is being performed. This unit is designed to perform thermocompression bonds and can be changed to make ultrasonic bonds.

Substrate Materials. Substrate materials which have been used successfully are:

1. Alumina—glazed and unglazed
2. Standard glass (Corning 0211 and 7093)
3. Epoxy glass—printed-circuit-board material
4. Beryllia

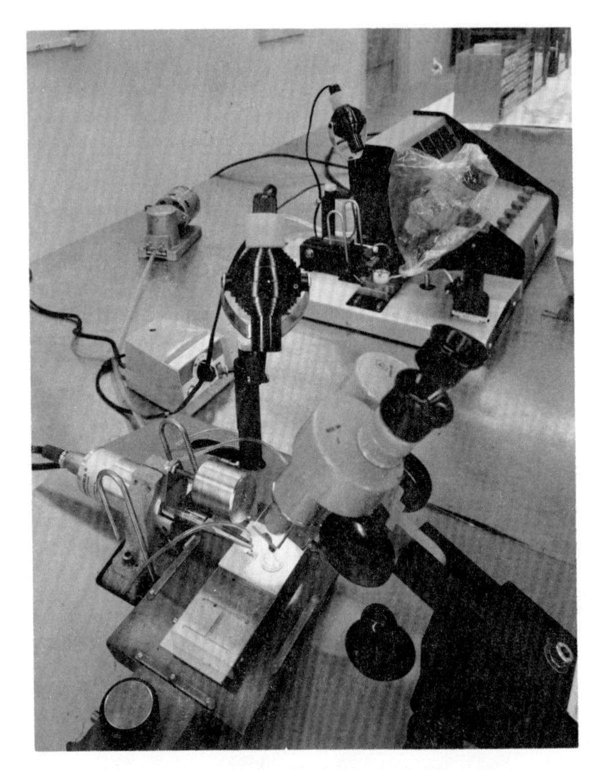

Fig. 45. Ultrasonic flip-chip die mounter. (*Hughes Aircraft Company.*)

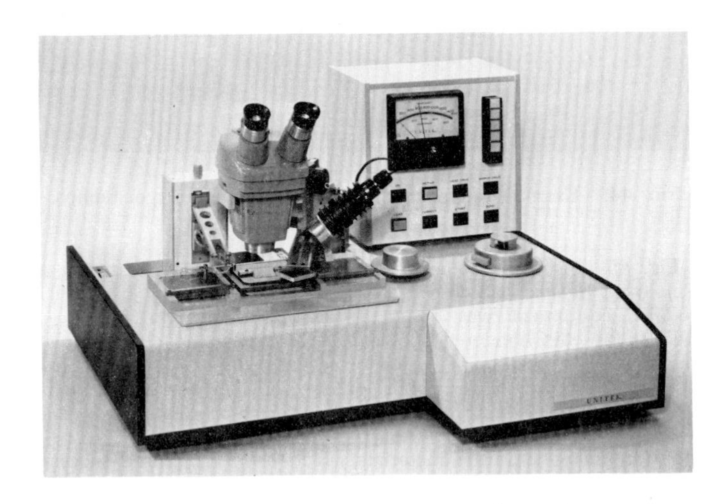

Fig. 46. Flip-chip bonder. (*Unitek Corp.*)

AUTOMATED WELDING

A basic requirement for automatic manufacturing is that the essential process variables be known and controlled. Many of the welding methods described in this chapter are easily automated. Some principles are given below to assist in decisions about automating a welding operation.

1. The quantity of the product must be high.
2. Flexibility of manufacturing equipment must be sufficient to accommodate a wide range of products easily.
3. Do not automate a human operation where the job is a natural human task and a difficult automation task. These operations can frequently allow human error without causing a product error, i.e., transport, start, stop, monitor, load, unload, etc.
4. The cost of the product must be substantially lower than by handmade methods when considering reasonable amortization of design and equipment costs.
5. Capability to schedule and to change the schedule on a quick-reaction basis is required.
6. Testing and/or quality control must be included in the items.
7. Automatic stop must be provided for equipment not within desired limits to prevent high rejection rates.
8. Diagnostic facility is needed to ascertain product degradation and factory equipment faults.
9. Automate by degrees where possible so that the processes can be evolved in digestible groups.

Cordwood Modules. The automatic welding of cordwood modules has been developed, as well as the automatic welding of matrix positioners. The automatic manufacturing of cordwood modules depends more upon the development of a system to assemble parts automatically than to weld.

Cordwood Module Welding. The welding process requires a one-piece premanufactured coordinated intraconnect layer for each side to effect high rate welding. The elimination of hand routing of ribbon removes most of the cost of intraconnection. Hand routing of ribbon can be effected if needed without requiring drawing changes. The welding operation can be automated by employing a table with x-and-y-positioned servo-controlled welding heads mounted in appropriate positions that will permit welding many modules simultaneously. An operator can guide one weld head while the balance are slaved, or a numerical control can effect all welding operations. Weld schedules can be automatically coordinated with each lead position so that optimum schedules are used and no missed welds are possible. Equipment to provide this facility is shown in Figs. 47 and 48. Approximately 10 modules can be welded simultaneously with this equipment by the addition of heads and power supplies. If necessary, hand welding can be employed with the same module drawings and parts; the major differences in processes are reliability and time. Many advances are contemplated in welding equipment, intraconnect devices, and materials which should be incorporated as applicable.

Intraconnection. The tubelet intraconnection design (Fig. 42) permits production of low-cost premanufactured positioners. Costs are low because the process uses known methods, permits preparation of large batches simultaneously, and employs photoetching techniques. Costs as low as ½ cent per tubelet are typical. Dimensions of tubelet positioners are governed primarily by the access for welding of 0.050 in. in one direction only. This permits the assembly of transistors and integrated circuits without lead spreading. Figure 48 shows a tubelet board being automatically welded. Tubelets are made of weldable nickel where the intraconnecting paths are deposited continuously with the metal making the tubes. A few (four or five) tubelet sizes will span all part lead diameters from 0.015 to 0.044 in. and will admit most lead materials presently used in welded modules.

Automatic Welding Equipment. The assembled modules are mounted on a Lear* or similar table which is controllable in x and y movements (see Fig. 47). Over the

* Trademark, Lear Siegler Corp.

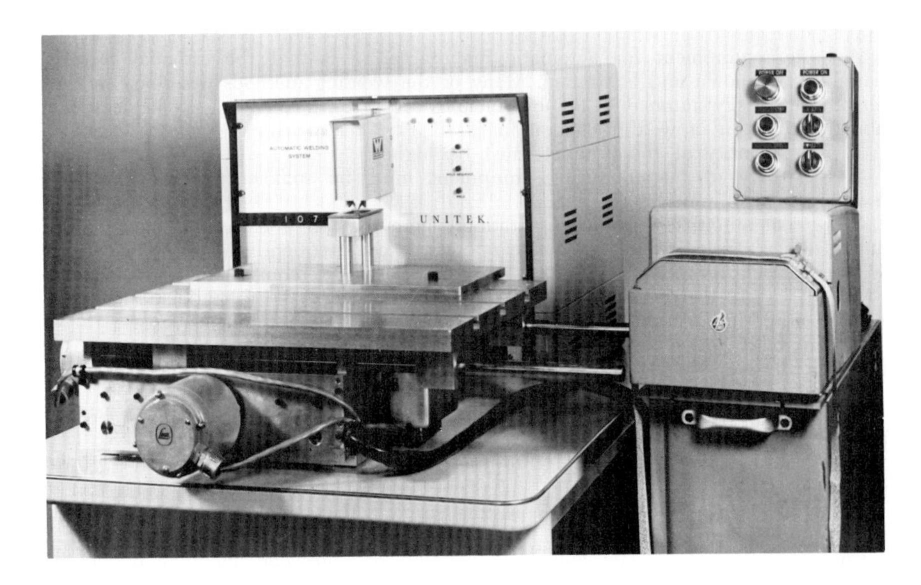

FIG. 47. Automatic welding with numerically controlled capacitor-discharge welding machine. (*Unitek Corp.*)

FIG. 48. A cordwood-tubelet module being welded automatically. (*Unitek Corp.*) Tubelet is patented by Litton Industries.

table are mounted a number of welding heads. The heads are electrically set for pressure. Each head has its own associated power supply. A numerical control system is provided which takes sequential information from a prepunched tape (one for each module surface) and commands each action of the welder. Thus the welds may be completed by all automatic actions and the various actions monitored to assure their proper completion. The welding operation described can be manual to the extent that an operator can weld one module which will cause identical operations on all the others, thus providing a saving in welding time even by pure manual techniques. This manual operation mode does not employ the numerical control mechanism. In addition to the welding controls provided in the automatic welder, certain other checking features are included integrally. Using automatic "self"-checking features will make it nearly impossible to produce a badly welded assembly. The self-checking operations include (1) proper weld schedule, (2) proper position, and (3) fault isolation and indication.

Typical equipment specifications for a resistance welding machine are as follows:

1. Capacitor-discharge-type power supplies with voltage regulation
2. Watts per second range—2 to 80 watts/sec
3. Head pressure range continuously variable by electrical control from 1 to 15 lb
4. Ten heads and associated power supplies mounted in position over an x-and-y-positioning table
5. Table overall location accuracy of 0.001 in. under electrical servo control
6. Control of all required table and setting operations by numerical tape program
7. Head vertically retractable by command; accuracy of return, 0.005 in.
8. Clearance for welding, 0.050 in.
9. Automatic electrode tip dressing adjustable each 100 welds
10. Recycle time each weld, 1.5 sec
11. Quick-change features of heads or electrodes
12. Welding current leads to each head shall be solid and not via slip rings. Routing shall be fixed and identical for each head. If head rotation is employed, flex leads may be employed at the head.
13. Weld condition monitor for each head to measure factors which determine the weld acceptability. Nondestructive methods are feasible to determine each weld quality accurately.
14. Output sensor for actual table position in x and y to accuracy of 0.001 in. to be used to obtain automatic check of table position.
15. Automatic shutdown circuit for any or all of 10 heads and table motion by external signal.
16. Service life of 10 years with normal maintenance, assuming 168 hr/month of operation. Availability of spare parts—five days for any replaceable item.

Printed-circuit Boards. Automatic welding of flat leads to traces on printed-circuit boards may be done by using parallel rollers which carry the weld current. The welding is parallel-gap type and the rollers are frequently flat at each step. There is difficulty in using round or curved electrode surfaces for this application. Automatic welding of flat-pack leads is effectively accomplished by a conventional parallel-gap electrode configuration lifted and stepped to each lead.

Laser Welding. The laser welding technique provides a means of welding in a location that is inaccessible except for an unobstructed beam line between laser head and the work. The beam can be quite long, so that automatic welding may be carried out by x and y coordinate control only. A tape-controlled table and the laser schedule setting commands are sufficient to effect automatic laser welding.

Electron Beam. The electron beam welder is ideally suited for automatic welding since the beam can be programmed to any pattern desired. A flying spot scanner can be coupled with the welding system to permit welding to templates photographically produced. There is no longer a requirement for beam welding in a vacuum (see discussion on beam welding), so that the flexibility of this welding technique has good potential for automated welding.

Other Automatic Welding. The combination of a welding technique with appropriate automatic loading, sorting, handling, and monitoring has been done for butt welding, lead welding, and assorted other electronic hardware processing. Primary consideration should be given to the automatic machinery for proper treatment and to the presentation of lead wires for joining, utilizing as a unit those basic elements of welding machinery where new automatic welding is indicated.

INSPECTION

The requirements of the welding operation are readily defined, notwithstanding the many factors requiring control. Welding machinery is available from many makers which, when properly used, will yield predictably acceptable joints. The parameters to be controlled and a method of effecting their control are described in the section on welders. The individual equipment types selected will determine the establishment of the specific values for equipment qualification and calibration.

Weld Process Proofing for Each Weld Schedule. Weld process proofing is the most important step in the welding operation since data are obtained to assure a proper weld schedule; since reliability predictions are obtainable; since the individual weld station variations, one to the other, are accounted for; and since control chart data are obtained.

Sample Size. The proofing tests require a sample size of 775 (5 for metallurgical examination) for a probability estimate of 3 welds in 1,000 outside the minimum stated at 90 percent confidence. When other probability and confidence limits are desired, a new computation of sample size is necessary. Data collected during normal production via control charts may be used to increase the sample size and thus obtain more accurate probability estimates. Such extensions of the process data must not include any kind of changes in the processes which are not included in the acceptable original limits.

Strength Requirements. The minimum allowable strength is established by a distribution-free statistical method for precision. The lowest strength in the sample should be above 50 percent of $\bar{X}$ chosen in the weld schedule development. If this value is not achievable, (1) the weld schedule is not good enough, (2) the welding process is not acceptable, or (3) the lead combination should not be used.

Selection of Sampling Conditions. The samples made shall be from the same materials used in the weld schedule development. Alternately, separate additional tests on material variations are necessary to determine changes of $\bar{X}$ which are due to material selection alone. A minimum of three operators on a minimum of three weld stations shall be used with a maximum of stations and operators so that the smallest sample weld lot shall be 50. All sample welds shall be tested on one pull tester all at one time, by one operator, and at the same pull test rate used in the weld schedule development.

Computations and Compilation. For the required sample size the necessary control data shall be recorded or computed:

$$\bar{X} = \frac{\Sigma x}{n} \qquad S = \sqrt{\frac{\Sigma(x_i - \bar{x})^2}{n-1}} \qquad \text{and } X_{\min}$$

The value of $X_{\min}$ shall be compared with the requirement, and production shall not proceed unless the requirement is met. Each equipment shall be computed separately in addition to the combined results so that the control limits can be established.

Acceptance Criteria. The welding process should be stopped and the causes investigated and corrected for any of the following out-of-control conditions: (1) any $X_{\min}$ is lower than the control limit; (2) any two consecutive $\bar{X}$ plots are out of the control limit; (3) any two consecutive strength-range measurements are out of the control limit; (4) any single value of x is below 50 percent of the value of $\bar{X}$ determined in the weld process proofing.

Schedule Proofing. It is necessary that the schedules be proofed using operation and equipment variations which are normal to the production operation. The *very-high-reliability* requirements of a joint used in electronic production suggest that a considerable effort be made to measure and record the expected strength and strength variations of each weld and then apply them routinely. The proofing tests for any distribution where precise measurements in the probability density function tails are required can result in gross errors if the estimation method is not designed for this statistic. Distribution-free methods are precise, and even though large sample sizes are needed, the extra complication is more than justified.

DISTRIBUTION-FREE STATISTICS. For situations in which very little is known about the distribution of the basic variable or for which it is known that the distribution is not of the required type, it is necessary to develop methods that do not depend on the particular form of the basic frequency function. A number of methods of this type have been designed. The only assumption that is needed for most of these methods is that the frequency function be continuous. A few, however, require that the frequency function possess low-order moments.

Since the methods of interest here are not concerned with testing or estimating the parameters of a frequency function of a given type, they are usually called nonparametric methods. Such methods are also called distribution-free methods because they do not require a knowledge of how the basic variables are distributed. Since neither name is strictly correct for all the methods usually listed under these names, the second name is used here arbitrarily.

The uncertainty associated with "goodness-of-fit" tests is caused by controlling only one error (the probability of making a wrong decision if the actual distribution is *identical* with the "fitted" distribution), whereas the error that is really of concern (the probability of making a wrong decision if the actual distribution is *different* from the fitted distribution) is dependent on the actual distribution and is, in principle, unobtainable. One way around this difficulty is to consider only distribution-free estimates of the percentage points of the true population. The error in these estimates is given by the confidence level, and at, say, the 90 percent confidence level, the probability (fiducial) that the true population percentile is less than the estimate (lower 90 percent confidence limit) is *exactly* 0.10. This assurance of an exact error more than compensates for the drawbacks inherent in the method. One drawback is that large sample sizes are required to meet high reliability requirements. This is not serious since large sample sizes are obtainable. Another drawback is that estimates are only available at specific points that are not determinable in advance, i.e., the points (pull strengths) where breaks actually occur in the sample. Provided that this fact is considered in specifying weld-strength requirements, it is also of minor importance.

Adoption of distribution-free estimates does not mean that other criteria are not useful. For example, statistical control is necessary in order to predict population parameters from sample characteristics. Standard criteria (e.g., $\bar{X}$, R control charts) can be used to provide assurance that the welding process is in control.

An example of the distribution-free estimates of the population percentiles is as follows:

With a sample size n let y_i represent the breaking load of the ith member of the sample. Then order the sample from smallest to largest and relabel the members from X_1 to X_n such that $X_1 \leq X_2 \leq X_3 \leq \ldots \leq X_n$. Call R_p the point in the parent population which cuts off p of the population on the lower tail (e.g., if $p = 0.003$ then 3/1,000 of the population is less than $R_{0.003}$). Then

$$P(X_r < R_p) = \sum_{i=r}^{n} \binom{n}{i} p^i (1-p)^{n-i}$$

where $\binom{n}{i}$ is the binomial coefficient, n things taken i at a time.

$P(X_r < R_p)$ is set equal to the desired confidence level. Given X_r and either n or p, the equation can be solved for the remaining parameter. In particular, if an estimate is desired at the lowest break strength X_1, at a confidence level of, say, 0.9, then,

$$P(X_1 < R_p) = \sum_{i=1}^{n} \binom{n}{i} p^i (1-p)^{n-i} = 0.9$$

If, in addition, p is fixed and equal to, say, 0.003 (3 out of 1,000), then,

$$P(X_1 < R_{0.003}) = \sum_{i=1}^{n} \binom{n}{i} (0.003)^i (0.997)^{n-i} = 0.9$$

and solving for n, $n = 770$.

Weld Station Calibration. The station calibration constitutes a check of each of the specified factors of the welding and weld position equipment. The calibration of electrical parts is done by conventional test equipment, while some special gauges may be required for mechanical tests. The purpose of this test is to assure that all parts, meters, and equipment are within the manufacturers' and/or users' requirements.

Weld Station Control. A continuous measurement of each weld station is desired to permit revisions in the production work station as early in the out-of-tolerance condition as can be detected. A control chart which has measurements and limits for $\bar{X}$, R, and X_{min} provides an inexpensive and suitable control method. The operator becomes a part of the system and each combination of operator, machine, and material requires separate recording charts. Automatic inspection by use of proper sensors and adjusted for each weld combination provides the best control by monitoring during the weld cycle.

Operator Qualification. The welding operation is not difficult nor does it require extensive special training. The operator effects are noticeable and important in the module weld quality. To maintain high quality standards, operator training and monitoring are needed. The welding operation is the most important, but other operations require training and control as well.

Prior Training. The inspector should assure that all operators have had prior training and that their proficiency is acceptable.

PERIODIC VERIFICATION. The operator should occasionally be provided with new information and given an opportunity to discuss problems associated with the operation. New equipment and procedures which are expected to be introduced should be taught, and operator proficiency should be rechecked to prevent a gradual decay of high-quality work.

OPERATOR CHARACTERISTICS. The operator characteristics listed below are either necessary or desirable characteristics specifically needed for the cordwood module welding operator. Other usual characteristics for electronic production workers are necessary but not detailed.

1. Must know how to read and understand manufacturing instructions via verbal and audio-visual aids.
2. Must know welding equipment used and settings permitted.
3. Must know three-dimensional welded module manufacturing techniques.
4. Must know electrode maintenance requirements.
5. Must know and recognize welding and polarity requirements.
6. Nice to know how to read and interpret color codes on electronic parts.
7. Nice to know how to read and interpret electronic part polarity markings.
8. Nice to know and recognize electronic parts by name, i.e., resistor, capacitor, transistor, diode, etc.

Visual Inspection. Visual inspection of each weld is not recommended because a good weld cannot be separated from a bad weld by visual criteria. Spending the large amount of inspection time necessary to examine each weld quantitatively is not economically advisable. Visual inspection is necessary on a sampling basis to assure that the processing procedures and controls are providing the desired results. The following items can be checked visually and are indicative of weld quality. Inspection of welds should be made at magnifications in the range of 10 to 20.

Missed Weld. A missed weld may result from omission altogether by an operator or from a power-supply misfire. Some lead deformation may be evident on a misfire, but no deformation is associated with an omission. In either case the inspector needs to ascertain whether the two pieces are attached, which *may* require light pressure on the lead wire or bus.

Incorrect Interface Area. A simple visual check of welding will reveal any welds made where one lead or bus is too short to overlap the other fully. Weld energy is carefully determined for each weld combination by the weld schedule development and is directly related to the volume of material between the electrodes as well as the area of contact. Any changes in volume of the material to be welded, such as insufficient material overlap, will cause weak or broken welds.

Misalignment. Any misalignment will degrade the weld; most misalignment can be readily seen under the specified magnification. Misalignment of electrode tips causes nonuniform lead deformation with some sharp discontinuities where lap welds are made in a bus. Misalignment has the same effect as mentioned in *Incorrect Interface Area* above. The degree of misalignment tolerated is difficult to specify quantitatively; however, the conditions which are usually typical of joints made for proofing tests and weld schedule development should be used for comparisons.

Angularity Requirements. The joints within a module weld should either form a right angle or be parallel. If other angles of joining are ever required, separate weld schedules are needed for each such special case. Weld schedules and proofing tests are made requiring a $\pm 10°$ tolerance on angularity, which makes it necessary that the same tolerance be maintained for production joints if the reliability estimates are to be precise. Overlap tolerances of parallel welds should be ± 10 percent for equivalent control of these welds unless the weld schedule and proofing tests were done with wider tolerances.

Bus Installation. The bus installation should be welded in place so that the sequence does not result in residual stresses. Improper installation is evidenced by bent component leads and bowed bussing where three or more welds are in straight sequence.

General Appearance. Generalizations about the appearance of weld conditions cannot be made because each weld combination differs. Standards about deformation are impractical because the only important condition is the strength achievement. Comparisons of welds with pictures and conditions established in the weld schedule development are useful in judging and resolving welding problems. Comparisons can be made on expulsion, splatter, gas holes, discoloration due to overheating, and other factors to aid in trouble detection rather than rejection.

REFERENCES

1. Larson, R. B.: Microjoining Processes for Electronic Packaging, *Assembly Eng.*, September, 1966.
2. Owczarski, W. A., and Palmer, A. J.: Percussive Welding, *Am. Machinist/Metalworking Mfg.*, June 12, 1961.
3. ITT Cannon Co.: Technical Supplement no. 32, 1963.
4. Epperson, J. P.: Laser Welding in Electronic Circuit Fabrication, *EDN*, October, 1965.
5. Miller, K. J., and Ninnikhoven, J. D.: Laser Welding, *Machine Design*, Aug. 5, 1965.
6. Conti, R. J.: Thermo Compression Joining, *Symp. Phys. Nondestructive Testing*, Sept. 27–30, 1966.
7. Gellert, R.: Ultrasonic Joining, *Machine Design*, Dec. 23, 1965.
8. Sawyer, H. F.: Q.F. vs E.T.P. Material Study, General Dynamics/Pomona Div., internal rept., Feb. 24, 1965.
9. Olson, D., and Purdy, K.: Welding Study Comparing Solder Coated Oxygen Free Copper vs Tough Pitch Copper, Allen-Bradley Co., technical rept.
10. Sawyer, H. F.: 3D Welded Module Design and Manufacturing Control Parameters, *4th Intern. Electron. Circuit Packaging Symp.*, Aug. 14–16, 1963.
11. Sawyer, H. F.: R & D for 3D Welded Circuit Packaging Design Requirements, 4th rept., pt. II, U.S. ERDL Contract DA-36-039 SC-90754, June, 1963.
12. General Dynamics/Pomona Div.: 3D Welded Module Design Standards, TM-314-P20.
13. Telfer, T.: The Welded Wire Matrix, *2d Intern. Electronic Circuit Packaging Symp.*, Aug. 16–18, 1961.
14. Amphenol-Borg Electronics Corp., R/A Div.: Intercon.
15. Sawyer, H. F.: A Model of an Automatic Manufacturing Facility for 3D Modules, *Natl. Electron. Packaging Conf.*, June 9, 1965.
16. Katzin, L.: Complex Interconnecting of Flat Packs without Multilayer Boards, *Natl. Electron. Packaging Conf.*, June 8–10, 1965.
17. Stearns, L. B.: Weldable Preformed Circuitry, *9th Welded Electronic Packaging Association (WEPA) Symposium.*
18. Meyer, F. R.: Ultrasonic Welding—Its Principles, Requirements and Uses, *Assembly Eng.*, March, 1966.

19. Hamilton Standard, Advanced Development, Electronics Dept.: Electron Beam Processes in Microelectronics, June, 1962.
20. Preston, D.: Welded Connections, *Product Eng.,* Aug. 5, 1963.
21. The Perfect Bond, *Electronics,* Jan. 9, 1967.
22. Warner, R. M., Jr., and Fordemwalt, J. N.: "Integrated Circuits—Design Principles and Fabrication," McGraw-Hill Book Company, New York, 1965, pp. 346, 347.

Chapter 5

DEPOSITIONS FOR MICROELECTRONICS

By

G. R. HEIDLER

General Electric Company
Missile & Space Division
King of Prussia, Pennsylvania

INTRODUCTION

In the past few years, circuit packaging has partly shifted from the conventional methods of assembling and intraconnecting discrete components into circuits to a technology where multiple components and their intraconnections are produced as an integral part of the component-manufacturing operation. This shift has also been accompanied by a trend toward what has been termed *miniaturization* and *microminiaturization.* It is sufficient to state that this new technology has been accompanied by a reduction in the size of a packaged circuit and a reduction in cost in many cases. It has also resulted in increased reliability because of reduced connections of all types. The packaging engineer, as a result, has become a component and process engineer along with being the engineer of the other technologies he is required to know. This chapter, then, will concentrate on the technologies which produce components and their intraconnections by means of depositions which result in assemblies that are smaller than their conventional component counterparts.

There are two basic types or styles of circuitry which result from various deposition techniques. The first is the "monolithic" or "integrated" circuit, as it has become known, in which active and passive components are manufactured simultaneously and then their required intraconnections are produced to form a completely functional or operational circuit. The second technology is the hybrid circuit, in which the passive components and intraconnections are produced on a substrate and the active devices added to form a complete circuit. The integrated-circuit technology is a highly specialized field in which only a few companies are engaged compared with the large number engaged in the manufacture of hybrid circuits. This chapter will discuss the types of depositions which are used in the production of hybrid circuitry and the primary characteristics of each.

The deposition technology encompasses vacuum deposition, screening techniques, and chemical depositions. Because of the variety and depth of deposition methods, this chapter will concentrate primarily on the depositions that have proved to be practical in that they are in use. Laboratory curiosities will not be discussed except where it appears that they may be important in the future. Finally, no attempt will be made to discuss all the possible deposition combinations that are being used. The number is staggering. Some of the more common combinations will be illustrated. Before proceeding, Table 1 contains a set of definitions for those who may not be familiar with some of the terms.

TABLE 1. Terms and Definitions

Active device. An electronic component which possesses gain, such as a transistor or diode.

Aspect ratio. The ratio of the length to the width of a two-dimensional resistor, or L/W.

Cermet, frit, ink, paste. Interchangeable terms used to designate the raw material for screened and fired components.

Cofired. A term denoting the firing of two or more SAF materials in one operation such as resistors and conductors.

Component, device. A term used to refer to any electrical component, resistor, capacitor, diode, transistor, choke, etc.

Deposition. A deposit of any substance or material on the surface of a substrate.

Drift. The rate of change in value of a passive component. Drift is usually specified in percent per 1,000 hr (%/1,000 hr) at a specified stress level.

Electroless plating. Plating of metal without the use of an electrical current—electrolessly.

Hybrid circuit. Any process in which the passive and active components are manufactured separately and joined together to yield a complete circuit in a later assembly operation.

Immersion plating. Plating based on an ion-exchange principle.

Integrated circuit (I/C). Any process in which the passive and active components are manufactured simultaneously to form a complete circuit.

Interconnection. Electrical connections between complete circuits.

Intermediate film. Any deposition between 10,000 and 127,000 Å (0.5 mil).

Intraconnection. Electrical connection between components within a single or functional circuit.

Large-scale Integration (LSI). A term which has not been precisely defined. It pertains to the incorporation of more than one functional circuit within a "chip." The term has also been used in describing "chip and wire" arrays of single I/Cs and in conjunction with multiple hybrid circuits on a single substrate.

Load life. A term used interchangeably with drift, but actually referring to the data obtained to establish drift rates, i.e., load-life data.

Noise. A term used to describe electrical noise generated by a passive component and usually expressed in decibels per volt (db/volt).

Oversquared. A resistor whose length is less than the width, so that the aspect ratio is less than 1.0.

Passive device. An electrical component without gain—resistor, capacitor, or inductor.

Resistor stress. The applied power in watts per square inch of film area.

SAF. Screened and fired.

Sheet resistivity. The resistance of the deposition in ohms per square.

Substrate. A base material upon which a deposition is made.

Temperature coefficient of capacitance. The unit change in capacitance per unit change in temperature, usually expressed in percent per degree centigrade (%/°C).

Temperature coefficient of resistance. The unit change in resistance per unit change in temperature, usually expressed in parts per million per degree centigrade (ppm/°C).

Thick film. Any deposition greater than 0.5 mil in thickness.

Thin film. Any deposition less than 10,000 Å thick.

Thixotropic. A fluid which changes viscosity as a function of the rate at which it is sheared.

Vapor plating. A method of deposition where a deposit is made on a substrate from a material in the vapor state.

Voltage coefficient. The unit change in a passive component per unit of dc voltage applied. Voltage coefficient applies to both resistors and capacitors and is generally expressed in percent per volt (%/volt dc).

TYPES OF DEPOSITION

There are three broad areas of depositions that are of interest in microelectronics: vacuum, screened and fired (SAF), and chemical depositions. In these three areas, various depositions and combinations thereof are used to produce resistors, capacitors, inductors, and conductors, to which active devices are added, or these components are added to active devices. In the true hybrid circuit, passive components and conductors are produced to form resistor, capacitor, inductor, and conductor networks. To these networks, capacitors and inductors, which cannot be produced by deposition, are attached to the substrate, along with the active devices, thus to form complete circuits or functions. In another form of hybrid circuitry, only the resistors and intraconnecting wiring are formed, with all capacitors added in "chip" form. In still another form of hybrid circuitry, resistors and conductors are deposited on the oxidized surface of a silicon wafer which contains active devices. In this case, resistors and conductors are added to a substrate which contains the active devices rather than attaching the active devices to the substrate. This method of hybrid-circuit fabrication is not in popular usage, however. The techniques of assembling hybrid circuits are discussed in Chap. 10.

To acquaint the reader with the various terms and terminology related to thickness and measurements thereof, Table 2 lists the common conversions between the English and metric systems. For example, surface finish is generally given in terms of microinches, while vacuum-deposited film thickness is measured in angstroms, thick films are in mils, and plating thicknesses in millionths. For the purposes of this chapter, a thin film will be considered as a deposit of 10,000 Å or less. Thick

TABLE 2. Thickness Conversion and Terminology

Inch (in.)	Common oral expression	Millimeters (mm)	Microns (μ)	Angstroms (Å)
1.000	. . . .	25.4	25,400	
0.100	1 tenth	2.54	2,540	
0.010	10 mils	0.254	254	
0.001	1 mil	0.0254	25.4	254,000
0.0001	1/10 mil or 1 tenth	0.00254	2.54	25,400
0.00001	10 millionths 10 microinches	. . . . . .	0.254	2,540
0.000001	1 microinch	. . . . . .	0.0254	254

1 micron = 10,000 angstroms
1 millimeter = 1,000 microns
1 millimeter = 10,000,000 angstroms

Thick films
0.5 mil ________________ 127,000 Å
Intermediate films
Approx. 40 millionths ________________ 10,000 Å
Thin films

films are those of 0.5 mil thick and up. The area from 10,000 to 125,000 Å has been facetiously called thick-thin films or thin-thick films. Being neither thin nor thick, they will be called *intermediate* films in this chapter. These divisions are arbitrary but should be acceptable to most readers.

Vacuum Depositions. There are two basic types of vacuum deposition: evaporation and cathodic sputtering. In evaporation, the material to be deposited is heated in a source in a vacuum chamber at 10^{-5} torr or lower, until the material vaporizes and deposits on the cool walls and surfaces within the chamber, including the substrate which is to be coated. A vacuum system is shown digrammatically in Fig. 1. The source can be a filament, boat, or crucible, as shown in Fig. 2, or an electron beam, as shown in Fig. 3. Each type of source supplies heat to the material which is sufficient to allow it to evaporate in a vacuum.

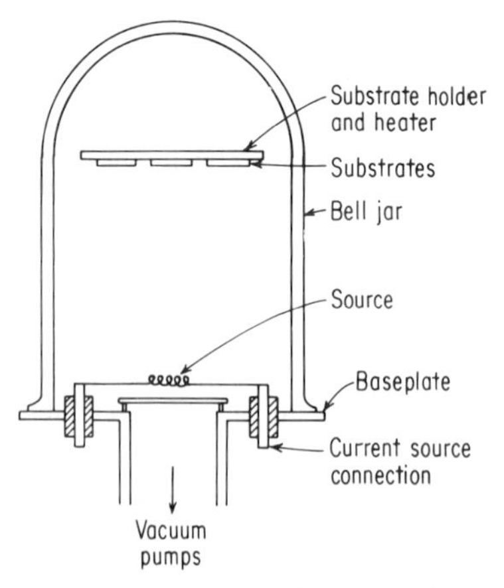

FIG. 1. Schematic diagram of a vacuum-deposition system.

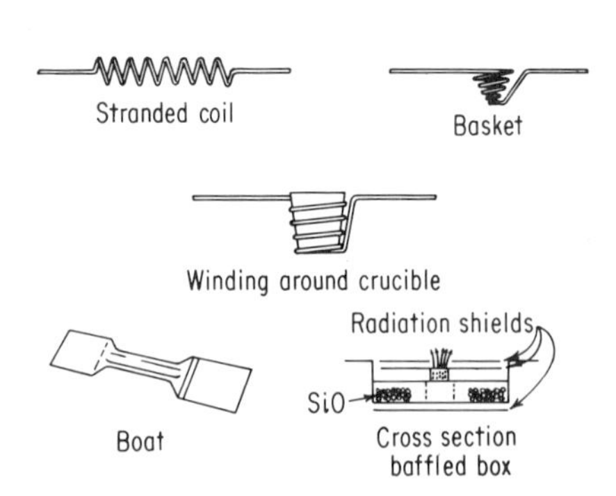

FIG. 2. Evaporation sources for vacuum deposition.[17]

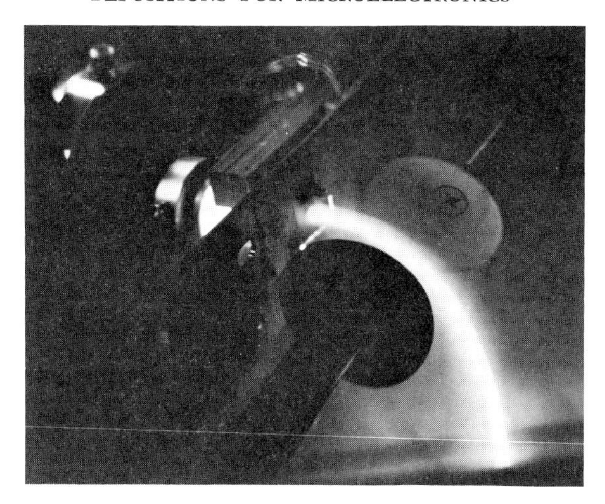

FIG. 3. Electron-beam source focused on material to be evaporated. (*Consolidated Vacuum Corp., Rochester, N.Y.*)

Cathodic sputtering is another form of vacuum deposition, but it is quite different from evaporation in that the deposition is made in a controlled partial vacuum, usually in the range of 1 to 100 μ pressure and most commonly at a partial pressure between 20 and 25 μ. In general, the chamber is pumped down to about 10^{-6} torr and backfilled with a known gas or mixture of gases. Argon is ordinarily used with some partial pressures of nitrogen or oxygen, which alters the properties of the film. Its simplest form is shown in Fig. 4, where sputtering is accomplished by applying a high dc potential between anode and cathode, which causes the gas to

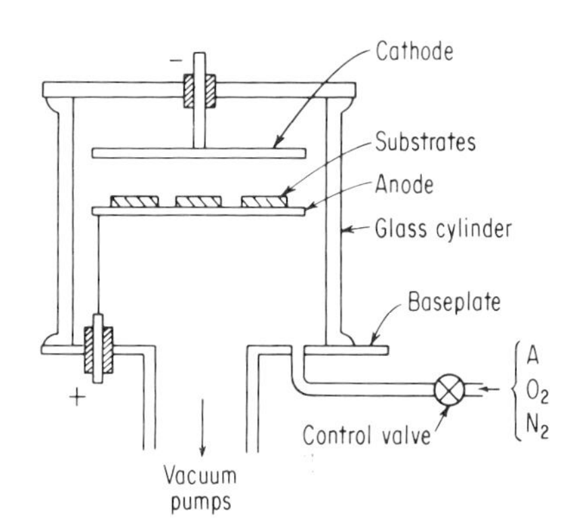

FIG. 4. Schematic diagram of a cathodic-sputtering system.

ionize and bombard the cathode. Particles are dislodged from the cathode and deposited on the anode and surrounding apparatus. Ionization and deposition can also be accomplished by using a radio-frequency field to energize the gas. In this manner a cathode, which is an insulator, can be sputtered when it could not be deposited by evaporation.

These two methods are used to produce deposits in one of two basic ways. The first method is to use metal masks in contact or close proximity to the substrate so that various materials are sequentially deposited through openings in the mask and onto the substrate to produce resistors, conductors, capacitors, and crossovers. This method is used in evaporation because masks distort sputtered depositions and are generally unsatisfactory. A second method is to deposit material over the entire surface of the substrate and, by using the photolithographic techniques described in Section 1, selectively etch material from the substrate, a network of resistors and conductors thus being left. This method is used for both evaporated and sputtered films. The third method is reverse photolithography. In this method, the deposition is made over a previously developed photoresist. After deposition, the photoresist is washed away, leaving the desired pattern. These methods are shown in Fig. 5.

Conductors. Most metals are relatively good conductors even when the deposited film is only a few hundred angstroms thick such as is found in vacuum-deposited

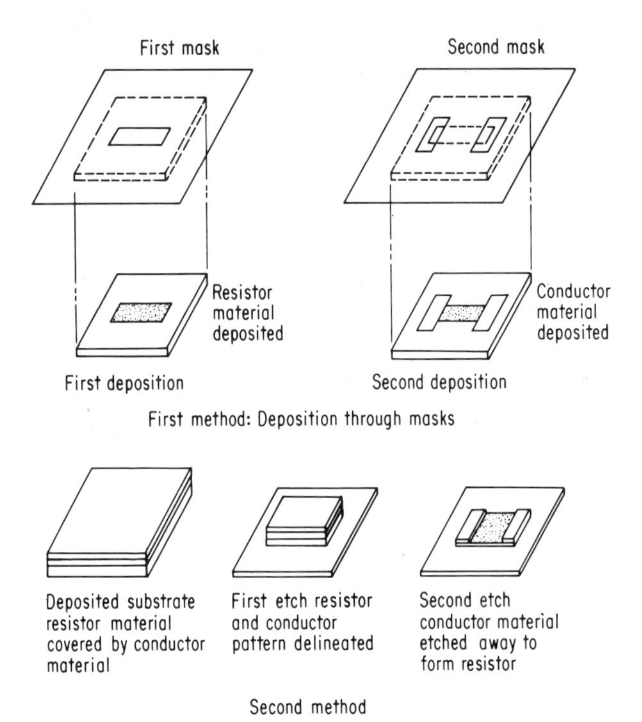

FIG. 5. The two most common methods of delineating components.

films. While most metals are conductors, only a few are useful. The metals most commonly used are aluminum, gold, copper, nickel, tin, and lead.

ALUMINUM. This is the most common vacuum-deposited metal because of its basically good properties. It is easily evaporated from a filament, boat, or crucible but almost impossible to sputter. Its frequent use is attributable not only to the ease of evaporation, but also to the fact that it is almost completely insensitive to the evaporation conditions. Aluminum has very good substrate adhesion whether the substrate is heated or cold. Conduction is very high, with resistance ranging from 0.05 to 0.001 ohm/sq, depending on thickness. Resistance normally averages 0.001 ohm/sq.

Aluminum does, however, leave something to be desired because it cannot normally be soldered with tin-lead solders. It can be soldered with zinc-tin solder. If lightly overcoated with a material such as tin or nickel, tin-lead solders can be used but an extra deposition is required. The secondary deposition provides a solderable material over the aluminum which will accept solder and prevents the under conductive layer of aluminum from oxidizing, which prevents aluminum from soldering. Once aluminum is solder-coated, there is little problem.

COPPER. Copper is easily sputtered or evaporated and slightly more conductive than aluminum. Unlike aluminum, copper has "marginal" adhesion on glass substrates and fair adhesion on unglazed alumina. It is not obvious whether the marginal adhesion on glass is a function of substrate cleanliness or other factors. Even when the substrate is heated, adhesion is not assured. The adhesion problem is generally eliminated by an undercoat of 50 to 200 Å of a metal with good adhesion such as nickel, chromium, tantalum, titanium, or tungsten. Nickel and chromium are most commonly used. While copper is solderable, it is somewhat soluble in solder so that a deposit of 1,000 Å or more is necessary if any great amount of resoldering is to be done. For this reason, copper is not used as frequently as aluminum.

GOLD. Gold is also easily evaporated or sputtered and is quite similar to copper in both adhesion and conductivity. It is different from copper in that it is highly soluble in tin-lead solder so that an undercoat is necessary before this solder can be used. Lower-temperature indium solders can be used and resoldered many times provided that indium-alloy solders are acceptable for the application.

NICKEL. Nickel can be both evaporated and sputtered. It is used primarily as an undercoat or base for copper and gold, because it has 10 to 100 times the resistance of copper or gold. Nickel has excellent adhesion itself and provides an excellent base for other metals. It is readily soldered and can be resoldered many times.

TIN AND LEAD. These are quite easily evaporated; so it is seldom that they are sputtered even though they are readily deposited in this manner. Solder, a combination of tin and lead, is also easily coevaporated from a single source. Tin and lead are not commonly used because they are soft and easily scratched. Both elements have been used extensively in thin-film cryogenic applications because both exhibit superconducting properties. Unfortunately, cryogenic microelectronics has not been too practical to date, and so the use of the individual metals has been rather limited. Solder, however, has found application in the fabrication of "flip-chip" transistors and diodes, where solder is evaporated on the surface of semiconductor wafers.

Resistors. Many metals exhibit properties which are suitable as resistors when deposited in thin layers in the range of 50 to 2,000 Å. Many of these have one or two disadvantages that limit their usefulness so that few metals are useful for thin-film resistors from a practical standpoint. Tungsten is a prime example. Thin films of tungsten have poor temperature coefficients of resistance. While it is not mandatory, most good resistor materials produce oxides whose molecule is larger than the molecule of the original metal so that the surface, when sufficiently oxidized, effectively seals the remaining metal film from further oxidation and drift. Hence, few materials have proved to be satisfactory—chromium, nickel-chromium (Nichrome*), and titanium.

Thin-film resistors of these materials have excellent load-life stability, drifting less than 1 percent/1,000 hr and exhibiting TCRs in the range of ±200 ppm, low noise, and virtually no voltage coefficient. Because metal-resistor films are thin, ranging from 50 to 2,000 Å, they are limited to a maximum of 500 ohms/sq, with 200 ohms/sq more generally used. Above 500 ohms/sq, resistors are generally unstable and difficult to reproduce. Because of this, a resistor of 500,000 ohms is a practical limit although 1-megohm resistors have been produced, but they consume a tremendous amount of area on the substrate. Even a resistor of relatively low value, such as 10,000 ohms, must be long and narrow, or it must be folded or serpentined to fit into a more reasonable area. Properties of thin-film resistors are shown in Table 3.

CHROMIUM. As a resistor material, chrome is not new, but until recently it was not well known because it was not a popular material. Its properties are excellent, and it produces stable resistors with specific resistances up to 500 ohms/sq. TCRs

* Trademark of Driver-Harris Co.

TABLE 3. Properties of Thin-film Resistors [9,22]

Material	Sheet resistivity, ohms/sq	TCR, ppm/°C	Drift, %/1,000 hr at 10 watts/in.² loading
Chromium	50–500	±100	0.4
Nickel-chromium	50–400	+25 to +200	Less than 0.1
Tantalum	25–200	−100 to +200	Less than 0.5 on glazed aluminum
Tantalum	25–2,000	−70	0.15 on 7059 glass
Tantalum nitride	25–200 is practical	−70	Less than 0.1 on glazed alumina, 3.0 on 7059 glass
Titanium	1–2,000; 1–500 is practical	−1,000 to +1,000 0 at 50 ohms/sq	0.5% at 100 ohms/sq
Noise	Negligible		

are normally ±100 ppm/°C with a drift of less than 1 percent/1,000 hr for a loading of 10 watts/in.² of film area. Chrome is not difficult to evaporate, but it does require considerable heat and, of course, it can be sputtered. It can be deposited through masks, but it is easily etched so that overall deposition is common. The circuits in Figs. 6 and 7 were produced by using overall deposition with a subsequent nickel-gold chemical deposit.

NICKEL-CHROMIUM. The combination of nickel and chromium commonly called

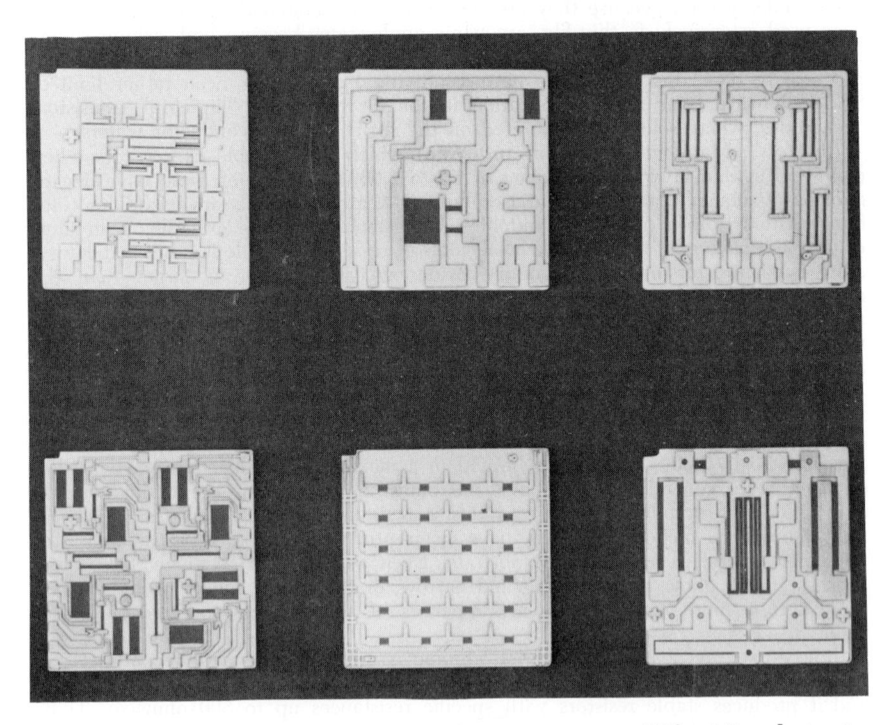

FIG. 6. A variety of etched evaporated chromium circuits. (*Film Microelectronics, Inc., Burlington, Mass.*)

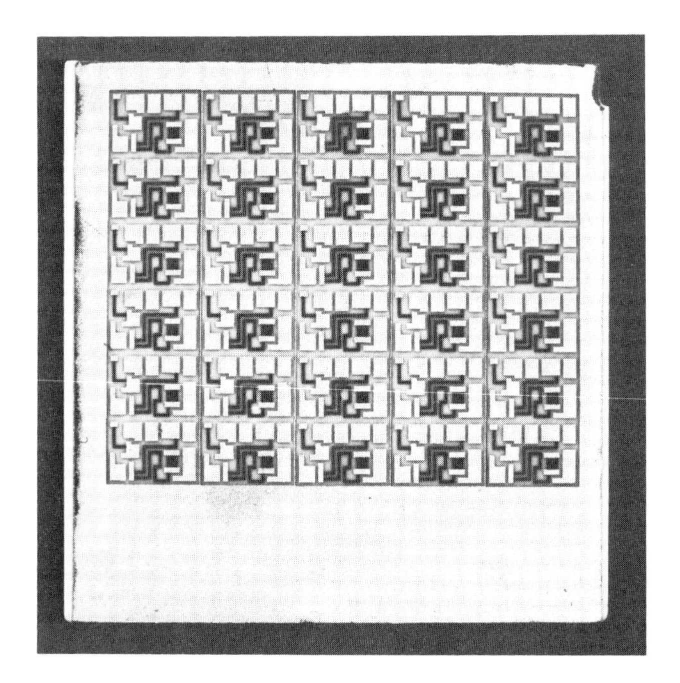

FIG. 7. Multiple circuits etched from evaporated chromium. (*Film Microelectronics, Inc., Burlington, Mass.*)

Nichrome is easier to evaporate and hence more widely used than pure chromium. Unlike the pure material, Nichrome films are frequently protected by an overcoat of silicon monoxide, although high-temperature stabilization is sufficient to yield very low drift rates. Nichrome is typically deposited from 50 to 200 ohms/sq and can be deposited up to 400 ohms/sq, but with a tolerance of ± 30 percent. TCRs can be tailored by regulating substrate temperature, but $+100$ to $+200$ ppm is typical, with a drift rate of 1 percent/1,000 hr at a loading of 10 watts/in.^{2} of film area. Figure 8 illustrates a multiple aluminum, Nichrome, SiO_x circuit deposition which includes capacitors, as well as conductors and resistors. The circuits are then cut apart and assembled as shown in Fig. 9, with an assembly close-up shown in Fig. 10.

TANTALUM. Unlike the two preceding materials, tantalum is not easily evaporated except with an electron beam. It is easily sputtered, and this is the general method of deposition. Because it is sputtered, the entire substrate is generally coated, for deposition through masks results in shadowed deposits. Tantalum coating is followed by nickel or copper and then gold. Since the entire surface of the substrate is coated, selective etching techniques are used to delineate resistors and conductors. Tantalum is different from other materials in that it is generally stabilized after the pattern is formed, either by baking the entire substrate or by anodizing each resistor to value. This serves to stabilize and trim the resistors in one operation. Tantalum has properties similar to chrome and nickel-chrome. Stable resistivities are to 200 ohms/sq, and TCRs are $+100$ to $+200$ ppm, which can be tailored by substrate temperature control during deposition. Tantalum resistors have low noise and voltage coefficients and drift about 1 percent/1,000 hr at 10 watts/in.^{2} of film area.

Tantalum films can also have high-sheet-resistivity values by reactively sputtering with nitrogen or oxygen. Nitrogen and oxygen both increase the value and stability

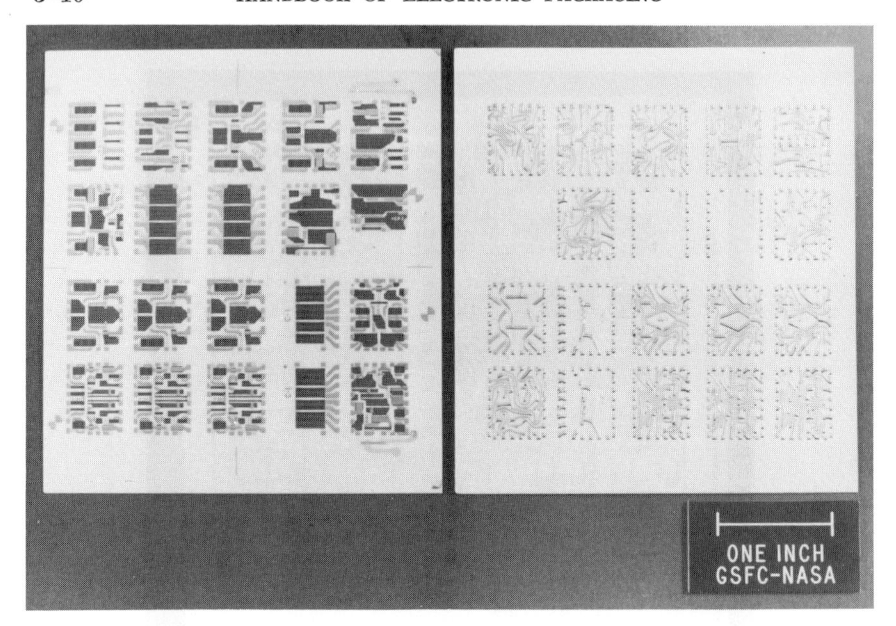

Fig. 8. A multiple-circuit substrate deposited by vacuum evaporation using nickel-chromium and SiO_x for insulation, capacitors, and crossovers. (*Radiation Systems, Inc.*)

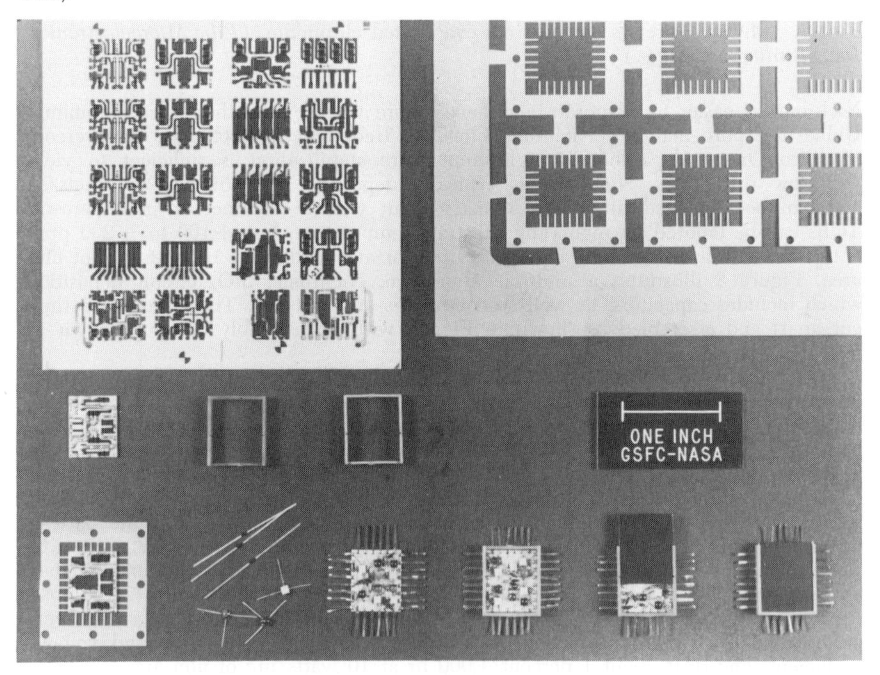

Fig. 9. Steps in the assembly of the substrate shown in Fig. 8. (*Radiation Systems, Inc.*)

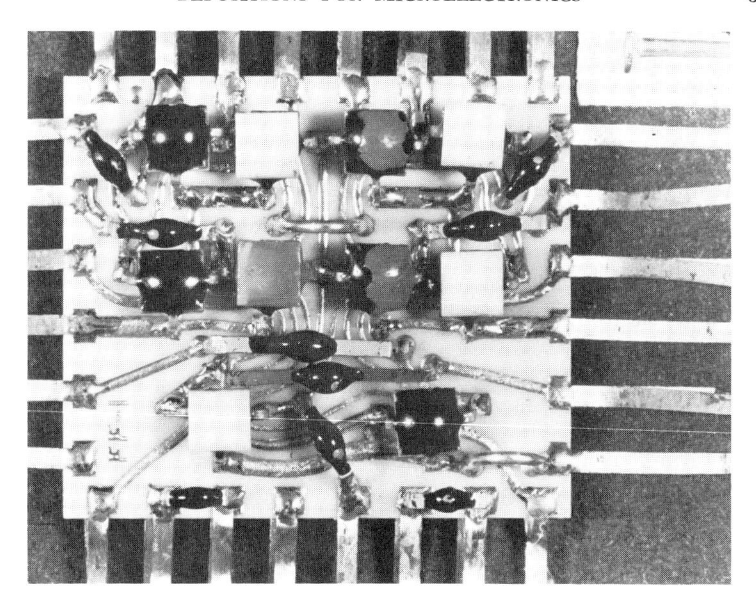

FIG. 10. Close-up of the assembly shown in Fig. 9. (*Radiation Systems, Inc.*)

of the films. Stable films of 2,000 ohms/sq can be produced. The essential problem with these films is the poor TCRs, generally about −1,500 ppm at the higher resistivities so that low-sheet-resistivity sputtered tantalum nitride films are more commonly used.

Figure 11 illustrates a multiple-circuit substrate produced by sputtered tantalum as opposed to the evaporated substrate shown in Fig. 8. The entire substrate has

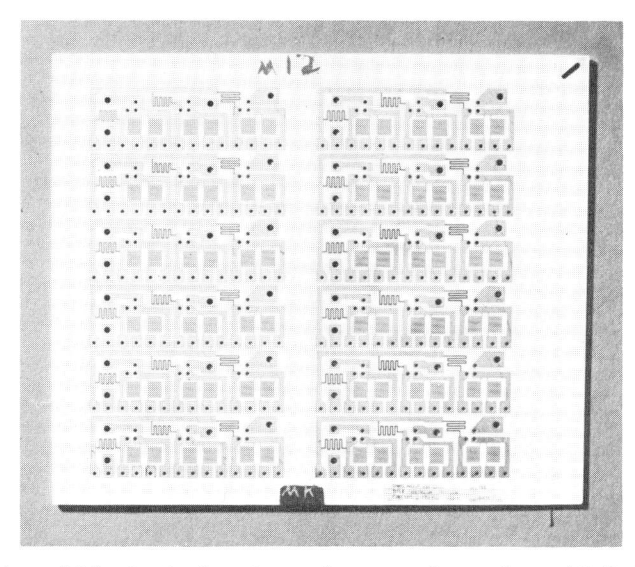

FIG. 11. A multiple-circuit deposition of sputtered tantalum. (*Bell Telephone Laboratories, Inc.*)

a conductor deposition on it, apparently electroless, and then the circuit is delineated. Finally, individual circuits are separated. Then, pins and transistors are added to complete the assembly as shown in Fig. 12.

TITANIUM. This is one of the easier materials to evaporate, with sheet resistivities of 1 to 5,000 ohms/sq and with a TCR from —1,000 to +1,000 ppm/°C. A more practical limit is 10 to 500 ohms/sq, with a 0 TCR at 50 ohms/sq. Drift is about 0.5 percent/1,000 hr. Titanium is not commonly used, but similar to tantalum, it is stabilized by both baking and anodizing.

Insulators. The properties of thin-film insulators are shown in Table 4. These materials are used to provide insulation layers for crossovers and, in most cases, can also be used as the dielectric material for capacitors.

SILICON MONOXIDE. Silicon monoxide is the only material that is normally used as both an insulator and capacitor dielectric in vacuum evaporation, as illustrated in Fig. 8. While many materials have been tried, SiO is easily evaporated with excellent adherence. SiO is somewhat of a misnomer, for the materials are not pure, so that some pure silicon, silicon monoxide, and silicon dioxide are all present within a deposition, the aggregate being between SiO and SiO_2. This accounts for the frequent use of the term SiO_x. Regardless of the actual composition, when evaporated from an indirect source, it is pinhole-free and is an excellent insulator for crossovers and capacitors. Capacitances of 50,000 pf/in.^{2} at 25 volts and 25,000 pf/in.^{2} at 100 volts are typical. Capacitors can be deposited to within ±10 to ±25 percent of value, with the SiO_x ranging from 5,000 to 10,000 Å thick. Capacitors 0.3 by 0.3 in. are a practical upper limit, although larger-sized capacitors have been fabricated. The T.C.C. is typically +200 to +400 ppm/°C, with no voltage characteristic and less than 3 percent dissipation factor. Capacitance is flat to 70 Mc, where series resistance increases the dissipation, although it is not an upper limit.

ANODIZED TANTALUM. Anodized tantalum also is used for capacitors, although it is more restrictive than SiO_x as a capacitor material and is not recommended for

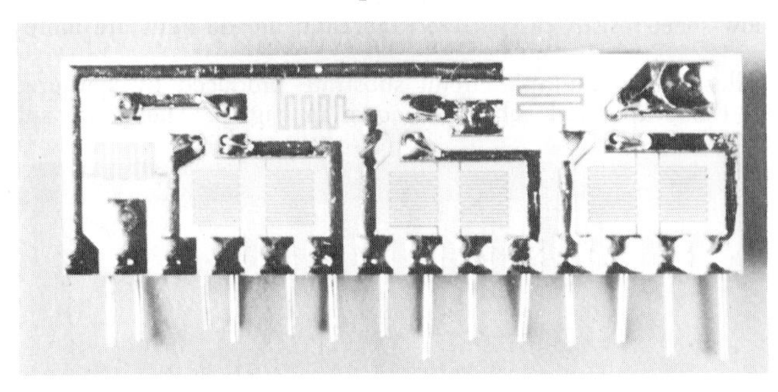

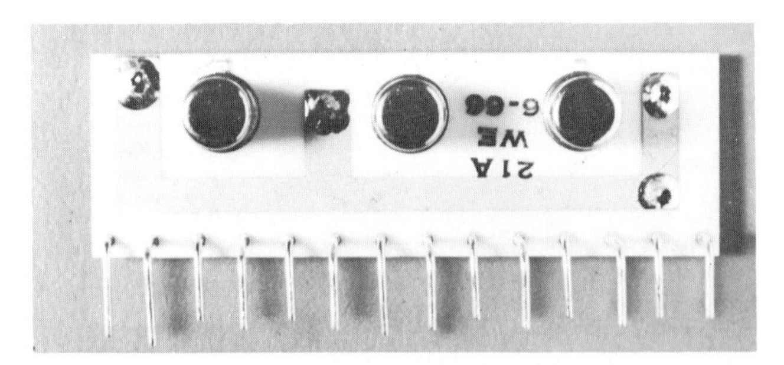

FIG. 12. Front and rear views of an assembled substrate illustrating the use of standard transistor packages with film depositions. The substrate is one of the units separated from the multiple shown in Fig. 11. (*Bell Telephone Laboratories. Inc.*)

crossovers. When anodized, tantalum pentoxide (Ta_2O_5) is formed, which has a dielectric constant of approximately 25. But because it is basically a resistive material in thin films, series resistance limits its usefulness. Second, because it must be anodized, a counterelectrode must be deposited, which requires two separate depositions compared with SiO_x, where the counterelectrode is deposited sequentially in a single pump-down, rather than two.

GLASSES. Some low-melting-point glasses have been evaporated experimentally but were found lacking as insulators, mainly owing to cracking and fissures. Common glasses are rf-sputtered to provide a seal for semiconductors, which are used in flip-chip applications. The entire semiconductor wafer is coated with about 1 mil of glass. Holes are etched through the glass and solder is vacuum-deposited, as previously discussed, to provide a means of connecting from the semiconductor chip to the substrate via a 5-mil-diameter sphere.

ALUMINUM. When anodized, forms Al_2O_3 with a dielectric constant of 10. A heavy deposit of aluminum provides sufficient conductivity so that capacitors made from anodized aluminum do not suffer from the frequency limitations of tantalum and titanium capacitors. Aluminum, however, has a lower capacitance per unit area because of its lower dielectric constant.

TITANIUM DIOXIDE. Titanium dioxide has a dielectric constant of 100 when formed by anodizing titanium films. It is easily evaporated so that a spot of titanium can be applied where capacitors are desired even though it requires an extra deposition, if it is not used as the basic resistor material. Hence, a spot of titanium over aluminum can yield the capacity for higher dielectrics without the consequences of a high series resistance to limit frequency response.

Special Materials. Many of the materials already mentioned have shown unusual properties in special applications. Aluminum, silicon monoxide, tin, lead, and gold have all been used in the vacuum deposition of cryogenic devices. As previously stated, cryogenic electronics does not appear to be too practical at this time, so that most research has stopped. Permalloy, however, is a nickel-iron alloy that has shown interesting properties as a magnetic material.

PERMALLOY. This is an alloy of approximately 80 percent nickel and 20 percent iron. When deposited in the presence of a uniform magnetic field, it exhibits a square hysteresis loop as shown in Fig. 13 so that the deposited films can be used

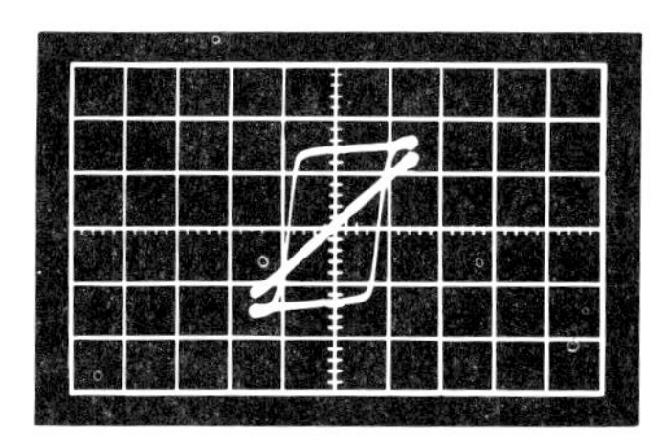

FIG. 13. Typical square and linear hysteresis loops of vacuum-deposited thin-film memory material. (*Burroughs Corp.*)

as a magnetic storage device similar to magnetic ferrite cores. Magnetic arrays, such as shown in Fig. 14, have been used for small high-speed memories of up to 512 words and for full-scale memories of 4,096 words and larger as shown in Fig. 15. Magnetic films have been sought for economic reasons since numerous substrates can be deposited in one operation. The substrates, after overall coating, are etched into arrays and tested. Since yields are fairly good, the cost of the magnetic devices is considerably cheaper than the more time-consuming method of fabricating and testing individual magnetic-memory cores. Magnetic thin films are wired by sandwiching them between printed circuits for a final electronic test and then permanently wiring them into a memory. Again, this is a relatively inexpensive operation. Cores must be wired into arrays. Even though this has been semiautomated, rework is frequently required to replace a defective core so that magnetic-film arrays appear to be cheaper to manufacture and assemble into complete memories.

Vacuum-deposited magnetic films have their drawbacks in that they require higher drive currents and have lower outputs, so that the electronics for them is more

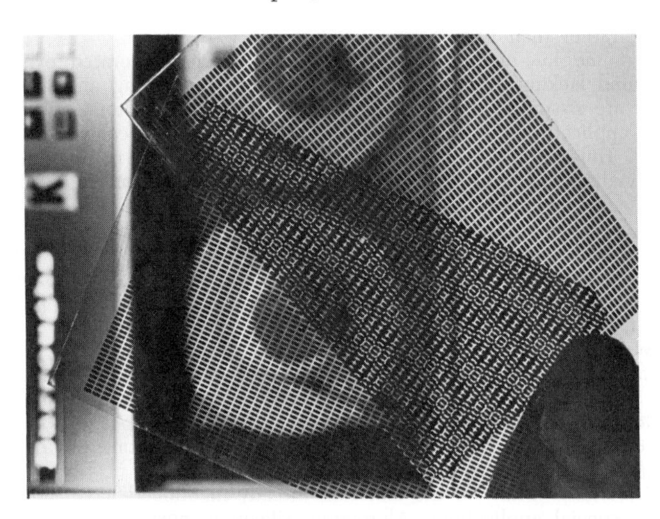

FIG. 14. Thin-film magnetic-memory array of 3,072 bits etched from an overall deposition of magnetic material. (*Burroughs Corp.*)

FIG. 15. Thin-film magnetic memory in the final stages of fabrication. Memory is composed of 40 of the arrays shown in Fig. 14. (*Burroughs Corp.*)

expensive. Hence, overall cost between any two systems appears to be nearly equal, neither system showing an economic advantage at this time. Since the two methods are being compared, it might be mentioned that the controversy over whether cores or films would produce the fastest memory system raged for many years. It is now obvious that both are limited by the time delay in the read-write loop so that memory-array packaging is the critical factor in determining the speed of an array, whether it be core or film.

ACTIVE DEVICES AND INTEGRATED CIRCUITS. While permalloy has been primarily used as a storage element, it has also been used to fabricate active devices in the sense that they possess gain. Permalloy has been used to fabricate flip-flops as logic devices, parametric amplifiers, sensitive magnetometers, and delay lines. Vacuum-deposited active devices have long been sought in research, and with some success, but they have not been as stable as or as repeatable as semiconductor active devices, so that few are available and most research has stopped. Certainly, no discussion of vacuum deposition would be complete without mention that these techniques are used to fabricate semiconductors and integrated circuits. Aluminum, gold, and other metallizations are commonly used as conductors in the fabrication of wafers. The reader is referred to any one of several good books which are available and will give more detailed information on semiconductor fabrication.

System Wiring. Thin films were initially developed to produce resistor, conductor, and capacitor networks for hybrid circuits to which active devices were attached. Later, thin-film resistors and conductors were deposited on silicon wafers which contained active devices which formed a sort of reverse hybrid circuit. Of course, metallization of semiconductors had long been known and used so that extension of the basic vacuum-deposition and lithographic techniques to form component (both active and passive) intraconnections was to be expected. Now, thin films are being used in techniques called *large-scale integration* (LSI).

Two forms of thin-film LSI have been developed. The first is the use of thin films to form the interconnection wiring between "good" integrated circuits on the SiO_2 layer of a single silicon wafer. The second approach is the deposition of conductors and crossovers with deposited "bumps" about 40,000 Å thick. Integrated circuits are then ultrasonically bonded to the bump to join the circuits into a small subsystem. Figures 16, 17, and 18 illustrate the vacuum-deposited bumps and attached chips. Figure 19 illustrates the application and potential for LSI where an entire memory array is wired on a single glass substrate. The memory is a chemically deposited magnetic array which will be discussed later.

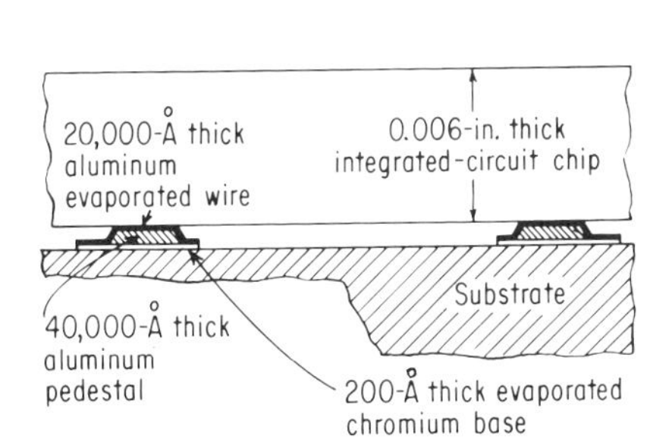

FIG. 16. Cross-sectional diagram of evaporated conductors and bumps. (*Sperry Rand Corp., Univac Division.*)

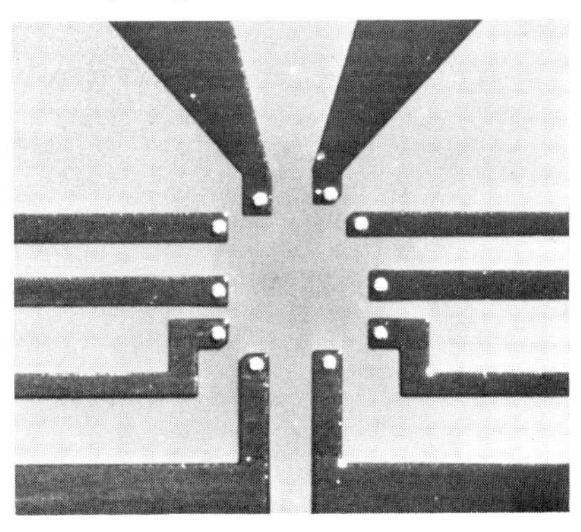

FIG. 17. Evaporated conductors and bumps as deposited on a glass substrate. (*Sperry Rand Corp., Univac Division.*)

Screened and Fired Depositions. Compared with thin films, screened and fired (SAF) depositions for hybrid circuits are the most economical method of high-volume microelectronic production. Screened and fired circuit technology extends from simple resistor-conductor networks to resistor, conductor, and capacitor networks with crossovers. The various "inks" or "pastes," as they are called, can be screened and fired on substrates of alumina, beryllia, and a variety of titanates. Screened deposits can be applied to one or both sides of a substrate, depending on the substrate and the application. Technically, the inks are *frits* or *cermets,* a combination of materials, mainly noble metals and metal oxides mixed with powdered glass, organic binders, and organic solvents to form a thixotropic paste. Upon firing, the residual solvents and binders vaporize, or burn off, at about 400°C. Continued firing at higher temperatures produces melting of the glass, which somewhat "glazes" the resistor deposits and accounts for the occasionally used phrase *glazed resistors.* The amount of glaze depends on the metal-to-glass ratio. For most resitive inks, a chemical reaction occurs during the high-temperature portion of the firing cycle, resulting in oxidation of the metals as well as glazing.

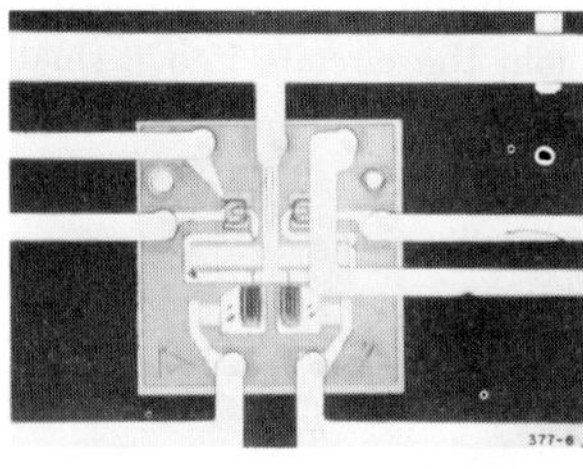

FIG. 18. Integrated-circuit chips ultrasonically bonded to evaporated conductors and bumps as viewed through the substrate. (*Sperry Rand Corp., Univac Division.*)

Conductor Materials. There are a variety of conductor materials which serve different purposes and functions. All the primary information available is presented in Table 4. Individual manufacturers should be consulted for specific details on each paste, its firing curve, and its properties. While conductor inks do not have the sensitivity of resistor inks, their solderability and adhesion are affected by firing temperature. Properties are shown in Table 5.

GOLD. Numerous conductive gold inks or pastes which fire between 800 and 1000°C are available. Some gold-conductor frits provide the capability of direct eutectic bonding, thermocompression, and ultrasonic bonding. Gold conductors have the best conduction properties available with one-fourth the resistance of any other material, except silver. Most gold conductors can be soldered only with indium-base alloy solders, although one paste is available which can be soldered with eutectic tin-lead solder. Gold is the second most expensive conductor material available. It appears to be compatible with the resistive inks, although definite data are not available. There is a marked difference in characteristics between palladium-silver resistors fabricated with gold terminations and those produced with other conductor materials as terminations. Figure 20 illustrates a typical application of gold conductors, including insulated crossovers. Here, four I/Cs are eutectically bonded to the pads and wired to the conductors by thermocompression bonds to form a four-stage counter. Hence, SAF techniques readily lend themselves more to customized LSI than vacuum deposition; i.e., screening techniques can be used to produce customized I/C arrays of any size or number of circuits and without the high cost of mask fabrication.

PLATINUM-GOLD. Several platinum-gold formulations are available, and they are one of the more popular conductor materials. Platinum-gold is used because of its high adhesion and good solderability even though it is the most expensive conductor material available. Platinum-gold has fairly good adhesion, scratch resistance, and pull and peel strength. It fires in the range of 850 to 1000°C, with a resultant resistance of 0.05 to 0.1 ohms/sq. In general, platinum-gold has very good soldering properties with the capability of resoldering several times. Platinum-gold conductors are compatible with all resistive inks known at present and can be cofired wth the platinum resistive inks. Direct eutectic bonding of silicon dice appears to be marginal. Gold-silicon solder preforms are being used more often than not for the

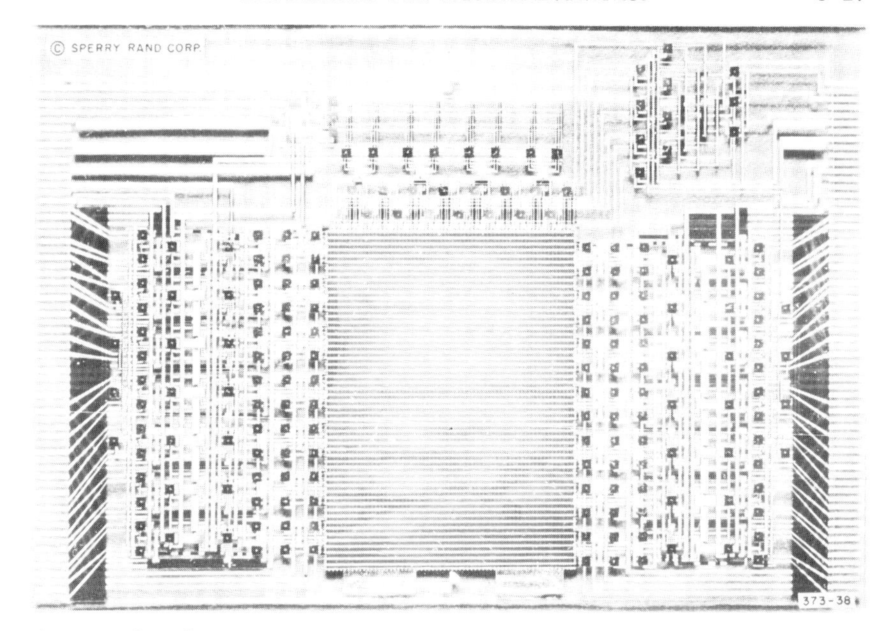

FIG. 19. Complete magnetic memory using a chemically deposited magnetic array, evaporated conductors and insulating SiO_x. All circuits are integrated chips which are face-bonded to the evaporated conductors. (*Sperry Rand Corp., Univac Division.*)

TABLE 4. Properties of Thin-film Insulators [29]

Dielectric material	Cap./ in.², μf	Dissipation factor at 1 kc, %	Operating potential, volts	Breakdown potential, volts	Temp. coef., ppm/°C	Dielectric constant	Counter electrode	Freq. limit, kc	Cap. range, μf	Amb. temp. limit, °C	Fabrication process
Silicon monoxide	0.09	0.1	35	50	110	6	Alum. or copper	†	10^{-6}–10^{-1}	125	Vac. dep.
Silicon dioxide	0.03	0.1	35	50	*	4	Alum. or copper	†	10^{-6}–10^{-1}	‡	Vac. dep.
Tantalum pentoxide	1.00	1.0	20	25	300	25	Tant. and alum.	10	10^{-6}–1	‡	Wet chem.
Titanium oxide	1.00	1.0	35	45	*	100	Tit. and alum.	1,000	10^{-6}–1	125	Wet chem.
Aluminum oxide . . .	0.55	1.5	10	15	*	10	Alum.	†	10^{-6}–10^{-1}	‡	Wet chem.
Alumina silicate . .	0.064	0.3	100	150	300	6–7	Alum.	500	10^{-6}–10^{-1}	150	Vapor plat.

* Accurate figures have not been established.
† Actual limits have not been established. Theoretically this material can be used in a circuit that operates at thousands of megacycles.
‡ Temperature limit has not been established, but it is probably not much higher than 125°C.

TABLE 5. Properties of Thick-film Conductors

Material	Relative cost	Adhesion	Resistivity, ohms/sq	Resistor compatibility	Interconnectability: Electron beam bonding	Interconnectability: Thermocompression	Interconnectability: Ultrasonic bonding	Solderability	Firing temp.
Gold (Au)	45	1	0.005–0.01	Yes, with some pastes, but blistering	Yes	Yes Gold wire	Yes Gold wire Alum. wire	With indium solder only	875–1,000
Pt-Au	60	2	0.05–0.1	Yes—5–10 mils overlap	Marginal solder preform recommended	Yes Gold wire	Yes Gold wire	Yes	900–975
Pd-Au ...	35	3	0.05–0.1	Yes—5–10 mils overlap	Not much No	Yes Gold wire	Yes Gold wire	Yes	900–975
Pd-Ag ...	13	4	0.01	Yes, with chemical reaction	No	Yes Alum. wire	Yes Alum. wire	Yes	680–780
Silver (Ag)	1.5	5	0.005–0.01	Not recommended	No	Yes Alum. wire	Yes Alum. wire	Yes	550–800

bonding. Thermocompression and ultrasonic bonding work well with either gold or aluminum wires.

PALLADIUM-GOLD. This is one of the least used conductor materials even though it has about the same properties as platinum-gold at about one-half the cost. It is compatible with the resistor pastes, can be thermocompression-bonded, and can be ultrasonic-bonded to gold wire. It fires at 850 to 1000°C.

PALLADIUM-SILVER. Palladium-silver is the most common silver-bearing conductor ink in use with the palladium-silver resistor inks. It is about one-fourth the cost of platinum-gold and has good adhesion and fair solderability. It fires at the same temperature as most of the palladium-silver resistor inks (760°C), so that cofiring is possible; i.e., resistors and conductors may be screened one after the other and fired together. During the firing process, a chemical reaction and/or diffusion occurs between the resistor and conductor inks which lowers the resistor value. This is particularly noticeable on resistors which are less than 80 mils long. Aluminum wires can be thermocompression- and ultrasonically bonded to palladium-silver. While palladium-silver can be soldered, the results are not as good as platinum-gold

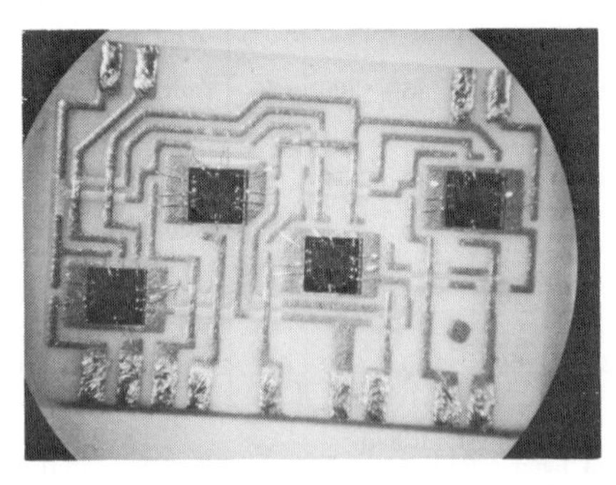

FIG. 20. Four-stage counter intraconnected via screened and fired gold conductors, including crossovers.

and the solder should contain at least 2 percent silver to prevent leaching silver out of the conductor during tinning.

SILVER. Silver conductors are the cheapest material available. They possess the best adhesion and highest conductivity (equal to or better than gold) and are readily soldered. But silver has its disadvantages. When this material is fired with resistors, the area adjacent to the termination is almost entirely "robbed" of metal material, leaving a very high resistance strip across the resistor, which very rapidly fails under load. This phenomenon has been termed "cleavage." Aside from this problem, silver conductors have been plagued by the common notion that they are unreliable owing to silver migration which will eventually cause short circuits between conductors. Silver migration occurs only in the presence of high humidity and where there is a voltage potential between the two conductors. Under most normal conditions for transistorized electronic equipment, migration does not occur. A simple glass overcoat on silver conductors drastically reduces the possibility of silver migration. Any coating which will prevent a continuous water film will suffice. Finally, the low potentials used in modern transistor circuitry are not of the magnitude to cause migration at an advanced rate.

MOLY-MANGANESE. This is a common term for molybdenum-manganese, a combination of two metallic powders mixed with organic binders. It is fired in a reducing or neutral atmosphere at about 1500°C. Above 400°C it will completely oxidize in a matter of minutes. Unlike the other conductor materials, moly-manganese is used primarily as an underlayer metallization in the fabrication of packages with "hermetic" seals. It is readily gold-plated so that it is used frequently as a metallization for ceramic hermetic packages rather than as a microelectronic conductor material, such as the frits previously discussed.

EPOXIES AND ADHESIVES. Numerous conductive epoxies and conductive adhesives are available which are used for a variety of purposes in microelectronic applications. Unlike the glass-base compositions, these materials are a combination of metal particles and epoxy resins which are "cured" rather than fired. Curing occurs in air at a general temperature of 150°C and seldom exceeds 200°C. Elevated temperatures are used only to accelerate the cure, although most of these materials would cure at room temperatures given sufficient time. Other epoxy-based materials are specifically formulated to cure at room temperature. A few epoxy pastes have been developed which will withstand the temperatures of soldering, and some are themselves solderable.

Copper, silver, and gold are the three metals normally mixed into epoxy resins to obtain conductive epoxies. Bonding strength, adhesion, cure temperature, and other properties are a function of the metal content and the epoxy system used, so that the combinations are almost limitless. Therefore, specific information from individual manufacturers is recommended, although most conductive epoxies have the general properties shown in Table 6.

TABLE 6. General Properties of Epoxy Conductors

Resistivity	0.1–0.001 ohm/sq
Heat resistance	150°C continuous duty min. 200°C continuous duty max. Soldering temperatures—short time
Thermal shock	−65 to 150°C min. −65 to 200°C max.
Tensile strength	500–5,000 psi
Compressive strength	10,000–30,000 psi

The conductive epoxies have been found useful in making low-temperature connections and replacing fired conductors in crossovers and capacitor counterelectrodes.

The materials can be applied to most plastics, ceramics, glasses, and metals by brush, hypodermic, dipping, and screening.

Resistor Materials. Similar to the conductor materials, resistor pastes consist of conductor or semiconductor particles mixed with ground glass, organic binders, and solvents which, when combined, form a thixotropic paste. The paste can be screened and fired to produce resistors of 0.1 ohm to 10 megohms although 1 ohm to 1 megohm is a more practical limit.

There are two basic resistor materials which are commercially available and numerous proprietary resistor compositions that are used in commercial products. The products are available to the public, but the inks are not. Some commercial products use resinates and carbides. Indium oxide and other glaze materials have been reported in the literature but are not at present being used in products, nor are the inks sold. Thus, discussion will be confined to commercial resistor pastes, of which there are two basic varieties with the general properties shown in Table 7.

PALLADIUM-SILVER. Unlike the conductor materials, palladium-silver resistor compositions are extremely sensitive to the firing profile. Three varieties are available. One fires at 690°C and the other two at 760°C. All require between 10 and 15 min at firing temperature to obtain complete firing. In addition, it is essential to maintain the time-temperature product under the curve constant so that the chemical reaction which occurs between palladium-silver and oxygen is maintained or reproduced. The chemical reaction is too complex to present here, especially since unknown dopants are present in the pastes, but the basic reaction is available in the literature for those who are interested. Not only are some of these compositions sensitive to firing, but also the value and tolerance of the resistor are dependent on the conductor material. As already indicated, there is a chemical interaction between Pd-Ag resistors and Pd-Ag conductors. There also appears to be a chemical interaction between platinum-gold and gold conductors and palladium-silver resistors. Gold conductors appear to have the least effect on palladium-silver resistors, with better temperature coefficients, slightly higher values, and a lower scatter of values.

From a packaging standpoint, most of these resistors must be overcoated with a low-temperature glass or other material which will seal the resistors from the adverse effects of hydrogen. Otherwise, when encapsulated with epoxy, the hydrogen released by the encapsulant reduces the "oxidized" resistor and lowers the value. This is also an important consideration when these resistors are placed on a heat column for die attachment or wire bonding. At bonding temperatures, the hydrogen

TABLE 7. Properties of Thick-film Resistors

System	Firing temperature, °C	Sheet resistivities, ohms/sq	TCR, ppm/°C, −55 to +125°C
Pd-Ag	760	0.2–17 K	Varies from +700 at 0.2 ohm/sq to +400 at 10 K/sq
Pd-Ag	690	1–20 K	+200 to −300 at 1 ohm/sq, −75 to −175 at 100 ohms/sq, −200 to −300 at 20 K/sq
Pd-Ag	760	1–100 K	+300 to −300 over complete resistance range
Pd-Ag	760	10–100 K	Most values from +150 to −150, complete data not available
Pt	980	300–1,000 K	Less than +150 from −55 to +175°C

Typical noise figure: −20 db/volt, 50 ohms/sq
+40 db/volt, 50 ohms/sq

in "forming gas" reduces the oxidized products in the resistors and lowers the value.

PLATINUM RESISTORS. These are relatively new but appear promising. The basic material is platinum, with other base metals mixed to form a thixotropic paste which fires at 980°C for 15 min. The basic resistance-forming mechanism appears to be similar to that of conductors: simple metallic contact with little or no chemical reaction occurring during firing so that the profile and maximum temperature are not critical. The resistors can either be postfired after the conductors or cofired with platinum-gold or gold conductors. Subsequent firing at lower temperatures does not appear to affect the resistors so that capacitor counterelectrodes and crossovers may be fired after resistor firing. Characteristics of the material are shown in Table 7.

Insulators and Dielectrics. Similar to the conductive and resistive pastes, insulator and dielectric pastes are available which can be screened and fired. Subsequent application of lower temperature firing conductive pastes produces crossovers or capacitors. The insulator pastes are primarily ground glass and the usual organic solvents and binders. The dielectric inks are typical of the conductor and resistor inks except that dielectrics such as titanium dioxide, strontium titanate, and barium titanate are combined with the glass rather than conductive or resistive materials. The insulator and dielectric materials fire at intermediate temperatures so that lower-temperature conductors may be screened on top of them and subsequently fired without the layers melting and diffusing. General properties are shown in Table 8.

Epoxies, varnishes, and silicones are also used as insulator coatings. Here the variety is too voluminous even to generalize on properties. However, the RTV silicone rubbers have found use and acceptance as an overall insulating material for microelectronics. Chapter 7 contains more detail on some of these materials and should be consulted.

Active Devices. Some work has been done recently on the fabrication of active devices by SAF techniques. Some success has been experienced in both diodes and insulated gate field effect transistors. While the initial results indicate that there are problems to be solved, these developments may be the breakthrough to truly low-cost circuitry for a variety of applications for industrial and commercial equipment.

Thick-film Circuits. The ease with which circuits can be fabricated and the variety in which they are manufactured are illustrated in Figs. 21 and 22. Figure 21 shows the individual steps in producing a circuit. Conductors are screened and fired, followed by resistors. The circuit is solder-coated, and active devices and leads are attached. Figure 22 illustrates a variety of circuits.

Chemical Depositions. Chemical depositions offer the largest variety of deposition techniques available, but those which will be discussed the least in this section for their techniques are either relatively familiar to most readers or nearly unheard of.

TABLE 8. Properties of Thick-film Insulators and Dielectrics

Dielectric constant K	Firing temp., °C	Application	Dielectric strength, volts/mil
6–7	550–575	Insulation	500
13	750–850	Crossovers Insulation Capacitors	1,000
50 ± 10%	800–950	Capacitors	NA
100 ± 20%	800–950	Capacitors	NA
150 ± 30%	800–950	Capacitors	NA
300–450	NA	Capacitors	1,000

NA = not available.

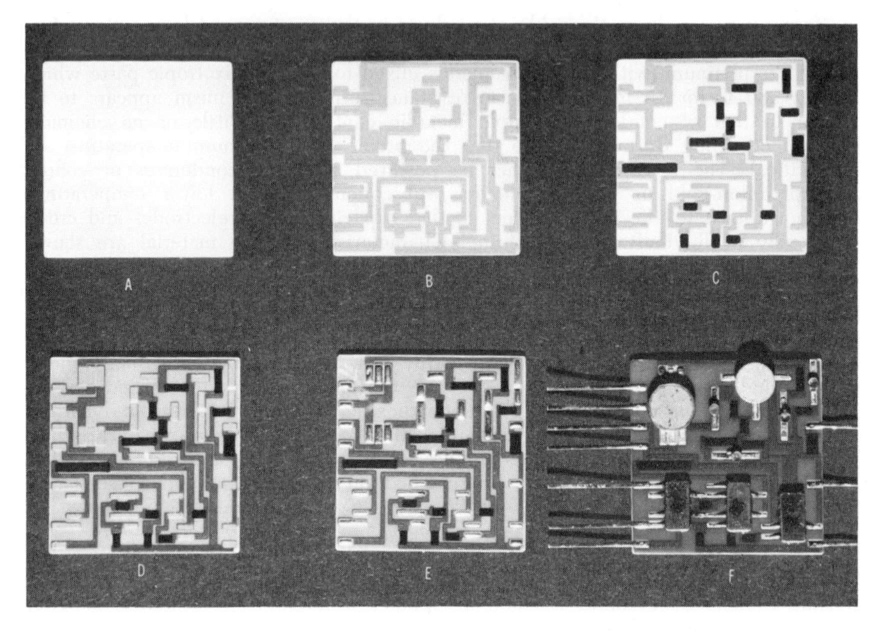

Fig. 21. Steps in the thick-film manufacturing process. (A) Blank substrate; (B) conductors screened and fired; (C) resistors added; (D) glass overcoat added over conductors and resistors; (E) uncoated conductors are soldered; (F) active devices are added.

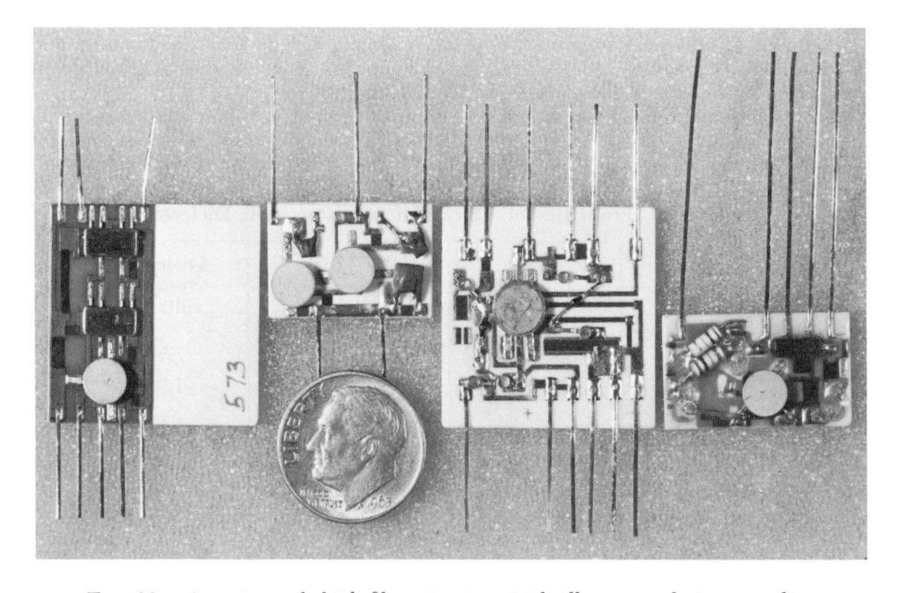

Fig. 22. A variety of thick-film circuits which illustrates their versatility.

As with most deposition techniques, the methods of application are in the literature, but the deposition secrets are seldom if ever discussed or included. Hence, discussion will concentrate on the types and properties of the depositions rather than the method of application.

There are four basic types of depositions: vapor plating, electro- and electroless plating, chemical-salts reduction, and chemical reactions. Of these, the vapor plating technology may be relatively unknown to the average reader, whereas plating methods are almost universally known. The chemical-salts deposition techniques are not well known, but most readers are familiar with many of the chemical reactions used in the fabrication of hybrid substrates.

Vapor Plating. As the name implies, vapor plating results in the plating of materials from the vapor state onto a substrate. In this deposition method, volatile metal halides decompose as they pass a heated substrate and a film deposit is produced. Such reactions take place at room temperature to 1200°C. Depositions are usually made at room temperature to 300°C. In the second form of vapor plating, a liquid halide solution is sprayed onto a heated substrate, where the decomposition and plating occur. Halides for doping, such as indium or antimony, are mixed into the solutions to tailor the physical properties to those which are desired. Films up to 20 μ thick can be deposited by using these methods.

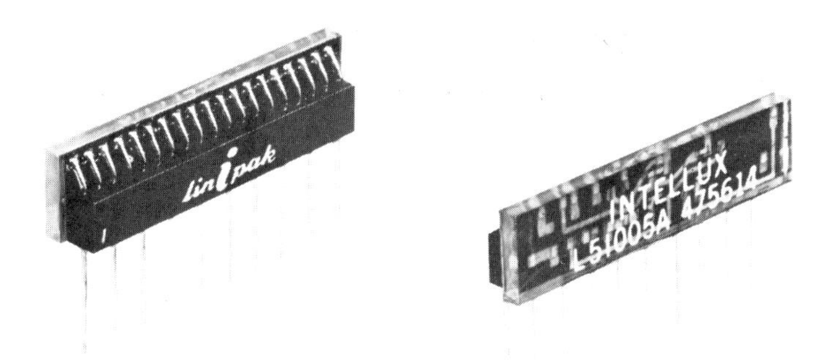

FIG. 23. Deposited tin oxide resistors as seen through the glass substrate. Note the vertical-mounting method for next assembly integration. (*Intellux, Inc.*)

EPITAXIAL SILICON. Epitaxial silicon is grown on the surface of a single-crystal semiconductor wafer by using vapor-phase plating. While single-crystal silicon is prepared from the liquid phase, epitaxial silicon is deposited at temperatures considerably below the melting point. The epitaxial growth or layers can be of either the same or the opposite conductivity type of silicon. For more complete information, books are available which cover the entire semiconductor-fabrication process.

TIN OXIDE. Tin oxide is deposited by the spray method of vapor plating. Tin oxide, doped with varying amounts of indium or antimony, produces highly stable resistive films with sheet resistivities ranging from 50 to 4,000 ohms/sq. Resistance values from 5 ohms to 3 megohms can be achieved by etching techniques. The TCRs range from near 0 to —1,500 ppm/°C over the range of sheet resistivities. Vapor-deposited tin oxide has excellent adhesion to both glass and alumina substrates and is easily plated with conductor materials to form resistor-conductor networks. The circuits in Figs. 23 and 24 clearly show the deposition areas through the glass substrate. These figures also clearly illustrate two distinctive methods of assembling the substrates for integration into the next level of assembly.

INSULATORS. Insulators can also be deposited by vapor plating, just as well as resistor and semiconductor materials. Both silicon monoxide and silicon dioxide can be deposited on metallic and nonmetallic surfaces to provide insulation.

Electroplating. This is a method of coating a conductive object with another metal 1 millionth of an inch thick to several mils, usually with a very tight grain structure and good adherence. In electronic packaging, electrically deposited platings are used extensively in the production of connectors, transistor headers, platings on leads, substrates, and printed circuits. Most metals can be electrically plated, such as copper, tin, nickel, silver, gold, solder, chromium, palladium, and rhodium.

Electroless Plating. As the name implies, this is the deposition of platings without the use of an applied potential as required in electroplating. Hence, some platings can be applied to nonconductive objects such as glass, epoxies, and ceramics when properly prepared by cleaning and sensitizing solutions such as tin chloride and palladium chloride. There are two basic types of electroless depositions. The

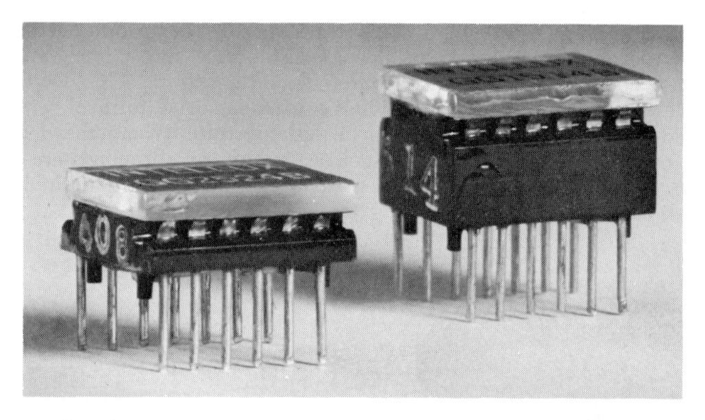

FIG. 24. Another form for next level integration. Note the transition from the large pins to the ribbons which attach to the substrate. (*Intellux, Inc.*)

first is the true electroless deposition, where plating thicknesses of one to five hundred millionths can be made, primarily depending on the deposition time. The second type is immersion plating, which operates on the basis of ion exchange. In this instance, plating stops when ions of the plating metal and the metal which is to be plated can no longer exchange. Thus, plating thickness is limited to between fifty and one hundred millionths. There is one more significant difference between the two. Electroless platings can be applied to conductive and nonconductive materials alike, although they will plate selectively on electrically platable materials when the object has not been sensitized. Immersion platings will plate only on metals.

CONDUCTORS. Obviously, all the materials are metals, and most are conductors. Copper, nickel, gold, tin, and other metals can be plated. While the solutions can be mixed in any well-equipped chemistry laboratory, prepared solutions can be purchased from several sources, essentially ready to use. For the average user, this is probably the best way.

RESISTORS. Resistors can be produced by the electroless deposition of a nickel alloy containing 10 to 12 percent phosphorus. They can be deposited on G-10 epoxy such as shown in Fig. 25 or on glass and ceramic substrates. Deposits to 5 percent tolerance are possible with sheet resistivities from 10 to 50 ohms/sq. The TCR is +80 ppm/°C, with drifts of 2 percent for 5,000 hr at 2 watts/in.^{2} loading on G-10 glass epoxy. Higher loadings and lower drifts have been observed when coated on ceramic, as shown in Table 9. After the initial deposition, overall plating and differential etching provide resistor-conductor networks. This process provides the additional advantage of coating both sides and plating through holes if they are in the substrate.

MAGNETIC MATERIALS. Magnetic alloys can be deposited by both electro- and electroless deposition with a variety of properties. Magnetic alloys are plated on "wire" to form magnetic rods which are fashioned into memories. Alloys have been deposited to form logic circuits, the *waffle-iron memory*, and the memory shown

TABLE 9. Properties of Chemically Deposited Resistors [22]

Material	Sheet resistivity, ohms/sq	TCR, ppm/°C	Drift, %/1,000 hr at 10 watts/in.² at 100°C
Titanium	50–2,000	+120 at 50 ohms/sq 0 at 1,300 ohms/sq −100 at 2,000 ohms/sq	Less than 1
Nickel-phosphorus ..	1–500	Depends on substrate G-10, +80 Alumina, +160	Less than 1 at 2 watts/in.² Less than 0.5
Tin-oxide	10–5,000 practical	±250 maximum	0.5

in Fig. 19. In general, the plated memory arrays have higher outputs and higher bit densities than their vacuum-deposited counterparts. Plated memories usually require lower drive currents and produce higher output signals. Production yields are somewhat lower, but equipment is considerably cheaper, so that the cost per bit is nearly the same, with the cost of the electronics being considerably lower because of the higher output signal, which results in lower overall system cost.

Chemical Salts. The reduction of molten chemical salts has been used to deposit films of metallic titanium on alumina. Substrates surrounded by titanium sheets

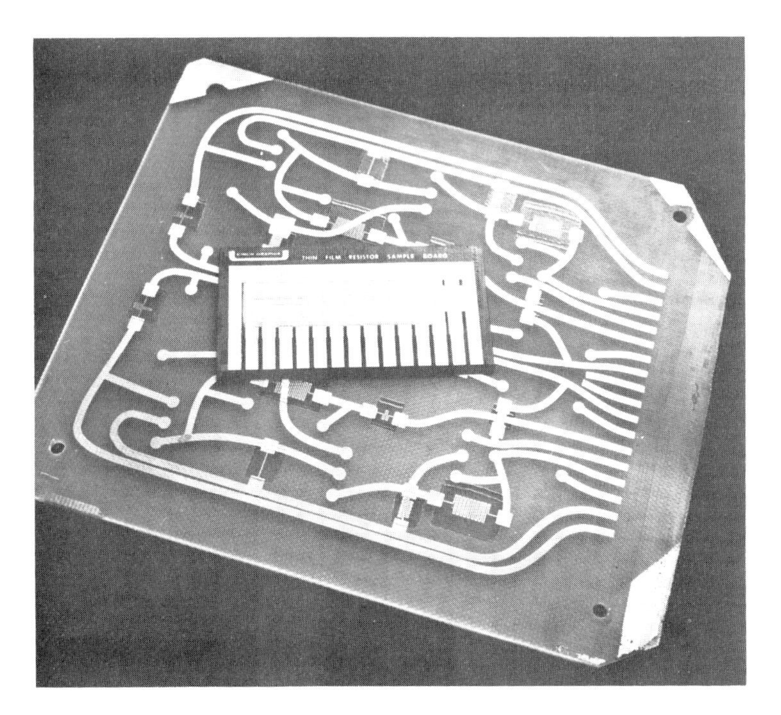

Fig. 25. Nickel-phosphorus resistors deposited on G-10 glass epoxy as part of a regular printed-circuit board. (*Cinch-Graphic.*)

and chemical salts are heated until the salts melt. Titanium goes into solution and comes to equilibrium. Upon cooling, titanium deposits back on the original titanium sheet and the substrates. This is shown diagrammatically in Fig. 26. The deposited substrates are then electrically plated with a conductor such as copper and resistor-conductor networks delineated by the photolithographic methods previously discussed. Resistor stabilization and trimming are by anodization or baking. Titanium exhibits a TCR of +120 ppm at 50 ohms/sq to —100 ppm at 2,000 ohms/sq, with a 0 TCR at 1,300 ohms/sq. Drift rates are less than 1 percent/1,000 hr at a loading of 10 watts/in.^{2} of film area and approximately 0.1 percent/1,000 hr at 5 watts stress level. Sheet resistivities of 50 to 2,000 ohms/sq can be obtained which are stable. Since Ti is adjusted in value by anodizing, titanium dioxide (TiO_2) is formed which has a dielectric constant of 100. Capacitance ratings of 60,000 pf/in. per square inch at 50 volts dc with a 1 percent dissipation factor are typical. The TCC is approximately +800 ppm.

Chemical Reactions. Chemical reactions, as the heading implies, are not depositions in the sense that something is always added to or deposited on the substrate. Nonetheless, reactions are important techniques that should be considered in microelectronics, for they change the characteristics or compositions of many of the

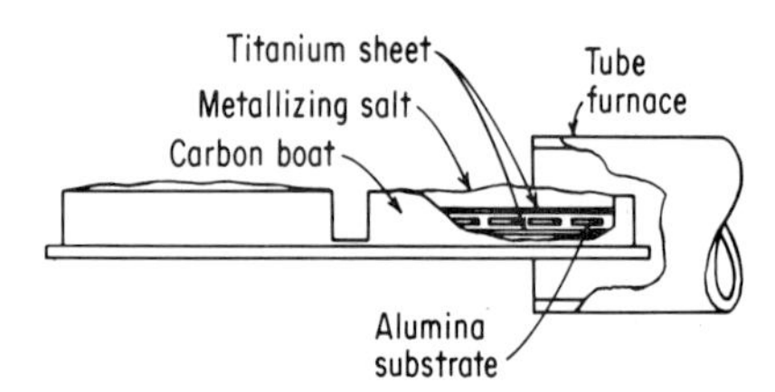

FIG. 26. Schematic diagram of the titanium-deposition process.[15]

deposits already there. For example, anodizing and baking are used to stabilize resistors and provide capacitor dielectrics. In such a case, metal forming the resistor element reacts with oxygen to form an oxide that stabilizes the resistor. In other cases, as has been discussed, the anodized films of Ta, Al, and Ti are used to form capacitor dielectrics. Etching of resistor and conductor metals are reactions that remove unwanted material from the substrate and can be accomplished differentially in most cases. As just discussed, decomposition is the chemical reaction involved in vapor plating. Finally, polymerization is involved in the conversion of photosensitive lacquers which serve as etching resists. Some photosensitive lacquers have proved to be effective insulators and have been used to produce crossovers in vacuum-deposited thin films and electroless plating.

System Wiring. As has been done with vacuum deposition, system wiring for LSI, including insulated crossovers, can be produced entirely by chemical depositions using electro-, electroless, and vapor-plating techniques. While this combination has not been published, it is an obvious alternative approach to the problem of providing LSI and system wiring without the use of expensive vacuum systems and with a larger capacity than a vacuum system can provide.

Technique Combinations. The possibilities in combining various deposition methods to achieve results that could not be achieved by a single deposition technique are limited only by the imagination of the user and the practical consideration that the combinations employed produce a useful result. Many examples could be cited, but they are too numerous to give here. A few will illustrate the results that can be achieved.

Certainly, one of the prime examples is the use of vapor-plating tin oxide and electrically plating the conductors. The circuits in Figs. 6 and 7 illustrate the use of vacuum deposition combined with electroless and electroplating. The beam lead transistor in Fig. 27 illustrates the combination of many disciplines: chemical reactions, diffusions, electroplating, and vacuum deposition.

SUBSTRATES

Although many materials have been used for substrates, ceramics and glasses are by far the most widely employed. Other substrate materials which have been used are dielectric-coated metals, glass-ceramics, plastics, and single-crystal materials such as quartz, sapphire, and aluminum oxide–sapphire. Their applications are specialized and relatively minor. Glasses and ceramics predominate because they alone possess the best combination of important properties: temperature, chemical and vacuum stability, high resistivity, good thermal conductivity, strength, flatness, smoothness, and, most important, relatively low cost. The choice between glass and ceramic depends on the best balance of such properties for each application.

Glass and Ceramic Surface Profiles. A comparison of some surface profiles is shown in Fig. 28. Here it is apparent that the glasses exhibit surfaces which are far smoother and flatter than the as-fired ceramics. These irregularities are of the

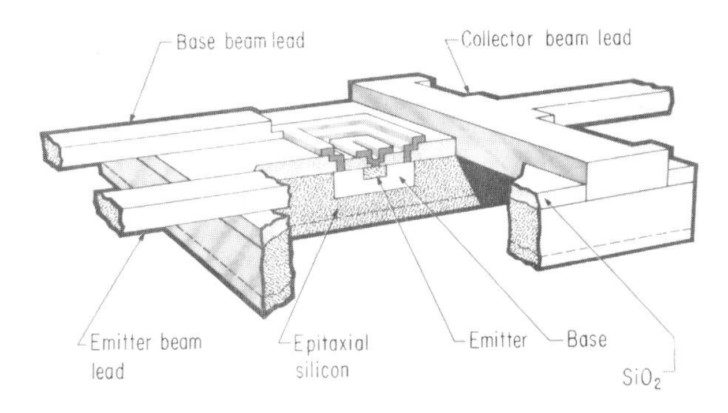

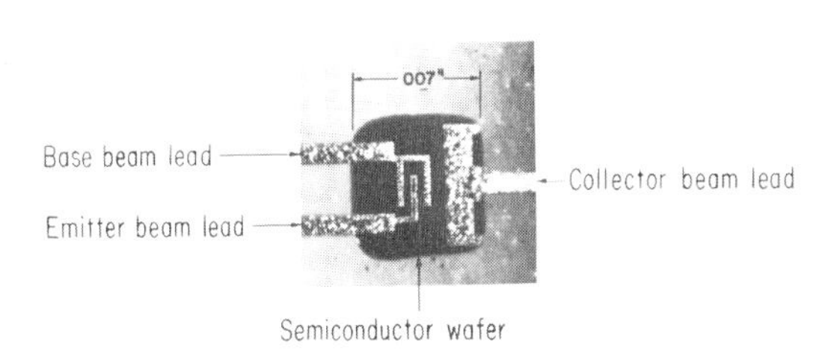

FIG. 27. Pictorial diagram and photograph of a beam lead transistor. (*Bell Telephone Laboratories, Inc.*)

order of 200 Å for some ceramics. The actual degree of smoothness required depends on the application. The only films which are nearly insensitive to surface smoothness are the screened and fired noble-metal–glass cermets and chemically deposited nickel-phosphorus. On the other hand, vapor-deposited high-value thin-film capacitors are extremely sensitive to the degree of substrate smoothness. This arises because of the large stresses at which these capacitors operate: approximately 10^6 volts/cm.

Obviously, the substrate must be able to withstand the processing temperature involved in the film-application method. In the case of the cermet systems, this temperature can be as high as 980°C, and the parts may be heated to lower temperatures several times during successive firing cycles. The substrate must have a high surface resistivity to avoid leakage paths or shunting of high-value resistors. A low, uniform dielectric constant is desirable to reduce capacitive effects.

A high degree of chemical durability is required because of the rugged chemical cleaning and processing steps, such as etching and anodizing, that are employed in some circuit fabrication methods. Another chemical requirement now recognized is that the substrate must be free of ionic species which can migrate under the thermal and electrical gradients associated with resistors. These species are alkali metal ions, especially sodium, which react with the film material and alter its properties often to the point of failure. The negative terminal corrosion encountered on soda-lime glass microscope slides is a result of alkali-ion migration. Figure 29 illustrates the percentage resistance change, with time, of resistors deposited on substrates of different sodium content. The loading of these resistors was such that those on glass were running at about 250°C, a temperature high enough for sodium-ion migration to occur in an electric field.

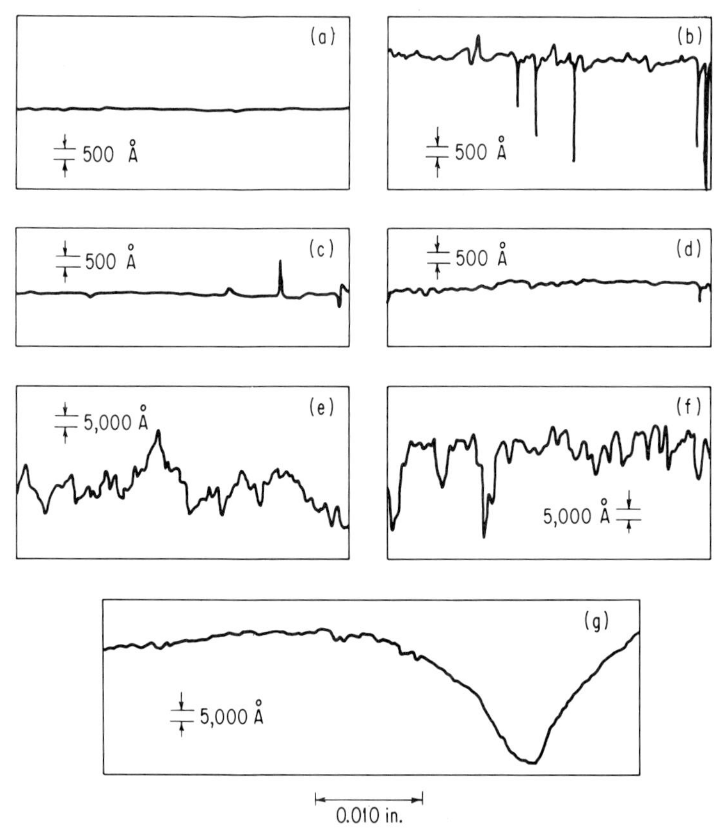

Fig. 28. Surface profiles of substrate materials.[31] Surface profiles observed on the following substrates: (*a*) drawn soft glass sheet; (*b*) high-density alumina, ground and polished; (*c*) photosensitive glass, polished; (*d*) photosensitive glass ceramic, ground and polished; (*e*) high-density alumina, as fired; (*f*) photosensitive glass, processed; (*g*) photosensitive glass ceramic, as fired. The distance scale along the surface (the abscissa) is the same in all figures, but the surface height scale (the ordinate) is larger in (*e*), (*f*), and (*g*). Note: 125 Å is equivalent to 0.5 μin.

Glass Substrates. One of the most commonly used substrate materials is glass, often in the form of microscope slides, since they are smooth, convenient in size, and readily available.

The properties of the more important substrate glasses are shown in Table 10. The designation "alkali-free" referring to glass type means that no alkali is intentionally added to the glass and that raw materials of extremely low alkali content are used. The maximum service temperature will be near the annealing point. The thermal conductivity of glass at 25°C ranges from 0.002 to 0.004 cal/(cm)(sec)(°C) and increases by about 10 percent to 100°C rise in temperature. Dielectric strength is measured as 1-min life at the cited voltage for a 2-mm sample at a frequency of 60 cycles.

The surface smoothness of glass depends primarily on the manufacturing process. Values less than 0.4 μin. (100 Å rms) are easily achieved either in the hot-forming process or in fine polishing. Ideally, the surface should be formed "free" by drawing or some similar process, with care taken to avoid contact of the smooth surface by tools or abrasive materials. Care must also be taken to prevent dust particles from touching the glass while it is in the hot plastic condition. Code 7059, the

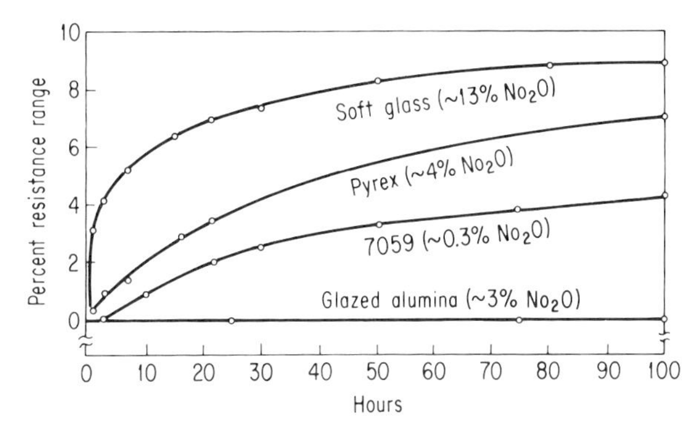

FIG. 29. Overload testing of various glasses.[31] Overload testing of tantalum nitride resistors on various substrates (60 watts/in.^{2} of film area, room ambient).

barium aluminosilicate glass listed in Table 10, has been drawn very carefully so as to preserve the smooth, virgin surface and is, therefore, especially suited to substrate applications.

Ceramic Substrates. Ceramic materials are used for substrates for film devices chiefly because of their high thermal conductivity, the ease of obtaining holes, and their excellent strength and electrical properties. Table 11 lists some typical properties of ceramic materials used for substrates. Attention is called to the high softening temperatures and thermal conductivities, which are of the order of 20 times those of glass for alumina and about 200 times for beryllium oxide (beryllia).

Processing. There are two methods of forming ceramic powders into ware, pressing and tape processes. Both forming methods have certain requirements in common in regard to the preparation of the ceramic composition for forming. The chemical purity, particle size, distribution, and thoroughness of mixing of the components must all be rigidly controlled. These factors are important because they, coupled with the sintering cycle, affect the microstructure of the final body, and nearly all properties of the substrate are structure-dependent. Dielectric properties in particular are sensitive to the ceramic microstructure. The nature of the surface is also a function of microstructure, and the surface is, of course, of great importance in thin-film work.

PRESSING. Pressing consists in compacting the ceramic powder with binders in metal dies at high pressures. The degree of powder compaction (pore-volume per-

TABLE 10. Properties of Some Common Glass Substrate Materials [19]

Glass type	Soda lime	Alkali zinc borosilicate	Lime alumino-silicate (alkali-free)	Lime alumino-silicate (alkali-free)	Barium alumino-silicate (alkali-free)	Alkali boro-silicate	96% silica	Fused silica
Annealing point, °C	512	542	866	710	650	565	910	1050
Softening point, °C	696	720	1060	910	872	820	1500	1580
Thermal-expansion coefficient, $10^{-6}/°C$	9.2	7.2	3.5	4.6	4.5	3.25	0.8	0.56
Thermal conductivity, cal/(cm)(sec)(°C) at 25°C	0.0023	......	......	0.0032	......	0.0027	0.0038	0.0034
Density, g/cm^3	2.47	2.57	2.48	2.63	2.76	2.23	2.18	2.20
Dielectric constant (1 Mc) at 25°C	6.9	6.6	5.9	6.4	5.8	4.6	3.9	3.9
Loss tangent (1 Mc) at 25°C	0.01	0.0047	0.0024	0.0013	0.0011	0.0062	0.0006	0.00002
Log volume resistivity (ohm-cm) at 250°C	6.4	8.3	13.6	14.1	13.5	8.1	9.7	11.8
Dielectric strength at 25°C, kv (rms)	0.35	2.0	>10	>10	>10	2.0	7.0	>10
Weatherability, g/cm^2	>5.0	0.05–0.25	<0.01	<0.01	<0.01	0.05–0.25	<0.01	<0.01
Chemical durability:								
5% HCl_2, 24 hr, mg/cm^2	0.02	0.03	0.10	0.4	5.5	0.005	0.001	0.001
5% Na_2OH, 6 hr, mg/cm^2	0.5	2.0	1.2	0.3	3.7	1.1	1.1	0.7
0.02 NNa_2CO_3, 6 hr, mg/cm^2	0.1	0.1	0.15	0.1	0.3	0.1	0.03	0.03

centage), and hence firing shrinkage, is dependent on the pressing force, up to some limiting value, on the order of 10,000 to 20,000 psi. Care must be taken to ensure that all portions of the pressed part attain the same unfired density; otherwise different shrinkages will occur during sintering and result in a warped piece. Pressing is particularly suitable for pieces which are essentially three-dimensional. Although some rather large, complicated shapes are pressed, the process as it relates to substrates is used chiefly for small parts of high precision. Surface finishes of pressed as-fired parts made of high-alumina ceramics will commonly be in the range of 30 to 50 μin., a range entirely suitable for screened and fired resistors.

TABLE 11. Properties of Ceramic Substrate Materials [19]

Ceramic type	Dense alumina 85% Al_2O_3	Dense alumina 94% Al_2O_3 + CaO + SiO_2	Dense alumina 96% Al_2O_3 + MgO + SiO_2	Dense beryllia 98% BeO	Dense beryllia 99.5% BeO
Softening temperature, °C	1100	1500	1550	1600	1600
Thermal/expansion coefficient, 10^{-6}/°C	6.5	6.2	6.4	6.1	6.0
Thermal conductivity, cal/(cm)(sec)(°C) at 25°C	0.060	0.073	0.084	0.50	0.55
Density, g/cm^3	3.40	3.58	3.70	2.90	2.88
Dielectric constant (1 Mc) at 25°C	8.3	8.9	9.3	6.3	6.4
Loss tangent (1 Mc) at 25°C	0.0058	0.0018	0.0028	0.0006	0.0006
Log volume resistivity (ohm-cm) at 300°C	10.7	12.8	10.0	13.8	>14
Dielectric strength at 25°C and 60 cps, volts/mil	230	230	230	255	260

TAPE PROCESS. In the tape process, the ceramic material is produced in a thin sheet or tape using plastic binders. The desired parts are then produced from the tape, much as mica insulating parts are stamped, and then fired. This process is especially suited for what may be considered two-dimensional parts, i.e., thin, flat sheets. The tape process is the least expensive method for obtaining a multiplicity of holes in a large, thin ceramic sheet held to close dimensional tolerances, such as shown in Fig. 30. One large substrate is an outstanding example of the current state of the art. This piece is over 16 in.^{2} in area and contains over 400 holes. Tape-processed substrates are produced in thicknesses of 15 to 40 mils, with an as-fired surface of 15 to 40 μin., which is too rough for the direct deposition of some thin films. Several papers have been published in which the effect of surface morphology on film properties has been reported. As a result of poor surface morphology and because of the desire to obtain the benefits of the thermal conductivity of the ceramics, glazes are frequently applied so that thin film may be deposited on ceramics.

PROGRESS IN CERAMICS. Progress in supplying ceramic substrates of better quality has been more spectacular than that of glass by comparison; the ceramic industry lagged in this respect. Recent history has been one of promise as the market potential of substrate materials has broadened and made its impact on corporate planners. There are intensive programs being carried on to improve ceramic substrates in several areas. One of these has been to control processing to allow reduction in the physical tolerances, especially in the larger sizes made by the tape process. Length and width tolerances have been cut from 1 to ½ percent. Problems in flatness still predominate. Bowing is evident on large and small pieces alike. The larger substrates are usually cut into subunits, which will all be well within the normally specified 0.004 in./in. bow, but which are still not as good as many users would like.

USE OF UNGLAZED CERAMICS. Some investigators in thin-film work regard the use of glazed ceramics as an interim measure and consider that eventually unglazed ceramics will be used almost exclusively. Some take this position on the premise that ceramics are becoming better, others on the basis that the film itself can be made less susceptible to the roughness of the substrate, and still others on the basis that the cost of ground substrates, which provides an adequate microstructure for thin films, is no longer economically prohibitive. To confine our remarks here to the ceramic and high-alumina ceramic, considerable progress has been made in controlling the microstructure in order to have a reproducible surface texture, as well as in improving the smoothness. Measurements since 1965 have shown that

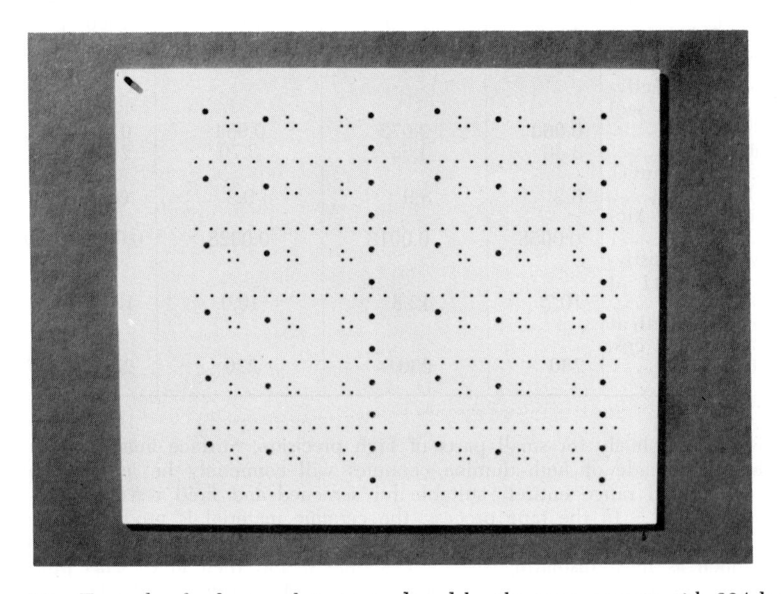

FIG. 30. Example of a large substrate produced by the tape process with 334 holes. The substrate is approximately 3 by 4 in. (*Bell Telephone Laboratories, Inc.*)

the surface finish of tape-process substrates has improved from 20 μin. to about 12 to 14 μin., based on Talysurf readings. Experimental bodies have been produced with a surface of 5 to 10 μin., which is acceptable for the direct deposition of some thin films. The present trend in the search for smoother surfaces appears to be toward the higher-alumina-content bodies; these materials also have slightly higher thermal conductivities and better dielectric properties as added benefits.

ECONOMICS. The present trend in ceramic substrates is toward the processing of larger substrates which can be separated into smaller units after processing. Hence, the equivalent of 4 to 10 substrates can be processed in the initial manufacturing process for approximately the same cost as a single substrate. Second, the ceramic cost itself is lower because the same economics apply to the production of the substrate itself. The problem has been one of separating these large substrates into individual units. Two methods have been attempted so that costly diamond sawing can be eliminated. One method is the "postage-stamp approach," where rows of holes serve as perforations for a break line. The second approach is the machining of V-shaped grooves on the reverse side, which acts as a break line.

The latter approach has been the more successful and appears more likely to be adopted by the industry.

TITANATES. The tape process was originally used to produce high-dielectric ceramics termed *titanates*. The tape process was adapted to the production of high-alumina-content substrates as a result of increased sales about 1962. Titanates are

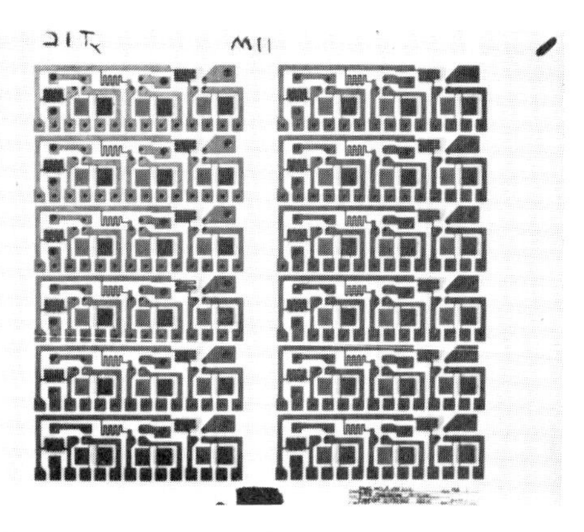

FIG. 31. Multiple deposition array which can be separated into individual circuits. (*Bell Telephone Laboratories, Inc.*)

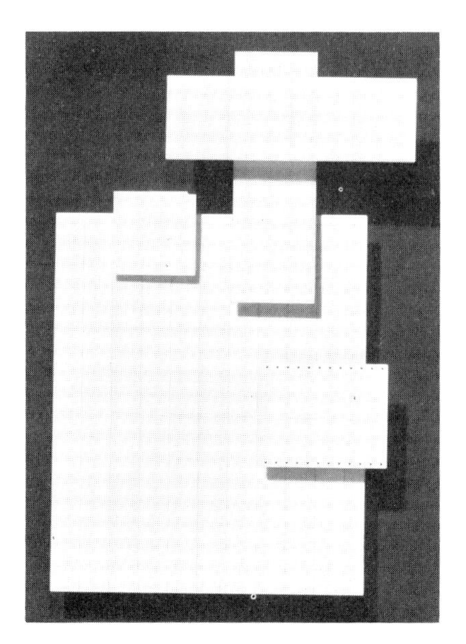

FIG. 32. Example of technique for separating smaller substrates out of large substrate. (*Coors Procelain Co.*)

ceramics, but the metal oxides possess high dielectric constants. These ceramics are usually composed of mixtures of titanium dioxide, barium titanate, strontium titanate, and barium-strontium titanate mixed with binders in various proportions to achieve varying dielectric constants, temperature coefficients, voltage coefficients, and Curie points.

Titanates have been used for microelectronic applications for many years because of their dielectric properties. They have been used since 1957 primarily to form resistor-capacitor networks such as shown in Fig. 33. In this *RC* network, the high dielectric constant of the substrate is used to form capacitors with cermet resistors formed on one or both surfaces. Because literally hundreds of compositions exist with dielectric constants ranging from 50 to 1,000, manufacturers should be consulted for individual properties.

RESISTOR DESIGN

Two-dimensional resistor designs can be achieved by means of simple theoretical calculations or more sophisticated methods in which compensation factors for tolerances, termination resistance, and power compensation are employed.

Basic Design. In designing a two-dimensional resistor, the first task is to determine the resistor area required for its dissipation, based on the film dissipation factor. All films have a specific dissipation factor, expressed in watts per square

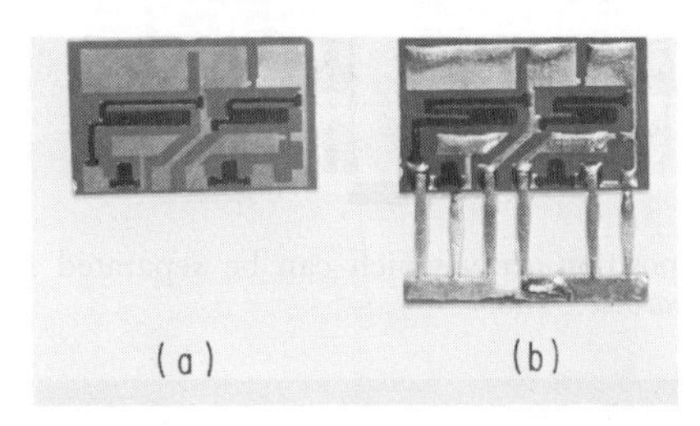

FIG. 33. Resistor-capacitor network screened on a titanate substrate. (*Philco-Ford Corporation, Microelectronics Division.*)

inch of film area. To simplify design, it is desirable to convert this figure into a power dissipation constant *C*, expressed in square mils per watt, as follows:

$$C = \frac{1{,}000{,}000}{\text{watts/in.}^2}$$

For example, the power-dissipation constant is 50,000 mils^2/watt for a film whose specific dissipation is 20 watts/in.^{2}

Next, the required area (in square mils) for the resistor is obtained by multiplying the power-dissipation constant by the resistor wattage. A resistor stressed at 20 watts/in.^{2} of film area requires 5,000 mils^2 area to dissipate 0.1 watt (50,000 $\times$ 0.1).

The area *A*, which has been calculated, is equal to the product of the length *L* and width *W*, or $A = LW$. Furthermore, for any given sheet resistivity *SR*, the area can be expressed as a function of *L* or *W*. However, neither is necessary. A constant *N* (number of squares) can be calculated as follows:

$$N = \frac{\text{R (ohms)}}{SR\ \text{(ohms/sq)}}$$

Thus $$L = NW$$
and $$A = NW^2$$
or $$W = (A/N)^{1/2}$$

For example, assume a resistor value of 2,000 ohms and sheet resistivity of 100 ohms/sq. Then,

$$N = \frac{2{,}000}{100} = 20$$

$$W = (5{,}000/20)^{1/2} = 15.8 \text{ mils}$$

Rounding off the value to 16 mils yields a length of 320 mils.

Nomographic Solution. While the preceding analytical method is relatively quick and easy, the nomograph in Fig. 34 greatly simplifies resistor-design calculations. To obtain the width and length for a specific resistor, only two calculations are necessary.

Using the previous resistor example—2,000 ohms at 0.1 watt, dissipation of 20 watts/in.^{2}, and a sheet resistivity of 100 ohms/sq—enter the nomograph on the extreme left at 0.1 watt and 20 watts/in.^{2} Draw a line, as shown, to intersect the area bar. The intersection falls at 5,000 mils^2. Next calculate N, the number of squares required, as before (for this example, $N = 20$). From the intersection on the area bar, draw a line through 20 on the N scale, to intersect the W^2-W scale. The answer reads slightly over 16 mils, which is sufficiently accurate when rounded off. The length is calculated as before.

Form Factor. Once the length and width have been determined, the form of the resistor becomes important. For many applications, a resistor 0.320 in. long is out of the question. Such a resistor must be folded back on itself to fit into the space which is available on the substrate.

There are three ways to fold or serpentine resistors. The first is the use of shorting bars shown in Fig. 35. The alternative method is to use the corner correction

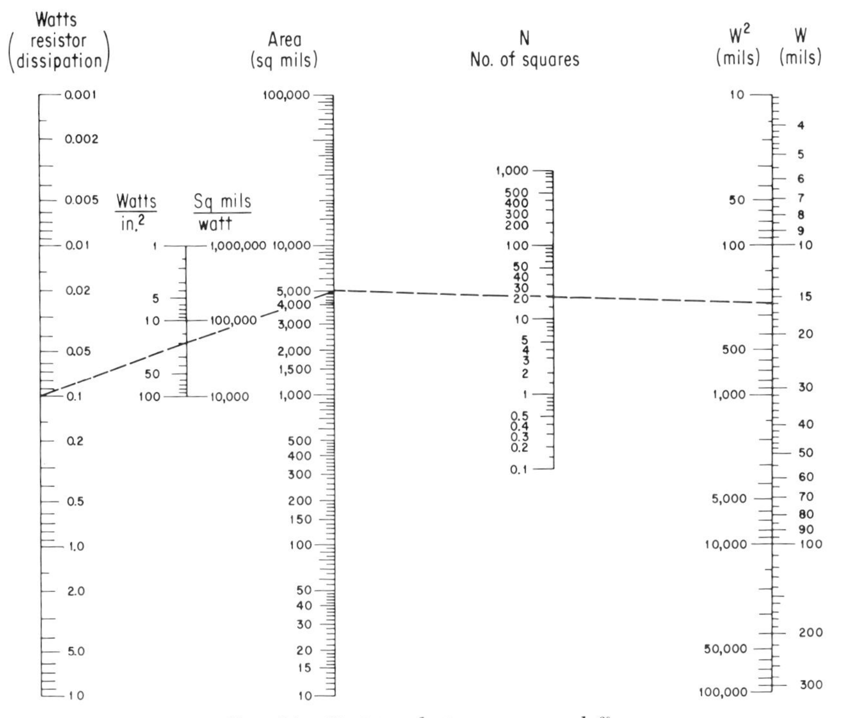

Fig. 34. Resistor design nomograph.[20]

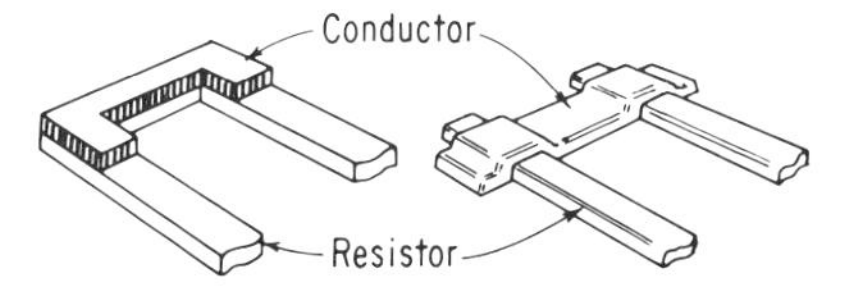

Fig. 35. Shorting bars used to serpentine resistors.[20]

factors shown in Table 12. It will be noted that a right-angle corner is 0.53 of a square, but high current densities occur which produce a local hot spot and a potential failure. The semicircular fold is more difficult to lay out in artwork preparation but cannot be used in vacuum evaporation because a mask fabricated for evaporation will not generally maintain contact with the substrate. It is useful in etched or screened networks, however.

Deviations from Theory. This method of design is theoretically correct but does not work from a practical point of view. First, depositions are not uniform. There are tolerances in both width and length, and the current path between conductor and resistor materials produces a physical shift of the actual electrical termination which establishes a geometric effect within each resistor. In some resistor-conductor systems, there is a noticeable contact resistance which further complicates design.

Mechanical Tolerances. The effect of mechanical tolerances is obvious; a small percentage deviation in either width or length results in a direct deviation of the value of each resistor. On a substrate where several resistors are made simultaneously, values can be off 10 to 15 percent simply as a result of mechanical tolerances. The effect of dimensional tolerances on electrical value becomes more severe as the value between resistors spreads.

Dimensional deviations occur in all types of resistors. In etched networks, some undercut is almost certain to occur during etching, and it is not necessarily a con-

TABLE 12. Corner Correction Factors for Resistors

Number of squares: The column gives the resistance of the area to the right of the broken line in terms of equivalent linear squares.

Power-density ratio. The column gives the peak power-density ratio in the area to the right of the broken line in comparison with current density in the linear portion of the resistor.

Geometry	Number of squares	Power density ratio
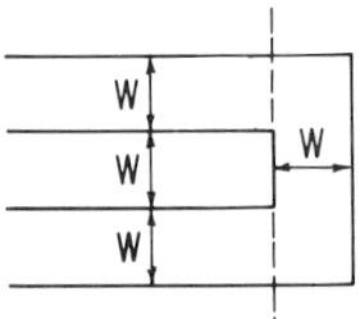 Square corner (W, W, W; W)	2.06	
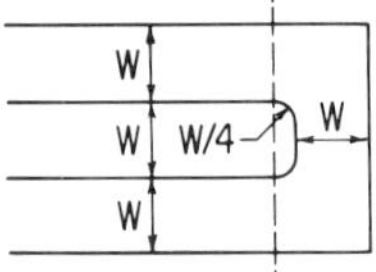 Rounded corner, radius W/4 (W, W, W; W)	2.68	3.6
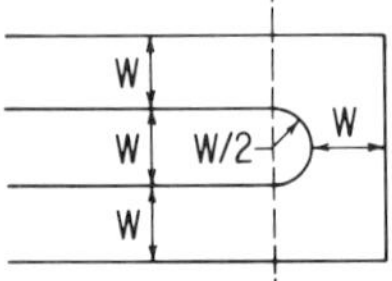 Rounded corner, radius W/2 (W, W, W; W)	2.65	3.4
Semicircular fold, radius D/2 (W, D, W; W)	D/W	
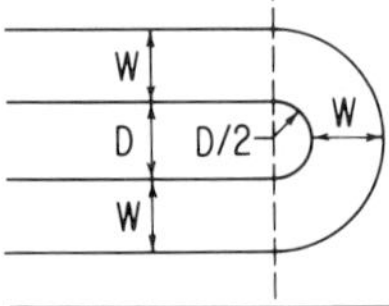	1/2 2.19	6.9
	1 2.92	3.2
	2 4.84	1.6

sistent undercut, so that the technique of designing the resistor wider and longer to compensate for undercutting during etching is limited. Essentially, the same is true for networks deposited through masks, for the mask openings vary from design, with the same result. Another effect occurs in evaporation in that most evaporation sources tend to be somewhat directional. It is best to design evaporated thin-film networks so that all resistor elements are parallel to each other, rather than having some elements at right angles. This will avoid the effect of edge shadowing of the deposition in one direction. This is of no consequence in other deposition methods.

Diffused resistors are subject to two effects: possible undercut during the etching of the SiO_2 on the wafer surface and then a spread of the resistor line width during the diffusion process itself. Screened resistors appear to be the worst from the standpoint of dimensional tolerance. Not only do the X-Y dimensions of the resistor vary as a result of inaccuracies in screen fabrication, but screening parameters effect the deposition as well. Finally, the thickness varies as a function of the X-Y dimensions, so that the value of the resistor is a composite of all three dimensions.

Termination Effects. For most resistor designs, termination resistance, or the effect of the current path between the resistor and conductor, is ignored, although it should not be. In Fig. 36, the current path does not occur precisely at the

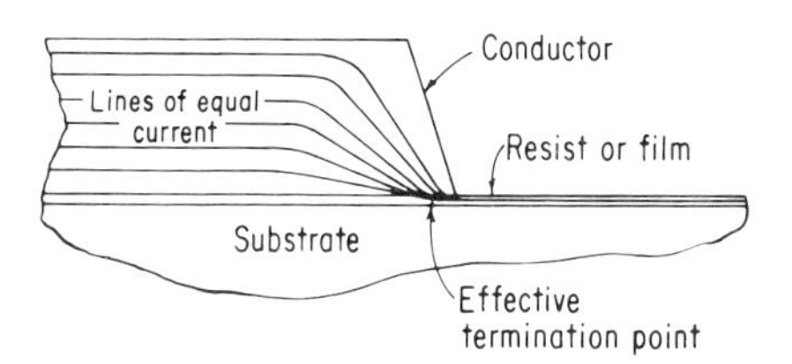

FIG. 36. Resistor-termination current path.

resistor-conductor interface, so that the resistor is effectively longer than designed. Second, should there be a contact resistance associated with the interface, the effect would be to change the value from design and cause the value function shown in Fig. 37. For a positive contact resistance, the value of a given resistor will change as a function of width.

This can be proved mathematically by assuming that the total resistance is equal to the sum of the three resistances,

$$R_T = R_t + R_r$$

where R_T = total resistance
R_t = contact resistance of two terminations combined
R_r = resistor value between terminations

If it is assumed that the contact resistance is constant, then a geometric effect is obvious. For example, assume a sheet resistivity of 500 ohms/sq and a contact resistance of 50 ohms/termination. If a constant resistor width is assumed, 1 square would be 600 ohms; 2 squares, 1,100 ohms, etc. If the width of the resistor is doubled, the contact resistance is one-half, or 25 ohms/contact, since one contact now represents two contacts in parallel. Such a resistor would be 550 ohms instead of 600 ohms for 1 square. Conversely, a resistor one-half the width would be 700 ohms in value for 1 square, 1,200 ohms for 2 squares, and 450 ohms for ½ square. It is significant to note that, when a contact resistance exists, the theoretically calculated value is low compared with what is obtained. Only positive resistances have been discussed. A negative contact resistance is theoretically impossible, but a mathematically negative resistance does occur when palladium-silver resistors are cofired with palladium-silver or silver conductors. There is a diffusion or chemical reaction between the conductor material and the resistor material during firing which lowers the resistor value and yields a mathematically negative contact resistance. The shorter the resistor, the more pronounced the effect.

The value of any resistor can be calculated from the following formula when the contact resistance is constant:

$$R_T = SR \times AR + 2\,R_t/W$$

where R_T = total resistance
SR = sheet resistivity
W = width, mils
R_t = termination resistance, ohms/mil

For example, assume an SR of 500 ohms/sq and an R_t of 5,000 ohms/mil. Using our previous example for a resistor 100 mils wide, we obtain

$$R_T = (500)(1) + (2)\left(\frac{5{,}000}{100}\right)$$
$$= 500 + 100 = 600$$

For 2 squares we obtain

$$R_T = 1{,}000 + 100 = 1{,}100$$

For ½ square

$$R_T = (500)(½) + (2)\left(\frac{5{,}000}{200}\right)$$
$$= 250 + 50 = 300$$

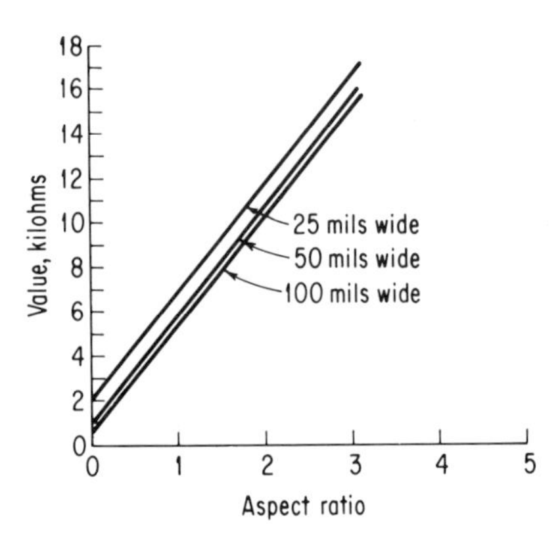

FIG. 37. Effect of termination resistance on value.

Complex Design. Certainly many techniques have been developed to circumvent these effects. In etched networks, or masks for evaporation, resistor line widths are generally made slightly larger than designed to compensate for the undercut which occurs during etching of the deposited film or evaporation mask. In etching TFs, there is some deviation from substrate to substrate which cannot be avoided and which results in minor tolerances even with compensation. Incremental design is a technique which maintains all resistor line widths constant by paralleling two or more resistor lines to obtain wider resistors where needed. In addition, the number of terminations within the resistor are maintained in approximate proportion to the resistor value. Not only does this technique compensate for dimensional variations, but it also compensates for termination effects and eliminates the necessity of precise photography and photolithography. While incremental design does compensate for these problems and eliminates the necessity for precision photography, it is cumbersome in design and time-consuming in layout, so that it is seldom used. In networks where the desired value is obtained by some form of trimming, all these effects may be essentially ignored unless they drastically influence the initial resistor value to such an extent that more than one-third the resistor must be removed or the resistor is above its initial value.

With all these factors possibly working together, resistor design could become extremely complex. The simplest and easiest method of avoiding this is to run a set of geometry experiments so that the value and tolerance of the process are known for each resistor width. Such a chart is shown in Fig. 38. For many resistor materials, these charts are not too beneficial, but they are almost essential in the design of screened resistors.

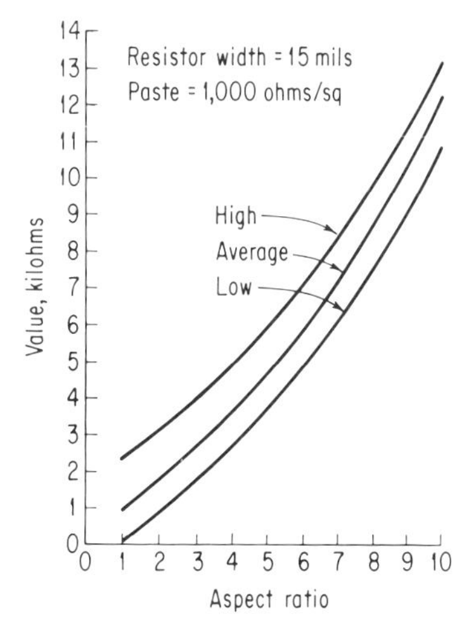

FIG. 38. Design chart for screened resistors.

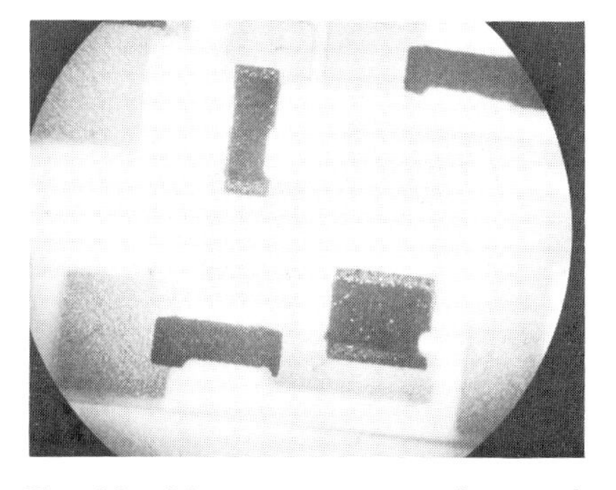

FIG. 39. Microscopic view of several abraded cermet resistors.

Power Compensation. Power dissipation is generally based on life-test data and is used itself as the basis for the determination of resistor area. As previously discussed, a single power-dissipation factor is generally chosen for the design of all resistors on a substrate, regardless of the ratio of the circuit-power dissipation to substrate area. Where miniaturization is truly required, substrate area can be reduced as a function of power dissipation to substrate area so that substrate area is optimized; i.e., the resistor stress level (watts per square inch of film area) can be increased, where the total power dissipation is relatively low, while the substrate size is reduced and the reliability of each resistor maintained. It requires some development for a particular resistor system to determine the stress level which is safe.

Practical Considerations. The foregoing has primarily been a theoretical discussion of resistor design, with little or no consideration of practical aspects. A great deal of the theoretical can be ignored for most of the resistor systems in use. Thin films, evaporated through masks, can be deposited to ±1 percent of value by using some of the design techniques described. Other systems employ trimming so that values within 40 percent of the desired value are acceptable.

Trimming. Trimming can be accomplished by abrading, baking, anodizing, and by use of the laser, or electrical pulse. In abrading, sandblasting on a miniaturized scale simply cuts the resistor away as shown in Fig. 39. Resistors are easily trimmed to ±1 percent in value, but must generally be trimmed one at a time unless expensive automated abrading stations are used. Where volume is sufficient, the expense of such a station is acceptable. It is, however, an untidy technique in that the fine abrasive scatters easily. Baking is a techniqe relying on high-temperature oxidation (300 to 500°C) of an entire substrate or sheet of substrates so that all resistors simply "drift" up to value. This technique is not generally used because baking times can be as long as 36 hr. Second, the original deposition must be extremely uniform and the processing precise so that all resistors arrive at value together. This seldom happens, and hence the technique is little used. For networks of tantalum or titanium, each resistor can be individually "anodized" to value. This trims and stabilizes the resistor in one operation. Third, the electrical-pulse method of trimming depends on the short-term overload of each resistor, which simply heats it to a temperature where it drifts. This can be accomplished at room temperature or on a heat column to reduce the time and energy input required. The problem with this system is the electronic switching systems involved and the fact that the last pulse can overshoot the desired value if the operator is not careful. Electron-beam milling has also been used to pattern thin-film networks in a matter of minutes, while providing complete feedback. Here, the drawback is the necessity of entering a vacuum system with its subsequent pump-down time. The latest trimming method is the use of CW lasers, which simply vaporizes a spot on the resistor. This technique appears promising for all resistor materials, for it permits continuous-value monitoring, it is not untidy, and it avoids most of the drawbacks of the other methods.

Top-hat Form Factor This has been used to obtain high-value screened resistors with high-value sheet resistivities which are always erratic in firing. Such a design is shown in Fig. 40, where trimming is accomplished simply by a straight abrasion cut up between the resistor legs.

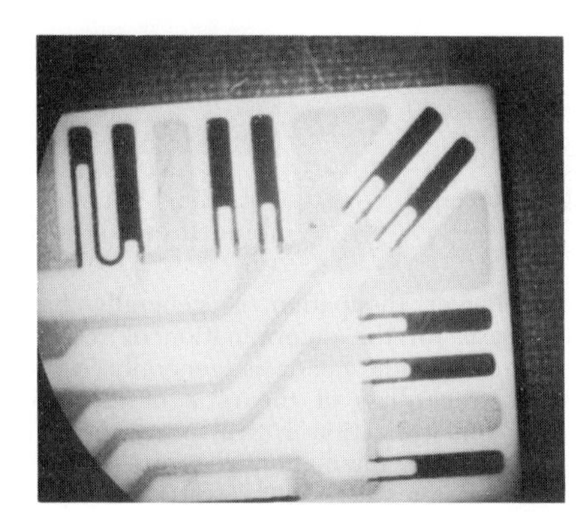

FIG. 40. Top-hat resistor form factor for large trimming ratios.

Oversquared Design. Resistors with aspect ratios of less than 1 are called *oversquared.* They are used almost exclusively in SAF networks; i.e., they are seldom used with thin films because high-value SRs are required to obtain resistors of reasonable value. The oversquared design is used frequently because of the ease with which it is trimmed.

CAPACITOR DESIGN

Capacitor design is similar to resistor design because both suffer from similar manufacturing problems. For example, a capacitor's value is a function of width, length, and thickness, just like a resistor's, so that dimensional tolerances have an obvious effect. In evaporated or anodized capacitors, thickness control does not appear to be nearly as critical as in screened capacitors, where deposition thickness is a function of the capacitor geometry. The formulas that follow have been derived from the general formula for parallel-plate capacitors and are accurate to ±10, for they neglect fringing effects and the effects of the capacitor physically being on a substrate with a definite dielectric constant. The general formula is

$$C = 0.0885K \frac{(N-1)A}{t}$$

where C = capacitance, pf
K = dielectric constant
N = number of plates
A = area, cm^2
t = thickness, cm

Thick-film Design. The above formula has the obvious drawback that it is in centimeters. Upon converting to easier terms for thick films, the following is obtained, one pair of plates being assumed, i.e., a single-layer capacitor:

$$C = 225K \frac{A}{t} \times 10^{-5}$$

where C = capacitance, pf
K = area, $mils^2$
t = thickness, mils

A handier version for design is as follows:

$$A = \frac{Ct}{225K} \times 10^{-5} \text{ mil}^2$$

As shown in the discussion on Resistor Design, for each dielectric and process, a design chart based on actual data is easier to use, faster, and more accurate.

Thin-film Design. The same general formula applies to thin-film capacitors except that the units are different. By converting the above formulas, the following formulas are obtained:

$$C = 57.097K \frac{A}{t}$$

$$A = \frac{Ct}{57.097K}$$

where C = capacitance, pf
A = area, $mils^2$
t = thickness, Å

INDUCTOR DESIGN

Small-value inductors can be and have been screened with a degree of success. They have been used as delay lines and as chokes for ultrahigh frequencies and microwave inductors and transformers. The general formulas used to calculate inductance values are highly inaccurate for deposited inductors since the formulas available are based on an air dielectric. The additional capacitance of the substrate increases the reactance so that the value of a deposited inductor is 10 to 100 percent higher than the calculated results. For a single-turn inductor, the following formula may be used for high frequencies:

$$L = 0.00508l\left(2.303 \log \frac{4l}{d} - \theta\right)$$

where L = inductance, μh
l = the perimeter, in.
d = line width, in.
θ = a constant for various figures, as follows:
Circle: $\theta = 2.451$
Square: $\theta = 2.853$

For multiple-turn square, flat inductors, the following formula may be used:

$$L = 0.02032n^2s\left(2.303 \log \frac{s}{nD} + 0.2235 \frac{nD}{s} - 0.726\right) - 0.02032ns(A + B) \quad \mu\text{h}$$

where n = number of turns
s = average length, in.
D = pitch between turns, in.
A, B = constants derived from Tables 13 and 14, respectively

These formulas, while helpful, produce an inaccuracy because the conductors are more rectangular than round. In general, it is easier to produce an inductor and measure its value than to calculate it. Again, a design chart based on actual data is extremely useful.

TABLE 13. Constants for Multiturn Inductors [18]

d/D	A	d/D	A
1.00	0.557	0.45	—0.242
0.95	0.506	0.40	—0.359
0.90	0.452	0.35	—0.493
0.85	0.344	0.30	—0.647
0.80	0.334	0.25	—0.833
0.75	0.269	0.20	—1.053
0.70	0.200	0.15	—1.340
0.65	0.126	0.10	—1.746
0.60	0.046	0.05	—2.439
0.55	—0.041		

TABLE 14. Values of Correction Term B [18]

Number of turns, n	B
1	0.000
2	0.114
3	0.116
4	0.197
5	0.218
6	0.233
7	0.244
8	0.253
9	0.260
10	0.266

SUBSTRATE LAYOUT

There is no substitute for experience in substrate layout. While substrate layout is similar to the layout of printed circuits, many of the advantages of printed-circuit layout are not available in substrate layout. In printed circuitry, resistors, diodes, capacitors, and even transistors can be used to effect conductor crossovers in both single- and double-sided circuitry. Although both surfaces of the substrate have been used, it is not common practice and is not recommended except in special cases where the advantages offset the added processing costs. Therefore, most layouts are basically confined to a two-dimensional layout on a single surface, where resistor and capacitor positions cannot be used as conductor-crossover points. Conductor-crossover capability is even further reduced as flip-chip or chip and wire semiconductors are used rather than individually sealed devices. From our previous discussion, it is obvious that physical crossovers or the addition of jumpers can be accomplished, but this requires extra steps in processing, which in turn lead to increased costs. Hence, layouts which eliminate crossovers are preferred to those which require them. Additional layout time is therefore advisable to eliminate crossovers if a reasonable number of circuits is to be produced.

As previously stated, there is no substitute for layout experience, but the approach to the hybrid system is about the same for most persons. First, the system is divided into functional or operational sections in a manner that minimizes the number of interconnections between substrates. Second, the schematic for each function or operation is rearranged to eliminate the necessity of jumpers to provide crossovers or the necessity of making them physically. The method by which crossovers can be avoided depends on what types of components are to be attached to the substrate. Obviously, individual diodes, transistors, capacitors, and chip capacitors provide ample versatility to route conductors around and between termination points so that crossovers and jumpers can be avoided. When flip-chip and chip and wire active devices or ICs are used, this versatility is somewhat reduced, although one or two conductors can be jumped by a wire bond from a transistor or diode chip to a termination pad.

Once the schematic has been rearranged so that as many jumpers and crossovers have been eliminated as possible, the passive-component dimensions are determined as described in component design. Next, a substrate size is chosen and a rough 10× layout is attempted in the style of the rearranged schematic, by using the dimensions of the passive components with conductor overlaps for terminations, conductor runs for intraconnections, and pad areas for the attachment of active devices and chip capacitors, if required. While general rules for pad sizes for each type of component could be specified, sizes vary so much in actual practice because of differences in processes that no recommendation will be made here. In general, pad areas vary depending on the area available on the substrate for each particular circuit, on the processes being used, and mainly on common sense. Conductors and connection pads should be made as large as possible, especially conductors. Even though the specific resistance of most conductors is relatively low (0.1 to 0.001 ohm/sq), long, narrow conductors can have resistance which can be detrimental to a circuit in some instances. Overlaps for SAF resistors should be standardized to correspond to the results obtained in development, basically 10 to 15 mils. In etched networks, overlaps are dependent on the factors previously discussed and on the etching and registration capability. Essentially, the same is true for substrates deposited through masks, with the additional factor of the etching tolerances involved or the registration tolerances within the vacuum system. In all systems, registration to 1 mil is easily achieved. Closer registration is quite possible and has been achieved.

SUBSTRATE TO SYSTEM INTEGRATION

Simply fabricating a number of passive components on a substrate does not produce or package an entire microelectronic system. Many things must be considered. The first is whether to use sealed or unsealed active devices. If individually sealed devices are used, then the quantity of electronics to put on a substrate is not too critical, for individual devices are easily replaced. The size and shape of the substrate are not too critical either, for there are a variety of sizes and shapes available which will satisfy most packaging needs where a small or modest quantity of circuits is required. Table 15 lists the sizes of substrates available that are standard with most manufacturers. For larger quantities of circuits, special tooling for a particular size of substrate, with or without holes, is economically feasible. The substrate shown in Fig. 30 is an example of a specially tooled substrate. Tooling typically costs $1,000 to $3,000 depending on complexity and holes.

When individually sealed devices are used, leads from the substrate are attached by one of two methods. The first of these is shown in Fig. 22, where ribbon or bus wire is simply soldered to pads on the edge of the substrate. The second method is the use of pins through holes in the substrate, as shown in Figs. 41 and 42. The pins can be loose fitted with an upset head which is soldered to a pad

TABLE 15. Standard Substrate Sizes*

0.25 × 0.25	0.375 × 0.440	0.50 × 0.50 0.50 × 0.75 0.50 × 1.00	0.75 × 0.75 0.75 × 1.00 0.75 × 1.50	1.00 × 1.00 1.00 × 1.50	1.50 × 2.00	2.00 × 2.00

Standard Tolerances

Thickness	10–40 mils ± 10% (tape process) No limit on pressed parts
Camber	5 mils/in. standard 4 mils/in. generally specified
Width and length	±0.5%
Surface	20–30 μin. rms for tape process 30–50 μin. rms for pressed parts

* Available without tooling charges.

surrounding the hole. A second method is to swage a copper pin into a hole so that it is mechanically secured prior to soldering. If uncased active devices are to be used, substrate size, hermetic package availability, and many other problems must be considered.

There is a tendency to use uncased devices for two basic reasons: (1) connections are eliminated by wiring the active devices directly to the intraconnections on the substrate, and so reliability is increased, and (2) the substrate area required for the active device is much smaller than when it is sealed in a package of its own so that the substrate size is reduced. These are the reasons commonly cited, but both may be pure rationalizations. While connections are eliminated, the ones eliminated, solder joints of active device leads, are generally highly reliable compared with the themocompression bonds used internally in active devices and on the substrates themselves. Second, by the time a substrate is sealed in a package, the package size is generally about the same size and volume as a substrate which uses individually sealed devices so that the size advantage is not necessarily a

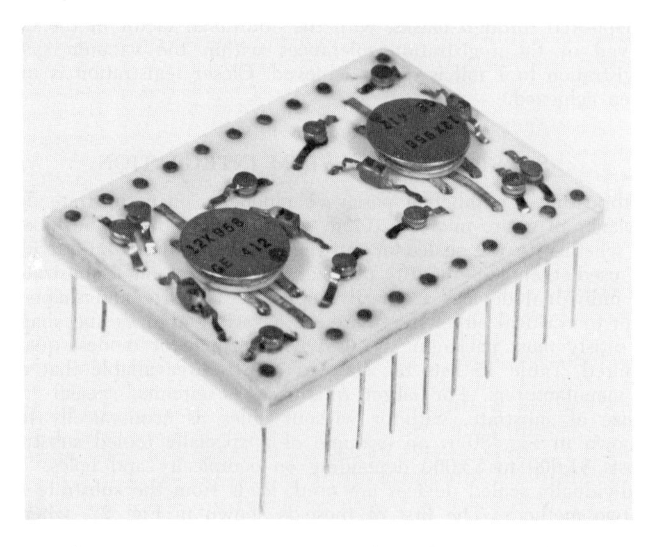

FIG. 41. Top of a cermet circuit using a perforated substrate for pin connections, active devices.

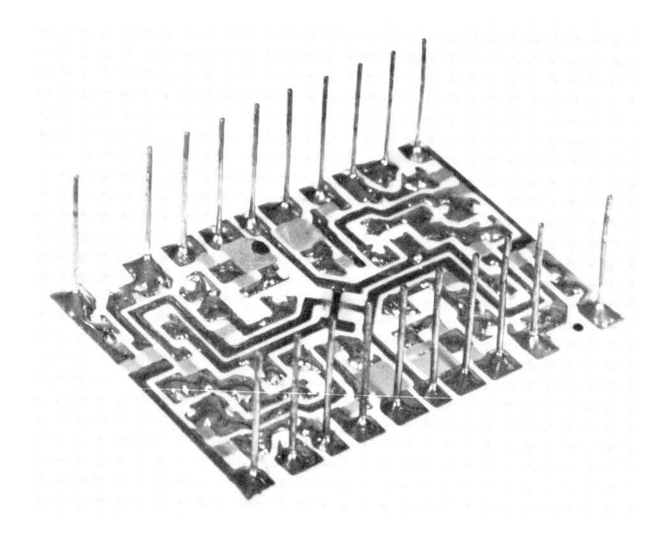

FIG. 42. Bottom view of the circuit shown in Fig. 41.

valid argument. Nonetheless, uncased devices are being used with the entire circuit sealed, even though this presents problems in circuit assembly and yields. For this method of circuit assembly, the internal and external fabrication processes must be considered as an integral part of the basic design.

Of all the methods of sealing hermetic modules, only one appears to be of any serious consequence to previous processes. Heat sealing or soldering can elevate the temperature of the entire package, with potentially detrimental effects to the semiconductors or previously processed solder joints, if they have been used. With heat sealing, it appears to be desirable to maintain the melting point of all joints above the temperature of the seal. Common melting temperatures are listed in Table 16.

TABLE 16. Melting Temperatures of Eutectic Alloy Solders

Solder	Melting temperature
Gold-silicon	372°C
Gold-germanium	356°C
Gold-tin	282°C
Tin-antimony	252°C
Tin-lead	183°C

It is quite obvious that it is not desirable to connect the substrate to the external leads by tin-lead solder if the package is to be sealed wih gold-tin solder. Hermetically sealed packages add difficulties other than processing problems. The selection of packages is rather limited, and the number of substrates available to fit these packages is even smaller. Little can be said about the package types available. The two basic kinds are shown in Figs. 43 and 44. The flat pack and the pin package are familiar to all in the electronics industry. Because of lack of adequate or available packages, integrated-circuit flat and pin packages, as shown in Fig. 45, are being used for hybrid circuits. Less familiar is the trend toward using the substrate as the base of the package, as shown in Fig. 46. This approach may become more popular as processes and products become available which lend themselves to this approach.

Certainly, the latest trend is toward the use of plastic encapsulants to avoid the cost of hermetic packages entirely. One method, shown in Fig. 47, uses a glass or

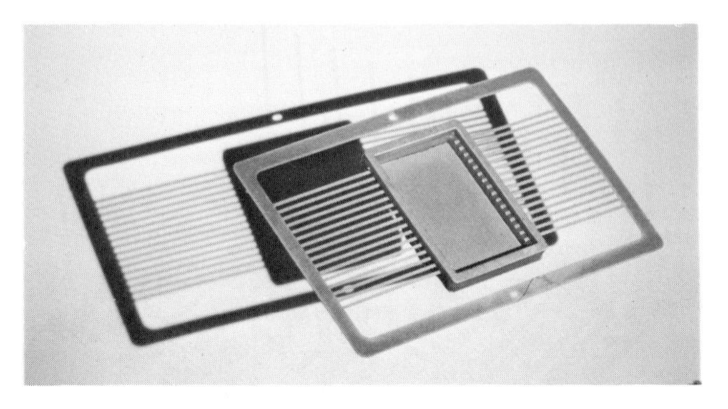

FIG. 43. Large-sized flat package used to seal hybrid circuits fabricated with unsealed active devices. (*The Bendix Corporation, Electrical Components Division.*)

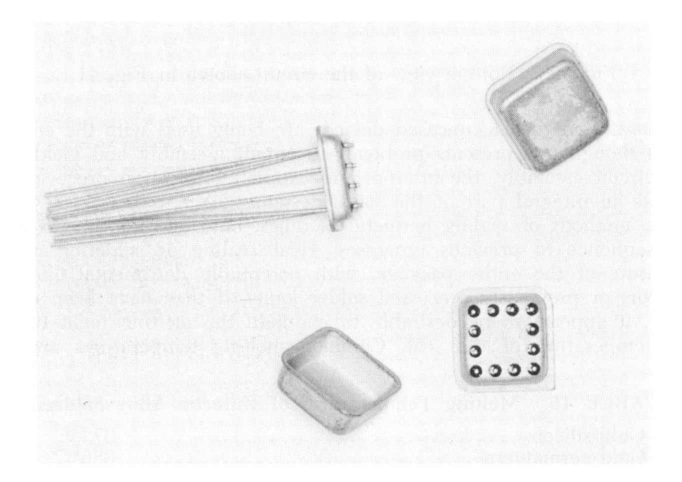

FIG. 44. Large-sized hermetically sealable pin package. (*The Bendix Corporation, Electrical Components Division.*)

ceramic cover over a layer of encapsulating material. Detrimental materials, such as water, cannot diffuse into this laminated packge. Finally, an even simpler approach is the direct encapsulation of the entire circuit in plastic. It would appear that this approach will eventually supplant the majority of all hermetic seals in microelectronic assemblies, even for military applications, once the reliability of the technique is proved.

ECONOMIC CONSIDERATIONS

One of the most important factors in microelectronic fabrication is the yield of the fabrication process. Certainly, a process is not economical if the processed substrate yield is 1 percent. And all processes have a yield factor at each step so that the effect of multiple yield factors should be understood.

The general formula for yield is

$$X = Y_1 Y_2 Y_3 \cdot \cdot \cdot Y_n$$

where X = overall yield
Y = operation yield for each process

This formula is based on a statistically random distribution of defect units in each operation. From this formula, it is obvious that, the greater the number of operations required, the lower the yield becomes, because few operation yields are 1.0, or

FIG. 45. A variety of packages designed for integrated circuits that have been used to seal hybrid circuits fabricated with uncased semiconductor devices. (*Glass-Tite Manufacturing, a Division of GTI Corporation.*)

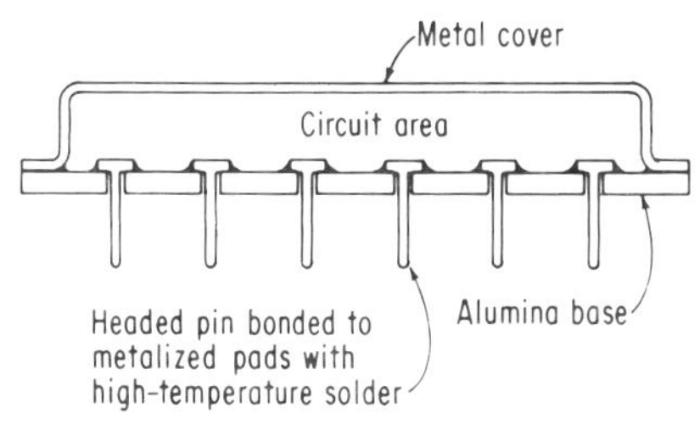

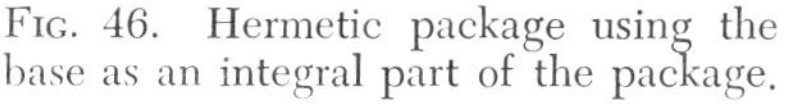

FIG. 46. Hermetic package using the base as an integral part of the package.

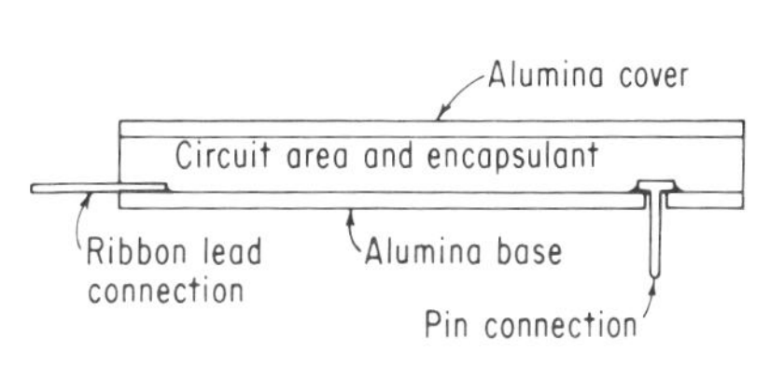

FIG. 47. Cover-layer substrate package with encapsulant.

100 percent. If this formula is applied to a single substrate rather than a process, it is also obvious that, as the number of components increases, the substrate yield is reduced. Thus, overall substrate yield is a composite of each component and process yield. For example, if the yield of each process is 0.99 for each 10 process steps, overall process yield will be 0.90, or 90 percent. This is derived from the formula $X = Y^n$, since all operations were assumed at 0.99. The labor of calculat-

ing individual or overall yields is greatly reduced by use of the chart in Fig. 48. The chart is almost self-explanatory and quite graphic. Overall yield falls drastically for even a small reduction in process yield.

Cost Considerations. The pros and cons of various deposition techniques have been argued long and loud, with proponents equally adamant in their respective basic views. Certainly, there is merit in the arguments on all sides and for all techniques. In some cases, there may not be an alternative to what has been done or will be done. In other cases, alternatives prevail, so that the choice of which deposition method to use is not academic, but a matter of pure economics.

This section will concentrate on the costs involved in establishing different types of facilities and their potential output for each set of equipment, in substrates per unit time, for no other method of comparison is really valid. The final cost of a processed substrate is quite dependent on local labor prices, the number of operations in the process, process yield for each operation, and the number of times the substrate must be handled in these operations. Costs would be very low in a highly automated installation compared with a prototype operation where every operation is essentially performed by hand. Therefore, specific cost figures cannot be attained without an intimate knowledge of the process and the equipment used. In all the following discussion, all numbers are based on an 8 hr/day operation.

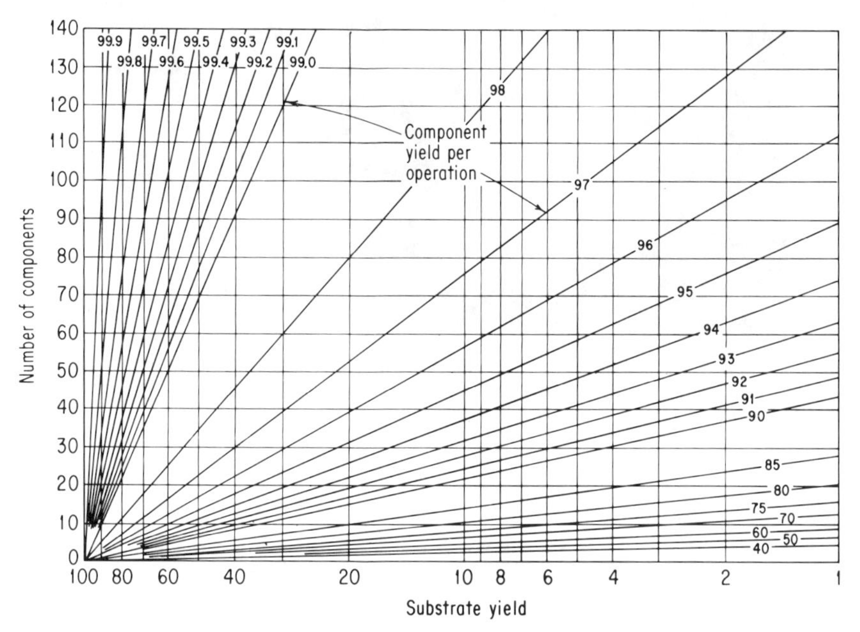

Fig. 48. Substrate yield vs. component yield.[16]

Vacuum Equipment. The least expensive vacuum system, such as shown in Fig. 49, when installed with all internal mechanisms, costs between $12,000 and $15,000. When depositions are made through masks, two to four deposition runs per day are attainable. For overall surface deposition, vacuum-system cost is reduced to approximately $10,000 per system, and four to eight depositions of about 160 in.^{2} of substrate surface can be deposited per vacuum-system cycle. The substrates must then be double-etched to obtain a resistor-conductor network. Each substrate must be handled individually even though batch operations are possible. When depositions are made through masks, one operator can produce 60 to 100 complete substrates per day that do not require any further operations or trimming.

By overall deposition and etching, two operators can fabricate 700 to 1,000 substrates per day exclusive of manpower for trimming.

Screen Fabrication. Equipment requirements for screened and fired resistor networks vary from one furnace and one printer to three or more furnaces and two or three printers. Belt furnaces, as shown in Fig. 50, cost \$7,000 to \$15,000 each and vary in capacity from several hundred substrates to several thousand substrates per day. Printers cost about \$2,500 for a hand-operated model such as shown in Fig. 51 to approximately \$7,500 for a production printer such as shown in Fig. 52. High-speed rotary printers are also available which will print 60 substrates/min or 3,600 substrates/hr. Even with a manual printer, 1,500 substrates/day is easily achieved. A rotary production unit is capable of handling about 25,000 substrates/day or more. As a basis for comparison, three operators using two hand printers and two furnaces can produce between 1,500 and 2,000 substrates/day exclusive of any trimming operation. Using self-feed automatic screeners, two operators can produce 5,000 to 7,000 substrates/day. This fact alone accounts for the increasing popularity of the screening process as a method of producing resistor-conductor networks. Screened resistors almost always require trimming. A basic abrader costs less than \$1,000. By using this unit, the abrading station in Fig. 53 is available at a cost of approximately \$15,000.

Chemical Depositions. Equipment for chemical depositions varies from a few beakers and hot plates to plating tanks and power supplies to accurately controlled furnaces for titanium-salts depositions and pyrolytically deposited tin oxide. Equipment can cost anywhere from \$200 to several thousand dollars, depending on volume requirements and the process itself.

FIG. 49. Vacuum system complete with vacuum pumps and controls. (*Consolidated Vacuum Corporation.*)

FIG. 50. Belt furnace used in the firing of cermet materials. (*BTU Engineering Corporation.*)

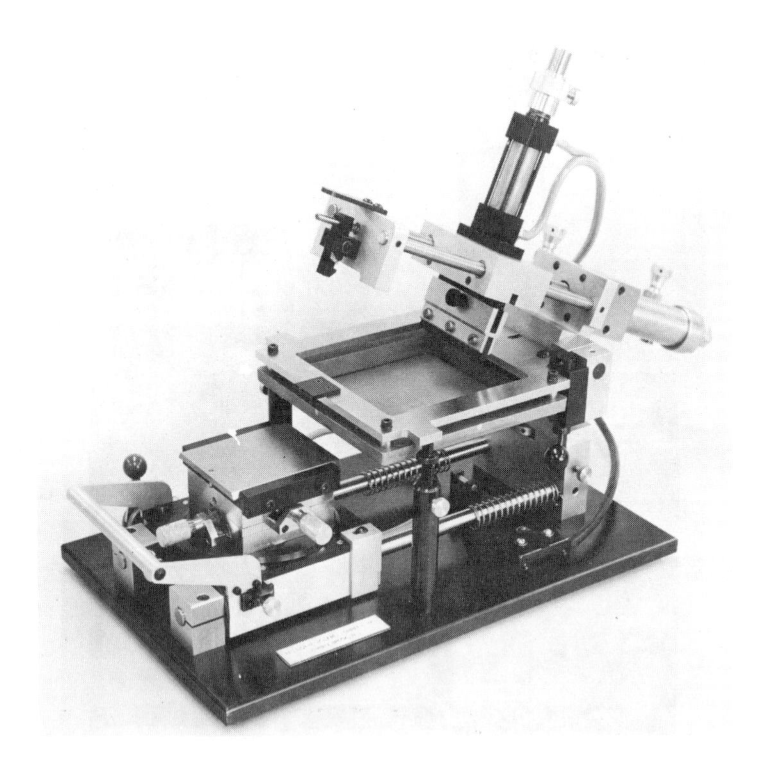

FIG. 51. Laboratory-model screen printer used in the fabrication of SAF hybrid-circuit substrates. (*Precision Systems Company, Inc.*)

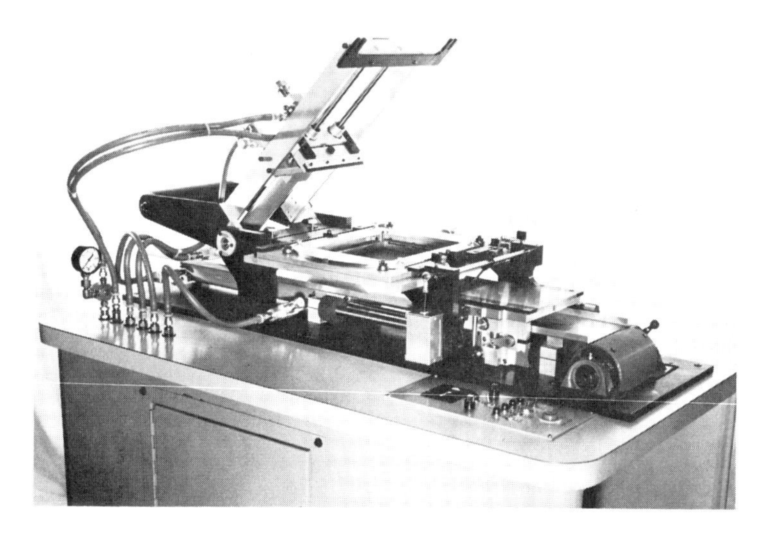

FIG. 52. Production-model screen printer. (*Precision Systems Company, Inc.*)

Capital-equipment costs are relatively low for all varieties of immersion and electroless deposition. Equipment consists of heated beakers or tanks. Two operators can deposit and etch about 800 nickel phosphorus substrates/day, using beakers and hot plates with simple etching equipment. Three operators could double this. Electroless and immersion plating of all sorts of small parts is done daily in large heated tanks. These cost no more than a few hundred dollars per tank where literally thousands of small parts can be plated by one operator.

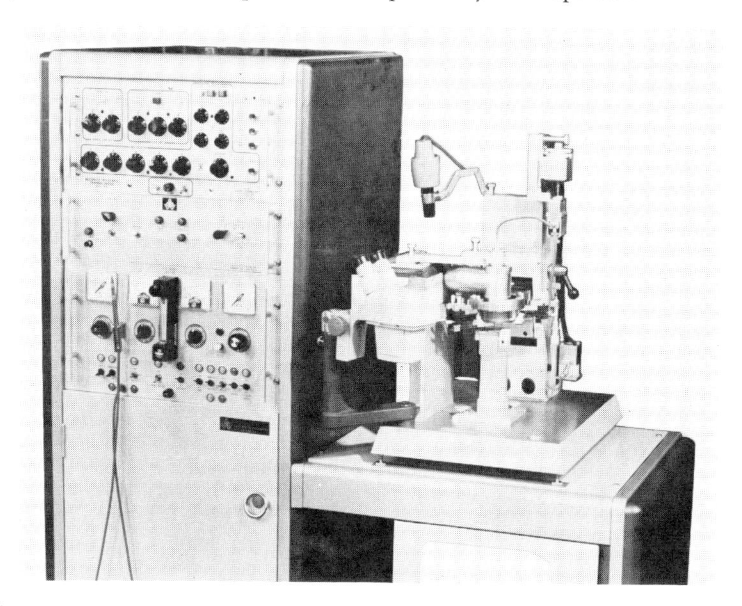

FIG. 53. Semiautomatic abrading system used to trim resistors to value. (*S. S. White Company, Inc.*)

In electroplating, tanks and power supplies cost $1,000 to $5,000 per plating operation depending on the size of the operation. Using barrel-plating techniques, one operator can plate thousands of parts per day through several different plating operations. This is the reason why electroplating is one of the most popular methods of metallic deposition: it is very economical. Second, the technology is relatively simple, with a minimum number of "tricks of the trade." Titanium-salts and pyrolytic tin oxide depositions require considerably more complicated equipment. While these processes are used, little is known about operator requirements. Since both depositions are used and are competitive with other methods, it must be assumed that they are sufficiently economical to warrant use or provide an advantage that makes any cost differential justifiable.

Finally, capital-equipment costs for chemically deposited magnetic films is certainly much lower. Again, equipment is no more elaborate than beakers or tanks and a magnetic field. While the composition and characteristics of chemically deposited films are different from vacuum-evaporated magnetic films, the production rate is considerably greater because the vacuum-system time requirements are eliminated. Little information is available about yields for either process. All companies are reluctant to state process yields. Since the films are being repeatedly produced and used, yields must be sufficiently high, for the costs of magnetic arrays are considerably lower than those of core arrays.

Integrated Circuits. As previously stated, integrated circuits cannot be considered a true deposition technique. They have been included because of the basic interest in them and the meteoric rise in their use in the past few years. It has been reported that it costs $250,000 to establish a small I/C capability and $500,000 to staff and operate it the first year. These costs are extremely high in comparison with the costs of establishing a hybrid-circuit facility and account for the specialization of only a few companies in I/Cs as compared with the vast number of companies engaged in all forms of hybrid circuity. This is so because equipment costs and the technical capability of both equipment and personnel are less stringent for hybrid circuits than integrated circuitry. Most companies developed I/Cs as an extension of their basic semiconductor capability. This alone has limited the number of entries into the field of I/Cs.

There is certainly no point in starting anew the debate on which technique to use. It is obvious that there must be an advantage to each technique, or the producer would switch to another method, as many have done. Certainly each technique offers distinct advantages, and these advantages prescribe one method over another. Thick films offer a distinct economic advantage where high-volume production is required and where component characteristics are not critical. This, of course, covers the bulk of present and potential microelectronic applications. It includes computers, radios, television, washing machines, machine tools, and a vast number of miscellaneous industrial and consumer products that will find economic advantage in hybrid circuitry. Thin films, on the other hand, have proved themselves in respect to extremely high reliability, improved component characteristics, or characteristics that can be tailored for special applications. Etched thin films are more economical than any other method of obtaining a few circuits or fast prototypes. With final processing equipment commercially available, anyone can fabricate his own circuits in small quantities (100 to 200 or less) and be economically ahead of the game, since deposited substrates can be purchased along with the equipment for the final processing. This eliminates all the deposition and control problems of the user and permits him to concentrate on the problems of fabrication and application. It also reduces the capital-equipment investment of the user to a minimum. Vacuum depositions are also useful where certain component characteristics are required or where other methods may not produce the same results. This is the case for some forms of flip-chip bonding and magnetic thin films. Chemical depositions and plating techniques are used extensively in electronic fabrication and in microelectronic assemblies for the reasons already discussed. Chemical depositions are used because of their ease of application, diversity of types available, and, for small users, the ridiculously low capital investment for some types of deposits. It is difficult to find a lower capital investment, for resistor-conductor networks, than two hot plates, two beakers, and $200 worth of chemicals.

Integrated Circuits. The use of I/Cs is growing rapidly. Originally, only digital circuitry was attempted in integrated form because of the large usage of a few circuit types. Now analog and operational circuitry is being produced in integrated form for special purposes where the volume is sufficiently large to warrant development. Just as hybrid circuitry is rapidly replacing individual component assemblies, even in small-volume applications, integrated circuits will eventually replace many hybrid applications when the cost is reduced to or below the point of economic competition and as I/C technology develops circuitry to replace hybrid applications. Hybrids, however, will continue to be used because of their special properties. Hybrid circuitry, with its method of producing intraconnections and components, will not fade from the scene, but one may well expect a marriage of these techniques in the near future and growth in the use of both as electronics continues to expand into consumer markets.

REFERENCES

1. Stein, S. J., and W. F. Ebling: Some Practical Considerations in the Fabrication of Printed Glaze Resistors and Circuits, *Proc. Electronic Components Conf.*, 1966, pp. 8–16.
2. Merz, K., C. Huang, C. Mabie, and R. Murphy: Tungsten Carbide–Tungsten Resistive Glazes, *Proc. Electronic Components Conf.*, 1966, pp. 1–7.
3. Peek, J. R.: Prediction of Temperature and Power Handling Capabilities of Thin Film Resistors and Circuits, *Proc. Electronic Components Conf.*, 1966, pp. 68–78.
4. Cubert, J. S., S. Markoe, T. Matcovich, and R. P. Moore: Face-down Bonding of Monolithic Integrated Circuit Logic Arrays, *Proc. Electronic Components Conf.*, 1966, pp. 156–167.
5. Block, M. L., and A. H. Mones: Properties of Indium Oxide Glaze Resistors, *Proc. Electronic Components Conf.*, 1966, pp. 191–196.
6. Farrell, J. P., and C. H. Lane: Cermet Resistors by Reactive Sputtering, *Proc. Electronic Components Conf.*, 1966, pp. 213–224.
7. Crowder, J. R., and L. W. Nail: Advanced Thin Film Capacitor Processes, *Proc. Electronic Components Conf.*, 1966, pp. 313–334.
8. Saunders, R.: Thin-film Hybrids—A Practical Approach to Integrated Circuitry, *Proc. Electronic Components Conf.*, 1965, pp. 63–69.
9. Kuo, C. Y., J. S. Fisher, and J. C. King: Thermal Processing of Tantalum-Nitride Resistors, *Proc. Electronic Components Conf.*, 1965, pp. 123–128.
10. Hoffman, C. C., V. L. Buchetta, and K. W. Fredrick: Adhesion of Platinum-Gold Glaze Conductors, *Proc. Electric Components Conf.*, 1965, pp. 381–386.
11. Driear, J. R.: Observations on Formation of "Cleavage" at Interface between Glaze Resistor and Termination, *Proc. Electronic Components Conf.*, 1965, pp. 387–392.
12. Melan, E. H., J. M. Greenman, F. J. Pakulski, and P. C. Reichert: Hybrid Integrated Circuit Delay Line, *7th IEC Electronic Circuit Packaging Symp.*
13. O'Connell, J. A.: Thick Film Technology, *Proc. NEP/CON*, 1966, pp. 112–129.
14. Gale, E. H., Jr.: Thermal Conductance of Thin Film Resistors, *Proc. NEP/CON*, 1966, pp. 130–140.
15. Grossman, J. J., and W. D. Fuller: Fabrication Techniques in Thin Film Circuit Technology, *Proc. NEP/CON*, 1965, pp. 140–156.
16. Saunders, R., and G. R. Heidler: An Analysis of Digital Hybrid Packaging and Connections, *Proc. NEP/CON*, 1964, pp. 303–314.
17. Warner and Fordemalt (eds.): "Integrated Circuits," McGraw-Hill Book Co., New York, 1965.
18. Terman, F. E.: "Radio Engineers' Handbook," McGraw-Hill Book Co., New York, 1943.
19. Editorial, Substrates for Microelectronics, *Electronic Packaging and Production*, March, 1966, pp. 86–99.
20. Heidler, G. R.: Simplifying Design of Two-dimensional Resistors, *Electronic Packaging and Production*, September, 1963, pp. 34–35.
21. Heidler, G. R.: Incremental Design of Two-dimensional Resistors, *Electronic Packaging and Production*, January, 1966, pp. 22–25.

22. Jones, D. E. H.: The Electrical Properties of Vacuum and Chemically Deposited Thin and Thick Resistive Films, *Microelectronics and Reliability,* vol. 5, pp. 305–321, January, 1966.
23. Hebb, E. L.: Microcircuitry by Chemical Deposition, June, 1962, Diamond Ordnance Fuze Laboratories, ASTIA AD281 843.
24. Morton, J. A.: The Microelectronics Dilemma, *International Science and Technology,* no. 55, July, 1966, pp. 35–44.
25. Devidse, P. D., and L. I. Maissel: RF Sputtering of Insulator Films Offers Many Advantages, *Insulation,* April, 1966, pp. 41–43.
26. Reimherr, G. W.: New Thin Film Magnetics Show Application Potential, *Electronic Design,* July, 1965, pp. 48–50.
27. Hoffman, L. C.: Precision Glaze Resistors, *Am. Ceram. Soc. Bull.,* September, 1963, pp. 490–493.
28. Short, O. A.: Silver Migration in Electric Circuits, *Tele-Tech and Electronic Industries,* February, 1956.
29. Schenkel, F. N.: Thin Film Capacitance Elements: Which Is Best for Your Purpose? *Electronics,* Jan. 25, 1965, pp. 67–72.
30. Sihovonen, Y. T., S. G. Parker, and D. R. Boyd: Printable Insulated Gate Field-effect Transistors, *J. Electrochem. Soc.,* January, 1967.
31. Schlabach, T. D., and D. K. Rider: "Printed and Integrated Circuitry," McGraw-Hill Book Co., New York, 1963.
32. Stetson, H. W.: Ceramic and Glass Insulation in Data Processing Equipment, *Proc. Electrical Insulation Con.,* September, 1965, pp. 175–178.

Chapter 6

CONNECTORS AND INTERCONNECTION DEVICES

By

RALPH SAUNDERS

Burroughs Corporation Laboratories
Defense, Space and Special Systems Group
Advanced Development Organization
Paoli, Pennsylvania

INTRODUCTION

Historically, an electrical connector was generally considered only a hardware item and, as such, was often the last item considered in the design and packaging of a piece of electrical or electronic equipment. Requirements were unsophisticated and were met by a number of standard, off-the-shelf connectors. These were reasonably satisfactory for most early applications of electronic equipment, in which signal voltage and current levels were relatively high. Thus, the selection and application of connectors posed no real problems.

Power connectors and cable connectors for signals between associated pieces of equipment were the first requirements for easily removable connections. In many cases, soldered connections and screw-type terminal boards were used to interconnect various subassemblies.

With the rapid growth in the field of electronics, equipment became more complex, and more and more connectors were needed to interconnect electronic functions in a practical, modular form that was both manufacturable and maintainable. *In today's packaging concepts, connectors have become a very vital link in forming or making up a complete electronic system, and as such, they are in fact a very important component part of that system,* rather than just being merely another item of necessary hardware. Thus, as with any component part, connector requirements must be evaluated, and connectors must be selected just as carefully as are the other components before a package design is frozen, rather than *after* the fact when volumetric considerations can dictate a compromise connector selection based almost entirely on size alone, which could be detrimental to reliability—or on the resultant need for a "special" that could be overly costly.

Consideration must be given to all electrical, mechanical, and environmental stresses to which the connector will be likely to be subjected in use, as well as to the compatibility of physical form and dimension with the intended packaging concept of the equipment in which it will be used.

Not only will consideration of all available application information, followed up by careful connector selection, help to prevent misapplication; it should, in addition, result in functionally satisfactory and reliable interconnections.

This chapter is not intended to be a catalog of available connectors, nor is it intended to be highly technical in content. Rather it is meant to be an introduction to the very extensive and sometimes confusing field of connectors. It is a practical approach to the connector problem——selection, application, reliability—rather than a series of charts, graphs, and technical data concerning connectors.

It would be impossible in a single chapter to give all the available technical data and design specifications on all of the multitude of types, sizes, and varieties of connectors that are presently available. Also, connector technology and requirements are both advancing so rapidly that whole new connector concepts are constantly being developed by the connector manufacturers to meet the constantly advancing needs of the user; therefore, the content of this chapter on connectors will be limited to an introduction to and a discussion of the various connector families; broad engineering parameters; and design considerations that should provide the design engineer with adequate guidelines for the intelligent selection of connectors and interconnection devices. It cannot be stressed too heavily that for specific applications, manufacturers' literature, technical data, and engineering services should be consulted by the user.

GLOSSARY OF CONNECTOR TERMS

The glossary in Table 1 contains the most used and generally accepted terms and definitions relating to connectors of all types. It should be useful for communicating between users and manufacturers when discussing and specifying specific connector requirements.[1] (See Table 1.)

TABLE 1. Terms and Definitions

Accordion. A type of printed-circuit connector contact in which the spring is given a Z shape to permit high deflection without overstress.

Bayonet coupling. A quick-coupling device for plug and receptacle connectors, accomplished by rotation of the two parts under pressure.

Bifurcate. Describes lengthwise slotting of a flat spring contact, as used in a printed-circuit connector, to increase the flexibility of the spring and provide redundant points of contact.

Blade contact. A flat male contact, used in multiple-contact connectors, designed to mate with a tuning fork or a flat formed female contact.

Body. Main, or largest, portion of a connector to which other portions are attached or inserted.

Boot. A form placed around the wire termination of a multiple-contact connector to contain a liquid potting compound. Also, a protective housing usually made from a resilient material to prevent entry of moisture into a connector.

Cavity. The lengthwise opening in a printed-circuit edge connector that receives the printed-circuit board.

Closed entry. A design which limits the size of mating parts to a specified dimension, usually used in reference to pin and socket contacts.

Connector assembly. A mated plug and receptacle.

Contact. A name given to the element in a connector which makes the actual electrical contact. Also, the point of joining in an electrical connection.

Contact area. The actual area in contact between two contacts or a conductor and a contact, permitting flow of electricity.

Contact engaging and separating force. The force needed to either engage or separate randomly picked pins and sockets when they are both in and out of connector inserts. Values are generally established for maximum, average, and minimum forces.

Contact plating. The plated-on material applied to the basic metal of an electrical contact to provide for required contact-resistance and/or specified wear-resistance characteristics.

Contact resistance. Electrical dc resistance of a pair of mated contacts.

Contact retention. Defines the minimum axial load in either direction which a contact must withstand while remaining firmly fixed in its normal position within an insert.

Contact size. The diameter of the engagement end of a pin contact; also related to the current-carrying capacity of a contact.

Crimp. The act of compressing (deforming) a connector barrel around a wire (conductor) in order to make an electrical connection.

Dip-solder terminal. The projecting terminal on a connector designed to be inserted into a hole provided in a printed-circuit board, to be soldered.

Environmentally sealed. Provided with gaskets, seals, potting, or other devices to keep out moisture, dirt, air, or dust which might reduce its performance.

Extraction tool. A tool used for removing a removable contact from a connector body.

Feed-through. A connector or terminal block, usually having double-ended terminals which permit simple distribution and bussing of electrical circuits.

Ferrule. A short tube. Used to make solderless connections to shielded or coaxial cable. Also molded into the plastic inserts of multiple-contact connectors to provide strong, wear-resistant shoulders on which contact retaining springs can bear.

Frame. In the case of a multiple-contact connector having a removable body or insert, the frame is the surrounding portion (usually metal) which supports the insert and provides a means for mounting the connector to a panel or a mating connector half.

Grommet. A rubber seal used on the cable side of a multiple-contact connector to seal the connector against moisture, dirt, or air.

Guide pin. A pin or rod extending beyond the mating face of a two-piece connector and designed to guide the closing or assembly of the connector to assure proper mating of contacts, and to prevent damage to these contacts that might be caused by mismating of the connector halves.

Hermaphroditic connector. A connector in which both mating members are exactly alike at their mating face.

Hermetic seal. Hermetically sealed connectors are usually multiple-contact connectors in which the contacts are bonded to the connector by glass or other materials and permit a maximum leakage rate of gas through the connector of $0.1\ \mu/(\text{ft}^3)/(\text{hr})$ at 1 atmosphere of pressure.

Hood. An enclosure, attached to the back of a connector, to contain and protect wires and cable attached to the terminals of a connector. A cable clamp is usually an integral part of the hood.

Insert. That part which holds the contacts in their proper arrangement and electrically insulates them from each other and from the shell. See Body; also referred to as a *panel.*

Insert cavity. A defined hole in the connector insert into which the contacts are inserted.

Insertion tool. A tool used to insert removable contacts into a connector.

Insulation resistance. Resistance measured between any pair of contacts and between any contact and the metal shell.

Interfacial seal. Sealing of a two-piece, multiple-contact connector over the whole area of the interface to provide sealing around each contact. This is usually done by providing a soft insert material on one or both halves of the connector which are in compression when mated.

Jackscrew. A screw attached to one-half of a two-piece, multiple-contact connector and used to draw both halves together and to separate them.

Key. A short pin or other projection which slides in a mating slot or groove to guide two parts being assembled.

Keyway. The slot or groove in which a key slides.

Modular. A modular connector is one in which similar or identical sections can be assembled together to provide the best connector size for the application.

Pin contact. A round male-type contact, usually designed to mate with a socket or female contact.

Plug. The part of a connector which is normally "removable" from the other, or permanently mounted, part; usually that half of a two-piece connector which contains the pin contacts.

Polarize. To design the two mating halves of a two-piece connector such that only a particular combination of halves can be assembled, thus preventing accidental assembly of unrelated circuits.

Potting. Sealing of the cable end of a multiple-contact connector with a plastic compound or material to exclude moisture, prevent short circuit, provide strain, relief, etc.

Quick disconnect. A type of connector shell which permits rapid locking and unlocking of two connector halves.

Receptacle. Usually the fixed or stationary half of a two-piece, multiple-contact connector. Also the connector half, usually mounted on a panel and containing socket contacts.

Shell. Outside case, usually metallic, into which the insert (body) and contacts are assembled. Shells of mating connector halves usually provide for proper alignment and polarization as well as for protection of projecting contacts.

Socket contact. A female contact designed to receive and mate with a male contact. It is normally connected to the "live" side of a circuit.

Solder cup. A tubular end of a terminal in which a wire conductor is inserted prior to being soldered.

Solder eye. A solder-type terminal provided with a hole at its end through which a wire can be inserted prior to being soldered.

Tape cable. A form of multiple conductor consisting of parallel metal strips embedded in insulating material. Also called *flat flexible cable.*

Taper pin. A pin-type terminal having a tapered end designed to be impacted into a tapered hole to form a connection.

Taper tab. A flat terminal having tapered sides, designed to receive a mating tapered female terminal.

Tuning-fork contact. A U-shaped female contact, either stamped or formed, so called because it resembles a tuning fork.

Umbilical connector. A connector used to connect cables to a rocket or missile prior to launching, and which is ejected from the missile at the time of launching.

Wiping action. Action of two electrical contacts which come in contact by sliding against each other.

Wire wrap. Method of connecting a solid wire to a square, rectangular, or V-shaped terminal by tightly wrapping or winding it around with a special tool.

Zero-force connector. A connector in which the contact surfaces do not mechanically touch until it is completely mated, thus requiring no insertion force. After mating, the contacts are actuated in some fashion to make intimate electrical contact.

RACK AND PANEL CONNECTORS

Strictly speaking, rack and panel connectors are connectors that have one half, usually the female side, mounted in a panel or back plane, and the other half mounted on the face of a drawer or in a module in such a manner that when the drawer is slid into the rack or the module is plugged in, the connector is engaged. Rack and panel connectors are generally considered to be rectangular, multipin connectors, although round connectors can be and are used for rack and panel applications also.

Actually, the greatest usage of rack and panel connectors is probably not in backplane applications, but rather in cable-to-fixed-receptacle and cable-to-cable applications.

In the typical and representative types of rack and panel connectors that are illustrated below, both the basic bare rack and panel types and the cable types are shown to illustrate the variety of hoods, locking devices, cable clamps, and associated hardware that are common to this family of connectors.

In a true rack and panel application the connectors must be considered a precision component, and the connector alignment and connector mating characteristics must be carefully considered to ensure proper functioning of the equipment.

A very basic requirement, and one that is often overlooked, is that in *all* connector applications at least one connector half *must* be floating with relation to the other half in order for the contacts and other connector parts to align themselves properly. This happens automatically in cable and printed-circuit-board applications but it must be designed in for back-panel and drawer applications. In this type of application it can be accomplished by the use of floating bushings, with the addition of guide pins, for instance (Fig. 1).

Mating Characteristics. The following is a list of mating characteristics which must be considered when designing rack and panel equipment.[2]

1. *Connector mounting.* Does the connector have float mounting provisions to aid connector mating? Float mounting will allow some latitude in installation tolerances.

2. *Connector restrictions.* Are wires to the connectors restricting the float mounting characteristics due to the rigidity of the wire or the clamping of the cable? Are the screws in float mounting bushings torqued too tightly, resulting in crushed float bushings?

3. *Guide pins.* Heavyweight installations require guide pins to put the plug-in equipment and connector into the proper mating plane and also to take the strain off connector housing and contacts during mating.

4. *Shell configuration for alignment.* Does shell design (tapered entry, for instance) aid connector mating?

5. *Maximum mounting and misalignment tolerance.* To ensure positive mating, racks and fixtures must be designed to assure that equipment upon mating does not allow connectors to fall outside this tolerance. Damage to contacts, inserts, and housings results if they are outside allowable tolerances.

6. *Minimum and maximum contact travel.* This will help ensure proper contact mating and connector functioning.

7. *Connector mating forces.* When mating forces begin to exceed easy handling

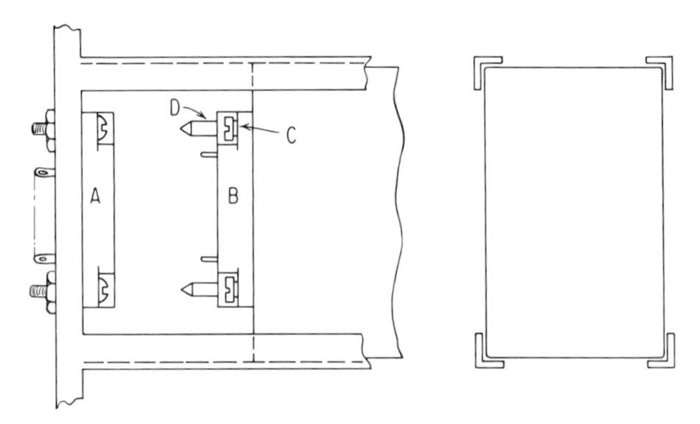

Fig. 1. Methods of floating and guiding connectors; drawer and panel application. With panel (female) connector A mounted solid, male connector B should be mounted to face of drawer with shoulder screws C that will allow a little sidewise movement of connector so that guide pins D can align connector before contacts mate. Also, drawer should have clearance inside track, both vertically and horizontally, to allow drawer to be aligned also by guide pins.

limits (which varies with equipment weight and size), mechanical assists are required.

8. *Mounting structure.* The mounting structure should be of sufficient thickness, strength, and rigidity to prevent deflection during mating of the structure on which the connectors are mounted. Deflection of the structure may prevent the connectors from mating and functioning properly.

9. *Connector sealing.* Employ proper gasketing to ensure integrity of environmentally sealed units, if required. It is very difficult to seal connectors with float mounting provisions, and special consideration must be given to such applications.

10. *Environmental requirements.* These must be considered when selecting connectors.

11. *Support hardware.* Support hardware requires space and necessary mounting provisions.

12. *Applicable connector specifications.* These are prescribed in the detail specification.

Available Types of Rack and Panel Connectors. There is an almost endless variety of types, sizes, and variations of rack and panel connectors available today. To simply list them all would require a whole book, rather than just a chapter, and would be of little help to the designer. What was considered a miniature connector yesterday has been largely replaced by the subminiature connector of today—which, in all probability, will be replaced by the microminiature connector of tomorrow. Contact density, contact numbers, and overall connector size are all relative to the state of the art from year to year, and in a handbook such as this it would be almost meaningless to attempt to catalog *all* the presently available connectors in all the multitude of types, styles, sizes, pin number and configurations, etc., because new specific designs and constantly changing equipment requirements would rapidly obsolete many of today's connectors in favor of tomorrow's required concepts.

Many of the various miniature rack and panel types are available already as subminiature connectors. If size and weight are significant factors, these subminiatures can be very useful. However, it is well to keep in mind that they are more fragile and cannot stand rough handling. Installation, wiring, and maintenance all require careful training of personnel, as with other subminiature component parts.

As a typical comparison of the size reduction that can be achieved in these subminiature designs, Fig. 2*a* shows the dimensional outline and pin spacing on a miniature 50-pin rectangular connector and Fig. 2*b* shows the comparable dimensions on a subminiature with a like number of pins.

Typically, the breakdown voltage of the connector in Fig. 2*a* would be about 2,800 volts ac rms with a contact rating of 7.5 amps. The connector in Fig. 2*b* would have a breakdown voltage of about 1,800 volts ac rms with a contact rating of only 3 amps.

Examples of Typical Rack and Panel Connectors. There is a wide variety of standard, miniature, and subminiature rack and panel connectors available—from rugged heavy-duty types to very-high-density, light-duty types. Associated hardware, such as hoods, cable clamps, locking devices, and various mounting methods, are available for most connectors. All of them can be used as true rack and panel connectors, or as cable-to-receptacle or cable-to-cable connectors. Contact ratings are dependent on contact size (see Table 2).

Heavy-duty Rectangular (Fig. 3). Illustrated are several examples of typical rugged, heavy-duty connectors especially suitable for heavy sliding-drawer applications. Terminals are available for solder and for taper pins. Crimp/removable contacts are also available as standard parts.

Miniature Rectangular (Fig. 4). This family is one of the most widely used of the rack and panel connectors. It is readily available as a plain rack and panel connector with polarizing guide pins (Fig. 4*a*); as a cable-to-receptacle connector with hood and cable clamp, either utilizing plain guide pins or with jack screws for ease of mating and locking (Fig. 4*b*); and with the male half as a hermetic seal connector for bulkhead mounting (Fig. 4*c*).

Solder and taper terminals as well as crimp/removable contacts are all standard.

Miniature Rectangular with Center Jack Screw (Figs. 5 and 6). Various shapes, from square to rectangular to long and narrow, are available, with a great variety of contact sizes and numbers.

Environment resistant (Fig. 7). Several shapes, sizes, and contact arrangements are available. Sealing is accomplished by interfacial contact of resilient insert moldings in one or both connector halves; by a pressurized seal around each contact and a grommet seal around the wires at the termination end; and by a lip seal barrier between the mated shells. Crimp/removable contacts with a closed-entry feature are available.

Rack and Panel Coaxial (Fig. 8). Various of the rack and panel shell types are available with contacts for standard and miniature coaxial and shielded cable. The individual coaxial connectors can be isolated by the use of an insulating insert, or the shields can all be brought to a common potential by the use of metallic inserts.

General-purpose Rectangular (Fig. 9). This is one of the oldest of the rack and panel connectors, but it is still generally useful. It is available with several combina-

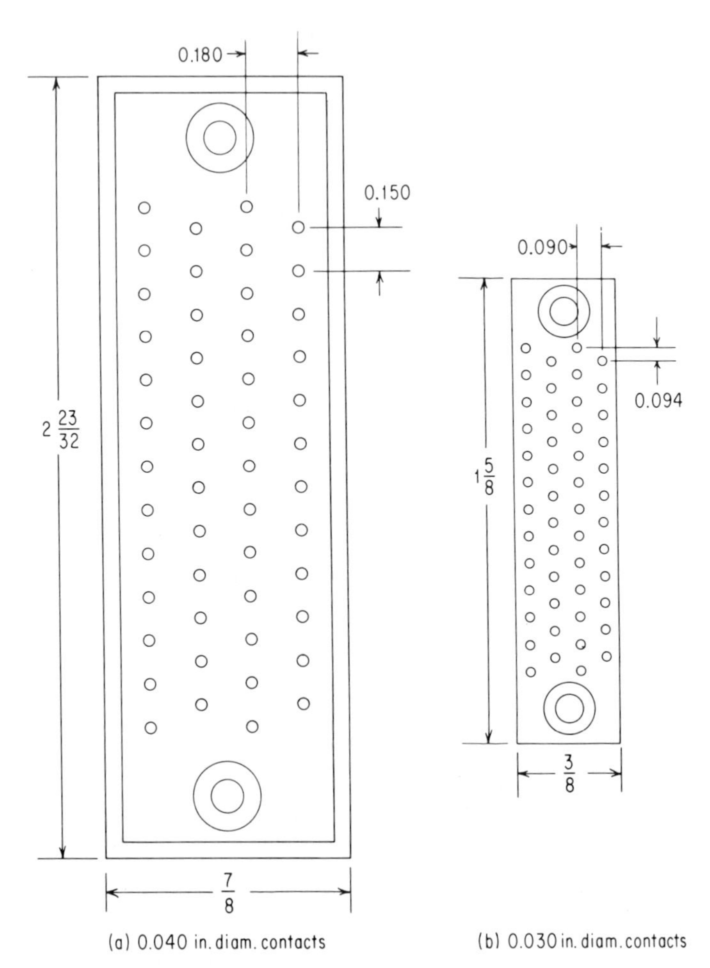

FIG. 2. Dimensional comparison of miniature and subminiature rectangular connectors.

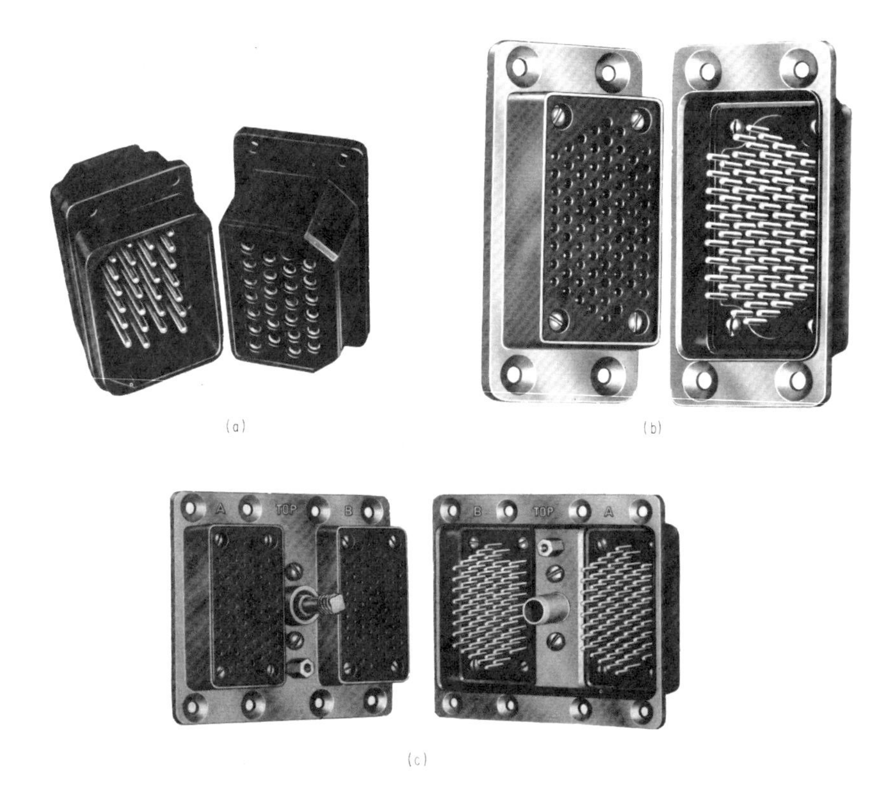

FIG. 3. Typical heavy-duty connectors. (*a*) Polarization accomplished by cut-corner shape of metallic shells. Male contacts protected by the shell enclosure. (*b*) Polarization accomplished by contact arrangement. (*c*) Two inserts in single housing. Center screw for ease of mating and positive locking. (*Burndy Corp.*)

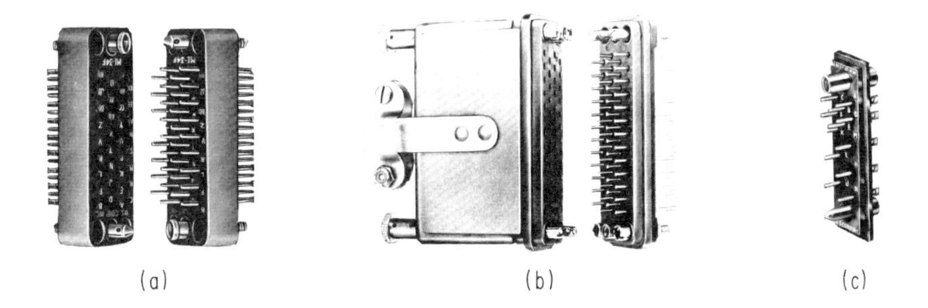

FIG. 4. Miniature rectangular connectors. (*U.S. Components, Inc.*)

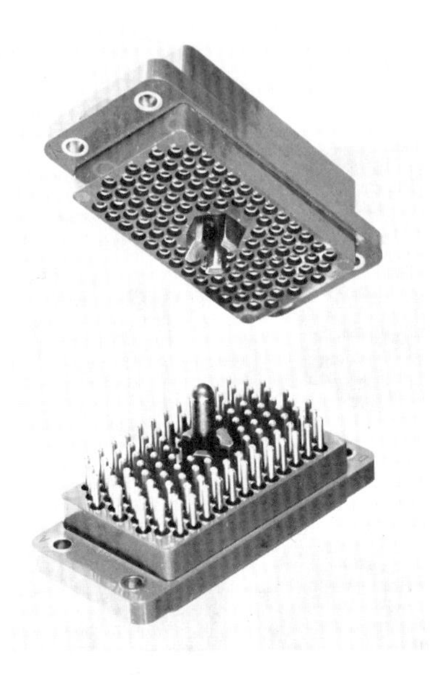

FIG. 5. Miniature rectangular connector with center jack screw. This patented PolarHex center jack screw assures alignment and provides for polarization where several similar connectors are used in the same application. (*Hughes Connecting Devices.*)

FIG. 6. Conventional rectangular connector with center screw. Here the center jack screw mates and unmates connector halves. Polarization and alignment provided by guide pins. (*U.S. Components, Inc.*)

FIG. 7. Typical environment-resistant rack and panel connector. (*Amphenol-Borg Electronics Corp.*)

tions of no. 12, no. 16, and no. 20 contacts which makes it very suitable for both power and signal circuits in a single connector block.

Contacts are available molded in, with solder terminations, or removable, with crimp termination. Hoods and cable clamps are provided.

Jones Connectors (Fig. 10*a*). This family of connectors has been an industry standard for over a generation. They are designed for medium or heavy-duty power applications, or for any application where a very rugged connector is required.

The male contacts are blade-type and the female contacts are "knife-switch" type that provides a solid contacting area on both sides of the male contact. They are available in from 2 to 33 contacts with solder terminations, and are rated at from 4.5 to 15 amps, depending on the series.

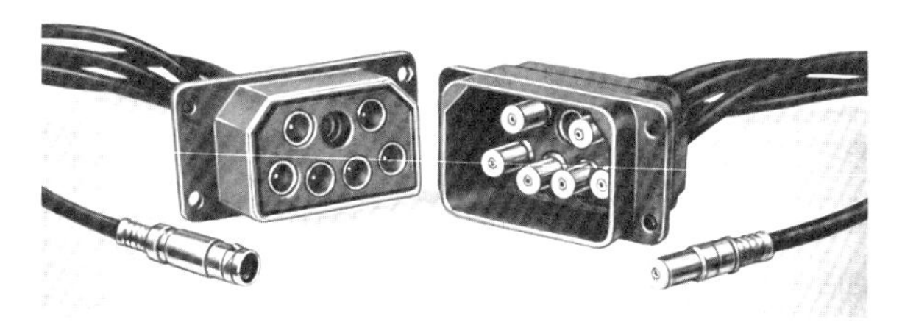

FIG. 8. Rack and panel coaxial connector. (*Burndy Corp.*)

FIG. 9. General-purpose rectangular connector. (*U.S. Components, Inc.*)

Mini-Jones Connectors (Fig. 10*b*). The "Mini-Jones" connectors have the same contact types as the standard Jones connectors, but are somewhat smaller and feature crimp-on, snap-in contacts that have a rating of 4.0 amps. They are available in from 19 to 68 contacts.

In both the Jones and Mini-Jones families, male and female contacts are available in either the plugs or the receptacles. Associated hardware includes hoods, straight-through and right-angle cable clamps, locking devices, and surface and flush mounting receptacles.

D Subminiature Series Connectors (Fig. 11). These are available in arrangements of from 9 to 50 contacts in size 20, and also with various combinations of coaxial and shielded connector contacts. The associated hardware includes floating mountings, locking devices, hoods, and cable clamps.

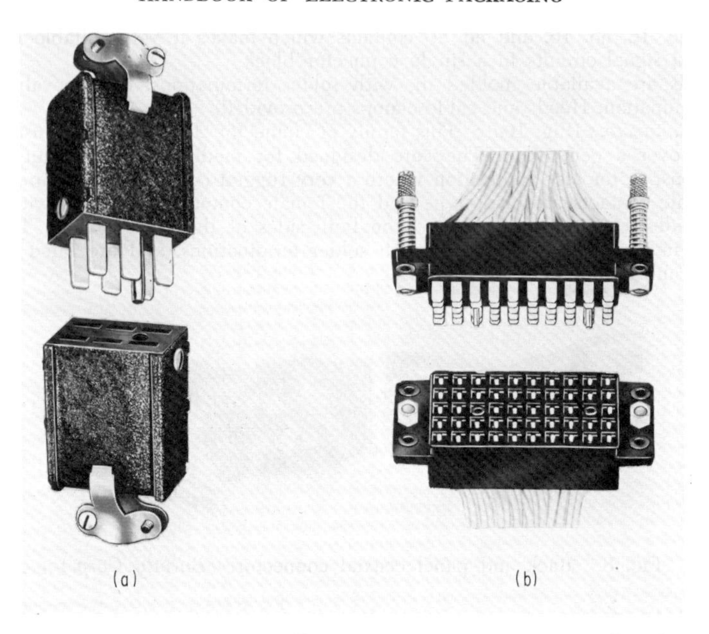

FIG. 10. (*a*) Jones connector; (*b*) Mini-Jones connector. (*Cinch Mfg. Co.*)

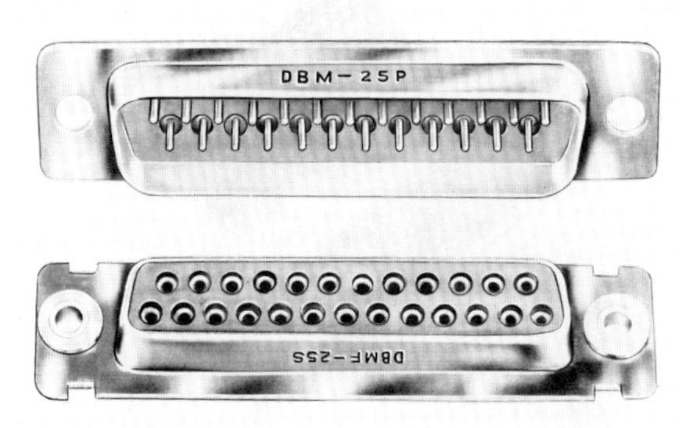

FIG. 11. D subminiature series conductor. (*Cinch Mfg. Co.*)

Blue Ribbon * *Connectors* (Fig. 12). This family of connectors is an extremely rugged, nonenvironmental, general-purpose connector, featuring easy insertion and extraction. Contacts are ribbon-type with generous contacting surfaces that are self-wiping and self-cleaning.

A variety of configurations and styles are available, including the barrier polarization type without shells (Fig. 12*a*); the pin polarization type without shells; the barrier polarization type with latching-type keyed shells (Fig. 12*b*); and the barrier polarization type with plain keyed shells. Straight-through and right-angle hoods with cable clamps are available, as are floating mountings for panel-mounted receptacles. The hardware is nonmagnetic, and they are available with 8, 16, 24, and 32 five-amp contacts.

* Trademark, Amphenol-Borg Electronics Corp.

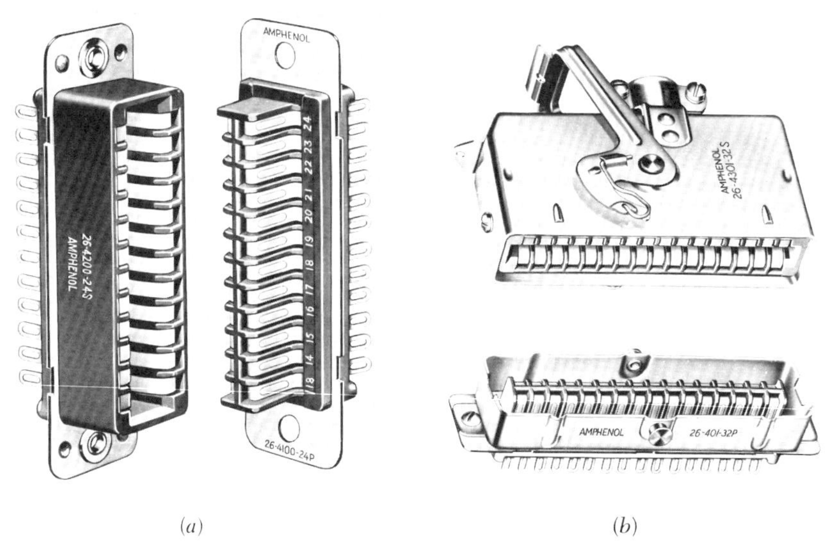

(*a*) (*b*)

FIG. 12. Blue Ribbon connectors. (*Amphenol-Borg Electronics Corp.*)

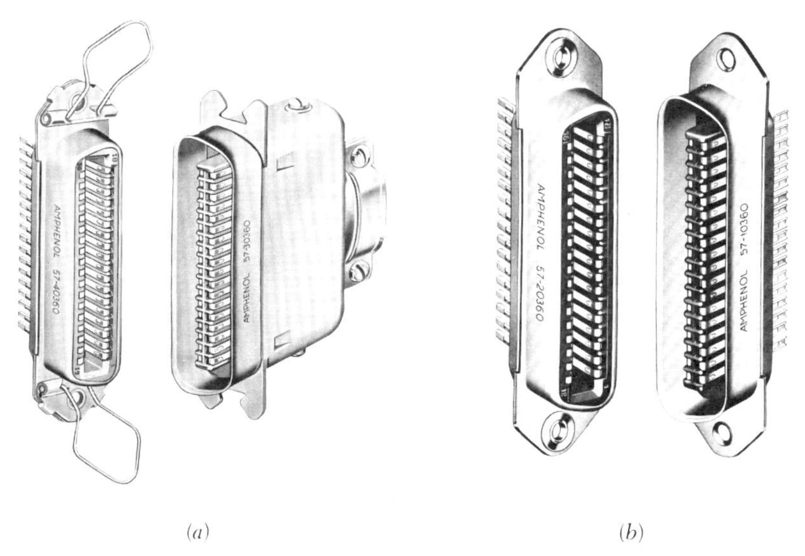

(*a*) (*b*)

FIG. 13. Micro-Ribbon connectors. (*a*) Cable-to-chassis; (*b*) rack and panel. (*Amphenol-Borg Electronics Corp.*)

Micro-Ribbon * *Connector* (Fig. 13). This family of connectors is nonenvironmental. Contacts are ribbon-type with solder terminations. The trapezoidal shape of the metal shells polarizes the connector halves. It features smooth, easy insertion and withdrawal, with self-wiping, self-cleaning contacts. It is used extensively in telephone

* Trademark, Amphenol-Borg Electronics Corp.

communications equipment. It is available in cable-to-chassis style (Fig. 13*a*), cable-to-cable style, and as a plain rack and panel connector (Fig. 13*b*). Hoods, latching devices, and floating mountings are available. The hardware is nonmagnetic, and it is available with 14, 24, 36, and 50 five-amp contacts.

Miniature Rectangular with "Floating" Molded Inserts (Fig. 14). This family features rugged die-cast frames in which the molded inserts containing the contacts are mounted in such a manner as to allow relative float. Mechanical stresses of mounting, engaging, and disengaging are carried by the frame rather than by the molded inserts. Guide pins align the contacts, and jack screws provide force for engaging, disengaging, and locking.

FIG. 14. Miniature rectangular connector with floating molded inserts. This example has three 50-pin inserts within a single housing. It is also available in single and double combinations. (*U.S. Components, Inc.*)

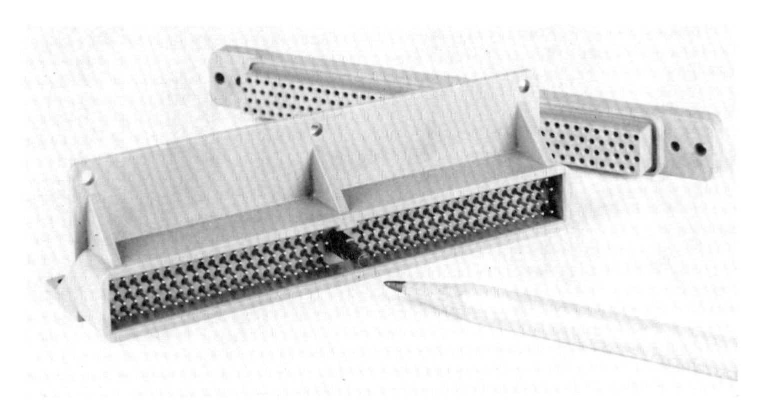

FIG. 15. High-contact-density—miniature rectangular connector. (*Amphenol-Borg Electronics Corp.*)

High-contact-density—Miniature Rectangular (Fig. 15). The connector illustrated is typical of a whole new family of miniature rectangular connectors now available. Contact spacing is 0.100 in. or less. The contacts are available in crimp/removable and solder termination types, and also with dip-solder tails either straight out or bent at right angles. The bodies are molded of high-impact-strength plastics.

Other Molded Rack and Panel Configurations (Fig. 16). A variety of connector types and sizes are now being completely molded from high-impact-strength plastic materials. They contain no metallic parts, except for the contacts. Mounting ears, guide pins, polarizing devices, and shells are all a part of the basic insert molding.

Functionally, similar connectors manufactured by different manufacturers are usually quite similar, but there are some design differences in hoods, shells, clamps, etc. Although it is possible to mate connector halves of similar types that have been produced by two different manufacturers, especially those manufactured to MIL specifications, it is generally not considered good practice because of sometimes incompatible manufacturing tolerances between manufacturers, and in the advent of problem areas there is no single responsibility to go back to.

With the exception of the ribbon contact connectors shown, all the other types and families illustrated are available with either factory-assembled contacts or crimp/removable contacts. The terminations available vary from manufacturer to manufacturer, but generally speaking, the terminal styles available include solder, wire-wrap, taper-pin, Termi-point,* and dip-solder. (Specific contact information will be found under the heading "Contacts and Tooling" in this chapter.)

FIG. 16. Typical examples of completely molded rack, panel, and cylindrical connectors. (*Amphenol-Borg Electronics Corp.*)

FIG. 17. Mix or match contacts. Four different wiring applications all within a single connector block: (*a*) stranded wire crimped in machined pin and socket contacts; (*b*) stranded wire crimped in sheet-metal formed contacts; (*c*) shielded wire in miniature coax contacts; (*d*) twisted-pair in miniature coax contacts. (*Burndy Corp.*)

One of the interesting features of many of today's rectangular rack and panel connectors employing removable contacts is the ability to instantly intermix various sizes of pin and socket contacts as well as miniature and subminiature coax contacts within the same connector block (Fig. 17). This is a practical and economical way of using a single connector to accommodate voltage, logic, and signal circuits.

* Trademark, AMP Incorporated.

Another advantage is that in development and prototype work it is extremely helpful to have the ability to change from single-wire leads to twisted pairs or coax if noise is a problem, as well as to accommodate circuit changes or additions conveniently.

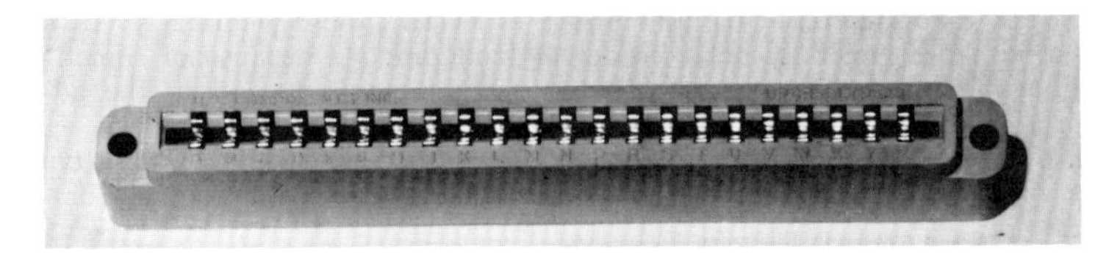

FIG. 18. Typical double-row printed-circuit edge receptacle with taper pin socket terminals. (*U.S. Components, Inc.*)

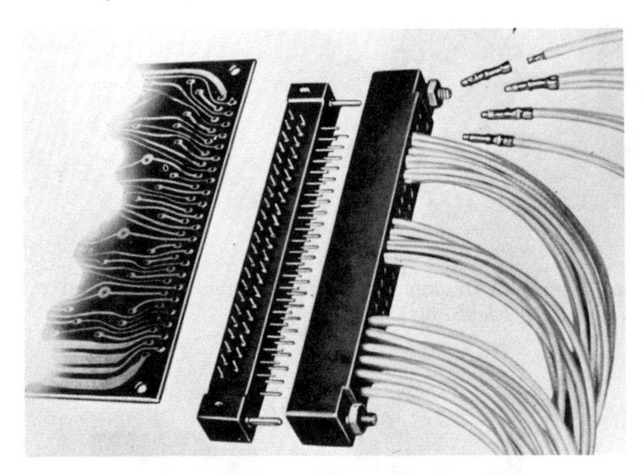

FIG. 19. Two-piece pin and socket printed-circuit connector. (*Burndy Corp.*)

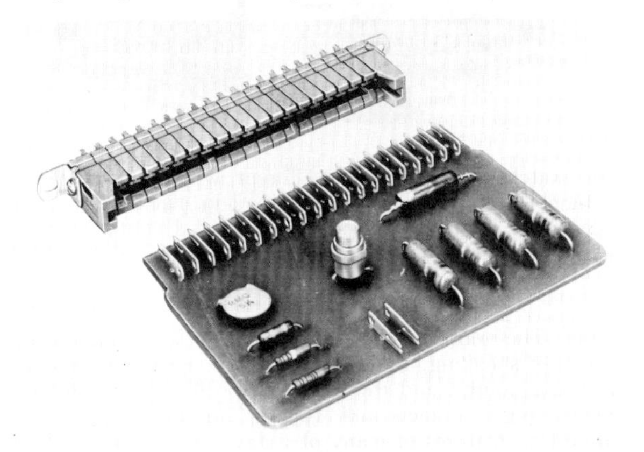

FIG. 20. Two-piece printed-circuit connector with individual board contacts. Conventional receptacle. Mating Varicon (trademark, Elco Corp.) contacts staked and soldered into circuit board. Individual contacts (shown) supplied mounted in carrier strip for ease of assembly. Board contacts also available (*a*) molded into board; or (*b*) in molding with dip-solder terminals and provision for fastening molding to circuit board. (*Elco Corp.*)

PRINTED-CIRCUIT CONNECTORS

Printed-circuit connectors are divided into two main types: (1) the one-piece "edge" connector, a receptacle containing female contacts designed to receive the edge of a printed-circuit board on which the male contacts are etched or printed (Fig. 18); (2) the two-piece male and female connector, one piece (usually the male contact) of which is attached to the printed-circuit board to allow the two connector halves to mate as do conventional connectors (Fig. 19). A variation of the two-piece printed-circuit connector is one in which the female contacts are contained in a conventional molding and the individual male contacts are fastened to the printed-circuit board in such a fashion as to form the mating half of the connector (Fig. 20).

Edge Receptacles. Standard connectors of this type are available with a single row of contacts (to mate with contacts on one side only of a printed-circuit board) or with a double row of contacts (to mate with individual contacts on both sides of a circuit board, thus allowing twice the contact density of the single-row receptacle in the same length). An illustration of a typical double-row connector is shown in Fig. 18. Figure 21 illustrates a typical miniature edge connector.

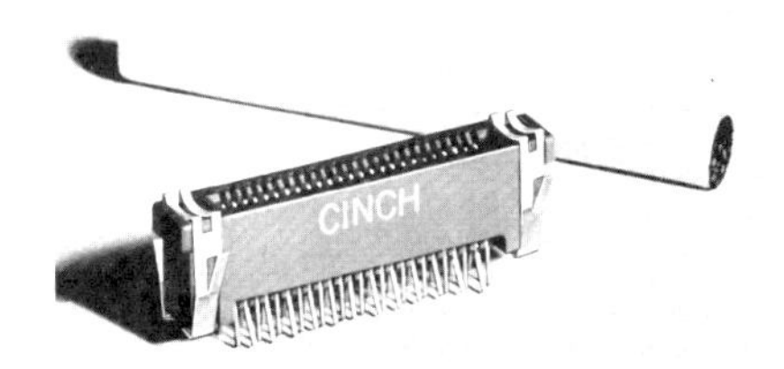

Fig. 21. Miniature printed-circuit edge-connector Tykon (trademark, Cinch Mfg. Co.). An example of 0.050-in. contact centers printed-circuit edge connector. Available in both single- and double-row contacts for 0.031- and 0.062-in.-thick printed-circuit boards. Contact terminals available for solder termination for wire and for dip-soldering into printed-circuit boards. Conventional and clip mounting. (*Cinch Mfg. Co.*)

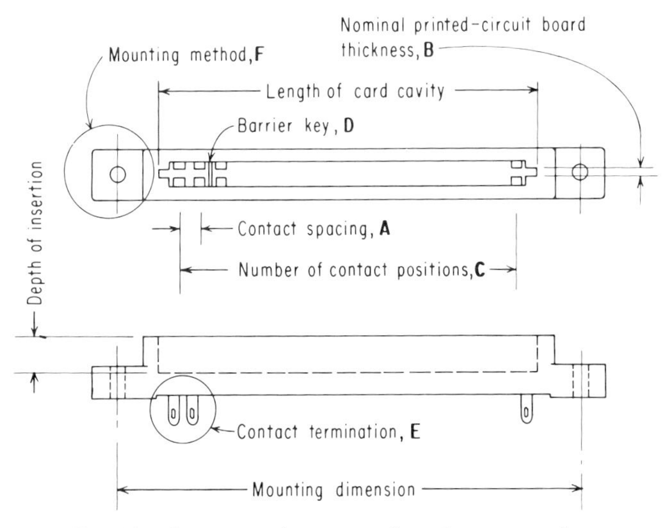

Fig. 22. Important dimensions for edge receptacles.

Figure 22 shows the important and necessary dimensions required in a given application. The technical data, following, refers to Fig. 22 and the prefixed letters refer to the applicable design dimensions and features indicated in the figure.

Standard Contact Spacing Ⓐ (inches).

0.200 0.156 0.125 0.100 0.078 0.075 0.050 0.025

Nominal Printed-circuit-board Thickness Ⓑ (inches).

0.125 0.093 0.062 0.031 0.025

Number of Contact Positions Ⓒ (X2 for total number of contacts in double-row receptacles). A variety of contact positions from 6 through 100 are readily available. Many manufacturers are now using modular molding techniques so that rather than the formerly fixed number of contact positions (6, 10, 15, 18, 22, etc.) it is possible today to obtain almost any required number of contact positions. Manufacturers' literature should be consulted for specific application requirements.

Barrier for Keying or Polarization Ⓓ. Press-in barrier slugs are available to key specific printed-circuit boards to desired locations in multiple connector applications, and to polarize boards in specific connectors (Fig. 23). There are two types, depending on the manufacturer: (1) the type that presses into slots between contact positions and (2) the type that is inserted into one of the contact positions. This type means the loss of a contact position.

Contact Terminations Ⓔ. Quite a few types of contact terminations are available. In connectors with 0.100-in. or greater spacing, the terminals illustrated in Fig. 24 are available as standard with most manufacturers.

Contact Configurations Ⓔ. Several basic types of contact configurations, and variations of these, are available. All of them can be good, reliable contacts, assuming proper materials and proper design. The bellows type, plain and modified, including bifurcated, is probably the most common type in use today, although all the other types also are in use depending on the manufacturer and his philosophy and also on the cost picture of the end product in which they will be used. Figure 26 illustrates several of the basic configurations.

Molded Body Configurations Ⓕ. There are three basic connector body configurations and mounting methods for edge-type printed-circuit receptacles. Figures 27 to 29 illustrate these three types.

Molding Materials, Contact Materials, and Platings. These are discussed under the heading Connector Materials in this chapter.

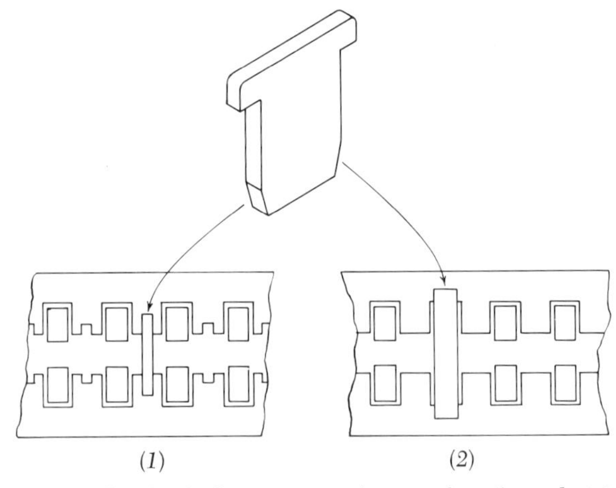

FIG. 23. Sketch of edge-connector barrier slugs for polarizing.

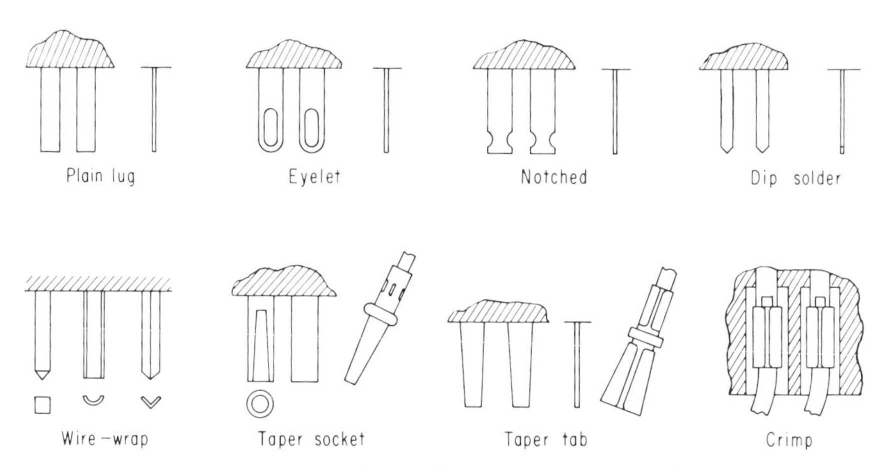

FIG. 24. Types of printed-circuit contact terminals.

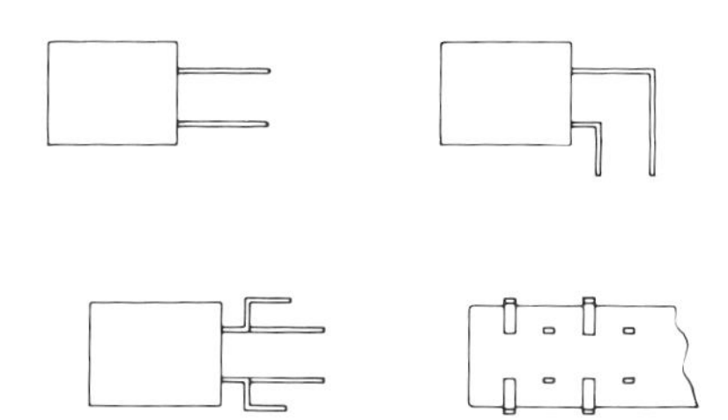

FIG. 25. Configuration of formed solder terminals.

In connectors with contact spacing of less than 0.100 in., the variety of contact termination types is limited basically to plain flat or round wire, due to the limitation on the physical size of the contact material. These terminations can be brought straight out or can be formed to make them more easily adaptable for wiring or dip soldering into a printed-circuit board. Figure 25 shows straight-out terminals and terminals formed in two different ways.

Edge-receptacle Applications. The most important design consideration in the application of printed-circuit edge receptacles is that there must be a floating relationship between the receptacle and the printed-circuit board. One or the other would normally be rigidly mounted, but both must *never* be. The float must be such that it allows the printed-circuit board to be properly aligned and centered in the receptacle by the pressure of the contacts themselves. Either the receptacle or the board must be free to mate with the other.

In most applications the receptacle is rigidly mounted in a frame or back plane, and the printed-circuit board is guided into the receptacle cavity freely, by hand, by means of mechanical guides on the equipment structure, or by means of built-in guides on the connector itself. A large majority of printed-circuit edge-receptacle failures can be attributed directly to misapplication in which there is no floating relationship between the board and the receptacle.

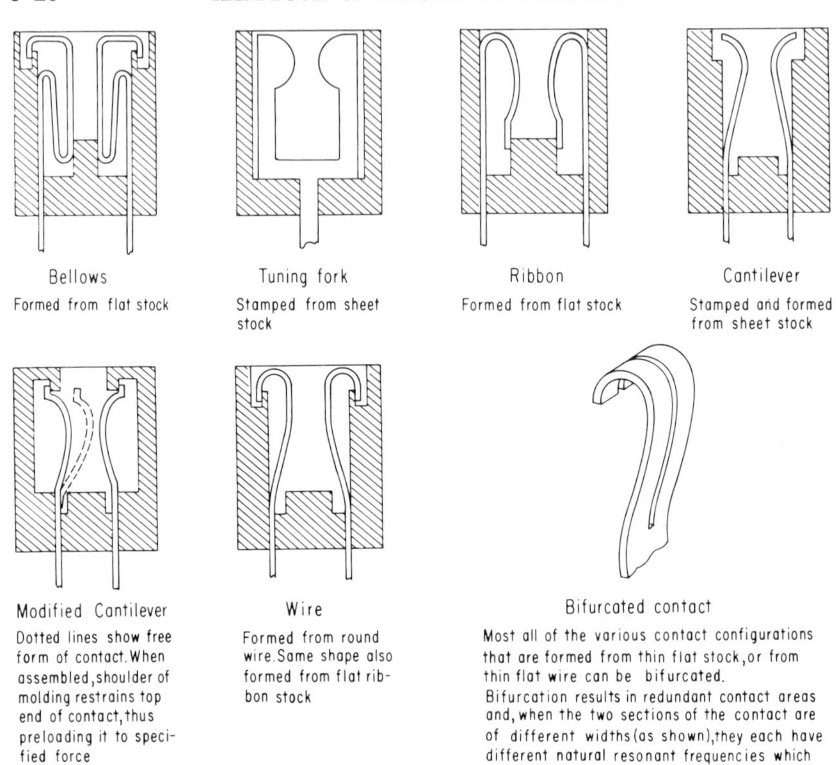

Fig. 26. Basic edge-receptacle contact configurations.

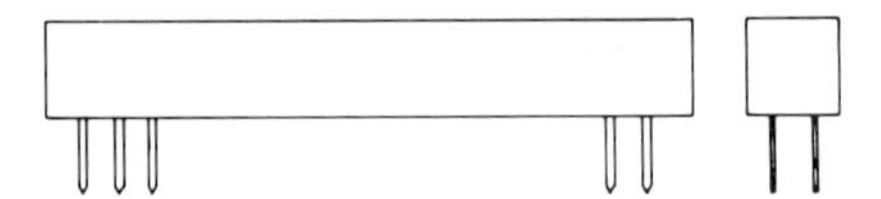

Fig. 27. Edge-receptacle body type with no mounting holes. Secured to a mother printed-circuit board by means of the terminations soldered to the board. Used primarily for small plug-in module boards. Not recommended for large boards.

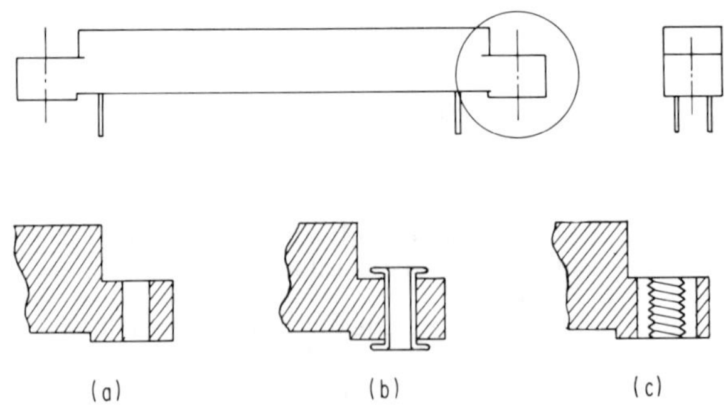

Fig. 28. Edge-receptacle body types with (*a*) straight-through holes to accept screws or rivets; (*b*) floating bushings to accept screws; (*c*) threaded inserts.

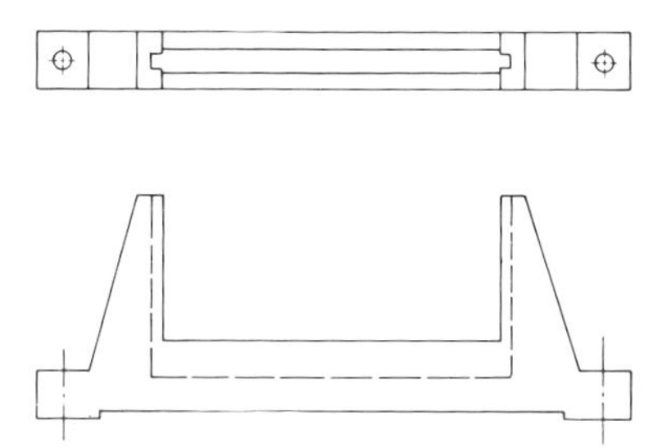

FIG. 29. Printed-circuit connector body type with built-in board guides. Used on both edge receptacles and two-piece connectors. Card guides available in various heights up to the entire length of the board. Mounting methods are the same as those shown in Fig. 28.

Printed-circuit-board thickness tolerance is another important design factor. Most edge receptacles are designed to accept boards with a ±0.007-in. thickness tolerance. This does not normally take into consideration board warpage, and in order to ensure a reliable relationship between the board and the receptacle a ±0.005-in. maximum tolerance over the nominal board thickness is recommended, especially in high-reliability applications. Board warpage should also be held to a reasonable minimum.

Contact width and spacing, overall dimensions, and tolerance build-up across the connector tab end of the printed-circuit board must be carefully considered and matched to the dimensions and tolerances of the chosen edge receptacle so that in a worst-case tolerance situation there will still be adequate mating of the contacts in the connector with the contact tabs on the printed-circuit board. Fairly long edge receptacles usually have a strengthening barrier across the board cavity, somewhere near the center, to prevent the long side walls from bowing out due to the build-up of pressure on the contacts with the board inserted. This barrier can be used in conjunction with the necessary clearance notch in the board as a zero reference point for properly locating the board in the receptacle, thus effectively cutting the overall tolerance build-up in half. This is especially important in a long connector with many contacts (Fig. 30).

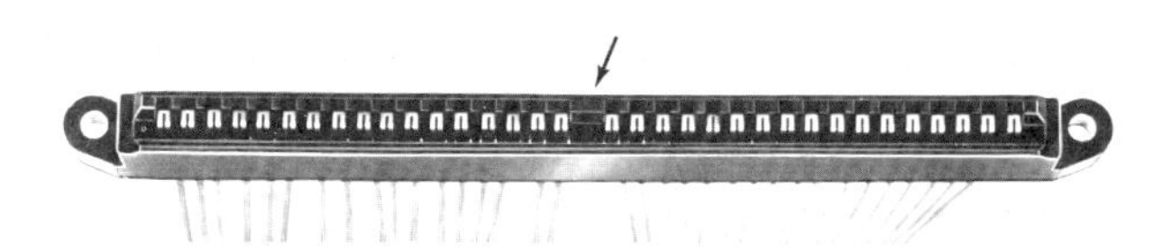

FIG. 30. Edge receptacle with molded-in center barrier; single-row contacts. (*Continental Connector Corp.*)

To assist in easy insertion of the printed-circuit board into the receptacle, as well as to help prevent damage to contacts and plating, the connector-tab end of the board must be chamfered and the tab pattern set back from the board edge, as shown in Fig. 31.

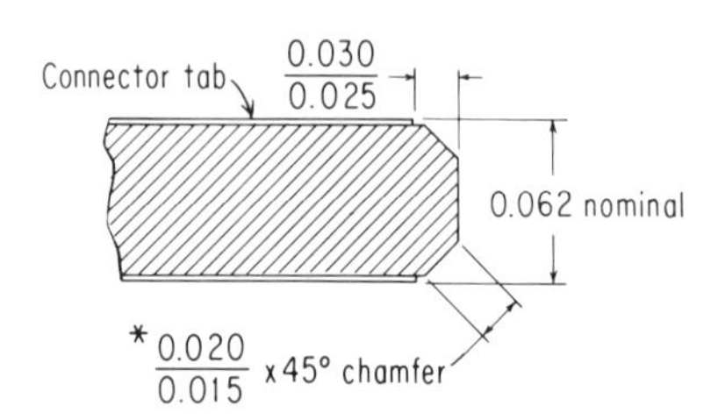

FIG. 31. Typical printed-circuit-board end configuration for insertion into edge receptacle. Dimension should be scaled up or down for thinner or thicker boards.

Always keep in mind that the printed-circuit board, when used with an edge receptacle, is actually the male half of a connector, and as such, must be designed to match the connector receptacle with the same care that any male connector half must be designed for reliable performance.

Two-piece Printed-circuit Connectors. This family of connectors is, essentially, a variation of the familiar two-piece rack and panel connector. One piece, usually the female half, is mounted to a frame or back plane, and the other piece is mounted to the circuit board.

This type of printed-circuit connector is generally more expensive, both initially and installed, than edge receptacles because there are two pieces rather than one, and one of the pieces must be affixed to the printed-circuit board, usually by soldering. However, the advantage of this type of connector is that the contact mating characteristics are controlled by a single manufacturer, as opposed to what is actually a joint effort between a connector manufacturer and the user with the edge type, thus reducing potential reliability problems that can be introduced in edge-receptacle applications if the manufactured receptacle and the printed-circuit board are not wholly compatible. In many applications where extreme vibration is anticipated, the two-piece connector would be desirable. Figure 19 is an illustration of a typical two-piece printed-circuit connector.

Body Configurations. Figures 20, 32, and 33 show the three basic types of two-piece printed-circuit connectors.

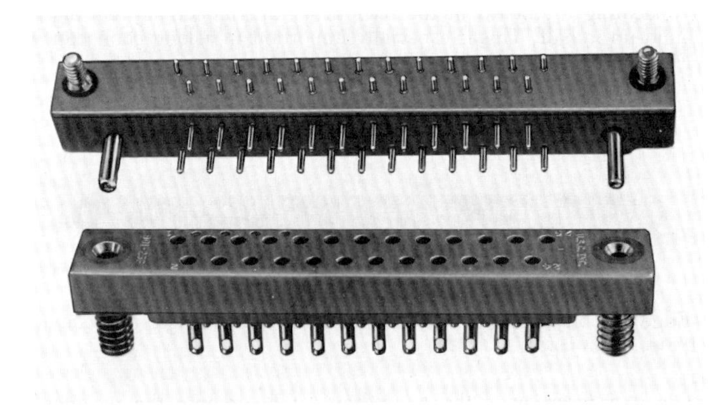

FIG. 32. Typical two-piece pin and socket printed-circuit connector. (*U.S. Components, Inc.*)

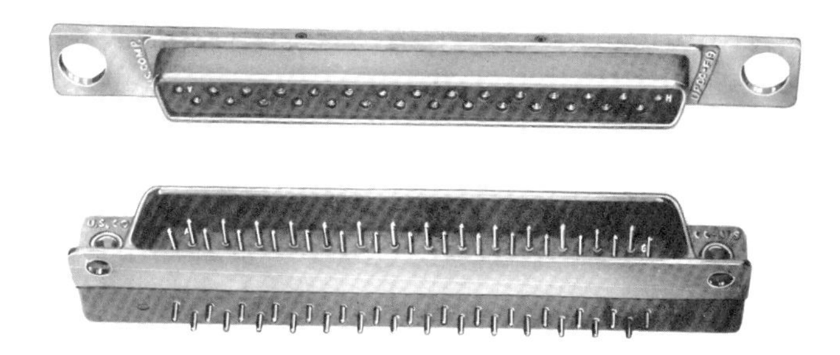

FIG. 33. Two-piece pin and socket printed-circuit connector with shell (NAS type). (*U.S. Components, Inc.*)

Contact Size, Spacing, and Density. There is a wide variety of contact sizes, spacing, and densities available in this type of connector because it is possible to have several layers or rows of contacts, as in rack and panel connectors. Manufacturer's literature should be consulted for those details most compatible with the intended package design.

Number of Contacts. The number of contacts available in a single connector of this type ranges from 10 to at least 75, arranged in one, two, or three rows. Figure 34 illustrates some typical arrangements.

Contact Configurations. Three basic types or configurations are available, as shown in Figs. 20, 32, and 33.

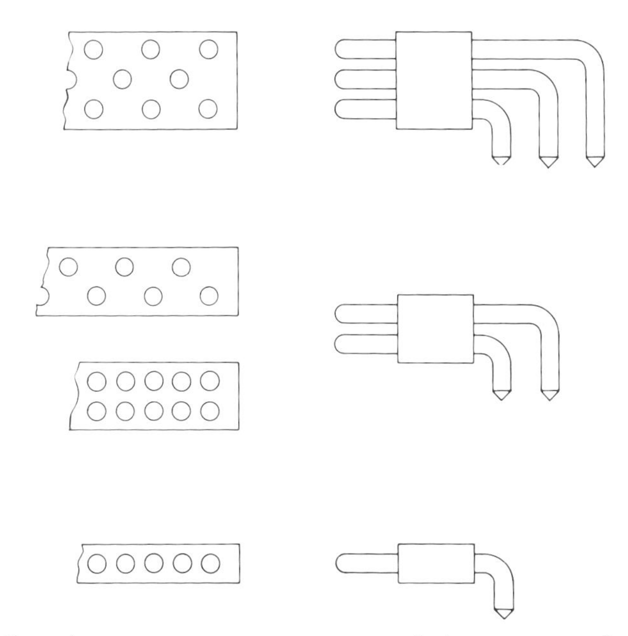

FIG. 34. Typical arrangements in one-, two-, and three-row printed-circuit connectors.

Contact Terminations.

CHASSIS-MOUNTED (FEMALE HALF). Same types as available for edge receptacles and for rack and panel connectors.

PRINTED-CIRCUIT-BOARD HALF. Dip-solder pins; solder cup and slotted eyelet for wiring from connector to board terminals.

Materials and Platings. Discussed under the heading "Connector Materials" in this chapter.

General Comments. Two-piece printed-circuit connectors are, in general, rugged and durable. They are specified in many military and space applications where shock and vibration are prime considerations. The type with the metal shell and plastic insulator insert, which was originally developed for the military, are very rugged and particularly adaptable for hostile environments, and can be environmentally sealed, if required.

For terminal pin layouts and hole sizes in printed-circuit boards, as well as for general mounting directions, manufacturer's data sheets should be consulted.

Cam-actuated Printed-circuit Connectors (Zero Force and Side Entry). In some specialized applications where it is desirable to engage a great many contacts at one time, or where it is necessary to slide a printed-circuit board lengthwise into a re-

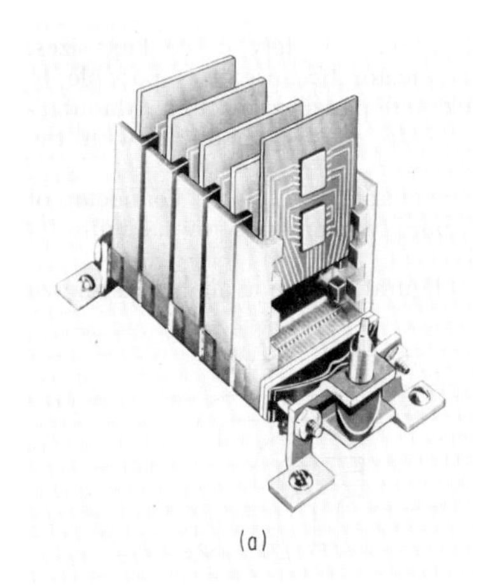

(a)

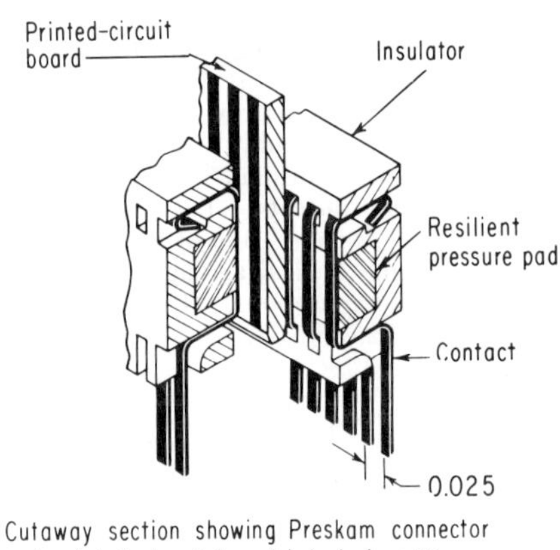

Cutaway section showing Preskam connector and printed-circuit board in locked position

(b)

FIG. 35. Preskam (trademark, Cinch Mfg. Co.) interconnection system; cam-actuated. Available in a variety of forms and sizes. (*Cinch Mfg. Co.*)

ceptacle (as in cases where it is necessary to have contacts along two edges of a board), a few special connector concepts have been developed. The design concept is the same with all; it is that the functional part of the connector containing the contacts can be opened or separated so that the printed-circuit boards can be inserted without actually engaging the contacts. This can be done either in the conventional manner or with the board sliding lengthways into the receptacle until it rests in position with opposing contacts in the correct operating position. The receptacle is then closed so that the contacts touch, and apply pressure to, the terminal pads on the printed-circuit board. Examples of this type of connector are illustrated in Figs. 35 and 36.

CYLINDRICAL CONNECTORS

Introduction. Cylindrical connectors are generally used on functional equipments that need relatively frequent disconnection for inspection, check-out, maintenance, or repair. They are used primarily to connect cables together and to interconnect equipment "black boxes." The greatest usage, by far, is in military and aerospace applications, although they are also widely used commercially for interconnection between computer cabinets and associated peripheral equipment and in all types of communication equipment.

Cylindrical connectors are basically rugged and, by the nature of their shape and construction, will withstand rough handling and generally hostile environments. There are many hundreds of types, sizes, contact and insert arrangements, polarizations, and associated hardware variations of cylindrical connectors. To attempt to list, classify, and describe them all would require a book in itself. In a handbook such as this, space does not permit more than a general introduction and description of typical classifications of this family of connectors.

The very large majority of cylindrical connectors are designed and manufactured to the specifications of the several military specifications covering this family of connectors. As a guide to the various types available, the appropriate military specifications will be listed in the chronological order of their development, starting with the original AN type on through to the latest types of miniature cylindrical connectors

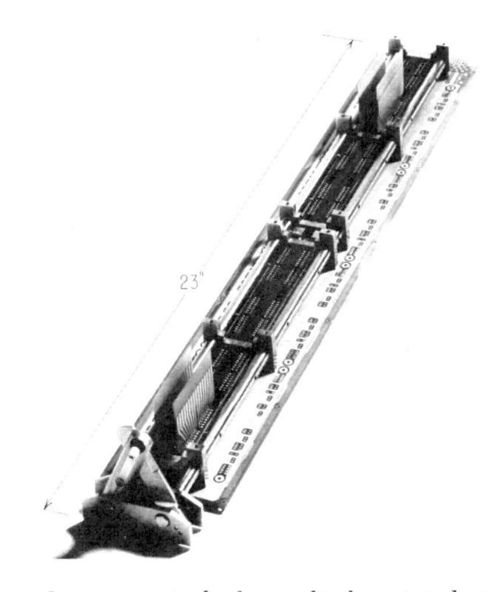

FIG. 36. Special zero-force receptacle for multiple printed-circuit boards. Connector is mounted to multilayer printed-circuit back plane. Two rows of double contacts accept a group of 16 printed-circuit boards containing a total of 512 contacts. A similar version allows printed-circuit boards to be inserted lengthwise into the connectors, after which the contacts are closed against the board terminals. (*Burroughs Corp. and U.S. Components, Inc.*)

developed for high-altitude and aerospace applications. Typical examples of each type will be illustrated and described. For specific details and application information the applicable military specifications and manufacturers' literature should be referred to.

Selection Factors. The nature of cylindrical connectors, particularly because many types include sealing grommets around the terminations and wires or cable, requires that several factors be considered in their selection, some of which apply to all connectors and some of which are especially applicable to specific applications of these connectors.[3]

1. Number of circuits involved (determines minimum size).
2. Voltage and current of circuits (determines contact sizes and spacing, also shell size).
3. Environmental considerations.
4. Type of wire termination:
   a. Solder.
   b. Crimp; including required tooling.

5. Wire characteristics:
   a. Type of wire—basic material and construction.
   b. Diameter—wire outside diameter must be compatible with sealing grommets.
   c. Shielded wires—shield termination techniques must be considered.
6. Area and location of mounting: In inaccessible areas special assists may be required for mating and unmating.
7. Support hardware: If special cable support hardware (clamps, caps, etc.) require space, mounting provisions must be made.

In all the various types, polarization is accomplished by the combination of insert arrangements and the rotational position of the insert in relation to key and keyway on the connector housing.

In Table 2 the various applicable MIL specifications are tabulated, and the mechanical and electrical characteristics of the contacts for these connectors are given.

Cylindrical Connector Types per MIL Specifications.[3]

MIL-C-5015—Connectors, Electric AN Type (Fig. 37). This type is an extremely rugged, versatile family of connectors, the "workhorse" of cylindrical connectors. The 5015 design is available in seven classes:

Class A: Solid shell, general use.

Class B: Split shell. Same as Class A except shell is split longitudinally to give accessibility to the solder connections.

Class C: Pressurized receptacles intended for use on walls and bulkheads of pressurized compartments. Good for leak rate of 1 in.^{3}/hr at 30 psi.

Class E: Environment-resisting; intended for use where the connector is subjected to heavy condensation and rapid changes in temperature and/or where the connector is subjected to high vibratory conditions.

FIG. 37. MIL-C-5015 connector with crimp/removable contacts. (*Amphenol-Borg Electronics Corp.*)

Class K: Fireproof; for use where connector must maintain electrical continuity for a limited time even though the connector is subjected to continuous flame (2,000°F). High-temperature wire should be used with this class of connector.

Class R: Environment-resisting (light-weight); intended for use where shorter overall length and lighter weight are required.

Class RC: Environment-resisting; identical to Class R except connector is supplied with strain relief clamp.

CONTACTS. See Table 2. Thermocouple contacts are available in various material combinations. The number of contacts ranges from 1 to 104, and they can be had in all the same size or in combinations of sizes. Typical examples are shown in Fig. 38.

MATERIALS. The shells and coupling rings are aluminum alloys. The finish is cadmium plate, dull olive drab in color, and is electrically conductive. Inserts are hard plastic or resilient material. Sealing is accomplished by interfacial mating of resilient materials or by the use of O rings when hard inserts are used. Terminations are sealed by potting or by grommet seals.

SHELL STYLES. There are many plug and receptacle shell types available to suit a

TABLE 2. Mechanical and Electrical Characteristics of Contacts[3]

Connector specification	Applicable contact specification	Contact plating	Contact size	Accept wire gauge (AWG)	Contact terminal type	Contact rating, amps†
MIL-C-5015		Gold	16		Solder*	22
		over	12			41
		silver	8			73
			4			135
MIL-C-22992		Gold	0			245
MIL-C-26482		Gold	20		Solder	7.5
		over	16			13.0
		silver	12			
	MIL-C-23216	Gold	20	24	Crimp	3.0
		over		22		
		silver		20		7.5
			16	20		7.5
				18		
				16		13.0
			12	14		17.0
				12		23.0
MIL-C-26500	MIL-C-26636	Rhodium	20	24	Crimp	3.0
		over		22		5.0
		silver or		20		7.5
		nickel	16	20		7.5
				18		16.0
				16		22.0
			12	14		32.0
				12		41.0
NAS 1599	NAS 1600	Gold or	20	24	Crimp	3.0
		silver or		22		5.0
		nickel		20		7.5
			16	20		7.5
				18		15.0
				16		20.0
			12	14		25.0
				12		35.0
MIL-C-25955		Gold	20	30	Crimp	
		over		24		
		silver		22		
				20		7.5
MIL-C-27599		Gold	20		Crimp	7.5‡
		over	16			13.0
		silver				

* Available with crimp/removable contacts under manufacturer's part numbers.
† Maximum contact rating for individual contact.
‡ Hermetic contacts rated at 5.0 and 10.0 amps, respectively.

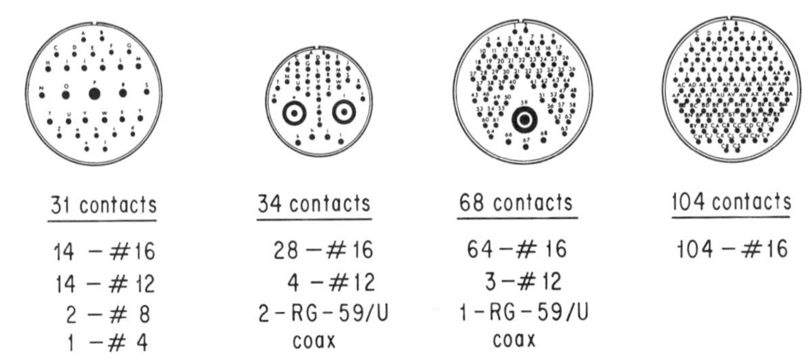

Fig. 38. Examples of cylindrical connector contact layouts and combinations. (*From MIL-C-5015.*)

variety of physical and environmental application requirements. Figure 39 illustrates the shell styles that are available for this series connector, that are typical of the shell styles presently available in many of the various families of cylindrical connectors, and that will probably be available eventually for some of the newer MIL specification types.

MISCELLANEOUS HARDWARE. Figure 40 illustrates the miscellaneous hardware available for many cylindrical connector families: dust caps (metal and plastic), conduit fittings and ferrules, reducing adapters, cable clamps, telescoping bushings (to adapt wires to cable clamps).

MIL-C-22992—Connectors, Electrical, Waterproof, Quick-disconnect, Heavy-duty Type. This specification covers heavy-duty, multicontact, waterproof electrical connectors, with either solder-type or crimp/removable-type contacts and with a maxi-launching equipment, aircraft ground equipment, and ground radar. The contacts and inserts are basic MIL-C-5015 design. They are available in three basic classes:

Class R: Environmental

Class C: Pressurized

Class J: Pressurized with grommet

Shells and coupling rings are made of aluminum alloys and are available with either conductive or nonconductive finishes. Sealing is accomplished by the use of interfacial interference of resilient inserts and also by gaskets and O rings. Shell types and miscellaneous hardware similar to MIL-C-5015 are available.

MIL-C-26482—Connectors, Electric, Circular, Miniature, Quick-disconnect (Fig. 41). This specification covers environment-resisting, quick-disconnect miniature connectors, with either solder-type or crimp/removable-type contacts and with a maximum operating temperature of 125°C. They are available in five classes:

Class E: Grommet seal; contains multihole grommet and follower for moisture sealing.

Class F: Grommet seal; same as Class E except it is provided with a strain-relief clamp.

Class P: Potted seal; the connector is supplied with potting mold for retention of potting compound.

Class H: Hermetic seal; for receptacles where leakage rate in the order of 0.1 $\mu/(\text{ft}^3)/(\text{hr})$ is required.

Class J: Gland seal; connector supplied with compression clamp for moisture-proofing multiconductor jacketed cables.

CONTACTS. (See Table 2.) The number of contacts available is from 2 to 61 in various combinations. Coaxial contacts are not covered in the specification but are available under manufacturer's part number. The shells and coupling rings are made of aluminum alloys, and the hermetic shells are made of material suitable for soft soldering.

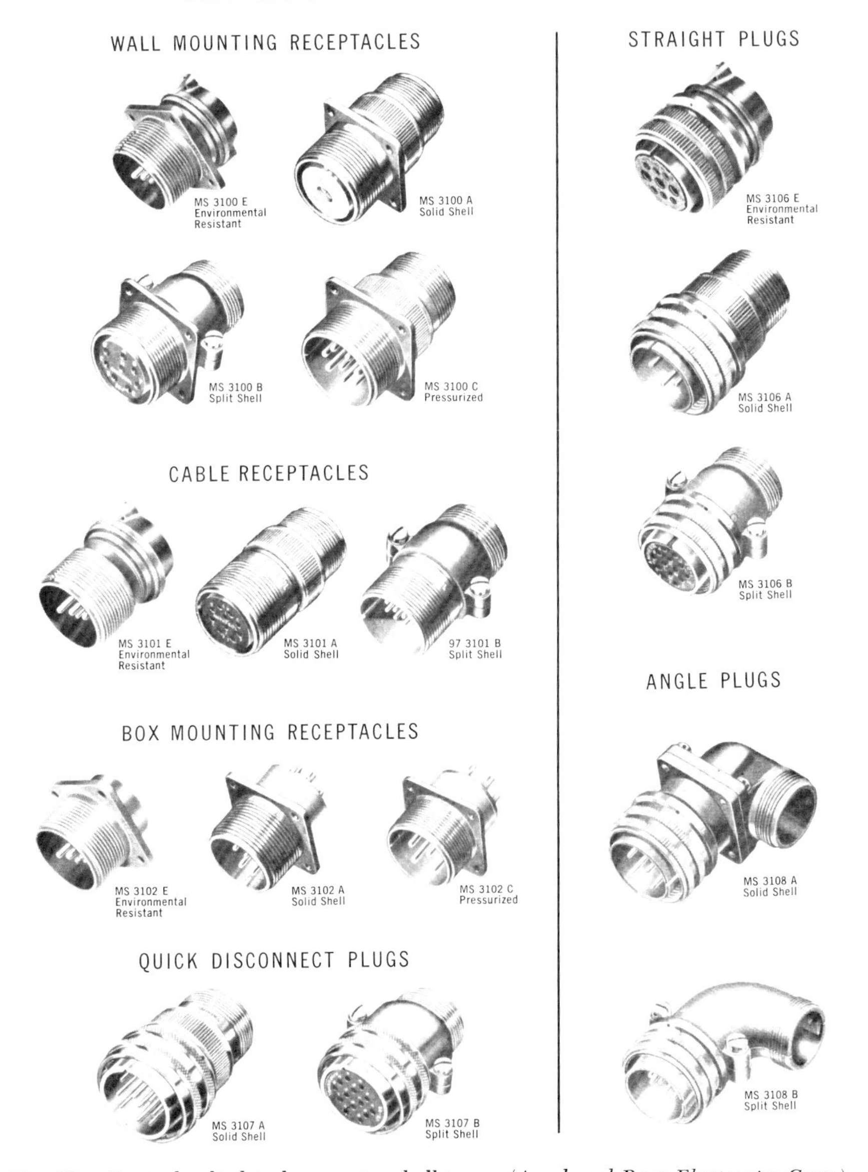

FIG. 39. Typical cylindrical connector shell types. (*Amphenol-Borg Electronics Corp.*)

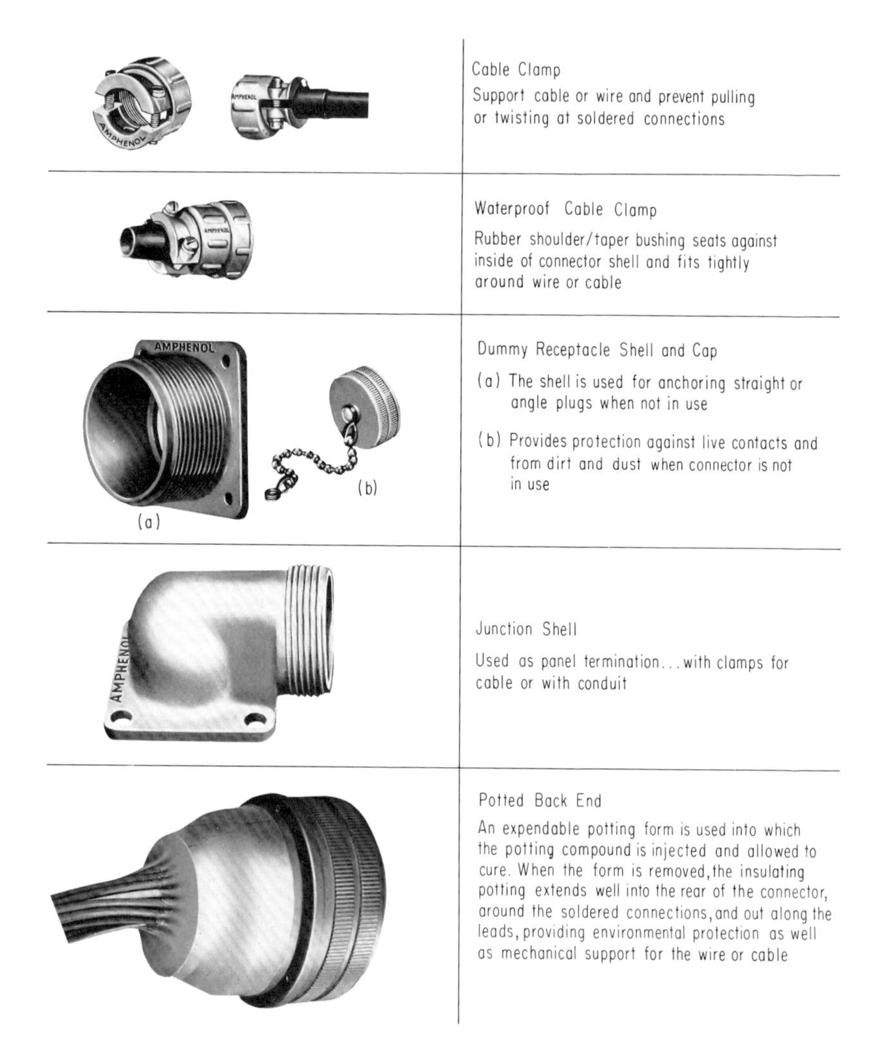

FIG. 40. Typical miscellaneous hardware for cylindrical connectors. (*Amphenol-Borg Electronics Corp.*)

MIL-C-26500—Connectors, General-purpose, Electrical, Miniature, Circular, Environment-resisting, 200°C Ambient Temperature (Fig. 42). This specification covers an environment-resisting family of miniature circular connectors designed essentially voidless and to meet the higher-altitude and higher-temperature requirements of missiles and space vehicles. There are three basic classes:

Class R: For use in environment-resisting applications.

Class H: Hermetic; for receptacles where leakage rates are in the order of 0.01 $\mu/(ft^3)/(hr)$.

Class G: Grounding, environment-resisting; for use where grounding to mounting structure is required, the anodized coating is removed to provide a conducting path.

These connectors are available with threaded couplings (Type T), bayonet coupling (Type B), and push-pull coupling (Type Q). The shells are aluminum alloy, stainless

FIG. 41. MIL-C-26482 plug and receptacle (wall mounting). (*Elco-Webster.*)

FIG. 42. MIL-C-26500 connector. Screw coupling is shown. Also available with bayonet and push-pull couplings. (*Amphenol-Borg Electronics Corp.*)

steel, or a material suitable for soldering or brazing for Class H. Sealing is accomplished by interfacial interference of the inserts as well as an O-ring seal which provides shell sealing before mating is accomplished. The terminals are grommet-sealed.

SHELL STYLES.

Square flange receptacle
Single-hole mounting receptacle
Straight plug
Solder flange receptacle (hermetic only)

CONTACTS. (See Table 2.) Coaxial contacts available.

MIL-C-38300. This specification covers an upgraded version of the MIL-C-26500 design and was primarily established to cover closed-entry design contacts.

NAS 1599—Connectors, General-purpose, Electrical, Miniature, Circular, Environment-resisting, 200°C Maximum Temperature. These connectors are much the same as the MIL-C-26500. The contacts are designed for rear insertion and removal. The sockets are closed-entry. See Table 2.

The connectors are capable of continuous operation between the temperature limits of −55 and +200°C. Hermetic receptacles are available to perform to the same temperature requirements as the environment-resisting construction. They are available in two basic classes:

Class R: Environment-resisting.

Class H: Hermetic; receptacles intended for use in applications where pressures must be contained by the connectors across the walls or panels on which they are mounted. Good to 0.01 $\mu/(ft^3)/(hr)$ leakage.

The design requires that the combination ambient temperature and contact current flow not allow the temperature of the connector assembly to exceed 200°C.

MIL-C-25955—Connectors, Electrical, Environment-resisting, Miniature, with Snap-in Contacts. This specification covers miniature connectors with threaded coupling provided with holes for safety wiring. They are available in two classes:

Class E: Environment-resisting

Class H: Hermetic receptacle

CONTACTS. Specification covers size 20 contacts (see Table 2). Size 16 and 12 contact arrangements are available under manufacturer's part numbers.

MIL-C-27599—Connectors, Electrical, Miniature, Continuously Shielded, Quick-disconnect. This specification covers the requirements for one type of bayonet-locking miniature connector designed to provide shell-to-shell mechanical orientation and electrical continuity of shells prior to mating of contacts. The pin and socket contacts are so located that they cannot be damaged during mating or unmating operation. The design is such that plug contacts cannot make contact with receptacle contacts, regardless of the angle of entry, until proper orientation is achieved. Spring fingers in the shell provide electrical continuity to the receptacle during mating and unmating. The connectors are available in two classes:

Class T: General-duty

Class H: Hermetically sealed receptacles

This connector has been designed for use with atomic weapons. For contact information see Table 2.

MIL-C-55181 and *MIL-C-12520.* These are power connectors. They are polarized and waterproof, with self-sealing cable clamps on the plugs. The cable plugs feature a center locking screw for positive engagement. Both plugs and receptacles may be specified with pin or socket contacts.

CONTACTS. There are five different arrangements in the MIL-C-12520:

4, 9, 14, 19, and 30 contacts (Fig. 43)

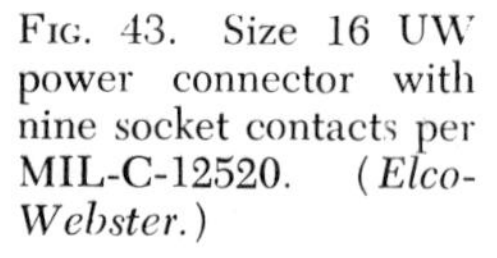

Fig. 43. Size 16 UW power connector with nine socket contacts per MIL-C-12520. (*Elco-Webster.*)

Fig. 44. High-density environmental connector. Amphenol's Astro/348 miniature cylindrical connector. Designed to meet requirements of MIL-C-81511 (Navy). (*Amphenol-Borg Electronics Corp.*)

The MIL-C-55181 connector is available with both removable and fixed contacts and with three contact arrangements: 4, 9, and 18 contacts.

MIL-C-81511. This is a miniature, environmentally resistant, very-high-contact-density cylindrical connector family. It features a monoblock internal construction that eliminates air voids between contacts, damage-proof mating with pins recessed beyond the reach of shells, closed-entry hard inserts for the socket contacts, and prod-proof socket contacts. There is a grommet seal on the terminal end that will accommodate a range of wire diameters from 0.020 to 0.054 in.

Environmental sealing is done internally by means of interfacial and shell O-ring seals. The mounting of the receptacle is single-hole and may be done on the front or rear panel. This type is available in six shell sizes with from 4 to 85 contacts. There are grounding springs in the receptacle that mate prior to electrical engagement of contacts. Shielding is provided for EMI and RFI protection (Fig. 44).

MIL-C-10544 and MIL-C-55116 Audio Connectors. The former is a 10-contact connector and the latter is a 5-contact audio connector. Both are widely used in Signal Corps ground communication equipment. They are both bayonet-locking and are available with either rigid contacts or nonrigid spring-loaded contacts (Fig. 45).

Determination of Part Numbers. As can be seen from the foregoing information, a wide and diversified selection of cylindrical connectors is available to the engineer and designer. The large majority of cylindrical connectors are manufactured to meet the mechanical and performance requirements of the various MIL specifications,

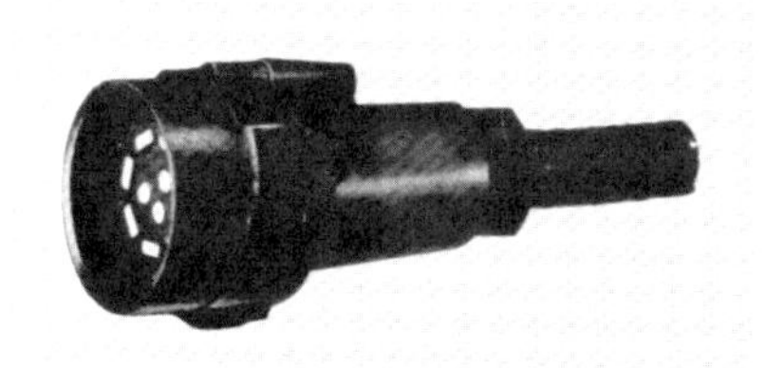

FIG. 45. Ten-contact U-161/U audio connector per MIL-C-10544. (*Elco-Webster.*)

although there are many differences in the same functional connector from manufacturer to manufacturer. It is most important that proper part numbers be designated when specifying and ordering these connectors because of the many contact arrangements, shell sizes, associated hardware, etc. Both the MIL specifications and the manufacturers' literature call out the information necessary to determine the particular part number for the selected connector. Table 3 shows required part number information, typical part numbers, and how they are derived for the MIL-C-5015 connector series. The same general method applies to all the various families of these cylindrical connectors.

COAXIAL CONNECTORS

Background. For the past 25 years, most coaxial connectors have been manufactured to conform to MIL specifications, with the possible exception of some of the microminiature ones recently developed. Continued requirements for different sizes, shapes, and electrical characteristics led to a confusing array of specifications that eventually covered over 15 types of coaxial connectors and a grand total of over 500 documents, specifications, and MS drawings to cover all the sizes, shapes, impedance characteristics, materials, and variations.

Mainly, these documents called out dimensional requirements, materials, and overall electrical parameters rather than actual performance requirements and standard, reproducible test requirements. The result of this multitude of documents was twofold:

1. No particular technical competence was required to manufacture the connectors to the dimensional and material requirements of the MIL specifications and MS drawings, with the result that connector performance of supposedly identical connectors

TABLE 3. How to Select MIL-C-5015 Connectors (According to Cataloged Information)

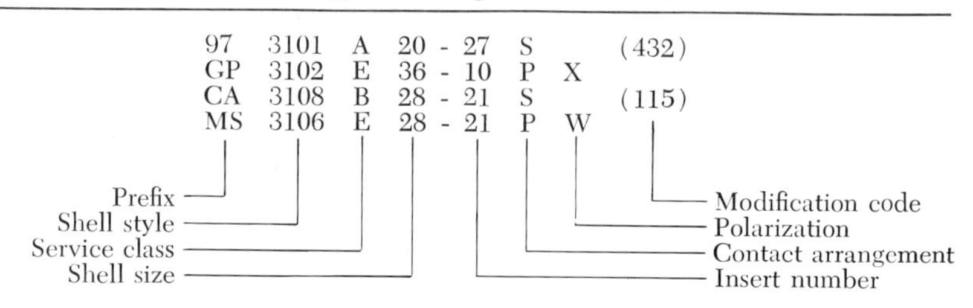

Prefixes

AN	Replaced by MS
CA	Cannon
CT	Cannon E series
FC	Flight connector
GP	Bendix gold contacts
MS	Military standard
SC	Bendix A type
SF	Bendix modification to E type
SG	Bendix modification to E type
SB	Bendix modification to E type
97	Amphenol

Shell style

3100	Wall receptacle
3101	Cable receptacle
3102	Box receptacle
3103	Wall receptacle for potting
3106	Straight plug
3107	Quick disconnect
3108	90° angle plug
25183	Straight plug for potting
25183A	Straight plug for potting; with ground plug

Steps to determine part number

1. Select number of contacts.
2. Select contact type and sizes.
3. Select shell type and size.
4. Select alternate position for inserts (polarization).
5. Select service class.
6. Add proper prefix.

SOURCE: "The Encyclopedia of Connectors," 2d ed., Spacecraft Components Corp.

Service class

A	Solid shell
AF	E type with threaded endbell and special coupling nut
B	Split shell
C	Pressurized
E	Environmental
ER	Cannon potted type
ES	Cannon potted type
F	E type with threaded endbell
K	Firewall
M	Replaced by E
P	Bendix potted type
PR	Replaced by C
R	Environmental with O ring under coupling nut.

Shell size

Outside diameter of mating portion of receptacle in 1/16-in. increments.

Insert number

Obtained from illustrated layouts in manufacturers' catalogs.

Contact arrangement

P for pin (male); S for socket (female)

Alternate positions

Rotational position of insert in shell in relation to key and keyway; used to obtain polarization.

Modification code

Used for changes in plating of shells or contacts, different insulation material, different types of contacts, etc.

manufactured by several different manufacturers could vary greatly due largely to mechanical tolerances involved.

2. The dimensional specifications did not permit the manufacturers to make design simplifications to improve the parts and/or reduce costs.[4]

The net outcome to the user was, in many instances, extreme confusion about what connector to use in a given application, and disappointment in performance of the selected parts.

The situation became acute as performance requirements for advanced electronic equipment such as radar, microwave communications, data transmission systems, and aerospace applications became more and more critical. In 1960 the American Standards Association formed a committee to study the problem and to suggest corrective action. The result of the study and recommendations of the committee was a coordinated *tri-service* specification, MIL-C-39012, which is a general specification covering several types of coaxial (rf) connectors. This new specification specifies an rf connector by means of all applicable performance parameters, tested in an appropriate and consistent manner, and specifies only envelope and mating face dimensions for a given connector.[5]

The electrical requirements for connectors qualified to MIL-C-39012 include insulation resistance, dielectric withstanding voltage, rf high potential, contact resistance, rf leakage, voltage standing wave ratio (VSWR), and insertion loss. The mechanical parameters include engaging force, cable retention force, coupling mechanism force, mating characteristics, and contact durability. Physical and environmental requirements are also specified. In addition, this new specification requires rigorous qualification testing, with periodic requalification.

Two classes of connectors are specified under this new specification, Class I and Class II. In general, the requirements for connectors under these classes are as follows:

Class I. A connector intended to provide superior rf performance at frequencies up to 10 GHz and for which all rf characteristics are completely defined. Mechanically it will mate with a Class II connector.

Class II. A connector intended to provide mechanical connection within an rf circuit and to provide reasonable rf performance.

Both Class I and Class II connectors will mate mechanically with connectors made to previous specifications but with some degradation in performance characteristics, particularly at the higher frequencies.

The two great advantages of this new specification are that a design engineer will be able to specify and use with confidence any connector that has been qualified to this specification, and connector manufacturers will have considerable latitude to improve the connectors and/or reduce cost.

Application. Probably the largest percentage of coaxial connectors in use today are not in critical rf applications, but rather in applications such as computer transmission lines where good shielding, low line resistance, low capacitance, and a reasonable impedance match to the coaxial cable used are important.

For such applications at the lower frequencies there is a multitude of coaxial connectors available in many sizes and shapes, from the standard sizes on down to the newer microminiature sizes, and with reasonably good electrical characteristics and fairly effective EMI and RFI protection. Also, there are shielded coaxial contacts available to fit multicontact rectangular and cylindrical connectors (Fig. 8).

When these connectors were first used, the upper frequency limit was about 300 MHz, and the fact that many of the connectors had nonconstant impedance, or that some did not meet their specified impedance, was not a real problem since there is no serious contribution by the connector until the connector length becomes about one-twentieth of the wavelength. (A wavelength of 300 MHz is 1 m.)[6]

Today, however, there are many applications at extremely high frequencies that require superior rf performance, and when you consider the very minute wavelengths of the very high frequencies, it becomes obvious that connector design and performance become critical in order that catastrophic discontinuities in the traveling signal not be introduced (Fig. 46).

Assuming a theoretically optimum design, very slight dimensional differences, well within normal manufacturing tolerances, can raise the impedance match to a greater and greater degree as the frequencies go higher and higher.

It follows, then, that for satisfactory results, coaxial connectors, especially for very high frequencies, should be procured only from reliable connector (not hardware) manufacturers who have the engineering know-how to design precision connectors properly, the manufacturing capability to produce them, and also the testing facilities required for adequate qualification testing and quality control.

There are some precision coaxial connectors available today that are qualified for the very high frequencies encountered in microwave applications, such as Amphenol's new type N connector which has a specified frequency range up to 10 GHz (Fig. 47).

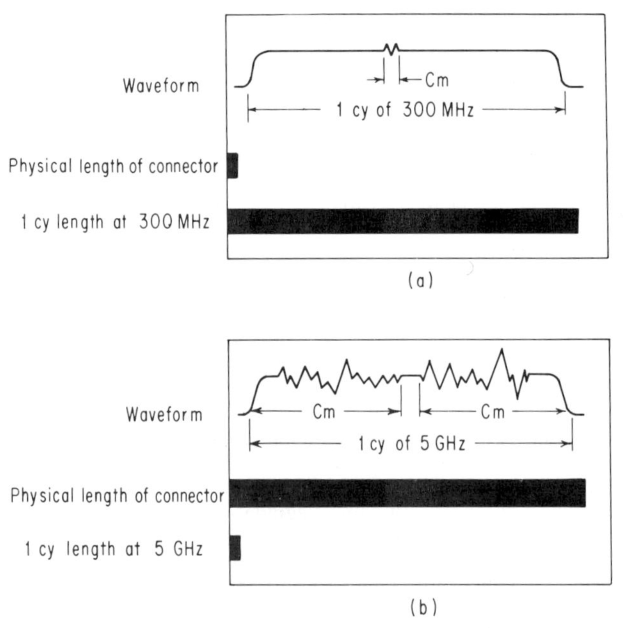

FIG. 46. Graphic illustration (not to scale) showing possible relative effect of impedance mismatch between connector and cable and/or EMI and RFI leakage through connector. In (*a*) Cm represents reflection caused by connector at low frequencies, and in (*b*) Cm represents catastrophic reflections caused by connector at higher frequencies.

For these very-high-frequency applications, it is most important that the designer have adequate information from the manufacturer regarding a proposed connector design. The new MIL-C-39012 specification will be of considerable help in this area because whether or not the application is for MIL specification equipment, coaxial connectors that have been qualified to this specification can be used with a high degree of confidence that they will meet the specified performance requirements.[5]

Selection Considerations. In general, selection of coaxial rf connectors should be based on the following considerations:

1. Electrical performance (VSWR, voltage rating, capacitance, etc.)
2. Environmental considerations, particularly shock and vibration
3. Physical interchangeability
4. Ease of assembly to cable without impairing performance

The last factor, 4 above, is extremely important. Most coaxial connectors are delivered to the user as a bag of parts, and time-consuming assembly of the connector to the cable can affect not only costs but also performance. A sloppy assembly job can completely destroy the connector's usefulness as a matched impedance device.

Most manufacturers offer complete instruction sheets on the preparation of the cable and on assembly of the cable to the connector. These sheets should be used in training assembly personnel, with careful attention being given to all of the steps and details outlined for a particular connector. Figure 48 is an example of a manufacturer's instruction sheet for the assembly of miniature coaxial connectors.

As an alternate to in-house assembly of coaxial connectors and cables, many manufacturers will provide connectors assembled to cables of lengths specified by the user. This can be technically and economically advantageous because their assembly people are well trained and expert in this tedious but all-important job.

Description of Coaxial Connector Types. Most manufacturers now offer coaxial connectors in which the shielding is crimped to the connector. Properly and carefully done, this appears to be a much more uniform method of attaching the braid shielding than the clamp nut method.

Crimping of the center conductor contact to the wire is also common practice, but there is considerable debate in the industry regarding the advisability of this method

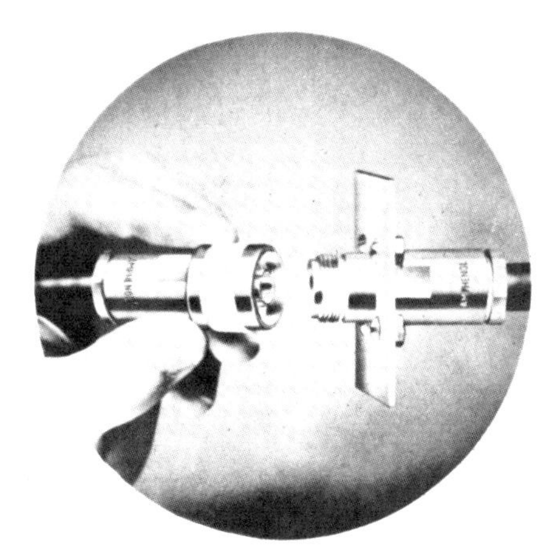

FIG. 47. Type N connector. Developed during World War II as microwave connector and with recent refinements is now usable to 18 GHz. (*Amphenol-Borg Electronics Corp.*)

for applications in the very high frequencies, as claims are made that the distortion of the center contact may cause serious impedance mismatch for reasons stated earlier in this section. Soldering of the center conductor requires more operator skill, but it does achieve a smooth transition from conductor to contact. Investigation in this area will no doubt result in significant data that can be evaluated by the user for his particular applications. In either method, the manufacturer's assembly method recommendations should be followed carefully.

Coupling Methods. Two-point bayonet coupling offers quick connect and disconnect; however, in some instances the connectors tend to rock during vibration, causing noise. Three-point bayonet types are generally considered to be free of rocking effects. Threaded types are generally considered also to be free of the effects of vibration. They provide a firmly seated connection and are used on most of the higher-frequency connectors. Push-pull snap-lock coupling, used mainly on subminiature connectors, is convenient but may be a potential reliability risk. Each manufacturer has his own approach to snap-on-type couplings, with the result that the subminiatures of different manufacturers generally cannot be mated.

Physical and Electrical Characteristics.

Size and Configuration. Coaxial cables are made in a size range (outside diameter)

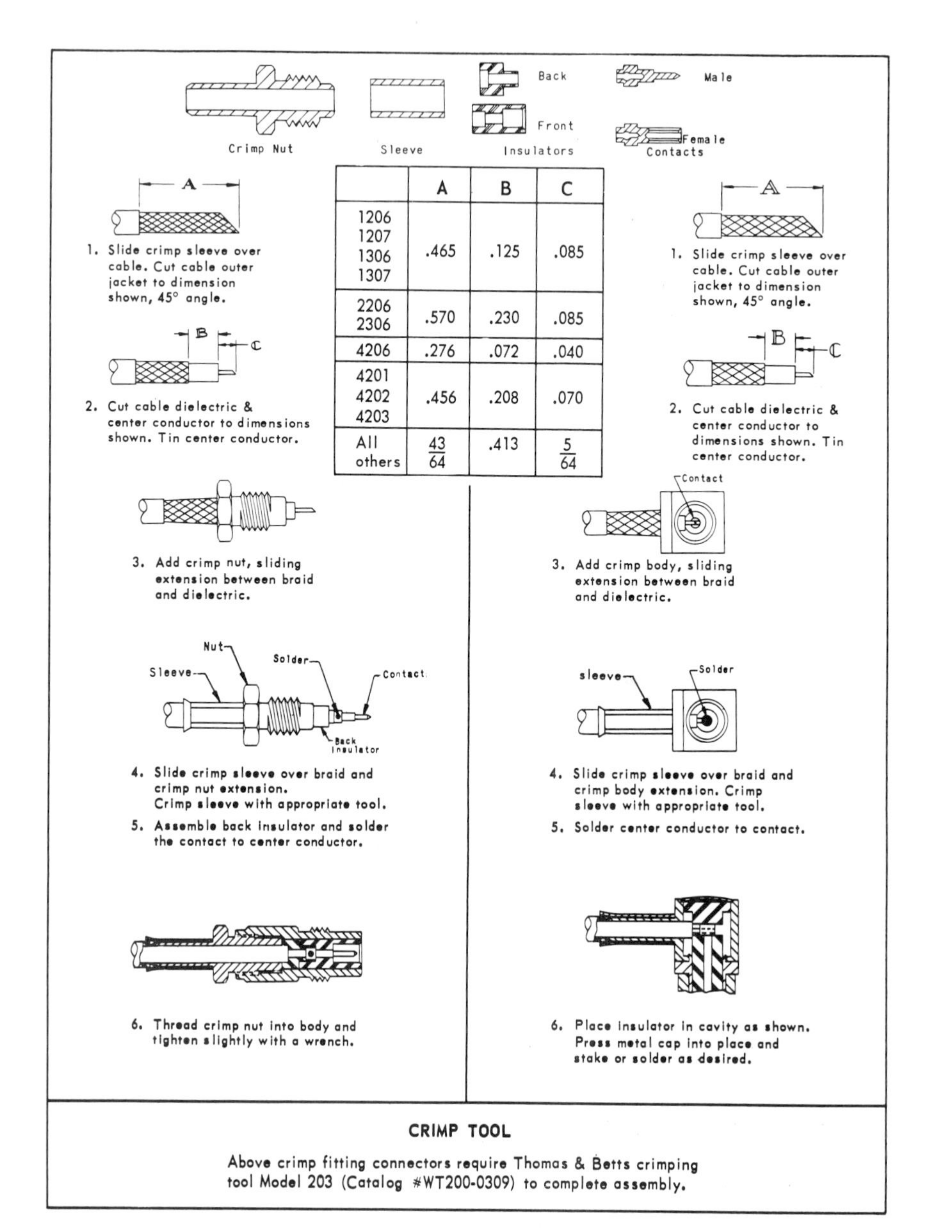

	A	B	C
1206 1207 1306 1307	.465	.125	.085
2206 2306	.570	.230	.085
4206	.276	.072	.040
4201 4202 4203	.456	.208	.070
All others	43/64	.413	5/64

FIG. 48. Typical coaxial connector assembly instruction sheet. (*Micon Electronics, Inc.*)

from about 1⅛ in. (rated at 14,000 volts rms) down to a subminiature size of 0.045 in. (rated at 500 volts). Physical sizes of coaxial connectors are compatible with the various cable sizes and range from a little over 2.0 in. across the coupling nut to as small as ⅛ in.

The physical configurations of coaxial plugs and jacks (receptacles), as well as those of associated hardware, are much the same, no matter what the series (type) or size. Figures 49 to 53 illustrate the various typical configurations, types of mounting, and associated hardware that are available in most series. The illustrations show the three types of coupling methods although not all are available on all series connectors. Detailed specifications for a given connector will indicate the coupling(s) available.

Voltage Rating Range. Voltage ratings range from 5,000 volts peak down to about 350 volts. As with any connectors, voltage rating is mainly governed by the length of the creepage path between current-carrying members or between current-carrying members and ground (shell). Materials for center insulators are a factor, but in general, the higher the voltage rating, the larger the connector.

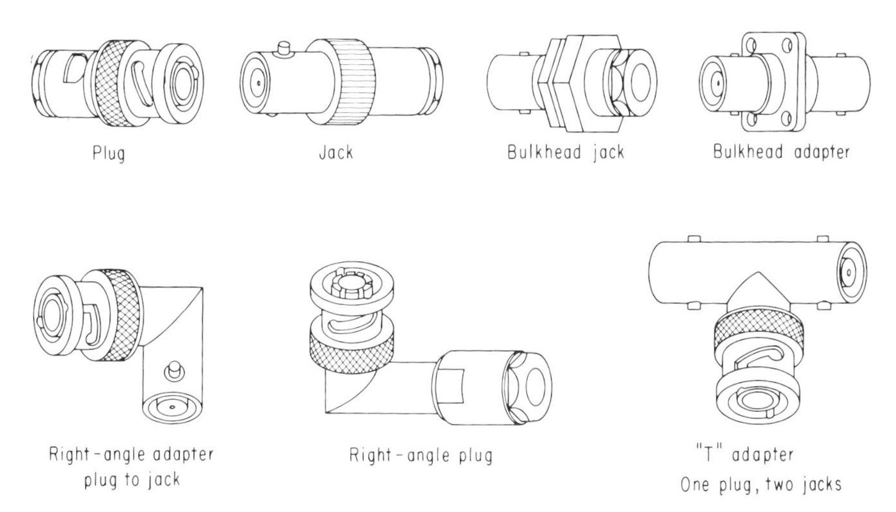

FIG. 49. Typical configurations: bayonet-lock coaxial cable connectors.

Thermal Limits. Although operating temperatures are usually limited by the temperature rating of the cable, the material of the center insulator is usually the limiting factor in the temperature rating of the connector. The temperature limits of the commonly used materials are as follows:

Materials	*Temperature limits*
Polyethylene	−67 to +185°F
Rubber	−67 to +250°F
Teflon *	−67 to +390°F
Rexolite †	−67 to +185°F
Kel-F ‡	−67 to +185°F
Ceramic	−67 to +1,800°F
Polystryrene	−67 to +185°F
Mica-filled Bakelite §	−67 to +300°F

* Trademark, E. I. du Pont de Nemours & Co., Inc.
† Trademark, Brand-Rex Div., American Enka Corp.
‡ Trademark, Minnesota Mining and Manufacturing Co.
§ Trademark, Union Carbide Corp.

Metallic Elements. Die-cast zinc, brass, and copper are used for connector bodies (shells) and associated hardware.

Brass is used for the male pin contacts and beryllium copper or phosphor bronze for the female contacts.

Platings. Shells and associated hardware are usually plated with 0.0002 minimum thickness silver. In some instances, 0.0001 gold over 0.0001 silver is used. In this family of connectors, precious metal platings are used throughout to ensure good electrical contact between all mating parts. Contacts are either gold-plated over the metal or gold-plated over silver.

Impedance Ratings. Coaxial connectors are specified to have either nonconstant impedance or with impedances to match the impedance of the cables that they can accommodate. The nominal impedance, in ohms, ranges from the common 50 ohms up to as high as 185 ohms. Unfortunately, many factors such as dimensional tolerance

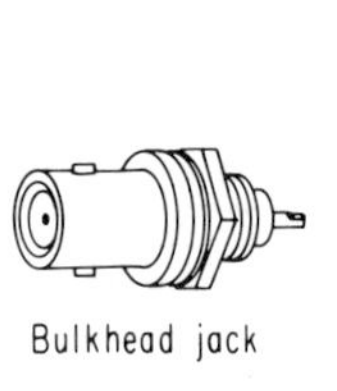

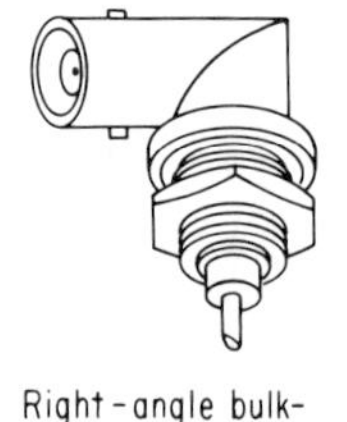

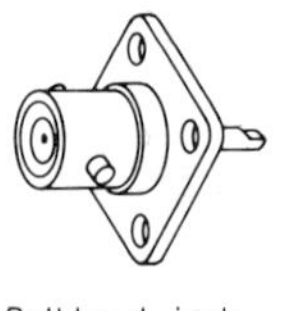

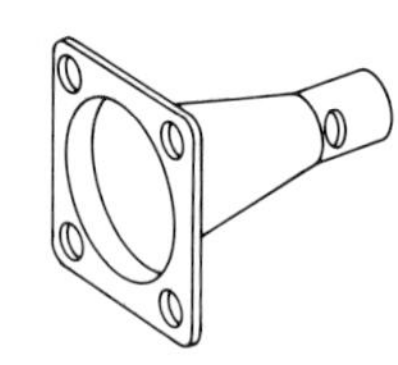

FIG. 50. Typical configurations: bayonet-lock, chassis-grounded jacks.

and construction details enter into the actual impedance of a given connector. A variation of plus or minus 20 percent from the specified nominal impedance is not uncommon on an installed connector. If the application is critical, the impedance match should be checked. The impedance match of right-angle coaxial connectors is traditionally poorer than the straight-through types because of the increased possibilities of discontinuities introduced by the nature of the right-angle configuration.

Although there are many reliable coaxial rf connectors on the market today, the new MIL-C-39012 specification will be of considerable value in helping to ensure that coaxial connectors perform as specified.

Selection Guide. There are many types of coaxial connectors available to the user today. For a given application, manufacturer's literature and the manufacturer's engineering people should be consulted in order to make an optimum selection.

As a guide in the selection of coaxial connectors, a checklist such as the following is suggested.[7]

1. Determine the VSWR of a connector and relate it to your application. Ascertain the amount of impedance matching necessary for the particular cable to be used, etc.
2. Determine the required peak voltage. (Do not overspecify. Remember that larger creepage distances may be required for high voltage requirements which magnify discontinuities.)
3. Determine the atmospheric conditions at which voltage is applicable. Take into account temperature, pressurization, etc.
4. Determine procedure for assembling cable to connector. (Analyze number of steps, number of parts, tools necessary, etc. Evaluate durability, reliability, uniformity of cable assembly.)
5. Equate costs against factors in item 4 and other guideposts. Check on field problems and requirements.
6. Determine physical requirements (mounting, size, weight restrictions, coupling means). Use a standard connector whenever possible.

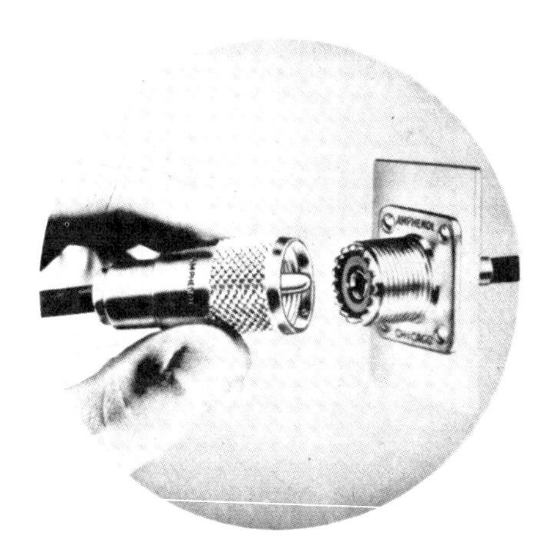

FIG. 51. UHF series connector; the original low-cost coaxial connector. Good up to 300 MHz, which was considered UHF before World War II. (*Amphenol-Borg Electronics Corp.*)

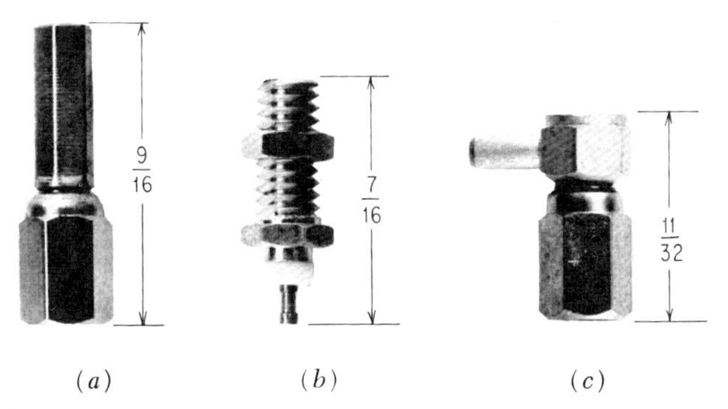

FIG. 52. Ultraminiature coaxial connectors for 50-ohm cable. These have screw coupling; also available with push-on coupling. Typical dimensions are shown. (*a*) Straight plug; (*b*) receptacle, rear mount; (*c*) right-angle plug. (*Microdot, Inc.*)

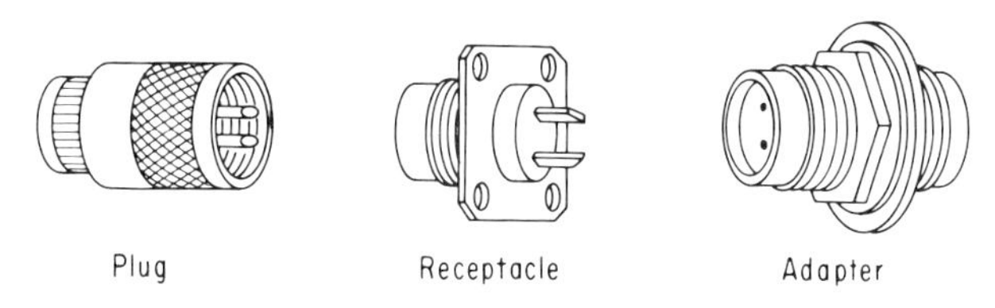

FIG. 53. "Twin" coaxial connectors.

TAPE CABLE CONNECTORS

Background. Tape (flat flexible) cable was first introduced about 10 years ago and although its many advantages became apparent over the years, its usage has been relatively limited, in part, because of the fact that there were no easily applied, good, reliable connectors available for terminating it until the past two or three years. Another hindrance to both the use of the cable and the development of connectors for it was the lack of generally accepted guidelines or standards for the cable, thus making it difficult for the connector manufacturers to design connectors for the cable that would be more or less standard.

Finally, in 1963, National Aerospace Standard 729 was introduced which, in cooperation with the Institute of Printed Circuits, became the basic standard for continuous, flat flexible cable. Materials, conductor spacing, conductor sizes, and number of conductors were specified, making it possible for the connector industry, as well as the manufacturers of the flat cable, to develop various connector concepts that were compatible with the standard cables.[8]

Although considerable development work is still being done by the tape cable manufacturers, with new, improved, and different cables rapidly being introduced, the fact that at least there was finally a measure of standardization led to the acceptance and use of flat flexible (tape) cable.

In 1965, the Institute of Printed Circuits released a general Standard for Connectors for Flat Flexible Cable, IPC-FC-218, which was also of considerable value in giving impetus to both the development of connectors and the use of flat flexible cable wiring systems. Table 4 lists the standard cable configurations as per NAS 729.

TABLE 4. Standard Flat Flexible Cable, NAS 729 (1963)*

Conductor spacing, in.	Conductor sizes, in.	Cable width per conductor		
		1 in. per no. of conductors	2 in. per no. of conductors	3 in. per no. of conductors
0.050	0.002 × 0.025 0.003 × 0.026	17	37	57
0.075	0.002 × 0.025 0.003 × 0.026 0.003 × 0.046	12	25	38
0.100	0.003 × 0.062 0.005 × 0.063	9	19	29
0.150	0.005 × 0.100	6	12	19

* This table was included in the first issued standard for flat flexible cable and is basic. However, with the rapid advancements in microelectronics, and the need for more and higher-density interconnections, custom cables have been fabricated up to 10 in. wide with conductors on 0.100-in. centers.

As a design note, it is suggested that, rather than using such a wide cable as this, when the application requires very large numbers of interconnections, it would best be handled by stacking smaller cables, or by a multilayer cable.

Two problems were apparent in developing methods of terminating the tape cable. One was the problem of exposing and cleaning the copper conductors embedded in the insulation of the cable, and the other was the problem of providing a good, practical, reliable interconnection between the conductors of the cable and another more or less conventional connector so that the cable could be used in a system.

Two concepts of flat flexible cable termination are presently being used in the various available connectors. One is simply a rather conventional style in which the contact terminals are attached to the cable conductors, and the other is a type in which the cable conductors themselves are used as the contacting surfaces in the connector.

Application. Details of terminating flat flexible cable are given in Table 5. Comments on the various methods are included.

TABLE 5

Type	Method of exposing conductor	Termination method	Comments
1	Chemical or mechanical stripping to remove all insulation in specified areas	Contacts are attached by solder, welding, or crimp.	Individual conductors are exposed to possible damage. Integrity of cable is lost due to complete removal of supporting insulation from conductors. Not recommended.
2	Mechanical removal of insulation from one surface only of conductors by the use of high-speed fiber-glass abrading wheels, or by skiving	Conductors are used as connector contacts, or contact terminals are joined to conductors by solder or welding.	Copper conductors are thin and soft, and care must be taken not to damage them during insulation removal. Available precision equipment does a good job of insulation removal with negligible damage to conductors if properly used. If conductors are to be used as contacts, they must be plated after insulation removal to prevent oxidation and to provide low contact resistance.
3	Cables prefabricated, leaving contacting areas exposed; no insulation to be removed	Same as type 2.	Best method for exposing conductors. No likelihood of damage to conductors. Most expensive because cables cannot be fabricated in continuous lengths and cut to size for application; must be fabricated to predetermined length with specified bare conductor areas. Must be plated as in type 2.
4	Insulation piercing	Contact terminals contain sharp points that are forced through insulation and into conductors to make interconnection.	Small contact area. Subject to possibility of insulation film between terminal point and conductor. Also possibility of cutting through conductor in such a fashion as to reduce strength and current-carrying capacity.
5	Melting of insulation	Welding.	Solid metal junction with conductor exposed only in weld area. Very reliable method. However, as with all welded connections, inspection is difficult. Weld schedule and control critical.
6	Flat flexible wiring hardness prefabricated with through-holes and solder pads exposed	Solder.	Must be custom designed to fit connector terminal configuration. May be one or several individual layers, or may be multilayer with plated-through holes. Very reliable but not applicable to continuous cable cut to length.

Types of Termination. Connectors for flat flexible cable applications can be grouped into two basic functional types:

1. Terminations at the ends of the cable (Fig. 54)
2. A right-angle tap connection to the cable (Fig. 55)

Either of the two types can be used in three basic ways, according to application requirements:

1. Flat cable to round wires (Fig. 56)
2. Flat cable to printed-circuit board (Fig. 55)
3. Flat cable to flat cable (Fig. 57)

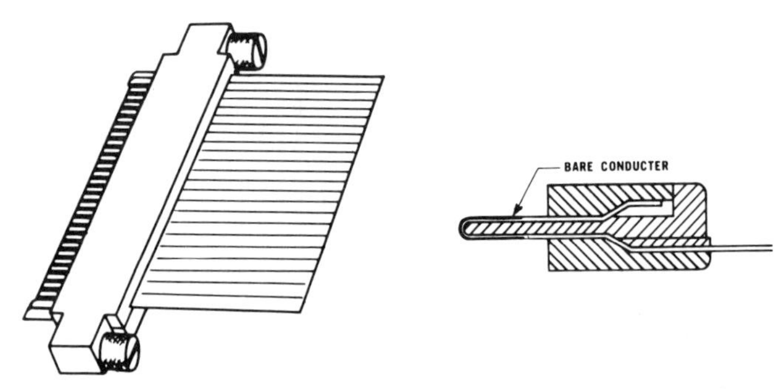

FIG. 54. Male connector for flat conductor cable. Flat conductors, selectively gold-plated, are used for contact with a female connector. The gold-plated contact area of the flat cable is folded and locked into the male connector. (*Advanced Circuits International.*)

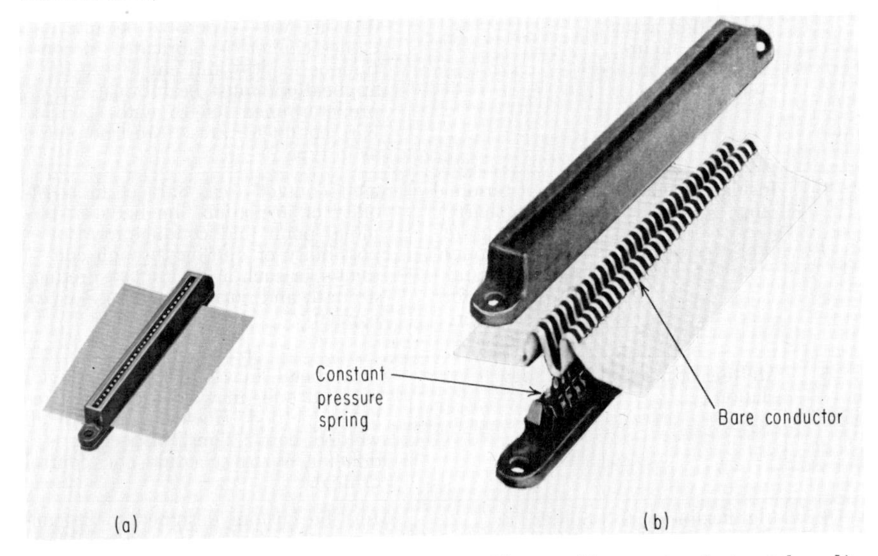

FIG. 55. Right-angle mid-span tap (cable-to-cable or cable-to-printed-circuit-board). (*a*) Assembled connector on tape; (*b*) exploded view showing construction. The selectively gold-plated contact area of the cable is formed and locked into the connector housing. A beryllium copper spring exerts pressure on each conductor, assuring positive electrical contact of each conductor with a mating piece. The spring carries no current; its function is only mechanical. (*Advanced Circuits International.*)

Terminating Methods. Referring to Table 5, the various types and categories are described and illustrated below:

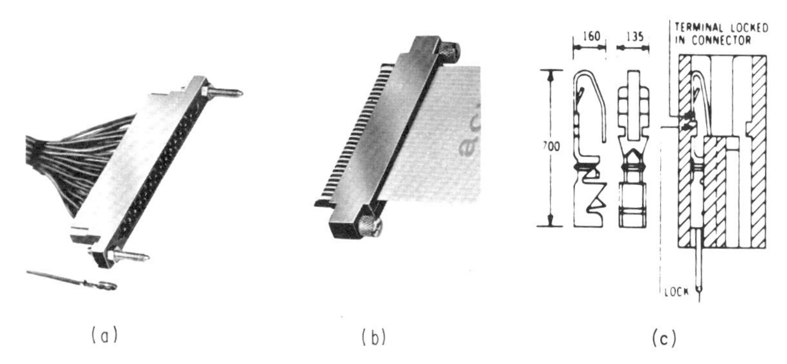

FIG. 56. Round-wire-to-flat-cable connectors. (*a*) Round wire connector receptacle; (*b*) flat conductor cable plug; (*c*) section of receptacle showing contact. (*Advanced Circuits International.*)

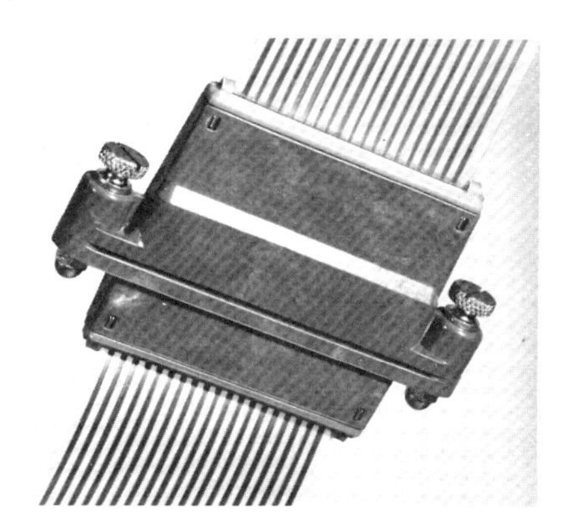

FIG. 57. Flat-tape-cable-to-flat-tape-cable connector with cable support and locking screws. (*Amphenol-Borg Electronics Corp.*)

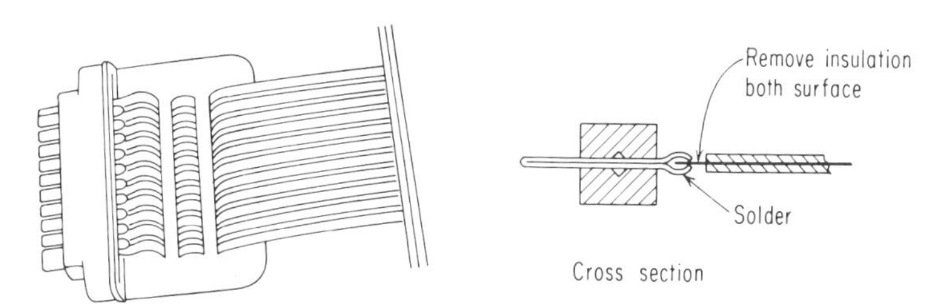

FIG. 58. Tape-cable connector with cable support. Blade-type contacts are used. A similar type is available for insertion into an edge-type printed-circuit receptacle.

Type 1: Cable conductors are stripped bare and are joined to connector contact terminals. The cable conductors are no longer supported by the insulation, and therefore the terminal junctions must be potted, or a firm cable clamp or support must be used (Fig. 58).

Type 2: Insulation is removed from one surface of the conductors, at the ends and/or at some point(s) along the length of the cable. Connector terminals are then

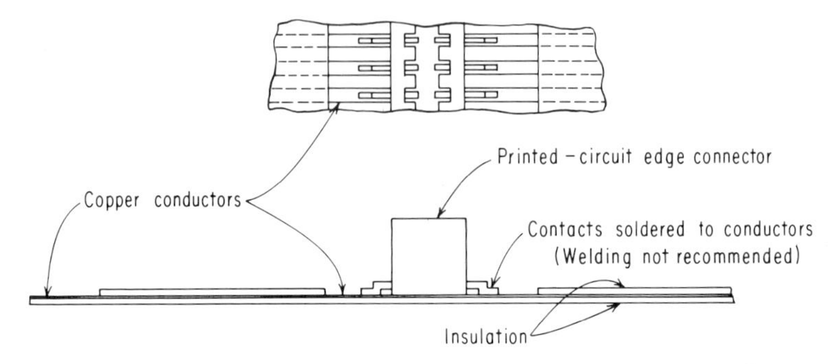

FIG. 59. Sketch of end and center stripping of tape cable with mounted connector.

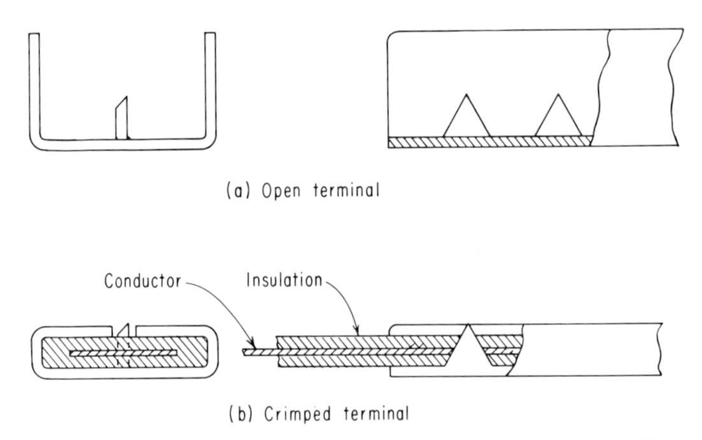

FIG. 60. Sketch of crimp-type insulation-piercing terminal.

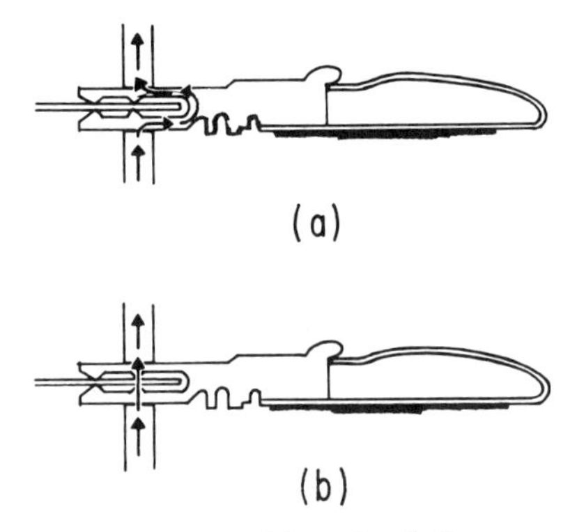

FIG. 61. Welded contact termination without insulation removal. (*a*) Shunted welding current heats terminal—melts insulation; (*b*) insulation melted, current shorts through conductor, completing weld.

joined to the bared conductors (Fig. 59), or the cable is formed and retained by spacers and molded carriers in such a manner as to result in flat connector contacts (Fig. 55). Cable must be supported either by the connector moldings themselves or by external cable clamps. Molded connector housing can be custom made to fit any cable width, with a variety of conductor spacings and sizes.

Type 3: Same as type 2.

Type 4: Contacts are applied to the cable, either individually or assembled in a connector, by crimping the contact to the cable conductor, thus causing the contacting points of the terminals to pierce the insulation of the cable and make intimate contact with the copper conductors by being forced into and through them (Fig. 60). Alignment of contacts to conductors is critical.

Type 5: Contacts are either held tightly to the surface of the cable insulation or crimped to the insulation. In both instances the welding electrodes provide additional pressure, as specified in the appropriate welding schedule. The welding current is shunted around the contact terminal until the heat melts the cable insulation, at which time the weld between the contact terminal and the conductor is made (Fig. 61). Because this is a highly specialized welding application, the manufacturers of this type of connector also supply welders which are electrically and mechanically compatible with the connection system.

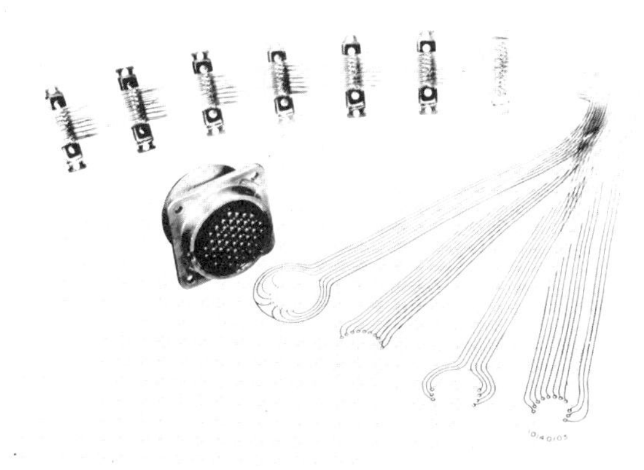

FIG. 62. Typical application of custom-made FREE-FLEX flexible circuitry. Four-layer circuit shown uses nineteen 15-pin miniatures, one 25-pin miniature, and two 51-pin AN-type cylindrical connectors. Circuit layers are positioned on connector terminals and soldered. Approximate size of complete circuit is 50 × 1 in. (*Arthur Ansley Mfg. Co.*)

Type 6: Connectors for this type of application can be any of the conventional connectors having dip-solder-type pin terminals. The flat flexible cable harness is designed with the conductor terminal pad geometry compatible with the connector(s) to be used. The pin terminals of the connectors are positioned through the holes in the pads and soldered using conventional techniques (Fig. 62).

Recent developments in connector concepts combined with low-cost assembly methods have made the use of flat flexible cable attractive and economical. Many manufacturers insist on supplying, and all prefer, their connectors fabricated to the cable. Both from an economical and a reliability viewpoint this is desirable in many instances. The technical know-how and experience of the manufacturer, combined with the advantage of a single responsibility, result in an overall saving to the user. Perhaps an exception to this philosophy is in the case of the type 6 method. In this instance, connectors and techniques are both fairly standard, which makes assembly of the connectors to the flexible circuits compatible with the printed-circuit assembly methods of many users.

PLATE CONNECTORS

Description. The plate connector is so named because it consists basically of a metal baseplate on which contact assemblies and/or connectors are precisely mounted in very accurately positioned holes located in a predetermined grid pattern (Fig. 63).[9]

The principal advantage of this concept is that, without any special tooling, a plate connector can be designed to accommodate almost any conceivable combination of plug-in modules, cable connectors, printed-circuit boards, and patch cords. These can be keyed in a number of ways, including variations in group patterns, variations in terminal orientation, and the use of stand-off keys and washers. In addition, visual keying may be achieved by means of color variation in the plastic insulators used.

Rather than a simple connector, the plate connector is, in fact, a packaging system with great flexibility. A major constraint upon this flexibility is that the contacts must necessarily be confined to one of the several grid systems for which tooling exists, both for the plate and for the contact assemblies. However, the fact that all the contacts are on a predetermined grid pattern makes possible the use of automatic machinery to wire the contact terminals as a back plane. Wire wrap and Termi-point * are two of the most common and economical methods of wiring, although solder, weld, and crimp-type terminals are available.[9]

Four basic connector contact types are available: the conventional pin and socket type, the blade and fork type, the printed-circuit edge-receptacle type and the Elco Varicon (Fig. 64). Bus strip contacts and terminal studs are also available.

The *contacts* are available individually, loose, and assembled in insulators; in modules containing two or four contacts assembled in insulators; or as complete connectors designed to mate with specific printed-circuit connector boards. The individual contacts can also be arranged to accommodate module headers or conventional mating connectors (Fig. 65).

At assembly, the insulators are pressed firmly and securely into the holes in the metal baseplate, with the terminations extending out the opposite side. Where contacts are to be grounded to the metal plate, metal bushings are substituted for the insulating bushings. These metal bushings make intimate contact between the individual contact and the metal plate, thus providing a solid ground connection right at the contact.

Application. The size and shape of the basic plate is dependent on its final usage. The most commonly used plate material is an aluminum alloy with a chromate finish; however, other materials and finishes are also used in some instances.

The baseplate must be a rigid, self-supporting structural member that can withstand, without deformation, the forces required to insert and withdraw mating connecting assemblies, whether they be printed-circuit boards, module packages, or mating connectors.

Contact patterns are usually arranged in rows and groups, allowing for unpunched areas in between, which act as strengtheners and, on very large plates, permit the use of additional structural support to the plate.

The holes in the plate must be distributed in a pattern that is compatible with the automatic wiring equipment to be used. Deviation from true location of any terminal, which is a combination of hole location and the perpendicular attitude of the terminal post in relation to the plate, is limited by the tolerance specification set by the manufacturer of the terminating machinery and must be considered in the plate layout and overall design.

The basic plate design is, of course, dependent on the overall packaging concept, and is the responsibility of the user. All the manufacturers of plate connectors will supply engineering assistance in the design of the plate, especially in the specific details associated with the component contact parts and assemblies. Although the same functional designs are available from several manufacturers, the component parts and design details are not quite the same and are not interchangeable.[10]

Manufacturer's data sheets should be used as references in working out the basic layout and design of the plate, after which it can be submitted to one or several manu-

* Trademark of AMP Incorporated.

FIG. 63. Typical plate connector assemblies. (*Elco Corp.*)

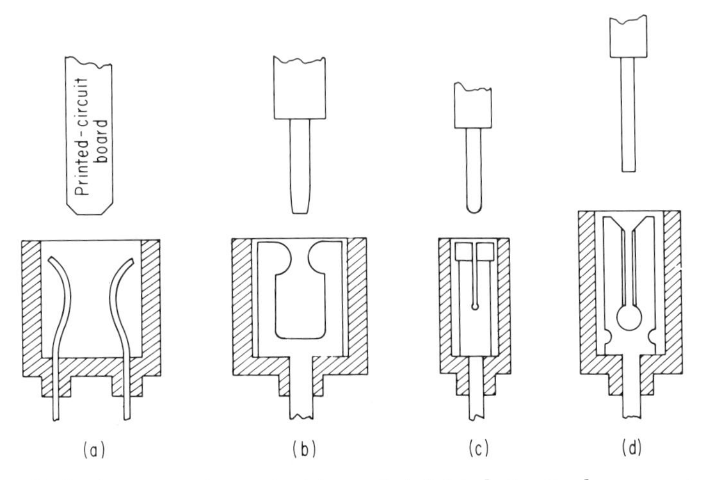

FIG. 64. Basic plate connector contact types. (*a*) Printed-circuit edge-connector type, single- and double-sided; (*b*) blade and fork type; (*c*) pin and socket type; (*d*) hermaphroditic *(Elco Varicon)*.

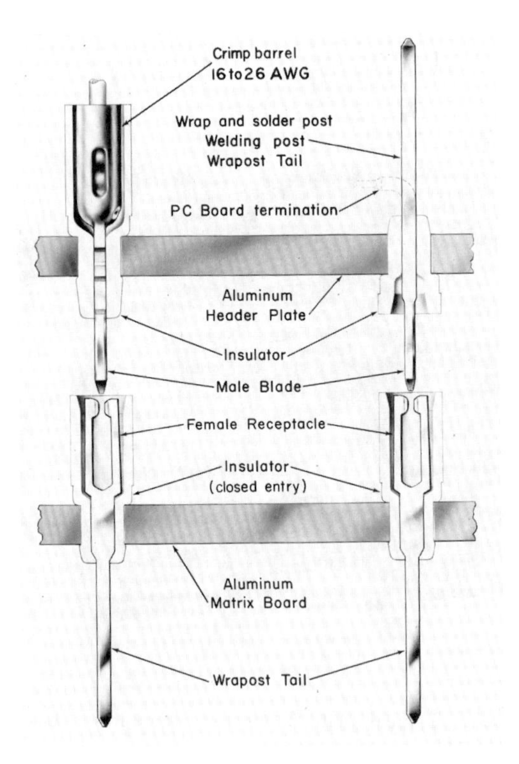

FIG. 65. Basic plate connector concept illustrated by Malco's WASP construction. (*Malco Manufacturing Co., Inc.*)

facturers for their proposals. At present, all the manufacturers prefer to fabricate the plates and assemble the contacts to it. They have the necessary tools and equipment to do the job, and an important advantage to the user is that there is a single responsibility for the finished assembly. All associated hardware, such as studs and terminals, should be assembled at the same source and the finished plate delivered ready to be wired and installed. Many of the manufacturers are also equipped to do the wiring, with automatic programmed machinery where applicable.

After the manufacturer has been selected, details of component parts, dimensions, and tolerances can be worked out as a joint effort. Before proceeding with the fabrication job, the manufacturer will supply the user with his engineering drawing of the details and assembly of the plate, based on the user's layout, for final checking and approval. A typical manufacturer's drawing is shown in Fig. 66.

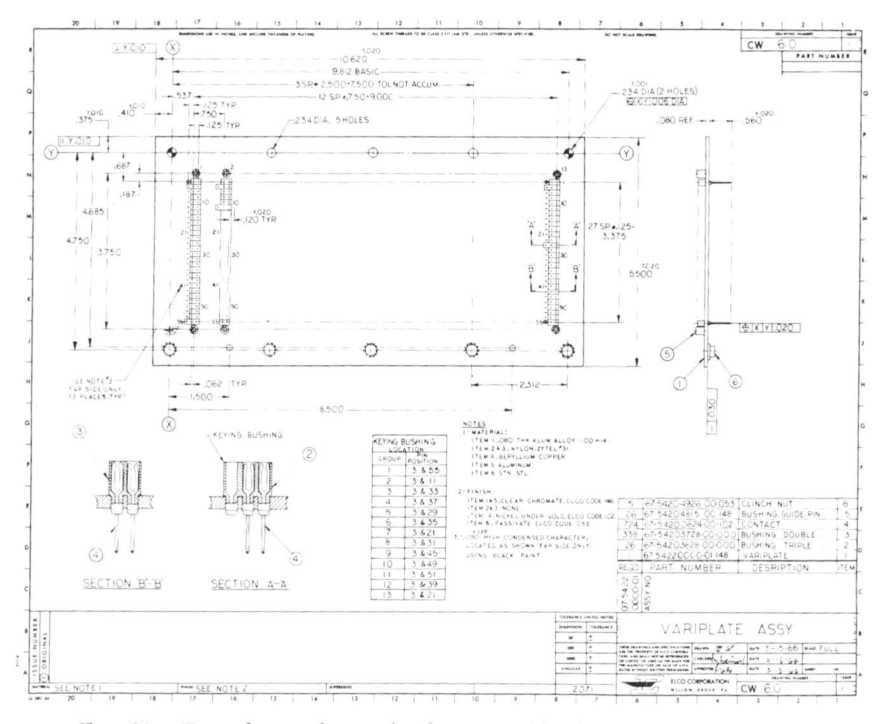

FIG. 66. Typical manufacturer's plate assembly drawing. (*Elco Corp.*)

Engineering Data.

Plate Size. Back-plane plates have been made as large as 24 × 48 in.; the only restrictions are that the size must be one that is practical to handle and must not exceed the capacity of the available fabrication tooling. Plates of just a few inches in area can be made for individual module headers and for mating connector bases (Fig. 67).

Plate Thickness. At present, contacts and insulators are available for plate thicknesses of 0.080 and 0.125 in. Future developments will undoubtedly allow for other plate thicknesses.

Hole Grid Patterns. Presently available grid spacings, compatible with automatic wiring machinery, are shown in Fig. 68. A hole tolerance location of 0.003 in. nonaccumulative can be held.

Contact Materials and Finishes. Phosphor bronze contacts with various platings are readily available. The manufacturer should be consulted for particular application.

Insulator Materials. Nylon and polycarbonate. Many colors are available for identification and programming.

Assuming a flat back-plane application, the plate connector concept is very versatile and readily adaptable to high-density packaging. Examples are illustrated in Figs. 63 and 69.

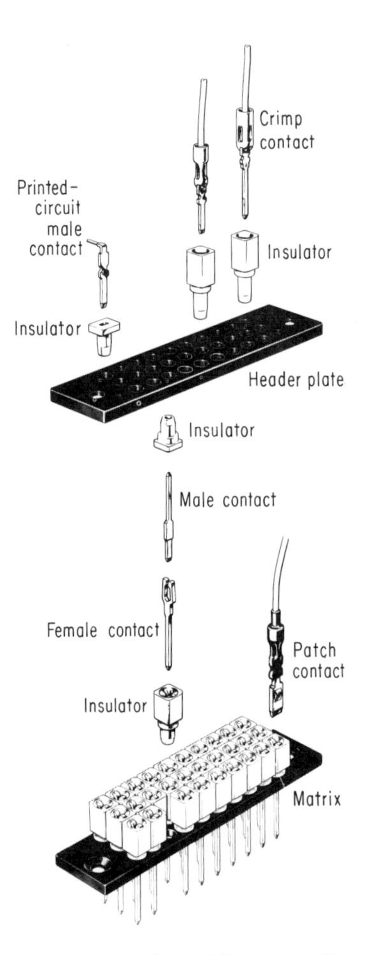

FIG. 67. Details of plate concept used to fabricate individual connectors. (*Malco Manufacturing Co., Inc.*)

HERMETICALLY SEALED CONNECTORS

Definition. A hermetically sealed connector is one that offers a gas- or airtight interconnecting junction through a wall or bulkhead of a "black box" electronic package, or wherever it is necessary to conduct electrical energy or signals between ambient air pressure and either a pressurized or a vacuum chamber or container.

A practical definition: A hermetic seal must be gas-tight and be able to conduct an electrical current into a sealed container with minimum disturbance to the circuit. The performance must be maintained for extremes of pressure, temperature, humidity, thermal and physical shock, vibration and corrosive atmospheres.[11]

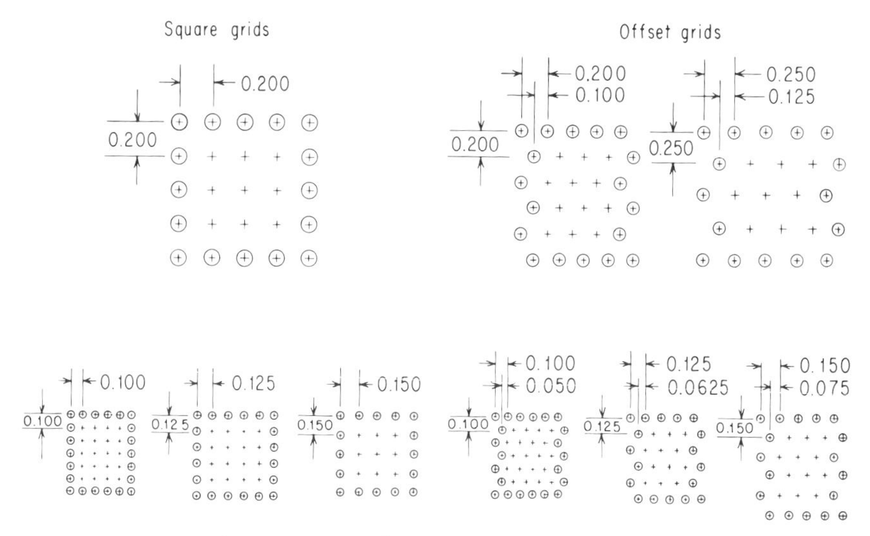

FIG. 68. Basic plate connector grid spacings.

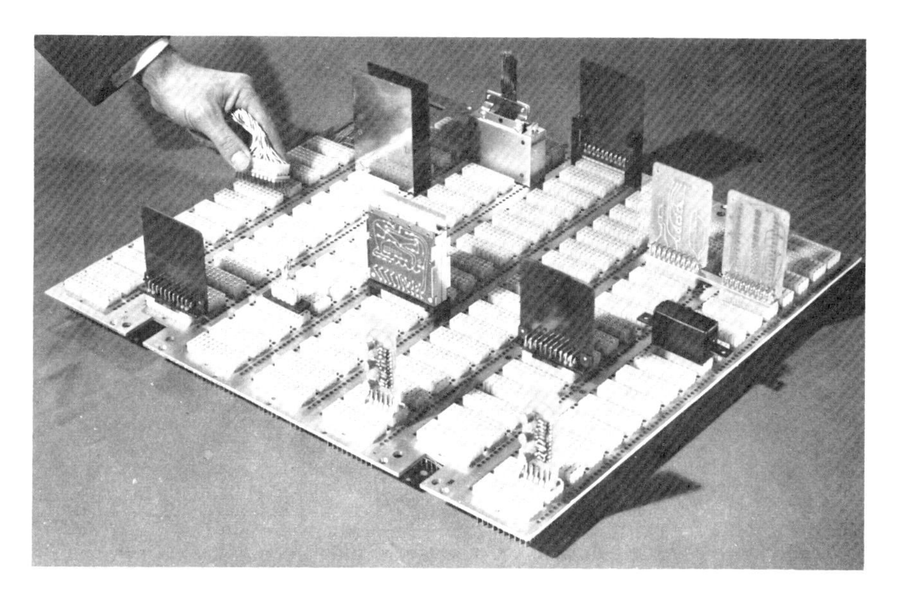

FIG. 69. Plate connector system showing how design versatility enables interconnection of printed-circuit boards, modules, and cable connectors all in one plate. (*Malco Manufacturing Co., Inc.*)

Under these conditions, the connectors must withstand rated current and voltage and must maintain high insulation resistance. They should produce no corona, nor have excessive shunt capacitance or dielectric losses.

Description. Simply described, a hermetically sealed connector is one in which individual contacts are mounted in a metal connector body and insulated from it either with separate glass beads surrounding each individual contact, or with all the contacts sealed in a larger piece of glass. The metal body is then attached to the container or chamber bulkhead by brazing or soldering, or in some instances of vacuum applications, it is held in place by bolts and sealed to the bulkhead with an O-ring-type gasket (Fig. 70).

The hermetic seal in a connector depends on a bond between the glass insulator and the individual contacts and between the glass and the metal body of the connector. This bond, or glass-to-metal seal, is produced in three ways: a soft-glass frit, matched glass, and compression glass. Although each type depends on a bond between a metal and the glass, the method of obtaining the bond differs.[11]

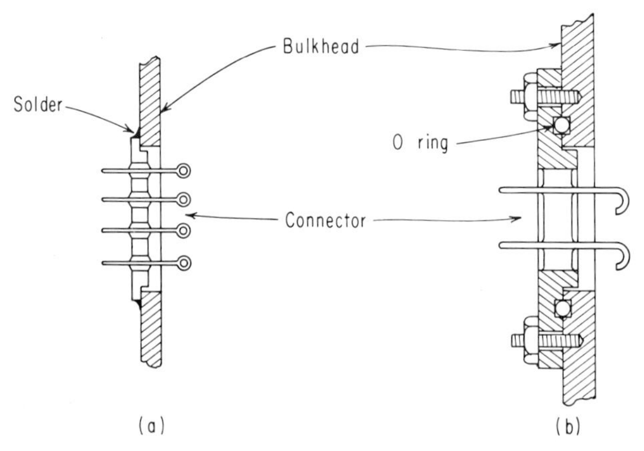

Fig. 70. Cross sections showing two methods of mounting hermetically sealed connectors. (*a*) Individual bead seal construction and eye terminals. Applicable to either rectangular or round connectors. (*b*) Single glass seal and hook terminals. Primarily suitable only for round connectors.

The soft-glass type uses an enamel-like glass bonded to mild steel. It makes an economical seal that is useful over a limited temperature range.

The matched-glass type uses low-expansion metal alloys, such as Kovar, Rodar, or Therlo, and glasses with matched expansion coefficients. This type is used for withstanding thermal shock and wide temperature ranges.

The compression-glass type has a steel outer shell shrunken into the glass insulator so that the glass is under compression throughout the operating temperature range. The thermal coefficient of the glass and the metal are deliberately mismatched, which aids in maintaining a seal over a wide temperature range. It is especially useful in miniaturized and multiple-contact connectors.

A routine temperature-range requirement for hermetically sealed connectors is from −65 to +400°F, although some connectors are available that will operate in temperatures from −200 to +600°F.

A variety of special glasses is used for hermetic seals, and along with the mild steel and special alloys already mentioned, several of the austenitic stainless steels are also used where mechanical stresses are great.

A variety of round, flat-pierced, formed, and hollow terminal types are available for soldering and welding connecting wires. Wire-wrap terminals are also available, but crimp-type are not.

The two most common types of hermetically sealed connectors are the rectangular rack and panel type, where the panel-mounting male half is the sealed half (Fig. 71), and the multipin cylindrical type with the seal in the male bulkhead-mounting half (Fig. 72). The mating female connector halves are the conventional types. A "special" printed-circuit type of hermetically sealed connector header is shown in Fig. 73.

As in all connector selection, complete application requirements should be supplied to the manufacturer who is to supply or design a hermetically sealed connector, because of the normally critical specifications of most such applications.

Fig. 71. Miniature rectangular hermetically sealed connector (male). (*U.S. Components, Inc.*)

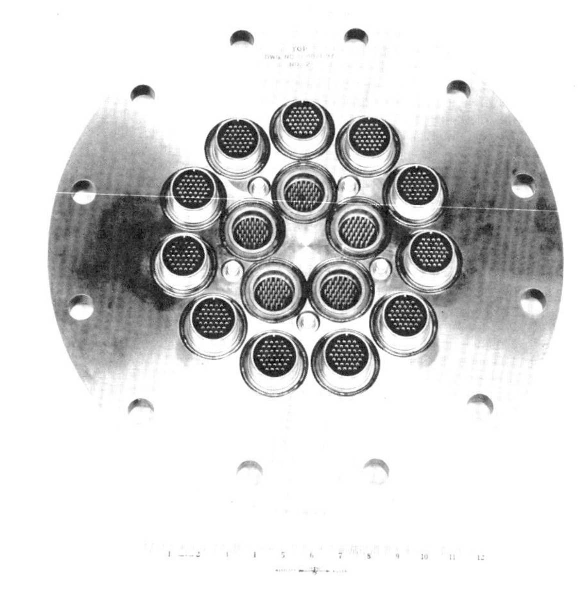

Fig. 72. Application of cylindrical hermetically sealed connectors. Bulkhead plate contains 11 female and 5 male connectors, 37 contact cylindrical connectors, and 5 coaxial connectors. (*Amphenol-Borg Electronics Corp.*)

MISSILE CONNECTORS

Missile and rocket development, both for military use and for space exploration, brought forth the need for a whole new family of connectors. In the general group of missile connectors, there are two functional types. One is the breakaway or umbilical, whose function is to provide electrical power and signal interconnection between the missile (and where applicable, a satellite or space vehicle) and ground power and control equipment, prior to and up to the moment of launch. At launch, the connector halves must separate cleanly and reliably, thus breaking the physical and electrical contact. The separation is accomplished by means of mechanical, electromechanical, or percussion devices built into the connectors.

Another type is the interstage connector, whose function is to provide electrical communication between stages of a missile after launch and up to the time when the stages are separated in flight. In most cases no release mechanism is required, as the force of stage separation causes the unmating. These connectors are either adaptations of existing connectors or, more often, special connectors that have been designed to meet the stringent performance criteria of the specific application.

FIG. 73. Special hermetically sealed printed-circuit card connector. Special connector/header contains 50 glass-sealed blade contacts. Structure holds printed-circuit cards on each side. Formed can is soldered to flange, resulting in a rugged, environmentally sealed package. (*Amphenol-Borg Electronics Corp.*)

FIG. 74. Breakaway-type missile connector. (*Amphenol-Borg Electronics Corp.*)

A breakaway-type connector is illustrated in Fig. 74. The contacts are butt-type and spring-loaded; they include 39 contacts of size 16, 20 amps; 3 contacts of size 12, 35 amps; and 2 coaxial contacts. The release mechanism is lanyard plunger/collet.

FIG. 75. Interstage missile connector. (*Amphenol-Borg Electronics Corp.*)

FIG. 76. Breakaway-type missile connector; twin cable. (*Amphenol-Borg Electronics Corp.*)

Figure 75 is an interstage connector. The contacts are pin and socket: 24 contacts of size 16, 4 contacts of size 12, and 2 contacts of size 2, with arc-suppressing feature. No release mechanism is required.

Figure 76 is a twin cable breakaway connector containing a variety of butt-type contacts ranging from size 0 (200 amps) to size 16, and including a pair of thermocouple contacts. The release mechanism is solenoid/mechanical and waterproof. Because of the variety of contacts, a multitude of electrical requirements can be met in breakaway situations.

MODULE HEADERS

A module header is essentially a connector half, usually the male, that not only provides connections from a circuit into a system but also provides the supporting structure for the circuit components.

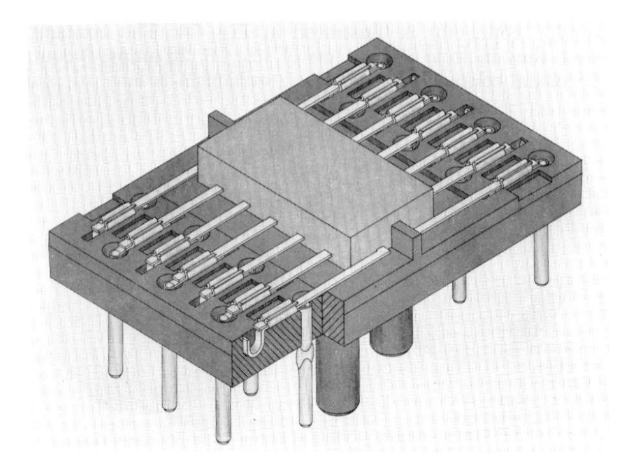

(*a*) Header.

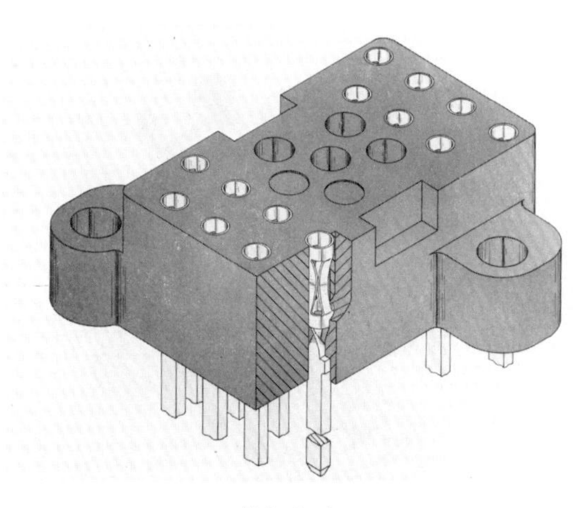

(*b*) Socket.

FIG. 77. Crimp-type flat-pack header. All leads are crimped simultaneously in this unique header. Mating sockets are available, or header can be dip-soldered into printed-circuit board. (*AMP Incorporated.*)

For many applications, headers are custom-designed to fit a particular packaging concept, although there are a few types that are available as standard products from some manufacturers.

In microelectronic packaging, in particular, the use of high-contact-density headers contributes both to overall high-density component packaging and to high operational speeds because of the close proximity of the contacts to components. This proximity results in the shortest possible lead lengths to and through the interface with the next larger logic block, whether it be a "mother" board or a back plane. The use of module header packaging is especially compatible with the plate-connector concept or other fixed-grid back-plane systems.

Several examples of microminiature circuit headers are shown, as well as a special header designed to accommodate two miniature printed-circuit boards environmentally sealed in an inert atmosphere (Figs. 77 to 82).

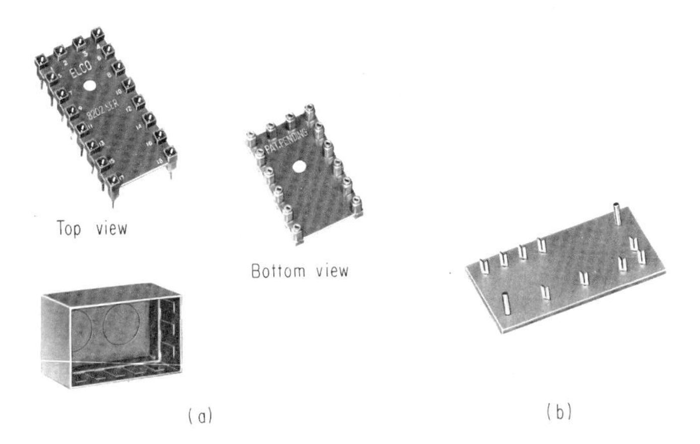

FIG. 78. Module header with cover. (*a*) This header concept, the Modu-Con (trademark, Elco Corp.) features Microcon (trademark, Elco Corp.) contacts recessed in individual cavities in the molded header base. (*b*) Mates with stand-up contacts on mother board. (*Elco Corp.*)

FIG. 79. Hermetically sealed relay header and socket. (*U.S. Components, Inc.*)

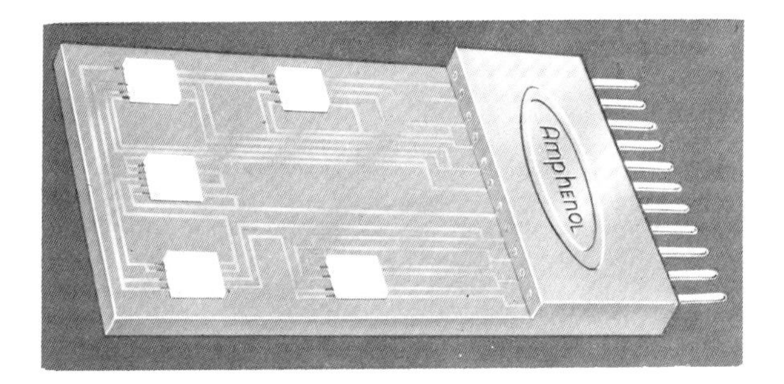

FIG. 80. Precision molded ceramic substrate with integral connector header. Such substrate/headers can be molded into almost any configuration of various materials including alumina, glass, or steatite. (*Amphenol-Borg Electronics Corp.*)

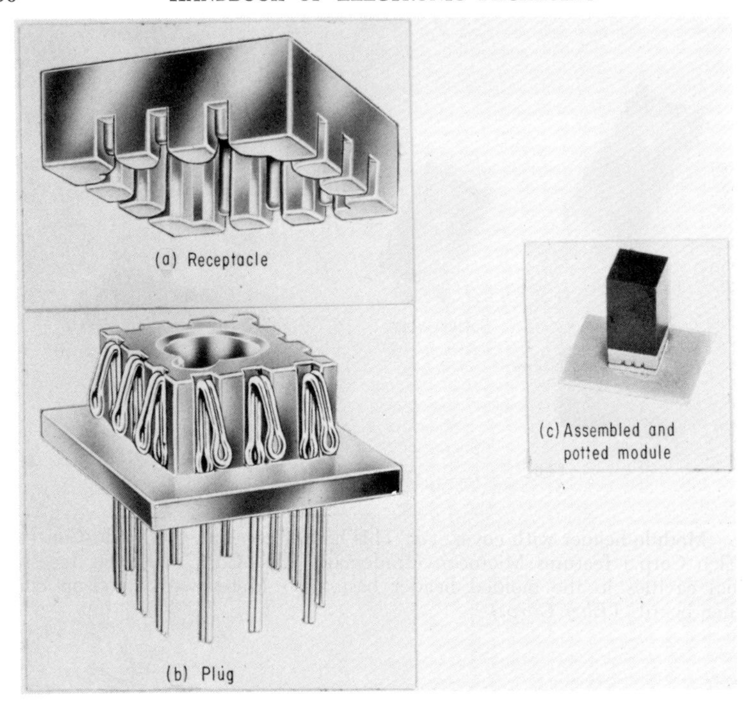

FIG. 81. Micromodule plug and receptacle. (*Amphenol-Borg Electronics Corp.*)

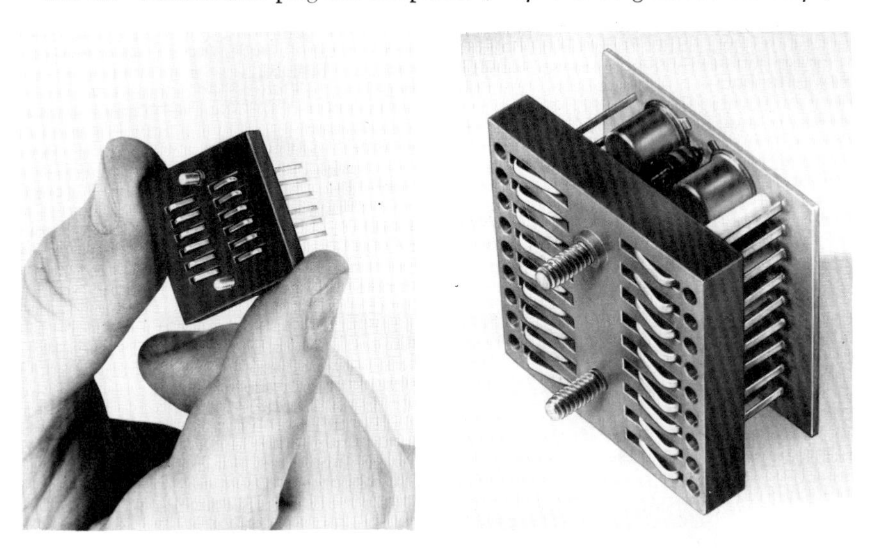

FIG. 82. Printed-circuit-board module header. Spring contacts permit solderless connection to conductive circuit lines on mother board. Center bolt provides high contact pressure that withstands shock and vibration, and also allows for easy disassembly for maintenance. (*Burndy Corp.*)

CONNECTOR MATERIALS

Contact Materials. The choice of contact material has a direct bearing on all other connector design considerations and an important influence on the electrical characteristics of the connector. The contact is the heart of any connector, and thus it must not only function as an electrical conductor, but must be adequately able to withstand all the projected mechanical and environmental conditions to which the connector will be exposed in service. Some commonly used contact materials will now be discussed.[12]

Beryllium Copper. Beryllium copper is an especially suitable spring contact material for connector applications because of its combination of good mechanical properties, electrical conductivity, thermal conductivity, and resistance to wear and corrosion. It is the best electrical conductor of any spring alloy of comparable hardness, and it is stronger and more resistant to fatigue than the other copper-base alloys.

The corrosion resistance of this alloy is about the equivalent of that of pure copper in most environments. It has a higher operating temperature than other copper alloys and retains good spring properties up to 149°C. It can also be used at subzero temperatures.

Beryllium copper springs are readily formed in the cold-worked and annealed condition. A relatively simple subsequent heat treatment can double their strength values. Practically all internal stresses caused by rolling, drawing, or forming are relieved during the heat treatment. The material also exhibits low mechanical hysteresis.

Phosphor Bronze. Phosphor bronze, specifically alloy grades A and C, is a widely used copper alloy because it has good corrosion resistance and fair conductivity, and it is easily formed. Alloy C is usually preferred where moderately high mechanical properties are required. Alloy A is used because of its lower cost where strength is not an important design factor.

Phosphor bronze should not be used at high stress levels at temperatures exceeding 107°C, nor should it be used in salt atmosphere even though it has generally good corrosion resistance. It is a good, general-purpose spring material for limited applications.

Spring Brass. Spring brass is especially useful in low-cost electrical applications where high temperature or repeated flexing at high stresses is not a consideration. Although it has relatively low spring properties, it is often used in conjunction with beryllium copper spring members which provide the electrical contact pressure. It is readily crimped, electroplated, welded, brazed, and soldered.

Low-leaded Brass. This material is used primarily as rod stock for male (pin) contacts. Its machineability characteristics make it very suitable for high-speed, high-production screw machine work. It is therefore an ideal material for high-volume rack and panel pin contacts. It has good electrical properties and good resistance to general corrosion and stress corrosion cracking. It has relatively low spring properties, but it is often used in conjunction with beryllium copper spring members, which provide the electrical contact pressure.

Plastic Molding Materials. The choice of the plastic insulator material to be used in a particular connector application is governed not only by the related design and size of the part (i.e., wall thickness variations, type of inserts, size, etc.,) but also by the electrical, mechanical, thermal, and chemical resistance requirements of the application. The chemical resistance is a requirement that is many times overlooked but is most important where chemical cleaning of solder joints on terminals or printed-circuit boards is involved.

From the cost standpoint, the chosen material should offer the lowest material cost per cubic inch of component part, and also the shortest molding cycle while still satisfying all the application requirements.

Molding materials, whether for use in commercial or military grade connectors, are generally designated by the connector manufacturer according to type, as classified in the specification MIL-M-14, "Molded Plastics and Molded Plastic Parts, Thermosetting."

Table 6 should be helpful to the user in determining the applicability of cataloged connectors in regard to molding material, or in making a determination of a suitable molding material for a special.

TABLE 6. Plastic Molding Materials

Basic resin	Type per MIL-M-14	Description and comments
Phenolic	CFG	General purpose, wood-flour-filled compound intended for applications requiring good electrical properties with mechanical properties better than the acceptable minimum. Moldability is excellent.
Phenolic	CFI-5	Moderate-impact, cotton- or paper-filled compound intended for use where good all-around mechanical properties are required. Impact strength approx. 0.6 ft-lb/in. notch.
Phenolic	CFI-10	Medium-impact, cotton-rag-filled compound that provides good finish. Impact strength is approx. 1.0 ft-lb/in. notch.
Phenolic	CFI-20	High-impact, rag- or cotton-filled compound that provides good finish. Impact strength is approx. 2.0 ft-lb/in. notch.
Phenolic	CFI-40	Highest-impact strength grade of cotton-filled phenolic compound. Impact strength is approx. 4.0 ft-lb/in. notch.
Phenolic	MFE	Low-loss, high dielectric strength, low-water absorption, mineral-filled compound intended for applications requiring the best dielectric properties for a phenolic material.
Phenolic	MFG	General-purpose, asbestos-filled compound intended for applications requiring good mechanical and heat-resistant properties.
Phenolic	MFH	Mineral-filled compound intended for applications requiring highest heat resistance. Mechanical properties are relatively low.
Phenolic	MFI-10	Heat-resistant, medium-impact, asbestos-filled compound. Impact strength approx. 1.0 ft-lb/in. notch.
Phenolic	MFI-20	Heat-resistant, high-impact-strength, asbestos-filled compound. Impact strength approx. 2.0 ft-lb/in. notch.
Phenolic	GPI-100	Glass-fiber-filled compound with high impact strength and good electrical properties. Impact strength approx. 10.0 ft-lb/in. notch.
Melamine	CMG	Cellulose-filled compound with high impact strength and good electrical and mechanical properties, for use where good arc resistance is required.
Melamine	CMI-5	Cellulose-filled, moderate-impact compound with good all-around mechanical properties, for use where resistance to arcing and moderate impact strength are required.
Melamine	MME	Mineral-filled compound for use where good dielectric properties and arc and flame resistance are required. Most dimensionally stable of all the melamine compounds.
Melamine	MM-5	Glass-fiber-filled compound with lower impact strength and higher dielectric constant and dissipation factor at 1 Mc than Type MMI-30. Good moldability. Impact strength approx. 0.5 ft-lb/in. notch.
Melamine	MMI-30	Glass-fiber-filled compound with high impact strength for use where heat resistance, arc resistance, and flame resistance are required.
Polyester	MAG	Mineral-filled compound for use where good dielectric properties and arc resistance are required.
Polyester	MAI-60	Glass-fiber-filled compound for use where high impact strength, good dielectric porperties, and arc resistance are required.
Alkyd	MAI-30	Mineral-filled, glass-fiber-reinforced compound having excellent handling and molding characteristics. It is an arc-resistant, flame-resistant, high-impact compound with good mechanical and excellent electrical characteristics.
Diallyl phthalate	MDG	Mineral-filled compound for use where good dielectric properties and low shrinkage are required.
Diallyl phthalate	SDG	Glass-filled compound of low loss, high dielectric strength, low shrinkage, and good moisture resistance. Relatively low impact strength.
Diallyl phthalate	SDI-5	Acrylic polymer fiber-filled compound of low loss, high dielectric strength, low shrinkage, very good moisture resistance, and high impact strength.
Diallyl phthalate	SDI-30	Polyethylene terephthalate fiber-filled compound of low loss, high dielectric strength, low shrinkage, very good moisture resistance, and high impact strength.
Silicone	MSG	Mineral-filled compound of low loss, high dielectric strength, and excellent heat resistance.
Silicone	MSI-30	Glass-fiber-filled compound with high impact strength and heat resistance. Somewhat poorer electrical properties than Type MSG.
Polycarbonate (Lexan)		High impact resistance, good dimensional stability, good heat resistance, self-extinguishing, good electrical properties.

Miscellaneous Materials. Connector shells and hoods are mainly made from formed sheet aluminum and die-cast aluminum alloys, although in a few cases, cold-rolled steel and stainless steel are used. Locking devices and cable clamps are made from cold-rolled steel, spring steel, and stainless steel. Guide pins and miscellaneous screws and nuts are made from brass, cold-rolled steel, and stainless steel. Jack screws are made from cold-rolled and stainless steel.

Protective coatings used are anodizing, chromate, tin and cadmium plating, and paint. If a bright silverlike finish is desired for appearance, a clear chromate finish over cadmium is used.

Contact Plating. Plating of the basis metal of connector contacts is normally required to prevent deterioration of the mating surfaces of the contacts, mechanically or chemically. Such deterioration eventually results in the inability of the mated contacts to perform their required function in a given application.

Cost per contact of a given connector can be greatly affected by the plating; however, so many variables affect plating that it is difficult to make meaningful comparisons. Variations in plating materials, plating thickness, contact shape, and combinations of all these parameters are all significant factors.

Contact plating should be considered on an individual basis for a particular application, and it is recommended that the experience and technical know-how of the connector manufacturer be utilized in determining the plating to be used.

The two main problems to be considered are wear and chemical environment. The plating should be adequate to cover and protect the basis contact material in the "worst-case" conditions of both these factors.

Hard gold plating on socket contacts and soft gold on pin contacts is used where numerous insertions and withdrawals are anticipated. This combination of hard and soft plating results in a burnishing action that improves wear resistance and also actually improves the contact resistance factor.

An overplating of gold with an underplating of a less precious metal is often used for specific environments as an added protective factor in the event that the gold overplating wears through.

A compilation of presently used plating materials, thicknesses, and combinations of materials for underplatings and overplatings would indicate well over 50 different specifications that have been requested of manufacturers by users. This is not only a confusing situation but an expensive and needless one. Actually the great majority of these many specifications are merely slight deviations from commonly accepted and satisfactory industry standard platings.

There is no one optimum contact plating, but considerable study and evaluation work is presently being done by many of the connector manufacturers as well as by various industry associations with the hope in mind that eventually hard facts and figures will be available that will take the guesswork and choice by personal preference out of connector contact plating and put it on a scientific and engineering basis.

For the present, Table 7 lists some of the most commonly used plating specifications with comments regarding their use.

CONTACTS AND TOOLING

The majority of connectors being used today, especially the multiple-contact rectangular and cylindrical types, employ insertable/removable crimp-type contacts that are assembled to wires, cables, or harnesses and are then inserted into the connector dielectric blocks or inserts at assembly of an equipment. Their use permits connector flexibility that has proved to be an economical advantage for both manufacturing and maintenance.

Advantages of Crimp/Removable Contacts.[13]

1. Inspection and testing of the termination of a wire to the contact can be made prior to assembly into the connector, and without the space restrictions of a normally wired solder-type connector.
2. Wiring errors can be easily corrected by removal and reinsertion of contacts.
3. Damaged contacts can be easily replaced, at assembly or in the field.
4. Contacts can be installed on wires individually by the use of hand tools, or at fast production rates by the use of automatic assembly equipment.

TABLE 7. Platings for Contacts

Material and thickness, in.	*Remarks*
0.0003 hard or soft gold over 0.0002 (min.) silver	Suitable for most most crimped contact applications. Provides low contact resistance for signal circuits and good wear resistance. Corrosion resistance is good unless excessive wear exposes the silver underplating which is then sensitive to sulfide atmosphere. Porous gold or thickness less than the specified 0.0003 will also result in sulfide contamination.
0.00005 hard or soft gold over 0.0002 (min.) silver	Same as above, but much more resistant to wear, with subsequent higher resistance to chemical deterioration.
0.00005 gold over 0.0002 nickel . .	Excellent for wear resistance and hostile environments.
0.00005 gold over copper flash . . .	Excellent for low-level circuit applications with low to medium insertion and withdrawal requirements, as in many data processing and computer applications.
0.0005 to 0.001 silver	Suitable for power contacts with relatively high contact forces.
0.0003 electrotin	For use in low-cost applications where few disconnects are anticipated.

5. Electrical and mechanical characteristics of the termination are uniform and reproducible.

6. Insulation support at the contact terminal is an aid to reliability.

7. Normally, the design of the molded insulator is such that the contact terminals are captured in individual cavities, thus eliminating the possibility of random contact between bare metallic portions of the terminals or wires. This feature also provides a very long creepage path between contact terminals, even on extremely high-density connectors.

Contact Designs.

Basic Retention Systems. Although there are many different trade names for crimp-type insertable/removable contacts, there are only a very few basic design concepts for the contact retention systems.

For printed-circuit edge connector and some blade-type contacts, there is a single basic method of retaining the contact in the connector molding. The body of the contacts contains a simple protruding tine that is compressed as the contact is inserted into the molding and then extends out again and is retained by a shoulder inside the molding. The contacts are inserted by simply pushing them into the molding by hand pressure or with the aid of a simple push-type tool.

Removal is accomplished by the use of a blade-type tool that is pushed into the contact cavity to depress the tine, thus allowing the contact to be pushed or pulled from the molding (Fig. 83).

For pin and socket contacts there are two basic retention methods:

1. A cylindrical collar or clip containing one or more protruding tines is fabricated to the basic contact or, in some cases, the tines are fabricated as part of the contact, especially in the case of formed contacts.

The contacts are inserted into a restrictive cavity hole in the molding, and when they are in position, the tines expand over a retaining shoulder which locks the contact into position.

Insertion is accomplished either by hand pressure or with the aid of an insertion tool that applies pressure against the contact, thus forcing it into position in the molding.

Removal is accomplished by the use of a hollow cylindrical tool that fits over the outside diameter of the contact and, with pressure, compresses the tines, thus allowing the contact to be removed from the cavity in the molding. Some types of tools simply compress the tines, thus allowing the contact to be pulled from the molding. Other types have a plunger, either manually operated or spring-loaded, that exerts force on the contact after the tines have been compressed, thus expelling the contact from the molding (Fig. 84).

Note: Some connectors merely have molded retaining shoulders in the connector inserts or moldings, while others have a molded-in insert, usually of stainless steel.

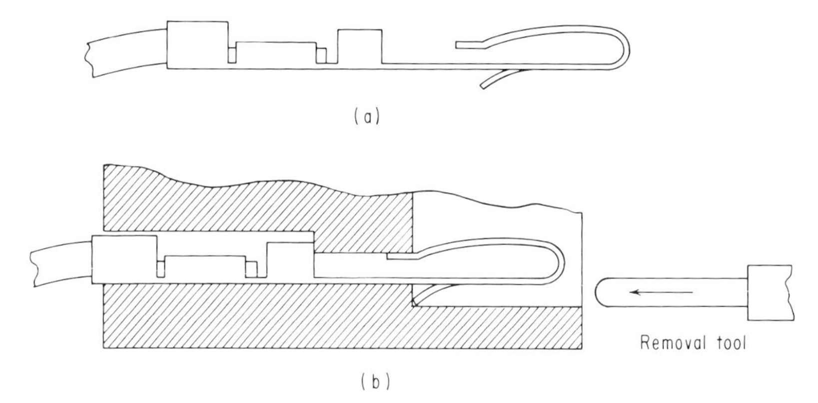

FIG. 83. Simple single-tine locking system.

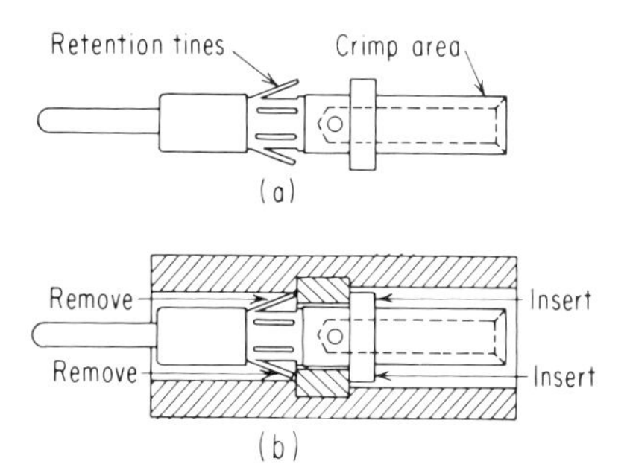

FIG. 84. Typical crimp/removable pin and socket contact. (*a*) Details of basic construction; (*b*) contact inserted and locked in connector molding. Arrows on the right indicate where pressure is applied by insertion tool, and arrows on the left indicate where hollow cylindrical removal tool is inserted to compress retention tines, thus allowing contact to be removed. Basic construction is same for socket contact.

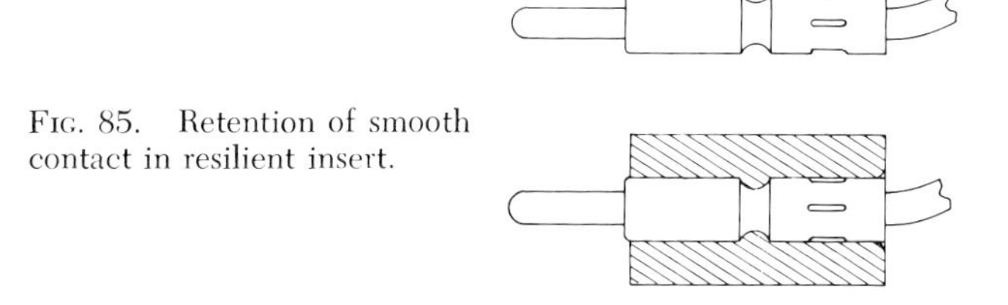

FIG. 85. Retention of smooth contact in resilient insert.

The molded-in insert is more durable than the plastic molded shoulder and allows for many more insertions and removals without degradation of the retaining shoulder.

2. The contact has a circumferential groove that retains the contact when it is pushed into a resilient insert. Simple push-type insertion and removal tools are required to insert and remove this type of contact (Fig. 85).

Other Contact Retention Systems. A unique retention system is that of the patented REMI* contact. In this concept, retention sleeves in the connector molding are employed to receive the contacts, which are snapped into the sleeve rather than being retained by the shape of the molded contact cavity. During insertion or removal, all mechanical stresses are confined between metallic elements rather than between metal and plastic (Fig. 86). The contacts snap in with finger pressure and are removed with a tool provided by the manufacturer. An additional feature of this concept is that

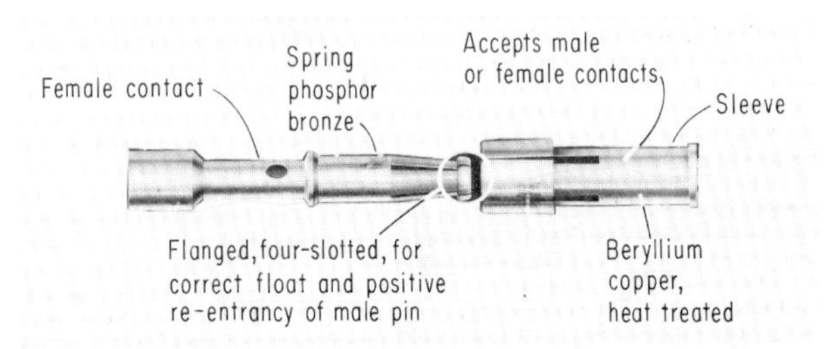

FIG. 86. REMI contact concept. Mechanical stresses confined between *metallic* elements. (*U.S. Components, Inc.*)

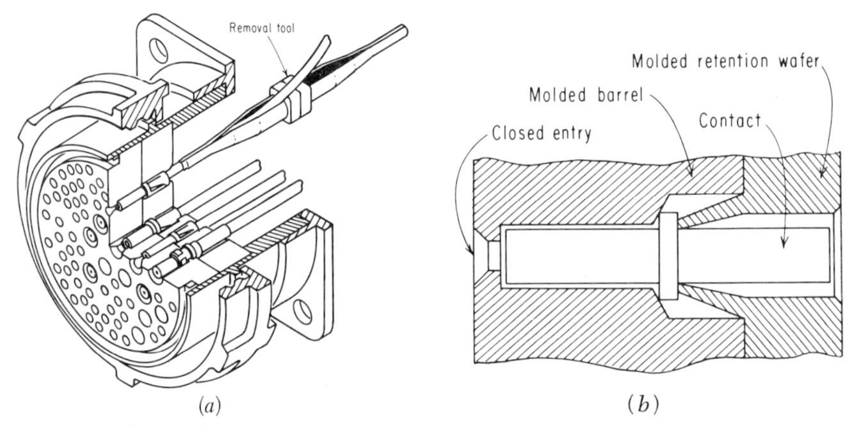

FIG. 87. The Little Caesar (trademark, ITT Cannon) contact retention system. Contacts are inserted from rear of connector and held by cone-shaped retainers. Removal is accomplished by use of a simple plastic removal tool that fits around terminal end of contact, expanding retaining cone in insert and thus allowing contact to be removed. (*a*) Sectional sketch of connector incorporating the Little Caesar rear release contact retention assembly; (*b*) detail of barrel, retention wafer cone, and female contact. (*ITT Cannon Electric, a division of International Telephone and Telegraph Corporation.*)

either male or female contacts can be inserted into the same sleeve. The retention sleeve is so designed that it provides the closed-entry feature.

Another unique contact retention system is one in which a single-piece, molded wafer of resilient material (Lexan† or nylon) contains integrally molded, cone-shaped contact retainers which lock behind the rear shoulder of the contact. The front of the shoulder butts against a barrel insulator, thus locking the contact into the connector body (Fig. 87). The contacts are inserted or removed with a simple, inexpensive plastic tool provided by the manufacturer.

* Trademark, U. S. Components, Inc.
† Trademark, General Electric Co.

Closed-entry contacts are provided in many pin and socket connectors. The closed entry is a solid ring of some sort at the entrance of a female contact that limits the diameter of a pin contact that can be inserted into the female contact to a predetermined maximum dimension, thus preventing overstressing of the female contacting spring members (Fig. 88).

The closed-entry feature assists in the proper mating of a contact pair if the pin contact is slightly bent or otherwise misaligned, within the limits of the location and float tolerances of the contacts. Also, the closed entry protects the female contact from damage caused by improper insertion of a test probe.

Contact spring pressure to provide intimate contact between the male and female contacts is accomplished in several different ways:

1. The female contact is essentially a thin-walled tube that has one end split into two or four sections. These sections are bent inward slightly, to provide spring pressure against the inserted male pin contact (Fig. 89).

2. The female contact is a hollow cylinder containing a flat formed spring that provides the contacting pressure (Fig. 90).

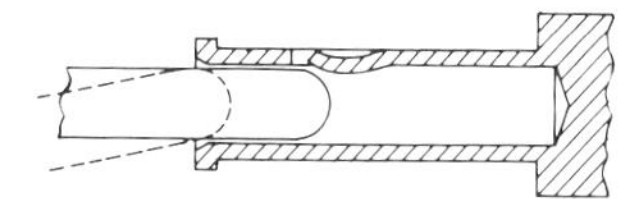

FIG. 88. Sketch of closed-entry contact concept.

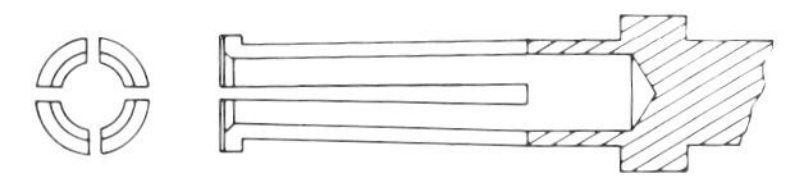

FIG. 89. Sketch of open-ended split-socket contact.

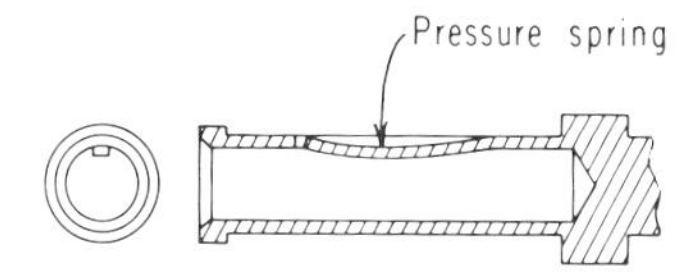

FIG. 90. Sketch of cylinder and spring female contact. Pressure member is either cantilever type or an inserted bow spring.

3. The female contact is a tube and the male contact is constructed of formed wires that have a slightly greater outside diameter than the inside diameter of the tubular female contact. When the formed wire pin is inserted into the female contact, the wires compress and provide contacting area along each of the wires, thus providing a highly reliable connection due to the redundancy of contacting areas (Fig. 91).

There are two main advantages to the concept of contact spring pressure. One is that it is possible to manufacture these contacts in very small diameters, thus allowing for extremely high contact density when used in a multicontact connector. The other is that even though the pin diameter is very small, the fact that it is composed of several flexible spring wires rather than being a solid, though fragile, pin makes it a very durable contact that is not subject to damage as are like-diameter solid pins.

These contacts have been incorporated in a number of very high-density connectors by several manufacturers (Figs. 15 and 16). In some of them the closed-entry feature is effectively accomplished by having the pin contacts recessed in cavities in the connector molding that are only slightly larger than the outside diameter of the female contact and also by having a generous chamfer inside the female contact that guides the flexible pin into it.

Crimp Technique. Compression crimping is a method for joining an electrical conductor (wire) to another current-carrying member (contact or terminal). The method

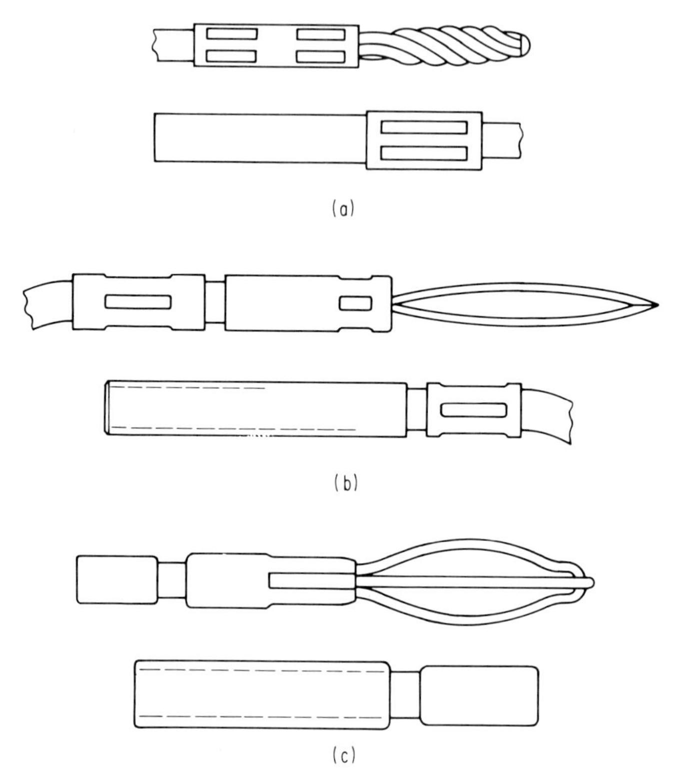

FIG. 91. Formed wire contacts. (*a*) Micropin (trademark, ITT Cannon) and Microsocket (trademark, ITT Cannon). (*ITT Cannon Electric, a division of International Telephone and Telegraph Corporation.*) (*b*) Bow Pin (trademark, Cinch Mfg. Co.) pin and socket. (*Cinch Mfg. Co.*) (*c*) Wire-Form (trademark, Amphenol-Borg Electronics Corp.) pin and socket. (*Amphenol-Borg Electronics Corp.*)

includes the use of a tool to compress the member tightly onto the conductor wire. The compressed juncture is called the *crimp joint.* A proper crimp joint is electrically sound and mechanically strong. External heat is not utilized, so there is no damage to insulation or to small conductors.[14] Degradation of the crimp joint due to shock or vibration is practically eliminated by using insulation grip/support as an integral part of the contact terminal.

Crimp joints are quickly made, no special skills are required, and all guesswork and operator variables are eliminated.

A good crimp termination is defined as one in which the mechanical connection of the wire and the contact do not break or become distorted before the minimum specified tensile strength is reached. Where there are no specified tensile values for specific crimped termination applications, it is common practice to set the requirement at 75 percent of the minimum allowable tensile strength of the wire (Table 8).

Crimp Configurations. There are a variety of crimp shapes that are used for contact and terminal crimping. Figure 92 illustrates some of the more common ones. Although hand or semiautomatic crimping tools are used to make the crimps, it is important to remember that good, repeatable crimp joints are the result of engineering the crimping

TABLE 8. Tensile Strengths of Wire Sizes in Crimped Terminations

Wire size	*Tensile strength, lb* *
26	7
24	10
22	15
20	19
18	38
16	50
14	70
12	110
10	150

* Values taken from MIL-T-7928.

tool, the crimping dies, and the contact together, along with the proper combination of tool, contact, and wire to be crimped. Crimped joints can be undercrimped or overcrimped, as well as correctly crimped. Undercrimping will result in loose connection between the terminal and the wire, whereas overcrimping will cause damage to both the wire and the terminal; either case results in a basically poor and unreliable connection (Fig. 93).

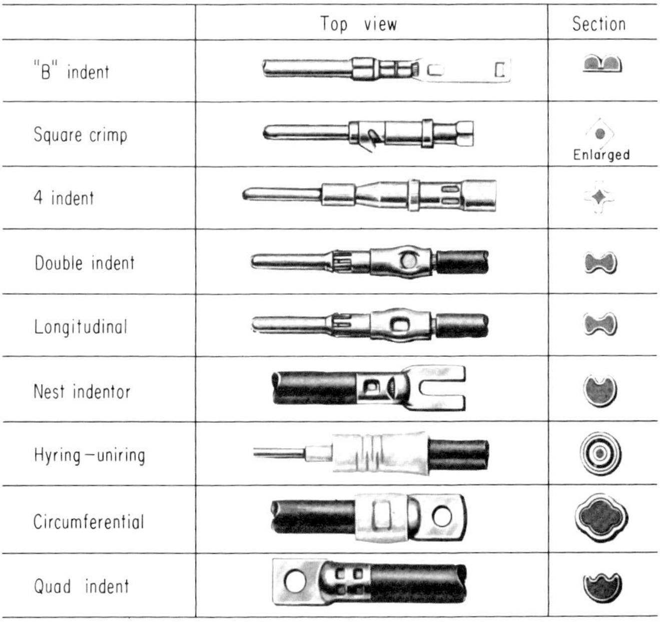

	Top view	Section
"B" indent		
Square crimp		Enlarged
4 indent		
Double indent		
Longitudinal		
Nest indentor		
Hyring—uniring		
Circumferential		
Quad indent		

FIG. 92. Typical and common crimp configurations. (*Burndy Corp.*)

Crimping tools for contacts are specified by the contact manufacturer, along with instructions for their use. In most cases of MIL specification connectors and contacts, the crimping tool specified will be one conforming to the requirements of MIL-T-22520A (WEP), and detailed in MS Drawing 3191 (Fig. 94).

In other instances the crimping tools specified will be ones of proprietary design of individual manufacturers or ones that are commercially available. The tools, dies, depth of crimp, and other pertinent information supplied by the manufacturers are the result of engineering work done in their laboratories for their products, and the instructions and engineering information should be followed carefully (Fig. 94).

Types of Crimping Tools. The fundamentals for good crimped connections are the same whether the tool to be used is a simple plier type or a semiautomatic machine. Descriptions of the several types and their intended uses follows:[14]

Simple Plier Type. This is the earliest type of crimping tool and it still is used for repair operations or where there are only a very few crimps to be made. They are similar in construction to ordinary mechanic's pliers except that the handles are usually longer and the jaws are specially shaped to perform the crimp. Many of these tools have no safety feature to prevent partial crimping and are fully dependent on the operator to perform the crimp properly by closing the pliers until the jaws butt together. (This type is not recommended.)

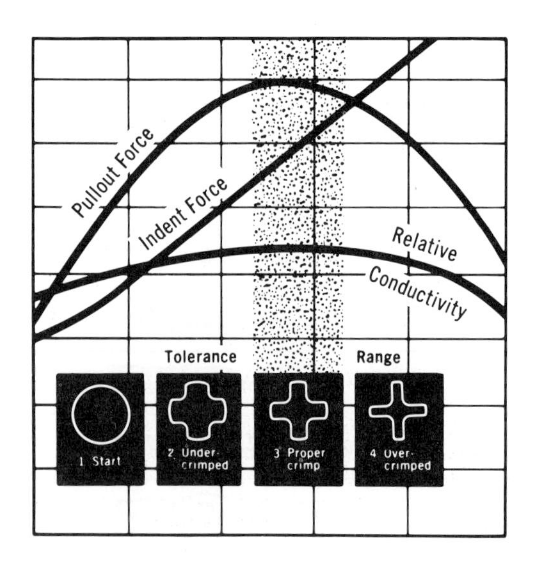

FIG. 93. Depth in indention; laboratory work curves. (*Burndy Corp.*)

How to Use M-S3191-1 Hand Crimping Tool

Setting up Tool

1. Put tool in open position by squeezing handles to their maximum position to trip ratchet, then releasing them.
2. Loosen latch locking screw and pull latch to open position.
3. Pull positioner release all the way down against force of spring and insert or remove positioner.
4. Select proper positioner for contact size. Positioners are color-coded and stamped for size. Be sure flat on flange mates with the flat in handle. Positioner flange must be flush with handle before positioner latch assembly and locking screw can be fully closed and locked.
5. After positioner is in place, push latch to closed position and tighten latch locking screw. Tool is now ready to crimp.

Crimping

1. Insert prepared contact and wire through the indenter opening into positioner.
2. Squeeze handles together until positive stop is reached. Tool will then release and return to fully open position. Remove crimped contact and wire.

Procedures for the other crimping tools are similar. See the applicable manual.

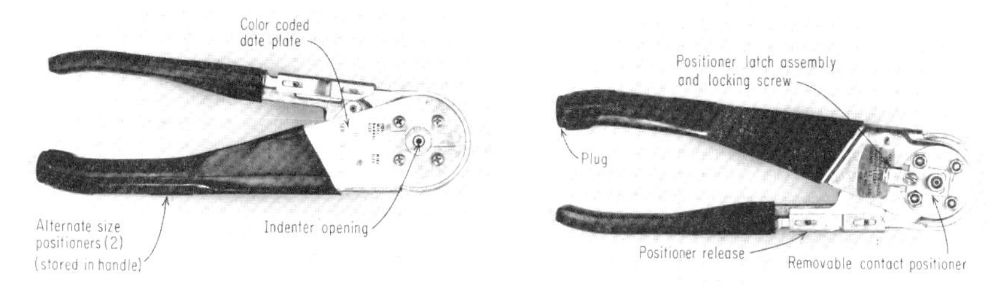

Insert or Remove Individual Contacts at Will

Insertion

1. Cradle the crimp end of contact in round end of tool, making sure tip of tool rests against contact shoulder at base of crimp.
2. Carefully direct mating end of contact into appropriate wire hole in grommet assembly.
3. Push contact into grommet assembly until contact is seated with a positive stop. When shoulder on insertion bit reaches grommet face, insertion is complete.
4. Withdraw tool, keeping it at right angles to grommet face during withdrawal.

Removal

1. Select proper removal tool for size of contact.
2. Insert bit (*into* mating end of female contacts, *over* mating end of male contacts) and push the contact out.

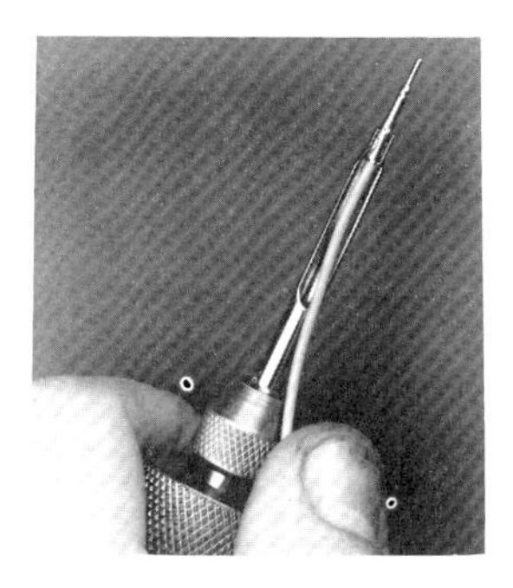

Cradling contact in tool.

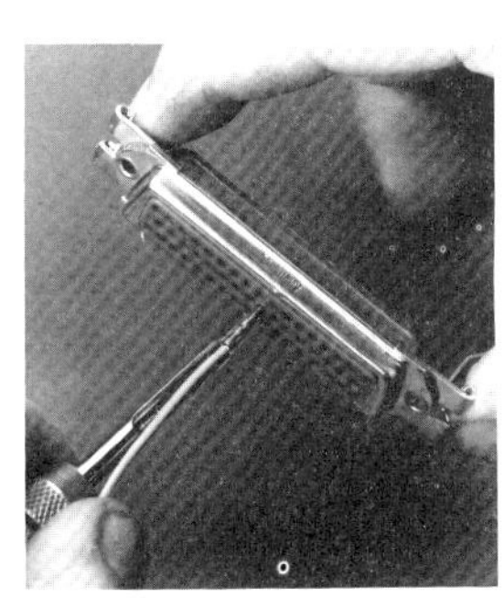

Inserting contact.

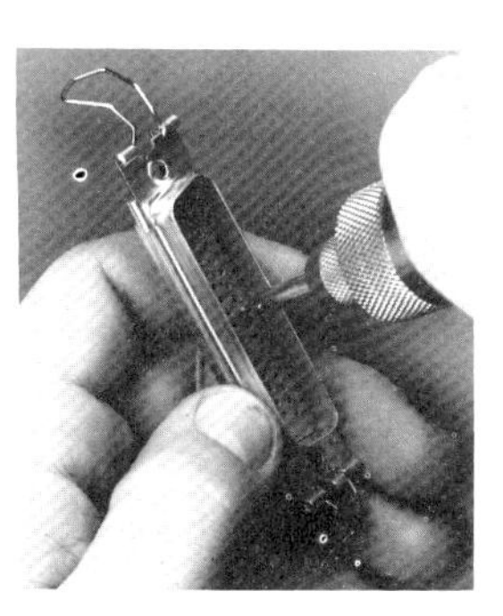

Removing contact.

FIG. 94. Typical manufacturer's instruction sheet on crimping, inserting, and removing contacts. (*Amphenol-Borg Electronics Corp.*)

An improved version of the simple plier tool has a safety ratchet that prevents opening the tool and removing the contact before the jaws have closed completely and performed a proper crimp (Fig. 95).

This type of tool is made to crimp specific terminals and cannot be adjusted, nor are the crimping jaws interchangeable.

The mechanical advantage of this type of tool is not great and the high hand force required to make some crimps is one of the chief limitations of this tool.

Ratchet-controlled Hand Tools with High Mechanical Advantages. This type of hand tool is designed with linkage or cam mechanisms connecting the handles to the crimp dies so that great crimp force can be applied to the crimping dies with minimum hand force. A ratchet mechanism prevents opening the tool and removing the contact before the crimp has been properly completed.

There are two basic versions of this tool. In one the crimp dies are not interchangeable but may contain two or three positions for crimping different-sized terminals (Fig. 96). The other type allows for interchanging dies for crimping a wide variety of different terminals (Fig. 97). The MIL specification tool (Fig. 94) falls into this category. The interchangeable die feature makes the tool very versatile in both production and field usage.

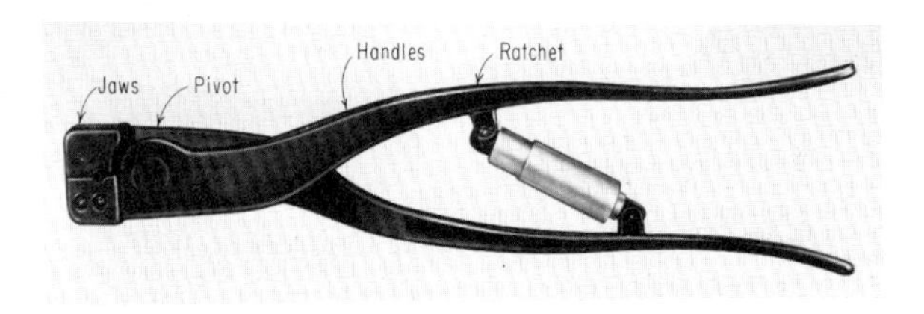

Fig. 95. Hand-plier crimping tool with ratchet. (*Burndy Corp.*)

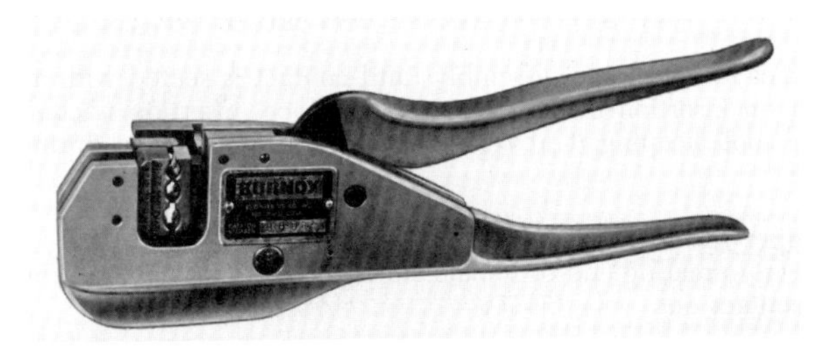

Fig. 96. Hand crimping tool; fixed dies in three sizes. (*Burndy Corp.*)

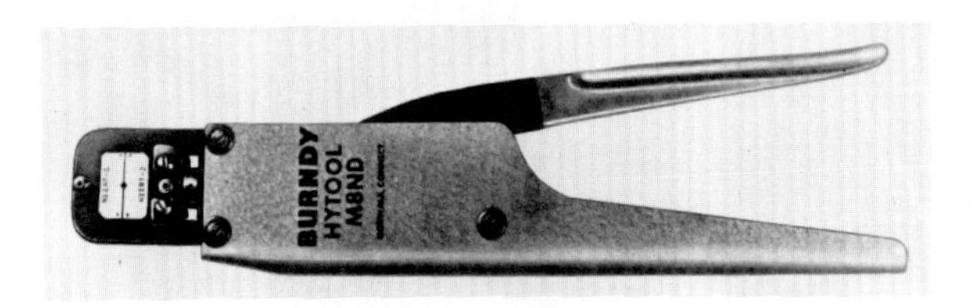

Fig. 97. Hand crimping tool; removable dies. (*Burndy Corp.*)

Air-powered Tools. Crimping tools of this type are available in both hand-held and bench-mounted styles. An air cylinder, operating off normal shop air lines, does the work of forming the crimp; the operator merely inserts the terminal and wire into the tool and pulls a trigger. This type of tool uses the same interchangeable dies as the tool in Fig. 97. It is especially useful for medium-volume production. The air-powered, bench-mounted tool shown in Fig. 98 is actuated by a foot pedal. These air-powered tools contain a full cycling mechanism which ensures full closure of the crimping dies.

Magazine-loaded Tools. For somewhat larger production rates than are possible with the single-terminal tools, air-powered tools which hold a magazine load of

Fig. 98. Bench-mounted, air-powered crimping tool. (*Burndy Corp.*)

Fig. 99. Magazine-loaded, hand-operated, air-powered crimping tool with removable dies. (*Burndy Corp.*)

terminals or contacts are available (Fig. 99). The terminals or contacts are supplied loaded in strips or carriers, and with the tool loaded, crimps can be made as fast as the operator can insert the wire.

Semiautomatic Crimping Machines. These machines are available for high production rates of installing terminals or contacts on wires (Fig. 100). They are available for either continuous strips of formed contacts on carrier strips, or machined contacts mounted on continuous carriers. Depending on size, reels of from 1,000 to 5,000 terminals or contacts can be loaded into the machines, and they operate as fast as an operator can insert prestripped wires into the crimp area. They are either foot-pedal- or electric-trip-actuated. In the electric-trip type, the operator inserts the wires into the crimp area, and when they are in the proper position, the wire strands contact an electrical trip plate, causing the machine to operate and form the crimp. Installed terminal rates of from 1,000 to 4,000 per hour can be achieved, depending on the size and type of wire and terminal being installed. Interchangeable dies can be installed in the machines, allowing installation of a wide variety of sizes and types of terminals and contacts.

In all cases where insulation support is included as part of the termination, the crimping action of the tooling crimps both the insulation and the wire conductor at the same time.

Contact Insertion and Removal Tools. Several types of tools are used for inserting and removing contacts, terminated to wires, from the connector moldings or inserts. Some of the tools are essentially tubes that require only hand pressure to insert or remove a contact, while others have various arrangements of tubes and spring-loaded plungers that exert pressure on the contacts.

Each manufacturer of connectors specifies and supplies tools specifically designed for his particular contacts. In order to properly insert or remove contacts, it is extremely important that only the specified tool be used for a specific contact.

(a)

(b)

(c)

FIG. 100. Typical high-production crimping machines. (*a*) The AMP-O-MATIC (trademark, AMP Incorporated) machine is pneumatically powered, bench-mounted, and operated by a foot pedal. Capable of making up to 3,000 uniform terminations per hour, it can easily be moved from one location to another. (*b*) The AMP-O-LECTRIC (trademark, AMP Incorporated) Model C operates on 110 volts, bench-mounted. Connected to any standard factory outlet, it can be operated by either a foot pedal or an electric switch in the crimping area. It is faster than AMP-O-MATIC, with rates as high as 4,000 per hour, and installs a wider range of terminals in a wider range of wire sizes. (*c*) The automatic stripping terminator (stripper crimper) is an electrically operated bench press that strips a circuit wire, then attaches a terminal to it in one machine cycle. Actuated when the wire is thrust into the die area, it can strip and terminate up to 2,000 circuit wires an hour. Stripped-wire particles are removed by compressed air. (*AMP Incorporated.*)

Insertion and removal tools are normally not interchangeable between one manufacturer's contact and another, and damage to contacts and/or moldings can very easily result if the proper tools are not used. Figure 94 illustrates typical insertion and removal tools.

Other Solderless Termination Methods.

Wire Wrap. Wire-wrap terminal posts are available on some contacts and connectors, especially printed-circuit connectors and plate connectors where the contacts can be located on a predetermined grid pattern and wired by automatic, programmed machinery. Terminal posts are either square or rectangular in shape, with length enough to allow for at least three wire-wrapped connections, depending on wire size.

A wire-wrap connection is a reliable, gas-tight connection that produces large contact areas and a high-pressure, low-contact-resistance joint. The wrapped connection can be achieved by hand-held tools, by hand- or power-operated hand tools, or for high production, by programmed, automatic wire-wrap machinery manufactured by the Gardner-Denver Company. Continued development of this type of termination is constantly reducing both the wire size that can be used and the density and spacing of the wire-wrap terminals (Figs. 101 and 102).

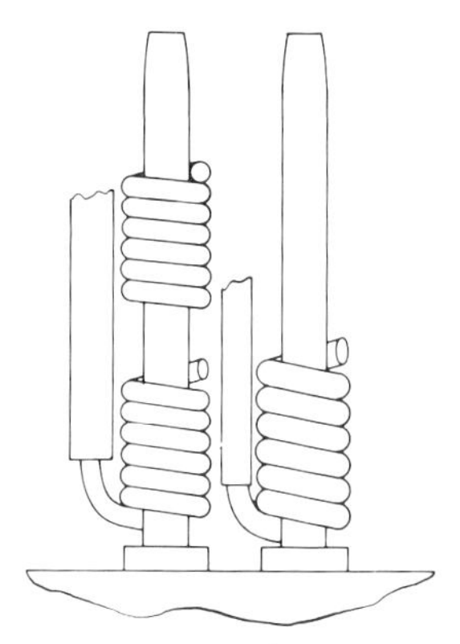

FIG. 101. Wire-wrapped connection sketch.

Taper Pins and Taper Tabs. Taper pins and taper tabs can be installed on wires and cables in the same ways as crimp-type terminals are installed. Impact-type installation tools drive the taper pin into a tapered hole in the contact terminal or onto a tapered-tab terminal of a contact (Fig. 103). The impacted pin or tab may be removed a limited number of times to allow for circuit testing or circuit changes.

Termi-point. A fairly recent development is the Termi-point termination. In this method, by the use of hand-operated, semiautomatic or programmed automatic machinery the connecting wire is stripped and fastened to a rectangular contact terminal by means of a small, formed clamp that exerts pressure between the wire conductor and the terminal (Fig. 104).

Welding. Welded terminations are achieved in two basic ways: by lapping wires or flat conductor cable over a flat terminal and performing the weld, or by butt-welding the conductor to the end of a contact terminal designed for the application (Figs. 65 and 105).

CONNECTOR SELECTION

It cannot be stressed too heavily that connectors are a vital part of an electrical or electronic system, for no printed-circuit board or black box can perform any better than the characteristics of the connector through which it is made into the system.

Even in the simplest connector applications, consideration must be given to all the

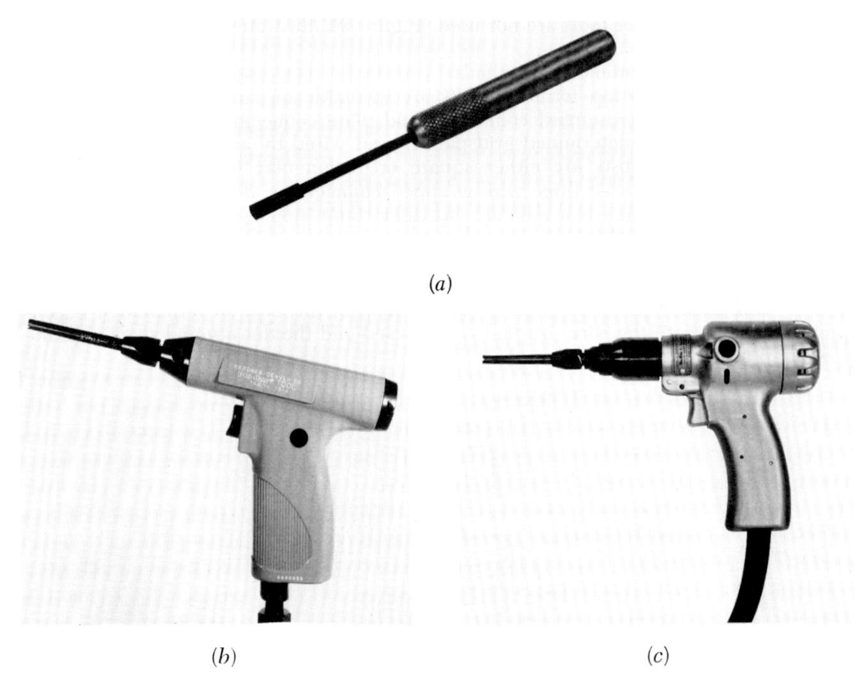

FIG. 102. Wire-wrap hand tools. (*a*) Hand-wrapping tool; (*b*) air wire-wrap tool; (*c*) electric wire-wrap tool. Stripped solid wire is hand-fed into the nose of the wrapping bit. The nose of the tool is then placed over the wire-wrap contact terminal, and the bit is rotated—by hand, using the tool in (*a*), or by depressing the trigger in (*b*) or (*c*)—thus completing the connection.

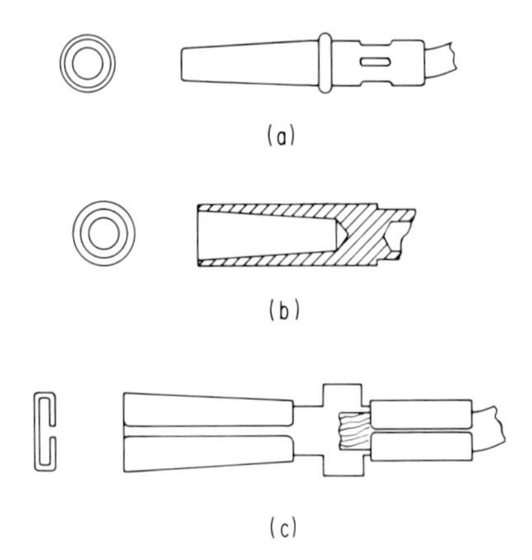

FIG. 103. Taper terminations. (*a*) Taper pin wire termination; (*b*) section through taper socket wire termination; (*c*) taper tab wire termination.

electrical, mechanical, and environmental stresses that it is anticipated that the connector will be subjected to in end use. Also, physical form, size, and compatibility to the intended package concept must be considered.

A common pitfall in connector selection is that the equipment in which it will be used is "only commercial," and therefore any connector will do as long as the price is right. Nothing could be further from the truth!

No matter what the equipment may be—whether it be an electronic garage door opener, a radio, or an electronic computer for use in a small business—it must perform its intended function reliably. Otherwise the user will rapidly become disenchanted

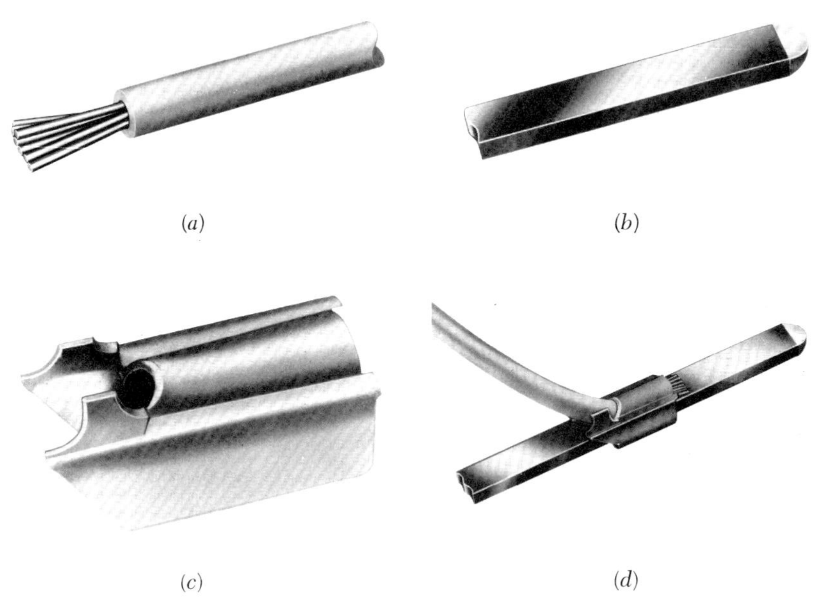

(*a*) (*b*) (*c*) (*d*)

Fig. 104. Termi-point (trademark, AMP Incorporated) termination. Applied with hand-operated or programmed automated tooling that strips the wire and applies the clip in a single operation. (*a*) *The conductor:* Bulk or precut, stranded or solid wire normally employed in point-to-point wiring as well as other wires such as tinsel wire and flexible flat cable. (*b*) *A-MP terminal post:* Rectangular, properly chamfered, and specially plated for a low coefficient of friction and high conductivity. Has been designed specially for Termi-point clips. The post, although specifically designed for Termi-point clip applications, can be wrapped, welded, or soldered. (*c*) *The Termi-point clip:* Utilizes the "memory" of fine-grain spring hard phosphor bronze to provide the retention force necessary to give the termination long-life characteristics. A high-pressure gas-tight contact area between the conductor and post surface is held firm by the integral rolled contact pressure springs. (*d*) *The resulting termination.*

with the product and with the manufacturer who produced it. The lives of people may not be directly affected by the malfunction of a home appliance or a business machine, as opposed to the many lives that may depend on the proper functioning of a computer in a military or space installation. However, the very business life of a manufacturer may depend on the reliable operation of his products.[15]

The selection of connectors and connecting devices for any and all applications should be just as carefully considered as is the selection of all of the other component parts.

Selection Guide. In considering, specifying, and eventually buying connectors, the following general suggestions should prove useful:

1. Provide for a few more contacts than initially anticipated to allow for changes, modifications, or otherwise unforeseen circuit requirements.

2. Choose terminal spacings and density that are industry standards, and choose termination types that will allow for the possibility of an alternate method of wiring that might possibly be dictated by economic or unforeseen physical limitations, especially in the engineering and prototype stages of product development.

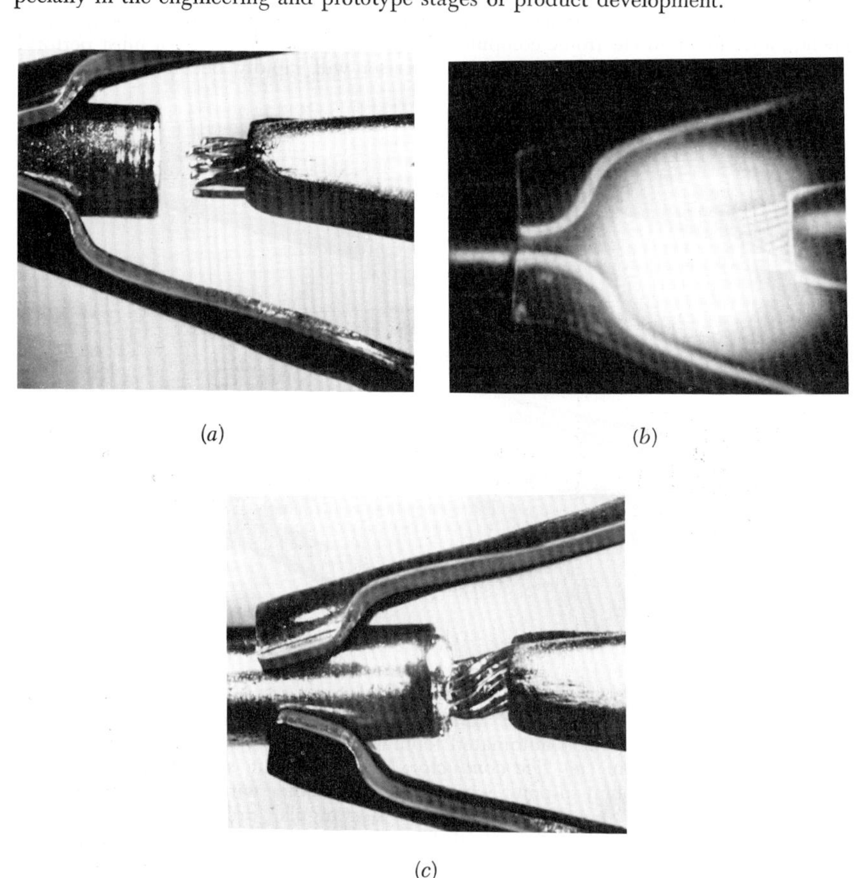

(a) (b)

(c)

FIG. 105. The arc/percussive phenomenon. A burst of rf energy ionizes the air gap (*a*) between two materials to be joined; immediately followed by a discharge (*b*) from a capacitor bank. This heats the surfaces into a molten state as an actuator forges the wire and contact terminal into fusion (*c*), completing the weld. (*The Sippican Corp.*)

3. Consider the broad connection and connector requirements at the initial stages of product development, with an eye toward incorporating available items into the contemplated package design, rather than attempting to adapt conventional connectors into an unconventional package concept and eventually having to have a special connector designed that will be compatible with the resultant packaging concept.

4. Consider the anticipated number of insertion and withdrawal cycles that the connectors might be subjected to in the production package. The basis contact material and the plating required are directly related to this parameter, and a realistic evaluation of projected usage can be a significant economic as well as reliability factor.

For example, full hard brass is entirely satisfactory in many applications, such as a household appliance socket where a very few insertions and withdrawals are anticipated, whereas an application where a great many insertions and withdrawals are to be expected requires a material such as heat-treated beryllium copper that will withstand considerably more mechanical stressing, as well as possible rough handling, than most other nonferrous contact materials.

The conductivity of such material is very nearly as good as that of brass, but contact cost can be as much as 5 times that of hard brass, due to both material cost and tool cost and maintenance. Where a great many connectors and contacts are involved, as in a computer application, connector cost can be as high as 20 percent of the total material and component cost and is therefore a very important economic consideration.

5. Discuss the anticipated connector requirements in as much detail as possible with potential suppliers. The more information the connector manufacturer has to work with regarding the specific application, the better prepared he will be to make a proposal. Lack of complete information regarding the intended end use in a connector application not only puts the potential supplier at a disadvantage, but can actually make it difficult to obtain the most suitable connector for a specific application.

Quite often a user is reluctant to disclose complete information to the potential supplier because of the need for secrecy concerning competitive product development. A good, workable rule of thumb to use when dealing with a manufacturer is whether

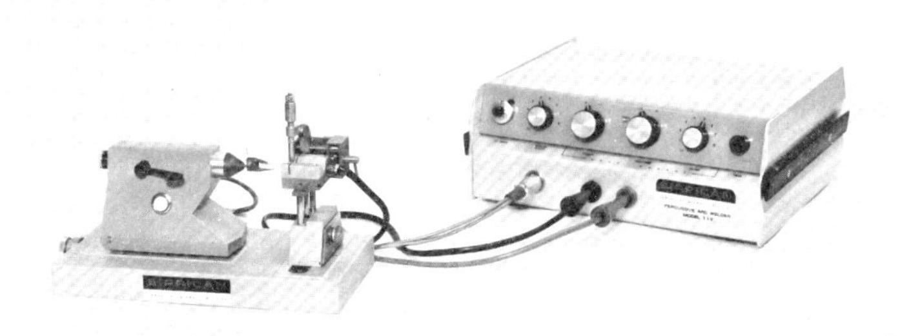

Fig. 106. Sippican's Model 720 bench welding fixture with Model 520 power supply. (*The Sippican Corp.*)

or not he discusses specific projects he is working on with other users. He may show samples of particular designs that he is working on, for instance, and talk in generalities (which is almost necessary in order to demonstrate his capabilities), but if he talks specifically about work he is doing for other customers, you can be very sure that he will discuss your product with others.

Required Specification Data. Certain basic facts and information must be determined by the user and conveyed to a potential supplier of a connector for a given application. In some instances this is done by a questionnaire supplied by the manufacturer which delineates the basic design considerations for the benefit of that manufacturer's engineering group. It is a very helpful tool for both parties because the omission of any pertinent data, no matter how seemingly unimportant, can prove to be quite costly in both time and money.

Connectors and connector design are the full-time job of the connector manufacturer, and his experience and technical know-how should be utilized to the full by the user.

A typical questionnaire would include the following information requirements:[16]

Type of Connector Application. The types available include printed-circuit, rack and panel, cylindrical, tape cable, and cable-to-receptacle, etc.

Number of Contacts Required. Assuming that connector selection is made very early in package and equipment development, a very practical idea is to provide a few more contacts than are anticipated initially. Also, total contact requirements can be of help in determining the optimum number of connectors required for the most reliable operation.

Applicable Specifications. If connector must conform to a particular MIL, Federal, commercial, or other specification, this should be noted.

Contact Considerations.

CURRENT-CARRYING CAPACITY REQUIRED. This is especially important where supply voltages and signals are to be carried by the same connector.

In many types of connectors it is possible to mix contact sizes and types in order to accommodate both high power requirements and low-level signals in a single connector block (i.e., no. 20, no. 16, and coax contacts).

CONTACT RESISTANCE. If significant, the maximum contact resistance that can be tolerated in specific circuits should be noted, both initially and after a specified number of insertions and withdrawals.

CONTACT AND TYPE TERMINATION. The overall application requirements are very important when selecting suitable contacts and terminations. For instance, a heavy package that will be used in a blind insertion application, or one that is expected to receive rough handling, would require sturdy contacts, quite possibly much heavier than those dictated by the actual power and signal requirements.

Molded-in or factory-inserted solder contacts are quite satisfactory for short-run or prototype applications, but for large-quantity production runs with a lot of cabling involved, crimp-type removable contacts would probably prove to be more economical, from both a manufacturing and a maintenance viewpoint.

Contact placement and spacing, as well as termination type, must be carefully considered if some method of automatic wiring is to be used for production.

Dielectric Considerations. Materials that would be suitable in the application must be considered on the basis of physical, chemical, and atmospheric environments to be encountered.

Mechanical Considerations. Physical size limitations, if any, must be specified. General application, mounting considerations, and type of equipment should be specified. Materials and finishes required should be specified, or these might be dictated by environmental requirements. Cable or wiring strain relief, hoods and hood locking devices, keying and polarizing provisions, etc., should be specified, if required. Environmental sealing, such as gaskets or inner-facial seals, should be specified if it is necessary to isolate contacts from hostile atmosphere. Hermetic sealing should be specified if no air leakage into a package can be tolerated. For printed-circuit connector applications, nominal board thickness and tolerance must be specified.

Miscellaneous Information. It is helpful and necessary to a manufacturer to have an idea of total requirements and projected quantity usage. Also, any specific test requirements to be performed by the manufacturer or by the user should be spelled out in detail.

Vendor Selection. There are well over 300 connector manufacturers in the United States today. Some of them offer a wide variety of connector families and types, while others specialize in a single family, such as printed-circuit connectors, coax, or AN types.

Some manufacturers are strictly "copiers," or hardware builders, with little or no engineering and design capability.

Fortunately, many manufacturers have excellent engineering and design staffs, adequate research and testing facilities, and competent field engineering personnel. A very important but often overlooked consideration in connector selection is that the selected supplier actually becomes a member of the engineering and production teams of the user. In a word, vendor selection is important.

Common Mistakes in Connector and Vendor Selection. The following list of common mistakes engineers make when selecting connectors and vendors was compiled from a recent survey of connector manufacturers. Some of the points have been made previously but are worth repeating. All of them should be helpful when selecting connectors and/or connecting systems.[17]

1. Not properly evaluating the integrated capability of the vendor, i.e., (*a*) staff—technical, production, and quality; (*b*) in-house manufacturing and laboratory facilities; (*c*) outside vendors employed by the connector manufacturer.

2. Not evaluating vendor's past experience in a particular type of connector design (i.e., circular crimp/removable and printed-circuit-card edge connectors require different approaches and experience).

3. Not evaluating present volume capability and backlog for new requirements.

4. Not remembering that the connector is only a part of any interconnection. It may be a perfectly good connector and pass certain tests, yet not be satisfactory when used in a particular application. This could be the result of any number of problems—assembly, crimping, soldering, plating, etc.

5. Making an inadequate facilities survey.

6. Placing prime consideration on cost instead of quality and reliability.

7. Waiting until a project is near completion before considering interconnection devices.

8. Being unwilling to consider new ideas and to look to new manufacturers in the field.

9. Not fully assessing the producibility of proposed designs.

10. Not considering connector design from a systems standpoint.

11. Overlooking services, such as field support.

12. Specifying requirements in some areas not actually required for system performance, thus forcing the connector manufacturer to make design compromises in other areas to meet those specific requirements.

13. Ignoring basic interface requirements—such as selecting the wrong connector for the cable to be used.

CONNECTOR RELIABILITY

General Discussion and Definition. The whole subject of reliability as applied to electrical connectors, as well as to all electrical and electronic components and extended on into complete systems, is one that is discussed freely and authoritatively by many learned people—but all too often on a strictly theoretical level rather than as a practical, applied science.

It is not within the scope of this chapter to present a lengthy treatise on the subject of reliability as such. Therefore, the subject of connector reliability will be presented on a practical level, with the basic premise that *a connector that is properly defined, well-engineered, and correctly used will be basically a reliable connector.*

In the electronics industry, reliability has been defined in many ways by many people. The definition given here is a simple and easily understood statement of what reliability really is, as applied to components in a system:[18]

> *Reliability is the probability that a given device will perform without failure to a given set of requirements for a specified length of time.*

A common misconception, particularly in the case of connectors, is that because a particular connector appears on a military Qualified Products List (QPL), or because a manufacturer claims that it is manufactured to a particular MIL specification and conforms to that specification, it is therefore a reliable connector for all applications. Impressive figures may be quoted, based on unrealistic but nevertheless honest testing on the part of the manufacturer, that bear little, if any, similarity to the actual application conditions in which the user may use the connector.

Too often, laboratory evaluation of connectors normally does not duplicate field usage, and individual interpretation of basic tests as applied to end-use similarity can sometimes be catastrophic.

Another generally overlooked contribution to connector reliability—or unreliability—is the handling of the connectors during installation and maintenance. Mishandling, physical damage particularly to contact pins, dirt, improper or careless attachment of wire or cable, improper mounting, etc., can result in reliability problem with an otherwise normally good, reliable connector.[15]

Reliability Rules. A few simple rules and suggestions are listed, following, that should be of aid to the design engineer concerned with reliability in the selection and application of connectors:[19]

1. Unlike other components, a connector does not contribute to the function or characteristics of an electrical or electronic circuit. Its function is a mechanical one, to

provide a means of engaging and disengaging electrical circuitry that provides energy or routes intelligence from one point to another. Therefore, of prime consideration is that it be of sound mechanical design and construction, and that all materials used—basis contact material, platings, insulating, and structural materials—be adequate and compatible with the physical and environmental stresses to which it will be subjected.

2. Be sure that connectors are properly mounted. If mounted in a panel or frame there must be sufficient rigidity so that any mechanical stresses on the mounting will not be transmitted to the connector. Proper float relationship between connector halves must be maintained. Make sure that wiring or cabling is properly dressed and supported so that it does not put undue strain on contacts (especially crimp/removable types), and also so that it does not inhibit connector float. Consider the mounting of connectors in relationship to their availability for maintenance and servicing.

3. Connectors must be installed and serviced by trained, capable personnel. Mishandling, from installation and wiring through maintenance, is probably the greatest single cause of connector failure. Be sure that all terminations are properly made using proper tools and processes. Be sure that miscellaneous hardware items are properly applied.

4. Make sure that the supplier or manufacturer has complete specifications for the application. The competitive situation is such that a manufacturer must naturally try to keep costs down and still supply what he feels will do a satisfactory job. It is possible that one manufacturer's price for a particular connector might be significantly lower than all others; if this be so, insist that he prove comparable quality and performance. Remember: you can always buy an item cheaper—and get less!

5. Know the qualifications of the connector manufacturer, and buy from a *connector manufacturer*, not a hardware manufacturer or assembler. Proper connector engineering is a most important initial step, and adequate quality control during manufacturing is of prime importance. Quality and reliability are synonymous.

6. Make sure that reliability claims made by a manufacturer are valid and applicable to the actual use requirements. Do not *assume* that a connector that has been qualified to a specification in the laboratory will necessarily perform as well under normal service conditions. A qualification test only demonstrates the connector's capability of meeting specified requirements under closely controlled laboratory conditions. Rough handling, for instance, is normally not a part of testing. Be sure that any anticipated physical stresses and the actual environmental conditions are incorporated into test and evaluation procedures. For example, a large-scale computer is normally installed in an air-conditioned room on a solid foundation, which would be an ideal environmental atmosphere. However, keep in mind that it must be transported from the manufacturer to the customer, during which time it can be subjected to appreciable shock and vibration conditions that are not applicable to the end-use environment. It must be capable of withstanding these conditions if it is to arrive at its permanent location in good working condition.

7. When measurements are made during testing, especially in low-level work, they must be extremely accurate if they are to be meaningful, either pro or con. Also, it is important that personnel involved in test work be trained in the handling and terminating of connectors.

8. Keep in mind that indicated current ratings do not necessarily mean that *all* contacts in a connector can be operated simultaneously at full rated current. Check the connector specifications carefully, with the manufacturer if necessary.

9. Determine whether the stated connector thermal limit is under operating or non-operating conditions. Some connectors, rated as high-temperature units, would burn up if operated at the stated ambient temperature.

10. Avoid a radical and unproved design unless the manufacturer can prove that this design will *best* meet your particular requirements fully.

11. Another significant cause of connector failure is misapplication. A broad rule of thumb is that a connector should be physically compatible with package size, structure, and handling requirements. For example, a delicate microminiature connector used in an otherwise rugged external cable application would most certainly be the weakest link in the system and, as such, be subject to physical damage and

eventual failure. A rugged application *dictates* a rugged connector, whereas in a light, protected application, a microminiature connector can very well be utilized.

Reliability and Cost. Reliability is a relative matter and is closely tied in with economics. Our economy is geared to cost. Equipments are designed and sales are made with a particular price in mind, and in many cases, the cost determines the degree of reliability required.[20] The higher the degree of reliability required or specified, the higher the component cost in general and thus the higher relative cost of equipment.

For high-reliability applications, a considerable part of the cost of a connector is in the testing required of the manufacturer. Compared to most other components, connector testing and evaluation are very costly because, if test information is to be of value on a statistical basis, large numbers of connectors and contacts must be tested under a great number of conditions, with many physical and electrical parameters considered. For example, a test lot of only 50 connectors, each containing 50 contacts, would be a total of 2,500 contacts, but would still be only 50 items (connectors) as compared to hundreds of thousands of resistors or capacitors normally considered an adequate test size. The cost of these 50 connectors could easily run from $500 to as much as $2,500 for some of the more exotic military types, whereas the cost of 500 or 1,000 resistors or capacitors would only be perhaps $50 to $250. Required measurements on the 2,500 contacts in the 50 connectors could be 25,000 or more, which would of course be expensive initially; it would also be expensive to collect and reduce the data obtained. The user or his customer must pay for this testing and data reduction, either directly, or in the cost of the connectors involved. Unfortunately, testing does not *make* a reliable connector or improve it. It is merely an indication of the probability of the connector's performing its required function.

A connector is a very vital component part of a system. It is a mechanical component, and if proper consideration is given to all of the many mechanical facets involved—design; construction; quality control during manufacturing, mounting, and installation termination; and compatibility with the mechanical and environmental application—a high level of confidence can be expected initially which, when backed up by a minimum of reliability testing, should result in solidly reliable connections.

REFERENCES

1. Ruth, S. B.: Connectors and Terminations, *Electron. Ind.*, p. 56, April, 1963.
2. Economon, T.: Rack and Panel Connectors—Part 1, *Electromech. Design*, p. 28, October, 1965.
3. Economon, T.: Connectors—Part 1—Cylindrical, *Systems Designer's Handbook*, p. 19, January, 1966.
4. Blair, N. M.: Time to Take Stock, *Electron. Forum*, Amphenol-Borg Electronics Corp., vol. 3, no. 2, p. 15, 1967.
5. Bowman, D. S.: MIL-C-39012—What It Means to RF Connector Users, *Electron. Forum*, Amphenol-Borg Electronics Corp., vol. 2, no. 4, p. 12, 1966.
6. Anderson, T.: How and Why of Coaxial Connectors at Microwaves, *Electron. Forum*, Amphenol-Borg Electronics Corp., vol. 2, no. 4, 1966.
7. Lippke, J. A.: Subminiature RF Connector Designs Proliferate, *EEE*, p. 42, September, 1963.
8. Amphenol-Borg Electronics Corp.: "Flex-1 Connectors Manual."
9. Cromer, E. G., Jr.: The Plate Connector, *Electron. Packaging Production*, p. 146, March, 1966.
10. Elco Corp.: "Catalog MP966—VARIPLATE ™ Connectors," 1966.
11. Stasch, A.: The Hermetically Sealed Connector and Its Capabilities, *Electron. Ind.*, p. 72, May, 1964.
12. Burndy Corp.: "Catalog RP—Rack and Panel Connectors," p. 5, 1963.
13. Amphenol-Borg Electronics Corp.: Product Development Release, No. 45, March, 1958.
14. Anderson, J. D.: "Crimping Manual for the Electronics Industry," Burndy Corp., 1965.
15. Ganzert, A. E.: "Preventing Multiple-contact Connector Problems," *Assembly Eng.*, p. 22, January, 1967.

16. Ruehlemann, H. E.: How to Specify Connectors," *Electron. Products,* March, 1963.
17. Connectors—1967, *Evaluation Eng.,* p. 22, January/February, 1967.
18. Amphenol-Borg Electronics Corp.: "Reliability Statistics for Electro Mechanical Devices."
19. Connectors and Reliability—1964, *Evaluation Eng.,* p. 18, September/October, 1964.
20. Nitschke, N. E.: How IBM Selects Components, *Evaluation Eng.,* p. 32, May/June, 1967.

Chapter 7

MATERIALS FOR ELECTRONIC PACKAGING

By

CHARLES A. HARPER

Westinghouse Electric Corporation
Aerospace Division
Baltimore, Maryland

INTRODUCTION

Knowledge and proper use of engineering materials are two of the most important factors in optimum design of an electronic package for a given function. This section will cover the important data and design information for the various classes of materials most commonly used in electronic packages.

Material Classifications. Materials may be classified in many ways, and the choice of classification depends primarily on the purpose for which they are being classified. Basic classifications and practical classifications may be used. All materials have a basic chemical structure, and hence the basic classification is the classification according to chemical structure. A simplified, basic chemical classification of materials is shown in Fig 1. Organic materials are materials based on a carbon atomic structure, and inor-

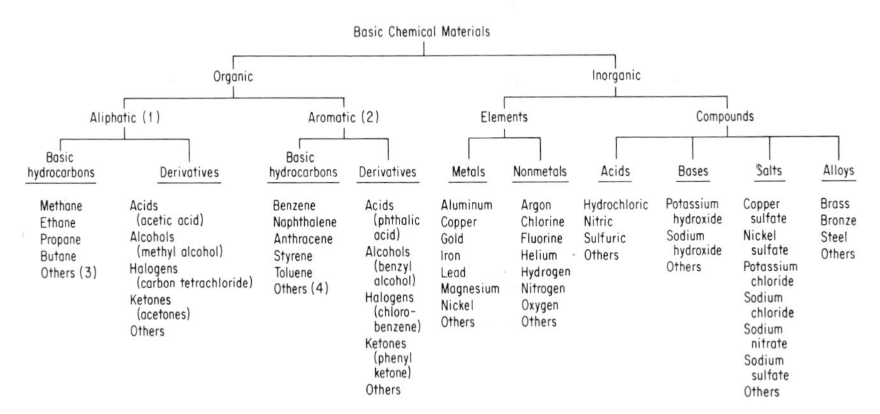

FIG. 1. Simplified classification of materials according to basic chemical structure. (1) Aliphatic organic chemicals are denoted by their basic straight-chain carbon structure, –C–C–C–C. (2) Aromatic organic chemicals are denoted by their basic ring (or benzene) carbon structure, which is usually simply indicated as —⟨⟩— or a hexagon symbol with carbon atom –C– at some of the points. (3) The list continues to grow with each addition of one carbon atom and two hydrogen atoms to the molecule. (4) Addition and rearrangement of basic benzene ring structure.

ganic materials are those with a chemical structure based on any atom other than carbon.

A widely used practical classification is simply metallic and nonmetallic materials. A simplified classification using these two categories and incorporating the materials discussed in this section is shown in Fig. 2. Using this type of classification, everything that is not metal, whether organic or inorganic, would be included in the nonmetal category. Actually, most of the materials listed in Fig. 1 are not used directly in electronic packaging. Some exceptions are metals, some organic chemicals which are used as solvents and cleaners, some inorganic acids and bases which are used as metal cleaners and etchants, and some inorganic salts which are used in water solution for electroplating and electrochemical operations. Most plastics are made from organic chemicals, and thus most plastics and plastic products are organic materials. The same is true for rubbers and elastomers, paints, insulating coatings, adhesives, oils, greases, and fluids. Exceptions to all of these are silicone rubbers, fluids, greases, etc., which have a chemical structure based on silicon rather than carbon. Most silicones, however, also have some carbon atoms in their polymer structure —but the basic chemical structure is silicon and oxygen. Thus, silicone materials are not considered to be organic materials.

Many of the various materials can also be classified according to form and shape, whenever this method of classification is propitious, which is mostly for application purposes, of course. Further, many of the materials are available as composites, blends, mixes, compounds, formulations, etc., which, again, are usually for application purposes. Such classifications are, of course, very helpful to the end user. It is the other material categories and properties which are, however, most helpful to the design engineer who is designing around properties rather than shapes. Needless to say, the design of an electronic package for optimum manufacturability and economy also requires that the design engineer know available shapes and forms, as well as processing techniques.

Terms and Definitions. Important terms and definitions, useful for understanding materials for electronic packaging, are given in Table 1.

TABLE 1. Terms and Definitions

Accelerator. A chemical used to speed up a reaction or cure. For example, cobalt naphthanate is used to accelerate the reaction of certain polyester resins. The term *accelerator* is often used interchangeably with the term *promoter.* An accelerator is often used along with a catalyst, hardener, or curing agent.

Acid. A substance that gives hydrogen ion in solution or that neutralizes bases, yielding water, e.g., hydrochloric acid.

Adhesive. Broadly, any substance used in promoting and maintaining a bond between two materials.

Aging. The change in properties of a material with time under specific conditions.

Aliphatic Hydrocarbon. See Hydrocarbon.

Alkali. A chemical that gives a base reaction. (See Base.)

Alloy. A mixture of elements, not necessarily chemically bound together, that constitutes some characteristic physical structure.

Ambient Temperature. The temperature of the surrounding cooling medium, such as gas or liquid, which comes into contact with the heated parts of the apparatus.

Anion. An ion which is negatively charged.

Anode. The electrode at which current enters, or electrons leave, the solution; the positive electrode in electrolysis, the electrode at which negative ions are discharged, positive ions are formed, or other oxidizing reactions occur.

Anodizing. Anodic treatment of metals, particularly aluminum, to form an oxide film of controlled properties.

Arc Resistance. The time required for an arc to establish a conductive path in a material.

Aromatic Hydrocarbon. See Hydrocarbon.

B Stage. An intermediate stage in the curing of a thermosetting resin. In this stage, resins can be heated and caused to flow, thereby allowing final curing in the desired shape. The term *A stage* is used to describe an earlier stage in the curing reaction, and the term *C stage* is sometimes used to describe the cured resin. Most molding materials are in the B stage when supplied for compression or transfer molding.

Barrel Plating (or cleaning). Mechanical plating (or cleaning) in which the work is processed in bulk in a rotating container.

Base. A substance that gives hydroxide ion in solution or that neutralizes acids, yielding water, e.g., sodium hydroxide.

Blowing Agent. Chemicals that can be added to plastics and that generate inert gases upon heating. This blowing or expansion causes the plastic to expand, thus forming a foam. Also known as *foaming agent.*

Bond Strength. The amount of adhesion between bonded surfaces.

Bright Dip (nonelectrolytic). A solution used to produce a bright surface on a metal.

Bright Plating. A process which produces an electrodeposit having a high degree of specular reflectance in the as-plated condition.

Brush Plating. A method of plating in which the plating solution is applied with a pad, or brush, within which is an anode and which is moved over the cathode to be plated.

Capacitance (capacity). That property of a system of conductors and dielectrics which permits the storage of electricity when potential difference exists between the conductors. Its value is expressed as the ratio of quantity of electricity to a potential difference. A capacitance value is always positive.

Cast. To embed a component or assembly in a liquid resin, using molds that separate from the part for reuse after the resin is cured. (See Embed and Pot.)

Catalyst. A chemical that causes or speeds up the cure of a resin, but that does not become a chemical part of the final product. Catalysts are normally added in small quantities. The peroxides used with polyester resins are typical catalysts.

Cathode. The electrode through which current leaves, or electrons enter, the solution; the negative electrode in electrolysis. The electrode at which positive ions are discharged, negative ions are formed, or other reducing reactions occur. In electroplating, the electrode which receives the deposit.

Cation. An ion which is positively charged.

Chlorinated Hydrocarbon. An organic compound having hydrogen atoms and, more importantly, chlorine atoms in its chemical structure. Trichloroethylene, methyl chloroform, and methylene chloride are chlorinated hydrocarbons.

Coat. To cover with a finishing, protecting, or enclosing layer of any compound (such as varnish).

Coefficient of Expansion. The fractional change in dimension of a material for a unit change in temperature.

Cold Flow (creep). The continuing dimensional change that follows initial instantaneous deformation in a nonrigid material under static load.

Compound. Some combination of elements in a stable molecular arrangement.

Contact Bonding. A type of adhesive (particularly nonvulcanizing natural rubber adhesives) that bonds to itself on contact although solvent evaporation has left it dry to the touch.

Conversion Coating. A coating produced by chemical or electrochemical treatment of a metallic surface which gives a superficial layer of a compound of the metal (e.g., chromate coatings on zinc and cadmium, oxide coatings on steel).

Copolymer. See Polymer.

Crosslinking. The forming of chemical links between reactive atoms in the molecular chain of a plastic. It is this crosslinking in thermosetting resins that makes them infusible.

Crystalline Melting Point. The temperature at which crystalline structure in a material is broken down.

Cure. To change the physical properties of a material (usually from a liquid to a solid) by chemical reaction, by the action of heat and catalysts, alone or in combination, with or without pressure.

Curing Agent. See Hardener.

Curing Temperature. The temperature at which a material is subjected to curing.

Curing Time. In the molding of thermosetting plastics, the time it takes for the material to be properly cured.

Dielectric Constant (permittivity or specific inductive capacity). That property of a dielectric which determines the electrostatic energy stored per unit volume for unit potential gradient.

Dielectric Loss. The time rate at which electric energy is transformed into heat in a dielectric when it is subjected to a changing electric field.

Dielectric Loss Angle (dielectric phase difference). The difference between ninety degrees (90°) and the dielectric phase angle.

Dielectric Loss Factor (dielectric loss index). The product of dielectric constant and the tangent of dielectric loss angle for a material.

Dielectric Phase Angle. The angular difference in phase between the sinusoidal alternating potential difference applied to a dielectric and the component of the resulting alternating current having the same period as the potential difference.

Dielectric Power Factor. The cosine of the dielectric phase angle (or sine of the dielectric loss angle).

Dielectric Strength. The voltage which an insulating material can withstand before breakdown occurs, usually expressed as a voltage gradient (such as volts/mil).

Dissipation Factor (loss tangent, tan δ, approximate power factor). The tangent of the loss angle of the insulating material.

Elastomer. A material which at room temperature stretches under low stress to at least twice its length and snaps back to original length upon release of stress. (See Rubber.)

Electric Strength (dielectric strength or disruptive gradient). The maximum potential gradient that a material can withstand without rupture. The value obtained for the electric strengh will depend on the thickness of the material and on the method and conditions of test.

Electrode. A conductor of the metallic class through which a current enters or leaves an electrolytic cell, at which there is a change from conduction by electrons to conduction by charged particles of matter, or vice versa.

Electrodeposition. The process of depositing a substance upon an electrode by electrolysis. Includes electroplating, electroforming, and electrorefining.

Electroforming. The production or reproduction of articles by electrodeposition upon a mandrel or mold that is subsequently separated wholly or partly from the deposit.

Electroless Plating. Deposition of a metallic coating by a controlled chemical reduction which is catalyzed by the metal or alloy being deposited.

Element. A substance composed entirely of atoms of the same atomic number, e.g., aluminum or copper.

Embed. To encase completely a component or assembly in some material—a plastic, for current purposes. (See Cast and Pot.)

Encapsulate. To coat a component or assembly in a conformal or thixotropic coating by dipping, brushing, or spraying.

Exotherm. The characteristic curve of a resin during its cure, which shows heat of reaction (temperature) versus time. Peak exotherm is the maximum temperature on this curve.

Exothermic. Chemical reaction in which heat is given off.

Faying Surface. The surface of an object that comes in contact with another object to which it is fastened.

Filler. A material, usually inert, that is added to plastics to reduce cost or modify physical properties.

Film Adhesive. Thin layer of dried adhesive. Also describes a class of adhesives provided in dry-film form with or without reinforcing fabric, which are cured by heat and pressure.

Flexibilizer. A material that is added to rigid plastics to make them resilient or flexible. Flexibilizers can be either inert or a reactive part of the chemical reaction. Also called a *plasticizer* in some cases.

Flexural Modulus. The ratio, within the elastic limit, of stress to corresponding strain. It is calculated by drawing a tangent to the steepest initial straight-line portion of the load-deformation curve and calculating by the following equation:

$$E_B = \frac{L^3 m}{4bd^3}$$

where E_B = modulus — L = span in inches
b = width of beam tested
d = depth of beam
m = slope of the tangent

Flexural Strength. The strength of a material in bending, expressed as the tensile stress of the outermost fibers of a bent test sample at the instant of failure.

Fluorocarbon. An organic compound having fluorine atoms in its chemical structure. This property usually lends stability to plastics. Teflon is a fluorocarbon.

Gel. The soft, rubbery mass that is formed as a thermosetting resin goes from a fluid to an infusible solid. This is an intermediate state in a curing reaction, and a stage in which the resin is mechancally very weak. *Gel point* is defined as the point at which gelation begins.

Glass Transition Point. Temperature at which a material loses its glasslike properties and becomes a semiliquid.

Glue Line Thickness. Thickness of the fully dried adhesive layer.

Hardener. A chemical added to a thermosetting resin for the purpose of causing curing or hardening. Amines and acid anhydrides are hardeners for epoxy resins. Such hardeners are a part of the chemical reaction and a part of the chemical composition of the cured resin. The terms *hardener* and *curing agent* are used interchangeably. Note that these can differ from catalysts, promoters, and accelerators.

Heat-distortion Point. The temperature at which a standard test bar (ASTM D-648) deflects 0.010 in. under a stated load of either 66 or 264 psi.

Heat Sealing. A method of joining plastic films by simultaneous application of heat and pressure to areas in contact. Heat may be supplied conductively or dielectrically.

Hot-melt Adhesive. A thermoplastic adhesive compound, usually solid at room temperature, which is heated to a fluid state for application.

Hydrocarbon. An organic compound having hydrogen atoms in its chemical structure. Most organic compounds are hydrocarbons. Aliphatic hydrocarbons are straight-chained hydrocarbons, and aromatic hydrocarbons are ringed structures based on the benzene ring. Methyl alcohol, trichloroethylene, etc., are aliphatic; benzene, xylene, toluene, etc., are aromatic.

Hydrolysis. Chemical decomposition of a substance involving the addition of water.

Hygroscopic. Tending to absorb moisture.

Impregnate. To force resin into every interstice of a part. Cloths are impregnated for laminating, and tightly wound coils are impregnated in liquid resin using air pressure or vacuum as the impregnating force.

Inhibitor. A chemical added to resins to slow down the curing reaction. Inhibitors are normally added to prolong the storage life of thermosetting resins.

Inorganic Chemicals. Chemicals whose chemical structure is based on atoms other than the carbon atom.

Insulation Resistance. The ratio of the applied voltage to the total current between two electrodes in contact with a specific insulator.

Micron. A unit of length equal to 10,000 Å, 0.0001 cm, or approximately 0.000039 in.

Modulus of Elasticity. The ratio of stress to strain in a material that is elastically deformed.

Moisture Resistance. The ability of a material to resist absorbing moisture, either from the air or when immersed in water.

Mold. To form a plastic part by compression, transfer, injection molding, or some other pressure process.

Monomer. See Polymer.

NEMA Standards. Property values adopted as standard by the National Electrical Manufacturers Association.

Noble Elements. Those elements that either do not oxidize or oxidize with difficulty, e.g., gold and platinum.

Organic. Composed of matter originating in plant or animal life, or composed of chemicals of hydrocarbon origin, either natural or synthetic. Used in referring to chemical structures based on the carbon atom.

Permittivity. Preferred term for dielectric constant.

pH. A measure of the acid or alkaline condition of a solution. A pH of 7 is neutral (distilled water), pH values below 7 are increasingly acid as pH values go toward 0, and pH values above 7 are increasingly alkaline as pH values go toward the maximum value of 14.

Pickle. An acid solution used to remove oxides or other compounds from the surface of a metal by chemical or electrochemical action.

Plastic. An organic resin or polymer. (See Resin and Polymer.)

Plasticizer. Material added to resins to make them softer and more flexible when cured.

Polymer. A high-molecular-weight compound (usually organic) made up of repeated small chemical units. For practical purposes, a polymer is a plastic. The small chemical unit is called a mer, and when the polymer or mer is a crosslink between different chemical units (e.g., styrene-polyester), the polymer is called a copolymer. A monomer is any single chemical from which the mer or polymer or copolymer is formed. Styrene is the monomer in a styrene-polyester copolymer resin. Polymers can be thermosetting or thermoplastic.

Polymerize. To unite chemically two or more monomers or polymers of the same kind to form a molecule with higher molecular weight.

Pot. To embed a component or assembly in a liquid resin, using a shell, can, or case which remains as an integral part of the product after the resin is cured. (See Embed and Cast.)

Pot Life. The time during which a liquid resin remains workable as a liquid after catalysts, curing agents, promoters, etc., are added; roughly equivalent to gel time. Sometimes also called *working life.*

Power Factor. The cosine of the angle between the voltage applied and the current resulting.

Promoter. A chemical, itself a feeble catalyst, that greatly increases the activity of a given catalyst.

Refractive Index. The ratio of the velocity of light in a vacuum to its velocity in a substance. Also the ratio of the sine of the angle of incidence to the sine of the angle of refraction.

Relative Humidity. The ratio of the quantity of water vapor present in the air to the quantity which would saturate it at any given temperature.

Resin. High-molecular-weight organic material with no sharp melting point. For current purposes, the terms *resin, polymer,* and *plastic* can be used interchangeably.

Resist (noun). A material applied to a part of a cathode or plating rack to render the surface nonconducting; a material which will resist specific etchants.

Resistivity. The ability of a material to resist passage of electrical current either through its bulk or on a surface. The unit of volume resistivity is the ohm-centimeter, and the unit of surface resistivity is the ohm.

Rockwell Hardness Number. A number derived from the net increase in depth of impression as the load on a penetrator is increased from a fixed minimum load to a higher load and then returned to minimum load. Penetrators include steel balls of several specified diameters and a diamond cone penetrator.

Rubber. An elastomer capable of rapid elastic recovery.

Shore Hardness. A procedure for determining the indentation hardness of a material by means of a durometer. Shore designation is given to tests made with a specified durometer.

Solvent. A liquid substance which dissolves other substances.

Specific Heat. The ratio of a material's thermal capacity to that of water at 15°C.

Storage Life. The period of time during which a liquid resin or adhesive can be stored and remain suitable for use. Also called *shelf life.*

Strain. The deformation resulting from a stress, measured by the ratio of the change to the total value of the dimension in which the change occurred.

Stress. The force producing or tending to produce deformation in a body, measured by the force applied per unit area.

Surface Resistivity. The resistance of a material between two opposite sides of a unit square of its surface. Surface resistivity may vary widely with the conditions of measurement.

Thermal Conductivity. The ability of a material to conduct heat; the physical constant for the quantity of heat that passes through a unit cube of a material in a unit of time when the difference in temperatures of two faces is 1°C.

Thermoplastic. A classification of resin that can be readily softened and resoftened by repeated heating.

Thermosetting. A classification of resin which cures by chemical reaction when heated and, when cured, cannot be resoftened by heating.

Thief. An auxiliary cathode so placed as to divert to itself some current from portions of the work which would otherwise receive too high a current density.

Thixotropic. Describing materials that are gel-like at rest but fluid when agitated.

Throwing Power. The improvement of the coating (usually metal) distribution ratio over the primary current distribution ratio on an electrode (usually a cathode). Of a solution, a measure of the degree of uniformity with which metal is deposited on an irregularly shaped cathode. The term may also be used for anodic processes for which the definition is analogous.

Vicat Softening Temperature. A temperature at which a specified needle point will penetrate a material under specified test conditions.

Viscosity. A measure of the resistance of a fluid to flow (usually through a specific orifice).

Volume Resistivity (specific insulation resistance). The electrical resistance between opposite faces of a 1-cm cube of insulating material, commonly expressed in ohm-centimeters. The recommended test is ASTM D257-54T.

Vulcanization. A chemical reaction in which the physical properties of an elastomer are changed by causing it to react with sulfur or other crosslinking agents.

Water Absorption. The ratio of the weight of water absorbed by a material to the weight of the dry material.

Wetting. Ability to adhere to a surface immediately upon contact.

Working Life. The period of time during which a liquid resin or adhesive, after mixing with catalyst, solvent, or other compounding ingredients, remains usable. (See Pot Life.)

PLASTICS AND POLYMERS

Plastics and polymers represent one of the most important classes of materials used in electronic packaging. This section will present a detailed discussion of these materials. It will cover types of plastics, fabrication processes, forms and shapes, and application products such as molded parts, embedding resins, organic protective and insulating coatings, and adhesives.

Introduction to Plastics and Polymers. The definitions of plastics, polymers, and plastic resins are given in Table 1. For engineering property design purposes, the terms can be considered interchangeable. Practically stated, and omitting the fine points of definition, a plastic is an organic polymer, available in some resin form, or some form derived from the basic polymerized resin. These forms can be liquid or pastelike resins for embedding, coating, and adhesive bonding; or they can be molded, laminated, or formed shapes.

The number of basic plastic materials is large and the list is growing. In addition, the number of variations and modifications to these basic plastic materials is also quite large. Taken together, the resultant quantity of materials available is just too large to be completely understood and correctly applied by anyone other than those whose day-to-day work puts them in direct contact with a diverse selection of materials. The practice of mixing brand names, trade names, and chemical names of various plastics only makes the problem of understanding these materials all the more troublesome.

Another variable that makes it difficult for the packaging engineer to understand and properly design with plastics is the large number of processes by which plastics can be produced. Fortunately, there is an organized pattern on which an orderly presentation of these variables can be based. Such an explanation is the subject of this section.

While there are numerous minor classifications for plastics, depending upon how one wishes to categorize them, nearly all plastics can be placed in one of two major classifications. These two major plastic material classes are thermosetting materials (or thermosets) and thermoplastic materials, as shown in Fig. 2. Although Fig. 2 shows

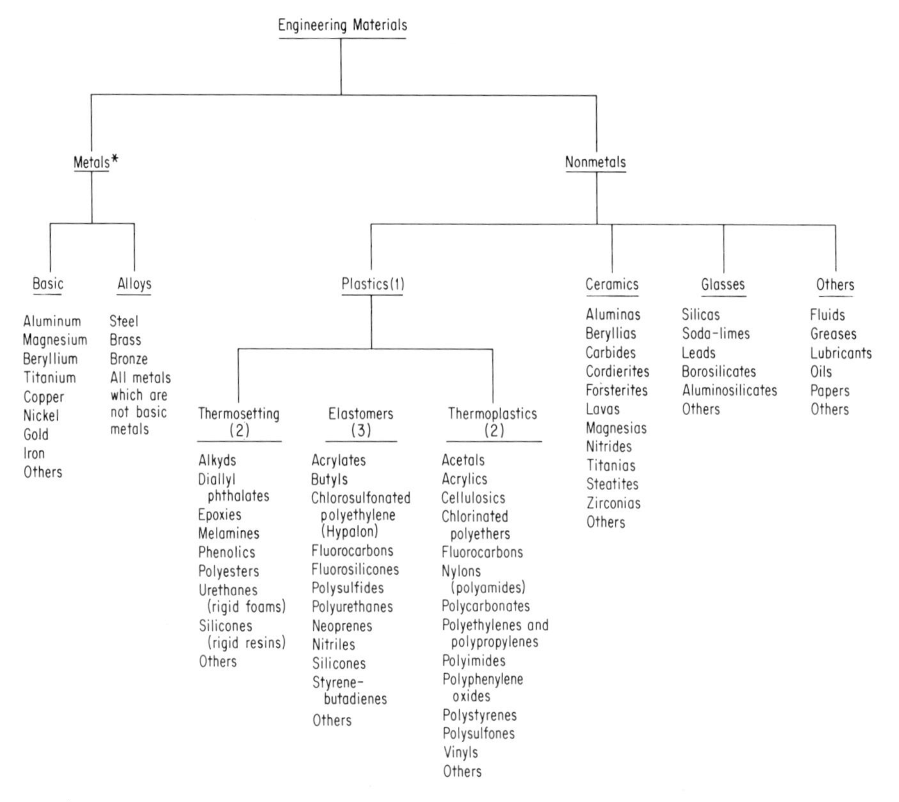

FIG. 2. Simplified classification of materials into metal and nonmetal categories. (1) Most organic paints, insulating coatings, embedding resins, and adhesives are also basically plastic or polymer formations. (2) Detailed discussion of these plastics is given in subsequent parts of this chapter. (3) Natural rubber is an organic compound but not a plastic.

elastomers separately—and they are a separate application group—elastomers, too, are either thermoplastic or thermosetting, depending on their chemical nature. Likewise, foams, adhesives, embedding resins, and other application groups can be subdivided into thermoplastic and thermosetting plastics.

THERMOSETTING MATERIALS.[1] As the name implies, thermosetting materials or thermosets are cured, set, or hardened into a permanent shape. This curing is an irreversible chemical reaction known as crosslinking, which usually occurs under heat. For some thermosetting materials, however, curing is initiated or completed at room temperature. Even here, however, it is often the heat of the reaction, or the exotherm, shown in Fig. 3, which actually cures the plastic material. Such is the case, for instance, with a room temperature curing epoxy, polyester, or urethane compound.

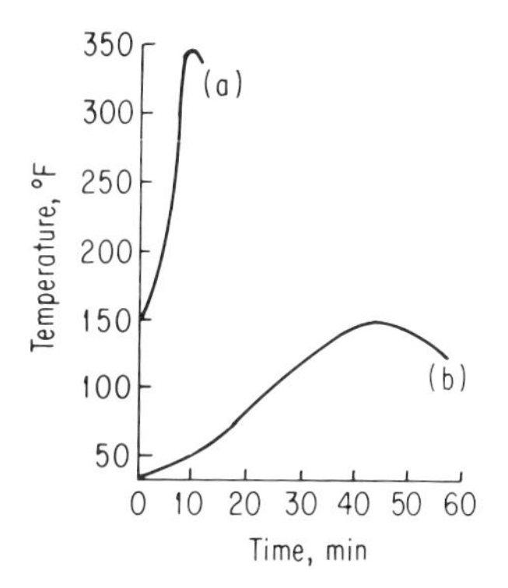

FIG. 3. Comparison of the exothermic curves for two polyester resins.[1] (*a*) 50-gram sample catalyzed with 0.5% benzoyl peroxide and cured at 180°F. (*b*) 50-gram sample catalyzed with 0.5% MEK peroxide, promoted with 0.25% cobalt naphthenate, and cured at room temperature.

The crosslinking that occurs in the curing reaction is brought about by the linking of atoms between or across two linear polymers, with this crosslinking of atoms making a three-dimensional rigidized chemical structure. One such reaction is shown in Fig. 4. Although the cured part can be softened by heat, it cannot be remelted or restored to the flowable state that existed before the plastic resin was cured. A tabulation of most of the thermosetting plastic materials is shown in Fig. 2.

THERMOPLASTIC MATERIALS.[2] Thermoplastics differ from thermosets in that thermoplastics do not cure or set under heat as do thermosets. Thermoplastics merely soften, when heated, to a flowable state in which under pressure they can be forced or transferred from a heated cavity into a cool mold. Upon cooling in a mold, thermoplastics harden and take the shape of the mold. Since thermoplastics do not cure or set, they can be remelted and rehardened, by cooling, many times. Thermal aging, brought about by repeated exposure to the high temperatures required for melting, causes eventual degradation of the material and so limits the number of reheat cycles. A tabulation of common thermoplastic materials is shown in Fig. 2.

Thermosetting Plastics. The nature of thermosetting plastics, or thermosets, was described above. Design data and information on the various thermosets will be discussed in this section. A tabulation of the major thermosetting plastics, along with the primary design and application considerations for each, and typical commonly used names for these materials, is shown in Table 2. More detailed data for each type of thermosetting material are given below and in other following sections. Likewise, plastic fabrication processes are discussed in a following section.

It will be seen that fillers and reinforcements are widely used, and of great importance, in thermosetting materials. Fillers and reinforcements will also, therefore, be covered in a later separate section.

Reaction A

One quantity of unsaturated acid reacts with two quantities of glycol to yield linear polyester (alkyd) polymer of n polymer units

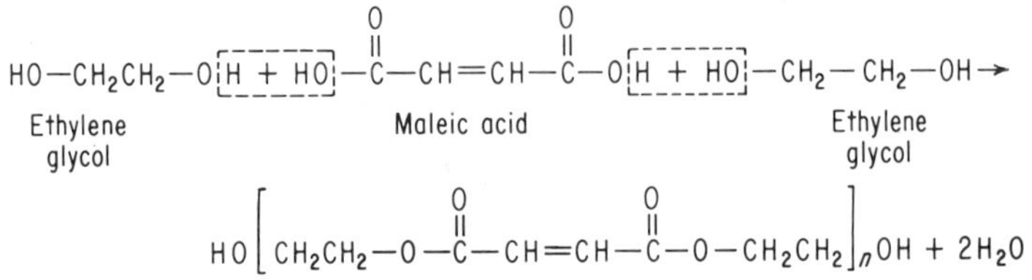

$$HO-CH_2CH_2-O\boxed{H + HO}-\overset{O}{\overset{\|}{C}}-CH{=}CH-\overset{O}{\overset{\|}{C}}-O\boxed{H + HO}-CH_2-CH_2-OH \rightarrow$$

Ethylene glycol — Maleic acid — Ethylene glycol

$$HO\left[CH_2CH_2-O-\overset{O}{\overset{\|}{C}}-CH{=}CH-\overset{O}{\overset{\|}{C}}-O-CH_2CH_2\right]_nOH + 2H_2O$$

Ethylene glycol maleate polyester

Reaction B

Polyester polymer units react (copolymerize) with styrene monomer in presence of catalyst and/or heat to yield styrene-polyester copolymer resin or, more simply, a cured polyester. (Asterisk indicates points capable of further crosslinking)

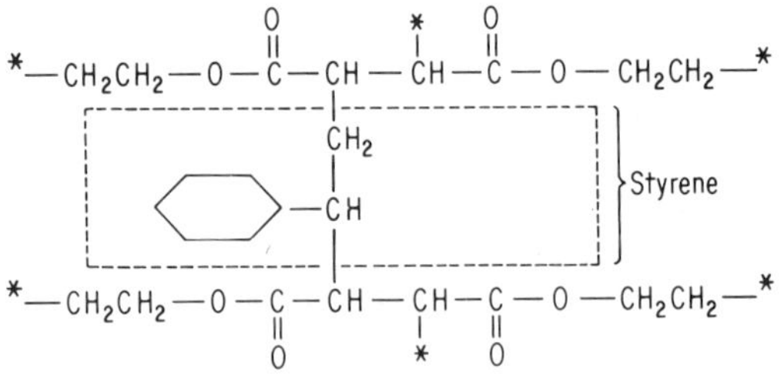

$$*-CH_2CH_2-O-\overset{O}{\overset{\|}{C}}-CH-\overset{*}{\overset{|}{C}H}-\overset{O}{\overset{\|}{C}}-O-CH_2CH_2-*$$

CH_2

C_6H_5-CH (Styrene)

$$*-CH_2CH_2-O-\underset{O}{\underset{\|}{C}}-CH-\underset{*}{\underset{|}{C}H}-\underset{O}{\underset{\|}{C}}-O-CH_2CH_2-*$$

Styrene-polyester copolymer

FIG. 4. Simplified diagrams show how crosslinking reactions produce polyester resin (styrene-polyester copolymer resin) from basic chemicals.[1]

Alkyds. Alkyds are widely used for molded electrical parts where the general application considerations given in Table 2 apply. They are chemically somewhat similar to polyester resins. Normally, the term alkyds is used to designate molding compounds, and the term polyester is used to denote liquid resins. Alkyd molding compounds are commonly available in putty, granular, glass-fiber-reinforced, and rope form. The properties of these forms are shown in Table 3.

Alkyds are easy to mold and economical to use. Molding dimensional tolerances can be held to within ±0.001 in./in. Postmolding shrinkage is small, as shown in Fig. 5. Their greatest limitation is in extremes of temperature (above 350°F) and in extremes of humidity. Silicones and diallyl phthalates are superior here, silicones especially with respect to temperature and diallyl phthalates especially with respect to humidity. The electrical insulation resistance of alkyds decreases considerably in high, continuous humidity conditions, as shown in Fig. 6.

Aminos. Amino molding compounds can be fabricated by economical molding methods. They are hard, rigid, and abrasion-resistant, and they have high resistance to deformation under load. These materials can be exposed to subzero temperatures without embrittlement. Under tropical conditions, the melamines do not support fungus growth.

Amino materials are self-extinguishing and have excellent electrical insulation characteristics. They are unaffected by common organic solvents, greases and oils, and weak acids and alkalies. Melamines are superior to ureas in resistance to acids, alkalies, heat, and boiling water, and are preferred for applications involving cycling between wet and dry conditions or rough handling. Aminos do not impart taste or odor to foods.

Addition of alpha cellulose filler produces an unlimited range of light-stable colors and high degrees of translucency. Colors are obtained without sacrifice of basic material properties.

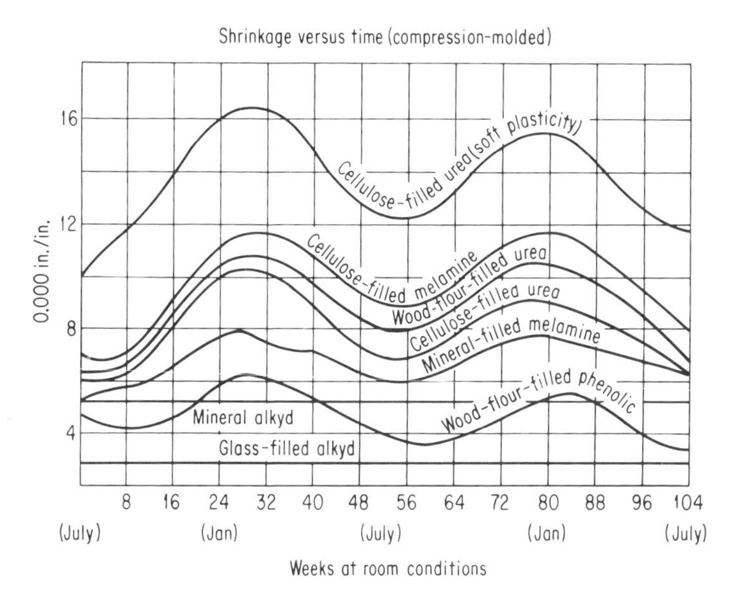

FIG. 5. Postmolding shrinkage variation of several molding compounds over a period of weeks, and showing stability of alkyds.[3]

Melamines and ureas provide excellent heat insulation; temperatures up to the destruction point will not cause parts to lose their shape.

Amino resins exhibit relatively high mold shrinkage, and also shrink on aging. Cracks develop in urea moldings subjected to severe cycling between dry and wet conditions.

Prolonged exposure to high temperature affects the color of both urea and melamine products.

A loss of certain strength characteristics also occurs when amino moldings are subjected to prolonged elevated temperatures. Some electrical characteristics are also adversely affected; arc resistance of some industrial types, however, remains unaffected after exposure at 500°F.

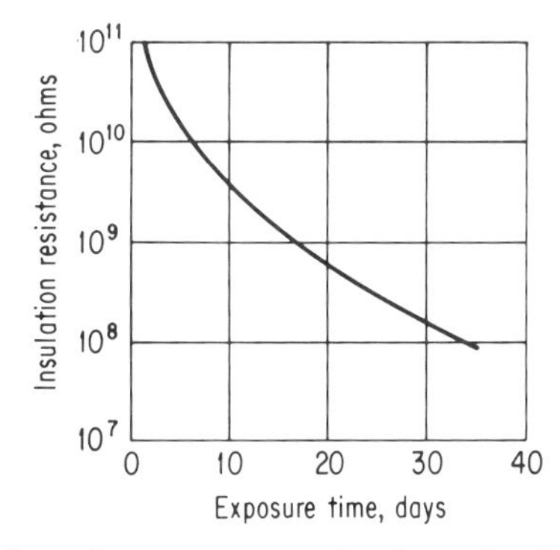

FIG. 6. Effect of 95% humidity at 60°C on the 500 volts dc insulation resistance of an alkyd molding resin.[4]

TABLE 2. Application Information for Thermosetting Plastics

Material	Major application considerations	Common available forms	Typical suppliers and trade names*
Alkyds	Excellent dielectric strength, arc resistance, and dry insulation resistance. Low dielectric constant and dissipation factor. Good dimensional stability. Easily molded.	Compression moldings, transfer moldings	Allied Chemical Corp. (Plaskon); American Cyanamid Co. (Glaskyd)
Aminos (melamine formaldehyde and urea formaldehyde)	Available in an unlimited range of light-stable colors. Exhibit hard glossy molded surface, and good general electrical properties, especially arc resistance. Excellent chemical resistance to organic solvents and cleaners and household-type cleaners.	Compression moldings, extrusions, transfer moldings, laminates, film	Allied Chemical Corp. (Plaskon); Monsanto Co. (Resimene); American Cyanamid Co. (Cymel for melamine; Beetle for urea)
Diallyl phthalates (DAP) (allylics)	Unsurpassed among thermosets in retention of properties in high humidity environments. Also, have among the highest volume and surface resistivities in thermosets. Low dissipation factor and heat resistance to 400°F or higher. Excellent dimensional stability. Easily molded.	Compression moldings, extrusions, injection moldings, transfer moldings, laminates	FMC Corp. (Dapon); Allied Chemical Corp. (Diall)
Epoxies	Good electrical properties, low shrinkage, excellent dimensional stability and good to excellent adhesion. Extremely easy to compound, using nonpressure processes, for providing a wide variety of end properties. Useful over a wide range of environments. Bisphenol epoxies are most common, but several other varieties are available for providing special properties.	Castings, compression moldings, extrusions, injection moldings, transfer moldings, laminates, matched-die moldings, filament windings, foam	Shell Chemical Co. (Epon); Jones-Dabney Co. (Epi-Rez); Dow Chemical Co. (D.E.R.); Ciba Products Co. (Araldite); Union Carbide Corp. (ERL); 3M Co. (Scotchcast)

Phenolics	Among the lowest-cost, most widely used thermoset materials. Excellent thermal stability to over 300°F generally, and over 400°F in special formulations. Can be compounded to a broad choice of resins, fillers, and other additives.	Castings, compression moldings, extrusions, injection moldings, transfer moldings, laminates, matched-die moldings, stock shapes, foam	Union Carbide Corp. (Bakelite); Hooker Chemical Corp. (Durez)
Polyesters	Excellent electrical properties and low cost. Extremely easy to compound using nonpressure processes. Like epoxies, can be formulated for either room temperature or elevated temperature use. Not equivalent to epoxies in environmental resistance.	Compression moldings, extrusions, injection moldings, transfer moldings, laminates, matched-die moldings, filament windings, stock shapes	Pittsburgh Plate Glass Co. (Selectron); American Cyanamid Co. (Laminac); Rohm & Haas Co. (Paraplex)
Silicones (rigid)	Excellent electrical properties, especially low dielectric constant and dissipation factor, which change little up to 400°F and over. Nonrigid silicones are covered in elastomers and embedding material sections.	Castings, compression moldings, transfer moldings, laminates	Dow Corning Corp. (DC Resins)
Urethanes (rigid foams)	Low-weight plastics. Excellent electrical properties, which are basically variable as a function of density. Easy to use for foam-in-place and embedding applications. Flexible urethane foams and nonrigid high-density urethanes are covered in sections on foams, elastomers, and embedding materials.	Castings, coatings	Nopco Chemical Co. (Nopcofoam); Hooker Chemical Corp. (Hetrofoam); Emerson and Cuming, Inc. (Eccofoam)

* This listing is only a very small sampling of the many possible excellent suppliers. It is intended only to orient the nonchemical reader into plastic categories. No preferences are implied or intended.

TABLE 3. Typical Properties of Alkyd Molding Compounds [3]

	ASTM test method	Putty	Granular	Glass-fiber reinforced	Rope
Electrical					
Arc resistance, sec	D495-58T	180+	180+	180+	180+
Dielectric constant	D150-54T				
1 Mc	....	5.4–5.9	5.7–6.3	5.2–6.0	7.4
1 Mc	....	4.5–4.7	4.8–5.1	4.5–5.0	6.8
Dissipation factor	D150-54T				
60 cycles	....	0.030–0.045	0.030–0.040	0.02–0.03	0.019
1 Mc	....	0.016–0.022	0.017–0.020	0.015–0.022	0.023
Dielectric strength, volts/mil	D149-59				
Short-time	....	350–400	350–400	350–400	360
Step by step	....	300–350	300–350	300–350	290
Physical					
Specific gravity	D792-50	2.05–2.15	2.21–2.24	2.02–2.10	2.20
Water absorption, 24 hr at 23°C, %	D570-57T	0.10–0.15	0.08–0.12	0.07–0.10	0.05
Heat resistance max., °F					
Long periods (continuous)	....	250	300	300	300
Short periods (0–24 hr)	....	300	350	350	350
Short periods (0–1 hr)	....	325	375	400	400
Heat distortion temp., 264 psi, °F	D648-56	350–400	350–400	>400	>400
Coefficient of linear thermal expansion/°F	D696-44	$10–30 \times 10^{-6}$	$10–30 \times 10^{-6}$	$10–30 \times 10^{-6}$	20×10^{-6}
Thermal conductivity, (g)(cal)/(sec)(cm²)(°C/cm)	....	$15–25 \times 10^{-4}$	$15–25 \times 10^{-4}$	$8–12 \times 10^{-3}$	10×10^{-4}
Flammability	D635-56T	Nonburning	Self-extinguishing	Nonburning	Self-extinguishing
Mechanical					
Impact strength, Izod, ft-lb/in. of notch	D256-56	0.25–0.35	0.30–0.35	8–12	2.2
Comprehensive strenth, psi	D695-54	20,000–25,000	16,000–20,000	24,000–30,000	28,800
Flexural strength, psi	D790-59T	8,000–11,000	7,000–10,000	12,000–17,000	19,500
Tensile strength, psi	D651-48	4,000–5,000	3,000–4,000	5,000–9,000	7,100
Modulus of elasticity, psi	D790-49T	$2.0–2.7 \times 10^{6}$	$2.4–2.9 \times 10^{6}$	$2.0–2.5 \times 10^{6}$	1.9×10^{6}
Barcol hardness	....	60–70	60–70	70–80	72
		MIL-M-14F Type MAG	MIL-M-14F Type MAG	MIL-M-14F Type MAI-60	

Ureas are unsuitable for outdoor exposure. Melamines experience little degradation in electrical or physical properties after outdoor exposure, but color changes may occur.

Typical physical properties of amino plastics are shown in Table 4, and typical mechanical and electrical properties are shown in Table 5.

Diallyl Phthalates. Diallyl phthalates are among the best of the thermosetting plastics, with respect to high insulation resistance and low electrical losses, which are maintained up to 400°F or higher, and in the presence of high humidity environments. Also, diallyl phthalate resins are easily molded and fabricated.

There are several chemical variations of diallyl phthalate resins, but the two most commonly used are diallyl phthalate (DAP) and diallyl isophthalate (DAIP). The primary application difference is that DAIP will withstand somewhat higher temperatures than will DAP. Typical properties of DAP and DAIP molding compounds are shown in Table 6. The retention of insulation resistance properties after humidity conditioning is shown in Table 7.

The excellent dimensional stability of diallyl phthalates has been mentioned above. This is demonstrated in Fig. 7, which compares diallyl phthalates to other plastic materials, at various temperatures.

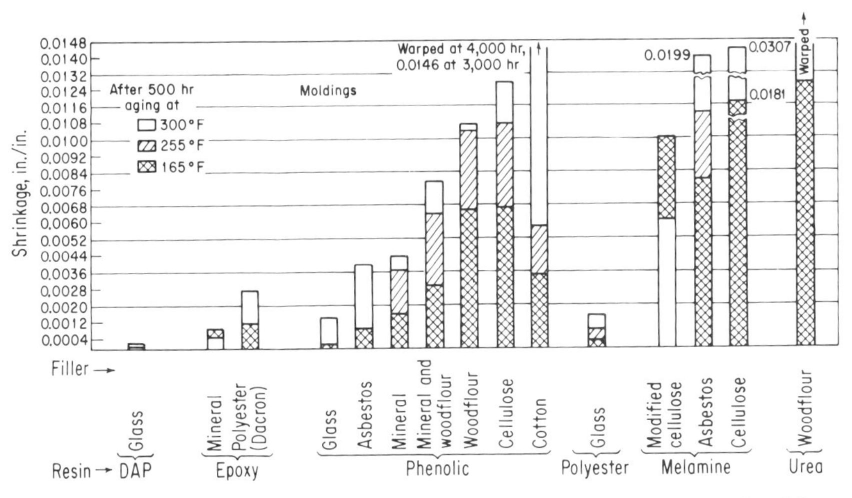

Fig. 7. Shrinkage of various thermosetting molding materials as a result of heat aging.[7]

Likewise, the excellent electrical properties of diallyl phthalates have been stressed. The effect of frequency and temperature on dielectric constant is shown in Fig. 8, and the effect of frequency and temperature on dissipation factor is shown in Fig. 9. Diallyl isophthalate, DAIP, is also especially good in retention of dielectric strength, as shown in Fig. 10. Further, the resistivity of diallyl phthalates stands high among plastics and insulating materials, as shown in Fig. 11.

Epoxies. Epoxies are among the most versatile and most widely used plastics in the electronic packaging field. This is primarily because of the wide variety of formulations possible, and the ease with which these formulations can be made and utilized with minimal equipment requirements. Formulations range from flexible to rigid, in the cured state, and from thin liquids to thick pastes in the uncured state. Conversion from uncured to cured state is made by use of hardeners and/or heat. The largest applications of epoxies, in electronic packaging, are in embedding applications (potting, casting, encapsulating, and impregnating) and in laminated constructions such as metal-clad laminates for printed circuits and unclad laminates for various types

TABLE 4. **Typical Physical Properties of Amino Molding Compounds** [5]

	Urea		Melamine					
	Alpha cellulose	Wood flour	Alpha cellulose	Wood flour	Alpha cellulose, modified	Rag	Asbestos	Glass fiber
Specific gravity	1.5	1.5	1.5	1.42	1.43	1.5	1.78	1.94–2.0
Density, g/in.^{3}	24.6	24.6	24.6	23.8	23.5	24.6	29.2	31.8–32.8
Hardness, Rockwell E	94–97	95	110	94	....	100	90	
Shrinkage,* in./in.								
Molding	0.006–0.009	0.006–0.014	0.008–0.009	0.007–0.008	0.006–0.008	0.003–0.004	0.005–0.007	0.002–0.004
Postmold	0.006–0.012	0.006–0.012	0.009–0.011	0.004–0.007	0.001–0.002	0.004–0.008	0.002–0.003	0.002–0.005
Deflection temperature, °F, at 264 psi	266	270	361	266	266	310	266	400
Heat resistance, continuous, °F	170†	170	210†	250	250	250	300	300
Coefficient of thermal expansion, per °C $\times 10^{-6}$	22–36	30	20–57	32–50	34–36	25–30	21–43	12–25
Thermal conductivity, (cal)(cm)/(sec^2)(cm)(°C) $\times 10^{-4}$	10.1	10.1	10.1	8.4	....	10.6	13.1	
Water absorption, 24 hr, at 23°C, percent	0.4–0.8	0.7	0.3–0.5	0.34–0.6	0.3–0.6	0.3–0.6	0.13–0.15	0.09–0.3
Color possibilities	Unlimited	Brown, black	Unlimited	Brown	Brown	Limited	Brown	Natural, gray

* Test specimen: 4-in. diam. × ⅛-in. disk.
† Based on no color change.

TABLE 5. Typical Mechanical and Electrical Properties of Amino Molding Compounds [5]

	Urea		Melamine					
	Alpha cellulose	Wood flour	Alpha cellulose	Wood flour	Alpha cellulose modified	Rag	Asbestos	Glass fiber
Mechanical:								
Tensile strength, 1,000 psi	5.5–7	5.5–10	7–8	5.7–6.5	5.5–6.5	8–10	5.5–6.5	5.9
Compressive strength, 1,000 psi	30–38	25–35	40–45	30–35	24.5–26	30–35	25-30	20–29
Flexural strength, 1,000 psi	11–18	8–16	12–15	6.5–9	11.5-12	12–15	7.4–10	13.2–24
Shear strength, 1,000 psi	11–12	....	11–12	10–10.5	11.4–12.2	12–14	7–8	13.0–15.6
Impact strength, Izod, ft-lb/in. of notch	0.24–0.28	0.25–0.35	0.30–0.35	0.25–0.38	0.30–0.42	0.55–0.90	0.30–0.40	0.5–6.0
Tensile modulus, 10^6 psi	1.3–1.4	....	1.35	1.0	1.0	1.4	1.95	
Flexural modulus, 10^6 psi	1.4–1.5	1.3–1.6	1.1	1.0	1.1	1.4	1.8	2.4
Electrical:								
Arc resistance, sec	80–100	80–100	125–136	70–106	90–120	122–128	120–180	180–186
Dielectric strength, volts/mil								
Short time								
At 23°C	330–370	300–400	270–300	350–370	350–390	250–340	410–430	170–370
At 100°C	200–270	....	170–210	290–330	140–190	110–130	280–310	90–350*
Step by step								
At 23°C	220–250	250–300	240–270	200–240	200–250	220–240	280–300	170–270
At 100°C	110–150	....	90–130	190–210	90–100	60–90	190–210	60–250*
Slow rate of rise								
At 23°C	250–260	....	210–240	240–260	280–290	210–240	270–290	170–210
At 100°C	120–170	....	90–120	170–200	90	70–80	170–190	70–90
Dielectric constant								
At 60 cps	7.7–7.9	7.0–9.5	7.9–8.2	6.4–6.6	7.0–7.7	8.1–12.6	10.0–10.2	7.0–11.1
At 10^6 cps	6.7–6.9	6.4–6.9	7.6–8.0	5.6–5.8	5.2–6.0	6.7–6.9	5.3–6.1	6.6–7.9
At 3×10^9 cps	....	....	....	....	....	....	4.9	5.5
Dissipation factor								
At 60 cps	0.034–0.043	0.035–0.040	0.052–0.083	0.026–0.033	0.192	0.100–0.340	0.100	0.14–0.23
At 10^6 cps	0.029–0.031	0.028–0.032	0.026–0.030	0.034–0.035	0.044–0.12	0.036–0.041	0.039–0.048	0.013–0.016
At 3×10^9 cps	....	....	....	....	....	....	0.032	0.040
Dielectric loss factor								
At 60 cps	0.28–0.34	0.24–0.38	0.44–0.78	0.17–0.22	0.90–2.4	2.0–5.0	0.5–1.0	1.5–2.5
At 10^6 cps	0.19–0.21	0.18–0.22	0.20–0.33	0.20–0.21	0.19–0.28	0.24–0.26	0.21–0.31	0.09–0.19
Volume resistivity, ohm-cm	0.5–5.0 × 10^{11}	....	0.8–2.0 × 10^{12}	6–10 × 10^{12}	6 × 10^{10}	1.0–3.0 × 10^{11}	1.2 × 10^{12}	0.9–20 × 10^{11}
Surface resistivity, ohms	0.4–3.0 × 10^{11}	....	0.8–4.0 × 10^{11}	0.3–5.0 × 10^{12}	1.7 × 10^{12}	0.7–7.0 × 10^{11}	1.9 × 10^{13}	3.0–4.6 × 10^{12}
Insulation resistance, ohms	0.2–5.0 × 10^{11}	....	1.0–4.0 × 10^{10}	1.0–3.0 × 10^{11}	2.0–5.0 × 10^{9}	0.1–3.0 × 10^{10}	1.0–4.0 × 10^{10}	0.2–6.0 × 10^{10}

* At 50°C.

TABLE 6. Typical Properties of Unfilled Diallyl Phthalate Molding Compounds [6] (DAP is diallyl phthalate and DAIP is diallyl isophthalate)

Property by ASTM procedures	DAP [a]	DAIP [b]
Dielectric constant		
At 60 cycles, 25°C	3.6	3.5
At 10^3 cycles, 25°C	3.6	3.3
At 10^6 cycles, 25°C	3.4	3.2
At 10,000 Mc, 25°C	....	3.0
At 10,000 Mc, 200°C	....	3.1
Dissipation factor		
At 60 cycles, 25°C	0.010	0.008
At 10^3 cycles, 25°C	0.009	0.008
At 10^6 cycles, 25°C	0.011	0.009
At 10,000 Mc, 25°C	....	0.014
At 10,000 Mc, 200°C	....	0.031
Volume resistivity		
Ohm-cm at 25°C	1.8×10^{16}	3.9×10^{17}
Ohm-cm at 25°C (wet)	1.0×10^{14} + [c]	
Surface resistivity		
Ohms at 25°C	9.7×10^{15}	8.4×10^{12}
Ohms at 25°C (wet)	4.0×10^{13} + [c]	
Dielectric strength, volts/mil at 25°C	450 [d]	422 [d]
Arc resistance	118	123–128
Moisture absorption, %—20 hr at 25°C	0.09	0.1
Tensile strength, psi	3,000–4,000	4,000–4,500
Rockwell hardness (M)	114–116	119–121
Barcol hardness	43	52
Specific gravity	1.270	1.264
Izod impact, ft/in. notch	0.2–0.3	0.2–0.3
Heat distortion temperature		
°C at 264 psi	155 (310°F)	238 [e] (460°F)
°C at 546 psi	125 (257°F)	184–211 (364–412°F)
Comprehensive strength, psi	22,000–23,000	21,200–24,000
Flexural strength, psi	7,000–9,000	7,400–8,300
Refractive index at 25°C	1.571	1.569
Modulus of elasticity in flexure, psi	0.6×10^6	0.5×10^6
Chemical resistance		
% gain in weight after 1 month in:		
Water at 25°C	0.9	0.8
Acetone at 25°C	1.3	—0.03
1% NaOH at 25°C	0.7	0.7
10% NaOH at 25°C	0.5	0.6
3% H_2SO_4	0.8	0.7
30% H_2SO_4	0.4	0.4
Heat-resistant properties [f]		
Weight loss after aging 6 weeks at 177°C, %	7.6	1.2
Dielectric constant after aging 6 weeks at 177°C, measured at 60 cycles, 25°C	3.7	3.6
Dissipation factor after aging 6 weeks at 177°C, measured at 60 cycles, 25°C	0.01	0.006
Volume resistivity after aging 6 weeks at 177°C, ohm-cm at 25°C	2.7×10^{15}	7.1×10^{15}
Surface resistivity after aging 6 weeks at 177°C, ohms at 25°C	1.4×10^{13}	1.1×10^{14}
Flexural strength, psi at 260°C (glass cloth laminate)	....	15,400 [g]

[a] Diallyl phthalate.
[b] Diallyl isophthalate.
[c] Tested in humidity chamber after 30 days at 70°F (158°) and 100% relative humidity.
[d] Step by step.
[e] No deflection.
[f] 50% silica filled.
[g] 12-ply, 181 glass cloth laminate (resin = 40%).

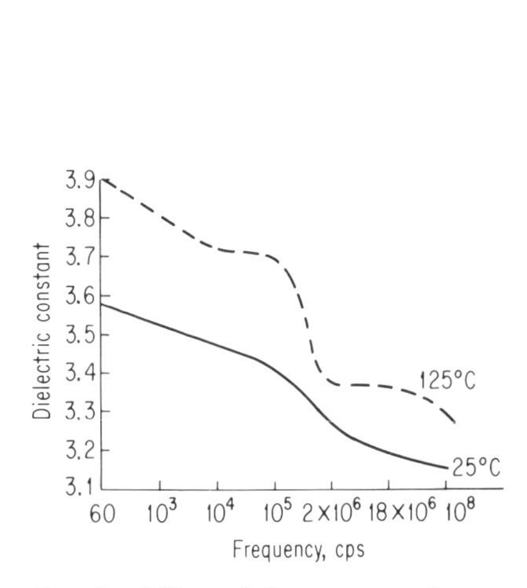

FIG. 8. Effect of frequency and temperature on the dielectric constant of unfilled diallyl phthalate.[8]

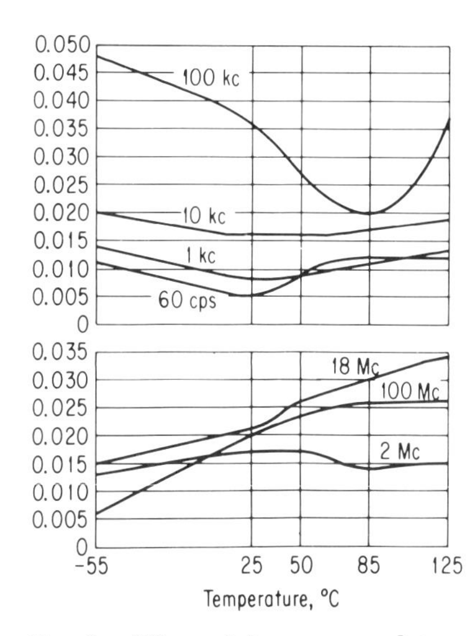

FIG. 9. Effect of frequency and temperature on the dissipation factor of unfilled diallyl phthalate.[8]

of insulating and terminal boards. Hence, the detailed properties of these embedding compounds and laminates are given in those parts of this section which describe laminates and embedding materials.

Basically, epoxies are available as liquid or solid resins, and as powdered molding compounds. It is the liquid resins which find the broadest use in embedding applications and in the fabrication of laminate boards. The molding compounds, while broadly used for embedment of electronic assemblies by the transfer molding technique, are also used for transfer and compression molding of many other types of electrical parts. Typical properties of glass-fiber-filled and mineral-filled epoxy compounds are shown in Table 8.

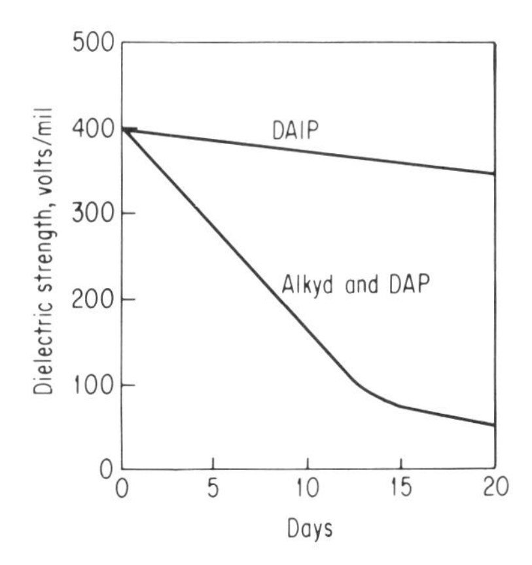

FIG. 10. Effect of heat aging at 400°F on the dielectric strength of diallyl phthalate (DAP), diallyl isophthalate (DAIP), and alkyd molding materials.[9]

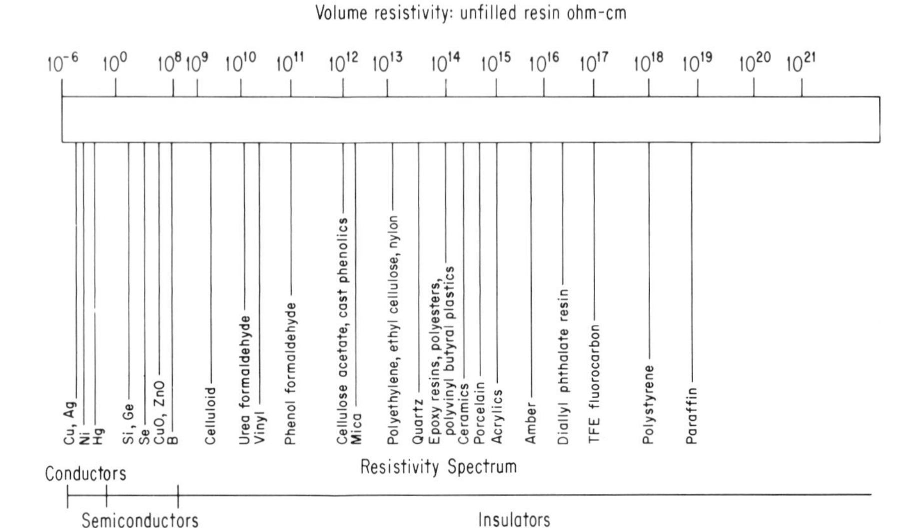

FIG. 11. Resistivity spectrum for a variety of engineering materials. Note that diallyl phthalate, TFE fluorocarbon, and polystyrene rate highest among insulating materials.[6]

TABLE 7. Retention of Insulation Resistance Properties of Several Plastics during Humidity Conditioning [6]

Material	Insulation resistance in megohms		
	As molded (conditioned 2 days at room temp. in the desiccator)	After 30 days at 100% RH and 80° F before removing moisture film	After 30 days at 100% RH and 80° F and after removing moisture film
Compound A, (mineral-filled alkyd, type MA 1-60, MIL-M-14E)	5,000,000	45	50
Compound B (Orlon-filled diallyl phthalate, type SDI-5, MIL-M-18794) ...	5,000,000	5,000,000	5,000,000
Compound C (glass-fiber-filled diallyl phthalate, type GDI-30, MIL-M-19833)	5,000,000	3,000,000	5,000,000
Compound D (glass-fiber-filled diallyl phthalate, type SDG, MIL-M-18794)	5,000,000	2,000,000	5,000,000
Compound E (mineral-filled melamine, type MME, MIL-M-14E)	1,000,000	15	90
Compound F (mineral-filled electrical-grade phenolic, type MFE, MIL-M-14E)	1,000,000	100	100
Compound G (mineral-filled silicone, type MSG, MIL-M-14E)	2,000,000	100	100
Compound H (nylon, MIL-P-17091)	1,000,000	11	20
Compound I (Spec. AMS-3651), TFE fluorocarbon.	5,000,000	5,000,000	5,000,000

In addition to their versatility and good electrical properties, epoxies are also outstanding in their low shrinkage, their dimensional stability, and their adhesive properties. Their shrinkage is often less than 1 percent, and the as-molded dimensions of an epoxy part change little with time or environmental conditions, other than excessive heat. Because of the low shrinkage and good strength properties of epoxies, cured epoxy parts resist cracking, both upon curing and in thermal shock, better than most other rigid thermosetting materials. Based on the excellent bonds obtained with epoxy resins to most substrates, epoxy formulations are broadly used as adhesives. Adhesives will be discussed separately in a subsequent portion of this section. Even when epoxies are not specifically used as adhesives, their bonding properties often provide a better seal around inserts, terminals, and other interfaces than do most other plastic materials.

Phenolics. Phenolics are among the oldest, best-known general-purpose molding materials. They are also among the lowest in cost and the easiest to mold. An extremely large number of phenolic materials are available, based on the many resin and filler combinations, and they can be classified many ways. One common way of classifying them is by type of application. Typical properties for some of these common classifications are shown in Table 9.

While it is possible to get various grades of phenolics for various applications, as shown in Table 9, phenolics, generally speaking, are not equivalent to diallyl phthalates and epoxies in resistance to humidity, shrinkage, dimensional stability, and retention of electrical properties in extreme environments. They are, however, quite adequate for a large percentage of electrical applications. Further, the glass-filled, heat-resistant grades are outstanding in thermal stability up to 400°F and higher (some are useful to 500°F).

Polyesters. Polyesters are versatile resins, which handle much like the epoxies. They are available in forms ranging from low-viscosity liquids to thick pastes or put-

TABLE 8. Typical Properties of Epoxy Molding Compounds

	Glass-fiber filler	Mineral filler
Tensile strength, psi	14,000–30,000	5,000–7,000
Elongation, percent	4	
Tensile modulus, 10^5 psi	30.4	
Compressive strength, psi	25,000–30,000	18,000–25,000
Flexural strength, psi	20,000–26,000	10,000–15,000
Impact strength, Izod, ft-lb/in. of notch	8–15	0.25–0.45
Hardness, Rockwell	M100–M108	M101
Specific gravity	1.8–2	1.6–2.06
Thermal conductivity, (cal)(cm)/(sec)(cm^2)(°C)	$7–10 \times 10^{-4}$	$7–18 \times 10^{-4}$
Specific heat, per °C	0.19	
Coefficient of thermal expansion, per °C $\times 10^{-5}$	1.1–3	2.4–5
Heat resistance, continuous, °F	330–500	300–500
Heat-distortion temp., °F	400–500	250–450
Volume resistivity, ohm-cm	3.8×10^{15}	9×10^{15}
Dielectric strength, ⅛-in. (volts/mil)		
Short time	360	330–400
Step by step	340	350
Dielectric constant at 60, 10^3, and 10^6 cps	4–5	4–5
Arc resistance, sec	125–140	150–180
Burning rate	Self-extinguishing	Self-extinguishing
Water absorption, percent*	0.05–0.095	0.1

* Test performed on ⅛-in.-thick piece immersed in distilled water for 24 hr at room temperature.

TABLE 9. Typical Properties of Several Common

ASTM test method	Property	Phenolics	
		General-purpose	Shock-resistant (cellulose fiber)
D792	Specific gravity	1.33–1.45	1.33–1.43
	Density, lb/in.^{3}	0.048–0.052	0.048–0.051
D955	Mold shrinkage, in./in.	0.004–0.009	0.004–0.009
D635	Flammability	←	
C177	Thermal conductivity, (cal)(sec)(cm^2)/(°C)(cm $\times$ 10^{-4})	4–7	4–7
D570	Water absorption, 24 hr, ⅛-in. thickness, 73°F, percent increase in weight	0.3–0.7	0.3–0.7
D696	Coefficient of thermal expansion, 10^{-5} in./(in.)(°C)	3.0–4.5	3.0–4.5
D648	Heat distortion, temperature, °F	260–340	260–340
	Maximum recommended intermittent service, temperature, °F	300	300
	Maximum recommended continuous service, temperature, continuous loading, °F	250	250
D638	Tensile strength, psi (average)	6,500–10,000	6,500–10,000
D638	Tensile modulus, psi, at 2 in./min	8–12 $\times$ 10^5	8–13 $\times$ 10^5
D790	Flexural strength, psi	8,000–12,000	8,000–12,000
D790	Flexural modulus, psi	8–12 $\times$ 10^5	8–13 $\times$ 10^5
D695	Compressive strength, psi	22,000–35,000	22,000–36,000
D732	Shear yield strength	9,600	....
D638	Elongation, percent (ultimate) at 2 in./min	0.4–0.8	0.4–0.8
D256	Impact strength, Izod, 73°F (notched), ft-lb/in. of notch	0.24–0.33	0.34–8.0
D785	Hardness (Rockwell M scale)	95–120	95–120
D257	Volume resistivity, ohm-cm (50% RH and 23°C)	10^9–10^{13}	10^9–10^{13}
D150	Power factor 24 hr in H_2O, 23°C conditioning		
	60 cycles	0.05–0.30	0.05–0.30
	10^6 cycles	0.03–0.07	0.03–0.07
D150	Dielectric constant		
	60 cycles	5.0–12.0	5.0–12.0
	10^6 cycles	4.0–6.0	4.0–6.0
D149	Dielectric strength, step by step, ⅛-in. thickness, volts/mil at 23°C	200–400	200–400
D495	Arc resistance, sec	20–50	20–50
	UV resistance	←	
D543	Resistance to acids	← Resistant to weak	
D543	Resistance to bases	← Resistant to weak	
D543	Resistance to solvents	←	
	Relative material costs (cents/in.^{3})	1.01–1.06	1.22–1.38

Grades of Phenolic Molding Compounds [10]

Phenolics				
Heat-resistant (mineral-filled)	Glass-filled	Low-loss	Chemical-resistant	Cast phenolics
1.50–1.95	1.7–1.83	1.50–1.95	1.24–1.50	1.3
0.054–0.070	0.061–0.066	0.054–0.070	0.045–0.054	0.047
0.001–0.004	0.001–0.004	0.001–0.005	0.005–0.009	0.005
——— Self-extinguishing ——→				
8–22	2–4	10–14	4–7	4
0.1–0.5	0.05–0.5	0.01–0.05	0.01–0.4	0.02
1.5–4.0	1.0	1.9–2.6	3.0–4.5	5
300–400	375–600	230–350	300–350	95–212
400	450	350	300	
350	400	300	250	
5,500–7,500	7,000–11,000	6,000–7,000	4,000–7,000	3,000–10,500
$10–20 \times 10^5$	3×10^6	$30–50 \times 10^5$	$1–15 \times 10^5$	
8,000–12,000	16,000–22,000	8,000–12,000	7,000–9,000	2,000–15,000
$10–20 \times 10^5$	3×10^6	$30–40 \times 10^5$	$7–15 \times 10^5$	
20,000–35,000	16,500–32,000	25,000–30,000	25,000–33,000	4,000–33,000
	15,000			
0.18–0.50	<1	0.13–0.5	0.2–0.4	
0.27–3.5	0.6–15.0	0.30–0.38	0.20–0.30	0.25–0.6
95–120	90–120	95–120	95–120	20–120
$10^9–10^{13}$	$10^{12}–10^{14}$	$10^{12}–10^{14}$	$10^{11}–10^{14}$	10^9
0.1–0.3	0.03–0.07	0.03–0.05	0.10–0.20	
0.4–0.8	0.010–0.10	0.005–0.20	0.03–0.06	
7.5–50.0	4.5–6.8	4.5–6.0	6.0–13.0	4.0
5.0–10.0	4.2–6.8	4.2–5.2	4.5–6.0	5.0
200–350	300–400	350–450	225–300	400
30–70	70–180	30–70	20–50	
——— Darkens ——→				
acids, attacked by strong acids ——→			Slightly attacked by strong acids	Good to weak; decomposes with strong acids
alkalies, attacked by strong alkalies ——→			Slightly attacked by strong bases	Poor
Resistant at room temperature ——→				None to poor
1.21–1.93	2.11	......	1.80	3.5

ties. The liquids are used for embedding applications and laminated products, much like the epoxies, and the pastes are used for molding applications. Polyester properties and embedding applications will be discussed in a later portion of this section, devoted to embedding. While both epoxies and polyesters are available in formulations for room-temperature cure and in formulations for heat cure, the chemical curing mechanism, or polymerization mechanism, is different for the two types of resins. Likewise, of course, the basic resins are chemically different. It is their physical forms and application forms which make them similar.

The major advantages of polyesters over epoxies are lower cost and lower electrical losses for the best electrical-grade polyesters. Some important disadvantages of polyesters, as compared to epoxies, are lower adhesion to most substrates, higher polymerization shrinkage, a greater tendency to crack during cure or in thermal shock, and greater change of electrical properties in a humid environment.

Polysulfides. Polysulfides are basically liquid resins which are converted to rubbery products, and are normally classified as elastomers. As such, they will be further discussed in the portions of this section which deal with elastomers. Likewise, polysulfides are used in certain embedding applications, notably for potting connectors. This aspect of their use will be further covered in the portions of this section which cover embedding materials. Polysulfides are thermosetting polymers, however, and can be included in any classification which strictly classifies plastics as thermosetting or thermoplastic. Synthetic rubbers and elastomers, like all other plastics or synthetic polymers, can be classified as thermosetting or thermoplastic—depending upon whether they are remeltable (thermoplastic) or permanently set (thermosetting).

Silicones. Like polysulfides, silicones are thermosetting polymers which can be either classified as elastomers, when the cured product is rubberlike, or classified as embedding materials, when the basic plastic form is a castable liquid (either rubberlike or rigid) or a low-pressure transfer molding compound. Further, silicone polymers are widely used in electronic packaging for other applications, including laminated products, molding compounds, and adhesives. Silicone elastomers and silicone embedding materials will be further discussed in subsequent portions of this section which cover those material applications. However, silicones play such an important role in electronic packaging that additional discussion is in order at this point.

In addition to the application areas mentioned above and discussed in other parts of this section, basic silicone molding compounds are used for many of the molded products required in special electronic packaging applications. These molding compounds are normally either mineral-filled or glass-fiber-filled, and the properties of these two classes of molding compounds are shown in Table 10.

The most important properties of silicones, for electronic applications, are excellent electrical properties which do not change drastically with temperature or frequency over the safe operating-temperature range of silicones. Further, silicones are among the best of all polymer materials in resistance to temperature. This is shown in Fig. 12, which compares the relative thermal stability of a group of plastic materials. Useful temperatures of 500 to 700°F are available in silicone materials. Hence, silicones are broadly used for high-temperature electronic packaging applications, especially those applications requiring low electrical losses. Additional data in this respect are presented in the embedding and elastomer parts of this section.

Mechanical properties of silicone molding compounds are affected by temperature, as shown in Fig. 13. Hence, mechanical properties are not as stable with increasing temperature as are electrical properties. Generally, most silicone properties are stable in extreme environments such as humidity and vacuum. Thus, silicones are widely used both for military and for space applications.

Urethanes. Urethanes are basically liquid resins which are converted to rubbery products, and are normally classified as elastomers. As such, they will be further discussed in the portions of this section which deal with elastomers. Likewise, urethanes are used in certain embedding applications, notably for potting connectors. This aspect of their use will be further covered in the portions of this section which cover embedding materials. Urethanes are thermosetting polymers, however, and can be included in any classification which strictly classifies plastics as thermosetting or thermoplastic. Synthetic rubbers and elastomers, like all other plastics or synthetic

TABLE 10. Typical Properties of Silicone Molding Compounds [10]

ASTM test method	Property	Silicone resins: Mineral-filled	Silicone resins: Glass-fiber-filled
	Specific gravity	1.80–1.95	1.88
	Density, lb/in.^{3}	0.065–0.101	0.068
	Mold shrinkage, in./in.	0.007	0.0005
	Flammability	Self-extinguishing	Self-extinguishing
	Thermal conductivity (at 500°F), Btu/(hr)(ft^2)(°F)(ft)	2.0–2.1	1.2
D570	Water absorption, 24 hr, ⅛-in. thickness, 73°F, percent increase in weight	0.05–0.20	0.10–0.12
D696	Coefficient of thermal expansion, 10^{-5} in./(in.)(°C), 25 to 250°C		
	Perpendicular	2.5–6.0	Not isotropic 1.0
	Parallel	....	12.1
D648	Heat distortion, temperature, °F	340–900	>900
	Maximum recommended intermittent service temperature, °F	500–600	700–750
	Maximum recommended continuous service temperature, continuous loading, °F	500–750	700–750
D638	Tensile yield strength, psi (average) at 73°F	2,500–3,500	3,500–6,500
	Tensile modulus, psi, at 2 in./min		
D790	Flexural strength, psi, at 77°F	6,500–8,500	18,000–21,000
D790	Flexural modulus, psi, at 77°F	1.4×10^6–1.6×10^6	2.5×10^6
D695	Comprehensive strength, psi, at 77°F	11,000–18,000	10,000–12,500
	Shear yield strength	———	———
	Elongation, percent (ultimate) at 2 in./min.	———	———
D256	Impact strength, Izod, 73°F (notched), (ft)(lb)/in. of notch	0.30–0.34	10–24
D785	Hardness (Rockwell M), 73°F	70–95	83–88
D257	Volume resistivity, ohm-cm (50% RH and 23°C)	2×10^{13}–5×10^{15}	Dry, 1.4×10^{14}–9×10^{14}
D150	Power factor		
	Dry 10^6 cycles	0.002	0.004
	24-hr immersion 10^6 cycles	0.003	0.005
D150	Dielectric constant		
	60 cycles	3.4–6.3	4.35
	10^6 cycles	3.4–6.3	4.26–4.28
D149	Dielectric strength, 500 volts/sec rate of rise, ⅛-in. thickness, volts/mil	300–465	275–300
D495	Arc resistance, sec (tungsten electrodes)	190–420	230–240
	UV resistance	Excellent	Excellent
	Resistance to acids	To mild acids, excellent	To mild acids, excellent
	Resistance to bases	To mild bases, fair to good	To mild bases, fair to good
	Resistance to solvents	Fair to excellent	Fair to excellent
	Relative material costs (cents/in.^{3})	26–41	24

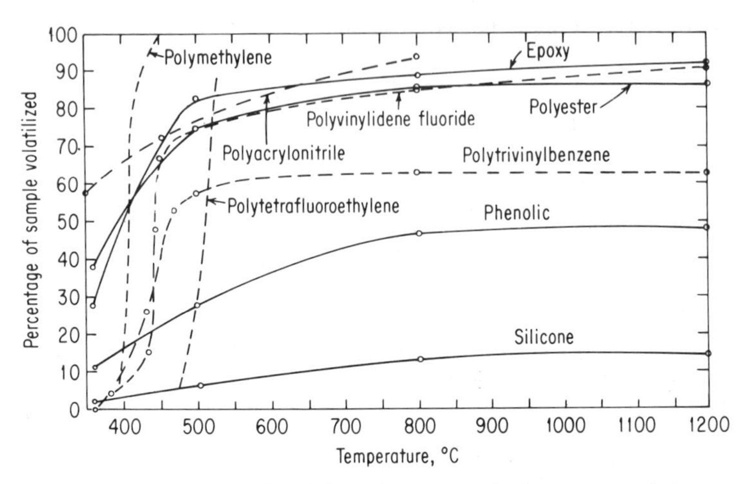

FIG. 12. Thermal stability of a group of plastic materials.[11]

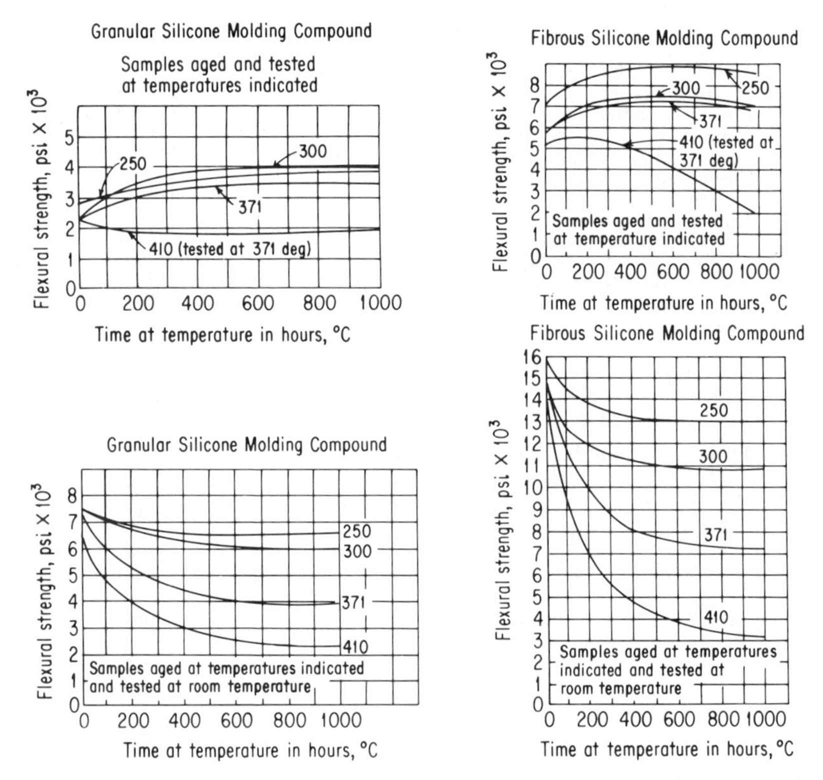

FIG. 13. Change in flexural strength of silicone molding compounds as a result of heat aging.[10]

polymers, can be classified as thermosetting or thermoplastic—depending upon whether they are remeltable (thermoplastic) or permanently set (thermosetting).

In addition to the above applications of high-density urethanes, low-density rigid urethane foams are used for embedding applications. These are discussed in the part of this section which deals with embedding materials. Also, urethanes are used as humidity-resistant coatings for printed-circuit boards, and this use, too, is further discussed in a subsequent part of this section.

Thermoplastics. As mentioned earlier in this section, thermoplastics differ from thermosets in that thermoplastics do not cure or set under heat, as do thermosets. However, thermoplastics will melt to a flowable state upon heating, whereas thermosets do not melt. Essentially all thermoplastics are processed by heating them to a soft state and applying pressure, such as injection molding, extruding, and thermoforming. Both thermosets and thermoplastics soften upon heating, to varying degrees. However, fully cured thermosets cannot be pressure-formed upon softening.

Design data and information on the various thermoplastics will be discussed at this point. A tabulation of the major thermoplastics, along with the primary design and application considerations and the typical commonly used name for each, is shown in Table 11. Basic physical and mechanical data on these thermoplastic materials are shown in Table 12, and basic electrical data are shown in Table 13. A more detailed discussion on each is given below. In addition, further strength data are given in a subsequent portion of this section which covers plastic reinforcements and plastic strength properties.

ABS Plastics. ABS plastics are derived from acrylonitrile, butadiene, and styrene. This class possesses hardness and rigidity without brittleness—at moderate costs. ABS materials have a good balance of tensile strength, impact resistance, surface hardness, rigidity, heat resistance, low-temperature properties, and electrical characteristics.

MECHANICAL PROPERTIES. The most outstanding mechanical properties of ABS plastics are impact resistance and toughness. Impact resistance does not fall off rapidly at lower temperatures. The Izod impact strength at 75°F is in the range of 3 to 5 ft-lb/in. of notch. This figure is gradually reduced to 1 ft-lb/in. of notch at −40°F. When impact failure does occur, the failure is ductile rather than brittle. Modulus of elasticity versus temperature is shown in Fig. 15.

ELECTRICAL PROPERTIES. While ABS plastics are used largely for mechanical purposes, they do have good electrical properties that are fairly constant over a wide range of frequencies. These properties are little affected by temperature and atmospheric humidity in the acceptable operating range of temperatures.

The dielectric strength of ABS plastics is about 350 volts/mil. The approximate dissipation factor of the best electrical grades is: 0.004 at 60 cycles; 0.005 at 1,000 cycles; and 0.009 at 1 million cycles.

Dielectric constants of these resins are also quite low (2.84 to 3.17) and are relatively independent of frequencies between 60 and 1 million cycles.

Acetals. Acetals are a class of thermoplastic materials which resemble nylon in appearance and have a natural whitish color.

MECHANICAL PROPERTIES. The most outstanding properties of acetals are high tensile strength and stiffness, resiliency, good recovery from deformation under load, and toughness under repeated impact. Acetals have low static and dynamic coefficients of friction and are usable over a wide range of environmental conditions. A fluorocarbon-fiber-filled acetal (Delrin AF*) is available and offers even better low friction and resistance properties.

The tensile yield strength of acetals is compared with that of some other thermoplastics in Fig. 14. Figure 15 shows the modulus of elasticity as a function of temperature for acetals and several other thermoplastic materials. The deflection under load for Delrin acetal is compared to that of other thermoplastics in Fig. 16.

ELECTRICAL PROPERTIES. Acetal resins are good insulators having relatively low dissipation factors and dielectric constants over the operating temperature range for these materials. The electrical properties (Table 13) of acetals are largely retained under exposure to high humidity and water immersion.

* Trademark, E. I. du Pont de Nemours & Co.

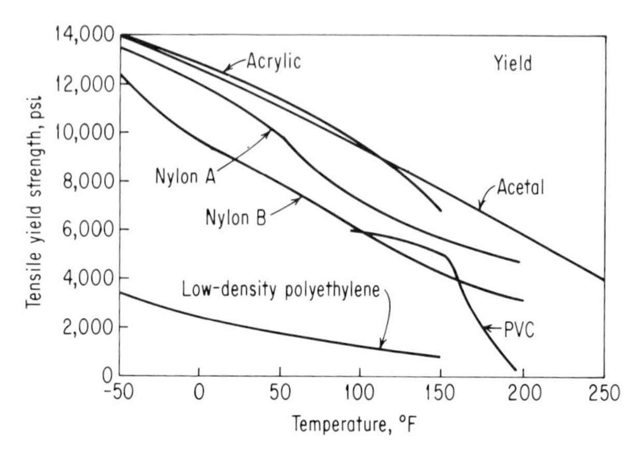

FIG. 14. Tensile yield strength of several thermoplastic materials as a function of temperature.[13]

Acrylics. The general properties of acrylics are given in Tables 11 to 13. Their optical clarity, good electrical and mechanical properties, and ready availability in a wide range of shapes and forms give them some special application areas.

MECHANICAL PROPERTIES. Acrylics, also called polymethylmethacrylates, are primarily known for their exceptional clarity and excellent light transmission. These materials are strong, rigid, and resistant to sharp blows. Their physical properties are not affected by outdoor weathering, and they do not become exceptionally brittle at low temperatures. Acrylics exhibit high tensile strength which is comparable to that of acetals (Fig. 14). However, the strength of acrylics falls off above 150°F. Modulus of elasticity is shown in Fig. 15.

ELECTRICAL PROPERTIES. Acrylics have no tendency towards arc tracking. This high arc resistance and excellent tracking characteristic makes acrylics a good choice for certain high-voltage applications such as circuit breakers. The dielectric strength is 450 to 500 volts/mil. Acrylic is one of the few plastics that exhibit an essentially linear decrease in dielectric constant and dissipation factor with increase in frequency, as shown in Fig. 17.

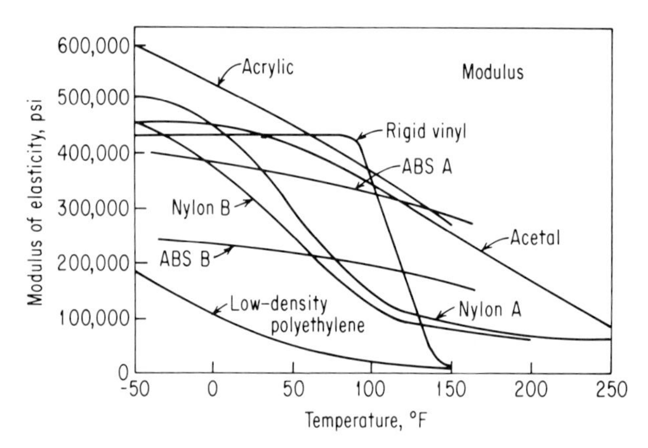

FIG. 15. Modulus of elasticity of several thermoplastic materials as a function of temperature.[13]

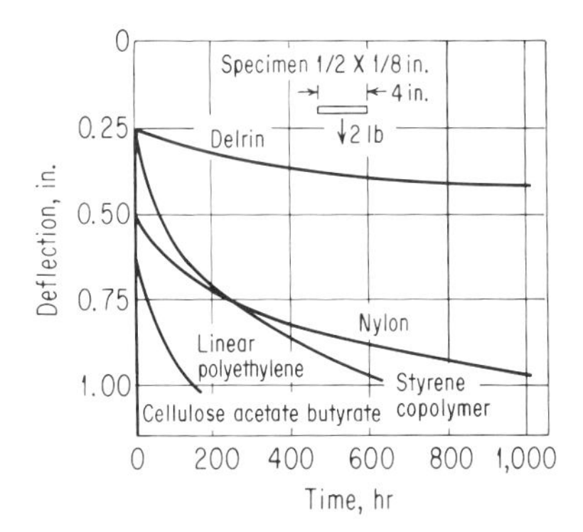

FIG. 16. Deflection of several thermoplastic materials as a function of time at 90% RH and 150°F.[14]

Cellulosics. Cellulosics have basically good electrical insulating properties, which are affected to varying degrees depending on the type and formulation. Cellulosics are not as resistant to extreme environments as many other thermoplastics which are more often used in electronic packaging applications. Toughness and transparency at moderate cost are the outstanding characteristics of cellulosics. Cellulose acetate is generally useful in the approximate temperature range of -25 to +170°F. The generally useful temperature range for cellulose propionate is —40 to +220°F, and for ethyl cellulose it is —40 to +200°F.

Chlorinated Polyether. Chlorinated polyether is better known for its use in piping

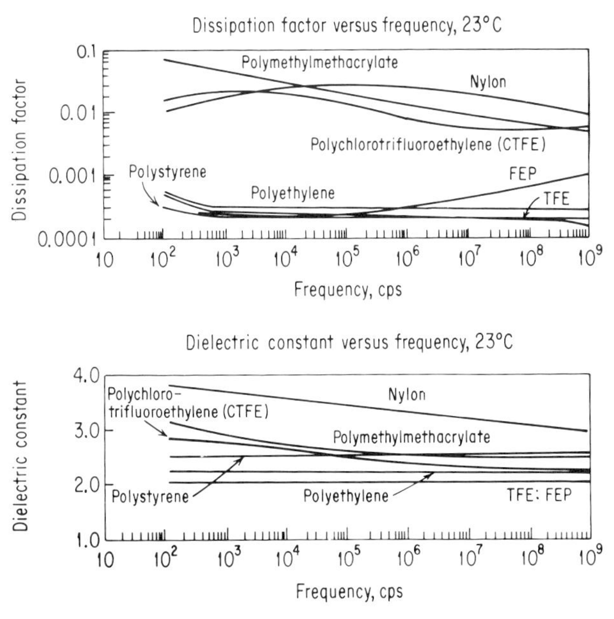

FIG. 17. Effect of frequency on dielectric constant and dissipation factor of several thermoplastic materials.[15]

TABLE 11. Application Information for Thermoplastics

Material	Major application considerations	Common available forms	Typical suppliers and trade names *
ABS (acrylonitrile-butadiene-styrene)	Extremely tough, with high impact resistance. Can be formulated over a wide range of hardness and toughness properties. Special grades available for plated surfaces with excellent pull-strength values. Good general electrical properties but not outstanding for any specific electric applications.	Blow moldings, extrusions, injection moldings, thermoformed parts, laminates, stock shapes, foam	Borg-Warner Corp. (Marbon Cycolac); Monsanto Co. (Lustran); Goodrich Chemical Co. (Abson)
Acetals	Outstanding mechanical strength, stiffness, and toughness properties, combined with excellent dimensional stability. Good electrical properties at most frequencies, which are little changed in humid environments up to 125°C.	Blow moldings, extrusions, injection moldings, stock shapes	Du Pont, Inc. (Delrin); Celanese Corp. (Celcon)
Acrylics (polymethyl-methacrylate)	Outstanding properties are crystal clarity, and resistance to outdoor weathering. Excellent resistance to arcing and electrical tracking.	Blow moldings, castings, extrusions, injection moldings, thermoformed parts, stock shapes, film, fiber	Du Pont, Inc. (Lucite); Rohm and Haas Co. (Plexiglas)
Cellulosics	There are several materials in the cellulosic family, such as cellulose acetate (CA), cellulose propionate (CAP), cellulose acetate butyrate (CAB), ethyl cellulose (EC), and cellulose nitrate (CN). Widely used plastics in general, but not outstanding for electronic applications.	Blow moldings, extrusions, injection moldings, thermoformed parts, film, fiber, stock shapes	Eastman Chemical Co. (Tenite); Dow Chemical Co. (Ethocel-EC); Celanese Corp. (Forticel-CAP)
Chlorinated polyethers	Good electrically, but most outstanding properties are corrosion resistance and good physical and thermal stability by thermoplastic standards.	Extrusions, injection moldings, stock shapes, film	Hercules Powder Co. (Penton)
Ethylene-vinyl acetates (EVA)	Excellent flexibility, toughness, clarity, and stress-crack resistance. Somewhat like a tough synthetic rubber or elastomer. Not widely used in electronics. Comparatively low resistance to heat and solvents.	....	U.S. Industrial Co. (Ultrathene); Du Pont, Inc. (Alathon); Union Carbide Corp. (Bakelite EVA)

Fluorocarbons *a.* Chlorotrifluoroethylene (CTFE)	Excellent electrical properties and relatively good mechanical properties. Somewhat more stiff than TFE and FEP fluorocarbons, but does have some cold flow. Widely used in electronics, but not quite so widely as TFE and FEP. Useful to about 400°F.	Extrusions, isostatic moldings, injection moldings, film, stock shapes	3M Co. (Kel-F); Allied Chemical Corp. (Plaskon CTFE)
b. Fluorinated ethylene propylene (FEP)	Very similar properties to those of TFE, except useful temperature limited to about 400°F. Easier to mold than TFE.	Extrusions, injection moldings, laminates, film	Du Pont, Inc. (Teflon FEP)
c. Polytetrafluoroethylene (TFE)	Electrically one of the most outstanding thermoplastic materials. Exhibits very low electrical losses, and very high electrical resistivity. Useful to over 500°F and to below −300°F. Excellent high-frequency dielectric. Among the best combinations of mechanical and electrical properties but relatively weak in cold-flow properties. Nearly inert chemically, as are most fluorocarbons. Very low coefficient of friction. Nonflammable.	Compression moldings, stock shapes, film	Du Pont, Inc. (Teflon TFE); Allied Chemical Corp. (Halon TFE)
d. Polyvinyl fluoride	Mostly used as a weatherable, architectural facing sheet. Not widely used in electronics.	Extrusions, injection moldings, laminates, film	Du Pont, Inc. (Tedlar)
e. Polyvinylidine fluoride (PVF_2)	One of the easiest of the fluorocarbons to process. Stiffer and more resistant to cold flow than TFE. Good electrically. Useful to about 300°F. A major electronic application is wire jacketing.	Extrusions, injection moldings, laminates, film	Pennsalt Chemicals Corp. (Kynar)
Ionomers	Excellent combination of toughness, solvent resistance, transparency, colorability, abrasion resistance, and adhesion. Based on ethylene-acrylic copolymers with ionic bonds. Not widely used in electronics.	Film, coatings, injection moldings	Du Pont, Inc. (Surlyn A); Union Carbide Corp. (Bakelite)
Nylons (polyamides)	Good general purpose for electrical and nonelectrical applications. Easily processed. Good mechanical strength, abrasion resistance, and low coefficient of friction. There are numerous types of nylons; nylon 6, nylon 6/6, and nylon 6/10 are most common. Some nylons have limited use due to moisture-absorption properties. Nylon 6/10 is best here.	Blow moldings, extrusions, injection moldings, laminates, rotational moldings, stock shapes, film, fiber	Du Pont, Inc. (Zytel); Allied Chemical Corp. (Plaskon); Union Carbide Corp. (Bakelite)
Parylenes (Polyparaxylylene)	Excellent dielectric properties and good dimensional stability. Low permeability to gases and moisture. Produced as a film on a substrate, from a vapor phase. Such vapor-phase polymerization is unique in polymer processing. Used primarily as thin films in capacitors and dielectric coatings. Numerous polymer modifications exist.	Film coatings	Union Carbide Corp. (Parylene)

TABLE 11. Application Information for Thermoplastics (*Continued*)

Material	Major application considerations	Common available forms	Typical suppliers and trade names*
Phenoxies	Tough, rigid, high-impact plastic. Has low mold shrinkage, good dimensional stability, and very low coefficient of expansion for a thermoplastic. Useful for electronic applications below about 175°F. Useful in adhesive formulations.	Blow moldings, extrusions, injection moldings, film	Union Carbide Corp. (Bakelite Phenoxy)
Polyallomers	Thermoplastic polymers produced from two monomers. Somewhat similar to polyethylene and polypropylene, but with better dimensional stability, stress-crack resistance, and surface hardness than high-density polyethylene. Electronic application areas similar to polyethylene and polypropylene. One of the lightest commercially available plastics.	Blow moldings, extrusions, injection moldings, film	Eastman Chemical Products, Inc. (Tenite)
Polyamide-imides and polyimides	Among the highest-temperature thermoplastics available, having useful operating temperatures between about 400°F and about 700°F or higher. Excellent electrical properties, good rigidity, and excellent thermal stability. Low coefficient of friction. Polyamide-imides and polyimides are chemically similar but not identical in all properties. They are difficult to process, but are available in molded and block forms, and also as films and resin solutions.	Films, coatings, molded and/or machined parts, resin solutions	Du Pont, Inc. (Vespel fabricated blocks, Kapton film, and Pyre-M.L. Resin)
Polycarbonates	Excellent dimensional stability, low water absorption, low creep, and outstanding impact-resistance thermoplastics. Good electrical properties for general electronic packaging application. Available in transparent grades.	Blow moldings, extrusions, injection moldings, thermoformed parts, stock shapes, film	Mobay Chemical Co. (Merlon); General Electric Co. (Lexan)
Polyethylenes and polypropylenes (polyolefins or polyalkenes)	Excellent electrical properties, especially low electrical losses. Tough and chemically resistant, but weak to varying degrees in creep and thermal resistance. There are three density grades of polyethylene: low (0.910–0.925), medium (0.926–0.940), and high (0.941–0.965). Thermal stability generally increases with density class. Polypropylenes are generally similar to polyethylenes, but offer about 50°F higher heat resistance.	Blow moldings, extrusions, injection molding, thermoformed parts, stock shapes, film, fiber, foam	Du Pont, Inc. (Alathon Polyethylene); U.S.I. Chemical Co. (Petrothene Polyethylene); Allied Chemical Corp. (Grex H.D. Polyethylene); Hercules Powder Co. (Hi-Fax

			H.D. Polyethylene); Hercules Powder Co. (Pro-Fax Polypropylene); Eastman Chemical Co. (Tenite Polyethylene and Polypropylene)
Polyethylene terephthalates	Among the toughest of plastic films with outstanding dielectric strength properties. Excellent fatigue and tear strength and resistance to acids, greases, oils, solvents. Good humidity resistance. Stable to 135–150°C.	Film, sheet, fiber	Du Pont, Inc. (Mylar)
Polyphenylene oxides (PPO)	Excellent electrical properties, especially loss properties to above 350°F, and over a wide frequency range. Good mechanical strength and toughness. A lower-cost grade (Noryl) exists, having somewhat similar properties to PPO, but with a 75–100°F reduction in heat resistance.	Extrusions, injection moldings, thermoformed parts, stock shapes, film	General Electric Co. (PPO and Noryl)
Polystyrenes	Excellent electrical properties, especially loss properties. Conventional polystyrene is temperature-limited, but high-temperature modifications exist, such as Rexolite or Polypenco crosslinked polystyrene, which are widely used in electronics, especially for high-frequency applications. Polystyrenes are also generally superior to fluorocarbons in resistance to most types of radiation.	Blow moldings, extrusions, injection moldings, rotational moldings, thermoformed parts, foam	Dow Chemical Co. (Styron); Monsanto Co. (Lustrex); American Enka Corp. (Rexolite); Polymer Corp. (Polypenco Q-200.5)
Polysulfones	Excellent electrical properties and mechanical properties to over 300°F. Good dimensional stability and high creep resistance. Flame-resistant and chemical-resistant. Outstanding in retention of properties upon prolonged heat aging, as compared to other tough thermoplastics.	Blow moldings, extrusions, injection mold thermoformed parts, stock shapes, film sheet	Union Carbide Corp. (Polysulfone)
Vinyls	Good low-cost, general-purpose thermoplastic materials, but not specifically outstanding electrical properties. Greatly influenced by plasticizers. Many variations available, including flexible and rigid types. Flexible vinyls, especially polyvinyl chloride (PVC), widely used for wire insulation and jacketing.	Blow moldings, extrusions, injection moldings, rotational moldings, film sheet	Diamond Alkali Co. (Diamond PVC); Goodyear Chemical Co. (Pliovic); Dow Chemical Co. (Saran)

* This listing is only a very small sampling of the many possible excellent suppliers. The listing is intended only to orient the nonchemical reader into plastic categories. No preferences are implied or intended.

TABLE 12. Typical Physical and Mechanical Properties of Thermoplastics [12]

Resin material	Impact strength notched Izod, ft-lb/in., ½" bar	Tensile strength, psi × 10^3	Tensile modulus, psi × 10^3	Elongation, %	Flexural strength, psi × 10^3	Compressive strength, psi × 10^3	Compressive modulus, psi × 10^3	Heat distortion temperature, °F, 264 psi	Heat resistance, continuous, °F
Acetal	1.1–1.4	8.8–10	400–410	12–75	13–14	18	410	230–255	185
ABS	1.3–10.0	4.5–8.5	200–450	5–200	5–13.5	5–11	120–200	180–245	160–235
Acrylic	0.3–0.4	8.7–11.0	350–450	3–6	14–17	14–17	350–430	167–198	130–195
Acrylic high impact	0.5–2.3	5.5–8	225–330	23–38	8.5–12	7–12	250–360	169–190	140–195
Cellulose acetate	0.5–5.6	2.3–8.1	....	10–70	2.2–11.5	2.0–10.9	....	111–209	140–175
Cellulose acetate butyrate	0.4–11	2.6–6.9	....	40–88	1.8–9.3	2.1–9.4	....	113–227	140–175
Cellulose propionate	0.7–10.7	1.8–7.3	....	30–100	2.8–11	2.4–9.6	....	119–250	140–175
Chlorinated polyether	0.4	6	160	60–160	5	....	130	185–210	250–275
Ethyl vinyl acetate	No break	20–40	3.0–15	500–1,500	....	....	....	....	120–170
Chlorotrifluoroethylene	3.5	6	150–190	60–190	8–10	6–12	180	160–170	390
Fluorinated ethylene propylene	No break	2–3.2	60–80	250–350	....	....	70	124	400
Polytetrafluoroethylene	No break	2–5	50–100	75–400	....	4–12	70–90	132	500
Nylon 6	0.9–4	9.5–12.4	200–450	25–300	9–16.6	4–11	347	150–175	250
Nylon 6/6	0.9–2	11.2–13.1	410–480	60–300	14.6	5–13	400	200	250
Nylon 6/10	0.8–3	7–8.5	160–280	50–300	10.5	4–6	....	145	220
Polyallomer	1.5–12	2.9–4.2	100–170	400–650	4–5	....	....	124–133	250
Polycarbonate	2–3	8–9.5	345	60–110	11–13	12.5	350	265–290	250
Polyethylene, low-density	No break	1–2.4	14–38	20–800	....	....	....	....	140–175
Polyethylene, medium density	No break	1.7–2.8	50–80	80–600	....	....	....	....	150–180
Polyethylene, high-density	0.5–23	2.8–5	75–200	10–800	1–4	0.8–3.6	50–110	110–125	180–225
Polyethylene, high molecular weight	>20	2.3–5.4	102	525–600	3.5	2.4	110	120	180–225
Polyimide	0.8–1.1	5–14.0	....	6–7	7–14	12–24	....	680	500–600
Polypropylene	0.5–15	3.2–5	150–650	3–700	4.5–8	6–10	....	140–205	250
Polystyrene	0.25–0.40	6–8.1	400–500	1.5–2.5	9–15	11.5–16	300–560	160–215	150–190
Polystyrene, high-impact	0.7–3.5	1.9–4	200–430	10–75	5.5–12.5	8–16	....	160–205	130–180
Polyurethane	No break	4.5–8	1–3.7	400–650	0.7–1	>20	85	....	190
Polyvinyl chloride (flexible)	Varied	1–4	....	100–450	....	....	....	....	150–175
Polyvinyl chloride (rigid)	0.4–22	6–9	200–600	5–40	8–15	10–11	300–400	140–175	160–165
Polyvinyl dichloride (rigid)	1.5–7.0	7.5–9.0	360–450	10–65	14.2–17	13–22	....	212–235	195–210
Styrene acrylonitrile (SAN)	0.3–0.50	8–12	500–600	1–3.2	17	15–17.5	650	200–218	170–210
Ionomer	5.7–14	3.5–5.5	28–40	300–450	....	....	....	....	140
Phenoxy	1.5–12	8–9.5	350–410	50–100	12–14.5	....	325	175–188	
Polyphenylene oxide	1.5–1.9	11	380	50–80	15	15	380	375	250
Polysulfone	1.3	10.2	360	50–100	15.4	15.4	370	345	300

TABLE 12. Typical Physical and Mechanical Properties of Thermoplastics (*Continued*)

Resin material	Coefficient thermal expansion, (in./in) (°C × 10^{-5})	Thermal conductivity, (cal)/(cm²)(sec)(°C)(cm) × 10^{-4}	Water absorption, 24 hr, %	Rockwell hardness	Flammability, (in./min) 0.125 in.	Specific gravity	Mold shrinkage, in./in.	Clarity	Price range per lb
Acetal	0.25	1.6	0.25	M94, R120	1.1	1.410–1.425	0.022	Translucent to opaque	$0.65
ABS	3–10.5	4–9	0.2–0.5	R80–120	1.0–2	1.01–1.07	0.003–0.007	Opaque	$0.33–0.43
Acrylic	....	1.4	0.3	M84–97	9–1.2	1.18–1.19	0.002–0.006	Transparent	$0.455–0.75
Acrylic high impact	6.5–10.5	4.0	0.2–0.3	M20–67	1.1–1.2	1.11–1.18	0.004–0.008	Translucent to opaque	$0.525–0.70
Cellulose acetate	8–18	4–8	1.7–4.4	R7–122	0–2	1.22–1.34	0.001–0.008	Transparent	$0.40
Cellulose acetate butyrate	11–17	4–8	0.9–2.2	R17–113	0.5–1.5	1.15–1.22	0.003–0.006	Transparent	$0.62
Cellulose propionate	11–16	4–8	1.2–2.8	R15–120	0.5–1.5	1.16–1.23	0.001–0.006	Transparent	$0.62
Chlorinated polyether	8	3.13	0.01	R100	Self-extinguishing	1.4	0.004–0.006	Semitranslucent to opaque	$2.50
Ethyl vinyl acetate	10–20	8	<0.01	R3–7	Slow burning	0.93–0.95	0.01–0.02	Transparent	$0.2775–0.3575
Chlorotrifluoroethylene	5–7	4–6	Nil	R85–112	Nil	2.09–2.14	0.010–0.015	Transparent to opaque	$4.70
Fluorinated ethylene propylene	8.3–10.5	5.9	<0.05	D55	Nonflammable	2.16	0.03–0.05	Transparent to opaque	$5.60–9.60
Polytetrafluoroethylene	5.5 (25–60°C)	6	0.01	D60–65	Nonflammable	2.13–2.18	0.02–0.06	Transparent to opaque	$3.25
Nylon 6	4.6–5.8	5.9	1.5	R107–119	Self-extinguishing	1.13–1.14	0.007–0.011	Transparent to opaque	$0.86–1.19
Nylon 6/6	8.1	5.8	1.3	R118–123	Self-extinguishing	1.13–1.15	0.007–0.015	Translucent to opaque	$0.84–0.875
Nylon 6/10	10	5.5	0.4	R111	Self-extinguishing	1.07–1.09	0.015	....	$1.26
Polyallomer	8–11	2–4	<0.05	R50–85	Slow burning	0.90–0.906	0.01–0.02	Transparent to opaque	$0.28
Polycarbonate	6.7–7	4.6	0.15	M70, R112	Self-extinguishing	1.2	0.005–0.007	Transparent	$0.90–3.05
Polyethylene, low-density	10–20	8	<0.05	R10	Slow burning	0.910–0.925	0.01–0.03	Transparent to opaque	$0.1525–0.29
Polyethylene, medium-density	10–20	8	<0.05	R15	Slow burning	0.926–0.940	0.01–0.035	Transparent to opaque	$0.17–0.235

TABLE 12. Typical Physical and Mechanical Properties of Thermoplastics (*Continued*)

Resin material	Coefficient thermal expansion, (in./in) (°C × 10^{-5})	Thermal conductivity, (cal)/(cm²)(sec)(°C)(cm) × 10^{-4}	Water absorption, 24 hr, %	Rockwell hardness	Flammability, (in./min) 0.125 in.	Specific gravity	Mold shrinkage, in./in.	Clarity	Price range per lb
Polyethylene, high-density	10–20	1.9–3.3	<0.01	R30–60	Slow burning	0.941–0.965	0.01–0.04	Translucent to opaque	$0.18–0.32
Polyethylene, high molecular weight	13	8	<0.01	R55	Slow burning	0.93–0.94	0.03	Translucent to opaque	$0.26–0.50
Polyimide	......	....	0.32	R85–95	......	1.43	......	Opaque	
Polypropylene	3.8–9	2.8–4	<0.01	R45–99	Slow burning to nonburning	0.90–1.24	0.008–0.025	Transparent to opaque	$0.19–0.55
Polystyrene	6–8	8	0.03–0.05	M65–80	0.5–2.5	1.05–1.06	0.002–0.006	Transparent	$0.145–0.245
Polystyrene, high-impact	6.5–8.5	1–3	0.05–0.10	M25–69	.5–2.5	1.04–1.06	0.003–0.005	Translucent to opaque	$0.16–0.27
Polyurethane	10–20	7.4	0.60–0.80	M26, R90	Slow to self-extinguishing	1.11–1.26	0.009	Translucent to opaque	$1.19–1.60
Polyvinyl chloride (flexible)	7–25	3–4	0.15–0.75	......	Self-extinguishing	1.15–1.80	0.002–0.004	Transparent to opaque	$0.16–0.455
Polyvinyl chloride (rigid)	5–10	3–5	0.07–0.40	R100–120	Self-extinguishing	1.33–1.58	......	Transparent to opaque	$0.21–0.42
Polyvinyl dichloride (rigid)	7–8	3–4	0.07–0.11	R118	Self-extinguishing	1.50–1.54	0.006–0.007	Translucent to opaque	$0.50–0.53
Styrene acrylonitrile (SAN)	7	3	0.23–0.28	M30–83	0.4–0.7	1.07–1.08	0.003–0.004	Transparent	$0.26–0.30
Ionomer	12–13	5.8	0.1–1.4	D60–65	0.9–1.1	0.94–0.96	0.001–0.005	Transparent	$0.47–0.49
Phenoxy	3.2–3.8	....	0.13	R113–118	Slow burning, self-extinguishing	1.17–1.34	0.003–0.004	Transparent to opaque	$0.75–1.00
Polyphenylene oxide	5.2	....	0.06	R120	Self-extinguishing	1.06	0.006–0.008	Transparent to opaque	$1.15
Polysulfone	3.1–10^{-5} in./(in.)(°F)	1.8 Btu/(hr)(ft)(°F)(in.)	0.22	M69, R120	Self-extinguishing	1.24–1.25	0.0076	Transparent to opaque	$1.00–1.25

TABLE 13. Typical Electrical Properties of Thermoplastics [12]

Resin material	Volume resistivity, ohm-cm	Dielectric constant, 60 cycles	Dielectric strength, ST,* ⅛-in. thickness, volts/mil	Dissipation or power factor, 60 cycles	Arc resistance, sec
Acetal	$1–10^{14}$	3.7–3.8	500	0.004–0.005	129
ABS	$10^{15}–10^{17}$	2.6–3.5	300–450	0.003–0.007	45–90
Acrylic	$> 10^{14}$	3.3–3.9	400	0.04–0.05	No tracking
Acrylic high impact	$10^{16}–10^{17}$	3.5–3.7	450–480	0.04–0.05	No tracking
Cellulose acetate	$10^{10}–10^{12}$	3.2–7.5	290–600	0.01–0.10	50–130
Cellulose acetate butyrate	$10^{10}–10^{12}$	3.2–6.4	250–400	0.01–0.04	
Cellulose propionate	$10^{12}–10^{16}$	3.3–4.2	300–450	0.01–0.05	170–190
Chlorinated polyether	1.5×10^{16}	3	400	0.01	
Ethyl vinyl acetate	1.5×10^{8}	3.16	525	0.003	
Chlorotrifluoroethylene	10^{18}	2.65	450	0.015	>360
Fluorinated ethylene propylene	$> 10^{18}$	2.1	500	0.0002	>165
Polytetrafluoroethylene	$> 10^{18}$	2.1	400	<0.0001	No tracking
Nylon 6	$10^{14}–10^{15}$	6.1	300–400	0.4–0.6	140
Nylon 6/6	$10^{14}–10^{15}$	3.6–4.0	300–400	0.014	140
Nylon 6/10	$10^{14}–10^{15}$	4.0–7.6	300–400	0.04–0.05	140
Pollyallomer	$> 10^{16}$	2.3	500–1,000	0.0001–0.0005	
Polycarbonate	6.1×10^{15}	2.97	410	0.0001–0.0005	10–120
Polyethylene, low-density	$10^{15}–10^{18}$	2.28	450–1,000	0.006	Melts
Polyethylene, medium-density	$10^{15}–10^{18}$	2.3	450–1,000	0.0001–0.0005	Melts
Polyethylene, high-density	$6 \times 10^{15}–10^{18}$	2.3	450–1,000	0.002–0.0003	Melts
Polyethylene, high molecular weight	$> 10^{16}$	2.3–2.6	500–710	0.0003	Melts
Polyimide	$10^{16}–10^{17}$	3.5	400	0.002–0.003	230
Polypropylene	$10^{15}–10^{17}$	2.1–2.7	450–650	0.005–0.0007	36–136
Polystyrene	$10^{17}–10^{21}$	2.5–2.65	500–700	0.0001–0.0005	60–100
Polystyrene, high-impact	$10^{13}–10^{17}$	2.5–3.5	500	0.003–0.005	60–90
Polyurethane	2×10^{11}	6–8	850–1,100	0.276	
Polyvinyl chloride (flexible)	$10^{11}–10^{15}$	5–9	300–1,000	0.08–0.15	
Polyvinyl chloride (rigid)	$10^{12}–10^{16}$	3.4	425–1,040	0.01–0.02	
Polyvinyl dichloride (rigid)	10^{15}	3.08	1,200–1,550	0.018–0.0208	
Styrene acrylonitrile (SAN)	10^{15}	2.8–3	400–500	0.006–0.008	100–150
Ionomer	$> 10^{16}$	2.4–2.5	1,000	0.001	
Phenoxy	$2.75–5 \times 10^{-5}$	4.1	404–520	0.0012–0.0009	70
Polyphenylene oxide	10^{17}	2.58	400–500	0.00035	75
Polysulfone	5×10^{16}	2.82	425	0.008–0.0056	122

* Short-time.

and hardware in the chemical processing industry than for its use in electronic packaging. Nevertheless, it does have properties that might be useful for certain electronic packaging applications. Tables 11 to 13 outline the major physical, electrical, and thermal properties of this material at room temperature.

MECHANICAL PROPERTIES. Mechanical creep properties are particularly good: the percentage creep remains below 4% after over 10,000 hr of a sustained 2,000-psi load at 75°F; a creep percentage of slightly over 4 percent exists after 5,000 hr of a sustained load of 1,000 psi at 280°F. For most cases up to the conditions just mentioned the percentage creep increases very little after the first 200 to 400 hr of sustained loading.

ELECTRICAL PROPERTIES. The electrical properties of dielectric constant and dissipation factor do not change much in the frequency range of 60 to 5 $\times$ 10^7 cycles, nor are they much affected by immersion up to 20 hr in boiling water. Regarding the effect of temperature, however, the dissipation factor does increase considerably in the range of 73 to 250°F. The 73°F properties are shown in Tables 11 to 13. Dielectric constant, however, increases only to about 3.5 at 250°F.

Fluorocarbons. Fluorocarbons are very important in electronics due to their excellent electrical properties, which are relatively unaffected by most extreme environments encountered in electronic packaging. Some fluorocarbon classes are more used than others, of course. The most widely used fluorocarbon, perhaps, is polytetrafluoroethylene, or TFE fluorocarbon. This was the original fluorocarbon, and is still known to many as Teflon. Correctly speaking, however, Teflon* is the trade name for Du Pont TFE and FEP fluorocarbons. Other fluorocarbons exist, however, and there are now other suppliers of TFE fluorocarbon. Table 11 clarifies the fluorocarbon terminology, and Tables 12 and 13 give the general properties of these materials. Due to the wide use of these materials, additional important discussion and data are given below.

MECHANICAL PROPERTIES. Like all plastics, especially thermoplastics, mechanical properties of fluorocarbons vary with temperature. Some of these properties and their relations to temperature will be given at this point. In addition, since there are countless data on the many properties, the references listed for this subject area are especially recommended.

TFE fluorocarbon, the most commonly used fluorocarbon in electronic packaging, is a semisoft plastic which exhibits some cold-flow properties. TFE and FEP are unique in their ability to retain a useful balance of flexibility and strength over a wide tem-

* Trademark, E. I. du Pont de Nemours & Co.

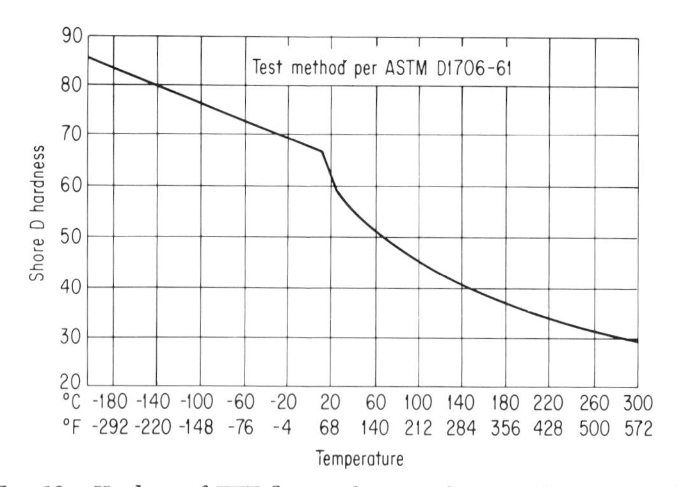

FIG. 18. Hardness of TFE fluorocarbon as a function of temperature.[16]

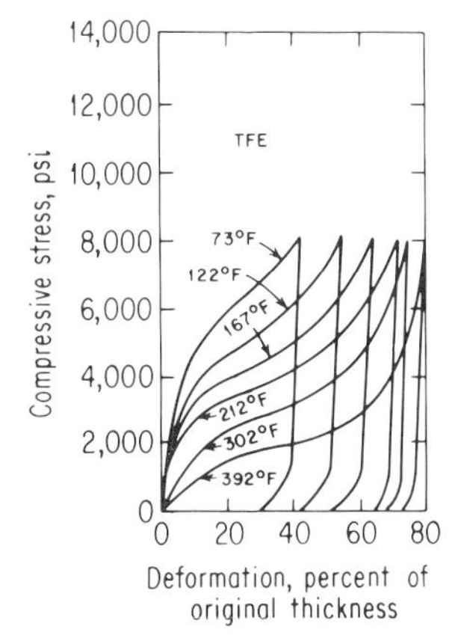

Fig. 19. Effect of temperature on deformation and recovery of TFE fluorocarbon.[17]

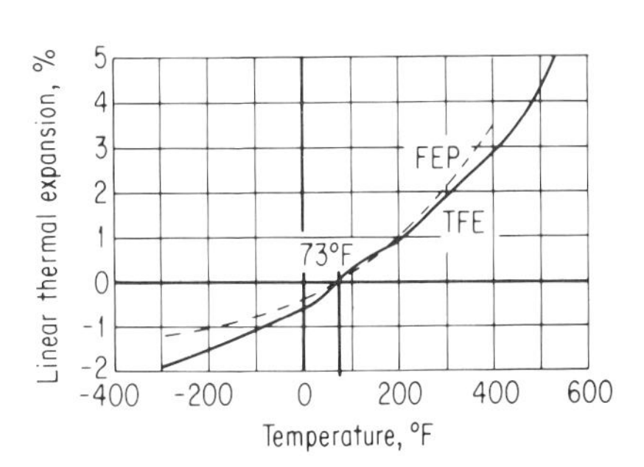

Fig. 20. Linear thermal expansion of TFE and FEP fluorocarbons as a function of temperature.[17]

perature range. The hardness of TFE as a function of temperature is shown in Fig. 18, and the deformation and permanent-set characteristics are shown in Fig. 19. Linear thermal expansion of TFE and FEP increases with temperature, as shown in Fig. 20. Stress-strain relationships for TFE, FEP, and CTFE at various temperatures are shown in Fig. 21, and the changes in various mechanical properties up to 500°F are given in Table 14. Mechanical properties of fluorocarbon plastics (as well as other thermoplastics) can be improved by use of various fillers. This will be further discussed later in this section. It might be mentioned that while thermosetting plastics normally always have fillers in the molded product, thermoplastics do not. However, fillers do improve many thermoplastic properties, and they are coming to be more widely used for critical applications.

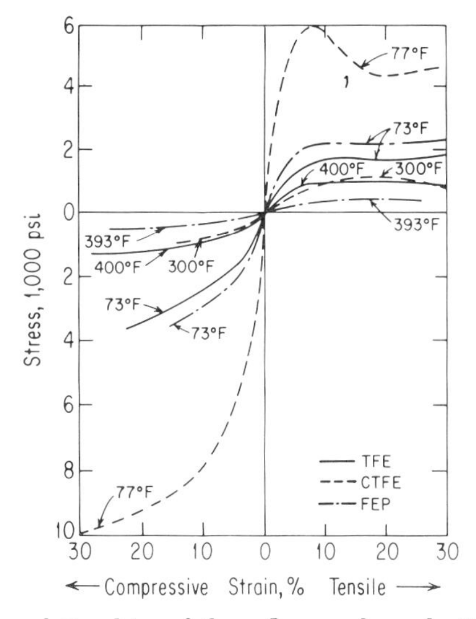

Fig. 21. Stress-strain relationships of three fluorocarbon plastics at several temperatures.[18]

TABLE 14. High-temperature Mechanical Properties of TFE Fluorocarbon Resins [14]

Property	Temperature *			
	72°F (23°C)	212°F (100°C)	400°F (204°C)	500°F (260°C)
Ultimate tensile strength, psi	3,850	2,500	1,500	900
Yield strength, psi	1,050 at 2%	400 at 2%	200 at 4%	200 at 8%
Ultimate elongation, %	300	>400	360	360
Flexural modulus of elasticity, psi	80,700	28,700	....	6,500
Flexural modulus of elasticity (35% glass-reinforced), psi	208,700	113,000	....	25,900
Compressive stress, 1% strain, psi	700	290	100	60
Compressive stress, 5% strain, psi	1,850	800	430	260
Compressive stress, 5% strain (15% glass-reinforced), psi	2,600	1,300	600	320
Linear expansion, in./(in.)(°F)	6.90×10^{-5}	6.90×10^{-5}	8.40×10^{-5}	9.70×10^{-5}
Linear expansion, %	0.0	0.9	2.7	4.3
Linear expansion (35% glass-reinforced), %	0.0	0.15	....	1.0
Coefficient of friction	0.04 over temperature range 80°F to 621°F for static loads			

* These values are typical of those for "Teflon" TFE-fluorocarbon resins in general. Variations may be expected from the values shown depending on the exact type of TFE resin used, methods of molding, and fabrication techniques employed.

ELECTRICAL PROPERTIES. Fluorocarbons have excellent electrical properties. TFE and FEP fluorocarbons in particular have low dielectric constants and dissipation factors which change little with temperature or frequency. This is shown in Figs. 17 and 22, which also show data for some other thermoplastics.

The dielectric strength of TFE and FEP resins is high, and does not vary with temperature and thermal aging. Initial dielectric strength is very high as measured by the ASTM short-time test (Fig. 23). As with any material, the value drops as the thickness of the specimen increases.

Life at high dielectric stresses is dependent on corona discharge (Fig. 24). The absence of corona, as in special wire constructions, permits very high voltage stress without damage to either TFE or FEP resins. Changes in relative humidity or physical stress imposed upon the material do not diminish life at these voltage stresses.

Surface arc resistance of TFE and FEP resins is high, and is not affected by heat aging. When these resins are subjected to a surface arc in air, they do not track or form a carbonized conducting path. When tested by the procedure of ASTM D495, they pass the maximum time of 300 sec without failure.

The unique nonstick surface of these resins helps reduce surface arc phenomena in two ways: (1) it helps prevent formation of surface contamination, thereby reducing the possibility of arcing; (2) if an arc is produced, the discharge frequently cleans the surface of the resin, increasing the time before another arc.

Volume resistivity ($> 10^{18}$ ohm-cm) and surface resistivity ($> 10^{16}$ ohms/sq) for both FEP and TFE resins are at the top of the measurable range. Neither resistivity is affected by heat aging or temperatures up to recommended service limits.

Ionomers. Outstanding advantages of this polymer class are combinations of toughness and transparency and combinations of transparency and solvent resistance. Ionomers have high melt strength for thermoforming and extrusion coating processes, and a broad processing-temperature range.

Limitations for the current products include low stiffness, susceptibility to creep,

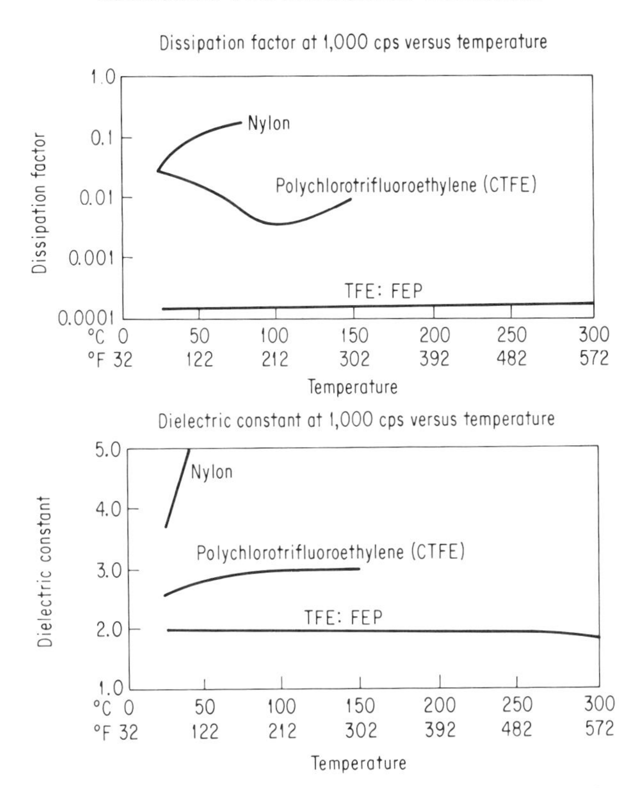

FIG. 22. Effect of temperature on dielectric constant and dissipation factor of fluorocarbons and nylon.[20]

low heat-distortion temperature, and poor ultraviolet resistance unless stabilizers are added.

Most ionomers are very transparent. In 60-mil sections, internal haze ranges from 5 to 25 percent. Light transmission ranges from 80 to 92 percent over the visible region, and in specific compositions, high transmittance extends into the ultraviolet region. The refractive index of ionomers is 1.51.

MECHANICAL PROPERTIES. Current commercial ionomers are nonrigid plastics but contain no plasticizers. Generally good flexibility, resilience, high elongation, and excellent impact strength typify the ionomer resins.

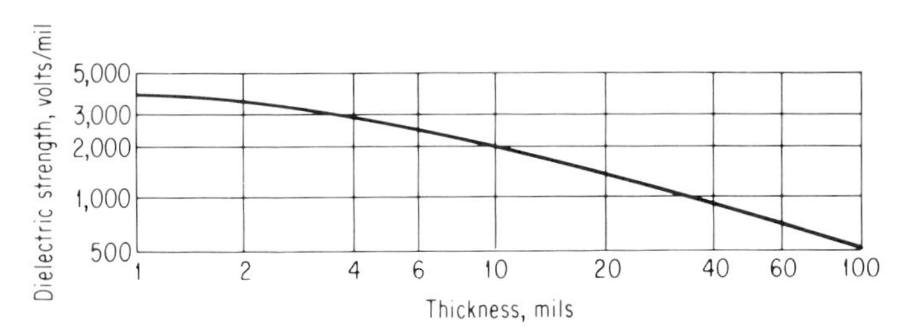

FIG. 23. Short-time dielectric strength of TFE fluorocarbon versus thickness of test sample.[21]

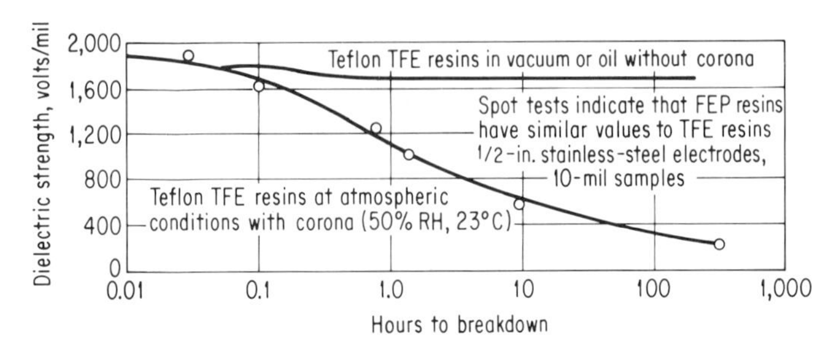

FIG. 24. Insulation life versus continuously applied voltage stress for TFE and FEP fluorocarbons.[21]

Deterioration of mechanical and optical properties occurs when ionomers are exposed to UV light and weather. Some grades are available with UV stabilizers that provide up to 1 year of outdoor exposure with no loss in mechanical properties. Formulations containing carbon black provide UV resistance equal to that of black polyethylene.

ELECTRICAL PROPERTIES. Most ionomers have good dielectric characteristics over a broad frequency range. The combination of these electrical properties, high melt strength, and abrasion resistance qualifies these materials for insulation and jacketing of wire and cable.

Nylons. Also known as polyamides, nylons are strong, tough thermoplastics that have good impact, tensile, and flexural strengths from freezing temperatures to 300°F, excellent low friction properties, and good electrical resistivities. They are not generally recommended for high-frequency, low-loss applications. Also, since all nylons absorb some moisture from environmental humidity, moisture-absorption characteristics must be considered when designing with these materials. They will absorb anywhere from 0.5 to nearly 2 percent moisture after 24-hr water immersion. In many cases, however, their moisture absorption properties do not have to be limitations in their use, especially for the lower moisture-absorption grades.

MECHANICAL PROPERTIES. Tensile strengths of nylons are compared with several other thermoplastic materials in Fig. 14. Figure 15 compares similar materials in modulus of elasticity. Deflection is shown in Fig. 16.

One special process exists wherein nylon parts are made by compressing and sintering, thereby creating parts with exceptional wear characteristics and dimensional stability. Various fillers such as molybdenum disulfide and graphite can be incorporated into nylon to give special low friction properties. Also, nylon can be reinforced with glass fibers to give it considerable additional strength. These variations are further discussed in a later part of this section dealing with reinforced thermoplastic materials.

Nylon, like all thermoplastic materials, exhibits some creep when subjected to stress.[22] Extent of creep depends on stress level, temperature, and time. A part that is subjected to long-time stress must be designed accordingly so that deformation with time is not excessive for the application and so that fracture will not occur. Typical values of creep versus time under various stress levels for nylon and polycarbonate are shown in Fig. 25. For nylon, most cold flow takes place during the first 24 hr. This is a useful checkpoint for testing parts under load.

The creep values shown in Fig. 25 are at relatively low stress levels compared to published tensile-stress values. However, these are typical working stresses. In metals, it is common practice to use yield stress or endurance limit and apply a safety factor to arrive at design stress. The same procedure does not apply for plastics. Stress levels must be kept at a fraction of published strength values to ensure proper performance. Published tensile strength of nylon, for example, is 12,000 psi. Stressing a part to 1,800 psi—a safety factor of 6.7—still results in 16 mils/in. cold flow.

ELECTRICAL PROPERTIES. The effect of frequency on the dielectric constant and dissipation factor of nylon is shown in Fig. 17. The effect of temperature on these two

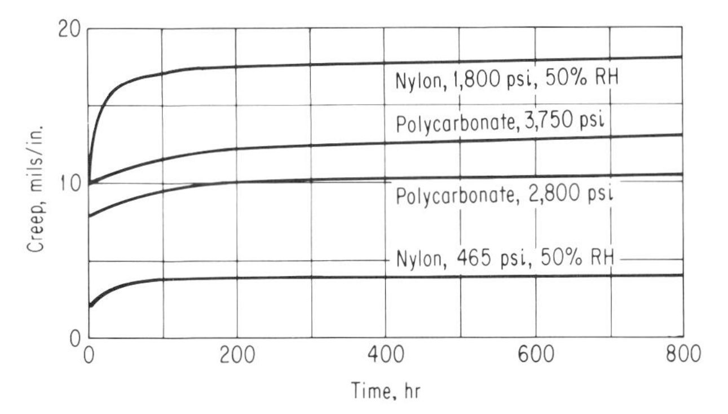

FIG. 25. Creep versus time at room temperature for nylon and polycarbonate.[22]

electrical properties is shown in Fig. 22. Nylons do not compare favorably with the good electrical thermoplastics such as fluorocarbons or polystyrene in these two electrical parameters. Nylons are good general-purpose electrical materials, however, as noted in Table 11.

Parylenes. Parylene is the generic name for members of a thermoplastic polymer series developed by Union Carbide's Plastics Division.[23] The basic member of the series, called Parylene N, is poly-para-xylylene, a completely linear, highly crystalline material.

Parylene C, the second member of the series, is produced from the same monomer modified only by the substitution of a chlorine atom for one of the hydrogen atoms on the ring. Parylene C, or poly(monochloro-p-xylylene), is somewhat less crystalline than Parylene N.

The parylenes are produced by vapor-phase deposition in a variety of forms. By effecting polymerization in an aqueous system, the material is obtained in particulate form. It can also be deposited onto a cold condenser, then stripped off as a free film, or it can be deposited onto the surface of an object as a continuous, adherent coating in thicknesses ranging from 0.2μ (about 0.008 mils) to 3 mils or more. Deposition rate is normally about 0.5μ per minute (about 0.02 mil). On cooled substrates, the deposition rate can be as high as 1.0 mil/min.

The material can be used at both elevated and cryogenic temperatures, although long-term (10-year) service in air is limited to 140 or 175°F, depending on the type of parylene involved. Its dimensional stability is said to be better than that of polycarbonate, and its barrier properties to most gases are reported superior to those of many other barrier films.

Parylenes are good dielectric materials, with Parylene N having a dielectric constant of 2.65 and a dissipation factor which increases from 0.0002 to 0.0006, increasing over the range of 60 cps to 1 megacycle. The chemical modification of Parylene C raises the dielectric constant to 2.9 to 3.1, and the dissipation factor to 0.012 to 0.020 over the above frequency range.

Phenoxies. The general properties of phenoxies are given in Tables 12 and 13. Phenoxy resins are high-molecular-weight thermoplastics that exhibit high rigidity, toughness, tensile strength, and elongation.

Principal advantages of phenoxy resins are excellent processability, low mold shrinkage (0.003 to 0.004 in./in.), and excellent dimensional stability and creep resistance. Regrind can be processed without loss of properties. The chief limitation is their heat distortion temperature of 188°F. Recommended maximum continuous-use temperature is approximately 170°F. Phenoxies are not widely used in electronic packaging.

Polyallomers, Polyethylenes, and Polypropylenes. The general properties of these three groups of thermoplastics, also sometimes known as polyolefins, are given in Tables 12 and 13.

Polyethylenes, which are among the best-known plastics, come in three main classifications based on density: low, medium, and high. These density ranges are 0.910 to 0.925, 0.925 to 0.940, and 0.940 to 0.965, respectively.

Polypropylene is chemically similar to polyethylene, but has somewhat better physical strength at a lower density. The density of polypropylene ranges from 0.900 to 0.915.

MECHANICAL PROPERTIES. All polyethylenes are relatively soft; hardness increases as density increases. Generally, the higher the density the better are the dimensional stability and physical properties, particularly as a function of temperature. Thermal stability of polyethylenes ranges from 190°F for the low-density material up to 250°F for the high-density material. Thermal stability of polypropylene ranges from 250°F to over 300°F.

The effect of temperature on the tensile length of polyethylene is shown in Fig. 14. Temperature effect on the modulus of elasticity is shown in Fig. 15, and deflection is shown in Fig. 16.

ELECTRICAL PROPERTIES. The electrical properties of polyethylenes and polypropylenes are similar and equally excellent over the operating temperature range of the materials. The electrical properties of dielectric constant and dissipation factor also remain very low over a wide range of frequencies (Fig. 17).

POLYALLOMERS. The general properties of polyallomers are given in Tables 12 and 13, and application information is given in Table 11. Polyallomers are crystalline thermoplastic polymers produced from two or more different monomers, such as propylene and ethylene, which would produce propylene-ethylene polyallomer. As can be seen, the monomers, or base chemical materials, are similar to those of polypropylene or polyethylene. Hence, as would be expected, many properties are similar, especially electrical properties. Polyallomers offer selective mechanical advantages such as rigidity, impact strength, and resistance to abrasion flexural fatigue, as in plastic hinges. Also, some selective processing advantages exist in properties such as flow characteristics, softening point, stress-crack resistance, and mold shrinkage.

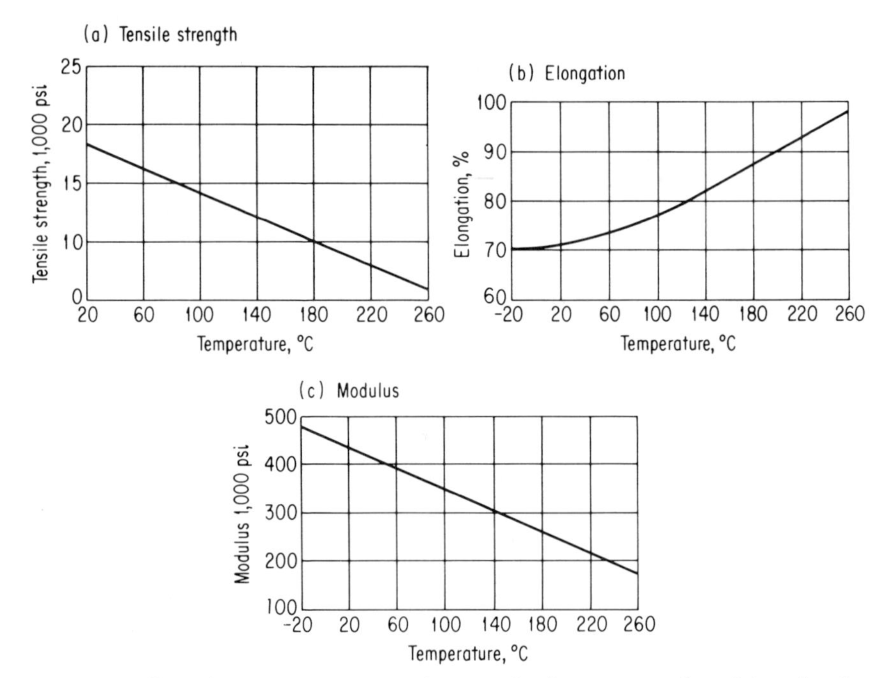

FIG. 26. Effect of temperature on tensile strength, elongation, and modulus of polyimides.[26]

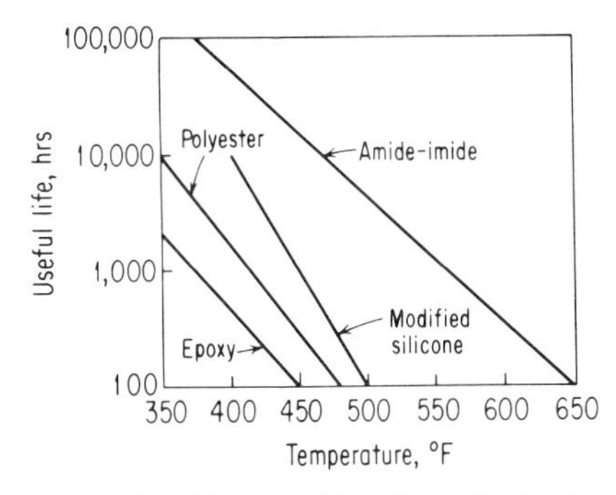

FIG. 27. Thermal life of amide-imide and polyimide wire enamels compared to other wire enamel materials.[27]

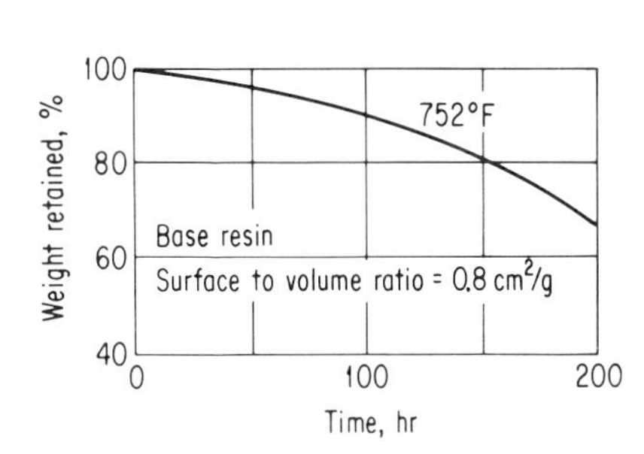

FIG. 28. Weight loss of polyimide resin at 752°F.[28]

CROSSLINKED POLYOLEFINS. While polyolefins have many outstanding characteristics, they, like all thermoplastics to some degree, tend to creep or cold-flow under the influence of temperature, load, and time. In order to improve this, and some other properties, considerable work has been done on developing crosslinked polyolefins—especially polyethylenes. The crosslinked polyethylenes offer thermal performance improvements of up to 25°C or more. Crosslinking has been achieved primarily by chemical means [24] and by ionizing radiation.[25] Products of both types are available. Radiation-crosslinked polyolefins have gained particular prominence in a heat-shrinkable form. This is achieved by crosslinking the extruded or molded polyolefin using high-energy electron beam radiation, heating the irradiated material above its crystalline melting point to a rubbery state, mechanically stretching to an expanded form (up to four or five times the original size), and cooling the stretched material. Upon further heating, the material will return to its original size, tightly shrinking onto the object around which it has been placed. Heat-shrinkable boots, jackets, tubing, etc., are widely used. Also, irradiated polyolefins (or irradiated polyalkenes) are important materials for certain wire- and cable-jacketing applications, which are further discussed in Chap. 2 of this handbook.

Polyamide-imides and Polyimides. These two somewhat related groups of plastics have some outstanding properties for electronic packaging applications. This is due to their combination of high-temperature stability, good electrical and mechanical properties which are generally stable to temperatures exceeding those for most plastics, and dimensional stability (low cold flow) in most environments. While their electrical properties are not so good as those of TFE fluorocarbons, polyamide-imides and polyimides do have very good electricals, and are better than TFE fluorocarbons in mechanical and dimensional stability properties. This provides advantages in many high-temperature electronic packaging applications. Polyamide-imides and polyimides, as well as fluorocarbons, are useful in the extreme environments of space and temperature—low negative temperatures as well as high positive temperatures.

MECHANICAL PROPERTIES. Although there are property variations dependent upon the particular material being used, the mechanical properties of polyimides are perhaps representative, and will be detailed further at this point. The effect of temperature on tensile strength, elongation, and modulus is shown in Fig. 26. The relative thermal stability of polyimides and several other materials, in the form of insulating coatings on magnet wire, is shown in Fig. 27. The weight loss of polyimides at an elevated temperature is shown in Fig. 28, and the dimensional changes of polyimides with temperature are shown in Fig. 29.

Molded polyimides have good wear and low friction properties, which can be improved even more by addition of fillers such as graphite.[28] The use of such fillers slightly degrades some other physical properties, however.

A potentially weak point of amide-imides and polyimides is moisture absorption

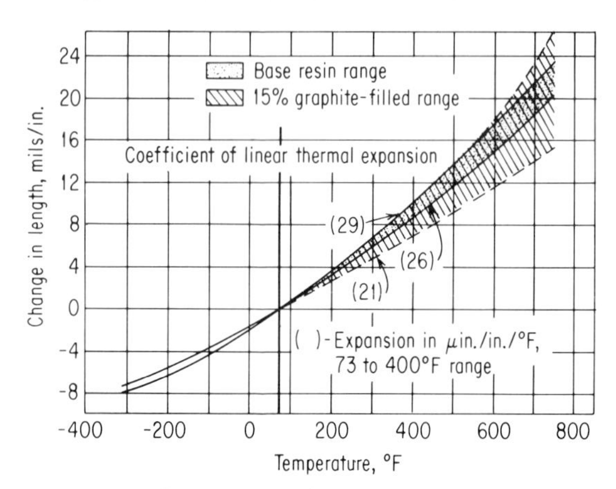

Fig. 29. Dimensional changes of polyimide resin versus temperature.[28]

in high humidities and attendant effects on physical and electrical properties. This varies with the material and condition, and should be investigated for a given application. Some amide-imides are chemically etchable, especially in caustic solutions, and this presents design opportunities for etchable amide-imide films in applications such as flat cabling.

ELECTRICAL PROPERTIES. The electrical properties of dielectric constant, dissipation factor, dielectric strength, and resistivity, as a function of temperature, are shown in Fig. 30. Both dielectric constant and dissipation factor begin increasing rapidly, at some frequencies, above 200 to 250°C. The dielectric strength decreases with increase

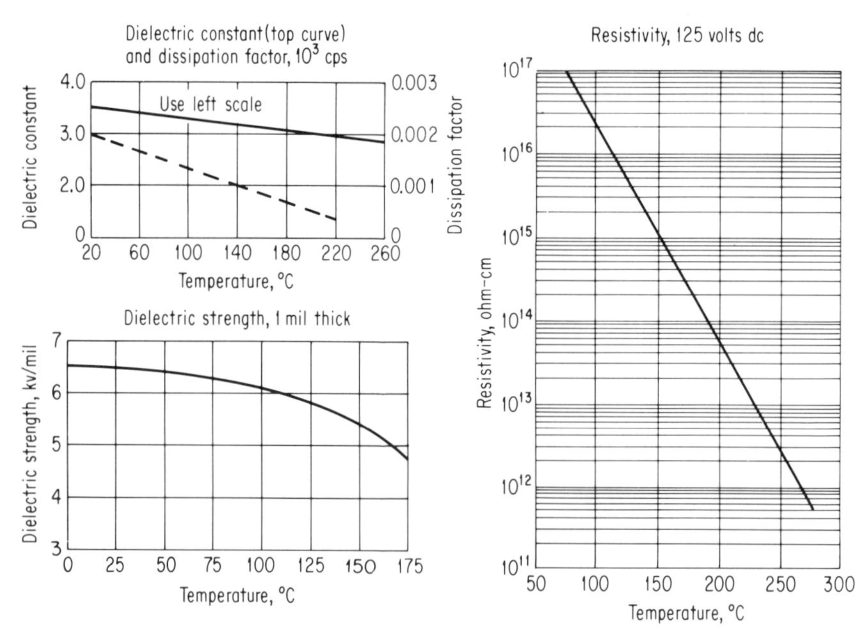

Fig. 30. Important electrical properties of polyimides as a function of temperature.[26]

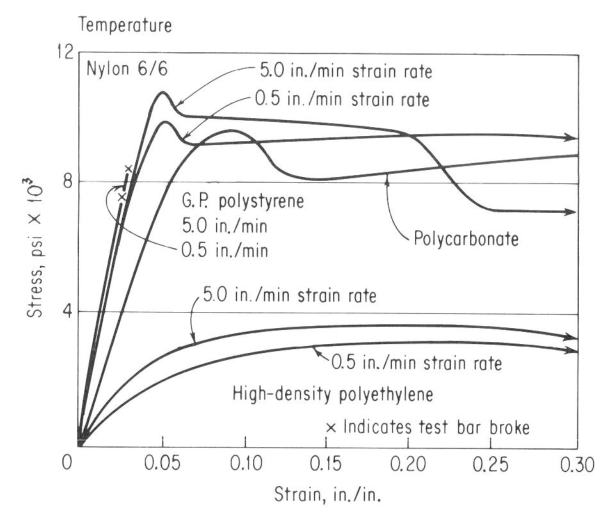

FIG. 31. Stress-strain curves for several thermoplastics.[30]

in thickness of section, being about 900 volts/mil at 30-mil thickness, and under 500 volts/mil at 120-mil thickness. This dielectric strength decrease with thickness increase is, of course, characteristic for insulating materials. Resistivity values are basically very good, but do decrease with temperature, as shown.

Polycarbonates. The general properties of polycarbonates are given in Tables 12 and 13. General application information is given in Table 11. Polycarbonates have an excellent combination of properties for use in electronic packaging applications, as described further below.

MECHANICAL PROPERTIES. Polycarbonates are especially good in impact strength, heat resistance under load, dimensional stability, creep resistance, outdoor weatherability, and low-temperature strength.[29] The Izod notched impact strength is about four times better than that of nylons or acetals. Much strength is maintained at very low negative temperatures. The useful temperature limits of polycarbonates are 250°F or higher, slightly exceeding those of acetals and exceeding those of nylons by a larger margin. Creep versus time for polycarbonate and nylon plastics is shown in Fig. 25. The stress-strain curves for several plastics are shown in Fig. 31, and the reduction of tensile yield strength and modulus with temperature is shown for polycarbonates in Fig. 32. The dimensional changes of several thermoplastics as a function of absorbed moisture are shown in Fig. 33. This is a design characteristic of plastics which is often overlooked.

ELECTRICAL PROPERTIES. The basic electrical properties of polycarbonates are very good. The power factor and dielectric constant, as a function of temperature, are shown in Fig. 34. It can be observed that there is no major change up to about 150°C. Both begin to increase beyond the 150 to 175°C temperature area, however. The dielectric constant value remains at about 3 up to 10^8 cycles or higher, whereas the power factor or loss factor does increase somewhat in this frequency range. Electrical properties remain relatively stable in high-humidity environments, which also helps in many design areas.

Polyphenylene Oxides. These materials represent another class of plastics which are very useful for electronic packaging. There are two groups of this family of plastics, namely, PPO and Noryl.* Noryl is the lower-cost member of this family, at a sacrifice in some properties. Polyphenylene oxides have good all-around mechanical and electrical properties, with some of the PPO materials being especially good in their

* Trademark, General Electric Company.

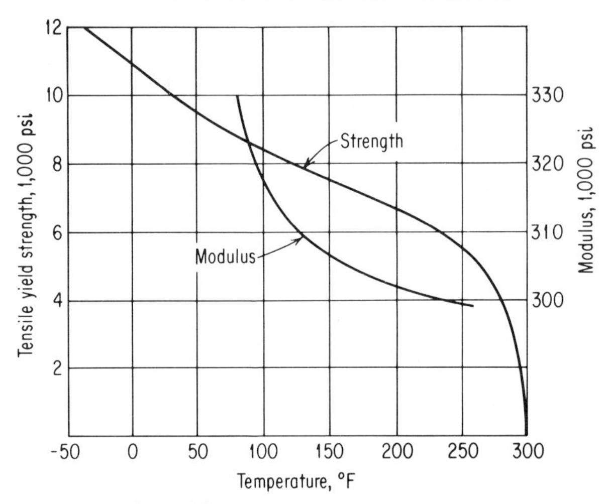

FIG. 32. Tensile yield strength and modulus of polycarbonate.[29]

combination of mechanical properties and high-frequency electrical properties. For application purposes in electronic packaging, polyphenylene oxides might be compared with acetals, polycarbonates, polysulfones, and nylons (polyamides). Some of the important comparative data are given below.

MECHANICAL PROPERTIES. The heat-deflection temperature of the polyphenylene oxides ranks high among the more rigid thermoplastics, as shown in Fig. 35. Polyphenylene oxides also rank well in tensile strength versus temperature (Fig. 36), tensile modulus (Fig. 37), and tensile creep (Fig. 38). Water absorption is also relatively low (Fig. 39), as in the attendant change in dimensions and weight associated with water absorption. Impact strength is not as good as that of polycarbonates at elevated temperatures, however, as is shown in Fig. 40.

ELECTRICAL PROPERTIES. Polyphenylene oxides have excellent electrical properties which remain relatively constant with frequency and temperature over their safe operating-temperature range. The effect of frequency on dissipation factor is shown in Fig. 41 and the effect of temperature on 60-cps frequency is shown in Fig. 42. Like PPO, Noryl also has a relatively flat dissipation factor with temperature, with typical 60-cps values being 0.0004 at 73°F, 0.0006 to 140°F, and 0.0008 at 220°F. The dielectric constant of the high-frequency grade of PPO, with frequency and temperature, is shown in Table 15. The dielectric constant of Noryl is also low and relatively stable up

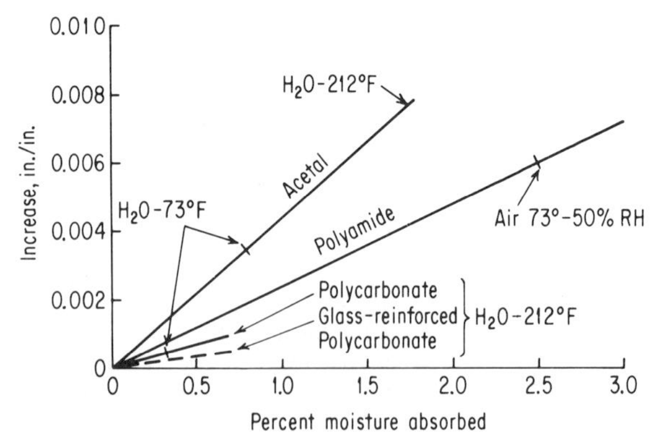

FIG. 33. Dimensional changes of several thermoplastics due to absorbed moisture.[29]

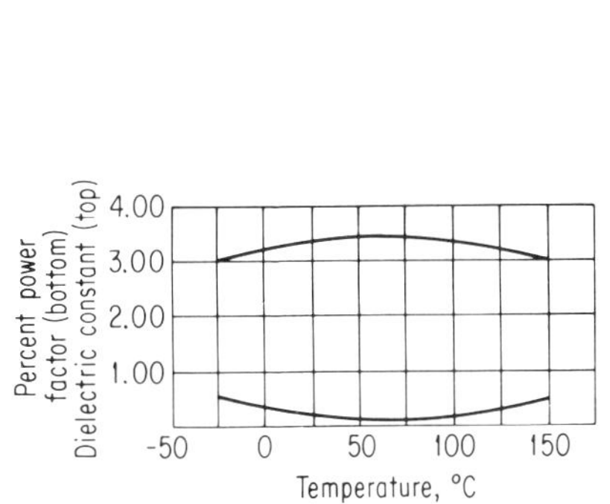

FIG. 34. Effect of temperature on power factor and dielectric constant of polycarbonates.[29]

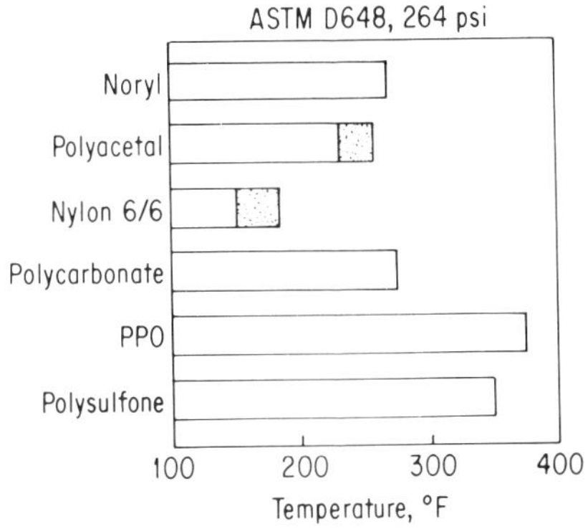

ASTM D648, 264 psi
Noryl
Polyacetal
Nylon 6/6
Polycarbonate
PPO
Polysulfone
100
200
300
400
Temperature, °F

FIG. 35. Heat-deflection temperature of several thermoplastics.[31]

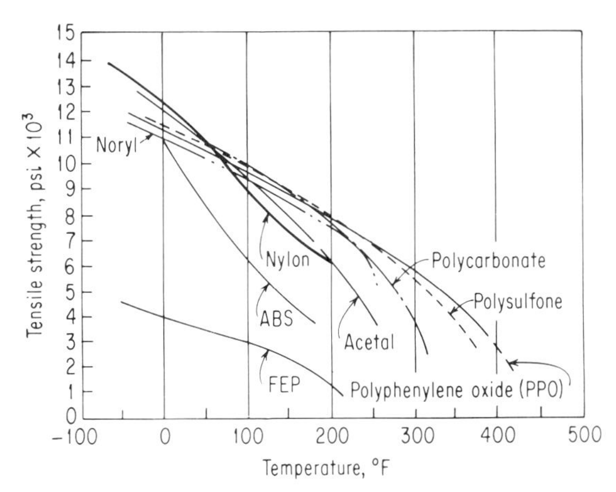

FIG. 36. Tensile strength versus temperature for several thermoplastics.[31]

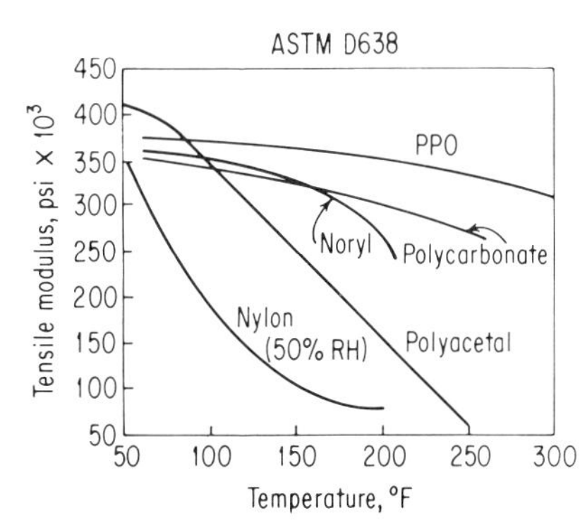

FIG. 37. Tensile modulus versus temperature for several thermoplastics.[31]

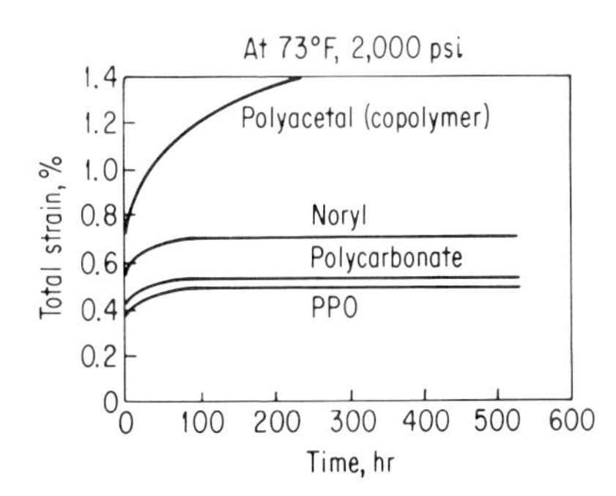

FIG. 38. Tensile creep for several thermoplastics at 73°F and 2,000 psi load.[31]

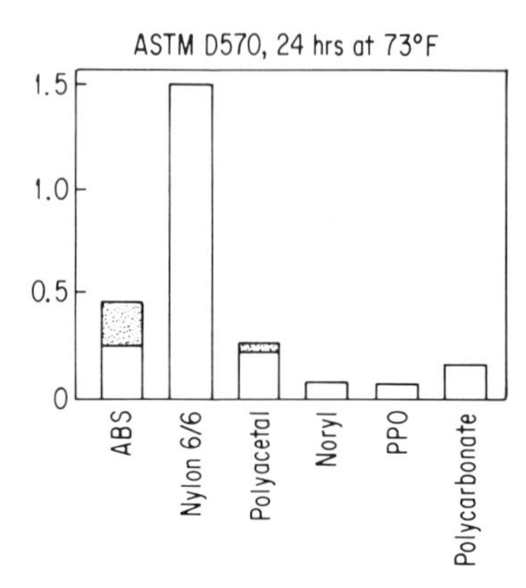

FIG. 39. Water absorption of several thermoplastics.[31]

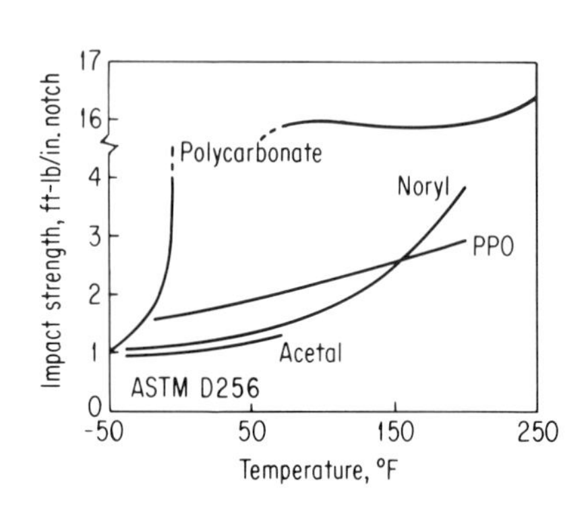

FIG. 40. Impact strength versus temperature for several thermoplastics.[31]

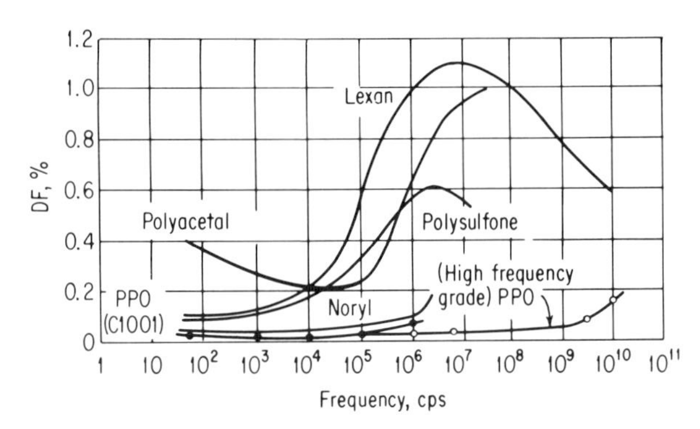

FIG. 41. Dissipation factor versus frequency for several thermoplastics.[32]

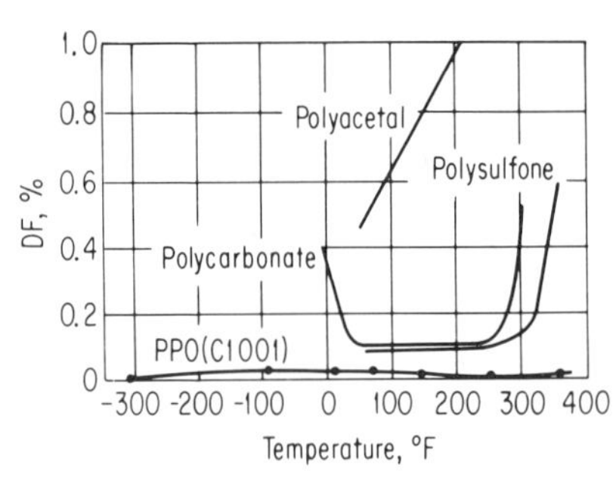

FIG. 42. Dissipation factor at 60 cps versus temperature for several thermoplastics.[32]

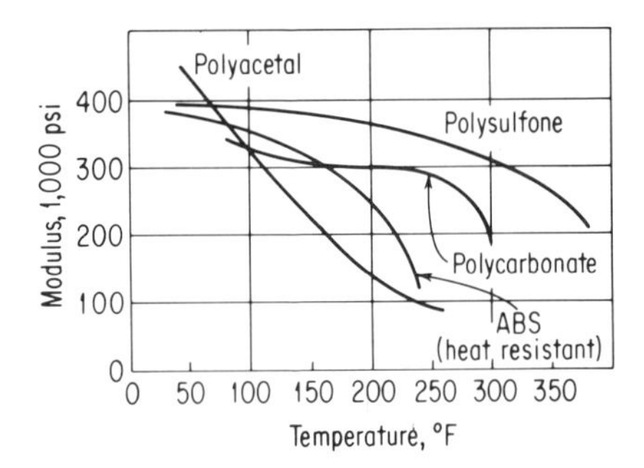

FIG. 43. Flexural modulus versus temperature for several thermoplastics.[33]

TABLE 15. Dielectric Constant of PPO Polyphenylene Oxide at Various Temperatures and Frequencies [32]

Frequency, cps	*Dielectric constant*
60	2.56
10^3	2.55
10^4	2.55
10^6	2.55
10^9	2.59
Temperature, °F	***Dielectric constant at 60 cps***
360	2.52
190	2.54
73	2.56
20	2.55
–320	2.57

to the 200 to 225°F range. Dielectric strength of both polyphenylene oxide classes is 500 to 550 volts/mil in 1/8-in. sections. Electrical properties of both materials are also relatively unaffected by humidity.

Polystyrenes. Polystyrenes represent an important class of thermoplastic materials in the electronics industry because of their very low electrical losses. Mechanical properties are adequate within operating-temperature limits, but polystyrenes are temperature-limited with normal temperature capabilities below 200°F. Polystyrenes can, however, be crosslinked to produce a higher-temperature material (see Table 11).

MECHANICAL PROPERTIES. The effect of stress and time on the deflection of polystyrenes is compared with that of some of the other thermoplastics in Fig. 16. The strength properties of polystyrene drop off rapidly at 100°C or perhaps lower.

ELECTRICAL PROPERTIES. The exceptionally low dissipation factor and dielectric constant for polystyrene as a function of frequency are shown in Fig. 17. The dielectric constant of some polystyrenes increases rapidly above 10^9 to 10^{10} cps. This low dissipation factor coupled with the relative rigidity of polystyrene compared to polyethylene and TFE gives polystyrene advantages in many electronic packaging applications requiring material hardness and extremely low electrical losses, particularly at high frequencies. The crosslinked polystyrenes are especially useful here.

The dielectric strength of polystyrenes is excellent, and their resistivity properties are outstanding (Fig. 11). These properties, coupled with the other above-mentioned excellent properties, make polystyrenes most useful in high-frequency electronic packaging applications which are not temperature-limited.

Polysulfones. Polysulfones are very useful thermoplastics for electronic packaging, having excellent strength vs. temperature properties, good electrical properties (though not outstanding for high frequency), and outstanding strength retention over long periods of aging up to 300°F or over. Generalized data and information are given in Tables 11 to 13. Important properties will be detailed below.

MECHANICAL PROPERTIES. The heat-deflection temperatures of polysulfones are compared with those of several other thermoplastics in Fig. 35, flexural modulus vs. temperature with those of other thermoplastics in Fig. 43. The tensile strength properties of polysulfones are generally similar to those of polycarbonates and polyphenylene oxides, which are shown in Fig. 36, and the dimensional changes of polysulfones due to absorbed moisture are also similar to those of polycarbonates, shown in Fig. 33. Creep versus time is low, as shown in Fig. 44. The retention of

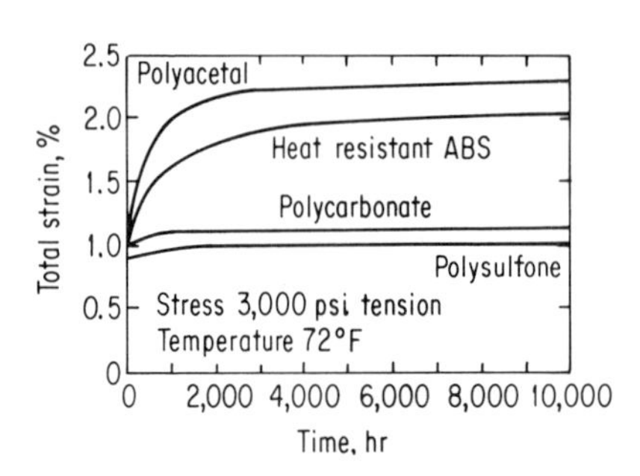

FIG. 44. Creep behavior of several thermoplastics at 72°F in air and under 3,000 psi tensile stress.[33]

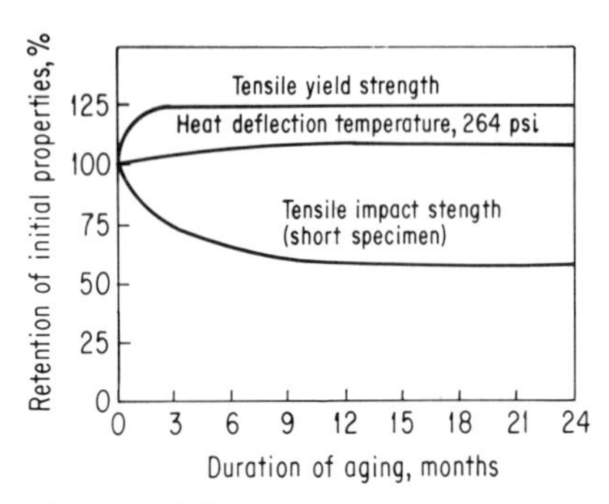

FIG. 45. Effect of heat aging at 300°F on three properties of polysulfones.[33] (All property tests made at room temperature on ⅛-in.-thick specimens.)

strength properties upon prolonged heat aging, mentioned above as perhaps the outstanding feature of polysulfones, is shown in Fig. 45.

ELECTRICAL PROPERTIES. The dissipation factor versus frequency is shown in Fig. 41, and the dissipation factor versus temperature, at 60 cps, is shown in Fig. 42. The dielectric constant of polysulfones is approximately 3.1 up to 10^6 cps, and decreases slightly at 10^7 cps. The other important electrical properties of polysulfones are also good, and are satisfactory for most electronic packaging applications. The electrical properties of polysulfones are maintained to approximately 90 percent of their initial values after 1 year or more of exposure at 300°F. Also, the basic electrical properties are generally stable up to about 350°F, and under exposure to water or high humidity.

Vinyls. Vinyls are good general-purpose electrical insulating materials, but are not outstanding from the electronic packaging viewpoint. Perhaps their widest applications in electronic packages are as hookup wire jacketing and as sleeving and tubing. There are many grades and types of vinyls, among which are some special electrical grades, which should be considered for any electronic packaging applications. Due to the many variations available, consultation with leading supplies is especially important. In addition to various basic vinyl classifications, vinyls may be rigid, flexible, or foamed. Further, they may be filled in many ways, alloyed with other plastics, plasticized with various plasticizers. Some vinyls are particularly outstanding in their resistance to corrosive chemicals, such as plating baths. A convenient-to-use form of vinyls is the polyvinylchloride dispersion, available in liquid form, which solidifies upon application of heat. Hence, these dispersions can be used for casting, potting, and dip-coating-type applications—somewhat as RTV silicones are used, though with considerably different properties, of course. Properties favor silicones, and costs favor the vinyl dispersions.

A good general reference on vinyls is the article by Bulkley.[34] Wire insulation and tubing are covered in another chapter of this handbook.

Elastomers and Rubbers. At one time, when natural rubber and a few synthetic rubbers constituted the primary type of rubberlike materials in use, the term rubber was predominantly used to describe this group of materials. However, with developments in the field of polymer chemistry, numerous other rubberlike materials have been developed whose chemical composition bears no resemblance to the chemical composition of the natural or the early synthetic rubbers. Also, these newer materials often exhibit vast improvements over the early rubbers in many respects while still being basically rubberlike or elastic in character. Therefore, the term elastomer came to be used to encompass the broadened range of rubberlike materials. The ASTM definition of an elastomer is: "A material which at room temperature can be stretched repeatedly to at least twice its original length and upon immediate release of the stress will return with force to its approximate original length." Currently, there are over a dozen recognized classes of elastomers, a number of which are useful in electronic assemblies.

Nature of Elastomers. Although elastomers are perhaps not so widely used in electronic packaging as are thermoplastic or thermosetting plastic materials, elastomers do have application in many instances. They are often used for cushioning materials, for vibration damping, for gasketing and sealing, and in many other applications where rubberlike properties coupled with some selected combination of mechanical, electrical, or fluid-resistant properties are required.

Elastomers are sometimes known by their popular name, sometimes by their chemical name, and sometimes by the ASTM standards designation or some other previously used symbol. Table 16 is a cross reference of these identifications. Also commonly used

TABLE 16. Designations and Application Information for Elastomers[95]

Elastomer designation		Chemical type	Major application considerations
ASTM D-1418	Trade name or common name		
NR	Natural rubber	Natural polyisoprene	Excellent physical properties; good resistance to cutting, gouging, and abrasion; low heat, ozone, and oil resistance. The best electrical grades are excellent in most electrical properties at room temperature.
IR	Synthetic natural	Synthetic polyisoprene	Same general properties as natural rubber; requires less mastication in processing than natural rubber.
CR	Neoprene	Chloroprene	Excellent ozone, heat, and weathering resistance; good oil resistance; excellent flame resistance. Not so good electrically as NR or IR. However, the combination of generally good electricals for jacketing application, coupled with all of the other good properties, gives this elastomer broad use for electrical wire and cable jackets.
SBR	GRS, Buna S	Styrene-butadiene	Good physical properties; excellent abrasion resistance; not oil-, ozone-, or weather-resistant. Electrical properties generally good but not specifically outstanding in any area.
NBR	Buna N, Nitrile	Acrylonitrile-butadiene	Excellent resistance to vegetable, animal, and petroleum oils; poor low-temperature resistance. Electrical properties not outstanding; probably degraded by molecular polarity of acrylonitrile constituent.
IIR	Butyl	Isobutylene-isoprene	Excellent weathering resistance; low permeability to gases; good resistance to ozone and aging; low tensile strength and resilience. Electrical properties generally good but not outstanding in any area.
IIR	Chlorobutyl	Chloro-isobutylene-isoprene	Same general properties as butyl.
BR	Cis-4	Polybutadiene	Excellent abrasion resistance and high resilience; used principally as a blend in other rubbers.

TABLE 16. Designations and Application Information for Elastomers (*Continued*)

Elastomer designation		Chemical type	Major application considerations
ASTM D-1418	Trade name or common name		
	Thiokol (PS) (Thiokol Chemical)	Polysulfide	Outstanding solvent resistance; widely used for potting of electrical connectors.
R	EPR	Ethylene propylene	Good aging, abrasion, and heat resistance; not oil-resistant. Good general-purpose electrical properties.
R	EPT	Ethyl propylene terpolymer	Good aging, abrasion, and heat resistance; not oil-resistant. Good general-purpose electrical properties.
CSM	Hypalon (HYP) (Du Pont)	Chlorosulfonated polyethylene	Excellent ozone, weathering, and acid resistance; fair oil resistance; poor low-temperature resistance. Not outstanding electrically, but has some special-application uses based on other properties.
SIL	Silicone	Polysiloxane	Excellent high- and low-temperature resistance; low strength; high compression set. Among the best electrical properties in the elastomer grouping. Especially good stability of dielectric constant and dissipation factor at elevated temperatures.
	Urethane (PU)	Polyurethane diisocyanate	Exceptional abrasion, cut, and tear resistance; high modulus and hardness; poor moist-heat resistance. Generally good general-purpose electrical properties. Some special high-quality electrical grades available from formulators.
	Viton (FLU) (Du Pont)	Fluorinated hydrocarbon	Excellent high-temperature resistance, particularly in air and oil. Not outstanding electrically.
ABR	Acrylics	Polyacrylate	Excellent heat, oil, and ozone resistance; poor water resistance. Not outstanding for or widely used in electrical applications.

are the ASTM-SAE application classifications for various elastomers, which are defined as follows:

Type R—non-oil-resistant
Type S—resistant to petroleum chemicals
  Class SA—very low volume swell
  Class SB—low volume swell
  Class SC—medium volume swell
Type T—temperature-resistant
  Class TA—resistant to high and low temperatures
  Class TB—resistant to hot air and oil

While elastomeric properties may be indicated in a manner similar to that commonly used for plastics, there are some properties of particular value that are widely used in the identification of these materials. One of these is durometer. Durometer, or hardness, is often the primary description used in identifying the characteristics of an elastomer. The hardness of an elastomer is related to its degree of vulcanization or cure and to the presence or absence of filler materials. The Shore durometer is widely

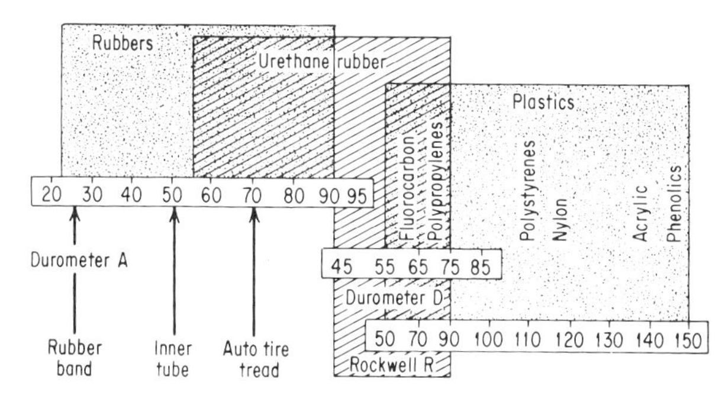

FIG. 46. Hardness comparison chart for various rubbers and plastics.

used for this measurement, and the shortened term durometer is often used to specify hardness: the softer the rubber, the lower the hardness or durometer. A comparative hardness guide is shown in Fig. 46.

Compression set is another property often given for elastomeric materials. Compression set, according to ASTM test D-395, is the residual decrease in the thickness of a test specimen which is observed after 30 min of rest following the removal of a specified compressive loading applied under established conditions of time and temperature.

Elongation and tensile strength are also often used to describe elastomers. Elongation is the amount the material is stretched at the moment of rupture. The amount of force necessary to rupture the material is the tensile strength. Tensile strength is normally expressed in terms of the original cross section of the specimen tested.

As with plastic materials, a broad range of properties can often be obtained with any given type of elastomer by compounding or modifying the basic material. Some representative data ranges for the various classes can nevertheless be shown, as in Table 17.

Design Considerations for Elastomers and Rubbers. Each elastomeric material has some basic characteristics which are important for general design purposes, as well as some distinct advantages and disadvantages. These are briefly described below for the

TABLE 17. Important Property Comparisons for Rubbers and Elastomers[35]

Material	Dielectric constant, 10^6 cps	Power factor $\times 10^2$, 10^6 cps	Volume resistivity, ohm-cm	Surface resistivity, ohms	Dielectric strength, volts/mil
Natural rubber	2.7–5	0.05–0.2	10^{15}–10^{17}	10^{14}–10^{15}	450–600
Styrene-butadiene rubber	2.8–4.2	0.5–3.5	10^{14}–10^{16}	10^{13}–10^{14}	450–600
Acrylonitrile-butadiene rubber	3.9–10.0	3–5	10^{12}–10^{15}	10^{12}–10^{15}	400–500
Butyl rubber	2.1–4.0	0.3–8.0	10^{14}–10^{16}	10^{13}–10^{14}	400–800
Polychloroprene	7.5–14.0	1.0–6.0	10^{11}–10^{12}	10^{11}–10^{12}	100–500
Polysulfide polymer*	7.0–9.5	0.1–0.5	10^{11}–10^{12}	......	250–325
Silicone	2.8–7.0	0.10–1.0	10^{13}–10^{17}	10^{13}	300–700
Chlorosulfonated polyethylene†	5.0–11.0	2.0–9.0	10^{13}–10^{17}	10^{14}	400–600
Polyvinylidene fluoride-*co*-hexafluoropropylene‡	10.0–18.0	3.0–4.0	10^{13}	......	250–700
Polyurethane§	5.0–8.0	3.0–6.0	10^{10}–10^{11}	......	450–500
Ethylene propylene terpolymer¶	3.2–3.4	0.6–0.8	10^{15}–10^{17}	......	700–900

* Thiokol, Thiokol Corp.
† Hypalon, Du Pont, Inc.
‡ Viton A, Du Pont, Inc.
§ Adiprene, Du Pont, Inc.
¶ Nordel, Du Pont, Inc.

important elastomer classes. Stalwart Rubber Company[35] is credited for many of the important points of this discussion.

Butyl. Butyl is best known for its outstanding air-retention qualities in products such as tire inner tubes. This impermeability extends to other gases, including nitrogen and oxygen.

In addition to its excellent impermeability, butyl provides outstanding resistance to ozone, oxidation, weathering, and acids and many other chemicals. Its superior resistance to abrasion and excellent flexing as well as dampening characteristics make it ideal for rugged industrial service.

Strict quality-control measures must be followed in compounding with butyl stocks to assure both dimensional and physical stability. Creep and cold-flow characteristics are also limiting factors.

Typical uses of butyl rubber parts include weather stripping, bumpers, shock absorbers, lining for bowling pits, and chemical tubing, as well as tubing that handles hot fluids.

Chlorobutyls. These materials have excellent heat resistance up to 400°F. They have good tear strength, low compression set, good flex life, and low permeability. Further, they have good resistance to aging, oxidation, and ozone, as well as acidic and alkaline materials and oxygenated solvents.

EP Rubber. Two basic types of EP rubber are available in commercial quantities: peroxide-cured ethylene-propylene and sulfur-cured ethylene propylene-terpolymer. Generally, comparable to NR and IR rubbers, EP rubbers promise attractive price advantages conducive to a growing volume of custom-engineered parts.

EP rubber is ideal for outdoor applications because of its excellent resistance to ozone, oxidants, and severe weather conditions. Other outstanding characteristics of EP rubber include excellent color stability, odorfree qualities, high heat resistance, and dielectric qualities. It offers many of the advantages of neoprene at a cost comparable to that of polyisoprene. Processed and cured with conventional rubber machinery, EP stocks are easier to fabricate and control than some of the newer elastomers.

EP rubber is slightly lower than both natural rubber and polyisoprene in resilience and tensile strength. It is not recommended for applications involving petroleum derivatives.

Low cost plus high performance make this elastomer ideal for a wide variety of molded and extruded rubber parts for the appliance and automotive industries: weather stripping, boots, seals, grommets, dust covers, sleeves, and mounts, to mention only a few.

Fluoroelastomers. One of the most promising members of the elastomer family, fluoroelastomers resist a wide variety of corrosive fluids at elevated temperatures while retaining their mechanical properties.

Resistance to solvents, acids, bases, fuels, oils, and hydraulic fluids, plus outstanding performance at elevated temperatures ranging from +450 to +600°F, are the most significant advantages of fluoroelastomers. Other important features include resistance to weathering, ozone, oxygen, and flame; good tensile strength; resilience; and low compression set. Unlike many of their counterparts, fluoroelastomers retain their basic properties at extremely high temperatures. Their low-temperature range is approximately 60°F.

Although fluoroelastomers are still undergoing extensive development, their excellent mechanical stability under severe operating conditions makes them an ideal component from any standpoint except price.

Hypalons. Like most synthetics, Hypalon* is superior to natural rubber in resistance to temperature, oil, and corona, and also has virtually total resistance to the effects of ozone.

Outstanding resistance to most chemicals, heat, and oil is among its most important characteristics. Further, it is flame-resistant, and offers excellent color stability and weather and abrasion resistance. Low moisture absorption, good dielectric qualities, and high abrasion resistance are among its other features.

* Trademark, E. I. du Pont de Nemours & Co.

Poor compression-set qualities limit the applications of Hypalon. Tensile strength is not particularly high, and resilience is lower than that of either natural rubber or polyisoprene (particularly at low temperatures).

Natural Rubbers. Only one synthetic—polyisoprene—comes close to providing the perfect balance and variety of properties produced from the Hevea Brasiliensis tree of the Far East.

High resilience, good tensile strength, and tear resistance are among natural rubber's chief advantages. In addition to notable wear resistance, natural rubber also offers low permanent-set characteristics plus good flexing qualities at low temperatures.

Natural rubber does not perform well when exposed to chemicals and petroleum derivatives, including petrochemicals. It is not recommended for outdoor applications where maximum resistance to sunlight, ozone, oxygen, or heat aging is a major factor.

Neoprene. Because neoprene provides the best all-purpose balance of properties, it more than pays for itself in outdoor applications involving extreme weathering and moderate oil exposure.

Neoprene is one of the best all-purpose elastomers where resistance to ozone, sunlight, oxidation, and many petroleum derivatives is of prime importance. In addition to inherent heat and flame resistance, neoprene rubber will not support combustion. Added advantages include good resistance to water and to many chemicals, plus good resilience characteristics and tensile strength properties.

Neoprene has few practical limitations. While it is slightly higher in price than general-purpose synthetic rubbers, its resistance to ozone, oil, and chemicals overrides the cost differential.

Nitriles. The nitrile-based elastomer is ideally suited for many automotive and aerospace components requiring resistance to gasoline and other petroleum derivatives.

In addition to its resistance to petroleum oils and aromatic hydrocarbons, NBR is highly resistant to mineral oils, vegetable oils, and many acids. It also has good elongation properties as well as adequate resilience, tensile strength, and compression set.

Because of its cost, this elastomer is not usually recommended for applications where oil resistance is not a major problem. Where oil resistance is required, the cost is more than justifiable.

Polyacrylics. The heat resistance of acrylic rubber is higher than that of any other rubber except silicone and fluoroelastomers. Further, it has excellent oil resistance at elevated temperatures. In addition to outstanding heat and oil resistance, acrylic rubbers offer excellent resistance to oxygen and ozone, plus heat aging and flex life.

Acrylic rubber parts are not suitable for applications where they would be exposed to acids, alkalis, steam, or water.

Polybutadienes. Physically similar to polyisoprenes and natural rubber, polybutadiene in recent years has been subjected to sophisticated compounding techniques to produce a useful, economical family of "stereo" rubbers.

Polybutadiene rivals rubber in many applications, providing many of the same physical properties at a lower cost per pound. It offers the user unusually good performance at low temperatures (−100°F), lower heat buildup, and higher resistance to wear and abrasion than natural rubber.

At the present time, its tensile strength and elongation characteristics are not comparable to those of natural rubber or polyisoprene.

Polysulfides. The relatively poor physical properties of polysulfides are compensated for by their unsurpassed resistance to petroleum products. Resistance to corrosive reagents, oil, solvents, and greases is the primary advantage offered by polysulfide rubber.

Polysulfide rubber has a very strong, objectionable odor. Polysulfide is low in tensile, tear, elongation, compression, and other important physical properties.

Polysulfides can be formulated in liquid, pourable compositions, useful for potting applications, as discussed in another part of this chapter under embedding materials and processes.

Polyurethane. Polyurethane is notable for its outstanding tensile strength and abrasion resistance, and also for its resistance to gasoline and kerosene, which exceeds that for either neoprene or nitrile rubbers.

For high-durometer materials, good elongation and high tensile strength are among the more outstanding characteristics of polyurethane rubber. In addition, it offers excellent abrasion and tear strength plus good resistance to ozone and oxygen. A low coefficient of friction makes polyurethane suitable for many other applications.

There are two types of urethanes, liquid and dry. The dry types, which can be readily fabricated on conventional rubber-making equipment, have slightly lower physical properties than the liquid types, which require special processing equipment. Cost is another limiting factor. Liquid urethanes are used in embedding electronic packages, and these materials and processes are reviewed in another part of this chapter under embedding.

Polyurethane rubber is not recommended for applications involving exposure to acids and alkalis, although its resistance to aliphatic solvents and cold water is quite good.

SBR Rubber. SBR is a copolymer created from butadiene and styrene. While its general properties are slightly below those of natural rubber, SBR is one of the lowest-priced polymers on the market, and has better uniformity than natural rubber.

SBR stock can be compounded to provide very fine abrasion, wear, and tensile qualities. It can be readily substituted for natural rubber in many applications with significant cost savings. Its resilience is about the same as that of natural rubber. SBR also bonds readily to many materials.

Like natural rubber, SBR offers little resistance to oils and chemicals. SBR must be specially compounded to provide resistance to ozone, sunlight, and heat.

Silicones. Silicone rubber is ideally suited for many electrical/electronic and aerospace custom rubber requirements where high resistance to both high and low temperature extremes is involved. Silicone rubber compounds are commonly divided into seven application categories: general purpose, extreme high-temperature service, extreme low-temperature service, low compression set, high tensile or tear strength, oil and fuel resistance, and food grade.

Some silicone rubbers are formulated in liquid, pourable compositions (RTVs), which are useful for embedding applications. These are reviewed separately in the part of this section which discusses embedding materials and processes.

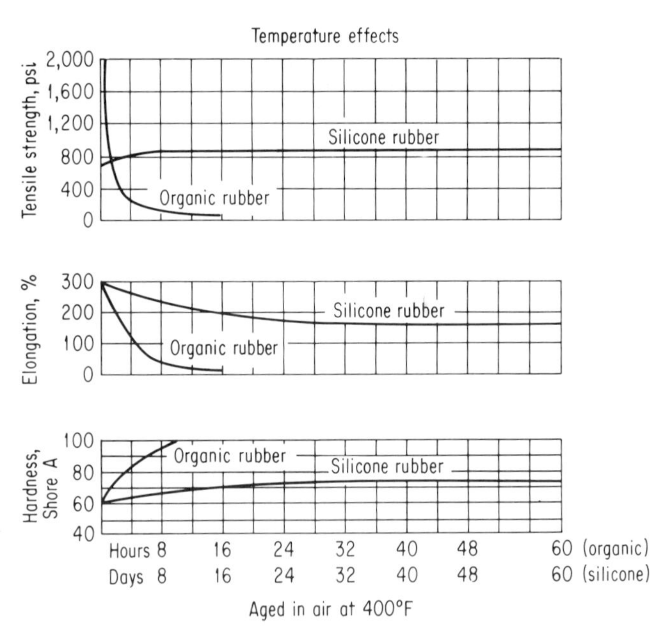

FIG. 47. Comparison of 400°F aging effects on important properties of silicone and organic rubber materials.[36]

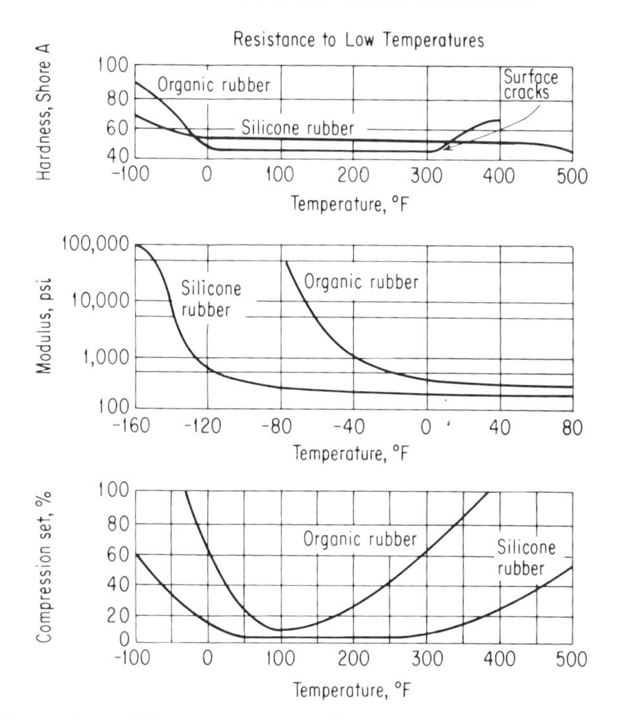

FIG. 48. Comparison of low-temperature effects on important properties of organic and silicone rubber materials.[36]

Temperature resistance for silicone rubber ranges from −160 to +600°F. Resistance to temperatures as high as 700°F for short periods is attainable. Tensile strengths as high as 1,800 psi are possible. Elongation characteristics as high as 800 percent are available in certain silicone compounds. In addition, silicone offers good resistance to oils, weathering, and compression set, as well as fatigue and flexing. The more exotic (and higher-price) fluorosilicones have unusual oil-resistance characteristics.

Cost continues to be a somewhat limiting factor where extreme temperature resistance is not essential. Unreinforced silicones also have lower tensile strengths than most other members of the elastomer family.

Comparisons of the aging effects of 400°F temperatures on silicone and organic rubbers, for several important physical properties, are shown in Fig. 47.

Comparisons for some low-temperature properties are shown in Fig. 48. Important electrical properties, as a function of temperature, are shown in Fig. 49.

Synthetic Rubber. Synthetic rubber elastomers closely approximate the chemical composition of natural rubber. The basic cost of these stocks is usually lower than that of natural rubber. Thus, better price stability and more uniform quality make this synthetic an attractive selection for many products with the same end uses listed for natural rubber.

In addition to better price stability, these synthetic rubbers are comparatively free from problems created by political or trade manipulations. Synthetic rubber offers outstanding resilience properties plus better resistance to temperature, heat, aging, and weathering. Color stability is another important characteristic. Compression set and tear resistance are slightly lower in synthetic than in natural rubber.

Synthetic rubber has lower viscosity, but its mixing cycles are more critical. Its tensile strength is in the range of 2,000 to 3,000 psi, or somewhat lower than that of natural rubber, which is approximately 4,000 psi.

Fillers and Reinforcements for Plastics. Fillers and reinforcements are widely

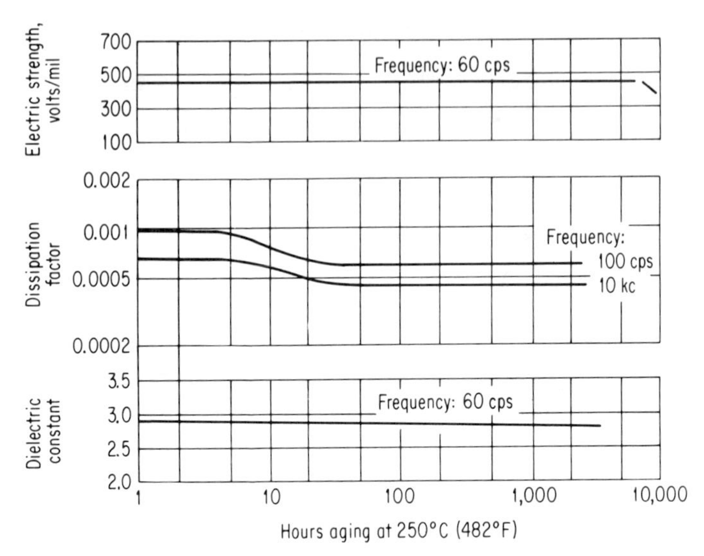

FIG. 49. Effect of temperature on important electrical properties of silicone rubber materials.[37]

used in the formulation of plastic compositions. Although fillers have always been widely used in thermosetting materials as part of the basic compound, they have not been so widely used in thermoplastic resin systems. Use of filled thermoplastic compositions is increasing, however, since (1) many thermoplastics offer certain design properties not attainable in thermosets and (2) application of thermoplastics for these unique design properties is limited unless other weak properties of thermoplastics are overcome. Thermoplastics do have a basic toughness and flexibility which allow their use without fillers in most cases, whereas many thermosets exhibit brittleness or other physical weaknesses which necessitate the use of fillers.

Applications of Fillers. Fillers may reduce plastic material costs or increase them, depending on the filler used. Even in cases where fillers reduce material costs, some property improvements usually exist. When compounded into thermosetting resins, the proper selection of filler or fillers can offer such material improvements as improved heat resistance, chemical resistance, physical strength, dimensional stability, lubricity, and thermal conductivity. Further, mold shrinkage may be reduced, as may thermal expansion. Also, crack resistance, thermal shock resistance, weatherability, and effects of adverse environments can be reduced. Proper selection of fillers can also offer similar advantages for thermoplastic materials. However, perhaps the greatest application of fillers to thermoplastics has been the use of fillers for increasing the strength of thermoplastics, especially the strength versus temperature properties, and the dimensional stability of thermoplastics.

Fillers can be classified as reinforcing or bulk, depending on their form factor or application. Also, they can be classified according to material type, such as organic or mineral. Organic fillers, such as wood flour, cotton, paper, cellulose, sisal, and jute, are natural organics. Also, certain synthetic organics such as nylon, Dacron,* or Orlon* are sometimes employed. Mineral fillers include asbestos, silica, glass, mica, clay, talc, carbonates, and metallic oxides (ceramics) such as alumina and magnesia. Certain other fillers also find special applications. Some such fillers are powdered metals, lubricating fillers such as graphite and molybdenum disulfide, and metal fibers or whiskers.

A tabular listing of fillers and reinforcements which are often used in molding compounds, and the property improvements for which they are used, is shown in Table 18.

* Trademark, E. I. du Pont de Nemours & Co.

TABLE 18. Applications of Fillers in Molding Compounds[38]

Filler or reinforcement	Properties improved *													
	Chemical resistance	Heat resistance	Electrical insulation	Impact strength	Tensile strength	Dimensional stability	Stiffness	Hardness	Lubricity	Electrical conductivity	Thermal conductivity	Moisture resistance	Processability	Recommended for use in †
Alumina, tabular	×	×				×								S/P
Alumina trihydrate, fine particle			×				×						×	P
Aluminum powder										×	×			S
Asbestos	×	×	×	×		×	×	×						S/P
Bronze							×	×		×	×			S
Calcium carbonate‡		×				×	×	×					×	S/P
Calcium metasilicate	×	×				×	×	×				×		S
Calcium silicate		×				×	×	×						S
Carbon black§		×				×	×			×	×		×	S/P
Carbon fiber										×	×			S
Cellulose				×	×	×	×		×					S/P
Alpha cellulose			×		×	×								S
Coal, powdered	×											×		S
Cotton (macerated/chopped fibers)			×	×	×	×	×	×						S
Fibrous glass	×	×	×	×	×	×	×	×				×		S/P
Fir bark													×	S
Graphite	×				×	×	×	×	×	×	×			S/P
Jute				×			×							S
Kaolin	×	×				×	×	×	×			×	×	S/P
Kaolin (calcined)	×	×	×			×	×	×				×	×	S/P
Mica	×	×	×			×	×	×	×			×		S/P
Molybdenum disulfide							×	×	×			×	×	P
Nylon (macerated/chopped fibers)	×	×	×	×	×	×	×	×	×				×	S/P
Orlon	×	×	×	×	×	×	×	×				×	×	S/P
Rayon			×	×	×	×	×	×						S
Silica, amorphous			×									×	×	S/P
Sisal fibers	×			×	×	×	×	×				×		S/P
TFE–fluorocarbon						×	×	×	×					S/P
Talc	×	×	×			×	×	×	×			×	×	S/P
Wood flour			×		×	×								S

* The chart does not show differences in degrees of improvement; calcined kaolin, for example, generally gives much higher electrical resistance than kaolin. Similarly, differences in characteristics of products under one heading, such as talc (which varies greatly from one grade to another and from one type to another), are not distinguished.
† Symbols: P—in thermoplastics only; S—in thermosets only; S/P—in both thermoplastics and thermosets.
‡ In thermosets, calcium carbonate's prime function is to improve molded appearance.
§ Prime functions are imparting of UV resistance and coloring; also is used in crosslinked thermoplastics.

Fillers for Thermoplastics. Use of fillers in thermoplastics has been steadily increasing due to increasing need for thermoplastics with higher strength, higher thermal stability, and higher dimensional stability. These needs have been increasing in electronic packaging, and hence, in many cases, filled thermoplastics should be considered. One of the largest growing classes of thermoplastics is the group of glass-filled materials. Thermoplastics which are commonly glass-filled include nylons, polystyrenes, styrene-acrylonitriles, polycarbonates, polypropylenes, polyethylenes, acetals, ABS materials, polysulfones, and polyphenylene oxides. Data on some important property improvements gained in several thermoplastics are shown in Table 19. The degree of improvement possible in the deformation properties of TFE fluorocarbon by use of certain fillers is shown in Table 20. This can be used as a guide for the type of dimensional stability improvements possible by use of fillers in soft thermoplastics. Typical filler effects on electrical properties of thermoplastics are shown in Table 21, using TFE fluorocarbon as the base material.

In addition to glass-filled thermoplastics, asbestos-filled thermoplastics are becoming increasingly popular. Asbestos-filled polypropylene has gained special usage, basically since it has a set of properties between those of unfilled polypropylene and glass-filled polypropylene, at a price generally also between those two.[41] Asbestos fillers are also gainfully used in other resins, such as fluorocarbons, silicones, phenolics, polyesters, and epoxies.[42]

Plastic Fabrication Processes and Forms. There are many plastic fabrication processes, and a wide variety of plastics can be processed by each of these processes or techniques. The fabrication process and the tooling determine the forms or shapes which are produced, of course. Fabrication processes can be broadly divided into pressure processes and pressureless or low-pressure processes. The former will be described at this point. Pressureless or low-pressure processes such as potting, casting, impregnating, encapsulating, and coating are reviewed separately in other parts of this chapter. Pressure processes are usually either thermoplastic materials processes (such as injection molding, extrusion, and thermoforming) or thermosetting processes (such as compression molding, transfer molding, and laminating). There are exceptions to each, however, as mentioned below.

Compression Molding and Transfer Molding. Compression molding and transfer molding are the two major processes used for forming molded parts from thermosetting raw materials. The two can be carried out in the same type of molding press, but different types of molds are used. Thermosetting materials are the materials normally molded by the compression or transfer process, but it is possible to mold thermoplastics by these processes since the heated thermoplastics will flow to conform to the mold cavity shape under suitable pressure. These processes are usually impractical for thermoplastic molding, however, since after filling the mold cavity to final shape, the heated mold would have to be cooled to solidify the thermoplastic part. Since repeated heating and cooling of this large mass of metal and the resultant long cycle time per part produced are both objectionable, injection molding is commonly used to process thermoplastics.

COMPRESSION MOLDING. The compression molding technique is shown in Fig. 50. In compression molding, the open mold is placed between the faces or platens of the molding press, the mold is heated, then filled with a given quantity of molding material, and closed under pressure, causing the material to flow into the shape of the mold cavity. The actual pressure required depends on the molding material being used

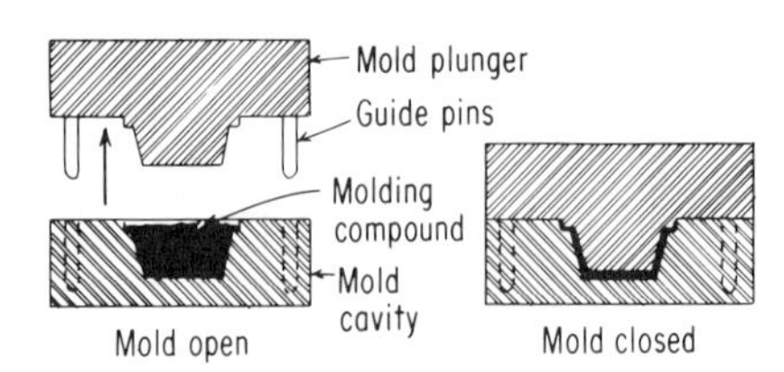

FIG. 50. Simplified illustration of compression molding process.[2]

TABLE 19. Effect of Glass on Physical Properties of Several Thermoplastics [39]*

Material	Percent glass loading	Tensile strength, 73°F, psi × 1,000 (D638)†	Elongation, %, 73°F (D638)	Flexural strength, 73°F, psi × 1,000 (D790)	Impact strength, notched Izod, 73°F (D256)	Heat distortion, temp., °F, 264 psi (D648)	Rockwell hardness (D785)	Specific gravity (D792)
Nylon, 6/10, raw	...	8.5	85–300	....	1.2	....	R111	1.09
Short fiber	30	17–19	3.0	22–26	1.4–2.2	400	E35–45, R118	1.30
Long fiber	30	19.0	1.9	23	3.4	420	E70–75	1.30
Nylon, 6/6, raw	...	9.0	60–300	12.5	1.0–2.0	150–186	R108–118	1.13–1.15
Short fiber	30	18.5–23	3.0	26.5–32	1.2–2.0	400–470	E50–55–R120	1.37
Long fiber	30	20	1.5	28	2.5	498	E60–70	1.37
Nylon, 6, raw	...	7.0	25–320	8	1.0–3.6	152–158	R103–118	1.12–1.14
Short fiber	30	17–24	3	22.5–32	1.3–2.0	400–420	E45–50, M90	1.37
Long fiber	30	21.0	2.0	27	3.0	420	E55–60	1.37
Polycarbonate, raw	...	9.5	60–110	13.5	2.5	265–280	M70–R118	1.2
Short fiber	20	12–18.5	2.5–3	17–25	1.5–2.5	285–295	M92–R118	1.35
Long fiber	20	14–18.5	2.2–5	18.5	2.5–3.0	295	H80–90	1.35
Polypropylene, raw	...	4.3	200–700	6	....	135–145	R85–110	0.90–0.91
Short fiber	20	6.0	3.0	7.5	1.0	230	M40	1.05
Long fiber	20	8	2.2	10	3.5	283	M50	1.05
Polyacetal, raw	...	10.0	15	14	1.4	255	M94–R120	1.425
Short fiber	20	10–13.5	2–3	14–15	0.8–1.4	315–325	M70–75–95	1.55
Long fiber	20	10.5	2.3	15	2.2	325	M75–80	1.55
Polyethylene, raw	...	1.2	50–600	4.8	0.5–16	90–105	(Shore) D50–60	0.92–0.94
Short fiber	20	6	3.0	7	1.1	225	R60	1.10
Long fiber	20	6.5	3.0	8	2.1	260	R60	1.10
Polysulfone, raw	...	10.2	50–100	....	1.3	345	M69–R120	1.24
Short fiber	30	16	2	21	1.8	360	....	1.41
Long fiber	30	18.5	2.0	24	2.5	333	E45–55	1.37

* Most favorable figures for short-fiber performance are based upon results with nominal ¼-in. fibers. Not included in the table are glass-reinforced styrene, SAN, ABS, polyurethane, and PPO, for which comparable data comparing short and long fibers are not available.

† ASTM test method.

SOURCES: Data for raw resin and long glass properties for all resins—Fiberfil Inc.; short glass polysulfone, polyethylene, polypropylene, all three nylons—Fiberfil and Liquid Nitrogen Processing Corp.; polycarbonate—Fiberfil, LNP, and General Electric Co.; polyacetal—LNP, Fiberfil, Celanese Corp., and Du Pont Co.; nylon 6/6—Fiberfil, LNP, and Polymer Corp.

TABLE 20. Effect of Fillers on Deformation Properties of TFE Fluorocarbon [40]

Property*		Un-filled TFE	15% glass fiber	25% glass fiber	15% graph-ite	60% bronze	20% glass 5% graph-ite	15% glass 5% MoS_2
% deformation at 78°F, 2,000 psi, 24 hr	MD‡	14.3	8.3	7.1	8.1	6.0	6.8	6.9
	CD	16.7	13.4	7.5	9.5	5.3	6.7	7.1
% permanent deformation†	MD	7.9	4.1	3.9	4.4	2.5	4.9	3.8
	CD	8.4	9.0	4.6	5.3	2.3	3.9	3.9
% deformation at 78°F, 2,000 psi, 100 hr	MD	16.3	12.6	8.9	10.1	6.1	7.6	7.8
	CD	18.7	14.9	9.4	11.5	6.4	8.7	8.1
% permanent deformation	MD	8.8	6.0	4.4	6.4	2.5	5.8	5.6
	CD	9.1	7.9	5.6	7.3	2.5	5.9	5.5
% deformation at 500°F, 600 psi, 24 hr	MD	30.1	16.6	10.6	16.0	10.6	11.3	9.6
	CD	32.8	27.7	27.8	15.4	8.4	12.2	10.9
% permanent deformation	MD	17.4	11.9	4.9	12.0	7.1	8.4	6.4
	CD	19.2	16.2	17.9	10.8	4.9	8.4	6.8

* ASTM D621-59 (modified).
† After 24-hr recovery.
‡ MD, parallel to molding direction; CD perpendicular to molding direction.

TABLE 21. Effect of Fillers on Electrical Properties of TFE Fluorocarbon [40]

Property	Un-filled TFE	15% glass fiber	25% glass fiber	15% graph-ite	60% bronze	20% glass 5% graph-ite	15% glass 5% MoS_2
Dielectric strength,* volts/mil							
Air	1,500	448	327	63	×§	63	690
Oil	....	921	866	69	×	187	932
Dielectric constant†							
60 cps	2.1	2.50	2.63	×	×	3.38	2.71
10^6 cps	2.1	2.35	2.85	×	×	3.25	2.68
Dissipation factor†							
60 cps	<0.0003	0.0753	0.0718	×	×	0.0761	0.0464
10^6 cps	<0.0003	0.0029	0.0028	×	×	0.0024	0.0061
Surface resistivity,‡ ohms	>10^{17}	10^{13}	10^{13}	10^4	×	10^{13}	10^{14}
Volume resistivity,‡ ohm-cm	>10^{16}	10^{16}	10^{16}	10^4	×	10^{13}	10^{14}

* ASTM D149a (20-mil sample).
† ASTM D150-54T.
‡ ASTM D257-57T.
§ Too conductive to be measured.

and the geometry of the mold. The mold is kept closed until the plastic material is suitably cured or hardened. Then the mold is opened, the molded plastic part is ejected, and the cycle is repeated. The mold is usually steel with a polished or plated cavity.

The simplest form of compression molding involves the use of a separate self-contained mold or die that is designed for manual handling by the operator. It is loaded on the bench, capped, placed in the press, closed and cured, and then removed for opening under an arbor press. The same mold in most instances (and with some structural modifications) can be mounted permanently into the press and opened and closed as the press itself opens and closes. The press must have a positive up-and-down movement under pressure instead of the usual gravity drop found in the standard hand press.

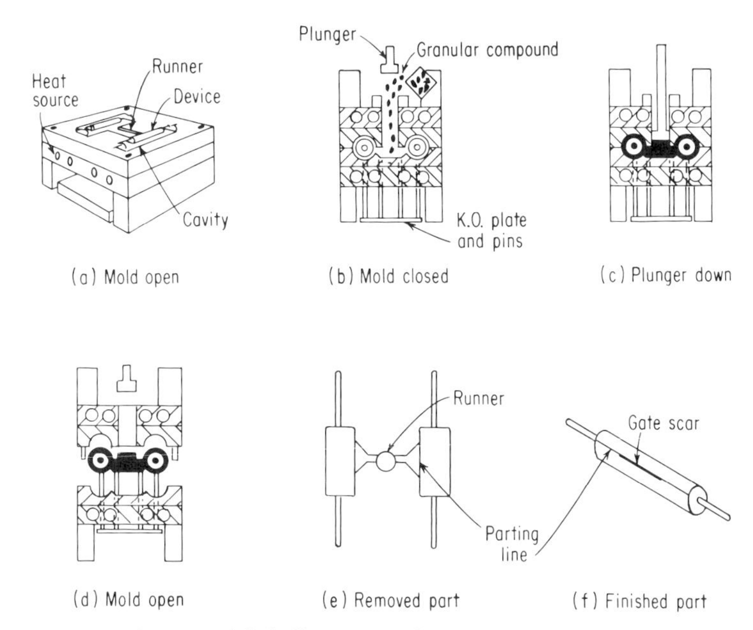

FIG. 51. Simplified illustration of transfer molding process.[2]

TRANSFER MOLDING. The transfer molding sequence is shown in Fig. 51. The molding material is first placed in a heated pot, separate from the mold cavity. The hot plastic material is then transferred under pressure from the pot through the runners and into the closed cavity of the mold.

The great advantage of transfer molding lies in the fact that the mold proper is closed at the time the material enters. Parting lines that might give trouble in finishing are held to a minimum. Inserts are positioned and delicate steel parts of the mold are not subject to movement. Vertical dimensions are more stable than in straight compression. Also, delicate inserts can often be molded by transfer molding, especially with the new low-pressure molding compounds.

Injection Molding. Injection molding is the most practical process for molding thermoplastic materials. The principle of operation is shown in Fig. 52. The operating principle is simple, but the equipment is not.

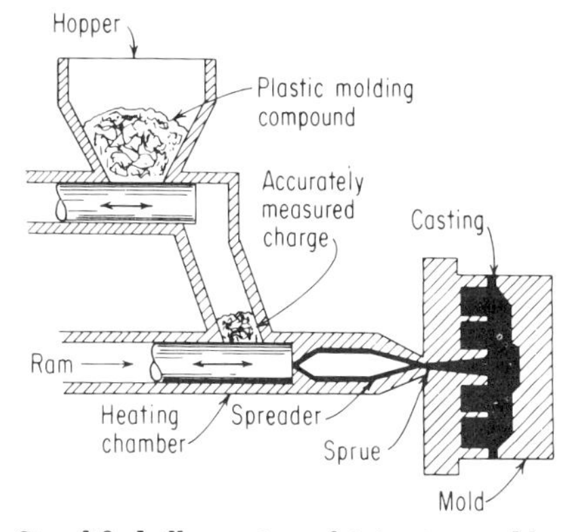

FIG. 52. Simplified illustration of injection molding process.[2]

A material of thermoplastic qualities—one that is viscous at some elevated temperature and stable at room temperature without appreciable deterioration during the cycle—is maintained in a heated reservoir. This hot, soft material is forced from the reservoir into a cool mold shaped to the desired form. The cool mold is opened as soon as the material has given up enough heat to hold its shape and allow repetition.

The speed of the cycle is determined by the rapidity with which the temperature of the material used can be reduced, which in turn depends on the thermal conductivity of that material. Acrylics are slow performers and styrenes are among the fastest.

The machine itself is usually a horizontal cylinder whose bore determines the capacity. Within the bore is a piston which, when retracted, opens a hole in the top of the cylinder through which new material can be added to replace the charge shot into the mold. The cylinder is heated by electric bands which permit temperature variation along its length. Inside the exit end of the cylinder is a torpedo over which the hot material is forced just before coming out of the nozzle into the channels leading to the cavities. This gives the material a final churning and provides thorough heating of all particles. The mold opens and closes automatically and the whole cycle is controlled by timers. An injection machine is rated by the number of ounces it will inject per stroke of the piston, and by the square inches of working area that can be clamped against the injection pressure.

Extrusion. The process of extrusion consists, basically, of forcing heated, melted plastic continuously through a die which has an opening shaped to produce a desired finished cross section. The process can be envisioned as similar to that of a "heated meat grinder." Normally it is used for processing thermoplastic materials, but it can, as described below, be used for processing thermosetting materials. The main application of extrusion is the production of continuous lengths of film, sheeting, pipe, filaments, wire jacketing, or other useful forms and cross sections. After the plastic melt is extruded through the die, the extruded material is hardened by cooling, usually by air or water.

Although not widespread, extruded thermosetting materials are used increasingly. The main object here is production of shapes, parts, and tolerances not obtainable in compression or transfer molding. Any thermoset, B-stage, granular molding compound can be extruded. Any type of filler may be added to the compound. In fiber-filled compounds, the length of fiber is limited only by the cross-sectional thickness of the extruded piece.

A metered volume of molding compound is fed into the die feed zone where it is slightly warmed. As the ram forces the compound through the die, the compound is heated gradually until it becomes semifluid. Before leaving the die, the extruded part is cured by a controlled time of travel through a zone of increasing temperature. the cured material exits from the dies at temperatures of 300 to 350°F and at rates of 3 to 6 in./min for phenolic compounds.

Thermoforming. Thermoforming is a relatively simple basic process, consisting of heating a plastic sheet, then forming it to conform to the shape of the mold either by differential air pressure or by some mechanical means. By this processing technique, thermoplastic sheets can be rapidly and efficiently converted to a limitless number of shapes whose thickness depends on the thickness of the film being used and the processing details of an individual operation. Although there are many variations of this process, they generally involve heating the plastic sheet and making it conform to the contour of a male or female form by air pressure or by a matching set of male and female molds.

Laminating. This process, generally well known, is used to manufacture laminates. Laminating is the process whereby multiple layers of material are bonded or welded together in such a way that the resultant part can be used as a single piece of material. Most commonly this process involves soaking sheets of paper or cloth in an uncured resin, stacking the soaked sheets in a pile of the desired thickness, and then curing these resin-soaked sheets so that they become a single, thicker sheet. Any resin can be used; however, most widely employed are thermosetting resins such as epoxies, polyesters, phenolics, silicones, and melamines. Thermoplastic resins are also used to produce laminates but not to the same extent as thermosetting resins. The sheet materials most commonly used are glass, paper, linen, and sometimes asbestos. While only a single-sheet material with a single resin material has been mentioned thus far, laminating also includes bonding different materials or different layers of material. For instance, copper-clad laminate is made of a given sheet stock and resin combination and a copper sheet.

In the actual manufacture of flat laminates, the sheets may be pressed together between flat plates in a compression molding press, or pressed together using little or no pressure. This is determined by the processing characteristics of the resin used and the properties desired of the end product. Resins such as epoxies or polyesters that are liquid at room temperature can usually be made into simple laminates with little or no pressure.

Forming of Thermosetting Sheets. Often it is desired to form resin-soaked sheets such as those used in laminating into some form other than flat sheets or standard rod-and-bar stock. This is done in a number of ways. One way is matched-metal molding, a process similar to compression molding. In matched-metal molding, matching male and female metal molds are placed in a compression mold and the resin-soaked sheets (the resin may be partially cured or liquid) are placed between the male and female halves of the mold. The press is closed and the part is formed to the mold contour when exposed to sufficient heat, pressure, and time.

A second common method of forming thermosetting sheets is vacuum forming or vacuum bagging. Here, one face of the resin-soaked sheets is placed onto a male or female form; the whole assembly interior is within a bag or cover that can be forced onto the other face of the resin-soaked sheets when a vacuum is applied inside the bag.

Again, as with flat laminates, forms can be made without pressure or vacuum when certain resins are used.

Laminates and Reinforced Plastics. Widely used among products made from plastic materials is that class known as laminates and reinforced plastics. These materials are made by impregnating paper, cotton, asbestos, glass cloth, or other base fibers with various plastic resins such as phenolics, melamines, silicone, and epoxies. The resin-impregnated layers are built up to the desired thickness, and the lamination is cured by heat and pressure which fuse the composite structure into a dense homogeneous material. Normally, if this built-up structure is a flat sheet, it is called a laminate; if it is fabricated in some other customized form, it is called a reinforced plastic. In addition, these composite structures can be formed in the shape of rods and tubes.

Much of the work which has been done on standardization and classification of these materials has been done by the National Electrical Manufacturers Association (NEMA). NEMA standards exist for most classes of flat laminates and for rods and tubes. Since these standards also cover copper-clad laminate constructions, the NEMA grade is often referenced in describing the base laminate of a copper-clad construction. Predominantly used for copper-clad constructions are laminates with a paper-phenolic or a glass-epoxy base.

Although special classes of laminates are made with thermoplastic resins, most laminated products are made with thermosets. This article primarily deals with thermosetting materials.

Another form of laminated material which should be mentioned is the classification known as vulcanized fiber. Vulcanized fiber is a regenerated cotton cellulose material made by passing layers of paper through a bath of zinc chloride. This bath vulcanizes the paper, causing individual fibers in the various layers to form together in a homogeneous sheet. The product that results, after finishing operations are completed, is a dense material of a fibrous nature in which the structure remains intact to varying degrees depending on the grade being processed.

Base Materials. Base materials, or reinforcing agents, used in the manufacture of laminates also run a wide range. Properties of the major base materials used today are summarized below.[43]

PAPER. Three types of paper are in common use as reinforcing agents. Kraft paper is used where good mechanical strength is an important requirement. Alpha paper is utilized where better electrical properties, good machinability, and uniform appearance are required. Rag paper combines improved strength characteristics with low moisture absorption and excellent machinability.

COTTON FABRIC. Cotton fabrics are used where mechanical strength is a prime requisite. These materials vary in weight; the heavier weights offer the highest strength, and the lighter, fine-weave fabrics offer improved machinability.

ASBESTOS. Asbestos is used as a base material in laminates requiring high heat resistance and flame resistance. It is used in a variety of forms including felts, paper, and fabric. For example, grades A and AA—the two most common grades of asbestos base laminates—are made of asbestos paper and asbestos woven fabric.

FIBROUS GLASS. Glass materials, both fabrics and mats, have become extremely valuable in laminates requiring superior electrical properties. They are also important in applications requiring low moisture absorption; high tensile, flexural, and compressive strengths; and heat resistance.

The newest base material and one of the recent developments cited earlier is a random glass-mat reinforcement manufactured on a modified paper-making machine. The material fills the gap between cellulose paper and glass fabric in both properties and price, and makes possible a laminate offering excellent mechanical strength and machinability in combination with the outstanding electrical properties usually found only in woven-glass laminates.

NYLON. Nylon fabric contributes many desirable properties to laminated materials. These include good electrical properties, very low water absorption, high impact strength, good abrasion resistance, and excellent resistance to chemical attack.

Resins. Resins used in the manufacture of laminates are usually of the thermosetting type because they must provide rigidity, dimensional stability, and good bond strength under a wide range of operating conditions. While a large number of resins can be employed, the four groups most extensively used today are the phenolics, melamines, silicones, and epoxies, discussed below.[43]

PHENOLICS. Phenolics, the most versatile resins, are still known as the "workhorse" of the industry. They are low in cost and possess nearly every desirable characteristic for wide application in many fields. They have excellent chemical resistance, water resistance, insulating properties, and heat resistance, and their mechanical strength is good.

SILICONES. Silicones are used primarily with glass-cloth fillers to provide laminates with a high degree of heat resistance and superior electrical properties under wide ranges of temperature and humidity. They resist temperatures up to 500°F and have arc resistance comparable to that of the melamines and moisture absorption comparable to that of the epoxies.

EPOXIES. Epoxies are resistant to all but the strongest acids, caustics, and solvents. Their extremely low moisture absorption makes them valuable in applications where the laminate must retain its electrical and mechanical properties under high humidity conditions. Epoxy resins also possess high tensile, compressive, and flexural strength; high impact resistance; and high heat resistance. Their bond strength is superior.

MELAMINES. Melamine resins are particularly noted for resistance to arcing and tracking, and are used primarily in electrical grades. They also possess high mechanical strength, good flame and heat resistance, and good resistance to alkalies.

Forms and Grades of Laminates. NEMA standard data are available on flat sheets and round tubes and rods.[44] The nature of the various product forms is as follows:

SHEETS. After impregnation and curing, the laminated material is cut into sheets which are stacked together between metal pressing plates and pressed under high temperatures and pressures to form laminated thermosetting sheets. During this operation, the resin passes from a fusible soluble stage into one which is practically infusible and insoluble. Temperatures of 270 to 350°F are commonly employed. For the silicone resins, temperatures up to 500°F may be required. Molding pressures of approximately 1,000 to 2,500 psi are common, but lower pressures are used for some resins such as polyesters.

ROUND TUBES. Tubes are formed by rolling impregnated sheet materials upon mandrels between pressure rolls. The material is then either oven-baked to get rolled tubes, or pressed in heated molds to produce molded tubes.

RODS. Molded rods are composed of laminations of impregnated sheet material which have been molded in cylindrical molds under high temperature and pressure and then ground to size. Rods machined from sheets are also made. In these rods, the laminations are parallel chords of a circular cross section. In general, their properties conform to the grade of sheet stock from which they are cut. Rods made in this manner are low in flexural strength when stress is applied perpendicular to the lamination.

MOLDED SHAPES. Molded shapes are composed of impregnated sheet materials cut into various sizes and shapes to fit the contours of a mold and then molded under heat and pressure. In special cases, depending upon the design of the piece, some macerated materials are used in combination with impregnated sheet materials. The requirements of these standards, particularly with regard to mechanical properties, should not be considered as applying to molded shapes, except for rectangular and square tubes, since the properties of such shapes depend to a considerable extent upon the design of the piece.

Design Data for NEMA Classified Laminates. While the properties of laminated thermosetting materials can be varied within quite a large range by varying the reinforcing fiber, the resin binders, manufacturing processes, and other factors, it has been found that a reasonable number of grades or classes suffice for most applications. Therefore, NEMA has standardized its classification based upon the type of reinforcing fiber used for the laminate and according to the application and the type of resin used with the laminate. A description of the chief characteristics of the various NEMA grades of laminated thermosetting products appears in Table 22.

Comparison of the important properties of most NEMA grade laminates is given in Table 2 of Chap. 1. Chapter 1 also has additional discussion on laminates, especially as used in printed-circuit applications. A cross reference of NEMA grades and military specifications is shown in Table 23.

In addition to the properties of various laminate materials discussed thus far, one of the other important engineering considerations in using these materials is the available dimensional tolerances. NEMA has formulated standard commercial tolerances for flat laminates and for tubes and rods. These tolerances are shown in Tables 24 to 26.

Laminates at Elevated Temperatures. Often it is desired to know the thermal endurance of laminate materials. Statistical averages of insulation life for five temperature classes, together with laminates appropriate to these classes, are shown in Fig. 53. IEEE standard procedures recommend approximate test durations of 350, 1,000, and 3,000 hr, for an estimated service life of 40,000 hr (about 5 years).

In any new insulation, whether homogeneous or composite, the mechanism of deterioration may change as temperature increases, and the life-curve slope may change also. This fact renders extrapolation from test data more difficult—hence the necessity for great care in extrapolating.

It will be noticed that the life curves for 105, 130, and 155°C classes have approximately equal slopes, and those for 180 and 210°C are somewhat steeper. The difference is probably attributable to the inability of the resins to withstand the higher test temperatures.

A high-temperature material, however, is not necessarily better for a given application than one with lower thermal properties. The latter may be preferred if it has superior mechanical strength, high resistance to vibration, and low moisture absorption. The data in Fig. 53 apply primarily to voltage endurance, but they are a generalized life-comparison guide. Specific thermal endurance properties should be compared for each individual case.

Higher-temperature laminates are rapidly being developed for uses at 600°F and over. The most widely used high-temperature laminates are glass-reinforced laminates using the highest-temperature-rated resins, such as phenolic, silicone, fluorocarbons, and diallyl phthalates. Of these, the latter three have especially good electrical properties for electrically critical electronic systems. Of these, silicones are perhaps the most broadly used in electronics.

Laminates of glass cloth impregnated with silicone resin have uniformly low dielectric constants and loss factors at both high and low temperatures, throughout a broad frequency range. Also, they retain their physical and electrical properties even when aged for long periods at high temperatures. These properties make silicone laminates highly desirable for high-frequency transmission and reception (as in radomes), because they have less tendency to absorb, reflect, scatter, and distort signals than do high-constant, high-loss materials.

Figures 54 and 55 compare dielectric constants and dissipation (loss) factors for

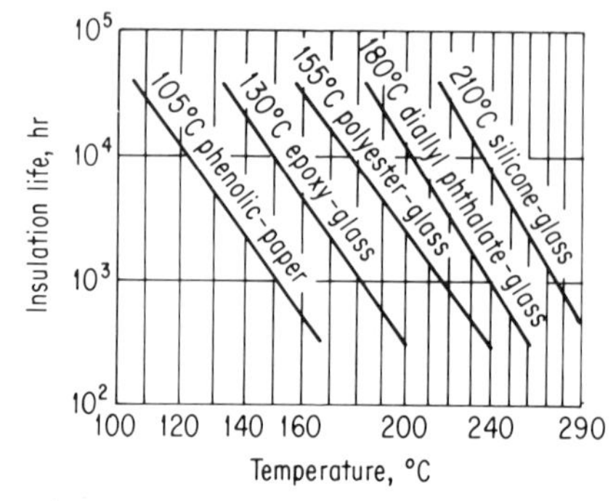

FIG. 53. Statistical averages of insulation life deterioration for laminates in five IEEE temperature classes (105°C, 130°C, 155°C, 180°C, and 210°C).[45]

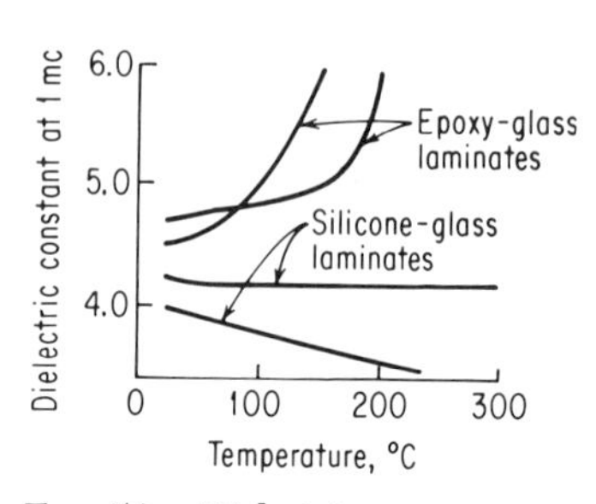

FIG. 54. Dielectric constant versus temperature for silicone-glass and epoxy-glass laminates.[46]

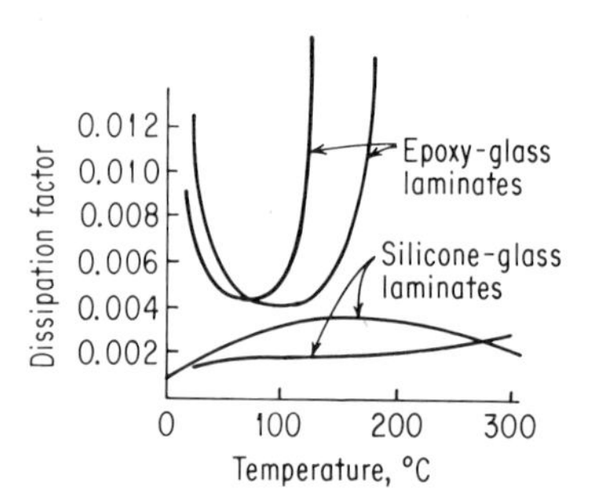

FIG. 55. Dissipation factor versus temperature for the laminates shown in Fig. 54.[46]

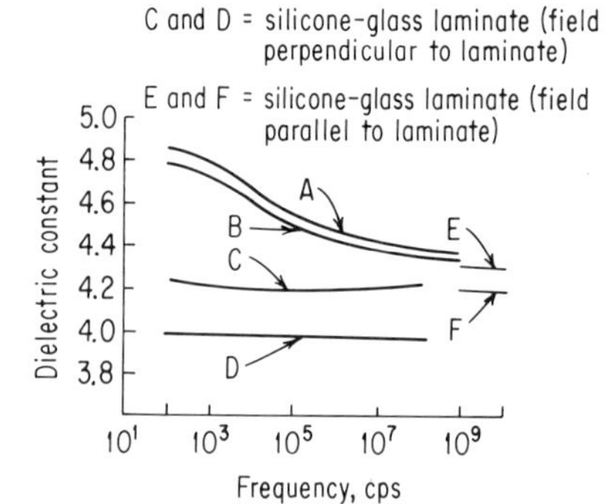

FIG. 56. Dielectric constant versus frequency at room temperature for various laminates.[46]

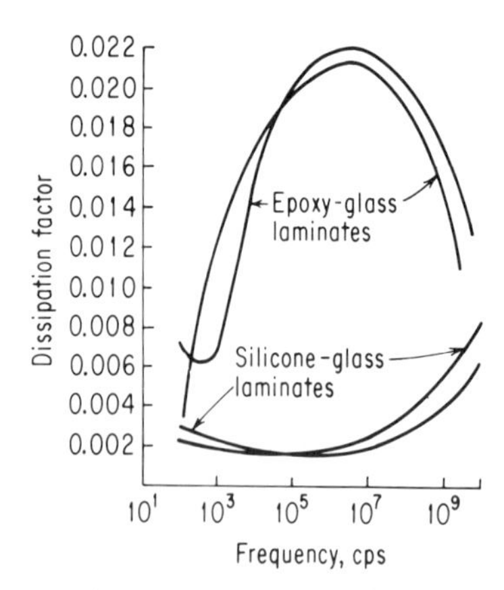

FIG. 57. Dissipation factor versus frequency at room temperature for four laminates.[46]

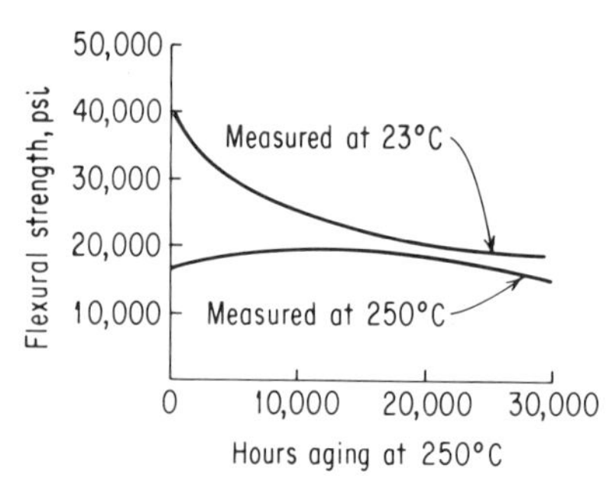

FIG. 58. Effect of long-term aging on flexural strength of silicone-glass laminate.[46]

TABLE 22. Application Characteristics of NEMA Laminate Grades[43]

Grade	Type and characteristics	Grade	Type and characteristics
C	Mechanical applications requiring toughness and impact strength. There may be several subgrades in this class adapted to specific end uses.	XXP	Better electrical and moisture resistance than XX. More suitable for hot punching.
CE	Electrical applications requiring greater toughness than XX, or mechanical applications requiring greater moisture resistance than C. Not recommended for primary insulation in applications involving commercial power frequencies at voltages over 600.	XXXP	High insulation resistance and low dielectric loss under severe humidity conditions. Better electrical properties than XXX and more suitable for hot punching. Falls between XXP and XX in punching characteristics.
MC	Good alkali and arc resistance. Made from purified cotton fabric.	XXXP-C	Similar to grade XXXP. Designed for punching and shearing at room temperature.
L	Mechanical uses where toughness requirements are lower than for C. Suitable for fine machining applications, particularly in thicknesses under ½ in.	G-3	General-purpose grade. High impact and flexural strength; good electrical properties under dry conditions; good dielectric strength perpendicular to laminations; and good dimensional stability. Made with continuous-filament cloth.
LE	Electrical applications requiring greater toughness than that of XX. Better machining properties and appearance than CE. Good moisture resistance. Not recommended for primary insulation in applications involving commercial power frequencies at voltages over 600.	G-5	Highest mechanical strength and hardest laminate grade. Good flame resistance; second only to silicone laminates in heat and arc resistance. Excellent electrical properties under dry conditions. Low insulation resistance under high humidities. Made with continuous-filament cloth.
A (paper)	More resistant to flame and heat than cellulosic-reinforced grades. Small dimensional changes when exposed to moisture. Not recommended for primary insulation in applications involving commercial power frequencies at voltages over 250.	G-7	Class H insulation. Dielectric strength perpendicular to laminations is best of silicone grades. Bond strength slightly lower than for G-5, but tensile and flexural strengths are higher. Made with continuous-fiber cloth.
AA (fabric)	More resistant to heat and stronger than A.	G-10	Extremely high flexural, impact, and bond strengths at room temperature. Good dielectric strength and loss properties under both dry and humid conditions. Insulation resistance under high humidity is better than for G-7.
X	Mechanical uses requiring strength and rigidity. Electrical properties secondary. Laminate should be used under dry conditions.	G-11	High mechanical strength retention at elevated temperatures. Will not support combustion.
XX	General-purpose electrical grade. Good machinability.	FR-3	Combines some of the superior performance characteristics of glass-epoxy laminates with fabrication ease of paper-phenolic laminates. Self-extinguishing.
XXX	Radio-frequency and high-humidity applications. Minimum cold flow.	N-1	Excellent electrical properties under high humidity. Good impact strength, but subject to flow or creep at temperatures higher than normal.
P	Hot punching grade. Heated to 220–250°F, thicknesses up to ⅛ in. can be punched. Sheets up to 1/16 in. can be punched cold. More flexible than and not so strong as X. Between X and XX in moisture resistance and electricals.		

NEMA grades G-10 and G-11 epoxy-glass laminates versus a low- and a high-pressure-molding silicone-glass laminate. Note that values for both properties of silicones are substantially lower and much more uniform than values for epoxies throughout the temperature range from 0 to 300°C.

The low, stable dielectric constant and dissipation factor of silicones throughout a wide frequency band is shown in Figs. 56 and 57. Figures 58 and 59 illustrate the essentially negligible effect of aging at 250°C on the flexural and compressive strength of these laminates. Often, of course, strength at particular temperatures is as important as strength after aging.

TABLE 23. Cross Reference of NEMA and Military Specifications for Plastic Laminates [2]

Resin and reinforcement	Classification		
	NEMA grade	MIL type	Military (MIL) specification
	Sheets		
Epoxy:			
Glass cloth	G-10	GEE	P-18177
Glass cloth	G-11	GEB	P-18177
Paper	FR-3	PEE	P-3115C
Glass (copper-clad)	G-10, G-11	GE, GB	P-13949B
Glass (copper-clad)	....	GF	P-13949B
Melamine:			
Paper	ES-1, ES-3	NDP	P-78A
Glass cloth	G-5	GMG	P-15037C
Glass (copper-clad)	G-5	GM	P-13949B
Phenolic:			
Paper	X, P		
Paper	XP	HSP	P-78A
Paper	XPC		
Paper	XX	PBG	P-3115C
Paper	XXX, XXP	PBE, PBE-P	P-3115C
Paper	XXXP, XXXPC	PBE-P	P-3115C
Paper	ES-2	NDP	P-78A
Paper	FR-1	PBG	P-3115C
Paper	FR-2	PBE-P	P-3115C
Cotton fabric	C	FBM	P-15035C
Cotton fabric	CF	FBG	P-8655A
Cotton fabric	L	FBI	P-15035C
Cotton fabric	LE	FBE	P-15035C
Asbestos paper	A, AA	PBA, FBA	P-8059A
Glass fiber	G-2		
Glass fiber	....	II-2	P-25515A
Glass cloth	G-3		
Nylon cloth	N-1	NPG	P-15047B
Glass (copper-clad)	XXXP, XXXPC	PP	P-13949B
Polyester:			
Glass fiber	GPO-1		
Glass fiber	GPO-2	1, 2, 3	P-8013C
PTFE:			
Glass cloth	GTE	GTE	P-19161A
Silicone:			
Glass cloth	G-6, G-7	GSG	P-997C
Glass (copper-clad)	G-6	GS	P-13949B
	Rods and tubes		
Phenolic:			
Paper	X	PBM	P-79C
Paper	XX	PBG	P-79C
Paper	XXX	PBE	P-79C
Cotton fabric	C	FBM	P-79C
Cotton fabric	CE	FBG	P-79C
Cotton fabric	L	FBI	
Cotton fabric	LE	FBE	P-79C
Glass cloth	G-3		
Melamine:			
Glass cloth	G-5	GMG	P-79C

TABLE 24. Standard NEMA Thickness Tolerances for Laminated Sheets[44]*

Nominal thickness, in.	X, XP, XPC, XX, XXP, XXPC, XXX, XXXP, XXXPC, FR-2, FR-3	C	CF, CE, A, MC	L	LE	AA	G-3, G-5, G-7, G-10, G-11	N-1
0.010	0.002	....	....	0.003	....	....	0.002	
0.015	0.0025	....	....	0.0035	0.0035	....	0.003	0.0035
0.020	0.003	....	....	0.004	0.004	....	0.004	0.005
0.025	0.0035	....	0.005	0.0045	0.0045	....	0.005	0.0056
1/32	0.0035	0.0065	0.0065	0.005	0.005	....	0.0065	0.0065
3/64	0.0045	0.0075	0.0075	0.0055	0.0055		0.0075	0.0075
1/16	0.005	0.0075	0.0075	0.006	0.006	0.018	0.0075	0.0075
3/32	0.007	0.009	0.009	0.007	0.007	0.018	0.009	0.009
1/8	0.008	0.010	0.010	0.008	0.008	0.020	0.012	0.010
5/32	0.009	0.011	0.011	0.009	0.009	....	0.015	0.011
3/16	0.010	0.0125	0.0125	0.010	0.100	0.024	0.019	0.0125
7/32	0.011	0.014	0.014	0.011	0.011	....	0.021	0.014
1/4	0.012	+0.030	0.015	+0.024	0.012	0.028	0.022	0.015
5/16	0.0145	+0.035	0.0175	+0.029	0.0145	0.034	0.026	0.024
3/8	0.017	+0.040	0.020	+0.034	0.017	0.038	0.030	0.032
7/16	0.019	+0.044	0.022	+0.038	0.019	0.044	0.033	0.040
1/2	0.021	+0.048	0.024	+0.042	0.021	0.048	0.036	0.048
5/8	0.024	+0.053	0.027	+0.048	0.024	0.058	0.040	0.054
3/4	0.027	+0.058	0.029	+0.054	0.027	0.068	0.043	0.058
7/8	0.030	+0.062	0.031	+0.060	0.030	0.076	0.046	0.062
1	0.033	+0.065	0.033	+0.065	0.033	0.086	0.049	0.066
1 1/8	0.035	+0.069	0.035	+0.069	0.035	....	0.053	
1 1/4	0.037	+0.073	0.037	+0.073	0.037	0.106	0.055	
1 3/8	0.039	+0.077	0.039	+0.077	0.039	0.124	0.058	
1 1/2	0.041	+0.081	0.041	+0.081	0.041	0.144	0.061	
1 5/8	0.043	+0.085	0.043	+0.085	0.043	....	0.064	
1 3/4	0.045	+0.0899	0.045	+0.089	0.045	....	0.067	
1 7/8	0.047	+0.093	0.047	+0.093	0.047	0.160	0.070	
2	0.049	+0.097	0.049	+0.097	0.049	....	0.073	

* Tolerances are ± unless otherwise indicated.

TABLE 25. Standard NEMA Tolerances for Tubing and Rods[44]

Rolled, molded tubing and rods, OD tolerances, in.		Rolled, molded tubing, ID tolerances, in.	
Size	Tolerance	Size	Tolerance
1/8 to 1 15/16	±0.005	1/8 to 23/32	±0.003
2 to 4	±0.008	3/4 to 1 15/16	±0.004
4 1/8 to 12 (rolled only)	±0.025	2 to 4	±0.008
		4 1/8 to 12 (rolled only) ..	±0.010

TABLE 26. Standard NEMA Tolerances for Sanded Laminated Sheets[44]

Condition	Tolerance, in.		
	Sheet width, in.		
	12	24	48
Sanded on one side	±0.002	±0.003	±0.004
Sanded on two sides	±0.001	±0.001	±0.001

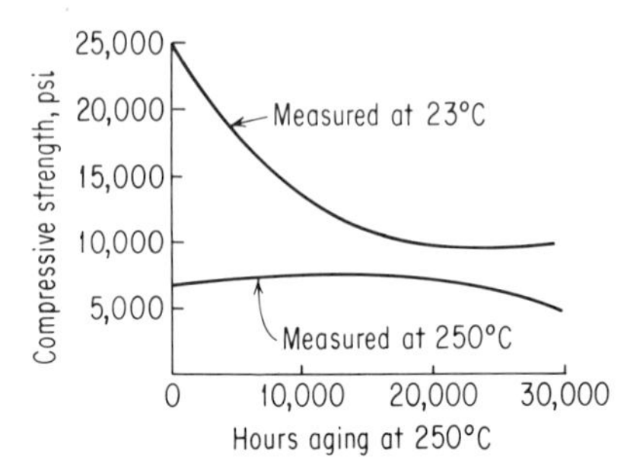

FIG. 59. Effect of long-term aging on compressive strength of silicone-glass laminate.[46]

Some of the newer high-temperature laminates being commercially developed are glass-reinforced polyimides, amide-imides, and polybenzimidazoles. The general properties of these high-temperature resins have been discussed earlier in this section. Flexural strength and modulus properties at 600°F for these and some other high-temperature laminates are shown in Fig. 60. Flexural strength versus temperature for three polyimide laminates is shown in Fig. 61.

Prepregs and Special Laminate Constructions. Prepregs are B-staged or partially cured laminate constructions. Here, the reinforcing materials, either fabrics or fibers, are impregnated with the plastic resin chosen and partially cured. In this state, they can be heat-softened, and thus can be formed into a permanent shape upon the application of heat or pressure. Prepregs are available using most resins, but materials based on epoxy, silicone, diallyl phthalate, phenolic, polyester, and melamine are prominent. While most forms of reinforcing fibers are used, particularly common are glass-fiber reinforcing materials. The availability of prepreg materials often makes possible, with simple techniques, the fabrication of many reinforced plastic shapes and forms in laboratories or model shops. Prepregs are very easy to work with since they do not involve messy handling of wet resins and the problems associated with using liquid resins. Prepregs are economical for short or pilot runs and can be conveniently used for making prototype assemblies. Prepregs are flexible and can be contoured around most forms to create the final shape desired. Heat and a minimum of pressure are then applied for molding into the final permanent shape. There are four major fabrication techniques or molding methods for forming end products from prepreg sheets. These four methods are shown in Fig. 62.

Plastic Films, Sheets, and Tapes. Thermoplastics can be formed into films or sheets by extruding, casting, calendering, or skiving techniques. The determination of when to use the term *film* or the term *sheet* is an arbitrary one, but a generally accepted practice is to use the term *film* for thicknesses under 10 mils, and the term *sheet* for thicknesses over 10 mils. Upon slitting and rolling, films become tapes. Tapes may be either adhesive-backed or plain. Numerous flexible foam tapes are also available on the market. Predominant here are urethane, vinyl, and neoprene foams.

Films and Tapes. Nearly every thermoplastic is available in film, sheet, and tape form. There are some thermoplastics which are basically available only in film form, due to processing limitations of the base-material systems. Perhaps the most common thermoplastic in this category is polyester, or PE terephthalate, such as Mylar* film.

As would be expected, films and sheets exhibit generally the same properties as the basic thermoplastic from which they are made. Any differences are those inherent to thin or thick sections of thermoplastic materials, such as greater flexibility and higher dielectric strength values in volts per mil for films than for thicker sheets or forms. Dielectric strength varies widely in tapes, depending not only on the material and the thickness, but also on the method of fabrication, particularly as the fabrica-

* Trademark, E. I. du Pont de Nemours & Co.

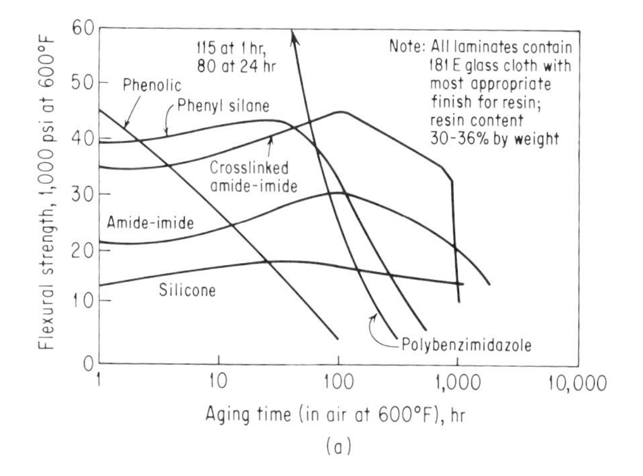

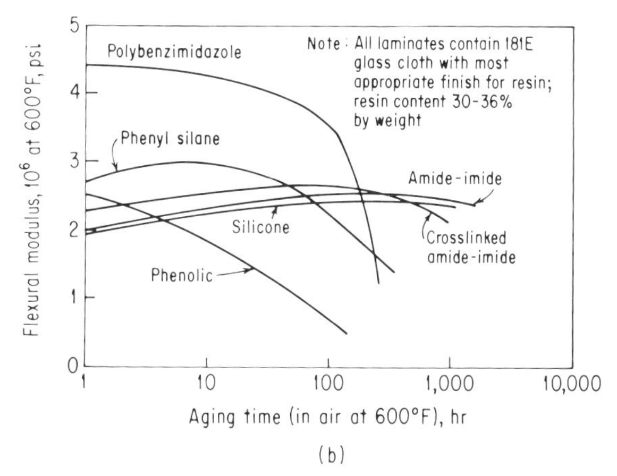

FIG. 60. Flexural strength and modulus of several high-temperature laminates at 600°F.[27]

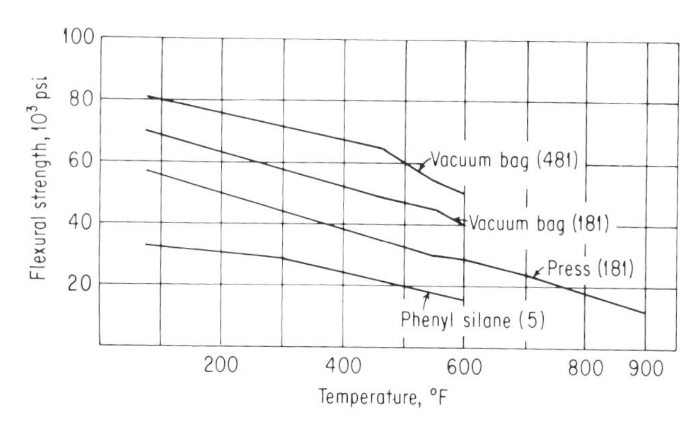

FIG. 61. Flexural strength versus temperature of vacuum bag and press-made polyimide laminates on 481 and 181 style E glass with A-1100 finish (13 plies).[47]

Vacuum Bag Molding.
Lay-up of prepreg in male or female mold is encased in a plastic bag and air evacuated. Assembly is then oven-cured

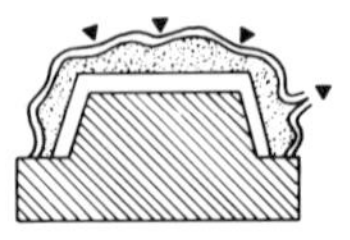

Pressure Bag Molding.
Lay-up of prepreg in female mold is sealed off with a diaphragm or bag and pressure up to 100 psi applied. Where a male mold is used or for pressures up to 200 psi, lay-up is placed in an autoclave

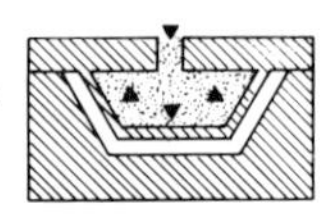

Matched Metal Die Molding.
Prepreg is shaped and cured between mated halves of metal mold at pressures up to 150 tons and temperature up to 400°F

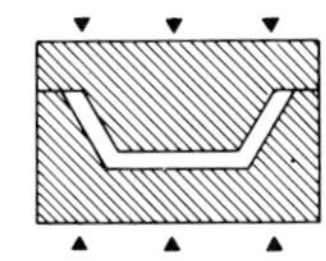

Filament and Tape Winding.
Unidirectional prepreg is wound on a rotating mandrel with the shape and inside dimensions of the finished part. Vacuum bag, pressure bag, autoclave, or shrinkage overwrap can be used to cure the wound structure to final, permanent shape

FIG. 62. Molding methods for prepreg laminate sheets.[48]

tion method affects porosity and uniformity of thickness. Notable in this respect is that cast or extruded films, especially multilayer films, are more free of these irregularities than are skived films. Skived films are inherently more irregular since they are essentially mechanically sliced from a thicker section. Needless to say, the thicker the film, the less chance there is for internal voids and surface irregularities to "line up" and cause functional problems. Among the films with the highest dielectric strength are polyester, TFE and FEP fluorocarbons, polyimides, polysulfones, and polycarbonates. Among the highest-temperature films are the fluorocarbons and polyimides. Table 27 shows the useful temperature range of a number of film materials.

Adhesive Tapes. While films are not normally affected by periods of storage, adhesive-back tapes are affected. This is due to the deterioration of the adhesive backing. Adhesive backings are either thermosetting or nonthermosetting. Typical thermosetting adhesive backings are natural rubber with crosslinking agents, thermosetting acrylics, silicones, and epoxies. Typical nonthermosetting adhesive backings are natural rubber, synthetic rubber, and acrylics. Tackifiers are often added to provide the necessary adhesive back. The usual effects of aging upon adhesive-backed tapes are difficulty of unwinding, dimensional distortion, and loss of bond strength. Electrical properties are usually not affected, since the film rather than the adhesive backing controls these properties. Storage at reduced temperatures of 30 to 60°F is the best means of minimizing the above-mentioned storage problems.

Plastic Foams. Plastic foams are cellular materials made from a variety of plastics by chemically or mechanically expanding the resins. This expansion can be part of the chemical reaction; for example, liberation of carbon dioxide causes expansion in urethane foams. Expansion can also be caused by gassing of a chemical blowing or gassing agent added to the resin mix. Also, expansion can occur by mechanical mixing of some type. The type of expansion used depends on the plastic

TABLE 27. Useful Temperature Range for Some Thermoplastic Films[12]

Film material	*Useful temperature range, °F*
Acrylonitrile butadiene styrene	−65–215
Acrylonitrile styrene copolymer	−80–185
Cellulose acetate	−15–200
Cellulose acetate butyrate	−30–180
Cellulose propionate	−30–200
Cellulose triacetate	0–400
Ethyl cellulose	−75–250
Fluorinated ethylene	−415–400
Polyamide	−100–300
Polycarbonate	−150–275
Polyester	−75–300
Polyethylene (low density)	−70–180
Polyethylene (medium density)	−70–200
Polyethylene (high density)	−70–250
Polyimide	−450–600
Polymethyl methacrylate	-190
Polypropylene, cast	−0–275
Polypropylene, balanced, oriented	−60–240
Polypropylene, balanced, oriented, coated	−60–225
Polystyrene (casting, extrusion)	−55–200
Polystyrene (oriented)	−80–175
Polytetrafluoroethylene	−450–500
Polytrifluorochloroethylene	−423–300
Polyvinyl alcohol	−40–420
Polyvinyl chloride (rigid)	−50–200
Polyvinyl chloride (flexible)	−50–200
Polyvinyl fluoride	−100–225
Regenerated cellulose (cellophane)	0–300
Rubber hydrochloride	−20–205
Vinyl chloride–acetate copolymers (rigid)	−50–200
Vinyl chloride–acetate copolymers (flexible)	−60–200
Vinylidene chloride–vinyl chloride copolymer	0–200 (dry) 300 (wet)
Vinyl nitrile rubber	32–200

and the desired end product, but the expansion technique must be controlled. Properly controlled, foams can be made from many plastics, with either open- or closed-cell structures, and in a variety of densities. For most foams, however, a fairly narrow density range is common, usually at low densities of 1 to 5 lb/ft^3. For some foams, notably rigid urethanes, densities up to the area of 15 to 20 lb/ft^3 are practical. Another type of low-density material, often classified as syntactic foams, is produced by adding hollow fillers such as glass or phenolic microballoons to the resin mix. Here, considerably higher-density materials can be produced.

Foams are used for design purposes such as thermal insulation, buoyancy, cushioning, low-weight packaging, acoustical insulation, and electrical insulation. They are available in rigid forms and in flexible forms. Most foams are available from suppliers in various forms such as blocks, sheets, rods, and molded shapes. Some foams can be foamed in place, and hence are especially useful for embedding electronic packages. Most common here are urethanes, epoxies, and silicones, especially the rigid foam-in-place materials.

Types of Foams. Plastic foams are often classified as rigid or flexible, and they may also be classified according to their chemical types. The most common plastic foams are urethanes, polystyrenes, vinyls, polyethylenes, polypropylenes, phenolics, epoxies, silicones, urea formaldelydes, cellulose acetates, and fluorocarbons. Many other plastics are capable of being foamed by one method or another. They can also practically be classified as thermosetting or thermoplastic; the thermosets are urethane, epoxy, silicone, and phenolic. All these classifications have a control on foam plastic properties—as with high-density plastic properties.

Urethanes can be made as either flexible or rigid foams, and they can be foamed in place, which makes them very practical for use in any laboratory or shop. High-temperature varieties, up to 400°F or so, as well as flame-retardant urethanes, are available. Epoxy foams are usually closed all-rigid or semirigid foams which can be obtained in forms and shapes or in formulations which can be processed in-house. They have the usual good chemical properties of epoxies, tempered in the areas where density is a controlling factor. Silicone foams are available in rigid or flexible formulations and, like epoxies, either in forms and shapes or as formulations which can be processed in-house. As with high-density silicones, silicone foams are outstanding in high- and low-temperature stability, and in electrical properties. Phenolic foams, rigid open-cell materials, have as an outstanding property good heat resistance, which also gives them good dimensional stability among foam materials. This is generally true, of course, for all the above-mentioned thermosetting foams, as compared to the thermoplastic foams discussed below.

Thermoplastic foams have the advantages of all thermoplastic materials, such as low-electrical-loss properties for fluorocarbon, polystyrene, and polyethylene foams, and lack of brittleness. They also have the usual thermoplastic disadvantages, most common of which is relatively weak thermal and dimensional stability. Fluorocarbon foams, however, do have good thermal stability, but, like high-density fluorocarbons, they suffer from cold-flow weakness. Polystyrene foams are available as rigid materials in standard and custom shapes and forms. Parts can be made in either integral foam form or expanded bead form. The expanded bead form is readily identifiable since the bead formation stands out, and individual beads can often be broken out of the foam. A major limitation of polystyrene foams is the low temperature at which they must be used, 150 to 175°F. Also, they are readily attacked by many organic solvents. Polyethylene and polypropylene foams are considerably tougher and harder than polystyrene, and are much more chemically inert. They, like polystyrenes, are available primarily in standard and custom-molded forms. Polystyrenes, however, are more practical to process and more widely used. Polystyrene beads can be conveniently used in-house by soaking them in some resin, such as epoxy, urethane, or silicone, which is drained off, and then acts only as a binder in the polystyrene bead matrix. Polyethylene and polypropylene foams will withstand temperatures as high as 200 to 225°F or higher. Polyethylene foam is used as the insulation in coaxial cabling, due to its good high-frequency electrical properties.

Vinyl foams are available as open- or closed-cell structures, in both rigid and flexible forms. The flexible foams are broadly used for cushioning and are resistant to many strong inorganic chemicals. They are attacked by many organic chemicals, however. Fluorocarbon foams, especially foamed FEP fluorocarbon, are widely used as insulation in coaxial cabling since they, like polyethylenes, have excellent high-frequency electrical properties. Fluorocarbons have much higher thermal stability, however, and can be used to 400°F or higher.

Other commercially available thermoplastic foams are urea formaldehydes and cellulose acetates. Numerous other thermoplastic foams, such as nylon, acrylic, ABS, and polyphenylene oxide, are being developed. At present, these are only available in limited quantities by direct negotiation with the developing companies.

Foam Properties. Some of the general foam properties have been discussed above. Many basic properties such as thermal stability and chemical resistance are not dependent on density, and hence are similar to the properties previously discussed for the specific high-density plastic-base material. Other properties, such as strength and electrical and acoustical characteristics, are at least partially controlled by density. Some typical density-sensitive properties will be shown at this point. One of these important properties is thermal conductivity. Figure 63 compares the thermal insulation ranges of several foam plastics with other normal thermal insulating materials.

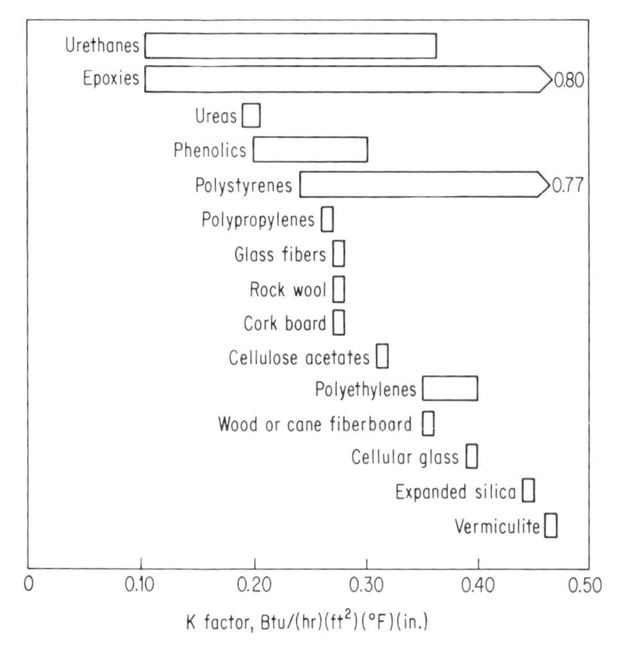

FIG. 63. Thermal insulation ranges for several foamed plastics compared with other insulating materials.[49]

Thermal conductivity is influenced by temperature, and typical examples of this are shown in Fig. 64.

Sound absorption of foams varies with the foam and with the frequency of the sound source. This is shown for four foams in Table 28. The air-filled cellular structure of foams gives low dielectric constants and dissipation factors for foams, as shown in Table 29. The values for these two characteristics increase with increasing foam density, often almost linearly to the values of the specific high-density plastic involved. These low values for foams can sometimes be misleading, however, in that the cell walls are still composed of the high-density plastic, and in some instances, especially as high-frequency power levels increase, breakdown of the cell walls occurs even though the averaged characteristic values for the foam are low. The breakdown mechanism is usually dielectric heating of the cell walls, which ultimately leads to melting, burning, or carbonization of the plastic wall.

Mechanical properties of foams are inherently low, due to the weak structure. Tensile and compressive strengths of 25 to 100 psi are common, and deformation under load can be high, especially as a function of increasing temperature, and especially with thermoplastics. Abrasion resistance of thermosetting foams and some

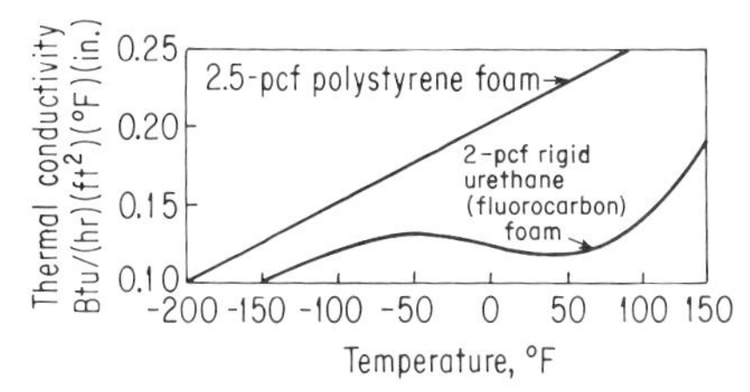

FIG. 64. Thermal conductivity versus temperature for two plastic foams.[49]

TABLE 28. Sound-absorption Properties of Four Plastic Foams Using ASTM C384 and ASTM C423 Test Methods[49]

Material	Frequency, cps	Sound-absorption coefficient*
Expanded polystyrene (2.5 pcf) ..	250	0.03
	500	0.05
	1,000	0.14
	2,000	0.49
	4,000	0.19
	NRC†	0.18
Rigid urethane (2 pcf)	250	0.18
	500	0.27
	1,000	0.19
	2,000	0.62
	4,000	0.22
	NRC	0.32
Flexible urethane (1.9 pcf)	250	0.02
	500	0.80
	1,000	0.99
	2,000	0.79
	4,000	0.88
	NRC	0.60–0.70
Phenolic (2–4 pcf)	NRC	0.50–0.75

* Mounting No. 7.
† Noise-reduction coefficient.

TABLE 29. Electrical Properties of Plastic Foams[49]

Foam	Density, pcf	Dielectric constant				Dissipation factor × 10^{-4}					
		10^3 cps	10^4 cps	10^5 cps	10^9 cps	10^3 cps	10^4 cps	10^5 cps	10^6 cps	10^8 cps	10^9 cps
Cellulose acetate	6–7	← 1.10–1.12 →				← 20–30 →					
Epoxy	5.0	← 2.0 →				← 50 →					
Polyethylene ..	32	1.48	1.49	1.49	1.49	3.3	3.3	3.3	3.3	3.3	3.3
	2.0	....	....	....	1.05						
	18	....	....	....	1.575	...	...	....	...	...	2.0
Polystyrene	1.5*	1.05	....	....	.....	<5.0	<5.0	....	...	...	33
	1.5*	← 1.017 →				← 1.0 →					
	9.8*	....	....	....	1.161	...	...	....	...	...	1.7
	4.3*	....	....	....	1.077‡	...	...	....	...	...	6.0 ‡
	1.9†	....	....	<1.05	<1.05	...	...	<4.0			
	2.8†	....	....	1.07	1.07	...	...	<4.0			
	4.3†	....	....	1.07	1.07	...	...	<4.0			
Silicone	3.5	....	....	1.09	.....	...	...	28			
Polyether urethane	9.3	....	....	....	1.193 ‡	...	...	....	...	...	38.0 ‡
	9.8	....	....	....	1.206 ‡	...	...	....	...	...	35.0 ‡
	1.5–3.0	....	1.05	....	.....	...	...	3–13			

* Molded.
† Extruded.
‡ Measured at 24 × 10^9 cps.

thermoplastic foams is quite low. Many of these properties depend heavily on both the material and the density, and hence, they must be evaluated almost on an individual basis.

Adhesives. Adhesives represent a widely used form of synthetic or plastic materials. Adhesives take many shapes and forms, and can be classified in several ways. One way is according to the chemical nature of the base material, that is, thermosetting, thermoplastic, or elastomeric. Also, they can be classified as structural or nonstructural, depending on whether their design use involves performance as a structural element or performance merely as a fastening and positioning element. The latter category can be the more critical in some electronic packaging applications, since positioning over wide ranges of environments can be more difficult than obtaining good structural strength over a limited range of environments. In addition to the above two classifications, other possible classifications are (1) air drying (or curing) and heat drying (or curing), (2) physical form, such as liquid, paste, solution, or dry film, (3) pressure bonding and nonpressure bonding, including contact adhesive bonding, and (4) solvent-based and non-solvent-based, i.e., 100 percent solids or completely reactive. There can be other classifications, of course, such as application method (brush, spray, ultrasonic, RF bonding). Because of this large number of possible classifications, no attempt will be made to indicate a preferential classification, as this depends so much on the application. Rather, prime application considerations and design data will be discussed.

Basic Application and Design Considerations. The general electrical, thermal, chemical, and physical properties of adhesive materials will closely parallel those of the basic plastic material being used. These properties have been discussed earlier in this section for each plastic. The adhesive or bond-strength properties, however, are unique for each class of materials, or in some cases, for the individual adhesive material. For instance, thermosetting adhesives will usually maintain better strength with temperature increase than will thermoplastics or elastomers, many of which soften or flow readily at elevated temperatures. On the other hand, many thermoplastic and elastomeric adhesives have considerably better peel strength at room temperature than do thermosets, due to their good bonding properties and lack of rigidity. It should be mentioned, however, that some thermosets, such as epoxies, can be flexibilized to overcome the brittleness problem.

THERMOSETTING ADHESIVES. Thermosetting adhesives are available primarily in three forms: liquid, paste, and solid. The liquid form is generally a free-flowing one- or two-part system which may or may not contain solid fillers and/or colorants. Such systems may be the so-called 100 percent solids systems (the contents of which are either already solid or completely reactable to a solid) or they may contain a nonreactive volatile solvent or dispersant which either is necessary for the existence of such a system or imparts certain application characteristics. Occasionally, one component of a two-part system (usually the curing agent) is a solid which requires that the two components be mixed at a temperature in excess of the melting point of the solid.

A distinction is made between the liquid form and the paste form (although both forms are fluids), because the flow behavior of the pastes is different from that of the unfilled or lightly filled liquids. Certain pastes are modified so that they are thixotropic. The practical advantage of such behavior is that an unrestrained mass of adhesive will not sag or flow unless forced to do so by mechanical action (for instance, spreading with a spatula), or that a mass of adhesive will not flow out of a vertical joint during assembly and cure of a structure.

Thermosetting adhesives are also available in supported or unsupported films of various thicknesses. Films are flexible and can be cut or punched to shapes which conform to a particular joint. Ease of handling and layup of these materials and the lack of cleanup which is required with liquids or pastes are their main advantages. Refrigeration of certain film-type adhesives, as with all one-part thermosetting materials, increases storage life. Refrigeration of certain other film types is mandatory, since they are compounded to cure at room temperature.[50]

Typical thermosetting adhesives are epoxies, polyesters, and phenolics, among others.

THERMOPLASTIC ADHESIVES. Thermoplastic adhesives are available in much the same forms as are the thermosets, with the possible exception of pastes. The liquid form of these materials may be either a solution and/or a dispersion of the base thermoplastic polymer and other compounding ingredients in a volatile vehicle, or they may be 100 percent solids systems containing liquid monomer, prepolymer, and catalyst, to which is added an accelerator which induces polymerization to a solid high-molecular-weight polymer.

Thermoplastic adhesives are also available in solid form (primarily unsupported or supported film), in granules, or in lengths of extruded, flexible, cordlike material wound into coils suitable for automatic application.

Thermoplastic film adhesives are converted into the fluid form by either solvent or heat activation. For instance, polystyrene can be dissolved into an adhesive solution by such solvents as ethylene dichloride, methylene chloride, xylene, and toluene, whereas fluorocarbons and polyethylenes are not readily soluble and must be heat-bonded. Solvent activation is only suitable in those cases where one or both adherents permit release of solvent, as by diffusion. Use of solvent-based adhesives must be avoided where solvent entrapment could cause blistering. Heat activation is suitable for use when the adherends are not permeable and are able to withstand the required temperatures. In compounds containing thermosetting resin, the heat also cures the resin. These activation techniques are also used to join substrates, one or both of which has been coated with an appropriate solvent-based adhesive and allowed to dry to a tack-free state.

Thermoplastic adhesives containing no heat-reacting resin, which are heated to render them fluid for joining, are called hot-melt adhesives. They are especially suitable for use with machinery for high-speed applications.[50]

Typical thermoplastic adhesives are polyethylenes, polypropylenes, FEP fluorocarbons, ionomers, vinyls, polystyrenes, and cellulosics.

ELASTOMERIC ADHESIVES. Rubber-base adhesives are widely used, and most common elastomers are useful as adhesives. Liquid, solvent-based adhesives are commonly the form of the more conventional "rubber-type" materials, such as nitriles, neoprenes, butyls, and SBRs. However, elastomers such as polyurethanes, polysulfides, and silicones are normally available and handled as two-component, 100 percent solid (solventless) adhesives; hence, they are processed similarly to epoxy adhesives. Polyurethanes and polysulfides are among the numerous materials which can be used to flexibilize epoxies.

The detailed properties of these elastomers were discussed above.

OTHER ADHESIVES. In addition to the plastic adhesives listed above, there are some types of adhesives which have advantages in special applications. Two of these are anaerobic adhesives and cyanoacrylate adhesives.

The anaerobic adhesives are unique in that they remain in liquid form as long as they are exposed to oxygen. If oxygen is removed—for example, by confining the material to a thin film between the threads of a metal nut and bolt—the material hardens. One series of one-part anaerobic adhesives is available which has the strength characteristics of the structural adhesives. These adhesives cure at room temperature in several hours, but their cure can be accelerated to a matter of minutes at slightly elevated temperatures.[50, 51]

Cyanoacrylate adhesives are also unique in their method and speed of cure. They cure in extremely short periods (sometimes 10 to 15 sec) if confined in a thin film between close-fitting parts. The mechanism of cure in this case is an anionic polymerization (induced by the presence of a base). The thin film of moisture which is usually present on surfaces exposed to normal atmosphere is generally sufficient to harden these materials if they are squeezed to a thin film.[50, 52] The materials with which strongest cyanoacrylate adhesive bonds can be made are vinyls, phenolics, cellulosics, polyurethanes, nylon, steel, aluminum, brass, copper; butyl, Buna N, SBR, natural rubber, most types of neoprene; and most woods. Among the weaker binding materials are polystyrene, polyethylene, and fluorocarbon plastics (sheer strengths up to 150 lb/in.).

Factors Affecting Performance of Adhesives. The performance of adhesives depends on many factors, and each application requires specific analysis. However, some general guidelines can be drawn. Certainly the nature of the joint is a major factor, as well as whether the forces are tensile, shear, or other. The larger the bondable area, the greater can be the bond-strength reliability. Also, mechanical aids such as locks and keys in the joint design will be helpful. The temperature of operation and the desired operating life at that temperature are certainly factors, as all adhesives are affected to some degree by heat aging. The plastic materials with the higher thermal stability are selected for higher-temperature operation, taking into account

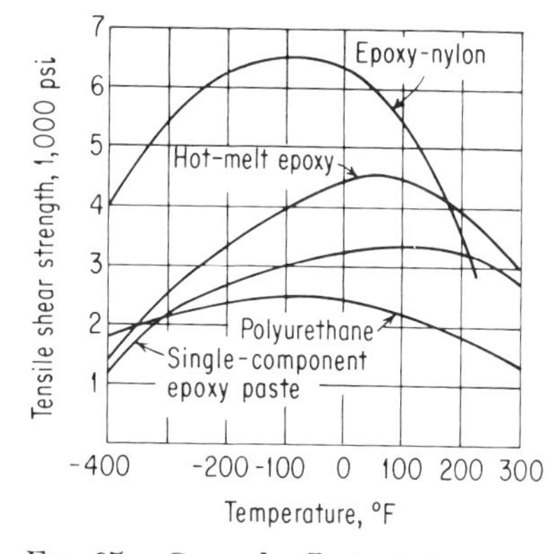

FIG. 65. General effect of temperature on strength of aluminum-to-aluminum joints bonded with various adhesives.[53]

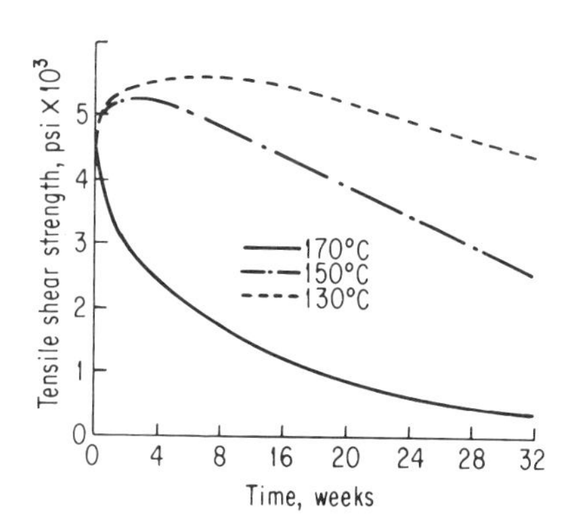

FIG. 66. Effect of elevated-temperature aging in air on an epoxy adhesive.[53]

whether an elastomeric, a thermoplastic, or a thermosetting adhesive—or one of the many alloys of these—was the best choice for the given application. The materials capable of providing higher-temperature capabilities are silicones among elastomers, FEP fluorocarbon among thermoplastics, and phenolics or high-temperature epoxies among thermosets. Typical strength versus temperature curves for two epoxies, an epoxy-nylon blended adhesive, and a polyurethane are shown in Fig. 65. Note the influence of low temperatures as well as high temperatures, indicating the usefulness of polyurethanes at low temperatures. Silicones, not shown here, are useful at both low and high temperatures. The effect of elevated temperature aging on a typical epoxy adhesive is shown in Fig. 66. For extremely high temperatures, over 300 to 400°F, some of the polyaromatic polymers (such as polybenzimidazoles, polyimides, and polyamide-imides) provide the best bond strength. Such a high-temperature bond-strength comparison is shown in Fig. 67.

Bond strength is also affected by oxygen, humidity, and other environments. Most adhesives will yield a considerably better thermal life and long-term, high-temperature bond strength if used in an inert atmosphere, such as nitrogen, as compared to heat aging in air. Also, humidity degrades most adhesive joints to varying degrees. Again, the basic nature of the polymer used in the adhesive can be a general guide.

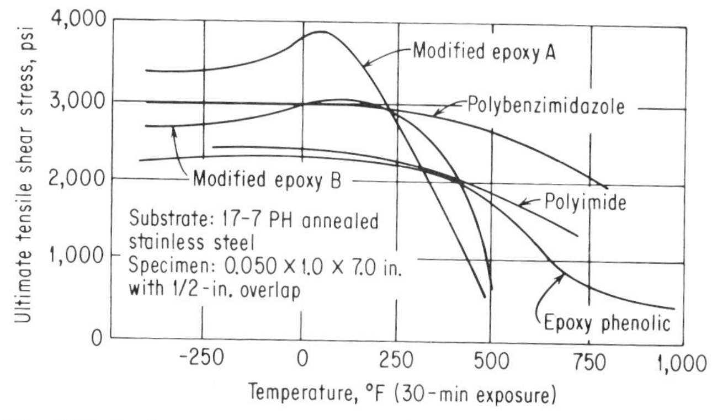

FIG. 67. Effect of temperature on several classes of structural adhesives.[54]

Certain plastics, notably thermoplastics such as fluorocarbons and polyethylenes, are difficult to bond with adhesives. They can be heat-bonded, but this is often impractical. To overcome this limitation, numerous surface treatments have been developed for these thermoplastics. While the material supplier is the best source of information regarding treatment of his particular thermoplastic, one widely used etching solution for pretreating fluorocarbons is sodium naphthalene. Many other plastics are difficult to bond without the use of primer pretreatment materials. Such primers, for instance, are commonly used to bond RTV silicone rubber materials to substrates. Here again, the material supplier is the best source of specific recommendations.

Even for materials which normally bond well, such as metals, thermosetting plastics, and elastomers, surface preparation before bonding is all-important to the final bond properties. As a minimum, all substrate materials to be bonded should be thoroughly cleaned and degreased. Vapor degreasing is excellent where it can be used, including with most metals and many thermosetting plastics. Degreasing solvents attack some plastics, however, notably elastomers and the less inert thermoplastics. Further, a mechanically abraded surface usually yields a more bondable surface and a better bond, where some form of abrading or surface roughening can be tolerated. Such abrading should precede degreasing and liquid cleaning, so that the loose particles resulting from abrading are removed before bonding. Another commonly used and very helpful surface preparation for bonding is chemical etching of the surface of the substrate to be bonded. Although plastic substrates can be, and often are, chemically etched for bonding (especially thermoplastics), many metal substrates lend themselves particularly well to chemical surface etching. These metal etchants are usually acids or alkalis, and the particular choice depends on the metal to be used. Selective chemical reactions are often involved, and the material supplier can best advise on etching specific metal substrates. A good guide for these is in the article by Sharpe.[50] Regardless of the bonding application, one universally accepted rule is that the best bonded joints are obtained when optimum surface preparation has been given to the substrates to be bonded. This rule should always be observed.

Plastic Coatings. Plastic, or organic, coatings are formulated from most plastic materials. Like other plastic products, they can be classified as thermosetting, thermoplastic, or elastomeric. Under such classification, many of the key end-product properties of coatings made from any given basic plastic will be similar to the properties of the base plastic. For instance, rigid thermosetting plastics will usually give harder coatings, and resilient or flexible thermoplastics or elastomers will give resilient or flexible coatings. Also, thermal aging properties, electrical properties, environmental properties, and many mechanical properties of coatings will be similar to the properties of the basic plastic from which the coating is made. For instance, low-electrical-loss coatings would be made from polystyrenes, fluorocarbons, diallyl phthalates, etc. Higher-temperature coatings would be made from silicones, polyimides, fluorocarbons, etc. Some of the important organic coatings are listed in Table 30. Properties of the basic thermosetting, thermoplastic, and elastomeric materials have been detailed in earlier parts of this chapter. There are, of course, properties of coatings which are basic to the use of the plastic as a coating, rather than or in addition to the primary plastic properties previously discussed. Some such properties are coatability, surface wetting, coating adhesion, and environmental chalking. For such properties, the obtaining of supplier advice is recommended since such properties are often affected by specific formulation, depending on whether the coating has been formulated for decorative, environmental, or insulation purposes.

Insulating Varnishes. Insulating varnishes are commonly used either as coatings or as liquid-insulating materials for impregnating or through-soaking of certain electrical structures such as transformer coils. Some varnishes are formulated from natural resins and others are derived from synthetics. Varnishes formulated from synthetic resins, however, are predominantly used in the electronics industry.

Varnishes may be either solvent-based, for viscosity reduction, or of a solventless or completely reactive type.

Lower-viscosity or thinner varnishes are usually possible with the solvent-based materials. They can, however, present the problem of entrapped solvent during the varnish-drying period, particularly if the varnish is used to impregnate deep sections or if a surface skin forms while the solvent evaporates.

TABLE 30. Application Information for Organic Coatings[55]*

Coating	*Distinguishing characteristics (max. continuous-use temperature, °F)*
Thermosetting coatings:	
Acrylics	Excellent resistance to UV and weathering. (250)
Alkyds	Most widely used general-purpose coating. (200–250)
Epoxies	Excellent chemical resistance and good insulation coating. Widely used as circuit-board coatings. (400)
Phenolics	Good chemical resistance against alkalies and good insulation coating. (350–400)
Phenolic-oil varnishes	Electrical impregnating varnishes. (250)
Polybutadienes	Good electrical insulation properties. (450)
Polyesters	Good electrical insulation properties. (200)
Silicones	Excellent high-temperature electrical insulation properties. (500)
Thermoplastic coatings:	
Acrylics	Excellent resistance to UV and weathering. (180)
Cellulosics	Fast-dry commercial lacquers. (180)
Fluorocarbons	Excellent chemical resistance, excellent electrical insulation properties, even at high temperatures. (400–500)
Penton chlorinated polyethers	Primarily used as chemically resistant coatings for equipment. Coatings have good electrical properties, however. (250)
Phenoxies	Low coefficient of expansion coatings. (180)
Polyimides	High-temperature coating having excellent insulation properties. Widely used as high-temperature magnet wire enamel. (600–700)
Polyurethanes	Good electrical insulation coatings. Widely used as circuit-board coatings. (250)
Vinyls	There are many varieties of vinyl coatings. Some vinyls are outstanding in resistance to inorganic and plating chemicals. Vinyl plastisols are convenient for dip coating of electrical parts. (150)

* In addition to the above basic types of coatings, there are many specialty coatings such as ablative coatings, thermal control coatings, flame-retardant coatings, fungus-resistant coatings, electrically conductive coatings, and magnetic coatings. In addition, many elastomers can be applied as coatings (e.g., neoprene). Most of the specialty coatings are specifically filled or otherwise modified variations of basic thermosetting, thermoplastic, or elastomeric coating polymers. Hence, the possible coatings available are as unlimited as the broad range of plastics and polymers.

Solvent-based varnishes are further classified as air-drying or baking. Obviously, air drying means drying or curing without heat, while baking, heat reacting, or polymerizing means that heat is required to dry or cure the coating. Air-drying coatings usually harden upon evaporation of the solvent, which is usually of a low boiling type. Among the important groups of insulating varnishes are alkyd, alkyd-phenolics, alkyd-silicones, silicones, phenolics, diallyl phthalates, and polyimides.

Solventless Varnishes. Solventless varnishes, such as polyester resins, epoxy resins, and urethane resins, are those which can be made to have relatively low viscosities at 100 percent solids. This low viscosity, however, is often only obtained by heating the resin. Being solventless, these materials do not incur the entrapment problems of the solvent-based varnishes. However, the solventless varnishes are usually two-component materials consisting of a resin and a curing agent. Once the curing agent is added, the resin varnish starts to react chemically by gradually thickening. Solventless varnishes, therefore, usually have a much shorter working life than do the solvent-based types. Depending on the application, this property could present certain processing problems.

In general, the functional properties of solventless varnishes are similar to those of the same class of resins when used for casting or potting applications. This is particularly true for the basic electrical properties. However, the coating thickness, as well as the basic resin used in the varnishes, can be a large factor in such properties as environmental resistance, chemical resistance, and mechanical characteristics, and where electrical losses are important, it can be important in electrical properties as well. Functionally, urethane coatings are tough and resilient; polyester and epoxy coatings can be formulated over a wide range of hardnesses and other end-property values.

Other Liquid Insulating Coatings. In addition to the varnishes used as coatings, there are numerous other types of insulating coatings useful for electronic packaging applications.

A number of the synthetic resins can be formulated into dip coatings, either as solvent solutions or as hot melts (which are actually melted plastic material). Also, many of the liquid resins, such as epoxies, can be formulated into thixotropic coatings and applied by dipping, spraying, or brushing. Any number of material combinations are commercially available and they can be readily procured from resin or coating suppliers. The term *thixotropic* implies a coating which will not run off the coated part.

Most epoxy-resin formulators have a series of liquid, thixotropic epoxy coating materials, either filled or unfilled, that can be applied to electronic assemblies as thick coatings. These materials are primarily used as conformal insulating coatings for electronic components and as conformal insulating coatings for printed-circuit boards.

Thixotropic coatings are particularly useful as conformal coatings for many electronic packaging applications. They can be used to obtain coatings of relatively controllable thickness and complete coverage, as compared to varnishes which are frequently too thin for effective coating of electronic parts. Varnish coatings are usually under 1 mil thick—not thick enough to cover surface irregularities—whereas thixotropic conformal coatings can range from 1 mil or so to 25 to 50 mils or higher.

Circuit-board Coatings. One area of electronic packaging where conformal coatings have special application is environmental protection of printed-circuit boards. Certain printed-circuit designs for military applications must conform to MIL-STD-275, which requires these coatings for low-conductor spacings as a function of voltage and altitude of operation, as shown in Table 31. Another design requirement of many printed-circuit boards, especially for military applications, is retention of insulation resistance properties in humid environments. Such coatings are specified in MIL-I-

TABLE 31. Minimum Allowable Spacings between Conductors on Printed-circuit Boards for Conformance to MIL-STD-275B; Uncoated and Coated Boards

Voltage between conductors, dc or ac peak, volts	*Minimum spacing, in.*
Uncoated boards—sea level to 10,000 ft	
0–150	0.025
151–300	0.050
301–500	0.100
Greater than 500	0.0002 (in./volt)
Uncoated boards—over 10,000 ft	
0–50	0.025
51–100	0.060
101–170	0.125
171–250	0.250
251–500	0.500
Greater than 500	0.001 (in./volt)
Coated boards—all altitudes	
0–30	0.010
31–50	0.015
51–150	0.020
151–300	0.030
301–500	0.060
Greater than 500	0.00012 (in./volt)

46058. Coatings conforming to this specification will provide retention of insulation resistance properties of printed-circuit boards, for a period of time, as shown in Fig. 68.

Powdered Coatings. In addition to liquid coatings, powdered coatings also frequently find use in electronic applications. Powdered coatings can be applied several ways, including electrostatic spraying and fluidizing. Fluidizing is perhaps most widely used, due to the simplicity and versatility of the fluidized-bed process. The coating is applied by dipping the preheated object into a fluidized bed of powdered plastic coating material. The coating of plastic is thereby melted onto the hot surface of the part being coated, except for electrostatic coating, where charged particles are sprayed onto a cold surface. The fluidized bed itself is a container partially filled with the powdered plastic coating material, and the powder is set in a fluid motion by an upward surge of low-pressure air. The fluidized powder has the appearance of a lightly boiling liquid and assumes flow characteristics similar to those of a liquid when a small air pressure is applied to the powder through a porous layer such as a filter. Figure 69 is an illustration of the fluidized-bed principle. In the process, heavy coatings can be applied in one dip, even over sharp edges. Normally, one dip in the coating powder will give a heavier film able to stand higher breakdown voltages than four to five dips in a conventional solvent-thinned insulating varnish.

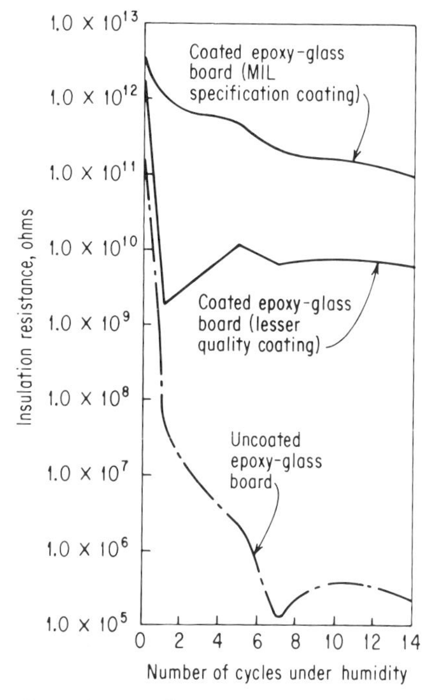

FIG. 68. Insulation resistance of coated and uncoated printed-circuit boards as a function of humidity cycling.[56]

Although the most widely used fluidizing powders for electronic packaging applications are the epoxy powders, other fluidizing powders are also used. Some of the other fluidized-bed coating powders available are fluorocarbons, silicones, chlorinated polyethers (Penton*), vinyls, cellulosics, nylons, polyethylenes, and polyesters.

Because of the manner in which this process is accomplished, certain variables become critical in determining the coating that will be actually obtained. For instance, parts that have a low heat capacity dissipate heat rapidly, and as they are being transferred from the oven to the fluidizing chamber they lose enough heat that a poor coating results. (For such parts, electrostatic coating is superior.) The part to be coated should have an adequate heat capacity and the operating conditions should be set for any given part. The temperature to which the part to be coated is preheated is also a factor in determining the ultimate coating thickness, as is the dipping time. The effect of preheat temperature and dipping time on film buildup, in coating a steel bar with an epoxy-fluidizing resin, is shown in Fig. 70.

After dipping, the coating is only melted—not cured. Therefore, the fluidized coating must have a final cure before the part is used. As in other situations involving the curing of a thermosetting plastic, cure time is a function of curing temperature.

Embedding Materials and Processes. Embedding processes and materials are perhaps among the most widely used plastics in electronic packaging. The materials used are predominantly rigid or elastomeric thermosetting liquids which can be hardened by use of a curing agent and/or heat to produce a hardened mass which

* Trademark, Hercules Powder Co.

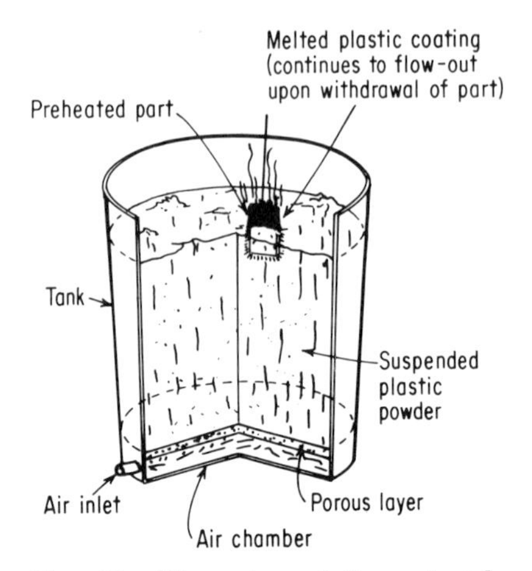

FIG. 69. Illustration of the principle of the fluidized-bed process.[57]

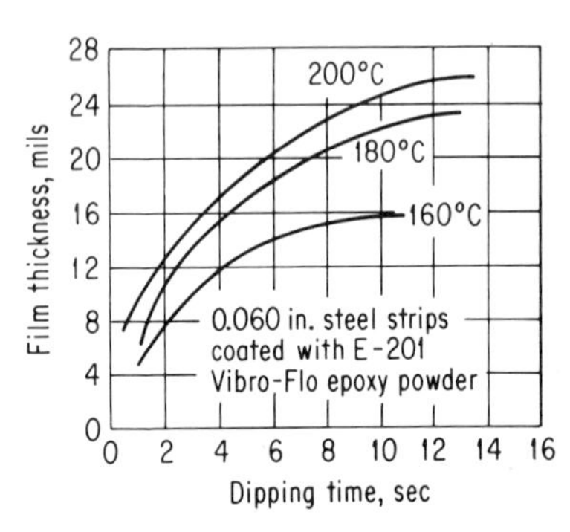

FIG. 70. Effect of preheat temperature and dipping time on film buildup in coating a steel bar with epoxy resin.[58]

embeds the desired electronic assembly, using the selected process or embedding technique. An exception to this is transfer molding. The important embedding processes and materials used in electronic packaging will be discussed below.

Processes and Terminology. The processes commonly used with embedding resins are potting, casting, transfer molding, encapsulating, and impregnating. Some of these terms can be used interchangeably, especially *embedding, potting,* and *encapsulation.* However, regardless of terminology, the utilization of low-pressure reacted resins is usually implied, as opposed to high-pressure molding resins. Embedding resins are those resins which can be converted from a liquid to a solid at room or slightly elevated temperatures, and at atmospheric pressure or low molding pressures. The conversion process is known as *curing* or *hardening.*

The low temperatures and pressures make these materials and processes attractive when compared with molding, which utilizes other plastic materials requiring more expensive molds and high temperatures and pressures. Pressure is perhaps the most important aspect, because processing costs are directly affected by the low-cost tooling that can be used with low-pressure processes.

EMBEDDING AND ENCAPSULATING. These are two of the most frequently interchanged words. Generally, *embedment* implies the complete encasement of a part or assembly to some uniform external shape. A large volume of the completed package is made up of the embedment material.

Encapsulation, on the other hand, is more correctly considered as a coating applied to a component or assembly. This usually involves dipping the part in a high-viscosity or thixotropic material to obtain a conformal coating on the surface—usually to a thickness of 10 to 50 mils or more. With encapsulation, the primary protection is sealing against atmospheric contaminants although, in some cases, a degree of mechanical strength is also gained. Problems in encapsulating are the need for surface wetting, control of resin runoff, and, in the completed product, variations in coating thickness and surface uniformity.

The terms *embedding* and *encapsulating* can be considered in another way. An embedding operation is usually performed by housing the part or assembly in a mold or case. The embedment material is then introduced, thereby completely surrounding the assembly. The mold or case confines the embedment material during its transition from a liquid to a solid state, at the completion of which the mold is removed. A smooth, uniform surface results, because the final product takes the shape of the mold. For an encapsulated coating, the final surface coat of the part is irregular (although it may be smooth), because it does not conform to the shape of a mold or case.

CASTING. For casting applications, the design of the mold and the design of the assembly should provide for minimum internal stresses during curing and for proper final dimensions after allowing for shrinkage.

In casting processes, the mold is cleaned thoroughly and a suitable release agent is applied, for many of the resins used normally adhere to the walls of the mold. The mold is then positioned around the part, and any points where leakage of the liquid resin from the mold might occur are sealed with a material such as cellulose acetate butyrate if they cannot be conveniently gasketed against leakage. Next, the resin and catalyst are mixed and slowly poured into the mold so as to avoid air entrapment during the pouring.

The entire assembly is allowed to cure either at room temperature under its own exothermic heat, or in an oven at some higher temperature, if such is the requirement of the resin-catalyst system. Finally, the part is separated from the mold. A typical process cycle for producing embedded assemblies by the cast-resin technique is shown in Fig. 71.

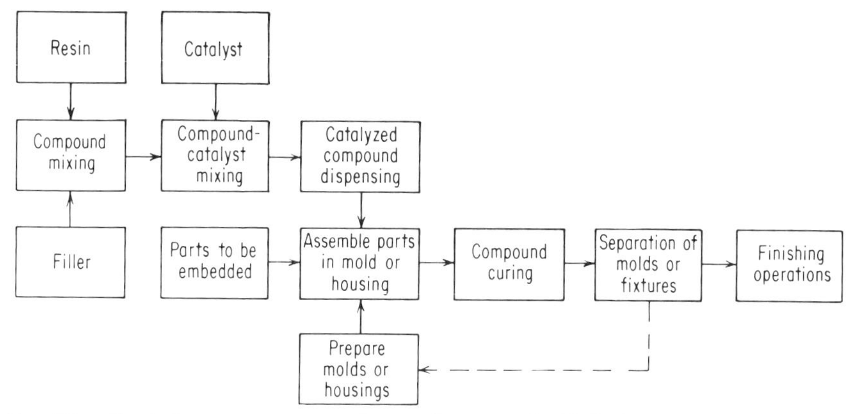

FIG. 71. Typical process sequence for producing embedded electronic assemblies by casting.[1]

The two main types of molds in common use are metal and plastic. The relatively high thermal conductivity of metal molds is an advantage where oven curing of the resin is used. The heat is rapidly and uniformly transferred into the resin. The insulating properties of plastics may require that parts processed in plastic molds be preheated before the liquid resin is poured. Fortunately, in many cases, curing is not really this critical, and preheating can usually be avoided. In fact, for room-temperature curing compounds, the insulating property of plastic molds can be advantageous because it retains the exothermic heat, thus permitting the resin to cure at room temperature. If a metal mold is used with a room-temperature curing compound, the exothermic heat may be dissipated through the mold, resulting in insufficient heat for the curing reaction.

If a part can be made in several ways, plastic tooling is often the best choice because it generally can be produced more economically, especially where large numbers of molds are required. The life of epoxy molds in particular is extremely good, and the cost is usually low enough to justify scrapping the molds when they fail.

Flexible plastic molds have the general advantage of allowing cured parts to be removed by squeezing and physically distorting the mold. However, for large runs and/or where dimensional control is important, flexible molds are often not suitable, because of their relatively short life and because they are easily distorted.

Where metal molds are required for assemblies with simple shapes, and where minimum cost is required, sprayed or dipped metal molds can be used. These are made by spraying molten metal on a mold or by dipping a mold into a molten metal, usually an alloy that melts at a low temperature.

If only small prototype molds are required, they can be machined from polyethylene or fluorocarbon flat or round stock. Polyethylene is more economical and is generally equivalent to fluorocarbon except where higher curing temperatures must be used. Neither provides close dimensional control, however, where oven cures are involved.

One useful technique for manufacturing large quantities of molds economically is to make them of polyethylene or polypropylene by injection-molding techniques. The same general limitations apply, however, to these types of molds as to machined plastic molds.

POTTING. The potting process is similar to casting, except that the assembly to be potted is positioned in a can, shell, or other container. Since the shell or can will not be separated from the finished part, no mold-release treatment is necessary. However, if the container is metal and the assembly is electrical, it is usually necessary to place a sheet of insulating material between the assembly and the can. The higher the voltage application, the more stringent is this requirement.

A problem sometimes occurs in potting, where adhesion to the case is poor, as with certain polyesters. A high-mechanical-strength package does not result if the can separates from the rest of the unit. Such is also the case with plastic shells, unless they bond well with the potting resin, or are prepared for potting by abrading or vapor-blasting the inside surface. An example of a good plastic-shell potting-resin system is a polystyrene or modified polystyrene shell used with a polyester resin. Here, the potting material and the shell will bond chemically.

A second problem with a potted unit is the difficulty in preventing spillage, overflow, and dropping of the resin onto the outside surface of the shell or can (the surface appearance is not really destroyed but it does become less neat). Solvent cleaning is unsatisfactory, because the solvents may affect both the shell and the potting material. Scraping is also unsatisfactory because it produces objectionable scratches.

Although, in one sense, housings might be considered tools, they are also an integral part of the final product. Selection of a housing is therefore an important design decision.

Housings for potted assemblies are made of metal or plastic. The same comparisons of these materials can be made for housings as for molds, with respect to curing and thermal conductivity. Since the housing remains an integral part of the embedded unit, however, some additional considerations are in order. For instance, corrosion resistance in humidity and salt spray can be an important factor, particularly if metal housings are used. Also, there may be standard metal containers available for certain types of products, such as transformers, which can reduce tooling costs. Thus, the choice between metals and plastics can be more important for potted units than for cast units, since the latter use only a temporary mold.

Probably even more important in the choice of housings is whether a metal or a plastic housing should be used. Metal provides good thermal dissipation to a heat sink where a high-heat-output device is embedded. But a metal container can cause shorting or arcing problems when electrical stress points are close to the metal wall. Where either metal or plastic will do, plastic housings are often more economical in mass-production operations.

Most economical, perhaps, are thermoplastic shells. However, thermoplastic shells have adhesion and temperature limitations. These usually can be overcome by using thermosetting shells or housings. Consideration of the relative importance of these factors must be made before a final choice of plastic housing is made. If a thermoplastic shell and a compatible resin system can be used, this combination can provide excellent adhesion. Such a combination would be polyester resin in a polystyrene housing. Here, the styrene in the polyester would effect welding to the polystyrene housing.

One other interesting consideration—for cylindrical embedded units—is the use of centrifugally cast shells for these cases. Tooling for centrifugal casting is reasonable, and precision cylindrical shells can be made from epoxy resins by this technique.

IMPREGNATING. In this process, the liquid resin is forced into all the interstices of the component or assembly, after which the resin is cured or hardened. This can be an independent operation, or it can be done in conjunction with embedding, encapsulating, casting, or potting. Impregnation differs from encapsulation in that encapsulation results in only a coating, with little or no resin penetration into the assembly. Penetration is most important for certain electrical parts such as transformers.

Impregnation is accomplished by submerging the assembly in the catalyzed resin and applying vacuum, pressure, or a vacuum-pressure sequence. The length of the impregnation cycle depends on the degree of impregnation desired, the desired freedom from entrapped air, the viscosity of the material, and the type of surface barrier through which impregnation is done.

Impregnation is sometimes accomplished by centrifugal means. The part is positioned in a mold, the mold is filled with resin, and the entire assembly is spun at a high velocity.

If both encapsulation and impregnation are required, as in some transformer applications, the encapsulation dip coating is usually applied first. A hole is left in the coating so that the low-viscosity impregnating resin can be forced in after the shell has hardened. This procedure provides a container, thus eliminating drainoff of the impregnating material during its hardening or curing cycle.

TRANSFER MOLDING. Although most resin-embedded electronic packages are produced by casting or potting techniques, the use of transfer molding is also widespread. Transfer molding offers advantages in economy and increased production rates for those assemblies that adapt to this technique, and that are produced in large quantities.

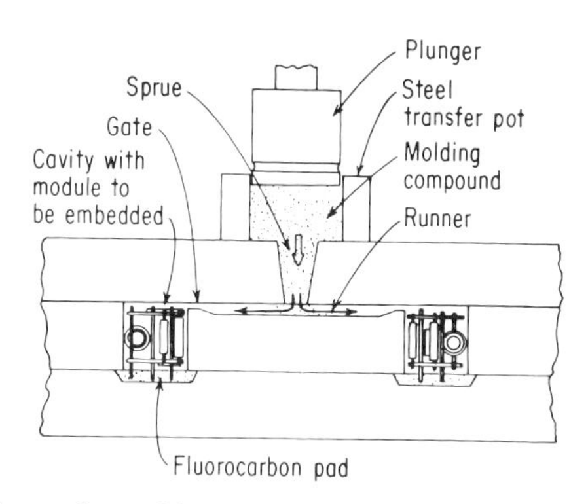

FIG. 72. Typical transfer molding assembly, showing molding compound flow.[1]

These advantages consist of a large reduction in processing steps (over casting or potting), and a shorter curing time of the embedding compound. Transfer molding materials cure in minutes; liquid casting and potting resins require hours.

Major limitations of the transfer molding process are: (1) The assembly must be able to withstand pressures of 50 to 250 psi. (2) The assembly must be able to take the curing temperatures of 250 to 350°F. (3) The production volume must be large enough to justify equipment expenditures.

In transfer molding, a dry, solid molding compound—usually in powder or pellet form—is heated in a molding press to the point of becoming flowable or liquid, at which time it flows (is transferred) under pressure into a mold cavity containing the assembly to be embedded (Fig. 72). The plastic remains in the heated mold for a short time until curing is completed.

The transfer mold shown in Fig. 72 embeds two similar parts, although a larger number of parts can be molded simultaneously. Multicavity molds are common in transfer molding, and represent one of the major points of economy for large-volume runs. Many cavities can be filled as rapidly as a single cavity, thereby reducing the cost per part. While mold cost increases as the number of cavities increases, total cost does not increase proportionately. Overall mold cost per part produced can be further reduced by incorporating cavities of different shapes into the same mold in proportion to the production volumes required, or by using mold inserts to vary the cavity configuration as required by changing production needs.

A gate scar remains on the finished part at the point where the molding compound goes out of the runners from the transfer cylinder into the cavity. The small scar is usually unobjectionable.

Processing Characteristics of Embedding Resins. Nearly all embedding resins have important basic processing characteristics of viscosity and exotherm. Each resin-curing agent-curing cycle system provides its own unique set of viscosity and exotherm curves, however. The nature of these two important processing characteristics will be explained below. It should be mentioned that even transfer molding materials, which are received as solids, go through a liquid phase in their cure, and that they have viscosity and exothermic characteristics equally important to those of liquid resin systems. While there are numerous chemical types of resins, each having different end properties, most liquid resins cure by heat and/or curing-agent influences, give off heat during the curing process, and are thermosetting. Resin viscosity or fluidity and the time-temperature curve (exotherm) for the exothermic reaction vary with each resin system, and are key properties describing the nature of the individual resin through the curing cycle.

VISCOSITY. Flow properties of resins are important because of the need for flow and penetration at atmospheric or low pressures. When the viscosity is too high, the formulation is difficult to pour and does not flow properly around inserts or components, thus allowing internal cavities to form. A high-viscosity resin is usually too thick to allow evacuation of entrapped air, which also causes cavity formations. High viscosity also makes mixing difficult. On the other hand, a resin whose viscosity is too low may cause problems of leakage through openings in the mold or container.

For most embedding applications, there is an optimum range of viscosity. For impregnating operations, extremely low viscosities—100 centipoises or less—are desirable because complete impregnation of the parts under vacuum is required. In practice, however, impregnation is often achieved with viscosities considerably higher—up to 1,000 centipoises. However, the higher the viscosity, the longer the cycling time and/or the higher the vacuum required for complete impregnation.

For embedment operations such as casting or potting, there is no limit to how low the viscosity can be, provided that the mold or container is tight enough to prevent leakage. Usually, however, if impregnation is not required and if the components are not packed tightly, viscosities in the range of 1,000 to 5,000 centipoises are satisfactory for casting and potting operations.

An encapsulation coating requires a thixotropic (nonflowing) material with an extremely high viscosity, because the part is dipped into the compound and the coated part is cured without the use of a mold or container. The coating must not flow off during the curing operation.

Viscosity usually can be lowered by heating the resin (Fig. 73) or by adding diluents, and it can be raised by addition of fillers. Not all resins exhibit as great a viscosity dependence on temperature as that shown in Fig. 73; silicones, for example, have a relatively flat viscosity curve.

EXOTHERMIC PROPERTIES. Most polymeric resins used for embedment have an exothermic reaction; that is, heat is produced as the reaction progresses. It is essential that the exothermic properties of a particular system be known and controlled. Too much heat can cause resin cracking during cure, and the heat generated can also affect heat-sensitive components.

Three specific values are commonly used for control measurements of these exothermic properties: gel time, peak exothermic temperature, and time to peak exothermic temperature. These characteristics are measured from a single graphic plot of exothermic temperature versus time for a given resin-catalyst-curing-agent system. A typical exothermic curve for a polyester resin is shown in Fig. 74. Although shapes of these curves vary widely from system to system, the curve for a given system should be closely reproducible. Exothermic curves also vary with mass of resin, as shown in Fig. 75.

Gel time is the interval from the time the exothermic reaction temperature reaches 150°F to the time when it is 10°F above the bath temperature (see Fig. 74). The reason

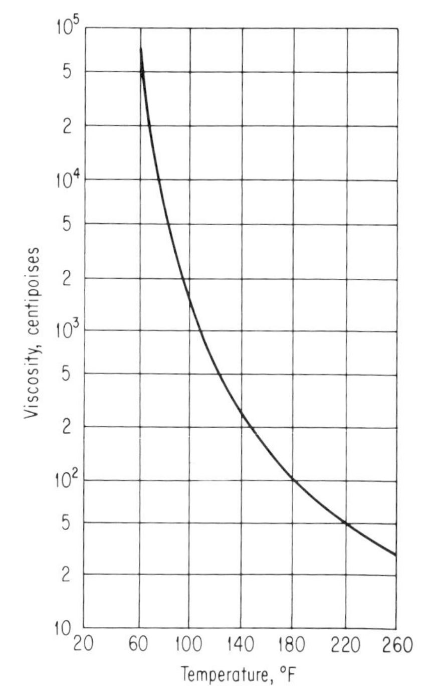

FIG. 73. Viscosity-temperature relationship for a bisphenol epoxy resin.[1]

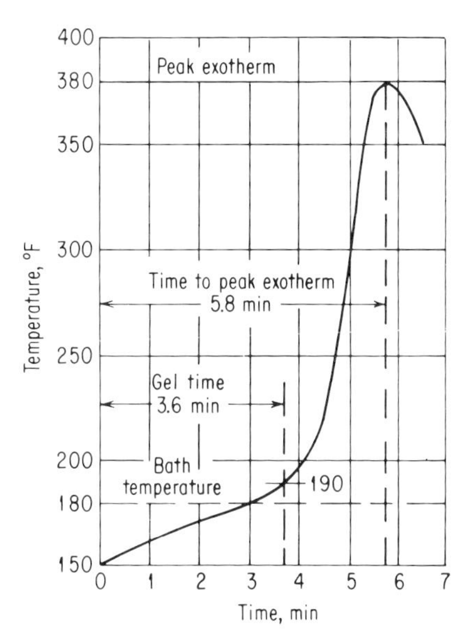

FIG. 74. Exothermic reaction curve for a typical polyester resin, using benzoyl peroxide catalyst and cured at 180°F.[1]

for starting the timing from a 150°F exothermic temperature rather than from the time at which the catalyst and resin are initially mixed is that it is not always practical to have the temperature of the ingredients precisely the same when the reaction starts, that is, during and immediately after the initial mixing. The common base point is used to assure better reproducibility. Gelation usually occurs by the time the exothermic temperature has slightly exceeded the bath temperature, for this type of resin system.

The exothermic temperature rise is much greater after gelation has occurred than

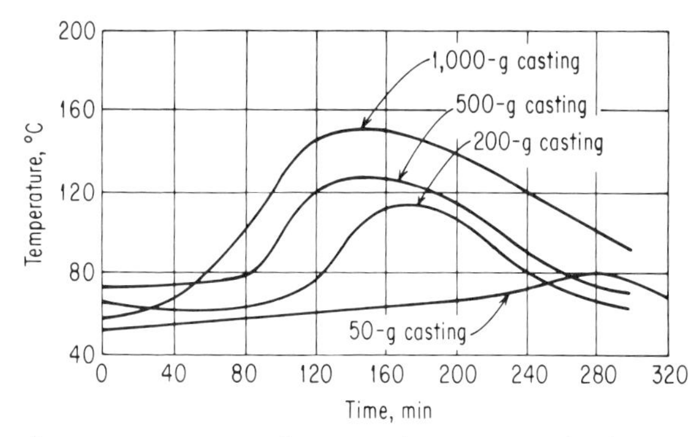

FIG. 75. Exothermic curves, as a function of resin mass, for bisphenol epoxy and 5% piperidene curing agent (catalytic type) cured at 60°C.[1]

before. This tendency is common for these materials. All resin-curing agent-curing cycle changes vary this curve, which can be nearly flat or very steep.

Resins for Embedding. Many chemical groups of embedding resins are available, and there are many variations in each group. The most important of these are discussed in the following sections. Typical mechanical, physical, thermal, and electrical properties of several of these classes are shown in Tables 32 and 33.

TABLE 32. Typical Mechanical and Physical Properties of Several Common Embedding Resins[59]

Material	Tensile strength, psi	Elongation, percent	Compression strength, psi	Impact strength, Izod, ft-lb/in. of notch	Hardness	Linear shrinkage during cure, percent	Water absorption, percent by weight
Epoxy:							
Rigid, unfilled	9,000	3	20,000	0.5	Rockwell M 100	0.3	0.12
Rigid, filled	10,000	2	25,000	0.4	Rockwell M 110	0.1	0.07
Flexible, unfilled	5,000	50	8,000	3.0	Shore D 50	0.9	0.38
Flexible, filled	4,000	40	10,000	2.0	Shore D 65	0.6	0.32
Polyester:							
Rigid, unfilled	10,000	3	25,000	0.3	Rockwell M 100	2.2	0.35
Flexible, unfilled	1,500	100	....	7.0	Shore A 90	3.0	1.5
Silicone:							
Flexible, unfilled	500	175	....	No break	Shore A 40	0.4	0.12
Urethane:							
Flexible, unfilled	500	300	20,000	No break	Shore A 70	2.0	0.65

TABLE 33. Typical Thermal and Electrical Properties of Several Common Embedding Resins[59]

Material	Heat-distortion temp., °C	Thermal shock per MIL-I-16923	Coefficient of thermal expansion, ppm/°C	Thermal conductivity, (cal)(cm)/(cm²)(sec)(°C)	Dissipation factor*	Dielectric constant*	Volume resistivity,* ohm-cm	Dielectric strength,* volts/mil	Arc resistance, sec
Epoxy:									
Rigid, unfilled..	140	Fails	55	4×10^{-4}	0.006	4.2	10^{15}	450	85
Rigid, filled	140	Marginal	30	15×10^{-4}	0.02	4.7	10^{15}	450	150
Flexible, unfilled..	<RT	Passes	100	4×10^{-4}	0.03	3.9	10^{15}	350	120
Flexible, filled....	<RT	Passes	70	12×10^{-4}	0.05	4.1	3×10^{15}	130	360
Polyester:									
Rigid, unfilled..	120	Fails	75	4×10^{-4}	0.017	3.7	10^{14}	440	125
Flexible, unfilled..	<RT	Passes	130	4×10^{-4}	0.10	6.0	5×10^{12}	325	135
Silicone:									
Flexible, unfilled..	<RT	Passes	400	5×10^{-4}	0.001	4.0	2×10^{15}	550	120
Urethane:									
Flexible, unfilled..	<RT	Passes	150	5×10^{-4}	0.016	5.2	2×10^{12}	400	180

* Dissipation factor and dielectric constant are at 60 cps and room temperature; volume resistivity is at 500 volts d-c; and dielectric strength is short time.

EPOXIES. The most used of the embedding resins are the epoxies, in many types and modifications. All classes of epoxies have certain outstanding characteristics which are important in electronic assemblies. Chief among these properties are low shrinkage, excellent adhesion, excellent resistance to most environmental extremes, and ease of application for casting, potting, or encapsulation.

Bisphenol Epoxies. The original class of epoxies—the bisphenols—is the workhorse of the electronics industry. Bisphenols are available as liquids over a wide viscosity range, and also as solids. The resins are sirupy, having viscosities in the range of 10,000 to 20,000 centipoises at room temperature. Their viscosity can be reduced, of course, by heating.

End properties of bisphenol epoxies, as well as of other epoxies discussed here, are controlled by the type of curing agent used with the resin. Major types of curing agents are aliphatic amines, aromatic amines, catalytic curing agents, and acid anhydrides (Table 34).

TABLE 34. Curing Agents for Epoxy Resins [1]

Curing-agent type	Characteristics	Typical materials
Aliphatic amines	Aliphatic amines allow curing of epoxy resins at room temperature, and thus are widely used. Resins cured with aliphatic amines, however, usually develop the highest exothermic temperatures during the curing reaction, and therefore the mass of material which can be cured is limited. Epoxy resins cured with aliphatic amines have the greatest tendency toward degradation of electrical and physical properties at elevated temperatures.	Diethylene triamine (DETA) Triethylene tetramine (TETA)
Aromatic amines	Epoxies cured with aromatic amines usually have a longer working life than do epoxies cured with aliphatic amines. Aromatic amines usually require an elevated-temperature cure. Many of these curing agents are solid and must be melted into the epoxy, which makes them relatively difficult to use. The cured resin systems, however, can be used at temperatures considerably above those which are safe for resin systems cured with aliphatic amines.	Metaphenylene diamine (MPDA) Methylene dianiline (MDA) Diamino diphenyl sulfone (DDS or DADS)
Catalytic curing agents	Catalytic curing agents also have a working life better than that of aliphatic amine curing agents, and, like the aromatic amines, normally require curing of the resin system at a temperature of 200°F or above. In some cases, the exothermic reaction is critically affected by the mass of the resin mixture.	Piperidene Boron trifluoride-ethylamine complex Benzyl dimethylamine (BDMA)
Acid anhydrides	The development of liquid acid anhydrides provides curing agents which are easy to work with, have minimum toxicity problems compared with amines, and offer optimum high-temperature properties of the cured resins. These curing agents are becoming more and more widely used.	Nadic methyl anhydride (NMA) Dodecenyl succinic anhydride (DDSA) Hexahydrophthalic anhydride (HHPA) Alkendic anhydride

Aliphatic amine curing agents produce a resin-curing-agent mixture which has a relatively short working life but which cures at room temperature or at low baking temperatures, in relatively short time. Resins cured with aliphatic amine usually develop the highest exothermic temperatures during the curing reaction; thus the amount of material which can be cured at one time is limited because of possible cracking, crazing, or even charring of the resin system if too large a mass is mixed and cured. Also, physical and electrical properties of epoxy resins cured with aliphatic amines tend to degrade as operating temperature increases. Epoxies cured with aliphatic amines find their greatest usefulness where small masses can be used, where room-temperature curing is desirable, and where the operating temperature required is below 100°C.

Epoxies cured with aromatic amines have a considerably longer working life than do those cured with aliphatic amines, but they require curing at 100°C or higher. Resins cured with aromatic amines can operate at a temperature considerably above the temperature necessary for those cured with aliphatic amines. However, aromatic amines are not so easy to work with as aliphatic amines, due to the solid nature of the curing agents and to the fact that some (such as MPDA) sublime when heated, causing stains and residue deposition.

Catalytic curing agents also have longer working lives than the aliphatic amine materials, and like the aromatic amines, catalytic curing agents normally require curing of the epoxy system at 100°C or above. Resins cured with these systems have good high-temperature properties as compared with epoxies cured with aliphatic amines. With some of the catalytic curing agents, the exothermic reaction becomes high as the mass of the resin mixture increases (Fig. 75).

Acid anhydride curing agents are particularly important for epoxy resins, especially the liquid anhydrides. The high-temperature properties of resin systems cured with these materials are better than those of resin systems cured with aromatic amines. Some anhydride-cured epoxy-resin systems retain most electrical properties to 150°C and higher, and are little affected physically even after prolonged heat aging at 200°C. In addition, the liquid anhydrides are extremely easy to work with; they blend easily with the resins and reduce the viscosity of the resin system. Also, the working life of the liquid acid anhydride systems is long compared with that of mixtures of aliphatic amine and resin, and odors are slight. Amine promoters such as benzyl dimethylamine (BDMA) or DMP-30 are used to promote the curing of mixtures of acid anhydride and epoxy resin. Thermal stability of epoxies is improved by anhydride curing agents, as shown in Fig. 76.

Novolac Epoxies. Excellent high-temperature properties are characteristic of novolac epoxy systems, especially using high-temperature curing agents. These resins contain more of the benzene ring or phenolic-type structure in the molecule and thus combine the excellent thermal stability of the phenolics with the reactivity (with curing agents and catalysts) and versatility of the epoxies. Because their average epoxide

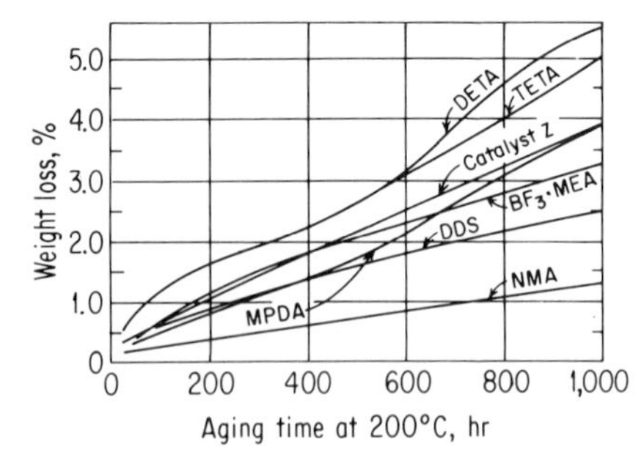

Fig. 76. Weight loss data at 200°C for Epon 828 cured with various curing agents. DETA and TETA are aliphatic amines, MPDA and DDS are aromatic amines, and NMA is a liquid acid anhydride.[60]

functionality* is greater than 3 or 4, tightly crosslinked structures are readily formed, producing cured masses that exhibit higher heat-distortion temperature, better chemical resistance, and better adhesion than do bisphenol epoxies having an epoxide functionality of about 2. The functionality of novolac epoxy ERR-0100, produced by Union Carbide, averages about 5 epoxy groups per molecule, and that of D.E.N. 438, produced by Dow, averages 3.3 epoxy groups per molecule, compared with a functionality of 2.10 and below for most conventional epoxy resins. The close-knit crosslinking of the cured novolac epoxy assures excellent retention of mechanical properties at high temperatures and thereby enlarges those application areas in which properties of epoxies are needed.

Viscosity of many novolac epoxies is originally high, but can be reduced by conventional ways: with solvents, with lower-viscosity resins, with diluents, with heat, or with low-viscosity hardeners. Hardeners or curing agents used with conventional bisphenol A epoxy resins are also used with novolac epoxies because the curing reaction is through the epoxy groups of the molecule. Postcuring is usually necessary to develop the maximum properties of the novolac epoxy resins.

TABLE 35. Properties of Epoxy-Polyamide Systems [61]

Epoxy-polyamide weight ratio	80:20	70:30	60:40	50:50	40:60	30:70	25:75
Heat-distortion temperature, °F. .	220	215	136	100	65		
Hardness	B70*	B66	B58	B50	B30	A70†	A40
Specific resistivity, ohm-cm	10^{15}	10^{15}	10^{14}	10^{12}	10^{10}	10^{9}	10^{8}
Moisture absorption, %	0.15	. . . .	0.20	. . . .	0.50	1.0	2.0

* Barcol M.
† Shore A.

TABLE 36. Properties of Epoxy-Polysulfide Combinations [62]

Epoxide/polysulfide weight ratio	1:0	2:1	1:1	1:2	1:3	0:1
Elongation, %	1	10	30	50	100	200
Hardness, Shore A . .	100	98	95	70	50	35
Specific resistivity, ohm-cm	10^{14}	10^{13}	10^{12}	10^{11}	10^{10}	10^{9}
Dielectric constant at 10^{6} cps	3.5	3.7	4.0	5.5	6.5	7.5
Loss tangent at 10^{6} cps	0.005	0.01	0.03			

Cycloaliphatic Diepoxides. These materials do not contain the phenolic rings which are associated with most epoxy resins. They offer unusual combinations of low viscosity, low vapor pressure, and high reactivity toward acidic curing agents such as polycarboxylic acids and anhydrides. A wide range of cured resin properties is possible depending on the selection of epoxide and hardener. Either rigid or flexibilized cured resins are possible. These formulations, which are free of aromatic structures, show outstanding resistance to the formation of carbon tracks under an electrical arc. They have excellent electrical properties. They also resist discoloration when exposed to ultraviolet light.

Flexibilized and Modified Epoxies. Four major flexibilizers used with epoxy-resin systems are polyamides, polysulfides, polycarboxylic acids, and polyurethanes. See

* Average epoxide functionality is the average number of epoxide or epoxy groups per molecule. Since crosslinking is through the epoxy group, a higher functionality means more crosslinking, more rigidity, and more resistance to thermal degradation. Unfortunately, it often also means more brittleness and tendency to crack.

TABLE 37. Effect of Polycarboxylic Acid Flexibilizer on Anhydride-cured Epoxy Resin [63]

	Formula 1*	Formula 2†
Hardness, Shore D, 10 sec	86	72
Percent weight loss after heat aging 64 hr at 150°C on 2-in. circle casting of ⅛-in. thickness	0.23	3.51
Percent weight increase after water absorption on immersion in boiling water for 2 hr on 2-in. circle casting of ⅛-in. thickness	+0.16	+1.40
Dielectric constant, 1,000 cycles at 25°C	3.45	4.23
Dissipation factor, 1,000 cycles at 25°C	0.0063	0.0258
Volume resistivity, at 25°C	1.18×10^{15}	1.41×10^{14}

* Formula 1 contained 100 parts liquid epoxy resin, 75 parts phthalic anhydride, and 0.1 part benzyl dimethyl amine. The cure cycle was 2 hr at 120°C and 16 hr at 150°C.

† Formula 2 contained 100 parts liquid epoxy resin, 53.6 parts phthalic anhydride, and 79.5 parts Harcure E. The cure cycle was 16 hr at 150°C.

TABLE 38. Properties of Epoxy-Polyurethane Systems [64]

Epoxy/polyurethane weight ratio	0:100	25:75	50:50	75:25	100:0
Ultimate tensile strength, psi	5,000	2,050	6,000	10,000	10,000
Ultimate elongation, %	420	350	10	10	10
Hardness, Shore A	90	60			
Hardness, Shore D	40	15	80	85	90
Heat-distortion temperature, °F	*	*	100	176	260

* Too low to measure.

Tables 35 to 38. Although flexibilizers improve thermal shock resistance and reduce internal stresses, some degradation in electrical properties results. However, it is usually possible to achieve a compromise between the electrical and physical properties desired. A large variety of flexibilizers and flexibilized resins is available from resin formulators. These suppliers should be consulted for final selection of flexibilized system.

Polysulfides also can be used alone for potting and coating assemblies. They provide a flexible end product with low moisture permeability, but with relatively low thermal endurance. Polyurethanes can also be used alone for many applications.

Another flexibilizing system for epoxy resins, linear polyazelaic polyanhydride (PAPA), offers several advantages. Epoxides flexibilized with PAPA do not increase in hardness, and do maintain their toughness with under 1 percent weight loss after aging for periods of up to 8 weeks at 300°F.[66]

While epoxy resins constitute the largest-volume usage for embedding electronic packages, other resins are also important. These include silicones, urethanes, polyesters, thermosetting hydrocarbons, thermosetting acrylics, and polysulfides. Also, foams and low-density resins are widely used for electronic packaging for weight-reduction purposes. These resins and their important characteristics are discussed below.

SILICONES. The silicone resins are convenient to use, they are available over a wide range of viscosities, and most of them can be cured either at room temperature or at low temperatures. Silicones maintain their properties over a wide temperature range, generally from approximately −65 to over 200°C and, in some cases, up to 300°C. A third advantage is their excellent electrical properties, particularly low loss factors and low dielectric constants, similar to data shown in Fig. 49 and Figs. 54 to 59. Their electrical properties are generally stable with temperature and frequency, an advantage over most other embedding systems.

Three classes of silicones are used for embedding applications: RTV silicones and flexible resins; silicone gels; and rigid, solventless resins.

RTV silicones and flexible resins are by far the most widely used silicones. These flexible materials have excellent thermal-shock resistance and low internal curing stresses. Some can be cured at room temperatures. While most are pigmented or colored, several clear, flexible resins are available. These materials are increasing in usage since they have most of the good properties of the pigmented materials in addition to their own optical clarity. The combination of flexibility and clarity facilitates cutting and repairing when needed.

The cure of some clear, flexible resins is inhibited when these resins contact certain materials. Notable inhibiting materials are sulfur-vulcanized rubber and certain RTV silicone rubbers. This problem can usually be overcome with a coating of a noninhibiting material on the inhibiting component.

Silicone gels, as the name implies, exist in a gel state after being cured. Although these materials are very tough, they are usually used in a can or case. The gel-like characteristic of a silicone gel allows for electrical checking of circuits and components. After the test probes are withdrawn, the memory of the gel is sufficient to heal the portion which has been broken by the probes.

Rigid solventless silicones are not used so widely as the other groups of silicones, because their resistance to thermal shock and cracking is not so good as that of the flexible materials, and because the rigid solventless resins are not so convenient to work with as are the room-temperature curing materials. However, where the general properties of silicones are desired and rigidity is preferred to flexibility, the rigid solventless silicone resins should be considered.

URETHANES. Cured urethane or polyurethane resins are very tough. Their tear strength is high compared with most other resilient or flexible materials. Polyurethanes have very good thermal shock resistance, and are convenient to work with. Their combination of toughness and resistance to cracking in thermal shock offers advantages in many applications. Cure is effected at room temperature or at low baking temperatures. Their electrical properties are excellent, and their adhesion to most materials is excellent—an advantage over silicones in some instances. Their operating temperatures—limited to below 300°F—do not compare with those of silicones, however. Because liquid polyurethanes react with moisture, careful drying of parts to be embedded is required to avoid bubbles. Many liquid polyurethane resin systems are available.

POLYESTERS. These materials were among the first of the liquid resins to be used for embedding electronic devices, and they are available in all degrees of flexibility. Most general-purpose polyesters are copolymers of a basic polyester resin and styrene monomer. However, other monomers are also used, and particularly good high-temperature stability is obtained in polyesters which are copolymerized with triallyl cyanurate.

Most polyesters have higher shrinkage, less thermal shock resistance, less humidity resistance, and lower adhesion than the epoxies. The electrical properties of polyesters are generally very good, and these materials are widely used in commercial applications because their cost is comparatively low.

THERMOSETTING HYDROCARBONS. Another class of liquid resins which can be used for embedding applications is the group called thermosetting hydrocarbons. Thermosetting hydrocarbons can be formulated for fairly wide ranges of flexibility or hardness.[67] Their primary attributes are excellent electrical properties and low moisture absorption. The cured resins are transparent, and the flexible types can be cut open easily for repair purposes. Their dielectric constant ranges from 2.5 to 3.1, their dissipation factor ranges from 0.002 to 0.010, and their volume resistivity ranges from 10^{15} to 10^{17}, at room temperature.

THERMOSETTING ACRYLICS. Another class of synthetic resins which have potential application for embedding of various assemblies is the thermosetting acrylics. An outstanding property reported for these materials is their good heat resistance up to 500°F, which approaches that of the best epoxy-resin systems.[68] Weight losses of under 5 percent are recorded after 1,000 hr at 200°C. Like the other resins discussed here, thermosetting acrylics are solventless and hence give off no by-products during the curing reaction. Like polyesters, they can be modified with monomers such as styrene. Their electrical properties are comparable to those of polyesters.

POLYSULFIDES. Polysulfides are available as liquids for use in potting of electrical connectors. The cured polysulfide rubber is flexible and has excellent resistance to solvents, oxidation, ozone, and weathering.[62] Its gas permeability is low, and its electrical insulation properties are good at temperatures between -65 and +250°F.

Polysulfide rubber resins are of the same chemical class as the polysulfide rubber resins used in modifying epoxy resins. Chemically, polysulfide rubbers are organic compounds containing sulfur; the sulfur groups in the polymer chain are known as mercaptan groups.

LOW-DENSITY FOAMS. Since most liquid resins can be made into low-density foams by addition of selective foaming or blowing agents, most of the resins discussed previously are available in formulations which can be foamed in place. Foams in general have been discussed earlier in this chapter. Epoxy and silicone foams are used in many embedding applications. However, the polyurethane foams are by far the most used of the foam materials. Urethane foams do not require blowing agents since gas is liberated during polymerization. These foams cure at room temperature or at a low baking temperature and are relatively easy to work with, particularly the prepolymer foams.

Generally, foams have lower electrical losses, lower dielectric strength, lower thermal conductivity, and less mechanical strength than high-density resins. Usually, changes in these properties are almost a direct function of foam density.

Fillers for Embedding Resins. Fillers play a most important role in the application of resins for embedding applications. They are additives, usually inert, which are used to modify nearly all basic resin properties in the direction desired. Fillers overcome many limitations of the basic resins. Through proper use of fillers, major changes can be made in important resin properties such as thermal conductivity, coefficient of thermal expansion, shrinkage, thermal shock resistance, density, exotherm, viscosity, and cost.

A tabulation of costs and effects on resin properties of the more commonly used fillers is given in Table 39. Because of the large number of materials and suppliers available, this listing is necessarily not comprehensive; however, it does provide basic information on the more commonly used fillers.

Although different fillers affect a given resin property to various degrees, the nature and direction of the effect are often similar regardless of the filler employed. That is, the magnitude of the effect a filler has on a given property seems to be closely associated with the amount of filler used. Figure 77 demonstrates this effect; the data shown are averaged for several fillers.

A major problem with most resin systems is their tendency to crack because of the difference in thermal expansion between embedded parts and embedment material. Figure 78 compares the thermal expansion of various materials with that of filled and unfilled epoxy resins. Addition of sufficient filler can reduce the thermal expansion coefficient of epoxies to the range of coefficients for metals. This is also true, to some extent, of other embedding resins. Although the trend is the same for most fillers, effects of specific fillers vary to some degree (Fig. 79).

The viscosity of resin compounds is also increased by addition of fillers, especially thixotroping fillers such as finely dispersed silica. This type of filler is often used to produce a thixotropic resin for encapsulation applications.

EFFECTS ON THERMAL PROPERTIES. In addition to the beneficial effects of reducing the thermal expansion characteristics of a resin, fillers also increase thermal conductivity (Fig. 80) and reduce the weight loss (during heat aging) of a resin. The more filler incorporated into a resin, the lower will be the weight loss of that system during heat aging. Although the resin portion of the system will degrade upon heat aging, overall performance is almost always improved by the incorporation of fillers. This results not only from reduced weight loss, but also from shrinkage reduction and thermal conductivity increase produced by incorporation of filler into the system.

Another beneficial effect of fillers on properties of a resin system is reduction in the exothermic heat of the system during the curing cycle. This effect, when com-

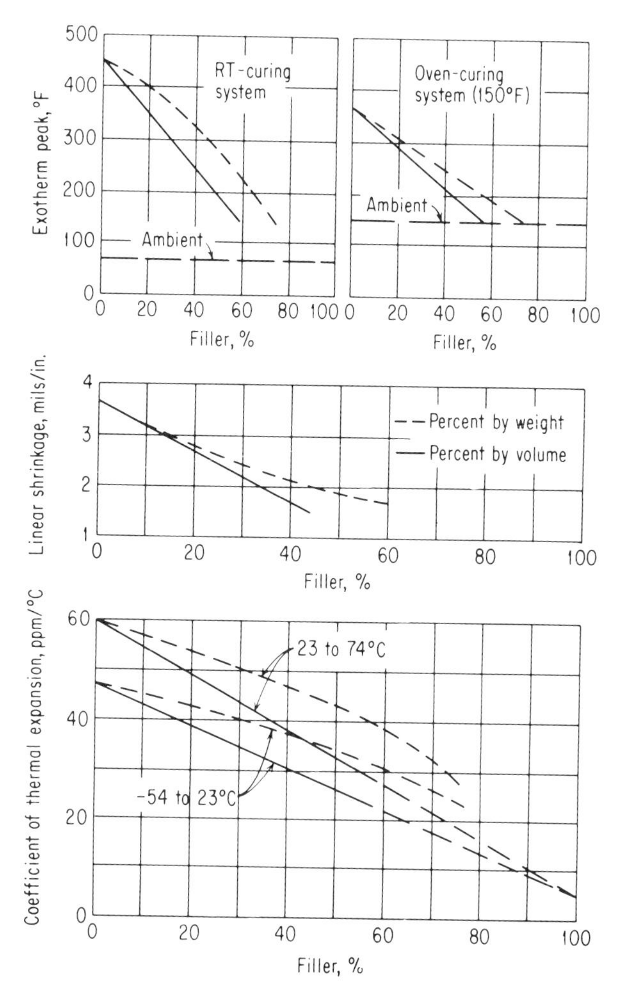

FIG. 77. Effect of filler concentration on exotherm, shrinkage, and thermal expansion of an epoxy resin.[69]

bined with reduced shrinkage and decreased thermal expansion, gives many resin systems minimal resin-cracking characteristics. Thus, the addition of filler can change a normally unsatisfactory system into a very usable system with respect to cracking.

Fillers can also increase the pot life of a resin system. The extension of pot life is related to the control of exothermic heat of the system.

Still another thermal property that can be improved by the use of fillers is fire resistance or burning rate. The burning rate is reduced considerably through the addition of a filler such as antimony oxide. Certain phosphates can be used to reduce the flammability of embedding resins.

EFFECTS ON MECHANICAL PROPERTIES. Fillers can have a major effect on mechanical properties of resins; however, the type of filler is also important. With respect to their effects on mechanical properties, fillers are classed as nonreinforcing (bulk, or nonfibrous) and reinforcing, as shown in Table 39.

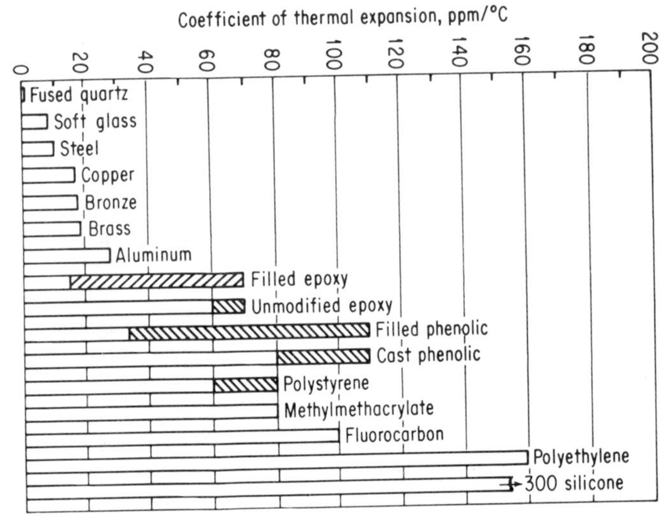

FIG. 78. Effect of filler content on coefficient of thermal expansion of epoxy resins compared with coefficients for various other materials. Ranges are shown cross-hatched.[70]

Hardness and machinability characteristics depend on specific fillers, but generally hardness is increased and machinability decreased by the use of fillers. Abrasive fillers such as silica and sand can produce particularly difficult machining problems.

Impact strength and tensile strength can be increased by the use of reinforcing fillers and are normally decreased by the use of bulk fillers. Milled or chopped glass fibers are especially good reinforcing fillers.

EFFECTS ON ELECTRICAL PROPERTIES. Although certain electrical characteristics of a resin system can be improved by the incorporation of fillers, these effects are minor compared with improvements in mechanical and thermal properties of the system. Dielectric strength may even be decreased if the filler has absorbed moisture or a contaminant. The dissipation factor and dielectric constant usually can be con-

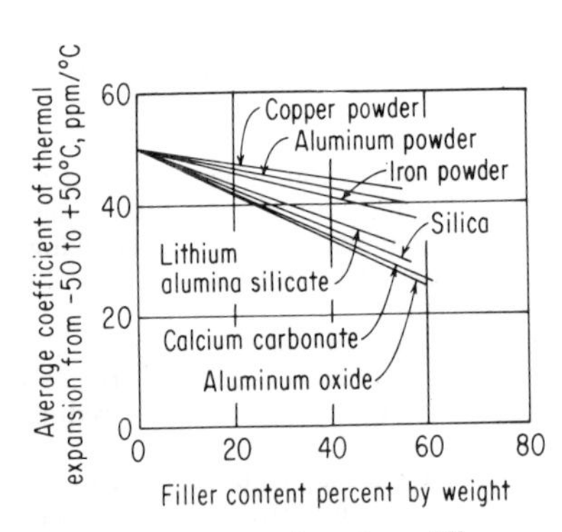

FIG. 79. Effect of various fillers on coefficient of thermal expansion of an epoxy-resin system.[70]

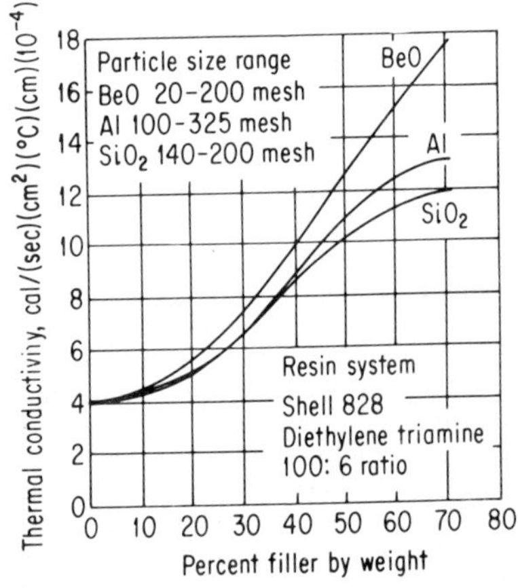

FIG. 80. Thermal conductivity versus filler content for three filler materials.[71]

TABLE 39. Costs and Effects on Resin Properties of Commonly Used Fillers[1]

Type of filler*	Approx. cost, cents per lb	Property increase							Property decrease					
		Thermal conductivity	Thermal shock resistance	Impact resistance	Compressive strength	Arc resistance	Machinability	Electrical conductivity	Cost	Cracking	Exotherm	Coefficient of expansion	Density	Shrinkage
Bulk:														
Sand	1	X			X				†		X	X		X
Silica	1–2	X			X				†		X	X		X
Talc	1–4						†		X		X			X
Clay	1–3		X		X		†		X	X	X	X		X
Calcium carbonate	0.5–5						†		X		X			X
Calcium sulfate (anhydrous)	2–4	X			X	†			X		X	X		X
Reinforcing:														
Mica	3–9		X	†					X	X				
Asbestos	2–5		X	†						X				
Wollastonite	2–3	X	X	†					X	X				
Chopped glass	45		X	†						X				
Wood flour			X	X			X		X			†		
Sawdust			X	X			X		X			†		
Specialty:														
Quartz	2–5	X		X	X					X		†		X
Aluminum	5–15	X	X		X						X	†		X
Hydrate alumina	3–6					†								
Li-Al silicate		X			X					X	X		†	X
Beryl		X			X					X	X		†	X
Graphite	6–30						X	†						
Powder metals		X	X	X	X		X	X	X	X	X	X		X
Low-density spheres	75–150												†	

* Particle size of fillers listed is 200 mesh or finer, except for sand, hollow spheres, and reinforcing fillers that depend on particle configuration for the desired effect.
† Most significant use of each filler listed.

trolled, however, by the use of low-density fillers and other selective fillers such as barium titanate.

Stresses in Resins. One of the mechanical effects which embedding resins have on components and critical circuits is the internal stresses created by either the shrinkage of the embedding resin during cure or the internal stresses created in the unit during thermal excursions caused by operating conditions, storage conditions, or testing to some set of specifications. Basically, embedding resins have a considerably higher coefficient of thermal expansion than do glasses and metals. This is shown graphically in Fig. 78, which compares thermal coefficients for a filled and unfilled epoxy resin with those for several other materials. Although the actual value will vary for different resin-curing-agent systems and for different measurement parameters, it can be observed that a wide range exists between unfilled and filled epoxy systems. The same holds true for resins other than epoxies. In general, the coefficient or expansion will depend on the original coefficient of the unfilled resin, the filler material used, and the concentration of filler used. These last two points are shown in more detail in Fig. 79 for a variety of commonly encountered fillers.

The object, of course, in designing with these data is to attempt to match the coefficient of expansion of the resin system with that of the construction material of the critical component to be packaged.

Much can be and has been written on the above-mentioned stress effects. Each situation is an individual one, and must be evaluated accordingly. Many components are unaffected by the highest stresses, while others are extremely sensitive. One generalization which can be stated, however, is that rigid, unfilled resins usually exhibit higher stresses than do soft resins, such as silicone rubbers, urethanes, and flexibilized epoxies. Silicone rubbers and urethanes are better at lower temperatures, since they maintain better resiliency or flexibility at these low temperatures. A general set of curves is shown in Fig. 81. Although there are many points of debate on protecting critical components by coating with a resilient coating, such a coating quite often gives improved yield and performance. One example of this is shown in Fig. 82.

CERAMICS, CERAMOPLASTICS, AND GLASSES

Ceramics, ceramoplastics, and glasses are inorganic, electrical insulating materials which offer some special design features which are frequently valuable to the electronic packaging engineer. Each of these material classes offers advantages over plastics, generally speaking, in higher-temperature capabilities and greater dimensional stability. However, they differ from each other in important engineering and application respects. Hence, ceramics, ceramoplastics, and glasses are discussed separately below.

Ceramics. Like plastics, ceramics are good electrical insulators, but they have much better thermal conductivity than plastics. This characteristic makes ceramics

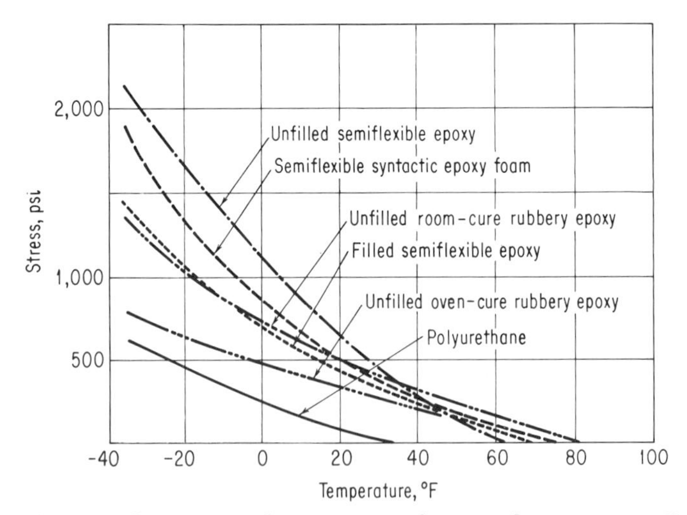

FIG. 81. Comparison of stress curves for several resin systems.[72]

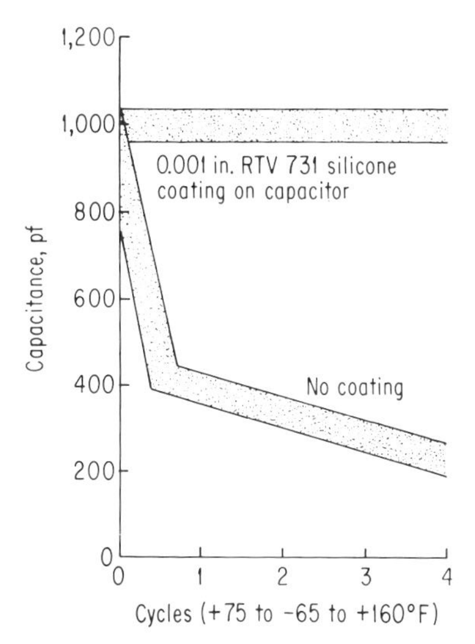

FIG. 82. Effect of silicone coating on behavior of capacitors potted in rigid, silica-filled epoxy.[73]

very useful in electronic package design requiring electrically insulated heat sinks. Ceramics do not, in general, have low dielectric constants and hence are not as good as many plastics in this respect.

In addition to their good thermal and electrical properties, ceramics offer greatly improved dimensional stability over most plastics—especially at slightly elevated temperatures. This is true in both thermal expansion and physical distortion properties, such as flatness and warpage. Although ceramics are generally harder and more brittle than plastics, they can often be machined, assembled, and fabricated with comparative ease, depending upon the differences that exist among the several types. Further, ceramics can be metallized or coated and thus fabricated into many hardware items useful in electronic packaging. They are widely used as substrates for deposited circuits, and for hermetically sealed electronic units, especially in packaging high-reliability semiconductor devices. In addition, certain ceramic materials, notably those based on titanates such as barium titanate, can be made to a controlled dielectric constant over a wide range, enabling usage for capacitors.

Classified in the broad range of ceramic materials are aluminas, steatites, forsterite, zircons, titania, titanates, cordierite, lava, magnesia, magnesium silicate, silicon carbides, zirconia, beryllia, porcelain, mullite, spinel, quartz, and boron nitride. Some of these ceramics are available in various compositions, degrees of purity, etc., so that the materials must be fairly well defined when making comparisons. Also, some of these ceramics, while good insulators, do not have properties suitable for high-reliability performance on some of the extreme environments (such as humidity) required of modern electronic packages. Hence, it is the electrical-quality ceramics, formulated to produce definite physical and/or electrical properties, which are of greatest interest to electronic packaging engineers.

Basic properties of some of the more important ceramics are shown in Table 40. Variations of dielectric constant and dissipation factor, with frequency and temperature, for one alumina and one beryllia are shown in Fig. 83.

Aluminas. Aluminas are very important in the ceramic field. They are harder, stronger, and more resistant to wear and are better electrical insulators, especially at higher temperatures and higher frequencies, than most other compositions. Aluminas are produced as dense ceramics, valuable primarily for their mechanical strength and great hardness, and for their excellent electrical properties. They are also produced

TABLE 40. Properties of Some Representative Ceramics[76]

Property	Steatite $MgO \cdot SiO_2$ Alsimag 665 L-533	Forsterite $2MgO \cdot SiO_2$ Alsimag 243 L-723	Zircon $ZrO_2 \cdot SiO_2$ Alsimag 475 L-514	Cordierite $2MgO \cdot 2Al_2 \cdot O_3$ $5SiO_2$ Alsimag 701	Lava (natural stone) grade A, aluminum silicate	Alumina Al_2O_3 Alsimag 753 L-724	Magnesia‡ MgO Alsimag 714	Beryllia§ BeO Alsimag 754 L-623
Water absorption, %	0	0	0	0.02–1	2–3	0	16–20	0
Specific gravity	2.7	2.8	3.7	2.3	2.3	3.85	2.4	2.88
Safe temperature at continuous heat	1000°C	1000°C	1100°C	1200°C	1100°C	1650°C	1600°C	1600°C
Hardness, Mohs scale	7.5	7.5	8	8	6	9	....	9
Thermal expansion, 25–300°C	6.9×10^{-6}	10×10^{-6}	4.3×10^{-6}	2.4×10^{-6}	3.3×10^{-6}	7.1×10^{-6}	11.2×10^{-6}	6×10^{-6}
Tensile strength, psi	10,000	10,000	12,000	....	2,500	28,000		
Compressive strength, psi	90,000	85,000	100,000	50,000	40,000	380,000	8,000	>185,000
Flexural strength, psi	21,000	20,000	22,000	15,000	9,000	50,000	2,000	25,000
Resistance to impact, in./lb	5.0	4.0	5.5	4.0	3.3	7.0		
Thermal conductivity, 300°C*	0.006	0.008	0.012	0.008	0.005	....	....	0.28
Dielectric strength, 60-cycle ac, volts/mil	230	240	220	225	80	230	50	240
Volume resistivity, ohm/cm								
25°C	$>10^{14}$	$>10^{14}$	$>10^{14}$	1.0×10^{14}	$>10^{14}$	$>10^{14}$	$>10^{14}$	$>10^{14}$
100°C	1.0×10^{14}	5×10^{13}	2×10^{13}	2.5×10^{11}	6×10^{11}	$>10^{14}$	$>10^{14}$	$>10^{14}$
Dielectric constant, 1 Mc, 25°C†	6.3	6.2	8.8	5.3	5.3	9.4	5.6	6.4
Dissipation factor, 1 Mc, 25°C†	0.0008	0.0004	0.0010	0.0047	0.010	0.0001	0.0019	0.0001
Loss factor, 1 Mc, 25° C†	0.0050	0.002	0.009	0.025	0.053	0.0009	0.011	0.0006

* Conversion factor. Figures are in cal/(cm) (sec) (cm^2), one of which equals 2902 Btu/(in)(hr)(ft^2)(°F)

† AlSiMag 243, 475, and 665 measured wet at 1 Mc, after immersion in water for 48 hr (MIL-I-10A).

‡ Thermal conductivity and dielectric strength values vary with the degree of compacting obtained in the user's swaging operation. Mechanical strength depends on the mutually agreed-upon hardness specifications.

§ In working with beryllia ceramics, personnel should avoid exposure to dust- or fume-producing operations, such as sawing and grinding, in moist atmospheres at high temperatures. Specialized equipment is necessary to prevent the dispersal of the dust and fumes into the air.

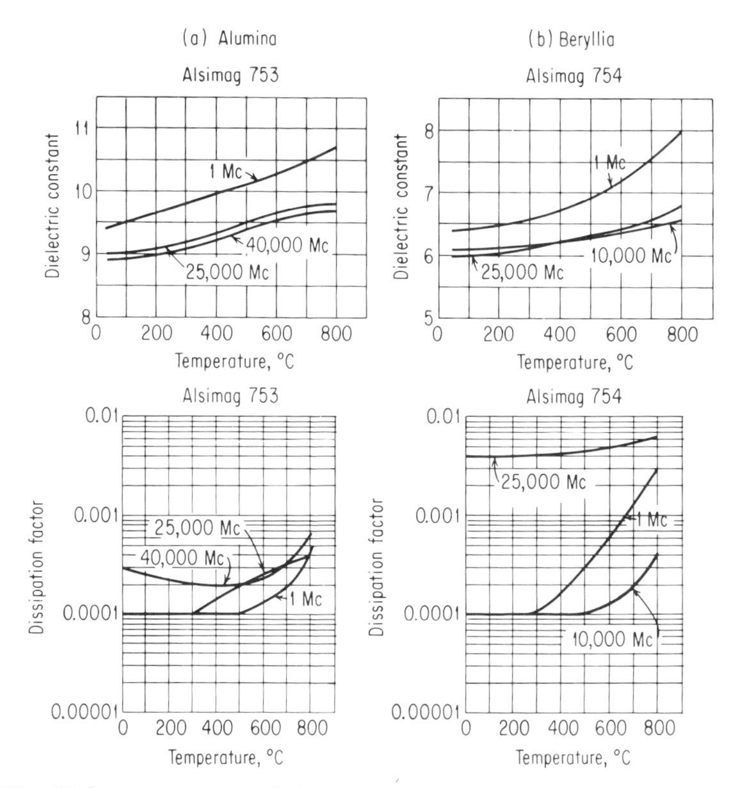

FIG. 83. Dielectric constant and dissipation factor of a representative alumina and a representative beryllia as a function of temperature and frequency.[76]

as porous ceramics which are easily outgassed and which find wide use as electron tube insulators.

Alumina powder (in a finely ground state having a particle size of 5μ or smaller) and fluxing agents such as silica, magnesia, and calcia are compounded together at high press.[77] They are then sintered to form a high-strength, dense, nonporous ceramic material. During the sintering or firing process, temperature, atmosphere, and time are carefully controlled to produce the desired polycrystalline structure of the particular ceramic part. After firing, the ceramic is composed mostly of the pure alumina crystals, carefully controlled as to size. The bond is achieved by the fluxing compound matrix—a tough glassy phase—and the interlocking crystals.

Thus, the apparent surface hardness is that of the pure alumina crystals, which compose the largest percentage of the surface. However, the small interstices between the crystals are filled with the relatively softer glassy phase matrix, and are therefore subject to erosion by ultrasmall particle impingement. When a surface is polished, the crystals may be perfectly flat, but the spaces in between may be lower, because the softer material is more quickly worn away, presenting a profile of plateaus and valleys reflecting the crystal size of the alumina crystals. The alumina crystals can be polished to smooth, flat surfaces of extreme hardness, but designers must consider the small "valleys" possible in all ceramics. Designers can turn these small valleys into advantages. For example: the valleys are ideal for lubrication reservoirs in bearing designs.

Ceramics are subject to attack by the fluoride ion, because of the minute amount of exposed glassy phase matrix at the crystal interstices. However, the alumina crystal remains completely inert.

The phenomenal compressive strength of alumina ceramics is due to the crystal formation. With most of the crystals lying against each other, the compressive strength of the ceramic approaches the strength of the pure crystal. Consequently, the compressive strength increases as the amount of the alumina content of the ceramic increases. The tensile strength of the ceramic is lower than the compressive strength, because it is derived from the combined strength of the glassy phase matrix and the mechanical strength of the interlocking crystals.

Temperature stability, even under repeated cycling, is achieved because the ceramic structure does not change until temperatures reach the softening point of the ceramic.

Steatites. Steatites are usually more readily formed than are aluminas, are fired at somewhat lower temperatures, and are more economical. For many years, steatites were the top performers in technical ceramics. They still have a wide field of usefulness and deserve study to see if they meet needed performance requirements; this may help avoid needless use of a more expensive ceramic.

Forsterite. Forsterite serves well where the primary requirement is for very-low-loss insulators. It is somewhat difficult to form and frequently requires grinding to meet close dimensional requirements. The high coefficient of expansion matches that of several metals but at a sacrifice in thermal shock resistance.

Zircons. Zircons have excellent electrical properties, moderate strengths, and good resistance to thermal shock.

Titania. Titania has high mechanical strength and great hardness. It is excellent for mechanical applications requiring wear and chemical resistance; it lends itself to close tolerance work and is often produced in controlled finishes measured in microinches. Normally an excellent electrical insulator, it can be processed to become a partial conductor to assist in control of static electricity.

Cordierites. Cordierites have low coefficient of expansion and excellent resistance to heat shock. They are used mostly in the extruded form for insulators in products such as heating elements and thermocouples. They also lend themselves to dry pressing.

Lava. Lava is mined natural mineral (aluminum silicate or magnesium silicate) which can be machined and then kiln-fired with little change in size. It has good electrical properties and good heat resistance. Lava is often used in prototypes or where small quantities of a technical ceramic are needed. Two types of silicates are available, as mentioned above.

Magnesia. Magnesia (magnesium oxide) ceramics are principally used as crushable ceramics for metal-sheathed heaters, range units, etc. Magnesias are available in medium purity for commercial uses or in extremely high purity to meet certain AEC specifications.

Beryllia. Beryllia, or beryllium oxide, is a material that insulates electrically as a ceramic does, but conducts heat as a metal does. Its conductivity is 62 percent of that of copper, compared with aluminum's 55 percent and steatite's 0.9 percent. Beryllia, which has all the electrical characteristics of a high-quality oxide ceramic, equals the best metals in thermal conductivity. Electrically, a component insulated with beryllia is isolated; thermally, it is the same as though the component were grounded. Uniquely among practical insulators—although the diamond exhibits this same combination of properties—beryllia prevents electrical leakage from a component while promoting conduction of heat away from it.[78] Thermal conductivity can vary considerably, if porosity or impurities are present. Care must be taken in machining beryllia, due to possible toxicity problems with beryllia dust. The supplier, however, can either machine for the user or advise on machining.

Boron Nitride. Boron nitride is another inorganic insulator useful for electronic packaging applications. While boron nitride is, in its raw form, a white powder, it is hot-pressed and compacted to form a dense, strong material which is easily machined and fabricated. Some typical values for important properties: thermal coefficient of expansion —1 to 3×10^{-6} in./(in.)(°F); thermal conductivity —10 to 20 Btu/(ft)(hr)(ft^2)(F).[79] Typical electrical properties of boron nitride are shown in

TABLE 41. Typical Electrical Properties of Boron Nitride[79] *

Frequency, cps	70°F		392°F		572°F		752°F		1067°F	
	K†	tan Δ‡	K	tan Δ	K	tan Δ	K	tan Δ	K	tan Δ
10^2	4.4	0.0005	4.45	0.0048	4.52	0.015	4.86	0.10	—	1.0
10^3	4.4	0.0003	4.43	0.003	4.48	0.0075	4.60	0.029	7.48	0.35
10^4	4.4	0.0002	4.42	0.0015	4.44	0.004	4.52	0.010	5.08	0.08
10^5	4.4	0.00075	4.41	0.0004	4.42	0.0015	4.48	0.004	4.60	0.02
10^{10}	4.4	0.0005	4.40	0.0006	4.40	0.0008	4.40	0.001	4.40	0.0015
Resistivity, ohm-cm	70°F		900°F		1830°F		2730°F		3630°F	
	1×10^{13}		5×10^8		1×10^7		2×10^3		1×10^2	

* Dielectric strength: 500–1,000 volts/mil.
† Dielectric constant.
‡ Dissipation factor.

Table 41. Boron nitride wafers are used for thermal conductivity mountings, coil forms, waveguide windings, etc. Though they are tough, they are soft enough to allow penetration of small edges into the ceramic body, which in turn allows intimate contact. There are variations in the moisture absorption of boron nitride materials, and the best grade should be chosen for a given design objective. Low-moisture-absorption grades (under 0.5 percent) are obtainable.

Fabrication of Ceramics. Knowledge of the principal fabrication methods for ceramics will be useful to the application packaging engineer in suggesting what is and what is not possible, and in judging the effect of processing on desired end properties. Most ceramics start out as finely divided powders; the forming method chosen is determined by the contour, the surface finish, the size, the quantity, and the dimensional tolerances required for the final part. Where alternate methods are equally feasible, the most economical one is chosen by the supplier.[76,77]

DRY PRESSING. Dry pressing is the least expensive production method for quantity production of a wide range of designs. Both mechanical pressing equipment and hydraulic pressing equipment are used to produce a range of designs. Hydraulic presses are used for more difficult designs.

EXTRUSION. Like squeezing toothpaste from its tube, extrusion produces definite lengths in a wide range of cross-sectional shapes. Internal extrusion pins can be added to form accurate openings within the cross section, to produce single-holed or multi-holed tubes of desired internal and external shapes. Extruded shapes are dried by several methods including electronic- and humidity-controlled tunnel driers. Dried (but unfired) shapes have sufficient strength to be readily handled in subsequent processing.

MACHINING BEFORE FIRING. Extruded and dried shapes can be sawed, turned, milled, drilled, tapped, etc., before firing. In general, any conventional machining process which can be applied to metal can be applied to dried ceramic parts before firing—and almost as easily. Further machine operations are performed on some dry-pressed parts before firing.

ISOSTATIC PRESSING. Isostatic pressing is called for by certain sizes and shapes. This process permits close control over large tubular forms. Where precise internal diameters are important and particularly where machining of internal diameters would be unduly difficult or expensive, this process may be the answer. Isostatic pressing employs hydrostatic compaction at selected pressures in the range of 5,000 to 20,000 psi to compact evenly from all directions.

INJECTION MOLDING. Advances have been made in the production of complex shapes once considered impossible or prohibitively expensive. Many of these are now practical through injection molding. This process is most apt to be economical when complex shapes and/or large quantities are required. The tooling costs are often substantial and must be given careful study. This is one of the few methods of forming where prototypes may pose problems because of tooling expense.

GRINDING AFTER FIRING. This requires both special equipment and techniques which have been developed by large-volume production over many years. Present equipment and know-how make it possible to measure almost any dimensional tolerance that is required.

Finishes of Ceramics. Finishes on technical ceramics are either "natural" or "applied." If natural, the several ranges of finishes achieved are closely related to the fabrication method used; if applied, the finish may be either:

A modification of the natural surface obtained by grinding, lapping, polishing, tumbling, etc. In effect, a removal of the natural finish.

A modification obtained by the addition of a coating such as glaze, silicone, or wax to the natural finish.

A modification obtained by secondary operations or combinations of these various operations.

It is important that the user distinguish between applications where finish itself is the objective and those where dimensional accuracy and flatness are the prime objectives. In some cases, both finish and accuracy may be objectives. In the natural classifications, for example, pressed parts have a range of 25 to 100 μ/in. (arithmetic average) as measured on the Profilometer, and extruded parts have a range of 25 to 100 μ/in.

Ultrathin substrate ceramics offer a unique category of natural finish that is exceptionally smooth and uniform (10 to 30 μ/in. on the Profilometer). These ceramics set high standards of both flatness and dimensional accuracy in the "as-formed" state. Electrical puncture values are also exceptional.

In the applied group, an "as-ground" surface finish will register in the distribution from 20 to 40 μ/in. Lapped surfaces will be in the distribution from 15 to 30 μ/in. but with a high gloss, the figures being approximate. Rougher or smoother finishes and finishes with narrower ranges can be supplied at commensurate cost; special finishes for particular applications can be developed with proper consultation and experimentation.

Glaze is an applied coating frequently used: a glaze is essentially a glass film fused

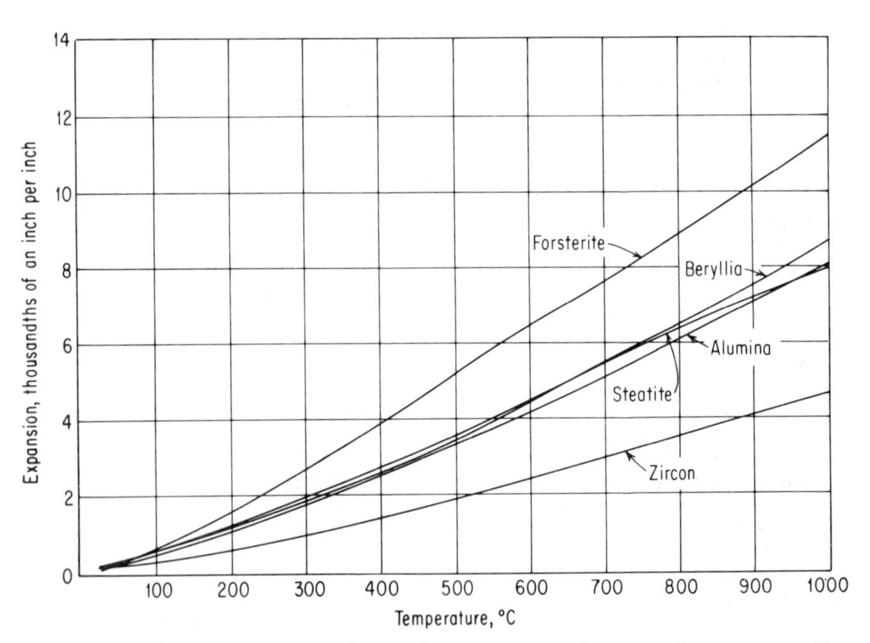

FIG. 84. Thermal expansion of typical ceramics as a function of temperature.[76]

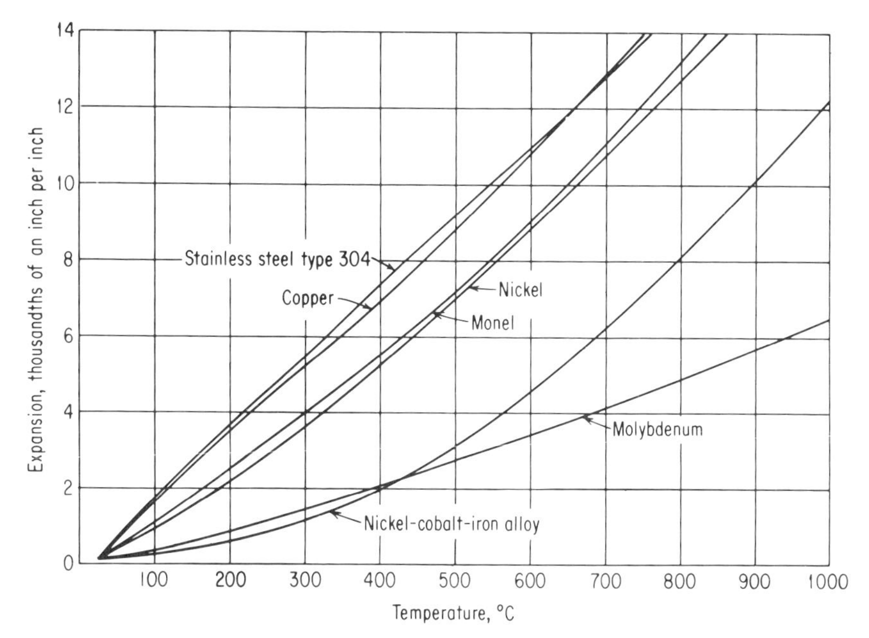

FIG. 85. Thermal expansion of some important metals used with ceramics.[76]

onto a compatible base ceramic. It is used to promote ease of cleaning and to maintain high resistivity across a surface under humid conditions. Glazes are normally applied to the fired ceramic by either spraying or screening and are then matured in a separate kiln operation. Glaze may, if desired, cover the entire ceramic shape (except for a single surface) or it may be limited to specific areas. Some glazes are available in certain colors.

On flat, thin substrates, especially those for thin-film deposition, an ultrasmooth glaze may be applied to obtain a finish of 1 μ/in. or less as measured on the Taylor-Hobson Talysurf instrument. The Talysurf is a graphic recorder; while it is primarily a research and development tool, it can supply charts at vertical scale amplifications up to 50,000 for certain critical needs.

Metallized Ceramics. The importance of metallizing insulators is rapidly increasing in the electronics industry as more and more applications of metallized insulators —both plastic and ceramic—are being found. As with plastics, much effort has been devoted to metallizing of ceramics, and many varieties of metallized ceramics are possible. In the case of ceramics, metal-to-ceramic assemblies are used for hermetic seals as well as for conductors, deposited interconnections, shielding, heat sinking, etc. Hermetically sealed devices are usually supplied by the manufacturer, but the bulk of the other metal-to-ceramic applications offer a wide range of design opportunities for the electronic packaging engineer. For this reason, some of the important information on the subject is presented here.

Since ceramic materials vary in their ability to be metallized, certain ceramic formulations are usually recommended for metallizing. The ceramic supplier should be consulted in this respect. In this way, best metal-to-ceramic bond strengths can be achieved, as well as the best matching of thermal coefficient of expansion. The latter point is very important, and Figs. 84 and 85 show coefficient of expansion for several important metals and ceramics as a function of temperature, which is the critical variable. Some metallizing processes are discussed below.[76]

THE ACTIVE-METAL PROCESS. This process is based on the use of a metal such as titanium. When brought to high temperature in a vacuum atmosphere, titanium will melt with silver, nickel, or other transition metals and react with most oxide ceramic materials to form a strong chemical bond. Unlike the silver-glass frit or the refractory-metal-powder processes, the active-metal process requires that flanges, pins, or other hardware be assembled and jigged in position during the one-shot, vacuum-firing operation. In the vacuum furnace the braze material becomes molten and quite fluid. Since commercial gases are not pure enough to provide a sufficiently protective atmosphere for this operation, it is performed in a vacuum of 10^{-4} torr or better. Vacuum brazing has other advantages: it is essentially inert and thus causes little change in the properties of any of the components except for an annealing of the attached metal hardware.

THE REFRACTORY-METAL-POWDER PROCESS. This process, sometimes referred to as the moly-manganese process, uses powdered molybdenum, or its oxide, alone or combined with catalytic materials such as manganese, iron, nickel, titanium, chromium, silica, calcia, or glass frit. A wide variety of mixes, when properly selected, will bond well with the appropriate high alumina or beryllia ceramics to provide parts with good mechanical and electrical properties at elevated temperatures.

The process steps in the refractory-metal-powder method are quite different from those in the active-metal process. In the refractory-metal-powder process, the mix is suspended in liquid organic materials which serve both as temporary binder and as a vehicle for the mix. This paste is then applied to the ceramic by one of several methods. The coating is dried and then sintered in a protective atmosphere of hydrogen, disassociated ammonia (75% H_2 to 25% N_2), or similar gas. The sintered coating, now firmly bonded to the base ceramic, is at this point ready for electroplating. The electroplating is usually nickel. This is sintered in a protective atmosphere at a lower temperature to improve adherence and reduce any tendency to oxidize. This sintered plating facilitates bonding with solders or brazes. Other metals may be electroplated to the base coat or they may be applied over the nickel as an addition.

Refractory-metal-powder coatings are applied only to selected unglazed areas of the ceramic part. A high-temperature glaze may be applied to other areas, so long as the glaze is capable of withstanding the temperatures and atmospheres necessary for metallizing and assembly operations, and so long as the glaze and the metallizing coats are separated by a gap of at least 1/16 in. A glaze reduces flashover and corona tendencies in service and provides a surface which may be readily cleaned from time to time.

LOW-TEMPERATURE METALLIZING. Four techniques of low-temperature metallizing are as follows:

1. A silver-glass frit process can be used for glazed alumina or steatite and unglazed steatite. Electroplating with copper, nickel, or tin is essential prior to the soldering operation. This process is applicable with low-melting (60:40 or 50:50) solders. Hermetic seals can be obtained by this method.
2. A direct-soldering silver-glass frit composition can be used on glazed alumina or steatite and unglazed steatite. Silver-bearing soft solder should be used.
3. Another process uses a platinum-gold, glass-frit composition. The fired coating can be soldered directly using 60:40, 50:50, or 61 percent tin to 35 percent lead to 3 percent silver solders. Therefore, electroplating usually is not used.
4. A fourth process is a modified silver-base process used primarily for unglazed alumina ceramics. Metallizing patterns cannot be applied over glazed surfaces. After kiln firing it is usually electroplated with nickel; it is then suitable for making hermetic seals using solders such as 60:40, 50:50, 95 percent lead to 5 percent indium, or 97½ percent lead to 2½ percent silver.

Ceramoplastics. Ceramoplastics are inorganic materials which can be molded or processed as plastics, but which have properties closer to those of ceramics. The useful application temperatures for ceramoplastics lie between those of plastics and ceramics. The processing advantage of ceramoplastics allows their use in some application requirements that are not as easily met with ceramics. The most common form of ceramoplastics is glass-bonded mica.

Glass-bonded mica is composed of finely powdered mica, natural or synthetic, bonded with special glasses. Modifications, to achieve certain properties, may include ceramics, glasses, and inorganic fibers or fillers. Molding of parts is performed at relatively high temperatures, commonly in the 1000 to 1500°F range. Close tolerances can be held, however, even in complex shapes or with inserts. Glass-bonded mica is available in custom-molded shapes or in sheet rod-and-bar stock. The materials are covered in MIL-I-10A.

Properties of Glass-bonded Mica. The important properties of some representative machining grades and some representative molding grades of glass-bonded mica are shown in Table 42.

The outstanding characteristics of glass-bonded mica are as follows:

1. High thermal endurance.
2. Arc resistance.
3. Radiation resistance.
4. Low thermal expansion.
5. Moldability with delicate insert inclusion.
6. Encapsulation of products.
7. Excellent electrical values.
8. More machinability than ceramics, with no firing after machining.
9. Corona-free parts.
10. Moldability with true hermetic seal.
11. High dimensional stability.
12. Plated circuitry may be applied.
13. Stamped or chemically etched circuitry may be molded in place.
14. High dielectric constant formulation products may be molded or fabricated.

TABLE 42. Properties of Some Representative Glass-bonded Micas[80]

Machining grades, sheets and rods			Property	Custom injection molded grades			
Mykroy 750	Mykroy 1100	Mykroy 1116		Units	Mykroy 761	Mykroy 777	Mykroy 1001
			General:				
3.3	3.2	3.2	Specific gravity	....	3.6	3.9	3.1
Nil	Nil	Nil	Moisture absorption	....	Nil	Nil	Nil
750	1100	1100	Max. continuous temp.	°F	750	790	1100
5.6	5.2	5.2	Coefficient of expansion	$\times 10^{-6}$ °F	6.0	5.1	5.0
			Electrical:				
10^{13}	10^{14}	10^{14}	Volume resistivity	ohm-cm	10^{14}	10^{14}	10^{13}
350	400	400	Dielectric, ⅛ in.	volts/mil	350	350	400
7.1	7.50	7.4	Dielectric constant	1 Mc	8.2	9.2	6.8
0.0014	0.0012	0.0019	Dissipation factor	1 Mc	0.0011	0.0016	0.0011
0.0099	0.016	0.014	Loss factor	1 Mc	0.009	0.015	0.008
10^{16}	10^{16}	10^{16}	Surface resistivity, dry	ohm	10^{15}	10^{15}	10^{15}
10^{10}	10^{11}	10^{12}	Surface resistivity, wet	ohm	10^{9}	10^{10}	10^{9}
			Mechanical:				
9,000	8,000	7,500	Tensile strength	psi	7,000	7,000	6,000
21,600	13,900	11,000	Flexural strength	psi	13,000	13,000	13,000
36,000	30,000	30,000	Compressive strength	psi	28,000	27,000	28,000
10×10^{6}	7×10^{6}	7×10^{6}	Modulus of elasticity	psi	6×10^{6}	6×10^{6}	7×10^{6}
1.7	1.3	1.2	Impact strength (Izod)	in./lb	0.7	0.7	0.6

High-frequency characteristics*

Glass-bonded mica	Frequency, Mc	*K*	Loss tangent
Mykroy, grade 750	1	7.06	0.0014
	1,000	7.33	0.0018
	3,000	7.32	0.0014
	8,500	7.54	0.0022
Mykroy, grade 1100	1	6.70	0.0014
	1,000	7.65	0.0013
	3,000	7.66	0.0010
	8,500	7.67	0.0016

* The data were measured at room temperature. Additional tests indicate that there will be less than a 10% increase for Mykroy 750 when measured at 250°C and for Mykroy 1100 when measured at 360°C.

One of the especially useful properties of glass-bonded micas is their relatively low linear coefficient of thermal expansion, especially among moldable products. There are various electrical grades of glass-bonded micas, and some grades have relatively stable values of dielectric constant and dissipation factor up to an 8,500-Mc frequency or higher and up to 250 to 350°C. Their resistivity, like that of plastics, does decrease with temperature. This is shown in Fig. 86.

Glasses. Glasses are produced from inorganic oxides, and one of the most important ingredients is usually silica, or sand. Some chemical compounds, notably oxides of silicon, boron, and phosphorus, are capable of being processed into glass products. Glasses are usually not single-chemical compounds but rather mixtures of inorganic oxides. The proportions of the different constituents may be varied freely within certain limits.[81] If these limits are exceeded, the formation of glasses becomes difficult or impossible. For silicate glasses alone, an infinite variety of compositions can be produced, and some hundreds of glass compositions with distinguishable differences in properties are melted more or less regularly. The glasses discussed below

TABLE 43. Approximate Compositions of Commercial Glasses[81*]

No.	Designation	Percent								
		SiO_2	Na_2O	K_2O	CaO	MgO	BaO	PbO	B_2O_3	Al_2O_3
1	Silica glass (fused silica)	99.5+								
2	96% silica glass	96.3	<0.2	<0.2	....	....	....	....	2.9	0.4
3	Soda-lime—window sheet	71–73	14–15	....	8–10	1.5–3.5	....	....	....	0.5–1.5
4	Soda-lime—plate glass	71–73	12–14	....	10–12	1–4	....	....	....	0.5–1.5
5	Soda-lime—containers	70–73	Na_2O, K_2O 13–16		CaO, MgO 10–13		....	....	....	1.5–2.5
6	Soda-lime—electric-lamp bulbs	73.6	16	0.6	5.2	3.6	....		....	1
7	Lead silicate—electrical	63	7.6	6	0.3	0.2	....	21	0.2	0.6
8	Lead silicate—high lead	35	....	7.2	....	....	....	58		
9	Aluminoborosilicate (apparatus)	74.7	6.4	0.5	0.9	....	2.2	....	9.6	5.6
10	Borosilicate—low expansion	80.5	3.8	0.4	....	....	....	....	12.9	2.2
11	Borosilicate—low electrical loss	70.0	....	0.5	0.1	0.2	....	Li_2O 1.2	28.0	1.1
12	Borosilicate—tungsten sealing	67.3	4.6	1.0	....	0.2	....	....	24.6	1.7
13	Aluminosilicate	57	1.0	....	5.5	12	....	....	4	20.5

* In commercial glasses, iron may be present, in the form of Fe_2O_3, to the extent of 0.02–0.1% or more. In infrared-absorbing glasses it is in the form of FeO in amounts from 0.5 to 1%. The numbers listed in this table may be identified with commercial glasses as follows: 2, Corning glass 7900, 7910, 7911; 6, Corning glass 0080; 7, Corning glass 0010; 8, Corning glass 8870; 9, Kimble glass N51a; 10, Corning glass 7740; 11, Corning glass 7070; 12, Corning glass 7050; 13, Corning glass 1710, 1720.

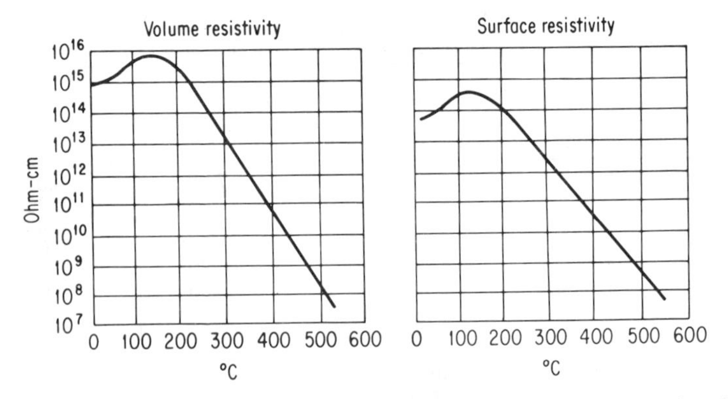

FIG. 86. Resistivity as a function of temperature for a representative glass-bonded mica.[80]

include types of glasses which are of importance in industry and technology. Representative chemical compositions are listed in Table 43. The properties of these glasses are listed in Table 44.

Silica Glass (Fused Silica). This glass may be made by melting crushed silica quartz, or sand, without other constituents. Other processes can be used in special cases. Silica is difficult to melt because of the high temperatures required and because it is so viscous when molten that the gas bubbles formed in melting do not readily free themselves from the molten mass. It becomes rigid so quickly that the operations of forming the glass into articles of various shapes are severely restricted.

Ninety-six Percent Silica Glass. This glass approximates fused silica in many of its properties. It is made by an ingenious process in which a composition relatively high in fluxing oxides is melted and formed to the desired shape, but somewhat oversize. After a heat treatment, these fluxes are practically all removed by acid leaching. The article is then fired at high temperature to consolidate the remaining porous silica structure, which causes the article to shrink to its finished size.

This glass has the same general field of application as fused silica. Its softening temperature is somewhat lower and its coefficient of expansion slightly higher. Greater ease of fabrication extends its uses materially beyond those of fused silica.

Soda-Lime Glass. The earliest glass made and still the most common glasses melted today are of this type. Soda, Na_2O, and sometimes small quantities of potash, K_2O, are added as fluxing agents, which reduce the viscosity greatly below that of the silica and thus permit the use of lower melting temperatures, which improve the "firing" qualities of the glass. The lime, CaO, and often small amounts of alumina, Al_2O_3, and magnesia, MgO, improve the chemical durability of the glass. The proportions of these constituents will differ slightly depending upon the particular uses for which the glass is intended. In Table 43, several compositions of soda-lime glasses are included.

Soda-lime glasses are commonly used for bottles, jars, window-sheet and plate glass, electric lamp bulbs, and ophthalmic (sight-correcting) lenses and for a wide variety of miscellaneous articles. This type of glass is economical both to melt and to fabricate.

Lead-Alkali Silicate Glasses. By replacing the lime of the soda-lime glass with lead oxide, PbO, another type of glass is formed. While the lime in the soda-lime glasses must usually be limited to 15 percent, the lead oxide constituent can be increased to 80 percent or more. The lead oxide acts as a fluxing agent and tends to lower the softening point of the glass still lower than for the soda-lime type.

Borosilicate Glasses. In addition to being a glass-forming material, boric oxide, B_2O_3, also serves as a fluxing agent for silica. Boric oxide has less effect than soda in lowering the viscosity of silica and in raising the coefficient of expansion. In general, the borosilicate glasses require higher melting temperatures than both the soda-lime and the lead-alkali silicate glasses. They are generally somewhat more difficult to fabricate. Borosilicate glasses of low expansion were developed well within the present century.

Because of their lower coefficients of expansion, which permit the glasses to be used at higher temperatures and to resist more severe thermal-shock conditions, and also because of their greater resistances to the corrosive effects of acids, these borosilicate glasses are used for many industrial and technical purposes. Among these may be included laboratory glassware, industrial glass piping, high-temperature thermometers, large telescope mirrors, household cooking ware, gauge glasses for steam boilers, enclosures for incandescent lamps, and electronic tubes of high wattage.

Although many compositions of borosilicate glasses have been developed to meet special requirements, only four examples of these have been included in Table 43.

Aluminosilicate Glasses. Glasses containing 20 percent or more alumina, smaller amounts of CaO or MgO, and sometimes relatively small amounts of B_2O_3 as a flux, but very limited amounts of soda or potash, are known as "aluminosilicate" glasses. They are usually more difficult to melt and to work than the borosilicate glasses. They are characterized by high softening temperatures and relatively low coefficients of expansion which make them particularly suitable for higher-temperature uses such as high-temperature thermometers, combustion tubes, water-level

TABLE 44. Properties of Commercial Glasses [81]

No.	Designation*	Viscosity data: Strain point, °C	Anneal point, °C	Soft point, °C	Flow point, °C	Coefficient of expansion per °C, 0–300°C	Density, g/cm³	Refract. index, sodium D line	Electrical properties: Log_{10} vol. res., ohm-cm: 250°C	350°C	Dielectric property, 1 Mc, 20°C: Power factor	Dielec-tric con-stant	Young's modulus, psi
1	Silica glass (fused silica)	1070	1140	1667	....	5.5×10^{-7}	2.20	1.458	12.0	9.7	0.0002	3.78	10×10^6
2	96% silica glass—7900..	820	910	1500	....	8×10^{-7}	2.18	1.458	9.7	8.1	0.0005	3.8	9.7×10^6
2a	96% silica glass—7911..	820	910	1500	....	8×10^{-7}	2.18	1.458	11.7	9.6	0.0002	3.8	9.7×10^6
3	Soda-lime—window sheet	510	545	730	920	85×10^{-7}							
4	Soda-lime—plate glass..	515	550	735	920	87×10^{-7}	2.46–2.49	1.510–1.520	6.5–7.0	5.2–5.8	0.004–0.011	7.0–7.6	10×10^6
5	Soda-lime—containers..	510	545	730	920	85×10^{-7}							
6	Soda-lime—elect. lamp bulbs	478	510	696	880	92×10^{-7}	2.47	1.512	6.4	5.1	0.009	7.2	9.8×10^6
7	Lead silicate—electrical	397	428	626	850	91×10^{-7}	2.85	1.539	8.9	7.0	0.0016	6.6	9.0×10^6
8	Lead silicate—high lead	398	429	580	720	91×10^{-7}	4.28	1.639	11.8	9.7	0.0009	9.5	7.6×10^6
9	Aluminoborosilicate—apparatus	540	575	795	....	49×10^{-7}	2.36	1.49	6.9	5.6	0.010	5.6	
10	Borosilicate—low expansion	515	555	820	1075	32×10^{-7}	2.23	1.474	8.1	6.6	0.0046	4.6	9.8×10^6
11	Borosilicate—low electrical loss	455	490	....	910	32×10^{-7}	2.13	1.469	11.2	9.1	0.0006	4.0	6.8×10^6
12	Borosilicate—tungsten seal	461	496	703	900	46×10^{-7}	2.25	1.479	8.8	7.2	0.0033	4.9	
13	Aluminosilicate	672	712	915	1090	42×10^{-7}	2.53	1.534	11.4	9.4	0.0037	6.3	12.7×10^6

* See Table 43 for compositions.

glasses for high-pressure steam boilers, and for household cooking ware intended to be used directly over flames or heating units. A glass of this type is included in Table 43.

Solder Glasses. Solder glasses have lower melting points than the common, soda-lime, or borosilicate glasses.[96] They are used as a sealing or bonding agent for higher-melting materials. Sealing is accomplished in the range of 400 to 700°C, depending on the glass, in time periods of about 1 hr. The density (4 to 7 g/cm^3) is greater than that of common glass (2 to 3 g/cm^3). These glasses have lower chemical durability, resulting from the specialized chemical composition required to achieve the low melting characteristic. Solder glasses are provided in a wide range of thermal expansions (40 to 120 $\times$ 10^{-7}/°C) to match the expansions of a variety of materials with which they can be used. Glass-glass, glass-metal, metal-metal, ceramic-glass, ceramic-metal, and ceramic-ceramic seals can all be made.

Other uses include films and coatings for glass, ceramic, and metal materials. The compositions may be used also for glass articles which are formed by pressing, molding, casting, or drawing. Powdered metallurgy techniques (pressing and sintering) using solder glass powders will result in a variety of shapes. By the use of proper additives, they can become electrically conducting, semiconducting, or heat-sensitive-conducting, or can be made to possess magnetic properties.

Solder glasses are available in two general classifications:

1. Vitreous type
2. Devitrifying type

Vitreous solder glasses resemble ordinary glasses because (1) they are relatively transparent, and (2) they can be reworked repeatedly while maintaining the same characteristics and properties. They have relatively high electrical resistivities and low dielectric constants. The thermal expansion range of these solder glasses is 45 to 100 $\times$ 10^{-7}/°C.

Devitrifying types crystallize during the initial curing cycle and form a two-phase glass crystalline material which is usually translucent or opaque. Upon reheating (even to temperatures above the initial curing cycle), these solder glasses remain stable. Heating to a temperature above the softening point causes these glasses to flow. As the heating is continued, crystals nucleate and grow. The rate of this crystal growth is dependent upon the temperature, with higher temperatures producing faster crystal-growth rates. Complete initial curing time schedules must be followed so that crystal growth will proceed properly. Devitrifying solder glasses are usually stronger than the vitreous types, and they have lower electrical resistivities and higher dielectric constants. The thermal expansion range for these solder glasses is 40 to 120 $\times$ 10^{-7}/°C. Conductive solder glasses are formed by combining a devitrifying type with selected additives. These glasses are quite temperature-sensitive. Overheating will reduce or destroy electrical conductivity. Underheating will produce seals which may leak.

Properties of Glasses. The general properties of representative commercial glasses are shown in Table 44. In addition, there are some properties which are of sufficient importance to electronic packaging engineers to be additionally detailed. These will be briefly discussed at this point.

The linear coefficient of expansion has particular significance in many electronic packaging applications. This characteristic is shown for several glasses, as a function of temperature, in Fig. 87.

Glass has mechanical properties corresponding to those of crystalline solids. It has elastic properties and strength so that it returns to its original shape after the release of applied forces which deform it. Glass does not exhibit the property of plastic flow, common to metals, and consequently has no yield point. Fracture occurs before there is any permanent deformation of the body, and failure is always in tension. The elastic constants of glass are shown in Table 45.

The electrical properties of glass are, of course, very important to electronic packaging engineers. In glasses, current is carried by the migration of ions, as in electrolytes, rather than by free electrons, as in the case of metals. For this reason, mobile ions, such as sodium ion, have a significant influence on the conductivity or

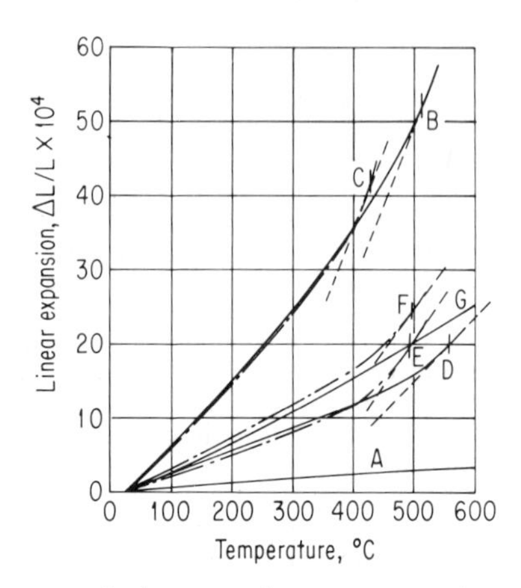

Fig. 87. Linear expansion of glasses with temperature.[81] Light broken lines show increased rates of expansion at annealing points. *A*, 96% silica glass; *B*, soda-lime bulb glass; *C*, medium lead electrical; *D*, borosilicate, low-expansion; *E*, borosilicate, low electrical loss; *F*, borosilicate, tungsten sealing; *G*, aluminosilicate.

TABLE 45. Elastic Constants of Glasses at 20°C[81]

Glass	Elastic constants, room temp., ultrasonic data		
	*E**	*G*†	*F*‡
Silica glass	10.0	4.3	0.17
96% silica glass	9.6	4.1	0.18
Soda-lime (plate)	11.0	4.4	0.23
Soda-lime (bulb)	10.0	4.1	0.24
Medium-lead glass	9	3.7	0.22
High-lead glass	8.5	3.4	0.23
Borosil, low-expansion	9.3	3.9	0.20
Borosil, low electric loss	7.3	3.0	0.22
Aluminosilicate	12.7§		

* Young's modulus—psi $\times 10^{-6}$.
† Shear modulus—psi $\times 10^{-6}$.
‡ Poisson's ratio.
§ Measured by deflection under static load.

resistivity of a glass. Conductivity tends to increase, and resistivity tends to decrease with an increase in the soda content of a glass, as shown in Fig. 88. Resistivity is also affected by temperature, as are nearly all dielectrics (shown in Figs. 88 and 89). Other important electrical properties of glasses, such as dissipation factor and dielectric constant, are also affected by temperature, as shown in Figs. 90 and 91, respectively. Voltage breakdown characteristics, as a function of thickness, are shown in Fig. 92.

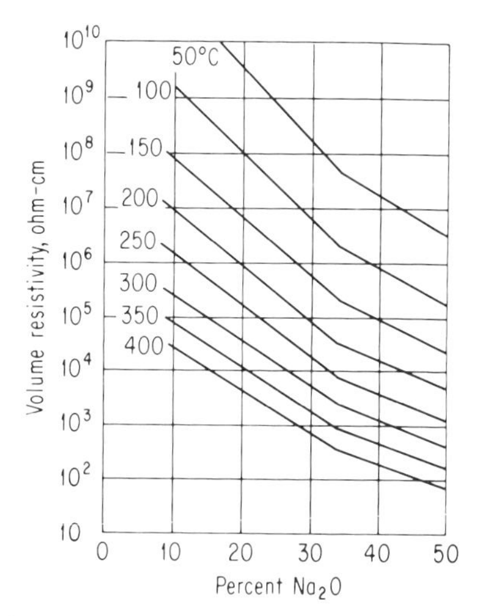

FIG. 88. Effect on volume resistivity of increasing the amount of sodium oxide (Na_2O) in a soda-silica glass.[81]

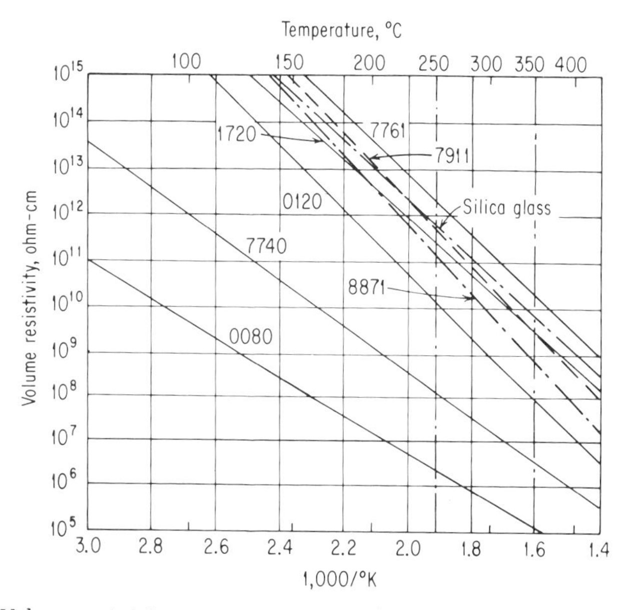

FIG. 89. Volume resistivity versus temperature for commercial glasses (dc values).[81]

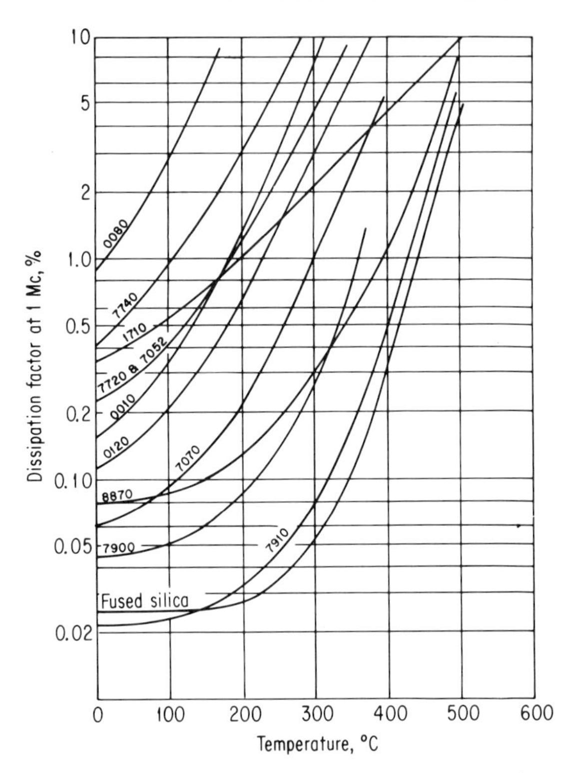

FIG. 90. Dissipation factor of commercial glasses at 1 Mc as a function of temperature.[81]

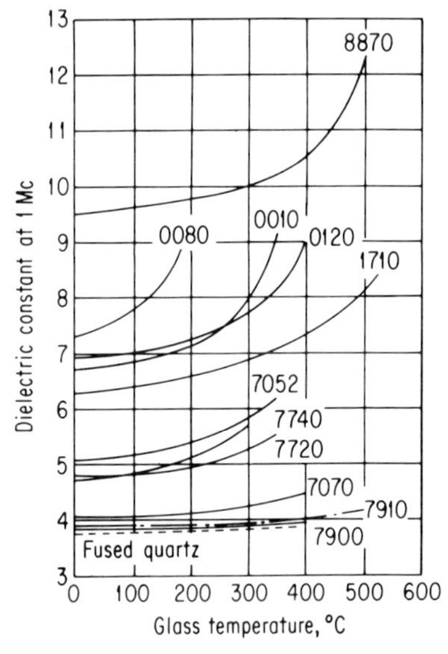

FIG. 91. Dielectric constant of commercial glasses at 1 Mc as a function of temperature.[81]

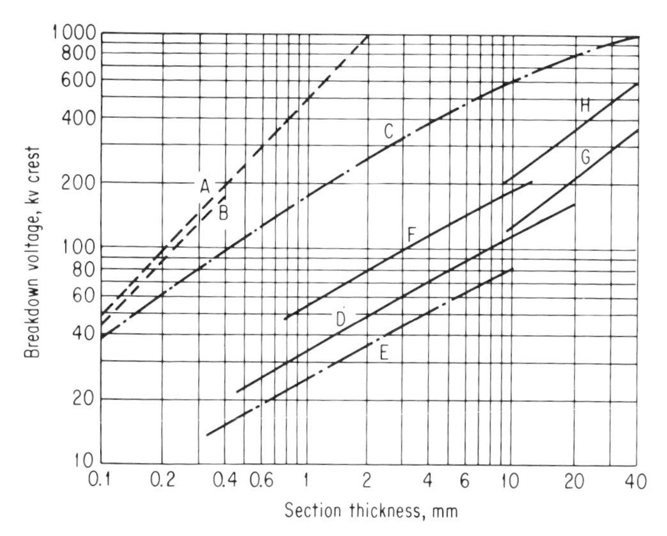

FIG. 92. Breakdown voltage versus thickness of glass for different conditions at room temperature.[81] *A*, intrinsic dielectric strength of borosilicate glass; *B*, intrinsic dielectric strength of soda-lime glass; *C*, highest test values available for borosilicate glass; *D*, borosilicate glass plate immersed in insulating oil; *E*, soda-lime glass plate immersed in insulating oil; *F*, borosilicate glass plate immersed in semiconducting oil; *G*, borosilicate glass power-line insulator immersed in insulating oil; *H*, borosilicate glass power-line insulator immersed in semiconducting oil.

METALS

The field of metals is extremely broad, and many complete publications exist in this subject area. Three such recommended publications are an issue of *Machine Design*,[82] a book by Hoyt,[83] and a book by Horger.[84] However, while it is not within the scope of this section to cover metals in any detail, brief, basic information will be given in the three areas of (1) basic discussion of steels and nonferrous metals, including the distinguishing properties or design considerations, (2) corrosion considerations, and (3) plated, metallic, and inorganic finishes. In addition, important electrical properties of metals are reviewed and tabulated in Chap. 10.

Basic Considerations for Metals. Basically, metals can be classified as ferrous or nonferrous. Ferrous metals are those that have iron as their primary element, and nonferrous metals are those that have other elements as their primary element. Irons and steels are ferrous metals, and all the others are nonferrous metals. Of the ferrous metals, steels are the most commonly used in the electronics industry. Many of the nonferrous metals find special uses in the electronics industry.

Steels. Steels can be categorized in many ways, each with its advantages for a particular situation. One breakdown of general and specialty steels is shown in Table 46, which also indicates the distinguishing property and general uses for each steel type.

Nonferrous Metals. As mentioned above, nonferrous metals are, for practical purposes, all metals other than irons and steels. Primarily, these metals are: elementary or alloyed aluminum, beryllium, cobalt, copper, lead, magnesium, nickel, zinc, and two groups of metals known as precious metals (gold, palladium, platinum, rhodium, silver) and refractory metals (tungsten, tantalum, titanium, molybdenum, niobium or columbium, zirconium). The important considerations and distinguishing properties for these metals are given in Table 47.

TABLE 46. Distinguishing Property and Uses of Some General and Specialty Steels [85]

Designation	Distinguishing property	Uses
Carbon steels, AISI C1030–C1080	Hardenable	Crankshafts, linkages, screw fastenings, cutting blades, springs
Carbon steels, AISI C1015–C1022, C1117–C1118	Carburizable	Structural shapes, wear-resistant machine parts, cams, bearing journals, gears, piston rings
Carbon steels, 135, N, EZ	Nitridable	Shafts, cylinder liners, cams, rollers, gears
Carbon steels, classes 60,000–100,000	Inexpensive, generally applicable, hardenable	Structural members, welded frames, electromagnet cores, transformer cores
High-strength steels, classes 45–60, D-6A, MX-2, 300 M	Workable, machinable, and weldable	Structural shapes, rivets, bolts, engine frames, turbine casings, pipes, and high-pressure tubing
Free-cutting steels, AISI B1111–B1113, C1211–C1213	Easy-machining and case-hardenable	Fastening devices, washers, spacers, shafts
High-speed steel, hi-molybdenum-M1–M44	Highly hardenable	Cutting tools, surfaces useful at moderate cutting speeds but can handle high cutting pressures
High-speed steel, hi-tungsten, T1–T15	Very hard	Cutting surfaces, instrument-bearing journals
Tool steels, chrome H10–H19, tungsten H20–H26, molybdenum H41–H43	Hardenable, tough	Forging dies, striking platens, plastic molds
Metallurgical powders, ferrous F series, stainless SS series	Magnetic, corrosion-resistant	Coil cores, armatures, magnetic linkages, light-duty gears, other machine parts
Austenitic stainless steels, AISI 200 and 300 series	Weldable, corrosion-resistant	Forged shapes, weldments, tubing, plumbing fittings
Martensitic stainless steels, AISI 400 series, except ferritic types	Hardenable, corrosion-resistant	Valves, impellers, ball bearings, springs, cutting devices, gears
Ferritic stainless steels, AISI 406, 430, 446	Free-machining, corrosion-resistant even at elevated temperatures	Chemical containers, tubing, flame tubes, nozzles
Age-hardenable stainless steels, wide variety of types, such as W, 17-7PH, and AM350 series	High strength, corrosion-resistant	Structural members, pressure vessels, chemical processing equipment

TABLE 47. Distinguishing Properties of Some Nonferrous Metals[86]

Metal	*Distinguishing Properties*
Aluminum and aluminum alloys	Aluminum is a versatile metal with combinations of light weight and tensile strength that range to 125,000 psi. Its excellent resistance to corrosion in many environments is made possible by the protective, highly adherent oxide film which develops in air, oxygen, or oxidizing media. High-strength aluminum alloys generally are not as corrosion-resistant as the high-purity or moderate-strength aluminum alloys. Aluminum is available in a great number of alloys and tempers. The metal can be fabricated economically by all common processes and can be cast by any method known to foundrymen.
Beryllium	Beryllium provides high stiffness-to-density ratios, high strength-to-density ratios, and excellent dimensional stability. Modulus of elasticity of 44×10^6 psi and density of 0.066 lb/in.^{3} are common to all forms of beryllium. High thermal conductivity and low thermal expansion coefficient contribute to dimensional control with temperature variations.
Beryllium-copper	Beryllium copper combines strength, wear resistance, electrical and thermal conductivity, and ease of fabrication properties. Age-hardened beryllium copper alloys can provide tensile strengths up to 215,000 psi with endurance strengths (at 10^8 cycles) of approximately 40,000 psi. A Rockwell hardness range up to C-45 broadens the alloy's usefulness in applications requiring resistance to wear. Electrical conductivities range up to 50% 1ACS, with a thermal conductivity range of 775–1,600 Btu/(hr) (ft^2) (°F) (ft). Beryllium-copper alloys can first be fabricated or machined in the unhardened state. Desired properties can then be imparted to the finished product by a simple low-temperature thermal treatment.
Beryllium-nickel	Beryllium nickel is the highest-strength nickel alloy at temperatures from room to 900°F. After precipitation hardening, tensile values over 270,000 psi with 230,000 yields can be obtained in wrought alloys. Like wrought forms of beryllium copper, beryllium nickel is normally supplied to a fabricator in relatively soft condition. The alloy can be formed using conventional methods. The parts can then age by a simple cycle in an ordinary furnace to achieve outstanding mechanical properties. The alloy exhibits the general corrosion resistance of nickel.

TABLE 47. Distinguishing Properties of Some Nonferrous Metals[86] (*Continued*)

Metal	*Distinguishing Properties*
Cobalt and cobalt alloys	The retention of hardness and strength at high temperature is a distinctive feature not only of cobalt-base alloys but of other cobalt-containing alloys. Materials are available for use under stress up to 2000°F and, at no load, up to 2400°F. Cobalt is the element with the highest Curie temperature (2050°F). The metal is used by itself or in alloys (cobalt-nickel, cobalt-phosphor, cobalt-nickel-phosphor) for memory and other magnetic devices. It is an important alloying element in a series of permanent magnets as well as soft magnetic materials. Cobalt and its alloys are remarkable for their low coefficient of friction and nongalling characteristics. They can often be used as high-temperature bearings without lubrication.
Copper and copper alloys	Copper has the highest electrical conductivity of any metal except pure silver. Copper alloys are easily fabricated and set the standard for nonferrous alloys in most fabricating operations. The brasses are ideally suited to cold-forming operations such as deep drawing, bending, spinning, and stamping. The ready solderability of copper is vital in electrical conductivity. The generally excellent corrosion resistance in natural environments accounts for the wide use of copper and copper alloys. These metals tarnish superficially in moist air, but the pleasing colors of the surface films developed after further exposure often are a plus factor. Copper and its alloys are resistant to corrosion by most natural waters—fresh as well as brackish and salt.
Lead and lead alloys	Lead is one of the most easily formed metals. At room temperature it is approaching the plastic state (melting point: about 620°F) and is easily rolled, extruded, cast, or often shaped by hand. Lead is resistant to most active chemicals. It is quite stable as a metal or in compounds, because of its chemical family and high atomic weight. Lead, with a density of 0.41 lb/in.^{3}, is the lowest-cost high-density material. The element has the rare, balanced combination of neutrons and protons that makes it an excellent shield against gamma rays and x-rays.
Magnesium and magnesium alloys	Magnesium, with a specific gravity of 1.74, has long been recognized as the lightest structural metal. Magnesium is well suited to modern die-casting processes. The metal flows easily and readily fills complex dies and thin sections. The most common magnesium alloys in regular use are in the magnesium-aluminum-zinc system for all cast and wrought forms. Magnesium-zinc-zirconium alloys, both cast and wrought, provide improved properties. Magnesium-thorium-zirconium, magnesium-thorium-zinc, and magnesium-thorium-manganese are the alloy systems used for retention of usable properties at elevated temperatures. Magnesium-lithium alloys are the lightest commercial magnesium alloys. A magnesium-lithium alloy, with specific gravity only 1.35, can be obtained in various mill forms. Magnesium-lithium alloys have improved cold-workability. They are fusion-weldable by the inert-gas shielded arc and also can be electrical-resistance spot-welded. They are receptive to the same chemical treatment as other

TABLE 47. Distinguishing Properties of Some Nonferrous Metals[86] *(Continued)*

Metal	*Distinguishing Properties*
	magnesium alloys. Although magnesium is sometimes used unpainted, protective and decorative finishes are applied in most cases. Application of electrolytic coatings is sensitive and requires good controls.
Nickel and nickel alloys ...	Nickel alloys, in general, are stronger, tougher, and harder than most nonferrous alloys and many steels. Their most important commercial mechanical property, however, is their ability to retain strength and toughness at elevated temperatures. The range of alloys based on nickel as a major constituent is wide enough to combat a great variety of corrosive environments. Generally, reducing conditions retard, while oxidizing conditions accelerate, corrosion of nickel. Nickel-base alloys often have the excellent corrosion resistance of elemental nickel, as well as that of other elements they contain.
Zinc and zinc alloys	Resistance to corrosion of wrought and cast zinc products is generally retained throughout the product's life. As coating, zinc's protection continues when the underlying material (generally steel) eventually becomes exposed. Corrosive attack is then directed to the zinc rather than the base metal. Zinc's low melting point facilitates low-cost production in cast, wrought, or coating form. While the melting point is in the soldering temperature range, zinc alloys have physical and mechanical properties suitable for many structural applications. Most of the finishes applied to other metal products can also be applied to zinc die castings. These include (1) mechanical—buffing, polishing, brushing, and tumbling; (2) electrodeposited—copper, nickel, chromium, brass, silver, and black nickel; (3) chemical—chromate, phosphate, molybdate, and black nickel; (4) organic—enamel, lacquer, and varnish; (5) plastic.
Precious metals	Precious metals are platinum, gold, palladium, iridium, rhodium, osmium, ruthenium, and silver. The precious metals are highly resistant to many corrosive environments either in their pure form or in alloys. Platinum is the most generally applicable but the others of the group are widely used. Silver is tarnished by sulfide environments. Gold and silver do not oxidize to form a scale at elevated temperatures, but silver does dissolve considerable oxygen and this must be considered in its use. The basic precious-metal alloy system used for structural purposes at elevated temperatures is platinum-rhodium. Pure platinum may be used in cases where the hot-strength requirement is not high. Where the strength requirements are higher than can be attained with platinum-rhodium alloys alone, platinum-rhodium alloys are used to sheath higher-strength materials such as molybdenum, which does not have adequate oxidation resistance. The stability and wide range of high electrical properties make the precious metals useful in a number of areas. Among the useful properties are stable thermoelectric behavior, high resistance to spark erosion, tarnish resistance, and broad ranges of resistivities and temperature coefficients of electrical resistance.

TABLE 47. Distinguishing Properties of Some Nonferrous Metals[86] *(Continued)*

Metal	*Distinguishing Properties*
Tungsten	Tungsten is stronger than any other common metal at temperatures over 3500°F. Tungsten's melting point of 6170°F is higher than that of any other metal. The electrical conductivity of tungsten is approximately one-third that of copper, much better than the conductivity of nickel, platinum, or iron-base alloys. Conversely, the resistivity in fine wire form has been exploited in many lamp and electronic applications in which tungsten wire serves as a light-emitting or electron-emitting filament.
Molybdenum	On a strength basis, pure molybdenum is generally considered the most suitable of all refractory metals at temperatures between 1600 and 3000°F. Small amounts of other refractory metals with molybdenum form alloys that have much greater strength-to-weight ratios at higher temperatures. Thermal conductivity of molybdenum is more than three times that of iron and almost half that of copper. Abrasion resistance of molybdenum is generally outstanding at high temperatures.
Tantalum	Tantalum is the most corrosion-resistant of the major refractory metals. It closely duplicates the corrosion resistance of glass. Tantalum is an easy-to-fabricate metal. Spinning, deep drawing, and severe bending can be performed without tears, cracks, or excessive peeling. Ductile, nonporous welds can be made easily by using TIG, resistance, or electron beam welding. Tantalum finds important use in electrolytic capacitors because it forms tantalum oxide (Ta_2O_5). This provides a high dielectric constant and good dielectric strength.
Columbium	At temperatures from 2000 to 3000°F (in vacuum or inert atmosphere), columbium alloys give the best performance on a strength-to-weight basis among metals that exhibit ductile welds. Columbium has excellent corrosion resistance, including resistance to liquid alkali metals. Columbium is the major component in high-field, superconducting alloys for use at cryogenic temperatures. Columbium has good nuclear properties, important in such applications as fuel-element cladding for nuclear propulsion reactors. The material is quite ductile, but is susceptible to air and hydrogen embrittlement at elevated temperatures. While columbium is easily worked when pure, it becomes difficult to work when highly alloyed. At high temperatures, columbium absorbs oxygen, nitrogen, and hydrogen and becomes brittle. It must be alloyed for high-temperature applications. Columbium is a reactive metal and accordingly reacts with coatings at high temperature.
Titanium	Titanium, with a density of 0.161 lb/in.^{3}, is classed as a light metal, being 60% heavier than aluminum (0.10 lb/in.^{3}), but 45% lighter than alloy steel (0.28 lb/in.^{3}). Titanium-base alloys are extremely strong, with ultimates starting at about 30,000 psi and

TABLE 47. Distinguishing Properties of Some Nonferrous Metals[86] *(Continued)*

Metal	*Distinguishing Properties*
	reaching in the neighborhood of 200,000 psi Strengths of 160,000 psi are attainable in the general-purpose (Ti6A1–4V) grade. The strength of titanium alloys is accompanied by excellent ductility. Titanium is virtually immune to corrosion from the atmosphere. Its corrosion resistance is excellent in most oxidizing environments and many mildly reducing environments. Titanium-base alloys provide excellent fatigue properties. Titanium is the only known structural metal with a corrosion-fatigue behavior in salt water practically identical to that in air.
Zirconium	Zirconium displays excellent resistance to many corrosive media. The metal resists both acids and alkalies, in particular alkali-and-chloride solutions, some inorganic acids, and chlorine-saturated water. Resistance covers a wide range of temperature and concentration. Zirconium offers a low-absorption cross section for the slow thermal neutrons necessary to sustain a chain reaction. The material is therefore used for reactor components. Zirconium is lighter than steel and heavier than titanium. When impure, it is very hard and brittle. When pure, it is soft, malleable, and ductile. Its mechanical properties resemble those of mild steel, although they change rather rapidly as temperature is elevated. For example, Zircaloy-2 has a yield strength of 65,000 psi at room temperature, but drops to 35,000 psi at 700°F.

Corrosion of Metals. Broadly speaking, corrosion can be defined as materials deterioration which is caused by chemical or electrochemical attack, in the presence of electrically conductive solutions, called electrolytes. Such electrolytes are common to some degree in nearly all environments, except perhaps extremely arid environments. When, for example, a steel part is in contact with another metal in the presence of an electrolyte such as salt in water, an electric current is generated that induces corrosion.

In a desert climate, absence of moisture will prevent corrosion even though contaminants are present; distilled water is an insulator and is not corrosive, but water found in nature contains enough contamination to make it conductive, and therefore corrosive, in the presence of a dissimilar metal junction. Salt used on winter streets to melt ice and snow produces an extremely corrosive electrolyte.

Degree of attack is also influenced by the position of the dissimilar metal in relation to steel in the galvanic series shown in Table 48, with the least noble at the top. This order is for usual types of electrolytes met in outdoor environments; with other conditions, such as specific types of chemical exposure, the order may vary.

A less noble metal (higher in the list than steel) will be consumed when in contact with steel in the presence of an electrolyte. A magnesium rod, for example, will corrode and protect the steel lining in a hot-water tank. Zinc, cadmium, and aluminum as plated coatings offer sacrificial protection to steel and also serve as barriers to moisture. Even when these coatings are porous, the steel underneath is not attacked.

More noble metals such as tin, nickel, copper, and chromium will cause steel to corrode. When used as a plated coating, these metals must offer a complete barrier to moisture; a pinhole or crack in the coating will cause formation of rust in the presence of moisture.

TABLE 48. Galvanic Series of Metals under Atmospheric Conditions [87]

Corroded end
(anodic, or least noble)

Magnesium
Magnesium alloys
Zinc
Aluminum 1100
Chromium
Cadmium
Aluminum 2017

Steel or iron

Stainless, ferritic (active)
Stainless, austenitic (active)
Lead-tin solder
Lead
Tin
Nickel (active)
Inconel (active)
Brass
Copper
Bronze
Copper-nickel alloys
Silver solder
Nickel (passive)
Inconel (passive)
Titanium
Stainless, ferritic (passive)
Stainless, austenitic (passive)
Silver
Graphite
Gold
Platinum
Monel

Protected end
(cathodic, or more noble)

Corrosion between dissimilar metals can be prevented by one of three means: (1) excluding moisture with an impervious barrier, (2) using a less noble or sacrificial metal, and (3) electrically insulating the dissimilar metal.

Electroplated and Chemically Deposited Coatings. Chemically deposited coatings constitute another of the groups of materials vitally important to the electronic packaging engineer. While used primarily with metals, chemically deposited coatings are finding increased usage with plastic materials. The primary reasons for such coatings on metals are decorative and protective requirements, and reflective and thermal surface requirements for aerospace applications. When used with plastics, chemically deposited metallic coatings often provide shielding for critical electrical circuits.

Electroplated Metallic Coatings. An electroplating solution is an electric field in a conductive liquid solution (electrolyte) with current entering the solution at the anode and emerging at the cathode. The cathode consists of the parts being plated. The anode and cathode can be considered uniform potential electrodes in an electrostatic field, with the anode positive and the cathode negative. If the field is plotted between the electrodes for different configurations, high potential gradients can be seen at projections and low potential gradients at recesses. Since current density is proportional to potential gradient, projections will receive high current density and recesses will receive low current density. In most solutions, the rate of deposition is proportional to current density. Therefore, on irregularly shaped parts, plating thickness may vary considerably. To have adequate thickness in recesses, average thickness may have to be two to three times the minimum specification thickness.

A discussion of individual electroplates is given below.[88]

CADMIUM PLATING. Electrodeposits of cadmium are used extensively to protect steel and cast iron against corrosion. Because cadmium is anodic to iron, the underlying ferrous metal is protected at the expense of the cadmium plate, even though the cadmium becomes scratched or nicked, exposing the substrate.

Cadmium is applied usually as a thin coating (less than 1 mil thick) intended to withstand atmospheric corrosion. It is seldom used as an undercoating for other deposited metals, and its resistance to corrosion by most chemicals is low. It is frequently used to coat parts and assemblies that are made up of dissimilar metals because of its ability to minimize galvanic corrosion. Its excellent solderability is advantageous in many electrical applications.

ZINC PLATING. Zinc is anodic to iron and steel and therefore offers more complete protection when applied in thin films (0.3 to 0.5 mil) than do similar thicknesses of nickel and other cathodic coatings. Because it is relatively cheap and readily applied in barrel, tank, or continuous plating facilities, zinc is often preferred for coating ferrous parts when protection from atmospheric and outdoor corrosion is the primary objective. Normal electroplated zinc without subsequent treatment becomes dull gray in appearance after exposure to air. Bright zinc that has been subsequently given a bleached chromate conversion coating or a coating of clear lacquer (or both) is sometimes used as a decorative finish. Such a finish, although less durable than heavy nickel-chromium, in many instances offers better corrosion protection than thin coatings of nickel-chromium, and at much lower cost. Plating of zinc on gray iron and malleable iron presents serious operational difficulties; cadmium is usually preferred.

COPPER PLATING. Copper electrodeposits are widely employed as underplates in multiple-plate systems, as stop-offs, for heat transfer, and in electroforming. Although copper is relatively corrosion-resistant, it is subject to rapid tarnishing and staining when exposed to the atmosphere; consequently, electrodeposited copper is rarely used alone in applications where a durable and attractive surface is required. In some applications, bright electrodeposited copper, protected against tarnishing and staining by an overcoating of clear lacquer, is used as a decorative finish.

NICKEL PLATING. Nickel plate, with or without an underlying copper strike, is one of the oldest protective-decorative electrodeposited metallic coatings for steel, brass, and other basis metals. Unless polished occasionally nickel plate will tarnish, taking on a yellow color during long exposure to mildly corrosive atmospheres or turning green on severe exposure. The introduction of chromium plate overcame the tarnishing problem and led to a great increase in the use of nickel as a component of protective-decorative coatings in various combinations with copper and chromium. The use of nickel in these coating systems is discussed in a following section.

CHROMIUM PLATING (DECORATIVE). Decorative chromium plating is a protective-decorative coating system in which the outermost layer is chromium. The layer is applied over combinations of plated coatings of copper and nickel. The function of this system is twofold: (1) to provide the basis metal with protection against corrosive environments, and (2) to maintain, in service, an appearance conforming to an agreed-on standard.

CHROMIUM PLATING (HARD). Hard chromium plating (commonly known as "industrial" or "engineering" chromium plating) differs from decorative chromium plating in the following ways:

Hard chromium deposits are intended primarily to restore dimensions of undersized parts or to improve resistance to wear, abrasion, heat, or corrosion, rather than to enhance appearance.

Hard chromium normally is deposited to thicknesses ranging from 0.1 to 20 mils (and for certain applications, to considerably greater thicknesses), whereas decorative coatings seldom exceed 0.1 mil.

With certain exceptions, hard chromium is applied directly to the base metal and is usually ground to a find dimension; decorative chromium is applied over undercoats of nickel or of copper and nickel, and is either buffed or used in the as-plated condition.

Whereas decorative coatings generally contain pores, the heavier hard coatings are more likely to be smoother, but they may contain microcracks.

TIN AND TIN-LEAD PLATING. Electrodeposits of tin are corrosion-resistant and nontoxic, possess excellent solderability, and are noted for their softness and ductility. Although tin plate is quite tarnish-resistant, it does not offer sacrificial protection in the air when plated on steel. To provide corrosion protection of a substantial nature, tin deposits must be thick enough to be virtually nonporous. Tin-lead electrodeposits are widely used in the electronics industry for solderability purposes, particularly on printed-circuit boards. A tin-lead electrodeposit is, of course, essentially a solder coating.

SILVER PLATING. Silver electrolytes of the cyanide type are the most widely used commercially, both for decorative and for engineering plating. With proper control of electrolyte composition and plating procedures, smooth, dense, fine-grained, strongly bonded silver plate ranging from less than 1 mil to more than ¼ in. thick can be deposited from cyanide baths on most metals and alloys. The smoothness and brightness of cyanide silver deposits can be controlled by adding to the bath small quantities of sulfur-bearing compounds such as ammonium thiosulfate and carbon disulfide. These additives also help to suppress burning and treeing in areas of high current density.

GOLD PLATING. Industrial uses of electroplated gold are centered primarily in the fields of electronics, control, and communications. Gold is ideal for these applications because it (1) resists tarnish and corrosion of various chemicals, (2) maintains a low electrical contact resistance, (3) resists high-temperature oxidation, and (4) is solderable after extended periods of storage.

Immersion and Electroless Coatings. Immersion and electroless coatings applied without the aid of an electrical current in the plating bath have become of increasing importance in the electronics industry. Numerous nickels are commonly deposited by one or both of these techniques. Perhaps most commonly applied are copper, nickel, gold, tin, and silver.

Two advantages of these coatings are: (1) application from a nonelectrolytic solution; variable potential gradient points do not occur on the part being plated and hence much more uniformity of deposit is obtained than from an electrolytic bath; and (2) nonconductors, such as plastics, can be given a basic nonelectrolytic metallic coating using electroless techniques, which can in turn be followed by electrolytic deposition of other metals onto the nonmetallic base material. The basics of immersion plating and electroless or catalytic reduction can perhaps best be described by describing their application in nonelectrolytic nickel plating.

The aqueous immersion plating bath is capable of depositing a very thin (about 0.025 mil) and uniform coating of nickel on steel in periods of up to 30 min. The coating is porous and possesses only moderate adhesion, but these conditions can be improved by heating the coated part at 1200°F for 45 min in a nonoxidizing atmosphere. (Higher temperatures will promote diffusion of the coating.)

ELECTROLESS PLATING. The electroless nickel-plating process employs a chemical reducing agent (sodium hypophosphite) to reduce a nickel salt (such as nickel chloride) in hot aqueous solution and to deposit various thicknesses of nickel on a catalytic surface. The deposit obtained from an electroless nickel solution is an alloy containing from 4 to 12 percent phosphorus and is quite hard. The hardness of the as-plated deposit can be increased by heat treatment. Because the deposit is not dependent on current distribution, it is uniform in thickness regardless of the shape or size of the plated surface.

Conversion Coatings. A conversion coating can be either a nonelectrolytic chemical treatment or an electrolytic chemical treatment that modifies a reactive metallic surface to a relatively inert inorganic film. Nonelectrolytic chemical treatments include phosphate, chromate, and oxide coatings. Typical of electrolytic chemical coatings are anodic treatments for aluminum- and magnesium-based alloys and other treatments such as anodizing for aluminum- and magnesium-based alloys. These conversion coatings then are primarily used to convert the surface of active metals such as magnesium and aluminum to a less chemically active or inert surface.

While chromate and oxide coatings are used predominantly on aluminum and magnesium, phosphate coatings are also used on steel and zinc metals.

Aluminum Coatings. Aluminum may be coated either by electroplating or by conversion coatings. Electroplating of aluminum requires particular precautions due to the high surface activity of the raw, fresh, clean aluminum surface. For this reason the common practice for electroplating on aluminum is first to apply a zinc coating onto the surface of the aluminum, a process known as zincating. Subsequent electrodeposited coatings can then be applied upon the zinc coating. A listing of the coatings which can be applied to aluminum and aluminum alloys is shown in Table 49. Because of the problems in electroplating aluminum, conversion coatings and anodized coatings are widely used on aluminum.

CONVERSION COATINGS ON ALUMINUM. In recent years there has been considerable development in coatings for aluminum. At the present time there are many proprietary compounds on the market (such as Alodine,* Iridite,† and Kenvert ‡). Their chief advantages are corrosion resistance, paint base, and ease of application. The corrosion resistance imparted to aluminum by using these dips has been reported by manufacturers of these compounds to be excellent, and in some cases better than chromic acid anodizing. As a paint base these materials are excellent, and it is recommended that one of the proprietary compounds be used in order to obtain better adhesion. The ease of application of these coatings is a very important factor in replacing anodizing, and because of this the cost is one-third to one-half that of anodized coatings. Usually these coatings have yellow to brown color (in some instances green) depending on the thickness of coating applied. There are on the market at present several compounds which are colorless or nearly so.

ANODIZING. Anodizing is an electrochemical process by which the natural oxide film present on aluminum can be increased in thickness.[90] This oxide is an inherent corrosion-resistant film, which normally has a thickness in the untreated form of 0.00000052 in. Subsequently, after an anodizing treatment the thickness of this oxide (Al_2O_3) will be increased by 500 to 2,000 times the original thickness.

The general procedure for the formation of this heavy oxide consists of making the aluminum part to be anodized the anode in a direct-current circuit. The current passes from the cathode through a solution of sulfuric acid, chromic acid, or other electrolyte to the anode. This procedure results in the liberation of nascent oxygen at the anode which reacts with the metal to form a film of aluminum oxide. This film forms at the surface of the metal and in effect grows outwardly.

There are two general types of anodizing normally employed. The first is the sulfuric acid process, the second chromic acid anodizing, and both are discussed below.[91]

SULFURIC ACID ANODIC TREATMENT FOR ALUMINUM. The sulfuric acid anodizing process for aluminum surfaces is probably the most widely used anodizing process in this country. It employs dc, although ac may also be used. It is the most economical process in use for anodizing aluminum. The coating varies from a clear, transparent film to one that is opaque or translucent; it usually allows some of the metallic sheen of the base metal to remain. The differences in color depend on the particular alloy, the length of treatment, and variations in composition of the bath. The film thickness ranges from 0.0001 to 0.001 in. However, for anodized coating on aluminum, the minimum amount of coating is specified by weight instead of thickness. When unsealed, the coating is porous, absorbent, and very hard; in order to give maximum corrosion protection (with some sacrifice in hardness) it must be made nonabsorbent by sealing. Sulfuric acid anodized coating is thicker and more porous than chromic acid anodized coating and easily lends itself to coloring when sealed with dichromate solution instead of hot water. It provides very good corrosion and abrasion resistance and is a very good paint base.

CHROMIC ACID ANODIC TREATMENT FOR ALUMINUM. The chromic acid method of anodizing is one of the two best-known processes for anodizing aluminum in this country, the other being the sulfuric acid process. The coating produced by the chromic acid process is opaque and gray in color, varying from light gray for purer aluminum to greenish gray on certain alloys. The greenish tinge, when present, is said to be owing to reduced chromium compounds. The thickness of the film ranges from 0.00003 to 0.001 in., and the average is about 0.00005 in. The thickness can be varied with bath conditions and length of treatment. Actually, as mentioned, the amount of coating is specified by weight.

The coating formed by the chromic acid process is not as porous or as absorptive as that formed by the sulfuric acid process. This makes it less desirable for dyeing purposes, but its protective qualities are equally as good and it is preferable in some instances, such as assemblies. Unlike the results of sealing the sulfuric acid anodic coating, sealing a chromic acid process coating does not increase, but sometimes may even decrease, its corrosion resistance. To begin with, the film produced with

* Trademark, Amchem Products, Inc.
† Trademark, Allied Research Products, Inc.
‡ Trademark, Conversion Chemical Corp.

chromic acid is not so porous. This is because the higher temperature and longer treatment time of the process favor the formation of the more nearly pore-free, hydrate crystal aluminum oxide coating. Second, hot-water sealing leaches out some of the protective chromate absorbed from the bath, resulting in reduced protection. A coating that is dyed, however, is still sealed in nickel acetate or hot water to protect the color.

HARD ANODIC COATINGS. The thickest anodic coatings formed on aluminum are produced by the hard anodic coating techniques.[92] Proprietary processes include Alumilite,* Duranodic,* Martin,* Sanford,† and Hardas.‡ Because a thickness of 1 mil or more is one of the primary requirements of a hard anodic coating, an electrolyte which enables the growth process to continue is essential. The other primary requirement is hardness. This term is somewhat ambiguous in describing this coating, however, because the aluminum oxide of the so-called "hard" anodic coating is actually no harder than that of conventional anodic coatings. The electrolyte and other conditions are such, however, that the thick aluminum oxide coating produced has a dense structure and therefore resists abrasion and erosion better than the conventional anodic coatings. This high density results from anodic oxidation at high current density in an electrolyte which, owing to its composition and low temperature, has a minimum solvent effect on the aluminum oxide formed. The electrolyte in general use is sulfuric acid (low solvent action) with or without additions of oxalic acid. Oxalic acid additions result in even less solvent action; consequently, denser coatings are produced.

Conventional anodic coatings produced in sulfuric acid (15 percent by weight) electrolytes at 70°F use current densities of about 12 amp/ft^2. Hard anodic coatings are formed in electrolytes of 15 percent or less concentration, in a temperature range of 25 to 50°F, at current densities of 24 to 36 amp/ft^2. All of these conditions favor high coating ratios, even on alloys that normally produce more porous anodic coatings.

Parts with thin sections or sharp edges must be treated with extreme care to produce satisfactory hard anodic coatings. Such areas, particularly in the copper-bearing alloys, promote "burning" or dissolving of the base metal. Sharp corners and edges produce high-current-density areas in conventional anodic-coating methods, and this condition is accentuated by the higher current densities used in hard anodic processing.

Magnesium Coatings. Magnesium, even more than aluminum, has a very reactive surface which necessitates close control of plating and finishing operations. Nevertheless, much work has been done in this field, particularly by the Dow Metal Products Co. Some of the more important chemically deposited coatings and coating processes are treated here.[93]

ELECTROPLATING. The Dow electroplating process for magnesium consists of the initial application of an immersion zinc coating followed by copper striking and electroplating in standard plating baths. With this process any metal which can be electrodeposited can be applied to magnesium. The success of the process depends almost entirely upon the adhesion and uniformity of the zinc immersion coating. This coating is approximately 0.0001 in. thick. Electron diffraction analysis shows the coating to be pure metallic zinc.

Copper- and silver-plated surfaces have been joined by soft soldering without disrupting the adhesion. Deposits have been hammered and bent without failure. Heavy chromium plates have been applied to articles for wear resistance without a tendency for peeling of deposits. General corrosion tests (salt spray, high humidity, interior and exterior exposure) over a period of years indicate good performance. Service conditions usually determine the entire plating cycle required in plating on any base metal; thus, for exterior use, service testing to determine the proper plating thickness on magnesium parts is recommended.

In many current electronic applications, hot flowed tin, gold, or silver electro-

* Trademark, Aluminum Company of America.
† Trademark, Sanford Process Company, Inc.
‡ Trademark, Hardas, Ltd.

plates are used over either a copper or a copper-nickel plate. Standard tin, gold, or silver electroplating baths are used. Such coatings are used primarily for rf grounding or hermetic sealing.

ELECTROLESS NICKEL PLATING. The Dow electroless nickel process for magnesium permits the chemical deposition of a nickel coating directly on magnesium with no corrosive attack on the base metal surface. The nickel plate applied by this process can be used as a final coat or as a strike coating (0.0001 to 0.0002 in.) over which standard plates (bright nickel, chromium, tin, cadmium, zinc, etc.) can be deposited. No electric current is required. The coating can be built up to any desired thickness. The process coats the surface wherever the solution touches it, and covers all surfaces evenly. This uniform deposition rate gives closer dimensional tolerances and better maintenance of detail than electroplating. The auxiliary anode and special racking commonly required to "throw" the plate inside blind holes and recessed areas are not needed with electroless nickel plating. If maximum adhesion and corrosion resistance are needed, the zinc immersion plus copper strike should be used prior to the electroless nickel plating.

Special Coatings and Processes

SELECTIVE PLATING. This process is a form of localized plating sometimes known as *brush plating* or *Selectron.** It is electrolytic solution plating using very small quantities of plating solution applied by means of a small brush or stylus which carries an electrical current. Many metals can be electrolytically deposited by this process; its use allows either the plating of small assemblies or the repair of plated parts which are not practical to replate in a tank electroplating bath.

ELECTROFORMING. Electroforming essentially consists of electroplating onto a master form or mandrel which is removed after plating, thus leaving an electroplated shell or form whose contour takes the shape of the mandrel or form on which it was plated. Low-melting metals which subsequently can be melted out of the electroformed part are commonly used for mandrels.

Electroformed objects are normally somewhat thicker than electroplated parts. Electroforming is, of course, most widely used for the fabrication of metal parts where this form of fabrication offers unique advantages. Electroformed parts can be made from many metals using a variety of electroplating solutions. Since electroformed parts usually have a considerable thickness buildup (compared with plating) electroforming involves more plating time than simple electroplating.

Some of the key advantages of electroforming include controllability of the metallurgical properties of an electrodeposited metal over a wide range of selection of metals, adjustment of plating bath composition, and variation of conditions of deposition. In addition, with the proper choice of matrix material, parts can be produced in quantity with a very high order of dimensional accuracy; reproduction of fine details is of a high order; there is almost no limit to the size of the end product; and shapes can be made that are not possible using any other fabrication method.[94]

Among the disadvantages, on the other hand, is the cost, which is often relatively high. Also, the production rate may be comparatively slow, sometimes measured in days. There are, too, limitations in design: the electroform cannot have great or sudden changes in cross section or wall thickness unless these can be obtained by machining after electroforming. In addition, because of the exactness of reproducibility, scratches and imperfections in the master will also appear in all pieces processed from it.

* Trademark, Selectron, Ltd.

TABLE 49. Chemically Deposited Coatings for Aluminum and Aluminum Alloys [89]

Treatment	Purpose	For use on	Operation	Finish and thickness
Zinc phosphate coating	Paint base	Wrought alloys	Power spray or dip. For light to medium coats, 1 to 3 min at 130 to 135°F.	Crystalline, 100 to 200 mg/ft².
Chromium phosphate coating	Paint base or corrosion protection	Wrought or cast alloys	Power spray, dip, brush, or spray. For light to medium coats, 20 sec to 2 min at 110 to 120°F.	Crystalline, 100 to 250 mg/ft².
Sulfuric acid anodizing	Corrosion and abrasion resistance, paint base	All alloys; uses limited on assemblies with other metals	15 to 60 min, 12 to 14 amp/ft², 18 to 20 volts, 68 to 74°F. Tank lining of plastic, rubber, lead, or brick.	Very hard, dense, clear. 0.0002 to 0.0008 in. thick. Withstands 250- to 1,000-hr salt spray.
Chromic acid anodizing	Corrosion resistance, paint base; also as inspection technique with dyed coatings	All alloys except those with more than 5 pct Cu	30 to 40 min, 1 to 3 amp/ft², 40 volts dc, 95°F, steel tanks and cathode, aluminum racks.	0.00002 to 0.00006 in. thick, 250-hr min salt spray.
Chromate conversion coating	Corrosion resistance, paint hesion, and decorative effect	All alloys	10 sec to 6 min depending on thickness, by immersion, spray, or brush, 70°F, in tanks of stainless, plastic, acid-resistant brick or chemical stoneware.	Electrically conductive, clear to yellow and brown in color, 0.00002 in. or less thick, 150- to 2,000-hr salt spray depending on alloy composition and coating thickness.
Chemical oxidizing	Corrosion resistance, paint base	All alloys, less satisfactory on copper-bearing alloys	Basket or barrel immersion, 15 to 20 min, 150 to 212°F.	May be dyed, 250-hr min salt spray.
Electropolishing	Increases smoothness and brilliance of paint or plating base	Most wrought alloys, some sand-cast and die-cast alloys	15 min, 30 to 50 amp/ft², 50 to 100 volts, less than 120°F, aluminum cathode.	35 to 85 RMS depending on treatment.
Zinc immersion	Preplate for subsequent deposition of most plating metals, improves solderability	Many alloys, modifications for others particularly regarding silicon, copper, and magnesium content	30 to 60 sec, 60 to 80°F, agitated steel or rubber-lined tank.	Thin film.
Electroplating	Decorative appeal and/or function	Most alloys after proper preplating	....	Same as on steel.
Chromium..	....	....	Directly over zinc immersion coat, 65 to 70°F, 6–8 volts, 200–225 amp/ft². Transfer to bath at 120 to 125°F if copper, or copper and nickel have been applied.	
Copper	....	....	Directly over zinc, or follow with copper strike, then plate in conventional copper bath.	
Brass	....	....	Directly over zinc, 80 to 90°F, 2–3 volts, 3–5 amp/ft².	

TABLE 49. Chemically Deposited Coatings for Aluminum and Aluminum Alloys (*Continued*)

Treatment	Purpose	For use on	Operation	Finish and thickness
Nickel	....	....	Directly over zinc, or follow with copper strike, then plate in conventional nickel bath.	
Cadmium ..	....	....	Directly over zinc, or follow with copper or nickel strike, or preferably cadmium strike, then plate in conventional cadmium bath.	
Silver	....	....	Copper strike over zinc using copper cyanide bath, low pH, low temperature, 24 amp/ft^2 for 2 min, drop to 12 amp/ft^2 for 3 to 5 min; plate in silver cyanide bath, 75 to 80°F, 1 volt, 5–15 amp/ft^2.	
Zinc	....	....	Directly over zinc immersion coating.	
Tin	....	....	Directly over zinc immersion coating.	
Gold	....	....	Copper strike over zinc as for silver, then plate in conventional bath.	

REFERENCES

1. Harper, C. A., "Electronic Packaging with Resins," 1st ed., McGraw-Hill Book Company, New York, 1961.
2. Harper, C. A., "Plastics for Electronics," 1st ed., Kiver Publications, Inc., Chicago, 1964.
3. Allied Chemical Corp., Plastics Division, Technical bulletin entitled "Plaskon Plastics and Resins."
4. Parry, H. L., et al., High Humidity Insulation Resistance of Resin Systems, *SPE J.*, October, 1957.
5. Sunderland, G. B., and A. Nufer, Aminos, *Machine Design Plastics Reference Issue*, June, 1966. Copyright 1966.
6. FMC Corp., Technical bulletin entitled "Dapon Diallyl Phthalate Resin."
7. Chottiner, J., Dimensional Stability of Thermosetting Plastics, *Mater. Design Eng.*, February, 1962.
8. Chapman, J. J., and L. J. Frisco, A Practical Interpretation of Dielectric Measurements up to 100 MC, *J.H.U. Dielectrics Lab. Rep.*, December, 1958.
9. Hauck, J. E., Heat Resistance of Plastics, *Mater. Design Eng.*, April, 1963.
10. Staff Report, Thermosets, *Design News*, May, 1964.
11. Madorsky, S. L., and S. Straus, Stability of Thermoset Plastics at High Temperatures, *Mod. Plastics*, February, 1961.
12. Staff Report, Tables of Plastics Properties, *Plastics World*, Cahners Publishing Co., Inc., 1967.
13. Patten, G. A., Heat Resistance of Thermoplastics, *Mater. Design Eng.*, May, 1962.
14. Barkan, H. E., and A. E. Javitz, Plastics Molding Materials for Structural and Mechanical Applications, *Elec. Mfg.*, May, 1960.
15. Riley, M. W., Selection and Design of Plastics, *Mater. Methods*, June, 1957.
16. Koo, G. P., et al., Engineering Properties of a New Polytetrafluoroethylene, *SPE J.*, September, 1965.
17. Du Pont, Inc., Plastics Department, Technical booklet entitled "Mechanical Design Data for Teflon Fluorocarbon Resins."

18. Staff Report, Fluorocarbon Plastics, *Mater. Design Eng.*, February, 1964.
19. Staff Report, Designing with Teflon Resins at High Temperatures, *J. Teflon,* April, 1965.
20. Doban, R. C., C. A. Sperati, and W. B. Sandt, "Properties of Polytetrafluoroethylene," *SPE J.,* November, 1955.
21. Du Pont, Inc., Plastics Department, Technical booklet entitled "Teflon for Electrical and Electronic Systems."
22. Ehner, W. J., Thermoplastic Parts, *Machine Design,* August, 1963. Copyright 1963.
23. Union Carbide Corp., Plastics Div., Technical bulletin entitled "Parylene Product Data."
24. USI Chemicals Co., Technical bulletin entitled "Petrothene Cross-linkable Polyethylene Compounds."
25. Raychem Corp., Technical bulletin entitled "Raychem Irradiated Polyolefins."
26. Staff Report, The New Polyimides, *Plastics Technol.,* December, 1962.
27. Freeman, J. H., New Amide-Imide Plastics at 600°F, *Mater. Design Eng.,* October, 1966.
28. Du Pont, Inc., Plastics Department, Technical bulletin entitled "Vespel Polyimide Resins."
29. General Electric Co., Technical bulletin entitled "A Designer's Profile of Engineering Plastics."
30. Staff Report, Why Plastics Don't Line Up to Design Data, *Prod. Eng.,* April, 1966.
31. General Electric Co., Technical bulletins entitled "Noryl Thermoplastic Resin" and "PPO Product Data."
32. Gowan, A., and P. Shenian, Properties and Applications for Polyphenylene Oxide Plastic, *Insulation,* September, 1965.
33. Union Carbide Corp., Plastics Div., Technical bulletin entitled "Polysulfone."
34. Bulkley, C. W., Vinyls, *Machine Design Plastics Reference Issue,* June, 1966. Copyright 1966.
35. Stalwart Rubber Co., Technical bulletin entitled "Stalwart Rubber Selector."
36. General Electric Co., Technical bulletin entitled "Silicone Rubber for Wire and Cable Application."
37. Dow Corning Corp., Technical bulletin entitled "Engineering with Silastic Silicone Rubber."
38. Seymour, R. B., "Fillers for Molding Compounds," *Mod. Plastics 1966 Encyclopedia Issue.*
39. Staff Report, Can RTP Meet Your Product Requirements, *Mod. Plastics,* March, 1966.
40. O'Rourke, J. T., Tailoring the Properties of Teflon with Fillers, *J. Teflon,* September–October, 1962.
41. Staff Report, Asbestos Filled Polypropylene, *Mod. Plastics,* August, 1966.
42. Cryor, R. E., Asbestos Reinforced Plastics Resist Heat and Chemicals, *Mater. Design Eng.,* April, 1966.
43. Widdop, H., and G. A. Ebelhare, Laminates, *Mod. Plastics 1967 Encyclopedia Issue.*
44. National Electrical Manufacturers Association, Publication entitled "Industrial Laminated Thermosetting Products."
45. Lee, R., Insulation Fundamentals Are Practical Design Tools, *Electron. Design,* July, 1965.
46. Herberg, W. F., and E. C. Elliott, Reinforced Silicone Plastics, *Plastics Design Process.,* January, 1966.
47. Courtright, J. R., and S. Graves, New Polymers for High Temperature Use, *Proc. SPI 23d Ann. Conf.,* Palm Springs, Calif., April, 1966.
48. Minnesota Mining and Manufacturing Co., Technical bulletin entitled "Prepregs."
49. Hauck, J. E., Plastic Foams, *Mater. Design Eng.,* May, 1966.
50. Sharpe, L. H., Assembling with Adhesives, *Machine Design,* August, 1966. Copyright 1966.
51. Loctite Corp., Technical bulletin entitled "Loctite Anaerobic Adhesives."
52. Eastman Chemical Products, Inc., Technical bulletin entitled "Eastman 910 Adhesive."
53. Burgman, H. A., Adhesive Technology Updated, *Westinghouse Res. Labs. Scientific Paper 63-131-353-P2,* August, 1963.
54. Staff Report, Polyaromatic Adhesives, *Design News,* September, 1965.

55. Licari, J. J., and E. R. Brands, Plastic Coatings for Commercial and Military Applications, *Machine Design,* May, 1967. Copyright 1967.
56. Beccasio, A. J., L. S. Keefer, and A. Z. Orlowski, The Effects of Conformal Coatings on Printed Circuit Assemblies, Insulation, November, 1965.
57. Westinghouse Electric Corp., Benolite Div., Technical data bulletin entitled "Insulating Varnishes and Finishes."
58. Lee, M. M., Application of Electrical Insulation by the Fluidized Bed Process, *Electro-Technol. (New York),* October, 1960.
59. Hinkley, J. R., Resins for Packaging Electronic Assemblies, *Electro-Technol. (New York),* June, 1965.
60 Lee, M. M., and R. D. Hodges, Heat Resistant Encapsulating Resins, *Proc. 15th Ann. Tech. Conf., Soc. Plastics Engrs.,* New York, January, 1959.
61. General Mills, Inc., Technical data bulletins entitled "Versamids."
62. Thiokol Chemical Corp., Technical data bulletins entitled "Thiokols."
63. Wallace & Tiernan, Inc., Harchem Div., Technical data bulletins entitled "Harcures."
64. Du Pont, Inc., Elastomer Chemicals Dept., Technical data bulletins entitled "Adiprenes."
65. Union Carbide Corp., Technical data bulletins entitled "Cycloaliphatic Diepoxides."
66. Black, R. G., Linear Polyazelaic Polyanhydride as a Converter for Epoxy Resin Systems, *Proc. 20th Ann. Tech. Conf., Soc. Plastics Engr.,* Atlantic City, N.J., January, 1964.
67. Enjay Chemical Co., Technical data bulletins entitled "Buton Resins."
68. Fekete, F., New Family of Thermosetting Acrylic Resins, *Plastics Technol.,* March, 1963.
69. Formo, J. J., and R. J. Isliefson, Producing Special Properties in Plastics With Fillers, *Proc. Soc. Plastics Engrs. Regional Tech. Conf.,* Fort Wayne, Ind., May, 1959.
70. Lee, H. H., "Epoxy Resins," McGraw-Hill Book Company, New York, 1957.
71. Wolf, D. C., Trends in the Selection of Liquid Resins for Electronic Packaging, *Proc. Natl. Electron. Packaging Conf.,* New York, June, 1964.
72. Minnesota Mining and Manufacturing Co., Technical data bulletins entitled "Scotchcast Electrical Insulating Resins."
73. Johnson, L. I., and R. J. Ryan, Encapsulated Component Stress Testing, *Proc. 6th Elec. Insulation Conf.,* Chicago, September, 1965.
74. Beach, N. E., "Government Specifications and Standards for Plastics," Plastics Technical Evaluation Center, Picatinny Arsenal, Dover, N.J. Document available from Clearinghouse for Federal Scientific and Technical Information Center, Springfield, Va., 22151.
75. Staff Report, Government Specifications and Standards, *SPE J.,* November, 1966.
76. American Lava Corp., Technical data bulletins entitled "Alsimag Ceramics."
77. Coors Porcelain Co., Technical data bulletins entitled "Coors Alumina and Beryllia Properties."
78. Hessinger, P. S., How Good Are Beryllia Ceramics, *Electronics,* October, 1963.
79. Union Carbide Corp., Technical data bulletins entitled "National Boron Nitride."
80. Molecular Dielectrics, Inc., Technical data bulletin entitled "MyKroy Glass Bonded Mica."
81. Shand, E. B., "Glass Engineering Handbook," 2d ed., McGraw-Hill Book Company, New York, 1958. Copyright 1958 by Corning Glass Works.
82. *Machine Design,* Annual Metals Reference Issue, September, 1966. Copyright 1966.
83. Hoyt, S., "ASME Handbook: Metals Properties," McGraw-Hill Book Company, New York, 1954.
84. Horger, O. J., "ASME Handbook: Metals Engineering—Design," 2d ed., McGraw-Hill Book Company, New York, 1965.
85. Katz, I., A Reconsideration of Metals, *Electromechanical Design,* June, 1966.
86. Wigotsky, V. W., Nonferrous Metals, *Design News,* September, 1965.
87. Bronson, H., Dissimilar Metals, *Prod. Eng.,* June, 1963.
88. American Society for Metals, T. Lisman (ed.), "Metals Handbook—Heat Treating, Cleaning and Finishing," vol. 2, 8th ed., 1964.
89. Staff Report, How to Get More for Your Metalworking Dollar, *Iron Age,* June, 1956.
90. J. W. Rex Co., Technical bulletin entitled "Chemical Finishing of Metals."

91. MIL-HDBK-132 (Ord), "Military Handbook on Protective Finishes."
92. Vandenberg, R. V., Characteristics of Hard Anodic Coatings on Aluminum, *Machine Design,* March, 1962. Copyright 1962.
93. Dow Metal Products Co., Technical data booklets entitled "Magnesium Finishing."
94. Graham, A. K., Design and Application Factors in Electroforming, *Machine Design,* November, 1962. Copyright 1962.
95. Mathews, J. P., Elastomers, *Machine Design,* August, 1965. Copyright 1965.
96. Owens-Illinois Co., Technical data bulletin entitled "O-I Solder Glasses."

Chapter 8

PACKAGING WITH CONVENTIONAL COMPONENTS

By

H. G. CARTER

Mechanical Design & Development Engineering
Westinghouse Electric Corporation
Aerospace Division
Baltimore, Maryland

INTRODUCTION

Scope. This chapter is devoted to data, information, and discussion on design of electronic packages which contain conventional discrete components. In terms of component sophistication, this covers items up to and including transistors. The packaging of more advanced items such as microcomponents, thin film, integrated circuits, and hybrid assemblies is discussed in Chap. 9. The material in this chapter is concerned with the use of techniques and materials insofar as they apply to the packaging of standard conventional components; the specific details of the applicable techniques can be found in earlier chapters. Primarily the reader should learn from this chapter the various ways of packaging conventional components, the techniques and materials which can be used to fabricate these designs, the trade-offs to be made when applying a packaging concept to a particular problem, and the types of electrical components and hardware items available to implement the design.

Component Parts. The term *conventional component,* as used in this chapter, refers to standard parts—electrical or electromechanical—which perform a single limited function in a larger assembly. These parts are available on relatively short notice from one or more manufacturers, and data are available on their physical size and electrical characteristics. Individually cased transistors form the upper limit of the sophistication considered, and the types of standard configured resistors and capacitors normally used with these transistors form the lower limit on size. Table 1 gives the more common component parts grouped as to function plus the hardware items used in packaging them.

TABLE 1. Components Grouped by Function

Electrical		
Active	Passive	Inductive
Electron tubes Diodes Transistors Rectifiers	Resistors Capacitors	Transformers Chokes Special coils
Electromechanical		
Switches Relays	Meters Synchros	
Mechanical (hardware)		
Clamps and clips Sockets Connectors	Heat sinks Nuts and bolts Enclosures Shields	

Future Usage. Ever since the introduction of the transistor in the late 1940s, predictions have been made of the imminent total extinction of the electron tube, but millions of receiver-type tubes are still used in new designs each year. When semiconductor integrated circuits became a functional reality rather than a laboratory curiosity, predictions were again made that conventional discrete components would soon become obsolete, having been replaced by totally integrated systems. Although the trend is downward, it is obvious that conventional components are still very much in common usage, and will continue to dominate in such areas as high power, certain high precision, adjustment required, and low unit volume. Continued improvement in cost, reliability, and size, plus imaginative innovation, will undoubtedly result in a complementary coexistence of integrated circuitry and conventional components for some time to come. A recent survey of manufacturers, designers, and users of electronic equipment forecasts a 50 to 60 percent share of the market for discrete conventional components for many years in the future (Fig. 1).

Standards and Specifications. To an increasing extent, the design, fabrication, and testing of electronics parts, components, and systems are being controlled by standards and specifications. These standards, defining what are considered preferred practices for a given application, are prepared by professional societies, industrial organizations or groups, and various agencies of the government. Standards are available for most areas of engineering, and their use is encouraged in the interest of economy, simplicity, and quality; their use is often made mandatory, particularly in the execution

TABLE 2. Sources of Standards and Specifications [1]

Sources of industrial standards	
ASME, American Society of Mechanical Engineers 345 East 47th Street New York, N.Y. 10017	IPC, Institute of Printed Circuits 3525 West Peterson Road Chicago, Ill. 60645
ASTM, American Society for Testing and Materials 1916 Race Street Philadelphia, Pa. 19103	NEMA, National Electrical Manufacturers Association 155 East 44th Street New York, N.Y. 10017
AWS, American Welding Society 345 East 47th Street New York, N.Y. 10017	RTCA, Radio Technical Commission for Aeronautics 2000 K Street N.W. Washington, D.C. 20006
EIA, Electronic Industries Association 2001 Eye Street, N.W. Washington, D.C. 20006	UL, Underwriters' Laboratory, Inc. 207 East Ohio Street Chicago, Ill. 60611
IEEE, Institute of Electrical and Electronics Engineers, Inc. 345 East 47th Street New York, N.Y. 10017	USASI, United States of America Standards Institute 10 East 40th Street New York, N.Y. 10016
Sources of Federal standards	
ASEA, Armed Services Electro-Standards Agency Fort Monmouth Red Bank, N.J. 07703	NBS, U.S. National Bureau of Standards Washington, D.C. 20234
FCC, U.S. Federal Communications Commission New Post Office Building Washington, D.C. 20554	OTS, Office of Technical Services U.S. Department of Commerce Washington, D.C. 20230
GPO, U.S. Government Printing Office Washington, D.C. 20401	SBA, Small Business Administration Washington, D.C. 20416
GSA, General Services Administration Washington, D.C. 20406	USAF, U.S. Air Force Air Materiel Command Wright-Patterson Air Force Base Dayton, Ohio 45433
NASA, National Aeronautics and Space Administration Office of Technical Information 1520 H Street, N.W. Washington, D.C. 20546	

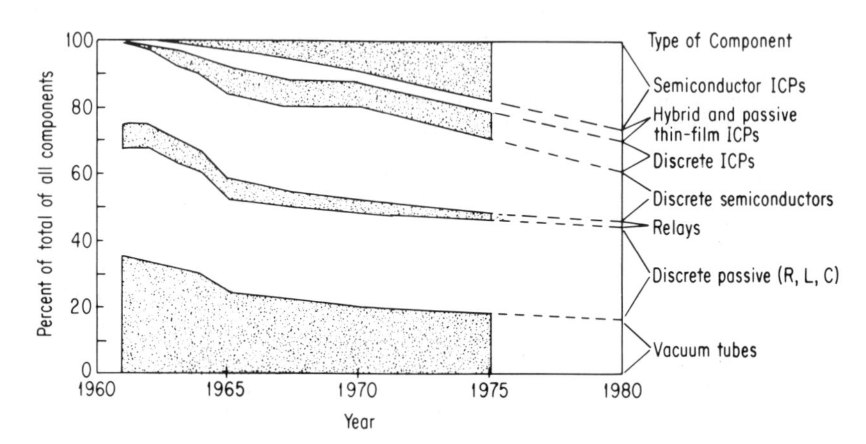

FIG 1. Share of electronic component market by type. (*IMA Incorporated.*)

of government contracts. Standards are revised from time to time to reflect a changing technology, and designs should incorporate the latest revision. Sources listing the more common standards and specifications are given in Table 2.

TERMS AND DEFINITIONS

The definitions of the major terms used in the discussions in this chapter are given in Table 3.

TABLE 3. Terms and Definitions

Active device. A device displaying transistance, that is, gain or control. Examples are transistors, diodes, and vacuum tubes.

Back panel. A relatively large piece of insulating material upon one side of which components, modules, or other subassemblies are mounted and on the other side of which interconnections between them are made. Also referred to as *motherboard.*

Chemical milling. The process in which metal is formed to intricate shapes by masking certain portions and then etching away the unwanted material.

Circuit. The interconnection of a number of components in one or more closed paths to perform a desired electrical or electronic function.

Component. An individual functional element in a physically independent body which cannot be further reduced or divided without destroying its stated function.

Component density. The number of components per unit volume (cubic inch or cubic foot).

Conventional component. The usual in form. This type of component is readily available from a number of suppliers and data are available on its physical size and electrical characteristics. The upper limit of sophistication is the transistor.

Cordwood technique. The arrangement of components within a module so that the component bodies are parallel and in close proximity to one another. Often referred to as three-dimensional packaging.

Derating. The reduction of the application stresses of components used in a design by using parts of higher manufacturer's rating than the design calculations call for; i.e., a ¼-watt resistor is "derated" to be used where calculations indicate that an ⅛-watt resistor is satisfactory.

TABLE 3. Terms and Definitions (*Continued*)

Device. An individual functional element in a physically independent body which cannot be further reduced or divided without destroying its stated function.

Diode. A semiconductor device which exhibits a high resistance to current flow in one direction.

Discrete component. Individual nonintegrated components, either active or passive.

Embedding. The complete encasement of a part or assembly in a plastic material to some uniform external shape. Often used synonymously with *encapsulation,* which is more properly considered a coating applied by dipping, brushing, or spraying.

Enclosure. A mechanical item which wholly or partly surrounds some electrical or electronic item or group of items and is an integral part thereof.

Header. A preformed part made of an insulating material and containing the input and output leads of a module. Usually forms a part of the outer case of the module.

Heat sink. A device used to absorb or transfer heat away from a heat-sensitive part.

Hybrid. Any assembly containing elements of different designs or employing different techniques in one package.

Interconnection. The physical wiring or connection between units in a system other than within modules. Often used synonymously with *intraconnection.*

Interface. The boundary surface between two different media.

Intraconnection. The physical wiring or connection between components in a module.

Module. A combination of components or devices which are contained in one package or common to one mounting and which provide a complete function.

MS. Military standard.

Packaging densiity. The number of components per unit volume in a working system or subsystem.

Passive device. A device which displays no transistance, that is, gain or control. Examples are resistors and capacitors.

Planar module. A packaging module in which the individual components are positioned and terminated flat or parallel to the plane of the substrate.

Planar packaging. Packaging method in which the components are arranged and mounted in essentially one plane. Often referred to as two-dimensional packaging.

Point-to-point wiring. The connecting together of the various function points and components with discrete conductors by the most direct route. May be wire or ribbon, soldered or welded.

Positioning film. Relatively stiff plastic sheets, usually Mylar, used to position components in a cordwood module. They are printed full size with component location, lead holes, and wiring runs. The films remain with the completed module.

Potting. A form of embedment in which a preformed case or shell is used and which remains with the assembly. Often used to indicate all embedment methods.

Printed circuit. A pattern comprising component parts and the conductors which connect them, all formed in a predetermined pattern on a common insulating base.

Printed-wiring board. A pattern of interconnecting wiring formed on a common insulating base. The pattern is usually formed by etching away the unwanted material. The components are mounted onto the board and connected to the wiring pattern, usually by soldering.

Resistance. The opposition that a device or material offers to the flow of direct current; the unit of measurement is ohms.

Resistance welding. The joining together of materials from the heat produced by their resistance to the flow of electric current. The materials are part of the current flow path and no filler material is used. The forcing together of the materials is required.

Rf. Abbreviation for *radio frequency.* A frequency at which coherent electromagnetic radiation of energy is useful for communication purposes.

Ribbon. A common intraconnection material, usually square or rectangular in cross section, used to form the wiring in a welded cordwood module.

Substrate. The physical material upon which a circuit is fabricated. It is used primarily for mechanical support, but may also provide thermal or electrical functions.

Transistor. A semiconductor device which by its atomic structure and fabrication permits the control of current flow.

Volumetric efficiency. The ratio of parts volume to total volume of a functional

TABLE 3. Terms and Definitions (*Continued*)

unit, expressed in a percentage. In a module, the ratio is based on the volume of the component bodies and the nominal outside dimensions of the module; sometimes called *packing factor.*

Wiring matrix. A method of intraconnection in which wires or ribbons are run at right angles to each other on either side of an insulating film. Welds are made from side to side through holes in the film to give the desired circuit pattern.

TYPES OF PACKAGING

Early in equipment development, determination must be made of the mechanical and packaging concept which will be used to ensure that the finished product will meet the requirements of function, environment, cost, and schedule. In order to make that determination, the designer must have a working knowledge of the packaging methods which may be suitable for the job. The component parts which make up the equipment can be divided into categories that require different packaging methods based upon their size, weight, or environmental attributes; a typical breakdown in this manner is shown in Table 4. Subassemblies may also require unique packaging

TABLE 4. Component Parts of a System Categorized as to Physical Characteristics [2]

Panel-mounted parts. Readouts in the form of meters and indicators, manual controls and adjustments, protective devices, switches, connectors, handles

Massive parts. Parts that are so large they must be mounted directly on the chassis at or near a stiff part of the structure; transformers, reactors, magnetic amplifiers, large capacitors, relays, servomotors and gearing, vacuum tubes

High-heat-emitting parts. Parts that emit so much heat they must be separated from other parts and mounted on plates or structures which conduct, radiate, and convect heat away

Rf components. Rf components are relatively large and require inflexible leads or waveguide for interconnection

Shielded subassemblies. Subassemblies which must be contained in a box or boxes to provide magnetic and/or electrostatic shielding

Circuit parts. Parts that are smaller in size and weight than any of the parts falling into the foregoing categories: carbon and film resistors, small capacitors, solid-state devices (transistors and diodes), small inductors. Generally, parts in this category weigh less than ½ oz and have axial leads suitable for mounting on circuit boards or in a small module

methods for electrical or functional reasons such as short propagation times, high power needs, or specific circuit pattern arrangements. Whatever the reason, a technique must be selected for each subassembly and an overall concept must be developed which is compatible with the various subassembly techniques.

It is not possible to describe all the packaging schemes and variations which are used in present-day equipments, for these are as numerous as ingenuity can provide. However, fairly detailed descriptions are provided for the more basic methods in use today. Thoughtful design can extend these techniques to the solving of almost any packaging problem. The assumption is made that all packaging schemes can be classed into three fundamental categories: planar packaging or two-dimensional; modular packaging or three-dimensional; and chassis-mount construction which might or might not contain some of the other types of assemblies.

Planar Packaging Methods. *General Considerations. Planar packaging* refers to that method in which the components are arranged together into functional testable groups and mounted in one plane in such a manner that they can be suitably intraconnected to form integral units. It is also referred to as *two-dimensional packaging* since the unit exists essentially in a single plane. The components do have a finite height off the mounting surfaces, but this dimension is small compared to the length and width outline; the completed unit is normally one component thick. A single planar assembly may make up the system, but usually the system is composed of a number of separate but interconnected planar units. The individual planar units may be conformally coated after electrical testing, to provide environmental protection.

PACKAGING TYPES. As illustrated in Fig. 2, there are basically two types of construction used in planar packaging: (1) Point-to-point wiring, in which the components are mounted parallel to the board upon posts or stand-off terminals and the circuit connections between them made with discrete wires which go from terminal to terminal by the most direct route. Due to the terminals, this type of unit tends to be large and is normally used for higher-power components, or where parts may be subject to frequent replacement. Most often these units are permanently fixed to the chassis. (2) Printed-wiring boards, in which the circuitry consists of a

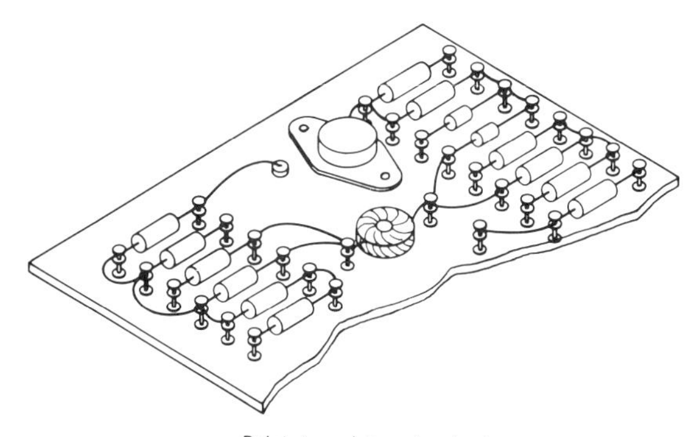

Point-to-point on terminals

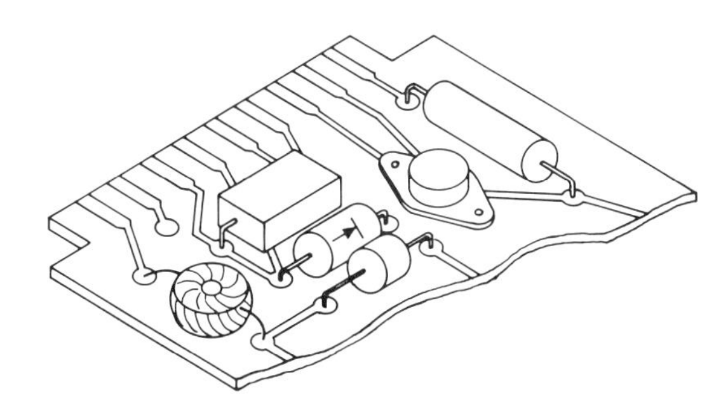

Printed-wiring board

FIG. 2. Two types of planar packaging.

pattern of electrical conductors formed on the common insulating base. The circuitry may be formed (as the name implies) by printing or by deposition, but more commonly is prepared by etching away the unwanted material from a metal-clad base so the remaining material forms the conductors of the circuit. Components are then mounted against the board surface and connected into the circuitry so as to give electrical continuity. Some advantages and disadvantages of printed-wiring boards are given in Table 5. This type of unit is by far the most commonly used planar packaging method, and the following discussions will be centered around that type of construction.

BOARD MOUNTING. If the boards contain relatively heavy components, require con-

TABLE 5. Advantages and Disadvantages of Printed-wiring Boards

Advantages

1. No wire jumpers are necessary between components and connectors
2. The printed-wiring board frequently uses a frame for support and could reduce the weight of the chassis
3. Dense packaging is possible since standoffs are not required, wire bundles are eliminated, and the leads of most plug-in-type components go directly to the board
4. Once it is properly etched the printed board is unlikely to be damaged unless it receives a direct blow. There are no free wires to catch on objects, tear away from terminals, or become frayed
5. The flat ribbons of metal, comprising the conductors, dissipate heat to the atmosphere better than round wire, thus tending to maintain lower temperatures
6. Since the printed-wiring board requires no forming, seldom needs machining operations other than drilling and trimming, and can be soldered in one dipping operation, it is faster and more economical than the hand-wired chassis
7. Wiring errors are eliminated once a circuit layout has been proved
8. Conductors are more readily traced and circuits are more readily checked than with wired circuits

Disadvantages

1. The surfaces of the conductors are exposed to the effects of humidity and to chips and dirt. It is generally necessary to clean and protect with a coating
2. The board is somewhat flexible and may increase the effects of vibration toward the center of the board
3. Components must be applicable to the board or made adaptable to board use. Some transformers and coils need repackaging before they can be used
4. High-frequency leads are difficult to shield and strong rf frequencies require special design consideration
5. Printed wiring is not as adaptable to design changes as is hand-wired construction
6. The planar characteristics accompanying boards and connectors may inherently result in inefficient use of multipin connectors in analog circuitry and pin-limited restrictions in digital circuitry

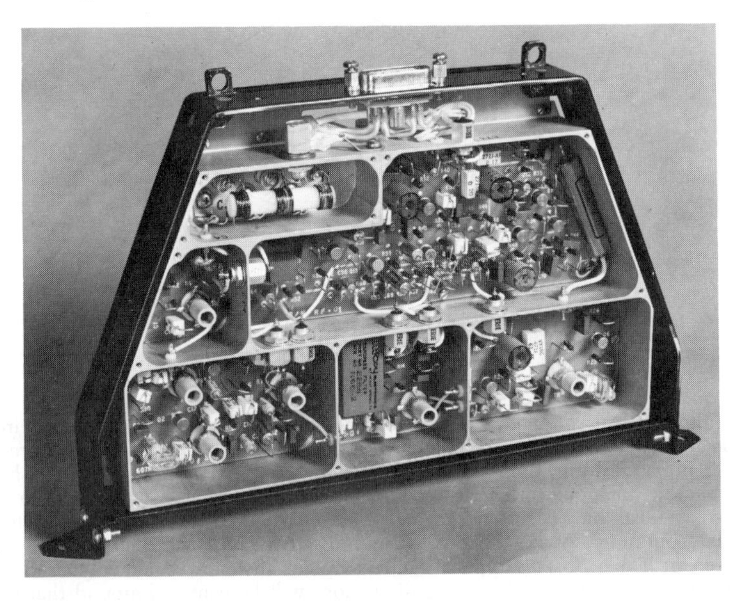

FIG. 3. Rigidly mounted printed-wiring boards. (*Westinghouse Electric Corp., Aerospace Division.*)

tinuous edge grounding, or will see extremely severe shock and vibration environments, they should be rigidly attached to the chassis. This type of mounting is illustrated in Fig. 3; here, small boards of irregular shape are mounted within individual shielding cavities. Subassemblies of smaller components which have no unusual functional or environmental parameters should be mounted so they are easily replaceable. This is most easily accomplished by providing (1) channels or guides into which the board edges slide and (2) a separate connector between board and chassis as shown in Fig. 4. The boards must be supported within a maximum of 1 in. of the edge on at least two opposite sides. The support should be sufficient to prevent fracture or loosening of the foil or breakage of the components or component leads resulting from flexing of the board. It is recommended that in addition suitable provision

FIG. 4. Plug-in boards and card guides. (*Card guides and frame by Calabro Plastics.*)

be incorporated in the design to provide for positive retention or clamping of the board in position after connector mating.

BOARD DEFLECTION. The primary cause of board failure under vibration loading is board deflection, or displacement, and not the actual g loads themselves. The repeated flexure destroys the solder joints and the adhesion of the circuit lines to the boards, and fatigues component leads until they break. Most components and board assemblies can accept the high g if excessive flexure is not allowed. The deflection under shock or vibration loading depends upon: (1) the weight of the board and components, (2) the method of supporting the board, and (3) the flexural rigidity of the board material and configuration. The basic formulas for determining board deflection and natural frequency are:

$$f_n = \frac{K}{2\pi}\sqrt{\left(\frac{Eg}{WL^3} \cdot \frac{Bt^3}{12}\right)}$$

$$f_n = 3.13\sqrt{\frac{1}{d_s}}$$

where f_n = natural frequency of vibration, cps
K = constant
E = modulus of elasticity of board, psi
W = weight of board and components, lb
L = board length, in.
B = board width, in.
t = board thickness, in.
d_s = static deflection of board, in.
g = gravitation constant = 386 in./sec²

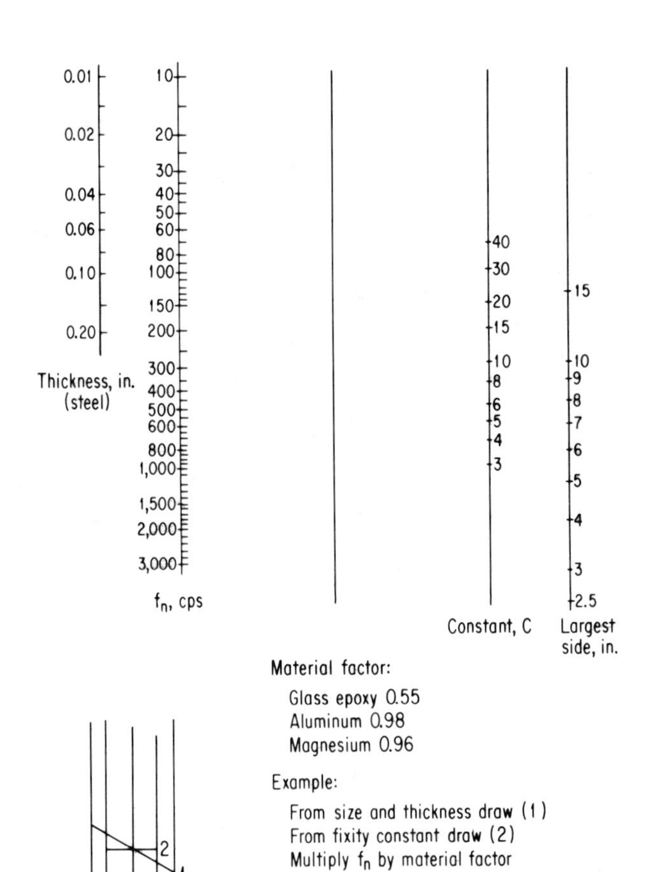

FIG. 5. Nomograph for determination of board natural frequency.[3]

From this it can be seen that to obtain the desired minimum deflection and maximum natural frequency, the length of the board should be a minimum and the thickness a maximum within the limits of the design. The board natural frequency should be made as high as possible and at least 50 percent greater than the resonant frequency of the complete package.

A nomograph which can be used to estimate the natural frequency of a printed-wiring board is given in Fig. 5. The fixity constant (not the same as K above) to be used with this nomograph can be obtained from Fig. 6; selection of the proper constants depends upon proper interpretation of the actual mounting conditions. Most printed-wiring boards which mount in side guides and mate with a connector on one end fall into the supported-edge category. Edge guides cannot develop bending moments in the boards; thus these appear to be free-hinged supports. The same is true of practically all connectors. The end opposite the connector can be supported or left free without drastically affecting the fixity constant. Boards which are screwed to a mounting frame at the corners, center, and edge centers are not in the clamped-edge category but are still in the edge-supported category. The screw in the center modifies the effective size rather than the fixity constant. The nomograph applies to bare boards only. Components tend to stiffen the board somewhat, often as much as 50 percent.

Design Factors. Factors to be evaluated when using printed-wiring boards are summarized in Table 6. These are all interrelated and the final decision will be

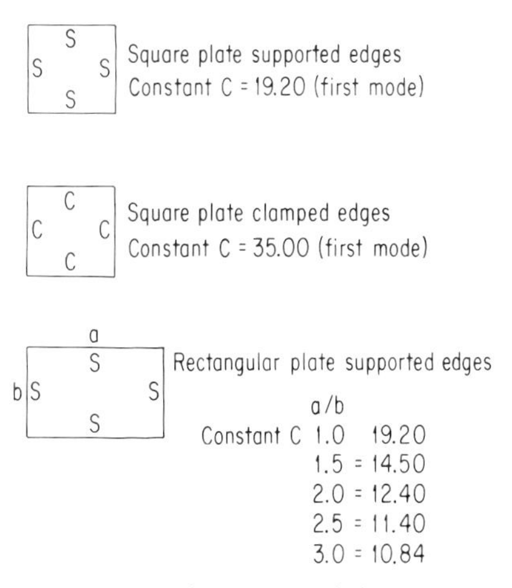

FIG. 6. Constants to apply to natural frequency nomograph.[3]

based upon a trade-off consideration. A further discussion of some of these factors is given below.

BOARD TYPE. In planning the detail design, it must be determined whether a single-sided or double-sided board is to be used; the basic considerations here are cost and size. A single-sided board with unsupported holes is the easiest to fabricate, and the least costly. However, complex circuitry in a single-sided layout occupies considerable area and the boards can become unmanageable; for higher component density with complex circuitry a double-sided board is desirable. It is generally considered acceptable to use a limited number of jumper wires to facilitate use of single-sided boards; as a rule of thumb, if the circuitry cannot fit into the designated size with a single-sided board plus five jumpers, then a double-sided design should be employed.

BOARD MATERIAL. The most commonly used base materials (substrates) are listed in Table 7. The grade XXXP paper-based phenolic is most popular for commercial application where conditions are not severe; grade G-10 or G-11 fiber-glass epoxy is

TABLE 6. Factors to Be Evaluated for Printed-wiring Board Design *

Factor	Considerations
Circuit needs and characteristics	Signal level, frequency, power, and impedance. Electrical interference, circuit couplings, shielding, grounding, and common lines. Circuit types and layouts: dc, ac, input, output, auxiliary connections
Circuit pattern ...	On one or both sides of a board. Complexity and packing density. Mechanical and thermal needs. Part accessibility, reliability, and comparative costs
Type of layout ...	Pad layout or strip layout. Component orientation and placement: coordinate, symmetrical, or nonsymmetrical
Conductor paths ..	Short and direct, evenly distributed. Minimum jumper leads and crossovers
Conductor sizes and spacings	Foil thickness and width according to current rating. Conductor spacings to suit peak voltages and environmental extremes
Conductor shapes ..	Smooth contours and free-flowing lines. No sharp corners, acute bends, or uneven distribution of copper
Board shape	To fit equipment, or conform with housing. Clearance for adjacent parts. Cutouts and notches for connectors, polarization, and special parts. Hole sizes, locations, registration, mating parts, and tolerances
Board size	External limitations and tolerances. Area needed to accommodate components, conductors, connections, and attachments. Overall projections and clearances
Board mounting ..	Fixed on chassis; permanent, semipermanent, or removable. Hardware: brackets, spacers, screws, and other fasteners. Plug-in types; guides, grooves, card rails, clips, locating pins, drawer pulls, and slides. Pressure fit between mating parts, clamps, or molded recesses
Components	Types, sizes, variations in styles and dimensions. Lead arrangements, mountings, heat sinks, and shields. Accessibility for adjustment, replacement, or repair. Special considerations (body sizes, shapes, lead diameters, etc.) for machine assembly
Component layout	Symmetrical layout, uniform spacings, and modular dimensions, especially for machine assembly. Nonsymmetrical part placement, to suit layout, with nonmodular dimensions
Terminations	Component terminals, eyelets, wiring holes, plated-through holes, test points, and edge connectors. Wired connections, lead forming, lead tipping, tab connectors, and leadouts. Hand soldering or machine soldering; flow dip, or wave. Welded, or wrapped, connections
Customer	Commercial or military, specification or control drawings. Known customer preference
Manufacture	Schedule and due date. Firmness of electrical design. Breadboard, small-quantity, or high-production rate. In-house capability. Funds available
Environments	Reliability level. Ground-based, shipborne or airborne, spacecraft. Maintainability. Acceptable throw-away cost. Type of cooling

* Adapted from Shiers.[1]

the most used in military applications. These materials may be single- or double-clad and are available in a wide range of thicknesses; 0.032 and 0.062 in., including the foil thickness, are the most widely used. Most boards are clad with copper although special claddings such as nickel or Kovar* can be obtained. Two weights (ounces per square feet) of copper cladding are in general use, 1 oz (0.00135 in.) and 2 oz (0.0027 in.); and other weights are available. Single-clad material, used for single-sided boards, has copper on one side only; double-clad material is used where a double-sided board is required. When plated-through holes are employed to connect

* Trademark, Westinghouse Electric Corp.

TABLE 7. Common Printed-wiring Board Substrates

Base material	Impregnating material	Comments
Paper	Phenolic	XXXP is most used grade in commercial application. Maximum temperature is 250°F. Poor arc resistance, tendency to moisture absorption, poor high-frequency loss factor. Low cost, relatively easy to machine, fair strength
	Epoxy	Slightly more expensive than paper. Relatively good electrical properties, low moisture absorption, strength not as good as that of glass cloth
Glass fabric	Teflon*	Extremely expensive. Electrically superior to other substrate materials. Good for microwave assemblies and high-temperature applications (500°F or higher)
	Epoxy	G-10 and G-11 (flame retardant) most used grades in military applications. High-strength operation to 340°F, low moisture absorption
	Melamine	Least expensive glass-cloth substrate. Good surface hardness. Difficult to machine. Good arc resistance
	Silicone	Lower loss factor than epoxy, more expensive, lower mechanical strength. Good for high temperatures (500°F or higher) and high frequency
Linen or cotton fabric	Phenolic	Electrical properties not too good. Stronger than paper-based phenolic. Easy to fabricate. Not good in high-humidity applications
	Melamine	Difficult to machine. Good arc resistance, good heat resistance, tendency to absorb moisture, not so strong as glass cloth

* Trademark, E. I. du Pont de Nemours & Co., Inc.

from one side to the other, a cladding thickness equivalent to approximately 1 oz/ft^2 (0.00135 in.) is added to all circuitry by the plating process.

BOARD SHAPE. The shape of a printed-wiring board is usually dictated to some extent by the equipment in which it will be used; a rectangular shape with a minimum of cutouts or configuration discontinuities is the most economical to manufacture. The standardization of board sizes is desirable but not always practical; efforts should be made to use one board shape and size within any one system or subassembly.

CONDUCTOR WIDTH AND SPACING. The basic conductor width and spacings are based upon the currents and potential differences in the circuitry; the usual precaution must be taken to isolate as far as possible the sensitive lines from the power and ac circuits. In all cases, finished conductor widths should meet the requirements of Fig. 7. Normally the 30°C maximum-temperature-rise curve is followed; Table 8 gives a summary of these line widths. In processing, line widths will be reduced about 0.002 in./oz due to undercutting, which must be compensated for in layout (i.e., an 0.62-in.-wide line in a 2:1 layout will yield an 0.27-in. final line width in 2-oz. copper). Conductor widths should be made as liberal as possible for manufacturing ease and tolerance leniency. In critical circuits, make power and ground leads large to ensure equal potential difference between points of usage. A 0.062-in.-wide conductor in 2-oz foil is approximately equal in cross section to a wire size of 27 AWG.

The spacing between conductors is a variable depending upon voltage potential between any two given points at a specified operating altitude, and depending upon whether or not environmental protection is provided. Recommended minimum spacings are given in Table 9; if the spacings in part C are used, complete conformal coating of the board in a material such as polyurethane, epoxy, or silicone is required. Also, since the spacings are minimum, a tolerance variation must be provided in artwork layout. This usually amounts to about 0.005 in.; i.e., for a 0.025-in. minimum

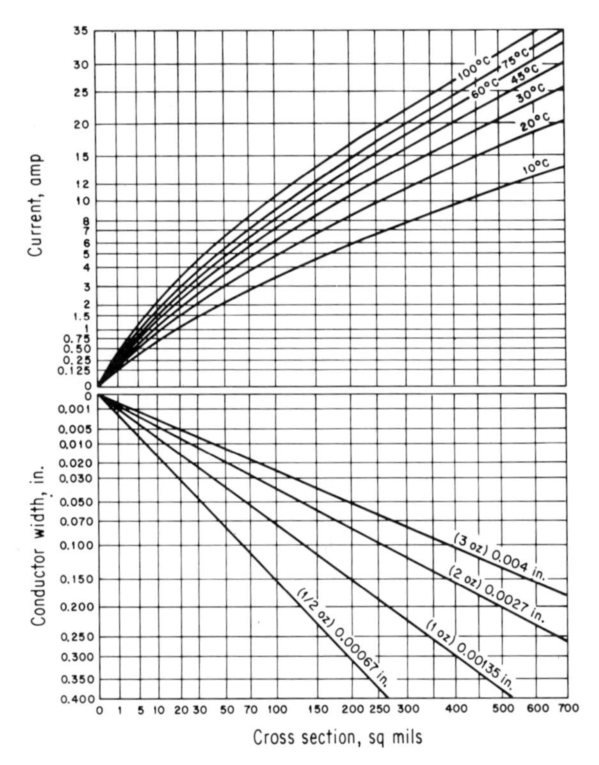

Fig. 7. Conductor thickness and width (for use in determining current-carrying capacity and sizes of etched copper conductors for various temperature rises above ambient).[4]

TABLE 8. Standard Conductor Widths [4] *

Finished conductor width, in.	*Current limitation, amp*
0.010	2.3
0.015	2.9
0.020	3.5
0.025	4.1
0.030	4.7
0.040	6.0
0.050	7.5
0.060	8.3
0.125	14.0

* Based upon use of 2-oz (0.0028-in.) copper thickness and 30°C temperature rise.

TABLE 9. Minimum Conductor Spacing for Printed-wiring Boards [4]

Part A Conductor spacing (sea level to 10,000 feet inclusive) (uncoated boards)

Voltage between conductors dc or ac peak	*Minimum spacing, in.*
0–150	0.025
151–300	0.050
301–500	0.100
Greater than 500	0.0002 (in./volt)

Part B Conductor spacing (over 10,000 ft) (uncoated boards)

Voltage between conductors dc or ac peak	*Minimum spacing, in.*
0–50	0.025
51–100	0.060
101–170	0.125
171–250	0.250
251–500	0.500
Greater than 500	0.001 (in./volt)

Part C Conductor spacing (any altitude) (conformal coated boards)

Voltage between conductors dc or ac peak	*Minimum spacing, in.*
0–30	0.010
31–50	0.105
51–150	0.020
151–300	0.030
301–500	0.060
Greater than 500	0.00012 (in./volt)

finished spacing, the separation should be 0.060 in. in a 2:1 layout. In certain applications, spacings may have to be increased due to capacitive coupling effects in parallel high-frequency runs, or in order to reduce risk of flashover in critical environments; these factors are discussed in Chap. 10.

BOARD SIZE. The board must accommodate not only the circuitry based on the data given above, but also the components required and the various hardware items used with the board. A listing of some items which should be considered when sizing boards is given in Table 10. A first approximation of required board area (and usually close enough for planning purposes) can be obtained by the following equation:

$$\text{Board area} = 1.75\,(HA + CA)$$

where HA equals the sum of the surface area required for the connector, card guides, test points, and other hardware items; and CA equals the sum of the areas required for the various components. Add 0.20 in. to the length of each component to allow for lead bending and insertion before determining area. For round components, assume they are square with the sides equal to the diameter.

TYPICAL PACKAGING DESIGNS. Figure 8 illustrates a reasonably complex type of printed-wiring board. It contains components mounted both parallel and perpendicular to the board, as well as larger items bracket-mounted with discrete wire connections. Boards of this type may contain components on both sides but usually have them on only one side, which is referred to as the *component side*. The other side is known as the *wiring side*, although both sides may contain printed wiring.

Figure 9 illustrates an assembly with components mounted on end normal to the surface of the board. This is acceptable if the components weigh less than 1/2 oz and if the height from the board surface to the uppermost point of the top lead is not greater than 1/2 in. The upper lead must have a straight run of about 0.05 in. before being bent over, and the minimum bend radius should equal two lead diameters. Since the parts are cantilever-mounted, protection with a conformal coating or potting may be needed if high-stress environments are encountered. Part density per

TABLE 10. Hardware Considerations in Sizing Printed-wiring Boards

Item	*Consideration*
Test points	May be printed pads on board, small connector, receptacle, or formed test jacks; pads occupy minimum room. Should be accessible with board in position; usually placed along top edge. Helpful to color-code or number
Card guides ...	Plastic or metal guides mounted inside both sides of card housing; edges of cards slide into these. Metal best for heat conduction; plastic lighter. Conformal coating should not be allowed in this area. Average guide requires 0.125 in. width each side for whole length of card
Connectors	Occupied area is size of block when pin-and-socket type used. Area with edge type is connector overlap, printed conductor lead-in, and board edge cutouts. For edge connectors, the leading edge of the board should be tapered. The leading edge of printed conductor contact may be tapered and should not extend to the edge to prevent circuit lifting. The printed contacts should be plated with a hard surface; typical are rhodium or gold over nickel, electrodeposited nickel-tin, or plain gold
Component mounting	Tubular components weighing less than ½ oz may be flush-mounted and adequately secured by their leads. Larger ones require supplemental support to remove vibration strain from leads. Clip-type holders mounted with rivets are usually used; they add approximately 0.05 in. to each side and about 0.1 to 0.2 in. to height. Adhesives are used to secure components if removal is not anticipated; this requires less area. If large or heavy components must be placed upon a board, they should be located near a supporting edge and provided with adequate securement other than their leads. Modules and other components having numerous relatively large leads designed for direct soldering into board holes may be mounted by these alone
Moisture entrapment	To prevent moisture entrapment, flat-base components should not be mounted flush on the board, regardless of whether single- or double-sided boards are used. They should be above board about 0.015 in.; small spacers can be used to provide this. On uncoated boards, the body of flush-mounted axial lead components should not bridge more than one conductor; this does not apply on conformally coated boards
Lead termination	Two leads should not be placed in one hole; provide multiple pads with holes. Lead termination may be accomplished by passing the lead through the termination hole and clinching over the lead end to make contact with the circuit pad, or it may be straight through. Complete termination is accomplished by soldering the lead to the pad. For clinched leads the cutoff should be of sufficient length to contact the pad but it should not extend beyond the edge of the terminal pad or its electrically connected conductor. For straight-through leads, the cutoff lead should be visible above the solder fillet; however, the length should not normally exceed 0.075 in. beyond the board
Heat sinking ..	Components in contact with board should have a surface temperature below 125°C in continuous operation; items over 2 watts should not be mounted onto board. High-power components should have finned heat sinks; these increase occupied component area. The board metal cladding can often be used for heat sinking; circuitry density is reduced. Silicon grease helps heat transfer
Card puller ...	Withdrawal forces from multicontact connectors can reach high levels (50 contacts, 4 oz/contact, is 12 lb). This force cannot be achieved with the fingers, and auxiliary leverage is required. Many varieties of jacking screws or camming devices are used; all require some board area. Simple removal device often used is formed double hook into grommets in each side at top of board; typical grommet occupies 0.06 in.[2]
Ground plane ..	Large metal-clad area on board to shield or provide special grounding conditions. Inhibits free component placement, occupies board area, and reduces circuitry density

unit area is increased here, but this may result in increased deflection and decreased natural frequency unless additional support is provided.

Figure 10 illustrates an assembly containing components mounted both parallel to the board and normal to it. If only one or two of the components are vertically mounted or have a large diameter, then the spacing between the boards in the system must be based upon these, and considerable packaging volume is lost if the other smaller-diameter components are not vertically mounted.

The same volume loss occurs on hybrid boards if only one or two cordwood modules are placed on each board. It would be advantageous from a packaging density standpoint to separate the larger components or modules onto a separate board if it were possible from a circuit consideration.

The nesting of components on two adjacent boards is shown in Fig. 11. In this unit, the single board with these components utilizing the required output pins would have been severely connector-pin-limited; the arrangement shown provides an additional connector in little more space than that occupied by one board. The

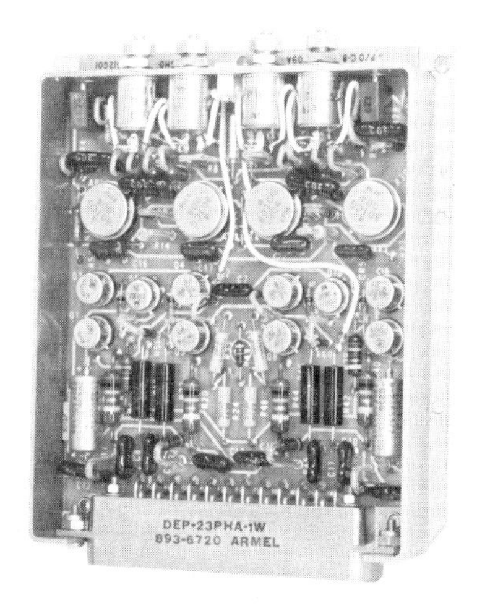

FIG. 8. Typical printed-wiring board. (*Westinghouse Electric Corp., Aerospace Division.*)

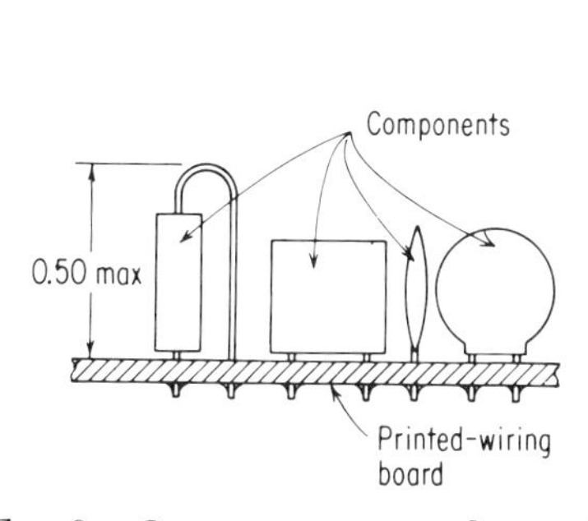

FIG. 9. Components mounted normal to board.

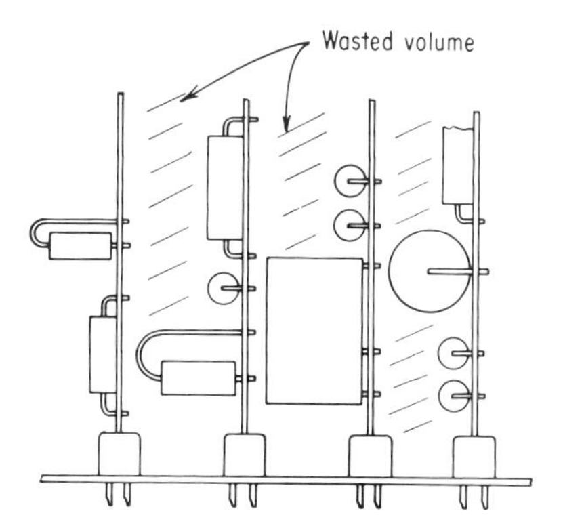

FIG. 10. Wasted volume in assembly of boards with mixed mounting.

FIG. 11. Nesting of boards containing conventional components. (*Westinghouse Electric Corp., Aerospace Division.*)

nesting of board-mounted modules is shown in Fig. 12. With all modules on one board, interconnection would be impossible without a multilayer board. This technique provides four wiring planes and ready access to all connections in almost the same height as a single board. If a pressure-sensitive adhesive is applied before the two boards are finally joined, a single rigid truss structure is formed which is resistive to severe stress inputs.

Modular Packaging Methods. *General Considerations. Modular packaging* refers to the method in which the components are arranged together in small func-

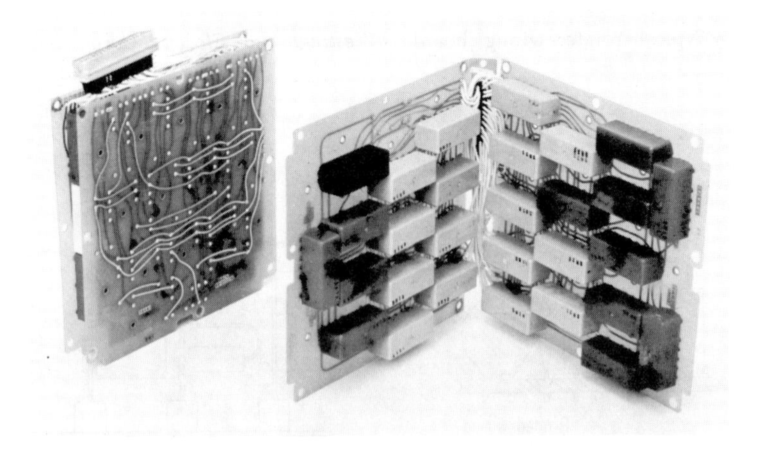

FIG. 12. Nesting of modules. (*Motorola, Military Electronics Division.*)

tional groups and electrically intraconnected to form discrete testable modules before being placed into the next level of assembly. It is also referred to as three-dimensional packaging since the components are stacked and wired so as to utilize a finite height as well as the length and width outline of the planar method. Most often modules are conformally coated or encapsulated for environmental protection after electrical check-out. The resulting miniature package is a high-density assembly with components occupying a very large percentage of the space within the package outline.

MODULE TYPES. There are two basic types of modules:

(1) Cordwood modules, as shown in Fig. 13, in which the piece parts used in a particular circuit are grouped together with their bodies touching, or in close proximity, and their leads parallel. Components are arranged so as to occupy the least possible space consistent with their electrical and thermal characteristics. Parts too long for the specified height of the assembly may be placed at right angles to the other parts with the leads modified to be parallel with the others.

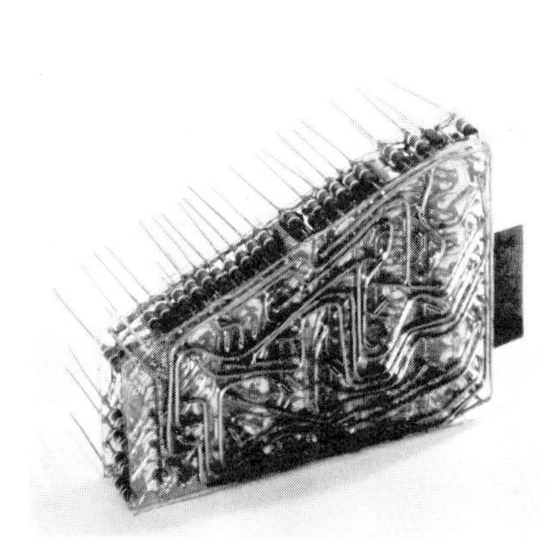

FIG. 13. Cordwood module. (*Westinghouse Electric Corp., Aerospace Division.*)

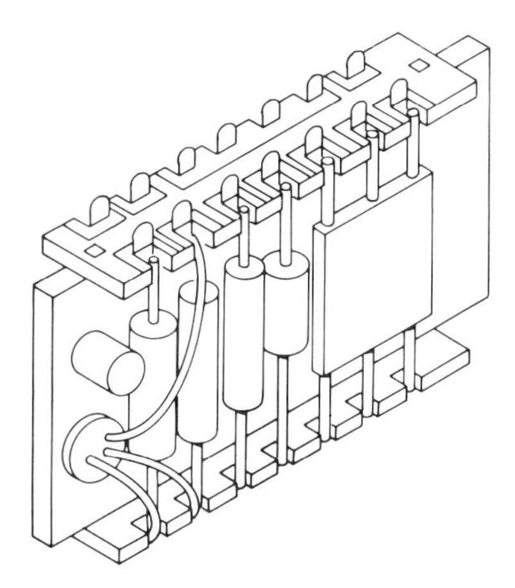

FIG. 14. Typical planar modular unit. (*ITT Telecommunication Systems.*)

(2) Planar modular, in which the components are arranged in small functional groupings in essentially one plane; the completed assembly is only one or two components thick. This is illustrated in Fig. 14. The highest packaging density obtainable with this module is achieved by mounting it in the next assembly so that the components are perpendicular to the mounting surface. Due to its relatively open construction, the planar module is more easily fabricated than the cordwood type.

OUTPUT PINS. In component modules the output leads can exit either parallel or perpendicular to the components as shown in Fig. 15; the choice is based upon the height available above the mounting surface. Usually the output leads are simple round pins which are attached to the intraconnecting circuitry at appropriate points. Extremely soft material should not be used for output leads since they may suffer handling damage. It is possible to use component leads as output leads if they are of the correct size and material. Preformed headers containing the output leads are often used to accommodate specific output pin size, shape, material, or special packaging considerations. A typical header is shown in Fig. 16. In any case, the output leads should exit on a regular grid pattern so as to simplify module fabrication and the next level of assembly.

MODULE SHAPE. The shape which occupies the least mounting-surface area for the occupied volume is the cube; for maximum packaging density this form factor should be approximated. However, since most discrete components available today result in modules between 3/4 and 1 in. thick, a cube of this dimension would contain only a limited number of components; a rectangular shape is a reasonable compromise. It is usually advantageous, when packaging systems with a large quantity of modules, to set standard shapes and design all units to this, with variations made only in certain

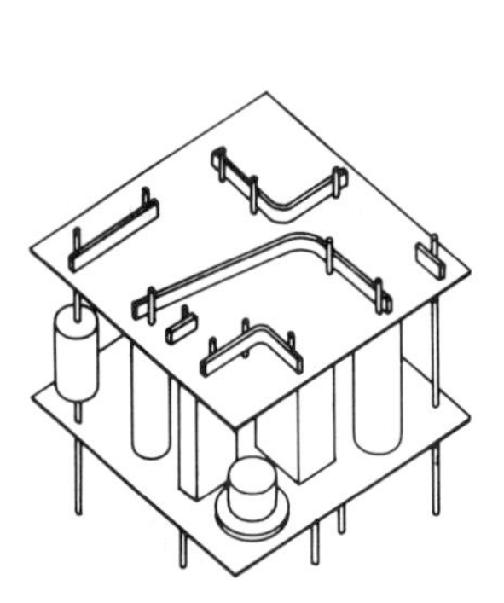

Output leads parallel to components. Requires minimum height, maximum mounting area.

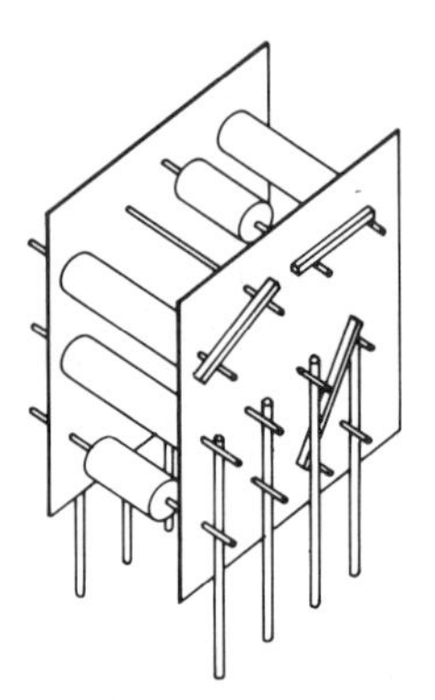

Output leads perpendicular to components. Requires maximum height, minimum mounting area.

FIG. 15. Parallel and perpendicular lead exits.

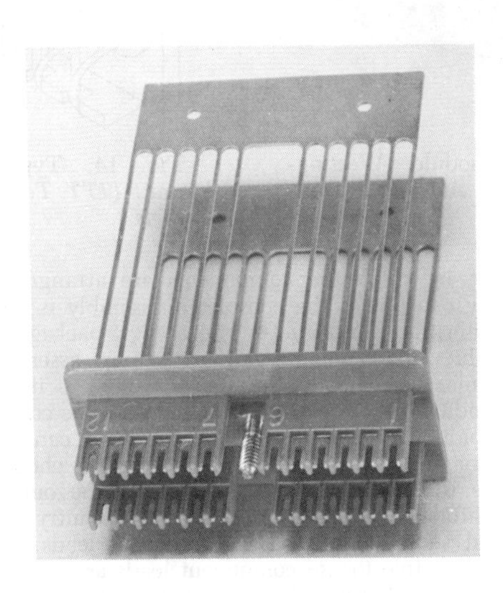

FIG. 16. Typical preformed module header. (*Jet Propulsion Laboratory.*)

preestablished increments. This standardizes tooling and reduces fabrication costs. If absolute maximum packaging density is required, then modules should be sized on a space-available basis and shaped as necessary to fit the space, as shown in Fig. 17. Cordwood modules are well suited to this technique and permit the utilization of space for electronic circuitry which might otherwise be vacant.

MODULE SIZE. The size required for a module can be determined by using the *squared-component-area method.* This is based upon assuming that all component bodies are touching each other and have a square or rectangular shape as shown in Fig. 18; the side of the square is equal to the diameter of the component. Since the height of the assembly is set by the length of the longest component, if one or two are considerably longer than the rest, they should be laid upon their sides. The

FIG. 17. Modules shaped to fit space available. (*WEMS, Inc.*)

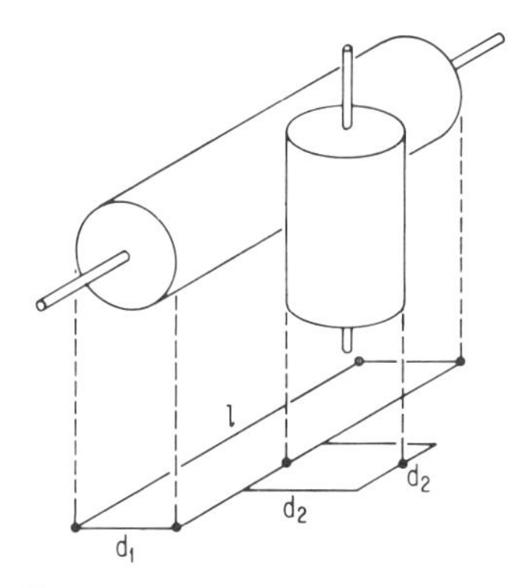

FIG. 18. Square-component area.

packaging area of these components then becomes the length of the part times its diameter. To determine the module size, use the following procedure:

Component area (*CA*) is found by totaling the areas occupied by the individual components when stacked side by side. The components' dimensions should be rounded off to the nearest high whole number for the grid system being used. For example, in a module containing 8 transistors, 0.225 in. in diameter; 12 resistors, 0.090 in. in diameter; 4 diodes, 0.100 in. in diameter; 1 inductor, 0.130 in. in diameter by

0.97 in. long; and 3 rectangular capacitors with radial leads such that the mounting face is 0.100 in. by 0.250 in., and using a 0.050-in. grid system, the component area is:

$$CA = 8(0.25 \times 0.25) + 12(0.10 \times 0.10) + 4(0.10 \times 0.10) + (0.15 \times 1.0) + 3(0.10 \times 0.25)$$

$$CA = 0.885 \text{ in.}^2$$

Layout area (LA) is the minimum area allocated for component placement. In designing a module some additional area in addition to the component area is needed for terminal and jumper locations, for making intraconnections, and to prevent shorting. Also, the use of a grid system is not 100 percent efficient in space utilization. An empirical value of 0.2 of CA can be used for these factors:

$$LA = 1.2CA$$

Module size (MS) is determined after the module shape has been established. Normally modules require a conformed coating or embedment for protection and structural support. The area needed for this, when added to the layout area, determines the final module size. It is found by adding an additional amount to each dimension of the module after the shape is determined from the layout area. The use of 0.040 to 0.060 in. of embedment material per side has been found adequate in most cases:

$$MS = (W + 0.10)\,(L + 0.10)$$

where W = established width from layout area
L = established length from layout area

The above determinations are based upon modules designed on standard grid systems and having round wire output leads. If other spacings, special output leads, or preformed leaders are used, then adjustments must be made.

Module height determinations are covered under specific module types described later in this section.

It should be pointed out that a cordwood module tends to become wasteful of space if too few components are placed in it. In this case, the percentage of the occupied volume that is consumed in intraconnection and structural protection is large compared to the volume consumed by actual active component elements. Although no rigid rule can be established since function often dictates module size, this packaging concept tends to become inefficient with fewer than 20 components.

INTRACONNECTION METHODS. The intraconnection between components within the module may be accomplished by welding with discrete wires and ribbons, or through the use of standard soldered printed-wiring board techniques. A comparison of the two methods is given in Table 11.

If welding is desired, then components with weldable leads must be used. These are obtainable from most manufacturers, but it is well to specify the need for weldable leads on the purchase orders. Also, the ribbon used for the intraconnection must of course be weldable. The most commonly used materials for this purpose are shown in Table 12. Copper is very difficult to weld and is seldom used. Nickel A, Dumet,* and Kovar* are the most commonly used materials. If the module is to be used in an application requiring a high degree of magnetic cleanliness, then one of the nonmagnetic materials should be used insofar as possible. In the welding process, the leads must be free from all nicks, scratches, or potting and marking materials in the weld area. These materials in the area of a solder joint usually result in poor joint, but for a weld joint they usually cause high localized heating resulting in a blown (ruptured) component lead.

Welded Cordwood Modules. Modules of the cordwood type have been made using various methods of welding to attach the interconnecting lead to the component leads. These include laser beam, electron beam, percussive arc, and resistance

* Trademark, Westinghouse Electric Corp.

TABLE 11. Comparison of Welding and Soldering of Modules

Advantages	*Disadvantages*
Welding	
1. No danger to heat-sensitive components by short weld pulse	1. Joints made one at a time—slower, more costly
2. Joints can be made closer to component body—volume reduction	2. All components must have weldable leads—procurement more difficult
3. Joint has inherent high reliability	3. Bad weld usually results in scrapped component
4. Cordwood modules more easily repaired if welded	4. Weldable material is usually magnetic
Soldering	
1. All joints made in one operation—less costly	1. Danger of damage to internal module connections if soldering at the next assembly level
2. No special equipment (i.e., welding machines) required. No new skills needed	2. Possibility of flux contamination
3. Noncritical requirements as to time, temperature, lead material, form factor	3. Not good for operation above solder melt temperature

TABLE 12. Commonly Used Materials for Electronic Welding*

Type	Name	Composition	Plated with	Clad with
Basic metal	Nickel A	99% nickel		
	Tough pitch copper†	Copper	Solder or tin	
	OF copper†	Copper		
Alloy	Kovar or Rodar‡	17% Co 29% Ni 53% Fe	Gold	
	Alloy 90†	12% Ni 88% Cu		
	Alloy 180†	21% Ni 79% Cu		
	Alloy 42	42% Ni 58% Fe		
	Alloy 45†	45% Ni 55% Cu		
	Dumet	42% Ni 58% Fe	Gold	Copper
Clad	Copper weld	Steel core		Copper
	Nickel-clad copper	Copper core		Nickel

* Adapted from NASA material.[5]
† Nonmagnetic.
‡ Trademark, Wilbur B. Driver Co.

spot welding, techniques which are described in detail in Chap. 4. By far the most used welding method is the resistance spot weld, in which the flow of electric current is used to generate the heat required for welding. The materials to be welded are placed between two electrodes which exert pressure on the materials. A high-intensity current is passed through the electrodes and the materials for a controlled length of time; the heat generated by the resistance of the materials to the current flow produces the weld. Modules made with this technique are usually classified by the method used to form the intraconnection pattern: point-to-point, wire matrix, or preformed wiring.

POINT-TO-POINT INTRACONNECTION. As the name implies, the point-to-point method utilizes a ribbon or wire which is routed between component leads by the most direct route. The components are usually stacked cordwood fashion between two insulating films, usually Mylar,* in such a manner that the leads extend through holes in the films. The component outlines and the intraconnecting circuit pattern are preprinted onto the insulating films. This makes the routing of the wiring and the later inspection verification a much easier task.

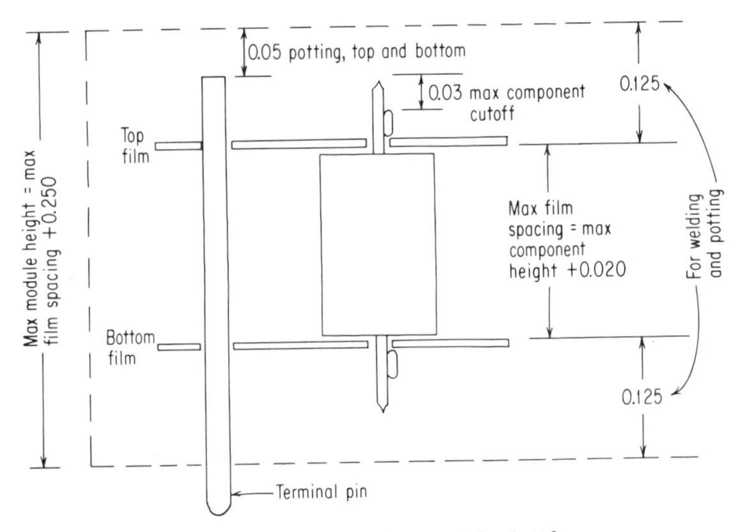

FIG. 19. Point-to-point module height.

Normally the ribbon used for the intraconnection is not insulated, so that crossovers in the routing cannot be made. When the wiring density is too great to be placed on one layer at each end, then additional layers of wiring can be made by placing additional layers of insulating film between circuit layers. More than two layers of circuitry at each end of the module are seldom used.

It is possible to place the components randomly in the assembly with all body parts touching, and this placement will certainly give the minimum module volume. However, since this type of procedure does not lend itself to standardization, the tooling and fabrication costs are greatly increased. Most often the module is built upon a standard 0.100- or 0.50-in. grid pattern with all component leads being placed, insofar as possible, on the grid intersections. In the interest of organizing the next level of assembly, output pins should also be placed on grid intersections.

Modules of this type are sometimes made with only one film to provide component support and positioning, and occasionally designs are utilized which eliminate the films entirely; this can be done by the use of some removable holding fixture or a potting-before-welding technique. However, modules made by the double-film method are in far more general usage than the other types.

The module height (*MH*) considerations for point-to-point modules are shown in

* Trademark, E. I. du Pont de Nemours & Co., Inc.

Fig. 19. The maximum component height is determined from a survey of the parts; care should be taken to add the plus tolerance on height. It is of course possible in layout to place certain components having radial leads so that they are two high between the films with their leads protruding from opposite films; if this is done, the stacked height may establish the maximum component height. The maximum film spacing then becomes the maximum component height plus a reasonable manufacturing tolerance. The module height is:

$$MH = \text{film spacing} + 0.250 \text{ in.}$$

The 0.125-in. dimension at each end includes the thickness of the film, the maximum height of the welding ribbon over the positioning film (assuming a 0.030-in.-high ribbon), the maximum allowance for component lead cutoff, and the minimum embedment allowance. If it is necessary to utilize added layers of welding ribbon for intraconnection, an allowance of 0.075 in. per additional layer for ribbon height and component cutoff should be made. As noted earlier under module shape, a standard module height for a quantity of modules is preferable. The welded assembly within the standard height may vary somewhat, resulting in reduced packaging efficiency, but the gain made in manufacturing more than offsets this loss.

In order to reduce the number of layers of ribbon and insulating films necessary when many circuit crossovers are encountered, small-diameter insulated wire may be used for the intraconnection; this is the form of wire referred to as *magnet wire.* Since the wire itself is insulated, crossovers can be made where desired without fear of shorting. The wire must be weldable—the most commonly used materials are nickel and OF copper. The insulating coating is selected so that it melts under relatively low heat without a great amount of char; polyurethane (also solderable) and polyvinyl-formal are typical materials. In the welding process, one of the electrodes is heated, or a separate heat source is applied in the electrode area. This of course melts the insulation away in the immediate area of the joint and allows a normal weld to be made.

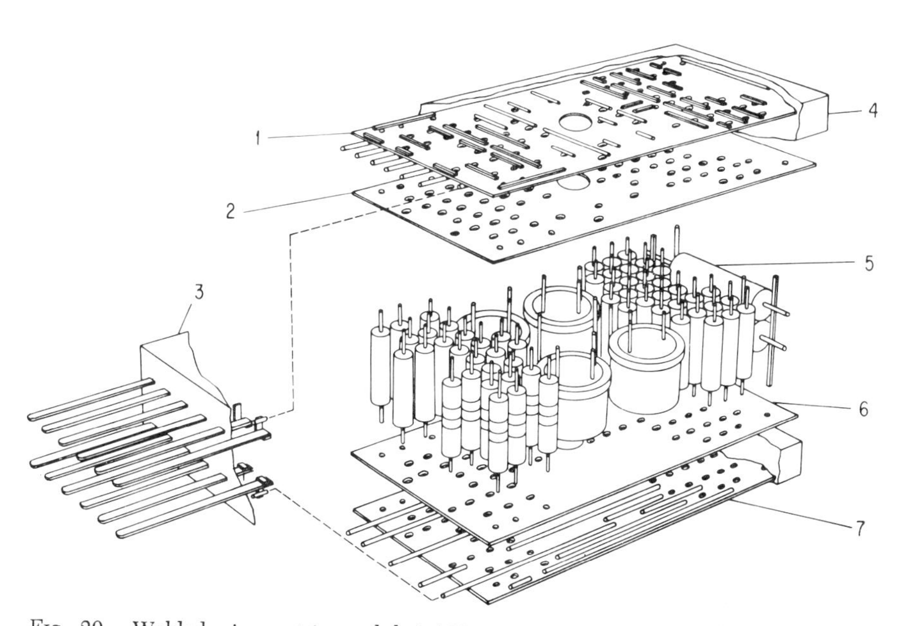

FIG. 20. Welded-wire matrix module.[6] (1) Top matrix film. (2) Top positioning film. (3) Header assembly. (4) Encapsulant. (5) Piece parts. (6) Bottom positioning film. (7) Bottom matrix film.

WIRE MATRIX. The wire matrix method, as shown in Fig. 20, utilizes a preassembled wiring pattern consisting of two layers of parallel wires at right angles to each other and separated by a thin insulating sheet such as Mylar. Since the wires are bare, the insulating film prevents shorting between the two layers. The film usually has the interconnecting pattern of the wiring printed upon it. Holes are punched in the film wherever one of the wires in the upper layer must be connected to a wire in the lower layer; holes are also made in the film for the component leads. The two layers of wires are positioned at right angles in a fixture of some sort, the film is inserted between them, and the two layers are welded together at the designated points through the holes provided. Unwanted sections of wire are then clipped out to give the desired wiring pattern. Where relatively long lengths of wire are run without being connected to a cross wire, a short section should be welded to the opposite side to act as an anchor. This anchor tie prevents the lead from moving and shorting to a parallel run; the maximum unsecured length should be 0.5 to 0.6 in. The matrix assembly can now be tested for continuity and welded to the components.

The components are placed through the punched holes in the film so that their leads are adjacent to the correct matrix wires, and the interconnection is made by welding; positioning films are often used in addition to the matrix layers. Compli-

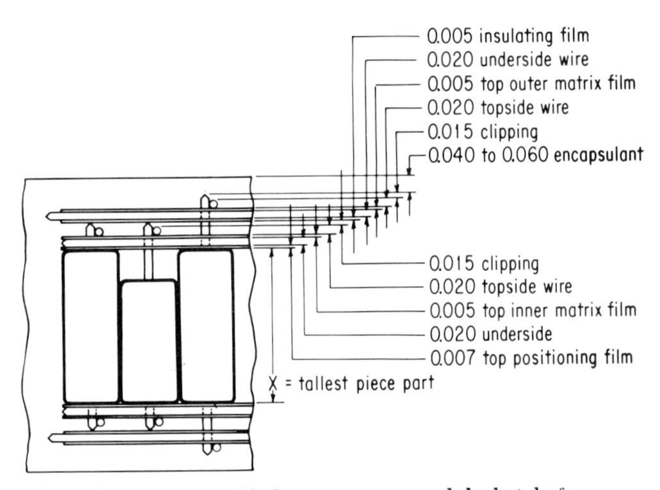

FIG. 21. Welded-wire matrix module height.[6]

cated circuits with a high packaging density may require two or more matrix layers per side. Output pins may be applied in the same manner as the components or may be placed in the plane of the matrix grid wires. As with the point-to-point wiring method, the components should be located on grid pattern intersections to simplify fabrication. A comparison between point-to-point and matrix modules is given in Table 13.

The module height (MH) may be determined as indicated in Fig. 21. Again the maximum film spacing is obtained from the maximum component length (including the component-plus-length tolerance), plus a reasonable manufacturing tolerance, usually 0.020 to 0.030 in. For modules with single matrix layers at each end,

$$MH = \text{film spacing} \times 0.364 \text{ in.}$$

The 0.182 in. at each end is made up as shown in the sketch, and a 20-mil-diameter wire is assumed. If more matrix layers are required, a height allowance of 0.08 in. per additional layer must be made.

PREFORMED WIRING. In the preformed-wiring method of module construction,

TABLE 13. Comparison of Advantages of Point-to-point and Wire-matrix Modules

Wire matrix

1. Wiring can be tested before components are applied
2. Complicated bending of welding ribbon is not required
3. Inspection is simpler since done in two stages
4. Manufacturing time is somewhat less

Point-to-point

1. Fewer welds are required
2. Repair or component replacement is easier
3. Module assembly is somewhat easier
4. Maximum flexibility in layout design
5. Intraconnection runs are more direct and shorter

etching or plating techniques are used to preform a wiring pattern into which the components are inserted and welded. Three basic types are in general use: (1) raised tab, (2) formed tubelets, (3) chemically milled and unsupported. The use of these preformed parts is generally restricted to large production runs so that the initial tooling costs can be reduced on a per-module basis. If the quantity of a single module design to be produced is in the thousands, then this type of construction should be considered since reduced assembly and welding costs can result. In addition, the controlled conductor length and fixed regular position of the component leads make this type of unit suitable to automated production. On the other hand, this fixed positioning of components tends to increase the module size. Also, the module circuit should be firmly established before this method is considered, since any changes could require the scrapping of considerable inventory and the preparation of new tooling. Module repair or component replacement is almost impossible with this type of unit due to the problem of cutting through the carrier boards and the equally difficult task of replacing any circuitry removed in the process.

In the raised-tab construction, panels similar to printed-wiring boards, with holes for the component leads and raised weldable tabs as part of the printed conductors, are used. A panel of this type is shown in Fig. 22. The tabs project out at right angles to the surface of the panel adjacent to the location of each component lead. The components are assembled as in an ordinary cordwood between the panels with the leads extending through the holes; the leads are then welded to the projecting tabs. This type of circuitry can also be obtained on a flexible backing which allows a module of slightly curved geometry to be made. Most cordwood modules are formed between a top and bottom wiring layer since this allows the maximum interconnection area with the minimum component-occupied volume. However, if some compromise can be made in component density within the module, then it is possible to prepare a module with only one wiring layer. One lead of each component with axial leads is bent over parallel to the component body so that all components essentially have radial leads. All interconnection must be done on one surface and the module size must increase to accommodate the additional leads. The raised-tab type of preformed circuit is well suited to this approach, and the cost of the required panels is cut in half.

The formed-tubelet construction employs a panel similar to the tab type, but the printed conductor terminates in small tubelets at right angles to and integral with the printed circuitry. A panel of this type is shown in Fig. 23. Holes for the component leads are placed in the carrier panel beneath each tubelet. The inside diameter of a tubelet is formed so as to be 0.010 to 0.015 in. larger in diameter than the component lead designated for that location. After the components are placed cordwood fashion between the panels, a weld is made between the tubelet and the component lead; the tubelet deforms and a joint is formed on each side of the component lead.

Chemically milled circuitry can be used on modules that have relatively simple

FIG. 22. Raised-tab preformed circuit. (*Amphenol Connector Division.*)

interconnection patterns. The circuitry is etched from thin metal sheets and is unsupported, that is, not attached to any backing material. Welding tabs, which are a part of the circuit runs, are provided at each component location for lead attachment. To provide some strength to the circuit pattern during module fabrication, tie bars and/or a perimeter frame are provided when the etching is done; these must of course be removed once the module is completed.

MOLDED SPACERS. Instead of insulating films to support the components during the module assembly and welding, preformed spacers may be used. These are usually molded of plastic insulating material and designed in the form of an egg crate or a honeycomb; a module of this type is shown in Fig. 24. The components are inserted into their respective cavities in the spacer, and interconnection between leads is made

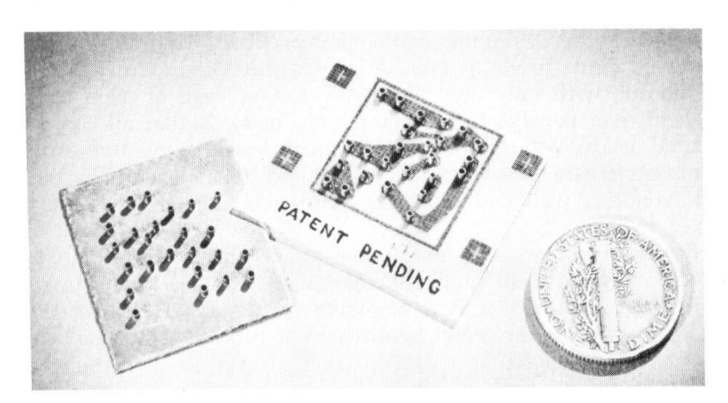

FIG. 23. Formed tubelet, preformed circuitry. (*Litton Industries, Advanced Circuitry Division.*)

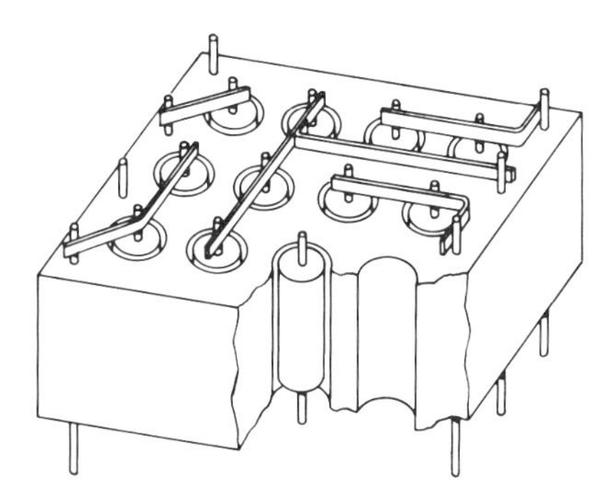

Fig. 24. Module with plastic honeycomb support.

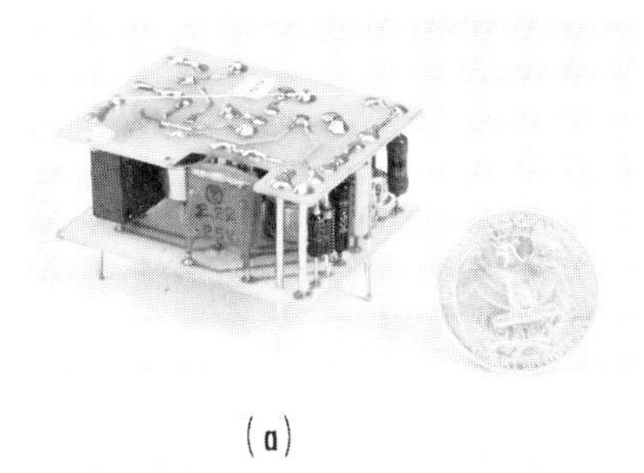

(a)

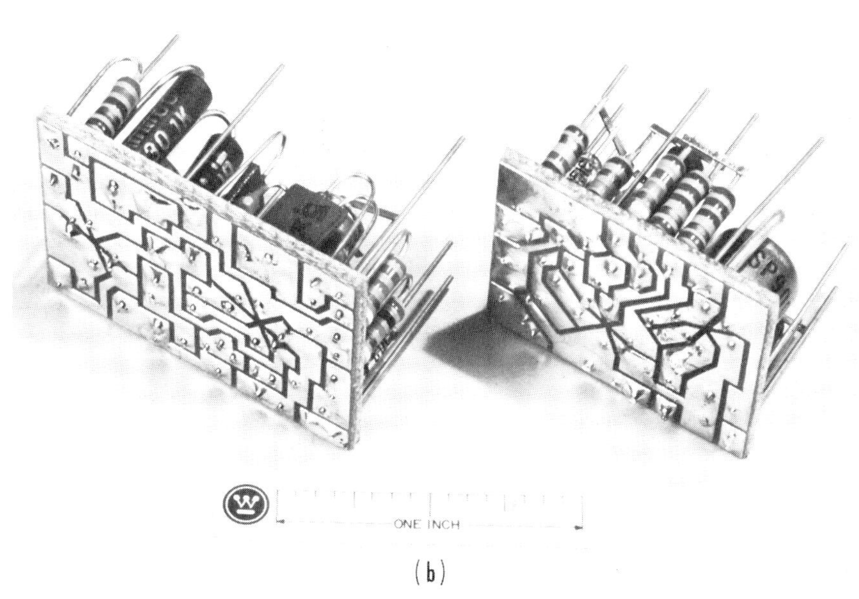

(b)

Fig. 25. Soldered cordwood modules. (*Westinghouse Electric Corp., Aerospace Division.*) (*a*) Standard version uses two printed-wiring boards. (*b*) Left-hand version uses one printed-wiring board and bent-over component leads; right-hand version is a hybrid with solder board on one end and welded ribbon on other.

by welding with a wire or ribbon, usually point-to-point. The spacer provides structural support for the module assembly and insulation between metallic body components and output or jumper pins. Encapsulation of this type of module is not normally necessary, but a conformal coating is advisable on the welded interconnection to give improved structural integrity. Repair of this type of module is relatively simple.

Soldered Cordwood Modules. Soldered cordwood modules are similar in design to the welded cordwood modules previously discussed, except that small printed-wiring boards are substituted for the insulating films and intraconnecting ribbons. Three modules of this type are shown in Fig. 25. In the standard version, the components are placed between two boards, and all the connecting joints on each board are made at one time by wave or dip soldering; this reduces fabrication time over the welded version. The components used do not require weldable leads, so that procurement problems are much simplified. With double-sided plated-through-hole boards, the component density can approach that of welded modules. The soldered module is ideally suited for certain semiconductor devices which may be injured in the welding process due to shunting or back biasing by improper welding sequence. Repair of a two-board soldered module is difficult since it is almost impossible to remove components in the center of the assembly without removing one entire board. One solution to this is shown in the lower half of Fig. 25; for these modules, only a single board is used. In the left-hand version, the component leads are bent in

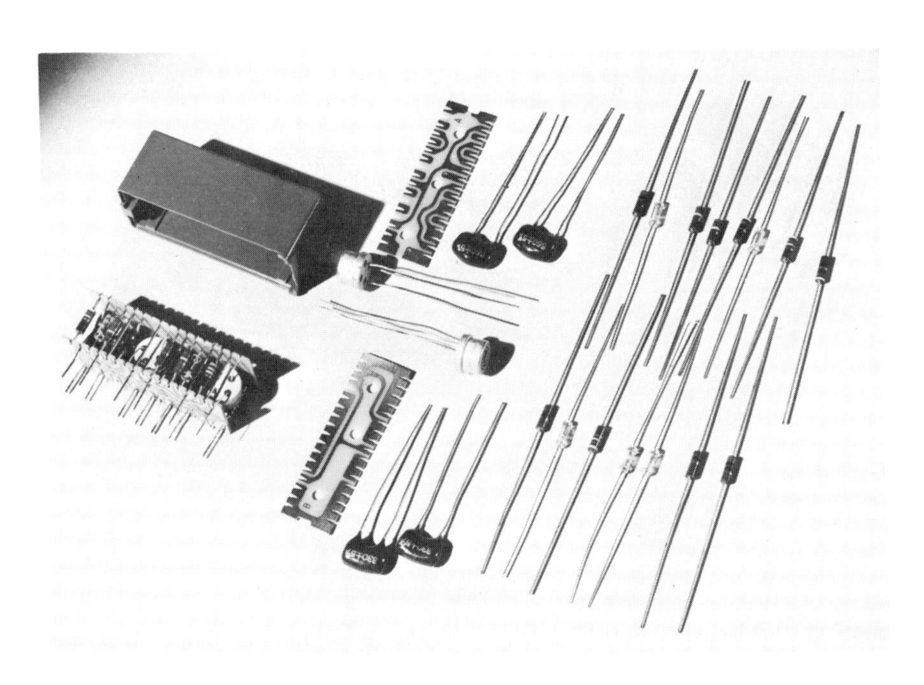

FIG. 26. Edge-loaded soldered module. (*Electronic Modules Corporation.*)

horseshoe fashion; this of course, requires greater board area. The right-hand module is a hybrid version which utilizes a board on one end and welded ribbon interconnects on the other. The greatest repairability is given by the edge-loaded version shown in Fig. 26. In this type of unit, the components are placed around the periphery of the printed-wiring boards. The boards may have holes, but for maximum ease of assembly, notches or slots are used. Automated assembly is easily incorporated here. This design tends to have a long narrow form factor, and is therefore generally limited to a module with a relatively small quantity of components.

Planar Modular Modules. *Planar modular modules* refers to the form in which the components are assembled in a two-dimensional (planar) array and then formed into a functional, testable module. A standard printed-wiring board assembly may be thought of in this manner, but a much smaller quantity of components is implied in the module version. Modules of this type may take many forms, but they all tend to be one component thick. Two forms of this module are shown in Fig. 27. In the version in Fig. 27*a* the interconnection is in the form of a welded wire matrix, and

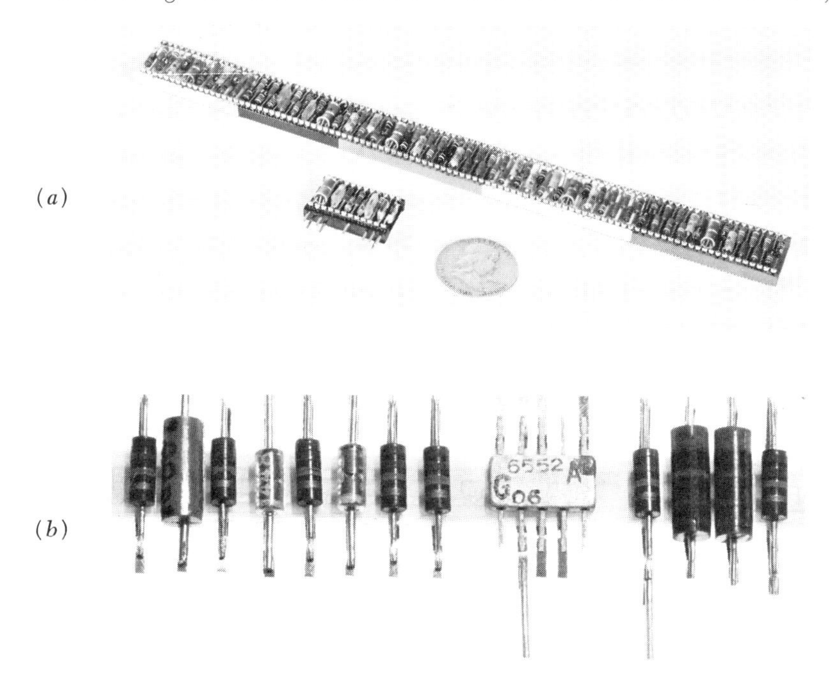

FIG. 27. Two planar-modular modules. (*a*) Wire matrix module. (*General Electric Company, Light Military Electronics Dept.*) (*b*) Stick model. (*Engineered Electronics Company.*)

slotted terminals are welded to the tranverse matrix wires so that the components can be soldered in place. In the version in Fig. 27*b* the interconnection is made with a series of etched metallic layers stacked with thin insulating sheets between them. Components are welded to the tabs which extend from the stack; layer-to-layer interconnection is also made by means of these fingers.

Stacked Wafer Modules. Stacked wafer modules are prepared by stacking small printed-circuit boards (wafers) upon which the components have been mounted by soldering. The boards are usually stacked vertically, and the interconnection between boards is made by vertical riser wires which are soldered or welded into formed grooves spaced around the periphery of the boards. A module of this type is shown in Fig. 28. The output pins are normally an extension of the riser wires, or a preformed leader may be used. Relatively small components must be utilized or the assembly becomes too high for its base area; components which are too high for one layer can be extended through a cutout in the board above. Repairability is poor here since all risers must be disconnected and a complete wafer removed to get to a specific component.

Embedment. Electronic assemblies must be able to operate over a wide range of environmental conditions such as temperature, humidity, vibration, acceleration, and shock. In general, due to their compact nature, modules tend to be resistive to these environments. However, to achieve maximum protection, they should be embedded or molded in a plastic or polymer after all electrical tests have been completed. This provides a unitized construction which is nonresonant and structurally rigid. At the same time, the assembly is sealed against the deleterious effects of such irritants as moisture, corrosive atmospheres, and fungi.

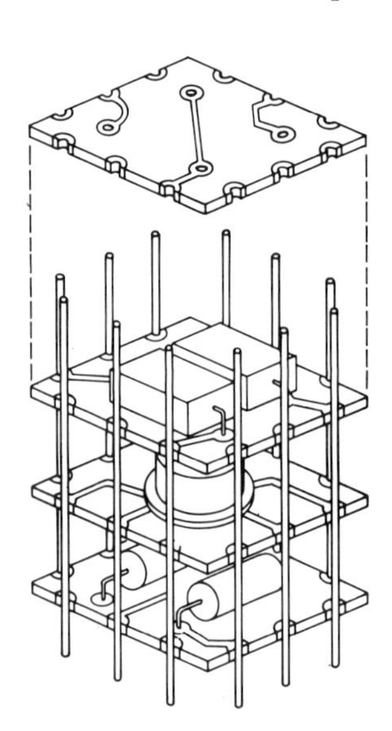

FIG. 28. Stacked wafer module.

The type of material and the process used for the embedment depend upon the module configuration, the functional requirements of the unit, and the environments to which it will be exposed. For instance, modules with low power dissipation can be embedded in foam material for low weight, while units with high power levels should have a filled epoxy for good thermal properties. A module with a voltage over 300 volts that is to be exposed to a low-pressure condition would best be embedded in a denser material such as unfilled epoxy or a silicone. A complete guide to the selection of the proper material can be found in Chap. 7. A quick method for estimating the weight of modules embedded in several common materials is found in Fig. 29 of this section. The weight of the components is included in the values shown.

Chassis-mount Construction. The basic building block in the design of electronic equipment is the chassis. This may take a wide variety of forms depending upon the size and function of the equipment, but in all applications it provides the surface or surfaces upon which the various components, subassemblies, or modular packages are mounted. It serves to give rigidity to the structure and provides a convenient attachment point for handles, connectors, and other hardware items. In addition, since it is usually fabricated from metal, it gives a good ground return and provides for electric, magnetic, and thermal shielding.

Chassis Sizing. Much equipment is most conveniently packed onto chassis which are sized according to standardized rack-and-panel assemblies for the type of service that the electronics will provide: ground-based, shipborne, or airborne. The basic consideration of size in each of these categories is given below:

GROUND-BASED ELECTRONICS. Ground-based electronics is often designed around panel mounting racks based on dimensional standards established ty the Electronic Industries Association (EIA) and MIL-STD-189. The standard panel width is 19 in. ($\pm 1/16$ in.), as shown in Fig. 30 along with the other major width dimensions and slot details. Panel height can be determined from:

$$\text{Height} = (1\tfrac{3}{4}n - \tfrac{1}{32}) \pm \tfrac{1}{64} \text{ in.}$$

where n is an integer. Dimensions for standard panel sizes are given in Table 14.

Standard cabinet racks are built to a height of 83 in., which gives a panel clearance of 77⅛ in.; other rack heights are available in increments of 7 in. with corresponding panel clearance. Although designs may vary, the most commonly used overall cabinet depths are 15¼, 18, and 24 in., which result in minimum inside clearances of 13¾, 16⅞, and 22 in. Overall enclosure widths may vary from 22 to 24 in. Panel mounting holes in the vertical sections of the cabinet are spaced as shown in Fig. 31.

The chassis used must of course fit within the outlines of the panel size selected; its depth is governed by the inside clearances of the cabinet and the dimensions of the cabling or output connectors at the rear. The chassis may be mounted either vertically or horizontally and attached to the panel, placed on a shelf behind the panel, or hung from drawer glides fastened to the vertical members of the enclosure. Various types of chassis arrangements are shown in Table 15.

SHIPBORNE EQUIPMENT. Shipborne equipment can be packaged in the same basic type of rack-and-panel assemblies as outlined for ground-based equipment if allowance is made for the restricted space and access doors. No unit of any equipment intended for interior shipboard installation should exceed 72 inch in overall height including any shock mounting. All units must be capable of passing through a doorway 26 in. wide by 45 in. high with 8-in.-radius rounded corner and through a hatch 36 in. long by 30 in. wide with 7½-in.-radius rounded corners. For submarine installation, units must pass through a doorway 20 in. wide by 36 in. high with 10-in.-radius rounded corners and through a tube 25 in. in diameter.

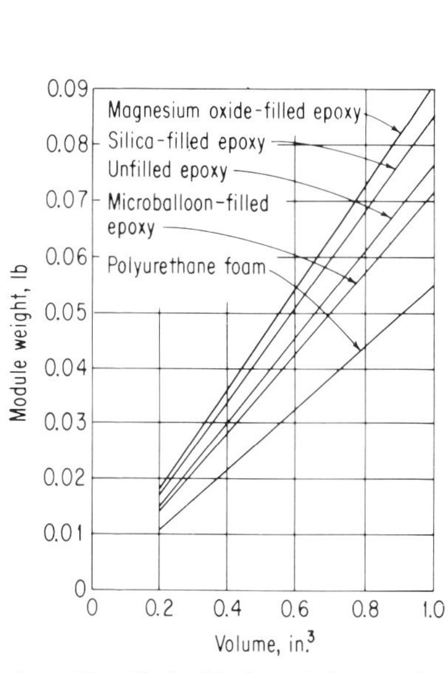

FIG. 29. Embedded module weight estimation.

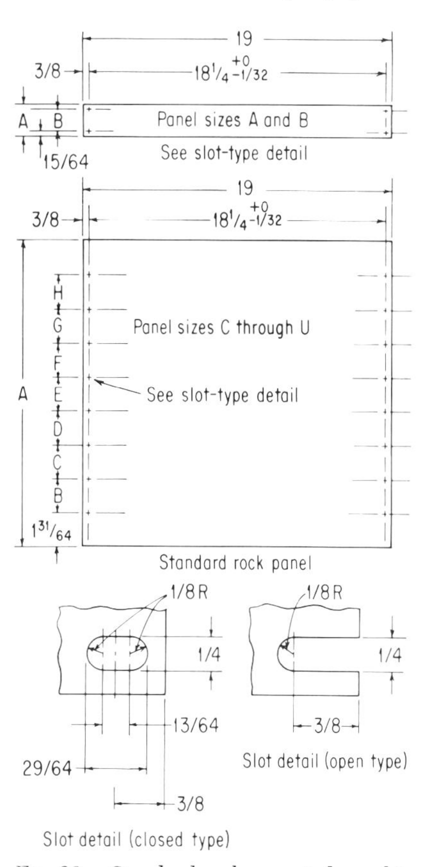

FIG. 30. Standard rack-mounted panel.[7]

The chassis should be mounted on channel guides with track and rollers or hinges to allow it to be moved toward the front for servicing. Automatic locks should be provided to secure the chassis in both the operating and servicing positions, and the chassis should be completely removable from the front of the enclosure.

AIRBORNE EQUIPMENT. Certain units of airborne equipment (such as fighter air-

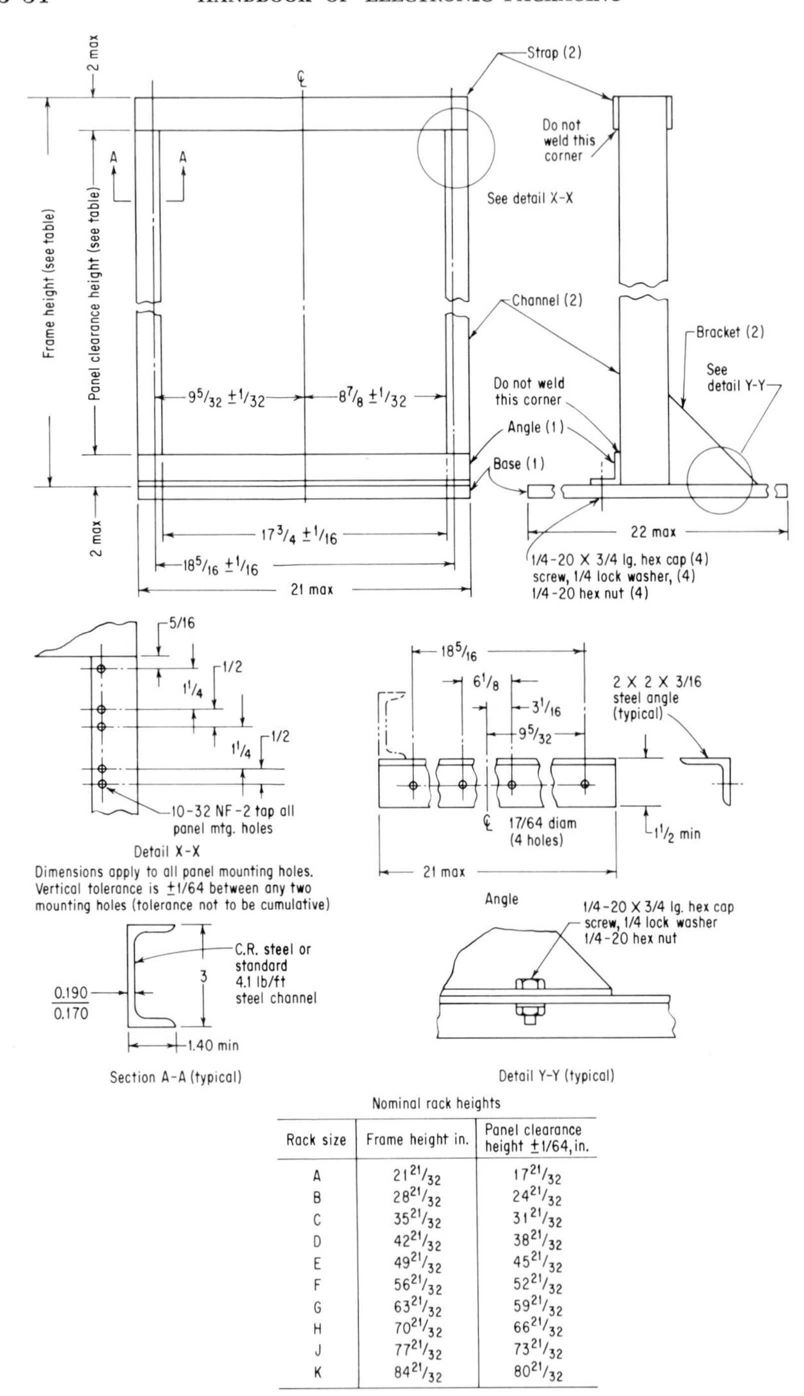

Nominal rack heights

Rack size	Frame height in.	Panel clearance height ±1/64, in.
A	$21\frac{21}{32}$	$17\frac{21}{32}$
B	$28\frac{21}{32}$	$24\frac{21}{32}$
C	$35\frac{21}{32}$	$31\frac{21}{32}$
D	$42\frac{21}{32}$	$38\frac{21}{32}$
E	$49\frac{21}{32}$	$45\frac{21}{32}$
F	$56\frac{21}{32}$	$52\frac{21}{32}$
G	$63\frac{21}{32}$	$59\frac{21}{32}$
H	$70\frac{21}{32}$	$66\frac{21}{32}$
J	$77\frac{21}{32}$	$73\frac{21}{32}$
K	$84\frac{21}{32}$	$80\frac{21}{32}$

FIG. 31. Dimensions of standard panel racks.[7]

TABLE 14. Dimensions of Standard Panel Racks [7] *

Panel size	Dimensions of panel, in.							
	A	B	C	D	E	F	G	H
A	1 23/32	1 1/4						
B	3 15/32	3						
C	5 7/32	2 1/4						
D	6 31/32	4						
E	8 23/32	1 3/4	2 1/4	1 3/4				
F	10 15/32	2 1/4	3	2 1/4				
G	12 7/32	1 3/4	5 3/4	1 3/4				
H	13 31/32	3 1/2	4	3 1/2				
J	15 23/32	3 1/2	5 3/4	3 1/2				
K	17 15/32	3 1/2	7 1/2	3 1/2				
L	19 7/32	5 1/4	5 3/4	5 1/4				
M	20 31/32	1 3/4	5 1/4	4	5 1/4	1 3/4		
N	22 23/32	5 1/4	1 3/4	5 3/4	1 3/4	5 1/4		
P	24 15/32	1 3/4	5 1/4	7 1/2	5 1/4	1 3/4		
R	26 7/32	1 3/4	7	5 3/4	7	1 3/4		
S	27 31/32	1 3/4	5 1/4	3 1/2	4	3 1/2	5 1/4	1 3/4
T	29 23/32	1 3/4	5 1/4	3 1/2	5 3/4	3 1/2	5 1/4	1 3/4
U	31 15/32	1 3/4	5 1/4	3 1/2	7 1/2	3 1/2	5 1/4	1 3/4

* Thickness: 1/8, 3/16, 1/4, or 5/16 in.

TABLE 15. Typical Chassis Arrangements [8]

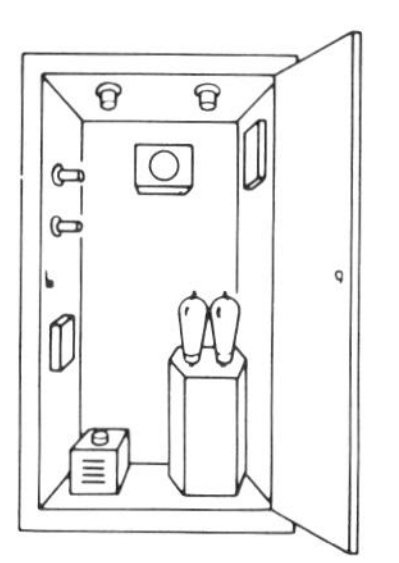

No removable chassis. Parts mounted directly on framework of cabinet. (Used for heavy, high-power units)

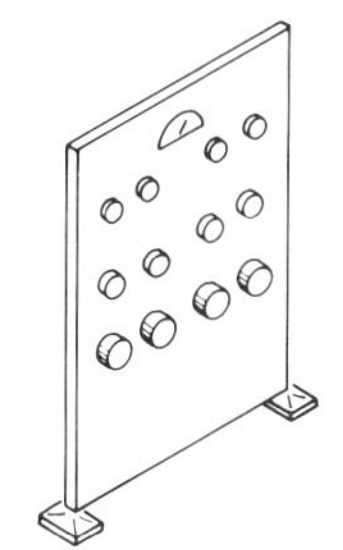

Parts mounted on single vertical panel

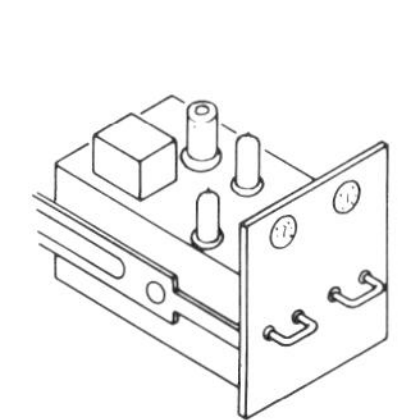

Horizontal chassis. Tubes, etc., mounted on top, other parts and circuitry mounted on bottom

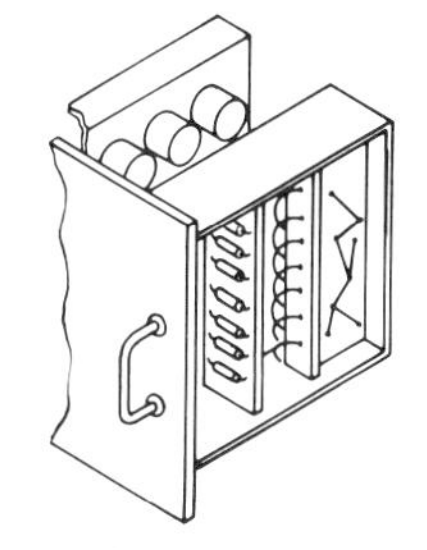

Vertical-mounted chassis. Tubes and large parts mounted on one side (inside) and small parts and circuitry mounted on other side (outside)

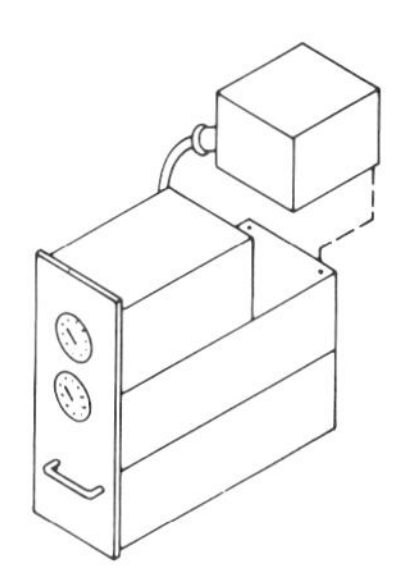

Subdivided chassis

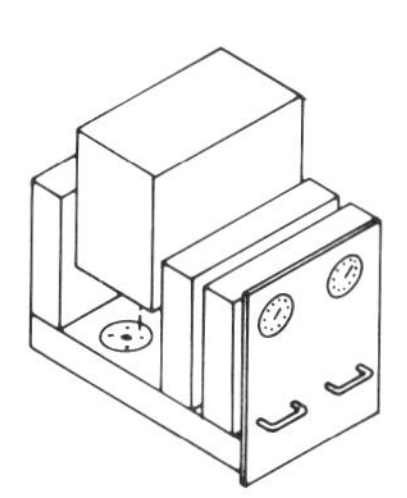

Plug-in modular units

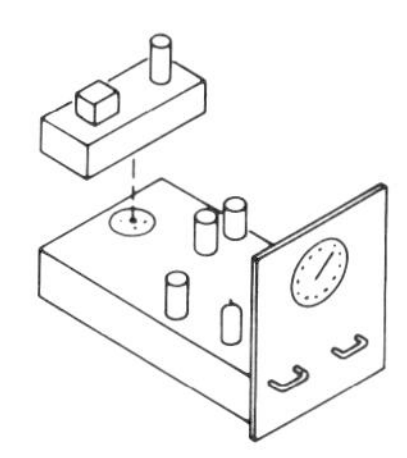

Plug-in subassembly

TABLE 16. Method of Retaining Plug-in Components [8]

Description	Advantage	Disadvantage(s)	Tools required	Operating time
Hinged clamp secured by wing nut	One device can retain several parts	Difficult to close with one hand. Possible loss of wing nut if not captive	None	Approximately 0.6 min
Retaining clamp pressed down on screw threads or ratchet device. Released by compressing spring	Support at top of part prevents cantilevering	Difficult to release at times. Possible loss of clamp	None	Approximately 0.2 min *
Screw through flange into tapped hole in socket or chassis	Strength comparable with other parts secured with screw fasteners without the need for soldering leads	Excessive time required to operate screws. Possible loss of screws or washers	Screwdriver	Approximately 0.8 min per screw fastener
Toggle-action clamp around base of part	No loose parts	Difficult to operate if too tight. Clamp at bottom of part permits cantilevering –undesirable on parts with high center of gravity	None	Approximately 0.2 min *
Spring clips hook over part	No loose parts	Two hands required to install parts	None	Approximately 0.1 min *
Miniature tube, heat-dissipating and -retaining shield	Easy, fast operation	Possible loss of retaining shield	None	Approximately 0.2 min *

* Based on average of estimates made by six engineers.

craft fire control radar) cannot conveniently be configured into a standard racking system and must be packaged on a space-available basis with the various chassis shaped to take best advantage of that space. However, much other electronics equipment can be packaged into the standard racking system specified for this service.

For smaller control packages which are to be cockpit-mounted, the size is given in MS 25212. The standard provisions of this are shown in Fig. 32.

More complex units are designed around standard form factors and case sizes to mount into air transport racks (ATR) in an electronics bay. These packages have rear locating dowels and rear-mounted "blind plug-in" connectors, and are removable from individual mounting cradles, usually shock-isolated. The smaller sizes are specified by MS 91402 and the larger sizes by MS 91403. The major provisions of these two specifications are shown in Figs. 33 and 34.

Chassis-mount Components. In most assemblies there are some components which by virtue of their configuration, mass, or heat emission are not adaptable for inclusion on printed-wiring boards or in modular packages. These components, or small subassemblies of them, must be separated from the other parts and mounted directly upon plates or structures tied to the chassis. Some considerations necessary with components of this type are given below:

PLUG-IN DEVICES. The category of plug-in devices includes tubes, relays, transformers, and other components equipped with a pattern of pins to plug into a socket mounted onto the chassis. The most used socket types are the standard octal (8-pin), miniature (7-pin), and naval (9-pin); details on these can be found later in this chapter. Since parts of this type tend to have considerable mass and a high center of gravity, they must be retained to ensure mechanical stability; Table 16 shows methods of retention. In addition, the tube-shield-type retainers act as heat dissipators to aid in proper thermal control of the tubes.

PARTS WITH LARGE MASS VOLUME. Heavy parts such as oil-filled capacitors, heavy-duty relays, and power transformers must be securely tied to the chassis. Also, these parts should be mounted near the chassis edge or beside a stiffener partition to prevent "oil-canning" or buckling of the structure. Typical retaining methods for large components are shown in Table 17.

SMALL COMPONENTS. Items which have a weight of less than 1/2 oz may be soldered to terminals, or standoff and thus secured to the chassis only by the wire leads. However, if the components weigh over 1/2 oz or have a high length-to-diameter ratio, additional mechanical support should be provided. Typical mounting provisions for items of this type are shown in Table 18. Other small components are designed to be secured to the chassis with screw fasteners and interconnected to the circuitry through solder terminals incorporated into the body of the component. Mounting provisions for this type of component are shown in Table 19.

HIGH DEGREE OF ISOLATION REQUIRED. Certain components require shielding from stray fields—electric or magnetic—which may cause performance degradation. A shield is a metallic barrier used to attenuate the rf energy passing through the barrier; since the chassis is usually metallic, it can often serve as a useful shield if properly designed. Electrostatic shields, used against electric fields, are made of high-conductivity materials such as copper, brass, and aluminum; low-frequency shielding against magnetic fields is based on ferromagnetic materials such as iron, nickel, and cobalt.

For electrostatic shielding, the attenuation afforded by a partition of some commonly used materials is given in Fig. 35. This is based on the formulas

$$A = 3.338\,\sqrt{f}$$

$$L = A\,\frac{1.72\mu}{R}$$

where A = attenuation of pure copper, db/mil
f = frequency, Mc
R = resistivity, μ ohms/cm^3
μ = magnetic permeability
L = attenuation of materials other than copper, db/mil

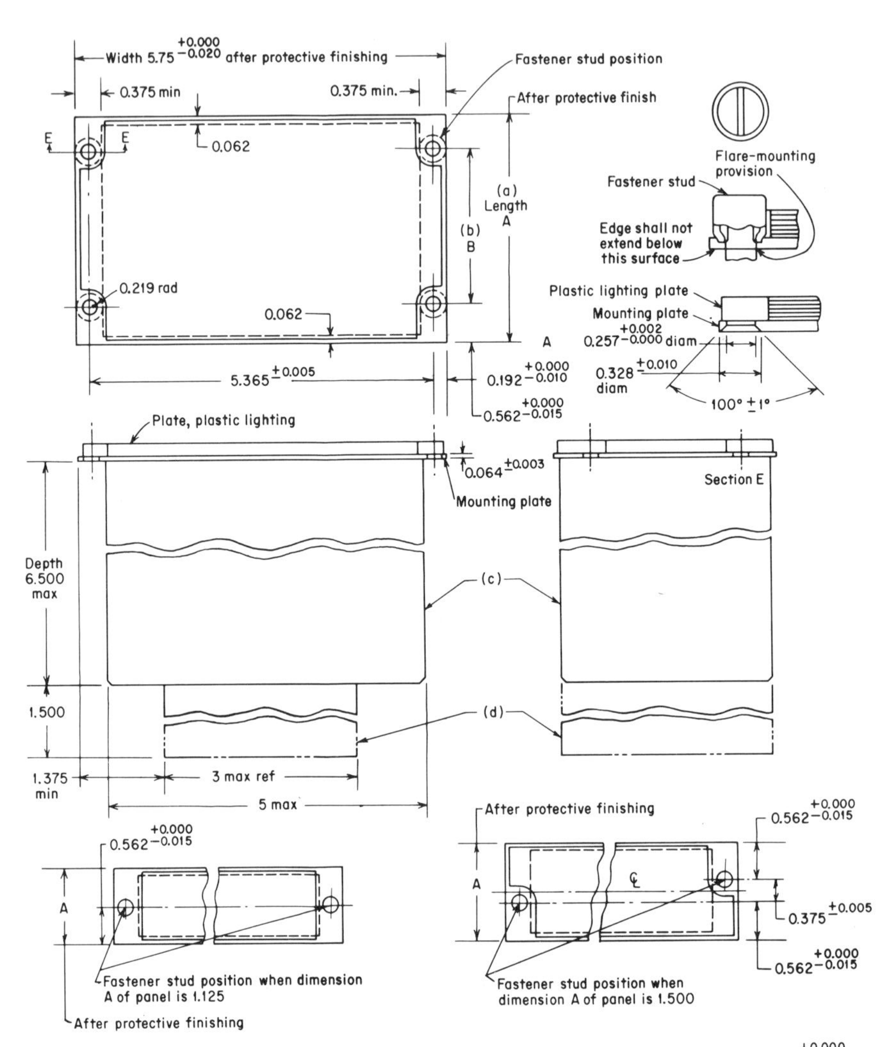

(a) Dimension A is a multiple of 0.375 (with minimum of 1.125 and a maximum of 9). Tolerance $^{+0.000}_{-0.015}$.

(b) Dimension B is 1.125 less than A (three multiples of 0.375). Distance between studs shall not exceed 4.875. Tolerance is ±0.005.

(c) Stage available for mounting of component controls, disconnect receptacle with mounting bracket and mating disconnect plug and cable clamp. When a dust cover is used, it shall be contained in this area.

(d) Maximum available space in the aircraft for the exit of cables, etc., from the control panel. Dressed cables must lie within this area and run in the direction of the control panel length.

Dimensions in inches.

FIG. 32. Cockpit-mounted control box outline (MS 25212).

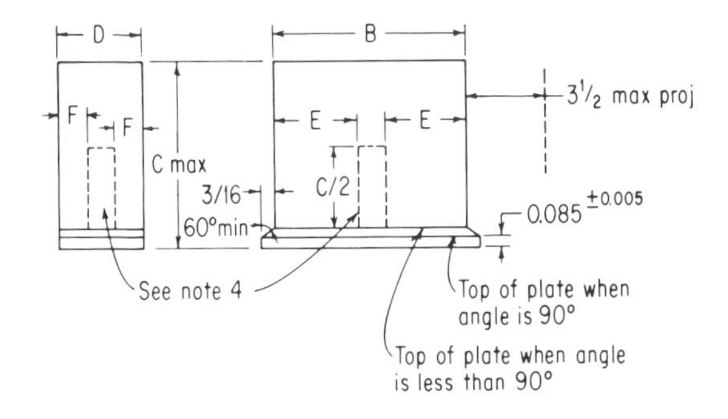

Part number	Dimensions, in.					Weight range, lb (See Note 5)
	B	C	D	E	F	
-S1	9 7/8	7 5/8	5 1/8	4	2	6 to 12
-S2	13 1/8	7 5/8	4 7/8	5 1/2	2	10 to 22

Notes:

1. All dimensions in inches
2. Unless otherwise specified, tolerances are ±1/32 on fractions
3. If front panel is used, it projects no more than 1/16 beyond the top and sides of the case and is considered part of the case
4. Center of gravity for cases used with class A mounting bases does not fall outside limits shown by dashed lines. The weight of the connectors and 1 foot of the attached cabling is considered for establishing location of the center of gravity
5. For use on mounting bases when a specific load is isolated, the weight range is the specific load for which the base is designed ±15 percent

FIG. 33. Small-size cases (MS 91402).

To determine the attenuation of copper, select the frequency on the f scale and read attenuation on scale A. For other metals locate resistivity on the R scale and permeability on the M scale and join these points with a line; at the intersection of this line with the T scale, locate a turning point. Construct a line from the turning point to the frequency value on the f scale. Read attenuation per mil on the L scale.

For magnetic shielding, the chassis itself is seldom used unless the field is of very low intensity. The highest magnetic attenuation is obtained from the material in its annealed state, and in this condition it is undesirable for a structural member. The field attenuation is usually achieved by a shield case around the source or the affected part. In securing these cases to the chassis, care must be taken that the annealed state is not destroyed by any cold working due to the mounting method used, and that the "soft" case will indeed remain in place under vibration and shock loads.

More information on isolation may be found in Chap. 10.

HIGH POWER AND HEAT DISSIPATION. Components which operate at high voltage or power levels are generally not suitable to "high-density" modular construction since they require large connection leads and wide spacings between different potential levels. Chassis mounting allows them to be spread.

Since the chassis is usually a metallic member, it acts as a good thermal control for components with high heat dissipation. The chassis may assist in the cooling by providing a greater surface for heat transfer through natural convection and radiation, or it may transfer the energy to the final heat sink by conduction. In any event, the mounting of the component should be such as to provide a good contact thermal inter-

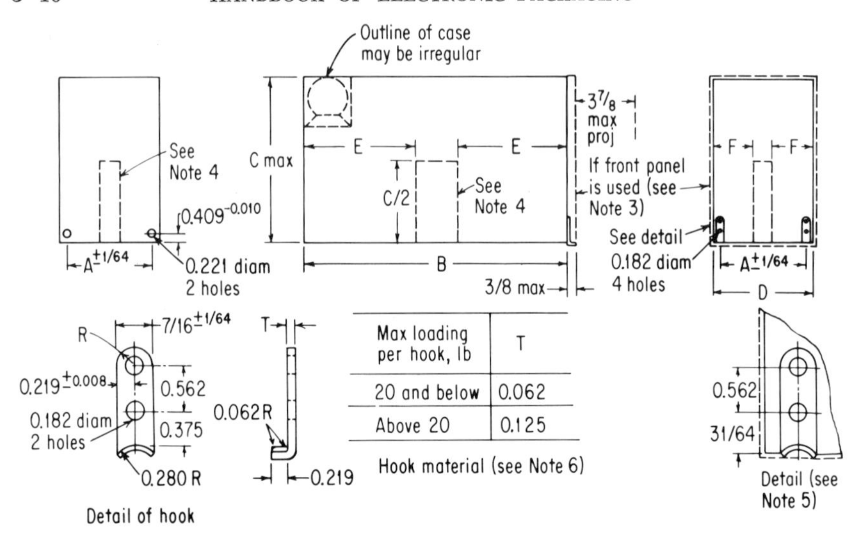

Part number	Dimensions, in.						Weight range, lb (see Note 7)
	A	B	C	D	E	F	
A1B	4 1/8	12 9/16	7 5/8	4 7/8	5 1/4	2	8 to 18
A1C	4 1/8	15 9/16	7 5/8	4 7/8	6 1/2	2	12 to 20
A1D	4 1/8	19 9/16	7 5/8	4 7/8	8 3/16	2	18 to 40
B1B	9 3/8	12 9/16	7 5/8	10 1/8	5 1/4	4 1/4	18 to 40
B1C	9 3/8	15 9/16	7 5/8	10 1/8	6 1/2	4 1/4	25 to 50
B1D1	9 3/8	19 9/16	7 5/8	10 1/8	8 3/16	4 1/4	40 to 80
B1D2	9 3/8	19 9/16	7 5/8	10 1/8	8 3/16	4 1/4	40 to 80
C1D	14 5/8	19 9/16	7 5/8	15 3/8	8 3/16	6 7/16	40 to 80
C2D	14 5/8	19 9/16	10 5/8	15 3/8	8 3/16	6 7/16	40 to 80

Notes:

1. All dimensions in inches
2. Unless otherwise specified, tolerances are ±1/32 on fractions and ±0.005 on decimals
3. If a front panel is used, it projects no more than 1/16 beyond the top and sides of the case and is considered part of the case
4. Center of gravity for cases used with class A mounting bases does not fall outside limits shown by dashed lines. The weight of the connectors and 1 foot of the attached cabling is considered for establishing location of the center of gravity
5. When 0.125 hook is used, the bottom of the hook is approximately flush with bottom of the panel, or 1/8 below the bottom of the case when panel is not used. When 0.082 hook is used, the bottom of the hook is approximately 1/16 above the bottom of the panel, or 1/16 below the bottom of the case when panel is not used. Additional hooks, when used, are located 2 1/16 on either side of case centerline. (See detail drawing)
6. Hook material is 410 corrosion-resistant steel, heat-treated to 126,000 psi
7. Rear travel of unit when front panel is used is limited by contact of back face of front panel with front of mounting base
8. For use on mounting bases when a specific load is isolated, the weight range is the specific load for which the base is designed ±15 percent

FIG. 34. Large-size cases (MS 91403).

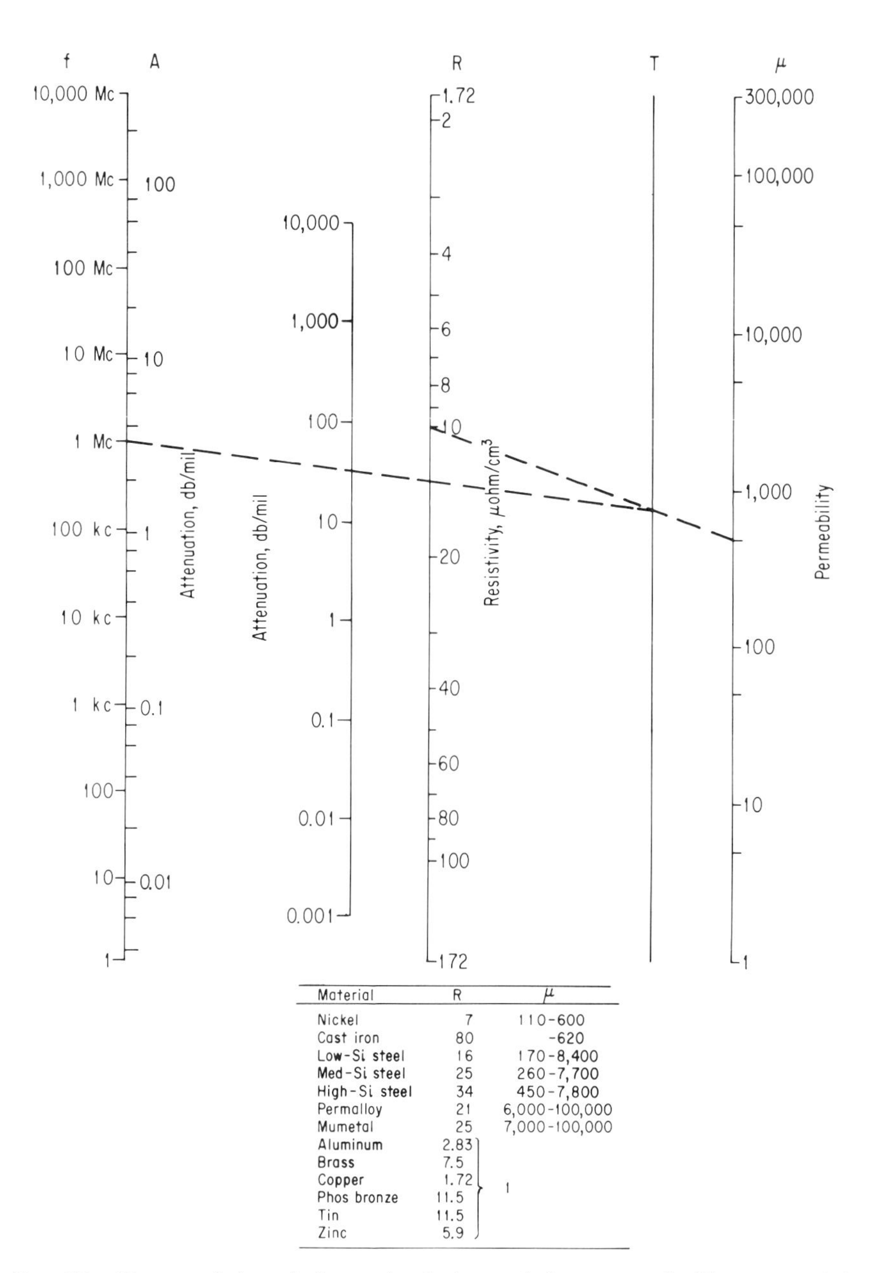

Material	R	μ
Nickel	7	110-600
Cast iron	80	-620
Low-Si steel	16	170-8,400
Med-Si steel	25	260-7,700
High-Si steel	34	450-7,800
Permalloy	21	6,000-100,000
Mumetal	25	7,000-100,000
Aluminum	2.83	1
Brass	7.5	1
Copper	1.72	1
Phos bronze	11.5	1
Tin	11.5	1
Zinc	5.9	1

FIG. 35. Nomograph for calculating the thickness of electrostatic shielding materials.[9]

TABLE 17. Retaining Methods for Large Components [8]

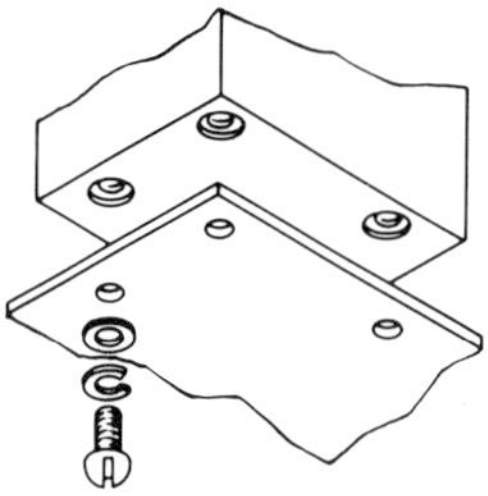	Description	Horizontal surface mounted. Screws with flat washers, and lock washers into tapped hole in part
	Advantage	Can usually be performed with one hand if necessary
	Disadvantages	Possible loss of washers or screws. Alignment of holes difficult
	Tools required	Screwdriver
	Operating time	Approximately 0.8 min per fastener
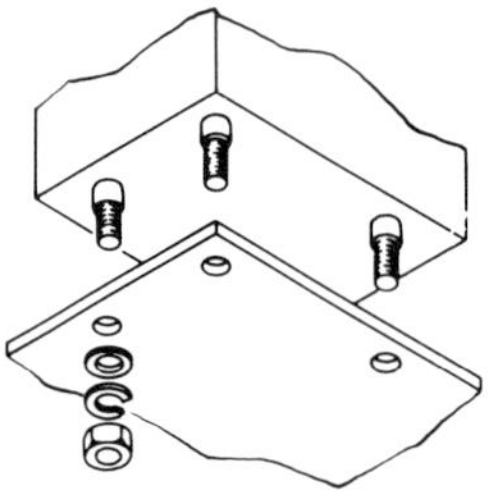	Description	Horizontal surface mounted. Studs through clearance holes with nuts, flat washers, and lock washers
	Advantages	One-handed operation. Studs act as locating pins
	Disadvantages	Possible loss of nuts or washers. Part must be lifted
	Tools required	Wrench or "spintite"
	Operating time	Approximately 0.7 min per fastener
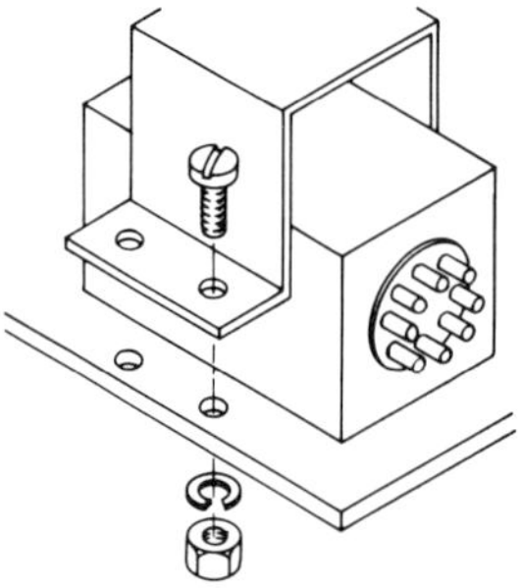	Description	Strap clamp with screws through clearance holes
	Advantage	Complete removal of clamp may not be required
	Disadvantages	Requires simultaneous access to both sides of chassis. Two-handed operation
	Tools required	Screwdriver and wrench or "spintite"
	Operating time	Approximately 0.8 min per screw fastener plus 0.2 min for positioning clamp
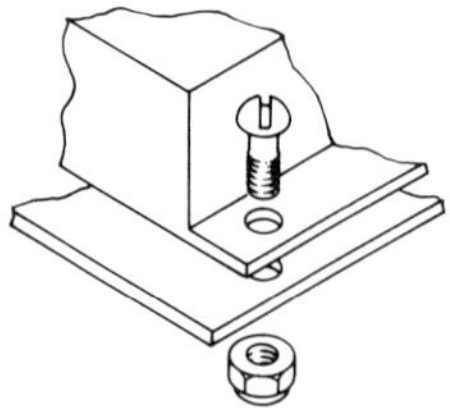	Description	Horizontal surface mounted. Screws with stop nuts through clearance holes
	Advantage	No washer required
	Disadvantages	Stop nut difficult to turn. Two-handed operation. Alignment of holes difficult. Possible loss of nuts
	Tools required	Screwdriver and wrench or "spintite"
	Operating time	Approximately 1.4 min per fastener
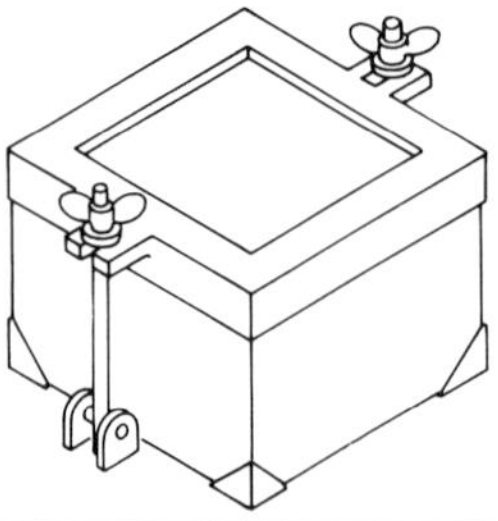	Description	Horizontal surface mounted. Clamping bracket held by clamp screws with captive nuts
	Advantage	Can be performed with one hand
	Disadvantage	Part must be lifted
	Tools required	None
	Operating time	Approximately 0.6 min per fastener *

* Does not include time to place clamp bracket in position and bring up clamp screws.

TABLE 18. Methods of Retaining Lead-mounted Components Requiring Additional Support [8]

Securing method	Maintainability considerations	
	Description Advantage Disadvantage	Clamp hooks over part Clamp need not be completely detached to remove part Requires use of screwdriver
	Description Advantage Disadvantages	Strap around part Part does not have to rest on chassis May be necessary to completely detach strap. Requires use of screwdriver
	Description Advantages Disadvantage	"Fuse-clip"-style holder Not necessary to detach clip from chassis. No tools required Sometimes difficult to remove part when clip has strong spring action
	Description Advantage Disadvantages	Mounting bracket soldered to part Part will not slip Replacement parts must be supplied with bracket attached. Requires use of screwdriver
	Description Advantages Disadvantage	Part held by spring fingers on sleeve No tools required. No loose parts Limited application

face; often the use of a silicone thermal compound will ensure a voidless contact between the two surfaces. To enhance the chassis conduction, all surfaces should be nonoxidized and all joints should be brazed or welded. If the component is a particularly high heat generator, then the use of an additional heat sink such as a tube shield or a finned device (Fig. 36) should be used.

The basic equipment-cooling method to be used can be chosen from the following guides: (1) When the density is less than 0.25 watts/in.^{2}, and with ambient pressure of 1 atmosphere, natural cooling (free conduction, convection, radiation) should be used. (2) For densities between 0.25 and 2 watts/in.^{2}, forced-air cooling should be used. (3) If the density is over 2 watts/in.^{2}, indirect liquid cooling should be used with metallic conduction paths from heat sources to the liquid coolant. A more detailed discussion of thermal dissipation can be found in Chap. 11.

BASIC CHASSIS-MOUNT CONSIDERATIONS. A listing of the basic rules to be followed in chassis-mount construction is given in Table 20.

RF COMPONENTS. Units of the rf type (typical examples shown in Fig. 37), by the very nature of their function, tend to be relatively large, heavy, and high-powered; these characteristics make it almost mandatory that they be chassis-mounted. Many of these components are strong sources for magnetic or electrostatic fields, and proper shielding must be provided for any sensitive equipment nearby. In addition to these packaging considerations, the interconnections between these devices present unique problems. The interconnection leads are relatively large and inflexible in themselves, and the terminations are made with connectors which require considerable space. Although the prime purpose of the lines is to transmit rf energy from one place to

TABLE 19. Small Components Secured with Screws [8]

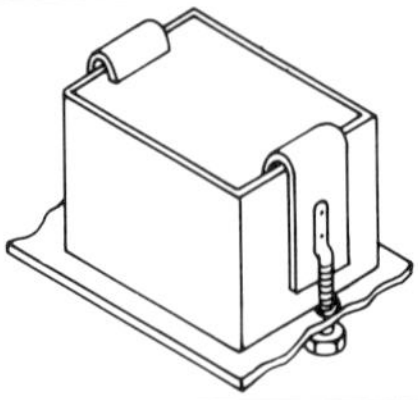	Description	Clamps hook over rim of part
	Advantages	Complete removal of clamp unnecessary in most cases
	Disadvantages	Requires simultaneous access to both sides of chassis
	Tools required	Wrench or "spintite"
	Operating time	Approximately 0.9 min to position one clamp and tighten nut
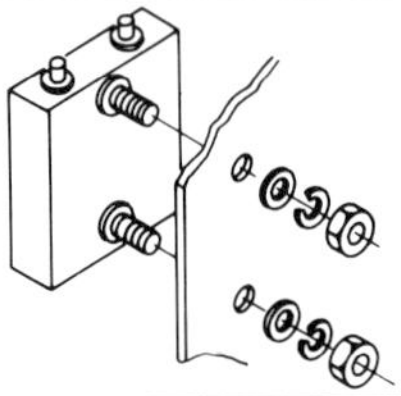	Description	Studs through clearance holes
	Advantages	Can be mounted with one hand if part is supported. Studs act as locating pins
	Disadvantage	May require one hand to support part while starting nut
	Tools required	Wrench or "spintite"
	Operating time	Approximately 0.7 min per fastener
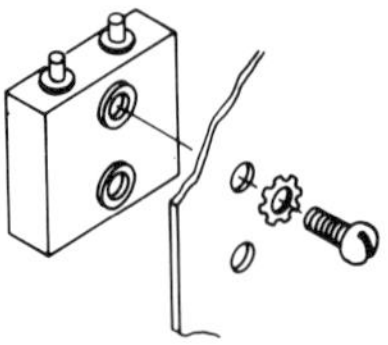	Description	Screws into tapped holes in part
	Advantage	No loose nuts
	Disadvantages	Requires simultaneous access to both sides of chassis. Two-handed operation
	Tools required	Screwdriver
	Operating time	Approximately 0.8 min per fastener
	Description	Nut on bushing through clearance hole
	Advantage	Can be performed with one hand if part is supported
	Disadvantage	May require one hand to support part while starting nut
	Tools required	Wrench or special purpose "spintite"
	Operating time	Approximately 0.7 min
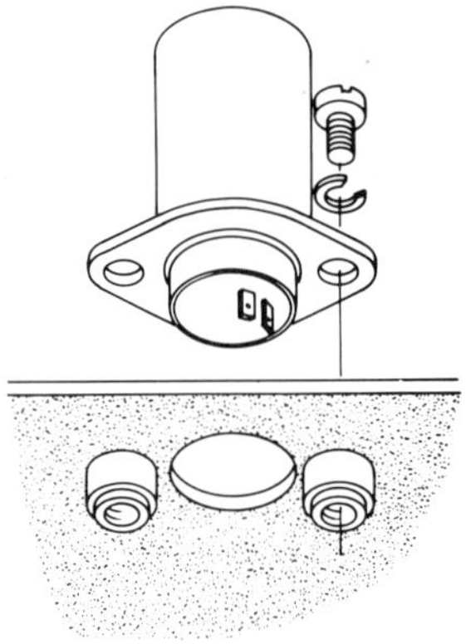	Description	Flange mounted with screws into tapped holes
	Advantages	One-handed operation. No loose nuts
	Disadvantages	Screws must be completely removed
	Tools required	Screwdriver
	Operating time	Approximately 0.8 min per fastener

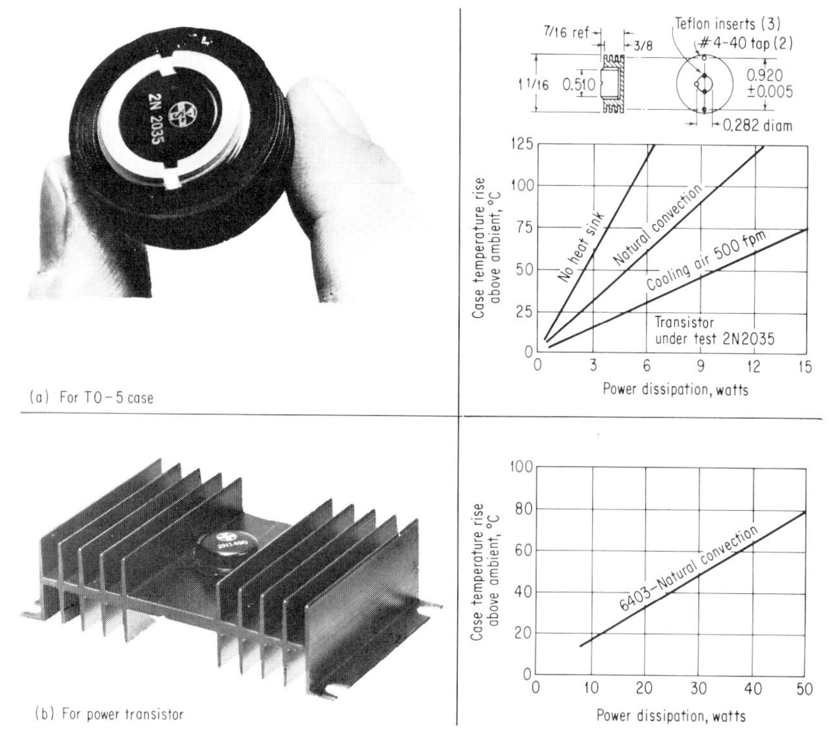

FIG. 36. Typical finned heat sinks. (*Thermalloy Company.*)

TABLE 20. Considerations in Chassis-mount Construction

Do	*Do not*
1. Provide grommets for wires passing through holes in all partitions less than ⅛ in. thick. Radius holes in panels over ⅛ in. thick to $\frac{1}{16}$ in. radius	1. Carry wires around or bend them over sharp edges or corners
2. Orient parts so that markings, polarity, etc., are visible	2. Apply more than three leads to each section of a terminal
3. Keep all rf circuit leads as short as possible	3. Anodize chassis to which components or lugs are to be grounded
4. Support cables to prevent fatigue failures at rigid terminations	4. Cantilever mount assemblies
5. Locate heavy item as close to load-bearing structures as possible	5. Use rivets to mount replaceable parts
6. Plate or protect metal parts from corrosion	6. Join leads without a support at their junction
7. Make deliberate considered effort in the thermal design	7. Mount components so another must be removed before the first is accessible
8. Design so interconnecting cabling can be prewired	
9. Furnish warning labels where danger exists	
10. Use standard components and parts whenever possible	

FIG. 37. Typical rf components. (*Microwave Associates, Burlington, Mass.*)

another, they are also used as circuit elements and as matching devices for ensuring that power may be transmitted with the least loss.

There are two basic types of devices used to transfer the power: transmission lines and waveguides. In general, transmission lines are smaller, lighter, and will conduct a wider range of frequencies than will waveguides. Transmission lines are of two types: the multiwire or twin-lead type, and the coaxial type in which a central conductor is surrounded by an outer conductor and separated from it by a dielectric material. Coaxial lines are obtainable in rigid, semiflexible, and flexible forms. Waveguides have greater power-handling ability and less attenuation and are considerably larger than coaxial lines. They are single-conductor devices that resemble a metal pipe or tube through which energy is transferred. Normally, both waveguide and coaxial cable are used in the frequency range between 1,000 and 10,000 Mc; for higher frequencies, waveguide is most practicable, and for lower, cable is employed.

Since the transfer elements also provide circuit and matching functions, not only are they available in a wide range of forms, but most systems will employ several different forms to obtain proper performance. This presents no great design problem since standardization in this field has been very good, so that adaptors and transitions are available for transferring from one transmission method to another. However, these transitions usually occupy considerable volume, which must be accounted for in the design. Also, most of the transfer elements require greater space than is normally considered for conventional wiring due to the size of the lines themselves, the large bend radii necessary, the connector size, and the length required for connector removal. Often additional length of run is required over a point-to-point distance in order to obtain a "matching" function. A typical equipment, illustrating the problems associated with rf "plumbing," is shown in Fig. 38; guides for the rf interconnection are given in Table 21.

Dramatic reductions in cost, weight, and size can be achieved through the application of printed-wiring techniques to rf circuitry packaging, as illustrated in Fig. 39. This process, known as the *strip-line technique*, uses a multilayer arrangement in

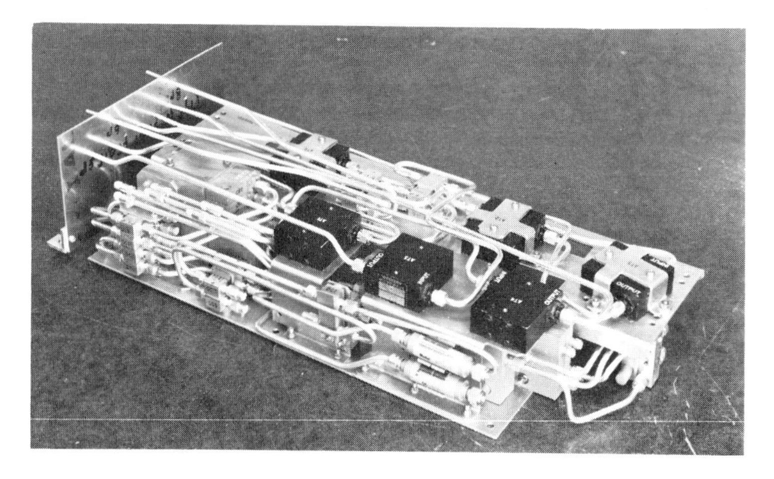

FIG. 38. Complex rf "plumbing," (*Westinghouse Electric Corp., Aerospace Division.*)

TABLE 21. Considerations for RF Interconnection [10]

Do

1. Separate or shield cables operating at low-power levels from those carrying high rf or control power to minimize interference
2. Select items from preferred or standard lists; the apparent advantages of a nonstandard item are generally offset by the maintenance of special fittings, test instrumentation, and so on
3. Use the least number of waveguide couplings possible; good preformed bends or flexible assemblies can contribute less to the overall system VSWR
4. Exercise extreme care in assembly and grounding of all fittings operating at high voltage to reduce corona and radiated noise; grounding should be done at several points for long runs
5. Use adjustable hanger straps or clamps to relieve strain on rigid lines; use additional resilient protection such as tubing or tape wrap for cable
6. Carefully remove all filings, loose solder, and similar foreign particles prior to assembly; cleanliness should be observed in all operations
7. Seal the ends of all lines and cables during storage to prevent the ingress of moisture or dirt; protect them from dents or bruises which can cause latent operating defects
8. Provide an adequate number of gas servicing vents for free circulation in pressurized systems; check for leaks periodically and make sure that the dehumidifier is operating adequately

Do not

1. Permit cable to be stored or installed in close proximity to "hot spots," such as heat dissipative tubes or resistors, steam or exhaust pipes
2. Assemble cables with magnesium oxide dielectric without discarding the first 2 or 3 in. and drying the ends thoroughly
3. Exert excessive forces in tightening fittings containing rubber or plastic as permanent deformation will result; occasional light retightening is preferred
4. Force flexible waveguides beyond their natural "stop" position; contact will be broken in the guide or at the flange
5. Subject ceramic insert pulse connectors to shock; they crack easily
6. Apply more heat than necessary in soldering, brazing, or welding connections; where possible, use crimped connections on cable braids to prevent distortion of the dielectric
7. Operate waveguides too close to their cutoff limits for high-power use; select a guide size so that the desired frequency will be close to midband

FIG. 39. Size reduction with strip-line techniques. (*Sanders Associates.*)

which the flat center conductor is located between two symmetrical ground planes. Strip line can be used to produce components or groups of interconnected components in one compact package by bulk processing techniques. Leads to and from these strip-line packages may still have to be conventional waveguide or coaxial cable, but the maze of interconnecting lines and the associated problems are considerably reduced.

Plug-in Capability. *General Considerations.* There are levels in all classes of electronic equipment at which a group of connections must be made or broken with the minimum amount of time, skilled labor, or specialized equipment. The mechanical device by which this is accomplished is called a connector. A connector consists basically of two halves: a plug and a receptacle. The plug is that half of the connector which is normally removed when the disconnection has been made. The receptacle is that half which is rigidly attached to a panel, chassis, or other supporting structure. Both halves of the assembly may be provided with either pin or socket contacts.

SAFETY. To minimize the possibility of electrical shock to personnel or the accidental shorting of contacts with resultant harm to components within an equipment, the following rules should be observed wherever possible:

1. One of the connector pins should be used for the grounding, not the shell of the connector. This is also good design practice from an operational point of view. The shell should be well grounded to the chassis.
2. The protruding, or male, contact pin should not contain the "hot" side of the circuit when it is withdrawn.
3. The socket, or female, contact should be used for the "hot" or power side of the circuit.
4. The socket contacts should be recessed as far as practicable into the connector insert.

POLARIZATION. Any connector used should have built into it a mechanism to prevent incorrect mating. The polarization is most effective when it assures that a plug and receptacle which are designed to be mated can be mated in only one position, and prevents the mating of the plug of one unit with the receptacle of the incorrect circuit even though both have connectors of the same size and configuration. This can be accomplished through the use of positioned keys and mating keyways, the shape of the protective shell, dissimilar-sized guide pins, or through a nonsymmetrical arrangement of contact barriers. Contact elements should never be used for purposes of alignment or polarization.

SPARE PINS. Military specifications require the number of spare pins or contacts shown in Table 22 to allow for any future circuit changes caused by modification; this

TABLE 22. Military Requirements for Spare Connector Contacts [8]

Total no. of contacts in connector	*Spare contacts*
Up to 25	2
26 to 100	4
101 and over	6

is of course not possible or necessary with items such as encapsulated modules which are considered not repairable or modifiable. In addition to pins left for modification, it is desirable to leave spare contacts for maintenance purposes. A good rule of thumb is to leave a minimum total of 10 percent spare contacts.

SHIELDED CIRCUITS. Signals at frequencies below 100 kc are not likely to experience crosstalk interference where the circuits are continued through connectors that do not contain coax contacts; connectors generally have a low capacitance between contacts. It is advisable, however, to space the signal leads and alternate with shield ground contacts. At frequencies over 100 kc, tests are recommended to determine whether the above procedure will be effective or whether a coax connector is advisable.

INSERTION AND WITHDRAWING FORCES. The force required to insert and remove connectors, although only of sufficient value to ensure proper low-resistance continuity on each contact, can reach very high values on large connectors or units with many contacts. Typical values for individual contacts on pin-and-socket-type (standard MS style) connectors are shown in Table 23. Printed-wiring-board connectors normally

TABLE 23. Contact Insertion and Withdrawal Forces [8]

Contact size	Force per contact, lb		
	Average	Maximum	Minimum
18	2.1	3	0.25
12	3.5	5	0.5
8	7	10	0.75
4	10.5	15	1.0
0	14	20	2

measure about 4 to 10 oz/contact for the smaller sizes and 16 to 20 oz/contact for the larger sizes. Many connectors contain some device such as lead screws or threaded couplings to apply mechanical advantage which will assist in overcoming the insertion and withdrawal force. If the connector selected does not have such a device, then jacking screws or board lifters should be provided. When neither of these is possible, consideration should be given to the use of two smaller connectors in place of the large one.

Printed-wiring Connectors. Connectors normally used on printed-wiring boards may be classified in one of the categories illustrated in Fig. 40 and described below.

PLUG-AND-RECEPTACLE CONNECTORS. Plug-and-receptacle connectors are probably the most used style. The pins may be round, bayonet-shaped, or a formed configuration, and they are secured into a plug which is attached to the board. The receptacle with mating sockets is mounted onto the chassis. The contact spacing here is usually 0.075 in. or greater.

EDGE-TYPE CONNECTORS. Edge-type connectors utilize printed contact surfaces on the board which are continuations of the wiring and which mate with contacts in the receptacle. The contacts may be single-spring ribbon, single fork (tuning fork), or double-spring ribbon as illustrated in Table 24. A single contact implies one that contacts either or both sides of a board and has one electrically common terminal on

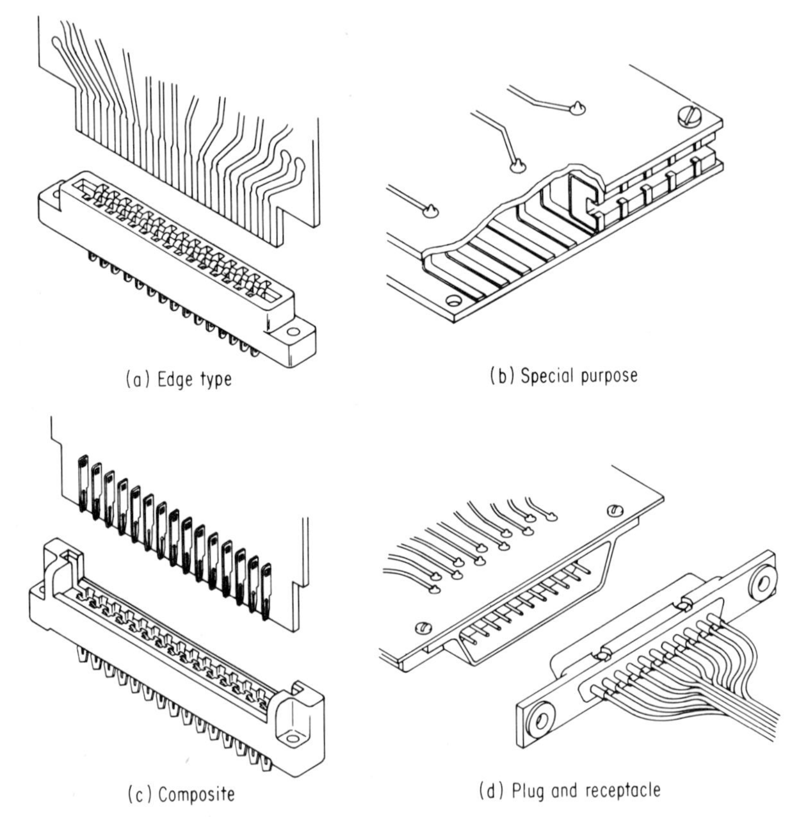

FIG. 40. Types of printed-wiring-board connectors.

TABLE 24. Comparison of Edge-connector Contact Types

Type	Advantages	Limitations
Tuning fork	1. Close spacing of contacts 2. High initial contact pressure 3. Double contact surface	1. Contact surface wears rapidly 2. Oversized board gives permanent set to contact
Spring ribbon	1. Large contact surface 2. Great resiliency, especially when bifurcated, results in contact under shock and vibration 3. Dual outputs per contact space possible 4. Insertion and removal forces low	1. Contacts are relatively fragile

the wiring end; a dual contact makes contact on both sides of the board and has two electrically separate terminals on the wiring end. Connectors of this type require very close control on overall board thickness and good adhesion of circuit to board. Dual-contact connectors allow circuit spacings down to 0.050 in. on standard models and closer on some special models. This style of connector tends to be less costly since no plug or plug-assembly operation is required.

COMPOSITE CONNECTORS. Also called *hermaphroditic connectors*, composite connectors consist of an adaptor or plug which mates with a receptacle, both of which have the same type of contact. The adaptor may be a set of individual contacts which are attached to a board, or an assembly containing the contacts which is attached to the board. Contact spacings here are generally 0.100 in. or greater.

SPECIAL-PURPOSE CONNECTORS. Connectors may be modified to fit an unusual situation or specially designed to perform a specific function. The connector illustrated has wraparound spring ribbon contacts and is used to interconnect two boards parallel to each other; both boards have printed-wiring terminations upon which the contacts press. This same style is available down to 0.050 in.

RETENTION OF BOARDS. Sufficient retention may be provided with card guides if the equipment will not experience high shock or vibration loadings. The guides also serve to lead the connector plug into the receptacle and provide a path for conductive heat transfer from board to case wall. Where high *g* loadings are involved, additional retention should be provided. This may be accomplished by elastomeric pads pressed upon the top edge of the board or by wedge (or cam) devices which press the board against a fixed surface.

CONNECTOR MOUNTING. The connector receptacle may be mounted above or below the chassis, but below mounting is preferred since this allows prewiring and cabling of the assembly. Sufficient float must be provided in the contacts and in the mounting to ensure that no mechanical stresses are placed on the mated contacts; this of course depends upon the tolerances allowed in connector and structure fabrication. If small components and high density are used, the required output contacts may "pin-limit" the board assembly; when subdividing the circuit into board groupings, this must be considered.

Module Connectors. To give an easily replaceable assembly, modules may be made pluggable; typical examples are noted below.

STANDARD MINIATURE CONNECTORS. These can often be used, as shown in Fig. 41*a*, with the connector body potted in with the rest of the components; the other half of the connector is mounted onto the back panel. Either half may carry the active contacts. The connectors are arranged in a set pattern, which may tend to inhibit the module design layout.

PREFORMED HEADERS. More versatility in module layout can be obtained with preformed headers since they may be designed for the specific package requirements. A typical example is shown in Fig. 41*b*. The use of a special header tends to increase cost.

ACTIVE PINS. With active pins used as the module output pins, the unit may be plugged into plated-through holes or eyelets in the back panel. An example is shown in Fig. 41*c*. Care must be taken to protect the active pins when handling a unit of this type. Modules designed with active pins mating with plated-through holes allow the closest output pin spacing and the most versatility in location of any plug-in module type.

STANDARD ROUND PINS. With correct spacing, a module with normal round pins may be plugged into active sockets which are soldered into the printed-wiring back panel; an example is shown in Fig. 41*d*. The output pins must of course be short and stiff enough so they do not deform when inserted; it is helpful to have the end of the pin radiused.

PREFORMED ASSEMBLIES. An assembly of this type is illustrated in Fig. 42. Units in this category allow considerable versatility in layout and module size, but tend to be uneconomical of volume when the modules are small.

RETENTION. Modules which are subject to high *g* loadings must be secured to the board other than by the connector. This may be accomplished with an adhesive, but then module replacement becomes difficult. Threaded fasteners are often employed

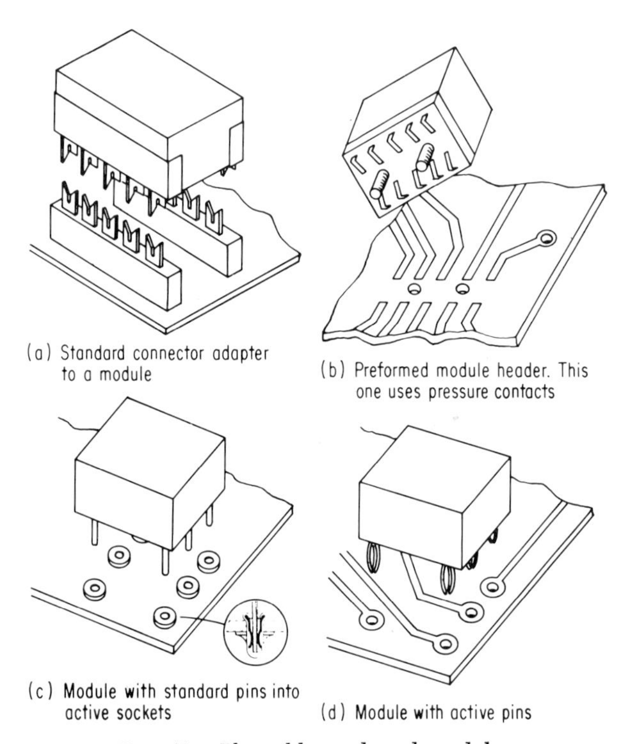

FIG. 41. Pluggable cordwood modules.

by molding studs or inserts into the module, or providing sleeves all the way through so longer screws may be used.

Chassis Connectors. Although chassis connectors may be obtained in many variations of construction and complexity, in usage they may be classified as rack-and-panel and cable-to-chassis connectors.

RACK-AND-PANEL CONNECTORS. Connectors of the rack-and-panel type are used for blind insertion of a chassis into a rack enclosure assembly; the chassis is most often mounted onto rails or other guides. As illustrated in Fig. 43, these connectors are usually rectangular in shape, with a shell which is used for polarization. One half of the connector is rigidly attached to the rack and the other half is firmly mounted to the removable chassis. Unless very close tolerance rails are used, relatively large guide pins should be placed in the rack to mate with sleeves in the chassis and provide lead-in positioning for proper connector mating.

CABLE-TO-CHASSIS CONNECTORS. One half of a connector in the cable-to-chassis category is firmly mounted to the chassis, and the other half forms the cable termination. The rack-and-panel type, with the addition of proper end bells on the plug half, may be used; or the so-called *MS series connectors* with a bulkhead-mounted receptacle can be employed. In either style, a cable clamp must be provided on the plug end to relieve strain on the wire to contact joint. A typical MS cable plug and mounted receptacle are shown in Fig. 44.

PACKAGING METHOD SELECTION

In any packaging design, the lowest level of sophistication that will satisfactorily accomplish the job should be used. This implies a complete understanding of the candidate packaging methods, the performance requirements of the package, and the

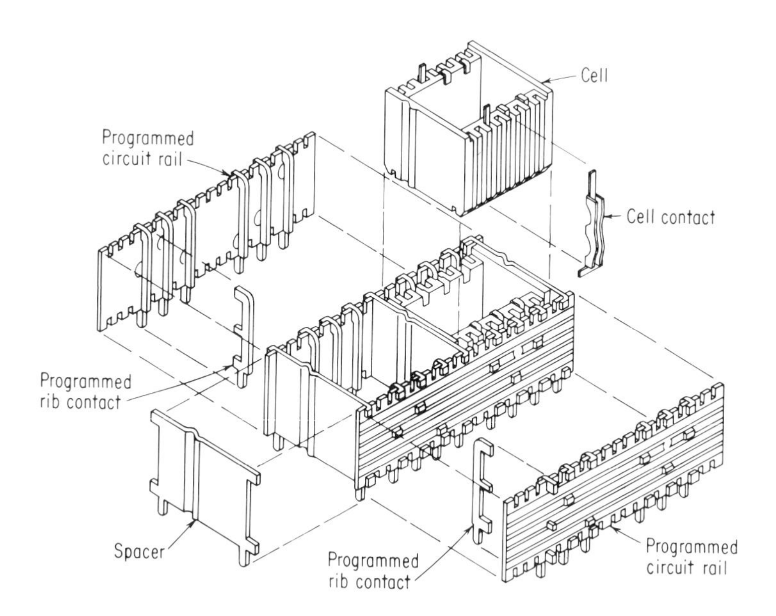

CELLS – Cells are modular units used to contain functional circuits. Length, width and height can be varied in increments of 0.100 in. to accommodate variations of modular assembly sizes. Cell walls are slotted in 0.100 in. increments to accommodate contacts which provide input-output connections to the circuits

CELL CONTACTS – Cell contacts are spring members which are retained in slots of the cell and soldered or welded to the input-output connections of the circuit. The cell contacts mate with the rib contacts when the cell is plugged into the receptacle formed by the circuit rails and spacers. The mated contacts, somewhat similar in design to the knife-blade switch, offer maximum reliability in that they consist of two independent spring systems with a total redundancy of four independent circuit paths

RIB CONTACTS – Rib contacts are metal stampings used to make connections between the baseboard, the cell contacts and the conductor circuit rails. Top and-bottom tabs of the contacts are used to secure the rib within slots at the top and bottom edges of the circuit rail. Rib contact tines are located on the same spacing as the circuit rail conductor paths and are programmed, in assembly, to suit interconnection requirements. The rib contact tine of the baseboard tine may be cut off if not required for interconnection

CIRCUIT RAILS – Circuit rails are held rigid by spacers and mounted on a printed-wiring board, in opposed fashion at 90°, to form two sides of a receptacle for the plug-in modular cells. Rail lengths are determined by the number and length of cells to be connected. Cell height or requirements for additional conductor paths for interconnections prescribe the rail height

SPACERS – Spacers are metal stampings used between cells and secured to circuit rails by bent-over tabs; together with the circuit rails they form receptacles for the individual cells. The spacers are equal in height to the rails and also act as rigidizing cross members

Fig. 42. Preformed module assembly. (*Amp-Meca.*)

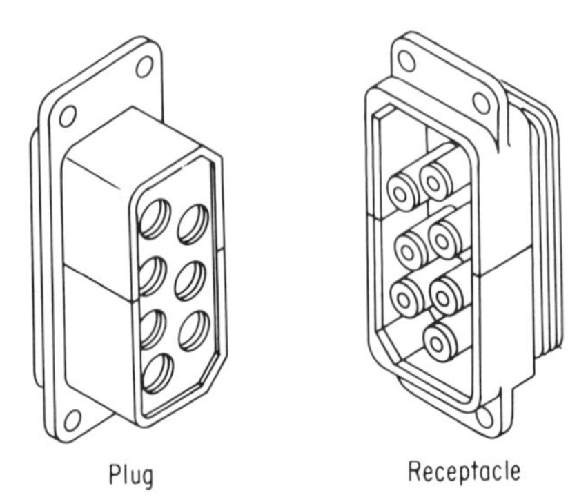

FIG. 43. Rack-and-panel connector.

method of integration of the package into the total system. To ensure that the designs are indeed practical and not unduly compressed as an exercise in miniaturization or compromised through some novelty technique, the design considerations as outlined below should be thoroughly explored in relation to the system being packaged. With this as background material, a merit-rating chart as described should be prepared to assist in arriving at an objective design decision.

Design Considerations. CUSTOMER SPECIFICATIONS AND PREFERENCES. The preferences and specifications of the customer must contribute to a large extent to selection of the packaging method. Specifications usually clearly define materials, component types, basic construction methods, and use environments to be encountered. If these guides cannot be followed, then it is necessary to establish the suitability of the substituted devices or method. In these cases, sufficient testing must be performed to provide proof data that the proposed property is indeed equivalent to that specified. In some cases it can be established that the specifications are general in nature and that certain provisions are not necessary to the unit function or may actually inhibit the design. Here waivers or exceptions to the requirements should be requested as soon as possible. It is of utmost importance that the nature of the requirements and restrictions be clearly understood. These should be established and outlined in a document agreed to by all persons concerned; this applies not only to

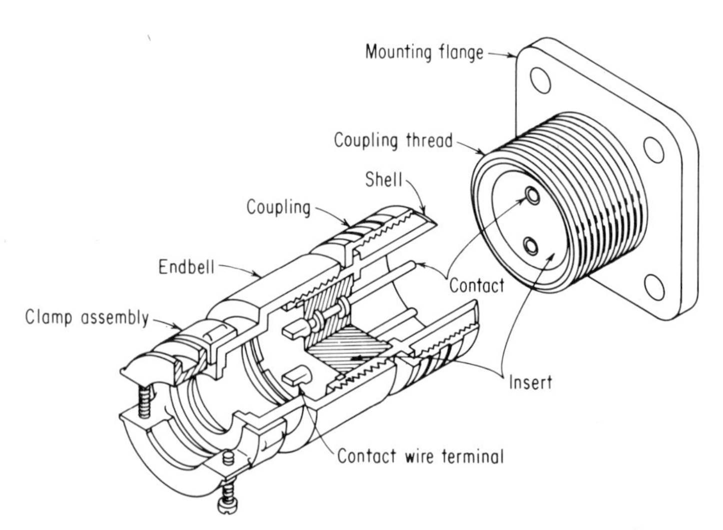

FIG. 44. Typical MS cable plug and mounted receptacle.

contacts with customers, but to work breakdowns between various groups within the supplying company as well. Time spent in initially defining the problem is well saved in reduction of misunderstandings and rework in later stages of construction.

PLANAR OR THREE-DIMENSIONAL PACKAGING. Fundamentally all packaging methods can be reduced to a variation of planar or three-dimensional packaging and a decision should be made as to the best type for the job. In terms of packaging sophistication, the simpler circuits should be placed planar on single-sided boards, and the most complicated ones designed to the three-dimensional concept.

PACKAGING DENSITY AND VOLUMETRIC EFFICIENCY. With a relatively normal conventional miniature component mix, densities close to those shown below can be obtained:

	Density, components/in.^{3}
Single-sided printed-wiring board	20–25
Double-sided printed-wiring board	25–30
Three-dimensional modules	40–50

These values are for individual unit assemblies of components, and cannot be applied on a system basis since they do not include volume for connectors, wiring, structure, etc. They must be divided by a factor anywhere between 2 and 10 to give reasonable overall system density; the divisor must be estimated based upon the amount of interconnection, the structural package required, the amount of accessibility desired, and like items. The actual density obtainable is dependent upon the component size and shape mix. If a great quantity of relays, crystals, or other large components are included in the circuitry, then modules are not practical and the board densities shown must be decreased. If massive rf components or power-supply cans are required, then planar chassis-mount construction should be used. Coax circuitry also tends to occupy a relatively large volume due to the minimum bend radii required with the cable and the space needed to engage and remove the coax connectors, as shown in Fig. 45. Circuitry with high power dissipation lends itself more easily to planar construction; a power density of about 1 watt/in.^{3} is the maximum that should be allowed in a cordwood module without special heat-sinking design. When the circuitry tends to be repetitive it is often advantageous to prepare and build a composite module, even though not all output pins and functions may be used in each application of that module.

Another measure of the component density obtainable on a system basis can be obtained from a determination of the system volumetric efficiency. The first step is to

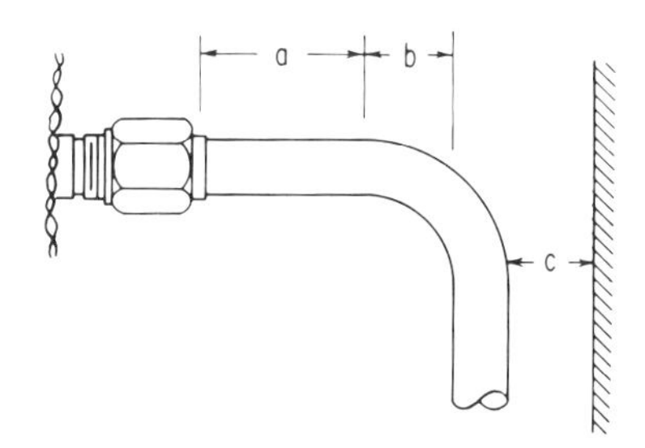

(a) Distance required before bend to protect cable-to-connector joint. Usually twice depth of nut.

(b) Bend radius, usually 5 to 10 cable diameters.

(c) Distance required to get complete separation of nut from fixed part. Approximately depth of nut.

FIG. 45. Space required to engage and remove coax connectors.

determine the volumetric efficiency of the first level of functional units such as cordwood modules and printed-wiring boards. Volumetric efficiency here is defined as:

$$\text{Unit } VE = \frac{\text{volume occupied by component bodies}}{\text{volume occupied by functional unit}}$$

Since some waste space is necessarily introduced in the process of assembling a functional unit, very rarely will this efficiency exceed 30 percent. Normally, cordwood modules will have an efficiency of about 20 to 25 percent, or one part electronic components plus three parts film, ribbon, and potting compound. Printed-wiring boards normally are somewhat lower in efficiency values.

In the next level of assembly, additional waste space is consumed in making a finished working system. The volumetric efficiency associated with this is:

$$\text{Integration } VE = \frac{\text{total volume occupied by functional unit}}{\text{total volume of system}}$$

Where the wiring is extremely complex, as in computer applications, or where open-type construction is used, as in high-power rack-and-panel assemblies, this efficiency may be as low as 10 percent. In other systems, such as receivers which do not require complex interwiring or which have minimum space between units and low power dissipation, the integration efficiency may get as high as 50 percent.

The overall system volumetric efficiency is then the product of the unit and integration volumetric efficiencies:

$$\text{Overall system } VE = \text{unit } VE \times \text{integration } VE$$

Even using high values of unit $VE = 30$ and integration $VE = 50$, the system VE is only 15 percent. In most practical cases, this is more likely to run about 2 to 4 percent for components on boards and about 5 to 7 percent for modules on boards.

INTERCONNECTION. Techniques that are normally used for connection within or among the various subassemblies are discussed in Table 25. Typical examples are shown in Fig. 46.

FIG. 46. Typical interconnection methods.

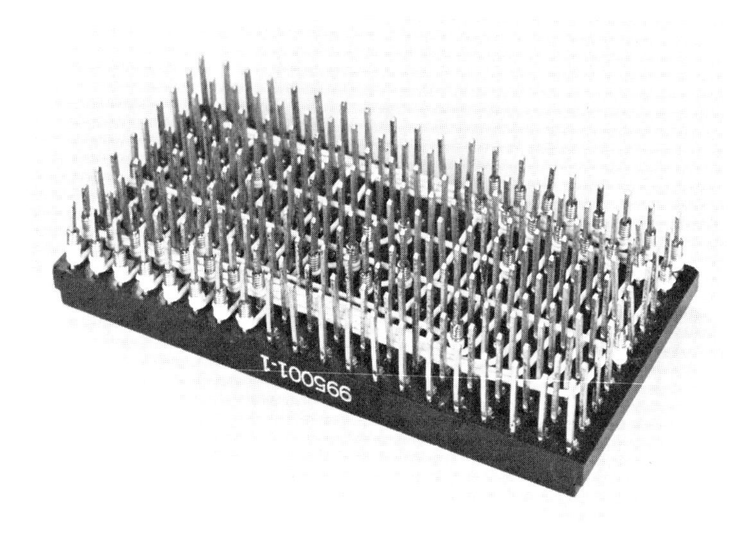

FIG. 46. Typical interconnection methods (*continued*).

TABLE 25. Interconnection Methods Available

Technique	*Characteristics*
Printed wiring	Usually used between components and small assemblies of components. May be single-sided, double-sided, or multilayer. Minimum standard grid spacing is 0.100 in., although 0.050 in. is fairly readily achieved. Components or modules replaceable. Minimum of tools required after board is made; artwork required for board manufacture. Circuit not easily changed
Soldered wire	Point-to-point connection; with insulated wire, crossovers are permissible. Most common method in use. Simple tools, medium skills. Spacing limited by size of wire and terminals used; with magnet wire to 0.050 in. Metallic junction; wire can be removed and reused to a limited extent. Quality depending upon operator if hand done; can be mechanized to minimize operator effect. Heat may cause damage. Relatively bulky. Circuit easily altered
Welded wire	May be point-to-point or matrix. Usual spacing 0.100 in., but often made on 0.050-in. centers. Requires special tools and skills; quality not as operator-dependent as solder. Fusion joint of parent metals; permanent joints, so repair is not readily made. High-temperature operation possible. Lightweight, small size
Flat cable	Replacement for cable harnesses. Lightweight, flexible, and small; may be soldered or welded at terminals. Can be designed to fit a space or bought in standard widths, conductor spacings, etc. The conductors are printed flat, or may be standard stranded wire
Solderless wrap . .	Multiple turns of wire around a square or rectangular post; contact through surface distortion and high interface pressure. Requires proper terminal design and special tools; may be automated. Quality largely controlled by tooling. Usual spacing 0.100 in., but can be performed to 0.075 in. Limited to solid wire; when removed, wire end cannot be reused
Crimped contact . .	Direct contact through deformation; used for wire terminations at lugs, pins, and removable contact elements. Requires special tools, terminals, and contacts; cannot be removed and reused. Joints usually preformed because of space required for tool. Quality depends upon tool use. One-at-a time connection. Relatively bulky. Spacings of contacts usually on 0.100-in. centers
Separable connectors	Pressure-type contact for single or multiwire connections. Wire termination to contacts may be solder, weld, crimp, or wrap. Maximum accessibility and interchangeability. Quality depends upon connector design and application. Contacts readily available to 0.050-in. centers. Assembly tends to be bulky
Screw terminals . .	Limited application to relatively large leads which are often removed. Pressure junction, subject to loosening under vibration. Easiest made, no tools; quality depends upon operator. Bulky

ELECTRICAL LIMITATIONS. The packaging methods considered must be compatible with system electrical requirements. Critical runs should be isolated by location, or introduced as shielded wires; high-frequency signals will require coaxial leads. These procedures require added volume for both the cables and their terminations. The wiring used must contain sufficient conductor material to carry the power levels involved; tables of allowable values are given in Chap. 2. In low-power circuits, long runs can result in line-voltage drops. If the ground return is not sufficiently large or properly applied, potential difference variations will occur between successive power taps and the ground. In many high-speed circuits, the propagation delays in the lines can exceed the delays in the processing elements themselves. A detailed discussion of the various electrical factors to be considered is given in Chap. 10.

HEAT DISSIPATION. A careful analysis should be made of the heat-removal characteristics of the techniques considered, and sufficient time should be allotted for this. Natural convection is effective if the cabinet volume is at least 6 in.^{3}/watt; over this, forced convection or heat exchangers must be used depending on wattage level.

In space environments, conductive cooling must be used and sufficiently large thermal conductors should be provided together with good thermal interfaces. In addition to determining the overall thermal level, the analysis should explore the suspected hot spots within the system to discover whether local heat sinks are required. Details of performing a thermal analysis are given in Chap. 11.

SENSITIVE COMPONENTS. Those items which cause or are susceptible to interference from if, rf, magnetic, or similar energy sources must be isolated by distance or by suitable shielding. In very dense packaging, separation is not possible and shielding must be used. This may take the form of grounding areas between lines on the surface of printed-wiring boards or complete shielding planes in multilayer circuitry In chassis construction, individual cans can be used, or a multicompartment box con taining and isolating the various stages, as illustrated in Fig. 47, is advantageous. Fo embedded modules, metallizing of the outside surface is an effective method.

FIG. 47. Multicompartment shielding case. (*General Electric Company, Light Military Electronics Dept.*)

Certain components can sustain a characteristic change in the potting process. Field-effect transistors are particularly sensitive in this regard; these should be precoated with some resilient or energy-absorbing material, checked for correct operation, and then subjected to the final encapsulation. Items containing nickel-iron cores often shift in performance due to the change in magnetic properties caused by the pressures generated during the resin-curing process; resilient prepotting is helpful here also. Subcircuits containing coils and delicately tuned for operation at high frequency suffer serious Q impairment when potted in the normal manner. Also, the Q value does not remain constant with temperature changes but increases as the temperature is increased. A lightweight (2 to 5 lb/ft^3) polyurethane foam can often be used to advantage for these circuits.

In the welding process, the ferromagnetic leads and intraconnection ribbon can become magnetized due to the induced field of the welding machine. If the units are sensitive in this regard, they must be degaussed after fabrication or made with nonmagnetic weldable lead materials. It is also possible to injure a transistor through back-biasing the junction during the welding process unless the welding machine is properly protected through insulated electrodes or shunting diodes.

BREADBOARDS. Although primarily intended to explore the circuitry characteristics, breadboards should also be usable to demonstrate the packaging characteristics as well. It is therefore advantageous to make the breadboard as close to the planned final configuration as possible so that the unknown areas of the design may be resolved. It may be necessary to package at a much lower density level and in many cases use substitute components so that the unit may be made rapidly and be easily reworked, but a form generally following that of the final package will save time later during design and fabrication. The time available and the level of confidence in the circuitry design will of course determine the degree to which this philosophy can be followed.

SCHEDULE REQUIREMENTS. If the lead time on the job is very short, it is best to use methods which are known and have been previously proved; new techniques and untried processes used on quick-reaction-type jobs can lead to disaster. Component availability must be balanced against time; remember that welding requires components with special leads, and items with specified characteristics or configurations always take longer than expected to obtain. If special tooling is required, sufficient time must be allowed for its fabrication.

QUANTITIES INVOLVED. Certain methods, such as printed-wiring boards, are suitable for both low and high levels of production using hand soldering for a few items and wave machines for high production rates. Other methods, which require elaborate tooling for first-piece fabrication, can be justified only by large quantities or unique packaging requirements. Also, repetitive operations which must be performed on a one-at-a-time basis tend to be more costly than operations that can be performed on a batch basis. If large quantities are involved, the method selected should readily lend itself to automatic assembly, connection, and testing.

Selection of Method. It would be ideal if a basic chart were available which listed all the possible packaging techniques and their level of suitability in numerical form, similar to a chart of physical constants of material. However, no single rating system can be realistically applied to all systems; the differences that exist between various system requirements, manufacturers' capabilities, and customer preferences make this impossible. It is possible to approach this with the development of charts based upon a specific set of requirements, capabilities, or preferences. With a good understanding of the characteristics of the basic techniques and sufficient knowledge of the system application, the following procedure can be used to develop merit-rating charts. This method will tend to give order to the selection process and will result in a more objective design decision. It should be noted, however, that the final choice is a personal one (through the assignment of ratings and weighting factors) based on an understanding of requirements, environments, customer, life, cost acceptability, and time available.

RATING CHARTS. From a study of the design considerations previously discussed, it is usually possible to discard techniques which are unsuitable for the application, and arrive at two or more techniques and variations which would satisfy most of the requirements. A table should then be developed to aid in the final choice of the one best suited to the application.

The table derives a single number representing the desirability of using the particular configuration; the higher the number, the greater the desirability. In arriving at this number, all factors which concern that particular design should be considered. Although only those items which are pertinent to the decision need be included in the chart, experience has shown that as a minimum the following items should be explored:

Design layout and drawing
Manufacturability
Repairability
Maintainability
Logistics
Physical characteristics
Environmental compatibility
Relative cost
Reliability

The ratings that these factors assume in any given system are composed of two parts:

1. Basic rating—the rating applied due to the inherent or inferred attributes of the technique under consideration. These attributes should be determined as objectively as possible after conducting the maximum research and study permissible in the time available.

Examples would be the better environmental capability of encapsulated modules over unencapsulated ones, and the lower cost of wave-soldered boards over hand-soldered ones.

2. Modifier—the rating applied based upon knowledge of the application and used to weight the basic ratings to suit the application. For example, capability to survive high-vibration inputs should not influence the selection as much for ground-based equipment as for spacecraft systems. A sample rating chart is shown in Fig. 48. Only

Rating factor applicable	Modifier (a)	Welded cord-wood modules		Soldered cord-wood modules		Printed-wiring boards	
		Basic rating (b)	Merit rating (ab)	Basic rating (b)	Merit rating (ab)	Basic rating (b)	Merit rating (ab)
Manufacturability	1	2	2	3	3	4	4
Repairability	2	2	4	1	2	4	8
Physical characteristics:							
Size, weight	3	4	12	3	9	2	6
Thermal	2	2	4	2	4	3	6
Cost	1	1	1	2	2	4	4
Reliability	3	4	12	3	9	3	9
Merit rating for application	...	...	35	...	29	...	37

Rating scale used:

Modifier—(a)		Basic rating (b)	
Helpful	1	Poor	0
Desirable	2	Fair	1
Important	3	Good	2
		Very good	3
		Excellent	4

FIG. 48. Sample packaging merit rating chart.

those rating factors applicable to the specific application were included. Based upon the scales selected and the ratings applied in this particular case, printed-wiring boards should be used.

RATING FACTORS. Table 26 gives discussions of the rating factors for use as a check-list in determining their applicability and merit when preparing a packaging method rating chart. In Table 27, the rating factors are listed in order of importance for four classes of application; different individuals may assign different levels of importance to some items, but this list serves as a representative order.

COMPONENT AND HARDWARE INFORMATION

The information presented here is intended to introduce the more common discrete components used and the hardware items normally employed in packaging them. There are so many types of components tailored to specific circuit functions and so many variations of each type that it is not practicable in this volume to describe them all. Many items are described in enough detail so that no further research is required; for those not so completely covered, the information herein contained will result in a more intelligent and fruitful search for the specific details desired. In general, the items described are of the types covered by military specifications since it is felt that these ensure the highest reliability and performance levels. In most cases, components of comparable values and similar characteristics but providing less en-

TABLE 26. Checklist for Preparation of a Packaging Rating Chart

Factor	Characteristic	Discussion of trade-offs
Design	Complexity	Do design specifications exist? Development required? Versatile enough to cover all items in system?
	Drawings	What are layout requirements? Special skills required? Drawing turnaround time? Separate manufacturing or tooling drawings needed? Can artwork be automated?
Manufacturability	Quantity	Model, limited quantity, or large production quantity to be made (this affects tooling and manufacturing method used)? Are joints made one at a time as in welded cordwood or all at one time as in printed wiring? Does quantity justify setting up a line operation, or is it possible in this design? Can design be broken into separate subassemblies? Do necessary skills and manufacturing facilities exist?
	Automation	Is design adaptable to automation? With standard equipment or especially developed equipment? How much tooling is required? Hard or soft?
	Components	Are components of nonstandard material, size, shape, or function required? What is component availability? Are weldable leads needed? How are components secured in equipment? Can components be set in with automatic assembly tools?
	Schedule	Can design be made in time available? What is parts procurement lead time? Manufacturing development required?
	Inspection and testing	Are assemblies inspectable? Does good process control make inspection unnecessary? Are built-in tests included? Is special test equipment required? What are test levels?
Repairability	Philosophy	Is repair done at component, module, or subassembly level? How complicated? Done by operator, repair shop, or return to manufacturer?
	In-house	Easy repair and rework at all levels during manufacture lower cost, increase yield, and help meet schedules
	In use	How complicated are steps required if a component fails? Are special skills required to repair? Are complicated instructions needed? Does repair lessen environmental protection?
Maintainability	Modularity	What is the replaceable unit level—the complete assembly or smaller subassemblies? How difficult is fault isolation? Are parts and modules easily accessible? How difficult is module removal?
	Requirements	Are any special skills or tools needed? Is required procedure evident? Is a special controlled-environment maintenance area required? How much time is required to maintain equipment? How long between maintenance operations?
Logistics	. . . .	What is the lowest-level subassembly "throw-away" cost? Is this reasonable to end use? Are hardware and assembly items standard or special? Are subassemblies multiple-use items?
Physical characteristics	Size	Consider size of overall assembly, not just small piece. Remember that connectors and interconnection consume considerable volume. As a rough guide, components occupy about 5% of total volume
	Weight	Consider weight required in structure as well as electronics. Include weight to communicate with the outside world. Can one man handle it? A nominal density figure is 45 lb/ft^3
	Thermal	How efficient is heat transfer? Do components operate within acceptable limits? Is thermal design simple or complicated? Does design give a weight penalty?
Environmental protection	. . . .	Does design satisfy all basic use environments—minimal or better? Does protection hurt in other areas? What is time required to get this protection? Is it compatible with schedule? How well protected from handling damage, shipping shock?
Costs	. . . .	Is design time minimum? Are drawings made the least expensive way? Are raw materials and hardware items reasonable, or are costly assembly parts required? Are manufacturing costs recurring or one-time items? Are processes the least expensive? How complicated is testing?
Reliability	. . . .	Are components known to have high reliability—actual or calculated? How many electrical junctions (joints) in system, and how are they made? Does the design have structural integrity? Are components properly derated, shielded, etc.? How will unit operate in its intended environment?

TABLE 27. Order of Importance of Rating Factors for Four Classes of Equipment *

Missile and spacecraft	Shipborne and military ground	Airborne	Commercial
Reliability	Maintainability	Reliability	Cost
Physical characteristics	Reliability	Physical characteristics	Reliability
Environmental protection	Repairability	Environmental protection	Maintainability
Cost	Logistics	Maintainability	Repairability
Manufacturability ..	Cost	Cost	Logistics
Design	Manufacturability	Repairability	Manufacturability
Repairability	Design	Logistics	Design
Maintainability	Environmental protection	Manufacturability	Physical characteristics
Logistics	Physical characteristics	Design	Environmental protection

* Listed in descending order of importance.

vironmental protection are available in commercial quality at a lower cost. Very large or massive components are not discussed since it is assumed that the reader is basically interested in the so-called *high-density packaging techniques* earlier described. The highest level of component sophistication covered is the transistor; for more advanced components see Chap. 9.

It is obviously desirable from a cost, manufacturing, and maintenance standpoint to use standardized components and parts wherever possible. However, due to rapid changes and improvements in the state of the art, and in order to meet unusual performance or reliability requirements, deviations from the standards may be necessary. In the formulation of a selection policy for any equipment, the possible use of standard items should be given first consideration. In any event, regardless of whether standard, nonstandard, or a combination of the two is selected, the goal should be to minimize the number of different types of components used.

One important aspect of component selection is proper derating, or reducing the application stresses of the components used in the design; this can result in substantial improvements in reliability and performance. Once a circuit is formulated, each part should be examined to ensure that no rating is ever exceeded under any likely combination of circumstances. A good overall design derating policy is to derate parts by approximately 50 percent of the manufacturer's recommended maximum application rating; often in high-reliability systems, parts are derated by as much as 90 percent. Some examples of derating factors are given in Table 28. A further discussion of derating is given in Chap. 10.

As an aid in relating component lead diameter to standard wire size, the information in Table 29 will be useful. Table 30 shows a ready translation between numbers in exponential form and their name and prefix.

Component Information. Component parts which primarily perform an electrical function in the circuit are described below.

Passive Components. By definition, *passive components* are items which do not require any input other than a signal to perform their function. These are represented by resistors, capacitors, and inductive components.

RESISTORS. Both fixed and variable resistors, of the types most widely used in electronic equipments, can be grouped into one of three general types: composition, film, or wire-wound. The composition type is a mixture of resistive material and a binder which is molded to shape and the proper resistance. The film type is composed of a resistive film deposited on a substrate (a typical fixed film type is shown in Fig. 49*a*), and the wire-wound type (as illustrated by the variable resistor in Fig. 49*b*)

TABLE 28. Typical Component Derating Factors

Part type	Recommended derating factor		
	Power, %	Voltage, %	Temp., °C
Capacitors		33–50	
Resistors	50		
Semiconductors, germanium:			
Diodes, MIL-E-1 or MIL-S-19500	50*	50	60
Transistors, MIL-S-19500	50*	50	60
Semiconductors, silicon:			
Diodes, MIL-E-1 or MIL-S-19500	50*	50–80	100–125
Transistors, MIL-S-19500	50*	50–80	100–125
Transformers, MIL-T-27A	25		
Tubes, electron	25	.....	110

* Depending on heat-sink condition for power devices.

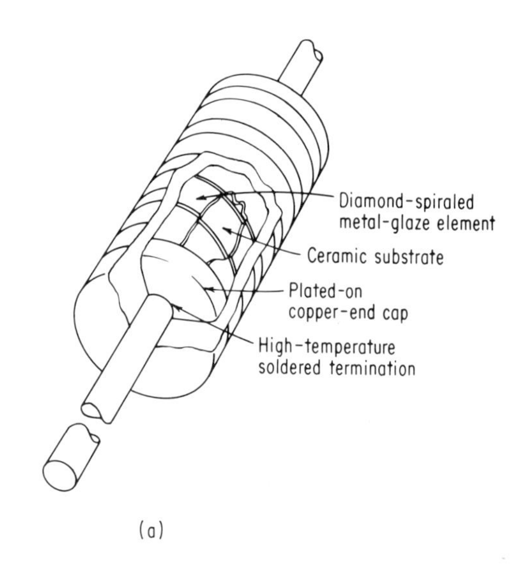

(a)

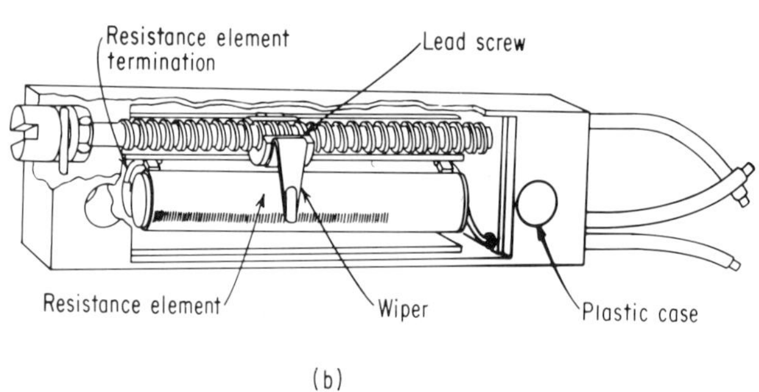

(b)

Fig. 49. Typical resistor construction. (*a*) Fixed-film type. (*IRC, Inc.*) (*b*) Variable wire-wound type. (*Amphenol Controls Division.*)

TABLE 29. American Wire Gauge Chart

Solid wire							
AWG	Diameter		Cross section, in.^{2}	Ohms/ 1,000 ft at 20°C*	Pounds/ 1,000 ft*	Feet/ pound*
	Inches	Mils				
8	0.1235	128.5	0.01297	0.6281	49.98	20.01
9	0.1144	114.4	0.01028	0.7925	39.62	25.24
10	0.1019	101.9	0.008155	0.9988	31.43	31.82
11	0.0907	90.7	0.00646	1.26	24.9	40.2
12	0.0808	80.8	0.00513	1.59	19.8	50.6
13	0.0720	72.0	0.00407	2.00	15.7	63.7
14	0.0641	64.1	0.00323	2.52	12.4	80.4
15	0.0571	57.1	0.00256	3.18	9.87	101
16	0.0508	50.8	0.00203	4.02	7.81	128
17	0.0453	45.3	0.00161	5.05	6.21	161
18	0.0403	40.3	0.00128	6.39	4.92	203
19	0.0359	35.9	0.00101	8.05	3.90	256
20	0.0320	32.0	0.000804	10.1	3.10	323
21	0.0235	28.5	0.000638	12.8	2.46	407
22	0.0253	25.3	0.000503	16.2	1.94	516
23	0.0226	22.6	0.000401	20.3	1.55	647
24	0.0201	20.1	0.000317	25.7	1.22	818
25	0.0179	17.9	0.000252	32.4	0.970	1,030
26	0.0159	15.9	0.000199	41.0	0.765	1,310
27	0.0142	14.2	0.000158	51.4	0.610	1,640
28	0.0126	12.6	0.000125	65.3	0.481	2,080
29	0.0113	11.3	0.000100	81.2	0.387	2,590
30	0.0100	10.0	0.0000785	104	0.303	3,300
31	0.0089	8.9	0.0000622	131	0.240	4,170
32	0.0080	8.0	0.0000503	162	0.194	5,160
33	0.0071	7.1	0.0000396	206	0.153	6,550
34	0.0063	6.3	0.0000312	281	0.120	8,320
35	0.0056	5.6	0.0000246	331	0.0949	10,600
36	0.0050	5.0	0.0000196	415	0.0757	13,200
37	0.0045	4.5	0.0000159	512	0.0613	16,300
38	0.0040	4.0	0.0000125	648	0.0484	20,600
39	0.0035	3.5	0.00000962	847	0.0371	27,000
40	0.0031	3.1	0.00000755	1,080	0.0291	34,400
41	0.0028	2.8	0.00000616	1,320	0.0237	43,900
42	0.0025	2.5	0.00000491	1,660	0.0189	52,800
43	0.0022	2.2	0.00000380	2,140	0.0147	68,300
44	0.0020	2.0	0.00000314	2,590	0.0121	82,600
45	0.00176	1.76	0.00000243	3,350	0.00938	107,000
46	0.00157	1.57	0.00000194	4,210	0.00746	134,000
47	0.00140	1.40	0.00000154	5,290	0.00593	169,000
48	0.00124	1.24	0.00000121	6,750	0.00465	215,000
49	0.00111	1.11	0.000000968	8,420	0.00373	268,000
50	0.00099	0.99	0.000000770	10,600	0.00297	337,000
51	0.00088	0.88	0.000000608	13,400	0.00234	427,000
52	0.00078	0.78	0.000000478	17,000	0.00184	543,000
53	0.00070	0.70	0.000000385	21,200	0.00148	676,000
54	0.00062	0.62	0.000000302	26,900	0.00116	862,000
55	0.00055	0.55	0.000000238	34,300	0.000916	1,090,000
56	0.00049	0.49	0.000000189	43,200	0.000727	1,380,000

TABLE 29. American Wire Gauge Chart (*Continued*)

Stranded wire							
Equivalent AWG No.	AWG nominal diameter	Number of strands per strand size (AWG)†					
		7	10	16	26	41	65
12	0.0808	. . .	. . .	. . .	. . .	28	30
14	0.0641	. . .	. . .	. . .	28	30	32
16	0.0508	24	26	28	30	32	34
18	0.0403	26	28	30	32	34	36
20	0.0320	28	30	. . .	34	36	
22	0.0253	30	32	34	36		
24	0.0201	32	. . .	36	. . .	40	
26	0.0159	34	36				
28	0.0126	36	. . .	40			
30	0.0100	38					
32	0.0080	40					

* Annealed copper wire.
† Sizes and strandings in common use.

TABLE 30. Numbers, Exponential Form, and Name*

Name of number	Exponential form
Googol	1.0×10^{100}
Vigintillion	1.0×10^{63}
Novemdecillion	1.0×10^{60}
Octodecillion	1.0×10^{57}
Septendecillion	1.0×10^{54}
Sexdecillion	1.0×10^{51}
Quindecillion	1.0×10^{48}
Quattuordecillion	1.0×10^{45}
Tredecillion	1.0×10^{42}
Duodecillion	1.0×10^{39}
Undecillion	1.0×10^{36}
Decillion	1.0×10^{33}
Nonillion	1.0×10^{30}
Octillion	1.0×10^{27}
Septillion	1.0×10^{24}
Sextillion	1.0×10^{21}

Name of number	Number	Exponential form	Symbol	Prefix
Quintillion	1,000,000,000,000,000,000.	1.0×10^{18}		
Quadrillion	1,000,000,000,000,000.	1.0×10^{15}		
Trillion	1,000,000,000,000.	1.0×10^{12}	T	Tera-
Billion	1,000,000,000.	1.0×10^{9}	G	Giga-
Million	1,000,000.	1.0×10^{6}	M	Mega-
Thousand	1,000.	1.0×10^{3}	k	Kilo-
Hundred	100.	1.0×10^{2}	h	Hecto-
Ten	10.	1.0×10^{1}	da	Deka-
Unit	1.	1.0×10^{0}		Deci-
Tenth	0.1	1.0×10^{-1}	d	Centi-
Hundredth	0.01	1.0×10^{-2}	c	Milli-
Thousandth	0.001	1.0×10^{-3}	m	Micro-
Millionth	0.000001	1.0×10^{-6}	μ	Nano-
Billionth	0.000000001	1.0×10^{-9}	n	Pico-
Trillionth	0.000000000001	1.0×10^{-12}	p	Femto-
Quadrillionth	0.000000000000001	1.0×10^{-15}	f	
Quintillionth	0.000000000000000001	1.0×10^{-18}	a	Atto-

* From *Design News,* October, 1966.

has a resistance wire wound on an insulating coil. The general characteristics and principal applications of these types are given in Table 31. The relative costs of the various types are listed in Fig. 50.

It is preferable to select resistors from lists of standard types and values. Standard values obtainable in the more common forms of resistors are given in Table 32. For cost purposes, closer tolerances than required should be avoided, but it must be realized that the normally specified tolerance describing the resistor is not the total

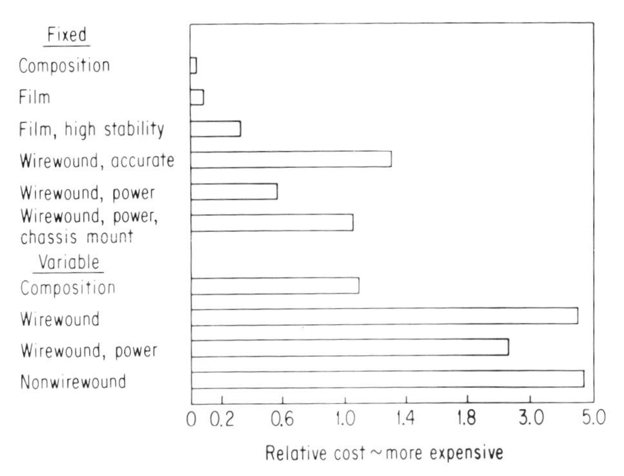

FIG. 50. Relative costs of standard resistor types.[12]

TABLE 31. Characteristics and Applications of Fixed Resistors [12]

Resistor types	Characteristics	Application
Composition	Small, inexpensive, good reliability when properly used. Poor stability, poor noise characteristics. Effective resistance reduced at higher frequencies	General-purpose uses where initial stability need be no better than ±5% and long-term stability no better than ±20% under rated conditions
Film	High stability, low tolerance, low environmental changes, low noise, excellent high-frequency characteristics. More expensive than composition types	Uses requiring high stability, where ac frequency requirements are critical, or for higher temperatures
Wire-wound	High stability, larger than others, good reliability, poor high-frequency characteristics above 50 kc	Used where ac frequency performance is not critical, where high cost and size are not important but stability is required

variation that can be expected in application. The specified tolerance is applied before the resistor is used; heat dissipation, impressed voltage, temperature, time, and moisture all contribute to a change in the resistive value, and may cause it to exceed the noted tolerance. Carbon resistors are the least stable (a ±5 percent unit may actually vary ±15 to 20 percent), deposited carbon resistors are quite stable, and wire-wound and metal film resistors are most stable. High-temperature operation is the largest contributor to variation; if extreme accuracy is required, operation well below specified temperatures is necessary. Do not fail to observe derating rules,

TABLE 32. Standard Resistor Values

Film, metal, or carbon, 1 to 10 decade						Carbon composition and power, wire-wound (±5%), 1 to 10 decade		
1.00	1.47	2.15	3.16	4.64	6.81	10	22	47
1.02	1.50	2.21	3.24	4.75	6.98			
1.05	1.54	2.26	3.32	4.87	7.15	11	24	51
1.07	1.58	2.32	3.40	4.99	7.32			
1.10	1.62	2.37	3.48	5.11	7.50	12	27	56
1.13	1.65	2.43	3.57	5.23	7.68			
1.15	1.69	2.49	3.65	5.36	7.87	13	30	62
1.18	1.74	2.55	3.74	5.49	8.06			
1.21	1.78	2.61	3.83	5.62	8.25	15	33	68
1.24	1.82	2.67	3.92	5.76	8.45			
1.27	1.87	2.74	4.02	5.90	8.66	16	36	75
1.30	1.91	2.80	4.12	6.04	8.87			
1.33	1.96	2.87	4.22	6.19	9.09	18	39	82
1.37	2.00	2.94	4.32	6.34	9.31			
1.40	2.05	3.01	4.42	6.49	9.53	20	43	91
1.43	2.10	3.09	4.53	6.65	9.76			

both for accuracy and for longer life under moderate temperatures. Derating curves for typical resistors are shown in Fig. 51.

The resistance value and specified tolerance of fixed resistors are indicated by color bands marked on the body of the part. The location and band designation are shown in Fig. 52. The color code used is shown in Table 33.

Data on typical fixed resistors are given in Table 34, and data on variable resistors are given in Table 35.

CAPACITORS. Energy-storage components used to accumulate energy through long time periods and to discharge the energy over longer or shorter periods are called capacitors. They consist essentially of two conductors separated by an insulator (dielectric), and are usually classified according to the dielectric used. The most widely used dieletrics are: (1) glass and mica, (2) electrolytic, (3) paper and plastic, (4) ceramic, (5) air, and (6) vacuum. Construction details of some of these are

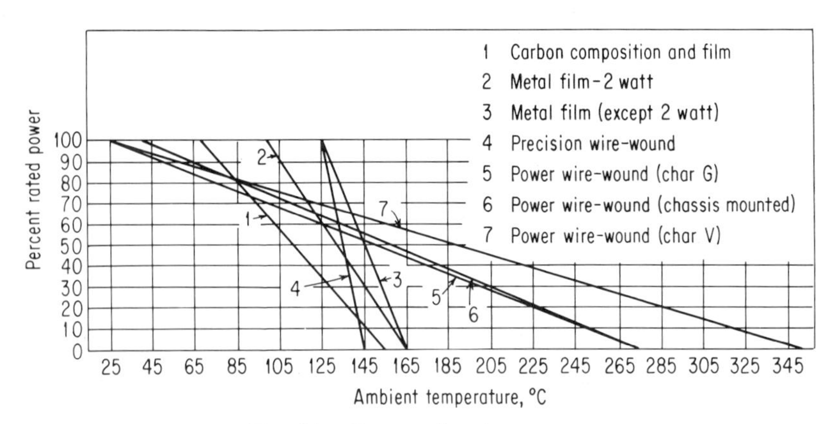

FIG. 51. Resistor derating curves.

TABLE 33. Color Code for Composition-type and Film-type Resistors (MIL-STD-221)

Band A *		Band B †		Band C ‡		Band D §		Band E ¶		
Color	First significant figure	Color	Second significant figure	Color	Multiplier	Color	Resistance tolerance, percent	Color	Failure rate level**	Terminal
Black	0	Black	0	Black	1			Brown ...	M	
Brown ...	1	Brown	1	Brown	10			Red	P	
Red	2	Red	2	Red	100			Orange ..	R	
Orange ...	3	Orange	3	Orange	1,000			Yellow ...	S	
Yellow ...	4	Yellow	4	Yellow	10,000	Silver ...	±10 (composition type only)	White ...	....	Solderable
Green ...	5	Green	5	Green	100,000	Gold	±5			
Blue	6	Blue	6	Blue	1,000,000	Red	±2 (not applicable to established reliability)			
Purple (violet)	7	Purple (violet) ..	7							
Gray	8	Gray	8	Silver	0.01					
White	9	White	9	Gold	0.1					

* The first significant figure of the resistance value. (Bands A through D shall be of equal width.)
† The second significant figure of the resistance value.
‡ The multiplier. (The multiplier is the factor by which the two significant figures are multiplied to yield the nominal resistance value.)
§ The resistance tolerance.
¶ When used on composition resistors, band E indicates established reliability failure-rate level. On film resistors, this band shall be approximatey 1½ times the width of other bands, and indicates type of terminal.
** M = 1, P = 0.1, R = 0.01, S = 0.001.

TABLE 34. Data on Typical Fixed Resistors[12]

Type	Styles available in standard	Power and max. voltage ratings	Resistance tolerance, ±%	Ohmic range, ohms	Temperature range, °C*	Max. body size, in.	Configuration
Composition (insulated)	RC05	1/8 w/150 v	5, 10	10 to 22 M	70–130	0.160 × 0.066	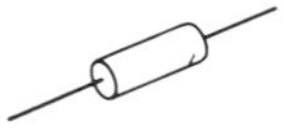
	RC07	1/4 w/250 v	5, 10	2.7 to 22 M	70–150	0.281 × 0.098	
	RC20	1/2 w/350 v	5, 10	2.7 to 22 M	70–130	0.416 × 0.161	
	RC32	1 w/500 v	5, 10	2.7 to 22 M	70–130	0.593 × 0.240	
	RC42	2 w/500 v	5, 10	10 to 22 M	70–130	0.728 × 0.336	
Film (high-stability)	RN50	1/20 w/200 v	1.0, 0.1	10 to 0.100 M	125–175	0.160 × 0.070	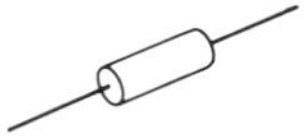
	RN55	1/10–1/8 w/200 v	1.0, 0.1	10 to 0.301 M	70–165; 125–175	0.281 × 0.140	
	RN60	1/8–1/4 w/300 v	1.0, 0.1	10 to 1 M	70–165; 125–175	0.437 × 0.165	
	RN65	1/4–1/2 w/350 v	1.0, 0.1	10 to 2 M	70–165; 125–175	0.656 × 0.249	
	RN70	1/2–3/4 w/500 v	1.0, 0.1	10 to 2.49 M	70–165; 125–175	0.875 × 0.328	
	RN75	1 w/500 v	1.0, 0.1	10 to 5.11 M	70–150; 125–175	1.124 × 0.437	
Film (power-type)	RD60	2 w/350 v	2	10 to 0.178 M	25–275	0.656 × 0.235	
	RD65	4 w/500 v	2	10 to 464 M	25–275	1.000 × 0.359	
	RD70	8 w/750 v	2	2.15 to 1.0 M	25–275	2.124 × 0.359	
Wirewound (accurate)	RB52	1/2 w/600 v	0.1, 1	0.1 to 1.5 M	125–145	1.062 × 0.406	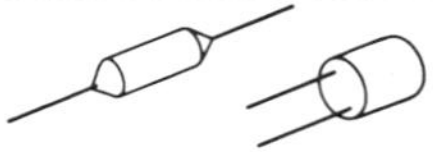
	RB53	1/3 w/300 v	0.1, 1	0.1 to 0.750 M	125–145	0.812 × 0.418	
	RB55	0.15 w	0.1, 1	0.1 to 0.226 M	125–145	0.562 × 0.281	
	RB71	1/8 w	0.1, 1	0.1 to 0.100 M	125–145	0.343 × 0.281	
Wirewound (power-type)	RW29	11 w	5, 10	0.10 to 5.6 K	25–350	1.812 × 0.500	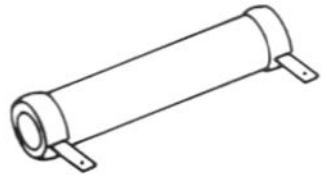
	RW31	14 w	5, 10	0.10 to 6,8 K	25–350	1.562 × 0.594	
	RW33	26 w	5, 10	0.10 to 18 K	25–350	3.062 × 0.594	
	RW35	55 w	5, 10	0.10 to 39 K	25–350	4.062 × 0.906	
	RW37	113 w	5, 10	0.10 to 82 K	25–350	6.062 × 1.312	
	RW38	159 w	5, 10	0.10 to 0.15 M	25–350	8.062 × 1.312	
	RW47	210 w	5, 10	0.10 to 0.18 M	25–350	10.562 × 1.312	
	RW55	7 w	5, 10	0.10 to 8.2 K	25–350	1.469 × 0.562	
	RW56	14 w	5, 10	0.10 to 15 K	25–350	2.094 × 0.562	
	RW67	6.5 w	5, 10	0.10 to 3.3 K	25–350	1.094 × 0.374	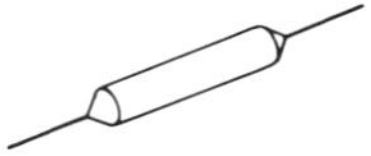
	RW68	11 w	5, 10	0.10 to 8.2 K	25–350	1.937 × 0.437	
	RW69	3 w	5, 10	0.10 to 820 Ω	25–350	0.562 × 0.249	
	RW70	1.5 w	0.1, 1	0.10 to 3.16 K	25–350	0.437 × 0.141	
	RW77	6.5 w	0.1, 1	0.10 to 38.3 K	25–350	1.094 × 0.374	
	RW78	11 w	0.1, 1	0.10 to 90.9 K	25–350	1.937 × 0.437	
	RW79	3 w	0.1, 1	0.10 to 10.5 K	25–350	0.562 × 0.249	
Film, insulated	RL07	1/4 w/250 v	2, 5	47 to 0.15 M	70–150	0.281 × 0.098	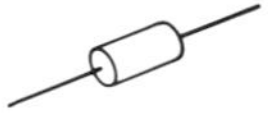
	RL20	1/2 w/350 v	2, 5	4.7 to 0.47 M	70–150	0.416 × 0.161	
	RL32	1 w/500 v	2, 5	10 to 1.0 M	70–150	0.593 × 0.205	
	RL42	2 w/500 v	2, 5	10 to 1.5 M	70–150	0.728 × 0.336	

Wirewound (power-type, chassis-mounted)	RE60	5 (Mtd on metal chassis)	1	0.10 to 3.32 K	25–275	0.662 × 0.677 × 0.351	
	RE65	10 (Mtd on metal chassis)	1	0.10 to 5.62 K	25–275	0.812 × 0.843 × 0.437	
	RE70	20 (Mtd on metal chassis)	1	0.10 to 12.1 K	25–275	1.124 × 1.125 × 0.593	
	RE75	30 (Mtd on metal chassis)	1	0.10 to 38.3 K	25–275	2.000 × 1.187 × 0.656	
	RE77	75 (Mtd on metal chassis)	1	0.10 to 34.8 K	25–275	3.531 × 2.843 × 1.761	
Composition (insulated), established reliability	RCR07	1/4 w/250 v	5	10 to 1 M	70–130	0.281 × 0.098	
	RCR20	1/2 w/350 v	5	10 to 1 M	70–130	0.416 × 0.161	
	RCR32	1 w/500 v	5	10 to 1 M	70–130	0.593 × 0.240	
Film, established reliability	RNR50	1/20 w/200 v	1	49.9 to 0.100 M	125–175	0.170 × 0.080	
	RNR55	1/10 w/200 v	1	10 to 0.301 M	70–165; 125–175	0.281 × 0.140	
	RNR57	1/8 w/200 v	1	10 to 0.200 M	70–165; 125–175	0.343 × 0.170	
	RNR60	1/8 w/250 v	0.1, 1	10 to 1.0 M	70–165; 125–175	0.437 × 0.165	
	RNR65	1/4 w/300 v	0.1, 1	10 to 2.0 M	70–165; 125–175	0.656 × 0.250	
	RNR70	1/2 w/350 v	0.1, 1	10 to 2.49 M	70–165; 125–175	0.875 × 0.328	
Wirewound (accurate), established reliability	RBR52	0.5 w/600 v	0.1, 1.0	0.1 to 1.2 M	125–145	1.062 × 0.406	
	RBR53	1/3 w/300 v	0.1, 1.0	0.1 to 1.75 M	125–145	0.812 × 0.406	
	RBR54	1/4 w/300 v	0.1, 1.0	0.1 to 0.850 M	125–145	0.812 × 0.281	
	RBR55	0.15 w/200 v	0.1, 1.0	0.1 to 0.525 M	125–145	0.562 × 0.281	
	RBR56	1/8 w/150 v	0.1, 1.0	0.1 to 0.350 M	125–145	0.406 × 0.281	
	RBR71	1/8 w/150 v	0.1, 1.0	0.1 to 0.270 M	125–145	0.344 × 0.281	
Wirewound (power-type), established reliability	RWR67	5.0 w	1	1.0 to 12.1 K	25–275	1.094 × 0.374	
	RWR68	10 w	1	1.0 to 38.3 K	25–275	1.937 × 0.438	
	RWR69	2.5 w	1	1.0 to 3.48 K	25–275	0.562 × 0.250	
	RWR70	1 w	1	1.0 to 1.21 K	25–275	0.437 × 0.124	
	RWR71	2 w	1	1.0 to 6.04 K	25–275	0.874 × 0.218	
Film (insulated), established reliability	RLR07	1/4 w/250 v	2, 5	47 to 0.15 M	70–150	0.281 × 0.098	
	RLR20	1/2 w/350 v	2, 5	10 to 0.47 M	70–150	0.416 × 0.161	
	RLR32	1 w/500 v	2, 5	10 to 1.0 M	70–150	0.593 × 0.205	
Wirewound (power-type, chassis-mounted), established reliability	RER65	10 w (Mtd on metal chassis)	1	0.1 to 5.62 K	25–275	0.812 × 0.843 × 0.437	
	RER70	15 w (Mtd on metal chassis)	1	0.1 to 12.1 K	25–275	1.124 × 1.124 × 0.593	
	RER75	30 w (Mtd on metal chassis)	1	0.1 to 38.3 K	25–275	2.000 × 1.187 × 0.656	

* Full-load ambient operating temperature and zero load temperature.

TABLE 35. Data on Typical Variable Resistors [12]

Type	Styles available in standard	Power rating, watts	Taper data	Nominal resistance range, ohms†	Temperature range, °C *	Mounting data	Maximum body size, in. (diameter × depth)	Configuration
Composition	RV4	2, 1	A, C	50 to 5M	70–120	Shaft and panel seal and locking bush.	1.156 × 0.750	
	RV6	1/2, 1/4	A, C	100 to 5M	70–120	Standard and locking bush.	0.515 × 0.468	
Wirewound (low operating temperature)	RA20	2.0, 1.1	A(linear); C(10% CW)	3 to 15K	40–105	Shaft and panel seal and locking bush.	1.313 × 0.703	
	RA30	4.0, 2.2	A(linear); C(10% CW)	3 to 25K	40–105	Shaft and panel seal and locking bush.	1.719 × 0.813	
Wirewound (power type)	RP06	6.25	Linear	1.0 to 3.5K	25–340	Standard bush. and locking bush.	0.906 × 0.751	
	RP10	25	Linear	2.0 to 5K	25–340		1.680 × 1.410	
	RP15	50	Linear	1.0 to 10K	25–340		2.410 × 1.440	
	RP20	75	Linear	2.0 to 10K	25–340		2.810 × 1.780	
	RP25	100	Linear	2.0 to 10K	25–340		3.190 × 1.780	
	RP30	150	Linear	2.0 to 10K	25–390		4.060 × 2.030	
Wirewound, semiprecision	RK09	1.5	Linear	10 to 50K	85–135	Standard bush. and locking bush.	0.570 × 0.650	
Wirewound (lead screw actuated)	RT12	3/4	Linear	10 to 10K	85–150	PC pin; mounting hole	1.260 × 0.200 × 0.330	
	RT22	3/4	Linear	50 to 10K	85–150		0.510 × 0.197 × 0.510	
Nonwirewound (lead screw actuated)	RJ11	1/4, 3/4	Linear	100 to 1M	See section	PC pin; mounting hole	1.282 × 0.300 × 0.374	
	RJ22	3/4	Linear	100 to 1M			0.510 × 0.197 × 0.510	
Wirewound (lead screw actuated), established reliability	RTR12	3/4	Linear	100 to 20K	85–150	PC pin; mounting hole	1.260 × 0.200 × 0.330	
	RTR22	3/4	Linear	100 to 20K	85–150		0.510 × 0.197 × 0.510	

* Full-load ambient operating temperature and zero load temperature, respectively. † M = megohms; K = kilohms.

given in Fig. 53. The letter designations of the standard types, basic construction descriptions, and military specification cross references are given in Table 36. In Table 37 are shown the principal applications of the various types, and Fig. 54 presents a measure of relative costs.

Capacitors are available in a wide variety of sizes and shapes governed by construction, use, rating, capacitance, and manufacturer. Several basic shapes are described in Fig. 55. Typical dimensions of several more common types of capacitors are given in Table 38. A schematic of a feed-through-type capacitor is shown in Fig. 56. These units are used when it is desired to pass low frequency through a chassis wall and to pass the rf current interference to ground.

It is necessary to indicate capacitance value, tolerance, working voltage, operating temperature, and desired case configuration when specifying capacitors. Do not forget to apply derating factors based upon environmental conditions; typical derating curves for temperature and frequency are given in Figs. 57 and 58. To determine the surface temperature rise of a capacitor, multiply the volt-amperes supplied to the unit by the power factor. This gives the watts lost in the capacitor.

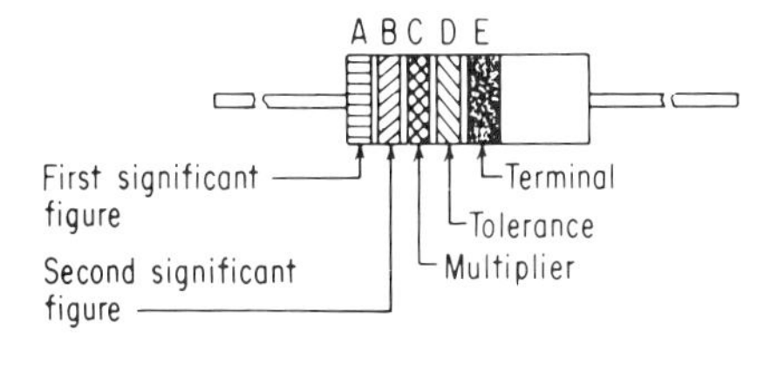

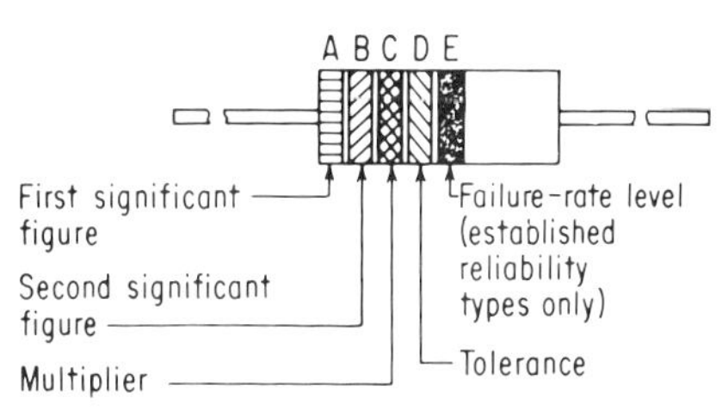

Fig. 52. Resistor value designation (MIL-STD -221).

Divide the watts lost by the capacitor surface area in square inches, and refer to Fig. 59 to get the approximate surface temperature rise.

INDUCTIVE COMPONENTS. Inductive components, also known as *magnetic components*, consist of one or more windings of wire on a core of magnetic material. The two basic classifications are transformers and inductors.

Transformers in their simplest form consist of two separate windings inductively coupled by being wound on a common magnetic core. When an ac voltage at one frequency is applied to one winding (primary) an ac voltage at the same frequency is induced in the second (secondary) winding. The magnitude of the secondary voltage depends upon the voltage applied to the primary, the turns ratio between the windings, and the flux coupling between the windings. The double-wound transformer also provides circuit isolation between the input and output circuits. Units are also made with only one winding, part of which is common to both the primary and the secondary. These are autotransformers and do not give circuit isolation.

Inductors are magnetic components which provide inductive reactance. Inductors for electronics systems are usually much simpler devices than transformers, consisting of a single winding with two terminals on magnetic core. Power inductors are used to smooth out ac ripple in dc power supplies (rectifier filter), and as charging inductors in pulse-charging circuits of radar modulators. Communication inductors are used as elements in filter networks or to limit ac while passing dc. The induc-

TABLE 36. Capacitor Type, Construction, and Military Specification Cross Reference Data [13]

Type	Construction	Military specification
CA	Fixed, paper dielectric, ac and dc, hermetically sealed in metal cans. Bypass, radio interference reduction	MIL-C-12889
CB	Fixed, mica dielectric, button style	MIL-C-10950
CC	Fixed, ceramic dielectrics, temperature-compensating. Tubular, disk, and rectangular styles; radial and axial leads	MIL-C-20
CE	Fixed, dry electrolytic, aluminum, polarized. Tubular, bathtub, and cylindrical case styles. Axial leads, solder lugs, or plug-in base	MIL-C-62
CG	Fixed or variable, vacuum dielectric	MIL-C-23183
CH	Fixed, paper or Mylar dielectric, metallized construction, dc, hermetically sealed. Tubular, bathtub, or rectangular case. Axial leads, solder lugs	MIL-C-18312
CHR	High-reliability version of above	MIL-C-39022
CJ	Fixed, dry electrolytic, ac, nonpolarized. Cylindrical case, solder lugs	MIL-C-3871
CK	Fixed, ceramic dielectric, general-purpose. Tubular, flat, disk, feed-through, and standoff styles. Axial and radial leads and mounting stud	MIL-C-11015
CKR	High-reliability version of above	MIL-C-39014
CL	Fixed, nonsolid electrolytic, tantalum (foil and sintered slug). Tubular, cylindrical, rectangular, and cup-style cases. Axial leads, solder lugs, and mounting stud	MIL-C-3965
CLR	High-reliability version of above	MIL-C-39006
CM	Fixed, mica dielectric. Rectangular, stack-mounting, or ear-mounting cases. Axial leads, solder lugs	MIL-C-5
CMR	High-reliability version of above	MIL-C-39001
CP	Fixed, paper dielectric, dc, hermetically sealed. Tubular, bathtub, and rectangular cases. Axial leads and solder lugs	MIL-C-25
CPV	High-reliability version of above	MIL-C-14157
CQ	Fixed, plastic or paper/plastic dielectric, hermetically sealed in metal, ceramic, or glass case. Tubular and rectangular case styles; radial leads, solder lugs	MIL-C-19978
CS	Fixed, solid electrolyte, tantalum. Tubular case, axial leads	MIL-C-26655
CSR	High-reliability version of above	MIL-C-39003
CTM	Fixed, plastic dielectric, dc, nonmetallic tubular case, axial leads	MIL-C-27287
CY	Fixed, glass dielectric. Rectangular and square cases, axial leads	MIL-C-11272
CYR	High-reliability version of above	MIL-C-23269
CZ	Feed-through, radio interference, reduction, ac and dc, hermetically sealed in metallic case. Tubular and bathtub cases; axial leads, screw terminals, cable connection	MIL-C-11693
CZR	High-reliability version of above	MIL-C-39011
EC	Covers a wide range of constructions and case shapes. See specification sheets	MIL-C-38102
CV	Variable, ceramic dielectric, trimmer	MIL-C-81
CT	Variable, air dielectric, trimmer	MIL-C-92
PC	Variable, glass or quartz dielectric and metal tuning pistons, tubular trimmer	MIL-C-14409

TABLE 37. Principal Applications of Capacitors[13]

Type	Application											
	Blocking	Buffing	Bypass	Coupling	Filtering	Tuning	Temperature compensating	Trimming	Motor starting	Radio interference reduction	Timing	Noise suppression
CA, paper, hermetic										×		
CB, mica, button			×	×		×						
CC, ceramic, temperature comp.			×	×	×	×	×					
CE, electrolytic, Al			×		×							
CH, paper or plastic, hermetic	×		×		×							
CJ, electrolytic, ac									×			
CK, ceramic			×	×	×							
CL, electrolytic, tantalum	×		×	×	×				×			
CM, mica, molded	×	×	×	×	×	×					×	
CP, paper, hermetic	×	×	×	×	×							×
CQ, paper, hermetic	×	×	×	×	×							
CS, electrolytic, tantalum			×		×							
CT, variable, air				×				×				
CV, variable, ceramic		×		×		×		×				
CY, glass	×		×	×		×						
CZ, feed-through										×		
PC, variable, glass, piston							×					

TABLE 38. Dimensions of Some Typical Capacitors [13]

Type	Style referenced in standard	Capacitance range, pf	Capacitance tolerance, ±	Dc working voltage, volts	Operating temp. range, °C	Max. body dimension, in.			Configuration
						Length	Width	Thickness or diameter	
Fixed, glass									
Fixed, glass dielectric	CY10	0.5–180	0.25 pf, 1, 2, or 5%	500	−55 to +125	0.391	0.203	0.109	CY10, 15, 20, 30; CY12, 16, 21, 31
	CY12	0.5–180	0.25 pf, 1, 2, or 5%	500	−55 to +125	0.406	0.219	0.187 or 0.250	
	CY15	220–470	1 or 5%	500	−55 to +125	0.516	0.297	0.156	
	CY16	220–470	1 or 5%	500	−55 to +125	0.515	0.312	0.187 or 0.288	
	CY20	560–4,700	1 or 5%	500	−55 to +125	0.796	0.469	0.187	
	CY21	560–1,000	1 or 5%	500	−55 to +125	0.812	0.484	0.187	
	CY30	5,600–10,000	1 or 5%	200 or 500	−55 to +125	0.828	0.828	0.187	
	CY31	1,200–3,300	1 or 5%	500	−55 to +125	0.823	0.843	0.187 or 0.288	
Fixed, electrolytic									
Fixed, tantalum (solid electrolyte), established reliability	CSR13	0.0047–330 (μf)	±10, ±20%	6–100	−55 to +125	0.317–0.817	....	0.151–0.367	CSR13
Fixed, paper and plastic									
Fixed, paper (or paper-plastic), dielectric, dc, established reliability	CPV07	0.0010–0.47	±1, ±5	200–600	−65 to −65	1.000–2.250	....	0.282–1.047	CPV07-09
	CPV09	0.001–1.0	±5, ±10	200–600	−65 to −125	0.845–2.657	....	0.275–1.040	

Fixed, ceramic								
Fixed, ceramic-dielectric, general-purpose	CK05	10–1,000	±10, ±20	200	−55 to +150	0.200	0.200	0.100
	CK06	1,500–10,000	±10, ±20	200	−55 to +150	0.300	0.300	0.100
	CK07	33–1,000	±10, ±20	50	−55 to +125	0.100	0.100	0.060
	CK08	1,500–3,300	±20	50	−55 to +125	0.150	0.150	0.060
	CK09	4,700–10,000	±20	50	−55 to +125	0.250	0.250	0.060
	CK10	15,000	±20	50	−55 to +125	0.350	0.350	0.060
	CK60	2.2–1,500	±10, ±20	500, 1,000	85, 125	0.310	.....	0.160
	CK61	2.2–2,200	±10, ±20	500	85, 125, 150	0.385	.....	0.160
	CK62	22–8,200	±10, ±20	500, 1,500	35, 125	0.590	.....	0.160
	CK63	680–10,000	±10, ±20	500, 1,000	85, 125, 150	0.690	.....	0.160
	CK64	680–3,000	±20	1,600	−55 to +85	0.770	.....	0.207
	CK65	4,700	±20	1,600	−55 to +85	0.830	.....	0.207
	CK66	5,600	±20	1,600	−55 to +85	0.930	.....	0.207
	CK67	6,800–7,500	±20	1,600	−55 to +85	0.990	.....	0.207
	CK68	10,000	±20	1,600	−55 to +85	1.090	.....	0.370
	CK69	15,000	±20	1,600	−55 to +85	1.150	.....	0.370
	CK70	10–1,500	±10, ±20	500, 1,000	−55 to +85	1.231	.....	0.327
	CK72	100–1,000	±20	1,500	−55 to +85	0.891	.....	0.391
	CK80	10–1,500	±10, ±20	500	−55 to +85	1.047	.....	0.327

CK05 through CK10
CK60 through CK69
CK70
CK72
CK80

tor also contains capacitance and resistance and this must be considered in design. As shown in Fig. 60, inductance and resistance vary with frequency. In high-frequency applications, location of the inductor is critical since nearby metal objects may affect the magnetic field and change the inductance. The basic military specification covering audio, power, and pulse transformers and inductors is MIL-T-27B, dated September, 1963. The method of classifying components under this

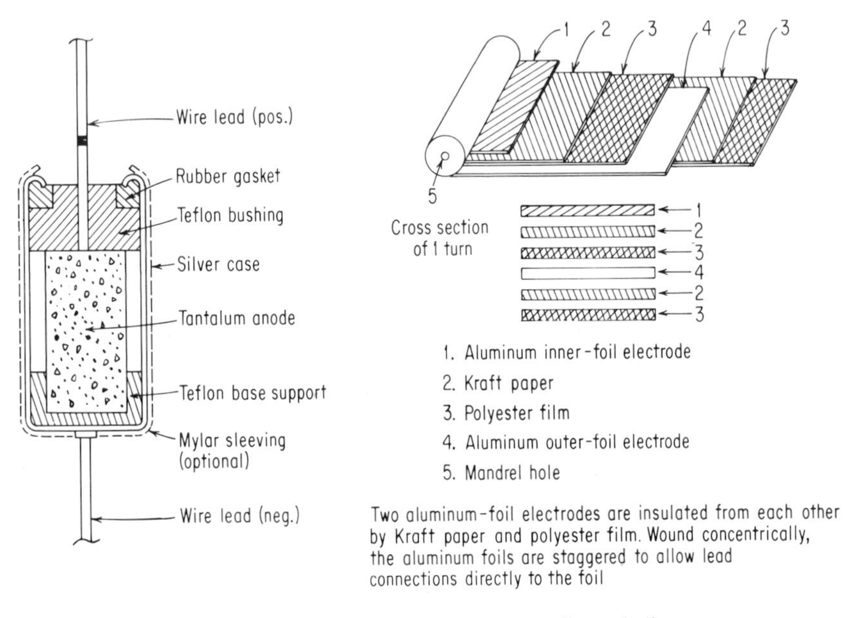

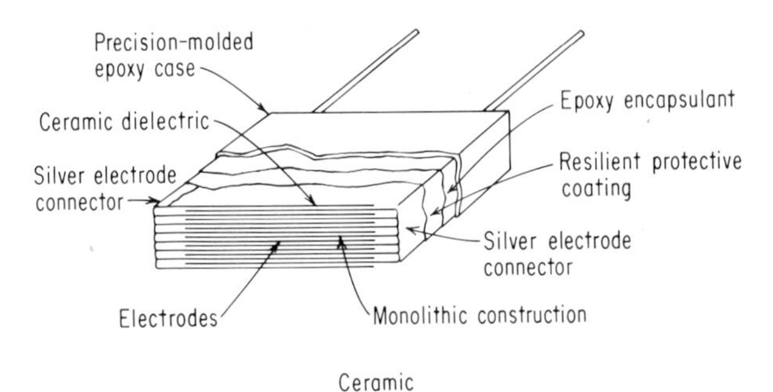

FIG. 53. Construction details of three capacitor types.

specification is explained in Fig. 61. Typical case sizes available under this specification are listed in Table 39.

Active Components. These are components which display transistance, that is, gain or control. Examples of this are the semiconductor devices (diodes, transistors) and the electron tube.

SEMICONDUCTOR DISCRETE COMPONENTS. Devices which take advantage of the unique electrical properties of semiconductor materials (usually germanium or silicon) to perform switching, amplification, rectification, and regulating functions are called semiconductor devices. A simplified version of the manufacture of a typical device, a silicon-controlled rectifier (SCR), is shown in Fig. 62. The basic device classifications and their applications are listed in Table 40. There are thousands of these device types in existence, and selection and identification of the proper one are difficult. Many manufacturers register their devices with the EIA. These registered devices are assigned a number 1n——, 2n——, or 3n—— based generally upon whether the device is single-junction, bipolar, or multijunction in construction, although this classification is not strictly followed. These numbers are of little value in identification since they do not indicate parameters, applications, or case size. In addition, many types are made which are not registered, and which carry only

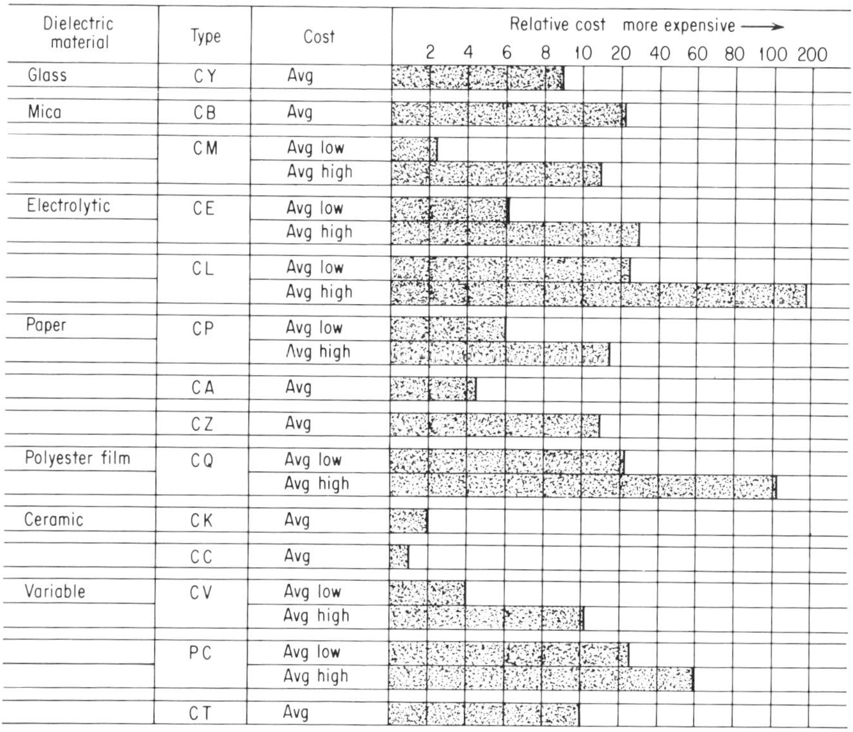

Dielectric material	Type	Cost	Relative cost more expensive ⟶ (2 4 6 8 10 20 40 60 80 100 200)
Glass	CY	Avg	
Mica	CB	Avg	
	CM	Avg low	
		Avg high	
Electrolytic	CE	Avg low	
		Avg high	
	CL	Avg low	
		Avg high	
Paper	CP	Avg low	
		Avg high	
	CA	Avg	
	CZ	Avg	
Polyester film	CQ	Avg low	
		Avg high	
Ceramic	CK	Avg	
	CC	Avg	
Variable	CV	Avg low	
		Avg high	
	PC	Avg low	
		Avg high	
	CT	Avg	

FIG. 54. Relative costs of capacitor types.[13]

manufacturers' in-house numbers. For identification, reference must be made to a handbook of data on semiconductors. There are also hundreds of case sizes and shapes available; often the device can be had in a choice of cases. The more common cases have been standardized through the EIA Joint Electron Devices Engineering Council and assigned a number of the form JEDEC TO-5. Reference must then be made to a data sheet to obtain specific dimensions. Several of the case styles in most common usage are shown in Fig. 63. The devices themselves are very small, being on the order of 5 to 50 mils on a side. The major part of the device container is consumed in giving environmental protection and providing a method of connecting the small device to the rest of the circuitry (see Fig. 64).

Diodes are devices which exhibit a very high resistance to current flow in one direction. They are all, including the so-called general-purpose type, tailored to meet

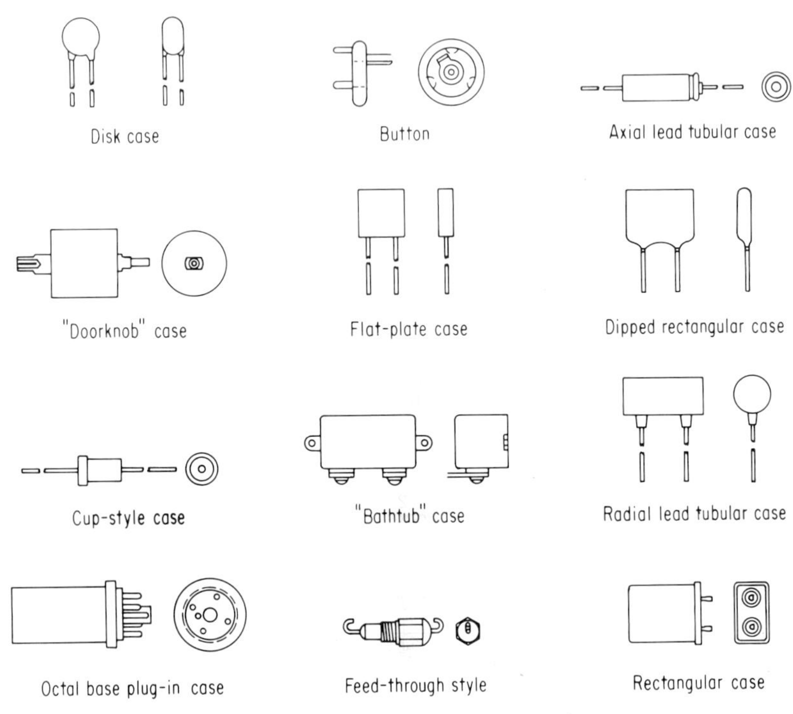

FIG. 55. Several basic capacitor styles.

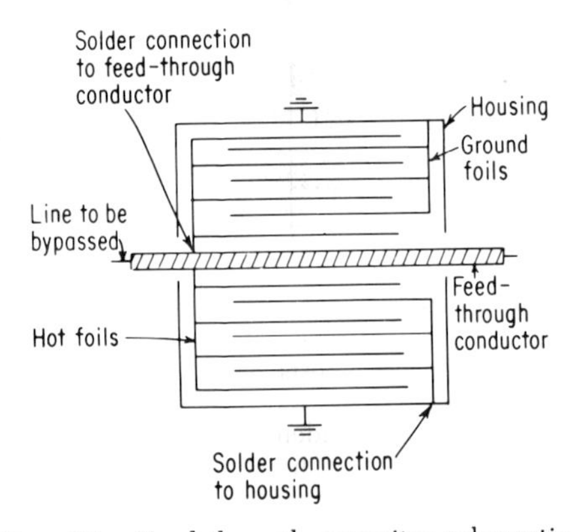

FIG. 56. Feed-through capacitor schematic.[13]

a specific class of application. Although they may be used with some success in other applications, it is probable that one or more of the parameters will be compromised. Selection should be made from comparisons of manufacturers' data sheets. Several typical diodes and SCRs are described in Table 41.

Transistors are semiconductor devices which by their atomic structure and fabrication permit the control of current flow. Several types are described in Table 42. Ratings are the limiting values assigned by the manufacturer which if exceeded may result in permanent damage to the device. These should not be used as design conditions. The characteristics are measurable properties of the device under specific operating conditions for which the unit will provide reliable performance. Insulated

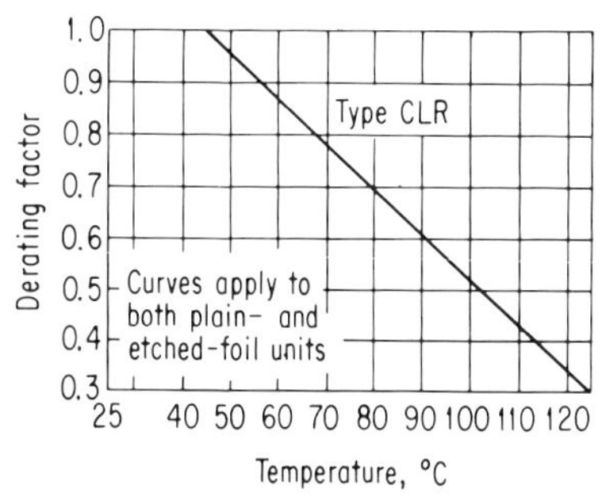

FIG. 57. Typical capacitor derating curve for temperature—type CLR.[13]

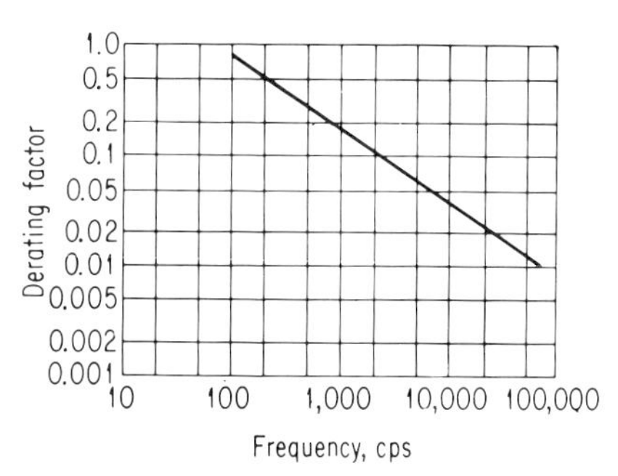

FIG. 58. Typical capacitor derating curve for frequency—type CLR.[13]

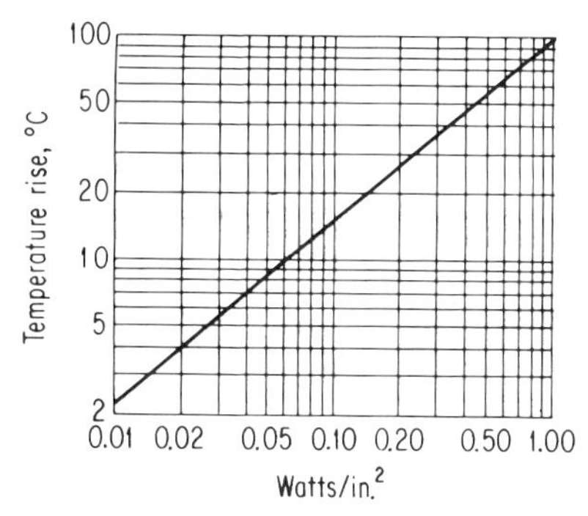

FIG. 59. Capacitor temperature rise as a function of power density.[13]

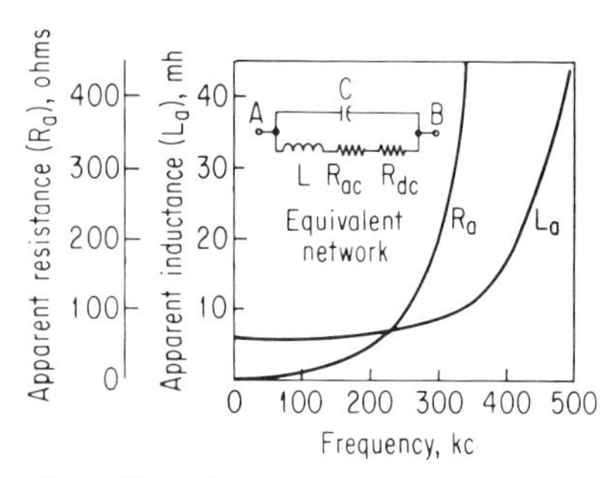

FIG. 60. Variation of apparent inductance and resistance of inductor with frequency.[10]

gate field effect transistors (IGFET or MOSFET), as illustrated in Fig. 65, due to their extremely high input resistance and small gate capacitance, are subject to potential damage from the accumulation of excess static charge. To avoid possible damage to the device when handling, testing, or in actual use, the procedure outlined in Table 43 should be followed.

ELECTRON TUBES. Tubes consist basically of an energized cathode which supplies electrons, and one or more additional electrodes, which control and collect these electrons; all are mounted within an evacuated envelope. The envelope may be a glass bulb or a metal shell. Tubes can be classified into three broad groups: receiving, microwave, and camera/display.

The type designation system prescribed by MIL-T-27B is:

TF	4	R	X	20	YY
(a) Component	(b) Grade	(c) Class	(d) Life	(e) Family	(f) Case

Explanation

(a) Component: The TF indicates the component is a transformer or inductor.

(b) Grade: Environmental level. Grades 1 and 4 are metal-encased and grades 2 and 5 are encapsulated.

Test	Grades 1	2	3	4	5	6
Sealing	×	×		×	×	
Temperature cycling	×	×	×			
Thermal shock				×	×	×
Immersion	×	×		×	×	
Moisture resistance	×	×	×	×	×	×
Vibration				×	×	×
Shock				×	×	×
Flammability		×			×	

(c) Class: Maximum operating temperature. No relation to insulation-type letter.

Q — 85°C V — 155°C
R — 105°C T — 170°C
S — 130°C U — 7170°C

(d) Life expectancy
X = 10,000 hr, minimum
Y = 2,500 hr, minimum

(e) Family: Intended application as listed below.

Family	Application
01	Power transformer supplying unrectified loads
02	Power transformer supplying rectified loads
03	Power transformer supplying both rectified and unrectified loads
04	Power inductor (single or multiple)
05	Vibrator transformer
06	Multiunit power transformer
07	Composite power transformer and inductor
08	Combination vibrator and line-input transformer
09	Miscellaneous single-frequency transformer or inductor

Family	Audio transformers: Primary dc	Primary impedance,* ohms	Secondary dc	Load impedance, ohms
10	No	Any	Yes or no	$>10,000$
11	Yes	$\leq 1,000$	Yes or no	$>10,000$
12	Yes	$>1,000$	Yes or no	Any
13	Yes	$>1,000$	No	$\leq 10,000$
14	Yes	Any	Yes or no	(Rf tube load)
15	Yes	$>1,000$	No	$>10,000$
16	No	Any	Yes or no	$\leq 10,000$
17	Yes	$\leq 1,000$	Yes or no	$\leq 10,000$

Family	Application
18	Audio oscillator
19	Multiunit audio transformer
20	Audio inductor (single or multiple)
21	Audio transformer, miscellaneous
22	Hybrid transformer
30	Pulse input; line to grid
31	Pulse interstage, plate or cathode to grid not drawing current
32	Pulse driver, plate or cathode to grid or diode drawing power
33	Pulse output: plate, cathode, or network to line
34	Pulse modulation: plate, line, or network to oscillator
35	Pulse oscillator, blocking tube
36	Pulse transformer, miscellaneous
37	Charging inductor
40	Saturable-core transformer
41	Saturable-core inductor
50	Carrier transformer } wide frequency range application
51	Carrier inductor } wide frequency range application
52	Sonar and ultrasonic transformer
53	Sonar and ultrasonic inductor
54	Computer applications

* This is the nominal value of impedance appearing across the primary terminals of the transformer when the primary is unterminated and all secondaries loaded with specified impedances.

(f) Case: This code identifies the case size. See Table 39.

FIG. 61. Classification of inductive components per MIL-T-27B.

TABLE 39. Standard Transformer Case Dimensions per MIL-T-27B

Case symbol	Case dimensions: A, in.	B, in.	C, in.	Mounting dimensions: D, in.	E, in.	S, in.	F, size	Stud length,[g] in.	Template: L,[h] in.
AF	3/4[a]	3/4[a]	1 1/8[b]	.....	.....	9/16	4–40	3/8	0.147
AG	1[a]	1[a]	1 3/8[b]	.....	.....	3/4	4–40	3/8	0.147
AH	1 5/16[a]	1 5/16[a]	1 3/4[b]	.....	.....	1 1/4	6–32	3/8	0.173
AJ	1 5/8[a]	1 5/8[a]	2 3/8[b]	1 3/16[d]	1 3/16[d]	...	6–32	3/8	0.173
EA	1 15/16[b]	1 13/16[b]	2 3/4[c]	1 3/8[d]	1 1/4[d]	...	6–32	3/8	0.173
EB	1 15/16[b]	1 13/16[b]	2 7/16[c]	1 3/8[d]	1 1/4[d]	...	6–32	3/8	0.173
FA	2 5/16[b]	2 1/16[b]	3 1/8[c]	1 11/16[d]	1 7/16[d]	...	6–32	3/8	0.173
FB	2 5/16[b]	2 1/16[b]	2 1/2[c]	1 11/16[d]	1 7/16[d]	...	6–32	3/8	0.173
GA	2 3/4[b]	2 3/8[b]	3 13/16[c]	2 1/8[d]	1 3/4[d]	...	6–32	3/8	0.173
GB	2 3/4[b]	2 3/8[b]	2 13/16[c]	2 1/8[d]	1 3/4[d]	...	6–32	3/8	0.173
HA	3 1/16[b]	2 5/8[b]	4 1/4[c]	2 19/64[d]	1 55/64[d]	...	8–32	3/8	0.199
HB	3 1/16[b]	2 5/8[b]	3 3/16[c]	2 19/64[d]	1 55/64[d]	...	8–32	3/8	0.199
JA	3 9/16[b]	3 1/16[b]	4 7/8[c]	2 5/8[d]	2 1/8[d]	...	8–32	3/8	0.199
JB	3 9/16[b]	3 1/16[b]	3 7/8[c]	2 5/8[d]	2 1/8[d]	...	8–32	3/8	0.199
KA	3 15/16[b]	3 3/8[b]	5 1/4[c]	3[e]	2 7/16[e]	...	10–32	1/2	0.228
KB	3 15/16[b]	3 3/8[b]	4 5/16[c]	3	2 7/16[e]	...	10–32	1/2	0.228
LA	4 5/16[b]	3 11/16[b]	5 9/16[c]	3 5/16	2 11/16[e]	...	10–32	1/2	0.228
LB	4 5/16[b]	3 11/16[b]	4 1/2[c]	3 5/16[e]	2 11/16[e]	...	10–32	1/2	0.228
MA	4 11/16[b]	4[b]	6[c]	3 11/16[f]	3[f]	...	1/4–20	5/8	0.2812
MB	4 11/16[b]	4[b]	4 15/16[c]	3 11/16	3[f]	...	1/4–20	5/8	0.2812
NA	5 1/16[b]	4 5/16[b]	6 13/16[c]	4 1/16	3 5/16[f]	...	1/4–20	5/8	0.2812
NB	5 1/16[b]	4 5/16[b]	5 1/2[c]	4 1/16	3 5/16[f]	...	1/4–20	5/8	0.2812
OA	5 1/2[b]	4 1/2[b]	6 3/4[c]	3 3/4[f]	3[f]	...	1/4–20	5/8	0.2812
YY[i]	All metal cases not included above								
ZZ[i]	Open type and encapsulated units								

Notes:

[a] Tolerance +0, −1/16 in.
[b] Tolerance +0, −1/8 in.
[c] Tolerance +0, −3/16 in.
[d] Tolerance ±1/64 in.
[e] Tolerance ±1/32 in.
[f] Tolerance ±3/64 in.
[g] Tolerance for studs 1/2 in. long and less, ±1/16 in. and longer than 1/2 in., ±1/8 in. Threaded inserts having 8 full threads (same size as stud threads) may be used instead of studs.
[h] Tolerance ±0.007 in.
[i] Corner radii may not exceed one-quarter of the smallest case dimension. Tolerances for cases YY and ZZ are: Dimensions not exceeding those of AJ case shall have same tolerance as AJ case, and for larger dimensions, the other tolerances apply.

*Dimension S is for two-stud mounting and dimensions D and E are for four-stud mounting

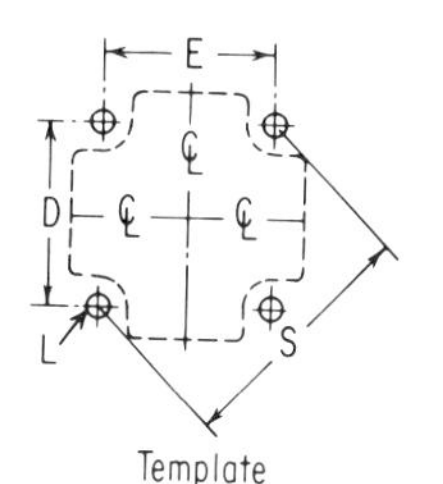

Template

Note: Threaded inserts may be used instead of studs

Dimensions of typical miniature transformers under case symbols ZZ and YY:

Transistor transformers

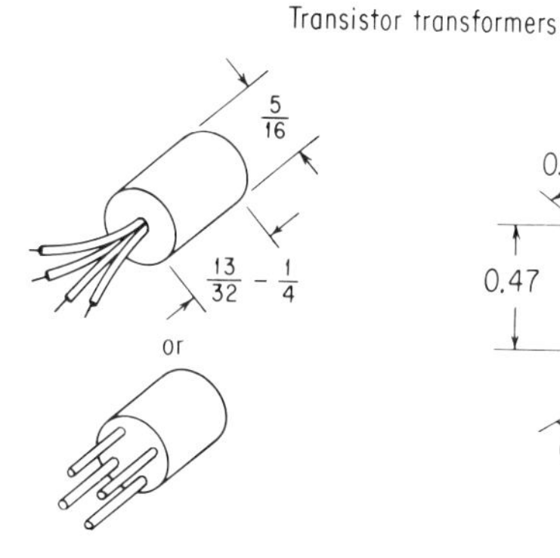

Symbol YY

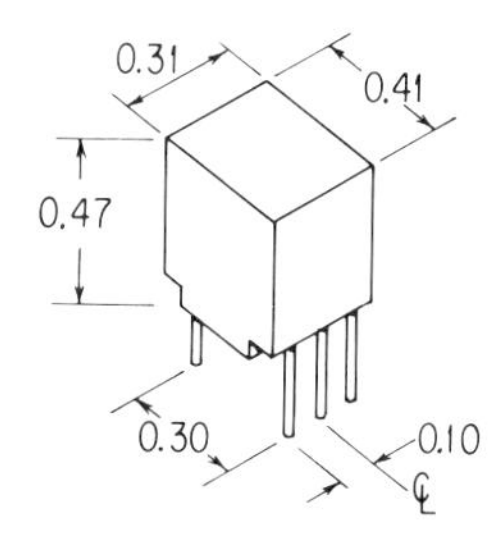

Symbol ZZ-printed-wiring-board style

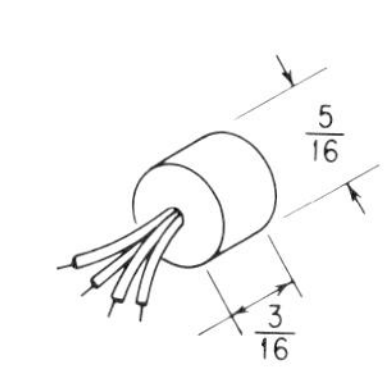

Symbol YY-audio inductor and pulse transformer

TABLE 40. Generic Information on Discrete Semiconductor Components[16]

Semiconductor device	Sketch	Schematic symbol	Functional analogy	State-of-art ratings * Not necessarily simultaneous	Applications	Electrical characteristics
Diode/rectifier		 Anode Cathode	Tube diode switch tube rectifier	2400 volts * 1000 amp	Voltage sensitivity switch, steering detection, rectification	Low forward resistance, high reverse resistance
Zener diode		 Anode Cathode	V.R. tube	250 volts 150 watts	Voltage reference, voltage regulator	Sharp zener knee, low dynamic impedance, stable temperature characteristic
Germanium bipolar transistor		 Collector Base Emitter	Triode tube switch	400 volts * 200 amp 1000 amp (in modules)	Amplification, oscillation, switching	High current gain, low input z, high output z
Silicon bipolar transistor			Triode tube switch	1200 volts (NPN) * 100 amp (NPN) 300 volts (PNP) 15 amp (PNP)	Amplification, oscillation, switching	High current gain, low input z, high output z
Germanium field-effect transistor		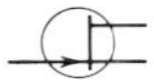 N-channel	Pentode tube	50 volts Gain -20,000	Amplification, oscillation, switching	High voltage gain, high input z, high output z
Silicon field-effect transistor		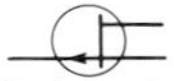 P-channel	Pentode tube	50 volts Gain -15,000	Amplification, oscillation, switching	High voltage gain, high input z, high output z
Unijunction transistor		 Base Base Emitter	None	50 ma	Long-time delay circuit, relaxation oscillator, triggering voltage sensing	Negative resistance
Silicon-controlled rectifier	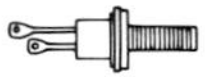	 Anode Gate	Thyratron	1800 volts 600 amp 1200 amp (dual pellet)	Relay replacement, variable voltage control, phase control	High voltage with high current
Bidirectional thyristor ("Triac")	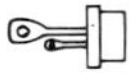	 Gate Anode 2 Anode 1	Thyratron	400 volts 15 amp	Switch (replaces two SCR's)	High voltage with high current
Trigger			Thyratron	200 volts	Switch	Fast switching low leakage

Receiving tubes have steadily decreased in usage since the introduction of the transistor, but they are still needed in many cases where impedance, linearity, noise, temperature, radiation, frequency, and power requirements are involved. Often, tube circuits require fewer total components. The construction of a miniature tube is shown in Fig. 66. A recent development, the Nuvistor, has a ceramic envelope in which all the electrodes are cylindrical and fit one inside the other; this promotes assembly by automatic means. For outline dimensions of the more commonly used tube envelopes, see Fig. 67.

Microwave tubes are designed for use in the highest frequencies (shortest wavelengths) of the radio spectrum. Characteristics and applications of the basic types are shown in Table 44. Construction details are illustrated in Fig. 68.

Camera and display tubes allow the collection, transmission, and display of visual information. The basic types are cathode-ray, storage tubes, and camera tubes; these are discussed further in Table 45. Construction details are shown in Fig. 69. A special type of display tube is the so-called *gas-filled cold-cathode tube.* These tubes are available in a wide variety of configurations and are used for digital

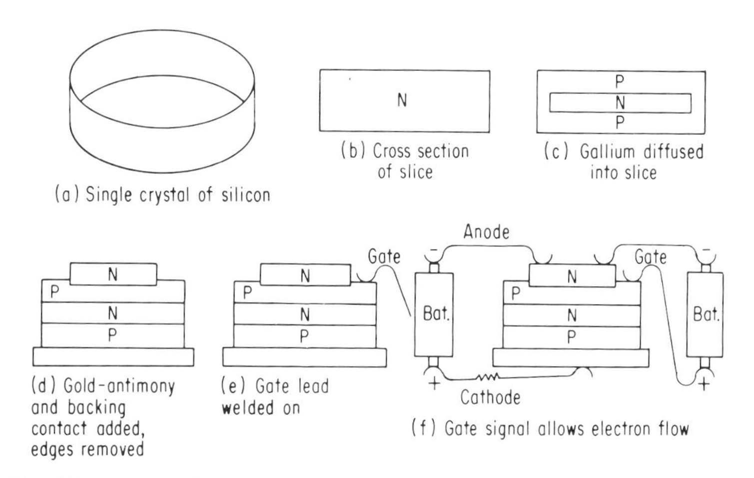

FIG. 62. How a typical power SCR is made. (*Reprinted from Iron Age, Jan. 26, 1967, copyright 1967. Chilton Company.*)

TABLE 41. Typical Diodes and SCRs

Device type No.	Material	Identification	Usage	Case
1 N914	Silicon	Signal diode	Medium-speed switch	DO-7
1 N2842	Silicon	Zener diode	Voltage regulation, 50 watts	TO-3
2 N690	......	SCR	....	TO-48
1 N3028B	Silicon	Zener diode	Voltage regulation, 1 watt	DO-13
1 N1198	Silicon	Rectifier	....	DO-5
1 N4645	Silicon	Zener diode	Reference	DO-14
1 N3611	Silicon	High conductance	Rectifier	

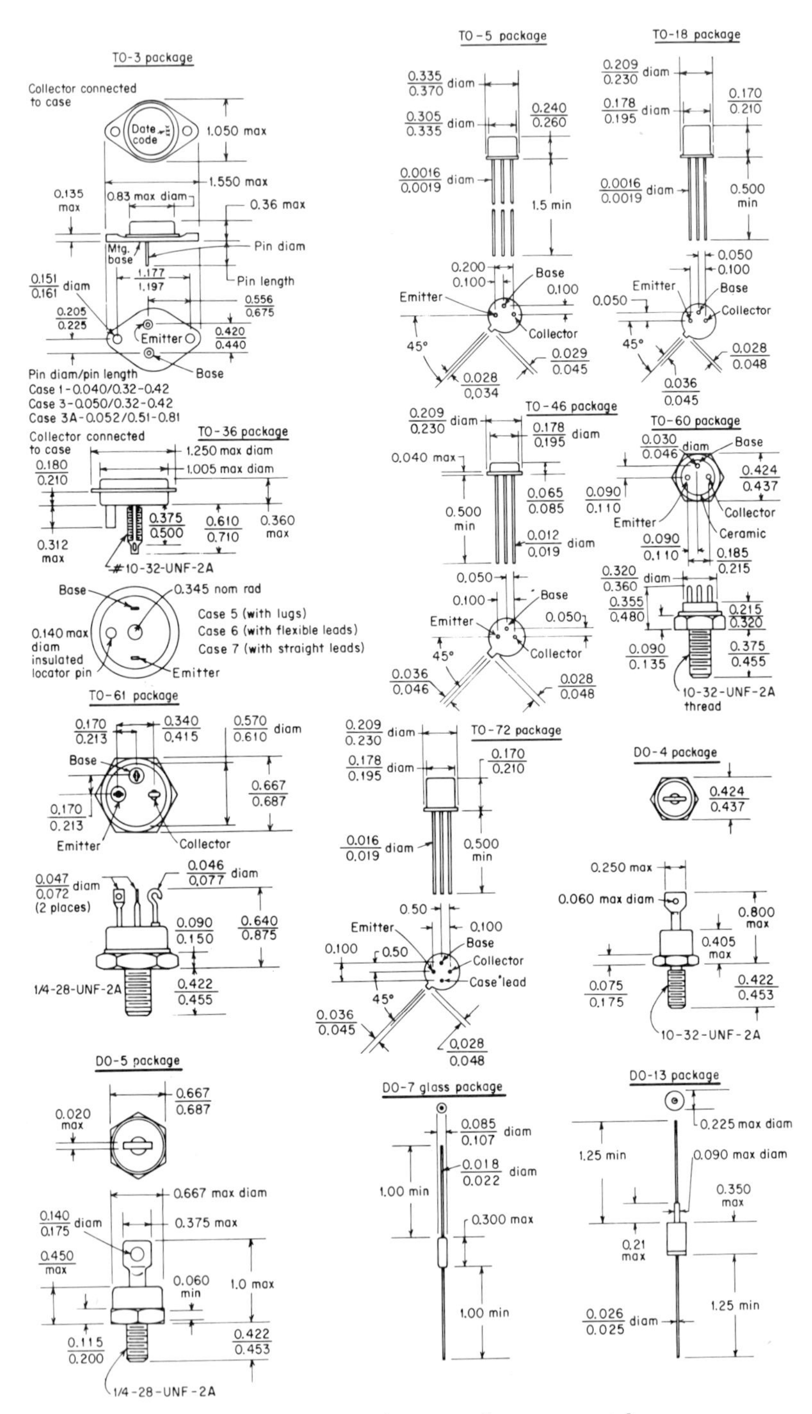

Fig. 63. Most used semiconductor case styles.

TABLE 42. Typical Transistors

Device type No.	Use	Case	Maximum ratings				Electrical characteristics						
			P_D,[a] @ 25°C	T_J,[d] °C	V_{CBO},[e] volts	V_{CE-},[f] volts	H_{fe}[j] Min.	H_{fe}[j] Max.	@ I_c[k]	$V_{CE\ (SAT)}$,[l] volts	@ I_c[k]	h_{fe}[m]	f_T[n]
2N1893	Gen. pur.	TO-5	80.0 mw[b]	200	120	100[g]	40	120	150 ma	5	150 ma	30	50 MHz
2N2906A	Switch	TO-18	1.8 watts[c]	200	60	60[h]	40	120	150 ma	0.4	150 ma	...	200 MHz
2N2946	Chopper	TO-46	400 mw[b]	175	40	35[h]	30	...	1 ma	...	.....	...	3 MHz
2N1016B	Power	TO-36	150 watts[c]	150	100	100[i]	10	...	5 amp	2.5	5 amp		
2N1724	Power	TO-61	50 watts[c]	175	120	80[h]	20	190	2 amp	1	2 amp	...	10 MHz
2N918	Rf amp.	TO-72	200 mw[b]	200	30	15[h]	20	...	3 ma	0.4	10 ma	...	600 MHz
2N3375	High freq.	TO-60	11.6 watts[c]	200	65	40[h]	10	100	0.25 amp	1	0.25 amp	...	400 MHz
2N3808	Diff. amp.	RO-52	0.5 watt[b]	200	60	60[h]	150	450	0.1 ma	0.2	0.1 ma	150	100 MHz

[a] Power dissipation.
[b] Ambient.
[c] Case.
[d] Operating junction temperature.
[e] Collector-base voltage.
[f] Collector-emitter voltage.
[g] Specified resistance.
[h] Base open.
[i] Voltage bias.
[j] Common-emitter dc short-circuit forward-current transfer ratio.
[k] Current flow out of collector terminal.
[l] Collector-emitter saturation voltage at specified collector current.
[m] Small-signal short-circuit forward-current transfer ratio (common emitter).
[n] Current gain–bandwidth product.

TABLE 43. Handling Considerations for Insulated Gate Field-effect Transistors [17]

1. The leads of the devices should remain wrapped in the shipping foil except when being tested or in actual operation to avoid the build-up of static charge
2. Avoid unnecessary handling; when handled, the devices should be picked up by the can instead of the leads
3. The substrate and case should be operated at the same or approximately the same dc potential as the gate
4. The devices should not be inserted or removed from circuits with the power on as transient voltages may cause permanent damage to the devices

FIG. 64. Transistor mounted on case header. (*Continental Device Corp.*)

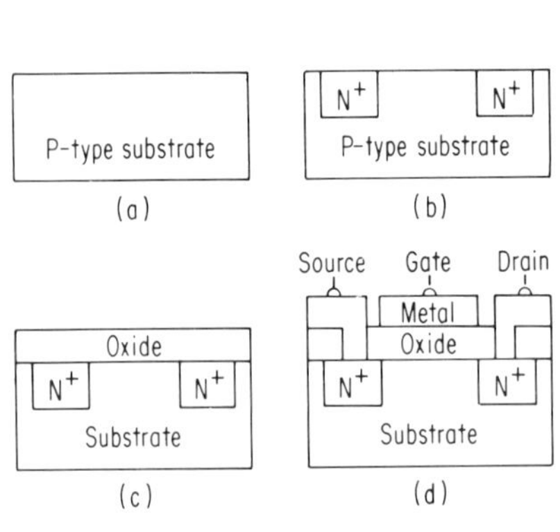

FIG. 65. Insulated gate field-effect transistor.

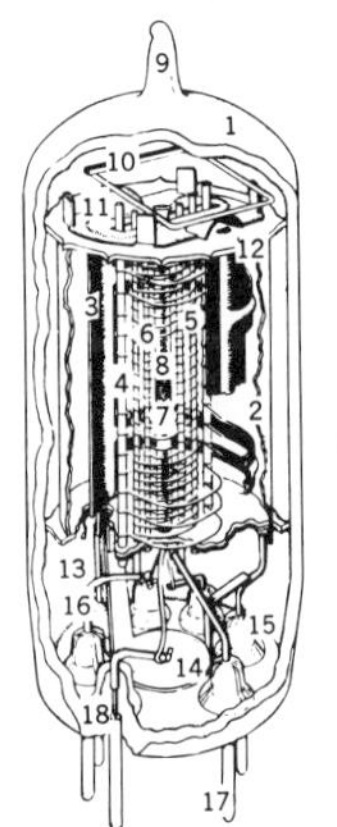

FIG. 66. Construction of a miniature tube. (*Radio Corp. of America.*)

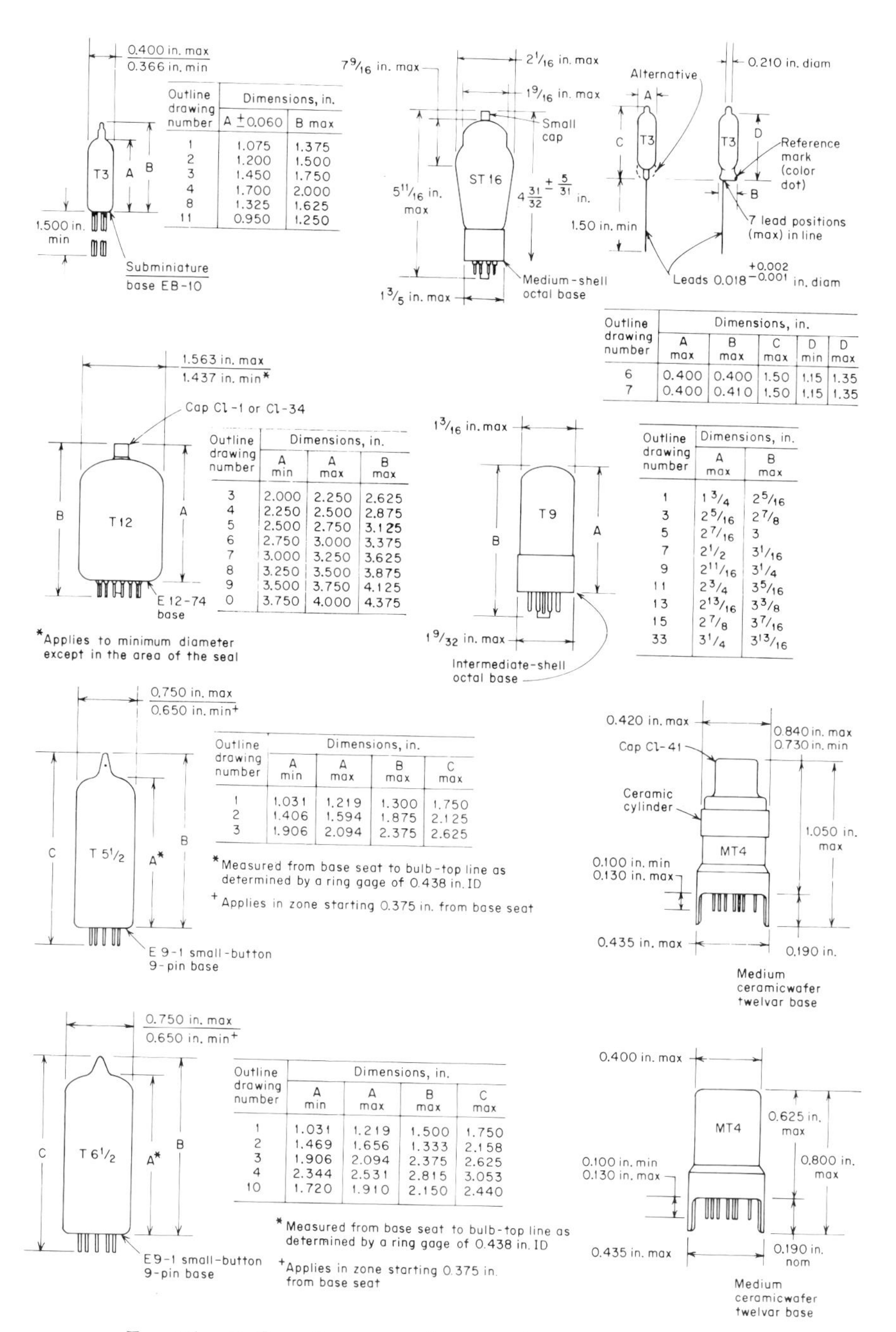

Outline drawing number	Dimensions, in.	
	A ±0.060	B max
1	1.075	1.375
2	1.200	1.500
3	1.450	1.750
4	1.700	2.000
8	1.325	1.625
11	0.950	1.250

Outline drawing number	Dimensions, in.				
	A max	B max	C max	D min	D max
6	0.400	0.400	1.50	1.15	1.35
7	0.400	0.410	1.50	1.15	1.35

Outline drawing number	Dimensions, in.		
	A min	A max	B max
3	2.000	2.250	2.625
4	2.250	2.500	2.875
5	2.500	2.750	3.125
6	2.750	3.000	3.375
7	3.000	3.250	3.625
8	3.250	3.500	3.875
9	3.500	3.750	4.125
0	3.750	4.000	4.375

Outline drawing number	Dimensions, in.	
	A max	B max
1	1 3/4	2 5/16
3	2 5/16	2 7/8
5	2 7/16	3
7	2 1/2	3 1/16
9	2 11/16	3 1/4
11	2 3/4	3 5/16
13	2 13/16	3 3/8
15	2 7/8	3 7/16
33	3 1/4	3 13/16

Outline drawing number	Dimensions, in.			
	A min	A max	B max	C max
1	1.031	1.219	1.300	1.750
2	1.406	1.594	1.875	2.125
3	1.906	2.094	2.375	2.625

Outline drawing number	Dimensions, in.			
	A min	A max	B max	C max
1	1.031	1.219	1.500	1.750
2	1.469	1.656	1.333	2.158
3	1.906	2.094	2.375	2.625
4	2.344	2.531	2.815	3.053
10	1.720	1.910	2.150	2.440

FIG. 67. Outline dimensions of commonly used tube envelopes.

readouts or in timers, counters, and scoring devices. A typical numeral-indicator type is shown in Fig. 70.

Crystals. Crystals are used to control the frequency of an oscillator. The functional part, or crystal plate, is a precisely cut slab of quartz crystal that has been lapped to final dimensions, etched to improve stability and efficiency, and coated with a metal on its major surfaces for connecting purposes. This is installed in a holder to provide support and protection and to allow for interconnection to the circuit. Typical holder dimensions are given in Table 46.

Switching Elements. Switching elements are electromechanical devices used to provide a positive (ohmic) type of circuit connect and disconnect; the two

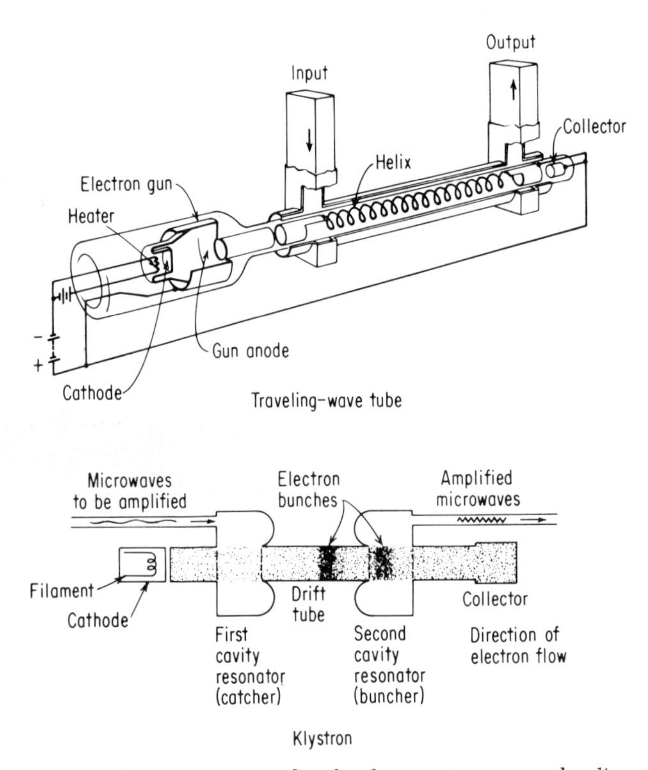

FIG. 68. Construction details of two microwave tubes.[14]

major devices used for this purpose are the relay and the switch. These are similar in that they both use the same basic nomenclature and symbols and are somewhat alike in construction.

RELAYS. Relays are probably the least standardized of any electronic components. This is undoubtedly because they are essentially devices designed to do a specific job, and many variations in type, characteristics, and construction have therefore evolved. Relays can be classified by many methods, as shown in Table 47. The constructions of two relay types are shown in Fig. 71. These categories are not mutually exclusive, however; a relay could, for example, be termed electromechanical, slow, interlock, waterproof, heavy-duty, polarized, or direct-current. The term *general-purpose* is often incorrectly used to define a type of relay. More properly a general-purpose relay is one which is not special-purpose or definite-purpose.

TABLE 44. Characteristics and Applications of Microwave Tubes

Type	Characteristics	Applications
Planar ceramic	Metal ceramic construction; tube elements flat and in separate parallel planes. Operates from dc to about 10 GHz and power outputs up to 300 watts cw. Radiation-resistant. Good linearity, power gain, power output, temperature stability, and gain bandwidth. Small size	Radar beacons, telemetry, mobile transmitters, ecm gear, radiation-exposed equipment, phased array radar
Klystron	Long tube with metallic cavities; reflex type uses reflector; electron beam focused magnetically or electrostatically. Electrons are periodically bunched; the resultant velocity-modulated beam is fed into resonator to sustain oscillations at desired frequency. Amplifies small signals, gives high power output. High reliability, long life. Reflex type is standard oscillator, operating at 1 GHz to *x* band at outputs from 2 to 5 mw. Two-cavity type operates to *Ke* band at 1.5 watts and *x* band to 250 kw. Amplifier types operate to *s* band at 5 mw and *x* band at 200 kw	Radar and communications systems, high-temperature plasma studies; local and master oscillators
Magnetron	A glass or ceramic envelope containing an anode having a number of coupled resonators surrounding a central cylindrical cathode; an external magnet provides a field parallel to the axis of the cathode. The flow of electrons is controlled by a combination of crossed steady electric and magnetic fields so as to produce rf power output. Operates from around 1 to 35 GHz at power to 5 mw at lower frequency and to 1,000 kw at higher frequency. Most power output for cost and weight. Poor stability; not good for coherent applications	Low-noise local oscillators, signal generators, telemetry, altimetry, parametric, pumping, ecm barrage jamming, broadband transmitters
Traveling-wave tube	An electron gun, wire helix, and a collector enclosed in a long glass envelope. The electron beam is directed down the length of the tube through the center of the helix. The helix slows down the speed of rf signal fed in for amplification so that it is synchronous with that of the electron beam. The beam is held in tight focus usually by a magnetic field from permanent magnets arranged periodically along the length of the tube. Operates from 500 MHz to *x*, *c*, and *s* bands, power to 10 kw cw. High gain, low noise, long life, broad bandwidth	Radar, countermeasures, wide-bandwidth amplifiers, point-to-point relays, telemetry, signal amplifies with varying frequency input

Table 48 lists standard relay contact nomenclature and symbols; the operating principles of several common types of relays are outlined in Table 49. Relays are obtainable in open construction (which is the least costly form), or in dust-tight or hermetically sealed enclosures (the most costly). Except for extreme environmental conditions, hermetically sealed units are not necessarily superior in performance to dust-tight units. Relay mountings include simple brackets, studs, clamps, and the terminals themselves; most manufacturers offer several mountings for each type of

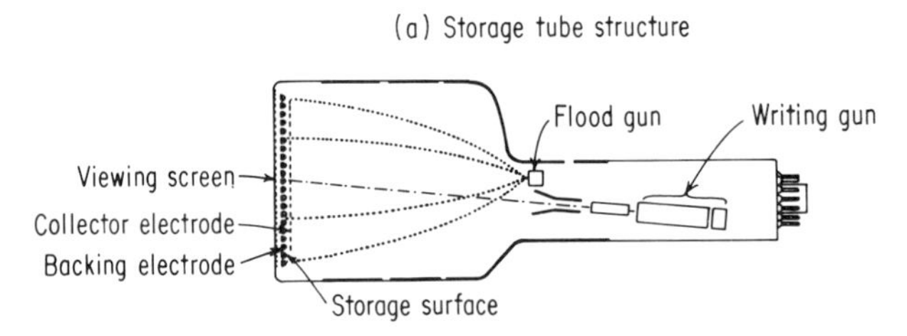

Elements of storage tube, right to left, are: writing gun and its deflection system, flood gun, collector electrode storage surface, backing electrode and tube face. Writing gun, deflection system and viewing screen are similar to conventional CRT

Simplified explanation of operation: Storage surface elements will become charged either positive, negative, or neutral, depending on energy of electrons from writing gun. If gun voltage is such that secondary emission ratio is greater than 1, element becomes positive. If ratio is less than 1, element becomes negative. If ratio is exactly 1, no charging occurs. Charge pattern remains, write gun turns OFF and flood gun turns ON. Low-energy flood gun electrons either are repelled and exit via backing electrode or are allowed to pass through to phosphor viewing screen depending on charge distribution. Erasing is accomplished by adjusting backing plate potential and flooding storage surface with flood gun electrons

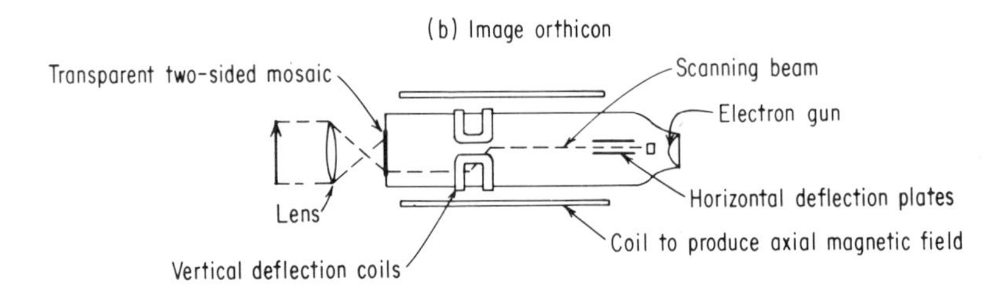

In this camera tube, a beam of low-velocity electrons scans a photoemissive mosaic that is capable of storing a pattern of electric charges

FIG. 69. Construction details of storage and image orthicon tubes.

relay. A wide variety of terminations are also available; the more common types are listed in Table 50. Almost any combination of type, mounting, and termination can be obtained, but considerable savings can be realized by using a standard type. Several of these are shown in Table 51.

SWITCHES. Switches are available in almost as many styles and functions as relays, and any classification system is overlapping in nature. Table 52 gives a grouping basically by actuation method. Examples of types often experienced in packaging are shown in Fig. 72; common switch mountings are described in Table 53.

Hardware Information. Items which primarily perform a mechanical function in an assembly are covered here. It is recognized that terminals do transmit an electrical signal, but they do not change its form or function, and so are considered basically as mechanical items.

Clamps and Clips. Clamps and clips are used where parts are massive or are not provided with suitable mounting provisions. On printed-wiring boards, components which weigh over ½ oz or have a large length-to-diameter ratio must be secured to the board; if the component is subject to replacement, clamping holders should be used. Figure 73 shows the dimensions of some typical types of clamps.

Sockets. Sockets are often used for tubes, relays, transistors, or crystals where the items are subject to frequent change. The more commonly used styles are illustrated in Fig. 74.

Printed-wiring Card Guides. Printed-wiring card guides serve the purpose of guiding the board into proper position for connector mating, securing the board from movement during shock and vibration, and forming a path for conductive heat transfer. Examples are given in Fig. 75.

Terminals and Lugs. Terminals and lugs are used to mount smaller components or as tie points and terminations for individual wires. Common used types are shown in Fig. 76. Data on connectors for termination of cables are given in Chap. 6.

Threaded Fasteners. Threaded fasteners are devices which have an externally threaded member designed for insertion into holes in assembled parts for mating with a preformed thread or forming its own thread. Only the smaller sizes likely to be used in the electronics package are described.

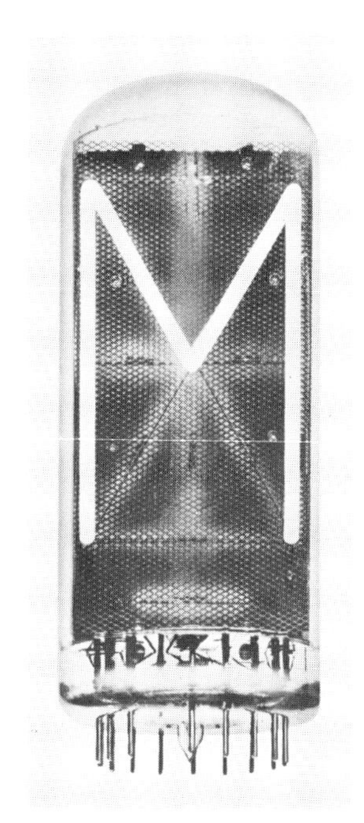

FIG. 70. Cold cathode numerical indicator tube. (*Burroughs Corp.*)

THREAD IDENTIFICATION. The method of thread identification is illustrated in Fig. 77. Sizes of typical screws are given in Table 54, and screw configurations are shown in Fig. 78; matching nuts are shown in Table 55. Tap drill sizes and screw clearance-hole diameters are provided in Table 56.

LOCKNUTS. Nuts having special internal means for gripping a threaded fastener or connected material to prevent rotation in use are called *locknuts.* Usually these have the same dimensions and mechanical characteristics as standard nuts, but with the locking feature added. Typical types are shown in Fig. 79. A locking feature

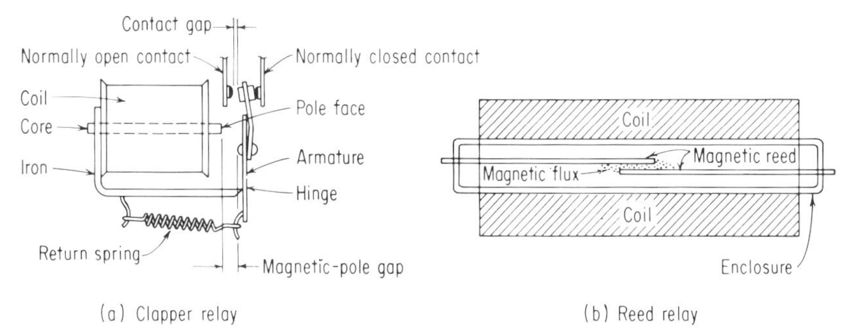

FIG. 71. Construction of two relay types: (*a*) Clapper relay. (*b*) Reed relay.[11]

Toggle Switch		L	D	H	h	Bushing thread
Standard	– SPDT···	1.3	0.625	1.0	0.480	15/32 - 32
	DPDT···	1.3	0.750	1.0	0.480	15/32 - 32
Miniature	– SPDT···	0.562	0.350	0.749	0.170	1/4 - 40
	DPDT···	0.562	0.560	0.749	0.170	1/4 - 40
Subminiature	– SPDT···	0.492	0.250	0.515	0.354	1/4 - 40
	DPDT···	0.492	0.450	0.515	0.354	1/4 - 40

Snap Switch (Momentary contact)	L	D	H
Standard ········	1.94	0.68	0.91
Miniature ········	1.09	0.41	0.63
Subminiature ····	0.78	0.25	0.38
Subsubminiature ····	0.50	0.25	0.38

Rotary Switch (Available in "make-before-break" and "break-before-make")

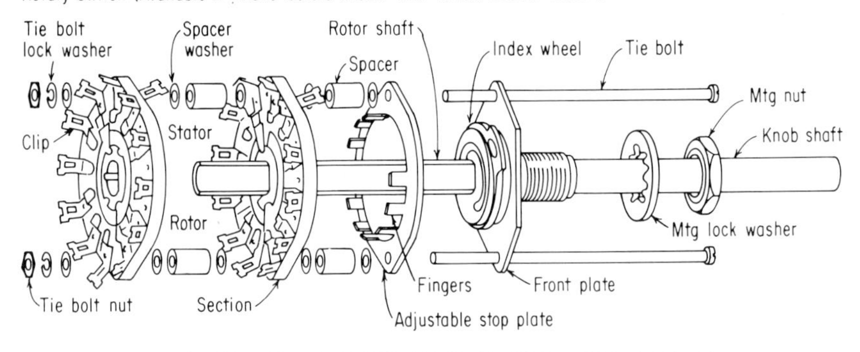

	Section diam	Bushing thread	Shaft diam		Positions per section
Standard ···	1.875	3/8 - 32	1/4	0.625	2 to 24
	1.500	3/8 - 32	1/4	0.558	2 to 24
Miniature ···	1.280	3/8 - 32	1/4	0.558	2 to 24
Subminiature	0.940	1/4 - 32	1/8	0.125	2 to 12

FIG. 72. Typical switch configurations.

TABLE 45. Characteristics and Applications of Camera and Display Tubes

Type	Characteristics	Applications
Cathode-ray tubes (CRT)	Electron source (cathode), accelerator, and control electrodes comprise "electron gun." This is housed at one end of glass envelope and aims electrons at other end, which is flat and coated with phosphor. Lit phosphor image visible on outside. Also available with fiber optic and conductive faceplates for direct graphic recording	Scopes, cockpit, displays, remote control, TV, strategic military displays
Storage tubes	Similar to CRT but has grid or storage surface mounted just behind faceplate which passes or blocks electrons in controlled manner. Allows variation in persistence, erase time, contrast, shades of gray, and selective erase	ATC and tactical displays of moving targets, slow-scan TV monitoring, medical diagnostic displays, controlled fading
Camera tubes	Long glass envelope with light-sensitive surface (photocathode) at one end, focusing or deflection coils along length, and collecting output at other end. Includes image orthicon, vidicon, and image dissector which transmits pinpoint parts of image. Orthicon is most stable, but larger, more complex, and less rugged than vidicon	TV cameras, space vehicles, military surveillance, document retrieval

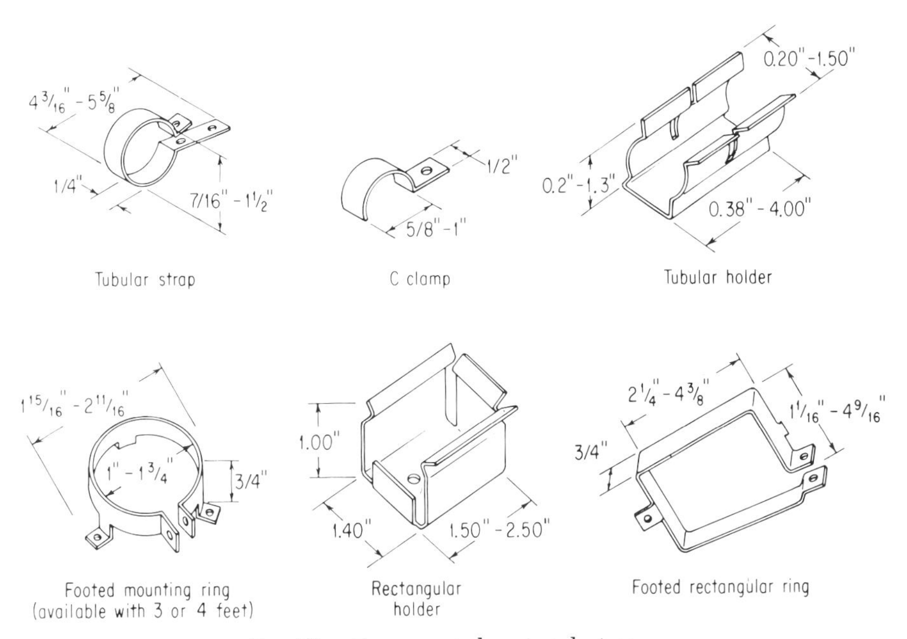

FIG. 73. Component-clamping devices.

is often added to the screw itself by placing an insert of nonmetallic material in the thread to provide a jamming action.

CAPTIVE NUTS. Captive nuts, also known as *retaining nuts,* are used for blind locations or for materials too thin to be tapped. Several types are shown in Table 57.

INSERTS. Inserts are used to provide threads in materials that ordinarily are not suitable for threading, such as plastic or light alloys. They also allow higher-strength threads than might be possible in the parent material. Several types are shown in Fig. 80. They are specified by standard screw size, and all require area in excess of a tapped thread.

Saddle type

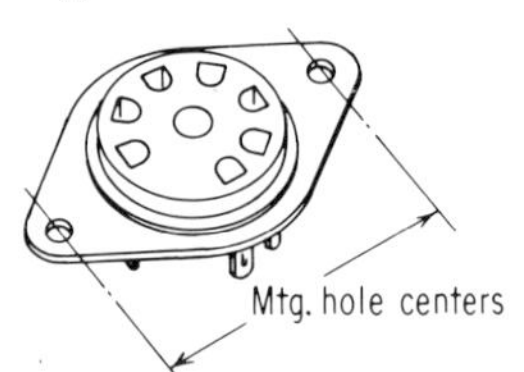

Available: top mount and bottom mount. 7-pin, 9-pin, and octal styles, with and without tube shield base, in standard and miniature sizes. Also octal style in 11, 14, 16 and 20 pins for relays.

Mtg. Mole Centers

7-pin = 7/8 in.
9-pin = $1^1/_8$ in.
Octal = $1^5/_{16}$ to $1^1/_2$ in.

Right-angle printed-wiring board (direct solder)

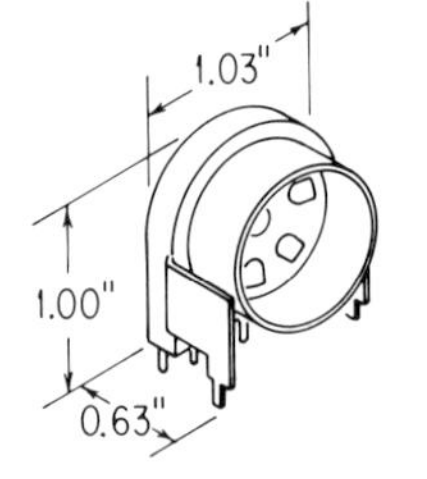

Available: With and without tube shield base; 7-pin and 9-pin styles. Also upright version

Subminiature tube or transistor

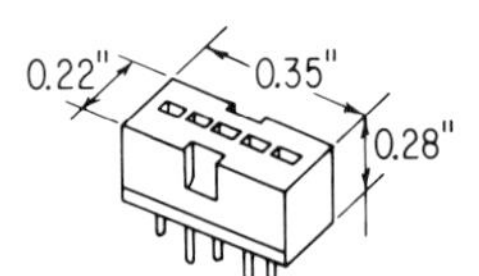

Blade contact for relay

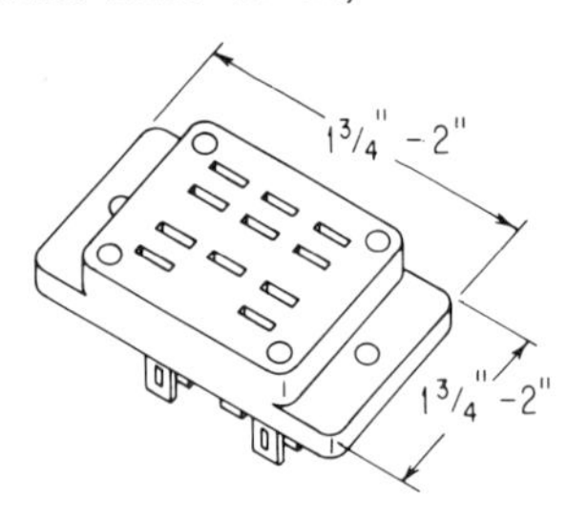

Transistor

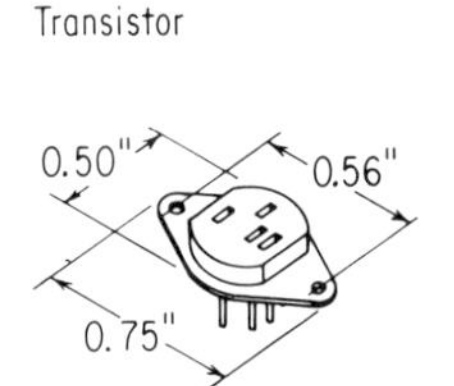

Crystal

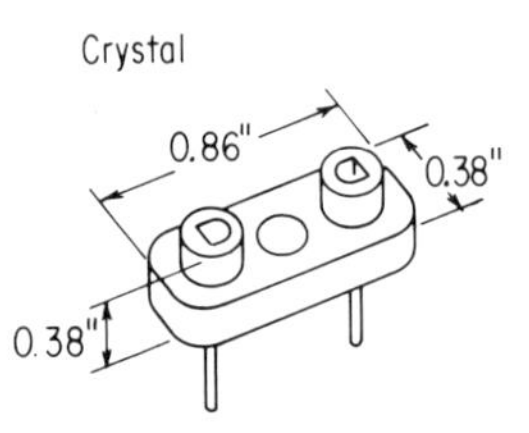

FIG. 74. Typical socket types.

WASHERS. Washers are used to give better bearing for nuts or screw faces, to distribute load, as locking devices, to prevent surface marring, and often to provide a seal. They are specified to fit a particular screw size. Typical types are shown in Fig. 81.

QUICK-OPERATING FASTENERS. Quick-operating fasteners are used where repeated access is required. Typical types are shown in Fig. 82. Many of these are not threaded but can replace threaded units.

TABLE 46. Dimensions of Typical Quartz Crystal Holders (MIL-H-10056)

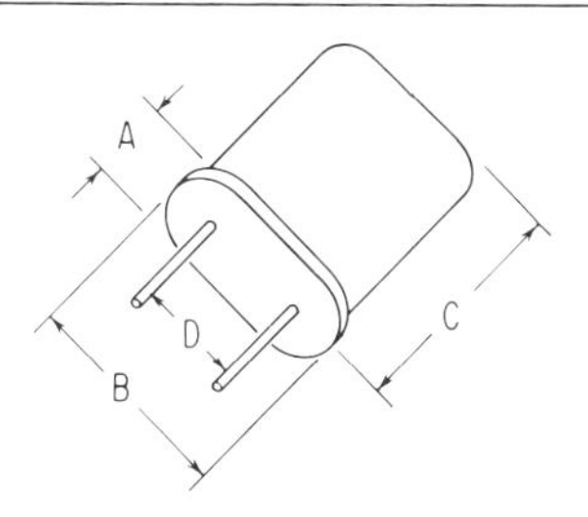

Holder	Dimensions, in.				
	A	B	C	D	Pin, diam.
HC-6/U	0.345	0.750	0.765	0.486	0.050
HC-11A/U	0.455	1.165	1.282	0.500	0.125
HC-13/U	0.345	0.750	1.516	0.486	0.050
HC-14/U	0.345	0.750	0.574	0.486	0.050
HC-17/U	0.345	0.750	0.765	0.486	0.093
HC-18/U	0.147	0.399	0.515	0.192	0.017
HC-25/U	0.147	0.399	0.515	0.192	0.040
HC-28/U	0.352	0.757	1.526	0.486	0.050
HC-32/U	0.202	0.530	0.640	0.275	0.040

Plastic

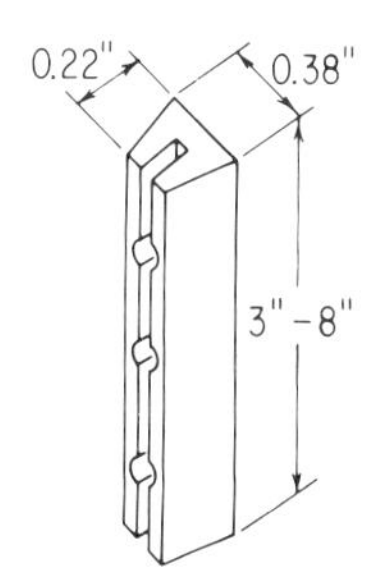

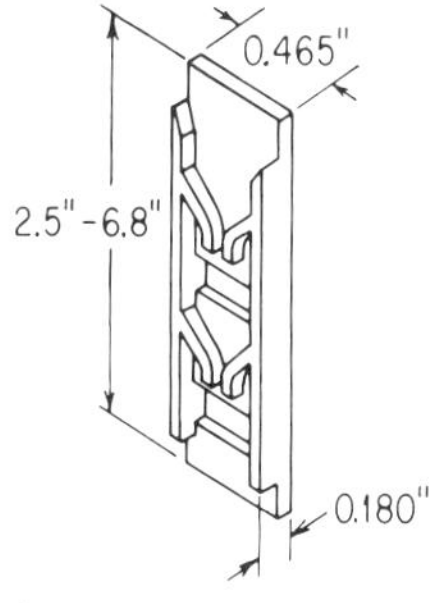

Metal

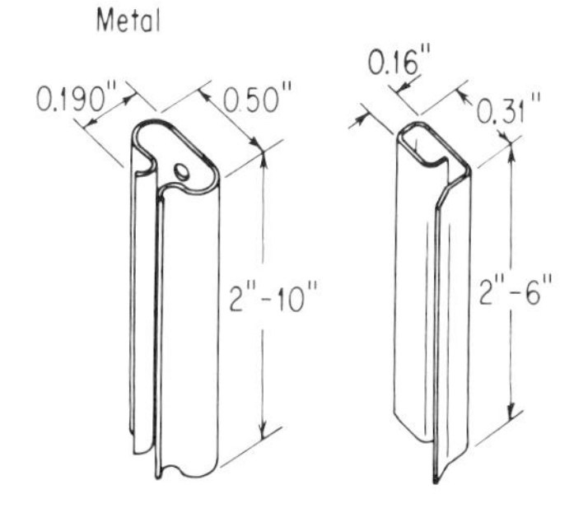

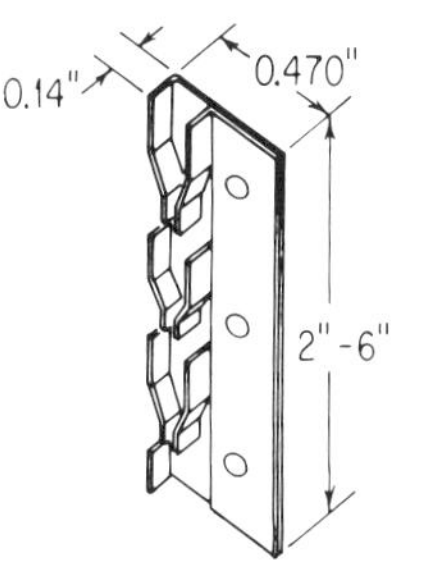

Fig. 75. Printed-wiring card guides.

Terminals

Normally used to mount smaller component or as tie points for wires. Available in screw-mounted, swaged-over, or press-in styles, and a wide variety of post configurations. They are ordered by style, size, and shank length to fit mounting board thickness. Terminal diameters range from 0.093 in. for subminiature sizes to 0.250 in. for the larger sizes

Mounting

Stud Screw Swaged Press-in

Typical styles

Turret

Insulated

Feed-through

Special Taper pin Test point

Solder lugs

Normally used for grounding leads to chassis. Ordered in standard screw sizes

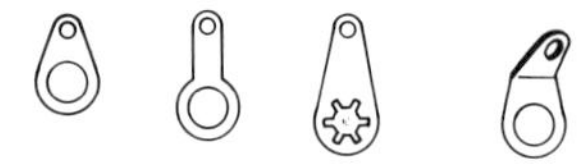

Wire-end lugs

Normally used to terminate individual wires to screw terminals or binding posts. Available in solderless (crimp) or solder styles, and with insulated or noninsulated shanks. Ordered for wire size used and connecting stud size

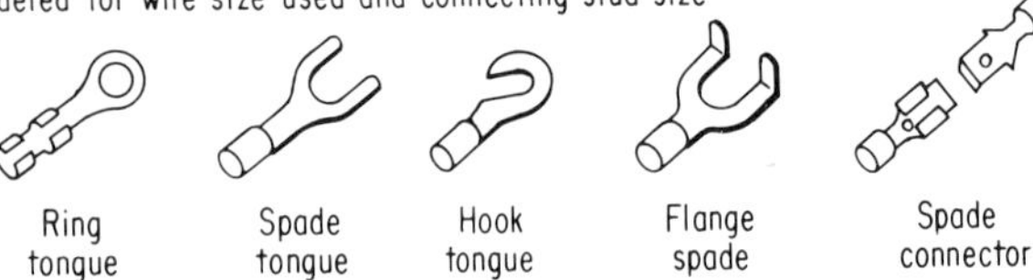

Fig. 76. Terminals and lugs.

TABLE 47. Relay Classification Methods [18]

By electrical control:
- Electromagnetic
  - Moving iron (plunger, clapper, ball, rotary)
  - Moving permanent magnet
  - Moving conductor
  - Electric coil
- Thermal
  - Bimetal drive
  - Pressure drive
  - Epansion drive
- Magnetostrictive
- Electric field
  - Piezoelectric
  - Electrostrictive
  - Electrostatic

By performance type:
- Marginal
- Fast
- Slow
- Sensitive
- Timing
- Latching
- Sequential
- Frequency-sensitive

By mechanical action:
- Two-position
  - Interlock
  - Latching
  - Ratchet
- Three-position
- Multiposition
  - Stepping
  - Coordinate
- Crossbar
  - Sequential

By enclosure type:
- Open
- Enclosed
  - Partially enclosed
    - Enclosed coil
    - Enclosed contacts
  - Fully enclosed
    - Dustproof
    - Waterproof
    - Explosionproof
    - Weatherproof
- Sealed
  - Partially sealed
    - Sealed coil
    - Sealed contacts
  - Gasket-sealed
  - Hermetically sealed

By duty:
- Light duty
- Medium duty
- Heavy duty

By electrical input:
- Direct current
  - Neutral
  - Polarized
- Alternating current

By load requirements:
- Direct current
- Power (heavy duty)
- Alternating current
  - Commercial frequencies
  - Radio frequencies
  - Higher frequencies

Typical Identification of Thread:

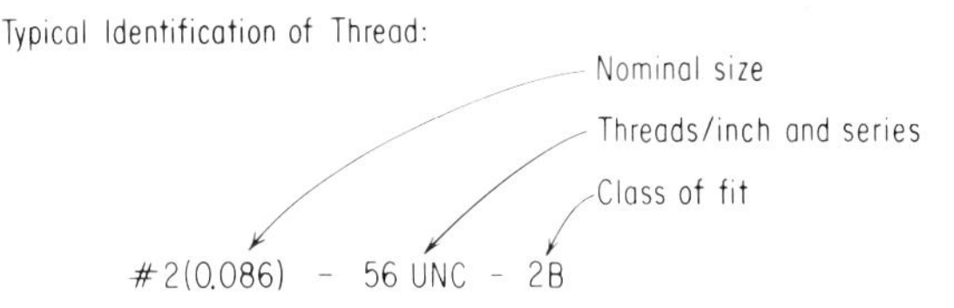

Explanation:

Nominal size - This is the major diameter of the screw or nut, and is usually given as the nominal fractional size or wire size; if a wire size, the decimal equivalent in inches should be shown in parentheses

Threads/inch - This is the number of threads per inch. This is not the pitch, which is the distance between threads in inches

Series - This identifies the thread as to form or configuration. Normally this will be stated as UNC for Unified National Coarse which has the fewest threads/inch or UNF which has the most. Use UNC for rapid assembly or soft materials, and UNF for high strength or limited thread engagement

Class of fit - The number here identifies the thread specification as to diameters (major, minor, pitch) and tolerances. The letter A identifies an external thread as on a screw; the letter B identifies an internal thread as in a nut. Class 2 is a medium loose fit, and is the standard for production of commercial fasteners. Class 3 is the closest fit, and used for critical applications. Class 1 is a very loose fit and is used only on larger sizes

FIG. 77. Thread identification.

TABLE 48. Relay Contact Nomenclature and Symbols[11] *

Form	Description	Symbol
A	Make or SPSTNO	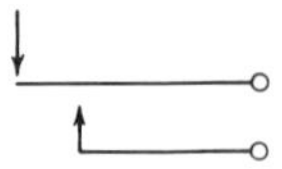
B	Break or SPSTNC	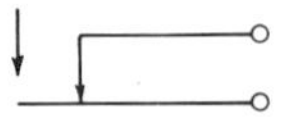
C	Break, make or SPDT (B-M)	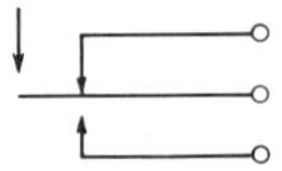
D	Make, break or SPDT (M-B)	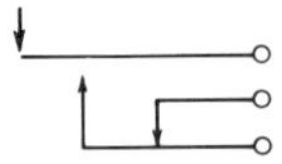
E	Break, make, break or SPDT (B-M-B)	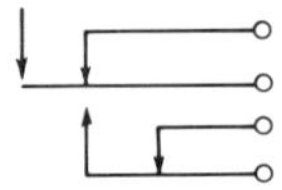
F	Make, make or SPST (M-M)	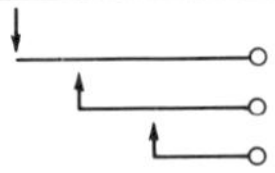
G	Break, break or SPST (B-B)	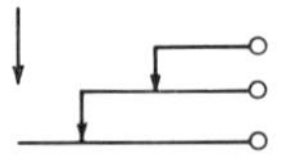
H	Break, break, make or SPST (B-B-M)	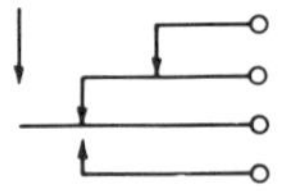
I	Make, break, make or SPST (M-B-M)	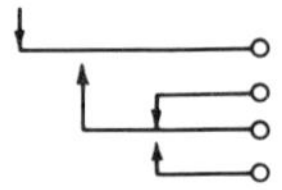
J	Make, make, break or SPST (M-M-B)	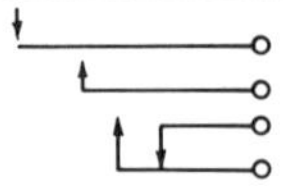
K	Single pole, double throw, center off or SPDTNO	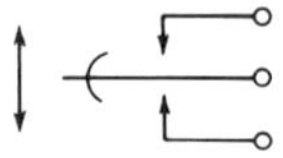
L	Break, make, make or SPST (B-M-M)	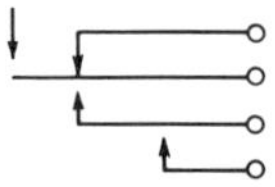
U	Double make, contact on arm or SPSTNODM	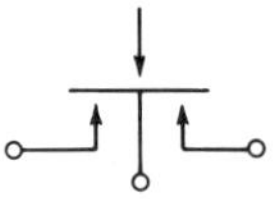
V	Double break, contact on arm or SPSTNCDB	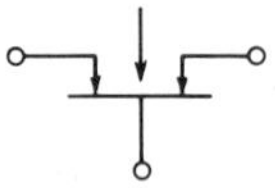
W	Double break, double make, contact on arm or STDTNC-NO(DB-DM)	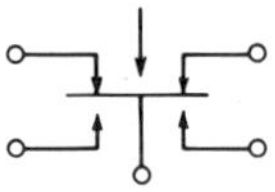
X	Double make or SPSTNODM	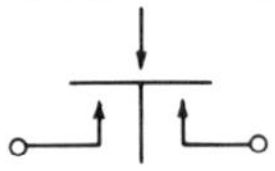
Y	Double break or SPSTNCDB	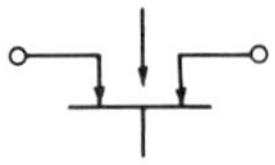
Z	Double break, double make or SPDTNC-NO(DB-DM)	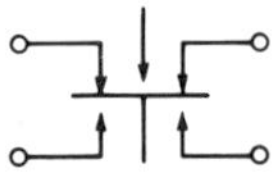

* The current-carrying parts of a relay which are used for making and breaking electrical circuits are available in various combinations of contact forms. Switching abbreviations are given in the following order: poles—single (SP), double (DP); throws—single (ST), double (DT); normal position—open (NO), closed (NC), double make (DM), and double break (DB).

Single-throw contact forms have a pair of contacts open in one armature position and closed in the other. The form in which the contacts are open in the normal or unoperated position of the relay is designated *normally open;* it is designated *normally closed* when the contacts are closed in the normal or unoperated position.

Double-throw contact combinations have three contacts, one of which is in contact with the second but not with the third in one position of the relay, and in the reverse connection in the other relay position. The basic double-throw contact form is the break-before-make (Form C).

Double-make and double-break contact forms have two independent contacts that are both connected to a third contact in one position of the relay. The contacts are designated *double-make* when normally open and *double-break* when normally closed.

Classifications of compounded contact arrangements such as for relays having more than two positions are designated by the symbol *MPNT*, where *M* signifies the number of poles and *N* the number of throws. Thus, 8P20T identifies a relay with 8 poles and 20 throws (as in a 20-step relay with 8 banks of contacts).

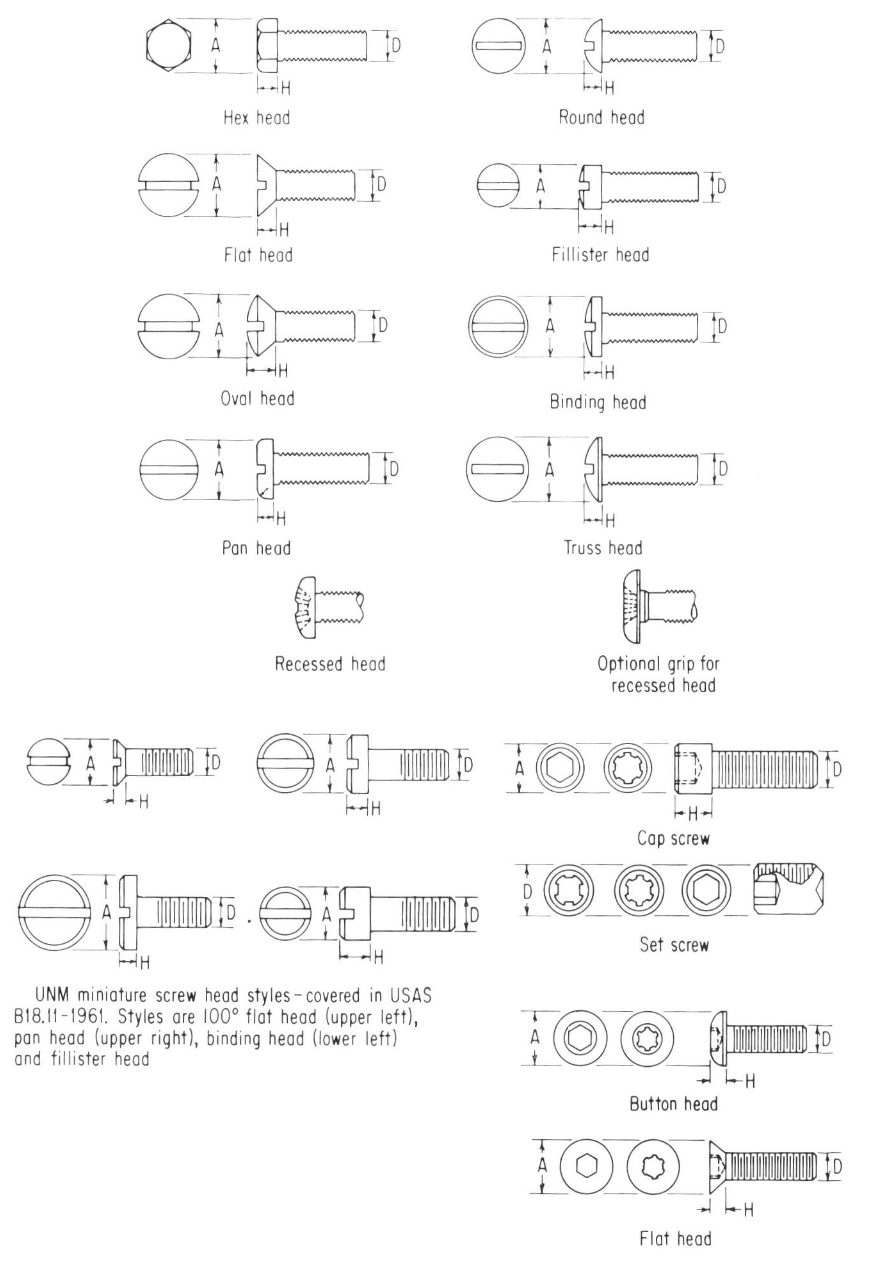

FIG. 78. Miniature screw head styles.[15]

TABLE 49. Description of Basic Operating Principles of Relays

Type	*Description*
Armature relay	Operation depends upon energizing an electromagnet which attracts a hinged or pivoted lever of magnetic material to a fixed pole piece. Normally refers to a *clapper relay*
Plunger relay	Operated by an electromagnetic coil or solenoid which when energized operates a movable core or plunger to which the movable contacts are attached. Used where considerable contact pressure is required; not normally used under conditions of shock and vibration
Rotary relay	Operates by the rotation on an armature to close the gap between two or more pole faces. Most have balanced armatures and are used primarily under conditions of vibration and shock
Meter relay	Also called *instrument relays.* Operate through the use of modified D'Arsonval meter movements as the actuator. Very sensitive devices; used in applications requiring critical current or voltage sensing
Reed relay	Operates by an electromagnetic coil which causes two thin, flat magnetic strips (or reeds) to be attracted to each other; the reeds serve as conductor contacts, springs, and magnetic armatures. Contacts are usually sealed in a glass capsule. May be dry-reed or mercury-wetted contact construction. Extremely fast operation and of small size
Thermal relay	Uses the heat produced by a resistance element to provide mechanical motion to a contact mechanism, often a bimetal strip. Since time is required to heat the strip, these devices are often used as time-delay relays

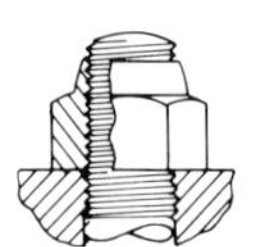

Slotted section of this prevailing-torque nut forms beams which are deflected inward and grip the bolt

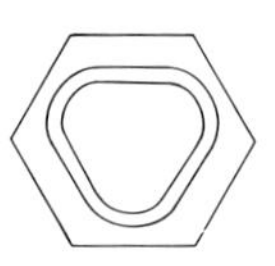

Three sectors of tapered cone, preformed inwardly, are elastically returned to circular form when the nut is applied

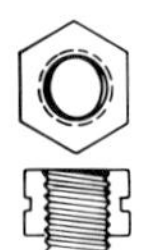

Nondirectional, out-of-round nut. Locking power comes from deflected threads in the center of the nut

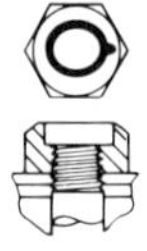

Threaded elliptical spring-steel insert grips the bolt and prevents turning

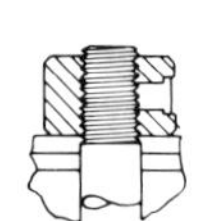

Nonmetallic plug insert grips the threads and causes a wedging action between bolt and nut

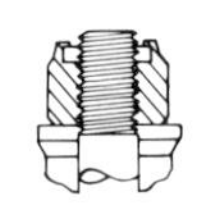

Nonmetallic collar clamped in the top of this nut produces locking action

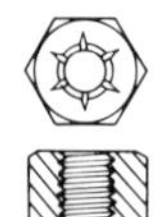

Deformed thread type. Depressions in the face of the nut distort a few of the threads

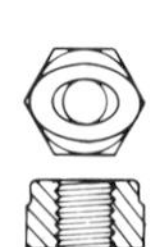

Out-of-round threaded collar above regular load-bearing threads grips the bolt

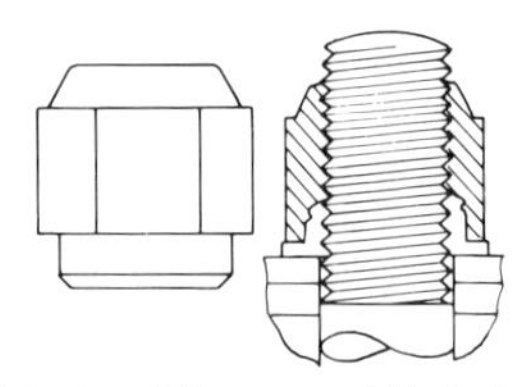

Nylon insert flows around the bolt rather than being cut by the bolt threads to provide locking action and an effective seal

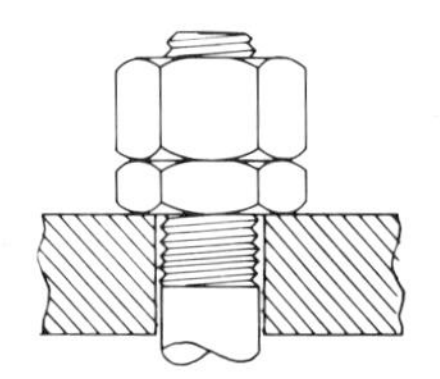

Jam nut, applied under a large regular nut, is elastically deformed against bolt threads when the large nut is tightened

FIG. 79. Several lock-nut types.[15]

TABLE 50. Common Relay Terminations

Termination type	*Description*
Lead wires	Relatively long wires to the relay coil only; normally not supplied on contact terminals
Printed-wiring-board terminals	Short straight pins arranged on a standard grid pattern (usually multiples of 0.010 in.) which are inserted into printed-wiring holes and soldered in place
Solder terminals	Includes eyelets, solder cups, stiff bent wire, flat pierced tabs, etc. Usually characterized by hook or hole used to secure lead as a strain relief for solder joint
Screw terminals	Provides a compression-type joint. Used where frequent change may be necessary
Quick connectors	Typical type is a flat tab on relay with flat spring tube on end of connecting wire
Plug and socket	Available in most standard tube sizes from miniature to octal. Facilitate replacement of relay, but relay must be retained under vibration or shock loads. In low-voltage circuits, added resistance of connector interface may be undesirable
Solderless wrap or clamp	Patented methods in which lead wire is wrapped or clamped against a square or rectangular relay contact post. Special tools required

TABLE 51. Data on Several Standard Relay Types

Hermetically Sealed Crystal Can Style
Size = 0.88 × 0.80 × 0.36 in. (approx.)

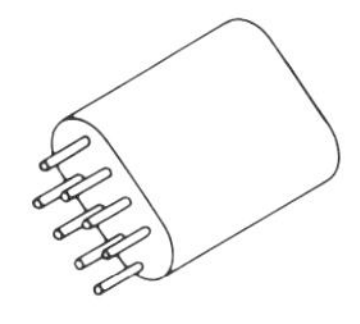

Plug-in-
no bracket

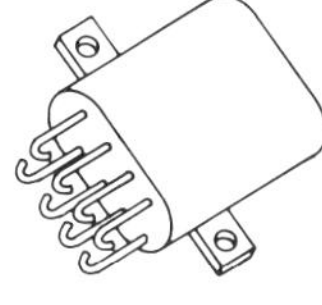

Solder hooks-
side bracket

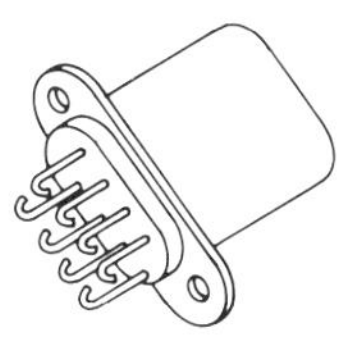

Solder hooks-
flange bracket

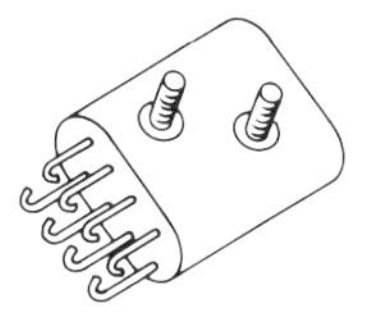

Solder hooks-
threaded studs

Hermetically Sealed
One-half Can Style

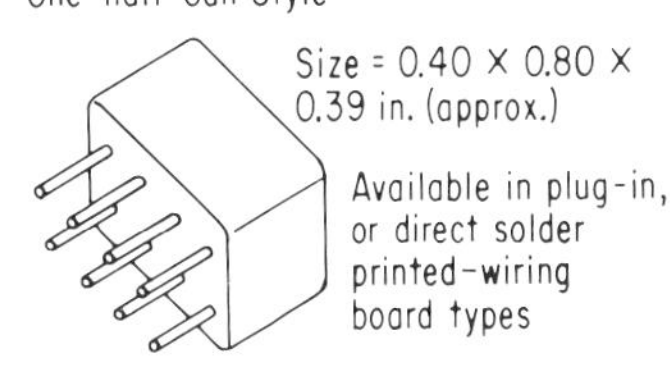

Size = 0.40 × 0.80 × 0.39 in. (approx.)

Available in plug-in, or direct solder printed-wiring board types

Dust Cover or Enclosed Style

Size = 1.38 × 1.38 × 2.13 in. (approx.)

8-pin octal, or 11-pin plug-in types

Reed style

L W H 0.13

Size (approx., single pole)

	L	W	H
Microminiature	1.17	0.28	0.28
Miniature	1.44	0.44	0.44
Standard	1.70	0.60	0.68

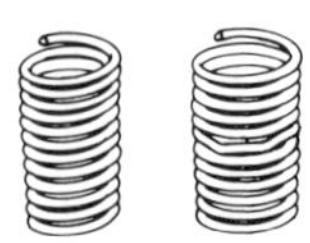

Diamond-shaped wire insert with or without self-locking configuration. Special driving tool used to install insert in most engineering materials. Driving tang can be broken off after installation. Insert diameter is slightly oversized and when wound into correctly sized tapped hole, the outward spring action of the coil locks it in place

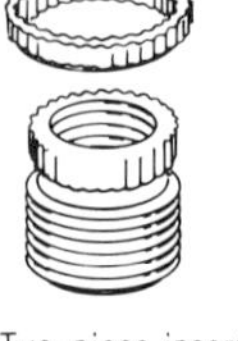

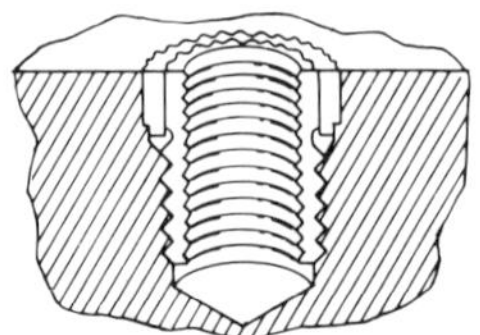

Two-piece insert with serrated locking ring. Tapped hole is counterbored to accommodate locking ring. Internal teeth on ring engage external teeth at top of insert while external teeth on ring broach counterbore wall when ring is driven into place. Insert and lock ring installed with special tools in any material that can be drilled and tapped. Available with split end for internal self-locking feature

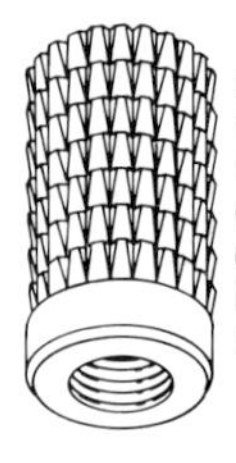

Knurled and serrated external surface for pressing into soft materials (plastic, synthetics, rubber, etc.) in the hot or cold condition. Flow of material into depressions in surface locks insert in hole. Installed with standard tools

Self-tapping external threads have portion cut out which acts as thread-cutting edge in drilled or cored holes. Internal threads are broached to receive a hexagon socket wrench for driving. Thread-cutting action produces interference fit to lock insert in place

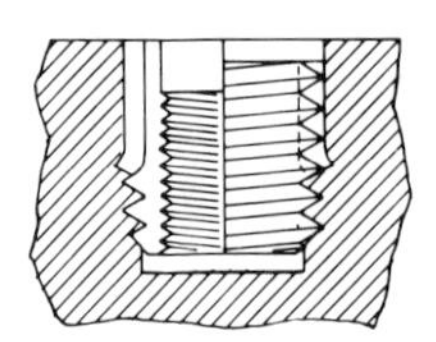

Two-piece insert for most materials installed with simple tool in counterbored tapped hole. Locking ring has projecting keys which mate with axial grooves in external threads. Ring is pressed into counterbore after insert is screwed into place and keys cut tapped threads to lock insert in place. Available with nylon self-locking collar, or beam-locking feature for internal threads

Top of insert has splines to prevent rotation. Remainder of insert is flanged. Insert is driven into hole and expanded so that sharp flange edges engage hole wall to lock insert in place

FIG. 80. Several types of inserts.[15]

Tooth Lock Washers

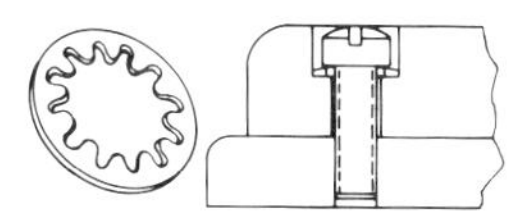

Internal – Used where projection of teeth beyond screw head or nut is objectionable. Presents a finished appearance, similar to plain washer. Standard sizes from No. 2 to 1 1/4 in. up to 2 3/8 in. available. Heavy washer made in 1/4 in. to 7/8 in. sizes

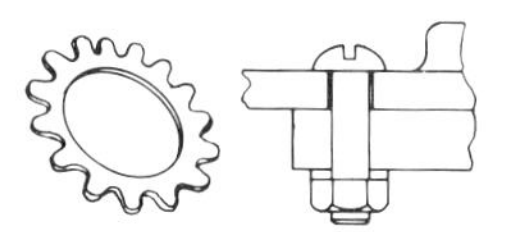

External – Most commonly used where security of assembly is critical. Teeth bite into standard screw heads and nuts to locks. Standard sizes from No. 3 to 1 in. up to 2 1/4 in. available

Flat Washers

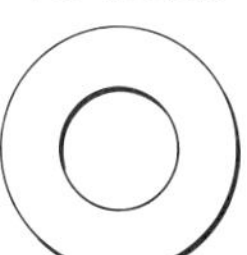

Plain, or flat, washers are used primarily to provide a bearing surface for a nut or screw head to cover large clearance holes, and to distribute fastener loads over a larger area, particularly on soft materials such as aluminum or wood

Sealing Washers

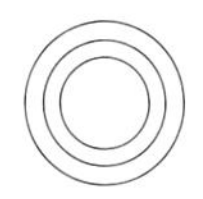

O ring

Bonded rubber

Laminated neoprene

Used to stop leakage or reduce moisture penetration

Helical Spring Washers

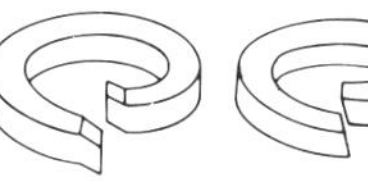

Nonlink-positive Plain

Spring take-up device to compensate for developed looseness. Not recommended for use on soft materials

Spring Washers

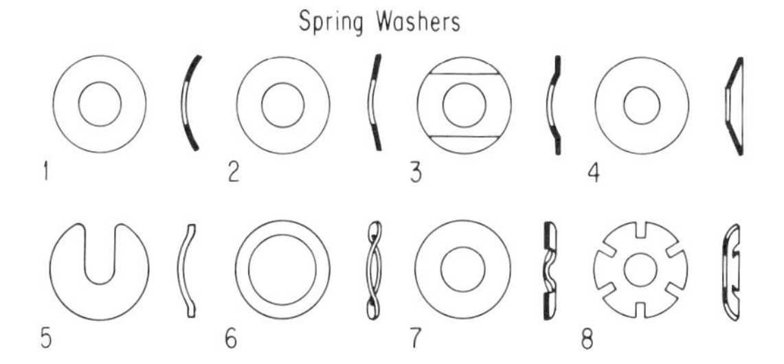

Used to produce predetermined pressure on adjacent members, as spring take-up devices, or to eliminate end play. Type 1 is the basic design and the most commonly used type

Fig. 81. Several washer styles.[15]

TABLE 52. Switches Classified as to Method of Actuation

Basic classification	Type	Characteristics	Applications
Manual (actuated by operator's hand)	Linear: Toggle	Most widely used form; often called "bat handle." Majority have maintained or momentary action and two or three operating positions. Positive position indication. Accidental actuation easy unless guarded	General-purpose use on central panels to start or stop a function, or to change parameters or inputs one at a time. Function limited to three conditions per switch
	Rocker	Modification of toggle with push action for actuation. Combines ease of operation with positive position indication	Same as above, but has lower profile on panel and better appearance
	Slide	Actuated by moving button (or sliding) from one position to another. Available in up to six stations. Danger of overshooting position stop. Positive position identification	Same as toggle, with lower profile, and can provide more function conditions
	Push-button	Actuated by in-line thrust; easy to operate. Must have method to indicate condition for position	Same as toggle, but usually limited to two function conditions
	Rotary	Actuated by twisting action of the hand. Rotation may be continuous or limited to 360°. Positional differences to 10° are available, but 30°, 45°, and 90° are more common. Momentary-contact versions usually confined to three positions. Shorting (make-before-break) or non-shorting (break-before-make) contacts are available. Multiple decks can be used to switch multiple signals simultaneously. Sliding or wiping contacts	Relatively low current levels. Selecting one or more circuits in a sequence. Test circuits, parameter variations changing function
Snap action		Small switch with closely spaced contacts snapped together or apart by spring action. Actuated by forces from 2 to 30 oz controlled within 1-oz spread and differential movements as small as 0.0005 in. Most are plastic-encased and mounted by two screws through holes in body	Door switches, travel limiters, timing controls, any application where only low force levels or small movements are desired

TABLE 52. Switches Classified as to Method of Actuation (*Continued*)

Basic classification	Type	Characteristics	Applications
Mercury tilt		Sealed glass or metal tube with stationary electrodes and pool of mercury; actuated by tipping tube to short electrodes with mercury pool. Temperature range from −35°F to over 400°F; reasonably rugged, particularly when encapsulated. Has low contact resistance, position sensitivity, and low force actuation	Motor controllers, ultrasensitive thermostats, gyro devices
Temperature (react to temperature changes)	Thermostats	May have force-limited, differential expansion, or liquid-filled elements. Temperature ranges from −40 to 1500°F	Thermal control of environment, high- or low-temperature control of equipment, measuring and indicating circuits
	Remote bulb	Gas-filled, vapor-filled, or liquid-filled elastic container. Movement of expansion transmitted to actuate contacts. Temperature ranges from −50 to 1500°F	
	Electronic	Resistance bulbs, thermocouples, or thermistors whose electrical characteristics change with temperature. Temperature −100 to 4500°F. Small size, accuracy to 1%, quick response	
Proximity (actuation by presence or absence of object without physical contact)	Rf inductive	Actuated by change in inductive coupling of coil in sensor. Metal sensing. Good sensitivity and resolution, operating frequencies from 10 kc to 1 Mc	High-speed counting and control, zero speed or motion indication, sensing ferrous machine-tool positioning control, material level control
	Magnetic-bridge inductive	Actuated by unbalancing a four-arm bridge when ferromagnetic material is near operating frequency from 60 to 500 Hz	
	Capacitive	Capacitance of probe to ground balanced in bridge circuit, change in ground plate area or dielectric medium unbalance bridge. Operating speed relatively low	
	Photoelectric	Actuated by interruption of beam of light between source and detector. Frequency of operation above 1,000 Hz	
	Ultrasonic	Requires two heads, one to transmit ultrasonic beam and other to receive it. Operates in poor environments	

Turn-operated

Standoff thumb screw

A polished screw and retaining device, which captivates the screw into a standoff. Widely accepted for use of electronic equipment. Attractive appearance. Can be fully disengaged without backing panel off door. Will operate in a tapped hole. Screw retracts completely for providing zero inside projection where panels move laterally with respect to each other

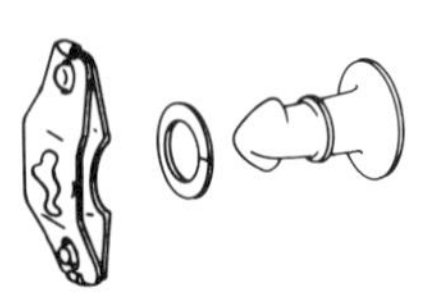

Quarter-turn fastener

A stud, locked into the door or cover panel by a retainer and engaging into a receptacle on the frame or chassis. Very fast actuation. Good resistance to vibration and shock. Light weight. Conserves space. Variety of head styles available

Lever-actuated

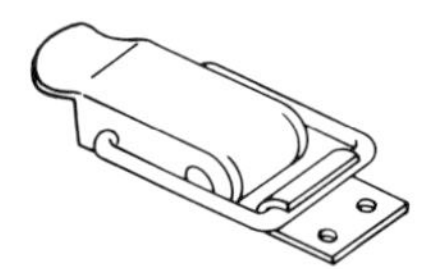

Draw-pull catch, bail type

A spring or drawhook assembly, with operating lever engaging a keeper or striker. Will handle edge-to-edge applications. Good leverage for pulling parts together. No parts inside assembly

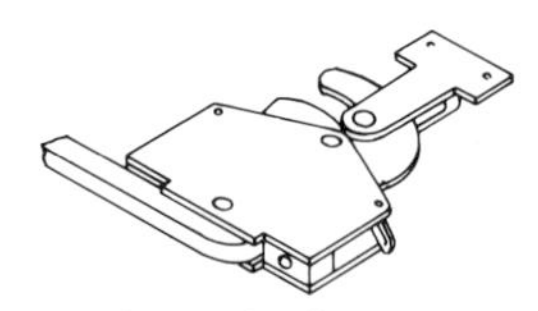

Cam-action fastener

A cam assembly, operated by a lever or handle, engaging a keeper. Quick disconnect (positive travel can be used to break or set up a circuit). Good resistance to shock, pull-out and vibration. Actuating lever can serve as carrying handle for subassembly

Slide-actuated

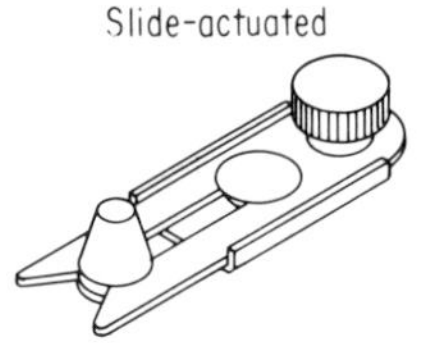

Snap-slide latch

A slide latch moving in a guide affixed to removable panel with a rivet stud. A locking stud is mounted on the fixed or frame member. Locking stud passes through hole in panel to be engaged by slide latch at shoulder

FIG. 82. Typical quick-operating fasteners.[15]

TABLE 53. **Common Switch Mountings**

Bushing mount	A threaded bushing, usually supplied as a part of the switch, is inserted through a single hole in the panel, and a nut is installed on the bushing from the other side. In rotary applications, a tab on the switch mates with a smaller hole in the panel, or a D hole is used to fit a flat on the bushing. This is the most used mounting method
Multihole mount	In addition to a hole for the button, lever, or shaft, additional holes are provided for mounting screws. Often these are blind-tapped from the rear so that the screw head does not show on front of panel
Spring mount	Usually a push-button style. Flat springs on the side of the switch case snap out and prevent removal when the switch is pushed through a single cutout in the panel
Multiple mount	Gangs of switches are preassembled and then inserted into a single-panel cutout. The matrix is secured with screws or other fasteners

TABLE 54. **Smaller Screw Dimensions**

Unified Miniature Screw Thread Dimensions *

Size designation, UNM	Threads per in.	Basic major diam., in.	Basic pitch diam., in.	Basic minor diam., in.	Minor diam. external threads, in.	Major diam. internal threads, in.	Lead angle at basic pitch diam.		Sectional area at minor diam. at D −1.28p,† in.^{2} × 10^{-4}
							Deg	Min	
0.30	318	0.0118	0.0098	0.0085	0.0080	0.0120	5	52	0.475
0.35	282	0.0138	0.0115	0.0101	0.0095	0.0140	5	37	0.671
0.40	254	0.0157	0.0132	0.0117	0.0110	0.0160	5	26	0.901
0.45	254	0.0177	0.0152	0.0136	0.0130	0.0180	4	44	1.262
0.50	203	0.0197	0.0165	0.0146	0.0138	0.0200	5	26	1.407
0.55	203	0.0217	0.0185	0.0165	0.0157	0.0220	4	51	1.852
0.60	169	0.0236	0.0198	0.0175	0.0165	0.0240	5	26	2.03
0.70	145	0.0276	0.0231	0.0204	0.0193	0.0281	5	26	2.76
0.80	127	0.0315	0.0264	0.0233	0.0220	0.0321	5	26	3.60
0.90	113	0.0354	0.0297	0.0262	0.0248	0.0361	5	26	4.56
1.00	102	0.0394	0.0330	0.0291	0.0276	0.0401	5	26	5.63
1.10	102	0.0433	0.0369	0.0331	0.0315	0.0440	4	51	7.41
1.20	102	0.0472	0.0409	0.0370	0.0354	0.0480	4	23	9.43
1.40	85	0.0551	0.0474	0.0428	0.0409	0.0560	4	32	12.57

Dimensions of UNM Miniature Screws ‡

Size designation, UNM	Threads per in.	Basic major diam., D	Head styles							
			Fillister		Pan		Flat		Binding	
			Max. head diam., A	Max. head height, H	Max. head diam., A	Max. head height, H	Max. head diam., A	Max. head height, H	Max. head diam., A	Max. head height, H
30	318	0.0118	0.021	0.012	0.025	0.010	0.023	0.007		
35	282	0.0138	0.023	0.014	0.029	0.011	0.025	0.007		
40	254	0.0157	0.025	0.016	0.033	0.012	0.029	0.008	0.041	0.010
45	254	0.0177	0.029	0.018	0.037	0.014	0.033	0.009	0.045	0.011
50	203	0.0197	0.033	0.020	0.041	0.016	0.037	0.011	0.051	0.012
55	203	0.0217	0.037	0.022	0.045	0.018	0.041	0.012	0.056	0.016
60	169	0.0236	0.041	0.025	0.051	0.020	0.045	0.013	0.062	0.014
70	145	0.0276	0.045	0.028	0.056	0.022	0.051	0.014	0.072	0.018
80	127	0.0315	0.051	0.032	0.062	0.025	0.056	0.016	0.082	0.020
90	113	0.0354	0.056	0.036	0.072	0.028	0.062	0.017	0.092	0.022
100	102	0.0394	0.062	0.040	0.082	0.032	0.072	0.019	0.103	0.025
110	102	0.0433	0.072	0.045	0.092	0.036	0.082	0.022	0.113	0.028
120	102	0.0472	0.082	0.050	0.103	0.040	0.092	0.025	0.124	0.032
140	85	0.0551	0.092	0.055	0.113	0.045	0.103	0.027	0.144	0.036

* As covered in USAS B1.10-1958.
† D = major diameter, p = pitch.
‡ As covered in USAS B18.11-1961.

TABLE 54. Smaller Screw Dimensions (*Continued*)

Small Socket Screw Dimensions ‡

No.	Max. body diam., D	Head styles						
		Set	Cap		Flat		Button	
		Basic body diam., D	Max. head diam., A	Max. head height, H	Max. head diam., A	Max. head height, H	Max. head diam., A	Max. head height, H
0	0.060	0.060	0.096	0.060	0.138	0.044	0.114	0.032
1	0.073	0.073	0.118	0.073	0.168	0.054	0.139	0.039
2	0.086	0.086	0.140	0.086	0.197	0.064	0.164	0.046
3	0.099	0.099	0.161	0.099	0.226	0.073	0.188	0.052
4	0.112	0.112	0.183	0.112	0.255	0.083	0.213	0.059

Class 2 Miniature and Small Inch Screw Threads §

Size No.	Threads per in.	External threads					Internal threads		
		Major diam.		Pitch diam.		Minor diam.	Pitch diam.		Major diam.
		Max.	Min.	Max.	Min.	Max.	Min.	Max.	Min.
0000	160	0.0210	0.0200	0.0169	0.0158	0.0140	0.0169	0.0181	0.0210
000	120	0.0340	0.0325	0.0286	0.0272	0.0232	0.0286	0.0300	0.0340
00	96	0.0470	0.0450	0.0402	0.0386	0.0334	0.0402	0.0418	0.0470
00	90	0.0470	0.0450	0.0398	0.0382	0.0326	0.0398	0.0414	0.0470
0	80 UNF	0.0595	0.0563	0.0514	0.0496	0.0442	0.0519	0.0542	0.0600
1	64 UNC	0.0724	0.0686	0.0623	0.0603	0.0532	0.0629	0.0655	0.0730
1	72 UNF	0.0724	0.0689	0.0634	0.0615	0.0554	0.0640	0.0665	0.0730
2	56 UNC	0.0854	0.0813	0.0738	0.0717	0.0635	0.0744	0.0772	0.0860
2	64 UNF	0.0854	0.0816	0.0753	0.0733	0.0662	0.0759	0.0786	0.0860
3	48 UNC	0.0983	0.0938	0.0848	0.0825	0.0727	0.0855	0.0885	0.0990
3	56 UNF	0.0983	0.0942	0.0867	0.0845	0.0764	0.0874	0.0902	0.0990
4	40 UNC	0.1112	0.1061	0.0950	0.0925	0.0805	0.0958	0.0991	0.1120
4	48 UNF	0.1113	0.1068	0.0978	0.0954	0.0857	0.0985	0.1016	0.1120
6	32 UNC	0.1372	0.1312	0.1169	0.1141	0.0989	0.1177	0.1214	0.1380
6	40 UNF	0.1372	0.1312	0.1210	0.1184	0.1065	0.1218	0.1252	0.1380
8	32 UNC	0.1631	0.1571	0.1428	0.1399	0.1248	0.1437	0.1475	0.1640
8	36 UNF	0.1632	0.1577	0.1452	0.1424	0.1291	0.1460	0.1496	0.1640

Miniature Screw Dimensions Based on UNC and UNF Threads

Size designation	Basic major diam., D	Hex		Round		Flat		Fillister	
		Max. head diam., A	Max. head height, H	Max. head diam., A	Max. head height, H	Max. head diam., A	Max. head height, H	Max. head diam., A	Total max. head height, H
0000	0.021	0.0781	0.033	0.038	0.019	0.040	0.011	0.035	0.019
000	0.034	0.0781	0.033	0.059	0.028	0.061	0.016	0.056	0.031
00	0.047	0.0781	0.042	0.085	0.041	0.089	0.024	0.078	0.043
0	0.060	0.0937	0.042	0.106	0.047	0.108	0.030	0.091	0.051
1	0.073	0.1093	0.055	0.130	0.055	0.136	0.038	0.111	0.062
2	0.086	0.1250	0.064	0.154	0.065	0.164	0.046	0.132	0.073

‡ As covered in USAS B18.3-1961.

§ Size Nos. 0 to 8 are from USAS B1.1-1960 "Unified Screw Threads." Sizes 00 to 0000 are not covered by published standards, but follow the Unified Thread formulation spelled out in USAS B1.1.

TABLE 54. Smaller Screw Dimensions (*Continued*)

Small Machine Screw Dimensions ¶

No.	Basic major diam., D	Flat		Oval		Round		Pan	
		Max. head diam., A	Max. head height, H	Max. head diam., A	Total max. head height, H	Max. head diam., A	Max. head height, H	Max. head diam., A	Max. head height, H (slotted)
0	0.060	0.119	0.035	0.119	0.056	0.113	0.053	0.116	0.039
1	0.073	0.146	0.043	0.146	0.068	0.138	0.061	0.142	0.046
2	0.086	0.172	0.051	0.172	0.080	0.162	0.069	0.167	0.053
3	0.099	0.199	0.059	0.199	0.092	0.187	0.078	0.193	0.060
4	0.112	0.225	0.067	0.225	0.104	0.211	0.086	0.219	0.068
6	0.138	0.279	0.083	0.279	0.128	0.260	0.103	0.270	0.082
8	0.164	0.332	0.100	0.332	0.152	0.309	0.120	0.322	0.096

No.	Fillister		Truss		Binding		Hex	
	Max. head diam., A	Total max. head height, H	Max. head diam., A	Max. head height, H	Max. head diam., A	Max. head height, H	Min. width across corners	Max. head height, H
0	0.096	0.059	0.131	0.037				
1	0.118	0.071	0.164	0.045				
2	0.140	0.083	0.194	0.053	0.181	0.050	0.134	0.050
3	0.161	0.095	0.226	0.061	0.208	0.059	0.202	0.055
4	0.183	0.107	0.257	0.069	0.235	0.068	0.202	0.060
6	0.226	0.132	0.321	0.086	0.290	0.087	0.272	0.093
8	0.270	0.156	0.384	0.102	0.344	0.105	0.272	0.112

¶ As covered in USAS B18.6.3-1962.

TABLE 55. Smaller Nut Sizes*

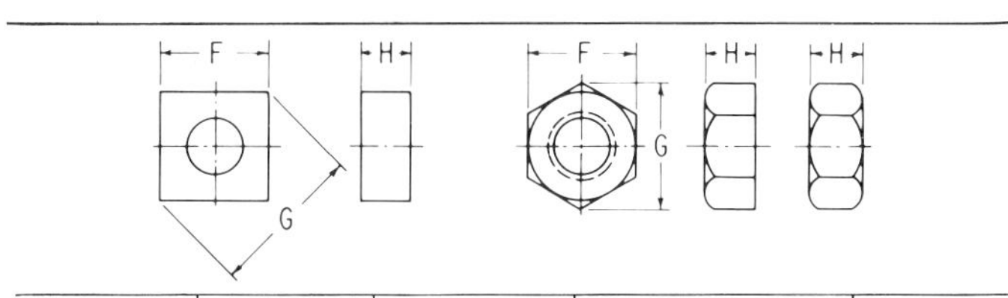

Size designation †	Major diam. of thread	Max. width across flats, F	Max. width across corners, G		Thickness, H
			Square	Hex	
0000	0.0210	0.0468	. . .	. . .	0.033
000	0.0340	0.0781	. . .	. . .	0.033
00	0.0470	0.0781	. . .	. . .	0.040
0	0.0060	0.156	0.221	0.180	0.050
1	0.0073	0.156	0.221	0.180	0.050
2	0.0086	0.187	0.265	0.217	0.066
3	0.0099	0.187	0.265	0.217	0.066
4	0.0112	0.250	0.354	0.289	0.098
6	0.138	0.313	0.442	0.361	0.114
8	0.164	0.344	0.486	0.397	0.130

* Modified from Ref. 20.

† Sizes 0 to 8 are covered by USAS B18.6.3–1962, "Square and Hex Machine Screw Nuts." Sizes 0 to 000 follow the same basic thread formulation. At present, there are no UNM standards for nuts.

TABLE 56. Tap Drill Sizes and Screw Clearance-hole Diameters

Screw No.	Diam., in.	Threads/in., UC	Threads/in., UF	Tap drill No.	Drill diam., in.	Recommended clearance-hole sizes, in.				
						Single screw	Two parts together with two screws			
							Clearance holes in both parts, tolerance between holes		Clearance hole in one part, tapped hole in other, tolerance between holes	
							±0.005	±0.020	±0.005	±0.020
0	0.060	..	80	56	0.047	0.063	0.063	0.073	0.067	0.089
1	0.073	64	72	53	0.060	0.079	0.079	0.089	0.082	0.106
2	0.086	56	64	50	0.070	0.094	0.094	0.102	0.098	0.120
3	0.099	48	..	47	0.079	0.104	0.106	0.113	0.120	0.141
3	0.099	..	56	45	0.082	0.104	0.106	0.113	0.120	0.141
4	0.112	40	..	43	0.089	0.120	0.120	0.128	0.125	0.147
4	0.112	..	48	42	0.094	0.120	0.120	0.128	0.125	0.147
5	0.125	40	..	38	0.102	0.136	0.136	0.147	0.141	0.161
5	0.125	..	44	37	0.104	0.136	0.136	0.147	0.141	0.161
6	0.138	32	..	36	0.107	0.147	0.147	0.156	0.147	0.172
6	0.138	..	40	33	0.113	0.147	0.147	0.156	0.147	0.172
8	0.164	32	..	29	0.136	0.172	0.172	0.180	0.180	0.196
8	0.164	..	36	29	0.136	0.172	0.172	0.180	0.180	0.196
10	0.190	24	..	25	0.150	0.196	0.196	0.213	0.203	0.228
10	0.164	..	32	21	0.159	0.196	0.196	0.213	0.203	0.228

TABLE 57. Types of Captive Nuts [15]

Plate or Anchor Nut. A nut with one or more lugs projecting from the base of the threaded body; it is attached by riveting or welding the lugs to the work surface. May be fixed or floating, and parallel or right-angle lug style

Caged Nut. A multiple-thread nut enclosed within a spring steel retainer. Attached by clips, welding, or deformed flanges. Usually in larger sizes. Studs in this style also available

Clinch Nuts. A solid nut having a pilot or other feature designed to be inserted into a preformed hole and permanently clinched to the parent material. Studs also available

Pilot collar

Completed clinch

Pilot hole

Self-piercing Nut. A one-piece metal nut with a formed hardened-steel body for punching its own mounting hole. The nut is installed by forcing the nut assembly into the parent material; deformed metal holds the nut in place

REFERENCES

1. Shiers, G.: "Design and Construction of Electronic Equipment," Prentice-Hall, Inc., Englewood Cliffs, N.J., 1966.
2. Franklin, H. G.: Packaging Electronic Circuits, *Machine Design*, Jan. 18, 1962.
3. Longmire, D. E.: Can Printed Circuit Boards Withstand High *G* Environments, *Proc. NEP/CON*, New York, 1964. Published by Industrial and Scientific Conference Management, Inc., Chicago.
4. MIL-STD-275B "Printed Wiring for Electronic Equipment," Sept. 7, 1960.
5. NASA: "Welding for Electronic Assemblies," NASA SP 5011, November, 1964.
6. Sippican Corporation, Report entitled "Electronics Packaging Concept, Typhon Missile," Marion, Mass., August, 1962.
7. MIL-STD-189 "Racks, Electrical Equipment, 19-inch and Associated Panels," revised Mar. 14, 1961.
8. NAVSHIPS 94324 "Maintainability Design Criteria Handbook for Designers of Shipboard Electronic Equipment," Fleet Effectiveness Branch, BUSHIPS, Naval Ships Engineering Center, Department of the Navy, March, 1965.
9. Carroll, J. M.: "Mechanical Design for Electronics Production," McGraw-Hill Book Company, New York, 1956.
10. Henney, K., C. Walsh, and H. Mileaf: "Electronics Components Handbook," 3 vols., McGraw-Hill Book Company, New York, 1957, 1958.
11. A Dictionary of Relay Types, *Machine Design*, Mar. 31, 1966.
12. MIL-STD-199A "Selection and Use of Resistors," July 15, 1965.
13. MIL-STD-198B "Selection and Use of Capacitors," Nov. 15, 1965.
14. A New Look at Microwave Tubes, *Elec. Design News*, October, 1966.
15. Special Reference Issue, Fasteners, *Machine Design*, March, 1965.
16. Special Report, Semiconductors, *Design News*, Oct. 12, 1966.
17. Motorola, "Semiconductor Data Book," 2d ed., Phoenix, Ariz., 1966.
18. Special Reference Issue, Electric Controls, *Machine Design*, Dec. 15, 1966.
19. Drummer, G. W. A., and J. W. Granville: "Miniature and Microminiature Electronics," John Wiley & Sons, Inc., New York, 1961.
20. Special Report, Fasteners for Packaging/Production, *Electron. Packaging Production*, September, 1966.
21. Keonjian, E.: "Microelectronics," McGraw-Hill Book Company, New York, 1963.
22. *Proc. NEP/CON*, Chicago, 1965, and *Proc. NEP/CON*, Chicago, 1966, Industrial and Scientific Conference Management, Inc.
23. "Advances in Electronics Circuit Packaging," Rogers Publishing Company, Englewood, Colo., vol. 4, 1964; vol. 5, 1965; vol. 6, 1965; vol. 7, 1966.

Chapter 9

PACKAGING OF MICROELECTRONIC AND HYBRID SYSTEMS

By

W. W. STALEY

Mechanical Design & Development Engineering
Westinghouse Electric Corporation
Aerospace Division
Baltimore, Maryland

INTRODUCTION

The packaging of electronic equipment has become a major factor in the design and manufacture of contemporary electronic systems. Newer and more advanced techniques are rapidly developing, and the packaging engineer is faced with the continuous task of keeping abreast of and utilizing these techniques to meet the ever-increasing demand for more systems capability and higher reliability in less space and at a lower cost. It is therefore essential that the packaging engineer become aware of and knowledgeable about the latest state of the art in the field of electronic packaging.

Materials, techniques, and processes used in applying microelectronic packaging techniques are discussed in other sections of this handbook. This section is devoted to the application and utilization of these materials, techniques, and processes in establishing the packaging concept that will best meet the systems requirements.

TERMS AND DEFINITIONS

Some of the terms and definitions peculiar to microelectronic and hybrid packaging are given in Table 1. Since this is a relatively new and growing field in packaging, terms and definitions are frequently vague or even conflicting. The military services and the various industry associations are continually working toward more commonality and standardization.

Background History of Microelectronic Packaging. The desire to make electronic equipment smaller probably started with the first system ever built. Through the years reduction in size has taken the form of more efficient circuit design as well as improved packaging concepts and improved component capability. The most significant steps include the changes in active devices from standard electronic

TABLE 1. Terms and Definitions

Active device. A device exhibiting transistance, e.g., gain or control.

Active substrate. A substrate which, by processing, is made to exhibit transistance.

Component part. The physical realization of an electrical property in a physically independent body which cannot practicably be further reduced or divided without destroying its function.

Device. A component part capable of affecting the behavior of an electronic circuit. Examples of component parts which are not devices are connectors, terminals, and fuses.

Discrete component circuit. A device in which separate discrete active and passive components, which have been fabricated prior to their installation, are mounted on a circuit board or substrate.

Electrical properties. The concept of basic electrical characterization. Basic electrical properties are resistance, capacitance, inductance, and transistance.

Hybrid integrated circuits. The arrangement consisting of one or more integrated circuits in combination with one or more discrete devices, or alternately, the combination of more than one type of integrated circuit into a single device.

Integrated circuit. The physical realization of two or more circuit elements inseparably associated on or within a substrate to form an electrical network.

Master slice. Silicon wafers containing many clusters of components. These elements can be interconnected with metallization paths to form the desired circuits. The wafer is then diced to form single circuits.

Micro. Prefix denoting a multiplication factor of 10^{-6}.

Microelectronics. The entire body of electronic art which is connected with or applied to the realization of electronic systems from extremely small electronic parts.

Microminiaturization. A technique for reducing the physical size of circuits, equipments, and component parts without sacrificing the intended function.

Module. A unit in a packaging scheme displaying regularity and separable repetition. It may or may not be separable from other modules after initial assembly. Usually all major dimensions are in accordance with a prescribed series of dimensions.

Monolithic circuit. A form of semiconductor microcircuit in which circuit elements are not readily identified as individual components.

Nano. Prefix denoting a multiplication factor of 10^{-9}.

Packaging. The process of physically locating, connecting, and protecting devices or components.

Packaging density. The number of devices or equivalent devices per unit volume in a working system or subsystem.

Passive substrate. A substrate which does not exhibit transistance.

Pico. Prefix denoting a multiplication factor of 10^{-12}.

Semiconductor integrated circuit. The physical realization of two or more circuit elements inseparably associated on or within a semiconductor substrate to form an electrical network.

Substrate. The single body of material upon or within which circuit elements are fabricated.

Transistance. The electrical property which affects voltages or currents so as to accomplish gain or switching action. Examples of transistance occur in transistors, diodes, voltage-controlled rectifiers, and electron tubes.

Wafer. A form of substrate, usually that on which semiconductor integrated circuits are fabricated.

tubes, to miniature tubes, to subminiature tubes, and finally in 1948 to semiconductor devices such as transistors and diodes.

The introduction of the new active devices was accompanied by many circuitry design changes and frequently a reduction in power, which allowed the use of smaller passive components with lower power ratings. This resulted in the introduction of the smaller ceramic and tantalytic capacitors and the ¼-, ⅛-, and ⅒-watt resistors. Even further size reductions were achieved with the introduction of uncased resistors in the form of small cylinders of carbon and uncased plastic-coated capacitors. Finally, even active devices in the form of uncased transistors and diode chips appeared on the market. Another significant advance was the approach introduced in the mid-1950s whereby the passive resistors and capacitors were produced in thick- and/or thin-film form. In the thick-film form, conductive, resistive, and insulating materials were screened or printed in proper physical form and proper sequence, and fired at elevated temperatures. In the thin-film form the conductive, resistive, and insulating materials were evaporated in proper sequence through appropriate masks to produce the desired component patterns. The total circuit functions were completed with the attachment of microminiature discrete active components.

In the meantime a more sophisticated approach to size reduction was taking place with the development of the semiconductor integrated circuit. This approach was a further extension of the transistor art whereby the passive and the active components were produced and interconnected on the same semiconductor chip or substrate. Completely new processing arts had to be developed, including vapor-depositing conductors and insulators, special microphotography masking, fine-line etching, and special chemical processing for diffusing and doping the semiconductor material. In addition, the art of solid-state circuitry design had to be developed to realize the full benefits of the semiconductor integrated circuit.

Results at first were frustrating, with low yield, poor reliability, and high cost. However, within a period of less than 10 years the semiconductor integrated circuit has become a proven, reliable, low-cost component and is now produced on a production basis by scores of electronic firms. Initial designs were simple and included such circuits as gates and flip-flops for digital-type applications. This has progressed to other digital designs as well as many higher-powered and more complicated analog-amplifier-type circuits.

The microelectronic evolution is now continuing in the area of packaging many bare integrated-circuit chips in a single container and interconnecting these into a multifunction unit using fine gold wires and prefabricated fine-line etched circuitry. Size reductions over the use of individually packaged integrated circuits range from 10 to as much as 100 times. More recently, development work is being conducted in the area of large-scale integration or what is frequently referred to as a *monolithic multifunction device*. In this approach the individual chip functions are produced on a single larger silicon substrate and interconnected using deposited metal conductors.

One of the most spectacular and promising microelectronic advances is in the area of the metal-oxide semiconductor. This device is reputed to be smaller in size and less costly than the more standard diffused integrated circuits. Size reduction claims of up to 50 times over the standard integrated circuit are being made. Needless to say, many of these newer concepts must be more thoroughly studied before their actual advantages and disadvantages can be determined and evaluated.

As this brief background history points out, frequent improvements and changes are constantly taking place in the field of electronic components and circuitry design. This in turn necessitates a thorough familiarity with all these areas by the packaging engineer if he hopes to have his designs remain competitive in the electronics industry.

Areas of Inclusion. Various components and/or circuits are available for use in building today's electronic equipment. A brief description of these components and circuits is included to help familiarize the packaging engineer with the materials available for his use. More detailed packaging information is given later in this chapter. Included are:

1. Microdiscrete components
   a. Pellet or dot
   b. Micro size
   c. Pico size
   d. Chips
2. Thin-film hybrid circuits
3. Thick-film hybrid circuits
4. Semiconductor integrated circuits
5. Metal-oxide semiconductor circuits (MOS)
6. Monolithic integrated circuits

Microdiscrete Components. Typical examples of microdiscrete components are the pellet or dot types which are available as both passive and active components (see Fig. 1). These components may be obtained with or without ribbon leads.

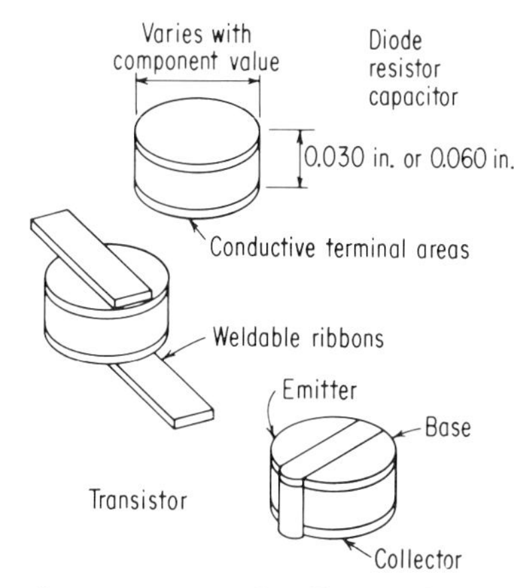

FIG. 1. Typical construction of pellet- or dot-type components.

Other available types of microdiscrete components fall into the categories known as micro and pico components. These components range from small lumps on a set of leads for the pico type to a somewhat larger size for the micro type (typical sizes range to less than 0.100-in. dimensions). These components normally are coated with a thin plastic material for environmental and handling protection.

A more advanced microdiscrete-type component is the semiconductor chip. Components of this type are becoming more widely used with the development of the multichip packaging concept. Typical component size is $0.020 \times 0.030 \times 0.010$ in. for the active components (transistors and diodes), while the passive types may be somewhat larger depending on the electrical values required. The chip components may come with either flat or raised terminal areas or with what is referred to as *beam leads*, for attachment to the interconnecting circuitry. This is discussed in detail in a subsequent section on the packaging of chip-type components.

Thin-film Hybrid Circuits. The thin-film hybrid circuit is usually produced on a glass or ceramic substrate using vacuum techniques to deposit various materials (see Table 2) in order to produce conductors, resistors, and insulators. Capacitors are produced with subsequent layers of conductive and insulating materials. Thin-film thicknesses are measured in angstroms ($1 \text{ Å} = 10^{-8}$ cm). Details of component designs are discussed in Chap. 5. Briefly, resistor values are a function of type of material, film thickness, and configuration. Capacitor values are normally a function of area and insulation or dielectric film thickness and dielectric constant. Thin-film circuits are larger (typical size might be 0.5×0.5 in.) than the equivalent semiconductor integrated circuits and are, therefore, better able to handle higher resistor power dissipation and can also contain larger capacitance and resistance values. However, active components are not readily available in thin-film form and are normally attached in microdiscrete form, thus creating a hybrid circuit. (Numerous companies are working experimentally with active thin-film devices, but no such devices are currently available on a production or large-scale basis.) Figure 2 shows a thin-film hybrid circuit and typical microdiscrete components used in construction.

TABLE 2. Materials Used in Making Thin-film Components and Circuits

Conductors	Resistors	Insulators
Aluminum	Nichrome	Silicon monoxide
Copper	Tantalum	Aluminum oxide
Gold	Titanium	Silicon dioxide
Nickel	Chromium	Titanium oxide
Silver		Tantalum oxide
Tin lead		

Thick-film Hybrid Circuits. The thick-film (thicknesses over 20,000 Å) hybrid circuit may be applied to a substrate (normally ceramic), using screening, printing, electroplating, electroless plating, or other similar techniques. The metallic materials are normally fired at relatively high temperatures (above 800°C) if they are screened or printed. Organic compounds are baked out at somewhat lower temperatures. Resistor and capacitor values are determined as they are with the thin-film approach. The active components are attached as microdiscrete components. The thick-film hybrid circuits are somewhat larger than the thin-film type because their dimensions and tolerances cannot be so well controlled. Figure 3 shows a thick-film network.

Semiconductor Integrated Circuits. The semiconductor integrated circuit is available in several basic forms. Included are the TO can configuration, the flat-pack configuration, the dual-in-line configuration, and the bare chip. The TO can config-

FIG. 2. Thin-film hybrid circuit and microdiscrete components. (*Westinghouse Electric Corp.*)

FIG. 3. Thick-film network. (*U.S. Army photograph.*)

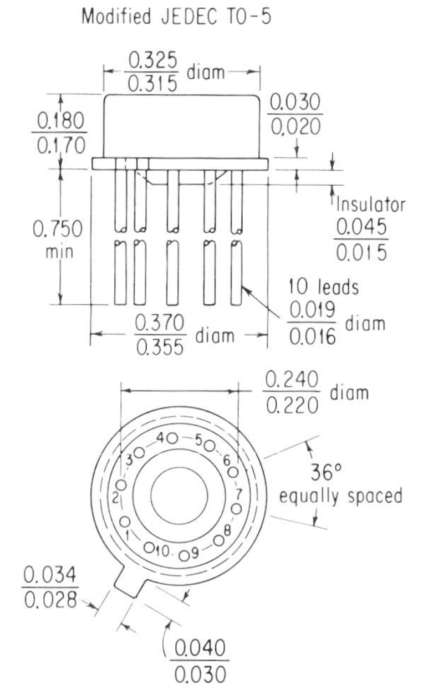

FIG. 4. Semiconductor integrated circuit—TO-5 can package.

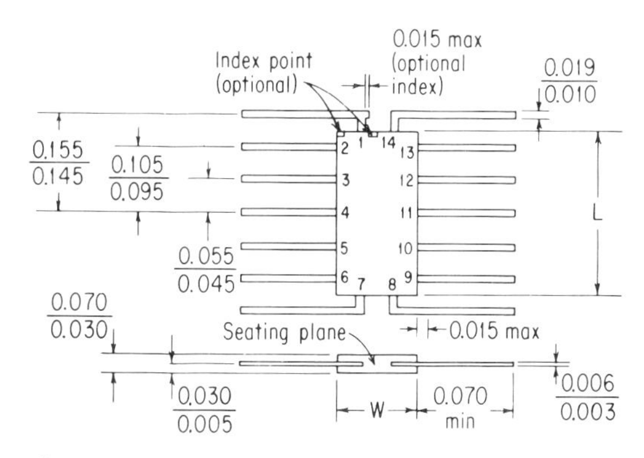

FIG. 5. Semiconductor integrated circuit—TO-84 flat package. No. of leads, 14; L min = 0.240 in., L max = 0.260 in., W min = 0.120 in., W max = 0.150 in.

uration is an outgrowth of the original TO transistor package. The most common configurations utilized for integrated circuits are the TO-5, TO-18, and TO-47 packages. Typical dimensions for the TO-5 package are shown in Fig. 4. The integrated-circuit chip is attached within the package and interconnected from terminal areas on the chip to the output terminals of the package using thermal-compression-bonded fine gold wires. The number of pins from the TO-5 package can range as high as 12, while the TO-18 and TO-47 packages can range up to 10. These packages provide hermetically sealed circuits.

The flat pack was specifically designed as a container for the semiconductor integrated circuit. It was designed to conform to the geometry of the chip it was to contain, which resulted in the flat rectangular shape. The leads pass through the walls of the package rather than through the base as with the TO cans. This allows the use of all four walls for interconnecting the package in addition to allowing more choices for mounting. The flat pack utilizes a matched glass-to-metal seal as with the TO can. Flat-pack sizes and the number of leads vary. Figure 5 shows details of the ⅛- × ¼-in. TO-84 flat pack with 14 leads.

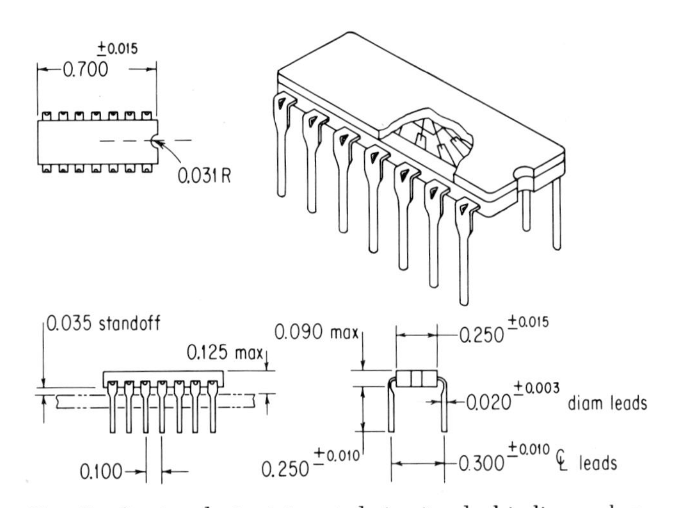

FIG. 6. Semiconductor integrated circuit—dual-in-line package.

The dual-in-line configuration is a more recent innovation in which the integrated-circuit chip is molded with a plastic material or mounted in a ceramic package. The lead configuration allows for easy insertion on a printed-wiring board with subsequent dip or wave soldering. Figure 6 shows the detail of the dual-in-line package. Although the package was developed primarily for commercial use, it also appears to be practical for some military uses.

As previously discussed with the discrete component, the semiconductor integrated circuit is also available in chip form. Chip size is generally larger; typical size might be 0.040 × 0.080 × 0.010 in. thick. Figure 7 shows a thimbleful of chips to illustrate their small size. Standardization of terminal area of location and spacing is becoming more critical with the trend toward use of the bare chips. Variation between manufacturers for the same device can no longer be tolerated. These chips are also available with raised terminal areas but to a lesser extent than transistor and diode.

Metal-oxide Semiconductor Circuits (MOS). A process comparison for the MOS and conventional integrated circuits is shown in Fig. 8. The MOS circuit is one of the newest and also most controversial devices to appear on the electronic packag-

FIG. 7. Thimbleful of semiconductor integrated-circuit chips. (*Sylvania Electronic Systems.*)

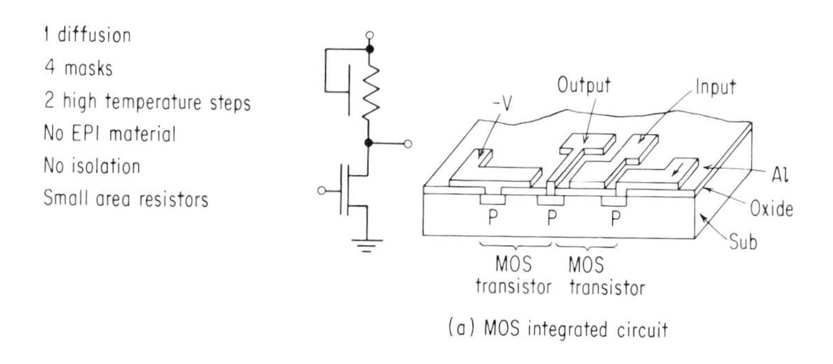

(a) MOS integrated circuit

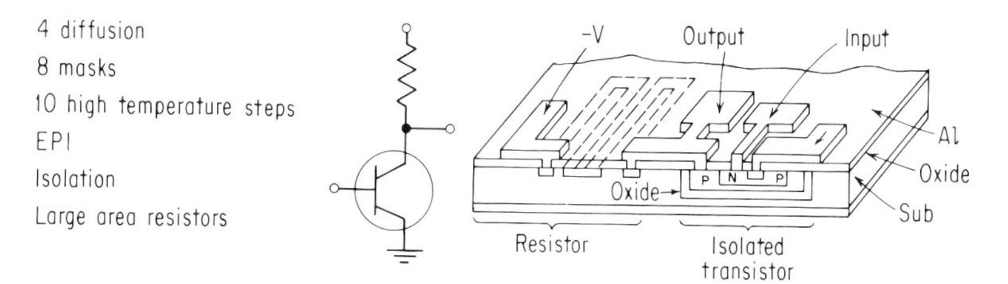

(b) Conventional double-diffused integrated circuit

FIG. 8. Comparison of MOS and conventional double-diffused integrated circuits. (*General Instrument Corp.*)

ing scene since the introduction of the transistor. It is claimed that the MOS can carry from 30 to 50 times more functions on a single chip of comparable size than the conventional double-diffused integrated circuit. It is also claimed that its cost is only one-tenth as much. These accomplishments are due primarily to a simpler process. The MOS is made with a single diffusion and 8 processing steps, compared to 4 diffusions and 13 processing steps for the conventional integrated circuit. Claims are that the MOS devices offer the same advantages over the conventional double-diffused integrated circuits as the transistor originally had over the tube, such as smaller size, lower cost, lower power, and improved reliability. In contrast, some people, who feel that the MOS is limited in its application, point to its speed, temperature, and radiation limitations. All things considered, the MOS devices must be seriously considered for use in most systems. With the increase in functions per chip, it is not surprising to find that the basic package must contain a larger number of leads than the conventional integrated circuit. Some possible ceramic packages for use with MOS circuits are shown in Fig. 9.

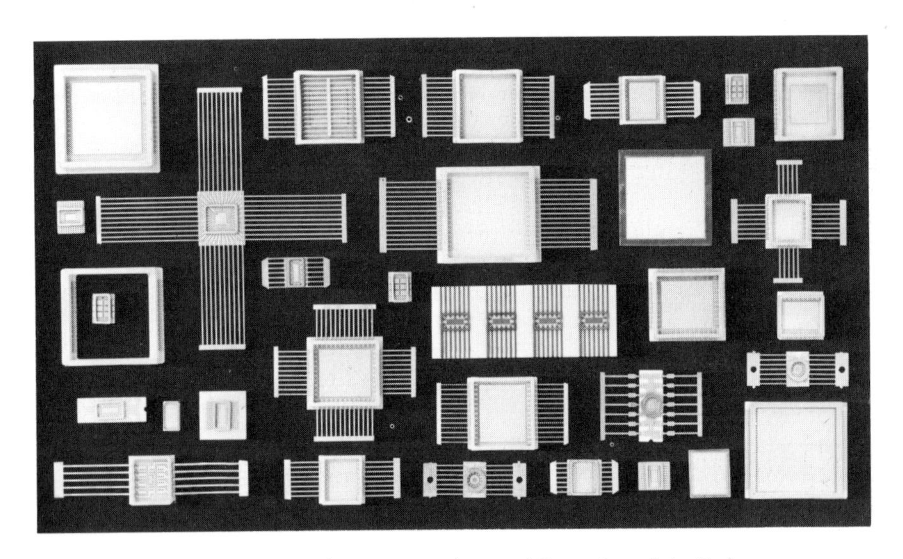

FIG. 9. Typical ceramic packages. (*Coors Porcelain Co.*)

Monolithic Integrated Circuits. Many semiconductor integrated-circuit manufacturers have been working toward producing more circuit functions on a larger chip. It was reasoned that reliability, size, and cost would all be improved if the integrated circuits could be interconnected into a multifunction array on a single larger chip. This was thought to be impossible in the early stages of the semiconductor integrated-circuit development because of poor yields. However, as materials and processes improved, yield improved; and many companies are now pursuing the monolithic multifunction package. Since 100 percent yield of, say, several hundred individual integrated-circuit functions on a single chip is virtually impossible, it becomes necessary to develop a selective technique for interconnecting the functions into the monolithic array. The functions may be interconnected in several ways.

The individual functions are tested, and the good circuits are marked. A custom interconnection pattern is then developed for interconnecting only the good circuits for that specific package. It is obvious that this can be an expensive process. However, with computer programming and automated pattern generation, the interconnecting layout labor cost can be reduced.

A more practical solution at present is to employ standard interconnecting patterns that allow for a certain percentage of defective circuits through the use of redundancy. This solution reduces the cost of interconnecting masks. The major disadvantage is that the number of circuits per unit wafer size is reduced.

Another approach is to determine the pattern of good circuits being produced on the standard production wafer. With this known, an optimum array of good circuits can be selected, and an appropriate interconnecting mask can be developed. The major objection here is that fewer circuits can be interconnected per single wafer.

The proper approach to interconnecting monolithic multifunctional integrated circuits is not obvious at this time. The device manufacturer will undoubtedly make his own decision. This will limit at least for a time what multi-arrays will be available. However, some monolithic devices are starting to become available and the packaging engineer must know how best to use them in his overall packaging

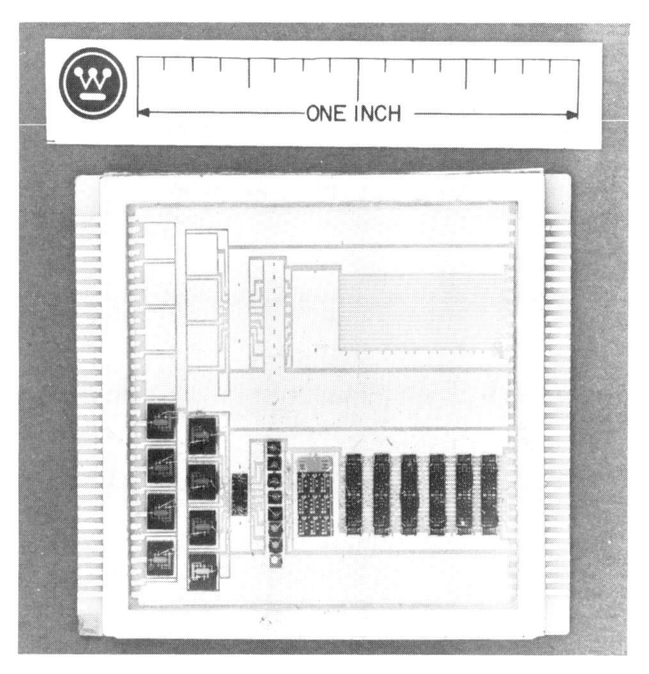

FIG. 10. Multifunctional package with 0.025-in. terminal spacings. (*Westinghouse Electric Corp.*)

concept. As with the MOS arrays, the input/output pin requirements have increased. Lead spacings have normally been limited to 0.050 in., but with the increased requirements, 0.025-in. spacings appear necessary. Some packages are available to meet these requirements (see Fig. 10).

Factors to Consider. Numerous factors must be taken into account when determining how to package a system. The importance of these factors is most frequently determined by customer requirements. One customer may want minimum cost while another may want maximum reliability. Frequently customers will want to emphasize a number of factors such as low cost, small size, and high reliability. Needless to say, frequently these factors can be incompatible with each other, and it is necessary to determine their order of importance with the customer. Once this is done, the packaging engineer must put each factor in its proper perspective and develop this overall packaging concept accordingly. Many of the factors that must be considered are obvious; however, many are also subtle. Table 3 shows the factors

TABLE 3. Factors to Consider when Determining Packaging Concept

Electrical (noise, speed, crosstalk, etc.)
Thermal dissipation (cooling available, ambient, component capability, etc.)
Size (weight, volume, shape)
Environment (vibration, shock, temperature, humidity, etc.)
Reliability (components, process, system, etc.)
Maintainability (field conditions, logistics, etc.)
Repairability (in the field, throwaway, etc.)
Interfaces (mechanical, electrical)
Cost (labor, material, testing, etc.)
Schedules (component availability, design, drafting, manufacturing capacity, etc.)
Customer preferences (experience, specifications, etc.)
Producibility (automation, in-house know-how, etc.)

to be considered. A more detailed discussion of these factors is given later in this chapter under design and special considerations.

Current Status of Microelectronic Packaging. Use of microelectronic and hybrid packaging techniques has varied considerably with the area of application, i.e., military, space, or commercial. Severe military environments such as shock, vibration, and temperature, along with increased systems complexity, have been responsible for the primary programs in microelectronic packaging. Many companies in the electronics industry have worked in coordinated teams with the different branches of the military to develop the current microelectronic packaging concepts available today. With the advent of outer space exploration, new environmental requirements such as radiation, temperature, and vacuum, along with the need for small size and high reliability, have necessitated even further microelectronic development.

Many military and most space electronic systems employ a large degree of microelectronic and hybrid packaging, and many commercial-type computers and data-processing equipments also employ these techniques. With the increased application of microcomponents and devices, cost per function has been greatly reduced. A number of companies manufacturing commercial electronic equipment have recognized this and are now starting to use these components and devices. Recently several portable television sets employing integrated circuits were placed on the market.

What the Future Holds. Predicting the future of microelectronic techniques is difficult because of the large steps that frequently take place with the introduction of the more significant discoveries. This was true in the past with the introduction of the transistor and the semiconductor integrated circuit. Now, with the greater effort being expended in the field of solid-state physics, we can expect even more rapid and more significant changes. Figure 11 is a predictions chart of future business in microelectronics. The predictions, if anything, are on the conservative side. More circuits are being converted to microelectronic construction every day. As cost goes down use goes up, and as use goes up cost goes down. New applications are opening up, including many new areas such as solid-state batteries, microwave devices, and solid-state cameras.

New commercial areas are certain to employ microelectronic techniques. This is especially true in the field of medicine, where such techniques will be useful in hearing aids, heart pacers, artificial eyes, etc. These are not blue-sky ideas but realities.

TYPES OF PACKAGING

All packaging, whether conventional, microelectronic, or hybrid, may be classified into two-dimensional or three-dimensional configurations. The two-dimensional configuration is frequently referred to as a *planar array*. The three-dimensional configuration is referred to as a *stacked array* or sometimes as a *module*.

The packaging of microelectronic and hybrid systems has generally followed the basic packaging configurations developed for conventional components. However, it has been possible to develop more regimented designs because of the fewer types of components and circuits with more standardized dimensions that fall into this category. Types of packaging are discussed and illustrated as various versions of planar and module arrays of the two basic circuit package forms, TO can and flat-pack configurations. This is followed by a discussion of the various techniques for packaging the bare chip components and circuits.

Planar Configurations of Packaged Components and Circuits. Planar configurations consist primarily of components and circuits mounted in a single plane on a panel or board and interconnected with some form of conducting lines. The panel or board consists of a dielectric base material clad with a conducting sheet of metal. The conducting lines are printed and etched to form the proper interconnecting pattern. Connection of the components to the interconnecting circuits is done primarily by soldering or welding. The following discussion illustrates some of the planar packaging concepts used to package the TO can and flat-pack circuits. It would appear that the flat-pack configuration of the integrated circuit would allow a closer board-to-board spacing than the TO can configuration because of its shorter height (typical 0.060 in. versus 0.180 in.). Unfortunately, this is not always obtainable since the board-to-board spacing may be dependent on the thickness of the

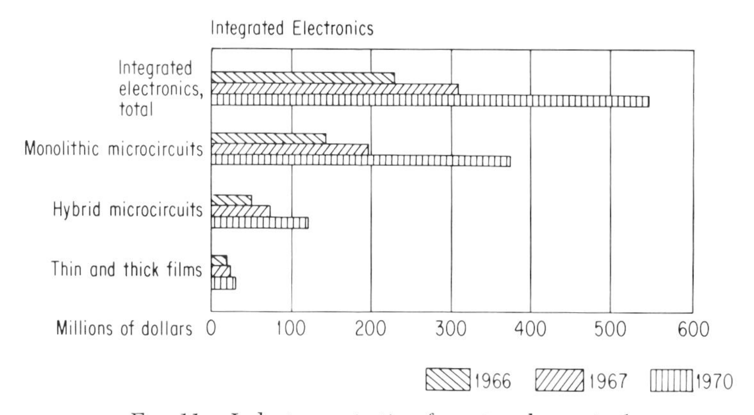

Fig. 11. Industry projection for microelectronics.[1]

connector. However, where connectors are not used or where some of the newer smaller connectors are available, the closer spacings can be realized. The volume required by these two basic device configurations then becomes dependent on the area that they require for proper layout. The following sections discuss this factor as well as other aspects of the various packaging concepts. It should be noted that the spacing between leads of most flat packs and some TO cans is too close for normal printed-wiring-board fabrication and spread patterns are used to avoid tight spacings. However, where volume efficiency is paramount the normal lead patterns are employed, and lower board yield and higher manufacturing cost are accepted. The following concepts are fundamental in nature, and many other variations may be developed by the packaging engineer.

TO Can with Spread Leads. The engineer has a choice of a number of lead pattern configurations depending on the complexity of his interconnecting circuitry and the grid to which he wants to work. Dimensions for lead forming and several basic lead patterns for a spread lead packaging concept are shown in Fig. 12. A major advantage of this concept is that a simple single- or double-sided printed-wiring board can be used. Engineering and manufacturing costs are therefore relatively low. The major disadvantage, however, is that overall package size is increased, since the pattern is spread out and added height above the board must be left for lead bending. Also, extreme care must be taken, when preforming the leads, not to allow bending stresses to reach the metal-to-glass seal. Since the patterns are normally repeated many times, a layout of the pattern is made up ahead of time, and it may be put down easily all in one operation. Intermixing of standard components is handled as described in Chap. 8.

TO Can with Multilayer Circuitry. Figure 13 shows the lead pattern employed for putting TO can devices on a multilayer printed-wiring board. It is not necessary to spread the leads for circuitry clearance when mounting TO cans on multilayer printed-circuit boards since the conductors are directed between holes on the various layers of the board (see Fig. 13). Circuitry is normally placed on the inner layers only, while the outer layers contain only terminal areas around the holes

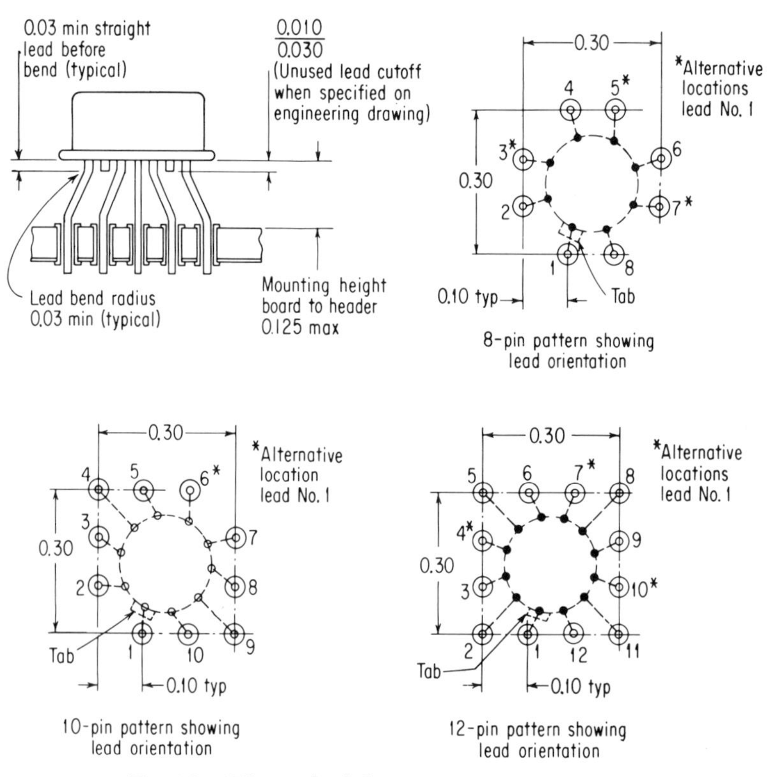

FIG. 12. TO can lead forming to spread patterns.

to allow for a maximum clearance between the terminal areas. This packaging concept results in smaller overall size. The TO cans may be spaced closer together and board-to-board spacing may be held to a minimum since there is no need to raise the cans off the board to allow for lead spreading. The major disadvantage of this concept is that the engineering design and manufacturing costs are relatively high. Also, many companies do not have multilayer fabricating equipment and must therefore purchase the boards outside.

TO Can Spread Spacing. A concept that is a compromise between the spread lead approach and the multilayer circuitry approach is the spread spacing of TO cans. In this concept the leads are not bent but are mounted straight into a single- or double-sided board as with the multilayer concept. The center-to-center spacings of the TO cans on the board are opened up to the point necessary to lay out the interconnecting circuitry between the basic lead patterns. Figure 14 illustrates this approach. Although the board area is necessarily larger with the spread TO can spacing, the board-to-board spacing is minimum since the leads are not formed and additional height above the board is not necessary. This concept is a little more critical to manufacture than the TO can spread lead concept, since smaller spacings are required for the straight-through lead patterns, and printed-wiring-board yield will probably be lower. However, the cost is considerably lower than that of the multilayer concept.

TO Cans Nested on Dual Boards. This concept employs the basic spread pattern technique but closely approaches the volume efficiency of the TO can with multilayer circuitry, at a much lower cost. The TO cans are placed in a spread pattern on two printed-circuit boards such that the TO cans will intermesh or nest when

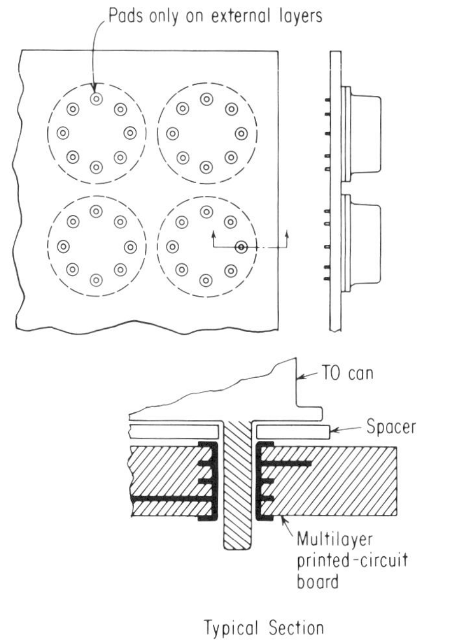

FIG. 13. TO can mounting to multilayer board.

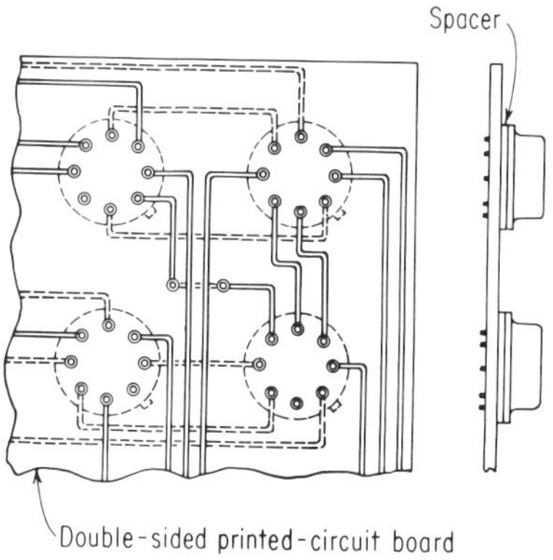

Spacer

Double-sided printed-circuit board

FIG. 14. TO can spread spacing.

brought together. Several types of connectors are commercially available for use with this concept. The details of construction are shown in Fig. 15. Interconnections between boards are possible through the connector or through a spacer connector, if employed. This concept, sometimes referred to as *board pairs,* has the same general manufacturing characteristics as the TO can spread spacing, but repair is more difficult after assembly and soldering.

It is also possible to employ this concept using a combination of spread TO can pattern and spread leads if normal printed-circuit spacings are felt necessary. This would result in a lower volume efficiency, but it would be an improvement over using the spread lead concept alone. This concept is also shown in Fig. 15. Both versions of this concept have good structural characteristics with their resulting boxlike configurations.

Flat Packs with Spread Spacing. The leads of the flat packs are not spread radially as with the circular TO can patterns. Rather they are extended away from the flat-pack body. This in turn requires spreading of the pack-to-pack spacing. A typical pattern is shown in Fig. 16. It should be noted that the flat pack can also be mounted with its leads extending through the board. The former approach is normally used when welding or resistance soldering (sometimes referred to as pulse soldering) is employed to interconnect the leads to the circuitry. The latter approach is used when wave or dip soldering techniques are employed. Again the patterns must be spread out as the complexity of the interconnecting circuitry between flat packs increases. This in turn increases board area used, and volume efficiency drops off.

Flat Packs with Multilayer Circuitry. The disadvantage of poor volume efficiency with the spread spacing of flat-pack patterns is avoided by using multilayer circuitry. Although the same basic pattern is normally used (since the 0.050-in. spacing is

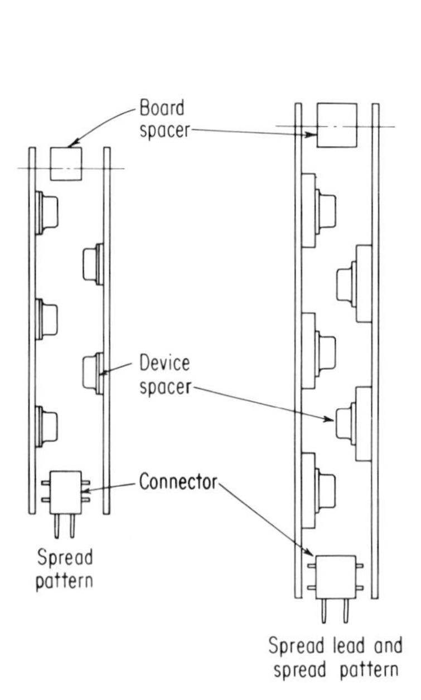

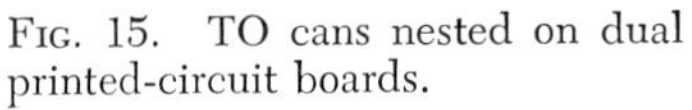
Fig. 15. TO cans nested on dual printed-circuit boards.

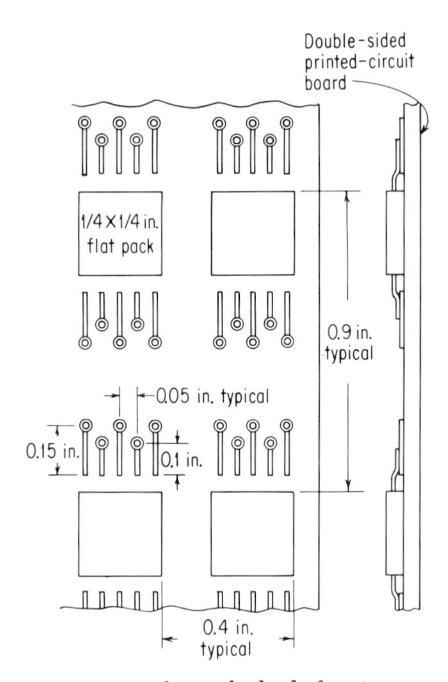

Fig. 16. Flat-pack lead forming to spread patterns.

beyond the limit of economical standard printed-wiring production), the pattern-to-pattern spacings are held much closer. This is possible since the interconnecting circuitry is distributed to the number of layers necessary to complete the wiring hookup. Figure 17 illustrates this concept.

Flat Packs with Plated-up Circuitry. Plated-up circuitry is used where volume efficiency is extremely important. The flat-pack leads are attached to matching plated-up circuitry without going through to the other side of the board. Figure 18 shows the details of this concept. The circuitry is not laminated as is normal printed circuitry, but rather the circuitry is built up of alternate layers of plating and insulating material. This technique allows a tighter interconnecting pattern than is possible with the laminated-type multilayer printed circuit. Unfortunately, only a few companies have developed this process and it becomes necessary for most companies to purchase these boards from outside sources. Welding and soldering techniques are both possible depending on the metals used for plating up the circuitry.

Commercial Planar Concepts. It is only natural to expect that some companies will devise and patent certain processes and packaging concepts. These companies frequently will market products based on the patents or sometimes sell the patent rights to other companies for their use. Several examples of these in the planar packaging field are discussed below.

Raised-tab Circuitry. The concept of raised-tab circuitry is basically a weldable interconnecting printed-circuit board for both standard and microelectronic components. Figure 19 shows the basic construction. The circuitry consists of a plated weldable metal on an insulating base material. The termination of a conductor at a component mounting hole is a tab bent or raised at right angles to the base material. This allows for easy attachment of the component lead to the conductor with opposed electrode resistance welding techniques. This concept is excellent

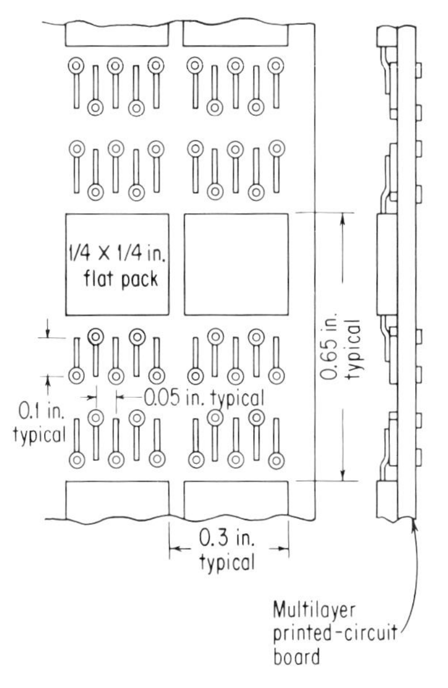

FIG. 17. Flat-pack mounting to multilayer board.

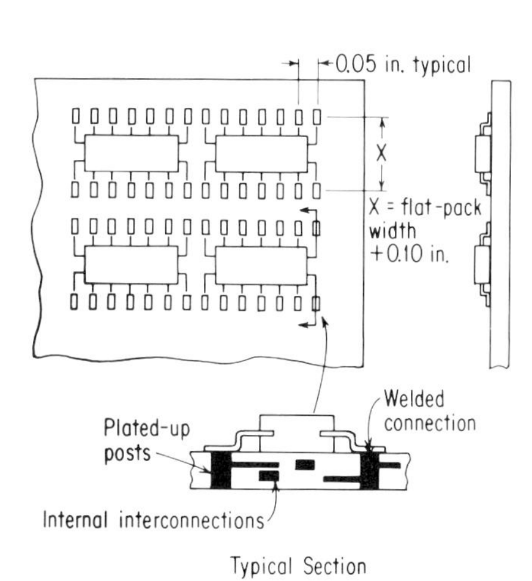

FIG. 18. Flat-pack mounting to plated-up circuitry.

from a cost standpoint when large quantities are required. Good volume efficiency is possible since close spacings can be obtained with photographic and plating techniques. Certain basic layout rules such as tab orientation are important and the manufacturer should be consulted before starting design.

Tube or Tubelet Circuitry. The concept of tube or tubelet circuitry is also basically a weldable interconnecting printed-circuit board for standard and microelectronic components. Figure 20 shows the basic construction. The circuitry is constructed by plating a weldable metal onto a pattern containing small posts at points of component termination. The circuitry is then transferred to an insulating base (prepunched). The resultant component interconnection points are weldable nickel tubelets. Orientation of component leads is not important with this technique since the lead is surrounded by the tubelet. Plating tolerances and conductor sizes and spacings should be discussed with the manufacturer before starting design.

Weldable Pin Circuitry. Weldable pin circuitry provides a weldable mounting base to which both components and interconnecting circuitry are applied. It is designed primarily for integrated-circuit flat packs. Figure 21 shows the detail of construction of this concept. Weldable pins (usually nickel) are molded into a fixed array for a number of integrated-circuit flat packs. A connector may also be included on one end of this assembly. In another version, also shown in Fig. 21, pins are molded into a strip which is then placed into slots on an insulating base. In both versions, the flat packs are welded to the pins on one side of the unit, and circuitry

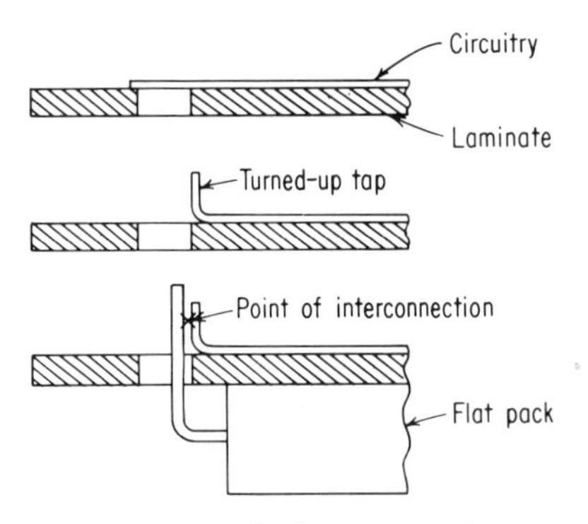

Fig. 19. Raised-tab circuitry. (*Intercon, trademark of Amphenol Borg Electronic Corp.*)

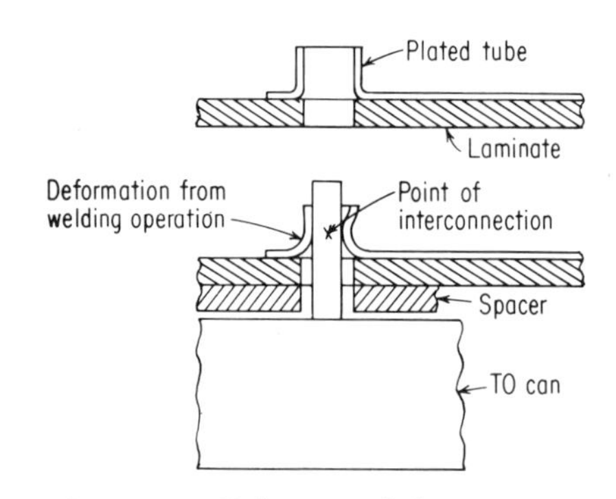

Fig. 20. Tube or tubelet circuitry. (*Polyweld, trademark of Litton Industries.*)

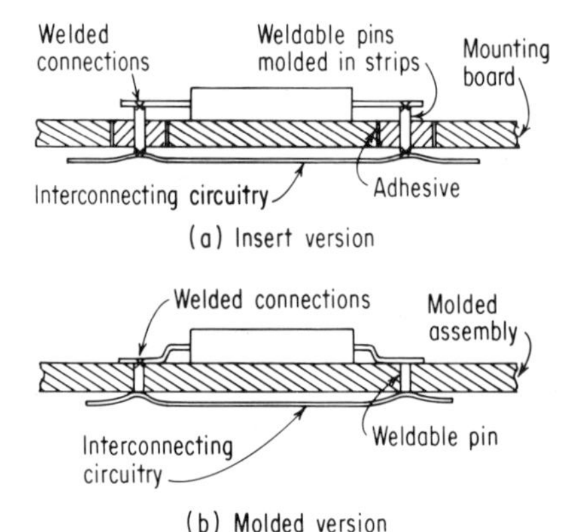

Fig. 21. Weldable pin circuitry.

is then welded onto the opposite side. This circuitry can be made either with welded insulated wire in a point-to-point array or with a weldable prefabricated circuit board (usually flexible).

This concept can be very useful in development-type programs when the point-to-point interwiring approach is preferred. Components and interconnecting circuitry are easily added or changed. Volume efficiency and cost are considered good for a development type of program. This concept is also useful on a large production system but the prefabricated interconnecting circuitry is preferred from a cost standpoint.

Module Configurations of Packaged Components and Circuits. These module configurations consist of components and circuits mounted in some form of three-dimensional array and interconnected with some form of conductors or conducting lines. These concepts are frequently, but not necessarily, embedded in some type of plastic. Connection of the component and/or circuit leads to the interconnecting circuitry can be done with soldering, welding, and even with deposited-metal techniques. Most of the hybrid concepts (integrated circuits mixed with conventional components) are a continuation of the cordwood module designs described in Chap. 8. The integrated-circuit flat-pack configuration generally allows for more varied and more efficient packaging concepts than the TO can configuration. The three-dimensional or module packaging concepts are usually the most efficient from a volume standpoint. Their main disadvantages normally are high cost and poor repairability. The following concepts illustrate the majority of the fundamental approaches. Many variations of the basic configurations are possible and frequently are seen as individual company packaging variations.

Hybrid Cordwood Module. The cordwood module, as discussed in detail in Chap. 8, is normally constructed by placing components vertically in cordwood style between two pieces of insulating film for welding, or between two printed-circuit boards for soldering. In the hybrid concept the integrated circuits (flat-pack and/or

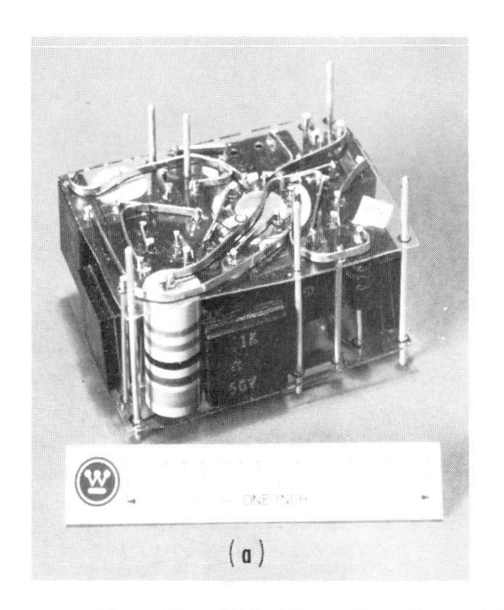

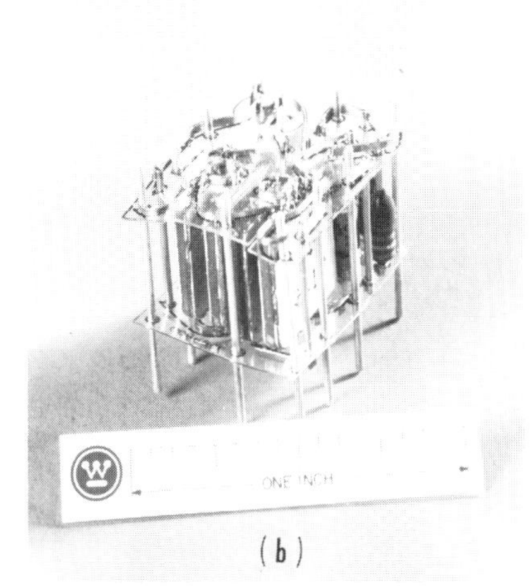

FIG. 22. Hybrid cordwood modules. (*Westinghouse Electric Corp.*)

TO can configuration) are intermixed in their proper circuit location with the more conventional components. If printed-circuit boards are used to interconnect the components, special spacing and conductor widths are necessary unless a spread lead approach is used. If welding techniques are used, care must be taken to place interconnecting ribbon properly for ease of welding. This is especially critical with the flat-pack configuration, and the leads are frequently twisted 45° to their normal orientation unless the spread lead concept is not used. Care must be taken in bending the leads to prevent the stresses from reaching the glass-to-metal seal. Figure 22 shows two welded designs.

Stacked Flat-pack Modules. Stacked flat-pack modules are designed primarily for the packaging of flat-pack integrated circuits, although it is not unusual to include some of the microdiscrete types of components. A number of variations on this concept are possible, but only one approach is discussed here. The basic concept is to construct a comblike structure as shown in Fig. 23. This structure contains input and output pins and through-connecting pins to aid in interconnecting complex wiring. The flat packs are assembled within this comblike structure (usually cemented into place) and interconnected with either welded-wiring or printed-wiring techniques. The basic comb structure may be embedded or used in its open form, depending on its application. The comb structure is normally made with plastic (usually a filled epoxy). Minimum tooling cost can be obtained by employing only a few basic comb sizes. It is probable that modules with various numbers of flat packs will be designed when the circuitry is divided into modules. It is a simple matter to select a basic comb size large enough for the maximum required number of flat packs and cut off the excess portion of the comb when not needed.

The stacked flat-pack module is normally designed with a comb tooth center-to-

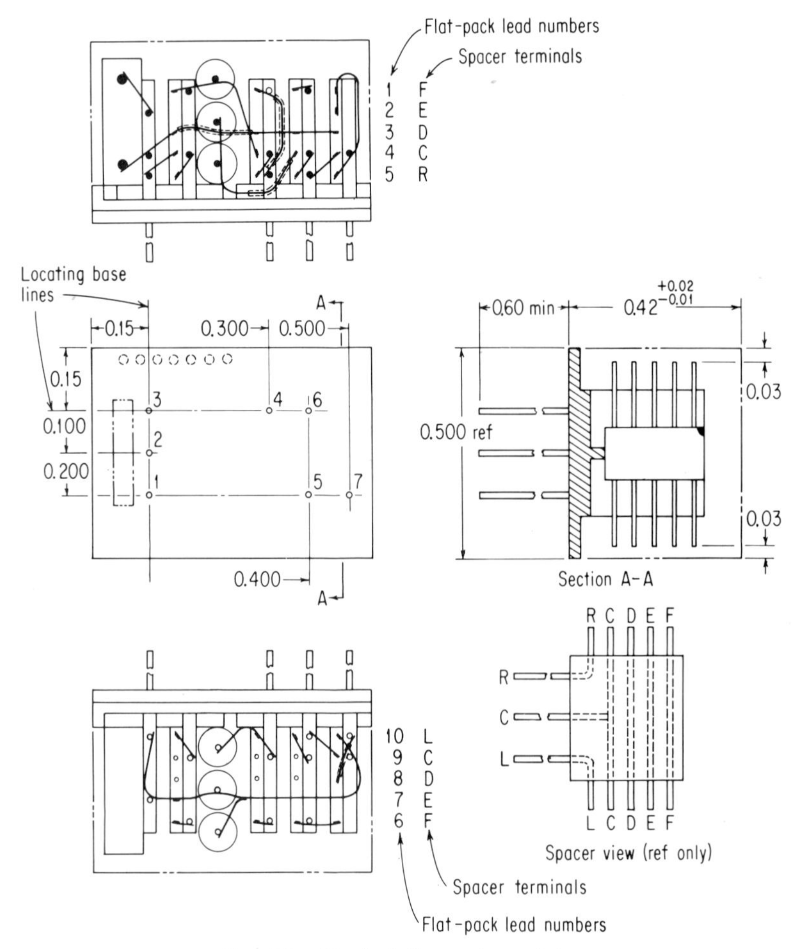

FIG. 23. Stacked flat-pack module.

center grid of 0.100 in. Individual teeth of the comb may be removed if greater center-to-center distances are needed. This concept can be relatively low in cost when interconnections are made with small printed-circuit boards. However, cost is much higher if the standard welded-ribbon techniques are used, since a 0.050-in. grid is employed and hand welding is difficult. Welded insulated wire techniques help to reduce the cost somewhat, since insulating the ribbon is not necessary and shorting out is less of a problem. Standard tooling and fixturing are possible since only a few basic dimensions are employed.

Compartmentalized Module. There are times when shielding is the principal requirement of the packaging design. The compartmentalized concept, as shown in Fig. 24, is an example of a packaging concept that will meet this requirement. The concept consists of a boxlike metallic structure with individual compartments to house the various groups of components and circuits to be shielded from each other and the surrounding circuitry. Normal fabricating techniques such as drilling and milling can be used to fabricate the structure, but cost will be high because of the small sizes and tight tolerances required. This concept is an excellent area for the application of photochemical milling techniques. The structural parts may be designed using printed-circuit layout technique, i.e., at a larger scale and photoreduced to the proper size for final fabrication. Metallic sheet material, such as copper, brass, and beryllium copper, in various thicknesses, such as 0.006, 0.008, and 0.010 in., may be used. Conventional photoresists are used. Parts are frequently imaged on both sides and etched with appropriate solutions. The resulting tolerances are primarily dependent on material thickness and etching control. After parts are

FIG. 24. Compartmentalized module. (*Westinghouse Electric Corp.*)

etched and cleaned, they are formed and soldered or brazed together. Gold plating is normally used to give better solderability.

Components and circuits are welded or soldered into their appropriate groupings. These subassemblies are then placed within the proper compartment, and interconnections between compartments are made. The total assembly may be embedded if desired.

Deposited Module. The deposited module makes it possible to overcome the problem of the high cost of interconnections in three-dimensional packaging. The processes required are critical, however, and good process control is important when this technique is used. Figure 25 shows the fabrication and assembly details of this concept. Components and circuits are aligned in proper location and embedded. The surfaces of the module to be interconnected are machined, exposing the component and circuit conductor leads. Using appropriate deposition processes, the surfaces are deposited with a conductive metal. The interconnecting pattern may be produced by depositing through a mask or by depositing on the entire surface and using photoetching techniques to produce the required conductor pattern.

This approach is an advanced concept, and only companies that have wide experience in deposition techniques will be able to fabricate units with satisfactory yields. Some of the major problems are the tight component lead location tolerances and adhesion of the deposited metal. This concept has good potential for a low-cost module, and a number of companies are actively pursuing the development of the concept as a production technique.

Multistacked Substrates. An example of the multistacked substrates technique is the Signal Corps Micromodule, one of the best-known packaging concepts ever developed. The microelectronic version is known as the *enhanced micromodule;* the basic concept is the same, but integrated circuits are employed rather than the more conventional types of components. Figure 26 shows the detail of this concept.

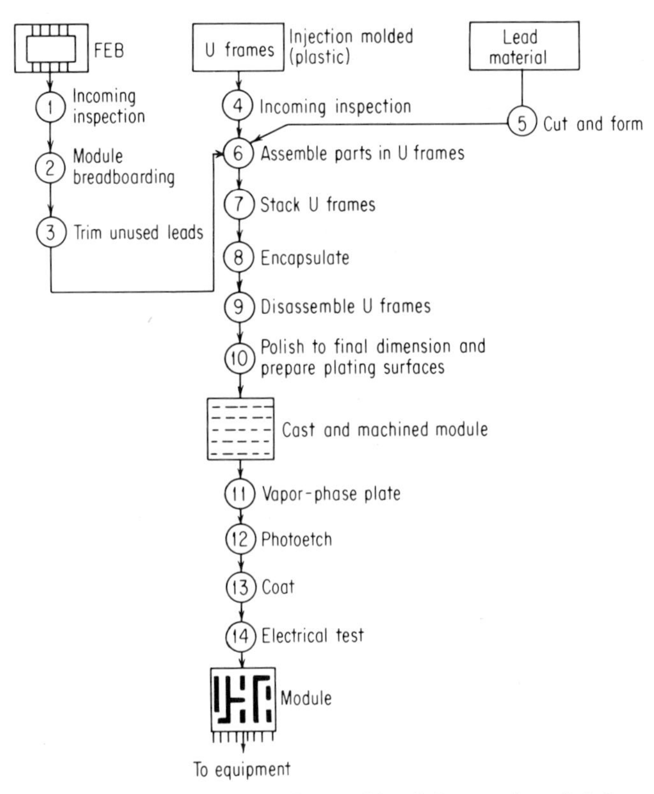

FIG. 25. Fabrication and assembly of deposited module.[2]

The basic substrate is 0.300 × 0.300 × 0.20 in. and is normally of alumina. Circuitry is screened or deposited, and components are attached to the appropriate circuitry on each substrate. The substrates are then put in an assembly fixture that establishes the proper substrate spacings. Interconnections are then made between substrates using riser wires. Soldering or welding techniques may be used. The finished module may be coated or embedded with a plastic, or hermetically sealed in a metal can if the application requires it.

The cost of this concept should be rather low since a number of companies are producing units on a mass-production basis.

Stick Module. The stick module is a three-dimensional packaging concept for integrated-circuit flat packs that has some of the advantages of a two-dimensional system. In this concept (see Fig. 27) the interconnecting circuitry is made up of a number of planar circuitry arrays. The individual array is made of a thin flexible insulating material clad with a conductive material (usually a weldable material). The circuitry is etched and the edges of the insulating material are etched back, resulting in extended weldable tabs. This circuit can be made with universal circuitry and punched out for the individual layers. When the layers are laminated together, the tabs are in proper position to allow easy alignment with the leads of the integrated-circuit flat packs. The flat packs are attached and the leads and tabs are welded together. The cost of this unit is very reasonable since tooling and processing are simple. Since each lead of the flat pack is automatically aligned with the proper tab (as in two-dimensional circuitry) welding is a simple low-cost operation. Circuitry changes or additions are very easily made. The unit may be embedded if desired. Volume efficiency is good since little volume is needed, even for complex circuitry. The long, thin configuration may be undersirable in some applications.

Fig. 26. Multistacked substrates. (*U.S. Army photograph.*)

Folded Module. Another three-dimensional packaging concept that has some of the advantages of two-dimensional systems is the folded module. In this concept the circuitry is made in a planar array, the components and circuitry are attached to this array, and then the array is folded into a three-dimensional package. See Fig. 28 for the details of this unit. The circuitry may be a series of parallel discrete wires (usually weldable ribbon) attached to an insulating flexible base. Circuit runs are achieved by welding component and/or circuit leads or jumper leads to the appropriate interconnecting wires and jumping to different levels using discrete wires where necessary. Interconnecting wires are also cut out where appropriate to complete the circuitry. In another version the interconnecting ribbons are embedded in a plastic film (Mylar,* Teflon,* etc.) and component leads are attached using a heated electrode and welding techniques. The cost of constructing this unit is very reasonable, and the system can easily be adapted to a continuous production line if large quantities are required. Overall volume will be greater than with the standard

* Trademark, E. I. du Pont de Nemours & Co., Inc.

welded module, however, since a greater module height is needed for interconnecting ribbons.

Commercial Module Concepts. As with the planar concept, a number of commercial module concepts are available. Both the turned-up-tab concept and the tubelet concepts may be used as the interconnecting medium with the hybrid

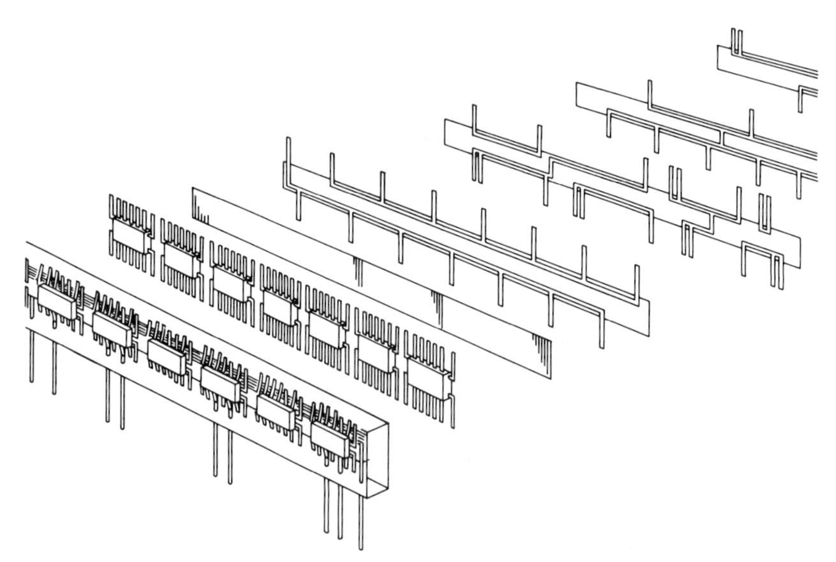

FIG. 27. Stick module. (*Engineered Electronics Co.*)

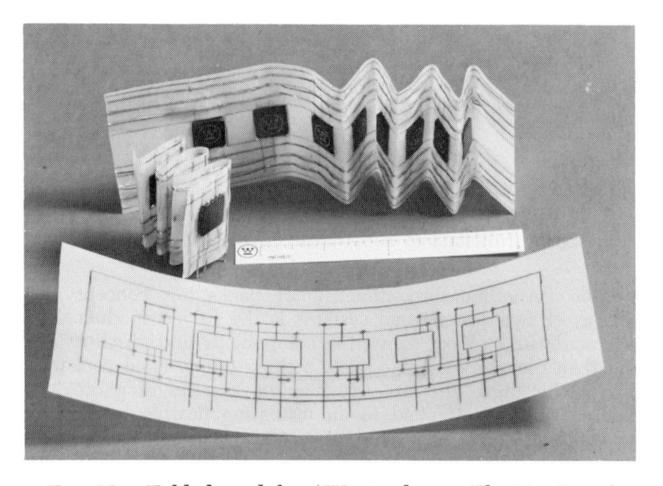

FIG. 28. Folded module. (*Westinghouse Electric Corp.*)

cordwood and the stacked flat-pack module concepts. Also, in the case of the multistacked substrates, the Signal Corps Micromodule is commercially available from a number of companies.

Another rather novel commercial module-packaging concept employs a cell or shell into which are assembled components and/or circuits. The cell contains a female connector arrangement which is plugged into a set of knifelike male connectors for the next level of interconnections. Details are shown in Fig. 29.

Component and Circuit Chip Packaging. The art of packaging component and circuit chips has various degrees of complexity. It might vary from a package containing a number of discrete component chips interconnected into a circuit function, to a package containing a single chip containing the entire circuit function in monolithic form. It might even vary from a package containing a number of monolithic chip functions interconnected into a larger array of circuitry, to a package containing a larger single chip with a number of circuit functions interconnected into a larger monolithic array of circuitry. The selection of any one approach is dependent on considerations such as cost, reliability, time to manufacture, and logistics. Each of the concepts and its variations are discussed in detail in the following pages.

Multidiscrete-component Chip Package. One of the earliest available semiconductor integrated packages consisted of discrete-component chips such as resistors, capacitors, diodes, and transistors mounted on a base substrate and interconnected with fine wire (0.0005- to 0.001-in. gold) bonding techniques (see Fig. 30). The container for the circuit function normally is either the TO can or the flat-pack configuration. Several companies also manufacture these packages in molded-epoxy flat packages. Note that bonded leads are necessary between the discrete components as well as to the input/output leads of the package.

FIG. 29. Pluggable cell module. (*Amp. Inc.*)

The multidiscrete-component chip approach is extremely useful for small quantities and for development-type jobs because of its low initial engineering and manufacturing tooling cost. However, if large production quantities or high reliability is required, a redesign to a monolithic integrated circuit is recommended.

Monolithic Semiconductor Integrated Circuits. The packaging engineer should have some knowledge of the processes and materials employed in the fabrication of monolithic semiconductor integrated circuits. This knowledge will allow him a

FIG. 30. Multidiscrete-component chip package. (*U.S. Army photograph.*)

better understanding of the restrictions and limitations these processes and materials might place on his packaging designs. Semiconductor integrated circuits are normally fabricated on a wafer of semiconductor material such as silicon approximately 1 in. in diameter and 0.017 in. thick. Since individual circuits are normally in the size ranges from 0.040 × 0.040 to 0.080 in. × 0.080 in., a large number of circuits are produced on a single 1-in.-diameter wafer. The final circuits are produced by performing a series of diffusing, etching, oxidizing, and evaporating steps on the wafer. The circuits are inspected and electrically tested in the large wafer form and then scribed and broken into the small individual circuit chips. Detailed steps of the process for producing a simple planar passivated silicon epitaxial integrated circuit are shown in Table 4.

TABLE 4. Process Steps for Producing a Simple Planar Passivated Silicon Epitaxial Integrated Circuit

1. Grow single silicon crystal.
2. Slice crystal 0.017 in. thick.
3. Lap and polish wafer.
4. Etch wafer.
5. Deposit epitaxial layer.
6. Grow oxide layer.
7. Coat surface with photoresist.
8. Expose pattern through suitable photographic mask.
9. Develop pattern.
10. Etch exposed oxide.
11. Remove photoresist.
12. First diffusion ($P+$ type dopant).
13. Grow oxide layer.
14. Repeat steps 7 to 13 using additional masks and $N-$ and $P-$ dopants.
15. Coat with photoresist.

16. Etch oxide and expose terminal pads.
17. Evaporate aluminum.
18. Coat with photoresist.
19. Expose interconnection pattern.
20. Etch interconnection pattern.
21. Test electrically.
22. Scribe and break wafer into chips (die).
23. Attach die to package header (TO can or flat pack).
24. Bond leads.
25. Seal package.
26. Test mechanically and electrically.

There are of course a number of variations on this basic process. Double-diffused devices, for example, delete the epitaxial growth and isolation diffusion steps. Other devices use multiple epitaxial growth cycles, buried layers, etc., to yield particular characteristics or capabilities. However, in all cases the operations are performed simultaneously on a large number of individual circuits producing significantly

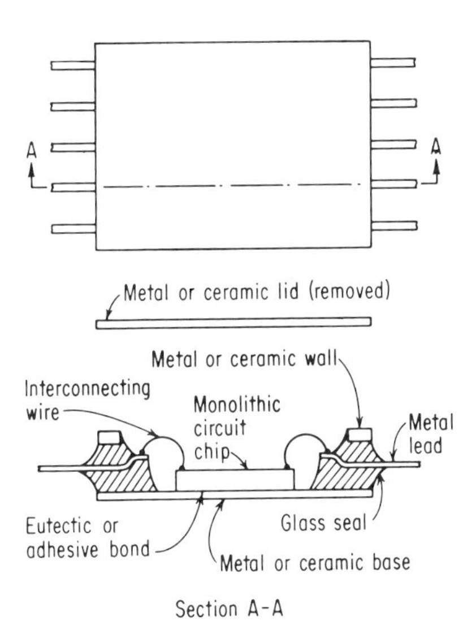

Fig. 31. Monolithic integrated-circuit chip construction (flat pack).

closer matched parameters and higher reliability than the multidiscrete chip circuits. Although initial tooling cost is high, this may easily be disregarded when large quantities are required.

The completed circuit chip is bonded to a base substrate (normally a TO can or flat-pack configuration) and interconnected to the appropriate input/output terminals as shown in Fig. 31. Note that in this package the only wire bonds necessary are those from the integrated-circuit terminal pads to the input/output lead of the package.

Master-slice Chip. A more sophisticated approach than the multidiscrete chip package for low-cost development-type jobs is the master-slice concept now offered by a number of integrated-circuit manufacturers. In this concept a basic chip is produced in monolithic form but designed with separate termination pads for each element. Various circuit functions are then produced from the same basic chip by using different interconnecting metallization patterns. The interconnecting patterns are designed and placed on the wafer before it is diced into individual chips. The result is a highly reliable monolithic integrated-circuit chip produced at a reasonably low cost. The major drawback to this approach is that the device manufacturer rather than the systems designer has control over the circuit to be manufactured.

General-purpose Chip. The concept of the general-purpose chip is similar to the master-slice approach in that a basic chip is also designed with separate termination pads for each element. The difference, however, is that the chip is available without the metallization interconnecting pattern, and the device user is able to make his own interconnections to obtain his desired circuit. Interconnections between elements are normally made by bonding fine gold wire to the termination pads. Figure 32 shows the layout of a typical general-purpose chip. This particular layout contains 10 transistors of four different types, 6 pairs of diodes, and 18 isolated, tapped, diffused resistors which can be divided into more than 100 resistors. The resistors range in value from 50 to 110,000 ohms.

Some companies use this concept to breadboard a system without using conventional discrete components at all. As the breadboarded circuit is checked out, a redesign to a completely monolithic package may be instituted. In some cases companies package the interconnected chips in TO cans or flat packs and use them in fabricating their finished equipment.

Multi-integrated-circuit Chip Package. It is only natural that the concept of the multidiscrete-component chip package would be extended to a multimonolithic integrated-circuit chip package. A simple but effective concept is one where a few chips are mounted in standard packages (TO cans and flat packs) and interconnected with bonded-wire techniques all of which are known state of the art. This concept may be extended to include a greater number of chips in larger packages, again using only bonded-wire interconnections. The larger package is generally a TO can or flat pack scaled up to ½ × ½ in. or even to 1 × 1 in. Input/output lead spacings are generally 0.050, 0.075, or 0.100 in. The approach is straightforward, and costs are reasonable. Reliability, however, is somewhat questionable, as with the multidiscrete-component chip package, and efforts have been directed toward an improved package and improved bonding techniques.

The design of a highly reliable multi-integrated-circuit chip package is quite complex. The major areas of design consideration are: (1) the basic package configuration, (2) the interconnecting circuitry, and (3) the bonding method. Each of these areas is discussed below, and the critical design considerations are pointed out in detail.

Basic Package Configuration. A number of basic package improvements have been developed, including better hermetic sealing capabilities and greater numbers of input/output terminations. Two typical package types are discussed; however, it should be remembered that many others also exist.

Figure 33 shows an all-ceramic package. The input/output terminals can be spaced as close as 0.025 in.; they are made with fired-on molymanganese for good adhesion and are then gold-plated for better conductivity. The desired terminal pattern and interconnecting circuitry are screened on with molymanganese. The ceramic ring metallized on one side is located unmetallized side down in proper position on glass frit and the unit is then fired at a high temperature. The molymanganese areas are then gold-plated. The package is enclosed with a ceramic lid, also metallized to match the ring pattern, normally using soft-soldering techniques. Solder preforms are frequently used in reducing the chance of voids in the solder. The final package results in an all-ceramic or glass-to-metal hermetic seal.

In the case shown in Fig. 34 (basically a metal package), the terminal pin locations are normally held to 0.075-in. minimum spacing because of glass-to-metal sealing limitations. The interconnecting area is fired on molymanganese and then gold-plated. This package is enclosed with a metal lid, using soft-solder techniques, for an all-metal-to-glass hermetic seal. The lid and terminal pins are normally made of Kovar* and gold-plated for better solderability. The lid is usually dimpled to increase its structural rigidity.

The increased number of functions per package necessitates a greater number of terminal pins per package. Theoretically with integrated circuits, terminal points

* Trademark, Westinghouse Electric Corp.

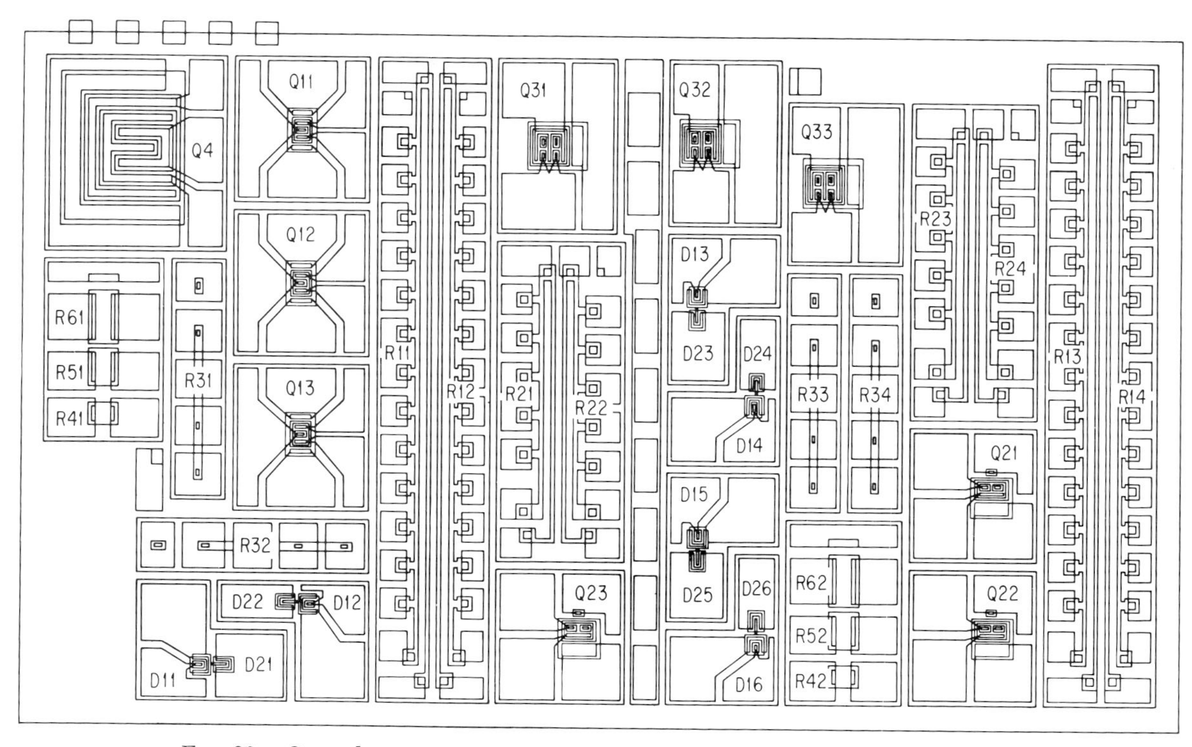

Fig. 32. General-purpose integrated-circuit chip. (*Westinghouse Electric Corp.*)

could be spaced on 5- to 10-mil centers. However, the capability of subsequent interconnecting techniques has been limited to 100-mil and, more recently, 50-mil centers. It is expected that the ultimate limit of interconnecting on the next level above the multifunction package will be 25 mils. Figure 35 shows the pin requirement versus the number of functions included in a single package. Note that the

FIG. 33. All ceramic multi-integrated-circuit chip package. (*Coors Porcelain Co.*)

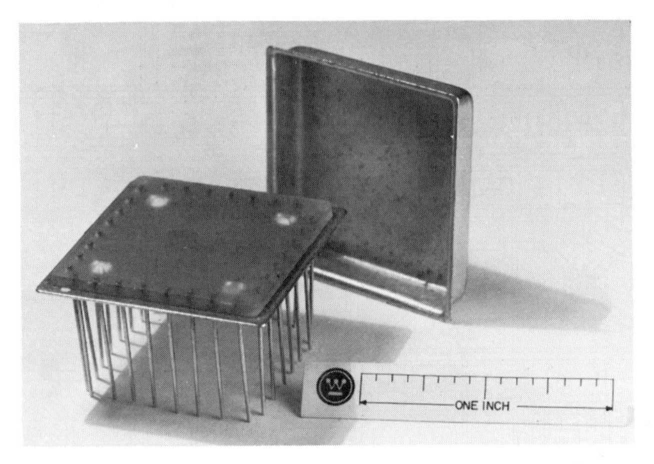

FIG. 34. Metal package for multi-integrated-circuit chips. (*Hughes Aircraft Co.*)

present systems require a greater number of terminal pins than that expected in the future (this is based on a projected improvement in electrical functional designs). Figure 35 also shows superimposed various basic flat packs and what can be expected in the way of maximum number of terminal pins and maximum number of functions for these packages.

Interconnecting Circuitry. The use of interconnecting deposited circuitry is one of the major contributors to the high reliability of the basic package. It is possible to limit the length of or even eliminate bonded wires which have been a major source of failure in semiconductor devices and circuits. The circuitry may be deposited in its final form or deposited as an interconnecting area and etched to its final form. If the latter technique is used, two etching steps are necessary—one for the gold and one for the molymanganese. The etching technique will produce finer lines and closer tolerances than the screening technique. (Note: Recent efforts by some companies have been in the area of aluminum conductors as a replacement for the molymanganese gold. This has some decided advantages in the elimination of the purple plague problem discussed in Chap. 10.)

The interconnecting circuitry of the multi-integrated-circuit chip package is a functional part of the circuit as well as the interconnecting medium. The conductors are functional elements having predictable resistive, capacitive, and inductive characteristics which contribute to the final circuit design. It is therefore necessary for the packaging engineer to become familiar with the electrical characteristics of his materials and of circuitry conductor geometry. The electrical characteristics include

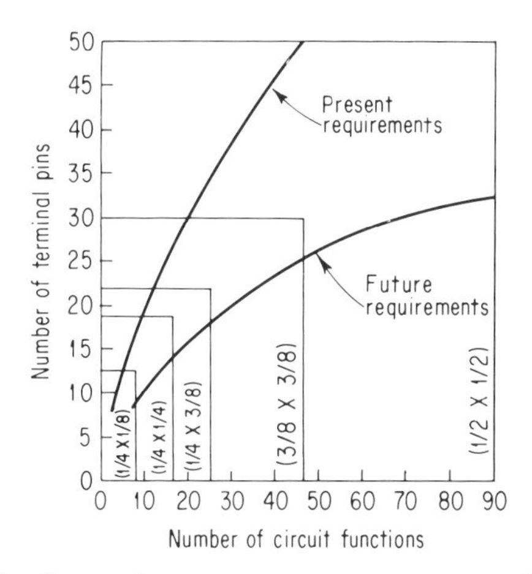

FIG. 35. Terminal pin requirements per circuit function.

propagation time, crosstalk, impedance, shielding resistance, capacitance, and inductance. The circuitry frequently must be designed and laid out with the same precision as is used with semiconductor circuitry. For each circuit interconnection conductor, physical characteristic, location, and relation to the circuit elements as a passive element contribute to the resulting circuit function.

Conductor line resistance should be minimized, which means making the interconnecting paths as short and as conductive as possible. Conductivity is a function of material resistivity, cross-sectional area, and temperature. Metallic films used for conductor materials have finite resistances in relation to thickness and are expressed in terms of ohms per square. (That is, for an area with equal square sides, a metallic film will have a specific ohmic value for a specified thickness regardless of the size of the square.) It is therefore possible to control the line resistance by controlling line width, thickness, and length. Figures 36 and 37 show the line resistivity of gold and aluminum, respectively, for various temperatures. In designing interconnecting circuitry, temperature must be considered. A thin metallic film will increase in temperature above the design ambient temperature from the power dissipated in the conductor itself. For a typical package operating at 150°C, the interconnecting conductor resistance for a 10,000-Å thickness will increase about 30 percent above the resistance of the conductor operating at room ambient.

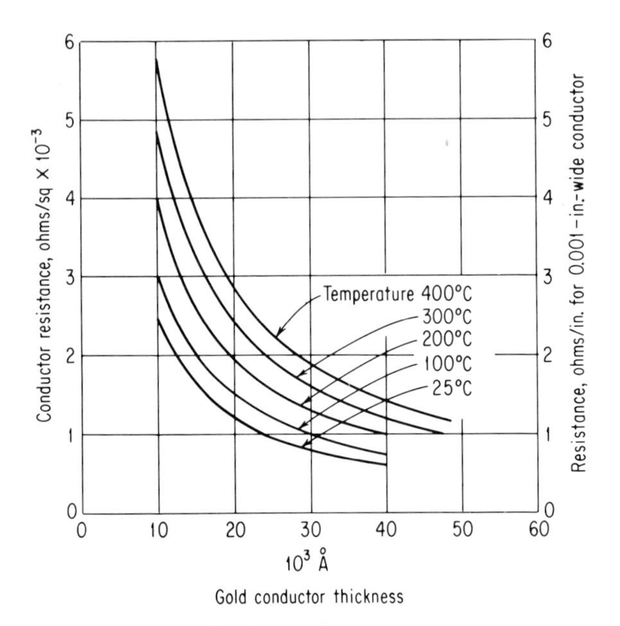

FIG. 36. Resistance of gold conductors at various temperatures.

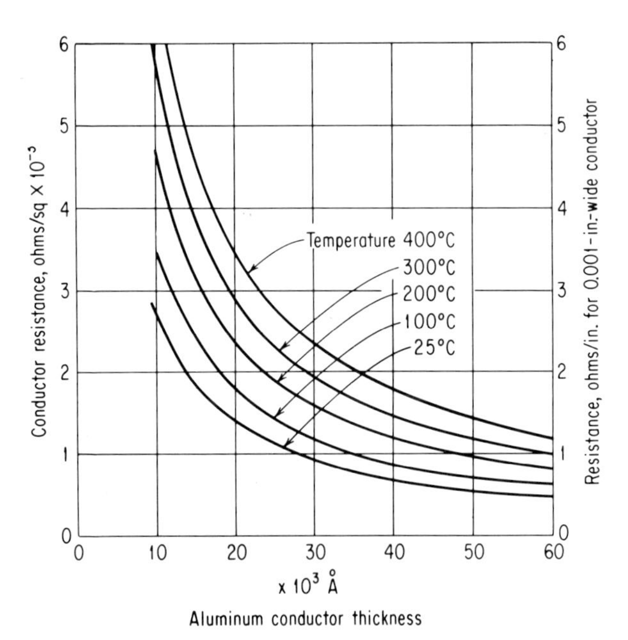

FIG. 37. Resistance of aluminum conductors at various temperatures.

The decrease in size resulting from the use of bare chips has brought signal lines extremely close together. It is essential therefore to control and minimize parasitic capacitance and inductance since they have a direct effect on the operation of the circuit. Capacitive and inductive coupling between conductor lines is a function of spacing, line length, and dielectric constant of the substrate material. These effects are controlled by minimizing the number of conductors routed parallel to each other and by limiting the conductor areas consistent with the line resistance. Capacitive coupling is also reduced with the use of ground planes, as shown in Fig. 38. Although capacitive coupling can be determined from the conductor and ground plane geometry, it is more frequently determined experimentally. Capacitance between conductors is difficult to determine analytically. It is made up of the capacitance between the sides facing each other and between the upper and lower faces, as shown in Fig. 38.

Relative effects of capacitance between two adjacent conductors for a specific set

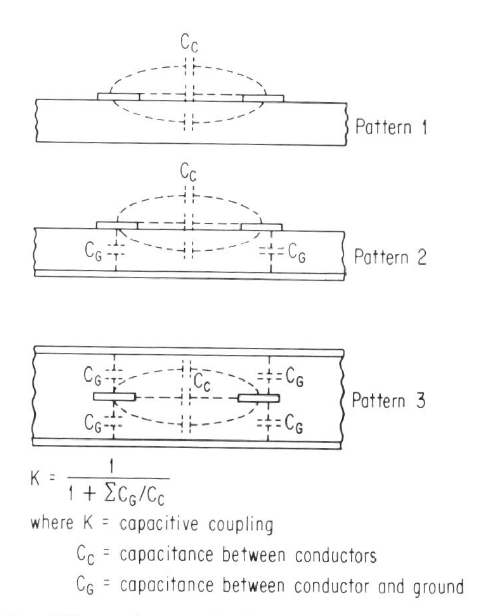

$$K = \frac{1}{1 + \Sigma C_G/C_C}$$

where K = capacitive coupling

C_C = capacitance between conductors

C_G = capacitance between conductor and ground

FIG. 38. Effect of ground planes on capacitive coupling.

of conditions (i.e., dielectric constant, dielectric thickness) are shown in Fig. 39. These curves are representative of the significance of using ground planes to reduce capacitive coupling.

Representative values for inductive coupling, a similar problem, are shown in Fig. 40. Inductance values are shown for various conductor center-to-center spacings for various conductor thickness-to-width ratios. The fundamentals of both capacitance and inductance are discussed in Chap. 10.

Voltage breakdown between conductors is normally not a problem with this type of packaging since the voltage potentials are usually very low. However, good design practice should minimize the voltage potentials between adjacent conductors.

Other electrical characteristics such as crosstalk, propagation time, and characteristic line impedance all must be considered in the final design of the multi-integrated-circuit chip package. These are a function of resistance, inductance, capacitance, and conductance, which are related to the physical properties of the structure, signal conductor width and thickness, conductor material, and dielectric constant of the supporting substrate. Their determination can be quite complex, and they are usually determined from closely controlled electrical tests performed during the breadboard stage of the electrical design.

Ideally, the interconnecting circuitry should be designed as a single-sided pattern. This will allow for minimum processing steps, thus enhancing cost and reliability. Unfortunately, interconnecting integrated-circuit chips frequently require a complex array of conductors, and the single-sided pattern will not solve the problem. Alternate solutions are possible, including double-sided circuitry and multilayer circuitry.

Double-sided circuitry may be constructed on a ceramic substrate as shown in Fig. 41. The through hole may be formed in the ceramic while it is in the green state (before high-temperature firing) or may be drilled after firing. The former technique is more economical and the latter technique yields closer tolerances. The

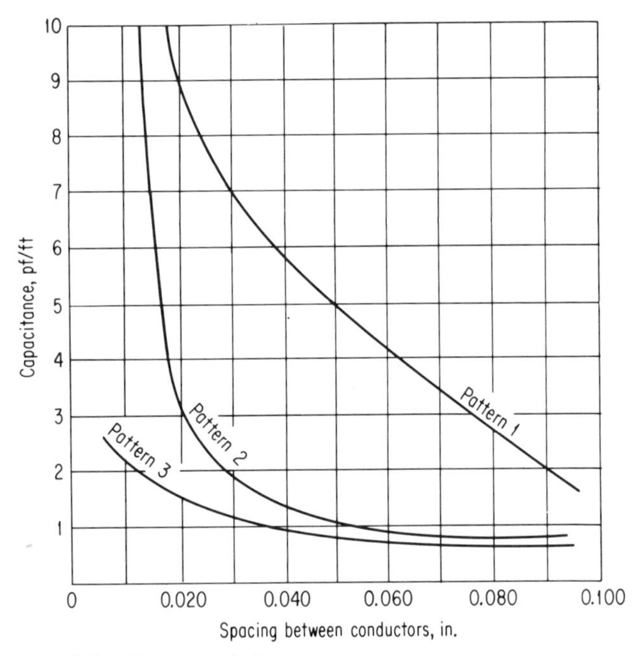

FIG. 39. Typical capacitance between two adjacent conductors (for a fixed spacing between conductor and ground plane).

advantage of this concept is that all processing, including the bonding of the chips and testing, can be performed in a single plane requiring only simple tooling. The substrate can then be mounted and soldered into the final package, shown in Fig. 34.

Multilayer interconnecting circuitry becomes a more complex problem. Additional materials and processes are required. A suitable dielectric material must be used between the various layers of circuitry. Problems of etching holes in the dielectric and of plating or depositing conductors on the dielectric must be solved. Materials that are compatible to process temperatures must also be used. However, if proper process controls are maintained, this concept can offer a reliable package. The multilayer circuitry and the chips may be placed in the final package, shown in Fig. 33. This concept eliminates the discrete wire interconnections to the output terminals, as

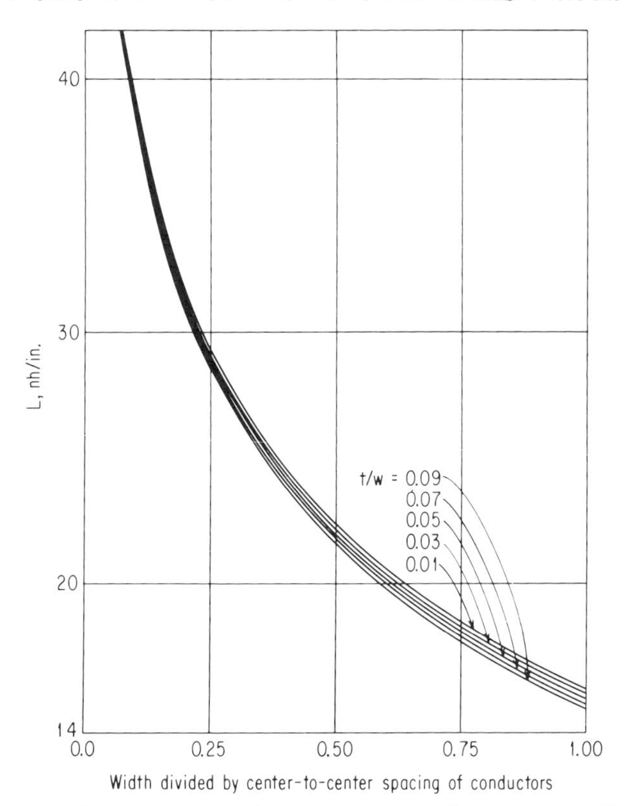

FIG. 40. Inductance curves. (*Institute of Printed Circuits Handbook.*)

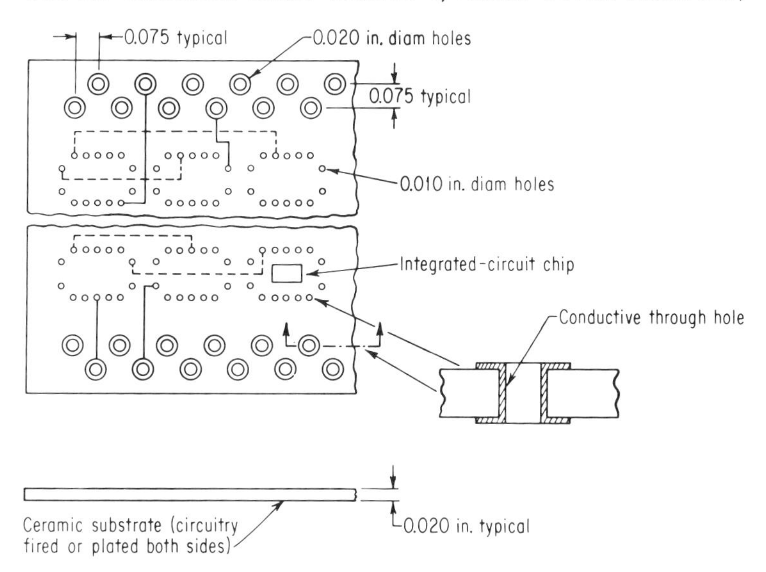

FIG. 41. Double-sided substrate for multichip interconnection.

well as giving minimum-length interconnections from chips to interconnecting circuitry. More than two layers of circuitry are possible if the layout requires it. The multilayer concept requires special design attention if electrical parameters are critical, since parasitic capacitance and inductance values become more significant.

Bonding Chip to Substrate. The bonding of the integrated-circuit chip to the substrate may be accomplished in several ways, such as with eutectics, organic adhesives, and inorganic frits. The most common technique used is eutectic bonding (a eutectic is a combination of two or more metals whose melting point is very much lower than any of the individual metals). The substrate contains areas of deposited gold or gold-plated molymanganese to which are mounted integrated-circuit chips. Heat, pressure, and movement are all necessary to achieve a satisfactory bond. The bonding operation may be done by machine or hand. Temperatures in the range of 420 to 500°C are used, and gold germanium eutectic preforms are sometimes used to aid in the bonding.

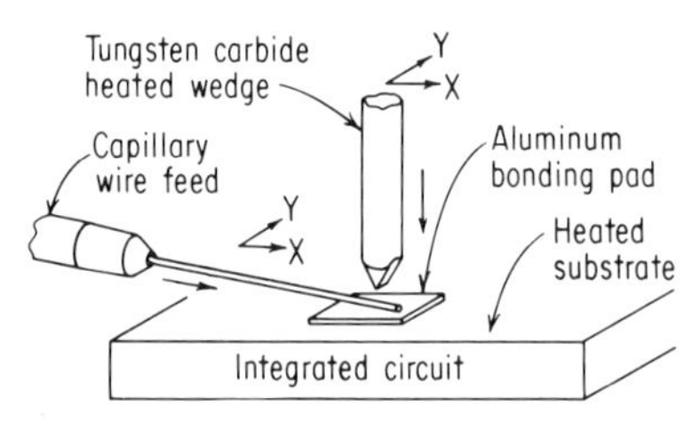

Wire: 0.0003 to 0.003 in. diam Al or Au

Substrate: 300°C

Capillary: 150°C

Gas: N_2

FIG. 42. Chisel or wedge bonding.[3]

Certain types of inorganic varnishes are also used as bonding materials. This technique is useful where electrical isolation from the substrate is necessary.

A third bonding method employs inorganic frits. They are usually low-melting-temperature glasses and are used either in a dry powder form or in a paste form with organic binders. The process is performed in an inert atmosphere with temperatures in the 500°C range. Crazing and cracking frequently occur but do not seem to be detrimental to the package reliability.

The bond, in all cases, must provide a good stable mechanical base for the integrated-circuit chips which are later bonded to the interconnecting circuitry by various techniques. Another consideration more subtle but just as important is that the bond must provide a good thermal-conductive path from the chip to the substrate. Voids between the chip and the substrate should be avoided since they will impede thermal conduction. This has been found to be quite detrimental in the case of certain types of linear circuits.

Bonding from Chip to Interconnecting Circuitry. There are many techniques for bonding from the chips to the interconnecting circuitry. The reasons for choosing one particular technique over another are varied and include such reasons as available equipment, cost, reliability, and chip configuration. Since the operation of bonding the chips to the circuitry is a major portion of the packaging design, it is necessary for the packaging engineer to have a reasonable understanding of all the available bonding techniques.

Two distinct bonding areas must be considered, i.e., those used with standard chip configurations and those used with special chip configurations such as flip chips and beam leads. Included in the techniques for the standard chip configuration are the following:

1. Thermal-compression bonding
   a. Chisel or wedge bonding
   b. Ball bonding
   c. Stitch bonding
2. Ultrasonic bonding

In the area of special chip configurations are the following:

1. Ultrasonic bonding
2. Welding
3. Diffusion
4. Soldering
5. Deposition

Detailed descriptions of the various bonding techniques follow.

Thermal-compression Bonding. The thermal-compression bonding technique is the one most commonly used in the manufacture of integrated-circuit assemblies. Its principle is to join two metals such as a wire and terminal area using heat and pressure in an inert atmosphere but without melting. The elevated temperature maintains the metal in an annealing state as the two metals join in a molecular metallurgical bond. The bonding process is rather involved, but generally speaking the softer the metal, the more readily it bonds. Aluminum and gold are the two metals most commonly used.

The most common and oldest form of thermal-compression bonding is known as *chisel or wedge bonding* (Fig. 42). This technique requires two separate alignments and is rather slow. However, it is very useful with small-diameter wires. It is also quite simple to produce two bonds with the same wire on the same pad, thereby improving joint reliability. Either aluminum or gold wire may be used with this technique.

A technique more suitable for higher production rates is known as *ball bonding* (Fig. 43). Only one alignment operation is necessary, allowing for faster bonding rates. Wire cutoff is achieved by a flame-off operation which produces the ball used

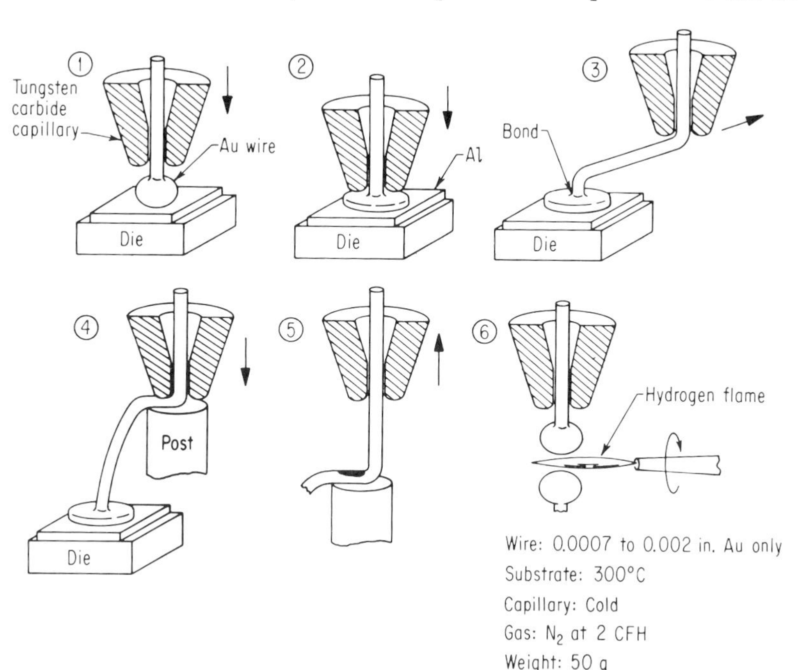

FIG. 43. Ball bonding.[3]

at the next termination. The disadvantages of this technique are that a larger bonding area is required and that aluminum is not usable since it will not ball properly during flame-off.

A technique frequently used as a compromise between wedge and ball bonding is *stitch bonding* (Fig. 44). In this technique a cutoff arrangement is used in place of the flame-off. This allows for the use of aluminum and gold wire and smaller bonding areas. The cutoff operation forms the wire for the next stitch bond.

Ultrasonic Bonding. A different concept (Fig. 45) of bonding has been developed which employs a rapid scrubbing or wiping motion in addition to pressure as the means of achieving the molecular bond. The scrubbing action removes any

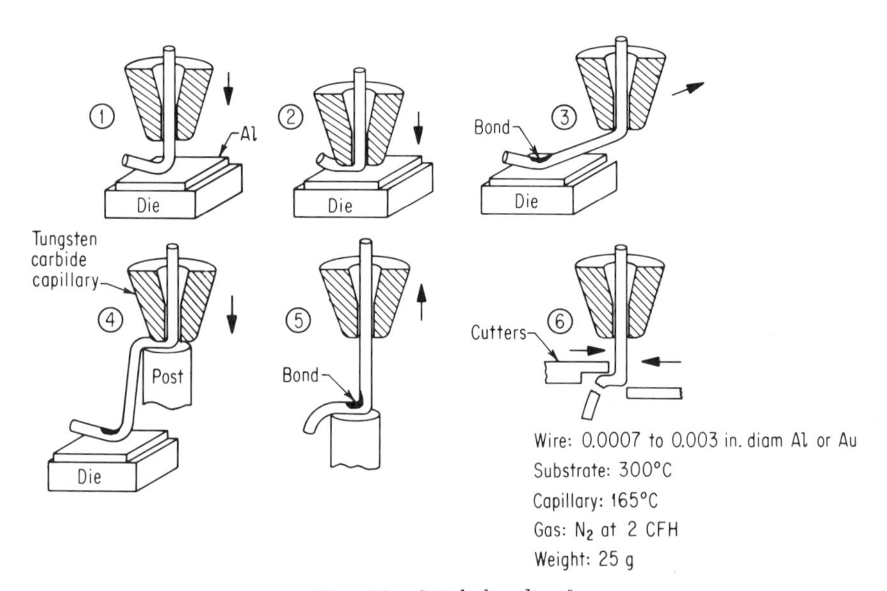

FIG. 44. Stitch bonding.[3]

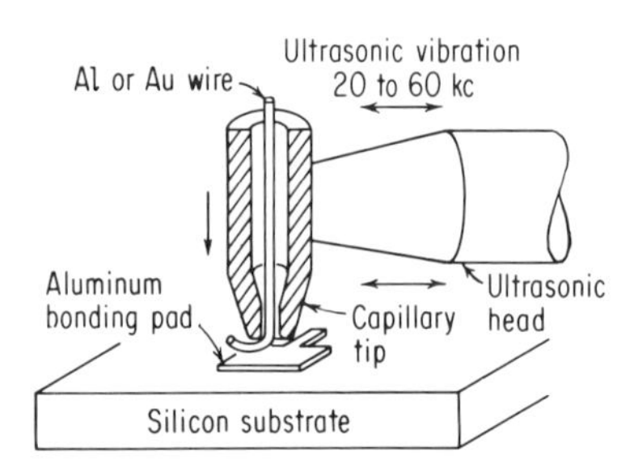

FIG. 45. Ultrasonic bonding.[3]

oxide films that might be present. Heat is not necessary in order to make the bond, and thus the hazard of "purple plague" is eliminated. This technique is compatible with either aluminum or gold wire. A slightly larger area of contact is necessary because of the scrubbing action. Extreme care must be taken not to damage the chip during the ultrasonic operation.

Face Bonding. In the past several years a number of different chip termination

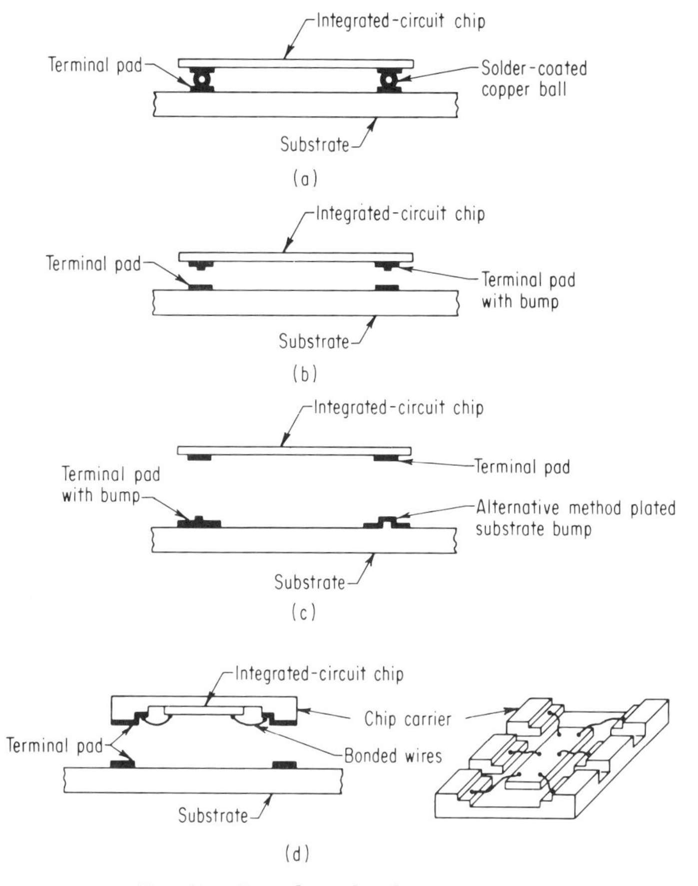

FIG. 46. Face-down bonding concepts.

areas have been developed as a means of improving the bonding operation from both a reliability and a cost standpoint. Most notable of these are the flip chip and the beam lead. The actual bonding operation used is popularly known as *face bonding* (Fig. 46).

The original flip-chip concept employs small solder-coated copper balls sandwiched between the chip terminal pad areas and the appropriate terminal pad areas of the interconnecting circuitry (Fig. 46*a*). The resultant soldered joints are made when the unit is exposed to an elevated temperature. Handling and placement of the small-diameter balls is extremely difficult, and the operation can be costly.

A more advanced technique is shown in Fig. 46*b*. In this concept a raised metallic bump or lump is developed on the chip terminal area. This is normally done on all terminal pads of all chips while they are still in the larger wafer form. The individual chip is then aligned to the appropriate circuitry on the substrate and bonded in place using thermal-compression or ultrasonic bonding techniques. Interconnection bonds between the chip terminal pads and the substrate circuitry are made simultaneously, thus drastically reducing fabrication costs. Since only a few integrated-circuit manufacturers are currently producing chips of this configuration, availability may be a problem.

A solution to the availability problem is found in using the concept shown in Fig. 46*c*. In this approach the bump or lump is produced as a part of the substrate or the interconnecting circuitry on the substrate rather than on the chip. Thus the equipment manufacturers are able to use any integrated circuits that are available in bare chip form. Bonding may be accomplished by thermal-compression or ultrasonic techniques. A diffusion bond may be achieved if the interconnecting circuitry is made of aluminum and thermal-compression techniques are employed. This concept is gaining in popularity because of its higher reliability potential and also because it eliminates chances of purple plague. A drawback to this concept is the problem with bump deformation if chip replacement is necessary.

Another solution to flip-chip nonavailability is illustrated in Fig. 46*d*. In this concept a formed ceramic carrier is employed as a mounting holder for the chip. The terminal pads of the chip are interconnected with contact or terminal pads on the holder using conventional bonded wiring techniques (thermal compression or ultrasonic). The holder is flipped and aligned to the appropriate circuitry on the substrate, and interconnection is then made with techniques such as ultrasonic bonding or soldering. This concept may be used with any bare chips that are available. It is also possible to establish a standard holder configuration or family of configurations. The major drawback to this concept is the extra substrate area required over the other flip-chip concepts.

Beam Lead Bonding. The beam lead concept is shown in Fig. 47. The lead is produced with plating techniques during the processing of the undiced wafer. The lead thickness is in the range of 0.0005 in. There are a number of advantages to this interconnecting configuration: alignment to the interconnecting substrate circuitry is simple compared to the flip-chip approach; a number of bonding techniques may be employed such as ultrasonic, thermal compression, and even welding. The major drawback of the beam lead device is its larger area requirement. This means that fewer devices are produced per wafer and also that fewer devices may be put on an interconnecting substrate.

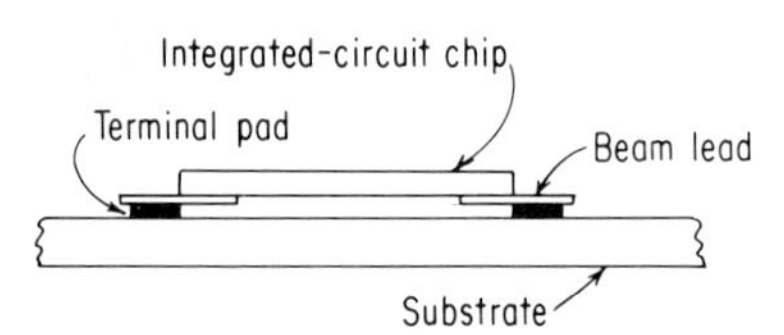

Fig. 47. Beam lead bonding.

Deposited Thin-film Interconnections. A promising advanced concept for interconnecting bare chips is the *thin-film deposition technique* under development by several companies. Two versions of this concept are shown in Fig. 48. It is necessary to provide a smooth uninterrupted surface between the circuitry on the substrate and the terminal pads on the chip. Figure 48*a* shows a concept in which the chips are held in recesses in the substrate with a filler material which also serves as a bridging material for the thin-film interconnection. Candidate materials must have coefficients of thermal expansion compatible with the substrate, the chip, and the metal film. Some promising materials appear to be certain high-temperature organic varnishes and certain low-temperature glasses. A simpler concept is shown in Fig. 48*b*. The chips are carefully located and placed on the bridging material on an uninterrupted substrate surface. The bridging material, if proper amounts are used, has a tendency to flow up the chip sides to a smooth fillet on top. The first concept requires more fabrication time since cavities must be made. The second concept requires more care in holding the chip in proper location. The thin-film deposited interconnection potentially is a low-cost–high-reliability interconnection system for integrated-circuit chips. Efforts to date have been very promising when extreme temperature changes are not required. When these large changes are necessary, more compatible materials with closely matched coefficients of thermal expansion must be developed.

Monolithic Multi-integrated-circuit Arrays. The most recent advance in higher circuitry density has been to put more interconnected circuits on a larger single chip. This may vary from several circuits to as many as several hundred. They may be the bipolar integrated-circuit type or the MOS type. In either case the term *large-scale integration* (LSI) has been used when over 100 circuits are interconnected in the monolithic form.

Current yields have not produced wafers with all good circuits and it is hardly likely that this will ever be the case. It is therefore necessary to limit the number of circuits to be interconnected or to develop a technique that interconnects only the good circuits. Another possibility would be to allow for some faulty circuits in the final interconnection pattern. Several promising techniques have been developed by the device manufacturers to achieve a practical multifunction array.

Optimum Arrays. A number of techniques may be used in mapping good and defective circuits on a wafer and determining the optimum number to interconnect. This may be done by studying yield patterns of good chips on the wafer. When optimum array patterns are established, standard interconnection patterns may be developed for these arrays. Figure 49 illustrates a technique of using overlays for establishing yield patterns for two- or four-circuit arrays. Figure 49 illustrates that the larger the array the lower the yield. For example, the 68 good single circuits yield

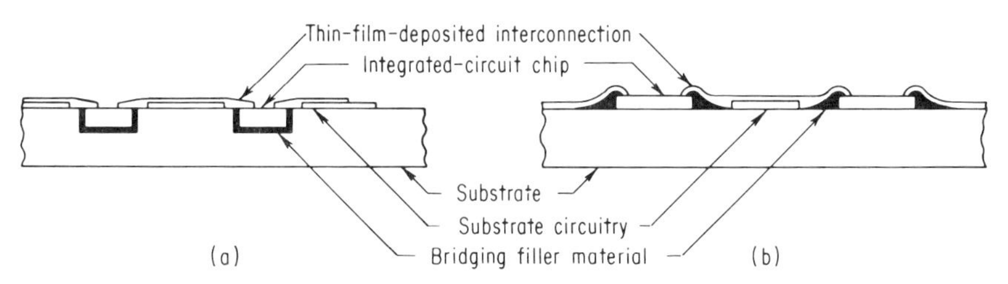

FIG. 48. Deposited thin-film interconnections.

only 27 two-circuit arrays or 54 good circuits, only 9 four-circuit arrays or 36 good circuits. The several curves shown in Fig. 50 indicate that as the wafer yield increases the array yield increases.

Selective Interconnections. The chances for LSI patterns with the optimum array technique are rather low at the current industry level of good circuit yields. It is therefore necessary to establish a technique for interconnecting circuits on a selective basis, that is, interconnecting only the good circuits on the larger wafer. The obvious—but costly—approach is to custom-design the interconnection pattern for each individual wafer. This requires the fabrication of new masks for each wafer, which might be justified under certain circumstances.

Another approach with greater flexibility, now under development by a number of companies, employs computer techniques. The basic principle uses a machine that scans the wafer and identifies the good circuits. The computer, having been given the interconnection instructions, decides on optimum layout (i.e., minimum area, isolation, speed, etc.). The actual interconnection pattern is then reproduced with a machine-controlled light beam or electron beam. This approach, although presently only in the development stage, will most likely be available for production use in one form or another in the near future.

Multilayer Interconnections. Some simple multicircuit arrays may be interconnected with the normal three levels of interconnections developed during the processing of the conventional integrated circuit: (1) the metalized interconnection, (2) the substrate used as a ground plane, and (3) the diffused components themselves, or intentionally placed diffused cross-under connections. More complex multicircuit arrays, however, require one or more additional interconnecting layers. After the first metalization step has been completed, another oxide layer is deposited; appropriate windows are opened; a second metalization layer is deposited, imaged, and etched; and so on for the total number of layers required. Unfortunately the process is critical because of the problem of the formation of aluminum oxides and because of the possibility that subsequent processing temperatures may damage prior metalization.

Recent developments by a number of companies have shown promise of overcoming

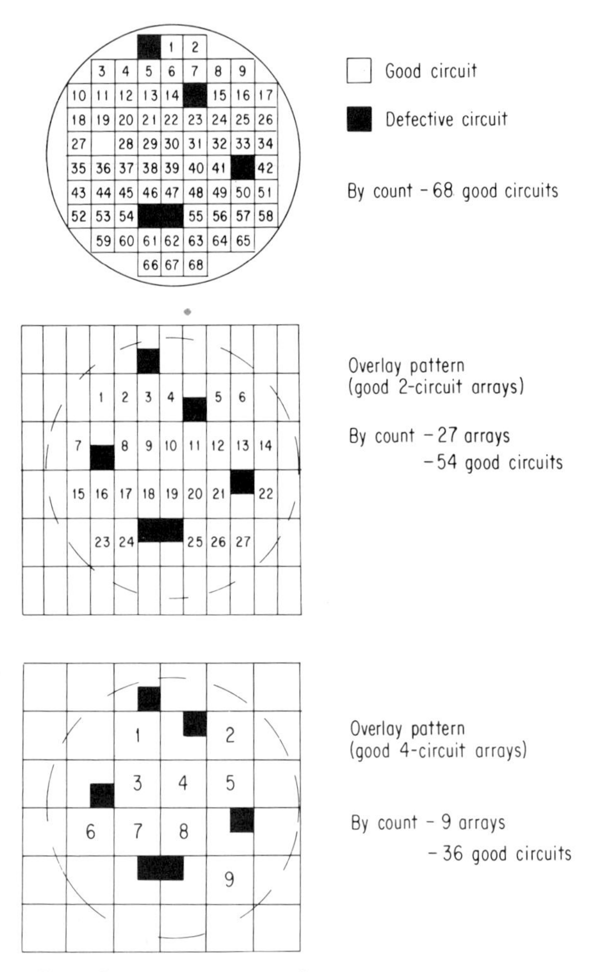

FIG. 49. Determination of optimum circuit arrays.[4]

these problems. A two-metal interconnection system seems to be a possible answer. The system employs a gold or platinum conducting layer on top of such materials as chromium, molybdenum, or titanium which provides the necessary adherence to the substrate dielectric. These two-metal systems allow the use of higher temperatures for the formation of the dielectric layers without the formation of surface oxides.

Thick- and Thin-film Circuits. Other useful techniques in the field of micro-electronics are the thick- and thin-film concepts. These concepts came into being about a decade ago with the development of thick- and thin-film passive components such as resistors and capacitors. These components were crude at first but very useful where close component tolerances were not required. As techniques, processes, and knowledge increased, more sophisticated and more useful components and circuitry were developed. Today thick and thin films are combined with integrated circuits and microdiscrete components to make up a significant part of the industry's micro-circuitry. A brief discussion of the art of thick- and thin-film circuitry follows. (A detailed discussion of these techniques may be found in Chap. 5 of this handbook.)

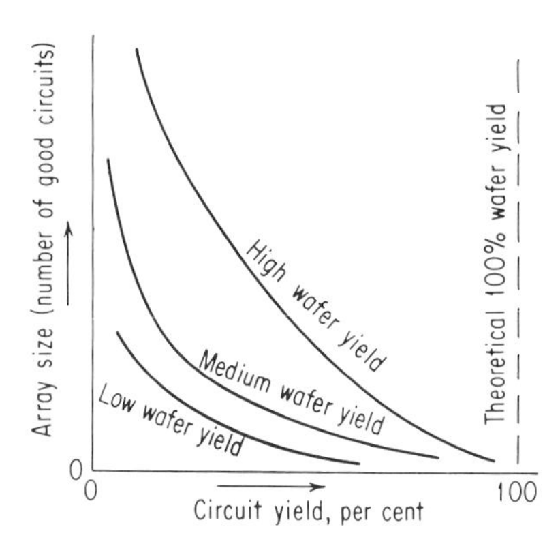

FIG. 50. Effect of array size on yield.

Thick Films. Thick films are produced primarily by screening resistive, conductive, and insulating inks or pastes in a predetermined pattern on a glass or ceramic substrate, usually followed by firing at elevated temperatures. Alternate disposition methods to screening are offset printing and transfer tape. Typical components produced with thick-film techniques include capacitors, inductors, potentiometers, resistors, and transformers.

Capacitors are produced by sandwiching an insulator between two conductive areas. Capacitance values are determined by the formula

$$C = 225K \frac{A}{t} \times 10^{-5}$$

where t = thickness
A = area
K = dielectric constant of the insulator

Inductors are produced by concentric paths of a conductive material. Inductance values are determined by the formula

$$L = 0.00508 \left(2.303 \log \frac{4}{d} - \theta\right)$$

where d = line width
θ = constant for various figures such as:
For a circle, $\theta = 2.451$
For a square $\theta = 2.853$

Resistors are produced by depositing resistive materials in a pattern in which the resistance value is based on the total number of squares in the pattern; i.e., if a line is 10 mils wide and 100 mils long, the number of squares is 10. This figure is then multiplied by the resistivity of the material, which is given in ohms per square. Resistance tolerance as screened can be controlled within ±10 percent. Closer values may be obtained by trimming by mechanical or chemical means.

Typical thick-film materials are shown in Table 5. Some ceramic substrate materials are shown in Table 13. Other ceramic and glass substrates are discussed in Chaps. 5 and 7.

TABLE 5. Thick-film Materials

Conductors	Resistors	Insulators
Gold Platinum gold Palladium gold Palladium silver Silver Molymanganese Copper-loaded epoxy Silver-loaded epoxy Gold-loaded epoxy	Palladium silver Platinum	Epoxy Silicone Glass

Thick-film passive components are frequently combined with microdiscrete components and packaged or bare chip integrated circuits to form a hybrid circuit. These circuits are stable, reliable, and capable of high-voltage, high-power dissipation. Attachment of the microdiscrete components and integrated circuits is accomplished with soldering, welding (parallel gap), or thermal compression or ultrasonic bonding. The thick-film substrate is frequently embedded in a shell form or just coated and used as is. It is also packaged in a number of other configurations, the most common of which are the TO can or flat pack. Input/output leads may be round wire or flat ribbon. Planar and stacked packaging concepts are then employed as with packaged integrated circuits, the major difference being that thick-film circuits are normally quite a bit larger, ranging from $1/2 \times 1/2 \times 0.100$ in. on up.

Thin Films. Thin films are produced in a variety of ways. Included are:

Vacuum deposition (both evaporation and sputtering)
Vapor plating
Electroplating
Electroless plating
Chemical salts
Chemical reactions

The dividing line between thick and thin films has been placed at 10,000 Å, although recently all vapor-deposited films have been considered thin films. Thin films have a number of advantages over thick films, including:

1. A larger selection of resistive and insulating materials is available.
2. The vacuum environment employed in vacuum deposition of thin films allows the exclusion of most contaminates.
3. Closer control of film thickness is possible, allowing closer component tolerances.
4. Closer line widths and spacings are possible, allowing greater component values.
5. Thin-film hybrid circuits are currently more readily available than thick-film hybrid circuits.

Vacuum deposition is the principal technique employed in the production of thin films. Two concepts may be employed: (1) Evaporation deposition is accomplished by heating the material to be deposited to its evaporation point (vacuum ranges of 10^{-4} to 10^{-6} torr are used). (2) Deposition by sputtering employs a partial gas pressure from 10^{-1} to 10^{-2} torr. A high voltage (1 to 20 kv) is applied to the anode, and the positive ions of the surrounding gas bombard the cathode material, which is knocked loose in a molecular form and deposited on the substrate. Since heat is not involved, sputtering is very useful in deposition of the high-temperature melting-point metals such as nickel and chromium. Table 6 is a listing of some of the more common materials employed for thin-film components and circuits.

A number of companies are working to perfect thin-film active components. Results to date have been very promising, and small quantities of experimental devices have been successfully produced.

Some of the materials used in making active components include cadmium selenide,

TABLE 6. Thin-film Materials

Conductors	Resistors	Insulators	Magnetic
Aluminum	Chromium	Anodized aluminum	Nickel-iron
Copper	Nickel-chromium	Anodized tantalum	
Gold	Tantalum	Anodized titanium	
Nickel	Titanium	Glasses	
		Silicon monoxide	

cadmium sulfide, tellurium, cadmium telluride, zinc oxide, tin oxide, and indium oxide.[5]

The electrical isolation of the insulating substrate is a decided advantage of the thin-film active device over the monolithic semiconductor integrated circuit, for it permits circuitry of greater complexity on larger substrates. Potentially lower costs are also possible as better fabrication techniques are developed.

Thin-film passive components are frequently combined with microdiscrete components and/or integrated circuits to make hybrid circuits. Attachment of the discrete components and/or circuits is done in essentially the same manner as with the thick-film circuits. Circuit package configurations are also the same, although it can be expected that the package will be smaller since the same value passive components can be made in smaller areas. Planar and stacked packaging concepts are also employed, as mentioned for the thick-film circuits.

SYSTEM PARTITIONING

The packaging engineer is faced with the decision of how best to subdivide the system into modules, submodules, and other levels of functional units. The choice is usually complex since many factors can have a significant effect on the decision. The choice can vary from a single circuit unit to a large complex unit of many hundreds of circuits. In general, small functional units permit a reduced number of unit types which in turn enhances spares logistics and manufacturing complexity. Small units are frequently of low enough cost to be considered throwaway items rather than repairable. The major disadvantage is the increased number of connectors, supporting hardware, and interconnecting wiring required. A large functional module has two advantages, ease of electrical diagnosis and inherent higher reliability with fewer joint interfaces, both of which enhance system maintainability. The final partitioning decision can be made only after a series of trade-off studies considering all the factors listed below:

Manufacturing cost
Reliability
Size
Repair or throwaway
Maintainability
Electrical requirements
Environmental requirements

The inherent high reliability of microelectronics has made possible the maintenance concept of logistic self-support. It is now practical to provide functional units as replacement units or even built in as redundant elements. A study of the diagnostic requirements will help determine the optimum functional unit size.

The choice between repairing and throwing away units has historically been based only on a flat functional unit cost. With the advent of microelectronics and higher reliability the military has recognized that reliability is also a determining factor in this decision: if a unit is very reliable it is frequently more economical to throw it away than to repair it. An MTBF of $1 per 10,000 hr is frequently used as the criterion for a throwaway unit. It is therefore easy for the packaging engineer to design the functional unit so that cost and reliability meet this requirement. Table 7 shows various sizes of functional units, permissible throwaway cost, and permissible cost of the integrated circuit assuming a failure rate of 0.003 percent per 1,000 hr. With an average integrated-circuit cost of approximately $3 to $5, the throwaway functional unit contains between 25 and 40 integrated circuits.

Generally the larger the functional units, the fewer the total input/output connector pins required. Also, the larger the functional unit, the fewer times it is repeated in the system. Table 8 illustrates these principles on a study of a system using 1,600 integrated circuits. Note that the number of connector pins per integrated circuit is reduced sharply as the number of integrated circuits in the functional unit goes up. Also note in the last column that the reuse factor drops off drastically above 40 integrated circuits. Two types of interconnection patterns are shown. In the register logic the maximum logical interconnections are made on the board; this approach minimizes the number of connector pins and level 2 wiring but normally reduces the reuse factor. The control logic brings all signal leads to the connector pins; this approach reduces the number of types required but requires more connector pins and more level 2 wiring.

TABLE 7. System Partitioning Based on Reliability and Cost, Based on $1 per 10,000 Hr [6]

N No. of integrated circuits per level 1 module (1)	$\frac{1{,}000}{N\lambda\ 10^{-2}} = \frac{10^5}{N\lambda}$ MTBF per level 1 module, hours * (2)	$\frac{\text{MTBF}}{10{,}000} = \frac{10}{N\lambda}$ Permissible † sell price per level 1 module (3)	$\frac{\$/\text{module}}{N} = \frac{10}{N^2\lambda}$ Permissible average price per mounted integrated circuit (4)
1	31,200,000	$3,120.00	$3,120.00
10	3,120,000	312.00	31.20
20	1,560,000	156.00	7.80
50	625,000	62.50	1.25
100	312,500	31.25	0.31
250	125,000	12.50	0.05
500	62,500	6.25	1.25

* λ = failure rate in % per 1,000 hr per connected integrated circuit = 0.003 + (20 × 0.00001) = 0.0032%/1,000 hr.

† The selection of $1 per 10,000 hr cannot be independent of the complexity of the throwaway module. Consider the case of a hypothetical 5,000-MIC (integrated-circuit) computer with an MTBF of 10 years (80,000 operating hr) and a cost of $50,000 ($10 per installed integrated-circuit package). With an MTBF equivalent to the probable usable life, it would appear reasonable to consider the computer a throwaway item, even at a $50,000 cost. This results in a throwaway cost of $6,250 per 10,000 hr ($0.625 per operating hour). This "system throwaway" approach eliminates the need for highly trained maintenance personnel, complex fault isolation and testing equipment, expensive maintenance manuals, and elaborate maintenance bases; such an approach could prove to be most cost-effective. Further study is warranted in this area.

Table 9 illustrates the effects of various size functional units. This is a study of a computer central processor, an in/out unit containing 3,689 integrated circuits. The smallest replaceable unit (SRU) quantity is the number of functional units required for the total system. It can be seen that the number of connector pins required increases rapidly as the functional unit size decreases. The total connector failure rate has been evenly divided between the functional unit and the level 2 wiring. It is significant to note that the system failure rate has only increased from 1.306 to 1.494 failures per year when the largest replaceable unit (LRU) count reduces from 3,689 to 9, while the number of connector pins required increases from 580 to 21,900.

TABLE 8. Number of Integrated Circuits versus Pin Requirements and Usage [6]

Integrated-circuit packages per module	Module input/output leads per circuit package		Module usage in 1,600-MIC system			Reuse factor
	Register logic	Control logic	Number of types	Total modules	% of total modules by type	
1	10	12				
5	5*	9*	10	320	3.2	96.7
10	4*	9*	15	160	9.4	90.6
20	3½*	9*	18	80	22.5	77.5
40	3*	7*	25	40	62.5	37.5
50	2*	2*	31	32	96.9	3.1
100	1*	1*	16	16	100.0	0
200	0.85*	0.85*	8	8	100.0	0

* Does not include power and voltage leads, two required.

TABLE 9. Functional Unit Size and System Reliability [6]

Failure rate analysis chart for various-sized modularity approaches

	SRU size, ICs	SRU quantity	System failures per year	System MTBF hours	λ per SRU *	λ back wiring *	Percent total λ in back wiring *	Connector pin total	No. of types of SRU
A. Single assembly	3,689	1	1.306	6,710	14.904	0.000	....	580	1
B. Large card	119	31	1.328	6,598	0.4867	0.070	0.46	5,199	26
C. Small card	9	410	1.494	5,865	0.0388	1.1385	6.7	21,900	15
D. Individually pluggable	1	3,689	1.692	5,178	0.004	4.558	23.6	44,848	15

Basis of calculation—failure rates assumed

Integrated circuit	0.004 % per 1,000 hr
Solder joints	0.000001 % per 1,000 hr
Plated-through joints	0.0000005 % per 1,000 hr
Connector pin friction joint	0.0001
Wire wrap, single	0.0000005 % per 1,000 hr

System failures per year = λt
MTBF = $1/\lambda$

* Percent per 1,000 hr.

INTERCONNECTION TECHNIQUES

Numerous interconnection techniques are available to the packaging engineer. These techniques may be used for level 1 interconnections (from component and/or circuit to component and/or circuit to form submodules or modules); for level 2 interconnections (from module to module to form subsystems); and finally for level 3 interconnections (from subsystem into system). Some of the techniques are common to all levels of interconnections, while some are peculiar to one level. Table 10 shows the various techniques available and the level of interconnections where they are normally used.

Printed Wiring. Single- and double-sided boards and multilayer boards are frequently used with microdiscrete components and integrated circuits when packaged in the flat-pack or TO can configurations. Details of the design of these circuits are covered in Chaps. 1 and 8.

Printed-wiring techniques may be employed with both the planar and module concepts of packaging. They may be solderable and/or weldable. The major differences in their use with microelectronic packaging as compared with conventional packaging is that closer spacing, smaller conductors, and tighter tolerances are required. Frequently

TABLE 10. Interconnection Techniques

Interconnection technique	Interconnection level		
	1	2	3
Printed wiring:			
Single-sided board	✓	✓	
Double-sided board	✓	✓	
Multilayer board		✓	
Flat cabling		✓	✓
Flexible printed wiring		✓	✓
Point-to-point wiring:			
Insulated wire soldered			✓
Uninsulated wire welded	✓	✓	
Insulated wire welded	✓	✓	
Deposited wiring:			
Plated	✓		
Screened	✓		
Vacuum	✓		
Special wiring:			
Tuned-up tabs	✓	✓	
Formed tablets	✓	✓	
Woven wiring	✓	✓	
Weldable pins	✓	✓	

conductor widths and spacings of 0.010 in. and hole spacings of 0.050 in. are employed. More recent efforts have been toward 0.005-in. conductor widths and spacings and 0.025-in. hole spacings. Thinner laminates are also employed, in the range of 0.015- to 0.030-in. thickness.

Multilayer Wiring. Tighter packaging of conventional components and, more recently, tighter terminal spacings of microelectronic devices have necessitated the development of multilayer techniques (see Fig. 51). The initial multilayer designs (still in use in many applications) employed what is referred to as the *clearance-hole technique.* In this approach, access to the next layer is accomplished by leaving a larger hole in the covering layer than the terminal pad in the lower layer. It can be seen that the number of layers had to be restricted if small size was to be maintained. Electrical shielding between layers is easily accomplished by inserting copper sheets (with clearance holes for through holes) between circuitry layers.

A more compact multilayer concept employs the plated-through-hole technique originally developed for double-sided printed-wiring boards. Each layer is fabricated with its circuitry etched for the required configuration. The layers are then stacked and laminated with heat and pressure to form a uniform single multilayer board. Through holes are drilled in the appropriate locations, and the holes are plated through (normally with copper, or copper nickel-gold). The plating makes contact with the various layers where the copper is exposed in the drilled holes. The ability to achieve close hole spacings is dependent upon how small a hole can be plated through. The rule of thumb is usually that the hole diameter should be no smaller than one-third the thickness of the board to be plated; however, special techniques have been developed to plate through smaller holes.

The most compact multilayer plated board is fabricated with a plated-up post. In this concept alternate layers of conductors and insulation are built up using plating and screening techniques. Connections between layers are accomplished by plating to previously plated circuitry (forming a post) through holes in the insulating layers. Since connections between layers are built up as a post from layer to layer, conductors exist only on these layers. Thus this space on the balance of the layers can be used for other conductors. Holes are only drilled where it is necessary to insert component leads. Flat-pack integrated circuits may be parallel-gap-welded to the circuitry. No holes are required, and thus greater circuitry density is provided.

Preformed Flexible Wiring. Several preformed flexible printed-wiring techniques are frequently used in level 2 and level 3 microelectronic packaging. Flat cabling is a good replacement for discrete round wire bundles. The conductors are parallel flat copper strips (they can be nickel for welding) embedded in a plastic. Conductor widths, thicknesses, and spacings can vary for different requirements. Terminations to these cables are more difficult than standard cabling but a number of satisfactory techniques do exist (refer to Chaps. 1, 2, and 6). A more sophisticated concept is flexible printed wiring. In this approach the cabling or interconnecting wiring is custom-designed to fit a specific pattern of terminations or connecting points, and maximum interwiring volume utilization is achieved. The conductors in this case are

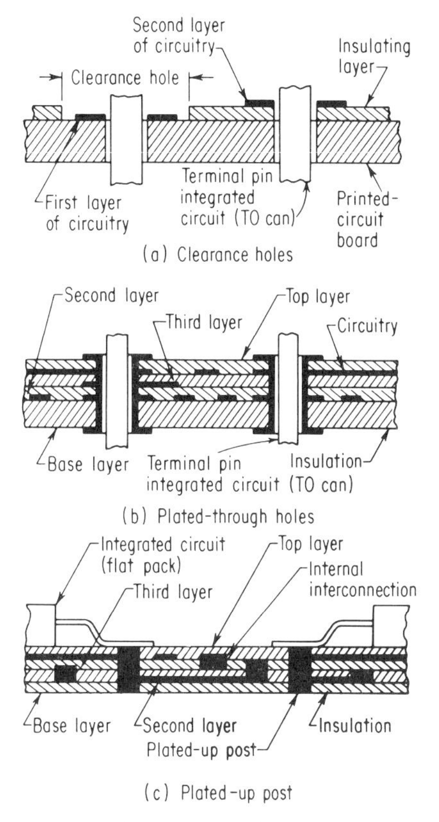

FIG. 51. Multilayer printed-wiring techniques.

normally copper as with the flat cable, but the pattern is more random to meet the specific layout requirements. Multilayer patterns may be made by stacking individual layers and laminating with heat and pressure. The initial design cost is high, but overall cost can be considerably lower than the cost of conventional wiring since assembly labor is reduced and troubleshooting for wiring errors is eliminated.

Point-to-point Wiring. Interconnection of electronic components with insulated discrete-wiring techniques has been in use since the electronic industry began. This concept meets military and commercial design requirements and is compatible with most existing hardware and fabrication techniques. The major drawback has been its relatively high cost. However, a number of automated machines have recently been developed to make point-to-point wiring economically more favorable to printed wiring. Terminations are made by wire wrapping, crimping, or other mechanical means with automated approaches. Wire wrap requires solid wire, while crimping techniques may use stranded wire. More details on these techniques are found in Chaps. 2 and 3.

Although soldering is generally used with point-to-point wiring, several welding concepts have been developed in recent years. One concept employs uninsulated nickel wire or ribbon and is used frequently with the stacked module for level 1 interconnections. (Some limited applications on level 2 interconnections for space systems also employ this concept.) Supporting or positioning insulating films are

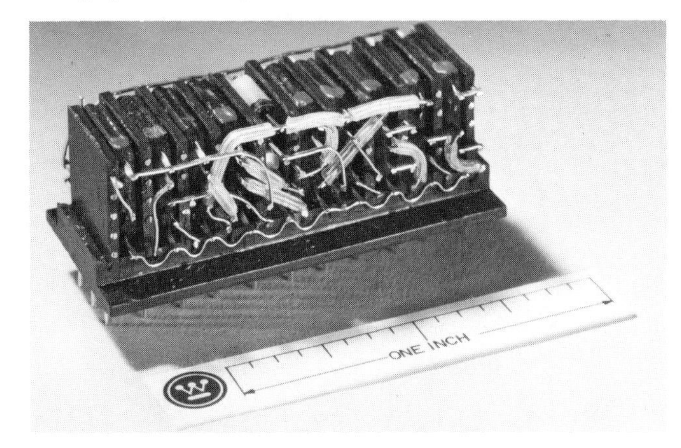

(a)

(b)

FIG. 52. Point-to-point welded interconnections. (*Westinghouse Electric Corp.*)

frequently employed to aid in properly locating the interconnecting wire or ribbon to the component lead. Care must be taken with this technique to preform and route the wire or ribbon to prevent any shorting. All crossover points must be insulated. Figure 52*a* illustrates this concept.

Another point-to-point welding concept more suitable to the close lead spacing required in microelectronic designs is the insulated wire approach. The wire is coated with an insulating covering. A heated electrode melts the dielectric prior to the welding pulse. This technique eliminates the need for carefully routing leads. The result is lower engineering design and fabrication costs. Figure 52*b* shows how the insulated wires may be routed in a random manner.

Deposited Wiring. There are numerous methods for depositing wiring, among them plating metals, screening conductive inks, and vacuum-evaporating metals. Deposited wiring is usually a low-cost operation since it is a batch-type operation with a minimum of labor. Level 1 interconnections are quite compatible with this technique, but levels 2 and 3 are less so.

Plating. Plating metals may be done by both electro and electroless techniques. The metal is usually plated onto an insulating dielectric surface. It may be plated as a complete layer and etched to the interconnecting pattern or plated through a mask for the final interconnecting pattern. This concept is illustrated in Fig. 53. Small conductor widths and spacings are possible with this approach; however, the small area of interconnection at the component and/or circuit lead interface makes the joint reliability critical to large temperature changes. A good match of thermal expansion coefficients of the metal being plated, the lead material, and the dielectric material is necessary to make this concept usable. Some supplementary bonding at the joint

FIG. 53. Deposited interconnections. (*Westinghouse Electric Corp.*)

interface such as laser welding or parallel gap welding will improve the joint reliability, but it adds greatly to the cost.

Screening. Screening conductive inks is another interconnecting technique available to the packaging engineer. Application is limited because of the high temperatures required to burn off inorganics used for the carrier material. The use of ceramic substrates for mounting bare integrated-circuit chips, however, presents a compatible temperature situation. Conductor widths and spacings are limited because of the limited tolerance capabilities of screening techniques. Line widths and spacings of 0.010 in. are possible, though, if screen mesh size and ink viscosity are carefully selected. The screened interconnection technique is a low-cost batch-type process, and a minimum of fabrication equipment is required.

Vacuum Deposition. Vacuum-deposited wiring is a sophisticated and relatively costly operation. The technique requires an initial costly investment for vacuum equipment. Tooling can be expensive and production rates are normally low. Materials that will not outgas at the vacuum levels encountered during the process must be selected. As with plating, a vacuum-deposited film may be deposited as a complete layer and subsequently etched to the interconnecting pattern, or deposited through a mask for the final pattern. Close control of film thickness and line widths and spacings is possible with this technique, which makes it very useful where extremely small dimensions are required. Its application is very compatible with the interconnection patterns used on integrated-circuit chips or with large-scale integration concepts.

Special Wiring. There are a number of commercially available interconnection systems that are worthy of mention. These techniques are proprietary and have the disadvantage of being available only from a single source. It is possible that licensing could be made in some cases.

Raised Tabs. One approach employs prefabricated circuitry with raised tabs at the points adjacent to the component and/or circuit lead holes. Details are shown in

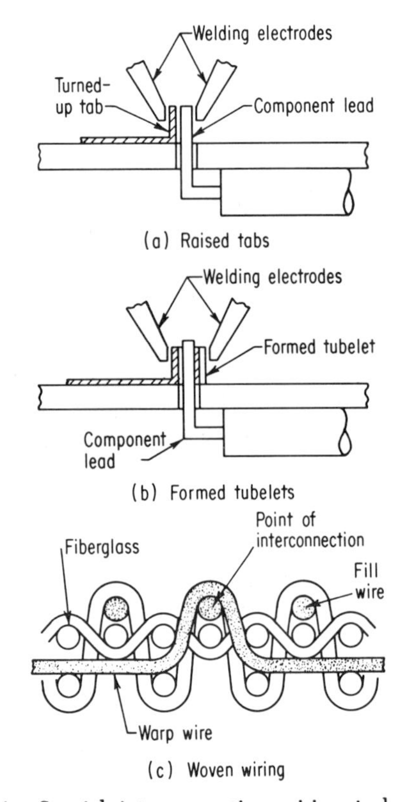

FIG. 54. Special interconnecting wiring techniques.

Fig. 54*a*. Close conductor widths and spacings are possible for microelectronic applications. The lead and tab are aligned for ease of welding and result in a significant reduction in welding labor. As with standard printed-wiring boards, this technique has a high initial design and tooling cost but this becomes less of a consideration as the quantity required goes up.

Formed Tubelets. Another, similar approach is the formed tubelet concept. In this case metal tubelets are plated as part of the preformed circuitry located at the point where the component and/or circuit leads project through the board (see Fig. 54*b*). The welding operation is actually easier than with the turned-up-tab concept since orientation of the welding electrodes is not required.

Woven Wire. A recent interconnection concept with automated possibilities has been developed that employs weaving techniques. Discrete solder-coated copper wires are woven with fiber-glass threads in both the X and the Y directions. Where a signal crossover from X to Y is required, the X wire is looped over the Y wire or vice versa. When the woven circuit is passed through a solder bath the points of wire crossovers are soldered together, forming the circuitry network. Wires may be left extended and exposed on one side of the circuitry to allow for interconnection of the components and/or circuits. Figure 54*c* shows the details of this concept. It can easily be seen, although the concept is relatively new, that with a little ingenuity the weaving looms could be programmed and numerically controlled to give a fully automated process.

METALS JOINING TECHNIQUES

Table 11 gives a breakdown of the various metals joining techniques available for

TABLE 11. Metals Joining Techniques

Soldering	Welding	Bonding	Mechanical
Hand Dip Wave Pulse	Opposed electrode Parallel gap Percussive arc Laser Electron beam	Thermal compression Ultrasonic Diffusion	Solderless wrap Crimp

interconnecting microelectronic systems. Some of the techniques are quite simple to perform and require little or no capital investment in equipment. Other techniques are quite complex and require a large capital outlay. These techniques are discussed only briefly since full details are given in Chaps. 3 and 4.

Soldering. Although soldering has been used since early civilization, it is still a very useful metals joining technique in the modern electronics industry. This technique joins two metals with a third metal. The basic 60-40 tin-lead eutectic solder is still used in most electronic applications. In some applications, metals such as indium, gallium, and silver are employed in special alloys for lower melting temperatures or for better metallurgical characteristics for certain interconnecting materials. Hand soldering in microelectronic applications is normally done with low-wattage irons (7- to 10-watt ratings) and solder is frequently applied in a preform configuration. Micro hand soldering is frequently used in the attachment of microdiscrete components to thin-film circuits. Dip and wave soldering are compatible techniques where printed-wiring boards are used. Pulse soldering is a technique recently developed for attaching flat leads to solder-coated printed-wiring boards. A resistance-type parallel-gap welding machine is used but at lower power levels than those used for welding. The resultant joint has been found to be very reliable with good repeatability since pressure, time, and energy are all machine-controlled. There is a minimum amount of heat applied to the printed-wiring board, and delamination problems are avoided.

Welding. The welding technique has had wide acceptance, especially in military applications, because of the ability to closely control the process variables. This technique joins two metals by raising their temperature to the point where fusion of the metals takes place. Although there have been many debates on welding versus soldering, recent reliability figures appear to favor welding (especially where severe environments are encountered). Another indication of higher reliability with welding is its widespread use in manned space systems. A number of welding techniques are available to the packaging engineer, as shown in Table 11. Some techniques were developed for special configurations while others use a different physical principle for making the welds.

Opposed-electrode Welding. Opposed-electrode welding, a resistance welding technique, is the original form of welding used as a metals joining technique for interconnecting electronic components and circuitry. It has been used primarily in cordwood or stacked module packaging. It is used where the two parts to be welded are located between the welding electrodes (this distance is normally limited to 1/4 to 1/2 in.). It is a very useful technique where weldable pins in a fixed array are used. First the component and/or circuit leads are welded to the pins, and then preformed thin flexible wires or insulated wires are welded to the other end of the pins to form the interconnecting circuitry (see Fig. 55*a*).

Parallel-gap Welding. Parallel-gap welding techniques were developed for use where it is impractical to place the parts to be welded between the welding electrodes. This technique is very useful where flat-pack integrated circuits are welded to printed-wiring boards (see Fig. 55*b*).

Percussive Arc Welding. Percussive arc welding is a very useful technique where component and/or circuit leads are welded to standoff leads. It is also useful where leads or terminal pins must be lengthened. A voltage is applied to the two parts in contact and then the parts are separated, setting up an electrical arc between the parts. The parts are then pushed quickly together, forming a sound metallurgical bond (see Fig. 55*c*).

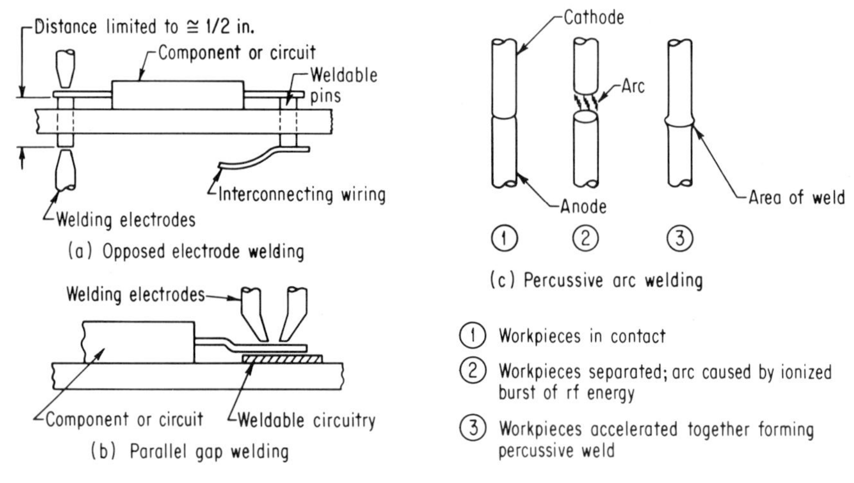

FIG. 55. Metals joining techniques (welding).

Several advanced metals joining techniques have been developed for interconnecting electronic circuitry. Included are laser welding and electron-beam welding.

Laser Welding. Laser welding is accomplished with a light source amplified to produce a power density capable of raising the materials to be welded to a temperature needed for fusing but below the vaporizing points. The heat input is relatively small and is delivered in a short period of time, thereby reducing the danger of thermal damage to the substrate material. A major advantage of laser welding is that all available lead materials can give satisfactory welds. The laser beam can be focused in an extremely small area (0.001-in. diameter) necessary for high-density microcircuit interconnections. Surface contaminants such as oxides and organic materials have little effect on the quality of the resultant welds.

Electron-beam Welding. Electron-beam welding is accomplished with accelerated electrons from a hot cathode emitter focused to a small beam of the shape necessary to achieve a welding operation. The entire operation takes place in a vacuum chamber. As with the laser operation, any metallic lead material will give satisfactory welds. Both the laser and the electron-beam welding techniques require expensive equipment, but both are compatible with automated techniques and should be very compatible with high production requirements. Both techniques are rather complex, however, and close control over processing procedures is necessary.

Bonding. Bonding is basically a welding operation and, as referred to here, may be any of several techniques employed in the interconnection of thin-film and semiconductor integrated circuits. Major categories of bonding include thermal compression, ultrasonic, and diffusion. Thermal-compression bonding employs heat and pressure to join two parts of the same metal. These concepts have been described in detail earlier in this chapter under chip packaging techniques (refer to Figs. 42 to 45).

Mechanical. Two major mechanical metals joining techniques, wire wrap and wire crimp, are frequently employed for level 2 and 3 interconnections with microelectronic systems. Figure 56 shows the details of these two concepts. For wire wrap, which employs solid wire, hand and automated equipments are already designed and available. Extreme care must be taken to prevent any nicking of the solid wire. Crimping techniques employ an intermediate piece of hardware between the interconnecting wire (usually stranded) and the post to be interconnected. The interconnecting post must be designed and fabricated specifically for the crimping hardware. This technique may also be performed with a hand gun or automated equipment. Both techniques are discussed in detail in Chap. 3.

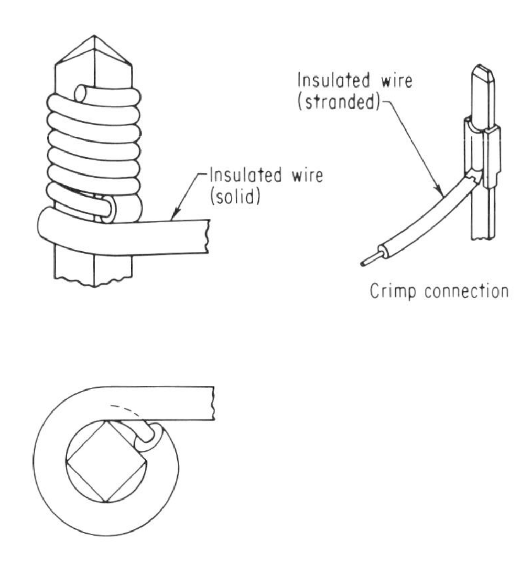

FIG. 56. Metals joining techniques (mechanical).

MATERIALS

Microelectronic packaging requires many special materials in addition to the materials used in conventional packaging. The packaging engineer should have a minimum familiarity with materials in general, plus detailed information available for his use on specific materials where needed.

Chapter 1 discusses in detail materials used in the construction of printed-wiring boards, multilayer boards, and flexible printed wiring. Chapters 3 and 4 discuss in detail materials used in soldering and welding as metals joining techniques. Chapter 5 discusses in detail the various materials used in the fabrication of thin- and thick-film circuitry. In addition, Chap. 7 covers the broader subject of materials for electronic packaging. Information needed for materials normally employed in standard packaging concepts for packaged integrated circuits may be found in these chapters.

The fabrication of multi-integrated-circuit chip packages requires more sophisticated materials. Because of the microminiature sizes utilized in these concepts, extreme care must be taken in materials selection for maximum compatibility of substrates, interconnecting conductors, and integrated-circuit chips.

Substrate Materials. Candidate materials for substrates include beryllia, forsterite, glass, alumina, steatite, and titanate. Coatings and glazes of various types are frequently used with the glasses and ceramics. Glasses and ceramics are most often used for substrate materials because of their compatible properties of temperature, chemical stability, high resistivity, high thermal conductivity, strength, flatness, and low cost. Alumina (Al_2O_3), which is a high-purity single-crystal sapphire, is a hard crystalline substance prepared by flame fusion techniques from high-purity starting material. This material has good dimensional stability with heat and is resistant to the temperature shock encountered during the soldering or welding operations. Table 12 shows the various properties associated with alumina and beryllia. NOTE: Special precautions must be taken for personnel safety when working with beryllia.

TABLE 12. Mechanical Properties of Ceramic Substrate Materials

	Dense alumina 85% Al_2O_3	Dense alumina 94% Al_2O_3 + CaO + SiO_2	Dense alumina 96% Al_2O_3 + MgO + SiO_2	Dense beryllia 98% BeO	Dense beryllia 99.5% BeO
Softening temperature, °C	1100	1500	1550	1600	1600
Thermal expansion, coefficient 10^{-6}/°C	6.5	6.2	6.4	6.1	6.0
Thermal conductivity, cal/(cm)(sec)(°C) @ 25°C	0.060	0.073	0.084	0.50	0.55
Density, g/cm³	3.40	3.58	3.70	2.90	2.88

SOURCE: American Lava Corp.

Design parameters necessary for specifying substrate material are shown in Table 13. Government or industry specifications and manufacturers' type or code numbers should be used when available.

TABLE 13. Parameters for Specifying Substrates

1. Material—government or industry specification, manufacturer's code number, or
   a. Chemical composition—includes permissible percent of contamination
   b. Density—lb/in.³
   c. Structure—grain size, porosity, internal defects, etc.
   d. Strength—modulus of elasticity, tensile strength, etc.
2. Size—face dimensions and thickness with tolerances, squareness
3. Surface finish—flatness and smoothness
4. Electrical properties—volume conductivity, surface conductivity, dielectric strength, etc.
5. Thermal properties—transverse thermal conductivity, maximum physically stable temperature, expansion characteristic, and maximum transverse temperature gradient

SOURCE: Electronics Industry Association.

Compatibility of the interconnection circuitry materials is a dual problem. The materials must closely match the substrate material from a thermal expansion standpoint for good adhesion and must also be compatible with the material used between the circuitry and integrated-circuit-chip terminal pads. Figure 57 shows how the thermal expansions for several substrate materials and several metals compare over a wide range of temperatures. Note the close correlation of the molybdenum and alumina materials, a popular combination.

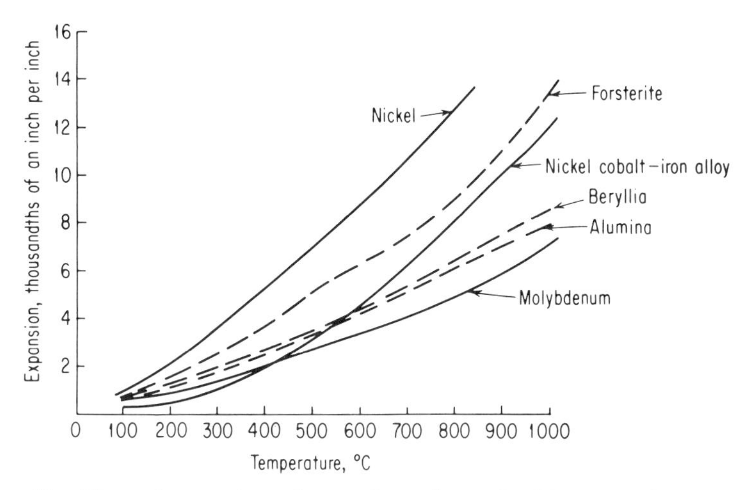

FIG. 57. Thermal expansions of ceramics and metals. (*American Lava Corp.*)

Conductor Materials. Table 14 shows a number of possible materials for interconnecting to the circuitry and to the integrated-circuit chip. This may be, as previously discussed, in the form of small-diameter wires, bumps on the integrated-circuit chip, or bumps on the substrate. The selection of the material must be based on low resistivity because of the low power levels involved. From Table 14 it would appear

TABLE 14. Materials for Interconnecting Integrated-circuit Chips

Material	Thermal conductivity, cal/(cm²)(cm)(°C)(sec)	Resistivity, μohm-cm
Copper	0.94	1.673
Aluminum	0.53	2.655
Gold	0.71	2.19
Chrome	0.16	13
Silver	1.0	1.59
Nickel	0.22	6.84
Molybdenum	0.35	5.17

that copper would be an ideal choice; unfortunately, however, copper poisons the silicon, thereby rendering it unusable. Another choice would be silver, but there are problems associated with adhesion and whisker growth. Gold also has adhesion problems but not if used in conjunction with molymanganese. Another material with promise is aluminum since the terminal pads on the integrated-circuit chips are aluminum. A number of companies are working on an all-aluminum interconnection system.

When plug-in terminal pins are used with the substrate, Kovar* is usually the material chosen. Its coefficient of expansion is compatible with the glasses used in making the hermetic seal around the pins. Kovar is normally gold-plated to improve its solderability.

Plastic Materials. Nonhermetic sealed packages are becoming more popular as demonstrated by the increased usage of the plastic embedded dual-in-line package. Plastic materials such as epoxy, silicone, urethane, and diallyl phthalate are the materials most frequently used.

* Trademark, Westinghouse Electric Corp.

Plastics are also employed for dielectric films between multilayer metallization. Polyimides and amide-imides are both used for their superior electrical and thermal characteristics. More recently, experimentation with photopolymers has shown promising potential for these materials as dielectric films.

COMPONENTS AND CIRCUITS

Only a few companies manufacture many types of components or circuits because of the low quantities used in the industry. This section will give some useful information about the components and circuits and what companies manufacture them. This

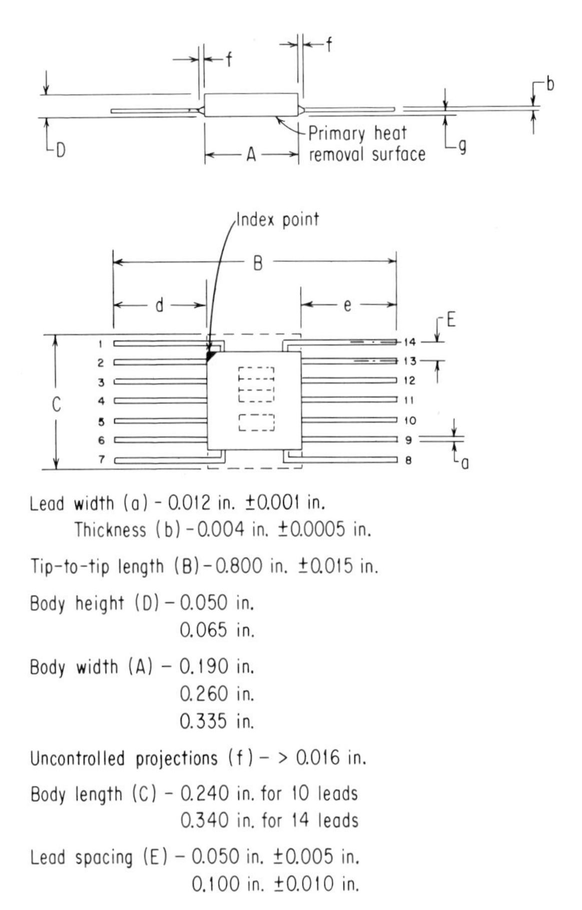

FIG. 58. Recommended standard packages. (*Electronics Industry Association.*)

information should be used only as a guide since it is subject to change when new companies form or old ones go out of business or change their product line. The information is broken down to include various component characteristics (both electrical and mechanical).

Microdiscrete Components. Discrete components of various forms and sizes are used in the fabrication of such microelectronic systems as those with thin-film circuits or with analog circuits where integrated circuits are not available. The components are micro in nature. They may come with or without leads. Table 15 shows micropackaged transistors, Table 16 shows micropackaged diodes, Table 17 shows microcapacitors, Table 18 shows microresistors, and Table 19 shows microinductors. As can be seen there are many sizes and shapes. Many fabrication and metals joining techniques are required to utilize these components in systems.

Several efforts have been made to develop more disciplined or uniform geometry for microcomponents. Most notable of these is the "pellet" or "swiss cheese" concept. These components are small cylinders of two basic nominal thicknesses (0.030 and 0.060 in.); diameter varies with component type and component value. Table 20 gives some useful information about the components and who the suppliers are.

A more recent effort has been put forth by the Electronics Industry Association (EIA) to develop a standard package or family of packages for microdiscrete components more compatible with the flat-pack configuration of integrated circuits. Figure 58 shows a sketch of the basic package shape and a table of recommended sizes. It is intended that the discrete component manufacturer package the component or components within the flat-pack container, which then becomes more compatible with the flat-pack configuration of the integrated circuit.

Packaged Integrated-circuit Configurations. The earliest package for the semiconductor integrated circuit was the TO configuration. This package had a long history of reliability with its use for packaging transistors. Since most semiconductor manufacturers had fabrication equipment for handling this configuration, it was logical that this package was first chosen. There are many variations in size, both in height and in diameter, and many leads to choose from. The more common ones include the 8-, 10-, and 12-lead configurations with the designations TO-70, TO-71, TO-73, and TO-74, as shown in Fig. 59. Complete information may be obtained for a small fee from the EIA headquarters in Washington, D.C.

Another configuration, developed later, was the flat-pack configuration. Again, configurations of many sizes and different numbers of leads quickly became available. Figure 60 shows a tabulation of the registered EIA flat packs. It also shows two configurations preferred by the users. Unfortunately they are conflicting, which is typical of the problems involved in standardization.

The most recent configuration is the dual-in-line package (DIP). Most major device manufacturers are now producing integrated circuits in this configuration either as a plastic encapsulated package or as a ceramic hermetically sealed package. A major advantage of the DIP is the 0.100-in. terminal spacing and its compatibility with dip or flow soldering techniques. A typical package with its dimensional detail is shown in Fig. 61; also included is a dimensional tabulation of DIPs manufactured by a number of different companies.

Uncased Components and Circuits. The most recent trend in component and circuit configuration is the uncased form. Components may be in the pellet or pill form, as previously described, or they may be in the thin- or thick-film or diffused chip form. They may be single components, or multiple components such as rc networks. The chip configurations usually come with metallized terminal pads for interconnections. In some cases they may be obtained with beam leads or in the flip-chip form with bumps. Table 21 shows some device manufacturers who will supply components and/or circuits in the uncased form. Again the reader is cautioned to use this table only as a guide since other companies not listed are also likely to supply the same product. When uncased components and/or circuits are used, the packaging engineer should recognize that the different manufacturers of the same component or circuit frequently use different chip sizes and different terminal pad locations. In fact, it is not unusual for a single manufacturer to revise or redesign his chip layout, to improve electrical performance or to increase fabrication yield, without changing the device designation. These changes must be recognized and taken into account in the interconnection pattern layout.

Several more formal uncased chip forms are available, namely the beam lead devices and the flip chips with bumps. Some efforts at standardization are being made with these components and circuits. Figure 62 shows the basic construction of a transistor with electroformed gold beam leads. The lead is approximately 0.0005 in. thick $\times$ 0.001 in. wide $\times$ 0.008 in. long.

TABLE 15. Micropackaged Transistors [8]

Manufacturer	Type	Package size	Package material	Lead configuration	Lead size
General Instruments	Switching, rf	Globular: 0.06 in. × 0.04 in. × 0.09 in. ht	Dip-coated	....	Radial 90° angle between
General Electric	....	0.08 in. × 0.06 in. × 0.03 in. ht	Au, Kovar base; chip is epoxy-coated	Perpendicular to base	0.001-in. diam
Fairchild	Rf, switching, etc.	0.05 in. × 0.05 in. × 0.025 in. ht	Molybdenum, Kovar,* or steel-base die is coated to base	Three parallel leads from one side	0.002-in. diam
Hughes	Switching, 12F, medium power	0.08-in. diam, 0.03-in. ht, pill	Ceramic hermetically sealed	Bottom is collector contact; top is divided for E and B	0.005 × 0.025 in. or without leads
Philco	....	Pill: 0.155-in. diam, 0.06-in. ht, similar to T0-51	....	Radial 90° between	0.03 × 0.05 in.
P.S.I.	....	Pico: 0.05 × 0.06 in. Micro: 0.08 × 0.05 × 0.14 in., T0-41	Encapsulated chip	Radial, C on one side, B and E other radial	C = 0.004 × 0.02 in.; B and E = 0.001 × 0.003 in.; Au platinum, Kovar * 0.004 × 0.02 in.
Sylvania	....	T0-46 T0-51	Hermetically sealed		
Texas Instruments	....	Pill: 0.215-in. diam, 0.06-in. ht	....	Radial 90°	0.025 × 0.003 in.
Transitron	Small-signal switching	Pill: 0.160-in. diam, 0.06-in. ht, similar to T0-51	Glass hermetically sealed	Radial 90°	0.025 × 0.003 in.

* Trademark, Westinghouse Electric Corp.

TABLE 16. **Micropackaged Diodes** [8]

Manufacturer	Type	Package size	Package material	Lead configuration	Lead size
General Instruments	....	Tube: 0.045-in. diam, 1-in. length; 0.045-in. diam, 0.08-in. length	....	Axial	0.004 × 0.019 in.
Hughes	....	Pill: 0.062-in. diam, 0.03-in. ht	Ceramic	With or without	0.025- × 0.003-in. ribbon, also 0.020-in. diam. axial
Western Electric	....	Pill: 0.04 × 0.016 in.	Glass	Top and bottom.	
Fairchild	....	0.05 × 0.05 × 0.025 in.	Kovar * or molybdenum base encapsulated die	Parallel, perpendicular to base	0.002 Au
Microsemiconductor	General-purpose computer	0.02 × 0.02 × 0.007, die only	None (surface passivated)	Top and bottom	
		Blob: 0.06-in. diam, 1-in. length	Blob encapsulated	Axial	0.018- × 0.03-in. ribbon
		0.035-in. diam, 0.08-in. length	Blob hermetically sealed	Axial	0.018 × 0.03 in.
Microstate Electronics	Tunnel	Pill: 0.12 × 0.05 in.	Hermetically sealed	Top and bottom	
PSI	Zener	0.13 × 0.055 in.	Blob	Axial	
RCA	Tunnel	Pill: 0.115-in. diam × 0.055 in. or × 0.075 in.	Ceramic	Top and bottom ribbon leads	0.115 × 0.003 in.
Texas Instruments	....	0.068 × 0.042 in.	Glass, hermetically sealed	Axial	0.012 in. diam
Transitron	Zener, switch, high conduction	Tube: 0.050 × 0.080 in.	Glass, hermetically sealed	Axial	0.004 × 0.015 in.

* Trademark, Westinghouse Electric Corp.

TABLE 17. Micropackaged Capacitors [8]

Manufacturer	Values	Size, in.	Voltage, volts	Other comments
Aerovox Corp. . .	3–60μf	1/8 × 1/2	6–50	Fixed, tubular electrolytic, series TTC
	10 pf, 702 μf	0.1 × 0.26	100	Ceramic dielectric, fixed tubular, series MC-70
	10 pf–0.1 μf	0.09 × 0.32	100	Ceramic dielectric, fixed, axial and radial, series C 80V
Air-O-Tronics . .	2–35 pf	5/16 × 7/16	160	Ceramic dielectric, disk trimmer, series 75 TRIKO
Astron Corp. . . .	0.1–330 μf	0.125 × 0.25	6–10	Tantalum dielectric, tubular electrolytic, series TES
Barco, Inc.	0.3–3 μf	1/8 × 7/16	3–70	Paper dielectric, hermetically sealed, aluminum foil electrolytic, series P
Bourns, Inc.	10 pf–0.12 μf	0.06 × 0.2	200	Fixed tubular, series 4230
Centralab	0.005–2.2 μf	0.29 diam	3	Fixed disk, series DA
Condenser Products, Inc.	0.001–2 μf	0.156 × 11/16	100	Mylar dielectric, fixed tubular, series MAW
Cornell Dubilier	0.0001–0.033 μf	0.225 × 5/16	100	Polyester film dielectric, fixed axial and radial, series MF
Electro-Cube Inc.	0.033–4 μf	0.33 sq × 0 1	100	Metalized Mylar, fixed rectangular, series 217A
Erie Electronics	5.6–1,200 pf	0.095 × 0.25	100	Molded, fixed tubular, series 390
Gudeman Co. . .	0.0068–10 μf	11/16 × 0.25 × 0.125	100–600	Metalized Mylar, fixed rectangular, series 363
JFD Electronics	5–62 pf	7/16 sq × 1/16	300	Ceramic dielectric, fixed rectangular, series UY01
E. F. Johnson . .	1.4–17 pf	0.433 sq	300	Variable, series 189
King Electronics	47–470 μf	0.1 × 0.03	100	Fixed disk, series DH
	0.012–0.1 μf	0.14 × 0.422	50–100	Fixed, epoxy molded tubular, series K1
Ohmite Mfg. Co.	0.1–150 μf	0.06 × 0.15	1.25–150	Tantalum dielectric, fixed tubular, series TW
Potter Co.	10 pf–0.08 μf	0.125 × 0.25	100	Epoxy encased, ceramic dielectric, fixed tubular, series 2042
Roanwell Cord	1–10 pf	1/4 × 9/32	. . . .	Glass dielectric, variable, series SG 11054/AG
San Fernando Electric Mfg. Co.	0.001–0.33 μf	0.3 sq × 0.1	100–200	Metalized Mylar dielectric, fixed rectangular, series 4M
	0.00001–0.056 μf	0.2 sq × 0.1	200	Ceramic dielectric, fixed rectangular, series RHO

TABLE 17. Micropackaged Capacitors (*Continued*)

Manufacturer	Values	Size, in.	Voltage, volts	Other comments
Sprague Electric Co.	5–36 μf	0.094 × 0.25	100	Ceramic dielectric, fixed tubular, series 252C
	0.02–2 μf	0.065 × 0.125	2–20	Tantalum dielectric, tubular electrolytic, series 160
Tansitor Electronics Inc.	3.9–68 μf	0.28 × 0.275 × 0.125	6	Tantalum dielectric, fixed rectangular, series LTS
	0.1–40 μf	0.065 × 0.15	1.5–35	Tantalum dielectric, fixed rectangular, series TS
TRW Capacitors	0.001–2 μf	0.156 × 0.5	100–1,000	Mylar dielectric, fixed tubular, series 663 US
	0.01–30 μf	0.125 × 0.12	6–50	Tantalum dielectric, fixed tubular, series 935
Union Carbide Co., Kemet Dept.	0.047–18 μf	0.085 × 0.25	6–50	Tantalum dielectric, fixed tubular, series Z
U.S. Semcor	0.15–250 μf	0.36 sq × 0.15	6–50	Tantalum dielectric, rectangular, electrolytic, series TSD
	0.6–6 μf	0.065 × 0.12	2–15	Tantalum dielectric, tubular electrolytic, series TSD
Vitramon	10 pf–0.01 μf	0.145 sq × 0.07 (radial)	200	Ceramic dielectric, flat rectangular, series VIC
		0.125 sq × 0.07 (axial)	200	
Voltronics	0.6–0.16 pf	0.167 diam	750–1,250	Glass and quartz dielectric, tubular trimmer, series QM

The flip-chip components with bumps are currently more popular and therefore more standardized. The EIA has recommended standard chip and bump sizes for both uncased transistors and dual diodes (see Figs. 63 and 64). In addition, some effort has been made toward standardizing integrated-circuit flip-chip bump locations (see Fig. 65). In any event, a component or circuit with prelocated terminal, whether beam lead or bump or terminal pad where it is to match bumps on the substrate, must be of standardized size and location if orderly and economical circuitry layout is to be obtained.

HARDWARE

The requirement for special hardware for microelectronic packaging beyond that used in conventional component packaging has been limited mainly to smaller size. Frequently, microelectronic concepts employ many of the same designs for brackets, slides, connectors, etc., as are employed in the more conventional concepts. Information on this hardware may be found in Chaps. 6 and 8.

Connectors. The packaging of more circuitry in a smaller space has necessitated the use of 0.050- and 0.025-in. connector contact spacings. It is not uncommon to need 100 connector contacts for a microelectronic subassembly. Many of the more conventional pin-and-socket connector concepts result in prohibitive insertion and withdrawal forces. The trend has been toward pressure- and cam-type connectors.

Another very useful approach is the edge-type connector which makes it possible to make contact with both sides of the card or board, thereby doubling the number of separate contacts in the same connector length. Care in both design and fabrication is needed when using the closer spacings or poor yields will result in unnecessarily high costs.

Individual component- or device-type connectors are also frequently used in microelectronic designs. Figure 66*a* shows an individual spring socket which is normally mounted in a printed-wiring board. The sockets are arranged in a fixed pattern to match the pin configuration of a module or submodule. Figure 66*b* shows a socket or connector for the TO can device configurations. These are available for the various

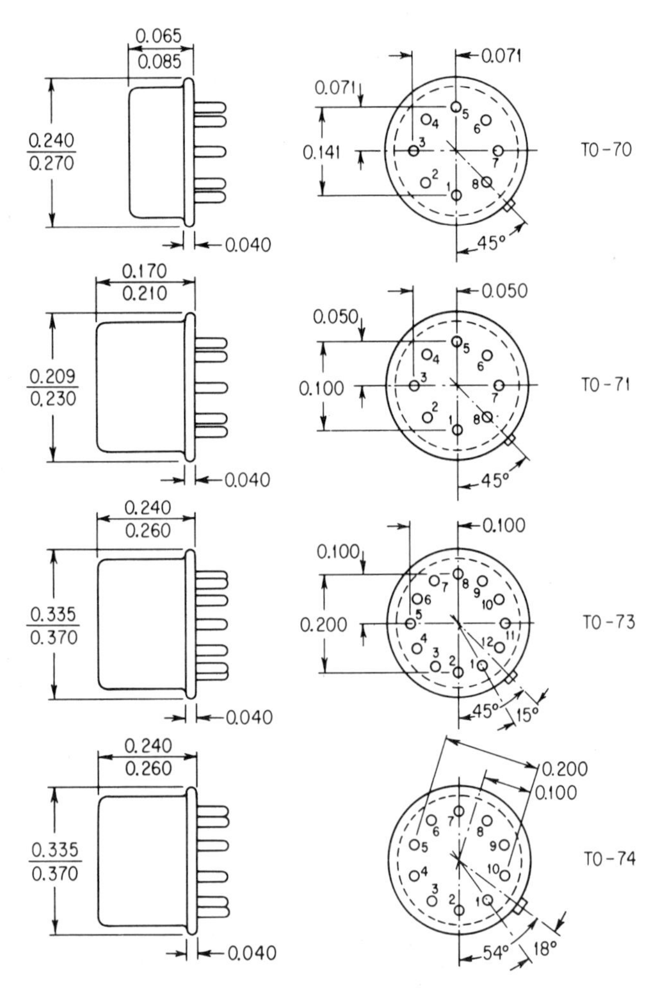

FIG. 59. Typical TO can configurations. (*Electronics Industry Association.*)

TO can pin configurations from 3 to 12 pins and are normally soldered into a printed-wiring board. Figure 66*c* shows a crimp-type carrier for the flat-pack device configurations. This configuration may also be obtained with weldable pins in place of the crimping arrangement. These carriers are normally plugged into connectors such as that in Fig. 66*a* or into fixed connectors soldered into a printed-wiring board. Figure

TABLE 18. Micropackaged Resistors [8]

Manufacturer	Values	Size, in.	Power, watts	Tolerance, percent	Other comments
Precision Resistor Co. Inc	0.1–300 kilohms	0.08 × 0.395	0.04–0.166	±1/100–5	Tubular, wire-wound, series SX
H. F. Priester Co.	47–220 kilohms	1/32 × 3/32	1/2	±20	Tubular, carbon film, series RSX-00
RCL Electronics Inc.	To 123 kilohms	0.1 × 0.26	0.02	±0.01–1	Tubular, wire-wound, series 7003
Reon Resistor Corp.	To 400 kilohms	0.15 × 0.25	0.1–0.25	TO ±0.01	Tubular, encapsulated, series TC
Resistance Products Co.	05–500 kilohms	0.25 × 5/16	1/2	±0.05–0.1	Epoxy encapsulated, wire-wound, series PB
	20–10 megohms	0.1 × 9/16	1/4	±0.5–20	Steatite, carbon film, series FAC
RHO Electronics	100 kilohms–2 megohms	0.25 × 0.475 × 0.125	1/10	±0.1–1	PC type wire-wound, series MP
	500–750 kilohms	0.15 × 0.26	1/10	±0.1–1	Tubular, wire-wound, series MA
Robinson Electronics Components, Inc.	1–2 megohms	0.5 × 0.25 × 0.125	1/4 and 1/8	±0.01–1	Flat, rectangular wire-wound, series 420 SP
	To 35 kilohms	0.1 × 0.2	1/10–1/4	±2	Axial and radial tubular wire-wound, series 200
Speer Carbon Co.	10–22 megohms	0.09 × 0.25	1/4	±0.5–20	Carbon composition, series SR
Sprague	2.5–500	0.098 × 0.25	1	....	Silicon-coated wire-wound, series 219E

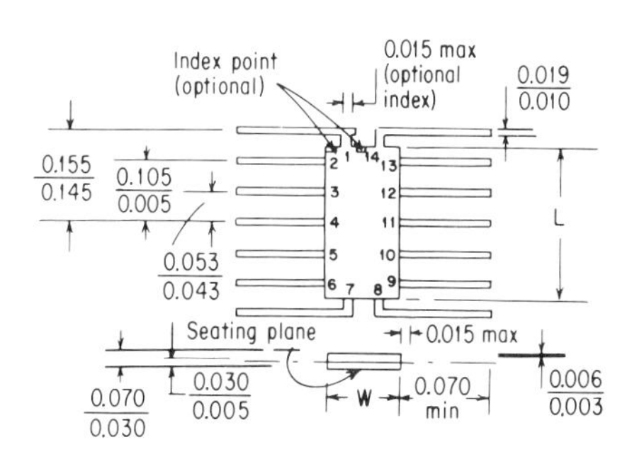

Outline No.	No. of leads	L Min	L Max	W Min	W Max	Remarks
TO-84	14	0.240	0.260	0.120	0.150	A, 1/4 X 1/8
TO-85	14	0.240	0.275	0.160	0.185	A, 1/4 X 3/16
TO-86	14	0.240	0.275	0.240	0.260	A, 1/4 X 1/4
TO-87	14	0.360	0.410	0.240	0.275	B, 3/8 X 1/4
TO-88	14	0.330	0.350	0.240	0.260	B, 3/8 X 1/4
TO-89	10	0.240	0.290	0.120	0.150	B, 1/4 X 1/8
TO-90	10	0.240	0.290	0.160	0.186	B, 1/4 X 3/16
TO-91	10	0.240	0.290	0.240	0.260	B, 1/4 X 1/4
TO-95	14	0.308	0.329	0.240	0.260	B, 1/4 X 5/16

A, offset end leads; B, no leads at end

Case type	Max leads	A max	B ±0.015	C max	D ±0.010	E max
1	14	0.185	0.800	0.345	0.260	0.050
2	14	0.260	0.800	0.345	0.335	0.065

A variety of flat-pack sizes have been accepted by the Electronic Industries Association's circuit-manufacturers' committee (left chart). The circuit-users' committee recommends only the two styles shown above

FIG. 60. EIA registered flat packs. (*Electronics Industry Association.*)

TABLE 19. Micropackaged Inductors [8]

Manufacturer	Values	Size, in.	Core material	Q	Other comments
Caddell-Burns Mfg. Co.	0.1–230 μh 0.25–70 μh	0.165 diam. 0.42 sq × 0.575	....	......	Phenolic inductor, series 426, variable coil
Piconics, Cu.	0.1–6,800 μh	0.25 × 0.097	Ferrite and powdered iron	5–40	Fixed, molded, series A
	0.1–5,800 μh	0.325 × 0.097		10–45	Tunable, molded type, series B
	0.1–6,800 μh	0.155 × 0.075	Ferrite and powdered iron	11–40	Unencapsulated, fixed type, series C
	0.1–5,800 μh	0.475 × 0.125	Ferrite and powdered iron	10–45	Tunable, chassis-mount-ing type, series F. NOTE: Transformers also available in each series
Torotron Corp.	0.5–200 mh	3/8 × 1/4	....	......	Epoxy-dipped toroids, axial, series TAS
Torotel, Inc.	1–250 mh	0.45 × 0.285	....	......	Epoxy metal-encased toroids
United Transformer Corp.	3–120 mh	7/16 × 1/4	....	......	Hermetically sealed inductor, series MM
Vanguard Electronics Inc.	0.01–10 μh	1/4 × 1/4	....	......	Variable inductor, series 53
	0.1–1 μh	0.2 sq × 0.1	....	......	Flat, toroidal, inductor, series 9100
Wabash Magnetics, Inc.	2.2 mh–27 h	17/32 × 3/4			

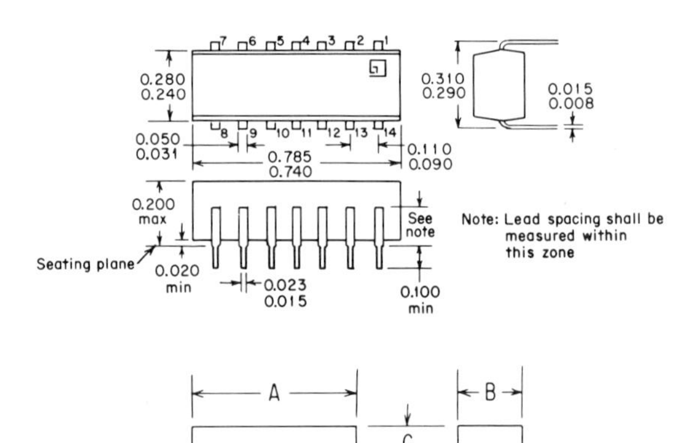

Manufacturer	A	B	C
Fairchild	0.785 max	0.280 max	0.180 max
ITT	0.775 max	0.350 max	0.200 max
Motorola	0.740 - 0.785	0.290 - 0.310	0.200 max (0.170)
Philco	0.595 - 0.605	0.380 - 0.420	0.260 max
Signetics	0.740 - 0.785	0.290 - 0.310	0.200 max
Sprague	0.740 - 0.780	0.300	0.125 max
Sylvania	0.700 ±0.015	0.250 ±0.015	0.100 max
Texas Instruments	0.770 max	0.250 ±0.010	0.180 max
Transitron	0.685 - 0.715	0.235 - 0.270	0.100 max
Westinghouse	0.600 - 0.660	0.245 - 0.255	0.185 max
Recommended	0.685 min - 0.785 max	0.240 min - 0.310 max	0.200 max

FIG. 61. Dual-in-line packages. (*Electronics Industry Association.*)

TABLE 20. Pellet or Dot Component Suppliers [9]

Component type	Manufacturer	Size, in.		Type	Value range
		D	T		
Resistors . . .	CTS, Berne, Ind.	0.050 0.050 0.100 0.100	0.030 0.062 0.030 0.062	Cermet	50 ohms–100 kilohms 100 ohms–200 kilohms 15 ohms–25 kilohms (special) 25 ohms–50 kilohms
Capacitors	Gulton Industries, Inc., Metuchen, N.J.	0.100 0.250	0.063 0.031	Ceramic	10–5,000 pf
	Scionics Corporation, Canoga Park. Calif.	0.050 0.250	0.063 0.030	Ceramic	To 4,150 pf
Diodes	Hughes Semiconductor Div., Newport Beach, Calif.	0.063	0.030	Silicon	Computer memory core driver
	Western Semiconductor, Santa Ana, Calif.	0.100	0.065	Silicon	General-purpose, switching zener

66*d* shows a connector for the dual-in-line device configuration. This connector is normally soldered into a printed-wiring board.

Spacers. The TO can configuration of the semiconductor integrated circuit is frequently packaged with a spacer between the device and the printed-wiring board. This may be done for the purpose of spreading the device leads to a fixed grid arrangement. It may also be done to provide clearance between the device and the printed-wiring board for better soldering or to provide a clearance between the base of the TO can and the printed wiring on the board. It may also provide a thermal heat sink when the spacer is made of a high thermally conductive material such as beryllia. Figure 67 shows two typical spacers.

Flat-pack Carriers. From the beginning the flat-pack configuration of integrated-circuit devices presented a critical handling problem. It was soon apparent that a carrier of some sort would be desirable from both a shipping and a testing standpoint. As with the device itself, standardization was not realistically carried out. Many individual device manufacturers developed carriers to meet their own needs, which resulted in many confusing configurations. Two major types evolved: a low-cost (15 cents or less) carrier to provide protection in shipping and handling, and a higher-

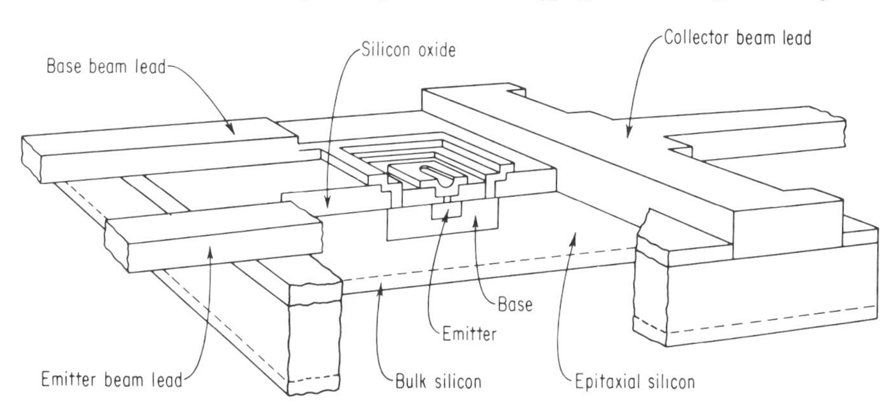

FIG. 62. Beam lead transistor structure.[7] (*Bell Telephone System.*)

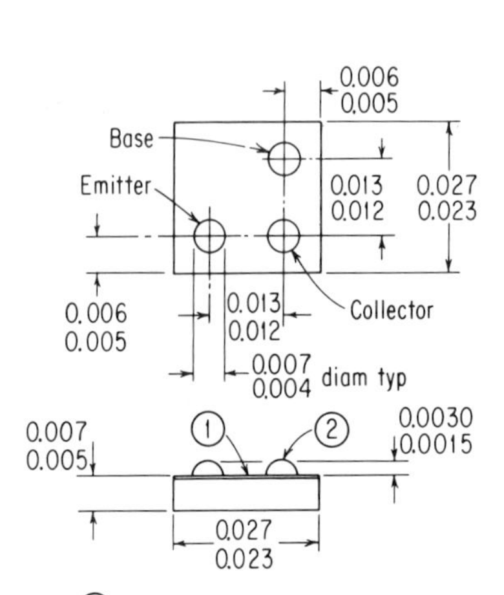

FIG. 63. Recommended standard—uncased transistor. (*Electronics Industry Association.*)

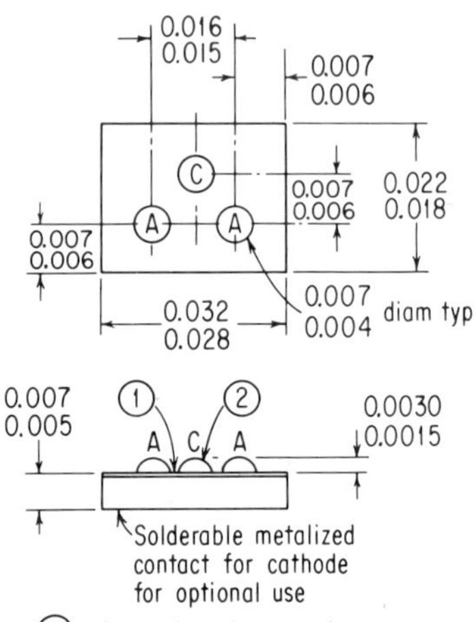

FIG. 64. Recommended standard—uncased dual diode. (*Electronics Industry Association.*)

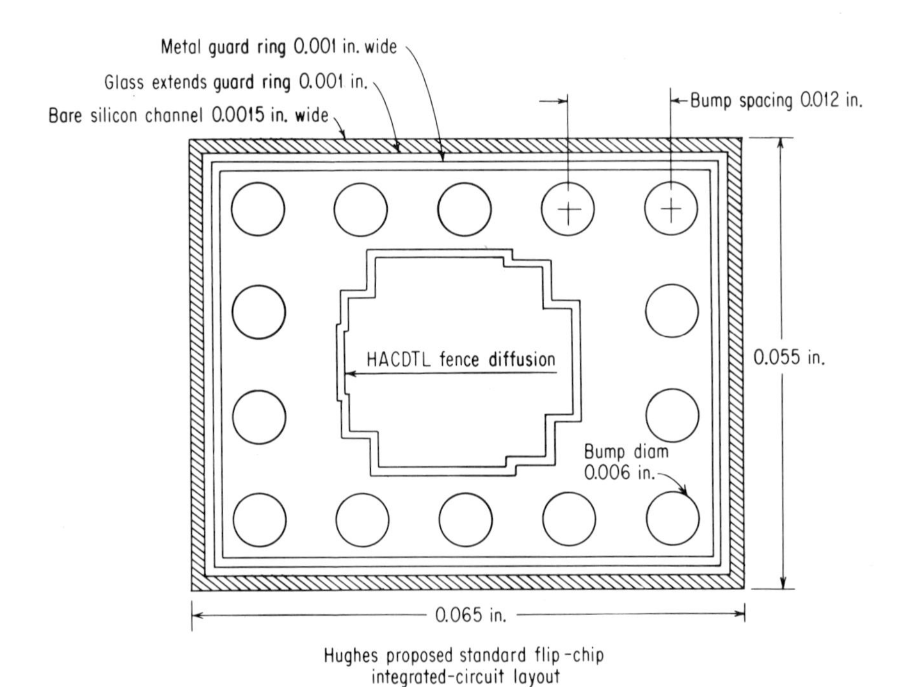

FIG. 65. Proposed flip-chip dimensions. (*Electronics Industry Association.*)

TABLE 21. Some Suppliers of Uncased Components and Circuits

Vendor	Type of device	Physical dimensions	Termination
Fairchild	Diodes, transistors, and integrated circuits	0.050 × 0.050 in., 0.060 × 0.060 in.	1-mil Au long, 2-mil Au wire, 5-mil Ni Au flush
Gen. Micro Elect.	Transistors and integrated-circuit chips	0.017- × 0.019-in. transistor; 0.035- to 0.050-in.-square integrated circuit	No leads
Sylvania	Single and multiple diodes, transistors, and integrated circuits	0.5-in.-square high-frequency transistors; 0.025-in.-square transistors; 0.040-in.-square integrated circuits	No leads or 1-mil Al wire
Texas Instruments	Transistors and diodes	0.040-in. square	Mounted on —5 headers with tall bonded term
Westinghouse	All semiconductors and integrated circuits produced	0.020- to 0.2-in. square	Au wires to Al circuit pads

SOURCE: Electronics Industry Association.

cost carrier ($2 or more) to provide for shipping, testing, and experimental circuit development. Figure 66*c* shows an example of the latter type.

The EIA has attempted to standardize flat-pack carrier configurations as shown in Fig. 68. The following 14 points are considered important requirements by the EIA's JS-10 Task Force on Microelectronic Device Carriers for evaluation of flat-pack carriers:

1. Protect the device from mechanical damage during shipping and handling, and maintain the device inside the carrier.
2. Provide mechanical clearances for ready access to positioned leads, and provide support for wiping action of test probes as well as provision for testing formed leads, and testing close to the body of the device.
3. Accommodate all existing registered peripheral lead outlines (this may require several standard carriers; one won't do the job alone).
4. Accommodate multiple axial and peripheral lead packages.
5. Should be easily groupable (e.g. stack, nest, horizontal strip, etc.) and yet permit easy removal of a selected sample.
6. Allow visibility for identification.
7. Accommodate flat-pack lead length from 0.070 in. to 0.500 in.
8. Support flat pack by the body.
9. Accommodate formed leads.
10. Allow physical access to the device from top and bottom.
11. Allow mechanical clearance for solder dipping.
12. Allow automatic insertion and removal of packs from carriers.
13. Provide means of indexing and orientation for automation and interchangeability.
14. Allow sufficient surface suitable for marking, coding, printing, or other identification.

FABRICATION

Microelectronic packaging may consist of the straightforward assembly and interconnection of packaged semiconductor integrated circuits such as the dual-in-line, the TO can, or the flat-pack configurations; it may consist of a more complex

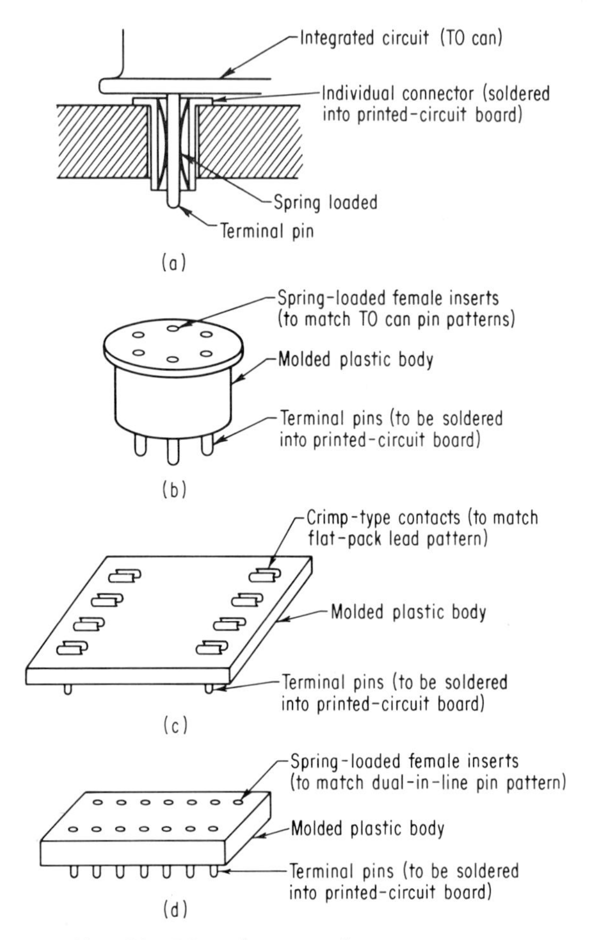

Fig. 66. Microelectronic device connectors.

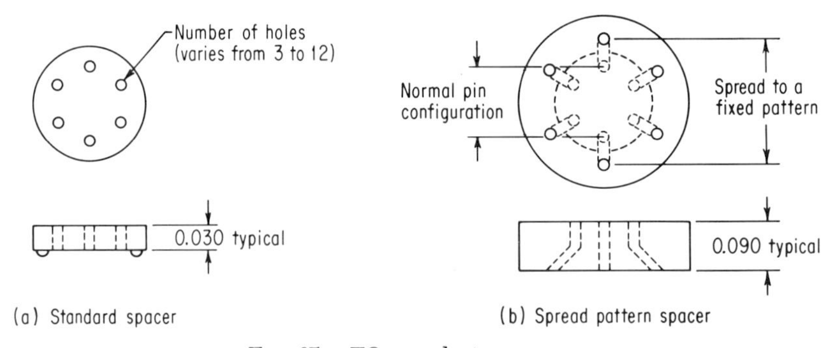

Fig. 67. TO can device spacers.

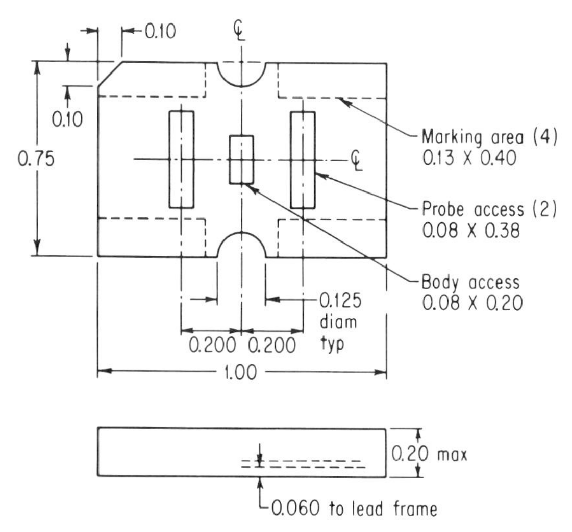

FIG. 68. EIA proposed carrier standard. (1) All dimensions in inches. (2) Details of body retention optional. (3) Additional index notches optional—may be used to identify carriers differing in body cavity size. (4) Body cavity should allow positioning to within 0.005 in. of true carrier center. (5) Probe access from underside permitted (if lid contains lead). (*Electronics Industry Association.*)

assembly of microdiscrete components and thin films; it may consist of an extremely complex assembly and interconnection of integrated-circuit chips; or it may consist of the ultracomplex interconnection of large-scale integration arrays. Each level of complexity requires a different level of fabrication capability. The packaged device designs may be fabricated with normal electronic manufacturing facilities with printed-circuit, soldering, and component welding capabilities. The other levels, however, require more elaborate facilities and more advanced fabrication capabilities. The following is a brief discussion of some of the areas that must be considered in order to fabricate the more complex microelectronic packaging designs successfully.

Facilities. An engineering laboratory development facility and a pilot line facility in addition to a full-scale production facility are desirable though not mandatory for producing microelectronic packages on a large scale. The engineering facility would develop processes and materials and test their feasibility for application to microelectronic packages. The pilot line would extend and modify these processes and materials to establish feasible, reliable, and economical fabrication capability. The full-scale production facility puts into practice the processes and materials developed, to produce microelectronic package hardware.

All the facilities should provide clean filtered air, controlled humidity and temperature, and proper lighting. Employees should wear lintfree gowns and caps. Critical operations should perhaps be performed in special clean-box arrangements. It is extremely important for management to motivate the worker on the importance of cleanliness.

Equipment. General equipment, such as microscopes, tweezers, and micromanipulating tools, is needed. In addition, three major categories of equipment are needed, deposition equipment, photo-imaging equipment, and bonding or metals joining equipment.

The deposition equipment may vary from relatively simple electroplating or electroless plating tanks to elaborate vacuum chambers. The photo-imaging equipment may include large, accurate reduction cameras for producing photomasks to photoequipment for projection of the image. Numerous exposure equipments may also be necessary, including special collimated light sources. The bonding equipment may include the more common resistance-type welders or microsoldering tools, it may include more sophisticated equipment such as laser and electron-beam welders, or it may include bonders of the thermal-compression and ultrasonic types.

There are many types of equipment available for fabricating microelectronic systems. The selection of equipment for any particular installation must be based on both what is to be fabricated and what quantities are to be produced. Equipment selection is critical to a successful microelectronic fabrication operation.

Careful study and evaluation of each process step should be made before any equipment is selected.

Inspection and Testing. It is generally conceded that inspection and testing make up a significant part of the cost of microelectronic systems. Every effort should be made to design the microelectronic packages for ease of inspection and electrical testing as well as for ease in troubleshooting malfunctioning units. Automatic test equipment, although costly, may be justified on full-scale production installations. Qualifying the process rather than the product is a relatively new philosophy now being accepted by the military. This greatly reduces the need for elaborate part inspection and testing. The control levels of the processes are established, maintained, and used as the means of verifying the product quality.

Training. Careful detailed training programs are necessary in microelectronic systems fabrication. Periodic reviews of the individual's capabilities should be maintained. Motivation is probably the one most significant ingredient necessary for maintaining a high level of product reliability. Management and line supervision must recognize that they hold the key to proper motivation.

DESIGN CONSIDERATIONS

The actual selection of a packaging concept for a particular application is a complex problem. Both technical and nontechnical requirements must be considered and weighed before the actual packaging concept is selected. The following is a discussion of these considerations.

Electrical. The packaging engineer and the electrical engineer must work closely together to establish critical electrical parameters before a design is established. When electrical parameters are not critical, a large number of packaging concepts may be acceptable. When they are critical, the design must accommodate them with proper shielding, line length, component isolation, etc., which limits the packaging concepts that can be used. Chapter 10 discusses electrical factors and their effect on electronic packaging design. It would be well for the packaging engineer to have a thorough understanding of the information given in Chap. 10 since electrical parameters are, not uncommonly, the controlling factor in microelectronic packaging design.

Thermal Dissipation. Although microminiaturized circuitry is normally accompanied by lower power dissipation, the power density (i.e., watts per unit volume) frequently is higher than that found with conventional circuitry. It is not uncommon to dissipate 100 to 150 mw from some microelectronic devices. Therefore, thermal design in microelectronic packaging must be given careful consideration. Techniques such as liquid and thermal-electric cooling must often be considered, in addition to the more conventional convective, conductive, and radiation techniques. Chapter 11 discusses in detail the various thermal techniques employed in electronic packaging design.

Volume and Weight. Volume figures can be confusing and deceiving unless they are carefully defined. For example, a circuit may occupy 1/10,000 in.^{3} as a semiconductor integrated-circuit bare chip, 1/1,000 in.^{3} as a semiconductor integrated circuit, 1/100 in.^{3} as a thin-film circuit, and 1/20 in.^{3} as a conventional discrete-component circuit. This size advantage can be maintained only if all the interconnecting wiring, hardware, etc., are also reduced similarly. This unfortunately has not been possible, and the actual relationships are more like that shown in Fig. 69. These figures are typical of digital applications where a major portion of the system may be microminiaturized. The volume advantage of microelectronic packaging would be somewhat less for analog systems where system requirements for components and component tolerances are either not available or difficult to obtain in thin-film or semiconductor integrated form.

Weight relationships generally follow the same trends. Typical packaged weights would be 1/100 to 1/10 gram per semiconductor integrated-circuit bare chip, 1/10 to 8/10 gram per semiconductor integrated-circuit flat pack, 8/10 to 8 grams per thin-film circuit, and 5 to 15 grams per conventional discrete-component circuit (the discrete-component circuit contains approximately 25 components).

Environmental. Environmental requirements may run from a laboratory to a deep space environment. Fortunately, the very nature of microelectronic packaging and its inherent low mass and high packaging density make it compatible with most environmental structural requirements such as shock, vibration, and acceleration. In most cases the circuits are either hermetically sealed or embedded, making them compatible with most humidity requirements. Temperature limitations for most microelectronic components and circuits are generally the same—in the range from −50 to

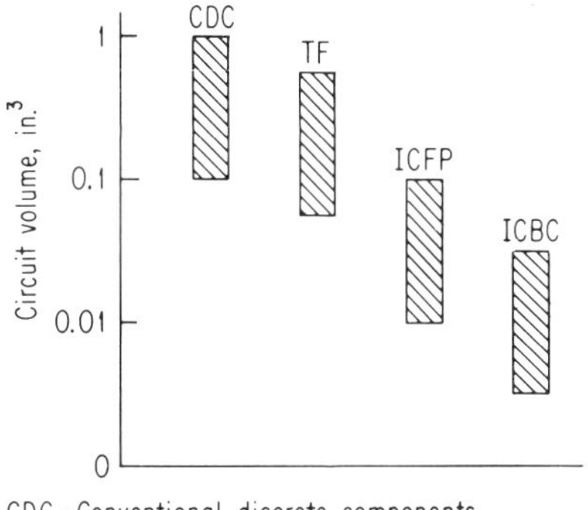

FIG. 69. Volume relationship for various microelectronic packaging concepts.

+125°C. A major environmental weakness in most electronic circuitry, whether conventional or microelectronic, is the degradation of performance of semiconductor devices when exposed to radiation. If this is a requirement, the system should be designed to allow for the expected degradation; otherwise, costly space- and weight-consuming shielding techniques must be employed.

Reliability. Microelectronic packaging is rapidly developing a reputation for high reliability. This is due in part to the high reliability of semiconductor circuits and other microcomponents as well as to the reduced number of interconnections needed. It is also due to the high reliability obtained from more highly skilled workers and more closely controlled fabrication processes and techniques. High reliability is reflected in the fact that a number of years ago a throwaway cost was limited by the military to around $50 per unit. Throwaway requirements are now based on dollars per hour of mean time between failures (MTBF) for a unit. With this set at $1 per 10,000 hr of MTBF, it is not uncommon to justify $100 to $200 throwaway unit costs.

Maintainability. Problems can arise with the maintainability of microelectronic systems. Frequently, connectors are avoided and units are hand-wired together, which may necessitate putting the maintainability point at a higher level of equipment module size. However, this may be justified as the higher throwaway cost is; the mean time between maintenance hours can be expected to be higher for microelectronic equipment than for conventional components and conventional packaging.

Repairability. A new philosophy of repair must be developed for microelectronic equipment. Depending on the extent of the microminiaturization, most of the units or submodules have to be repaired under the same conditions as when they were manufactured, i.e., clean room and special equipments. It is therefore necessary in many cases to consider microelectronics equipment nonrepairable in the field.

Interfaces. Microminiaturization is frequently employed in packaging only a part of a system, which creates a problem with interfacing to the rest of the system. It will be necessary to develop compatible electrical and mechanical interfaces. This necessity can create problems especially in the interwiring cabling and connector areas, and can also dictate restrictive equipment configurations with thermal and structural problems. The packaging engineer should consider the interface problems early in his design study.

SPECIAL CONSIDERATIONS

Many facets other than design types must be considered when selecting a packaging concept, including costs, quantities to produce, and delivery schedule. These considerations must also be thoroughly explored before proceeding with the design. A brief discussion of these other facets follows.

Cost. Microelectronic system packaging costs include component and/or circuit cost as well as fabrication cost. In small quantities, microdiscrete components are usually the most economical, while the semiconductor integrated circuits are the most expensive. For example, the initial cost of developing a new integrated circuit runs between $5,000 and $50,000 (the thin-film approach would cost $750 to $1,000), while the separate microdiscrete components could run less than $500. Approximate quantities, where costs cross over, are up to 25 circuits for separate microdiscrete components, from 25 to 1,000 circuits for the thin-film technique, and about 1,000 circuits for semiconductor integrated circuits. Needless to say, if a circuit is already developed (as with many digital types), the semiconductor integrated-circuit design will reflect the lowest total cost even for small quantities.

Schedules. Frequently a customer will allow only a short time period in the delivery schedule. Unfortunately this does not allow the full exploitation of the most appropriate packaging concepts, and frequently less than the ultimate must be used. Items such as component and/or circuit availability and in-house manufacturing capability greatly affect the packaging concept chosen.

Customer Preference. It is not uncommon for a customer to have a particular preference for a specific packaging concept or for certain fabrication techniques. His preference may be based on familiarity with the environmental requirements for his equipment or on past experience with good results obtained with a particular concept or technique. In any event it is wise for the packaging engineer to develop a close working relationship with his customer before he firms up his design concepts.

PACKAGE CONCEPT SELECTION

Occasionally the selection of a packaging concept is a straightforward process; this is possible only where the system requirements are not conflicting. More often the requirements are conflicting and the selection process is extremely complex, requiring a thorough study and evaluation of all the system requirements. A detailed approach to selecting a packaging concept is given in Chap. 10. In addition, a relative rating system has also been developed in Chap. 8. As pointed out, there is no really scientific approach to selecting the right packaging concept. The greatest assurance of the optimum selection is possible when the packaging engineer becomes thoroughly familiar with the art of packaging electronic systems, as well as with the systems requirements.

Microelectronics packaging concepts may vary from the relatively simple assembly of microdiscrete components to the complex interconnection of monolithic semiconductor integrated-circuit arrays. Table 22 includes basic evaluation information on the major categories of microelectronic packaging concepts. A further breakdown of

TABLE 22. Evaluation of Basic Microelectronic Packaging Concepts, on a Scale of 1 to 5 *

Evaluation factor	Basic packaging concept				
	Micro-discrete compo-nents	Thin films	Packaged integrated circuits	Bare integrated circuits	LSI
Engineering design	1	2	2	4	5
Drafting and artwork	1	1	1	3	5
Manufacturing complexity	1	1	1	4	5
Adaptability to automation	4	2	2	3	2
Quality-control complexity	2	1	1	5	4
Electrical test complexity	1	2	2	4	5
Size	5	5	4	2	1
Environmental resistance	4	3	3	2	1
Reliability potential	5	4	3	2	1
Maintainability	2	2	2	3	3
Repairability	1	2	2	4	5
Mechanical interface	2	2	2	4	4
Cost potential	4	3	3	2	1

* These ratings represent a hypothetical set of packaging parameters—in this case the greater the microminiaturization, the greater the reliability and cost potential even though more complex manufacturing and test procedures are required. Another set of parameters may result in a different conclusion.

these concepts should be developed by the packaging engineer so that he can realistically evaluate his particular problem.

EXAMPLES OF MICROELECTRONIC PACKAGING

The following examples are included to illustrate the actual application of the various basic microelectronic packaging concepts discussed in this section. Note that in many cases several concepts are employed in the same system. Also note that in some cases several basic concepts might be combined to develop a unique concept. Figures 70 to 81 are included as examples of typical microelectronic packaging.

Figure 70 shows part of the nondestructive readout (NDRO) memory of an airborne data processor. In this application the basic packaging concept employed is semiconductor integrated circuits in TO cans nested on dual planar boards. Note the use of the TO can spread spacing in addition to the spread pattern to allow for the nesting of the TO cans. The result is an easily maintainable economical system.

A good example of the stick-type module is shown in Fig. 71. The interconnecting pattern for the integrated circuitry (in this case in flat-pack form) is achieved in a series of layers of conductors stacked up to bring the interconnection points to the proper flat-pack lead position. The interconnection points are easily welded without any special positioning. Connections to the second level of interconnections are made with molded headers welded to the interconnecting stack as shown. The finished unit can be contained in a shell-like structure as shown or can be embedded in a more environmentally secure package.

Another type of flat-pack module is shown in Fig. 72. In this approach a molded header containing weldable pins is used to mount the integrated-circuit flat packs. The pins are spaced to fit the flat-pack lead pattern allowing direct in-line opposed electrode welding. The interconnection pattern is developed on the opposite side using point-to-point welded insulated wiring or prefabricated weldable circuitry techniques. Input/output connections to the second level of interconnections are made through the terminals shown. This module may be conformally coated or embedded for environmental protection.

FIG. 70. NDRO memory for airborne computer employing TO can integrated circuits with spread leads and nested board. (*Westinghouse Electric Corp.*)

FIG. 71. Stick module with weldable tabs. (*Motorola Micro-Harness.*)

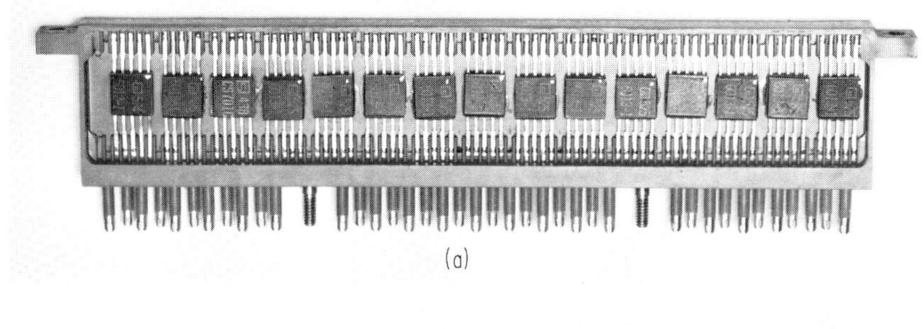

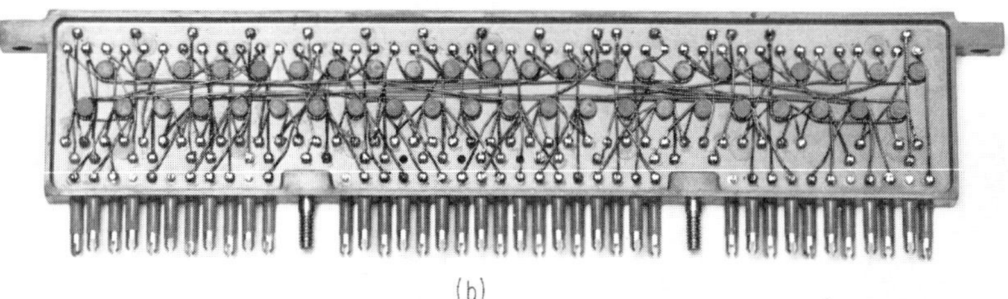

FIG. 72. Stick module with weldable pins interwired with insulated wire. (*Jet Propulsion Laboratory.*)

A dramatic example of size reduction possible through the use of microelectronic techniques is shown in Fig. 73. In this example the volume reduction was over 100 to 1. The equipment illustrated is the if amplifier section of an airborne radar system. The three large chassis in the upper portion of the photograph are constructed with conventional discrete-type components. The microelectronic version is shown in the lower portion. In this design integrated circuits in the flat-pack configuration are used with a number of microdiscrete components. The entire amplifier is contained in three compartmentalized metallic chassis mounted on an interconnecting printed-circuit board. Note also the use of micro-type coaxial connectors.

NORMOD,* shown in Fig. 74, is a three-dimensional module of integrated-circuit flat packs. In this design the flat packs are mounted on a frame which accurately locates the leads in a known grid pattern. The frames are stacked, laminated, and machined to expose the device leads. An interconnection wiring pattern is chemically milled for each side of the module, and attachment is made using parallel-gap welding techniques. This concept is efficient from a volume standpoint and is economical as well as reliable.

The equipment shown in Fig. 75 is a digital differential analyzer used for missile guidance and other digital control applications. A planar array of integrated-circuit flat packs is the basic concept employed. The planar interconnecting printed-circuit boards containing integrated-circuit flat packs are mounted on metallic frames which in turn are plugged into the second-level interconnecting panel. High-density welded discrete components are used in the interface circuit shown in the rear area of the analyzer.

Figure 76 shows Hi Pac,† a rather unique design with high thermal-dissipation capabilities. The metallic structure is constructed with holes as shown to allow the passage of air for cooling purposes. Ceramic is fired on the metallic structure to provide the necessary electrical isolation, and interconnecting circuitry is screened on and fired in place. The integrated-circuit flat packs are then welded or resistance-

* Trademark, Norden Div. of United Aircraft Corp.
† Trademark, Solitron Devices, Inc.

FIG. 73. Comparison of receiver unit before and after using microelectronic compartmentalized packaging. (*Westinghouse Electric Corp.*)

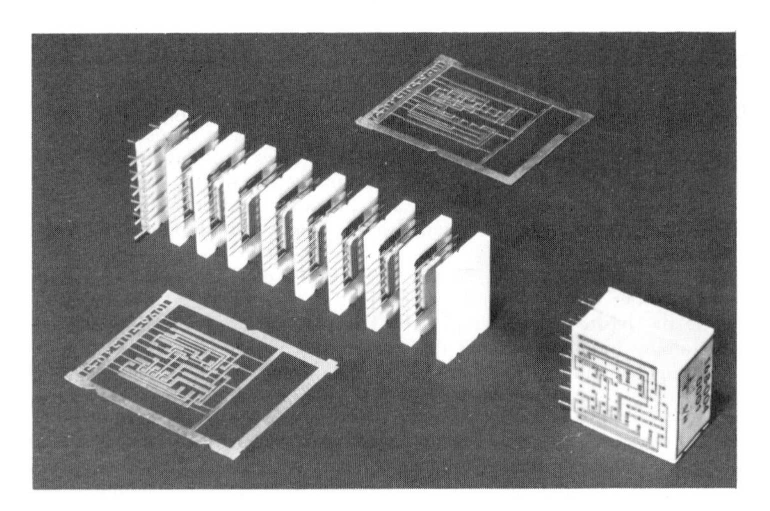

FIG. 74. Stacked flat-pack module using chemical milled interconnecting circuitry. (*Norden NORMOD.*)

soldered in place. The connector pin spacing is 0.050 in. The metal structure provides mechanical rigidity and shielding in addition to high thermal dissipation.

Figure 77 shows an all-glass-to-metal sealed container housing a monolithic analog-to-digital digital-to-analog converter containing 350 active components on a chip 0.062 × 0.104 in. This package illustrates the latest advancement in microelectronics—large-scale integration (LSI). The package contains 40 input/output leads which may be soldered into a printed-circuit board or be used with a 40-pin socket. The packaging engineer can expect to see many more of these larger electronic functions in small

FIG. 75. Digital computer employing planar array modules of integrated-circuit flat packs. (*Raytheon Co.*)

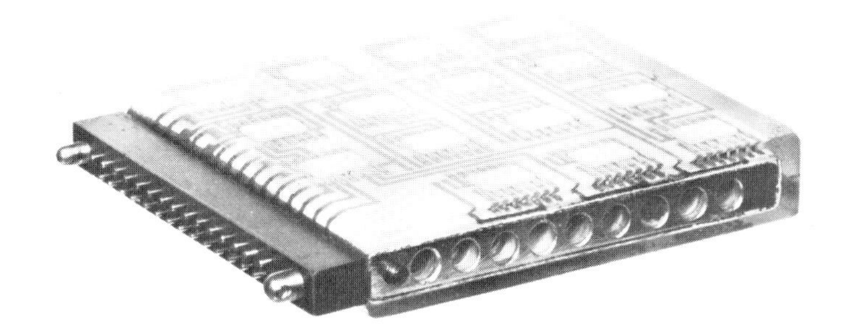

FIG. 76. Planar array of integrated-circuit flat packs mounted on metal interconnecting board for better heat dissipation. (*Solitron Hi-Pac.*)

packages becoming available at an ever-increasing rate as production yields are increased. Reliability is high with the great decrease in discrete interconnections, and cost is low considering the extent of the electronic functions performed.

Another of the recent advanced microelectronic packaging concepts is shown in Fig. 78. The multi-integrated-circuit chip package is not as advanced as the monolithic LSI; however, it is more versatile from the standpoint of immediate application without commitment to expensive processing and tooling. The concept is quite fundamental, and all the equipment and processes are readily available with many electronic manufacturers. The illustrated package is an all-alumina-to-metal hermetic sealed unit; the circuit is part of a radar altimeter and consists of flip-flops, gates, and rc network chips. The circuit shown is interconnected with, though not limited to, a single layer of circuitry.

Figure 79 shows the Ministick* system for interconnecting integrated circuits of the flat-pack configuration into a compact, economical planar array. The interconnections,

FIG. 77. An example of an advanced LSI device. (*General Instrument Corp.*)

FIG. 78. Multi-integrated chip package. (*Westinghouse Electric Corp.*)

which are built up in layers, bring the proper interconnection points to the appropriate flat-pack leads. This system is essentially the same as the stick-module approach except that larger arrays are used. Input/output terminations are located at one end of the frame. The flat packs are welded into position, and the system function is achieved by bolting a number of frames together and interconnecting the second level of circuitry through the input/output terminals. This concept is reliable, economical, and easily accomplished with a minimum of equipment and processing.

Large-scale integration is an outgrowth of the techniques employed in Fig. 80. This is a four-stage differential amplifier with vapor-deposited interconnections. The same techniques are used in LSI but on a larger scale. It is possible with this approach

* Trademark, The Johns Hopkins University Applied Physics Lab.

to have standard patterns of components, such as transistors and diodes, and/or electronic functions, such as gates and flip-flops, and interconnect them in different arrangements for various larger electronic functions. The problem to the packaging engineer is principally one of more leads with tighter spacings. Circuit configurations are essentially larger flat packs and larger TO or pluggable-type cans.

Hybrid thin-film circuits have been overshadowed with the use of semiconductor integrated circuits but they do have many areas of application. A typical circuit is

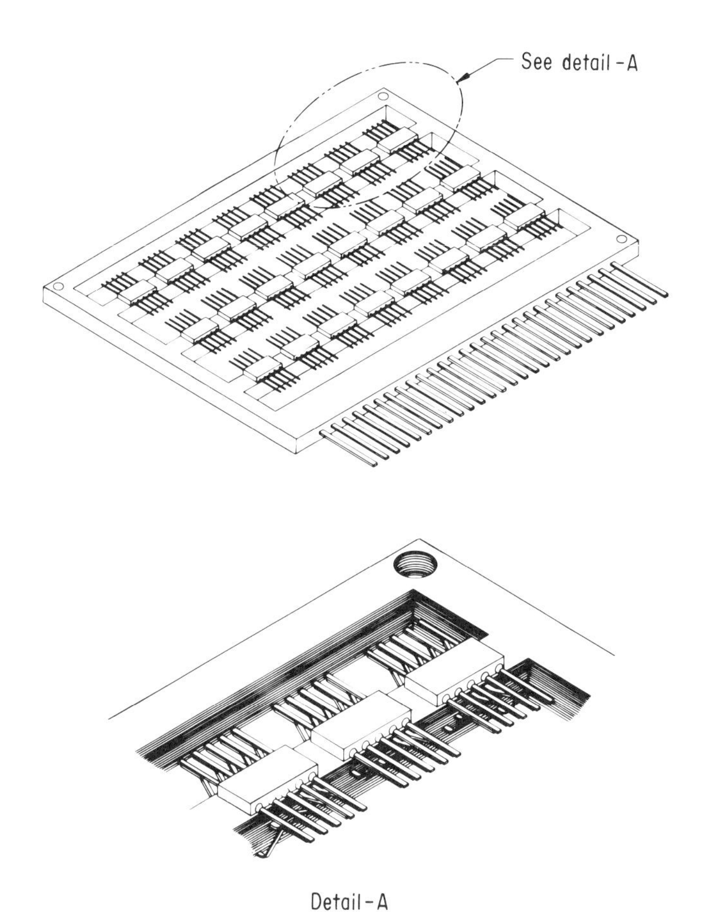

FIG. 79. The Ministick system—planar array of integrated-circuit flat packs with multilayer interconnections. (*The Johns Hopkins University Applied Physics Lab.*)

shown in Fig. 81. The smaller passive circuits (conductors, resistors, insulators, and capacitors) are normally vapor-deposited while the active and larger passive devices are soldered into place—for example, a small potentiometer and numerous transistors are attached as discrete components in the illustrated circuit. Circuit configurations can vary, but the flat pack (of various sizes) appears most frequently; however, the circuit shown is in a miniature plug-in arrangement similar to a plug-in printed-circuit card. Higher-power circuits are normally more available with the hybrid thin-film approach.

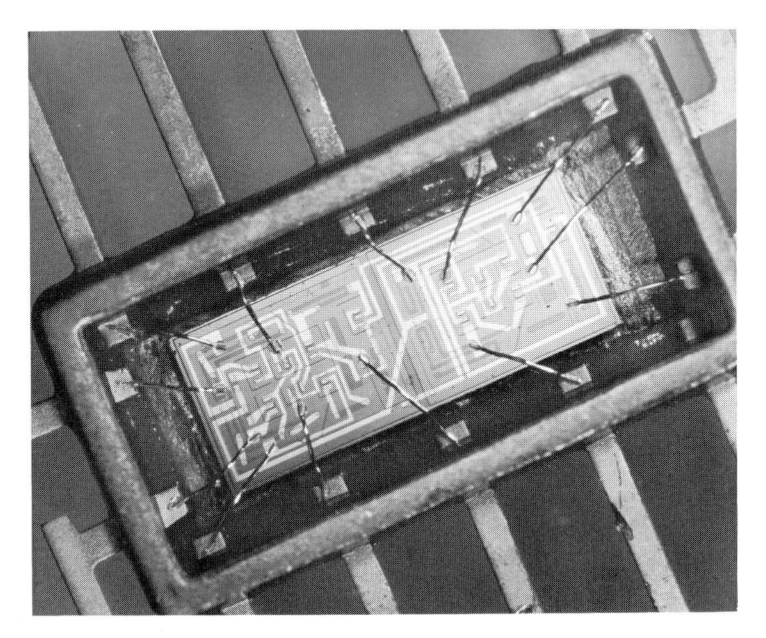

Fig. 80. Four-stage differential amplifier with vapor-deposited interconnections. (*Autonetics Division of North American Aviation.*)

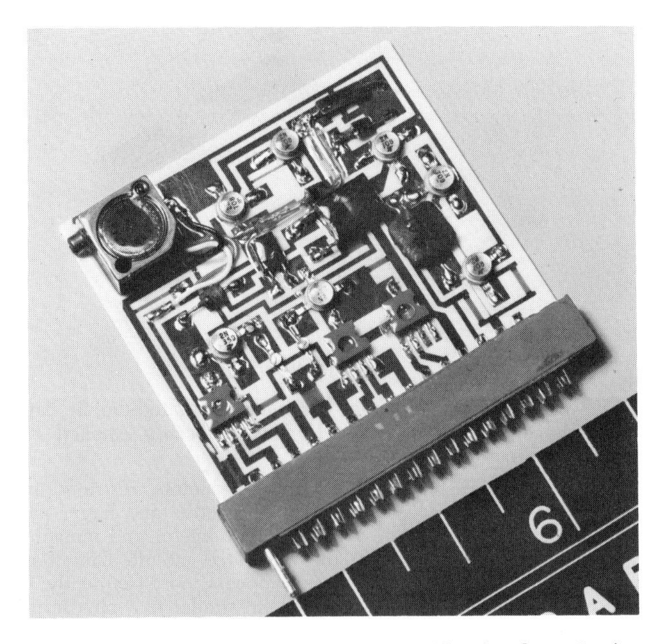

Fig. 81. Hybrid thin-film circuit assembly. (*Melpar, Inc.*)

REFERENCES

1. *Electronics,* Jan. 9, 1967.
2. Norden Division of United Aircraft Corporation: Report No. 1129R0204 entitled "Interconnection and Organizational Electronic Blocks," October, 1963.
3. "Integrated Circuit Engineering—Basic Technology," 4th ed., Boston Technical Publishers, Inc., Cambridge, Mass., 1966.
4. Sack, E. A.: More Complex Integrated Circuits for Digital System, *Westinghouse Engr.,* vol. 26, no. 3, pp. 88–92, May, 1966.
5. Weimer, P. K., et al.: Integrated Circuits Incorporating Thin Film Active and Passive Elements, *Proc. IEEE Integrated Component,* vol. 52, no. 12, pp. 1479–1486, December, 1964.
6. Staller, J.: Guidelines for Implementation of System Requirements into Microelectronic Mechanical Designs, IEEE Lecture Series, Microelectronics Comes of Age, Boston, Mass., November, 1966.
7. Lepselfer, M. P.: Beam Lead Technology, *Bell System Tech. J.,* vol. 45, pp. 233–253, February, 1966.
8. Westinghouse Electric Corporation: Technical report no. RADS-TR-65-61 entitled "Microelectronic Packaging Concepts," July, 1965; written by T. Hamburger et al. for the Reliability Branch of Rome Air Development Center, New York.
9. Sweany, L. P.: Workshop Session, National Electronics Packaging Conference, June 9–11, 1964.
10. Keonjian, E.: "Microelectronics Theory, Design, and Fabrication," McGraw-Hill Book Company, New York, 1963.
11. Boeing Company: Report no. D2-22311-1 entitled "Microminiaturization Study —Concepts and Effects," vol. 1B, sec. A6, Seattle, Wash., October, 1964; written by L. A. Hartley.
12. Loral Electronics Corporation, Manual of microelectronic techniques entitled "Micromin Digest," New York, June, 1963.
13. Fiderer, L.: "New Approaches to High Density Module Interconnections," 6th International Electronic Circuit Packaging Symposium, San Francisco, Calif., August, 1965.
14. Walker, J. S.: "Micro-harness: A New Concept in Interconnection for Integrated Circuits," 6th International Electronic Circuit Packaging Symposium, San Francisco, Calif., August, 1965.
15. Rupprecht, G., and E. Stubler: "Two Dimensional vs. Three Dimensional Packaging of Integrated Circuits," 6th International Electronic Circuit Packaging Symposium, San Francisco, Calif., August, 1965.

Chapter 10

ELECTRICAL FACTORS OF ELECTRONIC PACKAGE DESIGN

By

D. O. McCAULEY

Development Engineering
Westinghouse Electric Corporation
X-ray Division
Baltimore, Maryland

INTRODUCTION

Electrical design obviously influences mechanical design, but mechanical packaging design has a far greater effect on electrical parameters than the novice would suspect.

Purposes of the Chapter. The purposes of this chapter are to point out to the beginner the effect his packaging design has on system performance and to provide, for both beginner and experienced engineer, data which will be useful in minimizing the effect of packaging. This data is directed toward the packaging engineer concerned with electrical and mechanical interactions rather than toward the electronic specialist. Experienced packaging engineers recognize that electrical parameters are affected by packaging techniques; making them predictable is the key to good design.

Organization of the Chapter. Roughly, the material is arranged to follow the needs of the electronic packaging engineer in the sequence in which it will be used. The circuit designer usually provides a rough schematic and parts list; then the packaging engineer must follow through to satisfactory performance of the system in its ultimate usage.

Following the introduction, which includes terms, definitions, and some basic electrical formulas, a portion of this chapter is devoted to interfacing with the circuit designer, including steps that must be mutually taken to select components and arrive at a general packaging method. Next, circuit groupings, division points, and standardization are considered with emphasis on their effect on electrical characteristics, followed by selection of more detailed packaging techniques. Voltage distribution and grounding systems are covered, and considerable information is included on the important subject of shielding.

Following shielding, the bulk of this chapter is devoted to electromechanical data: properties and usage of materials, wire, printed-circuit boards, thin films, integrated circuits, etc. References are made to data contained in other chapters, but in some cases an overlap may exist where there seems to be some advantage to the reader in doing so.

Finally, electrical versus mechanical trade-offs are given. Although other chapters also discuss trade-offs, here they relate to electrical properties rather than to the overall advantages or disadvantages of any particular technique.

Terms and Definitions. One of the most important aspects of dealing with the electrical circuit designer is communicating with him. Some terms, such as *speed, loss,* or *leakage,* can mean different things. Practical descriptions are given which the packaging engineer can relate directly to his job of designing electronic gear.

Electronic Terms. Table 1 lists some of the more commonly used terms which the

(*Text continues on page 10–5*)

TABLE 1. Electronic Terms and Definitions

Arc resistance. The time required for an arc to establish a conductive path in a material.
BIT (built-in-test). Circuits included in operational equipment expressly for the purpose of providing for on-line automatic testing.
Breakdown. A disruptive discharge through insulation.
Burn-in. The practice of operating components, subassemblies, or systems for a limited period of time to cause short-term failures.
Capacitive coupling. Interaction of two or more circuits by means of the capacitance between them.
Characteristic impedance. The impedance a line offers at any point to an advancing wave of the frequency under consideration.
Corona. A luminous discharge caused by ionization of the air surrounding a conductor around which exists a voltage gradient exceeding a certain critical value.
Corona inception voltage. The voltage required to initiate corona in a specific set of circumstances.
Crosstalk. Electrical disturbances in a circuit as the result of coupling with another circuit.
Derating. The practice of using a component at less than maximum rating to avoid degradation under excessive physical conditions such as high temperatures.
Dielectric constant (relative permittivity). The ratio of the capacitance of a device using the dielectric material to the capacitance of the same device in a vacuum.
Dielectric strength. The maximum voltage gradient a material can withstand without rupture.
Dissipation factor. The ratio of power loss to power input for an insulation system (the same as the power factor for a normal insulation system).
Downtime. The interval of time which is required for repairs, maintenance, incorporation of changes, and other functions which result in loss of the equipment for its intended use.
EMI (electromagnetic interference). Replaces the radio-frequency interference (RFI) term, which restricts the spectrum of interest to radio communications frequencies.
Fall time (decay time). The time interval for a pulse to decay from 70.7 to 26.0 percent of its maximum amplitude.
Fan-in. The ability of a circuit to accept multiple inputs.
Fan-out. The output ability of a circuit to drive multiple circuits.
Fault isolation. Location of a failure to the level of replacement.
FIM (fault-isolation meter). Circuits included in operational equipment expressly for locating failures to the replacement level.
Flip-flop. A bistable multivibrator. The 0 and 1 sides are the digital logic circuit equivalent of a single-pole double-throw switch.
Floating ground. A level not tied into system ground, to which a group of circuits is referred.
Gate. A circuit having an output and a multiplicity of inputs so designed that the output is energized when and only when a certain definite set of input conditions is met.
Ground level. The difference in voltage level from the system ground reference point to a second point in the ground system.
Ground plane. A thin flat conductor at ground potential, to which circuit grounds may be connected.
Hard wiring (fixed wiring). Interconnecting wire that must be unsoldered or broken to disconnect and remove subassemblies.
Inductive components. Components which create an inductive field when voltage is applied, although that may not be their primary function.
Inductive coupling. Interaction of two or more circuits by means of either mutual inductance or self-inductance common to the circuits.
Input impedance. The impedance presented by the device to the source.
Leakage. Electrical loss resulting from poor insulation.
Leakage path. The path of least resistance for electrical loss.
Line drop. The voltage drop between any two points on a line.
Loss. Decrease in signal power in transmission from one point to another.

TABLE 1. Electronic Terms and Definitions (*Continued*)

LRU (line replaceable unit). A large complex functional assembly that may be easily replaced by a similar unit.

LSI (large-scale integration). Complex monolithic substrates containing multiple circuits.

Matching. Selection of two or more parts to provide characteristics when used together that similar parts selected at random would not provide.

MHz. Megahertz; 10^6 cps (same as megacycles, Mc).

MOS (metal-oxide semiconductor). A single-diffusion, small-area method of making semiconductor components that operate on the field-effect principle.

Nanosecond. 10^{-9} sec (1 sec = 1,000,000 μsec, 1 μsec = 1,000 nsec).

Noise. Any undesired electrical disturbance tending to interfere with normal circuit operation.

Noninductive components. Components which are designed to eliminate the inductive field which they would otherwise create.

Operating life. The period of time during which a device, subassembly, or system can be expected to operate satisfactorily, disregarding all nonoperating time.

Operating speed. The time required for a circuit to complete its specified function.

Overload. Current, voltage, or power level beyond the nominal operating level of a device.

Overload capacity. The current, voltage, or power level beyond which permanent damage occurs to the device considered. This is usually higher than the rated load capacity.

Permeability. The measure of a material's capability to provide a path for magnetic lines of force.

Pick-up. Interference from a nearby circuit or system.

Plug-in. That device or subassembly level at which replacement can be made by simply inserting one unit in the place of another. No soldering, welding, crimping, or other fastening means are required.

Propagation speed. The speed at which a wave travels through a medium.

Q factor. The relationship between stored energy and rate of dissipation: (1) For an inductor—the ratio of coil reactance to effective coil resistance at any given frequency. (2) For a capacitor—the ratio of susceptance to effective shunt conductance at any given frequency. (3) For a magnetic or dielectric material—2π times the ratio of maximum stored energy to energy dissipated in the material per cycle.

Reluctance. The measure of a material's resistance to the passage of magnetic flux.

Rise time. The time interval required for a pulse to rise from 26.0 to 70.7 percent of its maximum amplitude.

Shelf life. The period of time during which a device, subassembly, or system can be stored in nonoperating condition and still function properly when used.

Shield. Material used to suppress the effect of an electric or magnetic field within or beyond definite regions.

Shielding effectiveness. The reduction, expressed in decibels, of the intensity of an electromagnetic wave at a point in space after a shield is inserted between that point and the source.

Test jack. A terminal or jack designed to support a test lead.

Test pad. A conductor area (commonly on a printed circuit) provided for probing to assist in trouble shooting and engineering testing.

Throwaway. The subassembly or component level at which repair is considered unfeasible.

Time constant. The time required to rise to 63.2 percent or decay to 36.8 percent of maximum.

Tuning. Adjusting a variable resistor, capacitor, or other component that affects system operation.

Variable components. Components whose values can be changed (usually by screwdriver adjustment).

Voltage drop. The reduction in voltage level from the reference point to a second point. The percentage of voltage drop is the ratio of voltage drop to total voltage at the reference point.

Volume resistivity. The electrical resistance between opposite faces of a cube of insulating material, commonly expressed in ohm-centimeters.

electronic packaging engineer may find confusing at first. The definitions given are simple. In a specific environment they may be used somewhat differently, but the definitions are representative of USA standards, the *IRE (IEEE) Dictionary of Electrical Terms and Symbols,* and other references.

Basic Electrical Formulas. Table 2 lists some basic electrical formulas which are

TABLE 2. Basic Electrical Formulas

Ohm's Law for DC Circuits. The two basic formulas ($E = IR$ and $P = EI$) can be rearranged into the following twelve forms:[1]

To find	Given	Equation
E	I, R	$E = IR$
E	P, I	$E = P/I$
E	P, R	$E = \sqrt{PR}$
P	E, I	$P = EI$
P	I, R	$P = I^2R$
P	E, R	$P = E^2/R$
I	E, R	$I = E/R$
I	P, E	$I = P/E$
I	P, R	$I = \sqrt{P/R}$
R	E, I	$R = E/I$
R	E, P	$R = E^2/P$
R	P, I	$R = P/I^2$

where P = power, watts
R = resistance, ohms
V = voltage, volts
I = current, amp

Variations of Ohm's law can be solved quickly using the nomograph of Fig. 1.

Power in AC Circuits.

$$P = EI \cos \phi$$

where P = power, watts
ϕ = phase angle between E and I
$\cos \phi$ = power factor
E = voltage, volts
I = current, amp

Power Factor.

$$pf = \cos \phi$$

where pf = power factor
ϕ = phase angle between E and I, or
ϕ = arctan (X/R)
R = resistance, ohms
X = inductive or capacitive reactance, ohms

Inductive Reactance of a Coil.

$$X_L = \omega L$$

where X_L = inductive reactance
$\omega = 2 \pi f$
L = inductance of the coil, henrys
f = frequency, cps

TABLE 2. Basic Electrical Formulas (*Continued*)

Capacitive Reactance.

$$X_c = \frac{1}{2\,\pi\, fC}$$

where X_c = capacitive reactance, ohms
f = frequency, mc
C = capacitance, μf

Admittance of a Circuit.

$$Y = \frac{1}{\sqrt{R^2 + X^2}} \qquad Y = \frac{1}{Z}$$

where Y = admittance, mhos
R = resistance, ohms
X = reactance, ohms
Z = impedance, ohms

Resistors in Parallel.

$$R_T = \frac{1}{\dfrac{1}{R_1} + \dfrac{1}{R_2} + \dfrac{1}{R_3} + \cdots}$$

Resistors in Series.

$$R_T = R_1 + R_2 + R_3 + \cdots$$

Capacitors in Parallel.

$$C_T = C_1 + C_2 + C_3 + \cdots$$

Capacitors in Series.

$$C_T = \frac{1}{\dfrac{1}{C_1} + \dfrac{1}{C_2} + \dfrac{1}{C_3} + \cdots}$$

Inductors in Parallel.

$$L_T = \frac{1}{\dfrac{1}{L_1} + \dfrac{1}{L_2} + \dfrac{1}{L_3} + \cdots}$$

Inductors in Series.

$$L_T = L_1 + L_2 + L_3 + \cdots$$

Impedance; Resistance and Reactance in Series.

$$Z = \sqrt{R^2 + X^2} = \sqrt{R^2 + (X_L - X_c)^2}$$

where Z = total impedance, ohms
R = resistance, ohms
X = total reactance, ohms
X_L = inductive reactance, ohms
X_c = capacitive reactance, ohms

Impedance; Resistance and Reactance in Parallel.

$$Z = \frac{RX}{\sqrt{R^2 + X^2}}$$

where Z = total impedance, ohms
R = resistance, ohms
X = inductive or capacitive reactance, ohms

TABLE 2. Basic Electrical Formulas (*Continued*)

Characteristic Impedance of Twin-lead Cable.

$$Z_o = 276 \log \frac{2d}{r}$$

where Z_o = characteristic impedance
d = distance between leads
r = radius of conductors

Characteristic Impedance of Coaxial Cable.

$$Z_o = 138 \log \frac{d_2}{d_1}$$

where Z_o = characteristic impedance
d_1 = diameter of center conductor
d_2 = inside diameter of shield

Q Factor of a Capacitor.

$$Q = \frac{1}{\omega RC} \qquad (R \text{ and } C \text{ in series})$$

$$Q = \omega RC \qquad (R \text{ and } C \text{ in parallel})$$

where $\omega = 2\pi f$
R = resistance, ohms
f = frequency, Mc
C = capacitance, farads

Q Factor of a Coil when L and R Are in Series.

$$Q = \frac{\omega L}{R}$$

where $\omega = 2\pi f$
L = inductance, henrys
R = resistance, ohms
f = frequency, cps

Resonant Frequency of an L-C Combination.

$$f = \frac{1}{2\pi\sqrt{LC}}$$

where f = frequency, cps
L = inductance, henrys
C = capacitance, farads

Resistivity.

$$R = \rho \frac{l}{A}$$

where ρ = resistivity, ohm-cm
R = resistance, ohms
A = area, cm^2
l = length

Time Constant for an R-C Combination.

$$T = RC$$

where T = time, sec
R = resistance, ohms
C = capacitance, farads

TABLE 2. **Basic Electrical Formulas** (*Continued*)

Time Constant for an L-R Combination.

$$T = \frac{L}{R}$$

where T = time, sec
R = resistance, ohms
L = inductance, henrys

Gain or Loss when the Ratio of Power, Voltage, or Current is Known.

$$\text{db} = 10 \log \frac{P_1}{P_2} \qquad \text{db} = 20 \log \frac{E_1}{E_2}$$

$$\text{db} = 20 \log \frac{I_1}{I_2}$$

where db is the gain or loss in decibels, P_1 is the larger and P_2 is the smaller of the two powers, E_1 is the larger and E_2 is the smaller of the two voltages, and I_1 is the larger and I_2 is the smaller of the two currents. If P_1, E_1, or I_1 is the output power, voltage, or current, the answer will be decibel gain; conversely, if they are the input, the answer will be decibel loss.

Sinusoidal Conversions.

Known	Average	Rms	Peak	Peak-to-peak
Average	....	1.11	1.57	3.14
Rms	0.9	....	1.414	2.828
Peak	0.637	0.707	....	2.0
Peak-to-peak ..	0.32	0.3535	0.5	....

Multiply known values by the factors given under the proper heading to convert from one method of measurement to another.

Wavelength.

$$\lambda = \frac{300{,}000}{f}$$

where f = frequency, kc
λ = wavelength, meters

Permeability.

$$\mu = \frac{B}{H}$$

where μ = permeability
B = flux density, gauss
H = magnetizing force, oersteds

Reluctance.

$$\mathcal{R} = \frac{l}{\mu A}$$

where $\mathcal{R}$ = reluctance
μ = permeability
A = cross-sectional area, cm^2
l = flux path length, cm

commonly used by electronic packaging engineers. More specific formulas are given in portions of this chapter dealing with shielding, thin films, etc., and in other chapters where applicable. Variations of Ohm's law can be solved quickly using the nomograph of Fig. 1.

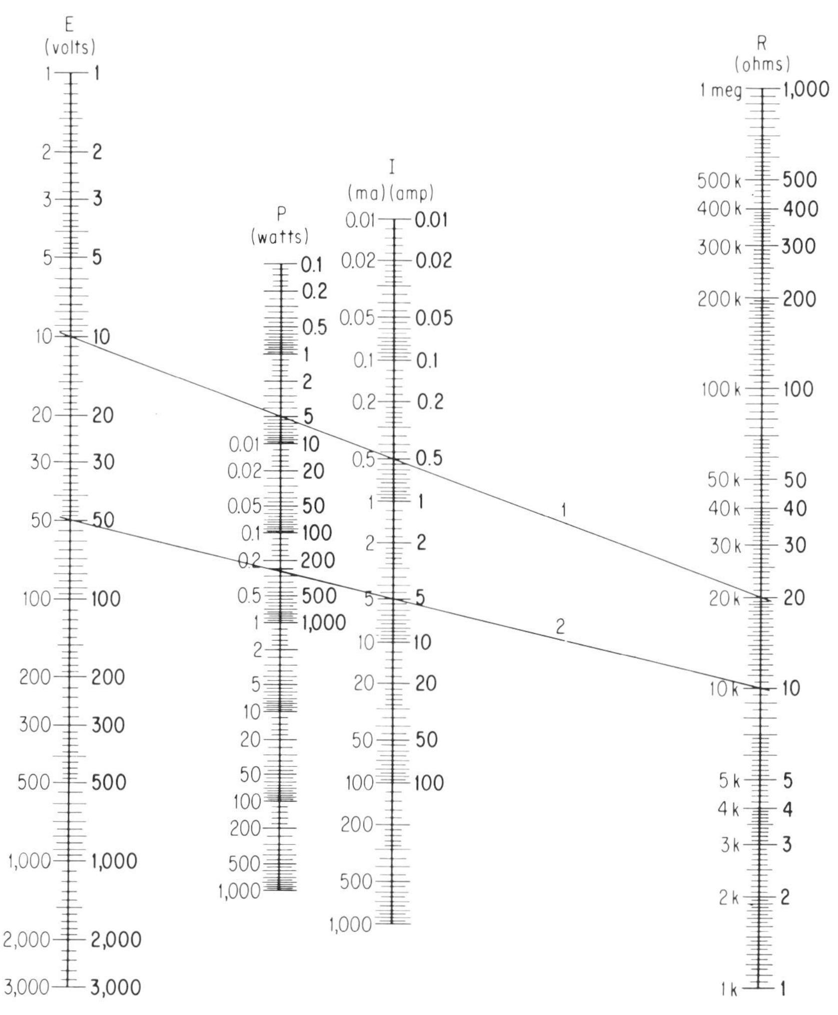

FIG. 1. Ohm's law nomograph.[1] Two scales are given on the nomograph. The boldface numbers on the right of each scale are for resistances for 1 to 1,000 ohms and currents from 0.01 to 100 amp. The lightface numbers on the left of each scale are for resistances from 1,000 ohms to 1 megohm and currents from 0.01 to 1,000 milliamp. *Always* use the same set of values for a given problem—that is, either all boldface or all lightface numbers.

INTERFACING WITH THE CIRCUIT DESIGNER

Learning to get all the necessary information from the circuit designer before the packaging design proceeds too far is probably one of the most difficult tasks for the untrained packaging engineer. This critical interplay of information determines how many changes or redesigns will be necessary later. All too often, repackaging is necessary because the packaging engineer did not find out in the beginning that one critical circuit had to be shielded from another, two parallel leads would cause excessive crosstalk, or other such problems.

Rough Schematic and Parts List. Schematics and parts lists received from the circuit designer may range from rough scratches on a memo pad to finished drawings. One "in-between" level is illustrated in Fig. 2. It seems fairly complete, but many things the packaging engineer needs are missing. For instance: What is the polarity of the transformer? What are the terminal numbers? Does the ¼-watt resistor note specifying ±10 percent hold for R12, the ½-watt resistor also? What kind of capacitor is C1? Is it polarized? These and other questions may not be answered before the rough schematic reaches the packaging engineer. In fact, when decisions have been made and changes are not permitted, he is definitely restricted in his design.

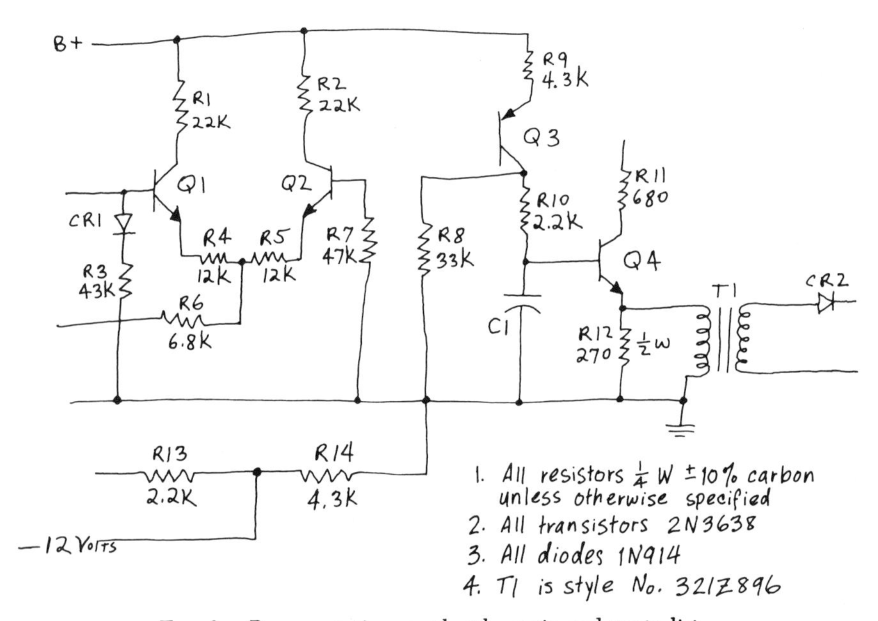

FIG. 2. Representative rough schematic and parts list.

Mutual decisions between circuit designer and electronic packaging engineer are an absolute must. Inputs from both are necessary to reach the best solution. A checklist of questions that help define the packaging task is given in Table 3. This represents a starting point only. Some questions will not apply to a particular system and addition of others may be necessary.

Selection of Components. Initial selection of components must be made before mechanical design can progress. Gross selection is the primary concern at this point—whether to use transistors or integrated circuits, carbon or wire-wound resistors, paper or tantalum capacitors, etc. The choice may be easy in a simple case, or difficult for a complex system, particularly if the "best" solution is desired.

Electrical Suitability. The most important criterion for components is performance of electrical function. Normally, the circuit designer will have built a breadboard, frequently with the components available to him in the laboratory. These parts may not be suitable in the final system. The checklist shown in Table 4 should be helpful in deciding whether components are electrically suitable. Further discussion and data are given later in this chapter as well as in other chapters to help the packaging

TABLE 3. Checklist—Interface between Packaging Engineer and Circuit Designer

1. System definition:
   a. Applicable specifications?
   b. Phase; engineering model, prototype, preproduction, production?
   c. Number of units to be built?
   d. Key schedule dates?
   e. Size and weight limitations?
   f. Cost restrictions?
   g. Reliability goals?
   h. Maintenance philosophy?
   i. Spares required?
   j. Allowable throwaway cost?
   k. Qualification tests required?
2. Environmental requirements:
   a. Temperature extremes?
   b. Altitude?
   c. Humidity?
   d. Vibration?
   e. Shock?
   f. Pressure?
   g. Vacuum?
   h. Radiation?
   i. Explosive atmosphere?
   j. Sand and dust?
3. Power:
   a. Dissipation in all circuits known?
   b. Air available?
   c. Conduction to heat sink?
   d. Power supply requirements?
   e. Voltage levels; drop allowed?
   f. Filtering necessary at subchassis level?
4. Grounding:
   a. Philosophy?
   b. Multiple grounds required?
   c. Ground symbols all indicate same ground?
   d. Change in level allowable?
5. Testing:
   a. Philosophy determined (100% component inspection vs. subassembly test, etc.)?
   b. Test points indicated?
   c. BIT included in circuits?
   d. FIM required?
6. Interference:
   a. Any circuits generate interference?
   b. Any circuits particularly sensitive to interference?
   c. Must certain circuits be isolated from others?
   d. Shielding required?
   e. Shielded wire or coax indicated?
7. Inputs and outputs all specified?
8. Breadboards built and tested? Temperature cycled?
9. Any temperature-sensitive circuits? Controlled temperature necessary?
10. All components specified (see Table 4)?
11. Does system or subsystem configuration restrict packaging?
12. Fan-in and fan-out considered in logic circuits?
13. Redesign likely at a later date?

engineer answer these questions. All the other factors that affect component usage in the system must be considered in conjunction with electrical suitability.

Environmental Suitability. Components are unusable if they will not perform adequately in the environment in which they are to operate. Environment may range from an air-conditioned room to a jungle or from under the sea to deep space.

TABLE 4. Checklist—Electrical Suitability of Components

1. Values all specified?
2. Tolerances all specified?
3. Tolerances based on published values or need?
4. Tolerances include effects of humidity, temperature, etc.?
5. Values and tolerances based on "worst case" or laboratory results?
6. Deratings considered?
7. Additional derating or advance to next rating necessary for maximum reliability?
8. Any components used incorrectly? Overstressed?
9. Voltage level, ripple, allowed at component level?
10. Matched components required?
11. Directional components (diodes, capacitors) indicated?
12. Noninductive components required?
13. Nonmagnetic materials required?
14. Isolation sufficient (integrated circuits particularly)?
15. "Dry" circuits require special components such as extended foil capacitors?

Equipment must be transported to the location where it will be used. If the equipment does not have to operate in the interim environment of its mode of transportation, the environment is important only in the effect it has on the equipment in normal service later. A good example is the system built in a normal shirt-sleeve environment, stored in a warehouse, shipped by truck and air, stored underground for a year or more, then fired in a missile. In some space systems, components must be stored in deep space for prolonged periods of time before the system is required to perform its final task.

Table 5 lists some of the questions that should be considered with respect to

TABLE 5. Checklist—Environmental Suitability of Components

1. Environmental conditions at the component location as opposed to the environment conditions for the system?
2. Voltage dissipation in all components?
3. Shipping conditions; rough handling?
4. Storage effects, shelf life?
5. Effect of humidity?
6. Change in carbon resistor value with humidity considered?
7. Vibration level allowable?
8. Shock component can withstand?
9. Temperature ranges; derating and performance deterioration?
10. Temperature shock requirements?
11. Effects of radiation known?
12. Vacuum performance; gases given off?
13. High-pressure application effect (crush)?
14. Effect of cutting leads at assembly?
15. Spread lead components weaken vibration mounting?
16. Support necessary for large components?

environmental suitability of components. Help in answering these questions is to be found throughout this handbook.

Mechanical Compatibility. Size and weight considerations usually come to mind when mechanical compatibility of components is mentioned. These considerations are important, but others are equally important in building a successful electronic system. Questions related to mechanical compatibility are listed in Table 6. One missing detail can result in an unsatisfactory system. For example, components with unweldable leads obviously cannot be used in a design that utilizes components welded in place.

Availability. Needless to say, no system can utilize a component that is not available, but the term *availability* means more than that. It implies "the best available," "the most reliable available," etc. No decisions can be made, regardless of how electrically suitable, environmentally suitable, or mechanically compatible a component is, until its availability is determined.

TABLE 6. Checklist—Mechanical Compatibility of Components

1. Configuration known?
2. Components compatible with interconnecting technique?
3. Component sizes consistent with overall size objectives of the system?
4. Changes to aid packaging allowable?
5. Can the whole circuit be made in thin-film or integrated-circuit form?
6. Are values specified possible in final form (for example, high-value thin-film capacitors)?
7. Is component lead spacing compatible with standard printed-circuit spacings?
8. Standard components specified? Can substitution be made if not?
9. Can parallel capacitors or series resistors replace unwieldy capacitors and resistors?
10. Adjustable components required? Can circuit changes eliminate the need?
11. Shielded components specified where needed?
12. Directional components marked?
13. Will embedding affect components?
14. Must individual components be replaced?
15. Are tantalum capacitors solderable only beyond the weld bead transition?
16. Permanent component markings? Positioned where they can be seen?
17. Are leads sturdy enough to support the component?
18. Special support required?
19. High-impedance circuits to be used on printed-circuit boards?
20. Electrically "hot" cases require insulation?
21. Heat sinking necessary during soldering?
22. Weldable leads required? Solderable?
23. Availability determined?
24. What type of drawings is required?

Selection of General Packaging Method. A clear definition of system requirements is absolutely essential. Ideally, these requirements should be defined before packaging design starts. Unfortunately, they seldom are, and assumptions have to be made, preliminary trade-offs considered, and design begun, in order to meet dates.

System Definition. Some characteristics of the system are nearly always predetermined. Whether it is continuously operated or used one time only may be obvious by definition. More subtle are questions of signal voltage levels, required signal rise times, operating speeds, mean time between failure goals, etc. No attempt will be made here to discuss all the characteristics which define a system. Computer applications are discussed in Chap. 12, military applications in Chap. 13, and space applications in Chap. 14. Other chapters define characteristics from their own particular viewpoint. This chapter is concerned with those electrical characteristics that restrict packaging design and those that may be affected by packaging design.

There is no clear dividing line between electrical and nonelectrical system requirements. Size requirements would appear to be a purely mechanical requisite, but when small size prevents effective isolation between components, or long leads result in signal propagation delays, size becomes an electrical factor. The system should be defined as completely as possible in the predesign stage.

Preliminary Trade-offs, Electrical versus Mechanical. Before design starts, trade-offs are considered, sometimes unconsciously, sometimes more formally. A relatively high percentage of trade-offs involve mechanical versus electrical characteristics. Many of these are discussed in the last portion of this chapter and should be considered when making the preliminary trade-offs. However, other trade-offs must be included. For a complicated system, it may be helpful to list the candidate packaging techniques in one direction, and the characteristics of interest in another, as illustrated in Table 7, giving each checkerboard square a weighted number. The weighted numbers should apply to the particular system being evaluated, since they will vary from system to system. An $\times$ may be used in a block to show that, at the time the chart is made, some characteristic is impractical using a particular packaging technique. Table 7 indicates that conventional components could not be used on printed-circuit boards because the resultant long leads caused excessive signal delays. Table 7

TABLE 7. Typical Chart of Packaging Trade-offs

Type of packaging	Characteristic *										
	Size	System cost	Throw-away cost level	Reliability	Electrical compatibility	Manufacturability	Automation possibilities	Thermal dissipation	Ease of maintenance	Spares and logistics	Development needed
Conventional components on boards	10	5	1	6	×	1	5	1	2	2	1
Soldered modules on boards	9	6	4	10	10	4	5	2	5	5	1
Welded modules on boards	8	10	4	8	10	5	10	2	5	5	1
Hybrid modules (with integrated circuits)	6	9	6	5	8	5	10	3	6	5	1
Hybrid compartmentalized	6	9	6	5	6	6	10	3	6	5	2
Etched circuits	5	8	5	6	7	8	2	2	4	5	5
Pluggable flat-pack modules	6	7	2	9	5	5	4	5	1	4	1
TO-5 can spread lead integrated circuits	5	2	2	4	3	2	2	2	2	2	1
TO-5 can integrated circuits on multilayer	4	4	2	4	3	4	3	3	3	1	3
Flat-pack integrated circuits on plated-through-hole board	3	3	2	4	2	3	4	2	2	1	1
Welded flat-pack integrated-circuit stack	3	8	8	4	2	4	9	6	8	7	2
Thin-film circuits	3	3	5	4	1	7	3	7	8	6	7
Bare integrated-circuit chips	2	1	9	2	2	2	8	9	9	9	×
Large-scale integration (LSI)	1	2	10	1	7	9	1	10	10	10	×
MOS devices	1	2	9	3	9	10	8	×	9	8	10

* 1 = best of group; 10 = worst of group; × = impossible at this time.

also reflects the fact that excessive development would be needed before bare integrated-circuit chips or large-scale integration (LSI) devices could be used, and that metal-oxide semiconductor (MOS) devices could not be adequately heat-controlled to maintain the necessary circuit tolerances.

CIRCUIT GROUPINGS

Although no design takes place in truly series fashion, once preliminary packaging trade-offs have been made and a general packaging method agreed upon, circuit groupings or divisions are required.

Functional versus Optimum Package. The packaging engineer constantly has two goals: one to make the package as good as possible from a packaging viewpoint, the other to make the package as functional as possible. *Functional* means more than operable. It means making the package easy to test, install, repair, replace, trouble-shoot, etc. Features must be added that make the package less attractive from a purely packaging technique viewpoint, but make it more functional.

Plug-in Level. *Plug-in level* and *throwaway level* are not synonymous. In fact, several plug-in levels may be utilized in one system. The confusing term *plug-in module* may refer to what is sometimes called a *line replaceable unit* (LRU, a complex functional assembly), to a subassembly consisting of several smaller subassemblies, to a printed-circuit-board assembly, or even to an embedded module which plugs into a printed-circuit board. The more generalized criterion for a plug-in unit is simply that, regardless of level, it be removable and replaceable by another unit without requiring soldering, welding, or crimping.

Selection of plug-in level or levels is important for several reasons:

1. Rapid replacement in case of failure
2. Interconnecting wiring complexity
3. Reliability
4. Subunit test capability
5. Repair time
6. Spare parts inventory
7. Standardization
8. Electrical isolation

These considerations are important enough to justify a short discussion of each.

RAPID REPLACEMENT. Fixed wiring would be adequate if every part of a system went together, worked properly, and never failed. This is seldom the case, although some space systems approach this condition, since once they are launched, no replacement or repair is possible. Replacement may take place at various levels. Aircraft electronics may be replaced in large functional boxes to avoid long aircraft downtimes. On the other hand, where open racks of boards make it easy to get at individual printed-circuit boards in land-based computers, replacement at the board level is adequate. When expensive individual components are stocked unassembled, it may be necessary to plug them in on an individual basis. Chapter 9 discusses several methods of plugging in individual integrated circuits.

INTERCONNECTING WIRING COMPLEXITY. Selection of circuit groupings and plug-in levels has a great effect on interconnecting wiring. In general, the greater the number of circuits per plug-in subassembly, the greater the number of output pins required. However, it is not a linear factor, and eventually the number of pins required decreases as indicated in Fig. 3. The reasons for the shape of the curve can be understood by considering an eight-wire system (five inputs, one output, one voltage, one ground) which has 1,000 circuits and complicated internal interconnections. While the curve of Fig. 3 is representative of one system, each system has its own characteristics. The optimum division, considering board size, number of pins required, throwaway cost, etc., varies widely.

The logic representation of Fig. 4 can be utilized to represent several methods of division. Each of the four flip-flops is in either the 0 or the 1 state, and the four are connected to the 16 gates in a manner which provides every possible combination, forming the digital representation of 0 through 15. Therefore, only one gate can be on at any one time, and only one driver is amplifying a signal at any one time. Table

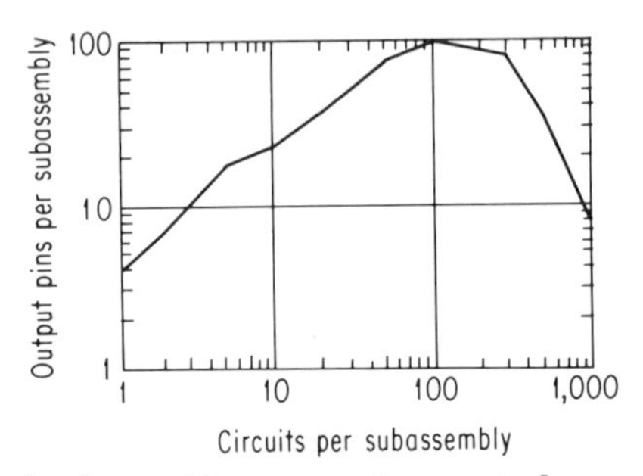

FIG. 3. Number of subassembly output pins required vs. number of circuits.

8 shows several circuit division possibilities, ranging from a subassembly (called a board here) with only one gate to a subassembly containing all 36 circuits. The number of pins required for each board is given and then combinations of boards are listed to indicate how many pins must be interconnected in the system. Board output pins per circuit range from 8 to 0.8. The total number of pins per system varies from 224 to 28.

Several factors other than interconnecting wiring may influence the final choice. In the last paragraph, the observation was made that only one driver was energized at a time. If each driver is located on a different board, 100 percent heat sinking will be required on each board, since there is no way of predicting which driver will be energized or for how long. If 4 drivers are on one board, they may be packaged so that only 100 percent heat sinking is required for the combination of all 4, and if all 16 are on one board, only 100 percent heat sinking is required for all 16. Standardiza-

TABLE 8. Pins Required for Various Circuit Combinations

Circuits per board	Inputs	Outputs	Voltage	Ground	Clock	Pins/board	Pins/circuit
1 gate	4	1	1	1	...	7	7
1 flip-flop	2	2	2	1	1	8	8
1 driver	1	1	2	1	...	5	5
4 gates	6*	4	1	1	...	12	3
4 flip-flops	8	8	2	1	1	20	5
4 drivers	4	4	2	1	...	11	2.7
8 gates	8*	8	1	1	...	18	2.2
8 drivers	8	8	2	1	...	19	2.4
16 gates	8*	16	1	1	...	26	1.6
16 drivers	16	16	2	1	...	35	2.2
1 flip-flop, 4 gates, 4 drivers	6*	4	2	1	1	14	1.6
4 flip-flops, 16 gates, 16 drivers	8	16	2	1	1	28	0.8

Combinations	*Total system pins*
Each circuit on an individual board	224
One board of 4 flip-flops, 4 boards of 4 gates, and 4 boards of 4 drivers	112
One board of 4 flip-flops, 1 board of 16 gates, and 1 board of 16 drivers (vertical division)	81
Four boards of 1 flip-flop, 4 gates, and 4 drivers each (horizontal division)	56
A single board for all	28

* Commons hooked together.

tion of board size may be important. One possible combination that lists 4 flip-flops on one board, 16 gates on a second board, and 16 drivers on a third board might require 3 entirely different sizes of boards and would, therefore, be impractical. Vertical division (similar circuits grouped) requires 3 different boards. Horizontal division grouping of 1 flip-flop, 4 gates, and 4 drivers per board requires 4 identical boards, and the total number of pins in the system is fewer.

While horizontal division seems to be a good compromise in this example, caution should be exercised in making a hasty decision in an actual case until other parts of the system are correlated. Drivers may be used elsewhere requiring entirely different considerations, and although wiring complexity is an important criterion, it will not be the only factor affecting a final decision.

RELIABILITY. Reliability of a system is difficult to evaluate properly. All too often, joint reliabilities are overlooked. Each soldered joint, each sliding contact, each welded joint, and any other method of interconnecting detracts to some extent from system reliability. Given the same set of conditions, fewer joints mean better system reliability. Total number of pins is not the same as total number of joints. Each "pin" may contain several joints—for example, a crimp, a sliding contact, and two solder joints in series.

TEST CAPABILITIES. Large systems should be built so that interim testing of subassemblies is possible. Critical signals should be brought out to a position where they can be analyzed. Test capabilities are easy to overlook but necessary. Testing may be done on many levels. Where automatic testing equipment is available, 100 percent testing of components may be practical. Conversely, economic studies sometimes show that no component testing should be done; statistically, rejects are small enough to allow throwing away small subassemblies with bad components. Sometimes subassemblies can be repaired less expensively than all components can be tested.

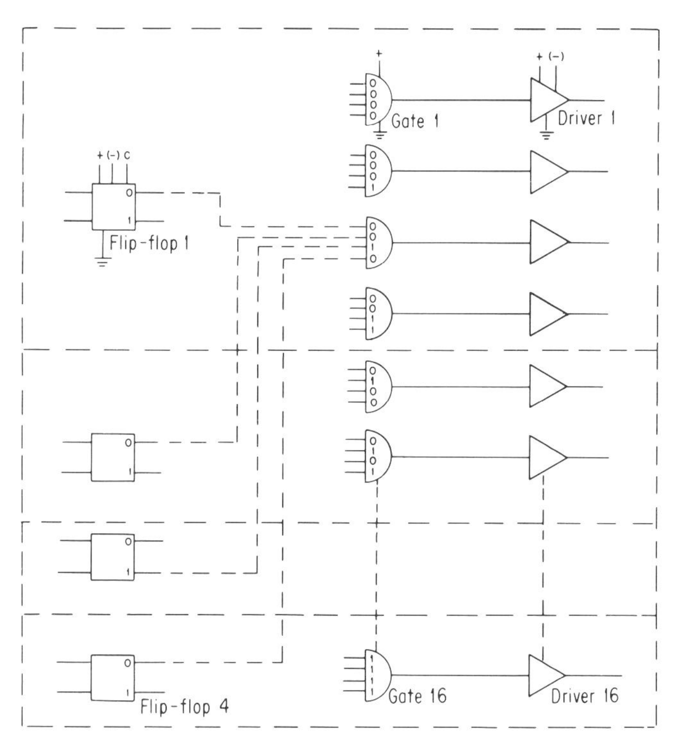

FIG. 4. Example of logic division.

Often, rework is impossible. Subassemblies which are embedded in plastic are commonly thrown away if they do not work properly. Molded subassemblies are generally tested before embedment to detect bad components, incorrect connections, or poor joints. They must be structurally sound to avoid damage during premold testing. Premold and postmold test results may be compared to ascertain changes due to molding.

REPAIR TIME. Practically impossible repair situations are sometimes built into equipment simply because repairability was not considered. Not only should repair be possible, repair time should be consistent with system usage. For example, where leads are welded in place, provisions should be left for a second weld. If welding equipment is not available in the field, repair may not be possible on location. Field techniques sometimes differ from original construction, and additional precautions may be necessary. When solder joints are substituted for weld joints, electrical characteristics may be affected if components are not heat-sunk to prevent damage.

SPARE PARTS INVENTORY. Packaging design has a great effect on spare parts. It is obviously impractical to stock resistors which are used inside embedded subassemblies. Even the embedded subassemblies may not be stocked if they are permanently attached to printed-circuit boards which can economically be considered throwaway units. Lack of standardization will lead to many spare parts problems.

STANDARDIZATION. Standardization makes a system easier to build, easier to stock parts for, and easier to test, replace, and repair. On the other hand, standardization may lead to more complex wiring, and to some degree it also limits flexibility in design.

ELECTRICAL ISOLATION. Breadboards built by the circuit designer are usually spread over a wide area. Those same circuits will frequently not operate in the same manner when brought together and packaged in a compact group. The effects of packaging on electrical characteristics is dealt with in detail later in this chapter. Where facilities are available, welded modules, printed-circuit boards, or other subassemblies should be built as nearly as possible according to the final design.

Throwaway Level. Throwaway units may be defined by customer price, by a price versus reliability relationship, or by other means. A nonrepairable subassembly automatically becomes a throwaway, but a repairable unit may not be repaired for economic reasons. An example of this is an inexpensive printed-circuit-board assembly. Test and repair of the board may be more expensive than a new board. Large complex LRUs are normally repaired at the bench level while replacements operate in the system. In this case, some smaller subassembly or sub-subassembly becomes the throwaway unit. Once the throwaway level is established, there is no need to add embellishments that make the unit easy to repair.

Fault Isolation. Providing test points is a compromise to optimum packaging design. Do not arbitrarily bring out leads which would normally be buried deep inside a subassembly; consider the effect on system performance. Additional electrical problems are nearly always discovered when test points are added as an afterthought.

Sophisticated systems require more than a few test points to locate failures. Complicated fault-isolation meter (FIM) circuits may be necessary, and the packaging design is correspondingly compromised. Design of the FIM circuits should relate to replaceable units and coincide with packaging design.

Built-in Test Circuits. Built-in test (BIT) circuits may be used to test a system in place. Initiation of the BIT circuit cycles a series of tests and presents some type of intelligence indicating that the equipment is working properly. In a fire-control system, for example, BIT may make the pilot's indicator presentation representative of the presentation expected in a real attack. Again, packaging design is compromised by the addition of what may be fairly complicated circuitry.

Standardization versus Optimum Package. Standardization is always a worthy goal, but when it restricts design, standardization must be compromised. The most compact, clean-cut design is a "one-of-a-kind" design which must be avoided. Two methods of circuit grouping are in common usage, general-purpose and specific-function subassemblies. Here, the subassemblies will be referred to as *boards* for the sake of simplicity, but the packaging form varies widely.

General-purpose Boards. All the inputs and outputs for each circuit are brought out on pins of a general-purpose board. Referring back to the logic circuit of Fig. 4, if four flip-flops were packaged on one type of board, four gates on another, and four drivers on a third, and if no leads other than voltages and grounds were connected together on the board, the resulting interconnecting wiring would be very complicated. The 112 pins shown in Table 8 for that particular configuration would rise to 152 since some of the inputs to the gates were considered common when the total number of pins was calculated. A real advantage results, however, because the basic flip-flop, gate, and driver boards are available for use throughout the entire system. Another advantage is that, with limited building blocks, a great variety of systems can be built and changes can be incorporated simply by changing wiring.

Whether a general-purpose board system is electrically acceptable or not depends on many things. Point-to-point wiring may be used to reduce pick-up, but systems built this way tend to be large in size. Lead lengths may limit signal propagation time and, therefore, operational speed.

Specific-function Boards. Circuit groupings chosen for a specific function usually require fewer leads but limit the flexibility of their usage. Referring again to Fig. 4 and Table 8, a system consisting of four boards, each containing one flip-flop, four gates, and four drivers, requires only 56 leads—roughly one-third of those required using general-purpose boards. Only one board type is required instead of three different ones. On the other hand, the particular interconnection pattern on this board may not be used anywhere else in the system. Looking at the problem from a system viewpoint, many more board types would be required using the specific-function approach than using the general-purpose approach.

Board cost, lack of flexibility, and long design time are disadvantages of the specific-function circuit grouping approach, but simplified wiring, smaller size, and electrical advantages of specific-function grouping may outweigh the disadvantages. Each system must be evaluated on its own merit.

Handling of Various Types of Circuits. Different types of circuits lend themselves more readily to one type of circuit division than to another.

Digital Circuits. General-purpose boards are most often used for digital circuits because a few basic circuits are repeated over and over again. Where smallest size, maximum speed, and highest reliability are design criteria, specific-function boards are sometimes used.

Linear Circuits. Since linear circuits are generally more complicated and less repetitive than digital circuits, circuit groupings tend to be functional in nature. Sometimes a repetitive amplifier may be packaged and used in the same manner as the general-purpose digital board.

Miscellaneous. High-power, high-frequency, and other special-purpose circuits are nearly always grouped by function. Electrical isolation of one circuit from another is a common reason for packaging functional groups together. Considerable space is devoted to shielding later in this chapter.

Effect of Circuit Grouping on Electrical Characteristics. Some of the effects circuit grouping can have on electrical characteristics are listed in Table 9. The examples indicate the kind of effect the packaging engineer should consider. Many of these examples are explored in more detail later in this chapter.

The effect of circuit grouping varies with packaging technique. Summing resistors of a high-impedance circuit, for instance, will have parallel impedance paths across the surface of a glass-epoxy board. If the same components are embedded in epoxy, the parallel paths may be beyond objectionable levels. High-impedance circuits should be grouped so that they can be dealt with more effectively.

SELECTION OF PACKAGING TECHNIQUES

After the packaging engineer and circuit designer have selected component types, arrived at a general packaging method, and made preliminary decisions about how the system should be divided into subassemblies, specific packaging techniques must be chosen. Some of the things that should be kept in mind while making the packaging techniques decisions will be discussed briefly.

TABLE 9. Some Effects of Packaging on Electrical Characteristics

Packaging technique	*Possible effect*
Tight packaging	Effective isolation impossible
No shielding between if stages	Interference between stages
Long leads	Increase propagation time
Parallel wires	Crosstalk
Low-level signal next to power line	Pickup on signal line
Parallel lines on printed-circuit board (same or opposite sides)	Mutual capacitance; crosstalk
High-impedance circuits on printed circuits	High-impedance shorts across the surface in humidity
Embedding of soldered or welded modules	Circuit changes caused by pressure and dielectric characteristics
Soldering components by hand	Component changes caused by heat
Grounding power and signals together	Signal interference through ground
Close interconnections on welded subassemblies	Blown semiconductors where electrodes touch
Grounding through anodized aluminum by use of toothed washer	Poor ground; opens
Use of component lead as output pin in soldered modules	Opening up of lead inside module when heat is applied to solder module on printed-circuit board
Use of nontwisted wire pairs	Mutual inductance; crosstalk
Large embedded subassemblies; leads from internal stages not brought out	Impossible to test once molded
No stress relief loop in wires or components	Intermittents and opens
No heat sinking on high-wattage components	Component failure
Sharp corners on high-voltage lead terminations	Corona
Basing subassembly design on overall power dissipation	Hot spots—component failure
Space not allocated for test points	Fault isolation very difficult
Plug-in of simple subassemblies	Many joint resistances in series
Series grounding	High ground level differences

System Requirements. Several extremely important system characteristics that are not directly electrical in nature influence packaging design. They are:

1. Cost
2. Size
3. Reliability
4. Manufacturability
5. Maintainability
6. Repairability

All these characteristics are explored in detail in other chapters and must be factored in with the requirements discussed here.

Speed. Wire length alone may easily be the limiting factor in a fast system. A rough rule of thumb is that 1 ft of bare copper wire causes a 1-nsec delay. Circuits are readily available with propagation delays of less than 10 nsec, and special circuits are much faster. System speed requirements may dictate tight packaging, microminiature parts, and point-to-point wiring, even though adequate space is available for more conventional packaging.

Where long leads are inevitable, the circuit designer and packaging engineer may elect to accept certain delays so long as the delays are equal in like branches. This is usually accomplished by making the leads equal in length even though the points are not equal distances apart.

High Voltage. High-voltage circuits normally form a small part of the overall system and require different packaging from the rest of the system. Clearances and materials are of prime importance. Useful data for high-voltage design of plastics and board coatings are included in Chap. 7, wires in Chap 2, printed circuits in Chap.

1, and connectors in Chap. 6. Corona and some design features are discussed in this chapter.

Power Dissipation. Few components can be operated at elevated temperatures without derating their electrical characteristics in some way. Total power dissipated per unit volume may be used as a guideline in establishing the cooling means (see Chap. 11), but hot spots will occur unless each component is individually questioned.

Power dissipation should be established for each component. Most of the very-low-dissipation components can be handled conventionally, but higher-dissipation components require special attention. It may be necessary to use heat sinks or to move components to more advantageous locations. The amount and type of cooling available, space for heat sinks, thermal paths, etc., must be traded off against deratings and effect on system performance. Frequently, a change in components will be required. The next-higher-power-rated resistor may be necessary to allow for additional derating.

Care should be taken in placement of components, proximity to other components, and handling of leads. Note in Fig. 5 that the center resistor of a string of five resistors dissipates only 40 percent of its rated wattage when the leads are not con-

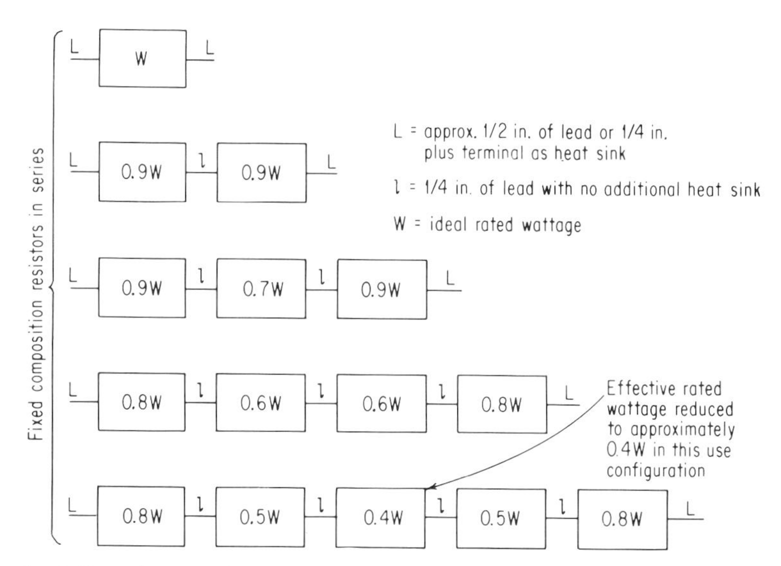

FIG. 5. Effect of use configuration on wattage ratings of fixed-composition resistors.[2] About three-fourths of the heat from fixed-composition resistors is dissipated through the leads. This accounts for the loss of effective wattage rating as this heat sink becomes less effective. Resistors of the RC-20, RC-32, and RC-42 sizes conform approximately to the above wattage-rating degradation.

nected to terminals. Most of the heat would normally be dissipated through the leads, but since the leads are attached to leads of another hot resistor, there is no place for the heat to flow.

Circuit Sensitivity. Recognition of sensitive circuits is half the solution of the problem. A core memory may require control of temperature variation to within a few degrees in order to maintain satisfactory electrical operation. Steps must be taken to heat, cool, or insulate so as to provide the required environment. Where only small temperature variations are allowed but the center of change is optional, higher temperatures are usually preferred because heating may be used with no cooling required.

Circuit sensitivity is most frequently related to pick-up of one kind or another. A circuit may be sensitive to high-frequency signals or to a magnetic field in its vicinity and may therefore require shielding. Shielding imposes packaging constraints. In the simplest case, an enclosure is placed around the circuit. Considerable data are included in this chapter on shielding.

Circuits or components which are sensitive to pressure, radiation, vibration, or shock may require special packaging design to isolate the sensitive areas. Sensitivity can take the form of degradation, temporary failure, or catastrophic failure. Classic examples of these forms are, respectively, changes in capacitor value, relay contact bounce at a particular resonant frequency, and diode whisker breakage.

Component Characteristics. The basic building blocks of a system are the components. The form of the building blocks places restrictions on packaging techniques.

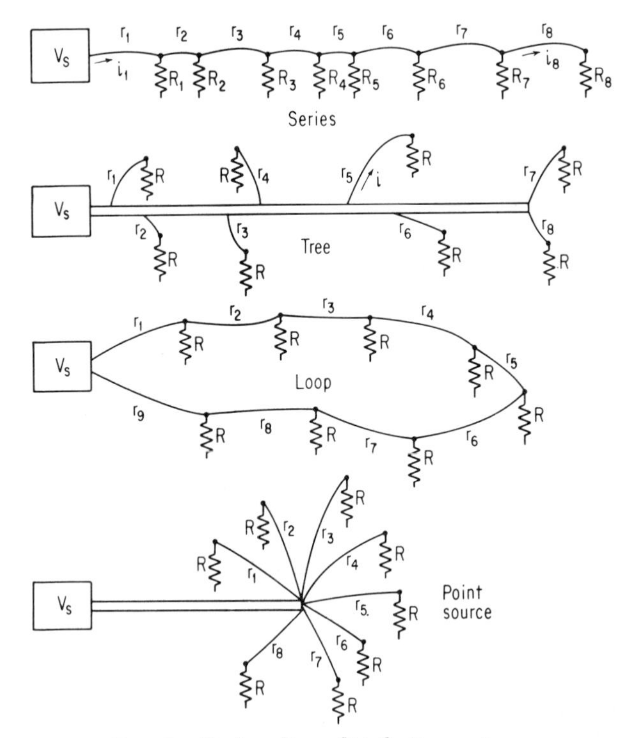

FIG. 6. Basic voltage distribution systems.

For example, TO-5 can integrated circuits cannot be used in the same manner as flat-pack integrated circuits; thin-film resistors are not directly interchangeable with ordinary carbon resistors; and solderable tinned copper leads cannot be used in welded subassemblies.

Component size, power rating, lead material, type of electrical connection, lead configuration, insulation, shielding, and environmental protection all influence the packaging design technique. A common mistake occurs when the circuit designer orders parts before the packaging engineer approves. As a result, components may not fit into the space allotted or the design may become unnecessarily complicated.

Packaging Techniques Decisions. Table 7 was used to illustrate how the advantages and disadvantages of packaging methods could be given weighted numbers. After considering circuit groupings and system requirements, weighted numbers can be assigned that apply to the major part of the system but probably not to special circuits. It is unwise to force the rest of the system into using the same technique if the requirements are vastly different. Several packaging techniques may be used within a single system.

System requirements which are not directly electrical in nature make a major contribution toward reaching a packaging technique decision. However, the characteristics concerned with electrical parameters are of more direct interest here. The rest of this chapter is specifically aimed at presenting data and detailed design aids of an electrical nature useful to the packaging engineer.

VOLTAGE DISTRIBUTION AND GROUNDING SYSTEMS

Each voltage level is a network underlying system wiring. Ground or separate grounds are usually even more complex.

Types of Voltage Distribution Interconnection Systems. Four basic types of voltage distribution interconnection systems are shown in Fig. 6. Each has advantages and disadvantages, some of which are listed in Table 10. Two fundamental considerations are current-carrying ability and voltage drop. To illustrate the differences in line

TABLE 10. Advantages and Disadvantages of Basic Voltage Distribution Systems

Type of voltage distribution system	Advantages	Disadvantages
Series	1. Easy to program and wire 2. Short leads between points	1. Joints in series depend one on the other 2. A broken lead affects multiple loads 3. Voltage drops are high 4. High current in the lead closest to the source 5. Two leads are required on each pin
Tree	1. Additive currents are in a hefty bar 2. A broken lead affects only one load 3. Low voltage drops 4. Low current in each line 5. Only one lead on each pin	1. Special busses are required 2. Space is required for busses 3. Not applicable to irregular systems and special applications
Loop	1. One broken lead does not affect operation 2. Easy to program and wire 3. Short leads between points 4. Voltage drop lower than series	1. Circulating currents 2. High current in one lead 3. Two leads are required on each pin
Point source . .	1. Low current in each line 2. Low voltage drops 3. Broken lead does not affect other loads	1. Long leads 2. Limited by number of leads possible on one terminal 3. Parallel leads and crosstalk are more likely than with other types 4. Unsoldering one lead affects the rest

current and voltage drop for the various methods, consider the simplified diagram of Fig. 6 where each of the loads R is equal and large with respect to the line resistances r, which are also equal. In addition, consider the buses in the tree and point source systems perfect conductors.

For the series type, the current $i_8 = V_s/R$ in r_8 but increases to $i_1 = 8V_s/R$ in r_1.

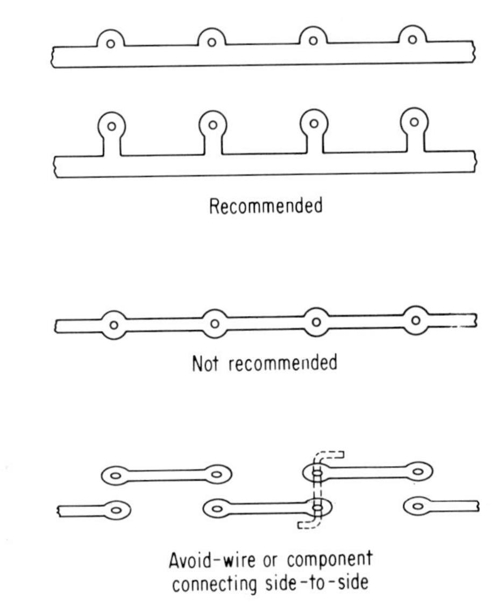

FIG. 7. Voltage distribution on printed-circuit boards.

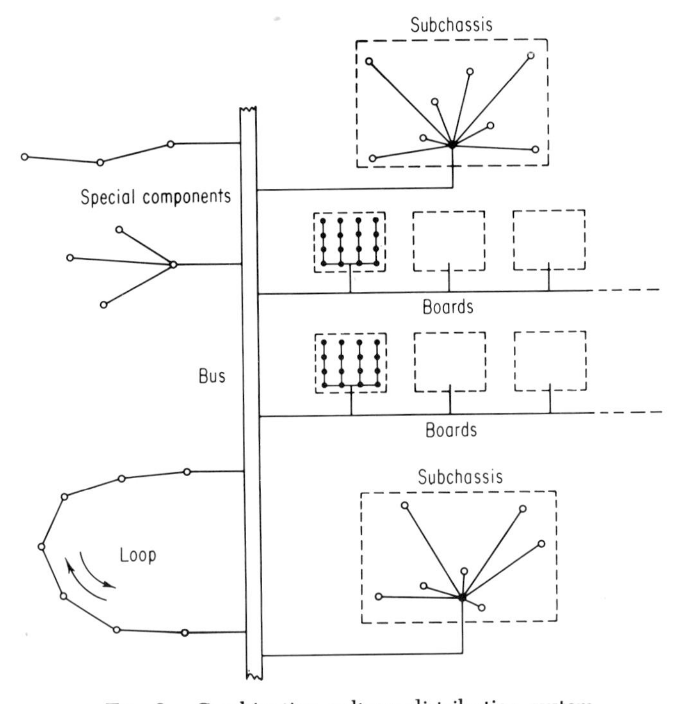

FIG. 8. Combination voltage distribution system.

The current in each line is only $i = V_s/R$ in both tree and point source systems. The voltage drop in the series line to load R_s is $v_s = 36rV_s/R$ but only $v = rV_s/R$ for any line in tree or point source system. Voltage drops become excessive rapidly using series wire connections. Series interconnections should not be used where currents are large, where the line resistance is an appreciable value compared to load resistance, or where long strings are required.

A complex system rarely utilizes one basic type of voltage interconnection system throughout, because of limitations in construction. For instance, wiring complexity normally prohibits use of single-point wiring on printed-circuit boards. Some of the disadvantages of series wiring, such as wire breakage or bad joints, can be avoided on printed-circuit boards if pads are offset from the current-carrying line as in Fig. 7. A hybrid system is illustrated in Fig. 8.

Types of Grounding Interconnection Systems. Basic grounding interconnection systems are similar to voltage distribution interconnection systems—series, tree, loop, and point source—with one major addition, the plane. Planes are rarely used for voltage interconnections but are commonly used for grounding. Chassis and printed circuits are the most commonly used ground planes. Metallic chassis are convenient conductors requiring no additional piece parts, but care must be taken in making contact with the chassis. Dissimilar metals in hardware may establish barrier corrosion that leads to a poor electrical joint.

Printed-circuit boards are easy to solder to, and permanent joints to ground can be flow-soldered in the same operation as the rest of the interconnections. Figure 9

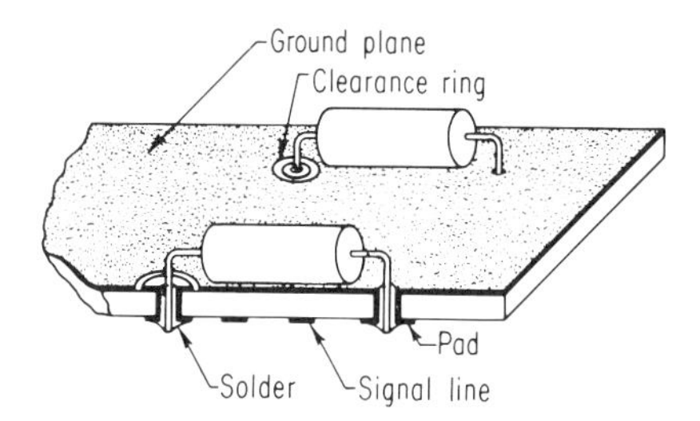

FIG. 9. Printed-circuit ground plane.

shows a typical double-sided plated-through-hole printed-circuit board with ground plane on the component side. The ground plane also doubles as a heat-conductive plane. Details of printed-circuit-board construction and use are given in Chap. 1.

Availability of multilayered printed-circuit boards has opened up even more sophisticated uses of ground planes. In the cross section of Fig. 10, two ground planes

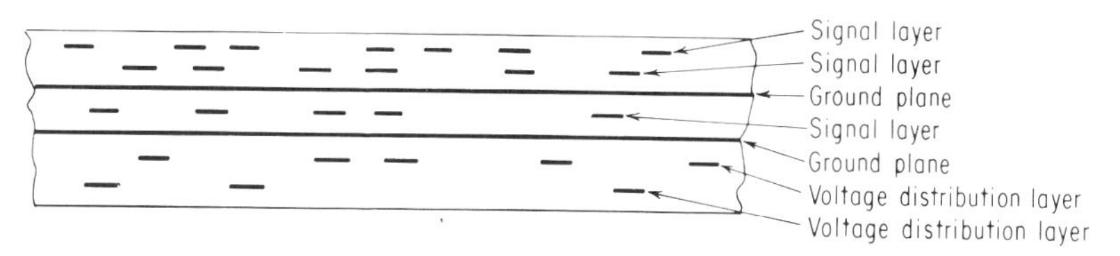

FIG. 10. Multilayer printed-circuit ground planes.

are used; one to separate voltage distribution layers from signals and another to separate two classes of signals.

Ground interconnections, like voltage interconnections, are seldom accomplished with a single technique. The grounding of a system may be similar to that in Fig. 8 with the addition of ground planes, but it is usually more complex.

Multiple Ground System. The more complex an electronic system gets, the

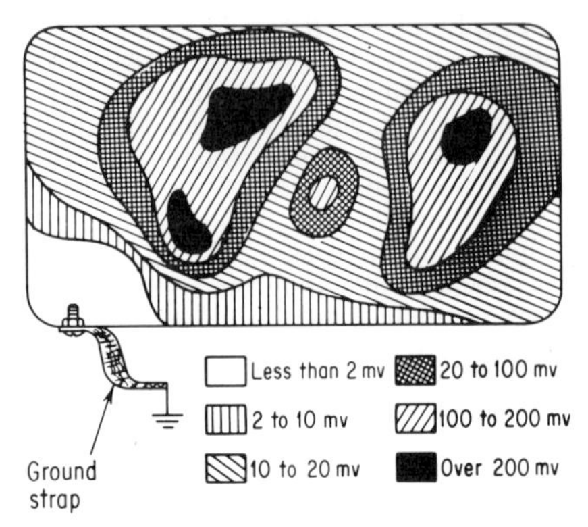

FIG. 11. Typical ground potential graph at a specific frequency. (*Filtron Co., Inc.*)

more unlikely it is that a single ground system will work. Low-power signals should not be grounded on the same line with high-power devices. Separation of grounds into three logical categories, signal, power, and chassis grounds, is one commonly used method. Signal grounds are those associated with sensitive signals, power grounds are those used for high current and relatively insensitive circuits, and chassis grounds are used mainly for tying together shield enclosures, component cases, etc. Different symbols must be used for each on the schematic, and in the equipment each type of ground is kept carefully isolated from the others except where they are joined at one point, usually the power supply. Floating grounds are infrequently used in electronic equipment. All circuits are usually referred back to the zero-level ground point.

Ground level at any particular point in any one of the three overlapping ground networks is not necessarily at ground zero level with respect to the point junction of the three grounds. Ground interconnections are susceptible to the same "drops" as the voltage distribution interconnection system. Where series grounding is used, the end of a long line may be appreciably above or below the ground reference point. Figure 11 shows how a typical ground plane deviates from ground zero at a specific frequency and loading. In this example, if low-level circuits were located in the black zone where the ground is over 200 mv above the ground reference, the design would probably be inadequate.

Electrical Considerations for Grounding and Voltage Distribution Systems. Interference generated by electrical and electronic equipment can degrade or make completely inoperable the primary function of an electronic system. Interference reduction should be an important part of the initial design of a piece of equipment. If it is overlooked and remedied after the equipment has been designed, it is much more

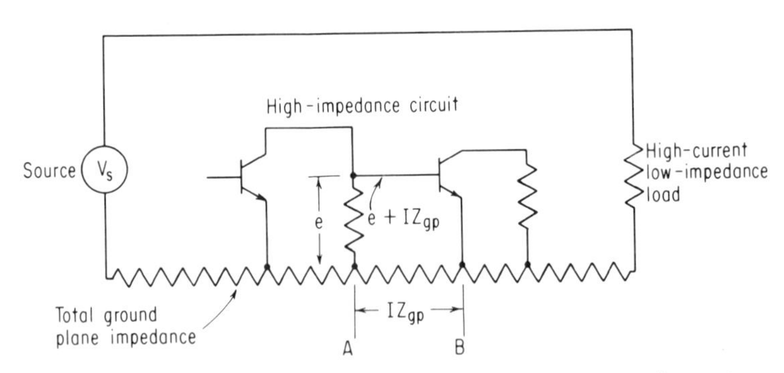

FIG. 12. Effect of circulating ground currents on a typical circuit.

difficult to incorporate. One of the most important aspects of interference reduction is establishment of grounding procedures to minimize interference. Shielding is covered in the next major portion of this chapter, following a discussion of grounding methods.

Electrical Circuits in Ground. An ideal ground should be a zero-impedance body to which all signals can be referred and all undesirable signals transferred for elimination. There should be no voltage difference between any two points, but unfortunately, this condition is never attained. Figure 12 shows the ground plane as an impedance path. Current flowing from the source through the load creates a voltage IZ_{gp} in the ground plane between points *A* and *B* which effectively changes the voltage level on the transistor.

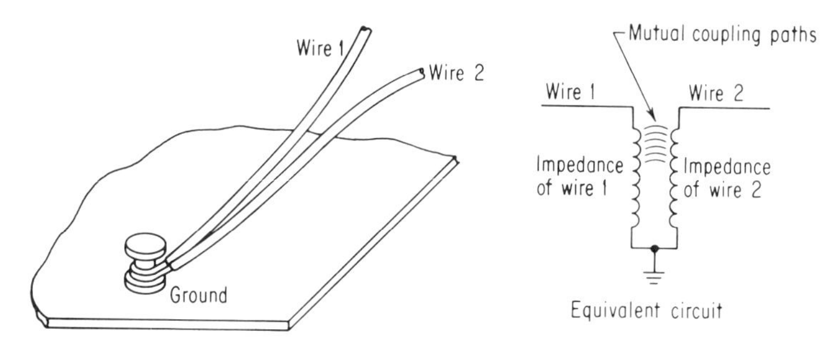

Fig. 13. Mutual coupling between two ground wires.

Two ground wires parallel to each other and attached to a ground terminal as in Fig. 13 are mutually coupled, as indicated, and a change in one will cause a change in the other. The ground leads should be as large as possible and as short as possible, and should run perpendicular to each other to reduce interference at high frequencies.

Where grounds run from black box to black box or cabinet to cabinet, special precautions are necessary. Figure 14*a* illustrates a typical method of connecting power and signal grounds. As the frequency increases, the inductance of the ground jumper can become appreciable, and if the power or signal circuit contains high-frequency interference currents, they may be conducted through the ground pin into the external wiring. In contrast, Fig. 14*b* shows the proper method of installing a ground to avoid conducting the interference through the connector.[3]

Long leads between cabinets provide good interference pickup possibilities. The ground potential effect can be canceled by isolating the circuits as in Fig. 15, using an isolation transformer. Although this method is very effective at low frequencies, the effectiveness decreases as the frequency increases because more coupling paths appear.

Cable Grounding. Whether cable shields should be grounded at only one end or at more than one point depends on the ratio of conductor length to wavelength (L/λ). Single point grounding is effective when $L/\lambda<0.15$, where λ represents the highest frequency expected. If $L/\lambda>0.15$, multipoint grounding at intervals of 0.15 L/λ should be used. Where large ground currents are present, multipoint grounding loses its effectiveness and should not be used.

Power Distribution. Simplified voltage distribution interconnection systems were shown in Fig. 6 to illustrate the effect the various methods of interconnection had on resistance alone. A more typical situation is shown in Fig. 16. A good technical discussion of how to design for all the elements of Fig. 16 is given in Gray's *Digital Computer Engineering*.[4] A complex problem exists that must be considered in design of high-speed circuits. The voltage at point *A*, the point where the circuit receives its power, changes for three reasons:*

* Harry H. Gray, "Digital Computer Engineering," copyright 1963. Reprinted by permission of Prentice-Hall Inc., Englewood Cliffs, N.J.

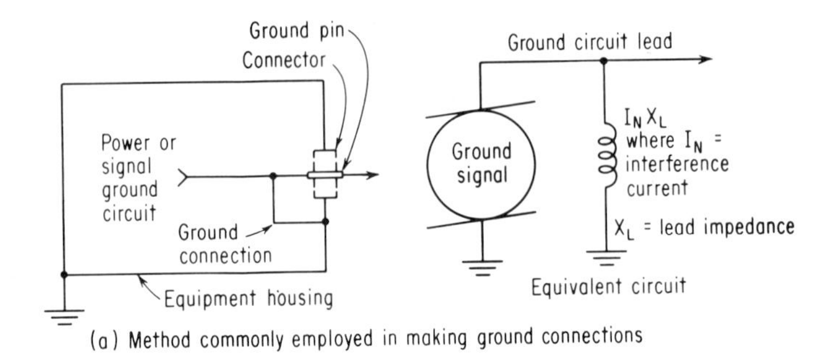

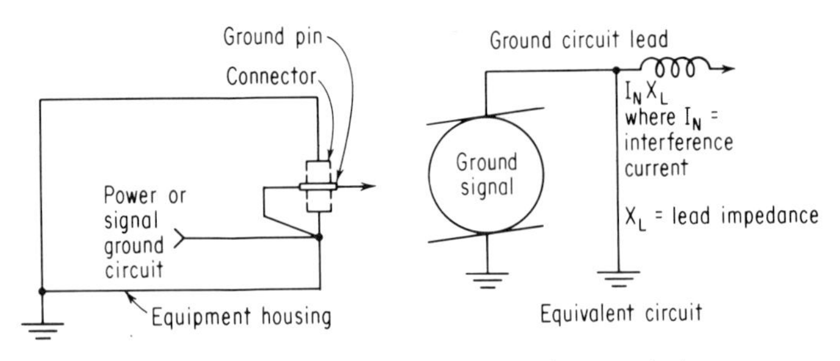

FIG. 14. Grounding between equipment housings. (*Filtron Co., Inc.*)

1. Static drops in the distribution system and power supply drift and regulation.
2. Transient changes because of the finite impedance of the distribution system.
3. Inductive crosstalk from signal leads and other dc leads into the dc distribution system.

Most of the equivalent circuit of Fig. 16 is present by virtue of the interconnection system itself, but capacitor C_1 at point B on the module is added to bypass spurious signal pickup. Since lines on the module occur after the bypass capacitor, they should be as short as possible and carefully located to avoid any more coupling than necessary. Some order-of-magnitude values for the equivalent circuit components of Fig. 16 are given in Table 11.

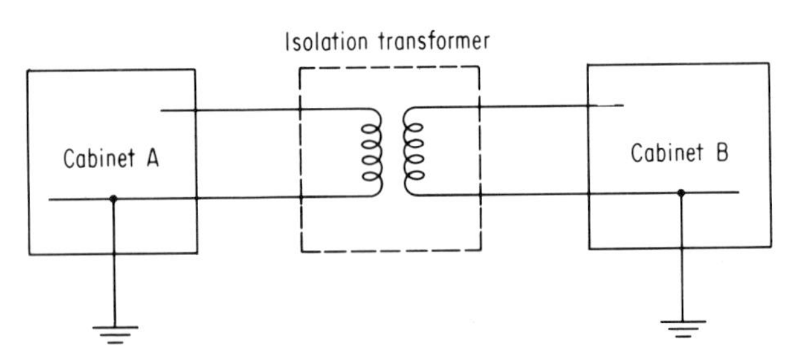

FIG. 15. Isolation transformer technique for minimizing ground potential. (*Filtron Co., Inc.*)

TABLE 11. Equivalent Circuit Values for a DC Power Distribution

Component	Description	Order of magnitude
L_1	Printed line on module	0.01 μh
L_2	Lead to module	0.1 μh
L_3	Dc bus (one of several)	0.001 μh
L_4	Power-supply leads	10 μh
l_1	Capacitor lead (C_1)	0.01 μh
l_2	Capacitor lead (C_2)	1 μh
C_1	Capacitor on circuit card	10 μh
C_2	High loss capacity	Farads
C_4	Capacitance of dc bus	1,000 pf
R_2, R_3, R_4, r_1, r_2	Lead resistances	$10^{-4} - 1$ ohm
R	Equivalent resistance of total load	60 milliohms

SOURCE: Harry H. Gray, "Digital Computer Engineering," copyright 1963. Reprinted by permission of Prentice-Hall, Inc., Englewood Cliffs, N.J.

Techniques Used for Grounding and Voltage Distribution. Unfortunately, no one type of mechanical hardware and joint combination is the best for every kind of system. Distance between points, frequency, power, signal proximity, and other criteria affect performance. Ideally, each voltage and each ground would consist of a single heavy, short, homogeneous conductor, but obviously this is impossible. Some of the methods used are described below.

Straps. Straps are used principally to tie two metallic structures together. A *bond* is the electrical union providing a low-impedance circuit between them. Bonds can be either direct or indirect. Direct bonds are permanent metal-to-metal joints like those provided by welding or brazing. Where direct bonds are impractical, indirect or mechanically held bonds, such as bolted or riveted joints, are used. Solder joints have appreciable contact resistance and should not be used for the bond between major structures. Solid straps are preferred to braid and should be flat to provide a large surface area. The length to width ratio should be less than 5:1. Figure 17 indicates the advantage of a thin, solid strap. Note that at a frequency of 10 Mc, the impedance is 10 times as high for the longer, thicker strap and over 100 times as high for the No. 12 AWG wire.

Wire. The most commonly used device for grounding or voltage distribution is wire. It may be the only usable means in many cases to get from a bus to a com-

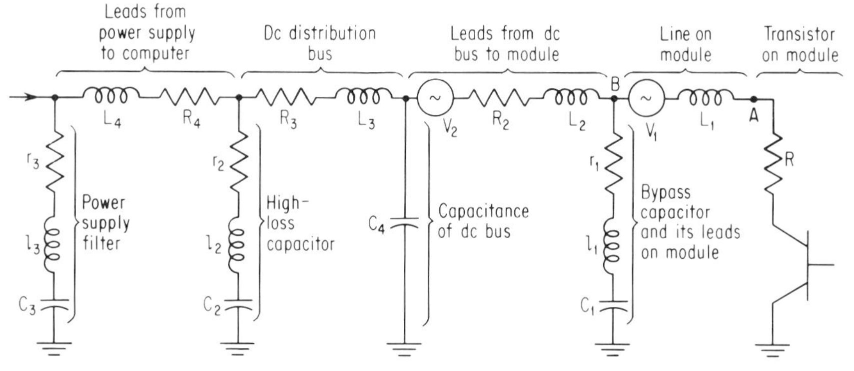

FIG. 16. Typical dc power distribution system. (*Harry H. Gray, "Digital Computer Engineering," copyright* 1963. *Reprinted by permission of Prentice-Hall, Inc., Englewood Cliffs, N.J.*)

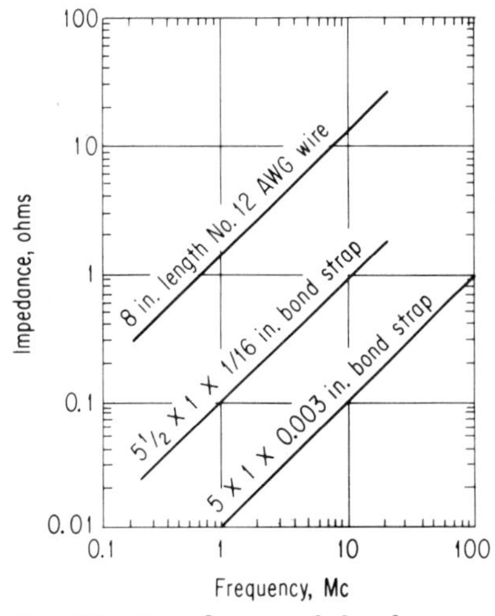

FIG. 17. Impedances of bond straps and No. 12 AWG wire. (*Filtron Co., Inc.*)

FIG. 18. Clothesline wire termination.

ponent, from a plane to a module, etc. Several techniques are used for terminating wire.

TERMINALS. The general category of terminals covers solder cup terminals, various types of turret terminals, and crimp, poke home, wire wrap, welded, and other types of mechanical joints, all of which are described fully in other chapters. All are limited in the number of wires they can accommodate. A single wire is most commonly used in a solder cup while a wire-wrap terminal post may accept four wires. The choice will depend not only on which gives the lowest-impedance joint, but on a combination of size, cost, and flexibility in satisfying overall system requirements.

CLOTHESLINES. Clotheslines like those shown in Fig. 18 are used as a point source radiating out to individual ground or voltage points to avoid series connections. The clothesline should be bowed to avoid straining the terminal supports, and each wire should be looped or arced to avoid straining the wire. While this method offers the obvious advantage of providing space for many wire terminations, there is a decided

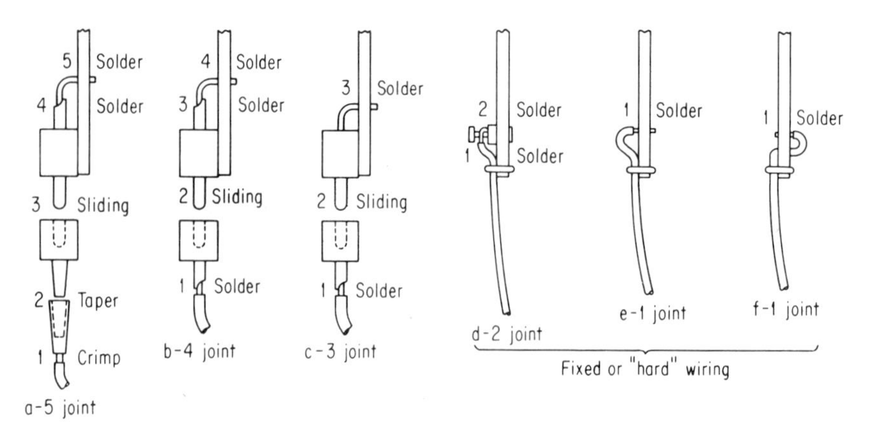

FIG. 19. Wire terminations to printed-circuit boards.

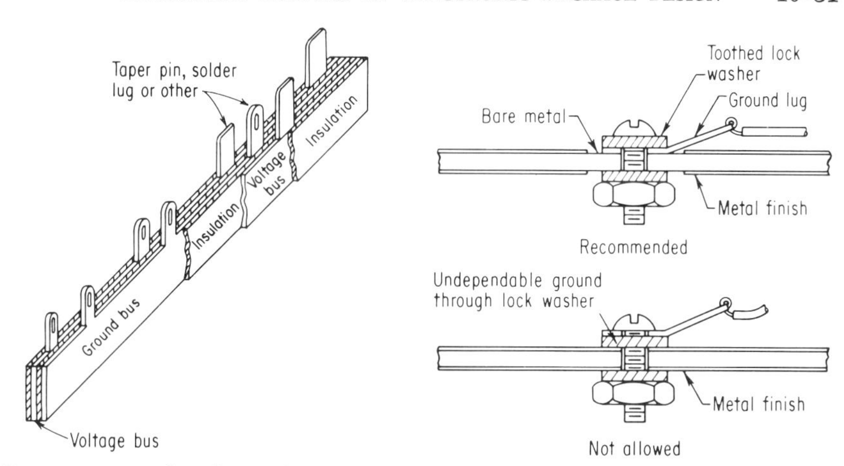

FIG. 20. Typical voltage distribution and ground bus.

FIG. 21. Grounding to a metal chassis.

disadvantage in soldering so many wires at one time or trying to unsolder one wire without affecting the others. Heat is dissipated rapidly, and a large soldering iron is required to soften the solder. Consequently, solder wicking and subsequent wire breakage may be a problem unless great care is exercised. Clotheslining is not permitted in some government specifications but has been successfully used for years on other military equipment.

PRINTED-CIRCUIT HOLES. Grounding and voltage distribution leads to printed circuits are commonly carried through connectors but, like the rest of the interconnections, are sometimes "hard-wired" to the board. Figure 19 shows several methods. Where reliability is more important than ease of replacement, wires can be soldered directly into plated-through or eyeleted holes to minimize the number of joints, but the leads must be supported to avoid strain on the wire or the printed-circuit hole. When connectors are used, guide pins should never be used for grounding since they are not designed to provide good electrical contact.

Buses. Buses, exemplified in Fig. 20, are designed to give minimum impedance and maximum number of interconnection points. The interconnection points may be solderable, tapered to accept a crimped taper tab, or designed for other types of termination. Designs that require different wire preparation on each end should be avoided. The characteristic impedance of a bus bar is the lowest of the commonly used voltage distribution systems (see comparison in Table 12), but system configuration does not always allow its use.

Chassis. Extreme care must be taken to make good electrical contact when chassis are used for ground planes. Anodizing (which is a good insulator), paint, or other coatings have to be removed to get a good metal-to-metal contact. A tooth-type washer will not consistently break through these coatings. The recommended method is shown in Fig. 21. Dissimilar metals will cause corrosion and a progressively degrading joint. Touching metals should be as close as possible in the electromotive-force series listed in Table 13. Since the higher metal in the electromotive-force series will be more affected by corrosion than the lower metal, choose hardware and other replaceable items such that, if corrosion occurs, it will be on the easily replaced part.

Printed-circuit Boards. Grounds and voltage distribution on printed-circuit boards have been discussed and are shown in Figs. 7, 9, and 10. Decoupling capacitors are commonly used on the voltage conductors because they take less space than the wide printed conductors which would be necessary to obtain a low characteristic impedance. Characteristics of these capacitors must be determined by the rise and fall time of the circuits rather than by system frequency. Transmitting a signal from a digital circuit with a rise time of 5 nsec is approximately equivalent to transmitting a

TABLE 12. Comparison of Three Power Distribution Techniques [5]

Type	Characteristic impedance *	Remarks
Two-wire line (d, h)	$Z_o \cong \frac{120}{\sqrt{\xi_r}} \log_e \frac{2h}{d}$	Has highest characteristic impedance of three types shown.
Single wire near a ground plane (d, h)	$Z_o \cong \frac{138}{\sqrt{\xi_r}} \log_{10} \frac{4h}{d}$ For $d << h$	Compromise between two-wire and bus bar. Can be fabricated with multilayer printed wiring.
Bus-bar (h, d)	$Z_o \cong \frac{377}{\sqrt{\xi_r}} \frac{d}{h}$ For $\frac{d}{h} < 0.1$	Lowest characteristic impedance of the three types shown; also the bulkiest.

* ξ_r = dielectric constant of the material surrounding the conductors. $\xi_r = 1$ for air.

100-MHz sine wave. Regardless of the time interval between circuit switches, the transmission line requirements do not change. Although decoupling capacitors are usually designed by a cut-and-try method, a technique for calculating their value is given in the article Cure Switching System Noise.[5]

Shielded Wire and Cables. Shielded wire can cause more coupling problems than common hookup wire if the shields are not grounded properly. Uninsulated shielded wire is especially difficult to handle since it may short to other shields anywhere along its length. Single-point and multipoint grounding have been discussed previously. The grounding is physically accomplished using several methods; the most common

TABLE 13. Electromotive-force Series of Commonly Used Metals [3]

Metal	*Electrode potential, volts*
Magnesium	+2.40
Aluminum	+1.70
Zinc	+0.762
Chromium	+0.557
Iron	+0.441
Cadmium	+0.401
Nickel	+0.231
Tin	+0.136
Lead	+0.122
Copper	−0.344
Silver	−0.798
Platinum	−0.863
Gold	−1.50

SOURCE: Filtron Co., Inc.

for single wires is shown in Fig. 22, for bundles of wires in Fig. 23. While physical contact may give good electrical contact in bundles initially, this method is not dependable over a long period of time and should not be used.

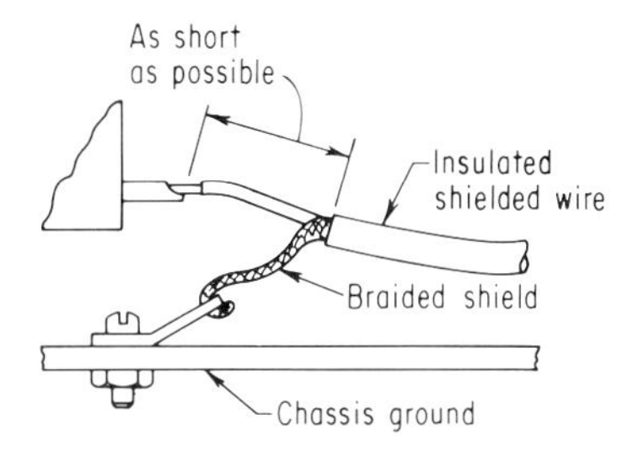

Fig. 22. Grounding of shielded wire.

SHIELDING

An excellent study of the whole problem of interference is given in the two-volume *Interference Reduction Guide.*[3] Control of interference is based on five principles:

1. Circuit design: Design for maximum desired signal and minimum spurious signal.
2. Component selection: Use, where possible, components which do not give off unwanted energy.
3. Placement: Locate sensitive components the maximum distance from spurious signals and route leads to prevent coupling, where possible.
4. Shielding: Use shields where the above methods are not effective.
5. Grounding. Establish a grounding procedure which will minimize interference.

In many cases, circuit design, component selection, and placement are not capable of solving interference problems, and shielding must be used.

Shielding may be used either to contain a source of interference or to protect equipment from an external source of interference. Physical separation or shielding, or a combination of the two, will attenuate an unwanted signal. Magnetic field intensity at any point is inversely proportional to the cube of the distance between that point and the field source, but adequate physical separation between components to attenuate the interfering signals to a harmless level is not always possible. Shields are divided into two general categories: magnetic shields in low-impedance, low-fre-

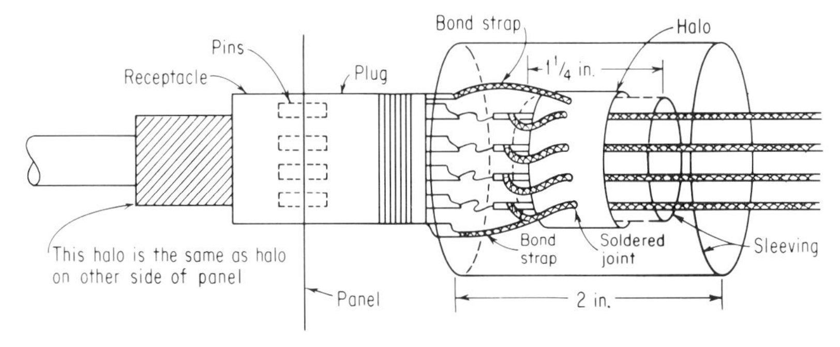

Fig. 23. Bonding ring or halo at connector for terminating shields in a harness. Bond strap may be connected as shown or with ¼-in. bond strap tied to structure or connector by means of eared washer. Halo is ¼ to ½ in. wide. (*Filtron Co., Inc.*)

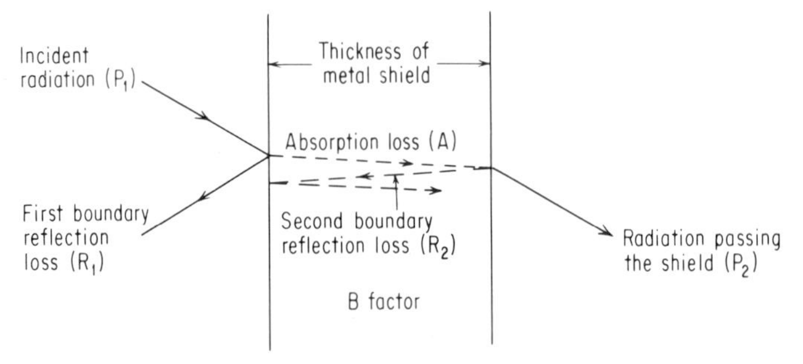

FIG. 24. Factors contributing to total shielding effectiveness. (*Filtron Co., Inc.*)

quency fields, and electric field shields in higher-impedance, higher-frequency fields.

Shielding Effectiveness. Shielding effectiveness, a measure of the total ability of a material to prevent propagation of electromagnetic energy, is represented by the wave in Fig. 24 and the formula:

$$SE = 10 \log (P_1/P_2) = R + A + B$$

where SE = shielding effectiveness, db
P_1 = incident power
P_2 = power passed through shield
$R = R_1 + R_2$ = reflected power, db, at the first and second boundaries
A = absorbed power loss, db, through the material
B = degradation factor, db, caused by multiple reflections in the material

The B term may be either plus or minus but may be neglected if A is greater than 10 db. Where A is less than 10 db, the shield is considered electrically thin and either a different or a thicker material should be considered. R may also be positive or negative but is always positive above 1 kc.

Since reflection losses for magnetic fields are small for most materials, magnetic shielding depends primarily on absorption losses. Electric fields, however, are easily reflected by metal shields, and reflected losses dominate. Since openings in the shield larger than half a wavelength decrease shielding effectiveness, great care must be taken in designing enclosures where high-frequency radiation is present. Table 14 shows the change in absorption loss from 60 cps to 10,000 Mc for copper, aluminum, and iron. The great range in absorption losses for different materials at one given frequency (150 kc) is shown in Table 15.

TABLE 14. Absorption Loss of Solid Copper, Aluminum, and Iron Shields at 60 Cps to 10,000 Mc [3]

Frequency	Copper		Aluminum		Iron		Absorption loss, db/mil		
	G	μ	G	μ	G	μ*	Copper	Aluminum	Iron
60 cps	1	1	0.61	1	0.17	1,000	0.03	0.02	0.33
1,000 cps	1	1	0.61	1	0.17	1,000	0.11	0.08	1.37
10 kc	1	1	0.61	1	0.17	1,000	0.33	0.26	4.35
150 kc	1	1	0.61	1	0.17	1,000	1.29	1.0	16.9
1 Mc	1	1	0.61	1	0.17	700	3.34	2.6	36.3
15 Mc	1	1	0.61	1	0.17	400	12.9	10.0	106.0
100 Mc	1	1	0.61	1	0.17	100	33.4	26.0	137.0
1,500 Mc	1	1	0.61	1	0.17	10	129.0	100.0	168.0
10,000 Mc	1	1	0.61	1	0.17	1	334.0	260.0	137.0

* Other values of μ for iron are: 3 Mc, 600; 10 Mc, 500; and 1,000 Mc, 50.
SOURCE: Filtron Co., Inc.

TABLE 15. Absorption Loss of Metals at 150 Kc [3]

Metal	G, relative conductivity	μ, relative permeability, at 150 kc	Absorption loss at 150 kc, db/mil
Silver	1.05	1	1.32
Copper, annealed	1.00	1	1.29
Copper, hard-drawn	0.97	1	1.26
Gold	0.70	1	1.08
Aluminum	0.61	1	1.01
Magnesium	0.38	1	0.79
Zinc	0.29	1	0.70
Brass	0.26	1	0.66
Cadmium	0.23	1	0.62
Nickel	0.20	1	0.58
Phosphor-bronze	0.18	1	0.55
Iron	0.17	1,000	16.9
Tin	0.15	1	0.50
Steel, SAW 1045	0.10	1,000	12.9
Beryllium	0.10	1	0.41
Lead	0.08	1	0.36
Hypernick	0.06	80,000	88.5*
Monel	0.04	1	0.26
Mumetal	0.03	80,000	63.2*
Permalloy	0.03	80,000	63.2*
Stainless steel	0.02	1,000	5.7

* Obtainable only if the incident field does not saturate the metal.
SOURCE: Filtron Co., Inc.

Magnetic Field Shielding. Typical sources of magnetic interference are:

1. Permanent magnets or electromagnets
2. High-current cables at dc or power frequencies
3. Motors and generators
4. Transformers, solenoids, and other coils

A magnetic shield is a ferromagnetic metal enclosure that surrounds the device as completely as possible. The attenuation g provided by the shield is the ratio of the field intensity outside the shield to the field intensity inside the shield, or $g = \frac{H \text{ out}}{H \text{ in}}$. Shielding effectiveness SE is related to attenuation and describes the efficiency of the shield in decibels as follows:

$$SE = 20 \log_{10} g = f(\mu)$$

This convenient form indicates the decibel drop from the interfering field on one side of the shield to the field on the other side of the shield.

Since shielding effectivenesss is directly related to configuration and permeability μ, the higher a material's permeability is, the better is its attenuation. However, permeability increases with field density only up to a point of saturation. Figure 25 shows how the permeability of a special nickel-iron alloy increases with flux density until it saturates; then as the flux density increases even further, the permeability decreases rapidly. Permeability also tends to decrease with an increase in frequency. Unfortunately, as permeabilities increase in alloys, their saturation levels decrease, so that the highest-permeability alloys have the lowest saturation levels. When a single material cannot be selected with high enough permeability to attenuate properly, considering its saturation characteristics, two or more layers of different materials may be necessary. The material closest to the high field should be a low-saturation, less efficient material which will reduce the field to a level which can be handled by the second-layer high-permeability alloy. When a device is to be protected from external

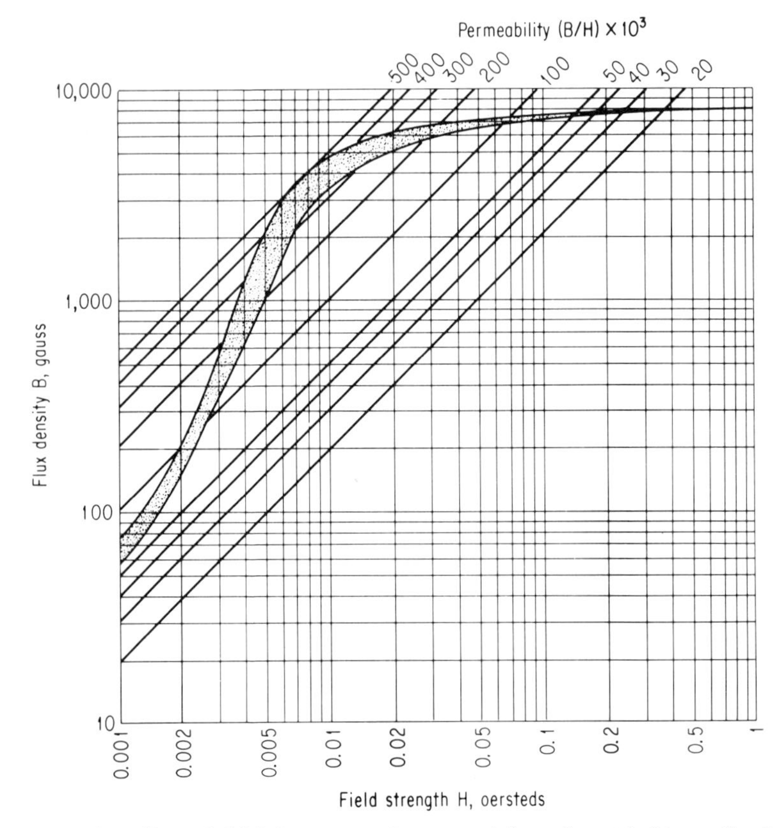

FIG. 25. Effect of field density on the permeability of a nickel-iron alloy.[6]

fields, the lower-permeability material goes on the outside and the higher-permeability material on the inside. Conversely, for shields used to prevent equipment from propagating interference, the lower-permeability material is placed on the inside. Normally, the layers are separated by an air gap, but laminations of magnetic and highly conductive layers may be used to extend magnetic and electrical shielding to higher frequencies.

Electric Field Shielding. The word *shielding* used in conjunction with electronic gear normally refers to rf shielding where higher frequencies (10 kc to 100,000 Mc) require only conductive material enclosures, but openings in the shield are critical. When openings are small compared with the wavelength of the energy passing through, the losses are similar to losses obtained with a waveguide below cutoff and are very great. Therefore, covering holes, seams, and other openings with mesh, knit wire, or conductive gaskets of various kinds can do an effective job of preventing an rf "leak."

Shielding Packaging Methods. Various packaging methods are used to enclose or exclude rf interference. All are based on getting a conductive enclosure between the unwanted signal and the circuit which would have been affected. The following paragraphs describe some of these methods, primarily with respect to rf shielding, but some magnetics information is included where a table or curve includes useful information on materials.

Cans. Insertion loss is the difference (in decibels) between readings taken between radiating and receiving antennas with nothing between them and with a given

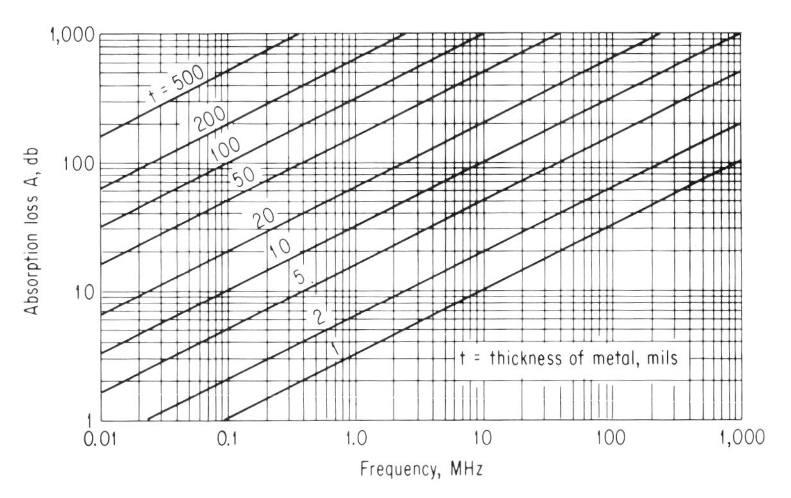

FIG. 26. Absorption loss for copper.[7]

material inserted between them. Insertion loss is equal to shielding effectiveness when the shield is thin, and for practical purposes the two terms are synonymous. Measured values of shielding effectiveness vary from the theoretical because of configuration discontinuities, particularly if openings at high frequencies approach half a

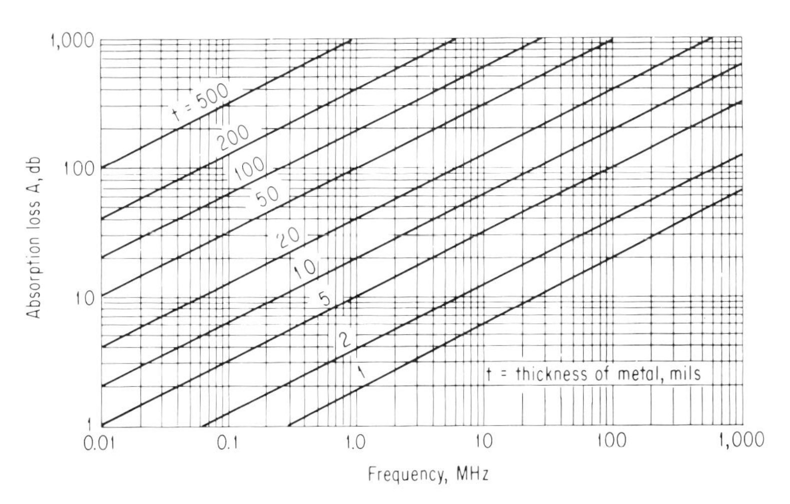

FIG. 27. Absorption loss for aluminum.[7]

wavelength, but theoretical absorption and reflection values for materials give useful approximations. Curves showing absorption and reflection losses for copper, aluminum, and Mumetal* are given in Figs. 26 to 35. Note that the far-field reflection loss lines

*Trademark, Allegheny Ludlum Steel Corporation.

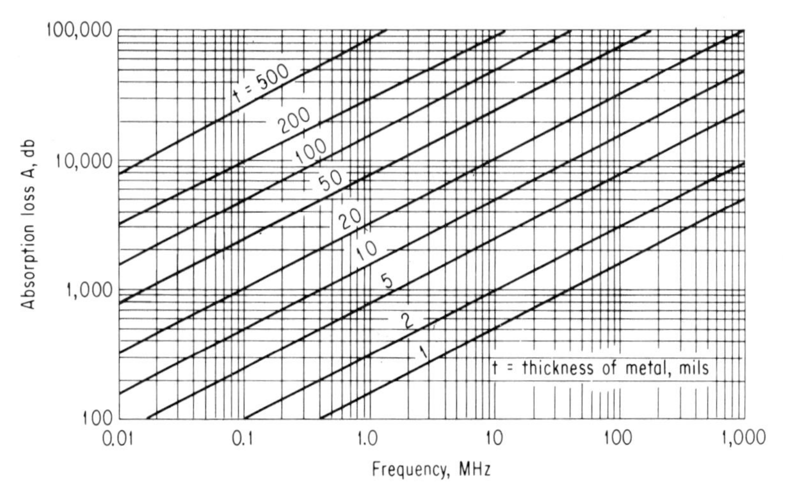

FIG. 28. Absorption loss for Mumetal.[7]

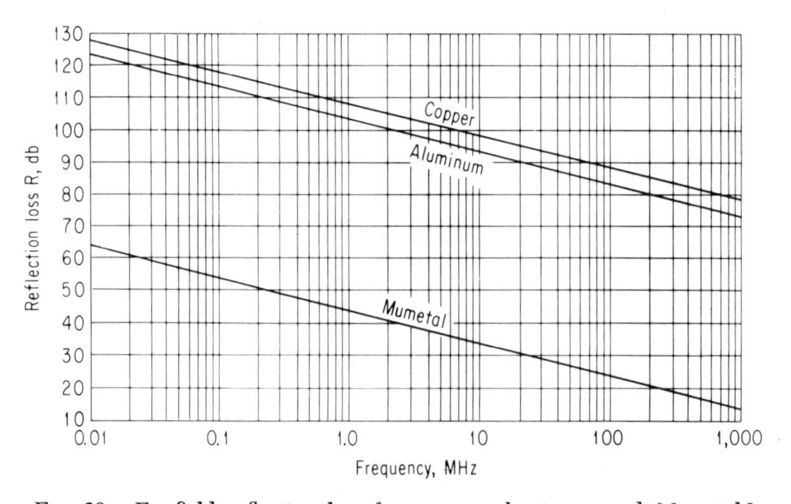

FIG. 29. Far-field reflection loss for copper, aluminum, and Mumetal.[7]

of Fig. 29 are repeated in Figs. 30 to 35, with near-field values added. Figure 36 shows the degradation factor B, due to low absorption loss, indicating how its value falls off and can be disregarded when the absorption loss is greater than 10 db. Using these curves, the shielding effectiveness for copper, aluminum, and Mumetal in varying thicknesses and frequencies can easily be arrived at, using the equation

$$SE = A + R + B$$

Figure 37 illustrates graphically the relationship between shielding efficiency, frequency, and the three types of fields—electric, magnetic, and plane wave—for a particular case of a source 12 in. from a copper or iron shield. Note that reflection is not dependent upon thickness, but the absorption loss curves are shown in decibels per mil of thickness. The B term is neglected in Fig. 37 but should be included for

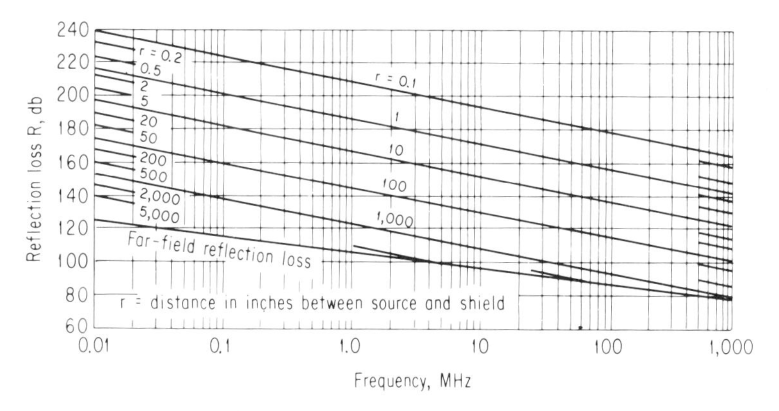

Fig. 30. Electric near-field reflection loss for copper.[7]

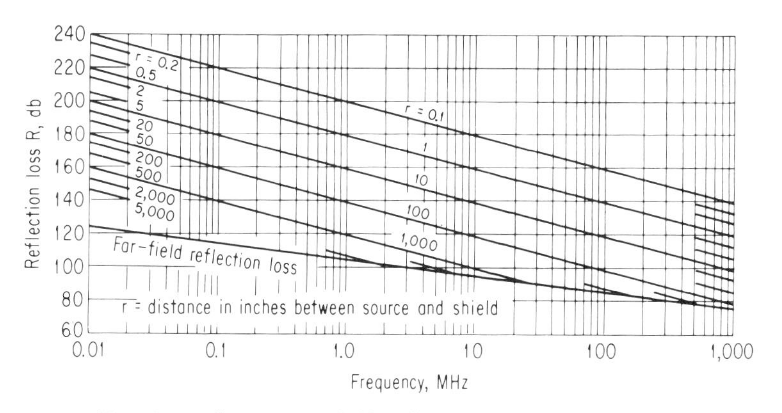

Fig. 31. Electric near-field reflection loss for aluminum.[7]

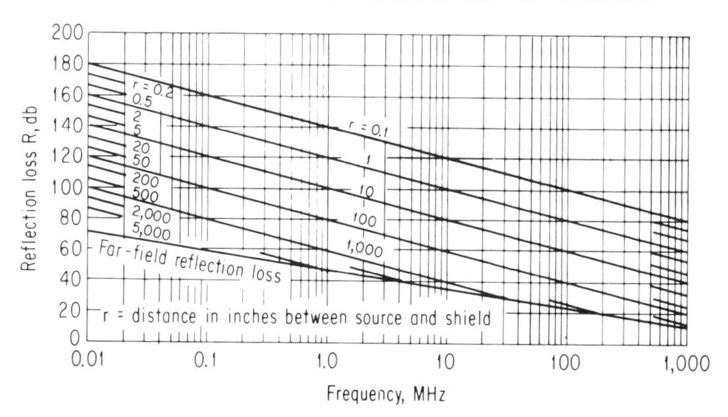

Fig. 32. Electric near-field reflection loss for Mumetal.[7]

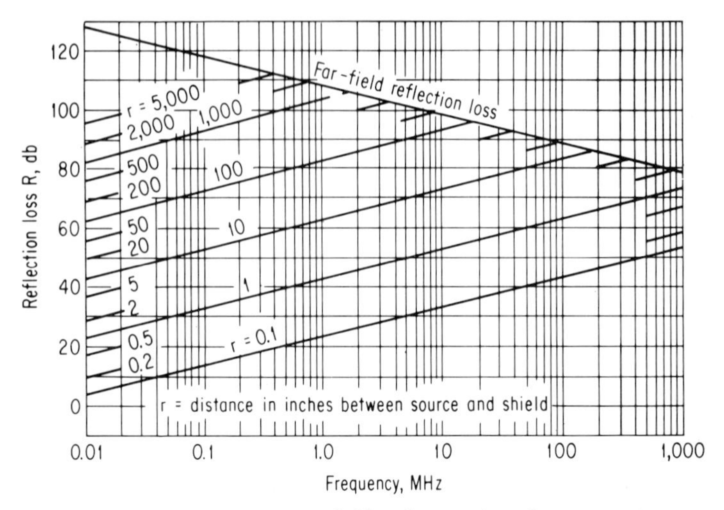

FIG. 33. Magnetic near-field reflection loss for copper.[7]

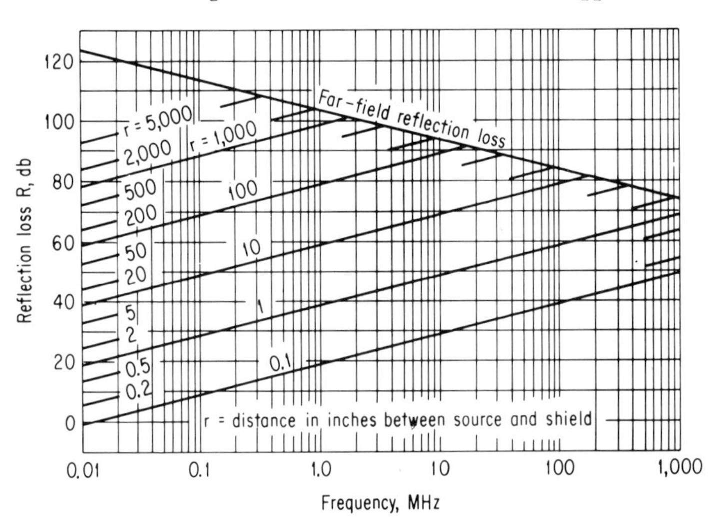

FIG. 34. Magnetic near-field reflection loss for aluminum.[7]

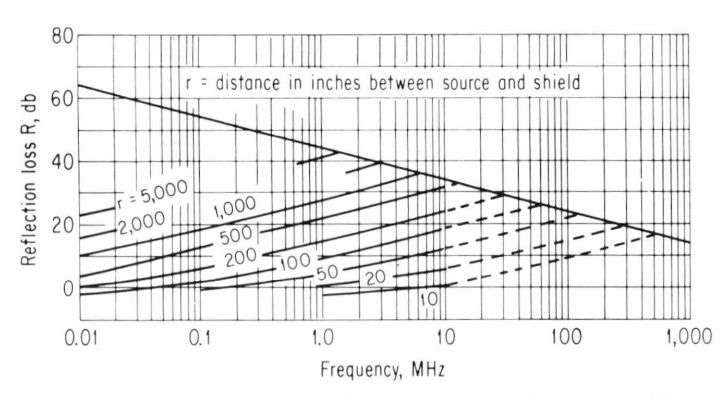

FIG. 35. Magnetic near-field reflection loss for Mumetal.[7]

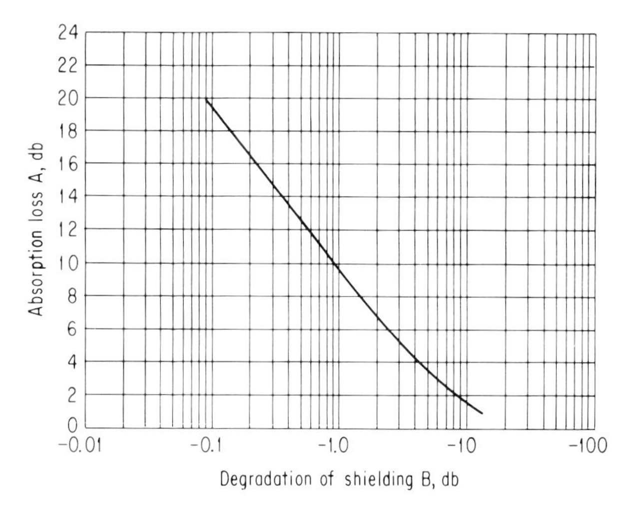

FIG. 36. Degradation in shielding effectiveness due to low absorption loss.[7]

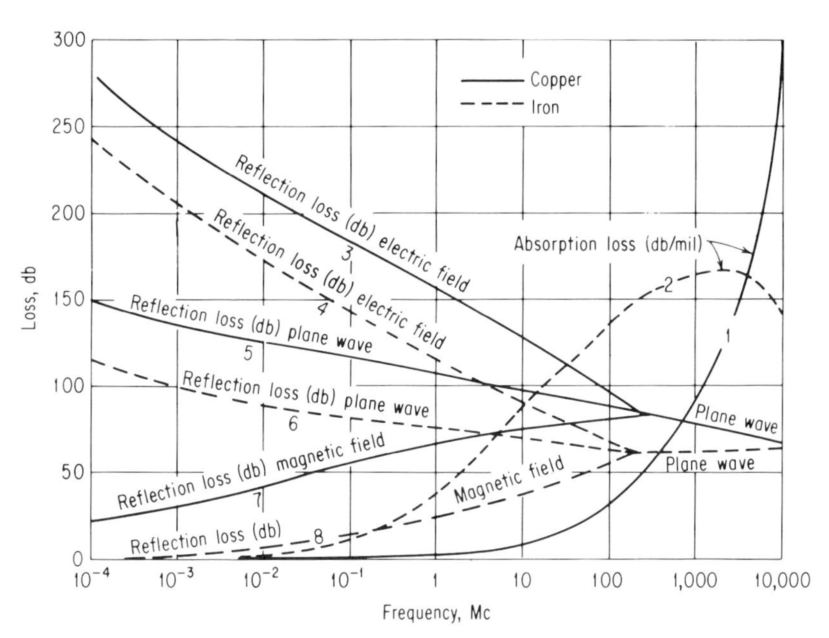

FIG. 37. Shielding effectiveness in electric, magnetic, and plane wave fields of solid copper and iron shields (for signal source 12 in. from shield at 100 cps to 10,000 Mc). (*Filtron Co., Inc.*)

electrically thin shields. Normally, any structurally sound material will be thick enough to shield electric fields effectively at any frequency.

Cans, boxes, or enclosures take many forms. The ideal continuously welded enclosure is almost never used because the equipment it encloses then becomes inaccessible; leads have to enter and exit, cooling air flow is necessary, etc. Some of these considerations and their effect on shielding effectiveness are discussed below.

Joints. Since openings larger than half a wavelength allow interference leaks, shield material should be joined either continuously or at close intervals consistent with the frequency. Sufficient spot welds, screws, or other fasteners should be used to obtain, as nearly as possible, continuous contact along the mating surfaces, but for minimum interference leakage, all mating edges should be bead-welded or continuously soldered.

Screens. Where large openings are necessary for air flow, various forms of screens are used to break the large opening into a series of small openings that act as waveguides below cutoff. To be most effective, the intersections between openings must be fused. Three commonly used devices are honeycomb, perforated metal sheet, and wire mesh screens. The smaller the openings and the smaller the percentage opening, the better the shielding effectiveness. Honeycomb is very effective and offers the additional advantage of low resistance to air flow. Figure 38 shows the attenuation ob-

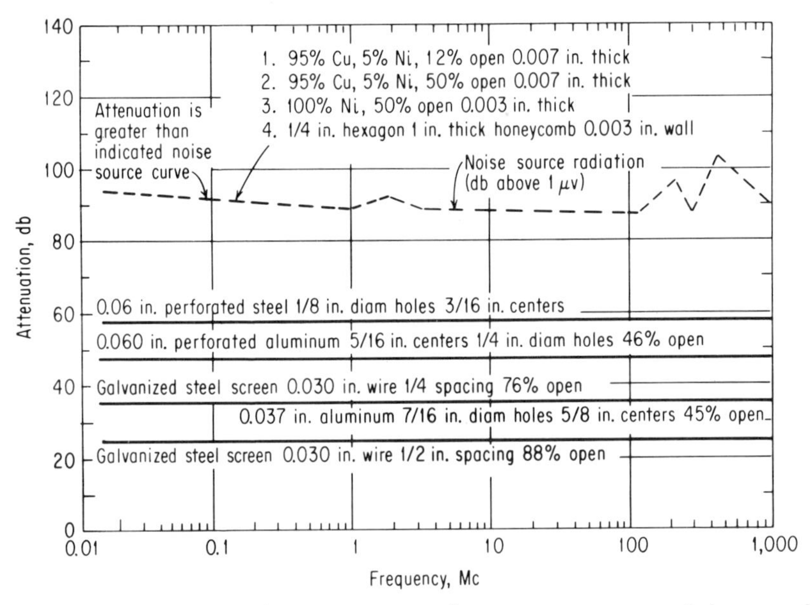

FIG. 38. Attenuation vs. frequency curves for various screens and honeycomb. (*Filtron Co., Inc.*)

tainable for various screens and honeycomb, and Fig. 39 indicates the effect of tightening the mesh on copper screening. Double layers of screens increase the shielding effectiveness, or more complex designs are available such as that shown in Fig. 40. The curve of Fig. 41 shows the change in shielding effectiveness from an open 15-in.-square hole to the same hole with the screen of Fig. 40 mounted over the hole.

Gaskets. To allow for servicing, most shielded equipment has a bolted-on cover of some kind. Spacing between fasteners must be much closer as the material gets thinner because the surfaces buckle with respect to each other. For maximum interference protection, rf gasketing is used to ensure short-interval metallic contact. Figure 42 illustrates the principle of rf gasketing. Figure 42*a* shows a section of two flanges bolted together with an opening of thickness *h* between the two bolts caused by buckling in the flanges. One of several types of conductive gaskets may be used to close the opening, but it must be thick enough and soft enough to fill in all irregulari-

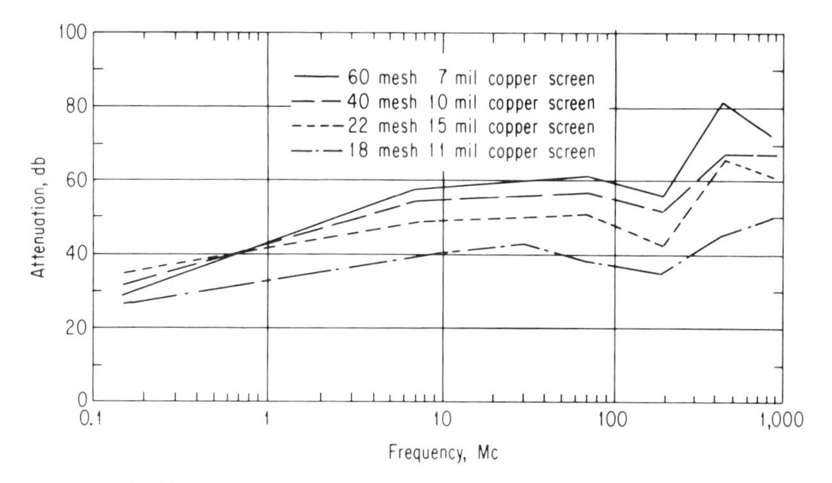

FIG. 39. Shielding effectiveness of various copper screens. (*Filtron Co., Inc.*)

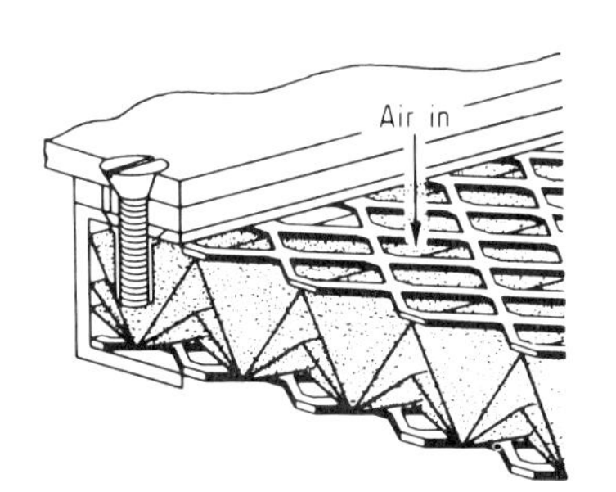

FIG. 40. A complex shielding screen.[8]

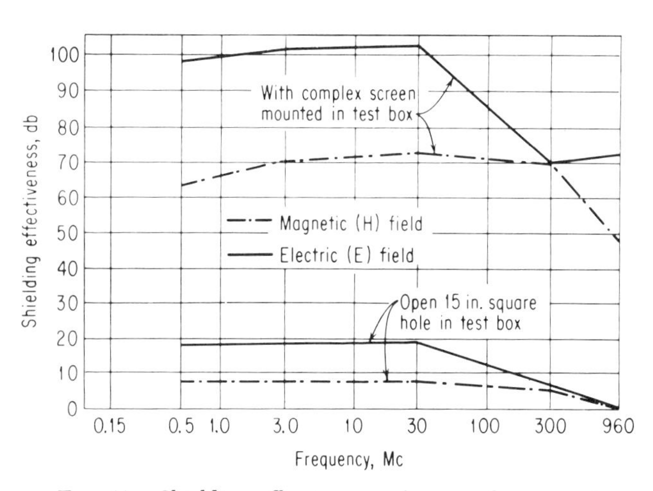

FIG. 41. Shielding effectiveness of a complex screen.[8]

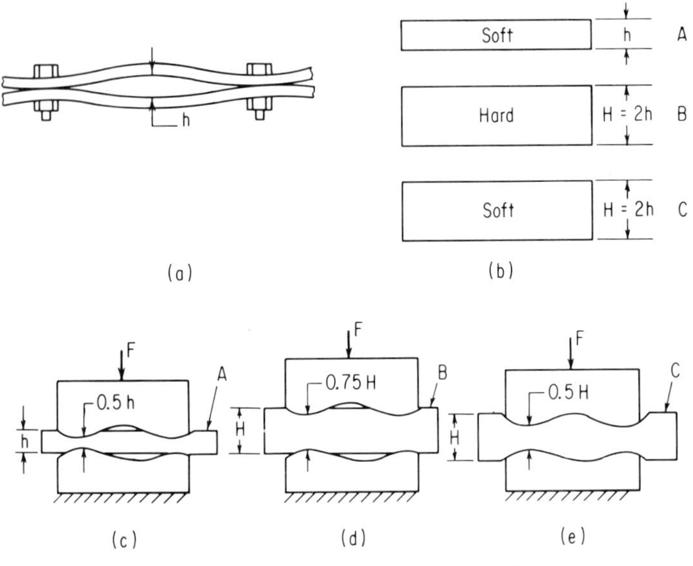

FIG. 42. Rf gasketing principles.[9]

ties. The three gaskets of Fig. 42*b* are shown in Fig. 42*c*, *d*, and *e* with the same force applied. Gasket *A*, while soft, cannot compress enough to fill the slot, which is its own height. Gasket *B* is thicker but twice as hard; therefore, it will compress only 0.5*h*. Gasket *C*, on the other hand, is both soft and twice the height *h*, and it will fill the void. Where covers are to be removed and replaced, the gasket must be capable of return because the irregularities may be displaced with respect to a "set" in the gasket. The pressure should be high enough to make adequate contact even when nonconducting corrosion is present. A typical insertion loss versus pressure curve is shown in Fig. 43.

Various gasket materials and configurations must be used to achieve best results for a variety of uses. Table 16 lists some of the advantages and disadvantages of the more common types. Most often, any one of a number of materials will electrically satisfy the requirements, but mechanical suitability is a problem. Table 17 gives a guide to the use of various rf gaskets.

Tape. Foils and pressure-sensitive conducting tapes may be used to build up the metal coating necessary for rf shielding. One particularly attractive use is for covering cabinet seams where the pressure-sensitive tape can be removed and replaced for access. One manufacturer claims a reduction of up to 45 db in rf leakage.

Plated Modules. Embedded electronic modules have been successfully plated to achieve the desired shielding. Masking of output pins and clearance around them is required, but the shield is usually plated directly onto one or more ground pins. These ground pins are subsequently grounded by wire or printed circuit to other module shields or the system shield ground.

Metal Sprays. Metal spray is another technique used to apply a conductive coat to the surface of an embedded module or other device.

Conductive Paints. Conductive paints or conductive epoxies are sometimes used where shielding is not so critical as to preclude their use. Volume resistivities are in the 0.001- to 0.003-ohm-cm range as compared to 1.7×10^{-6} ohm-cm for copper. One great disadvantage is that most of the best conductive paints have to be baked at from 150°C to several hundred degrees centigrade to get the best electrical characteristics. Most electronic gear cannot stand these temperatures.

Egg Crates. Egg crating is a method of packaging designed to separate circuits

TABLE 16. **Characteristics of Conductive Gasketing Materials** [3]

Material	Chief advantages	Chief limitations
Compressed knitted wire	Most resilient all-metal gasket (low flange pressure required). Most points of contact available in variety of thicknesses and resiliencies.	Not available in sheet form. (Certain intricate shapes difficult to make.) Must be 0.040 in. or thicker.
AEEL gasket or equivalent (expanded metal)	Best breakthrough on corrosion films.	Not truly resilient. Not generally reusable.
Armour Research gasket or equivalent (oriented wire in rubber)	Combines fluid and conductive seal.	Requires ¼-in. thickness and ½-in. width for optimum shielding.
Aluminum screen impregnated with neoprene	Combines fluid and conductive seal. Thinnest gasket. Can be cut to intricate shapes.	Very low resiliency (high flange pressure required).
Soft metals	Cheapest in small sizes.	Cold flows, low resiliency.
Metal over rubber	Takes advantage of the resiliency of rubber.	Foil cracks or shifts position. Generally low insertion loss yielding poor rf properties.
Conductive rubber	Combines fluid and conductive seal.	Practically no insertion loss giving very poor rf properties.
Contact fingers . . .	Best suited for sliding contact.	Easily damaged. Few points of contact.

SOURCE: Filtron Co., Inc.

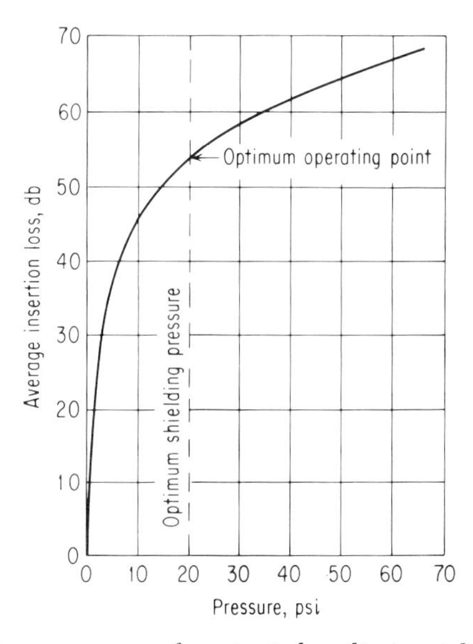

FIG. 43. Insertion loss vs. pressure for a typical resilient metal gasket. (*Filtron Co., Inc.*)

TABLE 17. Guide to Choice of Type of RF Gasket Based on Mechanical Suitability [3] [a]

	Mesh strips				Extrusion gasket	Combination strip	Formed rf gaskets	AEEL gaskets or equivalent	Combination gaskets	Woven aluminum and neoprene	Mesh over rubber	Conductive rubber[j]	Fingers
	○	□	○—	○—○									
Attachment method:													
Held in a slot or groove by tight sidewall fit	1	1	3	3	3	2	1	3	2	3	1	1	3
Nonconductive spot bonding[b]	2	1	2	2	3	2	1	3	2	2	1	1	3
Nonconductive bonding away from rf gasket portion[c]	3	3	1	1	3	1	3	3	1	3	3	3	3
Bond rf gasket portion with conductive adhesive[d]	2	1	2	2	3	2	1	3	2	2	1	1	3
Screw, spotweld, or rivet	3	3	1[e]	1[e]	1	3	3	2	3	3	3	3	1
Solder	3	2	2	2	3	3	2	2	3	3	3	3	1
Position by bolts through bolt holes	3	3	2	2	1	1	2	1	1	1	3	1	3
Pressure-sensitive adhesive backing	3	3	3	3	3	1	3	3	1	3	3	3	3
Other gasketing functions:													
Cooling air tightness	3	3	3	3	1	1	3	3	1	1	1	1	3
Raintightness	3	3	3	3	1[f]	1	3	3	1	1	1[f]	1	3
Pressuretight	3	3	3	3	3	1	3	3	1	1	3	1	3
Pressure available:													
0–5 psi[g]	1	1	1	1	1	1	1	2	1	3	1	1	1
5–50 psi	½	1	½	½	1	1	1	1	1	1	½	1	2
Over 50 psi	2	2	2	2	1	2	2	1	2	1	2	1	3
Total joint unevenness:[h]													
Less than 0.002	1	1	1	1	1	1	1	1	1	1	1	1	1
0.002–0.030	1	1	1	1	1	1	1	1	1	3	1	2	1
0.030–0.060	1	1	1	1	1	1	1	⅔	1	3	1	⅔	1
Over 0.060	1	½	1	1	2	2	½	3	2	3	1	3	1

Space available, width:													
Less than 0.060[i]	1	1	3	3	3	3	2	3	3	1	3	1	3
0.060–0.500	1	1	1	2	2/3	2	1	2	2	1	1	1	2
0.500–1.50	1	1	1	1	1	1	1	1	1	1	1	1	1
Space available, thickness:													
Less than 0.030	3	3	3	3	3	3	3	2	3	1	3	2	2
0.030–0.060	1	1	2	2	3	2	1	1	2	3	2	1	1
0.060–0.090	1	1	1	1	2	2	1	1	2	3	2	1	1
Over 0.090	1	1	1	1	1	1	1	3	1	3	1	1	1
Type of joint:													
Compression only	1	1	1	1	1	1	1	1	1	1	1	1	1
Combined compression and sliding	2	2	2	2	2	2	2	3	2	3	2	2	1
Sliding only	2	2	2	2	2	3	2	3	3	3	2	3	1

[a] 1, well suited; 2, can be considered; 3, not suited.
[b] Nonconductive spot bonding: a nonconductive adhesive can be used directly under an rf gasket if it is used only in 1/8- to 1/4-in. diameter spots, 1 to 2 in. apart.
[c] A nonconductive adhesive can always be used continuously if it is used under the attachment or rubber portion of combination strip and gaskets, but not under the rf gasket itself.
[d] A conductive adhesive can be applied continuously under an rf gasket.
[e] With backing strip over attachment fins.
[f] If mesh-over-rubber version of extrusion gasket is used.
[g] Evaluation is only for mechanical suitability. Pressure may be high enough to give sufficient insertion loss.
[h] Evaluation based on space being available to use thick enough gasket.
[i] Including space for attachment method integral to material considered.
[j] Evaluation is not based on electrical suitability of conductive rubber which is generally poorer than other materials listed.
SOURCE: Filtron Co., Inc.

electrically one from the other, as in the various stages of an if amplifier. This method is described in detail in Chap. 9, but is essentially the use of thin metal sheets to compartmentalize circuit groupings. A common form of construction utilizes etched metal patterns with tabs and slots which are assembled and furnace- or torch-brazed in place to form the compartments. Tabs are also used to fasten on removable lids to complete the enclosure. The location and number of tabs depend upon the frequencies being handled and the acceptable leakage.

Ground Planes. Printed-circuit ground planes are sometimes used for partial shielding where a complete enclosure is not required. Figure 44 illustrates how the ground planes of two printed-circuit boards and their metallic guide rails form a loose enclosure. Multiple ground planes within a single printed-circuit board are used to shield signals from power or signals from other signals, as illustrated in Fig. 10. Neither of these methods accomplishes true shielding, and where operation of a printed-circuit board is critical, an enclosure can is usually required.

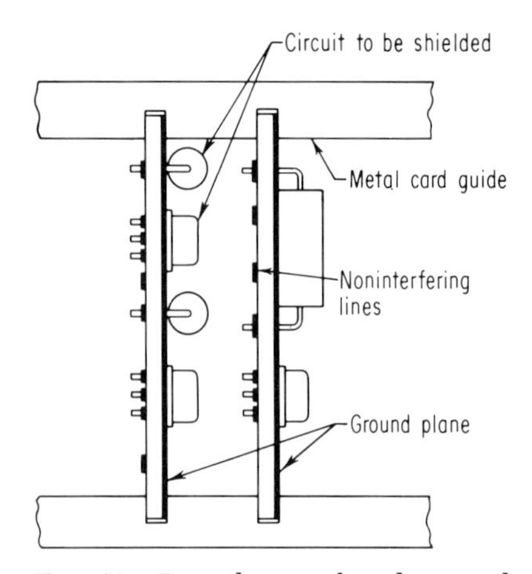

FIG. 44. Printed-circuit-board ground planes used to form an enclosure.

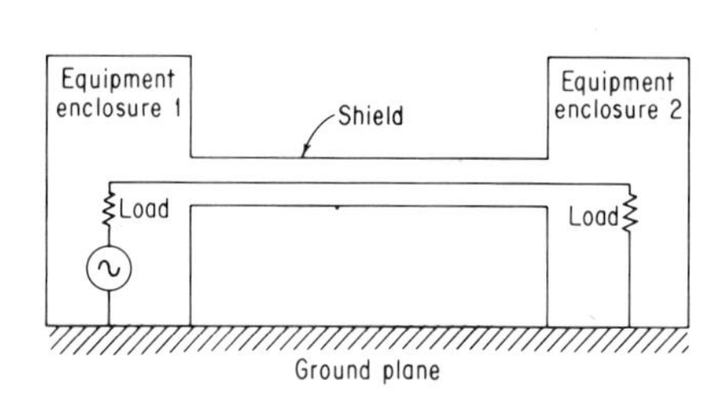

FIG. 45. Continuous equipment enclosure shield. (*Filtron Co., Inc.*)

Shielded Wire and Coax Cable. Shielding of wire between enclosures should be viewed as an extension of the enclosures, as shown in Fig. 45. Coax cable shields should be terminated so that no current loops exist inside the equipment. Figure 46*a*

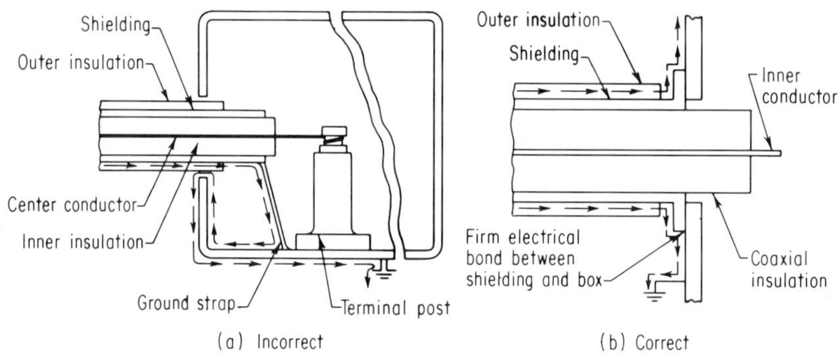

FIG. 46. Incorrect and correct methods of introducing shielded cable. (*Filtron Co., Inc.*)

shows a commonly used but incorrect termination, and Fig. 46*b* shows the correct method.

Shields of multiconductor cables are terminated at the plug leading to an enclosure or a subassembly within an enclosure. Figure 47 shows a typical system using the halo shield grounding technique detailed in Fig. 23.

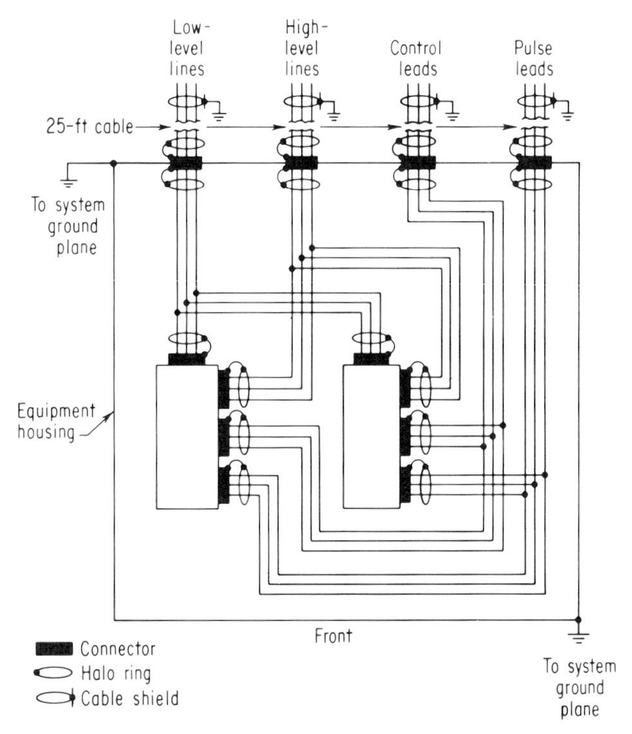

FIG. 47. Typical shielded cabling methods. (*a*) Modules bonded to cabinet. (*b*) All lines individually shielded and insulated. (*c*) Lines cross at right angles to each other to effect maximum interference reduction. (*Filtron Co., Inc.*)

Filtering. Radiated interference has been the principal interest so far, but interference can be conducted. Noise on a power line is the most commonly encountered conducted interference problem, and filtering is usually required. Like cable termination, mounting of the filter is important for maximum efficiency. The two methods shown in Fig. 48 illustrate how, in one case, the filter permits a bypass of radiated interference and, in the other, both direct and radiated interference are effectively blocked.

Shaft Feedthrough. Where switches, control knobs, and other devices must be accessible on the outside of an enclosure, the waveguide approach may be considered for shaft holes. The equation for a circular waveguide attenuator is

$$A = 31.95 \frac{D}{W} \sqrt{1 - \left(\frac{Wf}{6{,}920}\right)^2} \quad \text{decibels}$$

where D = depth of the waveguide, in.
W = inside diameter of the waveguide, in.
f = frequency, Mc

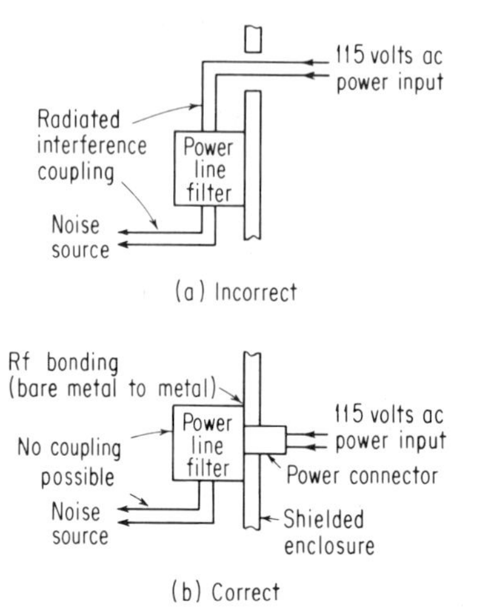

FIG. 48. Incorrect and correct rfi filter mounting.[10]

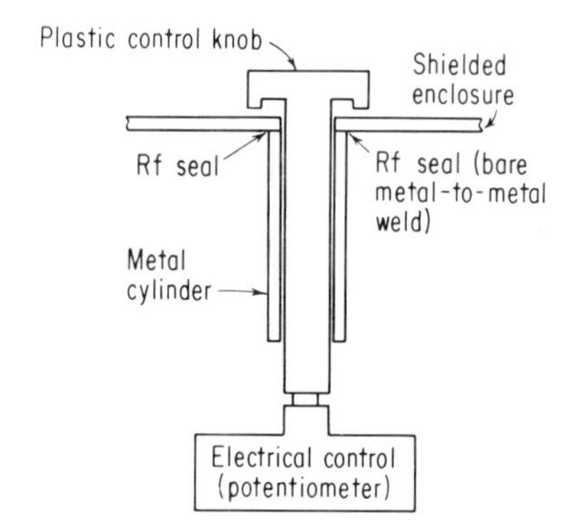

FIG. 49. Metal cylinder acting as a waveguide below cutoff blocks radiated interference.[10]

Figure 49 illustrates how a metal cylinder may be used with a nonconductive shaft as a waveguide below cutoff. Metal shafts may be used with contacts to the enclosure for even more effective interference control.

USEFUL ELECTROMECHANICAL DATA

The rest of this chapter contains miscellaneous electromechanical data the packaging engineer should find useful in designing electronic equipment.

Properties of Materials. Properties of materials are the subject of many books and the packaging engineer should consult them when necessary. However, some of the more frequently needed data are included here.

Metals. Resistivity, temperature coefficient of resistivity, and other physical properties of metals most commonly used in electronic packaging are given in Table 18.

Dielectric Values. Chapter 7 contains extensive data on various insulation systems and their physical and electrical characteristics. Resistivity and dielectric constant are two of the most useful electrical characteristics. Dielectric constant is defined as the ratio of the capacitance of a device using the dielectric material to the capacitance of the same device in a vacuum. Table 19 illustrates typical data for several kinds of epoxies which are used in packaging electronic subassemblies. Note that the dielectric constant varies with frequency.

Dissipation factor is the same as *power factor* for most insulation systems and is defined as a ratio of power loss to input power for an insulation system.

Dielectric strength is a measure of the maximum stress an insulation can take before breakdown. This value becomes very important when high voltages are present.

Insulation resistance, for a given configuration, or volume resistivity, for bulk material, is a required basic value, but one which may vary with humidity, temperature, or other environmental conditions. Volume resistivity values are given in Table 19 for epoxies, which are relatively stable, and insulation resistance is given for pressure-sensitive tapes in Table 20, but no such figures are given for the laminated and reinforced plastics of Table 28 because the values change with humidity.

Pressure-sensitive Tape. Typical properties of pressure-sensitive tape are given in Table 20.

Magnetic Materials. Properties of permanent magnet materials are given in Tables 21 to 24. The magnet steels of Table 21 are weak compared to the others, and since

TABLE 18. Properties of Metals at 20°C [11]

Metal	Specific gravity	Specific heat, cal/(g)(°C)	Melting point, °C	Resistivity, microhm-cm	Resistivity, ohms (mil-ft)	Temperature coefficient of resistivity per °C	Thermal conductivity, (cal)(cm)/(sec)(°C)(cm²)	Thermal coefficient of linear expansion per °C	Tensile strength, lb/in.²
Aluminum	2.71	0.214	660	2.828	17.00	0.00446	0.52	23×10^{-6}	24,000
Brass:									
Commercial bronze (90% Cu, 10% Zn)	8.80	0.092	1,045	4.66	28.03	0.00145	0.45	18×10^{-6}	95,000
Low brass (80% Cu, 20% Zn)	8.67	0.092	995	5.95	35.78	0.00114	0.34	19×10^{-6}	105,000
High brass (70% Cu, 30% Zn)	8.53	0.092	930	6.90	41.50	0.00098	0.26	20×10^{-6}	120,000
Copper:									
Annealed wire (100% conductive)	8.89	0.0921	1,083	1.7241	10.37	0.00393	0.92	17×10^{-6}	36,000–40,000
Hard-drawn wire (97.5% conductive)	8.89	.....	....	1.7683	10.63	0.00383	....	....	50,000–70,000
Gold	19.30	0.031	1,063	2.42	14.55	0.0034	0.71	14.2×10^{-6}	
Iron	7.86	0.107	1,535	10	60.14	0.0050	0.16	11.7×10^{-6}	50,000
Lead	11.3	0.031	327	22	132.31	0.0039	0.083	28×10^{-6}	1,800–4,000
Molybdenum	10.2	0.061	2,625	5.7	34.28	0.0033	0.35	5×10^{-6}	100,000
Nickel	8.9	0.105	1,452	7.8	46.91	0.006	0.14	14×10^{-6}	155,000
Platinum	21.45	0.032	1,773	10.60	63.75	0.00300	0.17	8.9×10^{-6}	
Silver	10.5	0.056	960	1.63	9.80	0.0038	1.01	18×10^{-6}	42,000
Steel (mild)	7.8	0.107	1,300–1,475	12	72.17	0.005	0.11	9×10^{-6}	50,000–70,000
Tin	7.3	0.054	232	11.5	69.16	0.0042	0.15	2.1×10^{-6}	4,000–5,000
Tungsten	19.3	0.032	3,410	5.52	33.20	0.0045	0.48	4×10^{-6}	490,000
Zinc	7.14	0.092	419	6	36.08	0.00347	0.27	28×10^{-6}	7,000–30,000

TABLE 19. Typical Physical, Electrical, and Thermal Properties of Epoxies [11]

	Molding compounds			Cast resins		
	Unfilled	Mineral filler	Mineral filler, glass-reinforced*	Unfilled	Silica-filled	Flexible
Volume resistivity, ohm-cm	10.5×10^{15}	$> 10^{15}$	$1.72–4.70 \times 10^{15}$	$10^{12}–10^{17}$	$10^{13}–10^{16}$	$10^{12}–10^{14}$
Dielectric strength:						
Short time, volts/min	400–433	345–387	304–364	400–550	400–550	235–400
Step-by-step, volts/min	384	400	340–350	380		
Dielectric constant:						
60 cps	4.43–5.0	4.4–5.56	5.82–7.17	3.5–5.0	3.2–4.5	3.0–6.0
10^3 cps	4.36–5.0	4.2–4.91	4.99–5.55	3.5–4.5	3.2–4.0	3.0–5.0
10^6 cps	4.37–4.8	4.1–4.56	4.75–5.18	3.3–4.0	3.0–3.8	3.0–6.0
Dissipation factor:						
60 cps	0.38	0.011–0.083	0.122–0.183	0.002–0.010	0.008–0.03	0.010–0.040
10^3 cps	0.020–0.240	0.019–0.140	0.117–0.150	0.002–0.020	0.008–0.03	0.012–0.050
10^6 cps	0.022–0.26	0.013–0.137	0.130–0.143	0.030–0.050	0.02–0.04	0.018–0.090
Arc resistance, sec	154–180	128–180	123–133	45–120	150–300	50–180
Specific gravity	....	1.6–1.9	1.79–2.06	1.11–1.14	1.6–2.0	1.15–1.25
Modulus of elasticity in tension, psi $\times 10^5$	....	....	....	3.0–3.5	3.5–4.0	0.01–3.5
Percent elongation	....	....	....	3–6	1–3	4–200
Flexural strength, psi $\times 10^3$	17	10–12	12.1–17.8	13–21	8–14	1–18
Compressive strength, psi $\times 10^3$	13–20	18	21.3–25.5	15–21	15–40	1–14
Rockwell hardness	....	M93	M99–M108	M80–M110	M85–M120	
Impact strength, ft-lb/in.	....	0.23–0.35	0.44–6.16	0.2–1.0	0.3–0.45	2–10
Heat distortion temp., °F, 264 psi	290	233–350	226–309	115–550	160–550	150–250
Thermal conductivity, cal/(sec)(cm²)(°C)(cm) $\times 10^{-4}$	....	11–18	....	4–5	10–20	
Thermal expansion, in./in. per °C $\times 10^{-5}$	....	2.5–5.0	1.5–2.6	4.5–6.5	2.0–4.0	2.0–10.0
Percent water absorption	0.5–0.1	0.1	0.1	0.08–0.15	0.04–0.10	0.27–0.50
Burning rate	Medium to self-extinguish	Slow to self-extinguish	Slow to self-extinguish	Slow to self-extinguish	Self-extinguishing	Slow

* Values vary due to different compositions (12 and 50% total weight).

TABLE 20. Typical Properties of Pressure-sensitive Electrical Tape [11]

Backing	Insulation class	Adhesion, oz/in. width	Tensile strength, lb/in.	Elongation, %	Thickness, mils	Insulation resistance, megohms	Dielectric strength, volts
Paper							
30-lb crepe paper, tan	105	40	20	12	9.0	20	2,500
30-lb crepe paper, black	105	40	20	12	9.5	20	3,000
Superfine creped cellulose fiber, epoxy-treated	105	35	30	10	8.0	30	2,000
3-mil rope flatback paper, tan	105	50	40	3	5.0	20	2,000
4-mil rope flatback paper, tan	105	50	60	3	5.5	20	2,500
4-mil rope flatback paper, black	105	50	60	3	5.5	20	2,500
Film							
2-mil acetate film, clear	105	50	25	10	3.5	300,000	6,000
2-mil acetate film, yellow	105	40	25	10	3.5	300,000	6,000
½-mil polyester film	130	20	10	100	1.0	>1,000,000	3,000
1-mil polyester film, orange-yellow	130	40	20	100	2.5	2,000,000	5,500
1-mil polyester film, clear	130	40	20	100	2.5	2,000,000	5,500
1-mil polyester film, yellow	130	50	20	100	2.5	2,000,000	5,500
1-mil polyester film, bondable	130	40	25	100	2.5	1,000,000	4,500
2-mil polyester film, orange-yellow	130	45	50	100	3.5	2,000,000	7,500
2-mil polyester film, clear	130	45	50	100	3.5	2,000,000	7,500
2-mil polyester film, shrinkable, transparent	130	40	40	100	3.25	1,000,000	7,000
1-mil polyester film, yellow, adhesive both sides	130	50	20	60	4.0	>1,000,000	6,000
Polyethylene, black	...	48	16	150	9.0	100,000	10,000
2-mil fluorohalocarbon film	155	50	15	150	3.5	1,000,000	8,500
1-mil polyimide film	180	25	35	100	2.5	>1,000,000	5,000
2-mil polyimide film	180	25	55	100	3.5	1,000,000	8,500
Polytetrafluoroethylene	180	20	20	100	3.5	2,000,000	8,500
Polytetrafluoroethylene	180	20	20	250	6.5	2,000,000	12,000
Polytetrafluoroethylene	180	25	45	300	11.5	2,000,000	16,000
Fully cured silicone rubber, unsupported	180	14	11	600	15.0	1,000,000	10,500
Vinyl, black	105	18	35	275	11.0	1,000,000	11,000
Vinyl, black, all-weather	...	20	20	250	8.5	1,000,000	10,000
Cloth							
White taffeta acetate cloth	105	40	45	10	8.0	100,000	3,000
Black taffeta acetate cloth	105	40	45	15	8.5	100,000	3,000
80 × 72 cotton cloth (nonthermosetting)	105	35	50	7	9.5	75	3,000
80 × 72 cotton cloth (thermosetting)	105	45	55	5	10.5	75	3,000
60 × 52 glass cloth	130	50	125	5	7.5	1,000	3,000
White glass cloth	155	25	125	5	8.0	500	4,000
60 × 52 glass cloth	180	20	110	10	8.0	1,500	4,000
Epoxy-resin-coated 0.002 in. glass cloth	155	40	100	5	5.0	3,000	5,000
Polytetrafluoroethylene-coated glass cloth	180	35	65	5	5.5	3,000	4,000
Silicone-varnished glass cloth	180	25	125	5	6.0	2,000	4,000
Mat							
Porous polyester mat	130	10	9	25	5.0		
Laminated							
Acetate film/acetate cloth	105	50	45	10	9.5	500,000	6,000
Rayon-filament-reinforced acetate film	105	40	200	10	9.5	100	6,000
Glass-filament-reinforced acetate film	105	45	130	5	7.0	1,000	5,000
Epoxy-bonded mica paper/glass cloth	155	40	140	8	7.5	1,000	4,500
Polyester film 2-mil rope paper	105	50	40	3	6.0	100	5,000
White polyester film/polyester mat	130	55	35	30	8.0	>1,000,000	4,500
Adhesive transfer film	...	40	...	...	2.3	500,000	

TABLE 21. Properties of Magnet Steels [12]

Material	Residual flux density B_r, gauss	Coercive force H_c, oersteds	Energy product, BH max $\times 10^6$, gauss-oersteds	Flux density, B_o at BH max, gauss	Density, lb/in.^{3}
0.65% carbon	10,000	42	0.18	6,500	0.283
1% carbon	9,000	51	0.20	5,900	0.282
5% tungsten	10,500	70	0.33	7,000	0.292
6% tungsten	9,500	74	0.33	6,500	0.294
1% chromium	9,500	52	0.23	6,500	0.282
2% chromium	9,300	60	0.26	6,300	0.282
3½% chromium	9,500	66	0.29	6,500	0.281
6% chromium	9,500	74	0.30	6,200	0.281
3% cobalt	7,200	130	0.35	4,280	0.278
6% cobalt	7,500	145	0.44	4,700	0.282
9% cobalt	7,800	122	0.41	5,100	0.286
17% cobalt	9,500	150	0.65	6,000	0.302
36% cobalt	10,000	240	1.0	6,300	0.296
40% cobalt	10,000	242	1.03	6,500	0.296

TABLE 22. Properties of Alnico [12]

Material	Residual flux density B_r, gauss	Coercive force H_c, oersteds	Energy product, BH max $\times 10^6$	Flux density, B_o at BH max, gauss	Density, lb/in.^{3}
Alnico 1	7,000	440	1.4	4,500	0.249
Alnico 2 (cast)	7,200	560	1.6	4,700	0.256
Alnico 2 (sintered)	6,900	520	1.45	4,300	0.243
Alnico 3	6,900	470	1.35	4,300	0.249
Alnico 4 (cast)	5,500	700	1.3	3,000	0.253
Alnico 4 (sintered)	5,200	700	1.2	3,000	0.250
Alnico 5 (sintered)	10,500	600	4.5	7,800	0.241
Alnico 5-NC1*	12,100	740	5.0	8,500	0.265
Alnico 5-NW1*	12,600	650	5.5	10,400	0.265
Alnico 5-NR1*	13,200	625	5.5	11,200	0.265
Alnico 5-PR1*	13,500	675	6.5	11,200	0.265
Alnico 5-OR1*	13,900	725	7.5	11,500	0.265
Alnico 6 (sintered)	8,800	800	2.75	5,500	0.241
Alnico 6-NS1*	10,500	750	3.7	7,100	0.268
Alnico 6-NS2	11,000	840	4.5	8,000	0.268
Alnico 7	7,500	1,100	3.0	4,300	0.259
Alnico 8†	8,050	1,810	5.76		
Alnico 9‡	10.500	1,450	8.5	7,700	0.265
Alnico 12	5,500	950	1.6	3,000	0.264

* Westinghouse grade identification and properties. Grades with similar properties are available throughout the industry.
† From Arnold Engineering Co. data published in *Electron. Prod.*, September, 1964.
‡ From Indiana General Bulletin 386, dated Mar. 23, 1964.

TABLE 23. Properties of Ceramic Magnets [12]

Material	Residual flux density B_r, gauss	Coercive force H_c, oersteds	Energy product, BH max $\times 10^6$	Density, lb/in.^{3}
Barium ferrite (unoriented)	2,000–2,000	1,400–1,700	0.8–1.01	0.17
Barium ferrite (partially oriented)	2,400–3,000	1,300–2,200	1.65–2.4	
Barium ferrite (oriented):				
High B_r	3,800–4,000	2,000–2,200	3.3–3.5	0.18
High H_c	3,000–3,300	2,300–2,600	2.4–2.8	0.17
Plastiform 1*	2,200	1,480	1.3	0.135
Strontium ferrite (oriented only):				
Westro Alpha (high B_r)	4,000	2,500	3.66	0.175
Westro Beta (high H_c)	3,600	3,200	3.1	0.171

* Rubber- or vinyl-bonded barium ferrite.
SOURCE: Leyman Corp.

TABLE 24. Properties of Other Magnetic Alloys [12]

Material	Residual flux density B_r, gauss	Coercive force H_c, oersteds	Energy product, BH max $\times 10^6$	Flux density, B_o at BH max, gauss	Density, lb/in.^{3}
Cunife I	5,400	500	1.3	4,000	0.311
Cunife II	7,300	260	0.78	4,700	0.311
Cunico I	3,400	700	0.8	2,000	0.3
Cunico II	5,300	450	0.99	3,400	0.3
Remalloy, Comol	10,000	230	1.1	6,900	0.295
Indalloy	9,000	240	0.9	6,000	0.288
Vicalloy	6,000–10,000	60–260	0.9	5,500–6,000	0.293–0.296
Silmanal	590	6,300	0.083	292	0.325
Platinum	5,830	1,570	3.07	3,400	0.559
Platinum (co.) ...	4,500	2,700	4.0	2,600	0.559
Platinax II	6,400	4,800	9.2	3,400	0.548
Nipermag	5,600	660	1.34	3,400	0.249
Bismanol	4,800	3,650	5.3	2,640	0.25
New KS	7,150	785	2.03	4,300	0.268

their magnetic properties are liable to deteriorate, they are seldom used. Alnico magnets yield maximum force in minimum space but can be machined only by grinding. Ceramic-type magnets are useful where nonconductive properties are required, but they are temperature-sensitive. Other alloys such as Cunife have been developed to be ductile and easily formed, others for high coercive force, and still others for lower cost.

Components. Discrete components and the data required to understand and use them are a large subject. Several chapters, notably Chap. 8, present data for a variety of components. Choice of components varies widely depending upon accuracy, reliability, size, cost, and other requirements.

Standard Envelopes, Ratings, and Tolerances. No attempt will be made here to list standard envelopes, ratings, and tolerances for resistors, capacitors, transistors, and other conventional or discrete components available to the electronic packaging engineer. There are, however, several ideas worth mentioning that are not always apparent on suppliers' data sheets.

ENVELOPES. MIL specification envelope sizes, or configurations, allow for a variety of manufacturers and methods, and this factor can lead to design limitations. For instance, a resistor built by a particular manufacturer may always be on the upper side of the diameter tolerance, never approaching the nominal value. Using a statistical approach to determine overall size leads to the wrong solution. Conversely, a diode manufacturer may always stay below the nominal diameter allowed, although the upper tolerance requires 0.2-in. instead of 0.1-in. spacing on a 100-mil grid printed-circuit board.

DERATINGS. Elevated temperature is the common reason for derating resistors and other components, but other criteria have an effect. Proximity to other heat-generating components, or grouping of resistors, must be taken into account as exemplified in Fig. 5. The fact that there are voltage limits in the application of fixed resistors is often overlooked. Both insulation breakdown and power dissipation must be considered in arriving at maximum voltages or maximum resistances for a given voltage. Figure 50 illustrates the trade-off between resistance value and maximum permissible applied voltage for several wattage value resistors.

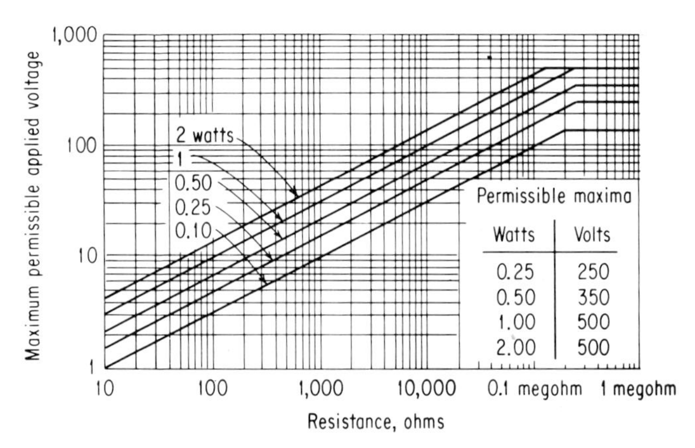

FIG. 50. Limitations to applied voltage on fixed-composition resistors (MIL-R-11C).[2]

Temperature derating curves vary enormously as illustrated by the resistor curves of Fig. 51. Many types of resistors can be used at 100 percent rating only up to 40°C; others have the ability to operate at several hundred °C at full rating. Where maximum reliability is required, additional derating is doubly important.

TOLERANCES. Component tolerances vary widely. Generally speaking, the closer the tolerance, the higher the price. Side effects appear; increased accuracy may mean an entirely different method of construction, larger size, less desirable environmental characteristics, etc. Generally it is electrically sound practice to design the circuit around standard component tolerances. The packaging engineer must monitor electrical design and suggest the possibility of redesign when unnecessarily close tolerance components are specified.

Inexperienced circuit designers are apt to fall into the trap of using manufacturers' component tolerances as the usage tolerance. Plus or minus 10 percent carbon resistors, for instance, may be manufactured within this tolerance, but exhaustive measurements show that a ±20 percent tolerance is not unusual after shelf life in a normal shop atmosphere. Packaging techniques in themselves may change component values. Pressure-sensitive components may change when embedded in an epoxy subassembly. Determination of the effect of usage is fully as important as initial tolerance.

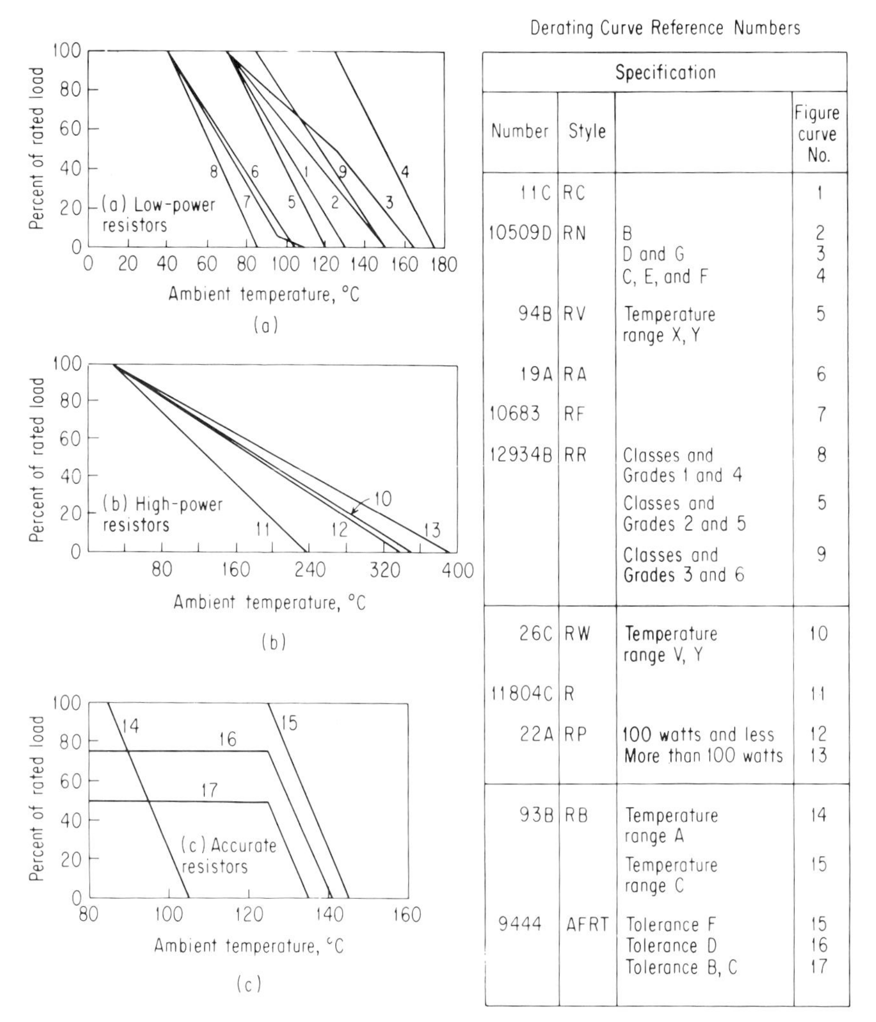

Derating Curve Reference Numbers

Specification			
Number	Style		Figure curve No.
11C	RC		1
10509D	RN	B	2
		D and G	3
		C, E, and F	4
94B	RV	Temperature range X, Y	5
19A	RA		6
10683	RF		7
12934B	RR	Classes and Grades 1 and 4	8
		Classes and Grades 2 and 5	5
		Classes and Grades 3 and 6	9
26C	RW	Temperature range V, Y	10
11804C	R		11
22A	RP	100 watts and less	12
		More than 100 watts	13
93B	RB	Temperature range A	14
		Temperature range C	15
9444	AFRT	Tolerance F	15
		Tolerance D	16
		Tolerance B, C	17

FIG. 51. Resistor derating for high ambient temperatures.[2]

Radiation. Where radiation is expected in space systems, the effect of radiation on components is important, not only because of the possibility of catastrophic failure, but because component operation may deteriorate beyond allowable limits. Degraded performance should be considered in initial design.

RESISTORS. It is generally accepted that carbon-composition resistors suffer greater degradation in a nuclear environment than any other type of resistor. The bar charts of Fig. 52 indicate the susceptibility of various resistors to radiation damage. Roughly, in order of increasing maximum safe dosage of nuclear radiation, carbon-composition resistors are followed by carbon film, metal film, and last, wire-wound resistors.

CAPACITORS. Due to the great variation in method of manufacture of capacitors, only generalities can be stated. Ceramic capacitors vary from less than 1 percent to 20 percent above initial values, but usually recover when removed from the radiation environment. Glass capacitors are more stable but do not return as well. Mica capacitors vary more than glass, and do not return well either. Paper capacitors are prone to

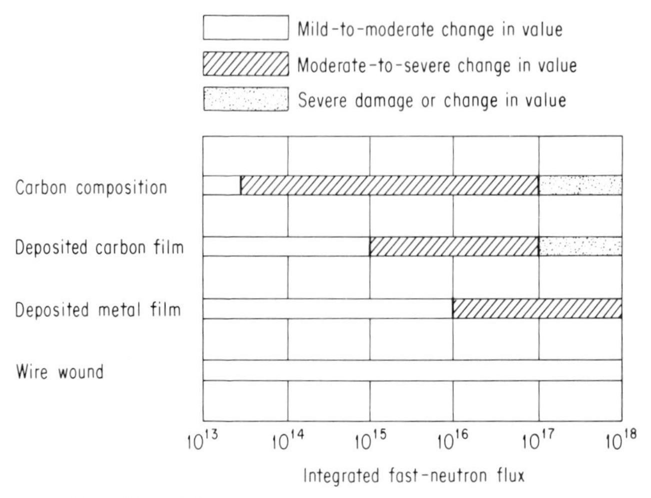

FIG. 52. Relative radiation sensitivity of resistors.

heavy damage, particularly the oil-filled type which tend to rupture or distort as the oil evolves gas. Plastic diodes are roughly 10 times worse than inorganic dielectric types.

Tantalum capacitors may be capable of surviving intense radiation, but unlike other types, they offer a biological hazard because thermal-neutron bombardment activates them. This secondary radiation has a half-life of 111 days as compared with 2.3 min for aluminum. The radiation sensitivity of capacitors is summarized in Fig. 53.

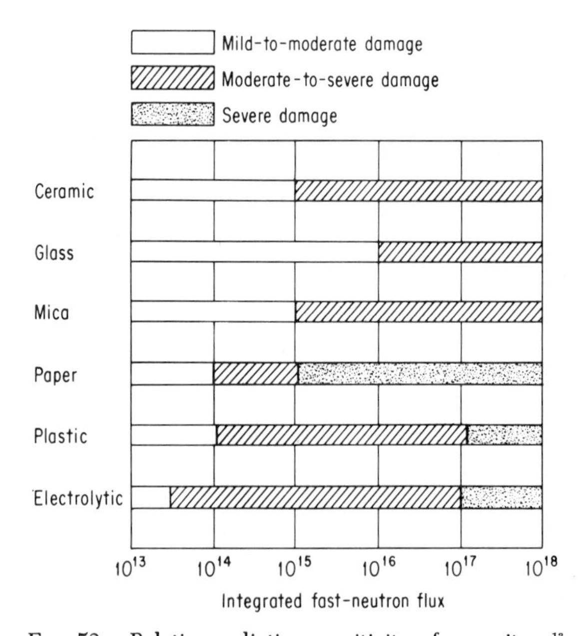

FIG. 53. Relative radiation sensitivity of capacitors.[13]

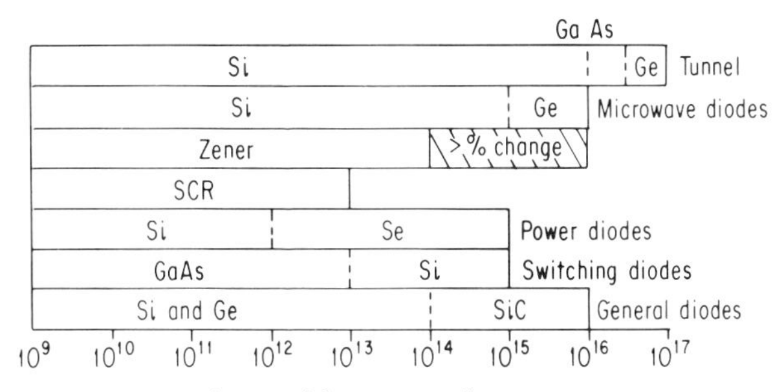

FIG. 54. Effect of radiation on diodes.[13]

DIODES. Nuclear radiation affects the lattice structure of semiconductors; diodes change characteristics. The level of exposure in which the device still operates within usable limits is indicated in Fig. 54. Note that germanium diodes are less sensitive to radiation than are silicon diodes.

TRANSISTORS. Like diodes, transistors change their characteristics in a radiation environment. The level of exposure in which transistors still operate within usable limits is indicated in Fig. 55. Note again that germanium transistors are less sensitive to radiation than are silicon transistors.

OTHER COMPONENTS. Most other components, such as transformers, relays, connectors, and indicators, are affected principally as the dielectrics from which they are made are affected. Magnetic materials, other than soft magnetic materials, may not be directly affected, but winding insulation, potting compounds, or insulating plastics may lose their dielectric qualities or rupture and damage the transformer.

Effect of Soldering or Welding. Chapter 3 is concerned with soldering techniques and Chap. 4 with welding and other metal-bonding techniques. From an electrical viewpoint, the packaging engineer should have a good feeling for what will happen to component characteristics during the assembly stage. Generally speaking, soldering has more effect on components than welding because the heating cycle is longer. Heat-sinking clips may be necessary on critical component leads to prevent excess heat from reaching the component body. Wave soldering creates a much more predictable level of heat than does hand soldering, which can only be partially controlled by iron size and operator training. On the other hand, direct electrical damage is possible in welding if the electrodes short to leads other than those being welded—a common occurrence with closely packed microelectronic subassemblies. Insulated electrodes are required in many cases.

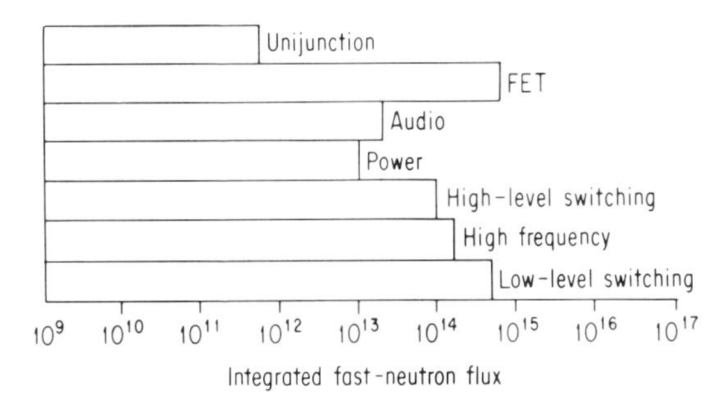

FIG. 55. Effect of radiation on transistors.[13]

Effect of Vibration and Shock. Components may be damaged by shock or vibration for two reasons: one, the component itself may be susceptible, or two, the mounting (or lack of it) may result in damage. Several other chapters of this book give rules and design criteria for packaging components. Physical damage and electrical damage may be one and the same. For example, diode whisker breakage is an electrical failure caused by a physical failure which is not apparent by looking at an opaque diode body. This type of failure is not usually classified with broken leads, cracked component bodies, torn cases, etc.

Some component types are more susceptible to damage than others. Certain types of tantalum capacitors have a suspended slug which will break loose under vibration long before anything happens to other tantalum capacitors. Choose components capable of surviving vibration and shock levels far above system requirements; amplification will occur in certain locations and at certain frequencies which are difficult to predict regardless of how much care is taken.

Reliability. Reliability is extremely important in electronic packaging. Components are at the heart of reliability but statistical component reliability figures should not be used to arrive at system reliability without considering proper usage of components. The most reliable component money can buy will fail instantly if used incorrectly. More subtle failures occur after a longer period of time, because something was not taken into account, such as a nearby hot spot, line transients, or statistically "impossible" duty cycles.

A commonly used technique employs "burn-in" at rated or above-rated power. Short-time failures are weeded out without noticeable effect on long-term life. Burn-in levels should be consistent with usage instead of with ratings. Analysis of failures during burn-in can provide useful feedback information on component characteristics and clues to better circuit design.

Inductive and Noninductive Components. Components intended for a particular purpose frequently exhibit another characteristic that ruins circuit performance. Wire-wound resistors, by the nature of their construction, add considerable inductance if the resistance wire is wound in one direction on a bobbin. Where this inductance is harmful to circuit operation, half of the wire can be wound in the opposite direction to cancel the inductive effect. Since component placement determines the effect of this type of inductive field, it is important to know what the component characteristics are and to design for them or specify special components where necessary. Components may be much more closely packaged in the final configuration than in the breadboard stage. Additional problems always crop up because of the close proximity

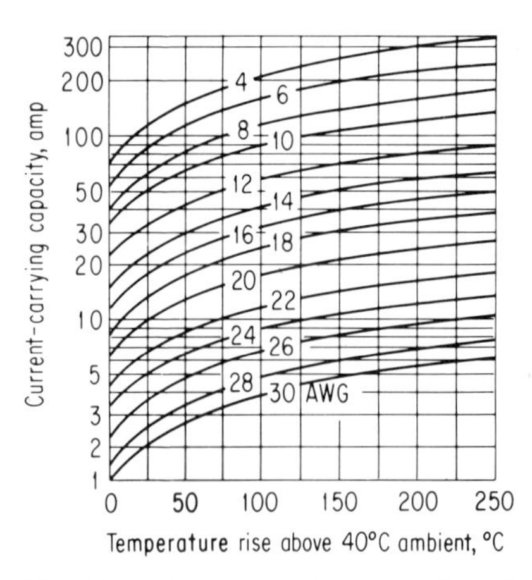

FIG. 56. Current-carrying capacity of copper wire vs. temperature rise.[40]

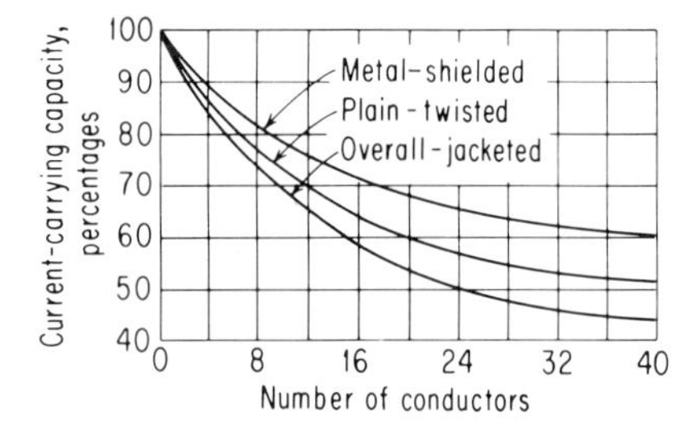

FIG. 57. Current-carrying capacity of multiconductor cables.[40]

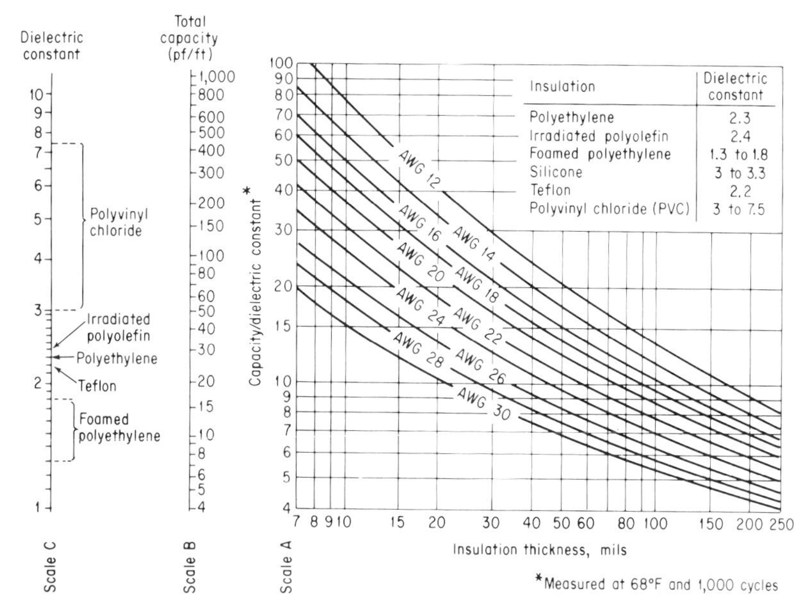

FIG. 58. Capacitance values for shielded wire.[24]

of components. Breadboards should, as nearly as possible, approximate the final configuration to avoid such problems.

Component leads may also add an unwanted electrical characteristic. Capacitive coupling is present between long parallel leads; weldable leads of magnetic metals upset magnetic cleanliness requirements, etc.

Wire. Chapter 2 is devoted entirely to wire and cable, but some of the electrical characteristics of wire will be considered here.

Current-carrying Capacity. Current-carrying capacity is basic to selection of wire size. The ability to handle current is dependent upon temperature, since IR^2 loss in the wire has to be dissipated in the surrounding environment. Figure 56 shows the current-carrying capacity of copper wire versus ambient temperature rise. These curves vary somewhat with different insulations and thicknesses but are useful design guidelines for low altitudes. At altitudes over 5 miles, there is no conduction of heat because there is no air, and an order of magnitude derating of 50 percent is necessary when transfer is by radiation alone. A companion limitation for the copper is the dielectric. The maximum current for copper obviously cannot be used when the resultant power loss results in a temperature above the safe temperature of the dielectric. Bundling also has a significant effect, as indicated in the curves of Fig. 57 for three typical wire bundling constructions. See Chap. 2 for additional detail.

Capacitance of Shielded Wire. The many variables inherent in ordinary shielded wire, such as insulation thickness, tightness and weave of the shield, additives and curing of the insulation, frequency, and temperature, preclude one accurate curve for capacitance, but usable results can be gained from Fig. 58. The intersection of wire size and insulation thickness curves gives the capacity/dielectric constant on the *A* scale. The capacity per foot in picofarads can then be determined either by multiplying this result by the dielectric constant or by using the convenient nomograph on the left side of Fig. 58. To use the nomograph, draw a line from the dielectric constant on scale *C* to the capacity/dielectric constant previously found on scale *A* and read the capacity per foot where the line intersects scale *B*.

For two-conductor twisted and shielded wire, the abbreviated data of Table 25 give

TABLE 25. Capacitance of Two-conductor Twisted and Shielded Wire, Picofarads per foot

Size, AWG	Conductor to shield		Between conductors	
	600 volts	1,000 volts	600 volts	1,000 volts
24	37.0	28.7	21.3	17.8
22	42.0	34.8	24.2	19.3
20	47.5	37.7	28.4	21.3
18	53.0	41.1	31.0	23.4

both capacitance between one conductor and the shield, and capacitance between the two conductors, in picofarads per foot.

Crosstalk. The effect one circuit has on a second because of their mutual capacitive and inductive coupling is called *crosstalk*. Equations describing complex wiring

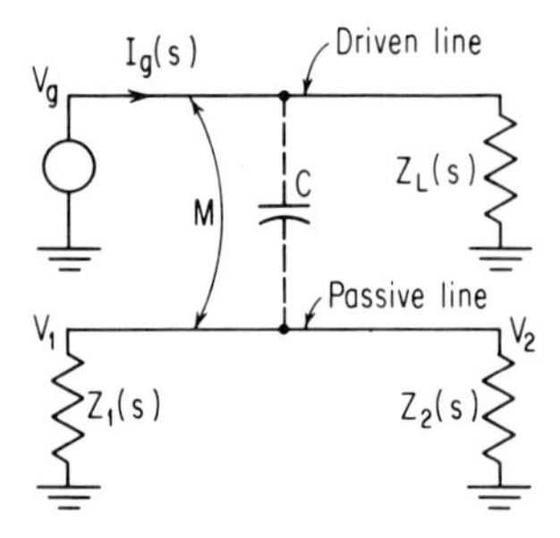

FIG. 59. Mutual capacitance and mutual inductance between a driven and passive line. (*Harry H. Gray, "Digital Computer Engineering," copyright* 1963. *Reprinted by permission of Prentice-Hall, Inc., Englewood Cliffs, N.J.*)

systems are not practical, but understanding an idealized system helps in designing a more complex one. Consider two circuits, a driven line and a passive line, illustrated in Fig. 59. Assume that the impedances are all resistive. The voltages across the passive line resistance are *

$$V_1 = C_m \frac{R_1R_2}{R_1 + R_2} \frac{dV_g}{dt} + M \frac{R_1}{R_1 + R_2} \frac{dI_g}{dt}$$

$$V_2 = C_m \frac{R_1R_2}{R_1 + R_2} \frac{dV_g}{dt} - M \frac{R_1}{R_1 + R_2} \frac{dI_g}{dt}$$

Each equation has two terms, one proportional to the mutual capacitance C_m, called the capacitive crosstalk component, and one proportional to the mutual inductance M, called the inductive crosstalk component. The capacitive crosstalk component is proportional to the rate of change of voltage in the driven line, and the inductive crosstalk component is proportional to the rate of change of current in the driven line. Mutual inductance varies with configuration, and mutual capacitance varies with configuration and dielectric.

MUTUAL CAPACITANCE. Mutual capacitance equations for various configurations are given in Table 26. The simplifying assumption is made that the wire size is small with respect to distances. Capacitive crosstalk is: *

1. Directly proportional to the signal voltage on the driven wires

* Harry H. Gray, "Digital Computer Engineering," copyright 1963. Reprinted by permission of Prentice-Hall, Inc., Englewood Cliffs, N.J.

TABLE 26. Mutual Capacitance for Several Lead Configurations [15]

Description	Configuration	Equation*
Inner and outer conductors of concentric coaxial cable	r_1, r_2	$C_m = \dfrac{k_1k_2}{\ln(r_2/r_1)}$
Single wire above a plane at ground potential	R, d	$C_m = \dfrac{k_1k_2}{\ln(2d/R)}$
Parallel wires on either side of a datum	d_1, d_2, R_1, R_2	$C_m = \dfrac{k_1k_2}{\ln \dfrac{4d_1d_2}{R_1R_2}}$
Shielded pair	d, D	$C_m = \dfrac{k_1k_2}{2\ln(6D/5d)}$
Any two of a shielded group of three or more wires	d, D	$C_m = \dfrac{k_1k_2}{2\ln(1.216D/d)}$

* Where C_m = mutual conductance in farads/cm
$k_1 = 0.2416$
k_2 = dielectric constant (= 1 for air, 4 for common insulating materials)
and r_1, r_2, R, R_1, R_2, d, d_1, d_2, and D are any units of length consistent within the equation.

2. Directly proportional to the distance over which the wires are near each other
3. Directly proportional to the dielectric constant of the medium
4. Inversely proportional to the signal rise times
5. Directly proportional to functions of the logarithms of the distances involved

Although the capacitive crosstalk is proportional to some function of the logarithm of distance, appreciable reduction in crosstalk can be obtained by dealing with spacings.

MUTUAL INDUCTANCE. Mutual inductances for various configurations are given in Table 27. The assumption is made that wire size is small with respect to distances. Making the same approximations for a twisted pair or inner and outer conductors of a coaxial line results in a mutual inductive coupling which is zero. This is never quite true because of end effects and other phenomena, but the use of either technique can be very effective. The top signal circuit in Fig. 60 is being disturbed by the magnetic and electric field from a third wire. The alternating current I_n produces a magnetic field which cuts both wires of the signal circuit, inducing a voltage in the signal loop. The induced emf is proportional to the magnitude and frequency of I_n and the area enclosed by the signal loop, but inversely proportional to d_1. Since d_1 and d_2 are unequal, the distributed capacities are also unequal. Their difference in magnitude causes the electric field to produce current I_x in the signal circuit which develops noise voltage e_x across the amplifier input resistance R_L.

If the conductors in the signal circuit are twisted (transposed) at regular intervals, as shown in the lower portion of Fig. 60, the distances d_1 and d_2 are approximately equal, and the area of the circuit loop is almost zero. In addition, twisting the signal

TABLE 27. Mutual Inductances for Several Lead Configurations

Description	Configuration	Equation *
Wires over a ground plane—current returns through ground plane	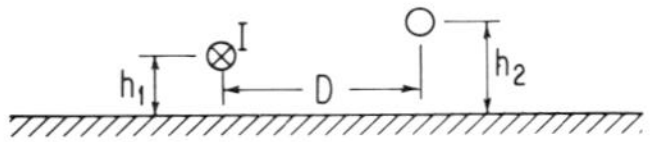 I, h_1, D, h_2	$M = 0.2 \log \sqrt{\frac{(h_2 + h_1)^2 + D^2}{(h_2 - h_1)^2 + D^2}}$
Wires between conducting planes—current returns through either or both planes	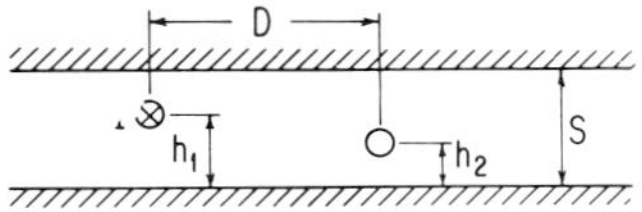 D, h_1, h_2, S	$M = 0.2 \log \sqrt{\frac{\cosh^2(\pi D/2S) - \cos^2[\pi(h_1 + h_2)/2S]}{\cosh^2(\pi D/2S) - \cos^2[\pi(h_1 - h_2)/2S]}}$
Parallel pairs in free space	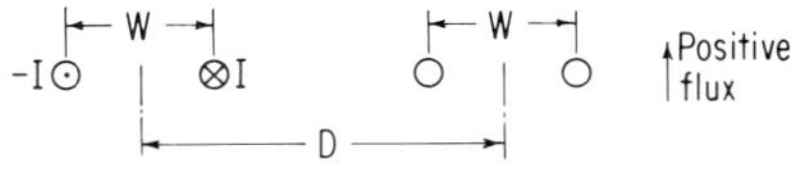 W, -I, I, W, D, Positive flux	$M = 0.2 \log \left[1 - \left(\frac{W}{D}\right)^2 \right]$
Parallel pairs near a ground plane	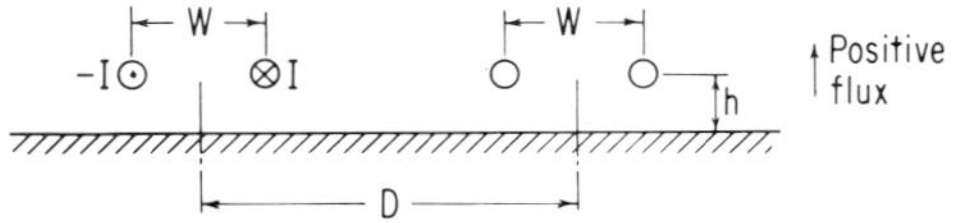 W, -I, I, W, h, D, Positive flux	$M = 0.2 \log \left[\frac{(D^2 - W^2)(D^2 + 4h^2)}{D^2\sqrt{(D + W)^2 + 4h^2}\sqrt{(D - W)^2 + 4h^2}} \right]$
Parallel pairs between two plane conducting surfaces	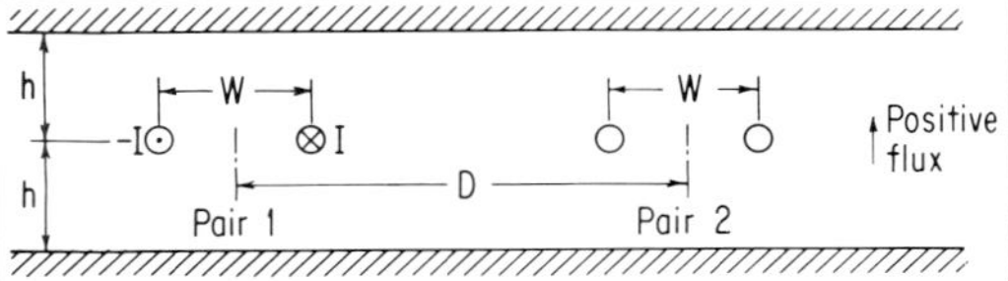 h, h, W, -I, I, W, D, Pair 1, Pair 2, Positive flux	$M = 0.2 \log \frac{\left(\coth \frac{\pi D}{4h}\right)^2}{\coth[\pi(D - W)/4h] \quad \coth[\pi(D + W)/4h]}$

* Where M = mutual inductance in microhenrys per meter and W, D, h, h_1, h_2, and S are any units of length consistent within the equation.

SOURCE: Harry H. Gray, "Digital Computer Engineering," copyright 1963. Reprinted by permission of Prentice-Hall, Inc., Englewood Cliffs, N.J.

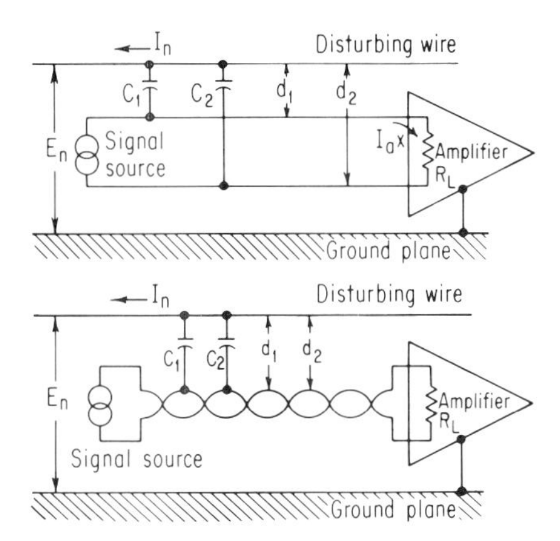

FIG. 60. Effect of twisting wires on induced crosstalk.[16]

wires has made C_1 approximately equal to C_2 and will reduce the effects of the electric field so that signal-circuit shielding becomes less of a problem.[16]

It is important to remember that if there is more than one driven line and all are driven at the same time, the crosstalk voltages produced by each in the passive line are additive.

Propagation Speeds. A rough rule of thumb is that there is a loss of 1 nsec/ft in bare copper wire. This is true only when the dielectric constant is unity. Propagation velocity is proportional to $1/\sqrt{k_2}$, where k_2 is the dielectric constant. Since dielectric constants of commonly used insulations average about 4, the speed is reduced to roughly 6 in./nsec. Since many circuits have switching times and propagation delays of less than 5 nsec, it is obvious that lead length can be the controlling factor. Some wire insulations have lower dielectric constants; Teflon * is 2.2, but the "foot per nanosecond" rule is a theoretical limit restricted to the speed of light which forces tight packaging and short leads.

Printed-circuit Boards. Chapter 1 covers the subject of printed circuits thoroughly, giving design criteria for line width, spacing, holes, pads, and other physical requirements as well as electrical properties. Some electrical design data are given here, but in far less detail.

Properties. A summary of physical and electrical properties of materials commonly used for printed-circuit boards is given in Table 28.

Current-carrying Capacity. The charts of Fig. 61 present an easy method for quick calculation of etched conductor size for variations of current and allowable temperature rise above ambient. Any intersection of current and temperature lines on the top chart yields a cross-section value on a vertical line. This cross-section value combined with the thickness desired (½-, 1-, 2-, or 3-oz copper) gives a required conductor width on the lower chart. These curves are only a convenient yardstick. Deratings and allowances for various manufacturing techniques are discussed in Chap. 1.

Propagation Speed. Signal propagation speeds in printed circuits are equivalent to those in wire. High-speed systems require short lead lengths which lead to dense packaging and multilayer boards. A conductor buried in a glass-epoxy board, which has a dielectric constant of approximately 4.5, conducts a signal at roughly 6 in./nsec. As discussed earlier, line delays may approach circuit delays. Figure 62 shows graphically the relationship between circuit and line delays. The solid lines represent logic delays of 0, 1, 2, 3, 4, and 5 nsec. The dotted lines represent percentages of logic delay: 10, 20, 30, 40, and 50 percent. For a required maximum total delay and a given logic delay determined by the circuits available, the maximum

* Trademark, E. I. du Pont de Nemours & Co.

TABLE 28. Properties of Laminated and Reinforced Plastics [11]

NEMA grade	NEMA minimum or maximum average values, sheets: Min. flexural str., 1/16 in., psi		Min. izod impact str., (ft)(lb)/in. notch, edge		Min. bond str., lb	Water abs., max., 1/16 in., %	Min. diel. str., kv	Max. diel. const., 1 Mc 1/32 in. or more	Max. diss. factor 1 Mc, 1/32 in. or more	Min. arc resist., sec	Typical values, sheets: Tensile str., psi		Compr. str., psi		Rockwell hardness M scale	Sp. gr.	Diel. str., perp. to lam., volts/min		Thickness range, in.		Base material	Resin
	LW	CW	LW	CW							LW	CW	Flat	Edge			Short time	Step by step	Min.	Max.		
X	25,000	22,000	0.55	0.50	700	6.00	..	...	.....	...	20,000	16,000	36,000	19,000	110	1.36	700	500	0.010	2	Paper	Phen.
XP	13,000	11,000	....	....	.....	3.60	40	...	.....	...	12,000	9,000	25,000	......	95	1.33	650	450	0.010	1/4	Paper	Phen.
XPC	10,000	8,000	....	....	.....	5.50	..	...	.....	...	10,500	8,500	22,000	......	75	....	600	425	1/32	1/4	Paper	Phen.
XX	15,000	14,000	0.40	0.35	80	2.00	40	5.5	0.045	...	16,000	13,000	34,000	23,000	105	1.34	700	500	0.010	2	Paper	Phen.
XXP	14,000	12,000	....	....	0	1.80	60	5.0	0.040	...	11,000	8,500	25,000	......	100	1.32	700	500	0.015	1/4	Paper	Phen.
XXX	13,500	11,800	0.40	0.35	950	1.40	50	5.3	0.038	...	15,000	12,000	32,000	25,500	110	1.32	650	450	0.015	2	Paper	Phen.
XXXP	12,000	10,500	....	....	.....	1.00	60	4.6	0.035	...	12,400	9,500	25,000	......	105	1.30	650	450	0.015	1/4	Paper	Phen.
XXXPC	12,000	10,500	....	....	.....	0.75	60	4.6	0.035	...	......	......	25,000	......	105	1.31	650	450	1/32	3/16	Paper	Phen.
ES-1	13,500	13,500	0.25	0.22	.....	2.50	..	...	.....	...	......	......	......	......	...	1.58	...	...	3/64	1/4	......	Mel.
ES-2	......	......	0.25	0.22	.....	....	..	...	.....	...	......	......	......	......	...	1.46	...	...	0.085	1/4	......	Phen.
ES-3	13,500	13,500	0.25	0.22	.....	2.50	..	...	.....	...	......	......	......	......	...	1.48	...	...	3/64	1/4	......	Mel.
C	17,000	16,000	2.10	1.90	1,800	4.40	15	...	.....	...	10,000	8,000	37,000	23,500	103	1.36	150	...	1/32	10	Cotton	Phen.
CE	17,000	14,000	1.60	1.40	1,800	2.20	35	...	.....	...	9,000	7,000	39,000	24,500	105	1.33	500	300	1/32	2	Cotton	Phen.
L	15,000	14,000	1.35	1.10	1,600	2.50	15	...	.....	...	13,000	9,000	35,000	23,500	105	1.35	150		0.010	2	Cotton	Phen.
LE	15,000	13,500	1.25	1.00	1,600	1.95	40	5.8	0.055	...	12,000	8,500	37,000	25,000	105	1.33	500	300	0.015	2	Cotton	Phen.
A	13,000	11,000	0.60	0.60	700	1.50	5	...	.....	...	10,000	8,000	40,000	17,000	111	1.72	225	135	0.025	2	Asb. paper	Phen.
AA	16,000	14,000	3.60	3.00	1,800	3.00	..	...	.....	...	12,000	10,000	38,000	21,000	103	1.70	...	...	1/16	2	Asb. fabric	Phen.
G-2	20,000	16,000	4.5	3.5	1,000	1.50	30	5.5	0.025	...	13,700	10,000	38,000	15,000	105	1.50	500	360	1/32	2	Staple gl.	Phen.
G-3	20,000	18,000	6.5	5.5	850	2.70	..	...	.....	...	23,000	20,000	50,000	17,500	100	1.65	700	500	0.010	2	Cont. gl.	Phen.
G-5	50,000	40,000	7.0 (to ½″)	5.5	1,570	2.70	23	7.8	0.020	180	37,000	30,000	70,000	25,000	120	1.90	350	220	0.010	3½	Cont. gl.	Mel.
G-7	20,000	18,000	6.5	5.5	650	0.55	32	4.2	0.003	180	23,000	18,500	45,000	14,000	100	1.68	400	350	0.010	2	Cont. gl.	Sil.
G-9	60,000	40,000	13.0	8.0	1,700	0.80	60	7.5	0.018	180	40,000	25,000	65,000	......	...	1.90	400	350	.....	...	Cont. gl.	Mel.
G-10	55,000	45,000	7.0	5.5	2,000	0.35	35	5.2	0.025	...	35,000	30,000	70,000	30,000	110	1.75	700	500	0.010	1	Cont. gl.	Epoxy
G-11	55,000	45,000	7.0	5.5	1,600	0.35	35	5.2	0.025	...	35,000	30,000	70,000	30,000	110	1.75	700	500	0.010	1	Cont. gl.	Epoxy
N-1	10,000	9,500	3.0	2.0	1,000	0.60	60	3.9	0.038	...	8,500	8,000	28,000	......	105	1.15	600	450	0.010	1	Nylon	Phen.
FR-2	12,000	10,500	....	....	.....	0.75	60	4.6	0.035	...	12,400	9,500	25,000	......	105	1.30	650	450	0.030	1/4	Paper	Phen.
FR-3	20,000	16,000	....	....	.....	0.65	60	4.6	0.035	...	12,000	9,000	28,000	......	...	1.45	600	500	1/32	1/4	Paper	Epoxy
FR-4	......	......	7.0	5.5	2,000	....	35	5.2	0.025	...	35,000	30,000	70,000	30,000	110	1.75	700	500	0.010	1	Cont. gl.	Epoxy
FR-5	......	......	7.0	5.5	1,600	....	35	5.2	0.025	...	35,000	30,000	70,000	30,000	110	1.75	700	500	0.010	1	Cont. gl.	Epoxy
GPO-1	18,000	18,000	8.0	8.0	850	1.00	40	4.3	0.03	100	12,000	10,000	30,000	20,000	100	1.5–1.9	400	...	1/16	2	Gl. mat.	Polyes.
GPO-2	18,000	18,000	8.0	8.0	850	0.9	40	...	.....	100	10,000	9,000	30,000	20,000	100	1.5–1.9	...		1/16	2	Gl. mat.	Polyes.

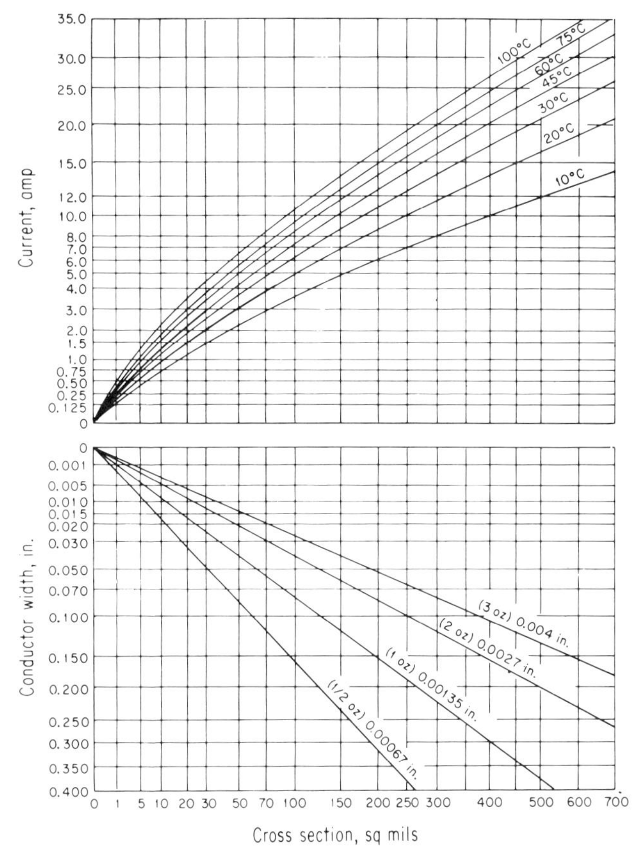

FIG. 61. Current-carrying capacity vs. etched conductor size for various temperature rises above ambient.[17]

lead length can be determined. The dotted lines can be used to determine line length needed to hold interconnection delay to a given percentage of logic delay.

Crosstalk. With conductors getting closer and closer together in multilayer boards, crosstalk becomes more important than ever. Consult the data in Chap. 1 for crosstalk values.

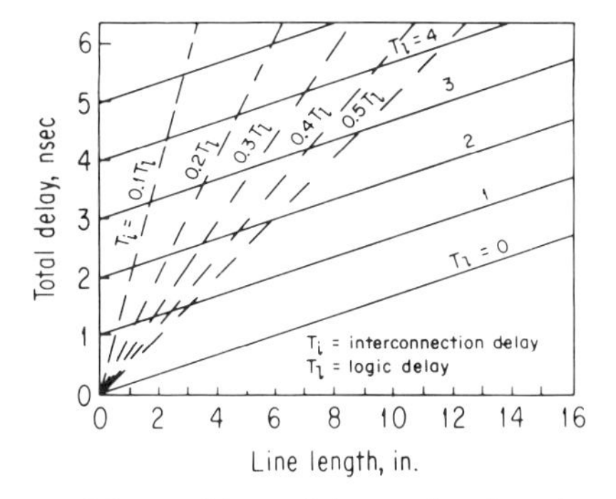

FIG. 62. Interconnection delay and logic delay vs. line length for a dielectric of 4.5.[18]

Environmental Characteristics. Temperature has an effect on current-carrying ability, but printed-circuit boards are affected more by humidity than by any other factor. The surface resistivity of glass-epoxy boards may decrease by a factor of 10 to 1 in a humid atmosphere. Volume resistivity, on the other hand, does not change as drastically. Effective insulation resistance between conductors embedded inside the epoxy (as in multilayer boards) is therefore much better than for conductors on the surface. In high-impedance circuits where insulation resistance may effect circuit operation, printed-circuit boards are not a preferred method of packaging, although for marginal cases, spacing on the surface should be greater than for spacings inside the epoxy. Normally, surface spacings at the solder pads may be a limiting design feature even if all conductors are buried.

For maximum insulation resistance, several precautions can be taken:

1. Store printed-circuit boards in a controlled dry atmosphere.
2. Dry printed-circuit boards in an oven immediately before soldering or welding components in place.
3. Coat the printed-circuit-board assemblies with a nonpermeable coating.

Flat Cable. Many electrical characteristics of flat cable are similar to those of printed circuits, particularly multilayer circuits. Although different materials may be used, the basic feature that makes them similar is the configuration and spacing of flat conductors. Unlike hook-up wiring cabling, electrical characteristics can be calculated and then maintained uniformly in production. Basic characteristics of flat cable are given in Chap. 1.

Crosstalk. Whenever two conductors run parallel to each other in close proximity, the capacitive coupling causes crosstalk. Flat cable can result in considerable crosstalk between leads, but fortunately the configuration of flat cable lends itself to effective shielding. A grounded shield bonded to one or both sides of the cable lowers the direct conductor-to-conductor capacitance, and some thought about lead placement and grounded leads has an additional effect. A summary of crosstalk measurements for several flat cable configurations is given in Chap. 1. Figures 63 to 65 show the difference in unshielded, single-side-shielded, and both-side-shielded flat cable for three different circuit lead configurations. Note that lead configuration alone may have over a 10:1 ratio effect on crosstalk, and shielding may have an additional 10:1 effect in reducing crosstalk.

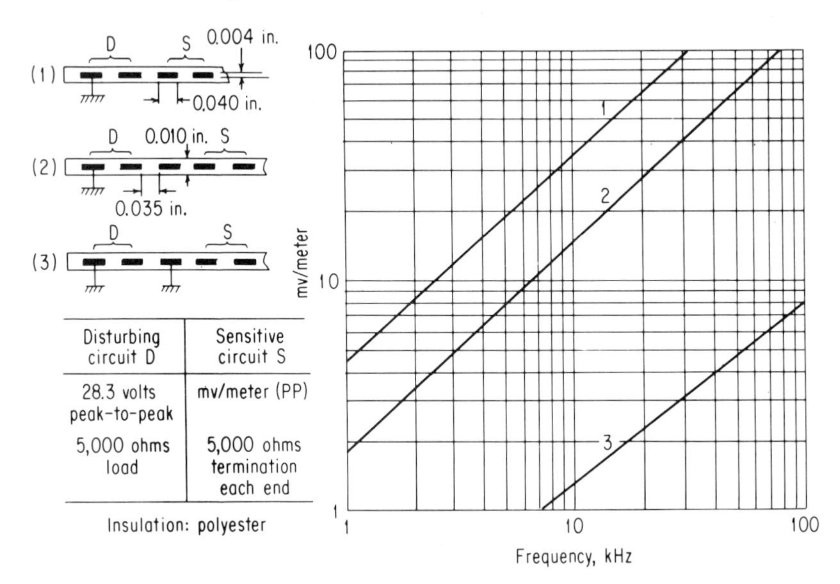

FIG. 63. Crosstalk in unshielded flat cable.[19]

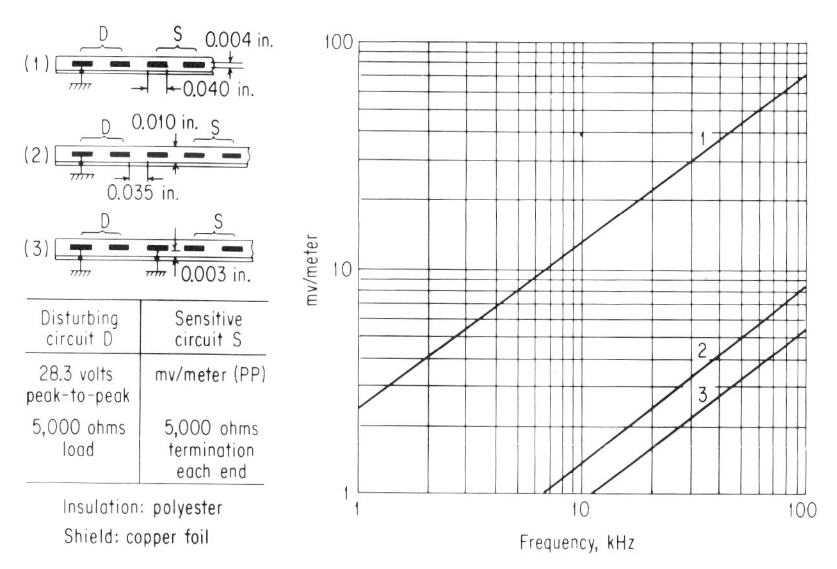

FIG. 64. Crosstalk in flat cable with one side shielded.[19]

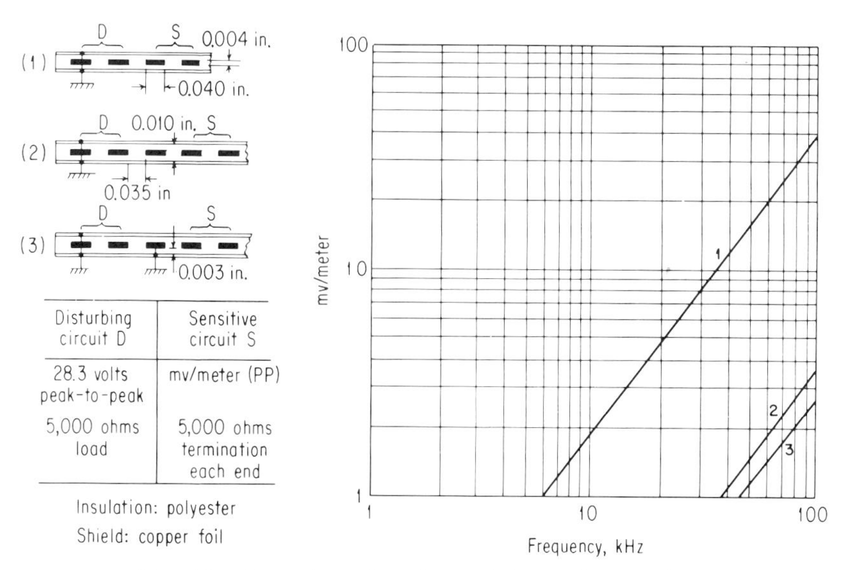

FIG. 65. Crosstalk in flat cable with both sides shielded.[19]

Thick and Thin Films. The large range of component values available to thin- or thick-film technology as opposed to integrated-circuit values is one major reason why they are used in hybrid systems.

Resistors. The basic equation for the design of thick-film resistors is

$$R = \rho(l/w)t$$
$$lw = P/D$$

where R = resistance, ohms
w = width, in.
l = length, in.
t = thickness, mils
ρ = surface resistivity, ohms/(sq)(mil)
D = allowable dissipation, watts/(in.^{2})(mil)
P = power rating, watts

Thin-film resistors are typically 200 ohms/sq, but nichrome and tantalum resistors of up to 400 ohms/sq are possible in production. Figure 66 shows how resistance varies with film thickness and illustrates why close tolerances become more difficult as film thickness decreases.

Capacitors. The basic equation for the design of thick-film capacitors is

$$C = 225\, k_2 A/t$$

where C = capacitance, pf
k_2 = dielectric constant (relative to air)
A = area of one side of one plate, mils^2
t = dielectric thickness, mils

Thin-film capacitors also vary considerably with materials. Representative thin-film capacitance materials and their characteristics including range of values are given in Chap. 5.

Inductors. The basic equation for thick-film round and square coils is

For circular coils

$$L = 0.126 a n^{5/3} \log 8 \frac{a}{c}$$

For square coils

$$L = 0.141 a n^{5/3} \log 8 \frac{a}{c}$$

where L = inductance, μh
n = number of turns
a = mean radius = $\dfrac{\text{OD} + \text{ID}}{4}$ in.
c = depth of winding = $\dfrac{\text{OD} - \text{ID}}{2}$ in.

ID, OD = inside and outside diameters of circular coil or the corresponding sides of square coil

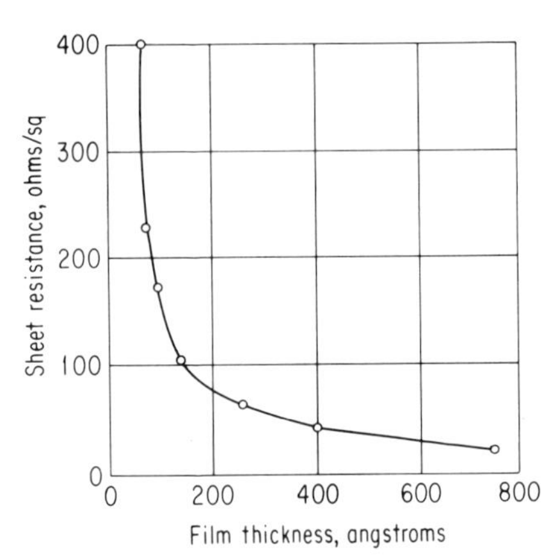

FIG. 66. Sheet resistance as a function of thickness for an NiCr film.[20]

TABLE 29. Comparative Characteristics of Several Logic Circuit Types [21]

Logic type manufacturer and component number	Speed		Fan-out		Noise immunity 25°C		Power dissipation		Temperature range, °C	Supply voltages, volts
	Gate delay, nsec (fan-out = 3)	Binary count, Mc	Gate	Buffer	Gate and binary, mv	Percent of logic swing	Single gate, mw	Binary element, mw		
DCTL Fairchild HSμL900	15	10	5	25	250	18	12	52	−55 to +125	+3
RCTL Texas Instr. Series 51	200	0.8	5	25	500†	17†	2	2	−55 to +125	+3
DTL Signetics SE100	30	10	5	25	500	14	7	16	−55 to +125	+4 and −2 or 0
RTL Fairchild MWμL 910	40	*	5	*	100	13	2	12	−55 to +125	+3
TTL Sylvania SUHL II	7	30	5	12	1,000	33	20	55	−55 to +125	+5
CML or ECTL Motorola MC300	7	40	26	*	320	40	35	52	−55 to +125	−5.2 −1.15 Ref.
MOSTL GME SC1174	50 (3-4 mw/node)	1	∞	∞	2,000	20	0.6	0.66	−0 to +70	−20

* Information not available.
† Depends on noise rise/fall time.

Integrated Circuits. The appearance of integrated circuits has brought about a whole new set of problems for the packaging engineer because they are not basically electrical or mechanical problems. They require a new vocabulary and familiarity with a great many new techniques. Integrated-circuit packaging methods are discussed in Chap. 9, and bonding techniques used in some of these packages are detailed in Chap. 4.

The processes for building, assembling, and packaging integrated circuits will not be discussed here (*Integrated Circuit Engineering—Basic Technology* [20] is an excellent source), but some of the basic types of circuits and some integrated-circuit characteristics will be covered briefly.

Logic Types. Several different logic types are available from a variety of manufacturers. Logic characteristics vary widely, and the type must be chosen carefully for the purpose intended. Speed, fan-out, and other characteristics are summarized in Table 29 for seven representative logic types from different companies. These characteristics should not be used in designing, but are provided as a means of comparing various types of logic circuitry.

Performance ranking of each logic type is given in Table 30. The logic type dis-

TABLE 30. Performance Ranking of Basic Logic Circuit Types (1 = best performance) [21]

Type of transistor logic	Speed	Fan-out	Noise immunity	Power level	Manufacturing cost
DCTL	3	7	7	5	1
RCTL	6	5	6	2	7
DTL	4	4	3	4	4
RTL	5	6	5	3	3
TTL	2	3	2	6	5
ECTL	1	2	4	7	6
MOSTL ...	7	1	1	1	2

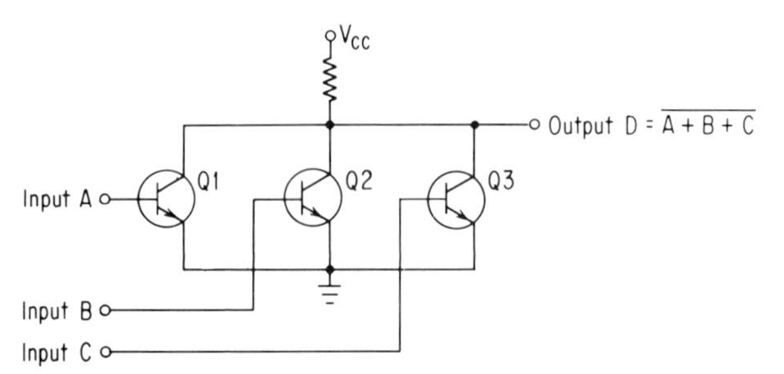

Fig. 67. Typical DCTL gate.[21]

playing the best performance for a given characteristic is given a value of 1 with values 2 through 7 assigned in order. A value of 7 is assigned to the logic type displaying the poorest performance in that particular characteristic. No one logic type is best in every characteristic, so trade-offs must be made to decide which type should be used for any system. Each type is discussed briefly below.

DCTL. The basic circuit element of the typical direct-coupled-transistor logic (DCTL) circuit of Fig. 67 is the transistor. Since resistors are used only for loads, there are no coupling resistors to slow down circuit speed. One major problem with DCTL is relatively low and variable fan-out characteristics caused by current hogging in some transistors with different characteristics than others.

RCTL. The resistor-capacitor-transistor logic (RCTL) of Fig. 68 costs more because the resistors and capacitors occupy more space than transistors alone, and integrated-circuit cost is proportional to area. The RC circuit prevents current hogging and helps fan-out. Some optimum value of capacitor can be chosen to trade-off speed versus noise immunity.

DTL. Figure 69 shows the diode-transistor logic (DTL) configuration. This form is more suited to integrated-circuit production than is RCTL logic because diodes require less space than resistors and capacitors. DTL circuits have greater fan-out, larger voltage swing, higher speed, and higher noise immunity.

RTL. The resistor-transistor logic (RTL) circuit shown in Fig. 70 is a modification of the DCTL circuit, with input resistors added to prevent current hogging. It is less complex than the RCTL type but still requires more space than DCTL or DTL, and the parasitic capacitances associated with the input resistors increase the propagation delay.

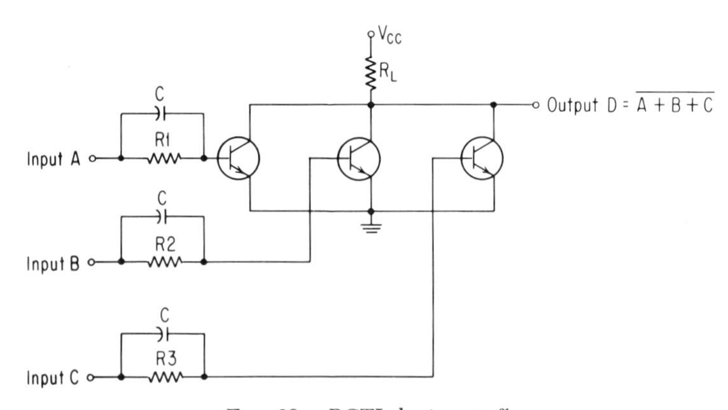

Fig. 68. RCTL logic gate.[21]

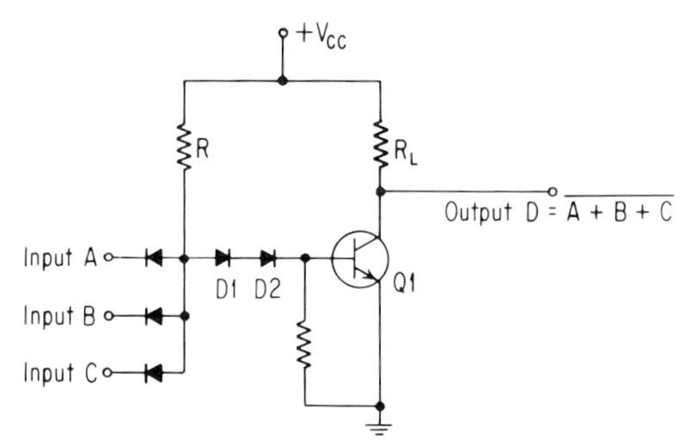

FIG. 69. DTL logic gate.[21]

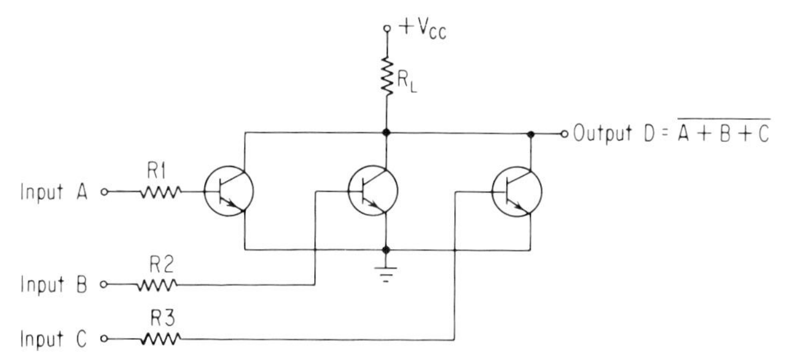

FIG. 70. RTL logic gate.[21]

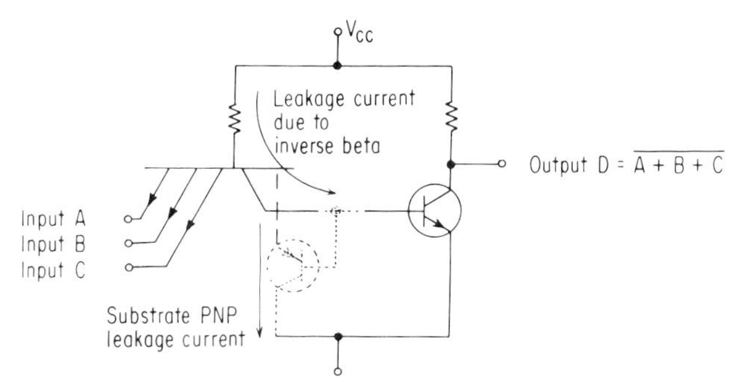

FIG. 71. TTL logic gate.[21]

TTL. Figure 71 shows the transistor-transistor logic (TTL) in its basic configuration. The input transistor has a number of emitters in place of the coupling diodes of a DTL gate. It is easily built in integrated-circuit form, since active components are used with a minimum of passive components. Unlike other circuits, the PNP transistor formed by the substrate isolation region (dotted lines in Fig. 71) conducts, resulting in a slower circuit.

ECTL. The basic emitter-coupled transistor logic (ECTL) circuit is shown in Fig. 72 along with two emitter-follower stages, one for each output. This type of logic exhibits high speed, high fan-out, and high capacitance-load driving capabilities, but it has low output voltage swings and requires three stable power supplies. ECTL is

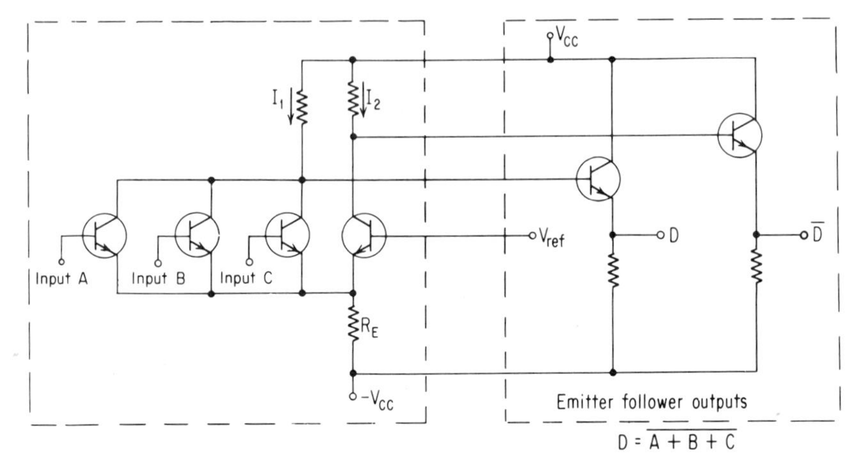

FIG. 72. ECTL logic gate with emitter-follower outputs.[21]

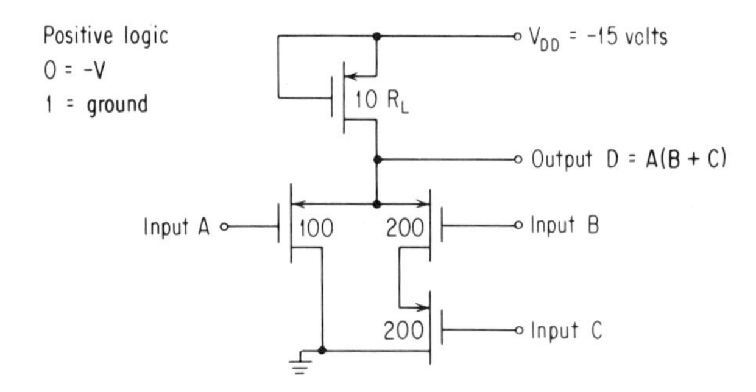

FIG. 73. Typical MOS logic gate.[21]

also referred to as current-mode logic (CML) because it is basically a current switch rather than a voltage switch.

MOSTL. The newest and most rapidly developing type of logic is metal-oxide-semiconductor-transistor logic (MOSTL). A typical MOS logic gate is shown in Fig. 73. MOS devices are easily made in integrated form because there are no passive elements, and only a single diffusion is necessary. They are extremely small compared with other devices. Fan-out is very large, but speed and operating temperature are limited. Long-term parameter stability problems are disappearing as design and processing improvements are made.

Isolation. Isolation has always been a problem in integrated circuits, and it has been handled in one of four general ways:

1. Diode isolation
2. Substrate diffusion isolation
3. Dielectric isolation
4. Physical separation

In order to save money, reduce interconnections, and increase reliability, the trend continues toward including more and more on a single monolithic substrate. Several variations in large-scale integration (LSI) packaging techniques are discussed in Chap. 9. Electrical limits are often reached long before processing limits because of isolation problems with critical circuits.

DIODE ISOLATION. Discrete components are physically separated by virtue of the fact that they are packaged in individual cases and isolation of a sort is automatically

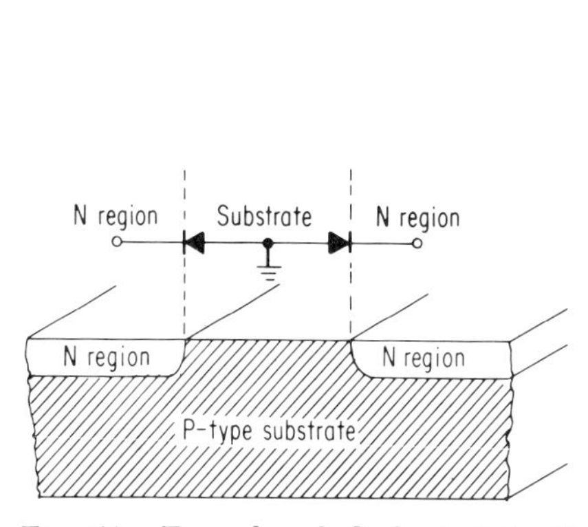

FIG. 74. Example of diode isolation.[20]

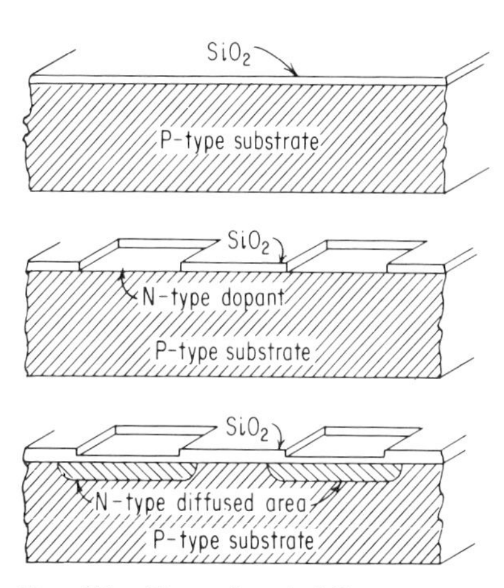

FIG. 75. Example of diffusion isolation.[20]

provided. Since the components of an integrated circuit are produced on a single substrate, isolation is anything but automatic. Pairs of reversed biased PN junctions like those in Fig. 74 are the most commonly used form of isolation in integrated circuits.

SUBSTRATE DIFFUSION ISOLATION. Diffusion is used to isolate regions from the substrate, as shown in Fig. 75. Diffused collector isolation was one of the first methods used for monolithic circuit production. A variation diffuses the entire substrate except at the collector (or other isolated region) in a process which is essentially the inverse of the diffused collector method. For certain applications, the substrate resistance is enough isolation, particularly if it contains enough impurities. Figure 76 shows the resistance isolation between two transistors. Figure 77 indicates the variation in the resistivity of silicon versus impurity concentration.

DIELECTRIC ISOLATION. The various methods used in attaining dielectric isolation will not be described here, but the end result is shown in Fig. 78. Each element is

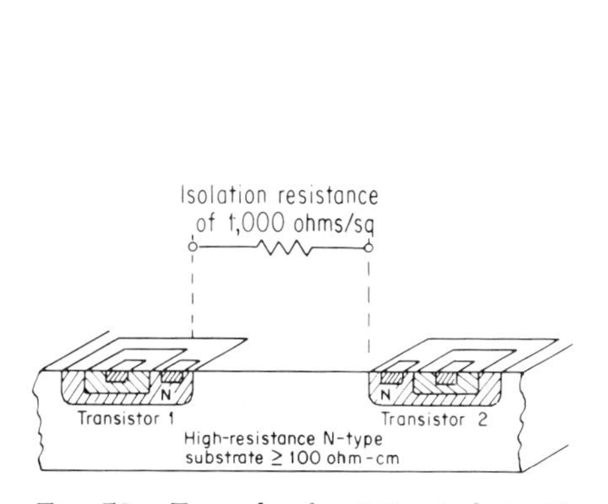

FIG. 76. Example of resistive isolation.[20]

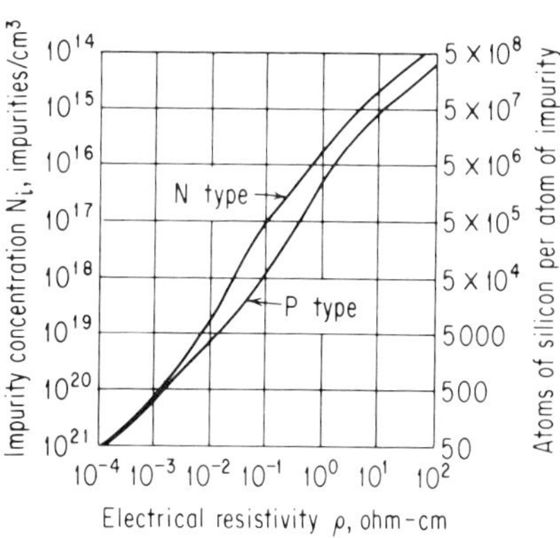

FIG. 77. Variation of electrical resistivity in silicon at 300°K as a function of impurity concentration.[20]

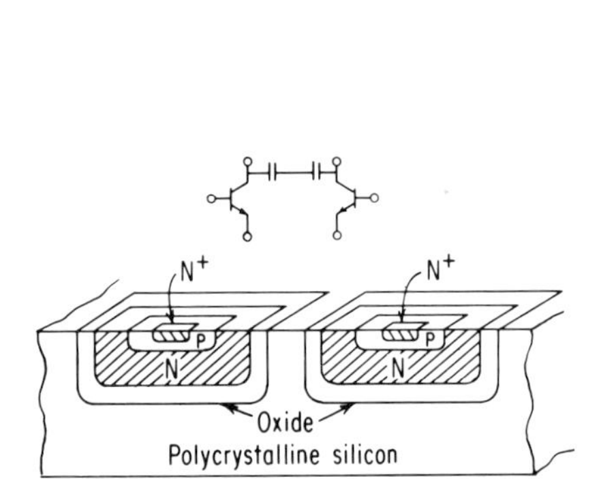

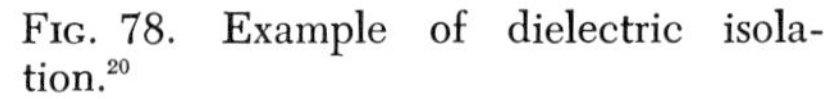
FIG. 78. Example of dielectric isolation.[20]

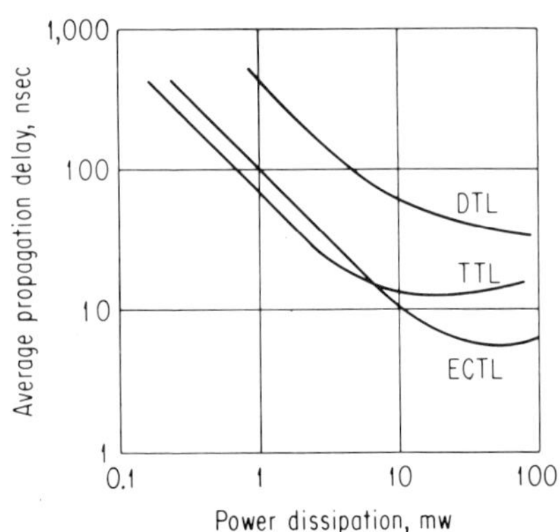

FIG. 79. Propagation delay as a function of power dissipation for various types of logic.[20]

separated from the other by a true dielectric layer, and isolation is much better than the inefficient isolation of diffusion.

PHYSICAL SEPARATION. Heavy leads may be built up on the substrate to connect circuit areas which must be isolated one from the other; then the substrate can be removed entirely under the lead area without disconnecting the circuits. More common methods include the interconnection of circuits on separate chips using standard wire-bonding techniques (see Chaps. 4 and 9).

Speed. In previous passages, propagation delays in various types of circuits have been discussed. Dissipated power affects the delay, as indicated in the curves of Fig. 79, for three types of logic. Figure 62 illustrates how lead propagation times are significant when circuits switch fast. Figure 80 indicates the wide variations in rise time caused by three different environments for an idealized 1-nsec rise-time circuit. In the ideal circuit at the top of Fig. 80, the switch represents the transistor. The lowest curve shows how the rise time varies with power when the driving and driven circuits are all on one monolithic chip; total load capacitance is 1 pf. The second curve shows what happens when driving and driven circuits are located on separate chips with an additional 5 pf appearing on the transmission line. The uppermost curve

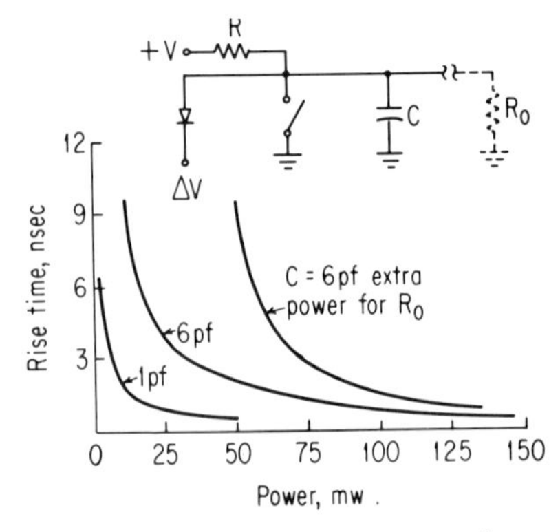

FIG. 80. Rise time vs. power for a 1-nsec circuit operating in three environments.[22]

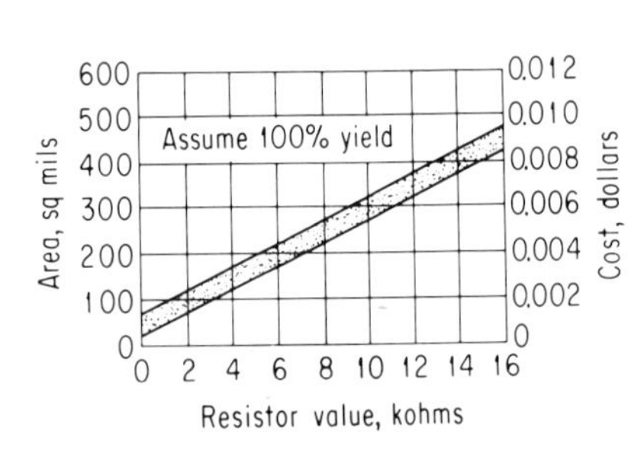

FIG. 81. Resistor value vs. area and cost for a single diffused resistor.[20]

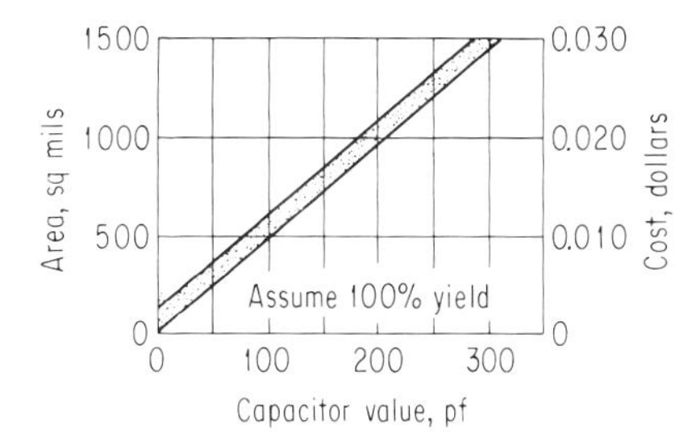

Fig. 82. Capacitor value vs. area and cost for an oxide capacitor.[20]

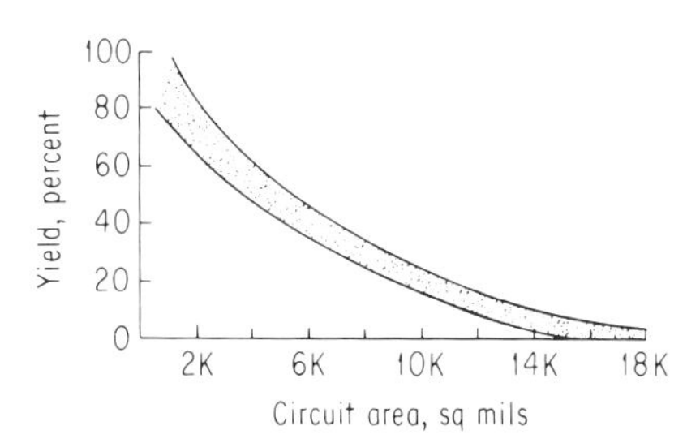

Fig. 83. Yield of silicon monolithic circuits as a function of die area.[20]

indicates the additional delay when driving and driven circuits are located more than 1¼ in. apart and a terminating resistor is required.

Component Ranges. For maximum utilization of integrated-circuit capabilities, discrete-component circuits should not be converted to integrated-circuit form, value for value; however, it is important to know what range of values can be made and the tolerances to expect. Resistor and capacitor values are dependent upon area as shown in Figs. 81 and 82, respectively. These curves assume a 100 percent yield, which is unrealistic. The curve of Fig. 83 shows how yield varies with circuit area; the larger the die area, the less the yield. The curve indicates that a yield of about 20 percent could be expected for a die 100 mils square. Yields are constantly being improved and these figures should not be taken literally; rather, they are indicative of the increased cost and decreased yield to be expected as larger-value components are specified.

Since all the components in an integrated circuit are made in one process batch, matching within a circuit will be much better than tolerance between batches. Figure 84 shows the distribution of absolute values and ratios for diffused resistors. Integrated circuits should be designed such that absolute values are not so important as the ratio of values within the circuit. Active components should be used instead of passive components wherever possible. Typical design tolerances for silicon monolithic integrated circuits are given in Table 31. Characteristics of integrated resistors, capacitors, and transistors are given in Tables 32, 33, and 34, respectively.

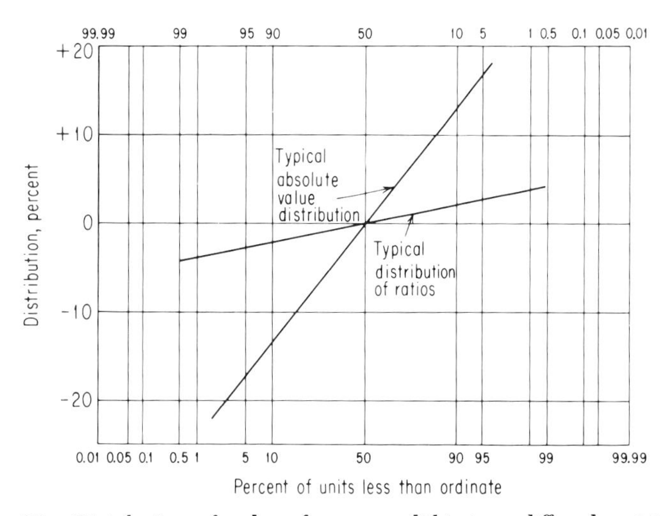

Fig. 84. Distribution of values for a monolithic-type diffused resistor.[20]

TABLE 31. Typical Design Tolerances for Silicon Monolithic Integrated Circuits [20]

Component	Symbol	Value	Tolerance, %
Epitaxial layer sheet resistance	R_s	150 ohms/sq	±15
Base diffusion sheet resistance	R_s	200 ohms/sq	±10
Emitter diffusion sheet resistance	R_s	2.5 ohms/sq	±30
Resistor matching	$R_1 : R_2$	....	± 5
Low current forward transfer ratio (gain) ..	h_{FE}	70	±50
Diode forward voltage drop	V_F	0.7 volts	± 3
Collector-substrate breakdown voltage	BV_{CSO}	80 volts	±25
Collector-base breakdown voltage	BV_{CBO}	45 volts	±30
Emitter-base breakdown voltage	BV_{EBO}	6.7 volts	± 3
Low current collector saturation voltage ...	$V_{CE(sat)}$	0.4 volts	±25

TABLE 32. Characteristics of Integrated Resistors [20]

Parameters	Si P— type	Si N+ type	NiCr	SnO_2	Cr-SiO (Cermet†)	Conventional molded carbon
Range, ohms/sq	50–250	2–5	40–400	50–500	100–10 K	
Max. value, ohms ...	25 K	250	100 K	100 K	200 K	22 meg
Min. value, ohms ...	25	5	20	25	50	0.24
Tolerance,* %	±10	±15	±10	±15	±20	±5
Max. power, mw/sq mil	3	3	2	2	2	2 watts total
Max. voltage, volts ..	±20	±5	±100	±100	±100	±750
Aging allowance, %	±2	±2	+3	To 4	+5	10
Temperature coefficient, ppm/°C	+50 to +2,500	+100	±100	To −1,500	±500	±1,500
Circuit applications:						
Multichip	Yes	Yes	Yes	Yes	Yes	
Si—monolithic ...	Yes	Yes				
Compatible	....	.....	Yes	Yes		
Thin film	....	.....	Yes	Yes	Yes	

* Unadjusted.
† Trademark, CTS Corporation.

TABLE 33. Characteristics of Integrated Capacitors [20]

Parameters	Junction	SiO_2	Al_2O_3-Si	Ta_2O_5	SiO
Max. value, pf	1,000	250	1,000	5,000	100
Pf/mil²	1.2	0.25	0.3	2.5	0.2
Transfer ratio (C/C shunt) ...	3	5	15	100	1
Voltage coefficient	$V^{-1/2}$	0	0	0	0
Breakdown voltage (BV).....	7	30	30	20	30
Temperature coefficient, ppm/°C	....	+100	+125	+400	±110
Dissipation factor at 1 Mc, %	....	0.7	0.4	0.3	0.7
Dielectric constant	13.7	3.78	≈10	25	≈6
Circuit applications:					
Multichip	Yes	Yes			
Si—monolithic	Yes	Yes			
Compatible	....	Yes	Yes		
Thin film	....	....	Yes	Yes	Yes

TABLE 34. **Characteristics of Typical Silicon Small-geometry Transistors** [20]

Parameters	Monolithic 0.1-Ω-cm collector	Monolithic 0.5-Ω-cm collector	Hybrid 2N834 type
Collector, substrate, BV_{CSO}, volts	80	80	
Collector, base, BV_{CBO}, volts	20	40	40
Base, emitter, BV_{EBO}, volts	6	7	5
Collector, emitter, BV_{CEO}, volts	12	20	30
Collector, substrate capacitance, pf	10	10	
Collector, base capacitance, pf	8	4	2.5
H_{FE} (I_C = 10 ma)	40	40	40
R_{SAT}, ohms	20	80	10
$V_{CE(SAT)}$ (I_C = 5 ma), volts	0.3	0.55	0.15
f_T (I_C = 10 ma, V_C = 5 volts), Mc	550	500	500

Purple Plague. Purple plague is a chemically caused failure possible only when gold, aluminum, and silicon are brought together, and it is accelerated with high temperatures. $AuAl_2$ has a characteristic purple color but does not cause the failure. A white powdery form of Au-Al-Si causes a high-resistance interface which effectively opens up the output leads. Purple plague is controlled in normal gold-aluminum integrated-circuit systems by controlling bonding temperatures and times. Large circuit packages requiring many bonds extend the bonding time at elevated temperatures, and to avoid purple plague, all-aluminum or all-gold systems may be necessary. Chapter 9 discusses a variety of packaging techniques including ultrasonically bonded subassemblies which do not require elevated temperatures.

High-voltage Circuits. Microminiature packaging carries the connotation of low voltage and low power; most of this handbook is directed toward that type of circuit. Many complex systems have a high-voltage section that cannot be handled in the same manner as the rest of the system. High-voltage packaging will be reviewed and some helpful design curves presented.

Corona. Corona occurs when high voltage causes a gas to ionize. Air is the medium most frequently present in electronic equipment, and corona causes a luminous discharge in it. Corona usually precedes arc breakdown in a uniform field, but some dielectrics will not deteriorate after being subjected to corona for long periods of time. Ac voltages are more critical than dc voltages by a ratio of approximately 1.6 to 1, although this figure is decreased if the direct current has a ripple on it. Corona can be decreased by increasing spacing, improving the configuration, providing shielding between breakdown surfaces, or eliminating the air gap.

Configuration Effect. Increasing distance between points where corona is likely to occur is an obvious possibility. In general, sharp corners permit corona at much lower voltages than smooth corners. Figure 85 shows the corona and breakdown levels for a 90° corner and an edge (0° corner) above a ground plane at 60 cycles. The effect of both distance and shape is apparent. Even fairly small changes in shape such as from a 13/16-in.-diameter sphere to a 1½-in.-diameter sphere affects corona level as evidenced in Fig. 86. Figure 87 shows the same type of information for a cylindrical shape at varying distances above a ground plane. Table 35 summarizes the effect of several configurations on corona level in air with a 1-in. gap. Rounding off all sharp corners on bare conductors is an important design rule.

Wire. Ionization of pockets or voids within wire insulation must be considered, as well as ionization of the air surrounding the insulation. Assuming a perfectly solid insulation, the air dielectric and insulation dielectric are in series. This is not true where the lead is terminated and the insulation is broken. Corona losses in wiring may be reduced in the following ways:

1. Increase distance.
2. Apply insulation tightly to prevent gas space between conductor and insulation. Solid wire can be handled in this respect more easily than can stranded wire.

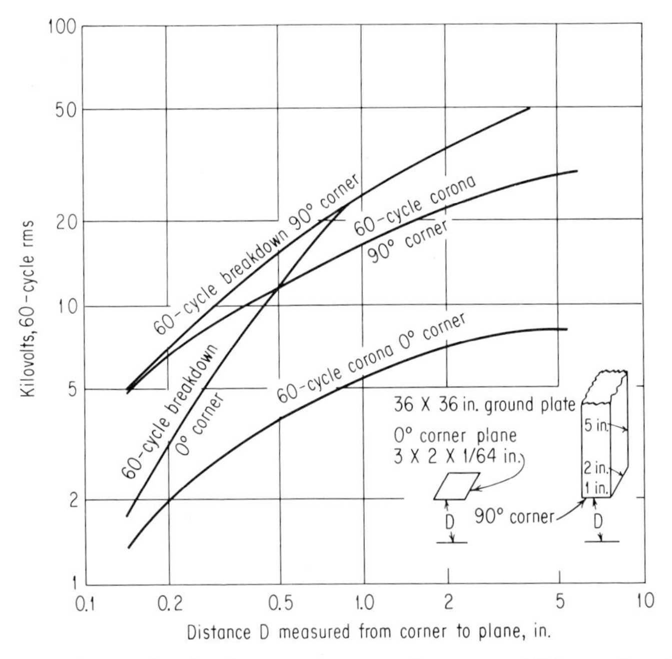

FIG. 85. Corona level, plane-to-corner configuration. (*Filtron Co., Inc.*)

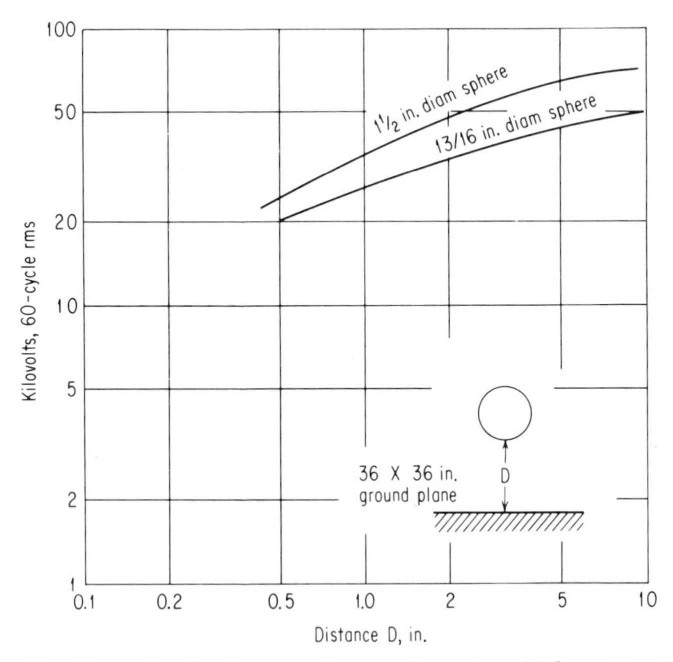

FIG. 86. Corona level, plane-to-sphere configuration. (*Filtron Co., Inc.*)

TABLE 35. Relationship of Geometry of Device to Corona Voltage [3]

Geometry	*Kv corona voltage (with 1-in. air gap)*
0° corner	5.4
90° corner	17
13/16-in.-diam sphere	26
1½-in.-diam sphere	35
½-in.-diam cylinder	35
1-in.-diam cylinder	42
Plane	51

SOURCE: Filtron Co., Inc.

3. Increase conductor diameter (see Fig. 87).
4. Round off all corners, especially at terminals and joints where conductors are bared.
5. Raise the wire insulation off ground. Figures 88, 89, and 90 show the corona levels for various diameter wires on ground, raised above ground 1/8 in., and raised above ground 1/4 in.
6. Pot or mold lead terminations, including wire and terminal, with a material that will bond well to the wire and leave no air pockets or voids.
7. Choose the insulation dielectric carefully. The ordinate of Fig. 88 is voltage times dielectric constant, indicating the effect of the dielectric. For a given air gap in series with an insulation dielectric, it is preferable to use a low dielectric material to minimize the stress on the air.
8. Electrically eliminate the air gap by coating the insulation over two voltage differential conductors with shielding conductors which are tied together.

Altitude. As atmospheric pressure decreases, the dielectric strength of air de-

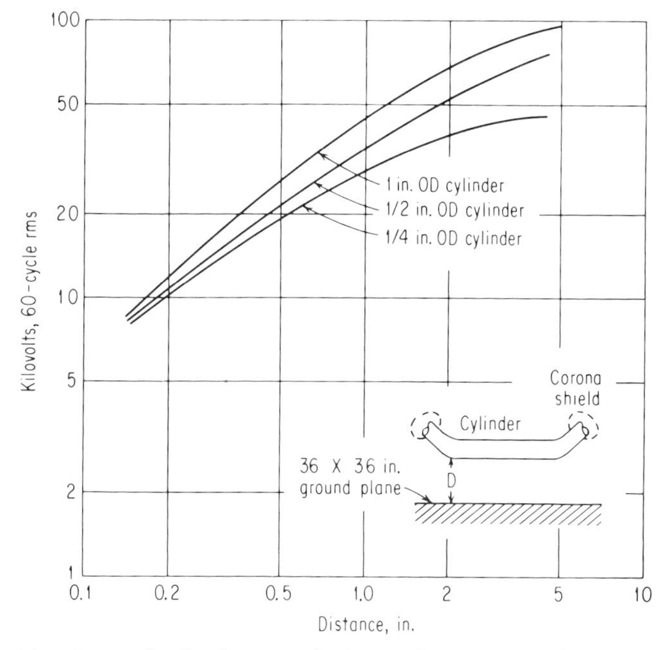

FIG. 87. Corona level, plane-to-cylinder configuration. (*Filtron Co., Inc.*)

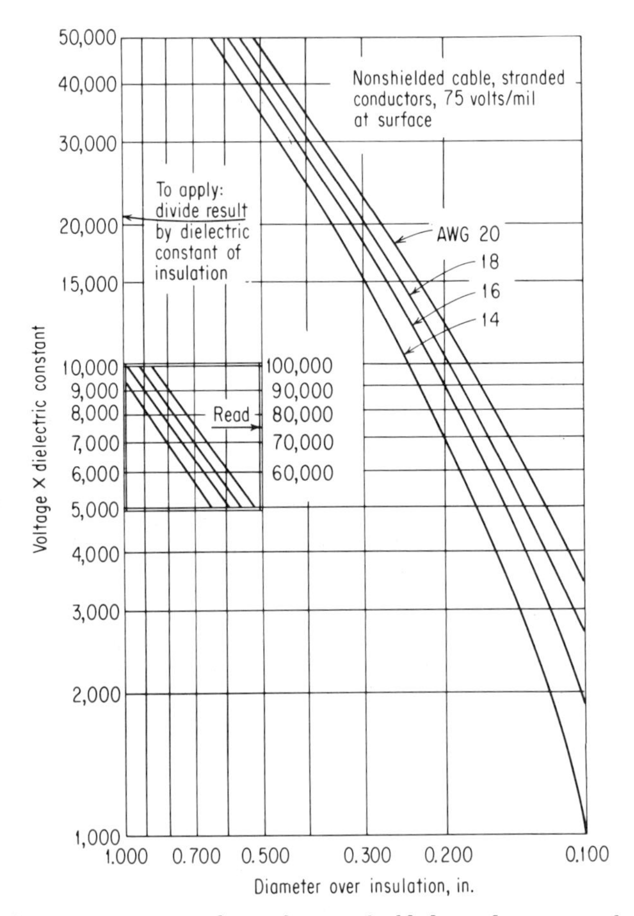

FIG. 88. Corona inception voltage for nonshielded conductors touching ground plane.[14]

creases until the pressure reaches about 1 mm Hg (about 30 miles). Beyond that point, the dielectric strength gradually increases again. The Paschen's curves of Fig. 91 show how the breakdown voltage of three gases varies with the product of pressure and distance. Note that no breakdown ever occurs below 300 volts in air. The critical region for normal conductor spacings at 1,000 volts or more is in the 10- to 50-mile altitude range. The safest approach is to rely entirely on insulating materials and allow no air space between conductors or between conductors and ground.

Even at sea level, the dielectric strength of air is variable. The normal dielectric strength of 75 volts/mil may decrease to 45 volts/mil under extreme conditions involving a hot, humid, and contaminated atmosphere.

Reliability. Reliability is extremely important in electronic systems, and the packaging engineer should consider it carefully. When interfacing with the electrical designer, the packaging engineer should consider three principal areas from an electrical reliability viewpoint: component usage, redundancy, and the contribution his packaging design makes to electrical parameters.

Component Usage. A large part of this chapter has been devoted to the proper usage of components in building an electrically compatible system. Reliable com-

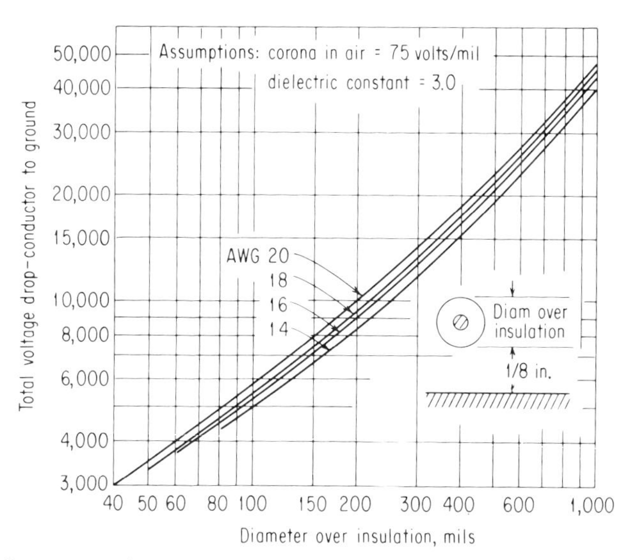

FIG. 89. Operating voltages without corona for nonshielded conductors raised ⅛ in. above ground plane.[14]

ponents do not make a reliable system unless they are used properly, although system reliability figures are all too often based solely on statistical component reliabilities. Refer to other portions of this chapter, to Chap. 8, and to Chap. 9, as well as to the reliability section for information in a particular area. Consider the possibility of long-term failures and design accordingly. Temperature deratings far below those normally required add considerably to component reliability, for example.

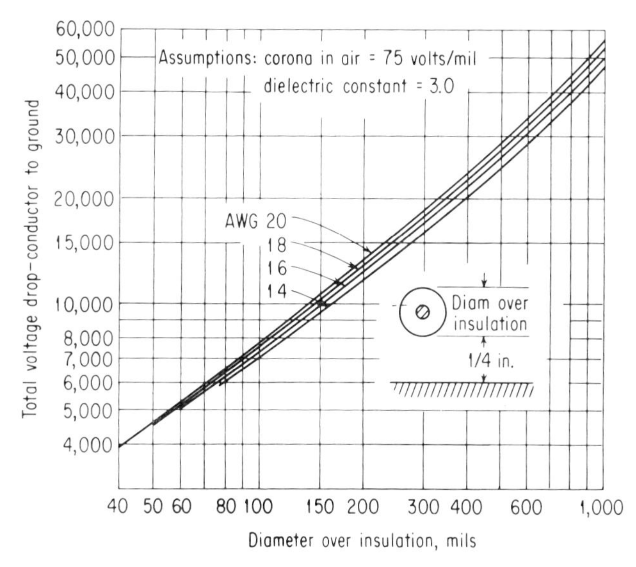

FIG. 90. Operating voltages without corona for nonshielded conductors raised ¼ in. above ground plane.[14]

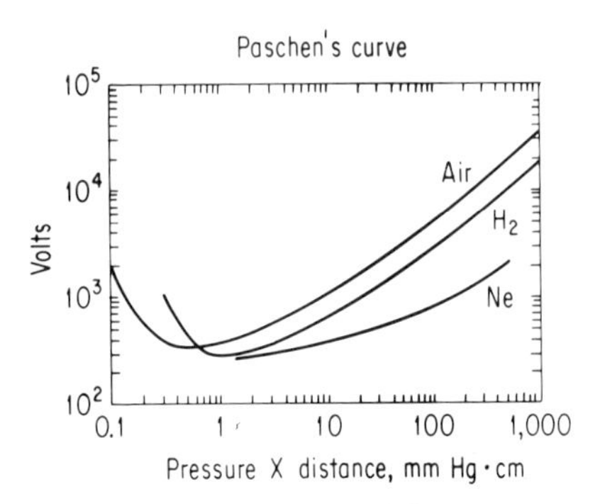

FIG. 91. Uniform field breakdown of gases.

Redundancy. Redundancy can be provided at any level from system to component. The effect on system reliability varies depending upon the method used. Straight parallel redundancy of X elements, all operating, is shown on the curves of Fig. 92. If only one element operates and the others are on standby, reliability is improved, as shown in Fig. 93, because the standby elements do not operate until necessary. A system is seldom that simple; parallel series redundancy is commonly used with the results shown in Figs. 94 and 95 for operating and standby conditions.

Paralleling components or assemblies is not as straightforward as a block diagram might imply. Additional circuits are usually required to interface between elements properly. The series of Figs. 96 to 98 illustrates this point. Figure 96 shows a simple two-stage logic system. Three logic A's cannot be paralleled without affecting system operation, and therefore a majority-vote unit is added between the A and B stages, as shown in Fig. 97. The A function becomes more reliable, but a failure may occur in M. Adding three majority voters as shown in Fig. 98 overcomes this drawback, but the total number of components and the wiring complexity have more than tripled. More intricate duplexing techniques have been worked out that use fewer components and give greater reliability than triplicated majority-vote logic.[23]

Contribution of Packaging Techniques. Interconnection techniques contribute

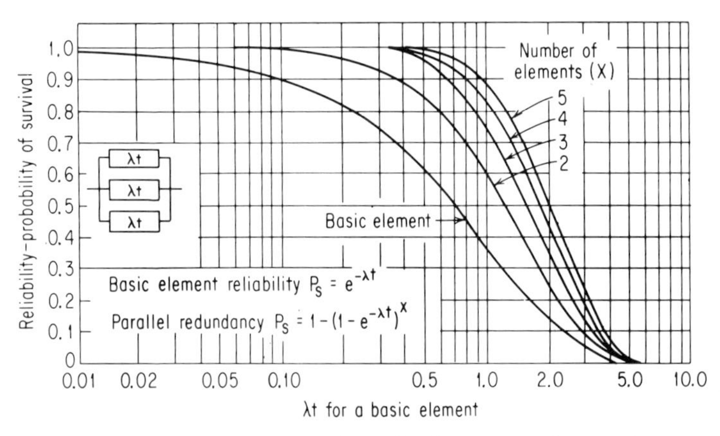

FIG. 92. Effects on reliability of parallel redundancy considering X operative elements.[2]

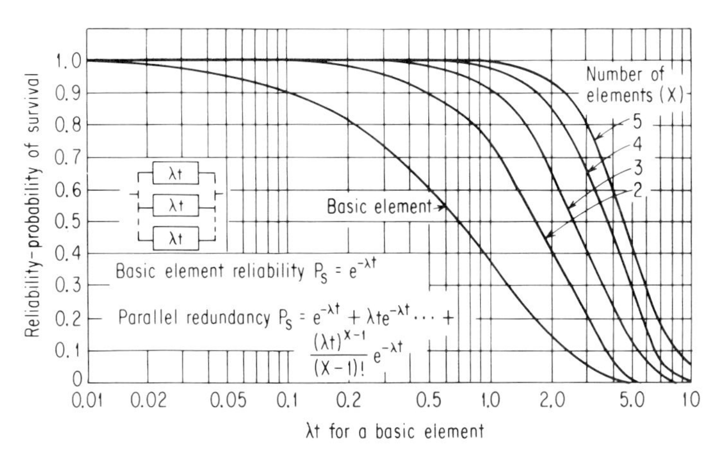

FIG. 93. Effects on reliability of parallel redundancy considering X standby elements.[2]

heavily to system reliability and should not be overlooked. Comparative reliability data for soldered, welded, ultrasonically bonded, sliding, and other types of joints cause much heated discussion, largely because the human element affects reliability. A soldered joint made on a tightly controlled wave solder machine in one plant is not the same as a soldered joint made by hand in another. Welded joints may be more reliable than soldered joints in a given location, but in another, the reverse may be true. The effect of packaging techniques on system reliability should be considered for the particular location and skill available.

Subassembly size and "break point," the division point at which a subassembly can be removed without loosening wires, may be in part determined by the reliability of the connectors or contacts being used. Given accurate reliability figures for the joints at various levels, divisions can be made consistent with maximum reliability and other tradeoffs such as maintainability. Where repairability is not required, as in a space system, "hard" wiring (no plugs) may be necessary for maximum reliability.

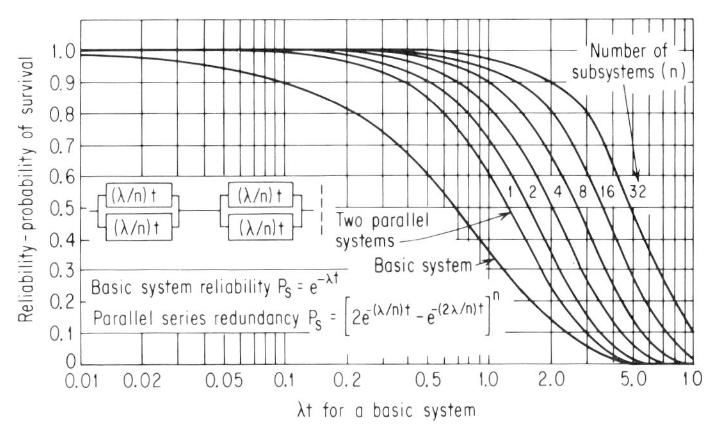

FIG. 94. Effects on reliability of parallel series redundancy using two systems and an operative redundant configuration.[2]

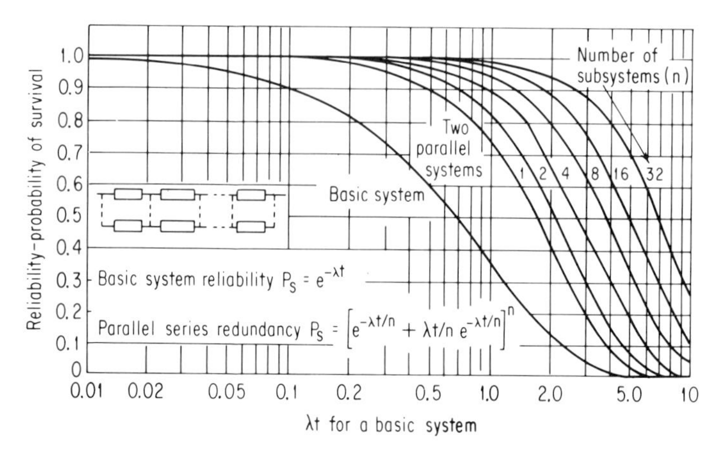

FIG. 95. Effects on reliability of parallel series redundancy using two systems and a standby redundant configuration.[2]

FIG. 96. Conventional nonredundant logic.[23]

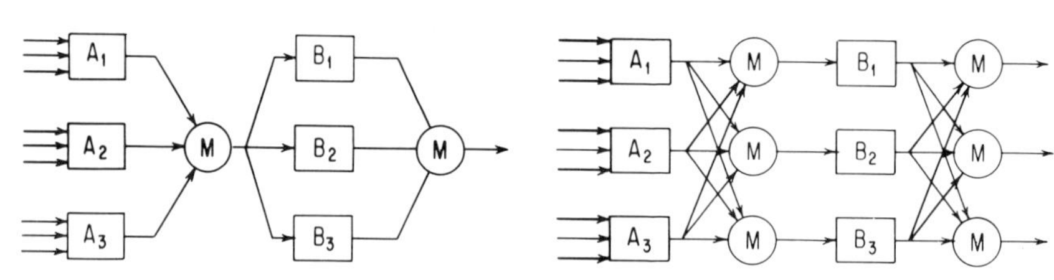

FIG. 97. Conventional majority-vote logic with one voting element per stage.[23]

FIG. 98. Majority-vote logic with triplicated voting elements.[23]

ELECTRICAL VERSUS MECHANICAL TRADE-OFFS

Overall system trade-offs should be made before any design job is advanced to where it would be costly to change. Many basic trade-offs can be made in the proposal or planning stage. A chart similar to Table 7 may be very helpful as suggested earlier. Table 36 summarizes some of the important electrical and mechanical trade-offs for various packaging techniques.

TABLE 36. Electrical versus Mechanical Trade-offs for Various Packaging Techniques

Technique	Electrical characteristics	Mechanical characteristics
Conventional components on printed-circuit boards	Longest interconnection propagation delays Isolation between circuits easy Compatible with high voltage and high power High-impedance shorts in humidity High-voltage drops and ground-level changes	Maximum interconnection complexity Fast reaction time Largest size Longest leads

TABLE 36. Electrical versus Mechanical Trade-offs for Various Packaging Techniques (*Continued*)

Technique	Electrical characteristics	Mechanical characteristics
Welded cordwood modules	Long interconnection propagation delays Isolation easy Circuit may change during embedment Final design can be breadboarded Compatible with high voltage and high power Shielding is simple	Size large in comparison with integrated circuits Throwaway once embedded Cost high in large quantities
Hybrid cordwood modules (integrated circuits and conventional components)	Long interconnection propagation delays Isolation easy Compatible with high voltage and high power Circuit may change during embedment Final design can be breadboarded Shielding is simple	Smaller size than conventional components alone, but not so small as all-integrated circuit Closer leads make welding more difficult Compatible with standard welded cordwood techniques Throwaway cost higher than for conventional cordwood modules
Hybrid compartmentalized modules (integrated circuits and conventional components)	Maximum isolation between i.f. stages Not applicable to high power or high voltage Unnecessary for common circuits which do not require shielding Circuit may change during embedment Final design can be breadboarded	Very expensive in large quantities Not as small as an all-integrated circuit system Compatible with other module packages High throwaway cost if embedded
Etched circuits	Electrical characteristics influenced by shape and proximity Especially applicable to isolation between i.f. stages Active components must be added Limited-value resistors, capacitors, and inductors Repetitive characteristics once established	Cut-and-try design method Inexpensive in large quantities Masks necessary for each circuit Fairly large size Complex assembly
TO-5 can integrated circuits on printed-circuit boards	Average interconnection propagation delay Good isolation possible High-impedance circuits marginal Mutual capacitance and cross-talk between lines likely Difficult to shield effectively	Good trade-off in size—fairly small but utilizes standard printed-circuit techniques Can be flow-soldered in large quantities Individual units can be replaced Multilayer boards required for maximum density Compatible with conventional components Assembly utilized standard printed-circuit techniques Good connection but poor heat conduction characteristics

TABLE 36. Electrical versus Mechanical Trade-offs for Various Packaging Techniques (*Continued*)

Technique	Electrical characteristics	Mechanical characteristics
Flat-pack integrated circuits on printed-circuit boards	Average interconnection propagation delay Good isolation possible Crosstalk on boards likely High-impedance circuits marginal Difficult to shield effectively	Several fastening methods available Two-dimensional fastening is much simpler than stacked flat packs Small quantities require printed-circuit artwork Spacing may be dictated by connectors Individual flat packs can be replaced Not compatible with three-dimensional parts Heat can be conducted away easily
Stacked flat-pack integrated-circuit modules	Average interconnection propagation delay Good isolation possible Requires welding care to avoid blowing semiconductors High-impedance circuits can be embedded Shielding easy	Throwaway once embedded Compact flat-pack subassembly Interconnections are very dense Very expensive in large quantities Compatible with other module forms and thick connectors
Thin-film circuits	Resistor and capacitor values can be higher than integrated circuit type Isolation better than integrated circuit Active devices not usually in thin-film form	Medium size Active devices added in separate operation Low cost in large quantities Long lead time—artwork required
Uncased discrete chip packages	Good isolation possible Medium high resistor and capacitor values possible Breadboard circuits possible Large-value resistors and capacitors require large area	Compatible with thin-film techniques Complex interconnections Costly in large quantities No complex masks needed Standard pieces allow fast reaction time
Uncased multi-integrated circuit chip packages	Isolation between circuits possible Short leads—minimum propagation delay Active devices easy to make Resistor and capacitor values limited Not applicable to high power or high voltage	Problems in buying and handling chips Use standard circuits—no complex masks Interconnections reduced but usually made point by point High-cost throwaway unit
LSI complex monolithic chips	Fast circuit and interconnection propagation speeds Isolation between circuits difficult Active devices easy to make Changes impossible Large-value resistors and capacitors impossible Not applicable to high power or high voltage	Interconnections built in Very small size Low yield Long development time for new masks Costly for small quantities High-cost throwaway unit

TABLE 36. Electrical versus Mechanical Trade-offs for Various Packaging Techniques (*Continued*)

Technique	Electrical characteristics	Mechanical characteristics
MOS (metal oxide semiconductor) devices	Slow circuit propagation speeds Parameters unstable Large fan-out No passive components necessary	Small size Heat dissipation difficult Interconnections large with respect to device Low cost on basis of small area Single diffusion process

REFERENCES

1. Lytel, A.: "Handbook of Electronic Charts and Nomographs," TAD Inc., 1965.
2. Department of Defense, MIL-HDBK-217, "Reliability Stress and Failure Rate Data for Electronic Equipment," 1962.
3. "Interference Reduction Guide," vols. 1 and 2. Prepared for the U.S. Army Electronics Laboratories, Fort Monmouth, N.J., by the Filtron Company, Inc., 1964.
4. Gray, H. J.: "Digital Computer Engineering," Prentice-Hall, Inc., Englewood Cliffs, N.J., 1963.
5. Feulner, R. J.: Cure Switching System Noise, *Electronic Design,* Aug. 16, 1966.
6. Westinghouse Electric Corporation, Metals Division, Blairsville, Pa., Technical bulletin entitled "Westinghouse Designer's Handbook: The When Why and How of Magnetic Shielding," 1966.
7. DiMarzio, A. W.: Which Shielding Materials and Why, *EDN,* October, 1966.
8. Technical Wire Products, Inc., Cranford, N.J., Data Sheets EMC-005 and EMC-007, copyright 1965.
9. Schreiber, O. P.: RFI Shielding Explained by Analogy, *Electron. Packaging Production,* January/February, 1962.
10. Cowdell, R. B.: Help Stamp Out EMI, *EDN,* Nov. 23, 1966.
11. *Insulation Directory/Encyclopedia Issue,* vol. 12, no. 6, May/June, 1966.
12. Cochrane, J. H.: Surveying the Field of Permanent Magnet Materials, *Machine Design,* Sept. 15, 1966.
13. Radiation Effects Information Center: REIC report no. 36 entitled "The Effect of Nuclear Radiation on Electronic Components, Including Semiconductors," Oct. 1, 1964.
14. Boston Insulated Wire and Cable Co.: private correspondence, copyright 1963.
15. Vogelman, J. H.: Transmitter-Receiver Pairs in EMI Analysis, *Electro-Technol.,* November, 1964.
16. Nalle, D. H.: "Elimination of Noise in Low Level Circuits," Clevite Corporation, Brush Instruments Division, Cleveland, Ohio.
17. Department of Defense, MIL-STD-275B, "Printed Wiring for Electronic Equipment," December, 1964.
18. Garth, E. C., and I. Catt: Ultra-speed I.C.'s Require Shorter, Faster Interconnections, *Electronics,* July 11, 1966.
19. The Institute of Printed Circuits: "Flexible Flat Cable Handbook," 1965.
20. Staff of Integrated Circuit Engineering Corporation: "Integrated Circuit Engineering—Basic Technology," 4th ed., Boston Technical Publishers, Inc., Cambridge, Mass., 1966. Copyright 1966 by Integrated Circuit Engineering Corporation, Phoenix, Ariz.
21. Mitchell, E. P., and K. D. Pope: Which IC Logic Form? *EDN,* Sept. 14, 1966.
22. Flynn, M. J.: Complex Integrated Circuit Arrays: The Promise and the Problems, *Electronics,* July 11, 1966.
23. Lowrie, R. W.: High-reliability Computers Using Duplex Redundancy, *Electron. Ind.,* August, 1963.
24. Smith, R. S.: Capacity of Shielded Wire, *EDN,* March, 1965.
25. Vacca, A. A., and A. G. Lambert: Selecting the Integrated Logic Circuit, *Electro-Technol.,* November, 1965.

26. Microcircuits—Digital and Linear, *EDN*, July, 1966.
27. DeGennaro, L. I.: Make Your Own Thick-film I.C.'s, *Electronic Design*, Sept. 13, 1966.
28. Schenkel, F. W.: Thin-film Capacitance Elements: Which Is Best for Your Purpose? *Electronics*, Jan. 25, 1965.
29. Sanders Associates, Inc.: "Flexprint Circuit Design Handbook," 1965.
30. Litton Industries, U.S. Engineering Co. Div.: "Design Manual for USECO Multi-layer Circuitry," 1962.
31. Sladek, N. J.: Electromagnetic-interference Control, *Electro-Technol.*, November, 1966.
32. Gale, N. H.: Shielding RF with Copper Foils, *Electronic Eng.*, September, 1966.
33. Bakker, W. F., and A. H. Cohen: "A Comparison and Survey of Flat Gasket Materials," Metex Corp., Clark, N.J.
34. American Standard Definitions of Electrical Terms, 1941; ASA C42.65, Communication, Jan. 24, 1965; ASA C42.70, Electron Devices, Oct. 10, 1956. American Institute of Electrical Engineers, New York.
35. "Van Nostrand's Scientific Encyclopedia," 3d ed., D. Van Nostrand Co., Inc., Princeton, N.J., 1958.
36. Federal Electric Corp.: Bureau of Ships Reliability Design Handbook, NAVSHIPS 94501. Prepared for Fleet Electronics Effectiveness Branch, Bureau of Ships, Department of the Navy, 1963.
37. Sarbacher, R. I.: "Encyclopedic Dictionary of Electronics and Nuclear Engineering," Prentice-Hall, Inc., Englewood Cliffs, N.J., 1959.
38. "IRE Dictionary of Electronics Terms and Symbols," Institute of Radio Engineers, Inc., New York, 1961.
39. Knowlton, A. E.: "Standard Handbook for Electrical Engineers," 9th ed., McGraw-Hill Book Company, New York, 1957.
40. Wong, H. M.: Wire Sizes and Insulation Material for High Temperature Applications, *Missile Design Develop.*, March, 1960.

Chapter 11

THERMAL-DESIGN CONSIDERATIONS FOR PACKAGING ELECTRONIC EQUIPMENT

By

JAMES R. BAUM

Aerospace Center
Government Electronics Division
Motorola, Inc.
Scottsdale, Arizona

CONTRIBUTORS

PAUL DICKERSON, *Motorola Aerospace Center*

VICTOR DUFFY, *Motorola Aerospace Center*

RAYMOND JIMEMEZ, *Universidad de Costa Rica*

JOHN WELLING, *Motorola Aerospace Center*

INTRODUCTION

The basic nature of the electronic equipment thermal design problem can be simply stated. First, all electronic components are sensitive to their temperature so that they have degraded performance above and below some limiting temperature range and will finally fail as the temperature greatly exceeds this range. Second, many parts generate heat internally which must be dissipated to the external environment. As a result of military and space requirements the general operating environments themselves have become increasingly severe. Space and weight are typically at a premium so that high power densities and marginal dissipating areas are the rule rather than the exception. In spite of all the foregoing constraints the electrical performance requirements and the reliability requirements become more demanding every year. Thermal design can no longer be an afterthought or done on a hit-or-miss basis.

This chapter will provide a broad overview of many of the aspects of thermal analysis and design in electronic equipment. It is organized so that the material is in approximately the same chronological order as used to approach an actual thermal design problem. Typically any thermal design effort is a close combination of analysis, hardware design and selection, and testing. The emphasis in this chapter is on the fundamental problems and the necessary analytical tools to handle these problems. No attempt is made to suggest a specific cooling configuration for any given equipment environment combination. In most cases the thermal design constraints are but a part of a multitude of constraints contributing to the overall design, and no handbook can cover all exigencies. Sufficiently detailed analysis techniques are provided to allow reasonable trade-offs to be made. This chapter is component-oriented, for after all, it is the parts which dictate the thermal design. The broad and important field of overall cooling system design is not covered in any depth since this type of work requires specialized knowledge and experience. Many good references are available.

Initially a brief review of heat-transfer fundamentals will be presented to provide a common discussion basis. Since the first step in any equipment thermal design study is to investigate the external environment as it affects the thermal design, the section on thermal environments points out the broad area of external effects and provides examples of how equipment design is influenced by the environment. The next step in any thermal design problem involves careful consideration of the internal thermal characteristics of the equipment; specifically, the design limits, heat dissipation, and heat-transfer characteristics of the components being used. This chapter will discuss the various types of electronic components from a thermal design standpoint.

Once the environment has been defined and the internal part requirements delineated it remains to provide for the efficient transfer of the dissipated heat within the equipment to some "ultimate sink." An efficient design is one which accomplishes this without exceeding allowable temperature limits and without unduly compromising other aspects of the equipment design, such as weight or cost. Therefore, some of the thermal-environment-control components which can be utilized to improve equipment heat transfer are discussed. Much of the preceding material is then correlated by presenting some examples of preliminary design calculations for entire equipments. Thermal testing is an important part of any thermal design program, and the important considerations involved are covered.

The entire chapter is oriented toward cooling, the major thermal design problem. For those few situations where heating is required the basic principles still apply and it is necessary to determine the required heater power and a satisfactory control technique.

There are very few commercially published reference books specifically on the thermal design of electronic equipment.[1] Most of the literature in print is in government-sponsored reports, papers given at various conferences, and articles in various periodicals. The attempt has been made herein to provide suitable references and bibliographies on the subject material.

Units. Traditionally all heat-transfer calculations and thermal property information in the United States utilized English units: British thermal units, feet, inches, hours, seconds, degrees Fahrenheit, etc. Electronic engineers, on the other hand, have dealt in a system with inches, watts, calories, and degrees centigrade. In order to facilitate communications between heat-transfer and electronic engineers a hybrid set of units has come into common usage. Basically, this consists of using the centigrade temperature scale, expressing heat-flow rates in watts, and using English system units otherwise. All calculations of basic heat-transfer coefficients and all thermal property data are handled in conventional English units. This approach will be utilized in this chapter. Although normally a consistent unit approach is desirable, in this case it would mean development of arbitrary conversion factors and make it difficult to use the published literature. However, care must be taken to maintain consistent units in utilizing the data and techniques given herein. The following list of nomenclature will aid in this regard.

a	Free-convection parameter, $c_p\rho^2\,\beta g/\mu k$	1/(ft³) (°F)
A	Area; A_o outside, A_i inside	in.², ft²
A_c	Cross-sectional area	in.², ft²
A_s	Surface area	in.², ft²
B	Constant for free convection	
c	Specific heat	Btu/(lb) (°F)
c_p	Specific heat at constant pressure	Btu/(lb) (°F)
C	Conductance	Btu/(hr) (°F), watts/°C
D	Diameter	in., ft
D_e	Equivalent diameter $4A_c/P$	in., ft
E	Torque	in.-lb
f	Fanning friction factor	Dimensionless
F_a	View factor	Dimensionless
F_e	Emissivity factor	Dimensionless
g	Acceleration due to gravity $(4.17)(10^8)$	ft/hr²
G	Mass velocity of fluid $= V\rho = m/A$	lb/(sec) (ft²)
h_c	Convection coefficient; h_c internal, h_{co} external	Btu/(hr) (ft²) (°F)
h_r	Equivalent radiation heat-transfer coefficient; h_r internal, h_{ro} external	Btu/(hr) (ft²) (°F)
I	Current	amp
I_{opt}	Optimum current	amp
k	Thermal conductivity; k_w wall	Btu/(hr) (ft) (°F)
k_b	Printed-circuit-board thermal conduction	
k_c	Copper-clad thermal conduction	
L	Length	in., ft

L_e	Length from beginning of tube	ft
m	Mass rate of flow	lb/min, lb/hr
m	Fin modulus $\sqrt{2U/k\delta}$	1/ft
n	Number of screws	
N_s	Specific speed	Dimensionless
N_{Gr}	Grashof number $gD^3\beta\Delta t\rho^2/\mu^2$ or $gL^3\beta\Delta t\rho^2/\mu^2$	Dimensionless
N_L	Load speed	Dimensionless
N_{Nu}	Nusselt number hD/k or hL/k	Dimensionless
N_{Pr}	Prandtl number $c\mu/k$	Dimensionless
N_{Re}	Reynolds number $DV\rho/\mu$ or $LV\rho/\mu$	Dimensionless
p	Pressure	lb/in.^{2}
Δp	Pressure drop or gain	lb/in.^{2}
P	Perimeter	in., ft
q	Rate of heat transfer	Btu/hr, watts
q_c	Convective heat-transfer rate; q_{ci} internal, q_{co} external	Btu/hr, watts
q_e	Internally dissipated power	Btu/hr, watts
q_i	Total internal heat-transfer rate	Btu/hr, watts
q_k	Rate of conduction heat transfer	Btu/hr, watts
q_r	Radiant heat-transfer rate; q_{ri} inside, q_{ro} outside	Btu/hr, watts
q_w	Heat transfer through wall	Btu/hr, watts
Q	Quantity of heat	Btu
$\dot{Q}$	Volumetric flow rate	ft^3/min
Q_s	Solar irradiation constant; 442 in earth orbit	Btu/(hr) (ft^2)
r	Radius; r_c chip, r_d dissipator, r_e equivalent	ft, in.
R	Thermal resistance	°C/watt, (°C) (in.^{2})/watt
R_{c-a}	Case to ambient thermal resistance	°C/watt
R_{c-p}	Semiconductor case to mounting, surface contact resistance	(°C) (in.^{2})/watt, °C/watt
R_i	Interface or contact thermal resistance	(°C) (in.^{2})/watt
R_{j-c}	Semiconductor junction to case thermal resistance	°C/watt
R_{sq}	Resistance per square for sheet material; R_{sqb} circuit-board material, R_{sqc} conductor material	°C/watt
R_t	Overall thermal resistance	°C/watt
t	Temperature	°C, °F
t_a	Temperature of air; t_{ai} inside, t_{ao} outside	°C, °F
t_c	Case temperature; cold-side temperature	°C, °F
t_f	Temperature change of fluid stream; film temperature $= (t_w+t_a)/2$	°C, °F
t_h	Hot-junction temperature; hot-side temperature	°C, °F
t_i	Initial stabilized temperature	°C, °F
t_{in}	Inlet fluid temperature	°C, °F
t_j	Semiconductor junction temperature; t_j maximum allowable case, maximum temperature	°C, °F
Δt_m	Log mean temperature difference, LMTD	°C, °F
t_o	Temperature of ambient air or/and surroundings; t_{oi}, initial surrounding temperature	°C, °F
t_{out}	Outlet fluid temperature	°C, °F
t_p	Plate temperature; mounting-plate temperature	°C, °F
t_s	Surface temperature; t_{si} inside, t_{so} outside	°C, °F
Δt_{1-2}	Temperature difference between points 1 and 2	°C, °F
T	Absolute temperature	°R
U	Combined heat-transfer coefficient; U_t overall, U_i internal, U_o external	Btu/(hr) (ft^2) (°F)
V	Velocity	ft/hr
$V\infty$	Free-stream velocity	ft/hr
w	Width	ft, in.
W	Fin spacing	in.
W	Weight	lb
x	Length or distance	in., ft
x_w	Wall thickness	in., ft
Δx_{1-2}	Linear distance between points 1 and 2	in., ft
y	Height	in.

α	Absorptivity for radiation	Dimensionless
α_s	Absorptivity for solar radiation	Dimensionless
β	Coefficient of volumetric expansion	$1/°F$
γ	Rate of change of environment temperature	°F/hr, °C/hr
Δ	Prefix indicating finite increment	
δ	Thickness	in., ft
ϵ	Emissivity; ϵ_1, of surface 1	Dimensionless
η	Fin efficiency	Dimensionless
θ	Time	hr
μ	Fluid viscosity	lb/(hr) (ft)
μ_w	Fluid viscosity at wall temperature	lb/(hr) (ft)
ρ	Density; ρ_{std}, referred to standard conditions; ρ_m, mean density	lb/ft^3
ρ	Reflectivity for radiation	Dimensionless
σ	Ratio of ρ to ρ_{std}, used in fan calculations	Dimensionless
σ_r	Stefan-Boltzmann constant; 0.171×10^{-8}	$Btu/(hr) (ft^2) (°R)^4$
τ	Transmittance for radiation	Dimensionless
ϕ	Altitude correction factor for free-convection coefficient	Dimensionless
ψ	Printed-circuit-board constant	Dimensionless

HEAT-TRANSFER FUNDAMENTALS

In order to provide a common basis for discussion of thermal design of electronic equipment, it is desirable to review briefly the fundamentals of the basic heat-transfer processes. Heat or thermal energy is transferred between two points as a result of a temperature difference between them. Heat may be transferred by conduction, convection, or radiation. Conduction is heat transfer within a medium by direct molecular communication. Convection is an energy-transport process combining conduction, mechanical mixing, and energy storage. It is most important as it defines the mechanism of transfer between a solid surface and a fluid. Radiation is the transfer of heat between two points across some space by electromagnetic radiation.

In general this chapter will be concerned with so-called *steady-state heat transfer,* which implies a constant rate of heat flow and constant temperatures. Transient or unsteady-state heat transfer, on the other hand, occurs as a result of time-varying heat fluxes or temperatures.

Conduction. *Conduction* is used herein to relate to the transfer of heat through solids. The simplest form of conduction heat-transfer equation is given by the following one-dimensional Fourier equation:

$$\Delta Q/\Delta \theta = -kA_c \Delta t/\Delta x$$

With steady-state heat transfer, the energy transfer per unit time, $\Delta Q/\Delta \theta$, is constant, and the most common form of the equation is obtained as

$$q_{1\text{-}2} = kA_c \Delta t_{1\text{-}2}/\Delta x_{1\text{-}2} \tag{1}$$

Figure 1 relates the equation geometrical parameters to a model.

The parameter k is the thermal conductivity of the particular solid. Table 1 lists thermal conductivity values of several materials used in electronic equipment. The term *thermal insulator* is associated with materials with low thermal conductivities, such as plastics, foams, epoxies, and fiber glasses. The metals are the best thermal conductors and also are electrically conductive. In general, thermal conductivity is a weak function of material temperature. For purposes of calculation the conductivity at the average temperature in the material is used. References 2 and 3 contain more complete property data.

Equation (1) shows that the temperature rise between any two points is inversely proportional to both the cross-sectional path area and the material thermal conductivity, and directly proportional to the path length. The ability of a particular geometry and material to conduct heat may be described by the parameter $kA_c/x = C$. This term C is often denoted the thermal conductance. A more common

TABLE 1. Thermal Conductivity and Specific Heat of Various Materials at 80°F

Materials	Conductivity k, (Btu) (ft)/(ft^2) (°F) (hr)	Specific heat c, Btu/(lb)(°F)
Silver	241	0.06
Copper	220	0.09
Gold	171	0.03
Beryllia, 99.5% pure	140	0.24
Aluminum, pure	125	0.21
Aluminum, 63S	116	0.21
Beryllia, 95%	90	0.24
Magnesium	91	0.23
Aluminum 6061-T6	90	0.23
Red brass	63.7	0.09
Yellow brass	54.6	0.09
Beryllium copper	47.8	0.09
Pure iron	43.2	0.11
Phosphor bronze	29.6	0.09
Soft steel	26.8	0.11
Monel	20.5	
Lead	18.9	0.03
Hard steel	14.8	0.11
Steatite	13.6	0.11
Alumina, 96% pure	17.0	0.20
Alumina, 90%	7.0	0.20
Pyrex	0.728	0.20
Grade A lava	0.683	
Soft glass	0.569	0.18
Water	0.380	1.01
Mica	0.341	0.21
Paper-base phenolic	0.159	0.40
Plexiglass	0.107	0.35
Cast epoxies	0.1–0.8	0.3–0.5
Polystyrene	0.061	0.35
Glass wool	0.023	0.19
Air	0.016	0.24

form of this parameter is the thermal resistance, which is simply the reciprocal such that $R = 1/C$. Typically the thermal resistance is given in units of °C/watt. Thus it can be thought of as the temperature rise between two points per watt of heat transferred between these two points. An excellent overall discussion of conduction can be found in Ref. 3.

Contact Resistance. A variation in the standard conduction heat transfer is the model of thermal conduction at an interface. This surface interface consists primarily

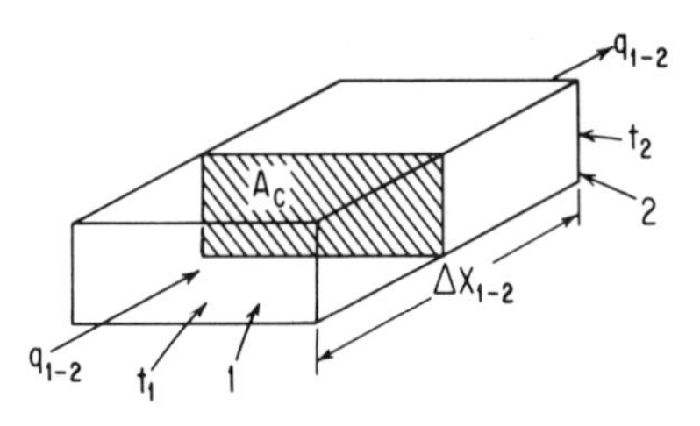

FIG. 1. One-dimensional conduction model.

of a finite discontinuity between the two surfaces with a few points of material contact. Heat transfer between the contact surfaces is by conduction through both the air gap and the material contact points. Each path is significant in the total interface thermal resistance. The most useful engineering data for interface thermal analysis is obtained empirically by tests of various interface models. These data are typically presented in the form of contact thermal resistance with units such as (°C)(in.^{2})/watt. Selected contact thermal-resistance data are presented in Table 2 (see also Fig. 15).

TABLE 2. Thermal-contact-resistance Preliminary-design Guidelines

Description	Environmental pressure	Approximate interface pressure, psi	R_i, (°C)(in.^{2})/watt
Small stud-mounted components (such as stud-mounted transistors)	Sea level	5,000 500	0.05 0.50
	High vacuum	5,000 500	0.08 0.80
Mounting feet of equipment with contact areas of about 1 in.^{2}	Sea level	1,000 100	0.5 1.0
	High vacuum	1,000 100	2.0 5.0
Large-surface contact areas	Sea level	100 10	1.0 3.0
	High vacuum	100 10	7.0 20.0

The four major variables affecting contact resistance are surface material, finish, interface pressure, and the presence or absence of some liquid or gas in the interface. The optimum interface model is one where there is no discontinuity of material at the interface. Up to some practical limit this is approached by increasing the interface pressure. Pressure on the interface surface tends to flatten the material high points and reduce the interface thermal resistance. Another technique for reducing interface resistance is to add an interstitial material at the interface. This material should have the characteristic of flowing into the interface surface hills and valleys, replacing the air in the voids. The use of an interface material requires that this material be thin enough that the decrease in interface contact resistance offsets the rise through the interface interstitial material. Typical interstitial materials used in electronic equipment are the silicon greases, thin silicon rubbers, and soft metals such as indium. In environments such as deep space where no air exists at the interface, thermal resistance can significantly increase. References 4 and 5 show data for particular interface models in air and in a vacuum.

Convection. *Convection* refers to the transfer of heat between a surface and the fluid surrounding the surface. This fluid may be any liquid or gas; however, in most electronic equipment the convection fluid is the atmospheric air. The most familiar form of convection takes place when a body at one temperature is placed in still air at another temperature. This process is known as *free-convection heat transfer* and relies on the buoyant force of the air at the surface to move the air over the body and thus remove heat from or transfer heat to the surface. Another form of convection heat transfer is forced convection, in which fluid motion is induced by external means.

The general form of the convection heat-transfer equation is given by:

$$q_c = h_c A_s (t_s - t_o) \tag{2}$$

Solution of this equation requires that the convection heat-transfer coefficient h_c be calculated or known.

Free Convection. Figure 2 shows the formulas for calculation of free-convection h_c values. Use of these formulas is only possible when the orientation of the convection surface is known, and further, the equipment surface temperature rise over the environment must be known or estimated. For the situation where the heat-balance equation is being solved for the surface temperature t_s, an iterative process is used. Typically t_s is assumed and h_c is calculated. The h_c value is then used in

Turbulent Heat Transfer

$$h_C = \varphi B (t_S - t_0)^{0.33}$$

where

$$aL^3 \Delta T > 10^8$$

① Top of a hot horizontal plate
② Vertical plate

Laminar Heat Transfer

$$h_C = \varphi B \left(\frac{t_S - t_0}{L}\right)^{0.25}$$

where

$$10^4 < aL^3 \Delta T < 10^8$$

③ Vertical plate or large diameter vertical cylinders (L = height)
④ Top of a hot horizontal plate (L = longest side)
⑤ Horizontal cylinder (L = diameter)
⑥ Bottom of a hot horizontal plate (L = shortest side)

Alt × 10^{-3} ft	0	10	20	30	40	50	60	70	80	90	100
φ	1.0	0.829	0.678	0.545	0.430	0.338	0.266	0.210	0.164	0.130	0.105

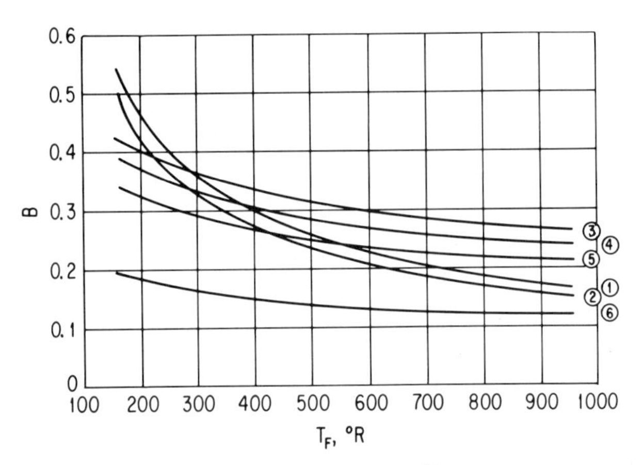

FIG. 2. Free-convection coefficients.

the heat-balance equation, and the equation is solved for t_s. This procedure is repeated until the t_s value used in the h_c calculation matches that resulting from the heat balance. Although h_c values are dependent on surface orientation, length, and temperature rise, when all equipment surfaces are lumped, overall free-convection h_c values in the range of 0.7 to 1.0 Btu/(hr)(ft²)(°F) are typical for the normal electronic equipment packages.

Forced Convection. Forced-convection heat transfer depends, in part, on the angle of impingement of the air with the surface. As such the heat-transfer coefficients are grouped according to the flow geometry. Flow parallel to a surface L long has h_c values given by the following:

For laminar flow, $LV_\infty \rho/\mu < 500{,}000$

$$h_c = 0.66 \left(\frac{c_p\mu}{k}\right)^{0.33} \left(\frac{k}{L}\right) \left(\frac{LV_\infty \rho}{\mu}\right)^{0.5} \tag{3}$$

For turbulent flow, $LV_\infty \rho/\mu > 500{,}000$

$$h_c = 0.037 \left(\frac{c_p\mu}{k}\right)^{0.33} \frac{k}{L} \left(\frac{LV_\infty \rho}{\mu}\right)^{0.8} \tag{4}$$

For the general case of forced-convection flow through ducts with equivalent diameter D_e, h_c may be the following:

For laminar flow, $D_eV\rho/\mu < 2{,}100$

$$h_c = 1.86 \left(\frac{c_p\mu}{k}\right)^{0.33} \frac{k}{D_e} \left(\frac{D_e}{L_e}\right)^{0.33} \left(\frac{D_eV\rho}{\mu}\right)^{0.33} \left(\frac{\mu}{\mu_w}\right)^{0.14} \tag{5}$$

For turbulent flow, $D_eV\rho/\mu > 10{,}000$

$$h_c = 0.023 \left(\frac{c_p\mu}{k}\right)^{0.33} \frac{k}{D_e} \left(\frac{D_eV\rho}{\mu}\right)^{0.8} \left(\frac{\mu}{\mu_w}\right)^{0.14} \tag{6}$$

Convection heat-transfer coefficients for flow over small bodies such as resistors can be roughly estimated in the range $100 < (D_eV\rho)/\mu < 10{,}000$.

$$h_c = 0.5 \frac{k}{D_e} \left(\frac{DV\rho}{\mu}\right)^{0.5} \tag{7}$$

The terms in the above equations, c_p, μ, k, and ρ, are each a function of fluid temperature to some degree. For usage of the equations, bulk temperature for the fluid should be estimated and can be confirmed on an iterative process once h_c and the final temperatures are estimated. Forced-convection coefficients associated with typical fans and blowers used in electronic equipment are in the range of 5 to 20 Btu/(hr)(ft^2)(°F).

The empirical convection heat-transfer relations found in published literature are most commonly in a dimensionless group equation or graphical form. As an example of this, Eq. (3) may be rewritten as

$$\left(\frac{hL}{k}\right) = 0.66 \left(\frac{c_p\mu}{k}\right)^{0.33} \left(\frac{LV_\infty \rho}{\mu}\right)^{0.5}$$

In this equation, each group in parentheses is dimensionless. This form of equation provides for interrelating dimensionless parameter groups from various test and analysis conditions. These groups are designated by specific names as shown in Table 3. In final dimensionless form Eq. (3) might then be found in published literature in the form of $N_{Nu} = 0.66\ (N_{Pr})^{0.33}\ (N_{Re})^{0.50}$.

TABLE 3. Dimensionless Groups

Designation	Name	Group
N_{Gr}	Grashof number	$\frac{gD^3\beta\,\Delta t\rho^2}{\mu^2}$ or $\frac{gL^3\beta\,\Delta t\rho^2}{\mu^2}$
N_{Nu}	Nusselt number	hD_e/k or hx/k
N_{Pr}	Prandtl number	$c_p\mu/k$
N_{Re}	Reynolds number	$D_eV\rho/\mu$ or $LV\rho/\mu$

Duct Pressure Drop. The flow of a fluid through a duct results in a pressure drop through the duct. In the original duct-flow design, provisions must be made for an adequate pressure head to maintain the required flow. The duct pressure-drop relationship is given by

$$\Delta p = \frac{0.1925\,(G)^2\left(\frac{1}{\rho_2} - \frac{1}{\rho_1}\right)}{g} + \frac{0.358f(G)^2\left(\frac{1}{\rho_m}\right)L}{gD_e} \tag{8}$$

where the units of pressure drop are in inches of water. The term f designates the duct friction factor and may be estimated by the use of Fig. 3. In addition to internal friction losses there are also losses due to entrance, exit, and bends or other flow restrictions in the duct geometry. Ordinarily these pressure drops may be minimized in a good duct design. Reference 6 presents detailed correlation for these and other types of duct pressure losses.

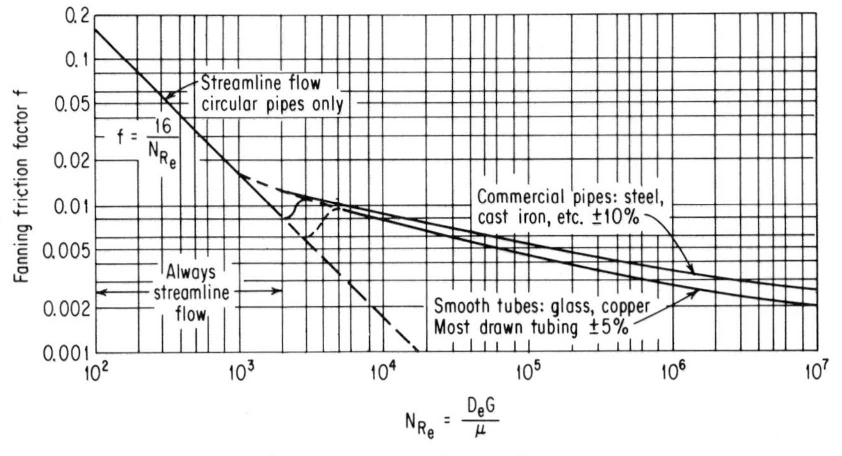

FIG. 3. Fanning friction factor.

Radiation. *Radiation* refers to the transfer of energy by electromagnetic wave propagation. The wavelengths between 0.1 and 100 microns are referred to as *thermal radiation wavelengths.* The ability of a body to radiate thermal energy at any particular wavelength is a function of the body temperature and radiating surface material characteristics. Figure 4 shows the ability to radiate energy for an ideal radiator, a blackbody, which by definition radiates the maximum amount of energy at any wavelength. The total energy radiated in Btu/(hr)(ft^2) by a blackbody at any particular temperature is given by the area under that temperature curve in Fig. 4 and is equal to $\sigma_r T^4$.

The incident radiation on any surface is partially absorbed, reflected, and in some cases transmitted through the surface. This situation may be expressed by $\alpha + \rho + \tau = 1$. Bodies such as metals and most plastics, which do not transmit radiation, are opaque, and the equation reduces to $\alpha + \rho = 1$. Glass and quartz are examples of materials which will transmit a portion of the incident radiation. Most materials encountered in electronic equipment packaging are opaque.

Materials or objects which act as perfect radiators are rare. Most materials radiate energy at a fraction of the maximum possible value. The ratio of the energy emitted by a nonblackbody to that emitted by a blackbody at the same temperature is called the emissivity ϵ. The emitted energy from a surface is given by $\epsilon\sigma T^4$. The

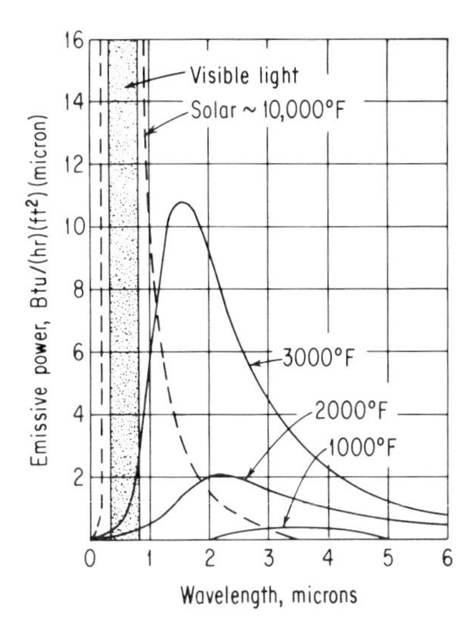

Fig. 4. Emissive power as a function of wavelength.

ability of a surface to absorb the total incident radiation on it is defined by the term *absorptivity* α. As in the case of emissivity, the absorptivity is also dependent on the material and temperature of the emitter, i.e., the wavelength of the incident radiation. Some materials, such as the dark-color paints, for example, exhibit the property of having $\alpha \approx \epsilon$ when the absorber and emitter are at temperatures $\pm$ 100°F of each other. These materials are defined as *gray bodies*. Table 4 lists some gray-body emissivity values.

TABLE 4. Emissivities of Surfaces at 80°F

Surface type	Finish	Emissivity ϵ
Paint	Black	0.91–0.97
Paint	Gray	0.84–0.91
Paint	White	0.95
Paint	White epoxy	0.91–0.95
Paint	Aluminum silicone	0.20
Metal	Nickel	0.21
Metal	Alodine on aluminum	0.15
Metal	Aluminum	0.14
Metal	Silver	0.10
Metal	Gold	0.04–0.23
Special surfaces:		
Metal	Aluminum (sandblasted)	0.41
Metal	Nickel (electroless)	0.17
Metal	Aluminum (machine-polished)	0.04
Metal	Gold (vacuum-deposited on aluminum)	0.04
Ceramic	Cermet (ceramic containing sintered metal)	0.58

The basic equation for radiation heat transfer between two gray surfaces is given by

$$q_{(1\text{-}2)} = 0.171\ F_e F_a A_s \left[\left(\frac{T_1}{100}\right)^4 - \left(\frac{T_2}{100}\right)^4 \right] \tag{9}$$

The term F_e is a function of the emissivity values of the two surfaces between which the radiation occurs and their relative geometries. The term F_a relates the "view" or amount of radiation interchange area seen by the radiating surface. Table 5 shows

TABLE 5. Radiation View and Emissivity Factors

Configuration	Area on which heat transfer is based	F_a	F_e
Infinite parallel planes	A_1 or A_2	1	$\dfrac{1}{\dfrac{1}{\epsilon_1} + \dfrac{1}{\epsilon_2} - 1}$
Completely enclosed body, small compared with enclosing body, subscript 1, enclosed body	A_1	1	ϵ_1
Completely enclosed body, large compared with enclosure, subscript 1, enclosed body	A_1	1	$\dfrac{1}{\dfrac{1}{\epsilon_1} + \dfrac{1}{\epsilon_2} - 1}$
Other configurations such as elements ΔA with various oriented surfaces and perpendicular and parallel finite planes	ΔA or A_1	See Ref. 3	

the procedure for calculating F_e and F_a.

When comparing the significance of radiation with other modes of heat transfer, the radiation heat-transfer coefficient concept is useful. This term is given by

$$h_{r(1\text{-}2)} = \frac{q_{r(1\text{-}2)}}{A_s(t_1 - t_2)} = 0.171\ F_e F_a \left[\left(\frac{T_1}{100}\right)^4 - \left(\frac{T_2}{100}\right)^4 \right] / (t_1 - t_2) \tag{10}$$

Typically, the h_r values for electronic equipment external surfaces vary between 0.9 and 1.5 Btu/(hr)(ft²)(°F).

Another class of materials, including most bare metals, includes selective absorbers and emitters. For these materials α is not equal to ϵ regardless of the emitter and absorber temperatures. A particular example of current interest which illustrates selective α and ϵ characteristics is the problem of determining the temperature of an equipment in deep space where the solar radiation representing a 10,000°F source is incident on the equipment. In this situation the solar absorptivity α_s of the exposed surfaces is important from a solar heating standpoint, while the emissivity is important from a dissipation standpoint. Table 6 lists various material α_s values. Neglecting planetary radiation to the equipment, a heat balance can be written for the equipment $q_{s(\text{in})} = \alpha_s Q_s A_s + q_e$, where Q_s is the incoming solar heat flux on the equipment. In a near-earth orbit $Q_s = 442$ Btu/(hr)(ft²). The heat transfer from the equipment is given by

$$q_{r(\text{out})} = 0.171\ \epsilon\ A_s \left[\left(\frac{T_s}{100}\right)^4 - \left(\frac{T_o}{100}\right)^4 \right]$$

TABLE 6. Solar Absorptivities

Surface type	Finish	Absorptivity α_s
Paint	Black	0.79–0.94
Paint	Yellow	0.54
Paint	Gray	0.49–0.57
Paint	Blue	0.48
Paint	White epoxy	0.34
Paint	White	0.19–0.33
Paint	Aluminum silicone	0.23
Metal	Alodine*	0.4
Metal	Nickel	0.33–0.45
Metal	Gold	0.30
Metal	Silver	0.25
Metal	Aluminum (sandblasted)	0.60
Metal	Aluminum (machine-polished)	0.35
Metal	Gold (vacuum-deposited on aluminum)	0.24
Ceramic	Cermet (ceramic containing sintered metal)	0.65

* Trademark, Amchem Products, Inc.

In deep space the sink temperature for energy radiated from the equipment is −460°F or absolute zero. The heat-balance equation becomes $\alpha_s Q_s A_s + q_e = 0.171 \epsilon A_s[(T_s/100)^4]$. For space equipment with little or no power dissipation, such as some antennas, mechanical devices, etc., $q_e = 0$, and a simpler relation is obtained.

$$T_s = 771 \left(\frac{\alpha_s}{\epsilon}\right)^{0.25} \tag{11}$$

Since α_s *and* ϵ are both characteristics of the equipment surface (or paint), the equilibrium temperature of the equipment is determined by the designer's selection of the equipment external housing material.

Combined Heat-transfer Mechanism. The preceding discussion has treated radiation, convection, and conduction heat-transfer mechanism as separate entities. In actual practice the three modes of heat transfer usually occur simultaneously in combined series-parallel thermal paths. The following illustrates one combined heat-transfer mechanism. Assume that an electronic equipment transfers heat by radiation and free convection to the equipment internal wall. Heat transfer from the external wall to the environment is by similar means. Figure 5 illustrates the example model.

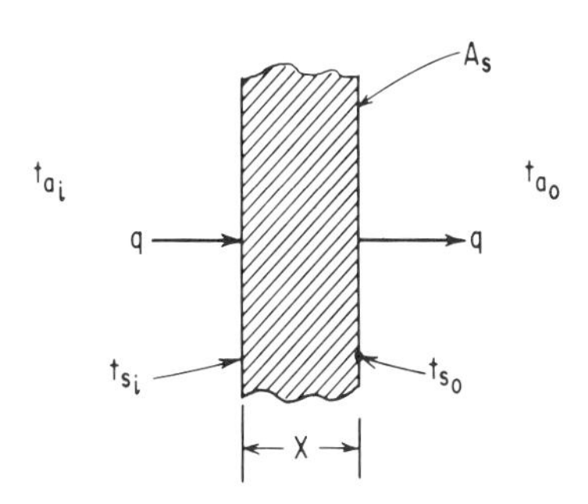

The internal heat-transfer equations are

$$\text{Radiation: } q_{ri} = h_{ri} A_s (t_{ai} - t_{si})$$
$$\text{Convection: } q_{ci} = h_{ci} A_s (t_{ai} - t_{si})$$

Therefore

$$q = (h_{ri} + h_{ci})(A_s)(t_{ai} - t_{si})$$

The heat transfer through the wall by conduction is

$$q = \frac{k_w A_s}{x_w} (t_{si} - t_{so})$$

As with the internal heat balance, the external heat balance is

$$q = q_{ro} + q_{co} = (h_{ro} + h_{co})(A_s)(t_{so} - t_{ao})$$

The internal wall and external heat balances may be combined to give a heat-balance equation which is in terms of the equipment internal and external environment temperatures. Suitable substitution results in

$$q = \frac{(t_{ai} - t_{ao}) A_s}{1/(h_{ro} + h_{co}) + x_w/k_w + 1/(h_{ri} + h_{ci})}$$
$$= U_t A_s (t_{ai} - t_{ao}) \quad (12)$$

where

$$U_t = \frac{1}{1/U_o + x_w/k_w + 1/U_i} \quad (13)$$

U_t denotes a combined heat-transfer coefficient relating the overall heat-transfer rate from inside to outside with the various conductance paths.

DC Electrical-flow and Heat-flow Analogy. The transfer of heat is governed by the equations:

$$\text{Conduction: } q_k = (kA/x)\Delta t$$
$$\text{Convection: } q_c = (h_r A)\Delta t$$
$$\text{Radiation: } q_r = h_r A)\Delta t$$

In physical terms each of the above equations relates the flow of heat q between some temperature potential Δt. In electrical terms the analogy is a flow of current I between a voltage potential E. The equation relating the current and voltage is $I = E/R$, where R is the circuit resistance. A comparison of the heat-transfer and electrical equations shows that if I were considered analogous to q and E to Δt then the electrical resistance would be analogous to the reciprocal of the thermal conductance, which is, as previously discussed, the thermal resistance.

The terms $1/hA$ and x/kA can then be referred to as the *thermal-resistance values* for the particular thermal model. The common thermal-resistance units used in electronics heat-transfer calculations are degrees centigrade of temperature rise per watt of power dissipation.

The result for the combined heat-transfer problem in the preceding section can be obtained in a direct form using the thermal-resistance concept. The equivalent electrical circuit would look as shown in Fig. 6.

The overall equivalent thermal resistance is

$$R_t = \frac{1}{(h_{ci} + h_{ri}) A_s} + \frac{x_w}{k_w A_s} + \frac{1}{(h_{co} + h_{ro}) A_s} = \frac{1}{U_t} \quad (14)$$

This form of equation is preferable for two reasons. First, resistance values for each path give a physical grasp for the effectiveness of each individual thermal path design. Second, the effect of a numerical change in an individual path thermal resistance on the total equivalent source to sink resistance is easily obtained. Thermal resistance is an extremely useful concept in working with heat transfer in electronic equipment. Typically we are concerned with the overall problem of transferring heat from a dissipating component to some ultimate sink. It is convenient to think in terms of an overall equipment thermal resistance consisting of a series-parallel path from the source to the sink. Each segment can be looked at separately and its effect on the total evaluated.

Transient Analysis. A transient thermal analysis requires a more complicated procedure for exact solutions than does the steady-state analysis since by its nature it involves another variable, time. Typically a solution of basic differential equations with appropriate boundary conditions by computer techniques is required. A simpler analysis procedure may be obtained by lumping the various heat-transfer-mode conductances and assuming that the thermal dissipation properties may be thought of as bulk equipment properties. To illustrate, consider electronic equipment where the package internal conductance U_iA_i (radiation, convection, and conduction) is large compared to the conductance from the equipment to its external sinks, U_oA_o. In this situation the time-temperature response of the equipment is primarily dependent on the package external-heat-transfer characteristics. Thus the package internal part temperatures will respond in time much as the housing of the equipment.

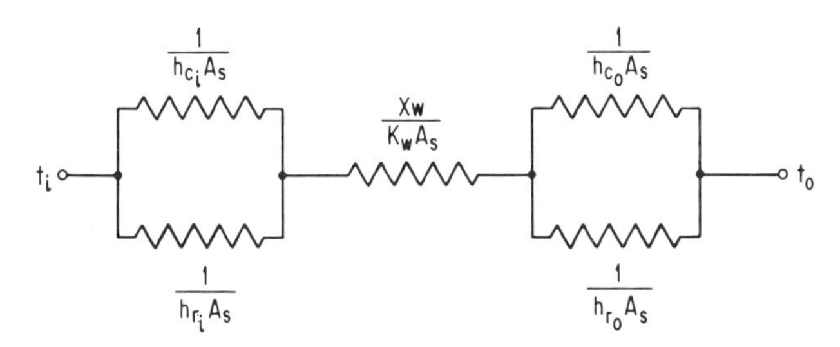

FIG. 6. Equivalent electrical circuit.

As an example of the use of this procedure, consider the following model. A sealed equipment with an initially stabilized temperature t_1 is turned on at time zero with q_e power dissipation. Further, the equipment operates in an environment t_o with a linear ambient temperature change in time. The environment temperature at time zero is designated as t_{o1} and the rate of change of the environment in time as γ. The environment temperature at any time θ is then given by $t_o = t_{o1} + \gamma\theta$. The solution to this model is given by the equation

$$t = t_1 e^{-\left(\frac{U_oA_o}{Wc}\right)\theta} + \left(\frac{q_e + U_oA_o\, t_{o1} - \gamma Wc}{U_oA_o}\right)\left[1 - e^{-\left(\frac{U_oA_o}{Wc}\right)\theta}\right] + \gamma\theta \tag{15}$$

Two lumped parameters must be calculated prior to solving this equation. These are U_oA_o and U_oA_o/Wc. The parameter U_oA_o is simply the product of the equipment's total equivalent external heat-transfer coefficient U_o and the equipment housing external surface area. The term Wc is the product of the equipment weight W and a bulk scientific heat c. The bulk specific heat should be representative of all of the equipment parts. However, as seen in Table 1, most materials have specific heat between 0.2 and 0.25 Btu/(lb)(°F), and the use of a composite specific heat is a reasonable procedure.

Heat Transfer in Change of Phase. The process which occurs when a material changes from one state to another (e.g., the change from liquid to gas, or evaporation) is called a *phase change.* The processes of sublimation, evaporation, and melting require that heat be added to the material, while the processes of condensation and solidification require that heat be taken away. Typically these are constant-temperature processes during the heat addition or subtraction. Change-of-phase processes are useful in a number of applications in cooling of electronic equipment. Usually the boiling and melting processes can be used in absorbing heat from a package. For short transients a fixed amount of material is used which absorbs heat and either melts or boils at some fixed temperature until the phase change for that amount of material is complete.

During the boiling process near a surface, extremely high heat-transfer rates to the fluid occur. This makes boiling heat transfer attractive for high-power-density equipment.

A discussion of applications of phase change to electronic equipment cooling can be found in Ref. 7.

Thermal Environments. The term *thermal environment* as used herein refers to all the external surroundings, configurations, and phenomena which in some way affect the thermal design of an electronic equipment. Environment includes all the obvious items such as ambient pressure, ambient temperature, external-thermal-radiation fluxes, and possible coolant-supply characteristics, as well as consideration of the type of equipment mounting, effects of adjacent equipment, duty cycle, etc.—items which are not normally thought of in these terms. As a general rule the external environment items are those beyond the immediate control of the equipment designer and therefore are design constraints.

The definition of the thermal environment for an equipment may result from interpretation of information from various sources. There may be a customer specification specifically detailing the items of interest. For many of the government agencies the equipment specification will reference or invoke the conditions covered in one of the general military environmental specifications such as MIL-E-5400 "General Specification for Aircraft Electronic Equipment."[8] A basic problem inherent in this type of specification is the lack of sufficient information for a given equipment application. For example, MIL-E-5400 merely defines a series of temperature altitude environments. An equipment mounted to a large structure of substantial heat-transfer capacity sees a considerably different thermal-design environment in a particular MIL-E-5400 condition than an equipment dissipating heat in the same condition but with minimum structural connections. It is necessary to completely investigate an environment situation beyond the apparent limits. The following represents a basic checkoff list for definition of a thermal-design environment. Not all items are necessarily applicable in every situation.

Ambient air temperatures
Surrounding structure temperatures
Mounting-structure temperatures
Method of attachment to structure
Ambient pressure variations
Incident radiation fluxes
Surrounding structure emissivities
Coolant properties
Flow rates of coolant
Temperatures of coolant
Coolant supply pressures
Allowable supply pressures
Time variation of external environment conditions

CLASSIFICATION OF THERMAL ENVIRONMENTS

Although almost every equipment thermal environment has some unique features, it is possible to categorize the environments or design conditions in two general ways. The most common classification is by type of use or installation. The typical use classifications are shown in Table 7.

TABLE 7. Electronic-equipment Use Classification

Application	Temperature range, °C	Features and remarks
Ground:		
Controlled environment	10 to 40	Primarily air-cooled in air-conditioned laboratories or blockhouses (see Fig. 7).
World-wide extremes	−55 to 71	Cooled by free-convection and radiation while exposed to the elements. Forced-air or -liquid cooling is employed in extreme cases of power density (see Figs. 7 and 8).
Shipboard	10 to 40 −55 to 71	Very similar to ground-equipment design (see Figs. 7, 8, and 9).
Aircraft:		
Low performance	−55 to 71	Extreme temperatures occur for short periods. Most of the operating life is in the moderate temperature range tolerated by man. Cooling design is primarily free-convection and radiation. Conduction to the vehicle structure is used at times (see Fig. 7).
High performance	−55 to 250	Primarily forced-air-cooled; some cabin-mounted equipment is exposed to relatively moderate environments (see Fig. 9).
Crew compartment	−10 to 40	
Remote	−55 to 150	Equipment remote from the cabin will see wide extremes in ambient air temperatures but is usually supplied with a more moderate (−20 + 50°C) forced coolant.
External equipment	−55 to 250	External equipment, at stagnation points, will see very severe temperatures but usually for short periods (see Fig. 9).
Missile	−55 to 425	Primarily designed to be free-convection and radiation-cooled during standby operation. Coolant supplied prior to launch. Thermal mass designed to withstand short transients of exposures to very hot skin temperatures (see Fig. 10).
Spacecraft:		
Controlled	−273 to 300 0 to 40	Most equipment is housed in bays controlled to moderate temperatures. Conduction to the mounting structure is the primary mode of heat transfer (see Fig. 9).
External	−273 to 300	Externally mounted equipment may see a radiation sink of −273°C and may rise to 300°C (see Fig. 11).

A second classification is by the predominant type of external heat dissipation from the equipment, e.g., conduction equipment, external-coolant-supply equipment. Some feature of the environment will always act as the sink for the dissipated heat from the equipment. The type of sink available is the major influencing factor in the overall thermal design.

The major classifications of external thermal sinks are described below. Note that the term *infinite sink* is often used in discussing the external environments. This concept refers to a point in the heat-dissipation path from an equipment which can absorb all the dissipated heat and maintain some temperature independent of that absorbed heat.

Conduction. A unit mounted upon a vehicle structure is a typical example of a conduction-cooled unit. The majority of the dissipated power is conducted to the structure which provides the infinite thermal sink. Spaceborne equipment which attaches to a vehicle-supplied cold plat represents conduction-cooled equipment.

Convection. A large percentage of the equipment being presently developed depends on convection for cooling. An equipment designed for typical sea-level environments employs natural modes of heat transfer. It dissipates at least half its power by free convection and the remainder by radiation. Forced-air convection becomes necessary as power densities increase, with forced-liquid-convection cooling being used on the very-high-power-density equipment.

Radiation. In space vehicles, radiation is the ultimate mode of transferring the dissipated power. For typical ground equipment, radiation is also an important mode of heat transfer since, as pointed out, a naturally cooled unit dissipates about half of its energy by radiation. Care must be taken to identify the surrounding structure to the equipment as either sinks or sources. Hotter surrounding equipment or solar flux will add to the heat energy which must be dissipated.

Overall Thermal-design Configurations. Figures 7 to 11 represent some of the general thermal-design configurations which may be utilized in the previously discussed environments.

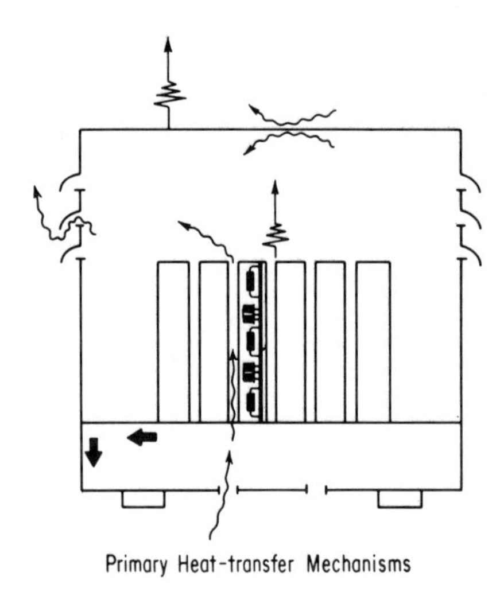

FIG. 7. Schematic of naturally cooled ground, shipboard, and airborne equipment.

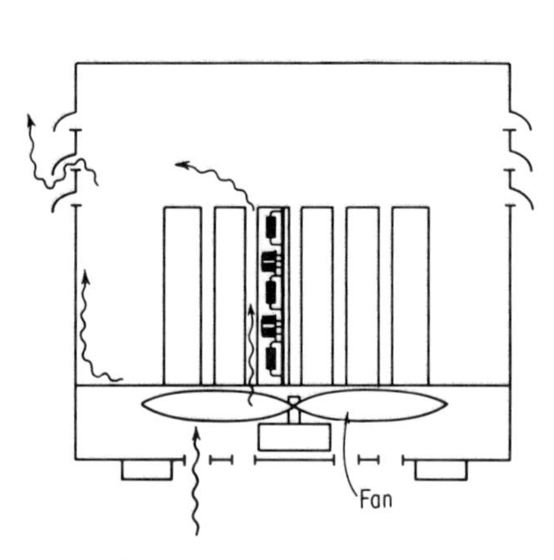

FIG. 8. Schematic of forced-convection-cooled ground, shipboard, and airborne equipment.

THERMAL CHARACTERISTICS OF ELECTRONIC COMPONENTS

Three types of information about an electronic part are required to provide a basis for effective thermal design. First, the limiting temperature for the part must be established. The location of the limiting "hot spot" on the part and the basis of selection of the maximum or minimum temperature are important auxiliary data. Second, it is necessary to know how much heat the part dissipates and under what conditions. Third, the internal and external heat-transfer paths and mechanisms for the part must be understood.

This section will discuss some of the most important broad classifications of electronic components with regard to their thermal characteristics. Since the limiting temperatures and the heat dissipation are a function of the specific application, these areas can only be covered in general terms. The internal thermal resistance, as applicable, and the external transfer are discussed in detail since this type of informa-

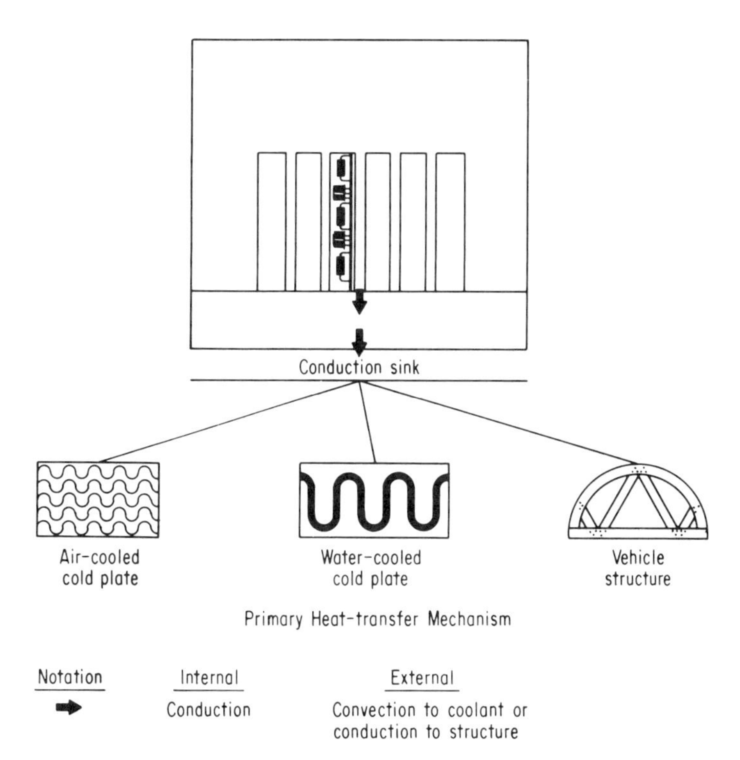

FIG. 9. Schematic of conduction-cooled shipboard, airborne, and space-vehicle equipment.

tion, which has not always been conveniently available in the past, has a major effect on the hardware design.

Semiconductors. Semiconductors constitute a general category of components which includes transistors, diodes, integrated circuits, silicon-control rectifiers, and varactors. These components are temperature-limited depending on the type of electrical-junction construction and the basic semiconductor material forming the device. Most semiconductors are made of germanium or silicon material. The typical maximum rating for germanium devices is a 100 to 110°C junction temperature. The maximum junction temperatures quoted for silicon devices range from 125 to 200°C. These values are stated in the semiconductor manufacturer's data sheets, specifically as the maximum allowable device operating-junction temperatures. Normally, these temperatures are not for catastrophic failures, but imply some nominal life value at this maximum temperature.

In many equipment designs, the allowable maximum component temperatures must be kept below the manufacturer's ratings to provide higher reliability. For example, in the latest space-flight equipment, the junction temperatures for silicon signal transistors are limited to 75°C although the maximum manufacturer's rating was 200°C. Reductions in allowable maximum junction-temperature ratings may also be necessary if a critical electrical parameter is temperature-sensitive.

Semiconductor Package Internal Heat-transfer Characteristics. The immediate thermal path from a semiconductor's power-dissipation region, the junction, is through the semiconductor chip, thence through a chip-to-case bond, and finally to the case surface.

In the past the internal thermal design was the responsibility of the manufacturer, and the user simply obtained the manufacturer's thermal data. Recently, however, more and more companies have acquired the facilities to fabricate their own individual integrated circuits. This has resulted from the wide diversification of electrical requirements for integrated circuits as opposed to other types of semiconductors.

Integrated circuits can take the form of monolithic, hybrid, or thin-film circuits. The monolithic circuit is described as a complete circuit on one semiconductor die.

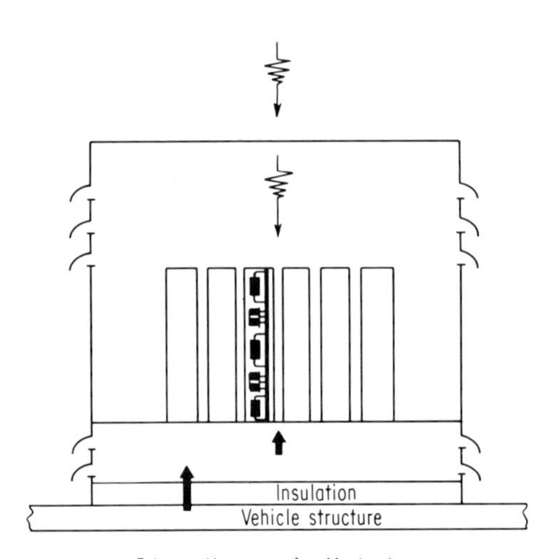

Primary Heat-transfer Mechanisms

Notation	Internal	External
	During standby operation:	
	Convection	Convection, see Fig. 7
	Radiation	Radiation, see Fig. 7
	Conduction	Conduction, see Fig. 7
	During flight transient:	
—ʌʌʌ→	Radiation	Radiation from the vehicle structure
➡	Conduction	Conduction from the vehicle structure

FIG. 10. Schematic of missile (transient) equipment.

It represents the most common type of integrated circuit. The hybrid makes use of multiple dies within a single case. In the case of the thin-film hybrid, the resistive and capacitative elements of the circuit are vapor-deposited on a wafer (die) with either transistor dies or monolithic integrated-circuit chips as the active elements.

Figure 12 illustrates the internal thermal paths for a typical semiconductor device. The total internal thermal resistance of the device, typically referred to as the junction-to-case thermal resistance R_{j-c}, is essentially made up of an intrachip, a bond, and an intracase thermal resistance in series.

INTRACHIP RESISTANCE (SPREADING RESISTANCE). The location and relative size of the dissipator(s) (junctions or resistive elements) on a chip define the intrachip or spreading resistance. In a qualitative sense, spreading resistance can be illustrated by the chip model in Fig. 13. Figure 13*a* represents a chip-dissipator thermal model in which the dissipating area covers the total area of the chip. This example is simply a model of one-dimensional heat transfer where the intrachip thermal resistance is given by $R = x/kA_c$. The model in Fig. 13*b* represents the situation in which the chip size is identical to that in Fig. 13*a* but the dissipator area is reduced. This reduction in the Fig. 13*b* dissipation area requires that heat transferred through the chip "spread" over a longer chip path, thus increasing the intrachip resistance. Figure 14 gives some spreading thermal resistance data for typical chip geometries. The entering parameters for this figure are the dissipator radius r_d and the chip radius r_c. For square or rectangular chips and dissipation areas an equivalent radius must be calculated. One technique that has been used successfully for estimating the equivalent radius is $\pi r_e^2 =$ length $\times$ width of the dissipator chip.

A study of Fig. 14 results in two major conclusions concerning chip-spreading thermal resistance. These are

1. For any fixed chip size (radius) the chip-spreading resistance increases as the dissipation area (radius) decreases.

2. For any fixed dissipating area there is a decrease in spreading resistance as the chip area is increased. However, the spreading resistance becomes nearly constant after some critical chip area (radius).

BOND RESISTANCE. The bond resistance is one of the largest variables in the total thermal-resistance path. The common gold-silicon solder preforms produce bond resistances of approximately 0.025 to 0.05 (°C) (in.^{2})/watt. Experimenters with aluminum-silicon solders have reported bond resistances of 0.004(°C)(in.^{2})/watt. Unfortunately, bond thermal-resistance data for typical solders still demonstrate wide variation of approximately 75 percent of the mean. References 9 and 10 provide more complete bond thermal-resistance information.

INTRACASE RESISTANCE. For external heat-transfer calculations, some reference point on the semiconductor case must be chosen to proceed from. The magnitude of the intracase resistance depends on where the particular reference point is chosen. For example, if the reference point is directly under the die then there may be no intracase resistance. If on the other hand the flange of a TO-5 device is chosen as the reference point then there is a very significant intracase resistance between the die mounting point and this reference. The intracase resistance of the semiconductor die mounted on the TO-5 case header is 40 $\pm$ 8°C/watt from the header to flange. The reference point for typical ceramic flat pack for integrated circuits lies on the case directly underneath the die. Therefore the intracase resistance is minimal.

Spreading thermal resistance within the case is of concern in hybrid-type integrated-circuit flat packs. There is a problem of minimizing chip-to-chip interheating effects in a typical hybrid integrated-circuit flat package. In this thermal model the dissipators are the chips, and the case of the integrated-circuit package provides the material into which the chip heat must spread. Note that this model is analogous to that discussed for the junction and chip, and a similar analysis approach is used. The usual design goal is a chip location pattern such that the chips are thermally independent of one another.

The foregoing discussion provides some background in the internal heat-transfer characteristics of semiconductor devices. Where actual semiconductor-device package designs are involved it is necessary to make detailed calculations to provide

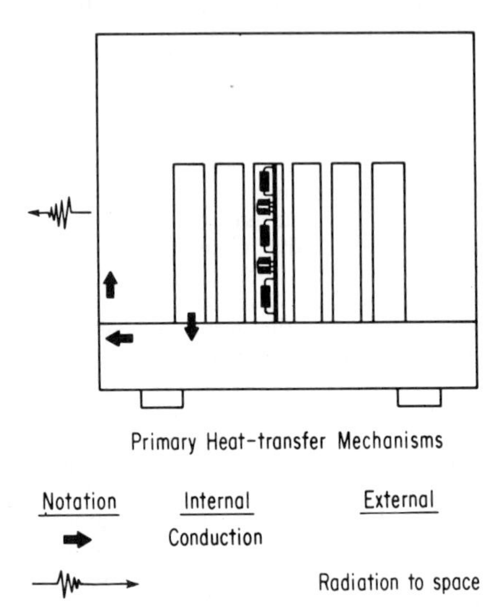

FIG. 11. Schematic of space-vehicle equipment.

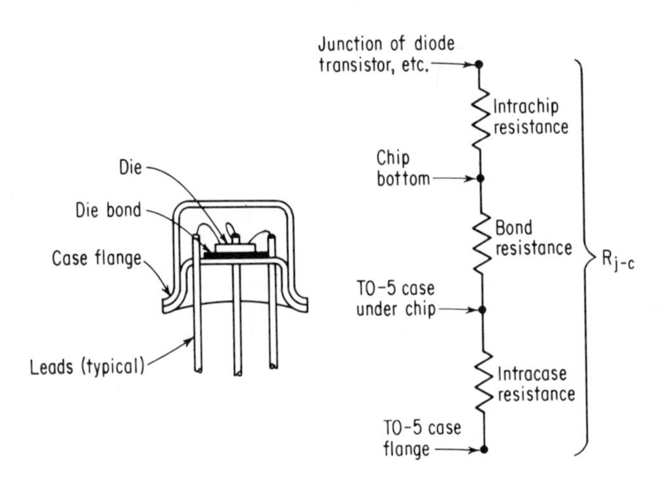

FIG. 12. Composite thermal-resistance paths for TO-5 package.

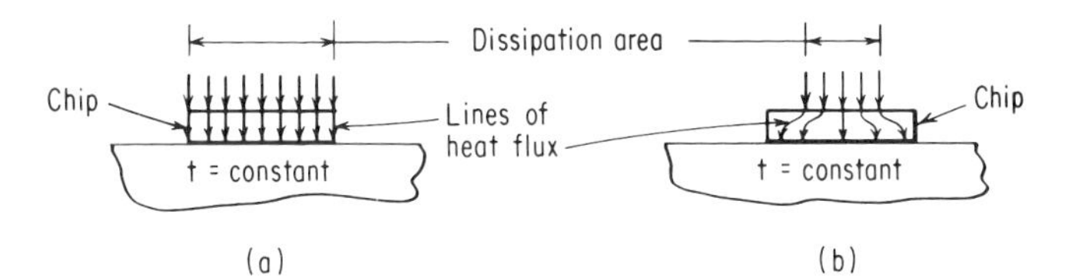

FIG. 13. Spreading-resistance models.

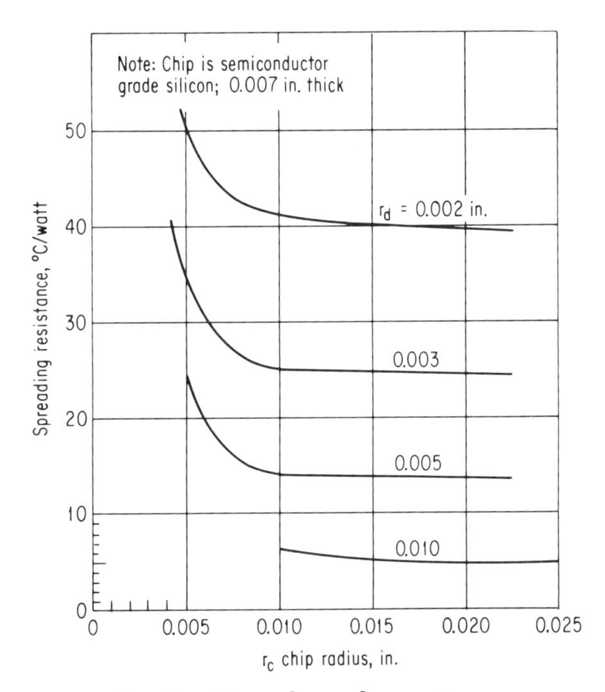

FIG. 14. Thermal spreading resistance.

the internal thermal data. However, in the vast majority of semiconductor design situations an off-the-shelf item is involved and therefore the user has no control of the internal thermal design. For standard lines of semiconductors the overall R_{j-c} value is determined by the device manufacturer and published in his data sheets. This resistance is obtained from the manufacturer's data in a number of ways depending on how the data is presented. It may be given directly (e.g., for a 2N3375, $R_{j-c} = 15°C/\text{watt}$) or it may be calculated from basic rating data as

$$R_{j-c} = \frac{t_{j\ \max} - t_c \text{ at } q_e}{q_e}$$

For example, for a 2N3445 transistor,

$$t_{j\ \max} = 200°C,\ q_e = 0 \text{ watts}$$

For

$$t_c = 25°,\ q_e = 115 \text{ watts}$$

$$\therefore\ R_{j-c} = \frac{200 - 25}{115} = 1.52°C/\text{watt}$$

A typical problem is to find the maximum allowable device-case temperature for a given power-dissipation level. The case temperature can then be used as a design value since it can readily be measured. For example, for a 2N3445 transistor dissipating 50 watts the allowable case temperature is

$$t_{c\ \max\ \text{allow}} = t_{j\ \max} - q\ (R_{j-c})$$
$$t_{c\ \max\ \text{allow}} = 200 - 50\ (1.52) = 125°C$$

Semiconductor Package External Heat Transfer. The actual mechanism(s) of heat transfer from the semiconductor case to the local sink depend(s) on the device configuration and the particular application. Two basic possibilities exist: predominant conduction or predominant convection/radiation. A conduction situation is one in which the majority of the device dissipation is conducted from the case of the device to some local mounting point. If the device is held down by a stud or screw mount, then a thermal contact resistance (R_{c-p}) exists. This can be calculated using Fig. 15 as a guide.

For example, on a 2N1016 transistor (stud mounting with mica washer), the recommended torque value for the 5/16-24 stud is 40 in.-lb. The mounting-face area for this transistor is calculated as 0.47 in.^{2} Using the interface pressure relationship in Fig. 15, we find that

$$p = \frac{5nE}{AD} = \frac{5(1)\ (40 \text{ in.-lb.})}{(0.47 \text{ in.}^2)(0.3) \text{ in.}} = 1{,}410 \text{ lb/in.}^2$$

Figure 15 shows $R_{c-p} = 0.38(°C)(\text{in.}^2)/\text{watt}$ (stud mounted with mica) and therefore at 50 watts,

$$\Delta t_{c-p} = 0.38(°C)\ (\text{in.}^2)/\text{watt}\ \frac{50 \text{ watts}}{0.47 \text{ in.}^2} = 40°C$$

Where a semiconductor device is isolated from a conduction sink, the predominant mode of transfer from the case is by convection and radiation to the local surrounding. This situation would be typified by a device mounted by its leads on a printed-circuit board. Although there may be some heat transfer by lead conduction, this may be neglected in calculations, thereby giving a somewhat conservative result. The case to surroundings thermal resistance R_{c-a} is tabulated in Table 8 for several common lead-mounted semiconductor devices.

Commercial Heat Sinks. There are two basic situations in which it may be necessary to provide a low semiconductor-case-to-local-sink thermal resistance. In one case there is no immediate conduction sink for heat dissipation. In the other there is a local conduction sink but it is necessary to improve the thermal path from the semiconductor to that sink.

CONVECTION/RADIATION SINKS. In situations where there is no intermediate conduction sink a low-thermal-resistance convection/radiation path must be provided to the local ambient sink. An example of this situation is a transistor mounted to a printed-circuit board and dissipating significant power. The board is a poor conductor and will transfer very little heat. Typically the transistor case has insufficient area to provide a low R_{c-a} value with only radiation and free-convection heat transfer from its surface. This problem can be alleviated by mounting the transistor to

TABLE 8. Transistor Case-to-ambient Thermal Resistance

Case type	R_{c-a}, *°C/watt*
TO-18 (transistor)*	380
TO-5 (transistor and IC)*	120–160
Alumina or beryllia IC flat pack (0.25 × 0.25 × 0.080 in.)†	140
Alumina or beryllia flat pack (0.4 × 0.4 × 0.075 in.)†	70

* Flange.
† Based on reference location on case, under chip mounting.

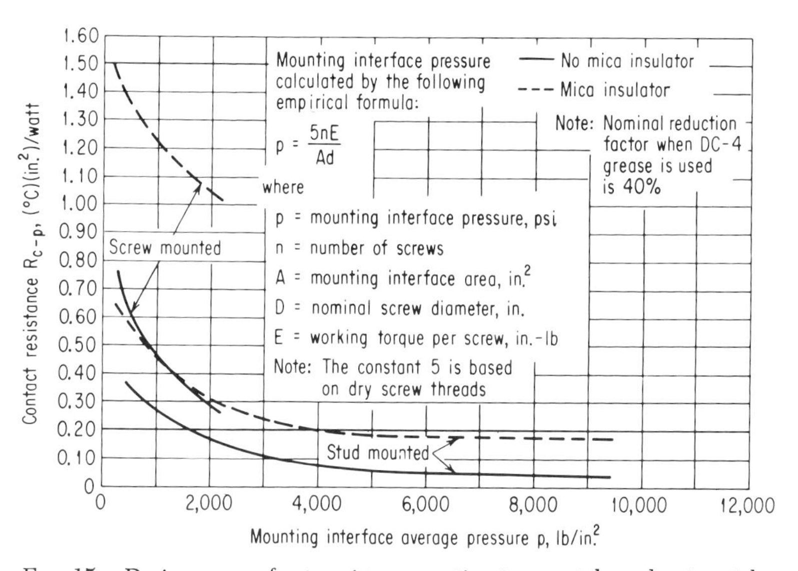

FIG. 15. Design curves for transistor mounting to a metal conduction sink.

any of a number of commercially available radiation/convection transistor heat "sinks." The term *sink* here is not quite accurate since these devices are simply designed to improve local heat transfer; however, the term is in common usage. These sinks typically have a large surface area and are made of a high-thermal-conductivity material such as aluminum. As such they distribute the semiconductor heat over the large intermediate area for dissipation to the ambient, thus minimizing the commercial sink to ambient thermal resistance. The manufacturers of these sinks provide sink-to-ambient thermal-resistance data for each of their products. All types of geometry are available. Reference 11 provides some basic comparison data on sink performance.

TABLE 9. Transistor Conduction Heat Sinks for TO-5 and TO-18 Cases

Sink description	No. 1 Beryllium oxide cup		No. 2 Aluminum oxide cup		No. 3 Screw-cap cup		No. 4 Copper-cup beryllium oxide assy.		No. 5 Spring clip	
Sink configuration										
Transistor case type	TO-5	TO-18	TO-5	TO-18	TO-5	TO-18	TO-5	TO-18	TO-5	TO-18
Sink-device manufacturer	National Beryllia		National Beryllia		Thermalloy		IERC		IERC	
Bonding material										
Transistor to cup	Eccobond 56C* epoxy		Eccobond 56C epoxy		N/A		Eccobond 56C		N/A	
Cup to mounting plate	Eccobond 56C epoxy		Eccobond 56C epoxy		N/A		N/A		N/A	
Sink requires mica for electrical insulation . .	No		No		Yes		No		Yes	
R_{c-p}, without mica, °C/watt										
Room pressure	2.4	5.8	7.5	12.5	5	17	9	14.5	18	30
High vacuum	7.3	. . . .	20	. . . .	. . . .	. . . .	. . . .	19		
R_{c-p} with 0.002 in. mica, °C/watt										
Room pressure	N/A	N/A	N/A	N/A	8	22	N/A	N/A		
High vacuum	N/A	N/A	N/A	N/A	. . . .	. . . .	N/A	N/A		

* Trademark, Emerson & Cuming, Inc.

With proper part layout in the preliminary design of an electronic package, such parts as an adequate metal bracket and a housing flange may be incorporated in the package for the specific purpose of providing a good heat-transfer path to the package surface. This practice will alleviate most requirements for the "hang-on" commercial radiation/convection sink.

COMMERCIAL CONDUCTION SINKS. A number of commercial low-thermal-resistance intermediate conduction sinks are available to provide conduction transfer. These are either permanently bonded to the case of the semiconductor device or mechanically clamped to it such that the resulting assembly can be readily conduction-mounted. Table 9 shows the characteristics of some representative conduction-mounting hardware.

As an example of a typical overall semiconductor mounting problem, consider a 2N2890 transistor to be mounted on a cold-plate-cooled chassis which has an expected temperature of 90°C. The dissipation of this device will be 2 watts. What mounting provisions are required? The manufacturer's data show that

$$R_{j\text{-}c} = 35°\text{C/watt} \qquad \text{and} \qquad t_{j\ \max} = 200°\text{C}$$

Therefore

$$\Delta t_{j\text{-}c} = (35°\text{C/watt})\ (2\text{ watts}) = 70°\text{C}$$

With a cold-plate chassis temperature of 90°C

$$\Delta t_{c\text{-}p\ \max\ \text{allow}} = 200 - (90 + 70) = 40°\text{C}$$

The case of this transistor is common to the transistor collector junction and therefore must be electrically isolated from the cold plate. From Table 9, mounting sink No. 4 has an $R_{c\text{-}p}$ value of 9°C/watt, which will produce $\Delta t_{c\text{-}p} = (9°\text{C/watt})\ (2\text{ watts}) = 18°\text{C}$. Therefore

$$t_j = 90 + 70 + 18 = 178°\text{C}$$

Since the junction temperature is below the rated maximum of 200°C, this design is adequate.

Lead-mounted Transistors. Transistors mounted as shown in Fig. 16 can conduct a significant portion of the dissipated heat to the printed-circuit board. Several case-to-plate thermal-resistance values are depicted in this figure for TO-18–type transistors. The combined heat-transfer mechanisms of conduction to the printed-circuit board, radiation to the environment, and convection to ambient air are included.

Resistors. Resistors are usually classified by power rating (½ watt, 1 watt) and by type (composition, film, or wirebound). This rating is based on the materials used and the type of resistor. It implies that the resistor will dissipate that amount of power without exceeding some limiting temperature, typically an external hot-spot temperature. Unfortunately there are no standard rating conditions or temperature definitions for resistors. The amount of heat that can really be dissipated by a resistor and not exceed the maximum hot-spot temperature depends upon how effectively the heat can be removed. This, in turn, depends upon the thermal path from the resistor to the sink. This thermal path may be via conduction, convection, radiation, or some combination of the three. Most of the common types of resistors have no specific mounting provisions other than their leads. Therefore, most of the manufacturer's thermal-design data are based on a conservative mounting such that the major heat transfer is by free convection and radiation with some minimum lead conduction. The data are presented in thermal derating curves for resistors at elevated-temperature operation and are similar to those given for transistors. Figure 17 shows a typical derating curve from MIL-R-11 [12] standard for a ¼-watt composition resistor, MIL Type RC07. The assumed lead length is 1 in.

A substantial reduction in resistor temperature can be made if the resistor is attached to a chassis or other member where the heat can be dissipated via a lower thermal resistance. Table 10 shows resistor surface-to-local-mounting thermal-

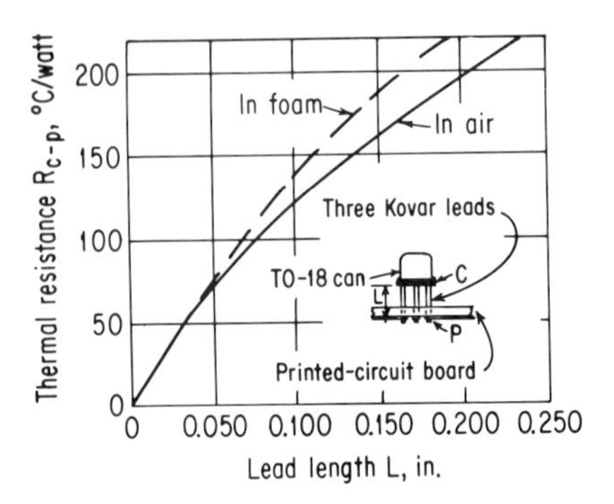

FIG. 16. Thermal resistance of lead-mounted transistors.

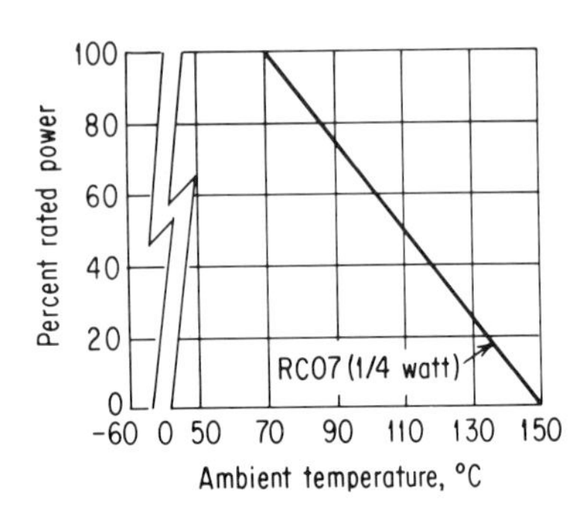

FIG. 17. Derating curve for ¼-watt composition resistor.

resistance values for a number of configurations. Note that values are given for both sea-level and vacuum-pressure conditions, the latter indicative of space-environment operation. Configuration C is similar to that considered typical for the derating-curve data of Fig. 17. It can be determined from Fig. 17 that the ¼-watt resistor RC07 has a thermal-resistance value from its surface to the ambient air of approximately $(150 - 70)/0.25 = 320$°C/watt. This is much larger than the 146°C/watt which is due in part to the longer lead length and in part to the inherently conservative approach of Fig. 17.

Some general comments on resistor mounting are, first, that anything other than intimate contact with a good conformal coating between the resistor and the mounting plate will significantly increase the thermal-contact resistance. Second, operation of the resistors in a vacuum results in a thermal-contact-resistance increase of about 1.4 times over that for atmospheric pressure conditions.

In some instances it is necessary to use forced-air cooling for resistors. This necessity might occur if several high-heat-dissipating resistors were mounted on a low-thermal-conductivity fiber-glass material. Forced-convection heat-transfer coefficients can be estimated using Eq. (7).

Example of Typical Resistor Cooling Problem. It is required to determine the maximum allowable power dissipation for a ¼-watt composition resistor that must operate in a vacuum environment. The resistor will be bonded to a chassis with conformal coating. The chassis temperature is 65°C. Reliability requirements necessitate that the maximum resistor surface hot-spot temperature not exceed 71°C.

Solution. From Table 10 it is found that a ¼-watt resistor staked to an aluminum chassis in a vacuum environment will have a hot-spot-to-chassis thermal resistance of 97°C/watt. For an allowable hot-spot-to-chassis temperature rise of 6°C $(71 - 65)$ the maximum power dissipation will be $q = \Delta t/R = 6/97 = 0.062$ watt.

Glass-envelope Vacuum Tubes. The glass-envelope vacuum tube represented the major active component in electronic equipment for many years. It also represented the largest source of heat in the equipment and the component most likely to fail. One major cause of failure was identified as temperature. Satisfactory cooling provisions are a necessary part of any design utilizing vacuum tubes. Tubes dissipate heat at the internal elements, specifically the filament and plate. This heat is predominantly transferred from the elements by radiation with some conduction to the glass envelope via the element supports and leads. The majority of the radiated heat is absorbed by the glass envelope, which is essentially opaque to thermal radiation in wavelengths of the radiated energy. The internal thermal design is be-

TABLE 10. Resistor Surface-to-mounting Thermal Resistance

Resistor-mounting configuration	Type watts	Thermal resistance, °C/watt	
		Sea level	Vacuum
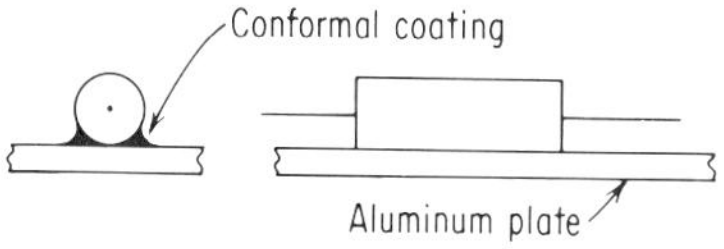 (a)	¼	87	105
	½	67	90
	1	43	73
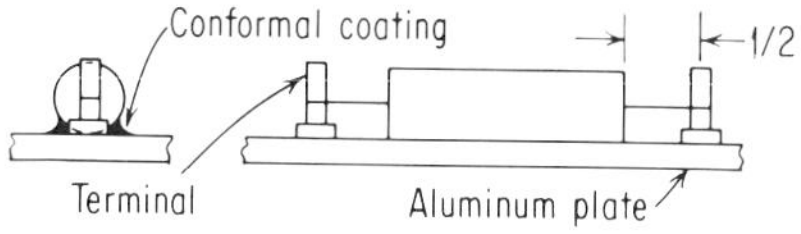 (b)	¼	79	97
	½	65	87
	1	40	58
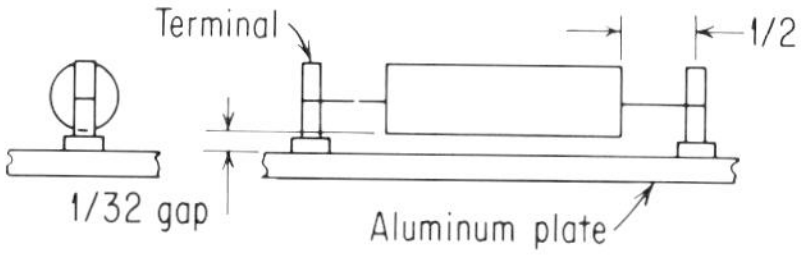 (c)	¼	146	214
	½	108	176
	1		
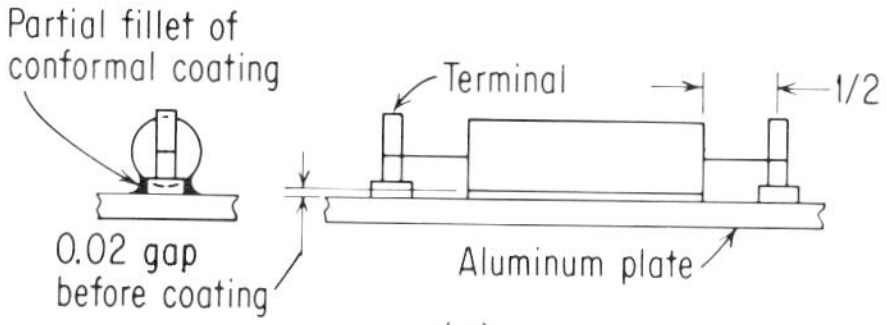 (d)	¼	135	192
	½	86	142
	1	64	112
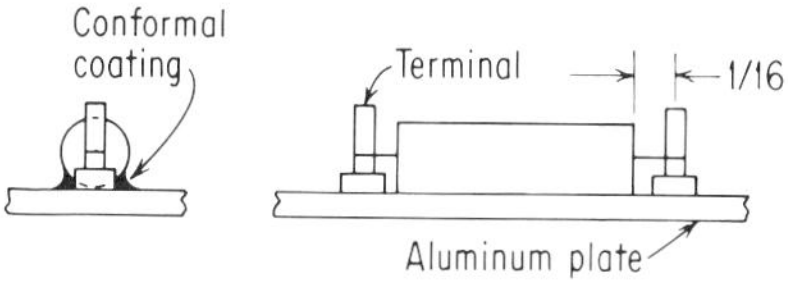 (e)	¼	78	95
	½	66	82
	1	39	57
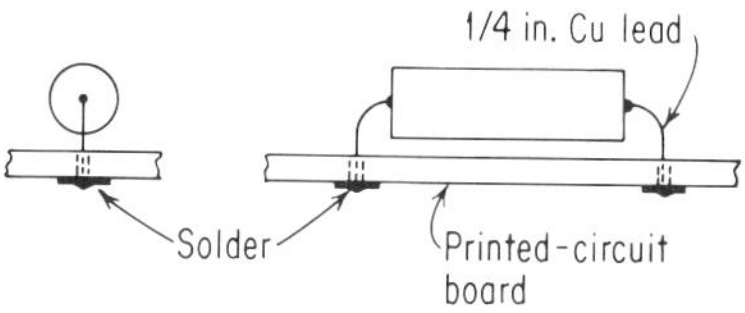 (f)	¼	120	
	½	90	
	1	50	

yond the control of the user, so that the application problem resolves itself into efficiently transferring heat from the tube envelope to the local sink.

Thermal-design characteristics and limits for tubes are based on envelope temperatures. As noted in MIL-HDBK-217A,[13] military-usage tubes are classified as either 165 or 200°C maximum bulb temperature. The effect of temperature on tube failure rate can be illustrated by noting, as an example, that the failure rate increases by a factor of 1.5 for a bulb temperature rise from 120 to 160°C.

Tube Cooling. There are a number of approaches to cooling tubes. Figure 18 shows the tube bulb temperature rise as a function of power density for various cooling methods. The simplest cooling technique is to utilize free convection and radiation directly with no auxiliary hardware. Curve A of Fig. 18 shows the temperature rise of the bulb over the local ambient for free convection and radiation cooling. The approximately linearized slope of this and any of the other curves on the figure may be calculated at any point to give a thermal resistance in (°C)(in.²)/watt.

Direct forced-air cooling may also be utilized with air flows both parallel and perpendicular to the tube axis. References 14 and 15 present detailed equations

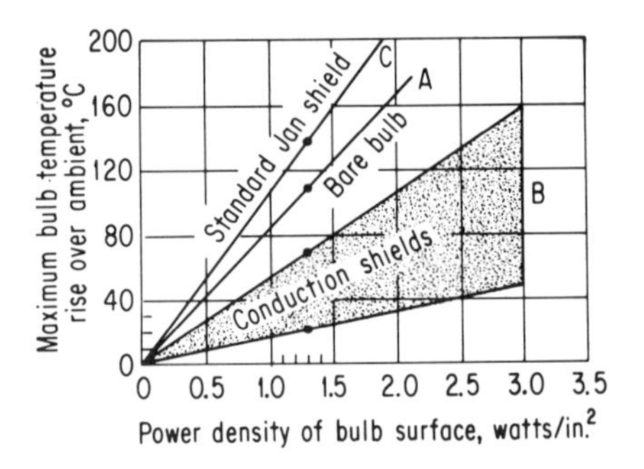

FIG. 18. Effects of tube shields on subminiature tubes.

for calculating h_c for various tube-mounting characteristics. A nominal value which may be used for estimation purposes is 10 Btu/(hr)(ft²)(°F).

The most efficient way to cool tubes is by the use of a clip or shield which makes contact with the tube envelope and acts as a conductive path to some local sink such as the chassis. A large number of special tube shields have been developed. Reference 16 presents some detailed performance data on various types. Curve B of Fig. 18 is typical of the performance for good conductive shields. Note the significant reduction in envelope temperature. Conductive shields also have the advantage of reducing the axial temperature gradients within the tube envelope, which forced convection alone may not do. The polished Jan-type tube shield which was used for electrical reasons is very poor thermally, as shown by curve C of Fig. 18. The polished shield effectively insulates the bulb.

Microwave Components. Low-power microwave components have been successfully cooled using a combination of free convection, radiation, and conduction-cooling methods. High-power traveling-wave tubes, klystrons, and magnetrons usually require forced-air cooling, liquid cooling, or mounting on a good conduction sink to prevent excessive temperatures. Since the construction of these devices varies widely, no general temperature limit or cooling technique can be specified. Typically, forced-air-cooled devices have a power dissipation of 10 to 100 watts; conduction-cooled devices, 10 to 1,000 watts; and liquid-cooled devices, 500 watts and up. In general, most of the power dissipation in a microwave device is concentrated in a local area, which can lead to very difficult cooling problems. For most applications, it is recommended that the device manufacturer be consulted on the specific cooling requirements.

Transformers. The two basic types of transformers are the iron-core and the foil-wound transformers. These transformers are rated on the basis of winding hot spot, insulating material, potting material, and electrical stress. Most transformer failures consist of breakdown of the insulating material, although occasionally open windings occur. The temperature classifications of insulating materials in accordance with AIEE Standard No. 1 are shown below:

Class	*Maximum temperature, °C*
0	90
A	105
B	130
H	180
C	No limit specified

Transformers are cooled by conduction, radiation, and forced-air or liquid methods. In each case, the individual transformer construction must be considered before the cooling method can be determined. For example, it might be necessary to bond the core of a transformer directly to the chassis of the equipment to prevent excessive temperatures. In the foil-wound transformer, heat dissipation can be as high as a kilowatt, and air or liquid cooling is definitely required in these cases.

Miniaturized transformers typically found in electronic equipment usually consist of an iron core wound with an insulated wire. In some cases, the core and winding are only impregnated with a thin protective coating prior to use. Generally, the impregnated core and winding are embedded in epoxy for additional protection. In a transformer of this type, the thermal path can be considered to consist of three series thermal resistances. The heat generated can be assumed to be dissipated at the core of the transformer. This generated heat must first pass from the core through the windings. A typical thermal-resistance value for this path is 1.5°C/watt. Second, the heat must pass from the surface of the winding through the epoxy encapsulating material. The temperature rise in this path can be determined utilizing the one-dimensional Fourier heat-transfer equation

$$\Delta t = 22.7\, q_e x / kA$$

where Δt = temperature difference through epoxy, °C
q_e = rate of heat generation, watts
x = epoxy thickness, in.
k = epoxy thermal conductivity, Btu/(hr) (ft) (°F) (typical value approximately 0.1)
A = surface area of winding normal to the direction of heat flow, in.^{2}

The third path, from the outer surface of the epoxy to the ultimate sink, depends upon how the transformer is mounted. For example, if the transformer were bonded directly to a metallic chassis which acts as a thermal sink, the resistance of the third path would be negligible. On the other hand, if the transformer were suspended by its leads, this thermal resistance could be significant (approximately 170°C for each watt per square inch of case area at sea level).

Capacitors. Capacitors utilizing air, paper, ceramic, mica, electrolytic, or glass as the dielectric normally dissipate little or no power. As such, these part types assume the temperature of the surrounding or local mounting point. Temperature limits for these components range from +40 to +150°C, depending on type. This information is specified by the manufacturer's data sheets. A capacitor may be the most thermally critical part in an equipment, since capacitors in general have low temperature limits. Normally, a capacitor does not fail catastrophically in temperature but degrades with respect to maximum operating voltage and/or frequency.

The varactor is a type of power-dissipating capacitor. It is basically formed of a semiconductor and should thermally be treated as such.

Subassembly Heat-transfer Characteristics. The practice of mounting discrete components and/or integrated circuits in cordwood modules and, in turn, mounting these modules on printed-circuit boards has become common. In addition to being complicated, the resulting thermal paths are at least partially through materials such as foam, epoxy, and laminated boards, which are normally considered to be insulators. Consequently, an analytical approach to thermal analysis of such equipment is exceedingly complex. However, some data, the results of approximations and measurements, are available.

Cordwood Modules. With the exception of rigorous computer-aided successive iteration analyses no accurate analytical approach is possible to the determination of the temperature distribution of a potted cordwood assembly of discrete components or integrated circuits. This is due to the extreme nonhomogeneity of properties, heat dissipation, and geometry. There are, however, two mathematical models which provide a gross method of bracketing the maximum internal temperature for preliminary calculations. A one-dimensional model of a cordwood module provides an upper limit by assuming uniform heat dissipation, uniform conductivity equal to that of the potting material, and insulation on five sides with an infinite sink on the remaining side, assumed to be the base. The most conservative estimate of the maximum internal temperature on this basis is

$$\Delta t_{max} = 11.4 \frac{q_e y}{kLw} \tag{16}$$

where Δt_{max} = maximum internal temperature rise over the surface, °C
q = total dissipated power (must be uniform), watts
y = module height, in.
k = potting compound conductivity, Btu/(hr)(ft)(°F)
w = module width, in.
L = module length, in.

The lower estimate of the maximum temperature is obtained by using a two-dimensional, infinitely long rectangular prism with uniform heat dissipation, uniform conductivity equal to that of the potting compound, and an infinite sink on four sides. The relation for maximum temperature in this case is

$$\Delta t_{max} = 5.8 \frac{q_e w}{kLy} \left(1 - \frac{1}{\cosh 1.57y/w}\right) \tag{17}$$

where Δt_{max}, q, k, L, w, and y are as before, and $y \geq w$.

Actual test data for various module configurations lie between the two limits. The equations shown are specifically for cordwood modules with uniform internal power dissipation. For nonuniform dissipation the hot-spot-to-surface and surface-to-ambient temperature rises are commonly two to three times the rises for uniform dissipation.

Printed-circuit Boards. Printed-circuit boards require computer-aided methods of analysis with considerable approximation required to model the sinuous thermal paths. The controlling factor in many printed-wiring-board heat-transfer situations is actually the copper circuitry pattern configuration on the boards. The configuration of the conductor as well as the total amount of conductor must be considered.

A concept useful for thermal analysis of plates or sheets of material such as printed-wiring boards is that of *conductance per square.* The equation for heat transfer by conduction in the plane of a sheet can be written as

$$q_e = (kw\delta/L)\,\Delta t = C\Delta t$$

where q_e = rate of heat flow
k = thermal conductivity
w = sheet width
Δt = temperature difference
δ = sheet thickness
L = sheet length
C = conductance

Note that if the sheet is square, so that $w/L = 1$, then the term $kw\delta/L$ reduces to $k\delta$. The conductance, then, of a square sheet of material is simply $k\delta$; hence the term *conductance per square.* The electrical analogy, previously explained, gives rise to the term *resistance per square* with similar logic. The resistance per square is, then, $R_{sq} = 1/k\delta$.

Figure 19 displays the results of computer analyses applied to a set of printed-wiring boards having copper-clad areas ranging from "no circuitry" to "100 percent circuitry," i.e., a solid conducting sheet. The figure provides an approximate bulk thermal resistance per square as a function of conductor configuration for any given printed-circuit board. In order to use Fig. 19, one must optically compare the circuit board to be analyzed with those shown in the figure to obtain an approximate resistance-per-square value.

The resistance-per-square numbers on Fig. 19 are valid only for printed-wiring boards having a conductor resistance per square of 35°C/watt and a board resistance per square of 5600°C/watt, 0.003-in.-thick copper and 0.029-in.-thick thermoplastic board. A resistance-per-square value for a printed-wiring board having different properties can be analyzed with the following procedure:

1. Read printed-wiring-board constant from Fig. 19.

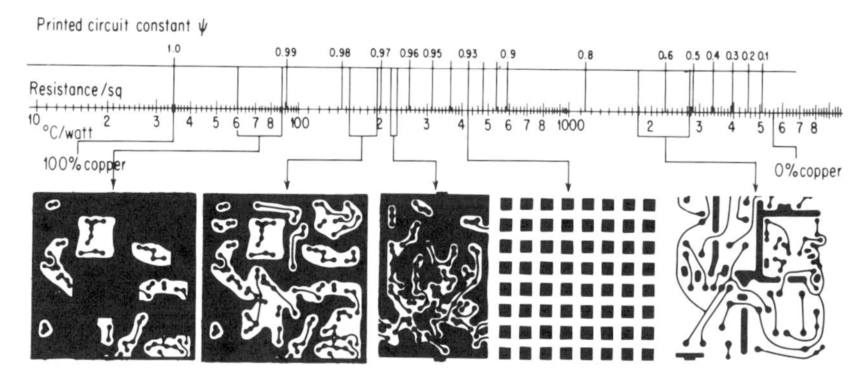

FIG. 19. Printed-circuit-board thermal resistance.

2. Substitute into the following equation:

$$R_{sq1} = R_{sqb} - \psi\,(R_{sqb} - R_{sqc}) \tag{18}$$

where R_{sq1} = desired resistance per square, °C/watt
ψ = printed-circuit-board constant for correct pattern (from Fig. 19)
R_{sqb} = resistance per square of actual circuit-board material being analyzed, °C/watt
R_{sqc} = resistance per square of actual conductor material being analyzed, °C/watt

There are two ways to determine the effective thermal resistance of double-sided printed-circuit boards. If (1) the two sides of the board have similar conductor geometry and (2) the resistance per square of the board alone is negligible compared to the effective resistance per square calculated for one side only, then the effective resistance per square of the double-sided board is approximately half that of the single-sided board. If, on the other hand, either (1) the conductors have grossly different geometry from one side to the next or (2) the resistance per square of the board is not negligible compared to the effective resistance per square, then the problem must be treated as two separate artificial one-sided printed-circuit boards, each having a board thickness of half the real double-sided board. The resulting two resistances when added in parallel approximate the resistance per square of the double-sided board.

The next step is the utilization of the resistance-per-square value to determine the expected temperature rise in a given board configuration. Printed-circuit boards are rectangular in most cases. They are usually attached to the supporting structure along opposite edges, as in commercial computing equipment, or along all four edges, as in spacecraft communication equipment. The effects of heat lost to the environment by convection from the board surfaces may or may not be important. If the bulk thermal resistance per square of the printed-circuit board is less than 200°C/watt, the effects of free convection will probably be negligible. If, on the other hand, the resistance per square is greater than 200°C/watt, the effects of free convection or conduction from the surfaces of the boards through even the best of insulators should be examined.

Figure 20*a* depicts a printed-circuit board with two insulated edges, two cooled edges, and insulated plane surfaces. The cooled edges of the board are at $t = 0$,

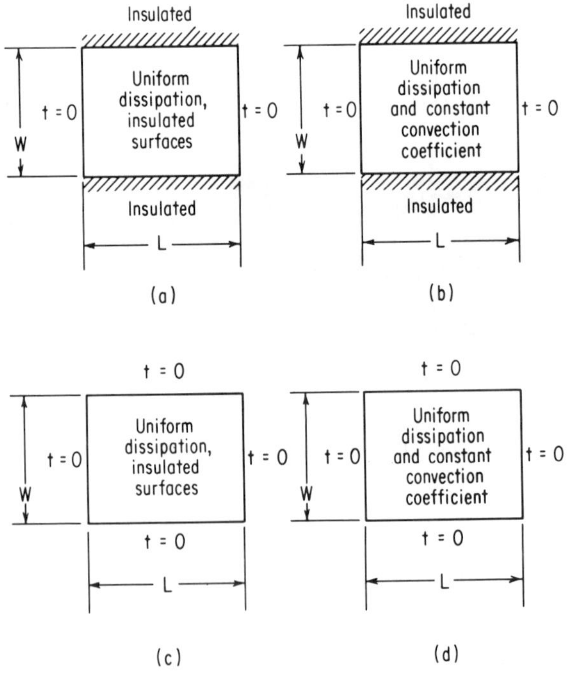

FIG. 20. Boundary conditions for printed-circuit-board analysis.

and the power is dissipated uniformly over the board. The maximum temperature rise is

$$\Delta t_{\max} = \frac{R_{sq}\, q_e L}{8w} \tag{19}$$

where $\Delta t_{\max}$ = maximum temperature rise over the sink, °C
R_{sq} = resistance per square, °C/watt
q_e = total power dissipated on the printed-wiring board, watts
L = total board length, in.
w = board width, in.

Figure 20*b* illustrates a printed-wiring board with two insulated edges, two cooled edges, and a convection coefficient h_c over the two plane surfaces. The sink for both the edges and the surfaces is at $t = 0$, and the power is dissipated uniformly over the board,

$$\Delta t_{\max} = 137\, \frac{q_e}{h_c\, Lw} \left\{ 1 - \frac{1}{\cosh\,[0.0426\, L\, (h_c R_{sq})^{1/2}]} \right\} \tag{20}$$

where $\Delta t_{\max}$, q, L, w, *and* R_{sq} are as before and h_c = convection coefficient, Btu/(hr)(ft²)(°F).

Figure 20*c* describes a printed-wiring board with all four edges held at $t = 0$ and both plane surfaces insulated. Power dissipation is uniform on the board.

$$\Delta t_{\max} = \frac{q_e\, R_{sq}\, w}{8L} \left\{ 1 - \frac{1}{\cosh\,[1.57\,(L/w)]} \right\} \tag{21}$$

where all parameters are as described before.

Figure 20*d* depicts a printed-wiring board with all four edges held at $t = 0$ and a convection coefficient h_c over the two surfaces. Power dissipation on the board is uniform.

$$\Delta t_{\max} = 137\, \frac{q_e}{h_c L w} \left\{ 1 - 0.785 \left[\frac{1}{\cosh\,\{[0.00725\, h_c R_{sq} + (\pi^2/L^2)]^{1/2}\, w/2\}} + \frac{1}{\cosh\,\{[0.00725\, h_c R_{sq} + (\pi^2/w^2)]^{1/2}\, L/2\}} \right] \right\} \tag{22}$$

HEAT-TRANSFER DEVICES

A portion of the section on thermal environments was devoted to the various types of overall thermal-design configurations. A number of specialized cooling techniques and hardware were indicated. Once the environment and internal part characteristics are known, the next logical step is to consider which configuration will best fit the application. It is necessary to know the characteristics of some of the basic cooling devices and techniques in order to implement their use in any design. The items to be discussed in this section include fans, heat exchangers, thermoelectric coolers, and fins, which represent the majority of those devices added for cooling purposes only.

Fans and Blowers. Fans may be used by the equipment designer for a variety of functions ranging from mild circulation of air in ground equipment racks to supplying air for a high-performance compact heat exchanger. Almost all applications can be classified either as a circulation function or as an application in which air velocity is increased to increase the convection coefficient. In the latter category are the heat-exchanger supply blowers and those used for direct impingement cooling of individual components.

Types of Fans and Blowers. A fan consists of an impeller, a housing, and a motor. In almost all electronic equipment applications the motor and impeller shaft are integral. There are a number of classifications of fans. The impeller configurations representing the major classifications used in electronic equipment are shown in Fig. 21.

The centrifugal blower, vane axial, tube axial, and propeller fans are most commonly utilized in electronic equipment cooling.

CENTRIFUGAL BLOWERS. The centrifugal or squirrel-cage blowers come in a number of configurations. The blading may be radial, forward-curved, or backward-curved. The forward-curved blades provide the most desirable combination of performance characteristics in the smaller sizes. The scroll or casing design controls the static-pressure-producing abilities of the unit with a tight scroll providing a higher pressure. This blower type is capable of delivering substantial quantities of air against a reasonable static head. In some cases, the inlet airflow passes over the motor, and in others, the motor is not in the airflow. The air inlet is normally at right angles to the discharge.

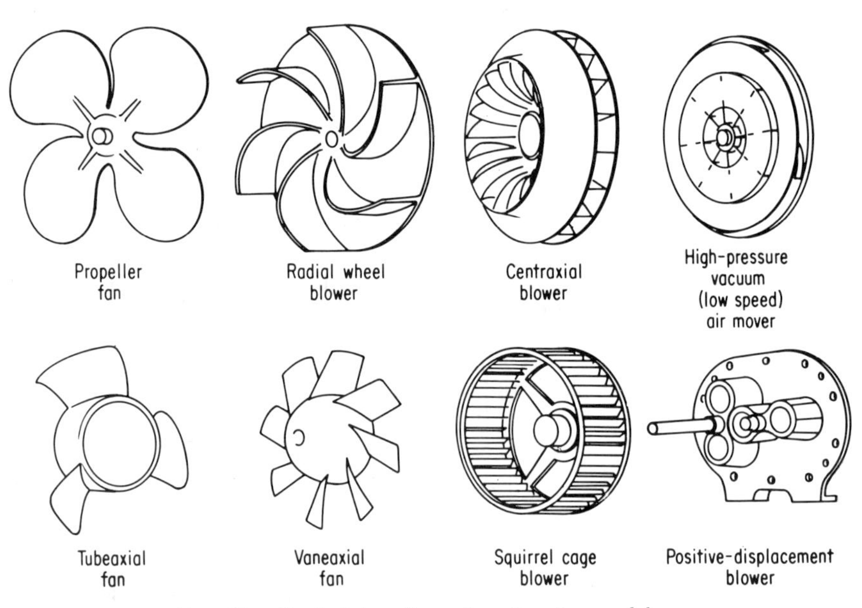

FIG. 21. Typical impellers of various types of fans.

VANE AXIAL AND TUBE AXIAL FANS. Vane axial and tube axial fans are designed to deliver considerable quantities of air against low-static heads. They are essentially ducted axial-flow fans with stationary guide vanes to increase their static-head-producing capabilities over those of unducted fans. These blowers usually have a higher efficiency than comparable centrifugal blowers. The drive motor is almost always integral with blower housing and rotor so that the motor wattage may have to be considered in the air requirements for the equipment. This type of blower is conveniently mounted to a front panel or directly in a duct. The static-head-flow characteristics of this type may indicate an unstable operating condition at high-static heads which must be investigated for the particular blower under consideration.

PROPELLER FAN. The propeller fan is essentially a high-volume circulation device. It delivers large quantities of air against negligible static pressure. However, for open-case construction, where the problem is just to provide sufficient airflow over the heat-dissipating parts or general circulation, the propeller fan may be quite satisfactory.

Fan Performance Curves. Constant-speed fan performance data are usually presented as a flow versus pressure drop characteristic similar to that shown in Fig. 22. The abscissa is typically flow rate in cubic feet per minute at some standard air density (ρ_{std}). The ordinate is the corresponding static pressure measured in inches of water. It can be shown that for any given density ρ, the volumetric flow rate $\dot{Q}$ is equal to m and the static pressure rise is equal to $\Delta p/\sigma$. These definitions allow interpretation of the fan performance at any density ρ from the standard density plot. See Ref. 17 for more information.

Blower Selection. SPECIFIC SPEED. In order to provide a common parameter to express the general performance characteristics of the various types of blowers, the specific-speed concept is used. The *"specific speed"* of a blower is defined as

$$N_s = (\text{rpm})(\dot{Q})^{0.5}/(\Delta p)^{0.75}$$

where N_s = specific speed
rpm = blower speed, rpm
$\dot{Q}$ = air delivery rate,* ft³/min
Δp = static pressure head,* in. of water

As might be expected since the characteristics of blowers of the same type are similar, the specific speeds are also similar. Each blower has a given range of specific speeds. In other words, there is a limited range of combinations of static pressures and air delivery rates at maximum efficiency for a blower type. Figure 23 shows the specific speed characteristics for the various blower types.

LOAD SPEED. If a relation similar to specific speed can be developed for the required *equipment* airflow characteristic, then an orderly blower-selection technique can be obtained by matching. The *load-speed characteristic* (N_L) is such a relation and is calculated by the same equation as used for N_s. However, the airflow and static pressure used are those for the required equipment operating point.

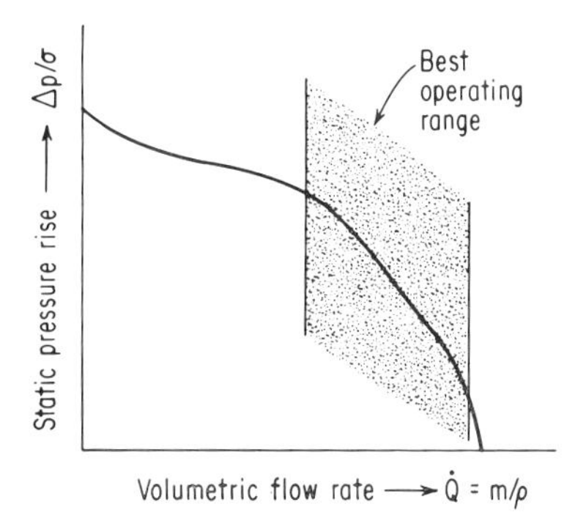

FIG. 22. Typical fan characteristic.

BLOWER SPEED. If an induction motor is used, the motor speeds are a direct function of the power frequency. For 60-cycle power, 1,750 and 3,450 rpm are typical values of actual operation, and for 400-cycle power, 3,700, 5,400, 7,200, 10,500, and 21,000 rpm are typical values. Dc power can provide a variety of speeds. In general, the higher the speed, the greater the noise level and the shorter the life. The usual situation with electronic equipment is a minimum size and weight requirement, so that 400-cycle high-speed induction motor-driven blowers are common. Typically, these motors maintain a reasonably constant speed during the change from

* At point of maximum efficiency, where the ratio of air horsepower to input power is a minimum. This occurs near or just after the midpoint of the characteristic.

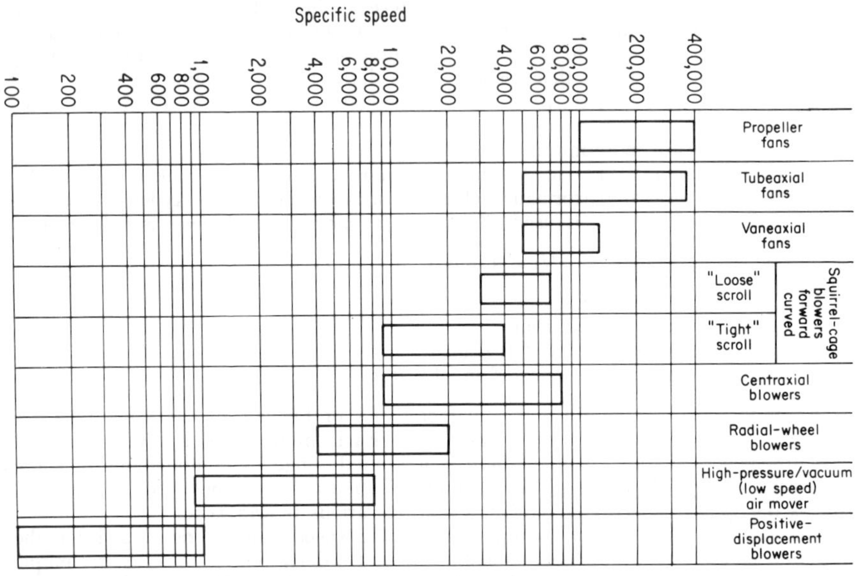

FIG. 23. Specific speed characteristics for various types of fans.[22]

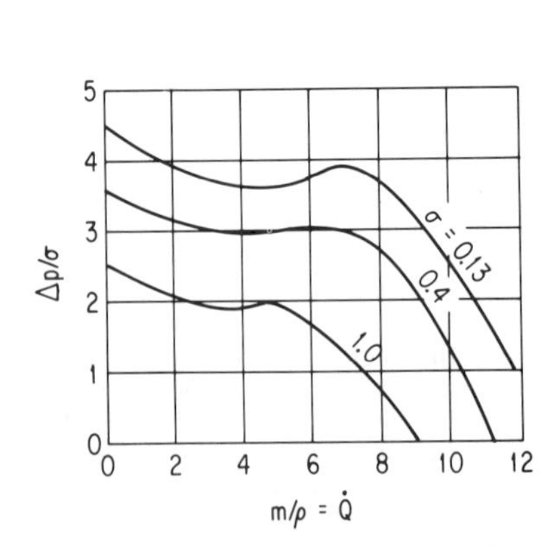

FIG. 24. Characteristic curves for variable-slip blower.

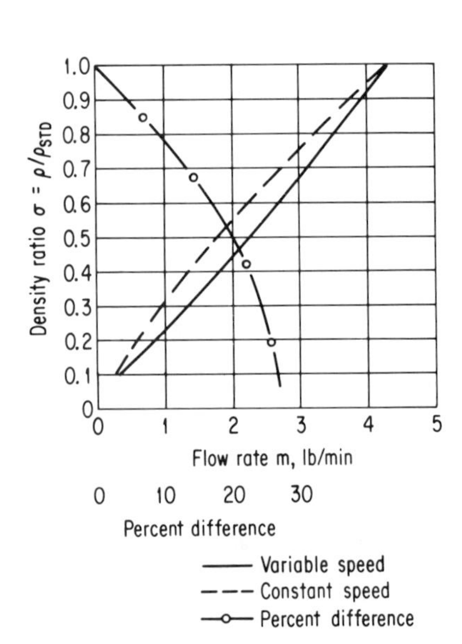

FIG. 25. Typical variable-speed vs. constant-speed blower performance.

full flow to zero flow at constant density. Since the volumetric flow is constant this results in decreased mass flow at low density. However, some motors are especially designed to increase in speed as the density of the air decreases, delivering more air at the low-density (high-altitude) operation than the standard constant-speed motors. A set of characteristics for such a variable-speed fan is shown in Fig. 24.

It should be noted that variable-speed fans can only increase speed over a limited range and therefore cannot fully compensate for density changes.

Figure 25 shows a comparison of the performance of a variable-speed and a constant-speed blower of similar basic characteristics.

If the load-speed characteristic is calculated for a specific application, the initial selection of blower type is made by matching this characteristic to the appropriate type in Fig. 23. The next step is to choose the particular fan, within those available of a given type, that will best match the equipment requirements. Note that these requirements must be defined in terms of a required flow and a corresponding pressure drop. At least one point on the pressure-drop–flow relationship for the equipment was required for the initial flow-speed determination. Reasonably complete data are now needed for final matching to the fan characteristics. In general the fan

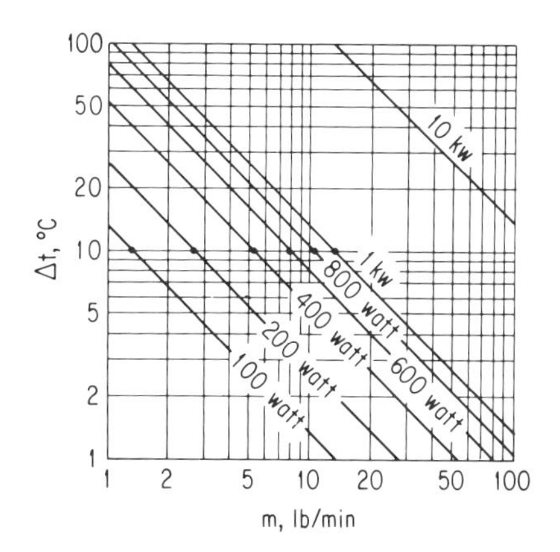

FIG. 26. Calculation of air-temperature rise as a function of heat input and flow rate.

should be selected such that the operating point is on the right decreasing portion of the fan characteristic. Calculations of flow resistance were briefly discussed in the section on fundamentals; no attempt will be made to cover this subject in more detail here. References 6 and 18 are recommended as guides for this purpose.

Example of Fan Selection. RACK CIRCULATION. Assume that a rack for electronic equipment will house numerous subassemblies which dissipate a total of 1 kw. It is desired to have a fan which will keep the internal rack temperature rise low. The subassemblies have been tested and are known to function reliably when operating in a local ambient of 55°C. The rack specification calls for operation in ambients up to 50°C. This means that the circulating air must not rise over 5°C. Experience indicates that similar racks operate with static pressure drops between 0.1 and 0.3 in. of water.

The equation for the sensible heat gain of a fluid is

$$q = mc_p\,(t_2 - t_1)$$

For convenience, this equation is presented in graph form in Fig. 26 for air.

From Fig. 26 it can be seen that 26.6 lb/min (350 ft^3/min) will be required to remove 1 kw with a 5°C rise in air temperature. Since 60-cps power is most readily available, 3,450 rpm is the maximum which should be considered for the fan. On this basis the load-speed characteristic is

$$N_L = \frac{(350)^{1/2}\, 3{,}450}{0.3^{0.75}} = 157{,}000$$

From Fig. 23 it is apparent that either a propeller fan or a tube-axial fan would be satisfactory. There are many propeller fans that fit the requirement. One manufacturer has a propeller fan which is 10 in. in diameter and 6 in. deep. It dissipates 75 watts under load and comes in a variety of 60-cps voltage combinations with a speed of 3,450 rpm. Therefore, a rack with this fan installed and with an adequate flow path will ensure that the subassemblies will not be exposed to ambients in excess of 55°C. No more detailed calculations than those shown are needed for this simple case.

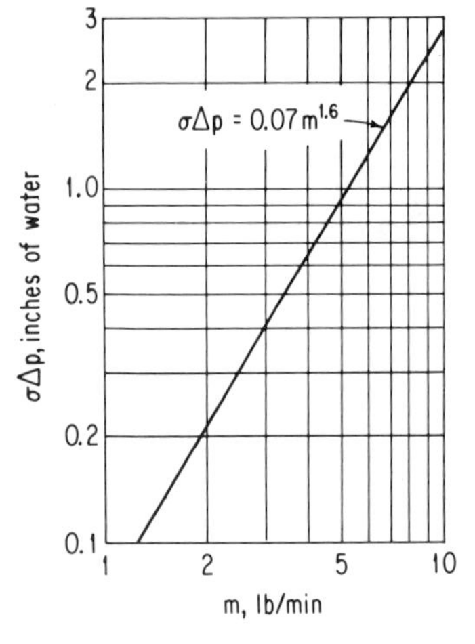

FIG. 27. Heat-exchanger flow-resistance curve.

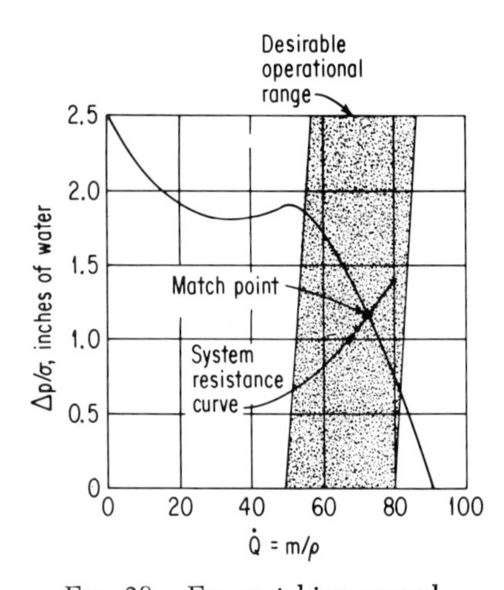

FIG. 28. Fan-matching example.

MATCHED FAN AND SYSTEM RESISTANCE. A more complex aspect of fan selection is the matching of the fan characteristic curve with a similar system resistance curve. Figure 27 represents a typical heat exchanger in which flow in pounds per minute is plotted against a corrected pressure drop $\sigma\Delta p$ in inches of water. This type of plot is initially developed from the analytic relationships previously discussed. If this is done the curve will have an equation of the form $\sigma\Delta p = C_1 m^n$, and n will be very close to 2. Once an exchanger or flow system has been fabricated, then a plot similar to Fig. 27 can be empirically obtained. From this figure, both the exponent n and the coefficient C_1 can be determined. The exponent is merely the slope of the line and is obtained by dividing the linear vertical displacement by the corresponding linear horizontal displacement. With the exponent determined, the value of the coefficient may be determined by solving the equation at any point on the curve. In this example, the exponent and the coefficient can be found to be 1.6 and 0.07 respectively. Typical heat exchangers have exponents in the range of 1.3 to 1.8 instead of the value of approximately 2 usually used in analyses.

For this example, initial heat-transfer calculations have indicated that the system weight flow must be at least 4.2 lb/min. A likely blower is chosen, and in order to find the fan/system operating point, a portion of the system plot is put on the fan characteristic (Fig. 28) to obtain the match point. Assuming the density at the operating point will be $\rho = 0.060$ lb/ft^3, then $\sigma = \rho/\rho_{std} = 0.060/0.075 = 0.8$. Determining the heat-exchanger resistance, from Fig. 27, for two arbitrary flow rates near the required amount of 4.2 lb/min will provide the first estimates of the match point on the fan curve (Fig. 28). For example, for $m_1 = 4.2$ and $m_2 = 4.8$ lb/min, the respective heat-exchanger resistances $\sigma\Delta p$ are 0.7 and 0.88 in. of water. However, since the fan curve is plotted as $\dot{Q}$ versus $\Delta p/\sigma$, the terms must be converted before plotting using the relationships $\dot{Q} = m/\rho$ and $\Delta p/\sigma = \sigma\Delta p/\sigma^2$. Therefore $\dot{Q}_1 = 4.2/0.06 = 70$ ft^3/min, $(\Delta p/\sigma)_1 = 0.7/0.8^2 = 1.09$ in. of water, $\dot{Q}_2 = 4.8/0.06 = 80$ ft^3/min, and $(\Delta p/\sigma)_2 = 0.88/(0.8)^2 = 1.38$. By plotting these two points on Fig. 28 and connecting them with a straight line, the match point for the system is obtained. The match point at 74 ft^3/min or 4.44 lb/min provides slightly more than the required airflow rate of 4.2 lb/min. The match point is also within the desired portion of the fan curve.

Heat Exchangers. Heat exchangers are used to cool electronic equipment when the heat dissipation from the equipment cannot be adequately removed by free convection, conduction, radiation, and/or direct forced-air impingement methods. Heat

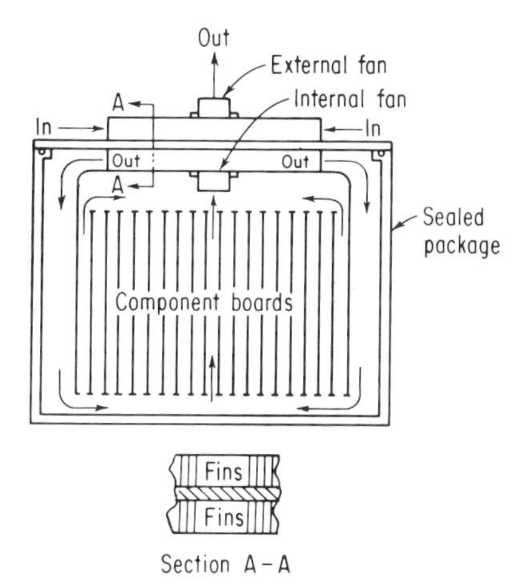

FIG. 29. Typical example of an air-to-air heat-exchanger application.

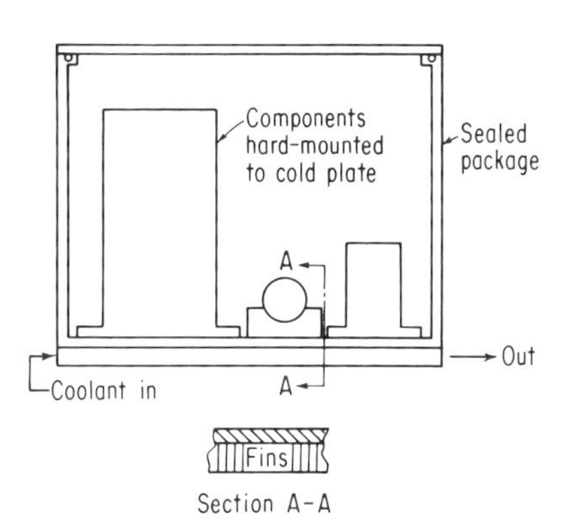

FIG. 30. Typical example of a cold-plate application.

exchangers can be divided into two basic types. The first, the convective-conductive-convective type, includes all liquid-to-liquid, liquid-to-gas, and gas-to-gas exchangers. In this type of exchanger heat is transmitted by convection from the hot fluid to the wall separating the fluid passages; by conduction through the wall; and by convection to the cold fluid. A typical example of an air-to-air exchanger is shown in Fig. 29. The second type, the conductive-conductive-convective exchanger, is commonly referred to as a *cold plate*. In a cold plate, heat is transmitted by conduction from the heat source to the plate, by conduction through the plate, and by convection to the cold fluid. This type of exchanger is widely used in electronic equipment cooling because of its packaging versatility. A typical example of cold-plate application is shown in Fig. 30.

Heat-exchanger Design. The design of a heat exchanger is typically a trial-and-error procedure. However, companies that specialize in the design and fabrication of exchangers can now optimize each exchanger configuration by utilizing digital computer techniques. Since the theory of exchanger design is too involved to be covered in a section of this size, only the basic concepts will be discussed. An excellent discussion of the design of heat exchangers is presented in Refs. 1 and 6.

The two main equations involved in exchanger design are the so-called *rate equation* and *energy equation*. The rate equation is given by

$$q = UA\Delta t_m$$

where q = the rate of heat transfer, Btu/hr
U = the overall heat-transfer coefficient from the hot side to the cold, Btu/(hr)(ft²)(°F)
A = the effective area upon which U is based, ft²
Δt_m = the log mean or average temperature difference, °F

This equation gives an indication of the size exchanger (UA value) and the temperature differential required to transfer a given amount of heat from the hot fluid (or hot surface for a cold plate) to the cooling fluid. It should be noted that the temperature differential used in this equation is the log mean temperature difference (LMTD). The LMTD can be expressed as

$$\Delta t_m = \frac{\Delta t_1 - \Delta t_2}{\ln (\Delta t_1/\Delta t_2)}$$

where Δt_m = LMTD, °F
Δt_1 = temperature difference between fluid streams or fluid and plate at one end of exchanger
Δt_2 = temperature difference between fluid streams or fluid and plate at other end of exchanger

The LMTD concept is necessary to account for the fact that there is a continually changing temperature difference between the hot and cold reference point as fluid stream(s) pass(es) through the exchanger. In most cases the arithmetic average temperature difference can be used for preliminary analysis.

The energy equation is expressed as

$$q = wc_p(t_{out} - t_{in})$$

where q = rate of heat transfer, Btu/hr
w = rate of coolant flow, lb/hr
c_p = specific heat of the fluid, Btu/(lb)(°F)
t_{in} = inlet fluid temperature, °F
t_{out} = outlet fluid temperature, °F

The energy equation shows that for a given flow rate, the temperature increase of the fluid is directly proportional to the amount of heat absorbed by the fluid. For a satisfactory design, the rate equation and the energy equation must be balanced. The amount of heat absorbed by the cold stream must equal that transferred by the exchanger, and in turn that amount must be equal to the heat dissipated on the cold plate or transferred from the hot fluid stream. This essentially means that both the energy equation and the rate equation must be satisfied. Another important point that must be remembered is that as the flow rate increases the pressure drop through the exchanger also increases. Therefore, there are actually three equations, for the rate, energy, and pressure drop, that must be "balanced" for a satisfactory design.

Preliminary Cold-plate Design Procedures. As a first trial a cold-plate design would typically utilize air cooling and a simple fin configuration. On a prototype model, for example, the fins might be milled as part of the chassis structure. The heat dissipation, maximum allowable cold-plate temperature, and maximum air inlet temperature are typically known. For a quick estimation, (1) a simple fin configuration, (2) a heat-transfer coefficient of approximately 5 Btu/(hr)(ft²)(°F), and (3) an airflow of 50 to 100 ft³/min can be assumed. This airflow is typical of miniaturized blowers used in electronic equipment packages. With the above-mentioned parameters established the calculation procedure would be as follows: First, use the energy equation to obtain the air-temperature increase through the exchanger. The average air temperature is then found by adding half this temperature rise to the inlet air temperature. Second, the assumed heat-transfer coefficient is used in the rate equation along with transfer area which is determined from the fin configuration to determine the estimated temperature difference between the cold-plate surface and the air. Third, this temperature difference is then added to the average air temperature to obtain an estimate of the cold-plate temperature. If the results are compatible with the allowable component limits, a more detailed calculation should be made utilizing the applicable forced-convection heat-transfer correlation for the exact flow-passage configuration. On the other hand, if the results of the quick calculation indicate that the mounting-plate temperature will be excessive, it might be necessary to consider a compact-core heat-exchanger design.

The compact-core heat exchanger typically consists of 0.006- to 0.008-in.-thick fins which range from 0.25 in. to approximately 0.5 in. high and have a spacing of 10 to 18 fins/in. This core material is usually dip-brazed to a mounting plate, forming a "sandwiched" configuration. The primary advantage of the compact-core exchanger over the simple fin configuration is the significant increase in heat-transfer area. The main disadvantage is an increase in airflow resistance, which results in an increase in pressure drop. Kays and London [6] present heat-transfer and pressure-drop correlations for a variety of compact-core configurations. The pressure drop obtained for the exchanger must then be matched to a coolant supply. A discussion on matching a system and blower is presented in the preceding section.

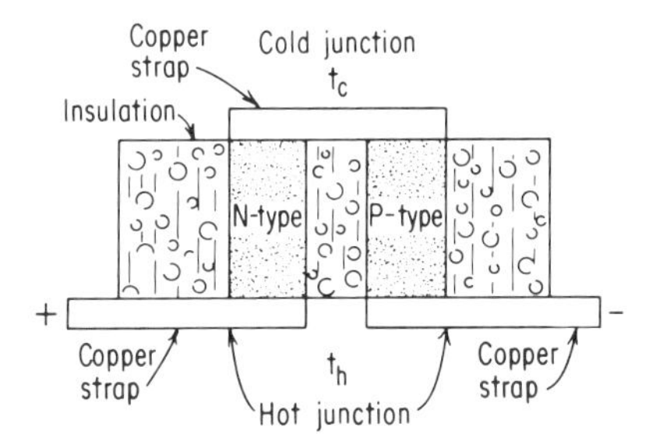

FIG. 31. Basic couple configuration.

If the cooling requirements cannot be met with an air-cooled exchanger, liquid cooling must be considered. By utilizing liquid cooling, a heat-transfer coefficient in the range of 100 to 200 Btu/(hr)(ft^2)(°F) can be expected.

Thermoelectric Coolers. Thermoelectric coolers are most effectively used in electronic equipment when it is necessary to provide spot cooling for a component, or to maintain the internal temperature of an electronic "equipment" below that of the ambient. Thermoelectric coolers offer several advantages over typical cooling systems. First and foremost is the ability to provide a temperature below ambient without the moving parts of a vapor compression compressor and without an expendable coolant approach such as liquid CO_2 or N_2. There is almost no minimum practical size limitation, so that pumping loads of 10, 50, or 100 Btu/hr can be handled with reasonable temperature differences. The size of the equipment to accomplish this is compatible with the task where conventional vapor cycle equipment would be excessively large. This section will not be concerned with the details of thermoelectric cooling theory (see Refs. 1 and 19 for a more detailed discussion) but will briefly review the concepts involved.

Figure 31 shows schematically the thermoelectric couple which is the basic design element for thermoelectric heat-pump systems. It consists of an N-type and a P-type semiconductor element (e.g., bismuth telluride, Bi_2Te_3, or antimony telluride, Sb_2Te_3) joined to conductive straps, usually copper. If a dc current is caused to flow as shown, heat will be absorbed at the cold junction and will be generated at the hot junction with a net heat transfer from the cold side to the hot side. A reversal in polarity will reverse the effects. The insulation minimizes conduction, convection, or radiation heat transfer between the hot and cold sides. The design of individual couples may be optimized for maximum heat pumping, maximum temperature difference, or maximum efficiency.

Typically commercial thermoelectric modules consist of some number of couples combined into a modular subassembly with appropriate internal electrical connections. These modules are the most practical basis for thermoelectric device design for any company that does not wish to fabricate the basic couples. Figure 32 shows three typical module configurations.

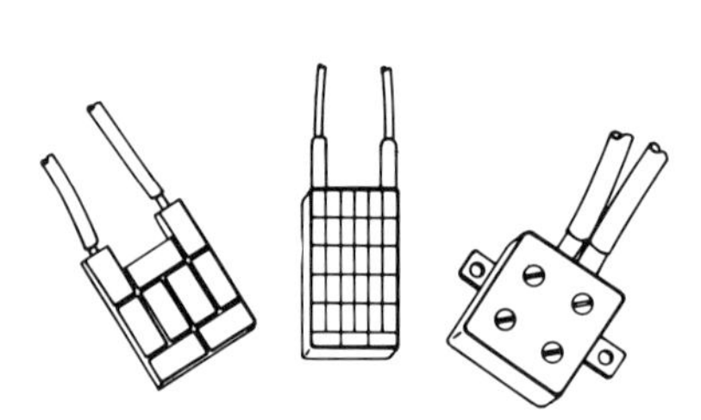

Fig. 32. Typical thermoelectric modules.

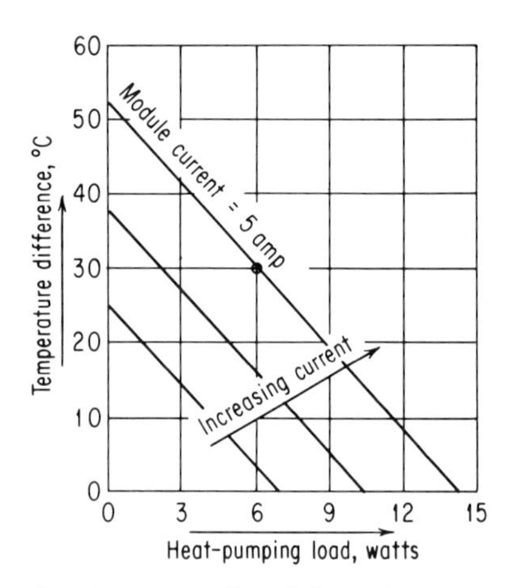

Fig. 33. Typical module performance curves.

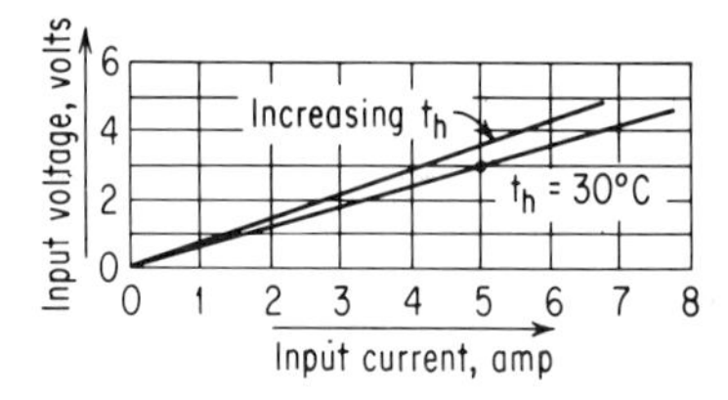

Fig. 34. Typical module current / voltage requirements.

Module Performance Characteristics. Figure 33 shows typical module performance data in terms of heat pumped, junction temperature difference, and input current for a particular hot junction temperature. For constant current, q pumped and Δt vary inversely. Increasing the current up to I_{opt} will increase Δt and/or q. For this reason no module is designed to operate at a current larger than I_{opt}. It may also be noted that the Δt is a maximum when the heat load is zero, and that the heat pumped is maximum when the Δt is zero. Figure 34 shows the relationship of input current to voltage for various hot junction temperatures. This plot provides the necessary information for the calculation of input power to the module.

The ratio of the actual heat pumped to the input power is called the *coefficient of performance* (COP) and is a measure of the "efficiency" of operation of the device as a heat pump. For the practical ranges of use for these modules this value is less than 1 and is typically close to 0.3. This means that it will require input power of at least three times the load to be pumped. As a basis of comparison typical COP values for residential vapor-compression systems are in the neighborhood of 3.

Thermoelectric modules are dc devices usually with low-voltage and high-current characteristics. Currents of 20 to 40 amp with voltages of 0.25 to 2.0 volts are common. Some modules with currents less than 8 amp and voltages of 3 to 8 volts are available. All modules require a relatively low ripple supply (less than 10 percent) for maximum COP. Since the thermoelectric module power requirements are not typical of most electronic components, special consideration must be given to the power supply.

Modules are usually supplied with electrically uninsulated top and bottom surfaces. This necessitates use of an electrical insulator between the module and the surface it will mount to if the latter is a conductor. A finite thermal resistance will be introduced at this interface which will decrease performance. Some manufacturers supply their modules with the top plates insulated from the couples. However, this merely means that there already is some loss incorporated into their performance.

The thermoelectric module absorbs the so-called *pumping load* at the cold junction and must dissipate it at the hot junction along with the electrical input power. The possible 3:1 or 4:1 multiplication of power between the cold and hot junctions poses definite problems due to the high power density on the hot side. Adequate provisions must be made to transfer the required heat at both junctions. Both the temperature rise between the cold junction and the cooled parts or air and the temperature rise between the hot junction and the ultimate sink subtract from the temperature difference provided by the module.

Example of Thermoelectric Cooling Design. The external case of a component must not exceed 1°C in a 25°C environment. The component dissipates 6.0 watts, and there is sufficient insulation so that negligible heat is absorbed from the environment. The total heat-pumping load is then 6 watts. A 1°C drop is expected between the component case and the cold surface of the thermoelectric element. It is estimated that the drop between the hot junction and the ambient air can be held to 5°C.

Using the performance curves of Figs. 33 and 34, the total heat to be removed from the hot side, and the COP of the module can be determined.

$$t_h = t_o + \Delta t_{h-o} = 25 + 5 = 30°\text{C}$$
$$t_c = t_s - \Delta t_{s-c} = 1 - 1 = 0°\text{C}$$
$$\Delta t_{\text{hot-cold}} = 30 - 0 = 30°\text{C}$$

For a total heat-pumping load of 6 watts and $\Delta t_{h-c} = 30°\text{C}$, it can be determined from Fig. 33 that a current of 5 amp will be required. Figure 34 indicates that for a 5-amp input current and a hot junction of 30°C, the voltage required will be 3 volts. It can now be determined that

$$q_{in} = \text{volts} \times \text{amp} = 3 \times 5 = 15 \text{ watts}$$
$$q_{\text{dissipated}} = q_{\text{pumped}} + q_{\text{input}} = 6 + 15 = 21 \text{ watts}$$
$$\text{COP} = \frac{q_{\text{pumped}}}{q_{\text{input}}} = \frac{6}{15} = 0.4$$

Note that provisions must be made to dissipate 21 watts at the hot side with a maximum temperature rise of 5°C.

Fins. Fins are used to add additional heat-transfer area to a surface. They may take various physical forms such as rectangular, parabolic, hyperbolic, and cylindrical (pin fins). The most common fin geometrical form is the vertical rectangular fin, the type discussed here. Heat transfer from the surface of a fin to the local ambient sink is given by the equation $q = \eta UA(t_s - t_o)$. The fin efficiency η is a measure of the fin's ability to maintain the fin outer tip at the temperature of the fin root. With a constant-temperature-profile fin the external heat-transfer equation reduces to the general form $q = UA(t_s - t_o)$. Analysis of any particular fin requires that both

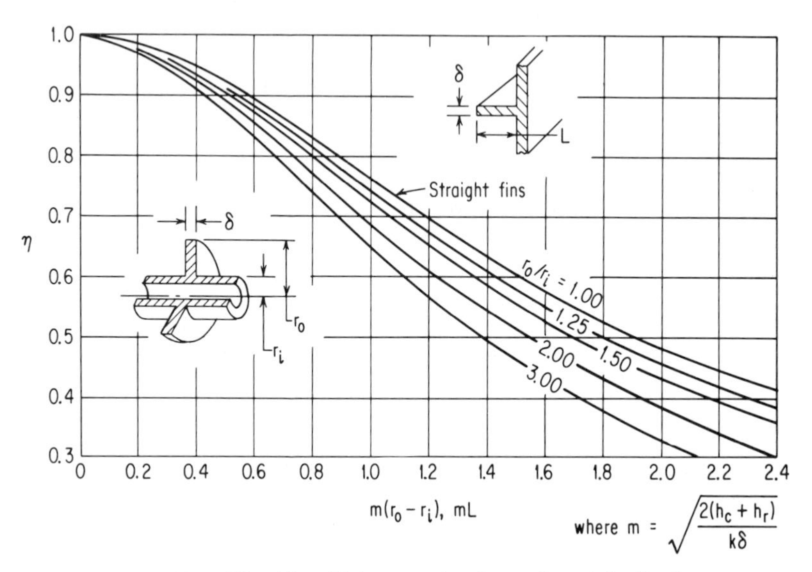

FIG. 35. Fin efficiency—circular and straight fins.[6]

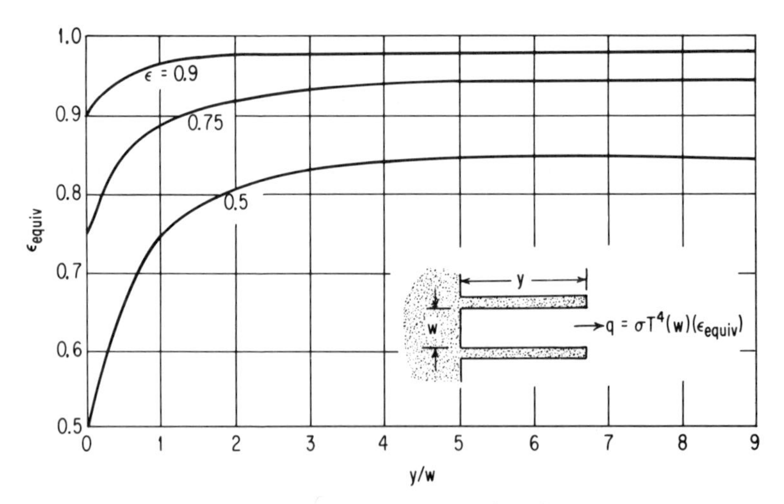

FIG. 36. Equivalent emissivity for finned surfaces.[23]

radiation and convection heat transfer be analyzed since both contribute to the overall coefficient U.

The fin efficiency may be calculated by utilizing the graph in Fig. 35. The use of this curve requires that the fin height, thickness, thermal conductivity, and overall convection and radiation coefficient be known or estimated.

The overall UA value for the fin is composed of both radiation and free-convection conductances and may be written as $UA = h_cA_s + h_rA_r$. The free-convection-coefficient value h_c is related to the fin height and spacing. As such, h_c is then related to the total fin-convection area A_s, thus indicating some desirable combination of h_c and A_s, that is, an optimum fin height and spacing for any given problem. The maximum attainable h_c value is that for a flat plate; therefore a general design rule is to space fins far enough apart that the various fin surfaces' convection flow

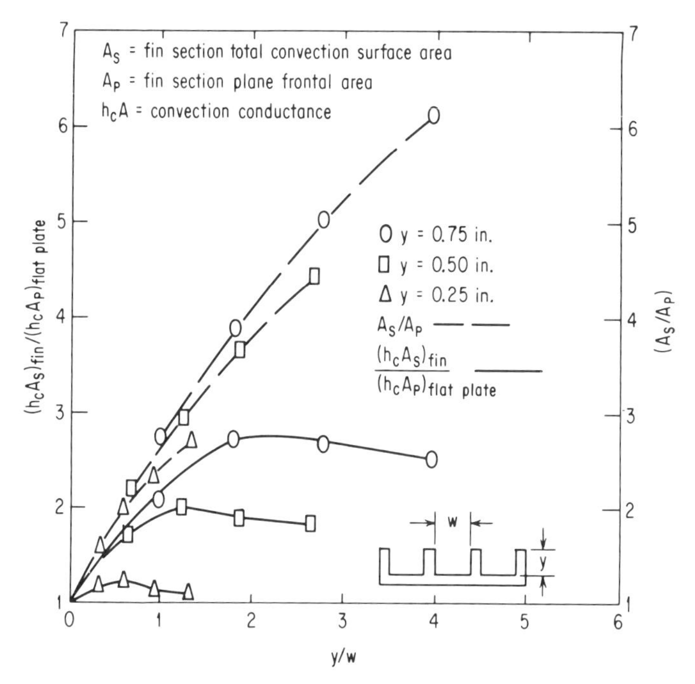

FIG. 37. Convection conductance data for 8-in.-long fins at a temperature of approximately 150°F.[20]

profiles are not affected by each other. Fin height-to-width (y/w) ratios in the range of 1:1 to 2:1 provide for good fin free-convection heat transfer. Reference 20 details the procedures for obtaining optimum fin free-convection geometries.

The addition of fins to a plane surface does not normally greatly increase the radiation conductance h_rA_r. Fins may be thought of as cavities or modified black-body radiators whose radiating area is the cavity face or space between the fins. Neglecting fin material thickness, the fin radiating area A_r is then the plane area of the finned plate. Fins tend to increase the basic plane area emissivity value, providing an "equivalent" emittance for the fin face. Figure 36 shows the correlations for calculation of fin radiation heat transfer.

The following example illustrates the techniques and procedures used in calculating fin heat transfer. The free-convection heat-transfer data used in the example are taken from Ref. 20. Assume that it is desired to obtain the maximum heat transfer from a set of 8-in.-long fins with a height of 0.75 in., a width w of 0.375 in., and a thickness of 0.080 in. Further assume that the finned plate has 18 fins and that a maximum fin temperature of 150°F is desired when placed in a 50°F ambient. Curves showing the optimum heat transfer for 8-in.-long fins with various y/w ratios are given in Fig. 37. This data shows that for a 0.75-in.-high fin, y/w is near optimum. Further, Fig. 37 also shows that for a given fin spacing an increase or decrease in the fin height from this optimum results in a decrease in the free-convection heat transfer. In terms of the free-convection heat-transfer coefficient, Fig. 38 shows that these fins have a coefficient of 0.8 Btu/(hr)(ft²)(°F). The fin convection area is

$$A_s = \frac{(0.75 \times 2 \times 8 \times 18) + (8 \times 18 \times 0.455)}{144} = 1.96 \text{ ft}^2$$

Assume that the fins are painted and have a flat-plate emissivity of $\epsilon = 0.75$. Figure 36 shows that for $y/w = 0.75/0.375 = 2$ the equivalent emittance value is 0.92. The fin radiation area A_r is nominally equal to the plane area of the finned surface, 0.46 ft². The radiation coefficient is then

$$h_r = \frac{(0.171)(0.92)(1)\,[(6.10)^4 - (5.10)^4]}{100}$$
$$= 1.11 \text{ Btu/(hr)(ft}^2\text{)(°F)}$$

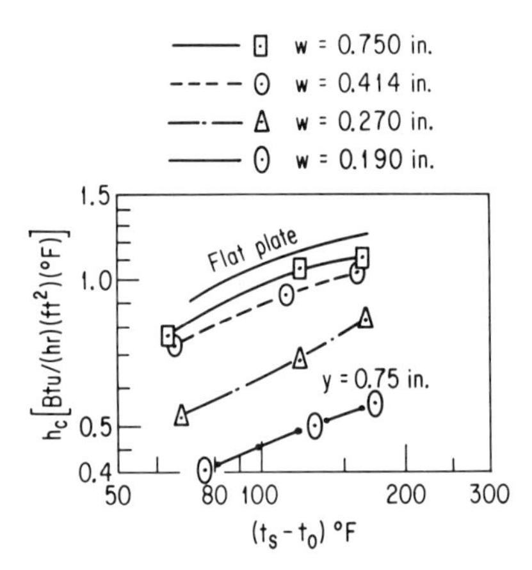

FIG. 38. Free-convection coefficients vs. temperature for finned surfaces.[20]

The total heat transfer is

$$q_{\text{total}} = (h_c A_s + h_r A_r)\ (t_s - t_o) = [(0.8)(1.96) + (1.11)(0.46)]\ 100$$
$$= 208 \text{ Btu/hr} = 61 \text{ watts}$$

Up to this point it has been assumed that the fins have an efficiency equal to 1. The data in Fig. 35 are used to evaluate the fin efficiency. The term

$$ml = \frac{y}{12}\sqrt{\frac{(2)(h_c + h_r)}{(k)(\delta/12)}}$$

in this figure is required as an entering parameter to obtain η_f. In the equation for ml, the terms h_c and h_r are required to be based on the total fin surface area A_s. Thus, h_r in the example must be multiplied by A_r/A_s to reflect this requirement.

$$h_r = [1.11 \text{ Btu/(hr)(ft}^2\text{)(°F)}]\ (0.46 \text{ ft}^2/1.96 \text{ ft}^2) = 0.26 \text{ Btu/(hr)(ft}^2\text{)(°F)}$$
$$h_c + h_r = 0.8 + 0.26 = 1.06 \text{ Btu/(hr)(ft}^2\text{)(°F)}$$
$$ml = \frac{0.75}{12}\sqrt{\frac{2\ (1.06)}{90\ (0.080/12)}} = 0.12$$

The efficiency $\eta \approx 1$, so that the assumption was valid.

THERMAL-DESIGN ANALYSIS TECHNIQUES

The previous sections have discussed the thermal environment considerations and the equipment characteristic information which influence thermal design. This section will illustrate the techniques used in making thermal analyses. Regardless of the type of equipment or the particular situation, the same general analysis procedure holds:

1. Define thermal environment in terms of temperatures, pressures, mounting provisions, and type of ultimate sink as previously discussed.
2. Specify equipment thermal-design characteristics. Both part dissipation and total dissipation must be determined. The types of components and their allowable temperature limits must be established. These allowable temperatures may be component manufacturer's ratings, they may be based on detailed reliability requirement studies, or they may simply result from experience. In any of the foregoing, it may be either a failure consideration or an electrical parameter that governs. In all cases, the maximum allowable component temperatures must be considered thermal-design requirements to ensure satisfactory equipment operation.
3. Provide for the necessary thermal paths from the dissipating components to the required dissipation sinks. This is the final thermal-design consideration. The heat dissipated by components follows complex paths to these sinks. It is convenient to use the thermal-resistance concept for consideration of the heat-transfer situation. In order to obtain a satisfactory design, the overall thermal resistance must be lowered to the value which will maintain allowable component temperatures while dissipating the heat. This overall component-to-sink resistance can be thought of as being the combination of individual resistances. Each represents an area for design effort. The actual detailed implementation of the analysis effort can range from a full-scale digital computer study to a very cursory estimate of overall temperature rise.

The mathematical tools available for the thermal analysis of real systems are varied in types and usefulness. Closed analytic solutions do not in general lend themselves to analysis of electronic packages since, in their most useful forms, they describe a homogeneous material with uniform properties. Electronic packages do not often satisfy this assumption. Further, the somewhat "ideal" boundary conditions assumed for the closed analytic solutions do not generally apply.

At the other end of the analytic scale is the simple one-dimensional conduction equation and various reduced empirical equations, as previously discussed and utilized in examples herein.

Between the closed analytic solutions and the simplified conduction correlations there is a wide range of numerical methods of analysis. In general, most of these numerical techniques involve subdividing the actual configuration into many small but finite subvolumes, typically called nodes. Both the thermal properties and the temperature are assumed to be uniform throughout the subvolume. It is further assumed that the heat transfer through each face of the subvolume is one-dimensional and can, therefore, be calculated using the simplest form of Fourier's conduction equation or suitably simplified convection or radiation equations. By noting that, for steady-state analyses, the sum of the heat entering a subvolume must equal the sum of heat leaving the subvolume, a simultaneous array can be derived consisting of one equation and one unknown per subvolume. Several approaches exist for approximating boundary conditions, heat dissipation, and transient conditions. This whole class of techniques is referred to as *finite difference solutions*. The theory and implementation of such solutions are thoroughly treated in the literature.[3]

With the advent of digital computing equipment, it became possible to handle large numbers of equations in short times with existing computer programs which utilize either successive iteration or matrix inversion methods. Successive iteration programs, while they are not exact and are slower than matrix inversion programs, are commonly utilized to solve equation sets with from 100 to 1,000 subvolumes. Matrix inversion programs, on the other hand, are exact and run rapidly, but the total nodes may be restricted due to the large storage capacity required.

The accuracy of such an analysis is, of course, only as good as the mathematical model; however, accuracies of ±1°C are not uncommon.

A great deal of the thermal design of electronic equipment is based on less sophisticated analysis techniques than those just discussed. One-dimensional assumptions and simplified estimating tools such as those shown in previous sections will satisfactorily handle a large variety of problems. The following examples will illustrate this type of general thermal analysis approach.

Example 1. An electronic equipment dissipates 84 watts and is required to operate in a worst-case sea-level environment of +55°C ambient temperature. The equipment is packaged in a 12- × 12- × 12-in. sealed housing. The part power-dissipation list for the unit is shown in Table 11.

TABLE 11

Part type	Number of like parts	Location in housing	Power dissipation per part, watts	Part mfg. max. temp., °C
2N2947 transistor . . .	4	Mounted to the housing bottom surface	10	175
1N4387 varactor	4	Mounted on aluminum housing bracket	7.5	175
2N3252 transistor . . .	4	Mounted in a BeO sink attached to the housing	2.5	200
2N2218A transistor . .	8	Lead-mounted on printed-circuit boards	0.25	175
¼-watt composition resistors	20	Staked to printed-circuit boards	0.1	150
Capacitors	15	Lead-mounted on printed-circuit boards	0	110

The ultimate thermal sink for the heat dissipated by the equipment is the ambient air and surroundings. The heat transfer from the housing is transferred via radiation and free convection. A simplified approximation of the housing temperature rise Δt_{so-ao} can be obtained from the curve in Fig. 39. The required entering parameter

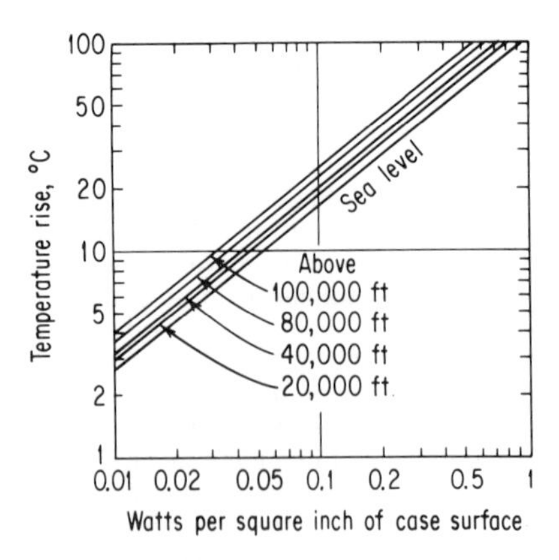

FIG. 39. Estimated surface temperature rise for typical closed electronic equipment cases with no external cooling provisions.

for the use of this curve is the equipment total dissipation per square inch of surface area calculated as (84 watts/864 in.^{2}) = 0.097 watts/in.^{2} The curve shows that $\Delta t_{so-ao} = 16°C$ so that $t_{so} = 55 + 16 = 71°C$. Figure 39 is valid for estimation of surface temperature rise for any closed equipment in any free-convection and radiation situation. Those equipment parts which are mounted with reasonable thermal paths to the housing and dissipate little or no power will be at the calculated housing temperature. The printed-circuit boards are poor thermal paths to the housing and for preliminary estimates may be considered to be at the housing internal air temperature. A conservative housing internal ambient temperature is obtained by assuming the same internal temperature Δt_{ai-si} as the external case had. Therefore,

$$\Delta t_{ai-si} = 16°C$$
$$t_{ai} = 55 + 16 + 16 = 87°C$$

For this temperature estimation the 110°C rated capacitors are at 87°C and are still below their rating.

The semiconductors (transistors and varactors) are to be mounted either to the printed-circuit boards, by a conduction sink, or to the housing. The semiconductor manufacturer's thermal resistance will be used to calculate the temperature rise of the semiconductor junction over the noted reference, i.e., either case or ambient. The junction-to-case thermal resistance is required for devices which transfer their heat by conduction to the equipment housing. The 2N2218A transistor's junction to ambient thermal resistance is used since these devices are mounted on poor thermal conductors and transfer their heat by radiation and free convection to the housing internal ambient. The junction temperature rise is obtained by multiplying the semiconductor R_{j-c} or R_{j-a}, as applicable, by the individual part power dissipation as tabulated in Table 12. The part case-to-local-sink temperature rise will be calculated

TABLE 12

Part	R_{j-c}	R_{j-a}	q, watts	Δt_{j-c}	t_{j-a}
2N2947	6°C/watt	N/A	10	60°C	N/A
1N4387	5°C/watt	N/A	7.5	38°C	N/A
2N3252	35°C/watt	N/A	2.5	87.5°C	N/A
2N2218A	N/A	188°C/watt	0.25	N/A	47°C

next. The local sink for the 2N2218A transistor is the equipment internal ambient. It has previously been shown that this internal ambient temperature will not exceed 87°C; therefore the 2N2218A transistor junction temperature is $t_j = 87 + 47 = 134°C$. The 2N2947 has a TO-3–type case with a mica insulator between it and the equipment housing. Figure 16, for contact resistance, shows that $\Delta t_{c-s} = (1.2°C/watt)(10\ watts) = 12°C$; thus $t_j = 55 + 16 + 60 + 12 = 143°C$. Applying the same analysis procedure for the stud-mounted mica-insulated 1N4387 varactor results in $t_j = 55 + 16 + 38 + 16 = 125°C$.

The 2N3252 transistor is mounted in a BeO cup with a silver-filled epoxy. The cup attaches by aluminum bracketing to the housing. Here $t_j = 55 + 16 + 87.5 + (9°C/watt)(2.5\ watt) = 181°C$. This semiconductor type may be considered as thermally most critical since it most closely approaches its maximum rated junction temperature.

The lead-mounted composition resistors mounted on circuit boards predominantly dissipate their heat to the housing internal ambient. The surface to ambient thermal resistance from Table 10 is 146°C/watt. Thus at 0.1 watts of dissipation these resistors will have a maximum surface temperature of $t_s = 55 + 16 + 16 + 15 = 102°C$, which is well below their 150°C rating.

All parts in this preliminary type of worst-case thermal analysis will operate with satisfactory part temperatures.

Example 2. An electronic equipment consists mainly of cordwood modules mounted on a printed-circuit board (see Fig. 40). The hottest component in the

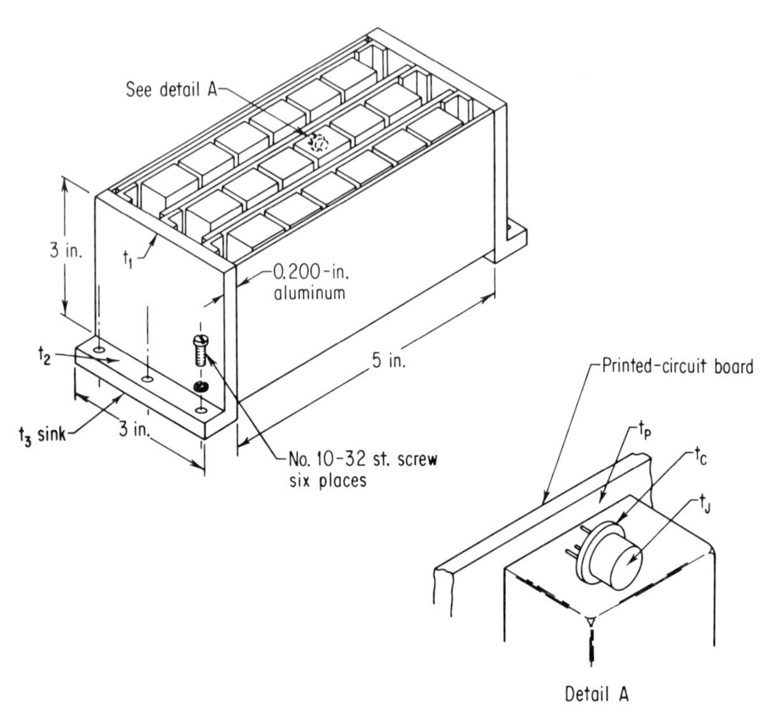

FIG. 40. Example 2 equipment modules on printed-circuit boards.

potted modules is expected to be a 2N918 transistor dissipating 30 mw. Otherwise, the power dissipated in each cordwood module is uniform at a total of 300 mw per module. The printed-circuit boards conduct heat from the cordwood modules to the equipment housing. The housing conducts the heat to a cold plate which is at a constant temperature of 48°C. Identical assemblies surround the equipment being studied so that no radiation heat transfer takes place. The equipment is mounted in a vacuum; hence, no convection takes place. Other salient details are:

Cordwood modules
- Total number, 18
- Length, L, 3.0 in.
- Width, w, 1.0 in.
- Height, y, 1.0 in.
- Potting conductivity, k, 0.12 Btu/(hr)(ft)(°F)
- Power per module, q_e, 300 mw

Printed-circuit board (see Fig. 41)
- Length, L, 5.0 in.
- Width, w, 3.0 in.
- Thickness, δ, 0.028 in.
- Conductivity, k_b, 0.14 Btu/(hr)(ft)(°F)
- Conductor thickness, δ, 0.003 in., copper

Cold-plate temperature, 48°C

The problem is to determine the junction temperature of the 2N918 transistor.

Solution. The ultimate sink for this equipment is the cold plate to which the equipment is attached. The first thermal resistance in the path from the cold plate to the transistor junction is the contact resistance between the cold plate and the equipment housing. Utilizing Fig. 15, the contact pressure in the immediate vicinity of each screw is

$$p = \frac{5E}{AD} = \frac{(5)\ (30)}{(0.25)\ (0.19)}$$

Hence,

$$p = 3{,}160 \text{ psi}$$

Then from Fig. 15, the contact resistance is approximately 0.09 (°C)(in.^{2})/watt. The thermal resistance at sea level is therefore 0.36°C/watt per screw. Then, multiplying by 2 since the environment is a vacuum, the thermal resistance at the interface is 0.72°C/watt per screw. It follows then that the temperature rise at the interface is

$$\Delta t_{3\text{-}2} = \frac{0.72°\text{C/watt per screw}}{3 \text{ screws}} \ 2.7 \text{ watts}$$

Therefore

$$\Delta t_{3\text{-}2} = 0.7°\text{C}$$

The temperature rise from point 2 to point 1 is that of one-dimensional conduction of heat from a plate with uniform dissipation, the equation of which is

$$\Delta t_{2\text{-}1} = 11.4 \frac{q_e L}{kA} = 11.4 \frac{(2.7)(3)}{(90)(3)(0.20)}$$

For the case in question, $\Delta t_{2\text{-}1} = 1.7°\text{C}$.

The printed-circuit boards are attached to the equipment housing in such a way that the contact resistance at this point is negligible.

Now proceeding to calculate the temperature rise from the housing to the hottest point on the printed-circuit board, $\Delta t_{1\text{-}p}$, the appropriate model is that shown in Fig. 20; hence, the governing equation is

$$\Delta t_{1\text{-}p} = \frac{R_{sq} q_e L}{8w} = \frac{(200)(1.8)(5)}{(8)(3)} \tag{19}$$

A required input to this equation, the quantity R_{sq}, can be determined by comparing the printed-circuit board in this example, Fig. 41, to Fig. 19. The printed-circuit board has a resistance per square of approximately 200°C/watt. Evaluating Eq. (19) reveals that the temperature rise from the edge of the printed-wiring board to the module surface, $t_{1\text{-}p}$, is approximately 75°C.

The temperature rise from the surface of the cordwood module to the hot spot can be bracketed by evaluating Eqs. (16) and (17).

$$\Delta t_{p\text{-}c} = 11.4 \frac{q_e y}{kLw} = 11.4 \frac{(0.30)(1.0)}{(0.12)(3.0)(1.0)} \tag{16}$$

$$\Delta t_{p\text{-}c} = 8°\text{C}$$

The conservative estimate of the surface-to-hot-spot temperature difference is therefore 8°C. The lower point of the bracket is

$$\Delta t_{p\text{-}c} = 5.8 \frac{q_e w}{kLy}\left[1 - \frac{1}{\cosh 1.57\ (y/w)}\right] \tag{17}$$

$$\Delta t_{p\text{-}c} = \frac{(5.8)(0.30)(1.0)}{(0.12)(3.0)(1.0)}\left[1 - \frac{1}{\cosh\,[(1.57)(1.0)/(1.0)]}\right]$$

$$\Delta t_{p\text{-}c} = 2.4°\text{C}$$

Empirical data typically fall roughly halfway between the bracket limits; hence, $\Delta t_{p\text{-}c} = 5°\text{C}$ is a reasonable approximation.

It is important to note here that, if the thermal conductivity of the adhesive used to attach the module to the printed-wiring board is equal to or greater than that of the potting compound, no contact resistance need be considered at the resulting interface.

Only the case-to-junction temperature rise now remains to be determined. The manufacturer's data sheet for this device lists the junction-to-case resistance at 582°C/watt. The temperature rise is therefore

$$\Delta t_{c-j} = (582)\,(0.030)$$

Hence,

$$\Delta t_{c-j} = 1.7°\text{C}$$

Now adding the various temperature increments:

Δt_{3-2}, cold plate to equipment	0.7°C
Δt_{2-1}, equipment to printed-wiring boards	1.7°C
Δt_{1-p}, printed-wiring boards to hottest cordwood module	75.0°C
Δt_{p-c}, module surface to component case	5.0°C
Δt_{c-j}, component case junction	1.7°C
Total temperature rise	84.1°C

Recalling that the cold plate is held at 48°C, the operating temperature of the transistor junction is established at 84 + 48 = 132°C.

FIG. 41. Example 2 printed-circuit-board pattern.

Example 3. A typical analysis requirement for space equipment is to define external α_s and ϵ characteristics such that the equipment housing temperature will not exceed some prescribed allowable value. As an example, consider an equipment which has package dimensions of 1 × 1 × 1 ft and dissipates 350 watts of electrical power. The equipment is oriented in space such that the only incident energy from the solar system is the flux from the sun incident on one surface. Other planetary emitted and albedo energy is negligible. Further assume that the equipment electronics design requires that the equipment bulk surface temperature not exceed 100°F.

The total energy to be dissipated by the equipment is given by

$$q_{\text{in}} = \alpha_s Q_s A + q_e$$

At equilibrium, the energy absorbed must equal the energy emitted to the sink space at 0°R. This emitted energy is given by

$$q_{r\ out} = \sigma \epsilon A_s T^4$$

Equating $q_{r\,in} = q_{r\,out}$, the equipment heat balance becomes

$$\alpha_s Q_s A + q_e = \epsilon \sigma A_s T^4 = 0.171 \epsilon A_s \left(\frac{T}{100}\right)^4$$

The above shows that the equipment equilibrium temperature is dependent on the values of α_s *and* ϵ. Both α_s and ϵ are characteristics of the equipment's external finish.

Reduction for the given example shows

$$\alpha_s[442 \text{ Btu/(hr)(ft}^2)]\ 1 \text{ ft}^2 + 350 \text{ watts } [3.416 \text{ Btu/(hr)(watt)}] = [0.171 \text{ Btu/(hr)(ft}^2)(°\text{R}^4)]\ (6 \text{ ft}^2)(5.60°\text{R})^4$$

$$\frac{\alpha_s}{\epsilon} \text{ (max allow)} = 0.62$$

Tables 4 and 6 show values for ϵ and α_s, respectively. As an example of an acceptable equipment external finish, the white paint given in these tables has a maximum α_s/ϵ value of $0.33/0.91 = 0.36$.

Although particular materials may be selected for their specific α_s and ϵ parameters, most materials are subject to degradation due to the ultraviolet (UV) portion of the solar flux. Much work has been done to find material coatings with stable α_s/ϵ values. Any aerospace coating (paint) supplier will furnish both undergraded and UV degraded α_s and ϵ data for his products.

THERMAL TESTING

The above portions of this chapter have presented the necessary data and references to accomplish the initial analytical portion of any thermal-design effort. Very soon after the beginning of any such program, thermal testing should assume an increasingly important role. The type of testing will typically follow the normal phases in any overall electronic thermal design. First there will be development tests, which are conducted to provide data to be used in the design or in the support of the design of a specific component or subsystem. They will typically start with individual component thermal-resistance tests where published data do not fit the configuration. As an example, a transistor case to local ambient temperature rise might be measured for a specific printed-circuit board mounting.

As actual hardware is fabricated and the design becomes finalized, more extensive testing of subassemblies is done. Almost all testing in the development phase involves extensive instrumentation of temperatures, flow rates, etc. The intent is to ensure that the analytical techniques were correct. Results are compared with predictions, and a final estimate of unit thermal performance is obtained.

Extensive use is made of partial and full-scale mockups during this initial design testing. The term *thermal mockup* implies that some portion of the configuration is not complete. Usually thermal mockup testing will proceed through a number of phases. Initially there will be the individual component mockups to obtain exact data on actual part-mounting schemes. Here actual parts and immediate mounting are used, and the part case to local mounting resistance is the item of interest. The next mockup phase would typically be a subassembly mockup to examine final temperature distribution and thermal resistances within modules and subchassis and on printed-circuit boards. Here, resistors may be used to simulate heat loads, but the actual geometry and materials should be utilized for one configuration of interest. Finally a full-scale equipment model may be built using all real hardware with the exception of the electronic parts. The dissipators may be simulated with resistors. This type of mockup will check over all heat-flow paths, airflow distribution, etc. It is used where the actual working electrical unit may not be available at a sufficiently early date to allow any thermal-design testing.

Much of the foregoing testing will be so-called *bench tests*. This typically involves running tests at room ambient conditions in such a manner that the results can be extrapolated to other environment conditions. One very important point to be realized is that the bench test data can only be extrapolated to other environmental conditions if the actual configuration is similar to that used during the bench test.

After the development phase, further thermal testing usually falls into one of two classifications: qualification testing or acceptance testing. Qualification tests are performed on items that are identical with production equipment, to demonstrate attainment of design objectives including margins of safety. Acceptance tests are conducted on deliverable items, to provide assurance that the item performance is within the limits of the design parameters.

Instrumentation. One of the most important aspects of any thermal test is proper temperature instrumentation. Usually an accuracy of ±2°C is desirable on most temperature measurements. Because thermocouples are sufficiently accurate and are easy to use, they are preferred as a temperature-monitoring device. In some cases, however, only an indication (within 5°C) of the temperature is required. In these cases paper thermometers or temperature-sensitive paints are sometimes satisfactory. When thermocouples are used to monitor temperature, the following guidelines are recommended:

The thermocouples should be of a size equal to or smaller than a 30-gauge wire. A welded-bead junction is preferred. Where the surface to be instrumented is basically flat, a bead and "collector-plate" configuration should be employed. To make the plate thermocouple, the bead of the thermocouple is inserted into a small hole in the collector plate and staked into place. An aluminum disk 0.25 in. in diameter with a thickness of 0.04 in. should be used for the collector plate. The collector plate should be bonded to the surface by means of an easily removable adhesive. Sufficient pressure must be used to produce a very thin film of adhesive.

Where the surface to be instrumented is curved to an extent which will prohibit the use of the collector plate, or where the addition of the collector plate may greatly influence the surface temperature field, the thermocouple bead should be bonded directly to the surface.

The air temperature immediately surrounding the unit should be monitored by the use of a 30-gauge bead thermocouple located within 3 in. of the center of each of the major surfaces of the equipment.

Test Methods. In general, the test method is dictated by the detailed equipment specification, which in turn is usually extracted in part from a military or general environmental specification. For these cases the step-by-step test requirements are delineated in some detail. In this type of test it is usually necessary to monitor most of the following items:

1. Surface temperatures of all critical components
2. Unit external surface temperatures
3. Test enclosure wall temperature
4. Air temperature in the local vicinity of the unit
5. Air temperature inside the unit
6. Air pressure in the local vicinity of the unit
7. Coolant inlet temperature
8. Coolant outlet temperature
9. Coolant flow rate
10. Coolant static pressure at inlet/outlet
11. Unit power input
12. Unit power output

Data from these tests will give a complete thermal profile for the unit at all operating conditions. This type of test is common during the qualification test phase.

In some instances it is desirable to make a complete thermal-performance-evaluation test of a unit. This type of test will define an envelope of thermal environments in which the unit will satisfactorily operate, and also the effectiveness of the thermal design. Specification MIL-T-23103 (WEP)[21] contains details on this type of testing.

REFERENCES

1. Kraus, A. D.: "Cooling Electronic Equipment," Prentice-Hall, Inc., Englewood Cliffs, N.J., 1965.
2. McAdams, W. H.: "Heat Transmission," 3d ed., McGraw-Hill Book Company, New York, 1954

3. Kreith, Frank: "Principles of Heat Transfer," International Textbook Company, Scranton, Pa., 1958.
4. Barzelay, M. E., K. N. Tong, and G. F. Hollaway: Effect of Pressure on Thermal Conductance of Contact Joints, NACA TN3295, May, 1955.
5. Bevans, J. T., et al.: Prediction of Space Vehicle Thermal Characteristics, Air Force Flight Dynamics Laboratory, Research and Technology Division Air Force Systems Command, Wright-Patterson Air Force Base, Ohio, AFFDL *Tech. Rept.* TR-65-139, October, 1965.
6. Kays, W. M., and A. L. London: "Compact Heat Exchangers," National Press, Palo Alto, Calif., 1955.
7. NAVWEPS 16-1-532 "Design Manual for Methods of Cooling Electronic Equipment," U.S. Naval Air Development Center, Johnsville, Pa.
8. MIL-E-5400 (ASG) "General Specification for Aircraft Electronic Equipment."
9. Baum, J. R., and R. J. Jimenez: Integrated Circuit Thermal Study Final Report, U.S. NADC Contract N62269-2393, Sept. 1, 1965.
10. Ruggierd, E. M.: Aluminum Bonding Is Key to 40 Watt Microcircuits, *Electronics,* Aug. 23, 1965.
11. Greenburg, R.: Factors Influencing Selection of Commercial Power Transistor Heat Sinks, *Solid State Design,* July, 1962.
12. MIL-R-11 "General Specification for Fixed Composition Resistors."
13. MIL-HDBK-217 (WEPS) "Reliability Stress Analysis and Failure Rate Data for Electronic Equipment."
14. Guide Manual for Cooling Methods for Electronic Equipment, *Bur. Ships, Navships* 900, 190.
15. Jones, C. D.: Techniques for the Cooling Design of Airborne Electronic Equipment, WADC *Tech. Rept.* 57-331.
16. McAdam, J.: "Heat Dissipating Electron Tube Shields and Their Relation to Tube Life and Equipment Reliability," International Electronic Research Corporation.
17. Berry, C. H.: "Flow and Fan," Industrial Press, New York, 1954.
18. Shoop, C. F., and G. L. Tuve: "Mechanical Engineering Practice," 5th ed., McGraw-Hill Book Company, New York, 1956.
19. Green, W. B. (ed.): "Westinghouse Thermoelectric Handbook," Westinghouse Electric Corporation, Youngwood, Pa., 1962.
20. Welling, J. R., and C. B. Wooldridge: Free Convection Heat Transfer Coefficients from Rectangular Vertical Fins, *Trans. ASME, Heat Transfer,* November, 1965.
21. MIL-T-23103 (WEP) "General Requirement for Thermal Performance Evaluation of Airborne Electronic Equipment."
22. "Cooling Equipment for the Electronic Industries," Rotron Manufacturing Company, Inc.
23. Sparrow, E. M., and J. L. Gregg: Radiant Emission from a Parallel-walled Groove, *Trans. ASME,* August, 1962.

Chapter 12

PACKAGING FOR COMPUTER APPLICATIONS

By

ROBERT M. KALB

Sperry Rand Corporation
Univac Division
Roseville, Minnesota

CONTRIBUTORS

Analog Computing Systems: EDWARD MASSELL, *Electronic Associates, Inc.*
Toroidal Core Memories: LEE R. CARLSON, *Rosemount Engineering Company*
Planar Thin Magnetic Film Memories: LARRY D. SLY, *Fabri-Tek Incorporated*

MAJOR TRENDS AND FACTORS

Governing Factors. For computers the packaging goal is to mount and interconnect the electronic elements in a suitable assembly. Suitability encompasses many factors whose relative importance determines what approaches are available to the designer in any specific application.

The cost, performance, physical characteristics, and practicality demanded for a particular computer design are usually stipulated in advance. Table 1 relates these to

TABLE 1. Design Factors and Related Computer Characteristics

Computer characteristics	Determining factors	Factors combine to establish
Performance	Speed Accuracy Power Reliability Storage capacity	
Physical attributes	Compactness Weight Ruggedness	Cost
Practicality	Manufacturability Maintainability Repairability	

several factors which can be controlled by packaging or which influence it. Selecting the best trade-offs among these factors is essential to good packaging design.

The nature of the electronic packaging choice for a computer design is dictated by type and speed requirements as well as by the intended applications. Digital computers, electronic analog computers, and network simulators each have particular characteristics, calling for packaging methods peculiar to them. In larger computer systems, packaging of various portions or modules differs according to the specific speed requirements and susceptibility to noise. In particular, the packaging of memories, since they are unique assemblies, has evolved its own practices. In other respects, the evolution of packaging practices in computers has followed the state of the electronic art, and of wiring and connection methods as they have come into vogue.

Special influences of military and space requirements for ruggedness and for enduring unusual environments are not considered in this chapter but are left to Chaps. 13 and 14. Consideration here is limited to packaging for general business and industrial applications.

Generations of Computers. Electronic computers are commonly referred to as belonging to generations defined by their types of active elements. Since each such generation exhibits some packaging practices peculiarly adapted to it, the generations as defined tend to segregate computers conveniently for packaging consideration. Four generations have been so defined to the present time.[1]

FIRST GENERATION. Computers constructed with vacuum-tube circuits. In the course of this generation, diodes made their appearance and cold cathode tubes were also used as circuit elements. All passive circuit components were discrete.

SECOND GENERATION. Computers characterized by transistor circuitry. The transition to this generation was marked by a few computers with ferromagnetic circuit elements. Discrete elements are used throughout.

THIRD GENERATION. Computers constructed with integrated circuits. These are the computers being produced today. In their most advanced forms, they employ hybrid and monolithic circuits.

FOURTH GENERATION. Projected design based on large-scale integration.[2, 3] This design concept was described[4] as far back as 1958: assembling an entire computer (or a large part) in a single process, using vacuum deposition of electrodes on blocks of pure silicon or germanium, and subsequently diffusing the electrode material into the block to form junctions. Such a process allows simultaneous fabrication of large numbers of transistors and creation of all their interconnecting wiring in one operation.

Table 2 lists several digital computers of each generation and shows the salient packaging features they typify. The succession of four generations ignores the early relay computers, which were few in number and were electrical rather than electronic. Most of them in the United States and abroad were built by telephone companies, using standard telephone practice in design and packaging.

Evolution of Storage. Concurrently with the several generations of computers, their *main memories* have been developed in a sequence less well defined in time. Successively appearing memories were packaged using relays, delay lines, magnetic drums, cathode-ray tubes, magnetic toroidal cores, and thin magnetic films. From the time that

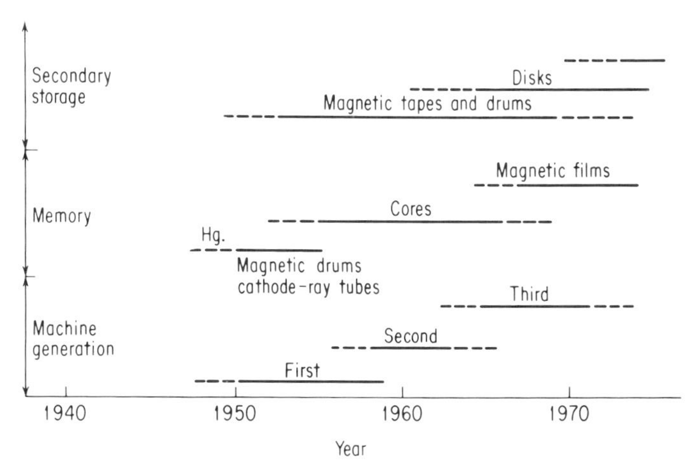

FIG. 1. Chronology of computer generations and types of storage.[1]

toroidal cores were first used in computers, they quickly outmoded other types for main memory and predominate today for that purpose. Special-purpose fixed memories, not alterable by the program, were first constructed with toggle switches; today they may use cores or capacitors. Smaller storage units, such as registers, employ all the schemes used for main memory, and in addition are often constructed from flip-flops or related electronic circuits. The advent of large-scale integration (LSI) promises the possibility of utilizing wafers of flip-flops to construct main memories. Steps in the evolution of memories are not uniquely identified with the generations of computers, as may be seen in Fig. 1 and Table 2.

Trends. The evolution of computers has, from its early days, been marked by a continual reduction in size and power demands, concurrent with increased speed and reliability. Advances in packaging techniques have made this possible despite the increasing component count. As discrete components give way to integrated circuits and large-scale integration, the component count has less meaning, but advances in packaging techniques promise to continue to govern the design possibilities of computers. Figure 2 shows the size decrease of circuit elements accomplished in the course of time; Fig. 3 shows how the packaging of these successive types of elements results in smaller circuit packages with less volume occupied by interconnections. Demand

TABLE 2. Evolutionary Steps in Packaging Digital Computers

Year introduced	Manufacturer	Computer	Logic circuit construction	Distinctive circuit elements	Internal memory
			First generation		
1950	National Bureau of Standards	SEAC	Pluggable modules and prewired panels of octal sockets	750 tubes, 10,500 diodes	Hg delay line (later CRT)
1950	Engineering Research Associates	1101	Assembled chassis	2,700 tubes, 2,400 diodes	Magnetic drum
1950	IBM	CPC	Single-tube pluggable chassis	Vacuum tubes	Mechanical counters
1951	Eckert-Mauchly	Univac I	Assembled chassis	1,900 tubes, 18,000 diodes	Hg delay line
1951	MIT	Whirlwind I	Two-tube pluggable chassis	5,000 tubes, 11,000 diodes	CRT
1953	National Bureau of Standards	SWAC	Multitube pluggable chassis	2,600 tubes, 3,700 diodes	CRT
1953	IBM	701	Eight-tube pluggable chassis	4,000 tubes, 17,000 diodes	CRT
1955	Bendix	G-15	Pluggable circuit boards	Vacuum tubes	Magnetic drum
1956	Univac	1103A	Multitube pluggable chassis	5,300 tubes, 9,400 diodes	Toroidal core
1956	IBM	705	Multitube pluggable chassis	Vacuum tubes	Toroidal core
			Second generation		
1956	Univac	Magnetic	Pluggable PC boards	1,500 magnetic amplifiers	Magnetic drum
1956	Bell Tel. Labs	Leprechaun	7- × 9½-in two-sided cards with central ground plane	976 transistors	Toroidal core
1957	Philco	S-1000	132 PC boards	7,000 transistors	Toroidal core
1957	Univac	Athena	PC cards in hermetically sealed cans	Ge transistors	Toroidal core
1958	Philco	S-2000	PC cards; back panel prewired on four levels	41,600 transistors	Toroidal core
1960	Packard-Bell	PB-250	PC cards; harness-wired back panel	375 transistors, 2,300 diodes	Magnetostrictive delay line
1960	IBM	7070	14,000 PC cards; wire-wrapped back panel	30,000 Ge transistors, 22,000 diodes	Toroidal core
1960	Univac	LARC	PC cards; taper pins wired point-to-point with steel-cored Cu wire	62,000 transistors, 174,000 diodes	Toroidal core
1961	IBM	7030	One- and two-sided PC cards; wire-wrapped back panel	Unsaturated transistors	Toroidal core
1962	Univac	1107	PC cards, inserted components	Discrete transistors	Toroidal core and planar thin film
1964	CDC	6600	Cordwood between 3 in. sq soldered PC cards	Discrete Si transistors	Toroidal core
1965	NCR	315-RMC	PC cards	Discrete transistors	Rod
			Third generation		
1965	IBM	360/50	Two-sided PC cards; ML backboard	Hybrid SLT modules	Toroidal core
1966	RCA	70/45	Two-sided and three-layer PC cards; ML backboard	Monolithic flat packs	Toroidal core
1967	Burroughs	B-8500	ML cards	Hybrid, dual-in-line modules	Planar thin film
1967	IBM	360/90	Two-sided PC cards; ML backboard	Hybrid ASLT modules	Toroidal core
1967	Univac	9300	Two-sided PC cards; wire-wrapped back panel	Monolithic dual-in-line modules	Plated wire
1968	Honeywell	4200	ML cards; wire-wrapped back panel	Monolithic dual-in-line modules	Toroidal core
1968	Honeywell	8200	ML cards; wire-wrapped back panel	Monolithic dual-in-line modules	Toroidal core
			Fourth generation		
1968*	Philco	AF guidance	CML	MOS wafer	MOS
1968*	Texas Instruments	AF special purpose	Bipolar TTL, discretionary wiring	LSI wafer	Integrated circuit
1968*	RCA	LIMAC	Bipolar ECL, 100% yield	LSI wafer	MOS

* Scheduled completion date for experimental model.

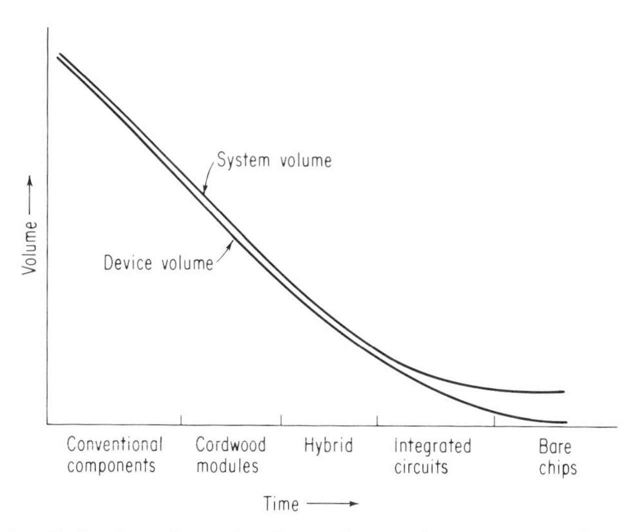

FIG. 2. Reduction of size by the evolution of component techniques.[101]

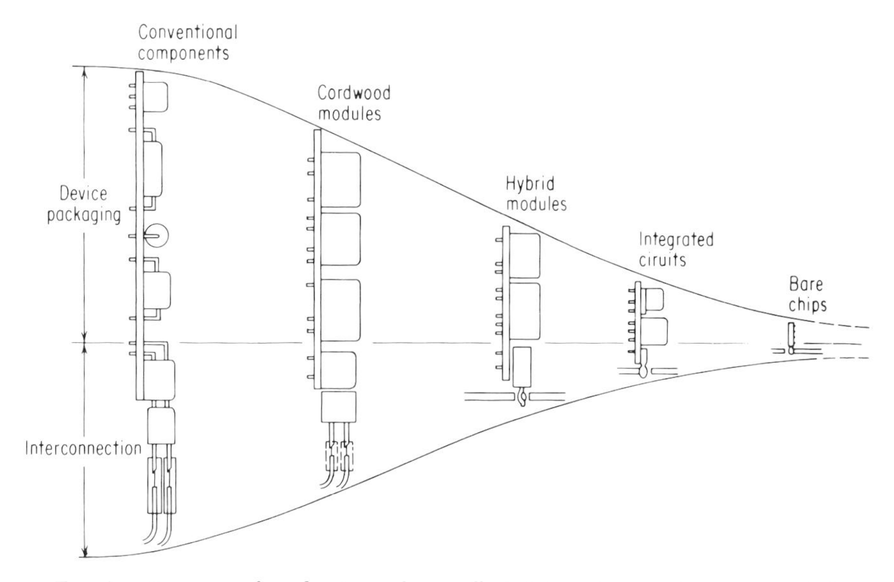

FIG. 3. Decrease of packaging volume effected by advances in components.[101]

for more speed in computers is met by these packaging advances and continues to make desirable any further compactness achieved through packaging.

As new packaging practices appear, the older ones do not disappear, but find an economic niche in slower parts of data-processing systems. The logic circuits and the fast memory sections of main memory or scratch-pad memory justify the costly practices of the most advanced packaging technology if reduced size results. The slower and cheaper approaches to design are used for peripheral devices. Past practices, obsoleted as solid-state modules replaced vacuum tubes, may again come into vogue if LSI again makes feasible circuits with large numbers of active elements. It is the logic and switching circuits of digital computers that change most rapidly with technology and utilize the latest packaging practices in large-scale machines. The higher levels of packaging change more slowly.

The gross structure of present-day large digital central processors is the same as that of the early vacuum-tube machines. In those, for instance, the chassis were mounted in parallel vertical planes to channel cooling air from the bottom of the cabinet, over the tubes and conventional discrete elements, toward the top where it carried away the heat they generated. Cards with integrated circuits today are packaged much more compactly, but they are mounted in guides in much the same fashion, and the cooling principle is the same.

The first level of packaging, the logic module, exhibits the greatest change in packaging technology. It is at this level that the most revolutionary changes seem to be imminent for the near future. The complexity of a first-level module is established by its component count and by the number of interconnections between components and to the external contact pins. Integrated-circuit technology is bringing many circuit elements and their interconnections into a single, fixed, small assembly until it is now possible to put more components on a single chip than practical design can use.[5] The larger the number of elements, the more specialized each chip becomes and thus the greater the variety required. There is a growing tendency to hold down the number of different first-level packages required, by utilizing multilayer boards for interconnection between "standard" packages, as well as with the next higher level. Another approach is to fabricate a complex assemblage of circuit elements, thoroughly interconnected on a single wafer, and then tailor it to specific applications by removing superfluous interconnections according to each application. Prevailing practice in current computer technology is to form first-level packages by assembling and interconnecting chips.

Mutual Impact of Computer Design and Packaging.

Evolution of Technology. For computers, as for most electronic equipment, the circuit design and the mechanics of packaging are closely interrelated. This is especially true of digital systems, in which the large component count can lead to excessive size unless close attention is given to packaging considerations. The quantity production of identical circuit boards quickly led to techniques for automatic insertion of components in early-generation computers. The transistor was invented a few years after the first digital electronic computer. Second-generation use of transistors reduced computer size and made practical parallel processing of data, which displaced serial processing, previously the common practice. Parallel processing multiplied the number of identical circuits used; as circuit speeds were increased, shorter leads and connections were required. The resulting widespread use of printed circuits and plated-through holes further reduced size. As smaller active elements and shortened interconnections compacted circuitry, the volume occupied by connectors between circuit mountings became important. This resulted in new smaller connections: card edge connectors, wire-wrap panels, multilayer boards, and similar connecting means with high volumetric efficiency. With the advent of monolithic elements, the cycle starts all over again: smaller elements, higher circuit speeds, more compact packaging, and further concern about the space taken by intermodular connections.

For the faster digital systems, the high-speed electronics required demands compactness; this in turn leads to concentration of energy in small volumes, resulting in high thermal density. As with most electrical equipment, the temperature rise places an ultimate limit on the design. In computers, this limit defines the maximum allowable energy density. Other limitations also result from high speed. The least circuit delay obtainable in any medium is fixed by the finite time it takes electricity to travel in that medium, and this is limited to the velocity of light (about 1 ft/nanosec). Recently it has been conjectured that the rising speed of computer circuitry may encounter a fundamental limitation dictated by quantum theoretical considerations.[6]

*Other Factors—Selection Dilemma.** The primary trends of electronic computer packaging are affected by many technical and economic factors, some unique to the computer field and some closely allied to those of other fields of application in electronics. Initially, when the computer field depended mainly upon other fields for components and application techniques, computer packaging was primarily an adaptation of that of the general electronic art. During subsequent growth of the computer field, its unique requirements have caused components and packaging technology to be specifically developed. Classic examples are the interconnection of large networks of logic circuits and the utilization of large-capacity storage devices.

The rate of change in computer design and packaging has kept pace with the accelerated development in the computer field specifically and in allied electronic fields in general. On the other hand, the overall time period for developing and marketing a computer product remains relatively long. A typical product program for a digital computer encompasses three years for development and a market life of five years or more. The complexity of development, manufacture, and application of computers, particularly in large-scale systems, tends to discourage early model change on the part of both supplier and user. The packaging engineer, as a result, is faced with the dilemma of committing for a long period of time designs chosen from a rapidly changing technology.

Selection of circuit and packaging techniques may be required some two years before the equipment first reaches the market. Automated processes for design, fabrication, and testing of computer hardware decrease the interval between technique selection and market introduction. But even if this portion of the development time is reduced substantially, most of the utilization still falls several years after packaging selection. To select from among the latest electronic packaging techniques in the laboratory those that can best survive this utilization period is a vital design problem.

The dilemma of rapid innovation and slow supersession of technical designs is perhaps unique to the computer field because of the unusual nature of the distribution and application of computer products. Factors such as marketing practices and software (programming) requirements affect selection of a design approach, both directly and indirectly. With major portions of the computer product on rental, the supplier has extraordinary responsibilities to provide maintenance and other supporting services, and has large incentives to recover investment through extended market life. The programming requirements unique to computer marketing make software development costs as high as or higher than those of hardware development. Software costs incurred by the supplier increase the problems of investment recovery; generally software cost has been prorated to hardware cost. Coverage of this economic area is beyond the scope of this book; reference is made only to indicate that unit cost considerations in the computer field include considerably more than basic hardware. The technical requirements of software operating systems are playing an increasingly important role in the specification of computer systems, with heavy impact upon their organization and modularity, and ultimately upon the packaging selection.

DEFINITIONS

Leading Concepts.

Computer Generations. As discussed in the introduction to this chapter, the four generations of computers and associated technology are defined as follows:

Generation	*Technology*
First	Vacuum-tube
Second	Discrete solid-state
Third	Integrated-circuit
Fourth	LSI

* Source: W. P. Burrell, Univac Division, Sperry Rand Corporation, Roseville, Minn.

Types of Computing Equipment. Electrical computers may be categorized in several ways, according to their construction or the nature of their intended application. For the purpose of this handbook, distinguishing them along the lines of structural characteristics is desirable. The two common types of computers are digital and analog; these differ markedly in principle and in the general nature of their circuits. A combination of both types, interconnected by analog-digital converters, forms a third type, the hybrid computer.

A special-purpose computer may be a computer of either type, designed for a specific purpose or a limited range of applications often in unusual environments. Most process control computers, many military computers, and all computers for airborne and spaceborne usage are of this class.

General-purpose computers are designed to be adaptable to a large and encompassing class of application problems, and consequently possess great flexibility. They operate in ordinary, habitable environments. In the following, general-purpose computers are discussed throughout because special-purpose ones merely omit some functional units and involve no new packaging principles other than environmental ones.

Analog Computers. Electrical analog computers are constructed to represent the

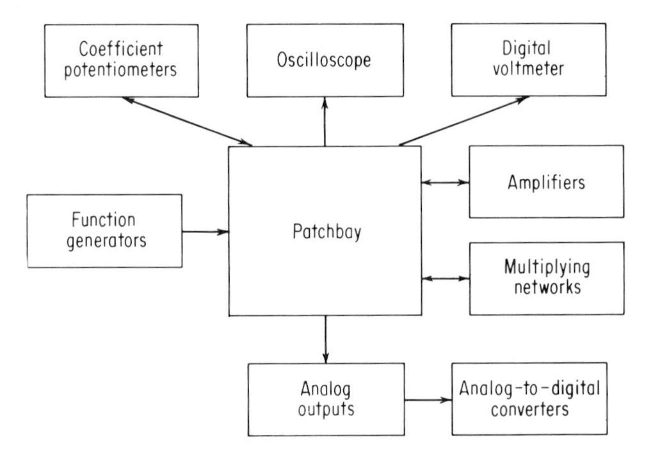

FIG. 4. Block diagram of an analog computer.

continuous mathematical variables and quantities by voltage or current values proportional to the physical quantities in the computation. The correspondence of the computer connections to the structure of the simulated system, and of the computer voltage values to the system parameters and variables, will be direct in the simplest analog computers.

Figure 4 shows the principal portions of an analog computer and their interrelationships. The various functional and operational portions of the computer cluster around a patch bay, to which they are connected, and by means of which connections can be arranged to create analogs of systems to be analyzed numerically.

Digital Computers. Digital computers comprise essentially a *central processing unit,* a hierarchy of *memories* which store information for reference and for use in the course of computation, and input and output *peripheral equipment* for entering data or information into the processor and memory and for receiving processed results. Input and output data are always in discrete digital coding. Figure 5 shows the major portions of a typical, large, general-purpose digital data-processing system and the relationships among them.

Memories. Memories of digital computers are also classified in various ways. They are called *volatile* if loss of electric power causes loss of information stored, and *nonvolatile* if the stored information is retained even in the absence of external power.

Another method of classifying memories is based on the manner in which they function:

NONDESTRUCTIVE READ-OUT (NDRO) MEMORY. Information stored can be sensed and read out without changing the data pattern of stored information.

DESTRUCTIVE READ-OUT (DRO) MEMORY. Information is destroyed upon being sensed; it must be captured during reading and restored afterward if it is to be retained in the memory for future reference.

READ-ONLY MEMORY (ROM). The contents cannot be altered by the computer, in contrast to NDRO and DRO memories.

Memories are most frequently typed according to the geometry of their construction (e.g., planar, matrix, rotary), or more specifically according to their physical nature. On the latter basis, the major types of such memories are the following:

CORE MATRIX: Array of ferromagnetic toroids.

PLANAR THIN FILM: Cells of thin magnetic film on a glass substrate.

WIRE: Conducting wire with thin magnetic plating.

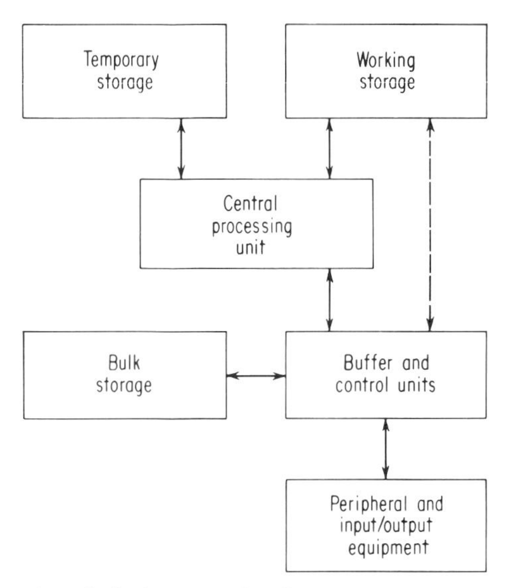

FIG. 5. Block diagram of a digital computer system.

ROD: A form of wire memory.

ROPE: Conductors threaded through or around toroids.

FLIP-FLOP: Standard two-state circuits, now solid-state only.

DRUM: Rotating cylinder with magnetizable coating.

DISK: Rotating plates with annular magnetizable coating on one or (usually) both sides.

MAGNETIC TAPE: Plastic ribbon with magnetic coating, supported on reels.

MAGNETIC CARD: Rectangles of magnetizable material; mechanically handled, individually, for reading and writing.

DELAY LINE: Originally mercury, acoustically excited into wave patterns in a closed tank. Now obsolete. Solid materials carrying acoustic waves now in use.

CRT: Cathode-ray tube with beam deflection controlled to spots on phosphor. (Also called *Williams tube*. No longer made.)

PHOTOCHROMIC: Color-alterable materials, controlled by light. In development.

CRYOELECTRIC: Circuits which sustain lossless electrical activity near absolute zero (experimental status).

Packaging Levels. There are three definite levels of packaging in electronic computers.[7] The first level interconnects the circuit elements, active and passive. These elements include discrete resistors and capacitors in the first two generations, vacuum tubes in the first generation, discrete solid-state elements (diodes, transistors) and a few magnetic logic elements in the second generation, and monolithic elements and integrated circuits in the third generation. The second level is the chassis or cards on which the first level components are mounted. The third level comprises the back plane and wiring which mounts and connects to packages of the second level; it may variously be a printed-circuit panel, a multilayer board, or wire-wrapped connectors. Higher levels comprise major assemblies and units of the computer, generally connected by cabling. Each level of packaging supports a number of units of the next lower level mechanically, and electrically through suitable connections which define the interface between the two levels.

Special Terms and Symbols.

Acronyms. Table 3 lists acronyms used in this chapter along with their translations.

TABLE 3. Acronyms and Their Meanings

A/D	Analog-to-digital
ASLT	Advanced solid logic technology
CML	Current mode logic
CRT	Cathode ray tube
CTL	Complementary transistor logic
D/A	Digital-to-analog
DCTL	Direct-coupled transistor logic
DRO	Destructive readout
DTL	Diode-transistor logic
ECL	Emitter-coupled logic
ECTL	Emitter-coupled transistor logic (same as ECL)
FET	Field effect transistor
LSI	Large-scale integration
ML	Multilayer
MOS	Metal oxide semiconductor
NDRO	Nondestructive readout
RCTL	Resistor-capacitor-transistor logic
RO	Read only
ROM	Read-only memory
R/W	Read-write
RTL	Resistor-transistor logic
SLT	Solid logic technology
TTL	Transistor-transistor logic
VTL	Variable threshold logic

Definitions of Terms. Definitions have been selected from current literature and from glossaries published by the Association for Computing Machinery, the Institute of Electrical and Electronics Engineers, the American Federation of Information Processing Societies, U.S.A. Standards Institute, Simulation Councils Incorporated, and the Bureau of the Budget. Selection, in some cases, was narrowed to those usages specifically applicable to packaging. These definitions are set forth in Table 4.

TABLE 4. Glossary

Access time. The time interval necessary to insert or extract data at a memory address.

Algorithm. A set of rules defining a fixed step-by-step procedure for solving a problem.

Analog computer. A computer that processes data which is continuously variable in nature.

Anisotropic. In magnetics, capable of being magnetized more readily in one direction than in a transverse direction.

Biax element (trade name). A magnetic memory cell with two orthogonal apertures.

Bit. Binary digit.

Cordwood. Parallel-stacked discrete components with axial leads extended through holes in circuit boards on which they are interconnected by soldering or welding.

Cryoelectric memory. A storage device based on persistent current flow in superconductors near absolute zero.

Digital computer. A computer that processes data which is variable in discrete, countable steps.

Discretionary wiring. Interconnection of integrated circuits on a wafer to avoid bad ones, thus creating an individual connective pattern for each wafer.

Domain. A region defined by its magnetic homogeneity within a magnetic material.

Fan-in. The maximum number of parallel inputs that a logic element can tolerate.

Fan-out. The maximum number of parallel loads that can be driven by a logic circuit.

Fixed wiring. A set pattern of interconnections among integrated circuits on a wafer.

Hybrid circuit. A circuit fabricated by depositing passive elements and attaching discrete components or integrated circuits.

Hybrid computer. A computer that includes both analog and digital stages, with means to convert the form of data passing between them.

Integrated circuit. A circuit fabricated by developing its elements and their interconnections inseparably from or on a common substrate.

Integrator. An operational amplifier with capacitive parallel feedback which makes its output the time integral of its input voltage.

Large-scale integration. The formation and subsequent interconnection of many integrated circuits on a single common substrate (wafer).

Logic. Circuitry for manipulating digital data.

Matrix memory. A storage unit composed of memory cells in a regular geometric array, with selection, driving, and sensing circuitry.

Memory cycle time. The shortest time permissible between successive references to a memory.

Memory dimensions (2D, 2½D, 3D). Nature of organization of a matrix memory, with respect to orthogonal electrical axes. The 2D arrangement has one word (address) dimension and one bit (data) dimension. The 2½D and 3D arrangements have two word dimensions (X and Y) and one bit dimension; the bit dimension is shared electrically with one of the word dimensions in the 2½D arrangement, but is distinct in the 3D arrangement.

Memory plane. An array of memory cells in a rectangular matrix, so wired that any cell can be selectively driven.

Memory stack. An assemblage of memory planes stacked and wired together to form a unit of memory.

Multilayer board. A connecting panel consisting of alternate layers of metallized wiring and insulation laminated or otherwise formed into a single structure. Connections to interior wiring are made from the surfaces via clearance holes or plated-through holes.

Noise immunity. Ability of a logic circuit to register a signal correctly in the presence of extraneous noise.

Nonvolatile storage. A memory device which retains its stored data in the absence of externally supplied power.

Operational amplifier. A stable dc feedback amplifier adapted to perform one of the elementary operations in an analog computer by choice of its input and feedback networks.

Patch bay. The interconnection portion of an analog computer, including the patch panel, switches, and connectable elements and networks.

Patch panel. A plug-in wiring board which can be pre-wired to establish the connections in an analog computer.

Substrate. The base material on which integrated circuits are formed, or into which their elements are diffused.

Transfluxor (trade name). A magnetic memory cell with two unequal apertures.

Twistor. A memory element comprising a conductor spirally wrapped with magnetic tape.

Volatile storage. A memory device which requires power continuously to sustain it.

Wafer. A substrate common to many integrated circuits.

Word. A set of bits which is stored or processed as a unit by a digital computer.

ANALOG COMPUTING SYSTEMS

Comparison With Digital Systems. The crucial difference between analog and digital computers lies in the form in which variables are represented when the mathematical operations (arithmetic, transcendental, analytic) are performed. In an analog computer the variables being operated upon are continuous and may take on all values between their maximum and minimum. In a digital computer they are discrete and may take on only a finite (though frequently very large) number of values designatable by numbers between their extreme values, usually at equal steps determined by the size of the increments used to represent the variable.

Analog variables are represented by values of voltage or current within the functioning circuits of the computer. They are thus very susceptible to changes in the values of the circuit parameters, and accurate determination of the value of the electrical variable governs the functional operation of the computer. Digital variables, in contrast, are represented by patterns of pulses, and traditionally only two values of voltage or current need be distinguished. Electrically, the determination of an elementary digital variable depends upon whether a voltage lies above or below a threshold value. Of necessity there is a region of uncertainty around the threshold value wherein electrical noise or disturbances may reverse an otherwise defined value. The circuits of the analog computer must be capable of preserving current or voltage values with great precision.

The analog computer usually represents and operates on a number of variables simultaneously; it integrates and differentiates directly with respect to time; it usually can generate solutions at a faster rate than the digital computer; it has limited accuracy. The digital computer usually performs only one computation at a time; it can change much more easily from one type of computation to another; it can be set for a new problem much more quickly; it can solve a much wider class of problems. Table 5 summarizes characteristic differences between analog and digital computers.

TABLE 5. Summary Comparison of Analog and Digital Computers

	Analog	Digital
Fundamental character	Continuous variables Voltage values indicative	Discrete, quantized variables Voltage presence indicative
Corollary attributes	Uncertainty of repeated solutions	Computations capable of exact repetition
	Limited precision	Unlimited precision of logical operations
	Integration direct	Integration by numerical approximations
	Error correction by triplication only	Threshold exact reconstitution, error correction
Secondary features	Parallel	Serial
	Wired programs	Electronically stored program; alterable
	Direct physical input/output	Transducers usually required
	Easier comprehension of physical simulation, and of modification of same	Faster change to new programs
	Faster data handling	Multiple function capability
Price range	\$200–2,000,000	\$2,000–20,000,000

Basic Limitations. In the near future, limitations of analog computing elements will be static accuracies of perhaps 1 part in 2,000. If the requirements of a computation include operating on variables to this accuracy and generating them continuously, then providing a solution every time the output varies 0.05 percent involves the following considerations:

At 10 kHz the variable changes at a rate of $2\pi \times 10{,}000 \times 1{,}000 = 62.8 \times 10^6$ parts/sec. A digital computer would therefore need to produce solutions at that rate to equal the performance of such an analog computer element. The large number of parallel elements usually available in an analog computer makes this magnitude still more impressive.

The foregoing discussion touches on both the strength and the weakness of the analog computer. If high frequencies are to be handled at high accuracies, only extremely small phase errors are tolerable, and unintended cross coupling—either via common impedance or via capacitive or inductive pickup—must be kept at extremely low levels. Frequently an operational amplifier is used with a 1-megohm feedback resistor. A 0.1 percent error because of crosstalk from an unwanted signal requires an impedance to the summing point of 10^9 ohms, which at 10 kHz is a mere 0.016 picofarad. General-purpose analog computers usually require relay contacts and patch cords for changing summing-point connections, making quite extensive the portions of the circuit which must be protected, but are difficult to protect, from crosstalk. Because of this, and also because electronic switching and resetting capabilities generate frequency components into the megahertz range, even for computers which have been in existence for several years, it can be seen that considerable cost is involved in approaching the limits of the state of the analog computer art.

Packaging Design Considerations. Leakage, stray capacitance, and noise degrade the accuracy of analog circuits, and therefore require the close attention of the packaging engineer to minimize their effects. These factors are least under control at the higher packaging levels where the precise placement of circuit interconnections is usually not attempted. Control of these factors at the lower packaging levels is by means of circuit element placement, generally following good circuit design practices, avoiding sneak paths and parasitic capacitances which may couple circuits in unplanned ways. Especially, high-level circuits should not be situated in proximity to low-level ones with high input impedance, or ahead of high-gain circuits. Since the impedance values in analog circuits may be varied over a wide range, the worst case for such unwanted coupling must be searched out and considered as part of the packaging design. Wide-band circuits are necessary to accommodate the rapid changes in fast analog computers, and operational amplifiers with gains over 60 db are not uncommon. Susceptibility to noise varies with the gain-bandwidth product, which may become the factor limiting speed. Noise may be introduced through coupling with other circuits, or by generation of electrical voltages at imperfect contacts, or through thermal effects or other disturbances within circuit elements. Deleterious effects resulting from improper grounding practices are discussed later in this chapter.

Packaged Analog Components. Analog computers are used in a wide variety of applications, including process control, guided-missile control, ground-support equipment, and individual test instruments. Operational amplifiers, multipliers, function generators, comparators, and electronic switches are manufactured in great variety by over a dozen manufacturers. Their methods of specification are not at all uniform. Table 6 is presented as an indication of the current state of the art with respect to operational amplifiers, with a warning to investigate specifications before purchase. There is a wide variation between maximum or minimum specifications and the typical ones. Most of those charted are typical. The chart is reproduced from one published for integrated circuits.[8] Note that the integrated circuits usually have differential input, sometimes have differential output, and usually require external frequency compensation resistors and capacitors for the particular application. Temperature ranges for operation and for storage are usually specified.

Analog Computers with Associated Digital Parallel Logic.

Interconnections to Digital Elements. In this category the designer may choose how the connections are made between the analog and digital domains. If the general-purpose analog computer has voltage levels compatible with those of the digital logic circuits, the logic lines may appear on the same patch panel as the analog signals, and may be patched via trunks to other logic panels. (For logic intended to run at much above 100 kHz, consideration must be given to proper termination, driving capa-

TABLE 6. Typical Characteristics of Operational Amplifiers [8]

Characteristics	Units	General-purpose	Transistor input: Wideband, fast response	Transistor input: High input impedance	FET input, high input impedance	Chopper-stabilized, one input	Parametric input	Integrated circuit
Gain, min at 25°	db	60–115	50–120	60–100	60–120	120–180	86–120	60–100
Rated output								
Volts	± volt	10–100	10–11	10–11	10–100	10–150	10–11	3.5–20
Current	ma	2–25	2.2–100	2.2–4	2.2–20	2.2–100	2.2–20	2.2–20
Frequency response								
Unity gain	MHz	0.5–20	10–300	0.5–300	0.5–300	0.2–100	0.075	0.5–30
Full output	kHz	2–150	80–5,000	3–5,000	5–100	10–1,500	1.1–500	2–500
Slew rate	v/μ s	0.1–5	30–300	1.5–2.5	0.3–20	1–200	0.06–0.3	0.2–100
Initial offset								
Voltage	± mv	0.3–10	0.3–10	0.3–15	0.5–20	0.03		0.5–20
Current	± na	2–150	0.05–300	1.5–50	0.001–20	0.01–0.1	0.001–0.01	0.15–5300
Input volt drift								
Max. vs temp (−25 to +85°C)	mv	0.5–20	0.02–11	0.1–12	0.1–10	0.02–1.1	2–6	0.1–6
Max. vs time	μv/day	10–50	1–100	20–100	20–50	1–15		10–50
Input current drift								
Max. vs temp (−25 to +85°C)	na	20–3000	0.05–3300	33–110	0.2–10	0.05–4.4	0.1	1–700
Max. vs time	na/day	1.5–100	30	1	0.003	0.001–0.1		0.003–5
Input impedance								
Differential	megohm	0.22–4	0.003–6	1–10^3	5–10^5	0.22–1.3	10^4–10^6	0.014–10^5
Common mode	megohm	1–100	0.1–500	500–10^3	5–10^6		10^4–10^6	0.5–2
Common-mode rejection ratio	v/v	300–50,000	300–50,000	300–50,000	300–3,000		10^6–10^8	1,000–150,000
Input noise voltage dc to 10 kc	μv, rms	1–2	1–20	1–20	1–20	1–30	1–20	1–20
Common-mode input voltage	± volt	5–11	3–10	3–11	7.5–10		200–300	2–10
Price (for 1 to 4)	$	15–175	47–235	35–98	85–198	90–295	198–242	10–180

bilities, and means of connection, to prevent deterioration of signal, reduction in noise immunity, confusion regarding fan-out capability, and logical errors caused by delay.) Alternately, the conversion from analog signal to patchable logic levels may be accomplished between two separate patch panels. This has the advantage of reducing the net amount of patching required, as well as the total length of critical leads. No matter which of the above choices is made, considerable care is required to avoid additional noise and error effects between the analog and digital domains.

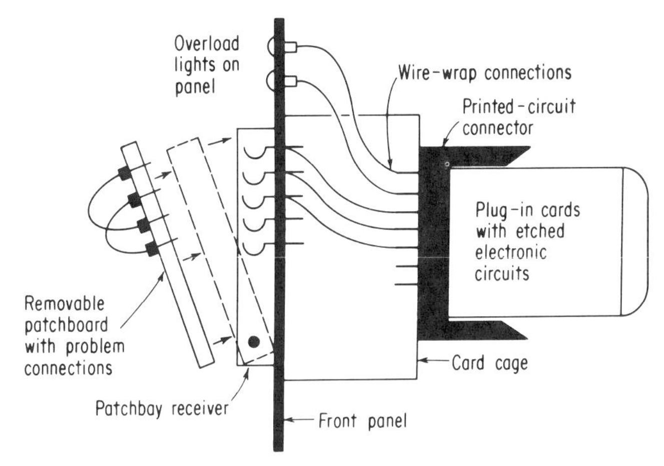

FIG. 6. Conventional mounting of circuit cards on patch panel.[9]

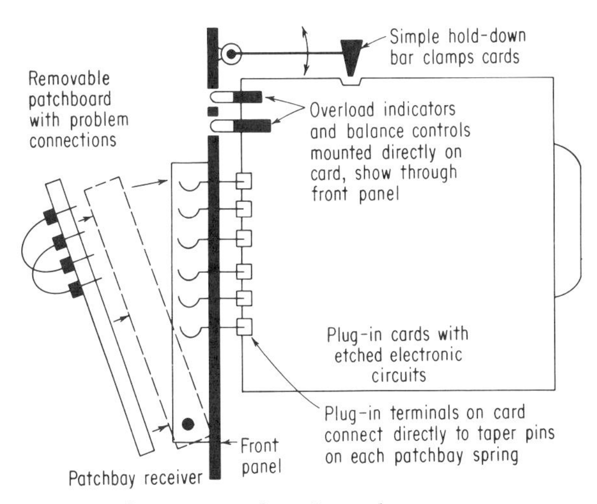

FIG. 7. Improved mounting of cards to eliminate wiring connections.[9]

The design of a small hybrid computer, the APE II, exemplifies the foregoing principles.[9] In its design, a conventional approach of mounting printed-circuit cards on a connector block, then individually wiring the connectors to the patch bay as shown in Fig. 6, was abandoned in favor of the direct method of connection shown in Fig. 7. This design eliminates the interconnecting wires and makes the connections between patch bay and cards as short as possible. In the design, all connections from the circuit cards, not just those to the patch bay, were made without intervening wiring. Figure 7 shows alarm lights and controls mounted through the panel and directly on the cards. Not shown is the power wiring, which was brought in through patch-bay connections.

The addition of parallel logic to analog computer systems imposes some additional requirements. Several of these, concerning grounding and shielding, are discussed in detail under that heading later in this part. Generally good practice is to shield the analog and digital elements separately.[10] The high-frequency noise energy of digital circuits must be kept away from those parts of analog circuits which are or may at times be at low level. This is best accomplished by minimizing the digital noise at its source.[10] Korn recommends decoupling filters on every digital circuit card and on every power-supply line of an analog element. He also recommends a decoupling filter on each line from a digital circuit to an analog relay, at the point it enters the analog computer. As a decoupling filter, a rf choke of around 100 μhenries in series with a shunt capacitor to ground is suggested.

Hybrid Computers. As analog computer users needed to utilize digital techniques to extend their overall computing capability or to obtain faster, more efficient setup, checkout, and documentation of analog computer solutions, the hybrid computer was developed. It must have an electrical linkage of some sort between the analog computer and the digital components or digital computer. It may consist, as a minimum, of an analog computer, a link in either direction, and digital components. A large hybrid system may have many linkages in both directions and may contain several large-scale analog and digital computers.

Links essentially include analog-to-digital (A/D) or digital-to-analog (D/A) converters. The resolution of the converters may be single-bit, or it may be multibit up to the full precision of the analog computer. A single-bit A/D converter may consist of a sensitive relay or a zero-crossing detector, while a single-bit D/A converter is typically an electronic switch or relay contacts. In the multibit case, analog multiplexers, sample-and-hold circuits, digital distributors, and complex control circuitry are usually involved.

The special packaging problems of hybrid computers, over and above those of all-analog computers, are closely associated with the inability to separate the two types of signals, since the converters themselves cannot provide complete isolation between input and output signals. For this reason some converters have utilized transformer coupling to permit dc isolation. The pressure to advance conversion rates into hundreds of kilohertz, with corresponding bit rates well into the megahertz region and rise times of only nanoseconds, makes the problems of eliminating noise spikes from the analog computer just about impossible to solve. The amplitude and duration of the noise spikes must be kept small enough that they do not affect an A/D conversion by more than the rated accuracy of the analog computer and converter, or cause greater than this significant error in analog components, especially those which may react to the noise pulses in a nonlinear relationship. The size of hybrid computers requires considerable physical extension of the interconnective leads, and since the operator must be able to change connections for different problems, the usual variable structure of the analog computer does not permit dual-signal leads. As a result, considerable common impedance is unavoidable in transmitting signals between the analog and digital computers.

Converters and Coupled Digital Systems. The problems associated with converters and digital computers differ in degree from the foregoing. Generally these equipments are large enough to require separate housing. The complications are similar to those of tying to other external equipment, such as cockpits or flight tables. In general, large systems are installed with a design attempt to limit the common impedances in signal paths, grounds, and power supplies; then empirical methods are used to reduce noise and crosstalk to tolerable values.

The General-purpose Analog Computer

Packaging Nature of the Computer. The general-purpose analog computer, for packaging considerations, may be described in terms of a housing with interconnections, a power-supply system, a control and indicator section, a patch bay, plug-in components, and peripheral equipment. The latter, including such equipments as displays, plotters, recorders, paper-tape readers and punches, typewriters, line printers, and analog-digital conversion equipment, may be packaged integrally with the computer. The functional units internal to the computer comprise resistance-capacitance networks or diode networks for transfer functions; potentiometers for coefficients and scale factors; operational amplifiers; relays or electronic switches; and other more specialized devices—comparator circuits, function generators, integrators and limiters. One essential device of this class is the patch board by which the analog configuration is set up for each problem. Patch boards are often made removable so that they can be prewired and programs interchanged almost instantly. Resolvers and servo-driven units such as gang-driven potentiometers were formerly used in analog computers, but their functions are provided by electronic circuits in today's models. The designer of a large general-purpose analog computer is usually engaged in a desperate struggle for equilibrium among the competing demands of operator access to controls and displays, maintainability, performance, and cost.

Korn and Korn [11] discuss many design considerations in detail. Some of the more recent major trends affecting packaging include the following:

Vacuum tubes are practically eliminated; the larger computers usually integrate a number of solid-state logic elements into the analog system.

More of the computing elements are placed directly behind the patch bay.

The read-out systems, at least as far as signal rates are concerned, have been decentralized and speeded up.

The size of the computers has increased, both in number of computing elements and in variety of functions performed (e.g., they are expected to communicate conveniently with digital computers, with control passing from one to the other according to the state of the system).

Standard air-conditioning equipment is used.

Rectifiers and preregulators are in separate locations.

Housing and Interconnections. Housing of analog computers is likely to compare more closely to that of large-scale test consoles than to that of digital computers. A large number of controls and output displays, and the patch panel, must be accessible to the operator in applications where simulation is used in conjunction with design synthesis or model building.

The much smaller size of transistors allows concentration of many more computing elements in the same volume. Thus, although transistors are much more efficient than vacuum tubes, the heat dissipation per unit volume is greater with high-voltage transistorized analog computers than with vacuum-tube computers. In addition, of course, the solid-state elements change characteristics or fail within more restrictive temperature ranges than do vacuum tubes. The 100-volt analog computers therefore require extreme care in provision for heat removal. Design for heat removal is covered elsewhere in this book, but it is worthwhile pointing out here that, unless compensated for, many of the specifications, such as integrator drift, integrator rate, multiplier accuracy, or converter offset, may be exceeded because of temperature variations of less than 0.5°F. Use of integral air conditioners is therefore advocated by some manufacturers.

Placement of Parts. The approach of one manufacturer to compact location of controls and displays is shown in Fig. 8. Table 7 shows the space allocations defined by the designer. The problem involved placing 20 ft^2 of primary and secondary operator controls and indicators above desk level on the main console, at the same time providing the operator with reasonable access to another 18 ft^2 of peripheral controls and displays, and also providing about 12 ft^2 of desk space. Maintenance controls were intentionally rendered less accessible, to avoid confusion as well as to conserve prime panel space.

In this instance the patch bays were located in the center of the main console, with the small (and lighter) logic panel located above the analog panel. The sloping panel at the left contains the main mode and analog read-out control and displays, and provision for transferring control to remote computers. The right-hand sloping panel contains the less frequently adjusted (and conceptually different) logic, function switch, interval timer, and manual function-generator setup controls. Above these two panels are the potentiometer manual controls. These may all be servo-set from the main control panel, but a number of these coefficient adjustors may be labeled with their function and more conveniently modified in parallel.

FIG. 8. Analog computer, showing disposition of controls and principal parts.

TABLE 7. Space Allocations in an Analog Computer

Control or display function	Area	
	In.^{2}	Percentage of total
Mode control	76	2.6
Readout control and display	102	3.5
Logic input-output	38	1.3
Function switches	56	1.9
Interval timer	46	1.6
Potentiometers	1,000	33.9
Analog and logic patch panels	1,260	42.7
Setup panel for manual diode function generators	38	1.3
Overload displays for analog components	168	5.7
Logic state indicators	168	5.7
Total	2,952 = 20.5 ft^3	100.0

Air is blown from an air conditioner through a channel enclosed by the base of the console, guided up through the heat-producing elements, and returned, in a system which is closed except for condenser-cooling water, through a duct above the console.

Some of the variations found in the products of several manufacturers include omission or modification of the integral cooling system, omission of manual access to servo-set potentiometer shafts, centralization of control and displays at a remote digital computer console, integration with static card readers or with conversion equipment, lower consoles to permit greater visibility from the seated position, and indicators and controls for the digital interface.

The housing must, of course, provide mechanical support and maintenance access for the computing elements and their supporting functions. In general, the considerations in these respects are similar to those of other electronic equipment. The analog computer is unique, however, in its requirements for extensive transmission of signals with frequency content from zero (direct current) to megahertz frequencies while maintaining a sensitivity to errors of 1 part in 1,000, or less.

Shielding and Grounding.[12] The principal causes leading to errors in analog computers, which can be controlled effectively by good packaging practices, are stated [12, 13] to comprise: coupling through common impedance, leakage, noise, distributed capacity, and parasitic oscillations. The principal means of packaging control are shielding and meticulous grounding. The proper use of shields reduces noise and fixes stray impedances and leakage paths; however, as pointed out by Stewart,[12] shielding controls distributed capacitance by adding to it.

Shielding is also effectively used to isolate portions of a system which may have separate grounds when they are connected through a transformer. In this case, the shield is built into the transformer (between the windings) so that the capacity between them is replaced by capacity from each to the shield. If, then, the shield is connected to the grounded side of the secondary as illustrated in Fig. 9, capacitive leakage from the primary is drained off. For complete shielding, both the primary and secondary can be wound with separate shields, each connected to the ground side of its winding to eliminate high-frequency leakage.

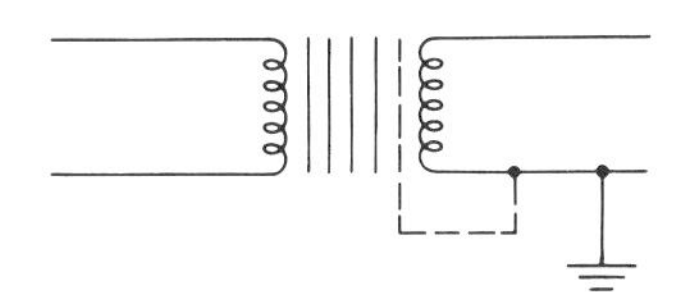

FIG. 9. Shielded transformer circuit.

When separately grounded systems are connected together through shielded leads, noise is minimized by connecting each end of the shield to the ground at its end. When a potential difference exists between the two grounds, the resulting ground currents in the shield may cause local currents to flow in the conductors. If this cannot be tolerated, the preferable practice is to omit grounding the shield at both ends and to float one end, at the expense of added noise.

Several types of grounds are identified in most electrical systems. Figure 10 shows the grounds common to analog computers. The signal ground is the low side of the circuits carrying the voltages and currents which represent the variables and other quantities of the computation. The shield ground is the system which connects the various shields of coaxial conductors and other parts of the shielding system together and eventually to a common ground. The power supply carries a separate ground of its own, which is also brought to the common ground. Power to the relays which control the action within the computer is normally brought to common ground separately from the power system ground. Also brought separately to this common ground are the ground leads from any external equipment connected to the computer system. The common ground is firmly attached to a point on the chassis or frame of the computer. It is denoted in Fig. 10 as the chassis ground. The "earth ground" in that figure represents the connection from the common chassis ground to a water pipe or an external buried grounding rod or plate.

It is essential that separate conductors be used for the several ground systems so that they are not coupled through a common ground wire impedance, which might cause the current flow in one grounding system to affect adversely the behavior of a sensitive signal circuit or to disturb a potential level from part of the computing system. The allowable resistance of common returns is easily calculated from the allowable dc and low-frequency cross-coupling and the net current flow through the return. At higher frequencies empirical methods are usually necessary.

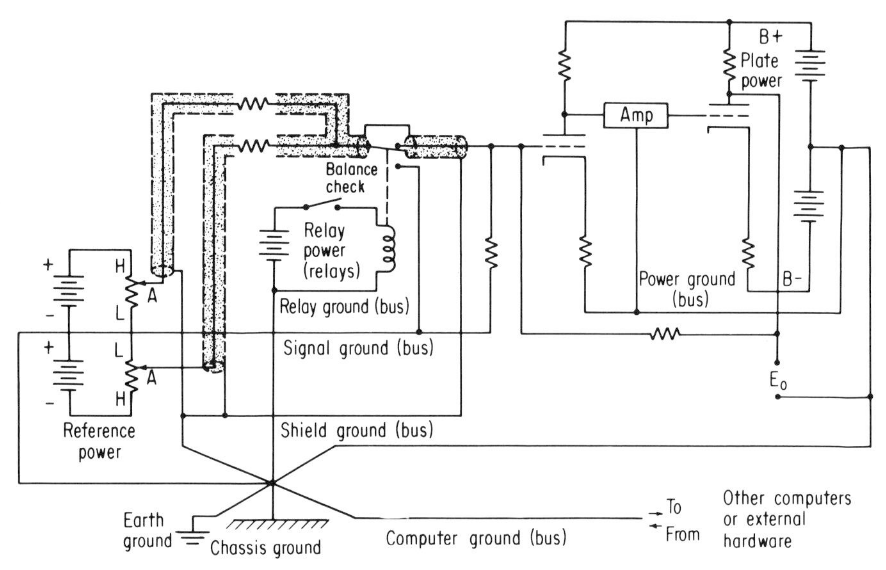

FIG. 10. Grounding system and power supplies for an analog computer.[12]

The plan of using a point ground as the connecting place for the several grounding systems can also be extended to parts of those systems which might mutually interfere. This is especially true in the relay ground system and in the signal ground system. Connecting through a point ground tends to lengthen ground leads but is good packaging practice. Short ground leads are most desirable, but where they cannot be obtained by means of a proper layout, heavier ground conductors may be the answer, to keep the ground potential of the circuit close to that of the chassis ground. It is not good practice to chain ground connections together along a ground conductor or to use a metallic chassis or frame as a ground conductor. An exception to this statement is found in circuit cards and multiple-layer boards which incorporate a ground plane as part of the structure and include its influence in the circuit design.

Power Supply. The general-purpose analog computer usually utilizes signal excursions of both polarities, balanced with respect to ground. Furthermore, the dc signal is more conveniently generated with several bias levels, and the amplifier offset necessarily shows some sensitivity to the voltage levels of the supply. All computing accuracy depends on an exact balance between the positive and negative reference supplies, lack of noise in the signal frequency range, and stability of the reference during a computing run.

Table 8 lists important factors that must be taken into account in planning the power supply. A low common source impedance minimizes coupling between loads via the power supply. Coupling between different units is eliminated by having each on its own power supply. Figure 10 shows three separate supplies, one for reference power, one for relay power, and one for amplifier power. Digital elements when

TABLE 8. Special Considerations Affecting Analog Computer Power Supplies

Physical location

Electromagnetic induction into signal circuits
Heat generation—effect on heat-sensitive computing elements

Loading

Up to hundreds of components
Tens of amperes
Wide variation during operation
Many different bias levels, both polarities
Coupling through common impedance of supply and leads

Classes of signals

Of limited bandwidth—physical simulation
Steps
Pulses from logic components

present would likewise be powered separately. Regulation for each supply is usually necessary. Requirements vary widely; typical values are given in Table 9.

The hum and other noise generated by the power supplies, as well as unwanted coupling between signals due to excessive impedance in the power supply or its leads, may be important factors up to quite high frequencies. Cross coupling may even result in parasitic oscillations capable of effectively disabling the computer. Electromagnetic fields generated by the supplies may also introduce hum directly into signal circuits.

Patch Bays. The patch bay provides means for interconnecting the various circuit elements and portions of an analog computer, resulting in its innate flexibility. The connections are made through patch cords, which generally are shielded coaxial cables with jacks on one or both ends. Removable patch panels are often part of the patch-bay assembly. With them, the wiring via the patch cords is done away from the computer; then the patch panel, one side of which is essentially a large connector in its construction, is attached to the patch bay. Figure 11 shows a patch panel being assembled to a patch bay.

The connections to the patch bay and the patch cords themselves will introduce cross coupling between circuits and set up unforeseen leakage paths, unless great design care is taken. This calls for shielding [14] and short connections throughout. The patch panel may be mounted on a large metal shield, or the individual jacks of the patch bay may be shielded as shown in Fig. 12. The patch bay is much more than an interconnection device; it contains circuit networks, switches, and often pluggable elements as well, which can be connected to produce the intended configuration of the computer. In designing the layout for these components, short noise-free connections and convenience to the operator are the paramount considerations.

TABLE 9. Specifications for Typical Analog-computer Power Supplies [11]

(Note that actual requirements vary widely)

Voltage change for a 5 percent line-voltage change	0.5 to 0.005 percent
Voltage change for a load-current change of 50 percent of rated load, and recovery time to 99.9 percent of steady state	0.5 to 0.005 percent; 250 to 2 μsec
24-hr drift, hum, and noise	0.5 to 10 mv
Source impedance	0.1 to 0.01 ohm (1 to 10 kc); shunt capacitance at higher frequencies

Fig. 11. Removable patch panel, being assembled to a patch bay.

Fig. 12. Patch-bay jack construction, depicting shielding. One plug is shown inserted.

One authority strongly advocates pluggable computing elements as follows:[15]

> To improve maintainability and convenient system expansion as well as computing bandwidth, analog computing elements ought to be plugged directly into the rear of a shielded analog-computer patch bay whenever it is at all possible. With reasonably designed transistor and integrated-circuit packages, such patch bay mounting of all analog computing elements is possible for ±100-V as well as for ±10-V computers, although the ±100-V machine will require careful management of forced-air cooling.

Circuit elements on the patch bay which must be kept at a uniform temperature for stability and to avoid errors in their circuit values are placed in temperature-controlled enclosures called ovens. This constant-temperature device often includes cooling as well as heating facilities.

Recent effort has been directed toward more efficient means of bringing connectors up to the rear of the patch bay. In designs now available, jack springs are mounted in boxed-in die castings at the front of circuit modules which assemble alongside each other to form the actual patch bay. Such assemblies can be seen in Fig. 11 directly behind the removed patch panel. The die castings are shaped to shield the springs as well as to hold them. Figure 12 shows the detailed construction, with one plug in place on the patch panel. This layout avoids the necessity of swinging cables and

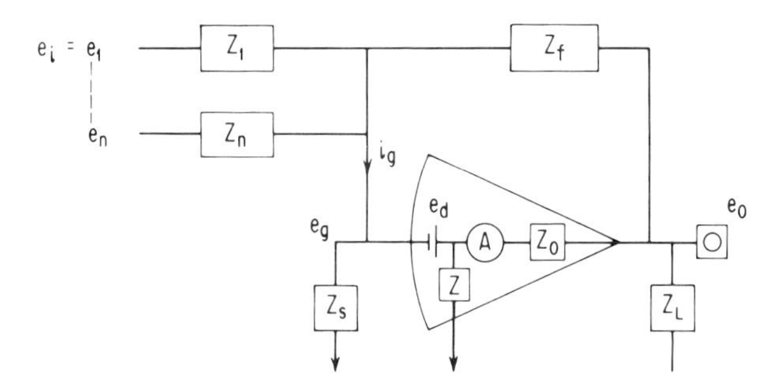

FIG. 13. Operational amplifier and networks.

eliminates difficult access to the patch bay rear. Most of the cabling is then at the rear of the console, with modules removable from the front.

Multiple-action springs are also used, capable of breaking normally made connections when the patch-cord tip is inserted, while others can connect the patch cord to two normally open leads. These effectively increase the number of functions which can be performed by a given size of patch bay.

Direct connection of digital printed-circuit cards to the patch bay was previously discussed in relation to Fig. 7. In that instance, all connections to the cards were on their front edges via the patch bay. Another practice is to build up the connections behind the patch bay from printed-circuit cards which plug into the rear of the patch bay and also contain provisions for side and rear connections. Computing elements plug into the rear of these, while flexible printed-circuit strips are used to connect up power supplies, control circuitry, and such signal leads as cannot be contained behind a single card.

Operational Amplifiers. For over two decades the dominant element in analog computation has been the operational amplifier. Its (ideal) ability to maintain a node at a fixed potential (usually zero) with respect to ground, without itself drawing current from the node, is the key to dc analog computation.

Referring to Fig. 13, the closeness with which Z and A can be made to approach infinity, and e_d and Z_o to approach zero, are measures of amplifier performance. Their approach to the ideal, as well as the frequency independence of the amplifier gain and impedances, are the subject of major design effort. In the ideal amplifier network combination, Z_s, Z, and A are infinite, and Z_o and e_d are zero, in which case e_g and $i_g = 0$. Then

$$e_0 Z_f + e_1 Z_1 + \ldots e_n Z_n = 0$$

$$-e_0 = \sum_{i=1}^{n} \frac{Z_i}{Z_f} e_i$$

In the most common usages the Z_i are resistive, while the Z_f are either resistive or capacitive. The former arrangement permits summation of input variables; the latter, summation and integration with respect to time.

The greatest sensitivity of performance to packaging lies in the parasitic capacitances appearing in the circuit at Z_s, in parallel with the input and feedback impedances. In applications where outputs must be transmitted long distances, stray capacitance in parallel with the Z_i may become significant. Crosstalk from unwanted signals most often occurs as a result of an emf feeding through such a stray capacitance C_i to the summing point. Such signals add directly to the desired signals.

ANALYSIS OF ERRORS IN ELECTRONIC INTEGRATORS. It is also true, of course, that Z_i and Z_f are never pure resistors and capacitors. Results of an analysis of the effects of imperfect amplifiers which particularly relate to packaging considerations of integrators are repeated here:[16]

> [Figure 14] shows an electronic-integrator model without many of the idealizing assumptions.
>
> 1. A suitably given expression for the amplifier gain A will account for effects of finite gain and can be written to account for the effect of the amplifier load.
> 2. The "capacitance" C is not a constant, but is given by
>
> $$C(j\omega) = C_{EFF}(\omega)[1 - jD(\omega)] \tag{1}$$
>
> to account for both resistive leakage and dielectric absorption.
>
> 3. The distributed capacitance C_R of the integrating resistor and the summing-point capacitance C_G are explicitly included. Resistor inductance can safely be neglected below 200 kHz.

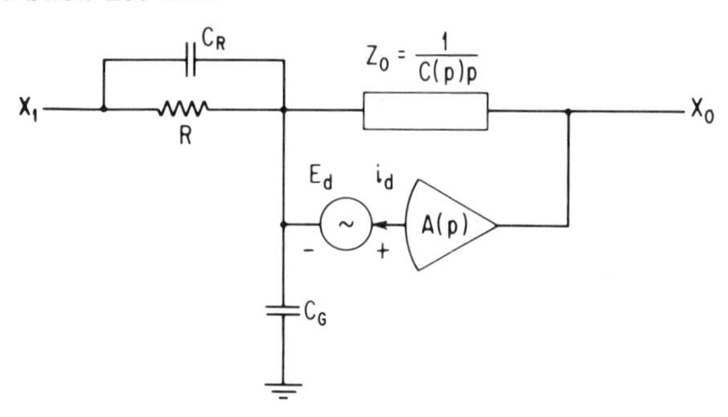

FIG. 14. Electronic integrator with finite gain, loading, parasitic impedances, and dc offset.[16]

> The purely additive effects of dc offset and noise [E_d and i_d in Fig. 14] will be considered separately. For $E_d = i_d = 0$ and reasonably small errors, the frequency response of a practical electronic integrator is given by

$$\frac{\overrightarrow{X_o}}{\overrightarrow{X_1}} \approx -\frac{1}{j\omega RC_{EFF}}\left\{1 + j\left[\frac{1}{\omega A_0 RC_{EFF}} + D(\omega) + \omega RC_R - \frac{\delta_A(\omega)}{A_0}\left(1 + \frac{C_G}{C_{EFF}}\right)\right]\right\} \tag{2}$$

throughout the integrator working frequency range. The resulting integrator phase error is

$$\delta_I(\omega) = \arg \frac{\overrightarrow{X_o}}{\overrightarrow{X_1}} - \frac{\pi}{2} \approx \frac{1}{\omega A_0 RC_{\text{EFF}}} + D(\omega) + \omega RC_R - \frac{\delta_A(\omega)}{A_0}\left(1 + \frac{C_G}{C_{\text{EFF}}}\right) \quad (3)$$

(phase lead for $\delta_I > 0$) and the absolute percentage error for sinusoidal components at the angular frequency ω is $|\epsilon_I(\omega)| = 100\delta_I(\omega)$. Equation (3) expresses the error as a sum of terms respectively due to finite amplifier gain (at dc or at the computer repetition rate), capacitor leakage and dielectric absorption, distributed capacitance of the integrating resistor, and amplifier high-frequency bandwidth limitations.

In practice, C_G/C_{EFF} is small for all but very fast integrators. The capacitor dissipation coefficient $D(\omega)$ is usually almost constant except at very low frequencies, where it is given by $1/\omega r_{\text{DC}}C(\infty)$. The phase shift caused by C_R precludes the use of 1-megohm integrating resistors above about 50 hz. . . . [Table 10] illustrates the relative importance of the various error sources with two numerical examples.

TABLE 10. Examples of Error Sources in Electronic Integration [16]

(For comparison purposes note that *unity-gain phase-inverter errors* are usually readily made less than integrator errors, while *errors due to potentiometer phase shift* might exceed the figures quoted for $100\omega RC_R$ percent by a factor of 10 without careful phase compensation)

Source	A typical "slow" integrator		A low-cost repetitive-computer integrator		A high-quality wideband (transistor) integrator	
Half-scale range	10 or 100 volts		10 or 100 volts		10 volts	
Calibration error	0.025 percent		0.5 percent		0.3 percent	
C, R	1 μF, 1 M		0.001 μF, 50 K		0.005 μF, 2 K	
$A_0, \delta_A(\omega)$	10^7, 10ω		10^4, $\omega/100$		10^6, $\omega/100$	
D	2×10^{-4}		$<10^{-3}$		$<10^{-3}$	
C_R	4 pF		2 pF		1 pF	
C_G	500 pF		20 to 100 pF		20 to 50 pF	
E_d, i_d	50 μV		200 μV, 20 nA*		100 μV, 10 nA*	
Frequency	0.01 cps †	100 cps	100 cps †	10 kc	1 kc †	100 kc
$\frac{100}{\omega A_0 RC}$ (percent)	1.6×10^{-4}	1.6×10^{-8}	3.2×10^{-2}	3.2×10^{-4}	1.6×10^{-3}	1.6×10^{-5}
$-100\frac{\delta_A(\omega)}{A_0}$ percent	-6.3×10^{-6}	-6.3×10^{-2}	-6.3×10^{-2}	-6.3	-6.3×10^{-3}	-0.63
$100D$ (percent)	2×10^{-2}	2×10^{-2}	<0.1	<0.1	<0.1	<0.1
$100\omega RC_R$ (percent)	2.5×10^{-5}	0.25	6.3×10^{-3}	0.63	1.25×10^{-2}	0.125
$\frac{E_d}{RC}\tau$ (per cycle)	5 m volts	0.5 μ volts	4 m volts	40 μ volts	10 m volts	100 μ volts
$\frac{i_d}{C}\tau$ (per cycle)	0	0	200 m volts	2 m volts	2 m volts	20 μ volts

* From electronic switch and/or transistor amplifier.
† One cycle at this frequency is one typical computer run with the R, C values quoted.

In addition to the dynamic errors discussed, integrators exhibit a constant-percentage calibration error mainly due to changes in the effective capacitance C_{EFF} with temperature and also with frequency. Finally dc offset voltage and/or current [Fig. 14] produce an additive error output

$$X_o = -\frac{\tau}{RC}(E_d + Ri_d) \quad (4)$$

where τ is the computing time. Effects of high-frequency noise are largely "integrated out."

Transistor amplifiers now are offered with E_d and i_d in the low microvolt and picoampere ranges.

The ideal operational amplifier, packaged perfectly, will perform as a summer or integrator only as well as the associated resistors and capacitors behave as two-terminal elements for which $I = E/R$ and $I = C(de/dt)$, with R and C independent of frequency, temperature, current, voltage, age, and previous history. The smaller the size of these elements, the more difficult it usually is to secure such independence, while increased size complicates the achievement of effects desired for similar reasons, in the rest of the computer.

Resistors and Capacitors. Resistors and capacitors are chosen for their low temperature coefficients and for stability with time. Noninductive wire-wound resistors stable to 25 ppm can be obtained with temperature coefficients of 2 ppm/°C. Metal film resistors are rapidly replacing carbon film resistors where higher resistance values (½ megohm and up) are required for computing elements in fast circuits. The metal film resistors have a lower temperature coefficient and excellent stability. Computing-quality capacitors are obtainable which have 0.01 percent accuracy over room-temperature range when compensated. Capacitors are subject to aging effects and exhibit dielectric hysteresis and the retention of residual charge. Anomalous effects from these causes can only be guarded against by carefully specifying and testing the capacitors chosen for computing elements.

Resistors or capacitors used to provide ratios or to divide computing voltages should be mounted adjacent to one another, so as to be subjected to the same environmental changes. Resistors and capacitors are commonly assembled as close as possible to each other, the mode-switching relays or electronic switches, the amplifier summing point and the patch bay to form a network. Considerable ingenuity is expended to achieve such close spacing, without permitting stray leakage resistance or capacitance to alter intended performance drastically.

Potentiometers. The next most common computing element is the coefficient potentiometer. It is required to introduce fractional values of initial conditions and coefficients, and usually is set by comparison with a decimal voltage divider, either manually with a digital voltmeter or by means of a servo system. The mechanical and wiring requirements imposed by the method of setting (or checking) are considerable factors in the system performance of potentiometers, since their output is commonly at a high impedance level.

More recently, it has become common practice to provide a number of voltage dividers with digital input controls, settable in anywhere from a few microseconds to several milliseconds. These are generally larger and more complex than potentiometers, and more remotely located. If they have a voltage output, they must be low-impedance (i.e., buffered by an operational amplifier) to avoid being affected by resistive or capacitive loading (although some methods of setting can compensate for a *constant* resistive load). A current output appearing on the patch panel tends to affect associated amplifier performance adversely.

In its higher-speed manifestations, the voltage divider becomes indistinguishable from a multiplying digital-to-analog converter, discussed elsewhere.

DYNAMIC ERRORS DUE TO POTENTIOMETERS.[17]

With increasing computing frequencies, the high-frequency phase shift in coefficient-setting potentiometers may greatly exceed phase shift in other computing elements and can introduce serious dynamic errors into computations. Potentiometer phase shift is mainly due to the capacitive load produced by computer circuits and wiring, and by the distributed capacitances of the potentiometer resistance, housing, and/or copper mandrel. With wire-wound potentiometer resistance elements wound on flat cards or small-diameter mandrels, the effects of potentiometer inductance are not ordinarily felt at computing frequencies below 1 MHz.

Potentiometer and load capacitances cause both lag and lead effects. While an ideal voltage divider would not introduce any phase changes at all, a wirewound potentiometer will introduce a lag at high potentiometer settings and a lead at low potentiometer settings.

Potentiometer phase shift decreases with the potentiometer resistance, but this cannot be reduced below a minimum value determined by amplifier power and, in wirewound potentiometers, by the required resolution. For fast (repetitive) analog computers which do not require setting stability below 0.1 percent, the best type of coefficient-setting potentiometer might be a type with a linear deposited-film resistance element with the slider positioned by a multiturn dial turning a lead screw. Such potentiometers have negligible inductance and distributed capacitance; note, though, that potentiometer construction can do nothing about the effects of capacitive loading due to computer circuits and wiring.

Compensating capacitor networks to offset phase shift in potentiometers can be designed on the basis of measurements or manufacturer's data of phase shifts versus displacement for a specific load.

Other Components. The other computing components are generally associated with the operational amplifiers, either being substituted for feedback or input elements or switching such elements. Multipliers and function generators use copious quantities of resistors, diodes, and/or transistors. Limiters and backlash and coulomb friction simulators utilize diodes, transistors, and variable resistors. The comparator uses precision resistors and a low-drift amplifier. The various digital-to-analog switches—mode control, electronic switches, track stores, converters—utilize field effect transistors (FETs) and chopper transistors to reduce the effects of the solid-state circuitry on computer precision. In all these components, the effect of leakage and stray capacitance is proportional to impedance level, while in solid-state elements the variation of voltage drop caused by ambient temperature is affected only by current level, which tends to be independent of computer reference level. The minimum resistance value is usually limited by the reference voltage; for example, 100 volts applied across a 100 K resistor generates 0.1 watt, enough to cause several degrees of temperature rise in a resistor of the physical size usable in large quantities in a general-purpose analog computer. It is advantageous when designing function generators to use high-stability metal-film resistors because of their low prices. Since their static accuracy is more affected by diode variations, temperature control or compensation of the diodes is necessary.

Table 11 shows a number of specifications, particularly with respect to independence of environmental factors, which may call for unusual rigor in analog computer applications.

Peripheral Auxiliaries. The peripheral equipment is used to speed setup and checkout of analog programs, in simplifying the modification of programs, in automating

TABLE 11. Susceptibility of Components to Environmental Factors

Components	Self-heating	Aging	Ambient temperature	Stray capacitance	Humidity	Voltage	Current	Barometric pressure	Leakage	Light sensitivity
Wirewound resistors	×	×	×	×	×	×				
Film resistors	×	×	×		×	×				
Adjustable resistors				×					×	
Precision capacitors		×	×		×	×		×	×	
Trimmer capacitors				×					×	
Diodes			×	×			×		×	×
Transistors—general	×					×				
Chopper transistors			×	×					×	
FETs			×	×	×	×	×		×	
Relays	×			×						
Printed circuits				×	×	×			×	

production runs, in recording results, and in displaying the results of computer runs. The problems these units pose for packaging are primarily those of location for convenience of the operator.

The analog computer usually uses peripheral equipment which has been designed for general-purpose use. The designer will usually go to some lengths to achieve uniformity of appearance and consistency of control functions. The high-sensitivity ranges of plotters, recorders, and large oscilloscopes may be dispensed with. The oscilloscope usually warrants a special design for the general-purpose analog computer because of the unusually high accuracy and convenience of multi-trace readability required, and its key place in the use of analog computers.

Analog computer technology has made extensive use of packaging techniques adapted from other electronic equipments, but has reached high design precision without developing unique technologies.

COMPUTER MEMORIES

The storage sections of a computer present packaging practices that are unique, in contrast to other parts of the computer which employ circuit and packaging technology common to other branches of the electronic art. Within a memory, in the logic to address their contents, the amplifier circuits which drive or sense memory elements utilize state-of-the-art circuit design practices. It is the storage cells themselves that are different, and this difference gives rise to novel packaging.

Hierarchy of Memories. Large digital computer systems contain several different categories of storage, dedicated to different purposes, and so require memories differing in capacities and speeds.[18] The principal categories, designated on the basis of their function, are the following:

TEMPORARY STORAGE. Very-high-speed memory for temporarily storing data being used by the central processing unit. Typical applications sometimes result in names such as *scratch-pad memory* or *control memory*. This class also includes registers—small, fast-acting memory units through which data is routed as the processor acts on it. A register often incorporates capability for altering the data it contains.

WORKING STORAGE. Main memory, for holding instructions and other information being operated on by the central processing unit.

MASS STORAGE. Collectively, the auxiliary memory units of a digital computer system which hold bulk data for potential reference. Copies of portions of this data are usually transferred in blocks to the working storage for processing.

BUFFER STORAGE. Small memory unit for temporarily holding data being transferred from one part of the computer system to another.

Table 12 lists for each of these categories the ranges of capacities and speeds, and

TABLE 12. Capacity and Speed Ranges of Storage

Category of storage	Capacity, bits	Memory cycle time	Type of memory
Temporary storage . .	10^2–10^4	10–100 nsec	Film Integrated-circuit
Working storage	10^3–10^6+	200–2,000 nsec	Film Wire Core
Mass storage	10^5–10^6	1–10 sec	Core, wire
	2×10^5–10^7	4–40 sec	Drum
	5×10^7–10^9	50–200 msec	Disc
	10^7–10^8		Removable cartridge disc
	10^7–10^8+	½ sec	Magnetic card
Buffer storage	1–10^4	1 sec or slower	Core

also shows the type of memory used. Those which store on movable media, mechanically driven, provide the largest capacities and also take the longest to complete a cycle; their use is limited to mass storage. Memories in all the other categories consist of arrays of memory cells, ferromagnetic or solid-state. The largest ferromagnetic arrays are practical for mass storage up to a million bits.

Two other functional classes of storage sometimes found in large systems are these: (1) *Fixed storage memory*—a RO device for holding material that must not be changed or written into. It may consist of an array of capacitors, cores, or other memory cells. (2) *Content-addressable memory* (also called *search memory* or *associative memory*)—a memory in which every word can be simultaneously compared against a given word, and the location or locations at which the given word is stored can be read out. In this sense the words it contains are not designated by addresses. (The previous specification of an address locates a word in the other kinds of memories described.) Memories of this class must be fast, and so utilize the techniques of temporary storage, differently arranged but with the same limitations on capacity.

Design Aims. The design of a memory must provide wanted capacity with adequate speed and at low unit cost. The capacity of a memory is fundamentally stated as the number of bits that it can store, and may range from a single bit for a buffer or register to hundreds of millions of bits for the largest mass-storage units. Speed is expressed in terms of the memory access rate—the number of times per second the memory can be referenced. However, the reciprocal of this quantity, the access time, is also frequently referred to as speed. The memory cycle time is the elapsed time from the moment the memory is addressed until a read-and-restore or clear-and-write operation has been completed and the circuits have settled so that it can be addressed again. This time is a measure of the speed of the entire memory, and is not to be confused with the cycle time of the computer. Unit cost, expressed in dollars or cents per bit, must be based on the complete memory, including sense, drive, and address logic circuits, to be meaningful in making comparisons, since the nature of these circuits and the number required changes markedly with the type and organization of the memory as well as with its speed and size. Table 13 relates the three factors of speed, size, and cost for the most prevalent types of memory arrays. The figures shown for each type are for the range where it predominates. Outside its range, each is overshadowed by other types. The cost figures in this table *do not* include the drive and access circuitry but serve only for comparing the arrays themselves.

TABLE 13. Compatible Ranges of Applicability for Principal Types of Memories

Type	Access time, nanosec	Size, bits	Element cost, cents per bit
Core	500 or more	10^6 and up	0.1–0.5
Wire	200–500	10^5	0.5–2
Planar film ...	50–200	10^4	2–10

Source: C. J. Kriessman, Ferroxcube Corporation, Saugerties, N.Y.

Through their effects on cost, packaging considerations are involved in selecting the type of memory. They also put an upper limit on its capacity, control its organization, and affect its speed through many intimate details. In any matrix memory, the cycle time is the sum of the propagation delays through the circuits and the recovery time following a write operation. In core memories, the magnetic elements give rise to series inductance and the wiring to shunt capacitance, so that the pulses travel along a lumped transmission line.[19] The attenuation and delay they suffer along this line limit the dimensions of the plane. Since the sensing circuits must respond to feeble excitation, they are particularly sensitive to noise generated by the much larger writing pulses. Packaging arrangements must be aimed at keeping this noise to tolerable levels by minimizing it at the source, by reducing the coupling through which it is induced into the sense lines, or by canceling it after it gets there. Application of these principles to a high-speed memory design is discussed by Bland [20] with particular relevance to the associated circuitry.

In addition to the design aims of speed and capacity, other influencing factors listed in Table 1 are of packaging concern—especially power, reliability, compactness, and manufacturability. Most kinds of storage elements are thermally sensitive and require provision for taking away the heat generated by them and by associated transistors and resistance. Conductive heat sinks may be necessary. Design must allow for repeated writing in all cells at once unless specifically inhibited. Overheating degrades reliability, as does excessive circuit noise. Volatility (loss of pattern if power is lost) may be incompatible with reliability requirements. Compactness is essential for high speed, less power, and lower noise, but it may impede manufacturability. Obviously, any design measures that lead to batch fabrication of arrays will greatly enhance their manufacturability.

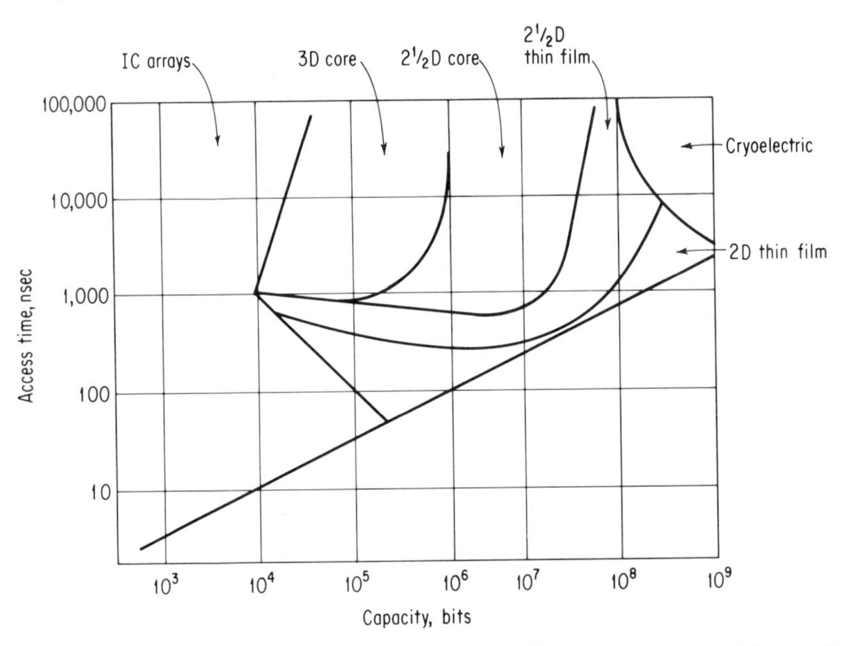

FIG. 15. Speed-capacity ranges for various types of matrix memories. (*George B. Strawbridge, Univac Division, Sperry Rand Corporation, Roseville, Minn.*)

Current and Future Technology. Figure 15 shows the present relationship among the available matrix memory technologies in terms of access time and memory capacity. For the smaller size ranges, it reflects the present availability of high-speed circuits using bipolar transistors in integrated-circuit arrays. For very large memories, cryoelectric techniques represent the only presently feasible approach. The large intermediate area of the figure, encompassing capacities between 10^5 and 10^8 bits, is the region served by magnetic arrays, ferrite cores, and thin-film planes or wires. In the future, this region will be encroached upon from the left, as larger arrays using metal-oxide-semiconductor (MOS) techniques in the slower speed ranges and bipolar devices in the faster ones become economically feasible. The sloping straight line bounding the areas at the bottom (fastest speeds) does not represent the theoretical speed-size limit set by the velocity of light, but represents the practical bound of today's technology; thus, it can be manipulated. If ways are found to reduce the physical size of the fastest memories of a given capacity, this boundary will be lowered.

Storage techniques presumably applicable to the principal categories of storage within a few years can be seen in Table 14. This shows memory cycle times and

TABLE 14. Storage Device Characteristics Anticipated in 1970 [21]

Type of storage	Registers and high-speed control memories		Main high-speed internal memories		Solid-state on-line auxiliary storage devices	
	Typical capacity, words	R/W cycle time, nsec	Typical capacity, words	R/W cycle time, μsec	Typical capacity, words	R/W cycle time, μsec
Integrated circuit arrays	256	50	0.01×10^6	0.2		
MOS arrays	512	250	0.02×10^6	0.7		
Planar thin-film	512	100	0.1×10^6	0.5	2×10^6	1
Laminated ferrite	512	150	0.1×10^6	1.0		
Plated wire	512	250	0.2×10^6	0.5	4×10^6	1
Magnetic core matrix	512	350	0.1×10^6	0.7	2×10^6	3
Etched Permalloy-sheet toroid			0.2×10^6	2.0	4×10^6	35
Continuous sheet cryogenic			2.0×10^6	2.0	20×10^6	5
Ferroacoustic					20×10^6	(Serial)

typical capacities anticipated in 1970 for a wide variety of storage types, including some presently in the development stage.[21] Integrated-circuit arrays are most promising for temporary storage applications; MOS arrays are promising for low-cost intermediate capacities, but their volatility is a disadvantage. Planar thin-film storage is promising for fast control memories and possibly for on-line auxiliary storage. Its use by that time for main internal memories is questionable. Plated wire is very promising in all categories. Magnetic core matrices are well established and are expected to be dominant for several years; they will eventually be outmoded because of batch fabrication techniques. The more "speculative" types of storage listed include laminated ferrite storage, not widely used. For it, reasonable yields are not proven for capacities over a few hundred words. Etched Permalloy * sheet storage, also not widely used, has potential for a very low unit cost, but the yield remains to be proven. The feasibility of a continuous sheet cryogenic memory is still unproved. Because of the high cost of tooling equipment, this type of memory is not economical for capacities below approximately 100 million bits. Ferroacoustic serial delay lines for mass storage are in early research stages, but feasibility is not proved. This concept is promising for low-cost block-oriented auxiliary storage.

Figure 16 outlines the ranges of capacities and speeds of the computer memory arrays being produced. All the boundaries are skewed with respect to cycle time, illustrating the reduction of speed entailed by the longer wires and more complex circuitry for a given type of memory as its capacity is increased. The utmost capacity for a given type of memory falls off with increasing speed, indicative of the limits of that technology. Although the ranges of capacity and speed overlap for the several

* Trademark, Western Electric Company.

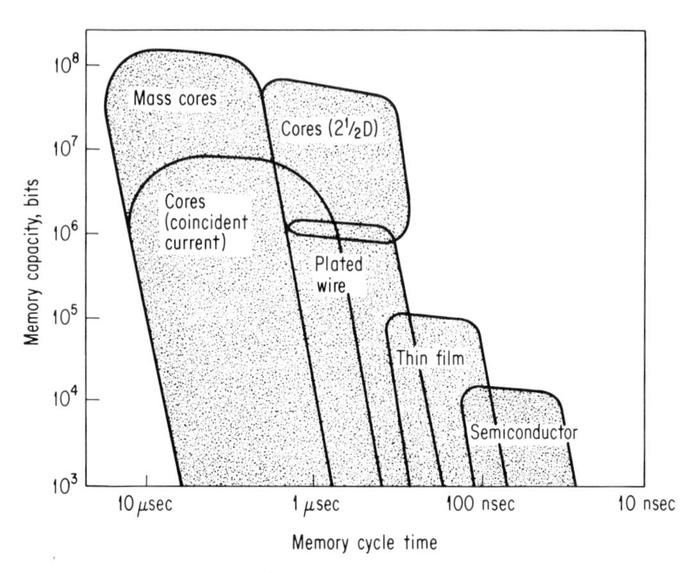

FIG. 16. Speed and capacity of core, thin-film, and semiconductor memories.[102]

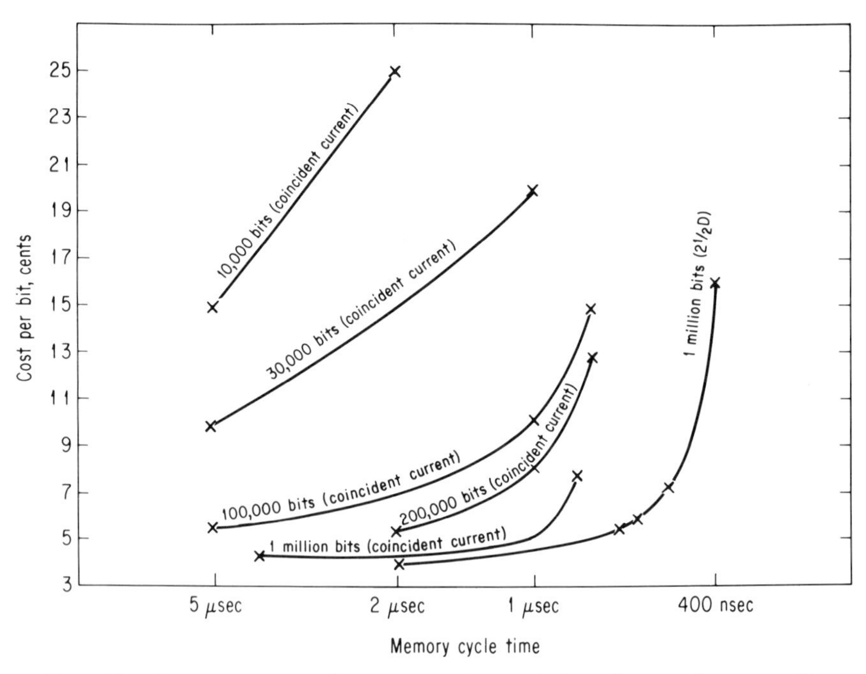

FIG. 17. Cost comparison of core memories, including driving electronics.[102]

types of memory, one type will be preferable for a given capacity and cycle time for economic reasons.

The unit costs for complete core memories of various sizes are shown as functions of cycle time in Fig. 17. The positions of the several curves illustrate the substantially reduced unit costs for the larger memories. Unit costs for a given size of memory begin to rise rapidly below a certain cycle time. For the higher-capacity memories this time is less but the rate of increase is more marked. The curves of this figure illustrate the trade-offs available in satisfying the design aims with the most widely used type of memory in its commonest forms of organization.

Toroidal Core Memories. Magnetic core memories use either single-aperture or multiple-aperture magnetic cores to store discrete bits (binary digits) of information. The most widely used core is the single-aperture toroid. It is used mostly for DRO memories, but it can be used also for NDRO memories. Multiple-aperture cores are used for NDRO memories, but generally their use is not extensive.

The electrical driving and sensing schemes chosen to produce either a DRO or NDRO memory directly affect the mechanical fabrication and packaging of both the planes and stack. Three common memory-organization schemes used for both DRO and NDRO memories are 3D, 2½D, and 2D. All utilize magnetic cores as discrete storage elements to store either a binary 0 or a binary 1, depending upon the particular data pattern at that instant. The physical and electrical arrangement of these discrete cores within a memory differentiates the three schemes.

Single-aperture DRO Organization.. 3D, 2½D, and 2D organization schemes utilize the phenomenon that a given switching current I_s is required to drive a core from one state of magnetic flux saturation to its opposite state of saturation. Figure 18 shows a hypothetical core and its accompanying hysteresis loop.

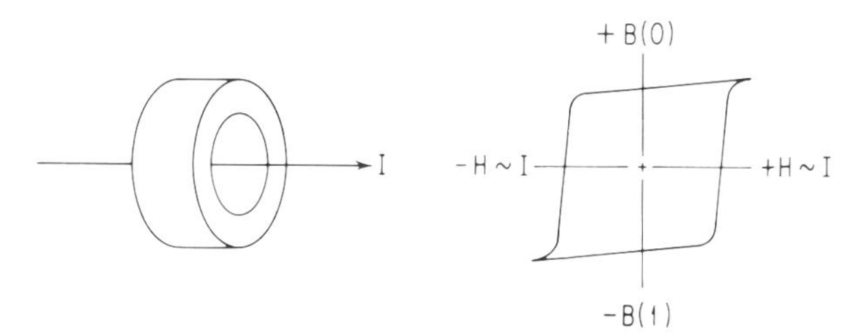

FIG. 18. Magnetic core and its hysteresis loop.

A core with good magnetic properties will be stable in either state of remanence. Arbitrarily, the positive state is defined here as a binary 0 and the negative state is defined as a binary 1. Current in the wire passing through the toroid creates the magnetic force necessary to switch the core from one state to the other. The magnetizing force H of the wire is directly proportional to the current flowing in it and to the resulting magnetic field. By reversing direction of the switching current, the magnetic state of the core can be switched from a 1 to a 0 or from a 0 to a 1. Once a 0 or a 1 state is established by the driving force H, the core will remain in that state even when the switching current I_s is turned off. The only means by which the core can revert to its opposite magnetic state is that of sufficient current in the reverse direction.

3D Organization and Operation.

DRIVE LINES. A *3D* coincident-current memory usually has either three or four wires passing through each core, taking advantage of the memory's geometric layout to perform part of the logic selection and to drive the selected core simultaneously. Figure 19 shows a 16-bit plane of a coincident-current memory. The plane has four X drive lines, four Y drive lines, an inhibit line parallel to the X drive lines, and a sense line strung diagonally through the core matrix. An inhibit and a sense line both pass through every core in a single plane. In some instances the inhibit and sense functions can be accomplished with a single wire, as shown in Fig. 20. Only three wires instead of four are needed when operating in this manner, but there is a sacrifice in speed.

In the four-wire plane of Fig. 19, each of the 16 cores in the plane is located at the

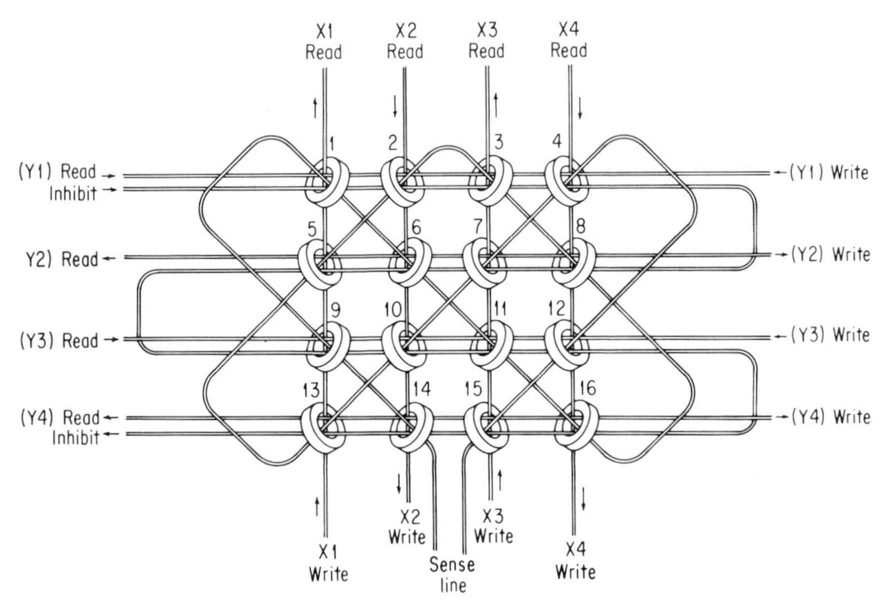

FIG. 19. A 3D coincident-current memory plane with four wires linking each core. (*Reprinted by permission from "Fundamentals of Core Memories," © 1962 by Sperry Rand Corporation.*)

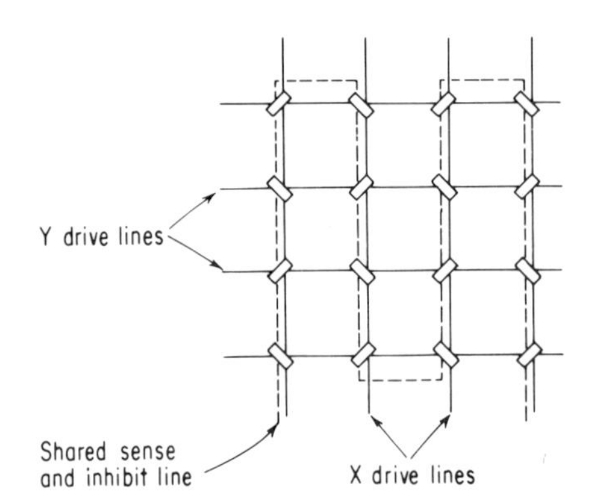

FIG. 20. A three-wire 3D coincident-current memory plane wiring plan.

intersection of an *X* and a *Y* drive line. Thus, each of the 16 cores in the plane is uniquely identified by one of 16 addresses defined by the coincident point of a given *X* and a given *Y* line. When a pair of specific lines is selected, half of the switching current is driven down each selected line, and the core at the point of coincidence is forced into the desired magnetic state because it receives the sum of the two half-currents on the *X* and *Y* drive lines. The other cores on the selected *X* and *Y* drive lines do not switch because they are excited by only half the current necessary to switch them.

Several planes are stacked on top of each other with the analogous X and Y drive lines each connected in a series fashion as shown in Fig. 21. The X and Y currents thus select and drive corresponding cores, one in each plane. In this manner, an address becomes three-dimensional; the core selected in each plane represents a bit in the word at the address defined by the point of X and Y coincidence. The number of planes in the stack equals the length of the word in bits. The number of intersections in each plane is the storage capacity of the stack, in words.

SENSE FUNCTION. To determine the contents of the memory, an X and a Y drive line are selected and driven (opposite state from write-in) to produce a binary 0 at the address or core selected. The presence or absence of a signal on the sense wire then indicates whether a 1 or a 0 was stored in the core. If a 1 was stored, the sense wire detects the change of state as the core switches from a 1 to a 0; no change of

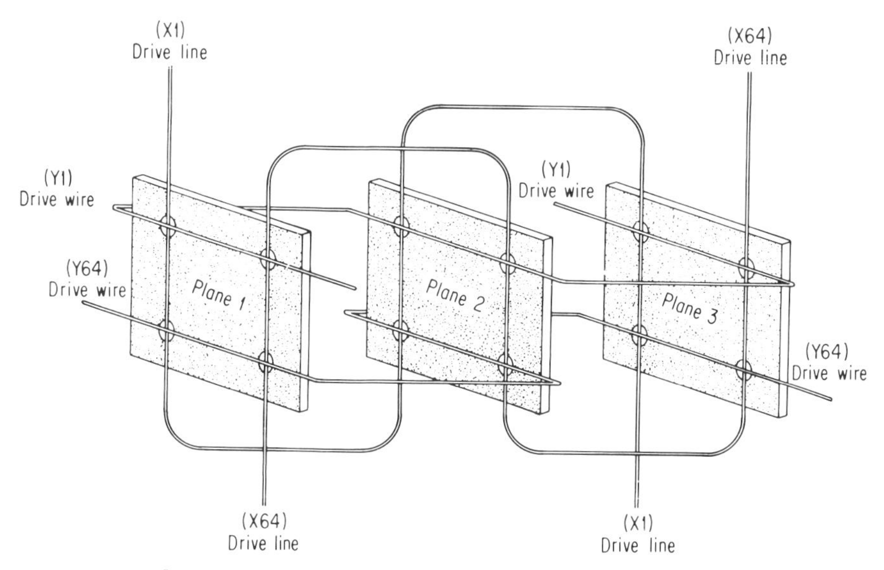

FIG. 21. Stacking arrangement of core planes, showing method of wiring drive lines from plane to plane. (*Reprinted by permission from "Fundamentals of Core Memories," © 1962 by Sperry Rand Corporation.*)

state indicates that a 0 was already present. The sense wire is connected to an amplifier which detects the small switching voltage, amplifies it, and stores the amplified 1 or 0 signal in a data flip-flop. The contents of a DRO memory are destroyed during a read-out operation, and it is therefore necessary to route the output to data flip-flops to prevent losing this information, and in some instances to re-store it in the core memory after use.

The diagonal sense winding configuration puts the sense line not parallel to the X drive, Y drive, or inhibit lines. If the sense wire were parallel to any of these lines, it would transmit to the sense amplifier the half-current drive signals produced by the cores on the parallel line. The sense wire is strung in such a manner that the partial switching noises cancel each other and eliminate the cumulative effect. Figure 19 shows a diagonal sense pattern, and Fig. 22 shows a "bow-tie" sense pattern, both utilized to achieve such cancellation. Each bit of a word must be individually sensed, and one sense amplifier can be common to the same bit of all words represented by the number of cores in that plane. The practical limit of a sense line's capability to sense core outputs is usually about 4,096 (4K) because of electrical noise considerations and attenuation of the signal. If there are more than 4,096 cores in a plane, the bit is sectioned off, and each group of 4,096 words is sensed by a wire connected to a preamplifier. The preamplifier then feeds the sense amplifier, which is not disturbed by the noise on the unselected preamplifiers.

READ-WRITE FUNCTION. Once a read operation has been initiated, the selected core driven to a 0, and the sense detection performed, it may be necessary to restore the contents of the address back to its original state. To perform the restore operation, the selected core is again driven to the 1 state with a half-current on each of the *X* and *Y* lines. However, if a 0 is to be restored, a half-current is turned on in the inhibit winding, opposite in direction to the half-write current in the *X* drive line. The inhibit current cancels the effect of the half-write current in the *X* drive line, leaving a half-current from the *Y* drive line at the point of coincidence, for no magnetic change. If a 1 is to be restored, the inhibit line for that plane is not turned on, and the *X* and *Y* half-write currents switch the core. The control of the inhibit line is a logic function, dependent on the stored data which was read out and stored in the data flip-flop.

The inhibit function is performed on a per-bit basis for all addresses. Therefore, it is possible to selectively write 1's and 0's in the bits of a given word, addressed by *X* and *Y* selection. As in the case of the sense wire, only one inhibit line is required per

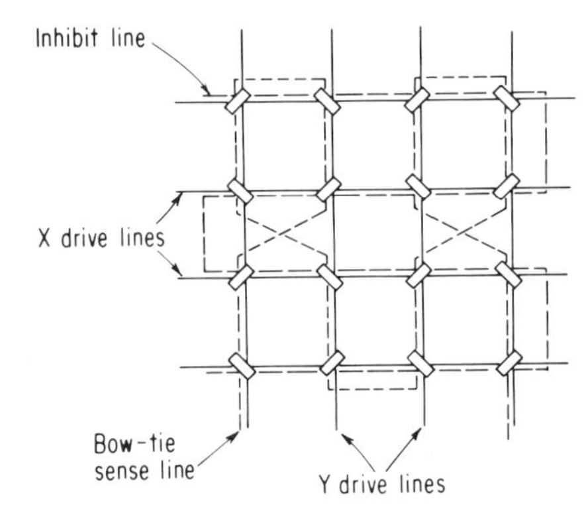

FIG. 22. An alternative four-wire 3D coincident-current memory plane wiring plan.

bit. Because of its length and resulting resistance, this wire can contribute significantly to stack heating when 0's are being written. The inhibit line is sometimes split and selectively driven to reduce stack heating and to limit line inductance. Selectively driving more than one inhibit line per bit is quite usual in memories of more than 4,096 addresses.

In Fig. 19 the inhibit line was chosen to run parallel to the *X* lines for the purpose of illustration, but it could run parallel to the *Y* lines. However, the inhibit line should not run parallel to the sense line.

2½D Organization and Operation. The 2½D coincident-current memory selects and drives cores much as does the 3D memory. As in the 3D memory, the *X* drive lines pass serially through all the bits of the entire memory. But, unlike the 3D memory, the *Y* lines do not pass serially through all planes. Instead, each ordinal bit of the memory has its own *Y* drive lines and *Y* selection scheme, as shown in Fig. 23.

Although the quantity of electronic circuitry required for driving and selection in a 2½D memory may be greater than that required by a 3D system of the same capacity, the 2½D memory is not necessarily more expensive. The ratio of *X* and *Y* drive lines may vary in different designs, to meet particular requirements and drive capability at optimum cost.

The memory mats are physically laid out much the same as they would be in a

3D system. Figure 23 shows twenty 64 × 64 mats arranged in a 4 × 5 matrix. During selection, only one of the 256 possible X lines is driven. Along the Y axis, though, each of the 5 mats is separated electrically into 4 sections, each representing a bit, for a total of 20 bits.

An address is selected by a combination of 1 of 256 X lines and 20 of 320 Y lines. The Y address circuitry selects the same one of 16 Y drive lines in each of the 20 bits. In this manner 20 cores, each at the same X address along each selected bit line, are driven.

The sense line in each plane runs parallel to the 16 Y lines but is transposed as shown in Fig. 24 to cancel partial switching noise.

Memories of this type have been built with only two wires instead of three, but at a sacrifice in speed. A two-wire, 2½D memory uses one of the drive lines as a sense line. When operating in this manner the sense amplifier must be able to detect the switching signal of the core superimposed upon the Y drive current. Thereupon, noise associated with drive currents becomes a very important factor. The advantage of operating in this manner is that only two wires are required. On large memories this can be a significant cost advantage.

2D Organization and Operation. Figure 25 shows a 2D linearly organized, or word-select, memory plane. This memory does not use coincidence to select and read, but does use it to write. During a read-out operation, one word line

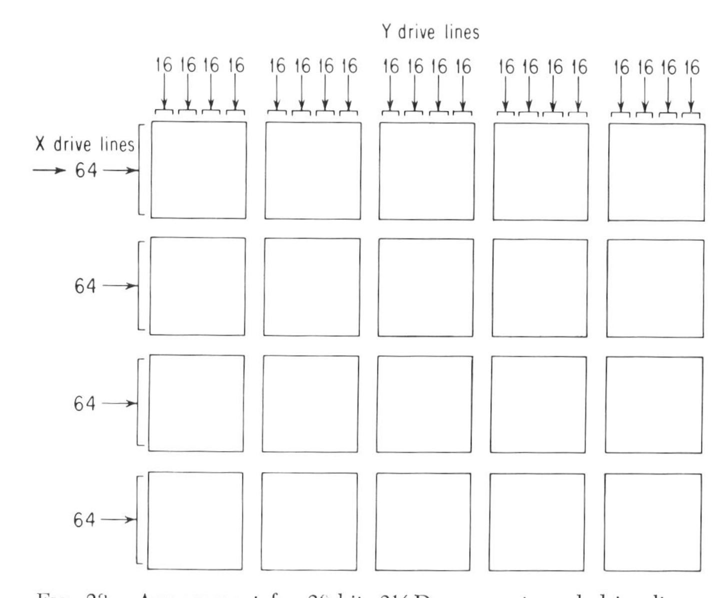

FIG. 23. Arrangement for 20-bit, 2½D core mats and drive lines.

is chosen, to drive all cores on that word line (each bit of the word) to 0. The individual sense lines associated with each bit then detect the presence or absence of the change of state in a core. The resulting signal is amplified, and the output is stored in a data flip-flop. The full switching current, or more, is driven down the word line. The combination of a short drive line and overdriving current results in a very fast read-out time.

To restore the contents of the memory or to perform the second half of a clear-and-write cycle, a half-select write current is driven down the word line to drive the cores toward a 1. The presence or absence of the other half-write current on each bit line will then enable 1 or 0 write-in at each core respectively.

Memory planes of the word-select type shown in Fig. 25 can be stacked up, with the bit and sense lines continued from plane to plane. However, each word line of the memory must remain independent.

3D, 2½D, and 2D Comparison. A comparison of the three DRO memory selection schemes yields the following general information:

1. The 3D coincident-current memory is the least expensive but the most difficult to operate in large capacities at high speeds. Its cost advantage results from its selection being performed once in the *X* and once in the *Y* dimension and carried through all the bits in the Z dimension by continuous drive lines. However, the

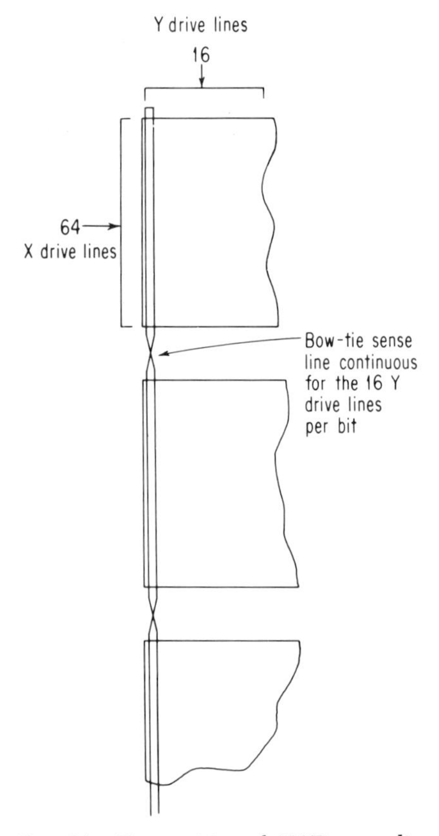

FIG. 24. Tranposition of 2½D sense line.

length of these lines increases the inductance, which tends to limit the memory's speed. Selection economy is enhanced by putting *X* and *Y* lines in a square configuration, and in binary multiples.

Longer word lengths (more planes) yield lower selection costs per bit. A 1,024-word × 1-bit memory, for instance, is not very economical compared to a 1,024-word × 16-bit memory. 3D memories are usually packaged in configurations of from 1,024 to 16,384 words. The larger-word-size memories often require more than one sense and inhibit line per bit because of noise and length (speed and drive voltage), respectively. The 3D memory can be of either three- or four-wire design; four-wire is the most common.

2. The 2½D memory requires more drive electronics than the 3D memory, but

the 2½D memory has the advantage of only two or three wires through the core arrays. With no diagonal sense wire and with less handling required, the stack cost is reduced. The three-wire 2½D memory is less expensive to produce for high-speed operation than a 3D memory.

3. The 2D word-organized memory is the fastest but also the most expensive of the three selection schemes. It is not used extensively, nor is it likely that it will be in the immediate future. As with a 2½D system, it can be operated with two wires but at a sacrifice in speed.

NDRO Organization. Each of the three DRO organization schemes previously discussed can also be used to produce an NDRO memory. The most common NDRO toroidal core memory is not electrically alterable; it is produced by stringing a restore wire through only the predetermined pattern of 1 cores in the memory matrix. When the memory is restored from the 0's created by the read operation, only the 1-designated and wired cores will receive the full current necessary to restore them to their 1 state. Since the restore wire skips the 0 cores, they will be excited by only a half-write current and will remain in the 0 state. One example of this type of memory is a 3D memory configuration with X and Y drive lines and sense lines, but with restore lines instead of inhibit lines.

Another less common type of NDRO memory is produced by passing an extra wire

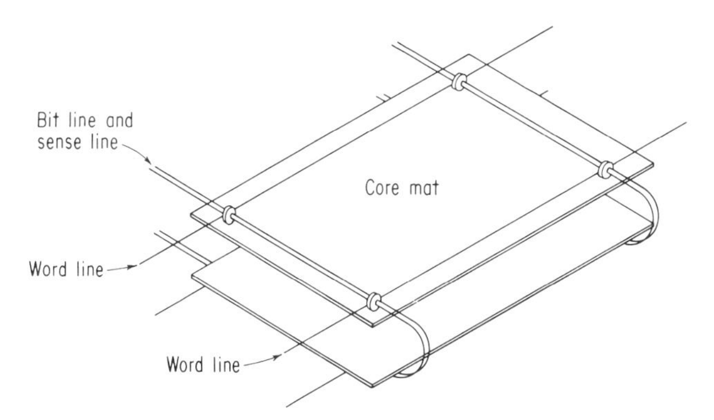

FIG. 25. 2D word-organized memory arrangement.

through the cores in such a manner that the NDRO current always sets up a predetermined magnetic state in each core. The NDRO wire supplies the full switching current and is strung so that the NDRO current sets each of the cores to the same state every time it is turned on. Operating in this mode, the information within the memory can be altered but is always available for restoration due to the wiring of the information into the memory.

Multiaperture Core NDRO Memories. Alterable NDRO memories can also be built with the use of multiaperture cores. Coincident-current organization schemes can be used with these devices, and information can be changed within the memory when desired. The distinction between these devices and toroids is that the read cycle does not destroy the information stored in the element.

Toroidal Core Array Fabrication. As with computers, the design trend in core memories is toward faster and more compact units with greater storage capacity. Fortunately for packaging engineers, the use of smaller cores facilitates this trend. During the past few years, core toroids have decreased from 0.080-in. outside diameter to 0.050, 0.030, and 0.020 in. Now 0.012-in. cores are being developed for production. Although these smaller cores benefit the packaging engineer in some ways, they also introduce many intriguing problems.

Arraying Technique. Smaller cores require more precision in making the core arrays. An etched metal plate mounted on a combination vacuum-vibration table is the most common method of making a core array. Holes for holding cores are etched in the plate in a pattern such as that shown in Fig. 26.

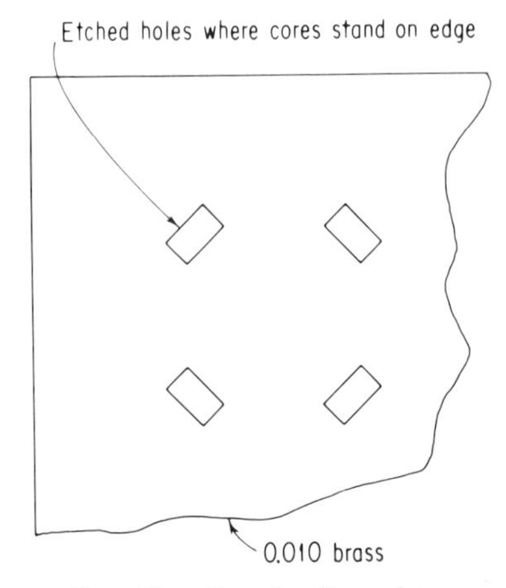

Fig. 26. Core loading plate.

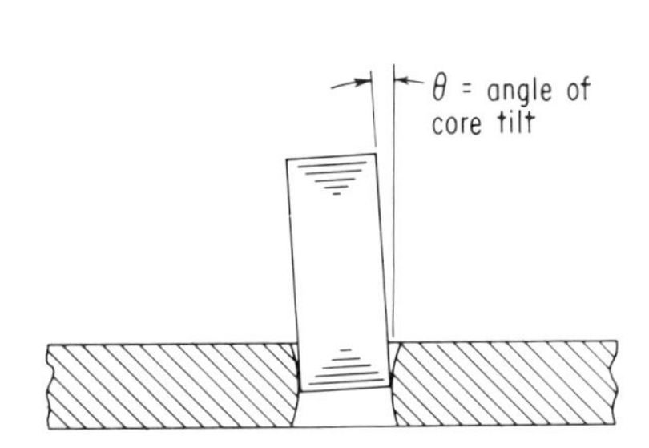

Fig. 27. Tilted core in loading plate.

To make the array, the cores are poured onto the etched plate, the vacuum is applied, and vibration is started. The combination of the vacuum on the underside of the etched plate and the vibration of the table causes the cores to fall edgewise into the etched slots, in their proper orientation. The vibration table is stopped, adhesive-covered tape is applied to the top of the core array, and the vacuum is turned off. The core array sticks to the tape and is lifted out of the etched holes as a complete core field.

As the cores and the core slots become smaller, the ratio of the etched hole size to the plate thickness decreases. This ratio of hole size to depth becomes critical when reduced, because the surface of the hole is etched more severely by the acid than the deeper sections of the hole.

Severe etchback at the surface of a plate may result in holes too large or too small. In addition, it creates nonrectangular holes, which may cause the cores to lean in the slots instead of standing vertically on edge, as in Fig. 27. When the cores are transferred to the tape in this position they are very difficult to string. Or, the cores in one section of the plate may be recessed deeper than in another; they then will not transfer to the tape, leaving a void in the array. In severe cases of nonuniform etching, cores may actually drop through some of the slots, making the plate useless. The problems of nonuniform etching and hole resolution can be solved by using plates made from a series of laminated layers.

Core Stringing. Most core stringing is done by hand. Some automatic equipment is being used, but designs where the number of wires exceeds two per core will probably continue to be strung by hand. The smaller cores of the future will require hand stringing. Most techniques require needles; others, depending on stiffened wire ends in lieu of a needle, exist but are not widely used.

Each smaller-size core development has generated stringing feasibility problems. However, the industry successfully made the transition in each instance with reasonable changes in stringing time. Typically, the stringing time for 0.020-in. cores is about 10 to 20 percent greater than that for 0.030-in. cores, although in some instances it is actually less.[22]

HOLE-SIZE LIMITATIONS. Core stringing is a meticulous task, and as cores get smaller, more refined stringing techniques must be developed. The smaller aperture size is actually further reduced because most of the wires pass through it at an angle.

Figure 28 shows that in such cases the core thickness reduces the available aperture. Decreased core and aperture sizes thus limit the size and number of wires which can be put through the core opening. However, smaller drive lines can not be used because their increased resistance degrades the performance of the faster memories.

Stringing difficulty can be reduced by eliminating one or two of the four wires in a 3D memory system. The two- or three-wire 2½D memory design does not require so many wires; further, it obviates the pattern generation necessary for a 3D diagonal sense winding, which can account for 50 percent of the stringing time in a plane. The decreased stringing time and the advent of low-priced integrated circuits has made the 2½D memory economically attractive. With fewer wires per core allowing smaller, faster cores, the high-speed 2½D memory has become technically more attractive.

Even though the smaller cores are more difficult to assemble, and even though they chip and crack more readily, they do provide a big electrical design advantage. As the physical dimensions of the memory decrease, so do the lengths of the drive and inhibit lines. Because the rise times of the drive and inhibit currents are proportional to the inductance and associated delay on the driven line, the shorter lengths decrease the rise times and allow faster cycle times. One test run on a memory using 0.030-in. cores on 0.030-in. centers showed that about 75 percent of the inductance on a drive

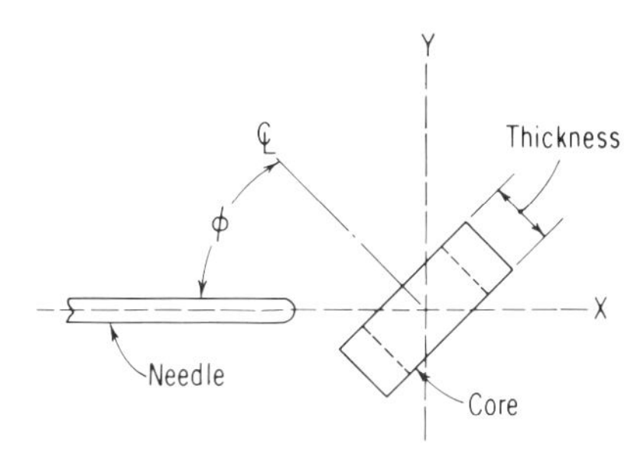

FIG. 28. Relation between stringing needle (or stiffened wire end) and core.

line was caused by the wire and about 25 percent by the cores. With so large a portion of the inductance contributed by the wire, decreases in wire length significantly increase memory speed.

Thermal Considerations.

HEATING. Increased core memory speeds introduce heating problems. When heating of the stack occurs, the increasing temperature raises the resistance of the drive lines and narrows the hysteresis loop of the core, making it less square. The heat comes from the $I_d^2R_d$ product of the drive lines, the $I_i^2R_i$ product of the inhibit lines, and the residual energy losses within cores caused by switching. The heat generated during each cycle by these three sources is proportional to their percentage of the total cycle time.

As the operating speed increases, either the drive currents tend to increase in amplitude, or the ratio of drive current time to total cycle time increases, or both. The result in either case is increased power dissipation. Also, because the density is greater with smaller cores, the packaging design must provide for the removal of greater amounts of heat energy per unit volume.

TEMPERATURE COMPENSATION. One method of solving the memory stack heating problem is to install a temperature-sensing device within the stack and vary the drive currents inversely to the temperature changes within the stack.[22] However, this technique assumes that stack heating is uniform, and imposes the restriction that temperature differences throughout the stack must be kept very low (which can be done with good heat sinking). This technique does not solve the individual core heating problem, which is discussed later.

A second approach is to heat the stack independently of operating conditions and to maintain it at a fixed temperature slightly above the maximum expected operating temperature. The drive currents are then adjusted for the established temperature and are independent of ambient temperature changes within the operational design range.

HEAT REMOVAL. With the development of higher-speed core memories, the problem of individual core heating has become more serious—with no apparent easy solution. Individual core heating occurs when the cores at one address are repeatedly switched at a high rate, and absorb energy from the magnetic fields of the drive lines. A single address then becomes a series of point sources of heat and creates differences of temperature within the stack.

The only means by which the heat can leave the core are (1) radiation, (2) convection of the surrounding fluid, and (3) conduction to an adjacent heat sink.

1. Radiation is insignificant in amount. For radiation to be effective, the difference of absolute temperatures $T^4_{source}-T^4_{sink}$ would have to be so large as to invalidate the basic assumption of small temperature differences within the stack.

2. Convection likewise requires too large a temperature difference between the air and the core being cooled. Further, increasing stack density makes air circulation both difficult and unreliable. Convection cooling with air necessitates either a closed system to prevent clogging the stack with dirt and dust, or else a reliable source of clean air. The former introduces heating within the loop, and the latter is difficult to achieve and maintain.

3. Conductive cooling is one rather obvious technique for readily cooling individual cores, but this is also difficult. Effectiveness depends on good contact between the core and the heat sink. Cores cannot be left to lie on the heat sink, because this method is unreliable, and large temperature differences between core and heat sink would still be possible. Bonding the cores to the heat sink with a thin layer of overspray is feasible, but significant temperature differences can still occur. The overspray must be carefully chosen so that it has a low dielectric constant, remains flexible over the operational temperature range, is inert to the cleaning solutions used during the remainder of the manufacturing operation, does not absorb moisture, has good adhesive action, and has good thermal conductivity. Obviously, these are very rigorous requirements for one product.

In addition to the difficulty of developing a good conductive heat-flow path with a heat sink and overspray, electrical degradation is introduced. Both the heat sink and the overspray increase the capacitance within the stack and slow the operating speed.

Cooling by immersion of the stack in a liquid involves both conduction and convection. This technique cuts down individual core heating, but requires hermetic sealing of the stack, including the connectors. The liquid must also be inert and must not attack the stack materials. Volumetric and pressure changes within the liquid, caused by the varying temperature, must be taken into account. The heat absorbed from the stack by the liquid must also be removed from the liquid in some manner.

Table 15 clearly shows the core heating problem for a 0.5-μsec cycle time memory driven at one address with a drive current of 0.8 amp. The advantages of intimate contact and of coolant are substantial.

TABLE 15. Experimental Data on Core Heating

Core situation	*Core temperature*
In open air	50°C greater than ambient
Lying on heat sink	50°C greater than ambient
Oversprayed on heat sink ...	20–25°C greater than ambient
Immersed in liquid	7–8°C greater than ambient

One other useful characteristic of a liquid is its ability to absorb energy and consequently damp out vibration. Sometimes the composition of a core is such that it rings at very high frequencies, preventing proper electrical functioning. In such instances, immersing the core stack in a liquid will damp out the ringing and at the same time provide the necessary individual core cooling. When using a liquid for cooling, little or no drive-line current compensation is required and only small temperature differences develop within the stack (including the cores). Higher operating temperature can be tolerated.

Core Planes and Stacks.

CORE PLANE ASSEMBLY. Once the core arrays have been established and the appropriate wires strung through them to produce the core mats, the next step is placing the wires in a frame to produce a memory plane. Two types of frames generally used for this purpose are: (1) printed-circuit board and (2) extended-pin frame. The wires going through the core mats are either soldered or welded to pads on the printed-circuit board, or to the extended pins of an extended-pin frame.

Several difficulties may arise which should be considered when designing the printed-circuit board. When a double-sided, copper-clad laminate has most of the copper etched away from one side while the other side remains nearly intact, the board tends to warp. Equal etching on both sides helps prevent warpage.

When solder plating is used as a resist during etching, solder slivers can result from undercutting by the etchant.[23] The heavier the copper laminate, the more severe the solder sliver problem will be.

Good mechanical design of memory planes includes attention to the following to avoid potential problems:

1. Tie-down areas are needed for splicing and repairs which occur during test and production. Without these tie-downs, splices may move around under vibration conditions and cause short circuits or other damage.
2. Heat sinking is needed to dissipate heat.
3. Shock and vibration integrity is needed. In many instances the heat sink can be utilized effectively to help provide shock and vibration resistance. Conformal coating and overspraying provide shock and vibration control and help dissipate heat. However, since all magnetic cores exhibit some degree of magnetostriction, complete encapsulation of a core is undesirable, especially with large operating temperature ranges. Even conformal coating or overspraying can be detrimental to very magnetostrictive cores.
4. Strain relief in some form is desirable for the lines connected to the frame and for the lines coming directly off the frame to the external circuits.
5. The frame must be easy to test electrically.

During fabrication of the frame it is important to guard against accidentally breaking the insulation on the wires going through the cores. An insulation break may not be readily apparent. It can cause high resistance or intermittent short circuits, or both. Either of the two conditions is very difficult to deal with, especially so if the defective frame is assembled into a stack before the defect is detected.

CORE STACK DESIGN AND ASSEMBLY. In a stack of printed-circuit-board frames, the plane-to-plane interconnections are usually accomplished by soldering wire risers on pads of adjacent frames. With the extended-pin type of frame the extended pins are either soldered or welded together. The extended-pin stack requires three interfaces in series—wire-to-tab, tab-to-tab, and tab-to-wire—to connect the drive line on adjacent frames. A printed-circuit memory frame requires four interfaces in series—wire-to-pad, pad-to-riser, riser-to-pad, and pad-to-wire—to connect the drive lines of adjacent frames. Thus, theoretically, the extended-pin stack interconnections should be intrinsically more reliable. When frames are interconnected, care must be used to get good solder joints or welds, to avoid dislodging pads with too much heat, to avoid short circuits caused by solder bridging, and to make sure all residual flux is removed after assembly to prevent corrosion at a later time.

Good stack design requires the following considerations:

1. Provision for heat sinking, a temperature-sensing device, and possibly stack heating.
2. Strain relief of X and Y drive terminations as well as the sense and inhibit lines which should already have been strain-relieved in the plane.
3. Provision for mounting the selection matrices on the stack if required.

4. Good electrical connection with the memory-driving circuitry.
5. Mounting hardware which is compatible with the mechanical interface to prevent galvanic corrosion.
6. Good repairability.

Planar Thin Magnetic Film Memories. Thin-film memories used in high-speed data processing computers extend the speed capability of magnetic memories to the 150- to 500-nanosec range. Core memories, on the other hand, are slower, with a minimum cycle time of about 375 nanosec. The magnetic characteristics of thin-film elements allow much faster switching time than can be achieved with cores, resulting in shorter cycle times. A typical film memory uses orthogonal drive lines and is word-organized.

The basic types of film memories are NDRO and DRO. From the packaging standpoint, the NDRO and DRO film memories are quite similar, with only differences in the magnetic alloy and in the sensing and driving electronics; hence, no further differentiation will be made between the two types.

Time delay in the circuit conductors is an important factor in high-speed memory design and affects cost, reliability, and performance. Typical delays run from 1 to 5 nanosec/ft of conductor length and directly affect memory speed. Large film memories may have conductors as long as 10 ft. However, it is very important to reduce delays by minimizing conductor lengths and packaging the associated memory circuits as close to the memory element as feasible.

Thin-film memory planes commonly have a matrix board attached to the magnetic element section, to contain a portion of the word-selection and drive electronics. The magnetic element section contains a film core array, an overlay, and a backup board. Figure 29 illustrates the relationship of these parts.

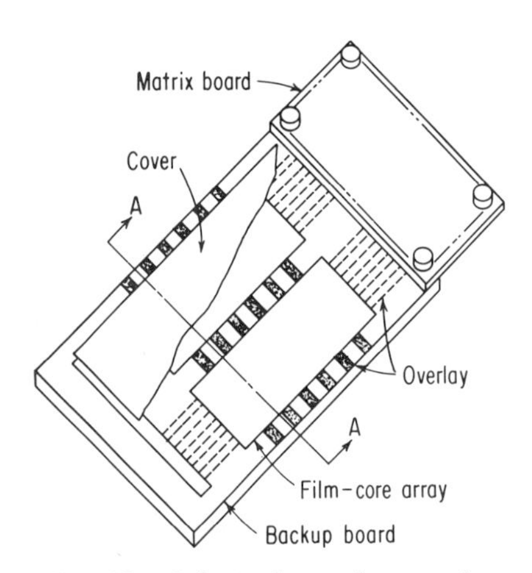

FIG. 29. Relationships of parts of a thin-film memory plane.

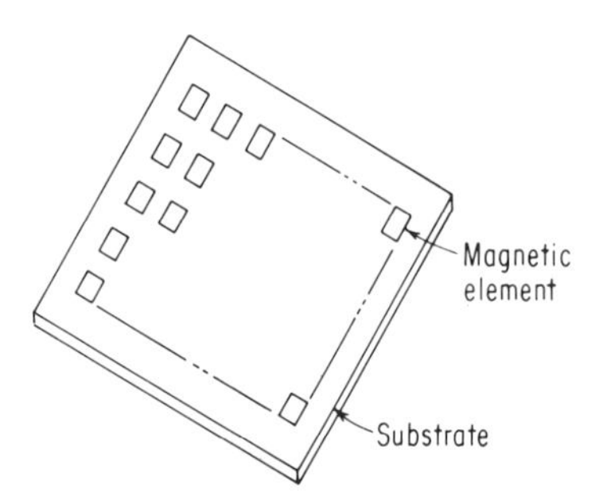

FIG. 30. Thin-film core array of magnetic elements.

Plane sizes vary greatly, depending on many variables such as computer and programming requirements, memory storage capacities, speed, volume, and cost. The number of word lines per plane varies, usually in multiples of 64. The word length, or number of bits per word line, is also variable, usually in increments of 12 or 25. For example, a typical small 3- × 4-in. plane might contain 64 words of 12 bits; a typical large 20- × 20-in. plane might contain as many as 256 words of 72 bits.

Packaging Considerations.

STABLE STRUCTURE. It is necessary to maintain the geometrical relationship of the film core array, overlay, and ground planes to a spacing tolerance of 0.001 to 0.002 in. for the specified environment.

REGISTRATION. The alignment of the film core array to the overlay is most critical and must be held to a tolerance in the range of 0.001 to 0.005 in. Also necessary is close control of the location of the connection tabs.

MAINTENANCE. Accessibility and repairability of the electronics and magnetic element section must be considered early in the design stages.

RELIABILITY. Proven printed-circuit and connection techniques should be used throughout the design.

QUALITY CONTROL. Good assembly processes and close inspection are necessary to assure an acceptable product. Since even dust particles can cause defects, cleanliness is one of the major problems.

Plane Makeup. The four major parts of the plane are described below: the film core array, the overlay, the backup board, and the matrix board.

FILM CORE ARRAY. Figure 30 shows a typical film core array of magnetic elements. Each magnetic memory element stores one bit of information. The magnetic elements, composed of an iron-nickel alloy, are vacuum deposited on the substrate. A typical element size is 0.030 × 0.040 in., approximately 1,000 Å (4 × 10^{-6} in.) thick. The element configuration and thickness can be varied to optimize the magnetic characteristics. Common element configurations are square, rectangular, round, and oblong. The most popular substrate material is glass; copper and aluminum substrates also are used. The substrate thickness can vary from 0.002 in. (glass) up to 0.125 in. (metal plate). Surface roughness must be in the 5- to 10-μin. range. A smooth surface reduces stress in the elements; stress causes skewed signals.

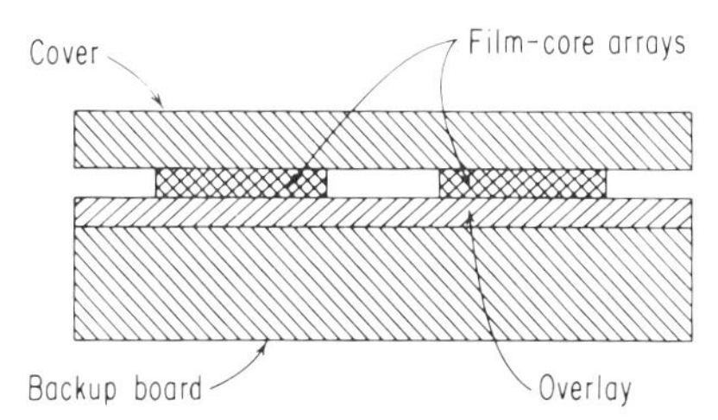

FIG. 31. Cross section of thin-film plane.

The difficult problem of handling thin glass limits the size of glass substrates to maxima of about 4 × 4 in. Metal substrates can be much larger. Element density is limited by the element size and configuration as well as by the disturbing effect of adjacent rows. A typical 3- × 3-in. substrate might contain 64 × 32 element rows.

OVERLAY. The overlay laminate consists of an insulator sandwiched between two copper layers. The drive lines or conductors are etched on each side.

There are basically two types of drive-line configurations: common sense-digit and separate sense-digit. Some designs incorporate a third layer of copper and locate the sense and digit lines on separate layers. The drive lines intersect at each element location when assembled in a film plane.

The copper thickness ranges from 0.0005 to 0.0028 in., depending on many parameters such as current capacity, impedance, time delay, conductor width, and overlay thickness. Either rolled or electrodeposited copper is used in the overlay laminate. The insulator usually is a 0.0005- to 0.002-in.-thick polyester film.

Artwork for the drive lines is cut on a coordinate plotter at high scale and reduced for printing and etching. Extremely accurate artwork is necessary to achieve good alignment between drive line and magnetic element. A 12- × 12-in. overlay is about the largest that can be readily etched and bonded to the plane assembly. Yields, difficulty in handling, and alignment problems are the limiting factors in the overlay size.

High-speed film memories require close control over the insulation or dielectric thickness. A ±0.0005-in. tolerance is common.

BACKUP BOARD. In the plane assembly the backup board is usually the main supporting member. Figure 31 is a cross section of the plane illustrated in Fig. 29, and shows the overlay bonded to the backup board. Some designs use a removable cover so that defective substrates can be replaced.

The thickness of the backup board varies from 0.031 to 0.187 in., depending on the plane size. The board's main function is to provide a flat, rigid base for the overlay and film core arrays. If the planes are stacked tightly together or potted in the stack assembly, thinner backup boards can be used. The main essential is that the film core arrays remain flat, so that the stress in the magnetic elements is minimized.

Epoxy-glass laminate is the most common backup board material, but other plastic laminates are used. Electrical, mechanical, chemical, and thermal characteristics must be considered when choosing a backup board material for a particular design.

In many film plane designs the backup board and cover are copper-clad to provide a ground shield on one or both sides of the substrate and overlay. The ground plane reduces electrical noise and provides the return path for the word, sense, and digit lines. This type of arrangement provides a transmission effect for the drive lines and results in a uniform and constant impedance in the substrate area.

MATRIX BOARD. Most film planes have some type of matrix or access board which contains a portion of the word-selection and driver circuitry. The matrix board is a printed-circuit board that attaches to the backup board as shown in Fig. 29, or is merely an extension of the backup board. This reduces the length of the drive conductors and minimizes electrical noise and connection problems.

The size of the matrix boards varies greatly, depending on the break-off point in the driver circuits. It is not uncommon to use multilayer matrix boards having as many as five copper layers for complex high-speed memories. Copper-clad epoxy glass is generally used for the matrix board material. Ground planes are used to maintain a constant impedance for the word-driver circuits. The driver circuits on the matrix board are attached to the word lines in the overlay by connectors, welded tabs, or soldered tabs, or by extending the overlay to the matrix board.

Plane Configurations. Four common film-plane configurations (coupled, closed-flux, wrap-around, and keeper) are described in the following sections. Most other plane designs are similar to these configurations with minor modifications. The unique packaging concepts for each type are illustrated in Figs. 32 to 35.

COUPLED FILM PLANE. An exploded cross section of a coupled film configuration is illustrated in Fig. 32. A pair of film core arrays sandwich the overlay containing the word and sense-digit drive lines. Ground planes on the backup board and cover are attached to each side of the assembly to complete the coupled film configuration.

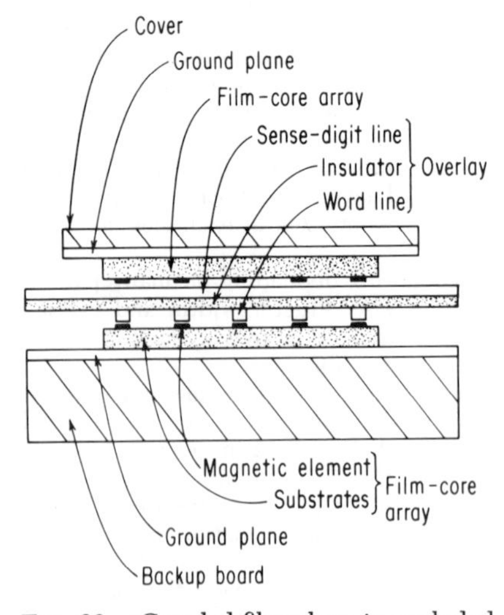

FIG. 32. Coupled film plane in exploded cross section.

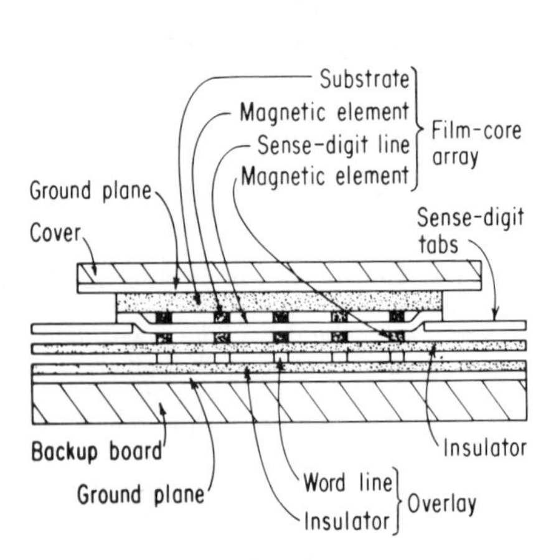

FIG. 33. Closed-flux film plane in cross section.

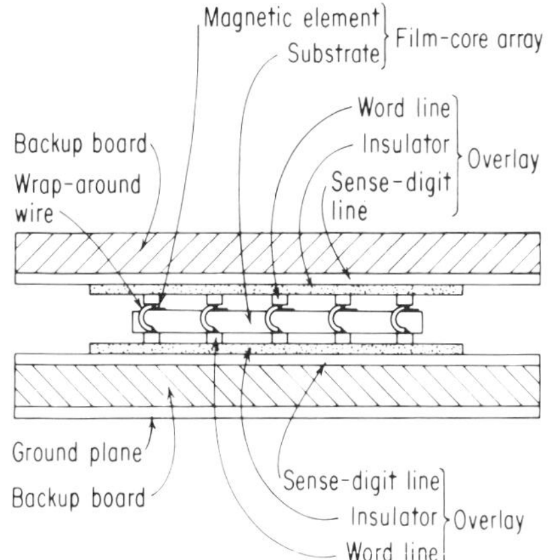

FIG. 34. Wrap-around film plane in cross section.

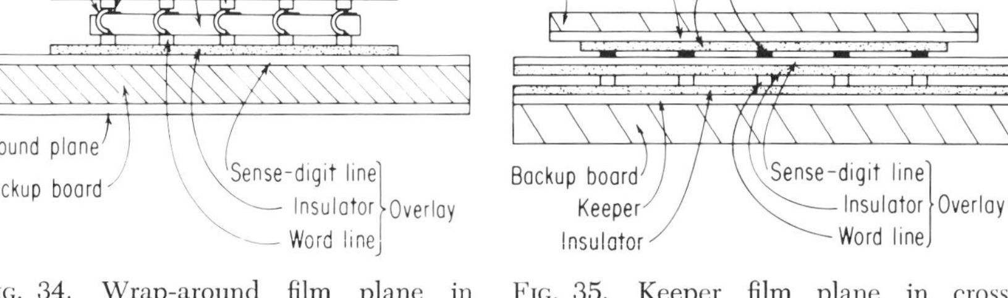

FIG. 35. Keeper film plane in cross section.

CLOSED-FLUX FILM PLANE. The cross section of a closed-flux plane is shown in Fig. 33. It is characterized by a multilayer film core array having successive vacuum-deposited magnetic elements, sense-digit lines, and another layer of magnetic elements. Elaborate deposition equipment and processes are required to produce the complex film core array. However, the closed-flux configuration shows much promise, and cost will decrease as better fabrication techniques are developed.

WRAP-AROUND FILM PLANE. Figure 34 illustrates the cross section for the wrap-around drive-line plane configuration. A word and sense-digit drive-line overlay is bonded to the backup board. Film core arrays are sandwiched between two of these overlay backup board assemblies. Wrap-around wires or cables are then attached at one end of the plane to make continuous word lines. Wires or cables are used to continue the sense-digit lines from plane to plane. Various crossover schemes are utilized at the sense-digit connections to provide noise cancellation. A remote ground plane is sometimes used to provide a circuit return path. The double overlay in the wrap-around plane improves the efficiency of the driving scheme over that of a single overlay. With this construction, it is quite difficult to replace film core arrays, as many connections must be unsoldered to open up the plane.

KEEPER FILM PLANE. A plane type utilizing a keeper is shown in Fig. 35. The keeper is of a high-permeability ferromagnetic material, providing a partially closed flux path around the drive lines for each magnetic element, to reduce the disturbing effect on surrounding elements. The keeper makes increased bit densities possible since with it the disturb threshold is higher than for some of the other plane designs. The keeper also minimizes the effect of ground plane current spreading. The keeper configuration is similar to the coupled-film version shown in Fig. 32. The main difference is that the keeper replaces one of the film core arrays and hence the drive lines are sandwiched between the keeper and film core array.

Thin Magnetic Film Memory Stacks. Usually, two to eight planes are assembled together in one of various possible configurations to form a thin-film memory stack. The planes are attached serially in the sense-digit direction; therefore, the length of the sense-digit line depends on the number and size of planes in the stack and can be as much as 10 ft. Speed, signal, noise, and circuit limitations are factors that must be considered when determining the allowable sense-digit line length. Stack sizes are referred to by the number of words and the word length. These are determined by the plane size and the number of planes in the stack assembly. For example, a 1,024-word × 36-bit stack contains 1,024 word lines each having 36 bits. A 128-word × 12-bit stack is considered small, whereas a 1,024-word × 72-bit stack is quite large.

Because it is relatively easy to tailor the stack size for a specific application, many stack sizes are in existence today.

Stack Packaging. When packaging thin-film stacks, three areas must be considered: configurations, connection techniques, and environmental factors.

STACK CONFIGURATIONS. Basically, the stack configuration is determined by the location of the sense amplifiers, planes, and digit drivers. Figure 36 shows a simplified version of an unfolded stack. Word current is driven down the selected word line. This current, added to the bipolar current driven by the bit driver, causes the magnetic film element to switch in one of two directions. In reading, this induces a current in the sense line; the polarity indicates whether a 1 or a 0 was stored.

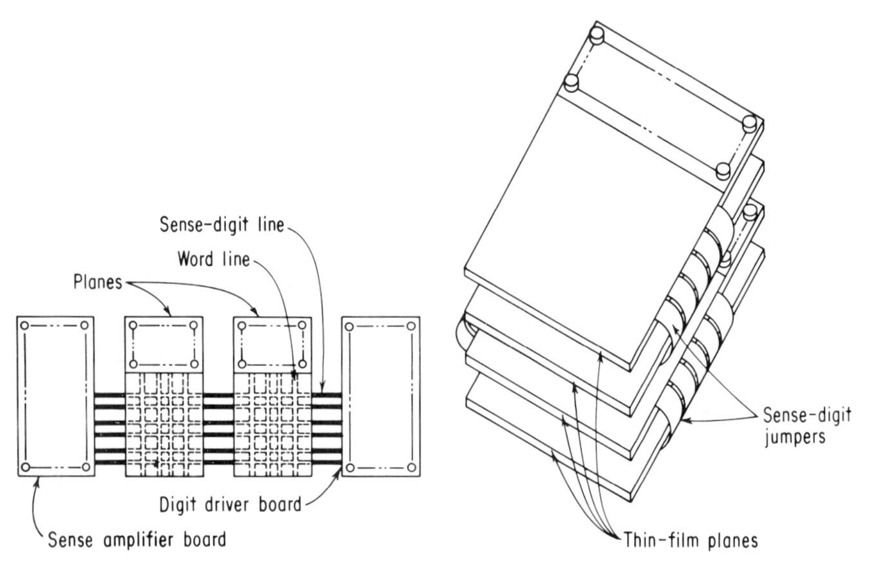

FIG. 36. Unfolded thin-film stack configuration.

FIG. 37. Folded thin-film stack configuration.

The simplest thin-film memory stack configuration contains only planes, as shown in Fig. 37. The sense amplifier and bit driver circuits are packaged on printed-circuit cards remotely located from the stack assembly. These cards are connected to the stack by some type of wire harness and pluggable connectors.

For high-speed film memories, it is desirable to mount the sense amplifiers and digit drivers close to the planes and to minimize discontinuities in the sense-digit line. Such a stack configuration is shown in Fig. 38, where the sense amplifier boards and digit driver boards are integral parts of the stack assembly. The sensed signal is amplified before it is sent to other circuits. Important reasons for close packaging of the bit drivers, sense amplifiers, and other electronics with the planes are to preserve rise times, reduce electrical noise, reduce delay times, and simplify connection problems.

CONNECTION TECHNIQUES. Designing a satisfactory connection technique for high-speed film stacks is one of the most imporant and difficult problems that must be solved. Connections in the sense-digit lines are by far the most difficult and critical. Connections or jumpers are necessary between planes and between planes and driver boards (see Fig. 38). The connection scheme must: be ultrareliable, minimize discontinuities and resistance variation, be repairable, and have reasonable cost. Some drive schemes require a ground plane that is continuous from plane to plane, and that maintains a constant impedance to the sense-digit lines. Frequently, connections must be made on 0.025-in. center lines.

Few connectors on the market can satisfy all these requirements. Consequently, many stack designs incorporate wires or flexible cable from plane to plane. The wires are soldered or welded to the drive lines on the overlay. This gives the best joint possible and satisfies most of the requirements that a connector fails to meet. Twisted pairs of wires are often used to carry the sense-digit lines. They are very reliable and afford an excellent connection; however, they are time-consuming to assemble, are difficult to repair, and have a large discontinuity between the flat conductors in the overlay and the round wires between planes. The flexible flat conductor cable comes closest to satisfying all the requirements. This cable is also soldered or welded to the drive lines on the overlay. A continuous ground plane can be laminated to the flat cable to maintain a constant impedance level and to minimize the discontinuity. Connections on 0.025-in. center lines can be handled quite easily. Difficulty in repairing or removing a plane from the stack assembly is the main shortcoming of the flexible cable method. Figure 38 shows how the sense-digit line weaves through the stack assembly. The wire and flexible cable techniques allow the stack to unfold so that repairs can easily be made to the matrix boards attached on the planes. Of course, when connectors are used to carry the sense-digit lines through the stack, they are

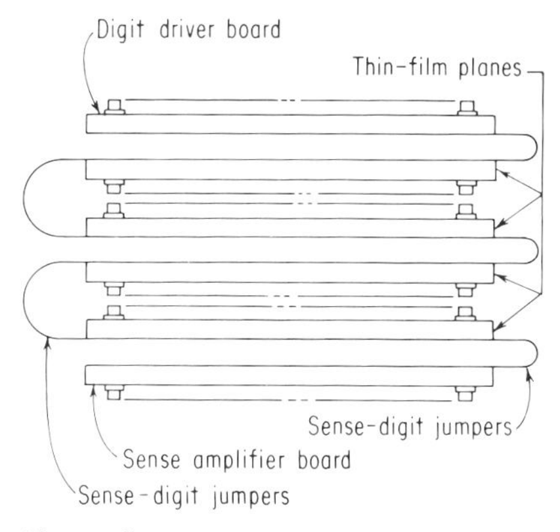

Fig. 38. Folded thin-film stack configuration incorporating sense amplifiers and digit drivers.

readily removable and make plane repair easy. In many designs it is necessary to cross the sense-digit lines to provide noise cancellation. This is done either on the plane or in the connection area.

ENVIRONMENTAL FACTORS. Shock, vibration, humidity, ambient air temperature, and stray magnetic fields are some of the more important environmental factors that should be considered when designing a film stack. An environmental specification is usually written for each application. Table 16 lists typical values for each factor.

TABLE 16. Typical Environmental Factors for Memories

Shock	5g random
Vibration	5–500 Hz 2–5g
Humidity	5–95 percent
Temperature (operating)	0–50°C
Magnetic field	0.8 gauss

Shock and vibration requirements for other than military or space applications are usually prescribed for shipping and handling. A well-designed shipping container will absorb most of the shock and vibration forces.

Cooling the stacks is generally accomplished by forced air or by conduction to cold plates.

In most film applications it is necessary to shield the stack from the earth's magnetic field as well as from fields generated by surrounding equipment. Nickel-iron alloys having a high permeability provide good magnetic shielding. As the material is most effective in the annealed condition, it is good practice to anneal the shield after forming, machining, or welding. The shields can be mounted between planes or to enclose the stack.

Cylindrical Thin-film Memories.

Wire Memories. A wire memory is constructed from conducting wires which are coated with a ferromagnetic material and situated between a pair of conductors orthogonal to them, as shown in Fig. 39.[24, 25, 26] Bits are represented along the wire by circularly magnetized cylindrical sections of the magnetic coating. These sections are not discretely defined but are formed, one under each word line, on each wire. They

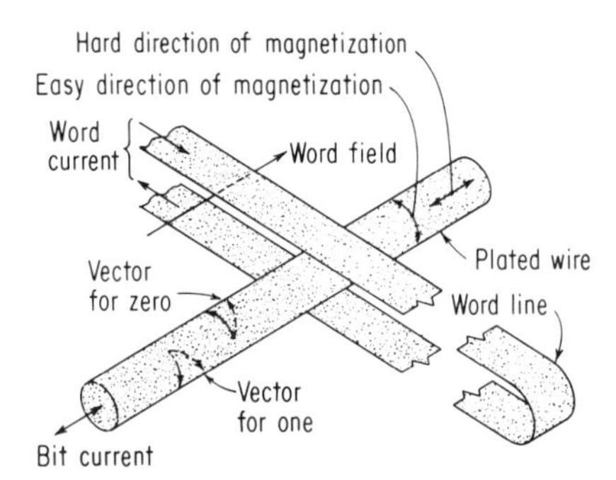

FIG. 39. Magnetization of plated wire in a memory.[26]

are analogous to the toroidal cores strung on driven wires. The plating process simplifies the packaging enormously by eliminating the stringing operation of the drive and sense wires. The magnetic coating itself is anisotropic; that is, it is easier to magnetize in one direction than in the direction perpendicular to the first. The coating is formed with the hard direction of magnetization along the axis of the wire and the easy direction circumferential. The bit current driven along the wire in one direction produces a circular magnetization in the magnetic coating; current driven in the reverse direction produces circular magnetization that is similar but reversed in polarity; thus 1's and 0's are distinguished. Writing is on a coincident-current basis, with the bit current small enough that it will temporarily diminish the circular magnetization, but not reverse it except in the presence of a longitudinal field produced by a word line. To read a word, current in the selected word line temporarily shifts the direction of the circular magnetization under it in each wire, as illustrated by the dashed vectors in Fig. 39. This shift develops, in each wire, a voltage whose polarity is dependent upon the direction of the circular magnetization that is shifted; this voltage is detected on each wire to identify the bit stored at the intersection with the energized word line. Thus the plated wire acts as a sense wire as well as a bit wire. When the word current is removed, the shifted circular magnetization returns to a circumferential course around the easy axis. Thus no restore current is required, and the device behaves as an NDRO memory.

The construction of a typical plated-wire memory can be seen in Fig. 40. The word lines are copper straps etched on an insulating substrate of epoxy glass. The wires are 36-gauge beryllium copper, plated with a thin coating of nickel-iron alloy. The tubular tunnels to contain the plated wires are formed by initially laminating the assembly around oversize pilot wires, which are then withdrawn and replaced by the plated wires. The plated wires are placed in alternate tunnels for even spacing, with the intervening tunnels empty except for one in each group associated with the sense amplifier. That one tunnel contains a noise-cancelling conductor connected to the amplifier. Word straps are typically on 0.06-in. centers, providing storage of more than 16 bits/in. of wire. Denser packing reduces signal output. Stronger output signals can be obtained from thicker platings. The thickness of the plated film is an order of magnitude greater than that of planar films but is still kept small enough to behave magnetically as a thin film. The spacing of the word straps is the principal factor controlling the physical size of the memory element.

In another construction of plated-wire memory, the word lines are insulated wires passing alternately over and under successively adjacent plated conductors, as shown in Fig. 41. The return portion of a word line passes likewise on the opposite sides of the plated wires to define a bit position. Spacer wires of similar construction are

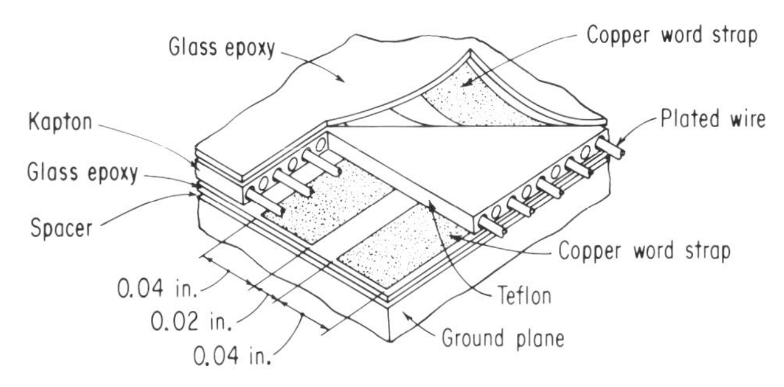

FIG. 40. Plated-wire memory construction.[26]

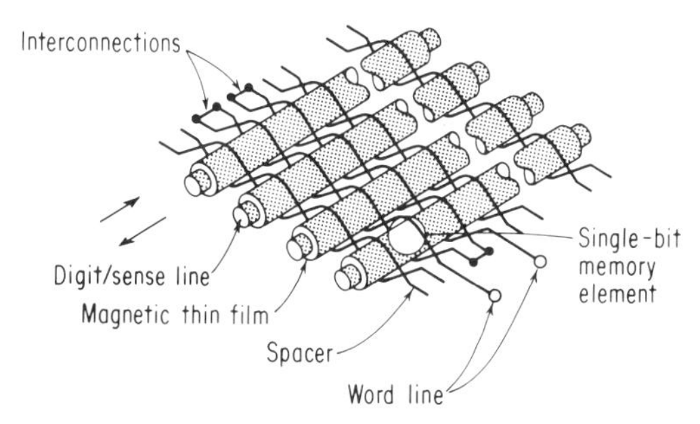

FIG. 41. Plated-wire memory matrix. (*Courtesy of General Precision Systems Inc., Librascope Group.*)

placed between adjacent word lines to maintain separation. Selected spacer wires are connected for noise-cancellation purposes. One design achieves extreme compactness by mounting plated-wire mats in pairs, one on each side of a printed-circuit board. An insulating spacer strip mounts and separates the bit wires as shown in Fig. 42 and supports the printed-circuit board in a raised channel. The printed-circuit board also holds the diodes which terminate the word lines.

Advantages of plated-wire memories, in addition to their compactness, include: easy cooling because of their low power dissipation; speed of switching due to the intimate coupling and the small energy involved at the bit levels, thus making large banks of fast memory feasible; and easy fabrication resulting from simple organization and the fact that strict orthogonality of the bit and word lines is not critical. Some plated-wire memories are manufactured in the United States and in Japan by weaving processes discussed later in this chapter.

The wires are plated in a continuous process; after early preparation stages, the nickel-iron alloy is deposited while a direct current flows along the wire as it passes through the plating bath. In the final stage of the process, the wire is automatically tested and then cut into segments of the proper length. Defective portions are automatically cut out under control of the tester, and then mechanically segregated since they are shorter in length than complete segments.

ROD MEMORY. The rod memory [27, 28] is a form of wire memory in which the easy

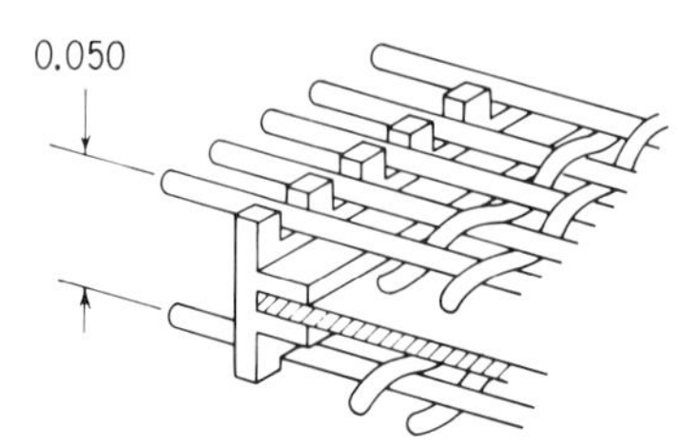

FIG. 42. Detail of matrix construction. (*Courtesy of General Precision Systems Inc., Librascope Group.*)

magnetization originally lay along the axis of the wire; a prewound single-layer solenoid surrounding each wire provides the axial magnetization. The plated wires, called rods in this memory, carry the bit currents and also serve as sense lines as in other forms of wire memory. The rods are beryllium copper wire; the coating is a nickel-iron alloy, about half as thick as that in the plated-wire memories but still four or five times as thick as planar thin films.

The low capacity occasioned by solenoids connected to a single word driver, and the mutual capacity which couples nearby rods, are both minimized primarily by circuit design rather than by packaging means. The preformed solenoids are potted within a molded frame to form a plane,[28] and the rods are inserted after the planes are stacked.

In an early form of this type of memory, the magnetic material was deposited on a silvered-glass rod which served as the conductive substrate.[29] Continuous plating processes and other advantages of a wire substrate have outmoded the glass rod, but the name persists. These other advantages include an easy axis of magnetization that is circumferential and a thicker magnetic coating; [30] the rod memory in these respects no longer differs from other forms of wire memories. In construction, long leads from the stack to associated circuits are avoided by mounting two-layer circuit cards vertically in a cluster around the stack on three sides, with one card edge toward the stack.

Woven Memories. Various approaches utilizing weaving techniques have been applied to the fabrication of digital memories. The advantage expected in all of these approaches is the use of well-established loom techniques to develop the configuration of a memory array. In one type of configuration a pattern of conductors is woven leaving apertures which receive ferromagnetic elements. In other arrangements some of the wires of the mesh are plated, either before or after weaving, with a ferromagnetic coating to provide the retentivity.

Adapting existing looms to the fabrication of memories introduces several problems.[31] There is a significant problem in weaving to dimensional tolerances with wires of different physical characteristics. This problem is compounded when it is necessary to protect insulating material on some of the wires. Extreme care must be maintained in weaving to minimize the damage to the surface of bare wires and to the insulation of others. When plated wires are being woven, mechanical strains imparted by the loom may result in magnetostrictive effects.[32] Despite these difficulties, memories are being successfully woven on modified Jacquard looms.

WOVEN MATS. In fabricating a memory of preplated wires, the plated wires are woven as the woof and the insulating conductors as the warp.[33] The woof (weft) runs across the output of the loom while the warp runs from the loom in a continuous length. Figure 43 shows an operating loom producing plated memory mats, two of which are visible in the foreground. On the frames in the background are the harnesses which lift selected conductors in the warp at appropriate times for the plated wires of the woof to be inserted beneath them; the conductors not lifted pass under the plated wire.[33] The mats are cut apart and mounted on glass-epoxy printed-circuit

FIG. 43. Weaving plated memory matrices on a loom. (*Courtesy of General Precision Systems Inc., Librascope Group.*)

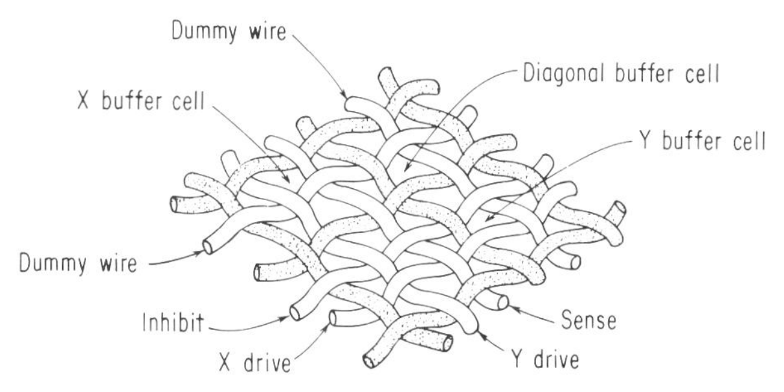

FIG. 44. Memory cell and buffer cells in a four-wire single-turn woven configuration.[31]

boards, where the wires are properly interconnected to form a plane. It is also feasible to form a memory stack by interconnecting the plated wires to form digit lines and folding the mats back and forth over each other across the warp conductors.

Another form of woven wire memory has separate sense and inhibit lines. Figure 44 shows the segments of wires which make up a single memory cell in this memory. The shaded wires are insulated, while the others are bare and are plated with a magnetic material after weaving.[31, 34] Afterplating obviates degradation of the magnetic material by mechanical strains during weaving and avoids the difficulties of bad spots in a continuous plating process. In Fig. 44, the active memory cell is at the lower center and three buffer cells are shown beyond it. Thus as this portion is replicated in the array, each active memory cell is surrounded by buffer cells which isolate it. The simple over-and-under weaving pattern of this arrangement is particularly adapted to fabrication on a wire loom and therefore not subject to some of the difficulties encountered in weaving wire on cloth looms.[31]

BRAID.[35] The braid memory is so named after the configuration of its loom-fabricated drive wire. The loom forms the word lines into a ladderlike array with square apertures, into each of which is inserted one leg of a U-shaped core as shown in Fig. 45. If a wire is in one rail of the ladder, it is coupled with the core; if in the other rail, it bypasses the core. The rungs of the ladder are formed by wires crossing between rails as they pass through or around successive cores. The configuration of the braid, once woven, fixes the stored pattern, making it thus a RO memory.

Separate sense wirings are wound on each core, as shown in Fig. 45. A sense winding is inhibited by short circuiting it with a transistor. The number of individual word lines forming the braid is an integral power of two. These lines are individually driven at one end and connected in groups to voltage or ground at the other. Ground return lines which bypass all cores are woven into the braid.

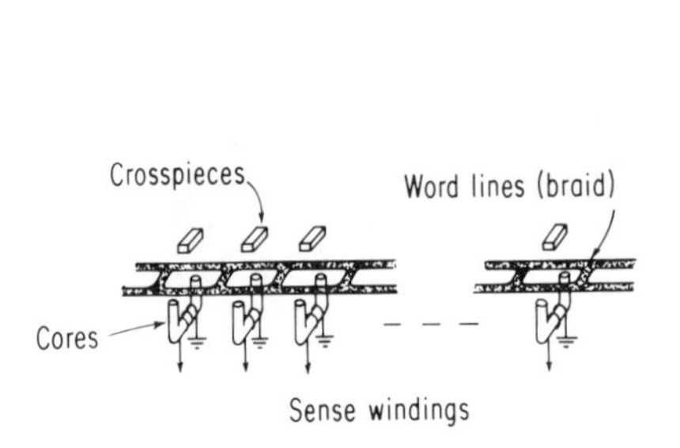

FIG. 45. Braid memory construction.[35]

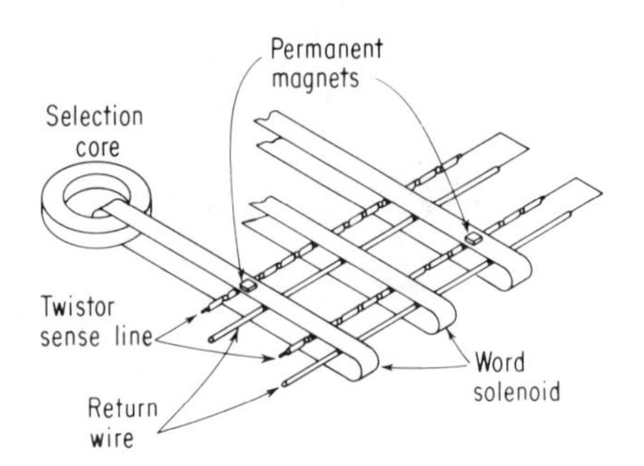

FIG. 46. Twistor memory configuration with permanent magnets.[37]

Voltage drop along a word line in the braid is kept low by making the area enclosed by the word line as small as possible, thereby reducing the self-inductance. The length of the braid is kept to a minimum for the required number of cores, since cycle time increases with braid length; it also increases with the number of word lines.

Various Ferromagnetic Memories.

Arrays of Ferromagnetic Elements. Although ferrite cores and thin magnetic films constitute most ferromagnetic memories, several other types have come into use for specific purposes. These types find their principal use in applications requiring RO memories whose contents are fixed permanently or are alterable only mechanically. The means for altering the memories introduce some novel packaging problems.

TWISTOR. The magnetic storage element in the twistor is a Permalloy tape spirally wound in a single layer on a conductor which forms the sense and bit line.[36, 37] The word lines are single-turn solenoids, usually individual straps of copper formed on a mounting board. Construction, illustrated in Fig. 46, resembles that of the plated-wire memory.

The twistor memory exists in several versions. In the one illustrated, which was developed to provide storage in an electronic telephone switching system, each word solenoid is a closed loop coupling a toroidal core through which pass X and Y selection wires and a bias wire.[38] In this version, small permanent magnets are located at certain intersections of bit word lines to inhibit switching at those points. By mounting these magnets in different patterns on removable cards, a mechanically alterable fixed memory can be constructed. In practice, small spots of a vanadium-iron-cobalt alloy are placed on an insertable aluminum card at every point which coincides with an intersection, and the spots are selectively magnetized to form the storage pattern. With this alloy, it is possible to remagnetize a card into a different pattern by inserting it in a suitable fixture. In some versions of the twistor, the sense lines are laid on top of the solenoid strap,[37] simplifying the construction when using etched or deposited word straps.

In assembling the twistor memory, mechanical problems are critical. The intersections must be precisely located, in order to register with the insertable magnet cards. However, there are advantages. A number of planes can be constructed along a continuous set of sense-digit lines and then folded across them to form a stack. The digit line and its associated return line are connected together at one end of such an array, and a sense amplifier is connected to the pair at the other end. The stack construction and the detail design are quite similar to those of the plated wire memory. In fact, the use of plating for applying the magnetic coating and of weaving to form the planes were both suggested with the original introduction of the twistor.[39]

FERRITE RODS. The ferrite rod memory is unlike the thin-film rod memory discussed earlier and must not be confused with it. In one form of ferrite rod memory [37] used in Great Britain, word and bit lines in an orthogonal array loop around one end of a plastic tube at each intersection and a like array loops the other ends of the tubes. A ferrite rod (half as long) is placed at one end of each plastic tube to establish a positive coupling at that end and, at the same time, zero coupling in the corresponding intersection at the other end. Since the flux paths close through air, a magnetic keeper is necessary at the center of every square, defined by four neighboring intersections, in order to prevent cross coupling.

Another kind of ferrite rod memory uses long ferrite rods (each wound with a sense winding) which go through a stack of single-sided printed-circuit cards. Each rod represents a bit, and each circuit card carries a word line which either encircles or bypasses a bit rod, as shown in Fig. 47. Construction is simple, since the circuit cards can all be made initially alike by printing word line paths which pass on both sides of each solenoid hole, and then severing the unwanted path at each position, as shown in Fig. 48. This ferrite rod memory is inexpensive but only moderately compact.

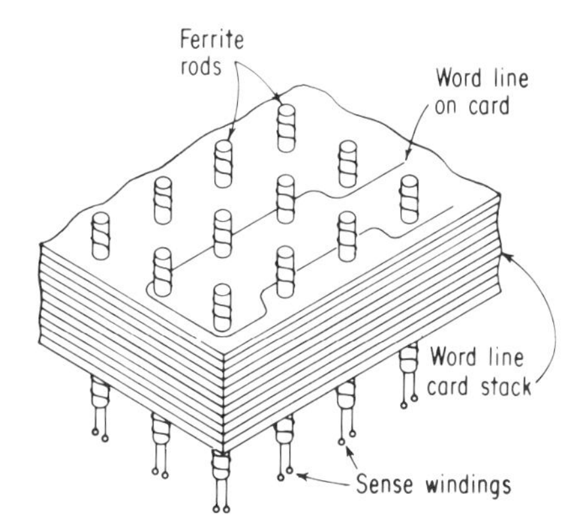

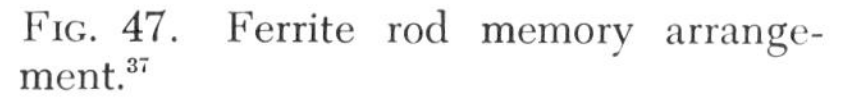
FIG. 47. Ferrite rod memory arrangement.[37]

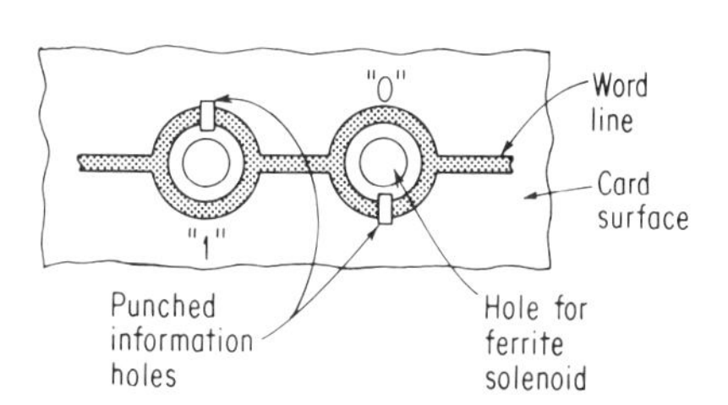

FIG. 48. Method of severing conductors to create permanent pattern in RO rod memory.[37]

TRANSFORMERS. The principle of this kind of memory is the coupling of drive and sense circuits through selected closed cores; this leads to large output voltages which may be sufficient to obviate the need for sense amplifiers.[37] This kind of memory is best suited for RO fixed-memory applications.

The topological arrangement of the windings and closed cores to obtain the desired information patterns poses a packaging problem. In one approach [40] the drive windings are formed on pairs of paper bobbins, in which the two U-shaped halves of the core fit, held together with a slight air gap. Before the cores are closed, a stack of as many as seven circuit cards is placed over a group of wound bobbins. Each circuit card has an array of holes, two for each core, around which formed wiring passes to provide the desired circuit configuration. The two legs of each core pass through a pair of apertures; thus the core is either linked or bypassed according to the configuration of the formed wiring. These wiring cards can be changed by removing half of each U-shaped core, making this type of memory mechanically alterable. Formed wiring on phenolic boards and etched copper wiring on Mylar* sheets have been used to produce the formed configurations. Windings on selected bobbins are driven for word selection and bit patterns are sensed from the wiring on the cards.

The *rope memory* is a transformer memory fabricated by passing read lines through or around closed cores.[41] Bundling the wires together as they pass through or around the cores of the array gives rise to the name. Forming the rope can be automated. The memory operates by means of a set line linking all the cores; it is energized simultaneously with inhibit lines of opposite polarity which link only the unselected cores. The contents of a rope memory cannot easily be changed, even by mechanical means.

Multiaperture Core Devices. A number of designs have been developed to simplify the packaging of magnetic memories or to provide wanted features in their operation by means of a more complicated geometry of the magnetic element. The packaging trade-off usually sought is primarily simplified wiring. These designs fall into two classes: one with two openings in a core, and the other with a multiplicity of openings in a larger magnetic element.

E CORE.[42] The magnetic element of an E-core memory consists of two E-shaped cores placed together to form a figure-eight magnetic element with two windows. The sense winding is on the common leg; word lines for 0 pass through one window and for 1 through the other window. This makes possible simple, straight, word lines, and also the assembling of the two halves of the core into a prewound sense winding. The word lines are formed on printed-circuit cards with holes to receive the center legs of the cores; these cards are assembled over the one half of an E core before the other half is put into place. Running two word lines of one polarity through one set of windows and two other word lines of opposite polarity through the other set of windows with a pair of return lines outside the core, and selecting these wires in pairs, reduces the number of lines required to store information.[42] The reduction is obtained by interpreting the logical combination of the selected pair of wires. The E core is a fixed memory with RO capability.

TRANSFLUXOR. Figure 49 shows the Transfluxor,† a ferrite core with two apertures, one larger than the other. The control winding causes a 1 or a 0 to be stored in the device by fully magnetizing the Transfluxor with a saturating pulse in one direction as shown (arbitrarily defined as the 0 state).

Once the magnetic state of the Transfluxor has been established, it can be interrogated via the input winding and the sense wire without changing its established binary state. Since current in the input winding disturbs the flux pattern only in the neighborhood of the small hole, the state of the core itself remains unchanged during reading.

To utilize the Transfluxor in a memory array, the control winding function can be performed by two coincident drive lines and an inhibit line just as in a conventional 3D organization. The input winding function can also be performed by two coincident drive lines passing through the Transfluxor between legs 2 and 3 (Fig. 49). A total of six wires is required to use the Transfluxor in the coincident current mode of operation.[43] Packaging such an array is beset with numerous difficulties.[44]

BIAX. The Biax ‡ element is another type of multiaperture core utilized for the fabrication of NDRO memories. This element is a block of ferrite material with two nonintersecting orthogonal holes, as shown in Fig. 50. The Biax element depends upon

* Trademark, E. I. du Pont de Nemours & Co.
† Trademark, Radio Corporation of America.
‡ Trademark, Aeronutronic Division, Ford Motor Company.

the interaction of flux in the common magnetic material between the storage and interrogate holes to achieve NDRO operation.

Biax elements, used in a word-select organization, depend upon signal polarity instead of signal amplitude to differentiate between 0's and 1's. For a read operation, only the word-oriented interrogation wire needs be driven. The outputs of the bits of each word are then sensed individually via the sense wire of each bit. To write, a word-write line and a bit-write line are required for random access. With this configuration 1's and 0's can be entered independently into the bits of a word.[45] Although the interrogate line always passes through the interrogate hole and the sense line through the storage hole, the word-write and bit-write lines may be interchanged between these two holes for different organization of Biax memories.[46, 47]

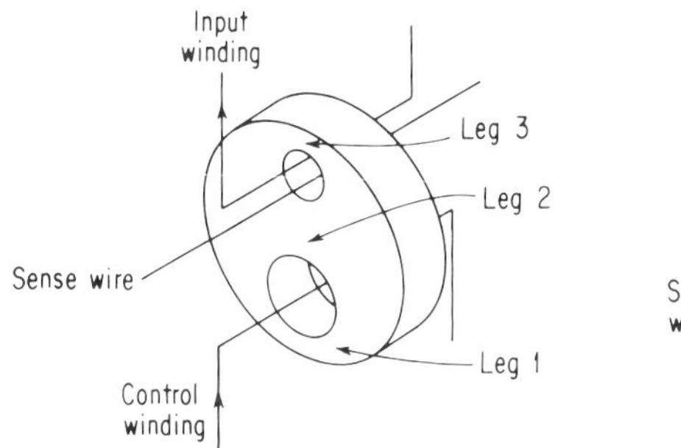

FIG. 49. Magnetic Transfluxor core device.

FIG. 50. Biax core element.

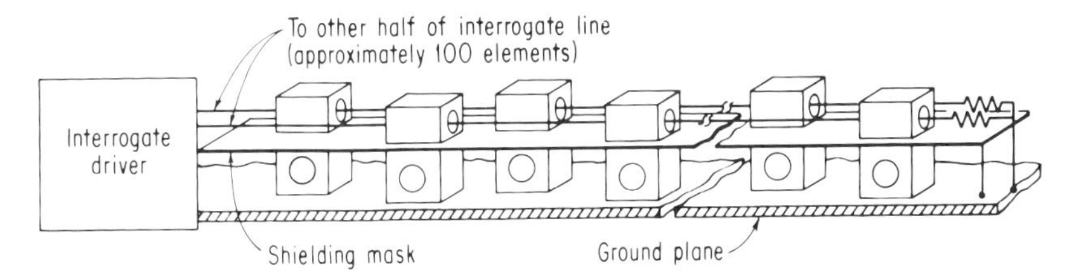

FIG. 51. Scheme for mounting Biax elements.[47]

For high-speed Biax memories, every signal path involved in the read operation, including interconnections, must be considered in terms of its transmission line characteristics.[48] A shield between the sense and interrogate lines, as shown in Fig. 51, is essential in high-speed applications to reduce the transmission line impedance.[47] This impedance, which is inversely proportional to the square root of the capacitance, can be further reduced by bringing the interrogate wire closer to the ground plane. As illustrated in Fig. 52, this can be accomplished by etching depressions in the copper ground plane to receive the Biax elements.

FERRITE PLATES AND DISKS. The multiapertured plate[49] is a sheet of ferrite with a large number of holes in rectangular array, developed to take the place of an array of strung ferrite cores. If the holes are spaced no closer together than their diameter, the area around each hole behaves magnetically as an independent toroid. The conductors through corresponding holes in a stack of plates are then easy to wire and can be connected in the coincident-current mode to form a memory.[50] Since the ferrite itself is an insulating material, windings in the plane of the sheet can be formed directly on it by plating and etching or by deposition.

A multiaperture ferrite disk has been devised as the basis of a RO memory.[37] In it, selected disks are driven by a current through a hole at the center, causing the magnetic disturbance to propagate outward toward the edge of the disk in expanding concentric circles. Lines of apertures in a spokelike configuration at regular angular intervals are threaded with sense windings. The configuration of each sense winding along a radius determines the pattern serially read out by each as a magnetic disturbance proceeds outward.

Nonmagnetic Memories.

Cryoelectric Memory. The cryoelectric memory, still in experimental status, retains each bit of information by means of a persistent current circulating in an element made superconductive by being maintained at a temperature very close to absolute zero. This superconductivity can be destroyed by a magnetic field; the cryotron, which is the computer element, is based on this principle.[51] Current in a superconductor has been shown to persist without diminution over a period of years.

The cryotrons are immersed in liquid helium to maintain the low temperature; thus a memory formed of them is volatile in the sense that loss of power causes loss of cooling and destroys the superconductivity. Because of the costly equipment required

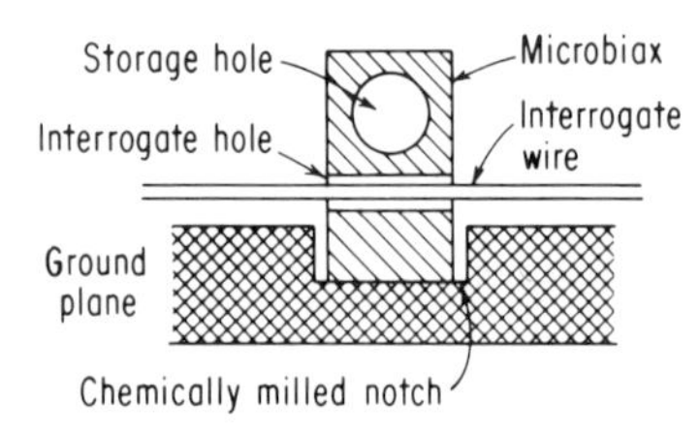

FIG. 52. Detail of mounted Biax elements.[47]

to maintain helium liquid, memories of this type are economically feasible only in very large arrays to bring the cost per bit within reason.

The packaging considerations involve problems caused by conduction of heat into the helium via the electrical conductors and heating of the helium by the resistive loses of the elements within it.[52] In calculating this heating effect, the low-temperature resistance of the conductors must be ascertained and used. A coincident-current cryoelectric memory of 16,000-bit capacity has been constructed on a 2-in.-sq substrate by vacuum deposition techniques.[53] Both 2D and 3D organizations are suitable for memories of this type.[52]

Impedance-coupled Matrices. Several kinds of RO memories utilize the principle of coupling orthogonal word and bit lines at selected intersections through an impedance. The pattern of coupled intersections determines the stored contents of the memory. Some types are mechanically alterable by inserted cards, but registration is critical. Using batch fabrication techniques, impedance-coupled fixed-memory arrays are made with inductive, capacitive, resistive, and diode coupling elements.

INDUCTIVE COUPLING. Coupling is provided by similarly shaped partial loops in the word and sense lines, in close proximity at each intersection. Proximity is achieved by placing the word and sense lines on opposite sides of a thin insulating sheet. For practicality in fabrication, a complete loop can be formed in the word line at each intersection and then severed on one side or the other to register a stored 1 or stored 0.[37] Mechanically alterable fixed memories can be made by fabricating the word and sense lines of an inductive memory on separate boards and inserting a metal plate between them with holes slightly larger than the inductive loops at selected intersections.[54] The plate acts as a shield and prevents coupling except at intersections with holes. Another type of inductive memory uses cards with closed loops at selected intersections. Eddy currents induced in the loops couple those intersections.[37] As before, cards may be fabricated with loops printed at all intersections and continuity of the loop broken at intersections where coupling is not wanted.

CAPACITIVE COUPLING. Capacitive arrays depend upon sufficient overlapped area of the word and bit lines at their intersections to form a suitable capacitor when the cards containing the two types of lines are mounted in close proximity.[37, 55] Placed between the cards is a grounded shield of thin foil with holes at the selected intersections to determine the pattern of the memory.

RESISTIVE COUPLING. In the resistive memory,[56] an array of film resistors is silk-screened onto a plastic card. All the resistors in a row are connected together at one end by a printed word line. The other end of each resistor terminates in a conductor printed around the periphery of a through hole. When the cards are stacked, each registered set of through holes is interconnected by injecting low-temperature solder to form a bit line. Wherever the word and bit lines are not to be coupled, the lead from the resistor to the through hole is severed before assembly.

DIODE COUPLING. Assembling a pattern of diodes to interconnect selected intersections of an array appears to be economical only if integrated arrays of diodes can be fabricated.[57] The attractive feature of this type of array is the ease of fabricating the encoders and decoders on the same substrate as the memory.[37]

*Integrated-circuit Memories.** Computers use integrated-circuit memories for temporary storage: registers, high-speed buffers, and scratch-pad memories.[18, 58] This type of storage generally uses arrays of interconnected flip-flops, rather than magnetic elements, to provide the memory function. Since such a memory is essentially an array of interconnected logic circuits, its design can be more closely compared to that of a central processor than to other present-day memory systems.

Semiconductor memories have several important advantages related to packaging. Their regular matrix geometry makes layout easy on printed-circuit boards and in integrated-circuit form. They do not require high-voltage, high-current drivers and sense amplifiers capable of detecting millivolt signals as do magnetic memories; in integrated-circuit memories both driving and sensing take place with logic-level voltages. These memories can be made physically quite small by use of packaging techniques for semiconductor chips. Integrated-circuit memory arrays are fabricated in batches, with a large number of circuits built at one time on a common substrate of silicon; with high yields this can be done at low cost.

Integrated-circuit memories are not without disadvantages. Stored information is lost in case of a power interruption, and this volatility may be their primary disadvantage. The problem can be overcome in low power memories using MOS devices, where it may be practical to supply a battery for auxiliary power. Large semiconductor memories, however, may require a great deal of power, since their power requirements are almost directly proportional to memory size, as in all memories dissipating power continually in the storage elements. The generally low voltages associated with integrated-circuit memories can result in very high current requirements which create problems in the bussing needed to distribute the power.

Two basic processes are used to build semiconductor memories, using either bipolar transistors or MOS transistors. The MOS transistor inherently tends to lead to lower speed because its gate is effectively a capacitor which must be charged through the high load resistor. MOS transistors appear to offer many advantages in the construction of memories. Such memories are easier to fabricate, and fewer processing steps are involved.[18] Memories based on this technique are emerging from the development stage.

Small memories are being assembled today from integrated chips with bipolar transistors, diodes, and diffused resistors.[58] Chips are packaged in a TO-5-type can or (usually) in a ceramic flat package.[59] These packages are arrayed on a printed-logic card, which interconnects them. One design, shown in Fig. 53, has an array of dual in-line packages containing the integrated-circuit storage elements, mounted and connected on a circuit card which also holds the sense amplifiers and driving circuitry, likewise packaged. Mounting of integrated associated circuitry on the substrate with

* Source: R. M. Englund, Univac Division, Sperry Rand Corporation, Roseville, Minn.

the film is also in use for magnetic memories to achieve compactness.[60, 61] Heat may be removed from the ceramic packages by air cooling and conduction. The conduction may be through the leads, or in addition to a heat sink via a plated-through hole thermally joined to the package by means of thermal joint compound or by soldering to a metallized bottom.[59]

DIGITAL LOGIC PACKAGING

In the central processing unit and other logic portions of a digital computer, the basis of operation is the gating and switching of pulses. From a packaging point of view, a digital computer is a large array of gates interconnected in a vast matrix that may involve scores of thousands of separate units.[62] Reliability, the paramount requirement for successful functioning of such a system, is inversely related to the number of

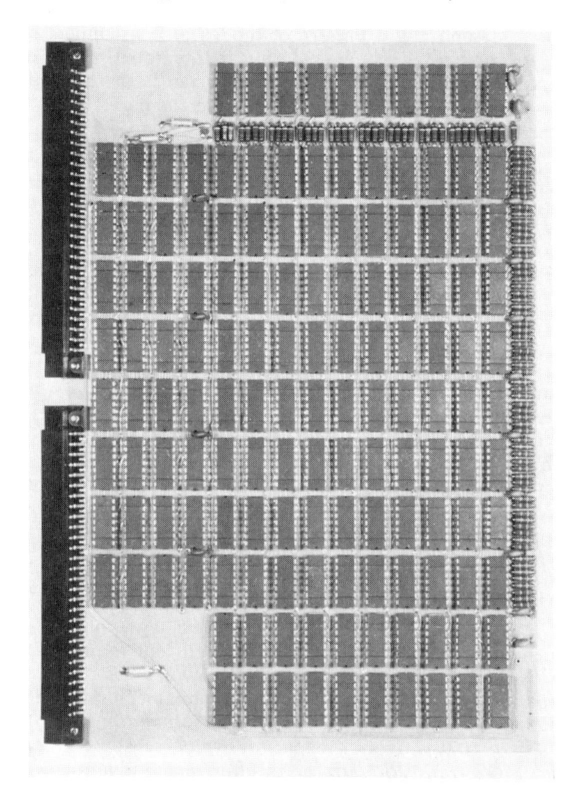

FIG. 53. 640-bit integrated-circuit storage with address and data drive circuitry, in dual in-line packages.

individual components and to the number of interconnections between them. Each separate circuit component is a potential source of failure, and each point of connection a possible source of trouble. Then, while high pulse rates and low voltages both diminish the reliability of circuit operation, both are essential to best performance. This is because their combination allows less energy to be involved in a gating or switching operation, thereby permitting high-speed performance in the computing circuits. Also, the power level is kept down, so that for a given packing density it is easier to design to stay below the maximum temperature ratings of the components. A thoroughgoing evaluation of factors governing reliability is essential to every computer packaging decision.

In an elementary processing operation, pulses advance through a chain of gates with a net gain of unity (no amplification or diminution). The output of a gate may "fan out" to reach several other gates, and as its input a gate may receive outputs from one or more other gates. So, it is advantageous to have input and output signal voltages the same for all logic elements (compatible operation). With a gain of unity through each element, the circuit design problem then becomes largely one of interconnection. It may, however, prove expedient to let the signal levels change through some gates in order to utilize logic best suited to the requirements at hand.

Types of Logic. Although it is the circuit designer who usually selects the type of logic to implement his circuits, the choice also concerns the packaging designer insofar at it influences the number of connections and of power leads, and determines the amount of heat to be removed. The common types of logic are listed in Table 17 with comparative evaluations of those properties of most interest to the packaging designer. Resistor-transistor logic (RTL), direct-coupled transistor logic (DCTL), and transistor-transistor logic (TTL) are the easiest to fabricate. Those types which require two or more power-supply voltages, such as diode-transistor logic (DTL), emitter-coupled logic (ECL), and variable threshold logic (VTL), call for more leads on the circuit board and may unduly complicate the packaging problem. RTL and DTL logic, used in discrete computer circuits, are giving way to TTL and ECL in integrated circuits; DTL is found in hybrid packages in third-generation computers.

TABLE 17. Comparative Properties of Logic Types

Form of logic	Speed capability	Power dissipation	Fan-out capability	Noise immunity	Fabrication in integrated circuits	Other characteristics
CML	Good	Moderate	High	Good	Difficult	Noise-sensitive; high fan-in
CTL	Good	Moderate	High	Good	Adaptable	Gain < 1
DCTL	Good	Moderate	Limited	Poor	Easy	Temperature- and noise-sensitive; limited fan-in
DTL	Fair	Low	Fair	Good	Complex	Often requires two power supplies; limited fan-in
ECL	Good	High	Very high	Good	Adaptable	Noise-sensitive; requires two power supplies; high fan-in
RCTL	Poor	Low	Limited	Fair	Complex	
RTL	Fair	Low	Limited	Fair	Adaptable	Temperature-sensitive; high fan-in
TTL	Good	Low	High	Very good	Easy	Wide tolerances
VTL	Poor to fair	Moderate to high	Limited	Good	Difficult	Requires two power supplies
MOS FET	Limited	Low	Very high	Excellent	Simple	Noise-sensitive; limited temperature range

An up-to-date résumé of the properties of each of the principal forms of logic presented here can be found in Ref. 63. Numerous variations of the basic types of logic exist and are separately named. For example, RCTL is a form of RTL modified to reduce power consumption at a sacrifice in speed, and DCTL is another form of RTL modified to reduce delay at the expense of additional power consumption. Component manufacturers offer lines of packaged logic and switching circuits based on RTL, DTL, TTL, and ECL. Often a manufacturer modifies the acronym for a type of logic by adding a letter or character to distinguish his particular version of it.

Fundamental Packaging Practices. In the design of the central processor and the other logic portions of a digital computer, packaging has its greatest influence on the chassis or card which holds and mounts the circuit components. In first-generation computers, discrete passive components were fastened by straps or their own axial leads to insulating boards, which in turn attached to the under side of a chassis holding sockets for vacuum tubes. The whole was interconnected by hand-soldered wiring, and connections from the chassis were via a connector, usually a tubular type. Such chassis frequently contained 10 to 20 vacuum tubes. Similar chassis, widely used in first-generation computers, had the discrete elements suspended by their axial leads between parallel insulating boards in tiers supported by posts at their corners. Smaller chassis evolved with a cluster of discrete elements in cordwood array or within a cage formed by the leads from the assembly. The whole assembly plugged into a connector or tube socket, and in some designs one or more tubes plugged into it opposite the connector. This type of construction was employed to provide modular replaceability and to add ruggedness; the caged elements were sometimes potted to withstand extreme environments.

As the second generation came into being, insertion mounting of both passive and active discrete components came into widespread use. For small production, the inserted components were connected by hand-soldered wiring. For volume production, single-sided printed-circuit boards were used to mount and connect the inserted com-

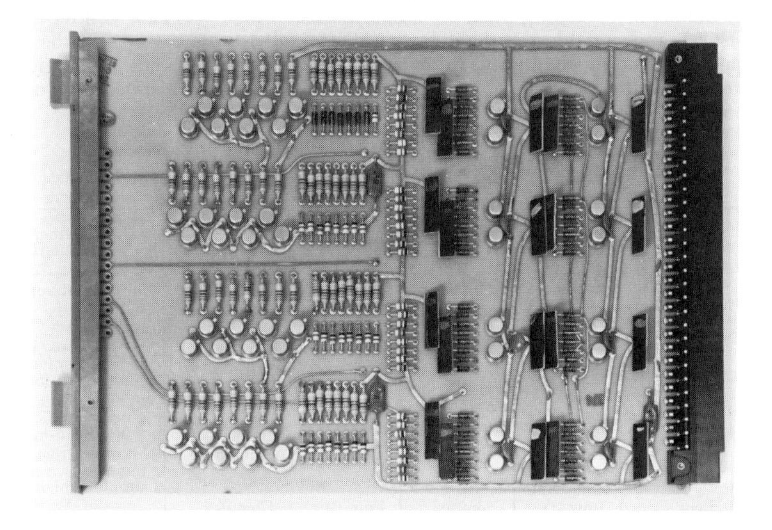

FIG. 54. Second-generation logic card featuring discrete components.

ponents. The boards were large, by today's standards, with dimensions of the order of 1 ft, tailored to the size of photographic plates used in the printing process. In one large-volume production line for computers, 4 to 16 printed-circuit cards, often not identical, were formed from one board and cut apart after it was completely fabricated. Flotation soldering with automatic card transport was used to complete assembly of the printed circuits. Faulty components were replaced and deficiencies in automatic soldering corrected by hand. Considerable cleaning, visual inspection, testing, and corrective hand soldering were necessary, which detracted from the savings and manufacturing advantages of the automatic insertion and soldering processes.

With the advent of subminiature resistors and capacitors and the TO-5 can for housing transistors, many more discrete components could be placed on a printed-circuit board than formerly was the case, and two-sided printed-circuit cards came into use for making the interconnections. Such a card, from a large-scale second-generation computer, is pictured in Fig. 54. On the side of the card shown, printed wiring is limited to common ground leads and connections are limited to test points. These connections and the leads from the components are carried via plated-through holes to the reverse side of the card; there, only power and signal wiring is located, and all connections to the 55-pin connector are made. Wave soldering completes all these connections at once.

This card mounts edgewise in a vertical position between channeled top and bottom guides and is secured by the connector at the rear. The flange on the card front is the same thickness as is the connector, and the guides are so dimensioned and placed that flanges and connectors on adjacent boards touch, to close the front and back openings. Therefore, they confine the cooling air as it passes upward over the components on several cards placed in vertical columns. The two protrusions on the front flange are shaped to accept a handle used to push the card connector firmly against its mating pins or to extract the card. Holes in the forward flange contain test probe insertion points, connected to the printed test and ground leads.

Figure 55 shows a printed-circuit card on which are mounted integrated circuits typical of the third generation. This card holds 10 flat packs, each 1/8 × 1/4 in., providing collectively eight gates with two inputs each. It is approximately 4 in. high × 3½ in. deep and is intended to mount vertically. Flat-pack connections are on its compo-

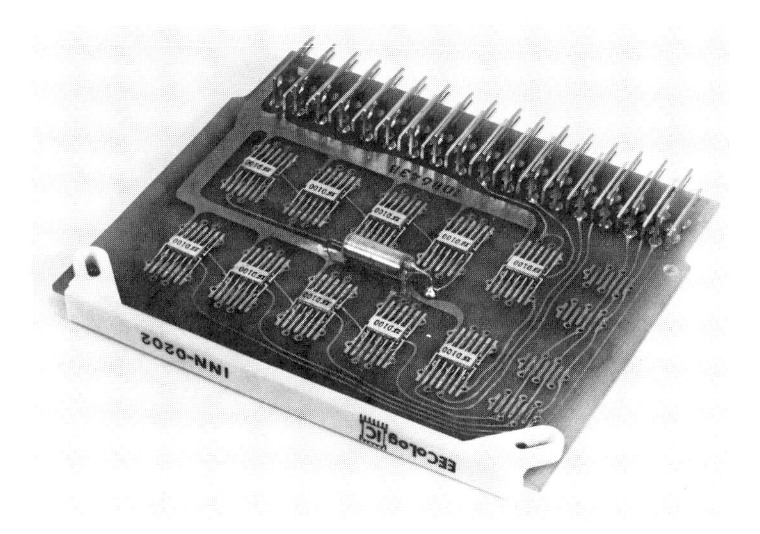

FIG. 55. Third-generation custom integrated-circuit logic card, mounting flat packs. (*Courtesy Electronic Engineering Company of California, Santa Ana, Calif.*)

nent side with most leads to the connector printed on the reverse side. Despite their small size, the leads from the flat packs are obviously quite wasteful of space. The layout of the card can accommodate two more flat packs, for which plated holes and printed wiring are in place. This is a standard card so designed that the printed layout can be used for more than one logic arrangement.

Minimum separation between adjacently mounted cards is governed by the 34-pin connector. The two rows of connector pins, staggered and one above the other, necessitate spacing between cards that suffices to allow for mounting discrete components, such as the decoupling capacitor on this card, and also provides separation between the handles on the card fronts. The plastic handle shown bows when the card is being extracted. The rounded ends on the side of the handle fit into notches in the guide and lock the cards in place when the handle lies flat. This has proved to be necessary only during shipment by air.

The foregoing illustrates the fact that fabrication of miniature components has developed faster than interconnection technology. In earlier generations the component count determined the size of the package; however, for printed-circuit logic cards of the third generation, the controlling factor determining their dimensions is the number of backboard connector pins required. One way around this difficulty is to interconnect the components via multilayer boards, with riser pins also serving as external connectors; then the number of connector pins is limited only by the board's area.

Basic Circuit Packages.

TO CANS. Discrete solid-state components are universally housed in a top-hat-shaped TO-5 can. This type of can, about ⅓ in. in diameter, has round leads extending from its bottom, equidistant from its axis, and suitable for insertion mounting on a printed-circuit board. When mounted staggered in parallel rows as in Fig. 54, these cans form a compact array suitable for heat removal by flowing air. For conductive heat removal, the cans can be inverted and inserted in holes in a heat-sinking board. In that case, the leads are bent over and connected to printed wiring on the base side of the board, or through clearance holes to wiring on the other side as shown in Fig. 56.

In the third generation, TO cans house monolithic integrated circuits or multichip ones formed by interconnecting discrete chips of resistors, capacitors, diodes, and transistors within the housing. These two forms of integrated circuits are described in Chap. 9 under the respective headings of Monolithic Semiconductor Integrated Circuits and Multidiscrete Component Chip Packages. The multichip technique affords flexibility and calls for minimum tooling. The monolithic circuit is better suited to the large quantities used in computers, since: (1) it is better adapted to automatic production techniques; (2) once tooled for production it offers high reliability and small size; and (3) it promotes cost savings when produced in large quantities.

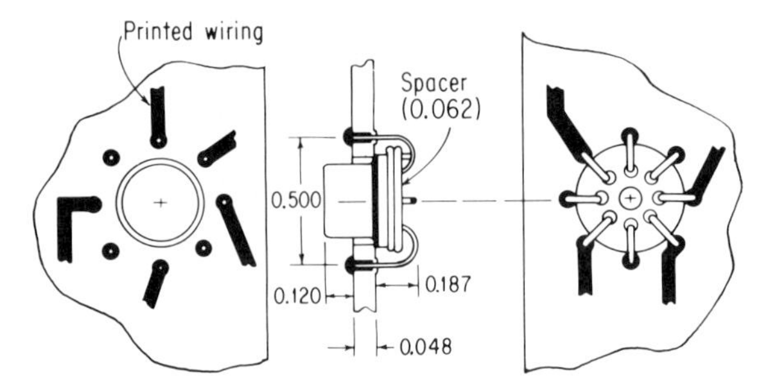

FIG. 56. Embedded mounting of TO can, for heat-sinking purposes.[103]

TO cans for integrated circuits generally have 10 leads, and 12 is the practical upper limit. Thus they are pin-limited for the more complex monolithic circuits or for interconnected monolithic chips within a single housing.

FLAT PACKS. Flat packs come in a variety of configurations and sizes. Standard ones generally available are listed in Table 18. The leads are flat ribbons on 0.050-in. centers, disposed equally along the two long sides; where space limitations make it necessary, some leads come out the ends in an L shape, as do those on the card in Fig. 55. In some computer applications, the leads on one side of the flat pack are bent up over it so that all can be connected from the same side in two planes. This arrangement improves compactness by freeing almost half the area otherwise required for interconnection. A flat-pack connector has been developed to receive packages with leads bent in this fashion, and to connect them to a multilayer board without soldering or welding.[64]

The maximum number of leads on standard flat packs is 14; special ones designed for more complex circuits in computers may exceed this number—the type of flat pack used in the ALERT computer has forty leads.[65] The limit of function complexity for one integrated-circuit package is not evident. Designing greater complexity into each package increases the number of specialized types of packages and hence means smaller production runs, but at the same time it decreases the number of external connections necessary.[66] Computers have been designed and built which used only one type of vacuum tube, or one type of diode, or had all gates identical; but third-generation computers normally use 3 to 10 different integrated circuits.[67] Further, it is expected that the diversity of circuit modules within a computer will increase greatly with time.

Flat packs are attached to printed-circuit cards by parallel-gap welding or reflow soldering, and there seems to be little reason to choose between the two methods.[68] One manufacturer of standard printed-circuit cards, who agrees with this premise, chose reflow soldering on the basis that the packages are more readily removed for replacement.

DUAL IN-LINE PACKAGES. The dual in-line package [68, 69] houses a monolithic integrated circuit within a rectangular ceramic case approximately ¾ × ¼ in. Fourteen flat leads, seven on each side, emerge from the sides of the case and are bent to extend downward at right angles, making insertion mounting easy. Each lead is wide near the case and has a shoulder which holds the insertion-mounted case some 0.035 in. from the board. This is the type of package used for circuit and storage elements of the integrated-circuit memory shown in Fig. 53. In volume it is not as compact as the flat pack, but it is suitable for two-sided printed-circuit layouts without necessitating plated-through holes, and it can be connected readily by dip or wave soldering.

Some lines of standard circuit cards are offered with either flat packs or dual in-line packages, the former for applications where superior speed is worth the extra cost. The printed-circuit card shown in Fig. 55 is paralleled by one of the same size and layout using dual in-line packages, at about two-thirds the cost.

ENCAPSULATED MODULES. Encapsulation of integrated circuits, although preferable to hermetic sealing from the standpoint of cost, nevertheless introduces difficulties of a mechanical nature,[70] occasioned by the strains resulting from thermal stresses and from pressures in the potting process. Changes of circuit values also may be introduced by contamination from the potting process. These difficulties are avoided by surrounding the circuits and their interconnections with a material which does not transmit stress to them or affect them adversely.[70, 71]

Solid logic technology (SLT) modules,[72] used in considerable volume to house hybrid circuits for computers, surround the circuit components with a silicone gel.[71] In SLT modules, connective wiring and thick-film resistors are deposited on a ceramic substrate, and chips containing transistors or diodes are soldered to the wiring pattern. In advanced solid logic technology (ASLT), capacitors are also tooled from a multilayer ceramic chip. This version of the module contains two ceramic substrates interconnected by posts, some of which protrude through a silicone rubber seal to provide the external connections. The circuitry is coated with glass, and covered on top and between substrates with the silicone gel, then housed in an aluminum case.

Custom-fabricated modules containing several flat packs are presently available

TABLE 18. Standard Flat Packs [68]

Outline no.	Nominal size, in.	No. of leads	Length, in.		Width, in.	
			Min.	Max.	Min.	Max.
TO-84	¼ × ⅛	14 A*	0.240	0.260	0.120	0.150
TO-85	¼ × 3/16	14 A	0.240	0.275	0.160	0.185
TO-86	¼ × ¼	14 A	0.240	0.275	0.240	0.260
TO-87	⅜ × ¼	14 B†	0.360	0.410	0.240	0.275
TO-88	⅜ × ¼	14 B	0.330	0.350	0.240	0.260
TO-89	¼ × ⅛	10 B	0.240	0.290	0.120	0.150
TO-90	¼ × 3/16	10 B	0.240	0.290	0.160	0.185
TO-91	¼ × ¼	10 B	0.240	0.290	0.240	0.260

* Elbow end leads.
† No leads at end.

under the name MICROSTICK.* Figure 57 shows a transparent sketch of an encapsulated stick containing seven flat packs, beside an expanded view of several layers of circuitry and conductors that compose this unit. Interconnections are formed by welding the flat-pack leads to teeth of the comb at various levels. The finished assembly before potting is shown in Fig. 58. The assembly is potted in black epoxy to form a stick suitable for mounting on a printed-circuit card.

LSI.† With uniformly higher yields of individual integrated circuits formed on wafers, it is anticipated that dicing into chips can be dispensed with. This leads to LSI, the interconnection of many circuit elements on a common substrate. *Discretionary wiring,* a technique which consists of connecting good circuits on a wafer to form a complex array, is one form of LSI. For it, the internal connections of functionally alike wafers are different, determined individually by a computer as a result of testing to find good circuits. The *fixed-wiring* technique uses sets of interconnection patterns which are fitted to semiconductor wafers to give the desired complex arrays. Obviously the fixed-interconnection scheme is more suited to high circuit yields, and the discretionary technique is practical where yields are relatively low. LSI is just coming into use for fabricating computer logic.

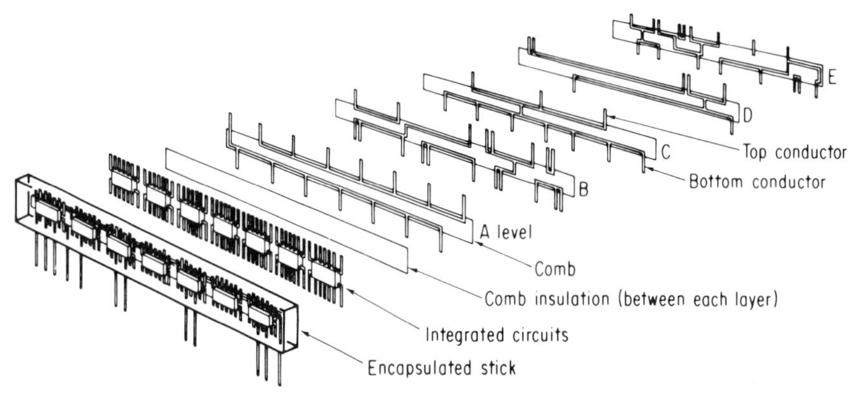

FIG. 57. Five-level custom interconnection of flat-pack assembly into an encapsulated module. (*Courtesy Electronic Engineering Company of California, Santa Ana, Calif.*)

Package Interconnections.

SINGLE- AND DOUBLE-SIDED PRINTED-CIRCUIT CARDS. The virtually universal use of integrated circuits in today's computer designs is outmoding the single-sided printed-circuit card as a means of mounting and interconnecting components. Two-sided cards make possible more conductors to interconnect the numerous leads from TO cans and flat packs, and provide a means for crossing over conductors. Even so, with miniature component packages making possible greater component densities at the same time that the number of leads to a package increases, the need for achieving greater interconnection density becomes acute. The number of circuits that can be placed on a single card is higher if the interrelationship between them is high and their need for connection to other logic areas is low. Thus the immediate design approach is to provide these interconnections via the printed-circuit-card wiring, not through the backboard connector.

One consequence of crowding more logic onto a single card is an increase in the variety of card types. Cards with few elements have simpler logic, which can be combined into more complex logic forms by interconnecting the cards through backplane wiring. Putting the more complex logic circuits on a single card proliferates the

* Trademark, Electronic Engineering Company of California.
† Source: R. M. Englund, Univac Div., Sperry Rand Corporation, Roseville, Minn.

number of card types and shortens the production runs. Small logic modules stacked or mounted in quantities on the printed-circuit card provide the solution to this problem—if they are removable. Microsticks, which when encapsulated can be mounted side by side over the area of a circuit card, are an example of this. SLT modules, likewise card-mounted,[73] are another example of this approach.

MULTILAYER CIRCUIT CARDS. Dense interconnection requirements can be met by additional layers of supported wiring in the form of a multilayer board.[7]

The number of layers required in a multi-layer board depends on the density of connection between devices mounted on it, as shown in Ref. 7. There are many approaches to multi-layer circuitry. The main difference between them involves inter-layer connection methods. Inter-layer connections can be formed with plated-through holes or by repetitive plating in a "build up" process. Multi-layer

FIG. 58. Welded flat-pack assembly before potting. (*Courtesy Electronic Engineering Company of California, Santa Ana, Calif.*)

boards with plated-through holes are composed of laminated layers. Copper pads are placed in registration on all layers to be interconnected, the laminated board is drilled, and connections between layers are formed by electroplating to the exposed peripheries of the holes in the copper pads. Built-up multi-layer boards are produced by alternately plating conductive pads and applying insulating compounds. Layers are interconnected by means of solid, plated copper feed-throughs.

The cost of multi-layer boards depends on their complexity and the number of layers; additional layers bring about a more-than-proportional increase in cost.

Multi-layer printed-circuit cards are particularly applicable to the high-density packaging requirements and sizable quantities required in production runs for computers.

Circuit cards for these computers are multilayer boards in almost all cases, as can

TABLE 19. Logic Packaging in Third-generation Computers

Computer system	Logic module			Circuit card			Back panel		
	Type and delay	Package	Leads	Capacity, modules	Construction	Pins	Capacity, cards	Description	External connections
IBM 360/30, 40, 50	DTL 30 nsec	SLT	12	6, 12, or 24	Two-sided PC in multiples of 1⅝ × 1½ in.	24	66	8⅜- × 12½-in. ML, supplementary wire wrap	20 flat cables with connector terminals
Honeywell 8200	TTL 15 nsec	Dual in line	14	50	5½- × 6-in. four-layer	80	20	PC, supplementary wire wrap and twisted pairs	Twisted pairs
IBM 360/70	DTL 10 nsec	SLT	12	6–24	Two-sided PC in multiples of 1⅝ × 1½ in.	24	66	8⅜- × 12½-in. ML, supplementary wire wrap	20 flat cables with connector terminals
Burroughs B-8500	CTL 7 nsec	Dual in line	14	56	6- × 7.3-in. four-layer	44	...	Wire wrap	
Univac 9300	DTTL 18 nsec	Dual in line	14	15	5- × 7-in. two-sided PC	55	84	10- × 50-in. wire wrap	Wire harness with wire-wrap terminations
IBM 360/90	ECL 5½ nsec	ASLT	16	72	4½- × 7-in. five-layer	96	18	8- × 12-in. ML	
RCA 70/45, 55	ECL 5 nsec	Flatpack	14	16	3- × 4-in. two-sided and three-layer	32–48	104	17- × 17-in. ML (five-layer) with supplementary wire wrap	26 wire harnesses with connector terminals

be seen from Table 19. Discrete and cordwood modules are mounted on single- or two-sided printed-circuit cards, but all hybrid and integrated-circuit first-level packages for the computer shown in the table are connected on multilayer cards. These cards are seen to be approximately square in all instances; this results from printed-wiring layouts which minimize the lengths of the longest possible paths.[74]

METHODS OF INTERCONNECTING BASIC PACKAGES. Combinations of conventional discrete components and of basic circuit packages can be connected together in almost any conceivable combination via printed-circuit cards. Figure 3 lists common arrangements, compared regarding size. Encapsulated cordwood modules of discrete components are being supplanted in today's computer designs by sealed hybrid modules containing integrated circuits of diodes and transistors connected with separately formed passive elements.[75] Resistors and low-valued capacitors are formed directly from the connecting strips of metallic film; capacitances of higher values must be attained by constructing separate, discrete elements.[76] Monolithic structures provide only transistors, diodes, resistors, and capacitors of limited size. Hybrid modules, or individual discrete components and basic integrated-circuit packages, are interconnected by soldering them to circuit cards in most instances. The circuit card thus becomes the basic field-replaceable package when provided with a suitable separable connector for attaching it to the computer wiring.

Exceptions to this practice may be found where, for practical reasons, it is desirable to attach circuit packages or elements directly to another assembly, as with preamplifiers for reading heads on magnetic drums or disks. As another example, it is becoming increasingly more common to construct memory planes with their drive and sense amplifiers and switching and decoding circuitry mounted on the same card as is the core matrix, wired directly to it and with interconnections printed on the card. External connections to the plane are made through connectors along its periphery or carried by flexible wiring to the next card in the stack. In this manner, many separable connections are eliminated and lead lengths are shortened. Even so, using discrete elements, the area they occupy on the plane may be as much as 20 times the area of the core matrix. The proportion of the plane area given over to interconnections is large even with unpackaged integrated circuits on chips, as seen in Fig. 59.

Other arrangements of circuit packages depend on mounting them in a volumetric assemblage rather than spread out on printed-circuit cards. Such arrangements generally depend upon stacking flat packs or ceramic substrates in layers and connecting them with vertical risers, or they comprise flat packs standing close together with their planes vertical, in order to gain more efficient utilization of volume. Computer applications of these arrangements have been limited mostly to military and aerospace computers where space was at a premium that would justify coping with the increased concentration of heat to be removed and with the assembly and accessibility problems introduced. However, some encapsulated cordwood-like modules of spaced flat packs are being introduced into commercial computer design. Volumetric modules of these types attach by insertion to printed-circuit cards by means of riser pins projecting from one side of the module, generally in one or two rows or in a rectangular pattern around all four edges of that side.

Another type of volumetric assemblage is the welded cordwood module consisting of a sandwich of circuit elements between two circuit boards on which the interconnections are made by means of conductors welded to the component leads. This construction is compact and mechanically strong, but does not lend itself to mass production techniques. Its principal use in computers has been for aerospace applications.[77]

Groupings of Interconnected Packages. An integral assembly of basic circuit packages, on a printed-circuit card or otherwise mounted, is generally referred to as a second-level package; the basic circuit packages themselves constitute the first level.[60] * In digital computer technology, best practice is to group first-level packages into a functional assembly at the second level.[80] By so doing, the second-level package becomes one that can be tested as a separate unit and easily substituted for another similar package. The packaging advantage of thus partitioning by function is a

* Usage of these terms is imprecise.[78] See also Levy[7] and Abbott.[79]

reduction in the number of connections needed at the interface with points beyond the second level. Since the interrelationships between components on a functionally integrated card are high and requirements for connection with other cards are low, the single functional card will accommodate a larger number of circuits than otherwise. More types of cards may become necessary under this procedure (especially cards for special functions), but since most functions of a digital computer are repeated many times, this circumstance does not pose a serious drawback. On the other hand, the reduction of connections to other cards eases one of the most critical problems of digital computer packaging: the number of such interconnections required, sometimes referred to as pin count.

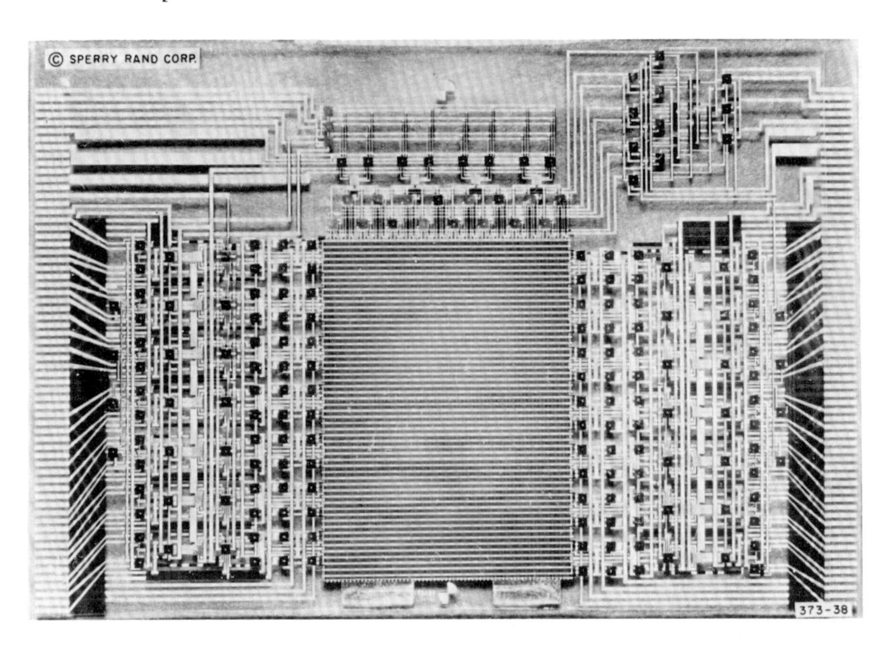

Fig. 59. 1,536-bit high-speed, low-power memory, comprising electroplated thin magnetic film and face-down bonded integrated chips on a common substrate, and vacuum-deposited Al and SiO wiring and insulation layers. (*Developed under contract AF 33(615)–1405 for the Air Force Materials Lab. Res. and Technol. Div., AF Systems Command, USAF.*)

Interconnective Logic Wiring.

Printed Wiring. The wiring of printed-circuit cards has to be planned subject to certain constraints of the electrical circuit; thus for the faster logic cards its configuration becomes a cooperative venture of the packaging engineer and the circuit designer.[81] The delay introduced by a signal path, plus the delay of the circuit element to which it connects, must not exceed the delays specified for that stage of logic. So, once the maximum delay in the signal path is known, its maximum allowable length is determined by the velocity of the signal it carries. This velocity, and hence the delay, must be determined for the lead in place on the card, as the speed of pulse travel will be found to be slower than its theoretical value because of the retarding effects of surroundings. Delays from one level of logic to the next must be independent of the route traveled, a fact which may prescribe minimum lengths for some wiring paths.

Besides meeting delay requirements, a signal path must have certain transmission properties which ensure its carrying a pulse of energy with sufficient fidelity to ensure circuit operation under the most adverse conditions of voltage and thresholds possible within the design tolerances. These properties are:

1. Low energy loss from attenuation and radiation
2. Propagation of all frequency components at approximately the same speed
3. Freedom from cross coupling with other circuits by which noise may be picked up or energy lost
4. Matched impedances at terminations and junctions to avoid reflections of energy

The implied transmission objectives are met by making the paths short, without sudden changes in direction or cross section, by avoiding exposure of long lengths or large areas to adjacent conductors, and by electrically designing the path as a distortionless transmission line with a characteristic impedance equal to the impedance of the load. Such ideal electrical design is seldom possible for these reasons: load impedances do not maintain the same values under all conditions, and electrical characteristics must be accepted within the limitations posed by feasible circuit layout and by the laminar thickness and dielectric properties of the supporting card. However, unless the attenuation is excessive, the line will be effectively distortionless. At all but the lowest pulse-repetition rates, design at this point becomes an experimental compromise to be worked out between the packaging and circuit design engineers.

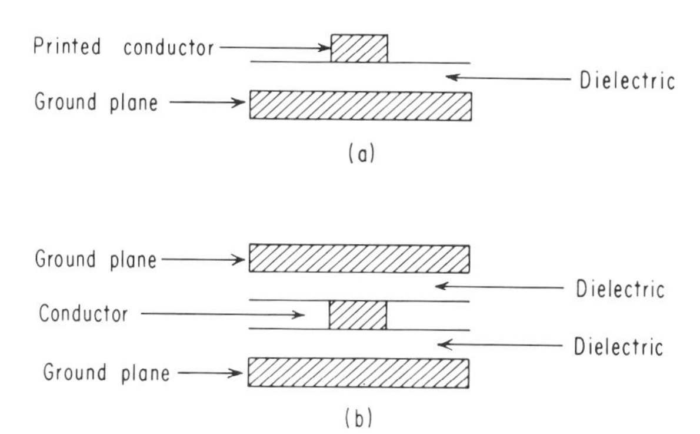

FIG. 60. Elementary transmission line cross sections. (*a*) Microstrip line, (*b*) strip line.

Factors which govern reflections and pickup of noise by cross coupling are well understood.[82, 83] Noise increases proportionately to the parallel lengths of coupled conductors, and decreases with the spacing between them. It also decreases with an increase in distributed capacity. The introduced noise is additive and so is proportional to the number of signal leads to which a conductor may be exposed.

To carry pulses satisfactorily with a rise time less than half their propagation time, a signal path must be designed as a transmission line.[83] The shortest rise time not requiring such techniques is ordinarily near 5 to 7 nanosec [73] in modern computer circuits. Transmission line characteristics are imparted to a signal path in printed wiring by its separation from a ground plane or other conductor by a thickness of dielectric, normally the circuit card.[84] Such an arrangement, illustrated by Fig. 60*a*, is termed a microstrip line. In multilayer boards, the signal wiring may be sandwiched between two ground planes in an arrangement called a strip line, as shown in Fig. 60*b*, affording it a measure of shielding. More complex configurations may result from the presence of power conductors or of other signal paths on different layers.[85] Transmission properties are governed by the width of the conductor as well as the thickness and dielectric constant of the medium which separates it from the ground plane. The values of these three quantities also establish the characteristic impedance, which it is desirable to keep in range of 70 to 100 ohms in logic circuits. For neighboring conductors, the distance separating their nearest edges (not the center-to-center spacing) is the figure of importance in transmission considerations.[86]

The physical layout of printed wiring is made subject to those of the foregoing considerations which are pertinent in a specific case, and subject to other considerations governing the placement of the circuit packages and removal of heat from them. Power leads are frequently brought in along one edge of the card, as in Fig. 55, and are made sufficiently wide to avoid excessive voltage drop in them. In two-sided cards, power and signal leads may be placed on one side and ground paths on the other. It is usually convenient to introduce the power at one or both ends of the connector so that power can be routed along the edge of an assemblage of cards or between adjacent assemblages on the back panel. Frequently, duplicate power pins are provided on both ends of the connector with a continuous printed voltage-distribution bus running around the periphery of the card from one of the pins to the other.[74] Avoiding crossovers in nonplanar circuits will require additional layers of signal wiring or a feed-through construction. The simplest arrangement of signal wiring in two layers has all horizontal interconnecting leads running on one side and all vertical leads running on the other side of a board.

Leads to the circuit packages may be depended upon to conduct some heat from them; however, they should not be designed primarily for this purpose, since increasing the width of a conductor raises its inductance and thereby introduces additional delay. The wiring configuration should be such that the packages form a regular array across which cooling air can pass with minimum turbulence.[74]

Back-plane Connections. Point-to-point wiring interconnects the connectors which mount the circuit cards. This wiring, referred to as backboard or back-panel interconnection, has evolved chronologically through soldered interconnections, taper pins, solderless contacts wrapped at first by means of a handgun and later by automatic machinery as well, and large multilayer boards. The development of new back-panel interconnection techniques has been slow in relation to advances in other electronic component packaging methods.

The connection interface on which the back plane is formed is made up of a compact grouping of the connectors to which the circuit cards attach. These connectors assemble precisely in a frame support, with the individual pins of the female connectors projecting to form an array to which the back-panel connections can be made. The types of connections that may be made are: [87]

1. Hand-soldered terminations
2. Welded terminations
3. Crimped terminations
4. Solderless wrapped wire attachment
5. Printed-circuit attachment

The last two of these methods are almost universal in commercial digital computers produced in quantity.

Point-to-point hand wiring still finds a place in small computers of low production volume. It is used to interconnect the encapsulated welded cordwood modules of aerospace computers with welded ribbon. In that instance, no true back plane is formed, as the multiplicity of interconnections is made on the end panels of the cordwood assemblies.

Two solderless wire-wrapping techniques, TERMI-POINT* and Wire-Wrap,† are in common use. Both are proprietary methods and each can be applied either with a hand-held wrapping gun or by an automatic machine which positions the back plane and places and connects the wires, one at a time. TERMI-POINT connections can be changed by removing a clip from the post to which the wire is attached and pushing the remaining ones down the post, thereby leaving room for adding a new clip at the top position to receive a new wire connection. A Wire-Wrap connection is changed by unwinding the wire from the connection and wrapping on a replacement. Satisfactory utilization of this replacement method is limited by damage to the post, because these solderless connections depend upon some deformation of its edges by the encircling wire when the connection is formed, and because solid-state diffusion subsequently occurs at the interface. Although back-panel connections are not ordinarily

* Trademark, AMP Incorporated.

† Trademark, Gardner-Denver Company.

disconnected and reconnected, easily made wiring changes are necessary to allow for corrections during manufacture and for additions and changes of wires made necessary by design modifications.

Connectors to form the back panel, having pins suitable for solderless wire wrapping, are available in a great variety of designs. The maximum number of pins in a single row of a connector is determined by the accumulation of interpin tolerances, since it matches the tolerance range which the head of an automatic wire-wrapping machine can accommodate. This limitation does not exist when wiring is done with a hand tool. Pins are regularly spaced on a grid to conform to the patterns set by the wire-wrapping machine. Grid centers normally are spaced 0.250 or 0.200 in. for 22- or 24-gauge wires, and 0.125 or 0.100 in. for 30-gauge wire. The machines will accommodate off-center tolerance variations of a pin amounting to about 0.015 in.[88, 89] Blocks of back-panel connectors are held together by a frame, die-cast or other metal, which will hold them so that all pins lie within these tolerances.

Paths between pins or between connectors are needed at intervals, to accommodate power distribution wiring and ground leads. Since the function of the back-plane wiring is to interconnect pins from the connectors, leads from the back plane can be made using some of these connectors. This may be done via flat flexible cable terminating in a male connector which fits the connector on the frame. One arrangement used * for this purpose utilizes printed-circuit cable cards upon which coaxial leads, twisted pairs, and other interconnecting wires can be terminated in mixed variety. These cable cards are shorter than the other circuit cards which plug into the back panel, thus allowing room for the cables connected to them. The cables are clamped to the corner of the frame for mechanical support.

Multilayer boards are finding increasing use as back panels in large commercial computers (see Table 19). They interconnect circuit cards with less waste of space than is the case with conventional connector assemblies. They will accommodate plug-in circuit cards [90] or encapsulated modules or an intermixture of both. These boards also effectively shield the circuit modules from the other side of the back plane by means of the ground planes which form one or more of the layers of a board.

With mass interconnections at increased density, the size of printed-circuit cards is determined more by the number of terminals needed on the connector than by the components mounted on them; this, in turn, results in an increase in capacity of the backboard. The demands, occasioned by increased speed, for diminished propagation delay and for avoidance of reflections in the backboard wiring, have shifted the emphasis on compactness to this part of the computer. This shift brings about a use of multilayer backboards for back-panel interconnections in the fast computers being produced.[91] The alternative, more complexity at lower levels of packaging to reduce the number of interconnections reaching the back plane, leads to more complex basic packages in a greater profusion of types.

Wiring Layout. Specifying how to run leads efficiently to make all the interconnections on a printed-circuit card or back plane—for any but the fewest number of connections—is a vastly complicated problem, generally handled by a computer program. The initial step is to define all the pairs of terminals which must be connected.[92] Succeeding steps, usually executed by a computer program, include: (1) assigning each connection to be made to one of the layers of wiring, (2) selecting for each layer the order in which pairs of terminals are to be joined, (3) tabulating all the paths by which they may be joined, and (4) selecting a single path to join each pair.

Some paths may be inadmissible because they exceed a maximum length and thereby introduce too much delay or because they encounter obstacles such as components, or leads which cannot be crossed over. Crossovers are admissible with insulated wires on wire-wrap panels; but on printed-circuit boards, where the number of crossovers in any path must be less than the number of layers, each such crossover requires a different layer. When the choice of number of layers is at the disposition of the packaging engineer, he can ignore crossovers while laying out the wiring, and then specify one more layer than the maximum number of crossovers resulting.

For initially assigning wires to one of the layers available, each pair of pins to be

* W. A. Merdinyan, Honeywell EDP Division, Waltham, Mass.

joined is assumed to be connected by a direct line; then those lines, representing tentative paths, are assigned to layers in such a way that the fewest lines intersect on each layer. Two algorithms [93, 94] provide a guide for selecting paths in an order which minimizes their average length. By applying either one, closest pairs are joined first; but pairs already joined, even by an indirect route, are not again joined. A method to determine the order for laying out wires to join pairs of points on a rectangular grid is to lay out a representation of the grid and draw on it a rectangle for each pair of pins to be joined. Each rectangle is drawn with the two pins of a pair at diagonally opposite corners of it; for pins in the same horizontal or vertical line on the grid, the rectangle degenerates to a straight line. Using the completed diagram, wires are chosen progressively: first those which connect pins that lie in the most rectangles, and thereafter in the order corresponding to fewer and fewer rectangles. This method tends to implement the rules of connecting closest pairs first.

Once the order for joining the pins has been thus determined, that order is followed in sequentially laying out paths to make the interconnections. Where two or more possible paths of equal length are indicated, an arbitrary choice may be made, subject

4	3	2	3	4	5	6	7	8	9	10	11
3	2	1	2	3	4		8	9	10	11	12
2	1	A	1		5		9	10	11		
		1	2		6	7	8	9	10	11	12
4	3	2	3		7					12	13
5	4				8	9	10			13 B	
6	5	6		10	9	10	11	12	13		
		7	8	9	10		12	13			
10	9	8	9	10	11		13				
11	10	9	10	11	12						

FIG. 61. Shortest path routing in a rectangular maze.[96]

to these factors: the proximity of other possibly interfering conductors, and the desirability of keeping the path as straight as possible, yet avoiding obstacles.

An algorithm by Lee [95] provides a straightforward means of finding the one or more best paths between two points on a grid which includes obstacles. The grid is divided into squares. Starting at the square occupied by one of the points to be joined, a 1 is written in each of the squares adjacent to the square it occupies to which a conductor may be routed. Then in each of the squares adjacent to those containing a 1 to which a conductor could be extended, a 2 is written. This process is continued until a square containing the other point to be joined is reached. A minimum path is determined by starting at that point and moving successively through squares each containing a number one digit lower until the initial starting point is reached. Squares which cannot be traversed because of obstacles are blanked out; no numbers are written in them. Such a layout is shown in Fig. 61. Of the two variations of the shortest path resulting there, the one which involves one less turn would normally be chosen. A modification of this procedure is available [96] to reduce the amount of memory storage required.

The algorithms and procedures are combined into computer programs which solve the sequence in layout and routing problems and prepare automated documentation for manufacture and test. This output can include tape or punched cards to control automatic wire-wrapping machines when such are used. Figure 62 is a scheme for such a program. Automated design and wire-wrapping procedures are being extended to peripheral units and communications auxiliaries for computer systems.

In wiring a back panel, the automatic machine attaches a wire to one grid and runs it between rows of pins to the next point of connection. The route traveled between pins may be horizontal, vertical, or diagonal. The pins can be in either a square or a staggered array, but for the square pattern the available diagonal conductor area is much smaller than it is for the staggered pattern, as illustrated in Fig. 63. In either configuration, the width of the diagonal avenue is less than that of a horizontal or vertical one. In the more advanced designs, where the dimensions between connector pins are reduced, constriction of this width becomes quite critical and the staggered array is much to be preferred. It has been found to result in shorter lead lengths. In multilayer boards the staggered pin arrangement has the further advantage of providing 50 percent more places for through-hole connections, as Fig. 63 shows.

Higher-level Packaging. Electronic packaging beyond the back-panel level concerns facilities for distributing timing pulses and power voltages into the various

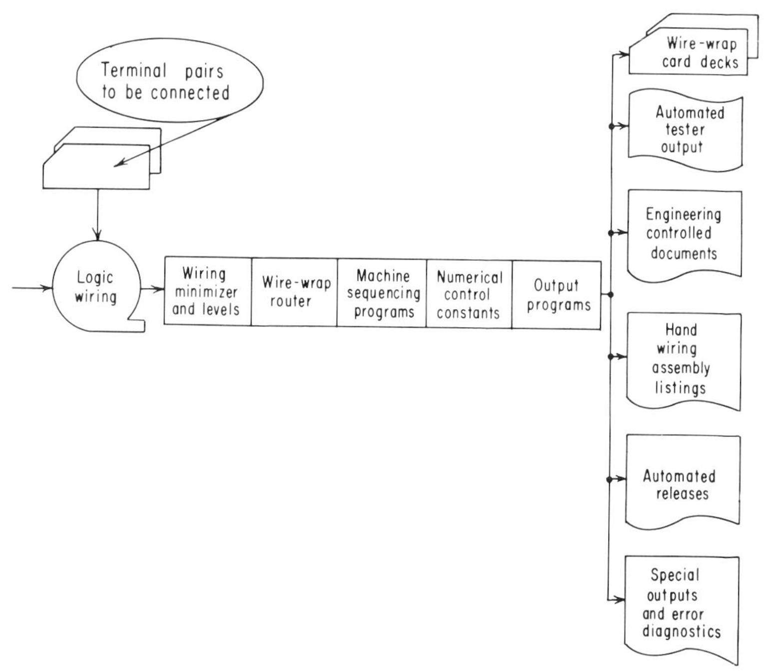

FIG. 62. Major program steps for computerized wiring design. (*W. B. Horsted, Univac Division, Sperry Rand Corporation, Roseville, Minn.*)

circuits and conveying signals between portions of the computer. It also involves a system of grounds and provision for shielding and for transfer of heat away from areas of power dissipation. External signal, power and ground connections to the computer are included. Many of the packaging problems arising are mechanical or thermal. Accessibility to the internal portions of the computer during manufacture and test, and subsequently for maintenance purposes, are also pertinent.

Distribution Systems. Single wires, twisted pairs, coaxial cable, or flexible flat cable are all used in computers to interconnect major modules and distribute power and timing pulses to them. For convenience in construction, bundles of wires can be made up into a preformed harness, which can be fastened into place by appropriate ties to the framework of the computer. The advantages of this construction include these: the wiring can be done apart from the equipment; the relationship of the wires to each other is the same from one harness to another; and lead ends fall near the points they connect to when the harness is put into place, thereby providing a measure of self-protection against wrong connections.

Coaxial cable is self-shielding and does not produce a field outside itself from currents it carries. It thus is ideally suited for transmitting the clock pulses. Terminations for coaxial cables are more difficult to make than those for other types of conductors. *Twisted pairs* also have some measure of immunity to external fields, but are subject to increased capacity to ground when run in the proximity of framework or grounded conductors. Twisted pairs are widely used for conducting signals and clock pulses and for distributing voltages. *Flat flexible cables* can be used for all these purposes. Flat cables hold the conductors in rigid relationship to each other and so maintain impedance values constant. Grounding the outside conductors or alternate conductors of

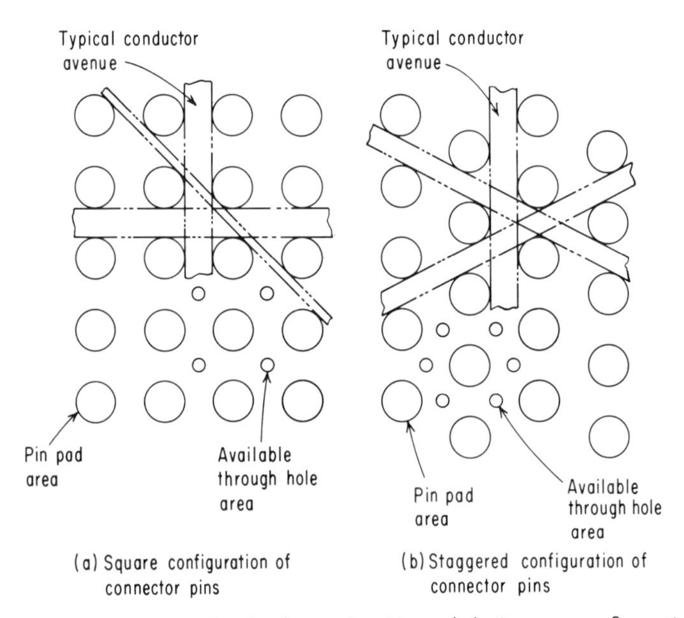

FIG. 63. Pin arrangements for back-panel wiring. (*a*) Square configuration; (*b*) staggered configuration. Staggered configuration allows more room for laying in wires than square configuration. (*Leon Schwartz, Univac Division, Sperry Rand Corporation, Philadelphia, Pa.*)

a flat cable and transmitting signals on the remaining conductors provides a good degree of freedom from external noise. This freedom can be increased by including a ground plane of foil or screen in the cable to provide a continuous shield. Such shielding will increase the capacity to ground of the signal conductors, but also helps to spread and dissipate the heat.

Ground conductors can be made an integral part of the cabling or can be run separately. The ground side of a signal path is preferably one side of a twisted pair or the outside conductor of a coaxial cable. Separate ground paths should be run for signals, for power, and for shielding; this avoids introducing noise or false threshold levels into the signals or changing delivered power supply voltages by drops caused by ground wire currents. The grounding principles for signals, power, and shielding discussed previously for analog computers apply in principle to digital ones. However, in analog computers, values are determined by precise voltages relative to a reference value, and any voltage drop in a ground lead introduces error; for digital signals, such a drop has no effect unless it is sufficient, when added to the signal, to push it beyond a threshold value.

Flat cables come in a great variety of arrangements as to number and size of conductors and are often fabricated in fixed lengths with connectors or special terminals at one or both ends. They can be obtained with retractable ability which causes them to roll up or fold up accordion style, yet allows extension so that modules which they connect can be moved (e.g., on a hinge or drawer) without disconnecting the cable and without danger of straining it. Although flat cables are widely specified on a customized basis, there is some attempt to standardize a few sizes and conductor spacings.[97]

The type of wiring, harness, or cable to be used is selected on a basis of its characteristic impedance and its capacitance to ground, its shielding properties, its resistance when carrying power, and certain mechanical considerations. The mechanical factors, except in cases of extreme environments, are principally the ability to withstand bending and flexing over the life of the equipment. When connected between hinged or movable sections of equipment, the cable should be so attached and given enough slack that individual conductors are not overstressed. Preformed cables should not be run directly across a hinge so that they are flexed in bending; rather they should be looped down along one side of the hinge and back up the other so that the cable twists only lightly and the individual conductors in it are stressed in torsion over a considerable length.

*Power Supply.** The much lower voltages and the higher currents required by solid-state components and integrated circuits, contrasted to first-generation technology, have placed increased emphasis on connections between the power supply and the load. The connections are a significant source of voltage drop and can offset the advantage of an otherwise well-designed power supply. The aggregate of circuits of a computer presents a nearly steady load to the power supply, since the switching rate of the circuits is much greater than the response time of the supply. Only under the unusual condition of switching a large section of circuits simultaneously will sizable changes in current demand arise and produce sizable variation in current flow from the supply. Except for such instances, it is only necessary that the power-supply system deliver voltages at suitable levels to the memory and logic elements of the computer, and this requires supplying wiring to accommodate a known, steady, voltage drop.

The packaging trend in computer power supplies has been one of slow evolution to smaller sizes, resulting from more efficient transistors, encapsulation of subcircuits, and improved construction with greater efficiency in heat removal. To obviate grounding problems intensified in the newer, faster computers, power supplies are being put closer to the loads. Recently there is a tendency to distribute power at a single voltage and convert to other voltage values in the neighborhood of the load. Voltage conversion units based on modern power transistor technology are built on circuit cards of the same type as logic cards, differing by having perhaps double thickness, mounted along with the logic cards in the computer frame.

Heat Transfer. Packaging arrangements for the electrical and electronic circuits of a computer must include facilities for removing the heat resulting from the total energy input. Leads, shields, and ground planes, increasingly effective in that order, conduct some of the heat from the circuits where it is generated. The remaining heat is transmitted into the envelopes and mountings of the components or circuits and must be carried away. Figure 64 shows the common way for accomplishing this. Circuit cards, as shown on the right, or packages of modules, as shown on the left, are racked so that blowers pass air over them vertically, exhausting at the top. The grill-like guides for the cards, and openings through the larger circuit modules, allow free passage of the cooling air. Commercial computers are designed to operate in rooms that are normally air-conditioned; the exhausted air adds a known amount to the cooling load of the room air-conditioning equipment. Cooling air is blown through stacks of memory planes much as through racks of logic cards.

* Source: L. E. Hansen, Univac Div., Sperry Rand Corporation, Roseville, Minn.

Power units or other particularly concentrated points of heat generation may be fitted with metallic heat sinks which conduct heat away from the elements to the framework or to radiating fins which are cooled by air. In rare instances, heat is removed from large commercial computers by circulating coolant. This technique is more common in computers for extreme environments, as in certain industrial control applications, or in computers for military applications, where a wider temperature range is required. Such computers, when built for airborne or space applications where size becomes important, often have their circuitry mounted on metal frames attached to an enclosing metal case which serves as well to equalize and dissipate the heat.

The purpose of heat removal is not alone to keep the temperatures of the elements

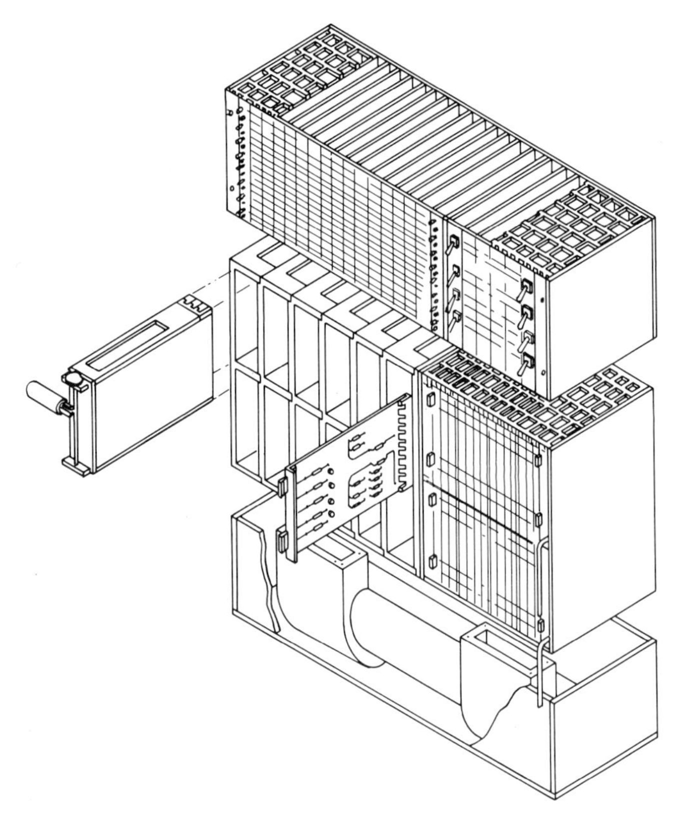

FIG. 64. Typical digital computer structure for convenient heat removal.[64]

below their safe temperature limits but also, since electrical characteristics generally change with temperature, to keep the temperature of each element always within limits which will ensure correct circuit values.

*Safety Specifications.** The provisions of the National Electrical Code [98] have been accepted *in toto* by most local governments. Compliance with it is prerequisite to many electrical installations. Most organizations dealing with electrical standards, codes, and specifications use it as a basis. The Underwriters' Laboratories, Inc., determines, defines, and publishes standards and specifications with a view toward reducing whatever hazards might be involved. Their work and published findings have been almost universally accepted in the United States. The principal document applying to computers [99] outlines the minimum requirements for electronic data processing

* Source: A. B. Poch, Univac Div., Sperry Rand Corporation, Roseville, Minn.

equipment. A very closely related Underwriters' Laboratories standard[100] is used to judge equipment not destined for use within computer rooms. These documents cover wiring, connections, and other packaging aspects from a safety standpoint. Counterparts to Underwriters' Laboratories, Inc., are the Canadian Standards Association in Canada and the International Commission on Rules for Approval of Electrical Equipment (C.E.E.) in Europe. The C.E.E. develops and publishes standards and test requirements based upon the national codes of the individual member countries.

Design Considerations. Factors which enter into the design of a computer and must be taken into account by the packaging engineer were set forth in Table 1. The relative importance of these factors as they influence the design will be dictated by the class of application for which it is intended. Comparison of the relative importance of these factors in four kinds of computer applications—business, industrial control, military ground service, and aerospace—will exemplify this. For *business applications,* cost will probably be paramount and the performance factors of speed, reliability, and power next in importance. The aim is to obtain as much performance as possible for a fixed amount of money dictated by the market price for which the design is intended. The physical attributes of compactness, ruggedness, and weight hold secondary place for this type of design. For *industrial control,* where the computer may operate in real time and on-line, reliability is most important and high speed will be required, probably with a lower limit of processing speed which must be attained. Cost will be the next consideration, subordinate to reliability and speed. If the location requires it, ruggedness will be an equally important matter. Power, compactness, and weight are minor considerations for this type of application. *Military ground-based systems* demand ruggedness as the first requirement, followed by performance factors. Of these, reliability is likely to be the most important, then speed and power, in that order. Cost can be considered only after the requirements for ruggedness and performance have been attained. Compactness and weight will be of lesser importance. In *aerospace applications,* reliability takes first place; weight, power, ruggedness, and compactness are all almost equally important. The stringent requirements for all these factors must be met before cost or speed considerations can be allowed to influence the design. Maintainability and repairability may warrant some increase in cost for business or control computers, but are matters of less importance for military and aerospace applications for which ruggedness and inherent reliability are relied on to obviate requirements for maintenance and repair.

An overall measure of the merit of a computer designed for any application is the ratio between performance and cost. When designing a computer for a *particular market,* the objective is to maximize the ratio of performance to cost. The packaging factors which affect performance can be traded off among themselves only to the extent that total cost is not increased in meeting this objective. For designs aimed at a *specific purpose,* it is easier to invert this ratio and consider the design objective to be minimizing the ratio of cost to performance. Trade-offs of disposable factors must be so made that the performance level is not disturbed from the value set for the specific purpose; final cost thus becomes a consequence rather than a fixed requirement.

In order to make the design problem tractable, it is well to proceed in distinct stages. The first stage is to choose the logic; the second, to choose the type of basic circuit package or module; the third, to determine the grouping of the basic packages on circuit boards or in equivalent assemblages; and the fourth, to design the interconnections among these assemblages. The type of logic may be decided prior to starting packaging design; at least, some types of logic will have been eliminated for circuit reasons. Packaging factors associated with particular types of logic should be brought to bear on the first stage before the final decision is reached on type. The choice of circuit package and planning of the circuit board should be done while keeping in mind the relatively long span in the design life of a computer from the time that its design is determined until its manufacture is discontinued.

Packaging practices for computers vary significantly among manufacturers in diverse ways, and even between lines of the same manufacture, with the result that no universal practices exist for implementing the design steps. A most useful tool is a figure of merit devised for the objective at hand. For example, in laying out a small circuit card, a figure of merit might be devised using the product of the clock rate and the number of logic levels as the numerator, and the product of cost and the delay through circuitry and wiring as the denominator.

Table 20 lists major packaging decisions along with factors influencing the particular choice in each case. A more detailed list of this kind may be made up and used as a guide to package the logic portion of a computer design.

TABLE 20. Digital Computer Packaging Considerations

Choice to be made	*Factors of most consequence*
Type of logic	Speed, noise susceptibility
Circuit package	Compactness, heat removal
Printed-circuit card size	External vs internal connections
Printed-circuit card mounting	Compactness, heat removal
Card interconnections	Delays, space required
Back-panel size	Maintainability, limits of wire-wrapping machine
Higher-level connections	Accessibility

REFERENCES

1. Nisenoff, N.: Hardware for Information Processing Systems: Today and in the Future, *Proc. IEEE,* vol. 54, no. 12, pp. 1820–1835, December, 1966.
2. Petritz, R. L.: Technological Foundations and Future Directions of Large-scale Integrated Electronics, *AFIPS Conf. Proc.,* vol. 29 (1966 Fall Joint Computer Conf.), pp. 65–87, November, 1966.
3. Amdahl, G. M., and L. D. Amdahl: Fourth-Generation Hardware, *Datamation,* vol. 13, no. 1, pp. 25–26, January, 1967.
4. Buck, D. A., and K. R. Shoulders: An Approach to Microminiature Printed Systems, *AFIPS Conf. Proc.,* vol. 14 (formerly *Proc. Eastern Joint Computer Conf., 1958*), pp. 55–58, July, 1959.
5. Baron, R. C.: Microelectronics and the Role of Circuit and System Designers, *Computer Design,* vol. 5, no. 9, pp. 16–18, September, 1966.
6. Ligomenides, P. A.: Wave-mechanical Uncertainty and Speed Limitations, *IEEE Spectrum,* vol. 4, no. 2, pp. 65–68, February, 1967.
7. Levy, A.: Interconnections, the Critical Link in the Microelectronic System, *Proc. Natl. Electron. Packaging Production Conf.,* pp. 686–701, June, 1965.
8. Packaged Operational Amplifiers, staff report, *Electro-Technol.,* vol. 79, no. 1, pp. 73–77, January, 1967.
9. Korn, G. A.: APE II . . . A Modern Analog/Hybrid Computer for Laboratory Instruction, *Simulation,* vol. 8, no. 2, pp. 97–104, February, 1967.
10. Korn, G. A.: Reduction of Digital Noise in Hybrid Analog-Digital Computers, *Simulation,* vol. 6, no. 3, pp. 151–152, March, 1966.
11. Korn, G. A., and T. M. Korn: "Electronic Analog and Hybrid Computers," McGraw-Hill Book Company, New York, 1964, pp. 455–464.
12. Stewart, E. L.: Grounds, Grounds, and More Grounds, *Simulation,* vol. 5, no. 2, pp. 121–128, August, 1965.
13. Huskey, H. D., and G. A. Korn: "Computer Handbook," McGraw-Hill Book Company, New York, 1962, secs. 4.4.6–4.4.8, pp. 4–30 ff.
14. Korn, G. A., and T. M. Korn: "Electronic Analog and Hybrid Computers," McGraw-Hill Book Company, New York, 1964, ch. 11, pp. 439–440.
15. Korn, G. A.: Progress of Analog/Hybrid Computation, *Proc. IEEE,* vol. 54, no. 12, pp. 1835–1847, December, 1966.
16. Korn, G. A., and T. M. Korn: "Electronic Analog and Hybrid Computers," McGraw-Hill Book Company, New York, 1964, pp. 106–110.
17. Korn, G. A., and T. M. Korn: "Electronic Analog and Hybrid Computers," McGraw-Hill Book Company, New York, 1964, pp. 87–88.
18. Hobbs, L. C.: Present and Future State-of-the-art in Computer Memories, *IEEE Trans. Electron. Computers,* vol. EC-15, no. 4, pp. 534–550, August, 1966.
19. Brown, J. R., Jr.: A Ferrite Core Memory Seen as a Transmission Line, *Computer Design,* pp. 44–54, January, 1966.

20. Bland, G. F.: Directional Coupling and Its Use for Memory Noise Reduction, *IBM J. Res. Dev.*, vol. 7, no. 4, pp. 252–256, July, 1963.
21. Hobbs, L. C.: The Impact of Hardware in the 1970's, *Datamation*, vol. 12, no. 3, pp. 36–42, March, 1966.
22. Ashley, A. H., E. U. Cohler, and W. S. Humphrey, Jr.: Temperature Compensation for a Core Memory, *Proc. Eastern Joint Computer Conf. 1959*, pp. 200–203.
23. Keister, Frank Z.: Beware of Solder Slivers, *Electronic Packaging Production*, vol. 6, no. 11, pp. 162–164, November, 1966.
24. McCallister, J. P., and C. F. Chong: A 500-nanosecond Main Computer Memory Utilizing Plated-wire Elements, *AFIPS Conf. Proc.*, vol. 29 (1966 Fall Joint Computer Conf.), pp. 305–314.
25. Noda, K.: Success Story: Japanese Originals, IV. Much Ado about Memories, *Electronics*, vol. 39, no. 13, pp. 93–100, June 27, 1966.
26. Fedde, G. A.: Plated-wire Memories: UNIVAC's Bet to Replace Toroidal Ferrite Cores, staff report, *Electronics*, vol. 40, no. 10, pp. 101–108, May 15, 1967.
27. Staff Report: Other Wires in the Plating Bath, *Electronics*, vol. 40, no. 10, p. 109, May 15, 1967.
28. Higashi, P.: A Thin-film Rod Memory for the NCR 315 RMC Computer, *IEEE Trans. Electron. Computers*, vol. EC-15, no. 4, pp. 459–467, August, 1966.
29. Meier, D. A., and A. J. Kolk: The Magnetic Rod—A Cylindrical, Thin-film Memory Element, in "Large-capacity Memory Techniques for Computing Systems," The Macmillan Company, New York, 1962, pp. 195–212.
30. Kaufman, B. A., P. B. Ellinger, and H. J. Kuno: A Rotationally Switched Rod Memory with a 100-nanosecond Cycle Time, *AFIPS Conf. Proc.*, vol. 29 (1966 Fall Joint Computer Conf.), pp. 293–304.
31. Davis, J. S., and P. E. Wells: Investigation of a Woven Screen Memory System, *AFIPS Conf. Proc.*, vol. 24 (1963 Fall Joint Computer Conf.), pp. 311–321.
32. Maeda, H., A. Matsushita, and M. Takashima: Woven Wire Memory for NDRO System, *IEEE Trans. Electron. Computers*, vol. EC-15, no. 4, pp. 442–444, August, 1966.
33. Maeda, H., M. Takashima, and A. J. Kolk, Jr.: A High-speed Woven Read-only Memory, *AFIPS Conf. Proc.*, vol. 27, pt. I (1965 Fall Joint Computer Conf.), pp. 789–790.
34. Howard, R. A., et al.: Investigation of Woven-screen Memory Techniques, in "Large Capacity Memory Techniques for Computing Systems," The Macmillan Company, New York, 1962, pp. 361–372.
35. Aldrich, W. H., and R. L. Alonso: The "Braid" Transformer Memory, *IEEE Trans. Electron. Computers*, vol. EC-15, no. 4, pp. 502–508, August, 1966.
36. Gianola, U. F., et al.: Large-capacity Card Changeable Permanent Magnet Twistor Memory, in "Large-capacity Memory Techniques for Computing Systems," The Macmillan Company, New York, 1962, pp. 177–193.
37. Lewin, M. H.: A Survey of Read-only Memories, *AFIPS Conf. Proc.*, vol. 27, pt. I (1965 Fall Joint Computer Conf.), pp. 775–782.
38. New Bell Telephone Switching System Based on Computer Technology, *Computer Design*, vol. 3, no. 12, pp. 18–22, December, 1964.
39. Bobeck, A. H.: A New Storage Element Suitable for Large-sized Memory Arrays—The Twistor, *Bell System Tech. J.*, vol. 36, no. 6, pp. 1319–1340, November, 1957.
40. Younker, E. L., et al.: Design of an Experimental Multiple Instantaneous Response File, *AFIPS Conf. Proc.*, vol. 25 (1964 Spring Joint Computer Conf.), pp. 515–528.
41. Kuttner, P.: The Rope Memory—A Permanent Storage Device, *AFIPS Conf. Proc.*, vol. 24 (1963 Fall Joint Computer Conf.), pp. 45–48.
42. Sidhu, P. S., and B. Bussell: Development of an E-core Read-only Memory, *AFIPS Conf. Proc.*, vol. 27, pt. I (1965 Fall Joint Computer Conf.), pp. 809–816.
43. Chamberlain, D. M.: *RCA Transfluxor Application Note SMA-28*, Radio Corporation of America, Camden, N.J.
44. Marlow, Lionel L.: Ferrite Memory Packaging for Improved Production Efficiency, *Proc. Natl. Electron. Packaging Production Conf.*, pp. 446–452, June, 1964.
45. Aeronutronic Division, Ford Motor Company, Technical Note: Biax Memory, *Aeronutronic Publ. 7-2042*, rev. March, 1964.
46. McAteer, J. E., J. A. Capobianco, and R. L. Koppel: Association Memory System Implementation and Characteristics, *AFIPS Conf. Proc.*, vol. 26, pt. I (1964 Fall Joint Computer Conf.), pp. 81–92.

47. MacIntyre, Robert M.: High-speed Biax Memories, *Computer Design,* pp. 54–61, June, 1966.
48. Pyle, W. I., T. E. Chavannes, and R. M. MacIntyre: A 10 Mc NDRO Biax Memory of 1024 Word, 48 Bit Per Word Capacity, *AFIPS Conf. Proc.,* vol. 26, pt. I (1964 Fall Joint Computer Conf.), pp. 69–80.
49. Rajchman, J. A.: Ferrite Apertured Plate for Random-access Memory, *Proc. Eastern Joint Computer Conf., 1956,* pp. 107–114.
50. Looney, Duncan H.: Magnetic Devices for Digital Computers, *Datamation,* vol. 7, no. 8, pp. 51–55, August, 1961.
51. Slade, A. E., and H. O. McMahon: A Cryotron Catalog Memory System, *Proc. Eastern Joint Computer Conf., 1956,* pp. 115–119.
52. Beesley, J. P., A. L. Leiner, and N. Rochester: Design of a Large-scale Cryogenic Memory System, in "Large-capacity Memory Techniques for Computing Systems," The Macmillan Company, New York, 1962, pp. 305–311.
53. Burns, L. L., D. A. Christiansen, and R. A. Gange: A Large Capacity Cryoelectric Memory with Cavity Sensing, *AFIPS Conf. Proc.,* vol. 24 (1963 Fall Joint Computer Conf.), pp. 91–99.
54. Endo, Ichiro, and Junji Yamato: The Metal Card Memory—A New Semipermanent Store, in "Large-capacity Memory Techniques for Computing Systems," The Macmillan Company, New York, 1962, pp. 213–230.
55. Takahashi, S., and S. Watanabe, Capacitance Type Fixed Memory, in "Large-capacity Memory Techniques for Computing Systems," The Macmillan Company, New York, 1962, pp. 53–62.
56. Lewin, M. H., H. R. Beelitz, and J. Guarracini: Fixed Resistor-card Memory, *IEEE Trans. Electron. Computers,* vol. EC-14, no. 3, pp. 428–434, June, 1965.
57. Lewin, M. H., H. R. Beelitz, and J. A. Rajchman: Fixed, Associative Memory Using Evaporated Organic Diode Arrays, *AFIPS Conf. Proc.,* vol. 24 (1963 Fall Joint Computer Conf.), pp. 101–106.
58. Potter, G. B., J. Mendelson, and S. Sirkin: Integrated Scratch Pads Sire New Generation of Computers, *Electronics,* vol. 39, no. 7, pp. 118-126, Apr. 4, 1966.
59. Catt, I., E. C. Garth, and D. E. Murray: A High-speed Integrated Circuit Scratchpad Memory, *AFIPS Conf. Proc.,* vol. 29 (1966 Fall Joint Computer Conf.), pp. 315–331.
60. Weilerstein, I. M.: Outlook for Computer Packaging, *Electron. Packaging Production,* vol. 7, no. 5, pp. 178–181, May, 1967.
61. Radio Corporation of America, RCA MF2100 MF2101 Monolithic Ferrite Memory Modules, RCA Memory Products Department, Needham Heights, Mass., November, 1965.
62. Maclay, W. R., et al.: Integrated Circuitry: A Study of a New Packaging Approach, *Proc. Natl. Electron. Packaging Production Conf.,* pp. 78–87, June, 1965.
63. Hugle, W. B.: Integrated Logic Circuits: A Comparative Evaluation, *Computer Design,* vol. 6, no. 1, pp. 36–47, January, 1967.
64. Joachim, O. C., and F. R. Metzger: A Miniaturized Packaging Technique, *Proc. Natl. Electron. Packaging Prod. Conf.,* June 1966, pp. 656–667.
65. ICs and Multilayers for "Alert," staff report, *EEE,* vol. 14, no. 3, pp. 40–44, March, 1966.
66. Martin, J. W.: What is Practical with Complex Integrated Circuits?, *Electron. Engr.,* vol. 25, no. 10, pp. 118–126, October, 1966.
67. Noyce, R. N.: A Look at Future Costs of Large Integrated Arrays, *AFIPS Conf. Proc.,* vol. 29 (1966 Fall Joint Computer Conf.), pp. 111–114.
68. Staller, J. J., and G. Sideris: The Packaging Revolution, Part II: Design and Manufacturing Overlap, *Electronics,* vol. 38, no. 22, pp. 75–87, Nov. 1, 1965.
69. Koeker, R. E.: Microcircuit Packaging Evolution, *EDN,* vol. 11, no. 11, pp. 52–65, Sept. 28, 1966.
70. Madland, G. R.: Plastic Encapsulation for Integrated Circuits?, *Electronic Prods.,* pp. 56–61, September, 1966.
71. Lloyd, R. H. F.: ASLT: An Extension of Hybrid Miniaturization Techniques, *IBM J. Res. Dev.,* vol. 11, no. 1, pp. 86–92, January, 1967.
72. Davis, E. M., et al.: Solid Logic Technology: Versatile, High-performance Microelectronics, *IBM J. Res. Dev.,* vol. 8, no. 2, pp. 104–114, April, 1964.
73. Springfield, W. K.: Multilayer Printed Circuitry in Computer Applications, *Electron. Packaging Prod.,* vol. 6, no. 12, pp. 111–117, December, 1966.
74. Karew, J., and T. Gilligan: Hybrid Circuits in Computer Design, *Electron. Engr.,* vol. 25, no. 9, pp. 90–100, September, 1966.
75. Saunders, R., and G. R. Heidler: An Analysis of Digital Hybrid Circuit Packaging and Connections, *Proc. Natl. Electron. Packaging Production Conf.,* pp. 303–314, June, 1964.

76. Saunders, R.: Thin Film Hybrid Approach to Integrated Circuits, *Electron. Ind.*, vol. 24, no. 6, pp. 34–37, June, 1965.
77. Linden, A. E.: Are Printed Circuits Passe, *Proc. Natl. Electron. Packaging Production Conf.*, pp. 260–269, June, 1963.
78. Staller, J. J.: The Packaging Revolution, Part I; Form and Function Interact, *Electronics*, vol. 38, no. 21, pp. 72–87, Oct. 18, 1965.
79. Abbott, M.: The System Designer's Role, *Electronics*, vol. 38, no. 22, p. 80, Nov. 1, 1965.
80. Heidler, G. R.: A Systems Approach to Fabrication and Miniaturization of Next-generation Digital Computers, *Proc. Natl. Electron. Packaging Production Conf.*, pp. 126–133, June, 1964.
81. Rhoades, W. T.: Guidelines in Implementing Logic Circuit Specifications into Packaging Concepts, *Proc. Natl. Electron. Packaging Production Conf.*, pp. 294–303, June, 1965.
82. Yao, F. C.: Interconnection and Noise Immunity of Circuitry in Digital Computers, *IEEE Trans. Electron. Computers*, vol. EC-14, no. 6, pp. 875–880, December, 1965.
83. Feller, A., H. R. Kaupp, and J. J. Digiacomo: Crosstalk and Reflections in High-speed Digital Systems, *AFIPS Conf. Proc.*, vol. 27 (1965 Fall Joint Computer Conf.), pp. 511–522.
84. Kaupp, H. R.: Characteristics of Microstrip Transmission Lines, *IEEE Trans. Electron. Computers*, vol. EC-16, no. 2, pp. 185–193, April, 1967.
85. Springfield, W. K.: Designing Transmission Lines into Multilayer Circuit Boards, *Electronics*, vol. 38, no. 22, pp. 90–96, Nov. 1, 1965.
86. Kaupp, H. R.: Pulse Crosstalk between Microstrip Transmission Lines, *7th Internatl. Electron. Circuit Packaging Symp. Record*, IECP 2/5, pp. 1–12, August, 1966.
87. Noschese, R.: Comparing the Backplane Wiring Techniques, *Electron. Industries*, vol. 24, no. 1, pp. 76–79, January, 1965.
88. Olds, W. L.: Packaging Techniques for Use with Automated "Wire-Wrap" Machines, *Proc. Natl. Electron. Packaging Production Conf.*, pp. 184–191, June, 1963.
89. Goodman, D. S.: The Role the Connector Plays in Modular Packaging, *Proc. Natl. Electron. Packaging Production Conf.*, pp. 47–54, June, 1964.
90. Ruth, S. B.: Backplane Wiring by Computer, *Electron. Engr.*, vol. 25, no. 12, pp. 78–81, December, 1966.
91. Messner, G.: Wiring Boards: The Types Available, *Electron. Packaging Production*, vol. 7, no. 12, pp. PC14–PC18, December, 1967.
92. Akers, S. B., Jr.: Some Problems and Techniques of Automatic Wire Layout, *Digest 1st Ann. IEEE Computer Conf.*, pp. 135–136, September, 1967.
93. Loberman, H., and A. Weinberger: Formal Procedure for Connecting Terminals with a Minimum Total Wire Length, *J. Assoc. Computing Machinery*, vol. 4, no. 4, pp. 428–437, October, 1957.
94. Ledley, R. S.: "Digital Computer and Control Engineering," McGraw-Hill Book Company, New York, 1960, sec. 23-4, pp. 773–782.
95. Lee, C. Y.: An Algorithm for Path Connections and its Applications, *IEEE Trans. Electron. Computers*, vol. EC-10, no. 3, pp. 346–365, September, 1961.
96. Akers, S. B., Jr.: A Modification of Lee's Path Connection Algorithm, *IEEE Trans. Electron. Computers*, vol. EC-16, no. 1, pp. 97–98, February, 1967.
97. Washtien, J. L.: "Flexible Flat Cable Handbook," The Institute of Printed Circuits, Chicago, 1965.
98. USA Standards Institute: "National Electrical Code," National Fire Protection Association, Boston, 1968.
99. UL Standard: "Standard for Electronic Data-processing Units and Systems," UL-478, Underwriters' Laboratories, Inc., Chicago, 1967.
100. UL Standard: "Requirements for Office Appliances and Business Equipment," UL-114, Underwriters' Laboratories, Inc., Chicago, 1965.
101. McCauley, D. O.: Systems Packaging Efficiency, *Proc. Natl. Electron. Packaging Production Conf.*, pp. 630–643, June, 1965.
102. Weniger, K.: Memory Systems Comparison, *Electron. Engr.*, vol. 26, no. 5, pp. 118–121, May, 1967.
103. Flat Pack or TO-5?, staff report, *EDN*, vol. 11, no. 8, pp. 27–43, August, 1966.

Chapter 13

PACKAGING FOR MILITARY APPLICATIONS

By

JACK J. STALLER

Microsystems Technology Corporation
Burlington, Massachusetts

CONTRIBUTORS

Military Specifications: E. KOVAL, *Sylvania*
Components for Military Programs: J. THOMPSON, *Sylvania*
Shock and Vibration: H. LAKE, *Sylvania*
Human Factors: L. BRICKER, *Grumman Aircraft Engineering Corporation*
Electromagnetic Interference Control: ARNOLD BUCKMAN, *Sylvania*

MILITARY PACKAGING REQUIREMENTS

Packaging for military applications differs from nonmilitary equipment packaging in the emphasis or priority placed on the various design factors and the need for strict adherence to the military specifications, standards, requirements, etc. Military equipment must be operated and maintained under conditions considerably different from nonmilitary equipment. It is subjected to environmental extremes of temperature, shock, vibration, humidity, radiation, submersion, and others. Space is frequently severely limited and of a unique form factor to meet the space availability of aircraft, projectiles, ships, or submarines. Weight is usually critical, since the equipment frequently must be transported by men or machines.

The equipment must be designed to meet military maintenance and logistics requirements. This involves special considerations of personnel skills, facilities, parts logistics, and instruction/maintenance manuals. Depending upon the application, the equipment may be required to be self-supporting in the field or to be sent to a well-equipped base for repair.

Military equipment, in general, demands shorter design and development cycles.

Because of changing technology and military requirements, new approaches are constantly sought. This results in fairly short-term obsolescence.

Reliability, or the ability to operate for extended periods without error or failure, is crucial to military equipment. Lives, equipment, and planning depend upon proper operation.

All aspects of military equipment design are regulated and governed by a tremendous family of military specifications, normally called *MIL Specs,* and a broad variety of supporting documents. These range from specific equipment definitions developed by a military organization to establish the requirements of a new system to be designed to routine definitions for components such as resistors and capacitors. Every move the packaging engineer makes is governed by the family tree of specifications that stem from the overall equipment specification. He must be completely familiar with them, understand their application, and know when and how deviations or variations from the specifications can be applied.

Examples of Documents Stemming from a General Equipment Specification. A military system specification for naval shipboard use designates MIL-E-16400F (NAVY), Amendment 1, Class 2, dated February 24, 1966. It is amended by Amendment 1, dated September 22, which is six pages long. Class 2 designates a temperature operating range of —28 to +65°C (ambient) ship exposed.

Specification MIL-E-16400F lists the following documents, of the issue in effect on the date of invitation for bids or requests for proposal that forms part of it:

Type	*Total number referred to*
Federal specifications	34
Military specifications	112
Federal standards	20
Military standards	20
Military handbooks	1
Bureau of Ships drawings	7
Bureau of Ships publications	7
Bureau of Naval Weapons publications	1
David Taylor Model Basin publications	1
National Bureau of Standards publications	1
Total	204

The documents listed may, in turn, designate other documents which add to the total specification requirement governing the system design.

This requirement for conforming to specifications, standards, design notes, and other material which the military services have found necessary is the major difference between nonmilitary and military equipment.

Military Documents. The military equipment packaging designer must work with a variety of documents from all branches of the military services and many government agencies. A later portion of this chapter covers the military specification system in detail. A summary of many of the primary working documents is presented below.

Detailed Equipment Specification. This is a specification which fully describes the requirements of a system or subsystem to be procured. It is issued by a government bureau or agency to define a system for which a quotation or offering is desired. In addition to defining all physical and performance requirements, the equipment specification will list other applicable specifications, which, in turn, may list further specifications.

Military Specifications. These are formal documents that define specific areas of design. A number of primary electronic equipment specifications are issued by the various branches of the military services covering the equipment areas of primary interest to their particular applications. Overall equipment specifications of particular interest to the military packaging designer are listed below. It should be noted that military specifications are subject to periodic revision and may be canceled or superseded. Therefore, the designer must be certain that he is utilizing the latest revision or, if specified, the particular revision called for in the contract.

The contractor is not normally required to respond to specification revisions after the award of a contract.

1. MIL-E-5400 "Military Specification—Electronic Equipment, Aircraft; General Specification for." This specification describes the general requirements for the design and manufacture of airborne electronic equipment for operation primarily in piloted aircraft. Five classes of equipment (1, 1A, 2, 3, 4) for various altitude and temperature ranges are covered. The individual equipment specification must define which one is to be applied.

2. MIL-E-16400 (NAVY) "Military Specification, Electronic Equipment, Naval Ship and Shore: General Specification." This specification defines the general requirements applicable to the design and construction of electronic equipment and associated and auxiliary electronic apparatus furnished as part of a complete system intended for naval ship or shore applications.

3. Electronics Command Technical Requirement SCL-6200 "Materials, Parts, and Processes Used in Military Electronic Equipment." This document describes the selection, application, and use of materials, parts, and processes in the design, construction, installation, and maintenance of ECOM-RD-D & L (Electronics Command Research and Development Directorate and Laboratories) and USAEL (United States Army Electronic Laboratories), including telecommunications, surveillance, their associated equipment and gear, radiac equipment, photographic, and other military electromechanical equipment.

4. MIL-P-11268 (SIG. C) "Parts, Materials, and Processes Used in Electronic Communication Equipment." This specification designates the selection, application, and use of parts, materials, and processes in the construction of electronic communication equipment. It appears to have been largely superseded by SCL-6200 above.

5. MIL-I-983 "Interior Communication Equipment, Naval Shipboard; Basic Design Requirements for." This specification covers the basic design requirements and test and operating conditions for interior communication equipment to be used in naval ships.

6. MIL-E-11991 "Electronic, Electrical, and Electro-mechanical Equipment, Guided Missile Weapon Systems; General Specification for." This specification describes requirements which are common to guided-missile weapon-systems electronic, electrical, and electromechanical equipment.

7. MIL-F-18870 "Fire Control Equipment, Naval Ship and Shore; General Specification." This specification includes the general requirements for the design and manufacture of surface, shore, antiaircraft, and underwater fire-control equipment and associated equipment furnished either as a complete system or as part of a complete system.

8. MIL-E-4158 (USAF) "Military Specification, Electronic Equipment, Ground; General Requirements for." This specification covers the general requirements for the design and manufacture of ground electronic equipment for the U.S. Air Force.

Military Standards. These, in general, provide guidelines for design rather than particular specifications. An example is MIL-STD-210 entitled "Climatic Extremes for Military Equipment."

Research and Development Material. These are periodic regulations published as guides for the development of designs to meet specific requirements. Two typical examples are: Army 705-8, NAVMAT 4600.5A, Air Force 80-18 MC 0 4610.14 "Criteria for Air-Transportability and for Transportability Program"; and AR-705-15 "Operation of Material under Extreme Conditions of Environment."

Military Standardization Handbooks. These are detailed handbooks, almost textbooks, covering in great detail a particular subject critical to military system design. An example is MIL HDBK-217A "Reliability Stress and Failure Rate Data for Electronic Equipment."

Bureau of Ships Publications. These are similar to the research and development material listed above. They give the results of studies and experience in the design of equipment for naval service. Typical examples are as follows:

1. NAVSHIPS 93820 "Handbook for the Prediction of Shipboard and Shore Electronic Equipment Reliability."

2. NAVSHIPS 250-423-30 "Shock Design of Shipboard Equipment, Dynamic Analysis Method."

3. NAVSHIPS 900,185 "Design of Shock and Vibration Resistant Electronic Equipment for Shipboard Use."

SCOPE OF THIS CHAPTER

It is the objective of this section to emphasize the packaging-design requirements that are particularly concerned with militarized equipment. General packaging-design technology will be covered in other chapters of this handbook.

Some definitions of terms to be used at different points in this chapter are given in Table 1.

TABLE 1. Terms and Definitions

Advanced development model. A model of the complete equipment or integral parts of an equipment for experimentation or tests, to demonstrate the technical feasibility of the design and its ability to meet existing performance requirements and to secure engineering data for further development. Dependent upon the complexity of the equipment and the technological factors involved, it may be necessary to produce several successive models, each containing additional objectives. The final advanced development model shall approach the required form factor and employ standard or nonstandard parts (which have been approved by the agency concerned). Serious considerations shall be given to military requirements such as reliability, maintainability, human factors, and environmental conditions.

Air Force Specialty Code (AFSC). A code consisting of a combination of digits and letters which is used to identify a given Air Force specialty such as AFSC 43250 Jet Engine Mechanic.

Anthropometry. Science of measuring the human body and its parts and functional capacities.

Benchmarks. Preferred packaging designs which serve as a reference base against which the effectiveness of all designs are compared.

Chassis. The physical structure which retains and electrically interconnects a group of modules which perform higher-level functions.

Commercial part. A part manufactured by a vendor primarily for nonmilitary end use or a part which is used by the military but has not been fully qualified to the applicable military specification, as in the case of new parts.

Control coding. Designing controls for optimum tactile, kinesthetic, or visual identification by taking into consideration such factors as shape, size, color, and mode of operation.

Control-display compatibility. A harmonious relationship between a display condition and the corresponding control action; a feature of equipment design that provides for compatibility between stimulus and response in a perceptual-motor task.

Criticality. Numerical weight assigned to subsystems or components on the relative effects of human malfunctions on system performance or mission success.

DESC. Defense Electronics Supply Center.

Design review. A technical review of the design approach by a group of impartial experts to assess its applicability and optimization. Design reviews may be performed at any identifiable point in the design cycle.

Discard at failure (throw away). An assembly of components or parts that is not capable of further subdivision for repair.

DOD. Department of Defense.

Enclosure. A combination of the external housing and the racks.

Exploratory development model. An assembly of preliminary circuits and parts, in accordance with commercial practices, to investigate, test, or evaluate the soundness of a concept, device, circuits, equipment, or system in breadboard or rough experimental form, without regard to the eventual overall design or final form.

Function analysis. The investigation of alternative man and machine capabilities which may be used to satisfy established requirements.

Generic parts. Parts of the same basic type but varying perhaps in only one or two parameters. An example of this would be solid tantalum capacitors. All solid tantalum units fall into one generic category. However, they differ in size, capacitance, voltage, leakage, package seal, etc.

Human engineering. The determination of man's capabilities and limitations as they relate to the equipment or systems he will use, and the application of this knowledge to the planning, design, and testing of man-machine combinations to obtain optimum reliability, efficiency, and safety.

Human factors. A body of scientific facts about psychological and biological characteristics in relation to complex systems. It includes human engineering, personnel selection, training, life-support requirements, job-performance aids, and human-performance evaluation.

IDEP. Interservice Data Exchange Program.

Individual equipment specification. An individual equipment specification is the detail specification covering a particular equipment.

Interface. The common boundary at which two elements must meet and be compatible in order to function properly. The elements may be anything from complete systems to components, and the boundary may be mechanical or electrical.

Link analysis. A procedure for determining the frequency and importance of functional connections or interactions between various elements of a system and types of communications involved, whether visual, auditory, or tactual.

Main external housing. The structure which mechanically supports the racks and which, together with the racks, forms the enclosure.

Military specification. Specifications are documents, intended primarily for use in procurement, which are clear, accurate descriptions of the technical requirements for items, materials, or services including the procedures by which it will be determined that the requirements have been met. Specifications for items and materials shall also contain preservation, packaging, packing, and marking requirements.

Military standard. Standards are documents that establish engineering and technical limitations and applications for items, materials, processes, methods, designs, and engineering practices.

Modular assembly. A replaceable assembly having outline dimensions which are integral multiples of a fixed set of modular dimensions.

Modularity. Partitioning or separating a system into successively smaller groups of components electrically and mechanically interconnected.

Personnel subsystem. Usually an Air Force term applied to a total human-factors program providing the human performance necessary to operate, maintain, and control the system.

Preproduction model. A model suitable for complete evaluation of mechanical and electrical form, design, and performance. It shall be in final mechanical and electrical form, shall employ standard or nonstandard parts (approved by the agency concerned), and shall be completely representative of final equipment.

PRINCE/APIC, Battelle, IDEP. These are government and private facilities that act as a reservoir for all generated test and application information on components. Companies can subscribe for a modest sum to these agencies and in return can participate in the sharing of each other's test information, thus saving duplication of testing on the part of the individual company.

Production model. A model in its final mechanical and electrical form of final production design made by production tools, jigs, fixtures, and methods. It shall use standard or nonstandard parts (approved by the agency concerned).

Rack. The mechanical support for the chassis, interconnecting cables, modules, front-panel performance-monitoring devices, and adjustment controls.

Replaceable assembly. An assembly that is capable of being easily removed and replaced as an integral item.

Second source. The requirement that a part or item be available from more than one supplier.

Serviceable main chassis. NEL designation equivalent to rack as used in this handbook.

Service test model. A model to be used for engineering or operational tests under service conditions for evaluation of performance and military suitability. It shall closely approximate an initial production design, shall have the required form, shall employ standard or nonstandard parts (approved by the agency concerned), and shall meet the standard military requirements such as reliability, maintainability, human factors, extreme environmental conditions, etc.

Spares logistics. This refers to the entire cycle required to maintain electronic equipment in the field. It includes maintenance of an adequate supply of spares at the operating site, the means of replenishing the supply, and the procedure for returning a subsystem or module for repair and replacement when required.

System effectiveness. A measure of the overall capability of a system to perform its assigned tasks. This includes balancing of all factors against one another to determine the combination that adds most significantly to the overall capability.

Unitized construction. A type of unit construction consisting predominantly of replaceable assemblies.

CATEGORIES OF MILITARIZED EQUIPMENT

Military equipment may be expected to operate under conditions to be found anywhere on earth and perhaps soon in environments other than on earth. The particular conditions under which equipment is expected to operate are normally defined in the overall system specification or the referenced military specifications. The environments defined are a result of the category and area of expected operation.

All military systems are given code letters which define their expected installation conditions, the type of equipment, and the purpose. The installation code listing provides a good definition of the variety of application categories that are encountered. Table 2 is a listing of the item indicator letters taken from Ref. 11.

TABLE 2. Item Indicator Letters

First letter—Installation	Second letter—Type of equipment	Third letter—Purpose
A—airborne (installed and operated in aircraft)	A—invisible light, heat radiation	A—auxiliary assemblies (not complete operating sets used with or part of two or more sets or sets series)
B—underwater mobile, submarine	B—pigeon	B—bombing
C—air transportable (inactivated, do not use)	C—carrier	C—communications (receiving and transmitting)
D—pilotless carrier	D—radiac	D—direction finder, reconnaissance, and/or surveillance
F—fixed	E—nupac	E—ejection and/or release
G—ground, general ground use (include two or more ground-type installations)	*F—photographic	G—fire control or searchlight directing
K—amphibious	G—telegraph or teletype	H—recording and/or reproducing (graphic meteorological and sound)
M—ground, mobile (installed as operating unit in a vehicle which has no function other than transporting the equipment)	I—interphone and public address	K—computing
P—pack or portable (animal or man)	J—electromechanical or inertial wire-covered	L—searchlight control (inactivated, use G)
S—water surface craft	K—telemetering	M—maintenance and test assemblies (including tools)
T—ground, transportable	L—countermeasures	N—navigational aids (including altimeters, beacons, compasses, racons, depth sounding, approach, and landing)
U—general utility (includes two or more general installation classes, airborne)	M—meteorological	P—reproducing (inactivated, do not use)
V—ground, vehicular (installed in vehicle designed for functions other than carrying electronic equipment, etc., such as tanks)	N—sound in air	R—receiving, passive detecting
W—water surface and underwater	P—radar	S—detecting and/or range and bearing, search
	Q—sonar and underwater sound	T—transmitting
	R—radio	W—automatic flight or remote control
	S—special types, magnetic, etc., or combinations of types	X—identification and recognition
	T—telephone (wire)	
	V—visual and visible light	
	W—armament (peculiar to armament, not otherwise covered)	
	X—facsimile or television	
	Y—data processing	

* Not for United States use except for assigning suffix letters to previously nomenclatured items.

A typical coding is Radio Set AN/GRC-5. The G defines general ground use, the R radio, and the C communications (receiving and transmitting).

General Categories. The major categories of operation are ground, water, air, and the outer atmosphere. Each of these may be further subdivided into applications as follows:

Ground

1. *Fixed installation.* Permanent buildings or other housings that are not normally moved from place to place.
2. *Ground transportable.* These may be transportable buildings which can be readily assembled and disassembled.
3. *Ground mobile—nonoperating.* These are customarily semitrailer vans or truck-mounted shelters which are designed for permanent equipment installation and transportation. The equipment is not customarily operated during transport.
4. *Ground vehicular.* The equipment is installed in vehicles designed for other purposes (such as tanks, Jeeps, etc.) and operates during mobility.
5. *Man pack or portable.* The equipment is intended to be transported by man or men.

Water

1. *Surface.* All shipboard installations fall into this class.
2. *Underwater.* All submarines fall into this class.
3. *Underwater projectile.* This includes torpedoes.

Airborne

1. *Aircraft.* Installed and operated in air-breathing aircraft.
2. *Helicopter.* Installed and operated in rotary-winged aircraft.
3. *Missile and rocket.* Installed and operated in air-breathing missiles and rockets.
4. *Parachute drop.* Designed to be dropped by parachute.

Outer Atmosphere

1. *Missiles.* Designed to operate beyond the earth's atmosphere for limited periods of time.
2. *Satellites.* Designed to operate beyond the earth's atmosphere for extended periods of time.

Environmental Conditions. Each of the categories listed above has its own set of environmental conditions, both climatic and dynamic, depending upon its area of application. These are discussed in general terms below:

General Climatic Conditions.[2] Six factors of the natural environment are considered for military equipment:

Thermal stress (hot and cold)
Humidity stress (high and low)
Precipitation (including snow load)
Wind (surface and altitude profile)
Penetration and abrasion (blowing sand, dust, and snow)
Atmospheric pressure

Seven spheres of operation requirements of military equipment are delineated in Ref. 2, as follows:

Operation, ground, worldwide
Operation, ground, arctic winter
Operation, ground, moist tropics
Operation, ground, hot desert
Operation, shipboard, worldwide
Operation, airborne, worldwide
Worldwide short-term storage

The particular equipment must be designed to operate under the worst combination of environments defined. In addition, it may have to be packaged for storage to meet particular storage conditions. Packaging (in this case, protection) equipment to meet military storage conditions is a specialty in itself. Typical environmental requirements are as follows:

1. Shock. This is normally specified as the ability to perform after a number of shocks of a specified value and impulse-time duration. The shocks are normally applied along each of the three mutually perpendicular axes. In general, the shock

levels for aircraft may be considered low, while the shock levels for shipboard are high.

2. Vibration. Vibration is normally specified by a frequency* in cycles per second coupled with a double amplitude in inches. Since the amplitude varies with frequency, the vibration requirements are normally given as a curve or series of curves. MIL-E-5400 provides three vibration curves (shown in Fig. 1) for aircraft equipment, for helicopter equipment, and for combined aircraft and helicopter equipment. Shipboard vibration is generally in the low (below 60 cps) range, aircraft in the medium (below 500 cps) range, and missiles in the high (below

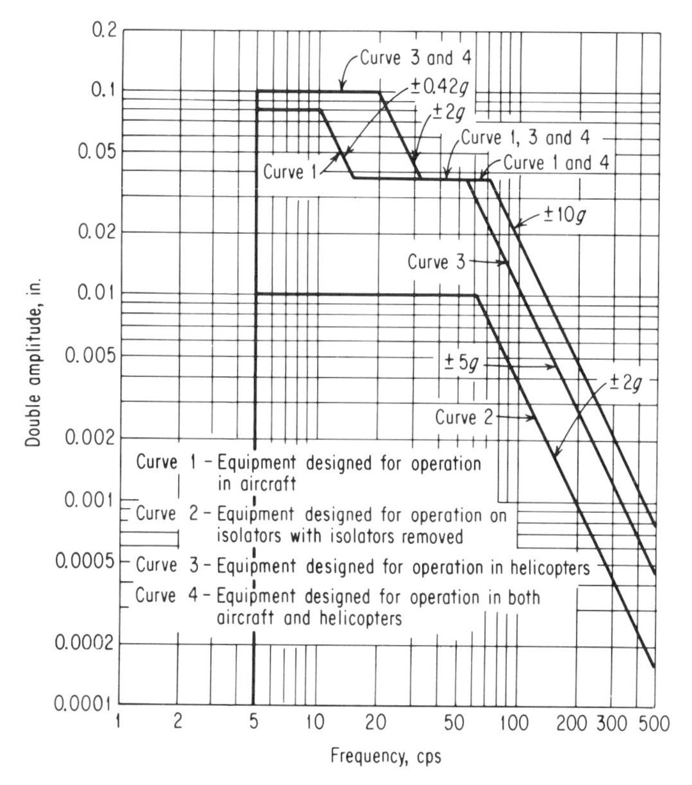

Fig. 1. Vibration requirements.

2,000 cps) range. The high-frequency vibration may also have random noise superimposed upon it.

3. Inclination. These are primarily for applications where the equipment is required to operate during rolling conditions such as in a ship or moving vehicle. A typical specification reads "The equipment shall be inclined at the rate of 5 to 7 cycles per minute in one plane to angles of 45° on either side of the vertical for a period sufficiently long to determine its characteristics under such motion or for a minimum of 30 minutes."

4. Bench handling. A typical specification is that any subassembly removable for servicing shall be removed from its enclosure, as for servicing, and placed in a suitable position for servicing on a solid 2-in. fir bench top. The test shall be performed as follows, in a manner simulating shocks likely to occur during servicing:

* For a constant acceleration (g) level.

a. Tilt up the assembly through an angle of 30°, using one edge of the assembly as a pivot, and permit the assembly to drop back freely to the horizontal. Repeat, using other practicable edges of the same horizontal face as pivots for a total of four drips.

b. Repeat (*a*) with the assembly resting on other faces, until it has been dropped for a total of four times on each face on which the assembly could practicably be placed during servicing.

5. Bounce (shock) test for equipment nonoperating. This simulates the shock environment during transport in a vehicle. A package tester for military equipment as made by the L. A. B. Corporation, Skaneateles, N.Y., or the equivalent is used for this test. The equipment on its shock mounts (if any) is secured to the actual or a simulated mounting plate and placed on the package tester. Wooden fences are provided to constrain the equipment from horizontal motion in a direction parallel to the axis of the shafts of the package tester by more than 2 in. and in a direction perpendicular to the axis of the shafts of a distance more than sufficient to ensure that the equipment will not rebound from fence to fence. An accelerometer is mounted on the adapter plate to record the shock transmitted to the equipment. The package-tester shafts, in phase, are adjusted to a speed such that random acceleration peaks of 5 to 10g are impressed on the accelerometer when measured at the output of a 100-cps low-pass filter for a total of 3 hr. After each ¾-hr period of the test, the adapter plate with the equipment constrained is rotated 90°, each time in the same direction. After completion of the test, the equipment is to meet the full specified performance.

6. Bounce test for complete assemblage nonoperating. This is a test of the completely loaded vehicle under simulated worst-case transportation conditions. Completed installations shall be capable of withstanding the following test without damage. A military vehicle with the van body or shelter mounted thereon shall be driven five (5) times over the following section of the Munson Test Course, at Aberdeen Proving Ground, Md., at the individual speeds (miles per hour) and in the order listed:

Course washboard (6-in. waves spaced 72 in. apart)	5
Belgian block	20
Radial washboard (2- to 4-in. waves)	15
Single corrugations (4- to 6-in. waves)	20
Any short sections between the above sections	20

The above-listed tests are not necessarily the precise tests or all the tests that will be needed on a specific piece of equipment, but they provide a measure of the types of environments that military equipment is expected to operate under and to be designed for.

MILITARY SPECIFICATIONS

General. The design of military equipment is guided entirely by the variety of documents encompassing a broad spectrum of specifications, standards, and formalized disciplines. Although this portion covers primarily the documents concerned with the technical aspects of design, it is also necessary to mention (1) documents concerned with engineering management, because all contract considerations are required to be made within the parameters of time, cost, and performance, and (2) documents concerned with engineering data, because these are firm requirements closely related to design in all contracts.

The Military-specification System. The military specifications, standards, handbooks, regulations, technical orders, directives, drawings, bulletins, and manuals comprise a system of documents which enables the Department of Defense (Military Services) to procure equipments and systems that satisfy military requirements. They include details on specified operations under various combinations of environmental and military conditions, together with operational reliability that will ensure the completion of assigned missions.

It is imperative that every project administrator and engineer concerned with the development of military systems and equipment have a thorough knowledge of the specification system. It is particularly important that engineers responsible for implementing design know the contractually invoked documents down to the smallest details. Failure to give these details adequate consideration can cause expensive and untimely redesign programs.

In recent years the Federal government and DOD have modified their policies on the minimal use of industry specifications and standards. Consequently, an increasing number of industry documents have been adopted by the military services for their use. However, caution must be observed and only those specifications or standards listed in approved DOD indexes or referenced contractually may be applied.

There are further precautions to be observed, for example, in the use of standard parts, materials, processes, tests, and documentation. It should be understood by the design engineer that, even though an item test or document can be obtained which outperforms in certain characteristics similar items covered by military specifications, if such an item does not fulfill exactly the requirements of a governing specification, he (the designer) must choose between processing a "nonstandard" item approval or applying for a waiver. Recognizing preferred, standard, and nonstandard areas in equipment specifications can prevent the indiscriminate use of items that may later create delays and complicate the development cycle.

Table 3 gives as its first part a list of contract items that will often be examined

TABLE 3. Index of Items Cross-referenced to Specification Approval
(Courtesy Electronic Design and Sperry Gyroscope Co.)

Item No.	Item name	Item No.	Item name
	Parts		Materials
A1	Adjusting devices using special threads	B1	Aluminum and aluminum alloys (chemical treatment)
A2	Batteries	B2	Antiseize compound
A3	Bearings, sleeve type	B3	Cable (application and treatment)
A4	Boots for push switches in watertight applications	B4	Castings; metals; alloys
A5	Capacitor, air dielectric	B5	Ceramics
A6	Capacitor, electrolytic	B6	Cotton fabric laminates
A7	Connectors	B7	Cotton- or wood-filled molding compounds
A8	Control of parts design	B8	Critical materials
A9	Control shafts and couplings	B9	Electrical insulation; solventless varnish
A10	Crystals	B10	Fiberboard
A11	Electron tubes (tube selection)	B11	Fungus-inert material
A12	Electron tube sockets	B12	Glass
A13	Fluorescent lamps	B13	Glass-bonded mica
A14	Fuseholders, clip type	B14	Insulated hookup wire
A15	Gears, fiber construction	B15	Insulating sleeving
A16	Knife switches	B16	Laminates (sandwich core material)
A17	Locking devices (tube thread)	B17	Magnesium; magnesium alloys
A18	Meters, high-sensitivity	B18	Nonslip surface coating
A19	Miniature parts	B19	Nonstandard material (unspecified material)
A20	Nonstandard parts	B20	Organic material
A21	Nuts, sheet spring	B21	Plastics (selection, color application)
A22	Power-plug locking devices	B22	Radioactive material
A23	Relays	B23	Rope
A24	Selection of parts; standard parts	B24	Rubber
A25	Self-tapping screws	B25	Solder and soldering flux
A26	Selenium rectifiers	B26	Standard material (material selection)
A27	Semiconductor devices; transistors	B27	Substitution of materials
A28	Set screws using other than hexagon socket heads	B28	Tape
A29	Shock mounts and vibration isolators		
A30	Terminal boards; terminal strips		
A31	Transformers		

TABLE 3. Index of Items Cross-referenced to Specification Approval—(*Continued*)

Item No.	Item name	Item No.	Item name
		C25	Preferred circuits
	Materials	C26	Power requirements
		C27	Processes; deviation from specified processes; nonstandard processes
B29	Thermoplastic material (rigid)		
B30	Toxic material	C28	Protection against corrosion (dissimilar metals)
B31	Unacceptable material		
B32	Wire (insulated hookup)	C29	Reliability
		C30	Ship's hull (chassis or enclosure for active circuits)
	Design, construction, process, finish		
		C31	Soldering
C1	Antijamming	C32	Special tools
C2	Control panel layout (operating controls)	C33	Threads in plastics
		C34	Through bolting or threading into watertight enclosures
C3	Cooling with heat exchangers		
C4	Design approval and design changes	C35	Tropicalization
		C36	Waveguides and waveguide assemblies
C5	Deviations from specific design requirements		
		C37	Working drawings
C6	Dial illumination		
C7	Dielectric strength and insulation resistance clearance		Documentation, tests, identification
C8	Drip-proof enclosures	D1	Circuit labels
C9	Electrical requirements	D2	Documentation; drawings; manuals
C10	Electronic tubes mounted horizontally	D3	Equipment errors
		D4	General requirements for identification
C11	Equipment weighing over 150 lbs		
C12	Factory adjustment controls	D5	Identification markings and labels
C13	Finishes and special finishes	D6	Preproduction (or design approval tests)
C14	General requirements		
C15	Impregnating, encapsulating, and embedding	D7	Printed wiring
		D8	Nameplates; nameplate facsimile; nomenclature
C16	Interchangeability		
C17	Interference reduction (filters)	D9	Test equipment provisions; test sets
C18	Maintenance		
C19	Mechanized production (including printed circuits)	D10	Test point plan
C20	Mock-ups	D11	Test procedures
C21	Modulator construction	D12	Through bolting in watertight enclosures
C22	Nonrepairable assemblies		
C23	Overload protection, equipment protection, and time delays	D13	Use of equipment subjected to shock test
C24	Parts and unit mountings	D14	Wire coding; external wiring

TABLE 3. Index of Items Cross-referenced to Specification Approval—*(Continued)*

Approval Requirements of Common Military Specifications

Item No.	MIL-I-983D(1)	MIL-E-4158C(2)	MIL-E-5400H	MIL-P-11268D (2)	MIL-E-11991B	MIL-E-16400F	MIL-T-17296D	MIL-T-18870C	MIL-T-21200F
A1						3.4.28.3			
A2						3.4.6		3.4.16.4	
A3						3.4.7.3			
A4	3.6.12.4								
A5		3.2.3*	3.1.3.1*		3.2.3.2.2.1*	3.4.9.2.1			3.1.4.1*
A6	3.6.4		3.1.3.2		3.2.3.2.1	3.4.9.1.1		3.4.16.13..	3.1.4.2
A7		3.2.16.8	3.1.5*	3.38		3.4.11*			
A8				3.3..					
A9								3.4.5.2	
A10							3.8.12		
A11		3.3.1.1.2	3.1.1.2.2	3.70..	3.2.3.38.2	3.4.31..*	3.8.22*	3.4.16.25.2	
A12						3.4.32			
A13								3.4.16.36.3	
A14						3.4.17			
A15					3.2.3.11	3.11.5.1	3.9.5.1		
A16								3.4.16.54.2	
A17	3.7.22.1					3.4.29.5			
A18						3.4.20.5			
A19		3.2.17				3.4.5			
A20	3.6.1.1	3.3.2..	3.1.1..		3.1.7	3.4.1..	3.8.1	3.4.16.2	3.1.1..
A21			3.1.27.5			3.4.28.15			3.1.27.5
A22	3.7.22.1.2								
A23	3.6.7.2			3.58	3.2.3.21	3.4.24	3.8.16		
A24		3.3.1			3.1.6		3.8.1	3.4.16.2	
A25				3.60.4		3.4.28.10		3.6.8.2	
A26					3.2.3.20.1				
A27	3.6.2*	3.3.1.1.2	3.1.1.2.2		3.2.3.38.1	3.4.31*	3.2.24*	3.4.16.48.5	
A28						3.4.28.12			
A29			3.2.3		3.2.3.24.4	3.11.8.4			
A30						3.4.11			
A31				3.67.1					
B1					3.2.1.10.1.1	3.6.3			
B2	3.7.11.5								
B3				3.7.1.2					
B4	3.4.8.3			3.17.2	3.2.1.13.3	3.11.1		3.7.19..	
B5	3.4.10				3.2.1.15				
B6	3.4.9.1								
B7	3.4.9.1								
B8							3.7.2		
B9	3.8.12..								
B10				3.14					
B11		3.4.4.1*	3.1.12*						3.1.13*
B12					3.2.1.16				
B13					3.2.1.16.1				
B14				3.22.2	3.2.1.21.1.2				
B15				3.33.7					
B16		3.5.7.2.1							
B17	3.4.8.2	3.4.12			3.2.1.10.2	3.5.7.2	3.7.6.3	3.7.19.10	
B18					3.2.1.18				
B19		3.4	3.1.1..		3.1.7				3.1.1..
B20								3.7.20	
B21	3.4.9.4							3.7.21..	
B22								3.7.9	
B23					3.2.1.9.6				
B24				3.20					
B25						3.11.3	3.7.4	3.7.10	
B26								3.7.1	

* Approval information will be found in another specification that is referenced in this paragraph.

TABLE 3. Index of Items Cross-referenced to Specification Approval—*(Continued)*

Approval Requirements of Common Military Specifications

Item No.	MIL-I-983D(1)	MIL-E-4158C(2)	MIL-E-5400H	MIL-P-11268D (2)	MIL-E-11991B	MIL-E-16400F	MIL-T-17296D	MIL-F-18870C	MIL-T-21200F
B27	3.42	3.4.1						3.7.7	
B28									
B29									
B30	3.4.6					3.5.3.1	3.7.4	3.7.10	
B31	3.4.3.1					3.5.3..			
B32				3.22.2	3.2.1.21.1.2				
C1			3.2.10						
C2				3.39.8		3.13.14		3.4.5.2	
C3		3.2.6	3.2.5					3.5.6.3	
C4	3.9.3.2			3.3.3					
C5		3.2							
C6	3.8.11								
C7	3.8.13								
C8									
C9				3.2.3..					
C10	3.7.22.1.1								
C11		3.2.32							
C12						3.13.10		3.4.5.3.1	
C13				3.26	3.2.1.12				
C14	3.2.3..				3.1		2.3		
C15	3.4.11.2					3.5.9.2			
C16						3.4.4..			
C17	3.8.4.2								
C18			3.2.12..						
C19		3.2.16..	3.2.1..						3.2.1.1
C20							3.9.2.2	3.11	
C21						3.3.2			
C22		3.2.16.4							
C23					3.3.10.2				
C24							3.4.6	3.5.14..	
C25						3.9.2			
C26									
C27	3.3..	3.5.1			3.1.7				
C28	3.4.15..	3.2.7*	3.1.8*		3.1.9*				3.1.8*
C29									
C30	3.8.5.								
C31	3.8.6.3..	3.5.5*	3.1.23.1						
C32		3.2.31.2	3.1.28		3.3.19.2	3.11.17	4.5..		
C33	3.7.11.7								
C34	3.7.7								
C35				3.31..	3.3.20*		3.7.5.1		
C36					3.2.3.35.1				
C37	3.9.3								
D1		3.6.9.2							
D2	3.9.3..								
D3	3.9.7.2							3.3.4.3.2	
D4						3.14..			
D5			3.1.19.2						
D6			4.3					4.3.1	
D7	3.6.17								
D8		3.6.1..	3.4.1		3.5.3			3.13*	3.4.1
D9					3.3.18.1		3.6.3	3.2.3.4	
D10		3.2.30.1.1							
D11			4.5						4.4
D12	3.7.7								
D13	4.4.18								
D14				3.3.11					

* Approval information will be found in another specification that is referenced in this paragraph.

for preferability. The list is separated into four sections: Parts; Materials; Design, Construction, Process, Finish; and Documentation, Tests, and Identification. A cross-reference number is listed against each entry in the first part of Table 3 and keyed to the second part of Table 3, which lists the approval paragraphs for each general specification. Note that paragraph numbers followed by two periods indicate that the approval requirements are detailed in a number of subparagraphs. The table may also be used in reverse to check the coverage of particular items by a general equipment specification.

Satisfactory operation of equipment designed to any standard is the responsibility of the contractor, and the use of standards specified by the military customer is in no way to be considered as a guaranty of system performance or of the acceptance of the completed product. Contemplated use of a nonmilitary item in a design is usually supported with a performance need. Nevertheless, trade-offs and changes in other areas should be considered before a decision is reached to propose the introduction of a new or commercial item into the military logistics system. Of course, the ultimate technical responsibility still remains with the designer. When an identical item is identified and described by military, industry, or possibly vendor documents, MIL-STD-143 is used to set the order of precedence. More is said later on component selection for military programs.

The trend toward miniaturization and more stringent reliability not only causes engineers to seek smaller components with closer tolerances but also accentuates the need for new design approaches and closer liaison with component suppliers who can make available sizable portions of equipment in modular form. The use of monolithic integrated circuits, for example, requires reexamination of existing military equipment and test methods specifications before application to a design situation. The effect of the emergence of solid-state electronics on equipment design has been profound. Expanded use of microelectronics will undoubtedly bring about increased numbers of military specifications and standards and guidelines, but where these do not exist, contractors will be expected to expand and exercise their expertise in coping with the modified system's problems.

It is important at the outset that an understanding in writing be reached between the design contractor and the military procurement activity on the necessary changes in and exceptions to the military specifications being invoked. These can be negotiated much more easily prior to final signing of the contract.

Indexes and Sources. The most comprehensive and widely used index of existing military specifications and standards is the Department of Defense Index of Specifications and Standards available from the Superintendent of Documents, Government Printing Office, Washington, D.C. 20402.

This index, which is updated periodically, contains:

Military specifications
Military standards
Federal specifications
Federal standards
Qualified products lists
Industry:
  AIA—Aerospace Industries Association of America, Inc.
  USA Standards–United States of America Standards Institute (formerly ASA, American Standards Association)
  ASTM—American Society for Testing and Materials
  AWS—American Welding Society
Military handbooks:
  Air Force—naval aeronautical standards
  Air Force—naval aeronautical specifications
Other department documents:
  Air Force—naval aeronautical bulletins
  U.S. Air Force specifications bulletins

The index lists only the unclassified Federal military and departmental specifications, standards, related standardization documents, and those industry documents which have been coordinated for DOD use. Qualified Products List information as well as the preparing activities is included.

Individual copies of the listed government documents may be obtained from the Naval Supply Depot, 5801 Tabor Avenue, Philadelphia, Pa. 19120.

In order to provide maximum responsiveness to design contractors' requirements, new and revised releases of those military and Federal specifications and standards (including Qualified Products Lists) which are to be listed in the Department of Defense Index of Specifications and Standards (DODISS) are available to industry on a subscription basis with automatic mailing upon payment of fees. Subscriptions will be accepted on a Federal Supply classification basis for a single class or for as many individual classes as the subscriber chooses. Available classes are listed according to subject (ex. under Group 47, Class 4710, Pipe and Tube) in the Cataloging Handbook H2-1, which can be obtained at no charge from the Director, Navy Publications and Printing Service Office, 700 Robbins Avenue, Philadelphia, Pa. 19111. Subscriptions may be forwarded at any time to the Director, Navy Publications and Printing Service Office, Building 4, Section D, 700 Robbins Avenue, Philadelphia, Pa. 19111, accompanied by a payment covering an annual subscription by class.

Copies of the above-listed industry documents when not available from the Naval Supply Depot can be procured from the following addresses:

ASTM:
American Society for Testing and Materials, 1916 Race Street, Philadelphia, Pa. 19103

AIA:
Aerospace Industries Association, 1725 DeSales Street, N.W., Washington, D.C. 20036

AWS:
American Welding Society, Inc., United Engineering Center, 345 East 47th Street, New York, N.Y. 10017

USASI:
United States of America Standards Institute, 10 East 40th Street, New York, N.Y. 10016

Performance and Equipment Specifications. A contractor must first analyze the pertinent detailed performance specification. This specification is centered in the operational and performance requirements of the equipment to be produced under the contract. Many other items such as size and weight limits, reliability requirements, and specific environmental conditions may be included. Each and every item must be given careful consideration in relation to all the other requirements, whether included directly or specified by reference and therefore existing as part of the specification.

The specialized requirements must be considered in relation to the general requirements specifications for equipment procured by the military services. These cover general performance and test requirements, parts and materials selection, design and construction considerations, workmanship, and other technical and procedural constraints. Examples of the more common general equipment specifications are given under Military Specifications earlier.

A comprehensive summary of representative general requirements specifications is contained in a report entitled "Digest of Military Specifications" issued by General Dynamics and available from this contractor or through IDEP (Interservice Data Exchange Program) *Rept.* 347.10.00.00-D2-02 (General Dynamics No. C-130-5588).

All general equipment specifications prohibit the use of certain material, parts, and practices. Table 4 shows items specifically prohibited by one or more military specifications as indicated by the pertinent paragraph number shown in the applicable column. Deviations are not to be permitted unless permission to waive is received from the cognizant procuring authority.

Among other parameters, the relationship of parts to systems and equipment specifications must be examined for environmental compatibility. Military equipment specifications prepared explicitly for specific missions invariably have more stringent requirements than specifications for the incorporated components. Thus, many available parts, components, and materials, manufactured and fabricated even to standard military specifications, are inadequate to meet these exacting equipment requirements. Therefore, contractors find a need to stipulate their own specifications for these items, invoking requirements as close as possible to the equipment environmental requirements. On occasion, some compensation can be realized in new equipment packaging techniques to meet the extended ranges of environment. Depending solely on the possibility of developing special items to meet the environmental differences can be hazardous, because of the reluctance of vendors to push the state of the art until volume sales can be assured. When specifications are prepared for such special items, either MIL-STD-100 or Defense Standardization Manual M200 must be used to meet contractual documentation requirements.

A condensed environmental comparison of representative equipment specifications is presented in Table 5.

General equipment specifications are supplemented with referenced documents covering design, construction, and testing. As an example, MIL-E-5400 references ANA *Bull.* 400, which lists the latest specifications, standards, and drawings to be used by the contractor when considering parts, components, and processes for his design. MIL-T-5422 supplements environmental testing covered directly. MIL-E-5272 is a comprehensive general-purpose equipment test specification. MIL-STD-202 is a similar specification but covers principally electronic parts. ANA *Bull.* 147 contains the nongovernment documents released for use in the construction of flight vehicles. This bulletin falls in the category of Special Lists as provided for in MIL-STD-143. MIL-STD-454 covers some of the common requirements to be used in military specifications for electronic equipment and is incorporated by reference in general equipment specifications. Although not referenced in any contractual documents or used in any way except as an index, DESC-E List 100 is a ready reference to the documents that are prepared and revised for which the Defense Electronics Supply Center is the agent for the Military Departments and is able to furnish information on specifications commonly referenced in equipment specifications.

Another area of prime importance in equipment design and electronic packaging is electromagnetic compatibility. Military procurement groups reference Military Specifications such as MIL-I-6181 as the governing interference-reduction document on their procurement of electronic equipment. This document is intended to resolve a problem that always faces the engineer concerned with rf interference. Anything less than complete conformity may render the design equipment useless when deployed. Redesign or field alterations are seldom effective and can be economically burdensome as compared with effective engineering during the earliest design stages. MIL-I-6181 requires that contractors submit design plans covering the design aspects insofar as electromagnetic compatibility is involved. Obviously, the cost of preparing the plan and design approaches must be considered at the outset. Similarly, a test plan is required. New military standards are continually being developed, and contractors should keep abreast of these.

In addition to prescribed tests, completed military equipments usually require an examination before acceptance by the procuring agency. A visual inspection of the equipment is made to verify that the materials, design and construction, necessary mechanical measurements, marking, and workmanship comply with the requirements of the contractual specifications. Parts and equipment are examined for workmanship, mechanical fit, loose nuts and screws, application of specified tropicalization treatments, and miscellaneous defects. Controls and fastening devices must be examined for mechanical operation. Wiring, soldered connections, ground connections, welds, finishes, etc., are examined for workmanship. Clearances, dimensions, and mechanical adjustments are measured. Labels and markings are inspected for conformity with approved facsimiles, when facsimiles are required. Examination also may be made for other visual or mechanical defects, similar to those described above, that are contrary to specified requirements for the equipment. Visual and mechanical

TABLE 4. Military Specification Prohibitions
(Courtesy Sylvania Electric Products, Inc.)

Prohibitions	Army		Navy	Air Force		MIL-STD-454
			Ship and shore	Aircraft	Ground	
	SCL-6200	MIL-P-11268	MIL-E-16400	MIL-E-5400	MIL-E-4158	
Materials:						
Flammable materials shall not be used (except when enclosed)			Para.*			Req. 3
No mercury or radioactive materials shall be used			3.5.3.4.1 3.9.12.5			
Fungus nutrient materials shall not be used	Para.		†	Para.		Req. 4
Cotton or linen-based laminated or molded plastic shall not be used for electrical insulation	3.3.14.1, table III, note B			3.1.15.7.2		Req. 11
Flammable material shall not be used in the construction of rf chokes, coils and transformers				3.1.14.1		
Glass-fiber materials shall not be used as the outer covering on cables, wire, or other components where they may cause skin irritation to operating or maintenance personnel or where there is any evidence of glass fibers protruding from surface						Req. 1
Materials as installed in the equipment and under service conditions specified in the specific equipment specification shall not liberate gases which combine with the atmosphere to form an acid or corrosive alkali, nor shall they liberate toxic or corrosive fumes which would be detrimental to the performance of the equipment or health of the equipment operators						Req. 1

No plastic material which softens within the equipment storage or operating temperature range shall be used						Req. 11
Component parts: Molded paper capacitors shall not be used				3.1.3.3		
Paper dielectric (nonmetallic cases) capacitors shall not be used			3.4.9.1.4			
Mica compression-type capacitors shall not be used			3.4.9.2.2	3.1.3.4		
MIL-C-25 Type D capacitors shall not be used				3.1.3.3		
Electrolytic capacitors shall not be used below −40°C			3.4.9.1.1			
Resistor types RN-10, RN-20, RN-25, and RN-30 shall not be used (MIL-R-10509), except in nonrepairable assemblies				3.1.22		
Wire leads on resistors (supported by their leads) shall be not less than 3/16 in. except on printed-circuit boards or nonrepairable items			3.4.36			
Wire leads on resistors (supported by their leads) shall be not less than 1/4 in. except on printed-circuit boards or nonrepairable items				3.1.22.3		
Relays containing mercury in any form shall not be used			3.4.24.1			
Metallic oxide rectifiers shall not be used				3.1.31		
Vibrators shall not be used			3.4.35			
Banana plugs shall not be used			3.4.11.1.4			

* Specifies requirement 3 of MIL-STD-454.
† Specifies requirement 4 of MIL-STD-454.

TABLE 4. Military Specification Prohibitions—(*Continued*)

Prohibitions	Army		Navy	Air Force		MIL-STD-454
			Ship and shore	Aircraft	Ground	
	SCL-6200	MIL-P-11268	MIL-E-16400	MIL-E-5400	MIL-E-4158	
Component parts (*Continued*)						
Meters containing radioactive self-luminous markings shall not be used			3.4.20.7			
The performance of equipment shall not be dependent on the selection of individual tubes or other parts			3.4.2.2			
Wax-coated ceramic parts shall not be used in electrical connection						Req. 11
A part shall not be subjected to any ambient hot-spot temperature voltage, current, or power dissipation exceeding that for which the part was designed, including derating curves		3.33.5		3.2.15	3.2.33.6	Req. 18
The temperature of any exposed parts of the equipment shall not exceed 60°C at 25° ambient			*			Req. 1
Hardware:						
Self-locking nuts of fiber binding type shall not be used			3.4.28.15.1			Req. 12
Thread-forming screws shall not be used	3.57.5	3.60.4	3.4.28.10			
Flathead screws shall not be used in sheet or thin materials less than 1½ times the height of the screwhead	3.57.4.1	3.60.3.1	3.4.28.13.1			
External tooth lockwashers shall not be used on front panels					3.2.25.5 2	

Roundhead screws shall not be used for panel mounting			3.4.28.16			
Cone-pointed set screws shall not be used except when the opposing surface has been countersunk to receive the point	3.4.47.4	3.60.5		3.1.27.3	3.2.25.2.4	Req. 12
Mounting parts: Threaded holes in ceramic materials shall not be used for assembly or mounting parts	3.5.18.1.5	3.29.3				
Failure of single rivet or screw shall not free a part completely	3.5.18	3.29.1			3.2.27	
Riveting shall not be used for mounting parts such as capacitors, resistors, transformers, and inductors	3.5.16	3.28			3.2.23	Req. 12
Friction shall not be used as the sole means of preventing fixed parts from rotating	3.5.18	3.29.1			3.2.27	
Nameplates, identification plates, and information plates shall not be mounted by means of rivets, self-tapping screws, or welding			3.14.6.1.2			
Lead washers shall not be used in fastening of brittle materials						Req. 12
Threaded devices securing parts mounted with pliable washers shall not depend upon lockwashers as a locking device						Req. 12
Enclosures: Ventilation openings shall not be located in the top of the equipment enclosure			3.12.3.1			
Air exhaust openings shall not be located on front panels			3.12.3.1			

* Specifies requirement 1 of MIL-STD-454.

TABLE 4. Military Specification Prohibitions—(*Continued*)

Prohibitions	Army		Navy	Air Force		MIL-STD-454
			Ship and shore	Aircraft	Ground	
	SCL-6200	MIL-P-11268	MIL-E-16400	MIL-E-5400	MIL-E-4158	
Wiring: Wires in a continuous run between two terminals shall not be spliced during the wiring operation	3.5.2.12	3.33.6				
Leads carrying potentials above or below ground shall not terminate in connector pins or other exposed contacts that may be accidentally touched or short-circuited	3.4.13	3.38		3.1.36.1	3.2.4	
Electrical connections shall not depend upon wires, lugs, terminals, and the like, clamped between a metallic member and an insulating pliable material	3.5.2.4	3.33.9.1	3.9.11.2	3.1.35.4	3.2.33.9	
Electrical connections shall not be made by clamping wires smaller than 14 between metal parts other than solderless terminals				3.1.35.2		
No more than 3 wires shall terminate at any one terminal			3.9.12.1			Req. 5
Primary power circuits shall not be directly grounded			3.9.7			
Self-locking nuts shall not be used for ground connections			3.4.28.15.1		3.2.25.3.3	
The bending radius of polyethylene cable shall be not less than 5 times the cable diameter				3.1.35.6		

Twine or tape shall not be used for securing wire and cable	3.5.2.12	3.33.4				
The shield shall not be depended upon for a current-carrying ground connection except with coaxial cables						Req. 1
Ground connection to shields and to other mechanical parts, except the chassis or frame, shall not be made to complete electrical circuits						Req. 1

TABLE 5. Condensed
(Courtesy Sylvania

Specification	Temperature range			Humidity, %	Pressure range	
	Operating		Nonoperating		Operating	Nonoperating
	Cont.	Iterm.				
MIL-I-983D (ships) Amend. 1 Sep. 16, 1965	−40 to 149°F		−40 to +167°F	100		
MIL-I-4158C (USAF) Amend. 2 July 9, 1964	0 to +52°C −40 to +52°C −54°C −40°C +71°C *a*		−62 to +71°C	100	30 to 20.58 in. Hg SL to 10,000 ft *c*	30 to 5.54 in. Hg SL to 40,000 ft *c*
MIL-E-5400H (ASG) Notice 1 Sep. 28, 1965 †	−54 to +55°C −54 to +55°C −54 to +71°C −54 to +95°C −54 to +125°C *a*		−62 to +85°C −62 to +85°C −62 to +95°C −62 to +125°C −62 to +150°C *a*	100	30 to 3.4 in. Hg 30 to 8.89 in. Hg 30 to 1.32 in. Hg 30 to 0.32 in. Hg 30 to 0.32 in. Hg *a*	30 to 3.4 in. Hg 30 to 8.89 in. Hg 30 to 1.32 in. Hg 30 to 0.32 in. Hg 30 to 0.32 in. Hg *a*
MIL-E-8189B (ASG) Amend. 2 July 1, 1964	*j*		*j*	100	*j*	*j*
MIL-E-11991B (MI) Dec. 10, 1964	*p*		*p*	*p*	*p*	*p*
MIL-E-16400F (Navy) Feb. 24,1966 *	−54 to +65°C −28 to +65°C −40 to +65°C 0 to +65°C *a*		−62 to +75°C −62 to +75°C −62 to +75°C −62 to +75°C	95	1,000 psi Hydrostatic	1,000 psi Hypostatic
MIL-T-17296D (WEP) Jan. 3, 1963 *	*j* 0 to +50°C		−55 to +65°C	95	Normal ground	Normal ground
MIL-F-18870C (WEP) Aug. 24, 1964 *	−54 to +65°C −28 to +65°C −40 to +50°C 0 to +50°C *a*		−62 to +75°C	95	700 psi Hydrostatic *j*	50,000 ft
MIL-E-19600A (WEP) Dec. 1, 1959 †	Low High −54°C +70°C or +85°C speci- +100°C fied +125°C limit		−62 to +85°C	†	20,000 ft 50,000 ft 70,000 100,000 ft +	
MIL-T-21200F Notice 1 Sep. 28, 1965 †	−54 to +55°C −40 to +55°C 0 to +55°C *a*	+71°C +71°C +71°C	−62 to +85°C −62 to +85°C −62 to +85°C	100	30 to 3.4 in. Hg 30 to 20.6 in. Hg 30 to 20.6 in. Hg *a*	50,000 ft

a Various values for different groups of equipment.
b Test required.
c SL—sea level.
d Details given—too involved to summarize.
e Type I vibration of MIL-STD-167.
f In accordance with MIL-S-901.
g In accordance with MIL-R-27055.
h In accordance with MIL-I-6181 or MIL-I-11748.
i In accordance with MIL-E-5400.
j In accordance with the individual equipment specification.

Environmental Comparison
Electric Products, Inc.)

Vibration, cps	Shock	Explosive atmos.	Fungus	Salt atmos.	Sand and dust	Radio interf.	Audible noise	Inclination, deg	Ice and snow, psf	Wind, knots	Dangerous radiation	X-ray radiation	Solar radiation	Gunblast	Depth charge	Radio interference	Insects and rodents	Rainfall
e	*f*			*b*		*m*	*d*	45–60						*b*	*b*			
10–20 20–55 for isolators *j*	*j*			*j*		*o*	*d*	10	*j*	*j*		*l*	*d*			*g*		
5–50	15*g* for equipment 30*g* for mounting base	*b*	*b*	*b*	*b*	*n*												
d	*c*	*b*	*i*			*n*												
p	*p*	*p*	*p*	*p*	*p*	*h*		10	*j*	*j*							*d*	*p*
e	*f*		*b*	*b*		*m*	*d*	45 90	4.5	75 oper. 100 non-oper.	*m*							
e	*f*		*b*	*b*		*m*	*d*	*f*										
e	*f*		*b*	*b*	*b*	*m*	*d*	45 90	4.5	75 oper. 100 non-oper.		*d*	*r*					
d †	*i* Except 50*g* in lieu of 15*g*		*b*	*b*	*b*	*q*												
d †	15*g* for equip. 30*g* for mtg. base	*b*	*b*	*b*	*b*	*n*												

k In accordance with MIL-E-4970.
l In accordance with MIL-STD-454.
m In accordance with MIL-I-16910.
n In accordance with MIL-I-6181.
o In accordance with MIL-STD-826.
p In accordance with MIL-STD-810.
q In accordance with MIL-I-6181 or MIL-I-26600.
r In accordance with MIL-STD-210.
* Specific testing outlined.
† Test in accordance with MIL-T-5422.

defects should be classified as major or minor in accordance with the definitions of MIL-STD-105.

COMPONENTS FOR MILITARY PROGRAMS

Component Selection. One of the most important factors to consider in designing high reliability into microelectronic circuits is the correct selection, application, and derating of components. Efforts must be made to select parts capable of a long, trouble-free life; and at the same time the parts must be approved by the contracting government agency. Unfortunately, government specifications and qualified products lists cannot be updated as fast as new components arrive on the market. This often results in a designer's need to utilize a nonstandard and in some cases an unproved part in his circuit.

Early in a military program, generally immediately after receipt of the contract, a parts-selection program must be developed. The following considerations should be included in this overall parts program.

Order of Precedence. Primary among the considerations for a parts program is the order of precedence for selecting components. Many times the government equipment specification or RFQ will designate the order of preferred component selection. However, owing to the large number of nonstandard parts existing in the microminiature field, the contractor generally has to modify and add to the requirements called for by the government.

One method of setting up a parts order of precedence is to categorize the various types of parts available. The list below gives an example of a typical order of precedence for a hypothetical naval microminiature program.

Category 1. Use parts which are listed in MIL-STD-242.

Category 2. Use any other military-approved parts for which a Qualified Products List exists.

Category 3. Use parts which have existing EIA, NEMA, or other industry standards.

Category 4. Use parts which conform neither to military nor to industry standards but which have been used successfully in past programs in a similiar application.

Category 5. Use new high-quality commercial parts.

The above component-selection precedence list is typical for government programs. Note that category 1 parts are preferred over category 2 parts, and so on. Normally, category 1 parts or parts specifically requested by contractual military specification do not require any further justification to the customer. Parts from category 2 on down require varying degrees of justification depending upon the program. In larger programs, a special parts engineering group is given the responsibility of generating a preferred-parts list. This list is merely a more detailed breakdown of the five-category listing shown above. The list could actually give various types of electrical and mechanical components, with comments to indicate their applicability, environmental capability, and category.

Types of Nonstandard Components. There are many different definitions of a nonstandard component. Perhaps the best is that it is a part which is not covered by an approved military standard or which is not listed as preferred (highest category) on the customer's or contractor's preferred-parts list. Some of the more common types of nonstandard parts are listed below. For purposes of illustration, we shall generally use electrical part examples although the same definitions hold true for mechanical and hardware items as well as for materials.

Nonpreferred Value. A part must be considered nonstandard if its value is nonpreferred as far as the government customer is concerned. All other parameters could be military-approved, but the nonpreferred value would result in the part being nonstandard. An example of this type of part would be an RC-20 military-approved composition resistor with a ± 5 percent tolerance and a value of 240 ohms. This resistance value and tolerance are both called out in the applicable military component specification (MIL-R-11), but let us suppose that the customer has a requirement that only 10 percent resistance decade values and tolerances are preferred. He might have this requirement if he did not desire to stock and keep on hand spares for an infinite number of parts. In other words, the customer's efforts at standardization often result in normally standard parts becoming nonstandard or at least nonpreferred for a particular program. Generally some sort of justification is required before a customer will approve a nonpreferred value.

Special Testing. A second type of nonstandard part is one where special testing, over and above the normal military testing, is required. Let us suppose that we have a requirement for a part which is going to be used in some unusual or special condition. In this case, it would be necessary to add a special requirement and possibly special acceptance testing to the purchase specification. Although this special testing over and beyond the military specification or standard enhances the reliability of the part, it still makes the part nonstandard and justification to the customer is required. Examples of special-type applications are:

1. Higher temperature requirement.
2. Other environments over and above the military-standard type such as vibration, shock, life, etc.
3. Space-radiation requirements.
4. Special electrical tests such as a pulse test or transient capability, or a more detailed test such as a transistor beta test at a different from normal collector current.
5. Tightening up of an electrical parameter such as a tighter dc leakage requirement on a capacitor.
6. Necessity of modified package. Perhaps a part is needed which is ¼ in. shorter than the military standard but in every other respect is identical to the standard part.

New State-of-the-art Parts. Another type of nonstandard part, especially important in the field of microelectronics, is the brand-new part for which no standard equivalent exists. These parts are generally components which offer something dramatically new such as a smaller-volume package, superior environmental and electrical performance, etc. Often a new design concept has to be based on this type of new component. The military-specification system cannot keep up with the large influx of new parts, especially microelectronic parts. New state-of-the-art parts have to be used in new designs, and the government is fully aware of this necessity. However, the contractor must be prepared fully to justify to the government the selection, use, and application of such parts and in many cases must include demonstrated reliability and life-test information.

Assemblies and Modules. Assemblies or modules are often considered as a part when one is responsible for designing a whole system. For example, a manufacturer may be building a complete system consisting of various subsystems further broken down into assemblies, subassemblies, and finally individual modules. In many cases these modules are composed of various microminiature components (semiconductor and capacitor chips, resistance elements, etc.). However, the module is potted, tested, and supplied to the customer as a throwaway item. In this case, the module would be considered a nonstandard part. Generally, the military customer does not require knowledge and further information regarding the components used within an encapsulated nonrepairable module, and thus the module can be considered to be "one nonstandard part." Occasionally, a semiconductor complement report is necessary covering briefly all the semiconductor discrete devices used within a given module.

Miscellaneous Mechanical and Hardware Items. The final general classification of a nonstandard part would be miscellaneous mechanical, material, and hardware items not covered in existing military specifications or MS data sheets. These hardware items are not too common in the microminiature design areas, and, in most cases, nonstandard-parts justification to the customer is not necessary. Again, most often you will find these items inside a module (see paragraph above) and therefore they can be considered a part of the final throwaway item. Exceptions to this would include epoxies and finishes used outside a module.

Component Applications. Of prime concern to the designer is the correct application of various components in his circuits. Unfortunately, not too much technical material on component circuit application is available. The majority of the information obtained comes from experience gathered by the specific designer. Many companies today have specialized engineering groups (standards, component, and part-application engineers) specifically trained in the component or materials areas. These specialists should be consulted whenever they are available. Other sources of component-application information are briefly outlined below.

Military Standards and Publications. The U.S. government has published and continues to maintain military standards on all basic generic electronic-component types. For example, MIL-STD-701 and 750 cover semiconductors, MIL-STD-199 covers resistors, and MIL-STD-198 covers capacitors. These standards cover various application information in addition to listing the various types available. With more and more present-day emphasis on miniature parts, these standards are periodically updated.

Vendor-applications Data. Technical papers and application guides published by the various component vendors still comprise over 75 percent of the available

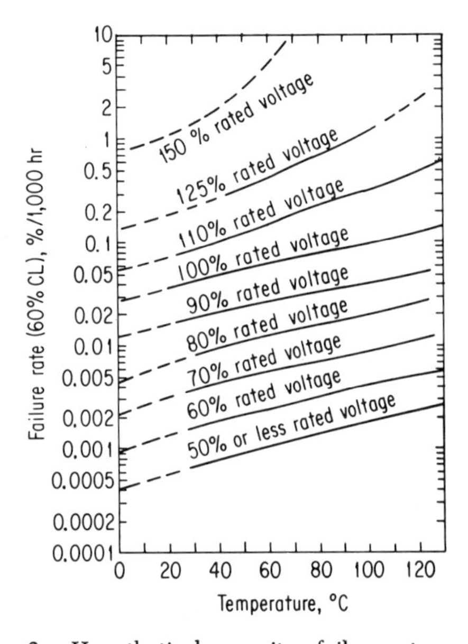

FIG. 2. Hypothetical capacitor failure-rate curve.

applications information pertaining to parts. Vendors are always up to date and are more than willing to supply you with all pertinent data. Many vendors will even go to the extent of running R & D test programs on a specific parameter of one of their components to satisfy a contractor's application requirement. An example of this might be surge or transient capabilities of a particular capacitor; or radiation resistance of a certain epoxy-cased transistor. In addition to vendor application notes, vendor sales-data sheets also contain much application information. Master industrial-company listings such as the EEM Radio Masters list the large majority of component manufacturers along with their addresses and local representatives.

Reliability Information. When one is specifically interested in reliability information such as failure-rate data or the probability of a particular component of surviving a certain period of operating time, reliability handbooks can be consulted. MIL Handbook 217A (updated in December, 1965) lists probable failure rates for all kinds of components under various operating and environmental conditions. RADC, Farrada, IDEP, PRINCE/APIC, and various other government and private agencies have similar handbooks and failure-rate data available. Most component vendors will also make reliability information available to potential users. Many vendors have a continuing reliability testing program in force at their companies and are thus able to offer proof of reliability backed up by actual component testing under power and temperature. Figure 2 shows a typical graph plotting the failure rate of a hypothetical capacitor as a function of applied power and temperature. Note that the failure curves seem to follow an exponential pattern.

Specifications for Components. It is extremely important that components purchased for military programs be adequately specified. Some type of company drawing or procurement specification should accompany each purchase order. In many cases, a referenced military specification, QPL, and/or MS sheet is more than adequate. However, when nonstandard parts are necessary, additional specification data are necessary.

The type of purchase specification depends mainly on the following factors:

1. Contractual reliability and test requirements
2. Number of components being purchased
3. End item use (breadboard, development model, or deliverable system)
4. Vendor rating (QC, and reliability rating of particular vendor).

Purchase specifications can vary from single sheets listing only general requirements to lengthy documents calling for reliability requirements, part screening, qualification, testing, and detailed acceptance inspection. The list below illustrates the normal necessary requirements for most military contract specifications for nonstandard components.

1. Scope (brief description of part).
2. Electrical requirements. All the general electrical parameters the designer needs, along with parameters necessary to prove the reliability and integrity of the parts. This would include voltage ratings, dielectric strength, power rating, insulation resistance, capacitance or resistance, etc.
3. Mechanical and physical requirements. Information such as weight, size, materials, finishes, marking, and special mechanical stresses should be included here.
4. Environmental requirements. All necessary pertinent environmental requirements such as ambient temperature, altitude, humidity, salt spray, vibration, shock, load life, solderability, etc.
5. Special requirements. Information such as radiation resistance, reliability proof, screening, and any other special requirements peculiar to the particular government contract.
6. Quality-assurance provisions. Some type of quality-assurance statement should be included on the specification. A QA plan for necessary qualification and acceptance inspection should be developed. The depth of QA testing is totally dependent upon the contract and the desire of the contractor to obtain some measure of proof as to the capability of the part to meet the specification requirements.

In addition to the above, a drawing should be included, and a true manufacturer's part number should be assigned. This helps to minimize any confusion as to the identity of the part when spares orders are placed many months after completion of the program. The nearest applicable component military specification should be referenced wherever possible and all efforts should be made to utilize available military standards.

Technical Considerations. The four real major considerations one must evaluate when making the final selection are: size, cost, electrical, and reliability. Actually a trade-off and feasibility study must be made, weighing each of the above considerations in hopes of reaching the most meaningful and appropriate compromise.

There are many examples of the above trade-off studies. Table 6 shows a hypothetical but typical listing of the various alternatives that may be encountered in making a final parts selection.

Vendor Selection. Once a decision is made about which nonstandard parts to use in a system, it then becomes equally important to select the best vendor. Some of the major factors involved in vendor selection are shown below, with appropriate comment.

Vendor Surveys. Visitations by user companies to potential vendors is one very good way to evaluate the technical and production capability of the latter. A most effective way of visiting a prospective vendor is for a technical team composed of persons from purchasing, quality control, and components and reliability divisions to inspect the company at the same time, each one concentrating on his own field of interest.

TABLE 6. Alternatives for Component Part Selection

Option	Vendor	Size, in.	Total cost, dollars	Electrical	Reliability	Remarks
1	A	4×4×1	2,000	Meets all req.	Poor	Low rel. would require extra testing
2	A and B	8×8×2	1,000	Meets all req.	Good	Large size would call for extensive repackaging
3	A	3×3×1	5,000	Meets all req.	Excellent	Part meets all req., but cost is considerably higher
4	C	4×4×1	2,500	Does not meet current req.	Good	Extra circuit needed since the part cannot perform all the necessary circuit functions

Testing. A second very important way of evaluating a vendor is actually to run sample parts through a testing program to assure that your requirements are being met. This often becomes expensive, but the vendor may be prevailed upon to share the cost of the testing, supply free samples, or actually run some of the tests at his own facility. It is also important to screen such existing test sources as PRINCE/APIC, Battelle, and IDEP, etc., to see whether or not testing has possibly already been run on the particular components.

Purchasing Input. The purchasing organization also has important contributions to make regarding the selection of the vendor. Purchasing generally obtains the necessary cost and delivery information. Moreover, it usually has available vendor rating lists, which show the past delivery performance of vendors. Often they obtain Dun & Bradstreet or other financial and organizational reports on the vendors.

Multisourcing. It is very important to obtain more than one source whenever possible. Efforts should always be made to keep away from designs based solely on one company's highly specialized part. It becomes even more important to have additional sources when large-quantity buys with short lead times are anticipated. Parts engineers should make every attempt to obtain additional sources at the outset of the program. It is never much more expensive to evaluate two or more vendors concurrently at the beginning of a program.

Typical Component Selection for a Military Program. Figure 3 lists the typical process for selecting components for a normal military program.

DESIGNING FOR SHOCK AND VIBRATION

In formulating the packaging design for any military electronic equipment, paramount consideration must be given to the environmental requirements.

Of the standard environments, i.e., temperature, humidity, altitude, etc., the most complicated are shock and vibration.

A general comparison of the environmental requirements imposed by equipment specifications of the Army, Navy, and Air Force is summarized in Table 7.

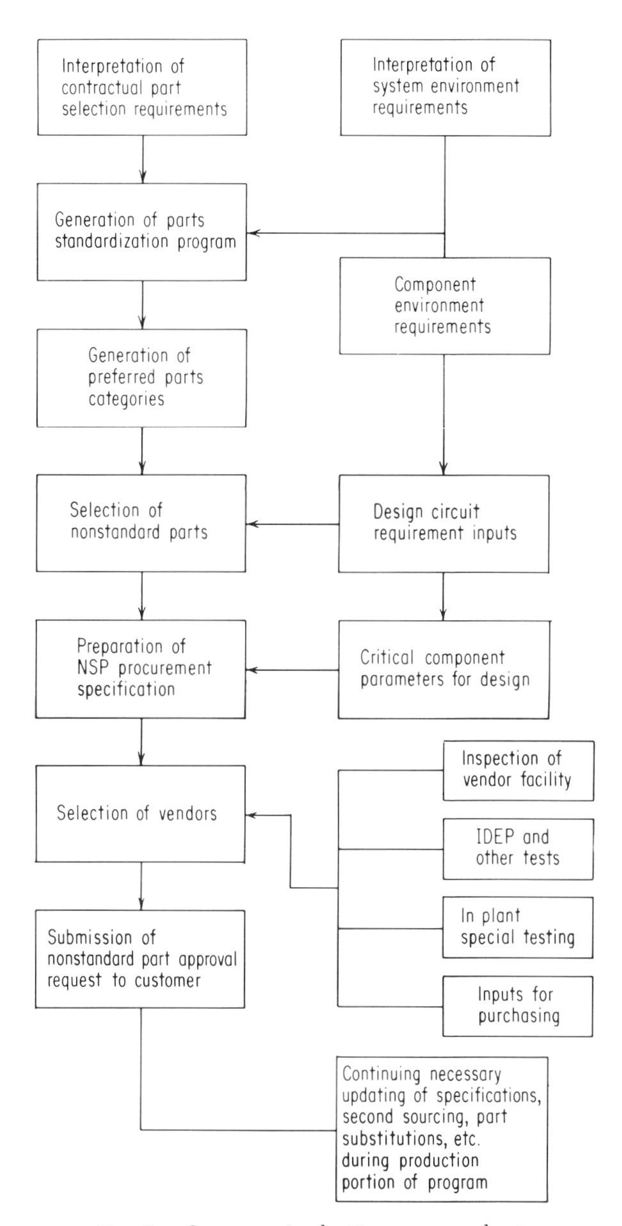

FIG. 3. Component-selection process chart.

TABLE 7. Summary of Environmental Requirements

Environmental tests	Navy	Army	Air Force
Temperature, °C: Operating Nonoperating	MIL-E-16400E −28 to +65 −62 to +75	Specified (typical) 0 to 50 −62 to +72	MIL-STD-810A −54 to +55 −62 to +85
Humidity, %	MIL-E-16400E 95 + condensation	MIL-STD-170 98	MIL-STD-810A 100 + condensation
Shock	MIL-S-901C 250 to 600*g*	Specified (typical) 10*g* random	MIL-STD-810A 20*g*, 10 msec
Vibration, cps	MIL-STD-167 5-15 at 0.060DA 16-25 at 0.040DA 26-33 at 0.020DA 2-hr endurance at resonance 3 planes	Specified (typical) 10-55 cps at 0.030DA 3 planes	MIL-STD-810A 5 to 2,000 cps (varies) 2 to 15*g* (varies) 3 planes
RF interference	MIL-I-16910C 14 kc to 10,000 Mc	MIL-I-11748B 14 kc to 10,000 Mc	MIL-I-26600 30 cps to 10,000 Mc
Altitude, ft: Operating Nonoperating		Specified (typical) 10,000 50,000	MIL-STD-810A 50,000 50,000

The extremes of shock as imposed by shipboard installations and vibration by space and airborne requirements will significantly influence or even dictate the packaging-design solution.

Many of the mechanical-design decisions relative to equipment shock and vibration considerations are commonly not made by specialists in these environments. However, complicated as these conditions are, they must frequently be treated by simple methods.

The following will introduce the packaging engineer to a basic understanding of the shock and vibration problem and present some of the practical approaches toward achieving solutions.

Shock is generally regarded as the application of a large force in a short time. Under such conditions we know from the theory of impulse that an object can momentarily maintain its position in space while the velocity of its parts changes appreciably. The dynamic description of the resulting motions and forces and the establishment of suitable structures safely to withstand these motions and forces constitute the shock-design problem.

While a shock imparts an impulse to a superstructure such as the hull of a ship, parts removed from the hull do not necessarily feel impulsive forces. Instead, the force exerted on any part by another depends upon the dynamic characteristics of the structure and generally can be described as some continuous function of time. There is great danger, however, in assuming that the force exerted by one part on another is independent of the system on which it acts as in a forced vibration. Under such conditions, infinite buildups of motion are possible; however, in shock, flare-ups of motion are possible, but, owing to the finite energy input, infinite buildups are not possible.

In the vibration problem, theoretical amplitudes of infinite measure are possible provided that the structure is tuned to the excitation. This can be illustrated by discussion of the magnification and transmissibility factors of a system due to forced vibration.

The amplitude ratio of the steady-state response to the static response of the system is known as the *magnification factor.* This factor is a function only of the frequency ratio ω/ω_n and the damping factor ζ and is plotted in Fig. 4,

ω = forcing frequency of the system
ω_n = natural frequency of the system, i.e., in the absence of a driving force
ζ = damping factor of the system, i.e., the ratio of the existing damping, C, as compared with that necessary to be critically damped, C_c

Critical damping C_c is defined as the least amount of damping required to cause a displaced mass to return to equilibrium, without oscillation.

When the forced frequency is extremely low (close to $\omega = 0$), the system will be deflected by the force to the amount of its static deflections only. Thus, this amplitude near $\omega = 0$ is nearly equal to unity. On the other hand, for very high forcing frequencies $\omega/\omega_n \gg 1$, the force oscillates so rapidly that the system simply

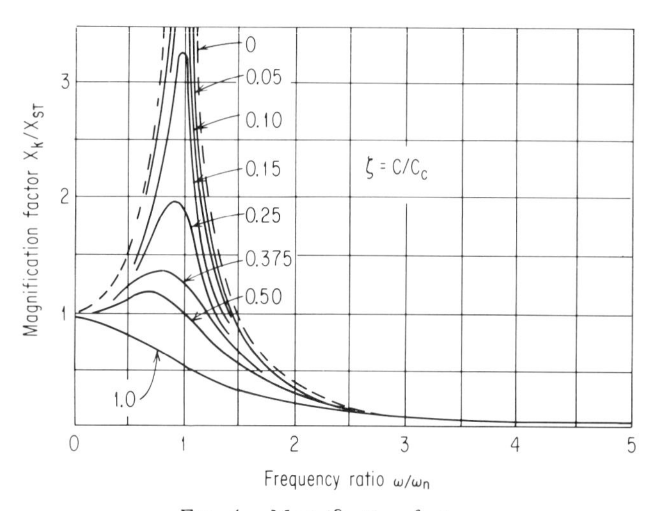

FIG. 4. Magnification factor.

has no time to follow and the amplitude is very small. At $\omega/\omega_n = 1$, the forced frequency coincides with the natural frequency; the amplitude becomes infinitely large. The force can then push the system always at the right time in the right direction, and the amplitude can increase indefinitely. When the natural frequency is equal to the forcing frequency, it is called the *resonant frequency;* and this important phenomenon, where a small force can make the amplitude very high, is known as *resonance.*

The ratio of the transmitted force to the disturbing force is called the *transmissibility* and is plotted in Fig. 5. It should be noted that force transmissibility and motion transmissibility are identical numerically.

Note that all the curves pass through $T = 1.0$ when $\omega/\omega_n = \sqrt{2}$. For $\omega/\omega_n < \sqrt{2}$ the transmissibility increases, and for $\omega/\omega_n > \sqrt{2}$ the transmissibility is less than 1. Therefore, for the isolation of the disturbing force, called *vibration isolation,* ω/ω_n should be greater than $\sqrt{2}$. Again note that, at $\omega/\omega_n = 1$, the resonant frequency of the system, infinite buildups of force are theoretically possible.

Design Considerations. The primary object of an adequate shock design is to transmit rather than absorb energy. In any shock design, the observance of a few relatively simple rules will go a long way toward achieving a good design. First and foremost is the aforementioned concept of trying to transmit rather than absorb shock energy. Second is the idea of attempting to make any structure stiffer than the structure supporting it and, thereby, achieving rigid-body response under shock. This rule, if followed, assures that subsystems will behave as rigid bodies and follow the motion of their bases. It does not assure low shock forces, but it does guarantee little relative motion in the subsystem. This rule, in essence, says that no additional shock forces of any significance are generated within a subsystem to cause flare-ups further down the line. In actual practice it is hard to achieve this situation, but nevertheless it represents a desirable goal. The *stiffness and lightness approach* is a very effective design criterion wherein the hardware should be as stiff as possible and as light as possible. This can be considered as the third rule for an efficient shock-resistant design and applies to both the structural design and the electronic equipment which it supports. Greater shock reliability is obtainable by the use of stiff supporting structures with natural frequencies above 35 cps than with flexible

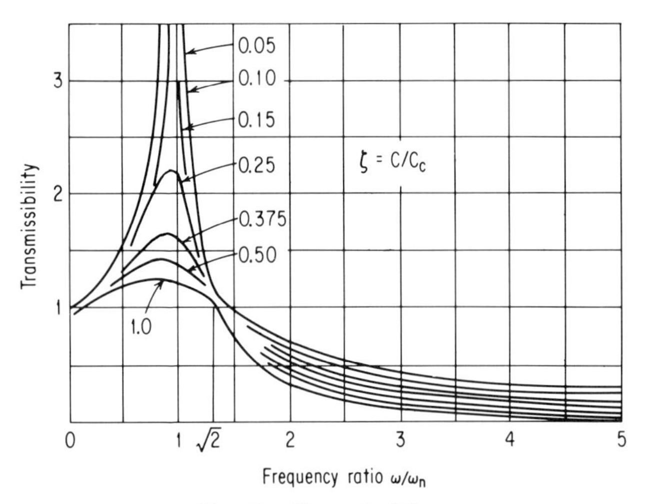

FIG. 5. Transmissibility.

structures, i.e., those having natural frequencies below 35 cps. This statement also applies to the structural design of the electronic components or parts supported by the equipment. This indicates that the addition of unneeded mass and additional spring flexibility in order to lower the natural frequency is undesirable. Instead, such changes usually weaken the supporting structure more than is gained by shock mitigation.

Shock and Vibration Mounts. A consideration of shock and vibration mounts usually arises in the design of equipment to meet the shock and vibration environments. Perhaps, too often they are accepted as a cure-all for the most severe conditions. A decision to use mounts can be made only with a complete understanding of both the environment that the equipment must be protected against and the ability of these devices to accomplish this objective.

Vibration and shock may be defined as the temporary storage and subsequent release of energy with a changed time relation. In most applications, isolation reduces the amplitude of transmitted force or motion.

Vibration isolation is obtained by controlled mismatch of the excitation frequency and natural frequency of the system. Of course, this is not always easily achieved, for there are other aspects such as the following which must be considered: The equipment may have unfavorable natural frequencies. This is generally the case in considering today's military equipment, which contains numerous components and subassemblies, each having their own natural frequency. Next, the flexibility required for vibration mounts may not be obtainable because of high stressing or space restrictions. The excitation function may be a complex wave of variable frequency, which complicates the solution.

Vibration isolation does not usually provide shock isolation to any great degree. The difference between a vibration mount and shock isolator is the time distribution of energy with which each must cope. An ideal shock-protected system is one with a zero natural frequency—the equipment suspended on an indefinitely flexible spring. An actual system does not have unrestricted shock excursions, nor does it have zero excursion; but it usually has some specified distance within which the motion must be confined. In attempts to optimize the arrestment, nonlinear shock-isolation systems are often devised. Generally, most shock isolators are tailored to each situation and are not available in catalogues.

The shock environmental requirements imposed by equipment specifications of the Navy, Army, and Air Force are defined by a shock pulse in (g's) of very short duration (in milliseconds), a condition which may seldom occur in service. Shock requirements are so defined for lack of good field data and the difficulties encountered in proper simulation of the actual shock spectra to which the equipment may be subjected. For this reason, shock isolators may be designed to help a system pass the shock specification tests and then hinder and help to destroy it under shock after it has been installed in service.

The experience of the U.S. Navy has indicated that the proper approach, in general, for shipboard electronic equipment is to eliminate the shock and vibration mounts and to use the stiffness and lightness approach previously discussed.

One popular early method of expressing a shock environment was to present a shock design number, in *g*'s, as a function of equipment weight and shock direction. A figure derived by multiplying the equipment weight times the *g*'s was then used for stress and deflection calculations. Shock theories such as the above, based on acceleration alone, cannot be accurate, since acceleration alone has never caused failure; instead, there must be relative motions within the equipment. Obviously, what has been omitted in acceleration shock theories is the important variable, time.

It must be recognized that all design, whether for shock or not, is an iterative procedure whereby certain configurations are checked for accuracy rather than designed in the true sense. However, the design cannot be assumed satisfactory until it is determined by either shock analysis or possibly by shock test. Analysis would certainly be the least expensive method to prove a design, especially if the equipment is complicated and costly.

The Navy dynamic shock-design analysis method, NAVSHIPS 250-423-30 "Shock Design of Shipboard Equipment," is a practical approach to shipboard-equipment design evaluation for shock environments.

This approach has been applied to several classes of naval surface vessels and is destined to play an increasingly important part in the design of future shipboard equipment. The shock-design method makes use of standard engineering and mathematical techniques. It is basically a normal-mode analysis method, with shock inputs empirically derived from underwater explosion tests on realistic ship and submarine installations. It assumes that an equipment and its foundation together make up a system which responds as a linear elastic structure to the input, which is described by a design shock spectrum. The design method, in brief, requires that the equipment be evaluated as a multimass spring system for which equations of motion are generated, and by use of the method of matrix iteration the normal mode shapes and frequencies (often called the *eigenvectors* and *eigenvalues*) are computed. Forces, displacements, and other criteria are established from which stress and deflection are calculated. The specification itself contains basic theory, calculated examples, and step-by-step procedures to follow for its use.

A similar analysis can be performed for vibration analysis of equipment. The most useful are based again on modeling of the equipment into a multimass spring system to form a mathematical model. Equations of motion using influence or stiffness coefficients are established and, by normal-mode theory of linear elastic structures,[11] the natural frequencies, or eigenvalues, and amplitudes, or eigenvectors, of the system are established. Determination of these parameters is essential for even the most minimal vibration analysis.

This portion cannot cover these prescribed methods of shock and vibration analysis in detail; this must be left to the list of referenced material.[11–22] However, in the use of any analysis method which requires the analyst to select a model of multimass to represent the mechanical system, experience is the best teacher.

A simplified model can eliminate many calculations and also reduce possible errors; a complicated model may produce more design information, but the analysis requires more time and money. Engineering judgment must be exercised to the fullest extent to achieve a proper balance between simplicity and accuracy.

Shock and Vibration Testing. The purpose of testing is manifold. It provides performance data for establishing reliability; it provides engineering data for use in design; and it provides a measure of acceptability for the buyer and the seller. From it, the designer can measure the performance built into the equipment and adjust his design accordingly.

A good test should impose conditions of greater severity than normal field conditions, thus allowing for service extremes, reserve potential, measurement error, and test variations.

The shock test establishes the response of the equipment to suddenly applied forces. It is one of the most difficult to define, measure, and perform because the duration of the shock pulse is usually very short and may seldom occur in service. Consequently, good field data are usually not available to define tests.

A given configuration of equipment becomes its own shock instrumentation. It displays, without recording or direct reading, the effects of the shock by its manner of response in terms of both observable motion and final condition. It is from repeated observation that a number of standardized shock-test machines have evolved.

There are three shock-test machines currently used to evaluate military equipment subject to severe shock environments. They are rated by the size of the equipment to be tested and are described along with test procedures in MIL-S-901C.

High-impact Shock-test Machine for Lightweight Equipment. The lightweight test machine is for equipment weighing 250 lb or less. It is a fabricated frame structure with two 400-lb weights to deliver the blows. One weight pivots on an arm about 5 ft long and may be lifted to the horizontal for maximum blow. The hammer strikes an anvil centered on the test platform when the arm reaches the vertical. The other weight may be lifted vertically to be dropped on the anvil on the top of the test platform. The test table, to which the specimen and mounting fixture are fastened, may be pivoted in a number of positions to develop side, back, and top blows. Some snubbers on the table contribute to the characteristics of the shock.

The test pattern is nine blows of 1-, 3-, and 5-ft heights along the three principal axes of the equipment. Equipment with two or more operating conditions must be tested with nine blows in each condition.

High-impact Shock-test Machine for Medium-weight Equipment. The medium-weight machine is for equipment weighing between 250 and 6,000 lb. The blow is struck by a pivoted hammer weighing 3,000 lb swinging on an arm of 60 to 72 in. length. The hammer strikes the underside of the 4,000-lb anvil plate at the horizontal with an upward blow. The table-mounted equipment is snubbed to permit vertical travel through a distance of 3 or 1½ in. as specified by test. This machine is rather large and is mounted in a pit on a huge block of concrete which is seismically supported on springs.

The mode of operating during the tests is given in the equipment specification. There are a minimum of six blows required consisting of three groups of two blows. Depending upon equipment and mounting-fixture weight, the hammer height and the anvil travel distance are specified. One blow of each group is with the equipment mounted in an inclined position. A series of standard mounting fixtures are described in MIL-S-901C for use on the anvils. They are described by equipment specifications, or they may be specially designed fixtures.

Floating Shock-test Platform for Heavyweight Equipment. The floating shock-test platform (large) is used for items over 6,000 lb and up to about 40,000 lb. It is a double-bottomed barge with high sides and a lightweight roofing to prevent the water column caused by the explosion from coming aboard. The equipment is mounted and, where possible, operating to design values on the deck. A series of shots suspended 24 ft below the water surface and at 60, 40, 30, 25, 20 ft offset provide the six-test shocks. The charge is 60 lb of explosive.

Other shock machines take on many forms and shapes. Some machines depend upon a free fall of the equipment into various media such as lead or plastic pellets or sand. Some machines attempt to reproduce the shock pulse electrically; this, of course, is limited to small equipment. Instead of the floating barge, a submerged test vehicle may be used to test submarine components.

The good features of the two-hammer shock-test machines include low cost, availability, reliability, standardization, and established acceptance by all agencies involved. Procedures and fixtures for most equipment have been worked out so that test costs are low. The machines have been in use for a long time and service results have substantiated the tests.

Vibration-test Equipment. Modern vibration-test instruments create vibration environments and measure the responses of objects attached to them. The field has expanded rapidly, and many hundreds of specialized instruments are available. Only the basic equipment necessary to reproduce sinusoidal environments will be discussed.

A vibration-test system consists of three major components: a vibration exciter, a power amplifier, and an audio oscillator with automatic signal-level control capabilities. The vibration exciter converts electrical energy to mechanical motion. The item to be tested is attached to the moving table of the exciter and thereby experiences the motion of the exciter table. The power amplifier is necessary to provide the large amount of signal power necessary to drive the vibration exciter. The signal source and control device supervises the vibration test by monitoring a vibration transducer mounted on the exciter. This transducer signal is used as a source of feedback for automatic control of vibration levels.

VIBRATION EXCITER. The exciter (commonly called a *shaker*) can be compared with the electrodynamic speaker; in both cases the cone or table is driven by a coil which is suspended in a dc magnetic field. When the coil passes an alternating current, it is attracted and repelled from the magnetic field with each alternation of the ac signal.

The exciter table is so suspended as to be more than self-supporting, allowing specimens of substantial weight to be mounted thereupon. A common suspension stiffness is 1,000 lb/in. The table is further endowed with special devices to restrain lateral motion and to detect table excursions beyond its maximum displacement limits.

The governing limitation of the exciter's frequency operating range is the mechanical resonance of the moving table (called *axial resonance*). Axial resonance generally occurs above 2,000 cps dependent upon the size of the exciter. Displacement and velocity limitations are common in the region of 1 in. D.A. and 70 in./sec, respectively. Electrodynamic exciters are available in force ranges of 25 to 30,000 lb. For higher force ratings, these exciters may be used in tandem. A single hydraulically operated exciter can produce up to 230,000 lb of force. However, its frequency range is considerably lower than that of the electrodynamic exciter.

The power amplifiers needed to drive such exciters vary up to 150,000 va. The requirements for power amplifiers to drive exciters are more stringent than for most high-fidelity or radio-transmitter amplifiers. The power range is very high, thereby making it more difficult to provide protective devices, sequencing, output transformers, etc. The exciter load is highly reactive, and thus high-plate dissipation ratings are required in the output tube stages. The voltage distortion must be minimized because of the irregular response of the exciter. Any distortion at a submultiple of the exciter axial resonance, for instance, will be magnified by the Q of that resonance. Output impedances of these amplifiers are kept as low as possible to minimize the loading effects of the changing impedance of the exciter.

1. Sinusoidal generator. A common source of sinusoidal signal is from the automatic vibration exciter control (MB/B&K model N-575). It is an electronic servo, an integral part of a special logarthmic wideband oscillator. It can maintain constant displacement or constant acceleration at the exciter with changes in frequency. Also provided is a completely automatic transfer of control function. By means of a vibration feedback signal from the exciter transducer, the oscillator output voltage is varied inversely to the monitored response changes, thereby maintaining a preset exciter level with changing frequency. A vibration meter is an integral part of this instrument; so the vibration level is under constant surveillance by the operator. A variety of manually and automatically selectable servo speeds are available to accommodate the many variances in tests. An automatic cyclic frequency-scanning capability is provided with a variety of speeds to suit various applications of sweep sine tests.

2. Accessories. There are a variety of accessories to expand system capabilities. Among those which are common in sinusoidal testing are: programming devices, integrator/amplifiers, horizontal test tables, automatic signal selectors, sine-noise discriminators, and portable vibration meters, to mention a few.

Typically, a specimen is placed on top of an exciter and subjected to vibratory motion in accordance with the requirements of a specification, e.g., MIL-E-5272C. To accomplish this, the control equipment must be adjusted to create the specified environment and maintain it throughout the course of the test. This process is called equalization. Equalization is accomplished by the adjustment of the signal input to the exciter inversely with the vibration response at the attachment point of the specimen to the exciter. In this way the resonant characteristics of the exciter and the reflected resonances from the specimen are compensated for at the motion input to the item under test. Sweep sine testing is the most common test technique in use today. Sweep sine tests are most frequently used to test those devices intended for use in land vehicles, ships, and aircraft powered by reciprocating engines, since the varying sinusoidal excitation activates all resonant modes sequentially and does so at moderate cost.

The impedance effects of structures cannot be ignored in the determination of the vibration control point. The servo control method of equalization is effective for controlling the vibration level at the location of the control transducer only. This is so because the controlling device has no feedback information relating the vibration levels at other locations on the test specimen. Consequently, the controlled input forces at a single point will travel through the specimen structure modified by the structure impedance. Thus, if measurements are made at a variety of locations on the specimen, they may easily vary by factors of 1,000:1.

Such conditions would undoubtedly subject some areas of the structure to damaging levels far in excess of the intended values relative to the test at hand or to the field vibration environment. To overcome this, it is important properly to correlate the location of input forces at the laboratory test relative to the input forces at the environment. In this way, the necessity for one or more exciters can be determined. In a similar fashion, the necessity for one or more control transducers for each exciter is also determined.

In conclusion, it can be accepted that, through scientific analysis of a component's resonant characteristics and relating this to the component's life properties at resonance, particular emphasis can be placed on the component design to increase its life in environments which excite the resonances.

HUMAN ENGINEERING

Meaning of Human Engineering. Human engineering is the application of knowledge of human limitations and capabilities to the design of systems and components in order to achieve maximum effectiveness at a minimum cost in the operation and maintenance of these systems. Human engineering attempts to minimize the difficulty of the operator's task by designing the equipment, procedures, and environment so that they are compatible with the capabilities and limitations of the personnel operating and maintaining the system.

The term *human engineering* as used here is not synonymous with "human-factors engineering." The latter term is an all-encompassing one which includes the physiological, psychological, sociological, and biomedical aspects of the total system in its special environment. Human-factors engineering is sometimes referred to as the *personnel subsystem.*

Requirements of Specifications. Human engineering can be specified in various ways and in several degrees of involvement in the request for proposal (RFP) or as a contract requirement. The Air Force Systems Command has issued a manual, AFSCM 375-5, which denotes the systems engineering management procedures to be used in the conceptual definition, acquisition, and operational phases of a major system. AFSCM 375-5 indicates 106 action blocks in a functional flow diagram that have to be performed in the development of a system. Of the 106 blocks, 95 require the involvement of human-factors engineers. When responding to an RFP, the bidder's human-engineering personnel should carefully read the statement of work and the contract data requirements list (CDRL). The CDRL specifies the deliverable outputs of the human-engineering tasks. The specification most often called for is MIL-H-46855, entitled "Human Engineering Requirements for Military Systems, Equipment, and Facilities." This specification establishes and defines the general requirements for applying the principles and criteria of human engineering to the concept formulation, definition, and acquisition of military systems, equipments, and facilities. A basic requirement of this specification is the issuance of a contractor's program plan which indicates the human engineer's participation in the systems analysis, hardware development and design, design reviews, and in the hardware and systems testing programs. The Appendix in MIL-H-46855 defines the additional task and data requirement that must be submitted when this specification is invoked on naval systems. MIL-H-46855 refers to MIL-STD-1472, entitled "Human Engineering Design Criteria for Military Systems, Equipment, and Facilities." Whenever this standard alone is called for, the requirement is applicable primarily to the actual equipment design in the areas of controls, displays, workspace design, maintainability and safety.

The Air Force usually designates the latter specifications in its contracts. Other military services sometimes designate the same specification also, but others may specify their own agency specification, which is usually very similar. The U.S. Army Electronics Command now usually refers to MIL-H-46855 entitled "Human Engineering Requirements for Military Systems, Equipment and Facilities." Many U.S. Army organizations will also refer to the following U.S. Army Human Engineering Laboratories (HEL) documents:

HEL Standard S-2-64 "Human Factors Engineering Design Standard for Vehicle Fighting Compartments"

HEL Standard S-3-65 "Human Factors Engineering Design Standard for Missile Systems and Related Equipment"

HEL Standard S-5-65 "An Evaluation Guide for Army Aviation Human Factors Engineering Requirements"

HEL Standard S-6-66 "Human Factors Engineering Standard for Wheeled Vehicles"

Similar documents exist for use on U.S. Navy projects and these usually will be military specifications. Various naval bureaus have their own specification, but in general most embody the same requirements and criteria.

Application of Human Engineering. The human-engineering effort begins in the conceptual phase of equipment design. In order to design equipment, we must know what the mission of the equipment is, i.e., what it must do, when it must do it, under what circumstances it must do it, and by whom it must be done. The procedure for developing this information is called the *systems analysis.*

Systems Analysis. The human engineer participates in the systems analysis by applying his knowledge of the human's capabilities and limitations to the system functions.

Men surpass machines in ability to:

1. Sense or detect small quantities of visual or acoustic energy
2. Recognize and interpret different arrangements of light and sound
3. Exercise judgment
4. Reason inductively
5. Improvise
6. Store information over long periods of time and apply the information when needed

Machines surpass men in ability to:

1. Respond rapidly to a change in control signals
2. Apply great force, precisely, smoothly, and continuously
3. Perform routine tasks with high reliability
4. Store information and erase it when no longer needed
5. Perform computational analysis quickly
6. Compare data accurately and quickly
7. Perform many complex tasks at once

These rules guide the human engineer in deciding which tasks should be assigned to the machine and which to the human in the loop.

Workspace Design. After the completion of the systems analysis an overall layout of the equipment is made. The layout may include many racks and consoles or just a few of each. In making the layout it is necessary to consider provisions for personnel passage for performing tasks. The workspace area is laid out to accommodate 90 percent of the military population, i.e., everyone between the 5th and 95th percentile. Some clearances are indicated in Table 8.

TABLE 8. Workspace Clearance Criteria

	In.	
	Nude body	With bulky clothing
Minimum overhead height for standing	76	78
Maximum allowable overhead reach	76	76
Maximum height required for crawling	31	34
Maximum allowable depth of reach	23	21
Minimum clearance for passing body width	23	27
Minimum clearance for passing body thickness	13	16
Minimum height for bending or kneeling	48	50

All racks shall have a kick space at the front of the rack that is 4 in. deep and 4 in. high

The distance from the front of the rack to any obstacle in front of it shall be not less than 42 in.

For standing operation, work surfaces should be 36 in. above the floor

For seated operation, writing surfaces should be at least 16 in. deep and 30 in. above the floor

The clearance required for leg room under a console is shown in Fig. 6. Figure 7 indicates the configuration for a wraparound console for the seated operator.

Equipment Design. The specific configuration of the controls and displays on the panels in a console or rack layout should be based not on aesthetics but on the functional utilization of the components. The designer should consider how the various components are used in each procedure and the criticality of the procedure. The more critical a procedure, the more important it is to have a proper sequential flow. Under stress more errors will occur in a nonfunctional layout. A link analysis which is a measure of the total movement of the arm from start to finish of a procedure can be utilized for comparison of total travel. The shortest distance traveled by the arm is usually also the least time-consuming. The angles of the panels on a console are critical and dependent upon arm reach, height and location of chairs, lighting, and precision required. For detailed dimensions of the human body refer to Wright Air Development "The Anthropometry of Flying Personnel."

Visual displays. A display must adequately convey sufficient message content to make its utilization worthwhile. The viewer must be able to understand the in-

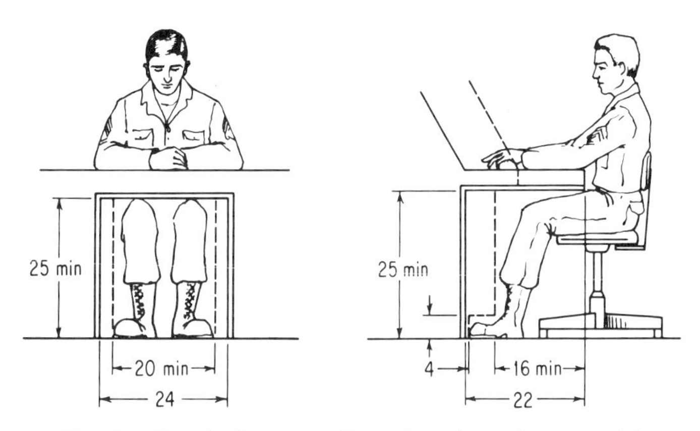

FIG. 6. Console leg-room dimensions (seated operator).[4]

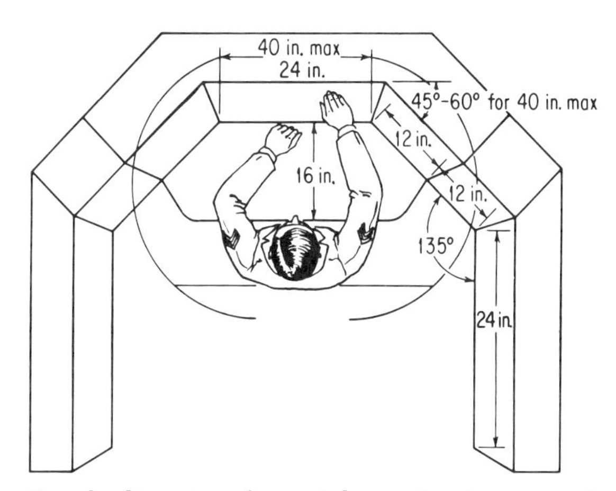

FIG. 7. Console dimensions for seated operator (wraparound panels).[4]

formation without any ambiguity and act upon the information seen. Some of the areas the designer should take under consideration are:

1. Viewing distance
2. Illumination (display and surrounding area)
3. Angle of view (see Figs. 8 to 11)
4. Proximity of other displays
5. Compatibility with associated controls
6. Method of use (qualitative and quantitative reading)
7. Purpose of the display (symbols, pictures, etc.)

Controls. There are four basic types of controls that can be utilized by a designer:

1. Continuous-adjustment rotary controls such as knobs, cranks, pedals, and thumb wheels
2. Discrete-adjustment rotary controls such as bar and pointer knobs
3. Continuous-adjustment linear controls such as joysticks, levers, and pedals
4. Discrete-adjustment linear controls such as push buttons, toggle switches, and detent levers

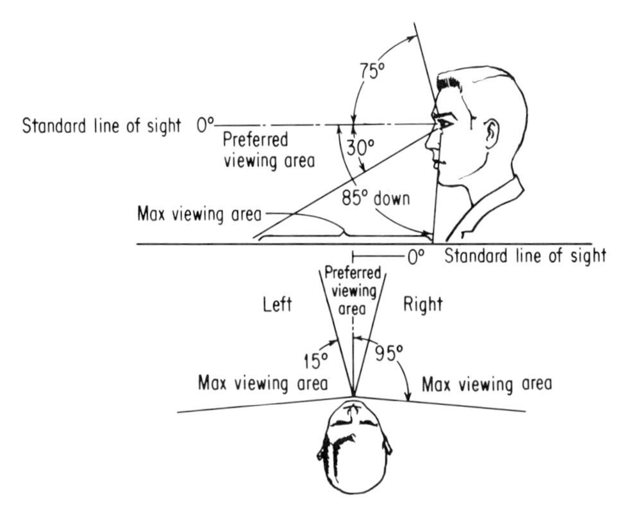

	Preferred*	Maximum*	
		Eye rotation only	Head and eye rotation
Up	0°	25°	75°
Down	30°	35°	85°
Right	15°	35°	95°
Left	15°	35°	95°

*Display area on the console defined by the angles measured from the standard line of sight

FIG. 8. Display viewing area.

Other considerations include:

1. Ease of movement
2. Size
3. Compatibility with associated display
4. Resistance
5. Type of utilization
6. Coding (shape) (see Tables 9 to 12)

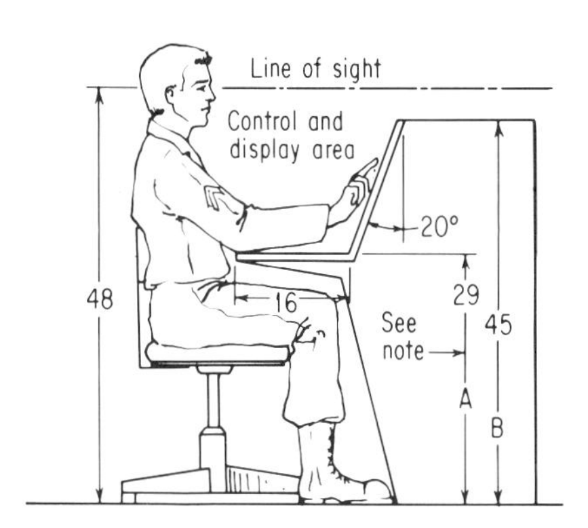

Note: Increase dimension A to 42 in. and B to 58 in. for standing operation with lookover requirement to maintain the same relationships

FIG. 9. Console dimensions for seated operators (simple panel).

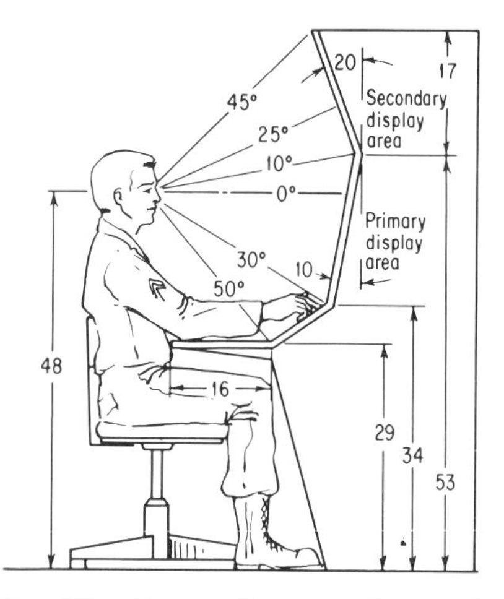

FIG. 10. Console dimensions for seated operators (compound panel).

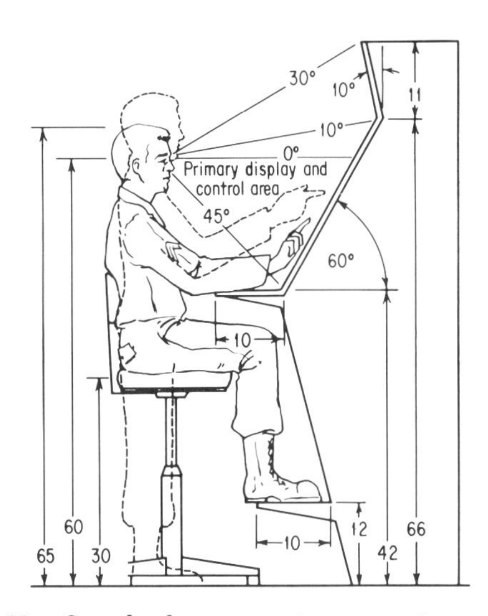

FIG. 11. Console dimensions for sit-stand operators.

TABLE 9. Coding of Simple Indicator Lights [5]

Size/Type	Color			
	Red	Amber	Green	White
½-in. diameter or smaller/steady	Malfunction; action stopped; failure; stop action	Delay; check; recheck	Go ahead; in tolerance; acceptable; ready	Functional or physical position; action in progress
1-in. diameter or larger/steady	Master summation (system or subsystem)	Extreme caution (impending danger)	Master summation (system or subsystem)	
1-in. diameter or larger/flashing (3 to 5/sec)	Killer warning (personnel or equipment)			

TABLE 10. Comparison of Display Coding Methods [5]

Code	Maximum number of items*	Evaluation	Comment
Color	11	Good	Little space required. Location time short
Numerals and letters	Unlimited for combinations of symbols	Good	Little space required if contrast and resolution are good. Location time longer than for color
Geometric shapes	~15	Good	Little space required if resolution is good
Size	5	Fair	Considerable space required. Location time longer than for color or shapes
Number of dots	6	Fair	Considerable space required. Easily confused with other coded items
Orientation of line	12	Fair	For special purposes
Length of line	4	Fair	Will clutter display with many signals
Brightness . .	4	Poor	Poor contrast effects will reduce visibility of weaker signals
Flash rate . .	4	Poor	Interacts poorly with other codes
Stereoscopic depth	Unknown	Fair	Requires complex electronic displays and special viewing equipment

* That generally will give overall accuracies of 95 percent or better.

TABLE 11. Relative Evaluation of Basic Types of Symbolic Indicators [5]

For	Moving pointer	Moving scale	Counter
Quantitative reading	Fair	Fair	Good (requires minimum reading time with minimum reading error)
Qualitative and check reading	Good (location of pointer and change in position are easily detected)	Poor (difficult to judge direction and magnitude of pointer deviation)	Poor (position changes not easily detected)
Setting	Good (has simple and direct relation between pointer motion and motion of setting knob, and pointer-position change aids monitoring)	Fair (has somewhat ambiguous relation between pointer motion and motion of setting knob)	Good (most accurate method of monitoring numerical settings, but relation between pointer motion and motion of setting knob is less direct)
Tracking	Good (pointer position is readily monitored and controlled; provides simple relationship to manual-control motion and provides some information about rate)	Fair (not readily monitored and has somewhat ambiguous relationship to manual-control motion)	Poor (not readily monitored and has ambiguous relationship to manual-control motion)
General	Good (but requires greatest exposed and illuminated area on panel, and scale length is limited)	Fair (offers saving in panel space because only small section of scale need be exposed and illuminated, and long scale is possible)	Fair (most economical in use of space and illuminated area; scale length limited only by number of counter drums but is difficult to illuminate properly)

TABLE 12. Advantages and Disadvantages of Various Types of Coding [5]

	Type of coding					
	Location	Shape	Size	Mode of operation	Labeling	Color
Advantages:						
Improves visual identification	X	X	X	...	X	X
Improves nonvisual identification (tactual and kinesthetic)	X	X	X	X		
Helps standardization	X	X	X	X	X	X
Aids identification under low levels of illumination and colored lighting	X	X	X	X	(When trans-illuminated)	(When trans-illuminated)
May aid in identifying control position (settings)	...	X	...	X	X	
Requires little (if any) training; is not subject to forgetting	...	...	...	...	X	
Disadvantages:						
May require extra space	X	X	X	X	X	
Affects manipulation of the control (ease of use)	X	X	X	X		
Limited in number of available coding categories	X	X	X	X	...	X
May be less effective if operator wears gloves	...	X	X	X		
Controls must be viewed (i.e., must be within visual areas and with adequate illumination present)	...	...	...	...	X	X

The controls should be so located as to make actuation relatively easy, with most critical controls located in the preferred control area (see Fig. 12). For a standing operator, the primary display and control areas are shown in Fig. 13. For several examples of maximum fingertip and reach control location see Fig. 14.

Maintainability. The equipment design must consider maintainability in the conceptual phase and throughout the design phase in order to minimize downtime, increase availability, and reduce cost. Good equipment design will include:

1. Modular packaging (throwaway type where practicable).
2. Replaceable units that can be interchanged and are independent.
3. Accessibility to test points and internal parts should be relatively easy (see Figs. 15 and 16).
4. Self-checking features or test points for auxiliary equipment should be provided.

Other prime considerations are:

1. Type of personnel that will perform the maintenance [in the Air Force it is known as the AFSC (Air Force Specialty Code)].
2. Weight-lifting capacity. One-man lifting capacity is shown in Fig. 17. For heavy equipment, provision should be made for two men to carry item (see Fig. 18) or provide lifting eyes so that mechanical means can be used.
3. Size and shape.
4. Lighting (available and auxiliary).
5. Labeling.
6. Type of test equipment (see Table 13).
7. Type of access opening required (see Fig. 19).

TABLE 13. Advantages and Disadvantages of Four Types of Test Equipment

Type	Advantages	Disadvantages
Built-in	Cannot be lost or damaged independently of prime equipment Requires no special storage facilities Does not need to be transported to prime equipment	May add appreciably to size and weight of prime equipment Will require greater total number of test units because there must be one for each prime equipment Calibration of each test unit may be difficult or inconvenient May increase complexity of and amount of maintenance needed on prime equipment
Go/no-go	Presents information that is clear and easy to read Simplifies decisions and tasks for maintenance man	Unique circuitry is usually required for each signal value to be tested Additional number and complexity of circuits often adds to cost of test unit, to time required for its development, and, later, to maintenance requirements Is usually of little help in checking common voltages or simple waveshapes, except in long sequences that must be checked quickly
Automatic	Can make rapid series of checks with little or no chance of omitting any steps	Is usually large, heavy, and expensive Is usually highly specialized with little versatility. Self-checking features almost essential to detect its own malfunctioning, which adds to cost and difficulty of maintaining it
Collating	Reduces number of indicators technician must read and so reduces checking time and errors Simplifies troubleshooting if it provides indication of which signal, if any, is out of tolerance	Disadvantages similar to those for go/no-go and automatic test equipments It merely indicates that all signals are, or are not, in tolerance, will not aid in troubleshooting

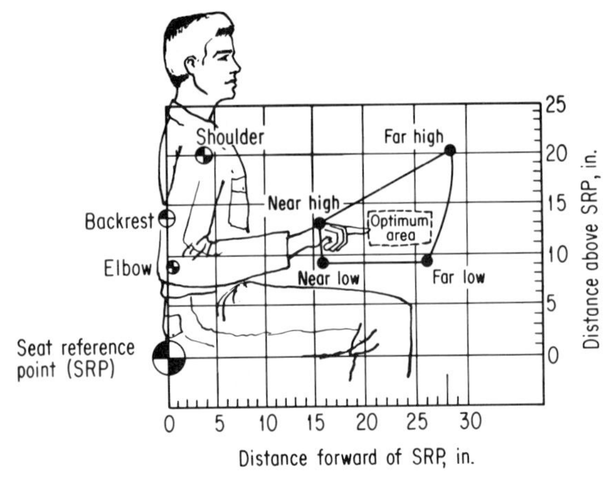

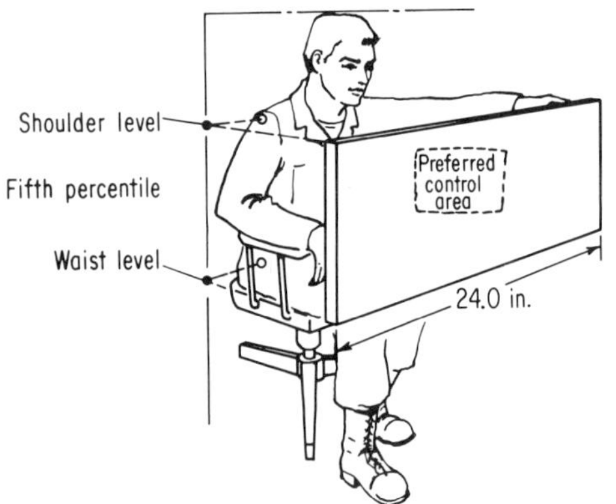

FIG. 12. Seated optimum manual control space.

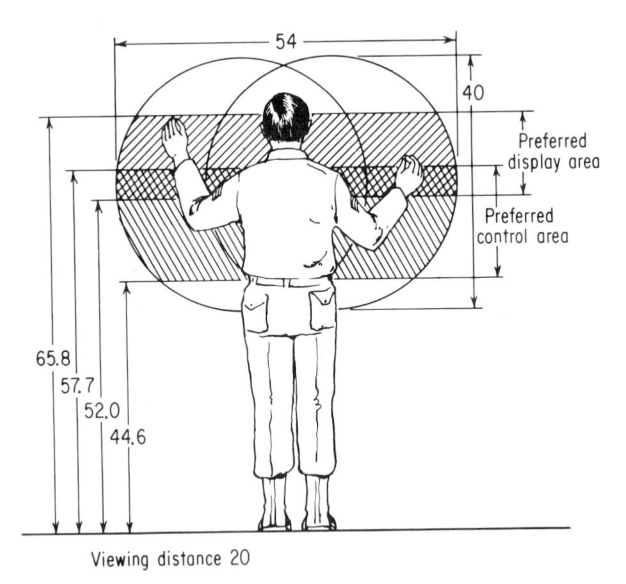

FIG. 13. Console dimensions for standing operators (flat panel).

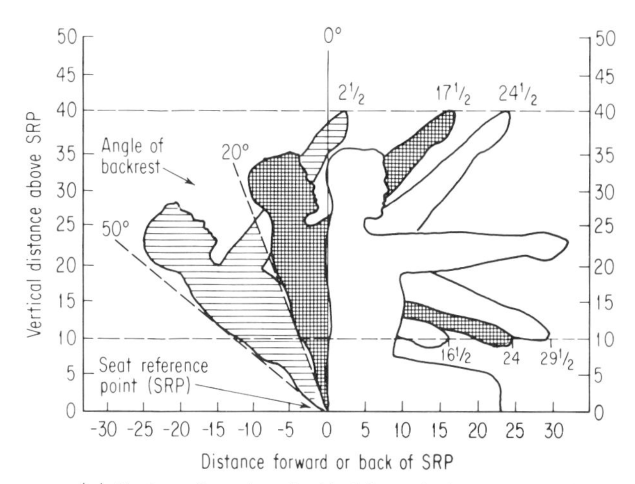

(a) Maximum forward reach with different backrest angles and vertical distances above the SRP, with 0° arm angle

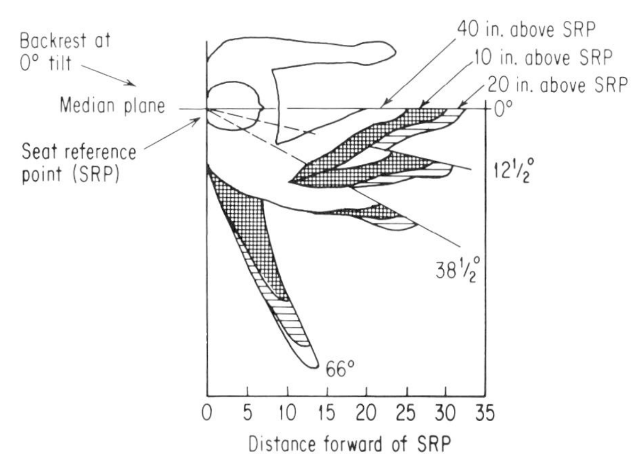

(b) Maximum forward reach with different arm angles and vertical distances above the SRP, with 0° backrest angle

FIG. 14. Examples of maximum fingertip reach.

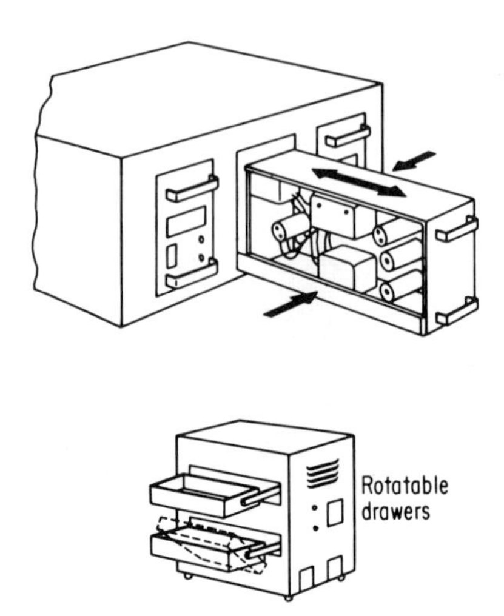

Fig. 15. Access features.

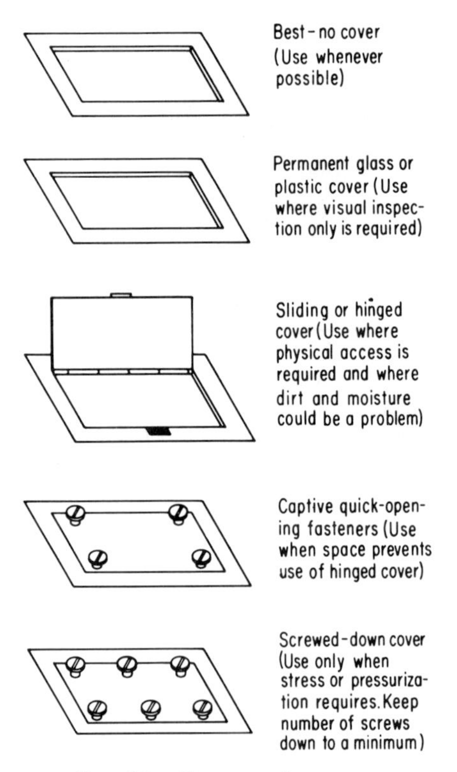

Fig. 16. Covers and accesses.

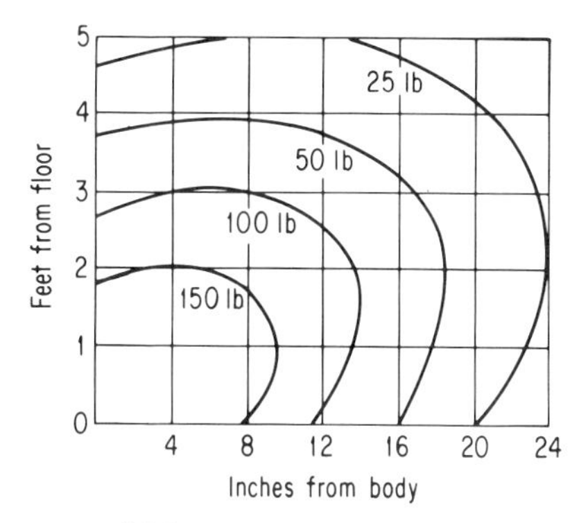

FIG. 17. Manual lifting capacity (using both hands).

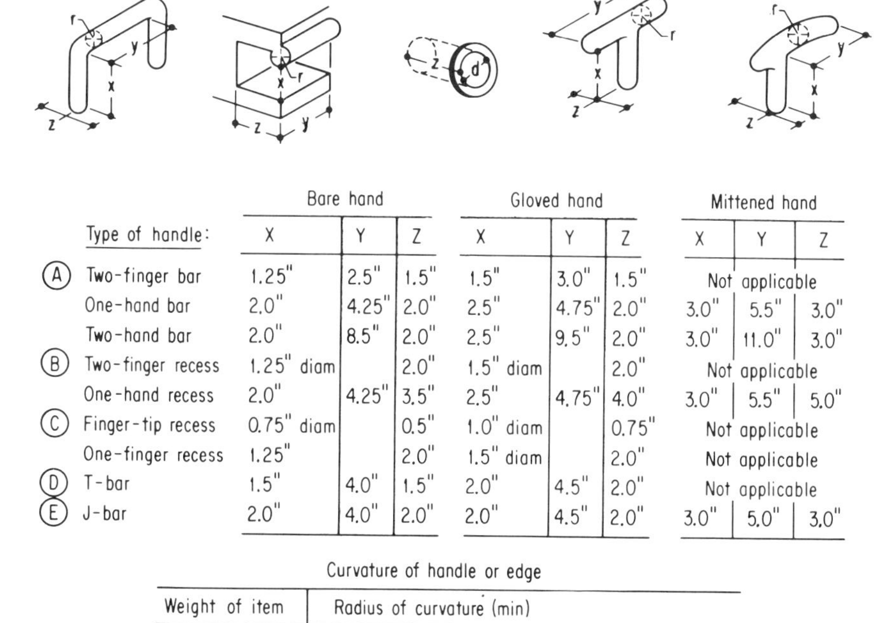

Type of handle:	Bare hand X	Bare hand Y	Bare hand Z	Gloved hand X	Gloved hand Y	Gloved hand Z	Mittened hand X	Mittened hand Y	Mittened hand Z
(A) Two-finger bar	1.25"	2.5"	1.5"	1.5"	3.0"	1.5"	Not applicable		
One-hand bar	2.0"	4.25"	2.0"	2.5"	4.75"	2.0"	3.0"	5.5"	3.0"
Two-hand bar	2.0"	8.5"	2.0"	2.5"	9.5"	2.0"	3.0"	11.0"	3.0"
(B) Two-finger recess	1.25" diam		2.0"	1.5" diam		2.0"	Not applicable		
One-hand recess	2.0"	4.25"	3.5"	2.5"	4.75"	4.0"	3.0"	5.5"	5.0"
(C) Finger-tip recess	0.75" diam		0.5"	1.0" diam		0.75"	Not applicable		
One-finger recess	1.25"		2.0"	1.5" diam		2.0"	Not applicable		
(D) T-bar	1.5"	4.0"	1.5"	2.0"	4.5"	2.0"	Not applicable		
(E) J-bar	2.0"	4.0"	2.0"	2.0"	4.5"	2.0"	3.0"	5.0"	3.0"

Curvature of handle or edge

Weight of item	Radius of curvature (min)	
Up to 15 lb	R-1/8 in.	Gripping efficiency is best if fingers can curl around handle or edge to any angle of 120° or better
15 to 20 lb	R-1/4 in.	
20 to 40 lb	R-3/8 in.	
Over 40 lb	R-1/2 in.	

FIG. 18. Minimum handle dimensions.

Minimal two-hand access openings
Reaching with both hands to depth of 6 to 25 in.:
Light clothing:
Width, 8 in. or ¾ depth of reach *
Height, 5 in.
Arctic clothing:
Width, 6 in. plus ¾ depth of reach
Height, 7 in.
Reaching full arm's length (to shoulders) with both arms:
Width, 19.5 in.
Height, 5 in.
Inserting box grasped by handles on the front:
½ in. clearance around box, assuming adequate clearance around handles
Inserting box with hands on the sides:
Light clothing:
Width, box plus 4.5 in.
Height,† 5 in. or .5 in. around box *
Arctic clothing:
Width, box plus 7 in.
Height,† 8.5 in. or .5 in. around box *

* Whichever is larger.
† If hands curl around bottom, allow an extra 1.5 in. for light clothing, 3 in. for arctic clothing.

Minimal one-hand access openings	Width	Height
Empty hand to wrist:		
Bare hand, rolled	3.75 in. sq	or diam
Bare hand, flat	2.25 in. ×	4.0 in. or 4.0 diam
Glove or mitten	4.0 in. ×	6.0 in. or 6.0 diam
Arctic mitten	5.0 in. ×	6.5 in. or 6.5 diam
Clenched hand, to wrist:		
Bare hand	3.5 in. ×	5.0 in. or 5.0 diam
Glove or mitten	4.5 in. ×	6.0 in. or 6.0 diam
Arctic mitten	7.0 in. ×	8.5 in. or 8.5 diam
Hand plus 1 in. diam object, to wrist:		
Bare hand	3.75 in. sq	or diam
Gloved hand	6.0 in. sq	or diam
Artic mitten	7.0 in. sq	or diam
Hand plus object over 1 in. diam to wrist:		
Bare hand	1.75 in. clearance around object	
Glove or mitten	2.5 in. clearance around object	
Arctic mitten	3.5 in. clearance around object	
Arm to elbow:		
Light clothing	4.0 in. ×	4.5 or 4.5 diam
Arctic clothing	7.0 in. sq or diam	
With object	Clearances as above	
Arm to shoulder:		
Light clothing	5.0 in. sq or diam	
Arctic clothing	8.5 in. sq or diam	
With object	Clearance as above	

Fig. 19. Access dimensions.[5]

Minimal finger access to first joint
Push button access, in.
Bare hand, 1.25 in. diam
Gloved hand, 1.5 in. diam
Two-finger twist access, in.
Bare hand, 2.0 in. diam
Gloved hand, 2.5 in. diam
Vacuum-tube insert (tube held as at right):
Miniature tube, 2.0 in. diam
Large tube, 4.0 diam

FIG. 19. Access dimensions (*continued*).

Safety. Major consideration should be given to safety factors to minimize personnel injury and/or equipment damage. Some of the considerations are:

1. Adequate warning labels should be provided for high voltages, hot equipment, etc.
2. Center of gravity should be marked for lifting ease and to prevent shifting of load.
3. Weight should be marked.
4. Control covers should be provided for switches that may cause damage if inadvertently moved.
5. Interlock switches for doors covering exposed voltage, heat, or rotation hazards.
6. Sharp corners and edges should be eliminated.
7. The result of a human-initiated error should be minimized. On the theory that anything will happen that can happen, controls should be designed so that accidental activation does not cause damage to human beings or equipment.
8. Alarms, audible and/or visible, should be provided to give adequate warning in the event an unsafe equipment situation develops.
9. Communication. Always provide some means of communicating with others in the event that the unexpected does occur.

PACKAGING FOR EMI CONTROL

Introduction. In military applications, electrointerference can cause the loss of communications, errors in data-processing equipment, and inaccurate presentations of radar displays. Such effects can destroy the success of a military mission, with all that such failure implies. This portion is devoted to practical implementation of electromagnetic interference (EMI) control techniques as required for compliance with military intereference control specifications.

Generation of Fields. Any circuit utilizing a time-varying voltage or current is capable of generating electromagnetic energy. With the advent of systems comprised of low-current solid-state devices, the magnetic-field radiation problem is not usually significant except in certain specialized applications. Generally, high levels of attenuation of low-frequency magnetic fields can be achieved only through use of high-permeability or very thick materials. At higher frequencies aluminum provides a degree of magnetic shielding, as can be seen in Fig. 20. Insofar as system compatibility and functional performance are concerned, magnetic fields dissipate very rapidly as a function of distance from the source in the near field. The primary source of interference is from electrostatic fields and plane waves. It is this problem which is of primary concern to the equipment designer.

The interference frequency spectrum generated by an operating unit extends beyond what the circuit designer considers as the operating range. The bandwidth and frequency response of high-speed digital circuits extends to many times the basic operating frequency.

Intraunit Wiring. *Wire Separation.* It is good practice to separate all wiring into a minimum of three categories. These are: (1) primary power; (2) secondary power; (3) control and signal. At any point where these wires must cross, the crossing should be accomplished at right angles.

Wire Shielding. When crossing at right angles is not practical, sufficient isolation can usually be obtained by wire shielding. However, if the wire shield is improperly terminated, it may be totally ineffective. The ideal termination would be a complete 360° connection to the shield and a low impedance connection to ground. Service loops on shield pigtails should be as short as possible.

Filtering. In order to comply with the conducted interference requirements of most military EMI specifications, line filters are usually required on the primary power lines. If the filters are to be effective, complete isolation of the input and output leads is mandatory.

Shielding. *Thickness and Material.* Theoretically, 5/1,000 in. of copper or aluminum provides a total shielding effectiveness of hundreds of decibels at most frequencies. The practical shielding problem is much more closely related to the treatment of the boundaries rather than to the material itself. Except when low-frequency magnetic shielding is required, the statement "If it's thick enough to hold together, it's thick enough as a shield" can be used as a general guide.

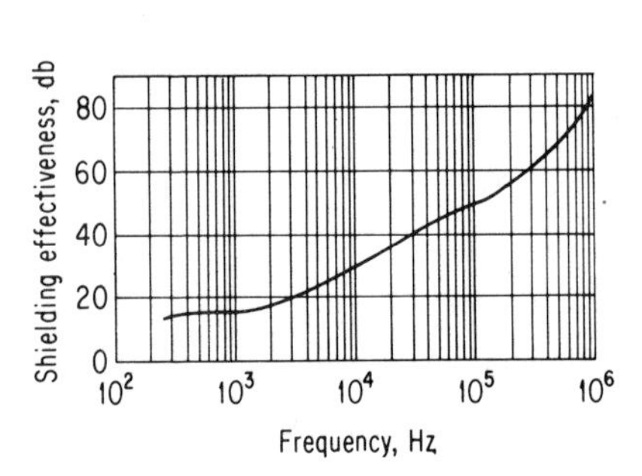

FIG. 20. Magnetic-shielding effectiveness of 3/32-in. aluminum.

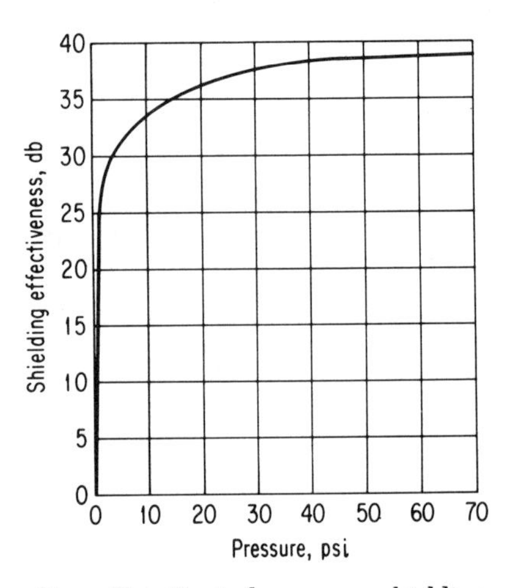

FIG. 21. Typical pressure-shielding effectiveness characteristics of knitted mesh gasket.

Removable Panels. If the two mating surfaces are perfectly flat and rigid, one surface resting on the other would provide an adequate rf seal. In a practical sense, all mating surfaces have imperfections and gaps which permit rf leakage.

1. Screw spacing and gaskets. One way to take up the imperfections is to use screws, closely spaced along the entire interface. As the number of screws is reduced, the gaps become greater and must be filled by other means. Knitted wire-mesh gaskets placed between the mating surfaces and compressed by a uniform pressure are the most common means of sealing joints. Figure 21 provides data on typical shielding effectiveness of this form of gasketing. The following precautions must be taken in using and installing knitted wire-mesh gaskets:

a. Pressure should not exceed about 30 psi in order to prevent the gasket's taking a permanent set. Positive stops should be employed so that the gasket is not compressed to less than two-thirds its original thickness.

b. When a weather seal is required in addition to an rf seal, various types of gaskets which combine the knitted wire mesh with some sort of rubber or Neoprene can be used. Highly conductive resins have been developed which are suitable as combination rf gaskets. These have the advantage of being more resilient and of higher conductivity than Monel mesh.

c. Installation. Rubber cement on the weather seal portion of the gasket is a simple means of holding the gasket in place without additional hardware. However, extreme caution must be exercised to ensure that the cement does not find its way to the conductive portion of the gasket. Figure 22 shows some typical ways of mounting gaskets. When gaskets must be joined end to end or prevented from fraying at the ends, a few stitches of ordinary nylon thread is suitable. Soldered gasket ends will harden and prevent proper compression.

2. Finger stock. Conductive metallic fingers, such as the type used on shielded room doors, are available in many sizes and shapes. When properly used, they are more effective than mesh gaskets in providing an rf seal. However, they are prone to breakage if subjected to any side thrust and should not be used except where mechanical linkages permit only compression. If they are overcompressed, they will take a permanent set and eventually break off.

3. Surface finishes. For any shield to be effective, the mating surfaces must be highly conductive. Lightly applied chromate (Iridite*) finishes are generally adequate, as are most pure metal platings. Anodize is an insulator and should never be applied to a surface which is part of an rf seal. Silver and aluminum "paint" are usually so lightly loaded with metal that the painted surface is nonconductive. Most paints and enamels are nonconductive.

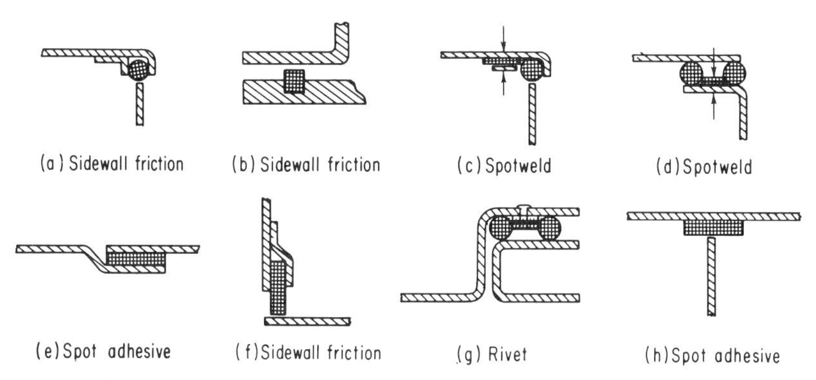

FIG. 22. Typical mounting methods for knitted mesh gasket.

Ventilation. When forced air cooling is required, openings must be constructed so that the shielding effectiveness of the enclosure is not degraded. Wire screen and perforated metal panels are commonly used to cover the ventilation openings. Metallic honeycomb can be used when the air-pressure drop is critical. For electric fields and plane waves, the approximate shielding effectiveness in decibels of a honeycomb panel is equal to 25 times the ratio of the cell depth to the cell diameter. Factors to consider in using honeycomb are:

1. Material. Honeycomb may be made out of any conductive material such as copper, brass, steel, aluminum, etc. The important point is the manner in which the cells are joined together.

2. Construction. Copper, brass, and steel honeycomb are almost always made with the cells soldered or welded. Aluminum is commonly assembled with an epoxy. Honeycomb made in this manner is useless as a shielding material. Certain types of "conductive" aluminum honeycomb are made by drilling through the aluminum sheets before they are stretched into the honeycomb shape. This creates a burr at each cell junction and provides very good conductivity but is suitable only if the

* Trade name, Allied Research Products, Inc.

honeycomb does not have to be treated for anticorrosive properties. The chemicals in the anticorrosive baths tend to seep into these burrs and insulate the cells. Other plating processes are more expensive so that, if a surface finish is required, only honeycomb with the cells permanently electrically bonded together should be used.

3. Mounting of the ventilation grill. For the grill to function as a shield, it must be electrically bonded to the main frame. This joint must be continuous around the entire periphery.

Grounding and Bonding of Hardware. *Structures.* As a rule, the equipment frame is the reference ground for all associated hardware. Any conductor which is not at ground potential can be a radiation pickup and/or emitter. Therefore, all conducting materials, including hardware which is not required to be above ground level, should be bonded to the main frame.

Front-panel Devices. Items such as meter movements, indicator lights, switches, and control shafts are discontinuities in the shielding integrity of the enclosure. By careful planning, these devices can be properly installed without the need for filters.

1. Meters. Metal-cased meters with conductive glass faceplates are satisfactory. In lieu of conductive glass, the feed-through terminals on the back of the meter case can be made with ceramic feed-through capacitors instead of plain terminals, and the meter case becomes its own shielded enclosure. The meter case must be bonded to the main frame.

2. Indicator lights. Most indicator assemblies are available with wire screens either behind or molded into the lamp cover. When these screens are properly grounded, the leakage path is essentially eliminated.

3. Switches.

a. Toggle. Toggle switches are available with copper springs which ground the toggle to the case.

b. Push button. Push-button switches can be effectively shielded by using waveguide-beyond-cutoff fittings and operating the switch remotely.

4. Control shafts. Control shafts should be mounted with wiper arms so that the shaft is grounded.

5. Brute-force approach. Any front-panel device can be shielded by building a shielded bulkhead behind it and pass all wiring through feed-through capacitors or filters. This essentially puts each device in its own shielded room.

By careful study of the circuit and the location of components, the number of filters can be significantly reduced.

Connectors. Whenever shielded cables are used, the shielded cables must be carried through to the mating surface. Before any connector is used in shielded-cable applications, the following factors should be reviewed:

1. The shield should be terminated around its entire periphery. Connector termination hardware should provide this connection without pinching the shield strands to the point where they break. Care should be taken to ensure that the strands are not broken by twisting during assembly or normal usage.

2. There must be a continuous path from the shield to the mating connector. Quick-disconnect types often use a spring with an insulated finish as the contact path. The replacement of the spring with one made of stainless steel represents a considerable improvement in the grounding of the shield. The screw-on type of connectors with conductive surface finishes are usually satisfactory.

THE TECHNICAL PROPOSAL

Most military system-design concepts start with a request for quotation (RFQ) or request for proposal (RFP) from the supply officer of a government agency or bureau. This document usually consists of two major items, the contractual details and the system specification.

Contractual Document. The contractual document defines the type of contract, the supplies or services to be provided, the date quotations are to be submitted, and the delivery dates required. Typical sections of an RFQ may be as follows:

Section 1.0. Supplies or services to be delivered and priced. The deliverable items are listed and described and the quantities indicated.

Section 2.0. Description or specification. The system specifications are identified and any special requirements delineated.

Section 3.0. Packaging. The method in which the equipment is to be prepared for shipment is defined.

Section 4.0. Deliveries. The delivery dates required, usually on an "on or before basis," are listed.

Section 5.0. Acceptance. The type and location of the final acceptance is defined. Preparation of invoices is also usually covered here.

Section 6.0. Referenced provisions. Any general or special provisions are listed here.

Section 7.0. Notice of offerers. This includes a variety of data required to carry out the bid, including the following:

Pricing breakdown
Government-furnished material
Delivery based on a contract award date
Type of contract
Cost-breakdown requirements
Technical proposal details

Section 8.0. Other provisions. Special considerations such as patent rights are covered here.

Technical Proposal Requirements. Frequently, the RFQ will include a listing of the detailed information expected to be contained in the proposal. This may include the following:

1. The prospective contractor's understanding of and design approach to the specification requirements
2. The prospective contractor's prior experience and demonstrated capability with the design, development, fabrication, and test techniques to be used
3. Details of the research, design, development, fabrication, and test facilities to be used in the work
4. A breakdown of the background and experience of the technical personnel to be applied to the program and the level of effort of each type
5. Identification of the major problem areas expected and the manner in which the problems are to be solved

System Specification. The system specification is normally made a part of the contractual document. It describes in detail the requirements of the system to be designed and the specifications that are invoked. Major sections of a typical document are customarily as follows:

1. Scope. General description of the system.
2. Applicable documents. Listing of specifications.
3. Requirements. Specific as well as generalized requirements are listed here. The general requirements may typically state that minimum size and weight, ease of maintenance, and improvements in reliability beyond the requirements of the specification are design objectives. The specific requirements define such items as maximum weight, maximum volume, minimum mean time to failure, interface with other subsystems, environmental performance, and specific electrical operating performance.
4. Quality-assurance provisions. The responsibility for inspection, the types of tests to be conducted, the test procedures to be prepared, and the approval rights of the customer are all defined in this section.
5. Preparation for delivery. The requirements for packaging and equipment for delivery are defined in this section.
6. Notes. Notes may be added defining specific data desired in the technical proposal, such as related work being carried on by the contractor, specific outline drawings of the equipment to be delivered, reports, handbooks, and drawings.

Technical Proposal Implementation. In almost every sense, a technical proposal requires all the same steps as carrying out a complete design program condensed into a short period of time. In general, a packaging design based on the requirements and specifications must be established. Various design approaches must be considered and an optimum approach established. This must be carried to a point where feasibility is established and sufficient detail is available to permit accurate estimating of design and fabrication costs. In cases where the design being proposed is new and untried, it may be desirable to build and test critical experimental units.

GUIDELINES TO THE CONVERSION OF SYSTEM SPECIFICATIONS INTO MECHANICAL SYSTEM DESIGNS

Introduction. The objective of this section is to emphasize those aspects of packaging that are unique to military systems. Previous sections of the handbook have covered specific packaging areas, most of which have applications to militarized equipment when properly specified and controlled. Military packaging places greater emphasis on the following design considerations:

Radio-frequency interference
Galvanic corrosion
Shock and vibration resistance
Environmental resistance
Modularization
Maintainability and spares logistics
Fault isolation
Interchangeability
Adherence to specifications
Thermal management

All militarized equipment must meet the requirements established in the basic system specification and the family tree of specifications that stem from it. This is the focal point of the design. The particular method of implementation will differ with the techniques available to the design organization and the characteristics of the design personnel involved. There is no one way to carry out an equipment design. The optimized design results from trade-offs and compromises. It should be neither overdesigned, with the penalties of increased cost, size, and weight, or underdesigned to the point where it will fail in operation. The optimized system fits into a narrow range between over- and underdesign.

Utilization of Microelectronics. Emphasis in this portion will be placed on packaging microelectronics into military systems, and the majority of the examples will use microelectronics. It is expected that the majority of future military systems will utilize microelectronics to the fullest extent possible to achieve size, weight, and cost reduction and reliability and performance improvements.

A paper entitled "Policies for the Use of Microelectronics in Military Systems and Equipment" addressed to the secretaries of the military departments and directors of all major equipment agencies dated Apr. 14, 1967, and issued by the Deputy Director, Defense Research & Engineering, contains the following:

> An ultimate objective in the area of military electronics is to provide equipment which satisfactorily fulfills the military need with a high probability of no failure for the entire lifetime of the equipment or system. The higher the equipment reliability, the higher becomes this probability and the simpler becomes the logistic support problem. The considerable improvement in reliability offered by microelectronics, the savings in space and weight and potential cost reduction make it most desirable to promote the widest possible, appropriate use of microelectronics in military systems. Further, the reliability of microelectronics circuits is sufficiently high to warrant packaging of several, or even many, such circuits into modules for which repair is neither practical nor effective. Such design modules, to be discarded upon failure, would reduce logistic support costs and further improve reliability. To maximize the potential benefits offered by microelectronics, certain changes in policy and philosophy are issued.

Other excerpts follow:

1. *Use of unitized or modular construction.* Microelectronic circuits shall be packaged into discrete replaceable modules of such cost and reliability that

disposal-on-failure rather than module repair is the most effective and economical logistic support action. Design complexity, reliability, system life, functional use, supply support, cost and equipment availability are typical trade-off factors to be considered in determining whether the module shall have several or many microelectronic circuits.

Hybrid construction using both microelectronics and discrete active and passive electronic devices should also be considered for modularization either for reason of (1) repair through replacement and disposal as in the preceding paragraph, (2) for better accessibility, or (3) for replacement and later repair. In the latter instances, at least microelectronics modules within a hybrid module shall be designed for replacement and disposal-on-failure although the ultimate goal is to package such hybrid construction into replaceable modules of such cost and reliability that disposal-on-failure is the most effective and economical logistic action for this type of construction also.

2. *Maintenance and logistic support.* Several concepts for maintenance and logistic support of microelectronics equipment and modules must be considered during the initial design and followed during the ensuing phases including the logistic support phase in order to achieve the duel objectives of high reliability and minimum support costs. One such concept, module discard-on-failure, was discussed in para. 1 above. Another concept, stemming from the reliability improvement of microelectronic devices is "logistic-self-support" wherein (*a*) built-in redundancy may substitute for replacement and repair or (*b*) replacement modules, accompanying the equipment, may be procured in sufficient quantities to support the total life expectancy of the equipment. Replacement of non-plug-in microelectronics modules shall be accomplished only at higher echelon special repair activities. All factors considered, the achievement of the ultimate objective of high reliability and long-life may lead to economical replacement of whole equipments upon failure.

3. *Intersystem standardization.* The requirement of and the design of standardized microelectronic circuits and modules for broad intersystem use is not favored since (1) this can lead to undesirable constraints on the design of each succeeding specific system, (2) the design and construction of each equipment can be optimized with little additional cost or logistic support difficulty, and (3) the need and value of standardization for logistic aspects disappears when, due to the life characteristic of microelectronics, logistics-self-support can be achieved as opposed to logistic support through supply inventory.

Limited intersystem standardization may be appropriate where the system designer of a new system selects previously acceptable microelectronics design for re-use or, where the DOD Components direct commonality of microelectronics design within two or more systems concurrently under development. Microelectronics commonality between concurrently developing systems should be a subject in each system's Technical Development Plan or other appropriate planning document.

4. *Intrasystem standardization.* Considering the present microelectronics technology along with the Maintenance and Logistic Support concepts outlined in para. 2 above, standardization of replacement modules within a specific equipment or system is a design objective secondary only to system optimization. This intrasystem standardization will enhance the achievement of logistic-self-support with a minimum number of replacement modules required to support the system/equipment for its total life span.

Steps in the Military Packaging Cycle. The generalized steps in carrying through a military packaging program are shown in Table 14. Figure 23 is a simplified flow diagram showing the major design decisions in a microelectronic system and the close relationship that must exist between the packaging engineer and the persons engaged in the other engineering areas, including the seimconductor manufacturer.

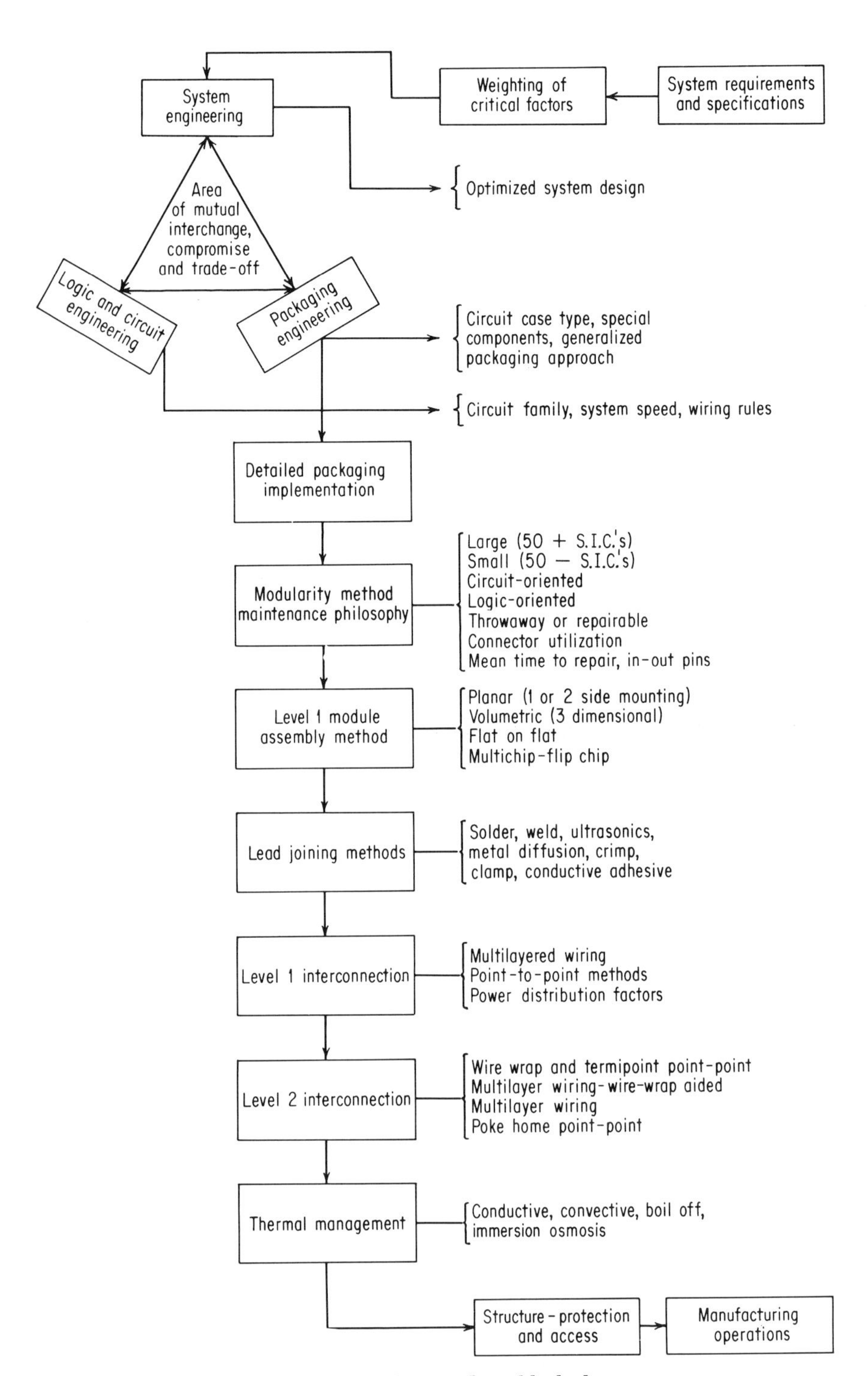

FIG. 23. Micropackaging flow, block diagram.

TABLE 14. Steps in the Military System Packaging Cycle

1. Review overall system specifications. Establish specification tree
2. Review contractual requirements and commitments
3. Establish performance schedule and milestone points
4. Determine circuit-design approach, special requirements, and constraints
5. Establish component requirements and component listing
6. Carry out system partitioning, and define system, subsystem, and module breakdowns
7. Establish and evaluate various packaging-approach alternatives. Finalize optimum design approach
8. Prepare experimental models of critical subassemblies. Conduct tests to define performance characteristics
9. Conduct manufacturing analysis
10. Carry out preliminary design review
11. Prepare detailed layouts of all major subassemblies
12. Carry out critical design review
13. Initiate preparation of detailed manufacturing drawings
14. Prepare breadboard, experimental, or engineering models as required
15. Conduct performance evaluation tests
16. Revise design and drawing set as required
17. Release to manufacturing, pilot or production run
18. Manufacturing support during fabrication and test
19. Revision of design and detailed drawings as a result of manufacturing and test experience
20. Field test programs
21. Revision of design and detailed drawings as a result of field test experience

MAINTAINABILITY CONSIDERATIONS

The approach to maintainability of a military system depends largely upon its expected use. A front-line equipment for ground use may require quickly replaceable discard-on-failure modules. A torpedo-guidance system will stress functional-item replacement whereby a large subsystem or system if identified as the defective unit is removed, a good unit substituted, and the defective unit sent back to a well-equipped ground base for repair and retest.

A number of military documents covering maintainability studies and guides have been prepared. A summary of the steps in designing for maintainability given in one of these documents[9] is provided as Table 15. A book entitled "Maintainability Design"[10] has been published as a result of the Fourth Electronic Industries Association Conference on Maintainability of Electronic Equipment.

The section of this chapter, Human Engineering, also provides guidelines to good equipment design for maintainability.

APPLICATION EXAMPLES—NAVAL ELECTRONIC PACKAGING PROGRAMS

Background. In May, 1967, the Headquarters, Naval Material Command, released NAVMAT P3940 entitled "Navy Systems Design Guidelines Manual—Electronic Packaging." This is a compendium of complementary electronic packaging approaches developed under the direction of naval facilities for the organization and construction of electronic systems. The objective is to reduce the proliferation of unique packaging hardware with advantages expected in cost, reliability, and logistic support.

The foreword of the manual states that "the use of this manual is mandatory in planning the development of future Naval electronic systems and equipment. All new projects, all projects that have not advanced past the planning stage as of this date, and all substantial redesigns, *will be reviewed* to ensure appropriate use of these standard modules and enclosures. The modules and enclosures described herein offer enough variety so that one approach or a combination can be found to fit the *majority* of Naval electronic applications."

TABLE 15. Summary of Steps in Designing for Maintainability

Stage of equipment design	*Steps in designing for maintainability*
Preparation of planning documents	1. Prepare realistic objectives for maintainability
Study of previous systems	2. Review experience on previous systems for application to maintainability of new system
Development of breadboard and experimental models*	3. Establish adjustment points and self-compensating features
Design of developmental model*	4. Determine maintenance levels
	5. Divide equipment into units
	6. Prepare maintenance procedures
	7. Determine requirements for maintenance supports
	8. Review and modify design for maintainability
	9. Translate design for maintainability into actual equipment and maintenance instructions
Testing of developmental model*	10. Make preliminary tests of maintainability, and modify design as necessary
Testing of service test model* (operational suitability test)	11. Make full-scale tests of maintainability, and modify design as necessary
Evaluation of prototype (preproduction) model*	12. Evaluate maintainability of the system
Engineering changes	13. Correct deficiencies in maintainability of production equipment
	14. Review engineering changes for effect on maintainability

* Models are as defined in MIL-STD-243 "Military Standard Types and Definitions of Communications—Electronics Equipment," Jan. 12, 1954.

These packaging designs are the first generation of preferred packaging approaches. They will be subject to revisions, modifications, and changes over the years. These will be necessitated by the continual changes in packaging requirements and technology.

Four packaging designs were selected from development under the following Navy programs:

Centralized Electronic Control (CEC), Naval Research Laboratory, Washington, D.C. (NRL)

Integrated Helicopter Avionics System (IHAS), Naval Air Development Center, Johnsville (NADC)

Standard Hardware Program (SHP), Naval Avionics Facility, Indianapolis (NAFI)

Electronic Packaging System (EPS), Navy Electronics Laboratory, San Diego (NEL)

The CEC program involves a unique approach to the solution of problems posed by the ever-increasing complexity of naval electronic systems. The concept is based on the coordinated control of all shipboard electronic functions by a central time and frequency system (CT & FS). Mechanization schemes for the CEC concept were a natural result when the functional performance of the concept was found feasible.

The IHAS program involves the development of a Central Computer Complex, which is one part of the overall integrated avionics system. The mechanization techniques developed for this computer complex are recognized by the very high packaging density achieved in using microelectronic circuits. This airborne system has to meet severe constraints in terms of space, weight, reliability, and maintainability.

The SHP program is aimed at reducing the number of noninterchangeable modules performing similar electronic functions. In this context both functional and mechanical interfaces are defined. This standardized approach to basic functional modules is flexible enough to be used in many naval systems. The reference base for development was guidance and fire control systems.

The EPS program is aimed at the development of a total package concept with primary emphasis on the physical resistance to adverse shipboard environments. Although no specific functional modules constitute a part of the program, hardware developments were accomplished for mechanical partitioning at all levels within an electronic system.

Details on the various packaging approaches have been extracted from the manual and presented here as typical military packaging approaches. Ground, shipboard, and airborne applications are considered. Additional data on the packaging approaches, design notes, and evaluation criteria for optimized package selection can be found in the Design Guidelines Manual.

Some specialized definitions of certain terms as used in the foregoing discussion on naval electronic packaging will be found in Table 1.

Module Approaches. *Centralized Computer Complex (CCC) Modules*

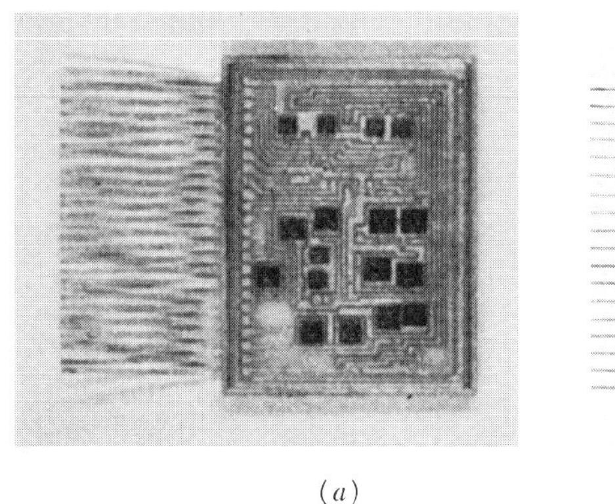

(*a*)

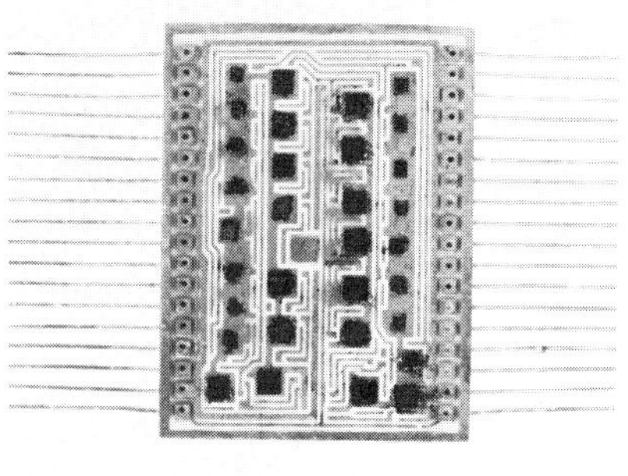

(*b*)

FIG. 24. Typical microelectronic modular assembly (CCC). (*a*) Single-ended MMA. (*b*) Double-ended MMA.

NADC. SUMMARY. The packaging concept for the Central Computer Complex (CCC) of the Integrated Helicopter Avionics System (IHAS) was developed for the Naval Air Systems Command. This concept is the only one of the six "benchmark" systems specifically designed to fulfill airborne requirements.

The Central Computer Complex is one part of IHAS. The packaging techniques developed for this computer complex are significant, particularly in the packaging of microelectronic circuits, in that they represent a major contribution to advancing the packaging state of the art.

The basic building block of the IHAS computer complex is a microelectronic modular assembly (MMA), which may contain up to 32 discrete integrated-circuit chips, each chip being a complex circuit in itself; each MMA therefore represents a complete functional assembly. Figure 24 shows a typical MMA, measuring 1 by 0.75 by 0.68 in. Note the integrated-circuit chips, the gold-plated, etched circuits, and the fine aluminum jumper wires attached.

Heat-removal considerations in the computer start at the MMA, where the basic substrate (alumina) is selected for its heat-dissipation characteristics as well as other electrical and mechanical properties. Heat is passed by conduction from the chips to the substrate to a submodule aluminum frame on which the MMAs are mounted.

A module in the computer consists of four subassemblies of MMAs, each subassembly consisting of five to six MMAs mounted on an aluminum frame. Subassemblies are held together by means of two removal clips mounted at each end of the module.

MECHANICAL CONSTRUCTION. The design concept of the CCC module evolves from the techniques used for integrated-circuit packaging. This technique is one of constructing functional packages of 0.75 by 1.00 in. which contain up to 32 integrated-circuit dice or chips. The dice package, called a *microelectronic modular assembly* (MMA), is a complete functional digital array. This package is one step beyond the individual dice (chips) discrete packaging level and is one step removed from the use of a single chip for multiple-use functions (such as adders and shift registers).

There are a total of 46 different circuit configurations of the MMA in the computer complex. Only six different dice are used in various combinations to achieve these different MMA configurations. (Multiple sources are providing these chips.) Each MMA may have either 24 or 36 external leads. The MMA shown in Fig. 24*a* is single ended with 24 leads. A 36-lead MMA is double-ended, i.e., it has 18 leads on opposite ends of the package, and is used in applications where fan-in and fan-out requirements can best be met by the double-ended arrangement, Fig. 24*b*.

The MMAs are manufactured with standard semiconductor techniques. Figure 25

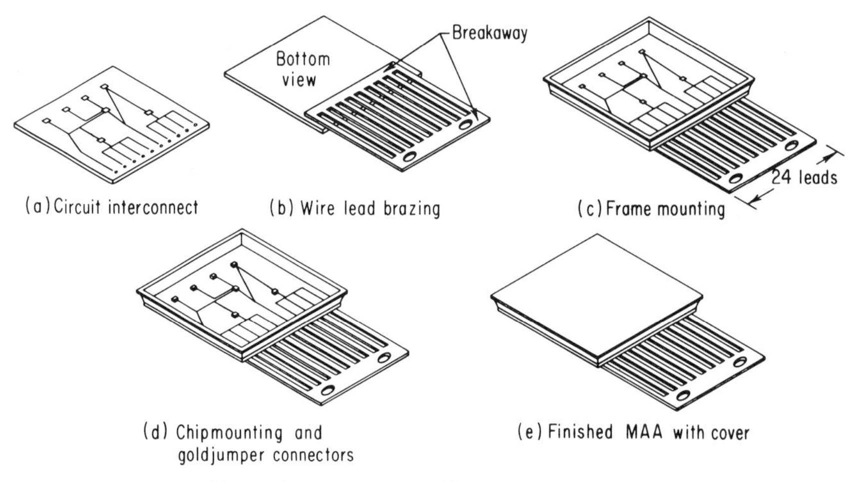

FIG. 25. MMA assembly process (CCC).

shows the order of manufacture used. An alumina substrate is metalized and then etched to form interconnect patterns. Leads are next brazed on, a frame then being welded to the substrate and tested for leaks. The chips are attached and aluminum jumpers added. The assembly is electrically tested, and a cover is welded on to complete the process.

Figures 26 and 27 show the MMAs mounted in a basic submodule. Figure 26 shows one design wherein six single-ended MMAs are mounted to an aluminum submodule frame. The MMAs are connected to a multilayer matrix board, which in turn is connected to the submodule connector. Soldering is used as the means for forming the connections between the MMAs and their matrix board and between the matrix and the connector. Figure 27 shows a five-MMA submodule, with the MMAs being oriented 90° to those of Fig. 26. These MMAs are double-ended, with the multilayer matrix board running between them to provide the running interconnection.

Submodules are used in groups of four to constitute a module. Figure 28 shows a module, with the four submodules held together by means of two clips, which slide over the ends of the outermost submodule frames to hold the module together. Each

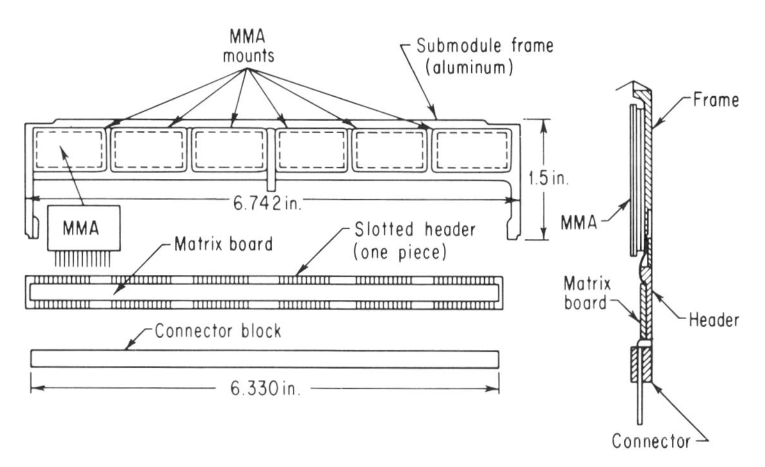

FIG. 26. Six-MMA submodule assembly (CCC).

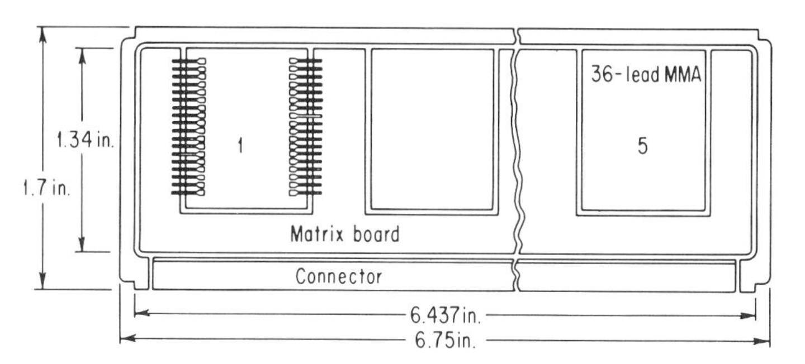

FIG. 27. Five-MMA submodule assembly (CCC).

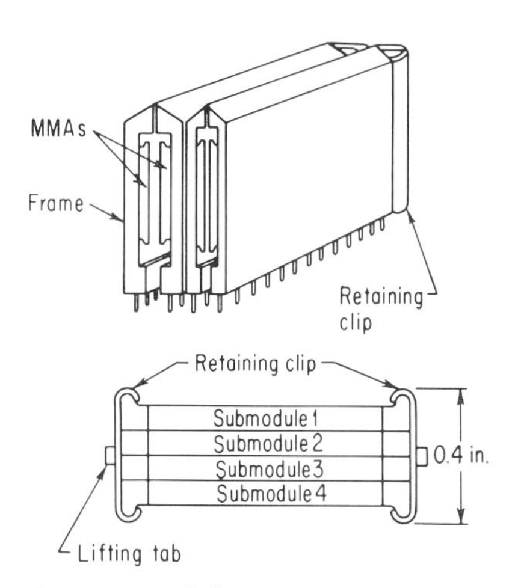

FIG. 28. Module construction (CCC).

module is approximately 0.4 in. in width and will therefore have either 20 or 24 MMAs as the basic functional unit.

The module design package provides a very compact lightweight and efficient package for the basic function building block. High-reliability assurance is reflected in unique design features and in the technologies used. Furthermore, a number of particular features of the module design contribute to effective maintainability. One such design feature is that the module is composed of four submodules, which are grouped or retained together by two clips. Submodules are basically identical except for the MMA complement and subsequent integrated-circuit usage.

The four submodule assemblies are joined to form a module as shown in Fig. 28. Each submodule is made up of five or six MMAs and the following items as illustrated in Fig. 29:

1. Frame
2. Header
3. Connector

For purposes of illustration, the six MMA submodules will be discussed in detail.

Item 1, the submodule frame, is the basic structural member as well as the MMA heat-transfer conductor. The frame is fabricated of aluminum alloy and finished in black anodize. It can be produced by casting or cold forming of wrought alloy. The frame includes the following design features:

Six individual bays for the mounting of the MMAs are maintained flat and finished to close tolerance. The ceramic substrates of the MMAs are bonded to the frame with the bonding agent filling the 1 to 2 mils air gap that may exist in areas under the MMA.

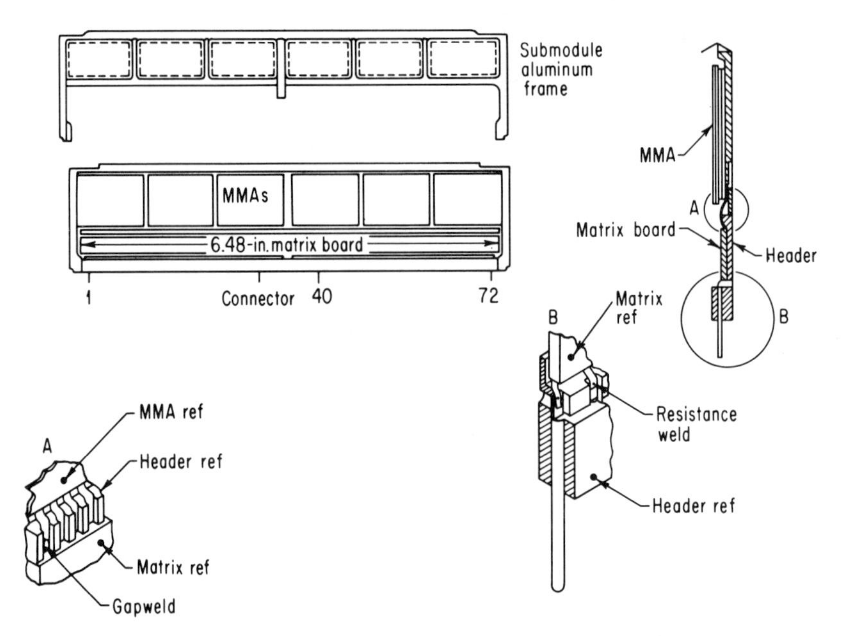

FIG. 29. Submodule assembly construction (CCC).

Between these bays, small rib sections provide rigidity to protect the MMA mounting and serve to reduce the temperature gradient across the light frame section.

At each end, the frame includes a full section "leg" that extends the structure to the connector portion of the module. This part of the frame also includes a slot for the module-retaining clip and an indexing notch that provides for "stepping" removal. This notch is differently indexed for each submodule in a module assembly as shown in Fig. 30.

The top edge of the frame is provided with a triangular cross section. The purpose of this is to provide lateral support along the length of the thin section module in vibration and shock. In addition, this surface is the heat-transfer interface to the heat exchanger, and the slope increases the effective area.

Weight of the frame is 8.5 gr. The back surfaces of the frame are recessed to reduce weight significantly while retaining adequate heat-transfer and structural characteristics.

Item 2 is the submodule header, which provides a continuation of the submodule structure in a dielectric material. The header is fabricated of epoxy glass laminate material and is bonded to the frame. Two notched ridges extend along the face of the header. The upper one is precisely notched to match the leads from the MMAs and the matrix. The lower ridge is notched to match the connector pins and the matrix. The purpose of the notches is to discretely position and retain alignment of all leads for the resistance-welding operation on assembly.

Item 3 is the submodule strip line connector. This connector provides the electrical interface with the housing-assembly motherboard connector. It has 72 pin positions, with one oversize pin used for polarization and alignment. The pins are

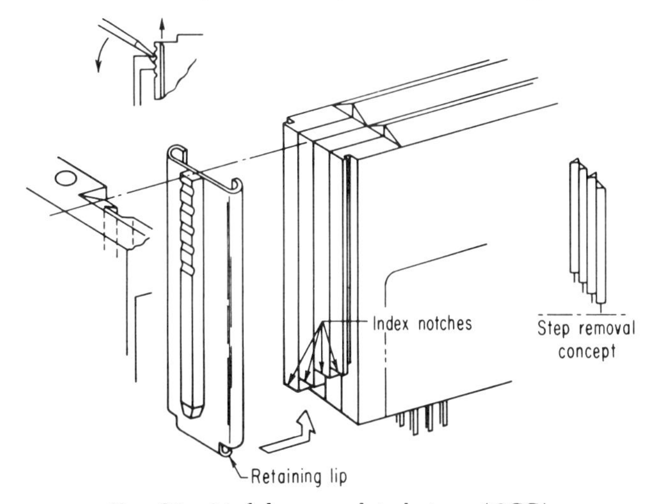

Fig. 30. Module-removal technique (CCC).

spaced on 0.075-in. centers, which allows a sufficient number of pins and a large enough pin (0.025 diameter or square) to assure reliability and durability. A large center gap in the pin arrangement is designed to span the large power bus planes within the housing motherboard. This connector is bonded to the submodule header. A reduced cross-section portion on the back ends of the pin fits into the header notched strip, as shown in Fig. 29.

The modules are designed for quick removal and replacement without the need for special tools. Maintainability is provided by the use of a connector and the minimization of fasteners. Module removal is accomplished by engaging two small screwdriver tips into the clip slots on each end of the module. By actuating the screwdrivers simultaneously in a downward direction, a lever action is exerted on the module. A series of slots provided on the clip allows the module to be evenly ejected in small increments without excessive force or a sudden disengagement (see Fig. 30).

The design of this arrangement incorporates "step removal," which, in effect, results in each submodule being disengaged from the connector individually in sequence. The module signal interconnections require a total of 288 pins disengaged. This would normally result in a fairly high removal force. Step removal reduces this force required by 75 percent.

Polarization is provided by two means. Initial visual orientation is accomplished by a color band at the top end of the module. Positive polarization for module installation in the housing unit is provided by discretely oversize guide pins on the submodule connectors.

Once the module is released from the unit, the clips may be slipped off and the module separated into its four submodule segments. These can be tested and replaced individually, or the complete module may be replaced.

The module retaining clip is a specially designed component that fulfills several functions. Essentially, its purpose is to retain the four submodules in a functional modular unit. It also provides keying for module installation and provides the unique step removal. Use of the clip eliminates the need for threaded miniature fasteners and greatly simplifies the submodule fabrication requirements. The clip, as installed, exerts a light spring tension to hold the submodules together.

INTERCONNECTIONS. A motherboard is used to provide a complete interconnect system for the modules. The complete motherboard includes the following components:

1. A multilayer interconnect board
2. Module receptacle connectors
3. Unit front panel and connector assembly
4. Power-distribution buses and filter capacitors

The motherboard is a multilayer interconnection matrix. Conductors and insulation layers as required by the interconnection complexity are laminated to form a relatively thin section board. All conductor layers are fabricated with complete continuity of individual runs. A typical cross section, representative of motherboard construction, is shown in Fig. 31. The number of interconnect layers is held to a minimum by judiciously grouping common and/or associated signals on the module pins and by making the maximum possible number of interconnections internal to the module.

Conductors for power distribution are necessarily large in cross section and are therefore treated separately, resulting in a thickened section down the central area of the board. Filter capacitors, one for each voltage and redundant for each module, are connected to special tabs on power planes.

Module connectors are retained on the motherboard by central mounting screws. On installation, additional mounting screws pass through the edges and retain both the connectors and the motherboard.

The module-connector receptacle provides phosphor bronze pins for connection to the respective motherboard pins. These pins pass through the complete laminate and align with tabs of the same dimension extending out of the board. A resulting 0.020-in.-square split pin is then wire-wrapped to complete the interface. The wire wrap is used exclusively to electrically couple the mating module receptacle and motherboard split-pin halves. There are no jumpers. All such interweaving is accomplished within the board laminates.

ENVIRONMENTAL CONSIDERATIONS. All functional electronics are located in MMAs, which are hermetically sealed from the ambient atmosphere. Degradation of this atmosphere in the form of contaminants will have no effect on their performance characteristics. Excessive heat, however, can have deleterious effects, and provisions are made for the rapid removal of heat energy from the MMAs.

The ceramic substrate is the heat sink for the silicon chip circuits. This substrate is placed in direct contact with the submodule frame and permanently bonded with a resilient bonding material having good heat-transfer characteristics. The heat is thereby conducted from the substrate to the submodule frame and then through the frame to the exterior of the module, where provisions must be made for its final removal. Such provisions may take the form of cold plates, forced air, natural convection, or any other suitable form, depending on the system requirements.

MAINTAINABILITY SUMMARY CHART

Item	*Discussion*
Accessibility	Modules can be removed by two screwdrivers step-lifting the module out of the chassis
Skill level to replace	Unskilled
Repairability	Standard modules are repairable items. Submodules may be throwaway items
Special tools	Two screwdrivers for module removal
Test point facility	Built-in test equipment (BITE) is used to identify faulty modules in the system. Special test equipment is used to isolate a faulty MMA in the module
Identification and keying provisions	Matching color bands may be used at the top of the module and on the side rail of the housing in which it mounts. Submodule connectors have oversize guide pins for positive polarization of module in housing
Logistics	Dependent on plans for maintenance support as evolved from maintenance engineering analysis for systems in accordance with NAVMAT INST 4000.20 "Integrated Logistic Support Planning Procedures"

Standard Hardware Program (SHP). SUMMARY. The Standard Hardware Program (SHP) is a functional electronic module development program coordinated by the Naval Avionics Facility, Indianapolis (NAFI), for the Naval Ordnance Systems Command. The program is an outgrowth of the Advanced Electronics Program sponsored by the Naval Air Systems Command Industrial Division (AIR 50211) as an industry-preparedness measure and the Advanced Packaging Techniques Program sponsored by the Naval Ordnance Systems Command Special Projects Office (SP-23) and the Naval Air Systems Command (AIR 5202).

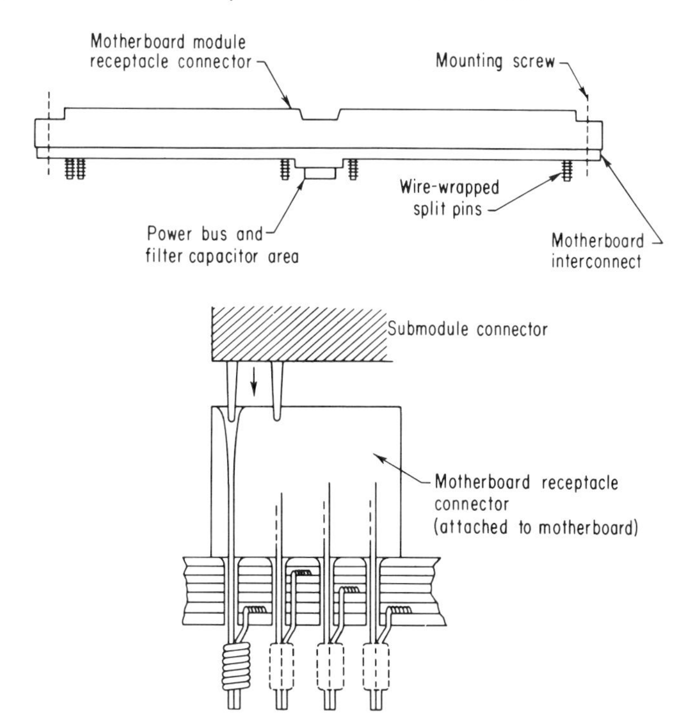

FIG. 31. Motherboard connection and construction (CCC).

The goals of the Standard Hardware Program are to select widely used electronic functions, a standard mechanical packaging design, and compatible electrical design parameters and, from these selections, to develop functional electronic building blocks which have general usage in a number of programs and projects. The purposes for establishing such goals are to increase military effectiveness, lower cost, reduce design and development costs, and ease the logistics support problem for present and future electronic systems.

The Standard Hardware Program receives the impetus to reach its goals because the program is not oriented to one system, leaves latitude for the introduction of new functional modules, restricts the use of proprietary designs, and selects and specifies functional modules in current use in such a manner that they will not become obsoleted owing to technological change.

The future success of the program is dependent on the wide usage of the modules and a broad industry base for module manufacture.

As of 1969, approximately 136 electronic functions had been selected for implementation, with many more under investigation. Of these, about two-thirds were potentially high-usage devices, and the remaining third were special functions developed for the need of a particular system. The quantity of these modules manufactured up to 1969 numbers in the thousands and shortly will be in the tens of thousands.

As will be discussed further, mechanical design for the module has been established, as well as reliability, quality control, documentation, and testing requirements.

Module and module piece-part vendors have been established and tooled and are currently manufacturing on a production-line basis.

For system applications requiring environments outside the range of conditions listed, the NAFI module design may be used as mechanically specified or modified to meet unique conditions. For example, the Poseidon guidance computer uses a NAFI-type module mechanically modified to optimize the extraction of heat by conduction through the side runners and electrically specified for a nuclear-radiation environment. It should be recognized that departures from the environments specified herein will result in modules that are not compatible electrically with the substantial quantity of SHP modules used interchangeably across a broad range of systems. To ensure appropriate identification and assignment of key codes in a range not to be confused with other modules in the NAFI format, coordination through the cognizant naval program manager is required. The standard modules should be used wherever possible when the NAFI format is acceptable, as the advantages of the large investment in design, production tooling, spare parts from other systems, and standardized test equipment are major economic, reliability, and logistic support factors.

MECHANICAL CONSTRUCTION. The basic SHP module configuration is the single-span single-thickness (1A) module increment, which is illustrated in Fig. 32. The key dimensions of such a module increment were derived from the following criteria: the amount of circuitry required to perform a specific electronic function, the maximum number of interface connections required, the size of the keying and retention mechanism, and the maximum tolerable cost for the module to be considered as a throwaway item. These considerations led to the development of a basic module with overall dimensions of 2.62 in. wide by 1.95 in. high by 0.290 in. thick.

In addition to the basic module configuration, SHP has provision for the development of modules of multiple span and thickness provided that such modules maintain standard growth increments. Such growth increments permit modules to increase in span by increments of 3.000 in. and increase in thickness by increments of 0.300 in. (see Fig. 33).

In the event of unavoidable module design requirements, SHP modules may deviate from certain design specifications in that the base of multiple-thickness modules may vary in depth as illustrated in Fig. 34. Some of the module configurations which have been used are shown in Fig. 35.

The SHP module is comprised of various component parts that are illustrated in Fig. 36. Each of these component parts is required by specification and serves functions pertinent to the mechanical integrity of the module. The following is a brief discussion of the form, fit, and function of such components.

1. *Module fin structures.* Each SHP module has configuratively controlled fin structures for the purposes of marking and extraction as illustrated in Fig. 37. These fin structures may be one and the same, provided that the requirements of both are suitably satisfied. In addition to the marking and extraction fins, multiple-increment modules are permitted to contain specialized heat-dissipating fin structures.

2. *Module pin skirts.* Module pin-skirt structures are protective shields for the module connector contacts and offer a marking surface for orienting and identifying the module.

3. *Module keying pins.* All SHP modules are uniquely keyed by the insertion and radial orientation of two specially configured keying pins, accomplishing the following objectives:

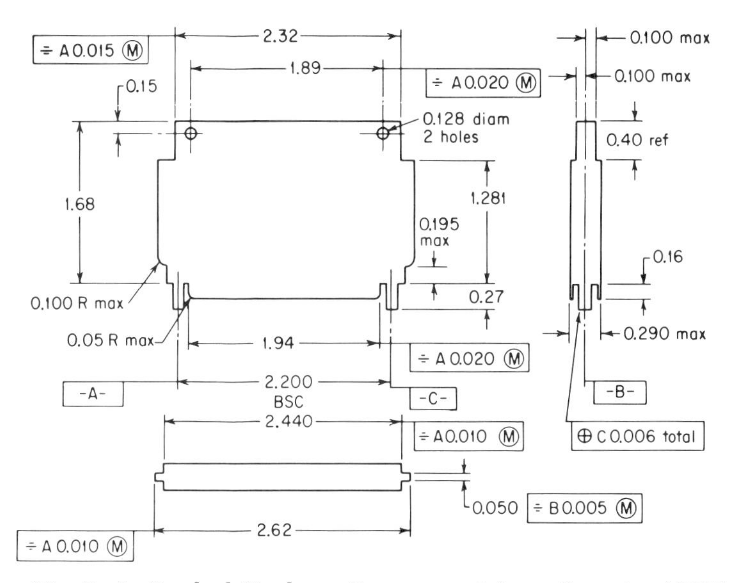

FIG. 32. Basic Standard Hardware Program module configuration (SHP).

a. Minimizing the possibility of a module being plugged into system in the wrong location

b. Preventing modules from being plugged into a system reversed

c. Assuring that the module male connector contacts do not mate with the female connector contacts in the event of (*a*) and (*b*) above

d. Assuring physical protection for the connector contacts of the module

4. *Module contacts.* The portion of the module male contact protruding from the module header surface is configuratively specified so as to assure proper engagement of the module and its interface mounting structure. The contact configuration within the module body is to be determined by the module developer in order that the total contact configuration may be selected which best suits the specific interconnection requirement and technique.

5. *Module connector.* Module male contacts are arranged in rows of 20 contacts each on a 0.100-in. grid system to form module connector increments. The basic module configuration has a maximum of 40 contacts or a minimum of 20 contacts. Multiple-increment modules may also have a maximum of 40 contacts per module increment; however, a minimum of 20 contacts per total module must be

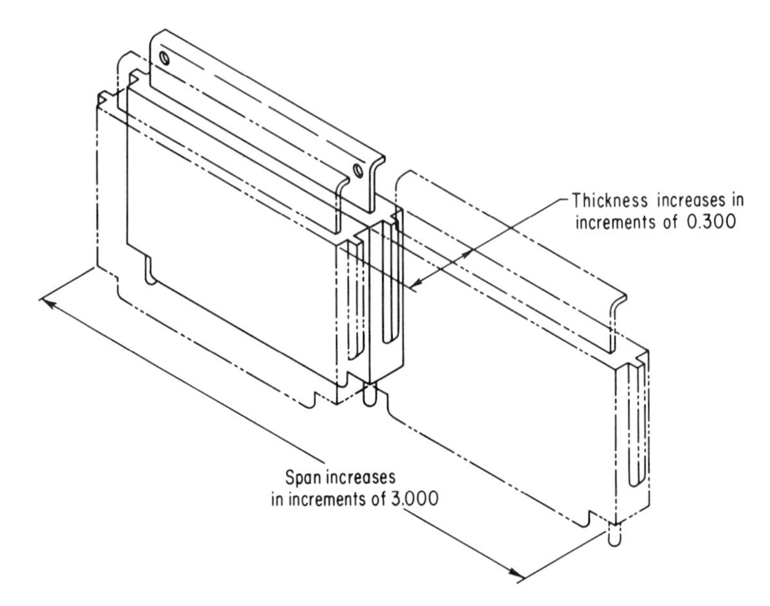

Fig. 33. Module incremental growth (SHP).

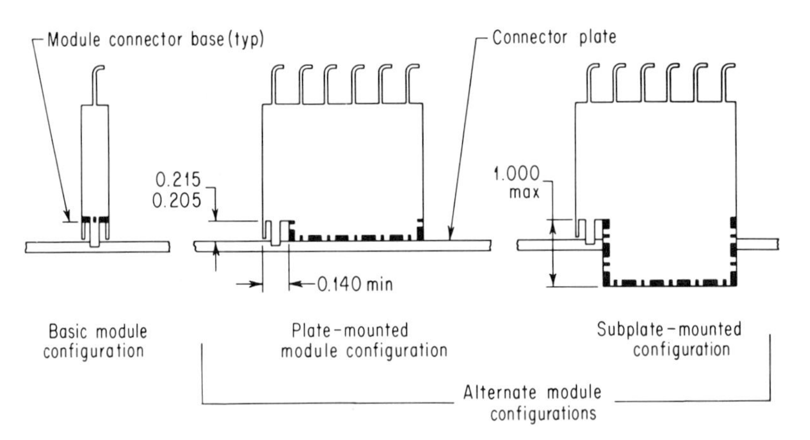

Fig. 34. Alternative module configuration (SHP).

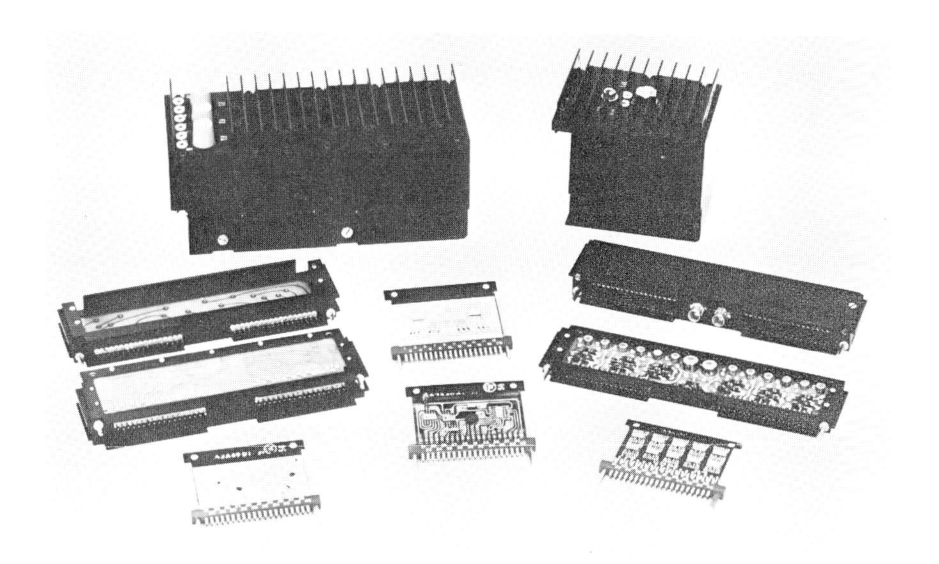

FIG. 35. Typical modules (SHP).

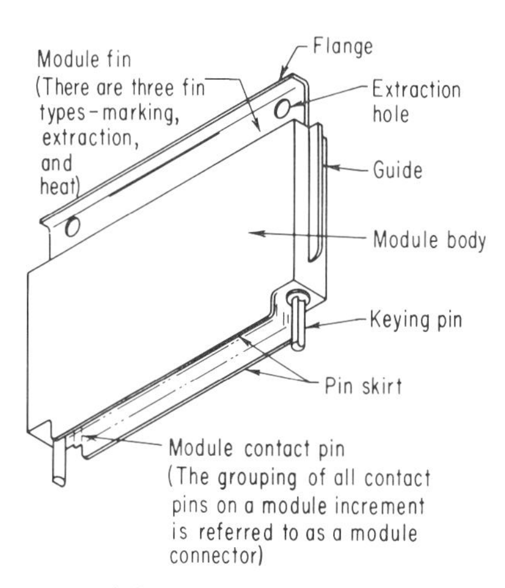

FIG. 36. Module component identification (SHP).

maintained. Each connector must maintain a full complement of 20 contacts even if they are not fully utilized for signal transmission.

6. *Module guides.* Each SHP module increment has guide structures at each end of the module. These guides are required to assist the proper mating and insertion of the module connector into its interfacing connector and mounting structure. The top surface of this structure is normally used to retain the module within a module card cage structure. Module guides, constructed of appropriate materials, provide an excellent method of conducting heat from the module and into the mounting structure.

INTERCONNECTIONS. SHP has been closely involved with the development and implementation of new automatic back-panel wiring techniques. In fact, such wiring

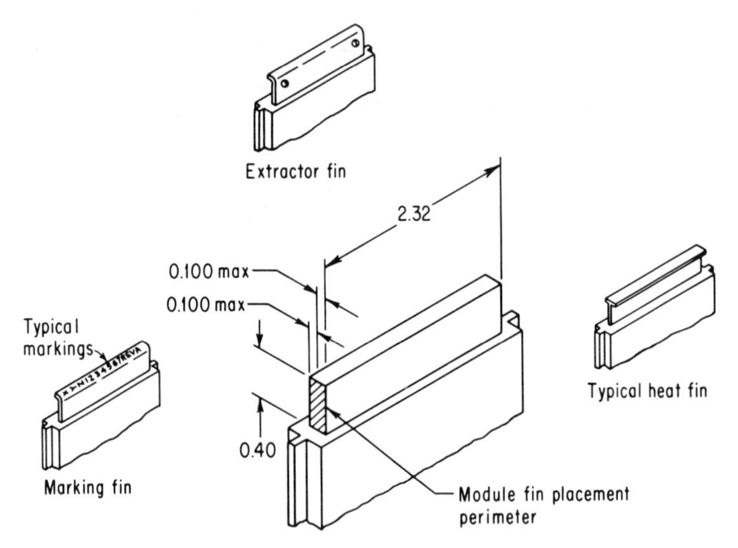

FIG. 37. Module fin structure (SHP).

techniques constituted an active influence in the development of such SHP module requirements as module contact-pin configuration, contact-pin location, and grid spacing. The SHP module mating interface was primarily established as a result of hardware application with 0.100-grid wire-wrap contacts. In order to alleviate commercial proprietary designs and offer a diversity of interconnection techniques, the SHP specifies only the actual female contact interface area and its mating insulator bushing outline. With this information, the systems user may employ the back-panel connection technique and hardware components as his requirements warrant. Back-panel wiring techniques that have been used with SHP modules are as follows:

1. Wire wrap
2. Crimp (Termi-point)
3. Percussion-welded
4. Multilayer printed circuitry

To date, wire wrap has been predominantly used for interconnection of SHP modules at the back-panel level. Wire wrapping is a technique for mechanically connecting a solid wire to a terminal by wrapping a specified number of turns of wire, under tension, around a terminal with two or more sharp edges. The average pressure between the wire and terminal at the corner contact stabilizes to 29,000 psi. This value is far above the requirements for a gastight connection and is capable of a life of 40 years in shore-based equipment.

Advantages of wire-wrapped connections are:
1. It is capable of high-density point-to-point wiring.
2. Manual or automatic machine application may be used.
3. It has a reliable gastight connection.
4. Its cost is low.
5. The connection is mechanically and electrically stable.
6. It is readily repaired or modified.
7. It withstands environmental tests and vibration.
8. It avoids failures incurred from application of heat.
9. Production quality control is simple.

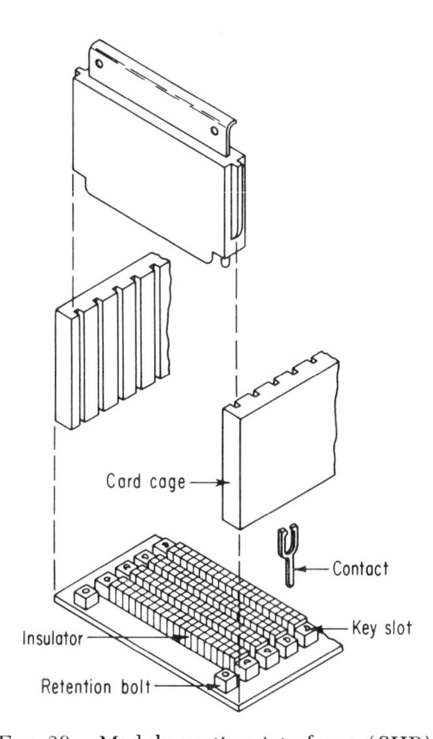

FIG. 38. Module mating interfaces (SHP).

CONNECTORS. The connector for the basic SHP module contains 40 pins. The pins are arranged in two rows of 20 pins each and are located on a 0.100-in.-square grid.

The male connector pins are the bayonet or blade type and mate with a female tuning fork (Figs. 38 and 39). Properties of the connector include the following: (1) the pins are easily fabricated and dimensionally stable; (2) they possess high strength and good corrosion resistance; (3) they are easily solderable to interface with the circuit substrates; and (4) the interface between the connector pin and connector receptacle has a low contact resistance.

For the male pins consideration was given to beryllium copper, phosphor bronze, and one-half-hard brass. All three materials are dimensionally stable, but the half-hard brass was selected, since it is more workable than either of the other two metals and is also easily gold-plated. The thin leaf portion of the circuit end of the contact is easily formed and will readily adapt to various printed-circuit or thin-film substrates.

In addition to the blade contact connector discussed above, the NAFI module also incorporates rf-type connectors for use on modules in rf applications. Production drawings, test results, and application notes for NAFI modules so equipped are available.

ENVIRONMENTAL CONSIDERATIONS. The entire spectrum of environmental requirements for equipment used in ground-based, shipboard, airborne, and missile systems has been investigated. In an effort to standardize, it has been necessary to reduce

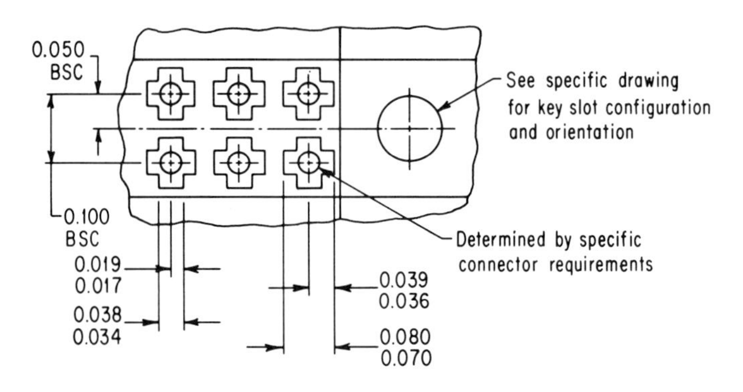

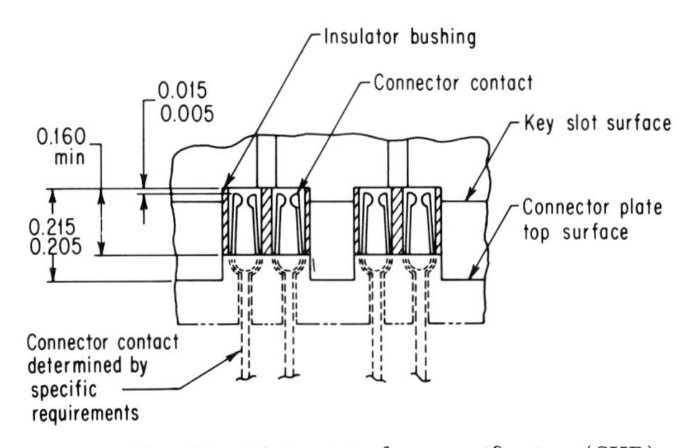

FIG. 39. Mating-interface specification (SHP).

this spectrum of environments to two basic independent classes, each class consisting of requirements for the following parameters:

1. Module fin temperature
2. Humidity
3. Vibration
4. Shock

An analysis of the range of military environments revealed that modules falling into either of the two classes are applicable to most military system applications. These classes will, if adhered to, assure the developments of modules utilizable in a large number of systems and locations.

It is the prerogative of the initial developer of a module to select the environmental class to which a new functional module is developed. It is recommended, but not a requirement for acceptability as a standard, that each new module be designed for operation in the more stringent environmental class so that the module may be used in a larger number of systems.

TABLE 16. Limits of Environmental Classes

Environmental characteristics	Class I limits		Class II limits
Fin temperature range (operating)	0 to +60°C		−40 to +100°C
Humidity	95^{+0}_{-5} % relative humidity at 44°C for 96 hr		
Vibration	Frequency range, cps	Table amplitude (plus or minus inch)	5–2,000 cps from 0.20 in. double amplitude to ±30*g* peak, sinusoidal; and 50–2,000 cps at 0.2g^2/cps, random
	5–15	0.030±0.006	
	16–25	0.020±0.004	
	26–33	0.010±0.002	
Shock	½ sine pulse for 11 msec at 50*g*		

Table 16 lists the limits of the two environmental classes to which modules may be designed. It is emphasized that each of these two module environmental classes is independent of one another; i.e., each is a separate and distinct entity. If it is decided to develop a module to meet the requirements of class II, all the requirements of class II must be met without exception. A module, for example, which meets the temperature requirements of class II and the vibration requirements of class I will be classified enivronmentally as a class I module.

The environments stated heretofore are sufficient for the majority of systems users. It is realized, however, that, for unique systems applications, there are other types of environmental requirements which modules will have to meet, such as radiation, RFI, altitude, etc. The majority of these other environmental requirements may be taken care of by the packaging of the overall system, controlled system ambient conditions, system circuit redundancy, etc. If, however, the system designer imposes additional environmental requirements on the module, e.g., radiation hardening, this requirement will be unique to the particular system and will not preclude the functional and mechanical interchangeability of this radiation-hardened module in other systems not specifically requiring immunity from radiation.

MAINTAINABILITY SUMMARY CHART

Item	*Discussion*
Accessibility	Depends primarily on rack and enclosure. Modules may usually be removed from the mounting structure without the need for removing any other module. Extraction force is approximately 12 oz contact
Skill level to replace	Unskilled
Repairability	Majority of standard modules are throwaway. A few are repairable on site, and a small minority are returned to a repair and replacement activity
Special tools	Module extraction tool (Fig. 40)
Identification and keying provisions	Module is identified on top flange of heat sink (see Fig. 41). Keying is such that a large number of unique combinations is possible
Logistics	Dependent on plans for maintenance support as evolved from maintenance engineering analysis for systems in accordance with NAVMAT INST 4000.20 "Integrated Logistic Support Planning Procedures"

Electronic Packaging System (EPS) Modules (NEL).

SUMMARY. The primary emphasis of the EPS packaging concept is on resistance to shipboard environments. Features of the packaging concept include: (1) packaging of all electronic functions in modular format; (2) forced-air cooling through each module; (3) advanced connectors, interconnections, and shielding.

There are two types of modules in the EPS packaging concept, a printed-circuit-board type and a shielded type.

No specific functional modules constitute a part of the program. The packaging concept is intended to be applicable to a number of different naval programs rather than to a specific system. However, this does not preclude the use of modules developed for a particular program in other, following systems.

MECHANICAL CONFIGURATION. The EPS module packaging concept utilizes two types of construction, printed-circuit board and shielded.

The printed-circuit module is a conventional circuit card designed for use in conjunction with a specially designed high-reliability high-contact pressure-cam type

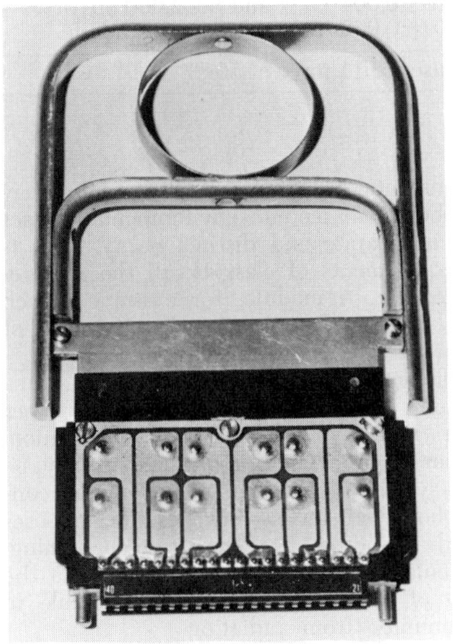
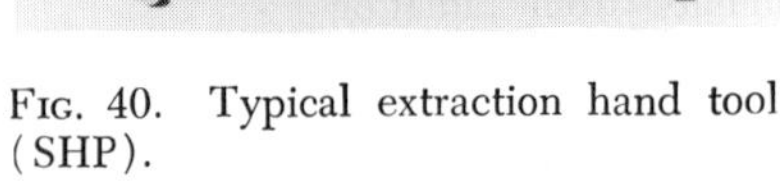

Fig. 40. Typical extraction hand tool (SHP).

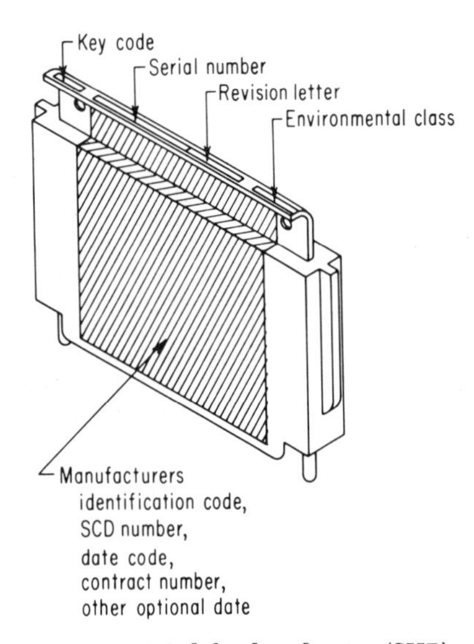

Fig. 41. Module identification (SHP).

of rack-mounted connector and is intended for use with circuitry that does not require shielding. This type of module is variable in width and depth and has a fixed vertical dimension of 4 in. The A dimension of the module can vary from 0.2 to 1.2 in. in 0.2-in. increments. The C dimension including connector can vary from 3.8 to 7.6 in. in 0.8 increments, with the 3.8 or 7.6 being preferred. The module can be keyed by slotting between contacts on the edge of the board. A typical example of a PC type of module is shown in Fig. 42.

The shielded module is designed primarily for the analog and rf type of circuitry and features specially developed high-contact pressure single- and multiple-contact connectors. The metal cover of the module fits over the electronic circuitry and attaches to the connector, which may contain several combinations of conductor contacts. The module cover is slotted top and bottom to interface with cooling ducts. The module covers also protect internal circuitry against atmospheric contamination and rough handling. The shielded module is dimensionally variable in all three directions. The A dimension can vary from 0.8 to 16.8 in. in 0.8-in. increments. The

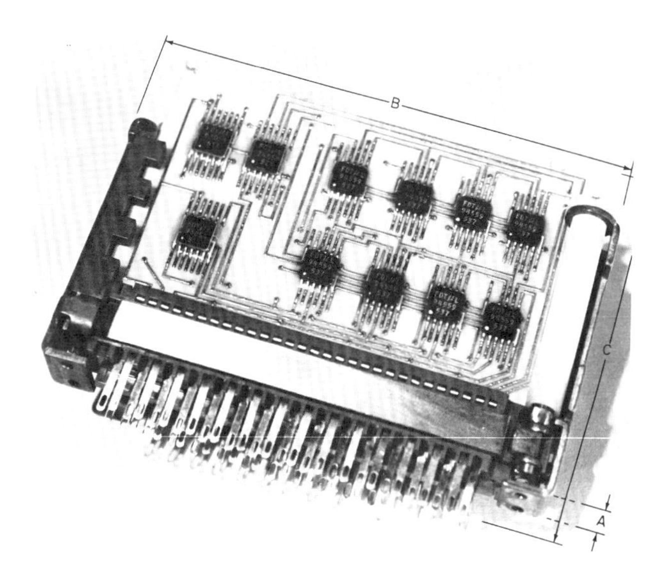

Fig. 42. Typical printed-circuit module (EPS).

Fig. 43. Typical shielded module (EPS).

B dimension can vary from 4 to 12 in. in 2-in. increments. The C dimension can vary from 3.8 to 7.6 in. in 0.8-in. increments. Each module contains the male plug of a collet-pin connector. The module is held to the serviceable main chassis by a locking screw and, when required, a dummy connector. A shielded module is shown in Fig. 43.

CONNECTORS. Two types of module connectors are used in the EPS packaging concept. One type is used for printed-circuit (PC) board-type modules, and the other is used on shielded modules. The printed-circuit card connector operates on the pressure-cam principle and the shielded-module type operates on the pin-collet principle. Both connectors can be unlocked for module replacement with minimal module replacement force.

The PC type of connector is 4 in. long with 0.100- or 0.050-in. pin spacing. Each connector has a total of 58 pins, or twice this for the 0.050-in. spacing. A special tool is used to lock and unlock the connector for insertion or removal of a circuit board. A thin-shanked screwdriver can be substituted for the special tool. Extremely

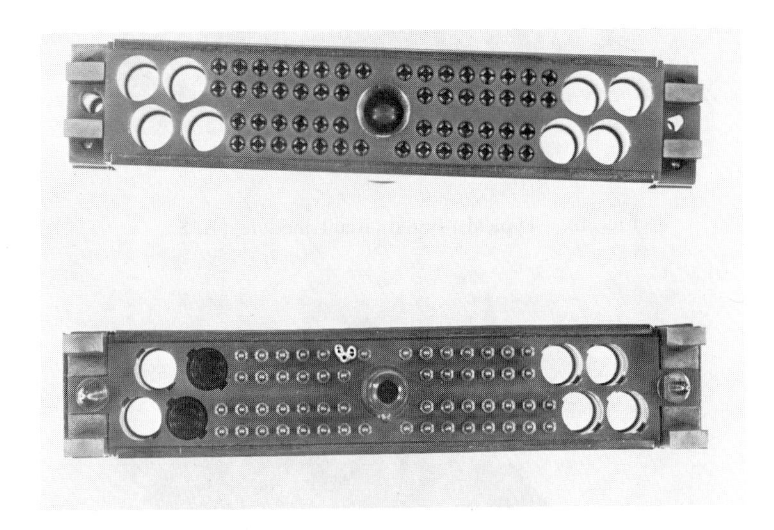

FIG. 44. Connector (EPS).

high contact pressures are realized when the connectors are in the locked position, and no additional mechanical support is required to hold the boards in place.

The shielded-module type of connector contains a mixture of single-conductor contacts and coaxial connections. Four combinations comprise the family of connector headers. Each header contains a different number of holes for insertion of contacts. This configuration allows insertion of a contact only when needed and of additional contacts resulting from equipment-revision requirements or test and check-out requirements. Extremely high contact pressures are obtained in this type of connector. One example of this type is shown in Fig. 44. In addition to contact pins and guide pins the connector is equipped with contacts for test probes. Test probes may be inserted in any of the coaxial-sized holes. The connector can be locked or unlocked with a standard allen wrench. As is the case with the PC type of connector, the shielded module can be removed or inserted with minimal force after the connector has been unlocked. The single-contact connections are spaced on ⅛-in. centers. Coaxial contacts are spaced on 0.28-in. centers. A dummy connector is provided for module support when required.

The five connector types in the EPS family of connectors allow for a wide flexibility as to both electrical and dimensional requirements. Each connector type can be obtained in a number of sizes and in a number of contact current ratings and is available either from multiple sources on a nonproprietary basis or on a licensing arrangement.

ENVIRONMENTAL CONSIDERATIONS. The EPS module is intended for use in an atmosphere which has been cooled and filtered by external means (see section on Enclosures). In the case of the shielded modules, cooling air flows directly through the cover. For printed-circuit types, air flows over both sides of the circuit board.

Provisions are made for shielding modules when necessary, and these provisions provide a high degree of RFI isolation between modules.

MAINTAINABILITY SUMMARY CHART

Item	*Discussion*
Accessibility	All modules and cable assemblies are readily accessible and easily removed for repair or replacement
Skill level to replace	Shipboard technician
Repairability	Dependent upon system and type of test equipment (off-line) available
Special tools	Dual cam actuating tool used to remove and replace PC-1 module. This tool protects the module. The module may also be removed with a plain screwdriver
Identification and keying provisions	Keying bars are inserted into slots provided in the PC-1 module connector, to match slot(s) in edge board connector. Arrangement of contacts which are individually inserted base cavities of the SC-1 shielded module may be altered
Logistics	Dependent on plans for maintenance support as evolved from maintenance engineering analysis for systems in accordance with NAVMAT INST 4000.20 "Integrated Logistic Support Planning Procedures"

Centralized Electronic Control (CEC) Modules (NRL). SUMMARY. The Centralized Electronic Control (CEC) packaging concept was developed by the Naval Research Laboratory. This particular packaging concept has evolved through three major design changes.

The primary characteristics of the CEC packaging concept are as follows: (1) each module is a complete functional unit; (2) modules have common electrical and physical interfaces; (3) advanced connectors and shielding techniques are used throughout the system; (4) provisions are made in each module for a fault-locating indicator.

Within the limits of the characteristics given above, module design is very flexible. Internal configuration and selection of circuitry type are left to the module supplier as long as electrical and physical interfaces are met. The basic module design is compatible with conventional thin-film, semiconductor integrated and hybrid circuitry.

MECHANICAL CONFIGURATION. The modules used in the Naval Research Laboratory's CEC system are expandable in selected increments, in width, depth, and height. The height-dimension multiples for the modules are governed by the width of the cooling interfaces. A representative range of typical module sizes is shown in Fig. 45. All the sizes shown conform with the summary of module sizes which is tabulated in Fig. 46. Commonly used module dimensions are 0.75 by 1.450 by 5.938 in., 1.490 by 1.450 by 5.938 in., and 2.980 by 2.950 by 5.938 in.

Typical module and connector configurations which are compatible with the CEC packaging concept are shown in Fig. 47. That shown in Figure 47*a* is the simplest and least expensive and would normally be used whenever the number of pins and component mounting area which it provides are sufficient. That shown in Fig. 47*b* provides three times the component-mounting area of that provided in Figure 47*a* and includes a total of 284 pins as compared with 142 pins for the connector of Fig. 47*a*. The construction of Fig. 47*c* is shown in exploded view in Fig. 48. This example provides eight smaller circuit boards, which typically mount eight integrated circuits on each board. The eight pressure-cam connectors are mounted on a multilayer printed-circuit board; this provides circuitry interconnection

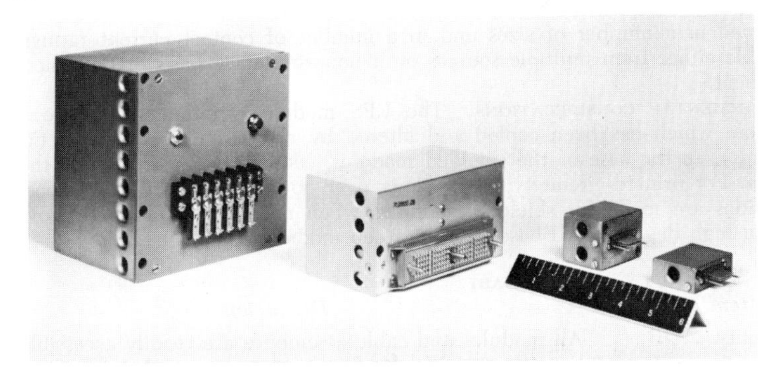

FIG. 45. Typical centralized electronic control modules.

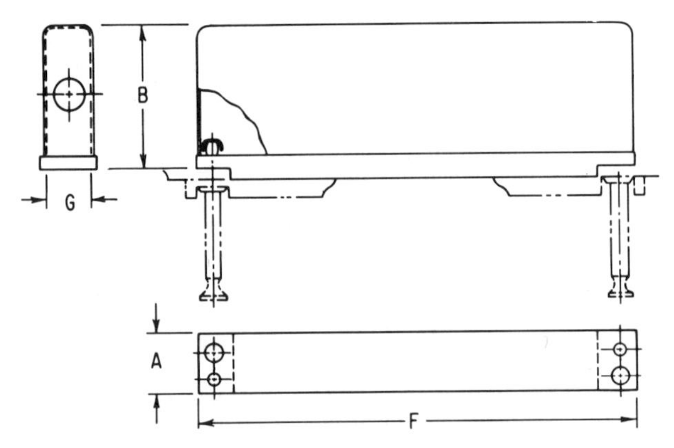

Module width			Module depth		Module height	
Size	*A*	*G*	Size	*B*	Size	*F*
A	0.745	0.690	B	1.450	F	2.328
2A	1.490	1.440	2B	2.950	2F	5.938
3A	2.230	2.180	3B	4.450	3F	13.159
4A	2.980	2.930	4B	5.960	4F	20.380
5A	3.720	3.670	5B	7.450	5F	27.601
6A	4.470	4.420	6B	8.950	6F	34.822
7A	5.220	5.170	7B	10.450	7F	41.943
8A	5.970	5.920	8B	11.950	8F	49.164
9A	6.710	6.660	9B	13.450	9F	56.385
10A	7.450	7.400	10B	14.950		
11A	8.200	8.150	11B	16.450		
12A	8.940	8.890	12B	17.950		
13A	9.690	9.640				
14A	10.440	10.390				
15A	11.190	11.140				
16A	11.940	11.890				
17A	12.690	12.640				
18A	13.440	13.390				
19A	14.190	14.140				
20A	14.940	14.890				

FIG. 46. CEC module dimensions.

and connections between the circuitry and module connector, which is a larger pressure-cam type.

A typical configuration used for some analog-type modules is shown in Fig. 49. The metal module cover provides RFI shielding as well as handling protection and cooling air ports. This type of construction provides for internal shielding compartments as well as a high order of protection from mutual interference between modules.

Suppliers of CEC modules must adhere to specified dimensions, satisfy cooling and mounting interface requirements, and select connectors from the types specified for CEC use. Beyond these requirements suppliers can configure modules in any manner which is compatible with electrical requirements.

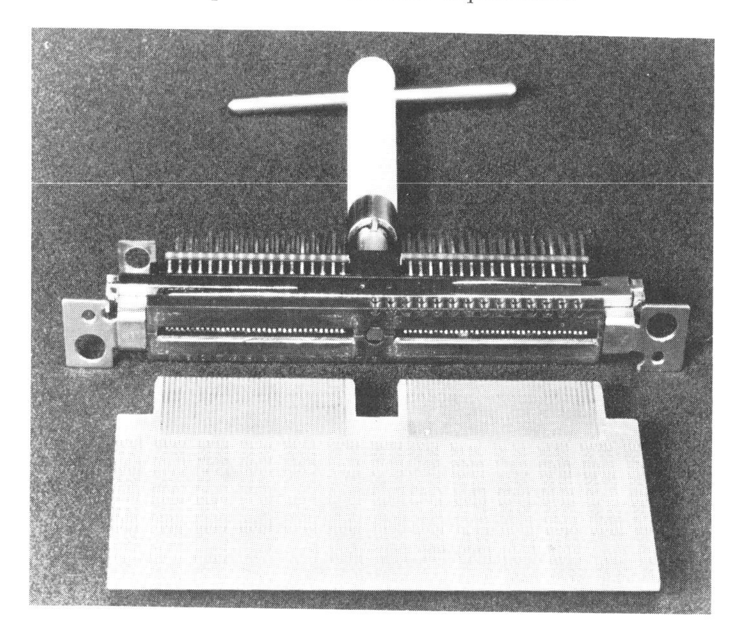

(*a*)

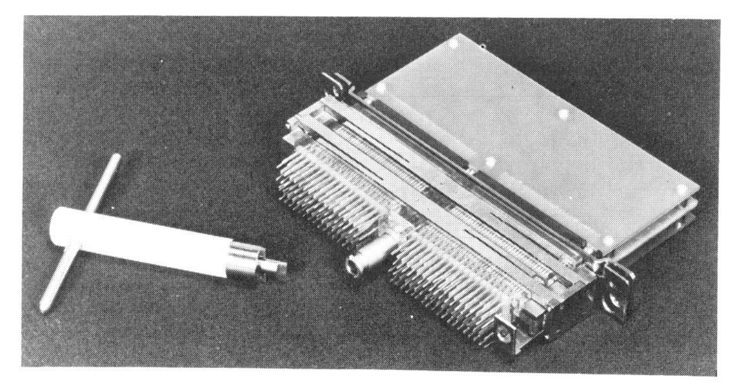

(*b*)

FIG. 47. CEC module and connector configuration.

The cooling and mounting interfaces are illustrated in Fig. 50. The nylon fasteners which hold the module to the rack connector are an experimental type and may be replaced with snap-in, DZUS, or some other type of fastener. The connector fastening force must be sufficient to hold most modules, even under shock and vibration loadings. Regardless of the module design, the supporting structure will include mounting holes and cooling-air intake and exhaust ports at fixed intervals. The module design will determine how many of these provisions are utilized.

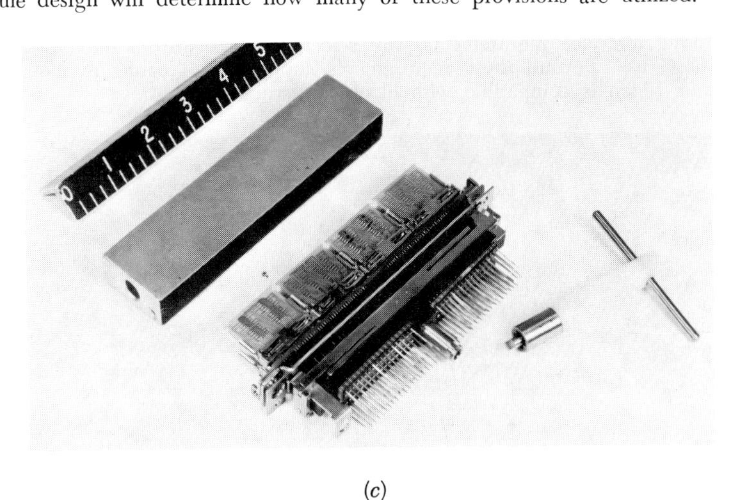

(*c*)

FIG. 47. CEC module and connector configuration (*continued*).

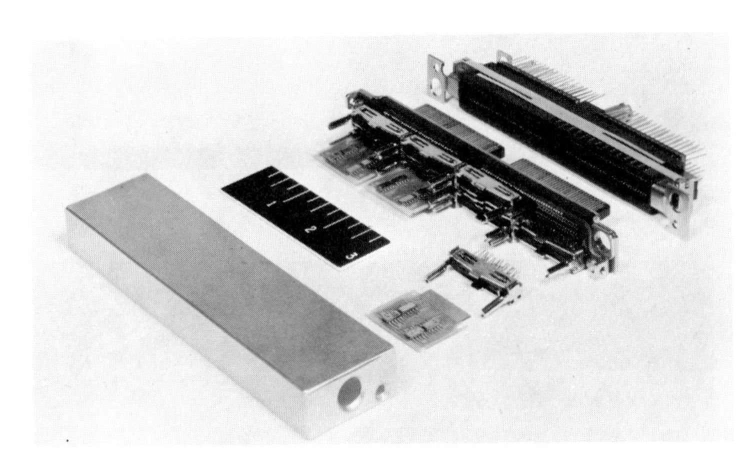

FIG. 48. Exploded view of a typical module.

CONNECTORS. Several types of experimental connectors have been developed for use with the CEC system. These connectors feature an extremely high reliability level of electrical contact, in addition to improved electrical, mechanical, and maintainability characteristics. Connector designs include single conductor contacts, coaxial contacts, and a combination of single and coaxial contacts. Planar-type connections for use in printed-circuit-board applications have been developed.

Fig. 49. Typical analog module (CEC).

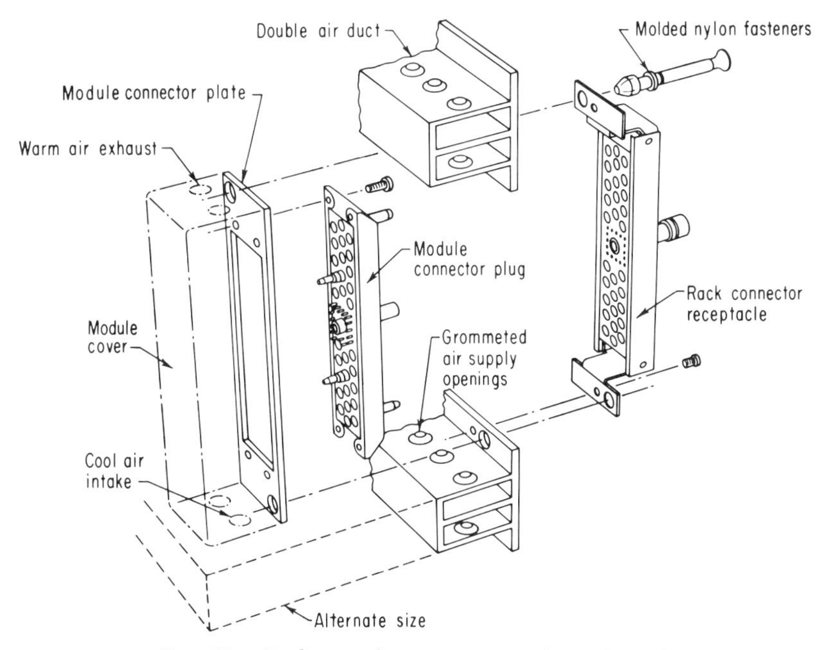

Fig. 50. Cooling and mounting interfaces (CEC).

The single-contact and coaxial conductor contacts are special multicontact connectors providing individual spring loading of each contact.

Each contact element, both single-conductor and coaxial, in the mounted connector receptacle is an individual collet. The module connector plug contains a mating pin for each collet. In mating the module plug to its collet, the collet is compressed around the pin by a tapered spring-loaded sleeve surrounding the pin. A jack screw in the connector body produces the mating force required to obtain the high contact pressures. The contact configuration of the connector plug is illustrated in detail in Fig. 51. The connectors provide extremely high contact pressures, resulting in low electrical contact resistances which are maintained even under high shock and vibration loadings.

The connector known as the Nike Zeus Junior type has been developed in a variety of sizes and contact combinations. For example, a connector for use on the standard ¾-in. module can be supplied with 76 single conductor contacts, 36 single conductor contacts and 10 coaxial contacts, 16 single conductor contacts and 25 coaxial contacts, etc. Some examples of the Nike Zeus Junior type of connector are shown in Fig. 52.

A connector system for interconnection of planar circuitry such as etched circuit boards provides for extremely high contact density as well as improved electrical, mechanical, and maintainability characteristics and is intended primarily for use with digital circuitry. Circuit-board interconnection is achieved by means of an ex-

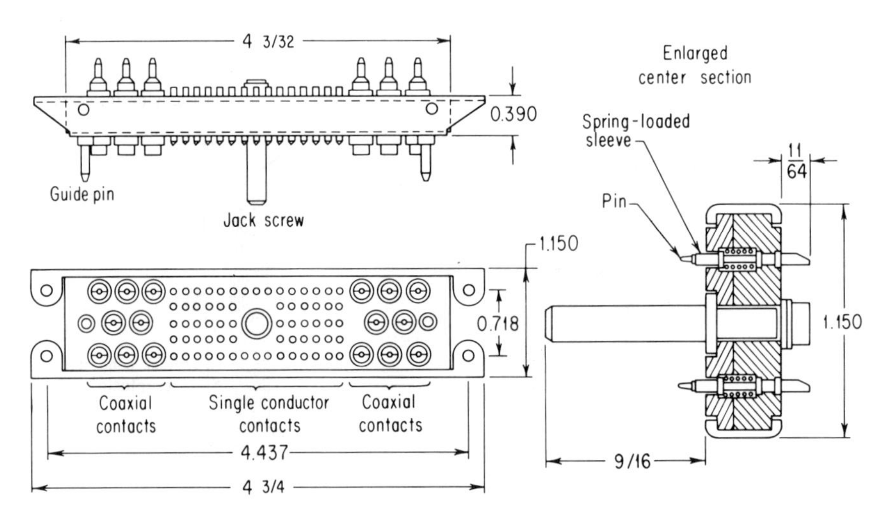

FIG. 51. Detail of connector receptacle (CEC).

ternally actuated insulated member that provides contact pressure to the boards by clamping after the board is inserted. A silicone rubber pressure pad, recessed in the insulator behind the connector contacts, as shown in Fig. 53, applies equal force over the entire length of the contact surfaces while keeping each contact flat against the board circuitry. This method not only provides more contacting area between male and female members but also eliminates the insertion/withdrawal forces and contact "drag" inherent in many conventional printed-circuit (PC) connectors, while upgrading resistance to shock and vibration.

The connector design also promises higher densities in equipment packaging design. The connectors currently produced have contacts which are able to mate with printed boards having 0.025-in. contact centers. Similarly the achievement of 0.100-in. clearance between laterally stacked boards is possible.

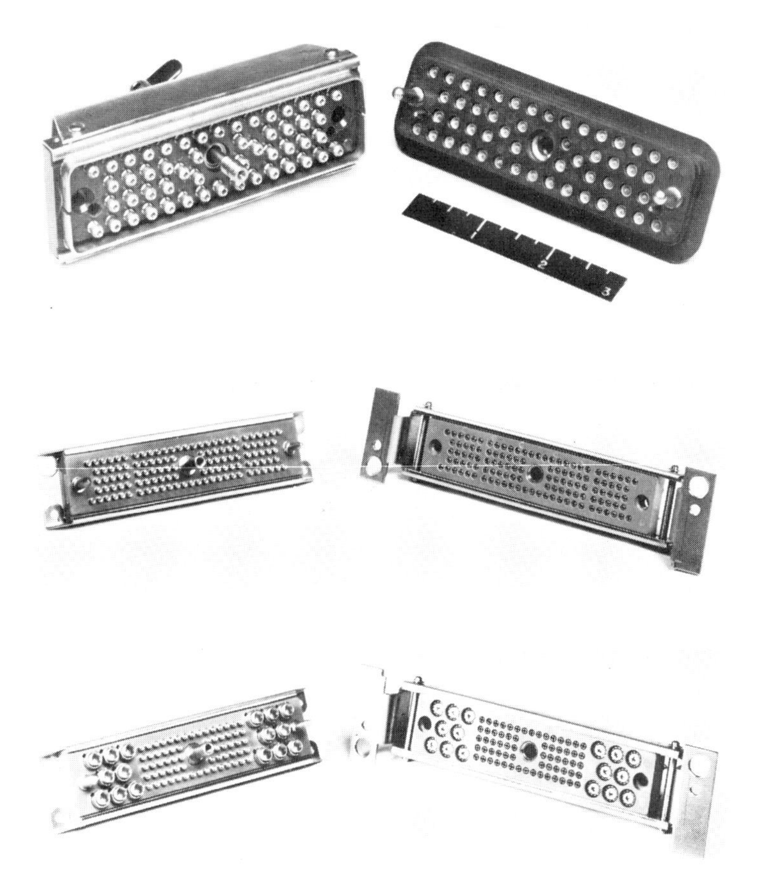

Fig. 52. Some representative CEC connectors (CEC).

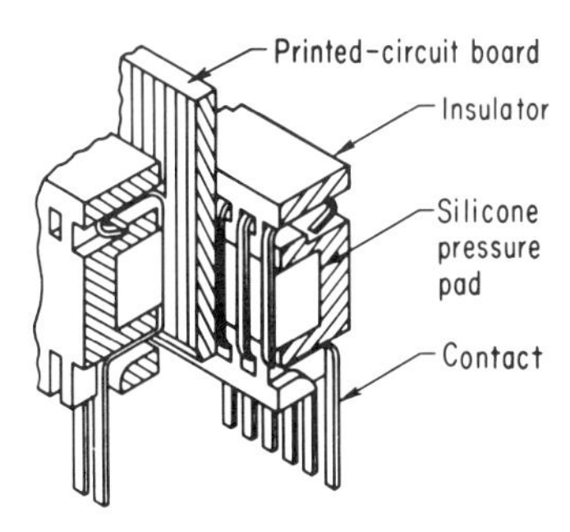

Fig. 53. Detail of pressure-cam mechanism (CEC).

The connectors are sometimes referred to as pressure-cam types because essentially a cam is used to release the spring pressure for opening or closing the connector.

A digital module has been provided as a result of high contact densities and the pressure cam's high contact pressures. This module (Fig. 54) has provisions for eight daughter (PC) boards, each of which is plugged into individual receptacles. The boards are double-sided with 34 contacts per side on 0.025-in. centers and inserted or removed from the receptacle with the use of a tool that relieves the clamping pressure from both sides of the normally closed receptacle.

The eight receptacles, totaling 544 contacts, are interconnected through a single multilayer motherboard. This in turn has 142 leads that go to an external plug-in header. Each digital module, therefore, consists of as many as eight daughter boards, eight individual receptacles, and one multilayer motherboard terminating in the plug-in header. The modules themselves then plug into another locking-type receptacle which has wire-wrapped terminations on the other end. A tool is used to relieve the pressure on the module contacts when a disconnect is desired.

FIG. 54. Typical digital module (CEC).

The CEC Nike Zeus Junior connectors also contain switch contact elements on the receptacle half. These contact elements are activated when the connector plug is fully seated. The contact elements also control indicator circuits which show that the connector is mated and the contacts are under pressure.

ENVIRONMENTAL CONSIDERATIONS. The CEC modules are designed to withstand certain environmental conditions. Of prime importance are provisions to control cooling, RFI shielding, and atmospheric and shock and vibration protection.

Cooling of CEC modules is accomplished by using positive forced air. Each functional module is provided with intake and exhaust air ports, and the forced air travels through each module by flowing through each intake port and out of the exhaust port (Fig. 50). The rate of air flow through each module may be controlled by incorporating only the necessary number of air-supply and -return holes in the module case and by governing their diameter.

The CEC modules are designed to give high order of protection from mutual RFI between modules.

MAINTAINABILITY SUMMARY CHART

Item	*Discussion*
Accessibility	Modules are accessible by releasing slide-fastener screw and connector jack or pressure cam
Skill level to replace	Faulty modules may be located and replaced by unskilled personnel
Repairability	Optional types of module construction. Some modules will be throwaway, and some will be repairable by removal of small circuit boards
Special tools	A special tool is required for opening the connector to remove and insert modules
Identification and keying provisions	Identification marking can be located on the exposed module face
Logistics	Dependent on plans for maintenance support as evolved from maintenance engineering analysis for systems in accordance with NAVMAT INST 4000.20 "Integrated Logistic Support Planning Procedures"

Enclosures. *Introduction.* Enclosures, with the associated rack, chassis, and cooling provisions, provide interconnection, support, and environmental protection of the functional modules of the system. The major design requirement of an enclosure is its resistance to shock and vibration. Beyond this requirement are criteria which influence the selection of a particular enclosure for specific application. Included among these criteria are accessibility, volumetric and gravimetric efficiency, cooling capacity, drip-proofing, and RFI-EMI shielding.

The two "benchmark" enclosures discussed in this section provide the flexibility for use of any of the benchmark modules contained in the preceding section. As previously mentioned, selection of either type will depend on engineering evaluation and trade-off analysis for the specific systems being developed. The basic philosophies of enclosure design have been maintained in each concept; however, the differences in design approach are such that each complements the other.

Electronic Packaging System (EPS) Enclosures (NEL). SUMMARY. The NEL Integrated Packaging Program embraces an enclosure consisting of a basic external housing in combination with a number of alternative rack configurations. With these, a high degree of versatility is possible. A single enclosure can protect not only modular elements, but also equipment which cannot be modularized for one reason or another.

An additional advantage of the NEL enclosure is its ability to accept modular concepts other than those developed at NEL. The enclosure has been used with modules developed by Control Data Corporation and is also usable with the SHP modules. Additional studies are now under way to determine commonality at the module level for avionic applications.

A major weakness of previous packaging concepts has been the inability to meet environmental requirements. It is for this reason that the primary emphasis of the EPS packaging program has been on resistance to shipboard environments. The resulting enclosures are designed to protect installed electronic equipment in a shock and vibration environment when deck-mounted only, and without the use of shock and vibration mounts. Other characteristics of the enclosure include the following: Cooling air flow is the same whether the enclosures are in the operate position or are opened for maintenance; modules are readily accessible for replacement; advanced connectors, interconnections, and shielding are used; the concept is equally applicable to single-enclosure equipment or multienclosure equipment.

The family of enclosures consists of four types, with each type available in various heights. The four types differ from one another primarily in the configuration of the pullout racks. The pullout racks can be essentially the rack, the double rack, or drawer-type racks arranged in various combinations. In the closed position, the enclosure isolates the installed electronics from the ambient atmosphere, including closure through RFI gasketing. Each enclosure can include its own cooling unit, or cooling air can be supplied from a central cooling unit.

There are no chassis as such in the EPS packaging concept. However, the pullout

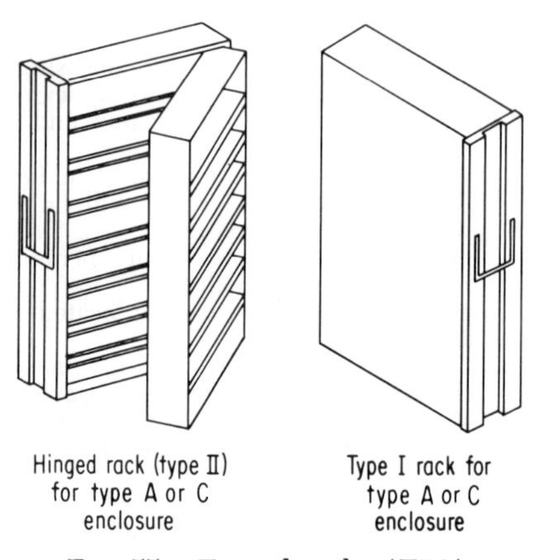

FIG. 55. Typical racks (EPS).

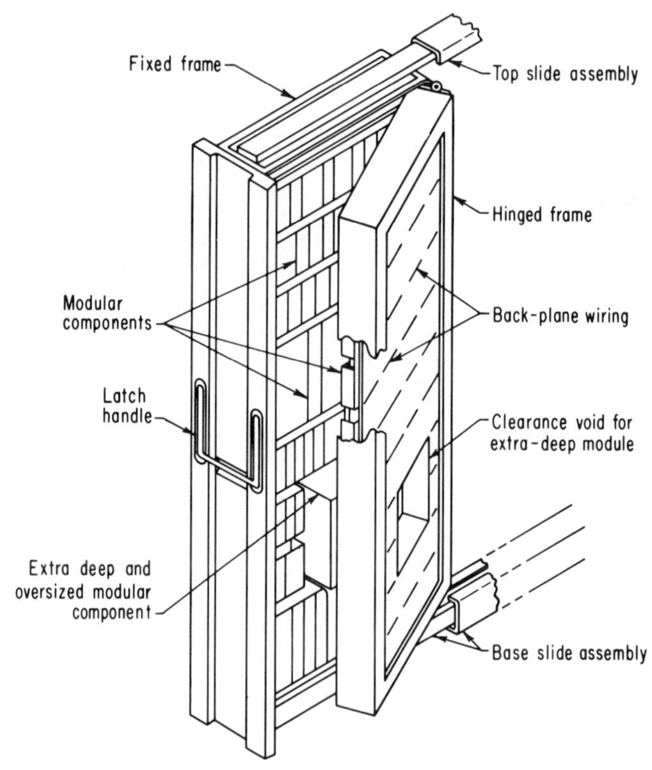

FIG. 56. Typical vertical type II rack, front view (EPS).

racks will mount standard 19-in. chassis, or, with adapters, chassis of other dimensions.

GENERAL CONFIGURATION. Each enclosure is a composite of the basic external housing, which is 24 in. wide by 25 in. deep by 72 in. high, and any compatible combination of main serviceable chassis (herein referred to as racks for brevity). These racks are generally one of four types:

Type I rack. A vertical half rack 7.88 in. wide. This rack is shown in Fig. 55.

Type II rack. A vertical half rack with two 3.88-in.-wide frameworks, one of which is hinged to allow complete access to both sides of both frameworks. This rack is illustrated in Fig. 56. Note that provisions are made for the accommodation of oversized modules or of equipment that cannot be profitably modularized.

(*a*)

(*b*)

FIG. 57. Type B enclosure (EPS). (*a*) Rack latched in place. (*b*) Rack extended in maintenance position.

Type III rack. A vertical full-width rack (17.55 in. wide), shown as part of the enclosure in Fig. 57.

No-rack designation. A drawer or horizontal rack shown as part of the enclosure in Fig. 58.

Each rack of types I and II is designed to accommodate loads up to 400 lb. The type III rack can carry loads up to 800 lb, and future developments should increase this to 1,000 lb. These racks can be extended up to 8 in. beyond the front face of the enclosure, providing easy access to the interior of the cabinet for maintenance of water hoses and interconnecting cables, and they will also lock in intermediate positions should any potentially dangerous situation arise that would cause maintenance personnel to inadvertently release the front handle while seeking access to the enclosure. This feature is intended to minimize danger under conditions of violent or unexpected ship motion.

As shown in Fig. 59, horizontal module supports are bolted to the rack at intervals determined by the module height. The horizontal module supports serve two basic purposes: they mount the receptacles for the modules, and they also contain ducts for cooling air.

The total enclosure (rack and housing) has one of four designations which depend primarily on the rack configuration. These are as follows:

Type A enclosure. This enclosure is preferred for modular applications and is shown in Figs. 60 and 61. It is composed of the standard external housing and any combination of type I and type II racks.

Type B enclosure. This configuration utilizes the standard external housing and a type III rack (Fig. 57).

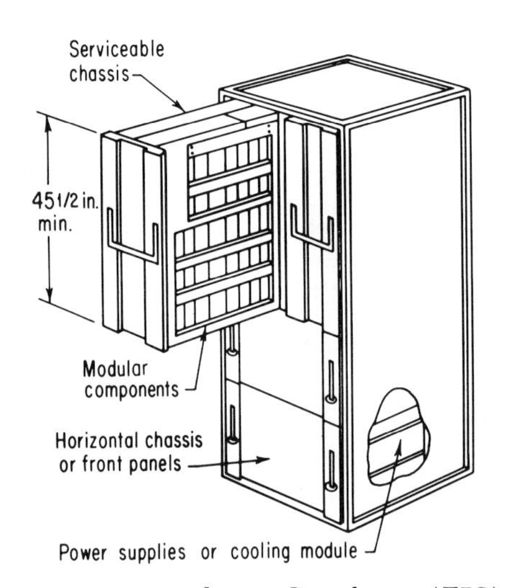

FIG. 58. Proposed type C enclosure (EPS).

Type C enclosure. This enclosure utilizes any combination of type I and type II racks in conjunction with one or more horizontal racks (Figs. 58 and 62). When used in this configuration, the vertical racks may have a minimum height of 45½ in., increasing in nominal increments of 7 in. up to the maximum of 72 in. Horizontal-rack (drawer) dimensions are compatible.

Type D enclosure. This enclosure is composed of the basic external housing and a number of horizontal racks. Figure 63 is a drawing of the proposed configuration.

Types B, C, and D embody modifications that are intended to accommodate electronic equipment that cannot be modularized or auxiliary equipment such as heat exchangers (see Fig. 58, for example). A possible, multiple-enclosure system utilizing several of these modifications is shown in Fig. 64. In a multiple installation of this type, the enclosures bolt together to form a single structural unit. Interconnecting wiring and cooling air are routed between enclosures through the tops or rear of the enclosures.

ENVIRONMENTAL CONSIDERATIONS. The enclosures were designed to have a minimum natural frequency of 40 cps when loaded to specifications, which is above the maximum shipboard vibration frequency. This applies to a single, base-mounted enclosure without top supports, side braces, shock mounts, or other form of external support.

Environmental requirements of MIL-E-16400 are satisfied, including such factors as drip-proof, spray-proof, and dustproof. The cabinets are also RFI shielded in the closed position by means of RFI gasketing.

The EPS cooling system provides not only cooling to the electronics, but atmospheric protection as well. Cooling air flows in a closed loop within the enclosure so that entrapped moisture and dust can be removed at the appropriate point in this cycle, thus providing a completely controlled environment for the electronics. Although cooling is not interrupted by opening the enclosure for maintenance, portions

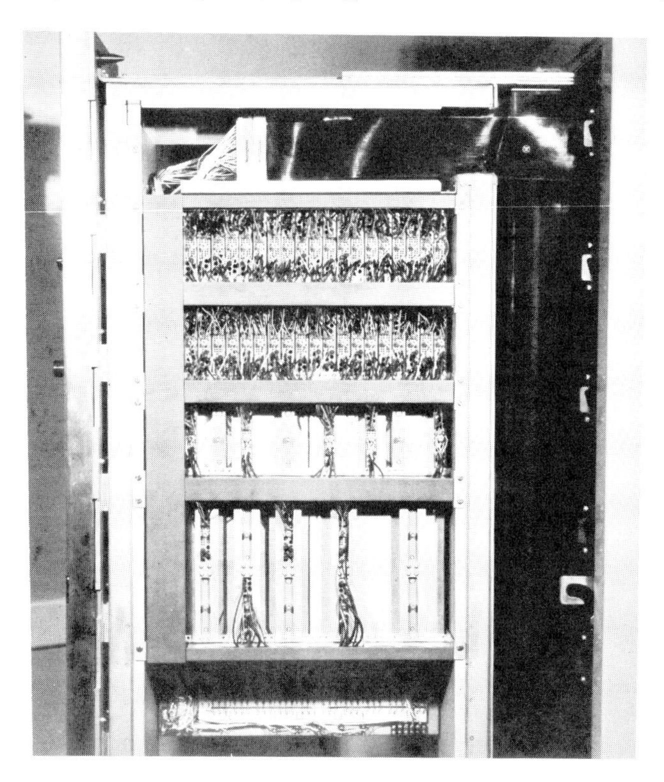

FIG. 59. Module mounting interfaces (EPS).

of the electronics are exposed to contaminants. These, however, are removed when the enclosure is returned to normal operating position.

The EPS cooling concept utilizes ship-supplied fresh water to cool a refrigeration unit, which, in turn, is used to cool the air circulated through the electronic modules.

The refrigeration unit with its blower and associated controls is called the *thermal control module*. The thermal control module consists of a vapor-compression-cycle refrigeration unit, a blower, and a control unit. The control unit is mounted in a separate chassis which mounts just below the thermal control module.

Three thermal control modules are under development. The three modules differ in cooling capacity and in configuration. The smallest unit has a cooling capacity of 1,300 watts and is designed to be installed in a vertical, serviceable main chassis. The next-sized unit has a capacity of 2,600 watts and is designed for installation in the bottom drawer of a type C enclosure. The largest unit has a cooling capacity of 5,000 watts and is designed to mount in a type B enclosure. Multiples of the large unit can be installed in a type B enclosure for cooling several cabinets in a systems installation.

The thermal control modules are designed for an outlet air temperature of 70°F±5°. They are designed to operate with ship-supplied cooling water at temperatures up to 105°F. The cooling water enters at the rear of any enclosure containing a thermal control module. Flexible water hoses run from the serviceable main chassis or drawer to quick-disconnected couplings located at the rear inside of the rack. The flexible base allows uninterrupted flow of water when the rack is in the extended position for maintenance.

FIG. 60. Type A enclosure, operating position (EPS).

FIG. 61. Type A enclosure, maintenance position (EPS).

Cooling air flows in a closed loop through or over the electronic modules. The cooling air leaving the thermal control module is supplied to a hollow, vertical, structural member of the serviceable main chassis. From the vertical member the air flows to the top half of the dual, horizontal, structural members. The air then flows vertically through the electronic modules and into the bottom half of the upper, horizonal, structural member. The air flows from each horizontal member into the second vertical member and is then returned to the thermal control module. In the case of shielded modules the cooling air flows directly through the module cover. For the printed-circuit card type of module the air flows over both sides of the circuit board.

Cooling air from the thermal control module mounted in the fixed, serviceable main chassis is supplied to the hinged chassis through a bellows type of flexible air duct which allows uninterrupted air flow when the chassis are in the normal operating position, in any internal position, or in the maintenance position.

INTERCONNECTION TECHNIQUES. Several features of the interconnection arrangement include the following: (1) back-panel point-to-point wiring; (2) conventional cabling from electronic modules to an overhead connector mounted on the rack; (3) flat cabling from the overhead connector to the interior of the enclosure; (4) armored cabling between enclosures.

(*a*)

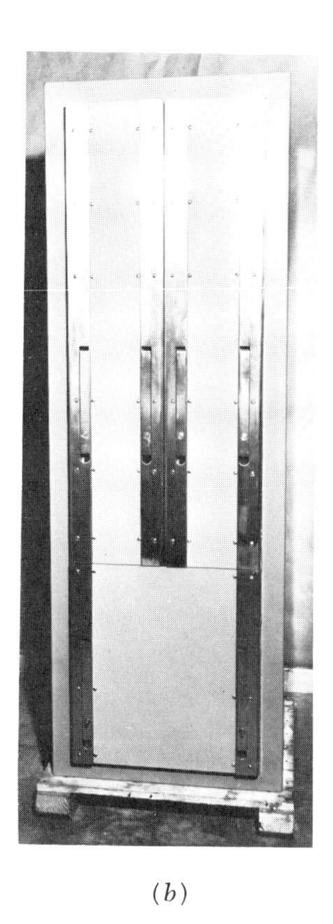

(*b*)

FIG. 62. Typical type C enclosure (EPS). (*a*) Drawer in maintenance position. (*b*) Drawer and type II rack latched closed.

The flat cables are flexible, to allow uninterrupted operation when the serviceable main chassis are in the maintenance position.

The general internal wiring arrangement is illustrated in Fig. 65. Conventional cabling is routed along a vertical structural member to an overhead connector which adapts the conventional cabling to a flat type of cable. This arrangement applies to both the fixed and the hinged rack.

Three connector types are available for use in implementing interconnections: a chassis type of connector is used for connections between back-panel wiring and cables coming into the rack; an enclosure connector is used for connections between the enclosure and ship's cabling; a front-panel connector is used to provide connection between back-panel wiring and front-panel indicators.

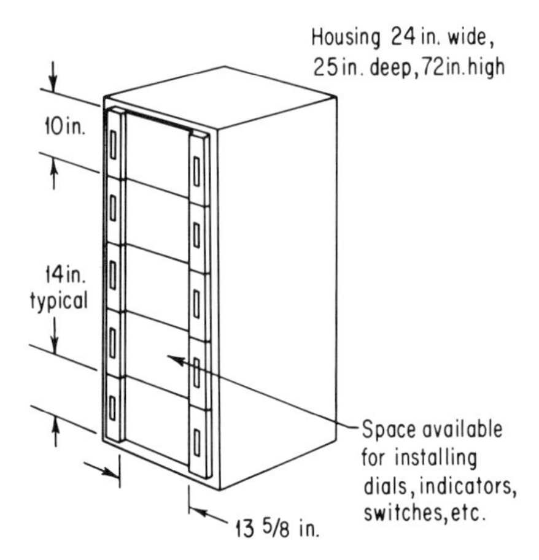

FIG. 63. Housing dimensions of proposed type D enclosure (EPS).

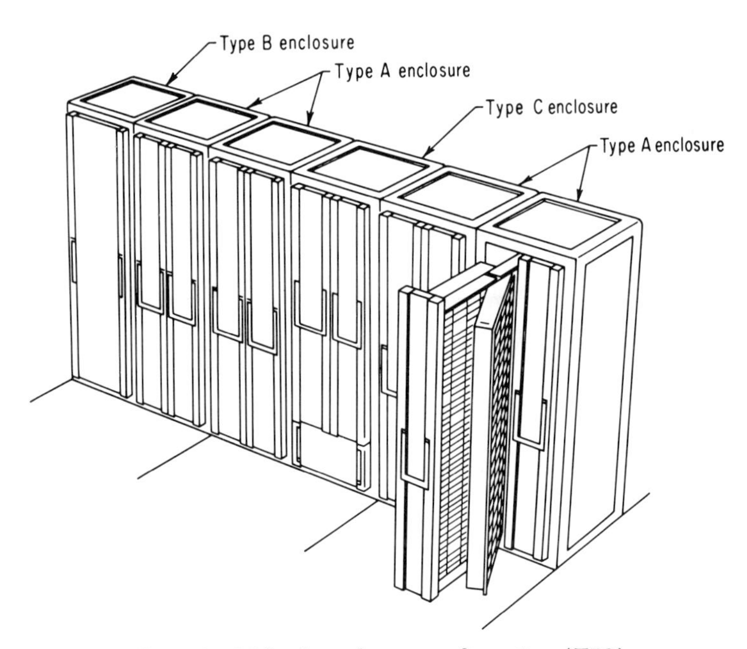

FIG. 64. Multiple-enclosure configuration (EPS).

These connectors, which are all available from multiple sources on a nonproprietary basis or on a licensing basis, allow a wide flexibility of both electrical and dimensional parameters. Each type can be obtained in a number of sizes and current ratings.

EXTERNAL-REQUIREMENTS SUMMARY CHART

Item	*Discussion*
Power ducting	Designed to accept any external power, particularly shipboard armored cable (upper rear of enclosure is preferred for cable entry)
Signal ducting	Same as above
Cooling	Shipboard fresh water (max. temp. 105°F) from 2.5 to 5.0 gpm for each type A (1,300-watt size) cooling module (2 type A chassis per enclosure)
Power requirement ...	Not defined (dependent upon system requirements)
Mounting structure ...	Enclosure cabinets are designed to be structurally sound when solid-base-mounted only (without the use of shock and/or vibration mounts)
Mounting area	Approximately 2- by 2-ft deck space plus access area in front of enclosure and cable-junction connector area in rear of enclosure
Volume	24 ft^3 per enclosure of maximum height
Location	Limitations relative to location in ship not yet determined
Shock and vibration supports	Not required
Enclosure dimensions ..	Approximate dimensions of enclosure are 24 in. wide by 25 in. deep by 72 in. high

INTERNAL ENVIRONMENT SUMMARY CHART

Item	*Discussion*
RFI shielding	A combination RFI and moisture-proof gasket is provided for each chassis or access opening in the enclosure
Atmospheric protection in open position	MIL-E-16400 compatible
Atmospheric protection in closed position ...	MIL-E-16400 qualified
Shock and vibration ..	MIL-E-16400 qualified
Cooling provisions, open position	Filtered air at 70°F ± 5° is capable of dissipating total of 1,300 watts of heat in max. size type A chassis or of dissipating 0.5 watt/$in.^3$ of electronics (2 type A chassis per enclosure)
Cooling provisions, closed position	Filtered air at 70°F ± 5° is capable of dissipating total of 1,300 watts of heat in max. size type A rack or capable of dissipating 0.5 watt/$in.^3$ of electronics (2 type A racks per enclosure)

Centralized Electronic Control (CEC) Enclosures (NRL). SUMMARY. Primary enclosure characteristics of the CEC packaging concept are: (1) the use of advanced connectors and shielding techniques throughout the system; (2) the provision in each rack of fault-locating indicators; (3) the elimination of false decking, ship's ducting, or overhead support for equipment groupings; (4) increasing the accessibility of all modules without interrupting or degrading cooling or interconnections; (5) the capability of cabinet sections, up to and including 11-in. widths, to pass through 25-in. circular tubes for submarine installation; (6) the inclusion of circuit-breaker protection for power wiring in each cabinet section.

Chassis as such do not exist in the CEC concept; modules are mounted directly to racks. However, the functional equivalent of a chassis may be obtained, if desired, by interconnecting a group of modules with a modular interconnection assembly.

The racks are slide-out assemblies and are enclosed in cabinet sections which isolate the racks from the room atmosphere. The structural members of the rack are used as cooling-air passages for routing of the air to modules in parallel paths. Heat-

FIG. 65. Typical interconnections (EPS).

FIG. 66. Centralized electronic control equipment arrangement.

exchanger cabinet sections are located as needed to transfer heat from the cooling air to ship-supplied fresh water.

GENERAL MECHANICAL CONFIGURATION. A possible arrangement of CEC equipment in a ship's compartment is illustrated in Fig. 66. An operator's console and a number of cabinet assemblies can be grouped together, as shown, in any arrangement suited to the compartment dimensions. The cabinet assemblies may be mounted against a bulkhead or arranged in rows. No rear access or overhead support is required.

An exploded view of a CEC cabinet group is shown in Fig. 67, and photographs of a mechanical demonstration model are shown in Fig. 68. Each cabinet group consists of a number of sections, including a heat exchanger, a ship's wiring inter-

FIG. 67. CEC cabinet group (exploded view).

face section, and two end panels. The number of heat exchangers in any cabinet group is governed by the maximum heat dissipation of the complete group. The wiring interface section is located at one end of each cabinet group, where it provides for interconnection with power inputs and interassembly signal cables.

The tops of the cabinet sections contain power buses, signal cables, and ducting. The bottom casting of each cabinet section serves to route the cooling air. Provision is also made on each cabinet section for indicators. The arrangement illustrated in Fig. 67 shows a group of indicators on the front panel of each slide-out rack. Four indicators could show information such as rack secured and locked, air temperature within tolerance, air pressure within tolerance, and electrical fault location. A summary of this information would also be presented on the operator's console.

The cabinet configurations illustrated in Figs. 67 and 68 show that the cabinet sections can be standardized in height and depth and that width may be varied in fixed increments. Figure 68 shows how this arrangement results in module dimensional flexibility in all three directions.

Some pertinent features of the CEC cabinet and rack configurations are as follows: No false decking or ship's ducting is required within an equipment grouping. Shielding and fireproofing is provided between cabinet sections, including front-panel closure through RFI gasketing. The assemblies are drip-proof. The racks automatically latch in any extended position when an operator releases the handle. Flexible bellows are not required for input of cooling air to racks when in the maintenance position.

ENVIRONMENTAL CONSIDERATIONS. The cabinet sections and racks used in the CEC packaging concept provide environmental protection for the enclosed modules.

FIG. 68. CEC demonstration model.

Cooling provisions, RFI shielding, and atmospheric, shock, and vibration protection are major environmental conditions which must be controlled. The manner in which these requirements are accomplished is discussed below.

In the CEC cooling concept, forced air cooling is provided in parallel paths to the electronic modules. The cooling system provides positive cooling of each module whether the rack is open or closed. Hollow structural members of the racks are utilized as passages for the cooling air. The complete air path, which is a closed loop, is shown in Fig. 69. Cool air travels to each rack through the cabinet base casting to the front, vertical, structural member which acts as a supply header feeding air to the upper half of each of the dual, horizontal, structural members. The air then flows vertically upward through each module, then horizontally through the lower half of the air duct to the return header, through the ducting located at the top of the cabinets, and is then returned to the heat exchanger. Single air ducts are used at the top and bottom of each rack for the cooling-air distribution. Tapered air passages made of reinforced plastic may be used as horizontal duct members in the racks to provide more efficient air distribution to the modules. The plastic material provides thermal insulation between the cool input air and the warm exhaust air.

When the rack is opened for maintenance, the air flow through the rack section is reversed. The rear vertical member becomes the supply header, and hot air is exhausted into the ship's compartment through the front member instead of being returned to the heat exchanger. Filtered makeup air is then introduced into the heat exchanger from the compartment.

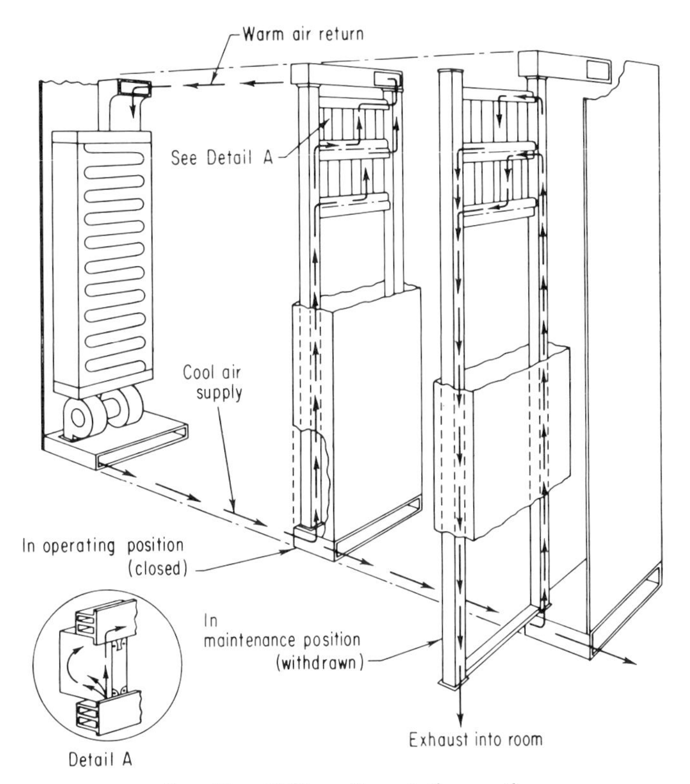

FIG. 69. CEC cooling-air-flow path.

Ship-supplied fresh water is used to cool the thermoelectric heat-exchanger unit, which in turn is used to cool the air circulated through the electronic modules. The heat-exchanger unit consists of a number of removable thermoelectric modules, a blower, and a control unit which is mounted as a removable assembly. The unit itself has a 2,500-watt capacity and is mounted in an 11-in. CEC cabinet section. It is designed to supply 350 cfm of cooling air at 70°F and at a 4-in. water column. Flexible water hoses connect from the rear of the cabinet section to the rack frame. These are of the guide-disconnect type and allow uninterrupted flow of water when the rack is opened for maintenance.

An exploded view of the rack ducting is shown in Fig. 70. Single horizontal air ducts are located at the top and bottom of each rack, and dual air ducts can be installed as needed to accommodate the standard CEC module sizes. A damper is provided for each horizontal and front vertical air duct so that the air flow can be balanced with heat dissipation.

Air temperature and flow rate can be monitored and malfunctions signaled by indicators located on the front panel of the cabinet sections.

RFI shielding is accomplished between cabinet sections, including front-panel closure, by use of RFI gasketing. Metal module covers for some modules accomplish this requirement by providing internal shielding compartments and a high order of protection from mutual interference between modules.

Atmospheric protection is accomplished by using drip- and spray-proof cabinets and a thermoelectric heat exchanger and by circulating cooling air in a closed loop. When cabinets are in the normal closed position, the equipment is protected from

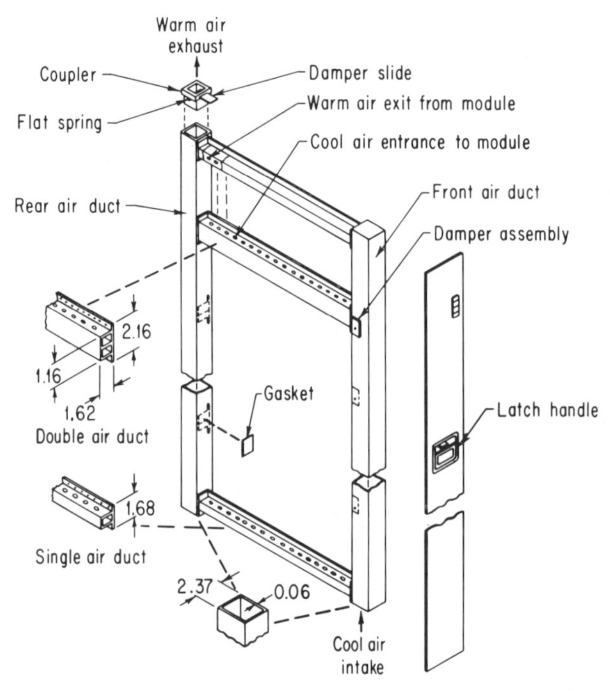

FIG. 70. CEC cooling-air-rack ducting (exploded view).

atmospheric contamination. The heat exchanger provides temperature control for the cooling air, and reinforced plastic air passages provide thermal insulation between the cool input air and the warm exhaust. When a rack is in the extended position for maintenance, cooling air from that rack is exhausted into the ship's compartment, and makeup air is drawn into the system through filters. For a cabinet group containing a number of racks, the decrease in fungus and humidity protection for the modules would be negligible. When a rack is opened for maintenance, the back-panel wiring is exposed to ambient air. However, except in the case of simultaneous multiple failures or a failure in the wiring itself, the rack would be open only for a few minutes at a time.

Shock and vibration requirements are accomplished by designing cabinet sections as base-mounted units not requiring shock mounts when installed in a system configuration. The dynamic forces felt by installed electronic equipment depend on the number of cabinet sections used to make up the cabinet group. When the cabinet group is made up of only a few sections, the transmitted forces can be high. However, an increase in the number of sections results in an increase in rigidity and a reduction in the level of transmitted forces.

INTERCONNECTION TECHNIQUES. The CEC interconnection concept is illustrated in Fig. 71. The electronic modules are interconnected by a back-panel wiring assembly, one or more flexible cables which connect from the midsection of the rack to the central portion of the overhead, and three overhead interrack wiring provisions (raw power, control wiring, and coaxial cables). The back-panel wiring assembly may be interconnected by individual wiring assemblies. The back-panel wiring assembly can be of any appropriate type; it may be a frame type of assembly with wire wrap, solder cup, percussive welding, or any other suitable type of pin, or, in

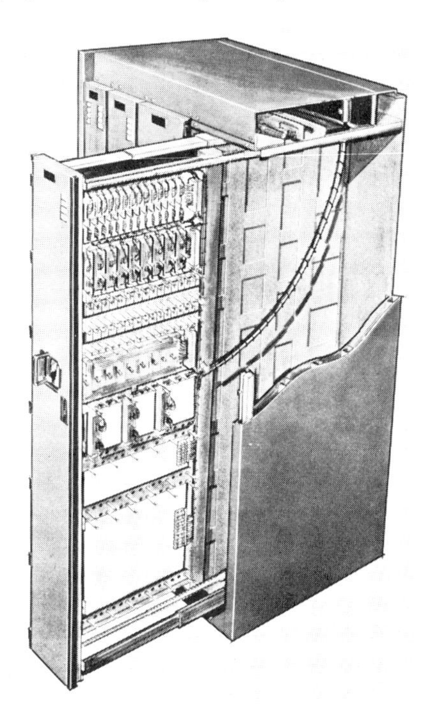

FIG. 71. CEC cabinet.

some cases, it may be a multilayer printed-circuit board. A connector may be located at one end of each back-panel wiring assembly for interconnection of wiring assemblies and connection with the flexible interrack cables. The flexible cables are also modular; each cable can be disconnected and replaced as a unit. The modular nature of the interconnecting concept provides an excellent maintainability and updating capability.

The bus-bar assemblies shown in Fig. 72 are used for distribution of power within a cabinet assembly. Each assembly consists of bus-bar and insulation sections plus a Nike Zeus type of connector receptacle. The bus bars of individual cabinet sections join together in an in-line overlap fashion and are connected to ship-supplied primary power through the wiring interface rack. The flexible cable in each cabinet section mates with the connector on the bus-bar assembly and carries power to the rack.

Power circuits, which connect to the racks in each cabinet section, are protected by circuit breakers mounted as part of the bus-bar assembly (Fig. 73). The circuit breakers are connected in series with the Nike Zeus connector mounted as part of the bus-bar assembly.

A "roll-up" cable is being developed to interconnect the power circuitry from the bus-bar assembly to the rack, i.e., it can be removed from the cabinet assembly as a unit.

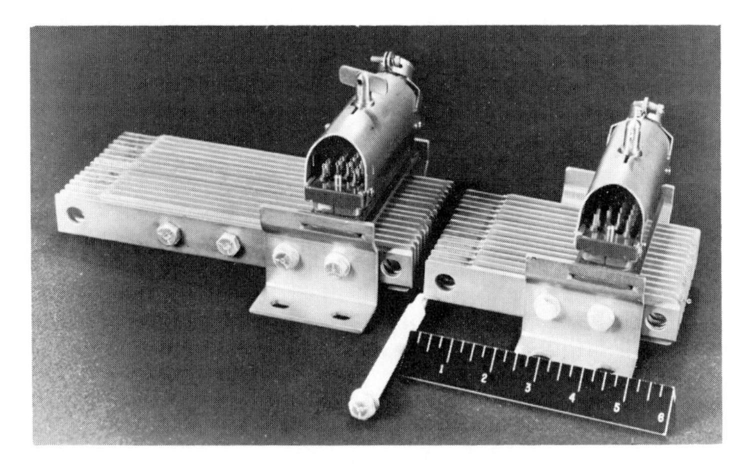

FIG. 72. CEC bus-bar assemblies.

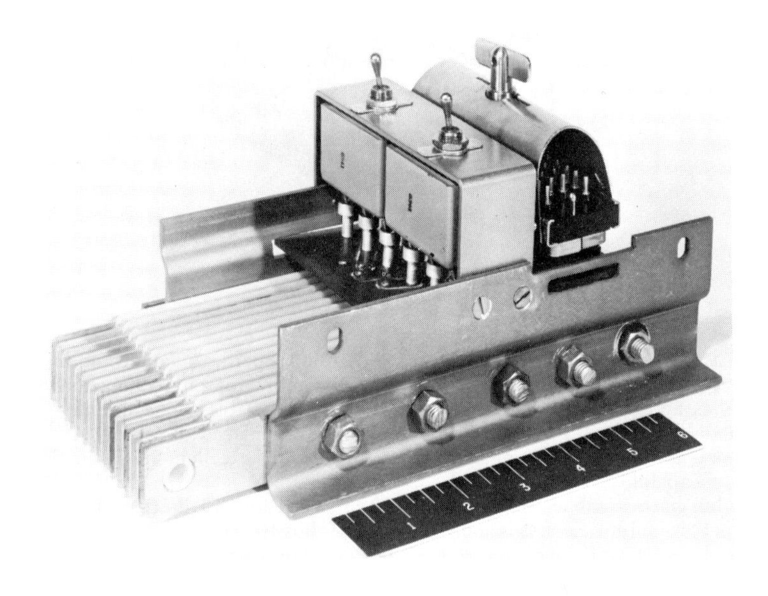

FIG. 73. CEC bus-bar assembly with circuit breakers.

INTERNAL ENVIRONMENT SUMMARY CHART

Item	*Discussion*
RFI shielding	Racks are enclosed in cabinets equipped with RFI gasketing
Atmospheric protection in open position	Decrease in fungus and humidity protection negligible for cabinet assembly containing a number of racks
Atmospheric protection in closed position	Cabinets are drip-proof and spray-proof for maximum protection in operation position
Shock and vibration	Racks contain no shock and vibration mounts
Cooling provisions, open position	In open position air is exhausted into ship's compartment, and makeup air is drawn in through filters
Cooling provisions, closed position	Cooling system is a closed loop. Module inlet and outlet area is variable; module internal flow area may vary within limits to be determined by the system designer

REFERENCES

Reference Number — **Title or Description**

1. MIL-STD-243A "Types and Definitions of Models for Communications—Electronics Equipment."
2. MIL-STD-210 "Military Standard—Climatic Extremes for Military Equipment."
3. Sperry Gyroscope Co.: Reduce Paperwork on MIL Designs, *Electron. Design Mag.*, vol. 14, no. 28, Dec. 6, 1966.
4. HEL Standard S-3-65 "Human Factors Engineering Design Standard for Missile Systems and Related Equipment," Human Engineering Laboratory, U.S. Army, Aberdeen, Md.
5. AFSCOM 80-3 "Handbook of Instructions for Aerospace Personnel Subsystem Designers."
6. McCormick, E. J.: "Human Factors Engineering," 2d ed., McGraw-Hill Book Company, New York, 1964.
7. The Packaging Revolution, parts 1, 2, and 3, *Electronics Mag.*, Oct. 18, Nov. 1, 1965.
8. Staller, Jack J.: "Guidelines for Implementation of System Requirements into Microelectronic Mechanical Designs," Sylvania Electronic Systems, Needham, Mass.
9. Guide to the Design of Electronic Equipment for Maintainability, *Wright Air Develop. Center Tech. Rept.* 56-218.
10. "Maintainability Design," F. L. Ankenbrandt, Engineering Publishers, Elizabeth, N.J.
11. Harris, C. M., and C. E. Crede: "Shock and Vibration Handbook," McGraw-Hill Book Company, New York, 1961.
12. Timoshenko, S. P.: "Vibration Problems in Engineering," D. Van Nostrand Company, Inc., Princeton, N.J.
13. Jacobson, L. S., and R. S. Ayre: "Engineering Vibrations," McGraw-Hill Book Company, New York, 1958.
14. Pipes, L. A.: "Applied Mathematics for Engineers and Physicists," 2d ed., McGraw-Hill Book Company, New York, 1958.
15. "Dynamic Shock," Bishop Engineering Co., Princeton, N.J.
16. Pipes, L. A.: "Matrix Methods for Engineering," Prentice-Hall, Inc., Englewood Cliffs, N.J.
17. O'Hara, J. J., and P. F. Cunniff: Elements of Normal Mode Theory, *Naval Res. Lab. Rept.* 6002.
18. NAVSHIPS 250-423-30 "Shock Design of Shipboard Equipment," part 1, Dynamic Design Analysis Method.
19. MIL-S-901C "Shock Tests, High Impact; Shipboard Machinery, Equipment and Systems, Requirements for."
20. Crede, C. E.: "Vibration and Shock Isolation," John Wiley & Sons, Inc., New York.
21. Forkois, H. M., and K. E. Woodward: Design of Shock and Vibration Resistant Equipment for Shipboard Use, *Naval Res. Lab. Rept.* 4789.

22. O'Hara, G. J.: Shock Spectra and Design Shock Spectra, *Naval Res. Lab. Rept.* 5386.
23. "R.F.I. Design Guide Catalogue," Technical Wire Products Inc., Cranford, N.J.
24. "Handbook on Radio Frequency Interference," Fredrick Research Corp., Wheaton, Md.
25. NAVMAT P 3940 "Navy Systems Design Guidelines Manual—Electronic Packaging," Headquarters, Naval Material Command, May, 1967.
26. OD 30355, rev. 2, "Standard Hardware Program—Data Handbook," vols. I–IV, Naval Avionics Facility, Indianapolis, Ind.
27. "Interference Reduction Guide for Design Engineers," prepared by Filtron Co. Inc., New York, issued by USAEL, Fort Monmouth, N.J.
28. "Integrated Packaging Manual for Shipboard Electronic Systems and Equipment," Navy Electronics Laboratory, San Diego, Calif., 1967.
29. "NRL Modular Electronic Packaging," Naval Research Laboratory, Washington, D.C., 1967.

Chapter 14

PACKAGING FOR SPACE ELECTRONICS

By

W. D. FULLER

Lockheed Missile & Space Company
Sunnyvale, California

INTRODUCTION

The purpose of this chapter is to illustrate the essential elements in the systems approach to the packaging of electronic equipment used in space missions. Systems packaging, like systems engineering, is concerned with design, i.e., packaging design and its management. It is implied that the output of systems packaging design is a set of specifications that detail the construction of a functioning system which will meet the system performance requirements within the environmental constraints of the space mission. The range and severity of these environmental conditions, in which space electronics must function, demand a rigorous, systematic approach to the integration of materials and processes into a system configuration that represents an optimum design among the trade-off parameters of performance, size, weight, producibility, maintainability, and cost. The success of a space mission is dependent upon how well system designers have been able to cope with the mission environment from concept to mission completion. The management of systems packaging interrelates highly with the other program organizations as shown in Fig. 1, and the success of these interrelationships is directly related to the success of accomplishing systems packaging within schedule and cost requirements. Documentation for systems packaging of space electronics, as might be required by the NASA 500 Series, the USAF 375 Series, or internal program management, is another important status and control function in systems packaging.

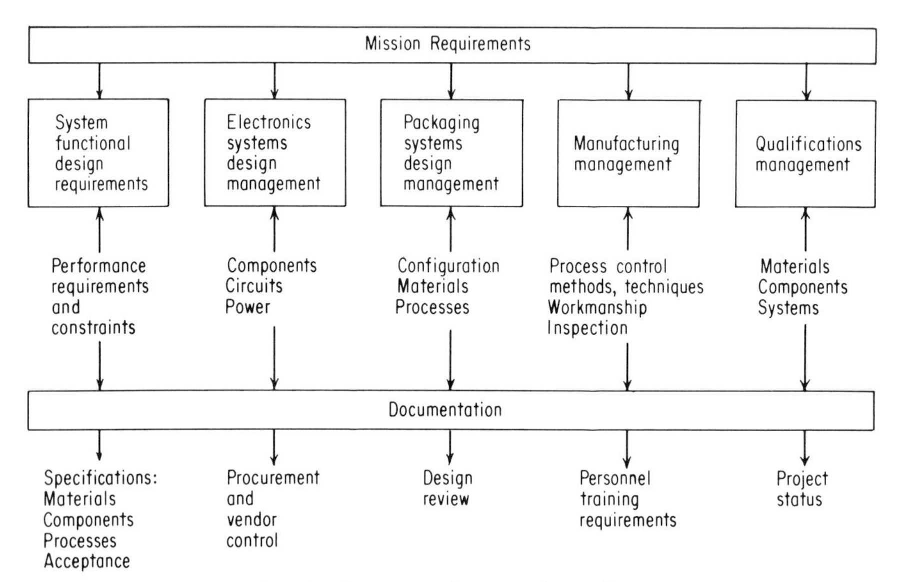

FIG. 1. System packaging relationships.

The range and scope of mission environmental factors require design techniques different from the conventional—different in materials selection, different in configuration, and different in workmanship. The differences in space missions result in a difference in specifications among the various types of missions. In this chapter, then, the types of missions are described, the environmental factors are examined, examples of packaging for space electronics are given, some characteristics of components and materials relating to space environments are tabulated, and some general comments on manufacturing and testing are made. Reference material is liberally cited, and references are made to applicable design information in other chapters.

SPACE MISSIONS

The primary objective of space missions is precise definition of the space environ- and eventually travel to the planets and beyond. Space missions encompass tasks of ment and its effects, so that man may confidently venture into space, utilize its volume, investigation, exploration, utilization, and control. These tasks have been expressed as national goals,[1] more definitively as: planetary exploration, biological research, astrophysical research, and man-in-space laboratory studies. The first Russian satellite was placed in orbit on October 4, 1957, and followed on January 31, 1958, by the United States Explorer I satellite. In the following decade approximately 700 space missions have been successfully accomplished, including the manned missions of the Mercury and Gemini series.

A partial listing of United States space vehicles with a description of their missions is given in Table 1. The configuration of the very successful Surveyor is shown in Fig. 2, and the Apollo Lunar Module Vehicle is shown in Fig. 3.

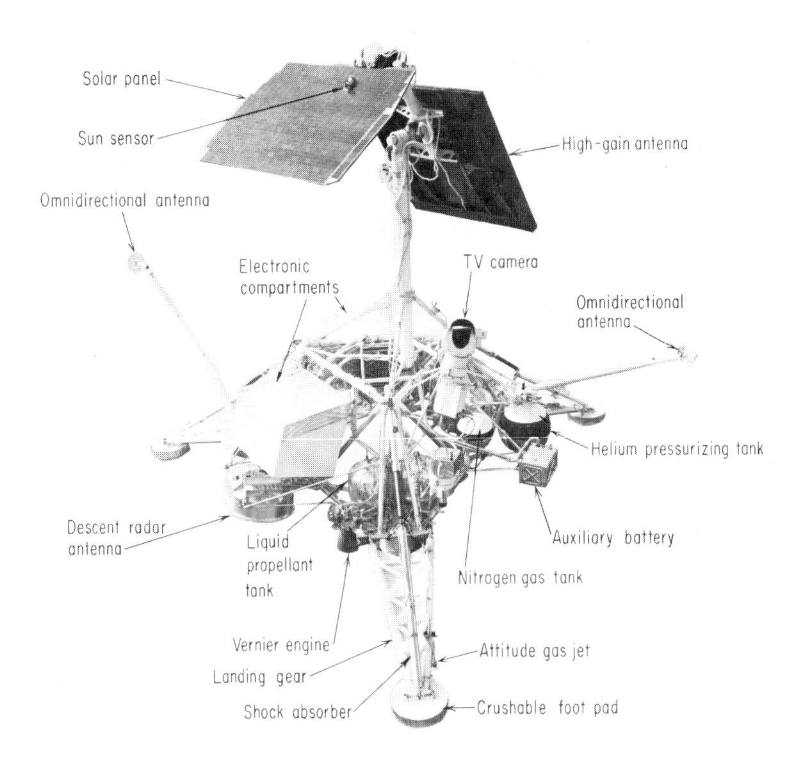

FIG. 2. Surveyor configuration. (*Photograph courtesy of Hughes Aircraft Company.*)

TABLE 1. Typical Space Missions [2]

Vehicle	*Mission*
Investigation:	
Orbiting observatories	
OSO	Detect and measure electromagnetic radiation from sun and other celestial bodies; data leading to solar-flare prediction system.
OGO	Look at earth, sun, and space, simultaneously for correlated studies of particle fields, within earth's atmosphere, magnetosphere, or circumlunar space.
OAO	Optical observations and mapping, mainly in a uv spectrum, some x-ray.
Explorer series	A series of unmanned satellites to study various aspects of space such as radiation belts, atmosphere, micrometeoroid, and the ionosphere from orbits ranging from near earth orbit to lunar orbit.
Pegasus	Use of modified boiler-plate models of Apollo as bases for meteoroid-impact detection surfaces.
Biosatellites	Instrumented, reconvertible capsules to conduct biological tests during 3 to 30 days of earth orbit.
Mariner series	A series of unmanned deep space probes to flyby Mars and Venus and send back en route data on particles and fields as well as TV pictures and other data during the flyby.

TABLE 1. Typical Space Missions (*Continued*)

Vehicle	*Mission*
Ranger	Hard impact on the moon after transmitting TV pictures from 900 miles altitude down to about 1,000 ft.
Pioneer	Small craft to transmit data on interplanetary particles over long distance from the earth in orbits from 0.2 to 1.2 astronomical units from the sun (1 astronomical unit $\cong$ 93×10^6 miles).
Surveyor	Soft-land 30-day-life payload on moon; transmit TV pictures and data on bearing strength of lunar surface as part of Apollo site-selection program; prove-out soft landing technique for L.M.
Exploration:	
Mercury	A series of two suborbital and four orbital flights with a one-man capsule which revealed human pilot capabilities and verified reentry and landing designs for manned orbital flight.
Gemini	Two-man spacecraft to demonstrate 14-day orbital mission, develop rendezvous and docking techniques, and explore feasibility of extravehicular activity.
Apollo	The three-man Apollo spacecraft's main mission is to reach the moon by 1970. There is also an extensive plan for earth and lunar orbital missions.
Utilization:	
Communication	
Telstar	Experimental low-orbit communications satellite, providing first transatlantic television.
Relay	Active communications satellite in medium-altitude elliptical orbit. Relay 2 had longer-life solar array.
Syncom	Active satellite in synchronous (stationary) orbit demonstrated precise altitude and period control, which led to synchronous technique for commercial satellites.
Early Bird	Synchronous commercial satellite to provide 240 duplex voice channels over North Atlantic.
Comsat	Global commercial communications using synchronous orbit. The ultimate global Comsat system, featuring multiple access, will have up to 1,200 duplex voice circuits.
Weather observations:	
Tiros	To develop global weather-observing techniques from low-altitude satellite, making cloud-cover photographs and measurements.
Nimbus	Large, modular craft to test advanced techniques of photographing cloud cover and making other meteorological observations.
Miscellaneous	
Geos	Active and passive satellite for gravimetric geodesy measurement leading to more accurate location of points on earth.
ATS	Large, multipurpose testbed for communications, weather, and navigation tests, including gravity, gradient stabilization at synchronous altitude.
Control:	
Samos	Survey earth from 100- to 300-nmi polar orbits using optical, ir, and electronic sensors. Advance types have data processing.
NDS	Carries x-ray, gamma-ray, and neutron sensors for detecting nuclear explosions in space.
MOL	Two-man, 30-day orbital flights with modified Gemini B capsule plus cylindrical laboratory to establish military usefulness of man in space.

Space vehicles are not homogeneous entities but composites of many systems or subsystems, each a critical part of the whole. Electronic assemblies make up a major portion of the mission payload and fall into the following functional categories:

Scientific instruments
Engineering housekeeping
Telecommunication
Command and control
Surveillance
Life support
Power
Navigation and guidance
Data processing

Table 2 lists some of the typical scientific-mission payloads and some of the missions on which they were used. The flight-control packages for Surveyor are shown in Fig. 4 as an illustration of exposed assemblies in an unmanned space probe. In Fig. 5 is a block diagram of an integrated payload for a typical exploratory space mission. Very few of the electronic systems in a space mission incorporate components or operating principles that are completely new. The problems are not in operational techniques, but in designing the equipment to achieve a specified performance within the constraints imposed by the spacecraft and the mission environment.[10]

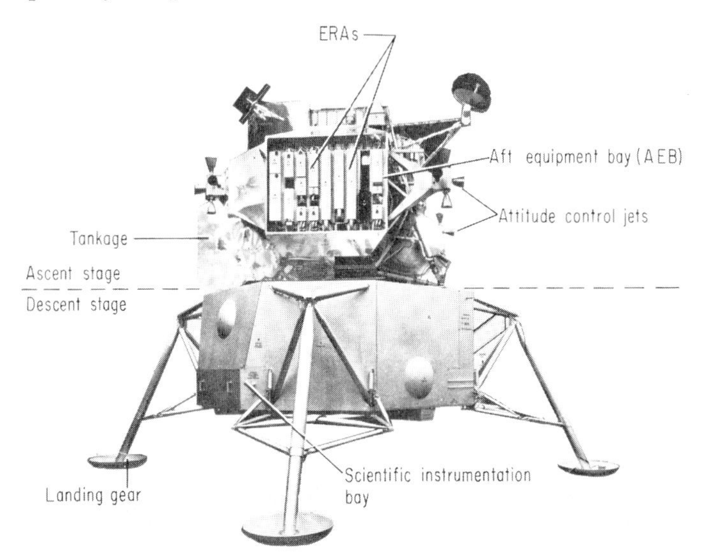

FIG. 3. Apollo Lunar Module Vehicle, showing aft electronics bay with Electronic Replaceable Assemblies (ERAs) exposed. (*NASA/MSC.*)

MISSION ENVIRONMENT

The environment in which space electronic equipment must operate varies widely throughout the mission, and knowledge of the constituents of this environment is a prime requirement for persons concerned with design and construction of satellites and spacecraft as well as the packaging of its equipment. Mission environment may be divided into natural and induced parts, where *natural environment* is defined as an environment which exists whether or not a man-made object is present. *Induced environment* is defined as an environment caused directly by a man-made device or due to an interaction between the device and the natural environment. The natural

TABLE 2. Typical Scientific-Mission Payload [3–8]

Measurement	Instrument	Vehicle
Cosmic-ray fluxes	Scintillation detector	OGO
Solar-proton concentrations	Electrostatic analyzer	OGO
Proton flux and energy spectrum	Faraday cup plasma probes	OGO
Positrons	Double gamma-ray spectrometer	OGO
Observation of trapped radiation and auroral particles	Scintillation detector	OGO
Study of galactic cosmic rays and isotopic abundance	Cosmic-ray telescope	OGO
Low-energy galactic radiation, protons, and other nuclei at high energies	Charged-particle telescope	OGO, IMP-I, Mariner IV, Solar Probe
Omnidirectional intensities of electrons	Geiger tubes	OGO
Electron energy	Spectrometer	OGO, Gemini
Magnetic-field fluctuations	Coil magnetometer	OGO
Dynamic radio spectra of solar bursts	Rf detector	OGO
Concentration and energy distribution of charged particles	Spherical-ion trap	OGO
Concentration and energy distribution of charged particles in the low-energy or thermal range	Planar-ion trap	OGO, Solar Probe
Positive-ion composition	Ion-mass spectrometer	OGO
Spatial density, mass distribution, and velocity of dust particles	Impact detector, adhesion to films	OGO, Voyager
Terrestrial and other emissions in the very-low-frequency range	Rf detector	OGO
Magnitude and direction of magnetic fields	Rubidium-vapor magnetometer	OGO, Solar Probe
Lyman-alpha and interplanetary geocoronal and interplanetary medium	Uv ion chamber	OGO
Ultraviolet, green, and infrared regions	Gegenschein photometer	OGO
Electron concentrations	Beacon to radiate linearly polarized signals toward earth	OGO
Obtain a fix on the star Canopus	Canopus star tracker	Voyager
Determine bearing of sun from spacecraft	Course-acquisition sun sensors	Voyager
Mass and flux of interplanetary gas	Mass spectrometer	Solar Probe
Three components of magnetic field	Flux-gate magnetometer	Solar-Probe Gemini, IMP-I
Monitor solar flares	Solar-flare x-ray ion chamber	OSO-1, Solar Probe
Solar radiation	Lyman-alpha detector	OSO-1, Solar Probe
Ionosphere characteristics	Lyman probe	Solar Probe
Intensity and polarization of light scattered by coronal electrons	White-light corona meter	Solar Probe
Dielectric constant	Rf resonant cavity	Voyager
Temperature	Thermocouple	Voyager
Pressure of atmosphere	Barometer	Voyager
Gas analysis of atmosphere	Gas chromatography	Voyager
Gas density	Accelerometer (during entry)	Voyager
Organism detection	Culture growth	Voyager
Surface features of planet	Microwave radiometry	Voyager

TABLE 3. Environmental-design Checklist

Environment	Storage/ Handling	Test checkout	Trans- portation	Launch	Flight	Landing
Thermal	...	×	...	×	×	×
Radiation	...	...	...	...	×	×
Particles	...	...	...	...	×	
Acceleration ..	×	×	×	×	...	×
Pressure	...	×	...	...	×	
Shock	×	×	×	×	...	×
Vibration	...	×	×	×		
Weathering ..	×	×	×	×		
RFI/EMI	...	×	...	×	×	×

environment of space includes extremely low pressure, electromagnetic radiation, high-energy particle radiation, dissociated and ionized gases, and high-velocity solid particles. Induced environmental factors result from activating a system and include acceleration, shock, heating, vibration, and acoustic excitation as well as the generation of toxic gases and organic contamination. These natural and induced environments occur throughout the mission regime as shown in Table 3, which includes storage and handling, testing and checkout, launch and reentry, orbit, and planet landing and relaunch.

FIG. 4. Flight control package for Surveyor which includes electronics, sun sensors, Canopus sensor, and inertial reference unit. (*Photograph courtesy of Hughes Aircraft Company.*)

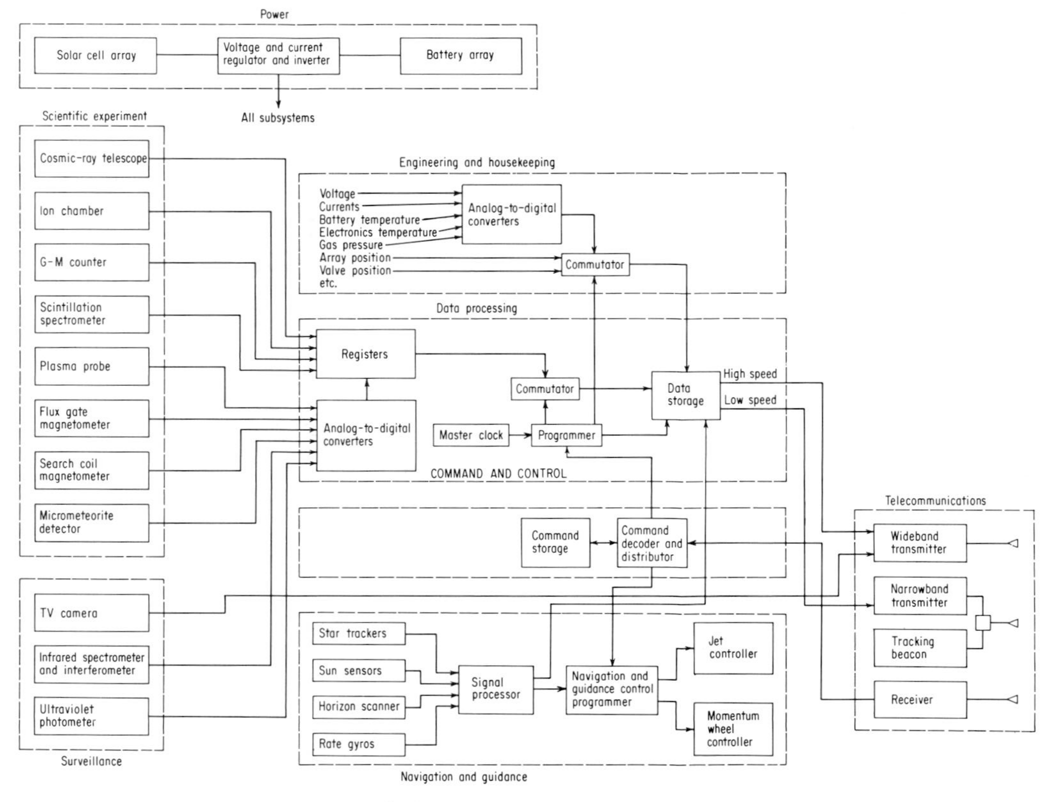

FIG. 5. Block diagram—space-mission payload.

Natural Environments. Natural environments exist independently of the aerospace vehicle system. Various natural environments in relationship to altitude from the earth's surface are illustrated in Figs. 7 and 8.

Ambient Temperature. The *United States Standard Atmosphere 1962* * generally describes the temperature gradient in the atmosphere. In space, however, *temperature* has no meaning. Temperature is an index of average translational molecular energy in a gas. When an individual molecule acquires translational energy, its kinetic temperature increases because of its greater velocity. In a tenuous atmosphere the heating effect of a relatively few molecules with a high kinetic temperature on a mass the size of a satellite is negligible.

Absolute Pressure. The earth's atmosphere is assumed to be a continuous medium of a gas in static equilibrium, and the pressure at any altitude is described by the

FIG. 6. Military communication satellites in burn-in test facility used to verify quality and reliability.[9] (*Photograph courtesy of Philco-Ford Corporation, WDL Division.*)

barometric law. The pressure or vacuum reaches 10^{-6} torr at approximately 100 miles altitude, and 10^{-13} torr in cislunar space. At 200 miles, 99 percent of the heat capacity of the atmosphere has been lost, and the only significant means of thermal-energy transfer is by radiation. In the direction away from the sun, space is an infinite heat sink with a blackbody temperature of 4°K.

Acceleration of Gravity. The acceleration of gravity varies as a function of the square of the distance from the earth's center and reaches a value of 0.01 ft/sec^2 in cislunar space. Weightlessness is typically achieved in orbital vehicles above 100 miles. "Zero gravity" has both harmful and beneficial effects on various components of a space system, but has no known effects on the properties of materials.

* U.S. Government Printing Office.

Geomagnetic Field. The magnetic field of the earth can be approximated by a simple dipole with a magnetic moment of 8×10^{25} cgs units near the center of the earth. This field, approximating 0.5 gauss at the earth's surface, has many fluctuations due to solar effects and is presumed to be terminated at 8 to 10 earth radii on the sun side due to solar winds, as shown in Fig. 9.

Solar Radiation. Solar radiation contains both electromagnetic radiation and corpuscular emission. The major part of the electromagnetic solar energy lies between the wavelengths of 0.3 and 2 microns, with a peak occurring around 4,550 Å as shown in Fig. 10. At altitudes of 100 to 120 miles a body is exposed to the full radiation spectrum, which has a thermal energy of 1,400 watts/m^2. Ultraviolet radiation below 2,000 Å represents a small part of the total solar energy, but has sufficient energy per quantum to initiate reactions, such as ionization, dissociation, and photoelectric effects, and to reduce sputtering thresholds. Corpuscular solar emission results from solar flares and solar winds. Flare events vary from minutes to several days in duration with fluxes in the range of 10^4 protons/ (cm^2) (sec) with energies up to 30 Mev. Total dose per flare may range from 10 to 10^3 r. Shielding against the high-energy protons is necessary in manned flights. The region below the Van Allen belt is well shielded from solar flare emission because of the magnetic field.

The solar wind is a plasma flowing from the sun with a flux of 3 to 100×10^7 particles/(cm^2)(sec) with energy varying between 200 and 8,000 ev. Prolonged exposure to solar winds can produce damage to exposed optical elements and thermal-control surfaces.

Cosmic Radiation. Cosmic radiation consists of nonelectromagnetic, charged atomic particles originating from outside the solar system, principally protons (90 percent) with velocities of 80 to 90 percent of the speed of light. The normal quiescent

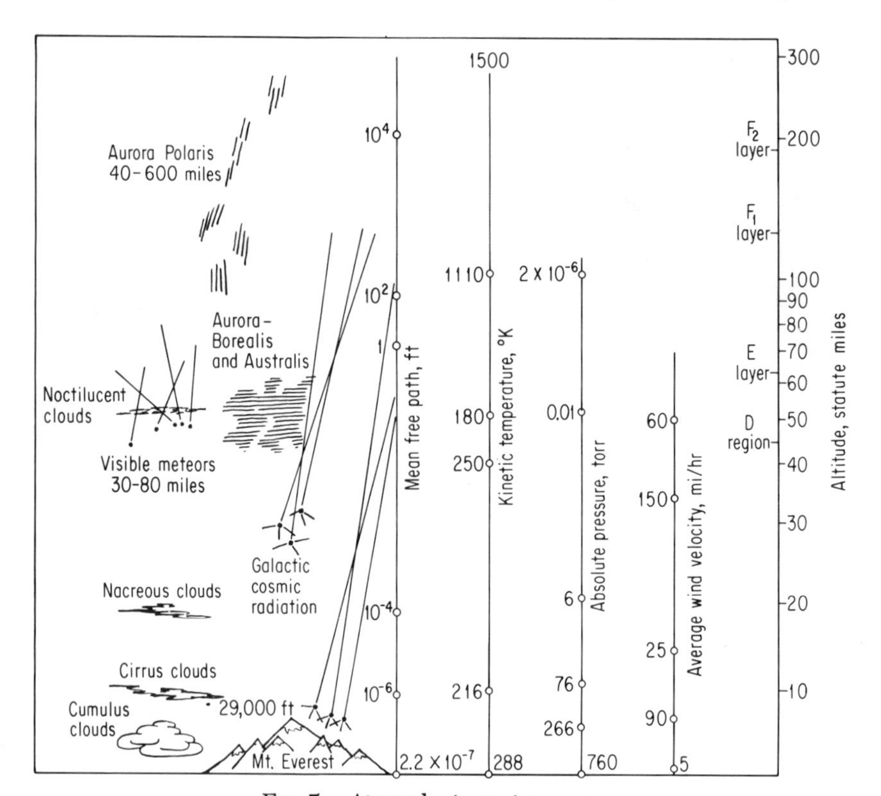

FIG. 7. Atmospheric environments.

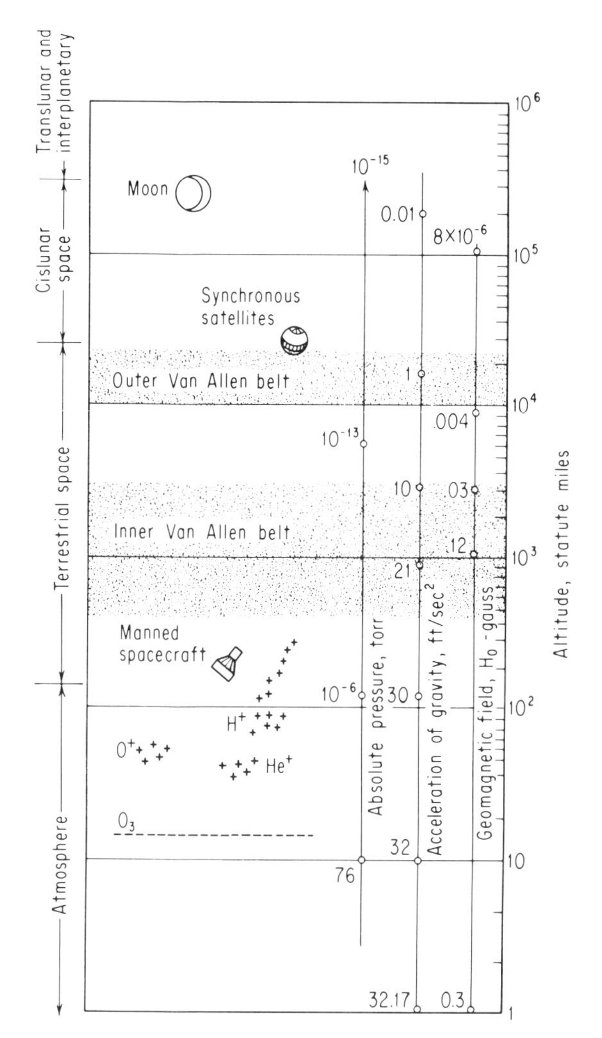

Fig. 8. Space environments.

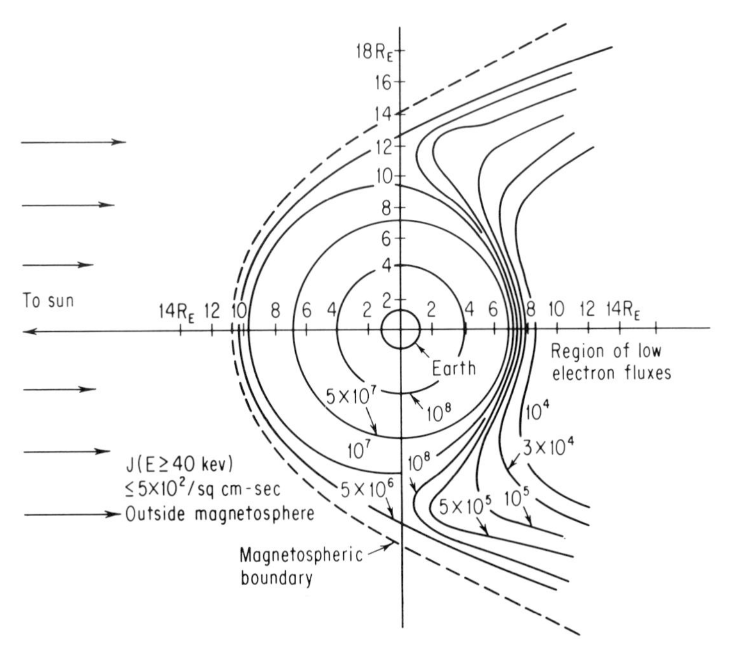

FIG. 9. Electron isointensity contours as modified by the solar wind. Quasistationary contours of constant omnidirectional flux of electrons ($E = 40$ kev) in the magnetic equatorial plane as measured with Explorers II and XIV.[12]

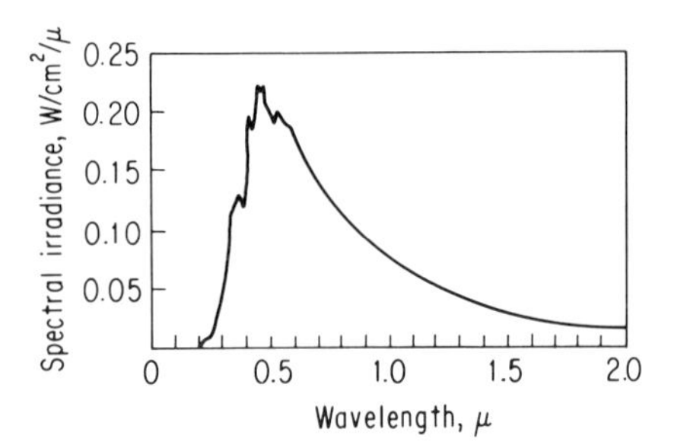

FIG. 10. Distribution of energy in solar radiation incident upon the earth's upper atmosphere.[11]

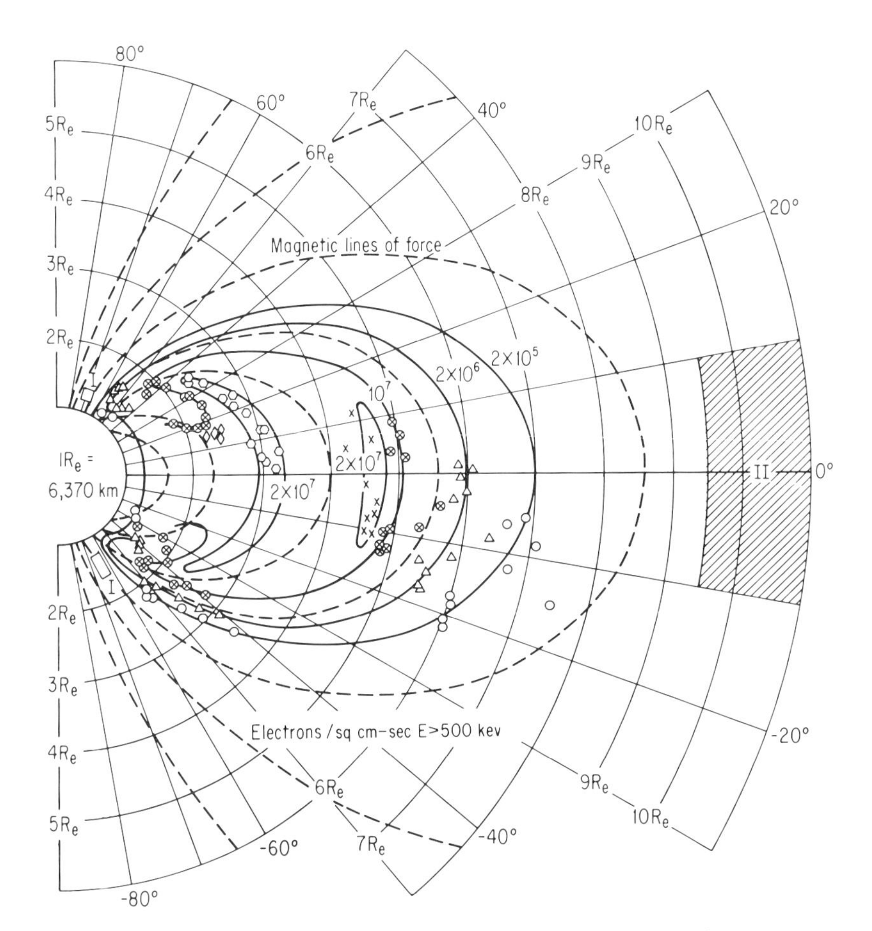

FIG. 11. High-energy electron isointensity contours.[11]

flux is approximately 2 particles/(cm^2)(sec) with energies ranging from 10^9 to 10^{19} ev. The distribution of these particles is isotropic and uniform in space, but inhomogeneities in the distribution are developed near the earth due to the geomagnetic field. Cosmic rays interact with the atmosphere to produce secondary particles and electromagnetic radiation. Negligible material damage results from the low effective ionization rate of 10^{-4} r/hr. A direct collision of a particle having high specific ionization with an electronic component could produce ionization, atomic displacement, or transmutation, although the probability of such a collision is small. Shielding is not economically effective due to the high penetrating characteristics of these near-relativistic ions and only results in the formation of secondaries and a correspondingly higher dose rate.

Auroral Radiation. The presence of auroral radiation fluxes are visually noticeable in the aurora borealis and the aurora australis at between 65 and 70° north and south

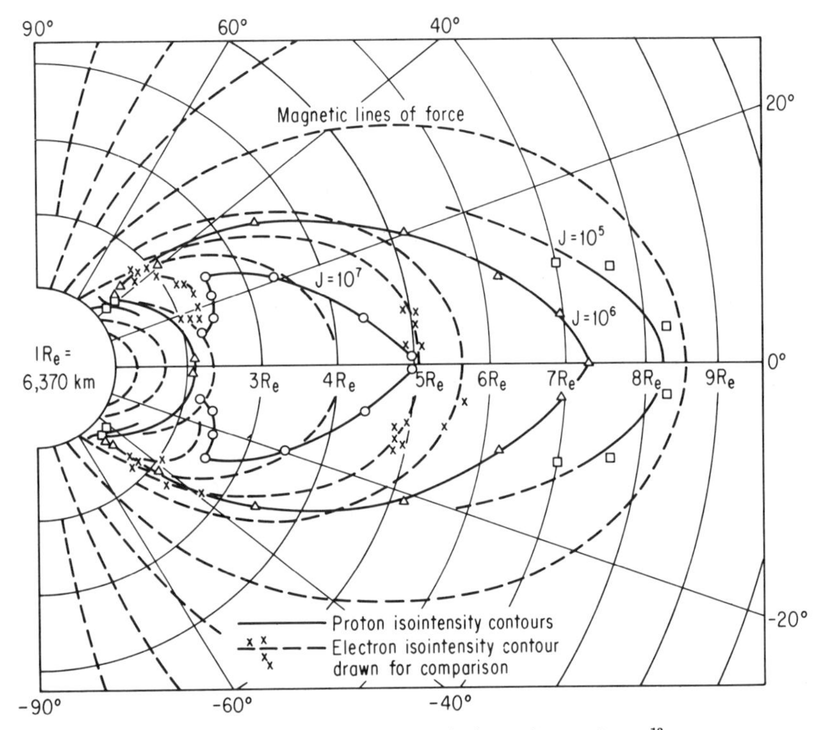

FIG. 12. Inner-zone proton isointensity contours.[12]

magnetic latitudes. Electron fluxes during auroral storms may be as high as 10^{11} electrons/(cm^2)(sec) with energies generally less than 50 kev, resulting in a surface dose rate of approximately 10^8 r/hr. Auroral proton flux is about 10^5 protons/ (cm^2) (sec) with energies ranging to approximately 650 kev, resulting in a surface dose rate approximating 500 r/hr. This intense surface dose rate over long periods of time will cause damage to thermal-control surfaces, optical surfaces, and exposed dielectric materials. Transient changes in the surface resistance of insulators will also occur. A vehicle in a polar orbit will traverse these auroral regions four times per orbit.

Van Allen Radiation. Two zones of radiation exist in the vicinity of the earth in the form of concentric rings. This radiation consists of protons and electrons trapped by the earth's geomagnetic field and displaced along the magnetic lines of force from pole to pole. The inner zone or belt, centered at approximately 2,200 miles, is charac-

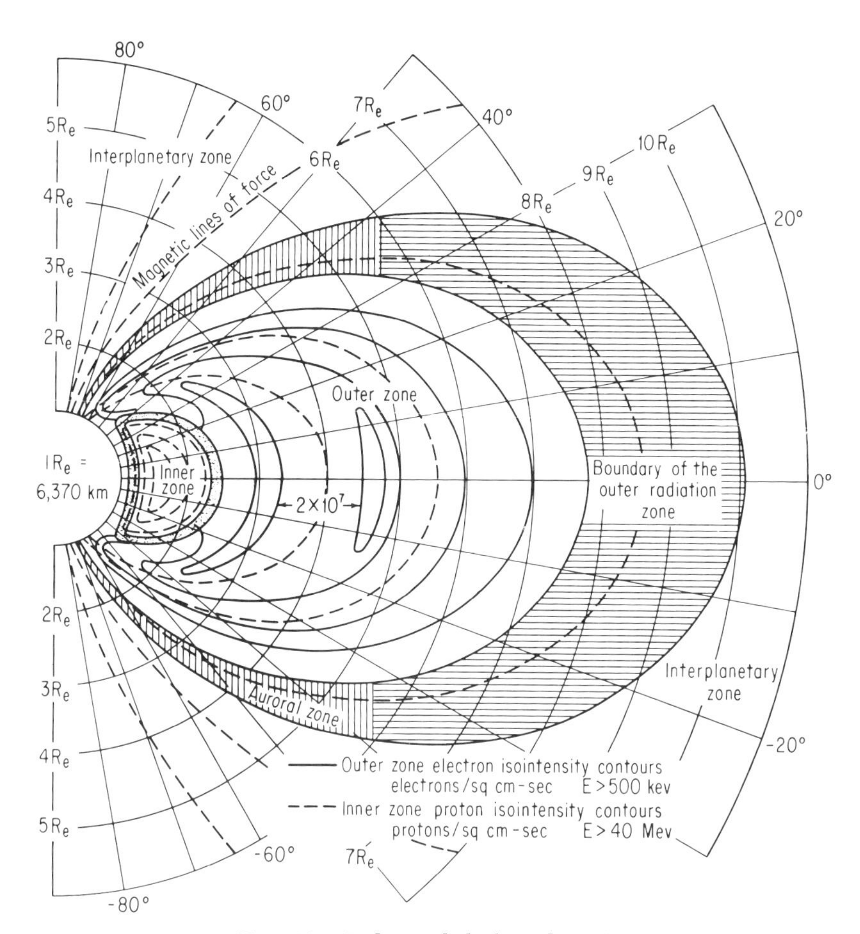

FIG. 13. Radiation-belt boundaries.[12]

terized by fluxes of high-energy penetrating protons and low-energy electrons and protons. The outer zone, broadly covering the altitude range from 5,000 to 20,000 miles, is characterized by high-energy electron flux. The radiation belt boundaries and the proton and electron intensity characteristics are shown in Figs. 11–13. Detailed discussion of penetrating radiation is found in Ref. 11. The inner zone has fluxes up to 8×10^4 protons/ (cm^2) (sec) with most electrons having energies greater than 30 Mev, which could produce a maximum dose rate of 100 r/hr. In the outer zone fluxes up to 10^8 electrons/(cm^2)(sec) exist, with most electrons having energies less than 1 Mev, which could produce a maximum dose rate of 10^5 r/hr. The radiation in these two belts can damage thermal-control surfaces, optical surfaces, solar cells, and electronic components. Shielding against this radiation is required in manned spacecraft.

Artificial Electron Belt. A radiation belt of high-energy electrons was created by high-altitude nuclear explosions. It has fluxes up to 10^8 electrons/(cm^2)(sec) for electrons with energies of 0.5 Mev. The center of this belt is at approximately 1,000 miles altitude, and a dose rate of 10^5 r/hr can be expected.

Ozone. One of the most active gases in the atmosphere is ozone, which is formed when ultraviolet energy of less than 2,400 Å in the solar radiation spectrum ionizes molecular oxygen. The maximum concentration is found at approximately 15 miles altitude. Ozone is a strong oxidizing agent that affects organic materials.

Ionized and Dissociated Gases. The absorption of solar radiation with wavelengths shorter than 1,850 Å can cause dissociation and ionization of oxygen into atomic oxygen at altitudes greater than 60 miles. Photodissociation of water vapor and methane can yield atomic hydrogen at altitudes above 50 miles. Ionized gases are not hazardous to electronic components, but dissociated gases such as atomic oxygen can react readily with metallic materials to form harmful oxides.

Micrometeorites. Solid particles in space orbiting the sun vary over a wide range of size, density, flux, and velocity. Approximately 90 percent of this meteoric material has a cometary origin, with densities ranging from 0.5 to 3.5 g/cm^3. The remainder are assumed to have asteroid origin with densities ranging up to that of iron (8 g/cm^3). The velocity of these particles ranges from 11 km/sec, the escape velocity of earth, to 72 km/sec. The flux mass varies from 10^{-12} to 10^{-17} g/(cm^2)(sec), with 1 to 10^{-6} impacts/ (m^2) (sec) as altitude increases from 60 miles to beyond 6,000 miles. Few particles of vehicle-skin-puncturing size ($>100\mu$) exist, so that the probabilities of a penetration's occurring in an exposed area of 100m^2 is only once in approximately 5 years. A large meteor could destroy a space vehicle, but such an event has an extremely low probability. The primary problem from the solid-particle environment of space is in erosion of exposed materials. A detailed discussion of the particle belts of the earth is found in Ref. 14.

Induced Environments. When a space system or any of its components are put into operation, environmental changes occur. These changes are associated with checkout on the bench and in the vehicle prior to launch, during the launch cycle and the subsequent mission phases.[15, 16] Packaging engineers have familiarity with most of these types of induced environments, but the magnitudes of these environmental conditions require different design techniques.

Thermal. The operating temperature of an electronic package is a function of the internally generated heat, the efficiency of the transfer of this heat to a sink, and the characteristics of that heat sink. Heat transfer normally occurs by a combination of convection, conduction, and radiation. For spaceborne electronic equipment, radiation is the only external mode of heat transfer and the heat from the internal equipment must be conducted to the skin of the vehicle for radiation into space. The characteristic of the heat sink varies from that of open bench tests to that of free space, 4°K, which includes the reverse heat flow or thermal shock that occurs during the several minutes of aerodynamic heating in the launch phase. The control of this thermal environment is an important design function and is unique for each equipment-vehicle configuration.

Acceleration. Electronic equipment in space vehicles is subjected to linear and angular acceleration. During the launch phase short-time linear accelerations approaching 20g may be experienced. Flight-acceptance testing may range as high as 75g, or even up to 10,000g, which might be encountered in hard landings. Angular acceleration results from spin or rotation imparted to the payload prior to separation from the last propulsion stage, to ensure stabilization of the vehicle axis in its orbit. Spin rates in the order of several hundred revolutions per minute have been used.

Acoustic Noise. Peak acoustic noise levels are encountered during vehicle launch, created by the reverberation of the engine field from the earth, and during the transonic flight phase, created by aerodynamic noise. These acoustic noise levels external to the spacecraft can reach more than 140 db (re 0.002 dynes/cm^2) within a spectrum of 10 to 10,000 cps.

Vibration. Excitation of mechanical oscillatory modes in electronics packages occurs primarily from engine ignition shocks, engine acoustic pressures, aerodynamic forces, and stage-separation shocks. The magnitudes of these driving forces are dependent upon the characteristics of the vehicle. The duration of these driving forces is in the order of 10 sec during launch, and 30 to 40 seconds during the transonic phase. The typical vibration spectrum is random but may have one or more narrow-band spectral peaks in the range of 10 to 2,000 cps. Typical excitation levels at the payload-booster interface may reach 20*g* rms, but decreases with heavier payloads on the same vehicle. Again, the vibration excitation is dependent upon the propulsion and the structural characteristics of the booster vehicle.

Shock. Shock levels depend upon the characteristics of the vehicle, the mission profile, and the handling and transportation environment. Transportation may result in many random shocks in the order of 2 to 10*g*. Shock excitation is typically in the order of 20*g* at launch, but may reach 200*g* on stage separation and 10,000*g* in hard landings.

Reactor Radiation. Fast neutrons and gamma radiation will be encountered from on-board nuclear reactor power sources and eventually nuclear propulsion reactors. A 1-Mw reactor may be expected to yield leak fluxes approximating 10^{16} fast neutrons/sec and 10^{17} gamma photons/sec with energies in the order of 1 Mev. Radiation from propulsion reactors is expected to exceed these fluxes greatly, but the duty cycle will be far lower than that of the power reactor. In either event, these radiation levels represent a severe environment for operating electronic equipment.

Sterilization. Biological contamination control is necessary if the planetary probes searching for extraterrestrial life are not to act as transplanters of earthborne life which could confuse the life sensors or alter or destroy the structure of planetary life. Biocontamination control[17, 18] requires assembly of the lander in clean rooms, enclosing the lander within a bacteriological barrier to maintain cleanliness, subjecting the lander to approved sterilization procedures, and not opening the enclosure, after sterilization, within any portion of the earth's atmosphere. Typical decontamination and sterilization procedures involve chemical treatments with gas composed of 12 percent ethylene oxide and 88 percent Freon 12 followed by thermal treatments at 135°C for times approaching 100 hr. Different numbers of cycles of these environmental conditions are used in the various phases of equipment prelaunch life to assure meeting the requirements of a probability of only 1 chance in 10,000 of the lander's carrying a single living microorganism. Equipment packaging materials and components must be selected with care to withstand these severe environmental requirements.

Cabin Contamination. Avoiding the release of toxic substances into the closed environment of a space capsule is imperative. Major hazards which can produce toxic effects include: leakage in the refrigeration system; failure of the air-filter system; overheating of electrical and mechanical equipment; boil-off of noxious vapors from paints, plastics, oils, and metals; and the interaction of biological matter with spacecraft materials. The compatibility of spacecraft materials with the vehicle's occupant is also a problem for the packaging engineer. Reference 19 contains an introductory discussion of this subject.

Summary. The environmental conditions for electronic equipment used in space missions vary greatly during its lifetime, and particularly very rapidly during the start of a space mission.

Prelaunch. The equipment has been subjected to all the simulated environmental conditions that can be produced, as well as to the quasi-predictable environments of contamination during manufacture; shock and vibration during transportation; shock during installation and final assembly; and the miscellaneous abnormalities in temperature, salt spray, sand, dust, and humidity encountered after leaving the controlled manufacturing environment.

Launch. At the start of the flight mission the equipment is subjected to high-level acoustic excitation; mechanical vibration; mechanical shock; acceleration; and indirect electrical, mechanical, and hydraulic transients that might result from these basic factors.

At 50,000 Ft. At this height a large multistage vehicle is approximately 30 sec into its flight, corona and arc-over are encountered due to reduced air density, ozone (O_3) concentration is increasing, and the atmosphere has lost approximately 80 percent of its thermal capacity. Acceleration and velocity of the vehicle are still increasing and the skin temperature of the vehicle is rising rapidly. The acoustic excitation has decreased approximately a thousandfold from its peak launch value.

At 200,000 Ft. At this height the vehicle is approximately 150 sec into its flight mission. The aerodynamic pressure peak has been encountered, acceleration has peaked, the outer skin has reached its maximum temperature ($\sim$700°F), the air pressure is less than 1 torr, and the atmosphere has lost 99 percent of its heat-absorbing capacity. Stage separation occurs with at least two shock peaks, and acceleration drops to a lower value.

At 600,000 Ft. At this height the vehicle has been maneuvered into a circular orbit for experiments or a parking orbit coast for an earth-escape mission. The second-stage cutoff has occurred with its subsequent stage-separation shock excitation. A 0-g condition exists, the air pressure is in the order of 10^{-6} torr, and the vehicle is subjected to the full solar radiation spectrum. The temperature of the vehicle will be the result of the thermal balance between radiation of internally generated heat and absorbed solar radiation. The vehicle is subject to cosmic radiation but is shielded from corpuscular solar emission by the magnetic field surrounding the earth. If it is in a polar orbit it will be subjected to auroral radiation.

ELECTRONIC SYSTEMS PACKAGING

Packaging in general is discussed in Chaps. 9 to 13; the materials, components, and devices are referenced in Chaps. 1, 2, and 5 to 7; and fabrication processes and testing are described in Chaps. 3 and 4. The information in these previous chapters is generally applicable to packaging of space electronics, with the understanding that more stringent requirements are placed on system configuration, thermal and power management, reliability and manufacturing processes by performance requirements and life environment of a space mission.

Systems packaging is a systematic approach to the integration of material and processes into a systems configuration that is the most cost-effective among the many alternates that can be synthesized through trade-off analyses of the factors affecting life performance, size and weight, producibility, and maintainability. As shown in Fig. 14 the procedures of systems packaging are a series of creative iterative operations constrained by applicable technology, specifications, and rigorous design review, in which documentation is the key to control from concept to completion.

Systems packaging is divided into three parts: design management, design development, and design review, which encompass the four time-sequential phases: problem definition, design synthesis, mechanization, and verification.

Design Management. Management makes systems packaging a productive activity as it places under single-point cognizance all the resources and skills required to accomplish the project. The function of design management is to translate the complicated and interrelated requirements of the space mission into design policy, procedural constraints, evaluation criteria, schedules, and cost limitations that are understandable by the designers. Design management also has the important responsibilities of anticipating the transition between the phases of systems packaging and of reallocating the project skills and resources compatibly with the phase requirements.

Design Development. The synthesis of an electronic systems configuration moves from concept to final design through a series of trade-offs among materials, components, processes, and configurations, as constrained by costs, life performance, and environmental requirements. These trade-offs are based upon experience, mathematical analyses, experiments, or combinations of these, with the objective of maximizing the system effectiveness for a fixed cost or, conversely, of minimizing the cost for a fixed system effectiveness. The use of computers for design analysis through simulation is becoming an important tool in systems packaging. (See Chap. 15.)

There is a wealth of information in this handbook and in the published literature covering these trade-off studies and the techniques used by packaging engineers.[20] In the packaging of space electronics there are no widespread standards, as the space age is still in the exploratory era and there are very few similarities among the many

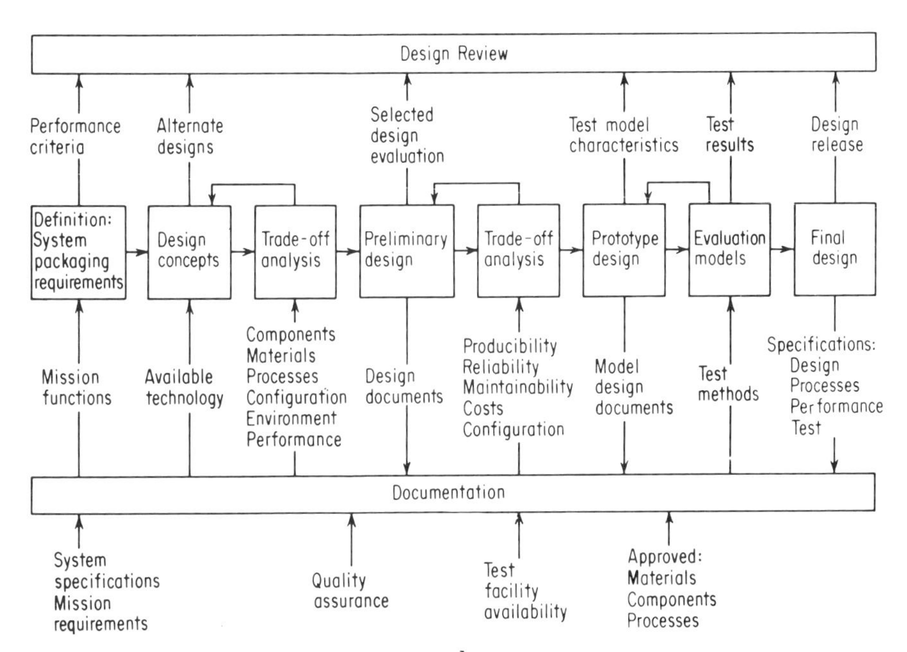

FIG. 14. Systems packaging management.

missions in requirements or payload complement. For example, the flight path establishes the scope and severity of the environmental exposure, while the mission length affects the life-reliability requirements. The requirements for extreme reliability in space missions have made this factor the pivotal point around which design decisions are carefully made.

Design Review. Assurance of the system design is based upon periodic audits from concept to completion. These audits focus on the design proposal to question the effectiveness of the selection of materials, components, processes, and configuration within the established program requirements, and also to probe for sensitive parameters within the design. The formal review serves the important function of establishing communication among the specialists whose disciplines are involved in the program, and of stimulating interdisciplinary contributions to the overall system effectiveness. Such interdisciplinary review can initiate alternate designs which are higher in value than those being evaluated. Again the mix of disciplines represented in a design review must necessarily change as the design progresses from concept to completion. Typically these reviews are held several times during the concept stage, several times during the preliminary design stage, again whenever significant changes are made from the preliminary design, at the scheduled design-freeze time, and finally, when the design has been finalized. Informal reviews and consultation with organizations concerned with value engineering (VE)[21] or systems effectiveness engineering (SEE)[22] may occur throughout the project.

Packaging Concepts. In the design of contemporary spaceborne equipment it has not been found possible to implement any system using only a single component technology, although the advent of new classes of thin-film and semiconductor integrated circuits has made it possible to reduce electronic subsystems significantly in size. Conventional components and assembly techniques are still used in space electronics because of the necessity of using components with proven reliability and the desirability of using proven manufacturing processes, involving existing equipment, facilities, and trained personnel.

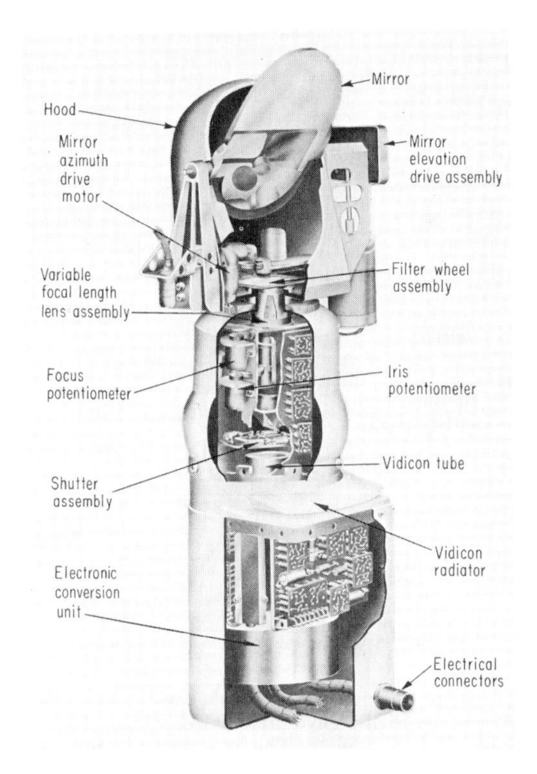

FIG. 15. Optomechanical and electronic assemblies of the Surveyor slow-scan television camera. (*Courtesy of Hughes Aircraft Company.*)

Surveyor Lunar Landing Camera.[23] The slow-scan television camera used in Surveyor I is shown in the cutaway drawing of Fig. 15. Six major subassemblies have been integrated in this design: the mirror, lens, shutter, filter wheel, vidicon, and electronic circuits. The camera through the mirror mechanism has a 360° azimuth coverage; a 100° elevation coverage; an optical field of view of approximately 6.5 or 25°, depending upon the focal length selection of the lens; and a focus capability from 1.23 m to infinity. The image focused on the vidicon is scanned in either a 600-line or a 200-line frame. In the 600-line mode one frame is provided in 3.6 sec, and in the 200-line mode one frame is provided in 60.8 sec. The electronics comprises five functional groups: drive circuits for lens and mirror mechanical positioning, video amplifier, horizontal and vertical sweep circuits, synchronization, and power conversion and control. The electronic packaging is based on discrete components of proven reliability arranged in cordwood-type modules with printed-circuit-board connection planes.

Lunar Module I.[24] The packaging concept illustrated in Figs. 16 to 18 applies to approximately 70 percent of the Lunar Module systems electronics. The equipment is located in the aft equipment bay of the vehicle's ascent stage, as shown previously in Fig. 3. This equipment is unpressurized. The electronics is packaged in a number of structures termed *Electronic Replaceable Assemblies* (ERAs), which are part of a mutually dependent structural system that includes the vehicle equipment bay, fuel tanks, and control jets. The ERAs are a long, narrow box-beam structure system (as shown in Fig. 16) consisting of two mounting flanges, two end plates, and top and bottom covers. The mounting flanges serve as the primary thermal path from the electronics to the cold plate as well as the support member of the electronic packages, as shown in Fig. 17. These full-width electronic packages act as bulkheads in the box-beam ERA structure, adding to its strength. Electronic subassemblies in the ERAs

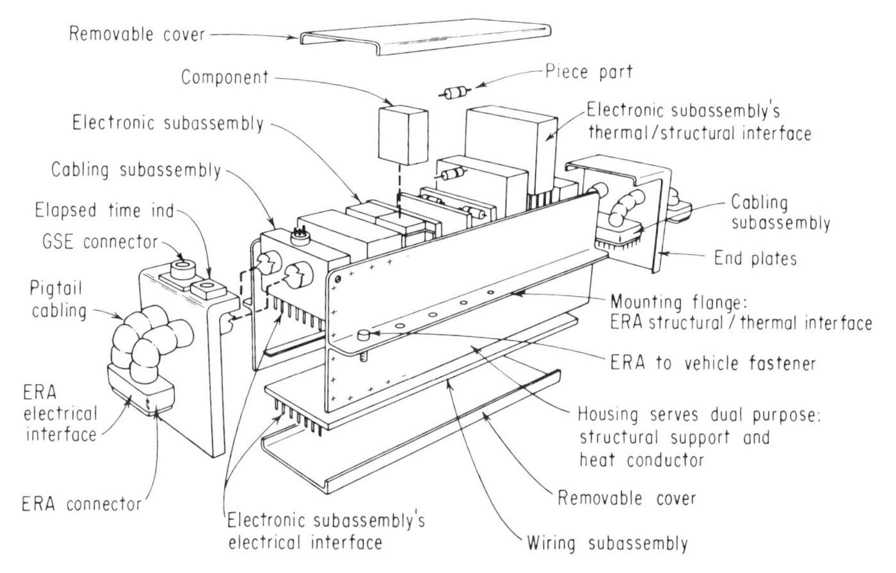

FIG. 16. Subassembly details of an Electronic Replaceable Assembly used in the Apollo Lunar Module Vehicle. (*NASA/MSC.*)

have been packaged as soldered or welded cordwood modules with some flat-pack integrated circuits on printed-circuit boards. Connections between electronic modules and wiring subassemblies are welded, soldered, or wire-wrapped split pins.

Lunar Module II.[25] Another packaging concept used in the Lunar Module is the frequency multiplier shown in Fig. 19. This unit produces approximately 0.5 watt of x-band power from a dc input of 25 watts. The package volume is 50 in.^{3} and its weight is 2 lb. The unit dissipates approximately 1 watt/in.^{2} into the Lunar Module structure through a 25-in.^{2} mounting area. The electronics comprises an oscillator-buffer amplifier, a power amplifier, and a series of five multipliers. The amplifier, as shown in Fig. 20, is assembled from discrete components mounted on a double-clad printed-circuit board. The components are arranged so that no rf shielding is needed between the stages.

Inertial Reference Unit (IRU).[26] The basic design in this unit is based on *Electronic Component Assemblies* (ECAs), which are cemented into a chassis and interconnected with soldered leads to form a module. The modules are then mounted into a frame with the inertial components to create the IRU. The assembly of modules and

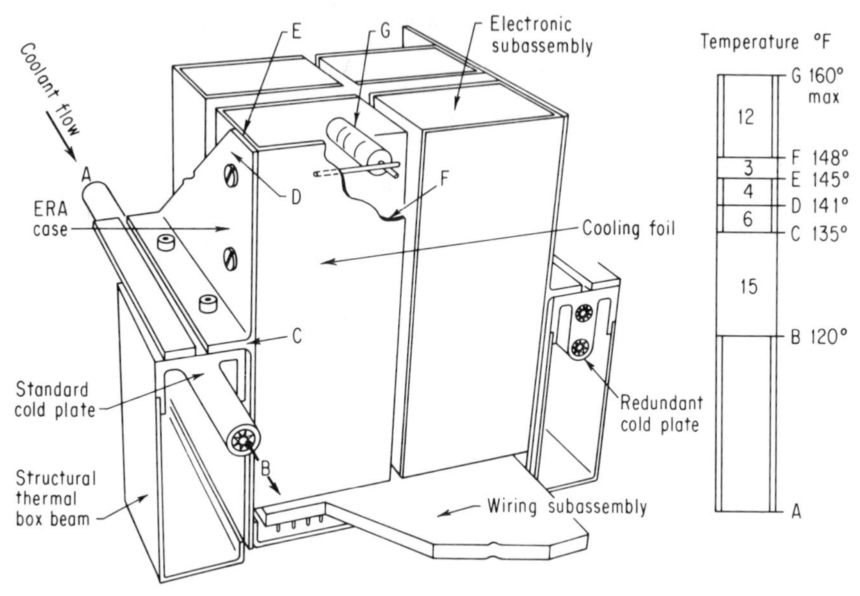

FIG. 17. Thermal-structural design showing the interface between the ERA and the redundant cold plate which is a box-beam structural member of the equipment bay. (*NASA/MSC.*)

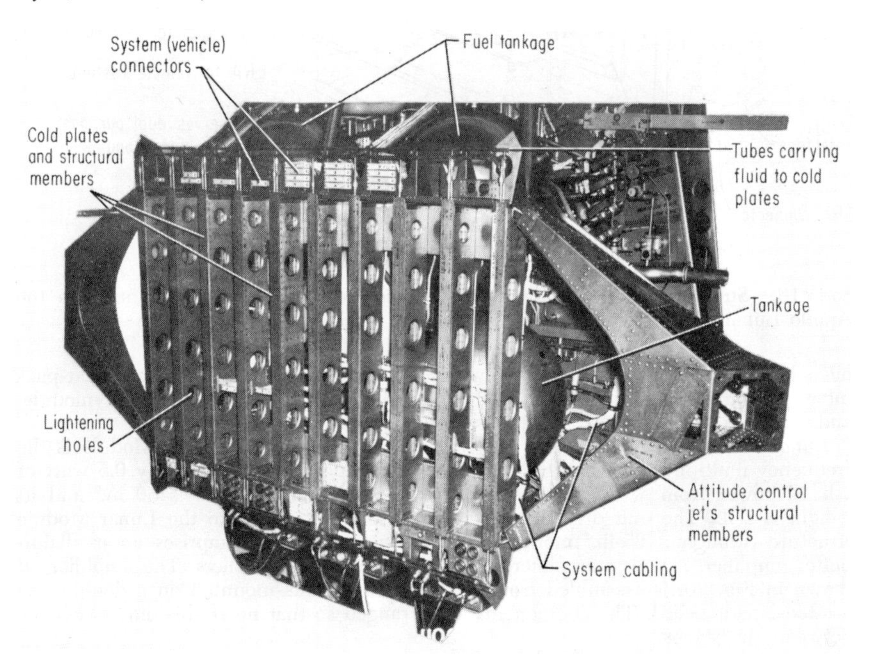

FIG. 18. Electronic equipment bay of the Apollo Lunar Module Vehicle. (*NASA/MSC.*)

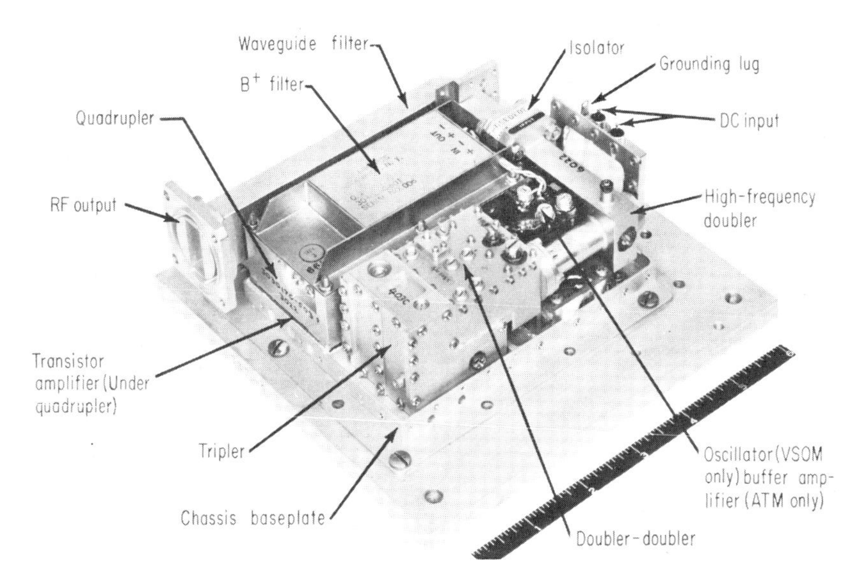

FIG. 19. LM frequency multiplier showing integrated electromechanical packaging. (*Reproduced from Ref. 25 by permission of IEEE. Photograph courtesy of Radio Corporation of America.*)

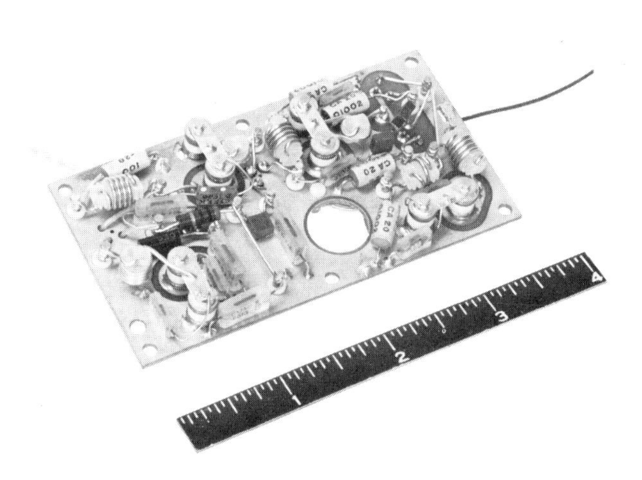

FIG. 20. Component layout of an rf amplifier on a double-clad printed-circuit board which acts as an rf shield. (*Reproduced from Ref. 25 by permission of IEEE. Photograph courtesy of Radio Corporation of America.*)

the intermodule cabling in the frame are shown in Fig. 21. The ECAs are based upon a honeycomb packaging construction, as shown in Fig. 22, and their assembly between rails in a module chassis is shown in Fig. 23. Over 70 percent of the electronic components in the IRU are tubular with axial leads as dictated by using a proven fabrication technique and approved components in the development of the IRU.

Infrared Radiometer.[27] The configuration of a wide-angle horizon sensor is shown in Figs. 24 and 25. This unit involves an optical system, a motor-driven chopper, a motor drive for the scanning optics, and the electronics. The unit is sealed in a pressurized housing for environmental protection. The circuits are assembled from pigtailed components mounted on single-sided printed-circuit boards. Four circuit boards are edge-mounted on vertical posts, underneath the mechanical base plate, to form a box-beam structure, and interconnections between boards are hard-wired with soldered joints.

FIG. 21. Inertial reference unit for the Lunar Orbiter, showing assembly of ECA modules into the IRU frame. (*Photograph courtesy of Sperry Gyroscope Company.*)

S-band Telemetry Transmitter.[28] Another approach to the packaging of a transmitter is shown in Fig. 26. This unit produces 10 watts of S-band power with a dc input of 50 watts. The package volume is 48 in.^{3}, the weight is 8 lb, and approximately 2 watts/in.^{2} must be dissipated into the vehicle structure through the 25-in.^{2} mounting base. The electronics is comprised of a vacuum-tube-cavity power oscillator, an AFC (automatic frequency control), an input power regulator, and a dc-to-dc converter. The AFC and power-supply circuits use semiconductor components where applicable and are packaged in plug-in modules, as shown in Fig. 27. These modules are based upon a plastic shell with integral long-blade contacts along two sides. Components are mounted cordwood-style within these shells and encapsulated. The shells have keyways molded into two ends which allow the modules to be mated or held in place with vertical keys.

Despun Antenna System.[29] In Fig. 28 is shown an assembly of conventional and integrated circuits that perform an antenna-pointing function in a satellite. The larger blocks are a power supply and a filter designed as cordwood-type assemblies of conventional components. The smaller block is the sun-sensor circuit assembly. The integrated-circuit portion of this system is in the 13 remaining modules. The details of the stacked integrated-circuit modules are shown in Fig. 29. This module consists of a militarized 20-pin header with a mating metal can in which integrated-circuit flat packs are assembled in a vertical stack and finally encapsulated, through a hole in the top of the can, with an epoxy resin. The flat-pack interconnection scheme is based upon nickel foil webs interleaved between the flat packs along with insulating shims to supply the crossover and jumper connections. The fingers of the webs are welded to the appropriate flat-pack terminals and risers from the header with procedures meeting the requirements of MSFC 271.* A thermal path from the module to the chassis is provided through Kovar† tabs on the can used for mounting the module to the chassis.

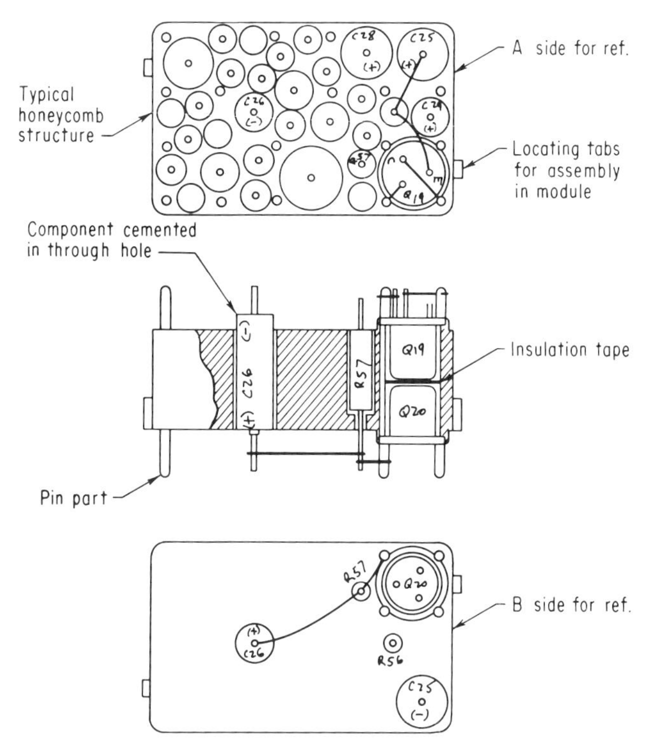

FIG. 22. Diagram of tubular components mounted in a molded honeycomb structure as used in the IRU. High density with good thermal conduction results from this packaging technique. (*Sperry Gyroscope Company.*)

Page Assembly.[30-32] Another approach to the packaging of assemblies of integrated circuits is shown in Fig. 30. Shown here is a page assembly comprising a page frame, two multilayered interconnection boards (MIBs), a 98-pin connector, and unit logic devices (ULDs). A ULD (as shown in Fig. 31) is fabricated by screening conductive and resistive materials on a 0.3-in.^2 ceramic substrate. Uncased semiconductor elements are attached to the circuit structure by solder reflow, and the entire unit is encapsulated. The MIBs are a 12-layered etched-circuit board with plated-through

* NASA–Marshall Space Flight Center.
† Trademark, Westinghouse Electric Corp.

holes for interlayer connections. The holes and conductors are located on an 0.020-in. grid. The MIB has sufficient mounting area for 35 ULDs located on 0.40-in. centers. The ULDs are attached to MIBs by solder reflow techniques, and after testing a protective coating is applied. The MIBs are bonded to the two faces of a page frame which is machined from a sheet of magnesium-lithium stock and serves to support the MIBs and to conduct heat from the ULDs to the main housing structure. The MIBs are bonded to the faces of the page frames. In the Apollo backup computer 71 page assemblies are used, and their layout within the main computer structure is shown in Fig. 32.

Integrated-circuit Module. Another technique for the assembly of integrated-circuit flat packs into a high-density assembly is shown in Fig. 33. The heart of this technique is the use of an etched-wiring circuit board molded with a shallow trough

FIG. 23. Electric Component Assembly (ECA) used in the IRU, showing soldered connections according to NASA/NPC 200-Y specifications for hand soldering. (*Photograph courtesy of Sperry Gyroscope Company.*)

approximating the width of a flat pack. A second thin board with square cutouts is layered over the formed substrate. Wiring traces are carried on both boards, and the terminals on the flat pack are formed to contact the traces at the appropriate level. The channel allows the crossover and jumper connections to be carried under the flat packs, eliminating the requirement for multilayer wiring boards. Interboard connections are made by point-to-point wiring with solder connections along the edges of board assemblies. A set of boards is then encapsulated to produce a module with single-side termination.

Universal Component Packaging System.[33] A modular packaging system designed to provide a high degree of flexibility in the choice of components and functional assemblies is illustrated in Figs. 34 to 36. This system has a standard enclosure fabricated from milled-aluminum sheet stock and assembled by dip brazing. These standard enclosures are designed to be stacked for larger volume requirements. Mounting rails are provided within the enclosure for attachment of module frames and the wire-wrap matrix. The module frame, made of cast aluminum, supplies mechanical support and heat-conduction paths for two double-sided etched-circuit boards mounted back-to-back on the frame. These frames are modular in thickness to accommodate the various component heights that might be placed on the circuit boards. The module frames, which are attached to the mounting rails by captive screws, complete the thermal path to the standard enclosure. The connector pins on the circuit boards and the wire-wrap matrix, as shown in Fig. 37, require slightly more volume than a multilayer board, but afford a high degree of flexibility for implementation of the changes that are characteristic of low-quantity-production space equipment. The main

FIG. 24. Optomechanical and electronic assembly in a wide-angle horizon sensor. Components are mounted on single-sided glass-epoxy etched-wiring board. (*Photograph courtesy of Barnes Engineering Company.*)

pin field contains 1,524 pins for interconnection to the circuit module. The pin field at the end of the matrix provides for connections between the matrix and the external connectors on the standard enclosure.

Components and Materials. Components and materials used in space electronics are required to perform their specific function in a wide range of space environments after exposure to the rugged conditions of manufacturing, testing, evaluation, prelaunch handling, launch ascent, and flight. Descent and impact may also occur. The choice of these materials and components is dictated by the specific mission requirements and availability of materials and components with proven stable characteristics.[34–42] In the typical spacecraft the component complement is on the average 70 percent passive components, 17 percent semiconductor diodes, and 13 percent transistors. In general the passive components are more resistant to space environments than are the semiconductors.

Of all the space environments, penetrating radiation constitutes the most important design constraint in the selection of materials and components. Semiconductors are particularily sensitive to this environmental factor. The magnitude of the radiation environment is illustrated in Tables 4 and 5, where the dose rate and annual integrated dose are shown for six different orbits and their radiation components. When the mission orbit and duration are known, the radiation environment can be determined; then the mission-life dosage can be determined, with factors for vehicle skin and chassis shielding, to establish a radiation dosage tolerance factor as a component-

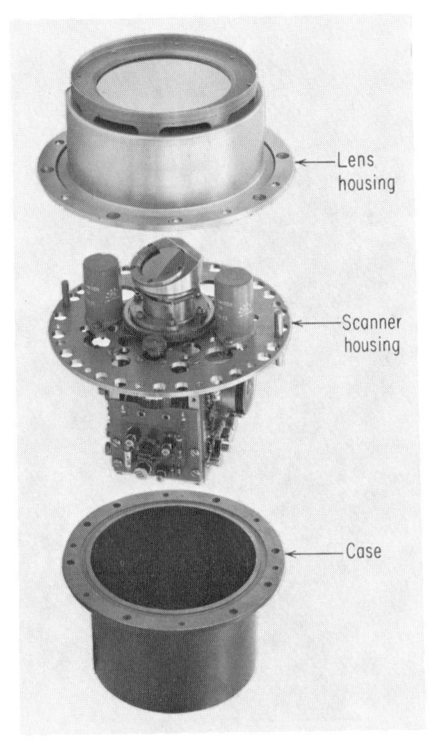

FIG. 25. Assembly of parts for the wide-angle horizon sensor, resulting in a hermetically sealed unit. (*Photograph courtesy of Barnes Engineering Company.*)

selection screening parameter. Mission reliability requirements supply another very necessary component-selection factor.

There is a voluminous amount of data in the literature[43] concerning radiation-induced effects in electronic materials and components. Most of these data have been obtained through irradiation in air with little or no control of the temperature conditions. These data can only serve the designer as a guide, as the radiation-induced effects may be radically different under the combined environments of space. One of the problems that exists is the rapid technological advances in electronics, which obsolete existing materials and components before space environmental-effects data with a high confidence level have been accumulated.

In Tables 6 and 7 some guideposts for the selection of general categories of components and materials have been listed relative to the missions described in Tables 4 and 5. These are not firm recommendations because of the uncertainties of the space

TABLE 4. Maximum Radiation Dose Rates for Selected Mission Orbits [11]

Sources of radiation	Radiation-dose rate,* r/hr											
	90°-inclination orbits (polar)								30°-inclination orbits			
	200 nmi (230 smi)		500 nmi (576 smi)		1,000 nmi (1,152 smi)		2,000 nmi (2,304 smi)		200 nmi (230 smi)		500 nmi (576 smi)	
	Surface	Internal	Surface	Internal	Surface	Internal	Surface	Internal	Surface	Internal	Surface	Internal
Cosmic primary protons	4×10^{-4}	4×10^{-4}	4×10^{-4}	4×10^{-4}	5×10^{-4}	5×10^{-4}	7×10^{-4}	7×10^{-4}	†	†	†	†
Trapped (Van Allen) protons	1.5	7×10^{-1}	2.5×10^{1}	1.5×10^{1}	1.5×10^{2}	8×10^{1}	5×10^{2}	2×10^{2}	2.5	1	70	40
Trapped‡ electrons (Van Allen and artificial)	6×10^{3}	5×10^{2}	7×10^{4}	8×10^{3}	1×10^{5}	1.5×10^{4}	9×10^{4}	5×10^{3}	9×10^{3}	7×10^{2}	1.5×10^{5}	2×10^{4}
Auroral protons	700	0	700	0	700	0	700	0	†	†	†	†
Auroral electrons	1×10^{7}	2	1×10^{7}	2	1×10^{7}	2	1×10^{7}	2	†	†	†	†
Solar-flare§ protons	300	100	300	100	300	100	300	100	†	†	†	†
Highest dose rate	1×10^{7}	5×10^{2}	1×10^{7}	8×10^{3}	1×10^{7}	1.5×10^{4}	1×10^{7}	5×10^{3}	9×10^{3}	7×10^{2}	1.5×10^{5}	2×10^{4}

* Given for vehicle surfaces and interiors (through 0.01-in. Al skin).

† These radiation sources have smaller intensities than those due to trapped electrons and will therefore not appreciably affect the total dose. When compared with doses for polar orbits, those for 30° inclination would be smaller approximately by a factor of 2 for cosmic rays and by a factor of 10 for solar flares, and would be essentially zero for auroral protons and electrons.

‡ Includes Van Allen electron and artificial electron belt data as of Jan. 1, 1963; relies principally on the data of McIlwain from Explorer 15.

§ Includes statistical maximum of one major flare per month and model flare data of Bailey.

TABLE 5. Integrated Radiation Doses for One-year Missions[11]

Sources of radiation	Radiation dose,* roentgens											
	90°-inclination orbits (polar)								30°-inclination orbits			
	200 nmi (230 smi)		500 nmi (576 smi)		1,000 nmi (1,152 smi)		2,000 nmi (2,304 smi)		200 nmi (230 smi)		500 nmi (576 smi)	
	Surface	Internal	Surface	Internal	Surface	Internal	Surface	Internal	Surface	Internal	Surface	Internal
Cosmic primary protons	4	4	4	4	5	5	7	7	†	†	†	†
Trapped protons (Van Allen)	30	10	3×10^3	1.5×10^3	1×10^5	5×10^4	5×10^5	2×10^5	50	15	8×10^3	4×10^3
Trapped‡ electrons (Van Allen and artificial)	1.5×10^5	1×10^4	1.5×10^7	1.5×10^6	1.5×10^8	1.5×10^7	8×10^7	4×10^6	2×10^5	1.5×10^4	3.5×10^7	3.5×10^6
Auroral protons	300	0	300	0	300	0	300	0	†	†	†	†
Auroral electrons	5×10^6	1	5×10^6	1	5×10^6	1	5×10^6	1	†	†	†	†
Solar-flare§ protons	3×10^3	1×10^3	3×10^3	1×10^3	3×10^3	1×10^3	3×10^3	1×10^3	†	†	†	†
Approximate total	5×10^6	1×10^4	2×10^7	1.5×10^6	1.5×10^8	1.5×10^7	9×10^7	4×10^6	2×10^5	1.5×10^4	3.5×10^7	3.5×10^6

* Given for vehicle surfaces and interiors (through 0.1-in. Al skin).

† These radiation sources have smaller intensities than those due to trapped electrons and will therefore not appreciably affect the total dose. When compared with doses for polar orbits, those for 30° inclination would be smaller approximately by a factor of 2 for cosmic rays and by a factor of 10 for solar flares, and would be essentially zero for auroral protons and electrons.

‡ Includes Van Allen electron and artificial electron belt data as of Jan. 1, 1963; relies principally on the data of McIlwain from Explorer 15.

§ Includes statistical maximum of one major flare per month and model flare data of Bailey.

FIG. 26. Modular assembly of space-rated telemetry transmitter. (*Photograph courtesy of Aerospace Products Division, EIMAC Division of Varian Associates.*)

FIG. 27. Molded component housing of subassemblies used in telemetry transmitter, showing integral electrical connecting devices. (*Photograph courtesy of Aerospace Products Division, EIMAC Division of Varian Associates.*)

TABLE 6. Electronic Components Acceptability Guide for a One-year Circular-orbit Space Mission [11]

Component	90°-inclination orbit (polar)								30°-inclination orbit			
	200 nmi‡		500 nmi		1,000 nmi		2,000 nmi		200 nmi		500 nmi	
	E*	I	E	I	E	I	E	I	E	I	E	I
Capacitors	All paper, ceramic, mica, plastic, tantalum, and oil-impregnated capacitors tested are suitable for all orbits.											
Resistors	All wire-wound, carbon, and composition resistors and potentiometers tested are suitable for all orbits.											
Electron tubes:												
Vacuum tubes	M†	S	U	S	U	U	U	M	M	S	U	M
Gas-filled tubes	S	S	M	S	U	S	U	S	S	S	S	S
Photomultipliers	U	M	U	U	U	U	U	U	U	M	U	U
Traveling-wave tubes	U	S	U	S	U	U	U	M	U	S	U	M
Camera tubes	U	S	U	U	U	U	U	U	U	S	U	U
Infrared detector cells	S	S	U	S	U	U	U	S	S	S	U	S
Diodes:												
Silicon	S	S	S	S	S	S	S	S	S	S	S	S
Germanium	S	S	S	S	S	S	S	S	S	S	S	S
Transistors:												
Silicon—thick base	M	S	U	M	U	U	U	M	M	S	U	M
Silicon—medium base	S	S	M	S	U	U	U	S	S	S	U	S
Silicon—thin base	S	S	S	S	S	S	S	S	S	S	S	S
Germanium—thick base	M	S	U	M	U	U	U	M	M	S	U	M
Germanium—medium base	S	S	M	S	U	U	U	S	S	S	U	S
Germanium—thin base	S	S	S	S	S	S	S	S	S	S	S	S
Miscellaneous:												
Quartz crystals	S	S	S	S	S	S	S	S	S	S	S	S
Differential transformers	S	S	S	S	S	S	S	S	S	S	S	S
Magnetic cores	S	S	S	S	S	S	S	S	S	S	S	S

* E = external; I = internal (refer to components located outside, or inside, the shielding of a satellite skin of at least 0.1-in. Al or equivalent).
† S = satisfactory for use; U = unsatisfactory; M = marginal value, more information needed.
‡ Nautical to statute miles: 1.15155.

TABLE 7. Electronics Materials Acceptability Guide for a One-year Circular-orbit Space Mission [11]

Component	90°-inclination orbit (polar)								30°-inclination orbit			
	200 nmi		500 nmi		1,000 nmi		2,000 nmi		200 nmi		500 nmi	
	E*	I	E	I	E	I	E	I	E	I	E	I
Potting compounds:												
Epoxy	S†	S	S	S	S	S	S	S	S	S	S	S
RTV silicones	S	S	U	S	U	S	U	S	S	S	S	S
Printed-circuit boards:												
Glass phenolic	S	S	S	S	S	S	S	S	S	S	S	S
Glass epoxy	S	S	S	S	S	S	S	S	S	S	S	S
Glass polyester	S	S	S	S	S	S	S	S	S	S	S	S
Coaxial-cable insulation:												
Irradiated polyolefin	S	S	S	S	S	S	S	S	S	S	S	S
RG11/U polyethylene	S	S	S	S	S	S	S	S	S	S	S	S
Suprenant No. 5561 FEP foam	S	S	U	S	U	U	U	S	S	S	U	S
Other insulators:												
Tenite II ‡	U	S	U	S	U	U	U	S	U	S	U	S
Lucite § and Plexiglas ¶	U	S	U	S	U	U	U	S	U	S	U	S
Phenolics:												
Unfilled and paper-filled	S	S	U	S	U	U	U	S	S	S	U	S
Mineral- and glass-filled	S	S	S	S	S	S	S	S	S	S	S	S
Polyesters	S	S	S	S	S	S	S	S	S	S	S	S
Polystyrene	S	S	S	S	S	S	S	S	S	S	S	S
Polyethylene	S	S	S	S	S	S	S	S	S	S	S	S
Irradiated polyolefin	S	S	S	S	S	S	S	S	S	S	S	S
Teflon §	U	S	U	U	U	U	U	U	U	S	U	U

* E = external; I = internal (refer to components located outside, or inside, the shielding of a satellite skin at least 0.1-in. Al or equivalent).
† S = satisfactory for use; U = unsatisfactory.
‡ Trademark, Eastman Chemical Products Co.
§ Trademark, E. I. du Pont de Nemours & Co., Inc.
¶ Trademark, Rohm and Haas Co.

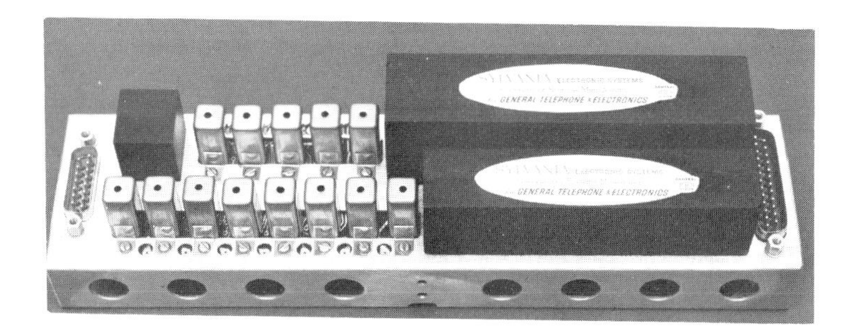

FIG. 28. Assembly of conventional and integrated circuits of the Despun antenna system for the NASA Application Technology Satellite. (*Photograph courtesy of Sylvania Electronic Systems, Sylvania Electric Products, Inc.*)

TABLE 8. Characteristics of Sealing Materials in Space Environments [11]

Class	Material types	Limiting dose for sealing properties, rads	Continuous-service temperature range, °F	Remarks
I	Viton A*	10^7	−65 to 450	Preferred for internal and external vacuum service (e.g., cabin seals, hatches, air locks) because of low permeability to gases, generally good vacuum stability, and resistance to moderate temperatures and radiation exposures
	Butyl	10^6	−65 to 250	
II	Teflon (TFE)*	5×10^4	−423 to 500	Satisfactory for internal and external vacuum seals but more permeable to gases than above types; use only at moderate temperatures and radiation doses
	Teflon (FEP)*	5×10^5	−423 to 450	
	Fluorobutylacrylate (IF-4)	10^7	0 to 350	
	Silicone	10^7	−65 to 450	
	Buna-N	5×10^7	−65 to 250	
	Fluorosilicone(LS-53)†	10^6	−100 to 400	
	Nylon	10^8	−423 to 300	
	Kel-F ‡	10^6	−423 to 350	
	Neoprene	5×10^7	−65 to 250	
	Buna-S (SBR)	10^6	−423 to 250	

* Trademark, E. I. du Pont de Nemours & Co., Inc.
† Trademark, Dow Corning Corp.
‡ Trademark, Minnesota Mining & Manufacturing Co.

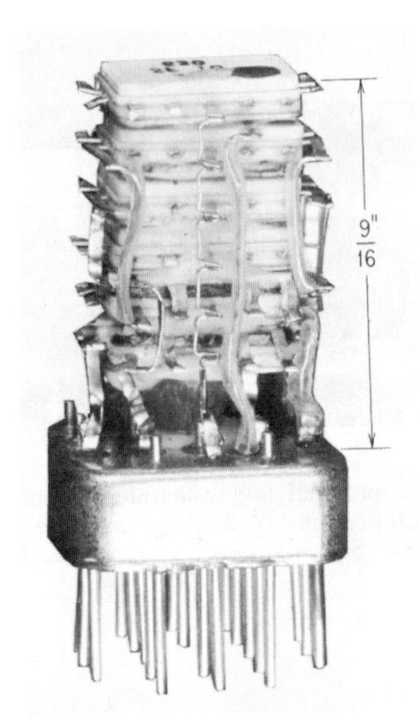

FIG. 29. Typical assembly of integrated-circuit flat packs in a vertically stacked module. (*Photograph courtesy of Sylvania Electronic Systems, Sylvania Electric Products, Inc.*)

FIG. 30. Page assembly of Unit Logic Devices (ULDs) used in the Apollo backup computer and the Saturn V computer. (*Photograph courtesy of International Business Machines Corporation.*)

TABLE 9. Characteristics of Adhesives in Space Environments [11]

Class	Chemical types	Representative commercial adhesives[a]	Radiation dose limit, rads	Temperature, °F: Maximum, Short time (½ hr)	Temperature, °F: Maximum, Continuous	Temperature, °F: Minimum
I	Epoxy-phenolic	422J,[b] HT-424,[c] 25-1,[d] EC-1469,[e] FM-47[c]	10^9	650	500	−423
	Vinyl-phenolic			300	250	−423
II	Nitrile-phenolic	AF-6,[e] EC-1245,[e] Plastilock 620,[f] Epon VIII,[c] A-1[g]	5×10^8	250	200	−423
	Epoxy (filled)			250	200	−320
III	Nylon-epoxy	Metlbond 406,[d] FM-1000,[c] Cycleweld C-3,[h] Metlbond MN3C,[d] Cycleweld C-14,[h] EC-1639,[e] HT-20[c]	10^8	250	200	−423
	Neoprene-phenolic ..			250	200	−100
	Nylon-phenolic			250	200	−100
	Modified phenolic ...			450	600	−100
IV	Silicones	....	5×10^6	650	500	−65
	General-purpose thermoplastic and elastomeric non-structural (low-strength) adhesives	EC-847,[e] EC-776[e]	5×10^6	225	180	−40

[a] Not necessarily limited to these formulations; other products may be satisfactory, but test data on radiation stability are lacking; consult manufacturers and/or conduct experimental evaluation if other materials besides those listed are to be used in structural applications where radiation dose exceeds 5×10^7 rads; use chemical types listed wherever practical.

[b] Shell Chemical Co.
[c] Bloomingdale Rubber Co.
[d] Narmco, Inc.
[e] Minnesota Mining & Mfg. Co.
[f] B. F. Goodrich Chemical Co.
[g] Armstrong Products Co.
[h] Chrysler Corp., Cycleweld Division.

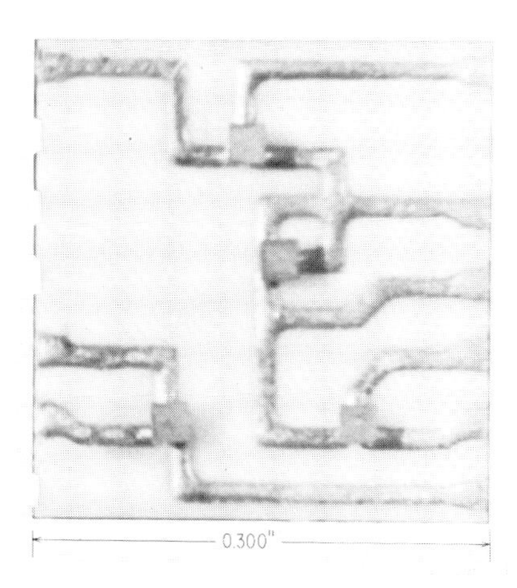

FIG. 31. Unit Logic Device showing thick-film circuit pattern and semiconductors prior to encapsulation. (*Photograph courtesy of International Business Machines Corporation.*)

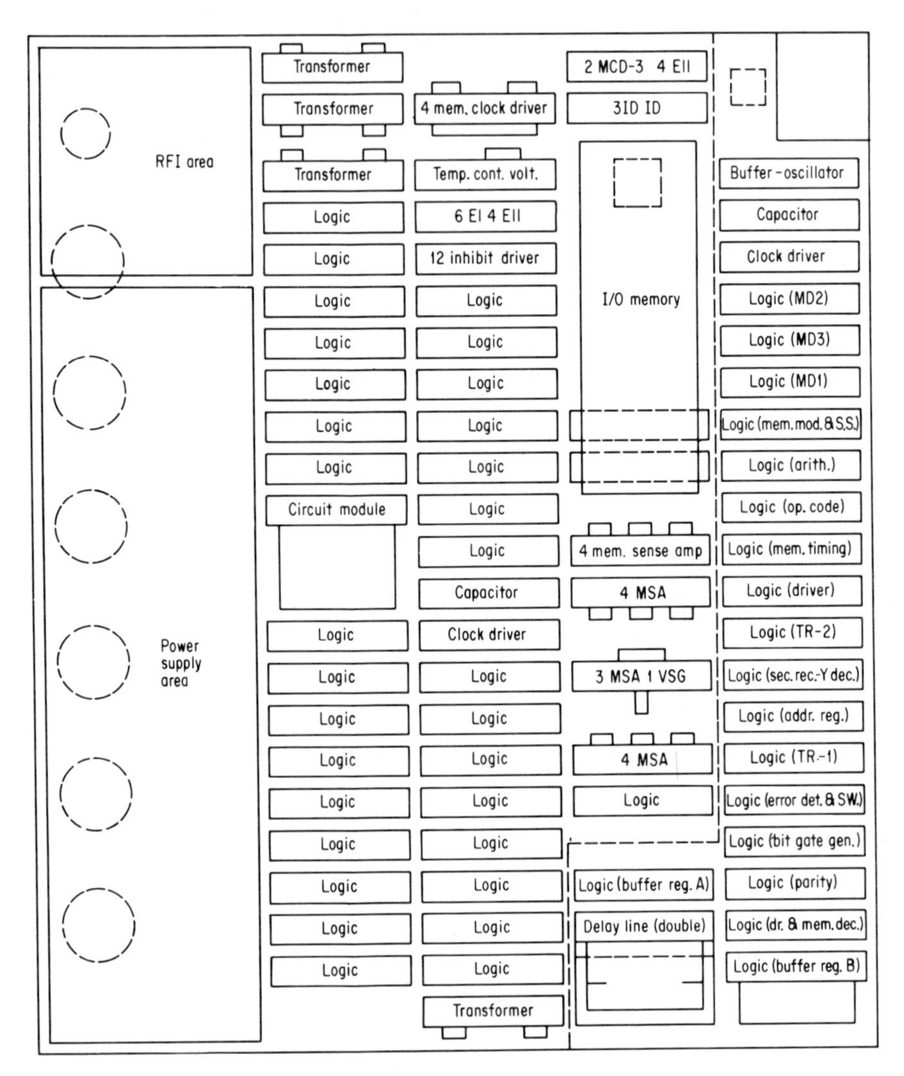

FIG. 32. Representative layout of the page assemblies in the main housing structure of the Apollo backup computer. (*International Business Machines Corporation.*)

TABLE 10. Characteristics of Films, Foams, and Fibers in Space Environments [11]

Class	Material types	Limiting dose for mechanical properties, rad	Continuous-service temperature range, °F	Ultraviolet radiation stability*	Vacuum stability	Remarks
I	Films:					Preferred materials for internal and external structural and related applications (e.g., inflatable space structures) particularly where long-term resistance to ultraviolet radiation (films) is required and moderate temperatures and radiation doses are encountered; good vacuum stability and mechanical-strength properties
	Polyimide (H-film)†	5×10^{8}	–423 to 600	Good	Good	
	Polyvinylfluoride (Teslar)†	10^{8}	–100 to 250	Good	Good	
	Mylar†	10^{8}	–423 to 300	Fair‡	Good	
	Foams (closed-cell, rigid-thermoset types):					
	Polyurethane (alkyd isocyanate)	10^{9}	–423 to 300	Not applicable if protected by sandwich skin	Good; (poor for open cell and some flexible types)	
	Polyether	10^{9}	–423 to 300			
	Phenolic	10^{9}	–423 to 450			
	Epoxy	10^{9}	–423 to 250			
	Silicone	10^{9}	–423 to 500			
	Fibers:					
	Dacron† (polyester)	10^{8}	–423 to 300	Fair‡	Good	
	Nylon, HT1†	5×10^{7}	–320 to 300	Fair	Good	
II	Films:					Not preferred for long-term exposures to ultraviolet radiation unless surface is metalized but may be used for shorter missions internally or externally at low to moderate exposures and moderate temperatures; good vacular stability
	Polyethylene (high-density)	10^{8}	–320 to 225	Poor	Good	
	Polypropylene	10^{8}	–423 to 275	Fair	Good	
	Kel-F	10^{6}	–423 to 350	....	Good	
	Teflon (TFE)†	5×10^{4}	–423 to 500	Good	Good	
	Teflon (FEP)†	5×10^{5}	–423 to 450	Good	Good	
III	Films:					Not recommended for external applications because of ultraviolet radiation susceptibility; may be used internally in applications where low-to-moderate temperatures are encountered
	Polyethylene (low-density)	10^{8}	–320 to 180	Poor‡	Good	
	Polyvinyl chloride (PVC)	10^{8}	–100 to 165	Poor‡	Poor	

* This is probably the most severe environment affecting the properties of thin films in external usage in space. It is beneficial to metallize (vacuum-deposit reflective coatings such as Al) the surface of such films whenever possible to prevent uv degradation. This practice also improves vacuum stability.

† Trademark, E. I. du Pont de Nemours & Co., Inc.

‡ Discolors and embrittles.

environment and the applicability of the tolerance data presently available, and it has been found that the manufacturing processes have a large influence on material and component radiation susceptibility.

In Table 8 is a short list of materials applicable for vacuum seals. In Table 9 are some adhesives and their constraints within space environments. In Table 10 are some limiting data concerning films, fibers, and foams that have some use in packaging. Finally, in Table 11 is some information concerning reinforced plastics that might be of interest to the packaging engineer.

It is expected that new products and processes will rapidly add to the lists of preferred materials and parts since the impetus for space exploration has accelerated research, development, and testing of materials and components for the space environment.

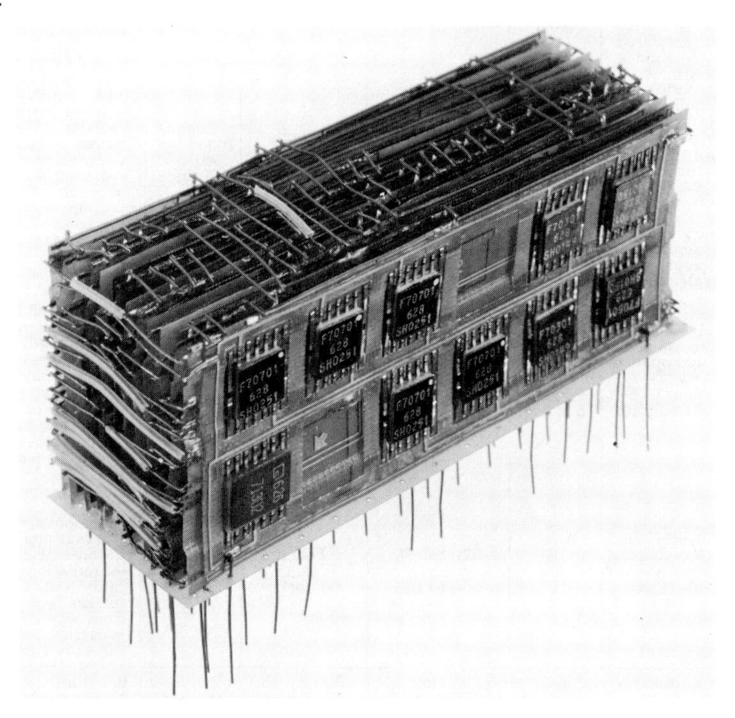

FIG. 33. Integrated-circuit module based upon channeled printed-circuit-board construction. (*Photograph courtesy of Philco-Ford Corporation, WDL Division.*)

Manufacturing Controls. Assuming that an approved equipment design has been generated and qualified parts have been selected and are available, the successful fabrication and assembly of that equipment to meet the documented requirements is dependent upon the control of the manufacturing processes and the motivation of all involved personnel to practice good workmanship, as exemplified by Zero Defects programs.

Good workmanship has five major facets: process control, work environment, personnel training, packaging and handling, and inspection with corrective action.[44] The equipment design has a large impact on the achievement of good workmanship in fabrication and assembly through the degree of design sophistication and the resulting demands on the manufacturing resources.

Process control is concerned with up-to-date documentation for each individual process and continual review of the process parameters and operating personnel for

TABLE 11. **Characteristics of Reinforced Plastics in Space Environments** [11]

Class	Material types	Limiting dose for mechanical properties, rads	Continuous-service temperature range, °F	Remarks
I	Phenolic resin–glass fiber†	10^{10}	–423 to 500	Preferred for general use in interior and exterior structural applications,* particularly where moderate to high temperatures and radiation exposures are encountered; excellent mechanical-strength properties and good stability to vacuum and uv radiation
	Epoxy resin–glass fiber	5×10^9	–423 to 250	
	Modified epoxy–phenolic–glass fiber (heat-resistant epoxy) ..	5×10^9		
II	Polyester resin–glass fiber	10^9	–423 to 250	Not preferred for general primary structural use because of lower strength properties, and slightly poorer stability to the space environment; recommended for external or internal electrical applications where optimum dielectric properties are required (e.g., radomes), particularly at moderate to high temperatures and moderate radiation exposures
	Melamine resin–glass fiber	10^9	–423 to 300	
	Silicon resin–glass fiber	10^9	–423 to 500	
	Triallylcyanurate resin (TAC)–glass fiber (heat-resistant polyester)			
III	Phenolic, polyester, epoxy resins and modifications filled or reinforced with organic fibers (e.g., Dacron and Orlon)	10^8	–100 to 250	Not recommended for structural applications but may be used in certain nonstructural internal applications such as dielectrics. These materials may be used instead of (I) and (II) only under exceptional circumstances, after thorough review of design and environment application. Relatively low mechanical-strength properties and temperature and radiation stability. Good vacuum and dimensional stability. Good electrical properties

* Many combinations of resins and reinforcement types and weaves available for specific structural applications. Directional-strength properties can be varied from unidirectional to orthotropic by choice of reinforcement type and laminate fabrication method.

† Other fibrous inorganic reinforcements, e.g., Refrasil quartz or asbestos, should be equally suitable for use in space but data are lacking to support recommending them. These reinforcements are generally used for more specialized applications such as thermal insulation, ablation, etc., where mechanical-strength properties are secondary to heat resistance.

deviations in performance. Design trade-offs must take full cognizance of the capabilities of these processes and the reliability of their control for producing acceptable results.

The work environment is concerned with the atmosphere,[45] the work space, and work protection in the manufacturing operation. The equipment design indirectly establishes the requirements for protection of the equipment from airborne and handling contamination, through the design features. Narrow spacings, for example, conductive lines on a printed-circuit board, can force assembly in a clean area with gloved handling; this will prevent contamination which, in combination with humidity and temperature in later usage, would have caused deterioration of insulation resistance.

Personnel training resulting in worker qualification is a key to good workmanship. Equipment designs that require new assembly processes for a one-of-a-kind equipment can impact the cost of the design through extensive worker training to assure good workmanship in this particular process or through multiplicities of work rejection.

Packaging and handling of piece parts and assemblies throughout the fabrication and inspection steps are a most important parameter in the production of acceptable equipment. The designer should be fully aware of the effects of fingerprints, scratches, air bubbles, and lead modifications on the reduction in reliability of the equipment design.

Inspection and corrective actions are complementary functions. Multiple levels of inspection normally occur in manufacturing to assure the production of defect-free equipment. The discovery of defects calls for immediate corrective action. Such action may be to improve process control or to provide refresher courses for personnel.

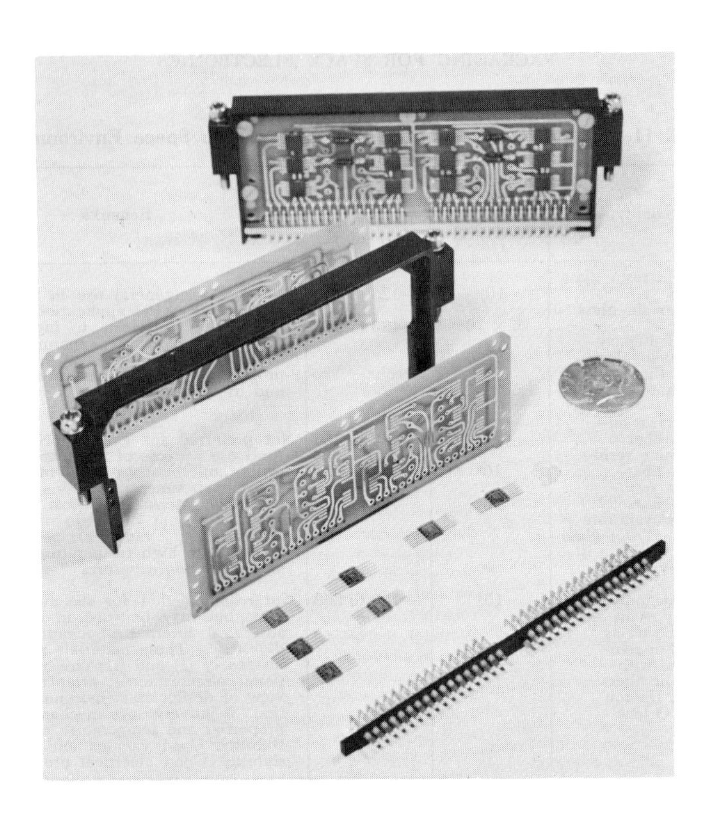

FIG. 34. Module component parts of the universal component packaging system. (*Photograph courtesy of Lockheed Missile & Space Company.*)

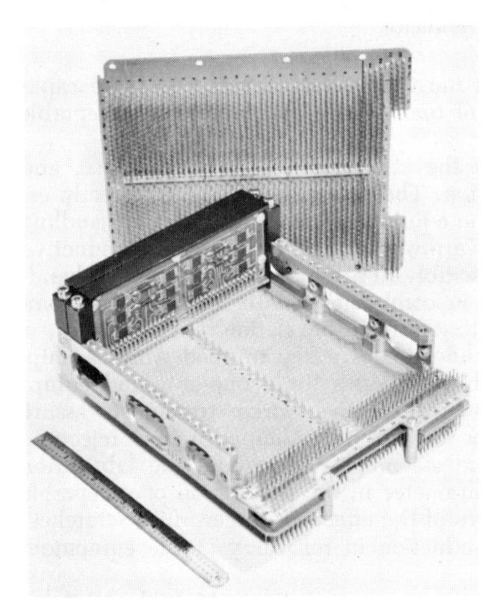

FIG. 35. Modules and wire-wrap matrix assembled on mounting rails prior to assembly into a standard enclosure. (*Photograph courtesy of Lockheed Missile & Space Company.*)

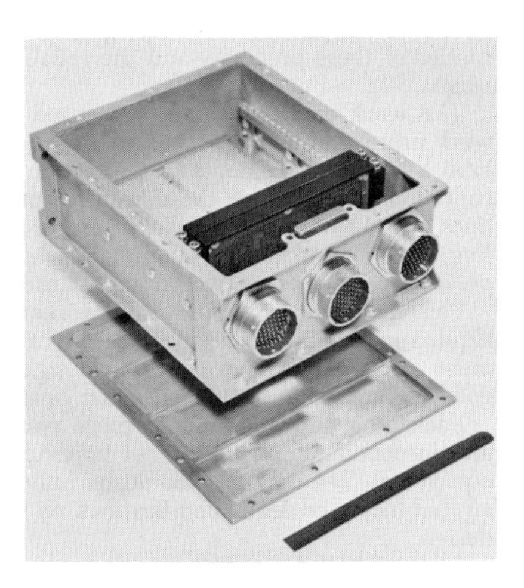

FIG. 36. Module and wire-wrap matrix mounted in a standard enclosure. (*Photograph courtesy of Lockheed Missile & Space Company.*)

A design feature, however, can also require a critical control of a process or an unusual worker skill, and in such a case the corrective action may well be to change a specific design feature. The equipment designer cannot remain aloof from the manufacturing disciplines and expect to develop cost-effective products.

Testing. The testing processes that develop data relevant to system performance are accomplished at all stages in system implementation from component to system qualification.[46–49] System packaging is highly dependent upon the results of these tests for assurance of defect-free designs. During design synthesis and mechanization, mock-ups will be constructed to test the concepts for configuration and assembly constraints; models will be fabricated to test for thermal performance; models will be fabricated to test shock and vibration characteristics; and finally a prototype will be

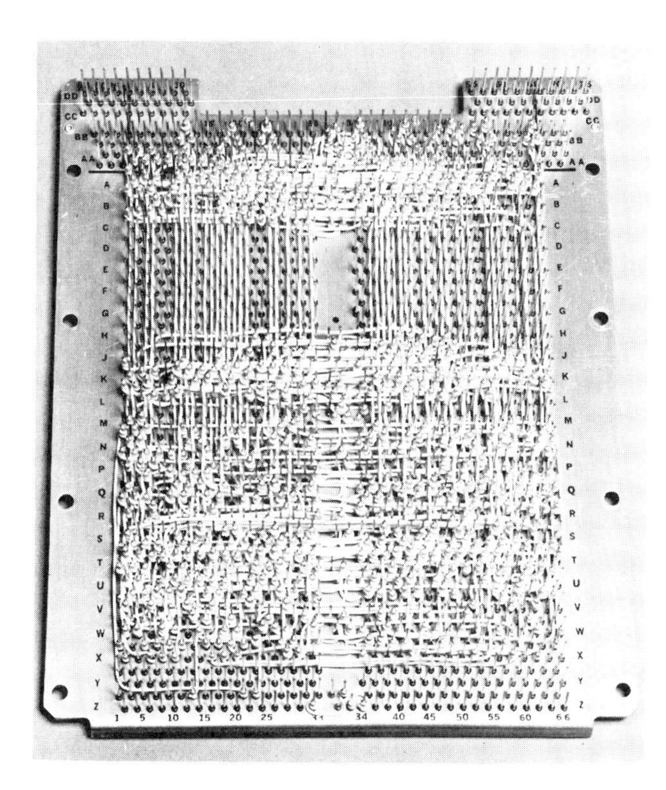

FIG. 37. The wire-wrap matrix of the universal component packaging system. (*Photograph courtesy of Lockheed Missile & Space Company.*)

fabricated using all the materials, processes, components, and configurations of the final design to test the manufacturability of the design. Type Approval Tests are accomplished on one of the first equipments fabricated and assembled by qualified personnel according to all documented specifications and procedures. This equipment is used to verify that the design is capable of meeting the mission performance requirements. In the extensive set of tests, stress levels considerably in excess of the mission environment will be used. When this equipment has acquired Type Approval it is subjected to a Life Test within an environment of vacuum and temperature cycling which simulates some of the mission environments profile. Considerable effort is under way to develop simulators that can more closely duplicate space environment and at the same time test entire spacecrafts for total environmental design assurance. One such simulator is shown in Fig. 38.

Fig. 38. A 39-ft-diameter earth-orbit simulator for testing full-size prototype equipment in an environment equivalent to that at 500 miles altitude. Shakers can be mounted on the internal bedplate to simulate vibration and shock conditions of flight. A solar-radiation simulator provides a 4-ft-diameter beam of energy, and the walls of the chamber can be cooled to −320°F or heated to 350°F. (*Photograph courtesy of McDonnell-Douglas Corporation.*)

REFERENCES

1. National Goals in Space, 1971 to 1985, *IEEE Trans. AES,* vol. AES-1, no. 6, p. 60, August, 1965.
2. Aerospace in Perspective, *Space/Aeron.*, vol. 47, no. 1, pp. 54–124, January, 1967.
3. Neumann, T. W.: The Automated Biological Laboratory, AIAA Unmanned Spacecraft Meeting, Los Angeles, March, 1965, pp. 224–229.
4. Martin, J. P.: Scientific Objectives and Instrumentation for a Solar Probe, *Advan. Astronaut. Sci.*, vol. 19, pp. 339–364, 1965.
5. Lassen, H. A., and R. A. Park: Deep Space Probes: Sensors and Systems, *Advan. Astronaut. Sci.*, vol. 19, pp. 365–395, 1965.
6. Gross, F. R.: Buoyant Probes into the Venus Atmosphere, AIAA Unmanned Spacecraft Meeting, Los Angeles, March, 1965, pp. 76–87.

7. Fosdick, G. E.: A Unique Mass Capture Probe and Its Relation to the Total Space Program, AIAA Unmanned Spacecraft Meeting, Los Angeles, March, 1965, pp. 104–116.
8. deMordes, C. A.: Mission Objectives and Design Considerations for a Scientific Solar Probe, AIAA Unmanned Spacecraft Meeting, Los Angeles, March, 1965, pp. 413–442.
9. Cohen, J. J.: Military Services' Satellites Will Ring the Earth, *Electronics,* vol. 39, pp. 96–99, May 2, 1966.
10. Johnson, R. L., et al: Some Electrical-Electronic Areas Requiring Emphasis in Preparation for Future Space Programs, *IEEE Trans. AES,* vol. AES-2, no. 1, pp. 121–136, January 1966.
11. Goetzel, C. G., J. B. Rittenhouse, and J. B. Singletary: "Space Materials Handbook," pt. 1, pp. 3–94, Addison-Wesley Publishing Company, Inc., Reading, Mass., 1965.
12. Machol, R. E.: "System Engineering Handbook," chaps. 5, 6, and 7, McGraw-Hill Book Company, New York, 1965.
13. Purser, P. E., M. A. Faget, and N. F. Smith: "Manned Spacecraft: Engineering Design and Operation," chaps. 7 and 12, Fairchild Publications, Inc., New York, 1964.
14. Moroz, V. J.: On the "Dust Envelope" of the Earth, "Artificial Earth Satellites," vol. 12, pp. 166–174 (translated from the Russian), Consultants Bureau, New York, 1963.
15. Demoret, R. B.: Titan III for Unmanned Space Exploration, *Advan. Astronaut. Sci.,* vol. 19, pp. 803–856, 1965.
16. Williams, F. L.: Description and Status of the Saturn IB Program, *Advan. Astronaut. Sci.,* vol. 19, pp. 857–909, 1965.
17. Lorsch, H. G.: Biocontamination Control, *Space/Aeron.,* pp. 82–91, November, 1966.
18. Lee, S. M., J. J. Licari, and R. O. Fewel: Sterilization Effects on Microelectronics, Report no. X7-79/501, North American Aviation, Inc.
19. Kammermeyer, K. (ed.): "Atmosphere in Space Cabins and Closed Environments," chaps. 7 & 8, Appleton-Century-Crofts, Inc., New York, 1966.
20. *Proc. Tech. Programs, Natl. Electronic Packaging and Production Conf.,* New York, June, 1964; Long Beach, June, 1965; Long Beach, February, 1967; New York, June, 1966.
21. Robinson, S.: Total Value Concepts in the Contract Definition Phase, *IEEE Trans. AES,* vol. AES-2, no. 4, pp. 402–408, July, 1966.
22. Sargent, K. N.: Insight into SEEing, *IEEE Trans. AES,* vol. AES-2, no. 5, pp. 506–510, September, 1966.
23. Montgomery, D. R., and F. J. Wolf: The Surveyor Lunar Landing Television System, *IEEE Spectrum,* vol. 3, no. 8, pp. 54–61, August, 1966.
24. Schwartz, J. L., and C. B. Tirrell: Electronic Packaging Concept for Apollo Lunar Excursion Module (LEM), *Proc. NEP/CON,* Long Beach, June, 1965, pp. 581–592.
25. Bliss, E., and M. Fromer: Design of Solid State Radar Transmitter Units for the Lunar Excursion Module, *Proc. Electronics Components Conf.,* Washington, May, 1965, pp. 257–263.
26. Walker, D. S.: Packaging for Lunar Orbiter Electronics, *Proc. NEP/CON,* Long Beach, June, 1965, pp. 538–549.
27. Barnes Engineering Company bulletins.
28. Varian Associates Telemetry Products Bulletin.
29. Flagg, W., and H. Lake: A Successful Approach to Vertical Stack Assemblies of Flat Packaged Microcircuits, *Proc. 1966 Electronics Components Conf.,* Washington, May, 1966, pp. 280–284.
30. Munroe, R. A.: The Use of Magnesium Alloys in Aerospace Electronic Packaging, *Proc. NEP/CON,* Long Beach, June, 1965, pp. 550–561.
31. Panaro, M. C.: IBM Apollo Computer Design Mechanical Packaging, *Proc. NEP/CON,* New York, June, 1966, pp. 544–569.
32. Kuehn, R. E., and F. J. Price: Reliability through Packaging Design, *Proc. NEP/CON,* New York, June, 1966, pp. 679–687.
33. Koenig, W. A.: Universal Component Packaging System, *Proc. 1965 Electronics Components Conf.,* Washington, May, 1965, pp. 230–236.
34. Goetzel, C. G., J. B. Rittenhouse, and J. B. Singletary: "Space Materials Handbook," chap. 19, Addison–Wesley Publishing Co., Inc., Reading, Mass, 1965.
35. Fewer, D. R., W. L. Gill, and J. R. Tomlison: Semiconductor Device Reliability

Evaluation and Improvement on Minuteman II CQAP, *Proc. 1966 Electronic Components Conf.*, Washington, May, 1966, pp. 389–406.
36. Yanikoski, F. F.: Some Factors Affecting Design, Construction and Testing of High Reliability Relays, *IEEE Trans. PMP*, vol. PMP-1, no. 1, pp. 79–83, June, 1965.
37. Michaelis, L. P.: Reliability Cost Effectiveness through Parts Control and Standardization, *IEEE Trans. PMP*, vol. PMP-1, no. 1, pp. 327–331, June, 1965.
38. Kirkman, R. A.: Failure Prediction in Electronic Systems, *IEEE Trans. AES*, vol. AES-2, no. 6, pp. 700–707, November, 1966.
39. Fulton, D. W.: Nonelectronic Parts Reliability Philosophy, *IEEE Trans. AES*, vol. AES-2, no. 2, pp. 169–174, March, 1966.
40. Adams, H. S.: Problems in Insulated Wire and Cable in Space Vehicle Systems, *Electro-Technol.*, September, 1963, pp. 133–136.
41. Doman, D. R.: Electronic Materials: Resistance to Radiation, *Electro-Technol.*, June, 1965, pp. 38–41.
42. Peck, D. S., and R. H. Shennum: Long Life Electronics, *Space/Aeron.*, vol. 46, no. 3, pp. 81–87, March, 1966.
43. *Radiation Effects Inform. Center Rept. Ser.*, 1957 to 1967, Battelle Memorial Institute, Chicago.
44. Farkas, L. L.: Meeting NASA Workmanship Requirements, *Electronics Packaging and Production*, vol. 6, no. 6, pp. 121–127, June, 1966.
45. Dye, T. G., and H. S. Jencks: Contamination Control in Packaging Design, *Proc. NEP/CON*, New York, June, 1964, pp. 436–445.
46. McGonnagle, W. J.: The Use of Penetrating Radiation in Failure Analysis, *Proc. 1966 Electronic Components Conf.*, Washington, May, 1966, pp. 361–368.
47. Carter, G. W.: A State-of-the-art Evaluation of Infrared in Heat Transfer Engineering, *Proc. NEP/CON*, Long Beach, June, 1965, pp. 33–43.
48. McCoy, H. L., and W. C. Hutton: Automation of Dynamic Testing, *Tech. Papers 11th ANE Conf.*, Baltimore, October, 1964, pp. 2.3.2-1 to 2.3.2-7.
49. Caren, R. P.: Environmental Simulation, *Space/Aeron.*, vol. 46, R&D issue, pp. 144–148, 190, July, 1966.

Chapter 15

COMPUTER DESIGN

By

W. R. COUCH

International Business Machines Corporation
Electronics Systems Center
Owego, New York

INTRODUCTION

In the design and fabrication of electronic packages, data-processing equipment can provide varying degrees of assistance. In addition, programs and computer-processable data developed for one purpose soon are proved useful for other purposes. As a consequence, computer-aided design and fabrication techniques soon become useful in virtually every facet of design, manufacturing, and testing of electronic systems.

Figure 1 shows a conceptual chart of the accumulation of data during a product cycle. Unfortunately, in practice, data accumulation is not such a smooth and orderly growth process. Development, which is normally thought of as a serial process, becomes iterative in practice. Products are developed by changes. Attempts to shorten the time cycle through parallel development of subsystems further complicate the smooth accumulation of valid data. Because it is essential that there be a prompt organization-wide awareness and response to changes, only data-processing equipment can efficiently handle the necessary data-organizing task and report preparation.

The key to the control of information is the early acquisition of data and the determination of a report production and distribution system, which is responsive to the complex operating requirements. In general, there is good correlation between engineering and manufacturing data, i.e., between the functional and the physical, or spatial, data. For example, an item detailed on one drawing may be used in

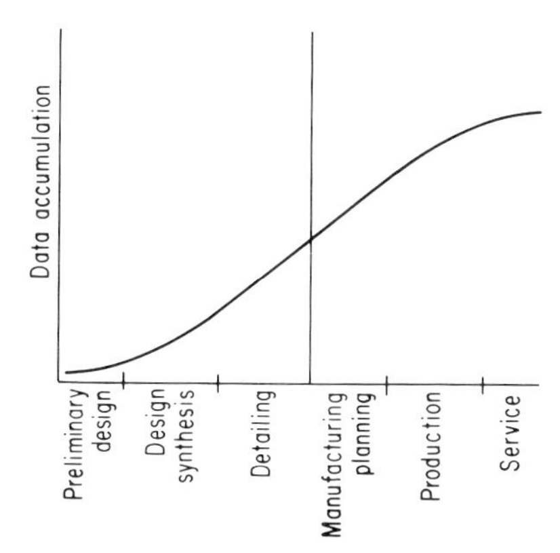

FIG. 1. Data accumulation during a product cycle.

multiple subassemblies, bills of material are accumulated from engineering data, or cabling for several distinct electrical systems can be assigned to the same bundle in the total system. Usually, a single unit of data generated in engineering can be found in many different places and formats in a sequence of documents within an organization. It is the pattern or arrangement of data that converts it to specific information useful for a particular operation. Data-processing techniques make possible the early acquisition of engineering data and provide the capability of expanding, correlating, and rearranging that data. The more successful systems for electronic packaging therefore consider the computer as the heart, not just a part, of the design and manufacturing system.

Form, fit, and functional data are accumulated within the computer system from the first breakdown of product end items. Level by level this occurs as the product requirements are broken down. It is mandatory, then, that changes or additions of pertinent design data are first passed through the computer. Stored computer programs are responsible for the maintenance of these data and will thoroughly check inputs for accuracy, compatibility, completeness, compliance with standards, or other constraints. Data required for analysis and synthesis of subsystems is extracted directly from computer files. Engineering decisions are recorded as generated. The data thus captured are immediately available to all planning and design functions. With the computer, central to the system, these data can be accurately and rapidly distributed to provide corrected and organized information in the format required to other functional responsibilities.

Although this chapter deals with what would appear to be separate computer use areas, it should be kept in mind that there is a requirement for the continuity of data through all phases of product manufacture from preliminary design to field maintenance. Many computer installations fail because of this lack of communications between computer programs of various use areas. The organizational structure of the user is frequently at fault. Each department or division has access to, and may utilize, the computer for its own specific job requirements. The data necessary for accomplishing the original responsibilities are passed along in traditional reports each in its own format. This occurs most often at the point where data processing could be of most value—the engineering release to manufacturing. Organization and budgetary boundaries make it difficult to break the traditional data-requirement routines for engineering releases consisting of drawings, specifications, and parts lists. Manufacturing, in turn, transcribes the parts-list data an average of 12, and has been known to go as high as 27, times. Schedules, inventories, routings, standards, purchases, usages, and many more files must be created and updated based on the released engineering data. This procedure is followed regardless of the expected manufacturing lot size. Only the number of copies and the care and accuracy of the transcription of details vary with lot size.

An integrated data-processing system ensures the prompt, error-free updating necessary for the profitable delivery of quality products.

A list of terms and definitions used in this chapter will be found in Table 1.

TABLE 1. Terms and Definitions

Data base. An integrated group of data sets.

Data field. A unique logical entity of data within a record; i.e., resistance of a component in a part-number record.

Data file (direct access). A device which provides the ability to randomly reach any given record area regardless of its physical location within the file.

Data file (logical). A set of relevant data.

Data file (physical). A device for storing data.

Data file (serial access). A device such as tape which records data in a sequential fashion and requires sequential processing of data.

Data record. A unique logical entity of data within a file; i.e., data defining a specific component within a part-number master file.

Data set. A logical data file.

Language. A method of notation to describe the series of specific operations to be performed on a specific data field to accomplish the desired processing within a computer.

Language, Cobol. A stylized Englishlike notation which can be translated by the computer into the stream of numeric instructions necessary to execute the function.

Language, Fortran. A contraction of *for*mula *tran*slation. A stylized algebraiclike notation which can be translated by the computer into the stream of numeric instructions necessary to execute the function.

Language, machine. The exact series of characters which, when accessed into the control circuitry of a computer, causes the appropriate switching within the data paths to allow the execution of an operation.

Language, PL/1. A stylized notation to provide the facilities of both Cobol and Fortran.

Program. A logically concatenated string of instructions designed to cause a computer to execute a planned function.

Program application. A computer program design to provide solutions or desired results in a specific problem area.

Program systems. A computer program designed to control the execution of application programs.

Program utility. A computer program design to assist in the use of specific computer functions or associated equipment.

System, computer. A group of interacting programs, hardware, and people organized to perform a broad functional requirement.

System, computer-aided design. Any use of the computer which assists the design process, but more specifically in reference to uses requiring direct communication between the designer and the computer.

System, design automation. A computer system encompassing a broad range of engineering requirements from design analysis through numeric-controlled manufacturing and testing.

System, engineering data-processing. The use of computers for maintaining and retrieving the large volume of information necessary to document a product.

STANDARDS

Need for Data Processing. There is a high degree of responsiveness between the early effort, which defines the overall product configuration with particular sizes and shapes of the component parts, and the work which must be done to define the final arrangement. The programming required both for computer-aided design and for automated fabrication, assembly, and testing equipment is highly dependent on the spatial relationships within the package. Lack of uniformity in the standards or lack of compliance with the standards would require excessive redesign and rework of both programs and equipment.

An intelligent standardization program can expand the creativity of the engineer. A flexible modular system of packaging allows the engineer to concentrate on the circuitry and variable design data. He does not have to take the time and effort to redesign the associated packages.

Properly established standards can service a wide range of market demands. Computer programs can be developed to provide maximum utility, and manufacturing equipment can be developed to provide economic returns even in a job-shop environment.

Modular Packaging. Automation is most readily accomplished where there is repetition; yet it is difficult to restrict the wide range of electronic systems requirements to a uniform package size. Both these constraints can be satisfied by development of a modular packaging system which is made up of integer multiples of the smallest desirable assembly. Hence, a standard for a printed-circuit card which describes the largest desirable configuration and provides for an aliquot division of that card into combinations of the smallest desirable configuration will satisfy a wide range of packaging requirements. Programs and manufacturing equipment designed to handle this large configuration can provide for segmentation to the smallest. A uniform repeatable pattern is established which requires only a single set of program logic for computer-aided design or servo-controlled manufacturing of electronic packages.

Figure 2 shows an array of connector sockets. The sockets or socket spaces are of uniform size. The module can be modified by changing either the number of rows or the number of socket spaces. The distance between pins or connectors is an integer multiple of the smallest increment. This includes the distance between pins on

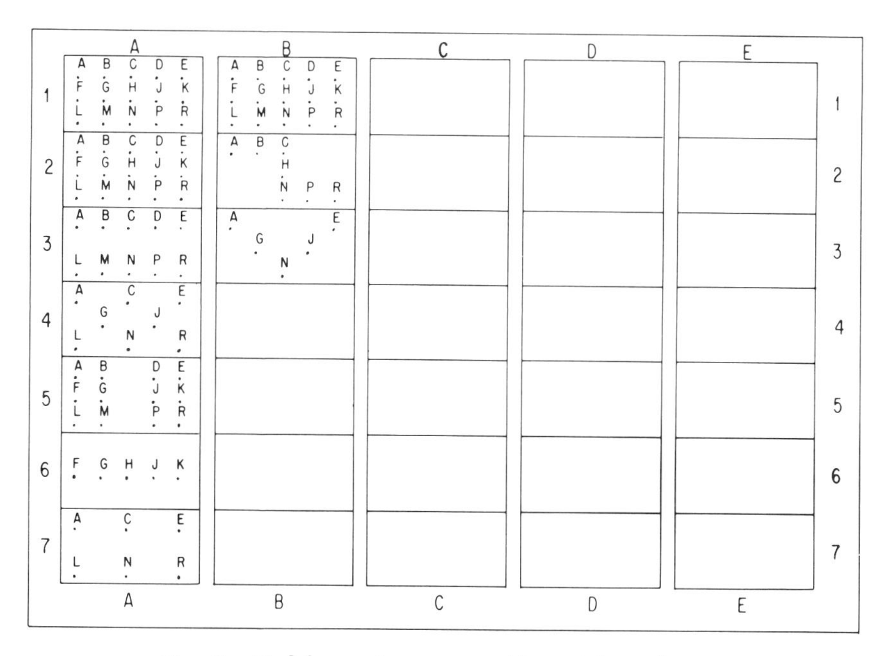

FIG. 2. Modular package array with spatial coordinates.

separate sockets; i.e., the distance between pin E on socket A1 (A1E) and pin A on socket B1 is an integer multiple of the distance between A1D and A1E. This, of course, should also be true of the distance between A1K, A1R, and A2E.

Nomenclature. Figure 2 also illustrates an area of standards which is rarely given enough attention—the selection and control of nomenclature. The successful application of computer aids to packaging requires that each and every point in the assembly be uniquely and uniformly identifiable. This identification should provide for a direct translation to a nonambiguous point in space.

In Figure 2 socket rows are identified by alphabetic characters, sockets within the rows by number. Pins or contacts within the socket are also identified by letter. In this example a three-character reference identifies an exact point on the plane. The chassis rack, or drawer number for this drawing, should further locate the point in space.

Important to this concept is the conformance to the uniform nomenclature when a particular item is omitted from a specific assembly. If within a given socket space the number and arrangement of pins or connectors vary, the positional notation should be maintained. No attempt should be made to assign contiguous identification. Socket locations B1 and A3 in Fig. 2 illustrate this nomenclature assignment. In practice, it may be necessary to develop nomenclature which is more dense than is required for any given element of assembly in order to provide exact positional notation to each space element used.

Additionally it is important that assignment of nomenclature always be made from a fixed reference point. In Fig. 2, if the rack frame or chassis space for the system is as shown for a particular unit and the first two socket rows are omitted, the numbering should begin with row C, A and B being omitted for that assembly.

The complexity, hence the cost, reliability, and flexibility., of both the programming system and the automated equipment is directly proportional to the number of exceptions within the nomenclature. The establishment and conformance to a simple, direct, and meaningful system of identifying points in space are essential to the development of a satisfactory computer-aided packaging system.

Control of Standards. A powerful monitor is now available to review individual compliances with standards. Each and every letter or symbol can be reviewed. Complex mathematical or logical functions can be carried out on a modular or complete systems basis to ensure proper usage of standardized circuits. This monitor is the computer.

The complexity and vagaries of the electronics art will require continuing compromises, but now each compromise can be flagged, documented, approved, and incorporated for future use.

Many benefits are now obtainable as an incentive for compliance with standards. Only with the practice of standard procedures and packages can the full capabilities of computer aids be realized—capabilities which can be developed to relieve the engineer from the routine of hand-sorting the many required parts into an orderly physical arrangement.

DATA BASE

The complexity of design, the short-term overlapping design-manufacturing cycle, and the high volume of inherent changes combine to create a number of data-handling problems. First, the downstream departments receive engineering data in a format unsuitable for their use and therefore must edit and transcribe the information into a more useful form. When this operation is done manually and by several different departments, errors often result. This copying is a significant problem in view of the number of data required to describe the thousands of components and interconnections in a complex electronic assembly. The problem is magnified further because of simultaneous design and manufacturing engineering activities, which produce continual changes. Each of these changes must be recorded, distributed, and controlled in many different areas.

Hence, the computer becomes extremely valuable in keeping track of the hardware items used and their functional relationship. In a properly designed system, this centralized data base can be used to provide accurate and current information to all concerned parties. The data captured for this vital function are virtually the same as those used by the engineer for packaging. Simple reporting programs can be devised to optimally organize this data for the engineer to assist in manual packaging efforts. Straightforward maintenance programs can then be used to incorporate the results of this effort back into the master data base.

Furthermore, the data base can be used directly as input to placement and routing programs. The outputs of these programs can be reported to the designer as well as added to the data base.

Data Base—Data Sets. *Part-number Master File.* The part-number master file represents the basic engineering and manufacturing data set. As such, it is one of the most important files within an engineering information system. This data set consists of a record for each part, component, subassembly, or end product which has been assigned a part number and is available for engineering consideration. At a minimum, each record would contain the assigned part number, a basic description, and the engineering change or variation level of the item. All additional generic data about the part relevant to planned data-processing applications should ultimately be contained in this record. Packaging applications, for instance, would require additional data fields such as number of leads or contacts, dimensions, and placement parameters. More sophisticated packaging analysis programs may require heat-sensitivity and dissipation parameters, power requirements, timing considerations, etc.

In any event, the data contained in this record should always be of a unique nature to the part, i.e., data which remain constant regardless of how, or how often, the particular item is used.

The importance of this file in the engineering-information systems requires that maximum concern be given the development of the necessary processing programs. These programs should be designed to provide:

1. Maximum retrieval efficiency
2. Unlimited record growth
3. Effective maintenance and protection
4. Ease of access by application programs

All applications which require specific data about a part number should be planned to utilize this master part-number data set. Likewise, programs designed to generate data about a part should be designed to recognize the engineering decision and utilize these data to update this file.

Product-structure File. The product-structure file contains one record for each unique component in an assembly. As such, it represents the parts list, or single-level bill of material for that assembly. The basic record consists of the component or subassembly part number, the assembly number, and the effect of the component on the assembly.

These records should also contain any additional data which are peculiar to the specific component or subassembly for this assembly. In many applications a single record per component part number is sufficient. In electronic assemblies, however, it is usually necessary to provide one record for each occurrence of a component. This requirement arises because of the uniqueness of each use of a component within a circuit. If for no other reason, interconnection data for each component require a distinct record of each individual component used. In addition, specific function, loading, timing, and heat-dissipation type data may be required for each logical occurrence of a component. Where this procedure is used, the part-number file key should be suffixed with the schematic-drawing reference part number to ensure proper identification of the part and its usage.

In applications such as cost estimating and materials planning, this file can be condensed so that it contains only one record for each unique part number contained in the assembly. The quantity per assembly in these cases is sufficient additional information to establish usage.

The inclusion in these records of references to both the component part number and the assembly number in which the part is contained provides the ability to resequence the file by component number in order to produce a where-used or used-on reference.

Bills of Material. The combination of the part-number master and the product-structure data sets constitutes a bill of materials or parts-list file. These files can be processed directly to provide single-level bills of material, level-by-level bills of materials explosion, or summarized bills of material for gross materials planning. Resequencing of these data provides where-used traces for components or assemblies to next higher-level or imploded where-used tracing of component usage to the end product.

These files can be successfully maintained and processed from virtually all types of computer-readable storage media: cards, tapes, disk, or direct-access devices. In fact, the bill-of-material maintenance and explosion have, for years, been a basic application of unit record or card processing installations.

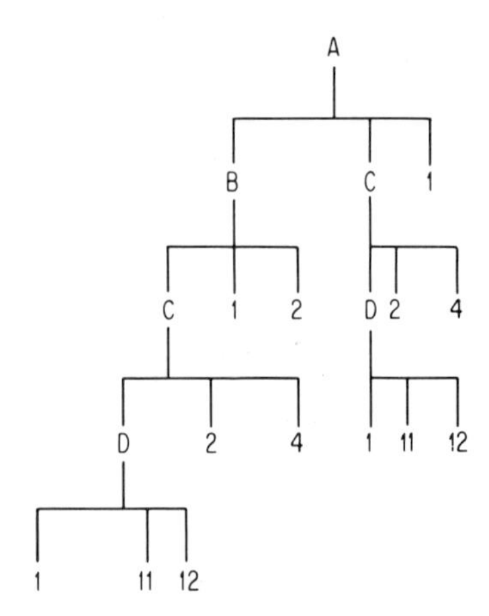

Fig. 3. Product-structure tree.

Figure 3 shows a classical Christmas-tree arrangement of a simple five-level product structure. Emphasis in this example has been placed on the reoccurrence of subassemblies and component parts. The fact that identical items can occur at various levels on multiple branches illustrates the requirement for random retrieval of the data records. If random retrieval is not available, the processing requirements are extended owing to the search requirements and/or the extensive duplication of data required.

Computer-controlled direct-access devices therefore provide the most efficient and effective method of maintaining these files. These two basic data sets can be maintained separately with a minimum of data redundancy. Retrieval programs can be developed to combine the required information in a random fashion to provide maximum retrieval efficiency.

Figure 4 illustrates a file organization which provides the interlocked cross referencing necessary to maintain these interdependent data sets. The part-number master record for assembly items contains, in addition to the unique information about the assembly, a single reference address to the first item record in its product structure. Each structure record, in turn, points to the next component in the assembly.

The pointing or linking of product-structure records can be accomplished within the computer in many ways. The simplest technique, assigning contiguous storage cells to all items in the assembly, also provides the most efficient retrieval for most applications. The primary disadvantage of this technique is that additions to the

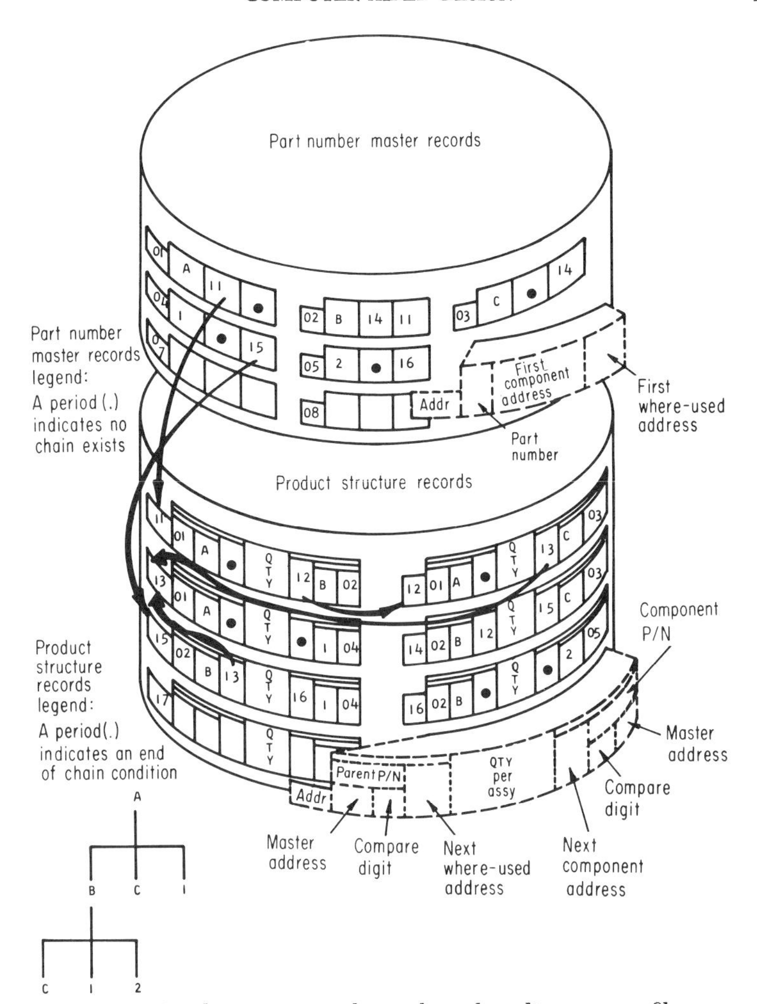

FIG. 4. Product-structure chains through a direct-access file.

number of items in the structures may require a complete reorganization of the data set. Hence, maintenance requirements may rapidly offset retrieval advantages.

The technique illustrated in Fig. 4 is known as *chaining*. Here each product-structure record carries a single reference address that points or directs the processing to the next sequential structure record regardless of its assigned location in the file. Changes in desired retrieval sequence can be made without reordering the data set. Additional items can be assigned to any available file space and the chain address modified to provide proper sequencing. This scattering of sequentially processed records will cause a degrading of the retrieval efficiency in most storage devices. For this reason, the data set is originally loaded into contiguous storage cells. Maintenance programs employ the chaining capability of the file to minimize routine updating process time. When the degeneration of the sequential organization causes a significant loss in retrieval efficiency, a single reorganization process can restore the sequential nature of the file.

Figure 4 also illustrates the chaining of item usage among the product structures. This chain provides the capability to retrieve all assembly data within which the item is used. Level-by-level processing can provide a detailed where-used trace from the lowest component level to the end products for which it is used—an invaluable aid in change analysis.

The universal requirement for parts-file capabilities has led many computer manufacturers to supply software packages to perform this function. The available programs vary widely in techniques, equipment requirements, functional capability, and efficiency. The complexity of an efficient system warrants a thorough evaluation of the available software and its customizing capabilities before undertaking a local development.

Interconnection Data. Interconnection data represent a third equally important data set for electronic design systems. Complete knowledge of the required connections within a product structure is necessary to analysis or to package the circuit effectively.

Two forms of interconnection data exist during a product development and must be provided for in the data base. Prior to packaging, interconnections exist only in logical form. As the circuits are designed, the desired connections are indicated as a collection of contacts to be connected in common. During packaging, each point in this net must be assigned a unique physical location and point-to-point wiring, or printed circuits are developed with associated routing details. These two forms of data orientation require separate techniques for computer processing.

Logical net representation requires the maintenance of some pseudo reference to identify individual occurrences of identical components. Classically components shown on schematics have been identified by labels R_1, R_2, . . . , R_n; T_1, T_2, . . . , T_n; etc. These references are then identified by part number and value in an accompanying parts list. This convention carried through to the computer-stored product-structure record will provide a nonambiguous base on which to build logical networks of interconnections.

In systems designed to provide a computer-generated schematic or logic diagram, unique identification of the component in the circuit is controlled by maintaining the coordinate position of the component on the drawing. Hence, instead of labeling a resistor R_1, it would be identified by the x and y coordinates at which it appears on the drawing.

To illustrate a typical file-processing system, Fig. 5 shows a two-level AND/OR gate. The schematic diagram has coordinates to provide positional data suitable for component identification and possible computer-generated reproduction. Figure 6 illustrates a document designed to replace the original manually generated parts list. Figure 7 illustrates the product-structure file and associated interconnection list generated by the maintenance programs for this input. (It is assumed that the master parts list was previously established within the system.)

In this example the engineer was required to supply two items of data not previously associated with a parts list: the coordinate location of each component symbol, and a net number or common copper reference number.

Other data normally associated with the parts have not been provided for in the manually generated list. Basic part description, component values, tolerances, and other generic data will be retrieved from the part-number master file. A formal printed parts list will be supplied from the computer processing. With proper programming, this data bank can be used to retrieve all pertinent data for packaging.

Figure 7 shows a file organization which would require a programmed search to retrieve the common copper points within a given net. Address chaining could, of course, be employed to improve the efficiency of net retrieval. Since nets normally consist of relatively few connections, the loss of efficiency in search procedures is usually justified by maintenance efficiency and storage savings.

Configuration Management. In order to provide engineering-change history and control proper effectiveness, an additional master data set must be added to the data base. This file, sequenced on engineering change or variation number, would contain generic information pertinent to the change. In addition, it should contain an anchor to a product-structure chain to provide for recovery of all items affected by the change.

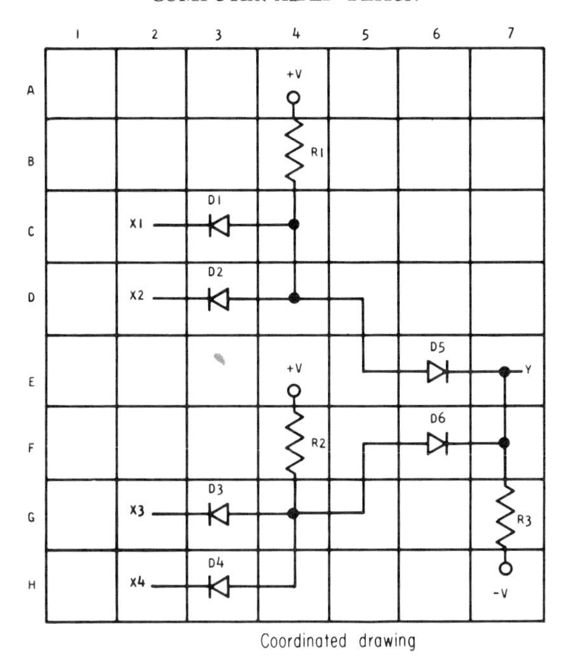

FIG. 5. Coordinated schematic drawing.

Circuit Description

Parts and Interconnection List

Circuit No.	Eng. change No.	Drawing No.	Circuit description	Date	Initials
1234	04	17B1694	Two level AND/OR gate	10/4	

Line No.	Dwg. Loc.		Label	Part No.	Net connections										Comments
	X	Y			A	B	C	D	E	F	G	H	J	K	
1	4	A	+V	8868	1										
2	4	B	R1	7986	1	2									
3	3	C	D1	5432	3	2									
4	3	D	D2	5432	4	2									
5	2	C	X1	8868	3										
6	2	D	X2	8868	4										
7	6	E	D5	5432	2	5									
8	4	E	+V	8868	6										
9	7	E	Y	8868	5										
10	4	F	R2	7986	6	7									
11	2	G	X3	8868	8										
12	3	G	D3	5432	8	7									
13	2	H	X4	8868	9										
14	3	H	D4	5432	9	7									
15	6	F	D6	5432	7	5									
16	7	G	R3	7987	5	10									
17	7	H	-V	8868	10										
18															
19															
20															

FIG. 6. Schematic-drawing data transcribed for computer input.

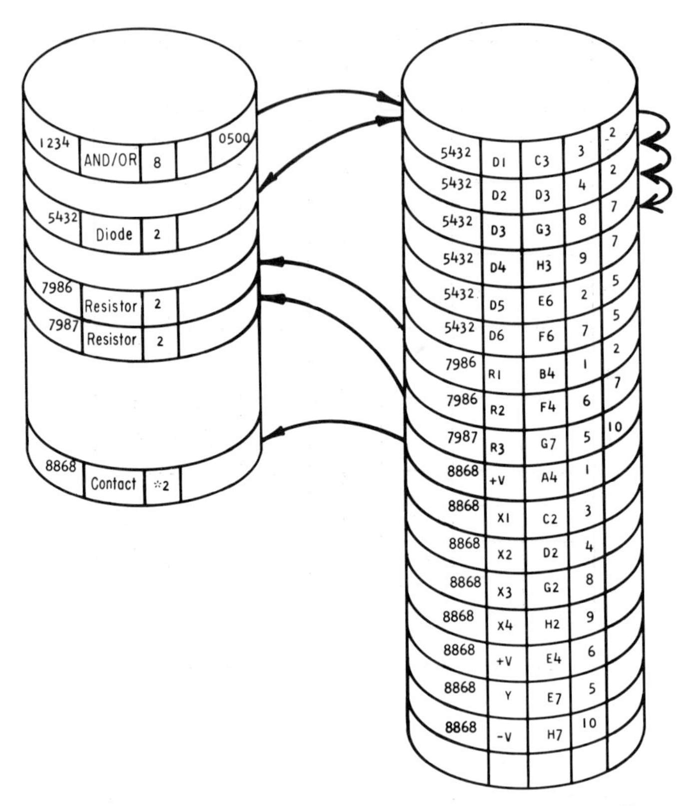

FIG. 7. Schematic-drawing data stored in direct-access file.

Figure 8 shows a file organization which incorporates basic change data. The interaction of these data presents a complex picture. It is admittedly difficult manually to thread through the interactions of this data base. Once considered and a commitment made to computer programming, the maintenance and retrieval of these data can be efficiently accomplished.

DATA-BASE MAINTENANCE AIDS

Run-activity Number. The run-activity control number has two purposes: (1) it can be an aid to reconstruction and restart procedures, and (2) it can be an aid to specialized retrieval functions. The master run-activity control number would be a field located in the part-number master file control record. At the beginning of any application program run, the program should:

1. Access the master run-activity control number
2. Update it by adding 1
3. Restore the master run-activity control number
4. Store the updated master number in the core for use during the program run
5. Display the run number on the operations log

During the program run, every part-number master record that is updated would have its own detail run-activity control number set equal to the value of the master for this run.

The first of the special retrieval functions employing the run-activity control number determines which components in the product-structure tree of one end item are common to those in the product-structure tree of a second end item. For in-

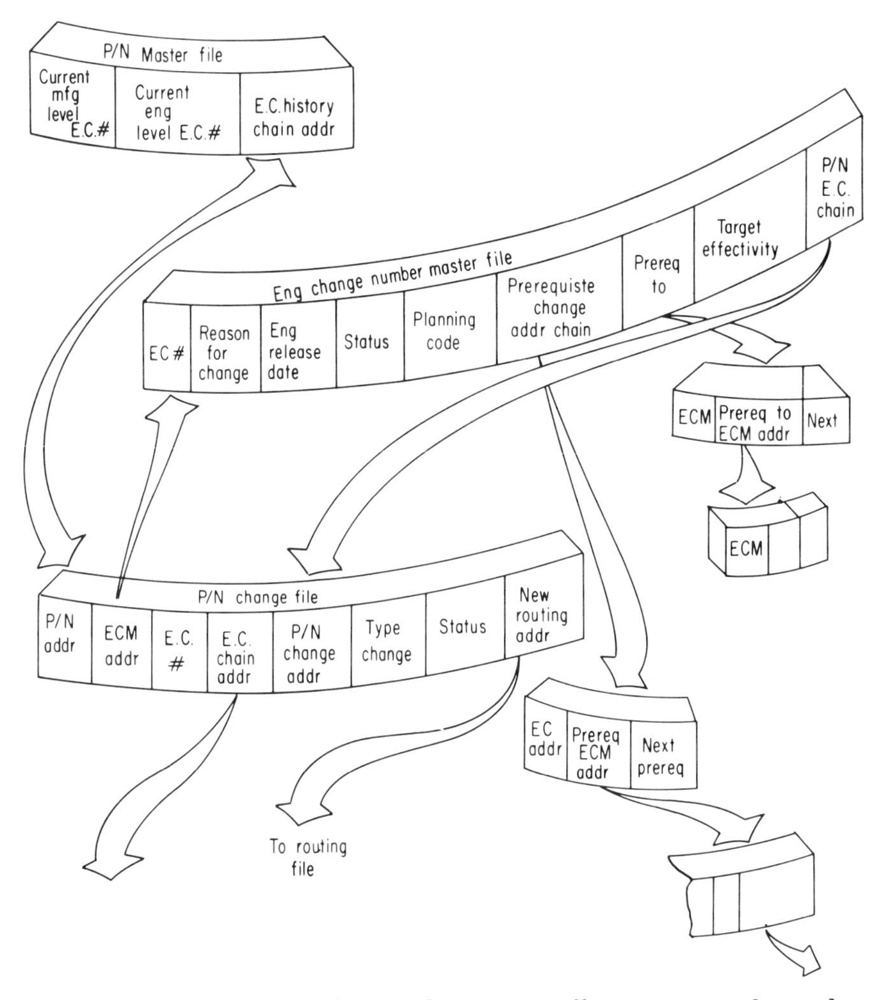

FIG. 8. Part-number master data, product-structure file, engineering change data chained for cross-referenced retrieval.

stance, assume that a summarized explosion inquiry is processed against the first end item in run 1. During a summarized explosion of the second end product in run 2, a test for run-activity control number equal to 1 is made against every part-number master record accessed. Those part numbers which meet the test must be common to both end items.

The second special retrieval function employing the run-activity control number deals with the question "What parts are unique to some product(s)?" As an example, if end product A is to be deleted from the product line, what component parts and assemblies may be deleted from inventory? The end products would be exploded in run 1. At low-level time for each part, a single-level where-used chain is traced. If all parent part-number master records accessed during this single-level where-used trace contain a run control number 1, these parts are unique to the end product(s) that was exploded in run 1. (Subassemblies that are not unique are exploded no further.)

Low-level Coding. Low-level codes can be used in the processing of retrieval programs and in checking assembly-to-subassembly continuity in the part-number master and product-structural files. A low-level code is a number indicating the lowest tier or level at which a particular part number is found in some structure tree. The low-level code is recorded in each part-number master record. Figure 9 depicts the product-structure trees of top-level assemblies *A* and *K*, showing a relative level number of each tier, starting with the top level defined as level zero. Figure 9 also depicts the low-level codes for each part number.

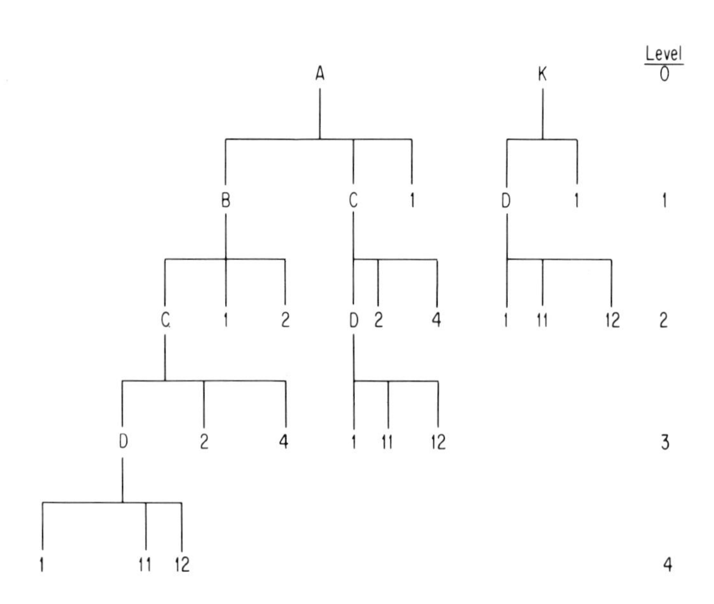

Part number	Low-level code
A	0
B	1
C	2
D	3
K	0
1	4
2	3
4	3
11	4
12	4

FIG. 9. Low-level coding.

The product-structure data-set creation and maintenance program should check the low level each time an assembly component is "added" to the file (during initial loading and subsequent adding). Low-level codes are updated, i.e., lowered, if necessary. Low-level codes need not be checked or updated when a component is deleted.

The lengthy processing time required to check and raise low-level codes, if necessary, is not warranted for deletions. The uses of low-level codes are not affected if they are lower, i.e., greater numerically, than the actual usages in the product-structure trees.

Summarized explosion techniques can utilize low-level codes to reduce processing time by eliminating the repeated explosion of a multiple-usage subassembly each time it is encountered in single- or multiple-product structure trees. If the basic summarized explosion is expanded into a complete gross-to-net requirements generation application, the low-level code takes on added significance. Accurate use of "netting" rules is facilitated when performed only once—at low-level time.

There are two other uses of low-level codes. Summarized implosion or where-used traces utilize low-level codes to reduce processing time in a manner similar to summarized explosion. The product-structure data-set creation and maintenance programs can use low-level codes to recognize a product-structure continuity error where an assembly contains itself as a component.

Product-structure Continuity Checking. Maintenance of a large volume of records requires that checks and controls be built into the system. One control required in the maintaining of product-structure records is the verification of continuity in assembly-subassembly breakdowns. This checking is done in two ways.

First, a complete product-structure file must specify that each assembly has at least one higher-level assembly usage, except an assembly defined as top-level and service (repair). This ensures that no missing connections exist in a product structure tree, thus leaving an assembly "stranded." The product-structure data-set maintenance-program modules should be designed to check automatically product-structure continuity whenever there are file additions.

Second, the product-structure data set must not contain an assembly that uses itself as a component directly or indirectly (through subassemblies). Figure 10 shows

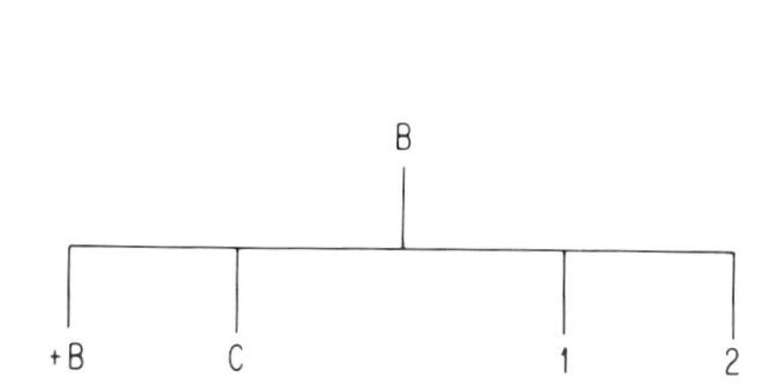

FIG. 10. Parent-assembly number identical to component part number.

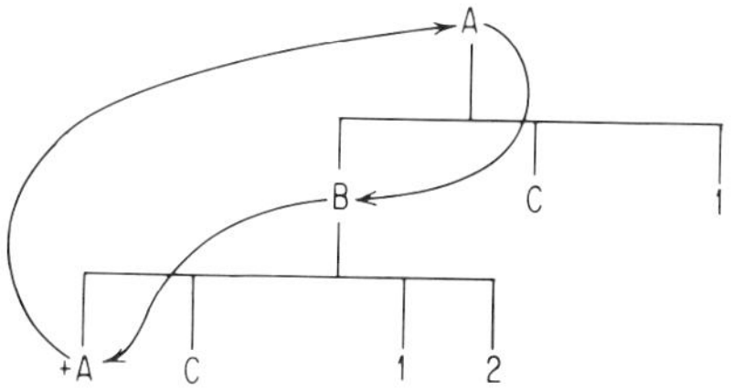

FIG. 11. Assembly contains itself through a subassembly, resulting in endless loop.

the addition of part number B to assembly B. This violates the assembly concept, since an assembly cannot be a component of itself. This error should be recognized by the product-structure file-maintenance program.

If Fig. 10 were altered to indicate the addition of part number A (instead of B) to assembly B, there would not appear to be any violation of the continuity rule. However, Fig. 11 further shows assembly B's position in the product-structure tree. It indicates that assembly A contains itself, one level removed, through subassembly B. Only by reflecting each assembly's position relative to other assemblies in the product-structure trees can an assembly containing itself one or more levels removed be recognized. This condition can be automatically recognized by low-level-code updating procedure. The logic in this procedure requires the low-level code of a component's part-number master record to be numerically greater than the low-level code of the parent assembly's part-number master record. If this is not true, the low-level code in the component part-number master record is made numerically one larger than the parent assembly's low-level code. If the component part number is a subassembly as well, the low-level-code assignment of each of its components must be checked by explosion. The process is repeated until the components of lower-tier subassemblies have proper low-level codes.

Adding a part number to itself directly or indirectly would place the low-level update logic in an endless loop, since the low-level codes are never large enough. Figure 11 shows the looping condition arising from adding part number A to assembly *B*. The product-structure file-maintenance program should recognize the loop and not allow a component causing this error to be added to the assembly.

DOCUMENTATION AIDS

Bills of Material. The part-number master and product-structure files readily provide for the three basic parts-list information forms.

Single-level Explosion. This is the most fundamental type of retrieval technique that uses product-structure data in component assembly sequence. The processing output is a bill of material. In the simplest form an assembly is exploded into its direct components and associated quantities per assembly. An example of a single-level explosion application that uses product-structure data as a framework for processing is the extended bill of material.

Indented Explosion. Indented explosion is a processing technique for completely breaking a top-level or other major assembly into its multiple subassembly levels or tiers. The term *indented* refers to the format of a printed output, frequently called an *indented parts list.* The incorporation of an extension (in which the order quantity of the starting assembly is multiplied by each of the quantity per assembly fields) gives the total quantity for each component needed.

Summarized Explosion. This is a retrieval technique that completely explodes a top-level or other major assembly into all its multiple subassembly levels and summarizes the quantities of all parts (subassemblies and simple parts) that are found in the entire product structure. Low-level codes are used to eliminate reexplosion of multiple-use assemblies. This type of retrieval provides the framework for performing cost estimating or net requirements applications.

Where-used List. When incorporated in the product-structure file, a where-used or used-on chain can provide three correlative reports.

Single-level Implosion. This is the most basic type of retrieval technique that uses product-structure data in where-used sequence. The output of the processing is a next assembly where-used listing.

Indented Implosion. Indented implosion is a processing technique used in programs that trace the usages of a given part number in an assembly and, in turn, the use of the assembly in higher-level assemblies up to the top assembly level. The term *indented* refers to the format of the printed output.

Summarized Implosion. This is a multilevel processing technique that accumulates the direct and indirect usages of a part number on all higher-level assemblies. The printed output of such programs indicates all the assemblies that directly or indirectly contain the part number, including the total quantity of the part number in each assembly. This type of retrieval technique can also be expanded to reflect the effect of incremental cost increases or decreases for simple parts or subassemblies on top-level products.

Interconnection List. Interconnection data contained in the file can be retrieved and readily processed into basic additional information forms.

Net List. The net list represents the basic form of interconnection information. Common copper connections are grouped in net-number–contact-number order. This basic listing provides information necessary for design of point-to-point wire connections.

Wire List. The wire list provides the primary point-to-point discrete wiring information. The list is sequenced by net number and interconnection or wire number. This list would contain a "from" and "to" connection specifying the exact physical terminus of each connection in the system. Additional data, such as "via" points or routing bend points, would also be shown where pertinent. Wire type, wire length, wrap level, and shield connection point should also be contained in this listing.

The wire list is designed to replace point-to-point wiring diagrams. It should contain the information necessary and sufficient to constitute a complete engineering design release of wiring data.

Checklist or Double-ended Wire List. Each wire appearing in the wire list is duplicated. The from and to contact location fields are reversed. The resulting records and the original wire-list records are sequenced together on the from contact field. The resultant listing provides a ready reference in connector sequence to all interconnections terminating at any connector. This list can significantly reduce the problem of locating the source of a wire when tracing "bugs" in a hardware model.

Installation Sequence List. Assembly processes rarely allow wiring to be installed in the sequence used for engineering documentation. Manual or machine efficiency must be considered. Wire-list data can be recorded against a set of established criteria to provide maximum installation efficiency.

Among the sequencing considerations are:

1. Installation level—in wire-wrap technology, all first-level wires must be installed before second-level wraps.
2. Wire length—short wires installed on top will hold longer wires in place.
3. Direction of wire routing—installation of all wires with major run lengths in similar directions tends to relieve buildup at intersections.
4. Economy of movement—minimize movement between successive connections.

Rework Net List. When engineering changes occur that affect wiring, it is desirable not only to produce a complete new wire list, but to rework instructions as well. For this purpose, with a properly maintained data base, an add and delete difference list can be developed between any two change levels of current or previous wiring configurations. In addition to adds and deletes, an associated resultant net giving common points after rework has been found useful for probing the reworked configuration. Where the resultant net technique is used, wires added by this change are contained as flagged items in the rework list.

Wire Labels. Labels of various types for wire identification can be produced on the printer directly from the wire-list data. Paper or fabric label forms, with or without adhesive, are available for continuous pin-feed printing.

Diagrams, Schematics, and Artwork. More sophisticated documentation can be produced from a properly established data base. The computer has been used in conjunction with a slightly modified line printer or a drafting machine to produce the necessary logical drawings, circuit schematics, component layouts, and printed circuit artwork. Figures 1 and 2 were both originally drawn for this text by a computer-controlled plotter. Several other figures were printer-produced, including the logic-diagram figure. More detail on these applications is contained in later sections of this chapter.

ENGINEERING AIDS

Design and Analysis. *Circuit Analysis.* The problem of assuring electronic-circuit performance faces most designers in the industry. Circuit performance is especially critical where extremely high reliability is desired. Throughout the industry the solution to the problem exists as a compromise between economy and product reliability.

The question to be answered in design of these circuits is "What is the value and tolerance required of each individual circuit component in order to obtain overall circuit performance within specific limits of error?" The relationship between individual component tolerance and overall circuit performance is not a direct one. Thus, it cannot be said that, to obtain a maximum of ± 1 percent variation of a given circuit output, all components must be ± 1 percent from nominal value. In fact, it may be that some components will require a ± 0.01 percent tolerance, while other tolerances may be as high as ± 10 percent or greater.

The analysis of a circuit, as a result of all the possible combinations of components, is a tedious, if not impossible, task to perform manually. It is not practical to build and test a large number of variations of any given circuit. Indeed with integrated circuits the tooling commitment demands a highly reliable design before even one circuit is produced. The digital computer provides a means of solving this widespread problem. Many programs exist for solving the various parts of this problem. They are available from computer manufacturers, consultants, service bureaus, users' groups, and sometimes from a user directly. Techniques and performances vary over a wide range depending upon the computer configurations and degree of solution required. Evaluation of these techniques and algorithms is beyond the scope of this chapter.

Logic Analysis—Simulation. The design of switching circuitry has become a highly complex art. An extremely large number of electrical paths is possible through the resulting switching maze. While this is particularly true in the design of digital computers, the problem is also prevalent in motor starting circuitry, machine-tool controllers, elevator controllers, etc.

It is difficult or impractical manually to trace the effects of all possible sequences of input pulses. Feedback and feed-through conditions frequently go unnoticed. Under certain untested combinations of conditions this results in elevators stopping between floors, strange behavior of machine tools, etc.

With the digital computer it is possible to set up truth-table models of the switching maze. This model can then be exhaustively exercised at microsecond rates to trace combinations of input sequences to determine the combination of paths which would be completed.

The interconnection file can be used to generate the required switching model.

Several techniques for simulation and diagnosis of logical failures have appeared in the literature. Among these techniques are the comparison of truth tables, the determination of the intersects of truth tables and complemented function truth tables, circuit tracing, and the D algorithm.*

Logic Reduction. The ability to express much of the switching-circuit art in the form of boolean algebra has led to the exploitation of this technique to reduce the number of actual switching elements necessary to perform a given function. Significant component reductions can frequently be made. The economic value of this reduction becomes extremely obvious for high production units.

The techniques developed to date have two basic weaknesses for many products:

Maintainability. The circuits developed by boolean manipulation are frequently too complex to effectively trace in the field. Troubleshooting and servicing may offset product cost savings.

Fail-safe techniques to ensure fail-safe design have not been too successful. Major attempts to date to provide this capability have produced only marginal results.

Component Selection. After circuit analysis it is necessary to select the appropriate components to implement the circuit. The data expressing the exact values and tolerances can be utilized to retrieve from the part-number master data set the component available to construct the circuit. Company part numbers, vendor lists, vendor part numbers, etc., should all be available through the part-number master data set. The new circuit can then be assigned a part number and entered into the part-number master data set and the product-structure data set. These necessary data are thus available for packaging of the circuit.

Electrical-load Analysis. Keeping track of fan-out and accumulated load on a given circuit becomes a major task. The use of the interconnection file as a data base for calculating and summarizing these loads can eliminate significant later troubles.

A system of integer load values can be developed to provide a simple high-speed approximation of the load on a network. This technique requires that each component pin, or contact to be used in the system, be assigned a positive or negative load factor proportional to a measured current value. An approximation of the network can be obtained by a simple summation of these values within a commonly connected net.

In more sophisticated systems, equivalent circuit equations can be used to develop a more accurate loading analysis.

Packaging. The areas of computer processing most closely associated with the packaging problem are partitioning, placement, and routing. All three of these

* The *IBM J. Res. Develop.*, vol. 10, no. 4, 1966, presents the D algorithm along with the mathematical derivation to prove that, if a test for any given failure exists, the D algorithm will compute such a test.

operations are clearly interdependent. Partitioning deals with the division of components for assignment to subassemblies. Placement covers the exact physical location of the components within the subassemblies, and routing is the assignment of connector paths for printed, etched, deposited, or discretely wired circuits.

Partitioning and Placement. Partitioning and placement obviously have a drastic effect on the routing of interconnections. A random arrangement of the components sets is unlikely to yield the nearly worst-case solution of the example Fig. 12*a*, but neither would it yield as good a solution as that of Fig. 12*b*. Even such a trivial

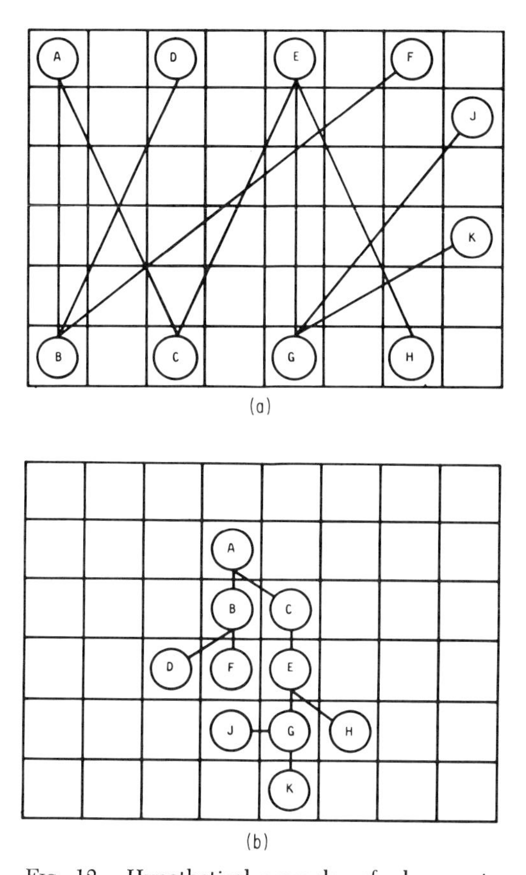

FIG. 12. Hypothetical examples of placement.

example helps to illustrate why packaging has become a special field in electronic design and why computer programs are highly desirable. The chassis of Fig. 12*a* contains 48 candidate positions for the 10 component sets of the net. There are, therefore,

$$\frac{48!}{38!10!} \times 10! \quad \text{or} \quad 25 \times 10^{16}$$

possible arrangements to evaluate. At a microsecond evaluation rate, almost 7 million computer-hours would be required to exhaust the combinations and to determine the optimum arrangement. Clearly, exhaustive analysis is out of the question even with the fastest computers available.

Both statistical and logical procedures have been used to address the problem and produce an economically feasible solution. The literature contains details of a number of approaches which can and are yielding production results at least as good as those of a capable packaging engineer. Generally these techniques can be divided into two classes: methods which employ successive trial arrangements, or heuristic techniques, and those which use a more direct algorithmic approach.

Successive trial-arrangements systems are based on a statistically controlled random ordering. Components to be placed are randomly distributed among the candidate positions, and the total length of interconnections thus created is summarized. The components are then rearranged, and the new interconnection length is compared with the best previous arrangement. The primary requirement for procedures of this type is the development of a technique which recognizes and bypasses lesser solutions in order to provide economic justification for "one more trial."

The algorithmic approach to partitioning and placement employs a technique of measurement of connectivity and disconnectivity between components. In this technique the connectivity of each net or electrically common function is established. In its simplest form, connectivity is a count of the number of separate contacts required for the function.

The component connectivity is then defined as the summation of the functional connectivities of all the contacts on the component. Because of their connectivity, components will tend to cluster into nearly disjoint sets. Weighting factors, based on circuit parameters such as frequency, can be added to the functional connectivity further to force this clustering and to constrain component placement by emphasizing the value of given functions.

This cluster definition and assignment are accomplished by first assigning the component with the highest connectivity to an exact chassis location. Then, successively, a single component is assigned to each additional chassis in the system on the basis of the highest remaining individual connectivity and maximum disconnectivity to previously assigned elements.

Each chassis, in this manner, will acquire a certain connective attraction by virtue of the component assigned to it. In turn, that component is guaranteed to have a high connectivity of its own. For the remaining unplaced components, an ambivalence of connectivity to more than one chassis is minimized and, at the same time, the total number of unplaced components which have high connectivity to the set of chassis is maximized. Thus, the total set of components can be distributed with a high probability of acceptance.

In both approaches, the number and the total length of connections are the primary consideration, but there are additional constraints which cannot be neglected. Each industry, each product has a unique set of constraints, but, cutting across the many packaging disciplines, five general parameter sets are common to all routing and placement problems:

Length of Interconnection. Reduction of total length implies a tangible savings in both cost and weight of the system. These savings are not, however, always inherent. Total minimization of systems which include special connections (for example, coaxial cable) may conceivably result in higher cost and weight. In systems of this type, a compromise must be made between total length and the length of these special connectors.

Position-dependent Components. Most systems include position-dependent components. These devices, for example, drivers, shields, delay lines, etc., are included because of the physical relationships of the systems components. Since they are usually heavier and more costly than a reasonable amount of interconnection material, special attention must be given to the creation of their requirement in placement and routing.

Reliability Considerations. The reliability of the system may be highly dependent upon the juxtaposition or separation of specific components. The most obvious case is the separation of heat-sensitive and heat-producing components and components creating electromagnetic interferences. Placement and routing procedures which are oblivious to such factors cannot achieve a reliable system.

Serviceability Considerations. In servicing a system, it is generally desirable to group those components which are performing similar or complementary functions as closely as possible. Similarly, components requiring frequent service must be easily accessible. Inasmuch as service over a period of years may represent a significant portion of the total systems cost, an understanding of this problem must be a fundamental consideration.

Effective Utilization of Space. In assigning physical locations to components it may, or may not, be desirable to utilize the available space completely. In the packaging of missile electronics, full utilization may be demanded. In most systems, however, it is desirable to provide unused space for later modifications and maintenance. To be effective, this available space should be distributed throughout the system.

The interactions of all these parameters are obvious. Mathematical optimization which is based on a single criterion seldom results in a practical design. The inclusion in a computer design program of a decision-making capability for all constraints complicates the programming task and increases computer process time. An economic trade-off, therefore, is generally made by permitting the engineer to exercise his experience and judgment manually to assign any location or connector prior to or during the processing cycle.

Interconnection Design. *Wire Assignment.* The assignment of discrete point-to-point wire segments from electrical or logical schematics is a time-consuming and error-prone task. Manually performed, it is generally impractical to calculate the minimal wire-length configuration for each of the thousands of nets required to interconnect an electronics complex. As a computer application, it is normally a straightforward operation. As such, it is frequently selected as a starting point for the development of design automation systems.

A wire assignment program requires two basic sets of information:

1. A precise definition of the pin or connector nomenclature as it relates to an exact xy physical location
2. A set of pins which are to be connected by "common copper"

Most electronic nets consist of a small number of contacts, under five. It is therefore practical to compute all of the $N(N-1)/2$ possible interconnections and select by exhaustive search the $N-1$ shortest segments required for common connection.

Figure 13 shows a six-pin example where the 15 possible segments are ranked by straight-line distance between the contacts. The classic procedure is to select progressively the topmost-ranked segment that does not violate constraints and is not redundant.

Unconstrained, this procedure results in a minimal configuration, Fig. 13*b*. It is not uncommon, however, to encounter a restriction on the number of contacts per connector. When the example is constrained to a maximum of two connections per pin, the first selection pass will result in the Fig. 13*c* configuration. As illustrated, this constraint may result in a less than minimal tree. Although it can be demonstrated that this condition will occur in an extremely small number of configurations, a simple correction technique has been devised and warrants consideration for inclusion in an automated wire assignment program:

> If the sum of the weights (or distances) of any two segments consecutively selected exceeds the sum of the weights of any two unused segments between the selected pair, a better solution is probable and a systematic substitution is warranted.

This is illustrated in the example by the first-pass selection of E5-G5 and G5-D7. Total length of this selection is 5.6 units. The total of unused segments C5-D7 and D7-E5 is 4.5 units. Systematic substitution results in the minimal tree of Fig. 13*d*.

It is assumed in this example that wire-routing constraints are such that the minimum physical separation will produce a minimum interconnection path. If shielding or special channeling void this assumption, special weighting factors must be imposed on the distance calculations prior to selection of the segments.

Wire Leveling. Wire-wrap technology requires that, when a wire is unwrapped

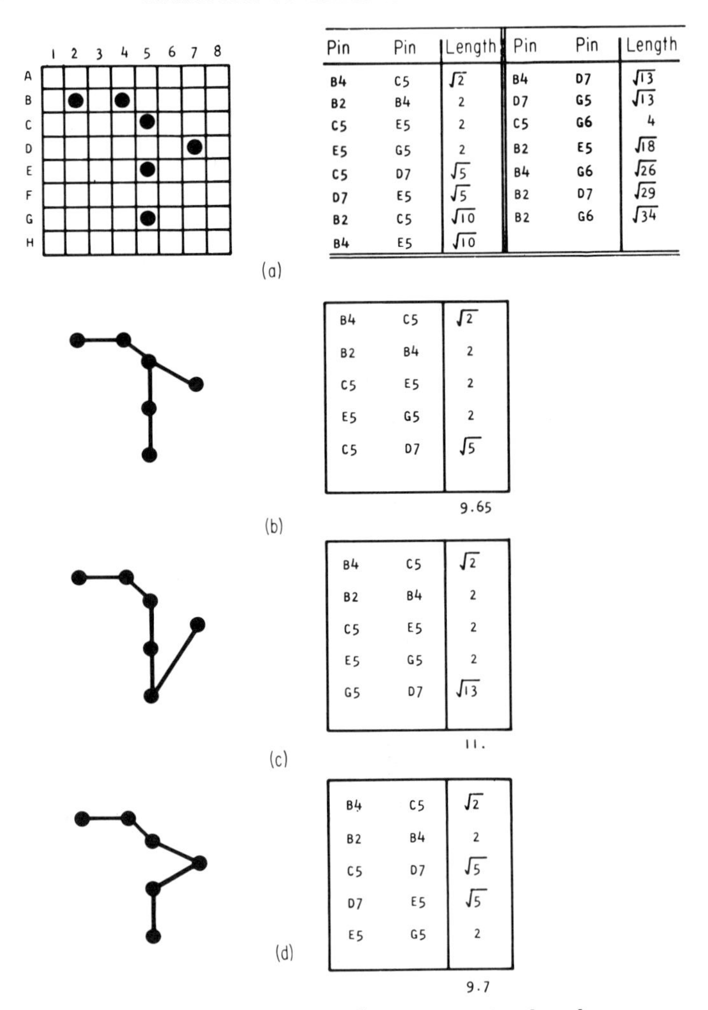

Pin	Pin	Length	Pin	Pin	Length
B4	C5	$\sqrt{2}$	B4	D7	$\sqrt{13}$
B2	B4	2	D7	G5	$\sqrt{13}$
C5	E5	2	C5	G6	4
E5	G5	2	B2	E5	$\sqrt{18}$
C5	D7	$\sqrt{5}$	B4	G6	$\sqrt{26}$
D7	E5	$\sqrt{5}$	B2	D7	$\sqrt{29}$
B2	C5	$\sqrt{10}$	B2	G6	$\sqrt{34}$
B4	E5	$\sqrt{10}$			

(b)

B4	C5	$\sqrt{2}$
B2	B4	2
C5	E5	2
E5	G5	2
C5	D7	$\sqrt{5}$
		9.65

(c)

B4	C5	$\sqrt{2}$
B2	B4	2
C5	E5	2
E5	G5	2
G5	D7	$\sqrt{13}$
		11.

(d)

B4	C5	$\sqrt{2}$
B2	B4	2
C5	D7	$\sqrt{5}$
D7	E5	$\sqrt{5}$
E5	G5	2
		9.7

FIG. 13. Minimization of interconnection lengths.

from a pin, it can never be rewrapped. It must be replaced. Improper initial leveling of the nets can result in extensive rework for a simple change.

Figure 14*a* shows a sawtooth wrapping arrangement of six wires. Should it be necessary to remove wire 2, wires 3 through 6, four additional wires, would also have to be removed and replaced.

Figure 14*b* shows the same interconnection arrangement, but both ends of each wire have been wrapped at the same level. Now if it is necessary to remove any wire, a maximum of two additional wires would be affected.

Assignment of wrap levels is a trivial addition to the wire-assignment program. The assignment is arbitrary as long as both ends of the wire are at the same level. Once assigned, the wrap level becomes the major sort field for an installation sequencing of the wire list.

Etched-wire Connections. Computation of a set of etched-wiring segments which satisfies the interconnection requirements is a demanding task. The process is somewhat analogous to designing wire routings for back panels in that a passable rout-

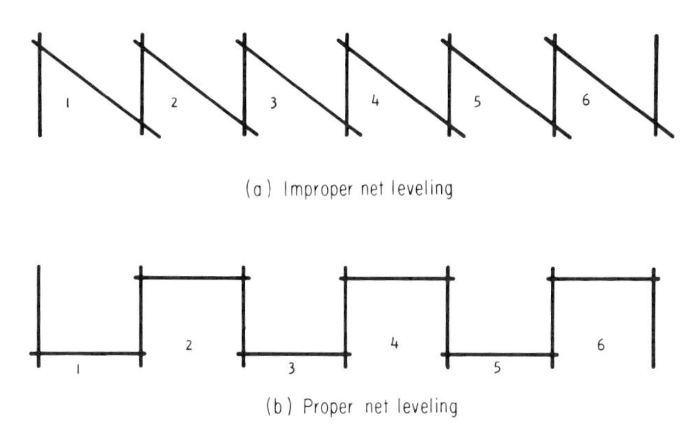

FIG. 14. Improper and proper net leveling for wrapped-wire connections.

ing must be determined between two points, the difference being that, in place of freestanding insulated wires, conductive paths are etched or deposited onto the surface or surfaces of a dielectric material. Conductors in nonelectrically common nets must be physically separated. This requirement demands the use of discrete wire jumpers or multiple surfaces even for relatively small circuit requirements. Cost and manufacturing considerations normally restrict the number of surfaces, hence severely restricting the solution and frequently causing the solution to become algorithmically indeterminate.

In order to compute the interconnections, it is necessary to create a model of

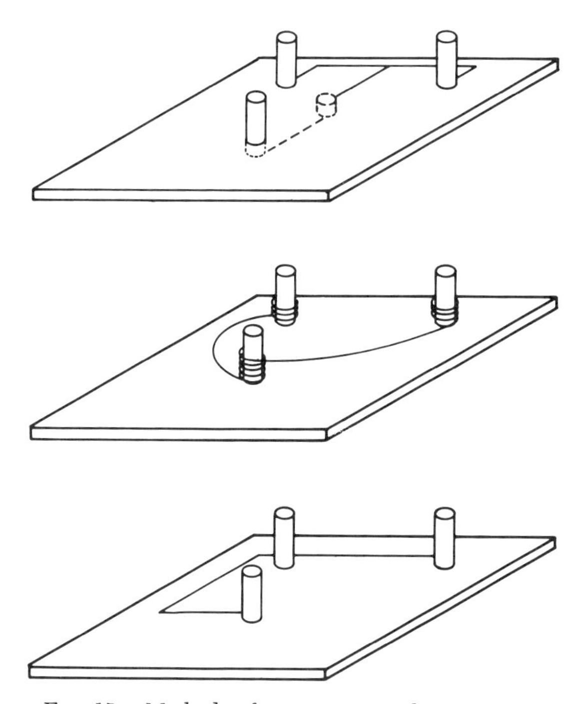

FIG. 15. Methods of connecting a three-pin net.

the surface within the computer. This model can be visualized as a grid or matrix superimposed on the surface, each cube or cell within the grid representing one unit of interconnection space on the surface. The grid spacing is determined by the unit of connector material allowed or the minimum separation between connectors. This grid can be represented in memory in a *list* form or more naturally in a matrix, where each cell is represented by a storage character.

The initial inputs to the routing programs supply the coordinates of path obstructions, feed-through hole patterns, and the location of component or module connections on the surface. These data are mapped into the grid pattern by encoding the corresponding character positions. This input is followed by a list of common connection points. As routing paths are determined, they are also mapped into the matrix-forming obstructions for subsequent connector paths.

While numerous procedures exist for determining paths within the surface maze, they are frequently highly package-oriented. The most universally adaptable algorithm published with variations under several names can best be visualized as a wavefront propagation. Starting from a terminal cell 1, the procedure is to move outward in all directions. Empty cells adjacent to the first are labeled 2, unlabeled cells adjacent to 2s are labeled 3, etc. This outward progression, like a wavefront in water, continues until the circuit is completed by the wavefront encountering a cell previously used for this common network or a desired terminus.

When a terminus is reached, a path back to the origin is traced along contiguous descending numbers so that only one cell of each numeric value is used.

Figure 16 shows a surface mapping. Alpha encoded cells represent path obstructions, including previously established circuit paths. A wavefront has been propagated from the upper right and has established contact with the desired terminus on the left. A path of contiguous numbers has been traced back to the origin. This path will now be encoded to block subsequent usage and establish this connection. All other wavefront cells will be cleared.

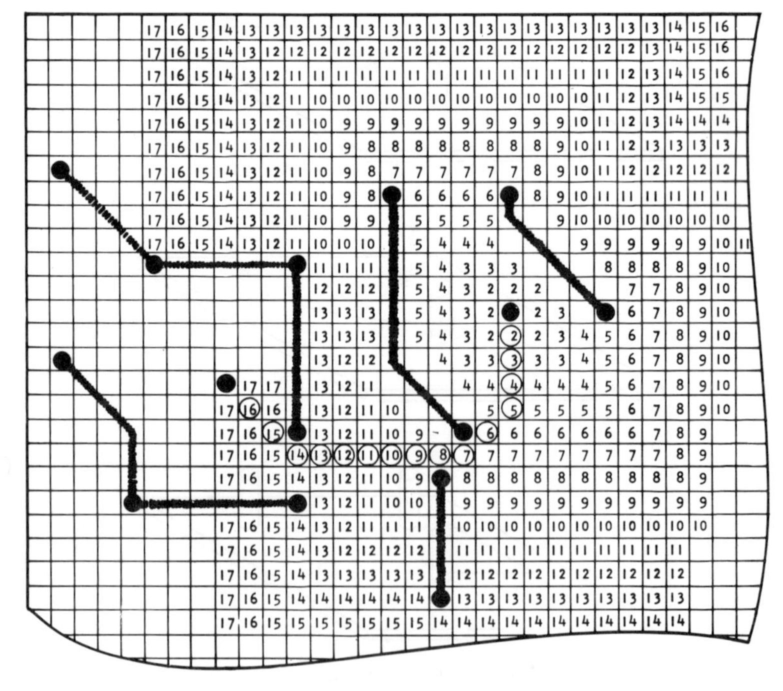

FIG. 16. Wavefront routing.

It is easy to note that circuit paths can be readily blocked. The sequence in which lines are laid is important to develop maximum effectiveness of the algorithm.

Integrated-circuit Design. Integrated-circuit design is a much more difficult task than the design of a conventional printed-circuit board. Much of the work consists of the geometric layout of diffusion and interconnection patterns required to fabricate the circuits within the base material. All elements must be arranged into a minimum area with proper allowance for interelement isolation such that interconnections can be routed without metallization crossovers. Errors are not likely to be detected until the first devices are made, and it is rarely possible to correct them.

This complexity has led the pioneers in this area to the development of man-machine interaction systems with direct graphic output. The circuit schematic representing all components and their connections is stored in the computer, either from a previous circuit-analysis application or from input supplied for this application.

The engineer selects and identifies a substrate by part number to the system. A preliminary layout is made either by manual assignment or by the computer. In either case, the system should automatically generate the connections implied in the schematic.

With the preliminary layout the operator can manipulate or shift the position of components to eliminate crossovers and ease manufacturing or other considerations. This manipulation has been performed with language inputs and plotter outputs. A more effective system employs the cathode-ray tube and a light-pen technique discussed later in the chapter. In either case, the layout is constantly checked against the schematic for accuracy and completeness. The final product is the master artwork prepared on a plotter for photographic processing.

DESIGN DISPLAYS

Design Output Techniques. An important part of a design automation system is the presentation of the problem solution. Vast numbers of data are captured in the computer storage. Effective economic systems must be employed to present information in a useful manner. Alphanumeric data such as bills of material, where-used traces, and other classical types of documents can be readily produced by high-speed printers.

Point-to-point wiring and pluggable-unit placement charts in alphanumeric list form have been proved to be useful and informative substitutes for functional schematics. In list form, these data can be resequenced, edited, and formatted for use by engineering during design, for quality control, for functional approval, and as actual fabrication and assembly documents or can even be converted to numeric controls for direct operation of wiring or drafting machines and circuit test devices.

Engineering changes can be rapidly updated and the documents generated and distributed to all areas of concern, a distinct advantage for the highly volatile data of electronic design. Lists also permit greater density of information per page than schematics, thus reducing total paper volume.

Some forms of data required for electronic systems presentation, however, do not lend themselves to alphanumeric listings. It is difficult to visualize the topological relationship of a circuit card from a data list. A graphic presentation is required. There are three basic types of output devices which can and have been used to produce pictorial output from computer-aided design systems: line printers, plotting or drafting machines, and cathode-ray tubes. All three devices require that a digitalized internal representation of the systems space be projected upon the planar surface of the output media.

Line-printer Diagrams. A line printer is a device designed to produce a series of images on paper at discrete intervals across a fixed span. The image may be developed by mechanical impact through an inked ribbon or by chemical deposition through static attraction or other nonmechanical phenomena. In any case, a fixed number of printing positions allow a fixed number of discrete characters to be developed along an image line. Computer installations, almost universally, have at least one line printer attached. Typically, these printers provide 132 characters/line at $\frac{1}{10}$-in. character spacing, with 10 to 12 lines/in.

While line spacing can be varied to some degree, the character spacing is normally fixed by the width of the mechanisms required to develop the character. Thus, the capability of producing scale drawings is extremely restricted. A significant part of the graphic requirements of a system, however, can be presented as nondimensioned, nonscaled diagrams. Schematics and logic diagrams, for instance, do not require scaling capabilities.

Diagrams of this nature can be produced directly from a model in the computer storage. The most direct method of producing these images is to establish a matrix of characters within the computer storage equal to the number of characters which can be produced on a single sheet of copy, the columns of the matrix being equal to the number of print positions per line and the rows equaling the number of lines per page. Each element of print, including blank spaces, has a unique one-for-one correspondence to a storage cell. The image to be printed is thus mapped into this storage with the character representations used which are required by the computer hardware to generate the desired character on the printer.

More sophisticated drawing programs utilize list structures or plex models to develop the image within the computer storage. These techniques allow larger images to be developed before printing and usually provide greater efficiency in processing the image. Character image mapping remains an effective technique and should be employed for initial attempts at graphic development.

Figure 17 shows a line-printer-produced logic diagram. These diagrams can be produced effectively through the development of a character image in the core by using the maze running techniques similar to those described for printed-circuit generation.

In procedures for generating printed diagrams of this type, the layout or placement of the major symbols is normally retained from the designer's original sketch. In transcribing the sketch for computer input, the coordinate location of the symbol is used to provide unique identity to the symbol, as well as to control its location for reproduction. By retaining the designer's layout, not only is the image-generating program simplified, but the designer's mental image of his idea is not deformed.

While Fig. 17 shows the use of rectangles for all switching circuits, other shapes may be desired. This can be accomplished by the development of macro images. That is, a set of characters is defined to represent a given function and is prestored in the system. This macroshape is then called up and mapped into the character image for each occurrence of the function.

While the aesthetic value of the logic diagram of Fig. 17 can be criticized, its economic value cannot. Diagrams of this type can be key-punched more rapidly and more economically from freehand sketches than they can be drafted. The data contained in the diagram, when committed to the data base, represent the total electronic design. Changes can be incorporated into the total design by merely inputting the change data. The data base is updated, and new documents are printed at the rate of over 1,000 lines a minute. The speed of printing, furthermore, is not affected by the density of the information on the page.

In addition to logic diagrams, pluggable-unit location charts, circuit schematics, cabling diagrams, charts, graphs, functional diagrams, and many other graphic displays, which do not require continuous scaled accuracy, can be satisfactorily produced on the line printer.

The speed of the line printer in producing the schematic type of sketches has led to the development of extended character sets. These character sets are constituted of image segments which can be combined to produce more pictorial representations than the normal alphanumeric characters.

Figure 18 shows a group of characters or image segments designed for graphic applications. These segments, when combined in a printing system with alphanumeric characters in both horizontal and vertical orientations, can provide an extremely variable graphic capability.

Plotters and Drafting Machines. Applications which require the generation of scaled, dimensioned drawings can be accommodated through the use of plotters or drafting machines. These devices are not restricted by the discrete print positions

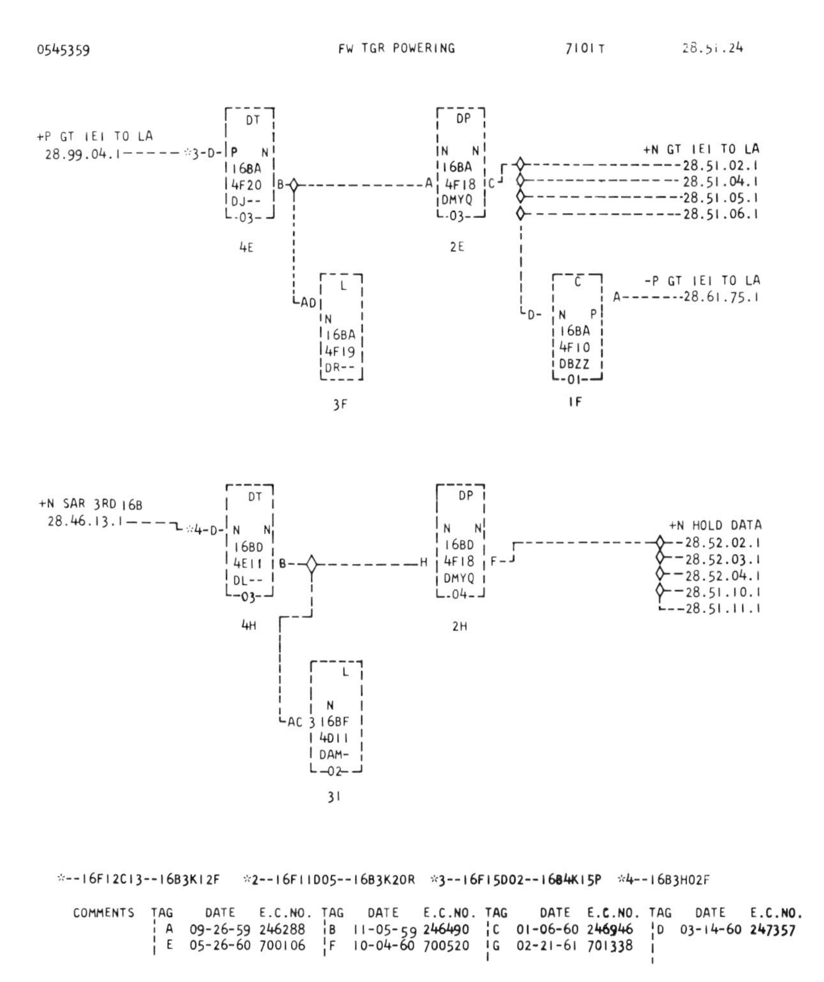

FIG. 17. Machine-printed logic page.

and limited character sets of the line printer. Instead, they utilize digital impulses generated by the computer to control a servo system. In this manner, a pen or stylus can be directed across the drawing surface to provide a continuous line.

A wide range of devices are available. Each has characteristics which require evaluation in light of the applications planned. Bed size, speed, accuracy, line width, and repeatability are factors which vary greatly. All these factors interact with the cost of the device, and thus price vs. performance of a particular machine must be evaluated on the basis of the application.

As in all other parts of the system, the total process must be kept in mind in considering graphic requirements. Aesthetics must frequently be sacrificed to economy. Artwork produced oversized and photographically reduced can be generated with less accurate equipment than would be required to produce exact sized masks. In all cases, the overall drawing completion time is a direct function of the density of data on the diagram. Alphanumeric data frequently must be generated as a series of discrete pen strokes, as with manual lettering. Extensive use of alphanumeric data on a document generated by a plotter of this type will greatly affect the eco-

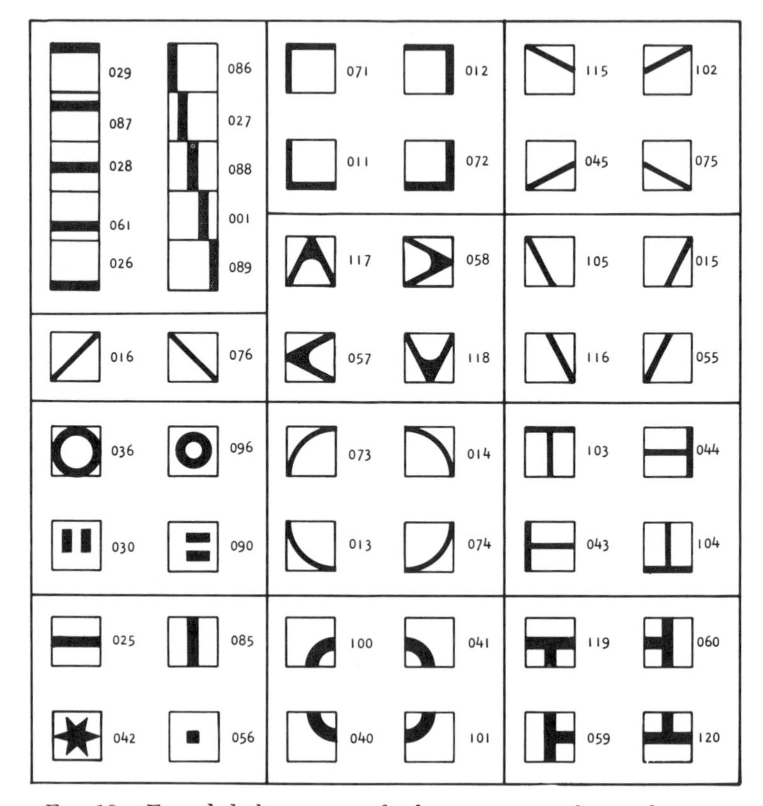

FIG. 18. Extended character set for line-printer graphic applications.

nomics of the system. Where possible, these data should be produced on the line printer as a separate document to accompany the plotter-generated drawing.

Normally the same techniques previously described for layout of printer output can be employed for generating the internal image for drafting machine or plotter output. Additional programming is required to convert this image to the numeric controls necessary to drive the specific plotting or drafting device.

In addition to their use for drawing output, similar devices have been employed directly in the manufacturing process. In these applications, a high-intensity light source is used to expose a photoresist on the circuit board to create customized circuit patterns. Line widths can be controlled mechanically by varying the focal distance of the light beam. In this manner, a large variety of circuit patterns can be created without the added time and expense of artwork creation and photo master generation.

Image Tubes. The third device available for graphic output from computing systems is the cathode-ray tube. The primary advantage of this device is the rapidity with which a numeric model can be displayed. With the tube coupled to a high-speed computer, images of electronic waveforms or mechanical components can be made to respond dynamically to mathematical manipulation of the numeric model. The calculated effect of parameter changes can be viewed as live action on the screen.

Circuit analysis programs, for instance, have been developed to depict the results of their calculations on the tube face such that the results would be identical in appearance to an oscilloscope image signal coupled to the physical circuit. The effect of individual changes in component parameters can be calculated and dynamically viewed as changes in the waveform.

Light Pens. The cathode-ray-tube facility has been further enhanced as a design tool by the development of the light pen, a pencillike device which, when held in close proximity to the face of the image tube, can be detected by the computer circuitry. Programs have been devised which allow this light pen to be used to "draw" images on the cathode display. As the light pen is moved across the surface of the tube, the computer program follows the motion. As image segments are drawn, a numeric model is created in the computer storage. This numeric model, in turn, is used to regenerate the image segments on the display continuously.

These devices offer the engineer the ability to view his design in detail as it is developed and, at the same time, create input data. In areas of topological arrangement, aesthetic design, and nonconvergent problem solution, it allows the man and machine to work in the closest possible manner, employing the most descriptive of languages—graphics.

Microfilm Recorders. Microfilm recorders provide a means of creating hard copy of image-tube outputs. Under computer or manual control, outputs displayed on an image tube can be photographed for the permanent-microfilm type of storage. Where visual display is also required, the microfilm recorder employs a second image tube slaved to the display tube. This tube is enclosed in order to ensure a quality film image.

Design Console. Design consoles made up of the typical components above provide an effective method of man-machine interaction for the solution of nonconvergent design problems. A full-design console consists of an image display tube, a microfilm recorder with a rapid film transport and developing process, a microfilm display screen, a light-pen capability, and two keyboard input devices. A standard typewriter keyboard is used for inputting alphanumeric data to the computer program and, hence, to the image. Additionally, a *program-function keyboard* is provided.

This keyboard provides identifiable signals to the computer program. The programs are written to respond to these inputs by initiating a specific function. Erasure of a line segment on the image tube, for instance, can be accomplished by pointing the light pen at the line and pressing a key programmed to initiate an erase function. Function keys can be programmed to augment the ability to draw mathematically straight lines, circles, parallel lines, etc.

The microfilm recorder is used in the design console to record intermediate decisions or promising solutions to the design problem. Developed and projected on the viewing screen, these intermediate solutions can be used for comparison with the active image. The film transport system allows any previously recorded image to be displayed on the screen. Thus, the design console represents an integrated facility for dynamic design. It facilitates the interaction of man's creative imagination with the high-speed mathematical logic capabilities of the computer. This combination reduces the all-important engineering time to decision to a minimum.

MANUFACTURING AIDS

Specialized Documents. In addition to the classically released documents, a computer-aided design system can provide many additional specialized documents and reports to a manufacturing facility. A few of these have already been described; installation sequenced wire, wire list, rework wire list, bill-of-material documents.

Additionally it has been found that performance in many fabrication and assembly operations can be improved with the generation of specially edited and formatted documents, documents which by manual methods have never been considered because of cost and possible error generation. These operations were forced to gather pertinent data scattered for their purpose in various areas of the standard document or drawing. High-speed printing and data manipulation drastically reduce the cost justification required to warrant a specialized data organization for individual operations.

Numeric Controls. Maximum value of an engineering design system can be obtained only when the data captured during the design phase is utilized directly for production of the product. The parts-list data base built up during design can be exploded directly into the manufacturing requirements planning; numeric descriptions of interconnections, hole patterns, cable details, etc., can be postprocessed into numeric controls. In fact, fabrication and assembly of electronic subassemblies can be completely controlled by the numbers. The design engineer can, in effect, build and test his product without further manual interference.

Fabrication and Assembly. The digitalized data base developed to describe package implementation can be processed directly into tool controls. Etched circuits can be drafted, hole patterns drilled, components inserted, and discrete wire connections made from the available data. Subsequently, these data can be used to direct extensive and accurate tests of the circuits produced. Short and continuity checking are readily accomplished from the interconnection data base. Functional and reliability testing can be developed from the logic of the interconnections.

The abundance of these data in numeric form provides for the capability of coupling the computer directly to the manufacturing facility. With on-line computers, it is possible to develop a continuous-process, semidiscrete manufacturing facility in which no two items would be necessarily identical. Basic material, as it entered the process, could be destined for a specific part number and change level subassembly. All processes in its fabrication and assembly could be directed by on-line computers.

Material Handling—Dispatching. The parts list developed in the data base is exploded and processed to net requirements. Daily build schedule is developed against a requirements forecast. Materials flowing into the line are dynamically destined for a specific subassembly. All materials are dispatched in accordance with stored alternative routings into the lowest available queue. Items rejected by on-line testing can be reinitiated consistent with build priorities.

Testing. Interconnection data developed as part of the computer-aided package design can be utilized to provide numeric controlled short circuits and continuity checking. Where multilaminate etched circuits are produced, it is important to ensure the adequacy of each circuit plane before it is sealed by the addition of the next laminate. Test equipment designed to input a current into a circuit and measure all other contacts for short circuits and continuity is being successfully employed.

Impedance testers to check the generation of impedance across independent circuits can also be controlled from data developed in the design system. Finally simulation of designed logic provides diagnostic input/output patterns to ensure proper functioning of the circuits.

On-line control of testing devices provides for dynamically following performance trends. Applications for testing drilled-hole patterns, for instance, have been developed such that the wear on each individual drill bit is followed in order to predict tool failure. An optical scanner is programmed to measure a series of radii on at least one hole from each drill for each laminate. These raw data are reduced by an on-line computer to compare the actual center of the hole to the design center. Burrs and the eccentricity of the hole can be measured. A statistical-trend line is maintained within the computer which will allow the machine to be shut off and maintenance ordered before rejects are created. Establishment of the trend will allow random failures to be passed without shutting down the facility. The trend also provides for a weighting factor to be applied to the sampling factor so that, as the bits wear, the number of tests per tool or per laminate can be increased. In addition to tool wear, such factors as table alignment can be measured and controlled.

Process Control. With micro miniaturization and integrated circuits, electronic packaging is becoming more and more dependent on chemical processes. Many of these processes are highly critical and require continuous monitoring of temperature, viscosity, pH, etc. Engineering limits established for these processes can be monitored and controlled by direct connection of the instruments through analog to digital converters to a process control computer.

While these data are not inherent in the design data base, the justification for an on-line computer can be increased by its ability to monitor and control these operations.

Engineering Change Control. Automation is the fastest method known to produce scrap. High-volume production based on obsolete engineering data must be reworked or scrapped. Hence, one of the most valuable aspects of engineering design data processing is the ability to develop and document required engineering changes rapidly and accurately. Where the manufacturing facility is geared to accept the engineering change data in numeric form directly into its systems, the change effectivity can be almost immediate. In multiplant or remote design control complexes, data-transmission facilities can be incorporated to provide rapid, accurate distribution of design changes.

Production-hardware Selection. In choosing automated fabrication, assembly, and testing equipment, consideration should be given to the control input requirements. Many times the real cost of procurement and usage of such equipment is hidden in the amount of effort required to provide control data. The more expensive equipment may embody control circuitry which could be readily generated by a general-purpose computer from the data base in the computer-aided design system. On the other hand, less expensive equipment may require extensive, complex programming in order to develop needed controls. Programming personnel familiar with the data base can provide significant insight into the compromising of this decision.

COMPUTER-HARDWARE CONSIDERATIONS

In considering a computer-aided design system there is no single "right" way. There is no single item of hardware which will do the job. Many decisions must be made concerning the ultimate and immediate extent of the system, the rate of development, the mode of operation. A reasonable job of wire-list maintenance, for instance, can be done with basic unit record equipment. Interconnection requirements can be transcribed to punched cards. Many necessary file maintenance operations can be performed by hand. Release and production listings can be made on readily available unit record sorting and tabulating equipment. Of course, the greater the automation of the system and its utility to the user, the greater the hardware/software demands. Many large producers have profitably employed multiple-interconnected computers with many access terminals to provide maximum utility for both design and manufacturing.

Operating Environment. Among the many important decisions to be made in the operating environment is the question "How will the computer be integrated into the design system?" Various degrees of utility can be provided to the individual user. Each increment of utility carries an appropriate price/performance compromise.

Batch Processing. The most common approach to data processing has been batch processing. This is a familiar concept and follows closely manual methods. A complete set of data is submitted by internal mail to a service group for processing, drafting, typing, etc. Jobs are processed serially, the execution of the first being completed before starting the next. This method provides an orderly approach to data processing with a minimum of hardware and manual intervention. Turn-around times can be extensive dependent on the queue and the efficiency of the mail service.

Remote Batch. This term has developed to describe a system wherein data are transmitted from a remote site directly into the centralized processing facility. The processing facility queues these tasks and executes them in a sequential batch mode. This technique reduces the delays of intermediate handling in mailrooms and by computer operating personnel. Since the data queue is maintained within computer-controlled storage, operating systems have been devised to provide complex priority scheduling and control. Significant improvements are frequently gained in central processor efficiency, in addition to turn-around improvements to the user.

Time Sharing. With various modifications and emphasis, time sharing reflects the ability of a computer to be switched from one job to another in extremely small time increments, always keeping track of where it has been and where it is going, always being able to come back to an interrupted task and continue execution from the point immediately following the last successfully executed operation. Hence, the available time on the computer can be optimized by instructing the computer to switch to another task when delays are encountered. These delays occur frequently, especially where large volumes of input or output are encountered. For design applications where manual input devices such as typewriter keyboards or light pens are employed, these relative delays can be extensive. Thousands of computer instructions can be executed between the key strokes of even the most accomplished typist. When these delays are augmented by the normal thought processes of the engineer developing a design, many terminals can be serviced by the computer in a manner that it appears to each user that he has complete and immediate access to the entire computer.

This technique, when coupled with a teleprocessing or communication facility which allows the individual terminals to be located at remote sites, provides extreme flexibility for design communication.

Operating Systems. Each of these environments requires a specialized operating system. The operating system is normally supplied by the manufacturer to control the sequencing of task within one or more of these operating environments. The function of the operating system is basically to provide the necessary data management, scheduling, and control of the computing facility. These systems vary from simple monitors, which recognize the end of one job and initiate the next, to large, complex systems, which divide the available time among many separate jobs, controlling multiple computers servicing a multitude of users on a time-shared basis. Balancing of the operating system to the requirements of the computer environment is most important for efficient use of the computer facility.

DATA-BASE STORAGE DEVICES

Cards. The well-known punched card provides utility for both data and program input/output communication. The unit-record nature of the card allows for manual sequencing and editing. Although many newer devices have challenged, and will, the supremacy of the card, it will undoubtedly remain in most systems for a long time to come. Inasmuch as the card is almost universally accepted as an input to computing systems, the effort expended in the early establishment of unit-record wire-list capability, for instance, is not lost when the system is converted for computer processing.

Punched Tape. Punched tape is used as an input/output medium for computers. Its continuous, permanent, sequential nature makes it awkward to create and extremely difficult to modify. It is therefore less desirable for transient data. Probably its most effective use has been as a numeric control medium. Sequential operating instructions generated in numeric form for the control of fabricating, assembling, or testing tools can be readily transferred to punched tape. The tape, either paper or plastic film for more permanent usage, provides not only a practical input medium to the tool controller but also a handy, economical storage device.

Magnetic Tape. Long a favorite of computer users, magnetic tape consists of a reel of ½- to 1-in.-wide high-strength plastic material coated with ferric oxide. Data are recorded by magnetic orientation of the ferric material. Millions of characters can be written on a single reel of tape. Data can be transferred between the tape and the computer at rates in excess of 100,000 characters/sec. Tape reels are easily exchanged on the driving mechanism for shelf storage.

Magnetic Disk. These devices consist of a stack of metallic plates similar in appearance to a commercial record player. The recording principle is similar to magnetic tape in that data are stored in the form of magnetic orientation within a ferric coating. Unlike tape, which must be processed sequentially from one end to the other, disk storage provides direct access of the read/write head to any record block. Where the processing job requires a random type of access to data, the use of disks for the storage of the basic data set will greatly enhance the performance of the system.

Disk storage devices are available in two basic configurations. The first provides a large capacity for data recording permanently maintained in the read/write mechanism. The second provides a smaller capacity for data recording on easily replaceable disk units.

Magnetic Drums. Drums are similar to magnetic disks except that the recording surface is on the periphery of a cylinder instead of the flat record surfaces. These devices tend to be mechanically more simple at the sacrifice of physical space.

Data Cells. Data cells were created in order to take advantage of the inexpensive nature of magnetic-tape recording media and to provide a degree of random accessibility of the data store. Magnetic-tape material has been cut into strips or sheets. These are suspended in a cartridge, many of which can be mounted on a mechanical handling device. This device can then select any individual section of tape and position it under a read/write head. Normally these mechanisms are considerably slower in access and data transfer time than disk units. They do, however, offer an economical method of mass data stored with a random access capability.

Microfilm. Microfilm output has long been available from computers. Data, either alphanumeric or graphic, can be photographed from the face of an image tube for microfilm storage.

Microfilm has also been used as an input medium. Microfilm images exposed to an optical scanner can be interpreted by a computer program. Graphic data scanned in this manner can be converted to a digital description for subsequent processing.

Microfilm filing cabinets have also been devised which, when connected to a computer, can access and copy any one of thousands of microfilm images in seconds.

SOFTWARE AND PROGRAMMING CONSIDERATIONS

Man-Machine Symbiosis. The use of the computer by the engineer is highly dependent on the ease with which he can communicate his problem to the computer. A practicing engineer who must run as fast as he can to remain abreast of his technology cannot be expected to remain proficient in the art of computer programming. In spite of the fact that the trend in a large portion of systems development work in recent years has been toward improving the relationship between the design engineer and the computer, it is to be expected that every installation must have a full-time staff of at least one competent programmer if full utilization of capability is to be realized.

Programming Languages. A job to be executed by a computer must be defined in minute detail. The computer itself is a numerically controlled tool. It is dependent on a string of detailed instructions of where to obtain, how to manipulate, and where to put data. These detailed instructions represent the *machine-language program.* Programming languages should not be confused with problem-oriented (or user) languages.

Assembly Language. Long ago it was recognized that the computer could assist in the translation of a program language more natural to the human programmer into the numeric stream required. Early developments along this line provided what is known as *assembly languages.* The assembly language is a sequenced step-by-step statement of the computer operations required. The data and computer operations, however, are given symbolic names. These names are much easier for the programmer to remember than a storage address. These symbolic instructions must then be passed through the computer under the control of an *assembly program* which translates the symbolic operations to numeric instructions and assigns data fields to specific storage locations. The resultant machine language or *object program* can then be executed on the computer.

Compiler Languages. The next step in automating the programming task is known as *compilation.* This facility allows the user to write programs in a highly stylized English (Cobol) or algebraic notation (Fortran, Algol).

This technique removed the programmer one step further from the machine-oriented instruction. The translation program required to convert these languages can be and has been implemented on a wide variety of computers. In many cases a program written in Fortran, for instance, can be compiled and executed without modification on a wide range of computers of competitive manufacturers.

Problem-oriented Languages. The term *problem-oriented language* refers to a technique which provides for direct execution of a series of preprogrammed modules based on a set of instructions and parameters native to the problem. It being recognized that each industry, each discipline has an individual semantic requirement, these languages must be tailored to a specific task. Frequently the preprogrammed modules may be used by different disciplines. In order to minimize the training of engineering users, they are given different titles. This language is then interpreted by the computer, and the required processing is performed by predefined program modules stored within the computer complex.

Graphic inputs represent the ultimate in this approach. Table 2 relates the various levels of programming languages as they have developed over the years.

TABLE 2. Programming Languages

Program type	Characteristics	Level	Examples
Object or machine language	Detail numeric step-by-step machine controls	Basic	
Assembly code	Symbolic operations step-by-step pseudo operations	Assembled or translated by the computer to machine language before execution	Autocoder, FAP, MAP, SAP, SOAP
Compiler languages	Stylized English algebraic equations	Compiled to assembly language by compiler program	Algol, Fortran, Cobol, PL/1
Problem-level languages	Parameter inputs only	Used normally to link precompiled or assembled modules	Ecap, Stress, Cogo

Application Modules. Preplanned and prepackaged application modules provide a wide range of utility for specific classes of problems. Mathematical and statistical functions are readily prepackaged in such a program and provide a facility for the solution of problems in any discipline.

More specific packages are also frequently available from user groups or supplied by the manufacturer. These programs provide significant generic portions of detailed data-processing applications.

User Libraries. Corporations which employ computers have, for a long time, recognized the advantages of a cooperative interchange. User groups are made up of representatives from installations employing similar hardware for similar problems. Standards are developed, and committees are formed to attack problems jointly and to pool resources for the development of large programmed systems. There is a relatively free interchange of programs and techniques. Programs created by individuals are submitted for distribution to any other member having a need or desire for them.

SYSTEMS CONSIDERATIONS

System Justification. The true value of a design automation system can rarely be measured in advance in actual dollar savings. The justification of such a system is necessarily involved in its ability to improve the product and deliver it sooner into the marketplace. This is not to say that dollar savings are not realized in the design area. While technical personnel are rarely replaced by such a design system, the output per man can be greatly improved. Routine drafting, calculation, and documentation can be drastically reduced. Additionally, many expensive breadboards and physical models can be eliminated through the use of mathematical simulation.

A very important cost savings which is frequently overlooked is material savings within a product. The overspecification and overdesign commonly practiced because of pressure on design time when manual methods are used can be eliminated or greatly reduced. It is not uncommon to justify the system in the cost-reduction area alone through improved material utilization as compared with past designs.

Organizational Considerations. A design automation system will eventually involve all segments of the manufacturing complex. While the computer-aided design aspects can and should be controlled under the engineering organization, provisions must be made for a cooperative and compromising effort between the engineering computing and the manufacturing requirements.

In a multiplant multilaboratory complex, serious considerations must be given to an organization which will ensure maximum coordination between the efforts of data-processing resources.

In short, it is relatively unimportant to whom the data-processing personnel report as long as the responsibility for the continual flow of data across organizational lines is recognized.

Cost Distribution. Viewed on the basis of the classical departmentalized budget, the free flow of data in the system produces an imbalance between cost and savings. Although major financial advantages can be shown throughout the system, the bulk of the savings will appear in the downstream operations, operations for which data are now available in a rigorously checked, immediately available, processable form. On the other hand, in some design areas, owing to the requirements for data input, the amortization of the cost of programming and data conversion may not appear to justify the expenditure. This frequently imposes an obstacle in the development of a comprehensive integrated system.

Several budgetary techniques have been used to redistribute data-processing expenses. One of the most satisfactory of these is the establishment of a *cost center*. The cost center represents a nonprofit service which is funded only for initial development. All users of the system are then charged for their outputs or use of the system. Charges are based on the relative value—not the *cost*—of the services performed.

This arrangement can provide a self-perpetuating expanding system with the development cost being spread over all users. Additionally it ensures against the perpetuation of unnecessary and unused reporting. Where a receiving department is forced to pay for a report, its usefulness will be continually reexamined.

Where to Start. There is no universally good starting point for the development of a design system. Although a good solid data base will be required, its original data set can be restricted to that which allows the solution of a particular problem The important consideration in initiating a design system is that data have no boundaries. Data generated for one problem area are always required for subsequent operations. All applications should be planned to work away from and into this common pool. With this in mind, the starting point can be chosen anywhere. The amount and rate of investment and the amount and rate of payoff must be measured as with any other sound capital expenditure.

The interconnection list represents one of the fastest, most profitable places to start. The wire list is extensive, detailed, and highly sensitive to engineering change. Highly satisfactory systems are obtained with a minimum investment in personnel or equipment. Hence, wire-list processing represents the normal starting point for most electronic design systems.

SUMMARY

Evolution—Not Revolution. The system of design and manufacture will always require people. These people must be trained in new techniques. Attempts to radically change the mode of operation in a design and manufacturing complex can be seriously hampered if the preparation of the personnel directly and indirectly affected is not properly completed. While the rate of evolution can be quite rapid, it should not approach the violence of a revolution.

Future. The rapidly developing technology of both electronic systems and design systems promises a future of inexpensive quality products beyond even the most lively imagination. The computer, itself a product of the dramatic evolution of electronic technology, holds the key to the continued development of that technology into a myriad of products too complex for the comprehension of an individual developer. Only with an understanding of the synthesizing abilities of the computer can these products be postulated.

REFERENCES

1. Pomeotale, T.: An Algorithm for Minimizing Backboard Wiring Functions, *Commun. ACM,* vol. 8, pp. 699–703, November, 1965.
2. Russell, S.: Algorithmic Concepts for Circuit Layout by Digital Computer, *Tech. Paper,* 1963 *IBM Workshop on Routing and Placement.*
3. Shannon, C. E.: Presentation of the Maze Solving Machine, *Trans. of the 8th Cybernetics Conf.,* Josiah Macy Jr. Foundation, New York, pp. 173–180, 1952.
4. Stone, A. J.: Partitioning of Logic into Physical Entities, *Proc. SHARE Design Automation Workshop,* 1966.
5. IBM, Data Processing Applications Manual E20-0119-0, entitled "Three Dimensional Placement and Routing."
6. Warshawsky, E. H.: Optimization Problems, 1964 *Proc. of Design Automation Seminar,* pp. 69–105, sponsored by MESA Scientific Corp., 2930 West Imperial Highway, Inglewood, Calif.
7. Weindling, M. N.: A Method for Best Placement of Units on a Plane, *Proc. SHARE Design Automation Workshop,* 1964; also Douglas Paper 3108, Douglas Aircraft Co., Santa Monica, Calif.
8. Chu, Y.: An ALGOL-like Computer Design Language, *Commun. ACM,* vol. 8, pp. 607–615, October, 1965.
9. Roth, J. P.: "Systematic Design of Automata," *Proc. FJCC,* pp. 1093–1099, 1965.
10. Stroupe, H. C., and A. I. Perlin: Standardized Topology—An Approach to Simplified Digital Circuit Design, *Solid-State Design,* vol. 5, pp. 41–45, August, 1964.
11. Special Issue on Automatic Testing Techniques, *IRE Trans. on Military Electronics,* vol. MIL-6, July, 1962.
12. Breuer, M. A.: The Formulation of Some Allocation and Connection Problems as Integer Programs, *Naval Research Logistics Quarterly,* vol. 13, pp. 83–95, March, 1966.
13. Evans, D. H.: Modular Design—A Special Case in Nonlinear Programming, *J. Operation Research Soc. of Am.,* vol. 11, pp. 637–647, July–August, 1963.
14. Gamblin, R. L., M. Q. Jacobs, and C. J. Tunis: Automatic Packaging of Miniaturized Circuits, in "Advances in Electronic Circuit Packaging," G. A. Walker (ed.), vol. 2, pp. 219–232, Plenum, New York, 1962.
15. Grim, R. K., and D. P. Brouwer: Wiring Terminal Panels by Machine, *Control Eng.,* vol. 8, pp. 77–81, August, 1961.
16. Burroughs Corp., Technical Report TR 60–40, June 28, 1965, entitled "Backboard Wiring Algorithms for the Placement and Connection Order Problems."
17. Algorithms for Backplane Formation in "Micro Electronics in Large Systems," pp. 51–76, Spartan Books, 1965.
18. Lee, C. Y.: An Algorithm for Path Connections and Its Applications, *IRE Trans. on Electronic Computers,* vol. EC-10, pp. 346–365, September, 1961.
19. Spitalny, A.: On-line Operation of CADIC, *Proc. of the SHARE-ACM Design Automation Workshop,* June, 1967.
20. Roth, J. P.: Diagnosis of Automata Failures: A Calculus and a Method, *IBM J. Res. Develop.,* vol. 10, no. 4, pp. 278–291, July, 1966.
21. Fisk, C. J., D. L. Caskey, and L. E. West: Topographic Simulation as an Aid to Printed Circuit Board Design, *Proc. of the SHARE-ACM Design Automation Workshop,* June, 1967.
22. Zell, C. C.: Manufacturing Engineering and Information, *Proc. of the SHARE-ACM Design Automation Workshop,* June, 1967.

INDEX